9th Textile Bioengineering and Informatics Symposium Proceedings (TBIS 2016)

AA006020

Melbourne, Australia
12-15 July 2016

Volume 1 of 2

Editors:

Yi Li
Rajiv Padhye

ISBN: 978-1-5108-2898-8

Printed from e-media with permission by:

Curran Associates, Inc.
57 Morehouse Lane
Red Hook, NY 12571

Some format issues inherent in the e-media version may also appear in this print version.

Printed by Curran Associates, Inc. (2016)

Textile Bioengineering and Informatics Society (TBIS)
Hong Kong Polytechnic University
ST733
Hung Hom, Hong Kong

Phone: (852) 2766 4206
Fax: (852) 2764 5489

tbis@tbisociety.org

Additional copies of this publication are available from:

Curran Associates, Inc.
57 Morehouse Lane
Red Hook, NY 12571 USA
Phone: 845-758-0400
Fax: 845-758-2633
Email: curran@proceedings.com
Web: www.proceedings.com

PROCEEDINGS OF TBIS 2016

TABLE OF CONTENT

Track 1 Textile Material Bioengineering

Track 2 Textile Biomechanical Engineering

Track 3 Clothing Bioengineering

Track 4 Textile Supply Chain Informatics

Track 5 Textile Thermal Bioengineering

Track 6 Textile Science and Engineering

Track 7 Textile Biomedical Engineering

TBIS

Textile Bioengineering and Informatics Symposium Proceedings

2016

ISSN:19423438

Edited by
Yi Li
Rajiv Padhye

The 9th Textile Bioengineering and Informatics Symposium
In Conjunction with The 6th Asian Protective Clothing Conference

July 12-15, 2016, RMIT University, Melbourne, Australia

Welcoming Message

Professor Rajiv Padhye
Director, Centre for Advanced Materials and Performance Textiles
School of Fashion and Textiles
RMIT University, Melbourne
Australia

We warmly welcome all of you to TBIS-APCC 2016 in Melbourne, the garden city, internationally famous for its restaurants, sporting events, beaches and beautiful weather.

TBIS conference is a brand in the field of textile bioscience that is attracting increasing attention from the International bio-scientific research community, which wants to present their research and interact with other world-leading researchers, students and industry from around the world.

The Textile Industry in Australia has been changing in recent years. Traditional manufacturing has moved offshore but industry is concentrating on producing new materials and designs for protective, sports, medical and smart textiles.

RMIT School of Fashion and Textiles is the only one-stop education, training and research institute for the entire supply chain of textile and fashion industry in Australia. The Centre for Advanced Materials and Performance Textiles (CAMPT) within the RMIT University School of Fashion and Textiles is involved in designing and engineering new products for various textile applications & has the state-of-the- art research and training facilities. CAMPT has strong collaborative links with Deakin University, CSIRO, Australian Wool Innovation and Defence Materials Technology Centre (DMTC) and consistently achieves internationally acclaimed research outcomes.

We invite you to attend TBIS-APCC 2016 Conference. Be our guests, and contribute to a memorable event in Melbourne! Once again we welcome you to Melbourne and know you will enjoy your stay with us.

Welcoming Message from TBIS

Prof. and Chair Henry, Yi Li
Professor of Textile Science and Engineering
School of Materials, The University of Manchester
Chairman, TBIS-APCC 2016 International Scientific Committee
Chairman, Textile Bioengineering and Informatics Society
Editor-in-Chief, Journal of Fiber Bioengineering and Informatics

With great pleasure, we are now collaborating with RMIT University to hold TBIS-APCC 2016, a joint conference between the 9th Textile Bioengineering and Informatics Symposium and the 6th Asian Protective Clothing Conference in July 12-15, 2016 in Melbourne, Australia.

With continuous success and enthusiastic support from authors and experts around the world, Textile Bioengineering and Informatics Symposiums (TBIS) have been recognized internationally as a brand of high quality conference with vision, creativity, excitement and global network. TBIS proceedings and JFBI (Journal of Fiber Bioengineering and Informatics) have been indexed by worldwide scientific and academic reputable databases such as Scopus and CPCI/ISI Web of Knowledge. Since 2008, TBIS has published 1553 research papers indexed in the CPCI/ISI Web of Knowledge database, becoming a leading academic conference organizer and publisher.

TBIS has established an effective platform for academic institutions and individual scholars, researchers and students to publish their research works and communicate with peers and experts in the field to get recognitions and increase impact such as citations and H-indexes. The TBIS symposium is a peer-reviewed conference aiming to set up a unique global stage and platform for people working in the whole supply chain from materials, fibers, textiles and apparel and technical textiles to relevant service industries such as biomedical industry, fashion industry and sport industry with common interests in providing high quality of products and services for protection, health, performance and comfort of human populations living in various environments and exposing to various natural and human disasters. To further enhance global networking, this year we have the pleasure to collaborate with RMIT University to organize TBIS-APCC in Melbourne, a beautiful and the world most and livable city located in the mainland of the Australian continent.

TBIS-APCC 2016 welcomes you and is looking forward to your active participation and contributions.

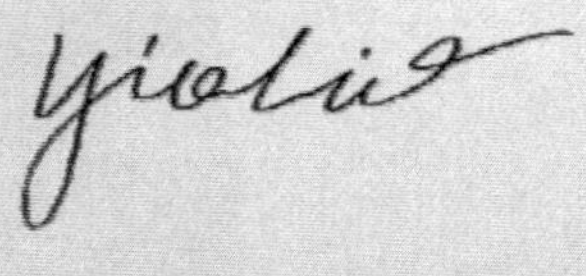

Welcoming Message from ASPC

Professor Kee Jong Yoon

President, Asian Society of Protective Clothing

Prof. of Dept. of Fiber System Engineering
Director of Personal Protective Equipment Center
Dankook University, Korea

Protective clothing is a relatively new field in Asia, but it is a field which requires immediate attention for the safety of people in diverse occupations such as fireman, policeman, car or motorcycle racers and industrial workers etc. The demand for protective clothing is expected to grow continually along with the advancement of the many developing countries in Asia.

When the Personal Protective Equipment Center of Dankook University organized the 1st Asian Protective Clothing Conferences (APCC) in 2010, interest was developing among the Asian countries but research on protective clothing was at an infant stage. However the delegates from Japan, Hong Kong, Taiwan, and China at this meeting all agreed on the urgent need for scientists and engineers to exchange ideas and collaborate in this area. Delegates of Japan from Shinshu University volunteered to organize the 2nd APCC in Ueda, Japan in 2011 and APCC started meeting annually. The 3rd APCC organized by the Taiwan Textile Research Institute was held at Taipei, Taiwan in 2012, and the 4th APCC organized by Soochow University was held in Suzhou, China in 2013. The 5th APCC is organized by the Hong Kong Polytechnic University in Hong Kong, China in 2014.

The international scientific committee hopes that many participants from across Asia participate in this conference to further the exchange and collaboration among engineers and scientists from academy and industry working in the protective clothing area.

윤 기 종

Welcoming Message from IFM

Professor Xungai Wang
Director, Institute for Frontier Materials
Deakin University

The Institute for Frontier Materials (IFM) welcomes TBIS-APCC delegates to Deakin University's Geelong Waurn Ponds Campus. As you all know, textiles and fibres play a vital role in society today – in our clothing, sport, the cars we drive, medicine, industry and much more. Geelong has historically strong connections with the fibres and textile industries and through the research carried out at IFM with many research and industry partners, these connections continue.

Our fibres and textiles research is carried out as part of the Australian Future Fibres Research and Innovation Centre (AFFRIC), a $103 million initiative supported by the Australian and Victorian Governments. AFFRIC is a collaboration between Deakin University and CSIRO, with research focusing on five core areas:

- Carbon fibres and composites
- Functional fibrous materials
- Green natural fibres
- Nanofibres
- Polymers

Another major collaboration is the new ARC Future Fibres Hub, a multi-million dollar collaboration funded by the Australian Research Council. The Future Fibres Hub brings together researchers from Deakin with partners in fibre research from around the world and a number of highly innovative industry partners.

The hub focuses on developing novel fibre technologies to facilitate more sustainable, advanced manufacturing of fibre materials and products that will help address significant global issues, including reducing our environmental footprint and improving public health and wellbeing. It's research is carried out under three broad themes: Advanced carbon fibre composites, high-value-added applications; and nanofibres.

IFM would like to thank you for your visit and hope that your tours of our facilities and the presentations of our work were of interest to you. As you leave our buildings and venture onto a small sample of our beautiful country and wildlife, we wish you well in your travels and a safe journey home.

TBIS-APCC 2016

Organizing and International Scientific Committees

Organizing Committee

Chairman:	Prof. Rajiv Padhye		
Co-Chairman:	A. Prof. Lijing Wang		
Vice Chairmen:	A. Prof. Jessica Bugg	Prof. Ruo-Mei Wang	Prof. Xin Zhang
Secretariat:	Ms. Mai Wang Li	Dr. Xin Wang	
Members:	A. Prof. Mohd Rozi Ahmad	Prof. Xue-Mei Ding	Dr. Jun-Yan Hu
	A. Prof. Rohana Kuruppu	Prof. Xiao-Nan Luo	Prof. Mangesh Teli
	Prof. Steven Westland	Prof. Xiong-Ying Wu	Prof. Wei-Lin Xu
	Dr. Lei Yao	Dr. Bo-An Ying	

International Scientific Committee

Honorary Chairman:	Academician Mu Yao	China		
Chairman	Prof. Henry Yi Li	UK		
Vice Chairmen:	Prof. Kanji Kajiwara (Japan)	Prof. Frank Ko (Canada)	Prof. Xungai Wang (Australia)	Prof. Kee Jong Yoon (Korea)
Members:	Prof. Seung Kook An (Korea)	Prof. Chi-Chung Bai (Taiwan)	Prof. Thomas Cassidy (UK)	Prof. B.K. Behera (India)
	Prof.Song-Mei Bi (China)	Prof. Ai-zheng Chen (China)	Prof. Guo-Qiang Chen (China)	Prof. Sheng-Fu Chiu (Taiwan)
	Prof. Joan Farrer (UK)	Prof. Jelka Gersak (Slovenia)	Prof. George Havenith (UK)	Prof. Emiel Den Hartog (USA)
	Prof. Lubos Hes (Czech Republic)	Prof. Morikawa Hideaki (Japan)	A. Prof. Simona Jesnik (Slovenia)	A. Prof. Hiroyuki Kanai (Japan)
	Prof. Eunae Kim (Korea)	Prof. Jun Li (China)	Prof. Marian McCord (USA)	Prof. Jiri Militky (Czech Republic)
	Prof. Xiu-Mei Mo (China)	Prof. Budimir Mijovic (Croatia)	Prof. Qing-Qing Ni (Japan)	Prof. Tetsuya Sato (Japan)
	Prof.Zhi-Juan Pan (China)	Prof. Elias Siores (China)	Prof. Guo-Wen Song (USA)	Prof. Yu-Chai Sun (China)
	Prof. Masayuki Takatera (Japan)	Prof. Rui-Jun Sun (China)	Prof. Kaoru Wakatsuki (Japan)	Prof. Qu-Fu Wei (China)
	Prof. Yong-Jin Wang (China)	Prof. Winnie Yu (Hong Kong, China)	Prof. Tochihara Yutaka (Japan)	Prof. Xian-Yi Zeng (France)
	Prof. Ke-Qin Zhang (China)	Prof. Ming Zhang (Hong Kong, China)	Prof. Feng Zhou (China)	Prof. Qing-Yong Zhu (China)
	Prof. Mei-Fang Zhu (China)	Prof. Malgorzata Zimniewska (Poland)		

Organizing and Supporting Institutions

Hosted by: RMIT University

Co-organized by:

Textile Bioengineering and Informatics Society

Asian Society Protective Clothing

Supporting Institutions

Deakin University
IFM-Institute for Frontier Materials

Aimer Human Engineering Research Centre

The University of Manchester

Xi'an Polytechnic University

Shaanxi Province's Research Center of Apparel Engineering and Technology

Sponsors

BEKAERT DESLEE

Bruck Textiles

TFS TECHNICAL FABRIC SERVICES AUSTRALIA

Acknowledgement to Reviewers - TBIS-APCC 2016 (Melbourne)

A	Abher	Rasheed		Hu	Jiyong	Q	Qiu	Chunting		Wang	Fa Ming
	Amin	Nabeel		Huang	Yu-Bing	R	Rajesh	Mishra		Wei	Bingqing
	Amirshahi	Seyed Hossein	I	Inui	Shigeru		Ratnapandian	Saminathan		Wei	Qu-fu
	An	Seung Kook		Ishimaru	Sonoko		Reischl	Uwe		Wardiningsih	Wiah
	Arumugam	Veerakumar	J	Jamdagni	Rishi		Ruan	Fangtao		Wu	Long
B	Barry	Blundell		Jevsnik	Simona	S	Salwa	Tashkandi	X	Xiao	Hong
	Behera	Bijoya Kumar		Jiang	Xiao Wen		Sato	Tetsuya		Xiao	Ping
	Bafekrpour	Ehsan		Jiang	Xiao Feng		Seyam	Abdel-Fattah		Xiaofeng	Yao
	Bai	Xiu-E	K	Kadole	PV		Shaw	Anugrah		Xin	Bin Jie
C	Cao	Mingliang		Kaleshnath	Chatterjee		Shang	Xiao-Mei		Xing	Xiao-yu
	Collie	Stewart Roger		Kajiwara	Kanji		Shao	Min		Xing	Xu
	Chen	Youfang		Kamijo	Masayhoshi		Shao	Xin-Ying		Xu	Fujun
	Chen	Zhige		Kanai	Hiroyuki		Shi	Hui		Xu	Shanqing
	Chen	Yanhua		Ke	Ying		Sikka	Monica		Xu	Pengjun
	Chen	Xiaogang		Khatri	Awais		Skenderi	Zenun		Xu	Jiahua
	Chen	Ai-Zheng		Kim	Gandhi		Song	Wenfang		Xu	Wei-Lin
	Chen	Jie		Kim	KyoungOk		Song	Guo Wen		Xue	Yuan
	Chen	Min-Zhi		Koptyug	Andrey		Soni	Sandip	Y	Yang	Jie
	Chen	Wei		Kotresh	TM		Strazdiene	Eugenija		Yang	Zhongkai
	Chen	Yan		Kuang	Cai Yuan		Stjepanovic	Zoran		Yee	Nee Elita Lam
	Chen	Ying		Kurniati	Nani		Sroma	Inese		Ying	Bo-An
	Cheng	Lan	L	Lee	Joo-Young		Srikulkit	Kawee		Yoshida	Hiroaki
	Choudhary	Awadhesh		Lee	Sun-Hee		Su	Junqiang		Yu	Winnie
	Cui	Hong		Lee	Young-A		Sui	Jianhua		Yuan	Yan
D	Dai	Hong Qin		Li	Li	T	Takatera	Masayuki		Yuan	Shi-ning
	Dai	Hong		Li	Zhi		Tausif	Muhammad	Z	Zeng	Wei
	Dai	Xiaoqun		Li	Gang		Tian	Li-Wei		Zeng	Xian-yi
	Deng	Yong Mei		Li	Jie		Tian	Ya-Nan		Zhang	Jing
	Ding	XueMei		Liang	Su-Zhen		Tian	Zhihong		Zhang	Ke-qin
	Du	ZhaoQun		Liu	XiaoXia		Tian	Mingwei		Zhang	Rui
	Duan	Xingyuan		Liu	Cheng-Xia		Tian	Wei		Zhang	TianJiao
	Dubrovski	Polona Dobnik		Liu	Yi		Tian	Bao-Zhong		Zhang	Xin
F	Fang	Xie		Liu	Li-fang		Tomoe	Masuda		Zhang	Yi
	Finn	Angela		Liu	Ya-Xia		Tong	You-Jun		Zhang	Li-yuan
	Fu	Ming		Liu	Xiao-song	U	Uttam	Devanand		Zhang	Tong-hui
G	Gersak	Jelka		Liu	Zheng-Dong	V	Vaclavik	Miroslav		Zhang	Zhaohua
	Gao	Shan		Louis	Kyratzis		Vinay	Midha		Zhang	Guangyu
	Gideon	Rotich K.		Lu	Ye Hu	W	Wakatsuki	Kaoru		Zhao	YuXiao
	Gu	ZhaoHui	M	Mao	Li-Li		Wang	Lijing		Zhe	Wang
	Guo	Xiaofang		Mao	Lei		Wang	Guohe		Zheng	Zi Jian
	Guo	Min		Mijovic	Budimir		Wang	Ruo-Mei		Zhou	Jie
	Guo	Yan		Militky	Jiri		Wang	Yongrong		Zhou	Feng
H	Harper	Catherine		Mitsuno	Tamaki		Wang	Zhao-Hui		Zhu	Chunhong
	Hartog	Emiel Den Hartog		Mo	Xiu-mei		Wang	Xin		Zhu	Chengyan
	Hashmi	Farina		Muthu	Subramanian		Wang	Xiao		Zhu	Qingyong
	He	Bin	N	Nawaz	Nazia		Wang	XiaoChun		Zou	Zhuanyong
	Hong	Jianhan		Nayak	Rajkishore		Wang	Guohe		Zuo	Baoqi
	Hong	Zheng-Lin		Nazarzadeh Za	Ehsan		Wang	Yanzhen			
	Hu	Jun-Yan	P	Purnomo	Ridwan		Wang	Lai-Li			

We sincerely thank every reviewer for providing valuable comments to the papers submitted to TBIS-APCC 2016 (Melbourne). We are looking forward to your continuing support in TBIS 2017 (China)

Waterproof Breathable Polyurethane Coating on Textile Substrate by Electrospraying

Amit Jadhav*, Lijing Wang, Rajiv Padhye

School of Fashion and Textiles, RMIT University, 25 Dawson St, Melbourne, VIC-3056, Australia

*Corresponding author: amit.jadhav@rmit.edu.au

Abstract

Electrospraying is a method of generating fine droplets through high electrostatic charging. It involves an electrohydrodynamic process for forming fine droplets. These droplets are highly charged that promotes self-dispersion which results in the uniform coating on the substrate. Electrospraying can be successfully utilised as a coating technique to coat textile substrates. Textiles substrate coated with electrosprayed polymer droplets have a large specific surface area, and it facilitates submicron range coating compared to conventional coating methods like roller coating or knife coating. This research study investigates the electrospraying of thermoplastic polyurethane as a coating technique on the textile substrate. The results showed that electrospraying facilitated the submicron size droplets deposition on textile substrates which imparted the waterproof property to the substrate without altering the moisture management and surface properties of the textile substrate.

Keywords: Electrospraying; Thermoplastic Polyurethane; Coating; Surface Roughness

1. Introduction

The use of coated textiles for protective clothing, shelters, covers, liquid containers, etc. dates back to antiquity. Historically, the earliest recorded use of coated textile was by the ancient people of Central and South America, who applied latex to fabric to render it waterproof [1]. Today, coated fabrics are essentially polymer-coated textiles. Advances in polymer and textile technologies have led to phenomenal growth in the application of coated fabrics for various end uses. Coated fabrics find an important place among technical textiles and are one of the most important technological processes in the modern textile industry.

The properties of a coated fabric depend on the type of polymer used and its formulation, the nature of the textile substrate, and the coating method employed. Most of the coated textiles are produced by conventional coating methods such as a knife or roller coating. There are problems associated with these coating methods, for example, impregnation of the polymer into the yarn/ fibre interstices, which alters the original properties of the fabric. Electrospray coating has the ability to deposit polymer droplets at submicron level due to electrostatic charging. The submicron level coating facilitates the retention of the fabric's original properties so that the fabric can deliver additional functional properties depending on the polymer characteristics.

The fine spray of polymer droplets results in very smooth and even coating. Surface coating of polymer with electrospraying is a promising technology for the near future in the field of coating of materials [2]. Electrospraying has opened new routes to nanotechnology. Electrospraying is used for micro– and nano–thinfilm deposition [3], micro-and nano-particle production [4], and micro- and nano-capsule formation. Thin films and fine powders are (or potentially could be) used in modern material technologies [5], microelectronics and medical technology [6,7]. This research work discusses the potential applications of electrospraying as a coating technique for textile substrates. A thermoplastic polyurethane was sprayed onto cotton fabric to impart waterproof-breathable functionality.

2. Experimental

2.1 Fabric materials

Fabrics produced from cotton or wool; natural fibres are mostly used for coating to enhance their properties and hence were chosen for this study. Mercerised and the bleached cotton fabric was received from Bruck Textiles, Australia. 100% wool was sourced from Macquarie Textiles, Australia. Both fabrics were used as received without any further chemical treatments. The fabric specifications are given in Table 1.

Table 1. Fabric specifications

Characteristics		Cotton	Wool
Weight (g /m^2)		200	230
Yarn count (Tex*)	Warp	48	36
	Weft	90	41
Ends/cm		34	36
Picks/cm		22	24
Weave		Plain	2/1 twill

*Tex - weight in grams per 1000 m of yarn

2.2 Chemicals

Thermoplastic polyurethane (TPU) of M_w 30,000 received in the form of chips from Pacific Urethanes (Australia). Tetrahydrofuran (THF) were received from Sigma-Aldrich Chemicals Co. (USA), The THF was used as a solvent for dissolving polyurethane because THF has a high evaporation rate. Its boiling point is 66°C, viscosity 0.55 cP and density 0.888 g/ml. All chemicals were used as received without any purification. Polyurethane solutions of different concentration were prepared by dissolving the polymer granules homogeneously in tetrahydrofuran for electrospraying.

2.3 Electrospraying

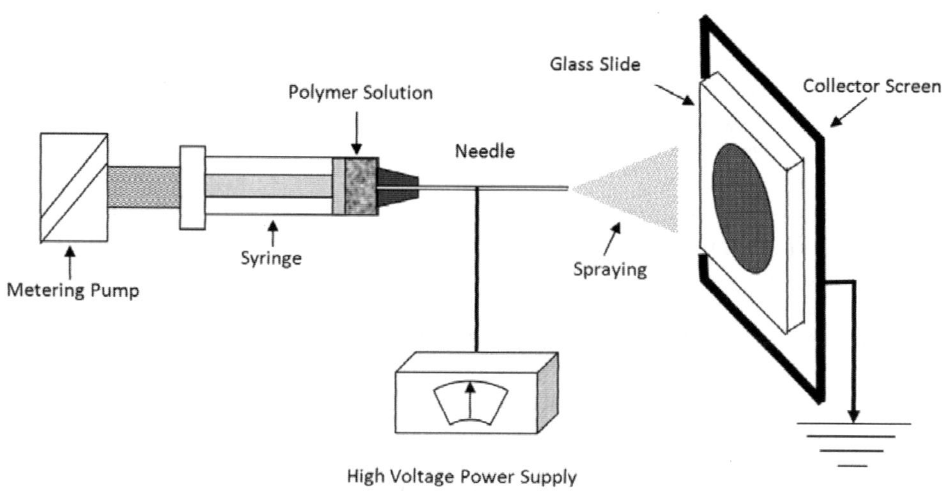

Fig.1 Equipment setup for Electrospraying Process [8]

The electrospraying apparatus consisted of a syringe which was filled with required polymer solution. The syringe has a capacity of 50 ml, and a specially designed needle is connected to the positive terminal of high voltage power source ES30P-5W from Gamma High Voltage Research (USA). This power source is capable of providing an output voltage from 0-30KV. The inner diameter of the nozzle is 0.5mm. A glass slide with fabric sample was placed on a metal plate was used as a collector to be coated with the polymeric solution. KDS 200 Digital Syringe Pump was used to deliver the polymer solution towards the tip of the nozzle.

A solution of 2.5% TPU was prepared by dissolving the polymer granules homogeneously in THF for the electrospraying. Woven cotton and wool fabrics were chosen for this study to enhance their properties. The polymeric solution was electrosprayed for 3 mins for each sample. Preliminary experimental trials showed that, 3 mins are enough to cover the fabric surface held at collector screen.

2.4 Scanning electron microscope (SEM)

Scanning electron microscopy (SEM) was used to study the coated surface morphology. SEM uses a focused beam of high-energy electrons to generate a variety of signals at the surface of solid specimens. The signals that derive from electron-sample interaction reveals information about the sample including external morphology (texture), chemical composition and crystalline structure. In most applications, data is collected over a selected area of the surface of the sample and a 2-dimensional image was generated that displays spatial variations in these properties. Areas ranging from approximately 1 cm to 5 microns in width can be imaged in magnification ranging from 20X to approximately 30,000X.

2.5 Contact angle measurement

Contact angle measurement is important for assessing the hydrophobicity of the textile material after polymer coating. The contact angle measurement was done using the PsS OCA 20 contact angle measurement instrument. The contact angle was measured by producing a drop of liquid on a solid surface. The contact angle was measured twice during the experiments, firstly 1 second after the droplet formed on the surface and secondly after 120 seconds. The results were compared before and after coating.

2.6 FTIR (Fourier transform infra-red spectroscopy)

FTIR was used to determine changes in the chemical structure by the effects of electrospray coating on various textile substrates. The FTIR spectra were obtained by a spectrometer (the PerkinElmer Spectrum-400) to determine the functional groups of the coated substrate. The number of scans per sample was 8 and FTIR spectra were obtained in the wave number region of 4000-650 cm^{-1}.

2.7 Moisture management testing (MMT)

To study the moisture management properties of TPU coated cotton fabric, MMT instrument was used. The MMT measures the liquid transfer in one step on a fabric sample in multi-directions. In the present study, the coated surface of the fabric samples is always a bottom surface (facing the bottom sensor) when the sample is tested, imitating the case where the uncoated surface is in direct contact with the skin.

2.8 Surface frictional properties

The KESFB4-A, Kato Tech Co. Ltd. Evaluation System was used to study the frictional properties of fabrics after electrospraying. The sensors measured the frictional coefficient and the roughness of the fabric surface simultaneously in three different areas within a fabric sample.

The evaluation of the fabrics was carried out having particulars that:

- The value of MIU ranges from 0 to 1 and its approaching value of 1 is interpreted as increasing friction and decreasing smoothness.
- The value of SMD ranges between 0 and 20 and its approaching value of 20 is interpreted as an increase in surface roughness and surface irregularities.

3. Results and discussions

3.1 Morphology of coated fabrics

Figure 2 shows the surface morphology of the TPU coated cotton fabric with the roller coating technique. As shown in Figure 2(a), the polymer is impregnated into the interstices of the yarns and fibres. All the yarns/ fibres were bonded together with the polymer. As seen in Figure 2(b), the cross-section of the coated fabric, the yarn crossover points are also bonded due to polymer impregnation. This is due to the fact that, in roller coating, the fabric is immersed into the polymer solution bath and then passed through the rollers to squeeze out the excess polymer. The polymer impregnation into the yarn interstices affects the original properties of the fabric including air permeability, moisture management, colouration, extension, tearing strength, etc.

The electrospraying technique could have an advantage over conventional coating technique due to submicron-level polymer droplets deposition. There is no mechanical action involved in the process. The deposition is taking place with the help of electrostatic charging. Figure 3 shows scanning electron images of uncoated and TPU-coated cotton fabric substrates. As shown in Figure 3(b) the tiny droplets of TPU polymer were deposited on the fibre surface and attached only to the fibre surface, not in the interstices of the fibres or yarns. Due to this, the original properties of the cotton fabric are unaltered, and the added advantages of TPU properties can be acquired. Also, the coating can be applied to only one surface so that the other surface will behave like 100% cotton. Also, different types of the functional polymer can be coated successfully on both sides to obtain dual surface functionalisation effects.

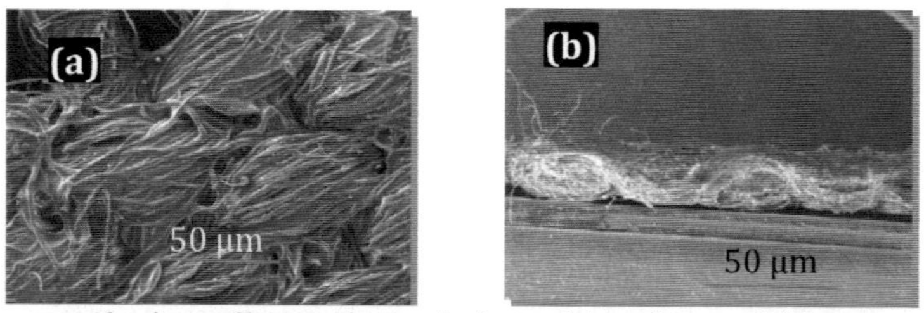

Fig.2 Cotton fabric coated using roller coating technique: (a) surface and (b) cross section

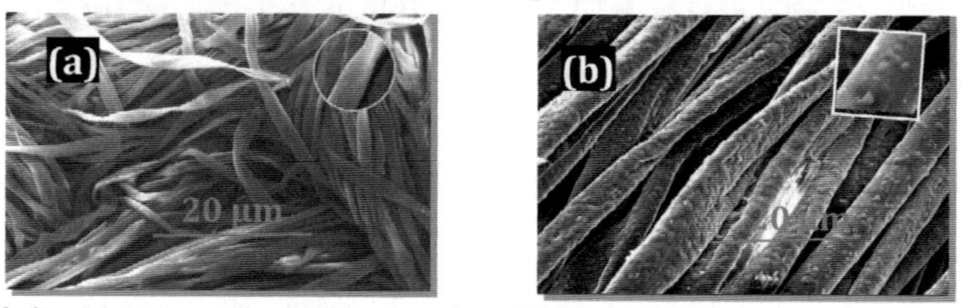

Fig.3 Cotton fabric: (a) uncoated and (b) coated at 20kV, 1mL/h flow rate, 10 cm nozzle-collector distance and 2.5% polymer concentration

Fig 4(b) shows the TPU submicron–sized droplets deposited on the wool fibre surface. This coating will give many added functional effects to wool substrates.

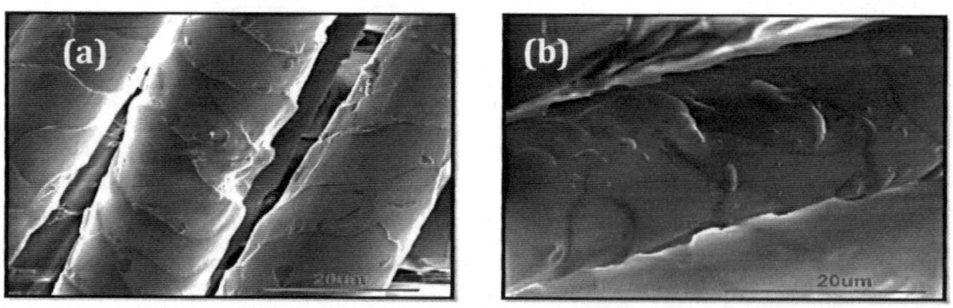

Fig.4 Wool fabric: (a) uncoated and (b) coated at 20kV, 1mL/h flow rate, 10cm nozzle-collector distance and 2.5% polymer concentration

3.2 FTIR analysis of TPU coated cotton fabrics

The FTIR spectra of the uncoated and TPU coated cotton samples were compared as shown in Figure 5, and their assignments are described in Table 2. The chemical formulation of polyurethane consist of isocyanates (-N=C=O) AND hydroxyl groups (-O-H) respectively. The reaction between isocyanates and hydroxyl group results in urethane (H-N-C=O). Figure 5 clearly shows a strong absorption at 1728 cm^{-1} (C=O) vibration, and hence, it is evident that the polyurethane is present on the substrate [9]. Also, a strong absorption is observed at 1222 cm^{-1} (C-O), 1414 cm^{-1} (C-N-C), 1530 cm^{-1} (N-CO) and 2943 cm^{-1}(C-N-O) and a decreased the intensity of the peak at 3330 cm^{-1} (N-H stretching).

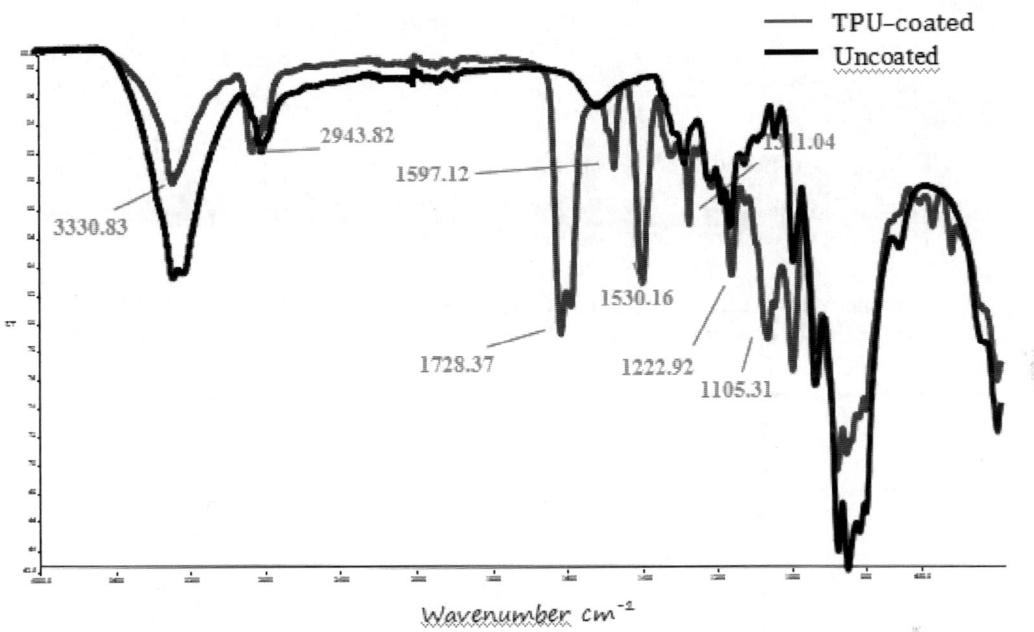

Fig.5 FTIR for uncoated and TPU–coated cotton fabrics

The presence of these peaks confirms that the TPU polymer droplets were bonded to cotton fabric substrates.

Table.2 FTIR bands of TPU coated cotton samples with assignments

Wavenumber (cm^{-1})	Vibrational assignments
1728	Strong C=O vibration
1414	C-N-C absorption
1530	Amide I N-C=O
2943	C-N-O vibration
3330	N-H stretching

3.3 Water contact–angle measurement

Water contact-angle measurements were made to observe any improvement in hydrophobicity of the coated fabric after coating. The PsS OCA 20 static contact angle measurement instrument was used to determine the contact angles. As shown in Figure 6(a) the water droplet spread on uncoated cotton fabric. This shows that the uncoated cotton fabric is hydrophilic in nature. Whereas water droplet, did not spread on the TPU-coated cotton fabric, as can be seen in Figure 6(b).

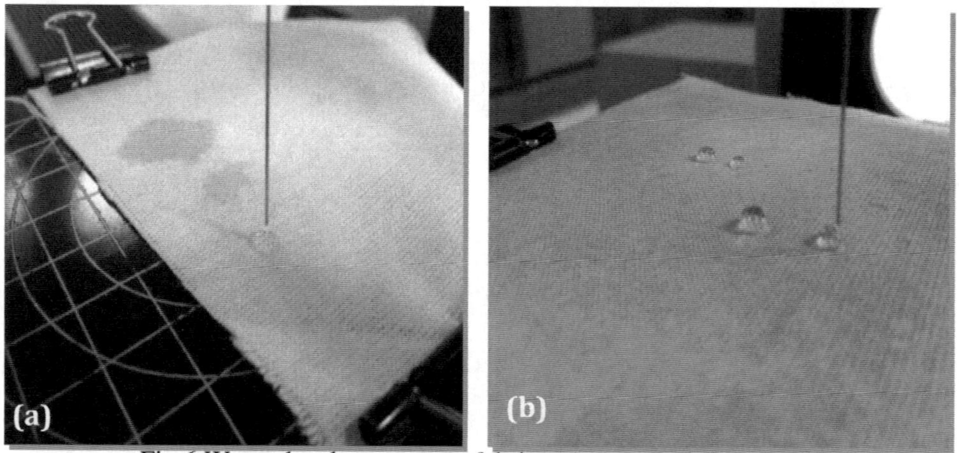

Fig.6 Water droplet on cotton fabric: (a) uncoated (b) coated

Figure 7 shows the contact angles after 1 second of droplet deposition. Contact angles of 66.3° for uncoated (a) cotton fabric and 117.4° for the TPU-coated (b) cotton fabric were observed. An almost twofold increase in contact angle was observed after TPU coating.

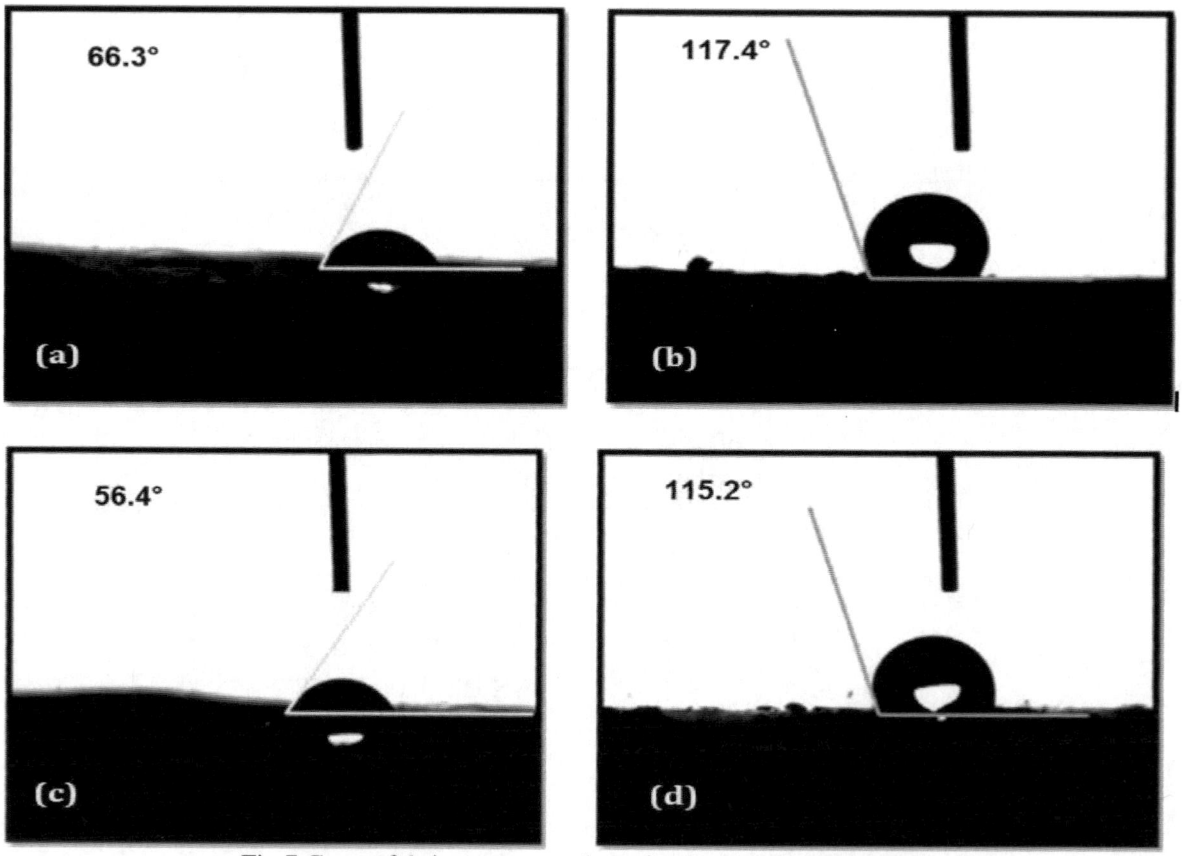

Fig.7 Cotton fabric contact angles: after 1 sec on (a) uncoated (b) coated; after 120 sec on (c) uncoated (d) coated

Contact angles after 120 seconds were measured, and it was observed that there was a 10° reduction in contact angle of the uncoated cotton fabric i.e. 56.4° as shown in Figure 7(c) and only a 2° reduction in the contact angle of the TPU coated cotton fabric i.e. 115.2° as shown in Figure 7(d). This is due to the submicron-sized droplets on the fabric surface, which held the water droplet and did not allow spreading. This shows that the hydrophobic property imparted to the cotton fabric.

Table 3. Contact angles for uncoated and coated fabrics

Fabrics	Contact angle after 1 sec	Contact angle after 120 sec
Uncoated	66.3°± 2°	56.4°± 2°
Coated	117.4°± 2°	115.2°± 2°

3.4 Moisture management testing

The moisture management properties of the cotton fabrics are given in Table 4. WTt and WTb are, respectively the time period in which the top and bottom surfaces of the tested fabric began to become wet after the commencement of the test. From Table 4, it can be seen that the uncoated and coated fabrics showed slow wetting times on the top surface and medium wetting times on bottom surface. Both coated and uncoated fabrics showed the same grade for wetting times. It can be seen from Figure 8.

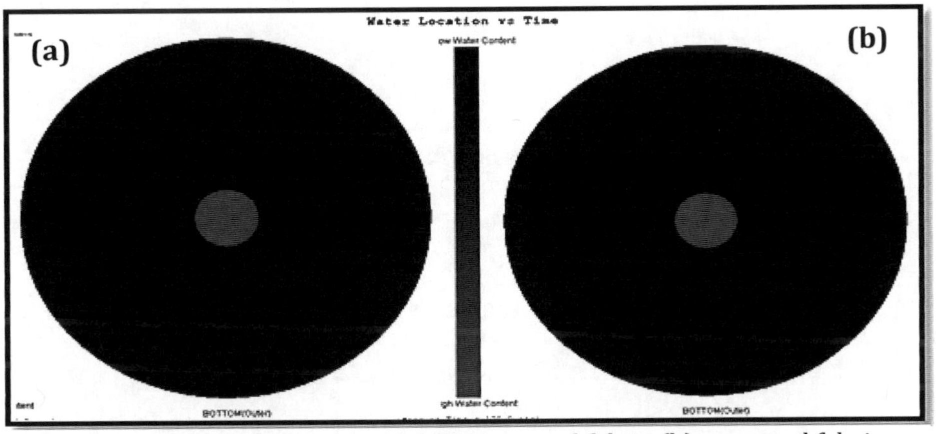

Fig.8 Water location vs time: (a) TPU–coated fabric (b) uncoated fabric

Table 4. MMT results in value of coated and uncoated fabrics

MMT Index	Fabrics	
	Coated	Uncoated
WTt	57.50	79.98
WTb	22.81	15.64
ARt	0.00	0.00
ARb	26.05	47.81
MWRt	0.00	0.00
MWRb	5.00	5.00
SSt	0.00	0.00
SSb	0.22	0.86
AOTI	266.21	357.00
OMMC	3.50	4.00

Due to the very fine coating, there is no significant difference in the wetting time index before and after coating. AR_t and AR_b also show similar grades for both fabrics. Considering the grades for the AOTI index, the uncoated fabric showed excellent and coated fabrics showed a very good grade.

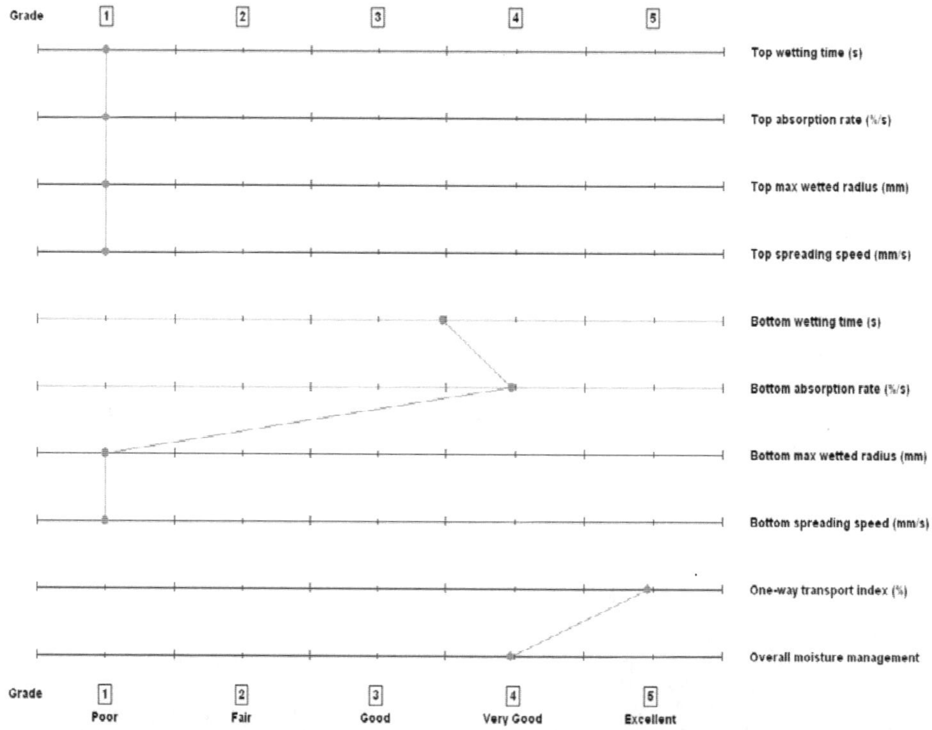

Fig.9 Fingerprints of uncoated fabric moisture management properties

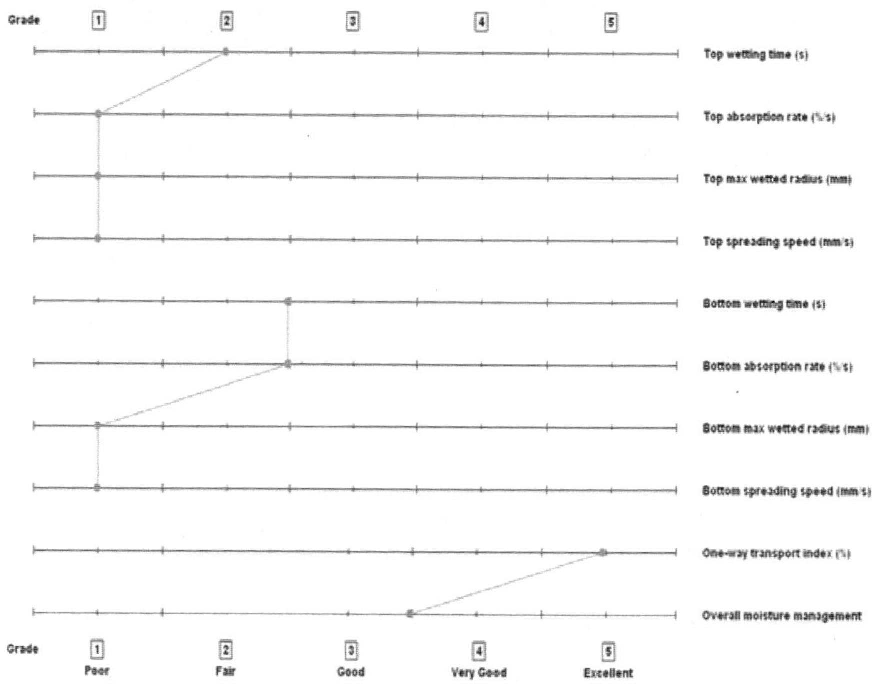

Fig.10 Finger print of coated fabric moisture management properties

The fingerprints of the moisture management properties of coated and uncoated fabrics are shown in Figures 9 and 10 respectively. It is evident that there were no significant changes in the moisture management properties of the fabrics duc to the coating of the TPU. This shows that these types of fabrics can serve as waterproof, breathable fabrics.

3.5 Surface roughness property

For the evaluation of fabric surface frictional properties, there are two main attributes to be considered: coefficient of surface friction (MIU) and geometrical roughness (SMD). In general, the frictional force is higher for fabric with a rough surface and lower for fabric with a smooth surface. Table 5 demonstrates the mean values of MIU and SMD for the coated and uncoated cotton fabric samples. It can be observed that there was no significant change in surface frictional properties. Figure 11 shows a graphical representation of the coefficient friction (MIU) for uncoated and coated fabric surface. There was no significant difference observed in MIU. Similarly, it can be observed from Figure 12 that there was no significant difference in geometrical roughness of the coated and uncoated fabric surfaces. This may be due to a very thin level of coating on the surface which has not created any rough patterns on the fabric surface.

Table 5. Surface roughness properties of coated and uncoated fabrics

Surface roughness Property	Uncoated Fabric	Coated Fabric
MIU	0.176	0.181
SMD	7.20	8.31

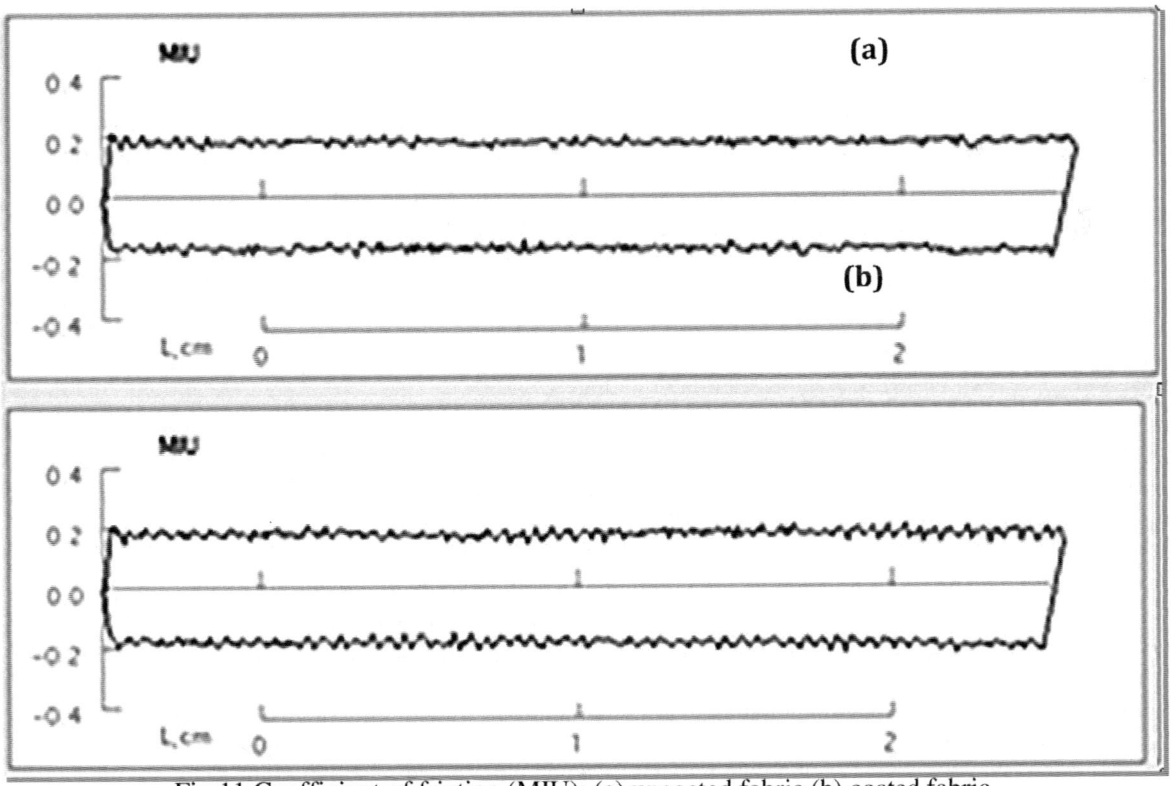

Fig.11 Coefficient of friction (MIU): (a) uncoated fabric (b) coated fabric

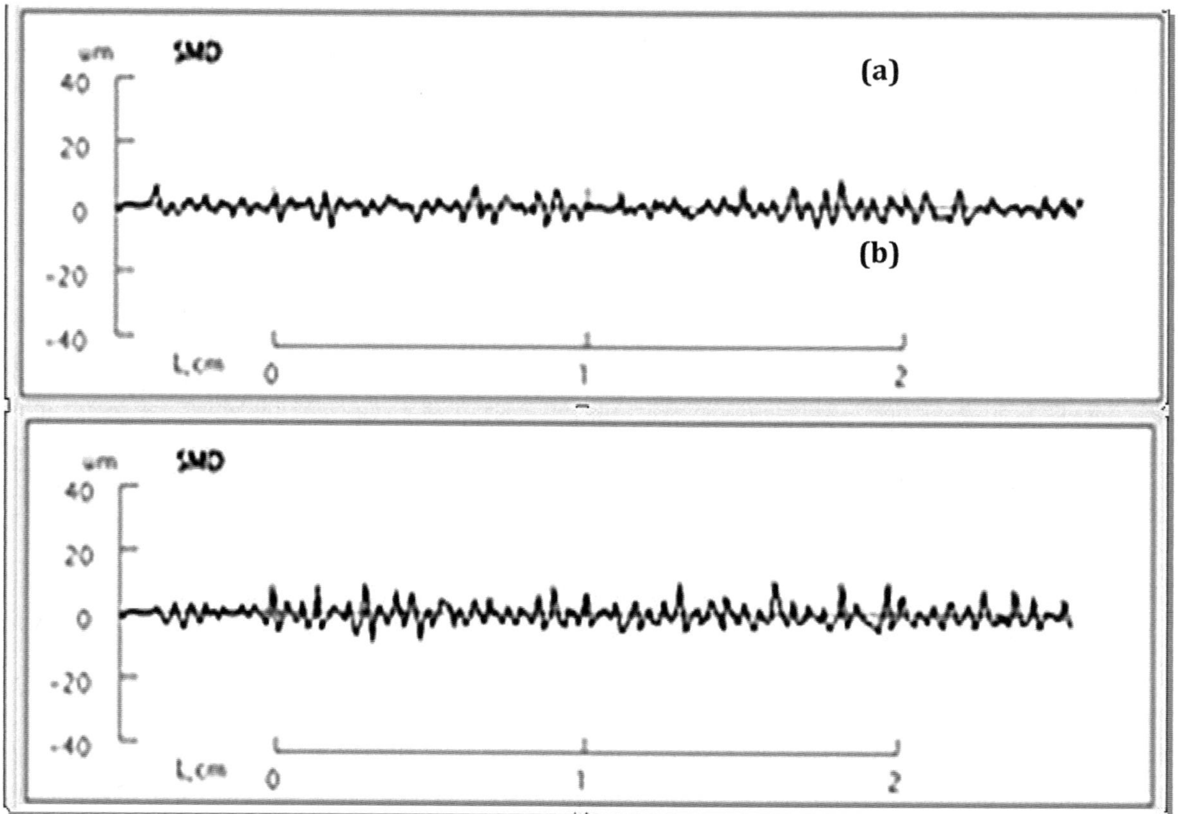

Fig.12 Geometrical roughness (SMD): (a) uncoated fabric (b) coated fabric

4. Conclusions

The electrospraying coating technique can effectively utilise as a textile coating technique. Textile substrates can be coated without losing their original property compared to conventional coating technique. From the results, it is observed that there was no significant change in surface frictional properties, no alternation in the moisture management properties of the fabrics. The water contact angle significantly improved after coating of TPU, which shows waterproof breathable property imparted to cotton fabric. This coating technique opens the door for novel textile coating applications.

References

[1] Sen AK. Introduction. In: Coated Textiles: Principles and Applications. New York. CRC Press: 2007; p: xix.
[2] Shin Y, Hohman M, Brenner M, Rutledge G. Electrospinning: A whipping fluid jet generates submicron polymer fibres. App. Phy. Let: 2001; 78: 1149.
[3] Arya N, Chakraborty S, Dube S, Katti DS. Electrospraying: a facile technique for synthesis of chitosan-based micro/ nanospheres for drug delivery applications. J Biomed Mater Res B Appl Biomater: 2009; 88: 17-31.
[4] Kumbar SG, Bhattacharyya S, Sethuraman S, Laurencin CT. A preliminary report on a novel electrospray technique for nanoparticle based biomedical implants coating: precision electrospraying. J Biomed Mater Res B Appl Biomater: 2007; 81: 91-103.
[5] Peltonen L, Valo R, Kolakovic, Laaksonen T, Hirvonen J. Electrospraying, spray drying and related techniques for production and formulation of drug nanoparticles. Exp opn on drug Del: 2010; 7: 705-719.
[6] Cavalli R, Bisazza A, Bussano R, Trotta R, Civra C, Lembo D, Ranucci E, Ferruti P. Poly(amidoamine)-Cholesterol Conjugate Nanoparticles Obtained by Electrospraying as Novel Tamoxifen Delivery System. J. Drug Del: 2011; 587-604.

[7] Tavares Cardoso MA, Talebi M, Soares PA, Yurteri CU, Ommen JR, Functionalization of lactose as a biological carrier for bovine serum albumin by electrospraying. Int. J. Pharm: 2011; 414: 1-5.
[8] Jadhav A, Wang L, Padhye R. Influence of Applied Voltage on Droplet Size Distribution in Electrospraying of Thermoplastic Polyurethane. Int. J. App. Mech. and Mat: 2013; 2: 64.
[9] Mishra AK, Chattopadhyay DK, Sreedhar B, and Raju KVSN. FT-IR and XPS studies of polyurethane-urea-imide coatings. Prog. in Org. Coat: 2006; 55: 231-243.

Self-cleaning Property of Cotton Fabrics Coated with Nano-sized Zinc Oxide

Chun-Hong Zhu[1*], Jian Shi[2]

[1]*Faculty of Textile Science and Technology, Shinshu University, 3-15-1 Tokida, Ueda, Nagano, 386-8567, Japan*
[2]*Faculty of Systems Science and Technology, Akita Prefectural University, 84-4 Aza Ebinokuchi Tsuchiya, Yurihonjo, Akita, 015-0055, Japan*

*Corresponding author's email: zhu@shinshu-u.ac.jp

Abstract

The visible light self-cleaning property of cotton fabrics coated with nano-sized zinc oxide (ZnO) was discussed in this study. After preparing different weight ratios of ZnO nanoparticle aqueous solutions, cotton fabrics were coated by the as-prepared ZnO solutions making use of dip-pad-dry-cure process. FESEM images were listed to show the morphologies of cotton fibers before and after treating. Dyestuff and coffee stained fabrics were used for investigating the photocatalysis effect by solar simulator for 6h, 12h and 24h irradiation. The results showed the ZnO treated fabrics had self-cleaning properties on organic material compared to untreated fabrics after 24h sunlight irradiation. Moreover, the visible light self-cleaning performance of 3wt% ZnO coated fabrics was found to be better than 1wt%, implying that the weight ratio of metal oxide was an important parameter influencing the sunlight self-cleaning property.

Keywords: Self-cleaning; Photocatalysis; Cotton Fabrics; Zinc Oxide; Solar Irradiation

1. Introduction

Energy use has become a critical problem in the last decades because of rapid increase in energy demand during industrial development. Moreover, environmental issues of traditional energy resources are more and more serious, such as climate warming which may cause natural disasters and diseases [1]. Therefore, it is necessary to exploit substitutable energy for the next generation. As the sunlight is abundance, clean and free, it is probably the most important source of renewable energy in the world. Many investigations have been made to extract solar energy in various kinds of fields, such as solar water heater, solar cooker, solar drier, solar pond, solar architecture, solar air-conditioning, solar chimney, solar power plants, and solar power generation and so on [1-3]. Associated with the development of functionalized textiles owing specific properties, solar energy has been widely used in the field of textiles, such as textile rechargeable battery [4]. Photocatalyst is often used in textiles for self-cleaning utilizing solar energy.

In the late 1960s, research on oxide semiconductors responding to light had been started by Akira Fujishima [5]. In 1977, Frank and Bard demonstrated the possibility of using titanium dioxide (TiO_2) to decompose cyanide in water firstly, and TiO_2 was considered as an ideal photocatalyst [6]. From then, metallic oxide used as photocatalyst has been researched for glass, plastics, and textiles and so on [7-9]. In recent years, more attention has been paid to the application of semiconductors, such as ZnO, Fe_2O_3, WO_3 and CdS as photocatalysts, for degradation of both organic and inorganic materials [10]. These metal-oxide semiconductors can help to break down carbon-based molecules and require only sunlight to start the reaction. The metal-oxide semiconductors work as follows [11]: they use light energy to catalyze oxidation-reduction reactions, producing electron-hole pair at material surface [5]. The generated electrons on the metal-oxide semiconductor surface have a high reduction potential, and the generated holes have a high oxidation potential [12-13]. Then aqueous reaction is catalyzed by electron-hole pairs, generating hydroxyl radicals and reactive oxygen species, which are directly responsible for the oxidative degradation of organic materials.

As a popular photocatalyst, TiO_2 has been widely used for photocatalyst because of its high photocatalytic activity, chemical stability, non-toxicity and low cost [9, 14]. Qi KH et al [15] investigated the decomposition of organic materials including colorants, red wine and coffee stains on cotton textile coated with TiO_2, and discussed the self-cleaning performance. Wang RH et al [16] reported a highly active Au/TiO_2/SiO_2 photocatalyst with a size of less than 30nm, and the self-cleaning durability of cotton was evidenced from the wash-fastness effect on decoloration of red wine and coffee stain. Karimi L et al [9] investigated on self-cleaning properties of TiO_2-coated cotton fabrics prepared by cross-link and non-cross-link methods. They achieved an optimal effect on the self-cleaning degree of treated fabrics.

As another kind of metal-oxide photocatalyst, ZnO, with band gap closing to that of TiO_2, is a highly useful and practical semiconductor material. It can be used including self-cleaning and anti-fogging surfaces, sterilization, gas sensing, energy production and environmental purification [17-18]. Moreover, ZnO materials are nontoxic and present little additional harm to the environment, contrary to most other methods of decontamination [11]. Furthermore, ZnO has the advantage of lower cost than TiO_2, which implies that ZnO should be a promising alternative to TiO_2 in photocatalysis [19]. There were many researches about the photocatalysis of ZnO. Zhang X et al [21] demonstrated the fabrication of different aspect ratios of ZnO nanorods with surface defects by mechanical-assisted thermal decomposition method. The results showed that ZnO nanorods with higher aspect ratio and surface defects have higher photocatalytic performances. Kenanakis G et al [22] investigated on highly c-axis oriented ZnO nanowires' arrays grown on glass substrates with temperature not exceed 95°C, and studied their photocatalysis and wetting properties upon UV light irradiation. Moore JC et al [11] measured the photocatalytic activity and stability of thin, polycrystalline ZnO films fabricated on sapphire substrates via direct current sputter deposition of Zn-metal films and described the growth parameters resulting in porous.

Even though it has been reported that ZnO shows better activity than TiO_2 in the photodegradation of some dyes in aqueous solutions since it can absorb more light quanta [20], until now, there are only a little researches on photocatalysis of ZnO for fabrics, which can be seemed as a kind of porous substrate. As a kind of metal-oxide photocatalyst, ZnO has low cost than TiO_2, which can be utilized in textile industry, and using solar energy for cleaning can be helpful for water saving and environmental protection. Therefore, in this study, the photocatalytic property of ZnO nanoparticles coating on cellulosic cotton fabric was discussed. The properties were investigated by different weight ratio (wt%) of ZnO nanoparticle aqueous solutions. After pretreatment and coating processes, the surface morphology of cotton fiber was observed by field emission scanning electron microscopy (FESEM). In order to demonstrate the self-cleaning effect under sunlight irradiation, both dyestuff and coffee strain were performed and the decoloration was observed.

2. Experiment

2.1 Materials

The cellulosic cotton fabric used in this experiment was plain woven, with the specification shown in Table 1. The pretreatment of cotton fabric was carried out to remove wax, grease, and other chemical residues from the fabrics. At first, pretreatment [23-24] was accomplished by immersing the fabric in a water bath of 80±2°C for 60min, twice. After air-dried, 1g fabric was pretreated by immersion in an 80±2°C water bath containing a 100g nonionic detergent solution (Polyxyethylene lauryl ether, Kao Corporation EMULGEN 108) with Emulgen 108/deionized water of 0.01g/100g at 80±2°C for 60min. Then as-pretreated fabric was rinsed in deionized water for several times. At last, the sample was air dried under constant conditions for 24 hours.

Table 1 Specification of fabric sample

	Weave	Warp weave density (cm^{-1})	Weft weave density (cm^{-1})
Cotton 100%	Plain	34.6	22.0

2.2 Preparation of different wt% ZnO nanoparticle aqueous solutions

The ZnO nanoparticle aqueous solution was purchased from Sigma-Aldrich Corporation, with dispersion of 50.1 wt%. The average particle size is less than 35nm. In the experiment, weight ratios of 1% and 3% ZnO nanoparticle aqueous solutions were prepared for coating.

2.3 Coating process

The as-prepared aqueous solutions can be applied onto the cotton fabrics by several techniques, such as dip-pad-cure, dip-coating, and spraying processes [16] and so on. In order to form a durable layer, in this study, the dip-pad-dry-cure process was employed. The cleaned substrate was dipped in the as-prepared ZnO nanoparticle aqueous solutions for 30 sec and then padded with a padder. The padded substrate was dried at 100°C for 30 min and finally cured at 130°C for 2min in a preheated curing oven (Asone Corporation, Japan, OFW-450B) to ensure particles adhesion to the substrate surface.

2.4 Study of photocatalytic activity

2.4.1 Morphology observation

The morphologies of coated fabric were investigated using field emission scanning electron microscopy (FESEM, HITACHI, Japan, S-5000).

2.4.2 Solar irradiation

Size of 2cm×2cm as-coated fabrics were cut and 10μL of dyestuff and coffee stain was dropped on fabrics separately at a height of 1cm. The dyestuff (Best color, Maxpoint Co.) used in this study is used for cotton, hemp and rayon fibers specially, purchased at the market. Coffee is a commercial one. In order to make a comparison, the untreated cotton fabric was performed in this investigation. Irradiations of all samples were carried out with solar simulator (SAN-EI ELECTRIC, Japan, XES-40S2-CE), with irradiance of $1000W/m^2$, shown in Fig.1. Both the dyestuff and coffee stained cotton fabrics were irradiated for 24hours. Processes during this period were recorded at 6h, 12h and 24h.

Fig.1 Solar simulator (SAN-EI ELECTRIC, Japan)

3. Results and discussion

3.1 SEM observations

The surface morphology of the obtained cotton sample is shown in Fig.2, which lists the FESEM images of treated cotton fabrics in comparison with untreated sample. In these images, the convolution of cotton fiber can be clearly observed. The cotton fiber showed dense paralleled macrofibrils and a clean and smooth surface at magnification of 7,000 times (Fig.2(a)). 1 wt% ZnO nanoparticles coated cotton fibers is shown in Fig.2(b). ZnO nanoparticles were wrapped spherical around the surface of cotton fibers. Because of lower wt%, fiber surface can also be observed obviously for the 1wt% treated fiber. 3wt% ZnO nanoparticles coated cotton fibers is shown in Fig.2(c). The presence of ZnO is also obviously observed around the cotton fibers. Along the longitudinal direction, compared with 1wt%, densely-wrapped spherical of ZnO nanoparticles are found around cotton fibers.

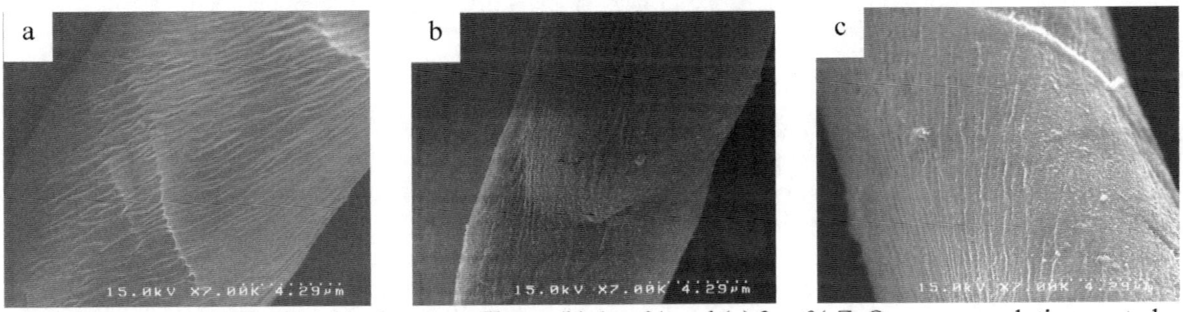

Fig.2 SEM images (×7000) of (a) blank cotton fibers, (b) 1 wt% and (c) 3 wt% ZnO aqueous solution coated cotton fiber.

3.2 Self-cleaning properties of light irradiation

The removal of original substrates such as dey and coffee is a key test to evaluate the self-cleaning properties of ZnO coating on cotton fabrics. Fig.3 shows the decoloration results of dey dropped on untreated and two weight ratios treated fabrics before visible light irradiation (0h), 6h, 12h and 24h irradiation separately. In this figure, for untreated fabric, there has only a little color change before and after sunshine. However, the metal oxide treated fabrics show color fading after 24h irradiation. This phenomenon does show the photocatalysis effect of ZnO. Moreover, it can also be observed that the decoloration of dye in ZnO nanoparticle of 3wt% was more significant after 24h of irradiation in comparison with both the 1wt% and untreated fabrics. It can be concluded that the self-cleaning property was determined by the weight ratio of ZnO nanoparticles. From the SEM images of Fig.2(b), there is some cellulose fibers unwrapped with ZnO nanoparticles at lower weight ratio, which weakened the photocatalysis effect.

Moreover, focus on images at 0h in Fig.3, the shape of dey stained on untreated and treated fabrics are different, and wetting area of untreated cotton fabric is different from treated fabrics. Liquid transfer on fabrics needs to be investigated in the future.

	0h	6h	12h	24h
Untreated				

Fig.3 The decoloration results of dey stains dropped on untreated cotton fabric, 1wt% ZnO treated fabric and 3wt% ZnO treated fabric before and after 6h, 12h and 24h solar irradiation.

As the purpose of metal oxide semiconductor utilization is to degradation organic substrate during daily life, it is necessary to investigate the self-cleaning property of ZnO on organic compound, such as coffee stains. Fig.4 shows the decoloration results of coffee stains dropped on untreated, 1wt% and 3wt% ZnO treated cotton fabrics. Focusing on the untreated fabric in Fig.4(a) and Fig.4(b), there is still residual coffee stains on fabric. However, for cotton fabrics (c), (d) and (e), (f) in Fig.4 treated with ZnO nanoparticles, the coffee stains seem to be unobvious, especially 3wt% treated fabric (e), (f) show a big chromatic aberration, of which the coffee stain can be difficult to be perceived.

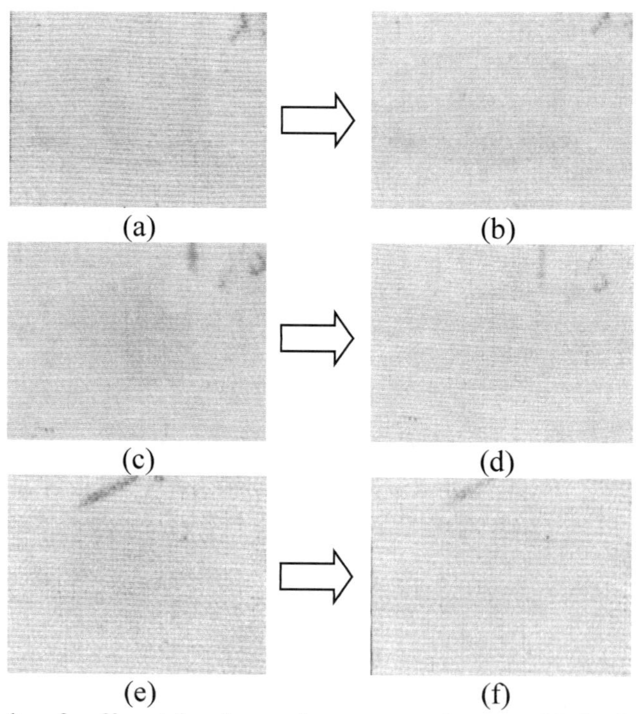

Fig.4 The decoloration results of coffee stains dropped on untreated cotton fabric, 1wt% ZnO treated fabric and 3wt% ZnO treated fabric before and after 24h solar irradiation. ((a), (c), (e) represent untreated, 1wt% and 3wt% ZnO treated fabrics before irradiation, (b), (d), (f) represent untreated, 1wt% and 3wt% ZnO treated fabrics after 24h irradiation separately)

From the above two investigations, it can be concluded that ZnO nanoparticles can be used for degradation of organic compounds on cellulose fibers. With different weight ratios of ZnO aqueous solutions, the self-cleaning properties are different, which can be implied that the weight ratio of coated ZnO nanoparticle is an important effect in this process.

4. Conclusion

In this study, after dip-pad-dry-cure process, 1wt% and 3wt% ZnO nanoparticles were coated on cotton woven fabrics. The photocatalysis of ZnO nanopartcles was investigated, using dye and coffee stained fabrics. The results showed that ZnO treated dey and coffee stained fabrics had self-cleaning properties on organic material compared with untreated fabrics after 24h sunlight irradiation. Moreover, the visible light self-cleaning performance of 3wt% ZnO coated fabrics was found to be better than 1wt%, implying that the weight ratio of metal oxide was an important parameter influencing the sunlight self-cleaning property.

As ZnO coated fabrics can degrade organic compound with irradiation of sunlight, which is a kind of clean, abundance energy in nature, textiles can be applied for various conditions, such as desert exploration and so on. Moreover, if the self-cleaning can be used for fabrics in the future, it will be another new usage of sunlight and domestic water can be saved effectively.

References

[1] Mekhilef S, Saidur R, Safari A. A review on solar energy use in industries. Renew Sust Energ Rev. 2011; 15: 1777-1790.

[2] Thirugnanasambandam M, Lniyan S, Goic R. A review of solar thermal technologies. Renew Sust Energ Rev. 2010; 14: 312-322.

[3] Parida B, Iniyan S, Goic R. A review of solar photovoltaic technologies. Renew Sust Energ Rev. 2011; 15: 1625-1636.

[4] Lee YH, Kim JS, Noh J, et al. Wearable textile battery rechargeable by solar energy. Nano Lett. 2013; 13: 5753-5761.

[5] Fujishima A, Honda K. Electrochemical photolysis of water at a semiconductor electrode. Nature. 1972; 238: 37-38.

[6] Fujishima A, Zhang XT. Titanium dioxide photocatalysis: Present situation and future approaches. C R Chim. 2006; 9: 750-760.

[7] Hashimito K, Irie H, Fujishima A. TiO2 photocatalysis: a historical overview and future prospects. Jpn J Appl Phys. 2005; 44: 8269-8285.

[8] You HC, Lin YH. Investigation of the sol-gel method on the flexible ZnO device. Int J Electrochem Sci. 2012; 7: 9085-9094.

[9] Karimi L, Mirjalili M, Yazdanshenas ME, Nazari A. Effect of nano TiO2 on self-cleaning property of cross-linking cotton fabric with succinic acid under UV irradiation. Photochem Photobiol. 2010; 86: 1030-1037.

[10] Mills A, Hunte SL. An overview of semiconductor photocatalysis. J Photochem Photobiol, A-Chem. 1997; 108: 1-35.

[11] Moore JC, Louder R, Thompson CV. Photocatalytic activity and stability of porous polycrystalline ZnO thin-film grown via a two-step thermal oxidation process. Coatings. 2014; 4: 651-669.

[12] Carp O, Huisman CL, Reller A. Photoinduced reactivity of titanium dioxide. Prog. Solid State Chem. 2004; 32: 33-177.

[13] Fujishima A, Rao TN, Tryk DA. Titanium dioxide photocatalysis. J. Photochem. Photobiol. C-Photochem Rev. 2000; 1: 1-21

[14] Xu J, Ao Y, Fu D, Yuan C. Low-temperature preparation of anatase titania-coated magnetite. J Phys Chem Solids. 2008; 69: 1980-1984.

[15] Qi KH, Daoud WA, Xin JH, Mak CL, Tang W and Cheung WP. Self-cleaning cotton. J Mater Chem. 2006; 16: 4567-4574.

[16] Wang RH, Wang XW, Xin JH. Advanced visible-light-driven self-cleaning cotton by Au/TiO2/SiO2 photocatalysts. ACS Appl Mater Interfaces. 2010; 2: 82-85.

[17] Zhao Q, Xie Y, Zhang Z, Bai X. Size-selective synthesis of zinc sulfide hierarchical structures and their photocatalytic activity. Cryst Growth Des. 2007; 7: 153-158.
[18] Moore JC, Kenny SM, Baird CS, Morkoc H, Baski AA. Electronic behavior of the Zn- and O-polar ZnO surfaces studied using conductive atomic force microscopy. J Appl Phys. 2009; 105: 116102.
[19] Hernandez-Alonso MD, Fresno F, Suarez S, Coronado JM. Development of alternative photocatalysts to TiO2: challenges and opportunities. Energy Environ Sci. 2009; 2: 1231-1257.
[20] Sakthivel S, Neppolian B, Shankar MV, Arabindoo B, Palanichamy M, Murugesan V. Solar photocatalytic degradation of azo dye: comparison of photocatalytic efficiency of ZnO and TiO2. Sol. Energy Mater Sol. Cells. 2003; 77: 65-82.
[21] Zhang X, Qin J, Xue Y, Yu P, Zhang B, Wang L, Liu R. Effect of aspect ratio and surface defects on the photocatalytic activity of ZnO nanorods. Sci Rep. 2014; 4: doi: 10.1038/srep04596.
[22] Kenanakis G, Vernardou D, Katsarakis N. Light-induced self-cleaning properties of ZnO nanowires grown at low temperatures. Appl Catal. A-Gen. 2012; 411-412: 7-14.
[23] Zhu C, Takatera M. Effect of fabric structure and yarn on capillary liquid flow within fabrics. J Fiber Bioeng Inform. 2013; 6: 205-215.
[24] Zhu C, Takatera M. Effects of hydrophobic yarns on liquid migration in woven fabrics. Tex Res J. 2015; 85: 479-486.

Construction and Mechanical Property of High Alignment Polyacrylonitrile Nanofiber Filaments using a Multi-needle Electrospinning Device

Tao-Yan[1], Zhi-Juan Pan[1,2*]

[1] *College of Textile and Clothing Engineering, Soochow University, Suzhou, 215021, China*
[2] *National Engineering Laboratory for Modern Silk, Soochow University, Suzhou, 215123, China*

*Corresponding author's email: zhjpan@suda.edu.cn

Abstract

In this article, continuous PAN nanofiber filaments were obtained for some hours by a special multi-needle electrospinning method. The influence of solution flow rate and applied voltage on the diameter of nanofibers, the morphology and the mechanical properties of the filaments were specially studied. The minimum average diameter of nanofibers was 258±51nm at the flow rate and applied voltage were 4.0ml/h and 17kV, respectively. With the increase of the spinning flow rate, the linear density of the filaments increased from 2.417tex to 4.735tex. The overall morphology of filaments first got better, and then worse. The maximum breaking stress and strain of the filaments were 5.96cN/tex and 4.74%, respectively. At the lower or higher flow rate, the mechanical properties suffered, and the beads were more. The preferable flow rates and applied voltages were 3.2mL/h at 16 kV and 4.0 mL/h at 17 kV.

Keywords: Electrospinning; PAN; Nanofiber Filament; Construction; Property

1. Introduction

Polyacrylonitrile(PAN) has a high carbon element content which is a direct precursor used to produce carbon fiber. Recently it was discovered that carbon nanofibr had excellent mechanical properties hence it was extensively studied. The preparation of nanofibers has been widely studied using the straightforward method of electrospinning, which can produce fibers made out of different compositions with nano-scale diameters through the action of electrostatic forces [1]. Current electrospun nanofiber yarn generally involves the deposition of fibers onto a substrate or a liquid bath. Various structures such as uniaxially aligned nanofibers filament, the nanofibrous twisted yarn, continuous aligned nanofiber bundles and short clips nanofiber bundle had been obtained using different mechanical collection devices.

Li et al [2]. demonstrated that continuous uniaxial nanofibers bundle filament made from electrospun nanofibers could be collected on the surface of a static liquid reservoir. The fluffy degree of the as-spun filament was low, while the orientation degree was high. However, the lower yarn production rate hindered its development. Liu et al [3]. prepared bundles through collecting electrospun nanofibers with a flowing water bath followed by stretching them in water. Through the stretching of bundles, the nanofibers diameter reduced to 56% of the original value, and the crystallinity and the crystalline orientation degree of PAN improved by 72% and 100%, respectively.

Wang et al [4] used a grounded needle tip to induce the self-bundling of polymer nanofibers at the beginning of electrospinning process. Through the needle tip guided the PAN yarn formation, and collected by drum winding. The experimental results revealed that the continuity of the yarn and the alignment of the separate nanofibers were enhanced with the increase of the conductivity of polymer solution. Moon et al [5]. obtained the yarn by a certain long of PAN nanofiber bundles twisting. The ultimate strength of these PAN yarns increased with increasing take-up velocity and was maximum at $9.8ms^{-1}$, and then decreased for higher take-up velocity. The strength was as high as 1GP in the optimum stabilization and carbonization condition. But the formation of the yarn was discontinuous through the spinning method .Yan et al [6]. used a pair of rotating metal tubes to collect the aligned electrospun nanofibers, and then twisted them into yarns. Between the metal tubes, a rotating plastic tube was inserted to wind the yarns. The average Young's modulus and strength of PAN nanofibers yarns were 9.18GPa and 180 MPa with twist angle for 35°, respectively. But this method could not be spinning stably for a long time.

Dabirian et al [7]. summarized the values of some parameters for quantitative comparisons. The nanofibers yarn was largely unoriented before the post-treatment, the values of E-modulus and breaking strength increased from 1.66 GPa and 58.08 MPa to 7.51 GPa and 171.84 MPa, respectively. Whereas the relatively lower breaking strain of post-treated yarn was significant surround 8.51%. Jad et al [8]. used a neutral surface to collect the nanofibers. The plate was placed in the middle of electric field, the half became positively charged and the other part charged negatively. The polymer solution coming from charged nozzles moved toward the part of plate with opposite charge. The nanofibers were collected onto this plate, and then twisted by rotating yarn around its axis. The experiment results showed that twists and heat treatment had a significant effect on capillary rise rate coefficient.

Xie et al [9]. collected nanofibers through a funnel and winded using the drum to manufacture yarn. The yarn showed excellent drawing performance, it could be drawn evenly up to 6 times without breaking. Meanwhile, they found that the drawing temperature and force had an effect on the drawing properties, the drawing treatment improved the yarn and nanofibers uniformity, polymer chain orientation within nanofibers, as well as breaking stress and modules of yarn, but declined the diameters of yarns and nanofibers and breaking strain. The highest strength and modules were 362±37 MPa and 9.2±1.4 GPa by 5 times its initial length, respectively.

However, various degrees of alignment of the nanofibers yarns or bundles could be achieved, the approaches invariably suffered from the drawback that all could not be continuous for a long time during the electrospining process and low output. A liquid collector may be a effective method to produce nanofiber yarn rapidly. In the present work, highly aligned electrospun PAN nanofiber filaments were continuously manufactured by a multi-needle electrospinning process. The influence of solution flow rates and applied voltages on the morphology and diameter of electrospun PAN nanofibers were investigated, and the optimal electrospinning conditions for continuously producing precursor PAN nanofiber filaments were finally obtained.

2. Experimental section

2.1 Experimental materials

PAN(Mw=150000) was purchased from Sigma-Aldrich Shanghai Chemical Reagent Co. Ltd.. Peregal O was obtained from Jiangsu Jiafeng Chemical Co. Ltd. N,N-dimethylformamide (DMF, be equal or greater than 99.5% of purity) were supplied by the Sinopharm Group Chemical Reagent Co.Ltd. (Shanghai, China). All chemicals were of analytical grade and were used as received without further purification. An electrospinning solution was prepared by dissolving the PAN powder in the DMF with continuous stirring vigorously using a magnetic stirrer for three hours at room temperature. The concentration of the PAN solution was 8 wt%. The bath was prepared by dissolving 0.5% Peregal O into deionized water at room temperature. Relative humidity and room temperature were controlled at less than 30% and 25±2℃ in the process of electrospinning.

2.2 Electrospinning nanofiber filament setup

A homemade multi-needle liquid bath electrospinning device with eight needles was used to continuously prepare the PAN nanofiber filaments as illustrated in Fig.1. Because of the electrostatic fields on the outer needles were larger than those on the inner needles [10], meanwhile, in order to decline the electromagnetic shielding, the central jet was removed. The distance between adjacent needles having 0.4mm inner diameter was 20mm, The PAN solution was loaded into a syringe, and the flow rate was controlled by the syringe pump (KDS 100, KD Scientific Inc). The spinneret was connected to the anode of high power supply by copper wires, and the cathode was inserted into the bottom of the bath reservoir, the diameter of copper wire was 0.4mm. PAN solution was turned into nanofibers under the high electrostatic field on the surface of the water, the distance between the needle tip and the surface of water was 9cm, then the as-spun nanofibers were initially assembled into a nanofiber filament in the bath with the guidance of a guide glass rod, the filament went through guide needle, heating device and stretching device, successively, during which the nanofiber filament was drafted to be necked, Finally, the as-spun filament was wound on the bobbin. The rotating speed of the winding device was kept constant at 1.75m/min. The length of drying device was 300mm with 350℃. The as-spun nanofiber filament was continuously manufactured by the drawing force caused by the rotating of the filament bobbin.

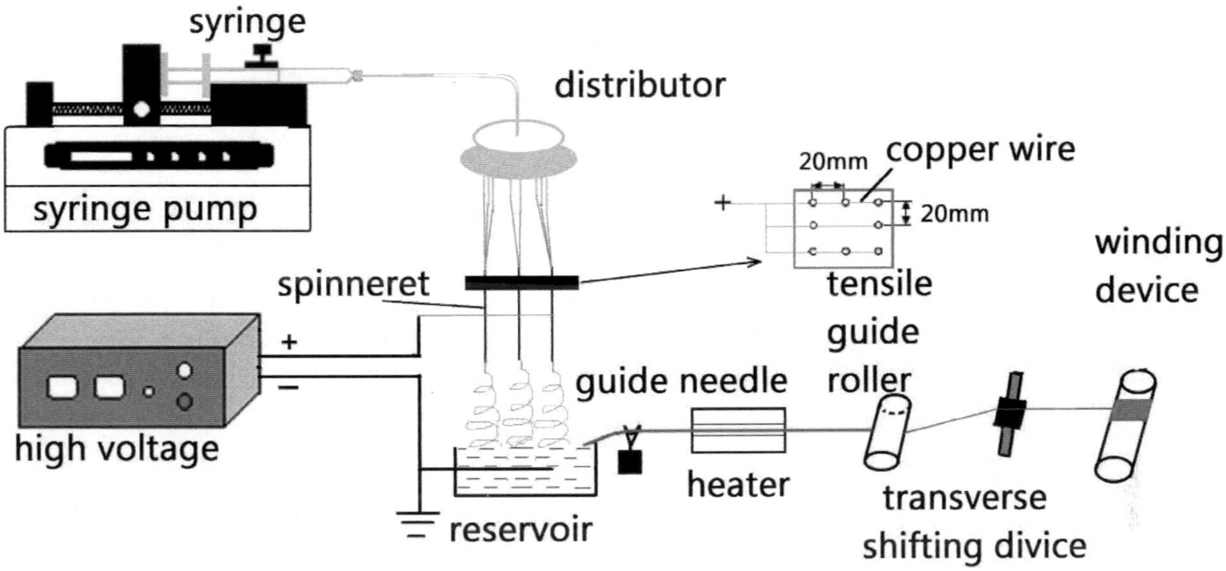

Fig.1 Schematic diagram of the electrospun nanofiber filament device

2.3 Characterization

The morphology of the nanofibers and filaments were examined using a cold field emission scanning electron microscope (SEM) (Hitachi S-4800), the samples were sputtered with a thin layer of gold prior to SEM analysis, and an accelerating voltage of 3kV was applied. The mean diameter of the nanofibers was calculated by averaging 100 measurements in SEM images of different areas of the filament, measured by Image Pro Plus 5.0 software.

The degree of orientation of the nanofibers in the filament was measured using a nonwoven fabric orientation analysis system (BD-11). This system was based on curvilinear integral algorithm. It is able to calculate the degree of the orientation of nanofibers from 1° to 180° in the SEM images. The orientation was obtained by averaging 10 SEM images of different areas of the filament, the orientation was calculated from 1° to 180°, the adjacent class interval angle and step value were 5° and 10, respectively. The final degree of orientation was calculated according to Eq.(1):

$$D = \sum_{\kappa=1}^{10} \frac{\sum A_\kappa}{10} \times 100\ \% \tag{1}$$

Where D was the degree of orientation of the nanofibers, ΣA1 was the sum value of 85° to 95°.

The breaking stress and strain of the nanofiber filaments were measured using an Instron 3365, the filaments specimens were maintained under standard conditions of (20±1)℃ and 65% relative humidity for 24h, ten samples (length = 1000mm) were randomly selected from each tested filament, the clamping distance was set to a fixed length of 50mm, the croosshead speed and the initial tension were 50mm/min and 0.1cN, respectively. The breaking strength and elongation resolution was set at 0.01 cN and 0.01 mm, respectively.

3. Results and discussion

PAN nanofiber filaments were fabricated using different operation voltages and spinning flow rates in this electrospinning process, and their effects were systematically discussed in this paper. For comparability, other conditions were fixed including the distance between the needle tips and the surface of the water and winding speed. The spinning solutions was kept the same at 8 wt%. The flow rates were 2.4ml/h to 4.8ml/h and interval of 0.8ml/h, the voltages were 15kV to 18kV and interval of 1kV.

3.1 Construction of Electrospun Nanofibers

Fig.2 shows that the morphology and the surface structures of PAN nanofiber filaments, the appearance of the filament was smooth. The process of electrospinning could last for a few hours. The electrospun PAN nanofibers within the filament had a rough surface as shown in Fig.2 (c), and even had porous structure on the surface of thicker fibers and beads as shown in Fig.2 (d) and (e). The formation of a rough surface and pores were ascribed to the vapor pressure of the solvent, i.e. the solvent evaporated from the nanofibers during electrospinning before the nanofibers were collected in the bath, but due to DMF was difficult to volatilize, there was still a part of solvent remaining in the nanofibers, and when this nanofibers were immersed in the water, the rough and porous structure was preserved after drying.

Fig.2 Structures of electrospun PAN nanofiber filament(a)Photograph (b)SEM of surface (c)as-spun nanofibers (d) thick fiber (e) beads

Fig.3 shows SEM images of PAN nanofiber filaments with different spinning flow rates at the same take-up velocity of 1.75m/min, spinning distance of 9cm and voltage of 15kv. The filaments thinned with the increase of spinning flow rates, caused by more solution extruded from the spinning jets. With the increase of the flow rate, the continuity of spinning increased, but the degree of orientation of the nanofibers along the axis of the filament decreased gradually from 72.67% to 53.17%, which was due to that the nanofibers in the filaments were unable to straighten enough and presented to be curved under the higher flow rate, limiting the orientation along the axis of the filaments. However, when the flow rate was overhigh (the flow rate of 4.8ml/h), the excessive solution could not be fully stretched and oriented by the lower electric field power as shown in Fig.3d, thus the beads began to increase, and the average diameter and diameter deviation of nanofibers also increased. Meanwhile, the producing process of nanofiber filament became difficult and poor continuity.

Fig.3 SEM images of PAN nanofiber filaments with different spinning flow rates(a)2.4ml/h (b)3.2ml/h (c)4.0ml/h (d)4.8ml/h

Fig.4 SEM images of PAN nanofiber filament with different spinning voltages (a)15kV (b)16kV (c)17kV (d)18kV

Fig.4 shows SEM images of PAN nanofiber filaments with different spinning voltages, at the same take-up velocity of 1.75m/min, spinning distance of 9cm and spinning flow rate of 4.0ml/h. As the electric field intensity increased gradually with the increase of applied voltages, the beads produced at a lower voltage, as shown in Fig. 4a, due to that the electric field force was not sufficient enough to overcome the surface tension of the fluid, and the excessive spinning solution (4.0ml/h) at even higher flow rate could not be fully stretched and oriented by the lower electric field power. Increasing the applied voltage made the Taylor cones more instable, thus worsening the alignment of nanofibers in yarns, the maximum degree of orientation of the nanofibers along the axis of the filament was as far as 76.52% with the voltage of 17kV. When the applied voltage was above 18kV, nanofibers were more dispersed in the surface of the bath, making which unfavorable to be drafted, meanwhile, the time of continuous electrospinning became shorter. However, it required a higher voltage with the increase of the flow rate to get a stable electrospinning condition.

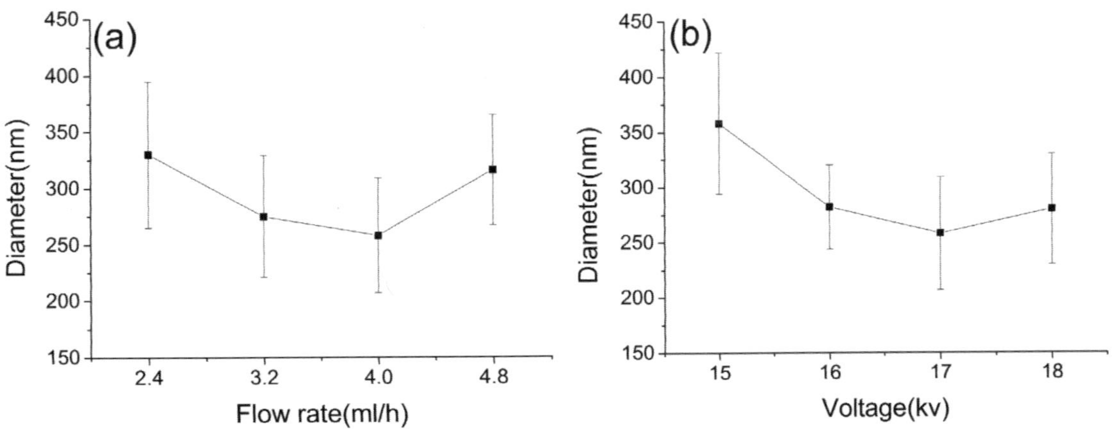

Fig.5 Changes of the nanofiber diameters (a) different flow rates with 17kV (b) different voltages with 4.0ml/h

Fig.5 shows that the changes of the nanofiber diameters at different voltages and flow rates. The nanofibers were found to have smaller average diameter and diameter distribution when the flow rate was 4.0ml/h at 17kV, the minimum average diameter of nanofibers was 258nm when the flow rate was 4.0ml/h, the whole spinning condition was also better.

Fig.5a shows that changes of the nanofiber diameter at the constant applied voltage of 17kV. The average diameter of nanofibers decreased initially, and then not increased until more than 4ml/h. The nanofiber diameter was determined by a pair of competitive factors: flow rate and electrostatic field force. Before the flow rate reached 4ml/h, the solution gradually became more with the increase of flow rate, the number of charges increased in the solution, the drafting force caused by electrostatic field increased, thus the diameter decreased. Meanwhile, the diameter of jet increased with increasing the flow rate, thus the nanofiber diameter increased. This phenomenon became more overwhelming when the flow rate was more than 4ml/h, which led to the diameter increased.

Fig.5b shows that changes of the nanofiber diameter with the constant flow rates of 4.0ml/h. When the spinning voltage increased, the nanofiber diameter decreased initially until 17kV, and then increased. The spinning solution could not be fully stretched with the lower electrostatic field force, but the electrostatic filed force increased with the increase of voltage, thus the nanofiber diameter decreased. However, when the voltage was higher than 17kV, the Taylor cone became instable, and there could be more solution to be drew out from the jets, thus the diameter increased.

3.2 Effect of voltage and flow rate on the mechanical properties of PAN nanofiber filaments

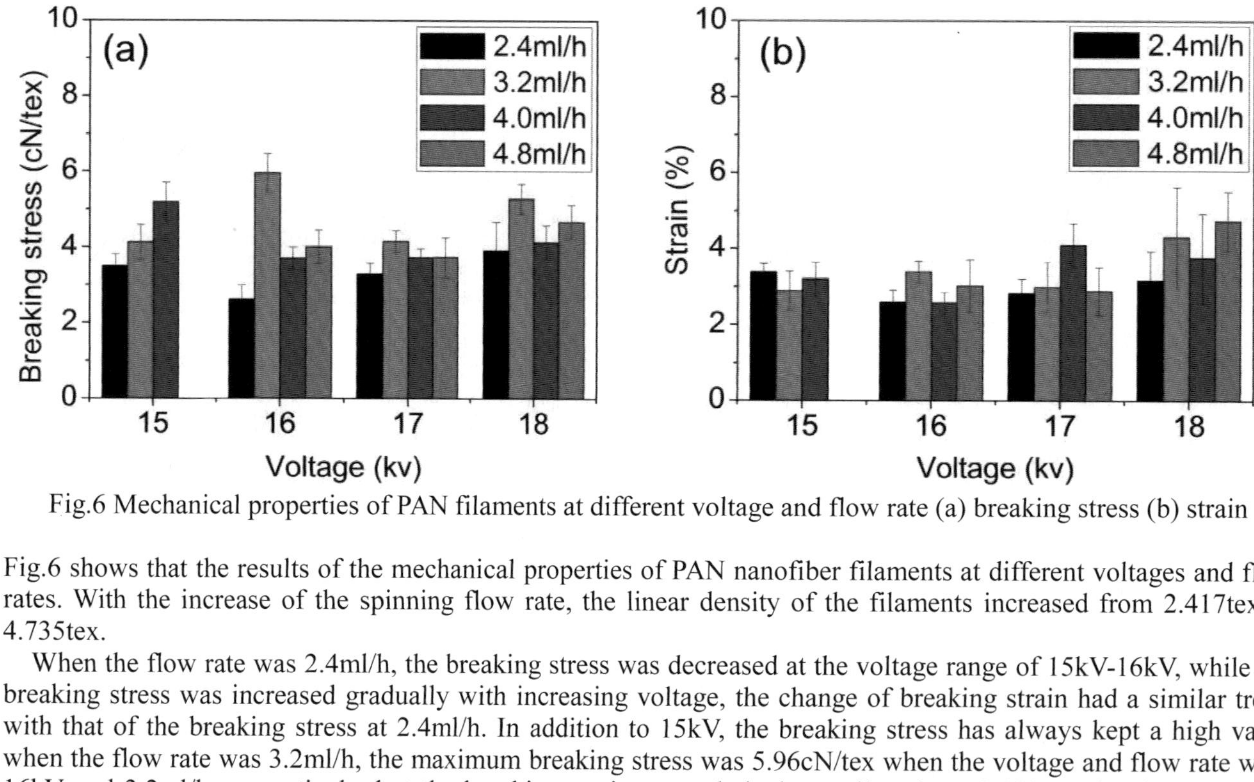

Fig.6 Mechanical properties of PAN filaments at different voltage and flow rate (a) breaking stress (b) strain

Fig.6 shows that the results of the mechanical properties of PAN nanofiber filaments at different voltages and flow rates. With the increase of the spinning flow rate, the linear density of the filaments increased from 2.417tex to 4.735tex.

When the flow rate was 2.4ml/h, the breaking stress was decreased at the voltage range of 15kV-16kV, while the breaking stress was increased gradually with increasing voltage, the change of breaking strain had a similar trend with that of the breaking stress at 2.4ml/h. In addition to 15kV, the breaking stress has always kept a high value when the flow rate was 3.2ml/h, the maximum breaking stress was 5.96cN/tex when the voltage and flow rate were 16kV and 3.2ml/h, respectively, but the breaking strain was relatively small, only at 3.39% under the condition. There was an unconspicuous trend of the breaking strain under the fixed flow rate.

The breaking stress had an unconspicuous trend with the increase of flow rate under the fixed voltage, the breaking strain kept a similar trend to the breaking stress under the constant voltages. However, the strain could maintain larger and the smaller deviation when the values of voltage and flow rate were 17kV and 4.0ml/h, respectively.

Due to the common influence of voltage and flow rate, the breaking stress and strain as the status of fluctuation, Hence as seen from Fig.5 and Fig.6, the perfect flow rates and applied voltages were 3.2mL/h at 16 kV and 4.0 mL/h at 17 kV. Meanwhile, the beads was less and the continuity of spinning was better under the two conditions.

4. Conclusion

The aim of this study was to optimize the applied voltage and flow rate of the multi-needle electrospinning system to produce PAN nanofiber filaments, the continuous time was as long as a few hours. The preparation of highly orientation filaments from electrospun PAN nanofibers were continuously manufactured by the multi-needle electrospinning process, the mechanical properties were investigated. Due to there was not enough solution to be stretched, the beads became more when the flow rate was 2.4ml/h, resulting in a decline in the overall mechanical properties of the filaments. As the excessive solution cannot be fully stretched at 4.8ml/h, the beads began to increase, the producing process of nanofiber filament became difficult and poor continuity. When the voltage Taylor cone instability increases nanofiber entanglement and makes alignment of nanofibers more difficult, especially nanofibers were more dispersed in the surface of the bath, and unfavorable to draft. The perfect flow rates and applied voltages were 3.2mL/h at 16 kV and 4.0 mL/h at 17 kV, the minimum average diameter of nanofibers was 258nm, the maximum breaking stress and breaking strain of the filaments were 5.96cN/tex and 4.74%, respectively. However, it required a higher voltage with the increase of the flow rate under a stable condition.

References

[1] Bhardwaj N, Kundu SC. Electrospinning: a fascinating fiber fabrication technique[J]. Biotechnology advances: 2010; 28(3): 325-347.

[2] Li J, Tian L, Pan N, et al. Mechanical and electrical properties of the PA6/SWNTs nanofiber yarn by electrospinning[J]. Polymer Engineering & Science: 2014; 54(7): 1618-1624.

[3] Liu J, Chen G, Gao H, et al. Structure and thermo-chemical properties of continuous bundles of aligned and stretched electrospun polyacrylonitrile precursor nanofibers collected in a flowing water bath[J]. Carbon, 2012; 50(3): 1262-1270.

[4] Wang X, Zhang K, Zhu M, et al. Continuous polymer nanofiber yarns prepared by self-bundling electrospinning method[J]. Polymer: 2008; 49(11): 2755-2761.

[5] Moon SC, Farris RJ. Strong electrospun nanometer-diameter polyacrylonitrile carbon fiber yarns[J]. Carbon: 2009; 47(12): 2829-2839.

[6] Yan H, Liu L, Zhang Z. Continually fabricating staple yarns with aligned electrospun polyacrylonitrile nanofibers[J]. Materials Letters: 2011; 65(15): 2419-2421.

[7] Dabirian F, Hosseini Y, Ravandi SAH. Manipulation of the electric field of electrospinning system to produce polyacrylonitrile nanofiber yarn[J]. Journal of the Textile Institute: 2007; 98(3): 237-241.

[8] Jad MSM, Ravandi SAH, Tavanai H, et al. Wicking phenomenon in polyacrylonitrile nanofiber yarn[J]. Fibers and Polymers: 2011; 12(6): 801-807.

[9] Xie Z, Niu H, Lin T. Continuous polyacrylonitrile nanofiber yarns: preparation and dry-drawing treatment for carbon nanofiber production[J]. RSC Advances: 2015; 5(20): 15147-15153.

[10] Liu Y, Guo L. Homogeneous field intensity control during multi-needle electrospinning via finite element analysis and simulation[J]. Journal of nanoscience and nanotechnology: 2013; 13(2): 843-847.

Granphene Fibre Review: from Spinning Methodologies to Applications

Yang-Yang Fan[1], Yi Li[2*], Xu-Qing Liu[2]

[1,2]*School of Material, Univeristy of Manchester, Oxford road, Manchester, M13 9PL, UK*

*Corresponding author's email: henry.yili@manchester.ac.uk

Abstract

Inspired by macroscopic high dimensional graphene materials, grahene fibres now have been fabricated in a range of methods, including drawing, solution spinning, thermal process, electrospinning, etc. Among them, most attractive and widely used methodology is wet spinning liquid crystal graphene oxide. Properties of different kinds of graphene fibres were discussed. Both neat graphene fibres and hybrid graphene fibres show remarkable mechanical and electrical properties, actuating properties, flexibility and weavability. Multifunctional properties provide graphene fibres potential in functional textiles, flexible and wearable sensors, supercapacitors, electrodes of supercapacitors, energy devices, and actuators etc. It is prospected that properties of graphene fibres will be further optimized, and their application will be more realistic and well developed.

Keywords: graphene fibres; spinning; mechanical properties; electrical properties; application

1. Introduction

1.1 Introduction to graphene

Graphene firstly found in 2014, by A.K. Geim and K.S. Novoselov [1], has been widely seen as a honeycomb structure of two-dimensional (2D) atomic crystals with high crystal quality and macroscopic continuity. Graphene can be seen as a single layer exfoliated from multiple layer graphite, a plat structure rolled out from single-wall carbon nanotubes or fullerene molecule [2,3]. The most attractive factor of graphene is its extraordinary properties, with an ideal tensile strength of 130 GPa, elastic modulus of 1.0 TPa, remarkable flexibility, electrical conductivity of 10^8 S/m, thermal conductivity of 5000 $Wm^{-1}K^{-1}$, transmittance of almost 97% and high temperature, chemical resistance [4]. Owing to these desirable properties, graphene was not only not only made into a neat structure, but also hybrided with other material.

1.2 Graphene based materials

1.2.1 Two dimensional (2D) and three dimansional (3D) graphene materials

First graphene films were prepared by continuous mechanical exfoliation of small mesas of highly oriented pyrolytic graphite. After this simple but reliable method, FLG films were found up to 10 μm in size. However, clearly, unrealistic and unfeasible problem, approach of making single layer graphene should be optimized or substituted. After micromechanical cleavage, several methods were developed and have been widely applied in making experimental graphene, e.g. acid oxidation, chemical vapour deposition etc [5,6]. Graphene oxide formed in some cases will then be reduced using thermal or chemical methods [7,8].

Graphene has demonstrated excellent mechanical, electrical, thermal, and optical properties [4,6-11]. It is assumed that this will bring graphene great potential in many high technical fields both experimental and practical because assembling 2D graphene nanosheets into 3D graphene structure is the most realistic and efficient approach to utilise its extraordinary properties [9]. Therefore, the most important issue of graphene is to transform and fabricate micro scale graphene into macroscopic structures and to make it functional for practical applications.

The assembly of graphene into macroscopic three-dimensional (3D) structures has been attracting intensive interest [10-12]. Till now, three-dimensional graphene aerogel, hydrogel, foams, sponges [9,13-17] and two-dimensional graphene film/ member/ paper/ nanosheets [5,18,19].

For 3D graphene structure, it is illustrated that 3D graphene achievements have been made in graphene 3D structure on their use in electrode of supercapacitor, supercapacitors, conductors with flexibility and biosensors etc. [13,20]

Other improvements in high-performance 2D films, paper, membrane etc. have spurred research efforts on their application in electric double-layer capacitors and pseudo-capacitors stretchable electrodes, biosensor, biomedical applications due to their acceptable mechanical strength, electrical conductivity, signal-to-noise ratio and potentially biocompatible[5,9,18,19].

1.2.2 One dimensional (1D) graphene fibres

In contrast, among macro scale graphene material, 1D graphene fibre was believed to emerge high electrical conduction, ultrahigh mechanical strength after stretching, functional possibility, unique flexibility and weavability compared with the other two kinds of graphene macroscopic materials. Hence, graphene fibre has outstanding potential in wearable smart textile, biosensor etc. Provided that structure of graphene fibre is continuous with 2D carbon monocrystals stacked together along fibre axial direction, graphene fibre is also designed with a purpose of overcoming the shortages of conventional carbon fibre PAN, e.g. vast grain boundaries and polycrystalline nature [21].

Nowadays, one-dimensional graphene fibres have already been prepared and confirmed with multiple superior properties compared with conventional large scale material. As a novel material, graphene fibre is fabricated to fill in gaps of one-dimensional graphene material, and to prepare multifunctional fibres with proper qualities in some specific areas.

In this review, we concluded current methodology with principle of fabricating macroscopic multifunctional graphene fibres. Thermal, electrical, mechanical properties of neat graphene fibres from different methodology were compared and discuessed. Strategies of doped graphene based fibres were also introduced. Finally, the current application of graphene fibres and their potential usage were rendered to lead to a positive prospect in fibre production field.

2. Graphene Fibres Spinning Methodology

In contrast to high-dimensional material have already been made such as areogels and membranes, graphene-based fibres were demonstrated to have remarkable mechanical, electrical properties alone fibre axis direction, and in some conditions can be knottable and weavable [22]. To fabricate a 2D graphene structure, common methods of polymer spinning, e.g. melt spinning, dry spinning, wet spinning, dry jet-wet spinning and electrospinning were taken into account.

Graphene, well-known as an extremely thermal stable crystalline material, has a melting temperature T_m around 4900 K (about 250 K higher than that of graphite, probably the highest in nature) [23]. In this respect, it is unrealistic to make neither neat graphene fibres nor graphene based composite fibres using melt spinning approaches.

In dry spinning, spinnability is influenced by both solution rheological properties and mass transfer rates. Solution reheological property is commonly known as one of the key factors to determine spinnability of dry spinning [24]. However, lacking of entanglement of 2D graphene leads to a relatively low solution reheological property. It highly restricts graphene fibre fabrication using continuous dry spinning. According to these natures of graphene, self-assemble graphene aqueous together with other methods were taken into consideration next section.

2.1 Drawing graphene

First neat graphene fibres were attempted to by drawing graphene macrosheets [25]. In this novel yet simple method, 2D graphene film prepared early by chemical vapor deposition was self-assembled into a 1D fibre structure. The short fibre was then dried showing a porous and crumpled morphology. However, only van der Waals force existed between graphene sheets after being dried indicating inferior mechanical properties.

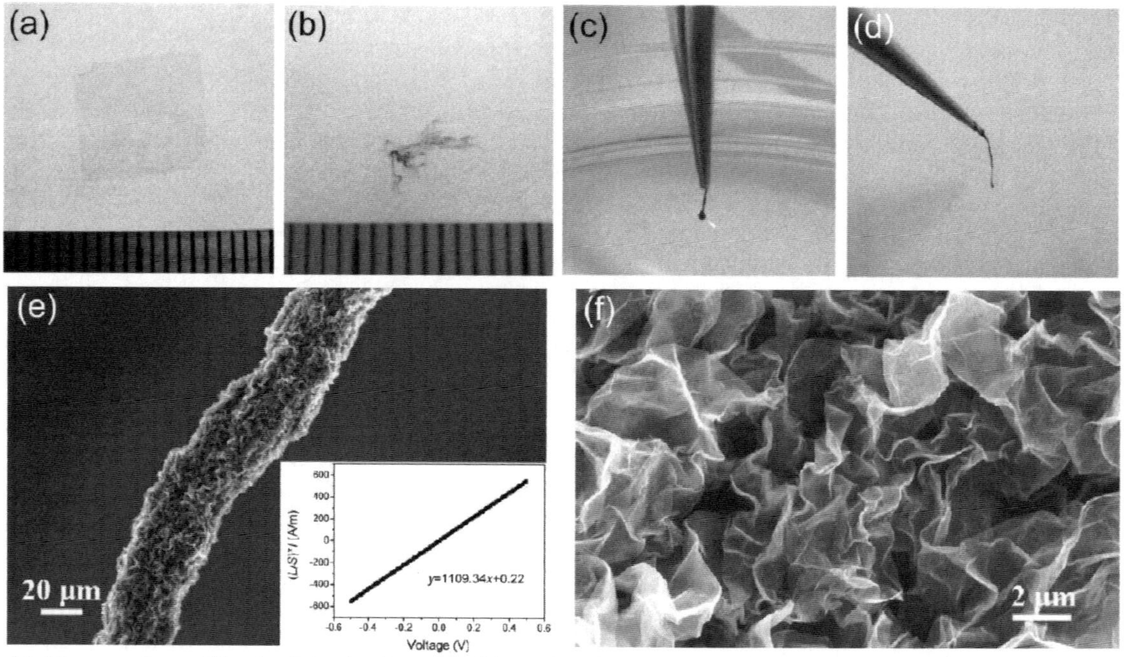

Fig.1 (a) Flat graphene film in deionized water. (b) Twisted and wrapped graphene film in ethanol. (c) Graphene fibre formed by drawing out of ethanol. (d) Graphene fibres with twisted structure. (e) SEM image of single graphene fibre with rough surface. (f) SEM images of single graphene fibre, showing porous structure

As the porous structure giving a potential in electrode material, manganese dioxide were added into graphene substrate. Following electrochemical characterisation confirmed that with incorporating manganese dioxide, graphene based fibre showed a tremendous improvement in electrochemical properties. Ever since, many other approaches of preparing macro scaled graphene were developed.

2.2 Solution spinning

2.2.1 Liquid crystal graphene oxide

Liquid crystal graphene (LC GO) is the essence of wet spinning. LC GO phase was firstly observed by Natnael et al when graphene was dissolved in chlorosulphonic acid. The superacid contributed to exfoliation and protonation of graphene, as a consequence of which induced repulsion between graphene layers. A spontaneous Liquid crystalline graphene phases appeared at high concentration dispersion (20 mg ml^{-1} to 30 mg ml^{-1}). After that, more LC GO were found and investigated in different other solutions [26]. Gao C.'s group firstly reported that soluble, chemically oxidized graphene or graphene oxide sheets can form chiral liquid crystalline (Fig.2) [27]. One of the most important applications of LCs is to make macro scale fibres using wet spinning. Gordon W.' group then investigated the quantitative relation between GO nanosheets size and GO concentration, liquid crystallinity, spinnability etc [28-31].

According to these researches, two key factors are demonstrated to form liquid crystalline graphene: aspect ratio and dispersibility or solubility. Aspect ratio was known as width/thickness ratio. For graphene nanosheets especially, the most critical is to maintain a high width of graphene in solution as unique thickness of graphene. Critical size of GO sheet size of forming LC GO, with a unique thickness of graphene 0.8nm and a concentration of 2.5 mg ml^{-1}, was established. When mean GO sheet size $\geq$ 1.5 µm, nematic phase was completely formed, i.e. LC GO was fully formed and capable to wet spinning. Mean GO sheet size is between 1.5 µm and 0.7 µm brought biphasic dispersions. Only discontinuous spun fibres were observed in coagulation. At mean GO sheet size < 0.7 μm, no nematic phase transition was found, the dispersion GO dope was totally isotropic, thus no processable.

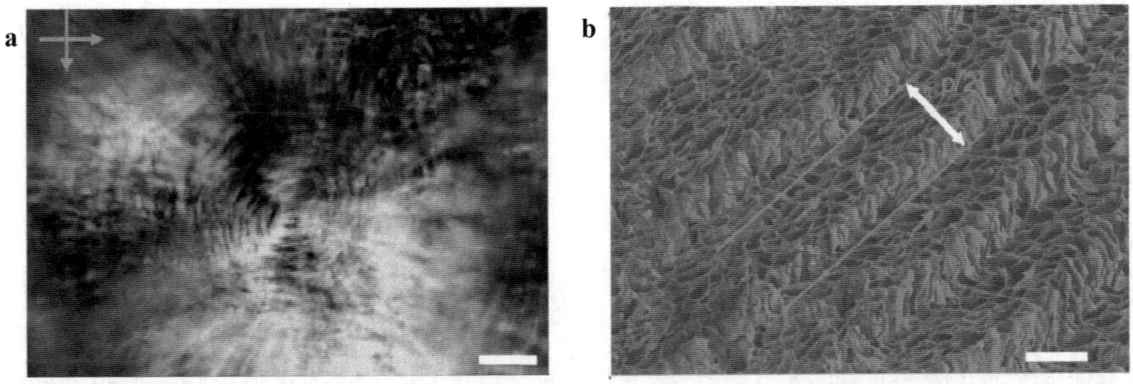

Fig.2 (a) Polarized optical microscopic images between crossed polarizers of GO CLCs in central. (b) SEM images of morphology of LC GO. The crosses are vectors (ns) into the paper and the arrows are alone the paper. (a, b) of LC GO, scale bar 200 μm [30]

Understanding that it is significant to have a sufficient dispersity of graphene, considering its poor interaction with most solvents, challenges remained were solved by using solvents, e.g. superacid or by chemical modifying. The former mentioned before has not been used for wet spinning yet. The latter, however, was commonly applied to form GO increasing dispersibility [21]. It was investigated that the rheological behavior of LC GO showed a non-Newtonian pseudoplastic fluid behavior. Continuous neat GO fibres combining conductive, mechanical properties were successfully spun at a minimum concentration of 0.75 mg ml-1. When Concentration is between 0.25 mg ml-1 and 0.75 mg ml-1 where existed a transition from biphasic to fully nematic phase, only partial discontinuous fibres were produced during wet spinning [28]. Hence the platform of spinning LC GO, containing relationship between GO sheet size, concentration and spinnability in different solvents, was established for carrying on wet spinning [20].

2.2.2 Wet spinning for Graphene fibres

First wet spun neat graphene fibre were developed in 2011 by Gao. C' group [27]. During wet spinning, LC GOs mentioned before, were injected into different coagulation baths to extract solvents and to form gel-like fibres. Gel-like GO fibres were then dried reduced into rGO fibres. As-spun fibres were slightly stretched by rotation movement in a rotating coagulation bath or drawing in a collection unit to maintain basic fibre structures during wet spinning (Fig.3) [22].

The formation of GO fibres was mainly based on orientation of LC GO in aqueous especially under sheer stress during uniaxial flow process. Strong attractions between GO sheets containing van der Waals and hydrogen bonds contributed to reinforce both strength along fibre axial direction and radial direction.

Coagulation bath and solvent are always considered to be key factors to wet spinning. The diffusional interchange between coagulation and dope can form solid or gel-like fibre directly, thus graphene fibre precipitations and crystalline are dramatically affected by choose of coagulation and dope. A series of coagulation bath formulations have been used as solvents, e.g. KOH, $CuCl_2$ or $CaCl_2$ in water, ethanol or isopropanol to investigate and evaluate fibre spinnability and properties in different coagulation, seen in Fig.4 [32]. After partly precipitation, as-spun graphene fibres were slightly drawn by a rotating coagulation or a collection unit to maintain a basic shape and further aligned to give better properties along fibre axis.

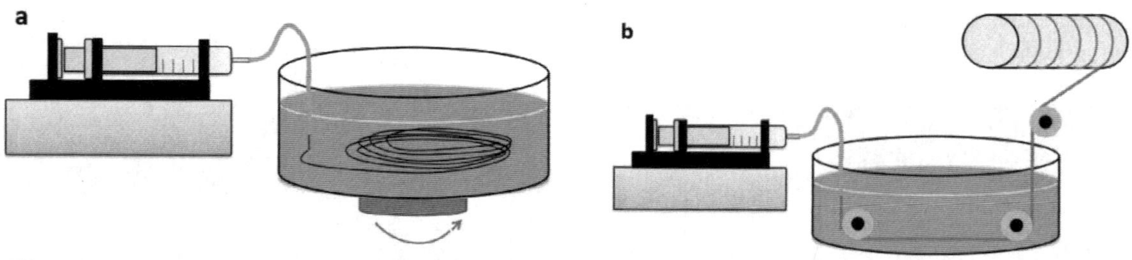

Fig.3 Sketch of wet spinning LC GO into graphene fibres. (a) Wet spinning with a rotating collector. (b) Wet spinning with a collector unit

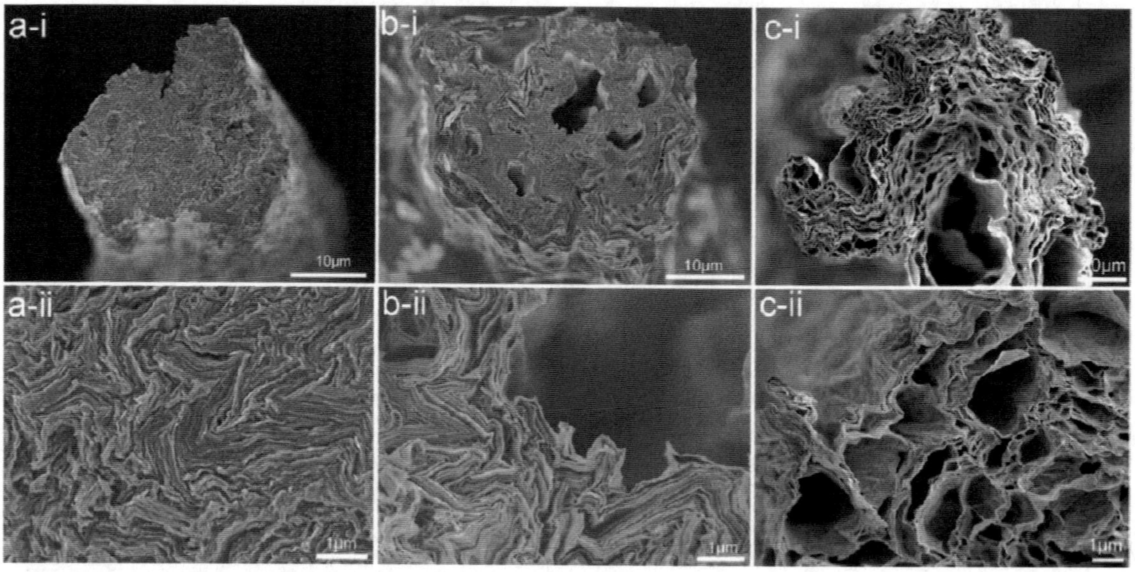

Fig.4 The evolution of structure in different coagulation baths was compared. From (a) to (c), the architectures of graphene fibres was becoming more porous and lower dense. (a-i) (a-ii) water was used as coagulation bath leading to a slow diffusion of water from LC GO. This resulted in a quite dense structure. (b-i) (b-ii) Acetone coagulation bath resuled in the higher rate of water extraction as a result of the difference in imbibition rate, consequently leaded to higher rate of solidification and porous fibre structure. (c-i) (c-ii) In acidic condition (pH ~3), diffusion of LC GO dopants further fasted bringing about the difference in imbibition rate, causing much higher water extraction rate, as a consequent of which resulting in an architecture with highest porosity [32]

2.2.3 Dry-jet wet spinning

Apart from impact from alignment of LC GO, structure of GO fibres was also dominated by sheer stress formed by drawing in the spinneret and stretching during rotating or rolling. Those two factors enhancing alignment of GO sheet finally leaded to transcendental mechanical properties. Although it is greatly improved by sheer stress and gel drawing, orientation of fibres was limited by cross-linked structure in coagulation bath resulted from diffusion of Ca^{2+} within GO sheets. To settle this problem, a widely used methods known as dry-jet wet spinning was applied with an aim of reducing effect of cross-links. LC GO was directly drawn by air gap after jetted from spinneret following a stage of gel-forming and cross-links forming (Fig.5).

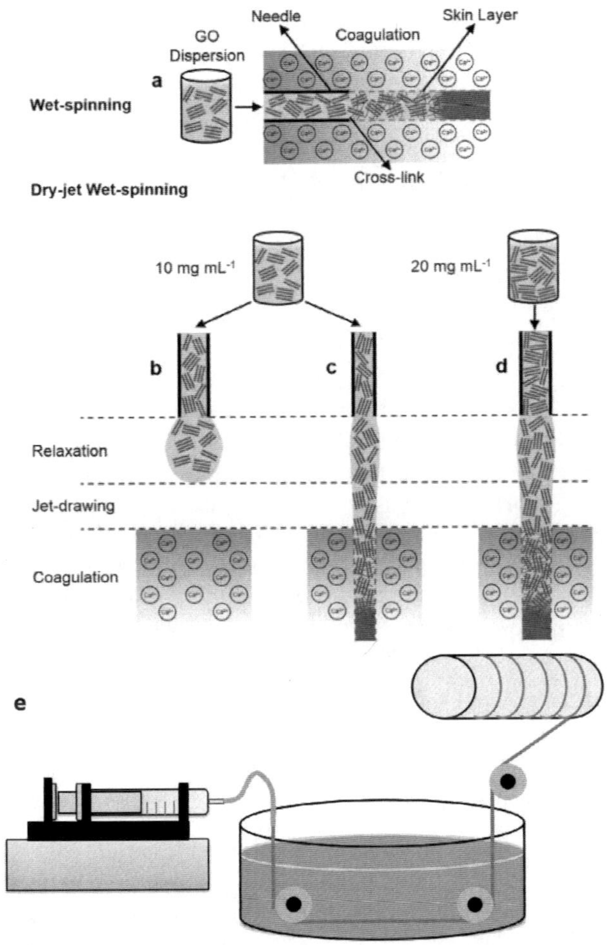

Fig.5 Schematics of wet spinning, dry-jet wet spinning stages and process [33]. (a) During wet-spinning, a skin formed at the early stage when fibres went into coagulation bath. Cross-linking structure was then formed internally inside graphene oxide fibres. (b) Lower concentration and higher diamter form a droplet at the spinneret tip. (c) Decreased needle diameter or (d) Increased GO concentration lead to continuous dry-jet wet spinning. (c,d) fibres were drawn at the early stage and cross linked in coagulation bath. (e) Wet spinning process

As the first solvent which generated LC GO, in chlorosulfonic acid as is previously stated, LC GO could be formed in a relatively high concentration of GO. Graphene oxide nanoribbons (around 4μm in length) instead of graphene nanosheets as basic fibre material successfully produced with a smooth fibre surface. Although it was reported that air gap may cause a droplet as a consequence of which would interrupt fibre spinning, with adding graphene oxide ribbons, hundreds meters of graphene fibre with smooth surface were successfully fabricated [33,34].

2.3 Hydrothermal process

As confirmed before, dry spinning of graphene nanosheets was not a practical choice, in spite of this, hydrothermal process, seen as analogy for dry spinning, was also developed to manufacture graphene fibre. Strategy of hydrothermal process is similar to wet spinning using LC GO while replacing coagulation bath with a sealed high-temperature pipeline. This high temperature environment not only removed solvents from LC GO forming fibres, but also reduced GO into rGO. As both dramatically precipitation from solvent and reduction of GO, the shrunk fibre diameter decreased by 80% [35].

As rapidly jetted into high temperature tub, for GO dope, there is not strict limit in forming liquid crystal phase. This one-step feasible process simplified the methodology of graphene fibres. Despite showing a lack ofsheer stressand tension compared with wet spinning, hydrothermal process shrunk behavior provided surface tension force and capillary force during heating reduction from gel-state to solid-state. In this case, as is reported the ratio of G-band intensity (parallel versus perpendicular) increased from 1:1 to 6:1 demonstrating that graphene nanosheets were aligned alone fibre direction [35]. Fe_3O_4 and TiO_2 nanoparticles were mixed into graphene suspension and fibres to make graphene fibre composite, rendering sensitive magnetic response and photocurrent response respectively [35].

2.4 Electrospinning

However, as is highly recognised as an efficient nano technology of nanofibre process, graphene based nanofibre composites were already fabricated [36-38]. Nanofibres mat of graphene/ polymer nanocomposites was electrospun applying an optical element in fibre lasers, seen in Fig.6a [38]. The fibre mat rendered wideband saturable absorbance for laser pulse shaping, and attain a larger modulation depth and smaller nonsaturable loss than single-walled carbon nanotubes. Except of graphene based material made by electrospinning, GO was also used for modifying electrospun nanofibres to make fibres with versatility [37,38]. Graphene was assembled directly onto the surface of electrospun polyamide 66 nanofibres (Fig.6b). Graphene on the surface is confirmed not to form a uniform coating, but an exudative network leading to a low surface coverage. Hence, conductivity and transmittance were measured to be promoted [36].

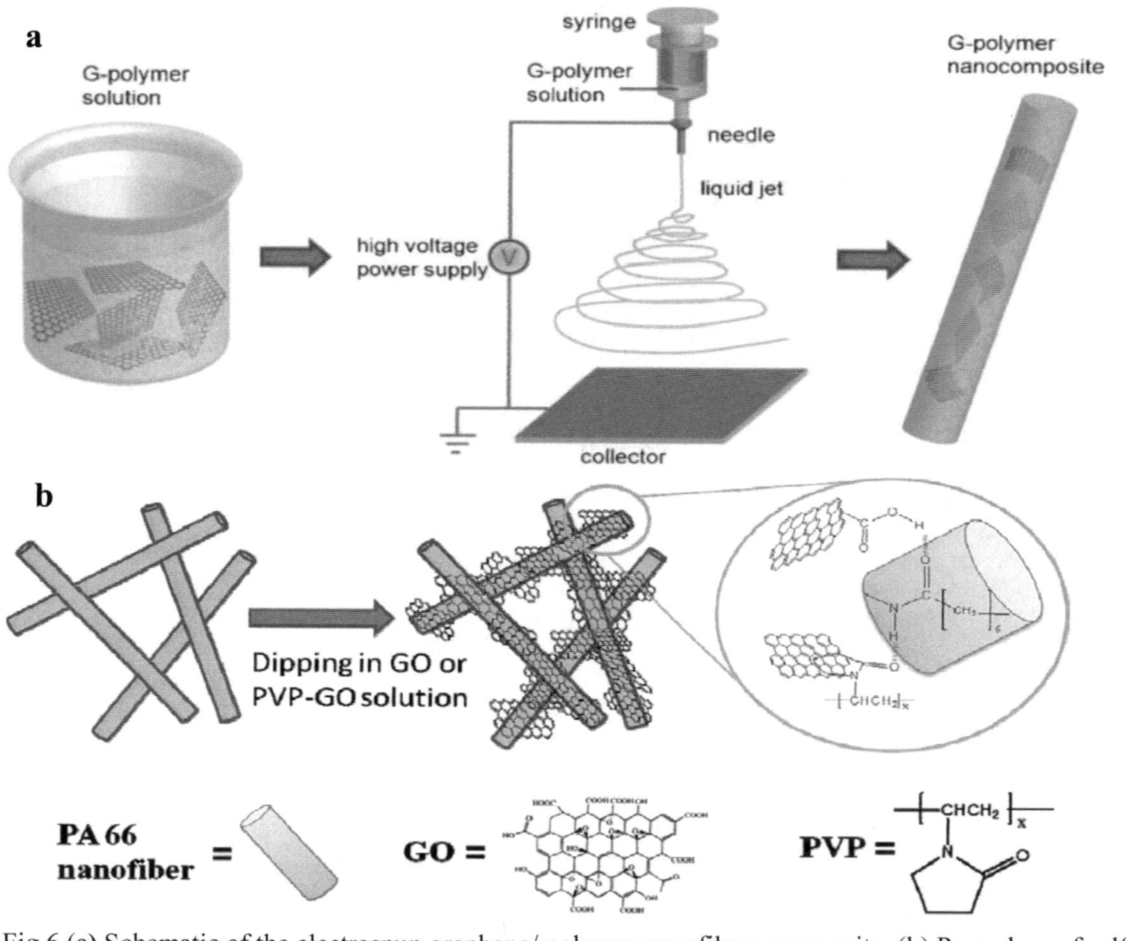

Fig.6 (a) Schematic of the electrospun graphene/ polymer nanofibres composite. (b) Procedure of self-assembled GO or PVP–GO on the surface of PA66 nanofibers

3. Properties of Graphene fibres

Making graphene fibres is actually a process of regularly aligning graphene sheets in a uniaxial direction along which the anisotropic merits of graphene are transformed into macroscopic properties. As graphene has numerous outstanding properties, it is highly expected that graphene fibres should have similar transcendental properties. Following to this aim, optimized or functional properties of graphene fibres were developed.

3.1 Mechanical properties

One of the initial purpose of developing graphene fibres is to have a better properties with fasile methodology compared with carbon fibres. First reported graphene fibre made by a simple drawing approach does not show any records about mechanical properties. Gao's group then firstly fabricated graphene fibres prepared by traditional spinning methods with a relatively poor tensile strength 140 MPa but decent flexibility. Although among all different methods of graphene fibre spinning, wet spinning gives best mechanical behavior, common wet spinning of neat GO or rGO fibres only gives a range of 41~182 MPa in tensile strength and a range of 2.9~8.7 GPa in Young's modulus, shown in Table 1. It is much lower than those of carbon fibres.

The factors dominate tensile strength are concluded as: 1) entanglement of twisted graphene sheets; 2) van der Walls force between graphene oxide or graphene interlayers; 3) hydrogen bonds between formed by oxygen groups in GO sheets; 4) chemical bonds between GO or rGO sheets and other infiltrate, additive [39-41]. Continuous GO fibres showed a typical plastic deformation under tension. It is predicted to arise from displacement of GO nanosheets and stretching of folded, crumbled GO nanosheets. The failure of graphene fibres was considered to be caused by defects and pull-out process under tension. According to these factors, three strategies were taken as to be effective [22]: 1) alignment of graphene nanosheets alone fibre direction; 2) eliminating defects inside fibres; 3) inducing interlayer interaction of nanosheets to improve tensile strength.

To keep a uniform, aligned and continuous formation of GO gel-state fibre, it is essential to have a higher moving speed of collector or rotator. Rotating collector was considered as a very convenient, small-amount instrument for collection. The limitation of it is thought as not precise and no strong, reliable drawing. Collection units, however, can precisely control the drawing rate during spinning by controlling speed of collection [42].

It was also demonstrated that coagulation with $CaCl_2$ gives better spinnability and better properties of graphene fibre [30]. The mechanism has already been discussed [28,43], Higher $CaCl_2$ concentration induced diffusion of $CaCl_2$ cations, facilitated forming crosslinking between GO sheets. When LC GO goes into coagulation, GO fibre gradually precipitated, meanwhile Ca^{2+}, Cu^{2+} from coagulation formulation goes inside GO fibres and between nanosheets forming cross-links. Mechanical properties cross-linked GO fibres were significantly enhanced. However, ultrahigh concentration, on the contrary, may cause an increasing of residual and did not show a further increasing in properties since cross-links were formed at the early stage of the diffusion.

The common main problem of graphene fibres made by different methods was lack of chemical bond of entanglement between graphene nanosheets even if they were well-aligned alone fibre, attracted by van der Waal force and hydrogen bond. One effective method was to form covalent bonds. Successively, graphene based fibres with ultrahigh tensile strength were fabricated. Gao' group combined polymers and nanoparticles into LC GO, then wet spun graphene based firbe composites were with a tensile strength of 652 MPa were fabricated. By Bonding with Ca^{2+} , and covalent bonding with 10,12-pentacosadiyn-1-ol (PCDO), The tensile strength of graphene based firbes was enhanced to as high as 842.6 MPa [44]. Recently, according to Zhen Xu, a breakthrough of graphene fibres has been made to fill in the gap in tensile strength. As is reported, graphene fibres The structure of graphene sheets were restored completely with thermal annealing at 3000 °C to remove defects inside fibres, therefore, the highest tensile strength was reported to be 1.45 GPa [45].

Although breakthrough was already been done to obtain superior properties compared with work before, there is still a huge gap in performance between ideal single graphene sheet and a single fibre. It is believed that, with offering novel approaches and optimization of spinning, graphene fibre can be dramatically improved.

Table 1. Mechanical properties of neat graphene fibres and graphene based fibres

Species	Ref	Diameter (μm)	Young's modulus(GPa)	Tensile strength(MPa)	Breaking strain(%)	Toughness (MJ m^{-3})
DrawingGF	[10]	20-50				
Wet spinning GOF	[27]		5.4	102	6.8-10.1	
Wet spinning rGOF	[27]		7.7	140	5.8	
H-rGOF	[35]	33		420	6	
Wet spinning assemble GOF	[46]	53	4.2	145		
Wet spinning assemble rGOF	[46]	43	8.7	182		
rGOF /epoxy	[46]		6.73		1.24	
HPM GO fibre	[47]		20.5			
HPM GO yarns	[47]	120	29.4			
GOF in KOH	[34]		2.9	41.2	3.6	1.0
GOF in $CaCl_2$	[34]		5.4	62.9	6.1	2.9
Coaxial wet-spun yarn GO@CMC	[48]			73-116	8~10	
Polymer grafted GOF	[49]		18.8	500		7.8
GGO-HPG-GA fibres	[41]		20.9	652	4.1	14
PCDO/GBF in $CaCl_2$	[44]			842.6		
U-GF	[45]	1.6	282	1450	0.5	
Carbon fibres PAN	[21]		230-290	3530-7000	1.5-2.4	
Carbon fibres Pitch	[21]		343-588	2740-4700	0.7-1.4	
Graphene single sheet	[4]		1100	130000		

3.2 Electrical properties

Low-dimensional materials are of great interest to both theorists and experimentalists, owing to their novel electronic properties which arise mainly because of a host of quantum confinement effects [1]. Hence, another most important properties of graphene fibres was electrical properties, including electrical conductivity, capacitive behavior etc.

1) Integration of chemical reduction and thermal annealing, even graphitization contribute to restack of graphene sheets [21];

2) Larger GO sheets give higher conductivity of rGO fibre [50];

3) Better alignment of GO sheets provide better [50];

4) Incorporating may bring in superior conductivity [51,52].

Jalili et al. fabricated neat wet spun rGO fibres (annealing at 220 °C) of which electrical conductivity is ultrahigh up to 2508 S m-1 following a sacrificing of mechanical properties. A high specific capacitance value was detected as 409 F g^{-1} at a current density of 1 A g^{-1}, much higher than best microsupercapacitors available now (265 F g^{-1}) and other types of multifunctional fibers [28].

As a limitation of defects and oxygen-containing groups in neat rGO fibres, doping strategy was applied with conductive dopants (alkali metals, acids) incorporating in rGO fibres. Doping with Ag ions has enhanced the conductivity to 9.4×10^4 S m^{-1}, and other may also possibly work well, as in the case of graphene transparent electrodes. Introducing Ag not only caused a improving in electrical conductivity but also enabled a high current capacity up to 7.1 × 10^3A cm^{-2} [53]. Besides, electrical conductivity of rGO fibres were also reported to be temperature-depended, and the current carrier density of rGO fibres were 440 A cm^{-2} [22,53].

3.3 Other properties

The thermal conductivity of ideal graphene nanosheet was ca. 5000 W $m^{-1}K^{-1}$ [4]. However, the free vibration of each graphene sheet was strongly limited by increased graphene layer, i.e. thermal conductivity of multilayer graphene was much lower compared with single sheet. The thermal conductivity of rGO fibres was determined to be over 1400 W $m^{-1}K^{-1}$, much higher than that of CNT fibres, 2D graphene film, implying usages in thermal management fields [54,55].

It is reported that graphene based fibres responded to voltage, and also have a magnetic property when incorporating Fe_3O_4. When mixing TiO_2 into graphene fibres, a fast repeatabile photocurrent response was found (Fig.7a-c) [35].

When only the upper surface of a GO fibre was reduced, the GO/rGO fibres showed reversible "fall down" and "stand up" between RH=10% and 80%. Further study of the humidity-actuating performance of a twisted GO fibre demonstrated that not only did the fibre-twisting angle reversibly change, but also stretch-out and draw-back in the axial direction were observed, as shown in Fig.7d [56].

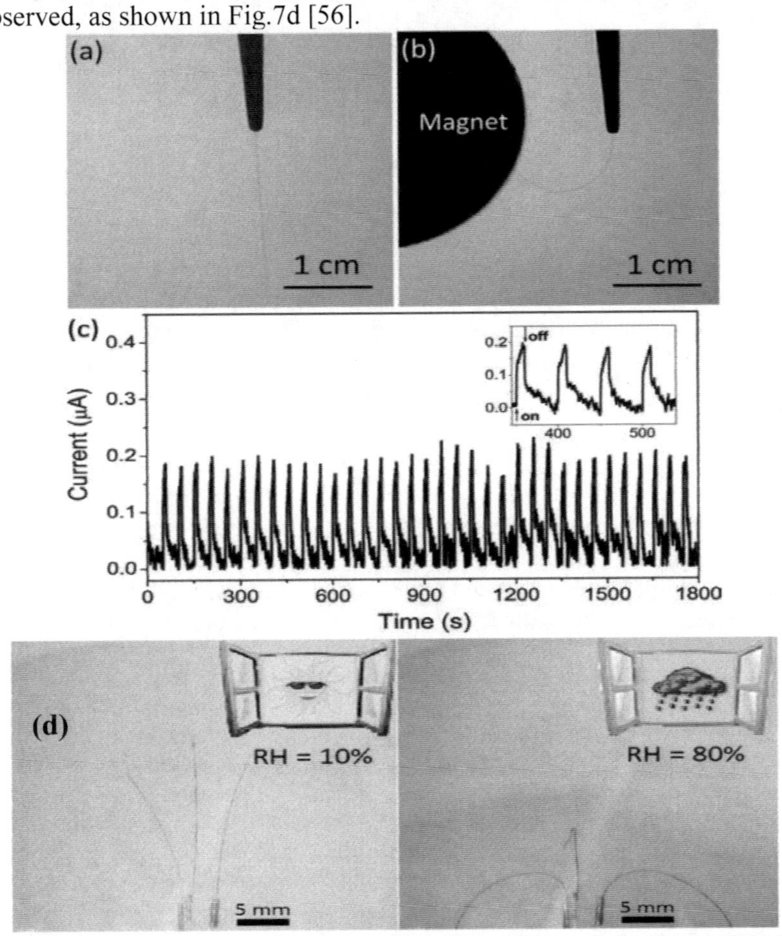

Fig.7 Magnetic and photoelectric response of functional graphene fibres. (a,b) the straight Fe_3O_4/ graphene fibre and the Fe_3O_4/ graphene fibre attracted to the magnet. (c) A representative photocurrent response for graphene/TiO_2 fibre upon exposure to on/ off light at 20°C. The inset shows an enlarged view for a small portion of the photocurrent response curve. The sample length is 1 cm. (d) Responses of laser-scribed GO/rGO asymmetric fibre under varied RH. [22,35]

4. Application of graphene fibres

Possessing the outstanding functional properties, graphene fibres were believed to have great potential in a wide range of application, e.g. functional textiles, flexible and wearable sensors, supercapacitors, electrodes of supercapacitors, energy devices, and actuators etc[47,52,57,58].

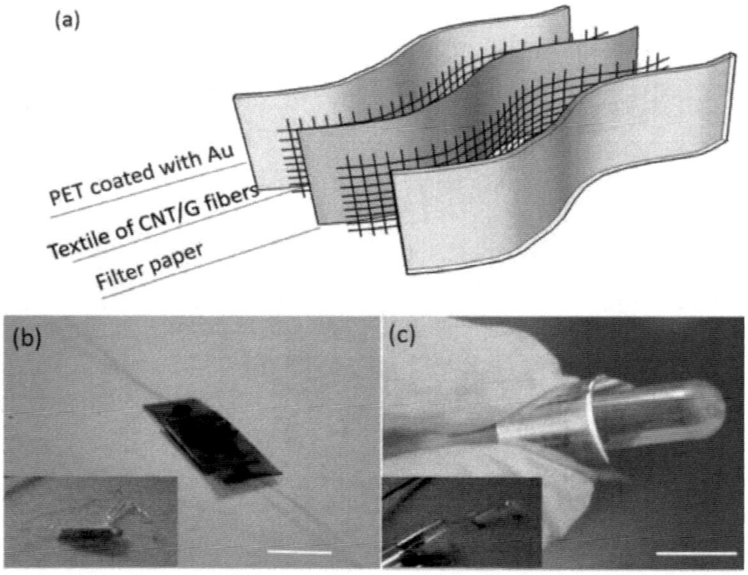

Fig.8(a) Schematic illustration of a multilayer textile supercapacitor using CNT/G fibres as electrodes. (b,c) Photos of fabricated textile supercapacitor showing a light-emitting-diode (LED) lighted by this textile supercapacitor in the flat and bending states. (b,c) Scale bar, 1 cm [59]

4.1 Electrodes of supercapacitors

High conductivity and flexibility facilitate graphene based fibres using in electrodes of supercapacitor. Fig.8 shows a schematic structure of the flexible supercapacitor using the CNT/G fibre textiles as electrodes (Fig.8a), and the final supercapacitors in the flat (Fig.8b) and bending (Fig.8c) states. A thin polyethylene terephthalate (PET) coated with Au was used as the supporting substrate and current collector. The separator is a filter paper soaked with 1 M Na_2SO_4 aqueous electrolyte. As shown in the insets of Fig.8b and c, the charged device can light a commercial LED in either flat or bending state, although the charge voltage is only 0.6 V [59].

4.2 Conducting wires

High electrical conductivity make graphene based fibre a candidate for conducting wires. Light weighted conductive GFs are good candidates for wires in cable and other fields. Ag/graphene hybrid fibres with ultrahigh electrical conductivity showed a good resistance after 50 cycles of stretching-relaxation test and a high breakdown current density 7.1×10^7 A/m^2 [53]. Integrating flexibility, relatively high mechanical properties, light weight and stretchable properties, graphene based fibres were fabricated into textile (Fig.9) [52]. Dong et al. prepared rGO/ poly (dimethylsiloxane) matrix textile [35], demonstrated hard to be deformed with stable conductivity.

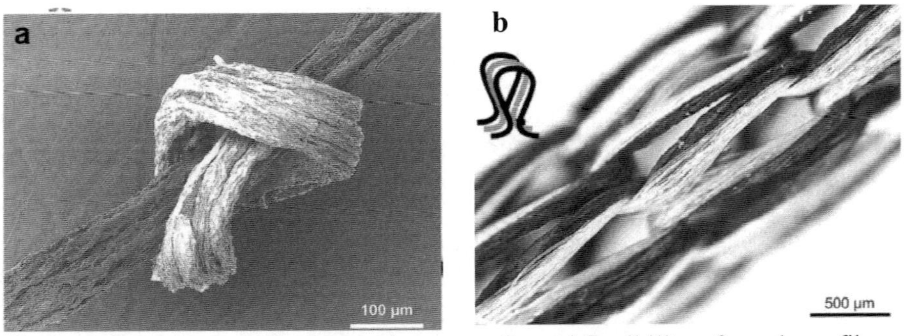

Fig.9 (a) SEM images of a knotted yarn confirmed flexibility of graphene fibres. (b) Optical microscopy images of a textile structure co-knitted with a commercial nylon yarn using double strand of GO fibre. As shown inset, black wires represent nylon, blue wires represent graphene fibres

4.3 Wearable capacitor

Considering high surface area, capacibility, weavability of graphene fibres, researchers have focused on its application in wearable capacitor. Jalili et al. fabricated neat wet spun rGO fibres (annealing at 220 °C) of which electrical conductivity is 2508 S m^{-1} with a sacrificing of mechanical properties [28]. After reduction, the specific surface area decreased from 2605 m^2g^{-1} to 2210 m^2g^{-1} remaining a high level. The charge storage capacity is 409 Fg^{-1} at 1 Ag^{-1} and rate capability is 56 Fg^{-1} at 100 Ag^{-1}, both of which indicated a usage in supercapacitor. According to Kou et al, two coaxial graphene based fibres were twisted and then coated with insulated PVA. Combining great capacity of graphene based fibres, coaxial yarn structure, together with flexible and scalable properties, coaxial wet-spun yarn was applied in cloth supercapacitor, shown in Fig.10. This graphene based fibre cloth supercapacitor showing a capacitance of 28 mF at the current of 10 mA, behaved well compared with commercial supercapacitors (25 mF) [60].

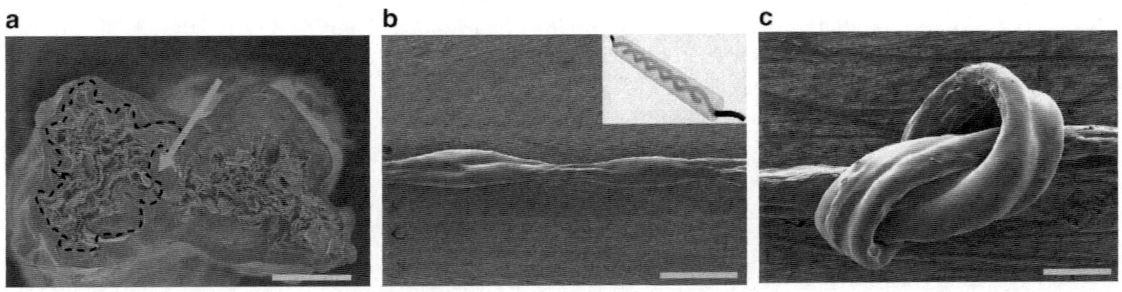

Fig.10. SEM images of cross-section of (a) and side (b) view of a two-ply yarn. The arrow area represents a PVA/H_3PO_4 electrolyte. Black area and red area are graphene firbes. Inset of b shows a schematic illustration of yarn. (c) SEM image confirmed that two-ply yarn can be knotted. Scale bars, 50 μm (a), 500 μm (b), 200 μm (c).

Apart from those applications, the temperature-dependent property in electro-conductivity gives graphene fibres potential usage in sensors. Graphene based fibres have also been illustrated to be applied in actuators, field emitters, solid-phase microextraction, springs, catalysis, etc [22]

5. Conclusion

In conclusion, as is motivated by outstanding properties, not only neat graphene fibres, but also hybrid graphene based fibres have been fabricated successfully. The methodology of graphene fibres spinning includes drawing, solution spinning, thermal process, electrospinning, etc. Up to date, as showing best properties and facile steps, wet spinning liquid crystal graphene oxide is the most attractive and widely used method. The properties of wet spun graphene have been optimized by increasing size of graphene sheets, choice of solvent and coagulation bath, etc. With incorporating functional groups, ions, etc, graphene based composite fibres have shown superior and abundant properties than traditional fibres. With optimizing graphene firbes, both neat graphene fibres and hybrid graphene fibres show remarkable mechanical and electrical properties (1450 MPa and 2508 S m^{-1}, respectively). Besides, graphene fibres also have desired actuating property, flexibility and weavability. Multifunctional properties provide graphene fibres potential in functional textiles, flexible and wearable sensors, supercapacitors, electrodes of supercapacitors, energy devices, and actuators etc. It should be pointed that despite of great progress of graphene fibres, properties of graphene fibres were still inferior to carbon fibres. However, according to this review, it has been realized that although graphene fibres were still in its infant stage, with great efforts from scientists and researchers, great potential of both neat and hybrid graphene firbes will lead progress in technology and science.

References

[1] Novoselov KS, Geim AK, Morozov SV, Jiang D, Zhang Y, Dubonos SV, Grigorieva IV, Firsov AA. Electric field effect in atomically thin carbon films. Science 2004; 306: 666-669.

[2] Novoselov KS, Geim AK, Morozov SV, Jiang D, Katsnelson MI, Grigorieva IV, Dubonos SV, Firsov AA. Two-dimensional gas of massless Dirac fermions in graphene. Nature 2005; 438: 197-200.

[3] Novoselov KS, Jiang D, Schedin F, Booth TJ, Khotkevich VV, Morozov SV, Geim AK. Two-dimensional atomic crystals. Proc Natl Acad Sci U S A 2005; 102: 10451-3.

[4] Zhu Y, Murali S, Cai W, Li X, Suk JW, Potts JR, Ruoff RS. Graphene and graphene oxide: synthesis, properties, and applications. Adv Mater 2010; 22: 3906-24.

[5] Kim KS, Zhao Y, Jang H, Lee SY, Kim JM, Kim KS, Ahn JH, Kim P, Choi JY, Hong BH. Large-scale pattern growth of graphene films for stretchable transparent electrodes. Nature 2009; 457: 706-710.

[6] Schniepp HC, Li JL, McAllister MJ, Sai H, Herrera-Alonso M, Adamson DH, Prud'homme RK, Car R, Saville DA, Aksay IA. Functionalized single graphene sheets derived from splitting graphite oxide. Journal Of Physical Chemistry B 2006; 110: 8535-8539.

[7] Fan X, Peng W, Li Y, Li X, Wang S, Zhang G, Zhang F. Deoxygenation of exfoliated graphite oxide under alkaline conditions: a green route to graphene preparation. Advanced Materials 2008; 20: 4490-4493.

[8] Pei SF, Cheng HM. The reduction of graphene oxide. Carbon 2012; 50: 3210-3228.

[9] Cao X, Yin Z, Zhang H. Three-dimensional graphene materials: preparation, structures and application in supercapacitors. Energy Environ. Sci. 2014; 7: 1850-1865.

[10] Castro Neto AH, Guinea F, Peres NMR, Novoselov KS, Geim AK. The electronic properties of graphene. Reviews of Modern Physics 2009; 81: 109-162.

[11] Wei N, Chen Y, Cai K, Zhao J, Wang HQ, Zheng JC. Thermal conductivity of graphene kirigami: Ultralow and strain robustness. Carbon 2016; 104: 203-213.

[12] Lee C, Wei XD, Kysar JW, Hone J. Measurement of the elastic properties and intrinsic strength of monolayer graphene. Science 2008; 321: 385-388.

[13] Sun H, Xu Z, Gao C. Multifunctional, ultra-flyweight, synergistically assembled carbon aerogels. Adv Mater 2013; 25: 2554-60.

[14] Sheng KX, Xu YX, Li C, Shi GQ. High-performance self-assembled graphene hydrogels prepared by chemical reduction of graphene oxide. New Carbon Materials 2011; 26: 9-15.

[15] Xu YX, Sheng KX, Li C, Shi GQ. Self-Assembled Graphene Hydrogel via a One-Step Hydrothermal Process. Acs Nano 2010; 4: 4324-4330.

[16] Wei W, Yang SB, Zhou HX, Lieberwirth I, Feng XL, Mullen K. 3D Graphene Foams Cross-linked with Pre-encapsulated Fe3O4 Nanospheres for Enhanced Lithium Storage. Advanced Materials 2013; 25: 2909-2914.

[17] Zhao JP, Ren WC, Cheng HM. Graphene sponge for efficient and repeatable adsorption and desorption of water contaminations. Journal Of Materials Chemistry 2012; 22: 20197-20202.

[18] Alwarappan S, Erdem A, Liu C, Li CZ. Probing the Electrochemical Properties of Graphene Nanosheets for Biosensing Applications. Journal Of Physical Chemistry C 2009; 113: 8853-8857.

[19] Chen H, Muller MB, Gilmore KJ, Wallace GG, Li D. Mechanically strong, electrically conductive, and biocompatible graphene paper. Advanced Materials 2008; 20: 3557-+.

[20] Jalili R, Aboutalebi SH, Esrafilzadeh D, Konstantinov K, Moulton SE, Razal JM, Wallace GG. Organic Solvent-Based Graphene Oxide Liquid Crystals: A Facile Route toward the Next Generation of Self-

Assembled Layer-by-Layer Multifunctional 3D Architectures. Acs Nano 2013; 7: 3981-3990.
[21] Xu Z , Gao C. Graphene fiber: a new trend in carbon fibers. Materials Today 2015; 18: 480-492.
[22] Meng FC, Lu WB, Li QW, Byun JH, Oh Y, Chou TW. Graphene-Based Fibers: A Review. Advanced Materials 2015; 27: 5113-5131.
[23] Zakharchenko KV, Fasolino A, Los JH, Katsnelson MI. Melting of graphene: from two to one dimension. Journal Of Physics-Condensed Matter 2011; 23.
[24] Griswold PD, Cuculo JA. An experimental study of the relationship between rheological properties and spinnability in the dry spinning of cellulose acetate–acetone solutions. Journal of Applied Polymer Science 1974; 18: 2887-2902.
[25] Li XM, Zhao TS, Wang KL, Yang Y, Wei JQ, Kang FY, Wu DH, Zhu HW. Directly Drawing Self-Assembled, Porous, and Monolithic Graphene Fiber from Chemical Vapor Deposition Grown Graphene Film and Its Electrochemical Properties. Langmuir 2011; 27: 12164-12171.
[26] Behabtu N, Lomeda JR, Green MJ, Higginbotham AL, Sinitskii A, Kosynkin DV, Tsentalovich D, Parra-Vasquez ANG, Schmidt J, Kesselman E, Cohen Y, Talmon Y, Tour JM, Pasquali M. Spontaneous high-concentration dispersions and liquid crystals of graphene. Nature Nanotechnology 2010; 5: 406-411.
[27] Xu Z, Gao C. Graphene chiral liquid crystals and macroscopic assembled fibres. Nat Commun 2011; 2: 571.
[28] Jalili R, Aboutalebi SH, Esrafilzadeh D, Shepherd RL, Chen J, Aminorroaya-Yamini S, Konstantinov K, Minett AI, Razal JM, Wallace GG. Scalable One-Step Wet-Spinning of Graphene Fibers and Yarns from Liquid Crystalline Dispersions of Graphene Oxide: Towards Multifunctional Textiles. Advanced Functional Materials 2013; 23: 5345-5354.
[29] Jalili R, Aboutalebi SH, Esrafilzadeh D, Konstantinov K, Razal JM, Moultona SE, Wallace GG. Formation and processability of liquid crystalline dispersions of graphene oxide. Materials Horizons 2014; 1: 87-91.
[30] Xu Z , Gao C. Graphene chiral liquid crystals and macroscopic assembled fibres. Nature Communications 2011; 2.
[31] Khan AA, Bin-Kamarudin MA, Kidambi PR, Hofmann S, Wilkinson TD, Qasim MM. Graphene and chiral nematic liquid crystals: a focus on lasing. Rsc Advances 2015; 5: 57437-57443.
[32] Aboutalebi SH, Jalili R, Esrafilzadeh D, Salari M, Gholamvand Z, Aminorroaya Yamini S, Konstantinov K, Shepherd RL, Chen J, Moulton SE. High-performance multifunctional graphene yarns: toward wearable all-carbon energy storage textiles. ACS nano 2014; 8: 2456-2466.
[33] Seyedin S, Romano MS, Minett AI, Razal JM. Towards the Knittability of Graphene Oxide Fibres. Scientific Reports 2015; 5.
[34] Xiang CS, Behabtu N, Liu YD, Chae HG, Young CC, Genorio B, Tsentalovich DE, Zhang CG, Kosynkin DV, Lomeda JR, Hwang CC, Kumar S, Pasquali M, Tour JM. Graphene Nanoribbons as an Advanced Precursor for Making Carbon Fiber. Acs Nano 2013; 7: 1628-1637.
[35] Dong ZL, Jiang CC, Cheng HH, Zhao Y, Shi GQ, Jiang L, Qu LT. Facile Fabrication of Light, Flexible and Multifunctional Graphene Fibers. Advanced Materials 2012; 24: 1856-1861.
[36] Huang YL, Baji A, Tien HW, Yang YK, Yang SY, Ma CCM, Liu HY, Mai YW, Wang NH. Self-assembly of graphene onto electrospun polyamide 66 nanofibers as transparent conductive thin films. Nanotechnology 2011; 22.
[37] Nardecchia S, Carriazo D, Ferrer ML, Gutierrez MC, del Monte F. Three dimensional macroporous architectures and aerogels built of carbon nanotubes and/or graphene: synthesis and applications. Chemical Society Reviews 2013; 42: 794-830.
[38] Bao QL, Zhang H, Yang JX, Wang S, Tong DY, Jose R, Ramakrishna S, Lim CT, Loh KP. Graphene-Polymer Nanofiber Membrane for Ultrafast Photonics. Advanced Functional Materials 2010; 20: 782-791.
[39] Zhang YY, Li YC, Ming P, Zhang Q, Liu TX, Jiang L, Cheng QF. Ultrastrong Bioinspired Graphene-Based Fibers via Synergistic Toughening. Advanced Materials 2016; 28: 2834-2839.
[40] Kou L, Gao C. Bioinspired design and macroscopic assembly of poly(vinyl alcohol)-coated graphene into kilometers-long fibers. Nanoscale 2013; 5: 4370-4378.
[41] Hu XZ, Xu Z, Liu Z, Gao C. Liquid crystal self-templating approach to ultrastrong and tough biomimic composites. Scientific Reports 2013; 3.
[42] Xiang CS, Young CC, Wang X, Yan Z, Hwang CC, Cerioti G, Lin J, Kono J, Pasquali M, Tour JM. Large Flake Graphene Oxide Fibers with Unconventional 100% Knot Efficiency and Highly Aligned Small Flake Graphene Oxide Fibers. Advanced Materials 2013; 25: 4592-4597.

[43] Xu Z, Sun HY, Zhao XL, Gao C. Ultrastrong Fibers Assembled from Giant Graphene Oxide Sheets. Advanced Materials 2013; 25: 188-193.
[44] Hu X, Xu Z, Liu Z, Gao C. Liquid crystal self-templating approach to ultrastrong and tough biomimic composites. Scientific reports 2013; 3.
[45] Xu Z, Liu Y, Zhao X, Peng L, Sun H, Xu Y, Ren X, Jin C, Xu P, Wang M. Ultrastiff and Strong Graphene Fibers via Full - Scale Synergetic Defect Engineering. Advanced Materials 2016.
[46] Cong HP, Ren XC, Wang P, Yu SH. Wet-spinning assembly of continuous, neat, and macroscopic graphene fibers. Sci Rep 2012; 2: 613.
[47] Aboutalebi SH, Jalili R, Esrafilzadeh D, Salari M, Gholamvand Z, Yamini SA, Konstantinov K, Shepherd RL, Chen J, Moulton SE, Innis PC, Minett AI, Razal JM, Wallace GG. High-Performance Multifunctional Graphene Yarns: Toward Wearable All-Carbon Energy Storage Textiles. Acs Nano 2014; 8: 2456-2466.
[48] Kou L, Huang T, Zheng B, Han Y, Zhao X, Gopalsamy K, Sun H, Gao C. Coaxial wet-spun yarn supercapacitors for high-energy density and safe wearable electronics. Nat Commun 2014; 5: 3754.
[49] Zhao X, Xu Z, Zheng B, Gao C. Macroscopic assembled, ultrastrong and H(2)SO(4)-resistant fibres of polymer-grafted graphene oxide. Sci Rep 2013; 3: 3164.
[50] Liu Q, Zhang M, Huang L, Li Y, Chen J, Li C, Shi G. High-Quality Graphene Ribbons Prepared from Graphene Oxide Hydrogels and Their Application for Strain Sensors. ACS nano 2015; 9: 12320-12326.
[51] Seyedin S, Romano MS, Minett AI, Razal JM. Towards the Knittability of Graphene Oxide Fibres. Sci Rep 2015; 5: 14946.
[52] Cheng HH, Dong ZL, Hu CG, Zhao Y, Hu Y, Qu LT, Chena N, Dai LM. Textile electrodes woven by carbon nanotube-graphene hybrid fibers for flexible electrochemical capacitors. Nanoscale 2013; 5: 3428-3434.
[53] Yoon SS, Lee KE, Cha HJ, Seong DG, Um MK, Byun JH, Oh Y, Oh JH, Lee W, Lee JU. Highly Conductive Graphene/ Ag Hybrid Fibers for Flexible Fiber-Type Transistors. Sci Rep 2015; 5: 16366.
[54] Balandin AA GS, Bao WZ, Calizo I, Teweldebrhan D, Miao F, Lau CN Superior thermal conductivity of single-layer graphene. Nano Letters 2008; 8: 902-907.
[55] Jalili R, Aboutalebi SH, Esrafilzadeh D, Shepherd RL, Chen J, Aminorroaya-Yamini S, Konstantinov K, Minett AI, Razal JM, Wallace GG. Wet-spinning of multifunctional graphene fibers using graphene oxide liquid crystals. 2013.
[56] Cheng H, Hu Y, Zhao F, Dong Z, Wang Y, Chen N, Zhang Z, Qu L. Moisture-Activated Torsional Graphene-Fiber Motor. Advanced Materials 2014; 26: 2909-2913.
[57] Cruz-Silva R, Endo M, Terrones M. Graphene oxide films, fibers, and membranes. Nanotechnology Reviews 2016; 0.
[58] Cheng H, Liu J, Zhao Y, Hu C, Zhang Z, Chen N, Jiang L, Qu L. Graphene Fibers with Predetermined Deformation as Moisture-Triggered Actuators and Robots. Angewandte Chemie International Edition 2013; 52: 10482-10486.
[59] Cheng H, Dong Z, Hu C, Zhao Y, Hu Y, Qu L, Chen N, Dai L. Textile electrodes woven by carbon nanotube-graphene hybrid fibers for flexible electrochemical capacitors. Nanoscale 2013; 5: 3428-3434.
[60] Kou L, Huang T, Zheng B, Han Y, Zhao X, Gopalsamy K, Sun H, Gao C. Coaxial wet-spun yarn supercapacitors for high-energy density and safe wearable electronics. Nature communications 2014; 5.

A Study on Bonding of Nanofibers over Nonwoven Substrates (Air Filtration)

Zuhaib Hassan[1,2*], Osman Bulent Ozipek[2]

[1,2]*Faculty of Textile Technologies and Design, Istanbul Technical University, Inonu Street No. 65, Istanbul, 34437, Turkey*

[1,2*]Corresponding author's email: zuhaihassan@itu.edu.tr

Abstract

This article presents a comparative study on nanofibers over nonwoven substrate. Here we used two different polymers named as thermoplastic polyurethane (TPU) and polyvinylidene fluoride (PVDF) for comparing their filtration efficiency and Dimethylformamide used as solvent and DMF polypropylene PP as substrate in the electrospinning machine. Air filtration requires efficient removal of large and small particles at high air flow rates. Fibers in the nano range, in comparison with larger ones, are well known to provide better filter efficiency at the same pressure drop in the interception and inertial impaction regimes. After making the materials, tests are evaluated by different testing machines. The results showed the air permeability differences, air filtration efficiency of the polymers. In this paper, we suggest solvent bonding method for nanofibrous filters, as an alternative for the traditional thermal and pressure bonding methods used until now.

Keywords: Electrospinning; Nanofibers

1. Introduction

Nanofibers exhibit structural features, such as high specific surface area, such as low weight, high permeability, small pore size and good pore's interconnection, low weight, high permeability and potential for incorporation of nanoscale functionality. Their application in filter media has been investigated since the 'Petryanov filter' in 1936. Nowadays, their production is the interest of many companies which try to lower their energy consumption, increase their filtration capacity, make their lifetime longer, and to facilitate their maintenance [1].

Nanofibrous mat thickness over the substrate is one of the factors for obtaining high filtration efficiency, as shown in PVA nanofiber mat deposited by electrospun over cotton substrate [2]. However, a study of multi-layered sandwiched filters, composed of PAN nanofibers and activated carbon fiber (ACF) mats where the nanofibers where collected changing the deposition directions, indicates that different fiber orientations benefits the filtration performances [3]. Nanofiber solidity also influence filter efficiency and drop pressure as more important criterions for evaluating filters [4]. The higher the fiber solidity the better the filtration efficiency, which means that less particles penetrate into the filter media. Nanofiber mat density is another factor that influence filter efficiency [5]. However, improved filter properties are mostly achieved by multi-layering, or distribution of sparsely nanofiber mats through stacking up multiple filters rather than a single nano-layer with the same density.

High performance filtration media are formed by lamination of one or more nanowebs and a substrate bonded together into a multilayered structure, as shown in Fig. 1. Additional layer(s) may be added in order to improve filtering performances and/or to protect the nonwoven mat during handling and any other subsequent process like pleating, corrugation, gathering or otherwise consolidated structural configuration [6,7]. Adhesive layers can also been added in order to improve the delamination resistance of the media [8]. Nanofiber mats can be positioned upstream before the substrate in order to enhance surface filtration performance, or downstream of the substrate in order to improve depth filtration by capturing particles within the body of the media. They can also be applied on the both sides of the substrate with different gradient density or sandwiched between two substrate layers in order to provide different filtration options [7,9]. The adhesion of the nanofiber layer with the substrate (and others layers) in the filtration media plays important role in the filtration media' performances and lifetime.

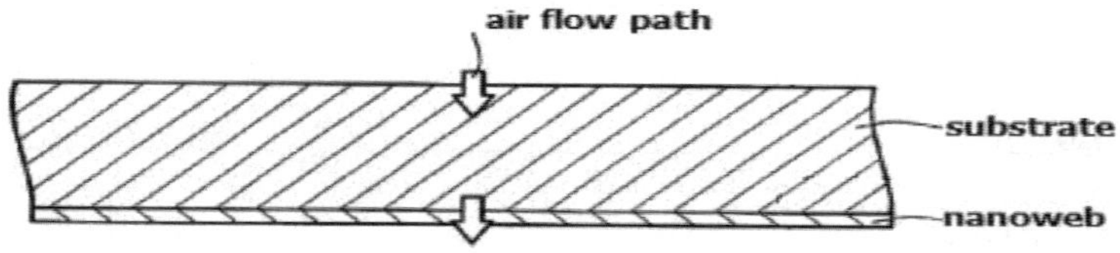

Fig.1 Cross sectional illustration of a multilayered filter media [6]

Nanofiber mats are usually directly deposited on the substrate during their formation by electrospinning, electro-blowing or melt-blowing method [6,8,10]. When the nanofibers come in contact with the adjacent substrate's fibers they bond through the residual solvent in the fibers from the spinning process – a step called solvent bonding. They may be directly incorporate into the filter media without any further bonding or the layers may firstly be further integrated by thermal or pressure bonding. Nonwoven nanofibrous filtration mats with filtration efficiency than HEPA filters have been produced by electrospinning PVC and PVDF nanofibers without further additional bonding [11].

When thermal bonding is applied the layers are heated together at temperature which is between the glass transitions and melt temperature of one of the layers. While during pressure bonding the layers pass through a pair of heated or not-heated calendaring rollers or heated dies [12]. Heated forming die method uses steam or hot air to heat the layers to a temperature sufficient to permanently bond them into a single structure element into a specific shape. The layers may also be integrated into filtration media by mixing of the nanofibers with the coarse fibers of the substrate by folding them into multiply folds and then pressure bonded into a desired thickness [10].

The effect of pressure bonding was observed on three-layered filter media [7]. Polyethylene oxide (PEO) nanofibers were electrospun over needle-punched polypropylene non-woven substrate and sandwiched with another needle-punched substrate layer by calendaring at 50ºC. It showed lower pore size, higher efficiency and pressure drop which did not change drastically after cycling compression in comparison to un-calendared two-layered filter media. Calendared filter media also did not delaminate after cycling compression, but it showed mixture of nanogranules and fibrillar structure as result of phase changes in the material during calendaring.

Pressure consolidation of a filter media with solvent based PU adhesive layer between the polyester substrate and the PA6.6 nanoweb increase the delamination resistance of the media compared to a same filter media but without adhesive layer consolidate under pressure at temperature higher than that of nanofibers' Tg [8].

In this paper, we suggest solvent bonding method for nanofibrous filters, as an alternative for the traditional thermal and pressure bonding methods used until now.

2. Experimental

2.1 Materials

Thermoplastic polyurethane (TPU) and poly(vinylidiene fluoride) (PVDF) were used as polymers for producing nanofibrous mats. Dimethylformamide (DMF) was used as a solvent for both polymers, and as a bonding agent. Polypropilene (PP) nonwoven mat was used as a substrate.

2.2 Methods

When Producing nanofibrous filter media by deposition of electrospun nanofibers on a substrate wetted with solvent in order to achieve stronger bonding between the nanofibrous mat and the nonwoven substrate. Three samples, using two different polymers were prepared in order to compare bonding effect and filter efficiency.

1st Sample:

13 wt% TPU solution was prepared by stirring at 80°C during 5h. As prepared solution was electrospun during 3h in a flow rate of 0.65ml/h; with 24kV applied voltage and 15cm working distance. PP nonwoven substrate sprayed with DMF was used as a collector of nanofibers.

2nd Sample:

10 wt% PVDF solution was prepared by stirring at 80°C during 2h. As prepared solution was electrospun during 2h in a flow rate of 1.5ml/h; with 24.5kV applied voltage and 15cm working distance. PP nonwoven substrate sprayed with DMF was used as a collector of nanofibers.

3rd Sample:

For comparison of the bonding effect and filter properties, 10 wt% PVDF was electrospun under same conditions as Sample 2, on a PP substrate without any solvent sprayed on it.

Fig.2 Electrospinning used during the experiments

2.3 Testing

Air permeability

Air permeability or, the velocity of an air flow passing perpendicularly through a test specimen under specified conditions of test area, pressure drop and time was tested in order to examine filter efficiency of the samples produced. The pressure drop on the filter is a linear function of gas velocity that indicates the viscous flow of air through the filtration mat. This value has significance with respect to the usage area.

PROWHITE AIRTESTER-II air permeability tester was used to test the samples, under EN ISO 9237 Special 400Pa standard parameters with pressure drop of 100 Pa and sample area of 20mm^2.

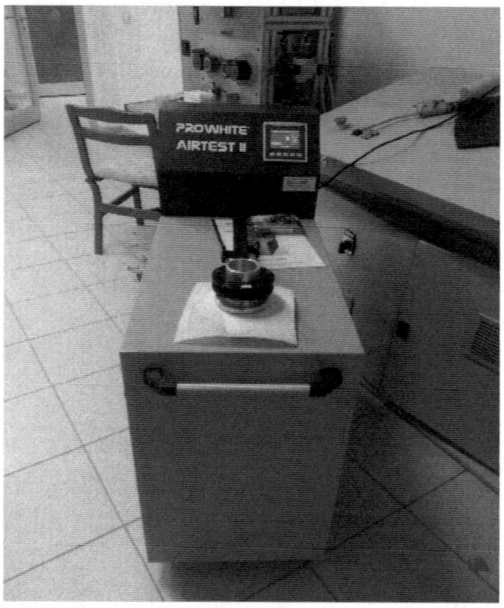

Fig.3 PROWHITE AIRTESTER-II air permeability tester

Peeling strength is a measure of the bond strength between two materials. Peel strength is the average load per unit width of bond line required to separate bonded materials where the angle of separation is 180 degrees.

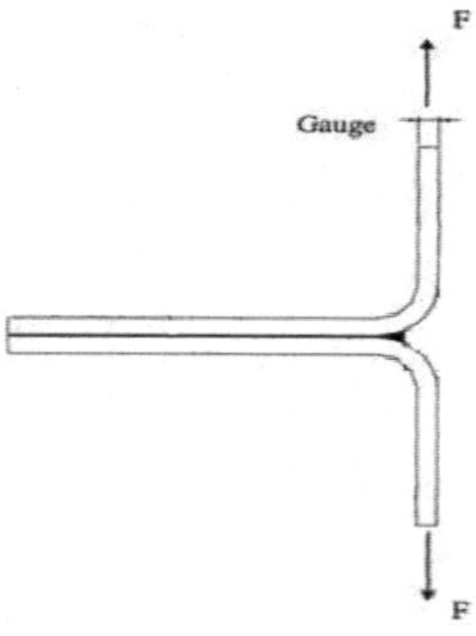

Fig.4 Peel Testing

3. Results and Discussion

Table 1. Air permeability results per sample are given below

	Sample 1	Sample 2	Sample 3
test 1 [mm/s]	12	9	28
test 2 [mm/s]	80 #	10	12
test 3 [mm/s]	10	10	17
test 4 [mm/s]	32		
test 5 [mm/s]	6 #		
average [mm/s]	**28 (18)**	**9,7**	**19**

From the two samples electrospun on wet substrate, Sample 1 is likely to provide greater filtration efficiency compared to sample 2. The large differences in air permeability values of Sample 1 may be attributed to the un-homogenous nanofiber mat obtained after electrospinning or the uneven application of the solvent on the substrate. The lower air permeability values of the Sample 2, compared to Sample 3, may be due to film forming at the beginning of the electrospinning process as consequences of the wet substrate.

Peel strength tests were not performed by standard tester due to the un-uniform mats obtained during electrospinning. From the visual peel tests performed by us can be conclude that Sample 1 (TPU on DMF) showed stronger bonding effect, followed by Sample 2 (PVDF on DMF) and Sample 3.

4. Conclusion

PVDF and TPU nanofibers were electrospun on microfiber nonwoven PP substrate wetted with DMF in order to achieve stronger bonding between the two, nano- and microfiber layers. Bonding forces of the samples electrospun on wetted substrate resulted stronger than those of the sample in which no solvent was applied. Although this method improve bonding effect, it decrease filter efficiency, as shown from the results obtained during air permeability tests.

However, deposition of electrospun nanofibers on a substrate wetted with solvent in order to achieve stronger bonding between the nanofibrous mat and the nonwoven substrate is a promising bonding method for nonwoven filter medias. By optimizing production parameters, such as electrospinning parameters, amount of solvent sprayed on substrate or maybe by heating the collector better results may be obtained.

5. Acknowledgement

The authors wish to acknowledge the Ass. Prof Ali Kilic who gave permission for using his TEMAG Lab at Istanbul Technical University. Without his permission, it was not possible.

References

[1] Barhate RS, Ramakrishna S. "Nanofibrous filtering media: Filtration problems and solutions from tiny materials," J. Membr. Sci., vol. 296, no. 1–2, pp. 1–8, Jun. 2007.

[2] Wang H, Yang ZY, Liu JY, Zheng GF, Chen LM. "Explore on Pressure-Drop Performance of Nanofibrous Filtration Membrane," Adv. Mater. Res., vol. 236–238, pp. 1885–1888, May 2011.

[3] Zhang Q, Welch J, Park H, Wu CY, Sigmund W, marijnissen JCM. "Improvement in nanofiber filtration by multiple thin layers of nanofiber mats," J. Aerosol Sci., vol. 41, no. 2, pp. 230–236, Feb. 2010.
[4] Wang J, Kim SC, Pui DYH. "Investigation of the figure of merit for filters with a single nanofiber layer on a substrate," J. Aerosol Sci., vol. 39, no. 4, pp. 323–334, Apr. 2008.
[5] Leung WWF, Hung CH, Yuen PT. "Effect of face velocity, nanofiber packing density and thickness on filtration performance of filters with nanofibers coated on a substrate," Sep. Purif. Technol., vol. 71, no. 1, pp. 30–37, Jan. 2010.
[6] Green TB, Li L. "Filter having melt-blown and electrospun fibers," US8172092 B2, 08-May-2012.
[7] Patanaik A, Jacobs V, Anandjiwala RD. "Performance evaluation of electrospun nanofibrous membrane," J. Membr. Sci., vol. 352, no. 1–2, pp. 136–142, Apr. 2010.
[8] Schelling A, kohli A. "Process for forming a laminate of a nanoweb and a substrate and filters using the laminate," US8679216 B2, 25-Mar-2014.
[9] Mei Y, Wang Z, Li X. "Improving filtration performance of electrospun nanofiber mats by a bimodal method," J. Appl. Polym. Sci., vol. 128, no. 2, pp. 1089–1094, Apr. 2013.
[10] Green TB. "Integrated nanofiber filter media," US20120255662 A1, 11-Oct-2012.
[11] Lackowski M, Krupa A, Jaworek A. "Nanofabric nonwoven mat for filtration smoke and nanoparticles," Pol. J. Chem. Technol., vol. 15, no. 2, pp. 48–52, 2013.
[12] Ward BC, Stoltz GM. "Multi-Layer, Fluid Transmissive Fiber Structures Containing Nanofibers and a Method of Manufacturing Such Structures," US20100206803 A1, 19-Aug-2010.

Development of 16.5 Tex Green Biomass Chitin/ Algae/ MulberryAntibacterial Yarn

Cai-Xin Qu[1], Gui-Cui Chen[1,2*], Hua Liu[1*], Shu-Dong Wang[1]

[1]*Yancheng Institute of IndustryTechnology, Jiangsu,Yancheng,224005, China*
[2]*Soochow University, Jiangsu,Suzhou, 215123, China*

Corresponding author's email: ycfylh@126.com, ycfycgc@126.com

Abstract

The green biomass chitin, alginate and mulberry fiber are used to produce the medical antibacterial functional yarn, The specification of the chitin, alginate and mulberry aredescribed. Then, the three kinds of fiber are used to spin the 16.5 Tex chitin/ alginate/ mulberry (40/ 30/ 30) blended antibacterial yarn. The technological design, technical measures and optimizing spinning process are studied. It is concluded that the antibacterial rate of the chitin/ algae/ mulberry blended yarn reaches 80%, which is significantly higher than that of the chitin/ algae/ cotton (40/ 30/ 30) blended yarn. From the above results, it shows that the chitin/ algae/ mulberry blended yarn has excellent antibacterial and antimicrobial properties for using antibacterial medical products.

Key words: Chitin Fiber; Algae Fiber; Mulberry Fibre; Blended Yarn; Process; Antibacterial

Introduction

Biomass is consisted of plants, animals, microorganisms and other organisms. Biomass fiber is made ofbiological or biological extracts [1], including natural animal and plant fiber, regenerated fiber and synthetic fiber derived from biomass. In this paper, Three kinds of biomass fiber such as chitin fiber, alginate fiber, mulberry fiber were used as raw materials to produce 16.5tex chitin fiber/ alginate fiber/ mulberry fiber 40/ 30/ 30 medical antibacterial blended yarn. The performance of three kinds of fiber was analyzed to develop reasonable measures of spinning process, guaranteeing the blended yarn quality, developingthe medical antibacterial yarn with high safety factor, It can not only improve the functionality of the product, also can use abandoned mulberry bark to produce mulberryfiber in order to reduce the production cost, and has broad social benefit and economic benefit, and is conducive to the comprehensive utilization of mulberry fiber.

1. Performance indicators of fiber and Pretreatment

1.1 Performance indicators of fiber

Chitin fiber contains natural moisturizing antibacterial ingredient, which is formed by the wet spinning process [2]. Physical performance indicators ofchitin fiber are: fiber length reaches 38 mm, fiber fineness reaches 2.2 dtex, fracture strength of fiber reaches 2.17 cN/dtex, moisture regain of fiber reaches12%, under the standard conditions (temperature: 20℃, humidity: 65%).

Algae fiber contains the extracted material from natural algae, which is formed by the spinning processing [3]. Because the raw material comes from natural algae, so that it has flame retardant, anti electromagnetic radiation, hemostatic and antibacterial properties, and it has very good development value in the field of medical dressing [4]. Algae fiber length is 38mm, fiber fineness is 2.78dtex, breaking strength of fiber is 2.36 CN/dtex, moisture regain rate of fiber is 18%.

Mulberry fiber is a kind of typical "green biomass fibers", which has good moisture permeability, warmthretention, good antibacterial properties, good luster, biodegradable and soft and so on [5, 6]. which belongs to the utilization of discarded and waste mulberry branchesand extensive source, turning waste into treasure. Average lengthof mulberry fiber is 22mm. Average fineness of mulberry fiber is 2.65dtex. Breaking strength of mulberry fiber is 4.8 cN/dtex. Moisture regain rate of mulberry fiber is 10%. Short fiber rate is 28.2%.Mulberryneps are 35 neps/g. Defects are 2.8%. Mulberry fiber longitudinal vertical linesare similar to the structure of hemp fiber, without convolution phenomenon of cotton fiber. The cross-sectional shape is oval, triangular and small irregular polygon asshown in fig.1.

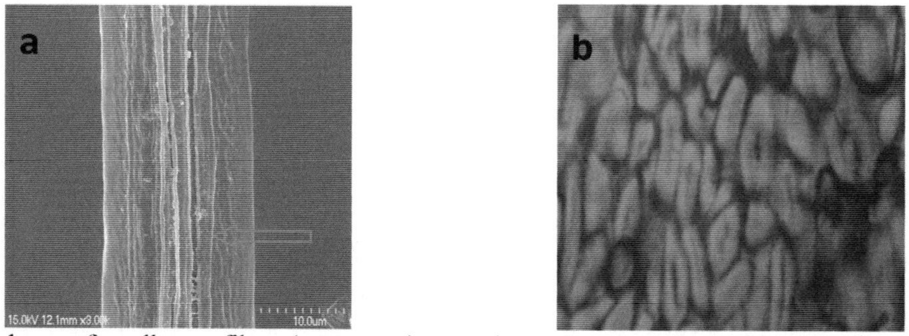

Fig.1 The morphology of mulberry fiber (arrow points to impurity) Rmarks: (a)Longitudinal scanning electron microscopephotograph, (b)Horizontal microscope photograph

1.2 Fiber pretreatment

Antistatic agent and emulsified oil are added into chitin fiber and algae fiber as spinning pretreatment in order to increaseits antistatic property, softness and fiber cohesion [4]. The antistatic agent is 0.3% of the feed rate, and the feed rate is 1.3%. Antistatic agent accounts for 0.3% of the feed rate, the proportion of oil accounts for 1.3% of the feed rate .

1.3 The preparation process of biomass antibacterial yarn

In order to solve that the pure mulberry fiber spinnabilityeasily forms the noil [5], chitin fiber, algae fiber and mulberry fiber usebales mixing method for spinning.in bales mixing process, chitin fiber wet weight mixing ratio accounts for 38%, algae fiber wet weight mixing ratio accounts for 28%, mulberry fiber wet weight mixing ratio accounts for 34%, in order to ensure thatblended ratio of the chitin fiber/ alginate fiber/ mulberry fiber is 40/ 30/ 30, production process as follows:

The chitin fiber/ alginate fiber/ mulberry fiber:

FA002 automaticplucker(×2)→FA121 metal impurity removal device→FA022 multi-mixer→FA106A comb needle roller opener (attached to the A045Bcondenser) → A062 electric distributing device (2-way assortingcottondevice) A092AST vibration type double hopper feeder (×2) (attached to the A045B condenser) → FA141single beater scutcher (×2)→FA201B carding machine→FA306 drawing frame (×3)→FA458 roving machine→FA507B spinning machine→NO.21c auto winding machine barrel removing.

2. Spinning technological measures

2.1 Picking process

The processing principle of picking process adopts "frequently picking and fewer picking, mixed evenly, more loose and less beating, falling early and less broken, less fiber damaged", in order to reflect the minute bale and uniform mixing, improving yarn evenness and reduce breaking strengthvariation coefficient of single yarn.

The main technical parameters: Dryweight of lap is 377.9g/m. Calculation length of lap is 33.187m. Actual length of lap is 34.38m. Elongation is 3.4%. Net weight of lap is 12.83kg, Combined beater speed is 921.6r/min, Roller speed of lap is 12.31r/min. Detection results show weight irregularity rate of lap is 1.2%.

2.2Carding process

The processing principle of carding process adopts “tightgauge, low speed, small tension, regularly card stripping". Tighter gauge is give between cylinder and flat to make full use of the comb the effectivenessof flatworking surface. “low speed, small tension ” can effectively reduce the damaged fiber, avoid excessive mulberry fiber noil, ensure the mulberry fiber blended ratio under the premise of ensuring the production. Cylinder and doffer need to be regularlycard stripped "to eliminate the inner fibers into the needle gap, improvethe surface carding efficiency and ensure the quality of the cardedweb.

The main technical parameters: Dry weight of sliver is 21.5g/5m. Cylinder speed is 330r/min. Doffer speed is30.4r/min. Taker-in speed is 800.05r/min. The five point guages(mm) between cylinder and flat are 0.20, 0.20, 0.15, 0.15, 0.20. The gauge between cylinder and taker-in is 0.19mm. The guage between cylinder and doffer is0.14 mm. Detection results show the weight of sliver is 1.2%, and the fiber neps are 43 neps/g.

2.3 Drawing process

The processing principle of carding process adopts “low speed, middle gauge, heavy pressure, normal draft”, “low speed” can avoid the too high speed and easily cause the phenomenon of winding roller. “Medium gauge" can avoid evenness deterioration and un-parallel-straightness, reduce thick knots and yarn defects, which is combined with the "heavy pressure" can ensure the gripping force is greater than the drafting force and effectively realizethe draft. “normal draft” is conducive to the extension of the hook fiber, in order to improve the quality of yarn.

It is seen from the main technical parameters as shown in Table 1. Conclusions showthe weight evenness of sliver is 1.2%, and the fiber neps are 36 neps/g.

2.4 Roving process

The processing principle of carding process adopts “lowspeed,heavypreasure,small winding, small tension”. The low speed and heavy preasure can reduce breakage and increasedraft. The small winding and small tensionavoid accidental elongation, improve evenness, overcome the lack of strength of yarn and is beneficial tospinning process for drawing.

Table 1. Drawing technical parameters

Programms		First drawing	Second drawing	Third drawing
Dry weight/g/5m		19.8	20.91	20.11
Sliver counts		6	6	8
Total draft	Mechanical	5.805	5.904	8.577
	Actual	5.64	5.73	8.33
Draft distribution	Front tension	1.02	1.02	1.02
	Front draft	5.215	5.399	7.466
	Back draft	1.03	1.06	1.15
	Back tension	1.02	1.02	1.02
Roller pressure /1×2×3×4×5/N		240x340x340 x340x370	240 x340x340 x340x370	240 x340x340 x340x370
Vehicle speed /m·min^{-1}		180	180	180

The main technical parameters: Dry weight is 5.0g/10m.Actual draft is 6.12, Mechanicaldraft is 6.02. Back draft is 1.15. Coefficient of twist is 66.02, Spindle speed is 625.7r/min. Front roller speed is 233.92r/min. Roller gauge is 10×25×35(mm). Detection results show the weight of roving is+1.4%, and the CV of evenness is 8.34%.

2.5 Spinning process

The processing principle of spinning process adopts "heavypreasure, low speed, medium back draft". "Heavypreasure" can ensure the roller jaw with enough gripping force, adapt to the change of the drafting force, so as to ensure the smooth progress ofdrafting. "Low speed"can prevent the fineyarn often brokenends, increase the production efficiency. Thebackzoneof spinning frames have poor control of fiber, and drafting ratio is only 1.14~1.50. Moderate drafting ratioof back zone is adoptedin draft rangeof the back zone, whichcan effectively share front draft task and is conducive to the draft and can improve evenness and strength.

The main technical parameters: Dry weight is 1.46g/100m. Moisture regain is 13.2%. Mechanical draft is 34.95. Actual draft is 34.25. Roller diameter is 25×25×25(mm). Model of ringis PG1/2 and diameter of ring is 42mm. Ring traveler model is CO and traveler number is 7/0. Back draft is 1.4. Actual twist factor is 348.631. Front roller speed is163.95r/min. Spindle speed is 12800r/min.

3. Testing and Analysis

3.1 Yarn quality test

Under the standard conditions (temperature: 20℃, humidity: 65%), the quality of blended yarn includes that twist factor is 347, breaking strength is 385cN, breaking elongation is 8.1%, single strength CV% reaches 9.7%, evenness CV% reaches 11.4%, thin nepsreaches 6 neps · 1000m^{-1},+ 50% thick nepsreaches 27neps · 1000m^{-1}. + 200% thick nepsreaches 73 neps · 1000m^{-1}, hairiness index reaches3.8, all of which can meet the requirements of weaving, Structure of the blended yarn is shown in Fig.2.

3.2 Test of antibacterial property

Antibacterial property of chitin/ algae/ mulberry 16.5tex 40/ 30/ 30 blended yarn is tested according to the method of literature introduction (FZ/T 73023-2006). First, theantibacterial property is tested by the oscillation method, then the strains are the gram negative bacteria (Escherichia coli) and gram positive bacteria (staphylococcus aureus). Secondly,the antibacterial property is characterized by its inhibitory rate. The results are seen from Fig.2. Inhibitory rate ofescherichia coli is (82.4±4.1)%, Inhibitory rate ofstaphylococcus aureus is (74.3±4.1%). Antibacterial test results are shown in Fig.3.

Fig.2 Structure of the blended yarn Rmarks: (a)SEM of Longitudinal, (b)SEM of Horizontal

Fig.3 Antibacterial testRmarks: (a)Escherichia coli test, (b)Staphylococcus aureus test

Inhibitory rates of two kinds ofstrains are respectively (82.4±4.1)% and(74.3±4.1)%. when inhibitory rate of sample is more than 70%, he sample is antibacterial according to GB/T20944.3-2008, so the blended yarn has good antibacterial properties. This is because the chitin is the only natural polymer material which is found in nature with unsaturated cationic groups, therefore,the negative charge of all kinds of harmful bacteria have a stongadsorption, so that it can inhibit the activity of harmful bacteria even loss activity, so as to achieve the purpose of antibacterial [2]. Moreover, algae fiber has also antibacterial effect [4], and the pectin of mulberry bark is removed (pectin can provide nutrients for the bacteria). The results analysis shows that chitin/ algae/ mulberry 16.5tex 40/ 30/ 30 blended yarnhas good antibacterial property.

4. Conclusions

Biomass fiber can effectively expand the medical antibacterial yarn sources of raw materials,make up for the lack of domestic textile resources, and also deal with oil resources are depleted day by day, therefore, it is an important means to realize the sustainable development of the textile industry [8]. Results are as follows by researching key technology of chitin fiber/ alginate fiber/ mulberry fiber antibacterial yarn:

1) Because the production of chitin fiber and algae fiber is easy to bring about static electricity. the chitin fiber and algae fiber need to add antistatic agent and emulsified oil to increase the fiber cohesion before spinning.

2) Strictly according to spinning technological measures to spinyarn,itreduces the card sliver neps and improve the clarityof the web, the structure of the web and the neps, therefore, 16.5tex chitin fiber/ alginate fiber/ mulberry fiber antibacterial yarn can meet the requirements of the production of weaving.

3) 16.5tex chitin fiber/ alginate fiber/ mulberry fiber 40/ 30/ 30 has good moisture absorption ability. E.coli antibacterial rate is more than 80%, staphylococcus aureusantibacterial rate is more than 70%, so it has excellent antibacterial, bacteriostatic and certain health characteristics and so on, and it is suitable for the raw material of antibacterial medical products and has a broad market prospect.

Acknowledgements

This research was financially supported by the Science and Technology Creative Team of Jiangsu Province, 2015, China (The main research direction is the preparation of biomass functional fiber). And this research was also financially supported by the Industry and University Research Project **(BY2015059-01)**. Morever, this research was also financially supported by the project supported by Scientific Research Fund of Yancheng Institute of Industry Technology (ygy1401) and science and technology project of Textile Industry Association, China (**2014055**).

References:

[1] Shen XY. Biomass and biomedical fiber. Proceedings of the National Symposium on polymer materials engineering and technology, HangZhou: 2010.
[2] Wong Y. Study on the structure and properties of chitin fiber. Modern textile technology, 2011; 6: 7-10.
[3] Wang SM, Wang Jk, Wang HT, Pan L. Effect of blending ratio on the properties of cotton algae fiber blended yarnCotton textile technology, 2011; 39(12): 28-30.
[4] Pan LQ, Wang MD, Zhang JW, Wang JK. Siro core spining yarn and process optimization of cotton spandex algaefiber. Cotton textile technology, 2012; 40 (5): 5-8.
[5] Zhang LF, Qu CX, Chen GC, Chen Y. The development of aloe viscose fiber/ mulberry fiber/ silver silk Siro Phil yarn.Shanghai Textile Science and technology, 2014; 42 (8): 42-44.
[6] Qu CX, Wang JM. Spinning and performance analysis of mulberry fiber blended rotor yarn. Cotton textile technology, 2012; 40 (5): 1-4.
[7] Zhuang CQ, Wu YS. Applied mathematical statistics. South China university of technology press, 2002; 4: 452-454.
[8] Cao J, Ding ZH, Shen XY. Present situation and development trend of new biomass fiber, China fiber inspection, 2012; 1: 82-86.i

Advanced in Smart E-Textile Apparel Technology

Yu-Xi Shi[1], Yi Li[1*], Zhi-Run Hu[2], Xu-Qing Liu[1]

[1]*School of Materials, The University of Manchester, Manchester, M13 9PL, UK*
[2]*School of Electrical and Electronic Engineering, The University of Manchester, Manchester, M13 9PL, UK*

*Corresponding author's email: henry.yili@manchester.ac.uk

Abstract

Functional wearable e-textile has high technology and shows widely potential application, attract numerous interests of fashion designer, costumers, venture capitals and scientists. Specifically, they are products where intelligent devices are integrated into fabrics and clothing and are expected to be high integrate, flexible, lightweight, comfortable, and aesthetic. However, there are still some research gaps of academy and technology, and we critically review the advancement of conductive materials, fabrication and integration technology, as well as input and output devices to tackle with these problems. In addition, this review discusses the limitations of current materials, fabrication techniques, devices manufacturability and performance, and finally provide our outlook over the field and a prediction for the future, especially the attractable prospects in graphene and hybrid conductive materials.

Keywords: E-Textiles; Smart Clothing; Wearable Systems; Apparel Technology; Graphene Conductive Materials

1. Introduction

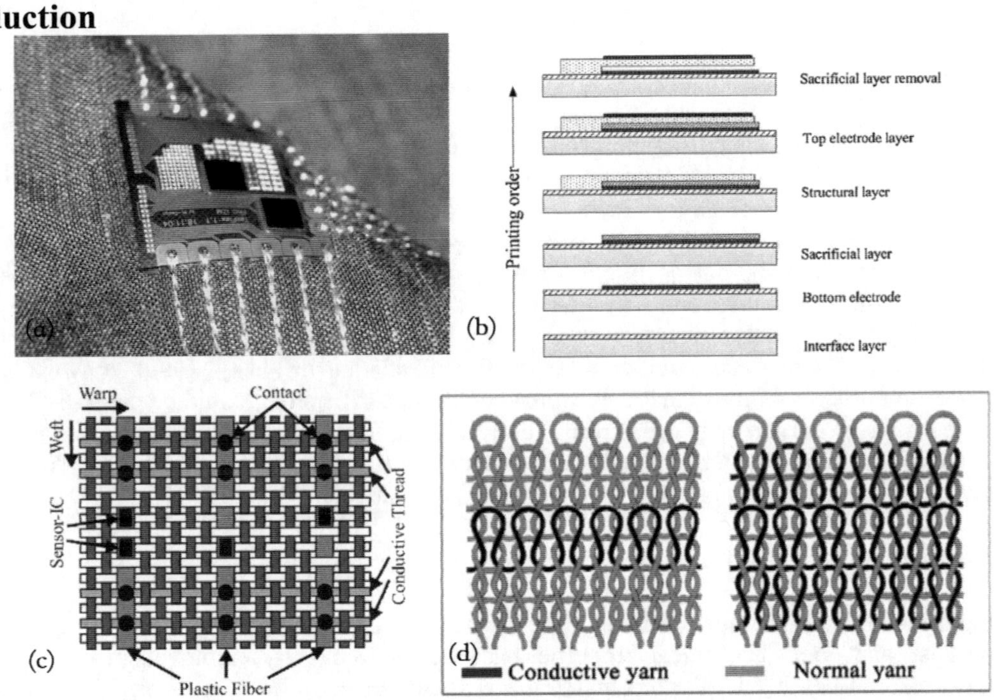

Fig.1 Different kinds of textile/fabric manufacturing and treatment
(a) Flexible substrate with metalized contact pads interconnected with embroidery [11]; (b) printing sequence of a capacitive cantilever through layer-by-layer screen printing process [12]; (c) a schematic of the woven textile with the conductive yarns [13]; (d) a chematic of the knitting textiles with the conductive yarns [14]

The term “E-textiles”, also known as smart clothing or intelligent textiles, refers to a broad field of studies and products where intelligent devices are integrated into the common fabrics and clothing. E-textiles are defined as textile products such as fibers and filaments, yarns can be integrated into the textile structure by embroidering [1], sewing [2], knitting [3], non-woven textile [4], weaving [5], printing [6], braiding [7], and chemical treatments [8] (Figure 1), which can build an intimate form that sense and respond to environment stimuli [9]. The environment stimuli conclude physical, thermal, mechanical and chemical responses e.g. changes in temperature, pressure, force, or others [10].

According to a different degree of sensing, reacting and integrating, e-textiles can be divided into three subgroups [10]: passive smart textiles, active smart textiles, and very smart textiles. Passive smart textiles can only able to sense the environment; active smart textiles can sense and react to stimuli from the environment; very smart textiles can sense, react and adapt their behavior to the given circumstances based on sensors and actuators.

E-textiles show widely potential application in wide areas, such as communications, medical monitoring, sports, military and aerospace where functional wearable e-textiles are designed to perform sensing and communication functions and support specialized and intelligent activities. It will also find applications in daily clothing on the aesthetic and functional aspects. Developing smart e-textile attracts numerous interests of fashion designer, costumers, venture capitals (CV) and scientists and also requires multidisciplinary approach involving textile technology, electronic engineering, fiber science, and human-computer interaction.

This paper reviews the definition and current status of functional wearable e-textiles. It covers conductive materials, textile fabrication technologies and smart textile input and output devices.

2. Conductive Materials

The conductive or semi-conductive materials are essential parts for functional wearable e-textiles. The requirements of materials in e-textiles are soft, flexible, high-conductive [15]. Conducting polymers, carbon-based micro/ nano materials, metallic nanoparticles/ nanowires are the most common conductive materials, in fiber form or as the coating. They have been used widely in sensors, actuators, transistors, flexible electronic devices, and field emission display in the textiles systems [16].

2.1 Conducting Polymers

Electrical conducting polymers are also called polyconjugated polymers, which are the organic materials that have single and double bonds in their chains. The commonly used conducting polymers are polyaniline (PANI), polypyrrrole (PPy), polythiophene (PTh), polyvinyl pyrolidone (PVP), poly (3,4-ethylene dioxy thiophene) (PEDOT), poly (m-phenylene diamine) (PMPD), polynaphthylamine (PNA), poly (p-phenylene sulfide) (PPS), poly (p-phenylene vinylene) (PPV), polyacetylene (PAC), polyfluorene (PF), polyphenylene, polypyrene and polynaphthalene (PN). There are different ways to prepare conducting polymers. For example, polyaniline nanofibers were synthesized by polymerization of aniline [17]; polypyrrole nanofibers were prepared in the presence of p-hydroxy-azobenzene sulfonic acid as a functional dopant [18]. Conducting polymers have conductivity and solution processability so that they were seen as ideal fibers that could be naturally integrated into functional wearable e-textiles during weaving, knitting, braiding, etc. However, the high cost, water solubility and low conductivity of conducting polymers limited their effective and wide usage in e-textiles. Thus, conducting polymers can only be used in components of electronic devices or solar cells [10].

2.2 Carbon-based Micro/Nano Materials

Carbon-based micro/ nano materials, due to the unique properties, such as high intrinsic carrier mobility, superior conductivity, low cost, and good mechanical properties, have been frequently employed in e-textiles for many decades. Nowadays, carbon-based micro/ nano materials (e.g. carbon particles (CP), carbon nanotubes (CNTs), graphene, carbon fibers, etc.) (figure 2), have been widely used in fiber-based sensors, transmission lines, microelectrodes and actuators. For instance, CP is a good commercial e-textiles material because of its high conductivity and low cost; CNTs can be used for flexible energy-storage because of the large specific surface area and the high electrical conductivity; CF are good substrates for industrial production because of the high conductivity [19]. Particularly, carbon fibers are good substrates for e-textiels because they can be woven into fabrics to acquire conductive fabrics. However, the black color of the traditional carbon materials may influence the aesthetics of e-textiles. Graphene, a new member of carbon material family, show high conductivity and high transparent.

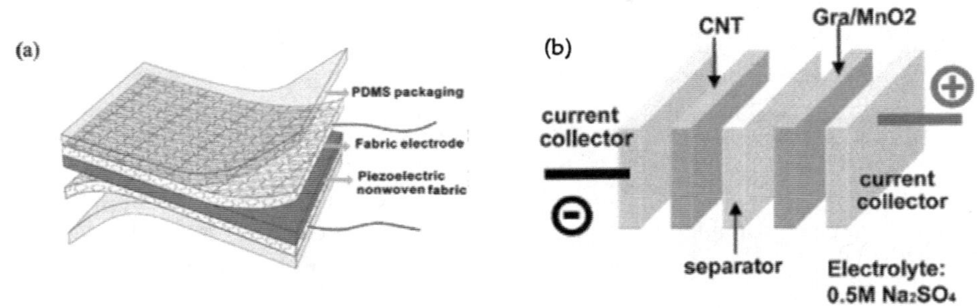

Fig.2 Carbon-based micro/ nano materials
(a) High durable piezoelectric nanogenerator based on PVDF/ NaNbO3 nanocomposite and conductive fabric electrodes [20]; (b) asymmetric (hybrid) ECs based on graphene/ MnO2-textile as positive electrode and CNT-textile as negative electrode [21]

Graphene is the first two-dimensional material, discovered by Prof. Andre Geim and Prof. Konstantin Novoselov at the University of Manchester in 2004. This new carbon nano material is one-atom-thick single layer, which is arranged in a hexagonal honeycomb lattice of pure carbon atoms. Graphene has high conductive, transparence, bendable, and high-stable properties [22] that make it show the potential application in e-textiles. For instance, graphene films can be used as solid-state dye-sensitized solar cells [23]-[24]; photovoltaic devices [25]-[26] tunable graphene antennas [27] gas sensors; pH sensors; biosensors; and heavy-metal sensing [28].

2.3 Metallic Nanoparticles/ Nanowire

Metal materials, such as gold, silver, copper, iron, aluminum, are common conductive materials in electronic devices. In e-textiles, the metal is directly used as fiber or coated in traditional fibers as nano-materials. Low dimensional nanostructures of metals, e.g. metallic nanoparticles and nanowires are introduced into functional wearable e-textiles because of the very high conductivity, low cost, lightweight, thin, and great mechanical flexibility. For example, Polyacrylonitrile (PAN) textiles are coated with magnetic nanoparticles in coating baths with Ni, Co [29]; metal conductors, such as Copper, fabricated by catalytic printing (MACP) [30]. So far, the metallic nano-materials are one of the most adaptable technologies [31].

3. Fabrication and Integration Technologies

3.1 Conductive Wires

Since the first wearable computer has been developed in 1955 [32], numerous technologies have been used in order to realize high integrate functional wearable e-textiles. Conductive yarns play important roles in textile integration since they form textile interconnect lines for electronic devices. They can be widely used in military garments [33], antistatic applications [34], medical apparel [35] and protective clothing [36].

The conventional process to produce metallic wires is drawing, a mechanical production process, which can be characterized by four drawing steps, named coarse, medium, fine and carding train. The wire drawing consists of a steel mount with a core out of ceramics, carbide and diamond.

Another fabrication method is coating conductive matter on the traditional fiber no matter natural or synthetic fiber. After coating, the fiber would maintain the flexibility and stretch-ability. Compared with other metallization methods, such as physical vapor deposition, chemical vapor deposition, and electrochemical plating, electroless deposition (ELD) has made its debut in making metal structures on substrates as a versatile and cheap tool, especially for plastic substrates. However, the most problem of those two methods lies in the low mechanical durability and low wash-ability of metallic wires [31].

3.2 Integration Technologies

Conductive fabrics are the basic foundations of e-textiles. Generally, various approaches to make conductive fabrics can be categorized into two groups. The first method is to integrate conductive yarns into textile structures by weaving, knitting and braiding. The other method is to attach conductive layers or patterns to garments directly by embroidering, gluing, welding and printing.

Weaving involves using a loom to interlace two perpendicular sets of yarns or threads at right angles to form a fabric. The longitudinal threads are named warp and the lateral threads are called weft. During weaving process, the warp threads are moved up or down using a harness. This process creates many different weave structures and provides various functions and performances of fabrics. In e-textiles, we can use conductive yarns to replace some of normal yarns and realize the conductive properties. In addition to those two-dimensional textiles, three-dimensional textiles, nowadays, become popular in e-textiles and other areas. Three-dimensional textiles consider the thickness (Z-axis) relative to planar (X-axis and Y-axis) dimensions, providing an endless number of possibilities for yarn spacing in a three-dimensional space and fabricating more complicated composite three-dimensional fabrics. For instance, Georgia Tech Wearable Motherboard used this method to reduce the need to connect wires during producing, cutting, and sewing to develop an uncut clothing [37].

Knitted fabrics have good elastic and stretch properties because of the looping structures. Knitting is a method by which yarns are manipulated to consecutive loops or stitches on the needle. Knitted textiles consist of a lot of consecutive rows of interlocking loops. By designing the sequence of knitting methods, we can hope for complex patterns in two-dimensional and three-dimensional structures. Furthermore, circular knitting technology allows us to create three-dimensional seamless tubular clothing [38].

This conductive fabric not only can acquire conductivity but also can provide a complex network to have multiple layers and spaces to insert circuits and sensors. The most noticeable advantage of this method is portability, flexibility and comfort when compared to traditional electronic devices. Woven fabrics are characterized by the duration, shape accuracy and have dense integration of electronic devices. Knitted fabrics and braided fabrics have good elasticity, thermal retention and humidity transport properties.

Another method to achieve a conductive fabric is to attach a conductive layer or pattern to the fabric directly like embroidering, gluing, welding, and printing.

Traditionally, embroidery technology was sewing decorations in or on textile fabrics or other surfaces (e.g. leather, plastic), through the application of threads, cords, sequins [10]. Gradually, embroidery technology was used in mechanical engineering [39], wound dressing [40] and e-textiles [41] because of the flexible, washable, comfortable, aesthetic properties. The normal process of embroidering in e-textile concludes four steps: conductive threads produce, computer design, digitization, and pattern embroidery [42]. The main points that will influence the results of e-textiles during embroidering are twist [43], density [44] and stitch direction [45]. Compared to knitting and weaving methods, embroidery technology in e-textiles can design high-accurate patterns and can be used to completed garments directly. For instance, Loughborough University used embroidery and related manufacturing techniques to design wearable antennas [46].

Functional wearable e-textiles can also be produced by printing technique (e.g. inkjet printing, screen printing and stencil printing). Printing is a process for copying text and images using a master form or template. In e-textiles, printing technology is used for reproducing electronic devices by using conductive solution. The conductive solution must contain an appropriate high conductive metal precursor, such as Copper, Silver and Nickel, to acquire conductive patterns. By printing specialized patterns using inkjet printing, screen printing, and stencil printing, functional wearable e-textiles can achieve high electrical conductivity, suitable viscosity, good adhesion to the substrate, and surface flexibility. Compared to other methods, printing offers the advantages of geometric design flexibility and the ability to simultaneously print various devices of the same or even different patterns [47]. For example, the active electrodes for electrocardiogram (ECG) monitoring have been fabricated by the screen and stencil printing [48]; sports-wear and work-wear have been prepared by inkjet printing of waterborne hydrophobic ink [49].

To maintain essential textile properties (e.g. durability, comfort), it is more recommended to integrate electronic devices at the fiber level rather than the fabrics level. On the other hand, however, integrating sensors and circuits in fiber level may meet some difficulties during the process of production, as we have to design the conductive parts of clothing before making it. Besides, the production process may destroy the performances and surface structures of conductive yarns by friction, bending, and high-temperature through sewing, knitting and weaving. All above problems will influence the functionality and usability of e-textiles.

4. E-Textile Input and Output Devices

Input and output devices can be used to generate textile systems. Typical examples are conductive textiles that can change their electrical properties as a result of the environmental impact such as sensors, wearable antennas, and output devices (Figure 3).

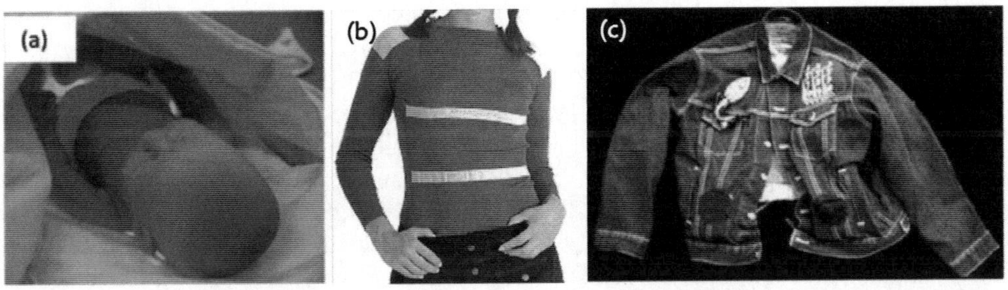

Fig.3 Smart textile input and output devices
(a) Mamagoose pajama developed by Verhaert [50]; sensorized shirt developed by Smartex within the European integrated project WEALTHY [50]; musical jacket comprising a fabric keypad on one side, a MIDI synthesizer on the other side, speakers behind speaker grills in the pockets and fabric buses visible inside the jacket [51]

4.1 Sensors

Sensors are the most mature and successful electronic devices that can measure physical quantities (e.g. temperature, pressure, force, etc.) and convert them into specific signals that can be stored and analyzed. Fibers may acquire their sensing signals via conductive materials, which as shown in part 2. In functional wearable e-textiles, sensors are often inserted into bio-monitoring clothing to measure physiological parameters in humans (e.g. blood pressure, heart rate, body temperature, etc.) or to determine changes in the ambient environment. Because of its application areas, the most important features of sensors are balanced sensitivity, durability and environmental stability.

4.2 Wearable antennas

Wearable antennas allow one to transfer electric information between garments and sensors. Wearable antennas are made of conductive yarns or conductive patterns and are integrated into the communication system [52]. When the distance of the different antennas is changing, the radio waves will change correspondingly. That is why wearable antennas can be used to wireless communications or health monitor functionality integratable into garments. To achieve good results, wearable antennas have to be thin, lightweight, flexible, high conformality, inexpensive, durable, washable and are subject to continuous stress caused by body movements [53].

4.3 Output devices

Output devices are used to communicate changes in the textile system with the ambient environment. In most cases, functional wearable e-textiles are interconnected with traditional display devices (e.g. colored LEDs, optical fibers, watches faces, etc.). In functional wearable e-textiles, output devices can also be transistors. Nowadays, various products to integrate transistors into textiles have been developed [51]. Furthermore, artificial muscles and artificial skins are other examples of textile output devices.

5. Future Perspectives

E-textiles represent an attractive method of electronic devices integrating into fibers and fabrics that are able to sense, compute, communicate and actuate. As various functionality and usage of electronic devices can be connected to clothing, e-textiles become more versatile, and can satisfy different demands and need from environmental changes and individual requirements.

In this paper, a critical review of conductive materials, fabrication technologies and smart textile input and output devices has been presented on the current status of e-textiles. The performance requirements of conductive materials, in particular conducting polymers, carbon-based micro/ nano materials, metallic Nanoparticles/ Nanowires and etc. are examined for developing e-textiles. Summarized descriptions of various fabrication methodologies can make e-textiles more feasible, reliable and advanced.

Critically, more fundamental research is needed to enable the next wave of e-textiles. Firstly, we are still far from fully utilizing the high-integrate and comfort form to fulfill the final application requirements. At present, the electronic devices are still separate to textiles or clothing and not intelligent enough. In the future, more and more high integrate electronic devices, which places a demand for high performance in terms of electronic functions, structural and functional integrity and stability during use, should be developed in fiber-based techniques by academic research labs. Furthermore, the e-textiles are far from automated, and large-area, low-cost and widely used e-textiles will only be achieved by seamless fabrication and integration technologies. Meanwhile, graphene and hybrid conductive materials can be attractable prospects in conductive materials because they have excellent transparency, electrical conductivity and good flexibility, together with high thermal stability and are easily processed in both water and organic solvents.

From the discussion above it becomes clear that there are still a lot of research gaps that need to be advanced before e-textiles become a mature technique. However, this fact also represents a golden opportunity to develop a new research area and expand the knowledge available to scientists. E-textiles research is a multiple discipline concerning electronic engineering, composite chemistry, polymer science, and textile technology. Thus, scientific understanding and proper design of conductive materials and electronic devices will lead to a new model for generating creative technologies for functional wearable e-textiles.

References

[1] Tsolis A, Whittow WG, Alexandridis AA, Vardaxoglou GC. Embroidery and Related Manufacturing Techniques for Wearable Antennas: Challenges and Opportunities. Electronics: 2014; 3 (2); pp: 314-338.

[2] Barbara J. Shape Shape: Sewing Clothing Patterns to Wear Multiple Ways. Booklist: 2012; 108 (15); pp: 13.

[3] Li L, Au WM, Li Y, Wan KM, Chung WY, Wong KS. A Novel Design Method for an Intelligent Clothing Based on Garment Design and Knitting Technology. Textile Research Journal: 2009; 79 (18); pp: 1670-1679.

[4] Kothari VK, Chakraborty S. Thermal protective performance of clothing exposed to radiant heat. The Journal of The Textile Institute: 2015; 106 (12); pp: 1388-1393.

[5] Stuart A, Monica L. Wright, Weaving Narrative: Clothing in Twelfth-Century French Romance. Medium Aevum: 2011; 80 (1); pp: 146.

[6] Dehghani A, Jahanshah F, Borman D, Dennis K, Wang J. Design and engineering challenges for digital ink-jet printing on textiles. International Journal of Clothing Science and Technology: 2004; 16 (1/2); pp: 262-273.

[7] Cho G, Jeong K, Paik MJ, Kwun Y, Sung M. Performance Evaluation of Textile-Based Electrodes and Motion Sensors for Smart Clothing. IEEE Sensors Journal: 2011; 11 (12); pp: 3183-3193.

[8] Havenith G, Den Hartog E, Martini S. Heat stress in chemical protective clothing: porosity and vapour resistance. Ergonomics: 2011; 54 (5); pp: 497-507.

[9] Van Langenhove L, Hertleer G. Smart clothing: A new life. International Journal Clothing Science and Technology: 2004; 16; pp: 63-72.

[10] Tao XM. Smart technology for textiles and clothing. Burlington: Elsevier Science: 2001; pp: 1-6.

[11] Linz T, Vieroth R, Dils C, Koch M, Braun T, Becker K, Kallmayer C, Hong S. Embroidered Interconnections and Encapsulation for Electronics in Textiles for Wearable Electronics Applications. Advances in Science and Technology: 2008; 60; pp: 85-94.

[12] Wei Y, Torah R, Yang K, Beeby S, Tudor J. Screen printing of a capacitive cantilever-based motion sensor on

fabric using a novel sacrificial layer process for smart fabric applications. Measurement Science and Technology: 2013; 24; pp: 1-11.
[13] Cherenack K, Zysset C, Kinkeldei T, Münzenrieder N, Tröster G. Woven Electronic Fibers with Sensing and Display Functions for Smart Textiles. Advanced Materials: 2010; 22; pp: 5178-5182.
[14] Zysset C, Cherenack K, Kinkeldei T, Troster G. Weaving Integrated Circuits into Textiles. IEE: 2010; 10(13); pp: 1-8.
[15] Mann DL, Acker MA, Jessup M, Sabbah HN, Starling RC, Kubo SH. Rationale, design, and methods for a pivotal randomized clinical trial for the assessment of a cardiac support device in patients with New York health association class III-IV heart failure. Journal of Cardiac Failure: 2004; 10(3); pp: 185-192.
[16] Jang SY, Seshadri V, Khil MS, Kumar A, Marquez M, Mather PT, Sotzing GA. Welded Electrochromic Conductive Polymer Nanofibers by Electrostatic Spinning. Advanced Materials: 2005; 17: 2177.
[17] Hatchett DW, Josowicz M. Composites of Intrinsically Conducting Polymers as Sensing Nanomaterials. Chemical Reviews: 2008; 108(2); pp: 746-769.
[18] Huang JX, Kaner RB. Nanofiber Formation in the Chemical Polymerization of Aniline: A Mechanistic Study. Angewandte Chemie: 2004; 43(43); pp: 5817-5821.
[19] Wang YH, Bian K, Hu CG, Zhang ZP, Chen N, Zhang HM, Qu LT. Flexible and wearable graphene/ polypyrrole fibers towards multifunctional actuator applications. Electrochemistry Communications: 2013; 35; pp: 49-52.
[20] Zeng W, Shu L, Li Q, Chen S, Wang F, Tao XM. Fiber-Based Wearable Electronics: A Review of Materials, Fabrication, Devices, and Applications. Advanced Materials: 2014; 26; pp: 5310-5336.
[21] Yu GH, Hu LB, Vosgueritchian M, Wang HL, Xie X, McDonough JR, Cui X, Cui Y, Bao ZN. Solution-Processed Graphene/ MnO2 Nanostructured Textiles for High-Performance Electrochemical Capacitors. American Chemical Society: 2011; 11; pp: 2905-2911.
[22] Chang HX, Wang GF, Yang A, Tao XM, Liu XQ, Shen YD, Zheng ZJ. A Transparent, Flexible, Low-Temperature, and Solution-Processible Graphene Composite Electrode. Advanced Functional Materials: 2010; 20; pp: 2893-2902.
[23] Wang X, Zhi LJ, Müllen K. Transparent, Conductive Graphene Electrodes for Dye-Sensitized Solar Cells. Nano letters: 2008; 8(1); pp: 323-327.
[24] Miao XC, Tongay S, Petterson MK, Berke K, Rinzler AG, Appleton BR, Hebard AF. High Efficiency Graphene Solar Cells by Chemical Doping. Nano Letters: 2012; 12(6); pp: 2745-2750.
[25] Balandin AA, Ghosh S, Bao WZ, Calizo I, Teweldebrhan D, Miao F, Lau CN. Superior Thermal Conductivity of Single-Layer Graphene. Nano Letters: 2008; 8(3); pp: 902-907.
[26] Kakavelakis G, Konios D, Stratakis E, Kymakis E. Enhancement of the Efficiency and Stability of Organic Photovoltaic Devices via the Addition of a Lithium-Neutralized Graphene Oxide Electron-Transporting Layer. Chemistry of Materials: 2014; 26(20); pp: 5988-5993.
[27] Filter R, Farhat M, Steglich M, Alaee R, Rockstuhl C, Lederer F. Tunable graphene antennas for selective enhancement of THz-emission. Optics Express: 2013; 21(3); pp: 3737-3745.
[28] He QY, Wu SX, Yin ZY, Zhang H. Graphene-based electronic sensors. Royal Society of Chemistry: 2012; 3; pp: 1764-1772.
[29] Akmana O, Kavas H, Baykal A, Toprake MS, Çoruh A, Aktaş B. Magnetic metal nanoparticles coated polyacrylonitrile textiles as microwave absorber. Journal of Magnetism and Magnetic Materials: 2012; 327; pp: 151-158.
[30] Guo RG, Yu Y, Xie Z, Liu XQ, Zhou XC, Gao YF, Liu ZL, Zhou F, Yang Y, Zheng ZJ. Matrix-Assisted Catalytic Printing for the Fabrication of Multiscale, Flexible, Foldable, and Stretchable Metal Conductors. Advanced Materials: 2013; 25(24); pp: 3343-3350.
[31] Liu XQ, Chang HX, Li Y, Huck WTS, Zheng ZJ. Polyelectrolyte-Bridged Metal/ Cotton Hierarchical Structures for Highly Durable Conductive Yarns. Applied Materials and Interfaces: 2010; 2(2); pp: 529-535.
[32] Thorpe EO. The invention of the first wearable computer. Proceedings of 2nd International Symposium on Wearable Computers. Digest of Papers: 1998; pp: 4–8.
[33] Winterhalter C, Teverovsky J, Wilson P, Slade J, Farell B, Horowitz W, Tierney E. Development of electronic textiles for u.s. military protective clothing systems. Studies in Health Technology and Informatics: 2005; 108; pp: 194-198.
[34] Palaniappan S, Ajit S. Preparation of processible polyaniline-formate salt via emulsifier free inverted emulsion

polymerization and its antistatic applications. Synthetic Metals: 2011; 161 (11); pp: 1029-1033.
[35] Ng SF, Hui CL, Wong LF. Development of medical garments and apparel for the elderly and the disabled. Textile Progress: 2011; 43 (4); pp: 235-285.
[36] Zhu FL, Li KJ, Zhu FL. Fractal Analysis of Effective Thermal Conductivity for Woven Fabrics used in Fire Fighters' Protective Clothing. Journal of Fire Sciences: 2011; 29 (1); pp: 3-20.
[37] Park S, Mackenzie K, Jayaraman S. The wearable motherboard: A framework for personalized mobile information processing (PMIP). Annual Design Automation Conference: 2002; 11(2); pp: 170-174.
[38] Yu W. Advanced in Women's Intimate Apparel Technology. Cambridge: The Textile Institute: 2016; pp: 53-66.
[39] Cui WG, Zhou JR, Li ZF. Dynamic analysis and optimization of beam of high-speed computerized embroidery machine. Manufacturing Automation: 2013; 17; pp: 56-62.
[40] Mecnika V, Hoerr M, Krievins I, Jockenhoevel S, Gries T. Technical Embroidery for Smart Textiles: Review. Rigas Tehniskas Universitates Zinatniskie Raksti: 2014; 9; pp: 56.
[41] Trindade IG, Spranger P, Martins F, Miguel R, Pereira M, Santos Silva M. Fully Integrated Embroidery Process for Smart Textiles. Nanotechnology: 2014; 3; pp: 65-68.
[42] Wang Z, Zhang L, Bayram Y, Volakis JL. Embroidered Conductive Fibers on Polymer Composite for Conformal Antennas. IEEE Transactions on Antennas and Propagation: 2012; 60(9); pp: 4141-4147.
[43] Choi J, Kim Y, Lee K, Chung YC. Various wearable embroidery RFID tag antenna using electro-thread. Antennas and Propagation: 2008; pp: 1-4.
[44] Ukkonen L, Sydanheimo L, Rahmat-Samii Y. Sewed textile RFID tag and sensor antennas for on-body use. Sixth European Conf. on Antennas and Propagation (EUCAP): 2012; pp: 3450-3454.
[45] Seager R, Zhang S, Chauraya A. Effect of the fabrication parameters on the performance of embroidered antennas. IET Microwaves, Antennas & Propagation: 2013; 7(14); pp: 1174-1181.
[46] Tsolis A, Whittow WG, Alexandridis AA, Vardaxoglou JC. Embroidery and Related Manufacturing Techniques for Wearable Antennas: Challenges and Opportunities. Electronics: 2014; 3(2); pp: 314-338.
[47] Wei Y, Torah R, Yang k, Beeby S, Tudor J. Screen printing of a capacitive cantilever-based motion sensor on fabric using a novel sacrificial layer process for smart fabric applications. Measurement Science and Technology: 2013; 24(7); pp: 13.
[48] Paul G, Torah R, Beeby S, Tudor J. Novel active electrodes for ECG monitoring on woven textiles fabricated by screen and stencil printing. Sensors and Actuators A: Physical: 2014; 221; pp: 60-66.
[49] Yu JC, Nierstrasz V, Seipel S. Inkjet printing of waterborne hydrophobic ink for functionalization of textile. Natural Sciences: 2015; pp: 32-37.
[50] Carpi F, De RossiD. Electroactive Polymer-Based Devices for e-Textiles in Biomedicine. IEEE Transactions on Information Technology in Biomedicine: 2015; 9(3); pp: 295-318.
[51] Stoppa M, Chiolerio A. Wearable Electronics and Smart Textiles: A Critical Review. Sensors: 2014; 14; pp: 11957-11992.
[52] Elham M, Karoliina K, Toni B, Lauri S, Jan MR, Jose MC, Yahya R, Leena U. Miniature Implantable and Wearable On-Body Antennas: Towards the New Era of Wireless Body-Centric Systems. IEEE Antennas and Propagation Magazine: 2014; 56(1); pp: 271-291.
[53] Kiourti A, Volakis J. Stretchable and flexible e-fiber wire antennas embedded in polymer. IEEE antennas and wireless propagation letters: 2014; 13; pp: 1381-1384.

The Flammability of Aerogel and PCM Coated Cotton Fabric for Firefighter's Garment

Abu Shaid[1], Lijing Wang[1*], Rajiv Padhye[1]

[1]*RMIT University, 25 Dawson St, Melbourne, VIC 3056, Australia*

*Corresponding author's email: lijing.wang@rmit.edu.au

Abstract

The use of phase change material (PCM) in fire fighter's protective gear is a growing interest. However the most suitable PCM used are petroleum derived and flammable. In current approach nanoporous silica aerogel was used in conjunction of PCM to improvise its flame protection capability. Fabric samples were coated with/ without aerogel or PCM and the ease of ignition of coated samples was investigated. It has been found that incorporation of aerogel provides better heat resistance, reduces ignition and slows down the flame propagation. Finally the air permeability of the samples was also analysed and found that Aerogel has negative impact on air permeability of the coated fabric.

Keywords: PCM; Aerogel; Ignition; Protection

1. Introduction

The phase change materials (PCMs) have long been used in clothing as a mean of body heat regulation for comfort enhancement. These materials absorb the extra heat generated by human body during action and release it when the body is cool. Including both natural and synthetic types, there are more than 500 types of PCMs [1-3]. Among all types of PCMs, paraffins are commonly used in clothing for thermal regulation as they have high heat storage capacity [4]. However, they are flammable and the binders used to apply the PCM on fabrics are also usually flammable [5]. In recent years, the concept of using PCM for firefighters protective clothing (FFPC) is gaining ground. Gao [6,7] investigated the effect of PCM on ergonomics of firefighters by introducing a cooling vest containing pockets of PCM. McCarthy and Marzo [3] incorporated PCM in FFPC by sewing different types of PCM between the batting and face cloth of the thermal liner. Rossi and Bolli [8] had also investigated the application of PCM in FFPC. They integrated PCM into foam for coating on a liner. There are few other studies available where mathematical models showed the good prospect of PCM incorporation in FFPC [9-12].

The application of flammable PCM in the FFPC field where the product may expose to heat and flame is a critical issue to consider. Rossi and Bolli [8] acknowledged that the burning behaviour of PCM containing layer in FFPC is a problem to be solved. Cardoso and Gomes [5] considered the flammability problem of PCM in case of FFPC and used melamine-formaldehyde to address the issue. They concluded that a conjugation of microcapsule protection and flame retardant finishing is appropriate to achieve flame retardant standard of a nonwoven lining inside FFPC [5]. Several other reports also used melamine-formaldehyde with PCM [13-15]. However, melamine-formaldehyde is not recommended to use in clothing which will come to direct skin contact due to their chronic allergic effect [16]. We have studied the thermal protection and comfort properties of aerogel and PCM-coated fabric for firefighter garment [17,18]. Current study aimed to present a resolution by using aerogel along with PCM to improve the flame protection.

Aerogel is an extremely light weight nanoporous material, derived from a gel by replacing its liquid component with a gas [19]. Among different types of aerogel, silica aerogel has several fascinating properties. Silica is nonflammable and thermal conductivity of silica aerogel is significantly lower than air under same condition [20]. It has high specific surface area (~1000 m^2/g), high porosity (> 90%), low thermal conductivity (~0.015W/mK), low bulk density (~0.1 g/cm^3), and hydrophobicity [21-27]. Aerogel insulations provides outstanding resistance to flame spread [28]. Hence silica aerogel has bright prospect to be used in fire protective clothing to enhance flame protection. The current study incorporated silica aerogel particle with form stable particulate phase change material to enhance flame protection of PCM containing layer. Then the ease of ignition was investigated. Additionally air permeability of the coated fabric was also analysed.

2. Experimental

2.1 Material

Rubitherm PX 52 heat storage powder from RUBITHERM GmbH was used as PCM material. Enova aerogel particle with particle size 7 to 11 µm was obtained from Cabot Corporation and used as received. Coating paste was purchased from Permaset, Australia. The fabric used was 100% cotton woven fabric.

2.2 Coating

Aerogel and PCM were coated on fabric by SV-MATIS laboratory coating machine using a coat-dry-cure method. Details of the coating formulation are shown in Table 1. As an example it can be stated here that for sample B, 5g of aerogel particles were slowly added with 150g of Permaset aqua binder while continuously stirring. To thin the paste, several drops of ammonium sulphate were added and a high speed blender was used to make the paste consistent. The paste was then coated on fabric specimens taking similar amount of coating for each sample. The coated fabric was dried at 60 °C, cooled and coated for second time, dried at 60 °C and finally cured at 160 °C. Samples A and C were also coated on similar fashion whereas in the case of sample D first coating was with PCM and then second coat was aerogel.

Table 1. Test samples description

Sample code	Coating description	Coating formulation / Coating ingredient
A	Coat on coat with binder	Binder only.
B	Coat on coat with Aerogel	5 g Aerogel in 150 g binder (paste A)
C	Coat on coat with PCM	5 g PCM in 150 g binder (paste B)
D	Aerogel coat on PCM coated fabric	Paste B for fast coat; Paste A for second coat

2.3 Scanning electron microscopy (SEM)

A FEI Quanta200 SEM was used to analyse the surface of the coated fabrics.

2.4 Ease of ignition

The ease of ignition of coated fabric samples were tested according to AS 2755.1 (ISO 6940) test method. It describes the ease of ignition through minimum ignition time which is defined as the 'minimum time of exposure of a material to an ignition source to obtain sustained combustion under specified test conditions' [29]. Coated fabric samples were vertically oriented and a defined ignition flame from a specified burner was applied. Then the time required to achieve ignition was determined as the mean of the measured times for ignition of that particular test fabric. The air movement was less than 0.2 m/s in the location and the samples were mounted at least 300 mm from any wall. Specimens were tested until at least five cases of ignition and five cases of non-ignition occurred. The flame height was adjusted to 40±2 mm as shown in Fig.1(a).

The burner was positioned perpendicular to the surface of the specimen as shown in Fig.1(b). Ignition was deemed to have occurred if the flame on the specimen persists at least 5 seconds after removal of the ignition flame. Timer setting was reduced by 1 second if the ignition occurred and timer setting was increased by 1 second when non ignition occurred. Thus the test continued until at least five ignitions and five non-ignitions had found. The result is expressed as *mean ignition time* which is calculated using Equation 1.

$$Mean\ ignition\ time = \frac{\Sigma(Time\ in\ second \times Number\ of\ cases\ in\ that\ specific\ time)}{Total\ Number\ of\ cases} \pm C \qquad (1)$$

(a) 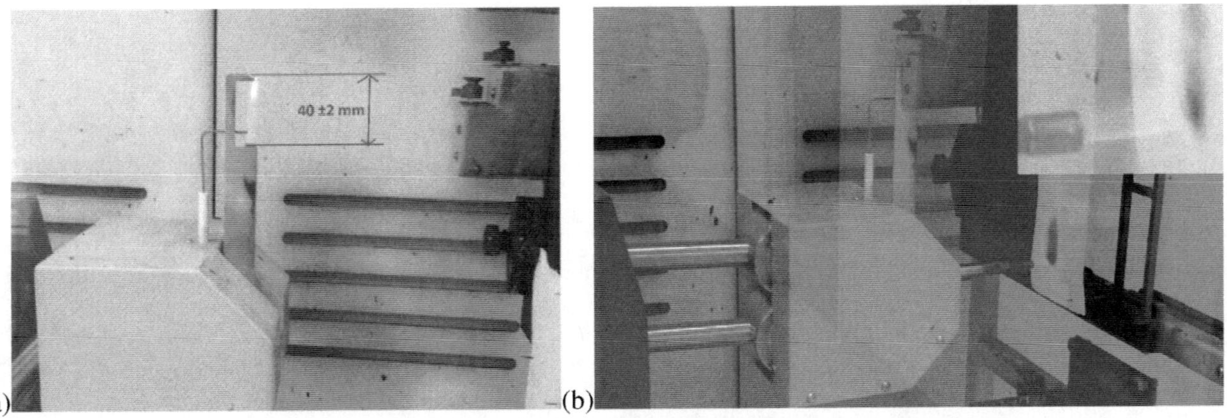

(b)

Fig.1 (a) Flame height; (b) Specimen exposed to ignition source

In Eq.1, the case is either ignition or non-ignition whichever occurs least according to appendix-B of the test standard AS 2755.1-1995. C is a constant value of 0.5 which is positive when non-ignition is least and negative when ignition is least. The resulting value is rounded to the nearest second and reported as the *mean ignition time*. As an example, the determination of *mean ignition time* for sample D is given here.

Total 12 specimens were tested in this case and the results are tabulated in Table 2. Based on the above tests, the summary of the results is shown in Table 3.

Table 2. Behaviour of test specimens when exposed to ignition flame at different time interval

Test Number	Time (second)	Result	Test Number	Time (second)	Result
1	4	Ignition	7	4	Ignition
2	3	Non-Ignition	8	3	Non-Ignition
3	4	Non-Ignition	9	4	Ignition
4	5	Ignition	10	3	Non-Ignition
5	4	Non-Ignition	11	4	Ignition
6	5	Ignition	12	3 (Not considered as five values for both ignition and non-ignition had achieved)	

Table 3. Ignition and non-ignition summary

Time (second)	Number of case of ignition	Number of cases of non-ignition
5	2	0
4	4	2
3	0	3

Table 3 indicates that non-ignition occurred least (i.e. there were six cases of ignition and five cases of non-ignition). Hence the mean of the recorded times at which non-ignition was observed is calculated according to Equation 1 as follows:

$$Mean\ ignition\ time = \frac{(2 \times 4s) + (3 \times 3s)}{2 + 3} + 0.5 = 3.9\ ,ie\ 4\ seconds \tag{2}$$

2.5 Air Permeability

The air permeability of the coated specimens was determined by using SDL air permeability tester according to the test standard EN ISO 9237:1995.

3. Result and discussion

3.1 SEM analysis

The SEM images in Fig.2 prove that coated fabric samples had reasonable amount of aerogel and PCM particles on their surfaces. However, the images also revealed that the surface coating significantly reduced the fabric porosity.

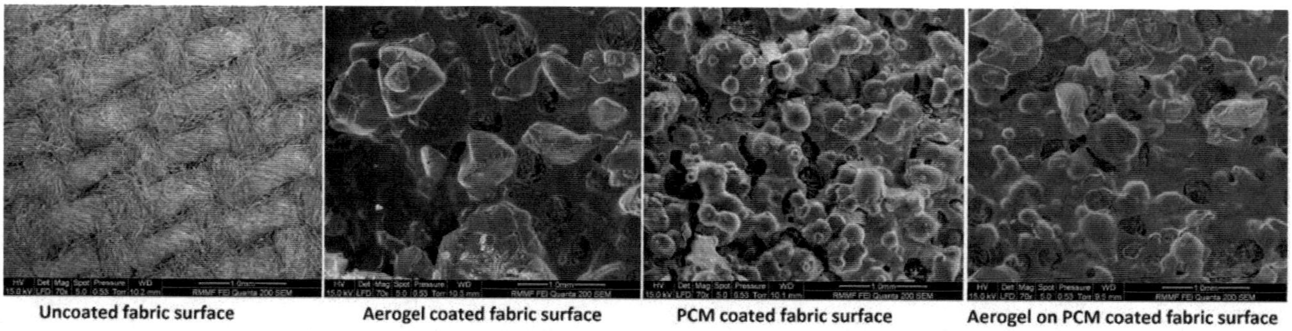

Fig.2 Microscopic image of coated and uncoated fabric surfaces

3.2 Ease of ignition

The *mean ignition time* for all the samples, rounded in nearest second, is shown in Table 4. However, the actual differences can be more precisely seen from the partial values as shown in Fig.3.

Table 4. Ignition and non-ignition case summary for all four types of samples.

	Sample A			Sample B			Sample C			Sample D		
	Time (s)	Number of cases of ignition	Number of cases of Non-ignition	Time (s)	Number of cases of ignition	Number of cases of Non-ignition	Time (s)	Number of cases of ignition	Number of cases of Non-ignition	Time (s)	Number of cases of ignition	Number of cases of Non-ignition
	4	3	0	5	1	0	4	2	0	5	2	0
	3	2	4	4	4	1	3	3	3	4	4	2
	2	0	2	3	0	5	2	0	3	3	0	3
Mean Ignition time	3.1			3.7			2.9			3.9		
Nearest round value	3			4			3			4		

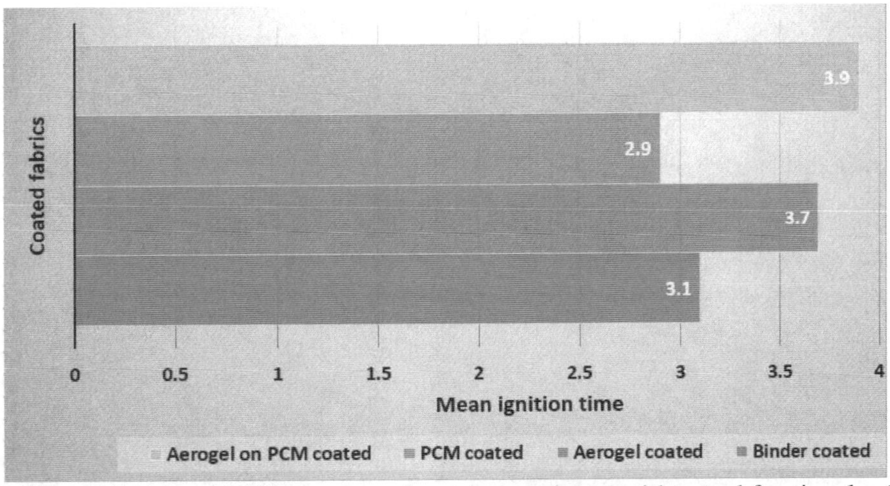

Fig.3 Ease of ignition of different coated specimens shown with actual fractional values

As seen in Table 4 and Fig.3, the incorporation of aerogel improved the flame resistance by showing higher *mean ignition time*. Sample C was coated only with PCM and it ignited comparatively more rapidly. From the standard test results in Table 4, it can be said that PCM coated fabric behaved similarly like the fabric which was coated only with binder. However, in reality sample C ignited even more rapidly than sample A, which can be clearly seen form Fig.3. On the other hand, aerogel coated fabric (sample B) showed the best resistance to direct flame exposure. The vital fact to notice in the test results is that the inferior flame resistance of PCM coated fabric (sample C) was improved when aerogel was incorporated with it (sample D).

During the test, the flame propagation was also observed carefully. PCM coated specimens started to burn more readily and in several cases it was found that PCM coated specimens started to burn from back surface first, instead of the front surface where flame was in direct contact and soon after flame was visible on face surface as well. This phenomenon was more observable when PCM coated samples were exposed for longer duration such as 5 seconds and 7 seconds or when the PCM content was increased. The results can be seen from Fig.4 where comparatively more flame propagation on back surface is evident.

To test the higher PCM content, another sample (sample E) was prepared by adding 30g PCM powder in the same amount of binder. Then it was found that the increase of PCM content among samples (sample E > sample C) even worsen flame propagation behaviour. Though both sample E and sample C burns at 4 seconds, in several cases the flam propagates more rapidly, also with that back burning issue, on the sample that contains higher PCM content as shown in Fig.4 (c) and (d). This phenomenon of back surface burning was never experienced for the specimens that contained aerogel. Hence from the test data of *mean ignition time* and flame propagation, it can be concluded that aerogel particles can be used as shielding material against direct flame exposure for the PCM containing fabrics which are intended for fire protective clothing, such as firefighter garment.

3.3 Air Permeability

The air permeability of the coated specimens was also tested. The results are shown in Fig.5. It can be seen that the incorporation of aerogel reduces the air permeability. Though this finding may have negative impact on comfort properties, lower air permeability can be taken as desirable property in protection perspective as it will reduce the convective heat transfer from external high heat sources.

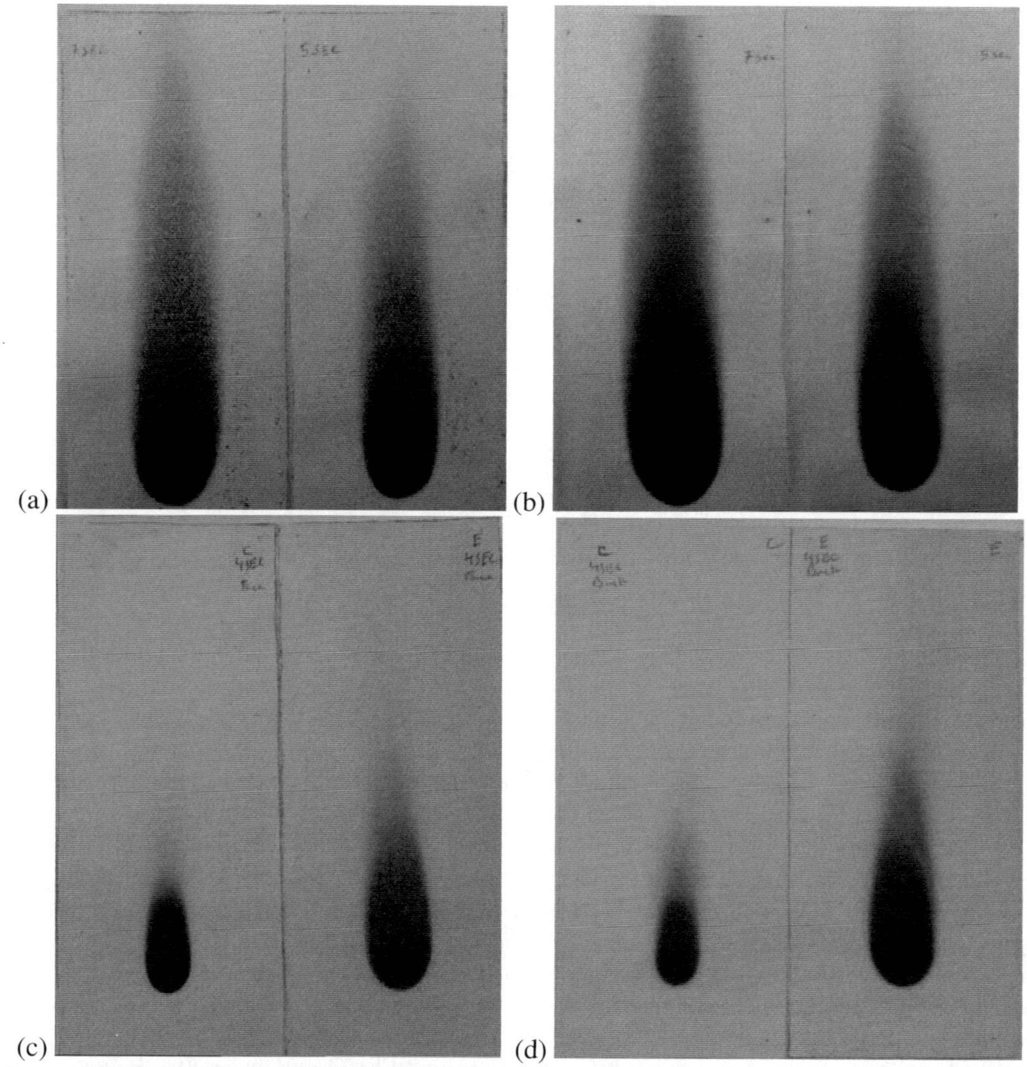

Fig.4 Ignition of PCM coated fabric: effect of time 5 and 7 seconds on - (a) face and (b) back surface; effect of PCM content on (c) face; (d) back surface

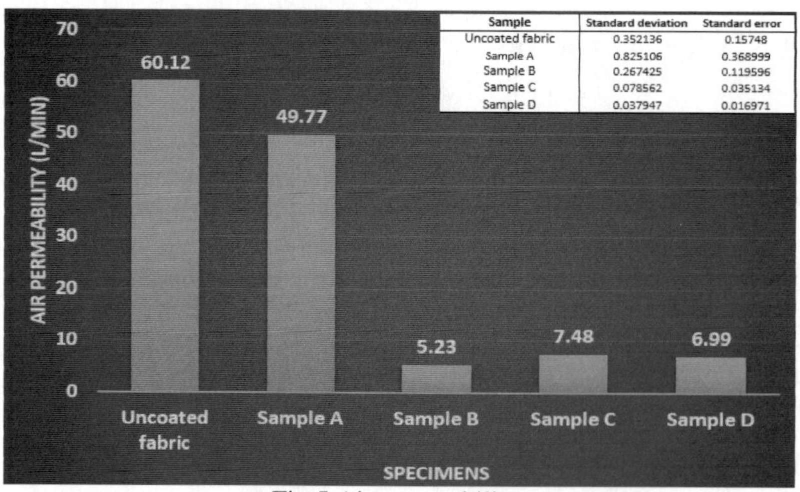

Sample	Standard deviation	Standard error
Uncoated fabric	0.352136	0.15748
Sample A	0.825106	0.368999
Sample B	0.267425	0.119596
Sample C	0.078562	0.035134
Sample D	0.037947	0.016971

Fig.5 Air permeability

4. Conclusion

The flame resistance of silica aerogel is well known. However its application in diversified areas has yet to be explored with full potential. Current study investigated the potential application of aerogel to improve the flame protection of PCM coated fabrics where the intended clothing has high chance of direct flame exposure in their application field, such as in fire fighter's garment. It had been found that aerogel can be successfully used to reduce the flammability of PCM in clothing. The use of aerogel decreases the risk of ignition by increasing the mean ignition time of PCM coated fabric. Thus the findings open a wide door of imminent further research to explore the extent of this type of flame protection and its efficient application in various fields.

References

[1] Hale DV, Hoover MJ, ONeill MJ. "Phase change materials handbook". Bradford Drive, Huntsville, Alabam: NASA and Lockheed Missiles and Space Co., 1971.

[2] Pause B. Nonwoven protective garments with thermo-regulating properties. J. Ind. Text. 2003; 33: 93-99.

[3] McCarthy L, Marzo M. The application of phase change material in fire fighter protective clothing. Fire Technology 2012; 48: 841-864.

[4] Fang G, Li H, Chen Z, Liu X. Preparation and characterization of flame retardant n-hexadecane/ silicon dioxide composites as thermal energy storage materials. Journal of Hazardous Materials 2010; 181: 1004-1009.

[5] Cardoso I, Gomes JR. The application of microcapsules of PCM in flame resistant non-woven materials. Int. J. Cloth. Sci. Technol. 2009; 21: 102-108.

[6] Gao C, Kuklane K, Holmer I. Cooling vests with phase change material packs: The effects of temperature gradient, mass and covering area. Ergonomics 2010; 53: 716-723.

[7] Gao C, Kuklane K, Holmér I. Cooling vests with phase change materials: The effects of melting temperature on heat strain alleviation in an extremely hot environment. Europ. J. Appl. Physiol. 2011; 111: 1207-1216.

[8] Rossi RM, Bolli WP. Phase change materials for the improvement of heat protection. Adv. Eng. Mater 2005; Vol. 7: pp: 368-373.

[9] Mercer GN, Sidhu HS. Mathematical modelling of the effect of fire exposure on a new type of protective clothing. ANZIAM Journal 2008; 49: C289-C305.

[10] Hu Y, Huang D, Qi Z, He S, Yang H, Zhang H. Modeling thermal insulation of firefighting protective clothing embedded with phase change material. Heat and Mass Transfer 2013; 49: 567-573.

[11] Mercer GN, Sidhu HS. Investigating the effectiveness of phase change material as a new type of protective clothing for firefighters. Chemeca 2008: Towards a Sustainable Australasia 2008: 1968.

[12] Prasad K, Twilley WH, Lawson JR. "Thermal performance of fire fighters' protective clothing: Numerical study of transient heat and water vapor transfer". US Department of Commerce, Technology Administration, National Institute of Standards and Technology, 2002.

[13] Mohaddes F, Islam S, Shanks R, Fergusson M, Wang L, Padhye R. Modification and evaluation of thermal properties of melamine-formaldehyde/ n-eicosane microcapsules for thermo-regulation applications. Appl. Therm. Eng. 2014; Vol. 71: pp: 11-15.

[14] Shin Y, Yoo DI, Son K. Development of thermoregulating textile materials with microencapsulated phase change materials (PCM). II. Preparation and application of pcm microcapsules. J Appl Polym Sci 2005; 96

[15] Choi K, Cho G, Kim P, Cho C. Thermal storage/ release and mechanical properties of phase change materials on polyester fabrics. Text. Res. J. 2004; 74: 292-296.

[16] De Groot AC, Flyvholm M-A, Lensen G, Menné T, Coenraads P-J. Formaldehyde-releasers: Relationship to formaldehyde contact allergy. Contact allergy to formaldehyde and inventory of formaldehyde-releasers. Contact Dermatitis 2009; 61: 63-85.

[17] Shaid A, Furgusson M, Wang L. Thermophysiological comfort analysis of aerogel nanoparticle incorporated fabric for fire fighter's protective clothing. Chemical and Materials Engineering 2014; 2: 37 - 43.

[18] Shaid A, Wang L, Padhye R. The thermal protection and comfort properties of aerogel and PCM-coated fabric for firefighter garment. J. Ind. Text. 2016; 45: 611-625.

[19] Thapliyal PC, Singh K. Aerogels as promising thermal insulating materials: An overview. Journal of Materials 2014; 2014: 10.

[20] Pierre AC, Rigacci A SiO2 aerogels. In Aerogels handbook. Springer, 2011; p: 21-45.
[21] Bheekhun N, Abu Talib AR, Hassan MR. Aerogels in aerospace: An overview. Advances in Materials Science and Engineering 2013; 2013: 18.
[22] Mulder C, Van Lierop J Preparation. densification and characterization of autoclave dried SiO2 gels. In Aerogels. Springer, 1986; P: 68-75.
[23] Pierre AC, Pajonk GM. Chemistry of aerogels and their applications. Chemical Reviews 2002; 102: 4243-4265.
[24] Hrubesh LW. Aerogels: The world's lightest solids. Chemistry and Industry (London) 1990: 824-827.
[25] Soleimani Dorcheh A, Abbasi MH. Silica aerogel; synthesis, properties and characterization. J. Mater. Process. Technol. 2008; 199: 10-26.
[26] Carraher CE. General topics. Polymer News 2005; 30: 386-388.
[27] Hrubesh LW. Aerogel applications. Journal of Non-Crystalline Solids 1998; 225: 335-342.
[28] Walkoff H. Space material meets industry. Chemical Engineer 2007: 24-25.
[29] Australia S. As 2755.1:Textile fabrics—burning behaviour, Part 1: Determination of ease of ignition of vertically oriented specimens (Vol. AS2755.1). Australia, 1985.

EMI Shielding Efficiency of Composite Materials Made of Conductive UHMWPE/PANI Composite Yarn

Jian-Han Hong[1,2*], Xiao Han[1], Zhi-Juan Pan[3,4], Mu Yao[5]

[1]*College of Textile and Garment, Shaoxing University, Huancheng West Road, Shaoxing, 31200, China*
[2]*Suzhou Institute of Trade & Commerce, Xuefu Road, Suzhou, 215009, China*
[3]*College of Textile & Clothing Engineering, Soochow University, Ganjiang West Road , Suzhou, 215001, China*
[4]*National Engineering Laboratory for Modern Silk(Suzhou), Renai Road, Suzhou, 215123, China*
[5]*Faculty of Textile & Material, Xi'an Polytechnic University, South Jinhua Road, Xi'an, 710048, China*

*Corresponding author's email: hongjianhan@sina.com

Abstract

The conductive UHMWPE/PANI composite yarn was continuously manufactured using a novel method based on in-situ polymerization, and electromagnetic shielding composite materials were prepared using the UHMWPE/PANI yarn in this paper. The effects of arrangement density and arrangement angle of UHMWPE/PANI yarns on the electromagnetic interference (EMI) shielding efficiency (SE) of composite materials were studied. The results indicated that the shielding efficiency increased with the increase of the density of the UHMWPE/PANI yarn. The SE could reach more than 10dB, when two parallel yarn layers with the distance of 1mm between the adjacent yarns in each layer were overlapped perpendicularly. The arrangement angle of two layers also affected the shielding effectiveness. The SE was the highest when the arrangement angle was 90 °, and the anti-electromagnetic radiation effect decreased gradually with the decrease of the arrangement angle.

Keywords: UHMWPE; Polyaniline; Conductive yarn; Composite materials; Electromagnetic interference (EMI); Shielding efficiency (SE)

1. Introduction

Ultra-high molecular weight polyethylene (UHMWPE) fiber has been widely used in fiber-reinforced composite materials because of its high strength, light weight and other excellent properties. However, the fantastic electric resistivity of UHMWPE fiber has been heavily hinders its further application in the fields such as antistatic composite materials, conductive materials and electromagnetic shielding materials. UHMWPE fiber was believed to be much more used in the fields not only including above mentioned but also in others, such as electrical and electronic, machinery, petrochemical and even military, aviation and space industry etc., in case of the obtaining of great conductivity.

There are many methods to manufacture conductive fibers, and the conductive components commonly used in conductive fibers are mainly inorganic conductive materials, such as carbon black, metals and metal oxides. Since the discovery of inherent conductive polymers (ICPs) in 1970s', the ICPs especially polyaniline (PANI) was believed to be promising in the producing of conductive fibers due to its excellent properties in conductivity and environmental stability [1]. Actually, researchers have produced conductive polymer/PANI composite fibers via solution spinning [2], melt spinning [3], electrospinning [4] and in-situ polymerization.

Among these methods, in-situ polymerization was believed to be a promising method for manufacturing conductive polyaniline composite fibers due to the simple manufacturing process, doing no harm to substrate fibers, high conductivity and environmental stability of the finished products [5]. So many kinds of conductive fibers such as polyamide/PANI [6], polyester/PANI [7], polypropylene/PANI [8] and polyurethane/PANI [9] were prepared by in-situ polymerization in recent decades.

However, preparing conductive yarns continuously using the traditional in-situ polymerization method is difficult because of its step-by-step process in solutions. In our previous work, a novel method based on in-situ polymerization to prepare conductive yarns continuously was presented and conductive UHMWPE/PANI composite yarns with high conductivity (~1 S/cm) were obtained, and the conductive UHMWPE/PANI composite yarn possessed good durability in the appropriate environment [10,11].

In this paper, the conductive UHMWPE/PANI composite yarn was manufactured continuously using the novel method based on in-situ polymerization, and electromagnetic shielding composite materials were prepared using the parallel conductive UHMWPE/PANI yarns. The structure of conductive UHMWPE/PANI composite yarn was studied, and effects of arrangement density and arrangement angle of UHMWPE/PANI yarns on the electromagnetic interference shielding efficiency of composite materials were also studied.

2. Experimental

2.1 Preparation and structure of conductive UHMWPE/PANI composite yarn

The UHMWPE filament yarn (400D/406f) was bought from Xiangsheng high strength fiber materials Inc. (Hangzhou, China). The preparation of conductive UHMWPE/PANI composite yarn was referred to our previous works [10,11]: the UHMWPE filament yarn was immersed in the aniline monomers and the reaction mixture of oxidant (150g/ammonium persulfate) and doping acid (1.75mol/L sulfuric acid) successively and extruded by a pressurizing roller to make the yarn contains aniline monomers, oxidant and doping acid on its surface. The yarn ran continuously (12m/min) under traction and then placed in a standard environment for at least 24h, so that the aniline reacts with the reaction mixture and finally a layer of conductive PANI was formed on the surface of UHMWPE yarn.

The surface morphologies of the UHMWPE and UHMWPE/PANI composite yarns were obtained using an S-4800 cold field emission scanning microscope (Hitachi Limited, Japan).

The chemical structure of UHMWPE yarn before and after conductive treatment was recorded using a NICOLET 5700 FTIR Spectrum (Thermo Nicolet Corporation, USA).

The electrical resistance of the UHMWPE/PANI composite yarn was tested using a ZC-90G high resistance meter (Shanghai Taiou Electronic Co., Ltd) and the electrical conductivity was calculated according to Eq.1:

$$\sigma = \frac{L}{R \cdot S} \quad (1)$$

where σ is the electrical conductivity (S/cm), R is the electrical resistance (Ω), L is the length (cm) and S is the cross section area (cm^2) of UHMWPE/PANI composite yarn, respectively.

2.2 Preparation of electromagnetic shielding composite materials

At regular intervals, the conductive UHMWPE/PANI composite yarns arranged in parallel and pasted onto a PVC membrane with special adhesive. The intervals were set as 6 mm, 4 mm, 2 mm and 1 mm, and two samples were made for each specification, and the samples were cut into circle with the diameter of 11.5 cm. And two samples with same yarn arrangement intervals were overlapped perpendicularly to form a composite material with grid structure, as shown in Fig.1.

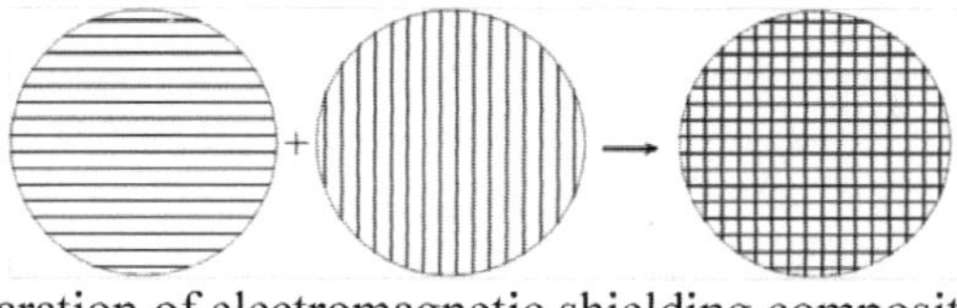

Fig.1 Preparation of electromagnetic shielding composite materials

2.3 Shielding effectiveness of electromagnetic shielding materials

The shielding effectiveness measurement of electromagnetic shielding composite materials was according to ASTM D4935-99 Standard Test Method for Measuring the Electromagnetic Shielding Effectiveness of Planar Materials. The major parts of testing device include an E5061A vector network analyzer (VNA) (Agilent Technologies Co., Ltd., USA) and a DN 1015A flange coaxial fixture for far field shielding effectiveness test device (EMC lab of southeast university, China), as shown in Fig.2. The frequency of plane wave tested in the range from 30 kHz to 1.5 GHz.

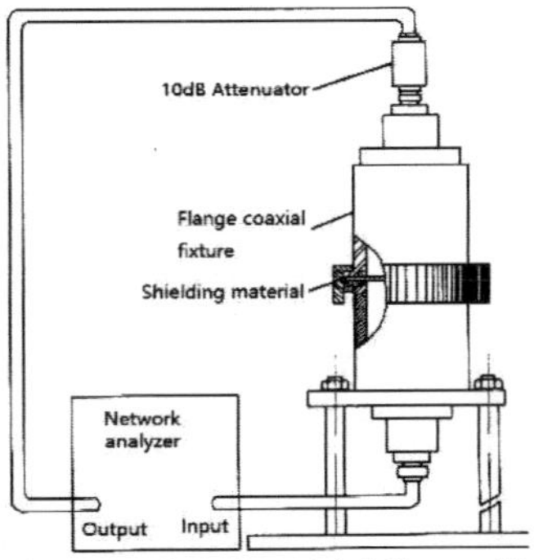

Fig.2 Testing device for shielding efficiency

The shielding efficiency was calculated according to Eq.2:

$$SE = 10\lg\frac{P_1}{P_2} = 20\lg\frac{V_1}{V_2} \tag{2}$$

where SE is the shielding effectiveness (dB), P_1 and P_2 are the respective received power with or without shielding material present (W), V_1 and V_2 are the respective received voltage levels with or without shielding material present (V).

According to the SE, the shielding ratio of composite materials to electromagnetic wave can be described by Eq.3:

$$\eta = (1 - \frac{1}{anti\log\frac{SE}{10}}) \times 100 \tag{3}$$

where η is the shielding ratio of composite materials to electromagnetic wave (%).

3. Results and discussions

3.1 Structure and conductivity of conductive UHMWPE/PANI composite yarn

The surface morphology of UHMWPE and UHMWPE/PANI composite yarn were shown as Fig.3. It can be seen from Fig.3a that the UHMWPE's surface was very smooth, and a complete layer of reaction product was adhered to the surface of UHMPWE fiber and the gaps between fibers were also filled with reaction product after conductive treatment, as shown in Fig.3b.

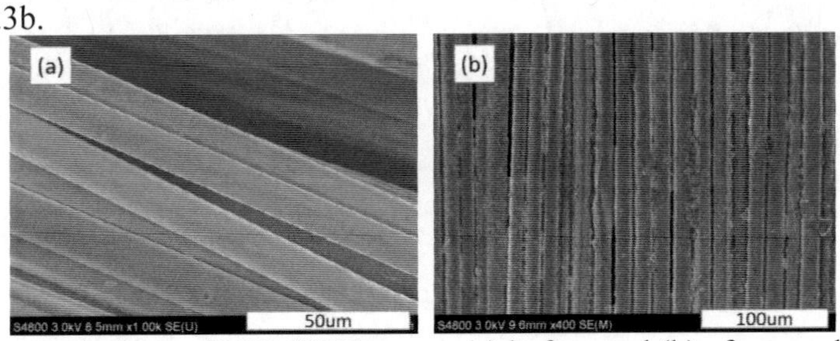

Fig.3 Surface morphology of UHMWPE yarns (a) before and (b) after conductive treatment

The chemical structure of reaction product on the yarn's surface was tested using FTIR spectra, as shown in Fig.4. Curve (a) corresponds to the FTIR curve of the UHMWPE fiber and the four absorption peaks at 2921 cm^{-1}, 2850 cm^{-1}, 1469 cm^{-1}, and 723 cm^{-1} were attributed to the methylene non-symmetrical stretching vibration, the methylene symmetrical stretching vibration, the methylene non-symmetrical changing angle vibration, and the methylene swing in-plane vibration, respectively [12]. After the conductive treatments, some new absorption peaks appeared, as shown in curve (b), such as the peak around 1587 cm^{-1} was attributed to the stretching vibrations of the quinine structure N=Q=N, and the peak around 1497 cm^{-1} was attributed to the benzene structure N–B–N in the PANI [13]. This proved that the reaction product on the yarn's surface was the emeraldine base PANI, both oxidized and reduced units existed in it.

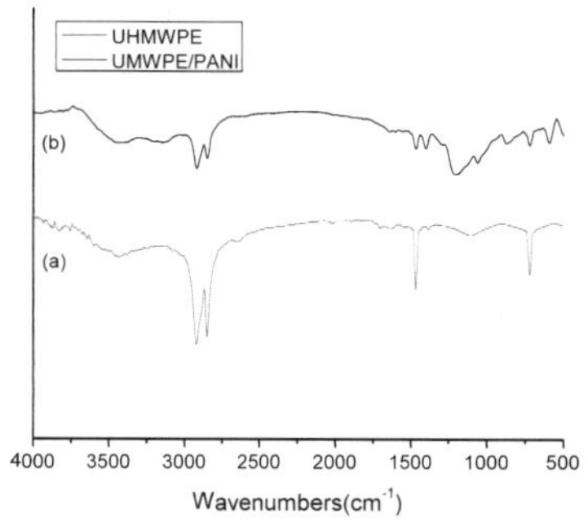

Fig.4 FTIR curves of UHMWPE yarn (a) before and (b) after conductive treatment

Fig. 5 shows electrical resistance for the UHMWPE/PANI conductive yarn within 100 meters in the length direction. It can be seen that the conductive yarn exhibited excellent electrical conductivity, the average resistance was $2.57\times10^3\Omega$/cm, and the conductivity was 0.88 S/cm calculated according to equation 1. The evidence suggested that conductive PANI composite yarn with high conductivity could be prepared using the continuous conductive treatment method based on in-situ polymerization.

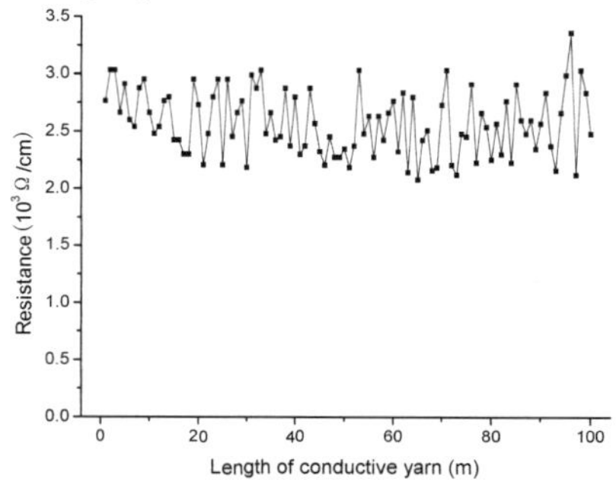

Fig.5 Electrical resistance of UHMWPE/PANI conductive yarn

3.2 Shielding effectiveness of composite materials

Fig.6 shows the EMI SEs of composite materials with different conductive yarn arrangement density. It can be seen that the composite materials possess great electromagnetic wave protection performance in low-frequency electromagnetic waves, the EMI SEs of all samples reached to more than 50 dB at 30 KHz. However, with the increase of frequency, the EMI SEs decayed rapidly, and approximately trend to be stable in the frequency up 50 MHz. The electromagnetic waves below 50MHz have small effect on human health, so only the EMI SEs of composite materials in the frequency range from 50MHz to 1.5GHz was discussed in this paper.

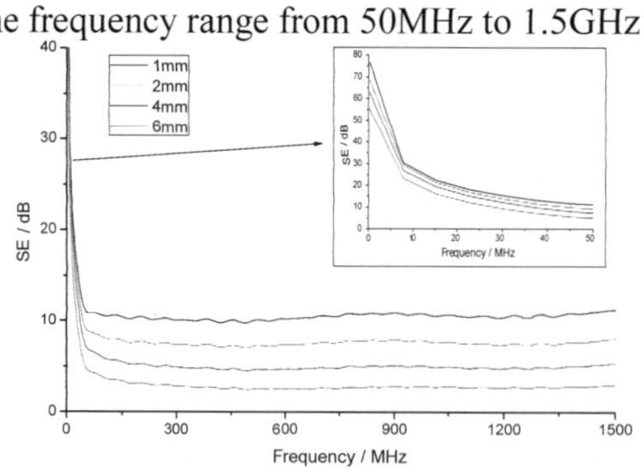

Fig.6 Shielding effectiveness of composite materials with different conductive yarn arrangement density

It can be seen from Fig.6 that, with the decrease of arrangement density of conductive yarn in the composite materials, the EMI SE increased gradually. When the interval of conductive yarn was 6 mm, the EMI SE was only ~3 dB, while the EMI SE of the sample with the conductive yarn interval of 1 mm was ~10 dB.

Fig.7 shows the effect of the conductive yarn interval on the shielding effectiveness of the composite materials, the values were the average at 194 frequency points from 50 MHz to 1.5 GHz. The shielding ratio to electromagnetic wave of composite materials with conductive yarn interval of 1 mm, 2mm, 4 mm and 6mm were 91.05%, 82.62, 68.38% and 47.88%, respectively, according to Eq.3.

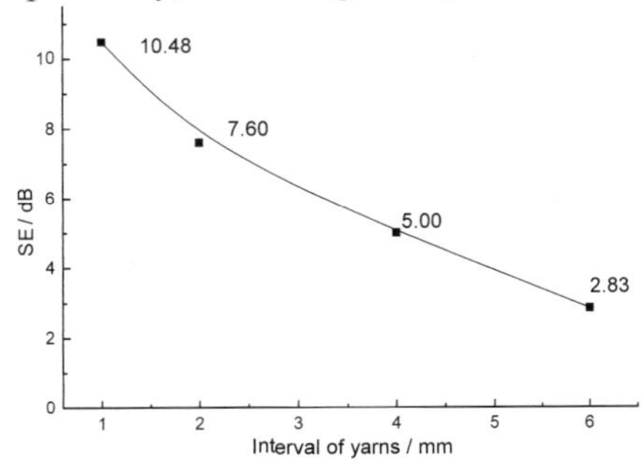

Fig.7 Effect of shielding effectiveness on the interval of conductive yarn

The effect of arrangement angle of two layers on the EMI SE of composite materials was also discussed, as shown in Fig.8, the arrangement angle were 90°, 45°, 30°and 0°, respectively, and the EMI SEs of samples with different arrangement angle in the frequency range from 50MHz to 1.5GHz were shown in Fig.9. The EMI SE was the highest when the angle was 90° (the average value was 10.48 dB), and with the decrease of arrangement angle, the EMI SE slightly decreased (9.47 dB when the angle was 0°). Although the conductive yarn interval of four samples were all 1 mm, the maximum space of each grid in composite materials was increased with the decrease of arrangement angle of two layers. The larger space benefits the transmission of electromagnetic energy to the composite materials.

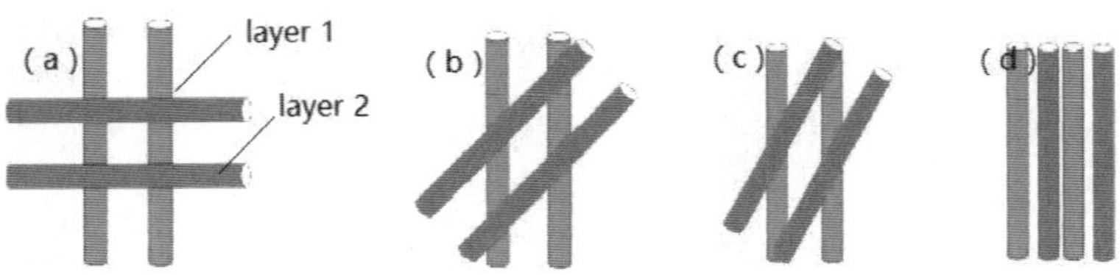

Fig.8 Arrangement angles between two layers of conductive yarn

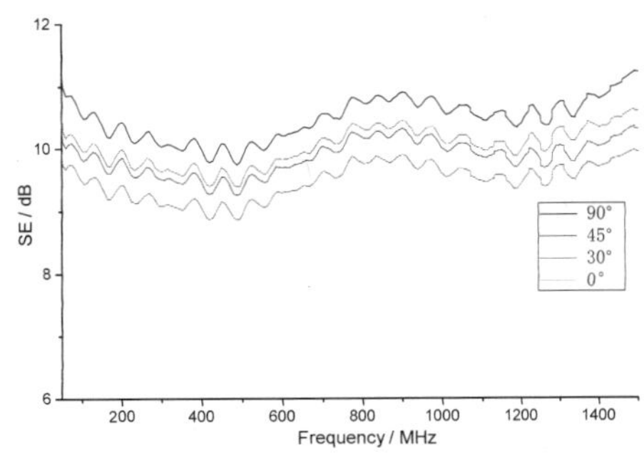

Fig.9 Shielding effectiveness of conductive yarn assembly with different arrangement angle

4. Conclusions

The conductive UHMWPEPANI composite yarn was continuously prepared using a novel method based on in-situ polymerization, and the electromagnetic shielding composite materials were prepared using the conductive yarn. The electromagnetic shielding composite materials showed a certain electromagnetic shielding effect. The EMI SE increased with the increase of the arrangement density of conductive yarn. The EMI SE could reach more than 10 dB, when two parallel yarn layers with the interval of 1 mm between the adjacent yarns in each layer were overlapped perpendicularly. The arrangement of two layers also affected the shielding effectiveness, and the EMI SE was the highest when the arrangement angle was 90°, and the anti-electromagnetic radiation effect decreased gradually with the decrease of the arrangement angle.

5. Funding

This work was supported by the Natural Science Foundation of Jiangsu Province（BK20150360, BK20141267）, the Cooperative Innovation Fund of Jiangsu Province （BY2014059-12）, the Technology Innovation Fund of Science and Technology Enterprises of Jiangsu Province （BC2014166）， the Science and Technology Support Program Funds of Suzhou （ SG201444 ）, and the Scientific Research Starting Foundation of Shaoxing University (20155036).

References

[1] Gordana CM. Recent advances in polyaniline research: Polymerization mechanisms,structural aspects, properties and applications. Synthetic Met 2013,177:1-47.
[2] Mattes BR, Wang HL, Yang D. Formation of conductive polyaniline fibers derived from hight concentrated emeraldine base solutions. Synthetic Met 1997,84:45-49.
[3] Passiniemi P, Laakso J. TEM and WAXS characterization of polyaniline/PP fibers. Synthetic Met 1997,84: 775-776.
[4] Repeker DH, Chun I. Nanometer diameter fibers of polyaniline, produced by electrospinning. Nanotechnology 1996,7: 216-223.
[5] Pang H, Xu L, Yan DX, et al. Conductive polymer composite with segregated structures. Prog Polym Sci 2014,39: 1908-1933.
[6] Chandran AS, Sunil KN. An elastomeric composite based on polyaniline coated nylon fiber and chloroprene rubber. Eur Polym J 2008,44:2418-2429.
[7] Liu WJ, Wu GM, Ma WH. Preparation and properties of the con-ductive composite fiber from polyaniline and PET fiber with sodium vanadateas oxidant. Polym Mater Sci Eng 1997,13:71–72.
[8] Zhang H, Li Y, Xu DZ. The structure and characters of polyani-line/polypropylene conductive fiber. Sh Tex Sci. Tech. 2005,6 :52–53.
[9] Li W, Zhuang QL, Ding X. Fabrication of electrically conductingPANI/Spandex composite yarns. J Donghua Univ 2004,12 :87–90.
[10] Hong JH, Pan ZJ, Yao M. Preparation and properties of continuously produced conductive UHMWPE/PANI composite yarns based on in-situ polymerization. Synthetic Met 2014,193:117-124.
[11] Hong JH, Pan ZJ, Tian L, et al. Continuous fabrication of conductive UHMWPE yarns based on in-situ polymerization with different doping acids. Synthetic Met 2015,209:512-520.
[12] Liu HJ, Pei YN, Xie D, et al. Surface modification of ultra-high molecular weight polyethylene (UHMWPE) by argon plasma. Appl Surf Sci 2010,256:3941–3945.
[13] Khalid M, Mohammad F. Preparation, FTIR spectroscopic characterization and isothermal stability of differently doped fibrous conducting polymers based on polyaniline and nylon-6,6. Synthetic Met 2009,159:119-122.

Polymer-matrix Catalytic Fabrication for Metal Nanoparticle Electroless Deposition

Chao Wang, Yi Li*, Miriam Caamaño Lago, Xu-Qing Liu

*School of Materials, University of Manchester, Oxford Road, Manchester, M13 9PL, UK

*Corresponding author's email: henry.yili@manchester.ac.uk

Abstract

Polymer-matrix catalysts have developed to be applied in electroless deposition (ELD) technique for metal nanoparticles. These long polymer chains on substrate play a role to capture catalysts selectively with high efficiency. More importantly, the polymer-matrix catalytic ELD provides a sample and effective approach to remarkably increase the adhesion between substrate and metal layer. This review focuses on the summary of this advanced method and a particular application of flexible electronics. An outlook on the future challenges and opportunities in this field is also provided.

Keywords: electroless deposition; polymer matrix; flexible electronics; printing methods

1. Introduction

In the past decades, the flexible and stretchable electronic devices have attracted tremendous attention owing to their excellent mechanical and electronical properties, comparing with conventional rigid and hard electronics. They have a wide range of applications, including robotic skins, point-of-care diagnostics, wearable displays and biological actuators [1-10]. One important step realizing these high-performance electronics is to fabricate conductive components. In the past few years, a large number of materials were studied as conductors, such as metal, conducting polymers, graphene sheets, carbon nanotubes and nanoparticles [2,10]. Nevertheless, metal is the best conductive material because it illustrates highly conductive, compatible, stabile and low-cost properties. For instance, metal materials can be flexible and stretched into suitable shapes visa optimizing their structures.

Many methods have been researched for fabrication of metal electronic materials, including electroless deposition (ELD), electrochemical plating [11-13], chemical vapor deposition [14,15] and physical vapor deposition [16]. In comparison with other techniques, ELD has attracted increasing interests as an efficient and low-cost method to form metal patterns or thin films on soft and flexible substrates, such as plastic, elastomer, fabric and some natural biomaterial [2,17-18]. Actually, ELD is a redox reaction for metal cautions to fabricate metal structures (Cu, Ni, Ag, Au etc.) on the surface of substrate. Furthermore, the synthesis situations are very gentle and high-temperature situation, expensive devices and conductive substrates are not needed [19-20].

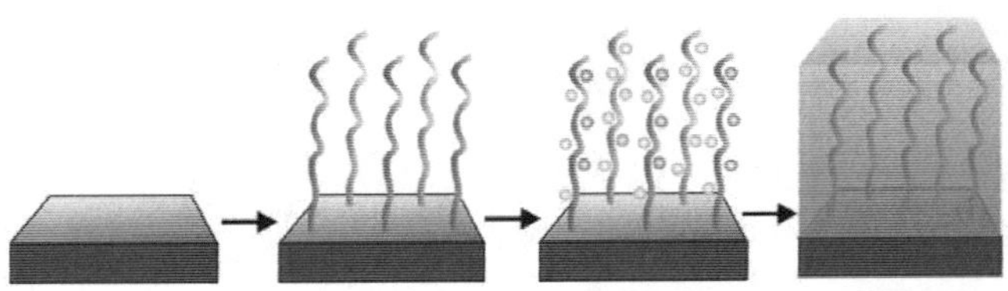

Fig.1 Three steps for process of typical polymer-matrix catalytic ELD [10]

In the recent years, polymer-matrix catalysts have developed to be applied in ELD technique. The polymer-matrix catalytic ELD provides a sample and effective approach to fabric metal nanoparticles on flexible and stretched substrates. The polymer layer plays a crucial role in this process, because it remarkably increases the adhesion between substrate and metal layer [21]. As shown in Figure 1, a typical process of polymer-matrix catalytic ELD involves three steps. First of all, the substrate is modified by grafted polymers. In addition, the ionic catalyst is immobilized on the anchoring polymer layer visa ion exchange. Finally, metal nanoparticles grow on the catalyzed area in an ELD bath [2,10].

2. Polymer Matrix and Catalytic Moiety

Several techniques of grafting polymers onto the target substrate have been employed in the past decades, such as polymer self-crosslinking, free radical polymerization (FRP), and surface-initiated polymerization [22-23], as illustrated in Figure 2. Meanwhile, some ELD studies about radiation-induced [24], plasma-induced [25] as well as UV-initiated [26] polymerizations have been reported. Among all the methods, surface-initiated atom-transfer radical polymerization is employed widely, because it is an efficient way to provide high-density polymer matrixes. But it still has some weaknesses which are the relatively slow polymerization speed and needing protection of nitrogen atmosphere. In comparison, the free radical polymerization permits that the polymerization occurs in the air environment. The thickness of the polymer layer ranges from tens to hundreds of nanometers [2]. The chemical or/and physical interaction ensures that the polymer matrix immobilized during the whole process of ELD. More importantly, the polymer chains with special ligands can capture the catalytic ions and provides powerful adhesion.

There exist various kinds of polymer matrixes (cationic, anionic, and nonionic) can be employed in polymer-matrix catalytic ELD as flexible electronics. The most common polymers reported in the field are poly [2-(methacryloyloxy) ethyl] trimethylammonium chloride (PMETAC) [27-31] and poly (acrylic acid) (PAA) [32-34]. Because they have both high feasibility in the process of synthesis and selection to special catalytic moieties.

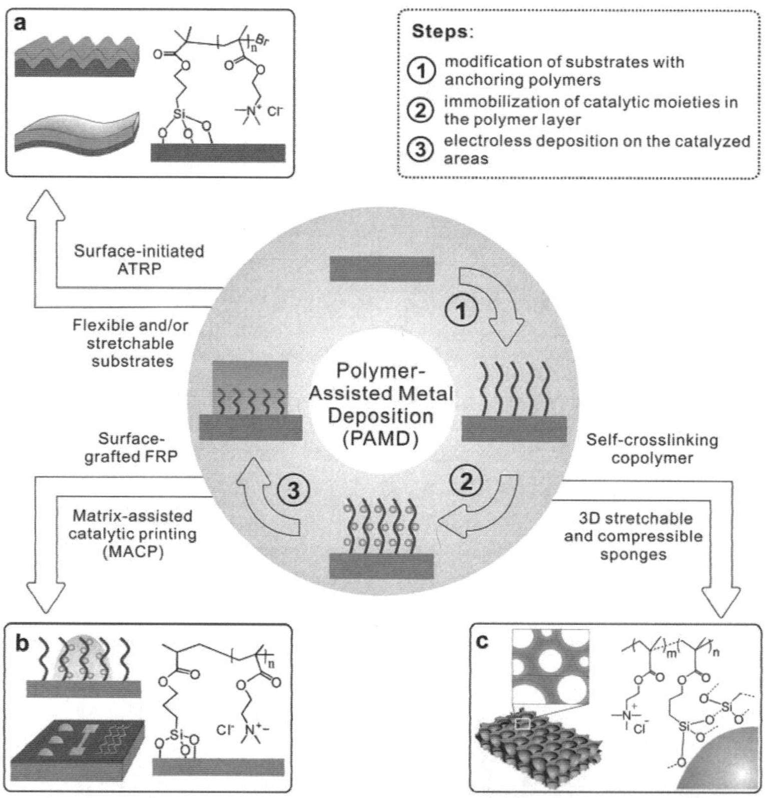

Fig.2 An illustration of polymerization methods of the polymer-matrix catalytic ELD. (a) Surface-initiated atom-transfer radical polymerization (ATRP) is applied on the flexible substrate. (b) In situ free radical polymerization (FRP) is employed visa various printing techniques. (c) Self-crosslinking copolymerization is used on a three-dimensional (3D) electronic [2]

2.1 Cationic Polymer Matrix and Catalytic Moiety

The cationic polymer matrix is a positively charged polyelectrolyte which has powerful affinity to the anionic catalytic moieties. For instance, in the aqueous solution, PMETAC is a cationic polymer matrix widely used to transfer ion species or capture the metal anions, including $[PdCl_4]^{2-}$ and $[AuCl_4]^{-}$. PMETAC firstly employed as a cationic polymer matrix in the field of ELD was reported by Huck and colleagues [27,35]. For example, the initiator-treated substrates (Au and Si) were modified with PMETAC via water-mediated surface-initiated atom-transfer radical polymerization. Then loading catalytic moieties were achieved by immersing the substrates with grafted PMETAC into the aqueous solution of $[PdCl_4]^{2-}$. The quaternary ammonium groups of PMETAC have a particular ability to immobilize $[PdCl_4]^{2-}$ species. Consequently, the catalytic areas were formed visa ion exchange. Then, the modified substrates were immersed into the copper electroless plating bath and the thin copper layer grew on the catalytic surface [27]. The amount of catalyst, which has a significant effect to the thickness of copper film, varies with the changing of concentration of $[PdCl_4]^{2-}$ species in the aqueous solution. While, PMETAC can control the catalyst amount directly. Besides, the method present both remarkable adhesion between metal film and substrate and excellent selectivity. Because the deposition only occurs in the area covered by catalyst.

2.2 Anionic Polymer Matrix and Catalytic Moiety

The anionic polymer matrix is a negatively charged polyelectrolyte that can immobilize the cationic catalytic moieties in the aqueous solution. The poly (acrylic acid) (PAA) [32-34] and poly (methacryloyl ethyl phosphate) (PMEP) are two common anionic polymer matrixes. The PAA brushes which can be negatively charged at a suitable PH exhibit a powerful binding affinity to metal cations including Ni^{2+}, Cu^{2+} as well as $[Pd(NH_3)_4]^{2+}$ [10]. PAA-matrix catalytic ELD was first studied by Kobersterin group [32]. In their report, a block copolymer polystyrene-b-poly(tert-butyl acrylate) (PS-b-PtBA) was employed. Firstly, the polystyrene blocks interpenetrated in a polystyrene substrate, while the PtBA blocks were explored in the UV radiation to be hydrolyzed to PAA blocks with the action of a photoacid generator. Then metal ions ($[Pd(NH_3)_4]^{2+}$ or Ni^{2+}) in the ELD bath were captured by the PAA brushes, as illustrated in Figure 3. One merit of the method is to decrease the cost because the price of metal cations is lower than that of metal anions [2]. Furthermore, it is a universal way to achieve metal ELD on many kinds of substrates. The crucial point is that the substrate and first polymer block should be the same.

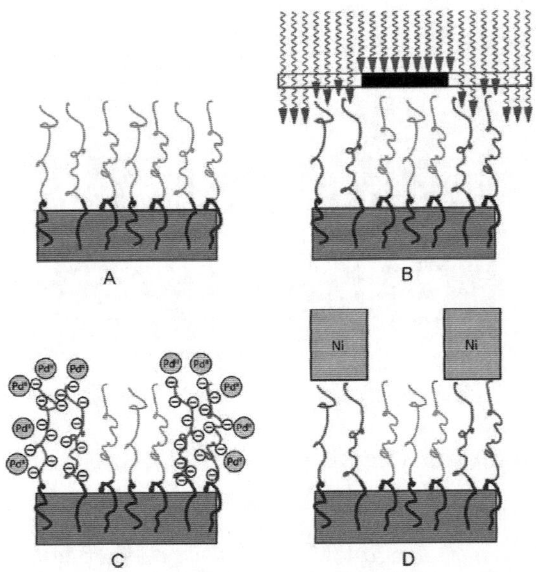

Fig.3 The process of ELD with diblock polymer PS-b-PtBA. (A) The copolymer grew on the PS substrate. (B) PtBA blocks were modified to PAA blocks under UV-radiation. (C) The metal ions were captured by PAA brushes. (D) Metal thin films were formed in the catalytic areas [32]

In the recent years, a technique, ligand-induced electroless plating (LIEP), has been introduced for formation of metal structures on substrates [36-38]. Through a GraftFast method, PAA grow on polymer substrates. In the GraftFast method, vinylic monomers in solution are initiated via radical polymerization in which the radicals are generated visa the reduction reaction of aryl diazonium salts with iron powder [39]. The PAA-modified substrate was immersed in a bath of alkaline solution containing Cu^{2+} ions and ammonia. Then the Cu^{2+} was captured by carboxylate groups of PAA and reduced by $NaBH_4$ so that a thin copper film fabricated on the polymer substrate, as illustrated in Figure 4. The remarkable advantage of LIEP technique is that the high-cost palladium catalyst is not necessary. Many various kinds of polymer substrates can be applied, such as polystyrene (PS), acrylonitrile-butadiene-styrene copolymer (ABS), polybutadiene (PB), poly(vinylidene fluoride) (PVDF), polyethylene terephthalate (PET) and polyacrylontrile (PAN).

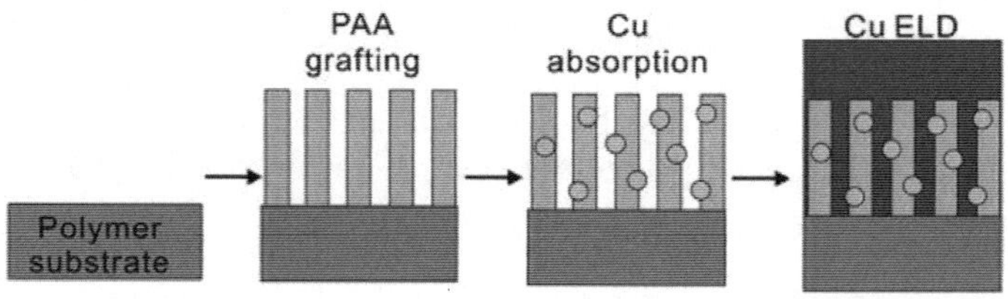

Fig.4 An illustration of PAA-matrix catalytic ELD [39]

Another example of anionic polymer matrix for ELD is PMEP. Zheng and his colleagues firstly reported the studies about PMEP for polymer-matrix catalytic ELD [40]. In their report, two opposite charged polymer brushes have strong affinity to different metal ions for site-selective ELD. Two polymer matrixes, PMETAC and PMEP, were synthesized through microcontact printing and surface-initiated atom-transfer radical polymerization. The quaternary ammonium groups of the positively charged PMETAC matrix could capture $[PdCl_4]^{2-}$ ions, while phosphate groups of negatively charged PMEP chains could immobilize $[Pd(NH_3)_4]^{2+}$ species. Then the modified substrate was immersed in a solution of copper and nickel ions for depositing two kinds of metal structures (Cu and Ni) (Figure 5). About this technique, it is very significant that various polymer brushes playing roles of two seeding templates to fabricate metal alloy on substrates.

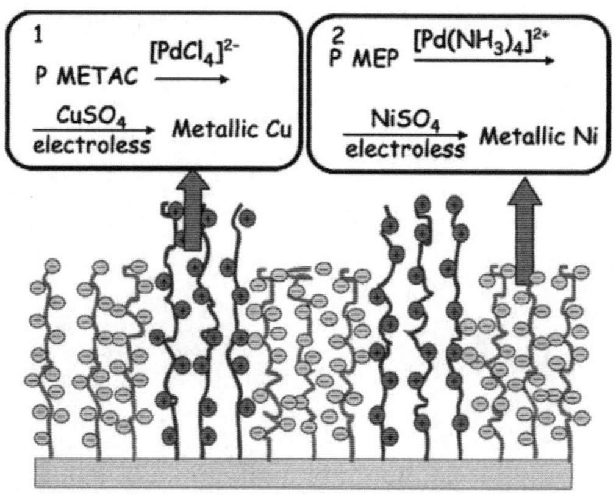

Fig.5 An illustration of mixed-polymer-matrix catalytic ELD. Positively or negatively charged polymers were used to deposit the copper and nickel films respectively [40]

2.3 Nonionic Polymer Matrix and Catalytic Moiety

Nonionic polymers is another effective polyelectrolyte showing a strong affinity to relative metal ionic catalysts for polymer-matrix catalytic ELD, including polyacrylonitrile (PAN), polyacrylamide (PAM), poly(vinyl pyrrolidone) (PVP), poly(2-vinyl pyridine) (P2VP) and poly(4-vinyl pyridine) (P4VP). Nonionic polymer matrixes have two particular features. 1) The metal catalytic moieties are captured to the binding sites visa coordination reaction in the present of free nitrogen electron pairs. 2) A large number of polymer can be applied as substrates in this technique visa different polymerization methods (radiation-, UV-, plasma-induced polymerization and surface-initiated atom-transfer radical grafting), such as polyethylene (PE), polypropylene (PP), polyimide (PI), poly (tetrafluoroethylene) (PTFE), fluorinated polyimide (FPI) and poly (tetrafluoroethylene-co-hexafluoropropylene) (FEP). Besides, the polymerization or copolymerization reactions of these nonionic polymers exhibit many strengths, including low-cost processing, high polymerizing rate, simple devices and potential applications in industry.

It is reported that, through postmodification of surface-grafted polymeric polyelectrolytes, N-containing nonionic polymer brushes can be polymerized. Some works in this filed were reported by Kang and co-workers [41-43]. Besides, the metal layers growing on these postmodified surface-grafted polymer brushes illustrated a powerful adhesion in relative tests [44]. For example, Chen et al synthesized the ELD gold film visa post-modifying PAA into amine-terminating layers [45]. As shown in Figure 6, firstly the PAA layer was grafted through UV-induced polymerization on poly (dimethyl siloxane) (PDMS) substrate and then the patterned PAA matrix with carboxylic acid groups was modified by ethylenediamine. After treatment, the resulting polymer with amine-terminating groups was employed to immobilize gold nanoparticles. There is an excellent adhesion between the gold thin film and PDMS surface, based on the testing consequences.

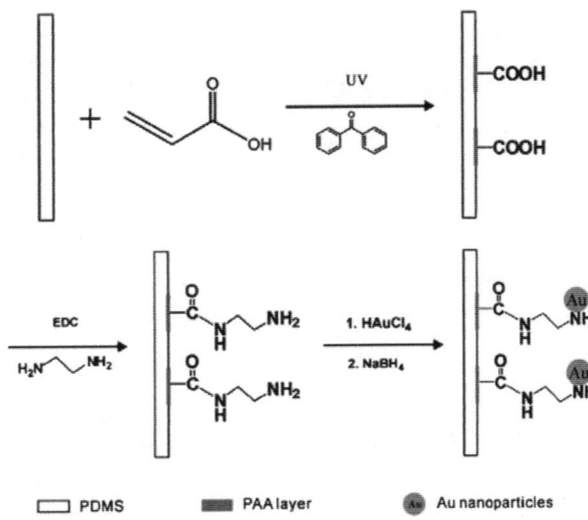

Fig.6 Process of ELD of a fold film visa the treatment of postmodification of surface-grafted PAA layer on PDMS substrate

3. Substrates and Application of ELD in Flexible Electronics

In the past decades, polymer-matrix catalytic ELD, as a high-efficient technique to fabricate metal electrodes, has been employed in the application of flexible and stretched electronics. Apart from the polymer matrixes and catalytic moieties, the substrate, affecting the property of the ELD electronics to a large extent [2], is also a significant part in this technique. Up to now, many different kinds of substrates has been used, including metals, glass, plastics, rubbers, cottons and some biomimetic materials [1].

3.1 ELD on Plastic Substrates

Polymer-matrix catalytic ELD is especially suitable for plastic substrates because of its gentile synthesis process, including room-temperature treatment, low-cost equipment and sample steps [19-20]. A suitable instance is a report in which a copper thin electrode was fabricated through ELD in the way of ligand-induced electroless plating [38]. PAA played a role of polymer matrix grafted on the surface of poly (ethylene terephthalate) (PET) substrate to immobilize catalytic moieties. Finally, a copper layer grew on the catalytic areas. The resulted copper electrode was stable and its electrical properties was closed to bulk values. More significantly, the metal structures on the flexible PET substrate could withstand mechanical deformation to some extent. The influence of this kind of deformation on resistivity of Cu structure on substrate was determined through bending PET around a cylinder (Figure 7). It showed that after thirty cycles the change of resistivity was quite small.

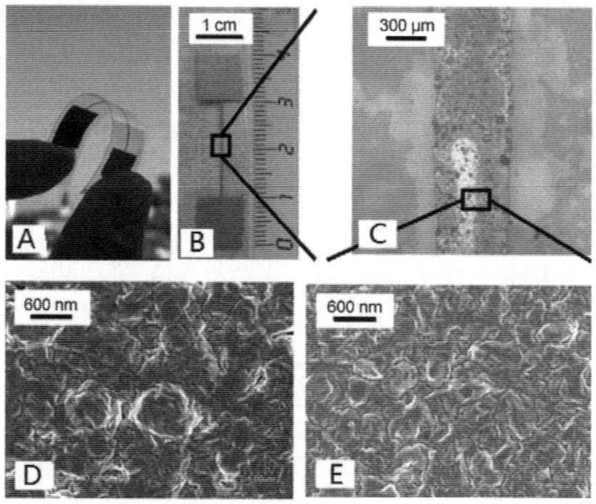

Fig.7 (A) (B) the photographs of PET sheet with copper film. (C) Optical image of the part in (B). (D) (E) SEM images of the copper structures on substrates of PET (D) and PVDF (E) [38]

3.2 ELD on Elastomer Substrates

Apart from plastic substrates, the elastomer substrates used for ELD also have particular merits in the application of flexible and stretchable electronics. The main challenge for employing elastomer in ELD is the poor adhesion because the substrate can be stretched but the metal layer is not stretched. This problem for formation flexible metal electronic was solved by Zheng group [28]. They reported that a copper thin film was fabricated on the prestrained elastomer substrate by ELD, resulting in an excellent stretchable conductor (Figure 8).

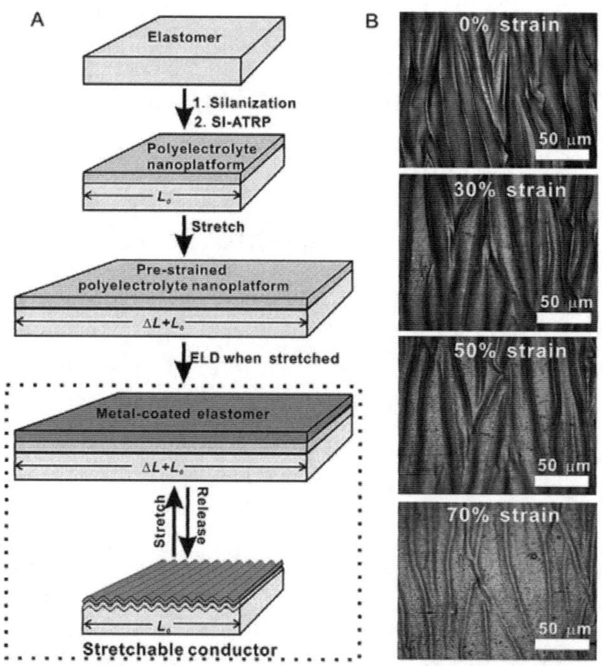

Fig.8 (A) An illustration of synthesis of stretchable conductor on elastomer substrate. (B) The optical microscopy images of copper layers of ELD on rubber at different tensile strains (from 0% to 70%) [28]

Firstly, PMETAC was grafted as a polymer matrix onto a rubber substrate. Then, after immobilizing catalysts, this modified rubber was stretched and immersed into a solution of copper ions. The ELD of copper thin coating was formed on the prestrain elastomer. When the tension was removed, the rubber sheet with copper conductor could recover to original shape. At the same time, when it was stretched again (maximum by 300%), the metal layer could keep good stability on conductive property without any cracking. Owing to the high binding strength between metal structures and elastomer substrate, the conductors can be stretched and relaxed reversibly staying highly stable conductivity, as illustrated in Figure 9. This method overcame the poor adhesion problem in previous reports to achieve a stretchable metal layer on rubber substrate.

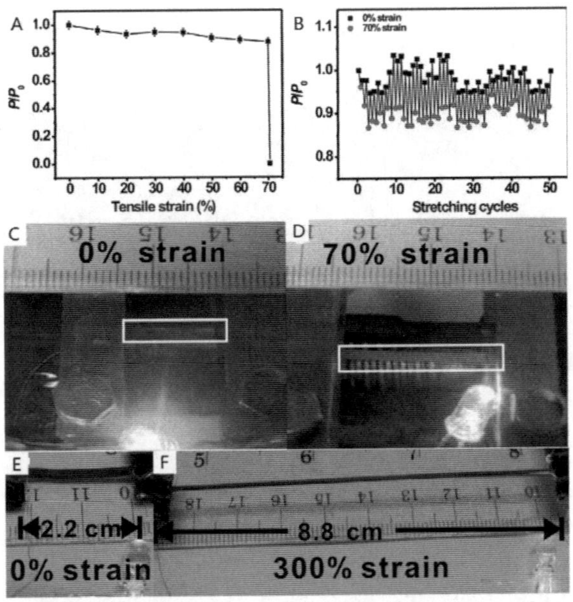

Fig.9 (A) The changes in conductivity of the high-stretchable conductor at various tensile strains. (B) The change of stability in conductivity with stretching cycles when the tensile strains are 0% and 70% respectively. (C) (D) The digital images of the high-stretchable conductor connected with a light emitting diode at 0% (C) and 70% (D) Tensile strains respectively. (E) (F) The digital images of the copper film with rubber substrate at 0% (E) and 300% (F) [2,28]

3.3 ELD on Cotton Substrates

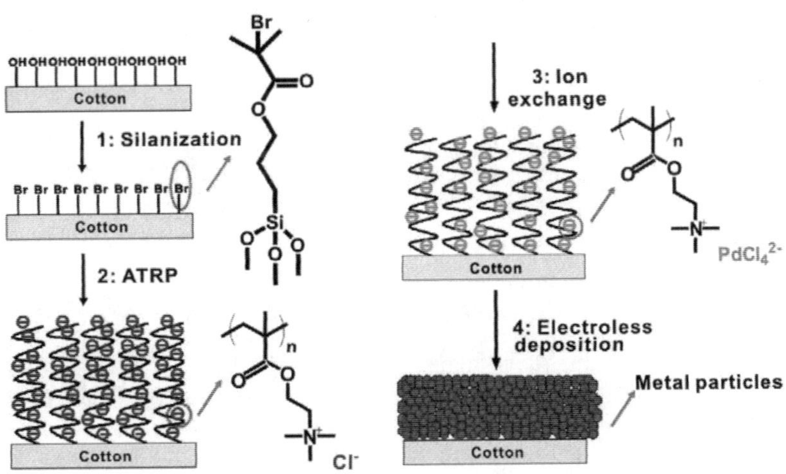

Fig.10 The illustration of fabrication of the metal particles on cotton substrate through ELD

Cotton is one of the most common natural fibers. Because of its excellent advantages, including low price and sample processing techniques, it can be applied in wide range of applications in daily life [30]. The usage of this natural fibers to be a soft and flexible substrate is more environmentally friendly than those man-made polymer materials. Zheng and co-workers designed a versatile method to fabricate metal conductors on cotton substrates [2,30]. At first, as shown in Figure 10, the textile substrate was surface-modified by the bromine-terminating silane which could act as a following reacted sites for polymer matrixes. Then the PMETAC matrix was grafted on the initiating points through surface-initiated atom-transfer radical polymerization. These long polymer chains with a high density of binding sites capture the $[PdCl_4]^{2-}$ ions in aqueous solution visa ion exchange action. Finally, the metal particles (Cu or Ni) were formed in the catalytic areas. The conductivity of copper film was tested with powder a light emitting diode (Figure 11). As a result, this copper structure exhibited not just excellent conductive and mechanical properties but also electrical stability when it was stretched, bended and even washed.

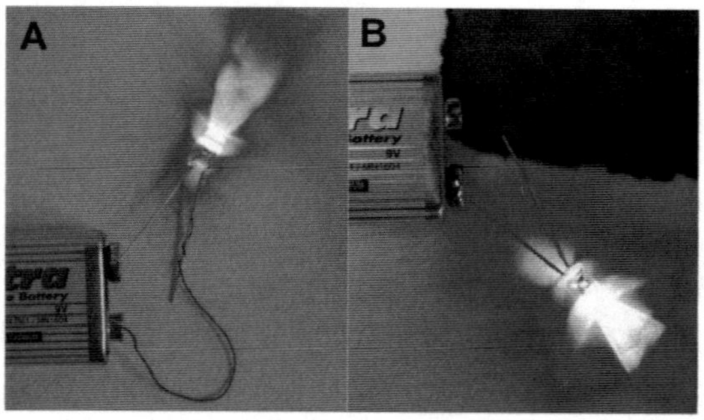

Fig.11 Digital images powdering a light emitting diode through an electrical wire (A) or through the conductive yarn (B)

3.4 ELD on Bio-Inspired Substrates

In the recent years, some bio-inspired materials was chosen as an omnidirectional substrate applied in ELD field, owing to the fact that natural materials are environmentally friendly, low-cost and easily to get [1,18]. For example, plant leaves exhibiting soft and stretchable structures are considered as proper substrates for synthesis metal conductor visa ELD. Recently, the vein-based transparent electrodes were fabricated by Zheng et al [1]. In this approach, the ELD of copper particles can be achieved in three steps (Figure 12). 1) The surfaces of natural veins are modified by copolymer P(METAC-co-MPTS). 2) These leaves with polymer matrix is immersed in a solution of $(NH_4)_2PdCl_4$ for some time and the $[PdCl_4]^{2-}$ ions were captured for the following deposition. 3) In the plating bath, copper particles were synthesized on the surface of vein substrates. This electrode is quiet conductive and transparent. As illustrated in Figure 13, based on the random determinations on different parts on the vein conductor, its bulk resistances are nearly 4 Ω, which is better than some commercial products in market [1]. Furthermore, the test of powering some electronical devices was taken. Consequently, the light emitting diodes fixed on the vein-based conductor could work normally when the vein was at different strains (Figure 14).

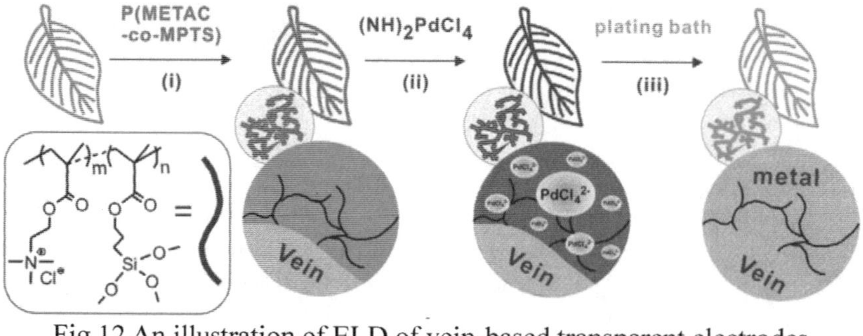

Fig.12 An illustration of ELD of vein-based transparent electrodes

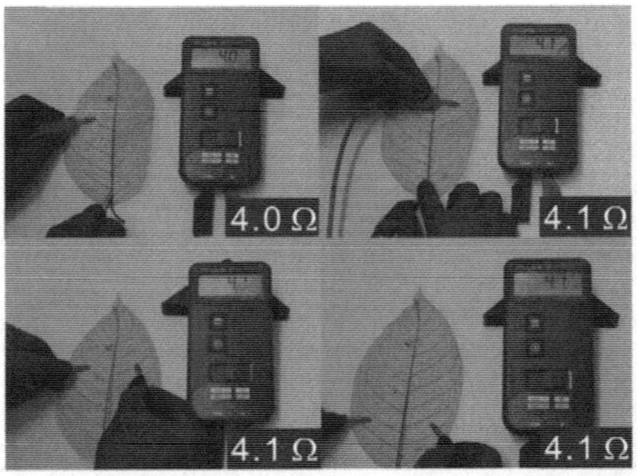

Fig.13 The random test about bulk resistances in various part of the vein conductor

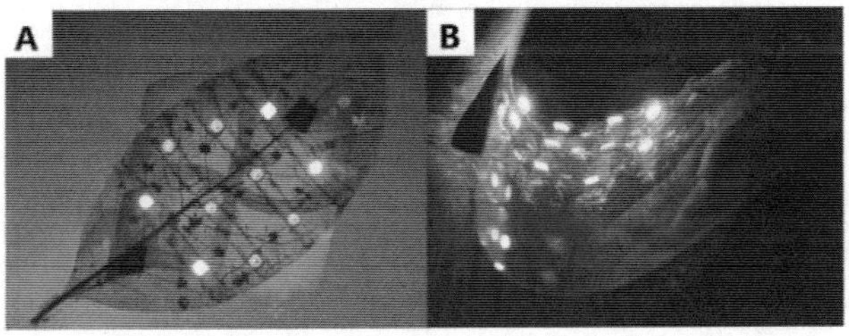

Fig.14 The LEDs glued on the modified vein worked well when the conductor stayed in different shapes

3.5 ELD on Three-Dimensional Sponges

In the past reports, most substrates applied in the polymer-matrix catalytic ELD is limited in two-dimensional sheets. While a three-dimensional sponge employed as the substrate for ELD was reported in 2014 by Yu et al [31]. They fabricated a flexible and stretchable three-dimensional conductor on the porous polyurethane (PU) sponges. The copolymer P(METAC-co-MPTS) was grafted visa the dip-coating and curing process instead of polymerization on the substrate surface directly. As shown in Figure 15, firstly, in the ammonia environment, bulk substrates were dip-coated by P(METAC-co-MPTS) at room temperature.

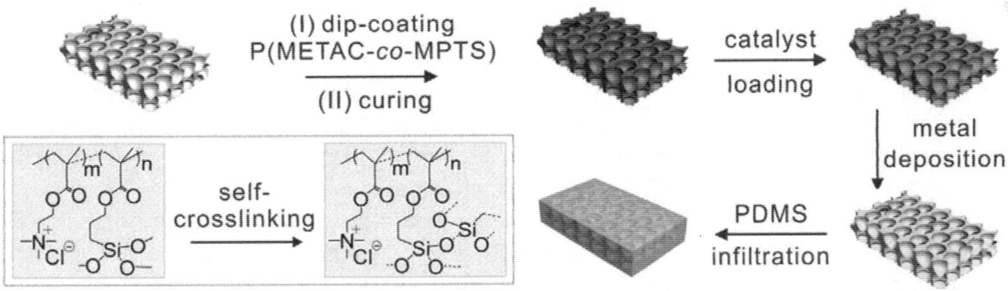

Fig.15 An illustration of three-dimensional metal conductor visa polymer-matrix catalytic ELD

A strong adhesion between polymer matrix and PU sponge was built, during the curing process, through forming a connected layer by silane groups in the PMPTS blocks. Then the surface-modified sponges were immersed in catalyst solution and then $[PdCl_4]^{2-}$ moieties were immobilized by PMETAC blocks. Finally, metal thin films (copper, copper/ silver, gold) are formed on the surface of PU sponges by ELD in the plating bath.

This three-dimensional metal-coated sponge with stable conductive property showed an excellent mechanical flexibility (Figure 16), such as stretching, bending, twisting and compressing properties. The outstanding advantage of the three-dimensional electrodes is that it can resist the surface damage more effectively than two-dimensional sheets, which can be explain by a 3D-network conductive pathway [31]. Even through it was polished by an abrasive paper or cut a half by a razor blade, the resistance did not change too much, as shown in Figure 17.

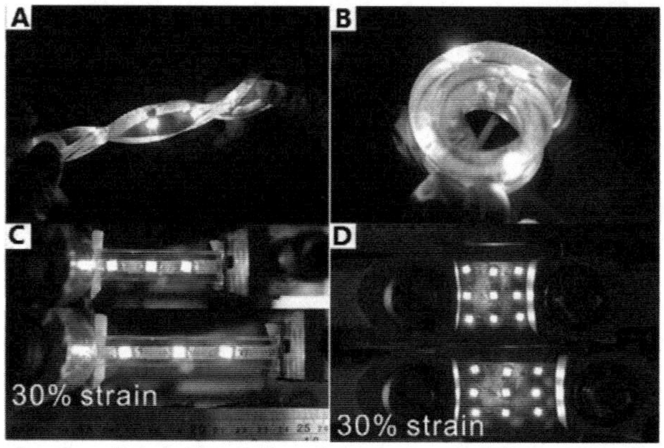

Fig.16 Digital images of conductive three-dimensional sponges at different strains with working LEDs on the surface

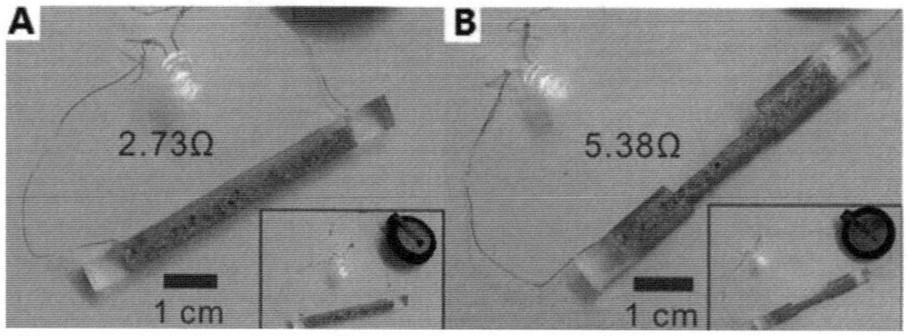

Fig.17 Demonstration of the conductive three-dimensional sponges before (A) and after (B) cutting some parts away

4. Printing Techniques for Polymer-Matrix Catalytic ELD

The catalytic moieties for polymer-matrix catalytic ELD technique usually is metal ions or nanoparticles [2]. The most common approaches to load of these catalytic moieties are immersing the substrates with polymer-modified surfaces into the metal-based salts solution and printing the ink containing catalytic metal compounds onto the modified surfaces. Even though the first method is widely used in the relative reports, due to the low cost and sample process, printing techniques still play a quite significant role to produce ELD electronical devices.

Accordingly, developing advanced printing approaches, through which low-cost and high-performance flexible ELD conductors can be fabricated, attracts increasing interests. Nevertheless, owing to the limitation in viscosity problem of catalyst ink, just inkjet printing was applied in the ELD patterning techniques, which has a negative influence for product output and efficiency. In order to advance new printing methods to synthesize ELD of metal layers, an effective way should be found first to solve the problem of low ink viscosity.

In the recent years, Zheng et al studied a novel strategy controlling the viscosity of the catalyst ink visa employing polyethylene glycol (PEG) as the delivering matrix [21]. As they reported, the matrix-assisted catalytic printing (MACP) method was taken advantage of to fabricate four metal conductors (copper, nickel, silver and gold). The catalytic moieties were printed onto the substrate surface through three printing techniques, namely dip-pen nanolithography (DPN), inkjet printing, and screen printing, rather than through immersing substrates directly in catalyst solution or traditional lithography printing. As shown in Figure 18, PMETAC working as the receiving matrix was grafted on the substrate surface. Then the ink containing catalyst solution and delivering matrix was printed by three various printing methods respectively. After ELD, the metal thin films were only formed on the printed areas, which exhibited great selectivity and good patterns transfer from ink to metal layer.

The delivering matrix (PEG) played a crucial role in these printing ways [2,21]. On one hand, the viscosity of ink could be controlled effectively by the amount of PEG. Especially for screen printing, the catalyst ink should be very viscous in comparison with other methods. On the other hand, the water-absorbing PEG kept the ink high moisture to enable that PMETAC had sufficient time to capture $[PdCl_4]^{2-}$ ions during the process of diffusion. Therefore, the ELD of metal films could be fabricated with various printing approaches. In addition, these flexible and stretchable conductors illustrated good mechanical property and stable electronical property. The minimum resistivity of four conductors (copper) is only 4.9 μΩ· cm [21]. The printing methods (such as screen printing) have a potential to synthesize high-throughput ELD conductors.

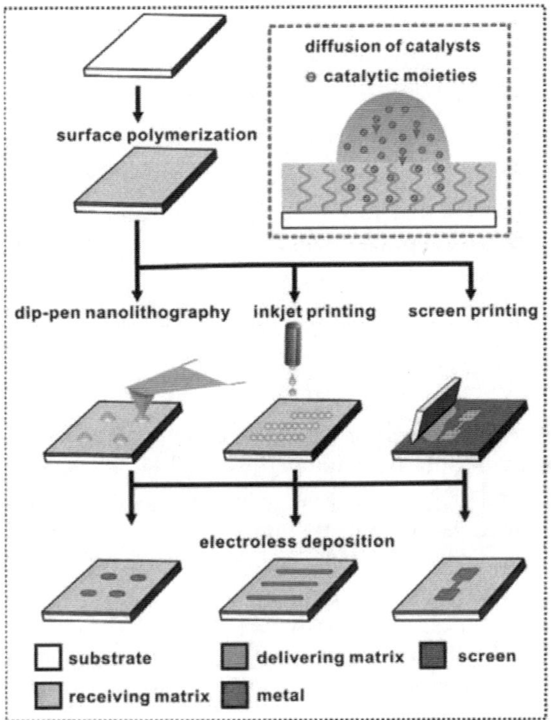

Fig.18 An illustration of ELD of metal films through dip-pen nanolithography, inkjet printing and screen printing

5. Conclusion and Perspectives

Table 1. Current results of polymer-matrix catalytic ELD [A]

Polymer matrixes	Polymerization method	Metal	Substrate
PMETAC [27-28,30,40]	SI-ATRP	Cu, Ni	Cotton, rubber, PDMS, spandex fibers, Au
PAA [32,36-38]	GraftFast, block copolymer	Cu, Ni	ABS, PET, PVDF, PAN, PB and PS
PVBVN [47]	Plasma grafting	Ag	PI
PMEP [40]	SI-ATRP	Cu, Ni	Au
PAA postmodified with ethylenediamine [45]	UV grafting	Au	PDMS
P2VP, P4VP [48-56]	UV, plasma, radiation grafting	Cu, Ni	PI, FPI, PTFE, SiLK, PVDF
PAM [56,57]	radiation grafting	Cu	PP
PVP [57]	radiation grafting	Cu	PE
PAN [44]	UV grafting	Cu	FEP
PHEA, PHMMAAm [41]	UV grafting	Cu	PTFE
PDDA [42,43]	oxidative grafting	Cu	PTFE, glass

[A] PDDA= Poly (diallyldimethylammonium chloride), PVBVN= Poly (1,1'-bis(4-vinylbenzyl)-4,4'-bipyridinium dinitrate), PHEA= Poly-2-hydroxyethyl acrylate, PHMMAAm= Poly-N- (hydroxymethyl) methacrylamide

The polymer-matrix catalytic ELD is an effect and efficient way to fabricate metal particles or thin films on substrates as the flexible and stretchable electronics, because of its outstanding advantages. Generally, this low-cost and convenient technique contains three steps to achieve metal conductors, including modifying substrates with polymer matrixes, capturing catalytic moieties onto the polymer-grafted surfaces and subsequent depositing metal layer in plating bath. The substrates applied for ELD can be chosen from a wide range of materials, such as plastics, elastomers, cottons, three-dimensional sponges and even natural leaves. Some previously reported substrates, polymer matrixes and polymerization methods employed for polymer-matrix catalytic ELD are summarized in Table 1 [10].

The polymer matrixes act a quite important role in the whole process of fabrication because interpenetrating networks are formed between the matrixes layer and metal structures. More significantly, this interpenetration provides a powerful adhesion to overcome the mechanical failures and instability, which ensures that the metal layers do not fall down from the substrate surface when the conductors are stretched or folded. At the same time, the metal films fabricated from polymer-matrix catalytic ELD is usually high-performance and have excellent conductive property. In addition, except traditional lithographic approaches [58-66] (including photolithography, scanning probe lithography and contact printing) some advanced printing techniques have been applied in ELD field. These effective printing methods, such as dip-pen nanolithography, inkjet printing and screen printing, have a potential to improve the throughput of patterned ELD electronics.

However, the researches about polymer-matrix catalytic ELD still at early stage [2,10]. There exist some challenges limiting the advancement of this technique. First of all, limited polymer materials can be used as matrixes. More kinds of polymers and polymerization methods need to be studied and employed in this field. Meanwhile, the limited types of metal layers are another question. The deposition process is full-solution in aqueous solution, which may be not suitable for those material sensitive to water and other salts. For instance, aluminum particles cannot achieved by polymer-matrix catalytic ELD. Besides, in the current reports, most ELD metal structures are only applied as simple conductive wires and electrodes. Many other fields, including robotic skins, point-of-care diagnostics, wearable displays and biological actuators, should be expected for the applications of ELD electronics. Further studies about polymer-matrix catalytic ELD is currently in process by relative researchers.

References

[1] Yu Y, Zhang Y, Li K, Yan C, Zheng Z. Bio-Inspired Chemical Fabrication of Stretchable Transparent Electrodes. Small 2015; 11: 3444-3449.

[2] Yu Y, Yan C, Zheng Z. Polymer-Assisted Metal Deposition (PAMD): A Full-Solution Strategy for Flexible, Stretchable, Compressible, and Wearable Metal Conductors. Advanced Materials 2014; 26: 5508-5516.

[3] Shang S, Zeng W, Tao Xiao. High stretchable MWNTs/polyurethane conductive nanocomposites. Journal of Materials Chemistry 2011; 21: 7274-7280.

[4] Benight S, Wang C, Tok J, Bao Z. Stretchable and self-healing polymers and devices for electronic skin. Progress in Polymer Science 2013; 38: 1961-1977.

[5] Hu L, Pasta M, Mantia F, Cui L, Jeong S, Deshazer H, Chio J, Han S, Cui Y. Stretchable, porous, and conductive energy textiles. Nano letters 2010; 10: 708-714.

[6] Yeo W, Kim Y, Lee J, Ameen A, Shi L, Li M, Huang Y. Multifunctional epidermal electronics printed directly onto the skin. Advanced Materials 2013; 25: 2773-2778.

[7] White M, Kaltenbrunner M, Głowacki E, Gutnichenko K, Kettlgruber G, Graz I, Major Z. Ultrathin, highly flexible and stretchable PLEDs. Nature Photonics 2013; 7: 811-816.

[8] Webb R, Bonifas A, Behnaz A, Zhang Y, Yu K, Cheng H, Yeo W. Ultrathin conformal devices for precise and continuous thermal characterization of human skin. Nature materials 2013; 12: 938-944.

[9] Kim D, Lu N, Ma R, Kim Y, Kim R, Wang S, Yu K. Epidermal electronics. Science 2011; 333: 838-843.

[10] Liu X, Zhou X, Li Y, Zheng Z. Surface-Grafted Polymer-Assisted Electroless Deposition of Metals for Flexible and Stretchable Electronics. Chemistry-An Asian Journal 2012; 7: 862-870.

[11] Brosteaux D, Axisa F, Gonzalez M, Vanfleteren J. Design and fabrication of elastic interconnections for stretchable electronic circuits. Electron Device Letters, IEEE 2007; 28: 552-554.

[12] Li X, Pei W, Tang R, Gui Q, Guo K, Wang Y, Chen H. Investigation of flexible electrodes modified by TiN, Pt

black IrO (x). Science China-technological Sciences 2011; 54: 2305-2309.
[13] Cho S, Kim S, Lee J, Lee N. Micro-scale metallization of high aspect-ratio Cu and Au lines on flexible polyimide substrate by electroplating using SU-8 photoresist mask. Microelectronic engineering 2005; 77: 116-124.
[14] Limb S, Gleason K, Edell D, Gleason E. Flexible Fluorocarbon Wire Coatings by Pulsed Plasma-Enhanced Chemical-Vapor-Deposition. Journal of Vacuum Science & Technology A 1997; 15: 1814-1818.
[15] Sun Y, Rogers J. Inorganic semiconductors for flexible electronics. Advanced Materials 2007; 19: 1897-1916.
[16] Silva R, Franchi D, Leone A, Pilloni L, Masci A, Pozio A. Surface conductivity and stability of metallic bipolar plate materials for polymer electrolyte fuel cells. Electrochimica Acta 2006; 51: 3592-3598.
[17] Liu L, Yu Y, Yan C, Li K, Zheng Z. Wearable energy-dense and power-dense supercapacitor yarns enabled by scalable graphene-metallic textile composite electrodes. Nature communications 2015; 6.
[18] Guo R, Yu Y, Zeng J, Liu X, Zhou X, Niu L, Zheng Z. Biomimicking Topographic Elastomeric Petals (E-Petals) for Omnidirectional Stretchable and Printable Electronics. Advanced Science 2015; 2.
[19] Zschieschang U, Klauk H, Halik M, Schmid G, Dehm C. Flexible organic circuits with printed gate electrodes. Advanced materials 2003; 15: 1147-1151.
[20] Cheng K, Yang M, Chiu W, Huang C, Chang J, Ying T, Yang Y. Ink-Jet Printing, Self-Assembled Polyelectrolytes, and Electroless Plating: Low Cost Fabrication of Circuits on a Flexible Substrate at Room Temperature. Macromolecular Rapid Communications 2005; 26: 247-264.
[21] Guo R, Yu Y, Xie Z, Liu X, Zhou X, Gao Y. Zheng Z. Matrix-Assisted Catalytic Printing for the Fabrication of Multiscale, Flexible, Foldable, and Stretchable Metal Conductors. Advanced Materials 2013; 25: 3343-3350.
[22] Edmondson S, Osborne V, Huck W. Polymer brushes via surface-initiated polymerizations. Chemical society reviews 2004; 33: 14-22.
[23] Pyun J, Kowalewski T, Matyjaszewski K. Synthesis of polymer brushes using atom transfer radical polymerization. Macromolecular Rapid Communications 2003; 24: 1043-1059.
[24] Deng J, Wang L, Liu L, Yang W. Developments and new applications of UV-induced surface graft polymerizations. Progress in Polymer Science 2009; 34: 156-193.
[25] Kang E, Zhang Y. Surface modification of fluoropolymers via molecular design. Advanced Materials 2000; 12: 1481-1494.
[26] Nasef M, Hegazy E. Preparation and applications of ion exchange membranes by radiation-induced graft copolymerization of polar monomers onto non-polar films. Progress in Polymer Science 2004; 29: 499-561.
[27] Azzaroni O, Zheng Z, Yang Z, Huck WT. Polyelectrolyte brushes as efficient ultrathin platforms for site-selective copper electroless deposition. Langmuir 2006; 22: 6730-6733.
[28] Wang X, Hu H, Shen Y, Zhou X, Zheng Z. Stretchable conductors with ultrahigh tensile strain and stable metallic conductance enabled by prestrained polyelectrolyte nanoplatforms. Advanced Materials 2011; 23: 3090-3094.
[29] MingáLau W. Grafting of polyelectrolytes onto hydrocarbon surfaces by high-energy hydrogen induced cross-linking for making metallized polymer films. Chemical Communications 2013; 49: 4658-4660.
[30] Liu X, Li Y, Zheng Z. Programming nanostructures of polymer brushes by dip-pen nanodisplacement lithography (DNL). Nanoscale 2010; 2: 2614-2618.
[31] Yu Y, Zeng J, Chen C, Xie Z, Guo R, Liu Z, Zheng Z. Three-Dimensional Compressible and Stretchable Conductive Composites. Advanced Materials 2014; 26: 810-815.
[32] Lancaster J, Jehani J, Carroll G, Chen Y, Turro N, Koberstein J. Toward a universal method to pattern metals on a polymer. Chemistry of Materials 2008; 20: 6583-6585.
[33] Garcia A, Hanifi N, Jousselme B, Jégou P, Palacin S, Viel P, Berthelot T. Polymer grafting by inkjet printing: a direct chemical writing toolset. Advanced Functional Materials 2013; 23: 3668-3674.
[34] Garcia A, Berthelot T, Viel P, Polesel-Maris J, Palacin S. Microscopic study of a ligand induced electroless plating process onto polymers. ACS applied materials & interfaces 2010; 2: 3043-3051.
[35] Azzaroni O, Moya S, Brown A, Zheng Z, Donath E, Huck, W. Polyelectrolyte brushes as ink nanoreservoirs for microcontact printing of ionic species with poly (dimethyl siloxane) stamps. Advanced Functional Materials 2006; 16: 1037.
[36] Garcia A, Berthelot T, Viel P, Mesnage A, Jégou P, Nekelson F, Palacin S. ABS polymer electroless plating through a one-step poly (acrylic acid) covalent grafting. ACS applied materials & interfaces 2010; 2: 1177-1183.

[37] Garcia A, Berthelot T, Viel P, Polesel-Maris J, Palacin S. Microscopic study of a ligand induced electroless plating process onto polymers. ACS applied materials & interfaces 2010; 2: 3043-3051.
[38] Garcia A, Polesel-Maris J, Viel P, Palacin S, Berthelot T. Localized ligand induced electroless plating (LIEP) process for the fabrication of copper patterns onto flexible polymer substrates. Advanced Functional Materials 2011; 21: 2096-2102.
[39] Mévellec V, Roussel S, Tessier L, Chancolon J, Mayne-L'Hermite M, Deniau G, Palacin S. Grafting polymers on surfaces: A new powerful and versatile diazonium salt-based one-step process in aqueous media. Chemistry of Materials 2007; 19: 6323-6330.
[40] Liu Z, Hu H, Yu B, Chen M, Zheng Z, Zhou F. Binary oppositely charged polyelectrolyte brushes for highly selective electroless deposition of bimetallic patterns. Electrochemistry Communications 2009; 11: 492-495.
[41] Wu S, Kang E, Neoh K, Tan K. Electroless deposition of copper on surface modified poly (tetrafluoroethylene) films from graft copolymerization and silanization. Langmuir 2000; 16: 5192-5198.
[42] Ma Z, Tan K, Alian A, Kang E, Neoh K. Electroless plating of copper on poly (tetrafluoroethylene) films modified by NH3 plasma and surface graft copolymerization with aniline. Journal of Vacuum Science & Technology A 2001; 19: 2471-2478.
[43] Chen Y, Kang E, Neoh K, Huan, W. Electroless metallization of glass surfaces functionalized by silanization and graft polymerization of aniline." Langmuir 2001; 17: 7425-7432.
[44] Yu Z, Kang E, Neoh K. Amidoximation of the acrylonitrile polymer grafted on poly (tetrafluoroethylene-co-hexafluoropropylene) films and its relevance to the electroless plating of copper. Langmuir 2002; 18: 10221-10230.
[45] Hao Z, Chen H, Ma D. Preparation of micro gold devices on poly (dimethylsiloxane) chips with region-selective electroless plating. Analytical chemistry 2009; 81: 8649-8653.
[46] Garcia A, Berthelot T, Viel P, Jégou P, Palacin S. 3D Amino-Induced Electroless Plating: A Powerful Toolset for Localized Metallization on Polymer Substrates. ChemPhysChem 2011; 12: 2973-2978.
[47] Sekitani T, Someya T. Stretchable, Large-area Organic Electronics. Advanced Materials 2010; 22: 2228-2246.
[48] Yang G, Lim C, Tan Y, Zhang Y, Kang E, Neoh, K. Electroless deposition of nickel on fluoropolymers modified by surface graft copolymerization. European polymer journal 2002; 38: 2153-2160.
[49] Wu D, Zhang T, Wang W, Zhang L, Jin R. Reflective and conductive surface-silvered polyimide films prepared by surface graft copolymerization and electroless plating. Polymers for Advanced Technologies 2008; 19: 335-341.
[50] Zhu Y, Kang E, Neoh K, Osipowicz T, Chan L. Plasma graft copolymerization of 4-vinylpyridine on dense and porous SiLK for electroless plating of copper and for retardation of copper diffusion. Journal of the Electrochemical Society 2005; 152: F107-F114.
[51] Li L, Yan G, Wu J, Yu X, Guo Q, Kang E. Electroless plating of copper on polyimide films modified by surface-initiated atom-transfer radical polymerization of 4-vinylpyridine. Applied Surface Science 2008; 254: 7331-7335.
[52] Yu W, Zhang Y, Kang E, Neoh K, Wu S, Chow Y. Electroless plating of copper via a Sn-free process on dielectric SiLK surface modified by UV-induced graft copolymerization with 4-vinylpyridine and 1-vinylimidazole. Journal of the Electrochemical Society 2002; 149: C521-C528.
[53] Wang W, Vora R, Kang E, Neoh K. Electroless plating of copper on fluorinated polyimide films modified by surface graft copolymerization with 1-vinylimidazole and 4-vinylpyridine. Polymer Engineering & Science 2004; 44: 362-375.
[54] Wang W, Vora R, Kang E, Neoh K. Electroless Plating of Copper on Fluorinated Polyimide Films Modified by Plasma Graft Copolymerization and UV-induced Graft Copolymerization with 4-Vinylpyridine. Macromolecular Materials and Engineering 2003; 288: 152-163.
[55] Wang W, Zhang Y, Kang E, Neoh K. Electroless deposition of copper on poly (tetrafluoroethylene) films modified by plasma-induced surface grafting of poly (4-vinylpyridine). Plasmas and polymers 2002; 7: 207-225.
[56] Ali A, Aal A. Conductive thin film formation onto radiation grafted polymeric surfaces using electroless plating technique. Polymers for Advanced Technologies 2009; 20: 729-735.
[57] Shojaei A, Fathi R, Sheikh N. Adhesion modification of polyethylenes for metallization using radiation-induced grafting of vinyl monomers. Surface and Coatings Technology 2007; 201: 7519-7529.
[58] Zhou X, Liu X, Xie Z, Zheng Z. 3D-patterned polymer brush surfaces. Nanoscale 2011; 3: 4929-4939.

[59] Zhou X, Wang X, Shen Y, Xie Z, Zheng Z. Fabrication of Arbitrary Three-Dimensional Polymer Structures by Rational Control of the Spacing between Nanobrushes. Angewandte Chemie International Edition 2011; 50: 6506-6510.
[60] Zhou F, Zheng Z, Yu B, Liu W, Huck W. Multicomponent polymer brushes. Journal of the American Chemical Society 2006; 128: 16253-16258.
[61] Gates B, Xu Q, Stewart M, Ryan D, Willson C, Whitesides G. New approaches to nanofabrication: molding, printing, and other techniques. Chemical reviews 2005; 105: 1171-1196.
[62] Nie Z, Kumacheva E. Patterning surfaces with functional polymers. Nature Materials 2008; 7: 277-290.
[63] Haq E, Liu Z, Zhang Y, Ahmad S, Wong L, Armes S, Weaver, J. Parallel scanning near-field photolithography: the snomipede. Nano letters 2010; 10: 4375-4380.
[64] Zhou XZ, et al. "Chemically functionalized surface patterning." Small 7.16 (2011): 2273-2289.
[65] Zhou X, Boey F, Huo F, Huang L, Zhang H. Controlled growth of single-walled carbon nanotubes on patterned substrates. Chemical Society Reviews 2011; 40: 5221-5231.
[66] Ginger D, Zhang H, Mirkin C. Zur Entwicklung der Dip-Pen-Nanolithographie. Angewandte Chemie 2004; 116: 30-46.

Fabrication and Characterization of Porous Chitosan/ Poly-Lactic-Acid Blend Membranes

Ye-Ping Xiang[1], MoYang[2], Li-Yun Xu[1], Mu-Wen Jiang[1], Dan-Jie Zhu[1], Rui-Yun Zhang[1*]

[1]*Key Laboratory of Textile Science & Technology, Ministry of Education, College of Textiles, Donghua University, Shanghai, 201620, China*
[2]*School of Economics and Management, Hebei University of Enginneering, Hebei, 056038, China*

[*]Corresponding author's email: ryzhang@dhu.edu.cn

Abstract

Chitosan(CS) and Poly-lactic-acid(PLA) both have excellent properties of biocompatibility and biodegradability. In this work, PLA solution is blended with chitosan solution to prepare the porous CS/PLA blend membrane by casting and lyophilization method. The mechanical properties, moisture regain, water vapor penetration and porosity rates of polymer blend membranes are estimated. And then the microstructures, morphology, chemical composition and thermal properties are also characterized by scanning electron microscopy (SEM), Fourier transform infrared spectroscopy (FTIR), and differential scanning calorimetry (DSC) respectively.

Keywords: Chitosan (CS); Poly lactic acid (PLA); Lyophilization; Porous membrane; Biodegradation

1. Introduction

Chitosan is converted by chitin deacetylation [1]. Chitosan is easy to film, has good transmission performance and physical and mechanical properties, and has strong alkali resistance and organic solvent resistance. Poly_lactic_acid (PLA), composed of dehydrated and condensed lactic acid, is a kind of completely biodegradable polymer. PLA fiber and its fabric waste can be processed under the suitable conditions. However, the disadvantages of PLA due to its chemical structure, like high crystallinity and low crystallization rate, low strength, poor toughness, low thermal deformation temperature (usually only 60 degrees centigrade), high cost and less species, limit the large-scale application in industry [2-4].

In this work, PLA solution was blended with chitosan solution to prepare the porous CS/PLA blend membranes by casting and lyophilization method, and then the blend membranes were characterized by such properties as mechanical properties, chemical composition and thermal properties, etc, The blend membranes can provide potential applications in the field of bio-medical, such as wound dressings, artificial skin, etc.

2. Experiment

2.1 Materials

CS (degree of deacetylation ≥90%) was purchased from Jinan Haidebei Marine Bioengineering Co. Ltd (China). PLA (Laboratory synthesis), acetic acid, acetone, chloroform, N, N- methyl amide two, two dimethyl sulfoxide (DMSO), sodium hydroxide(analysis of pure) were purchased from Boer Shanghai Chemical Reagent Co. Ltd.

2.2 Preparation of CS and PLA solution

CS was dissolved in 2% HAc solution to prepare a 3% solution. A certain amount of PLA was dissolved in DMSO solution to prepare 5% (w/w) PLA solution. Different weights of PLA with DMSO were added into the 3% chitosan solution to obtain CS/PLA weight ratios of (10:0), (8:2), (6:4), (4:6), and (2:8), and marked as CS100,

CS80, CS60, CS40 and CS20 respectively. The mixture solution was kept stirring for 1 hour until the PLA and chitosan completely reacted. The solution were then processed by the ultrasonic degassing.

2.3 Preparation of CS/PLA blend porous films

The solution was poured into glass molds for pre-frozen in the refrigerator for one night. After solution was frozen into solid, these samples were placed into the vacuum freeze-drying machine for several hours, until the porous CS/PLA blend porous films were formed.

3. Measurement and characterization

3.1 Moisture regain rate

The prepared CS/PLA blend membranes were cut into 20mm×20mm samples and then humidified (20±2 ℃ 65%±3%) for 24h. G_0 as the dry weight of the samples and G_1 as the wet weight of the samples. The moisture regain rate was calculated as Formula 1.

$$rate = \frac{G_0 - G_1}{G_1} \times 100\% \tag{1}$$

3.2 Water vapor penetration rate

The CS/PLA blend membranes were cut into 20mm diameter circular specimens, and then smoothly covered on the test tube with some proper deionized water. Then the assembly was placed into the environment of temperature of 20 ± 2 ℃, humidity 65% ±3% for 24h. The weight of films, liquid and test tube before and after process were weighed on an electronic balance. The water vapor penetration rates of films were then calculated from Formula 2.

$$rate = \frac{W}{\frac{1}{4}\pi D^2} = \frac{4W}{\pi D^2} \tag{2}$$

3.3 Porosity rates

The porosity rates of porous membranes were determined by dissolving the membranes in ethanol at room temperature. The cut-out porous membrane was placed in a scaled test tube with the media (ethanol) until the sample was fully immersed without air bubble on the surface. Denoting V_1 as the initial volume of ethanol, V_2 as the volume of ethanol when the porous membranes were immersed in the test tube, V_3 as the volume of ethanol after the membranes was removed, after each porosity rate experiment was repeated 3 times, the average value P was taken as the porosity rates, calculated as formula 3.

$$P = \frac{V_1 - V_3}{V_2 - V_3} \times 100\% \tag{3}$$

3.4 Mechanical properties

The mechanical properties of blend membranes were tested by using YG065 tester. The prepared CS/PLA blend membranes were cut into samples 20mm in width. The clamp length is 50mm, tensile speed is 50in/min and the pre-load is 2N.

3.5 Fourier transforms infrared spectra (FTIR)

Measured by NEXUS-670 Fourier transforms infrared spectra, the CS/PLA film samples were cut into particle whose size less than 40 um, KBr tablet compressed samples preparation.

3.6 Thermal performance

The DSC curves of blend membranes were measured in nitrogen atmosphere during the heating process. The heating rate was set 20 ℃/min while the temperature range was between 30-400 ℃.

3.7 Scanning electron microscopy (SEM)

The surface morphology of the films was analyzed by using the JSM-5600LV scanning electron microscope.

4. Results and discussions

4.1 Moisture regain rate

Fig.1 shows the moisture regain rates of CS/PLA porous membranes.

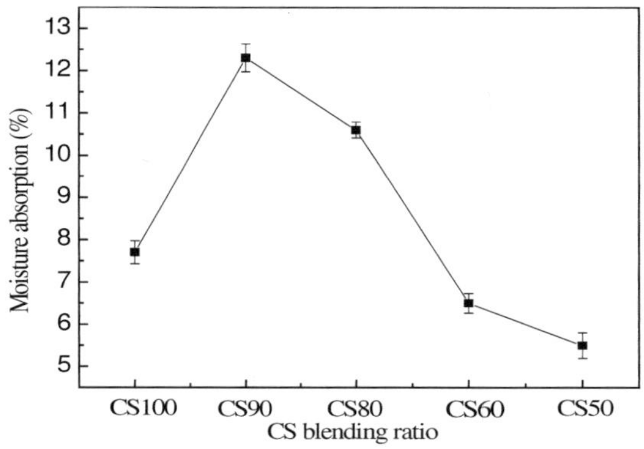

Fig.1 Moisture regains of CS/PLA porous membranes

It can be seen from Fig.1, with the increase of PLA content, moisture regain of CS/PLA blend films first increase and then decrease, and CS90 presents the best. This is partly because that in the CS/PLA system, the water molecules, penetrated into the blend membranes, mainly exist in the membrane pores, and only a small amount bond with the hydroxyl groups of CS. With the adding of PLA content, the porosity rate of the membrane first increase (this can be also proved by the following porosity rate value), so that the moisture property was improved. And in the meantime, because the molecular chain of PLA contains a lot of hydrophobic ester bonds, the further adding of PLA results in the decreasing of the hydrophilic ability of blend membrane. In addition, the increase of PLA content in the blend solution may cause a certain degree of phase separation, resulting in the gel precipitation of CS and PLA, the water absorption properties of blend membranes thus decrease.

4.2 Water vapor penetration rate

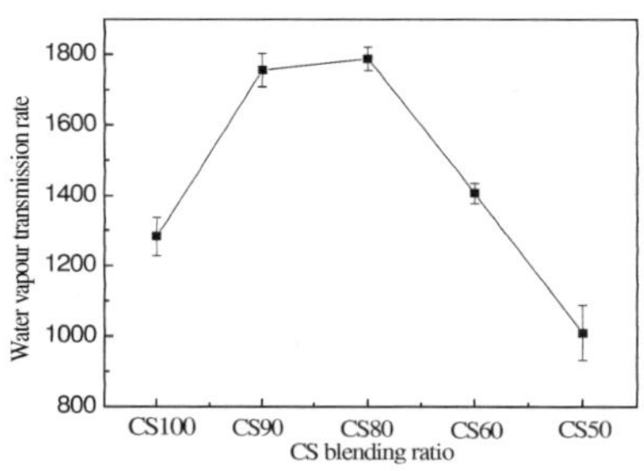

Fig.2 The water vapor penetration rate of the CS/PLA blend membranes

The changing trend of water vapor penetration rate of different mass ratio of CS/PLA blend membrane was shown in Fig.2.

With the increase of PLA content, water vapor permeability first increases and then decreases. The highest value reaches at 1810 when PLA content is 20wt% and the value is about 80% higher than that of CS50. That indicates the higher content of CS contributes the better water vapor penetration rate of the blend membrane. Because the water vapor penetration rate is closely related with membrane thickness, structure compactness and crystallinity degree, the thicker the blend membrane, the higher the structure compactness and crystallinity degree, while the lower the water vapor penetration rate, and vice versa. The interacting of CS and PLA in the blend results in difference of the density and crystallinity of different mass ratio of blend membranes, so that water vapor penetration rates show irregular change. In addition, the blend membrane porosity will also affect the permeability. The higher proportion of CS content, the higher porosity rate, thus the relatively higher water vapor penetration rate of the blend membrane.

4.3 Porosity rate

Fig.3 shows porosity rate of the different mass ratio of CS/PLA blend membranes.

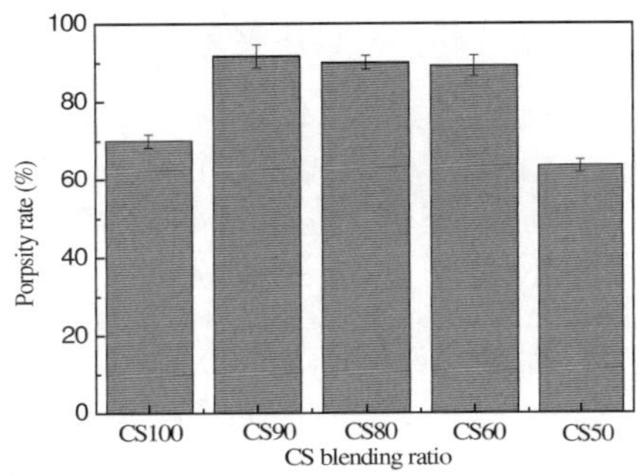

Fig.3 The porosity rates of the CS/PLA blend membranes

The CS50 porosity presents the lowest porosity value, because the high content of PLA results in phase separation of CS and PLA, in which the higher viscosity PLA solution floating in the upper layer makes the resistance force of water molecule movement larger. Aperture, formed on the blend membrane by freeze-drying process, is small, pore volume per unit volume is also small, but the porosity rates have no significant differences among other proportions of blend membranes. Therefore the mass ratio of blend membrane has a great influence on the porosity rate.

4.4 The surface morphology analysis

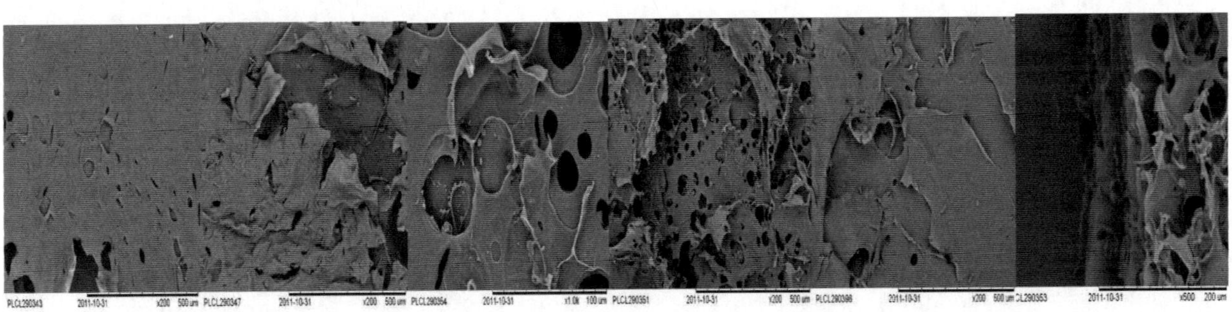

(a) (b) (c) (d) (e) (f)

Fig.4 SEM images of the surface of CS/ PLA blend membranes:
(a)CS90; (b)CS80; (c)CS60,×1000; (d)CS60; (e)CS50; (f)cross section

Fig.4 shows the SEM images of different proportions of CS/PLA blend membranes. In CS/PLA blend system, the surface morphology of the blend membranes have the obvious difference with the changing of CS content. From CS90 image, the whole surface is smooth, with some dispersed small and closed holes on the surface; In CS80, holes formed on the membrane surface are irregular, and some ripples emerge on the surface; In CS60, the pores on the membrane surface are isolated with bad connectivity among them, and all holes are distributed in uniform pore diameter with some holes in large open shape; In CS50, holes are not obvious, with a lot of small particles on the membrane surface. Fig.4 (f) is the image of the section structure of a membrane, presenting the lamellar structure of blend membrane. That maybe because during the freeze drying process, the temperature dropped to below freezing point, thus water molecules in CS solution got crystallized, resulting in phase separation of the two polymers. Phase separation process will be stop after water completely crystallized.

4.5 Mechanical properties

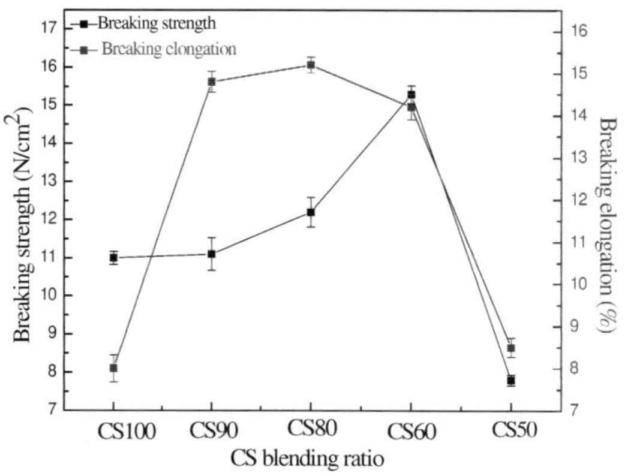

Fig.5 The mechanical properties of CS/PLA blend membranes

Fig.5 presents breaking strength and elongation of CS/PLA blend membranes in different blend ratio. The membrane fabricated with pure CS possesses the lowest tensile strength and the elongation is only 5.67%, indicating the frangibility of pure CS membrane. Experiments show that pure CS film is more difficult to be demold, and the film is easy to disintegrate. When the CS content decreases, the strength and elongation of the blend films first increases and then decreases, reaching the maximum value when mass ratio of CS and PLA is 3: 2. This is contributed to the interaction between CS and PLA molecules, which change the crystal structure of blend membrane and improve the mechanical properties., The decreasing strength with the continuously adding of PLA content contributes to that the inhomogeneity of different blend ratio of CS and PLA [10].

4.6 FTIR spectra

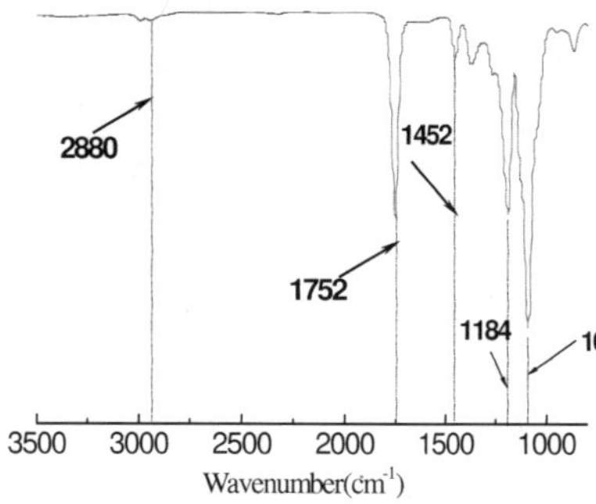

Fig.6 FTIR curve of the PLA membrane

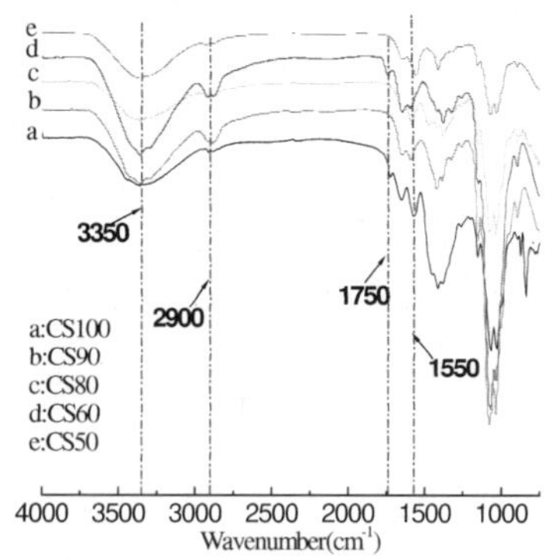

Fig.7 FTIR curves of the CS/PLA blend membranes

Fig.6 shows infrared spectrum of pure PLA film, while Fig.7 shows infrared spectra of CS/PLA blend membrane. The infrared absorption of the pure PLA film has stretching vibration of C=O near 1752cm-1 and stretching vibration of C-O near 1184cm-1, proving the existence of carbonyl. The wave number for the absorption peak near 1453cm-1 is bending vibration peak of -CH3, proving the existence of -CH3. The absorption peak at 2945cm-1 is the stretching vibration of -CH, proving the existence of -CH. And comparing with the previous reference [2], the position of the peak and intensity just has a little difference.

Refering to pure CS, there is a strong absorption peaks of amide in the region of 1625.4-1634.6cm^{-1}. The displacement of that peak for different samples is small, thus the peak can be used to determine CS deacetylation degree [11]. The crystallization peak of CS located at 1030cm^{-1} moves to the high range, indicating the changes of crystalline region of CS; 1084cm^{-1}, 1134cm^{-1}, 1184cm^{-1} are the peak of C-O-C stretching vibration, proving the existence of ester group [12]. 3000-3600cm^{-1} absorption peak attributes to the hydroxyl stretching vibration. Due to each hydroxyl of hydrogen bonds generated strong broad peak, from the figure, the width of the absorption peak of the blend membrane was gradually narrowed with the decrease of chitosan content, indicating that hydroxyl function enhanced. Meanwhile, the curve d appears characteristic peak at 1750cm^{-1}, i.e. carbonyl peak of PLA, proving the existence of PLA; But the carbonyl absorption peak of PLA disappeared in curves b, c, and e, indicating that after the blending of CS and PLA solution, the hydrogen bonds between them appears, which also proves the compatibility of the two polymers.

4.7 DSC analysis

Fig.8 shows the DSC curves of CS/PLA blend membranes.

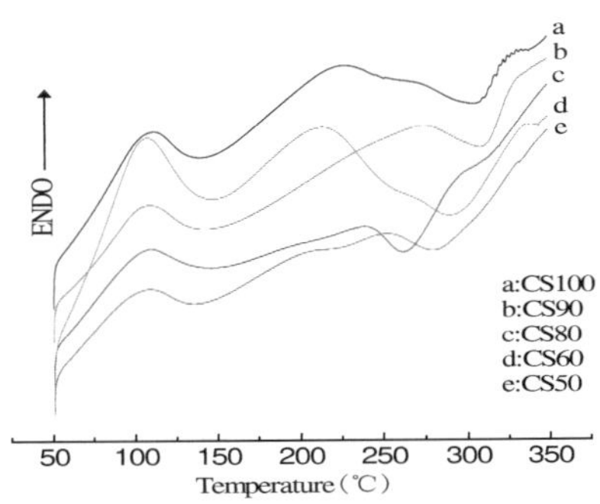

Fig.8 DSC curves of the CS/PLA blend membranes

From the DSC curves, the melting temperature of CS is 309.6 ℃ due to the dehydration of sugar ring and decomposition of polymer in acetyl and deacetylation. The melting temperatures of blend porous materials are lower than that of pure CS. And with the decreases of CS content, the melting temperature of the film first decreases and then increases. The melting temperatures of CS90 and CS80 are 307.3 ℃ and 261.9 ℃ respectively. While the melting temperatures of CS50 and CS60 are 291.15 ℃ and 279.1 ℃ respectively. These indicates that the two polymer solutions are not completely separated, a certain extent interaction force exists between them. The interaction force derives from the strong hydrogen bonds generated between the carbonyl of PLA molecules and -OH and -NH2 of CS molecules, and makes the connection among molecules of the two polymer solutions.

5. Conclusions

A series of porous membranes were prepared by freeze-drying the PLA and CS blend solution. After the performances of the series of blend membranes were analyzed and characterized, the main conclusions were obtained as follows:

1) moisture regain and water vapor penetration rate of CS/PLA blend membrane was higher than that of pure chitosan, which were mainly attributed to that the interaction between CS and PLA, disrupt the regular structure of the original CS. In addition, the structure of PLA itself, such as ester, also affect the performance of the blend films.

2) The blend membrane porosity was improved after freeze-drying, but the mass ratio of blend membrane has a

certain influence on porosity. The SEM images can clearly observe these changes.

3) As to mechanical properties of blend membranes, with the proportion of CS reducing, breaking strength first increases and then decreases, while the elongation decreases gradually. The reason may lie in the performance of PLA itself which changes the crystal structure of CS. In addition, the compatibility of two kinds of solution still need further studied.

4) the DSC curve analysis shows that with CS content decreasing, the decomposing temperature of the membranes decreases firstly and then increases. The reason may attribute to that there is some interaction force between the two solutions. The interaction force occur in the amorphous region of PLA, which inhibited the crystallization of CS and PLA phase and makes the components of melting point depressed.

5) The infrared spectra of CS/PLA results shows there is the formation of strong hydrogen bonding interaction between the two, which is consistent with the conclusion of DSC. Reducing the ratio of CS, the characteristic peaks of PLA appear in the group of them, indicating that the phase separation appeares, which is consistent with the conclusion of SEM.

6. Acknowledgment:

In this paper, we would like to give thanks to Shanghai Municipal Natural Science Foundation（14ZR1401000）and Shanghai University Students Innovation Project for providing support.

References:

[1] Huang GF, Qing SB, Li SB, Huang SQ. The Advance of Polysaccharide Biomedical Materials-Chitin and Chitosan. J Polym Bull: 2001; 03: 43-48.

[2] Yu RB, You HY, Xu XL, Li HX. Preparation and Properties of the Blend Film of PVA and Chitosan. J Mod Plast Process Appl: 2001; 13: 16-18.

[3] Liu JM, Lv YC, Jia LX. A feasible study of PLA fibers application to medical use, JSH Textile: Sci Technol: 2009; 07: 37.

[4] Xia M, Liu JC, Wan DY, Li SH, Chen YW, Wang LF. Advance on Modification of Environment-friendly PLA. J Eng Plast Appl: 2014; 03: 22-126.

[5] Guo C, Dong YS, Lin PH, Pu YP, Liu B. Study on Fabrication of Poly lactic Acid Porous. J SE Univ:Natl Sci Ed: 2004; 04: 495-499.

[6] Li RL, Li CY, Zhang W. Micro-structure of Composite Tissue Engineering Scaffold of Chitosan and Type Ⅰ Collagen with Scanning Electron Microscope. J SUN YAT-SEN Univ:Med Sci: 2006; 05: 557-561.

[7] Ding B, Kimura E, Sato T, Fujita S, Shiratori S. Fabrication of blend biodegradable nanofibrous nonwoven mats via multi-jet electrospinning. J Polym: 2004; 45: 1895-1902.

[8] Liu WJ, Lin ZH. Preparation and Mechanical Properties of Biodegradable Cellulose/Chitosan Blend Film. J China Plast Ind: 2003; 12: 44-46.

[9] Xue HT. Study on Chitosan function dressing. D Zhejiang Univ: 2006.

[10] Pearson FG, Marchessault RH, Liany CY. Infrared spectra of crystalline polysaccharides V chitin. J Polym Sci: 1960; 141: 101-116.

[11] Zislis T, Martin SA, Cerbas E. A scanning electron microscopic study of in vitro toxicity of ethylene-steri lized bone repair materials. J Oral Implantol: 1989; 1: 41-47.

[12] Domszy JG, RobertsG AF. Makromol, Evaluation of infraxed spectropic techniques for analyzing chitosan. J Macromol Chem Phys: 1985; 8: 1671-1677.

The Effect of Bio-inspired Polydopamine Coating on Polymer Fabric

Ting Wu, Yi Li, Xu-Qing Liu*

**School of Materials, University of Manchester, Oxford Road, Manchester, M13 9PL, UK*

*Corresponding author's email: xuqing.liu@manchester.ac.uk

Abstract

A simple surface modification method is introduced in this report. Polyester fabric, nylon fabric and cotton fabric was modified by polydopamine coating which can be obtained by oxidative self-polymerization in pH 8.5 Tris solution. By such surface modification, the polymerization of dopamine and hydrophilic property of these fabrics were investigated by UV spectroscopy, Moisture Management Tester. In order to improve the antibacterial property of fabrics, Ag nanoparticles deposited on the surface of fabrics, which was observed by XRD. The result showed that the PDA coating made repellent fabric to penetration fabric for both polyester fabric and nylon fabric. Ag nanoparticles deposited on the surface of fabrics from the data of XRD.

Keywords: Dopamine; PDA coating; polyester fabric; nylon fabric; cotton fabric; moisture management testing

1. Introduction

To fulfill the increasing demand of functional and smart materials, lots of surface modification methods have been employed, which can modify the surface of a material by changing physical, chemical or/ and biological characteristics different from the original one. Surface modification techniques can be applied to develop plenty of functional properties, including chemical, physical, electrical, electronic, magnetic, mechanical, wear-resistant, corrosion-resistance and hydrophilic-hydrophobic properties at the required substrate surface to a controlled extent [1]. Nearly all types of surfaces, including polymers, metals, ceramics and composite can be modified. Comparing with creating new materials, surface modification is a much easier and cheaper way which also can combine different properties of coating material and substrate together. According to current research, there are various surface modification approaches. For instance, chemical methods including oxidation [2], grafting [3], rare earth treatment [4], electrochemical treatment [5]; physical methods like plasma treatment [6], -ray irradiation [7]; and a combination of chemical and physical methods [8].

The mechanism of surface modification can be described as [9]:

- Removing the relatively weak bonding layer formed by some contaminant or other molecules absorption on the surface. This would impede the direct contact between fibre and matrix.
- Allowing matrix molecules diffuse into fibre to form interlocking between fibre and matrix.
- Promoting surface porosity so that matrix molecules can penetrate to form physical interlocking between fibre and matrix.
- Increasing the active sites on the surface forms chemical bonding.
- Using coupling agent can chemically connect fibre and matrix.

All these methods play important role in surface modification.

However, although these surface modification methods can enhance the properties at the required substrate to some extent, they also have some negative effects [9]. For example, using nitric acid modified the fibre would etch the bulk of fibre and damage the mechanical properties of fibres. Plasma surface modification is difficult to control and largely dependent on equipment. The extent of plasma is controlled by the power of plasma and the modification time. Moreover, the future of surface modification should be bio-degradable, biocompatibility, high-performance and low cost. The multifunctional properties of fabric are highly demanding, such as good physical and chemical performance, high durability, sustainability, good antibacterial performance, easy to dye, etc. In terms of antimicrobial applications, surface modification plays critical role. As bacterial infections affect many fields like medical, industry, food safety, resulting in serious threats to human health, the demand for production antibacterial surface is relatively urgent [10]. Therefore, how to find new surface modification method is an academic and industrial challenge. In terms of current research, some biopolymers are used for surface and interface modification, for example, PEG, DNA, protein, polydopamine and so on. Polydopamine coating has been a hot topic for surface modification due to its high performance.

Polydopamine (Fig.1) can be obtained from its monomer dopamine which was observed from the strong adhesion of mussels even on wet substrate [10]. Therefore, the origin of dopamine is very abundant. The most important property of polydopamine is that it can stick to almost all surface, organic, inorganic, metal, ceramic, which overcomes the sticky problem that most of surface modification methods [11]. Moreover, polydopamine can be obtained via oxidative self- polymerization at room temperature under basic condition. Such wild polymerization process makes polydopamine coating much more abstractive. The thickness of polydopamine coating layer can be controlled via changing the polymerization time with a maximum thickness of 50nm after 24 h [10]. Such coating also introduces functional groups on the substrate including amino group, carboxylic acid group, catechol or quinone functions; thus, it improves the chemical reactivity of inert surface without damaging the bulk material so that diverse substrates can be modified to construct hybrid materials [10]. In addition, polydopamine can bind to multivalent metal ions, for example, Mn^{2+}, Zn^{2+}, Fe^{3+}, and Cu^{2+} [12]. Except for metal ion binding, polydopamine can also reduce some metal ions like Au^{3+}, Ag^{+}, and Pt^{3+} under basic environments [13]. Polydopamine acts as a reducing agent in this condition. Such property of polydopamine makes surface modification to produce organic-inorganic hybrid materials, and polydopamine coating can carry with metal nanoparticle [10]. During the polymerization of dopamine, it can mix with graphene oxide sheets for large-scale production and stabilization of rGO, providing a new way to produce high electro-conductivity hybrid materials [14]. Besides, polydopamine provides a simple and inexpensive coating for antimicrobial applications. The reducing capability of polydopamine toward metal ions as well as its robust adhesion on different inert substrates has been employed for in situ fabrication of antimicrobial surface. The multi-properties of polydopamine make it a hot spot for surface modification for different materials.

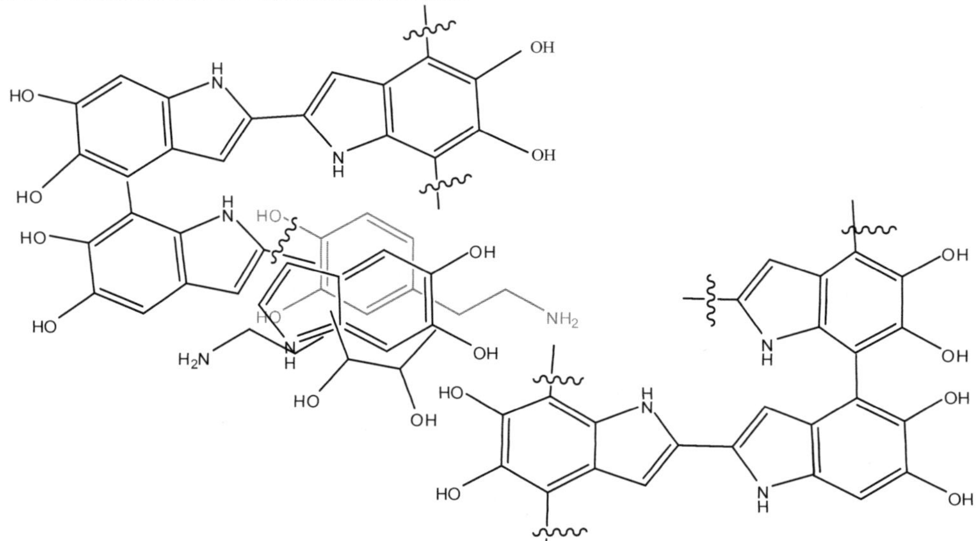

Fig.1 the structure of Polydopamine

In our report, polyester, nylon 66 and natural cotton fabric are coated by polydopamine for surface modification. PET fibre is widely used because of its tenacity, washability and wearability, dimensional stability and resistance to many chemicals. Those functional properties make PET fabric also popular for many years. However, there are some limitations for PET fabrics. It is hydrophobic inherently. Thus, it is necessary to improve the surface hydrophilicity for many end applications [15,16]. Many surface modification approaches have been employed, such as plasma treatment, chemical modification, grafting and so on. Comparing with polydopamine coating, these methods have many disadvantages due to high cost, damage on the surface, limited practical use. Nylon 66 also is one of the most widely used fabric because of its low cost, good resistance to shrinkage and abrasion. However, the inherent hydrophilicity property limits its applications in industry [17]. Similarly, many methods also have been used for surface modification, but many of them are as good as polydopamine coating which can improve the surface wettability greatly. In terms of cotton fabric, it has good hydrophily but bad antibacterial property [18]. Polydopamine coating provide a platform for improving the antibacterial property for cotton fabrics.

In this paper, we report our recent research efforts of polydopamine coating on PET fabric, nylon 66 fabric and cotton fabric to increase the surface moisture, which is showed in Fig.2. The coating process was under alkalinity solution, and dopamine polymerized on the surface of fabrics. Afterwards, polydopamine reduced metal ions to make metal nanoparticles deposit on the surface of fabrics. Moisture Management Tester, Ultraviolet-Visible Spectrophotometer and X-Ray Diffraction (XRD) were employed to analyze the surface moisture, the process of polymerization and surface properties.

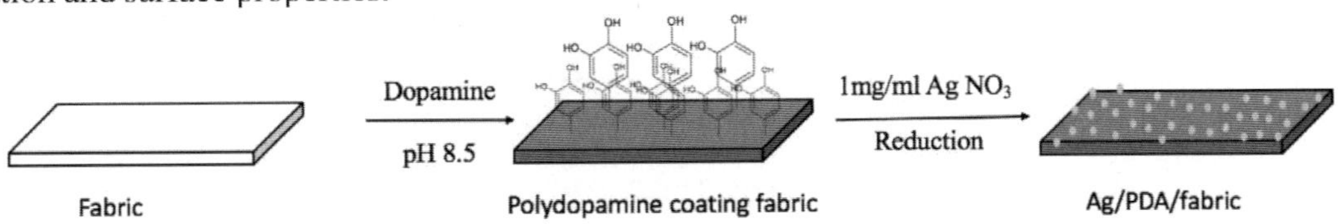

Fig.2 The process of

2. Experiment procedure

The surface modification of carbon fibre and Kevlar fibre.

Dopamine solution (2mg/ml) was prepared by dissolving dopamine in Tris-HCl buffer solution (10mM, pH 8.5). Subsequently, PET fabric, nylon 66 fabric and natural cotton fabric were immersed in dopamine solution art room temperature for 24h separately. A thin PDA layer was adhered on PET fabric, nylon 66 fabric and cotton fabric via the oxidative self-polymerization of dopamine. Then, the polydopamine coated fibers were taken out, washed with deionized water for several times until the solution was colorless and dried in a vacuum oven at 50 overnight.

The loading of AgNPs on the surface of polydopamine-coated fabrics

The polydopamine coating fabrics were immersed into silver nitrate solution (0.29 mol/L) at room temperature for 8 h. Then the samples were rinsed with flowing deionized water for 10 min and dried at 40 ° C in the vacuum oven for 6 h. The obtained samples were designated as polydopamine/AgNPs fabrics in the subsequent discussion.

3. Moisture Management Tester (MMT)

According to the basic moisture properties of fabrics, MMT was invented by Prof. Yi Li. Such testing technique gives many key terms for almost kinds of fabrics. It gives the wetting time, maximum absorption rate, maximum wetted radius, including top and bottom, spreading rate, all of them including the top surface and bottom surface; and one-way liquid transport index, overall moisture management index. By analyzing these key parameters, the moisture properties of a particular fabric can be determined quickly, such as water repellent fabric, absorbent and quick dry fabric and excellent moisture management fabric. Hence, in our experiment, the moisture property of PET fabric, nylon 66 fabric and cotton fabric before modification and after modification can be obtained.

4. Results and analysis

4.1 Ultraviolet-Visible Spectrophotometer

In the process of oxidative self-polymerization of dopamine, absorbance peaks for catechol group at 278nm, dehydro-dopamine at 300nm, and *o*-quinone at 405nm were observed by UV spectrum. A group of spectrum of absorbance against wavelength with time was obtained. It was obvious that the color change from colorless to dark for the reaction solution during the formation of PDA. From the spectrum, the absorbance at three wavelengths increased with time.

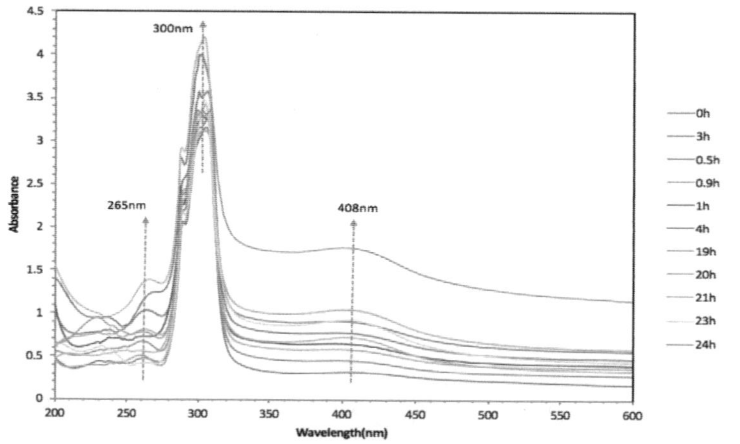

Fig.3 UV spectrum for polymerization of dopamine with time

4.2 Moisture Management Tester

To find out the change of hydrophilic property after PDA coating, the modified fabrics and unmodified fabrics were used for moisture testing. The unmodified polyester and nylon fabrics are hydrophobic, and the spreading rate from top surface to bottom surface is very slow. After modification, the hydrophilic property for fabrics increased a lot. According to the data of moisture state of the three different fabrics, the water content with time, the water location with time and finger print of moisture management properties were obtained. In terms of polyester fabric, after PDA-coating, the moisture state of polyester fabric was changed from water repellent fabric to slow absorbing, and its water location with time changed obviously. What's more, if increasing the layer of PDA coating, its hydrophilic property could be improved greatly. In terms of nylon fabric, after PDA coating, its hydrophilic property has improved from water repellent fabric to water penetration fabric. As for cotton fabric, it is a hydrophilic material inherently, after PDA coating, the moisture state has not been reduced, and it is still a water penetration fabric. Besides, the antibacterial property of cotton fabric can be improved by PDA coating.

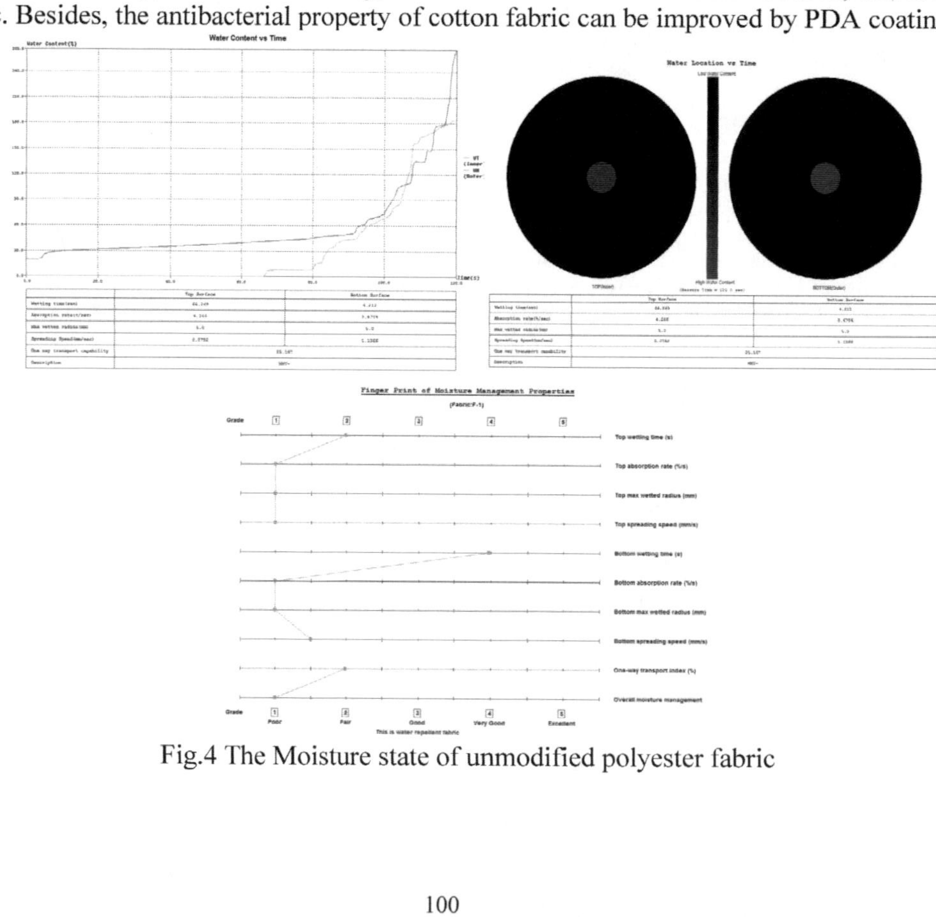

Fig.4 The Moisture state of unmodified polyester fabric

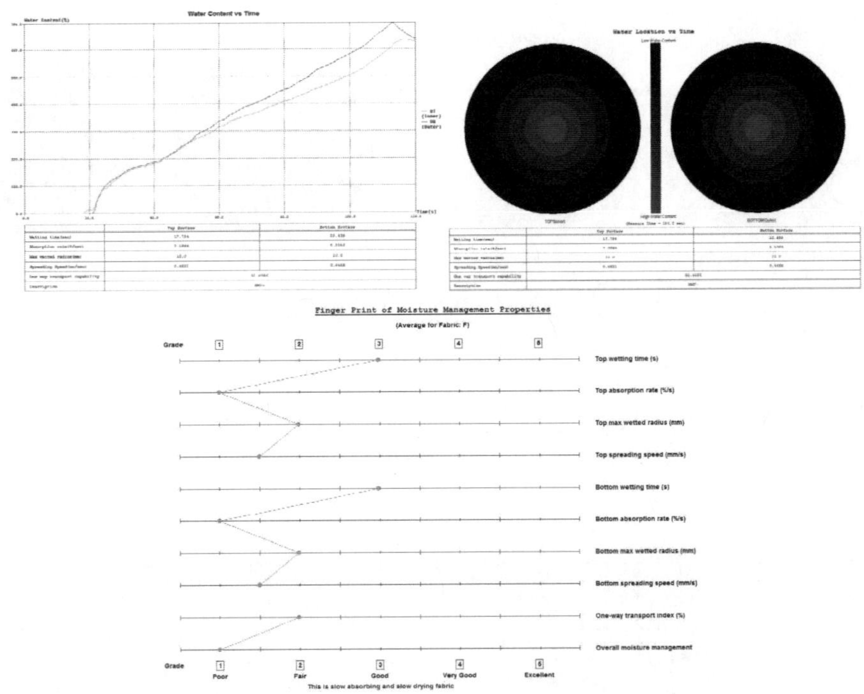

Fig.5 The moisture state of PDA coating polyester fabric

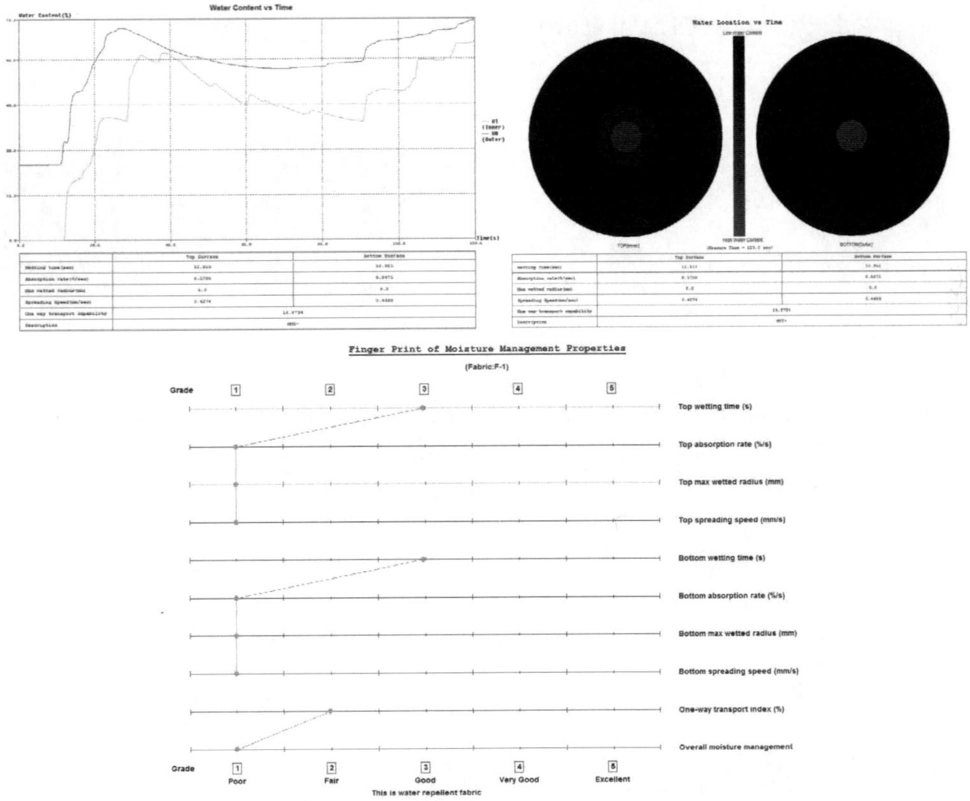

Fig.6 The moisture state of unmodified nylon fabric

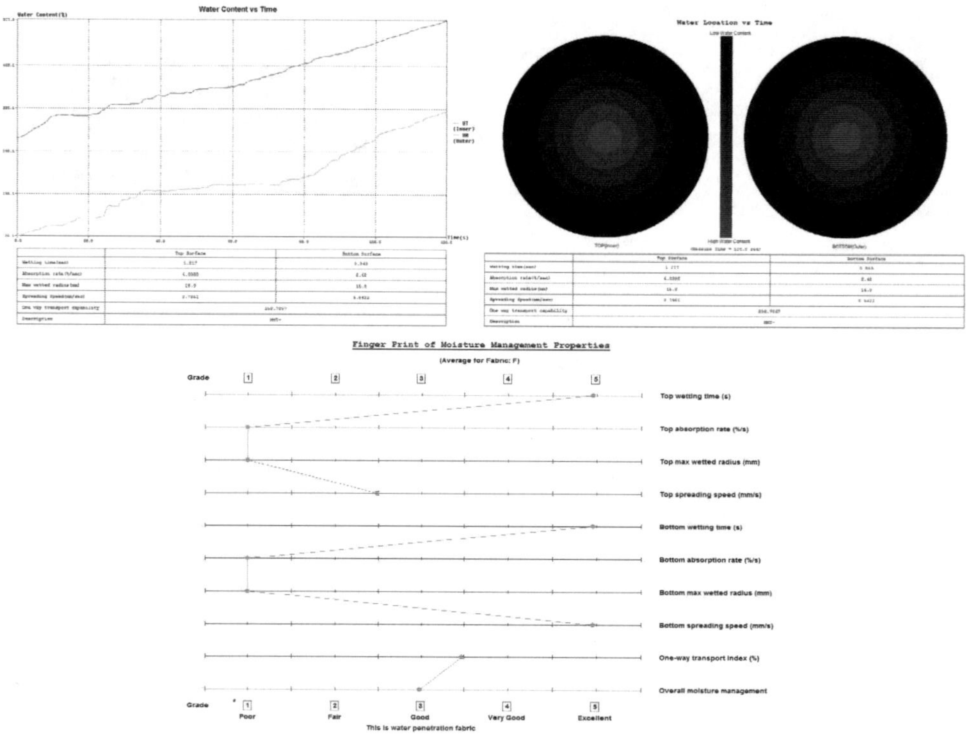

Fig.7 The moisture state of PDA-coating nylon fabric

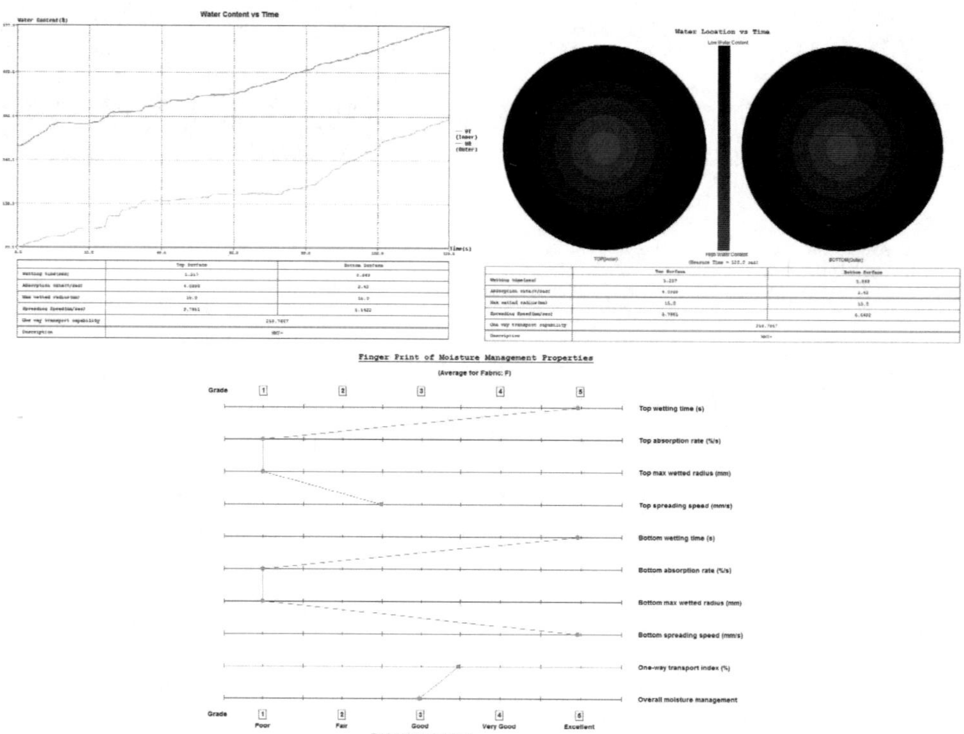

Fig.8 The moisture state of unmodified cotton fabric

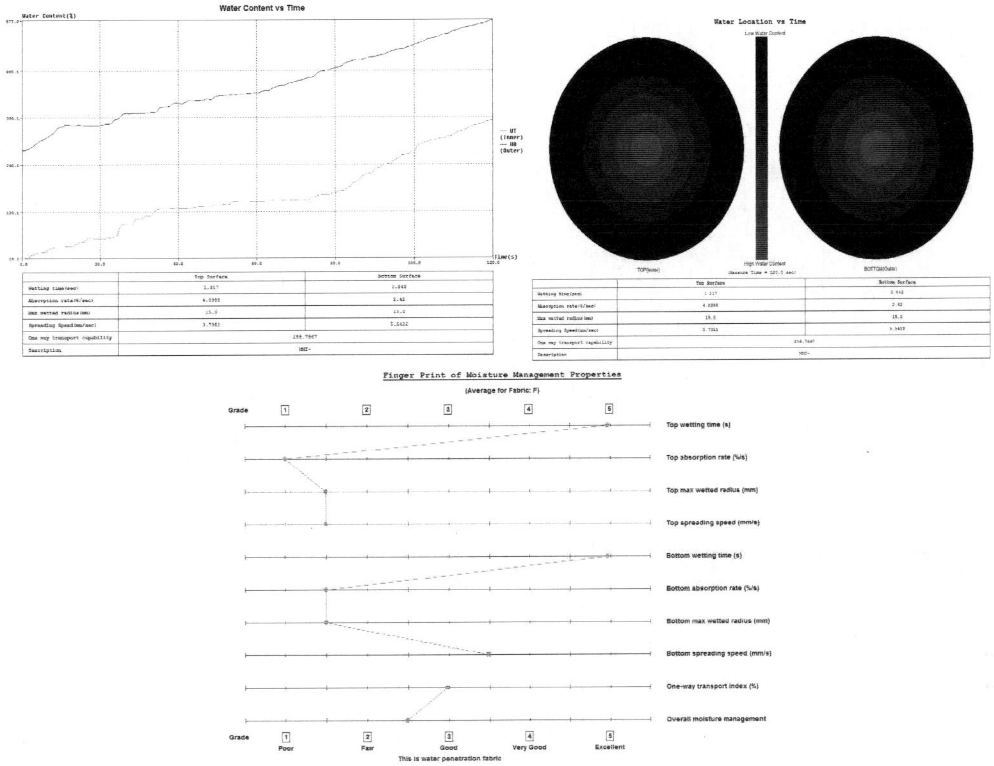

Fig.9 The moisture state of PDA-coating cotton fabric

4.3 X-Ray Diffraction

After finishing PDA coating, $AgNO_3$ was used for improving the antibacterial property of these fabrics. XRD was employed to characterize the crystal on the surface of modified cotton fabric. The 2 was range from 30 to 80. Three strong peaks on the XRD spectrum (Fig.9), 32.8, 38.1,45, and there were some small peaks which located at 64.4and 77.29, which are corresponding to the peaks for silver. However, the size of silver is relatively small and uniform.

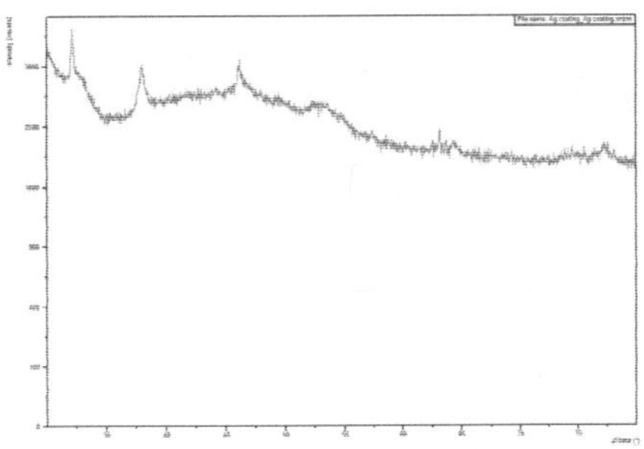

Fig.10 XRD spectrum for Ag

5. Conclusion

Polydopamine is a promising candidate for surface modification. It could be formed on the surface of different fabrics by oxidative self-polymerization in the Tris buffer solution. Polydopamine coating provide a new way to improve the hydrophilic property without damaging the surface or bulk material, and it provide a new method for metal deposition to improve the antibacterial property. The polydopamine modified cotton fabric was coated with Ag nanoparticles.

Acknowledgement

References

[1] https://en.wikipedia.org/wiki/Surface modification

[2] Tran MQ, Ho KKC, Kalinka G. Carbon fibre reinforced poly (vinylidene fluoride): Impact of matrix modification on fibre/ polymer adhesion, Composite Science and Technology: 2008; 68: 1766-1776.

[3] Fang CQ, Wu JX, Wang JL. Modification of carbon fibre via grafting with Meldrum's Acid, Applied Surface Science: 2015; 356: 9-17.

[4] Li J, Cheng XH. Effect of rare earth solution on mechanical and tribological properties of carbon fibre reinforced thermoplastic polyimide composite, Tribology Letters: 2007; 25: 207-214.

[5] Ishifune M, Suzuki R, Mima Y. Novel electrochemical surface modification method of carbon fibre and its utilization to the preparation of functional electrode, Electrochimica Acta: 2005; 51: 14-22.

[6] Tiwari S, Sharma M, Panier S. Influence of cold remote nitrogen oxygen plasma treatment on carbon fabric and its composites with specialty polymers, Journal of Material Science: 2011; 46: 964-974.

[7] Li JQ, Huang YD, Fu SY. Study on the surface performance of carbon fibres irradiated by - ray under different irradiation dose, Applied Surface Science: 2010; 256: 2000-2004.

[8] Liu TM, Zheng YS, Hu J. Surface modification of aramide fibres with novel chemical approach, Polym. Bull: 2011; 66: 259-275.

[9] Jang BZ. Control of interfacial adhesion in continuous carbon and Kevlar fibre reinforced polymer composites. Compo Sci Techno: 1992; 44(4): 333-490.

[10] Liu YL, Ai KL, Lu LH. Polydopamine and its derivative materials: synthesis and promising applications in energy, environmental, and biomedical fields. Chem. Rev: 2014; 114: 5057-5115.

[11] Lee H, Dellatore SM, Miller WM, Messersmith PB. Science 2007; 318: 426.

[12] d'Ischia M, Napolitano A, Pezzella A, Meredith P, Sarna T. Angew. Chem., Int. Ed: 2009; 48: 3914.

[13] Ball V, Nguyen I, Haupt M, Oehr C, Arnoult C, Toniazzo V, Ruch DJ. Colloid Interface Sci: 2011; 364: 359.

[14] Xu LQ, Yang WJ, Neoh KG, Kang ET, Fu GD. Macromolecules 2010; 43: 8336.

[15] Surface modification of polyester fabric with plasma pretreatment and carbon nanotube coating for antistatic property improvement. Wang CX, Lv JC, Ren Y, Zhi T, Chen JY, Zhou QQ, Lu ZQ, Gao DW, Jin LM. Applied surface science: 359; (2015): 196-203.

[16] Modification of Polyester Fabric Properties by Surfactant-Aided Surface Polymerization Ampornphan Siriviriyanun, [1]Edgar A. O'Rear, [2]Nantaya Yanumet, journal of applied polymer science: 2007; 103: 4059-4064.

[17] Surface modification of polyamide 66 fabric by microwave induced grafting with 2-hydroxyethyl methacrylate. Qian Zhao[1], Xiaoyu Gu[1], Sheng Zhang[*], Mingzhe Dong, Peng Jiang, Zhongwu Hu. Surface and coating technology 2014; 240: 197-203.

[18] The preparation and antibacterial effects of dopa-cotton/ AgNPs. Hong Xu[*], Xue Shi, Hui Ma, Yihang Lv, Linping Zhang, Zhiping Mao[*], doi: 10.1016/ j. apsusc: 2011.02.129.

Air Permeability of Polyacrylonitrile Electrospun Nanomembranes

Vinod Kadam[1], Amit Jadhav, Rajkishore Nayak, Lijing Wang, Rajiv Padhye*

School of Fashion & Textiles, RMIT University, Dawson st. Brunswick, VIC, 3056, Australia

*Corresponding author's email: rajiv.padhye@rmit.edu.au

Abstract

This research studied the effect of collector speed and deposition time on the air permeability of electrospun nanomembranes. Polyacrylonitrile nanofibres that had the diameter in the range of 291 – 877 nm were fabricated. It was observed that the collector speed did not affect air permeability significantly. However, the deposition time significantly influenced air permeability. The air permeability of all nanomembranes decreased drastically compared to the pristine fabric without nanomembrane. It is concluded from the results that the deposition time is a dominant factor compared to the collector speed influencing air permeability of nanomembranes. A good correlation was found between the theoretical porosity and experimental air permeability of nanomembranes.

Keywords: Electrospinning; Fibre Diameter; Nanomembranes; Air Permeability; Porosity;

1. Introduction

Electrospinning has gained a tremendous increase in research and commercial interest since beginning of the century [1]. Electrospinning is simple, easy and versatile process to produce polymeric nanofibres [2-4]. In a typical electrospinning process, a viscoelastic polymer fluid is electrically charged with high voltage power supply. The fluid then overcomes its surface tension and elongates in a jet form to deposit as nanofibres onto the collector [5, 6]. The nanofibre diameter is largely influenced by polymer fluid characteristics and electrospinning process parameters. The polymer fluid characteristics include molecular weight, solution viscosity and electrical conductivity, whereas the electrospinning process parameters include applied voltage, flow rate, needle gauge, needle to collector distance, collector speed and deposition time [7].

Electrospun nanofibres possess high surface area to volume ratio (specific surface area) and low pore size compared to microfibres [5, 6]. These properties of nanofibres have led to variety of applications by converting them into nanomembrnaes. Some of the important applications of electrospun nanomembranes include air filtration [5], tissue engineering and drug delivery [8], and protective textiles [9, 10]. Polyacrylonitrile (PAN) is one of the extensively used polymers in such applications due to its excellent mechanical and chemical properties [6, 11]. More attention has been given to optimize process parameters for PAN to get desired nanofibre diameter [11]. However, optimization of properties such as air permeability of nanofibre membrane is still in infancy.

Air permeability and porosity are the important parameters of nanomembrane, especially for air filtration applications. Air permeability is defined as the rate of air flow through a porous material under a specified differential pressure across the material [12], whereas, the porosity is a ratio of void in the nanomembrane over volume of fibre [2]. Electrospun nanofibres have good air filtration efficiency due to small fibre diameter and small pore size [13, 14]. However, attaining high filtration efficiency adversely effects on air permeability and porosity of the membrane. One major challenge in developing protective materials is to maintain low pressure drop and high air permeability while increasing the filtration efficiency. Many filtration media provide high efficiency by sacrificing pressure drop or air permeability [15]. Electrospun nanomebranes can be effective in resolving the challenge. Relatively fewer studies focused on air permeability of nanomembranes.

In this study, nanomembrnaes of PAN were fabricated by electrospinning using a laboratory scale electrospinning apparatus with indigenously designed collector assembly. The porosity and air permeability of nanomembrane were evaluated following standard procedures. The effect of collector speed and deposition time on nanofibre diameter, fibre distribution and air permeability was analyzed. The correlation between porosity, air permeability and nanomembrane thickness was established statistically.

2. Experimental

2.1 Materials

Polyacrylonitrile polymer (Mw 110 000 g/mol) was obtained from Goodfellow Cambridge Limited, England. The polymer was dissolved in N,N-dimethylformamide (DMF) solvent to prepare 8% (w/w) polymer solution. The solution stirred continuously using an electromagnetic stirrer at 70°C for 3 hours to obtain homogenous solution. The solution was stored at room temperature and all the electrospinning experiments were conducted at room temperature. The electrospun nanofibres were deposited on wool/polyamide fabric substrate as a support to nanomembrane.

2.2 Methods

2.2.1 Electrospinning

The polymer solution was filled in a 5 mL plastic syringe. The solution was pumped at a constant rate of 1 mL/hr using a metering pump (model KDS 100, K.D. Scientific Inc., Holliston, MA, USA) through a stainless steel needle of 18 gauge (950 µm internal diameter). The tip of the needle was grinded to make it circular. The high powered direct current (DC) voltage of 8 kV was supplied to needle tip. The needle tip to collector distance was 14 cm throughout all experiments. The collector was grounded to produce electric strength between needle tip and collector. The rotating collector drum was used to collect nanofibres in continuous form. The rotating collector drum assembly was designed indigenously. The collector drum, of 75 mm external diameter and 200 mm width, was connected to a variable speed motor drive. The speed of drum was adjustable within the range of 3–30 rpm. Nanomembranes were collected with three, six and nine rotations of collector drum per minute. Three deposition times of 20, 40 and 60 minutes were used for collecting the nanofibres.

2.2.2 Fibre diameter and diameter distribution

The nanomembranes were characterized for fibre diameter and diameter distribution using a scanning electron microscope (SEM) (Philips XL-30). The samples were fixed on a circular stub and sputter coated with gold. For fibre diameter measurement, one hundred fibres were counted from five different SEM images. The SEM images were analyzed using Image J® software and MINITAB 16® to obtain the average fibre diameter and fibre distribution.

2.2.3 Air permeability

The air permeability of nanomembrnaes deposited on fabric sample was measured according to IS EN ISO 9237:1996 standard at 65±2% relative humidity and 20±2°C. The samples were conditioned for 24 hours before the air permeability testing. The air permeability results were compared with pristine fabric sample. One way analysis of variance (ANOVA) was carried out to test the significance of collector speed and deposition time on air permeability of nanomembrane.

2.2.4 Porosity

The air porosity was determined by measuring the weight and volume of the nanomembrane using Eq. (1) [2, 16].

$$\varepsilon = \frac{V_o}{V_o + V_s} = \left(1 - \frac{\rho_{nanomembrane}}{\rho_{fibres}}\right) = \left(1 - \frac{w}{A.t}.\frac{1}{\rho_{fibres}}\right) \quad (1)$$

Where ε is Porosity, V_0 and V_s are volume of void and fibres in the nanomembrane respectively. $\rho_{nanomembrane}$ and ρ_{fibres} are the mass densities. ρ_{fibres} was taken an 1.184 gm/cc since the fibres produced from pure polyacrylonitrile. w, A and t are weight, area and thickness of nanomembrane, respectively.

3. Results and discussion

3.1 Nanofibre diameter and its distribution

Fig.1 presents the SEM images and fibre distribution of PAN nanofibres at collector speed of 3 rpm The average fibre diameter was in the range of 499 – 564 nm for deposition time of 20 – 60 minutes. The variation in fibre diameter with constant electrospinning parameters, except deposition time, indicates the difficulty in maintaining the uniformity in the fibre diameter. Although the average fibre diameter was in the same range, different values of fibre diameter for different deposition times during the electrospinning may be attributed to change in solution viscosity upon storage in a syringe. Fig.2 and 3 shows the SEM images and fibre distribution at collector speed of 6 and 9 rpm. Similar trend of fibre diameter was observed at different collector speeds and deposition times. During all variations in electrospinning, the minimum and maximum fibre diameter determined was 291 and 872 nm, respectively. It is evident that the fibres in all the nanomembranes were normally distributed (Fig.1 3). For all variations in the deposition time and collector speed the average fibre diameter range was 499 – 594 nm. It may be inferred that nanofibre diameter is independent to deposition time and collector speed, since no specific increasing or decreasing trend was observed. This is in agreement with previous study where collector speed not altered the average fibre diameter [2]. Park et al[12]. prepared nylon nanofibres at collector speed from 30 to 100 rpm. The average fibre diameter did not change. However, the distribution was shifted towards lower values. In another study, it was observed that the change in deposition time from 2 – 6 hours did not alter the average fibre diameter significantly [10].

Nanofibre diameter and its distribution are important parameter to control the pore size and air permeability [17]. An increase in nanofibre diameter reduced the number of pores and increased the pore size that resulted into higher porosity [17]. The nanomembrane with small fibre diameter resulted into reduced porosity and air permeability [18]. A small fibre diameter is desirable to achieve a small pore size which may be suitable for filtration. However, small fibre diameter resists the air flow. Hence to achieve maximum air permeability across the nanomembrane, fibre diameter and pore size optimization plays vital role.

3.2 Air permeability

Airflow through the nanomembrnaes can be analogous to the channel theory that assumes that porous media is a bundle of cylindrical tubes passing from one surface of the media to the other surface, and not necessarily perpendicular to the surfaces [19]. The air permeability of nanomembranes at 100 Pa air pressure is shown in Table 1. The air permeability of the pristine wool/nylon fabric, before nanofibre deposition, was found to be 693.85 mm/s which are significantly higher ($p < 0.01$) than all the fabrics with nanomembranes deposited (Table 2). This can be attributed to the resistance offered by highly entangled nanomembranes with low pore size. From the SEM images in Fig. 1-3, it is clear that when the fibre diameter is down to submicron size, the inter fibre space between the nanofibres is also reduced to micron scale. At the same air permeability, finer fibres are required to achieve the same coverage as that of the coarser fibres. More fibres in the same area result in smaller inter fibre space (pore size) [20]. It indicates a thin layer of nanofibres is sufficient to reduce the air permeability to a larger extent.

3.2.1 Effect of collector speed

The air permeability results of nanomembranes at different collector speeds are shown in Table 1. It can be observed that as the collector speed increased from 3 to 9 rpm for the same deposition time, the air permeability did not change significantly ($p > 0.01$) (Table 2). It means collector rpm variation from 3 to 9 rpm did not affect air permeability. It is in agreement with the previous study [2] that the air permeability of nanomembrnaes at different rotational speeds was comparable. It is important to note that the nanomembrane thickness for different collector speeds was also comparable.

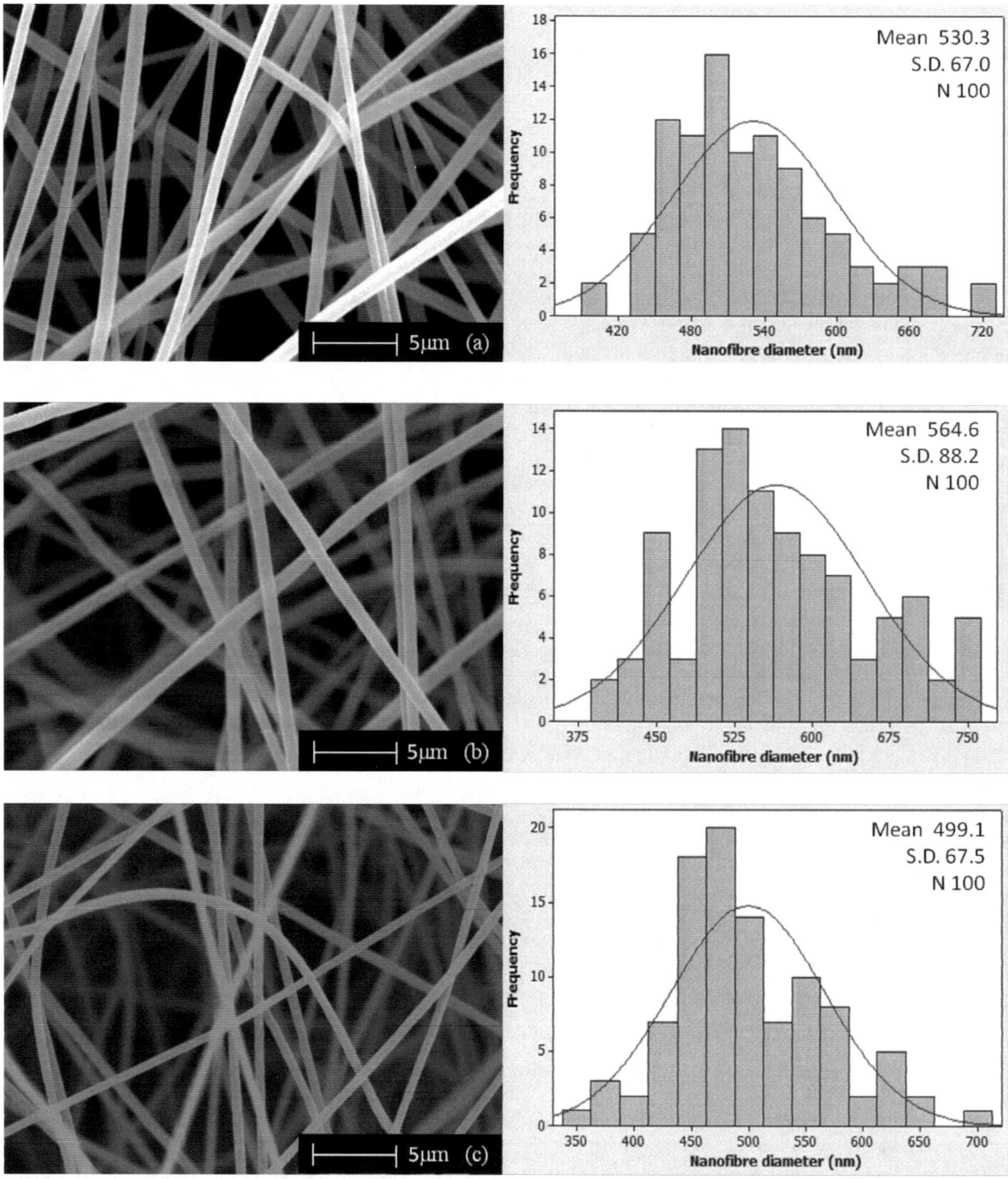

Fig.1 SEM images of 8% PAN nanofibres electrospun at 8 kV applied voltage with 14 cm collector distance from needle tip. The nanofibres deposited at collector speed of 3 rpm for (a) 20 minutes, (b) 40 minutes, and (c) 60 minutes, respectively. The corresponding fibre distribution is placed adjoining to the SEM image. SEM images captured at acceleration voltage of 10 kV with 10000 magnifications.

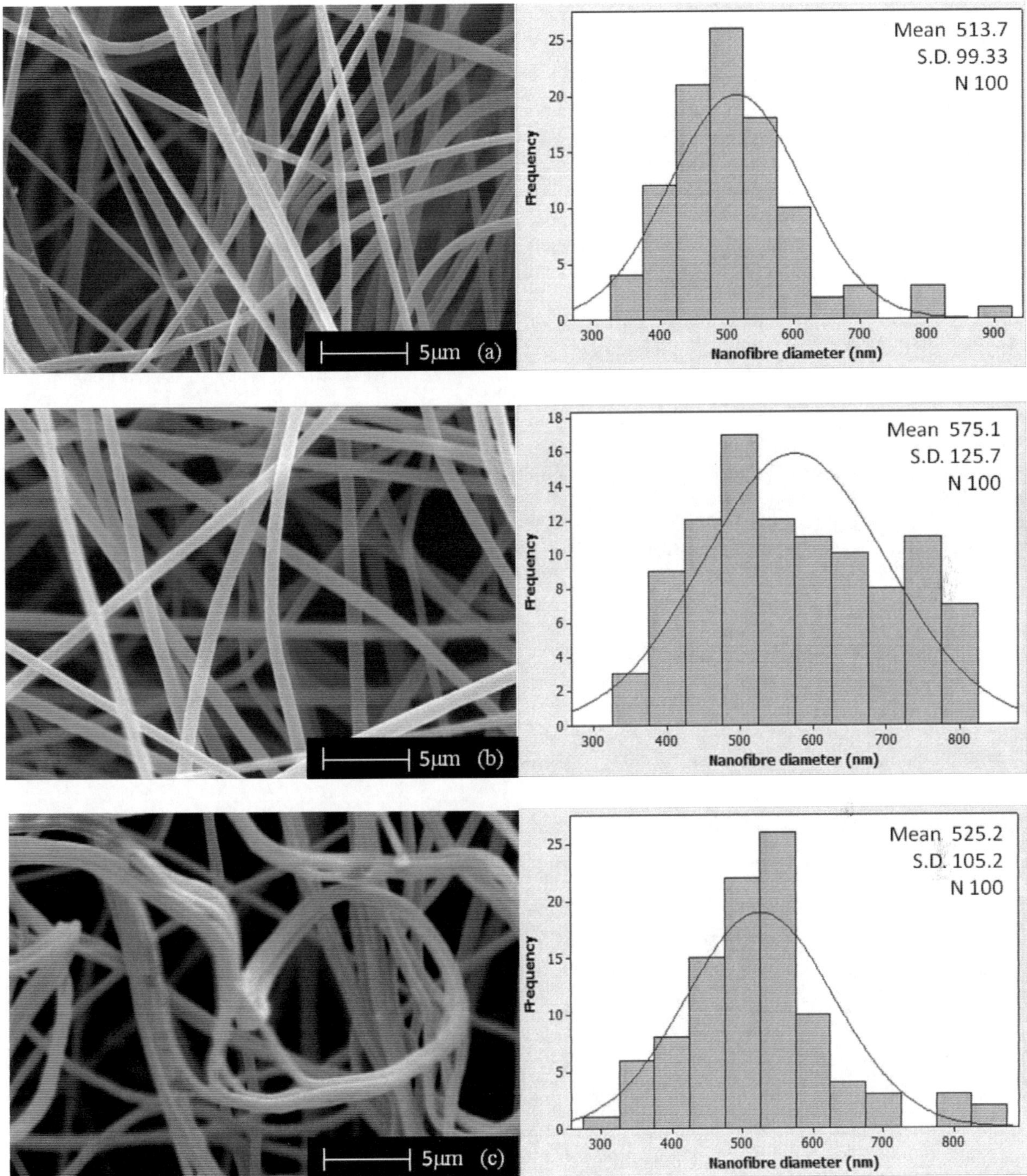

Fig.2 SEM images of 8% PAN nanofibres electrospun at 8 kV applied voltage with 14 cm collector distance from needle tip. The nanofibres deposited at collector speed of 6 rpm for (a) 20 minutes, (b) 40 minutes, and (c) 60 minutes, respectively. The corresponding fibre distribution is placed adjoining to the SEM image. SEM images captured at acceleration voltage of 10 kV with 10000 magnifications.

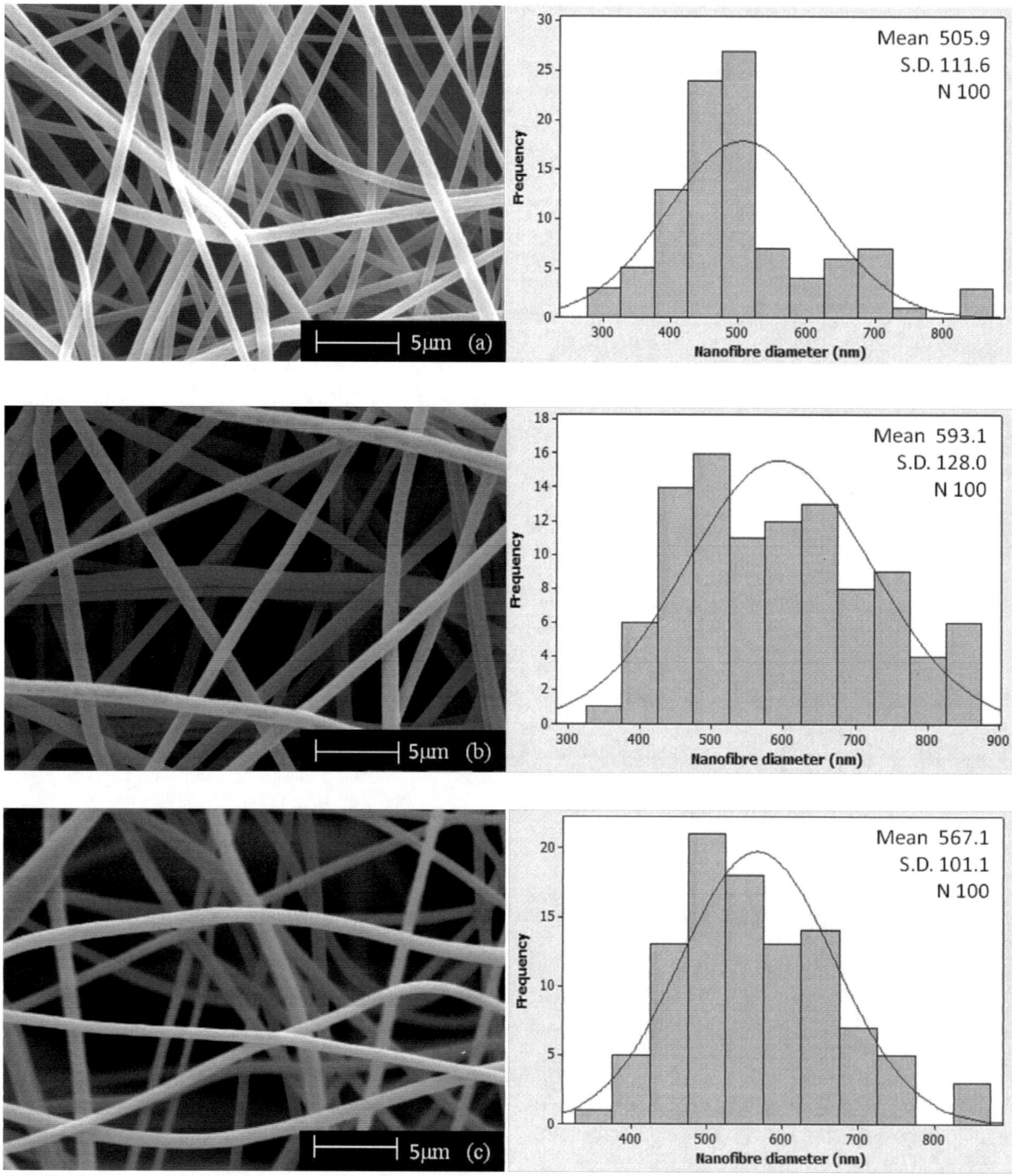

Fig.3 SEM images of 8% PAN nanofibres electrospun at 8 kV applied voltage with 14 cm collector distance from needle tip. The nanofibres deposited at collector speed of 9 rpm for (a) 20 minutes, (b) 40 minutes, and (c) 60 minutes, respectively. The corresponding fibre distribution is placed adjoining to the SEM image. SEM images captured at acceleration voltage of 10 kV with 10000 magnifications.

Table 1. Air permeability of nanomembrane samples

Collector speed (rpm)	Electrospinning time (minutes)	Air permeability (mm/s) at 100 Pa	Thickness (μm)	Porosity (%)
3	20	554.40 ± 53.29	41.928 ± 1.964	98.39
	40	346.92 ± 102.41	56.884 ± 5.468	98.34
	60	243.19 ± 42.46	69.148 ± 4.174	97.74
6	20	549.30 ± 61.08	45.503 ± 4.721	98.39
	40	306.11 ± 69.42	63.225 ± 7.082	97.06
	60	278.90 ± 53.89	71.263 ± 1.863	96.29
9	20	520.39 ± 66.20	42.328 ± 3.843	97.41
	40	314.61 ± 37.81	64.921 ± 5.216	97.66
	60	236.38 ± 24.64	74.453 ± 4.143	96.01

3.2.2 Effect of deposition time

The relation between air permeability and deposition time is not linear and equivalent to inverse of each other [10]. The deposition time significantly affects air permeability ($p < 0.01$) (Table 2). Increasing the deposition time from 20 to 60 minutes reduced the air permeability drastically, irrespective of the collector rpm. It may be due to the relatively higher thickness of nanomembrane with long deposition time (Table 1). The thickness of nanomembrane was increased by 27–32 μm when the deposition time varied from 20 to 60 minutes. Serbezeanu et al. [21] reported decrease in air permeability with increasing deposition time due to an increase in web thickness. It is worth to note that the nanofibres deposited on the fabric of same dimension for longer time increased the thickness of nanomembrane marginally. However, the marginal increase in thickness was sufficient to restrict the air flow to a great extent. The thickness of the upper nanofibre layer was increased with electrospinning time [22]. This resulted in reduced pore size and hence reduced air permeability. One of the reasons for the decrease in air permeability with the electrospinning time was the increase in the packing density of the filter media [23]. Thus, the air permeability of the nanomembrane can be governed by the deposition time. It is important to optimize the deposition time to achieve maximum air permeability without affecting filtration performance. It is reported that the presence of a thin layer of nanofibres on a microfibre substrate strongly increased the flow resistance of the filter [24]. The change in air permeability was attributed to different deposition time. The nanofibres deposited time was optimized to 15 minutes to achieve good air permeability and filtration efficiency for polysulphone nanomembrane with average fibre diameter of 300 nm [20].

Table 2. ANOVA results for air permeability of nanomembranes

Case	df	F value	p value
Use of nanomembrane	99	47.25	0.00
Change in collector speed (3-9 rpm)			
for 20 minutes	29	0.92	0.41
for 40 minutes	29	0.83	0.46
for 60 minutes	29	2.94	0.07
Change in deposition time (20-60 minutes)			
for 3 rpm	29	49.78	0.00
for 6 rpm	29	58.05	0.00
for 9 rpm	29	100.50	0.00

3.3.3 Correlation between air permeability, porosity and thickness

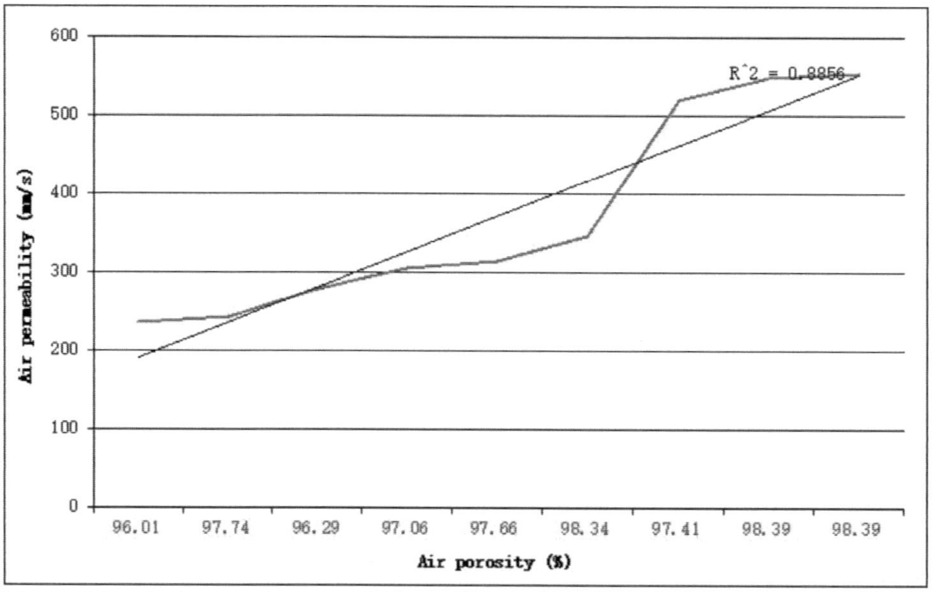

Fig.4 Correlation between air permeability and porosity of PAN electrospun nanomembrane

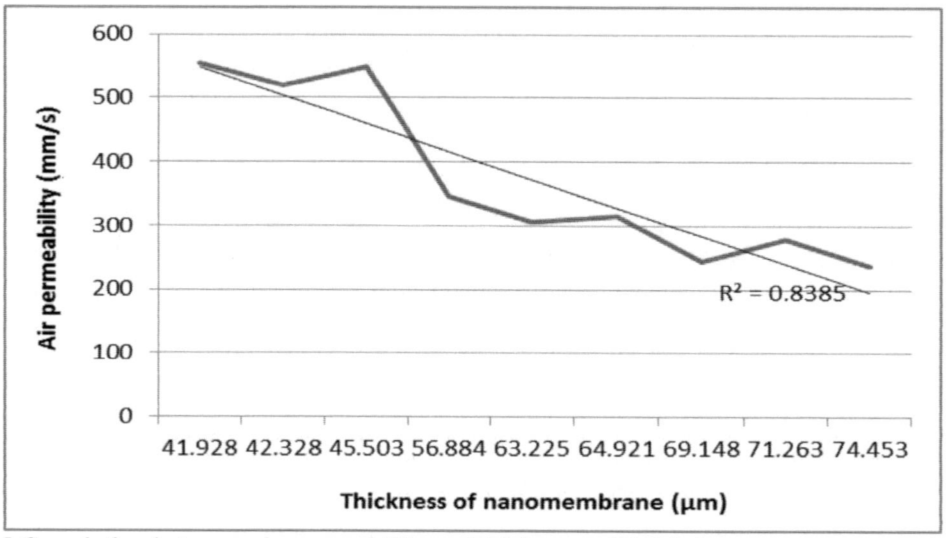

Fig.5 Correlation between air permeability and thickness of PAN electropsun nanomembrnae

The correlation curves for air permeability and porosity; and air permeability and thickness are graphically shown in Fig.4 and Fig.5, respectively. A strong positive correlation is found between air permeability and air porosity of PAN electrospun nanomembrane (Fig.4). The air permeability increased with an increase in porosity. The linear correlation coefficient of 0.8856 indicates a good agreement of the results obtained from air permeability and calculated porosity of nanomembranes. Furthermore, it was observed that air permeability is inversely proportional to the nanomembrane thickness (Fig.5). A negative correlation was noticed, which is in agreement with previously reported study [19]. Thus, the air permeability is strongly governed by the nanomembrane thickness, ultimately, by the deposition time rather than by the fibre diameter or any other variable in electrospinning process.

4. Conclusion

Air permeability of polyacrylonitrile nanomembrane deposited on wool/nylon textile fabric was presented in this research work. The collector speed and deposition time were the variables during the electrospinning. The nanofibre diameter determined was in the same range for different collector speed and deposition time. The change in collector speed at the constant deposition time does not alter the air permeability of nanomembrane. Electrospinning deposition time that derives nanomembrane thickness significantly affects the air permeability. A strong positive correlation was found between air permeability and porosity, whereas a negative correlation was noticed between air permeability and thickness of PAN electrospun nanomembrane. Air permeability of the nanomembrane can be governed by nanofibre layer thickness that depends on the deposition time. Thus, nanofibre deposition time is an important factor in controlling air permeability of the nanomembranes.

5. Acknowledgement

The first author wishes to acknowledge the scholarship support of RMIT University (Australia) towards PhD study. The authors are also thankful to Mr. Martin Gregory for his technical assistance in fabricating the collector assembly.

References

[1] Bhardwaj N,Kundu SC. Electrospinning: A fascinating fiber fabrication technique. Biotechnol Adv 2010; 28: 325-347.
[2] Barhate RS, Loong CK, Ramakrishna S. Preparation and characterization of nanofibrous filtering media. J Membrane Sci 2006; 283: 209-218.
[3] Chuang YH, Hong GB, Chang CT. Study on particulates and volatile organic compounds removal with tio2 nonwoven filter prepared by electrospinning. J Air Waste Manage 2014; 64: 738-742.
[4] Guibo Y, Yahong Z, Yin Y, Yumin Y, Qing Z. The electrospun polyamide 6 nanofiber membranes used as high efficiency filter materials: Filtration potential, thermal treatment, and their continuous production. J Appl Polym Sci 2013; 128: 1061-1069.
[5] Barhate RS,Ramakrishna S. Nanofibrous filtering media: Filtration problems and solutions from tiny materials. J Membrane Sci 2007; 296: 1-8.
[6] Huang ZM, Zhang YZ, Kotaki M, Ramakrishna S. A review on polymer nanofibers by electrospinning and their applications in nanocomposites. Compos Sci Technol 2003; 63: 2223-2253.
[7] Andrady AL, Science and technology of polymer nanofibers. 2008: John Wiley & Sons.
[8] Sill TJ,Von Recum HA. Electrospun materials for affinity-based engineering and drug delivery. J Phys: Conf Ser 2015; 646 IOP Publishing
[9] Schreuder-Gibson H, Gibson P, Senecal K, Sennett M, Walker J, Yeomans W, Ziegler D, Tsai PP. Protective textile materials based on electrospun nanofibers. J Adv Mater 2002; 34: 44-55.
[10] Gorji M, Jeddi A, Gharehaghaji A. Fabrication and characterization of polyurethane electrospun nanofiber membranes for protective clothing applications. J Appl Polym Sci 2012; 125: 4135-4141.
[11] Kirecci A, Özkoç Ü, İçoğlu HI. Determination of optimal production parameters for polyacrylonitrile nanofibers. J Appl Polym Sci 2012; 124: 4961-4968.
[12] Park HS, Park YO. Filtration properties of electrospun ultrafine fiber webs. Korean J Chem Eng 2005; 22: 165-172.
[13] Thavasi V, Singh G, Ramakrishna S. Electrospun nanofibers in energy and environmental applications. Energ Environ Sci 2008; 1: 205-221.
[14] Wang H, Zheng G, Wang X, Sun D. Study on the air filtration performance of nanofibrous membranes compared with conventional fibrous filters. in Nano/Micro Engineered and Molecular Systems (NEMS), 2010 5th IEEE International Conference 2010; London, UK: IEEE.
[15] Vitchuli N, Shi Q, Nowak J, McCord M, Bourham M, Zhang X. Electrospun ultrathin nylon fibers for protective applications. J of Appl Polym Sci 2010; 116: 2181-2187.

[16] Kok MDR,Gostick JT. Transport properties of electrospun fibrous membranes with controlled anisotropy. J Membrane Sci 2015; 473: 237-244.
[17] Abuzade RA, Zadhoush A, Gharehaghaji AA. Air permeability of electrospun polyacrylonitrile nanoweb. J of Appl Polym Sci 2012; 126: 232-243.
[18] Borhani S, Hosseini SA, Etemad SG, Militký J. Structural characteristics and selected properties of polyacrylonitrile nanofiber mats. J of Appl Polym Sci 2008; 108: 2994-3000.
[19] Hassan MA, Yeom BY, Wilkie A, Pourdeyhimi B, Khan SA. Fabrication of nanofiber meltblown membranes and their filtration properties. J Membrane Sci 2013; 427: 336-344.
[20] Li X, Gong Y. Design of polymeric nanofiber gauze mask to prevent inhaling pm2. 5 particles from haze pollution. J Chem 2015; 501: 460392.
[21] Serbezeanu D, Popa AM, Stelzig T, Sava I, Rossi RM, Fortunato G. Preparation and characterization of thermally stable polyimide membranes by electrospinning for protective clothing applications. Text Res J 2015; 0040517515576326.
[22] Kang YO, Im JN, Park WH. Morphological and permeable properties of antibacterial double-layered composite nonwovens consisting of microfibers and nanofibers. Compos Part B: Eng 2015; 75: 256-263.
[23] Karthick SA, Gobi N. Nano silver incorporated electrospun polyacrylonitrile nanofibers and spun bonded polypropylene composite for aerosol filtration. J Ind Text 2015; 1528083715622428.
[24] Dotti F, Varesano A, Montarsolo A, Aluigi A, Tonin C, Mazzuchetti G. Electrospun porous mats for high efficiency filtration. J Ind Text 2007; 37: 151-162.

Graphene based achieve moisture management on cotton fabric

Wen-Tao Chang, Xue-Ying Yu, Ting Wu, Yi Li, Xu-Qing Liu

The University of Manchester, Oxford Road, Manchester, UK, M13 9PL

*Corresponding author's email: xuqing.liu@manchester.ac.uk

Abstract

Moisture management is the most important property of textile area for smart wearable products, which influence the comfort, protection and functional performances for apparel, sports devices, and healthcare/ medical devices. Nowadays, the Graphene materials have a lot of advantages in textile area such as very thin film (only 1 atom thickness), light and have no effect on fabric hand. And the Graphene Oxide (GO) is hydrophilic while the Reduced Graphene Oxide (RGO) is hydrophobic. Thus, inside RGO coated (hydrophobic) layers and outside GO coated (hydrophilic) layers are combined to transfer moisture quickly from the body skin and evaporate it on the outer, which achieves the moisture management on cotton fabric and maintains the hand property of original fabric. In this project, using the GO coated on cotton fabrics and then using hydrazine hydrate for chemical reduction of GO to form RGO on the surface. And the results show that the RGO coated cotton fabric is a water repellent fabric while the GO coated cotton fabric is a hydrophilic fabric.

Keywords: moisture management; graphene oxide; reduced graphene oxide; hydrazine hydrate

1. Introduction

Moisture management is one of the key properties of textile materials for smart wearable products, which influence the comfort, protection and functional performances for apparel, sports devices, and healthcare/ medical devices. The inside hydrophobic layers and outside hydrophilic layers are combined to transfer moisture quickly from the body skin and evaporate it on the outer. This special functional structure of the material allows transference of moisture from the inside to the outside of the fabric to happen.

The 2D Graphene materials have a lot of advantages in textile area such as very thin film (only 1 atom thickness), light and have no effect on fabric hand. And the Graphene Oxide (GO) is hydrophilic while the Reduced Graphene Oxide (RGO) is hydrophobic.

In this project, by using the Reduced Graphene Oxide (RGO) and Graphene Oxide (GO) to form a coating around cotton fabric, we can get hydrophobic and hydrophilic properties of them. The aims of this project are shown below: 1) To achieve the moisture management of fabric by using Reduced Graphene Oxide (RGO) and Graphene Oxide (GO). 2) To improve the procedure of coating on fabric by RGO and GO. 3) To investigate moisture management properties of fabrics coating by RGO and GO.

2. Literature Review

2.1 Importance of the moisture management of textile

Generally, the moisture management could be understood to be the capacity of fabric materials to absorb liquid and gaseous moisture on the skin and transfer humidity from inter surface to the outer surface. In addition, the human being has a lot of ways to maintain the body temperature. For instance, when the human body feel very cold, the blood circulation in the body is going to be reduced by reason of minimizing the heat exchange with the nearby environment; when the body of human feel very hot, the blood circulation will be grew to release the surplus heat which results to sweat is transferred through the pores of the skin and then evaporated to readjust body temperature.

Obviously, clothing is thought to protect people from cold, heat, wind and other weather, which should achieve these functions without impeding the evaporation of moisture result from perspiration and not affecting with temperature regulation of human body. If moisture or water stay in textile or clothing and are not transferred to outside surface to evaporation, the temperature of skin cannot be cooling which means the body warms up and even more perspiration is formed. After a period of time, when the temperature of body cools down, the moisture retained in the fabric will evaporates and makes the body feel cold, even if the body does not need to be cooled any more [1].

Thus, moisture management properties of textiles play an important role in the thermo-physiological comfort of the body and skin and it is related to the heat behaviour and moisture diffusion behaviour of cloting. With the aim of maintaining the thermal balance of the human body, the cloth should permit moisture which is sensible and insensible perspiration to be transferred from human skin to the atmosphere, which can also decrease the degradation of thermal insulation of the textile which is resulted from moisture build-up [2].

Nowadays, some sportswear companies have a lot of strategies to improve the property of moisture management. Firstly, in order to transfer moisture to outside surface as fast as possible to evaporate, the capillary force should be as large as possible which will be increased when distances between adjacent fibres grow into thinner. That is to say the better the fibres are, the smaller the gaps are, and the greater the moisture transfer. Another method is to evaporate the moisture as fast as possible, which depends on outside surface area of textile. That means that more fibres there are at the surface, the larger the surface is and the faster the humidity evaporation [3].

2.2 Introduction to Graphene

In general, graphene is one-atom-thick single layer which are arranged in a hexagonal honeycomb lattice of pure carbon atoms and graphene is the building-block of graphite, which is a single, tightly packed layer of carbon atoms. In terms of chemical bonds, it is an allotrope of carbon which has a structure of the pi-hybridisized carbon plane with sp2 bonded atoms.

All over the world, graphene is the thinnest material known (only 1 atom thickness), the lightest material discovered (~0.77mg/m2), the strongest material known (between 100-300 times larger than steel), the best conductor of heat and electricity at room temperature. The mainly problem for graphene development in commercial consumptions was that the formation of high quality graphene was not a cheap and simple process just like chemical vapour disposition (CVD) which would use some toxic chemicals to produce a single layer graphene by introducing Pt, Ni or Ti Carbide to ethylene or benzene at high temperature. In addition, previously, it is too difficult to grow a large amount of graphene layers by using crystalline epitaxy method on some substrate except metallic substrate, which critically restricted its development in commercial uses because the separation of graphene layers from the metallic substrate without damaging graphene is a really difficult challenge [4] .

The possibilities of what scientists and research institutes can achieve with the known materials and knowledge have been dramatically improved. Moreover, everybody in the word can freely to imagine wonderful prospective situations just like invisibility cloaks, longer standby time smart phones with extra-long battery life and super-small, foldable computers [5].

2.3 The application of graphene in moisture management

Nowadays, more and more researchers and institutes focused on the smart actuators which can convert outside stimulations just like moisture to visualized movements by fast, reversible and controllable structure/morphology changes [6]. Because of smart driving method, these actuators have received a large amount of research interests and already discovered great potential in developing applications [7]. Furthermore, these smart actuators can achieve the moisture management.

As we mentioned before, the graphene and its related materials (GO and RGO) are really suitable for smart actuators, because they have fantastic physical and chemical properties such as the amazing mechanical strength, thermal conductivity, biocompatibility and large surface area. Thus, it is easy to find that the graphene-based smart actuators were successfully proposed by scientists and were currently being developed for the mechanism and applications.

During the last several years, a lot of scientists have proposed their opinions and methods to achieve these smart actuators. Firstly, the Qu et al. have successfully manufactured the fibre-type smart actuators using method of laser reduction of GO fibres which both have GO and RGO in this actuator [8]. Recently, Han D D et al. have proposed a Self-Controlled photoreduction method to form GO and RGO two-layer actuators which is the moisture-responsive graphene paper using focused sunlight irradiation [9]. However, by using the scanning processing method, the GO and RGO two-layer are not uniform which will result to unstable response. And in general, the more complex response is almost impossible to achieve because the response of those actuators is always limited to the bend and straightening.

Recently, the researches on graphene actuators still focus on the innovative technologies to achieve simple, more efficient and large scale production of graphene which can complete complex movements.

Han D D et al. have found a simple preparation of moisture-driven graphene actuators using unilateral UV irradiation of graphene oxide papers and the surface performances of the graphene oxide paper were significantly changed after UV irradiation for several minutes because photoreduction effects removes majority of oxygen-containing groups on graphene oxide sheets. However, the thick graphene oxide paper would not be completely reduced as photoreduction gradient would be controlled by light transmission and thermal relaxation. As a result, a self-controlled photoreduction process should be simply reached because the bending processes of anisotropic graphene oxide and reduced graphene oxide two-layer would be driven by humidity. For instance, smart moisture-driven graphene actuators that imitated the cilia of respiratory tract have been significantly developed for controllable objects transport [10].

The photoreduction of graphene oxide is a much more efficient reduction strategy which was commonly used for the production of graphene because it allows pretty control over content and distributions of oxygen-containing groups on graphene oxide sheets. Previously, the researchers have focused on the radiation-induced graphene oxide reduction by using different kinds of radiation sources such as sun, UV, and laser, which were used for the elimination of oxygen-containing groups on graphene oxide sheets and inducing chemical conversion on graphene-related materials [11]. By measuring the temperature change, Kumar et al. have detected the photo-thermal effects during the radiation-induced chemical transformation [12]. By comparing the wavelength of incident light, it is easy to briefly classify the mechanism for photoreduction of graphene oxide into two types: photochemical reduction and photo-thermal treatment [13]. In addition, the Smirnov et al. have reported the threshold theory which stated that photon energies bigger than 3.2 eV could trigger graphene oxide reduction during the photochemical process [14].

3. Experimental Procedure

1) Polymer fabrics surface modification

The Graphene Oxide will be introduced to modify fibres and the reduction of GO to RGO process will be optimized.

2) Materials Characterization

Using the Scanning Electron Microscope (SEM) and Optical Microscope to characterize the morphology of fabrics; Using the X-Ray Diffraction (XRD), Fourier Transform Infrared Spectroscopy (FTIR) and Raman Spectrum Analysis to characterize the composition of fabrics.

3) Performance Investigation

Using the Tension Mechanical Machine to investigate the tensile strength and other properties of coated fabrics; Using the Moisture Management Tester (MMT) to investigate moisture management properties of coated fabrics.

4. Results

4.1 Scanning Electron Microscope Test (SEM)

Using the scanning electron microscope to characterize the surface morphology of the cotton fabrics which are coated by graphene oxide (GO).

And according to the SEM figure 1 below, it is easy to find that there are lots of graphene sheets coating randomly around the cotton yarns which formed the fabric. And there are different layers of graphene sheets coated around the cotton fabrics, but the thinner the sheets are the better the property is.

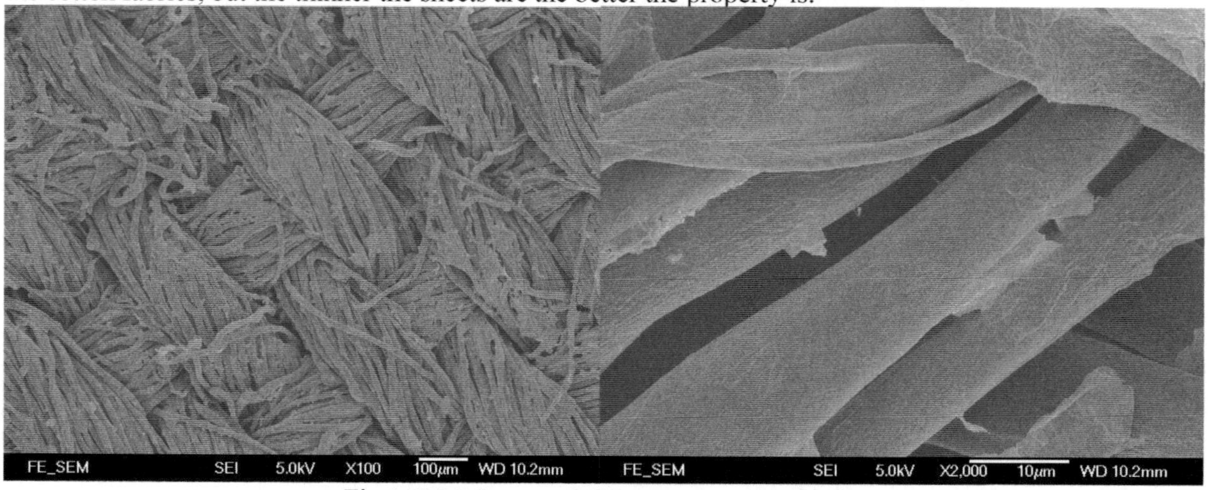

Fig.1 SEM figures of GO coated cotton fabrics

4.2 Fourier Transform Infrared Spectroscopy Test (FTIR)

Using the Fourier Transform Infrared Spectroscopy to characterize the functional group in the cotton fabrics which are coated by graphene oxide (GO) and reduced graphene oxide (RGO).

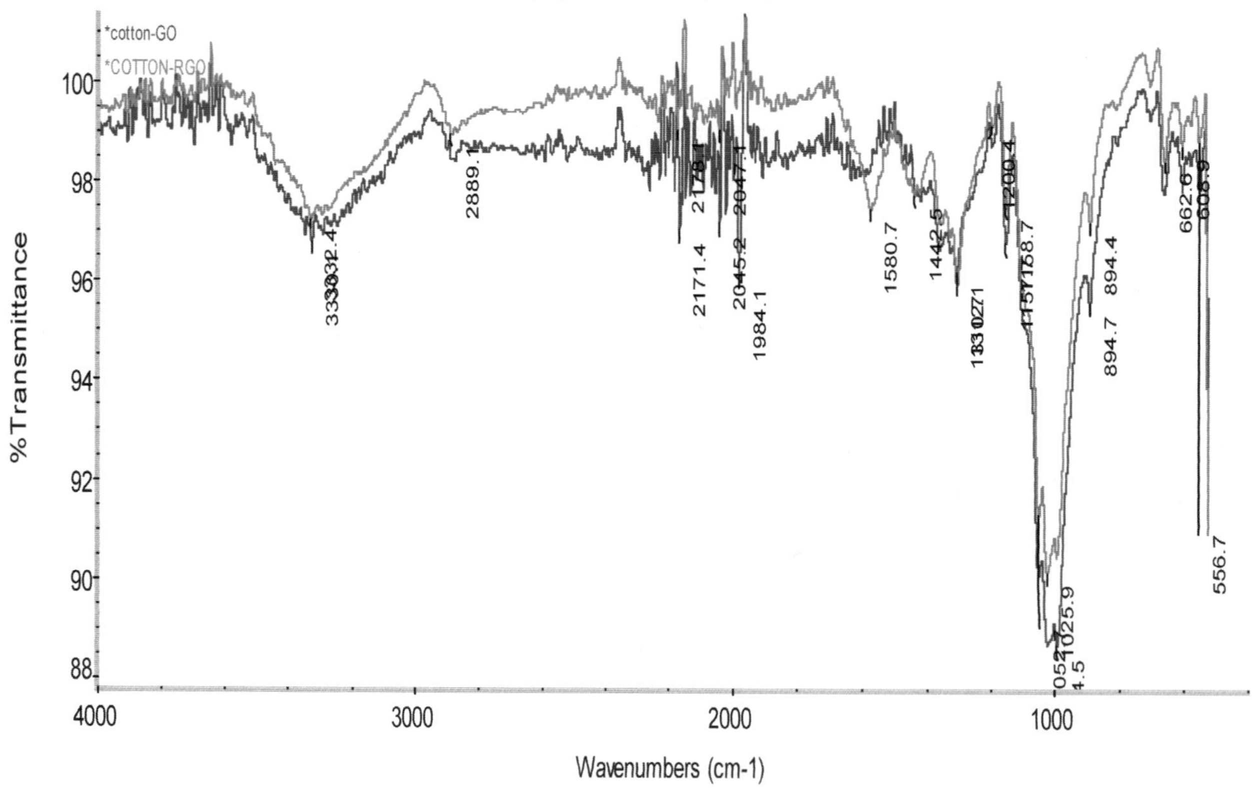

Fig.2 FTIR figure of GO and RGO coated cotton fabrics

And according to the FTIR figure 2 below, the peak around 1000 cm^{-1} is the hydroxyl (-OH) and it is easy to find that the intensity of peaks of hydroxyl for GO coated cotton is larger than that for RGO coated cotton, which is means that the chemical reduced process of GO is successful and most of GO is reduced to graphene sheets.

4.3 Moisture Management Tester (MMT)

Using the Moisture Management tester to investigate moisture management properties of GO and RGO coated cotton fabrics. And according to the MMT figures below, it is easy to find that GO coated cotton fabric is a slow absorbing and slow drying fabric while the RGO coated cotton fabric is a water repellent fabric. Thus, if the GO and RGO coated the two face of one fabric, this fabric will get the moisture management property.

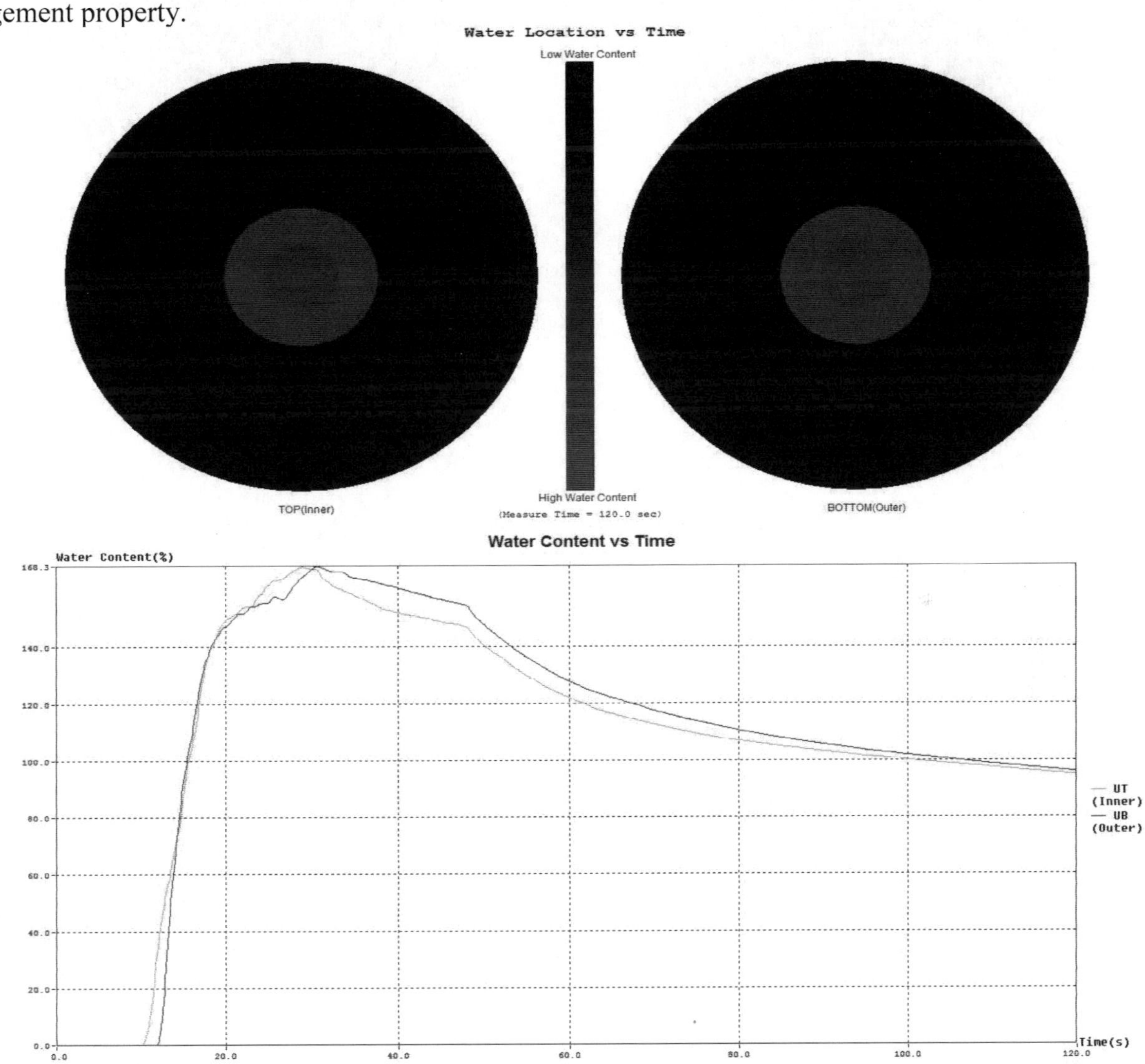

	Top Surface	Bottom Surface
Wetting time(sec)	10.109	11.7
Absorption rate(%/sec)	14.5015	15.5736
Max wetted radius(mm)	10.0	10.0
Spreading Speed(mm/sec)	0.8309	0.7524
One way transport capability	2.3832	
Description	MMT-	

Fig.3 MMT figures of GO coated cotton fabrics

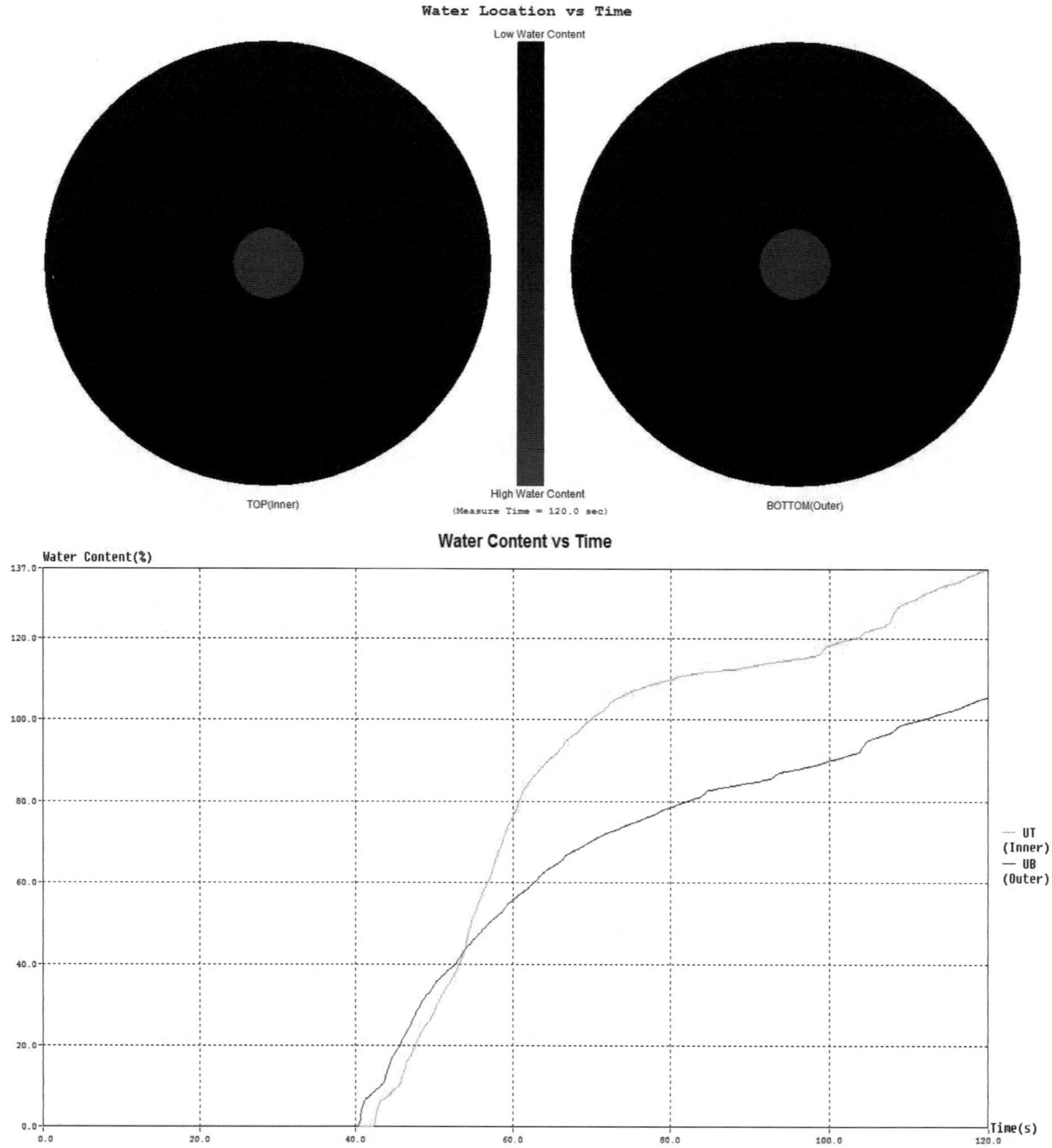

	Top Surface	Bottom Surface
Wetting time(sec)	42.214	40.249
Absorption rate(%/sec)	3.9852	4.1873
Max wetted radius(mm)	5.0	5.0
Spreading Speed(mm/sec)	0.1177	0.1234
One way transport capability	-14.4817	
Description	MMT-	

Fig.4 MMT figures of RGO coated cotton fabrics

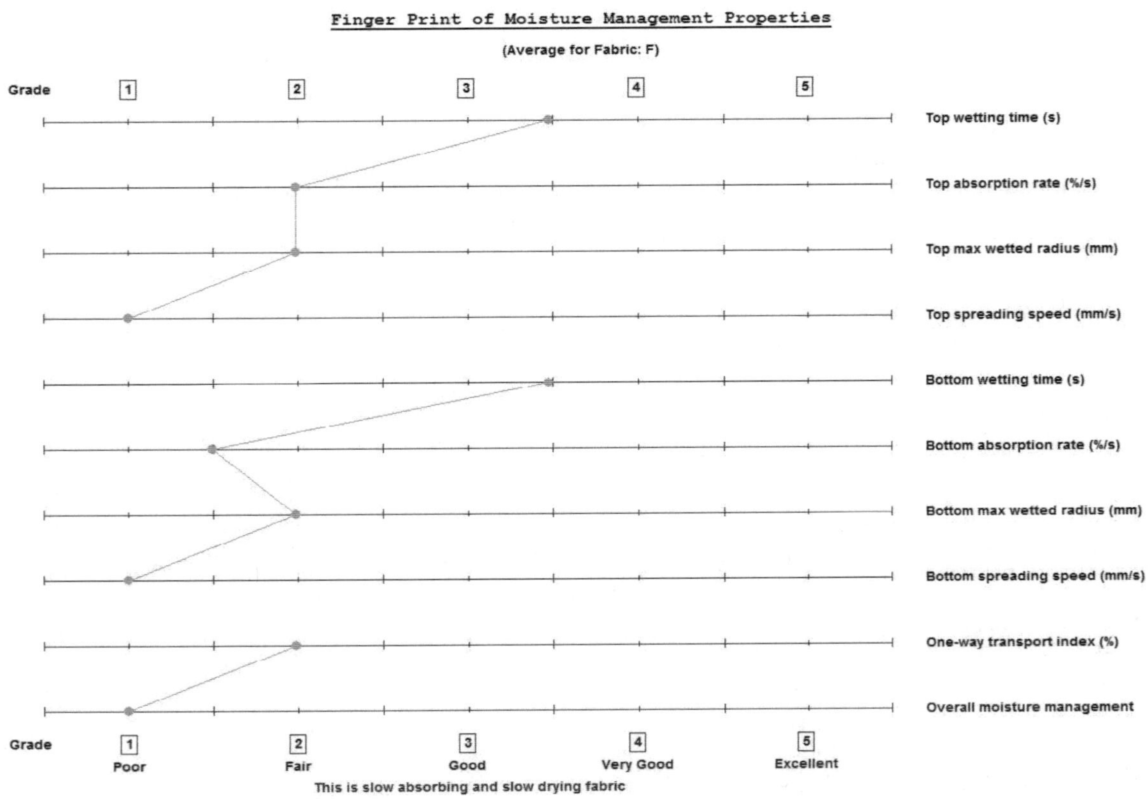

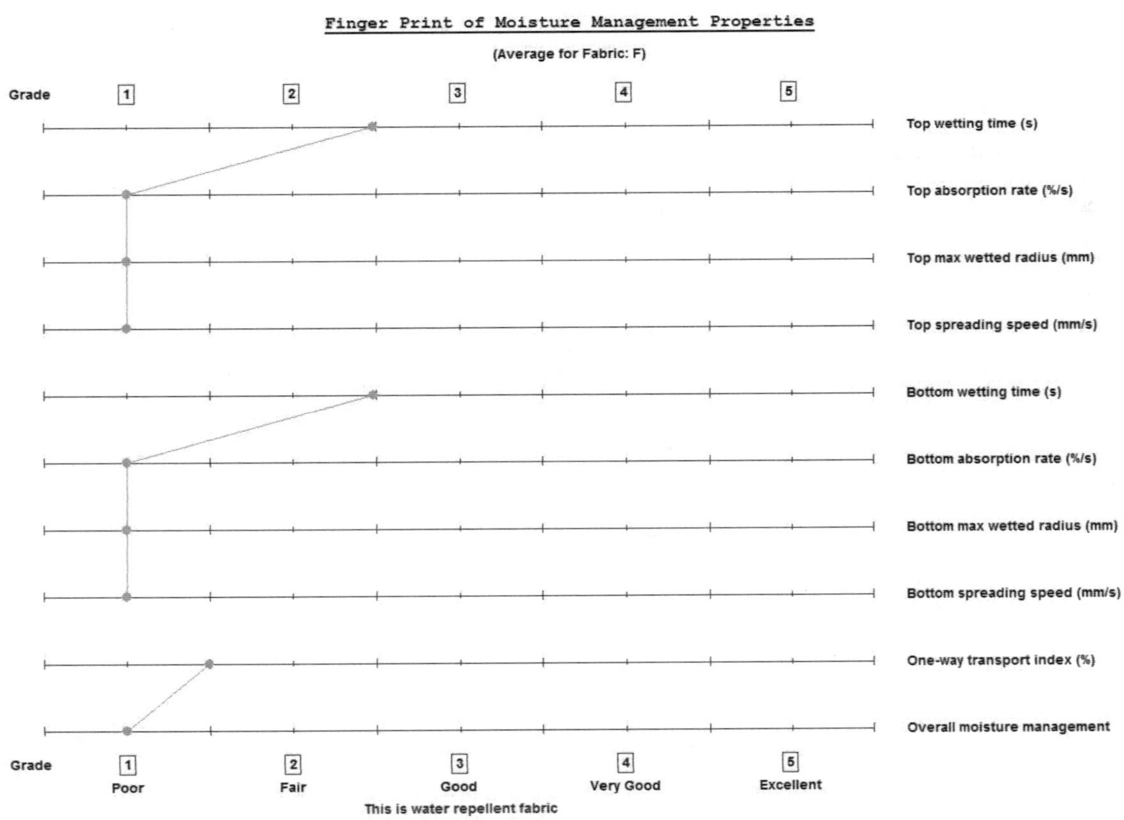

Fig.5 MMT figures of comparison of GO and RGO property

5. Conclusion

Nowadays, the Graphene materials have a lot of advantages in textile area such as very thin film (only 1 atom thickness), light and have no effect on fabric hand. And the Graphene Oxide (GO) is hydrophilic while the Reduced Graphene Oxide (RGO) is hydrophobic.

In this project, using the GO coated on cotton fabrics and then using hydrazine hydrate for chemical reduction of GO to form RGO on the surface. And the results show that the RGO coated cotton fabric is a water repellent fabric while the GO coated cotton fabric is a hydrophilic fabric.

Thus, inside RGO coated (hydrophobic) layers and outside GO coated (hydrophilic) layers are combined to transfer moisture quickly from the body skin and evaporate it on the outer, which achieves the moisture management on cotton fabric and maintains the hand property of original fabric.

6. Acknowledgment

Also, we would like to thank the support of the EU Horizon 2020 and University of Manchester through projects with project codes 644268 - ETEXWELD - H2020-MSCA-RISE-2014, AA14512 (UMRI project "Graphene-Smart Textiles E-Healthcare Network") and AA01906.

References

[1] Zhang P, Watanabe Y, Kim SH, Tokura H, Gong RH. Thermoregulatory responses to different moisture-transfer rates of clothing materials during exercise, J. Text. Inst: 2001; 92 (1): 372-378.
[2] Das B, Das A, Kothari VK, Fanguiero R, Arau´ jo M. moisture transmission through textiles part I: processes involved in moisture transmission and the factors at play. AUTEX Res J 2007; 7: 100-110.
[3] http://www.drpetry.de/index.php?id=48&L=1.
[4] Graphene: synthesis and applications[M]. CRC Press: 2011.
[5] http://www.graphene-info.com.
[6] Ji M, Jiang N, Chang J, et al. Near-Infrared Light-Driven, Highly Efficient Bilayer Actuators Based on Polydopamine-Modified Reduced Graphene Oxide[J]. Advanced Functional Materials: 2014; 24(34): 5412-5419.
[7] Cheng H, Hu Y, Zhao F, et al. Moisture-Activated Torsional Graphene-Fibre Motor[J]. Advanced Materials: 2014; 26(18): 2909-2913.
[8] Cheng H, Liu J, Zhao Y, et al. Graphene Fibres with Predetermined Deformation as Moisture-Triggered Actuators and Robots[J]. Angewandte Chemie International Edition: 2013; 52(40): 10482-10486.
[9] Han DD, Zhang YL, Jiang HB, et al. Moisture-Responsive Graphene Paper Prepared by Self-Controlled Photoreduction [J]. Advanced Materials: 2015; 27(2): 332-338.
[10] Han DD, Zhang YL, Liu Y, et al. Bioinspired Graphene Actuators Prepared by Unilateral UV Irradiation of Graphene Oxide Papers[J]. Advanced Functional Materials: 2015; 25(28): 4548-4557.
[11] Kumar P, Subrahmanyam KS, Rao CNR. Graphene produced by radiation-induced reduction of graphene oxide[J]. International Journal of Nanoscience: 2011; 10(04n05): 559-566.
[12] Kumar P, Das B, Chitara B, et al. Novel Radiation-Induced Properties of Graphene and Related Materials [J]. Macromolecular Chemistry and Physics: 2012, 213(10-11): 1146-1163.
[13] Zhang YL, Guo L, Xia H, et al. Photoreduction of graphene oxides: methods, properties, and applications [J]. Advanced Optical Materials: 2014; 2(1): 10-28.
[14] Smirnov VA, Arbuzov AA, Shul'ga YM, et al. Photoreduction of graphite oxide [J]. High Energy Chemistry, 2011; 45(1): 57-61.

Comfort Characteristics of Selected Textile Fabrics Proposed for Use as X-Ray Protective Apron Casing

Huda Ahmed Maghrabi[1,3], Rana Mahbub[4], Arun Vijayan[1], Pradip Deb[2], Lijing Wang[1*]

[1]*School of Fashion & Textiles, RMIT University, Brunswick, Victoria 3056, Australia*
[2]*School of Medical Sciences, RMIT University, Bundoora, Victoria 3083, Australia*
[3]*Department of Textiles and Clothing, Umm Al-Qura University, Mecca 24382, Saudi Arabia*
[4]*Clothing and Textile Department, King Abdulaziz University, Jeddah, Saudi Arabia*

[*]Corresponding author's email: lijing.wang@rmit.edu.au

Abstract

Lead aprons are typical protective clothing to protect radiographers from harm of radiation. As such, a good radiation shield must have a high Lead Equivalence to minimize the radiation dose of exposure. While most radiation shields in the market fulfil this requirement by using matrices of lead and other substances, most of them do not necessarily emphasize on the comfort properties. In the same context, the time of the examination takes longer than expected, which causes some discomfort to the person who wears this type of clothing. The discomfort can be caused by the heavy weight of these aprons or the smooth surface of the coated casing material. The poor fit and design of these aprons due to the stiffener of the fabric used is another issue that makes the final product uncomfortable to wear. In general, the apparent comfort characteristics of any textile fabric material are moisture management, weave structure, yarn types, fabric thickness and weight. The objective of this study is to use standard testing methods to characterize selected fabrics in terms of their X-ray shielding ability, physical, mechanical, and morphologic properties.

Keywords: Lead apron; X-ray shielding; Moisture management; Fabric comfort; Protective garment

1. Introduction

Ionizing radiation has two main uses in human medicine: to kill cancerous cells and to diagnose disease or injury [1]. X-ray is the most common form of radiation used in medical diagnosis. The radiation is passed from a source, through a specific part of the patient's body to generate images (radiography). During such procedures, the resulting ions affect normal biological processes and ecological balance [2]. To prevent unnecessary exposure of the radiologist and the patient, various protection measures have been standardized and are continually improved [1].

Radiation shielding is categorized into biological and thermal shielding. Biological shielding is designed to reduce radiation to a safe and acceptable level for living organisms. The danger of exposure is further classified into two: external and internal [3]. External radiation emanates from an outside source and can only be reduced by providing a safe distance, a protective shield, and a limit time of the exposure. Internal radiation is a medical and hygiene problem and does not involve any external shielding.

Lead aprons are the most common type of personal shielding in radiology departments and medical imaging industry in general. Lead is an ideal material for radiation protection, but it compromises on comfort for the wearer [4]. The issue of comfort becomes even more important for the lead apron to wear for a long duration. According to Van et al [5], the most common comfort issue when selecting a lead apron is the weight of the apparel. Lead is a heavy metal. Long hour use of heavy aprons and apparel especially in a catheterization laboratory or surgeries can lead to discomfort, constant back pain, and the risk of permanent back problems. Modern technology has resulted in ultra-lite and light-weight aprons but most of them are still heavy, about 3.8 kg [6].

Radiation shields are meant to be worn during radiological diagnostic/ interventional procedures and/ or oncological treatments. The shields are worn by clinicians, patients and physicians to selectively shield, isolate and protect particular parts of their body from radiation. The shields are reusable, meaning same shields may be used many times by different care providers and patients. For this reason, earlier shields were not finished to facilitate easy cleaning after use. These unsanitary shields may pose health risks to a person when they are used to cover parts of the body especially the gonadal region. The adoption of the hygienic use of the shields is one of the methods of tackling the problem [7,8]. Using an apron casing is another measure for the ease of cleaning.

Clothing insulation has a significant impact on thermal comfort, because it influences the heat loss and consequently the thermal balance. More layers of insulating clothing can avoid heat loss and help either keep an individual warm or lead to overheating. Medical protective clothing made of different materials should not induce any thermal discomfort. Such an uncomfortable sensation experienced by surgeons can decrease their psychomotor skills and at the same time, adversely influence the way an operation procedure is carried out [9]. In this context, if a radiologist attempts to do examinations for more than 60 min, thermal comfort becomes an issue and could negatively affect their performance even though they usually do this procedure in a conditioned room. The heavy weight of the aprons and different layers of clothes underneath as well the type of activities would have an impact on their thermal comfort sensation thus, may cause discomfort.

This research investigates the comfort performance of a specific range of textiles fabrics that could be used as alternative casing aprons. In ascertaining the comfort elements in the current line of aprons, the purpose of this research is also to provide a basis for proposing alternative enhancements to the fabric finishing, technology, or material used. These proposed alternatives should fulfil a balance between radiation protection and thermal comfort.

2. Experiment

All the fabric specimens were conditioned according to Australian Standard (AS 2001.1-1995, conditioning procedures) under 20±2 °C and 65±3 % relative humidity for at least 24 hours before testing.

2.1 Materials

The specifications of all materials are shown in Table 1. Two lead equivalent samples were used as the benchmark for comparison. A regular lead equivalency sheet made from rubber, polyvinyl chloride and lead powder, weight 2.4 kg/m^2 was used as inner layers. Similarly, a Lite Lead sheet made also from rubber and polyvinyl chloride and lead powder, weight 2.1 kg/m^2 was used. Both samples were obtained from Medical Concepts Australia Pty Ltd.

Both nylon and polyester fabrics are commonly used as casing materials for commercial lead aprons because of their strength, resilience, lightweight, and durable properties. The main fibre used as casing layer in this paper was nylon, polyester and Kevlar. Plain weave and weft knitting fabrics of 100% nylon, nylon/ wool, 100% polyester, cotton blend, 100% Kevlar, and Kevlar/wool were used for comparison.

2.2 X-ray measurement

Fabric and lead samples were tested for X-ray attenuation by using a medical X-ray machine (SHIMADZU X-ray system). The distance from the X-ray beam source to the specimen (SID) was 80 cm. Samples were exposed to X-rays at a tube voltage of 80 kVp, and at tube current and time of 12 milliampere-second (mAs). The shielding ability of each sample was evaluated by comparing their transmission doses with the measured transmission doses for reference lead samples. A 'Rad-Check Plus' dosimeter was used to measure the X-ray transmission. The fabric sample size was 15×15 cm. Five different positions, four different corners and the centre of each specimen, were exposed independently and the mean value for each sample was calculated.

2.3 Scanning electron microscopy (SEM)

Philips XL30 Field Emission Scanning Electron Microscope was used for the analysis of surface morphology. Samples were prepared by coating them with gold using a sputter coater in an argon-purged chamber for 60 sec. Then the specimens were observed in the XL30 SEM chamber at a high voltage of 30.0 kV and 10 mm WD. The magnification was set above 1500×.

Table 1. Material properties

Material code	Material composition	Construction	Density				Thickness [mm]	Mass per unit area [g/m²]
			No. of Ends/cm	No. of Wales/cm	No. of Picks/cm	No. of Courses/ cm		
RL	Regular Lead	Sheet (inner layer)	-	-	-	-	0.86	2600
LL	Lite Lead	Sheet (inner layer)	-	-	-	-	0.58	2200
N	100% Nylon	Plain woven (outer layer)	14.5	-	12.2	-	0.25	162
NW	Nylon/wool	Weft knitting single jersey (outer layer)	-	8.6	-	6.2	1.25	462
P	100% Polyester	Plain woven (outer layer)	7.4	-	9.8	-	0.60	154
PC	50%Cotton 24%Modal 24%Bamboo 2%Polyster	Weft knitting double jersey (outer layer)	-	10.6	-	8.2	0.81	228
K	100% Kevlar	Plain woven (outer layer)	8.6	-	9	-	0.33	213
KW	Kevlar/wool	Weft knitting single jersey (outer layer)	-	7	-	6.2	1.15	520

2.4 Material thickness

The material thickness was measured according to the Australian Standard, AS 2001. 2. 15-1989, determination of the thickness of textile fabrics. The arithmetic mean of five measurements was taken to find out the thickness.

2.5 Mass per unit area

The mass per unit area was measured according to the Australian Standard, AS 1587-1973, determination of mass per unit area and mass per unit length of fabrics. The sample dimensions were 100 mm × 100 mm. Fabric weight in g/m^2 was calculated from the weight of area measured as Eq.1.

$$g/m^2 = \frac{W}{A} \tag{1}$$

Where
W is the mass of specimen, g
A is the fabric area, m^2

2.6 Moisture Management

The Moisture Management Tester (MMT) measures the liquid management properties of fabrics according to AATCC TM 195: 2009, liquid moisture management properties of textile fabrics. Five fabric specimens measuring 80 mm × 80 mm were measured and the liquid moisture transport behaviour in different directions of the sample was assessed. The following grading was used for the evaluation:

● Wetting Time (sec) in top and bottom: 1) ≥120S No wetting; 2) 20-119S Slow; 3) 5-19S Medium; 4) 3-5S Fast; 5) <3S Very fast.

● Absorption Rate in (%/s) top and bottom: 1) 0-10%/s Very slow; 2) 10-30%/s Slow; 3) 30-50%/s Medium; 4) 50-100%/s Fast; 5) >100%/s Very fast.

● Max Wetted Radius (mm) in top and bottom: 1) 0-7mm No wetting; 2) 7-12mm Small; 3) 12-17mm Medium; 4) 17-20mm Fast; 5) >22mm Very fast.

● Spreading speed (mm/s) in top and bottom: 1) 0-1 mm/s Very slow; 2) 1-2 mm/s Slow; 3) 2-3 mm/s Medium; 4) 3-4 mm/s Fast; 5) >4 mm/s Very fast.

● One-way Transport Capacity (OWTC): 1) <-50 Very poor; 2) -50-100 poor; 3) 100-200 Good; 4) 200-400 Very good; 5) >400 Excellent.

● Overall Moisture Management Capacity: 1) 0-0.2 Very poor; 2) 0.2-0.4 Poor; 3) 0.4-0.6 Good; 4) 0.6-0.8 Very good; 5) >0.8 Excellent.

3. Results and discussion

3.1 X-ray measurements

Initial X-ray exposure was measured without any fabric and was taken into consideration as a control. Fabrics and lead sheets were measured for X-ray shielding at a medium level of radiation at 80 kVp. The results are shown in Fig.1. The value of the transmittance corresponds with the amounts of X-rays that penetrated through the specimen. It indicates the shielding effectiveness of the sample tested. A higher transmittance value corresponds to lower X-ray absorption, meaning that the sample is less effective for radiation protection.

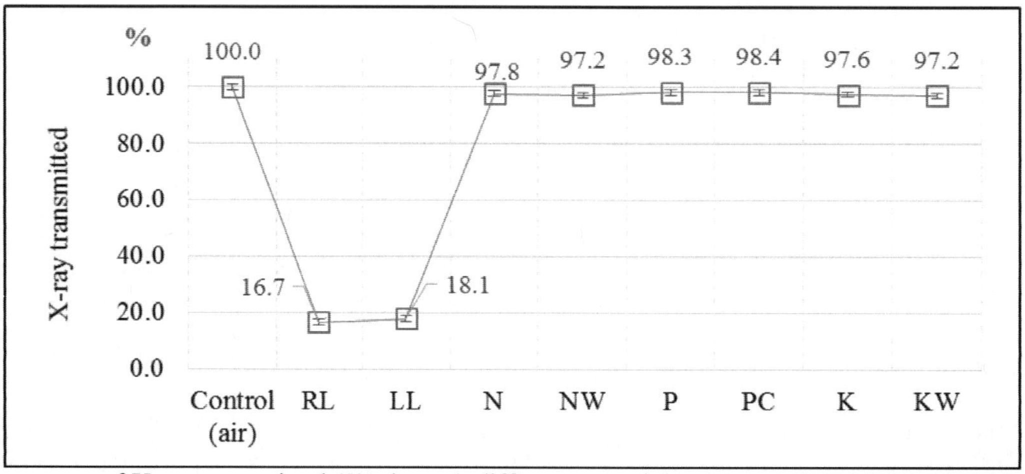

Fig.1 The amount of X-ray transmitted (%) through different materials (the error bars indicate standard deviation)

It can be seen from Fig.1 that the transmittance values for RL and LL were 16% and 18%, respectively, indicating that they can absorb the majority of the X-ray radiation. However, the control and the fabric samples transmitted almost all the X-rays and cannot be considered as having radiation shielding effect. Since the casing nylon or polyester fabrics have no radiation shielding effects though their structures are very tight as observed from the SEM images, the primary purpose of using these casing fabrics is to hold X-ray shielding sheets and facilitate cleaning. Hence, the casing materials should be light for apron weight reduction.

Table 1 shows the differences of proposed fabrics in thickness and weight. The RL and LL have a heavier weight of 2600 and 2200 g/m2 respectively because they were made from lead powder, which was clearly visible from their SEM images. On the other hand, the KW and NW weigh 520 and 462 g/m^2 respectively, indicating they are heavy as casing material. Samples PC, K and P are medium to lightweight fabrics. While very lightweight fabrics are not considered physically effective radiation shields, they would ideally offer the highest possible protection if the fabric was coated with a suitable radiation absorbent.

Fabric thickness affects the comfort properties of garment made from the fabric. It also determines the effectiveness of X-ray protection. Thickness tends to be related to fabric density. Table 1 shows samples N, P and K fell in the medium to low thickness range. They have relatively desirable (medium thickness) comfort characteristics while all the other samples fell in the upper thickness range. In context, the knitting samples NW and KW were the thickest (1.25 and 1.15 mm) among the samples even thicker than the lead sheets. They are likely to have the least desirable comfort characteristics because of their thickness. Hence, these fabrics may not be considered as the casing material.

3.2 Fabric moisture management

An individual wearing to RL and LL would, for example, experience a clammy or damp sensation because of the non-existent moisture management. Such lead sheets would be uncomfortable to wear for long even in a well-ventilated room. Most facilities that utilize radiation tend to prioritize on safety over ventilation. A fabric with low or non-moisture management properties would result in uncomfortable to the wearer even for a relatively short period.

Table 2 shows the moisture management results for all moisture management property indexes. Overall Moisture Management Capacity (OMMC) is an index that is used to determine the overall ability of the fabric to handle liquid moisture transport. This includes three performance aspects: one-way liquid transportability, moisture drying speed (bottom side maximum spreading speed), and the moisture absorption rate. The larger the OMMC value, the higher the moisture management ability of the sample fabric.

Table 2. Summary of fabric moisture management properties.

Sample code	Top surface				Bottom surface				Accumulative one-way transport [%]	OMMC
	Wetting Time [sec]	Absorption Rate [%/sec]	Wetted Radius [mm]	spreading Speed [mm/sec]	Wetting Time [sec]	Absorption Rate [%/sec]	Wetted Radius [mm]	spreading Speed [mm/sec]		
RL	3.95	32.6	5	1.2	120	0	0	0	-588.1	0
LL	9.4	40.6	5	0.5	120	0	0	0	-476.4	0
N	4.8	61.8	5	1	120	0	0	0	-913.5	0
NW	8.7	31.0	14	1.8	4.5	31.0	18	3.1	145.8	0.4
P	5.2	55.2	5	1	120	0	0	1	-850.4	0
PC	4.1	41.0	20	2.6	4.4	43.4	15	2.4	-5.9	0.3
K	4.7	51.7	5	1	120	0	0	0	-902.9	0
KW	6.0	90	10	1.6	7.0	118	21	3.7	548	0.7

The zero OMMC for the samples (RL, LL, N, P, and K) indicates that these samples have very poor overall moisture management ability. The radiation shielding using naturally dense materials such as pure lead sheets or heavy fabrics compromised thermal comfort. The NW and PC fabrics show poor OMMC. However, the KW fabric has a relatively high OMMC (0.7, very good) amongst all the samples. Regardless, it is essential to emphasize on radiation shielding, comfort characteristics on the fabric come second. The RL and LL samples are waterproof due to their very slow water absorption and spreading rate, as well as zero penetration and poor one-way liquid moisture transport.

Fig.2 shows the fabric max wetted radiuses with moisture absorption and spreading rate. Woven fabric samples N, P and K all have poor liquid moisture management properties with very slow water spreading rates and no wetted areas on the bottom surface. They also have negative or insignificant one-way transport capacities, which indicate that sweat cannot diffuse easily through the fabric. The sweat on the next-to-skin surface to the other side will thus accumulate on top of the surface and result in an uncomfortable wear for the user. On the other hand, the knitted fabric samples NW, PC and KW have medium to fast liquid absorption, penetration and spreading rate on the bottom surface of the fabrics. This is due to that the wool or cotton composition in blend would improve the fabric moisture management properties (see Table 1 and Table 2). Further research is suggested using knitted plated wool fabric as base material and coated with X-ray absorber material on the front side [10,11]. This would save half the weight of casing material for an apron.

As shown in Table 2 and Fig.2, the KW has the highest absorption rate on the top surface (next-to-skin) with 90 %; meanwhile, NW has an absorption rate of 31 % on the top surface. A higher absorption rate is related to better comfort characteristics, especially subjective perceptions of the sweat sensation. The K fabric had zero penetration and thus no one-way transports, slow spreading, and very slow absorption. The characteristic result of most of the samples is slow spreading and no OMMC, indicating that the wearer is likely to experience clamminess, i.e. moisture remaining either on the next-to-skin surface or within the fabric. This agrees with a report from Mahbub [12] on thermal comfort study for a similar type of fabric.

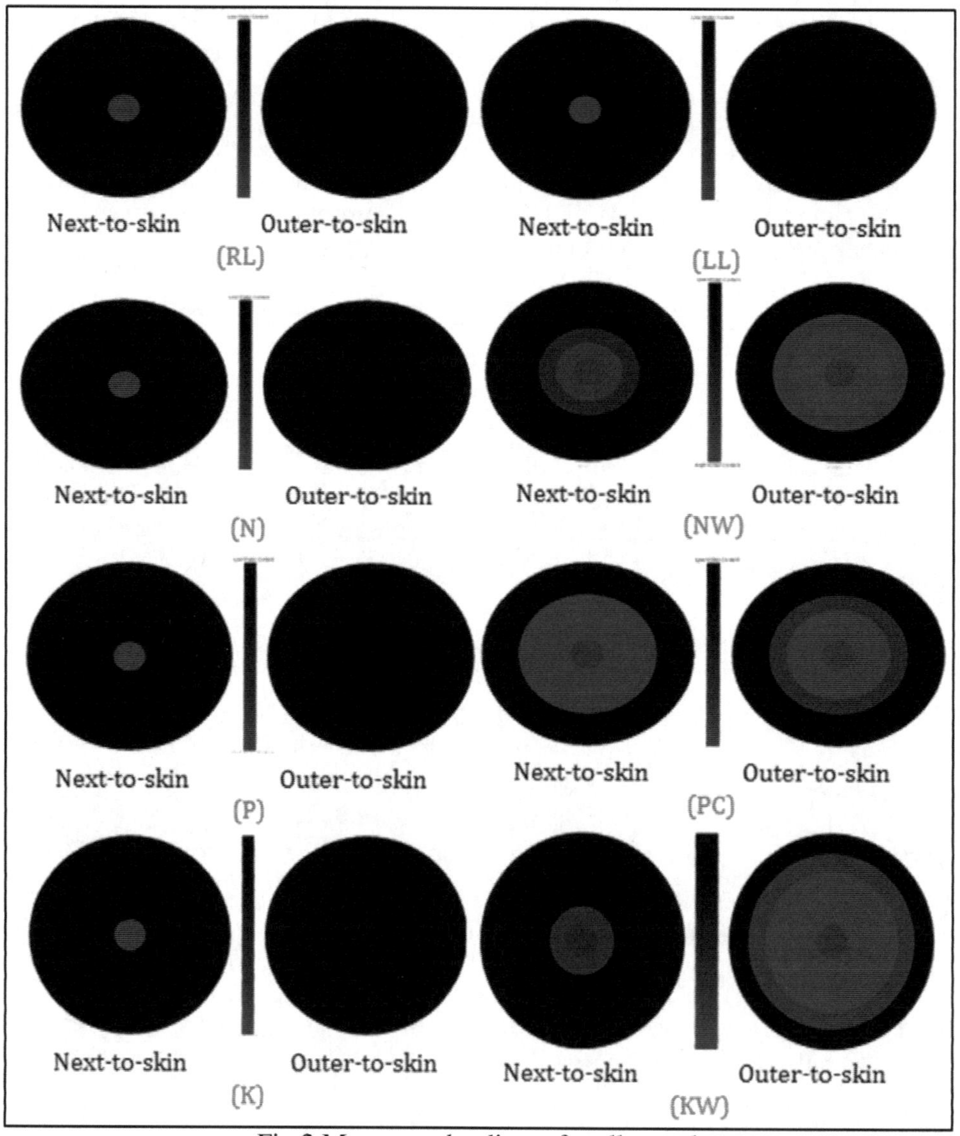

Fig.2 Max wetted radiuses for all samples

Fabric sample N is a very slow absorbing and very slow drying fabric which is characterized as poor one-way transport, very slow spreading without absorption of moisture, indicating that moisture cannot easily diffuse across the fabric and evaporate into the environment. The NW shows a high absorption rate on the wool surface. This suggested that the wool can absorb water faster than the nylon surface. Hence, the wool face next-to-skin is better for moisture management. Furthermore, the wet-out radius for the top surface was smaller than the bottom surface (Table 2) and the top surface has moderate ability to transfer the absorbed water to the outer surface. Also, from Fig.2 it can be seen that the wet radius was different between two surfaces for the NW. The NW fabric has a relatively large spreading rate (3.1 mm/s) and large wet-out radius (18 mm/s) on the bottom, indicating that liquid can speared quickly, transfer easily and dry quickly on the outer surface of the fabric. The fabric accumulative one-way transport capacity (OWTC) was 145.8%. Also, the overall moisture management capacity (OMMC) was 0.4. Therefore, the OMMC and OWTC grade was good, suggesting that the nylon wool knitted fabric has a moderate water penetration rate between both surfaces. Generally, samples N, P and K showed poor results for the moisture management properties.

4. Conclusion

Radiation protection is a critical aspect of the human application of different forms of radiation in medical contexts for diagnosis and treatment. The lead apron material was uncomfortable to wear because of its heavy weight, not permeable to air and inability to absorb and allow for evaporation of moisture in the form of sweat. The study evaluated the effectiveness of X-ray aprons to determine their protectiveness and comfort performance. Six fabrics and two commercial lead sheets were tested and evaluated for X-ray protection and comfort characteristic. The lead sheets have very poor comfort characteristics but excellent radiation shielding ability. On the other hand, the knitted fabric samples like NW, PC and KW show a poor absorption of the X-ray radiation, meaning that these samples do not have any effective shielding ability. Nevertheless, in term of comfort characteristics, they show good comfort properties particularly in term of moisture management. It is suggested that these fabrics could be used for apron casing to replace the woven fabrics which have poor comfort performance. The woven fabrics also showed ineffectiveness for radiation shielding. However, all fabrics showed enhanced comfort properties as compared to the lead sheets. The use of knitted fabric plated with wool will improve the comfort performance of the casing material and could be used as fabric substrate to hold the shielding absorber materials.

5. Acknowledgement

This work was supported by the Saudi Arabian Cultural Mission in Australia (SACM) on behalf of the Umm Al-Qura University, the Ministry of Education, and Higher Education of Saudi Arabia.

References

[1] Bushong SC. "Radiologic science for technologists: Physics, biology, and protection". (9th ed.). London: Mosby St. Louis, MO, 2008.

[2] Wagner LK, Eifel PJ, Geise RA. Potential biological effects following high x-ray dose interventional procedures. J. Vasc. Interv. Radiol. 1994; 5: 71-84.

[3] Lead Industries Association. "A guide to the use of lead for radiation shielding". 1984.

[4] Goldstein JA, Balter S, Cowley M, Hodgson J, Klein LW. Occupational hazards of interventional cardiologists: Prevalence of orthopedic health problems in contemporary practice. Catheter. Cardio. Inte. 2004; 63: 407-411.

[5] Van Veelen M, Nederlof E, Goossens R, Schot C, Jakimowicz J. Ergonomic problems encountered by the medical team related to products used for minimally invasive surgery. Surg. Endosc. 2003; 17: 1077-1081.

[6] Schlattl H, Zankl M, Eder H, Hoeschen C. Shielding properties of lead-free protective clothing and their impact on radiation doses. J. Med. Phys. 2007; 34: 4270-4280.

[7] Cadwalader CW. Slipcover for radiation shields, US Patent NO. 5, 523, 581: 1996.

[8] Huda Ahmed M, Pradip D, Arun V, Lijing W. An overview of lead aprons for radiation protection: Are they doing their best? Paper presented at the the 8th Textile Bioengineering and Informatics Symposium (TBIS 2015), Zadar, Croatia, 2015: 232-240.

[9] Bogdan A, Sudoł-Szopińska I, Szopiński T. Assessment of textiles for use in operating theatres with respect to the thermal comfort of surgeons. Fibres Text. East. Eur. 2011; 19: 65-69.

[10] Maghrabi HA, Vijayan A, Wang L, Deb P. Design of seamless knitted radiation shielding garments with 3D body scanning technology. Procedia Technology 2015; 20: 123-132.
[11] Maghrabi HA, Vijayan A, Deb P, Wang L. Bismuth oxide-coated fabrics for X-ray shielding. Text. Res. J. 2016; 86: 649–658.
[12] Mahbub RF, Wang L, Arnold L, Kaneslingam S, Padhye R. Thermal comfort properties of Kevlar and Kevlar/wool fabrics. Text. Res. J. 2014; 84: 2094-2102.

Thermal Stability of Nano Composite Phase Change Materials Microcapsules (NC-MicroPCMs) Incorporated With Nanoparticles of Different Surface Energy

Qing-Wen Song*, Shao-Feng Lu, Jia-Guang Meng, Huan-Huan Ren

School of Textile and Materials, Xi`an Polytechnic University, 19 Jinhua Road, Xian, Shaanxi, PC710048, PR. China*

*Corresponding author's email: 1143601883@qq.com

Abstract

NC-MicroPCMs samples by introducing P25 (Degussa) nanoparticles with different surface energy, and a control sample with the same specification were prepared via in-situ polymerization. SEM and TG (thermal gravimetric analysis) were applied to study the surface morphology, thermal stability and P25 loading rate. The results illustrated that: 1) surface morphology of the five microPCMs were different due to various "nanoparticles @polymer" combining state; 2) P25 loading rate in NC-MicroPCMs would increase as its surface energy decrease; 3) NC-microPCM had a maximum accommodation capacity for P25 nanoparticles, and surplus nanoparticles could form a reticular structure among microcapsules; 4) thermal stability of NC-MicroPCMs would increase as surface energy of P25 decrease; 5) weight losing features of NC-MicroPCMs were shown in 3 stages, which signified a substantial protection was given by composite shell polymer; while for control, the sample kept weight losing throughout the process. The half life temperature T_{50} of NC-MicroPCMs is 307°C, which is 94°C higher than the control sample (213°C).

Keywords: phase change material microcapsules (Micro-PCMs); Nano-composite; surface energy; thermal stability

1. Introduction

Energy shortage and environmental issues have gained considerable attentions recently. Energy conservation and raising its utilization efficiency become focus in the research area of energy consumption globally [1,2]. Microcroencapsulation is an important form in the applications of energy storage technology by means of encapsulating PCMs into a natural or synthetic polymeric shells which range in size from less than 1μm to 1000μm [3-5]; However, there are still some problems, such as low heat capacity, higher super-cooling, low thermo-stable temperature, liquid leakage, short service life, as well as formaldehyde release, and etc. Many scholars have done a lot of work and approaches have extensively explored by doping various functional nano materials into the core/ shell structure. In these studies, conventional MicroPCMs are often modified by introducing a composite structure that doping of nano materials, such as carbon nanotubes, silver nanopowder, graphite, ATO (Antimony Tin Oxide) powder, and etc [6-9]. Not only do these composite microcapsules have conventional functions of thermal regulating, but some extra properties, such as heat conduction strengthening, supercooling depression, and higher thermal stability are also exhibited. Xiaohua Liao etc [10] have reported that a newly MicroPCMs containing nano ATO powder (Antimony Tin Oxide) within polymer shell or core component, have both a photothermal conversion effect and thermal storage ability. As we have reported earlier that doping process could often induce physical structure changes of shell polymer, e.g., doping silver nano particles into polymer shell would increase degree of crystallinity (DOC) of shell polymer, and a better thermal stability of the NC-MicroPCMs could be achieved [11,12]. These structure modifications are usually prepared by a surface treatment for nano materials to make it compatible with a component of MicroPCMs. Processing techniques, such as quantity of nanoparticles, may have a great influence on MicroPCMs properties. However, researches on how to manipulate nanoparticles to different position of MicroPCMs, and further systematically researches on relationships between the loading rate of nano materials and the thermal performance of NC-MicroPCMs have rarely be reported.

In this study, titanium dioxide P25 (Degussa) was used as a doping substance for NC-MicroPCMs. Silane coupling agent octyltrimethoxy silane (OTMS) was applied for P25 surface modification; surface property of the P25 would be hydrophobic and the surface energy of it would decrease after the treatment. Contact Angle was chosen as an index to express the degree of hydrophobicity. In order to control the amount of P25 which had entered into the NC-MicroPCMs, four kinds of nanoparticles with different surface energy in terms of the contact angle of 51.5 ° (surface energy 50.81mN/m), 91.6 ° (29.57 mN/m), 109 ° (27.13 mN/m), and 120 ° (7.41 mN/m) were obtained respectively. The size of P25 was changed from 0.10μm to 0.13μm after the surface treatment. According to our earlier studies [13], melamine-formaldehyde resin was applied as shell material and n-hexadecane as a core. Four NC-MicroPCMs were prepared by using the four P25 nano powders with different surface energy, and a conventional MicroPCMs was also prepared as a control. These 5 kinds of samples were of the same specifications of core/ shell mass ratio of 5:1, formaldehyde/ melamine molar ratio of 2:1. The amount of P25 was chosen as 10% in the weight of the core material for NC-MicroPCMs. Relationships between the load quantity of nanoparticles and its surface energy were discussed in terms of P25 contact angle. The surface energy of the four P25 nano particles were calculated by means of "OWENS two liquids methodology" [14]. The final content of TiO2 particles which had been doped into the NC-Microcapsules were determined by thermogravimetric analysis (TGA) by comparing the remains of the control and NC-MicroPCMs. Thermal stability of the five samples were determined by means of T50 - the temperature at which a sample losing 50% of its original weight. The relationships between surface energy of P25 and its load quantity, as well as thermal stability of the corresponding MicroPCMs were studied accordingly.

2. Experimental

2.1 Surface hydrophobic treatment for P25 nano particles

Hydrophobic agent OTMS was diluted by anhydrous ethanol in advance. Adding prepared OTMS into a 250ml flask, 10% acetic acid was added to adjust pH to 5 under a constant speed mixing at room temperature for 10 minutes; put P25 nano particles into the flask and stir it for 10 minutes, raise temperature up to 60°C and keep for 60 min; then adding the final residue into a suction filter bottle, using deionized water washing and filtering three times to get filter cake. Put the filter cake into oven at 80°C drying to a constant weight, cooling and grinding the dried cake into a fine powder.

2.2 Contact angle measurement and determination of surface energy of P25 nanoparticles

Put 0.2g pre-treated P25 powder into DF-4 (Tianjin GangDong) tablet machine and press it into a tablet thickness of 1 mm; 6 tablets were needed for each sample. Then dripping a drop of pure water or n-tetradecane onto the surface of the tablet; measuring contact angle by JC2000C3 contact angle tester (Shanghai Zhongchen) respectively. Owens two liquids methodology are used to calculate the surface energy (mN/m) of the solid surface made by P25 nanoparticles.

2.3 Preparations of MicroPCMs and refine process

Based on our earlier studies, put core material emulsion into a 250ml flask with a sturdy stir, and a proper amount of nano particles was then added into the system; keep the process for 20minutes at room temperature; then adding prepared M-F pre-polymer droply, adjusting pH of the whole mixture to 5.0 by 10% acetic acid. Raise temperature to 70°C in 30min while keeping pH steady. The whole mixture was cured with a proper stir rate for 3 hours. Then adjusting pH to neutral to terminate the process by adding 10% sodium carbonate solution. Microcapsule suspension was then filtrated by a medium speed filter paper; wash the filter cake by distilled water three times, drying the powder at 60°C until the powder were completely dry.

A further refine process was needed before TGA measurement. Firstly, the dried MicroPCM powders were re-washed by anhydrous ethanol three times, then followed by pure water in a vacuum filter three times; the filter cake was then dried for 24 hours at 60°C and stored in a dryer. The refined MicroPCMs powders were ready for TGA test.

2.4 SEM observation

Surface morphology observation was taken by JSM-7500F FE-SEM. Put a thin layer of the refined MicroPCMs powder onto a conductive adhesive and then coating a thin layer of gold. The accelerating voltage was 5.0 KV.

2.5 Determination of loading rate of P25 nanopowder and thermal stability of MicroPCMs

STA 449 simultaneous Thermogravimetric Analyzer (TGA) was applied to analyze thermal stability of the five MicroPCMs samples and the P25 nano powders. TGA testing parameters were set as temperature range from 25°C to 850°C under nitrogen atmosphere at a rate of 10°C/min. The loading rate of the P25 could be determined by measuring the residue weight of the MicroPCMs sample at the ultra high temperature of 800°C.

"T_{50}" refers as "half-life temperature"; it is an index by which thermal stability of a McroPCM sample can be determined; a higher T_{50} means a better thermal stability of a sample has as shown in Fig.3.

In order to get an clear understanding of weight losing velocity for a MicroPCMs as a function of temperature, Differential Thermogravimetric (DTG) was applied that a sample`s TG curve was taken a derivative with respect to time in the study. DTG curve of the five MicroPCMs sample were obtained therein as shown in Fig.4.

3. Results and discussion

3.1 Surface morphology of the five MicroPCM samples

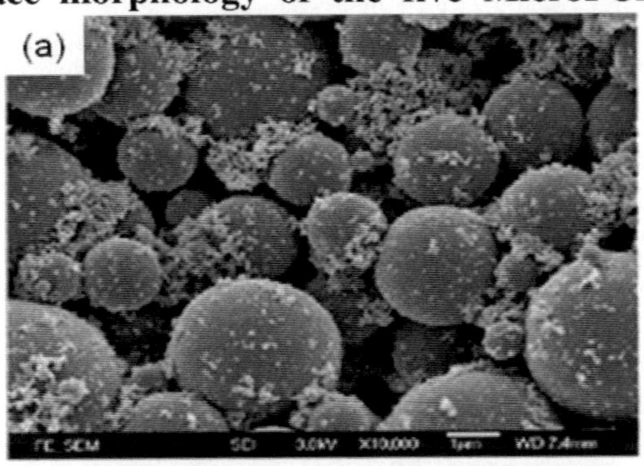

(a) EMP0, contact Angle 51.5°

(b) Sample EMP1, contact Angle 91.6 °

(C) EMP11, Contact Angle 120 °

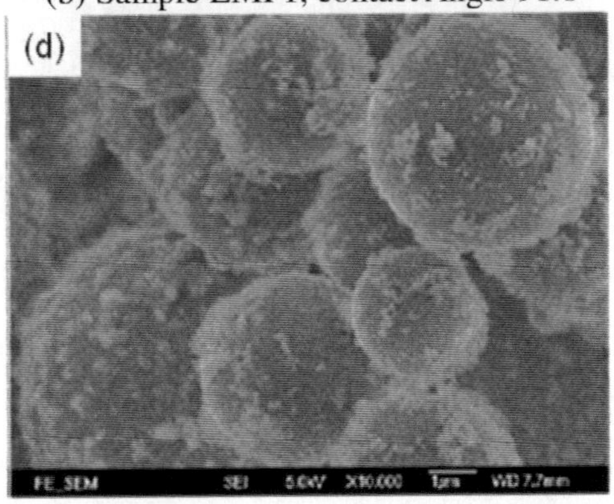

(d) EMP5, contact Angle is 109 °

(e) Con-MicroPCMs, no nanoparticle addition

Fig.1 SEM observation of the 4 NC-MicroPCMs samples and a Con-MicroPCMs (Control) with the same specification

Based on the SEM imagines named as a, b, c, d, e, the five samples have the following characteristics.

1) All the 5 MicroPCMs samples are sphere with size distribution of 2~5μm and in good condition.

2) The morphology features of the four NC-MicroPCMs doping of P25 are shown in Fig.1 (a) (b) (C) (d). For sample EMP0 (a), P25 was used as received; most P25 nanoparticles are spotted among microcapsule individuals in different degree of aggregation, rather than on its surface, which can be attributed to a higher surface energy of original P25 nanoparticles itself. Therefore, the affinity between naoparticles and shell polymer is not capable of forming a steady and compact structure.

3) After a hydrophobic treatment, the surface energy of P25 nanoparticles is decreased; this makes it more compatible with the polymer shell. Despite a degree of aggregation, nano particles are observed absorbing or entrapping into the shell material with a well dispersion; the P25 particles aggregation has been depressed as shown in Fig.1 (b, c, d).

4) As the contact angle is greater than 109° as sample EMP11, the bonding force between P25 nanoparticles and shell polymer reaches to the maximal point, and probably the shell polymer has a maximal accommodation capacity for nanoparticles; some surplus P25 nanoparticles have formed a reticular structure, as shown in Fig.1 (c).

5) Conventional MicroPCMs (e) are adhesive seriously between each other; some flakes like material that scattered on the surface of microcapsules are polymer shell residual as shown in Fig.1 (e).

3.2 Loading rate of P25 nanoparticles in NC-MicroPCMs and its surface energy

Table 1 shows data of contact angle, surface energy of the four P25 nanoparticles and residue weight of the five MicroPCMs at 800°C. The weight residues of the four NC-MicroPCMs are higher than the control sample of 1.42%. The weight residue of P25 nanoparticles remains stable in the whole testing process (80~800°C). Therefore, it can be speculated that the residue of these samples may probably be carbonizated substance and P25 nanoparticles. The P25 loading rate of the four NC-MicroPCMs can be calculated by the weight of residue subtract the weight of the control sample which no P25 added.

The loading rate of P25 of the four NC-MicroPCMs and its contact angle are positively correlated as shown in Fig.2; contact angle has a significant influence on P25 loading rate. The surface energy data shown in table 1 also indicates that the greater the hydrophobicity of P25, the lower the surface energy would be and the bigger affinity between P25 nanoparticles and polymer shell. Therefore, more P25 nanoparticles could enter into MicroPCMs to form a stable composite structure. But after it exceeds 109°, the shell polymer seem to has a maximum capacity in accommodating of P25, and some surplus nanoparticles are deposited orderly on the surface of the microcapsule shell as shown in Fig.1 (C).

Table 1. Thermal gravimetry test result for the 4 groups of Microcapsules incorporating with nano powders of different surface energy and a control sample

Sample	P25Nano-TiO_2 Contact angle°	Nano-TiO_2 Surface energy mN/m	800℃ weight residue %	P25 loading rate %	Mean T_{50} / ℃
EMP0*	51.50	50.81	2.21	0.79	224.33
EMP1	91.60	29.351	3.36	1.94	280.33
EMP5	109.00	27.18	5.07	3.65	307.67
EMP11	120.00	7.41	6.03	4.61	289
Con-MicroPCMs	/	/	1.42	/	213

*P25 used as received.

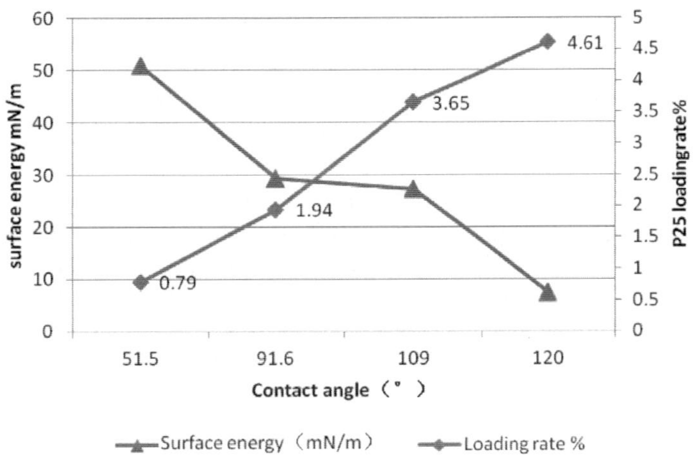

Fig.2 Relationships between contact angle, surface energy and loading rate of P25 nanoparticles

3.3 Relationship between thermal stability and surface energy of P25 integrated with the corresponding NC-MicroPCMs

Fig.3 indicates that there are significant differences among the five MicroPCMs in weight losing features, while the weight of P25 remain stable in the whole process (100~800°C). EMP5 exhibits the highest thermal stable T_{50}, while Con-MicroPCMs has the lowest. The surface energy of P25 nanoparticle decrease with the increase of its contact angle, and thermal stability (T_{50}) of the corresponding MicroPCMs will increase accordingly. However, as the contact angle exceed 109°C, a downward trend has been shown, indicating that thermal stability become worse.

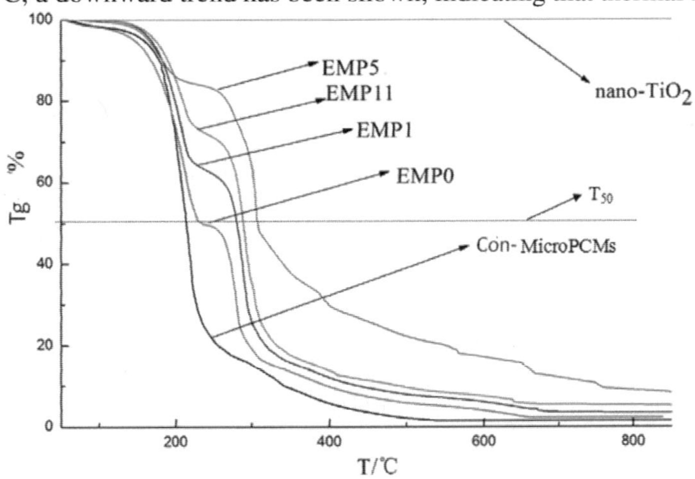

Fig.3 Thermal gravemetry analysis of the 4 NC-MicroPCMs, control MicroPCMs and P25 nano powder

Derivative thermogravimetric (DTG) curves of the five MicroPCMs can be obtained by taking a derivative with respect to time as shown in Fig.4, by which the maximum weight loss rate can be determined clearly. For NC-MicroPCMs samples, the weight losing features are detected in 3 stages that separated by "peak" and "plateau". The first weight losing stage may probably start from 100°C to 230°C, and there are about 50 ~ 70% weight remain at the end of the stage; it probably attributed to water, a small amount of core material hexadecane and some low molecular shell polymer volatilizing in the stage. Then followed by the second stage, at which MicroPCMs retains 10 ~ 20% of its original weight in a range of 260 ~ 380°C; most of core material have vaporized due to shell polymer has been collapsed and the weight losing rate is the highest. Ether bond, triazine ring and methylene bridges in aminoplast cross-linked shell polymer are started to collapse in sequence during the stage. The third stage starts from 380 to 500°C, the residual weight remains of 2~6% for NC-MicroPCMs samples, except for the control sample of 1.42% which has the lowest residue. This part of the residue in MicroPCMs may probably be identified as its carbonization. As temperature continues to increase, the weight residuals of MicorPCM samples are stable; weight losing rates trend to be zero in a range of 500°C to 800°C as shown in Fig.4.

The features in weight losing for sample EMP5 are in stages as shown in Fig.4, by which distinct weight losing peaks can be marked clearly; while in other temperature zone, weight losing rate trends to be zero in spite of temperature rising. On the contrary, the weight losing of Con-MicroPCMs continues at a certain speed in the whole process, except a relatively higher weight loss rate appeared at 216°C. These features in weight losing confirm that MicroPCMs with a nano composite structure have better thermal stable properties than that of MicroPCMs without doping of P25. Moreover, hydrophobic surface pretreatment makes it more compatible with polymer shell and the affinity between P25 and shell polymer would be strengthened therein due to the effects of small size, large specific surface area of P25 and etc. It is confirmed that a strong bonding force between the polymer substrate and pretreated nanoparticle makes the composites structure more toughened, by which the physical and mechanical properties of the polymer matrix can be improved. This conclusion supports our previous study of doping silver particles into NC-MicroPCMs had made shell crystallinity increase and thermal stability had been improved [6][11]. The principal mechanism to improve mechanical properties of polymer shell should be similar.

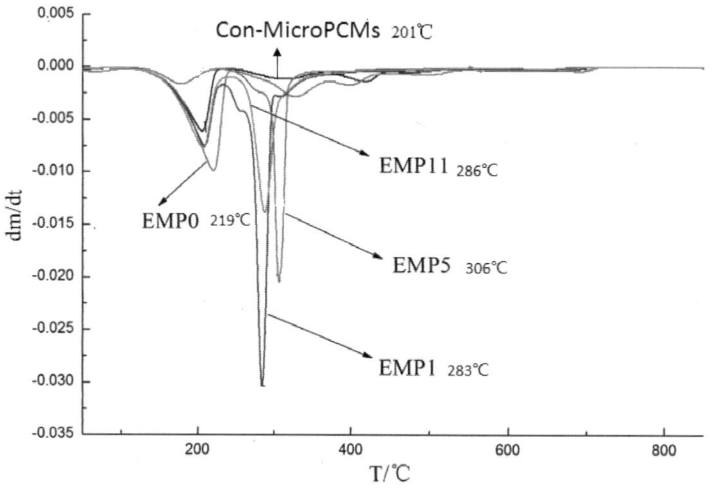

Fig.4 DTG curves of the five MicroPCMs with different P25 loading rate

The maximal weight losing temperature for samples of Con-MicroPCMs、EMP0、EMP1、EMP5、EMP11 are 201°C、219°C、283°C、306°C、286°C respectively, as shown in Fig.4. According to the analysis, this is the maximum temperature that a corresponding MicroPCMs could bear. As the hydrophobic property of P25 increase, more nano particles are compatible with shell or core material, that means a composite structure could be formed more compact and toughened, and thermal stable temperature are increased accordingly.

4. Conclusion

Based on the result and discussion, conclusions may draw as the following:

The dispersion state of P25 nanoparticles on the surface of microcapsules or among them are closely related with its hydrophobic property. The loading rate of P25 are positively correlated with its contact angle before 109°, then it will decrease as the contact angle increases and nanoparticles will agglomerate. With the reduction of surface energy of P25 after hydrophobic treatment, the affinity between TiO_2 nano particle and shell polymer increase, and more P25 are integrated with MicroPCMs. There is a maximal accommodation capacity for nanoparticles within MicroPCMs, as the P25 loading rate is higher than 3.65%, surplus P25 nanoparticles are observed forming a reticular structure that has no contribution on improving thermal stability of the NC-MicroPCMs.

Surface hydrophobic treatment of P25 reduce the surface energy of nanoparticle and makes it more compatible with polymer shell in forming a certain "nanoparticles @ shellpolymer" composite structure, by which thermal stability of NC-MicroPCMs can be enhanced as shown in sample EMP5. These features testify a conclusion that a “nanoparticle@shellpolymer” composite has the capacity to strengthen the aminoplast shell molecular structure and the NC-MicroPCMs are given a higher thermal stability.

5. Acknowledgments

The authors would like to express gratitude to the financial support of National natural science foundation of China with grant No.51403169, and the key laboratory of functional garment and fabric Shannxi province with grant No.2010JS006.

References

[1] Wang TY, Huang J, Zhu PP, et al. Fabrication and characterization of micro-encapsulated sodium phosphate dodecahydrate with different crosslinked polymer shells. Colloid and Polymer Science: 2013; 291(10); p: 2463-2468.

[2] Rathod MK, Banerjee J. Thermal stability of phase change materials used in latent heat energy storage systems: A review. Renewable and Sustainable Energy Reviews: 2013; 18; p: 246-258.

[3] Alkan, Cemil, Sarı, Ahmet, Karaipekli, Ali, Preparation. thermal properties and thermal reliability of microencapsulated n-eicosane as novel phase change material for thermal energy storage. Energy Conversion and Management: 2011; 52(1); p: 687-692.

[4] Khakzad, Fahimeh, Alinejad, Zeinab, Shirin-Abadi, Abbas Rezaee, Ghasemi, Maryam, Mahdavian, Ali. Optimization of parameters in preparation of PCM microcapsules based on melamine formaldehyde through dispersion polymerization. Colloid and Polymer Science: 2014; 292(2); p: 355-368.

[5] Li L, Liu S, Tao H. Characteristics of weaving parameters in microcapsule fabrics and their influence on loading capability. Textile Research Journal: 2012; 83(2); p: 113-121.

[6] Song QW, Xing JW, Yang JZ, Nano Composite Microcapsules-Crystallinity, Thermal Stability. Journal of Donghu University: 2012; 29(4); p: 321-325.

[7] Song QW, Li Y, et al. Supercooling of Silver Nano Composite PCM Microcapsules. Journal of Fiber Bioengineering and Informatics (TBIS): 2008; HongKong; 1(3); p: 239-248.

[8] Yu SY, Wang XD, Wu DZ. Microencapsulation of n-octadecane phase change material with calcium carbonate shell for enhancement of thermal conductivity and serving durability: Synthesis, microstructure, and performance evaluation. Applied Energy: 2014; 114; p: 632-643.

[9] Lu SF, Xing JW, Zhang ZH. Preparation and characterization of polyurea/ polyurethane double-shell microcapsules containing butyl stearate through interfacial polymerization. Journal of Applied Polymer Science: 2011; 121(6); p: 3377-3383.

[10] Liao XH, Shi HF, Li WW, You M. Facile Synthesis of Highly Photoactive ATO-Based Microcapsule for Solar Energy Harvesting. Advanced Science, Engineering and Medicine: 2013; 5(10); p: 1498-1503.

[11] Song QW, Li Y, Xing JW. Thermal stability of composite phase change material microcapsules incorporated with silve rnano-particles. Polymer 2007; 48; p: 3317-3323.

[12] Ren HH, Song QW, Lu SF, Meng JG. Relationships between surface energy of nano powder and its loading

capacity in composite microcapsules. New Chemical Materials 2015; 43(8); p: 143.
[13] Ren HH, Song QW, Xing JW. Study on hydrophobic treatment of nano-TiO2, in Textile Bioengineering and Informatics Society (TBIS): 2013; Xi-an China.
[14] Cwikel Dory, Zhao Q, Liu C. Comparing Contact Angle Measurements and Surface Tension Assessments of Solid Surfaces. LANGMUIR: Oct 5 2010; 26 (19); p: 15289-15294.

Investigation on Intra-Ply Shear Properties of 3D Weft Knitted Spacer Fabrics by Image Analysis

Veerakumar Arumugam[1], Rajesh Mishra[1], Maros Tunak[2], Jiri Militky[1], Dana Kremenakova[1], Mohanapriya Venkatraman[1], Vignesh Balaji Ramanisanthi Subramaniam[1]

[1]*Dept. of Materials Engineering, Technical University of Liberec, Liberec, Czech Republic, 461 17*
[2]*Dept. of Textile Evaluation, Technical University of Liberec, Liberec, Czech Republic, 461 17*

*Corresponding author's email: veerakumar27@gmail.com

Abstract

A fixture was designed to analyze the intra-ply shear properties of these fabrics. The nonlinear behavior of shear force versus shear angle and the deformation mechanism were analyzed. The curves for shear force versus shear angle and position of buckling for intra-ply shear test are recorded. Load-displacement curves of intra-ply shear tests are also analyzed. In addition to this, a program was developed in MATLAB using Hough transform to analyze the shear angle in the real-time image taken during displacement of specimen at various positions. These findings are important requirement for the further improvements in designing of picture frame fixture and to study the in-plane shear properties of 3D fabrics.

Keywords: Displacement; 3D spacer fabrics; Shear angle; Shear stress; Image analysis

1. Introduction

It is necessary to study the inter/ intra-ply shear behavior of 3D fabric because of their wide application in production especially in the case of forming process. The in-plane shear behavior of 2D fabric has been comparatively well investigated. Zhu et al [1]. carefully investigated the in-plane shear characterization of 2D fabric by experimental test, and found that the reduction of yarn was a key to wrinkling. Xiaohua et al [7]. studied the shear property of 2D fabrics using picture-frame test method, and pointed out that the shear results were sensitive to the tensions in the yarns during the experiment. The tensile force increased with the increase in shear angle. Lomov et al [2]. presented shear tests of unbalanced 2/2 twill glass/PP fabric on picture frame in three different pretension states and studied the influence of tensile load in the yarn direction on the shear resistance for the fabric and the repeatability of the test method. Lin et al [3]. established the finite element model based on the geometry of 2D fabric to simulate the in-plane shear deformation, the simulation results were identical with experiments. Cao et al [4]. compared the picture frame shear test results from seven different labs for developing a standard test setup and obtaining accurate and appropriate material properties. Chen et al [5]. developed a FEM model to predict in-plane and interlaminar shear properties of laminates. However, the in-plane shear behavior of 3D spacer fabric was rarely reported. Charmetant et al [6]. built a hyperelastic model to simulate the formability of 3D fabric.

In this research work, a picture frame shear fixture was developed and a careful study was made on its applicability in testing the intra-ply shear behavior of 3D spacer fabrics with different fabric density, thickness and structure. The shear force versus shear angle curves and position of wrinkles during in-plane shear test are recorded and analyzed. In addition to that suitable program was developed in MATLAB using Hough transform to analyze the shear angle during deformations. It can provide the foundation for investigating the performance of picture.

2. Materials and Methods

Six different types of spacer fabrics were developed using computerized Mayer & Cie, OVJA 1.6 E 3 WT knitting machine. These fabric samples were classified into two groups for convenient analysis of results, the first group has been developed using Polyester/ Polypropylene blend with three different proportions and second group with Polyester/ Polypropylene/ Lycra blend having another 3 different compositions. As a spacer yarn, three different types of 88 dtex Polyester monofilament yarn and Polyester multifilament yarns (167 dtex and 14.5 tex) were used. 14.5 tex Polypropylene and 44 dtex Lycra multifilament yarns were also used for the face and back side of the spacer fabrics (Table 1). The fabric characteristics such as areal density, stitch density, structure, thickness, are presented in Table 2.

Table 1. Fabric particulars

Fabric sample No.	Fabric layers	Technical face	Spacer yarn	Technical back	Fiber composition (%)
Group 1 - Without Lycra					
S1	**Type of yarns and linear density**	Polypropylene (PP) - 14.5 tex	Polyester monofilament (PES monofil - 88 dtex	Polypropylene (PP) -14.5 tex	58% PP, 42% PES monofilament
S2		Polypropylene (PP)- 14.5 tex	Polyester (PES) - 14.5 tex	Polypropylene (PP) -14.5 tex	45% PP, 55% PES
S3		Polypropylene (PP) - 14.5 tex	Polyester (PES) - 167 dtex	Polypropylene (PP) -14.5 tex	41%PP, 59% PES
Group 2- With Lycra					
S4	**Type of yarns and linear density**	Polypropylene (PP)- 14.5 tex Lycra - 44dtex	Polyester monofilament (PES monofil) - 88 dtex	Polypropylene (PP) -14.5 tex	55%PP, 39%PES monofilament, 6% Lycra
S5		Polypropylene (PP)- 14.5 tex Lycra - 44dtex	Polyester (PES) - 14.5 tex	Polypropylene (PP) -14.5 tex	42% PP, 52% PES, 6% Lycra
S6		Polypropylene (PP)- 14.5 tex and Lycra - 44dtex	Polyester (PES) - 167 dtex	Polypropylene (PP) -14.5 tex	39% PP, 55% PES, 6% Lycra

Table 2. Characteristics of spacer fabrics

Fabric sampl eNo.	Areal densit y g/m^2	Thickne ss mm	Density (kg/m^3)	Wales/cm	Courses/c m	Stitch Density (stitches/ cm^2)
Group 1 - Without Lycra						
S1	493	4.4	112	10	20	200
S2	443	2.62	169.1	10	15	150
S3	477	2.74	174.1	10	15	150
Group 2- With Lycra						
S4	632	4.4	144.8	14	25	350
S5	657	3.5	187.7	14	20	280
S6	695	3.4	205.4	14	20	280

2.1 Picture frame shear tester

Picture frame test as shown in Fig.1 is an effective way for characterizing intra-ply shear property of fabrics [8]. The picture frame test is preferred by many researchers for shear testing since it generates pure state of strain which can be imposed on the test specimen [9-11]. Shearing is induced by restraining the textile reinforcement in a rhomboid deformation frame with fibers constrained to move parallel to the frame edges. The frame is extended at diagonally opposing corners using simple tensile testing equipment. The 3D spacer fabrics for shear tests were prepared according to the size of the picture frame and the characteristics of samples are described in the Fig.2.

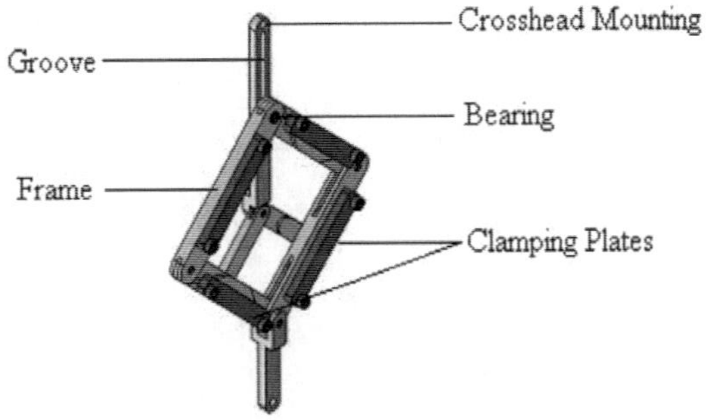

Fig.1 Picture frame shear fixture

Fig.2 Clamping of sample in fixture

Direct measurement of axial load and shear angle is possible through this following relationship [25,26].

$$F_s = \frac{F_x}{2cos\varphi} \tag{1}$$

Shear force (F_s) is determined by the axial force (F_x) frame rig length (L) and the frame angle (φ). Meanwhile frame angle can be determined directly from cross head displacement (d). Shear angle (γ) can be obtained from frame angle by using the following equations 2 and 3 [12].

$$\varphi = \cos^{-1}\left[\frac{L\sqrt{2}+d}{2L}\right] \tag{2}$$

$$\gamma = \frac{\pi}{2} - 2\varphi \tag{3}$$

2.2 Image analysis using MATLAB

Image analysis can aid in the determination of the shear angle and displacement at any particular point on the surface of fabric specimen. Grid pattern was applied on the specimen surface before the test and used as the reference points of image analysis. The complete displacement of the specimens during loading process was obtained by image analysis methods. The images were captured at certain regular interval of time using digital camera. A special program is developed in MATLAB 7.10 (R 2010a) using Hough's transform to find the angle between the lines on the specimens. The results of experimental methods and image analysis were compared with each other.

3. Results and discussions

3.1 Effect of fabric characteristics on in-plane shear behaviour of spacer fabrics

The picture frame test method is verified for the 3D spacer fabrics under consideration. Fig.3 shows the array of images for a 3D spacer fabric specimen captured during the loading process at each 5mm displacement for every 30 sec. In the beginning, (Fig.3a), there is no shearing. Subsequently the shear deformation resistance is mainly from the friction between the wale and course direction loops before reaching the limiting locking angle. The limiting locking angle is the ultimate shear deformation observed by means of local wrinkling. Based on the deformed configuration of the fixture, shear angle is calculated using the relationship shown in equation 3. A vertical displacement of 40 mm, which corresponds to a shear angle of around 50° was found to be the maximum displacement but the pre-buckling occurs after 20 mm displacement with shear angle range of 20-30° (Fig.3e). The local buckling is calculated using image analysis software. Fig.4 presents the linear fit curve between time and displacement.

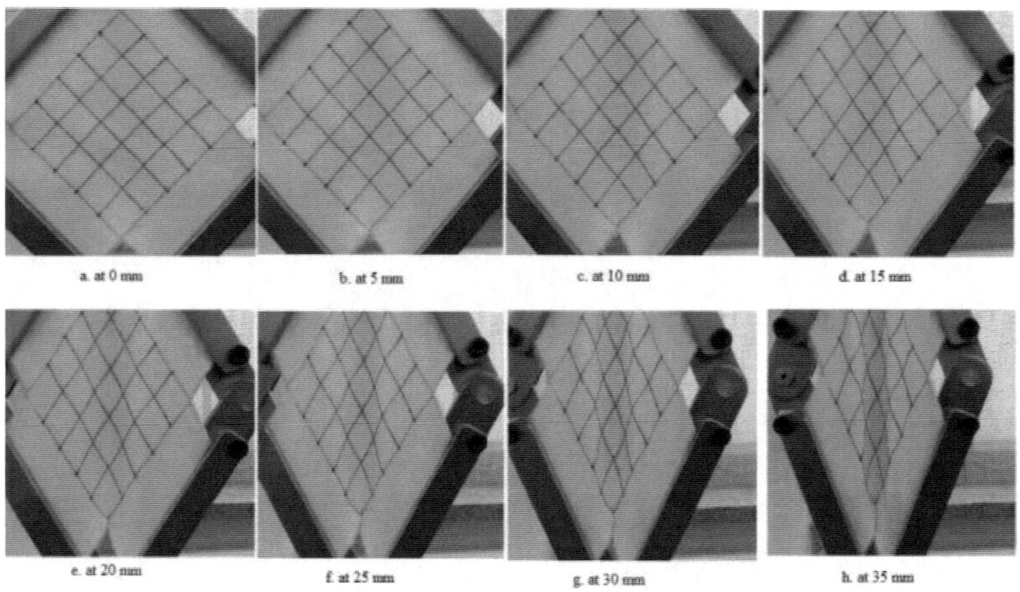

Fig.3 Shear deformation of specimen at different displacement levels

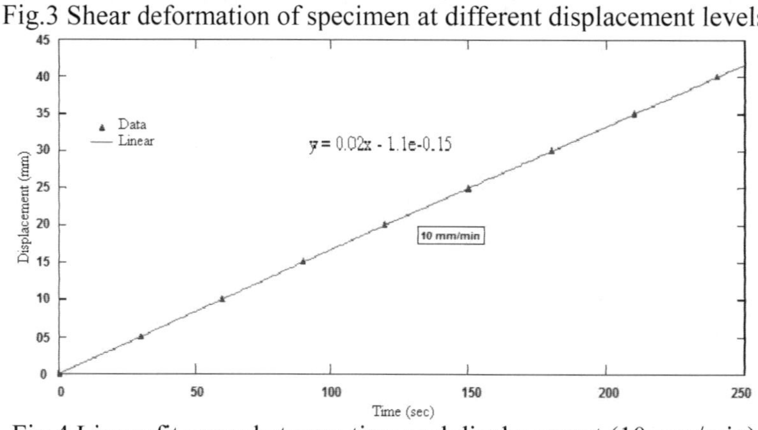

Fig.4 Linear fit curve between time and displacement (10 mm/min)

It was a challenge to maintain uniform rate of loading on the tensile testing device. Therefore, uniform displacement method was selected. In order to ensure that the frictional effects of the clamping device and picture frame do not influence the uniform displacement, the linear fit curve was plotted between time and displacement as shown in Fig.4.

The variation in shear angle mainly depends on the different fabric characteristics in samples like stitch density and thickness. The effect of lycra on in-plane shear behavior of spacer knitted fabric samples is not pronounced. The reason is that, lycra is used only as a spacer yarn with only a small contribution on the surface fabrics. Thus the influence of lycra on inter yarn friction, resisting the angular deformation is almost negligible. The typical load-deformation curves for the six types of 3D spacer fabrics with different stitch density and compositions are shown in Fig.5.

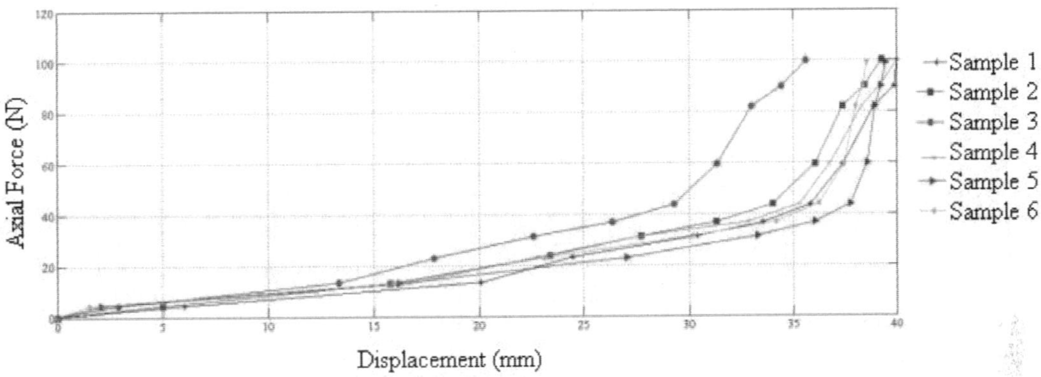

Fig.5 Axial force–displacement curves of specimens

The overall shear stiffness increases with increasing stitch density and the non-linearly for all curves increase in the whole shear process. In order to observe the wrinkling more distinctly, the curves of shear force– shear angles are plotted (Fig.6). There are significant differences between the six curves due to the difference of the fabric density and thickness. The standard deviation and CV% was calculated for all results. Less than 8% CV was obtained in all measurements. The effect of stich density on shear behavior was significant with less than 5% CV. In the case of 3D spacer fabrics, fabric width decreases rapidly during shear mechanism which results the spacer yarns to be compressed in to flat form. Thus the shear deformation is influenced also by the presence of spacer yarns in the z direction. It introduces obstacles to the in-plane yarns while shearing. It is apparent that fabrics using monofilament as spacer yarn generally have higher compression resistance than those using multifilament yarn because the behavior of monofilament yarns is similar to a linear spring when the fabric is compressed. The inherent resilience behavior of multifilament yarns is less than the monofilament. When the results for spacer fabrics using the monofilament and multifilament spacer yarn were compared, it was found that the shear angle of a sample is closely related to the types of spacer yarn as shown in Fig. 6. During the shear process, the gap between the yarns vanishes gradually and the adjacent yarns contact each other, but the width of yarn almost does not decrease. The width of yarn starts to decrease under lateral compression after the limiting locking angle during the large shear deformation, which can offer more space for the fabric to be sheared before wrinkling. The wrinkles occur in the central zone of the sample along the load direction due to the lateral compression from the left and right joints of picture frame. The intra-ply shear deformation of 3D fabrics is still predominated by the deformation mode.

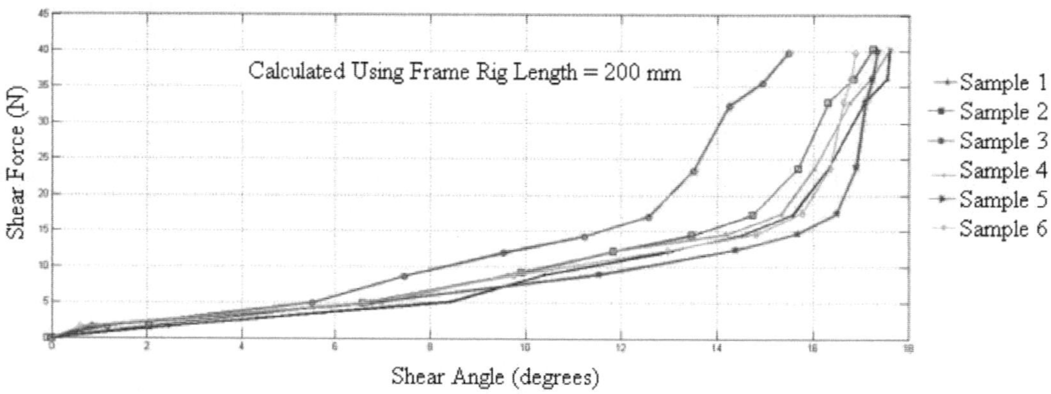

Fig.6 Intra-ply shear force –shear angle curves

3.2 Determination of shear angle using image analysis method

Fig.7 shows the image of a fabric specimen captured during the loading process without displacement. This image was analyzed to obtain shear angle and displacement of the specimen during the test. The image file string is examined and then the appropriate image reading function is called by MATLAB imread. The image is then converted to a gray image and then normalized to a matrix of values ranging from zero to one. This matrix is returned by the function in the variable I. Image analysis can aid in the determination of the shear angle and displacement at any particular point on the surface of fabric specimen. Grid pattern was applied on the specimen surface before the test and used as the reference points of image analysis. Around 16 points (25 square cells ($4cm^2$ each) with 90° angle at 4 points) can be chosen on a 100 mm x 100 mm specimen for image analysis to determine the displacements and shear angles at the chosen points. First, by manually choosing the points on a reference image, the X and Y coordinates of each point can be determined in pixel as shown in Fig. 8. The difference of X and Y coordinates of each chosen point between the reference image and the image chosen for analysis represents the displacements in X and Y directions.

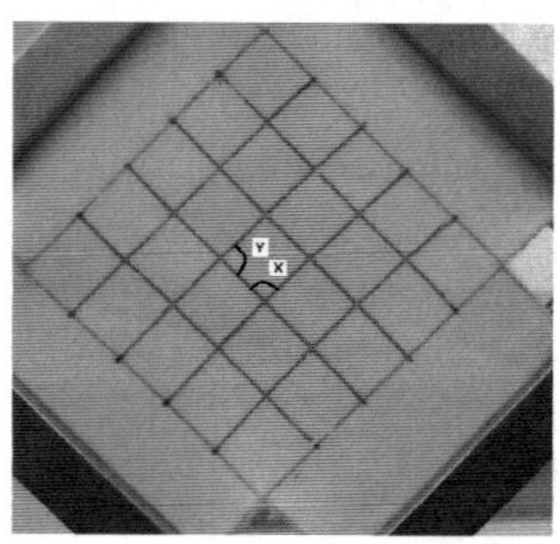

Fig.7 Determination of shear angle using image analysis

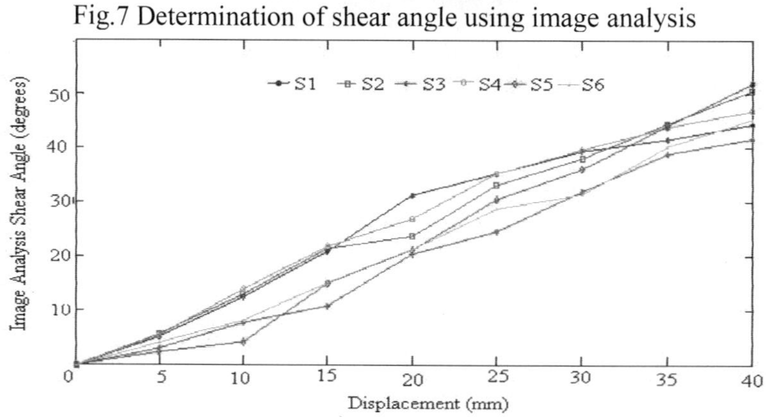

Fig.8 Shear angle using image analysis with respect to displacement

Fig.8 shows the comparison of shear angles between all six samples at different displacement levels using image analysis.

4. Conclusion

The shear is a complex mechanical behavior of fibrous structures. In woven fabrics, the frictional restraint between warp and weft plays the crucial role in deciding the shear rigidity. Whereas in case of knitted fabrics, the interlooping friction and the stitch density decide shear properties. Measurement of shear in knitted fabrics is relatively more challenging due to instability of normal weft knitted fabrics. However, warp knitted and weft knitted 3D spacer fabrics are relatively stable and can be prepared for shear testing. Clamping of bulkier spacer fabrics needs special care not to induce preloading and buckling before shearing action. The in-plane shear properties are partially influenced by the spacer yarns. The type of spacer yarn (monofilament or multifilament) has a strong influence on the the shear properties. It is found that fabrics using monofilament as spacer yarn generally have higher compression resistance than those using multifilament yarn because the behavior of monofilament yarns is similar to a linear spring when the fabric is compressed. The inherent resilience behavior of multifilament yarns is less than the monofilament. The effect of lycra on in-plane shear behavior of spacer knitted fabric samples is not pronounced as lycra is used only as a spacer yarn with only a small contribution on the surface fabrics. Thus the influence of lycra on inter-yarn friction, resisting the angular deformation is almost negligible. Further research needs to be carried out for development in fixture design in order to measure inter-ply shear properties of 3D spacer knitted fabrics.

References

[1] Zhu B, Yu TX, Tao XM. Large deformation and slippage mechanism of plain woven composite in bias extension. Compos Part A- Appl S 2007; 38: 1821–1828.

[2] Willems A, Lomov SV, Verpoest I, Vandepitte D. Picture frame shear tests on woven, textile composite reinforcements with controlled pretension. In: 10th ESAFORM conference on material forming, Parts A and B 2007: 999–1004.

[3] Lin H, Clifford MJ, Long AC, Sherburn M. Finite element modelling of fabric shear. Model Simul Mater Sci 2009; 17: 1–16.

[4] Cao J, Akkerman R, Boisse P, Chen J, Cheng HS, de Graaf EF, et al. Characterization of mechanical behavior of woven fabrics: experimental methods and benchmark results. Compos Part A- Appl S 2008; 39:1037–1053.

[5] Chen QQ, Boisse P, Park CH, Saouab A, Bréard J. Intra/inter-ply shear behaviors of continuous fiber reinforced thermoplastic composites in thermoforming processes. Compos Struct 2011; 93: 1692–1703.

[6] Charmetant A, Orliac JG, Vidal-Sallé E, Boisse P. Hyperelastic model for large deformation analyses of 3D interlock composite preforms. Compos Sci Technol 2012; 72: 1352–1360.

[7] Ye X, Hu H, Feng X. Development of the warp knitted spacer fabrics for cushion applications. J Ind Text 2008; 37, 213-223.

[8] Zhang Y, Sun F, Wang Y, Chen L, Pan N. Study on intra/inter-ply shear deformation of three dimensional woven performs for composite materials. Materials and Design 2013; 49: 151–159.

[9] Willems A, Lomov S, Verpoest I, Vandepitte D. Picture frame shear tests on woven, textile composite reinforcements with controlled pretension. In: 10th Esa form conference on material forming, Parts A and B 2007: 999–1004.

[10] Dash BP, Behera BK, Mishra R, Militky J. Modeling of internal geometry of 3D woven fabrics by computation method for structural composites. J Text I 2013; 104, 312-321.

[11] Mishra R. Impact tolerence of 3D woven nanocomposites: A simulation approach. J Text I 2013; 104: 562-570.

[12] Behera BK, Mishra R. 3-Dimensional weaving. Ind J Fiber Text 2008; 33: 274-287.

Research Progress on the Evaluation and Test Methods of Clothing Pressure Comfort

Ling Chen, Zi-Lan Li, Xiu-E Bai*

College of Textile and Clothing Engineering, Soochow University, Ganjiang East Road No.178, Suzhou, 215021, China

*Corresponding author's email: baixiue@suda.edu.cn

Abstract

In this paper, the formation theory of clothing pressure is firstly analyzed. The principle and characteristic of clothing pressure comfort's evaluation and test methods as well as the processing and analysis method of experimental data are emphatically elaborated. Researches have shown that the evaluation methods of clothing pressure comfort are divided into subjective evaluation method and objective evaluation method, and each has its advantages and disadvantages. Any single evaluation method is unable to fully reflect human pressure comfort. Only in the subjective-objective comprehensive evaluation method can we evaluate clothing pressure comfort fully and scientifically. It provides a scientific research mean or method for the study of clothing pressure comfort.

Keywords: Clothing pressure; Pressure comfort; Evaluation and test; Research progress

1. Introduction

With the improvement of people's living standard, the concept of healthy dressing is gradually taken root in people's hearts. People begin to pay more attention to clothing comfort when in the pursuit of its aesthetic and serviceability. Clothing comfort is determined by three main factors, namely thermal-wet comfort, tactile comfort and pressure comfort [1], and pressure comfort is one of the most important indexes to evaluate clothing comfort [2]. Clothing pressure plays a positive role in shaping body, sport protection, disease treatment and other aspects. However, too much clothing pressure will affect human body's normal movement, and even pose a threat to human health, so the evaluation and test of clothing pressure is an indispensable link for the study of clothing pressure comfort. Researches on the testing technology of clothing pressure and its application help to optimize clothing structure, enhance clothing added value and improve its fitness and health [3]. Clothing pressure test is a starting point for the study of clothing pressure comfort, and only by constantly improving the scientificity and accuracy of clothing pressure test can it provide a basis for the study of clothing pressure comfort.

2. The formation of clothing pressure

The pressure in physics is a kind of force that occurs on the surfaces of two objects in contact. It is due to the extrusion and deformation of two objects in contact with each other. Clothing pressure, as a form of pressure, is also the result of extrusion and deformation of clothing and human body in the process of contact (as shown in Fig.1) [4]. Specifically, when people wear a tight-fitting, clothing tensile effect causes the deformation of material and generates a component force perpendicular to human body surface. This component force together with the pressure caused by gravity form a physical mechanical signal and act on human skin(the physical process), and then the signal is transmitted to human brain through nerve terminals(the physiological process), finally the brain forms a feeling of pressure on the stimulus signals (the psychical process) [5].

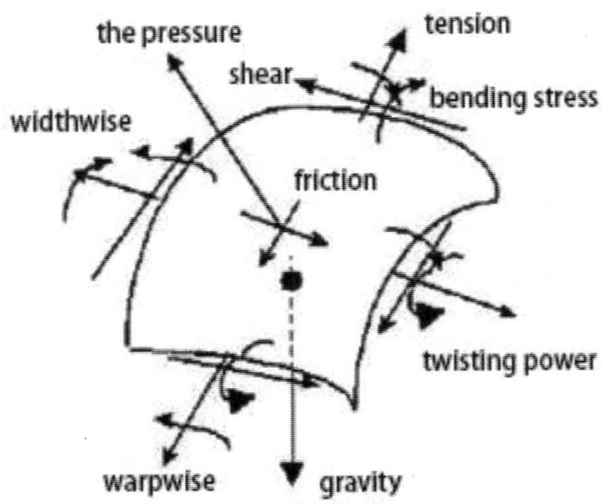

Fig.1 Clothing pressure forms [4]

According to the nature of pressure, clothing pressure can be divided into binding pressure and weight pressure. The pressure caused by the extrusion of clothing is called binding pressure [6]. In the process of dynamic contact between human body and garment, the changes in human movements or postures make clothing stretch, and in return, the clothing produces relative slippage and deformation to adapt to the changes in body shape. Due to the deformation of clothing material, various forms of interior stress are generated, including tension, compression, shear and bending stress, etc. The component force perpendicular to human body surface is the binding pressure. Researches have shown that appropriate binding pressure has a positive effect on human body, but too much binding pressure will pose a threat to human health. Take corsets for example, it uses the appropriate binding pressure to make breast adipose tissue more concentrated, finally achieve the effect of body shaping. But when the pressure is too large, it will produce discomfort for people, such as dizziness, hypoxia, dyspnea, and may even lead to splanchnodiastasis [7]. Weight pressure means a kind of power caused by the gravity of clothing [6]. Because the direction of gravity is always straight down, the horizontal contact point between human body and clothing is most susceptible to the weight pressure. This horizontal contact point is exactly the shoulder, so we can come to a conclusion that the pressure comfort of shoulder has a great relationship with clothing weight pressure. In order to improve the pressure comfort of shoulder, clothing designers should reduce clothing weight as possible as they can while ensuring clothing serviceability and other special features. For instance, in 2016-2017 autumn and winter Munich "Performance Days" fabric exhibition, either 3D structure fabric or Merino wool fabric has a common characteristic-lightness. Clothing made of these fabrics make people move freely without any sense of restraint. There is no doubt that they will be the perfect material for increasingly expanding market of luxury sportswear.

3. Evaluation and test methods of clothing pressure comfort

Clothing pressure test is the measurement of clothing objective pressure value with pressure instrument or the investigation on human subjective pressure comfort through questionnaire survey. Therefore, clothing pressure test can be divided into subjective evaluation method and objective evaluation method, and they are based on subjective pressure comfort and objective pressure value respectively. Moreover, objective evaluation method includes indirect measurement and direct measurement, and their difference lies in whether the clothing pressure value is obtained by directly measuring the human body.

Western countries have begun the research of clothing pressure test since the beginning of the 20th century. Over a century, new test technologies, means and instruments emerge in an endless stream, and the accuracy of clothing pressure test is continuously improved. In recent years, some experts in Asian countries, such as Japan, South Korea, Hong Kong and other countries or regions also conducted a number of studies in this area, especially the study of tight-fitting clothing pressure comfort.

3.1 The subjective evaluation method of clothing pressure comfort

The subjective evaluation method is also known as the sensory evaluation method, which is a way for people to evaluate things according to their own subjective feelings. It is a common way to evaluate clothing pressure in earlier research. The Weber law proposed by Ernst Weber in 1834, the Fechner law proposed by Fechner in 1860 and the Steven exponential law proposed by Steven in 1953 are the three theoretical bases for subjective evaluation [8]. Hollies used the psychological scale to get the feeling of subjects. Psychological scale mainly includes Hollies four-level scale, Hollies five-level scale (as shown in Table 1), Hollies mark sheet on subjective comfort, Fritz semantic scale and other evaluation scales [9]. Later, based on these studies, Hollies developed a wearing test technology to assess the clothing comfort. Specifically, the experimenter divides clothing pressure comfort into five or more levels, and then the subjects evaluate clothing pressure comfort by selecting corresponding level on the scale according to their subjective feelings [10].Now, the technology has been widely adopted and applied to the study of clothing pressure comfort by researchers. There are quite many foreign researches related to clothing subjective pressure comfort, especially in Japan. Some Japanese universities and research institutions perform noticeably in this field. As early as 1991, through a research on the pressure of girdle, Makabe [11] came to a conclusion that the comfortable pressure threshold of girdle was 4 to 5.33 kPa. Afterwards, some domestic researches in this field emerged. In 2002, Dr. You Fang [12] from Sun Yat-sen University evaluated the subjective pressure of different tight pants in the psychological scale method. The study showed that clothing pressure comfort was mainly determined by the feeling of pressure, at the same time it was also related to fabric handle, finally she came to a conclusion that it was a comprehensive sense. Therefore, there are many corresponding sensory indexes to evaluate it, mainly including the binding feeling, the compressive feeling, the dry feel, the prickle sensation, the soft feeling and the heavy feeling, etc.

Table1. Hollies five-level scale

Level of scale	Comfort Evaluation
1	Totally uncomfortable
2	Uncomfortable
3	Comfortable
4	Relatively comfortable
5	Totally comfortable

The feeling of clothing pressure is a complicated physiological and psychological reflection, so any single evaluation index can't fully reflect this complex feeling. At present, the probability theory and mathematical statistics, the fuzzy theory and the gray system theory are widely used in the data analysis of subjective evaluation. The research object of probability theory is that "the large sample is uncertain". The research object of fuzzy theory is that "the cognition is uncertain". The research object of grey theory is that "less data is uncertain" [13]. For pressure comfort of tight-fitting clothing, its subjective evaluation is based on the wearing experiment, which is completely certain. However the feeling of comfort for subjects is a subjective psychological quantity, which belongs to the uncertain concept. It can be determined that part of the information is known, and the other part of the information is unknown in the entire experiment. So, the grey cluster analysis is the reasonable analysis method to classify and sort the subjective comfort of different tight-fitting clothing.

3.2 The objective evaluation method of clothing pressure comfort

The objective evaluation method is a way for people to evaluate things on the basis of the objective data. This method is based on the objective facts, so its reliability is relatively high. According to different test objects, the objective evaluation method can be divided into the indirect measurement and the direct measurement. The indirect measurement is a method to obtain the value of clothing pressure indirectly through empirical formula or human models. Its test object is the mannequin, so firstly it can save a lot of human resources. What's more, the precision of experimental data is relatively high. The direct measurement is a method to obtain clothing pressure value directly with pressure instrument. Its test object is the real human body, so it can reflect human complex physiological and psychological reaction, thus the experimental data have more practical significance. Further, the indirect measurement mainly includes the theoretical calculation method, gypsum method, soft manikin method, etc. The direct measurement mainly includes the hydraulic pressure method, the sensor method and so on.

At present, the most advanced and widely-used test technology in indirect measurement is the soft manikin method. Foreign scholars firstly began the development of soft manikin. Mr. Yamada, the Japanese expert, firstly put forward a way to test the clothing pressure on human model, and made three human models [14]. Wang further improved the intelligent soft manikin system. Eight pressure sensors were embedded into the body surface of soft manikin to monitor and record the size of clothing pressure [15]. In recent years, some domestic clothing colleges also began the work of improving soft manikin. In 2004, Professor Fan Jintu and Qian Xiaoming from Hong Kong Polytechnic University developed a kind of soft manikin. It was made up of a special skeleton, soft tissue, smooth skin and a pair of changeable artificial breasts, and it was the most mature clothing pressure test model at that time [16]. In 2008, Donghua University developed a posture-adjustable soft manikin according to the demand of pressure test in the study of clothing pressure comfort, which remedied the defect that previous soft manikins all were unmovable, and further promoted the development of soft manikin [17].

The sensor method is the main content of the direct measurement method, and it is also the most widely-used test method in the whole garment industry. At present, the United States, Britain and Japan are in the leading position in the development of clothing pressure test equipment. Some related equipment from Japan and US has already been put into use, such as the airbag pressure sensor developed by Japanese AMITECHNO Company, as well as the Flexiforce pressure sensor developed by American Tekscan Corporation. As early as 1998, Toyonori used the elastic fiber to measure the pressure caused by socks on human skin and analyzed the relationship between output voltage and pressure, finally found that their linear correlation was -0.99 [18]. At home, the development of clothing pressure test equipment is relatively backward, but in recent years, some relevant researches have been done with foreign equipment. In 2006, Professor Zhong Anhua used the hydraulic pressure method to measure clothing pressure. The study showed that there was a close relationship between clothing pressure and clothing ease [19]. In 2008, Li Qiaolian used AMI-3037 airbag pressure test system to measure clothing pressure, and drew a conclusion that the comfortable clothing pressure range was less than 1256.89Pa for people [20].

The frequency analysis, the correlation analysis and the regression analysis are widely applied currently for the data analysis of objective evaluation. Frequency analysis refers to grouping all data according to a certain standard and counting up the number of objects in each group. Correlation analysis is a statistical method to study whether there is some kind of dependency relation among phenomena, and to explore concretely the direction and degree of relevance between those phenomena. Regression analysis is a statistical analysis method to determine the quantitative relationship between two or more variables. As to the data analysis for the objective evaluation of tight-fitting clothing pressure comfort, the correlation analysis and regression analysis are generally used to find out the subjective and objective factors closely related to the pressure, and to establish the effective regression equation between those factors and the pressure value.

4. The comprehensive evaluation of clothing pressure comfort

As mentioned above, the subjective evaluation is based on human subjective feelings, reflecting human dominant position. Its experimental data are directly obtained through questionnaire survey, so this method is simple, quick and maneuverable. Compared with the objective evaluation, the subjective evaluation directly reflects human true feelings, so it can solve many problems that can't be solved by objective measurement. However, its shortcomings are also obvious, which mainly displays in three aspects: first, individual differences are distinct. Individual subjective feelings with a strong personal consciousness are influenced by personal life experience, social environment and other factors. Second, the subjective evaluation depends on personal subjective judgment. Individual subjective judgment on objects or things is not the same. As a result, it has a certain unavoidable deviation of subjective evaluation. Third, the experimental process can hardly be repeated. People's subjective judgment is an instant feeling, namely at different times or places their subjective feelings are different, so it lacks in reproducibility.

The objective evaluation is based on objective pressure value and this method has outstanding performance in reliability and scientificity. Its advantages are mainly manifested in three aspects: firstly, the experimental data is reliable. Compared to the subjective evaluation, the outstanding advantage of objective evaluation is its high reliability. The objective evaluation method is based on the objective data, so its reliability is relatively high.

Secondly, it can save a lot of manpower. With the continuous improvement of soft manikin, there is no need to employ experimental subjects in clothing pressure experiment in the future. Thirdly, this method is easy to combine with the computer technology. With the development of intelligent virtual instrument technology, the organic combination of computer software technology with the traditional measuring instrument enables the user to complete the control of the pressure test process and the data processing by operating the computer. However, the biggest drawback of the objective evaluation is that it's completely out of human subjective feelings, so it can't reflect human complex physical and psychological reaction.

Clothing pressure comfort is a comprehensive psychological and physiological reflection. It is affected not only by fiber kinds, yarn structure, fabric texture and other objective physical properties, but also by subjective physiological and psychological factors, including human curvature, human posture and motion, the mechanical characteristics of human skin and subcutaneous soft tissue and human pressure sensitivity. It is a complex problem involving many factors, such as physics, physiology and psychology. Therefore, any single evaluation method for the clothing pressure comfort is incomplete. Only the combination of the objective evaluation and the subjective evaluation, namely the subjective-objective comprehensive evaluation is the most scientific and comprehensive method. As the term suggests, the comprehensive evaluation has all advantage that single evaluation method possesses. Back in 2007, Song Xiaoxia [2] conducted a clothing objective pressure test on sports underwear with different sizes and different fabrics, and at the same time, she also carried out a questionnaire survey about the subjective pressure comfort. The subjective-objective comprehensive evaluation method was adopted to assess experimental underwear. The research showed that there was a linear relationship between the subjective evaluation value and the clothing objective pressure value, thus drawing a relatively scientific and comprehensive conclusion.

5. Conclusion

In general, the subjective-objective comprehensive evaluation is a mainstream method in the study of clothing pressure comfort in present and future. First of all, it has the advantages of subjective evaluation method, which is based on human subjective feeling, reflecting human dominant position. Secondly, it also has the advantages of objective evaluation method, which is based on objective data, and the reliability of the experimental results is relatively high. Through the subjective-objective comprehensive evaluation method, the unity of the experimental data and the real pressure is realized, providing a reliable basis for the study of the clothing pressure comfort. With the continuous development of computer technology, the combination of computer software technology with the traditional measuring instrument improves scientificity and accuracy of the clothing pressure test, and provides a more scientific and comprehensive research means for the study of clothing pressure comfort, finally guide clothing design and production to realize the unity of clothing functionality and clothing pressure comfort.

References

[1] Song XX, Feng XW. Relationship between garment pressure and human body comfortablenesss. Journal of Textile Research: 2006; 3: 103-105.

[2] Song XX. Relationship between the pressure of knitted sportswear and the comfort of human body. Knitting Industries: 2007; 4: 33-37.

[3] Yang XY, Wang GH. The Method of Constructing Garment Pressure Test System Based on LabVIEW. Journal of Zhejiang Textile and Fashion Vocational Technology College: 2013; 4: 32-36.

[4] Wang J. Study of the Influence of Garment Pressure on Human Physiological Comfort. Suzhou: Soochow University: 2008: 8-9.

[5] Meng XL, Zhang WY. Progress of study on pressure comfort of clothing. Journal of Textile Research: 2006; 7: 109-112.

[6] Wang Q, Chen DS, Wei QF. Situation and prospects of the researches on clothing pressure effects on body. Journal of Textile Research: 2009; 4: 139-144.

[7] Wang J, Dai XQ, Liu GL. Introduction to the clothing pressure and its influence on the human body. Foreign Silk: 2007; 5: 20-22.

[8] Duan XY, Yu WD. Research progress in pressure comfort of foundation garment. Journal of Textile Research: 2007; 11: 144-148.

[9] Xu ZY, Song XX. Research Progress in Relationship Between Human Body and Clothing Pressure. Journal of Shanghai University of Engineering Science: 2013; 6: 170-173.
[10] Zhang WY. Clothing comfort and function. Beijing: China textile press: 2011.
[11] Makabe H. A study of clothing pressure developed by the girdle. Jpn Res Assn Tex End Uses: 1991; 9:424-438.
[12] You F. The subjective assessment of pressure comfort for tight-fitness. Journal of Xi' an Institute of Engineering Science and Technology: 2002; 1: 13-16.
[13] Lu LN, Ren CJ, Xu J, et al. Gray cluster analysis on pajamas' wearing comfort subjective assessment. Journal of Xi'an Polytechnic University: 2010; 2: 176-179.
[14] Yue WX, Wang GH. Effect of clothing pressure on human body and testing method. Journal of Xi'an Polytechnic University: 2012; 4: 452-456.
[15] Wang YR, Cui YH, Zhang PH. A smart mannequin system for the pressure performance evaluation of compression garment .Textile Research Journal: 2011; 11:114-123.
[16] Yu W, Fan JT, Qian XM, et al. A soft mannequin for the evaluation of pressure garments on human body. Sen Gakkaishi: 2004; 2: 57- 64.
[17] Zhang WB, Bi YX. Research on the Soft and Posture-adjustable Mannequin for Clothing Pressure Test. Shanghai: Donghua University: 2008.
[18] Toyonori N, Kenichi O, Sadamu S, et al. Comfort pressure evaluation of men's socks using an elastic optical fiber.Textile Res J: 1998; 6:435 -440.
[19] Zhong AH, Zhang Q. A research on the pressing comfortability of elastic women's underwear. Knitting Industries: 2006; 4: 27-28.
[20] Li QL, Xie MD. Relation between cloth pressure and stretch rate of elastic knitted fabric. Progress in Textile: 2008; 5: 97-98.

Influence of Material of Protective Pad Integrated Into Hip Protective Garments on Their Pressure Delivery to the Hip of the Wearer

Wiah Wardiningsih*, Olga Troynikov

RMIT University, Dawson Street, Brunswick, Victoria, 3056, Australia

*Corresponding author's email: wiah.wardiningsih@student.rmit.edu.au

Abstract

This study aimed to investigate the influence of material of hip protective pad integrated into knitted fabric assembly intended for hip protective garments on the pressure delivery by the garment to the body of the wearer. Soft type commercially available hip protective pads and materials intended for hip protective pad were studied. A protective pad was integrated into an experimental knitted fabric sleeve assembly that resembled the hip section of a common hip protective garment. The pressures generated by the assemblies on a cylindrical hip model were determined. The pressure generation properties of the sleeve assemblies were measured using Salzmann pressure-measuring device MST MKV IV. The results were analyzed and the influence of the protective pad on the garment pressure delivery to a cylindrical model, and therefore potentially to the wearer's body, was determined. It was found that experimental pad made of warp knitted spacer fabric exhibited low-pressure delivery value. This was because the pads made of knitted spacer fabrics were flexible and conformed well to the cylinder contour. The protective pad that generates low pressure delivery is generally preferable for the hip protective garments.

Keywords: Protective pad; Pressure delivery; Pressure comfort; Ergonomic comfort

1. Introduction

Due to the increasing life expectancy, falls resulting in facture of hip in elderly people are posing a major health problem worldwide [1,2]. Hip fracture is a common problem in elderly people, occurring in around one in four women and one in eight men [3]. It is an important cause of morbidity and mortality in old age [4]. It is predicted that the number of hip fractures worldwide will increase if not addressed [5]. In 1996, there were approximately 15,200 hip fractures in Australia; by 2051, this number is predicted to be 60,000. Between 2000 and 2051, the number of Australians aged 65 and over will rise from 13% to 23% of the population [6]. Worldwide projections show depressing picture, with the total number of hip fractures forecast to increase from 1.26 million in 1990 to 4.5 million by 2050, assuming no increases in age-specific hip fracture rates [7]. The worldwide annual number of hip fractures could rise to 6.26 million by 2050 [8] .The number of hip fractures is expected to increase most dramatically in Asia, where aging populations are expected to contribute more than two million hip fractures annually by 2050 [9].

The consequences of hip fractures can be severe. One year after hip fracture, the mortality of patients with a hip fracture is about 17–33% [10] and another 25–33% of the patients are severely disabled or cannot walk at all [11]. Furthermore, the costs of hip fractures are very high, especially due to hospital admission and nursing home stay [12].

The use of hip protectors was one of several approaches that have been identified to prevent hip fractures [13]. Hip protective garment were defined as special underwear designed to dissipate kinetic energy from an impact to the soft tissue and muscle, as well as anterior and posterior of the femur [13]. Hip protective garment consists of protective pad attached to a protective garment. Existing hip protectors were developed on the basis of biomechanical research suggesting that passive protection such as hip protectors may prove to be more effective for preventing hip fractures than efforts aimed at slowing the rate of bone loss in older people [14]. Hip protective garment belongs to a category called Protech, which is one of the mainstream technical textiles [15].

The main problem of current hip protectors was wearer's compliance, the percentage of people wearing the hip protectors. Compliance rates for some products fall significantly with time [10]. The factors that may negatively influence the compliance are: hip protector experienced as too tight, too hot, uncomfortable in bed and necessitating assistance in toileting [10], and that the garment is too bulky or uncomfortable and is difficult to put on [16].

To increase the adherence and wearer's compliance to hip protector, its negative determinant has to be addressed. Being too tight is one of several negative determinants that frequently arise, therefore to provide pressure comfort, where the hip protective garment is not perceived as too tight by the wearer, the pad integrated into garment fabric should not add significantly high pressure to the hip section of the wearer.

The protective pad has to be flexible and to conform to the body contour of the wearer to permit his or hers adequate range of motion. Where the pad is considerably rigid it would not conform to the body contour of the wearer and as a result will generate considerable interface pressure to the underlying body of the wearer and would be perceive as too tight.

Generally ergonomic wear comfort or pressure comfort is determined by the fit of the clothing and the resultant freedom of movement it allows. The garment construction and the elasticity of the comprising materials are the primary determinants of ergonomic wear comfort which is defined as the fit of the clothing and the freedom of movement it allows. Garment fit and pressure comfort play an essential role in clothing comfort [17]. Concerning clothing construction, Zhang, Young [18] specified that garment pressure is closely associated with the space allowance between the clothing and the body of the wearer during their body movement. When measurement of a garment girth is smaller than the human body, it has a negative fit with the space allowance between the body and the garment being less than or equal to zero, and as a result the interface pressure is created. Skin is exceptionally sensitive to pressure and under general circumstances, skin displacement of less than 0.001mm can result in a pressure sensation or touch [19]. Denton [20] stated that the level of clothing pressure discomfort was found to be in the range of above 20 to 40 g/cm^2, depending on the individual and the body part concerned.

Pratt and West [21] specified that three aspects affected the pressure generated by garments: 1) Shape of the body parts where the greater degree of curvature, the greater pressure exerted; 2) Fabric type and 3) the garment design, construction and fit, where a poorly constructed garment may lead to the perception of heightened interface pressure. Therefore the interface pressure generated by the garment worn is affected by the shape of the wearer's body, garment construction and comprising fabric mechanical properties.

Present research aimed to determine the influence of materials of the protective pad on the interface pressure delivery. Several soft protective pads that commercially available on the market and soft materials that are intended for impact protection were studied.

2. Materials and methods

Table 1. Specification of experimental pads

No	Pads Code	Thickness (m)	Mass (kg)	Material
1	A	0.01612	0.0493	Polyurethane closed cell foam
2	B	0.01606	0.0412	Polyester knitted spacer fabrics (4 layers)
3	C	0.01888	0.0326	Polypropylene closed cell foam- segmented (2 layers)
4	D	0.01589	0.1115	Polyurethane viscoelastic closed cell foam
5	E	0.01448	0.0235	Polyester knitted spacer fabric -heat set (2 layers)
6	F	0.01274	0.0327	Polyester knitted spacer fabrics (2 layers)

Six materials were selected for this study (Table 1). Three of them were commercial hip protector pads (A, B and C) and the remaining three materials were materials intended for the use for impact protection.

Further, knitted fabric was sewn into tubular piece so that the piece became sleeve-like. Then a protective pad was integrated into a knitted fabric sleeve-like assembly that resembled the hip section of a hip protective garment. The experimental sleeves were then placed onto a rigid cylinder and the interface pressure generated by the sleeves between the cylinder and the sleeve was measured. The experimental rigid cylinder was a 0.5 m in circumference and 0.30 m in length. 0.5 m in circumference was regarded as circumference of thigh or leg section extending from hip (Fig.1) [22]. The length of the cylinder was determined by the length of the sensor array used for the interface pressure measurement.

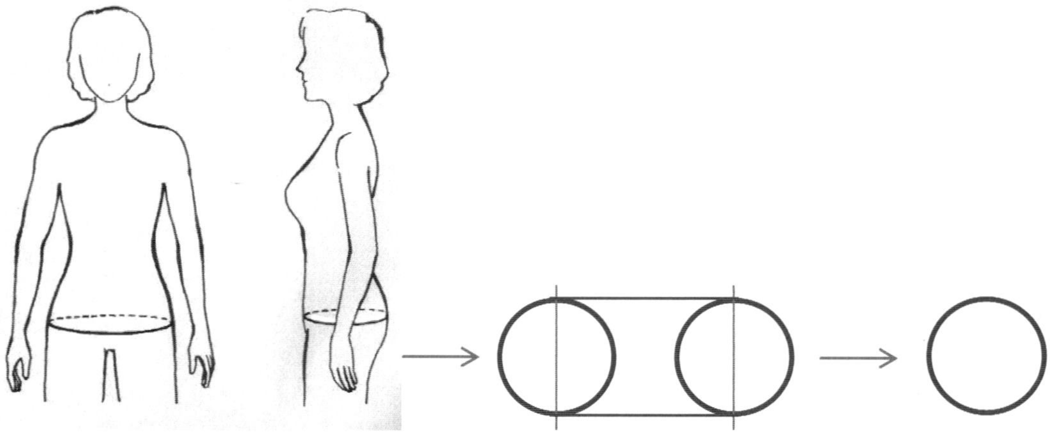

Fig.1 Simplified hip model

The sleeves were made from knitted fabric suitable for a hip protective garment. The fabric size was 0.45 m in length and 0.327 m in width, so that the sleeve provided 10 % practical extension in fabric course direction (in the width direction when worn on the cylinder). The seam allowance of 0.02 m was added. The fabric was sewn into sleeves. The pocket for the protective pad was made from the same fabric and was incorporated into the sleeve by seams. The protective pad was manually inserted into the pocket. All pads and sleeves were conditioned and tested in the standard temperature atmosphere of 20 ± 2°C and 65 ± 2% relative humidity.

The resultant interface pressure generated by pad and sleeves was measured using Salzmann pressure-measuring device MST MK IV and Salzmann MST 2007 software. Salzmann sensor array consisted of a thin plastic probe of 0.04 m wide and 0.005 m thick, with four paired electrical contact points, connected to an air pump and a pressure transducer. The probe was placed between the cylinder and the experimental assembly. The air pump inflated the probe until the contacts opened as a result of the inner pressure exerted by the air being just above the external pressure generated by the experimental assembly. When the contacts open, the transducer read the pressure at each measuring point and the pressure was displayed digitally with 1-mmHg resolution [23].

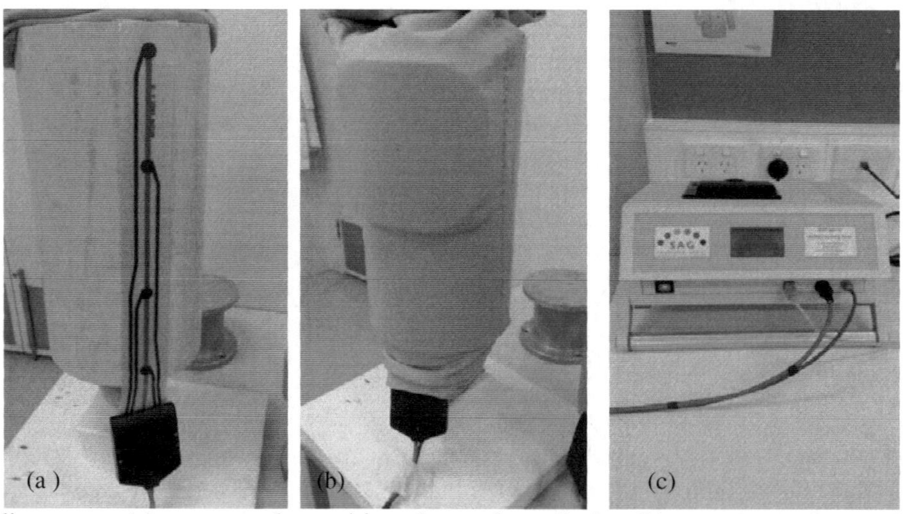

Fig.2 Pressure delivery test: (a) sensor probe positioned over the cylinder; (b) experimental assembly; (c) Salzmann MST MK IV portable device

The pressure measurement on the cylinder was conducted as follows: the cylinder was positioned vertically on its holder without allowance for movement (Fig.2). The sleeve was carefully positioned on the cylinder over the Salzmann sensor and was spread evenly so there were no folds kinks, or air gaps between the cylinder, the sensors and the fabric. The results were read on the MST portable device. The measurement of pressure induced by the sleeve on the cylinder was repeated three times. The sleeve was removed and re-positioned again before the next measurement was taken and the mean values calculated.

The experimental pads used in this study had different thickness, so that it was necessary to normalize the measured resultant pressure delivery.

$$Pn = (Pp-Ps) / T \qquad (1)$$

Where Pn : Pressure delivery of pad normalized by thickness, units; Ps : Pressure delivery of sleeve, units; Pp : Pressure delivery of pad inserted into sleeve pocket, units; T : Pad thickness, units.

To determine the significance of the difference between the pressures generated by experimental pads, statistical analysis was carried out where one way analysis of variance (ANOVA) test was conducted to determine the equality variance. The null hypothesis (H_0) for the test was that all population means (level means) were the same. The alternative hypothesis was that one or more population means differ from the others.

3. Results and discussion

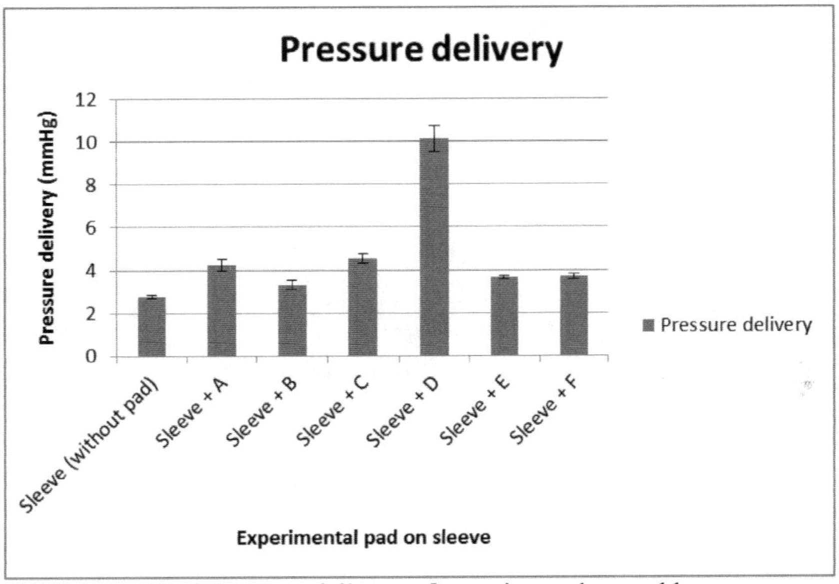

Fig.3 Pressure delivery of experimental assembly

The addition of experimental pad inserted into the fabric sleeve generated higher interface pressure delivery than the sleeve without a pad (Fig.3). The pressure delivery by the fabric sleeve without the pad and sleeve with inserted pad generated significantly different pressure delivery value (p value < 0.05): therefore protective pad inserted into the fabric sleeve influenced the resultant pressure delivery value.

Sleeve with inserted experimental pad D generated the highest pressure delivery value among all of experimental pads. When hip protective garment with experimental pad D is worn, it is valid to expect that this hip protective garment would generate high-pressure delivery, and would be perceived as being too tight and uncomfortable to the wearer. Other experimental pads generated low pressure delivery. The pressures generated by experimental pads A, B, C, E, and F were almost similar.

After the pressure delivery was normalized (Fig.4), it could be seen that experimental pad D generated the highest pressure delivery among all the experimental pads. Pad A and C had similar pressure delivery value. Despite Pads A, C and D being all made of closed cell foam, produced a different pressure delivery values. This was probably because pad D had the highest mass: the pressure was influenced by the mass of the pad as could be seen from (2):

$$P=F/A=(m.g)/A \quad (2)$$

Where P: pressure, unit; F: normal force, unit, A:the area of the surface of contact, unit, m:mass of what, unit and g:gravity, unit.

(2) implies that P was equal to m, and therefore if the pad had high mass, the generated interface pressure would be also high. As a result protective pads with the high mass are to be avoided. In addition to this it would be reasonable to expect that high mass of protective pad would also negatively influence the decision of the wearer to wear the hip protective garment.

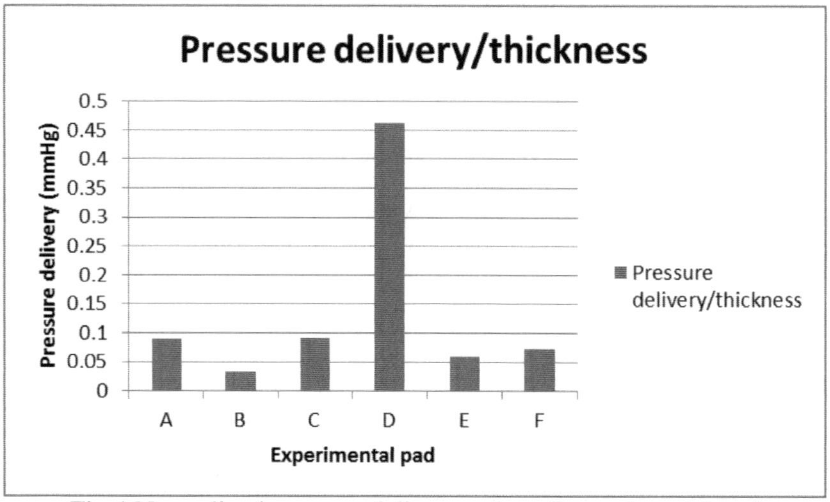

Fig.4 Normalized pressure delivery value of experimental pad

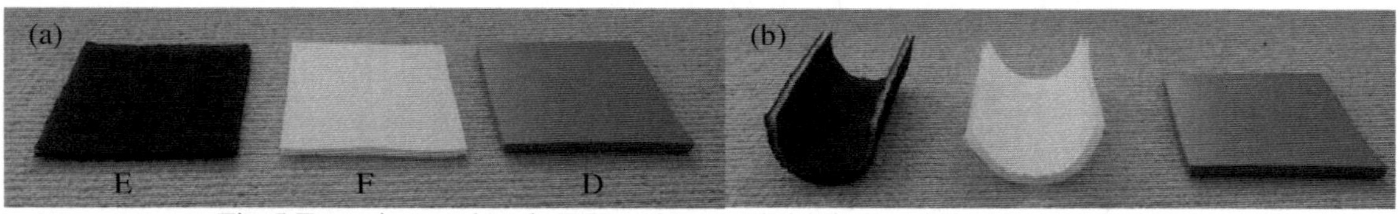

Fig.5 Experimental pads E,F and D: (a) initial shape; (b) shape after being bent

Higher pressure delivery value generated by pad D also probably due to the pad D being rigid and not flexible, in comparison to other pads (E, F) although pad D was made of "soft" foam. When pad D was bent, it did not assume the curved shape as did pads E and F, but remained flat due to its rigidity (Fig.5). When pad D was inserted into a pocket of the sleeve and was positioned over the cylinder, Pad D did not adjust to the cylinder contour.

Experimental pads made of knitted spacer fabric (B, E and F) exhibited low-pressure delivery values, as compared to soft foam pads A and C. This was because the pads made of knitted spacer fabrics were flexible (Fig.5) and conformed well to the cylinder contour. The protective pads that generate low interface pressure delivery are preferable for hip protective pads due to the wearer's comfort considerations.

4. Conclusion

Incorporation of a protective pad into the garment produced different outcomes in terms of the resultant interface pressure. The resultant pressure delivery revealed a dependence on the comprising materials of the pads. The addition of experimental pad inserted into the fabric sleeve significantly influenced the pressure delivery value of the pad-containing assembly: protective pad inserted into the sleeve generated significantly higher pressure delivery value than the sleeve alone. When hip protective garment with high-pressure delivery pad would be worn, it would be reasonable to expect that it would generate high-pressure delivery on the wearer body and would be perceived as too tight and uncomfortable by them. High mass of protective pad tends to produce high-pressure delivery. The protective pad has to be flexible so that it could conform to the body contour. Pad that generated low-pressure delivery would be preferable for the use in hip protective garments. Knitted spacer fabric could be used as material alternative to the foam for protective pad as it is flexible and generated low pressure delivery value. It is hoped that compliance of the wearer to wear hip protective garment will increase if the garment was not too tight, heavy and comfortable to wear.

These results are of significance for design and engineering of hip protective garments. The designer should consider that the incorporation of non-flexible pads into a garment will influence pressure delivery generated by them, so that the design of the garment will have to be adjusted accordingly.

References

[1] Minns R, et al. Are hip protectors correctly positioned in use?. Age and ageing: 2007; 36(2); p. 140-144.
[2] Spierings A, Derler S. Assessment of hip protectors and corresponding hip fracture risk using stress calculation in the femoral neck. Medical engineering & physics: 2006; 28(6); p. 550-559.
[3] Parker MJ, Gillespie WJ, Gillespie LD. Hip protectors for preventing hip fractures in older people. Cochrane Database Syst Rev: 2005; 3.
[4] Villar MTA, et al. Will elderly rest home residents wear hip protectors?. Age and ageing: 1998; 27(2); p. 195-198.
[5] Walker A. Active ageing in employment: Its meaning and potential. Asia-Pacific Review; 2006; 13(1); p. 78-93.
[6] Sanders KM, et al. Health burden of hip and other fractures in Australia beyond 2000. Projections based on the Geelong Osteoporosis Study. The Medical Journal of Australia: 1999; 170(10); p. 467-470.
[7] Gill KS. Correcting For Non-Adherence In A Randomized Study Of Hip Protectors To Prevent Fractures. in Department of Epidemiology of the School of Public Health: 2010; University of North Carolina.
[8] Kannus P, et al. Epidemiology of hip fractures. Bone: 1996; 18(1); p. S57-S63.
[9] Gullberg B, Johnell O, Kanis J. World-wide projections for hip fracture. Osteoporosis international; 1997; 7(5); p. 407-413.
[10] van Schoor NM, et al. Acceptance and compliance with external hip protectors: a systematic review of the literature. Osteoporosis International: 2002; 13(12); p. 917-924.
[11] Magaziner J, et al. Recovery from hip fracture in eight areas of function. The Journals of Gerontology Series A: Biological Sciences and Medical Sciences; 2000; 55(9); p. M498-M507.
[12] De Laet C, et al. Incremental cost of medical care after hip fracture and first vertebral fracture: the Rotterdam study. Osteoporosis international: 1999; 10(1); p. 66-72.
[13] Dubey A, Koval K, Zuckerman J. Hip fracture prevention: a review. American journal of orthopedics (Belle Mead, NJ): 1998; 27(6); p. 407-412.
[14] Lotz JC, Hayes WC. The use of quantitative computed tomography to estimate risk of fracture of the hip from falls. J Bone Joint Surg Am: 1990; 72(5); p. 689-700.
[15] Higgins L, Anand S. Textile materials and products for activewear and sportswear. Technical Textile Market: 2003; 52; p. 9-40.
[16] O'Halloran PD, et al. Factors affecting adherence to use of hip protectors amongst residents of nursing homes-a correlation study. International journal of nursing studies: 2007; 44(5); p. 672-686.
[17] Wong AS, Li Y, Zhang X. Influence of fabric mechanical property on clothing dynamic pressure distribution and pressure comfort on tight-fit sportswear: 2004; 60(10); p. 293-299.
[18] Zhang X, Yeung K, Li Y, Numerical simulation of 3D dynamic garment pressure. Textile Research Journal: 2002; 72(3); p. 245-252.
[19] Schiffman HR. The skin, body and chemical senses. Sensation and perception: 1995; p. 70-96.
[20] Denton M. Fit, stretch and comfort. Textiles: 1972; 3; p. 12-17.
[21] Pratt J, West G. Pressure garments: a manual on their design and fabrication. 2014: Butterworth-Heinemann.
[22] Heitmann BL, Frederiksen P. Thigh circumference and risk of heart disease and premature death: prospective cohort study. Bmj: 2009; 339; b3292.
[23] Flaud P, Bassez S, Counord JL. Comparative in vitro study of three interface pressure sensors used to evaluate medical compression hosiery. Dermatologic Surgery: 2010; 36(12); p. 1930-1940.

Characterizing Tensile Strength of Flame Resistant Fabrics after Multiple Radiation Exposures

Ye-Hu Lu[1,2,3*], Shu-Min Jiang[2], Qiang Gao[3*]

[1]*Laboratory for Smart and Protective Clothing, National Engineering Laboratory for Modern Silk, Soochow University, Suzhou, 215123, China*
[2]*College of Textile and Clothing Engineering, Soochow University, Suzhou, 215021, China*
[3]*Key Laboratory of Science and Technology of Eco-Textiles, Ministry of Education, Jiangnan University, Wuxi, 214122, China*

[*]Corresponding author's email: yhlu@suda.edu.cn, gaoqiang@jiangnan.edu.cn

Abstract

In this study, the tensile strengths of three typical aramid fabrics after exposure to multiple radiations were investigated. Nine scenarios with different exposure steps and durations were selected. Radiant heat flux of 21 kW/m^2 was simulated by a radiant protective performance (RPP) tester. The results indicated that the radiation exposure decreased the fabric tensile strength. Both the exposure step and individual duration affected the tensile strength. The more exposure steps with the same individual duration caused lower tensile strength. Under condition of the same cumulative exposure duration, the longer individual exposure duration resulted in lower tensile strength. It can be concluded that the effect of radiation exposure type on the tensile strength of firefighters' protective clothing should be considered during firefighters' protective ensemble selection and maintenance.

Keywords: Firefighters' protective clothing; Radiation exposure; Tensile strength; Thermal degradation

1. Introduction

During fire extinguishing and emergency rescue, firefighters will encounter multiple hazards, including thermal, chemical, biological, radiation and physical hazards. Firefighters' protective clothing (FPC) is essential to protect their health and safety [1]. NFPA 1971 Standard (2007) provides numerous performance requirements for a new piece of protective ensemble. During the lifetime, the performance of thermal protective clothing, especially the outer layer fabric, will change after exposure to these hazards. Therefore it was important to know how environmental conditions affect the fabric performance over time. NFPA 1851 Standard (2008) gives many indicators to identify the deterioration, but fails in providing quantitation of degradation.

The effect of thermal aging on the degradation of firefighters' protective clothing has been widely investigated. The intensity of incident heat flux or temperature, duration of exposure, and frequency of exposure affected the thermal decomposition [2,3]. Day et al [4]. investigated the impact of high temperature exposure on the mechanical and thermal protective performance of fireproof fabrics. Although all outer shell specimens showed significant reduction in tear strength, the flame-resistant properties did not change considerably. The changes in structure and properties of single layer aramid fabrics depended on both temperature and cumulative duration of exposure [5-7]. Davis et al [8]. found that the mechanical properties of the polyaramid and polybenzimidazole fabrics decreased dramatically due to ultraviolet radiation. Cui et al [9]. found that the intensity of heat flux had significant effect on the tensile strength and tear strength of X-fiper® fabric, however, the thermal protective performance did not change considerably. The performance of fabric combinations of firefighters' protective clothing was also evaluated after radiant exposures. Thorpe [10] evaluated the performance of in-use firefighters' protective clothing under exposure to continuous 5 to 30 kW/m^2 radiant heat and demonstrated that both the radiation heat flux and exposure time affected the performance. Rossi et al [11]. investigated the change of thermal protection and mechanical properties provided by fabrics combination under exposure to 40 kW/m^2 radiant heat or 80 kW/m^2 convective heat. The thermal protection for the second exposure decreased, especially under higher heat fluxes. The tensile strength and tear strength of the outer layer were measured after exposure for the duration equal to the discoloration time. The results showed three types of behaviors for the tested specimens.

With regards to actual wear and use, FPC will encounter repeated exposures of thermal hazards. However, previous studies mainly focused on the continuous exposures for different durations, there were few studies considering the effect of multi-stage exposures on fabric performance. In this study, the impact of multi-step periodical radiant exposure on the degradation of FPC was investigated. The research finding will provide technical suggestions on the selection and maintenance of FPC.

2. Methodology

2.1 Materials

Three typical flame resistant fabrics A, B and C used as the outer layer of FPC were selected. Fabric D as the moisture barrier and fabric E as the thermal liner layer were also used to compose different three-layer fabric assemblies, i.e., A+D+E, B+D+E and C+D+E. The physical properties are listed in Table 1.

Table 1. Physical properties of the tested fabrics

Sample	Fabric content	Weave structure	Weight (g/m^2)	Thickness (mm)
Outer layer A	50% m-aramid and 50% flame retardant viscose	Twill	204.5	0.5
Outer layer B	98% m-aramid and 2% p-aramid	Twill	193.7	0.49
Outer layer C	93% m-aramid, 5% p-aramid, 2% antistatic fibre	Twill	208.9	0.67
Moisture barrier D	100% m-aramid fleece laminated with PTFE membrane	Laminated	108.3	0.85
Thermal liner E	100% m-aramid fleece with woven m-aramid facecloth	Needle punched nonwoven	200.0	0.72

2.2 Test apparatus

2.2.1 Radiant protective performance tester

To simulate the multiple radiation exposures during daily firefighting activities, a standard radiant protective performance (RPP) tester (Precision Products LLC, Richmond, VA, USA) was used. The thermal radiation is produced by several quartz tubes and the radiation intensity can be adjusted. A shutter plate is installed to avoid preheating of the test sample. A specimen holder plate and sensor holder plate are located in front of the shutter plate. When the test begins, the shutter plate will be removed pneumatically, and the thermal radiation will reach the surface of the test sample. The sensor behind the fabric sample will record the temperature rise and heat flux profile during the test. The protective performance (H_{12}: time for 12 °C temperature rise and H_{24}: time for 24 °C temperature rise) and absorbed energy can be calculated.

2.2.2 Tensile strength tester

Before and after thermal exposure, tensile strength of the outer layer fabric was measured using an Instron 3365 test machine (Instron Corporation, Canton, MA, USA) in accordance to the ASTM D 5035 strip method. The space distance is 100 mm and the tensile speed is 50 mm/min.

2.3 Test protocol

To accelerate the thermal ageing of flame resistant fabric, the radiation intensity of 0.5±0.015 cal/cm²·s (21±0.6 kW/m²), i.e., an emergency condition was employed. The exposure time was determined by the protective performances of the selected outer layer fabrics. The H_{12} and H_{24} of the single layer fabrics were shown in Table 2. To ensure that there was pain sensation but with no 2nd degree burn during radiation exposure, minimal individual exposure duration was set to 15s. Considering the impact of exposure step and duration, there were nine exposure scenarios in total, as shown in Table 3. The tensile strength of fabric before and after exposure was measured respectively.

Table 2. Protective performance of the tested fabrics (s)

Fabric code	H12	H24
A	13.8	21.6
B	14.2	23.4
C	14.7	22.1

Table 3. Details of test scenarios

Scenario	Exposure step	Each duration (s)
E1	0	0
E2	1	15
E3	2	15
E4	3	15
E5	4	15
E6	1	30
E7	2	30
E8	3	30
E9	1	60

3. Results

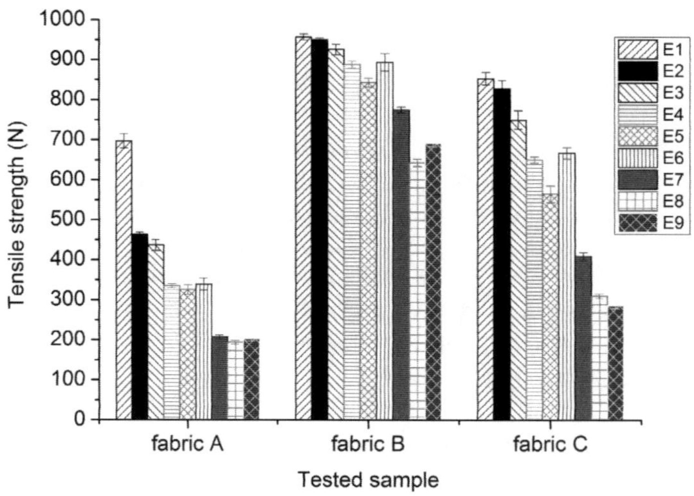

Fig.1 Tensile strength of fabrics in different scenarios

The tensile strength of flame resistant fabrics under multiple radiation exposure is presented in Fig.1. For all scenarios, the fabric B had the highest tensile strength, and the fabric A had the lowest tensile strength. It was clear that the radiation exposure decreased the fabric tensile strength to a certain extent. All three fabrics showed similar change trend but with different decrements. The tensile strength of fabric A decreased from 697.0 N to 194.2 N after E8, the tensile strength of fabric B changed from 957.5 N to 643.4 N after E8, and the tensile strength of fabric C dropped from 853.6 N to 288.1 N after E9. The maximal decrement of fabric A, B and C was 72.2%, 32.9% and 66.3% respectively. In all radiation exposure scenarios, the tensile strength of fabric A was much lower than 464 N, whereas the tensile strength of fabric B was higher than 643 N.

The impact of exposure step with individual exposure duration of 15 s on tensile strength is shown in Fig.2. The increase of exposure step decreased the tensile strength for all three fabrics. The tensile strength of fabric B and C decreased linearly, and the slope for fabric C was about three times than that for fabric B. The tensile strength of fabric A decreased exponentially, showing small decrement after the third exposure step. The effect of exposure step with individual exposure duration of 30 s on tensile strength is demonstrated in Fig.3. Generally, the change trend was similar to that with 15 s individual exposure duration, but showing higher change slope. It indicated that, for the same exposure step, the individual exposure duration also affected the tensile strength.

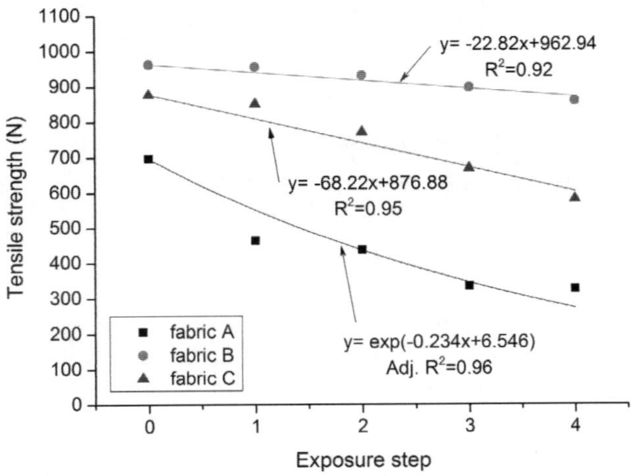

Fig.2 The impact of exposure step (exposure duration: 15 s) on tensile strength.

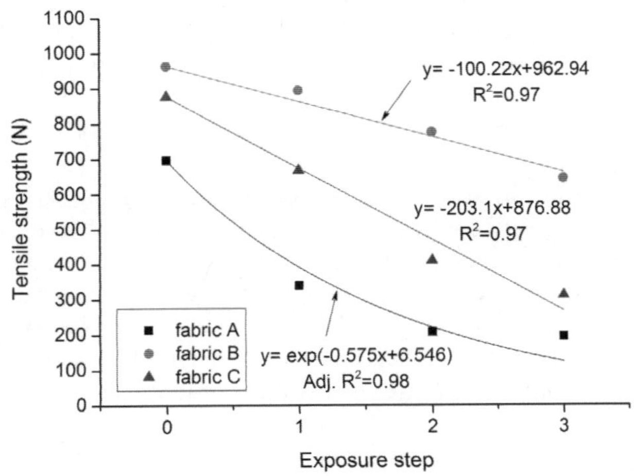

Fig.3 The effect of exposure step (exposure duration: 30 s) on tensile strength.

Fig.4 shows the effect of individual exposure duration on the tensile strength. It was observed that the tensile strength decreased with the increasing of individual exposure duration while keeping the same cumulative exposure duration (i.e., 60 s). When the individual exposure duration was 15 s, i.e., the fabrics were exposed in four steps, the tensile strength of fabric A was 46.8% of its original value, while it decreased to 29.1% after 60 s continuous exposure. For fabric B, the tensile strength decreased from 88.1% to 72.3%. For fabric C, the tensile strength changed from 66.2% to 33.8%. In general, the tensile strength of all fabrics decreased linearly with the increasing of individual exposure duration.

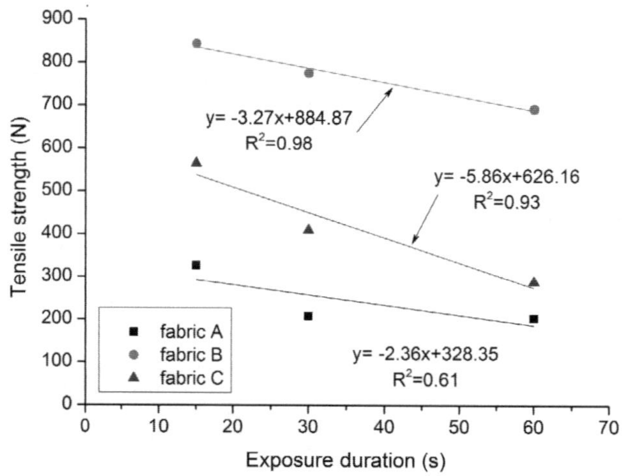

Fig.4 The effect of individual exposure duration on tensile strength.

4. Discussion

In this study, the tensile strength after multiple radiation exposures was investigated. It was observed that the radiation exposure decreased the tensile strength. The change rate varied in different fabrics, e.g., after three–step 30 s or continuous 60 s radiant exposure, the maximal decrement of fabric A, B and C was 72.2%, 32.9% and 66.3% respectively. In all radiation exposure scenarios, the tensile strength of fabric A was much lower than the minimum value (623 N) required in NPFA 1971, whereas the tensile strength of fabric B was higher than the required minimum value. For fabric C, in scenario E5, E7~E9, the tensile strength failed to meet the requirement. The decrease of tensile strength due to radiant exposure was in good agreement with previous studies [4,8,9,11]. It should be noted that the tested fabrics were composed of aramid fibers blend with other fibers, hence the characteristic peaks mainly included the stretching vibration of N–H bond, the stretching vibration of Amide I, of Amide II, the aromatic C–N stretching vibration [9,12]. The fabric temperature rose with exposure time, when it was higher than the glass temperature (T_g), the molecular chain obtained enough energy and movement space, the Amide I (amide C=O) and II (the N–H deformation and C–N stretching coupled modes) might be damaged, and thus reduced the tensile strength.

Secondly, our observation also demonstrated that the increasing of radiation exposure step decreased the tensile strength of aramid fabrics. The increase of exposure step enlarged the cumulative exposure duration, resulting in severer reduction in tensile strength. For fabric B and C, a linear empirical equation could be obtained, whereas a nonlinear correlation could be regressed for fabric A. Moreover, the tensile strengths of fabrics exposed to different exposure steps with individual exposure duration of 30 s were smaller than those with individual exposure duration of 15 s, which further confirmed previous finding that the tensile strength decreased with exposure time [5-7].

Furthermore, while keeping the same cumulative exposure duration, the tensile strength declined with the increase of individual exposure duration. Moreover, the tensile strength of all fabrics decreased linearly with the increasing of individual exposure duration. Those findings were consistent with Iyer and Vijayan's study [3]. They identified a new parameter, i.e., delta t (T), to describe the thermally induced effects on Kevlar 49 fibres. It was indicated that the individual exposure duration greatly affected the thermal degradation of Kevlar 49 fibres under exposures in different temperatures.

Finally, from application perspective, FPC will encounter various thermal conditions, and the tensile strength of outer layer fabric gradually decreases with exposure time. The current study demonstrated both the exposure step and individual exposure duration affected the tensile strength. The empirical equations for predicting the tensile strength in different exposure type should be established in the future study. A database of typical flame resistant fabrics used in engineering FPC should also be established to predict the lifetime of FPC during daily wear and give helpful suggestions on the maintenance and wear.

5. Conclusion

The tensile strengths of aramid fabrics under multiple radiant exposures were investigated in this study. It was confirmed that the fabric tensile strength decreased to a certain extent under the radiation exposure. Both the exposure step and individual duration determined the tensile strength. Under condition of the same individual duration, the more exposure step, i.e., longer cumulative exposure duration led to lower tensile strength. With regards to the same cumulative exposure duration, the longer individual exposure duration, i.e., less exposure step, caused lower tensile strength. The research findings provided technical bases for the selection and maintenance of garments that would encounter multiple radiant exposures during actual wear and application.

6. Acknowledgement

This study was financially supported by the Open Project Program of Key Laboratory of Eco-textiles, Ministry of Education, Jiangnan University (No. KLET1402), the Natural Science Research Project for Colleges and Universities in Jiangsu Province (project no. 14KJB540001), and Jiangsu Postdoctoral Science Foundation (1402108C).

References

[1] Song GW, Lu YH. Flame resistant textiles for structural and proximity fire fighting. In: Kilinc-Balci FS, editor. Handbook of fire resistant textiles. Cambridge: Woodhead Publishing Limited, 2013; P: 520-548.

[2] Rezazadeh M, Torvi DA. Assessment of factors affecting the continuing performance of firefighters' protective clothing: A literature review. Fire Technol 2011; 47(3): 565-599.

[3] Iyer RV, Vijayan K. Identification of a new parameter, delta t (T), controlling the thermally induced effects on Kevlar 49 fibres. Current Science 1998; 75(9): 946-951.

[4] Day M, Cooney JD, Suprunchuk T. Durability of firefighter's protective clothing to heat and light. Text Res J 1988; 58: 141-147.

[5] Iyer RV, Vijayan K. Decomposition behavior of Kevlar 49 fibres: Part I. At $T \approx T_d$. Bull Mater Sci 1999; 22: 1013-1023.

[6] Iyer RV, Sudhakar A, Vijayan K. Decomposition behavior of Kevlar 49 fibres: Part II. At T values $< T_d$. High Perform Polym 2006; 18: 495-517.

[7] Jain A, Vijayan K. Thermally induced structural changes in Nomex fibres. Bull Mater Sci 2002; 25: 341–346.

[8] Davis R, Chin J, Lin Chiao-Chi, PetitS. 2010. Accelerated weathering of polyaramid and polybenzimidazole firefighter protective clothing fabrics. Polym. Degrad. Stab. 95(5) : 1642-1654.

[9] Cui Z, Ma C, Lv N. Effects of heat treatment on the mechanical and thermal performance of fabric used in firefighter protective clothing. Fibers Text East Eur 2015; 23(2): 74-78.

[10] Thorpe P. Assessment of in-use firefighter's protective clothing. Dissertation : University of Saskatchewan, 2004.

[11] Rossi RM, Bolli W, Stampfli R. Performance of firefighter's protective clothing after heat exposure. Int J Occup Saf Ergon 2008; 14: 55-60.

[12] Villar-Rodil S, Paredes JI, Martínez-Alonso A, Tascón JMD. Atomic force microscopy and infrared spectroscopy studies of the thermal degradation of nomex aramid fibers. Chem Mater 2001; 13(11): 4297-4304.

Traction Pillow Model Design Based on Pressure Mapping and Material Analysis

Kasia Lee, Rong Zheng*, Yong-li Xing, Ying Wang

Aimer Human Engineering Research Centre, Beijing Institute of Fashion Technology, Beijing, 100029, China

*Corresponding author's email: yuki.zheng@gmail.com

Abstract

The purpose of this study was to design a pillow model that is comfortable and provides traction function. First, we analysed 3D shape attributes of the necks of 30 female college students. We then conducted a comparative study of five pillows purchased on the market. We investigated effects of pressure distribution and material selection on pillows' comfort level. Twenty female subjects evaluated the five pillows on 10 indices in a subjective questionnaire. We undertook pressure mapping tests of the head and neck on the five pillows. Consequently we were able to determine optimal pressure and material characteristics that correlated to greater subjective comfort. We found the functional relationship between pressure and material hardness and deformation to be Y=0.301P+29.807/H-1.495. Using Rapidform XOR 3D virtual modelling software, we created a 3D model of the traction pillow design.

Keywords: Traction pillow; Cervical pain; Pressure mapping; Material deformation; 3D modelling

1. Introduction

Cervical traction pillows were developed for traction treatment in the medical domain and have become increasingly popular in household healthcare products in recent years. It is believed that traction helps relieve muscle spasms and pain symptoms, increase intervertebral space and intervertebral foramen, and realign herniated discs and the fibrous ring tissue in the cervical intervertebral discs. In addition, it provides relief to compressed root nerves, increases blood circulation, improves joint function, and helps to restore general function of the spine [1].

From 1992 to 1995, Shan held a series of clinical trials comparing the effects between a traction pillow he developed and one available on the market [2]. He discovered that the pillow he devised had a greater success in treating musculoskeletal disorders, and suggested that patients with musculoskeletal disorders make use of traction pillows as an extension of traction treatment in the clinic, which may prevent flare-ups of past conditions and the onset of future disorders.

However, an international standard for ergonomic pillow design and evaluation has yet to be developed. While some studies focused on pillows' physical properties such as height, shape, type or composition, others took a more medical approach by evaluating the effectiveness of sleep quality or pain levels of the subjects [3-5]. It is difficult to compare the results of different studies as they assess different products based on disparate evaluation methods. Test duration and sleeping position are also important factors that might affect results directly or indirectly.

Moreover, most studies stop at proving the effectiveness of cervical pillows based on certain variables, such as pain, sleep quality, or comfort. Studies comparing pillows' physical properties often end with a generalised comment on the importance of a pillow with good shape, height, comfort, and firm support [3,5,6]. However, Perrson [6] also pointed out the difficulty in creating a one-design-fits-all pillow. He posits that individuals must test different pillows to get the best unloaded sleep position based on their respective head and neck configurations. A specially selected pillow can be an important part of the physiotherapy in patients with non-specific neck pain.

Poor lifestyle habits, such as incorrect posture and lack of exercise, have contributed to an increased incidence of non-specific neck pain and related musculoskeletal disorders in young adults [7,8]. Zhou et al. discovered that 41.02% of 78 university students spend more than 5 hours each day using electronic products [9]. For this study we decided to limit the target audience to asymptomatic female college students between the ages of 18-24 and investigated the students' 3D neck shape attributes. Concurrently, we obtained subjective and objective evaluation of five pillows purchased from the market, and derived the preferred material, pressure, and pillow dimensions from the experiment. Combining the above data, we explored and developed the model of an ergonomic traction pillow design using the Rapidform XOR (recently renamed as Geomagic Design X) 3D virtual modelling software.

2. Methods

2.1 Subjects

The study recruited 20 female students from online advertisements. Individuals with a history of dementia, neoplasm, or known infection in the cervical region were excluded. The subjects' heights ranged from 147.20cm-171.00cm with an average height of 159.99cm. The weight range was 44kg-66.5kg with an average weight of 51.82kg.

2.2 Pillows

Traction pillows can be subdivided into two main groups, namely material (memory foam, gel, or latex) and shape (flat, rectangular, convex, concave, S-shaped). The five pillows selected for this study represent different brands, materials, and shapes. All five pillows have favourable sales figures and positive reviews. Two pillows are under the Tempur brand, two are from Sinomax, and one is from De Rucci. The pillow brand names are: 1) TEMPUR® Original, 2) TEMPUR® Millennium, 3) De Rucci Latex, 4) Sinomax Icy Diamond, and 5) Sinomax 3D Space Tech. Illustrations of the pillows are shown in Figs.1 to 5. All units of measurement represent centimetres (cm).

TEMPUR is a patented technology in pillow design that contours to body shape in order to support the head and neck. The Millennium model is higher on the sides and concave in the centre, suitable for back sleepers who require traction while asleep. Both are made of memory foam. Tempur products are sized from S to XL. For this study we selected size S as it is more suitable for the average Chinese female.

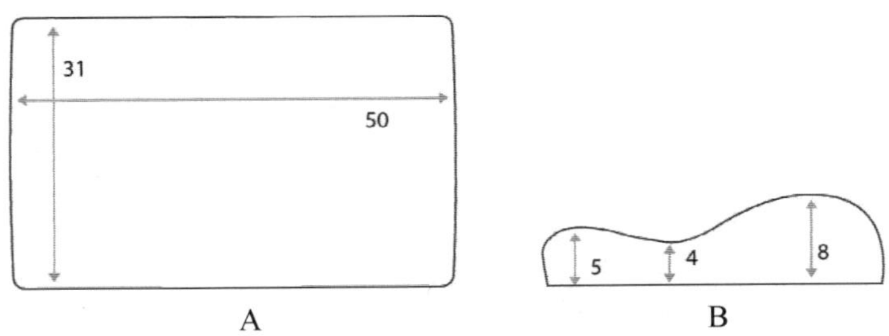

Fig.1 Top view (A) and lateral view (B) of TEMPUR® Original pillow

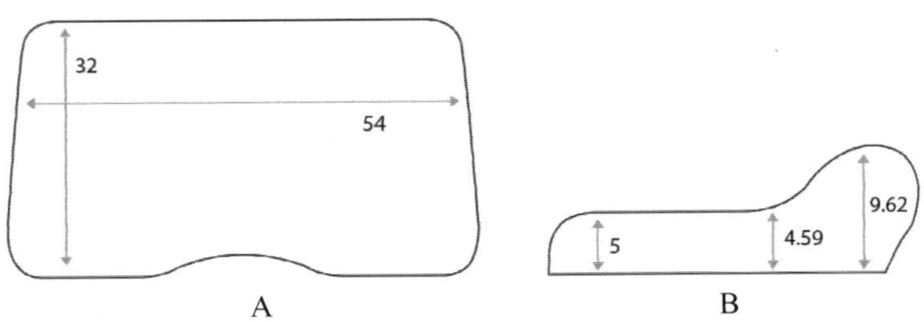

Fig.2 Top view (A) and lateral view (B) of TEMPUR® Millennium pillow

While made of different materials, the De Rucci Latex and Sinomax Icy Diamond pillows are similar in size. They have the same height and are both classified as high pillows.

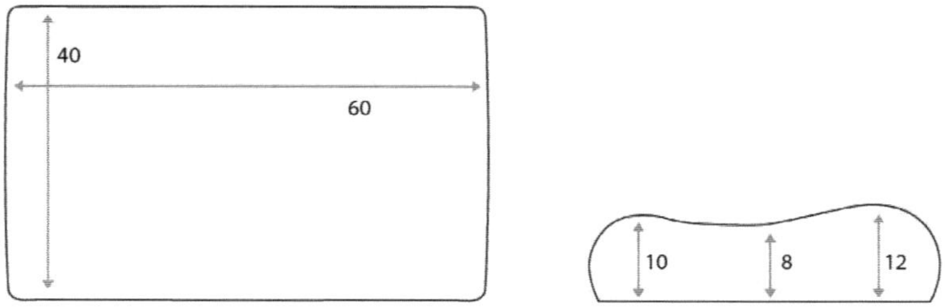

A B

Fig.3 Top view (A) and lateral view (B) of De Rucci Latex pillow

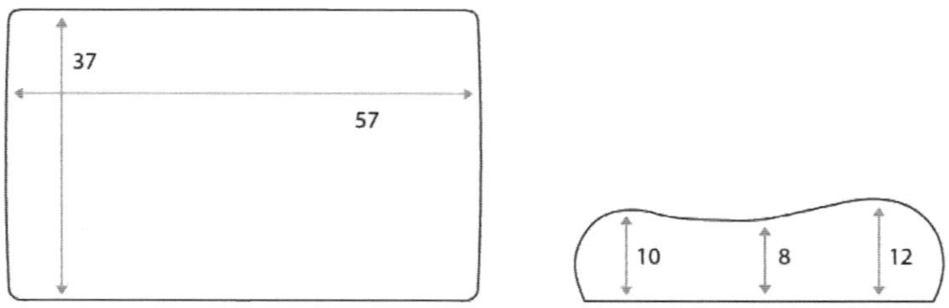

A B

Fig.4 Top view (A) and lateral view (B) of Sinomax Icy Diamond pillow

The Sinomax 3D Space Tech pillow is available in two sizes only, S and L. The small size is more suitable for women, with the following dimensions: 54cm length x 32cm width x 10/5cm height. The 3D Space Tech pillow is similar in shape to the Tempur Millennium pillow.

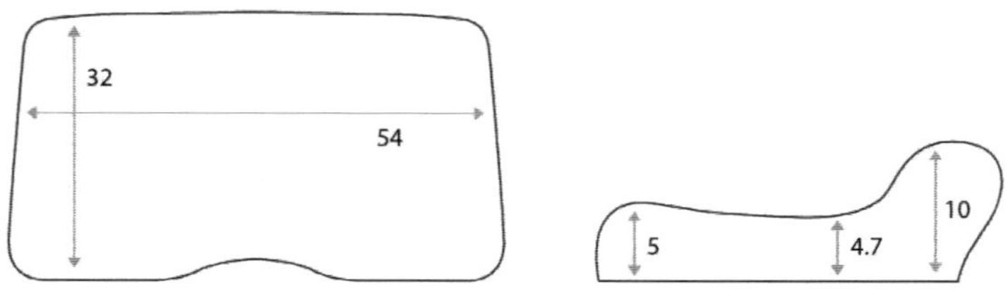

A B

Fig.5 Top view (A) and lateral view (B) of Sinomax 3D Space Tech pillow

During the experiment, labels and trademarks were removed and the pillows were wrapped in a white cotton pillowcase. The pillowcase of each pillow was marked with a code number A, B, C, D or, E.

2.3 Procedures

This study sought to design an ergonomic pillow that satisfies the human need for comfort and the function of traction. We thus designed an experiment that is divided into two sections: subjective evaluation and objective evaluation. In the subjective part of the evaluation, subjects responded to a satisfaction survey that encompassed basic personal information, general neck health, and subjective evaluation of pillows; the second part was the pressure test - an objective observation of the pressure value(s) of the subject's head and neck on the pillow after the body has stabilised in the supine position.

2.3.1 Satisfaction questionnaire

The 20 subjects were instructed to wear a standard test set of a long-sleeved shirt, long pants, and a test cap.

Subjects with long hair were asked to tie up and bun their hair to the top of the head so as not to interfere with the pressure measurement of the head and neck. Subjects were instructed to lie on the pillow in a supine position, in a natural relaxed state. The position of the head was in the middle of the pillow, as illustrated in Fig.6. Each pillow was tested for 20 minutes. Subjects rated the pillows when they had just lain down on the pillow, and at intervals of 10 and 20 minutes. The pillows were rated on 10 aspects, including pillow height, head and neck support, stiffness of the head and neck, feeling of stuffiness, and smell. After completion with one pillow the subject continued with another pillow until all five were completed. The test sequence for the different pillows was randomised.

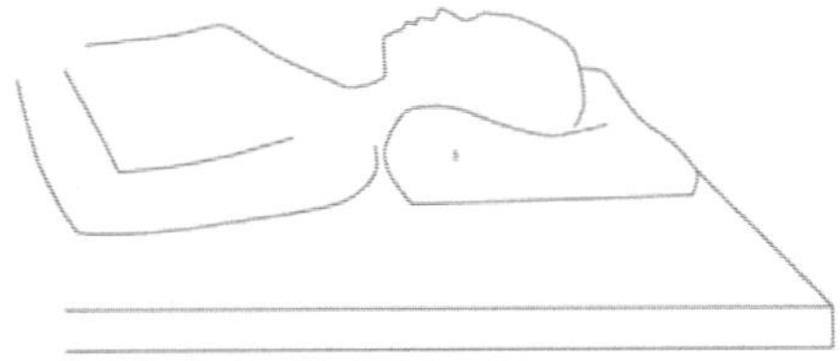

Fig.6 Position of the head and neck during the experiment

2.3.2 Pressure mapping experiment

Following the first part of the experiment, subjects were instructed to lie down in a supine position and breathe normally. We then proceeded with the pressure test using a pressure sensing mat, model no. UT1026-6030 manufactured by Hanweisikang (Fig.7). Data were recorded using the M-flex software (Fig.8).

Fig.7 Pressure sensing mat

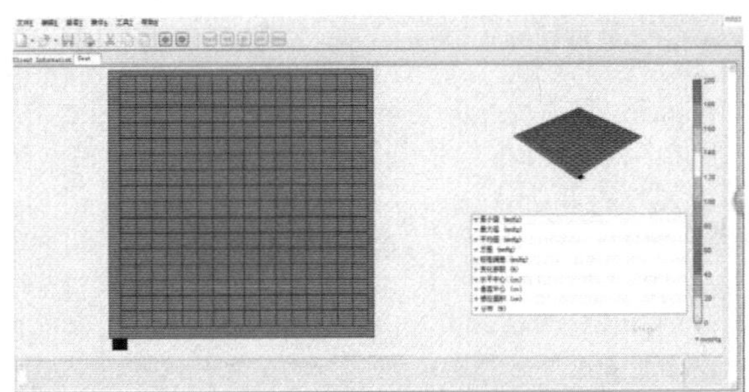

Fig.8 M-flex software

Subjects were instructed to lie on the pressure sensing mat, placed directly on the test pillow. The order of the test pillows was randomised. To ensure that the subjects maintained the same position in all of the tests, they were requested to lie in the centre of the pillow with their legs stretched straight out. Data were collected after 10 minutes of stabilisation and acquired in 10 seconds.

2.3.3 Material analysis

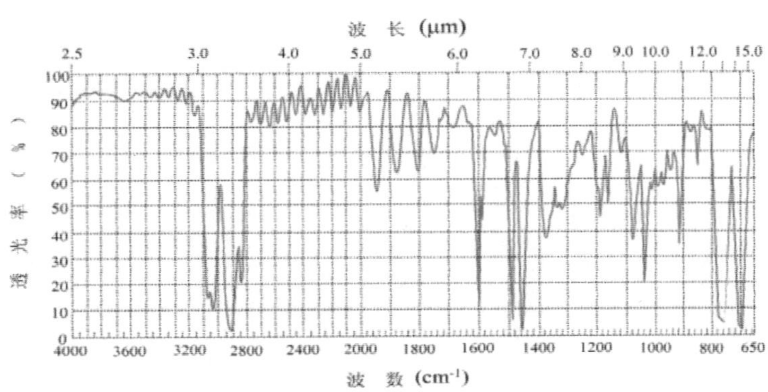

Fig.9 Infrared spectroscopy table

In pillow design, material selection is one of the most important factors to consider as it directly affects the comfort and function of the pillow. This study examined the polymer structure, hardness, and textile deformation characteristics of the different materials of the five pillows. Using the infrared spectroscopy method, the absorption of different wavelengths of infrared light of the material is observed. The structure of organic compounds was identified and inferred by comparing it to the standard table as shown in Fig.9 [10].

Material hardness was tested using a durometer - Teclock model FO GS-744G, manufactured by Teclock. Material deformation was tested according to the Chinese standard GB/T6669-2008, Flexible cellular polymeric materials - Determination of compression set [11]. The experiment was conducted at the material testing facility of Beijing Research Institute of Chemical Industries.

3. Results

3.1 Questionnaire results

To gain a better understanding of musculoskeletal disorders in the college student context, our questionnaires included the international standard Neck Disability Index (NDI). Among the 20 subjects, 50% of them sleep in a back sleeping position, 8 of them sleep on the right side of the body, and the remaining 2 sleep on the left side of the body. All 20 subjects prefer pillows of low (<7cm) and medium heights (7cm-12cm), with 10 subjects in each category, while none prefer high pillows (>12cm).

The subjects were instructed to rate the pillows on 10 indices - head height, neck height, head support, neck support, neck lean-back perception, neck stiffness, head stiffness, breathability, smell, and overall comfort - at three intervals: first when the subject has just started laying on the pillow, then 10 minutes later, and again 2 minutes later. The questionnaire questions are in the form of a scale from scores 1 to 9, 5 being the most suitable and the highest rating, and 1 and 9 being the lowest rating. The scores on the scale were adjusted for a real scale rating, with 5 adjusted to a score of 9 points, 4 and 6 into 7 points, 3 and 7 into 5 points, 2 and 8 into 3 points, and 1 and 9 into a score of 1. After adjustment of the values, a score system was derived. The higher the score, the better the results; the lower the score, the worse the evaluation.

First, the average ratings of the 10 indices provided by all 20 subjects for the same pillow were tabulated. Table 1 represents the scores of the different indices based on the first survey done just when the subject has started lying on the pillow. Values marked in red are the highest for the index across all five pillows, and values marked in green are the lowest.

Tables 1- 3 show the ratings of the different pillows at three different stages of the experiment - when the subject has just started lying on the pillow, 10 minutes later, and 20 minutes later. We can see that pillows A (TEMPUR® Original), B (TEMPUR® Millennium), and D (Sinomax Icy Diamond) rated consistently higher than pillows C (De Rucci Latex) and E (Sinomax 3D Space Tech). Table 4 shows the mean ratings over the three stages. Pillow A rated highest for 2 indices (head stiffness, overall comfort); pillow B rated highest for 4 indices (head height, neck height, neck support, neck stiffness); pillow C rated highest for 2 indices (neck lean-back perception, breathability). Pillow C rated lowest for 7 indices while pillow E had the highest rating for head support and the lowest rating for neck stiffness.

Table 1. Rating of pillow based on 10 indices - Just started laying on pillow

Just on pillow	A	B	C	D	E
Head height	7.81	7.90	6.00	6.60	7.40
Neck height	7.10	8.00	6.50	7.10	6.90
Head support	8.00	7.70	6.80	7.40	8.00
Neck support	7.21	8.05	6.37	7.74	7.42
Neck lean-back perception	8.00	7.56	8.67	8.89	8.16
Neck stiffness	7.78	8.38	7.22	8.44	7.95
Head stiffness	8.37	7.63	7.53	7.95	7.53
Breathability	6.60	5.90	5.70	7.30	6.40
Smell	9.00	8.80	9.00	9.00	8.90
Overall comfort	6.00	5.90	4.50	5.70	5.30

Table 2. Rating of pillow based on 10 indices - 10 mins

10 mins	A	B	C	D	E
Head height	7.50	7.90	6.20	7.10	7.50
Neck height	7.50	7.60	6.00	7.20	7.00
Head support	7.20	7.40	7.20	7.50	7.70
Neck support	6.68	8.11	6.11	8.00	7.32
Neck lean-back perception	8.11	7.56	8.56	8.60	8.44
Neck stiffness	7.89	7.67	7.44	8.00	7.00
Head stiffness	7.63	7.32	7.74	7.60	7.53
Breathability	4.90	5.32	4.30	5.30	4.30
Smell	9.00	8.80	9.00	8.90	9.00
Overall comfort	5.70	5.50	4.20	5.50	5.00

Table 3. Rating of pillow based on 10 indices - 20 mins

20 mins	A	B	C	D	E
Head height	7.90	7.90	6.30	7.00	7.60
Neck height	7.40	7.42	5.70	7.40	7.30
Head support	7.10	7.60	7.00	7.90	7.90
Neck support	7.40	6.79	6.47	7.21	6.79
Neck lean-back perception	8.10	7.33	8.33	8.44	8.44
Neck stiffness	8.10	8.21	7.22	7.39	7.33
Head stiffness	7.70	7.93	7.84	7.74	7.42
Breathability	4.40	4.30	4.10	4.70	4.20
Smell	8.90	8.80	9.00	9.00	9.00
Overall comfort	5.30	5.50	4.30	5.40	5.00

Table 4. Mean values of three test stages

Mean value (Just on pillow, 10mins, 20mins)	A	B	C	D	E
Head height	7.74	7.90	6.17	6.90	7.50
Neck height	7.33	7.67	6.07	7.23	7.07
Head support	7.43	7.57	7.00	7.60	7.87
Neck support	7.10	7.65	6.32	7.65	7.18
Neck lean-back perception	8.07	7.48	8.52	8.64	8.35
Neck stiffness	7.92	8.09	7.30	7.94	7.43
Head stiffness	7.90	7.63	7.70	7.76	7.49
Breathability	5.30	5.17	4.70	5.77	4.97
Smell	8.97	8.80	9.00	8.97	8.97
Overall comfort	5.67	5.63	4.33	5.53	5.10

Table 5. Ranking of 5 pillows

	Pillow				
	B	A	D	C	E
Lowest-rated indices in three test stages	4	2	2	1	1
Highest-rated indices in three test stages	12	8	9	3	3

Consolidating the results, we ranked the pillows, as shown in Table 5. Pillow B ranked first, followed by pillows A, D, C, and E.

3.2 Pressure mapping results

Results from this experiment are the measurements of the pressure of the head and neck on the five different pillows for every subject. We then processed the data and used the colouring method to arrive at the pressure distribution map of the human head and neck. In addition, the surface area of the pressure distribution was derived from the software.

Pressure measurements were taken after the subjects were stabilized in the supine position. The pressure distribution map is displayed on M-flex software (images on the left in Fig.10). The raw data were then imported into Excel and coloured according to the colouring rules in Fig.10. Here we arrived at the pressure measurement value distribution map (images on the right in Fig.10). Different colours represent different pressure intensity: red represents higher pressure values; blue represents lower pressure values. It is observed that pressure values are highest at the centre of the back of the head and decrease as we move outwards. To better differentiate the head and neck area, the researcher pressed down on the pressure sensing mat where the head joins the neck. The blue frame indicates the head area and the green frame indicates the neck area.

In general, we discovered that the distribution of pressure value is greater in the head than in the neck. The average pressure value is also higher for the head than the neck.

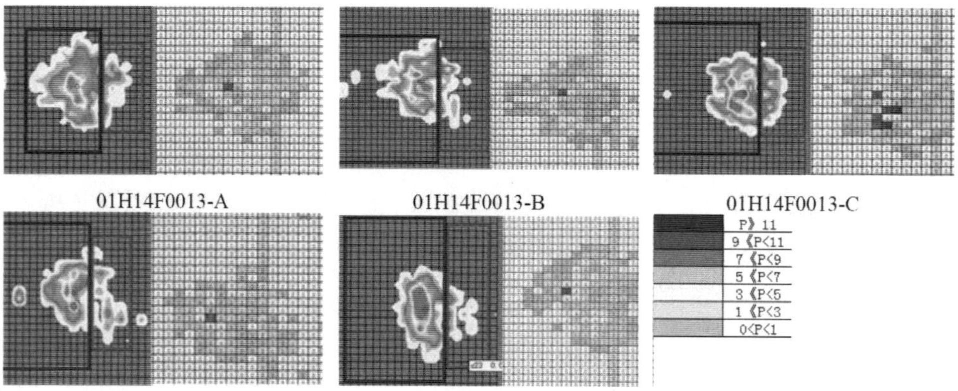

Fig.10 Pressure distribution maps of the five pillows

Table 6. Surface area (SA) where head-neck comes into contact with pillows & average pressure intensity

Pillow	Head SA [cm]	Neck SA [cm^2]	Head pressure intensity [kPa]	Neck pressure intensity [kPa]
A	125.19	39.27	1.95	0.91
B	130.25	43.34	2.06	1.14
C	123.82	72.53	2.66	1.49
D	104.37	50.56	2.29	1.06
E	131.81	46.17	2.04	1.22

Using M-flex pressure analysis software, we were able to determine the location of the area(s) where pressure is detected, and their corresponding surface areas. Through calculating the average surface area and pressure intensity of all 20 subjects for all five pillows, we arrived at the values shown in Table 6. From here we can see that the surface area where the head comes into contact with the pillow is far greater than that of the neck. The respective areas of contact for pillows A, B, C, and E are 125.19cm^2, 130.25cm^2, 123.82cm^2, and 131.81cm^2 - relatively close measurements. In comparison, the area of contact for pillow D is much smaller at 104.37cm^2. The surface area where the neck comes into contact with the pillow is the highest at 72.53cm^2. The pressure intensities for all five pillows are relatively close.

Based on the subjective and objective evaluation of the five pillows, we know that pillows A, B, and D have the highest ratings. To further understand the pressure properties of these three pillows, we performed differential analysis on their pressure values. We deduced that there is no significant difference between pillows A, B, and D.

We also observed that pillows receiving favourable ratings (A, B, and E) have similar pressure measurements. We took this into consideration in the design of the ergonomic pillow. We calculated and fitted the averages of the pressure values for the three pillows, and the results are shown in Fig.11.

As the human body is a symmetrical object, we selected the opisthion as the basal point, and used only data pertaining to the right side of the body. The measurements were averaged and adjusted for balance. First, pressure values for each of the three pillows for all 20 subjects were fitted and averaged. Then the averaged values of the three pillows were combined and fitted to arrive at the final head-neck pressure value of the three pillows. As shown in Fig.11, we can tell that the pressure value is highest at the centre, and decreases moving outwards. The highest value is highlighted in red -8.8kPa.

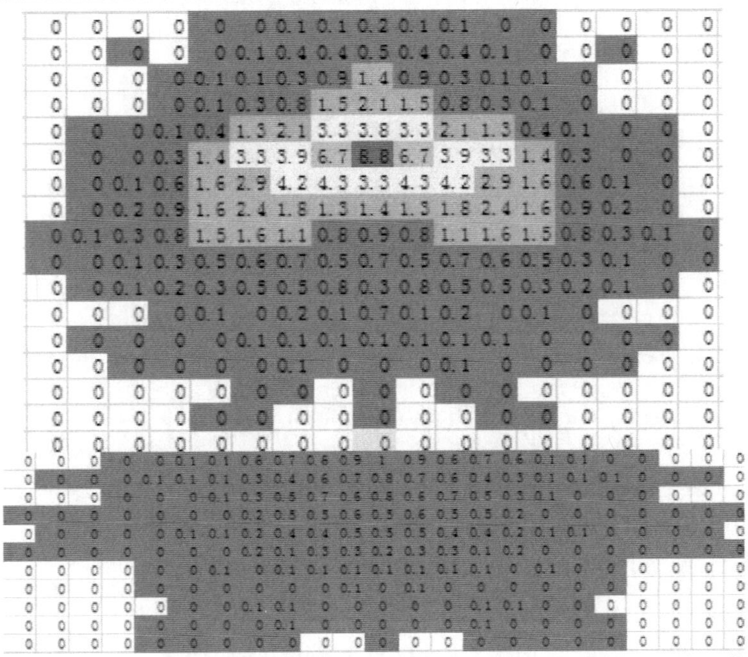

Fig.11 Distribution map of head-neck pressure values on pillow after fitting

3.3 Material test analysis

3.3.1 Infrared spectroscopy

Infrared spectroscopy revealed that the main composition of the five pillows is a polyurethane polymer instead of ether or other polymers.

3.3.2 Material hardness

We also compared the hardness of the materials of the five pillows. The greater the value, the harder or stiffer the material is; the smaller the value, the softer the material is. The material used for pillow C is the softest at 5.1, and is therefore not suitable for use in the design of an ergonomic pillow due to its lack of support. Pillow E uses the hardest material of all five pillows and this translates to a decrease in comfort compared to the other pillows. The material hardness results corroborrate the subjective and objective evaluations of the pillows. The hardness of materials used for pillows A, B, and D are 7.2, 8.6, and 9.5, respectively - values that are relatively close to one another, neither too soft nor too hard. We averaged these three values and obtained the new value 8.43, which we will use later in the design of the ergonomic pillow.

3.3.3 Material deformation

Using a durometer, we measured the deformation values of the materials of the five pillows under varying degrees of force. In Fig.12, the X-axis represents the pressure applied on the pillows in values of 1kPa, 3kPa, 5kPa, and 9kPa, respectively, while the Y-axis represents the corresponding numerical values of deformation. From the curve of the graph, we can see that the deformation value increases in relative to the pressure value. The graph indicates a steep incline from 1kPa and 3kPa for all five materials, and a more gradual incline after 3kPa. The curves for pillows A, B, and D are relatively close compared to those for C and E.

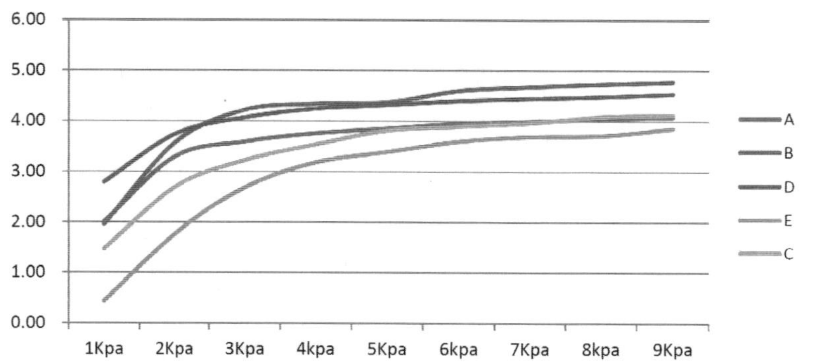

Fig.12 Relationship between pressure value and deformation value

3.3.4 Functional relationship between material deformation and pressure, and material hardness

Based on the findings from the subjective and objective evaluations of the five pillows, combined with the results of the material deformation and hardness tests, this study aimed to explore the relationship between the material deformation, pressure value, and hardness for pillows A, B, and D. Using the linear regression function in SPSS software, through the stepwise method and introducing variables, we arrived at the coefficient $R^2=0.717$, $p<0.05$.

The equation is as follows:

$$Y=0.301P+29.807/H-1.495 \quad (R^2=0.717)$$

Y represents the deformation value

P represents pressure value

H represents hardness of fabric

From the equation above we can tell that the deformation value is directly proportionate to the pressure value - i.e., the greater the pressure value, the greater the deformation value. It is, however, inversely proportionate to material hardness - i.e., the harder the fabric, the smaller the deformation value.

4. Design of pillow model

4.1 Modelling of head and neck shape and pillow

Previously, we had collected the 3D head and neck shape scan data, selected the relevant data, and developed a 3D model of the human head-neck shape. We then arrived at the shape of the traction pillow under compression of the head and neck. Combined with the functional relationship between material deformation, pressure, and hardness, we were able to develop a 3D mold of a traction pillow design with the material properties of pillows A, B, and D.

Using Rapidform XOR's Morphing function, we developed a head and neck model based on the collected results as mentioned above. We used anatomical landmarks on the upper body as reference points for the model (Fig.13). Blue lines indicate the contour lines along the mid-sagittal plane, and the head circumference, top neck circumference, base of neck circumference, shoulder point (SP1) contour, and axillary contour on the transverse plane. A total of 41 reference points were used, shown as red dots in Fig.13. They are distributed along the contour lines and include the vertex, glabella, pogonion, menton, front neck, top side neck, base side neck, shoulder, underarm, opisthion, and back neck. Midpoints of the contour lines on the sagittal plane were also taken (Fig.13C).

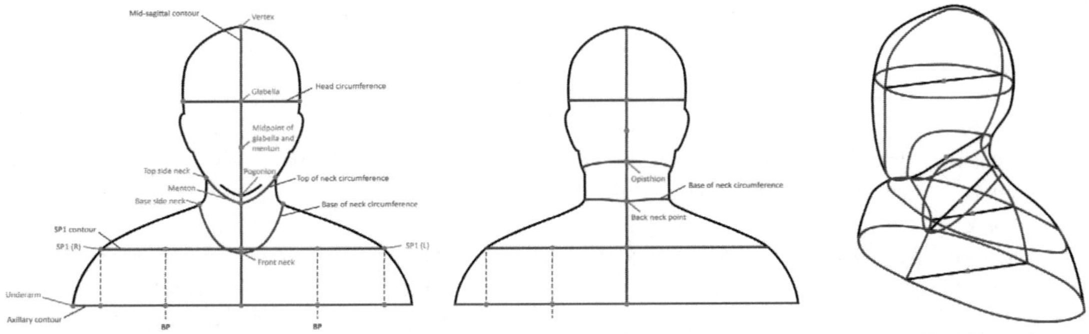

(A) Front (B) Back (C) 45° angle

Fig.13 Reference points for the Morphing model in different views

This study used forward modelling and Rapidform XOR software to create a 3D model of a pillow under the compression of the human head and neck. According to research by Cai and Chen [12], the width of the pillow is determined by the shoulder width, with adjustment allowances. The desirable pillow height is determined by the distance between the shoulder point and the point at the top of the head with extra adjustment allowances. Here, we set the pillow width at 40cm and and pillow length at 32cm.

4.2 Using deformation, presssure

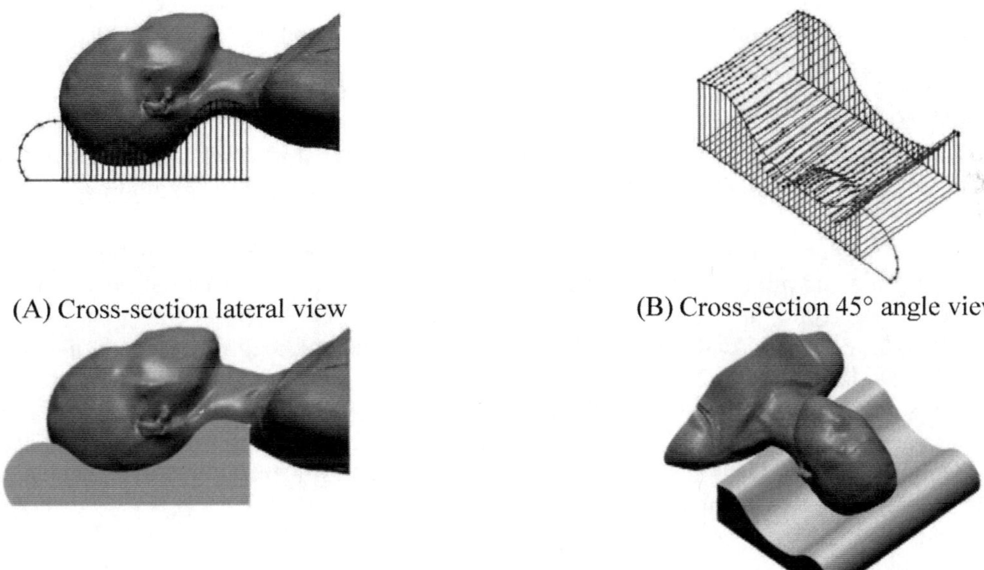

(A) Cross-section lateral view (B) Cross-section 45° angle view

(C) Full model lateral view (D) Full model 45° angle view

Fig.14 Stages of traction pillow design

From previous experiments we defined pillows A (TEMPUR® Original), B (TEMPUR® Millennium), and D (Sinomax Icy Diamond) as the highest rated and the basis on which we will design the new ergonomic form. Previously we calculated that the ideal material hardness is 8.43; the functional relationship between material deformation, pressure, and hardness is Y=0.301P+29.807/H-1.495 (R^2=0.717). Here we bring the head-neck pressure value and material hardness into the equation to arrive at the deformation value of the materials of the three pillows.

Based on the 3D post-traction head-neck morphology data, we were able to create a model of the traction pillow design (Fig.14). Adjustments were made to various parts of the head-neck area according to deformation values derived earlier before arriving at the final product.

5. Conclusion

The aim of this study was to provide a design solution that involves an ergonomic pillow that satisfies both comfort and the function of traction. The target group is female college students who are at risk of head and neck-related health implications due to unsatisfactory lifestyle habits.

In this study, we investigated the 3D shape characteristics of the neck of female college students. Combined with a consumer survey on the materials used for the products available on the market, we then explored a design method for the pillow. In the first part of the study, 20 female college students participated in a questionnaire survey and pressure mapping test. The questionnaire survey provided the subjective ratings of the pillows. From the pressure mapping experiment, we found the functional relationship between pressure and material attributes to be $Y=0.301P+29.807/H-1.495$. After further screening of the subjects' neck health condition and neck shape characteristics, we selected five candidates and used the corresponding sets of neck shape data to create a model of the pillow design. Using Morphing, we obtained the fitted data for neck shape in a comfortable sleeping position. Combining that with pressure value, material hardness and material deformation, we used Rapidform XOR 3D virtual modelling software to create a model of the traction pillow design.

References

[1] Su YX, Xu SL, Liu HY. A biomechanical study on the relationship between the incidence of vertebral disease and the height of pillow. Chinese J Trad Med Traum & Ortho 1998; 6(3): 5-11.

[2] Shan JP. A clinical study on pillows, pillow treatment & cervical spondylopathy - a report on 300 cases. J Cerv Lum 1997; 18(1): 55-56.

[3] Lavin RA, Pappagallo M, Kuhlemeier KV. Cervical pain: a comparison of three pillows. Arch Phys Med Rehabil 1997; 78: 193-198.

[4] Hagino C, Boscariol J, Dover L, LeTendre R, Wicks M. Before/after study to determine the effectiveness of the align-right cylindrical cervical pillow in reducing chronic neck pain severity. J Manip Physiol Therap 1998; 21: 89-93.

[5] Persson L, Moritz U. Neck support pillows: a comparative study. J Manip Physiol Therap 1998;21:237-240.

[6] Persson L. Neck pain and pillows - a blinded study of the effect of pillows on non-specific neck pain, headache and sleep. Adv Physio 2006; 8: 122-127.

[7] Zhang XL, Zhao X, Zhu QC. Analysis of the survey results of mental health status of female college students. Chinese J of School Health 2007; 22(4): 525-527.

[8] Kanchanomai S, Janwantanakul P, Pensri P, Jiamjarasrangsi W. Risk factors for the onset and persistence of neck pain in undergraduate students: 1-year prospective cohort study. BMC Pub Health 2011; 11(1): 566.

[9] Zhou HP, Zheng R, et al. A study on the factors impacting college students' cervical vertebrae health, Beijing: Beijing Institute of Fashion Technology, 2013.

[10] Hou JJ, Shen LM. A study on the design of a pillow for sleep comfort. Package Eng 2014; 35(8): 21-26.

[11] Standardization Administration of China. GB/T 6669-2008. Flexible cellular polymeric materials - Determination of compression set, 2008.

[12] Cai D, Chen HL. Ergonomic approach for pillow concept design. Appl Ergonomics 2016; 52: 142-150.

Characterization of Mechanical Properties of Basalt Hybrid Fabrics

Hafsa Jamshaid[1,2], Rajesh Mishra[1], Jiri Militky[1]

[1]*Technical university of Liberec, Liberec, Chezk Republic*
[2]*National Textile University, Faisalabad, Pakistan*

*Corresponding author's email: rajesh.mishra@tul.cz, jiri.mility@tul.cz

Abstract

Technical textiles are one of the imperative subjects in present time. Application of textiles in various protective garments application is increasing day by day. The aim of this work is to investigate the effect of fibre and structure on tensile and dynamic mechanical properties of hybrid and non-hybrid basalt structures. In this study, experimental investigations are carried out to check the mechanical behaviour of woven fabric Basalt/ Polyester (B/ PET) and Basalt/ Polypropylene (B/ PP) fibre hybrid woven and non-hybrid woven fabric. Three types of weaves are used i.e Plain (PW), Matt (MW) and 1/3 Twill (TW) for both hybrid and non-hybrid structures. Tensile testing of all the woven fabrics is carried out. The discussion of results is in terms of fibre effect and woven geometry. It was observed that plain weave has highest modulus in non -hybrid fabrics due to more interlacement. The exposure of fabrics to increase (DMA) shows an increase in the storage modulus indicating higher stiffness and lower damping ratio in the case of hybrid fabrics. These fabrics can be used for temperature resistant applications.

Keywords: Basalt; Technical textile; Hybrid, Woven; Storage modulus

1. Introduction

In this fast paced time, life is changing and standards are enhanced with every day passing by. People are becoming more demanding and well aware about their need. People are becoming specific with the level of comfort or specifications of textile material regarding end use. People are well aware with the need of having environment friendly products which is leading to invent of new environment friendly fibres. Basalt is a fibre that gives solution to need of environment friendly material. The origin of basalt fibre is rock of basalt that was known long ago for its thermal properties, durability and strength. This rock is used for production of basalt fibres. The extraction cost of basalt raw material is comparatively low and the production process for basalt is safe and simple. The methodology used for basalt production is simple and environment friendly i.e. no harmful emission or waste production during the production process, in addition to this it does not react with water. Basalt is termed as green industrial material. As basalt is made of natural material so it falls under sustainable material class. Basalt is harmless both environmentally and ecologically. It is free of carcinogens and other health hazard. Basalt has replaced asbestos, as it was potentially hazardous to respiratory system. Basalt can replace glass in many applications because of its advantage of possessing environment friendly [1-2].

Fabrics can be produced with different techniques with different fibres. The fabrics made from interlaced yarn falls under most common yet complex technique. It is most tradition method of fabric formation. Most common and famous example of interlacement of yarns to produce fabric is weaving in which two sets of yarns interlace and interweave to complete fabric production process. The force that holds the two sets of yarns together is inter-yarn friction. Woven fabrics have been utilized commercially for energy absorption, aerospace, defence, automotive parts, and structural applications because of their high strength and stiffness.

For special applications with affordable costs, hybrid structures are used. Two sets of yarns that are used in hybrid structures make it possible to form a material that may possess combined properties of both the material and it provides the flexibility to mitigate non desirable properties. It makes possible, tailoring of properties as per specific requirements.

In the research, basalt fibre with thermoplastic yarn is used. This research focuses on the mechanical properties of yarns made from these fibres and on the mechanical properties of woven fabrics made from these yarns in warp and weft direction. While designing fabrics for various industrial textile applications, the knowledge about the mechanical properties is very important. Strength, modulus, and elongation are the most important performance properties of fabrics governing their performance in high-end applications. These fabrics can be used for protective application like gloves. The dynamic mechanical properties were obtained to investigate the effects of adding basalt fibre on the mechanical performance of the fabrics based on the functional requirements of end applications.

2. Material &Methods

2.1 Material

The polyester (PET) and used in our study were available commercially. The Polypropylene (PP) yarn was received from Synthetic (Pakistan) and basalt yarn was used as received from company Kamenny Vek (KV) (Russia). Polyester fibres are the main synthetic fibres used in the industrial manufacturing sector and can be found in several areas of application. Polyester fibres (polyethylene terephthalate) dominate the world synthetic fibres industry. A very important property of polyester is that its mechanical properties in the wet state and under standard conditions are practically the same. PET fibres have excellent resistance to acids, alkalis and microbial attack. They also have good resistance to light degradation [3-4]. Polypropylene is presently one of the fastest growing fibres for technical end-uses where high tensile strength coupled with low cost are essential features.

2.2 Method

2.2.1 Production of Fabrics

The research was focused on the mechanical properties of hybrid fabrics with Basalt (B) in warp/ weft and, Polypropylene (PP) and Polyester (PET) yarns in warp/ weft . First basalt is used in warp and polyester and polypropylene yarns are used in weft and then basalt in weft and polyester and polypropylene in warp. The raw materials to be tested were Plain weave (PW), Matt weave (MW) and 1/3 Twill weave (TW). Pure fabrics were made for the comparison with other fabrics. All 21 fabrics were made on the CCI sample loom with the same density for all fabrics, 12 threads/cm in warp and 8 threads/cm in weft. All the fabric variants were measured, according to standardized procedures.

2.2.2 Yarn Tensile Properties

Yarn tensile properties are measured. Tensile properties of yarns used were measured according to standard CSNENISO 2062. For each sample, 20 measurements were done.

2.23 Fabric Static Mechanical Properties

Fabric stress and strain of all samples in warp and weft direction were measured on a TIRATEST universal tensile tester. The tensile testing speed was kept at 100 mm/min. This test was performed according to standard EN ISO 13934-1.The specimen size for tensile testing was 20 cm × 5 cm as per standard. For each typology, 20 samples were tested.

2.24 Fabric Dynamic Mechanical Properties

DMA facilitates in material selection for specific end-use applications: The task of evaluating new materials and projecting their performance for specific applications is a challenging one for engineers and designer. Forced Vibration (Non-resonant) variable frequency apparatus applies a constant amplitude (stress or deformation amplitude). The storage and loss moduli are then measured as function of temperature using dynamic mechanical analysis (DMA). Tensile mode is ideal for investigating thin specimens, such as films/fibers, thin materials in the low-to medium modulus range. The sample is clamped at its ends with sufficient pretension applied to avoid sample buckling and to keep the sample under tension during the whole experiment. The sample is clamped into a frame of measurement head and is heated by the furnace. The sample in the furnace is applied the stress from the force generator via probe. To make the strain amplitude constant, the stress is applied as the sinusoidal force. This frequency is one of the measurement conditions. The deformation amount generated by the sinusoidal force is detected. Viscoelastic values such as elasticity and viscosity is calculated from the applied stress and the strain and plotted as a function of temperature or time. The DMA was performed on a DMA 40XT RMI equipment. The samples were tested using tensile mode at a frequency of 1 Hz in temperature scan mode. The DMA test was executed in the temperature range of 27 to 100 °C at a heating rate of 3C/min. This test was performed according to EN ISO 6721-1. All of the samples investigated only in the warp direction for each sample, five measurements were done.

3. Result &Discussion

Mechanical properties describe the resistance of material against external forces. The response of the textile material depend on the material properties, type of loading, and direction of loading. Tensile test is fundamental test to measure the mechanical properties of any material. Tensile tests apply forces directly the material sample, usually by using clamps to securely grip two opposite ends of the sample and then pulling the ends away from each other. As the force is slowly increased the stress on the material slowly increases and the sample elongates until it reaches its maximum strain and the material break. Measurement of tensile stress–strain properties is the most common mechanical measurement on fabrics. It is used to determine the behaviour of a material while under an axial stretching load. The tensile properties of woven fabrics has many complexities as fabric is anisotropic in nature and has initial modulus which changes with strain. Anisotropy is a characteristic of most fabrics, especially woven; the impact of the direction of loading on tensile properties can be enormous and is frequently examined, in two principal directions - warp and weft , in which yarns and majority of fibres are oriented [5-12].

3.1 Tensile properties of yarn

The strength of a fabric depends on the strength of the constituent yarn thus tensile properties (i.e. breaking force and elongation) of all used yarns were measured and for a better comparison, breaking tenacity of yarns was calculated. It was established that basalt yarn is the strongest and can withstand the most stress per linear density. Furthermore, the Load -elongation curves were elaborated for each yarn in Table 1 and Fig.1 , where the load - elongation behaviour of the used materials can be observed. There are various ways in which fibres in yarn ultimately undergo failure or fatigue. In basalt fibre brittle elastic failure occur i.e. the fibre fails to extend due to its high modulus and rapid failure occurs due to crack initiation at a surface fault. .In PP and PET ductile crack growth occurs with slow formation of numerous cracks followed by opening to a large V-shape subsequent to catastrophic failure [13].

Table 1. Properties of fiber and yarn

Properties	Basalt	Polypropylene	Polyester
Diameter of fibres [micron]	12	34	22
No. of filaments	890	300	900
Linear density of yarn [Tex]	295	292	250
TPM [Twists/m]	20	30	24
Modulus[MPa]	9,378	721	1069

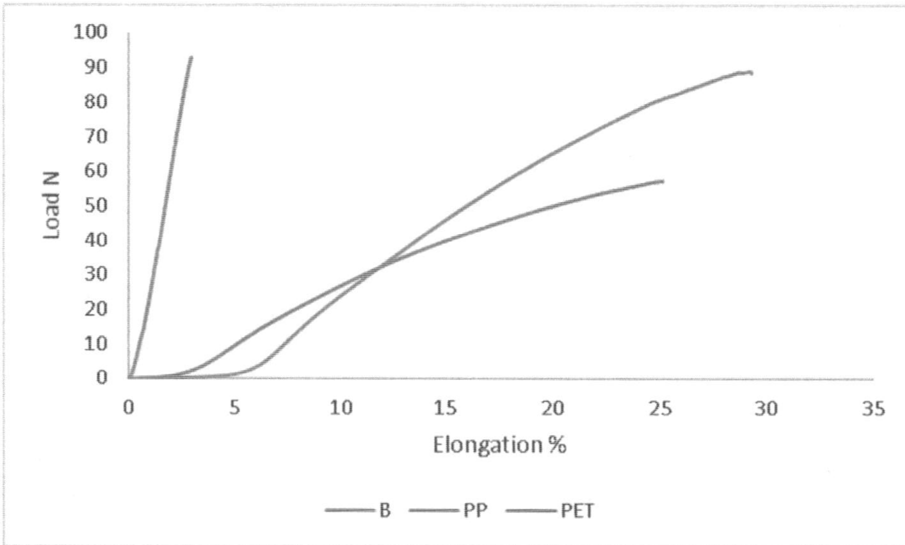

Fig.1 Load - Elongation % of yarns

3.2 Static Mechanical Properties

Some tensile properties of the fabric samples as tensile stress, strain, and elastic modulus are shown from fig.2 and 4. In warp direction, B/ B has highest tensile stress value as basalt is strongest yarn. With the use of basalt in warp/ weft stress/ strain behaviour changes a lot .Tensile moduli of all fabric increased. In hybrid structures B/ PP has highest stress value in warp direction. As PP yarn has higher strength than PET, the results are evident. It is quite clear that effect of yarn tensile properties have significant influence on fabric behaviour. With the use of PET in weft direction modulus in warp direction increase for all weaves .With the use of basalt in warp direction improves moduli more as compared to when used basalt in weft direction.

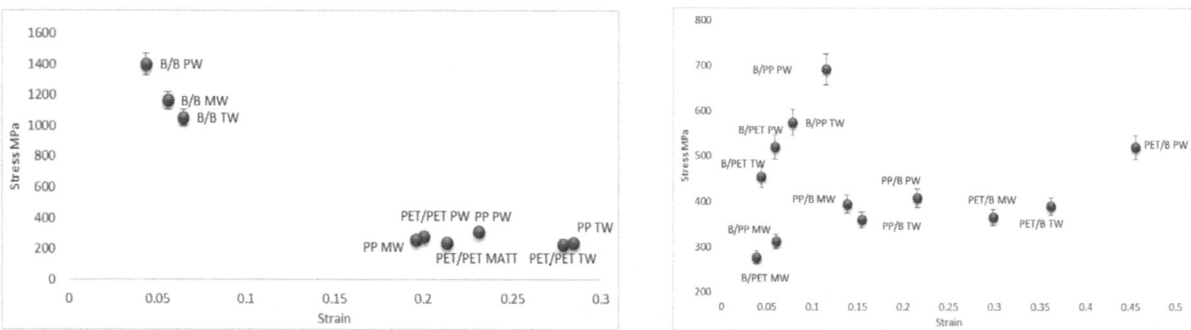

Fig.2 Stress-strain of Non-hybrid and Hybrid in warp direction

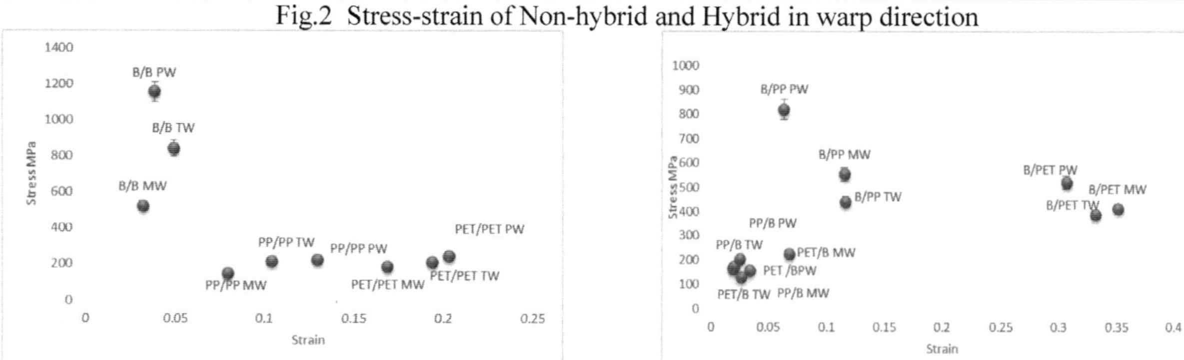

Fig.3 Stress-strain of Non-hybrid and Hybrid in weft direction

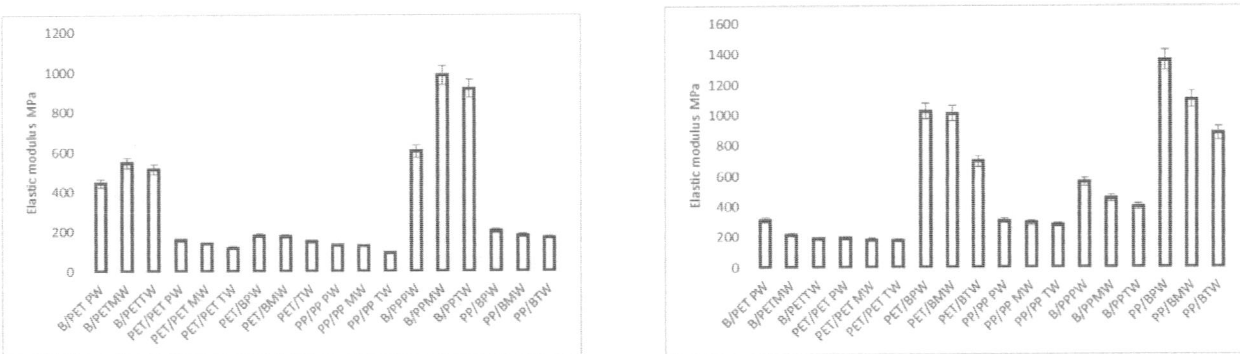

(a) Warp direction (b)Weft direction

Fig.4 Elastic Moduli in warp and weft direction

A comparative account of the effect of weave structure on tensile properties of fabrics is also shown in fig.2-4. It is also established that the type of weave has a great influence on the tensile force of fabrics in warp direction. It is obvious that in non- hybrid structure plain has highest value followed by matt and twill. The highest stress and elastic modulus is observed, in plain weave, because of the maximum number of interlacing points resulting in higher friction between yarns so they can sustain at high loads. In hybrid fabrics when basalt is used in warp, the highest moduli is obtained in matt weave (groups of yarns have woven together), due to it cohesiveness and because of two yarn leads to increase contact point so friction between warp and weft is high, and consequently also higher moduli in warp. Plain weave has lower value, as basalt is brittle yarn and during weaving a lot of weaving stresses acted on it so fibre rapture occur and also due to high crimp of plain weave. During tensile test, breaking elongation of warp yarn is low as yarn already extended so leads to low residual stress. The differences in the same weave depend considerably on the material used in weft. The lowest tensile elongation/ strain is observed in fabrics woven in basket weave. So matt weave has less elongation/deformation but high load bearing capacity. Maximum stress level in matt weave is low because of excessive frictional restraint catastrophic failure occur.

It was also found out that different properties of yarns have influence on the tensile elongation of fabrics in warp direction. When basalt is used in warp direction, tensile elongation values are lower as basalt is brittle yarn with low elongation. By using basalt in warp direction, fabrics tensile stress of all fabric increased. B/ PP has highest value followed by B/ PET respectively.

In weft direction, in all structures plain weave has highest moduli followed by matt and twill. When using basalt in weft direction, moduli is not affected much as compared to non- hybrid fabrics. During weaving of filament yarn with basalt i.e brittle yarn, it damage the warp yarn so not affected the moduli. Moduli of PP/ B and PET/ B is quite higher in weft direction due to the reason described above.

3.3 Dynamic Mechanical Properties

The mechanical properties of fibers depend on their molecular structure, where macromolecules can be arranged in crystalline (specific arrangements of molecules) or amorphous (coincidental arrangements of molecules) structure. The macromolecules are orientated mostly along the fibre axis and are connected to each other with intermolecular bonds. When a force is applied, the macromolecular structure starts changing [14]. The rate of molecular rearrangement depends on the load level and on the mobility of the macromolecules, which again is governed by the physical and chemical structure and arrangement of the molecules (i.e., side group, bonding, etc.). Temperature also accelerates the rearrangement of the macromolecules as a result of increasing mobility, coupled with a higher free volume. The dynamic mechanical analysis (DMA) is mostly employed for investigating the structures and viscoelastic behaviour of polymeric materials for figuring out their relevant stiffness and damping characteristics for various applications. It has considerable practical significance when determined over a range of temperature and frequencies [15]. Classical DMA outputs are storage modulus (E'), loss modulus (E'') and mechanical damping factor (tan θ). The storage modulus represents the stiffness of a viscoelastic material and is proportional to how much energy a material can absorb, whereas the loss modulus shows how much energy(heat) the material releases The mechanical damping factor or internal friction in a viscoelastic system can be utilized to show the impact resistance of the material. A high tan θ value is indicative that a material has a high, non -elastic strain component, while a low value indicates, that it is less.

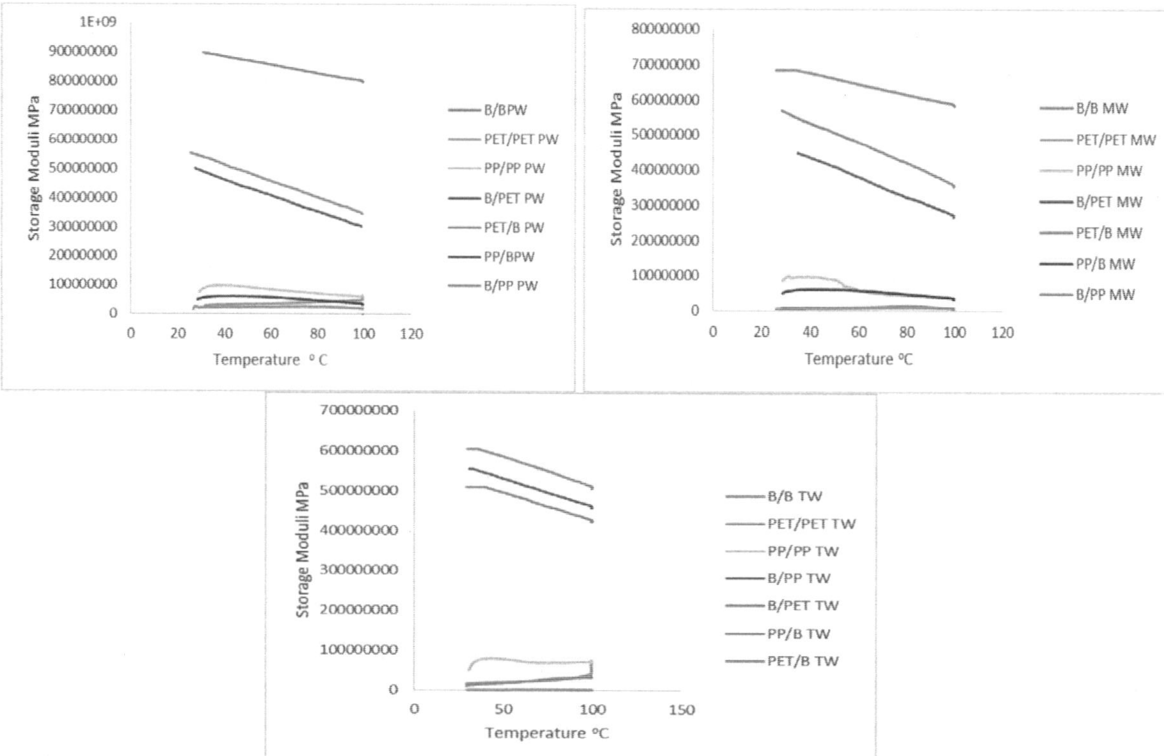

Fig.5 Storage Moduli of different weaves

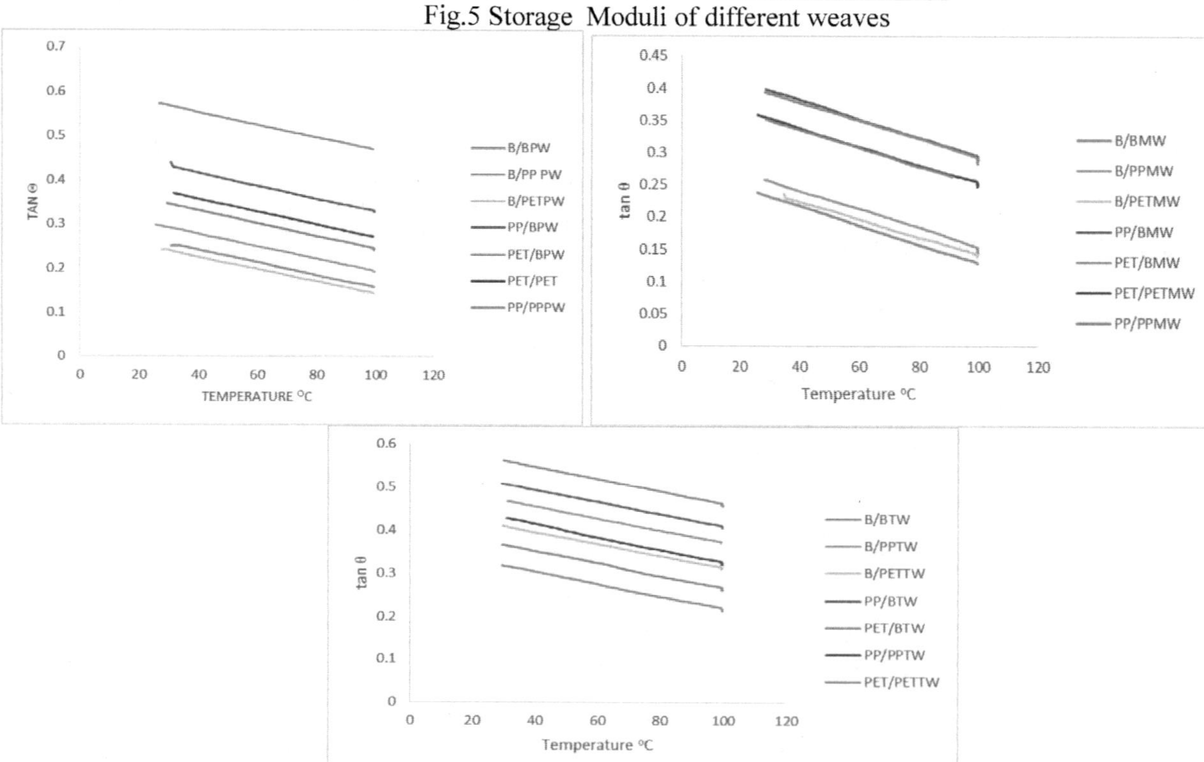

Fig.6 tan θ of different weaves

Storage modulus, loss modulus, and tan δ are described as a function of temperature in Fig.5-6 .All the hybrid fabric exhibited higher storage modulus (higher stiffness) than the non- hybrid fabrics. Tensile strength of the fabric showed the same behaviour as shown in fig.5. At each temperature, hybrid fabric exhibit higher storage modulus values than the non –hybrids fabrics. Storage modulus of hybrid fabrics increased significantly with the incorporation of basalt yarn especially when it is used in warp direction. This condition is primarily attributed to the increase in the stiffness of the structure due to the nature of basalt fibre that allows more stress to bear. As seen from the figure, there is a decreasing trend in the storage modulus (decrease in stiffness)_over the entire experimental temperature range. B/B structures are least effected by the temperature as compared to other materials due to the stable property of basalt (inorganic) fibre over wide range of temperature. Thermoplastic polymers are semi crystalline so they have high degree of orientation which strongly affect their mechanical properties. At low temperatures, the molecules are so immobile that they are unable to resonate with the oscillatory loads and therefore remain stiff. The macromolecular segments cannot change shape, especially through rotation about C–C bonds, and so the molecular entanglements act as rigid crosslinks. At elevated temperatures, the molecular segments become readily mobile and have no difficulty resonating with the load. When the timescale of molecular motion coincides with that of mechanical deformation, each oscillation is converted into the maximum-possible internal friction and non-elastic deformation. The entanglements more or less remain firmly in place, but may occasionally slip and become disentangled. The loss modulus, which is a measure of this dissipated energy, also reaches a maximum [16].

In PET structure below the glass transition temperature (Tg) modules decrease is not significant. It is important to mention that modulus in the glassy state is determined primarily by the strength of the intermolecular forces and the way the polymer chains is packed. At higher temperature mobility increase and lose their packing arrangement Near the glass transition, the storage modulus drops off significantly. All chain segments cannot store much more energy for a given deformation than a rubbery segment which is free to move. Thus, every time a stressed frozen in segment becomes free to move, excess energy is dissipated and consequently there is an abrupt fall near the glass transition region.

The loss tangent or tan θ is a ratio of the loss modulus to the storage modulus and is measured as the mechanical loss or damping factor. The damping properties of the material act to balance between the elastic and viscous phases in a polymeric structure. Material with higher tan θ suggests that more heat was produced and more deformation could not recover when outside force was removed. In nature, tan θ is the response of inner friction force .The variation of tan θ of non- hybrid and the hybrid as a function of temperature is represented in Figure 6. It can be observed that there was a decreasing trend in the loss tangent over the temperature range. This can be justified by the restriction of the motion of polymer molecules resulting from the incorporation of rigid fibres In this case, the incorporation of basalt fibres for hybrid fabrics, which act as barriers to the mobility of polymer chain, led to a lower degree of molecular motion and consequently lower damping characteristics. Matt weave structure shows lower value of tan θ showing fabrics has better impact resistance.

4. Conclusion

The results reveal that the hybridization of basalt with polypropylene and polyester in different weaves leads to significant improvement in the static and dynamic mechanical properties of fabrics. When choosing a weave design for a particular product, one must consider the aesthetic appearance of the weave as well as the desired properties. Structure of weave and fibre type have strong influence on static and dynamic mechanical properties of fabrics. Hybridization of basalt yarn in warp direction has more significant effect on mechanical properties. In hybrid structure matt weave have better properties as compared to other weaves.

References

[1] Jamshaid H, Mishra R. A green material from rock: basalt fiber- a review. J. Text. Inst. 2015. DOI 10. 1080/ 00405000: 2015; 1071940.

[2] Kamenny V. Advanced basalt fiber, Basfiber: 2009. retrieved from http://www.basfiber.com.

[3] McIntyre JE. Synthetic fibers: nylon, polyester, acrylic, polyolefin: 2005. Woodhead Publishing Limited, Boca Raton, FL.

[4] Matsuo T. 2008. Fibre materials for advanced technical textiles. Text. Prog: 40: 87-121.
[5] Dai X, Li Y, Zhang X. Simulating Anisotropic Woven Fabric Deformation with a New Particle Model. Textile Res. J: 2003; 73; 12: 1091-1099.
[6] Hu J. Structure and mechanics of woven fabrics. 2004;102. Woodhead Publishing Ltd. ISBN0-8493-2826-8.
[7] Kilby WF. Planar stress-strain relationships in woven fabrics. J. Text. Inst: 1963; 54; T: 9-27.
[8] Kovar R. Structure and properties of flat textiles (in Czech). 2003: 142 TU of Liberec, ISBN 80-7083-676-8.
[9] Kovar R, Gupta BS. Study of the Anisotropic Nature of the Rupture Properties of a Woven Fabric. Textile Res. J: Vol 2009; 79; 6: 506.
[10] Lo MW, Hu JL. Shear Properties of Woven Fabrics in Various Directions, Textile Res. J: 2002; 72; 5: 383-390.
[11] Pan N, Yoon MY. Structural Anisotropy, Failure Criterion, and Shear Strength of Woven Fabrics. Textile Res. J: 1996; 66; 4: 238-244.
[12] Postle R, Carnaby GA, de Jong S. The Mechanics of Wool Structures: 1988 Ellis. Horwood Limited Publishers, Chichester. ISBN 0-7458-0322-9.
[13] Tyrone Lv. Textile Processing And Properties Preparation, Dyeing, Finishing and Performance. 1994: 350. Elsvier. ISBN: 978-0-444-88224-0.
[14] Geršak J. Mehanske in fizikalne lastnosti tekstilnih materialov. Fakulteta za strojništvo, Laboratorij za oblačilno inžinerstvo ter fiziologijo in konstrukcijo: 2006; ISBN 86-435- 0754-7, Maribor.
[15] Pothan A. Laly Oommen, Zachariah, Thomas, Sabu. Dynamic mechanical analysis of banana fiberreinforced polyester composites. Composites Scien. Techn: 2003; 63; 2: 283–293.
[16] Joseph D. Menczel R. Bruce Prime.Thermal Analysis of Polymers: Fundamentals and Applications: 2009. Wiley Limited Publisher ISBN: 978-0-471-76917-0.

Effect of Woolen Knitted Stiffness and Draping on Modeling

Chun-Yan Qu, Yong-Mei Deng*

College of Apparel & Art Design, Xi`an Polytechnic University, No.19 Jinhua Rd (S), Xi'an, Shaanxi, 710048, China

*Corresponding author`s email: dengymxpu@sina.com

Abstract

Fabric performance can affect apparel modeling design. This is particularly true for the properties of knitted woolen fabrics as they have a direct impact on apparel modeling design. This study uses the method of literature study, experimentation, subjective evaluation and statistical analysis, etc. to understand the effect of woolen knitted stiffness and draping on modeling. First, test the fabric's stiffness and draping, then the experimental data is analyzed systematically and comprehensively. Compared to 100% cotton woven fabrics, the draping and stiffness of knitted woolen fabric is smaller, and when combined with selected different thickness, different clothing can be designed to have a specific shape effect.For instance, lightweight and thin fabrics are suited for creating a delicate shape, whereas thick fabrics can increase the sense of volume. Research indicates that 45 °oblique fabrics is the optimum to sculpt the wave shaping of garments.

Keywords: Knitted woolen fabrics; Fabric stiffness; Fabric draping; Modeling design; Subjective assessment

1. Introduction

Stiffness and draping is considered to be one of the key properties of fabric, because it is the main contributory factor in model designing. The subject of fabric stiffness and draping have been studied by a number of researchers, but only drape knitted fabrics or stiffness has been considered. Few studies have been conducted in regards to how stiffness can combine with drape, especially when stiffness is combined with the draping of woolen knitted fabric [1-5]. This paper presents the results of an examination on the effects of the stiffness and draping of the selected knitted woolen fabrics, then these fabric design are used to model and evaluate the performance.

2. Specifications of the test sample

Since each of the specifications of knitted fabric, such as thickness, weight, yarn density and stitch density, etc, can influence the stiffness and draping of fabric, knitted fabrics were mostly used for summer wear. It is particularly suitable for women's dress as it was required for a beautiful draping contour [6,7]. Recently woolen knitted fabrics are popular in fashion week. Compared with woven fabrics, the specifications of woolen knitting fabrics can create different shape performance. This is the reason why knitted woolen fabrics were selected with different specifications of thin and light structure.

Their specifications are show in Table 1. All fabrics were tested three times for the thickness and weight of square according to GB/ 3820-1997 and GB/ T 24218.1-2009.

3. Test and analysis of fabrics stiffness and draping

3.1 Testing bending stiffness and draping

3.1.1 Testing stiffness

Testing instrument: FY-207 fabric stiffness tester.

Testing conditions: standard atmosphere.

Test result of the processing.

During the test, the average of multiple measurements are taken to increase accuracy. Both aspects are measured five times according to GB/ T18318-2001 [8,9].

Table 1. Sample Specifications

Type	Main Content(%)		Thickness (mm)	Weight of Square meter (g/m^2)	Course wise density/5cm	Wale wise density/5cm
1*	W/ T55/45	Ave	0.76	215.39	110	85
		SD	0.06	0.51	-	-
2*	W/T 40/60	Ave	0.81	250.29	108	85
		SD	0.05	0.97	-	-
3*	W/T 35/65	Ave	1.03	300.42	78	55
		SD	0.07	0.66	-	-
4*	W/T22/78	Ave	1.34	420.48	60	55
		SD	0.09	0.64	-	-
5*	W/A/T22/38/40	Ave	2.03	480.77	60	55
		SD	0.08	0.89	-	-
6*	W/R/T30/30/40	Ave	1.93	530.76	35	40
		SD	0.06	0.87	-	-
7*	C 100%	Ave	0.26	125.69	190	230
		SD	0.05	0.61	-	-

Illustrate: "-"means there is no value;"Ave" means Average;"SD" means Standard Deviation

Calculation of bending length.

Bending length indexes and their formula:

$$C = L/2 \quad (1)$$

$$C^Z = \sqrt{C_T \cdot C_W} \quad (2)$$

L- the length of the elongation;

C- the bending length;

C_T- the average wale bending length, cm;

C_W- the course bending length, cm;

C_Z- the total length of the total average bending, cm.

Table 2. Bending length results (cm)

wale and course Type		C_T	C_W	C_Z
1*	Ave	1.60	1.68	1.64
	SD	0. 06	0.10	-
2*	Ave	1.67	1.79	1.73
	SD	0.05	0.06	-
3*	Ave	1.74	1.98	1.86
	SD	0.08	0.12	-
4*	Ave	2.11	2.27	2.19
	SD	0.05	0.06	-
5*	Ave	2.45	2.70	2.57
	SD	0.07	0.09	-
6*	Ave	2.69	2.97	2.83
	SD	0.09	0.05	-
7*	Ave	2.95	3.01	2.98
	SD	0.06	0.07	-

Illustrate: "-"means there is no value; "Ave" means Average;"SD" means Standard Deviation

Calculation of bending stiffness.
Bending stiffness indexes and their formula:

$$B = G \cdot C^3 \cdot 10^{-1} \quad (3)$$

$$B_Z = \sqrt{B_T \cdot B_W} \quad (4)$$

B-the bending stiffness of the fabric;
G-weight of fabric, g / m^2;
B_T-the average bending stiffness of the fabric's wale direction , mg · cm;
B_W-the average bending stiffness of the fabric's course direction, mg · cm;
B_Z-the average bending stiffness of the fabric.

Table 3. Bending stiffness results (mg · cm)

wale and course Type		B_T	B_W	B_Z
1*	Ave	88.06	118.54	102.17
	SD	0.55	0.67	-
2*	Ave	116.44	253.38	171.76
	SD	0.75	0.89	-
3*	Ave	158.04	232.87	191.84
	SD	0.73	0.98	-
4*	Ave	394.54	491.27	440.26
	SD	0.59	0.74	-
5*	Ave	705.89	944.78	816.65
	SD	0.59	0.84	-
6*	Ave	1031.60	1388.40	1196.78
	SD	0.66	0.74	-
7*	Ave	320.90	340.88	330.74
	SD	0.97	0.77	-

Illustrate: "-"means there is no value. "Ave" means Average;"SD" means Standard Deviation

3.1.2 Draping testing of fabric

Table 4. Draping statistical results(Drape coefficient%)

		OA	OB	OC	OD	Average
1*	Ave	41	37.5	39	39	40
	SD	0.52	0.56	0.35	0.66	-
2*	Ave	43	42.8	43.8	40.2	42
	SD	0.46	0.89	0.46	0.44	-
3*	Ave	45.2	43.9	41.5	44.5	44
	SD	0.56	0.70	0.53	0.75	-
4*	Ave	52	50	51	50	50
	SD	0.75	0.70	0.85	0.78	-
5*	Ave	52.9	52.6	54.5	52	53
	SD	0.70	0.80	0.60	0.90	-
6*	Ave	56.3	55.3	54.9	53.9	55
	SD	0.80	0.98	0.50	0.75	-
7*	Ave	84.8	81.9	84	83.9	83
	SD	0.52	0.26	0.90	0.60	-

Illustrate: "-"means there is no value

Testing instrument:

Testing instrument: YG (L) 811-DN fabric Static and Dynamic drapemeter.

Size of fabrics: circular specimen with the diameter of 24cm.

Testing conditions: standard atmosphere.

Statistical results

All fabrics were tested three times for the drape according to GB-T 23329-2009. The average of the drape coefficient was calculated by Rounding Method as shown in Table 4.

3.2 The analysis of draping and stiffness result

3.2.1 Summary of experimental data

Thickness: Every fabric was tested according to GB/T3820-1997, Pressurized area 50mm^2, pressing pressure 20KPa, 100CN, a summary of the experimental data is shown in the Table 5:

Table 5. Summary of draping and stiffness result

	Thickness (mm)	Bending length (cm)	bending stiffness (mg·cm)	Drape coefficient(%)
1*	0.76	1.64	102.17	40
2*	0.81	1.73	171.76	42
3*	1.03	1.86	191.84	44
4*	1.34	2.19	440.26	50
5*	2.03	2.57	816.65	53
6*	1.93	2.83	1196.78	55
7*	0.34	2.98	330.74	83

3.2.2 Analysis

Table 1, 2 and 2 shows the experimental results. It can be noted that with the increase of the fabric's thickness, the bending length and drape stiffness also increases. The test results are summarized in table 5. With the increase of fabric thickness, its bending stiffness also increasing, among them 1*, 2*, 3* fabric bending stiffness is smaller than 7* cloth, and 4* 5* 6* fabric bending stiffness is greater than type 7*.

A calculation of drape stiffness, bending stiffness, draping thickness test was conducted for knitted woolen fabrics and cotton fabric, and the analysis of the results are the following:

1) In terms of drape stiffness, the drape stiffness of several different knitted woolen fabrics are smaller than cotton, the order is the following: 7*> 6*> 5*> 4*> 3*> 2* > 1*. When designing, designers should select different thickness of fabric according to the requirement of the clothing model, For example: smaller drape stiffness fabric should be selected rather than the bigger drape stiffness when designing body-fitted clothing in order to reflect women's curves [5];

2) Bending stiffness: As knitted woolen fabric weight increases, the bending stiffness also increases. Its order is: 6*> 5*> 4* >7*> 3*> 2*> 1*, the bending stiffness impact the overall shape, a bigger bending stiffness of the fabric increases the clothing's volume and sense of space;

3) Draping: The draping of several knitted woolen fabrics were tested and the results suggests that that they are better than cotton fabric. Among them the drapability of 1* is the best and 6* is the worst. Several knitted woolen fabric tested from best to worst its order is 1*> 2*> 3*> 4*> 5*> 6*, the drape performance impacts the garment's shape as it reflects the shape of skirt, designers can use the smaller drape fabric sculpture to create delicate folds, as to reflect the feminine beauty;

4) Aspects of thickness: The thinnest of the tested fabric is 1*, and 5* is the thickest, from thick to thin in proper order is: 5*> 6*> 4*> 3*> 2*> 1*.

4. Modeling performance and subjective evaluation

This part selected the same version of fabric`s modeling by using the different knitted woolen fabrics and cotton fabric which we selected above to observe the performance of their three-dimensional, Then, selected the clothing partial modeling results, photographed,observed and compared seven experiments fabric such as pleats, wave modeling and so on [10,11] to evaluating the performance.

4.1 Pleats modeling performance

4.1.1 Flat pleat

The size of 60cm * 60cm specifications fabric samples were selected, the type number is the same as the previous section. And then make flat pleat, analysis of the modeling results were conducted by quantitative analysis, flat pleat fold size is 15mm. Modeling results were recorded with the camera to take pictures, and Adobe Photoshop was used to make the appropriate treatment to obtain the modeling results [12].

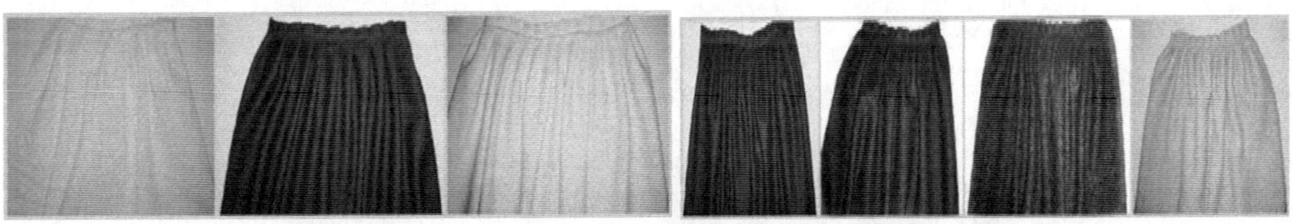

Fig.1 wale direction of flat pleat performance

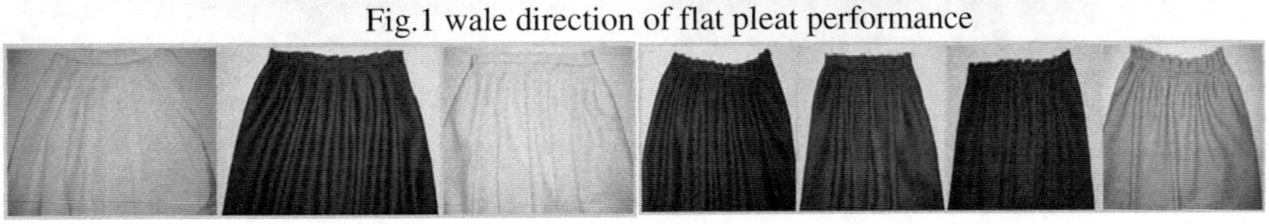

Fig.2 course direction of flat pleat performance

According to the experiment results of the modeling combined with the previous section results, the performance tests was analyzed and the conclusions drawn are the following:

1) Overall, the flat pleat of knitted woolen fabrics are more apparent, draping and stiffness can affect the silhouette of the flat. Different fabric thickness, stiffness and drapability can create a different fabric silhouette. For instance, light fabrics have a more delicate, smooth and tidy result whereas for thicker and heavier fabric, the flat plea is more rugged.

2) In regards to the stiffness aspect of flat pleat, the greater the drape stiffness and bending stiffness of the knitted woolen fabric, the more heavy the flat pleat shape.The smaller the drape stiffness and bending stiffness the more uniform the flat pleat;

3) based on the analysis of the drape, the bigger the drape coefficient of the fabric, the more heavy the flat pleat. The smaller the fabric drape coefficient, the more uniform the plat pleat will be.

4) The analysis results of the wale and course effect on the flat pleat is not clear, generally wale flat pleat is better than course shape.

4.1.2 Shirring

Shirring can be performed at the edge of the fabric or may be carried out inside the fabric [13]. This study has chosen the simpler one: shirring at the fabric edge. The fabric specifications and ordering are the same as the previous section, shirring to one-third of the original, then the effect was photographed. Experiments were performed along the wale and course directions [14].

The performance experiments were analyzed based on the test results of the modeling combined with the previous section. The following conclusions are summarized:

1) From the overall experiments results, the shirring effect of knitted woolen fabrics is different depending on the different thickness, stiffness and drape. In general the shirring effect of knitted woolen fabric is more ideal;

2) According to the analysis of the shirring effect of fabric's stiffness, drape stiffness and bending stiffness of knitted woolen fabric is bigger, which gives us the felling of formal, on the contrary small drape stiffness and bending stiffness can appear the feeling of fine and smooth ;

3) From analyzing the drape aspects, for good drape fabrics, the experimental effect is more ideal and shirring produces uniform pleats. For poor drape fabric performance, the results are not uniform;

4) From the wale and course direction of the shirring performance to analyze, course direction performance is better than wale direction performance.

4.2 Wave modeling results and subjective evaluation

4.2.1 Wave modeling results and analysis

The cutting pattern of a skirt shown is in Figure 1 where the waist length is 20cm and the length of skirt is 25cm. A 45o oblique direction cut was made on same size pieces along the sample wale direction and course direction as presented above. Then it was fixed on a model, and a photo was taken with an ordinary camera in a parallel direction. The photo was then processed by Photoshop. The performance of the skirt piece is shown in fig.2.to Fig.5.And the order of Fig 2 through Figure 5is 1*, 2*, 3*, 4*, 5*, 6*. Model specifications: half mannequin, B = 44 W = 34 H = 42.

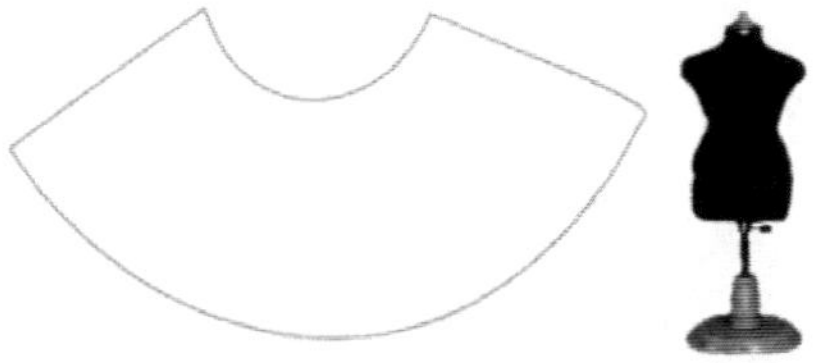

Fig.3 Cutting pattern and mannequins

Fig.4 wale direction of skirt wearing performance

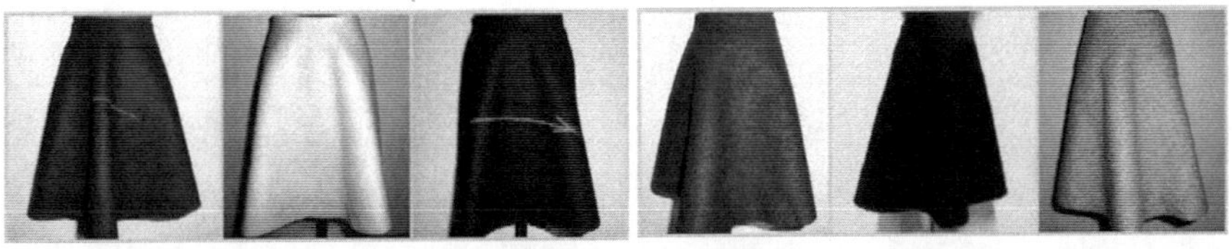

Fig.5 course direction of skirt wearing performance

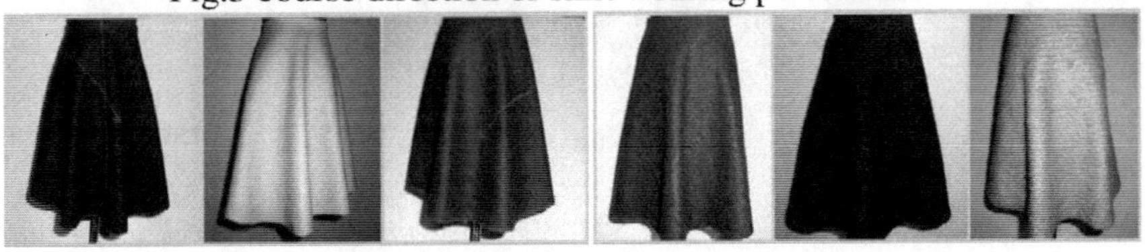

Fig.6 45 ° oblique direction of skirt wearing performance

From the research, the following conclusions are drawn:

1) Fabrics drapability and stiffness can influence the silhouette of wave performance:

2) Overall, the wave silhouette performance of knitted woolen fabric is more obvious, owing to the different properties of fabric, its wave shape gives a different subjective feeling.

3) From the wave silhouette comparison of the wale, course and diagonal direction of fabric, it is apparent that the wave silhouette performance were: oblique>wale>course.

4) From analysis of the drape stiffness and bending stiffness, the performance of wave silhouette is the same with the pleats and shirring in the previous section. Namely the size of the drape stiffness and bending stiffness can affect the performance of the wave silhouette.

5) From the analysis of silhouette performance based on drapability, the order is1*> 2*> 3*> 4*> 5*> 6*.

4.2.2 Subjective evaluation of wave silhouette performance.

1*, 2*, 3*, 4*, 5*, 6* were chosen for experimentation, and the wave silhouette was selected to evaluate the knitted woolen fabric. Subjective ratings is used to assess the Grade Evaluation of wave silhouette performance. In the Grade Evaluation, the best is 5 points and the worst is 1 point, as shown in Table 5. Evaluation can be achieved according to different type fabrics of the wale and course direction. The number of subjective rating is 50 people, and the professional fashion designer was selected as the object of subjective evaluation [13,14].

Table 5. The Grade Evaluation of wave silhouette performance

No.	Grade				
	Excellent 5	Good 4	Medium 3	Ordinary 2	Poor 1

Table 6. Part of subjective evaluation

No.	1*	2*	3*	4*	5*	6*
1	4/3	4/5	3/3	3/4	2/4	4/5
2	4/4	4/4	4/4	4/5	3/3	4/4
3	4/5	4/5	4/5	5/5	3/4	4/3
.	..	..	..	..	..	..
48	4/4	4/5	4/5	5/5	3/4	4/3
49	3/5	4/4	5/5	3/4	4/4	5/5
50	4/4	4/5	4/5	5/5	3/4	4/5

With the results of subjective evaluation analysis, the six experimented fabrics were put into 5 classes by the order of excellent, good, moderate, ordinary and poor, as shown in Table 7:

Table 7. Results of subjective evaluation

Rate	Type No.	Description of wave silhouette performance
Excellent,	1*wale2*3*course	Wave performance is very good, great pleats, the peaks and valleys of pleats is uniform, draping is good, ripple evenly distributed, the hem is flush.
Good	1*4*5*6*course 2*3*4*6*wale	Wave performance is better, great pleats, the peaks and valleys of pleats is relatively uniform, draping is good, ripple distributed, the hem is flush.
Moderate	5*wale	Wave performance is better, peaks and valleys of pleats is relatively uniform, draping is better, ripple relatively distributed, the hem is relatively flush.
Ordinary		Wave performance is not very good, peaks and valleys of pleats is relatively uniform, the hem is not flush anymore.
Poor		Wave performance, draping is poor, peaks and valleys of pleats is not very uniform, ripple unevenly distributed and the hem is not flush anymore.

From table 6 and Table 7 as well as Fig.4, Fig.5 and Fig.6, it can be seen that the different weight and material composition of knitted woolen fabric have different garment performances.Generally the performance of knitted woolen fabric is better, and they all are above average. In general, thick fabrics gives a sense of solemnity, and lightweight fabrics gives a sense of elegance.

Comparison drape stiffness and bending stiffness is shown in table 2. and table 3. The smaller the drape stiffness and bending stiffness, the easier it is to bend the fabric, and the greater the bending stiffness of thick fabric, the more thick and heavy the outline. As seen from the table, the stiffness of the wale and course knitted woolen fabric is different, the wale is smaller compared to the course, where the performance lines are softer;

Based on the comparison of table 4 it can be observed that draping, wavy curved radius increases as the drape coefficient of fabric and the hem of skirt becomes larger

5. Conclusion

This paper studies knitted woolen fabrics and the partial shape performance of clothing. It analyzes the influence of knitted woolen fabric's stiffness and draping on costume performance by the method of literature study, experimentation, subjective evaluation and statistical analysis etc. By testing the fabric's stiffness and draping, and conducting an analysis on the experimental data systematically and comprehensively as well as analyzing the performance by subjective evaluation and the evaluations of several experts, a conclusion was drawn on the best performance method of knitted woolen fabric which is shown as follows:

1) Stiffness and drape are the main factors of knitted woolen fabric that directly effects the performance of clothing;

2) Different thickness of knitted woolen fabric are able to create different shape performance.Thick fabrics can be designed apparent so matosensory performance to increase the clothing' s volume sense of space;

3) Knitted woolen fabrics have a unique advantage to created comfortable and body fitted clothing. Oblique fabrics is most appropriate to shape the performance of clothing, especially along 45° to cut the fabric, which can better reflect a women`s elegant beauty.

6. Acknowledgement

This paper is supported by HaoYunLai Innovation Fund(2014KJ-048)- Knitted woolen Dress Design Study (HYL201501); Science and Technology Planning Project of Shaanxi Province-Natural Science Foundation Research Program (Grant No:2015JM5167); Special Research Program of Education Department of Shaanxi Province (Grant No:12JK0571).

References

[1] Peiree FTJ. Text. Inst: 1930; 21: 377.
[2] Chu CC, Ciimmings CL, Texeira NA. Text. Res. J: 1950; 20: 539.
[3] Cusick GE: 1968; J Text. Inst: 59: 253.
[4] Hearle JWS, Grosberg P, Backer S. Structural Mechanics of Fibers, Yattis and Fabrics. Wiley- Inlerscience. New York. NY. USA: 1969; Chapter 12.
[5] Booth JE. Principles of Textile Testing. Heywood, London, 3rd edition: 1968.
[6] Liu CX, Zhang CQ, Chen WL. Effect of Bending Stiffness on Drape of Soft course-Knitted Fabric. Proceedings of the Second International Conference on Advanced Textile Materials & Manufacturing Technology: 2010: 521-524.
[7] Zhao QD, Yu WD. Analysis of Structure and Stiffness of Fabric by Three-point Bend in Principle. Proceedings of Textile Bioengineering and Informatics Symposium 2010: 1164-1170.
[8] Hearle JWS, Grosberg P, Backer S. Structural Mechanics of Fibers, Yattis and Fabrics. Wiley- Inlerscience. New York. NY. USA: 1969; Chapter 12.
[9] Ren YH, Fan LH. Textile performance test tutorial. X PU: 2014; 28-33: 94-101.
[10] Kenkare, Narahar, Lamar, Traci A. M. J Text. Inst: 2008; 99: 211-218.
[11] Zhu SW, Liu JW. Clothing Materials. China Textile Press: 2011. 79-100.
[12] Kenkare, Narahar, Lamar, Traci A. M. J Text. Inst. 2008.
[13] Li YL,Yang LH. 3D Modeling and Simulation of Fancy Fabrics in course Knitting. DHU: 2012; 4: 351-357.
[14] Zhu XX. Liberal Curve and Surface Modeling Techniques [M]. Beijing : Science Press: 2000: 138-168.

Bending Mechanical Properties of Textile Fiber and its Application

Mu Yao, Run-Jun Sun*, Kan Lai, Mei-Yu Chen, Cheng-Kun Liu

School of Textile and Materials, Xi'an Polytechnic University, Xi'an, 710048, China

*Corresponding author's email: sunrunjun2002@163.com

Abstract

Application of bending mechanical properties of textile fiber in production has a long history. Western violin, Chinese violin, and especially lyre have been used for 600, 1500, and 2500 years, respectively. The strings were all made by textile fibers, which were chosen and designed by professionals based on historical experience through repeated trials. Textile fibers have also been chosen to produce anti-noise ear plug for a long part of history. However, influence of fiber bending vibration mechanics has not been taken into account, though air flow resistance has been considered for a long time. In recent years, relationship between them has been explored theoretically, and validated through some experiments. Theoretical formulas have been deduced, and design methods have been investigated. Correction factor of non-circular cross section has also been supplemented in application. This paper will provide theoretical foundation and methods for material selection of musical instrument, test for the linear density, and material design of anti-noise ear plug. However, these contents have not been reported in textbooks of Science of Textile Materials both at home and abroad.

Keywords: Bending mechanical properties of textile fibers; String vibration; Simple beam supported by two ends; Vibration energy; Musical instrument; Tester of linear density; Ear plug

1. Introduction

For a long time, bending mechanical properties of textile fibers have been rarely reported in textbooks of Science of Textile Materials. Application areas and examples were also rarely mentioned. Theory, methods, and mathematical analytic equation were even not involved [1-4]. However, practical application history of textile fiber bending mechanics has been very long. Strings of many musical instruments were all made from or of textile fibers. Western violin and cello have been using these strings for 600 years, Chinese violin (Erhu, Jinghu, Banhu, or Sihu) used for 1500 years, Pipa used for 2000 years, and lyre used for 2500 years. Vibration of strings made by suitable textile fibers (considering linear density, tensile modulus, and bending rigidity) can produce a sound wave with a corresponding frequency on the condition of certain string length and preset tension through playing, rubbing, touching, and beating. But for a long time, selection and design of each kind of string were all performed by professionals through historical experience and repeated practices.

2. Basic Knowledge and Applications of String Vibration in Bending Mechanical Properties

2.1 Basic Knowledge of String Vibration

When length of textile fiber is many times larger than its cross-sectional size, and certain tension is applied between two end supports, string vibration occurs under the action of external force. As shown in Fig.1, first-order wave, second-order wave and multi-order wave can be produced between two end supports.

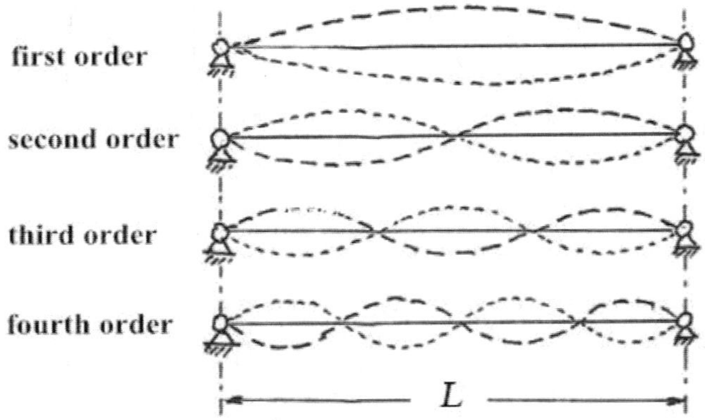

Fig.1 Multi-order bending vibration of two end simple beam

The theoretical equation for string vibration is as follows [5]:

$$f=[10^3/(4L)]\cdot\sqrt{F/N_t} \tag{1}$$

Where, f is the first-order frequency of string transverse vibration, Hz; L is linear distance between the two end supports, m; F is tension of the fiber, N; Nt is linear density of the fiber, tex.

2.2 Applications of Textile Fiber in String Vibration Musical Instruments

String vibration of textile fiber has been used by many musical instruments to produce music for a long time. Lyre in Zhou Dynasty, Pipa in Han Dynasty, Huqin (such as Erhu, Jinghu, Banhu or Sihu) in Tang and Song dynasty, as well as western violin, cello, and Chinese Guqin those were invented in the 15th century are all based on the string vibration of textile fiber to produce music through playing, rubbing, touching, and beating. Distance between the two end supports is usually over 300 mm. Cello requires 500-600mm. For ensuring the accuracy of music frequency, choice and design of linear density and string tension are very important. So fundamental Equation (1) are needed to be modified for it to work proficiently. Self-vibration frequency equation can be solved by string vibration differential equations [8].

Linear density of fiber can be solved by series equation obtained by quadratic equation. The corresponding derivation process can be seen in literature [9].

$$N_t=\frac{10^{10}}{4}\cdot\frac{F}{L^2\cdot f^2}+\frac{10^{18}E^{0.5}\cdot F^{1.5}}{8\pi^{0.5}\cdot\rho^{0.5}\cdot L^5\cdot f^2}+\frac{10^{25}E\cdot F^2}{128\pi\rho^8\cdot f^6}+\ldots \tag{2}$$

According to Formula (2), based on the selected type of material (considering modulus and bulk density) and the setting of frequency, tension and distance between the two end supports, linear density of textile fiber bundle can be calculated.

The above calculations has created a scientific and useful engineering condition for the design and production of string vibration musical instruments. Since a large amount of rosin on bow string is often used to form a friction condition between the bow string and vibration string due to a certain adhesion textile fibers used for bow string should be chosen and designed reasonably.

2.3 Methods and Applications for Test of Textile Fiber Linear Density by String Vibration

Due to the large area promotion of transgenic insect resistant varieties of cotton fibers, test for linear density and maturity of cotton fiber by using double compressed air meter was found invalid in America in 1990. International Standardization Organization also claimed that standard of double compressed air meter was suspended in 1997, and the micronaire was not linear density of cotton fiber multiplied by its maturity any more [10]. So a new, efficient method should be founded to measure linear density of cotton fiber. The testing of specific strength and specific modulus of a single fiber cannot be conducted if the middle part linear density is not measured accurately. Therefore, China began to develop the specific strength and specific modulus tester for a single textile fiber, and that means strength was obtained along with the middle-part linear density, which could be calculated by the measurement of vibration frequency through the use of fixed length and excitation. Formula (2) was generally used in principle. For polyester fiber, ρ=1380 kg/m^3, L=0.02-0.05m, E=1600 N/tex, when vibration frequency is f=1000-2000 Hz, and the first term in the right side of the equal sign is 10^{-2} order of magnitude, the second term is 10^{-3} to 10^{-4} order of magnitude, the third term is 10^{-5} to 10^{-6} order of magnitude, the fourth term is 10^{-6} to 10^{-7} order of magnitude. Therefore, when less than 1%, error is required the first and second term can be only taken in the right side of the equal sign in Formula (2), so Formula (3) is obtained as follows [9]:

$$N_t = \frac{10^{10}}{4} \cdot \frac{F}{L^2 f^2} + \frac{10^{18} \cdot E^{0.5} \cdot F^{1.5}}{8\pi^{0.5} \cdot \rho^{0.5} \cdot L^5 \cdot f^2} \tag{3}$$

When Formula (3) is used to calculate the middle-part linear density of a single fiber, not only distance of the two end supports (L), tension of the fiber (F), vibration frequency (f), but also the volume density in the middle part of fiber (ρ) and dynamic bending stiffness (EI) should be all considered. Therefore, cross sectional shape coefficient (η_f) should be introduced into the formula. η_f can be found in [1-3] for some conventional fibers, while for the special profiled fibers, such as oval, cross-shaped, five-leaf and hollow, η_f should be calculated specially. Under these conditions, the middle-part linear density of a fiber can be fast and accurately calculated, and then along with the test of its tensile strength, modulus of the fiber can be ultimately obtained through final adjustment. This kind of testing instrument named XD-1 vibration-type fiber fineness tester was initially made by Chinese Textile University. Meanwhile, there are many other corporations where this kind of instrument is supplied in China.

3. Basic Knowledge and Applications of Acoustic Vibration's Energy Absorption of Bending Mechanical Properties

3.1 Basic Knowledge of Acoustic Vibration's Energy Absorption

Textile fiber assemblies cross with each other. Every little part can be regarded as a simple beam that is supported by two ends or a cantilever beam. The simple beam that is supported by one end will swing, bend and vibrate due to violent vibration when faced with acoustic wave formed by air transport (shown in Fig.2).

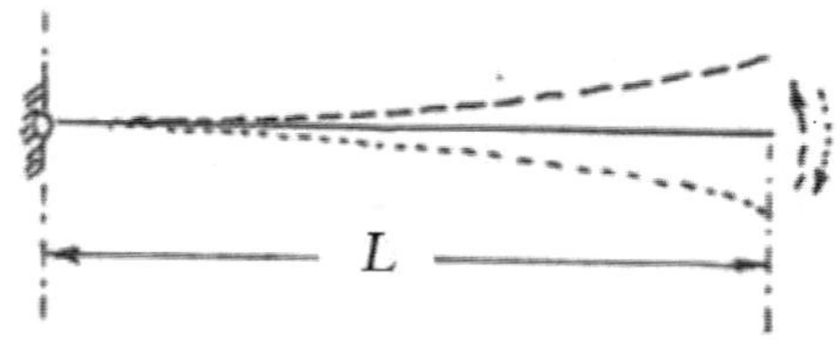

Fig.2 Bending vibration of cantilever beam with a single end support.

When the cantilever beam is very long, it not only has the first order vibration wave but also the second and the third. When cantilever beam is short, the first order vibration wave is dominant. Simple beam that is supported by two ends will also vibrate and bend when cross air transport acoustic wave as shown in Fig.1. These vibrations can exert the function of absorption of acoustic wave energy and precaution of the formation of noise through energy consumption and weakening of transport energy of acoustic wave.

In fact, this kind of vibration is the same as simple beam vibration of metal material, and mechanical principle is identical with mechanical vibration principle.

The basic equations of two-end simple beam are given as follows [12-15]:

Kinetic energy of vibration:

$$E_n=2\int_0^{L/2} \frac{1}{2}\rho \cdot A \cdot dndy^2=\frac{1}{2}(\frac{17}{35}\rho \cdot A \cdot L)y_o^2 \tag{4}$$

Equivalent mass of beam:

$$M_f=\frac{17}{35}\rho \cdot A \cdot L \tag{5}$$

First order frequency:

$$f = \frac{\pi^2}{L^2}\sqrt{\frac{E \cdot I}{\rho \cdot A}} \quad (6)$$

Concept of multi-order frequency:

$$\cos(kl)ch(kl)=1 \quad (7)$$

Coefficient of systematic mathematics: $(\frac{i\pi}{L})^2$ i=1,2,3,4...

Standard formulas of inherent frequency:

The first order:

$$f_1=\frac{3.730^2}{L^2}\sqrt{\frac{E \cdot I}{\rho \cdot A}} \quad (8)$$

The second order:

$$f_2=\frac{7.853^2}{L^2}\sqrt{\frac{E \cdot I}{\rho \cdot A}} \quad (9)$$

The third order:

$$f_3=\frac{10.996^2}{L^2}\sqrt{\frac{E \cdot I}{\rho \cdot A}} \quad (10)$$

The fourth order:

$$f_4=\frac{13.137^2}{L^2}\sqrt{\frac{E \cdot I}{\rho \cdot A}} \quad (11)$$

Where, f_i is resonance frequency, Hz, in order of the first order, the second order, the third order, and the fourth order; L is distance between the two end supports of simple beam, m; E is elastic modulus, N/m^2; I is moment of inertia of noncircular cross section, m^4; A is cross-sectional area of the beam, m^2; ρ is bulk density of the beam, kg/m^3.

$$I_o=\frac{\pi}{4}r^4 \quad (12)$$

$$I=I_o \cdot \eta_f \quad (13)$$

Where, I_0 is the moment of inertia of circular cross section, η_f is the correction coefficient of the moment of inertia of noncircular cross section [1].

Unit conversion is as follows:

Sectional area is translated to linear density of a fiber: tex is the mass in grams of 1000m long fiber.

$$A=\pi \cdot r^2 \quad (14)$$

$$N_t=A \cdot \rho \cdot 10^6 \quad (15)$$

E is obtained through E_0

$$E=E_o \cdot \rho \cdot 10^6 \quad (16)$$

Where, E_0 is initial modulus, N/tex.

Introduce (12), (13), (14), and (15) into (8), (17) can be obtained as follows:

$$f = \frac{n^2}{L^2}\sqrt{\frac{E_o \cdot \rho \cdot 10^6 \cdot I_o \cdot \eta_f}{\rho \cdot \pi \cdot r^2}} = \frac{n^2}{L^2}\sqrt{\frac{E_o \cdot \frac{\pi}{4} r^4 \cdot 10^6 \cdot \eta_f}{\pi \cdot r^2}} = \frac{n^2 \cdot r \cdot 10^3}{2L^2}\sqrt{E_o \cdot \eta_f} \quad (17)$$

Where, n is the equivalent constant of each order of resonance frequency.

3.2 Applications of Vibration Wave Absorption

When used in ear plug to absorb noise, length between the two ends of simple beam is very short. So first order string vibration is mainly existed.

Take circular cross-section polyester fiber for example:

$$E_o=11.074 \text{ N/tex}; \eta_f=0.91$$

r=0.1μm=10^{-7} m

$$f = \frac{3.730^2 \cdot 10^{-4}}{2 \cdot L^2}\sqrt{11.074 \cdot (10^{-7})^2 \cdot 0.91} = \frac{22083.13 \cdot 10^{-7}}{L^2} \tag{18}$$

r=1μm=10^{-6}m

$$f_{12} = \frac{3.730^2 \cdot 10^{-3}}{2 \cdot L^2}\sqrt{11.074 \cdot (10^{-6})^2 \cdot 0.91} = \frac{22083.13 \cdot 10^{-6}}{L^2} \tag{19}$$

r=10μm=10^{-5}m

$$f_{13} = \frac{3.730^2 \cdot 10^{-2}}{2 \cdot L^2}\sqrt{11.074 \cdot (10^{-5})^2 \cdot 0.91} = \frac{22083.13 \cdot 10^{-5}}{L^2} \tag{20}$$

Inherent frequency for different beams with different lengths can be calculated according to (18), (19), and (20), as shown in Table 1.

Table 1. First-order string vibration frequency for simple-beam polyester fiber

L(mm)	f(Hz)		
	r=0.1μm	r=1μm	r=10μm
0.5	8833.3	88332.5	883325.2
1	2208.3	22083.1	220831.3
2	552.1	5520.8	52207.8
3	245.4	2453.7	24536.8
4	138.0	1380.2	13802.0
5	88.3	883.3	8833.3
6	61.3	613.4	6134.2
7	45.1	450.7	4506.8
8	34.5	345.0	3450.5
9	27.3	272.6	2726.3
10	22.1	220.8	2208.3

According to Table 1, polyester fiber can absorb noises covering a majority of frequency in conventional air transport. So ear plug made by using loose fibers arranged randomly and compressed in a certain dense degree is effective. It is demonstrated that anti-noise effect of ear plug is not only related to air flow resistance, but also bending mechanical properties of textile fibers.

In addition, it is worth mentioning that there is a problem with anti-noise ear plug produced by textile fibers. Since 1980, in traditional test study of ear plug, noise intensity decay rate for sound wave has a frequency between 20-20000 Hz perceived by human hears have only been considered. So only this frequency interval was paid more attention during the study of fibers used in ear plug. But in some explosion environment. The worst damage to ear is in the initial stage. There is a serious instant negative pressure pulse after the positive pressure of the sound wave. In this occasion, tympanic membrane of both human and animals, especially for stape, incus, and malleus in internal auditory meatus will break due to this kind of negative pressure and make irreparable disability. Therefore, some thicker fibers (such as fiber with the diameter of 25 μm) should also be added into the ear plug, and ensure essential bulk density and stability with inner wall of ear canal. In the range of aural frequency, these fibers cannot act as energy absorption and noise protection, but in some special situations it can protect ear nerves from negative pulse caused by the start moment of explosion to a great extent.

4. Conclusions

Bending mechanical properties of fibers relevant with flattening, stiffening, bending, damaging, and breaking were conventionally paid more attention to. However, they also play a very important role in production of string vibration musical instruments and resonance absorption of air guided noise sound wave and protection. The corresponding theoretical basis and engineering implementation have also been formed.

References

[1] Yao M. Science of Textile Materials (Fourth edition). Beijing: China Textile Press: 2015; p: 261-266.
[2] Yu WD. Science of Textile Materials. Beijing: China Textile Press: 2006; p: 113-115.
[3] Li DG. Fiber Materials. Beijing: China Textile Press: 2006; p: 58-61.
[4] Jiang H, Zhu H. Textile Materials. Beijing: China Textile Press: 1979; p: 171.
[5] Sears FW. Translated by LIU JC. Exercises Answer for Second Book of University Physics. Beijing: People's Education Press: 1989; p: 144-145.
[6] Xia ZN. Dictionary. Shanghai:Shanghai Dictionary Press: 1989; p: 1370,1529,6,1698,1252,716.
[7] Wu LK. Sound Study of Erhu Fiddle and Teaching Exploration [M]. Beijing: People's Music Press: 2015; p: 50.
[8] Yao M, Zhou JF, Huang SZ, Shao LH, An RF, Fan DX. Science of Textile Materials (Second edition). Beijing: China Textile Press: 1990; p: 216.
[9] Lai K, Yao M. Influential factors of testing fiber linear density with disturbance resonator. Fiber Standards and Inspection 1996; (8): 5-7.
[10] Chen MY, Sun RJ, Lai K. Theoretic formula of Airflow Instrument for Cotton Fiber and Its Modification. J Tex Res 2011; (4): 118-122.
[11] China Textile University: Instructions for XD-1 Fiber Linear Density Tester. Shanghai: China Textile University: 1993.
[12] Yu KP, Zou JX. Structural Dynamics (Third edition). Harbin: Harbin Institute of Technology Press: 2015; p: 94-97.
[13] Wu TX, Hua HX. Mechanical Vibration. Beijing: Tsinghua University Press: 2014; p: 10,17-18.
[14] Zhang XK, Zan HP, Ling YW. Calculate Basic Physics. Xi'an: Northwestern Polytechnical University Press: 2015.
[15] Qing DT, Xie LY. Handbook of Modern Mechanical Design. Beijing: Chemical Industrial Press: 2013; p: 7,91.

Study on the Relaxation of Knitted Female Underwear Based on Pressure Comfort

Bo-Ni Zhang, Jun Xu*

Xi'an Polytechnic University, Xian, Shaanxi, 710048, China

*Corresponding author's email: xujun0127@sina.com

Abstract

Underwear fit and pressure comfort are of crucial importance to underwear design. This paper first used the Untouched 3D Body Scanner to scan the bodies of 110 young females, and extracted a total of 17 human body sizes related to underwear structure in all the directions including the horizontal direction, vertical direction and other directions. Through analyzing the data, this paper focused on that among the horizontal dimensions: waist circumference, abdomen circumference, hip circumference, and the thigh root circumference; for vertical dimension: height and waist high; and other sizes: lower trunk length, which has the highest variance and is the key part that affects underwear specification. Secondly, the 160/84A standard manikin was used as our study object. The importance of structure and style of underwear, abdomen, hips, thighs and fork length were studied and considered as the key parts of an underwear structure.

Moreover, this paper made the zero relaxation prototype with inelastic fabric by utilized vertical cutting. The negative loose quantity in several key parts is increased to make five different pairs of relaxation knitted female underwear. Eight eligible subjects are invited to try on the underwear to complete a series of dynamic testing, which also gives the subjects a chance to evaluate pressure comfort under different circumstances. Finally, through analyzing the data, this paper finds the measurements for knitted underwear that gives the most comfortable feeling: when the loose quantity of abdomen circumference and hip circumference is -17cm, that of thigh circumference is - 12cm, and that of fork is 13cm.

Keywords: The relaxation; Pressure comfort; Women underwear

Preface

At present, China's population reaches 1.4 billion people and there are around 670 million women. According to a survey, most of the women would buy 4 to 10 panties every year, therefore, there is great potential in the development of the female underwear market. With rapid developments in the society, modern female underwear does not only play a role in shading and sanitation, but also for comfortableness and health. The latter has not been studied or discussed as much in underwear design.

Loose quantity is one of the important factors contributing to the comfortableness of clothing. Previous studies have shown that the loose quantity is affected by many factors, including but not limited to basic physiological needs, style, aesthetics and motion. Among these factors, fabric texture also plays a significant role, for example, woven and knitted fabric has obvious differences in the usual loose quantity on the human body. Woven fabric is not very elastic, therefore loose quantity needs to be added to meet the physiological needs, Due to the flexibility of knitted fabric, a negative loose quantity can meet the demand.

Wearing comfort affects not only the wearer's psychological feelings, but also her physical health. Clothing pressure comfort is an important index for evaluating clothing comfort [1]. Since the underwear has direct contact with the human body, when the body is in motion, certain amount of pressure will be generated on the body, influencing wearing comfort. The design of a relaxing style can affect wearing comfort in a dynamic state, and it plays an important role in version design [2]. Since underwear fits your body and is in close contact with your skin, a more relaxed style will be more comfortable for the wearer, though it might affect the aesthetics. Relaxation alone is not enough, as it will produce large pressure, which affects the wearing comfort. Thus, it is important to analyze appropriate relaxation styles of underwear to exert the appropriate amount of pressure to the body.

1. Key factors of underwear specifications

1.1 Measuring equipment

Non-contact 3D body scanner is a high-tech human body measurement technology that uses modern optical technology and combines information processing technology, electronics, computer vision and computer image applications. At the same time, it can also measure the basic data of anthropometric measurements quickly and conveniently using the computer automatic measurement function that redefines the image information [3].This experiment uses the Non-contact 3D body scanner (model: JB-3DLS-BODY) located in the clothing engineering center of Xi'an Polytechnic University, as shown in Fig.1.

1.2 Body parts being measured

This paper uses the method of non-contact 3D body scanning (Brand: Human Solutions; Region: German; Type: VITUS Smart XXL) and chooses young women aged from 18 to 25 as experimental subjects. The effective sample number is 110. According to the law of the People's Republic of China national standard GB/T 16160-2008, there are currently 52 body measurement parts [4], and this paper chooses 17 related body parts for measurement. The woman's underwear reference data includes height, waist height, waist circumference and hip circumference in the Chinese standard [5], as shown in Fig.2.

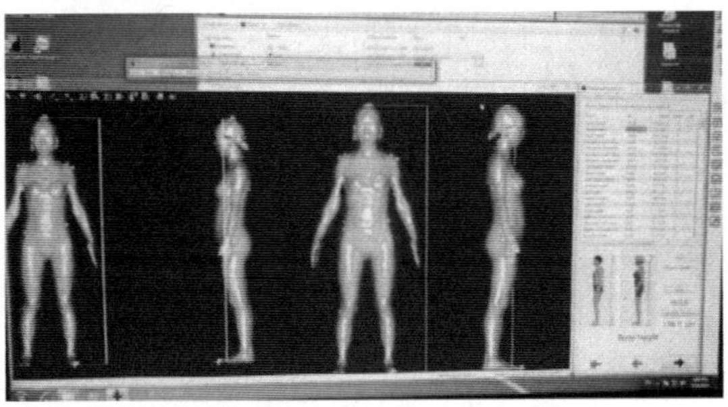

Fig.1 3D Body scanner

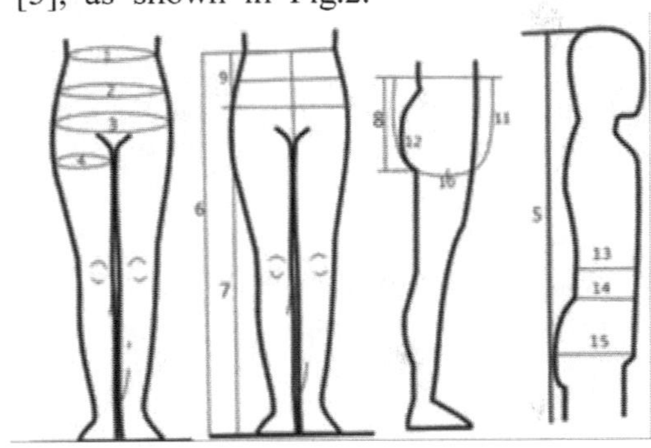

Fig.2 Figure at site

(Fig.2 Remarks: 1. waist girth; 2. abdominal girth; 3. hip girth; 4. thigh girth; 5. height; 6. waist height;7. hip height; 8. body rise; 9. waist to hips; 10. total crotch; 11. front crotch; 12. back crotch; 13. waist depth; 14. hip depth; 15. hip depth; 16. hip breadth; 17. waist breadth)

1.3 Measurement result analysis

Statistical methods were adopted to analyze the measurement results, as shown in Table 1.

Table 1 reflects the basic data of the measured object parts, including the maximum, minimum, average, standard deviation and variance. The average represents the average measurement of each part, variance and standard deviation reflect a basic deviation degree of the measured object data. The larger the value, the greater the degree of deviation.

Table1. Body part statistical analysis table

Part	Maximum	Minimum	Average	Standard	Variance
Waist girth	95.5	56	70.1	7.71	59.53
Abdominal girth	104.4	61	82.5	6.8 2	46.62
Hip girth	61	41.1	49.4	3.9 1	15.43
Thigh girth	104.8	79.5	89.4	4.7	28.52
Height	174.8	144	158.2	16.2	34.51
Waist height	108.7	87.2	96.77	4.38	19.16
Hip height	89.6	69.2	76.8	3.7	13.72
Body rise	33.9	25.7	28.6	1.74	3.01
Waist to hips	24.4	15.3	20.55	1.71	2.92
Total crotch	82.3	62.4	71.15	3.98	15.88
Front crotch	40.4	29.9	34.59	2.06	4.24
Back crotch	42.3	32	36.58	2.21	4.9
Waist thickness	29.7	15.3	20.17	2.65	7.02
Abdominal thickness	29.4	16	22.35	2.54	6.44
Hip thickness	30.5	19.5	22.91	1.87	3.5
Hip breadth	37.3	30.8	33.67	1.62	2.63
Waist breadth	32	21.5	25.4	1.62	6.17

1.4 Key parts of extraction

According to the law of the People's Republic of China national standard GB/T16160-2008, the human body size i s divided into three categories: a. horizontal dimension size (including surrounded degree); b. vertical dimension; c. other sizes. In the 17 sets of data of the human body associated with underwear, the parts with the greatest variance are picked after the statistical analysis: horizontal dimension: waist circumference, abdominal girt h, hip circumference, and thigh circumference; vertical dimensions: height, waist height; and other sizes: a long t runk

2. Fabric's performance test

2.1 Selection of measurement project

The fabric's weight, warp and weft density, and thickness can directly affect the size of the knitting clothing, and have an impact on wearing comfort. As the knitted fabric has elasticity and the underwear is close-fitting, the knitted underwear will adjust to the human skin. Influenced by clothing specification size and characteristics of design, fitting clothing should provide enough space for the skin. The elasticity of the fabric is an important factor of pressure comfort, which depends on the fabric's elastic recovery performance. Fabric elasticity is mainly measured by its elongation rate, tensile elastic elongation response rate, index of plastic deformation, and tensile stress relaxation [6]. The measurement of elastic performance of knitted fabric can provide a convenient and scientific basis for designing the sizes of underwear.

2.2 Measurement result analysis

The experiment was performed in a standard environment where the temperature is at (20 + 2)℃ and relative humidity at (65 + 2)%. In order to guarantee the accuracy of experimental results, each layout project was measured five times and the average of the five values was taken and presented as the final test results, shown in Table 2.

Table 2. Fabric performance schedule

	Weight (g/m^2)	Thickness (mm)	Density Warp /weft		Elongation Warp /weft (%)		Restoration ratio Warp /weft (%)		Deformation rate Warp /weft (%)	
Average	182	0.48	77	169	13.8	27.1	74.4	66.2	3.5	9.1
Standard deviation	3.03	0.0136	1.73	1.64	0.6	0.6	1.9	0.2	0.2	0.1

(Remarks: The instrument measuring fabric weight is MP120-1 ELECTRONIC BALANCE and data is measured according to the test standard BS EN12127.

The instrument used to measure fabric thickness is YG141D Digital fabric thickness tester and data is measured according to the test standard GB/T3820-1997.

The instrument used in the fabric tensile elastic recovery experiments is YC026PC-500 Electronic strengt h tester and data is measured according to the standard FZ/T70006.)

The variation in fabric's elasticity can be divided into high elastic fabrics, shot in fabric, and low stretch fabric, according to the elasticity of the clothing material. High elastic fabrics have good elongation and rapid rebound resilience, in which its elasticity is always in the range of 30% - 40%, and it can be used to make sportswear and elastic pants; that of shot in fabrics is between 20% to 30%, and it is suitable for making suits and underwear; that of low stretch fabrics is under 20%, and it is suitable for making trousers and coats. According to the tensile force experiment mentioned above, it is known that the fabric chosen for this study is a shot in fabric. Taking other performances into account, it can be found that this kind of knitting fabric is suitable for making underwear.

3. Version design

3.1 Style choice

The first principle of underwear selection is degree of comfort, followed by the degree of moisture, durability, fabric elasticity and fit front crotch [7]. Through investigation and analysis of the underwear market, it is found that waist boxer is most popular among young women due to its wearing comfort. This experiment chose to study the waist boxer style to cater to the public's need. However, in order to achieve better body shaping effect, an upward inclined angle will be adopted in the leg design (Specific design in Fig.3).

3.2 Obtaining the prototype

The main object of the study is young women with a height of 160cm, and this experiment used standard 160/ 84 A mannequin as the draping prototype. In order to study characteristics of underwear independently, inelastic fabric without any loose quantity was used as the draping sample cloth to eliminate effects of elastic fabric (Draping process in Fig.4).

Front Back

Fig.3 Low waist female boxer

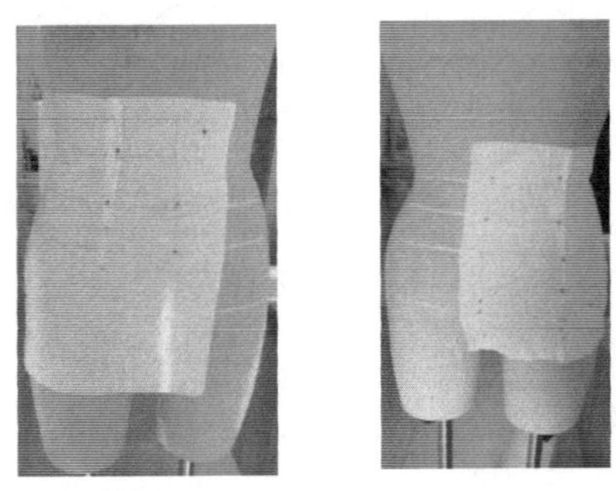

Front Back

Fig.4 Draping process

Considering the wear comfort and aesthetic requirements in bottom crotch in the process of draping, the crotch width was set to 9cm and length was set to 15cm under the condition of no loose quantity. Taking down the cutting sample, using the principle of dart merge, merging the front and back dart, then joining the crotch pieces together (as shown in Fig.5).

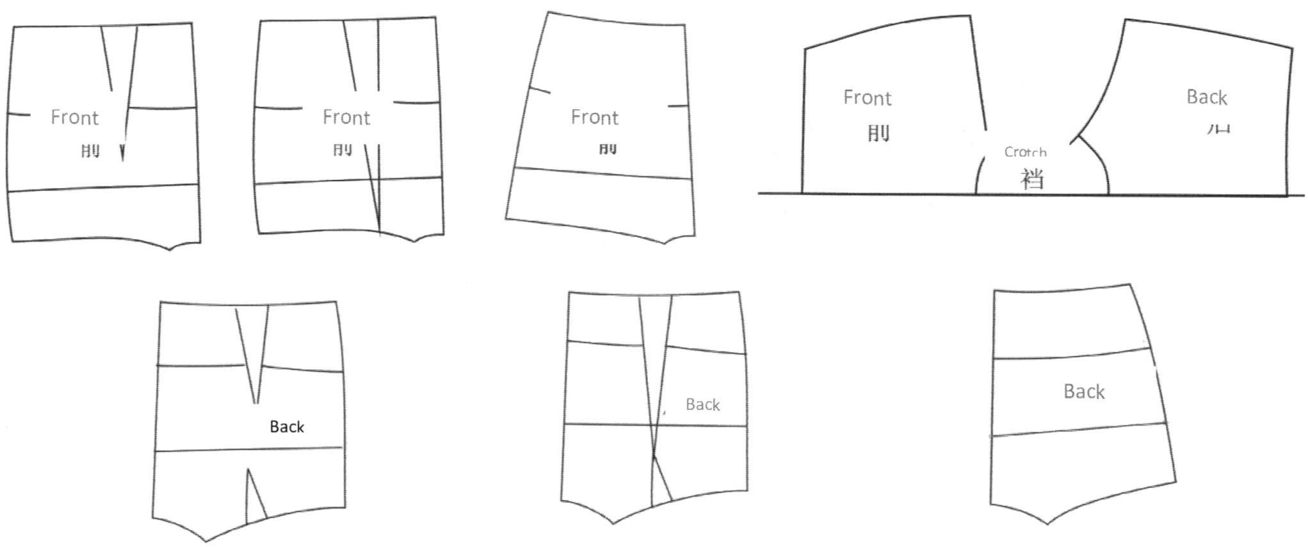

Fig.5 Underwear prototype

Since no relaxation was added to the underwear prototype, it can be assumed as a zero ease underwear. Abdomen and hips reduction was not addressed when the dart was transferred, as the knitting fabric has certain elasticity. Table 3 shows the prototype sizes of the zero ease underwear. Negative relaxation can be increased accumulatively based on the values in this table; the surround degree direction, abdominal girth, hip girth and thigh girth can be reduced by 4cm; the length direction and bottom crotch can be reduced by 0.5cm, in turn. A total of five underwear prototypes were produced and named as 1#, 2#, 3#, 4#, 5#, according to the order of relaxation increases (as shown in Table 4).

Table 3. Prototype specification unit: cm

Part / Size	Abdominal girth	Hip girth	Thigh girth	Side seam	Center front length	Center back length	Crotch length	Crotch width
Manneauin size	83	89	52	25	25.5	26.5	15	9
Prototype size	79	85	52	14.5	15.5	16.5	15	9
Difference value	4	4	0	10.5	10	10	15	9

Table 4. Underwear negative loose quantity table Unit:cm

sample	Abdominal girth	Hip girth	Thigh girth	Side seam	Center front length	Center back length	Crotch length
1#	79	85	52	14.5	15.5	16.5	15
2#	75	81	48	14.5	15.5	16.5	14.5
3#	71	77	44	14.5	15.5	16.5	14
4#	67	73	40	14.5	15.5	16.5	13.5
5#	63	69	36	14.5	15.5	16.5	13

4. Subjective pressure experiment design

4.1 Choice of tester

According to the national standard and prototype size requirements, in order to guarantee the accuracy of the test, this experiment chooses eight college female students as the experimental subjects (Specific information shown in Table 5).

Table.5 Tester information table

Age	Height(cm)	Waist girth(cm)	Abdominal girth(cm)	Hip girth(cm)	Thigh girth(cm)
22-25	158-164	62-68	81-87	86—92	52—57

4.2 Choice of experimental action

Five actions were selected: normal standing, walking, bending the leg by 90°, lifting the leg, and sitting. These actions are involved in the daily motions of the human body in everyday life (as shown in Fig.6).

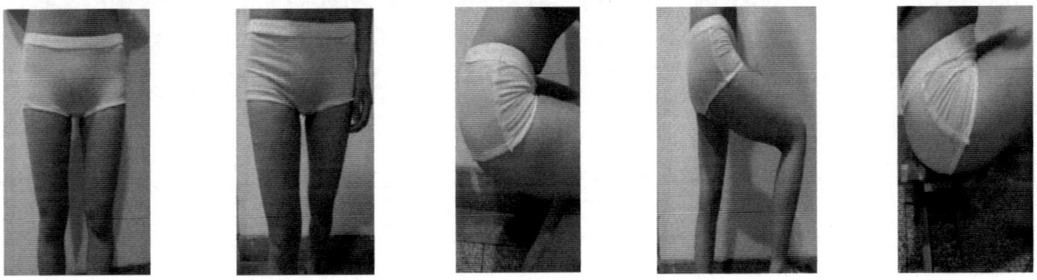

Normal standing Walking 90 ° bend the leg Lift leg Sitting

Fig.6 Choice of experimental action

4.3 Parameters to be studied and analyzed

This experiment chose to study the effects on the abdomen, hips, thighs, front crotch, back crotch and bottom crotch.

4.4 Data collection and analysis

From a psychophysical aspect, quantitative estimation is a simple and wide method of psychological scale. Likert scale is used to show the respondents' attitude, view, evaluation and intention for something. Likert scale includes a 5-level scale, a 7-level scale and a 10-level scale [8].

This experiment adopted the 7-level scale, and the pressure level increases from 0 to 6 (as shown in Fig.7).

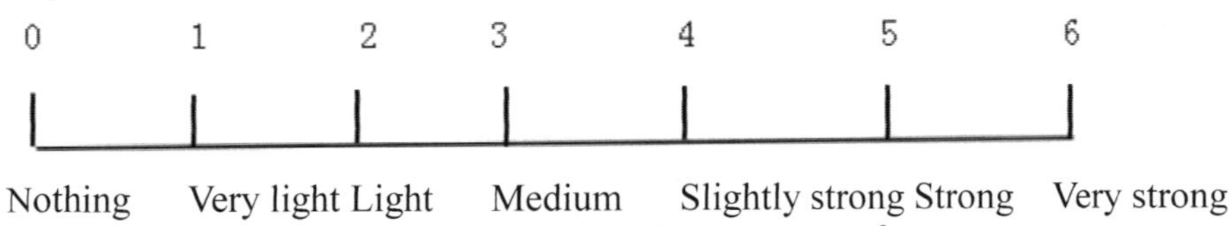

Fig.7 Evaluation pattern of pressure comfort

In the test, the eight subjects were asked to put on five different kinds of underwear and complete the five selected actions as described. After each action is completed, the subject must fill in a questionnaire within a time limit. Through calculating the average of values obtained from the eight subjects doing each motion, this study collected the statistics on the maximum and minimum level of pressure comfort (as shown in Table 6).

Table 6. Partial pressure comfort distribution of different loose quantity underwear

Serial number	Abdomen	Back hip	Front crotch	Back crotch	Thigh root circumference	Bottom crotch
1#	0.93-2.02	1.03-1.71	0.88-1.12	0.89-1.58	0.99-1.61	0.73-0.99
2#	1.32-2.18	1.45-2.47	1.34-1.58	1.89-2.78	1.45-2.45	1.43-1.63
3#	2.37-3.25	2.53-3.19	1.56-1.79	2.69-3.45	2.4-2.8	1.89-2.21
4#	2.47-3.65	3.25-4.53	1.73-1.99	2.43-2.93	2.86-3.76	2.31-2.57
5#	3.89-5.03	4.32-5.12	2.30-3.78	3.17-4.25	4.32-5.23	3.00-3.78

5. Conclusion

According to table 6, the general scope of pressure comfort generated from different relaxation underwear can easily be identified. For underwear 1# and 2#, the underwear were too loose on the subjects and did not fit their bodies, thus they did not experience much pressure from the underwear. On the contrary, underwear 5# was too small, exerting more pressure and uncomfortableness on the subjects. Since the underwear 3# and 4# were both within the reasonable scope of relativity, the mean value can be used as the optimal relaxation value, that is when abdomen pluses equals -17cm, the quantity of hip circumference equals -17cm, crotch size equals 13cm, and the quantity of reasonable thigh root circumference size equals -12cm. In addition, this test is mainly based on a subjective evaluation method, which may have some biases on the results. In the future, we hope to perform studies combining the subjective evaluation method with an objective pressure test to improve results of the experiments.

References

[1] Song XX, Feng XW. The relationship between Clothing pressure and human body comfort [J]. Journal of Textile Research: 2006; 3: 103-105.

[2] Jiang RF. The research and analysis of young men's underwear fit [D]. Xian: Xi'an Polytechnic University: 2013.

[3] Miehelle H, Demers, Jerfery D, Hurley, Riehard C, WulPern. Three Dimension Surface Capture for Body Measurement Projected Sinusoidal Patterns [Z]. SPIE: 1997; 3023.

[4] People's Republic national standard of China GB/T [S]. The measurement part of the human body and method for clothing 16160-2008.2008.

[5] Yin JR. Underwear design tutorial[M]. China's textile publishing: 2005; 23-24.

[6] People's Republic national standard of China FZ/T70006-2004 [S].The test method for Knitted fabric tensile elastic recovery rate: 2004.

[7] Chang JL, Yin JR. Underwear pattern design principle and instance[M].Shanghai:Shanghai Textile science and technology: 1995; 47-49.

[8] Jin LB. Likert scale statistical analysis and fuzzy comprehensive evaluation [J]. Journal of ShanDong science: 2006; 2:18- 24.

Tucking Clothing Dart in the Virtual World

Inui Shigeru[1*], Miyahara Noriki[2], Yosuke Horiba[2]

[1]*Interdisciplinary Cluster for Cutting Edge Research Institute for Fiber Engineering Shinshu University, 3-15-1 Tokida, Ueda, Nagano, 386-8567, Japan*
[2]*Faculty of Textile Science and Technology Shinshu University, 3-15-1 Tokida, Ueda, Nagano, 386-8567, Japan*

*Corresponding author's email: inui@shinshu-u.ac.jp

Abstract

Draping is one of the methods to design clothing, and it is possible to increase efficiency considerably by virtualizing draping. We constructed a prototype system for virtual draping, and the process to tuck a dart became reality. Cloth, hand and dress form were modelled. The movement or deformation was calculated mechanically in real time. The hand model was moved according to the movements of a hand in the real world detected by a sensor. An initial shape of the cloth model was made and the cloth model was handled by the hand model to tuck dart. It became possible to execute the fundamental process of draping in the prototype system.

Keywords: Virtualized Draping; dart; cloth model; Distance Function

1. Introduction

Production of clothing consists of many processes. Clothing design is different from other industrial products because it is based on design illustrations drawn by fashion designers. Pattern makers make paper patterns as design drawings of clothing from the design illustrations. To make paper patterns is one of the most important processes to design clothing, and there are two major methods to make paper patterns, planar drawing and draping. Textiles are cut along the outline of the paper patterns to make parts of clothes, and the parts are sewn together to make clothes.

In the draping processes, human body or dress form is utilized for the basis to make paper patterns. Cloth is cut to proper size according to the paper pattern to be made. The cut cloth is fitted to the surface solid shape of human body or dress form, and it is temporarily pinned at proper position as needed. When the cloth is fit to a curved surface such as human body surface, it is impossible that the cloth is completely fit to the surface as it's original shape is plane. Paper pattern is made to cut the excess part of the cloth by taking dart.

Draping has both advantages and disadvantages. It is better to make clothing fit to body shape of each person by draping than planar drawing. On the other hand, the method of draping takes more time and effort than the method of planar drawing. Paper patterns are made by draping for made-to-order. Made-to-order has to be high-cost because it requires much time and effort to run through the processes of basting and fitting.

The final goal of this study is to virtualize the whole processes of draping to bring drastic efficiency to the method of draping. And the purpose of this study is to virtualize the process to tuck dart as the part of draping processes. It is possible to make the processes of draping more efficient by virtualizing the draping processes. Basting prototype clothing in the virtual world is also possible if paper patterns are made by virtualized draping system. It takes much shorter time to baste prototype clothing in the virtual world than in the real world. If the data of human body surface shape is provided, it is possible to fit clothing and check defects in the virtual world in a very short period of time. For this reason, it is expected that the work efficiency is increased and the cost is controlled for draping.

Various studies to obtain pattern shape from three dimensional body shape have been carried out. In those methods, the garment form is represented by curves, a wireframe and surface, and changed by deforming the curves and wireframe to obtain the pattern. Cho et al [1]. proposed a pattern-making method for various body forms. The pattern of the garment is obtained by developing the contoured surface of the dress form model. Wang et al.

presented the method for making the garment pattern using two types of curve: contour curves and style curves [2]. First, the garment surface is modeled and contour curves are generated. The garment form is changed by deforming the curves. Such as darts and grain line are added to the garment surface by the style curves, and then two dimensional garment pattern is made. Huang et al. made wireframe-model by using a body features, and a garment form is made by deforming the wireframe [3].

Various studies on modelling of cloth behaviour and manipulation have been carried out. Harada et al [4]. presented real-time collision calculation between a cloth model and a complex object in virtual space. They modelled a cloth and an object with particles and springs, and defined collision detection and reaction between the cloth model and the object by the distance between particles in the cloth model and the object. Meng et al [5]. presented a system of virtual try-on simulation in real time. In this system, the user manipulates garment patterns and makes a rough wearing form before simulating the final try-on form. There are four types of manipulations such as moving and fixing. Igarashi et al. presented a set of interaction techniques for putting clothes on a 3D character. The techniques can put cloth models on the 3D character by marking on the cloth pattern and 3D character. Wibowo et al [6]. performed draping interactively using two special tools in virtual space. An operator cut cloth model by drawing lines with the tools on the surface and in the space. Paper patterns are made by draping and then an actual garment is made from the patterns.

In these studies, the cloth model manipulation was performed using a mouse. On the other hand, studies to manipulate and deform an object in virtual space by movement of hand have been also conducted. Satoh et al [7]. presented 3D CG modelling method by hand gesture. They detected movement of hand by Kinect and deform the object model with the movement. Liang el al [8]. proposed a system in which a non-rigid hyper touch interaction is used to detect the movement of hand and to manipulate an object in virtual space. As mentioned above, real-time virtual draping has been studied but the manipulation is not by the movement of hand. Mesuda et al [9]. studied a method to handle the cloth based on the movement of hand in real-time. Draping requires collision detection and reaction for various manipulations, such as cutting, attaching and cutting the cloth model.

2. System of Virtual Draping

The virtualized draping system consists of cloth model, hand model and dress form model. The shape of the cloth model is defined geometrically. The cloth model is formulated mechanically, and properties are set to the geometrically defined shape. The deformation of the cloth model by internal and external forces and boundary conditions can be obtained by mechanical calculation. The shape of the cloth is also deformed by the contact between the cloth model and the hand model and between the cloth model and the dress form model. The hand model is composed of geometrical forms. Movements of hand and fingers in the real world are detected and transferred to a computer by a sensor. The model of hand and fingers in the virtual world is moved according to the transferred information of the hand and fingers in the real world. The cloth model in the virtual world is handled with this human-machine interface.

2.1 Cloth Model

To model real cloth, geometrical shape of the cloth model has to be defined. The shape of the cloth model is limited to rectangle when it is not deformed. In draping method, paper patterns are made as the design drawing of the parts of clothing. To begin the process to make a paper pattern, rectangular cloth is cut the size of which is little bigger than the size of the part to make.

Then, cloth is mechanically formulated. There are two kinds of mechanical models for cloth, discrete model and continuous model. Here it is necessary to move the cloth model along with the movement of hand and fingers in the real world, so real time response is required for the cloth model. The mechanical calculation has to be executed in high-speed to react the movement of hand and fingers in real-time, and here complicated calculations can not be executed over time. Then, cloth is mechanically formulated with discrete masses and springs. Lattice is set on the cloth model, and springs are set between neighbouring lattice points. Some springs are along the directions of the lattice and some springs are along diagonal line of the lattice. An ideal property is given to the springs for tensile and compression, and the bending property between connected two springs is not given at this time. The value of the mass and the spring constant are set for the cloth model not with the value of measurement but the value which are proper to move the cloth model in real time.

2.2 Mechanical Calculation

The shape of the cloth model is determined by mechanical calculation. The cloth model consists of lattice points, and mass is given to each lattice point. The gravity force and the internal forces from the connected springs act on a lattice point of the cloth model. By the resultant of the forces, the time revolution of each lattice point of the cloth model is calculated by the motion equation. The motion equation is discretized and numerically integrated. Since the cloth model has to be moved along with the movement of the hand and fingers in the real world, it is necessary to adopt high speed calculation method. An explicit method is adopted which is less stable but high speed, but implicit methods which is stable but time consuming for numerical integration.

2.3 Geometrical Calculation

Geometrical calculation is also necessary for the cloth model. Moved or deformed cloth model sometimes reacts with the hand or the dress form model. The collision between the cloth model and other models has to be detected and reacted. The lattice points of the cloth model are considered to be spheres for collision detection and reaction, and a certain radius is set to the lattice points.

2.4 Human-Machine Interface

The cloth model in the virtual world is handled freely by the movement of hand and fingers in the real world. LeapMotion is utilized to detect the movements of hand and fingers in the three-dimensional space of the real world. LeapMotion detects the coordinates of the joints of hand and fingers and transfers the information of the coordinates to a computer in real time. Hand and fingers are modelled as the nodes of the joints in the virtual world. The hand model in the virtual world is moved according to the coordinate of the joints of hand and fingers in the real world.

Collision has to be considered between the cloth model and the hand model. A radius is set to the node of the joints of the hand model for detection and reaction of collision. When the hand model is shown on a display, the joints of the fingers are drawn as spheres. The radius of the finger joints is also utilized for the detection and reaction of collisions. The collision between the cloth model and the hand model is checked as the collision between lattice points of the cloth model and joint nodes of the hand model. If the distance between a node of a finger joint and a lattice point of the cloth model is shorter than the sum of the radius of the finger joint and the radius of the lattice point of the cloth model, a collision is detected. The collision is resolved to act a force on the lattice point of the cloth model. The direction of the force is the direction of a vector from the node of finger joint to the lattice point of the cloth model.

2.5 Dress form Model

In the real world, human body or dress form is used for draping. In the virtual world, solid object such as dress form model is also needed to fit cloth model in draping method. The surface shape of a real dress form (Kiiya, New Kypris 5AR-R) is measured by a three-dimensional shape measurement system (VOXELAN, LPW-2000FW). A dress form model is made from the coordinates of measured point clouds, then distance function is made from the dress form model. The distance function is a distance field to the surface of the dress form model. This distance function is used for collision detection between the cloth model and the dress form model. Here a discrete distance function is made from the dress form model. For the discrete distance function, a three-dimensional lattice is set in the cubic region containing the dress form model. The distance from a point to the surface of the dress form model is set for each lattice point. The distance of a point other than the lattice points is obtained by interpolation of the distances of the lattice vertices of a unit cubic lattice which contains the point. When a lattice point of the cloth model comes into the region of the distance function, the distance between the cloth model and the dress form model is calculated by the distance function. If the distance is shorter than a threshold, a collision is detected. Then, a vector from the surface of the dress form model to the lattice point of the cloth model is calculated, and a force the direction of which is same as the direction of the vector is applied to the lattice point of the cloth model. By this force, the collision between the cloth model and the dress form model is avoided.

2.6 Initial Shape of the Cloth Model

A dart is tucked when the cloth is temporarily pinned on the human body or dress form. In the real world, woven fabric consists of warps and wefts is mainly used for draping. For example, in the making processes of a skirt, cloth is fitted on the waist part of a dress form to make paper pattern. In this process, warps of the cloth are fitted to the vertical direction, and wefts are fitted to the horizontal direction. To make a skirt of four parts, the paper patterns of the right hand side are made, and then the parts of the left hand side are made from the right hand side patterns as the symmetrical shape. To make front right hand side paper pattern, one side of a rectangular cloth is fitted to the front centre line of a dress form and then the cloth is temporarily pinned along the line. To form a solid shape by cloth which is plane in a normal situation, darts have to be tucked over hip line as the circumference of waist is shorter than that of hip. Some warp lines are fitted in the process of making paper pattern. It means that a warp line is fitted to a vertical line on the surface of a dress form. The vertical line on surface means a line of intersection of a plane including vertical central axis of and surface. To tuck a dart, the warp lines of the both sides of the cloth model are fitted and temporarily fixed to vertical lines on surface. Some of the cloth is left in the cloth model over hip line of the cloth, cloth of the area is suspended above the surface of the . A dart is tucked to pick up the suspended area of the cloth by fingers. The dart is temporarily pinned and marked, and cut out when the cloth is opened.

In the virtual world, the shape of the cloth model which is temporarily fixed on the surface of the model is made to tuck a dart. Lattice is set on the rectangular cloth model. The lattice line in the lengthwise direction corresponds to warp direction, and the lattice line in the crosswise direction corresponds to the weft direction. The vertical central axis of the model is determined from the average coordinate value of the point cloud, as the model is made from the surface points of the in the real world. A vertical line of surface is the line of intersection of the plane including the vertical centre axis and the surface of the dress form model. Some of the lattice lines in warp direction of the cloth model are mapped to vertical lines of dress form surface to tuck a dart. In the case of front paper pattern, one of the sides of the cloth model is mapped to the front centre line of the dress form model. The upmost lattice point on the mapped lattice line of the cloth model is mapped to the position little over the waist line on the front centre line of the dress form model. Then, the length of the lattice pitch of the cloth model in warp direction is taken on the front centre line of the dress form model, and the second lattice point from the top on the mapped lattice line is mapped to this position. The lattice points on the mapped lattice line of the cloth model are mapped one after another on the front centre line of the dress form model. Hip line is one of the cross section of the dress form model. The nearest lattice point to the hip line is identified among the lattice points of the cloth model mapped on the front centre line of the dress form model. The lattice point neighbouring to the identified lattice point is mapped on the hip line. When the neighbouring lattice points in the horizontal direction are mapped one after another, one of the lattice line of the cloth model in the weft direction is mapped approximately on the hip line. The lattice points of the cloth model under the hip line are mapped in the same way as the lattice points on the hip line. Another side of the cloth model is mapped to a vertical line of dress form surface in the same way as the front centre line of the dress form model. The lattice points of the cloth model other than the lattice points which has already mapped on the dress form surface can not be mapped on the surface of the model, because the length of the cloth model in the weft direction is longer than the length of the model in the horizontal direction. The position of the lattice points which has not been mapped is determined so that the cloth model is suspended above the surface of the dress form model. Thus, the shape of the cloth model to tuck dart is determined in the virtual world.

2.7 Algorithm to Tuck Dart

Followings are the processes to tuck a dart in the real world. When a dart is tucked for a skirt, the right and left hand side of cloth is fixed, and the cloth between the both sides is suspended above the surface of a dress form. To tuck a dart, the suspended cloth is tucked from the left and the right hand sides to the centre, then the cloth is got together to the centre. In the case of skirt, the cloth is got together to the centre from the top to down below. Here we call the base of slit the bottom of a dart. A dart is tucked from the top to down below and then the position at which dart can not be tucked is determined to the bottom of the dart. When the position of a dart is determined, the dart is pinned or marked and then the cloth of the dart can be cut off.

An algorithm is determined to tuck a dart in the virtual world. The initial shape to tuck a dart described above is made and the cloth model can be tucked, as the lattice points of the cloth model are handled in the virtual world by the movement of hand and fingers in the real world. The position of dart has to be determined when the cloth model is tucked. Attention is paid to a lattice line in weft direction of the cloth model crossing the finger models that are tucking the cloth model. The lattice points at both ends of the lattice line are fixed on the vertical line of dress form surface. From the lattice points of both ends, the lattice points on the lattice line in the weft direction are explored one after another to the direction of the centre of the dart. When the distance between the surface of the dress form model and the lattice point of the cloth model is shorter than a threshold, the lattice point is judged to be on the surface of the dress form model. The lattice points on a dart are located inside of the fingers tucking the cloth model, and the distance from the points on a dart to the surface of the dress form model is longer than a threshold. The position of the dart is determined from the average value of the coordinate of the lattice points on a dart. The position of the bottom of the dress form is on the vertical line on the dress form surface which pass the position of the dart. A line is obtained as an intersection of a plane and the cloth model, and the plane contains the vertical centre axis of the dress form and the vertical line of the dress form surface. The bottom of the dart is the point on the line at which the distance between the surface of the dress form model and the lattice point of the cloth model. From the position and the bottom of the dart, dart part can be cut off from paper pattern.

3. Results and Discussion

The system described above is developed and executed on a computer. Microsoft VisualStudio 2013 is utilized for development of the system, and the hardware for the system is iMac (Core i5 3.5GHz, DDR3 8GB). The numbers of lattice in warp and weft direction are respectively 15 and 15. Fig.1 shows the situation to handle the cloth model in the virtual world by the movement of hand and fingers in the real world. It is possible to handle the cloth model in the virtual world in real time. Fig.2 shows the initial shape of the cloth model to tuck a dart. In the figure, the cloth model is fixed only on both vertical sides. Some part of the cloth model is suspended above the dress form surface which contacts on the dress form surface in usual situations to decrease vibration by the collision of the cloth model and the dress form model. Fig.3a shows the cloth model tucked by virtual fingers to tuck dart. Fig.3b shows the cloth model of Fig.3a, but the virtual fingers are not shown to see the shape of the cloth model. Spheres in Fig.4 indicate the position of dart on the cloth model. The sequence of the processes to provide the initial shape of the cloth model, to tuck by fingers, to tuck dart and to make a paper pattern becomes possible in the virtual world.

Followings are the future studies. As it is now, it is rather difficult to position a dart precisely as the lattice pitch of the cloth model is coarse. To execute the calculation in real time, the lattice pitch of the cloth model is constant while at work. On the other hand, to position a dart precisely, fine lattice pitch is desirable. While taking a dart, the cloth model is tucked on both sides, and the tucked part of the cloth model becomes narrow. The cloth model which is located outside of the finger model is fit to the surface of the dress form model, and that area of the cloth model does not move nor deform while the process to tuck dart progresses. It may possible to execute more precise calculation in real time to change dynamically the lattice pitch of the cloth model which has not been fit to the surface of the dress form model. Some times, some part of the cloth model vibrates when it is tucked by the model of fingers. In this case, it becomes difficult to position a dart precisely. The vibration happens when the cloth model and the model of fingers are colliding, and the collision between the cloth model and the dress form model is related to the vibration. The vibration may be resolved by decreasing dynamically the value to adjust the location as the reaction of collision when the vibration happens.

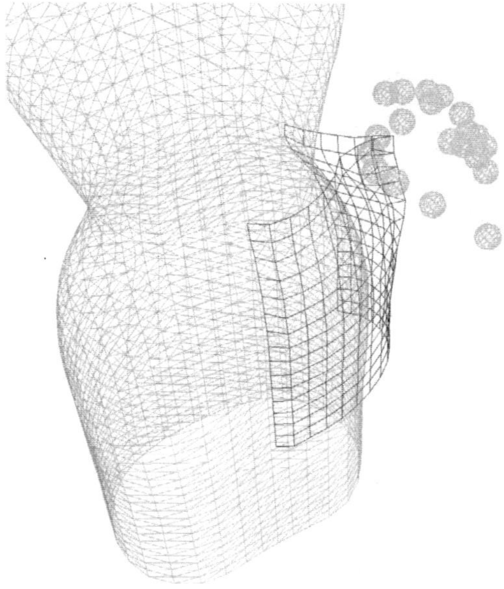

Fig.1 Handling the Cloth Model

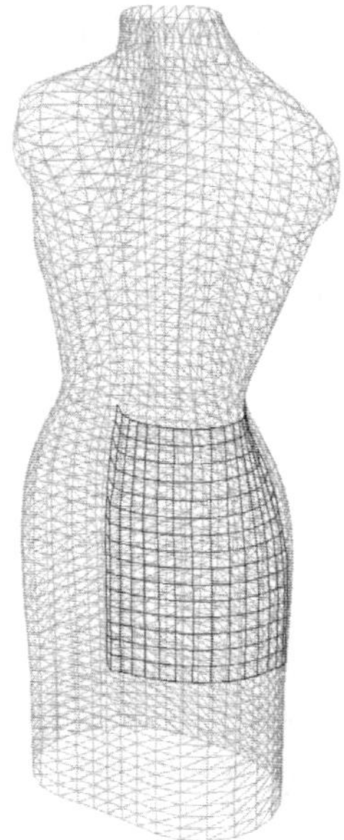

Fig.2 Initial Shape of the Cloth Model

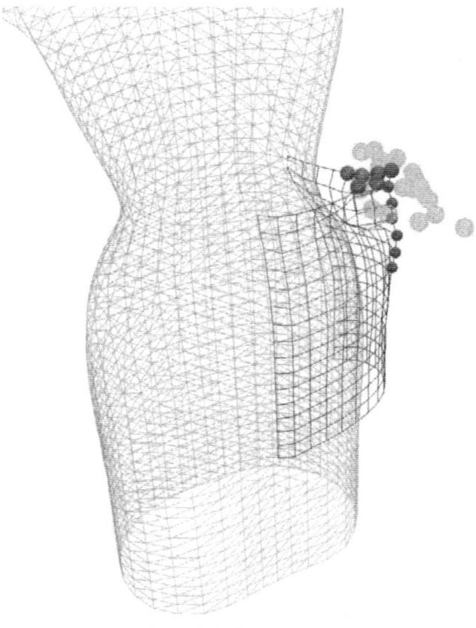

Fig.3a Cloth Model Tucked by Virtual Fingers.

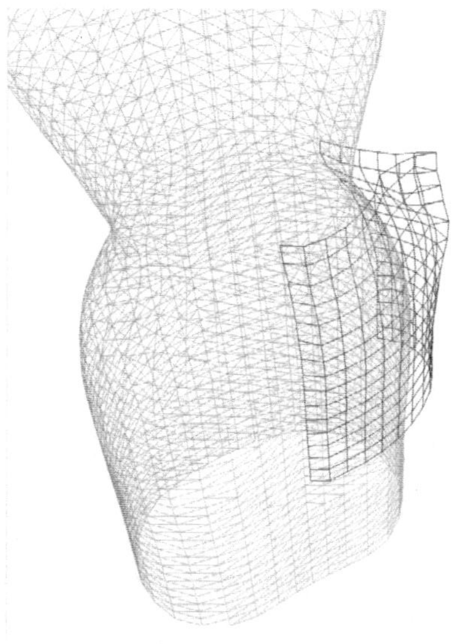

Fig.3b Tucked Cloth Model.

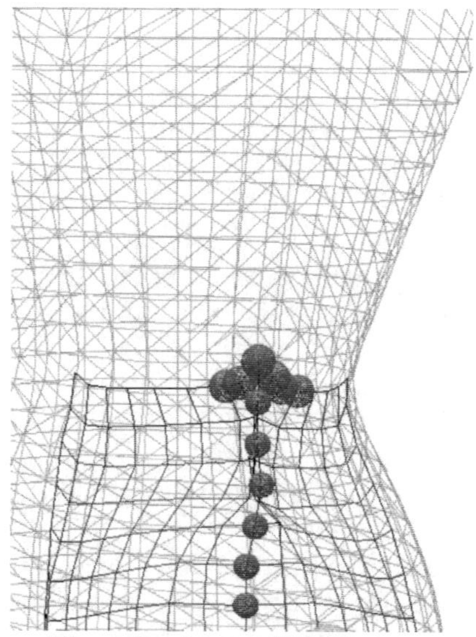

Fig.4 Position of Dart on Cloth Model.

4. Conclusion

A prototype system is constructed to design clothing in the virtual world by using the draping method through virtualizing composing elements of draping. Dart can be tucked by this virtual draping. Precision and stability of the system has to be improved for practical use.

5. Acknowledgement

This work was partly supported by Grants-in-Aid for Scientific Research (No. 23240100, 24220012) from the Ministry of Education, Science, Sports and Culture.

References

[1] Cho Y, Komatsu T, Inui S, Takatera M, Shimizu Y, Park H. Individual Pattern Making Using Computerized Draping Method for Clothing. Textile Research Journal 2006; 76: 646-654.

[2] Wang J, Lu GD, Li WL, Chen L, Sakaguti Y. Interactive 3D garment design with constrained contour curves and style curves. Computer-Aided Design 2009; 41: 614-625.

[3] Huang HQ, Mok PY, Kwok YL, Au JS. Block pattern generation: From parameterizing human bodies to fit feature-aligned and flattenable 3D garments. Computers in Industry 2012; 63: 680-691.

[4] Harada T, Koshizuka S. Real-time collision computation between cloth and highly-tessellated models. Transaction of Information Processing Society of Japan 2007; 48: 1829-1837.

[5] Meng Y, Mok YP, Jin X. Interactive virtual try-on clothing design systems. Computer-Aided Design 2010; 42: 310-321.

[6] Wibowo A, Sakamoto D, Mitani J, Igarashi T. DressUp: A 3D Interface for Clothing Design with a Physical Mannequin. Proceedings of the 6th International Conference on Tangible, Embedded and Embodied Interaction (TEI 2012), Ontario, Canada: Feb.19-22; 2012: 99-102.

[7] Satoh K, Yamamoto T, Matsuyama K, Konno K. Rokuro : Implementation and Examination of Gestural Modeling Interface. IPSJ Interaction 2012; 2012: 771-776.

[8] Liang S, Baciu G, Li RH, Jia JY, Zhang JH. Recommendation in Motion: Intelligent Hypertouch Garment Design. ADVANCES IN MECHANICAL ENGINEERING, Vol.610604.

[9] Mesuda Y, Inui S, Horiba Y. Virtual manipulations for Draping.International Journal of Clothing Science and Technology 2015; 27: 417-433.

The Application of Parametric Surface Model in Mould Head Design for Brassieres

Long Wu[1], Kit-Lun Yick[2*], Joanne Yip[2], Sun-Pui Ng[3]

[1]*Apparel & Art Design College, Xi'an Polytechnic Universigy, Xi'an, 710048, China*
[2]*Institute of Textiles and Clothing, The Hong Kong Polytechnic University, Hong Kong*
[3]*Hong Kong Community College, The Hong Kong Polytechnic University, Hong Kong*

*Corresponding author's email: kit-lun.yick@polyu.edu.hk

Abstract

The purpose of this paper is to establish a parametric surface model for characterizing the shape of mould head and moulded bra cups of various bra styles and cup sizes.

A parametric method for the design and grading of the master head cone is formulated. With this method, the new style of the mould head can be accurately and efficiently designed and/or graded into larger or smaller head sizes. In addition, parameterization based remeshing and a registration algorithm are used to characterize the 3D shapes of the convex surface of scanned mould heads. By using an example-based method, virtual master cones are constructed based on the 3D surfaces of the investigated bra cups after eliminating the style lines and size differences. Then, a new mould head design can be developed from the master cone by drawing the style lines in a virtual environment, different sizes can also be made by grading the wire frame model in a 3D manner.

The proposed method not only provides useful information for the selection of suitable foam materials and the determination of suitable moulding conditions, but also shortens the design time. It will allow a rapid response to the continuous changing needs of the intimate apparel industry.

The results from this study will provide a scientific, practical and effective solution for intimate apparel manufacturers so that they can improve the development process of mould heads.

Keywords: Mould head; Foam; Bra cup; Conformity

1. Introduction

Among the various types of apparel, brassiere design requires a particularly lengthy process that demands design creativity, precision pattern making skills and a detailed knowledge of fabric [1]. Mould head design is also a key process in producing brassieres. The process control parameters, such as the 3D shape characteristics of the mould heads, control of the moulding temperatures and dwell times, etc. determine the resultant shape and appearance of the final product. Conventionally, the moulded cup shape is visually inspected through a comparison with a plastic shot. The judgment of the quality and shape depends on the experience of the inspectors as seen in Fig.1. Geometric control of the mould head design is therefore a highly time-consuming process.

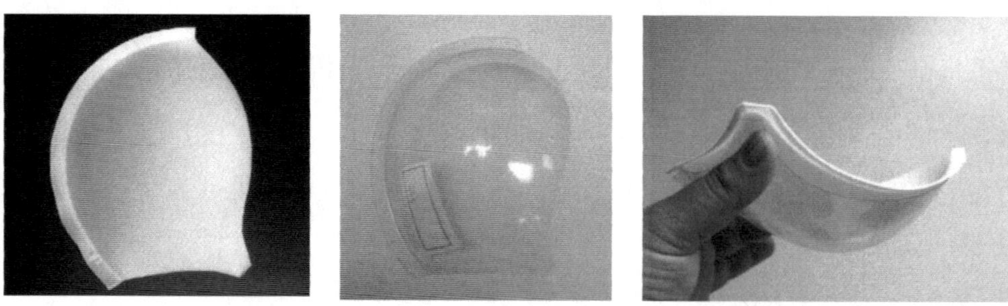

Fig.1 Foam cup sample is fitted into a corresponding plastic shot

The geometric feature and shape modification processes of developing new aluminum mould heads usually involve many "trial and errors". Problems have further increased in the new century due to the complex style of the moulded cups (such as full coverage, demi, triangle, etc.), extreme range of cup sizes (from AA to F) and softness requirements of bras introduced into the market. Researchers continue to explore rapid prototyping, body scanning and other innovative product development technologies to improve this development process (Liao and Lee, 2010). As there has been limited studies to provide accurate and reliable guidelines for the mould head design process. It is crucial to develop a parametric optimization system for the mould head design. The proposed method not only provides useful information for the selection of suitable foam materials and the determination of suitable moulding conditions, but also shortens the design time. It will allow a rapid response to the continuous changing needs of the intimate apparel industry.

2. Research situation

The general process for mould head development included the following key steps. The mould head design starts with a standard cup shape. The designer and/or engineer then designs the mould head by means of a master cone profile, manually creates a plaster mould, and casts the aluminum units with male and female moulds. When developing the aluminum mould heads, their material properties were considered based on the foam properties, foam shrinkage, compression rate, etc. Computerized numerical control (CNC) systems may be adopted by large-sized bra cup moulding manufacturers. In bra cup moulding, a wide variety of flexible PU foams can be used for different styles and end-uses of the products with providing adequate support, comfort and hand feel. A pair of aluminum male and female moulds with a conformed shape, which is similar to the anatomy of the female breast, is designed and made according to the specified cup shape given by the customer [2]. During the moulding process, the heated male mould stretches and compresses the originally flat foam sheets toward the female mould at a temperature over the softening point of the foam materials. A high temperature is used for about one minute to allow the heat to transmit through the foam sheets which are held in a space between the male and female moulds, so that the foam sheets are heat-set to the desired cup shape and thickness.

As a standard shape, the plastic master cone is used to compare against the shape of the moulded cup to assure 3D shape conformity of the moulded cup with the corresponding mould head design. When the moulded cup shape has been given approval, the pair of aluminum moulds, type of foam material and the optimal process parameters for bra cup moulding are then specified for mass production. Otherwise, the corresponding mould heads would be returned for further modifications.

Recently, reverse engineering technology has been adopted to assist rapid prototyping, complex geometric measurements, shape evaluation, etc [3-5]. In this research, by using a non-contact 3D digitizing apparatus, the surface coordinates of mould heads, cup samples and master cones can be accurately and quickly obtained. The key features of the surface will be constructed and revised from point clouds [6,7]. On the basis of a parametric design programme for master cones of bra cups, the 3D shapes of the convex surface of scanned bra cups can be characterized [8]. New styles of mould heads can also be accurately and efficiently designed and/or graded into different sizes.

3. Research method

3.1 Scan data

Recently, reverse engineering technology has been adopted to realize rapid prototyping, complex foam cup geometric measurements, cup shape evaluation processes, etc. In this research, by using a non-contact 3D digitizing apparatus, the surface coordinates of the cup samples, master cones, and mould heads can be accurately and quickly obtained. The key features of the surface will be constructed and revised from point clouds [6]. With this technique, new bra cups can be accurately and efficiently designed and/or graded into larger or smaller cup sizes.

To scan the 3D shape of bra cups, a 3D scanner is required. It is a device that analyzes a real-life object or environment to collect data on shape and possibly appearance (i.e. color). The collected data can then be used to construct digital 3D models. Many different technologies have been used to build these 3D scanning devices; each technology comes with its own limitations, advantages and costs. The collected 3D data is useful for a wide variety of applications. These devices are extensively used in industrial design, orthotics and prosthetics, reverse engineering and prototyping, quality control/inspection, documentation of cultural artifacts, etc [9-11]. There are a variety of technologies for digitally acquiring the shape of a 3D object which mainly includes digital photogrammetry, white light interferometry and 3D laser scanning.

3.2 Instrumentation

The equipment used in this research is a NextEngine desktop 3D laser scanner (NextEngine Inc., USA) mounted over a tripod. This scanner is based on multi-stripe laser triangulation (MLT): the device projects multiple laser-stripes and registers point positions with a charge coupled device (CCD) camera. The scanner comprises a compact box that is 22×28×9 cm in size, with twin arrays of four solid state lasers (red, 650 nm) and twin 3.0 megapixel complementary metal–oxide–semiconductor (CMOS) image sensors [12].

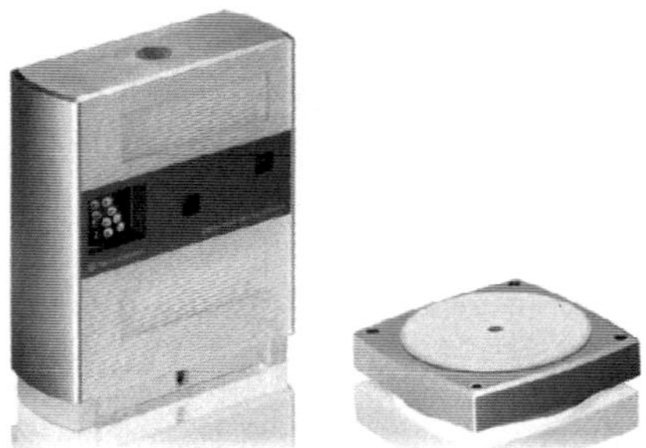

Fig.2. NextEngine desktop 3D laser scanner

The NextEngine desktop 3D laser scanner (Fig.2) uses a proprietary platform, called Scan Studio, which runs on Windows OS. This software, although necessary for the acquisition, is not suitable for the alignment of the meshes of complex objects. The average distance between the scanner and the object surface needs to be approximately 45-50 cm in order to properly acquire the highest percentage of the scanned object [13].

4. Results

4.1 Prototype model

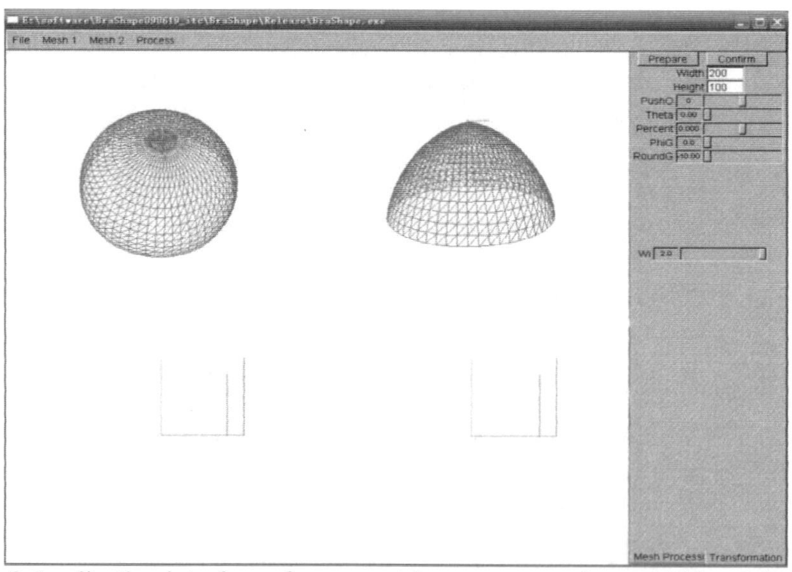

Fig.3 Application interface of a parametric programme for mould head design

The application interface of the parametric design programme for master cones is shown in Fig.3. A standard master cone (prototype) could be modified and improved through related functions in the programme, which is actually more convenient and faster than the existing 3D software. In my research, the programme was further revised and the master cone was changed to a mould head template to facilitate the mould head design process.

In the mould head design field, there is great uncertainty as to whether a new mould head design will actually do what is desired, because new designs often have unexpected problems. A prototype is often used as part of the mould head design process to give designers or clients the ability to explore design alternatives and confirm style and size prior to starting the production of a new mould head. The entire parametric design of the mould head by the programme is based on its master cone.

In the programme, the storage of the mould head prototype information is found in an object (OBJ) file format. It is a common format which supports lines, polygons, surface and free-form curves, etc. The lines and surfaces are defined by their control points and additional information attached to the curve types. This file format can be used to store modified mould head designs and also can be directly exported into existing 3D software for further processing. As shown in Figure 3, a 3D model of a master cone is displayed in the form of a mesh structure, and can also be displayed in the form of a surface mode.

The surface of the mould head or bra cups could be obtained as cloud points by a 3D scanner. Then, triangular meshes will be randomly generated. However, different grid structures and mesh density distributions bring forth inconvenience in assignment and deviation calculations. The function of remeshing could reconstruct different grid structures of mould heads or bra cups without altering their 3D shapes, by using the mesh structure of the master cone as reference. Furthermore, the registration function is used to align two mesh structures to a certain reference point, which is easier for deviation inspection and shape conformity calculation.

4.2 Parametric design of master cone

The functions include style design, remeshing, registration and 3D grading etc. The style design function can generate a new master cone for the mould head in terms of the specified width and height based on the prototype. In addition, there is a parametric conformity function for designated parts on the master cone.

The width and depth must be input into the dialogs at the top right corner of the interface before the master cone file is opened. The width represents the diameter of the underside circle of the master cone, while depth refers to the vertical distance from the highest point to the bottom face as shown in Fig.4. In defining the width and depth of the master cone, a standard master cone (in a circular shape) is automatically generated.

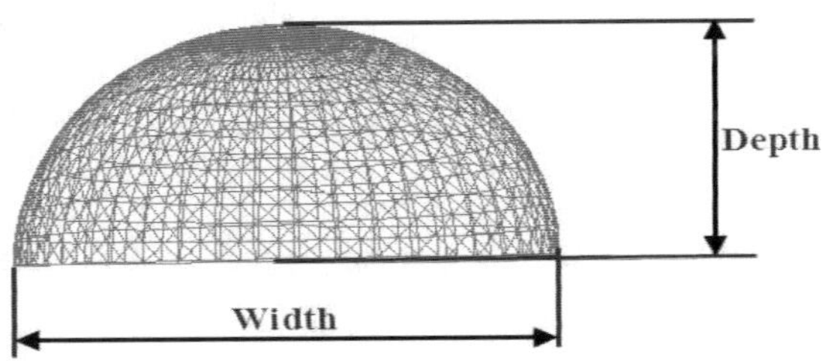

Fig.4 Schematic diagram of the width and depth of master cone

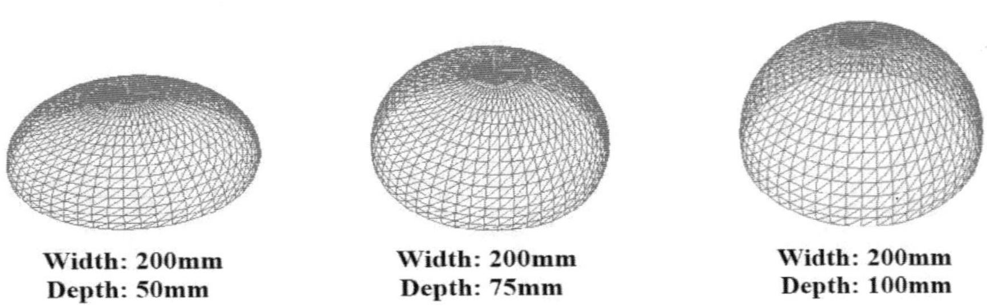

Fig.5 Master cones with three different depths

Fig.5 shows three different prototypes of the master cone in which their depths are set at 50, 75 and 100 mm, respectively, and all have a standard diameter of 200 mm.

To change the shape of the master cone, a number of regulating parameters can be adopted. These include:

PushO: mainly used to adjust the diameter of the master cone. As shown in Fig.6, the values of PushO can be set at -1, 0, 1, respectively. When PushO is set at "-1", it means only the upper region of the cone will be adjusted while the value "1" means only the lower part. The value "0" indicates that both two parts will be simultaneously adjusted.

Percent: represents the adjustment percentage in which the scope is limited by ±10%. Parameters PushO and Theta can be adjusted by the mould head designer so as to change the standard circular master cone to an oval shape.

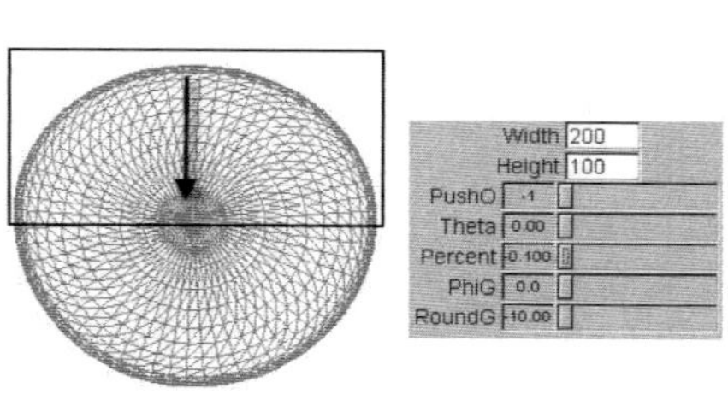

PushO: -1 Theta: 0 Percent: -0.1

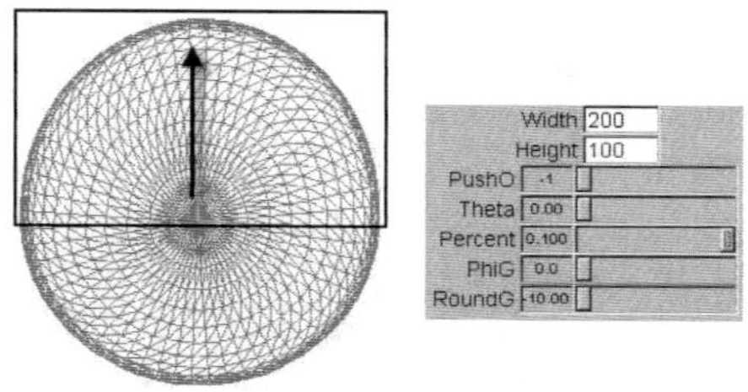

PushO: -1 Theta: 0 Percent: 0.1

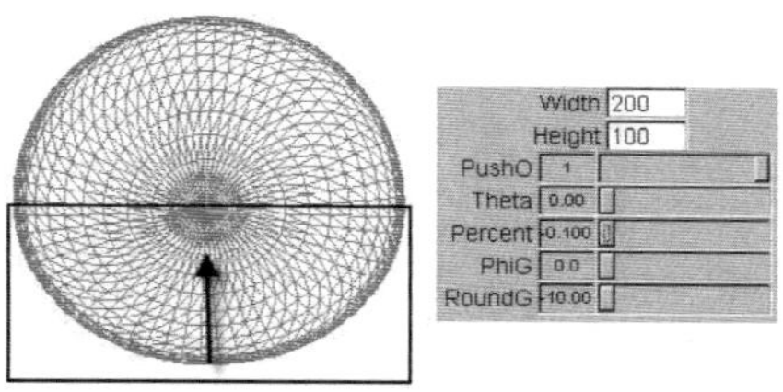

PushO: 1 Theta: 0 Percent: -0.1

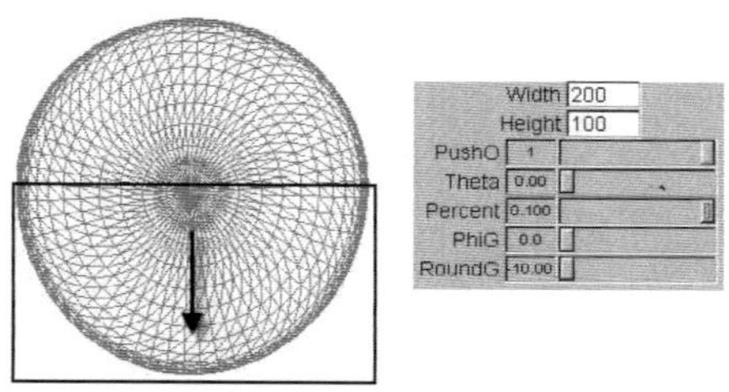

PushO: 1 Theta: 0 Percent: 0.1

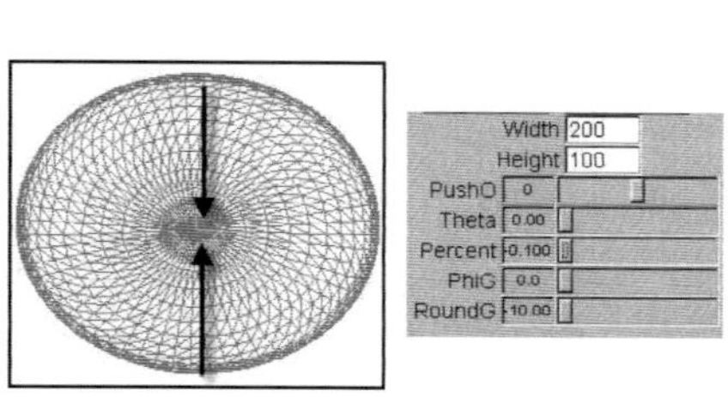

PushO: 0 Theta: 0 Percent: -0.1

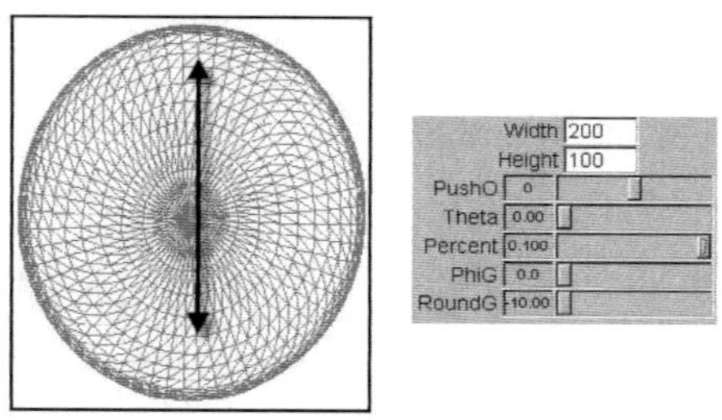

PushO: 0 Theta: 0 Percent: 0.1

Fig.6 Shape modification of master cone by using PushO and Percent parameters

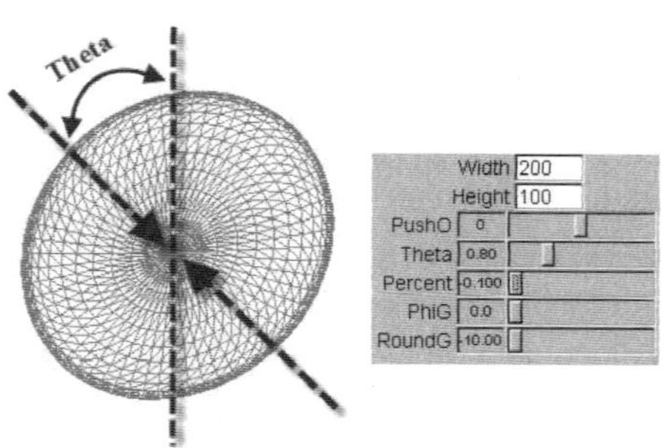

PushO: 0 Theta: 0.8 Percent: -0.1

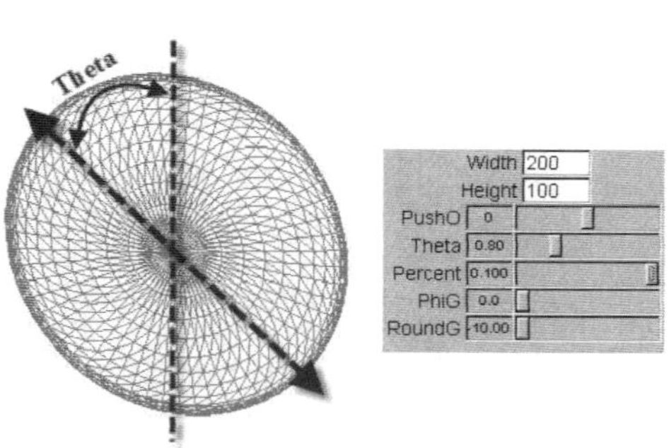

PushO: 0 Theta: 0.8 Percent: 0.1

Fig.7 Shape modification of master cone by adjusting Theta parameter

Theta: indicates the radian between the direction to be adjusted and the north-south axis, see Fig.7. The designer can therefore manipulate the master cone into different angles, which ranges from 0 to π (180°).

PhiG: defines the radian of the reserved area with the vertical axis from the side, see Fig.8. The radian ranges between 0 and 1.6 (90°) within which the mesh structure in the reserved sector area cannot be modified.

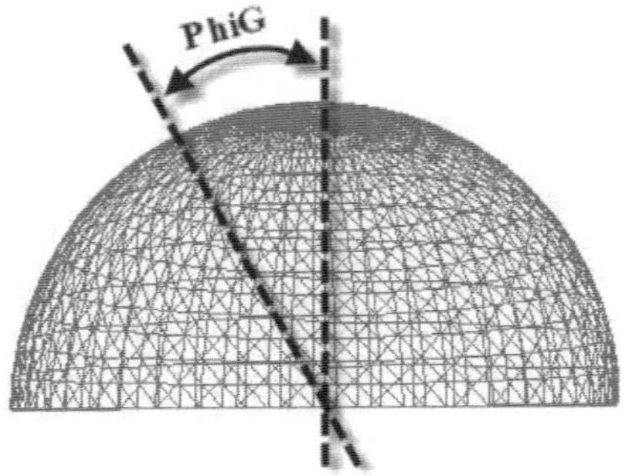

Fig.8 Schematic diagram of the PhiG parameter for modification of master cone

RoundG: means the round grade which modifies the curvature along the outside surface of the mesh structure except for the reserved sector region defined by the PhiG parameter. Fig.9 shows the shape modification of the master cone with RoundG from -10 to 10.

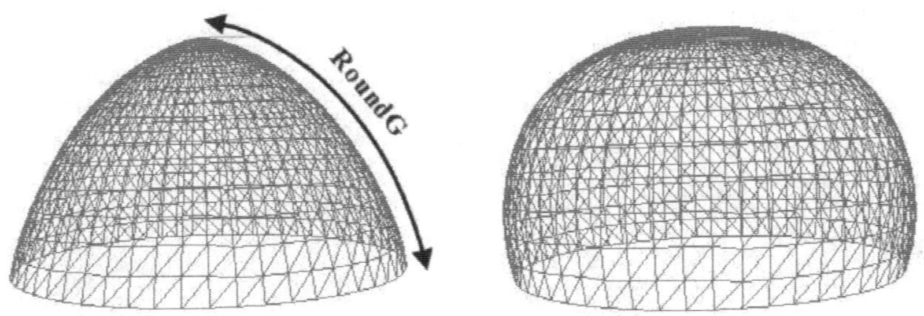

RoundG: -10 PhiG: 0 **RoundG: 10 PhiG: 0**

Fig.9 Shape modification of master cone with adjustments of RoundG when PhiG=0

4.3 Example-based design of cone shapes

By using an example-based algorithm, designers can merge the features of two different cone shapes in terms of various weightings. The example-based method aims to provide a tool for designers to produce new master cones with the aid of known samples. If each existing sample represents a typical modification in the design of a master cone, the new master cone can be conveniently modified by adjusting the corresponding weightings. An example is shown in Fig.10. Suppose S_1 is a push-up master cone and S_2 is a plunge style, respectively. Based on the original master cones of S_1 and S_2, new master cones (S_3, S_4, S_5) are generated by using different weightings as shown below.

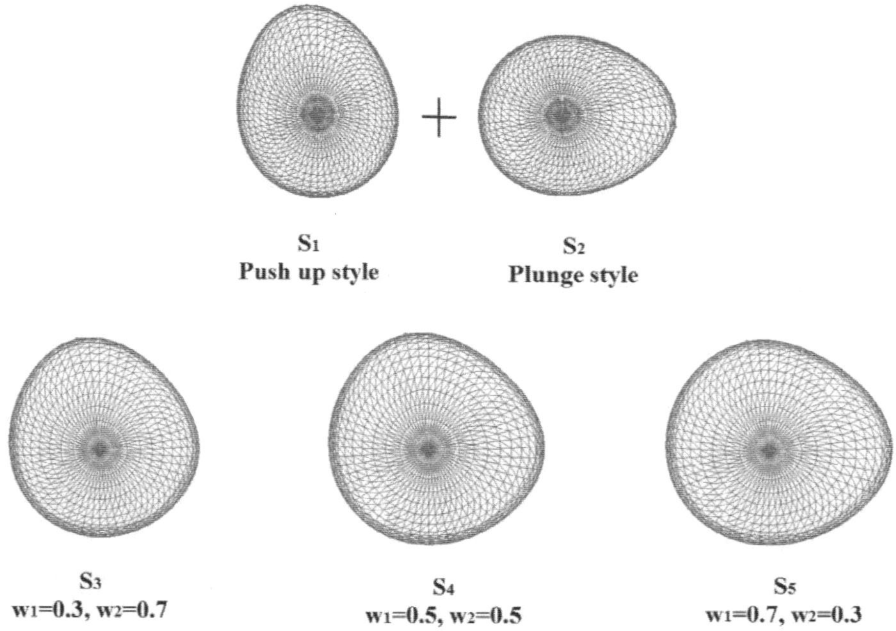

Fig.10 New master cones (S_3, S_4, *and* S_5) are generated by using different weightings of S_1 and S_2

As virtual master cones can be constructed based on 3D surfaces, designers need not modify the master cone and manually cut along the style lines for desired product design and the development process. The features of typical master cones could be created and normalized to a specific width and height by scaling. Modifications can also be made by adjusting the loading weight of each sample by using the following Equation (1), which is:

$$S = \sum_{i=1}^{n} w_i S_i \tag{1}$$

where S_i is the style of a typical example, w_i is the weighting of S_i and $\sum_i w_i = 1$. A larger w_i produces a master cone that is more similar to S_i.

4.4 Calculation of shape conformity

The 3D outer surfaces of moulded bra cups are firstly scanned and compared against a plastic shot. The highest bust point for the plastic shot is first defined. To measure the deviation of the corresponding points on the scanned cup surface, the parameterized-based remeshing and registration algorithm (PRRA) method is used to characterize the 3D shapes of the convex surface on the scanned bra cup [8]. The initial bra cup images with missing data after scanning are mended by reverse engineering software for curvature smoothing. Irregular triangular faces from cloud points when producing the surface are also rectified so as to generate a completed cup surface. The outer surfaces of the plastic shot and the bra cup are aligned and their 3D shape deviations are then calculated. In this process, the plastic shot is used as the reference mesh, the bra cup surface is converted to the structure of the reference mesh by rigid transformation, whereas the corresponding vertices are matched around the bust point [14]. The 3D geometrical data files would be the output for carrying out the subsequent processes. To produce a fitting surface for comparison, the following steps are carried out, see Fig.12.

For calculating shape conformity, remeshing and registration between the bra cups and master cone are first carried out with the parameterized-based remeshing and registration algorithm method. After scanning the 3D shape geometry of the plastic shot and the bra cups produced under different processing conditions, their fitting surfaces can be rebuilt based on the original 3D surfaces and then output into reverse engineering software through an OBJ file format.

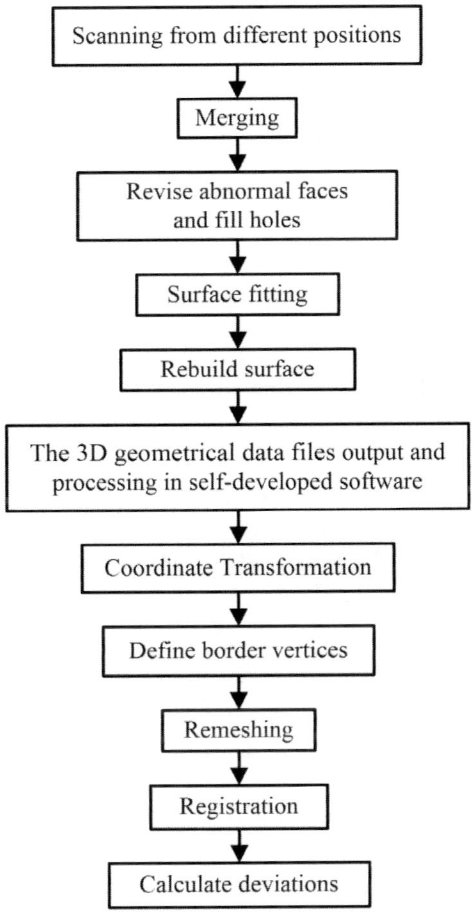

Fig.12 Processes of calculating deviations between plastic shot and bra cups

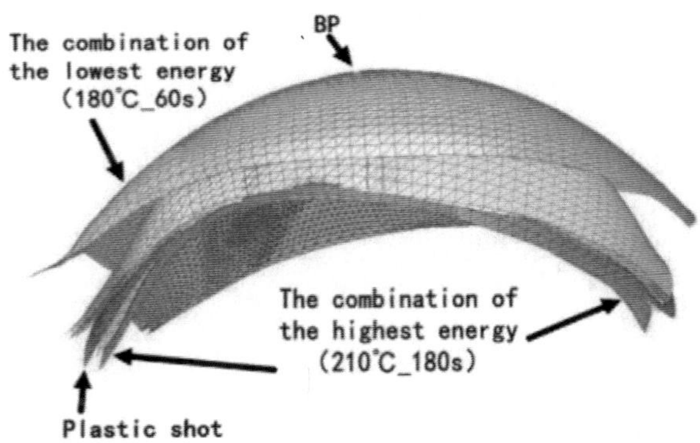

Fig.13 Profiles of moulded foam cups under the different combinations of energies

As shown in Fig.13, profiles of moulded foam cups under the different combinations of energies are aligned to the bust point through coordinate transformation. When the moulding temperature is lower with the shorter dwell time, viz., the combination of lower energy absorption, the shape conformity between moulded cup and plastic shot will be greater; when the temperature is higher with longer time, the shape conformity will be also greater.

The rigid transformation is computed by matching the corresponding vertices around the bra cup peak. By aligning the highest point (bust point) of the plastic shot that of the moulded cups, the coordinate differences of the same points between the two 3D surfaces can be quantified. According to industrial practices, the degree of shape conformity (T) between the moulded cups and the plastic shot at the cup peak region is evaluated by calculating the percentage of deviation at the corresponding points by using Equation (2).

$$T_{\mathrm{tip}} = \frac{f_1(0 < D \le 1)}{\sum_{i=1}^{10} f_i\left[(i-1) < D \le i\right]} \times 100\% \tag{2}$$

$$(i \in N),\ \text{and}\ \ D = \left|Deviation(X_j - M_j)\right|,\ \ (j \in R)$$

where T_{tip} is the percentage of frequency f_1 (which have deviations that are less than 1 mm) out of the total frequency within a surface area that cover a three centimeter radius from the cup peak. M_j is an arbitrary coordinate on the plastic shot and X_j is the corresponding coordinate on the moulded cup. The function Deviation (j) is used to calculate the absolute deviations between the plastic shot and moulded cup.

5. Discussion and conclusion

The style design function in a parametric design programme has been used to primarily solve the problems with style parameters, including style and size modification. This programme makes it easy to overcome these difficulties. In this study, an example-based method for designing mould heads or bra cups is presented. Based on the sample input by any existing sample, a master cone can be developed for a new style in any mould head design. The proposed wire frame based method provides a unique presentation of mould heads that facilitates the process of foam cup grading as well as its design and development processes. On the basis of the parameterization based remesh and registration algorithm method, it is possible to register the scanned bra cups, and shape conformity can also be evaluated and effectively quantified. The optimal moulding conditions for particular types of foam materials can also be easily determined. The current process of development shortens at least three or four cycle times of design and analysis compared with traditonal method. This not only facilitates the control of the bra cup moulding process and communication amongst different sectors of the intimate apparel industry, but also moves towards more sustainable production process for fast fashion trend in intimate apparel industry.

References

[1] Hardaker C FG. Communications: the bra design process: a study of professional practice. Int J Cloth Sci Technol 1997; 9: 311-325.

[2] Yip J, Ng SP. Study of three-dimensional spacer fabrics: physical and mechanical properties. Journal of Materials Processing Technology 2008; 206: 359-364.

[3] Sha L. The Application of Reverse Engineering in Innovation Design. Applied Mechanics and Materials 2014; 470: 479-483.

[4] Li D, Feng X, Liao P, Ni H, Zhou Y, Huang M, et al. 3D Reverse Modeling and Rapid Prototyping of Complete Denture. Frontier and Future Development of Information Technology in Medicine and Education: Springer: 2014; p: 1919-1927.

[5] Varady T, Martin RR, Cox J. Reverse engineering of geometric models-an introduction. Computer-Aided Design 1997; 29: 255-268.

[6] Tam KW, Chan KW. Thermoforming mould design using a reverse engineering approach. Robotics and Computer-Integrated Manufacturing 2007; 23: 305-314.

[7] Shimizu Y, Furukawa T, Maeda S, Konda A, Daiyasu K, Shimizu H. Representation Method of Form. Proceedings of the Symposium on Computer Graphics and Its Applications. Wako-shi Saitama at the Institute of Physical and Chemical Research (RIKEN): Tokyo KTK Scientific Publishers: 1992; p: 147-169.

[8] Yick K, Ng S, Zhou X, Yu W, Chan D. Wire frame representation of 3D moulded bra cup and its application to example-based design. Fibers and Polymers 2008; 9: 653-658.

[9] Bernard A, Fischer A. New trends in rapid product development. CIRP Annals-Manufacturing Technology 2002; 51: 635-652.

[10] Kai CC, Meng CS, Ching LS, Teik LS, Aung SC. Facial prosthetic model fabrication using rapid prototyping tools. Integrated Manufacturing Systems 2000; 11: 42-53.

[11] Koller D, Frischer B, Humphreys G. Research challenges for digital archives of 3D cultural heritage models. journal on computing and cultural heritage (JOCCH) 2009; 2: 7.
[12] Bradley BD, Chan AD, Hayes MJD. A simple, low cost, 3d scanning system using the laser light-sectioning method. Instrumentation and Measurement Technology Conference Proceedings, 2008 IMTC 2008 IEEE: IEEE: 2008; p: 299-304.
[13] Yang R, Cheng S, Chen Y. Flexible and accurate implementation of a binocular structured light system. Optics and Lasers in Engineering 2008; 46: 373-379.
[14] Su ZX, Cao JJ, Shi XQ, Liu FS. Surface deformation evaluation by parameterization of 3D raw data advances of applied and computational mathematics II. USA: NOVA Publishers: 2006.

Designing and Assessing New Firefighting Protective Clothing for Chinese Female Firefighters

Dan-Dan Lai[1], Fa-Ming Wang[1, 2*]

[1]*Department of Apparel Design & Engineering, Soochow University, 178 East Ganjiang Road, Suzhou, 215021, China*

[2]*Laboratory for Clothing Physiology and Ergonomics (LCPE), the National Engineering Laboratory for Modern Silk, Soochow University, 199 Ren'ai Road, Suzhou, 215123, China*

*Corresponding author's email: dr.famingwang@gmail.com

Abstract

In this study, a new set of firefighting protective clothing for Chinese female firefighters was designed. The two most important physical parameters (i.e., thermal insulation and evaporative resistance) to quantify clothing thermal comfort were determined using a thermal manikin under a variety of conditions (i.e. two levels air velocity: 0.4 m/s and 1.0 m/s; two walking speeds: 0 and 0.96 m/s). The existing firefighting clothing was used as a control. Results demonstrated that both the existing firefighting clothing and the new female firefighting clothing had a similar total thermal insulation. The total evaporative resistance of the newly designed female firefighting clothing determined on the standing manikin under the 0.4 m/s air velocity was 18.6% lower than that of the existing firefighting clothing. It was concluded that the moisture transfer ability of the newly designed female firefighting protective clothing has been much improved without compromise of thermal protection.

Keywords: Female firefighters; Firefighting protective clothing; Thermal manikin; Thermal insulation; Evaporative resistance

1. Introduction

Firefighting is one of the most perilous tasks, which may be involved with numerous physical activities in hazardous environmental conditions [1]. Generally, firefighters have a wide range of duties, including putting out fires, rescuing and caring the sick and the injured and responding to accidents where risks are posed to life and properties. Due to the use of new construction materials and improvements in building codes in modern building structures, fires in modern buildings are usually hotter and more dangerous than ever before [2]. Firefighting clothing serves as the most important personal protective equipment to protect firefighters, which may include a jacket, pants, boots, a hood, a helmet, and gloves. In the United States, the quality of firefighting clothing and equipment must meet the NFPA 1971 and 1981 standards to ensure firefighting clothing and equipment could protect firefighters against various hazards. The firefighting jacket and pants are normally composed of three layers: a durable and protective outer shell, a moisture barrier and a thermal liner [3]. Previous research work [4,5] has suggested that clothing design, fit and size in relation to the human body geometry play a vital role in determining body movement and thermal comfort. Therefore, wearing properly fitting firefighting clothing is critical not only to firefighters' safety but also to their job performance.

The population of volunteer and professional female firefighters in the world is gradually growing. Unfortunately, specially designed female firefighters' protective ensembles are still not available. Female firefighters normally wear exactly the same protective clothing that was specially pattern designed for male firefighters. Furthermore, the physique differences between male and female have been well documented. To improve the design of firefighting clothing for female firefighters, it is important to understand the differences between men and women in perception related to performance. Based on an online survey of 202 male firefighters and 314 female firefighters, Park et al [6]. reported that female firefighters showed significantly lower satisfaction with fit and functionality of the turnout ensemble in multiple areas (e.g., shoulder, chest, armhole, coat length, crotch, hip) than males. This study also pointed out that the cargo pockets and the straps were unfit for females. Boorady et al [7]. assessed the needs of female firefighters for firefighter turnout gear using focus group interviews centering on design issues and the sizing system. It was recommended that female firefighting clothing should be sized specifically for female wearers and the design should also be reconstructed. Hence, there is a great need to design better-fitting firefighting clothing for female firefighters to improve wear comfort.

The main purpose of this study was to develop new firefighting clothing for Chinese female firefighters based on the average body dimension of adult Chinese females. The thermal insulation and evaporative resistance of the existing firefighting clothing and the newly designed female firefighting clothing were determined using a thermal manikin under different test conditions. The differences in clothing total thermal insulation and evaporative resistance between these two firefighting clothing ensembles were also compared. In addition, the effect of zipper opening on thermal comfort attributes was examined.

2. Methodology

2.1 Design of new female firefighting clothing

2.1.1 Clothing design

To resolve the sizing, fit and thermal comfort issues in multiple areas for females, our new design is illustrated in Fig.1 and 2.

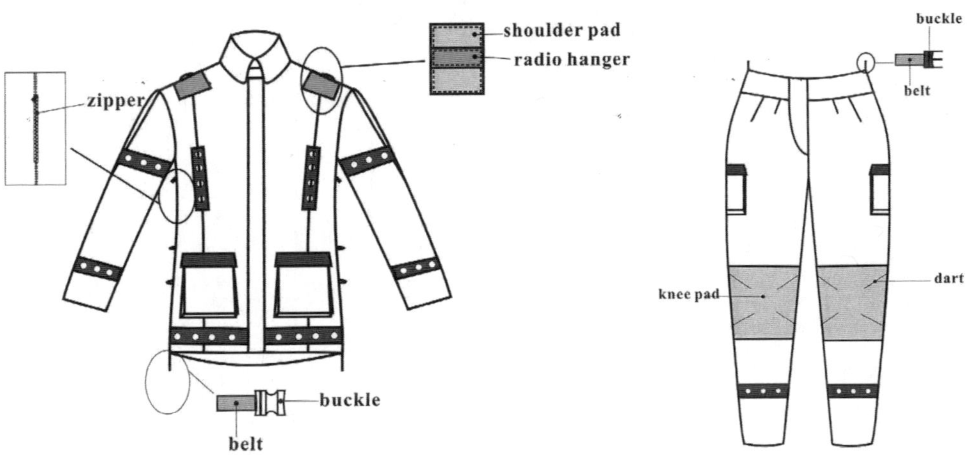

Fig.1 Prototype of the new female firefighting clothing (front view).

The design features of the new female firefighting clothing included:

Features to facilitate the fit: princess lines in the front and back of the jacket were introduced to add a tailored fit to the body shape of the females; the three-piece sleeve assembly design was introduced to eliminated the bulky folds in the underarm; pants with tapered legs were also introduced.

Features to improve comfort, utility and mobility: suspenders were replaced by belt buckles and were sewn into the side seam of the front and back lining. The belt buckles served the purposes of holding up the pants and eliminating pressures on the breasts brought by suspenders; foam shoulder pads were added to relieve the pressure on the shoulders caused by self-contained breathing apparatus; two 10cm long zippers were sewn into the side seam of the jacket to adjust the microclimatic ventilation; punched reflective straps (width: 5cm, diameter of the hole: 0.4cm, equal distance between two each adjacent holes was 2.5cm) were used to improve the air and vapor permeability; four vertical reflective straps were also introduced to the jacket to increase the visibility; tabs were sewed into the side seam of the jacket to help firefighters easily catch the necessaries; variegated length hem was designed to increase the range of motion; elbow darts and knee darts were also designed to improve the motion of joints.

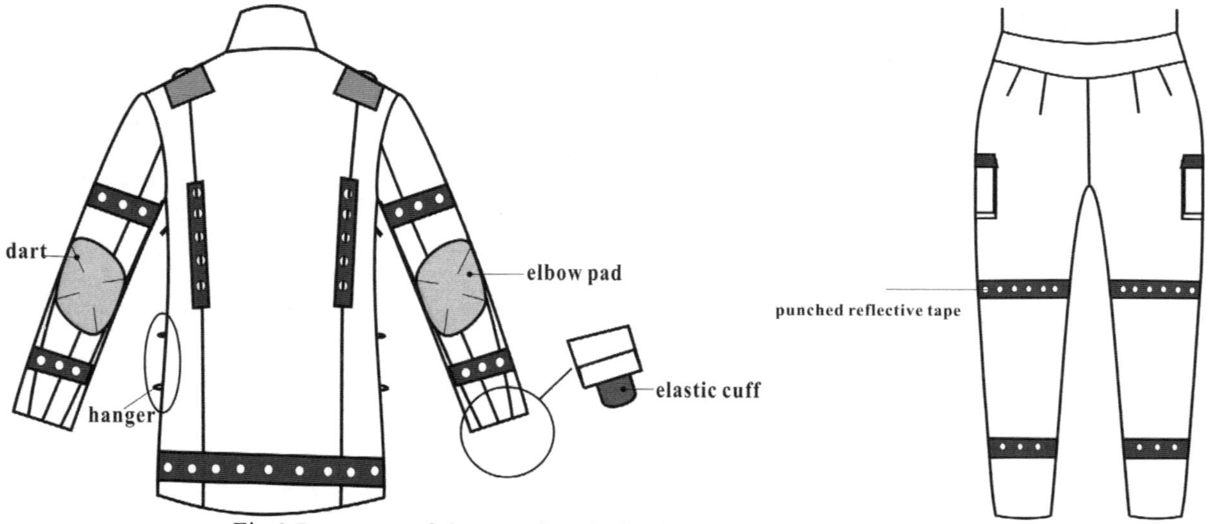

Fig.2 Prototype of the new female firefighting clothing (rear view).

2.1.2 Pattern design

Fig.3 demonstrates patterns of the existing and the new female firefighting jacket and pants. Black lines denote the existing pattern and the pink lines are the new pattern.

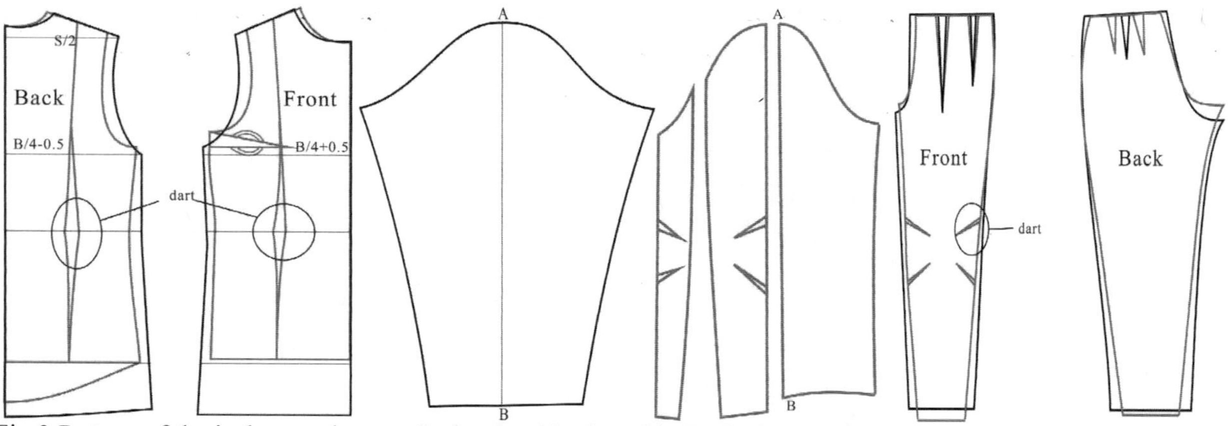

Fig.3 Pattern of the jackets and pants. S, the shoulder breadth; B, the breast circumference; A-B, the length of sleeve.

Measurements of the new firefighting jacket and pants at each location were based on the size of an average adult Chinese female with a height of 170cm female. The size design was aimed to provide maximum wearer comfort and mobility. In the existing design, the jacket was rather loose at the waist and shoulders. To adjust the jacket fit around these two sites, the waist circumference and shoulder breadth were reduced by 23cm and 7.5cm, respectively. Also, the length of the existing firefighting jacket was too long and thus, the length of center back (i.e., CB) and center front (i.e., CF) was reduced by 5cm and 11cm, respectively. Generally speaking, the crotch area was a major cause of impaired mobility in lower body movements due to a longer front-rise length. Hence, in the new design, the front-rise length was reduced by 5cm to improve the lower body mobility. The waist circumference of the pants was also reduced by 14cm in the new design. All measurements of the existing (size: 170A, model: SRO-RFM-B) and new firefighting clothing are provided in the Table 1.

Table 1. Measurements of the existing and the new type firefighting clothing

Measurement	Existing clothing (cm)	New type (cm)	Measurement	Existing clothing (cm)	New type (cm)
Shoulder breadth	52.5	45.0	Sleeve length	56.5	61.5
Chest circumference	122.0	112.0	Outseam	97.0	102.0
Waist circumference (jacket)	115.0	92.0	Waist circumference (pants)	93.0	79.0
Hip circumference (jacket)	114.0	108.0	Hip circumference (pants)	102.0	106.0
Center back length	80.0	75.0	Front-rise length	37.0	32.0
Center front length	78.0	67.0	Leg opening	48.0	41.0

2.2 Evaluation of thermal comfort properties of the existing and new firefighting clothing

2.2.1 Materials

a b

Fig.4 The Existing firefighting clothing (a) and the new female firefighting clothing (b)

Fig.4 displays the existing firefighting protective clothing ensemble (i.e., EXISTING) and the new firefighting clothing ensemble (i.e., NEW) designed for Chinese female firefighters. The jackets and pants were made of three layers to meet the Chinese Standard GA10-2014 [8]. The new female firefighting clothing ensemble used exactly the same fabric materials and accessories as those selected for engineering the existing firefighting clothing. A set of long underwear (81% cotton and 19% polyester) was used. Besides, a pair of firefighting gloves (size: M, type: SRO-RMS-MA, Shanghai, China) was selected. The firefighting boots and helmet were not included as the focus of the study was to determine the influence of design and construction on thermal comfort attributes of firefighting clothing.

2.2.2 Thermal manikin

A 34-segment 'Newton' thermal manikin (Thermetrics LLC, Seattle, WA) was used to determine the thermal insulation and evaporative resistance of the existing and new firefighting clothing ensembles. The segmental surface temperature of the manikin could be controlled individually. Wire sensors were embedded into each segmental surface to record segmental surface temperatures. In this study, the constant temperature mode was used and the surface temperature was set to 34.0°C. The experimental data were recorded by the ThermDAC® software (Thermetrics LLC, Seattle, WA).

2.2.3 Test protocol and test conditions

All manikin tests were conducted in a climatic chamber (ESPEC Inc., Osaka, Japan). Measurements of clothing thermal insulation strictly followed the test procedure defined in ISO 15831 (2004). In this study, the air temperature was 12.0±0.5°C and the relative humidity was 50±5%. Two levels of air velocity (0.4m/s and 1.0m/s) and two walking speeds (0m/s and 0.96m/s) were used to investigate the effect of zipper opening on the total thermal insulation and evaporative resistance of the newly designed firefighting clothing. For the measurements of clothing evaporative resistance, a pre-wetted stretchable fabric 'skin' was used to simulate sweating. Wet tests were performed in an isothermal condition (i.e., $T_{manikin}=T_a=T_r$), where the air temperature was 34.0±0.5°C, and the relative humidity was 40±5%. It should be noted that a relative humidity of 60±5% was used for the tests performed at 1.0m/s air velocity while the manikin was walking at 0.96 m/s to ensure the fabric 'skin' would not be dried out too quickly. Data were recorded at 1 min intervals and a 30-min steady state was required for data analysis. Each test scenario was repeated for 3 times.

2.2.4 Calculations

The total thermal insulation (I_t) was calculated by the parallel method as shown in Eq.1 [9].

$$I_t = \frac{(\overline{T}_{manikin} - T_a)}{\sum_{i=0}^{n}(a_i \times H_i)} \tag{1}$$

The total evaporative resistance (R_{et}) was calculated by the heat loss method as shown in Eq.2 [10].

$$R_{et} = \frac{P_s - P_a}{\sum_{i=0}^{n}(a_i \times He_i)} \tag{2}$$

where, I_t is the total clothing thermal insulation, °C·m²/W $T_{manikin}$ and T_a are the surface temperature of the thermal manikin and the air temperature, respectively, °C; a_i is the ratio of local surface area of the segment i to the total surface area (i.e., 1.697m²); H_i is the segmental dry heat loss, W/m²; R_{et} is the clothing total evaporative resistance, Pa·m²/W; P_s is the saturated water vapor pressure at the fabric 'skin' surface, Pa; P_a is the partial water vapor pressure in the air, Pa; He_i is the segmental evaporative heat loss, W/m².

2.2.5. Statistical analysis

Data were reported in mean±SD (standard deviation). One-way ANOVA analysis (SPSS v.22.0, IBM Corp., Armonk, NY, USA) was performed to determine whether the influence of design and construction on the clothing thermal insulation and evaporative resistance was significant or not. The Bonferroni post-hoc method was employed, and the significance level was set at $p<0.05$.

3. Results and discussion

3.1 Total Thermal insulation

Fig.5 shows the thermal insulation of the existing and the new firefighting clothing ensembles determined in four different test scenarios. The calculated thermal insulation was based on 26 segments excluding the head/face, hands and feet. Both EXISTING and NEW presented a similar total thermal insulation. For firefighting clothing, greater thermal insulation denotes better thermal protection (i.e., a better ability in preventing external heat flowing into the human body). It is obvious that both air and body movements could reduce the total thermal insulation, which is consistence with previous documented studies [11,12]. The total insulation of EXISTING, NEW and NEW(open) measured under the 1.0m/s air was reduced by 6.2%, 8.3% and 8.8% compared to those determined under the

0.4m/s air, respectively. At the 0.96m/s walking speed, the total insulation of EXISTING, NEW and NEW(open) was reduced by 2.1%, 2.5% and 5.1%, respectively, compared with those measured under the 0.4m/s air velocity on the standing manikin. It could be deduced that the percentage of reduction in the clothing thermal insulation of NEW(open) was slightly higher than NEW when the manikin was walking. This was mainly because the opening of the zippers promoted the microclimatic ventilation and thereby decreased the total thermal insulation.

At the 0.4m/s wind, the resultant thermal insulation of EXISTING, NEW and NEW(open) was reduced by 33.9%, 26.4% and 25.6%, respectively. Similarly, the percentage of reduction in the resultant thermal insulation measured at the 0.96m/s walking speed and the 1.0 m/s air velocity was 31.0%, 21.7% and 22.5%, respectively, compared with those measured while the manikin was in a standing posture at 0.4m/s wind. Apparently, the reduction in the total thermal insulation of NEW and NEW(open) caused by air and body movements was much smaller than that of EXISTING. Furthermore, statistical analysis indicated that the resultant thermal insulation of EXISTING was significantly lower than that of NEW and NEW(open) under the two levels of air velocity ($p<0.05$). One possible reason was that EXISTING was loose fitting. For loose fitting clothing, the air gap entrapped was greater than that in regular fitting clothing, hence, forced convection inside the clothing microclimate could be easier triggered by air and body movements.

Further, the body movement had a much greater influence on the total thermal insulation in EXISTING than that on NEW. Also, the air movement had a smaller impact on the reduction of clothing total thermal insulation. The main reason could be due to the airtight nature of the fabrics used for engineering firefighting clothing ensembles. Thus, it was difficult for the forced air to penetrate the clothing to promote the microclimatic convection.

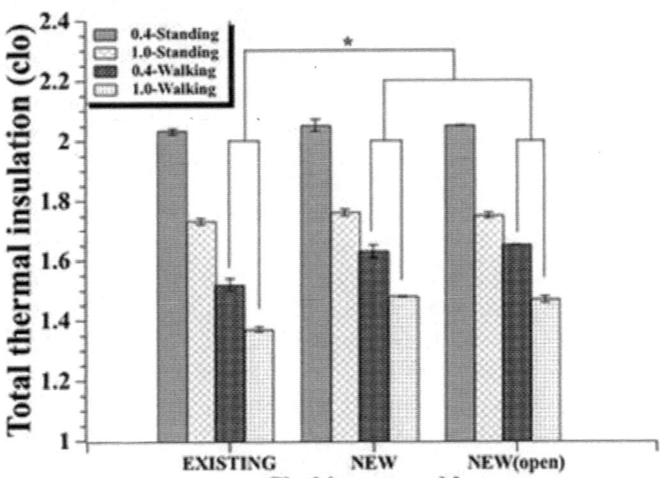

Fig.5 Total clothing thermal insulation of EXISTING, NEW, and NEW(open) determined in four different test scenarios.*, $p<0.05$.

3.2. Total evaporative resistance

Based on the research work of Wang et al [13], the uncorrected clothing total evaporative resistance was reported. Fig.6 shows the total evaporative resistance of EXISTING, NEW and NEW(open) determined under the four different test scenarios. It was found that the total evaporative resistance of NEW determined under 0.4m/s wind was 18.6% lower than that of EXISTING. For any type of clothing, a lower evaporative resistance is always preferred. Lower evaporative resistance denotes a better ability to permit water vapor transferring from the human skin surface to the ambient environment and thereby, a better evaporative cooling capacity.

When the manikin was standing, the evaporative resistance of EXISTING, NEW and NEW(open) determined under the 1.0m/s air was reduced by 12.62%, 10.43% and 13.98% compared to those determined under the 0.4m/s air velocity, respectively. This might be because the increased air velocity promoted the penetration of water vapor molecules through clothing and openings. At the 0.4 m/s wind, the percentage of reduction in the total evaporative resistance measured under the 0.96m/s walking speed was 49.05%, 34.39% and 35.53% in EXISTING, NEW and NEW(open) compared to those measured on the standing manikin, respectively. The greatest reduction in the total evaporative resistance was found in the test scenario of 1.0 m/s air velocity combined with a 0.96m/s walking speed. The total evaporative resistance was reduced by 56.47%, 39.29% and 40.72% respectively in EXISTING, NEW and NEW(open) compared to those determined under the condition of 0.4m/s air velocity on the standing manikin. It was not difficult to find that, the percentage of reduction in the total evaporative resistance in EXISTING determined under the three dynamic test scenarios (i.e., 1.0-standing, 0.4-walking, 1.0-walking) was higher than that in NEW and NEW(open). This was probably due to much smaller microclimatic air gaps existed in NEW and NEW(open) compared to those in EXISTING.

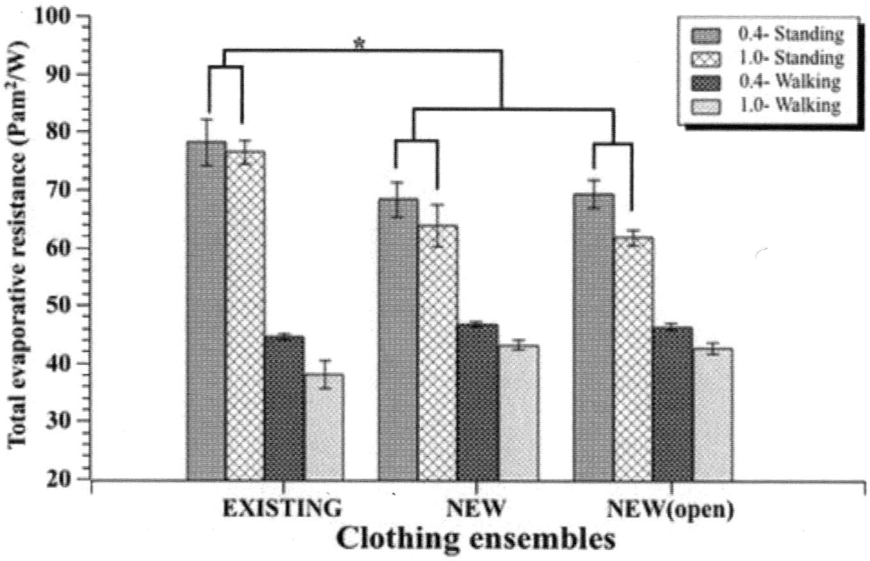

Fig.6 Total evaporative resistance of the EXISTING, NEW, and NEW(open) determined in four different scenarios. *, $p<0.05$.

Finally, the results reported in this study indicated that clothing construction design was one of the important factors in determining clothing thermal comfort attributes. Besides, it was found that walking speed had a much greater impact on both the thermal insulation and the total evaporative resistance than the air movement due to the relatively impermeable fabrics used for engineering protective firefighting clothing. On the whole, our study has presented evidence that the newly designed female firefighting clothing could greatly permit moisture evaporation and thus contributing to improving wear thermal comfort.

4. Conclusions

In this study, a new firefighting clothing ensemble was designed based on the average body dimension of the Chinese adult female firefighters. The thermal insulation and evaporative resistance of the newly developed female firefighting clothing and the existing firefighting clothing were determined using a thermal manikin. Results revealed that both EXISTING and NEW had a comparable total thermal insulation, whereas the evaporative resistance of NEW was significantly lower than that of EXISTING. In addition, the body movement had a much greater impact on both the thermal insulation and evaporative resistance than the air movement. It was finally concluded that the thermal comfort property (i.e., evaporative resistance) of the newly developed female firefighting clothing has been much improved without compromise of thermal protection.

References

[1] Coca A, Williams JW, Roberge RJ, Powell JB. Effects of fire fighter protective ensembles on mobility and performance. Appl Ergon: 2010; 41: 636-641.

[2] Boorady LM, Barker J, Lee YA, Lin SH, Cho E, Ashdon SP. Exploration of firefighter turnout gear; Part 1: Identifying male firefighter use needs. J Textile Apparel Technol Manag: 2013; 8: 1-13.
[3] Huang D, Yang H, Qi Z, Xu L, Cheng X, Li L, Zhang H. Questionnaire on firefighters' protective clothing in China. Fire Technol: 2012; 48: 255-268.
[4] Song G. Clothing air gap layers and thermal protective performance in single layer garment. J Ind Text: 2007; 6: 193-205.
[5] Tannie M, Song G. Investigation of the contribution of garment design to thermal protection. Part 2: instrumented female mannequin flash-fire evaluation system. Test Res J: 2010; 80: 1473-1487.
[6] Park H, Hahn KHY. Perception of firefighters' turnout ensemble and level of satisfaction by body movement. Int J Fashion Des Technol Educ: 2014; 7: 85-95.
[7] Boorady LM, Barker J, Lee YA, Lin SH, Cho E, Ashdon SP. Exploration of firefighter turnout gear. Part 2: Identifying female firefighter use needs. J Textile Apparel Technol Manag: 2013; 8: 1-12.
[8] GA 10-2014. Fireman's protective clothing for firefighting. Ministry of Public Security of the People's Republic of China, 2014.
[9] ISO 15831. Clothing-Physiological effects-Measurement of thermal insulation by means a thermal manikin. International Organization for Standardization, Geneva, 2004.
[10] ASTMF F 2370. Standard test method for measuring the evaporative resistance of clothing using a sweating manikin. American society for Testing and Materials, West Conshohocken, PA, 2015.
[11] Havenith G, Nilsson HO. Correction of clothing insulation for movement and wind effects, a meta-analysis. Eur J Appl Physiol: 2004; 92: 636-640.
[12] Lu Y, Wang F, Wan X, Song G, Shi W, Zhang C. Clothing resultant thermal insulation determined on a movable thermal manikin. Part I: effects of wind and body movement on total insulation. Int J Biometeorol: 2015; 59: 1475-1486.
[13] Wang F, Zhang C, Lu Y. Correction of the heat loss method for calculating clothing real evaporative resistance. J Therm Biol: 2015; 52: 45-51.

Adjustment of Individualized Dress Shirt Taking into Account Men's Shoulder Shape

KyoungOk Kim[1], Noriaki Innami[2], Masayuki Takatera[1*]

[1] *Division of Kansei and Fashion Engineering, Institute for Fiber Engineering (IFES), Interdisciplinary Cluster for Cutting Edge Research (ICCER), Shinshu University, Ueda, Nagano, 386-8567, Japan*
[2] *Graduate School of Science and Technology, Shinshu University, Ueda, Nagano, 386-8567, Japan*

*Corresponding author's email: takatera@shinshu-u.ac.jp

Abstract

To make more comfortable individualized shirt, we investigated relationship between characteristics of shoulder shape and men's dress shirt. To understand the common complaints of ready-to-wear dress shirts, wearing test of ready-to-wear shirts of 15 Japanese males were carried out. Among the problems, it was discovered that shoulder parts were not adjustable by current methods of body. Three-dimensional body shape of two subjects who had a problem in shoulder with ready-to-wear shirts was compared with one of a dummy designed to represent the average Japanese male. One subject's shoulder point and shoulder line was located in the forward than one of the average dummy. The other subject had smaller shoulder angle than one of the average dummy. Those were not possible to measure using simple method. Therefore, it became clear that new measurement method is necessary to adjust the pattern for forward thrust shoulder and square shoulder.

Keywords: Individualized shirt; Pattern adjustment metho; Men's shoulder shape

1. Introduction

In Japanese apparel size system [1], the shirt size is based upon the wearer's neck circumference and the distance measured from the back of the neck to the wrist through the shoulder point (sleeve length) in addition to body type sizing based upon the diameter of the chest and waist. Even if those measurements were fitted, it is not enough to satisfy the wearing comfort for various body types. To make a more comfortable shirt, it is necessary to understand the body shape and to obtain the more specific measurements.

In order to improve the satisfaction of wearing comfort, custom-made clothing service for individuals is a good way [2]. For custom-made shirts, more specific dimensions are necessary to distinguish the body characteristics. To make better fitted clothing, relationships between pattern and body measurements were investigated by many researchers. Makabe and Beppu [7, 8] investigated the relationship between the body measurements of Japanese females and the geometric qualities of the basic patterns that were made to fit the individuals. For shirts, A. P. Chan et al. [3, 4] showed the necessity of improvement of shirts pattern drafting methods by verifying current pattern drafting methods. They proposed a prediction method of pattern parameters from 3D body measurements and shirt patterns for different body sizes using multiple linear regression [5]. Although they proposed a prediction method of pattern parameters, it needs many measurements obtained by a three dimensional scanner which is expensive and not convenient for mass customization. More simple method will be necessary for the customization of a shirt using the Internet in particular.

The adjustment of a ready-made shirts is more effective for making fitted shirt than custom made using many measurements, because there are adequate kinds of ready-made shirt for the size of neck and sleeve including shoulder width. However, using current measurement method, it is difficult to obtain the necessary measurements regarding shoulder part.

Especially, shoulder comfort is a big problem for shirt making because of different shoulder shapes [6]. The characteristic shoulder shapes are divided into square or sloping, depending on the shoulder angle. It is also divided into forward and backward thrust shoulders, depending on the front-back direction of shoulder line.

Therefore, in this study, we investigated the dissatisfied point of ready-to-wear shirts by wearing experiments for Japanese males to confirm importance of the customization of the shoulder part. We also investigated the characteristics of forward thrust shoulder and square shoulder shapes by observing three dimensional body shapes.

2. Experimental

2.1 Dissatisfied points of ready-to wear shirt

To clarify the problem of ready-to-wear dress shirts, we investigated dissatisfying aspects of these shirts by wearing evaluation tests with questionnaire survey. Fifteen Japanese male subjects from 20~50 years old with various body shapes participated in the test as a volunteer. According to the subjects' neck circumference and sleeve length (back neck point-shoulder point-wrist), two expert pattern makers selected a ready-to-wear shirt.

Table 1 shows the measurements of subject and the shirt sizes. Subject wore an underwear (cotton 100%) under the shirt and a trousers of rubber waist band with a shirt. Hem of the shirt was placed in a trousers in a way that subject is usually worn. Fig.1 shows an example of shirt. They moved freely and watched their appearance on a mirror. We also took their pictures in front, side and back directions.

A survey was conducted related to comfort and appearance during the fitting. We asked subjects about their perception of the shirt on the questionnaire; 1) size, 2) comfort, 3) appearance, 4) ease of neck, shoulder, sleeve length and girth, and 5) the entire satisfaction out of 10 point scale. Experts checked the results of the questionnaire survey with the pictures. For the subjects who mentioned dissatisfaction on ready-to-wear shirts, experts adjusted the shirt pattern, and then their satisfaction was investigated.

Fig.1 Standard shirt for standard body

3. Results and discussion

3.1 Dissatisfied point of ready-to-wear shirt

Table 2 shows the questionnaire results of wearing tests. The entire satisfaction rate was 66%. The main reasons for dissatisfaction were tight neck, and incompatible of sleeve and shoulder, over 40%. Large ease of bodice also reason of dissatisfaction. Although it was not directly connected to the wearing comfort, it was found that it also affect the satisfaction of shirt. As a result, it was found that the tight neck, incompatible sleeve and shoulder were major problems of ready-to-wear shirts.

Table 1. subject measurements, shirt size and the satisfaction score

Subject No.	Subject measurement		Shirt size	
	Neck (cm)	Sleeve length (cm)	Neck (cm)	Sleeve length (cm)
1	33	74	36	74
2	34	85	37	84
3	36	82	38	82
4	38	81	40	80
5	33	78	36	78
6	35	83.5	37	84
7	36	83	38	82
8	34	79	37	80
9	34	84	37	84
10	40	78	42	78
12	41	84	43	84
13	39	82	41	82
14	40	84	42	84
15	43	87	45	86
17	41	84	43	84

*Sleeve length measurement: the distance from the back of the neck to the wrist through the shoulder point

The subjects' disaffection with neck was eliminated by wearing one large neck size of shirt. Disaffection with sleeve length was also eliminated by wearing one large sleeve length size of shirt. However, subjects 3 and 6 were not satisfied with the one size larger shirts. The experts were not able to adjust the patterns with the questionnaire results and pictures for the subjects 3 and 6. After checking their wearing and hearing subjects' comments, the experts adjusted the pattern and made modified shirts.

Table 2. Satisfaction score and reason of dissatisfaction

Subject No.	Satisfaction score	Reason of dissatisfaction					
		1)neck	2)sleeve length	3)Sleeve width	4)Shoulder width	5)Arm hole	6)Ease of bodice
1	8			wide			
2	8						large
3	7	tight	short	wide	tight		
4	6			wide		large	large
5	7	tight					small
6	4	tight			tight	tight	
7	6	tight		tight (upper arm)			
8	7			wide			
9	6	tight	long	wide			
10	7		long		wide		large
12	4	tight	short		tight	tight	
13	7						
14	6.5			tight (cuffs)			large
15	8	tight			wide		large
17	7.5	tight	long	tight (upper arm)	tight	tight	
Rate of dissatisfied (%)	34	53.3	33.3	53.3	40.0	26.7	40.0

Features of the each subject's body type are shown in Table 3. The experts distinguished the subject 3 is a forward thrust shoulder and subject 6 is a square shoulder. Experts modified the shirts pattern and made them. Using the modified shirts, the wearing evaluation experiment was also carried out with a questionnaire survey. The satisfaction score and a pattern adjustment method are also shown in Table 3. As shown in Table 3, both subjects showed over 8 score of satisfaction on the modified shirts which was higher than the ready-to-wear shirts. In addition, dissatisfaction points had been disappeared. Thus, it was found that the pattern adjustment methods by experts were reasonable. However, they couldn't adjust the pattern with conventional measurements such as sleeve length, shoulder width and neck circumference. Therefore, it was found that new measurement method is necessary to adjust the pattern for forward thrust shoulder and square shoulder.

Table 3. Features of body type and pattern adjustment by experts

Subject No.	Features of body type	Pattern adjustment	Satisfaction score
3	Forward thrust shoulders	Extend 0.5cm of shoulder part upward	8
6	Square shoulder	Add the ease 1cm around the shoulder line	9

3.2 Characterization of shoulder shape

Through wearing experiments and experts evaluation, we found two subjects of 3 and 6 who complain about shoulder part. One has a square shoulder shape and the other one has a forward thrust shoulder shape. To investigate the characteristics of those shoulder shapes, we compared the three-dimensional shapes of two subjects to one of a dummy of average Japanese men body size (MD-20A, NANASAI Co., Ltd, Japan). Three-dimensional scanner (Body Line Scanner, Hamamatsu Photonics, Japan) was used to obtain the three dimensional shape of subjects and the average dummy. We compared the silhouette and the cross-sectional shapes. Relationship between the adjustment of patterns in wearing test and the characteristics of shoulder shapes was investigated.

Fig.2 shows comparison of the shoulder angle of subject 3, 6 and average dummy. Comparing to average dummy (24 degree), subject 3's shoulder angle was 24.5 degree which was larger than one of the average dummy. Shoulder angle of subject 6 was 20 degree which was smaller than one of the average dummy. Fig.3 shows the comparison of the shoulder shape of subject 3, 6 and the average dummy. Comparing to the average dummy, subject 3's shoulder point is located forward and the subject 6 is located above. Fig.4 shows comparison of the armhole of subject 6 and the dummy. The armhole of subject 6 was located higher than one of average dummy.

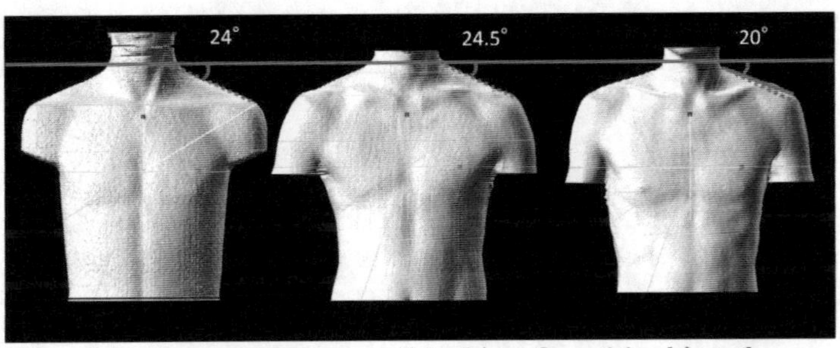

(a) average dummy (b)subject 3 (c)subject 6

Fig.2 Comparison of the shoulder angle of subject 3, 6 and average dummy (Front)

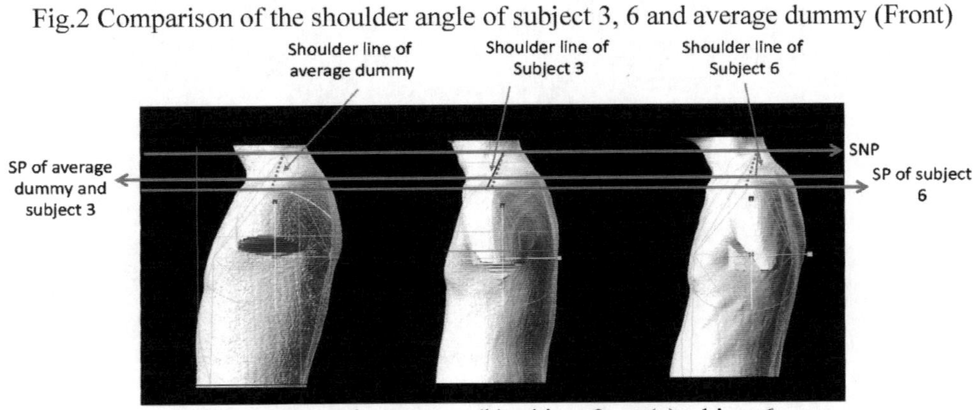

(a) average dummy (b)subject 3 (c)subject 6

Fig.3 Comparison of the shoulder shape of subject 3, subject 6 and average dummy (side)

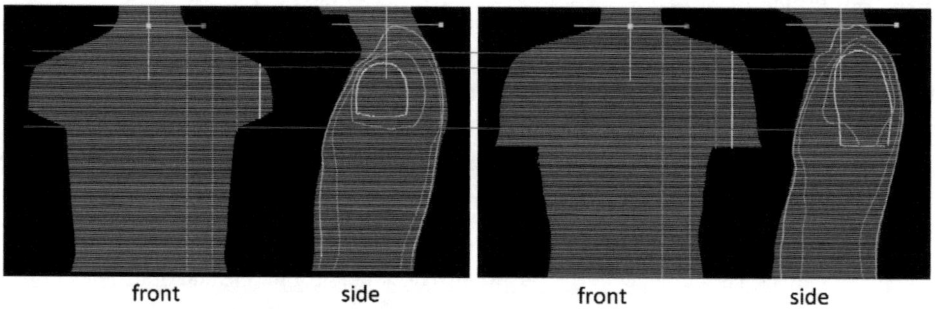

(a) average dummy (b)subject 6

Fig.4 Comparison of the armhole of subject 6 and average dummy

Fig.5 shows the seam lines of subject 3's shoulder line, shoulder seam line of ready-to-wear shirt and shoulder seam line of modified shirt. Shoulder seam line of ready-to-wear shirt were parallel to that the subject 3's shoulder line. By modifying the shirt patterns, shoulder seam line was moved to the shoulder line closely. Especially, the shoulder point of shirt moved to the subject 3's one.

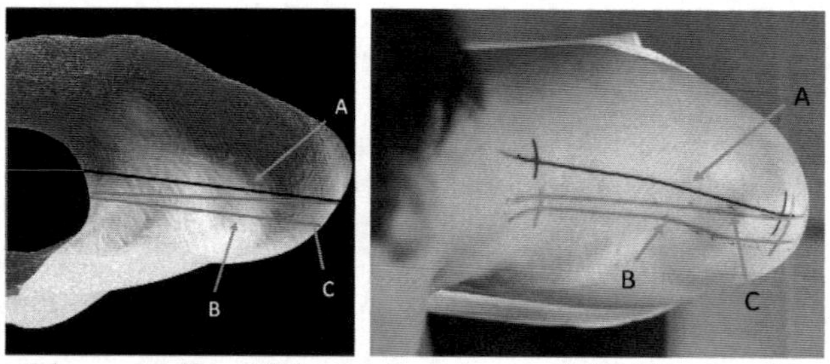

Fig.5 shoulder line (A), shoulder seam line of the ready-to-wear shirt (B), shoulder seam line of modified shirt (C) of Subject 3

4. Conclusion

By wear test of ready-to-wear shirt, it is clear that the dissatisfaction points of ready-to-wear dress shirt were tightness of neck, and incompatible of sleeve and shoulder. Experts could choose a shirt of appropriate size among the existing size other than the shoulder part, if they confirmed the fitting. The characteristics of shoulder shape for forward thrust shoulder and square shoulder were characterized by the three-dimensional body shape. In forward thrust shoulder, the shoulder point and shoulder line was located in the forward than one of the average dummy. The square shoulder showed smaller shoulder angle than one of the average dummy. Those were not possible to measure using simple method. Therefore, it is clear that new measurement method is necessary to adjust the pattern for forward thrust shoulder and square shoulder.

5. Acknowledgement

This work was supported by a Grant-in-Aid for Japan Society for the Promotion of JSPS KAKENHI under grant number 24220012.

References

[1] Japanese Standards Association, Japanese Industrial Standards hand book, Sen-I (2015)

[2] Ives, Blake, Gabriele Piccoli. "Custom made apparel and individualized service at Lands' End." Communications of the Association for Information Systems 11.1 (2003): 3.

[3] Chan AP, Fan J, Yu W. Men's Shirt Pattern Design: Part I: An Experimental Evaluation of Shirt Pattern Drafting Methods. Sen-I Gakkaishi, 2003, 59.8: 319-327.

[4] Chan AP, Fan J, Yu W. Men's Shirt Pattern Design: Part II: Prediction of Pattern Parameters from 3D Body Measurements. Sen-I Gakkaishi, 2003, 59.8: 328-333.

[5] Chan AP, J Fan, Yu WM. "Prediction of men's shirt pattern based on 3D body measurements." International Journal of Clothing Science and Technology 17.2 (2005): 100-108.

[6] COFFIN, David Page. Shirt making: Developing skills for fine sewing. Taunton Press, 1998.

[7] Makabe H, Beppu M. The construction of the basic system for the clothing pattern design (Part. 1) The indispensable measurement items of the upper body, The Japanese Journal of Ergonomics, 33. 1 (1997): 35-46.

[8] Beppu M, Makabe H. The construction of the basic system for the clothing pattern design (Part. 2) The indispensable measurement items of the lower body, The Japanese Journal of Ergonomics, 34 .1, (1998): 17-27.

Correlation between Fabric Property and Sewing Quality of Sportswear

Ling Chen, Xiao-Yu Han, Xiu-E Bai*

College of Textile and Clothing Engineering, Soochow University, Ganjiang East Road No.178, Suzhou, 215021, China

*Corresponding author's email: baixiue@suda.edu.cn

Abstract

In this paper, six kinds of common sportswear fabrics were selected, and then the fabric mechanical properties in warp and weft directions, the sewing shrinkage ratio, seaming strength and seaming efficiency were tested respectively. Finally the Pearson correlation analysis was performed with the SPSS software. The results showed that fabric property has a significant impact on the sewing quality. Specifically, fabric tensile property, shearing property, bending property at low stress level and the thickness have a significant correlation with the sewing shrinkage ratio, whereas fabric compression property has a relatively weak correlation with the sewing shrinkage ratio. Moreover, fabric tensile breaking property has a very strong correlation with the seaming efficiency, but it has a moderate correlation with the warp seaming strength and a very weak correlation with the weft seaming strength. The results also showed that the closer the tensile strength of fabric and sewing thread is, the higher the seaming efficiency is, and in this situation, the fabric and sewing thread mixed fracture is the main fracture mode at the seams.

Keywords: Sportswear fabric; Mechanical property; Sewing quality

With the improvement of people's living standards, their requirements for sportswear are becoming higher and higher. Meanwhile, people gradually pay more attention on clothing sewing quality [1]. Fabric property and sewing condition directly influence sewing quality and seaming efficiency, so the relevant research is of great value. Quality control is a very important part of garment production. The sewing quality is divided into appearance quality and inner quality. The appearance quality is manifested in the sewing flatness, and its evaluation indicators mainly include the sewing shrinkage, the displacement of upper and lower sewing materials, etc. Because of the serious effect of the sewing shrinkage and the displacement of sewing materials on the stitch appearance, they become the important indexes to evaluate the appearance of seam [2].The inner quality is mainly reflected in the performances of seam, including the seaming strength, elongation, fastness and the abrasion resistance of sewing thread and the seaming strength is the most important aspect for seaming performances [3-5]. Sportswear not only demands the smooth appearance but also requires the high strength, so the sewing quality is particularly important for it. Moreover, fabric, as the basis of clothing processing, directly affects clothing style and sewing quality. So the test on fabric performances and the study on the relationship between these performances and sewing quality are conducive to selecting appropriate fabrics, making better use of fabric performances and improving clothing quality. This paper mainly studied the relationship between fabric mechanical property at low stress level, tensile breaking property and the sewing shrinkage, the seaming strength, the seaming efficiency. The fabric mechanical property at low stress level, the tensile breaking property, fastness and fracture mode of the seam were tested respectively. The Pearson correlation analysis was adopted to analyze the experimental data. Finally, the influence degree of fabric mechanical property on the sewing quality of sportswear was obtained. It provided a certain reference for the rational use of the fabrics and the improvement of the sewing quality.

1. Experiment

1.1 Experimental condition

The experiment was conducted in the laboratory with constant temperature and humidity. The temperature is 20±2℃, the relative humidity is 65±2% during the experimental period.

1.2 Sample

The specifications of six kinds of common sportswear fabrics are shown in Table 1.

Table 1. The fabrics specifications

Fabric	Fabric type	Fabric structure	Thickness (mm)	Areal density (g/m^2)	Warp and weft density (number/10cm)
1	Cotton	Twill 2/1	1.31	314.5	260×190
2	Polyester	Plain	0.37	115.1	610×320
3	Polyester/cotton	Twill 2/1	0.54	132.4	460×330
4	Polyamide	Twill 2/1	0.15	80.6	730×510
5	Polyamide/cotton	Plain	0.36	114.4	620×320
6	Polyester/polyamide	Plain	0.37	16.3	560×380

1.3 The test on fabric mechanical property at low stress level

Fabric property, especially the mechanical property at low stress level is closely related to the sewing quality and clothing style. Garment sewing is very sensitive to the fabric mechanical property. Fabric bending, shearing and tensile properties at low stress level can cause fabric quality problems in the process of spreading, cutting and sewing [6]. In this paper, the KES fabric style instrument was used to measure the fabric thickness and areal density under $0.5N/cm^2$. The fabric tensile, shearing, bending, compression properties at low stress level were tested in warp and weft directions. The size of test sample is 200mm×200mm.

Table 2. The test items and indexes of fabric properties

Property	Instrument	Item	Index	Implication	Unit
Mechanical properties at low stress level	KES-FB1	Tensile property	LT	Tensile linearity	-
			WT	Tensile energy	gf/cm^2
			RT	Tensile resilience property	%
		Shearing property	G	Shear rigidity	N/cm
			2HG	Hysteresis value when the shear angle is 0.5°	N/cm
			2HG5	Hysteresis value when the shear angle is 5°	N/cm
	KES-FB2	Bending property	B	Bending rigidity	$N\cdot cm^2/cm$
			2HB	Bending hysteresis moment	Ncm/cm
	KES-FB3	Compression property	LC	Compression linearity	-
			WC	Compression specific work	$N\cdot cm/cm^2$
			RC	Compression resilience property	%
Fracture property	INSTRON-5967	Tensile breaking property	-	Tensile breaking strength	N

1.4 The test on fabric tensile breaking property

The fabric basic mechanical property includes tensile, tearing and bursting properties, etc. It directly affects fabric durability, and it is an important content of evaluating fabric inner quality. From the perspective of fabrics, in the study of the influence factors related to the sewing quality of sportswear, fabric tensile breaking property is the most important factor that affects the seaming strength. Tensile breaking strength is the maximum force when the fabric is stretched to break, and it is the direct indicator to test the seaming strength. This experiment used the INSTRON 5967 material testing machine, and consulted GB/T 3923.1-2013 Textiles-Tensile Properties of Fabrics-Part 1: Determination of Maximum Force and Elongation at Maximum Force Using the Strip Method.

The fabric properties and indexes measured by the KES fabric style instrument and the INSTRON 5967 material testing machine are shown in Table 2.

1.5 The sewing experiment

The flat-seam technology is adopted in this paper, which is widely used in the process of sportswear sewing. The sewer, sewing machine and sewing technology should always keep same during the sewing experiment. The sewing technology mainly includes five aspects: First, because of the 40S/2 polyester/cotton sewing thread's good abrasion resistance, high strength, small washing shrinkage and the suitability for high-speed sewing, it was selected in the sewing experiment; Second, the needle gage is 11 stitches/3cm, and the model of needle is 11#; Third, 6 pieces of samples (size: 200mm×200mm) were cut out of each fabric in both warp and weft directions; Fourth, every two samples must be placed on the operations area oppositely, and then sewed together. Fifth, the length of seam allowance is 1cm.

1.6 The test on the sewing shrinkage ratio

Problems of sewing shrinkage were often encountered during the sewing process. Sewing shrinkage is an important aspect needed to examine for fabric external sewing quality, and it is evaluated mainly through surface flatness and sewing shrinkage ratio. Sewing shrinkage ratio is an objective index to evaluate fabric surface shrinkage degree effectively. In the experiment, the sewing shrinkage ratios of fabric upper and lower layers were tested, and the comprehensive evaluation method is adopted to evaluate fabric sewing shrinkage ratio fully. The sewing shrinkage ratio is: $S_P' = \frac{|L_0 - L_2| + |L_1 - L_2|}{L_0} \times 100\%$ .Where L_0 is fabric length before sewed; L_1 is upper fabric length after sewed; L_2 is lower fabric length after sewed [7].

1.7 The test on the seaming strength

The test method of the seaming strength is the same as that of the 1.3. The fracture modes of the seam include sewing thread fracture, fabric breakage and the mixed fracture of sewing thread and fabric [8]. The seaming strength and seaming efficiency directly affect the fracture mode. The seaming efficiency is the ratio of the seaming strength to fabric strength in the seam [3].

2. Results and discussion

2.1 The correlation analysis on fabric property and sewing shrinkage ratio

Table 3. Test results of fabric property

Property			Fabric					
			1	2	3	4	5	6
Tensile property	LT	Warp	0.73	0.56	0.56	0.63	0.72	0.77
		Weft	0.63	0.63	0.64	0.64	0.79	0.74
	WT	Warp	4.77	6.80	3.15	4.53	9.20	6.32
		Weft	22.08	3.20	6.40	15.02	6.63	11.83
	RT	Warp	52.11	63.48	75.66	65.50	56.56	64.90
		Weft	42.95	88.54	64.59	42.57	57.88	53.54
Bending property	B	Warp	-0.59	0.02	0.16	0.03	0.03	0.03
		Weft	0.15	0.11	0.02	0.01	0.06	0.02
	2HB	Warp	0.47	0.01	0.05	0.04	0.04	0.03
		Weft	0.17	0.06	0.02	0.01	0.07	0.02
Shearing property	G	Warp	3.14	0.80	1.01	1.13	1.93	2.25
		Weft	3.39	0.66	0.98	1.31	1.87	2.29
	2HG	Warp	7.66	1.05	1.10	1.98	4.27	4.30
		Weft	6.41	0.38	1.01	2.52	3.35	4.34
	2HG5	Warp	10.04	3.37	4.33	4.70	6.81	4.28
		Weft	11.04	2.54	4.43	5.74	7.09	5.13
Compression property	LG	-	0.20	0.26	0.28	0.26	0.24	0.35
	WC	-	0.34	0.12	0.19	0.03	0.09	0.07
	RC	-	39.64	50.03	69.79	178.10	71.75	105.50

Table 4. Test results of sewing shrinkage ratio

Fabric		1	2	3	4	5	6
Sewing shrinkage ratio (%)	Warp	0.93	3.15	2.18	2.52	2.23	1.82
	Weft	0.8	0.7	2.43	3.23	0.97	1.07

Table 5. Correlation between fabric property and sewing shrinkage ratio

Correlative factor		Correlation coefficient	
		Warp	Weft
Tensile property	LT	-.824	-.471
	WT	-.141	-.726
	RT	.816	.835
Bending property	B.	-.871	-.750
	2HB	-.795	-.700
Shearing property	G	-.958	-.809
	2HG	-.969	-.876
	2HG5	-.812	-.772
Compression property	LG	.269	.122
	WC	-.513	-.044
	RC	.214	-.269
Thickness	T_0	-.702	-.259

Pearson correlation coefficient is used to measure whether two data sets are in the same line, further evaluate the correlation degree between two variables. The greater the absolute value of correlation coefficient is, the stronger the correlation is. More specifically, the closer the correlation coefficient to 1 or -1 is, the stronger the correlation degree is; on the contrary, the closer the correlation coefficient to 0 is, the weaker the correlation degree is. In general, when the correlation coefficient is 0.8-1, it indicates the extremely strong correlation; 0.6-0.8 illustrates the strong correlation; 0.4-0.6 represents the moderate correlation; 0.2-0.4 means the weak correlation; 0-0.2 denotes the extremely weak correlation or irrelevance [9].

Pearson correlations between fabric properties of six kinds of sportswear and the sewing shrinkage ratio are shown in Table 5. From the Table 5, it can be seen that fabric sewing shrinkage ratio and fabric mechanical properties at low stress level have a strong correlation as a whole. The correlations in warp and weft directions are different, and the correlation in warp direction is higher than that of weft direction. For samples sewed in warp direction, fabric sewing shrinkage ratio has a high correlation with fabric tensile property (LT, RT), bending property, shearing property and thickness. For samples sewed in weft direction, fabric sewing shrinkage ratio has a high correlation with fabric tensile property (WT, RT), bending property, shearing property, whereas it has a relatively weak correlation with fabric compression property.

2.2 The correlation analysis on fabric property and seaming strength

Fabric property, especially the breaking strength is the basis for deciding the seaming strength. The test results of fabric tensile breaking strength, seaming strength and the seaming efficiency are shown in Table 6. It should be specially explained that the samples sewed in weft direction are used to test the warp-wise tensile breaking property, whereas samples sewed in warp direction are applied to test the weft-wise tensile breaking property. Pearson correlations between the fabric tensile breaking properties of six kinds of sportswear and the seaming strength, the seaming efficiency are shown in Table 7.

Table 6. Test results of fabric tensile breaking strength, seaming strength and seaming efficiency

Fabric		1	2	3	4	5	6
Tensile breaking strength (N)	Warp	1302.07	711.43	1128.87	818.47	753.70	479.80
	Weft	589.00	808.77	539.00	462.80	594.57	484.37
Seaming strength (N)	Warp	273.80	272.83	313.47	259.43	294.83	247.43
	Weft	276.23	290.47	240.27	306.43	253.90	304.47
Seaming efficiency (%)	Warp	21.03	38.35	27.77	31.70	39.12	51.57
	Weft	46.90	35.91	44.58	66.21	51.34	62.86

Table7. Correlation between fabric tensile breaking strength and seaming strength and seaming efficiency

Correlative factors	Correlation coefficient related to seaming strength		Correlation coefficient related to seaming efficiency	
	Warp	Weft	Warp	Weft
Fabric tensile breaking strength	.534	-.109	-.961	-.862

From the Table 6, it can be seen that fabric tensile strength is larger than seaming strength. The fabric tensile breaking strength in warp direction as a whole is larger than that of the weft direction, whereas the seaming strengths in two directions are almost the same. It means sewing thread has a relatively significant impact on the seaming strength. The strength of experimental polyester/cotton sewing thread is 106N. Obviously, the strength of sewing thread is smaller than that of seaming strength, so the sewing thread fractures firstly in process of tensile experiment. This is caused by the fact that the strength of sewing thread is much smaller than fabric tensile breaking strength. This phenomenon can be verified in the test of the seaming strength. The greater the ratio between fabric tensile breaking strength and sewing thread strength is, the more easily the fracture of sewing thread happens. From the perspective of fabric utilization, the fracture of sewing thread is a relatively ideal failure mode. But the ideal seaming strength is this kind of situation that the sewing thread and fabric fracture at the same time, whereas in reality it is an relatively appropriate condition that the sewing thread fractures before the fabric is about to break. Therefore, the strength of sewing thread (mainly refers to the ratio of fabric strength to sewing thread strength) influences the seaming strength to a great extent, directly affects the seaming efficiency and fabric fracture mode. The experiment indicates that the closer the fabric tensile breaking strength and sewing thread tensile breaking strength are, the higher the seaming efficiency is, and in this situation, the mixed fracture of fabric and sewing thread is more likely to appear at the seams. Consequently, the sewing thread whose strength is close to fabric tensile breaking strength can obtain a high seaming efficiency. Moreover, the seaming efficiency of the tension in warp direction is generally much lower than that of the weft direction. This is determined by the fact that fabric tensile breaking strength in warp direction is larger than that of the weft direction.

From the Table 7, it can be seen that fabric tensile breaking property has a strong negative correlation with the seaming efficiency. It illustrates that fabric tensile breaking property is the root of the seaming efficiency. What is more, the correlation between the seaming strength and the tensile breaking property in warp direction is quite different from the correlation between the seaming strength and the tensile breaking property in weft direction. Specifically, the seaming strength has a moderate correlation with the tensile breaking property in warp direction, whereas it has an extremely weak correlation with the tensile breaking property in weft direction.

3. Conclusion

1) As a whole, fabric mechanical properties at low stress level have a strong correlation with seaming shrinkage ratio, and the correlation in warp direction is greater than that of the weft direction. Specifically, for samples sewed in warp direction, fabric sewing shrinkage ratio has a high correlation with fabric tensile property (LT, RT), bending property, shearing property and thickness, and for samples sewed in weft direction, fabric sewing shrinkage ratio has a high correlation with fabric tensile property (WT, RT), bending property, shearing property, whereas it has a relatively weak correlation with fabric compression property.

2) The closer the relationship between fabric tensile breaking strength and sewing thread tensile breaking strength is, the higher the seaming efficiency is, and in this situation, the mixed fracture of fabric and sewing thread is more likely to appear at the seams.

3) Fabric tensile breaking property has a strong correlation with the seaming efficiency. Concretely speaking, the seaming strength has a moderate correlation with fabric tensile breaking property in warp direction, whereas it has an extremely weak correlation with the tensile breaking property in weft direction.

References

[1] Yang XY, Shi Q. Brief introduction to the clothing international standardization. Quality & Standardization: 2014; 10:15.

[2] Li YM, Zhang WY. Research progress and tendency of fabric sewing compatibility. SHANGHAI TEXTILE SCIENCE&TECHNOLOGY: 2009; 7:1-4.

[3] Zhang WB. Garment Manufacturing. Beijing: China Textile Press: 2008.

[4] Lu X, Zhang S, Gu YF. Influence of knitted fabric properties on edge sewing of knitted apparel. Journal of Textile Research: 2012; 11:97-101.

[5] Ou Y, Zhao L. Effect of fabric character and sewing parameters on seam strength of chiffon fabric. International Textile Leader: 2014; 11:66-70.

[6] Liu K. The establishment of objective evaluation system for the grade of garment sewing flatness based on fabric mechanical properties. Shanghai: Donghua University: 2005.

[7] Wang JJ, Zhang HZ. Evaluation of Fabric Crease-Shrinkage in Sewing. Journal of Silk: 1998; 7: 43-44.

[8] Zou FY, Quan XF, Fang LY. Research on Relation between Seam Property and Sewing Condition for Shell Silk Fabrics. Journal of Textile Research: 2002; 1:53-56.

[9] Wu Q. Sewing Performances and Morphological Characteristics of the Fabric. Suzhou: Soochow University: 2004.

Foot Anthropometric Measurements of Hong Kong Elderly: Implications for Footwear Design

Pui-Ling Li[1], Kit-Lun Yick[1*], Sun-Pui Ng[2], Joanne Yip[1]

[1]*Institute of Textiles and Clothing, The Hong Kong Polytechnic University, HungHom, Hong Kong*
[2]*Hong Kong Community College, The Hong Kong Polytechnic University, HungHom, Hong Kong*

*Corresponding author's email: kit-lun.yick@polyu.edu.hk

Abstract

Foot deformities, such as hallux valgus, bunionettes and pes planus, are commonly found among Hong Kong older people, which may lead to changes in foot anthropometry. A decline in functional mobility and greater risk of falling are linked to foot deformities and footwear. This study therefore aims to evaluate the anthropometric measurements between healthy and deformed feet in order to determine the key foot measurements in relation to the deformed foot which can also act as indicators in current footwear sizing systems. By using a 3D hand-held scanner, 11 foot anthropometric measurements are captured and used to characterise the dimensions and foot shape between healthy and deformed feet. A total of 49 elderly people between the ages of 65-95 years old, including 41 women and 8 men (mean: 81.71; SD: 7.08) are recruited for this study. The results indicate that the foot characteristics of elderly people with foot deformities are different from those without deformities, especially in the degree of hallux valgus deformity and width of the ball for women, and instep height for men. The length of the foot and ball, width and girth of the ball, and degree of hallux valgus deformity are common predictors for differentiating between healthy and deformed feet. It is also found that the current footwear sizing systems fail to accommodate the foot dimensions of elderly people in both foot length and width, which may therefore lead to foot discomfort and even limit their daily life activities.

Keywords: Foot problems; Anthropometry; Footwear design; Elderly; 3D scanning

1. Introduction

Foot deformity is a prevalent issue among older people. With increased age, degenerative processes induce many physical changes in the body and feet, including pes planus, pes cavus, hallux valgus, bunionettes and enlargement of the width and height of the feet due to toe deformities or thickened nails [1]. A recent local study revealed that about 80% of the studied elderly people have at least one foot problem and/or various types of foot deformities [2]. Deformed feet may lead to a decline in functional mobility and add to a greater risk of falling. As compared to younger people, older people tend to have flatter, longer and wider (larger circumference) feet [3]. It is difficult for the elderly to find well-fitting shoes in the market since most shoe manufacturers utilise data from the feet of young adults for their shoe designs [4]. In traditional footwear sizing systems, customers are required to take the measurement of their foot length with a measuring tape or use the Brannock device to select a suitable size for footwear. However, the feet of older people are generally broader than the shoes available in their size. As a result, reports have indicated that around 80% of older men and women wear shoes that are too narrow and too short for their feet, which lead to lesser toe and hallux valgus deformities, corns on the toes and foot pain [5-6].

In light of the increasing population of older people and their increasing demands for footwear, foot anthropometry related studies have been conducted in Brazil, Australia, Thailand, Japan, etc [6-9]. As ethnic origin and the corresponding life style can influence foot shape, a study on the foot characteristics of older adults in each nation is infeasible. To the best of our knowledge, no previous investigations have been made on measuring the foot morphology of older people in Hong Kong in relation to their deformed foot. The aim of this study is to therefore characterise the dimensions and foot shape of the elderly in Hong Kong and compare the anthropometric measurements of healthy and deformed feet. The key foot measurements that determine the deformed foot in the elderly will also be determined. Since poorly fitting footwear is a known risk factor of falls for older people, their foot shape characteristics will also be compared with the current footwear sizing systems. The findings of this study are important for improving current understanding on anthropometric measurements of the deformed foot and providing more information that will improve the fit and comfort of footwear for older people.

2. Method

2.1 Participants

A total of 49 elderly people between 65-95 years old, including 41 women and 8 men (mean: 81.71; SD: 7.08) were recruited from a local self-care residence for this study. The inclusion criteria are people who are 65 or older, and able to walk independently across a distance of at least 6 m with or without the use of a walking aid [9]. Their body mass index (BMI) ranges from 14.13 to 34.11 kg/m^2 (mean: 24.2; SD: 3.9). Their foot size is a European 35 to 42 (mean size 38) for the women and 37 to 44.5 (mean size 40) for the men. Their foot conditions were assessed by a physiotherapist through a physical examination to classify them with healthy or deformed feet. Amongst the 49 participants (98 feet in total) in this study, 26 (26.5%) healthy feet are found, including 20 for the women and 6 for the men. There are 72 feet (73.5%) that are diagnosed as deformed, including 62 for the women and 10 for the men. All of the participants provided written informed consent before participation in the study.

2.2 Foot Anthropometric Measurements

Modern 3D surface scanning systems can obtain accurate and repeatable digital representations of the foot shape and have been successfully used in medical, ergonomic and footwear development applications [10]. In this study, all of the foot shape and characteristics are captured by using an Artec Eva 3D hand-held scanner (Fig.1). Eight markers were drawn on specified anatomical landmarks on each foot, which were visible to the laser of the scanner. These anatomical landmarks were used to characterise the external shape of each foot and facilitate the calculation of 11 anthropometric measurements (see Fig.2). During the scanning process, the subjects were given instructions to stand in their bare feet with their feet shoulder width apart (equal weight on each foot) and place their arms by their side to separately acquire the left and right foot images. An example of a 3D foot image that is taken by the Artec Eva scanner is shown in Fig.2.

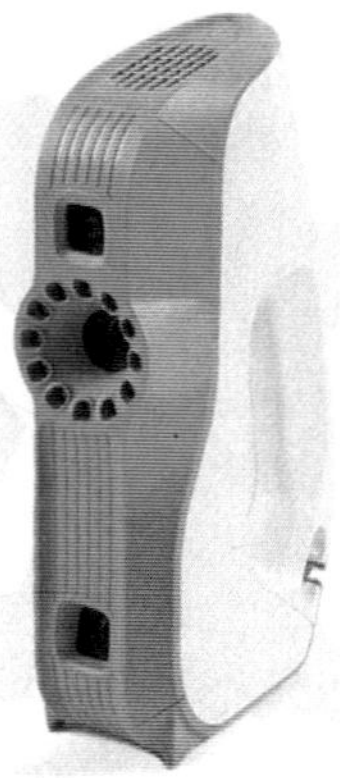

Fig.1 Artec Eva 3D hand-held scanner

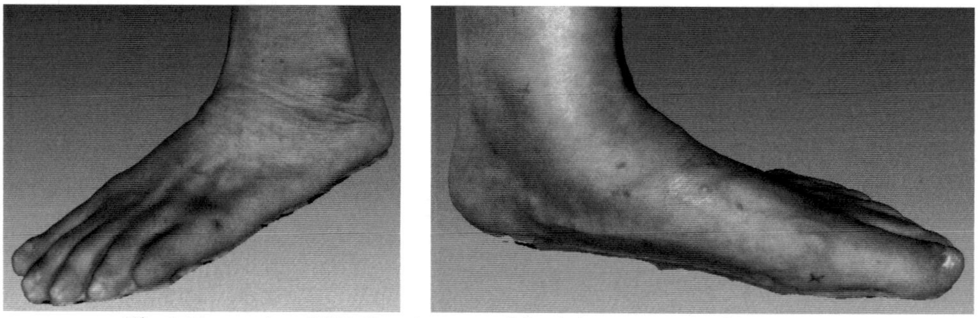

Fig.2 Example of 3D foot image with eight anatomical landmarks

2.3 Measurement Outcomes

Eleven foot anthropometric measurements were taken from the 3D foot images through the software of Geomagic Studio and SolidWorks, including the foot length (FL), heel length (HL), ball length (BL), foot width (FW) ball width (BW), bimalleolar width (BMW), ball girth (BG), instep girth (IG), instep height (IH), degree of hallux valgus deformity (HVD) and valgus index (VI), see Table 1 [9,11-12]. Geomagic Studio is used to align the feet with the floor and the XYZ coordinate in order to obtain the measurements in a more accurate manner. SolidWorks is also incorporated to obtain the angular and perimeter measurements.

Table 1. Details of 11 foot anthropometric measurements taken from 3D images

Dimension	Description
Foot Length (mm) 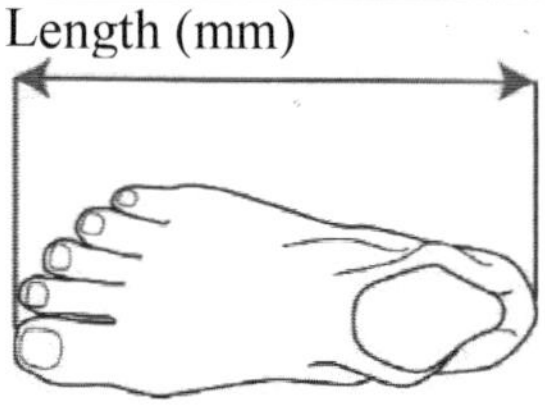	Distance between the pternion and the tip of the second toe, measured along the foot axis.
Heel Length (mm) 	Distance between the pternion and the line that joins the medial and lateral malleolus.
Ball Length (mm) 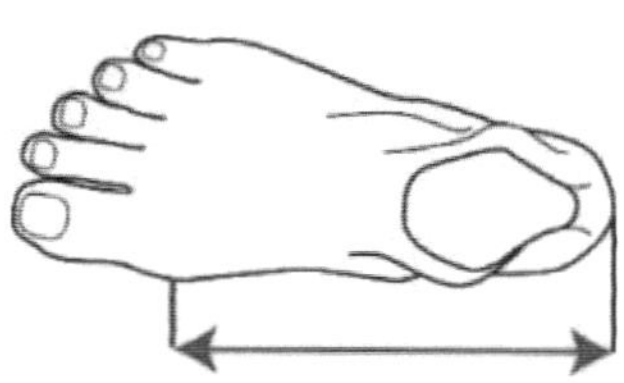	Distance between the pternion and the most medially protruding point of the first metatarsal head, measured along the foot axis.
Foot Width (mm) 	Distance between the most medially and laterally protruding points of the feet.

Ball Width (mm)

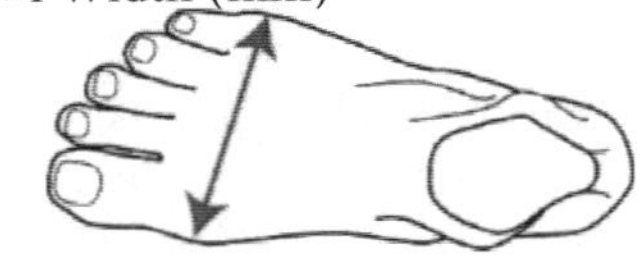

Distance between the most medially protruding point of the first metatarsal head and the most laterally protruding point of the fifth metatarsal head.

Bimalleolar width (mm)

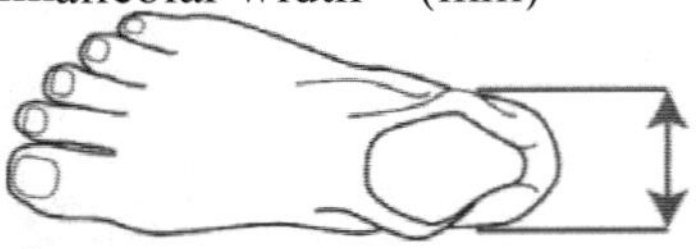

Distance between the most laterally and medially protruding points of the lateral and medial malleolus measured along a line perpendicular to the foot axis.

Ball Girth (mm)

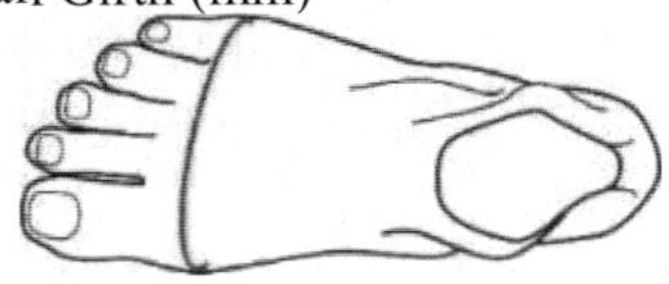

Measurement of the curve that passes from the first to the fifth metatarsal heads on the dorsum foot. The plantar area is not included.

Instep Girth (mm)

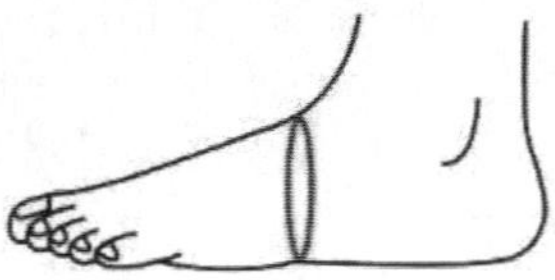

Measurement of the curve of the vertical section of the dorsum foot in the most prominent region of the navicular bone. The plantar area is not included.

Instep Height (mm)

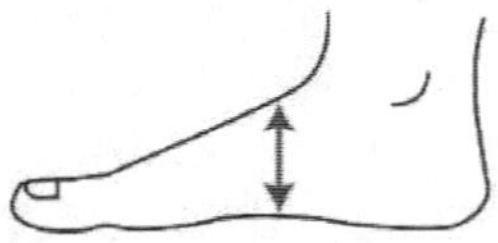

Distance between the highest point on the instep circumference and the floor.

Degree of Hallux Valgus Deformity (◦)

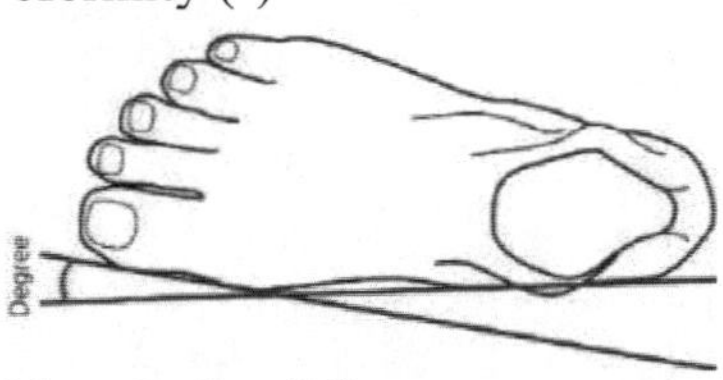

A base line is drawn from the medial heel width point to the widest point of the first metatarsal head and extends past the foot. Another line is drawn between the widest point of the first metatarsal head and the outer edge of the first toe joint. The angle between these two lines is calculated.

Valgus Index (%)

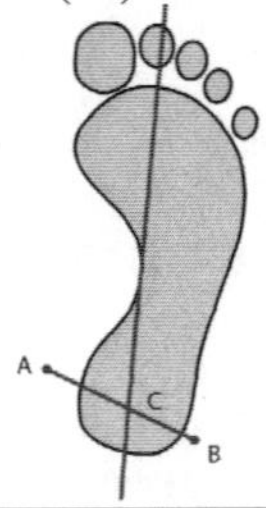

VI = 0.5AB-AC X (100/AB)

AB: The line that joins the two malleoli projection points, which is based on the actual position of each malleolus marked.

A second line is drawn from the centre of the heel to the tip of the second toe (foot axis). C is the intersection point of the two lines.

2.4 Statistical analysis

Independent T-tests were used to assess the statistical significance of the differences in the measurement outcomes between the healthy and deformed feet. The level of significance was set at 0.05. A discriminant analysis was used to determine the key foot anthropometric measurements of the deformed feet.

3. Results and Discussion

3.1 Differences between healthy and deformed feet

Table 2 shows that the male subjects with deformed feet have larger mean values for the FL, HL, BL, FW, BW, BMW, IG, IH, HVD and VI than those with healthy feet, except for BG. As compared to the healthy feet, the major mean differences in the foot dimensions are observed in the HVD (increase of 142.17%). However, a significant difference between healthy and deformed feet is only found for the IH (p=0.031). This may be because the mean value is exaggerated by some of the serious cases. For women, larger mean values are also found for deformed feet in the HL, BL, FW BW, BG, IG, HVD and VI as opposed to healthy feet, except for the FL, BMW and IH (. The largest mean value increment is 108.71% for the HVD and the least is 0.04% for BG. Only the BW and HVD show a significant difference between the healthy and deformed feet (p=0.040, p=0.000, respectively). It is found in this study that deformed feet tend to have larger mean values for some of the foot anthropometric dimensions in terms of the width, height and degree of hallux valgus deformity. These findings are similar to those in Mickle et al. in that foot anthropometric measurements of older people are affected by foot deformities, such as an increased width of the ball, girth of the ball, first toe angle, medial ball length, etc [9]. The male subjects have a significant difference in the instep height perhaps because pes cavus is more commonly found in men, while pes planus is more common in women [13]. The female subjects with a deformed foot tend to have a significant difference in the degree of hallux valgus deformity and width of the ball. These results are consistent with a study done by Saghazadeh et al. on older Japanese people in that the first toe angle is significantly greater in women due to the presence of hallux valgus, which occurs more frequently in women [8]. Moreover, in another study by Mickle et al., the older people with moderate-to-severe hallux valgus deformity have a significantly increased girth and width of the ball, medial and lateral ball lengths, heel bone and first toe angles [9].

Table 2. Anthropometric measurements of healthy and deformed feet amongst male subjects

Male	Healthy Foot (n=6)	Deformed Foot (n=10)	Mean
	Mean (SD)	Mean (SD)	Difference (%)
Foot Length (mm)	245.14 (12.43)	247.35 (14.46)	0.90
Heel Length (mm)	59.36 (8.94)	59.72 (4.80)	0.61
Ball Length (mm)	178.08 (9.34)	181.30 (10.57)	1.81
Foot Width (mm)	96.93 (9.82)	99.10 (5.36)	2.24
Ball Width (mm)	96.74 (7.87)	102.06 (6.67)	5.50
Bimalleolar width (mm)	66.05 (7.05)	68.93 (5.27)	4.36
Ball Girth (mm)	153.55 (13.80)	150.39 (10.48)	-2.06
Instep Girth (mm)	175.15 (21.12)	177.39 (15.46)	1.28
Instep Height (mm)	56.96 (8.13)*	68.63 (10.08)*	20.49
Degree of Hallux Valgus Deformity (◦)	7.02 (6.94)	17.00 (11.90)	142.17
Valgus Index (%)	-3.82 (3.37)	-1.00 (7.34)	73.82

*Significant difference at P<0.05 (2-tailed)

Table 3. Anthropometric measurements of healthy and deformed feet amongst female subjects

Female	Healthy Foot (n=20) Mean (SD)	Deformed Foot (n=62) Mean (SD)	Mean Difference (%)
Foot Length (mm)	230.39 (8.06)	229.29 (11.79)	-0.48
Heel Length (mm)	58.00 (5.59)	59.88 (4.62)	3.24
Ball Length (mm)	169.40 (5.97)	170.91 (8.64)	0.89
Foot Width (mm)	90.15 (6.14)	92.56 (5.16)	2.67
Ball Width (mm)	92.11 (6.12)*	94.95 (5.02)*	3.08
Bimalleolar width (mm)	64.52 (3.66)	64.49 (6.39)	-0.05
Ball Girth (mm)	142.66 (9.56)	142.72 (9.45)	0.04
Instep Girth (mm)	159.31 (9.84)	161.15 (11.20)	1.15
Instep Height (mm)	61.16 (4.06)	60.44 (5.49)	-1.18
Degree of Hallux Valgus Deformity (∘)	7.58 (3.67)*	15.82 (10.27)*	108.71
Valgus Index (%)	-1.67 (5.66)	-1.36 (7.35)	0.19

*Significant difference at P<0.05 (2-tailed)

3.2 Key foot measurements between healthy and deformed feet

The discriminant analysis indicated that the predictors below could be used to differentiate between deformed and healthy feet (p=0.000) and the overall effect size for the analysis (η^2=0.12) is small according to Cohen [14]. Table 4 presents the standardised function coefficients, which suggest that the FL, BL, BW and BG contribute most to differentiating feet with deformities from those without deformities through these predictors. The HVD tends to moderately contribute to the process of differentiating, but has relatively high correlation with the overall discriminant function. These findings are similar to those in Luximon and Goonetilleke [15] and could be explained by using their results in that the foot shape is modelled and predicted by using just length, width, height and a measure of the curvature of the metatarsalphalangeal joint, which has a mean accuracy of 2.4 mm.

Table 4. Differentiations in key foot anthropometric measurements

	Standardised function coefficient	Correlation between variables and discriminant function
Foot Length (mm)	-2.01	-0.10
Heel Length (mm)	-0.02	0.20
Ball Length (mm)	1.56	0.07
Foot Width (mm)	-0.22	0.19
Ball Width (mm)	1.13	0.30
Bimalleolar width (mm)	0.13	0.03
Ball Girth (mm)	-0.87	-0.09
Instep Girth (mm)	0.30	0.02
Instep Height (mm)	0.30	0.14
Degree of Hallux Valgus Deformity (∘)	0.43	0.61
Valgus Index (%)	0.00	0.08

3.3 Comparison between footwear sizing systems and foot dimensions

Table 5 shows a comparison of current footwear sizing systems, such as pressure measuring insoles (wide fit) by the Novel® Pedar system [16], the UK footwear sizing system [17], footwear sizing system of international brands that focus on indoor footwear (Mahabis® and UGG®) [18-19], and the foot characteristics of the subjects (with/without foot deformities and both genders) in terms of the dimensions of the foot length. The measured foot length of each subject is categorised in accordance with his/her shoe size. It can be seen that the foot length of the current sizing systems, especially for indoor footwear available in the market, is consistently longer than that obtained in this study by 3.5-8.0%. As compared to the Novel® Pedar system, the width of the ball of all of the subjects is broader than the insoles at around 6-12%. The findings are somewhat different from a previous study by Menz and Morris in which 23 subjects (13.7%) wore indoor shoes that were shorter than their feet and 136 subjects (81.4%) wore indoor shoes that were more narrow than their feet. [5] However, the results indicate that the indoor shoes available in the market overall fail to accommodate the foot dimensions of older people, which may cause discomfort or eventually injury to their feet. Moreover, the results in the present study indicated that the FL and BW, which are regularly used in traditional sizing systems, should not be the only key dimensions for choosing a correct size for feet with deformities. Dimensions such as the BL, BG and HVD should also be taken into consideration.

Table 5. Comparison of current sizing systems and foot length (mm)

Size	Novel® Pedar system	British Standard (A)	International indoor footwear brands (B)	Measured foot length (C)	Difference (%) A-C	Difference (%) B-C
37	235	233.5	229	221.2	5.6	3.5
38	245	240.1	238	227.0	5.8	4.8
39	245	246.8	244	231.0	6.8	5.6
40	260	253.5	252	236.6	7.1	6.5
41	260	260.1	258	240.8	8.0	7.1
42	275	266.8	265	249.5	6.9	6.2

4. Conclusion

To conclude, the foot characteristics of the deformed feet of the elderly are different from those of healthy feet. Shoe manufacturers should consider the gender and deformation differences in feet when designing the last measurements of footwear for older adults, especially in terms of the instep height for men, and the degree of hallux valgus deformity and width of the ball for women. Apart from the foot length and width of the ball, the length and girth of the ball and degree of hallux valgus deformity are common predictors used to differentiate between deformed and healthy feet. These findings could be used as indicators in footwear sizing systems as well as the design of geriatric footwear to provide better fitting and comfort for the elderly, such as heel height, heel counter stiffness and height, fixation, shoe length and width, forefoot height, tread pattern, sole hardness, sole flexion point, longitudinal sole rigidity [20-22]. For example, the findings on foot length imply that the shoe length for older people with foot deformities should be extended to provide extra space for the toes for enhancing comfort and fit. The findings of ball and hallux valgus measurements also imply that adjustments should be made when designing footwear for the elderly with hallux valgus. It is suggested to match the 'flex angle' of the forefoot with that of the anterior part of the shoe (e.g. ball width) so as to better accommodate the deformities in the forefoot. Moreover, footwear for older people with lesser toe deformities should include the toe box with increased depth to allow room for the increased heights of the first and fifth toes [9, 23]. Changes in the foot shape geometry in older people have exacerbated the problems of footwear fit and comfort due to inappropriate footwear design and sizing systems. In considering that there is an increasing population of older people and therefore a strong demand for geriatric footwear, the findings of this study will provide valuable foot anthropometric information to improve the design of footwear for older people and ultimately provide optimal fit and comfort.

5. Acknowledgement

We would like to thank the Innovation and Technology Fund (ITF) (ITF/Tier 3 Project ITS/359/14) for funding this research project.

References

[1] Dunn JE, Link CL, Felson DT, Crincoli MG, Keysor JJ, McKinlay JB. Prevalence of foot and ankle conditions in a multiethnic community sample of older adults. Am. J. Epidemiol: 2004; 159: 491-498.

[2] Lai TL, Chan KL, Chung SL. Foot problems among elderly people in Hong Kong. Asian J of Gerontology and Geriatrics: 2014; 9(2): 71-73.

[3] Scott G, Menz HB, Newcombe L. Age-related differences in foot structure and function. Gait Posture: 2007; 26: 68–75.

[4] De Castro AP, Rubens Rebelatto J, Rabiatti AT. The effect of gender on foot anthropometrics in older people. J Sport Rehabil: 2011; 20(3): 277.

[5] Menz HB, Morris ME. Footwear characteristics and foot problems in older people. Gerontology: 2005; 51 (5): 346–351.

[6] Chaiwanichsiri D, Tantisiriwat N, Janchai S. Proper shoe sizes for Thai elderly. The Foot: 2008; 18 (4): 186–191.

[7] De Castro AP, Rubens Rebelatto J, Rabiatti AT. The relationship between foot pain, anthropometric variables and footwear among older people. Applied Ergonomics: 2010; 41: 93–97.

[8] Saghazadeh M, Kitano N, Okura T. Gender differences of foot characteristics in older Japanese adults using a 3D foot scanner. J Foot Ankle Res: 2015; 8: 29.

[9] Mickle KJ, Munro BJ, Lord SR, Menz HB, Steele JR. Foot shape of older people: implications for shoe design. Footwear Science: 2010; 2(3): 131-139.

[10] Telfer S, Woodburn J. The use of 3D surface scanning for the measurement and assessment of the human foot. J Foot Ankle Res: 2010; 3: 19.

[11] Fritz B, Schmeltzpfenning T, Plank C, Hein T, Grau S. Anthropometric influences on dynamic foot shape: Measurements of plantar three-dimensional foot deformation. Footwear Science: 2013; 5(2): 121-129.

[12] Billis E, Katsakiori E, Kapodistrias C, Kapreli E. Assessment of foot posture: Correlation between different clinical techniques. The Foot: 2007; 17(2) : 65-72.

[13] Chaiwanichsiri D, Janchai S, Tantisiriwat N. Foot Disorders and Falls in Older Persons. Gerontology: 2009; 55(3): 296-302.

[14] Cohen J. Statistical power and analysis for the behavioral sciences (2nd ed.). Hillsdale: NJ. Lawrence Erlbaum Associates 1988.

[15] Luximon A, Goonetilleke RS. Foot shape modelling. Hum Factors: 2004; 46: 304-315.

[16] Novel.De. Pedar® Insole Catalogue. Novel GmbH: 2010.

[17] British Standards Institution. Footwear-Sizing-Conversion of sizing systems PD ISO/TS 19407. IHS: 2015.

[18] Mahabis Ltd. Sizing chart on http://mahabis.com/pages/sizing-chart.

[19] UGG Australia. Men and women sizing chart on http:// www.uggaustralia.com.

[20] Tencer AF, Koepsell TD, Wolf ME, Frankenfeld CL, Buchner DM, Kukull WA, LaCroix AZ, Larson EB, Tautvydas M. Biomechanical properties of shoes and risk of falls in older adults. J Am Geriatr Soc: 2004; 52: 1840–1846.

[21] Menz HB, Sherrington C. The Footwear Assessment Form: a reliable clinical tool for the evaluation of footwear characteristics of relevance to postural stability in older adults. Clin Rehabil: 2000; 14: 657–664.

[22] Menz HB, Auhl M, Ristevski S, Frescos N, Munteanu SE. Evaluation of the accuracy of shoe fitting in older people using three-dimensional foot scanning. Journal of foot and ankle research: 2014; 7(1): 3.

[23] Chantelau E, Gede A. Foot Dimensions of Elderly People with and without Diabetes mellitus – a Data Basis for Shoe Design. Gerontology: 2002; 48(4): 241-4.

Assessment and Analysis of Physical Beauty of Young Women in China based on 3D Scanning

Zhi-Peng Liu, Hao Zhang*, Lu Wang, Ying Wang

Aimer Human Engineering Research Centre, Beijing Institute of Fashion Technology, Beijing, 100029, China

*Corresponding author's email: zhhh1cn@126.com

Abstract

Since classical antiquity, assessment of physical beauty has been a popular topic studied by Chinese and international scholars. This study sought to assess physical beauty of 109 young women selected by means of random sampling. The study used 3D scanning software and subjective assessment software to evaluate formal characteristics of the human body, and then further selected subjects with certain representative physical characteristics of the population and statistical significance. Based on assessment scores and body size data of assessed samples, the study established a linear regression mathematical model to evaluate indexes of physical beauty encompassing shoulder-neck, chest, waist-abdomen-hips, legs, and the whole body.

Keywords: Physical beauty; 3D scanning; Assessment software; Regression model

1. Introduction

The bodily form is a unique aesthetic object, and physical beauty refers to the beauty of the bodily form as the aesthetic object. The perfect human body shall meet certain aesthetic indexes in terms of form, outline, position, and proportion, which will be modified with the change of age, social concept, and ethics of religion. Assessment of physical beauty has been a popular topic studied by Chinese and international scholars for centuries.

In the 6th century BC, an ancient Greek philosopher discovered the so-called Golden Ratio [1]. The proportional relationship of 0.618:1 is common in the ecological construction of nature and was gradually applied in the study of the physical beauty. In the 1960s, the Wacoal Human Science Research Center in Japan began studies of Japanese women's physical characteristics. Since then, Wacoal has released successive research results such as "Golden Section" (1965), "Beautiful Proportion" (1979), and "Golden Rule" (1994).

In recent years, researchers in the fields of social biology and evolutionary psychology have undertaken in-depth studies of physical beauty for different research purposes. Significant results have been achieved from the study of assessment of physical beauty. Currently, there are mainly three indexes for evaluating women's physical beauty: WHR [2,3], BMI [4], and VHI [5]. However, most indicators used to evaluate physical beauty are either subjective or objective, and the importance of combining the two is neglected.

2. Research method

The study combined subjective assessment and objective measurement to research physical beauty based on assessment indexes. Subjective assessment refers to the method that the assessor uses to evaluate the physical beauty of the assessed person; objective measurement refers to the method that the assessor uses to analyse the bodily form data of the assessed samples using the mathematical statistics method. First, the accurate and authentic bodily form data is obtained by 3D scanning; after 3D data is combined and repaired, the bodily form measurement data of samples is extracted. Second, sample assessment scores are obtained through a physical beauty assessment experiment; the assessment result is processed through normalisation algorithm [6], Euclidean distance detection, and independent-samples T-test, and proper physical samples are selected. Finally, the related regression equation is applied for multiple linear regression statistics and physical beauty measurement.

2.1 Data collection

For its assessment samples, the study used 3D-scanning body-size data of 109 childless women 20-29 years of age who were randomly selected from the 2009-2010 database of female bodies of BIFT-Aimer Human Engineering Research Centre. The sample selection meets sampling requirements for data collection of the Hong Kong Polytechnic University [7]. Data encompassed 147 items for use as measurements to investigate women's physical beauty in accordance with the actual research needs [8-11]. Of the 147 measured items, 100 were direct measurements that included 13 height measurements, 11 girth measurements, 28 thickness measurements, 11 surface measurements, 25 angle measurements, 9 chest detail measurements, and 3 volume weight measurements; the 47 indirect measurements included 8 high degree conversion items, 8 circumference transformed items, 13 width converted items, 5 thickness converted items, 5 flat ratio items, 2 exponential items, and 6 symmetrical shoulders and chest items.

2.2 Assessment of physical beauty

The study divided the assessment of physical beauty into two stages: 1) The first assessment of physical beauty was conducted for the body samples through the subjective assessment system of physical beauty developed by the Aimer Ergonomics Institute; upon calculation of assessment results, relatively beautiful samples were selected. 2) The same system was also used to conduct the second assessment on relatively beautiful samples; upon calculation of assessment results, the most beautiful samples were selected. As women's physical beauty is the topic of investigation, 40 professionals from the apparel industry evaluated the samples via the subjective assessment system; however, because assessors with different education and occupation backgrounds may make different assessments of physical beauty, the assessor sample group was divided into artistic and non-artistic groups based on their respective artistic professional backgrounds.

The following questions were put forward to assess the degree of beauty of the shoulder-neck, chest, waist-abdomen-hips, legs, and the overall body:

1) Degree of beauty of assessed person's shoulder-neck;
2) Degree of beauty of assessed person's chest;
3) Degree of beauty of assessed person's waist-abdomen-hips;
4) Degree of beauty of assessed person's legs;
5) Degree of beauty of assessed person's overall body.

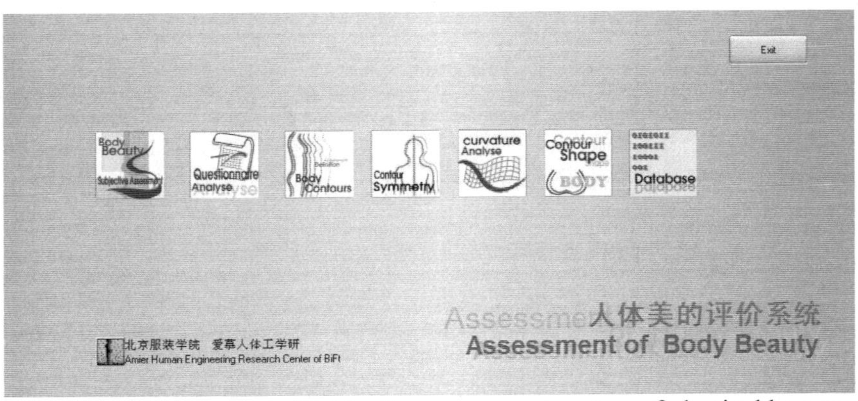

Fig.1 Navigation page of subjective assessment system of physical beauty

The questionnaire employed a graphic rating scale that indicates the viewer's response to the problem. A point was selected on a straight line and from this point the distance to both ends was measured. For example, in the evaluation of beautiful chest shapes, the five degrees of the scale were designed as 1, 3, 5, 7, and 9, with 1 representing the least beautiful, 9 representing most beautiful, and 5 representing average (as shown in Fig.2).

Fig.2 Scale of evaluation questionnaires

3. Data processing

3.1 Analysis of normalisation and differentiation of assessment scores

In order to reflect more factually a female's level of physical beauty, optimization theory was employed during calculation. Normalisation was adopted to convert the assessment scores of the artistic and non-artistic groups, calculate the accuracy of each assessor, and detect the outliers. Assessors of the artistic and non-artistic groups were numbered from 1 to 20, respectively. As shown in Fig.3, the Euclidean distance of assessors No.11 and No.17 in the artistic group, as well as that of assessor No.17 in the non-artistic group, falls into the scope of outliers. Therefore, their assessment scores were eliminated before recalculation of both groups' assessment scores by means of normalisation.

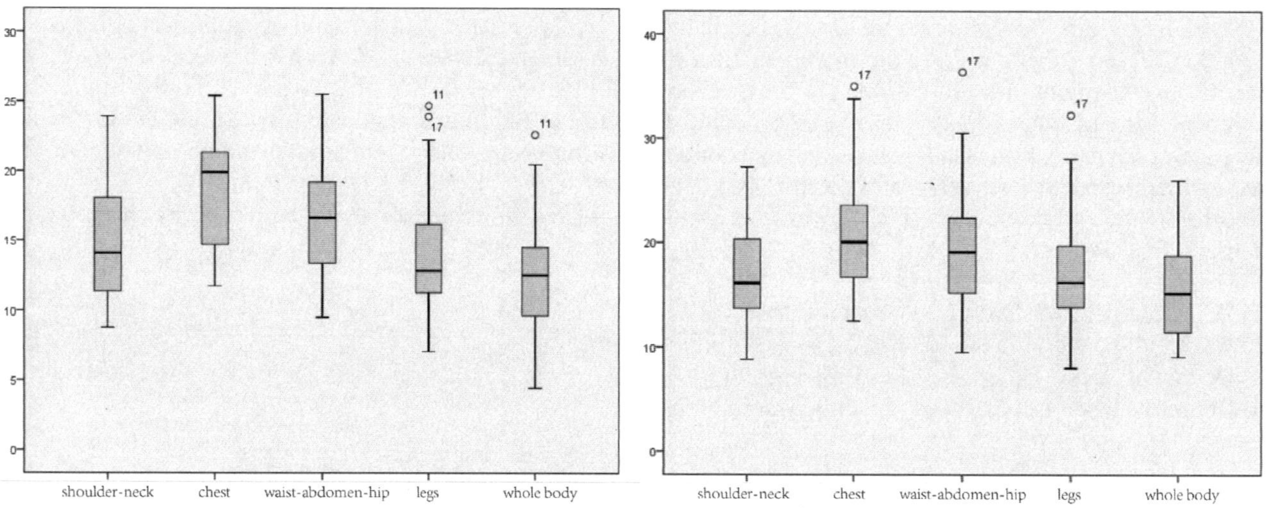

(a) Assessors of artistic group (b) Assessors of non-artistic group

Fig.3 Detection of assessor outliers

3.2 Differentiation analysis of both groups' assessment scores

Table 1. Independent-samples T-test results

Score	Levene test for variance equation		T-test for mean equation		
	F	Sig.	t	df	Sig. (both sides)
Shoulder-neck	2.881	.091	-.424	216	.672
Chest	.998	.319	-1.427	216	.155
Waist-abdomen-hips	3.095	.080	-1.148	216	.252
Legs	2.044	.154	-2.787	216	.006
The whole body	9.717	.002	-2.379	198.9	.018

Note: The significance level is 0.05.

Since the professional backgrounds of the assessors are different, it should be verified whether there exists any difference between the assessment of physical beauty by the artistic group and the non-artistic group, and independent-samples T-tests should be carried out on the scores of both groups, so as to determine whether the assessment scores may be combined.

Through comparison between the mean values of both groups (Fig.4), it is discovered that due to the special education and occupation backgrounds of the artistic group, their assessment of physical beauty features higher standards, resulting in lower assessment scores for shoulder-neck, chest, waist-abdomen-hips, legs, and the overall body compared with the non-artistic group.

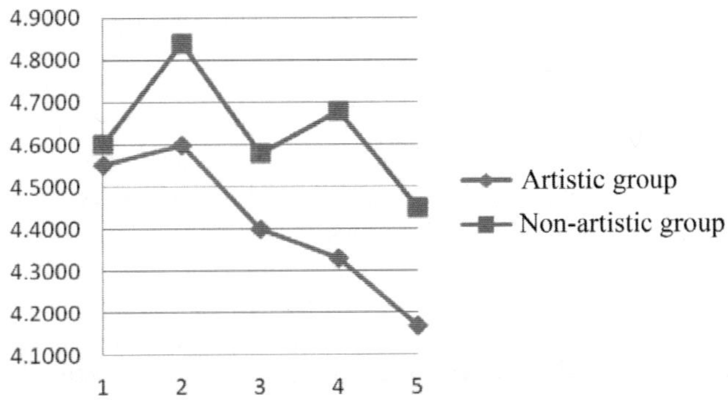

Fig.4 Comparison of mean values between both groups

3.3 Selection of relatively beautiful samples

Based on the analysis results of the independent-sample T-tests, there are no obvious differences in assessment scores of shoulder-neck, chest, as well as waist-abdomen-hips; therefore, assessment scores of the two groups were combined at the time of calculating such scores and the normalisation algorithm was used to calculate the combined assessment scores. There are obvious differences in the assessments of the legs and the overall body of the artistic group and the non-artistic group, so the normalisation algorithm was used to calculate the assessment scores of the two groups.

Through the Euclidean distance detection, it was found that there were abnormal samples in the assessment of shoulder-neck, chest, waist-abdomen-hips, and legs, thus they were eliminated at the time of selecting more beautiful samples and conducting follow-up regression analysis. Based on the analysis results of scores of four body parts and the overall body (as shown in Table 2), the correlation of scores of the four body parts and that of the overall body were ranked as follows: waist-abdomen-hips, legs, shoulder-neck, and chest. This shows that the score of waist-abdomen-hips has a highly positive correlation with that of the overall body, and that of legs and shoulder-neck also have a high correlation with that of the overall body. Therefore, the selection standard of relatively beautiful samples shall be subject not only to the comparison of the scores of the overall bodies, but also to the scores of waist-abdomen-hips, legs, and shoulder-neck.

Table 2. Pearson correlational coefficients for assessment of different body parts

	Scores of degree of beauty for:				
	Shoulder-neck	Chest	Waist-abdomen-hips	Legs	Overall body
Shoulder-neck	1				
Chest	.473**	1			
Waist-abdomen-hips	.670**	.331**	1		
Legs	.501**	.189*	.758**	1	
Overall body	.780**	.567**	.917**	.796**	1

** shows the significance level of correlation coefficient is 0.01.

For the overall body beauty assessment, 33 persons were regarded to be beautiful, with assessment results of waist-abdomen-hips, legs, and shoulder-neck taken into consideration. Ultimately, 23 persons were regarded to be beautiful in terms of overall body, accounting for 21% of the total population.

3.4 Establishment of mathematical model of assessment indexes of physical beauty

Upon correlation analysis at the time of selecting relatively beautiful samples, we sought to determine if assessments of the beauty of different body parts are correlated, how they are correlated, and the closeness of their correlation. The statistical method of regression analysis was further used to analyse exactly how assessment scores of different body parts and measurements correlate and to determine the casual relationship, and a mathematical model was used to express the specific relationship. The gradual multiple linear regression statistical method was used to conduct the regression analysis on the assessment scores of shoulder-neck, chest, waist-abdomen-hips, legs, and the overall body based on characteristics of 147 persons' bodily forms. We obtained the following multiple linear regression equations of corresponding body parts:

VHI, upper angle of chest and back, ratio of waist width and shoulder width, as well as angle of neck and shoulder were included in the regression model to evaluate the shoulder-neck. The regression equation for evaluating shoulder-neck beauty is $Y=8.597-0.11X_1-0.0320X_2-5.313X_3+0.027X_4$ (see Fig.5a for description of independent variable) and the coefficient of determination for goodness of fit is 0.444.

Width between two teats, lower chest circumference, ratio of waistline and bust, as well as perpendicular distance from BP to UBL were included in the regression model to evaluate the chest. The regression equation for evaluating chest beauty is $Y=12.93+0.418X_1-0.054X_2-15.179X_3+3.304X_4$ (see Fig.5b for description of independent variable) and the coefficient of determination for goodness of fit is 0.475.

VHI, ratio of waistline and thigh circumference, and ratio of BP and waist width were included in the regression model to evaluate the waist-abdomen-hips. The regression equation for evaluating waist-abdomen-hips beauty is $Y=14.189-0.34X_1-2.703X_2+2.003X_3$ (see Fig.5c for description of independent variable) and the coefficient of determination for goodness of fit is 0.630.

VHI, ratio of waist width and gluteal fold width, ratio of waist height and body height, as well as the ratio of chest height and body height were included in the leg assessment model of the artistic group; the regression equation of leg beauty is $Y=-9.406-0.257X_1+2.3X_2+12.671X_3+11.317X_4$ (see Fig.6a for description of independent variable) and the coefficient of determination for goodness of fit is 0.740. VHI, ratio of waist width and gluteal fold width, ratio of knee height and body height, as well as ratio of head height and body height were included in the leg assessment model of the non-artistic group; the regression equation of leg beauty is $Y=5.040-0.304X_1+2.322X_2+18.668X_3-14.969X_4$ (see Fig.6a for description of independent variable) and the coefficient of determination for goodness of fit is 0.714.

VHI, ratio of BP and waist width, flatness ratio of convex part of abdomen, ratio of chest height and body height, and difference of upper chest circumference and lower chest circumference were included in the overall body assessment model of the artistic group; the regression equation of overall body beauty is $Y=-8.364-0.155X_1+1.374X_2+2.083X_3+15.779X_4+0.048X_5$ (see Fig.6b for description of independent variable) and the coefficient of determination for goodness of fit is 0.576. VHI and ratio of BP and waist width were included in the overall body assessment model of the non-artistic group and the regression equation of overall body beauty is $Y=13.914-0.265X_1-4.589X_2$ (see Fig.6b for description of independent variable) and the coefficient of determination for goodness of fit is 0.540.

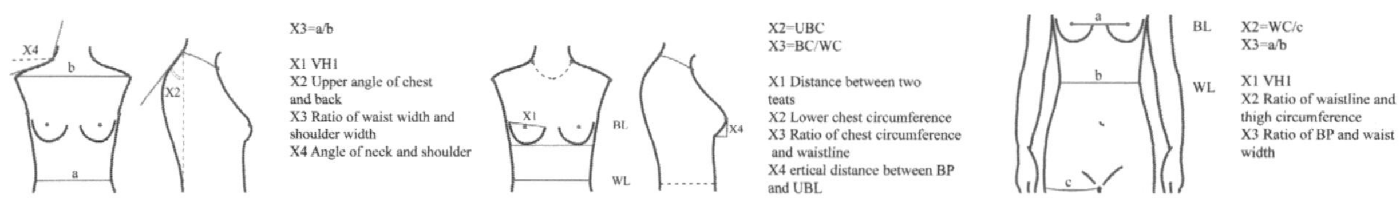

a b c

Fig.5 Independent variable description of regression equation for: (a) shoulders; (b) chest; (c) waist-abdomen-hips

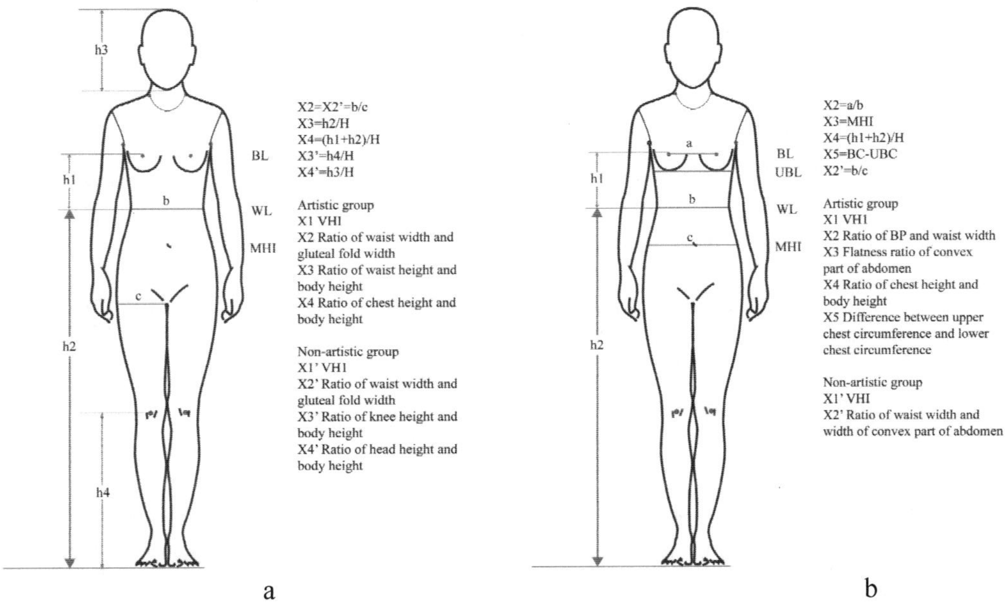

Fig.6 Independent variable description of regression equation for: (a) legs; (b) overall body

4. Conclusion

This study used three-dimensional scanning to obtain three-dimensional body shapes of 109 young women, and from direct and indirect measurement methods collected 149 human form data. Secondly, to obtain the sample evaluation, 40 fashion industry personnel used the human body evaluation system to evaluate the three-dimensional morphology of the human body. Then the normalisation algorithm, Euclidean distance detection, and independent samples of the T-test results were evaluated, and samples of the human form were selected and irrelevant data were eliminated. Lastly, an equation of the human modality form was obtained by using the multiple linear regression statistics to extract the sample data of the human modality form. The equation of the human body modality measurement is as follows:

Regression equation of shoulder-neck beauty is $Y=8.597-0.11X_1-0.0320X_2-5.313X_3+0.027X_4$ and the coefficient of determination for goodness of fit is 0.444. The regression equation of chest beauty is $Y=12.93+0.418X_1-0.054X_2-15.179X_3+3.304X_4$ and the coefficient of determination for goodness of fit is 0.475. The regression equation of leg beauty is $Y=-9.406-0.257X_1+2.3X_2+12.671X_3+11.317X_4$ and the coefficient of determination for goodness of fit is 0.740 for the artistic group; $Y=5.040-0.304X_1+2.322X_2+18.668X_3-14.969X_4$ and the coefficient of determination for goodness of fit is 0.714 for the non-artistic group. The regression equation of overall body beauty is $Y=-8.364-0.155X_1+1.374X_2+2.083X_3+15.779X_4+0.048X_5$ and the coefficient of determination for goodness of fit is 0.576 for the artistic group; $Y=13.914-0.265X_1-4.589X_2$ and the coefficient of determination for goodness of fit is 0.540 for the non-artistic group. With the change of the independent variables, as the dependent variable Y grows the more beautiful the human body's modality form will be.

References

[1] Chen L, Shi DW, An HJ. A study on the golden ratio in human morphology. Sichuan Journal of Anatomy: 2012; 20: 63-66.

[2] Singh D. Is thin really beautiful and good? Relationship between waist-to-hip ratio (WHR) and female attractiveness. Pers Indiv Differ 1994; 16: 123-132.

[3] Singh D. Ideal female body shape: the role of body weight and waist-to-hip ratio (WHR). Int J Eat Disord 1994; 16: 283-288.

[4] Bray GA. Definition, measurement and classification of the syndromes of obesity. Int J Obes 1978; 2: 99-112.

[5] Fan J, Liu F, Dai W. Visual perception of female physical attractiveness. Proc Biol Sci 2004; 271: 347-352.

[6] Zhang HB. Valuable book of SPSS. Beijing: Publishing House of Electronics Industry: 2007.
[7] Lim NY. The shaping effect of body shapers on female attractiveness. Hong Kong: Hong Kong Polytechnic University: 2007.
[8] Qian XN, Yin B. Research on subdividing of human body recognition based on 3D body measurement. Journal of Textile Research: 2011; 32: 107-120.
[9] Wang YL. A study of body dimension measurement based on 3D scanning. Techniques of Automation and Application: 2015; 34: 115-118.
[10] Zheng R, Hao Z, Huang HJ, Hu XD, Lee J, Duan XJ. A measurement and analysis method of the human body. CN Patent No. ZL200410009356.8: 2007.
[11] Yao T, Wang J, Meng LH. A study on the body proportions of female youths in northeast China based on non-contact measurement methods. Journal of Dalian Polytechnic University: 2014; 33: 226-228.

Optimization Study of Anthropometric Method for Trouser Design

Ya-Nan Tian , Li-Wen He , Xiao-Mei Shang*

College of Textile and Clothing Engineering, Soochow University, No. 1, Shizi Street, Suzhou City, Jiangsu Province, 215006, China

*Corresponding author's email: shangxiaomei@suda.edu.cn

Abstract

During the manual operation for anthropometric, many factors can influence the data accuracy. For trouser design, both crotch height and crotch circumstance are important data to consider. This paper mainly study how the measure method influence the two data accuracy and the potential influence degree. Using two methods, included "have reference" measurement and "no reference" measurement, obtained the data through experiment. Furthermore, by using mathematical analysis software, quantitatively analyzed the discrete degree for the data, then made the data do ANOVA test and statistical hypothesis test. Finally, we got the result that the two method significantly influenced the data accuracy, and the "have reference" method got the more accurate data.

Keywords: Manual anthropometry; Trousers design; Crotch height; Crotch circumference

1. Introduction

As the demand for suitable and beautiful garment increasing, the accuracy of the human body measurements should be improved especially for trouser design [1,2]. It's necessary to study anthropometric operation for designing more suitable and comfortable clothing [3]. At present, the forefront of anthropometry has focused on automation anthropometric technology [4]. The technology has been applied on the market, with the purpose to realize the clothing size standardization, and also to promote the development of digital garment customization system and P.MTM(Personalize Made to Measure) [5]. However, due to many factors, such as the clothing industry lacking of computer technology application talents, and capital investment, the industry is still widely using manual anthropometric [6,7]. Therefore this paper is still mainly in the service of manual anthropometric.

During the manual measure process, the measurers touch the human body directly, then obtain the data in accordance with the manual anthropometric principles and methods. Even though the measurers try their best to control the accuracy of the measurement data, but it still can't avoid that the system error make influences on the data accuracy [8]. The influence factors come from various aspects, but this study mainly focus on the influence that the measure method make on the measurement data.

Among the measurement data, both crotch height and crotch circumference are important for trousers design. Therefor, this paper mainly study on the two data accuracy and the measure method. Made use of two different measure methods, included "have reference" measurement and "no reference" measurement, to research whether the method could influence the data accuracy and the potential influence degree. Firstly, obtained the measurement data through manual anthropometric; Then made use of mathematical analysis software, calculated the arithmetic mean, standard deviation and AVEDEV function values to evaluate the data accuracy; Furthermore, combined with ANOVA test and statistical hypothesis test to evaluate the potential influence degree; Finally, according to the mathematical statistics results, did comparative analysis to judge whether the discrete degree was different and which one was better.

This study can optimize the anthropometric method, enrich the manual anthropometric experience, enhance the measurement data accuracy, improve the trousers suitable and comfortable, and will play a supporting role in improving automation of human body measurement technology.

2. Anthropometric Experiment

2.1 Sample selection

Measurer. Selected ten female measurers with certain measurement level randomly, aged from 20 to 25 years old, height from 155cm to 170cm. They have been trained unified measurement skills, and their operation skills and the measure accuracy have been up to the manual anthropometric standard.

Recorder. Selected ten female recorders with certain record level as the measurers' partner randomly.They have been trained unified record skills, and their record skills has been up to the manual anthropometric standard.

Subjects. Selected ten female subjects randomly, aged from 20 to 25 years old, height from 155cm to 170cm, and be numbered from NO.1 to NO.10.

2.2 Equipment and environment

Equipment. The experimental equipment included Martin-type Authropolometer, tape, graph paper, triangle, sitting ruler, record boards, pencils, special clothes, etc. All the measurers dressed in special white coat, and held measurement tools in hands; All the recorders dressed in special white coat as well, held pencils and record boards in hands; All the subjects dressed in special tights.

Environment. During each experimental process, decorated the same environment, the measurers used the same manual techniques. At the same time, ensured the consistency of the equipment specifications. In addition, the environment temperature should be around 27 ℃ and the humidity be around 60%, to ensure all the experiment participants feel comfortable; Light intensity was moderate, ensured clear readings; Wall and ground must be smooth, to avoid concave and convex by all means and avoid its effect on the measurement data accuracy.

2.3 Measure method

Crotch height: the vertical distance between the standing surface and the crotch. Crotch circumference: the curve length from the predecessor waist line midpoint via the perineum point to the back waist line midpoint. As we all know, the crotch point is just the perineum point.

The measurement methods just as shown in Fig.1, and as described below.

What is the subjects' posture? With the vertical scale of graph paper as the axis, subjects stood sideways and straight, feet together, kept binocular-as ahead, be relaxed and natural.

How did the measurer operate? The measurer stood on the side to the subjects, used right-angled triangle, one side cling to the vertical plane of graph paper, and one side cling to the horizontal plane of perineum point. The corresponding size of the triangle's horizontal plane was just crotch height; The subjects kept legs apart and the width of two legs as the shoulder breadth, then the measurer used one tape to measure, from the predecessor waist line midpoint via the perineum point to the back waist line midpoint, then the curve length was just crotch circumference.

How to distinguish the two measure methods? According to the different bench-marking method to make distinction. The one was "no reference" measurement, relied on the measurer's visual sense to determine where the perineum point was and the location of the waist line, as shown in Fig.1(a_1) and Fig.1(b_1). The other one was "have reference" measurement, fastened one standard reference before measuring, thus determined the perineum point and the location of the waist line, as shown in Fig.1(a_2) and Fig.1(b_2). However, the reference was different for the two measure data. For crotch height, in the middle of the subjects' two legs, put an appropriate size, thin but hard cardboard, then moved the cardboard's upper edge to the perineum's horizontal plane slowly, the contact point was just perineum point, as shown in Fig.1(a_2); For crotch circumference, fastened a 0.5cm wide elastic band with the subjects' waist, then the location of elastic band was just the location of waist line, as shown in Fig.1(b_2).

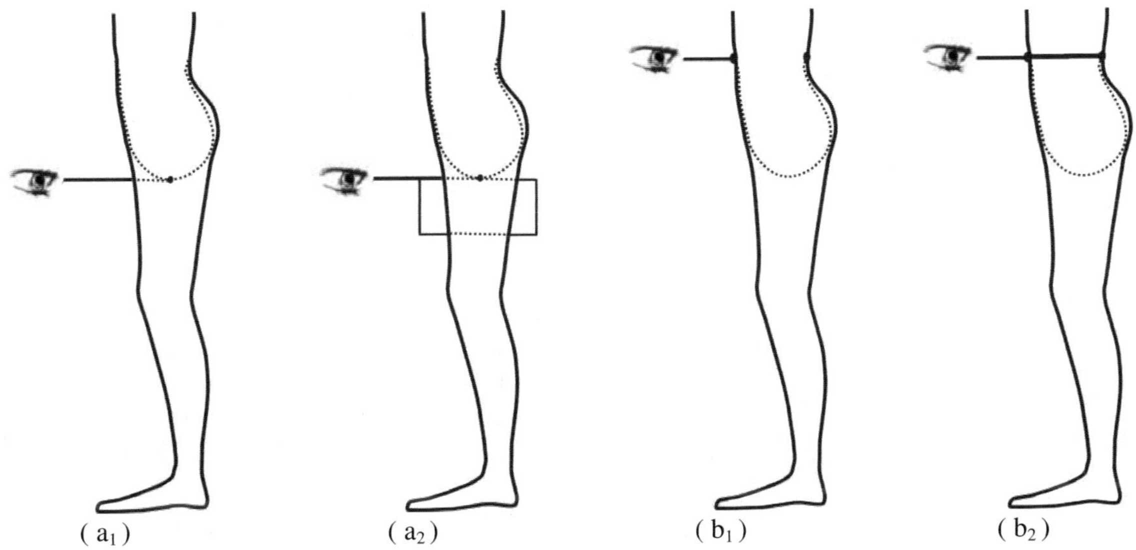

Fig.1 Schematic diagram for the measure method

2.4 Data Collection

According to the "have reference" measure method and "no reference" measure method, obtained the crotch height and crotch circumference of each subjects. Firstly, the ten measurers repeatedly measured every subjects for three times, then totaled 30 times for each one. Secondly, obtained the ten subjects' measurement data, then totaled 300 groups of measurement data. Thirdly, eliminated abnormal value and invalid data, then there still were 293 effective groups data for study. Finally, counted arithmetic mean of each group for further analysis.

3. Data accuracy analysis for "no reference" measurement

First of all, made the crotch height and crotch circumference as research object, chest circumference as reference object, preliminary analyzed "no reference" measurements, to judge whether the volatility of the data was beyond the scope of error allowed, and if the stability of the data met the standard required; Then for each group of data, made use of mathematical analysis software to calculate the arithmetic mean, standard deviation and AVEDEV(average deviation) function values. The one hand, the bigger standard deviation, the bigger difference between most of the values and the mean; the other hand, AVEDEV function refers to the average absolute deviation between a set of data and the average, to measure the discrete degree for the set of data, and the greater the AVEDEV function value, the greater the discrete degree for the set of data.

3.1 Stability analysis for measure data

Through human body shape cluster analysis, selected 3 subjects with different body shape as further research objects, from the 10 subjects. The 3 subjects included NO.1, NO.6 and NO.7, their shape size respectively were 160/84A, 155/82A and 168/90A (according to GB/T 1335). Based on the standard requirements for manual anthropometry accuracy, for chest circumference, crotch height and crotch circumference, the error range respectively were 2cm, 1.5cm and 2cm. Analyzed the measured values and obtained the fluctuation curve for "no reference" measurements, as shown in Fig.2. In the Fig.2, (I) represented chest circumference; (II) represented crotch height; (III) represented crotch circumference. And the measurements was corresponded with the subjects.

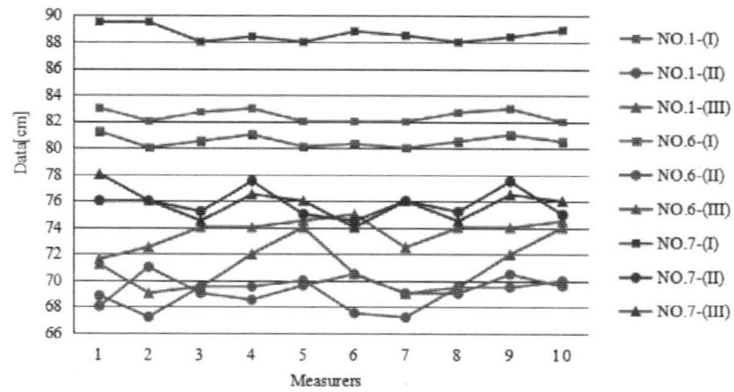

Fig.2 The fluctuation curve for "no reference" measurements

From the point of the fluctuation curves in Fig.2, we can concluded that, the three subjects' chest circumference data were fluctuating within 2cm of error allowed; while for crotch height and crotch circumference, the data were fluctuating within 4cm, but out of the range of allowable error. That's to say, through "no reference" measure method, for crotch height and crotch circumference, the data stability was not up to the standard.

3.2 Compare the discrete degree for each group of data

By using the mathematical analysis software, quantitatively analyzed the discrete degree for the measurements, and the statistical results as shown in Table 1.

Table 1. The statistics of discrete degree for each group of data

Measurements	Subjects	Arithmetic mean [cm]	Standard deviation	AVEDEV function values
Chest circumference	NO.1	82.49	0.38	0.44
	NO.6	79.91	0.93	0.33
	NO.7	87.81	0.65	0.46
Crotch height	NO.1	68.82	1.10	0.96
	NO.6	68.87	1.29	0.77
	NO.7	75.79	1.03	0.81
Crotch circumference	NO.1	73.66	1.08	0.88
	NO.6	72.37	1.49	1.10
	NO.7	75.80	1.18	0.88

The Table 1 indicated that as below:

The first point, all the standard deviation of chest circumference were less than 1.00, and all the AVEDEV function values were less than 0.50; Combined with the Fig.2, we got that the discrete degree of chest circumference was small, because that each group of data was fluctuating within 2cm of error allowed, so the measurements of chest circumference was accurate in fact; Therefore, we can make the standard deviation and AVEDEV function values of chest circumference as the standard discrete degree, provide the reference for crotch height and crotch circumference.

The second point, all the standard deviation of crotch height and crotch circumference were more than 1.00, and all the AVEDEV function values were more than 0.50; Reference the discrete degree of chest circumference, we speculated that the discrete degree both of crotch height and crotch circumference were bigger, and their measurements were not accurate enough; Therefore, we need to optimize the measure method for crotch height and crotch circumference, to effectively improve the accuracy of the measurements.

4. Data accuracy analysis for "have reference" measurement

First of all, finishing the measure data obtained by the two measure methods; Then for each group of data, made use of mathematical analysis software to calculate the arithmetic mean, standard deviation and AVEDEV function values, compared the discrete degree for each group of data, to judge whether the discrete degree was different and

$$MS_t = SS_t / df_t \quad (2)$$

$$MS_e = SS_e / df_e \quad (3)$$

$$SS_t = \sum_{i=1}^{k} n_i \left(\overline{x_{i_n}} - \overline{x_{...}}\right)^2 \quad (4)$$

$$SS_e = \sum_{i=1}^{k} \sum_{j=1}^{n_i} \left(x_{ij} - \overline{x_{...}}\right)^2 \quad (5)$$

which one was better; Furthermore, using mathematical analysis software, made the data do ANOVA test and statistical hypothesis test, to judge whether the measure method could make influence on the data accuracy and the potential influence degree.

4.1 Compare the discrete degree for each group of data

By use of mathematical analysis software, quantitatively analyzed the data discrete degree for the two methods, and the statistical results as shown in Table 2. In the Table 2, A_1 represented "no reference" measurement, and A_2 represented "have reference" measurement.

Table 2. The statistics of data discrete degree for the two methods

Measurements	Subjects	Methods	Arithmetic mean [cm]	Standard deviation	AVEDEV function values
Crotch height	NO.1	A_1	68.82	1.10	0.96
		A_2	69.28	0.32	0.28
	NO.6	A_1	68.87	1.29	0.77
		A_2	69.86	0.47	0.39
	NO.7	A_1	75.79	1.03	0.81
		A_2	77.88	0.45	0.38
Crotch circumference	NO.1	A_1	73.66	1.08	0.88
		A_2	71.11	0.73	0.54
	NO.6	A_1	72.37	1.49	1.10
		A_2	70.05	0.75	0.61
	NO.7	A_1	75.80	1.18	0.88
		A_2	77.49	0.28	0.21

Judging from the statistical results in Table 2, all the standard deviation and AVEDEV function values of A_2 method were lower than that of A_1 method, and there were obviously differences between the two arithmetic means in each group. That's to say, the "have reference" measure method effectively reduced the discrete degree, and improved data accuracy to a certain extent.

4.2 ANOVA test and statistical hypothesis test

The first step, did experimental design for two factors' variance analysis: the experiment investigated two factors, factor A represent the measure method, divided it into two levels. While factor B represent the measurer, divided it into ten levels. The combination of each group of level only had one observation value, which was the arithmetic mean obtained by 3 times repeated experiment, then totaled 20 observation values for each group. There were 3 subjects, 2 measurements, accumulative totaled 6 groups of experimental design and 120 observation values.

The second step, did ANOVA test for each group of experimental design, just as below:

$$F = MS_t^2 / MS_e^2 \quad (1)$$

The third step, did statistical hypothesis test: put up hypothesis for the whole sample, included null hypothesis H_0 and alternative hypothesis H_A; Determined the significant level as α=0.05 and α=0.01; Then from the point of null hypothesis H_0, calculated and tested statistics according to the information provided by sample; Finally inferred whether the null hypothesis was correct according to the probability value, so as to decide whether to accept or reject H_0.

For six groups of experimental design, made the data do ANOVA test and statistical hypothesis test as above, the results as shown in Table 3. Then compared the Sig.(which represent the significant level) with the significant level α, to judge the significant degree of each data set's difference.

Table 3. Two factors' ANOVA test and statistical hypothesis test results

Measurements	Subjects	Factors	Type III sum of squares	Mean square	F	Sig.
Crotch height	NO.1	A	6.160	6.160	25.734	0.001
		B	2.822	0.314	1.310	0.347
	NO.6	A	5.000	5.000	7.745	0.021
		B	0.930	0.103	0.160	0.994
	NO.7	A	21.840	21.840	39.121	0.000
		B	6.412	0.713	1.278	0.360
Crotch circumference	NO.1	A	32.512	32.512	43.463	0.000
		B	8.761	0.973	1.301	0.351
	NO.6	A	17.485	17.485	10.366	0.010
		B	9.981	1.109	0.657	0.729
	NO.7	A	14.280	14.280	20.484	0.001
		B	7.035	0.782	1.121	0.434

The Table 3 indicated that as below:

To factor B in the six groups experiment, the Sig. were all higher than 0.05, it showed no significant difference between groups. That's to say, the ten measurers' operation skills had no obvious difference.

To factor A of crotch height, both the Sig. of NO.1 and NO.7 were less than 0.01, it showed the influence that the method made on the measurements was most significant; while the Sig. of NO.6 was higher than 0.01, but less than 0.05, so it showed the influence was significant but not most significant.

To factor A of crotch circumference, all the Sig. of NO.1, NO.6 and NO.7 were less than or equal to 0.01, it showed the influence that the method made on the measurements was most significant.

To sum up, the measure method significantly influenced the data accuracy for crotch height and crotch circumference, the data obtained by "have reference" method showed lower discrete degree and higher accuracy.

5. Conclusion and discussion

This paper made crotch height and crotch circumference as mainly research subjects, made the "have reference" measurement and "no reference" measurement as two methods, then obtained the data through anthropometric experience. Then, made use of mathematical analysis software to calculate the arithmetic mean, standard deviation and AVEDEV function values, and combined with ANOVA test and statistical hypothesis test. Finally, the study found that, the two measure methods significantly influenced the data accuracy both for crotch height and crotch circumference, and the data obtained by "have reference" method showed lower discrete degree and higher accuracy.

This study can optimize the anthropometric method, enrich the manual anthropometric experience, enhance the measurement data accuracy, improve the trousers suitable and comfortable, and will play a supporting role in improving automation of human body measurement technology.

However, due to the limited experimental content and data analysis, there still be many more measurements to study for improving the trousers suitable and comfortable, even for other garments. Therefor, it is still necessary to go deep research into the measurements and the measure method.

6. Acknowledgement

The authors would like to thank all anonymous reviewers for constructive comments. The work described in this paper was partially funded by the Priority Academic Program Development of Jiangsu Higher Education Institutions.

References

[1] Li Q, Li PW, Hong ZL, Shang XM. Psychological Factors Affect Measured Data of Human Body. Jiangsu Textile: 2014; 1: 50-52.

[2] Jia SL, Tian YN, Ji JH, Shang XM. The effect of measurers' view angle deviation on measurement data. Advanced Materials Research, Materials Science and Engineering Technology: 2014; 2237-2242.

[3] He H, Wang JP, Zhi HH. Young female hip classification based on the pants fit. Journal of Donghua University Natural Science Edition: 2015; 3: 313-316.

[4] Xu H, Shang XM. Data acquisition of the human body for making pants. Foreign Silk: 2006; 4: 25-28.

[5] Zhang X, Wang HF. Based on 3D human body measurement of young female fork high end features discussed. Textile Journals: 2012; 33(7): 100-103.

[6] Jagir RK, Wane PE. Variation in measurements across different brands of same style ladies' pants in Swaziland. International Journal of Fashion Design, Technology and Education: 2011; 51-57.

[7] Shang XM, Wang ZF. Study on the Mathematic Model of Apparel Lapel Collar Pattern. Journal of Donghua University(English Edition): 2009; 03: 333-338.

[8] Marina AD, Gina RP. Evanesce. Investigating women's plus-size body measurements and hip type variation based on Size USA data. International Journal of Fashion Design, Technology and Education: 2012; 3-12.

[9] Jagir RK, Wane PE. Variation in measurements across different brands of same style ladies' pants in Swaziland. International Journal of Fashion Design, Technology and Education: 2011; 51-57.

[10] Shang XM, Lu C, Wang H. Human body circumference curve fitting model based on automatic measurement. Journal of Textile Research: 2010; 4: 128-131.

[11] Qian JL. A Research on the Relationship between Self-confidence in Characteristic Sports and Athletic Motives for the Students of Sports Schools. J Capital University of Physical Education and Sports: 2007; 19(5): 52-54.

[12] Na XF. Effect of Physical and Psychological Factors on the Short Track Speed Skaters' Competitive Power. J China Winter Sports: 2010; 32(2): 17-21.

[13] Shang XM, Xing XY, Gao PP. Hierarchical Analysis Method and Model Based Data of Controlling Parts of the Body for Clothing. J Fiber Bioengineering and Informatics: 2014; 7 (4): 479-494.

[14] Jiao J, Yao L, Li Y, et al. Thermal Physiology and Local Responses of Human Body During Exercise in Hot Condition. J Fiber Bioengineering Informatics: 2012; 2: 115-124.

[15] Sims RE, Marshall R, Gyi DE, et al. Collection of anthropometry from older and physically impaired persons: Traditional methods versus TC2 3D body scanner. International Journal of Industrial Ergonomics: 2012; 1: 65-72.

Evaluation of Fashion Design by Combining ANP and TOPSIS

Mohammad Adam Jerusalem*, Shi-Woei Lin

National Taiwan University of Science and Technology, No.43, Sec. 4, Keelung Rd., Taipei City 106, Taiwan

*Corresponding author's email: mohadamjerusalem@gmail.com

Abstract

Numerous studies evaluated the fashion designs using different criteria; however, these studies did not consider interrelation between criteria. In this study, we combined two approaches - the Analytical Network Process, and the Technique for Order Preference by Similarity to Ideal Solution - to evaluate various designs of fashion. We found that the new idea, comfort to the style and fashion forecast were the significant criteria for designing a fashion. The ideal design with the highest rank was Veracious. A fashion design company could consider using the method, framework, elements, and criteria in the daily fashion designing.

Keywords: Evaluation; Fashion design; Ideal solution; Analytical network process; TOPSIS

1. Introduction

Every fashion design introduces the concept of the clothing or accessories from the fashion design company. In order to meet customers' need for aesthetic design, fashion design company should consider various factors in designing a fashion (e.g, fashion trend, cultural, social assumptions, time and place) [1]. Designing a fashion scheme is a complex decision-making process for a company. Selecting an appropriate combination of materials, colors, patterns for specific target consumers also involves multifaceted and subjective judgments from fashion design experts. Hence, developing a multicriteria decision-making approach to evaluate fashion designs plays an important role in this field.

Some studies discussed how to evaluate fashion design schemes, and most of them focus on the criteria and methods to assess fashion designs. For example, Lin and Twu [2] proposed eight evaluation criteria in their decision framework (e.g, theme story, best-seller modification, new idea, fashion forecast, fashion event, opponent ability, brand image, and product position). Three criteria - fabric, color, and style - for analyzing a fashion design was implemented in the study of Goncu and Bayazit [1]. Harr suggested a different set of criteria including time management, concept, figure development, apparel development and interaction on figure, rendering, figure relationship, composition, quality, style, and presentation [3]. Kidd and Workman developed a model of evaluation that comprised of design creativity, aesthetic appeal, functionality, appropriateness and originality [4]. The criteria discussed thus far are internal factors of designing a fashion. On the other hand, Jones concentrated on external factors such as marketing when evaluating a fashion design. His criteria included consumer age, gender, demography, lifestyle, physical characteristics, psychographics, social class, value and attitudes, economic circumstances, and religion [5] .

Numerous studies stated that the main criteria, criteria, sub-criteria and alternatives are not independent each other in the real world situation [6-9]. In other words, in order to obtain valuable and reliable results, we should assess the magnitude of influence or dependence among elelments of the decision framework and take these assessments into consideration when deriving the final priorities or weights of the criteria. Therefore, we assumed that elements, criteria, and alternatives have interrelation among others in this study. This study developed criteria for evaluating fashion design and combined two approaches which can take the interrelation between elements of evaluation and the influence between criteria into consideration to evaluate different alternatives of fashion design schemes. The method and framework used in this study could be easily implemented in cases where evaluation criteria are interrelated with each other - cases that have not yet been well addressed in previous research.

2. Research Methodology

Numerous research not only constructed elements and criteria for evaluating fashion design schemes but also selected an ideal or the best fashion design alternative. Goncu and Bayazit [1] employed AHP to select fashion trend for a specific target group. Lin and Twu [10] presented eleven valuable criteria for fashion trend selection. They implemented fuzzy AHP to calculate the weights of the criteria. Lin [11] compared AHP with the consistent fuzzy preference relation (CFPR). He concluded that AHP was less efficient than CFPR. In his case, AHP needed 76 data in interval scale for fashion design selection among four schemes and eight criteria, while CFPR only needed 31 data.

Fuzzy Delphi Method and Fuzzy AHP were implemented by Lin [12] to determine criteria weights. In this study, fashion design expert first evaluated the criteria using FDM, and then Fuzzy AHP was used to calculate the criteria weights at the second step. Fuzzy AHP and Technique for Order Preference by Similarity to Ideal Solution (TOPSIS) were employed by Lin and Twu [2] to appraise the eight criteria weights of four fashion designs. The criteria weights were thenused by the synthetic evaluation method to derive the index value of each scheme. Lin and Twu also implemented the TOPSIS to analyze the scheme. The results indicated that TOPSIS perform better than synthetic evaluation method.

One of the issues in previous studies is that the elements and criteria were assumed to be independent to each other. This independent situation is too strong and is usually violated in the real world situation. Saaty [13] developed the Analytical Network Process (ANP) that simultaneously considers both the relationships of feedback and dependence. In addition, the ANP provides a more generalized model in decision-making without making assumptions regarding the independency of the higher-level elements from lower-level ones and also of the elements within own level [6]. On the other hand, selecting an alternative has its own advantages and disadvantages. A solution resulted by TOPSIS comes from the ideal of reference-dependent theory. Reference-dependent theory states that reviewers assess alternatives in term of gains and losses relative to a subjective reference point [7]. Therefore, we integrate the ANP and TOPSIS to take the advantages of each method. The ANP is a powerful tool to appraise the influence between criteria. The TOPSIS could determine a solution by considering the distance to the ideal solution. The model in this research provides a solution in evaluating fashion design that can take interrelation between elements and criteria into consideration.

2.1 Analytic Network Process

ANP is an extension of the Analytic Hierarchy Process (AHP). However, unlike AHP that assumes an uni-directional relationship between decision levels, ANP provides an effective approach to determine the priority of different alternatives in a system where interaction between elements or criteria are existing. ANP is a comprehensive tool in decision-making that could formulate interdependence between elements or criteria in real world situations. Four key steps of the ANP method are discussed as follows [6,7].

Step 1. Framework and problem structuring. Decision makers or experts construct the framework and problem through brainstorming or other suitable methods.

Step 2. Pairwise comparison matrices and priority vectors. Similar to AHP, ANP's pairwise comparison is made in the framework of a matrix. A local priority vector can be attained for measuring the relative importance associated with elements or criteria being compared using Eq.1.

$$A.w = \lambda_{max} \,.\, w \tag{1}$$

where A is the pairwise comparison matrix, w denotes the eigenvector, and the largest eigenvalue of A is represented by λ_{max}.

Step 3. Supermatrix formation. The local priority vector from step 2 is placed in appropriate cells in a supermatrix based on the direction of influence from one element or criterion to another. Eq.2 describes what the supermatrix looks like.

$$W = \begin{matrix} C_1 \\ \vdots \\ C_k \\ \vdots \\ C_n \end{matrix} \overset{\begin{matrix} C_1 & \cdots & C_k & \cdots & C_n \end{matrix}}{\begin{bmatrix} W_{11} & \cdots & W_{1k} & \cdots & W_{1n} \\ \vdots & \vdots & \vdots & \vdots & \vdots \\ W_{k1} & \cdots & W_{kk} & \cdots & W_{kn} \\ \vdots & \vdots & \vdots & \vdots & \vdots \\ W_{n1} & \cdots & W_{nk} & \cdots & W_{nn} \end{bmatrix}} \quad (2)$$

The supermatrix must be made into a stochastic form to ensure each column of the matrix sums to unity The resulting stochastic matrix is termed the weighted supermatrix. It needs to be stochastic to derive meaningful limiting priorities [13]. To get convergence of the importance weight, the weighted supermatrix should be transformed to the limit supermatrix by raising the power of k; where k is an arbitrarily large number as Eq.3. The final priorities of all elements can be achieved by normalizing each block of the limit supermatrix [14].

$$\lim_{k\to\infty} (W^{\alpha})^k \quad (3)$$

2.2 TOPSIS

In this proposed approach, the TOPSIS steps are the subsequent processes following the ANP. The final priorities as the result of limit weighted supermatrix will become an input in the TOPSIS steps. Proposed by Hwang and Yoon in 1981 initially and developed later by Chen and Hwan in 1992, TOPSIS approach provides any chosen alternative that have the shortest distance from the ideal solution and the farthest distance from the negative-ideal solution. The six key procedures compose the TOPSIS [15].

Step 1. Compute the normalized decision matrix, r_{ij}. Eq.4 is the formula to calculate the normalized decision matrix.

$$r_{ij} = \frac{f_{ij}}{\sqrt{\sum_{j=1}^{J} f_{ij}^2}}, j = 1, \dots, J; i = 1, \dots, n \quad (4)$$

where f_{ij} is the score of the ith criterion function for the an alternative; n is the number of criteria.

Step 2. Calculate the weighted normalized value, v_{ij}, using Eq.5.

$$v_{ij} = w_i r_{ij}, j=1,\dots, J; i=1, \dots, n. \quad (5)$$

where w_i is the weight of the ith criterion, and $\sum_{i=1}^{n} w_i = 1$.

Step 3. Determine the ideal solution (A^*) and the negative-ideal solution (A^-).

$$A^* = \{v_1^*, \dots, v_n^*\} = \{(\max_j v_{ij} | i \in I'), (\min_j v_{ij} | i \in I'')\} \quad (6)$$

$$A^- = \{v_1^-, \dots, v_n^-\} = \{(\min_j v_{ij} | i \in I'), (\max_j v_{ij} | i \in I'')\} \quad (7)$$

where $v_i^* = \max_j v_{ij}$, $v_i^- = \min_j v_{ij}$.

Step 4. Determine the separation measure by utilizing n dimensional Euclidean distance. The separation of every alternative from the ideal solution, D_j^*, and from the negative ideal solution, D_j^-, is given by Eq.8 and Eq.9.

$$D_j^* = \sqrt{\sum_{i=1}^{n}(v_{ij} - v_i^*)^2}, j = 1, \ldots, J. \quad (8)$$

$$D_j^- = \sqrt{\sum_{i=1}^{n}(v_{ij} - v_i^-)^2}, j = 1, \ldots, J. \quad (9)$$

Step 5. Ascertain the relative closeness to the ideal solution, C_j^*. The relative closeness of the alternative a_j with respect to A^* is defined as Eq.10.

$$C_j^* = \frac{D_j^-}{(D_j^* + D_j^-)}, j = 1, \ldots, J \quad (10)$$

Step 6. Rank the preference order of the alternatives.

3. Problem solution

3.1 Design, Element and Criterion

This study considered four fashion design alternatives, four promising design themes resulted from Indonesian Fashion Trend Forecasting 2015/ 2016 conducted by the Ministry of Tourism and Creative Economy of the Republic of Indonesia [16]. These four fashion design themes are Adroit, Alliance, Biomimetics, and Veracious. The characteristic of the four design themes could be seen in Table 1.

Table 1. Themes of Indonesian Fashion Trend Forecasting 2015/ 2016

Theme	Characteristic
Adroit	All limitations of distance, time, and space due to the rigors of competition in Jakarta have inspired the Adroit This theme describes urban strategy when response these limitations. By utilizing virtual world technology, the people of Jakarta have a way out of this pressure and still be able to recreation with joy.
Alliance	Inspired by the tug of war between tradition and modernization, Solo (Central Java) elected as local inspiration to the theme of the Alliance. This theme displays the grandeur and beauty of the ancestral tradition that combined with the spirit of modernity.
Biomimetics	Excessive natural exploitation that happened in Banjarmasin (Kalimantan) has been a local inspiration to the theme Biomimetics. This theme introduces a new idea and concept in an effort to save the world from extinction by making biological engineering that designed as similar as possible or even exceeds the original.
Veracious	Raja Ampat (Papua) with all the peace, beauty and authenticity of nature is become an inspiration of the Veracious. This theme emphasizes the importance of listening to natural changes and lives a more meditative concept. This is a factor that increasingly sought in happiness and applied in everyday modern life.

In order to evaluate the fashion designs, this study constructed an evaluation framework that consists of four elements and eight criteria (see Fig.1). This framework is based on the internal factors adapted from previous studies related to fashion scheme evaluation, and all the elements and criteria were also re-examined by domain experts. Element of innovation comprises of fashion forecasting and new idea (Lin and Twu, 2012; Lin, 2013). Element of style includes comfort of the style and the possibility of being combined with other garments or mix and match (Harr, 2004; Goncu and Bayazit, 2007). Color consists of sustainability for daily use and sustainability for body type and skin color [1]. The final element, fabric, covers easy to care and pattern of the fabric (Goncu and Bayazit, 2007).

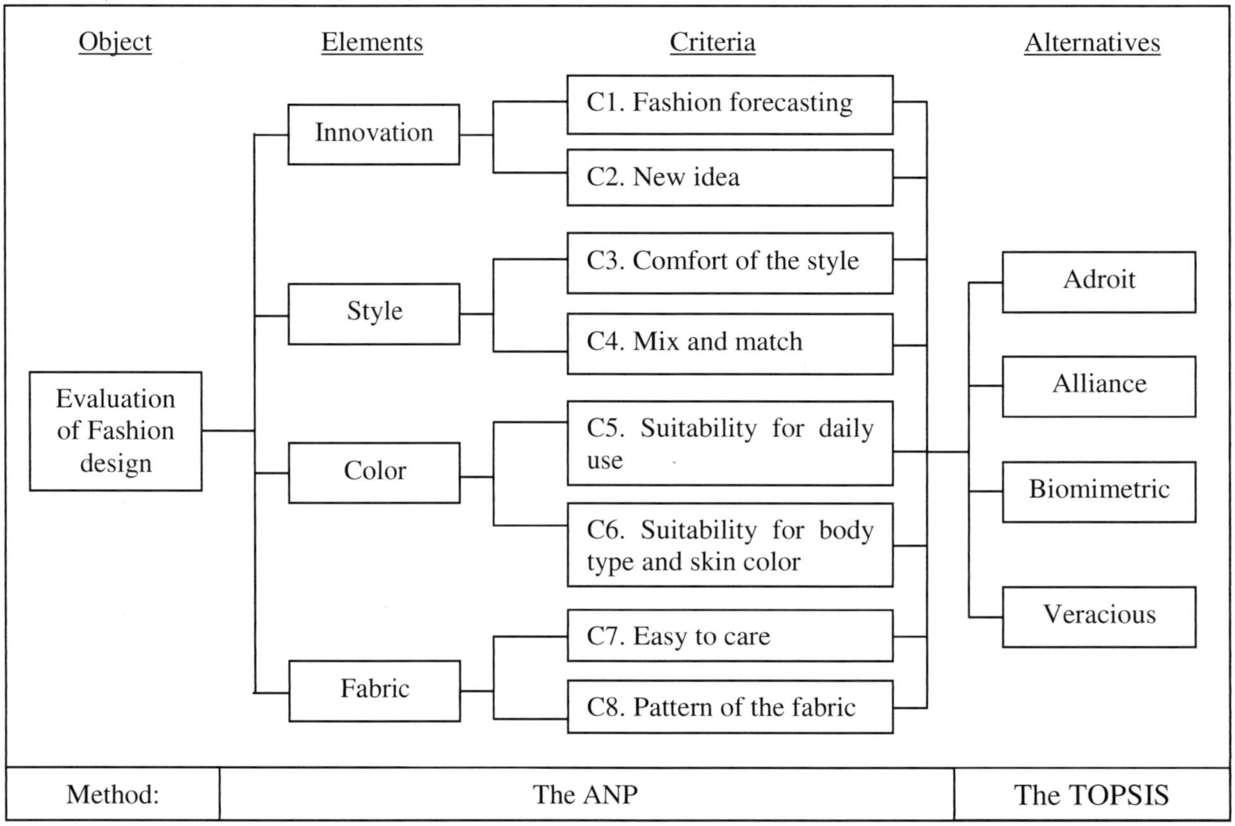

Fig.1 Hierarchical framework for evaluating fashion design

3.2 Proposed Method: Combining ANP and TOPSIS

Based on the four elements and eight criteria, this study evaluated four fashion designs in order to find the ideal or the best fashion design by considering inter-relation between elements and criteria. The proposed method to solve this problem is a combination of the ANP and the TOPSIS. The ANP was used to find the weights or priorities of the criteria, while the TOPSIS was used to rank the alternatives.

Ten experts who have more than 10 years of experience in the domain area of fashion were requested to make a series of pairwise comparisons and provide relative scores. By using Eq.2 and Eq.3, we can obtain unweighted supermatrix and limit weighted supermatrix as Table 2 and Table 3, respectively. Super Decision, a decision-making software which implments both AHP and ANP methods, were used in this study to calculate the supermatrices.

Table 2. Unweighted supermatrix

Criteria	Innovation		Style		Color		Fabric	
	C1	C2	C3	C4	C5	C6	C7	C8
C1	0.00000	1.00000	0.37888	0.37578	0.39099	0.35149	0.38042	0.37225
C2	1.00000	0.00000	0.62112	0.62422	0.60901	0.64851	0.61958	0.62775
C3	0.76487	0.82600	0.00000	1.00000	0.82788	0.83601	0.80361	0.83151
C4	0.23513	0.17400	1.00000	0.00000	0.17212	0.16399	0.19639	0.16849
C5	0.52358	0.57173	0.50224	0.50025	0.00000	1.00000	0.55476	0.50199
C6	0.47642	0.42827	0.49776	0.49975	1.00000	0.00000	0.44524	0.49801
C7	0.78369	0.76236	0.77988	0.76042	0.75616	0.76230	0.00000	1.00000
C8	0.21631	0.23764	0.22012	0.23958	0.24384	0.23770	1.00000	0.00000

Table 3. Limit the weighted supermatrix

Criteria	C1	C2	C3	C4	C5	C6	C7	C8
Weight	0.20305	0.24879	0.21048	0.10661	0.08108	0.07351	0.05620	0.02029

The ANP provides weights of the criteria that are useful for calculating the rank of alternatives. These derived weights can also provide useful information for a decision maker in this industry about the priorities of different criteria. Fig.2 points out that the priorities of the criteria could be categorized onto three groups. The first group consists of criteria with top three highest weights. They are new idea (the highest criterion), comfort to the style, and fashion forecasting, when listed in descending order. The second group comprises of mix and match, suitability for daily use, suitability for body type and skin color, and easy to care. The final group only consists of the pattern of the fabric. A fashion design company could consider this priority or group of priority when designing a fashion and evaluate it.

The results imply that according to the experts, innovation is the most important element. All criteria of this element are having higher weight than others. The findings reveal that the experts tend to consider innovation and style to be their top priorities when designing a fashion. In addition, we find that the experts are less concern about color and fabric; the mean scores of these elements are substantially lower than the scores of other elements.

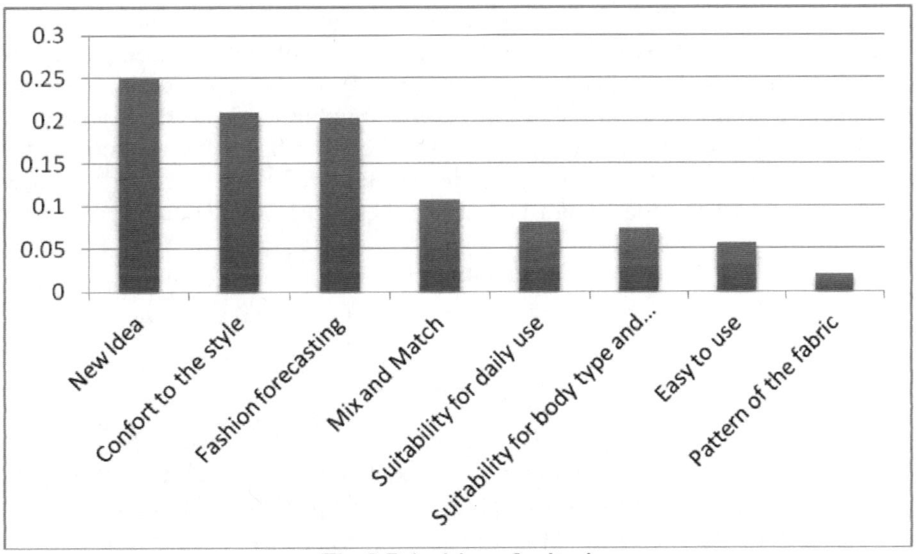

Fig.2 Priorities of criteria

Other than assigning a relative score for each pairwise comparison, the experts also appraised fashion designs based on the eight criteria (see Table 4). At the second step of TOPSIS, the weighted normalized value v_{ij} were calculated using the weights of the *i*th criterion derived from the limiting weighted supermatrix at the ANP phase. Passing the six steps of TOPSIS, the result shows the most ideal fashion design. The TOPSIS decided that Veracious exceed Alliance, Adroit, and Biomimetic, respectively (see Table 5).

Table 4. Fashion design score

Alternative of design	Criteria							
	C1	C2	C3	C4	C5	C6	C7	C8
Adroit	6.5	6.7	7.3	6.3	5.9	6.3	6.4	5.9
Alliance	6.9	7.3	6.8	6.5	6.6	7.0	5.9	7.4
Biomimetic	7.6	6.8	6.8	5.6	5.0	6.6	5.7	6.4
Veracious	7.2	6.9	7.0	7.7	7.0	6.8	6.8	7.4

By comparing each alternative to the ideal solution and negative ideal solution, the TOPSIS calculated the relative closeness to the ideal solution rank the four designs (see Table 5). Veracious style, which was inspired by the peace, beauty and authenticity of nature, was the best choice from experts' points of view.

Table 5. Ranking based TOPSIS approach

Fashion Design	$d*$	$d-$	$d* + d-$	$d*/d* + d-$	$C*_j$	Rank
Adroit	0.0246	0.0143	0.0389	0.6324	0.3676	3
Alliance	0.0170	0.0236	0.0406	0.4182	0.5818	2
Biomimetic	0.0318	0.0105	0.0424	0.7512	0.2488	4
Veracious	0.0061	0.0332	0.0393	0.1556	0.8444	1

Gaps between Veracious and others are large. In order to improve the quality of other fashion designs, a fashion company could improve begins from Biomimetic (0.2488), Adroit (0.3676) and Alliance (0.5818). This improvement should be based on the priorities of the criteria as the result of the ANP.

There are some advantages of integrating the ANP and the TOPSIS models in selecting the best fashion design. First, the compensatory approach of TOPSIS can help find the alternative that have shortest distance from the ideal solution by using the criteria weights derived from ANP, which takes interrelation among elements and criteria into consideration. Another advantage is the proposed approach need fewer data compare to the original ANP model. The original ANP also has capability to select the best fashion design but it needs 57 pairwise comparison matrixes (that comprise 150 items of relative importance question). On the other hand, the method used in this study only requires 28 pairwise comparison matrixes that consist of 48 items of relative importance questions, and one comparison matrix to evaluate the fashion design alternative that consist of 40 questions. Hence, the integrated method only need 88 questions in total, and thus is simpler and faster than ANP. Even need less pairwise comparison, this result is still valid since the experts confirmed it. The ideal fashion design is similar with the rank of fashion design score.

4. Conclusion

This study presented a hybrid method combining the ANP and the TOPSIS to select the most ideal fashion design according to expert perception. The proposed method implements the ANP to derive weights of criteria, and the TOPSIS to rank the alternatives and select the ideal fashion design. Four fashion designs from Indonesian Fashion Trend Forecasting 2015/ 2016 were evaluated. The priorities of criteria listed in descending order are new idea, comfort to the style, fashion forecasting, mix and match, sustainability for daily use, sustainability for body type and skin color, easy to care and pattern of the fabric, orderly. The best fashion design was Veracious. Thus, this study successfully developed a hierarchical framework of fashion design evaluation, combined two methods, generated priorities of criteria, and ranked the ideal fashion design.

This paper only considers internal factors of designing a fashion. A comprehensive model that takes external factors into consideration could be considered in the future. The approach proposed in this study assumed that all elements or criteria were related to each other without considering the level or magnitude of interrelation. Another future study could develop a method that considers the level or magnitude of interrelation to better formulate the influence between element or criteria.

References

[1] Goncu G, Bayazit N. An application of AHP in the selection of fashion trends for specific target groups. Design Principles & Practices 2007; vol.1, no.3: pp: 573-584.

[2] Lin C, Twu C. Combination of a fuzzy analytic hierarchy process (FAHP) with the Technique for Order Preference by Similarity to Ideal Solution (TOPSIS) for fashion design scheme evaluation. Textile Research Journal 2012; vol.82, no.10: pp: 1065-1074, 2012.

[3] Harr SJ. Development of assessment rubrics for fashion design projects. in 61st Annual Meeting of the International Textile and Apparel Association: 2004.

[4] Kidd LK, Workman JE. Assessment of Creativity in Apparel Design. Clothing and Textiles Research Journal 1999; vol.17, no.1: pp: 58-64.

[5] Jones SJ. Fashion Design. New York: Watson-Guptill: 2002.

[6] Tseng ML, Chiang JH, Lan LW. Selection of optimal supplier in supply chain management strategy with analytic network process and choquet integral. Computers & Industrial Engineering 2009; vol.57, no.1: pp: 30-340.

[7] Tzeng GH, Huang JJ. Multiple attribute decision making: methods and applications. Boca Raton, FL: CRC press: 2011.

[8] Lu MT, Lin SW, Tzeng GH. Improving RFID adoption in Taiwan's healthcare industry based on a DEMATEL technique with a hybrid MCDM model. Decision Support Systems 2013; vol.56, pp: 259-269.

[9] Yazgan HR, Boran S, Goztepe K. Selection of dispatching rules in FMS: ANP model based on BOCR with choquet integral. International Journal of Advanced Manufacturing Technology 2010; vol.49, no.5-8: pp: 785-801.

[10] Lin C, Twu CH. Fuzzy MCDM for evaluating fashion trend alternatives. International Journal of Clothing Science and Technology 2012; vol.24, no.2/3: pp: 141-153.

[11] Lin C. Application of CFPR to fashion design scheme selection. International Journal of Clothing Science and Technology 2014:vol. 26, no.4; pp: 316-329.

[12] Lin C. Application of fuzzy Delphi method (FDM) and fuzzy analytic hierarchy process (FAHP) to criteria weights for fashion design scheme evaluation. International Journal of Clothing Science and Technology 2013: vol.25, no.3; pp: 171-183.

[13] Saaty TL. Fundamentals of the analytic network process. Proceedings of the ISAHP 1999; pp: 1-14,.

[14] Hsu P, Kuo M. Applying the ANP Model for Selecting the Optimal Full-service Advertising Agency. International Journal of Operations Research 2011: vol.8, no.4; pp: 48-58.

[15] Opricovic S, Tzeng GH. Compromise solution by MCDM methods: A comparative analysis of VIKOR and TOPSIS. European Journal of Operational Research, 2004: vol.156, no.2; pp: 445-455.

[16] Ministry of Tourism and Creative Economy of Republik of Indonesia, Fashion Trend Forecasting 2015/ 2016: 2015.

A Study of Female Garments based on Female Pottery Figurines of the Northern Qi Dynasty

Yue Zhu, Hao Zhang*, Jing Guo

Aimer Human Engineering Research Centre, Beijing Institute of Fashion Technology, Beijing, 100029, China

*Corresponding author's email: zhhh1cn@126.com

Abstract

This study systematically investigated clothing data corresponding to 306 female pottery figurines from 15 Northern Qi Dynasty era tombs in Hebei, Shandong, Henan, and Shanxi regions. Through text description, corresponding photos, and sketch lines, the study analysed pottery figurines' classification, quantity, size, headwear, upper garments, bottoms, shoes, accessories, and other related features. Based on data findings, the study summarised dress features of female officials, waitresses, as well as housemaids in noble families during the Northern Qi Dynasty, and discussed reasons for and underlying cultural factors affecting dress transformation.

Keywords: Female garments; Northern qi dynasty; Pottery figurines; Burial system

1. Introduction

The Northern Qi Dynasty (c. 550-577 AD) was one of the north separatist dynasties of the Northern and Southern Dynasties in Chinese history. Although its duration was very short, Northern Qi integrated many cultures and successively generated many new cultures, which laid the foundation for the subsequent prosperity of Sui and Tang Dynasties. Shen Kuo, a famous polymathic scientist and statesman of the Northern Song Dynasty (c. 960-1127 AD), once wrote in his "Dream Pool Essays" that "Chinese clothes and headwear applied Hu dress since Northern Qi [1]." Kuo's observation shows the far-reaching influence of Northern Qi clothes on dress culture in later ages, and its status is indispensable in the history of Chinese clothing. Nonetheless, there is still a dearth of research pertaining to Northern Qi clothes.

Most authors of previous studies related to Northern Qi clothing were somewhat constrained by their specialised backgrounds in archaeology, museum science, and art design, which means most such studies are equally limited in their discussion of clothing culture, style, and aesthetics and do not present explicit or systematic data on Northern Qi clothes. Because of the many gaps in the micro research fields of Northern Qi clothes, the latter studies' findings cannot address problems faced by scholars whose specialty is clothing, nor can they be applied in a practical sense to such areas as Chinese traditional clothing restoration and costume design in movie and TV dramas.

Based on the above points and from the perspective of clothing design, this study gathered up existing clothing data pertaining to female pottery figurines from the Northern Qi Dynasty. The study summarises dress features of female officials, waitresses, and housemaids in noble families during the Northern Qi era, and discusses reasons for and underlying cultural factors affecting dress transformation. The study's findings may be a valuable resource for subsequent studies in fields such as clothing historiography, clothing structure, and fashion history. Moreover, the research findings may have a practical application for professionals involved in traditional clothes restoration, movie/ TV drama, stage, and travelling culture costume design, and Chinese fashion design.

2. Research content

2.1 Research methods

Among the extensive data related to clothing of China's Northern Dynasties (spanning from 386-581 AD), this study focused on clothing represented on female pottery figurines buried with the dead during the Northern Qi Dynasty (c. 550-577 AD). The female pottery figurines refer to three-dimensional, personified burial implements representing female images that are either shaped and/ or fired in mud, or carved from wood, stone, or bronze, and which first appeared in tombs of the late-period Shang Dynasty (c. 1600-1046 BC). During China's Warring States period (c. 475-221 BC), the use of wood carving figurines was popular in Chu tombs; the use of pottery figurines started to emerge in tombs since the Han Dynasty (c. 202 BC-220 AD), then faded during the Wei and Jin Dynasties (c. 220-589 AD), and became prosperous again during the Northern Dynasty. The burial pottery figurines system thrived a second time during the Sui and Tang Dynasties. The Northern Dynasty is a turning point in the course of burial pottery figurines development.

Although extant research on pottery figurines is indispensable, current studies of unearthed Northern Dynasty pottery figurines correspond mainly to text descriptions, which indicates that such studies did not sufficiently investigate the characteristics of pottery figurines. Moreover, due to the uneven knowledge levels of authors who wrote archaeology reports and because text descriptions may occasionally include errors or omissions, it is necessary to first take photos or create sketches when researching pottery figurines' garments.

Therefore, this research systematically gathered text descriptions and corresponding photos and sketches of pottery figurines' garments in archaeology reports on the Northern Qi Dynasty. The statistical information includes pottery figurines' classification, quantity, size, head clothing, face, upper garments, bottoms, shoes, accessories, and other related features. Based on its systematic gathering of data, this study summarises the dressing characteristics of female officials, waitresses, and housemaids of the Northern Qi Dynasty, and discusses reasons for and underlying cultural factors affecting dress transformation.

2.2 Research scope

The scope of the study encompasses female burial pottery figurines (that is the pottery figurines which have the word "female" in their category names) unearthed from tombs explicitly dated to the Northern Qi Dynasty era in Hebei, Shandong, Henan, and Shanxi. Table 1 shows the figurines' quantity and distribution.

Table 1. Quantity and distribution of female pottery figurines in Northern Qi Dynasty

Location	Tombs	Pottery figurines	Female pottery figurines	Proportion of female pottery figurines
Hebei	4	534	24	0.04
Shandong	1	33	8	0.24
Henan	2	290	36	0.124
Shanxi	8	1,474	238	0.16
Total	15	2,331	306	0.13

According to statistics, it is known that female pottery figurines occupy only 13% of the total amount of pottery figurines unearthed in 15 tombs, which indicates that the proportion of female pottery figurines is low in Northern Qi tombs. The types of female pottery figurines in these 15 tombs include figurines of female officials, waitresses, and housemaids; although the tombs also contained rare horsewomen figurines, this article will not discuss the latter due to their small quantity). The three kinds of pottery figurines primarily represent housemaids in noble families in Northern Qi, with only class and division differences between categories.

3. Results and analysis

3.1 Distribution of various female pottery figurines in Northern Qi Dynasty

Table 2 shows the distribution of various female pottery figurines in Northern Qi Dynasty; the figurines of female officials unearthed from tombs in Northern Qi are centralised in Shanxi, including 96 figurines bearing similar shapes of female officials from the Prince Dongan Tomb, Lourui Tomb, and Hanyi Tomb. However, the shape of 14 female-official figurines unearthed from Kudihuiluo Tomb is slightly different from the other female-official figurines. Because the nine figurines of female officials unearthed from the Kudiye Tomb did not wear the symbolic sarong hat, this study reclassified them into the ordinary waitress figurine category. Also, a waitress figurine unearthed from the joint burials of Heshaolong couples in Henan and five waitress figurines unearthed from Gaorun Tomb in Hebei are similar in shape to the dancing figurines in Northern Wei and Eastern Wei Dynasties, and thus have been removed from the category of waitress figurines.

Table 2. Distribution of various female pottery figurines in Northern Qi Dynasty

	Type of figurine		
Location	Female official	Waitress	Housemaid
Hebei		24	3
Shandong		3	3
Henan		33	4
Shanxi	119	118	8

3.2 Dress features of female-official figurines in Northern Qi Dynasty

"Female officials" refer to females with certain official status. People wearing sarong hats often assume official positions of authority, which is one of the bases for naming this kind of pottery figurine. During the Han Dynasty, only males wore sarong hats; they became popular in central China during the Wei and Jin Dynasties and then gradually in north China before becoming the major hat form of female officials in the Northern Dynasty.

a

b

Fig.1 Female-official figurines from: (a) Prince Dongan Tomb (445); (b) Kudihuiluo Tomb (55)

Data showed the following dress features of female-official figurines: a sarong hat and underwear with round collar; a bag-sleeve upper garment overlapping from right lapel as outerwear; a waistband tied at normal position of waistline; underwear long dress and inner wide-leg pants; and round toe shoes (see Fig.1a). The length of the longuette worn by female-official figurines unearthed from Kudihuiluo Tomb and their movements are different from those unearthed from other tombs (see Fig.1b).

To summarize, the clothes of female-official figurines in Northern Qi have strong aesthetic and hence social functions (sarong hat, upper garment overlapped from right lapel, and bag-sleeve), which are quite different from the clothes of ordinary waitress and housemaid figurines, that embody their status and daily work.

3.3 Dress features of waitress figurines in Northern Qi Dynasty

"Waitress" refers to female servants accompanying masters; their status is higher than housemaid but lower than female officials. Archeological materials show three clothes forms of waitress figurines in Northern Qi.

3.3.1 Waitress figurines' garment style 1

1) Fundamental form:

This type was unearthed in tombs in Hebei, Henan, and Shanxi; 7 are from Yuanliang Tomb, Fancui Tomb, and Hanyi Tomb, respectively, and 12 are from the joint burials of Heshaolong couples.

They wear either a crescent-shaped chignon (hat), double crescent-shaped chignon, or double spiral coil chignon; round-collar underwear; narrow-sleeve upper garment overlapping from right lapel as outerwear; waistband tied above normal place of waistline; long dress with inside underskirt and outer long dress; and round toe shoes (see Fig.2a).

a b

Fig.2 Waitress figurines from: (a) Yuanliang Tomb (CMMI: 64); (b) Zhanghaiyi Tomb (sample 6)

2) Fundamental subtype

This type only appears in the tombs in Taiyuan, among which 3 were unearthed from Zhanghaiyi Tomb, 1 from Hanyin Tomb, and 16 and 31 from the Prince Dongan Tomb and Lourui Tomb, respectively.

The differences between this type and the fundamental form are that they do not wear upper garments overlapping from right lapel and the long dress has shoulder straps (see Fig.2b). Relevant literature notes that this special "overalls dress" may be a variant of the waistcoat.

3.3.2 Waitress figurines' garment style 2

This type of pottery figurine was generally unearthed from the tombs in Hebei, Henan, and Shanxi: 6 come from Yuanliang Tomb; 1 from Yaojun Tomb; 4 from Gaorun Tomb; 9 from joint burials of Heshaolong couples and Kudiye Tomb, respectively; 3 from Kydiye Tomb; 34 from Prince Dongan Tomb; and 2 from Hebachang Tomb and Helouyue Tomb, respectively.

The dress features of these waitress figurines are: the type of headwear is uncertain; either a gold ingot shaped chignon (hat), topknot, crescent-shaped chignon, or double spiral coil chignon; round-collar underwear; bag-sleeve upper garment overlapping from right lapel as outerwear; waistband tied at normal place of waistline; long dress; and round toe shoes (see Fig.3a).

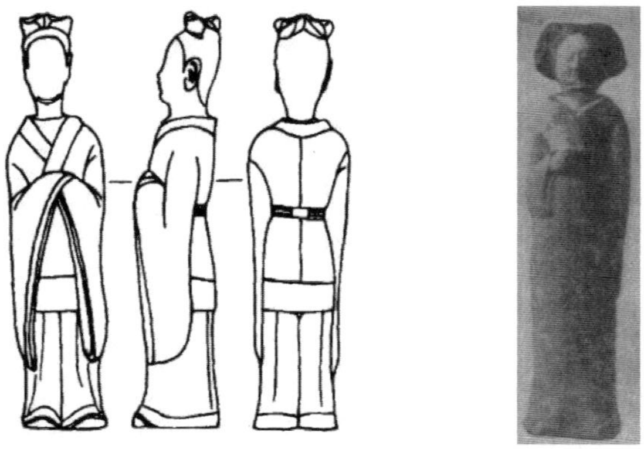

a b

Fig.3 Waitress figurines from: (a) Yuanliang Tomb (CMMI: 19); (b) Yaojun Tomb

3.3.3 Waitress figurines' garment style 3

This type of figurine was unearthed only from the tombs in Hebei, with 1 from Yaojun Tomb and 5 from Changwengui Tomb.

They wear a double bun or crescent-shaped chignon (hat), a narrow-sleeve upper garment overlapping from left lapel, long dress, and a waistband tied above normal place of waistline (see Fig.3b).

To summarize, the waitress figurine represents the most common form of female pottery figurines in Northern Qi and their clothes forms are relatively abundant. Although there are many differences in headwear, upper garments, as well as bottoms, it can be seen that the basic costume elements have certain inherent and combined features. With respect to clothing details, the first clothing type has aesthetic functions with complex apparel technology; the second type has aesthetic (bag-sleeve) and practical (wide leg pants) functions. Questions remain, however, regarding why this kind of women's dress appeared in Northern Qi combined both Hu and Han elements, and why its style of headwear is unconstrained. Such questions deserve further research. Compared with the first two types of clothing types, the third one is simpler in shape and only appears in the tombs in Hebei, which is speculated to be affected by regional culture.

3.4 Dress features of housemaid figurines in Northern Qi Dynasty

Housemaid figurines (also referred to as female work figurines or female sitting figurines) can be subdivided into female figurines holding basins and female figurines holding dustpans, based on the figurines' depicted movements and held objects. They represent the image of housemaids engaging in work and other domestic affairs. Housemaids' status is the lowest in noble families in Northern Qi.

a b

Fig.4 Housemaid figurines from: (a) Yuanliang Tomb (CMMI: 58); (b) Yuanliang Tomb (CMMI: 57)

Data showed the following dress features of housemaid figurines buried with the dead in Northern Qi: crescent-shaped chignon (hat), double crescent-shaped chignon, or double spiral coil chignon/ short chignon;

round-collar underwear and narrow-sleeve upper garment overlapping from left lapel as outerwear; waistband tied above normal place of waistline; and underwear long dress. Some individual pottery figurines tie the two upper-garment sleeves at the rear (see Fig.4a) or wrap silk cloth (see Fig.5b).

Fig.5 Housemaid figurines from: (a) Prince Dongan Tomb (521, 523); (b) Prince Dongan Tomb (524)

To summarize, compared with female official and waitress figurines, the clothes of housemaid figurines shows obvious practicality (narrow sleeves). In addition, the proportion of upper garment overlapping from left lapel among housemaid figurines is higher than that of the other two categories, which indicates that the direction of lapel may have a certain relation to social status.

4. Conclusion

Table 3 shows the statistical data of garment details of various female pottery figurines in the Northern Qi Dynasty.

Table 3. Garment Details of various female pottery figurines in Northern Qi Dynasty

	Type of chignon (hat)					
Type of figurine	Sarong	Crescent-shaped	Gold ingot shaped	Other shapes	Unrecognized	Total
Female-official	94					94
Waitress		87	49	19	8	163
Housemaid		13	1	4	15	33
Total	94	100	50	23	23	290

	Type of lapel				
Type of figurine	Right lapel	Left lapel	V-shaped lapel	Unrecognized	Total
Female-official	94				94
Waitress	65	21	55	22	163
Housemaid	1	18	3	11	33
Total	160	39	58	33	290

	Type of sleeves			
Type of figurine	Bag-sleeve	Narrow-sleeve	Unrecognized	Total
Female-official	94			94
Waitress	77	62	24	163
Housemaid	2	25	6	33
Total	173	87	30	290

	Type of bottoms			
Type of figurine	Dress	Trousers	Unrecognized	Total
Female-official	94			94
Waitress	96	63	4	163
Housemaid	22	10	1	33
Total	212	73	5	290

Most female pottery figurines from the Northern Qi Dynasty era represent images of female officials, waitresses, and housemaids in noble families. When conducting research on garments used during the Northern Qi Dynasty, we recommend that they be studied separately from data pertaining to noble females' clothing in order to avoid confusion. The clothing of female pottery figurines from the Northern Qi Dynasty have prominent contemporary characteristics; meanwhile, the pottery figurines unearthed from several regions or tombs also show strong regional characteristics. The clothing of female pottery figurines in Northern Qi show raised waistlines; however, this phenomenon only appeared in figurines corresponding to waitresses and housemaids who both had a lower status than female officials.

References

[1] Shen K. Dream pool essays. Liaoning: Liaoning Education press, 1997.
[2] Yang S. A study of female pottery figurines in northern dynasties. Beijing: Minzu University of China, 2011.
[3] Song BL. A study of non-religion garments in northern dynasties. Jinan, China: Shandong University, 2008.
[4] Shang CF. A primary study on garments of female pottery figurines in Luoyang Northern Wei. Huaxia Archaeology 2000; 3: 71-76.
[5] Wei QL. A study of pottery figurines in Eastern Wei and Northern Qi. Zhengzhou, China: Zhengzhou University, 2007.
[6] Hai Q. A study on headwears of pottery figurines in Shanxi northern dynasties. World of Antiquity 2015; 1: 21-27.
[7] Jia XZ. A study of ancient Chinese headwears. Shanghai: Donghua University, 2007.
[8] Wang RH. A study on female garments of fresco in Shanxi Northern Qi. Taiyuan, China: Shanxi University, 2010.
[9] Zhang ZY. Archaeological report: Yuanliang Tomb of Northern Qi in Hebei province. Archaeology 1997; 3: 33-39.
[10] Zhu QS. Archaeological report: Yaojun Tomb of Northern Qi in Hebei province. Cultural Relics 1984; 4: 16-22.
[11] Wang MZ. Archaeological report: Changweigui Tomb of Northern Qi in Hebei province. Cultural Relics 1984; 9: 39-42.
[12] Cultural Center of Ci County. Archaeological report: Gaorun Tomb of Northern Qi in Hebei province. Archaeology. 1979(3): 235-243.
[13] Su YQ, Yingju J. Archaeological report: Cui family Tomb of northern dynasties in Shandong province. Acta Archaeol Sinica 1984; 2: 221-244.
[14] Li XP, Gu Y. Archaeological report: Joint Burials of Heshaolong Couples of Northern Qi in Henan province. Cultural Relics 1987; 1: 8-18
[15] Henan Museum. Archaeological report: Fancui Tomb of Northern Qi in Henan province. Cultural Relics. 1972; 1: 47-57.
[16] Institute of Cultural Relics and Archaeology of Taiyuan City. Archaeological report: Hebachang Tomb of Northern Qi in Shanxi province. Cultural Relics 2003; 3: 11-25.
[17] Chang YM. Archaeological report: Helouyue Tomb of Northern Qi in Shanxi province. World of Antiquity 1992 ;3: 33-38.
[18] Wang KL. Archaeological report: Kudihuiluo Tomb of Northern Qi in Shanxi province. Acta Archaeol Sinica 1979; 3: 377-402.
[19] Li AG. Archaeological report: Zhanghaiyi Tomb of Northern Qi in Shanxi province. Cultural Relics 2003; 10: 41-49.

[20] Tao ZG. Archaeological report: Hanyi Tomb of Northern Qi in Shanxi province. Cultural Relics 1975;4:64-73.
[21] Institute of Cultural Relics and Archaeology of Taiyuan City. Archaeological report: Kudiye Tomb of Northern Qi in Shanxi province. Cultural Relics 2003; 3: 26-36.
[22] Institute of Cultural Relics and Archaeology of Shanxi Province. Archaeological report: Prince Dongan Tomb of Northern Qi in Shanxi province. Beijing: Cultural Relics Press, 2006.
[23] Institute of Cultural Relics and Archaeology of Shanxi Province. Archaeological report: Lorui Tomb of Northern Qi in Shanxi province. Cultural Relics 1983; 10: 1-23.

Study of Stitching Parameters and Functional Finishes of Suiting Fabrics on Seam Performance and its Quality

A. K. Choudhary[1]

[1]*Department of Textile Technology, National Institute of Technology, Jalandhar-144011, Punjab India*

*Corresponding author email: choudharyak@nitj.ac.in, akc1087@gmail.com

Abstract

The objective of this work is to come up with an ideal needle and stitch specification for various finished heavy suiting fabrics and to test the seam quality aspects especially seam strength and elongation of various combinations of needle and stitch specifications on various finished heavy suiting fabrics and derive the optimized combinations of stitching parameters for each of the finished fabric sample. With the increase in stitch density, breaking load of seam and seam elongation increases but after a particular value of stitch density, breaking load of seam and seam elongation decreases with increase in stitch density, which in turn affects the performance of the apparel. Also, with increase in needle size, the breaking load of seam and seam elongation decreases. So the suitable combination is of needle size with stitch density for best seam performance in heavy fabric with various finishes the best combination is needle size of 16 and stitch density of 13 SPI.

Keywords: Fabric finishes; stitch density; seam strength; seam elongation and sewing needle

1. Introduction

There are a number of factors on which quality fitness of clothing industry is based such as - performance, reliability, durability, visual and perceived quality of the garment. Quality needs to be defined in terms of a particular framework of cost and user's comfort. Here some of main fabric properties that are taken into consideration for clothing manufacturing:

- Overall look of the clothing.
- Right formation of the clothing.
- Feel and fall of the clothing.
- Physical properties.
- Colour fastness of the clothing.
- Finishing properties
- Presentation of the final produced clothing.

In the manufacture of clothing the main fabric is supplemented by a variety of trimmings whose number and complexity will depend on the type of the garment. All clothing employs sewing threads in some form or other. Trimmings may be for:

- Aesthetics reasons
- Performance reasons
- Both

Most trimmings involve both functions and are the result of various kinds of seams applied on the clothing. Functional - finishes involves: Fragrance Release, Protective Finishes, Skin Care Additives, Insect Repellent, Deodorising Fragrance, Antimicrobials, Flame Retardant Finishes, Cool Finish & Thermal Insulatory Finish, Water Proofing Finish and UV Stabilisers. Functional finishes represent the next generation of finishing industry, which, make textile materials act by themselves. This means that they may keep us warm in cold environments or cool in hot environments or provide us with considerable convenience, support, and even fun in our normal day-to-day activities.

Seams include two most important parameters i.e. needle size and the stitches per inch (stitch density) for best performance and best quality garment both in terms of aesthetic and performance. Seam qualities also have the effect of the fabric construction and the finished applied on the fabric, if any. This work, hence, focus on the effect of seam qualities (seam strength and seam elongation) on a heavy suiting fabric, with various finishes applied, with various sampled needle sizes and stitches per inch combinations.

Hence the objective of this experimental work is:

- To come up with an ideal needle and stitch specification for various finished heavy suiting fabrics.
- To test the seam quality aspects especially seam strength and seam elongation of various combinations of needle and stitch specifications on various finished heavy suiting fabrics and derive the optimised combinations of stitching parameters for each of the finished fabric sample.

1.1 Seam formation

It is discovered that almost all types of sewing stitch form the stitch interlock with the aid of the needle rise and the resulting yarn loop. However, this requires the sewing thread to not possess a high degree of stretch ability, as in this case the loop is unable to form and no loop interlock takes place. A highly stretchable sewing thread would seem to ensure the required sewing capabilities. In case of lockstitch (stitch type 301), with correct stitch position, the needle and the bobbin threads interlock in the centre of the fabric. As a result, the needle and the bobbin thread account for 50% of the thread used. Under these circumstances, the best possible seam elongation and the greatest seam strength are achieved. In case of double chain stitch (stitch type 401), in the standard stitch position the interlock of needle and bobbin threads takes place on the lower face of the fabric. This means that under customary conditions the proportion of needle thread used is lower compared to the bobbin thread supply. Further, 30% to 35% of the thread consumption on fine fabrics is needle thread and this is why the needle thread breaks first, not the looper thread - therefore the strength of looper thread is more or less important. This is why, the world over, the bobbin/ looper threads is always thinner than the needle threads. Only on a very thick material does the looper thread have an influence, when needle thread consumption is more than 50%. Fabric plies undergoes a complex sequence of loadings during sewing operations, and unless their properties are compatible, the strains developed will be quite different, which affects the balance, appearance, and stability of the seam. Incompatibility of the fabric components is not necessarily associated with stitching fabrics of different kinds. Owing to the orthotropic nature of the textile fabrics, i.e. anisotropic with two axis of symmetry, unless the seam is sewn along the same direction of both plies, their mechanical properties, except for the compression modulus, will be different even if they are cut from the same piece of fabric. In garment making, which require conforming a two-dimensional into a three dimensional surface, seams joining fabric plies along different directions are not avoidable. The clothing industry tends to use the finest thread that will provide adequate seam strength. There are a no of reasons for this:
The volume of the under thread bobbin on stitch machines are severely limited. The use of finest threads will allow a longer length of thread to be wound on the bobbin, thus giving longer running time between bobbin changes.

- Finer threads give less bulk and smoother chain stitch seams.
- Finer threads take up less space in the seam, causing smaller amounts of fabric distortion and a lower likelihood of seam puckering.
- Finer threads allow the use of finer needles, further contributing to a lack of fabric distortion and consequent seam puckering.
- Finer threads produce less obtrusive seams.
- Finer threads tend to bury themselves below the surface of the fabric, rendering the seam less affected by abrasion.

1) It must not be pull apart under the stresses of the service.
2) It must not be cockle or tight.
3) It must be as extensible as the fabric must or as needed by the movement demanded of each area of the garment.
4) The sewing stitches must not cut the fabric, break or crack on the extent of stretching.
5) The seam must not be grin.

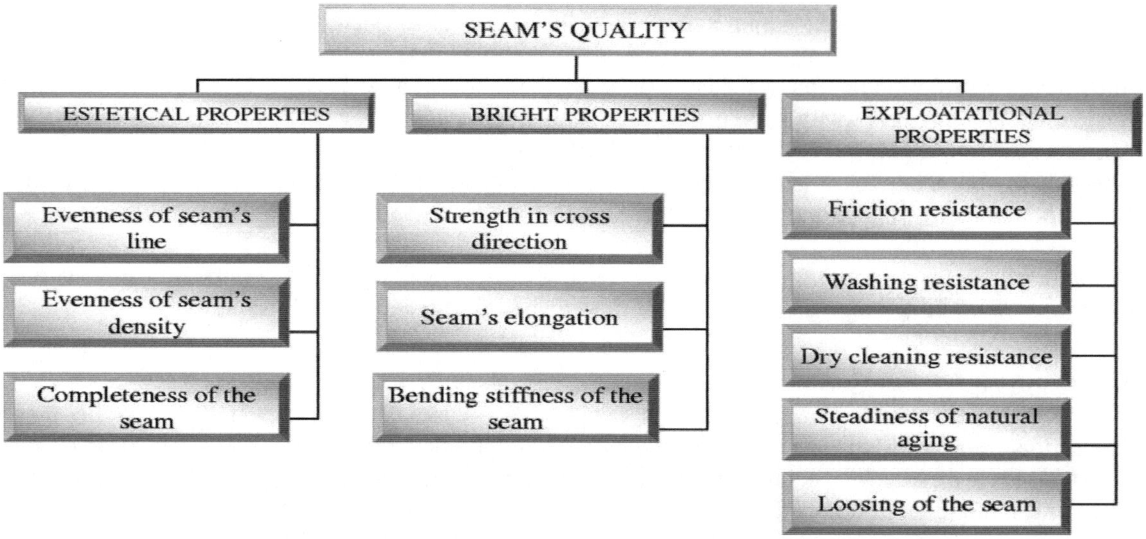

Fig.1 Desirable Properties of Seam

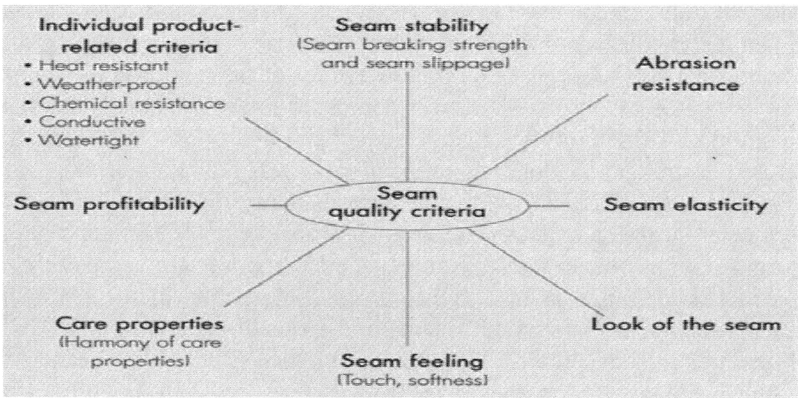

Fig.2 Seam quality criteria

2. Material and Methods

The samples of different functional finishes as well as the combination of these finishes are:

- Dyed
- Water repellent
- Stiff
- Wrinkle free
- Aroma
- Stain release
- Soft
- Flame retardant
- Stain release - Wrinkle free
- Water repellent - Wrinkle free
- Water repellent - Flame retardant

Fig.3 Prepared various finished Samples

The details of the various finishes used in experimental work are as follows

Type of finish	Main component	Chemical constitution	Mode of action
Water repellent	Phovotex JVA Olephobol CO	Oil,water and stain repellent based on dispersion of fluorochemical containing extender	Low surface energy finish – The low surface tension of fluorocarbons enables textile material treated with fluorochemical finishes to obtain low surface energy finish. The treated textile material repels water due to this property.
Stiff	Apprantan E M Apprantan MBX	Durable stiffening agents based on thermoplastic polymers-polyvinyl acetate	Coating – by formation of water insoluble films on fabric surface.
Wrinkle free	Knitex FEL	Based on modified dimethyloldihydroxy ethylene urea(DMDHEU)	Reactants (resins) chemically cross-link the cellulose polymer chains. Resins form covalent bonds with hydroxyl (-OH) groups of cellulosic materials.
Aroma	Lexroma	Silicon resin coated synthetic product	Application of silicone resin coated synthetic product on textiles gives fragrance by destruction of the micro – capsules through rubbing with hands or scrubbing rather strongly etc, or while worn.
Stain release	Oleophobol ZSR Ultraphil HSD	Based on fluorocarbons Oleophobol ZSR-non ionic/slightly cationic emulsion of a polymer, perfluorinated compound. Ultraphil HSD-durable stain release agents belongs to cationic surfactant group	Fluorocarbon-based compounds improve the surface energy of the fibres,making them more hydrophilic. Resultantly the water can more readily carry the detergent molecules into the fibre crevices and emulsify and remove the stain material. Hybrid fluorocarbons of the block copolymer are hydrophobic and oleophobic in air and hydrophilic and oil releasing during the laundering process. When dry, the fluorochemical blocks function as a conventional oil repellent finish. Under laundering conditions the hydrophilic blocks absorb water, causing the polymer backbone to flip. Now ,the fibre surface is hydrophilic and the coating functions as a stain release agent.
Soft	Recosoft	Cationic silicone softener	Reduces the coefficient of friction. Softeners function primarily on the surface of fibres. Small softener molecules, in addition , penetrate the fibre and provide an internal plasticization of the fibre-forming polymer by reducing the glass transition temperature (Tg). When silicone is applied, there is interaction between textile substrate (with characteristic functional group) and silicone molecule forming a strong covalent bond and thus modifying the surface subsequently. There is molecular interaction among the silicone molecules through the rreactive functional

			groups resulting in cross- linked network formed by self- polymerization. In cellulose, there are strong hydrogen bonds between the hydroxyl or amino groups of the fibres and the amino groups of the modified silicones. These bonds act as an anchor for the silicone, which forms an evenly distributed film on the fibre surface.
Flame retardant	Pyrovatex CP	Dialkylphosphonocarboxylic acid	Reacts with the hydroxyl groups of the cellulose. If flame or heat comes in contact with the flame retardant finished cellulose, a carbon scaffold is formed by dehydration of the cellulose.

2.1 Cutting stitching of samples

The samples mentioned above, were cut in dimensions of size - 4in. x 8in. This dimension was chosen in accordance with the requirements of the testing apparatus. The samples were stitched using JUKI lockstitch machine DDL8300N. Different parameters like needle size and stitch density i.e. stitches per inch of the sample were varied during stitching. According to the specifications for a medium heavy suiting fabric, standard parameters for better seam performance is Needle size-16 (Grozbecket) and SPI - 13. Apart from the standards 8 other combinations (of needle size and stitches per inch) were taken for sampling and testing.

Needle sizes - 14, 16, 18 and stitches per inch (SPI) -18, 13, 8.

Thus, the following nine combinations are prepared:

1) Needle size 14 and SPI 8
2) Needle size 14 and SPI 13
3) Needle size 14 and SPI 18
4) Needle size 16 and SPI 8
5) Needle size 16 and SPI 13
6) Needle size 16 and SPI 18
7) Needle size 18 and SPI 8
8) Needle size 18 and SPI 13
9) Needle size 18 and SPI 18

2.2 Testing of samples

Different tests were conducted on prepared samples to study the seam quality of various needle sizes and stitch densities. The tests were conducted for the samples to study the seam strength and elongation. Seam strength and elongation are determined with the help of ZWICK Tensile Testing machine as per the ASTM standard D-1683. The testing parameters considered are:

1) Preload
2) Specimen thickness
3) Specimen width
4) Loading rate
5) Gauge length

Testing is done for checking the seam qualities when needle size and stitches per inch are.

3. Result and discussions

3.1 Breaking Load

Table 1. Breaking strength of the samples

	Needle size - 14			Needle size - 16			Needle size - 18		
	SPI -18 (SD 1.5)	SPI - 13(SD 2)	SPI - 8 (SD 3)	SPI -18 (SD 1.5)	SPI - 13(SD 2)	SPI - 8 (SD 3)	SPI -18 (SD 1.5)	SPI - 13(SD 2)	SPI - 8 (SD 3)
DYED	246	199	82.8	300.5	172.5	90.05	199	111	60.7
AROMA	201.5	154.5	94.45	288.5	121.5	87.1	204.5	83.95	152
SOFT	377.5	167	76.25	254.5	150.5	92	283.5	167	58.45
STIFF	227.75	149	112	209.5	153	104.25	289	132.5	87.45
SR	229.5	184.5	100.15	330.5	147.5	111.5	320.5	76.45	97.2
WR	371.5	111.85	90.9	336	191	119.4	211.5	182.5	78.85
WF	266.5	197	103.8	260	177.45	132	138	200	124.5
SR+WF	145.5	226.5	107.65	289	196	116	306	208	91.55
FR+WR	323.5	127.8	117	296.5	176	114.6	240.5	144.5	95.35
WR+WF	291.5	217.5	110	274.5	205.5	116.1	253	139	87.95
FR	296.5	256.5	90.85	243	128.6	123	208.2	244	85.15

SR - Stain release, WR - Water repellent, WF - Wrinkle free, FR - Fire retardant

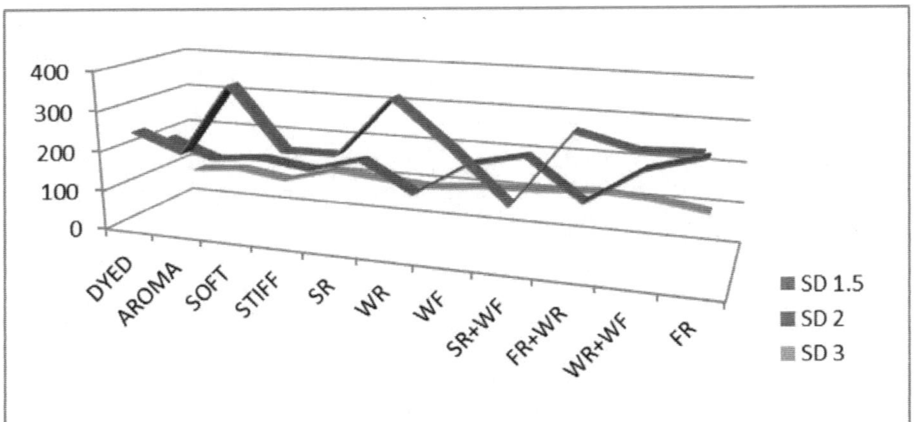

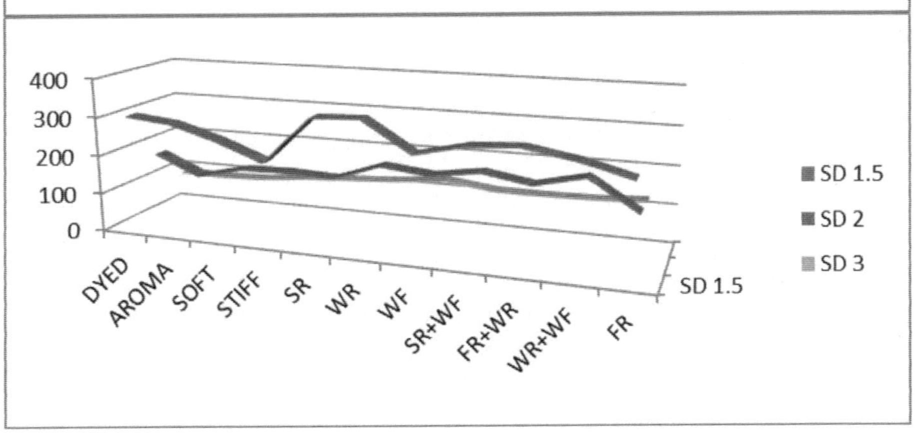

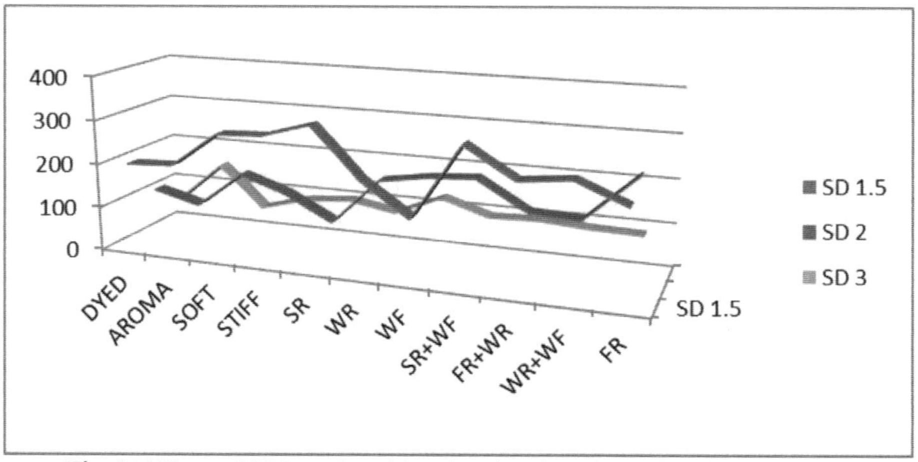

Fig.4 Effect of SPI on breaking loads of various finished samples

3.2 Breaking extension

Table 2. Breaking extension of samples

	Needle size - 14			Needle size - 16			Needle size - 18		
	SPI -18 (SD 1.5)	SPI - 13(SD 2)	SPI - 8 (SD 3)	SPI -18 (SD 1.5)	SPI - 13(SD 2)	SPI - 8 (SD 3)	SPI -18 (SD 1.5)	SPI - 13(SD 2)	SPI - 8 (SD 3)
DYED	10.3	8.4	6.4	10.85	12.5	8.6	9.6	10	8
AROMA	9.7	11.15	10.75	12.35	10.2	10.6	12.85	12.45	11
SOFT	12.6	13.75	12.5	14.55	12.15	14.1	14.5	12.2	12.85
STIFF	10.3	11.1	10.45	11.5	9.55	9.55	11.75	8.4	6.55
SR	10.95	12.6	10.25	13.85	11.8	10.85	13.7	13.1	10.3
WR	12.45	12.3	10.6	14.8	12.05	11.5	14.1	12.2	9.6
WF	12.4	11.4	11.5	15.55	11.4	10.45	16.55	10.95	11.6
SR+WF	12.3	12.4	10.65	13.65	11.15	11.05	12.7	11.4	9.9
FR+WR	8.45	11.6	8.05	12.95	12.4	11.6	12.85	11.6	9.7
WR+WF	10.65	11.55	10.75	14.9	10.6	11.75	14.85	11.45	7.8
FR	8.65	10.55	8.85	12.45	12.45	10.5	15.4	12	10.55

SR - Stain release, WR - Water repellent, WF - Wrinkle free, FR - Fire retardant

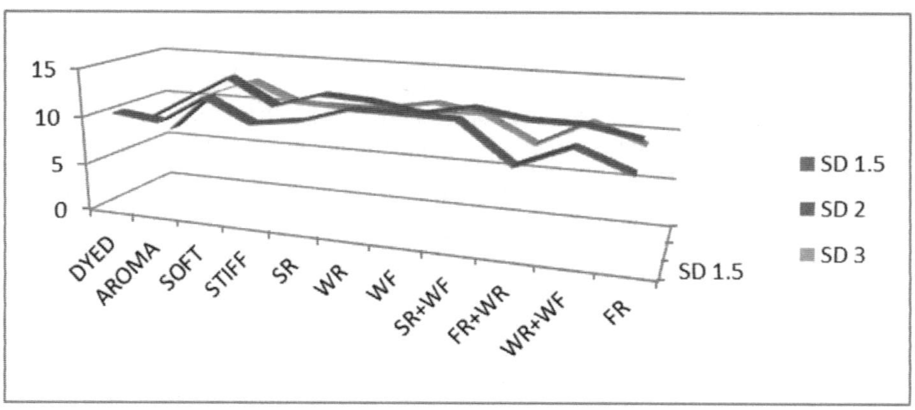

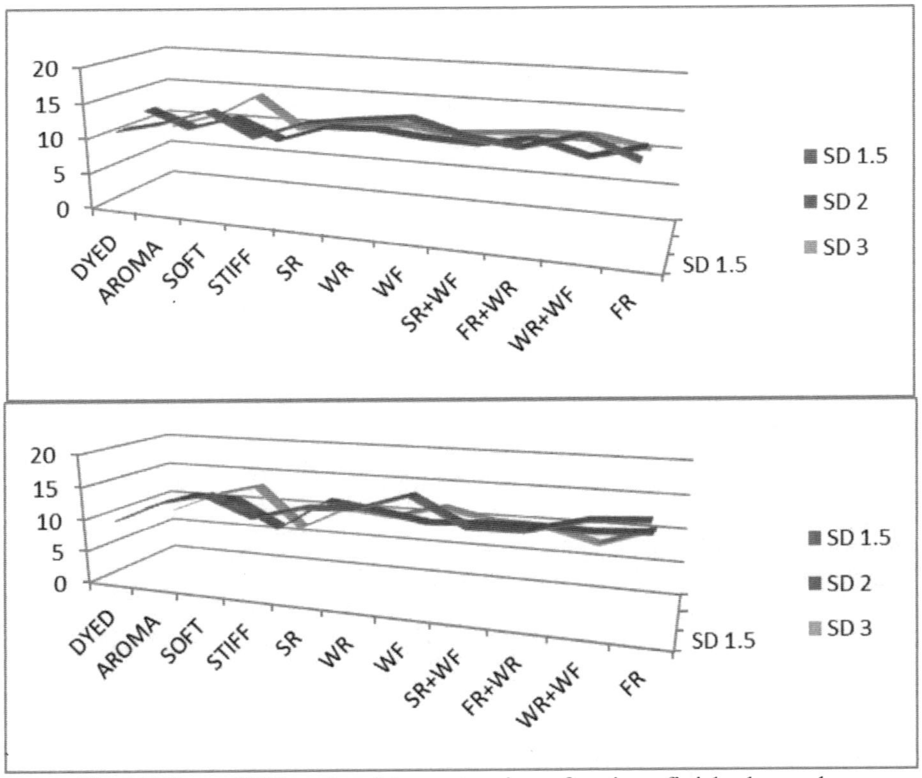

Fig.5 Effect of SPI on breaking extension of various finished samples

The following discussions are derived from the above table and graphs of both breaking load and extension:

1) As seen from the graph, as SPI (Stitches per Inch) increases, the breaking load of Seam and seam elongation will increase. It is due to more and tighter stitch formation per unit length of seam.

2) With the increase in needle size, breaking load of seam and seam elongation decreases. This is because as when needle size increases i.e. needle diameter increases so, the needle hole size of stitch is also increased and hence, the frictional contact of fabric with the sewing thread will be less so thread tension decreases and which lead to the decrease in seam strength.

3) In SR+WF (N-14) case, if the SPI is further increased from 13 to 18 then seam strength decreases, which is just opposite from the general trend. The reason for this is that the strength first increases upto a certain value and then it starts decreasing, same happens in this case that maximum point reaches earlier and further increase in SPI will lead to decrease in seam strength.

4) For the same sample, seam elongation also decreases if the SPI is further increased from 13 to 18 as the strength also decreases. This same trend is followed in other cases also.

5) In WF (18) sample, covalent bonds are formed which provide strength so increase in SPI from 13 to 18 will have effect on these bonds so strength will decrease.

4. Conclusion

It is concluded from the above result and discussion that with the increase in SPI, breaking load of seam and seam elongation increases but after a particular value of SPI, breaking load of seam and seam elongation decreases with increase in SPI. Also, with increase in needle size, the breaking load of seam and seam elongation decreases. So the suitable combination is of Needle size -16 with SPI -13 for best seam performance and its overall quality.

References

[1] Standard Test Method for Failure in Sewn Seams of Woven Apparel Fabrics1 Designation: D1683/D1683M - 11.

[2] Mukhopadhyay A, Ph D. Relative Performance of Lock stitch and Chain stitch at the Seat Seam of Military Trouser (J), Journal of Engineered Fibers & Fabrics, 2008; 3(1): 20-24

[3] Tensile Strength Test-Grab Test, ASTM D 5034-95 (Reapproved 2001), Standard Test Method for Breaking Strength and Elongation of Textile Fabrics (Grab Test).

[4] http://www.amefird.com/wp-content/uploads/2010/01/Minimizing-Seam-Puckering-2-5-10.pdf. 30/04/2011

[5] http://www.indiamart.com/canaren-india/fabric-protection-products.html. 29/04/2011

[6] http://www.arrowthreads.com/seamology files/seamproperties.html. 27/04/2011

[7] http://www.indiangarmentexporters.com. 30/04/2014

[8] http://ezinearticles.com/Quality-Control-Aspects-Of-Garment-Exports&id=373711. 26/04/2011

[9] http://www.fibre2fashion.com/industry-article/pdffiles/quality-control-aspects-of garmentexports.pdf?PDFPTOKEN=a9424826c1b99dd2862a3f66b3d6d24f3453da18|1305468547#PDFP. 30/04/2011

[10] http://www.am efird.com.cn/spi.html. 20/04/2011

[11] http://www.allfreecrafts.com/sewing/needle.html. 28/04/2011

[12] http://www.fabrics.net/SarahNeedleSelection.asp. 25/04/2011

[13] http://get-up-and-go.com/upholstery-forum/index.php,topic=6371.0st seam performance. 24/04/2011

Study on Military Uniforms of Northern Qi Dynasty based on Funerary Terracotta Warrior Figurines

Ting He, Rong Zheng*, Hao Zhang, Jing Guo

Aimer Human Engineering Research Centre, Beijing Institute of Fashion Technology, Beijing, 100029, China

*Corresponding author's email: yuki.zheng@gmail.com

Abstract

The Northern Qi Dynasty corresponded to a Cold War era in China. Its military apparel developed well, and it had a certain importance in history. However, scholarship in the study of military uniforms of this era is insufficient. The study is based on terracotta warriors in military uniforms from 22 tombs of the Northern Qi Dynasty in Henan, Hebei, and Shanxi province. The study proposed classification for military uniform patterns of the Northern Qi Dynasty with the perspective of clothing. The purpose of the study was to investigate the form and characteristics of military uniforms in the Northern Qi Dynasty.

Keywords: Northern qi dynasty; Funerary warrior figurines; Military uniforms; Form

1. Introduction

The Northern Qi Dynasty (550-577 AD) was a separatist regime during China's Southern and Northern Dynasties era. The Northern Qi Dynasty adopted the Northern Wei Dynasty's military system, from which the Sui and Tang Dynasties developed their respective systems for military uniforms. This study is based on research related to the military uniforms on unearthed Northern Qi terracotta warrior figurines, whose advantage is their three-dimensional nature. Currently, there has been more macro research on the clothing of the Northern Qi Dynasty based on terracotta figurines than on individual types of clothing. In addition, most studies undertaken by the clothing research community encompassed the entire Northern Dynasties or even the entire Wei-Jin and Southern and Northern Dynasties. Thirdly, the research vision of such scholars - who major mostly in specific disciplines such as archeology, museology, and fine arts - is usually limited to their respective fields of study [1-3].

This study extended previous research on the perspective of the Northern Qi Dynasty funerary terracotta figurines, and discusses characteristics and dress patterns of military uniforms of the Northern Qi Dynasty from a clothing perspective based on detailed classification of funerary terracotta figurines. Findings may provide a valuable resource for further specialized research on military uniforms of the Northern Qi Dynasty and may have practical application for theatre, film, television drama, animation, gaming, and other industries.

2. Research content

2.1 Research methods

We first extracted data from previous archeological reports pertaining to the Northern Qi Dynasty, and abstracted the terracotta figurines that met the study's requirements. We then enumerated the headwear, tops (collars, sleeves, shoulders, belts), pants, shoes, and other features of the figurines. During the report extraction, it was found that although some figurines look identical in pictures or photos, their text description is different, possibly because the reports were generated by varied organizations that did not have uniform naming standards. Therefore, with their specialized knowledge and clothing background, the authors of this study classified the terracotta figurines having the same clothing form and same or different names into a single category based on pictorial and text description, and proposed a new classification method for North Qi Dynasty military uniforms.

2.2 Research scope

In the current study, "terracotta warrior figurines" refer to terracotta figurines representing soldiers, private, guards, armoured soldiers, hercules, armoured riders, warriors, military officials, assistant military officials, and Generals, that are often depicted with accoutrements such as arrows, shields, knives, swords, armour, helmets, Kaijia armour, and/ or Jiazhou armour. In addition, depending on the circumstances, some ceremonial figurines and other related figurines described as weapon-holding figurines were also included in this study.

The authors studied and classified terracotta warrior figurines in military uniforms that were unearthed from 22 tombs of the Northern Qi Dynasty in Henan, Hebei, and Shanxi province. Among them were 6 Hebei tombs (with 171 figurines), 3 Henan tombs (with 56 figurines), and 13 Shanxi tombs (with 869 figurines). The subjects for this study comprised 634 military terracotta warrior figurines in military uniforms that represent the mainstream dress pattern, accounting for 58% of the total number.

3. Results and analysis

Northern Qi Dynasty tombs date from 553 to 576 AD. Although there is only a limited number of tombs spanning the dynasty's relatively short duration and the funerary figurines vary among the tombs in the same area, it is still possible to identify the forms and patterns of the figurines' military uniforms. According to the existing reports, the relevant names include: shield-pressing warrior figurines; shield-holding figurines; tomb-guarding warriors (or tomb-guarding figurines, tomb-guarding warrior figurines); horseback-riding warrior figurines; figurine guards, armoured figurines (or armour and helmet figurines, armour soldier figurines), and so on [4-24].

3.1 Characteristics of shield-pressing terracotta warrior figurines of the Northern Qi Dynasty

The tombs containing shield-pressing terracotta warrior figurines include Hebei Gao Run tomb, the tomb of Hebei Yuan Liang, Hebei Yao Jun tomb, Henan He Shao Long couple's tomb, the tomb of Shanxi Ku Di Hui Luo, and Shanxi He Ba Chang tomb. There are 11 shield-pressing terracotta warrior figurines in total. One shield-bearing figurine of Shanxi He Ba Chang tomb is the same as in the aforementioned figurines; nine warrior figurines unearthed from Hebei Cui Ang tomb, Henan Fan Cui tomb, four Northern Dynasty tombs in Hebei, Henan Gu An No. 2 tomb, and Shanxi He Lou Yue tomb are the same as well. There are 21 in total.

The basic form of shield-pressing terracotta warrior figurines show the figures wearing a pocket iron helmet, round Ming Guang armour protection, Bo on shoulder, belt on waist, hakama (a type of pant), and round toed boots (see Fig.1). "Pocket iron helmet" refers to the helmet worn by ancient warriors. The Shuowen, a 2nd-century Chinese dictionary during the Han Dynasty, also refers to pocket iron helmets as "Zhou," a word that later was used to mean "helmet." "Ming Guang armour" refers to two-piece plate-like chest armour that was popular in the period from the Northern and Southern Dynasties to the Tang Dynasty.

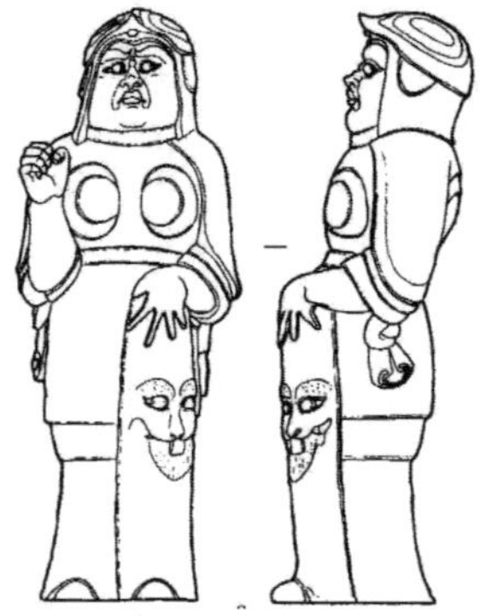

Fig.1 Shield-pressing warrior figurine in the He Ba Chang tomb

In spite of similarities in the above forms, there are still individual differences:

1) Headwear: The two shield-pressing warrior figurines from the Shanxi Ku Di Hui Luo tomb wear a gold dome helmet, with no ridges on the helmet, no protruding corners, and no protruding ear protection. one shield-pressing warrior figurine with pocket iron helmet decorated with beads was from Shanxi He Ba Chang T99HQH9 tomb; and three shield-pressing warrior figurines without bead decoration were from Hebei Yuan Liang tomb and Shanxi He Ba Chang T99HQH10 tomb.

2) Collar: Hebei Yuan Liang tomb warriors have front bows; there are two such figurines.

3) Belt: Shanxi Ku Di Hui Luo tomb figurines have front bows; there are two such figurines. The front portion of the belt of the figurine from Shanxi He Ba Chang T99HQH9 tomb has a narrow round form, while the back portion is flat and wide, and the waist is decorated with a red Sui band; there is one such figurine.

4) Sleeve: The Shanxi He Ba Chang T99HQH10 tomb warrior has bows on both arms; there is one such figurine.

5) Bottoms: The Hebei Gao Run tomb's shield-pressing warriors have legging skirts; there are two such figurines. The shield-pressing warrior figurines from Hebei Yuan Liang tomb and Hebei Yao Jun tomb have a leg wrap armour skirt; there are three such figurines.

3.2 Clothing characteristics of the Northern Qi period shield-holding terracotta warrior figurines

Tombs containing figurines holding shields include Hebei Gao Run tomb, the tomb of Hebei Yao Jun, Hebei Yuan Liang tomb, Hebei Chang Wen Gui tomb, four Hebei Wuqiao Northern tombs, Shanxi Ku Di Ye tomb, the southern Taiyuan suburbs tombs with the Northern Qi Dynasty murals, Shanxi Zhang Hai Yi tomb, Shanxi Xu Xian Xiu tomb, and Shanxi Di Zhan tomb. According to their characteristics, they were classified into four main forms:

3.2.1 Shield-holding terracotta warrior figurines (form 1)

These figurines wear head cloth, round neck narrow-sleeved shirt, bare right shoulder, coat draped over the left shoulder, right sleeve tied to the waist, belt, tied trousers, and round toed shoes (see Fig.2). Eight are from Hebei Gao Run tomb, three are from the tomb of Hebei Yao Jun, and 20 are from Shanxi Ku Di Ye tomb. Figurines with the same form but different names include Hebei Gao Run tomb's 49 guard figurines and Hebei Yao Jun tomb's four guard figurines. Shan Xi Ku Di Hui Luo tomb's 12 male guard figurines with folded coat edges and four bare-shouldered male guard figurines have the same form, except for the three ridged black crown. The form of Taiyuan southern suburbs mural tomb's one warrior figurine and Shanxi Xu Xian Xiu tomb's two warrior figurines is basically the same with this form, but they wear a dome hood with flipping cap ears pinned on both sides of the dome. There is a total of 99 form 1 figurines.

3.2.2 Shield-holding terracotta warrior figurines (form 2)

These figurines have a folded ear hood, tight sleeved underwear, round neck short-sleeve coat, shoulder armour, scarf ends hanging on the chest, leather lace, hakama, and round toed shoes (see Fig.3). Taiyuan southern suburbs mural tomb has six such figurines, Shanxi Zhang Hai Yi tomb has four, and Henan He Shao Long couple tomb has six warriors holding a shield that are basically the same as this form 2, the only difference being the headwear. There is a total of 16 form 2 figurines.

3.2.3 Shield-holding terracotta warrior figurines (form 3)

These figurines have a tri-ridged hood, narrow sleeved and neck wrapped Ming Guang armour, waist belt, tied trousers, and round toed boots (see Fig.4). Shanxi Ku Di Ye tomb has 26 such figurines, Shanxi Xu Xian Xiu has one, and the tomb of Shanxi Di Zhan tomb has two, for a total of 29 form 2 figurines.

3.2.4 Shield-holding terracotta warrior figurines (form 4)

These figurines have a hair scarf, short-sleeved lapel coat, narrow-sleeved underwear, belt with ends hanging on the back, tied trousers, and round toed shoes (see Fig.5). There are only seven form 4 figurines from Northern Hebei Wu Qiao tombs.

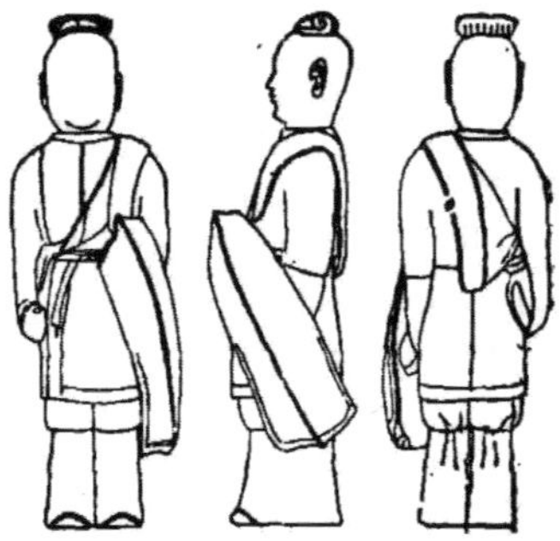

Fig.2 Yao Jun tomb's shield-holding figurine

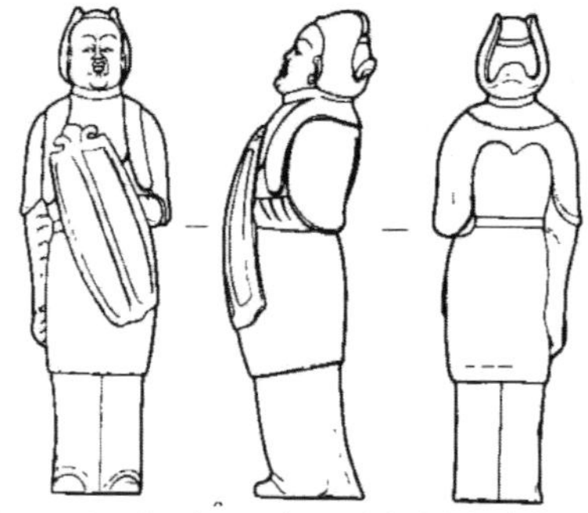

Fig.3 Zhang Hai Yi tomb's shield-holding figurine

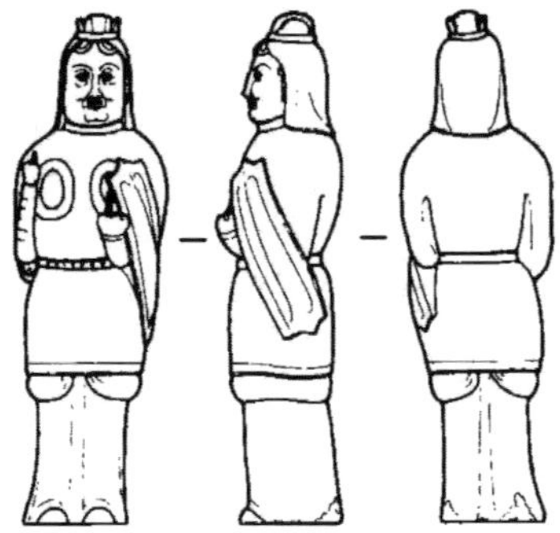

Fig.4 Xu Xian Xiu tomb's shield-holding figurine.

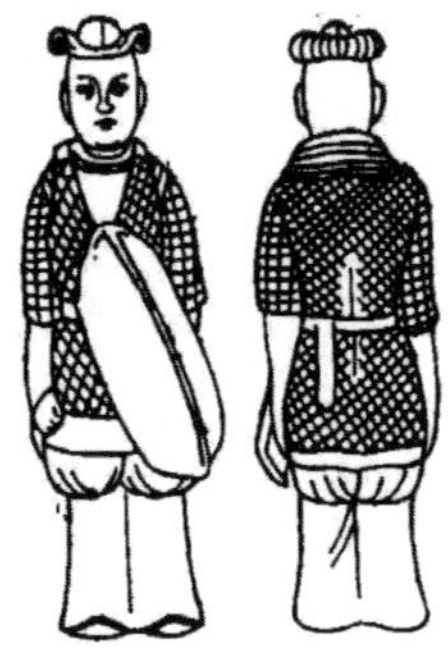

Fig.5 Wu Qiao tomb's shield-holding figurine

3.3 Clothing characteristics of tomb-guarding warrior figurines of the Northern Qi period

Tomb-guarding warrior figurines accounted for those named as "tomb warriors," "tomb figurines," and "Tomb Warrior figurines." The tombs containing the tomb-guarding warrior figurines include Shanxi Xu Xian Xiu tomb, Shanxi Zhang Hai Yi tomb, Shanxi Kai Hua village tombs, murals tomb in the southern suburbs of Taiyuan, Shanxi He Lou Yue tomb, tombs in southwest suburb of Shanxi, Shanxi Lou Rui tomb, the tomb of the Shanxi Ku Di Ye, and Shanxi Lou Rui tomb, all of which are in Shanxi Province. There are 13 tomb-guarding warrior figurines in total.

The basic form of the tomb-guarding warrior figurines includes pocket iron helmet, Ming Guang armour on upper body, shoulder armour, waist belts, tied pants on lower body, and round toed shoes (see Fig.6).

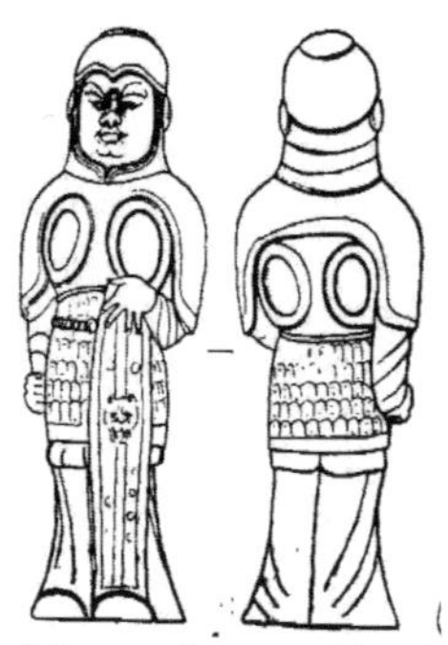

Fig.6 Lou Rui tomb's tomb-guarding warrior figurine

In spite of similarities in the basic forms, there are still differences:

1) Headwear: There are 11 figurines with dome-pocket iron helmets, two with steeple-pocket iron helmets, and two with high folded edged caps. The figurines' dress has a neck guard; four figurines have neck guard openings, and three have neck guards without openings. There are eight figurines with no neck guard on the dress, and five figurines have domes with ridges.

2) Shoulder-armour form varies: Some figurines may have two layers, but it may just be underwear folds. Some

figurines have a trim according to the text description, but it cannot be determined if there is trim based on the pictures.

3) Different ways for tying the belts, and different decoration on the belts. Nine figurines have beaded belts.

4) Eight figurines have scale Ming Guang armour, and one has tiger-skin Ming Guang armour.

3.4 Clothing characteristics of horseback-riding warrior figurines of the Northern Qi period

Tombs containing horseback-riding warrior figurines include Shanxi Lou Rui tomb, Shanxi Lou Rui tomb, and Shanxi Han Yi tomb. Based on their characteristics, they were classified into three main forms:

3.4.1 Horseback-riding warrior figurines (form 1)

These figurines have pocket iron helmet, oval ear protection on both sides, round-neck thin-sleeve underwear, Liang Dang on both chest and back, shoulder black leather armour, leather lace on waist, pants, and leather boots (see Fig.7). Shanxi Lou Rui tomb has 40 such figurines, Shanxi Lou Rui tomb has 40, Shanxi Han Yi tomb has three, all of which have the same basic form. The tomb of Hebei Gao Run has seven Jiaqijuzhuang (horseman with complete armour for both man and horse) with the same form, three of which have gold and red decoration; one Jiaqijuzhuang from Shanxi Zhang Hai Yi tomb is the same as the latter type, and three fully armoured horse-with-rider figurines from Hebei Wu Qiao tomb also have the same form. There are 94 such horseback-riding figurines in total.

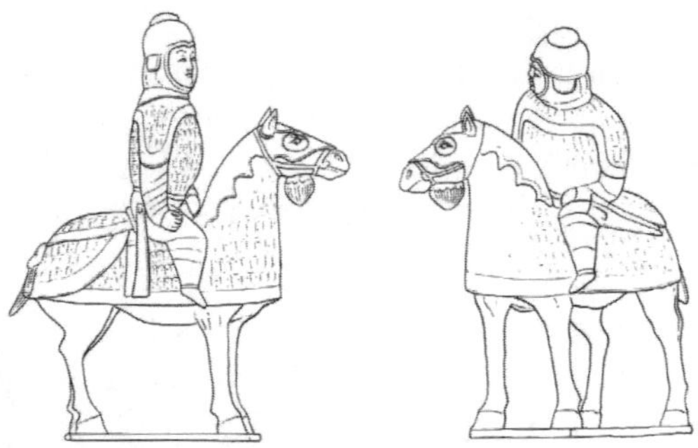

Fig.7 Lou Rui tomb's horseback-riding warrior figurines

3.4.2 Horseback-riding warrior figurines (form 2)

These figurines have a rolled edged hat, neck scarves, tight-sleeved round-neck short jacket, waist belts, breeches, and boots (see Fig.8a). Shanxi Han Yi tomb has three such figurines.

3.4.3 Horseback-riding warrior figurines (form 3)

These figurines have a high side hat with edges folded up, short jacket with round high collar and tight sleeves, waist belt, breeches, and riding boots (see Fig.8b). Shanxi Han Yi tomb has four such figurines.

a b

Fig.8 Han Yi tomb's horseback-riding warrior figurines: (a) form 2; (b) form 3

3.5 Clothing characteristics of bodyguard figurines of the Northern Qi period

Tombs containing bodyguard figurines include Hebei Gao Run tomb, the tomb of Hebei Yao Jun, Hebei Yuan Liang tomb, and Henan He Shao Long couple tomb. Based on their characteristics, they were classified into two main forms (originally there were three forms, one of which was later classified into the figurines holding shields form):

3.5.1 Bodyguard figurines (form 1)

These figurines wear a head cloth, cross-ren lapel and short-sleeve coat, round collar underwear, belt tied twice in the front, tied trousers, and round toed shoes (see Fig.9). Hebei Gao Run tomb has 11 such figurines, Henan He Shao Long couple tomb has seven, and Hebei Yuan Liang tomb has six. Some figurines have different names but the same forms. For example, of Henan He Shao Long couple tomb's seven shield- holding warrior figurines type-I, Henan Gu An No.2 tomb's A-type military officials figurines, D-type military officials figurines, and E-type military officials figurines, only one A-type figurine wears a crown, and three E-type figurines have a hair topknot. Henan Cui Fan tomb has five entourage figurines of type II, Hebei Yuan Liang tomb has four type-I arrow-bearing figurines and one type-II arrow-bearing figurine, and the Hebei Wu Qiao Northern tombs have four V-type warrior figurines with belts that are tied differently. There is a total of 50 form-1 bodyguard figurines.

3.5.2 Bodyguard figurines (form 2)

These figurines wear a head cloth, left ren long sleeved clothes, leather lace, tied trousers, and round toed shoes (see Fig.10). Henan He Shao Long couple tomb has nine such figurines. Figurines with different names but same forms include Henan Fan Cui tomb's five entourage figurines of type I, and Henan Gu An No.2 tomb's two C-type military official figurines. There is a total of 16 form-2 bodyguard figurines.

Fig.9 Gao Run tomb bodyguard figurine

Fig.10 He Shao Long tomb bodyguard figurine

3.6 Clothing characteristics of armoured warrior figurines of the Northern Qi period

Tombs containing armoured warrior figurines include Shanxi Ku Di Ye tomb and Shanxi Xu Xian Xiu tomb. Here the "armour figurines" from Shanxi Di Zhan tomb, the "armoured officials figurines" from Shanxi Zhang Hai Yi tomb, and murals tomb of the southern suburbs of Shanxi were classified into this category. Han Yi tomb's warrior figurines also fall into this category. Based on their characteristics, they were classified into two main forms:

3.6.1 Armoured warrior figurines (form 1)

These figurines wear a dome pocket iron helmet with ear protection, scale armour, thin and long sleeves, waist belt, jacket cape, cape lifted and exposed forming local twists drooping naturally, tied trousers, and round toed shoes (see Fig.11). Armoured warrior figurines have capes draped over the shoulders, and some have capes stacked on the neck. Shanxi Ku Di Ye tomb has three, Shanxi Xu Xian Xiu tomb has three, Han Yi tomb has 29, Shanxi He Lou Yue tomb has two warrior figurines with capes, and Shanxi southwest suburb tomb has 12 warrior figurines with capes, all of which have the same form. Shanxi Di Zhan tomb has three figurines with dome helmets with basically the same form; they were described as only wearing a short jacket, with no scale armour, but exactly the same contour profiles. There is a total of 52 form-1 armoured warrior figurines.

3.6.2 Armoured warrior figurines (form 2)

These figurines wear a dome pocket iron helmet with ear protection, scale armour, thin and long sleeves, waist belt, tied trousers, and round toed boots (see Fig.12). Shanxi Xu Xian Xiu tomb has 10 such figurines, Shanxi Di Zhan tomb has 10, Shanxi Lou Rui tomb has 64 with armour, all of which have the same form. Shan Xi Ku Di Ye tomb's 12 figurines with domed helmet have the same form. There is a total of 96 form-2 armoured warrior figurines.

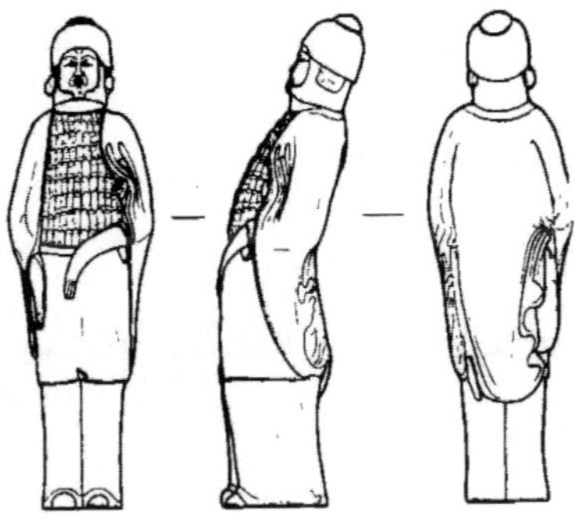

Fig.11 Xu Xian Xiu tomb's armoured warrior figurines

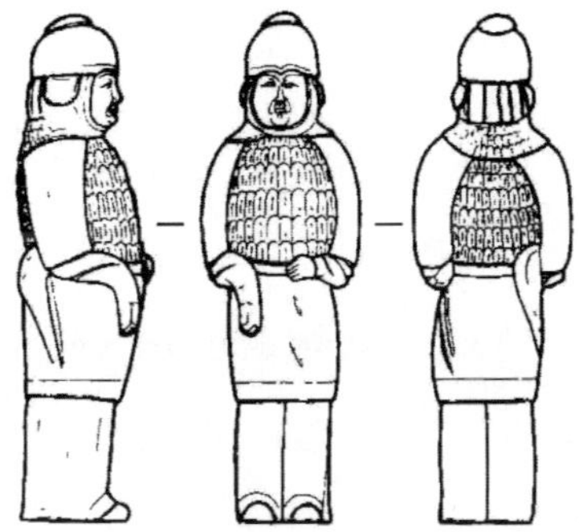

Fig.12 Xu Xian Xiu tomb's armoured warrior figurines

4. Conclusion

Based on the relevant literature, pictures, and other information of the tombs in Hebei, Henan, and Shanxi considered from a clothing perspective, this study classified the military uniforms of the Northern Qi Dynasty period into 13 major forms (see Table 1). It also generated each form's characteristics and dress pattern from the clothing perspective encompassing head wear, tops (collar, sleeve, shoulder, and belts), bottoms, shoes, and other relevant features (see Table 2). Previous researchers in archeological studies mostly named and classified the terracotta warrior figurines based on the figurines' posture, function, and identity, none of which is based on clothing. Moreover, because the archaeological reports were generated by different organizations, such studies had no uniform standard for naming. Under such circumstances involving complicated names and dress diversity, this study's findings resulted in more specialized data related to clothing forms. This is significant, because clothing form diversity amongst these figurines indicates military personnel's title and ranking; is climate-related (e.g., wearing a hood, cape to against high wind and severe cold of north); and is relevant to functional aspects of military uniforms (e.g., armour used for defensive purposes).

Table 1. Northern Qi Dynasty figurines by clothing category number and proportion

Type of figurine	No. of figurines	Proportion (%)	Head wear	Upper outer garment
Shield-pressing warrior	21	4	pocket iron helmet	Ming Guang armour
Shield-holding terracotta warrior (form 1)	103	20	head cloth	bare right shoulder
Shield-holding terracotta warrior (form 2)	16	3	folded ear hood	round neck short-sleeve coat
Shield-holding terracotta warrior (form 3)	29	6	tri-ridged hood	narrow sleeved Ming Guang armour
Shield-holding terracotta warrior (form 4)	7	1	hair scarf	short-sleeved lapel coat

Tomb-guarding warrior	13	3	pocket iron helmet	Ming Guang armour
Horseback-riding warrior (form 1)	97	19	pocket iron helmet	Liang Dang
Horseback-riding warrior (form 2)	3	1	rolled edged hat	tight-sleeved round-neck short jacket
Horseback-riding warrior (form 3)	4	1	high side hat with edges folded up	short jacket with round high collar and tight sleeves
Bodyguard (form 1)	50	10	head cloth	cross-ren lapel and short-sleeve coat
Bodyguard (form 2)	16	3	head cloth	left ren long sleeved clothes
Armoured warrior (form 1)	52	10	pocket iron helmet	scale armour/ jacket cape
Armoured warrior (form 2)	96	19	pocket iron helmet	scale armour

Table 2. Proportion of high frequency clothing of Northern Qi Dynasty figurines

Clothing	Clothing type	Proportion (%)
Head wear	pocket iron helmet	55
	hair scarf series	34
	folded ear hood	3
	tri-ridged hood	6
	rolled edged hat	1
	high side hat with edges folded up	1
Upper outer garment	Ming Guang armour	7
	bare right shoulder	20
	round neck short-sleeve coat	3
	narrow sleeved Ming Guang armour	6
	short-sleeved lapel coat	1
	Liang Dang	19
	tight-sleeved round-neck short jacket	1
	short jacket with round high collar and tight sleeves	1
	cross-ren lapel and short-sleeve coat	10
	left ren long sleeved clothes	3
	scale armour	29

References

[1] Liu YH. Chinese ancient military apparel. Beijing: Tsinghua University Publishing House, 1992; p: 90-106.
[2] Song BL. The study on common costume of Northern Dynasties. Jinan, China: Shandong University, 2008.
[3] Liu B. Wei Jin and Southern and Northern Dynasties. Taiyuan, China: Shanxi University, 2012.
[4] Tang YM, Wang YW. Investigation report of Hebei Pingshan Cui Ang tomb of Northern Qi Dynasty. Cultural Relics 1973; 11: 27-38.
[5] Wang MZ. Hebei Wu Qiao's four Northern Dynasty tombs. Cultural Relics 1984; 9: 23-38.
[6] Pan WB, Xue B, Pei T. A brief report on excavation of tomb no. 2 in Anyang County, Henan Province. Huaxia Archaeology 2007; 2: 30-38.
[7] Taiyuan Municipal Institute of Cultural Relics and Archaeology. Taiyuan Di Zhan tomb of Northern Qi Dynasty. Cultural Relics 2003; 3: 37-42.
[8] Zhang ZY. Hebei Cixian Yuan Liang's tomb of Northern Qi Dynasty. Archaeology 1997; 3: 33-39.
[9] Zhu QS. Hebei Cixian Dong Chen village Yao Jun's tomb of Northern Qi Dynasty. Cultural Relics 1984; 4: 16-22.

[10] Wang MZ. A brief report on excavation of Huanghua County Chang Wen Gui tomb of Northern Qi Dynasty. Cultural Relics 1984; 9: 39-42.
[11] Cixian Cultural Center. Hebei Cixian Gao Run's tomb of Northern Qi Dynasty. Archaeology 1979; 3: 235-243.
[12] Shanxi Institute of Archaeology, Taiyuan Municipal Institute of Cultural Relics and Archaeology. Taiyuan Xu Xian Xiu tomb excavation briefing of Northern Qi Dynasty. Cultural Relics 2003; 10: 4-21.
[13] Li XP, Yu G. Anyang He Shao Long tomb excavation of Northern Qi Dynasty. Cultural Relics of Central China 1987; 1: 8-16.
[14] Henan Museum. Excavation of the Fan Cui's tomb of Henan Anyang of Northern Qi Dynasty. Cultural Relics: 1972; 1: 47-57.
[15] Taiyuan Municipal Institute of Cultural Relics and Archaeology. Taiyuan He Ba Chang tomb of Northern Qi Dynasty. Cultural Relics 2003; 3: 11-25.
[16] Chang YM. Excavation of the Taiyuan Shentanggou He Lou Yue tomb of Northern Qi Dynasty. World of Antiquity 1992; 3: 33-38.
[17] Wang KL. Ku Di Hui Luo tomb of Northern Qi Dynasty. Acta Archaeologica Sinica 1979; 3: 377-402.
[18] Li AG. Taiyuan Zhang Hai Yi tomb of Northern Qi Dynasty. Cultural Relics 2003; 10: 41-49.
[19] Tao ZG. Shanxi Qixian Baigui Han Yi tomb of Northern Qi Dynasty. Cultural Relics 1975; 4: 64-73.
[20] Taiyuan Municipal Institute of Cultural Relics and Archaeology. Taiyuan Ku Di Ye tomb of Northern Qi Dynasty. Cultural Relics 2003; 3: 26-36.
[21] Shanxi Archaeological Research Institute. Northern Qi Dynasty tomb of Donganwang Lou Rui. Beijing: Cultural Relics Publishing House, 2006.
[22] Shanxi Archaeological Research Institute, the Tourism Bureau of Jinyuan District. Taiyuan Kaihua village cavern excavation of tomb of Northern Qi Dynasty. Archaeology and Cultural Relics 2006; 2: 7-12.
[23] Qu CF. The southern Taiyuan suburbs tombs with the Northern Qi Dynasty murals. Cultural Relics 1990; 12: 1-10.
[24] Shang TL, Zhou J, Li AG. The southwest suburb of Taiyuan, the Northern Qi Dynasty cave tomb. Cultural Relics 2004; 6: 35-46.

The Effect of Human Shoulder Slope on the Judgment of Acromion Points

Zheng-Lin Hong[1], Yan Yang[2], Shang Wu[1], Xiao-Mei Shang[1*]

[1]*College of Textile and Clothing Engineering, Soochow University, Suzhou, 215021, China*
[2]*Jiangsu Sunshine Garment Co.,LTD, Jiangsu Sunshine Group Co.,LTD, Wuxi, 214426, China*

[*]Corresponding author's email: shangxiaomei@suda.edu.cn

Abstract

The accuracy of obtaining acromion points will affect several measurements in anthropometry for garment. These measurements are important for pattern, so it's necessary to obtain accurate acromion points. There are three special shoulder slope among human body, such as sloping shoulder, flat shoulder, and high-low shoulder. Somebody have the shoulder with left and right shoulder point forward than ordinary being. In order to study the effect of human shoulder slope on the judgment of acromion points, 224 subjects were measured by experiments. Six body dimensions were measured. Through the analysis, it can be found that the shoulder slope doesn't influence obtainment of acromion points. Forward shoulder effects obtainment of shoulder point significantly. It's more difficult to obtain acromion points of forward shoulder body type than normal body type. It leads to less accurate measurements.

Keywords: Anthropometry; Shoulder slope; Acromion points

1. Preface

The size information of body parts is the foundation for tailor fashion. The Standard sizing systems for garments-Men has high coverage owing to a large number of human body measurement data [1]. The main factors to affect the accuracy of body measurement data for garment are skill level of the experimenters, measurement environment, difficulties to define datum points, motion range of the datum points, measuring path and so on. Difficulties to define datum points and range of the datum points' motion influence the accuracy of datum points' acquisition. Thus it leads to inaccurate measurement results [2,3]. The acromion points are relevant to six sizes in anthropometry, such as height of left shoulder, height of right shoulder, straight shoulder width, shoulder width, armscye girth and arm length [4]. Therefore, obtaining acromion points accurately is premise of obtaining accurate measurements of these sizes. These sizes will influence the determination of small shoulder, shoulder drop, sleeve length, arc length of armhole when clothing samples drawn.

China National Standards, Location and method of anthropometric surveys for garments, stipulate that acromion points are the most outward prominent points of lateral margin of acromion [5]. The subjects should keep standing posture with arms drooping naturally when experimenters estimate the acromion points. The experimenters stand in front of subjects to determine the lines that are respective tangent to shoulder slash and the arm contour line. The intersection of the angular bisector of the two lines with the shoulder is the acromion point. The judgment of acromion points is based on the shoulder slash. Therefore, this paper researches on the human body shoulder slope to analyze the acquisition of acromion points. In addition, the shoulder slash of special shape of forward shoulder is not easy to obtain. So it is also in consideration.

2. Anthropometric experiments

2.1 Experimental purposes and methods

This paper devoted to analyze how shoulder slope influences the judgment of acromion points. 224 subjects with different degree of shoulder slope and forward shoulder were measured. The relationship between the shoulder slope and the shoulder forward and the acquisition of acromion points can be found by analyzing the measurements.

2.2 Experimental design

The experimenters, subjects and recorders were all women in this anthropometry. The experiment had nine experimenters who are skilled and have rich measurement experience in anthropometry. Nine recorders were familiar with the measurement process. They have rich record experience and are seriously. The location was the special classroom which is used for anthropometry. The temperature and humidity were suitable. Measuring tools were tapes, Martintape (only used to measure shoulder breadth and side neck point width), set squares, rulers. In order to eliminate the tension of experimenters who were over attention in acromion points, the anthropometry took a normal process. That is the familiar process and body parts of experimenters. The experiment collected a total of 27 measurements parts size and body mass data [6,7]. Each subject was measured by arbitrary three different experimenters. The three experimenters measured the subject three times not continuously. The recorders ensured that every subject would be measured nine times validly. This experiment was carried out in two times. Experiment 1 measured 179 subjects who were first measured by the experimenters. Experiment 2 measured 45 subjects who have varying degrees of forward shoulder.

2.3 Experimental results

A total of 1611 pieces of human body measurement data had been obtained in the experiment 1, 537 pieces of data can be acquired by calculating the average of three measurements of each experimenter. Table 1 shows part data of six locations which are relative to acromion points. The locations are height of left shoulder (A), height of right shoulder (B), straight shoulder width (C), shoulder width (D), armscye girth (E) and arm length (F). The times mean that each subject was measured by three experimenters. Each measurement data in the table is the mean of one subject's three times measurement by the same experimenter. 405 pieces of data had been obtained in the experiment 2. 135 pieces of data can be acquired by calculating the average of three measurements of each experimenter.

The definition of height of left shoulder is that the subject stands upright with arms hanging and the experimenter measures the vertical height of left acromion point. The definition of height of right shoulder is that the subject stands upright with arms hanging and the experimenter measures the vertical height of right acromion point. The definition of straight shoulder width is that the subject stands upright with arms hanging and the experimenter measures the horizontal distance between left and right acromion points. The definition of shoulder width is that the subject stands upright with arms hanging and the experimenter measures the horizontal curve length between the left and right acromion points. The definition of armscye girth is that the subject stands upright with arms hanging and the experimenter measures the closed curve length goes through acromion point, anterior armpit point and posterior armpit point. The definition of arm length is that the subject stands upright with arms hanging and the experimenter measures the curve length goes through acromion point, radiale and ulna stylion.

Table1. Part of data of part locations

subjects	times	A/cm	B/cm	C/cm	D/cm	E/cm	F/cm
	1	124	123.3	38.7	44.5	33.2	50.5
001	2	124.5	125	38.5	46	33.2	50
	3	124	123.5	38.4	43.6	34.6	50.1
	1	126.4	126	38.5	42	32.5	53
002	2	125	125	39	44	32.5	53
	3	126.2	125.5	37.7	44.2	33.5	51.6

2.4 Experimental analysis

2.4.1 Figure analysis

Fig.1 Diagram of shoulder slope

Position change of acromion points of different shoulder slope were drawn according to the definition of acromion points. As the figure 1 shown, if the arm slope keep unchanged, the position of acromion points will move downward as the shoulder inclined more and more. Acromion points are intersections of the angular bisector of the lines which are respective tangent to shoulder slope line and arm contour line with the acromion. The intersection of the tangents of shoulder slope line and arm contour line will move downward with the increase of the degree of shoulder slope. The downward intersection causes the downward acromion points.

2.4.2 Data analysis

Variances of six parts' data of each subject were calculated. The six parts are height of left shoulder (A), height of right shoulder (B), straight shoulder width (C), shoulder width (D), armscye girth (E) and arm length (F). The experiment concentrated on the relation of angle of shoulder slope and judgment of acromion points. However, special tools didn't been used to measure angle of shoulder slope in the experiment. So a value was calculated to represent degree of shoulder slope. The value of height of side neck point minus height of left acromion point was denoted as X. The value of straight shoulder width minus width of side neck points was denoted as Y. Then the value of X divide half of Y was regard as representation of left shoulder slope. The representative value of right shoulder slope was calculated the same method as left shoulder slope. Table 2 shows the correlation of variance of the six parts and representative value of left and right shoulder slope.

The representative value of left and right shoulder slope are highly positive correlate with angle of shoulder slope [8,9]. Bigger the representative values are, bigger the angles of shoulder slope are. Conversely, smaller the angles are. Through the Table 2, it can be found that variance of shoulder width and representative value of left shoulder slope has significantly negative correlation at the 0.05 level (2-tailed). Variance of armscye girth and representative value of right shoulder slope has significantly negative correlation at the 0.05 level (2-tailed). Other variances are not significantly correlated with representative value of left and right shoulder slope. Variance characterizes the accuracy of measurement data. That is to say, variance is the system error of the different experimenters measuring body size. The greater the variance, the lower accuracy the measurement data are. Versa, the smaller variance indicates the higher accuracy of the measurement data. Variance of shoulder width and representative value of left shoulder slope showed a negative correlation. That means the larger angle the left shoulder slope is, the smaller the variance of shoulder width is and the higher the accuracy of measurement data are. Conversely, the accuracy of measurement data is lower. That is to say, it's easier to judge the left acromion point when angle of the left shoulder slope is larger. The size of the shoulder width is jointly determined by the left and the right acromion points. However, shoulder width and representative value of right shoulder slope was not significantly correlated. Therefore, we can't draw the conclusion that larger the angle of shoulder slope is, more accurate the shoulder width measurement data are. Variance of armscye girth and representative value of right shoulder slope showed a negative correlation. It means that the larger the angle of right shoulder slope is, the higher accuracy the armscye girth measurement data are. Conversely, the accuracy of measurement data is lower. That is to say, it's easier to measure the arescye girth and more accurate to judge right acromion point under the larger angle the right shoulder slope is. Right arm was measured to obtain armscye girth in the experiment. It's normal that the variance of armscye girth and representative value of left shoulder slope was not significantly correlated. Other variances and representative value of left and right shoulder slope were not significantly correlated. The different angles of shoulder slope have no significant effect on the judgment of acromion points.

Table 2. Correlation of variance of measurement data and representative value of shoulder slope

		Variance of A	Variance of B	Variance of C	Variance of D	Variance of E	Variance of F	RL	RR
Variance of A	Pearson Correlation	1	0.461**	0.103	0.003	-0.098	0.025	0.023	0.057
	Sig.(2-tailed)		0.000	0.169	0.967	0.191	0.737	0.759	0.447
	N	179	179	179	179	179	179	179	179
Variance of B	Pearson Correlation	0.461**	1	0.165*	0.004	0.094	-0.022	-0.002	0.028
	Sig.(2-tailed)	0.000		0.028	0.962	0.210	0.766	0.979	0.712
	N	179	179	179	179	179	179	179	179
Variance of C	Pearson Correlation	0.103	0.165*	1	-0.076	-0.035	-0.042	-0.069	-0.125
	Sig.(2-tailed)	0.169	0.028		0.311	0.645	0.576	0.360	0.095
	N	179	179	179	179	179	179	179	179
Variance of D	Pearson Correlation	0.003	0.004	-0.076	1	0.207**	0.173*	-0.160*	-0.111
	Sig.(2-tailed)	0.967	0.962	0.311		0.005	0.020	0.032	0.140
	N	179	179	179	179	179	179	179	179
Variance of E	Pearson Correlation	-0.098	0.094	-0.035	0.207**	1	0.182*	-0.130	-0.164*
	Sig.(2-tailed)	0.191	0.210	0.645	0.005		0.014	0.084	0.028
	N	179	179	179	179	179	179	179	179
Variance of F	Pearson Correlation	0.025	-0.022	-0.042	0.173*	0.182*	1	-0.073	-0.008
	Sig.(2-tailed)	0.737	0.766	0.576	0.020	0.014		0.334	0.911
	N	179	179	179	179	179	179	179	179
RL	Pearson Correlation	0.023	-0.002	-0.069	-0.160*	-0.130	-0.073	1	0.770**
	Sig.(2-tailed)	0.759	0.979	0.360	0.032	0.084	0.334		0.000
	N	179	179	179	179	179	179	179	179
RR	Pearson Correlation	0.057	0.028	-0.125	-0.111	-0.164*	-0.008	0.770**	1
	Sig.(2-tailed)	0.447	0.712	0.095	0.140	0.028	0.911	0.000	
	N	179	179	179	179	179	179	179	179

**. Correlation is significant at the 0.01 level (2-tailed),*. Correlation is significant at the 0.05 level (2-tailed). The representative values of left and right shoulder slope were respectively denoted as RL and RR.

In experiment 2, subjects have special shoulder which is more forward than normal shoulder. 45 subjects with different level of forward shoulder were measured in experiment 2. The variances of the measurement data of height of left shoulder, height of right shoulder, straight shoulder width, shoulder width, acromion girth and arm length were calculated as experiment 1. The variances are compared with corresponding parts of variances in experiment 1. The 179 sets of variances in experiment1 and 45 sets of variances in experiment 2 have been carried out by an independent T test of two samples with using SPSS. Experiment 1 was denoted as condition 1 and experiment 2 was denoted as condition 2. The analysis results are shown in Table 3 and 4.

Table 3. The variances of the measurement data

	Two Conditions	N	Mean	Std. Deviation	Std. Error Mean
Variance of A	1	179	0.4904	0.43932	0.03284
	2	45	0.6962	0.36855	0.05494
Variance of B	1	179	0.5103	0.47555	0.03554
	2	45	0.7591	0.39455	0.05882
Variance of C	1	179	0.7204	0.67615	0.05054
	2	45	0.7120	0.40666	0.06062
Variance of D	1	179	0.9457	0.88141	0.06588
	2	45	1.3918	0.57074	0.08508
Variance of E	1	179	0.8164	0.64596	0.04828
	2	45	0.8896	0.46428	0.06921
Variance of F	1	179	0.9363	0.78401	0.05860
	2	45	0.9967	0.45909	0.06844

Table 3 shows the basic statistics that are mean and standard deviation of the variances of the measurement data. The mean values show the overall size of the sample. The standard deviation shows the degree of dispersion of the sample. It can be seen in Table 3 that the degree of dispersion of the data in experiment 1 is greater than that in experiment 2. The mean values of variances of straight shoulder width, armscye girth and arm length in two experiments are very close. The mean values of variances of height of left shoulder, height of right shoulder and shoulder width in experiment 2 are bigger than that in experiment 1. Table 4 shows the results of independent sample T test of the same parts of the two experiments. In the equal variance assumed line, if the significant value is bigger than 0.05, which can be considered as the variance is equal, the results of the line are valid. Conversely, the results of the line of equal variance are not assumed valid. When the significant value (2-tailed) is bigger than 0.05, it shows that there is no significant difference in the two conditions. There is a significant difference when it was bigger than 0.01. And there was a very significant difference when it is smaller than 0.01.

According to the analysis results of Table 4, the variances of straight shoulder width, armscye girth and arm length have no significant difference in experiment 1 and 2. And variances of height of left shoulder, height of right shoulder and shoulder width have extremely significant difference. That is to say, the accuracy of measurement data of height of left shoulder, height of right shoulder and shoulder width had very significant differences in experiment 1 and 2. The measurement data of height of left shoulder, height of right shoulder and shoulder width in experiment 2 are more accurate than that in experiment 1 combined with the mean comparison in Table 3. This shows that the condition of forward shoulder will affect the judgment of acromion points and make it more difficultly to judge the position of acromion points. It goes through several datum points when measuring armscye girth and arm length. The measurement path is complex, so the accuracy of judgment of the acromion points influence the accuracy of the two parts' measurement low. According to the diagram of acquisition of acromion points, the horizontal movement is smaller than the vertical movement of the position of acromion points when acromion points move up and down in a small range. Therefore, the accuracy of straight shoulder width measurement data varies little in the conditions of forward shoulder and different degree of shoulder slope.

Table 4. Independent samples T test results of the variances of the measurement data in two conditions

		Levene's Test for Equality of Variances		t-test for Equality of Means						
									95% Confidence Interval of the Difference	
		F	Sig.	t	df	Sig. (2-tailed)	Mean Difference	Std. Error Difference	Lower	Upper
Variance of A	Equal variances assumed	1.975	0.161	2.895	222	0.004	0.20578	0.07108	0.06570	0.34585
	Equal variances not assumed			3.215	78.571	0.002	0.20578	0.06400	0.07837	0.33318
Variance of B	Equal variances assumed	1.197	0.275	3.239	222	0.001	0.24878	0.07681	0.09740	0.40016
	Equal variances not assumed			3.620	79.391	0.001	0.24878	0.06872	0.11200	0.38555
Variance of C	Equal variances assumed	8.974	0.003	-0.080	222	0.936	-0.00845	0.10538	-0.21612	0.19923
	Equal variances not assumed			-0.107	112.929	0.915	-0.00845	0.07892	-0.16481	0.14792
Variance of D	Equal variances assumed	2.433	0.120	3.226	222	0.001	0.44603	0.13827	0.17355	0.71851
	Equal variances not assumed			4.145	103.391	0.000	0.44603	0.10761	0.23263	0.65943
Variance of E	Equal variances assumed	4.549	0.034	0.714	222	0.476	0.07315	0.10243	-0.12871	0.27501
	Equal variances not assumed			0.867	91.868	0.388	0.07315	0.08439	-0.09445	0.24075
Variance of F	Equal variances assumed	12.748	0.000	0.495	222	0.621	0.06035	0.12193	-0.17994	0.30064
	Equal variances not assumed			0.670	116.670	0.504	0.06035	0.09010	-0.11809	0.23879

2.5 Experimental conclusion

The variance of shoulder width and representative value of left shoulder slope has significantly negative correlation at the 0.05 level (2-tailed). The variance of armscye girth and representative value of right shoulder slope has a significant negative correlation at the 0.05 level (2-tailed). And other variances of measurement data are not significant correlate to representative value of left and right shoulder slope. Therefore, different degrees of shoulder slope have no correlation to the judgment of acromion points. The variances of measurement data of height of left shoulder, height of right shoulder and shoulder width in different degrees of shoulder forward are significantly different with the variances in different degrees of shoulder slope. And the variances in condition of shoulder forward are bigger than that in condition of shoulder slope. It shows that condition of shoulder forward influences judgment of aromion points deeper and leads judgment of acromion points more difficult. Measurement error was larger in condition of shoulder forward.

3. Summary and prospect

There are many factors that affect the accuracy of the measurement results in the process of anthropometry for garment, such as environmental factors, psychological factors, skill levels, measurement path and so on. The paper concentrates on the acquisition of acromion points in different characteristics of the human body. The influence of shoulder slope and forward shoulder on the judgment of acromion points was studied. Experimental results show that degrees of shoulder slope don't influence the judgment of acromion points significantly. And degrees of forward shoulder influence the judgment of acromion points significantly. Therefore when the subjects have forward shoulder, the experimenters should measure them more carefully [10]. And it also need to add some measurement to improve accuracy of the measurement, such as measuring the arc length of two acromion points in the front of the subjects and the value that the left and right shoulder forward than normal people. These added dimensions also improves the accuracy of garment pattern [11,12].

Anthropomrtey for garment is one of the important ways to acquire the human body size. The accuracy of measurement data is very important. It has a great influence on the pattern drawing, garment manufacturing and data analysis.

References

[1] GB/T 1335.1-2008. The Standard Sizing Systems for Garments-Men [S].

[2] Shang XM, Xing XY, Gao PP. Hierarchical Analysis Method and Model Based Data of Controlling Parts of the Body for Clothing [J]. Journal of Fiber Bioengineering and Informatics, 2014; 7(4): 479-494.

[3] Hong ZL, Li PW, Shang XM. Psychological Factors Affect Measured Data of Human Body [J]. Textile Bioengineering and Informatics Symposium Proceedings, 2015: 122-127.

[4] GB/T 16160-2008. Location and Method of Anthropometric Surveys for Garments [S].

[5] International Organization for Standardization, Garment Construction and Anthropometric Surveys-Body Dimensions [S].

[6] Shen L. Pattern Adjustment Based on Special Body Type [J]. Tianjin Textile Science & Technology : 38-40.

[7] Wang FL. the Structure Balance of Shoulder Slope [J]. Melliand-China, 2005; 02: 82-83 +66.

[8] Pan CY, Lv J, Liang HE, Yao Y. the Measurement of Female Body's Shoulder Slope and the Deduction about Its Anti-trigonometric Function [J]. Progress in Textile Science & Technology, 2011; 06: 69-72.

[9] ISO 3635 1981.International Organizat ion for Standardization, Size Designation of clothes-Definitions and Body Measurement Procedure [S].

[10] Fang F, Zhang WY, Zhang WB, Lang J. Study on Standards of the Human Body Measurement [J]. Journal of Donghua University, 2005(1): 132~138.

[11] Jin J, Luo XJ, He WJ. Study on the Relationship Between the Shoulders's Modeling and Structure in Lady's Wear [J]. Advanced Textile Technology, 2010; 06: 12-14.

[12] Wang QM, Zhang WY. Comparision and Study on Datum Point of Anthropometry for Garment Industry [J]. Jiangsu Textile, 2006; 12: 39-41.

Uniform Design of Laifeng County Ethnic Primary School

Ye-Wen Wu, Wei Tian*

Xian Polytechnic University, No.19 South Jinhua Road, Xian, Shaanxi, 710048, China

*Corresponding author's email: 569790315@qq.com

Abstract

Nowadays, school uniforms have a special form of culture as it transmits the school's atmosphere, culture and spirits. This article is about taking which Laifeng County primary school of Tujia area as the object of the research, analyse the previous ethnic uniforms, and design ethnic uniform according to a series of characteristics of the Tujia Costume.

Keywords: Ethnic costume and decoration; Dress design; Chinese ethnic culture; School uniform

1. introduction

School uniforms minimise discrimination as you cannot tell the difference between students' economic backgrounds or family situations. Uniforms are unify and strengthen the overall image of the school and enhance the collective sense of honour; school uniforms are symbolic, reflecting the atmosphere of culture and spirit of a school.

In the early period of the 1911 Revolution, uniforms were introduced to China under the influence of Western culture. During the 1950s, a specific historical period, most school uniforms were blue or grey cadres outfits, Lenin outfits. Grass green color old uniforms became the most popular clothing in infrequent political movements in the 1960 and 70's. Since the reform and opening up, a variety of school uniforms were born: simple, loose cut sportswear of blue (green) and white with no sense of beauty or design was the most popular style, regardless of gender [4]. Therefore, modern school uniforms reflect the social situation at that time, and became the standard uniform for Chinese students. With the development of the society, school uniforms stand for a school's culture. In the 21st century, British style school uniforms are popular in private schools and prosperous, economically-developed areas, where the uniforms reflect the students' temperament and the students look more beautiful [6]. The men's style suits include trousers, skirts, and women's include suit, skirt, and stockings.

Schools require students to wear uniforms, which reflects the overall style of a school. It concerns a school's image and represents the students' youth, vitality and spirit. According to the particular culture of a school, different areas of the school uniforms will also reflect the local conditions and customs.

Some national school uniforms in ethnic minority areas show the national style and promote national culture. For example, (Fig.1) in Tawakkul township, Heitian country, Xinjiang province, students wear Uygur ethnic costumes, including the head wear Huamao, blue and white cloth stitching skirt with red and yellow patterned cloth. It combines strong ethnic characteristics with a sense of modern fashion style! Tujia ethnic school uniforms of Tianzishan Center School in Zhangjiajie city, Hunan province are red: jackets with a broad chest brimmed with black embroidery line segmentation, cuffs, the hems have embroidery broadside, and the black bottoms embroidery belt is the highlight. The happy students in Tongdao Dong autonomous country, Hunan province are sporting ethnic school uniforms which consists of a fake two piece shirt and pleated skirts. The students wear blue and white ethnic costumes of matching collar embroidery pattern, characteristic of th Tujia costume in Enshi Tujia and Miao autonomous prefecture, Hubei province. But, there still is no sense of wear ethnic uniform, few students wear ethnic costume in minority area.

Fig.1 National costume

In China, the majority of middle and primary schools, including ethnic schools, have are blue (green) and white sports school uniforms. (See Fig.2) Uniforms for these schools are only unified uniform dress. The definition of primary school uniform is not blue and white, not a work clothes, not a sports suit [4]. Sports uniform is different from structure line of cloth, or the name of school of back cross (运动校服的区别在于结构分割线或者衣服后背学校名称的不同). China is a country spanning vast territory and has a large population. There are 56 minority nationalities which have different local cultures, religions and costumes. Schools in each region have their own characteristics; not only the concept of the school motto, but also distinct district characteristics. Students' daily activities include running, jumping, squatting, and studying, so the choice of fabric must be comfortable, air permeable and elastic. At the same time, uniform design should also reflect schools' characteristics to act as a clear marker of identity.

Fig.2 Green (blue) and white sports school uniform

2. Tujia minority nationality

2.1 Tujia costume and decoration

Tujia people claim to be "Bizika", meaning native dweller. Nowadays, the Tujia people mainly reside in Hunan, Hubei, Chongqing and Guizhou provinces, such as Hunan Xiangxi Tujia and Miao Autonomous Prefecture, Hubei Enshi Tujia and Miao Autonomous Prefecture, Shizhu,, Xiushan, Qianjiang, Youyang and Yinjiang.

Young Tujia men often wear jackets with short collars, Pankou (a handmade button of made by cloth) with pockets on both sides and wide trousers. They often wear sandals made of straw and tie white scarves on their heads, while the older men prefer wearing black, and hangs smoking pipes on their scarves. Tujia women also wear scarves that are larger in size compared to the men's [1]. Tujia people live a simple life in the mountains, where the main colours of their clothes are black and blue.

2.2 Xi Lan Ka Pu

Xi Lan Ka Pu is a Tujia brocade. In Tujia language, "Xi Lan" means blanket and "Ka Pu" means flowers; the symbol of a Tujia flower blanket. Tujia girls learn to weave blankets using the picking weaving technique from their mother when they are young. Blue, black, red, white and other colored cotton threads are wrap of cloth. A variety of colors of silk, cotton, and wool weft fills the yarn on the wood loom. When a girl grows up she must have her own hand woven Xi Lan Ka Pu, which will be passed down from generation to generation [5].

[3] Hook patterns are a major characteristic of Xi Lan Ka Pu. Hook patterns such as the 48 hook (Fig.3 and 4), 24 hook, 12 hook, double hook and the single hook are used to illustrate Tujia folk tales, such as the Tujia's Valentine's day; the dance of Maogusi (a celebration of their ancestors); the Tujia hand-waving dance, a ceremony which is blind dateand mice's daughter weeding; pattern of animal plants, such as the tiger flower and the snake flower. The main colors of Xi Lan Ku Pu are red, yellow, blue, and black.

Fig.3 Patterns of the Tujia brocade

Fig.4 Patterns of the Tujia brocade

3. School uniform design

3.1 Design object

Laifeng County Ethnic Primary School is located in Laifeng County, Enshi Tujia and Miao Autonomous Prefecture, Hubei Province. Founded in 1951, it changed its name formally in September, 2000. When the school was first established, the name of the school illustrated the special nature of the region as well as reflecting national culture. In recent years, school uniform of ethnic schools have become standardized with schools in other areas in China - the blue (green) and white sportswear, changed the structure of cloth (调整衣服的结构线).

In 2003, (Fig.5) Laifeng country government call for all students to take part in the Hand Waving Festival. Students wear ethnic costume with low collars and Pankou (a handmade button of made by cloth). Men's clothing is symmetrical in design, with an embroidered pattern and a black line on the cuffs and trousers. Women's clothing style includes a blouse of right slant opening (右斜襟) and pants. (Fig.6) In 2013's Children's Day, students of Ethnic Primary School wore ethnic costumes. Women wore right slanted shirts and yellow A-line skirts with two embroidered hanging belts lined in black. Men wore embroidered black pants and black shirts with a small yellow patch that matched the women's.

Fig.5 2003 costume style

Fig.6 2013 costume style

3.2 Purpose design

In Laifeng County ethnic primary school, ethnic costumes are usually worn in festivals. While students are required to wear their blue and white sports uniforms in the weekly flag raising ceremony, they wear their own clothes on weekdays. In order to strengthen and carry forward the Tujia culture, the re-design of the school's uniform to replace the blue (green) and white sports clothes involve incorporating representative elements of the Tujia culture into the new design. The new uniforms should have distinct ethnic characteristics so that students can understand and contact their culture intuitively in the campus environment. This will play a positive role in protecting ethnic culture.

3.3 Design concept

Breathable and comfortable fabric is the basic requirement of school uniform design. Clothing designs based on Tujia traditional designs are combined with modern clothing features to ensure the clothing meets the needs of the students in their daily activities. Drawing from the clothing patterns of the Tujia's Xi Lan Ka Pu, clothing colors will reflect the characteristics of Tujia costumes.

3.4 Design description

The new ethnic uniform design differs for boys and girls and each has different designs for the summer and autumn seasons.

3.4.1 Summer wear

Laifeng County belongs to the subtropical monsoon mountain climate with four distinct seasons. Summer is hot and humid. With the development of modern textile technology, natural fiber has become the darling of summer. Breathable, moisture wicking fabrics such as silk, cotton, and hemp are ideal. The design of the women's shirt is of a T-shirt style tapered at the waist to accentuate female curves. Women's pants are longer than men's, but with a wider leg. The design of the clothes are of Tujia men and women wearing the traditional Tujia costume; the boy wears a head scarf and the girl has neatly combed hair. The uniform is mainly white because it is the least endothermic compared to other colors.

Fig.7 Blueprint of garments of summer wear

Fig.8 Design drawing of summer wear

3.4.2 Autumn wear

The Autumn uniform puts the classic Xi Lan Ka Pu patterns of "Yan Qiang flower" and "Tiger flower" into design. The "Yan Qiang flower" pattern originated from the flower of Tujia ancestors. The Yan Qiang flower is highly adaptable to its surrounding environment and symbolizes the rough character of the Tujia people [2]. As for the tiger flower, it is in fact a mythical plant that symbolizes good wishes. Man's is classic style of Tujia national costume. It is a fake two piece to reduce production costs. Sleeveless and embroidered, it is a versatile piece of clothing. The women's uniform on the other hand, sports a low collar. The Pankou is removed, and a modern is T-shirt design with a structure line and cloth stitch was chosen. Black, red, white are the classic colors of Tujia costumes. Men's wear is black and white, and women's clothes are red and black in design.

Fig.9 Effect drawing of autumn wear

Fig.10 Design drawing of autumn wear

Fig.11 "Suo Luo flower" pattern

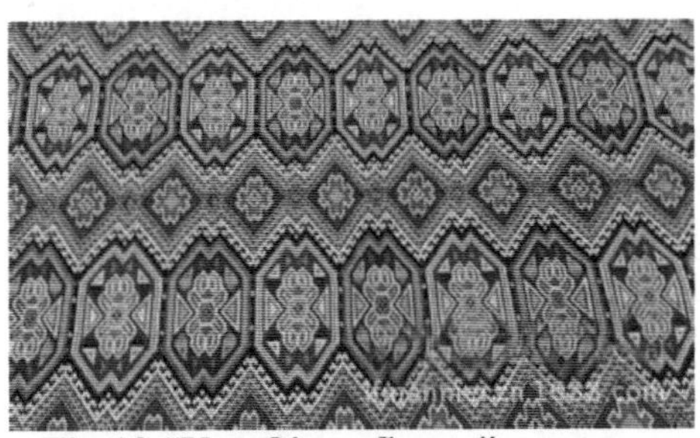

Fig.12 "Yan Qiang flower" pattern

4. Conclusion

The intangible cultural heritage of the Tujia's ethnic costume and Xi Lan Ka Pu is passed down from generation to generation. It is the product of a combination of material civilization and spiritual civilization. It gave a good meaning by combining things in real life, ethnic pattern and Xi Lan Ka Pu. It will promote and carry forward national culture, strengthen student's cultural connotation, and smriti the classical sense when ethnic culture is embodied in the school uniforms.

Due to limited time, the author was not able to investigate in school uniforms in most areasin China. And the time span is slightly larger, collect ethnic minority clothing materials is not sufficient., hence the analysis is not comprehensive enough. Therefore, the author analyzed the recent national costume of Ethnic Primary School. Meanwhile, design of the ethnic costume is just a draft. Tt will be better if the author makes a sample of the design, for the students to try on.

5. Comment

Fig.1.2.3.4.11.12. from the network, author take the picture of Fig.6, and drawing Fig.7.8.9.10. in CORELDRAW software.

References

[1] Tang Y. In Western Hubei Traditional Dress of Tujia in Modern Apparel Elements Applied Research. Master Degree dissertation of Wuhan Textile University 2012; p: 15.

[2] Zhang D. The Application of the Xi Lan Ka Pu Traditional Pattern to Modern Art Design. Master Degree dissertation of Shaanxi University of Science & Technology 2011; P: 22-23.

[3] Chen WW, Li Kaide. Tujia Brocade Treasure - Xi Lan Ka Pu. Journal of China Three Gorges University (Humanities & Social Sciences) 2010; 03: 32-2, p.6.

[4] Yu X. Study of Daily Protection Design and Its Development of Elementary School Uniforms in China. Master Degree dissertation of Dong Hua University 2014; p: 7-16.

[5] Wang XL. Grassroots power - Analysis Artistic Characteristics of Xi Lan Ka Pu of Tujia. Conference Papers 2006; p: 640-641.

[6] Zhou S. Study on the Design and Application of School Uniforms in Wuhan Area. Master Degree dissertation of South-Central University for Nationalities 2012; p: 5-12.

Analysis of Human Factors Inherited by Baiku Yao Costume Forms

Ling-Ling Xu, Hao Zhang*, Pan-Pan Li, Jing Guo

Aimer Human Engineering Research Centre, Beijing Institute of Fashion Technology, Beijing, 100029, China

*Corresponding author's email: zhhh1cn@126.com

Abstract

With the rapid development of global economies and increasing worldwide trade among nations, traditional costumes of many ethnic groups have disappeared. However, the Baiku Yao culture has maintained its primitive, ecological, and traditional forms of costumes, which merits further study. This study systematically investigated the development of Baiku Yao cultural costumes using qualitative research methods comprising a literature review and field investigation. The study used semi-structured interviews with 11 Baiku Yao people to explore human factors underlying traditional Baiku Yao costume forms from the perspective of physiological and psychological relationships of costumes and people.

Keywords: Baiku yao; Costume forms; Inheritance; Influencing factors

1. Introduction

Baiku Yao is an old branch of the Yao ethnic group, distributed mainly in the rocky mountainous area across the border region of Guizhou and Guilin. Baiku Yao people, who name themselves "Naogelao" or "Duonu", make a living from agricultural activities. The ethnic group name "Baiku Yao" originates from the knee-high white pants (pronounced as Baiku in Chinese) worn by Baiku Yao men throughout the year, and the two-piece Yao (or "karasilis") worn by Baiku Yao women during the summer. It is one of few ethnic groups whose name is derived from costumes. The Baiku Yao ethnic group is still a so-called "frying-pan" organisation (of people with the same ancestry who cook meals in the same pot) that is composed of families with shared kinships who unite together and help each other [1]. By far, Baiku Yao culture still maintains old cultural patterns characterised by widespread natural, spiritual, and ancestral worship practices.

In 1986, Wei recorded his observations during winter in Nandan region and described primitive folk scenes among the Baiku Yao that included playing with spinning tops, the sacrificial killing of bulls, and communal intoxication [2]. Wei's study vividly described Baiku Yao men's and women's clothing during winter: all men wear a black, homespun upper outer garment, a homespun belt around the waist, and knee-high five-finger print white pants; all women also wear a black homespun upper outer garment, but with a knee-high blue and white striped homespun skirt. Early in 1989, Yu researched Baiku Yao culture in a book titled Baiku Yao Society. The study provided a comprehensive analysis of the Baiku Yao ethnic group's history, religious beliefs, and culture and art using an anthropological research method, with the investigation of Baiku Yao costumes relying mainly on recorded observations. Yu's study also introduced categories such as wax printing and couching techniques for men's and women's costume embroidery [3]. In 2013, Sun Ying researched Baiku Yao costumes in the Nandan area in Guangxi on the basis of field investigation [4]. The study recorded categories of forms, colours, patterns, and technical characteristics of Baiku Yao costumes by means of illustration and summarised characteristics in costume shaping and colour matching.

Upon the review on the aforementioned literature, it is very noteworthy that the studies' research methods employed a combined investigation of literature and material objects. However, such research was mostly concerned with artistic and material characteristics of the garments and gave little consideration to the reasons for the continuous inheritance of costume forms.

2. Source of Baiku Yao costume forms

Baiku Yao costumes' functions include recording ancient events, honouring ancestors, and inheriting tradition, but the specific source of the costumes has not yet been confirmed [5]. The Baiku Yao ethnic group's own description essentially constitutes folklore akin to a fairy tale: a long time ago, Baiku Yao people wore no clothes and covered their bodies and kept out the cold with taro leaves; afterwards, an immortal gave directions and presented cotton seeds to them, from which they began to make clothes made with cotton cloth. Thus, this section focuses on tracing the Baiku Yao culture's costumes from relevant historical and archaeological materials.

2.1 Archaeological findings

In the 1980s, researchers investigated more than 30 ancient tombs in caves of Yao Village, Lihu Nandan, County, Guangxi and collected remnants of textiles (including clothes, pants, skirts, and scarves) buried with the dead [6]. Expert appraisal determined that nearly all of the textile artefacts were made of silk fabrics. Some of the pieces were produced during the Ming Dynasty (1368-1644), while others were undated. Among such pieces of cloth and silk buried with the dead, researchers identified pieces of Baiku Yao women's skirts [7]. We thus can infer from the archaeological results that the Baiku Yao people lived in Nandan, Guangxi at least as far back as the Ming Dynasty, so the history of Baiku Yao costume and silk culture also can be traced to that time.

2.2 Written records in ancient books

The Baiku Yao people do not have any written language and instead pass the history of their ethnic group orally from generation to generation or record it in their costumes with the language of imagery. Currently, Baiku Yao ethnic history and costume development can only be gleaned from existing Chinese books, mostly local chronicles and travel notes compiled by the ruling class. Such materials reveal that: 1) As Baiku Yao culture developed independently from its parent culture early on, the existing distinctive Baiku Yao costume developed gradually based on regional differences in living habits and other cultural characteristics; however, Baiku Yao costumes have similarities with certain Yao and Miao cultural costumes, or those of other branches of the latter parent cultures. 2) Currently available descriptions of costumes similar to those of Baiku Yao men can be traced back to the Qing Dynasty, and in case of women's costumes, to the Ming Dynasty; thus, it can be said that the characteristics of Baiku Yao men's black short gown and white pants and women's black gown, two-piece karasilis, and flowery skirt were first developed centuries ago.

2.3 Images

Table 1. Baiku Yao men's costumes

	A	B	C	D
Baiku Yao men's costumes:				
Timeline:	1763 Qing Dynasty	1786 Qing Dynasty	1939 Republic of China	2014
Sources:	Tributes to the Imperial Ch'ing [8]	Illustrations of Yi People [9]	Provided by museum	Photographed by author

Table 2. Baiku Yao women's costumes

	A	B	C	D
Baiku Yao women's costumes:				
Timeline:	1763 Qing Dynasty	1786 Qing Dynasty	1988	2013
Sources:	Tributes to the Imperial Ch'ing [8]	Illustrations of Yi People [9]	Baiku Yao in China	Photographed by author

Columns A and B in Tables 1 and 2 include images from Tributes to the Imperial Ch'ing and Illustrations of Yi People [8,9] respectively, while columns C and D show more recent photos. The images show that the earliest graphic illustrations relevant to Baiku Yao costumes are dated to the Qing Dynasty, and that Baiku Yao costumes to date have not changed much in form or in colour since at least the 18th century.

3. Analysis of inheritance of costume forms based on interviews

We employed an interview method to further analyse Baiku Yao costume forms. The specific purpose of interviews was to: 1) review the inheritance and evolution of costume forms from the perspective of Baiku Yao people; and 2) analyse the humanistic reason for the inheritance and evolution of costume forms from the perspective of the physiological and psychological relationship between costume and person.

3.1 Interview and interviewees

We employed a semi-structured interview method in which the researcher prepares an interview outline according to the research purpose and carries out the interview mainly around questions listed in the outline. However, interviewees can freely express their views and adjust the interview procedures at any time [10]. Each interview was carried out at the interviewees' respective homes to reduce any distractions from the interview environment. Interview lasted 30 to 60 minutes, and interviewees reviewed and consented to the use of sound recordings. All sound and written recordings were transcribed post-interview into words for statistical and analytical purposes. The interviews addressed the following three topics: 1) What are the inheritances and evolutions of ethnic costumes, and why is this so? 2) What rules govern the selection of costume structure and size? 3) What are your physiological and psychological feelings when you wear ethnic costumes?

The researcher selected 11 villagers from Yao Village, Lihu Nandan, County, Guangxi as interviewees, representing 4 men and 7 women, aged from 18 to 90, and average length of interviews was 41 minutes. Interviews were coded as FT001~FT011 according to their chronological order.

3.2 Interview coding and analysis

This part of the research data includes sound recordings, images, memos, and other field notes of interviews. The popular NVivo10 qualitative analysis software NVivo10 was used for data processing and analysis. Information points relevant to the research topic were coded and summarised to quantify the qualitative data as much as possible [11]. NVivo10 is coded mainly by means of nodes, which are divided into free nodes and tree nodes. The study coded written message, forming several free nodes and integrating such messages by determining the "axis" of nodes. Thus the coding involved three steps - open coding, axial coding, and selective coding. 1) Open coding involved the following steps: reading and sorting out the interview documents, annotating important messages, and extracting topics and coding the corresponding parts. 2) Axial coding was undertaken to categorize the 396 codes obtained in open coding, merging ones with same meanings and stratifying the open codes, which were then generalized into two grade-I nodes - costume development and correlation between costume and people - with their number of codes being 144 and 252, respectively. 3) Finally, selective coding involved sifting through the obtained codes; more general and explanatory topics were retained, and four grade-II nodes were achieved: inheritance, evolution, physiological level, and psychological level. The number of codes corresponding to inheritance is 126, including two grade-III nodes of form characteristics and size rules.

Upon review of interviewees' words, the core content was divided into two categories and coded in three rounds with nodes determined at the three grades. From Fig.1 and Table 3 it can be observed that the number of codes for each node implies the frequency of the node occurring in interviews. For example, the number of codes 72 for the node of form characteristics means the "inheritance" form characteristic had been mentioned 72 times in interviews. Node coverage rate refers to the percentage of the words in a specific node to the total documents, as shown in Fig.1.

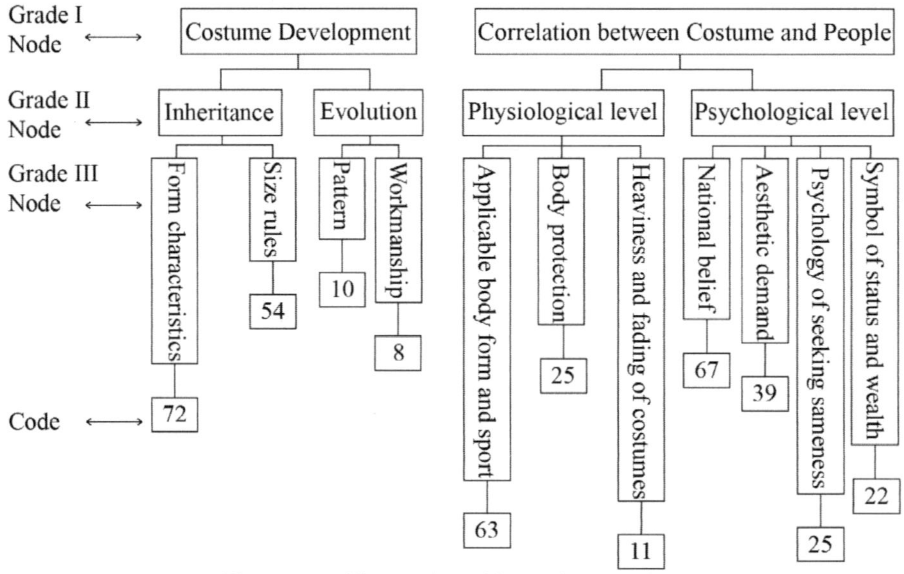

Fig.1 Tree-like nodes of interview coding

3.3 Analysis of results and discussion

It can be observed from the number and the coverage rate of nodes from Table 3 that: 1) the coverage rates of two nodes about inheritance reached 20.4% and 14.9%, and in contrast the coverage rates of two nodes about evolution stood only at 1.3% and 0.9%. In addition, according to interview documents, 10 out of 11 interviewees indicated that costume forms had not changed. Therefore, the conclusion is made that costume forms basically have not changed, and there are few changes in embroidery pattern and fineness of workmanship; and 2) regarding the relationship between costumes and people, the influence of people's physiological and psychological factors on costume development are roughly the same.

To explore the correlation among nodes, the study adopted a cluster analysis. After sorting out interview documents, two nodes were selected in turn with words under the two nodes entered. The correlation coefficient was calculated based on their similarities. Table 4 shows the results of the correlative analysis of "inheritance" and factors of three-level nodes.

Table 3. Coding of interview documents

Grade I Node	Grade II Node	Grade III Node	Coding reference points	Coverage Rate of Nodes
Costume Development	Inheritance	Form characteristics	72	20.4%
		Size rules	54	14.9%
	Evolution	Pattern	10	1.3%
		Workmanship	8	0.9%
Correlation between Costume and People	Physiological level	Applicable body form and sport	25	7.1%
		Body protection	63	17.1%
		Heaviness and fading of costumes	11	1.9%
	Psychological level	National belief	67	20.4%
		Aesthetic demand	25	8.4%
		Psychology of seeking sameness	22	12.9%
		Symbol of status and wealth	39	10.7%

Table 4. Correlation coefficient of "inheritance" and three-level nodes

Node A	Node B	Pearson Correlation Coefficient
"Inheritance" (Characteristics of Form)	Applicable bodily form and sport	0.931
	Belief of ethnic group	0.881
	Aesthetic demand	0.852
	Status symbol	0.821
	Body protection function	0.816
	Psychology of seeking sameness	0.802
	Heaviness and fading of costumes	0.118

By sorting the correlation coefficient, it can be seen that the highest correlation coefficient of costume inheritance and "applicable bodily form and sport" is 0.931, followed by "belief of ethnic group" (0.881) and "aesthetic demand" (0.852), both of which feature a high correlation because both correlation coefficients are greater than 0.85. "Heaviness and fading of costumes" is the negative feature of ethnic costumes, with a correlation coefficient of 0.118 with the costume inheritance; the two are nearly of no correlation, and thus are excluded.

The "Correlation Model" of "Exploration" tool in NVivo then was used to further explore the correlation of all factors. Fig.2 illustrates the correlation model (in which the solid-line double-headed arrow shows obvious interaction between nodes, while dotted-line double-headed arrows show low correlation).

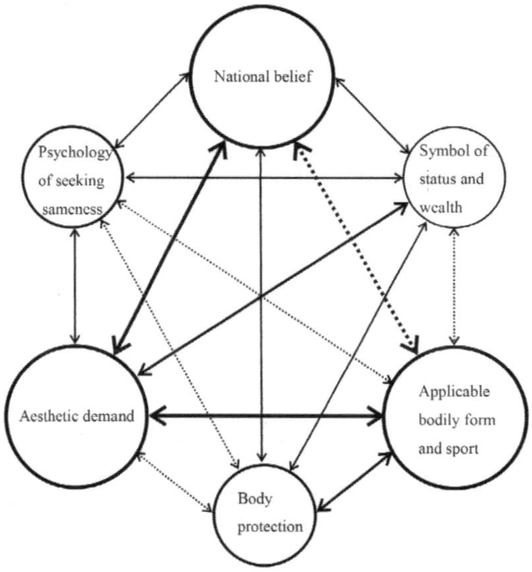

Fig.2 Correlation model of factors influencing inheritance of Baiku Yao costumes

Fig. 2 clearly shows the correlation relationship of all factors. Three major factors (i.e., "belief of ethnic group," "applicable bodily form and sport," and "aesthetic demand") interact with one another; in addition, the body protection function of costumes determined by the identification function of costumes, the strong psychology of seeking sameness due to the personality of the ethnic group, and the natural environment is also an important factor influencing inheritance.

In conclusion, the Baiku Yao ethnic group's strong "belief of ethnic group" prohibits the Baiku Yao people from violating ancestors' teachings, so the craftsmanship in costume handed down generation to generation is strictly preserved. Also, in the specific environment of the ethnic group, the costume forms conform to the aesthetic demand of the ethnic group and deliver a psychological feeling of comfort and freedom feeling. Thus, although Baiku Yao costumes are characterised by their heaviness, colour fading, and difficulty to clean, there is still no change in the characteristics of forms.

4. Conclusion

This study used qualitative research methods such as a literature review and field investigation to systematically investigate the development of Baiku Yao costume culture, analyse the characteristics of the costume form, and discuss human factors behind the inheritance of costume form. Findings reveal the following main conclusions:

1) Image-text records dated back to the Ming Dynasty and Qing Dynasty and related to Baiku Yao costumes show men's black short gown and women's black gown, two-piece "karasilis", and knee-high flowery skirt that all bear similar characteristics with existing Baiku Yao costume forms;

2) During the development process, there are much more inheritances than evolutions; there is no change in the costume form and there are few changes in embroidery pattern and fineness of workmanship;

3) Three factors have a significant influence on Baiku Yao costumes: "applicable bodily form and sport", "belief of ethnic group", and "aesthetic demand".

References

[1] Tan L. Value research on education of Baiku Yao silk culture. Chongqing, China: Southwest University: 2012.

[2] Wei JH. Customs of Baiku Yao. Ethnic Art 1986; 1: 216-219.

[3] Yu SJ. Baiku Yao society. Guilin, China: Guangxi Normal University Press, 1989; p: 108-109.

[4] Sun Y. A study of Baiku Yao costumes of Nandan Guangxi. Wuhan, China: South Central University for Nationalities: 2013.

[5] Lei WB. Interpretation of aesthetic anthropology for Baiku Yao costume culture in Nandan. J of Chongqing University of Arts and Sciences 2012; 5: 1-5.
[6] Zhang YM, He YD, Yu SJ. Research and preliminary discussion on cave tombs of Yao commune in Lihu Nandan Guangxi. J of Guangxi Normal University 1983; 3: 92-102.
[7] Zhang SQ, Peng SL, et al. Investigation report of cave tombs in Lihu Nandan Guangxi. Cultural Relics 1986; 11: 65-75.
[8] Fu H, et al. Tributes to the Imperial Ch'ing. Yangzhou, China: Guangling Publishing House: 2008.
[9] Chuai ZY. Illustrations of Yi people in Yunnan. Beijing: China Social Sciences Press: 2009.
[10] Babbie E. The practice of social research. Chengdu, China: Sichuan People's Publishing House: 1987.
[11] Feng DJ, Li P, Cao FL, et al. A qualitative study of nurses' psychological quality based on NVivo software. Nursing J of Chinese People's Liberation Army 2010; 15: 1124-1127.

Element Analysis of the Song Dynasty's Clothing for the Dynasty Symbol

Li-Ping He[1], Wei Tian[1*]

[1]*Xian Polytechnic University, No.19 South Jinhua Road, Xian, Shanxi, 710048, China*

[*]Corresponding author's email: 569790315@qq.com

Abstract

Based on the theory of semiotics, this paper analyzes Song Dynasty's clothing style pattern, color and symbolism, with the signifiers and the signified. It also summed up the elements which can be treated as the sign of the Song Dynasty from their choice in clothing. At last, the author made a point that puts forward these semantic elements into design elements, all of which are prevalent in modern fashion design.

Keywords: Semiotics; Signify; Signified; Clothing; Song Dynasty's symbol

1. Introduction

At present, there are a lot of scholars that have studied the garments of Song Dynasty from different perspectives in China. But the application of semiotics in the research of Song Dynasty fashion is not very common in China. This is due to the distance of the Song Dynasty compared to the present and the difficulty of preserving fabrics. In this paper, the author observed and studied ancient sculptures and painting works which highlighted the fashion of the Song Dynasty through the review of related literature materials. It has then been summed up through elements that can symbolize the Song Dynasty in a pragmatic and semantic manner. As a result, the meaning of these elements can be used in the field of clothing design.

2. Introduction to the theory of semiotics

2.1,The definition and relationship between the Symbol, Signifier and Semiotics

What is a symbol? And what can be served as symbols? Swiss linguist and founder of semiotics Saussure believed. "The symbol is a unity which consisted of "signifiers" (the function can be expressed in feeling directly) and the "signified" (which can be understood and speculated), and we can also say the symbol can also be a mixture of Signifiers and the Signified [2]. While the development of things can be seen as a symbolic process, which is the whole process beginning from a symbol that is first created by humans, given a symbolic form, hence creating a corresponding symbol that suggests meaning in a particular context, and then finally leading to symbolic behavior. Semiotics is the subject of studying symbols. Based on symbolic nature, semiotics believe humans created symbols from his understanding and expression of the physical world.

The signifier refers to the object itself, and the signified is the meaning it refers to in the specific context. One signifier can refer to a variety of signified meanings, but whether all that can be referred to as appropriate is usually judged by our understanding and consciousness at the time [5]. When the context or objects are different, a signifier will indicate a different meaning. For example: a picture of "Bloom mountain birds map" painted by Ji Zhao will have multiple meaning in different contexts, but the meaning is determined by the context of the painting [2]. In the Song Dynasty, the metaphorical imagery of birds of love illustrates the firmness of friendship and love in the world. In modern society, we can use it as a typical representative of Chinese traditional culture [6].

Therefore, "Context" is the important condition for a symbol to convey meaning and be understood correctly. However the method with which to create a symbol is by using one or a series of processes to determine the relationship between the signifier and the signified.

2.2 Construction of clothing as a symbol

Application in the field of semiotics in clothing include the construction of clothing as a symbol indicative of behavior. Symbols that indicate behavior is firstly the result of a specific event through expression, and secondly the communication that is completed through a number of actions. If symbolism is the metaphorical process of constructing symbols of clothes, indicating behavior is the symbol's behavioral process. Clothing symbolism is a process that merges signifiers and signified meanings in order to produce a symbol. In other words, it is a process to enrich and give meaning to make empty things substantial [4]. A noteworthy example would be the traditional Chinese "Dragon", where the golden dragon figure is the signifier, and the power, nobility, luck and success is the meaning that is signified. Its use in certain situations is notably embroidered on the garment, gives the piece of garment a sense of occasion for special events. The relationship between signify and signified makes the dragon a substantial symbol in Chinese culture.

3. The historical background and symbolic essence of the Song Dynasty

The Song Dynasty's symbol must be based on the premise of the Song Dynasty's historical background, and must have characteristics that is representative, symbolic, historic and diachronic (历时性) [5]. In this paper, the author selected the Song Dynasty's garments as the source of symbol extraction, whereupon the symbols themselves must be representative of widely known and recognized symbols.

3.1 The historical background of the Song Dynasty

The Song Dynasty includes the Northern Song Dynasty (A.D. 960-1127) and the Southern Song Dynasty (A.D. 1127-1279). In the Song Dynasty, literature, and science had both achieved fruitful results. And its culture is an abundant period in the Chinese cultural history. Some scholars in the Song Dynasty thought the destructive reasons of the Tang Dynasty is due to newfound cultural openness and the art of war that was prized above the art of peace. As a result, the Song Dynasty grew from the troubles of the previous dynasty and vigorously promoted the imperial examination system and the policy of governing the country by text. This helped to stabilize the formation of imperial power and feudal social ideology. Qian Si-ma in "Historical Almanac" wrote, "King changed family name and ordered, must be careful from the start, correct moon, change the color of garments, pushing performance of kings, Shun Cheng Jue Yi." (王者易姓受命，必慎始初，改正朔，易服色，推本天元，顺承厥意。) [11] A visible clothing system along with the physical changes of the dynasties meant it must contain the elements that can represent a dynasty's characteristics. The historical values of the Song Dynasty's that respected "literary talent above martial arts" (重文轻武) formed its own characteristics of clothing styles and clothing elements.

3.2 The essence of Song Dynasty garments

The main essence in the garments of the Song Dynasty are primarily function-based, symbolically significant and carries a high aesthetic value. Semiotics contain semantics, language structure and learning pragmatics. Semantics is the research of meaning in symbols and social thought, what is also known as the "signified". The learning of language is the study of the relationship between symbols. Learning pragmatics is the study of the relationship between symbols and users and the understanding of signifiers.

4. Analysis of the garments' style, pattern and color during the Song Dynasty by Syntax and Semantics

The best way to understand a man is to understand his background, just the same as studying a symbol. In this section, the author selected a representative sample of the Song Dynasty's clothing, and provided an analysis of the style, color and pattern. The semantics of Song Dynasty's costumes reflect a humanistic spirit and a strong aesthetic consciousness. The information obtained from these Song Dynasty's garments can be interpreted as symbolic elements of the Song Dynasty. There are many different kinds of Song Dynasty garments, including coronas and similar features that were used in the Tang Dynasty. In this paper, the main styles that were analyzed are square heart Song collar gowns, round collar robes, cross collar robes for men ,and Bei Zi (褙子)for women .

4.1 Analysis of men's tailoring in the Song Dynasty

The garments and styles for men in the Song Dynasty mainly consisted of robes, shirts, jackets, pleated shirts, crane cloaks and raccoon sleeves. In this section, we will analyze the Square and Curved Collar Robe, Collar Robe and Cross Collar Robe. The Square and Curved Collar Robe's position in the Song Dynasty is inferior to the Crown, where the emperor and the government official had a collar on. The book "Noted Examples"(《礼记集说》) record, "Today we have the Square and Curved Collar Robe, it's made from a fabric called White Luo, the size is two inches, it is embroidered on the round collar, and has a long ribbon round neck, to avoid collar surges . ("今朝有方心曲领，以白罗为之，方二寸许，缀于圆领之上，以系于颈后结之也。") [16] As shown in Fig.1 and Fig.2, the characteristics reflected in the Song Dynasty's clothing is both pragmatic and simple.

The signified meaning of the Square and Curved Collar shape refers to the theory of "a square earth and spherical heavens"("天圆地方") [11]. The top features a Curved Collar and underneath it is a square. In terms of its meaning, the number two is an even number and it's a Yin number (阴数), suggesting the coordination of Yin and Yang and life in endless succession. When we observe such details regarding its aesthetics, the shape of the square and circular collar provides the interpretation of a contrast between fluid transformation and a rigid framework of beauty. The Square and Curved Collar metaphorically suggests the symbol of heaven and earth along with the people within, which reflects the meaning of nature and the unity of nature and humanity. At the same time, the governor of the Song Dynasty borrowed the tailoring technique to portray the destiny of the royal family and the imperial power of authority. In this case, the round aspect represents the circular "Yin and Yang" destiny whilst the square represents the imperial power of authority.

Fig.1 The Square and Curved Collar Robe dressed by the emperor
("Chinese costume", Shanghai traditional opera school Chinese costume history of the team, Xuelin Publishing House)

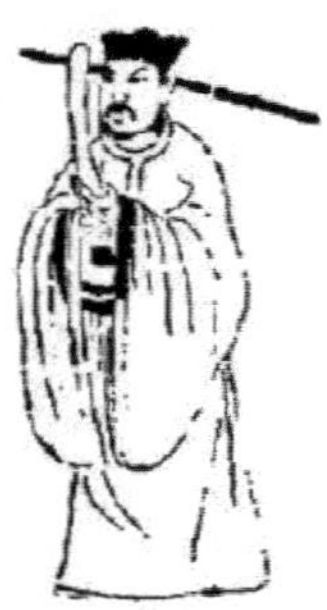

Fig.2 The Square and Curved Collar Robe dressed by civil official
("Fang Cheng County of Henan province unearthed in Song Dynasty stone workers", Cultural Relics,1983, p40-43)

The round collar robe of Song Dynasty is also referred to as Lan (襕衫), where it is used to acquire fame and Confucian literati dress codes. It is specifically named as a result of the circular collar and cuffs in addition to the armhole that provided great volume. Its more notable features include the Tang Dynasty to the Han Chinese traditional tight sleeves that is illustrated in Fig. 3. Regarding the gown-like tailoring and "Lan" structure, it quickly become the "splendid attire" ("盛服") of scholar-bureaucrats. Such a voluminous shape made the round collar robe become much more elegant. The Cross Collar robe contains Youren Cross Collar Dress (右衽交领长裙) and a Hand Over Collar (交领宽袖长袍) style. The shape and corona of Youren Cross Collar Dress is similar to the clothing of Taoism, with its straight collar cardigan mop that is worn outside robes as shown in Fig. 4. It reflects the beliefs of the Song Dynasty with high regards to Confucianism, Taoism and Buddhism, all of which directly affected the aesthetics of scholars during that period. The Hand Over Collar was the informal dress of people the seclusion, and it was usually tied around the waist with a piece of string. It is worn like a belt that creates the voluminous silhouette of a dress underneath a white coat with black piping around the edges. The characteristics of this style of clothing is consistent with the requirement that is both "convenient and fast" [19], qualities reflected by refined scholars during the Song Dynasty that is simple but elegant.

Fig.3 Li-Zong Song dressed the round collar robe
("The Chinese past dynasties dress art", Neng-Fu Huang[1],Juan-Juan Chen[2],China tourism publishing house)

Fig.4 Hui Zong Painted " TING QIN TU"
(www.Baidu.com)

4.2 Analysis of the clothing shape of women in the Song Dynasty

The garments of women in the Song Dynasty mainly consist of: Ru, coat, BeiZi（褙子）, robe, binding, and gowns. Most feature a straight collar double-breasted style without buckle, belt and neck edge sewing. Cropped sleeves are connected with the fabric because of the width in the back of the garment that is decorative. Clothes are patterned around the collar, sleeve edges, waist and hem using embroidered decorative patterns, Indian gold, embroidery and hand-painted motifs such as peonies, camellias, plums and lily patterns. In this section the author mainly focused on the garments of Hui and Bei Zi for analysis. The shape of the Hui garment is a long gown where its' signified meaning is "Female virtue"(女德专一) [26]. It's the suit of the Queen and Ming Fu (a noble woman in ancient China who was given a title or rank by the emperor) , equal to the Corona Suit which was used during conferring or sacrificial ceremonies. Hui clothes by the collar and cuffs, placket, the shoemaker's red cloth with flange, and compose with dragon patterns, teal fabric for the bottom pair of red and embroidered with birds patterns, refers to the "good in pairs" [29], red symbol in the right and noble, for most of the time to use at royal garments, beautiful bird means temperament and beauty. Birds were in close contact with nature dressed by human to show the thought of "the theory that man is an integral part of nature" as Fig.5.

Fig.5 The queen of Ren Zong wearing the Hui garment
("The Chinese past dynasties dress art",Neng-Fu Huang[1],Juan-Juan Chen[2],China tourism publishing house)

Fig.6 Soliders wearing the Bei Zi
("Dress aesthetics in China",Zi E Cai,HeBei Fine Arts Press,2000)

In the Song Dynasty , both women and men were permitted to wear the BeiZi (褙子), including scholars and soldiers. Despite the complex objects that serve as ornamentation it also has many kinds of tailored silhouettes. The basic shape is a narrow sleeve and robe with underarms open and armholes tucked behind the belt. The armholes with two front and rear skirts can be tied. However, in the Song Dynasty, the belt was only used as an decorative accessory to hang on the garments, like the traditional singlet (underwear) with cross form which symbolized the saying "Study carefully the ancient music."("好古存旧") [34]. The collar of the Bei Zi is in a straight collar double-breasted style, where the garment style features a maxillary collar type. In Fig.6 we can see two soldiers were wearing the Bei Zi, where the person on the left is wearing a half sleeve Bei Zi with a round collar, front skirt, cuffs and collar all lined in dark piping. The person on the right is also wearing the half sleeve Bei Zi, where the cuffs feature the dark piping, however, the back features a darker belt that is relatively rare.

In Fig.7, a Song Dynasty stone carving was exhibited at the National Museum of China. The exhibition further highlights the many varieties of this garment with the maids wearing a different length of BeiZi.

Fig.7 The carved stone of Song Dynasty
("Song Dynasty costumes and the application in modern fashion design",Jia Zi Chen Wang,2014)

The main function of Bei Zi is its convenience, which is also one of the main reasons why the Bei Zi was popular at that time. Compared with the two sides of the sutured robes, the two cut-parts were not sutured. The Bei Zi was very convenient for daily activities or horse riding. Its shape can be long or short and it can also be used by women as a dress in formal occasions. With the Southern Song Dynasty located in Hangzhou, the temperature is hot in summer hence the shape of the Bei Zi helps provide ventilation. The simple structure and fabrics reflect the simple and pragmatic style of the Song Dynasty, while its elegant natural shape reflects the belief that "man is an integral part of nature" [37]. Song Dynasty fashion is noticeable from its tailoring, and some of its style was influenced, "Mainly by the development of literature"("武随文制"), in which all " mental pursuits were placed above literature and martial arts", hence influencing the elegant, simple and pragmatic style of the clothing.

5. Analysis of the color of Song Dynasty clothing by Semantics

The color in Ancient Chinese clothing symbolized status, rather than today's populist and trend based preferences. The clothing color of the Song Dynasty was mainly inherited from the Tang Dynasty, which distinguished the official position by the color of their clothing. The History of the Song Dynasty "Yu Fu Zhi" (《宋史·舆服志》) records:

"The system of the Song Dynasty is rooted in the Tang Dynasty's system: people whose position were above the General of the Forward Army will wear the clothes that were purple, those under two levels wear red, the next two levels wear green, and the last grade wears cyan."("宋因唐制，三品以上服紫，五品以上服朱，七品以上服绿，九品以上服青。") [2]

Between the years of 1078 and 1085, purple was symbolic of the official position above the four, pink followed the next two positions, and green illustrated the official position above the ninth level. The ordinary dress color is given priority to with simple colors while white is reserved for the power of the unknown. The characteristics of the Song Dynasty's common clothing color is illustrated in Fig.8, also referred to as "Qing Ming Shang He Tu" (《清明上河图》) (Chinese Symphonic Picture Riverside Scene At Qing Ming Festival), painted by Ze-Duan Zhang. In the illustration, the dominating colors of the Song Dynasty clothing are gray, white and light gray. The color of Song

Dynasty clothing in comparison with the Tang Dynasty was much more subdued. It demonstrates that the Song Dynasty pursued a simple and elegant style that respected the belief of "valuing martial arts over the literary arts".

Fig.8 "Qing Ming Shang He Tu" painted by Ze Duan Zhang
(www.baidu.com)

6. Analysis of the patterns of Song Dynasty clothing by Semantics and Syntax

Song Dynasty clothing motifs were designed under its political influence and painting style, compared with the Tang Dynasty where flamboyant flower patterns and harmonious lines defined their main characteristics. Song Dynasty clothing motifs were mainly composed of birds and flowers, while intricate patterns were typical of the Song Dynasty with their realistic and vivid depictions of nature and elegant composition. Multiple tree branches illustrated using fancy patterns were juxtaposed with flower heads, branches and leaves of a single branch, and then depicted in their realistic and dynamic shapes and growth patterns. The common pattern of scattered tree branches was a symbolic gesture that suggested the static and dynamic formation of the state. The composition that takes heavy inspiration from natural elements gives a sense of quaint elegance as can be observed in Fig.9.

Fig.9 The clothing with Camellia Luo
("The Chinese past dynasties dress art",Neng-Fu Huang[1], Juan-Juan Chen[2], China tourism publishing house)

The fanciful patterns include flowers with birds, all of which are based on the patterns of phoenix birds across a variety of flowers along with the patterns of dragons around the flowers. The artist takes the opportunity to present a number of realistic fabric type flowers that are scattered in a certain distribution pattern. Through the organization of the tree branches, leaves, and vines, it is possible to stretch a single pattern that is closely linked, ultimately connecting each panel to the block diagram of the final Quartet. Smooth rhythmic lines combined with the realistic movement of the pattern formation provides contrast, vivacity and vibrant natural patterns. The color of the patterns are usually very subtle, making use of colors such as pale yellow, pink, silver, light green and light blue to highlight simple, clean and natural features. A string of branch patterns in the Song Dynasty also helped distinguish office positions.

In addition, Song clothing also utilize animal patterns such as dragons, phoenix, magpies, butterflies, deer, fishes and lions. Animals that featured strong geometrical patterns such as turtles were offset with Song Dynasty water lines. The Song Dynasty later introduced fruit patterns such as plums, orchids, bamboos, chrysanthemums and other literati aesthetics in line with a "Gentlemanly" themed flower pattern that helped reflect their ethical virtues. Also applied were auspicious patterns such as the phoenix peony pattern due to the combination of the phoenix's magnificent shape and the peony's auspicious meaning. Together they indicate a common Chinese saying. "Richly auspicious and infinite youth" [43]. The narcissus also connotes meanings such as "luck, beauty, purity and happiness ", while the Begonia combined with the peony suggests a prosperous family and so forth.

Whatever the plants, flowers, or animal patterns that were used, they were all used to convey the meaning of prosperity, all of which indicates the unity of "nature and humanity".

7. Conclusion

The styles, colors and patterns of the Song Dynasty are extremely rich. In this paper, the author studied the development of these symbols and elements whilst analyzing the Signified meaning of symbols in the Song Dynasty utilizing the style, color and pattern of their garments. The theory of semiotics was also explained by understanding and connecting the differences between the symbol, the Signifier and the Signified. By using the understanding of these symbols in fashion design, the traditional cultural values of China can be preserved, followed and practiced in the present day.

References

[1] Liu J. Taste Compare of Garment Pattern of Tang and Song.J.Silk: 2005: No.5.

[2] Qiao T. Research and characteristics of the Song Dynasty. J. Shandong textile economic publishing house: 2010: No.10.

[3] Pierre. ROM. An Introduction to Semiotics. M. The Sichuan people's publishing house: 1988; p: 13-30.

[4] Wang BFZ. Discussion of the Song Dynasty aesthetic characteristics and impact. J. Textile science and technology progress: 2009: No.5.

[5] Guo ZX. From the point of view of one kind of symbol of social and cultural phenomenon of semiotic interpretation. D.Guangdong people's publishing house: 2003; p: 30-75.

[6] Wu MC. Eco. Semiotics. Renmin. D. university of China publishing house: 1990; p: 47-68.

[7] Www.baidu.com.

[8] Huang NF, Chen JJ. The Chinese past dynasties dress art.D.China tourism publishing house: 1999; p: 20-25.

[9] Chinese costume.Shanghai traditional opera school Chinese costume history of the team. M. Xuelin Publishing House: 1984; p: 34-45.

[10] Huang Q. Chinese clothing painting. M. The yuan literature and art publishing house: 2007; p: 3-7.

[11] Wan JZC. Song dynasty costumes up.the application in modern fashion design: 2014; p: 32-40.

[12] Hua M. Dress psychology. M. China Textile Press: 2004; P: 13-19.

[13] Ye LC. Dress aesthetics. M.China Textile Press: 2001; P: 4-9.

[14] Li TZ. Chinese costume big dictionary. M. Shanxi People's Publishing House: 1992; P: 21-35.

[15] Lei SF, Yang XY. The ancient Chinese craft art history. M. Hubei Fine Arts Publishing House: 2002; P: 68-93.

[16] QiaoT. Research of the song dynasty clothing characteristics. J. Shandong textile economy: 2010: No.10.

[17] Shi L.The northern song dynasty city female entertainment research. D. Central China Normal University: 2013; p: 56-63.

[18] Zheng YP. Interpretation from the perspective of history and culture of the northern song dynasty summed up. D. Northwest University: 2005; p: 8-39.

[19] Guo MQ. The Collection of Poems of Yuefu. M. The bookstore of Hua Yashi: 2002.

[20] Zhu LB. Clothing and word aesthetic qualities. D. ECNU: 2012; p: 21-27.

[21] Qu JY. The Chinese and foreign clothing culture. M. Black long jiang fine arts publishing house: 1999.

[22] Gao Y. The theory of the research methods of semiology. D. Tianjin Polytechnic University: 2006; P: 14-30.

[23] Susanne K. Langer. Problems of Art. M. Harvard University Press: P: 144-145.

[24] Zhang Z, Zheng R. The fashion in one hundred.China Light Industry Press: 2001; P: 104-119.

[25] Suzanne Leng. Problems of art.M.China Social Sciences Publishing House: 1983; P: 28-29.

[26] Maurice merlot ponti. Symbol. M. The commercial press: 2003; P: 50.

[27] Di CZD. Introduction to the sociology of clothing. China Textile Press: P: 26.

[28] Xiao YJ. The customs in the works of the female Song Ci creators. D. JXNU: 2012; p: 7-19.

[29] Tang GZ. Full Song. M. Zhonghua Book Company: 2009; P: 1.

[30] Song YH. On the evolution of the Feminist characteristics of the Song Lyrics. D. Jilin University: 2012; P: 12-22.
[31] Li RH. The royal plaza image of ci poetry.D. Central China Normal University: 2013; p: 42-44.
[32] Liu F. Two song dynasty Beijing urban culture and literary production. D. Shanghai Normal University: 2008.
[33] Yang Y. Northern song dynasty city word study. D. Guangxi Normal University: 2011; P: 3-4.
[34] Zhang BB. The song dynasty han clothing research. D. Suzhou University: 2010; P: 20-31.
[35] Liu RB. The song dynasty folk culture aesthetic study. D. Shandong Normal University: 2015; p: 4-7.
[36] Xu J. The song dynasty women tire design research. D. Suzhou University: 2013; p: 17-24; p: 33-47.
[37] Tan JJ. Dress of Scholar-officials in Song Dynasty. D. Shandong University: 2008; p: 32-40.
[38] Men DH. On the connotations of the Song Dynasty plant decoration. D. QIQIHAR University: 2012; p: 35-60.
[39] Chen YJ. Song liao, study of dance dress art. D. Suzhou University: 2013; p: 20-23.
[40] Zhang FQ. Song Ming Neo-Confucianism economic justice thought research. D. JXNU: 2013; p: 20-22.
[41] Zhao M. In the tang and song dynasties women marriage clothing and modern fashion. D. Wuhan Textile University: 2014; p: 7-9.
[42] Wang DK. Chinese traditional clothes decoration Pattern and its Application in modern clothes design. D. Suzhou University: 2008; p: 17-19.
[43] Zhou HS. Chinese cuju during the song and yuan period. D. Xi' an Polytechnic University: 2009; p: 20-24.
[44] Xu QQ. Chinese clothing art. M. Shanxi education publishing house: 2001; P: 79.
[45] Wang HX. Art fabric. M. Jilin publishing group co., LTD: 2990: 43.
[46] Hong ZZ. Research on Chinese Han Nationality Female's Beizi in Song Dynasty. D. Fujian Normal University: 2014; p: 23-27.
[47] Pen M. Chinese costume aesthetic impact on western dress. D. Shanghai Normal University: 2008; p: 12.
[48] Shen CW. Study of Chinese Ancient Costumes. M. Shanghai Century Publishing Group: 2002; P: 11.
[49] Zhang WQ, Gao YQ. History of Chinese Civilization.M.Shandong people's publishing house: 2002; P:9.
[50] Xuan MY. Introduction to Chinese Traditional Culture. M. Fudan University Press: 2003; P: 15.

The Analysis of the Present Situation of Domestic Functional Sportswear Designs

Ye-Fu Zhang, Hao Zhang*, Hai Yu, Hai-Jing Liu

Aimer Human Engineering Research Centre, Beijing Institute of Fashion Technology, Beijing, 100029, China

*Corresponding author's email: zhhh1cn@126.com

Abstract

Along with the development of modern technology and the requirements of clothing in competitive sports, the fabric, function and comfort of contemporary sportswear has changed dramatically. The demand for functional sports clothing is also increasing. Therefore, in order to enhance the product competitiveness of our sports apparel brands, there is an urgent need to improve the status of functional sportswear.

This article analyzes the current status of the design of domestic functional sportswear through a questionnaire. The survey were filled out by consumers and enterprise designers, the results of the survey demonstrated that: 1) The designer's specific design points and the consumer's choices are consistent. The level of comfort of the clothing is one of the most important factors considered by both designers and consumers. Popular sport brands are the research objects of this article. Ex: brands such as Nike, Adidas and other foreign brands. 2) The results of the survey indicate that currently the Company designer's design is consumer-oriented, and their main goal is to be "in line with the market demands". The designer's orientation mainly refers to the sales department, 40.91% of the designs are in accordance with market conditions. 3) The technological content of domestic functional sportswear is low; the functional sportswear development team is too small; the designers have not paid enough attention to new technology and ergonomic factors; consumers have low satisfaction on the function of domestic sports brands. To improve the penetration rate of functional sportswear, enterprises should ensure an all-round development on the investment of scientific research, expand the scale of product research and development team, and pay more attention to the protection of product patent.

Keywords: Sportswear; Functional; Design; Domestic

1. Introduction

In China, along with the rapid development of the economy, the consumer's income levels grew continuously. Due to the popularity of keeping fit, it gave a broader space for the development of sportswear. " Sportswear made lifestyle like; everyday clothing made more athletic" has become one of the main features of modern sportswear [1]. After meeting people's basic necessities, they have the ability to pursue a more comfortable and qualified lifestyle. The lower-income consumers began to participate in the sports of higher-income groups, and their requirements for functional sports clothing are also increasing [2]. With the addition of the average consumers, the market of functional sportswear will be more promising, because the professional athletes are no longer the only sportswear sales targets anymore [3]. Under the development of modern science and technology, as well as the requirements of competitive sports, the fabrics, functions, and comfortableness of contemporary sportswear has changed dramatically. With the people's awareness of the importance of health, more and more people started to participate in recreational sports, so sportswear began to popularize. It developed from purely functional to fashion-oriented sportswear, and gradually into people's lives. Meanwhile, people are more concerned about competitive sports, in addition to their own physical fitness, sports clothing also became an important factor that affects athletic performances. An increasing number of sportswear is designed to adapt to different kinds of sports. Sportswear manufactured for competitive sports are developing towards a more professional direction [4]. Nowadays functional sportswear plays a more important role in people's lives. The Functions of the sportswear refers to its properties that can improve the performance of the athlete. This type of clothing needs technical support and ergonomic design. Its functionality includes applying pressure to specific muscles, improve blood flow and reduce wind resistance by utilizing the airflow principle [5].

The Chinese sport brands have a certain influence on the domestic market differentiated products and brand positioning are initiated [6]. The survey showed that the current development of sportswear design and production is developing toward the direction of diversification; style, color, fashion, younger integration and other design elements have become the mainstream of sports apparel [7]. However, there is still a large gap between the development of Chinese and foreign functional sportswear [8]. The domestic sportswear lacks beauty and uniqueness. On the use of style, patterns and the change of color to enhance taste aspects of sports apparel, there is still a disparity between domestic and foreign designs. Our constant goal is to combine fashion design with sport functionality, narrow the gap between foreign design sportswear and increase the added value of the sports apparel industry [9].

The development of functional sportswear will bring hope to Chinese sports apparel brands. Therefore, based on the consumer and the enterprise designer's bidirectional investigation, this paper will conclude a general understanding of the current status of sportswear companies, so to provide a reference for enterprises that specialize in functional sportswear design.

2. Research

2.1 The study sample

The samples of this study are divided into two parts, the consumer groups and the designer group. First of all, on account of the consumer group, this paper recovered 270 valid questionnaires, of which 41.85% are male and 58.15% female; Secondly, for the designer groups, aiming at several well-known sportswear company I conducted a survey of 10 designers, which includes 4 male designers and 6 female designers.

2.2 Statistical Analysis

2.2.1 Statistical analysis of consumer groups

The age distribution of consumers is as shown in Figure 1, 63.70% of the consumers are 20-25 years-old, and 15.19% are 26-30 years-old.

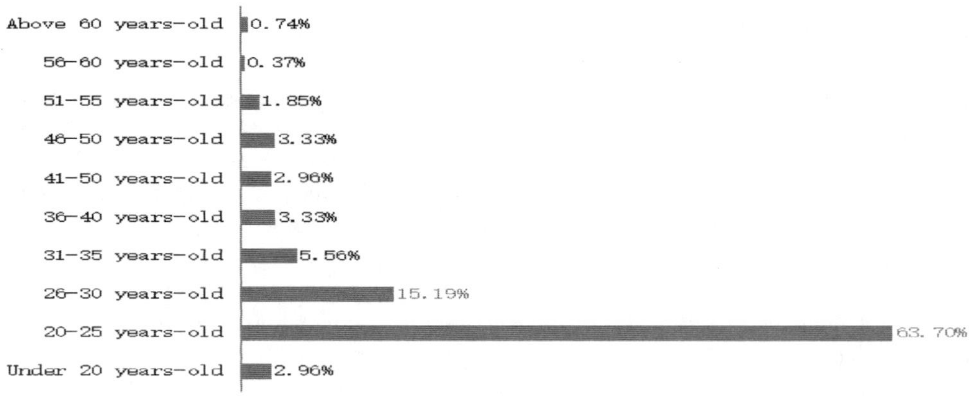

Fig.1 Consumer age

The main purpose of this questionnaire is to understand the exercise habits of consumers, as well as their buying behaviors of sports apparel and their cognition of functional sportswear. It includes the following three parts:

Part 1: The consumer's exercise habits.

In the survey of 270 questionnaires, 57.71% of the consumers exercises 1-2 times weekly, and only 14.34% accounted for never participate in any sports activity at all. The most participated exercise is jogging and badminton, accounted for 29.57% and 10.61%, respectively, in addition to swimming, fitness equipment, basketball and cycling.

Part 2: The investigation of the behavior of consumers while they purchase sportswear.

As shown in the chat below, the main factor considered by consumer is comfortableness (20.47%), in addition to the style, price and brand.

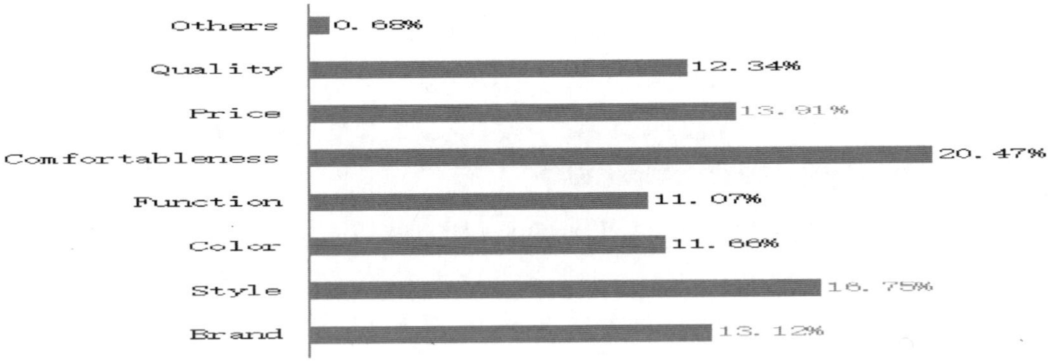

Fig.2 The main factors consumers consider while purchasing sportswear

When consumers evaluate a sportswear, comfortableness is the primary factor (35.38%), in addition to hardwearing (18.14%) and fashionable (15.44%), shown in Figure 3.

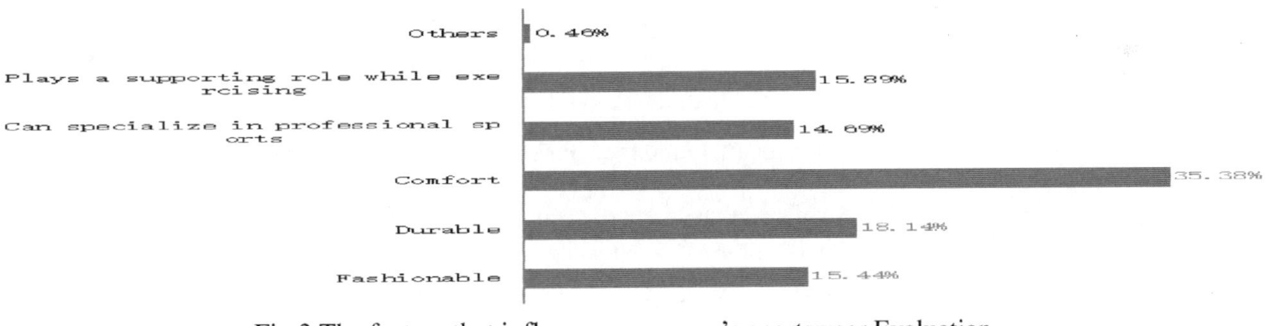

Fig.3 The factors that influence consumer's sportswear Evaluation

More than half of the consumers (52.69%) on some extent indicated that they purchase clothing according to brands, only 4.66% of the consumers weren't affected by brands, while 2.87% chose their attire entirely by brands. In the questionnaire, 25 of the most popular international and domestic sports apparel brand were cited. Nike (21.67%) and Adidas (20.17%) are the top two brands, followed by domestic brand Li Ning (7.60%), ranking third. Domestic sports brand Anta (5.35%), XTEP (4.48%) and 361° (3.74%) have high rankings, but compared to foreign brands, the purchase rate still lags far behind.

In addition, the consumer's preferences of purchasing sportswear is as follow: 67.38% of the consumers chose to purchase specialized sportswear according to their favorite sports; 73.48% of the consumers do not tend to purchase sportswear according to celebrities; Consumers that prefer gray toned sportswear (32, 92%) and high purity colored sportswear (31.97%) are accounted for the most; A majority of consumers prefer sportswear that has no pattern (29.81%), or ones with abstract designs (23.85%) .

Part 3: The investigation of functional sportswear from the consumer's perspective.

From the survey, 23.66% of consumers have never heard of functional sportswear before, but on the contrary only 32.62% purchased functional sportswear. 78.02% of the consumers that have purchased functional sportswear believe that it only play a supporting role while they exercise, and 33.69% of the consumers are dissatisfied with the functions of domestic sportswear brands.

From figure 4 we can be informed that consumers believe that three factors affects functional sportswear the most, they are comfort (33.24%), textile (22.72%) and the structure of the silhouette (17.81%). Thus, comfort is not only an important factor in consumer choice of the sportswear it is also a major part of the clothing's functionality.

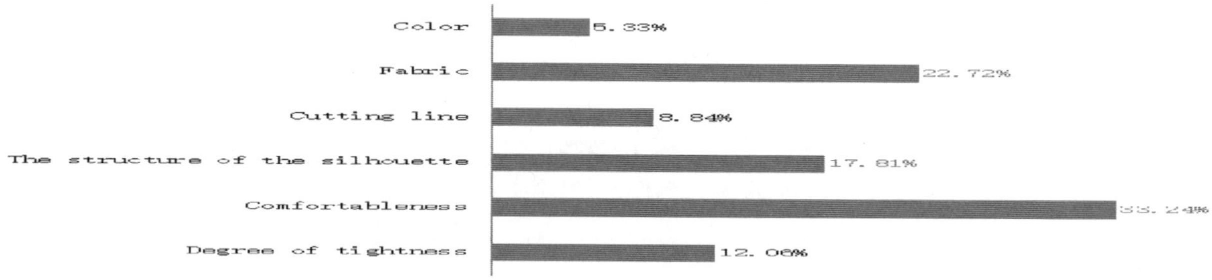

Fig.4 The main aspect of functional sportswear

As shown in figure 5, consumers believe that the main factors that effects comfort is the air permeability of that fabric (25.33%), the elasticity (19.67%) and the pressure of fabric on the skin (18.22%).

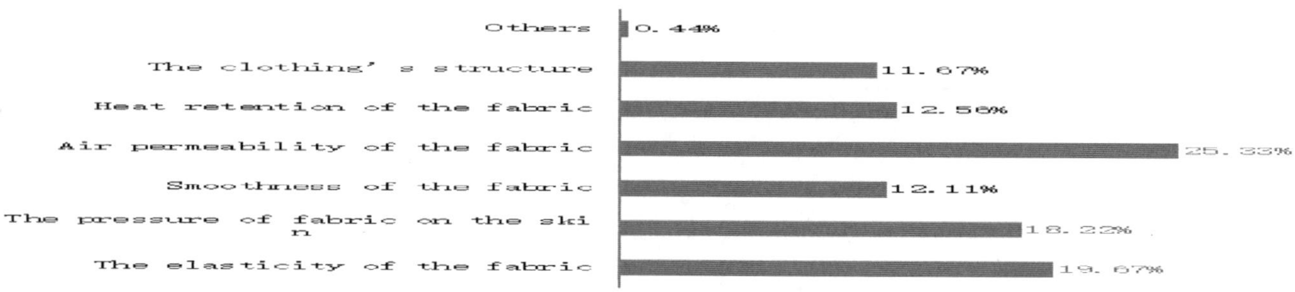

Fig.5 The main factors that maintain the comfortableness of the sportswear

As shown in the chart below, 19.17% of the consumers expect to buy functional sportswear that can maintain a comfortable temperature, and 18.26% needs sportswear that has good ductility.

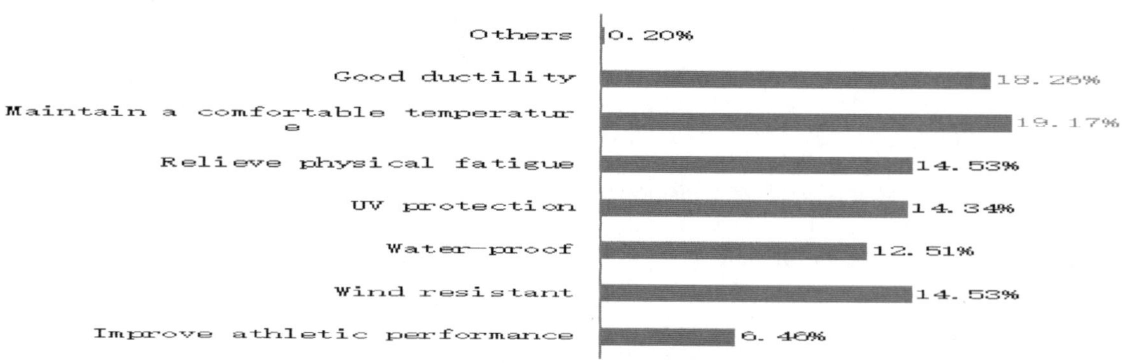

Fig.6 The factors that consumers desire in functional sportswear

As shown in figure 7, the requirements of functional sportswear is quite balanced among different sports, so we can concluded that the consumers desire to use corresponding functional sportswear.

Fig.7 Sports that demands functional sportswear

Through the analysis of the consumer's questionnaire, conclusions are drawn:

1) 57.71% of the consumers exercise 1-2 times per week. Running is voted the most popular exercise among consumers.

2) 20.47% of the consumers consider comfort as the most important factor when they purchase sportswear. In the factors that affect the degree of comfortableness, the breathability of the fabric is what the consumers concern the most.

3) In all of the sports brands, Nike and Adidas are the most frequently purchased brands. Consumers are more into neutral colors and bright colored sportswear, and garments that have no pattern or abstract designs are more desirable to consumers.

4) The popularity of functional sportswear is low, only 32.62% of the consumers have purchased functional sportswear. While 78.02% of the consumers believe that functional sportswear only play a supporting role as they exercise.

2.2.2 Statistical analysis of the designers

The main purpose of this questionnaire is to understand the current situation of the designs in sportswear companies, and the existing problems within its design methodology. The interviewed designers have 2-10 years of experience in sportswear garment design, and they are responsible for the company's product lines, which includes basketball clothing, football clothing, golf clothing, and swimwear etc. The designer's daily work content mainly include drawing, selecting materials, model fitting, design modifications, product audit, and a small number of designers will participate in the performance testing.

The results of the survey showed that among the companies 70% of them have a functional sportswear design team, the size of the team varies from 5 to 20 people. The team members includes planners, designers, pattern designers, craftsmen and quality control personnel, and only one of the company interviewed have a laboratory personnel. The manager or director of the design department in is position of the team's management.

In the stage of product designing this research focused on the designer's responsibility, the objectives of the design, design orientation, the source of inspiration and design evaluation. According to the survey, the target of the design is as shown in Figure 8, which is to conform to the kinesiology principle and explore new technology and fabric, accounting for 17.65%.

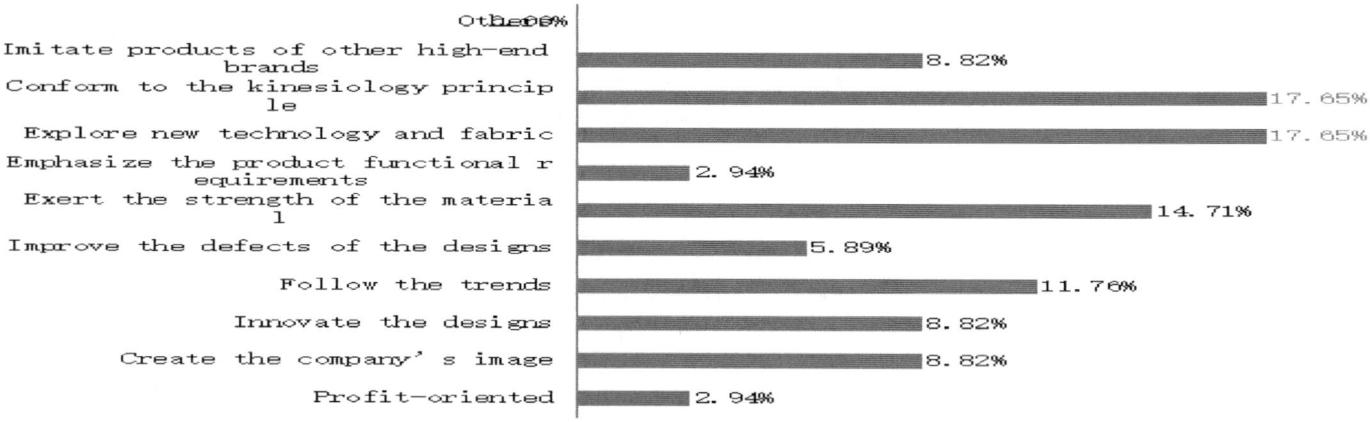

Fig.8 The designer's goal

40.91% of the designer's orientation is based on the advice of the sales department, followed by 27.26% of the designs that are based on the concept of innovation. The designer's orientations are as shown in Figure 9.

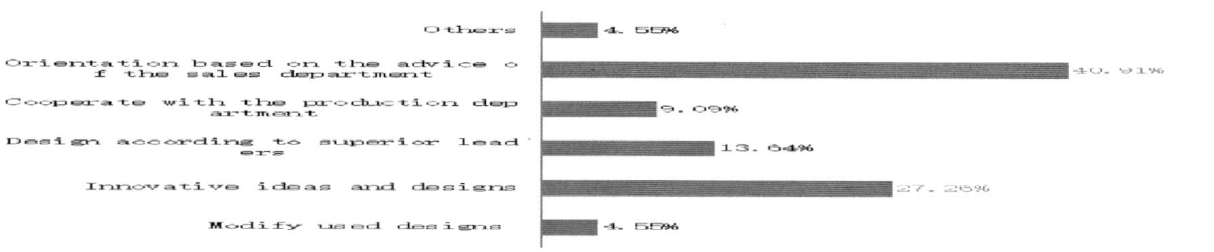

Fig. 9 Enterprise designer's orientation

Table 1. Enterprise designer's sportswear product review

	The Least Important	Unimportant	Generally	Important	Very Important	Average Score
Trendy	10%	10%	20%	60%	0	3.3
Level of Innovation	0	0	10%	60%	30%	4.2
Comfort level	0	0	0	30%	70%	4.7
Functionality	0	0	20%	50%	30%	4.1
Colour Matching	10%	0	0	90%	0	3.7
Structure Line	0	0	40%	60%	0	3.6
Pattern Design	0	0	40%	50%	10%	3.7
Accessory	0	0	30%	60%	10%	3.8

The development of new products is based on the designer's inspirations, consumer 's demands, and the existing product defected trends [10]. The designer's inspiration mainly comes from the fashion trend, market research and new products of competitive brands. A major part of their resources comes from the internet, research, magazines and other channels. While conducting the market research, designers choose popular brands such as Nike, Adidas, Puma and Converse as their research objects. The selection of fabric, aesthetic, color and style are the designer's main considerations. The evaluation of the quality of the sportswear is shown in Table 1. Comfort, innovation, and functionality are the main evaluated factors. Regarding the protection of patents, 50% of the companies applied for patents, 10% of companies apply occasionally, and 40% of the company's products have never been patent protected.

Regarding the design projects within the company, the research and development investigation, and the phase of performance testing, 30% of the company rely on other institution to do their testing, 50% of the company did the testing autonomously, while 20% of the company completed the testing partially outsourced.

The data analysis of the designer's questionnaire shows that:

1) The functional clothing development team is a small scale in our existing domestic sportswear company.

2) The main goal of the designers is to meet the market's demand.

3) The orientation of the designer's design depends on the situation of the sales department.

4) The market research regarding the designers is based primarily on foreign major sportswear brands. In the evaluation, the most important factor is the Sportswear's "level of comfortableness".

5) In the process of development, only half of the companies took the performance test solely, and the enterprise should be more aware of the importance of product patents.

3. Conclusion

Through the two-way investigation on consumers and enterprise designers, we can have a general idea of what the current situation of sportswear companies are like. By comparing the two questionnaires, we can draw to the following conclusions:

1) On the demand of sportswear, the focus point of both designer and consumer is consistent. For example, in the consumer product sportswear evaluation, comfort factors are accounted for the highest, which is 35.38%. 70% of the designers think that comfort is extremely important, 30% think it's quite important, these two had the highest proportion of all evaluated factors. The brands that are most popular among consumers were chosen as the objects of this study, such as Nike, Adidas and others international brands.

2) Currently, enterprise designer are mainly consumer-oriented, from the survey results accounted for 20.93% it is shown that its design objective to be "in line with market demand" is their most important goal. In their design's orientation, 40.91% of the designers will make reference to the sales department to design their designs in accordance to the market conditions. Designers can design clothes according to consumer's hobbies, for instance they can focus on the designs of functional sportswear for jogging.

3) At present, the development of functional sportswear is deficient. According to the 'designer survey', the results of the questionnaire shows that 30% of the companies do not have their own functional sportswear development team, and 30% of the company rely fully on other agencies to do their performance testing. The designers have not paid enough attention to new technology and ergonomic factors.

The designers rarely participate in any task that involves testing performance of the apparel. During the various stages of design and development, the step of testing the performance of the product was ignored. The results were reflected in the consumer questionnaire, 34.44% of consumers are dissatisfied with our domestic functional sportswear brands. In order to improve the penetration rate of functional sportswear Enterprises should focus more on the investment of scientific research, the expansion of the scale of product research and development team, and the protection of product patents.

Last of all, due to limited resources, the survey sample of designers is short, so the data analysis is limited. We are hoping to make up for this shortcoming in future studies.

References

[1] Wei TJ. The Discussion of the Sportswear Development Process and its Trends of Design. J Zhejiang Sport Science 2010; 32(3): 43-48.

[2] Li WP, Shi XD. The New Consumption Trends of Sports Apparel. J Sports Culture Guide, 2002; 3: 95-96.

[3] Fan J. The Marketing Cultural Study of Four Main Domestic Sportswear Brands. China University of Mining and Technology, 2014.

[4] Wu CR. Study of Modern Fashion Design Trend. J Sports Goods and Technology 2014; 2: 44-44.

[5] Deepti Gupta. Functional clothing-Definition and classification. J Indian Joumal of Fibre&Textile Research 2011; 36: 321-326.

[6] Zhang XZ, Chen LH. The Development Status and Trends of High-performance Sportswear. J China Textile Leader 2010; 5: 105-107.

[7] Lu XY. The Design of Fashion Oriented Sportswear. J silk 2008; 8: 18-20.

[8] Cao HM. The Research and Analysis of the Sportswear Market. J Qingfang Gongye Yu Jishu, 2014(01): 32-33.

[9] Chen YS. The Research of the Design of Modern Sportswear. Jiang Su: Soochow University, 2008.

[10] Tang JW. Study of Chinese Sportswear Companies Market Positioning. J China's Management of Information Technology 2015; 11: 97-98.

Emotional Evaluation of Special Dart in Women Fashion Design

Jiang Wu, Xing Zhang*, Qiao-Yun Tang

Xi'an Polytechnic University, 19 Golden Road, Xi'an, Shaanxi, 710048, China

*Corresponding author's email: Zhxing029@126.com

Abstract

Currently applied to some special dart, in the design of clothing, apart from complying to the human body surface form, meeting the demand of visual aesthetic is also important. To classify such a wide variety of dart in accordance with the specific operation and in the case analysis, a questionnaire was designed to measure different types of special dart in doing a gradient measurement technique of expression, from both the theoretical and practical aspects. Ultimately, through the questionnaire, the different types of special sensory effect of dart was studied.

Keywords: Special dart; Garments; Sensory effects

1. Introduction

1.1 Overview of special dart

Darts can be traced back to the 13th century, new cutting methods of clothing have been developed to get rid of the excessive difference between the chest and waist in three major areas, namely the front, back and side. Many triangular cloths are added when the sleeve is rooted to the side of the hem, and the irregular triangle between cloths that are arranged in many diamond space to remove the waistline is the dart [1]. Both Madeleine Vionnet who created a new way of cutting, known as the oblique cutting method [2], and Grès who is known as the "fabric sculptors" are skillful in using diagonal yarn to the characteristics of the fabrics in the 1930s [3], well coordinated and unified the relationship between fabric and the human body, showing the stereo aesthetic feeling of clothing. The birth of the cutting method led to the development of dart diversification, developed by three basic province shapes, namely bust dart, waist dart and armhole, to functional and decorative special dart [4].What we call special dart, just as its name implies, is different from the ordinary dart [5]. Its design is derived from that of basic dart, from the initial single fit development of a decorative design features of both [6]. Summary and analysis of contemporary fashion design, the use of unconventional dart manipulation can be roughly divided into segmentation, gather and comprehensive special dart [7].

1.2 Research status

The research of this subject matter is mainly based on dart technology and the view of clothing structure of pattern. For example, the history of dart was described in a few texts such as Apparel Sociology [8], Women's Clothing Structure Design and Application [9], and the Foundation Garment Design. These texts mainly introduced women's clothing structure design, which referred to the use of lines in designing costume details [10]. In Dress Aesthetics [11] and Susan B.Kaiser's Clothing Social Psychology, a comprehensive theoretical basis of the cause of darts and the scale of the aesthetic were provided [12].

Two aspects of different forms of special dart, namely image analysis and practical operation of a comprehensive experiment, were viewed, and through statistical data analysis, it is concluded that different expression of special dart generated the difference sensory effects. Further profound interpretation of different dart in clothing design and its aesthetic characteristics need to be carried out to understand the effects of different dart on dress sense, which will help designers grasp and choose the suitable dart for their designs.

2. Image analysis and practice operation

Through the analyses of a lot of pictures and hands-on practice, it can be summed up that special dart falls under the influence of different expression of visual rule. Details of the preparation and implementation steps are as follows:

1) Tools required: bust female model, cloths, beads needle, sewing kit, rulers, scissors, chalk

2) Design: design should pay attention to the aesthetic feeling of the clothing while applying common dart to the design.

3) Adjustment and manufacture: combined with prototyping and modeling to realize the design conception as much as possible, making ordinary dart transfers or disappears into the special dart.

4) Sample suture: suture the special dart, front and back piece.

2.1 Classification of special dart

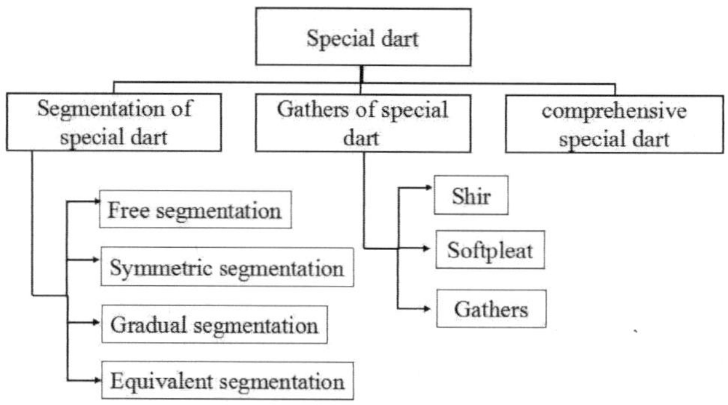

Fig.1 Classification of special dart

Segmentation of special dart: is a process of doing subtraction of the fabric, cutting off the excessive parts. The majority modeling characteristic is that this special dart has the opened dart, without dart ends (Fig.2).

Gathers of special dart: The modeling of stereoscopic relief effect, is to remove the redundant parts of the fabric by suture, is a process that doing addition to fabric (Fig.3).

Comprehensive special dart: due to its application in fashion design elements, sometimes they are used in complementary with other techniques. In fact, segmentation special dart and fold special dart for design are more commonly used (Fig.4).

Fig.2 Segmentation of special dart

Fig.3 Gather of special dart

Fig.4 Comprehensive special dart

2.2 Segmentation of special dart

Free segmentation principle: refers to the free expression in clothing that uses segmentation dart, which shall not be subjected to horizontal, vertical, oblique constraint. At the same time, the overall effect of dart must keep a balance after they are segmented, and during the design process in orderto maintain the integrity of visual effect,we need to transfer ordinary dart out from special dart. Free segmentation dart is included in free line segmentation and free curve segment (Fig.5,6,7).

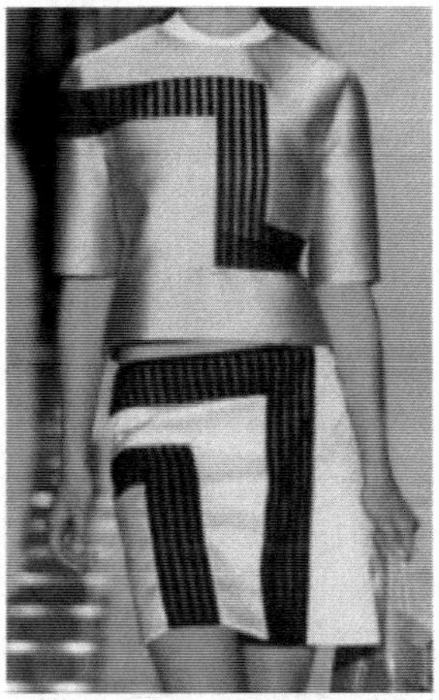
Fig.5 Fashion show

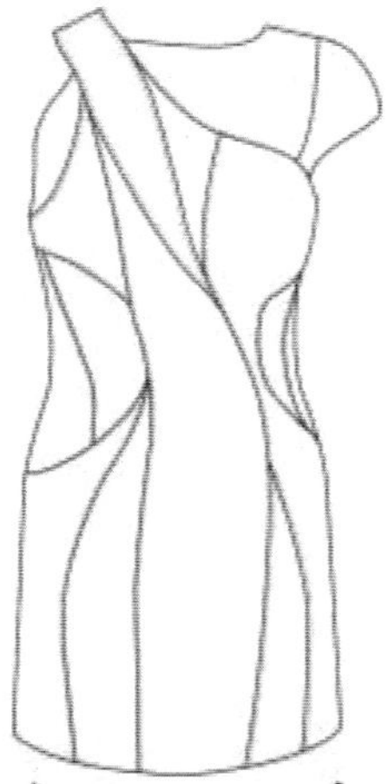
Fig.6 Structure chart

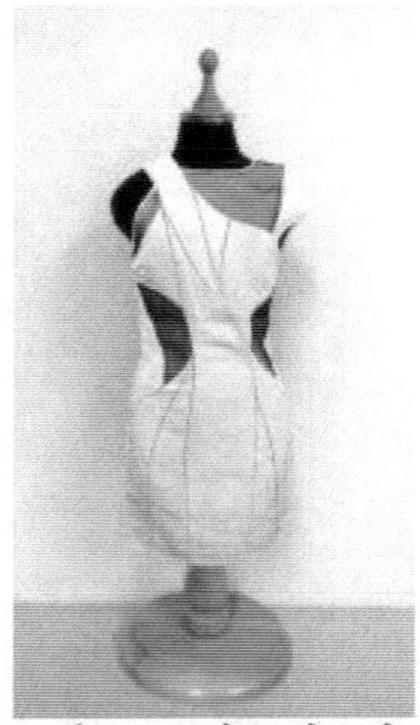
Fig.7 Design sketch

Symmetric segmentation principle: refers to clothing with either a vertical or horizontal center line as the symmetry axis, or one that uses the center symmetrical design, showing the effect of a mirror image in the divided structure (Fig.8,9,10).

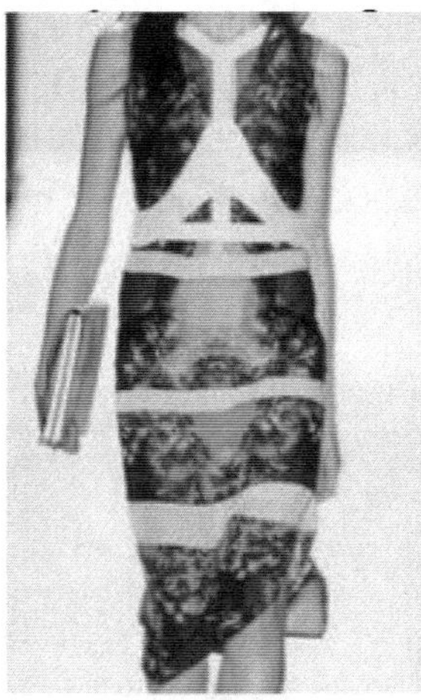
Fig.8 Fashion show

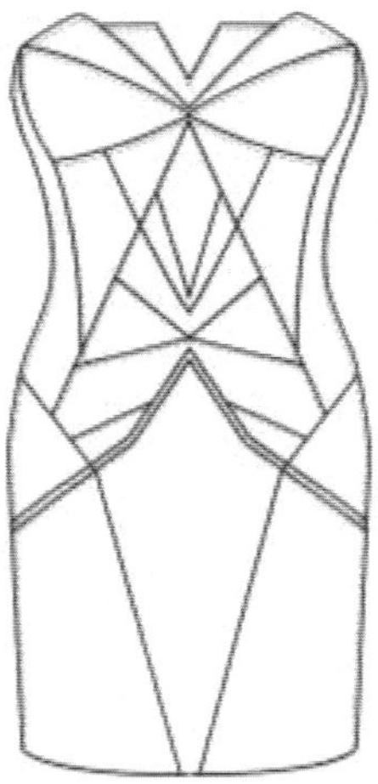
Fig.9 Structure chart

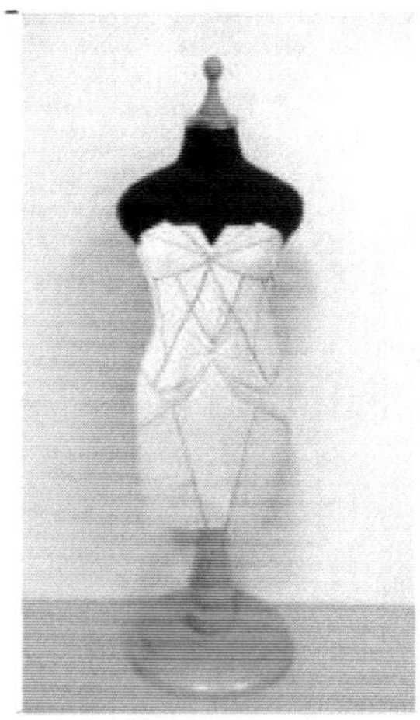
Fig.10 Design sketch

Gradual segmentation principle: it is a performance enhancing or decreasing performance method that is used in clothing in accordance with a certain distance. The special dart gradually increases or decreases in the form of a vertical, corrugated or spiral shape [13] (Fig.11,12,13).

Fig.11 Fashion show

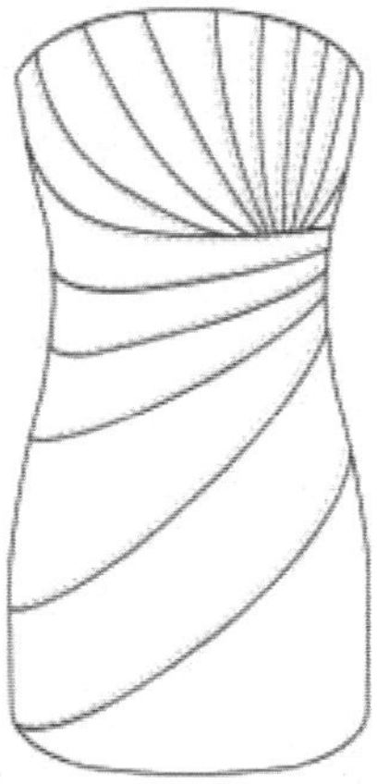

Fig.12 Structure chart

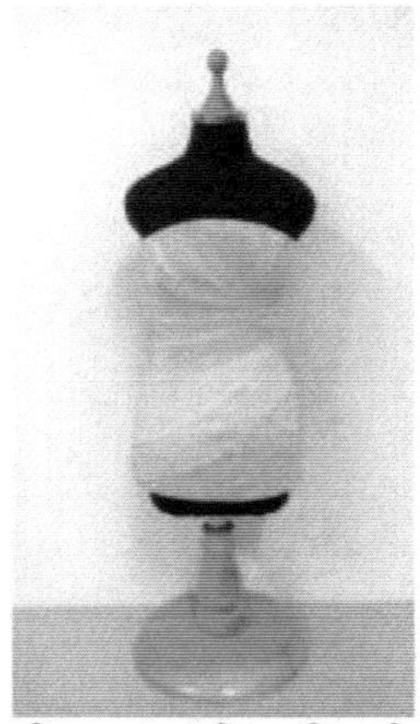

Fig.13 Design sketch

Equivalent segmentation principle: refers to clothing that can be different when used with shape segmentation in dart, but is equal in the visual area. The equality here is relative and not absolute. It is the character of Equivalent segmentation that the fabric shape is different from each other after segmentation, creating more diverse shapes. Since they are equal to each other in the area, they have been visually regarded as the same form full of balance and stability [14] (Fig.14,15,16).

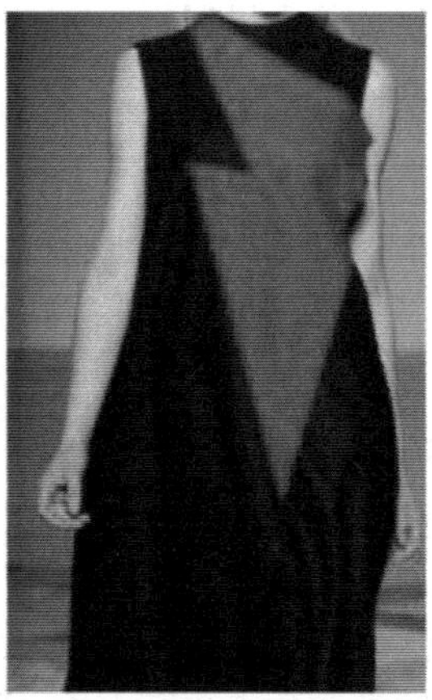

Fig.14 Fashion show

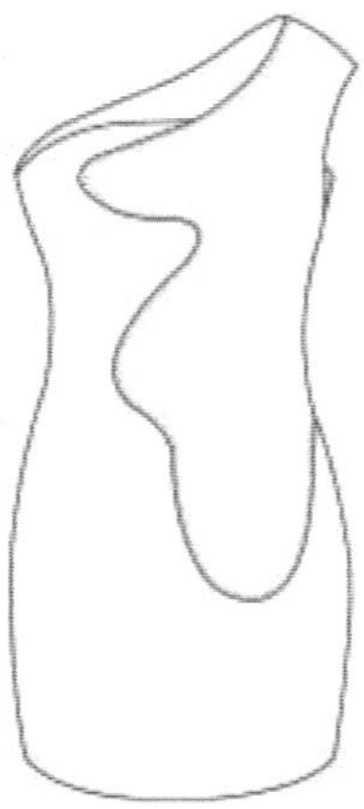

Fig.15 Structure chart

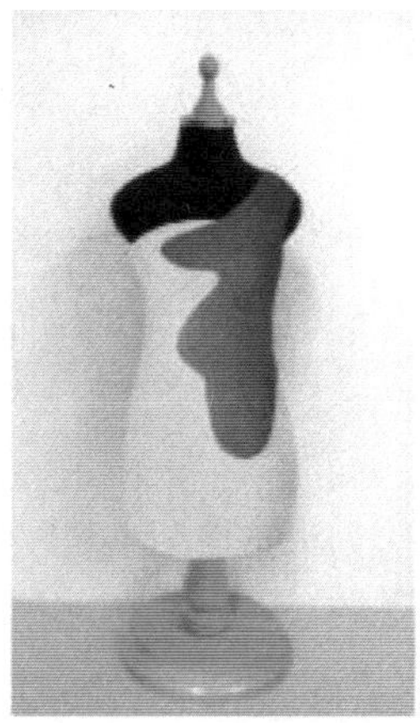

Fig.16 Design sketch

From figure 5 to 16, it can be concluded that different forms of free segmentation of dart give a person different feelings: polyline style brings a serious and stable feeling, while curve style gives a flexible and innovative feeling. The common method for segmentation dart is to open up a new location of the dart and then suture. According to structure design we can use multiple shear, which we call it elimination method of dart.

2.3 Gathers of special dart

Shir principle: refers to pleated sewing techniques in certain interval and depth from different direction of the fabric planned gathers. It does not change along with the shape of the human body in the process of using the gathers. It belongs to gathers category.

Fig.17 Fashion show

Fig.18 Structure chart

Fig.19 Design sketch

Softpleat principle: is to fold the fabric, then fixed on both ends of the position, without suturing the middle with free state in the process of using the gathers. It is usually used in concave and convex parts of the body, such as chest, waist and hip.

Fig.20 Fashion show

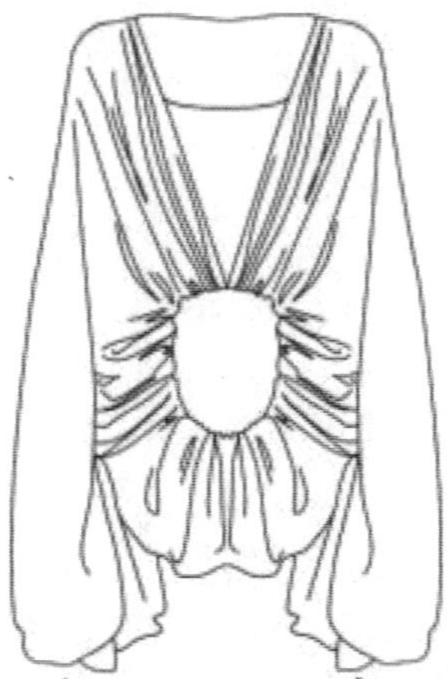
Fig.21 Structure chart

Fig.22 Design sketch

Gathers principle: means making the ribbon fixed on the fabric shrink to form folding. In modern women's fashion design, the fabric is usually transformed out the redundant parts of it in the use of the method of the fold, forming abundant modelling in the fabric [15].

Fig.23 Fashion show

Fig.24 Structure chart

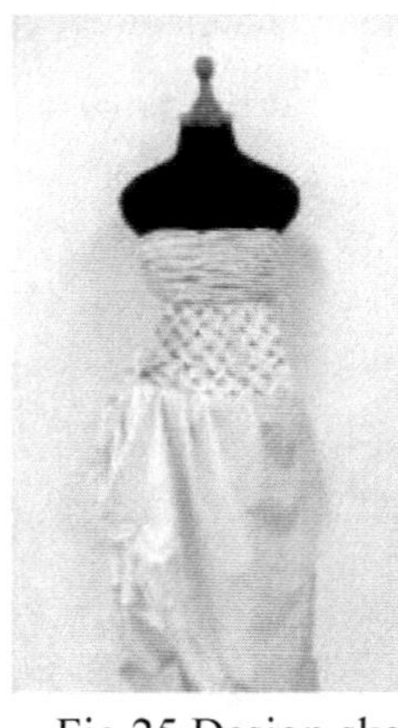
Fig.25 Design sketch

From figure 17 to 25, it is shown that different forms of free segmentation of dart create different feelings; polyline style brings a serious and stable feeling, while curve style gives a flexible and innovative feeling. Increasing pleat method is suitable for the transfer of pleat dart. The method is to eliminate the amount of pleated distribution spread to parts of the body where is needed, through the design of pleats in clothing structure design, which we call it elimination method of dart.

2.4 Comprehensive special dart

The principle of comprehensive special dart: Both segmentation special dart and gathers special dart are usual expression technique in the process of using segmentation dart. The effects of both methods are dependent on each other.

Fig.26 Fashion show

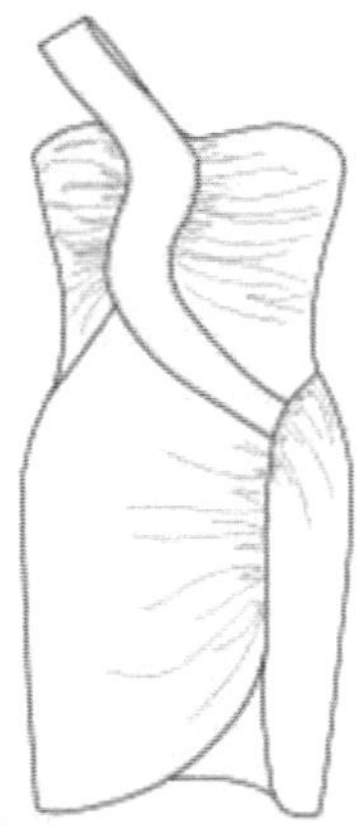

Fig.27 Structure chart

Fig.28 Design sketch

3. Questionnaire and data analysis

3.1 The questionnaire survey

This questionnaire was designed according to the research method of kansei engineering. Through analyzing the related retrieval data, 10 groups of perceptual words are finally selected, for each group set up by one to five, with a total of five levels of gradient. The closer to both sides of level the feeling of sensory effect is stronger, which is stated as A to E in the survey. Choose figure (7,10,13,16,19,22,25,28) as a reference object, and through the issuance of questionnaires, for different types of special dart to do a gradient measurement technique of expression, which measures different types of special dart sensory effects. This questionnaire was distributed to 100 girls in a specific field of Xi`an Polytechnic University Institute of Art and Design, increasing the credibility of the results. While filling out the questionnaire, every respondent must carefully observe each image and choose the appropriate vocabulary at the same time.

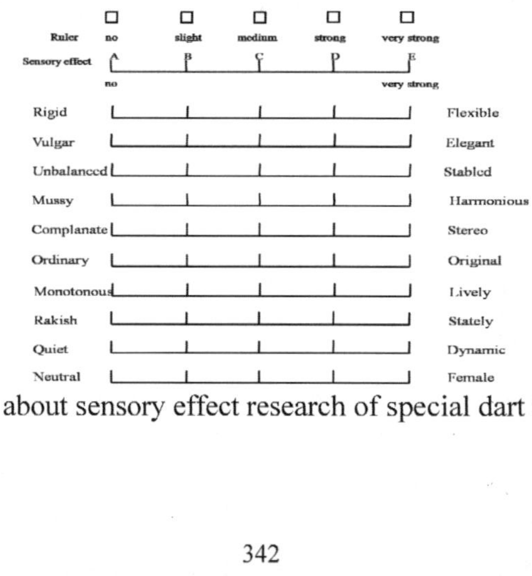

Fig.29 Questionnaire about sensory effect research of special dart in women fashion design

3.2 Data analysis

According to the above questionnaire, random sample survey of 100 copies of questionnaires, of which 98 effective questionnaires were retrieved. The following conclusions were drawn from the statistical analyses and results of the effective questionnaire. In the case of words categorized under rigidity and flexibility, the statistical conclusions of the 98 valid questionnaire data are as follows:

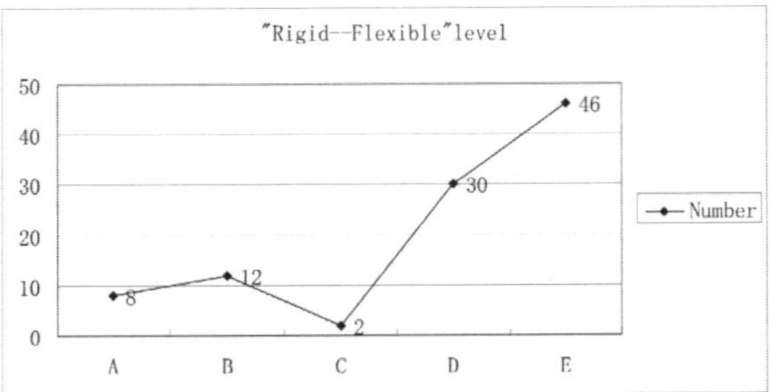

Fig.30 The number of "Rigid—Flexible" level

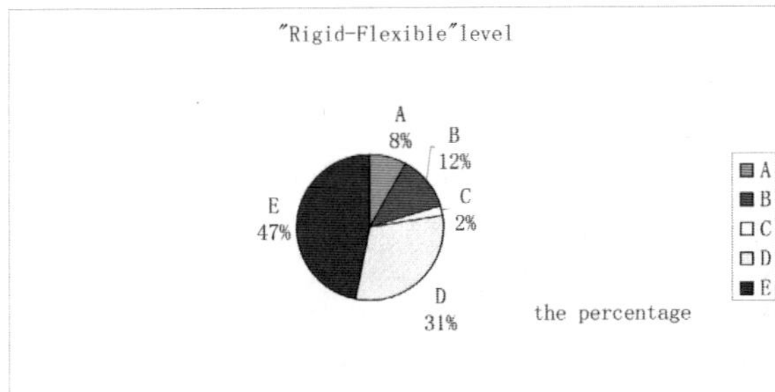

Fig.31 The percentage of "Rigid—Flexible" level

Conclusion from Fig.30 and 31: we learned that 57 people selected category E, accounting for 47% of all the people surveyed, meaning that most people voted for freedom segmentation dart, and they thought that it gives a person athletic and flexible sensory effects. The rest of the category of special dart also referenced the experiment methods above, as shown in the following Table 1.

Table 1. Statistical analysis about sensory effect research of special dart in women fashion design

Category	Modeling characteristics	Sensory effects
Free segmentation	Line and shape divided randomly There are rules to follow the Division	sense of movement, flexible Stiff and rigid
Symmetric segmentation	With the center line as the axis, the symmetry reversal	balanced and stable, solemn, neutralization
Gradual segmentation	The divider is increasing or decreasing trend	the feeling of novel and acceleration
Equal segmentation	The split shape is the same The segmentation shape is different, the area is relatively equal	drab and hard sense of balance and stability
Shir	At a certain interval and depth, from the vertical and horizontal fold	sense of sculpture, static feeling
Softpleat	Fixed on both ends of the pleat, free state in the middle	flexible, sense of sculpture
Gathers	With the stripe on fabric pleating	sense of sculpture, feminine

4. Conclusion

Special dart design has been regarded as a kind of indispensable way of fashion design. Based on the available data, three common expressions are summarized: Segmentation of special dart, Gathers of special dart and Comprehensive special dart. Combininh the technique of expression with the clothing styles can be applied to different position of the garment. In the process of practice, polyline segmentation dart is found to give a person a sense of neutral and serious feeling; curve segmentation dart gives people a sense of speedy and flexible feeling; and pleat dart produces a different feeling dependent on the number of folds, location, treatment method and the state of the human body. Therefore, in the future, analysis of clothing design and research on the special dart must be considered carefully and further research should be carried out, instead of reapplying existing fabrication approaches. At the same time, we should pay attention to its decoration and the visual influence. In conclusion, it will be helpful to a fashion designer when he or she understands and grasps the visual effects of different dart design.

5. Acknowledgment

I would like to express my gratitude towards the teachers and classmates who have helped me in any way.

References

[1] Zang YC. The comparison of women's clothing modeling between Chinese and Western. China Light Industry press: 2001.
[2] Song YL, Wang X. The research on the technique of bias cut of fashion modeling design. Shandong Textile Economy, 2010; 10: 53-54.
[3] Li DQ.A history of costume in the west. China Higher Education Press: 2005. 07.
[4] Tian BY, Wang LX. The relation of irregular dart and garment style. Progress in Textile Science & Technology. 2008; 4: 77-78.
[5] Jiang R. Research on contemporary women's garment with unconventional structural line. Donghua University: 2009.
[6] Tian BY. Research of garment with unconventionality structure line based on symmetry and wire direction characteristic. Donghua University: 2009.
[7] Wang L. The research on the design of special dart in modern well-fitting dress. Donghua University: 2007.
[8] Hua M. Costume sociology. China Textile & Apparel Press: 2005. 03.
[9] Wu J. Women's clothing structure design and application. China Textile & Apparel Press: 2000. 04.
[10] Li XG, Cui YM. Basic clothing design. Donghua University Press: 2010. 12.
[11] Ban CSL,Li JS. Dress aesthetics. Shanghai People's Publishing House: 1986. 06.
[12] Kaiser SB, Li HW. The social psychology of clothing. China Textile & Apparel Press: 2000. 03.
[13] Chen GH. A dictionary of world costumes. Shanghai Far East Press: 1996.
[14] Li TZ. Dictionary of Chinese costumes. Shaanxi People's Publishing House: 1992. http://www.ellechina.com/
[15] Wang L. The research on the design of special dart in modern well-fitting dress. Donghua University: 2007.

Analysis on Structure and Technology of Ujimqin Mongolian Robes

Xiao-Yan Ma, Li-Ying Huo, Xiao-Fang Guo*

The College of Light Industry & Textiles, Inner Mongolia University of Technology, Hohhot, 010051, China

*Corresponding author's email: alisa1996@imut.edu.cn

Abstract

By the means of fieldwork and literature review, the aim of this research is to study the structure and technology of Ujimqin Mongolian robes. The results showed that due to the different layers, Ujimqin Mongolian robes can be divided into five kinds. The type, pattern making method and sewing technology of all the robes are generally same, which are typical and traditional. This research is helpful to the heritage of the ethnic clothing and can provide reference for the related study.

Keywords: Ujimqin; Mongolian robe; Structure; Pattern; Sewing technology

1. Introduction

Ujimqin Mongolian robe is the typical clothing of the Mongolian robe, and is also the best existing traditional robes. Ujimqin is an ancient tribe in the history of China. The meaning of "Ujimqin" is "the person living in the grape mountain or the grapes" in Mongolian. This ethnic group resided in Altai Mountains originally, then surrendered to Genghis Khan, and finally settled in the grassland in Xilingol. The daily clothing of Ujimqin Mongolian was robes. As the complex history of this tribe, the robe shape was developed in Ming dynasty and solidified in Qing dynasty [1].

Nowadays "Ujimqin" is an administrative zone name in China, where most of Ujimqin Mongolian lived in. Ujimqin Banner includes East Ujimqin Banner and West Ujimqin Banner, located in Xilingol League in Inner Mongolian in China, the area of northern temperate continental climate. And the weather characteristics can be described as spring is arid and windy, summer is hot and short, autumn is cold and windy, winter is harsh and long. Overall, the amount of the gale days and sunshine duration are large throughout the year [2].

Based on the climate, seasonal robes were developed. And Ujimqin Mongolian robe can be classified into five categories, including the single-layer robe, lined robe, cotton robe, fur-lined robe and smoked-fur robe, named by their layers [1,3]. In addition to single-layer robe and smoked-fur robe, fabrics of other robes are generally same, focusing on satin and brocade, shown in Table 1 [4-7].

2. Structure of Ujimqin Mongolian robes

The shape of all kinds of Ujimqin Mongolian robes is basically same. Specifically, the design elements are fold-over collar or mandarin collar, wide border, Chinese opening (left front opening edge, named "right ren" in Chinese), straight silhouette, connected body sleeve, horse-hoof cuff, wide hem, side opening [1,8]. A typical type of Ujimqin Mongolian fur-lined robe is shown in Fig.1.

The structure of Ujimqin Mongolian robe belongs to "cross" flat structure, which is cross-shaped, planarity, cut by the whole piece of fabric and dressed like a cylinder. The center front and center back of the robe are the axis of symmetries [9-11]. There is no difference between the female and male robe. The net pattern can be used directly without being added seam amount, because of the large ease of the robe.

Table 1. The type of Ujimqin Mongolian robes

Name	Season	Wearing Month	Number of layers	Materials of the robe layers			
				1st	2nd	3rd	4th
Single-layer robe	Summer	6-7	1	Silk	/	/	/
Lined robe	Summer	6-7	2	Satin/ Brocade	Cotton	/	/
Cotton robe	Spring/Autumn	4-9	3	Satin / Brocade	Cotton	Cotton cloth	/
Fur-lined robe	Autumn /Winter/ Spring	10-12/1-3	4	Satin / Brocade	Cotton	Cotton cloth	Woolskin
Smoked-fur robe	Autumn /Winter/ Spring	10-12/1-3	1	Woolskin	/	/	/

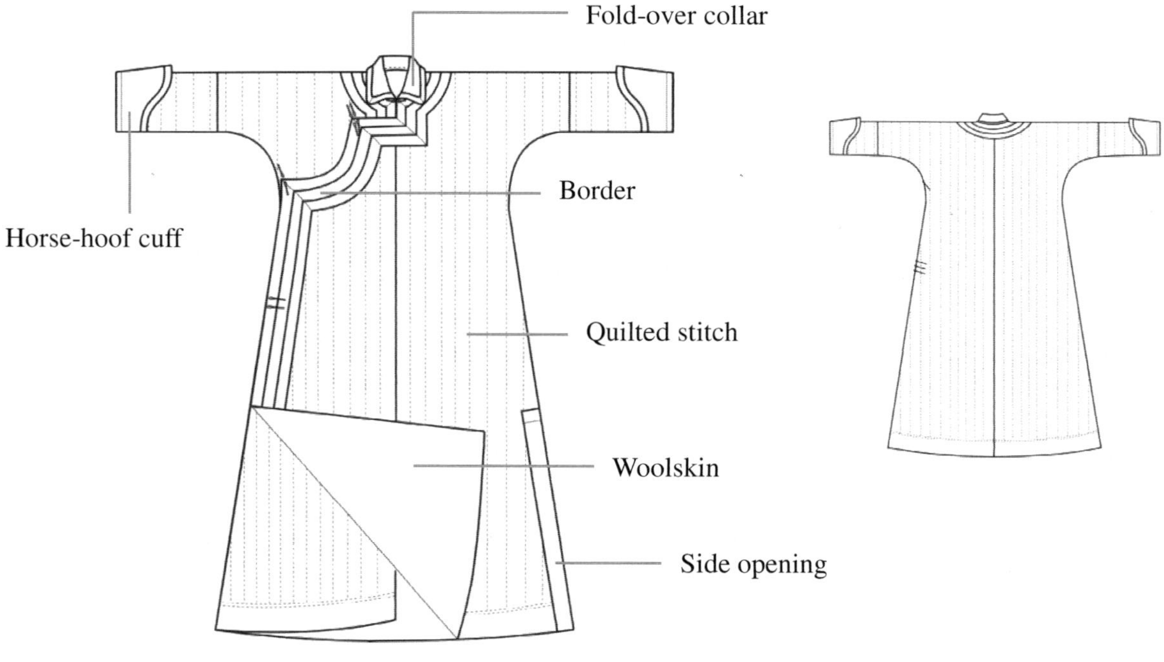

Fig.1 The shape of Ujimqin Mongolian fur-lined robe

The pattern of Ujimqin Mongolian robe is drawn directly on the fabric based on the dimension obtained by measurement and experience, which inherits the traditional methodology and differs from the pattern making method of western clothing [12]. This method was named fixed size method and commonly used in ancient China. The structure of Ujimqin Mongolian robe is shown in Fig.2. There are eleven key points and lines in the pattern, including length, whole sleeve length, front neck point, back neck point, side neck point, shoulder width (including the actual shoulder and part of arm), arm hole circumference, arm hole line, bust girth, cuff width and hem width [12,13]. They are marked in the pattern, and the full names of the markers are shown in Table 2. Among these dimensions, there are four measured parts including length, whole sleeve length, shoulder width, cuff width. Besides these, others are based on experience of the designer. The designer used hand to measure the size. The name of this method is "zha" in Chinese, which means the distance between the thumb and the middle finger.

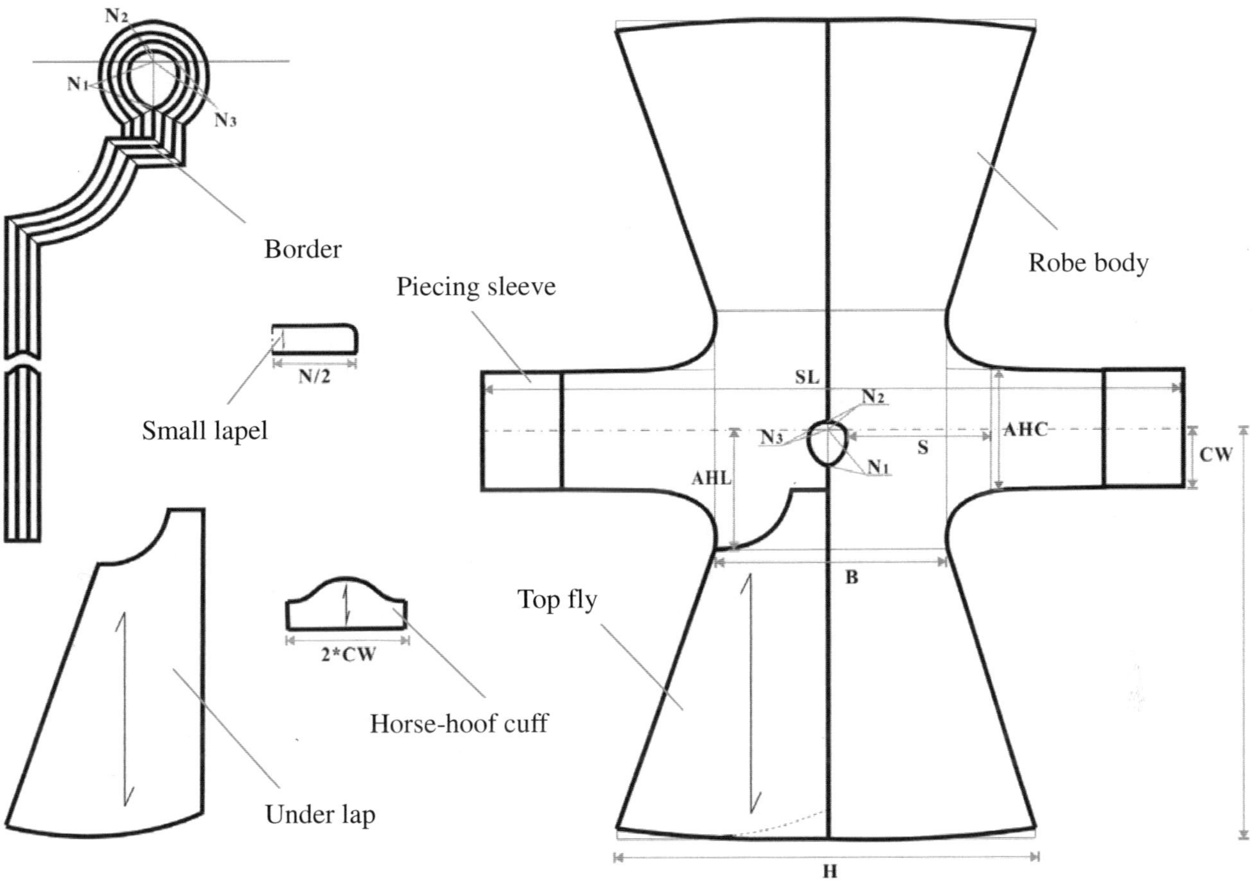

Fig.2 The structure of Ujimqin Mongolian robe

Table 2. The full name of markers in structural lines of Ujimqin Mongolian robe

L	B	SL	CW	S	AHC	AHL	H	N	N_1	N_2	N_3
Length	Bust girth	Whole sleeve length	Cuff width	Shoulder width	Arm hole circumference	Arm hole line	Hem width	Neck girth	Front neck depth	Back neck depth	Neck width

The Chinese opening almost covers the front robe body, and the opening (top fly) with the back body is in one piece. The sleeve in the robe belongs to the one-piece raglan sleeve. As the width of robe fabric is usually not enough to match the actual length of the sleeve, another same fabric named piecing sleeve is cut in remnants to be jointed. The shoulder line is straight, and the angle of armpit is relative to 90 degrees. However, this sleeve structure is unsuitable for the three-dimensional surface morphology of human body. The fabric in the armpit is easy to be accumulated, when the wearer's arm is laid down.

3. Sewing technology of Ujimqin Mongolian robes

As many steps in the sewing process are complex and can only be made manually. In the past, robes were made by the female herdsmen, and today one robe still can only be made at one time [14,15]. In addition to the different technology of the robe layers, sewing technology of the lined robe, cotton robe, fur-lined robe is similar. In this research, the sewing technology of Ujimqin fur-lined robe was investigated.

3.1 Sewing tools and materials

Sewing tools: Wooden ruler, hand needle, sewing machine, iron, flour, water
Fabrics: Silk, satin, brocade.
Filler: Cotton
Lining: Woolskin, lambskin

3.2 Sewing technology

Pieces of Ujimqin fur-lined robe include four main parts which are body pieces (left piece, right piece, top fly piece, under lap piece and piecing sleeve), collar pieces (outside collar piece and under collar piece), horse-hoof cuff pieces (outside cuff piece and under cuff piece) and border. The sewing technology of fur-lined robe was introduced based on these four parts, shown in Fig.3 - Fig.5.

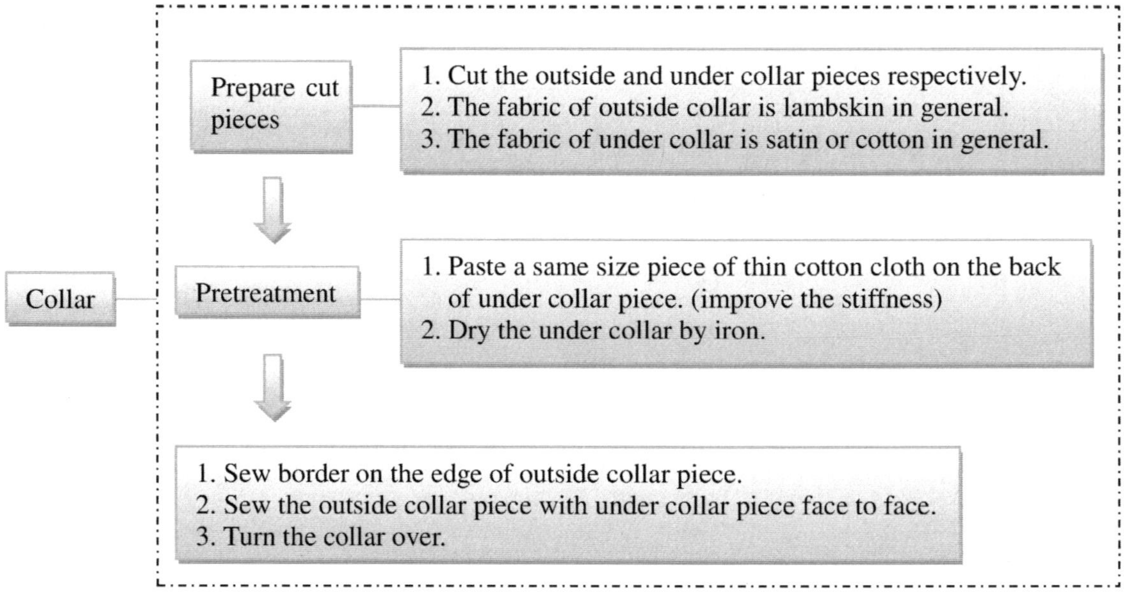

Fig.3 The sewing technology of the collar in fur-lined robe

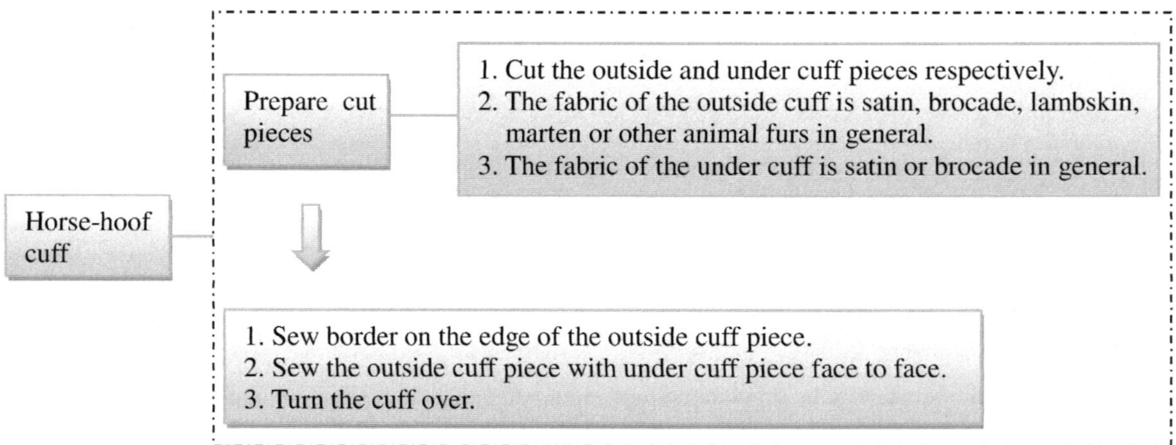

Fig.4 The sewing technology of the horse-hoof cuff in fur-lined robe

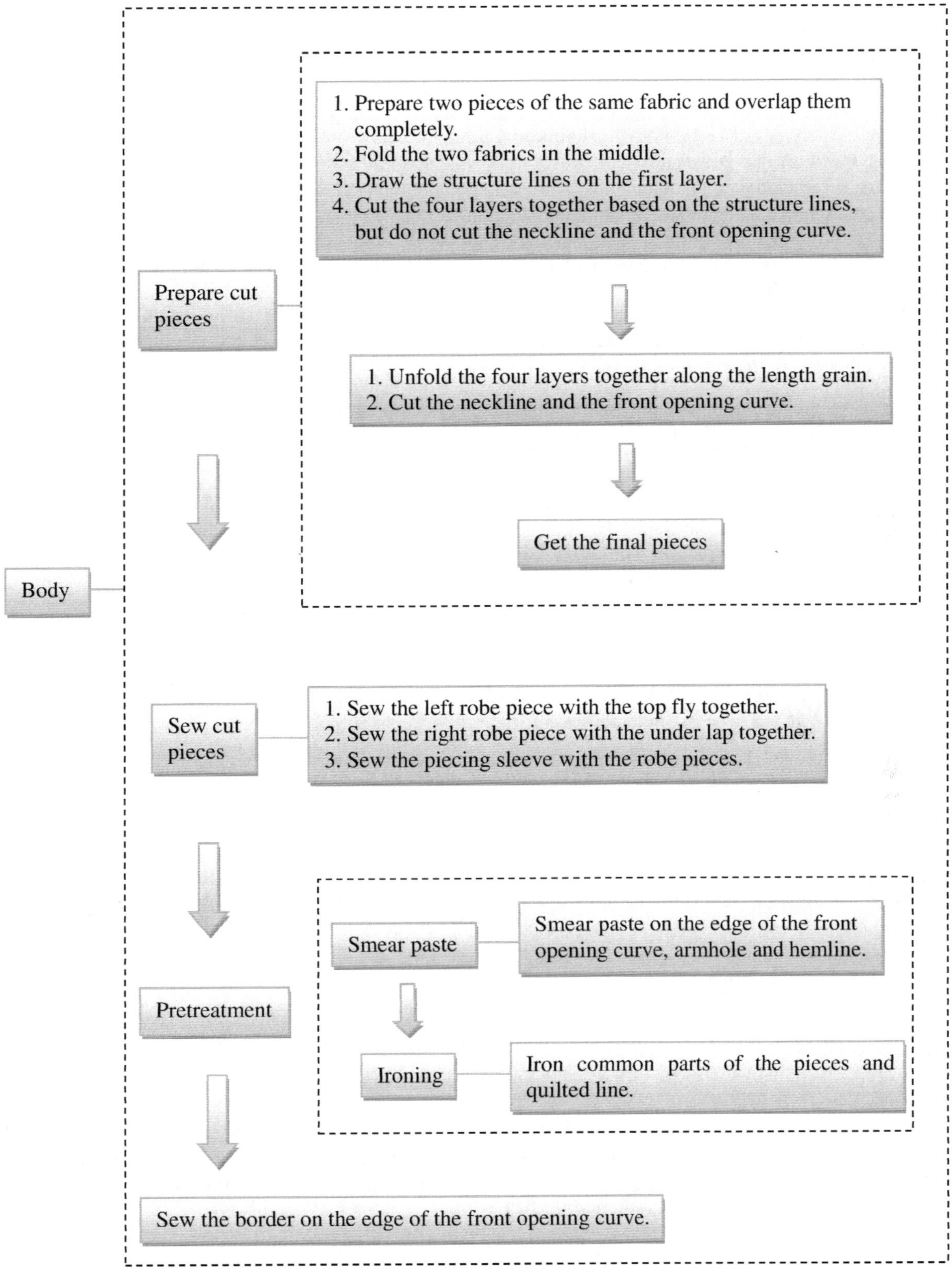

Fig.5 The sewing technology of the body in fur-lined robe

The robe can be completed by sewing the body with the collar and horse-hoof cuff. As the border technology is complex and based on the designer experience totally, it has not been analyzed deeply in this study and can be investigated further in the future [1].

4. Conclusion

In this study, the structure and technology of Ujimqin Mongolian robes were investigated. Major results of the present study can be summarized as follows:

1) Based on the weather characteristics, there are five kinds of Ujimqin Mongolian robes, which are single-layer robe, lined robe, cotton robe, fur-lined robe and smoked-fur robe, and their type is consistent.

2) The pattern making method of the robe is fixed size method, and most of the sizes are determined by designer. There is no difference in male and female basic pattern. But the pattern is unsuitable for the morphology of human body, especially the armhole.

3) The robe technology focused on the manual sewing, and little parts can be sewn by machine. The cut pieces include body, collar, horse-hoof cuff and border. Some special traditional processes are used to improve the robe quality, such as smearing paste and quilting.

Ujimqin Mongolian robe is the most typical robe of Mongolian robe. This study is highly significant to the heritage and protection of traditional clothing.

5. Acknowledgements

The authors wish to acknowledge the Nation Natural Science Foundation Research Project of China (No.51366011) and The Young and Middle-aged Teacher Training Program of Inner Mongolia University of Technology.

References

[1] Guo XF, Ma XY. Research on Composition and Culture Inheritance of Ujimqin Mongolian Robes. Journal of Silk: 2015; 52(8): 62-69.
[2] Bao G. Analysis on changes of temperature and precipitation in Inner Mongolia in recent 30 years. Journal of Inner Mongolia Normal University(Natural Science Edition): 2012; 6: 668-674.
[3] Na B, Dao C. A magical land: Ujimqin. Hohhot: Inner Mongolian Education Press: 2011; p: 1-4.
[4] Saina. The field work of the Mongolian smoked leather gown made in Ujimqin grassland of Xilingol League of Inner Mongolia: Take smoked leather as the main investigation content. Journal of Art College of Inner Mongolia University: 2013; 10(2): 97-103.
[5] Zhu SW, Liu JW. Fashion fabric. Beijing: China Textile & Apparel Press: 2010; p: 116.
[6] Zhang YQ. Analysis on woven structure and materials of ancient golden brocade - from ancient calligraphy and paintings of Zhang Daqian's collection. Art and Design Research: 2013; 4: 40-47.
[7] Shang G. Nasisher in China. Southeast Culture: 2003; 8:54-64.
[8] Liang Y. Mongolian Costume. Hohhot: Inner Mongolian People's Publishing House: 2013; p: 69.
[9] Shen CW. Study of ancient Chinese clothing. Beijing: Commercial Press, 2011; p: 646-647.
[10] Wang W. Research on structure and technology of the Zhuang women clothing in black. Hunan Normal University: 2013; p: 18.
[11] Zhao M. Line cutting and doublet structure: Study on the structure of the Chinese ethnic costume. Art & Design: 2012; (1):110-112.
[12] Liu RP. The principles and practices of pattern design. Beijing: Textile Press: 2008; p: 348-397.
[13] Zhang WB. Pattern making for fashion design. Beijing: China Textile Press: 2006; p: 25-40.
[14] Zhang YC. The research on trimming of women clothing of late Qing dynasty and its innovative application. Beijing Institute of Fashion Technology: 2014; p: 12-10.
[15] Ding YH. Study on the process of the late Qing dynasty and the early years of the republic of horse face skirt. Beijing Institute of Fashion Technology: 2013; p: 28-40.

Research on Knowledge Representation of Clothing Pattern Design Based on Ontology

Xin Zhang, Wen-Can Zhou*, Bo-An Ying, Ming-Ming Wang

School of Apparel & Art Design, Xi'an Polytechnic University, Xi'an, 710048, China

Corresponding author's email: zhouwencan123456@126.com

Abstract

Pattern methods, formulae and empirical data are three basic elements of clothing pattern. They are also known as the domain knowledge of clothing pattern. The unification, sharing and reusing of the domain knowledge will greatly improve the efficiency of designing clothing pattern. Computerized knowledge of clothing pattern will be more effective in promoting intelligent pattern, especially in semantic pattern. Ontology is important for knowledge sharing and reusing, and knowledge management. It has contributed tremendously to natural language process, information extraction, semantic web and so on. In this paper, we first analyzed then classified the knowledge of clothing pattern, followed by abstracting and representing the classified results based on ontology. Lastly, the concepts and relationships between clothing pattern knowledge will be computerized and formalized. The invisible knowledge can be identified and exploited by the intelligent clothing CAD system. It will provide a certain basis for the research of knowledge sharing and reuse in clothing pattern.

Key words: Clothing Pattern; Ontology; Knowledge Representation; Relations;

1. Introduction

In recent years, ontology has gained increasingly more applications in the field of knowledge engineering and artificial intelligence. It has been an important approach in the field of computer. Ontology is a formal, explicit specification of a shared conceptualization [1], which can promote resource sharing, improve interoperability of heterogeneous system, and be used to solve issues in resource sharing and reusing [2].

There have been a lot of local and foreign research performed on ontology-based knowledge representation. The most representative one is performed by the Institute of Applied Informatics and Formal Description Methods (AFIB) in Germany. It is universally acknowledged as the authority of research on ontology. It has constructed knowledge portals and semantic web portals based on ontology. On the other hand, the Knowledge Systems Laboratory (KSL) of Stanford University in the United States is constantly researching on knowledge sharing technologies. Ontology has been developed by an American corporation, MCC, based on ontology knowledge [3-5]. In China, the How Net has been constructed by Computer Language Information Engineering Research Center of Chinese Academy of Sciences, which has also completed research on large-scale knowledge acquisition associated with National Knowledge Infrastructure (NKI).

In addition, framework knowledge description language was designed by NKI, which has done countless experiments on knowledge representation in many different disciplines, such as medicine, military, mathematics and information science [6-11].

For ontological research in garments field, a distributed search engine system for apparel was designed by Gefu Zhang and Qi Xu based on semantic web services; a knowledge base of ontology for textile and garment was constructed by Yuqing Liang; a bra product knowledge model was constructed by Bei Cao based on the concept of ontology knowledge modeling; and a garment retrieval system was constructed by Aili Wang based semantic web [12- 14].

Clothing pattern is not only an extension and development of the style, but also the preparation and foundation of the design process. It also plays a connecting role in the whole garment design cycle. Nonetheless, there has been limited studies on clothing pattern ontology in the field of garment. In this paper, a Knowledge Representation method for Clothing Pattern Design based on ontology has been put forward.

2. Classification of clothing pattern knowledge

2.1 Analysis of characteristics of clothing pattern knowledge

The clothing pattern design process is visually expressed in the form of a directed graph. Taking some certain brand suit as an example, the directed graph is shown as Fig.1.

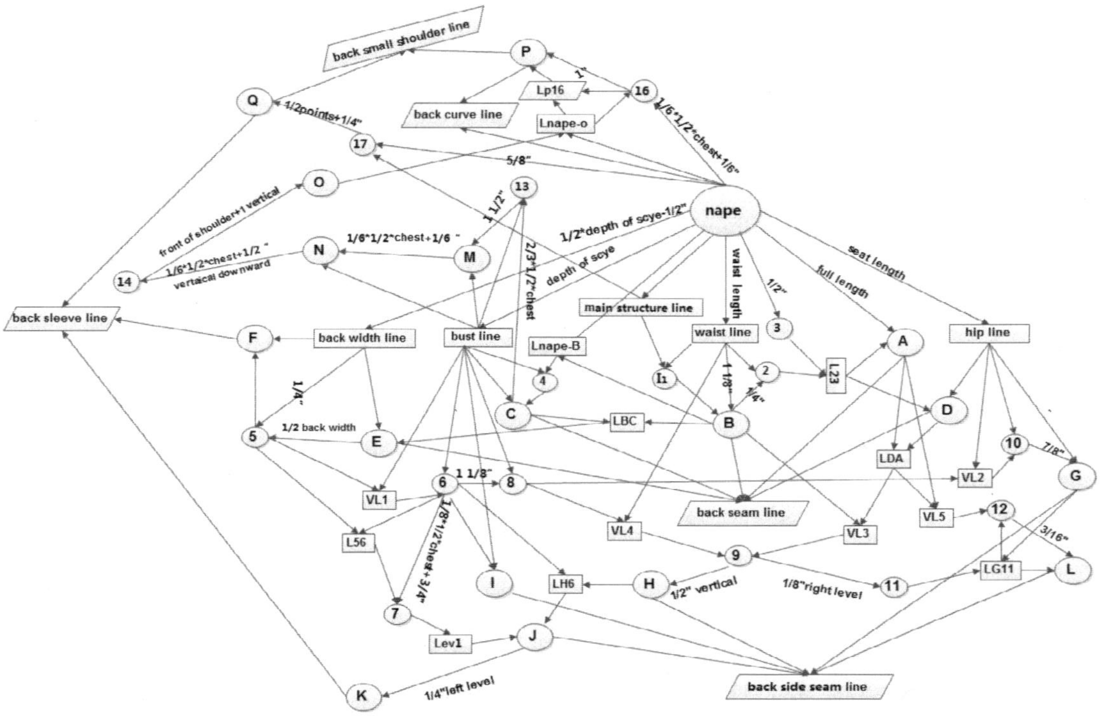

Fig.1 The directed graph of suit

Notes of directed graph Fig.1: ○ represents a point; ▭ represents a line; ▱ represents a curve line; Lev represents a parallel relationship; VL represents a vertical relationship; ⟶ represents the direction of line; the number and formula represent the basis data what the formation of point and line need.

From Fig.1, we can see the essential elements of clothing pattern design are points, line, data and direction, which are also known as the clothing pattern knowledge. There are also parallel, vertical relationships, deviation or others between the formation of points and line. Data, direction and relationship become the bridge connecting the points and line. The name of points and line are stationary and obvious, thus they are called explicit knowledge. Data is determined by the size of the human body and experiences of the pattern maker, thus it is called invisible knowledge.

Based on the analysis, the characteristics of clothing pattern knowledge can be summarized into four points: relevance of knowledge; restriction of knowledge; parameter and knowledge; and uncertainty of knowledge.

Relevance of knowledge: clothing specifications are determined by the dimensions of the human body; the size of each part of clothing pattern is determined by the clothing specifications; point coordinates is determined by the size of each part of the clothing pattern; line position is determined by coordinates of different points; styling and form of the pattern are determined by the line and point; garment fitness is determined by the styling and form of the pattern. The clothing pattern knowledge are interrelated in the whole cycle.

Restriction of knowledge: a parallel restrictive relationship exists among waist line, hip line, bust line and other parts of the body. Vertical restrictive relationship exists between the center line and the bust line, the center line and the waist line, the center line and the hip line, and so on. The lower-right pocket point is located by a deviation of 7cm from the lower-left pocket point in the right direction. The deviation and length relationship exists between the lower-right pocket point and the lower-left pocket point. There are many relationships such as the parallel restrictive relationship, the vertical restrictive relationship, the deviation and length relationship, and the angle relationship between the processes of the clothing pattern, which embodies the restriction of knowledge.

Parameter and knowledge: adjusting garment size also changes the size of the bust and the waist. They are called specification parameters. The ease decides the space between the body and the costume, and affecting the costume's comfort and modeling. It is decided based on previous experiences of the pattern maker. Ease is one of the custom parameters. In the pattern, collar = collar stand + 1, lapel collar angle = collar stand $\pm$ 0.5, turn-over collar angle = lapel collar angle-0.5, lapel collar width = 2, and lapel collar angle + 1.5. The size of lapel collar angle decides the width of the lapel collar, which will then change the styling of the collar. This is called a styling parameter. The parameterized design of clothing pattern refers to the dependence of pattern and garment style on the size of the garment. Parameterized design is also the theoretical basis of automatic pattern design for clothing.

Uncertainty of knowledge: ease is decided based on previous experiences of the pattern maker, features of the fabric and the sewing method. Different body shapes generates different data for the pattern. These data belong to a type of invisible knowledge, which is uncertain.

2.2 Classification of clothing pattern knowledge

Even though the characteristics of clothing pattern knowledge are applied to automatic pattern of clothing, there are still many obstacles in using them flexibly on the computer. The knowledge of clothing pattern must be classified to summarize the common laws between different patterns, which can be better communicated and understood by the computer.

The knowledge of clothing pattern is divided into two major categories: fundamental pattern knowledge and processing pattern knowledge. Fundamental pattern knowledge contains information on style, fabric and size specifications. Style information includes four-piece, six-piece and others. Fabric information includes cotton, wool, polyester, ramie, chemical fiber textiles and so on. Size specifications information include 160/84A, 165/88A, 170/92A, 180/100A and others. Processing pattern knowledge contain information on point, line and plane. Point information includes human body points, numerical moving points, vertical points, intersection points, extension points, equal diversion points and auxiliary points. Numerical moving points consist of direction and data information. Line information includes basic line, structure line and auxiliary line, in which auxiliary line includes parallel lines, vertical lines and angles lines. Plane information includes welt pockets, big pockets and diamond darts. The classification is too diversified, thus in order to present them in a clearer manner, a mindmap is created, as shown in Fig.2: (+ represent the project has the partial contents, maybe one or more).

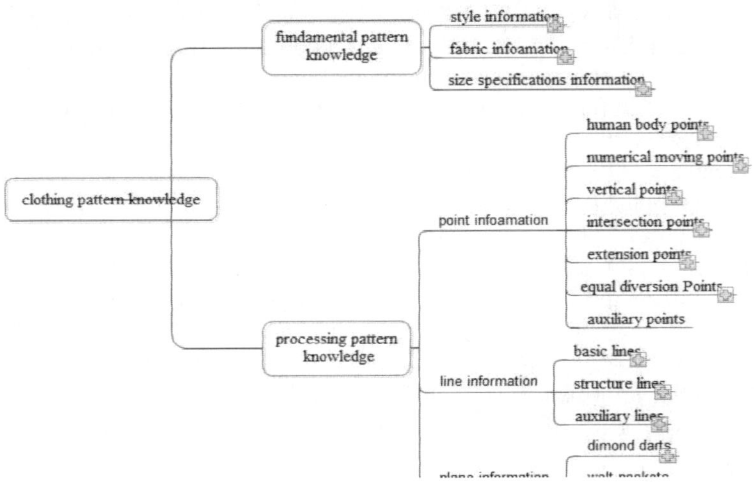

Fig.2 Mind map of clothing pattern knowledge classification

3. Knowledge representation of clothing pattern based on ontology

3.1 Class abstraction based on ontology relations

In the classification of clothing pattern knowledge, data, direction and start point consist of numerical moving points. Both data and direction can be classified as independent class-data. According to dart shape, the dart can be divided into three classes: sealing plane type, half sealing plane type, and separating line type, which overlaps in both plane and line. It can also be classified as an independent class-dart class. The example of the overlapping categories that dart belongs to is very common in the classification of clothing pattern knowledge. Thus, class simplification is necessary to reduce confusion. In this paper, ontology relationships are put forward to help class simplification.

Ontology relationships consist of part of, kind of, attribute of, and instance of. Part of represents the relationship between the whole and the part among concept. Kind of represents an inheritance relationship, like that seen in parent-and-child relationships. One concept can be regarded as the property of the other, which represents the meaning of attribute of. Instance of represents the relationship between concept and instances [15].

According to the characteristics of clothing pattern knowledge, class simplification must conform to three principles. The first principle is that the class being simplified must be the tree nodes of pattern knowledge classification. Class is a collective concept in which items belong to a particular class share the same property. Its subordinate floor at least shares common features with the original class. The subordinate concept is called a tree node and the concept that does not branch off to any subordinate concepts is called a terminal node. Only a tree nodes can be considered a class. For example, nape, bust point (BP) and side shoulder point are simplified to human body point, but they do not have any subordinate collective concepts. They can be called terminal nodes. The human body point is a tree node, as well as a class. Secondly, the terminal concept nodes possess the properties and instances of the parent concept. When the parent concepts are simplified to the last floor, the common property will be not found in this floor. Some scattered concepts are neglected. They are terminal nodes and regarded as instances. For example, the bust line is the instance of the level basic line. The start point, direction and data are the instances of the directional numerical point. Thirdly, some certain nodes can be under a different class in pattern knowledge classification. For instance, fabric information, although style information and size specification information are part of the fundamental pattern knowledge, they can be an independent class as well.

The simplification method is to tag part of, kind of, attribute of, and instance of in sequence like that in the mind map first, followed by the application of the ontology relationship to pattern knowledge classification. The relationship will be transmitted among concepts of clothing pattern, and the relationships need to be screened to ensure the uncrossed relations. If the uncrossed relations exist, we must modify until the relationships are correctly identified.

In this paper, two class As are simplified. They are fundamental pattern knowledge and processing pattern knowledge. The class A - fundamental pattern knowledge is further branched off into three class Bs. Stating the number of Class B - fabric, style and size specifications - is too complex, so the numbers are replaced with n. The class A - processing pattern knowledge is subdivided into six class Bs, 23 class Cs, and 2 class Ds. Class D contains points and lines of a certain style. Different style has different points and lines. Thus, the specific number of class D cannot be predicted and is replaced with n. The specific classes simplified are as followed:

Class A : fundamental pattern knowledge {style information, fabric information, size specification information};

Class A : processing pattern knowledge {point information, line information, plane Information, data information, direction information, dart information};

Class B: point information {human body point information, numerical moving point information, vertical point information, intersection point information, extension Point information, auxiliary point information};

Class C: human body point information {nape, BP, side neck point, shoulder endpoint, front neck point, elbow point};

Class C: numerical moving point information {directional numerical point, line direction numerical point};

Class C: intersection point information {line an line intersection, line and curve line intersection, curve line and curve line intersection};

Class C: equal diversion point information {point of bisection, three equal points, n equal points};

Class D: auxiliary point information{auxiliary point 1, auxiliary point 2, auxiliary point3, auxiliary point 4, auxiliary point 5, auxiliary point 6,auxiliary point n......}

Class B: line information {basic line, structure line, auxiliary line};

Class C: basic line information {horizontal basic line, vertical basic line};

Class C: structure line information {straight structure line, curve structure line};

Class C: auxiliary line information {parallel line, ordinary line, vertical line, angle line};

Class D: horizontal basic line information {waist line, bust line, hip line......}

Class D: curve structure line information {neck line, armhole line, side seam line, vest hem line, underarm curve line......}

......

3.2 The formalized representation of clothing pattern knowledge based on Pretégé 4.3

Ontology is a formal, explicit specification of a shared conceptualization. Formalization is one of the typical characteristics. In this paper, the ontology editing tool Pretégé 4.3 is used to simplify the class, which will contribute to knowledge sharing and refuse of clothing pattern. Pretégé 4.3 is an ontology editing and knowledge acquisition software developed by Stanford University based on Java language. Pretégé 4.3 is the newest edition at present. It is also an open source software that the editor based on knowledge and belonging to the open source software. The software is mainly used for the construction of ontology in semantic web, which is the core development tool. The 4.3 version is a relatively new version. Pretégé 4.3 provides ontology concept classes, relationships, attributes and instances for building. It uses specific ontology description language and help users build a domain ontology model on the conceptual level. The formalized representation interface of Pretégé 4.3 is shown in Fig.3 and Fig.4.

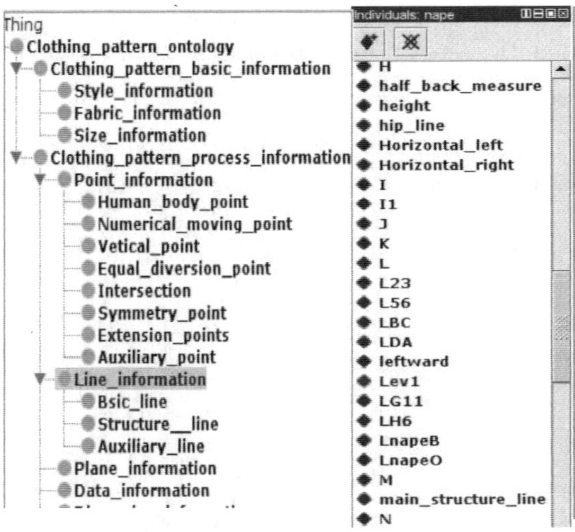

Fig.3 The knowledge representation of clothing pattern in Pretégé 4.3

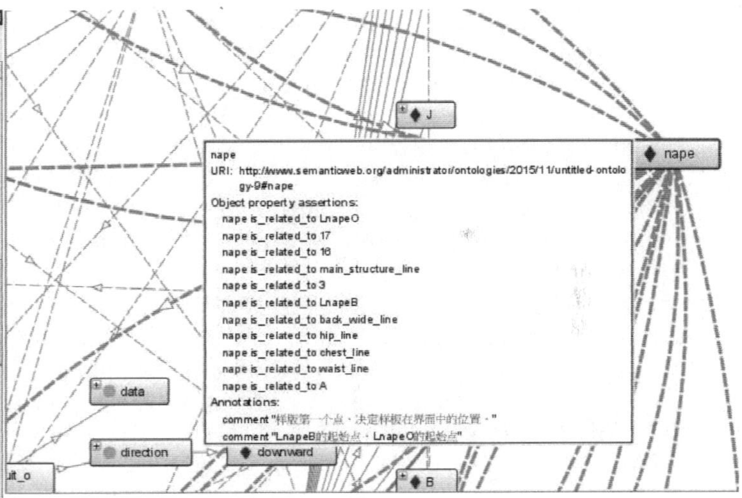

Fig.4 The relationship of nape in Pretégé 4.3

Notes of Fig.3 and Fig.4: Fig.3 represents the classes, from class A to class B, and the individuals (instances) of the class.

Fig.4 shows the relationships between the instances.

As shown in Fig.4, around 11 lines are issued from nape, which shows that there are 11 elements related to nape. In addition, the nape point coordinates will cause the related point coordinates. Each line represents the relationship between two components. This is a network of relationships on clothing pattern. As the saying says, pull one round and move the whole body. The linkage of changes in point and information are implemented as the major difference between Intelligent Clothing CAD and common Clothing CAD. It is confirmed that formalized representation of clothing pattern knowledge has great meaning and function.

4. Conclusion

In this paper, the characteristics of clothing pattern knowledge are summarized into four points: relevance of knowledge, restriction of knowledge, parameter and knowledge, and uncertainty of knowledge. The categories of clothing pattern knowledge are summed up and represented in a mind map. This paper set a principle of abstraction and abstract method on the classes of clothing pattern knowledge. This paper has simplified around two class As, nine class Bs, n class Cs and n class Ds. After the categories are simplified, the concepts and relationships of clothing pattern become cleare. Through further formalization by Pretégé 4.3, clothing pattern knowledge can be recognized and utilized by the computer, in which invisible knowledge can be transformed into visible knowledge. Pretégé 4.3 contributes to knowledge sharing and refuse, and the realization of semantic pattern in clothing.

5. Acknowledgment

The authors would like to express their gratitude towards the Apparel & Design College of Xi'an Polytechnic University.

References

[1] Studer R, Benjamins VR, Fensel D. Knowledge Engineering, Principles and Methods. Data and Knowledge Engineering: 1998; 25(1- 2): 161-197.

[2] Hua L. Research on Ontology-based Emergency Domain Knowledge Expression and Reuse. PHD of Tianjing University: 2008; 5: 15-16.

[3] Perez AG, Benjamins VR. Overview of Knowledge Sharing and Reuse Components: Ontologies and Problem-Solving Methods. In: Stockholm V R, Benjamins B, Chandrasekaran A, eds. Proceedings of the IJCAI-99 workshop on Ontologies and Problem-Solving Methods (KRR5) : 1999: 1-15.

[4] Mike Uschold, Martin King, Stuart Moralee, Yannis Zorgios.The Enterprise Ontology. The Knowledge Engineering Review: 1998: 13.

[5] Gruber TR. Towards Principles for the Design of Ontologies Used for Knowledge Sharing. International Journal of Human Computer Studies: 1995; 43: 907-928.

[6] Qing TZ, Cun GC, Yuefei M.Research on Ontology-based Mathematical Knowledge Acquisition and Knowledge Heritance Mechanism,microelectronics and computer: 2003; 20(9): 19-27.

[7] Qing TZ, Cun GC, Gai GT, Yi FS, Hua WL. Methods for mathematical concept knowledge representation, acquisition and management with in NKIMath. In: Second International Conference on Knowledge Economy and Development of Science and Technology, Beijing, SePtember17-19: 2004: 27-35.

[8] Dan HZ, Cun GC. Acquiring knowledge of Chinese cities from the encyclopedia of China. In: Proceedings of ICYCSPZOOI,Beijing: 2001: 1191-1193.

[9] Qing TZ, Cun GC, Hua D, Yan QL. NKIMathE2A multi2Purpose knowledge Management environment for mathematical concepts.In:Lang J,Lin F and Wang J. (Eds.).KSEM2006, LNAI4092: 2006: 625-636.

[10] Qing TZ, Hua D, Hong MY, Yan L, Yong QL. Applying research on domain ontology for knowledge processing. Information Learned Journal: 2006; 25(6): 713-719.

[11] Si JX, Cao CG, Wang HT, Gu F, Feng QZ, Zhang CX, Zeng QT, Mao WT, Zheng YF. An Environment for Multi-domain Ontology Development and Knowledge Acquisition Engineering and deployment of cooperative Information systems: 2002: 104-116.

[12] Ge FZ, Qi X. Design of distributed search engine system for apparel based on semantic Web services. Computer Applying: 2009; 49(2): 45-48.

[13] Yu QL, Dao LL, Guo TS. Establishment and Application of Textile and Clothing Domain Ontology Knowledge Base.Modern textile technology: 2013; 6: 21-24.

[14] Bei C. Study on the construction of bra product knowledge model based on ontology. Master thesis of xian polytechnical University: 2013; 2: 3.

[15] Maedche A, Staab S. Ontology for the semantic web. Intelligent Systems, IEEE: 2001; 16(2): 72-78.

Discussion on the Reconstruction of Fabrics in Creative Design of Clothing

You-Jun Tong, Xin-Yan Lu

Xiamen Academy of Arts and Design Fuzhou University, XiaMen, Fujian, 36000, China

*Corresponding author's email: tyjart@126.com

Abstract

Taking fisher-woman dress in the coastal environment as an example, to analyze the effect of the natural ecological environment on dress, the functional effect of Marine natural environment on headwear and dress, the effect of the environment on clothing material and color, and the effect of humanistic ecological environment on clothing decoration and accessories under the background of Marine culture. it is concluded that regional dress features, from the structure to the color, occurred, developed and evolved in a particular ecological environment. It has close relationship with the environment. the difference of the natural environment is one of the important factors of diversification of clothing design. It comes from nature, and also beautify the natural. This is the charm of Oriental clothing art and the clothing art beauty important features.

Keywords: Creative design; Fabric reconstruction; Art performance

Introduction

Clothing is an important part of the fashion, and the fabrics is the main physical media for clothing, in the process of clothing design, selection and design of fabric play an important role. Relative clothing style in terms of style, fabric change rhythm is often faster, stylist the visual acuity of fabrics and the secondary design ability determine and reaction costume design the ultimate success or failure. In the clothing design, how to effectively select and transform first-hand material, the designer make it more visual feeling, texture, and rich clothing modeling, function and visual effect, fabric design as a "barometer" lead the fashion trend in clothing.

1. Clothing creative design fabric physical properties of art

Fabric reconstruct on the premise of the design need, to enhance the clothing art effect as the goal, using various technological means to secondary existing fabric design process. Fabric of physical treatment is to point to only change the original appearance of the material form, in the skin texture, form, or texture change of fabric reconstruction, such as flat single become or have the appearance of the fabric surface texture, or there are ups and downs, or design, or superposition, or hollow out, or are joining together, or with embroidery and so on, Recreation give materials to new look, new features, new style, personality rich original fabrics, expand the use of the fabric, add the visual aesthetic feeling of clothing.

1.1 The same fabric material for reconstruction of the physical quantity of art

Physical quantity reengineering art creative fabrics, mainly refers to the original fabric quantity and two aspects of reducing, added amount in increasing the volume of a garment, overall and local increase, through the material various reconstruction technique to achieve the visual effect of garment size. The fabrics of the same material by superimposing, composition, techniques, such as repeatedly because of material texture is unified, no matter how complex level, is able to mediate on the vision, repeating the same elements, extremely easy to order and unity. The volume change of fabrics of processing is one of the commonly used gimmick recreation, through a variety of craft technique, make the fabric itself the original plane state produces change, fold the fabric into different shapes, present a certain sense of body modeling method, the purpose is to cause the fabric surface turning, ups and downs and various spatial change and combination, the fabric of the original fabric structure basic remain unchanged. For example, through the smoke plait, folds and contraction joint, weaving technique, the original surface morphology change fabrics, in does not destroy the basic structure of the premise, the person gives on the vision with level change, forming fabrics relief effect and stereo feeling, enhance the volume of a garment [1]. Fabric art, of course, reengineering and process characteristics must be based on the fabric itself to implement creative design, all kinds of fabrics and have specific attributes and characteristics of the production process, when the fabric art redesign, should try to play to the fabric and craft technique specialty, show the most suitable for artistic effect.

And add more relative to the decrement of reengineering art, is the original fabric decoration processing, reducing fabric characteristics such as basic material, the color remains the same, only change the physical component. Such as drawnwork, hollow out, cut, cut, burn, burnt-out, tear, abrasive cleaning and so on, three-dimensional space visual effect. Recreation is closely related to the clothing modeling, the structure is, uneven, often overlapping space change garments' integral modeling sculpture is more stereo feeling. Recreation, unlike a design, one of its basic characteristics is not out of fashion design itself and isolated, reengineering is in understanding the original fabric performance and characteristics, to ensure its comfort, functionality, security, etc, on the basis of will consider including of the basic elements of fashion design, reasonable use of the technology method, stress individual character art connotation, highlight the aesthetic feeling and decoration of a design [2].

1.2 Use of fabrics made of different materials for reconstruction of the physical art

To add a different decorative material of the original fabric of is another common way of recreation, designers through the use of sewing, embroidery, nail, stick, compose, hang wait for all sorts of adornment gimmick. Add design on the existing fabric, compare the different material, the performance of different body feeling, the visual effect of rich fabrics, craft technique with filling, Mosaic, printing, embroidery, beading, patch, ribbon, lace, hot stamping, etc [3].

Comparison of different material is creative fabrics commonly used gimmick, traditional embroidery in the design of process reengineering is different materials the most traditional, usage, a design technique, the most flexible embroidery material and color can be according to different design needs to change. Traditional embroidery craft culture deep, adornment effect is only beautiful rich, exquisite craft technique, design patterns, such as he, stick cloth embroidery, beads embroidery, carving embroidery, ribbon embroidery, etc., as a carrier of the embroidery, the material with different texture, through different embroidery techniques to achieve the artistic effect of material re-creation. With the development of new materials, new technology, the design technique and technology of the "embroidery" skills constantly change, a variety of material mixture of comprehensive technology of mature, and conventional materials, new processing method, which makes the embroidery craft of further innovation. such as disorderly needle embroider, it breaks through the traditional method of flat embroidery, stitch interweave, flexible, with all kinds of thread the iterated complex admixture, finished products rich color level, have a strong stereo feeling. Anyway, regardless of the change embroidery materials, or change the way the embroidery, or change the style of the design, the new embroidery effects are to the designers, the wearer to the design of the new feelings and experiences (Fig.1, Fig.2).

Fig.1 Embroidery1

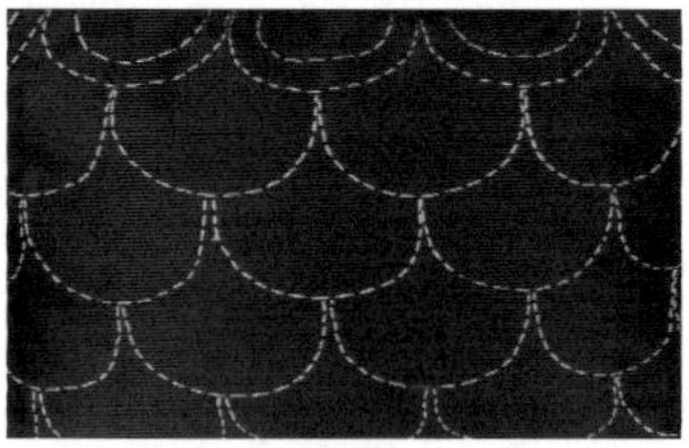

Fig.2 Embroidery2

1.3 Creative art of hand-painted fabrics

Hand-painted performance art design is a creative fabrics, fabric hand has a long history, now science and technology are development, Hand-painted art is a common method of creative fabrics (Fig.3). theme selection is flexible, such as animals, plants, character, geometry, landscape, architecture, open text, symbols, and so on. The composition forms of hand-drawn fabrics is various, horizontal or oblique, or simple or complex, scatter or dense, majestic or low-key reservation; Color performance can be pure and fresh quietly elegant, hand-painted fabrics can also be more brilliant; Hand-painted fabrics can draw and spray, flat coating and cancel, carefully carves and bold splash-ink, render large fine and painting in a small range. of course, Hand-painted can also be combined with other methods to achieve the perfect fabrics adornment effect, such as printing paint combination, dye paint combination, effect is often better especially with embroidery, nail bead and hot drilling technology [4]. hand-drawn parts of fabrics can be before or after, on or under, shoulder or chest, sleeve or skirt. with the constant improvement of the people and literacy conception, hand-painted clothing will increasingly favored by consumers, offering people increasingly rich and colorful life.

Fig.3 Hand-painted

2. Reengineering art of Creative fabric with chemical change

2.1 Chemical dyeing processing

Secondary printing and dyeing is the usually methods in fabric color respect. garment or the partial or whole of piece is dyed again to present a unique and rich color effects, such as gradient, local coloring. It has a strong hand lasting appeal and very original with manual operation [5]. The special manual dyeing include tie-dye, pour dye, batik dyeing and so on. The tie-dye is convenient and easy, has lively artistic characteristic and nature pattern, fresh and elegant. in the second fabrics design the printing and dyeing can be national, fashion, modern and traditional. the batik process is relatively complex, and must use reserving agent. people draw the pattern on cloth dipping into liquid wax with copper knife, pen or brush pen. After solidification, some subtle cracks generated. after putting in cylinder dyeing, waxed part can not be colored, presenting themselves. without wax and wax cracking appear patterns because of dye penetration natural dyeing. After remove wax with the boiling water and washing, patterns appear like drawing, having a unique ice grain effect. Above several kinds of dyeing can be successively to be hand-drawn, freehand brushwork in traditional Chinese painting, realism for needs.

2.2 Chemical treatment of fabric reconstruction

Burnt-out printing processing is also called corrosion coking printing. it is the intertexture of two fiber interweave, using the different acid and alkali resistance of two fibers. the effect depends on fiber decay degree mainly. The visual effect of burnt-out mainly reflects in fabrics do old, and make the fabric form hollow-out decorative pattern to realize space visual effect (Fig.4).

Burnt-out can also be used with printing at the same time. adding the appropriate tolerance in printing paste dye, while decaying certain rot fiber with reactive dyes or pigments on burnt-out products whole or part pattern to get burnt-out color effect [6]. water processing art achieve the visual effect of a natural do old through enzyme washing Mainly. enzyme is a kind of cellulose enzyme. it can degrade fiber structure in a certain PH and temperature, and make the cloth fade gently. cloth surface will produce a layer of soft cream white fluff with stone mill after washing. the fabric looks like fade effect and outdated feeling.

Fig.4 Manual dyeing

3. Achieve artistic effect with The combination of fabric reproduction and science technology

3.1 Photoelectric effect artistic effect of fabric reproduction

With the development of new textile materials science and technology, and some new composite materials with hi-tech appear constantly. Designers use it in innovation design of fabric reproduction. Technologies such as sound, light, electricity combine with stereo modeling, structure decorative sewing of fabric to perform themes and create atmosphere better. common metal coatings, plastics synthetic, glass cutting, metal silk, change of glow diode electricity, heating techniques are used to change the material of a part of the attributes and functions, and the local face fabric, to realize different clothing form and give people different feelings.

3.2 Digital printing reproduce fabric design art

Digital printing is an effective technique to realize the fabric pattern and color. digital printing can create color combination flexibly, and quick change color visual effect. It break through the traditional printing tinted on craft process limits. Fabrics digital printing pattern design maximum possibly rich fabric color vision, and realize the design and creation of photo images and details. Based on the change of pattern design and creation, we can expand the visual field of the effect of fabric design. The changeful special pattern effect is developed, Such as 3D effects, exercise effects, mixing effect, light effect, acid effect, tie-dye effect, watercolor effect, texture effect, simulation effect, neon effect, etc. Digital printing promote the popularity of the fabric pattern innovation effect and increase the interestingness and vitality of clothes [7].

3.3 3D printing technology and art

3D printing is a kind of rapid prototyping technology, and unlike digital printing digital printing. Digital printing is 2D. 3D printing is a three-dimensional display technology, not only can print high simulation 3D printing design, but also can select material to print model. 3D printing fabric and accessories have been used in the innovative design of clothing or accessories design, even the leather and fur. 3D printing is a revolutionary breakthrough for fabric 3D effect, 3D printing technology change garment life and fashion way in the future [8]. The explore of the combination of 3D printing and new materials brings new attraction and mystery for clothing creative design, and the possibility of seamless garment.

4. The conclusion

In the process of fashion design, both of the art reconstruction of fabric and costume design influence and make up each other. the art reconstruction of fabric change and enrich the clothing style and model. Fabric art reproduction integrate fabric design and fashion design in certain level. the combination and reproduction of fabric gradually become a new breakthrough of fashion design, and become an important means of adding garment products value. for the designers, Fabric reproduction offer a wider choice of space.

References

[1] Wu XQ, Wu J. The use of the recreation of textile art design of clothing. Textile science and technology progress: 2007; 01: 85-86.
[2] Liang HE, Liu S. Simple discussion fabric art design principle and the design program. Jiangnan university journal: 2008; 10: 147-148.
[3] Hhi NN. Introduction to hand-painted clothing decorative and personality. Design: 2015; 11: 112-113.
[4] Wu QH. Introduction to hand-painted clothing decorative and personality. Art Panorama: 2009; 10: 102-103.
[5] Ma XM. The diversity of fabric art and creative design. China textile leade: 2012; 04: 93-94.
[6] Zhang XX. Burnt-out technology research and product development. Tianjin Textile Science & Technology: 2014; 3: 25-27.
[7] Ni M. Analyses the influence of digital printing on the fabric pattern design. The fine arts education research: 2015; 08: 82-84.
[8] Zhou LL. 3D printing instead of digital printing in the future development trend of garment design. Meiyuan: 2015; 06: 54-55.

Study of Knit Woolen Fabric Pattern Design and Processing based on Database Building

Man-Yi Zhang, Yuan Xue*, Jin Duan, Bo-Ni Zhang

Xi'an Polytechnic University, Xi'an, Shaanxi, 710048, China

*Corresponding author's email: xyuan98@hotmail.com

Abstract

This paper attempts to study application of pattern in knit woolen clothing based on weaving technology. Knit clothing is getting more and more popular in modern society as it is casual and comfortable. When compared with woven garments, it has more limitations on modeling but can be more innovative in terms of color and pattern. In this forward-looking context, abstract painting is selected as an example in this paper to study into pattern designs suitable for knitting woolen fabric combined with weaving technique characteristic. Several experiments were done, and some pattern design requirements about pretreating, size processing, color processing and cycle processing could be summarized which make the patterns produced more conviniently. Additionally, the results can also provide some references for establishing pattern database of knitting woolen fabric.

Keyword: Knit woolen fabrics; Pattern design; Drafted pattern; Pattern database

1. Introduction

Knit woolen clothing with good elasticity is loved for the comfort it provides - it is one of the most important categories in modern apparel. Recently, consumers both globally and in Mainland China have become increasingly interested in knit woolen goods [1]. However, the supply of single-colored knit woolen clothing has been unable to meet consumers' demand for personalized clothing - there is a need for variety of pattern elements used in knit woolen fabric is very necessary.

However, many domestic knit woolen enterprises are not reacting to this trend. The fabric pattern is mainly designed and developed by factory workers, not specialized designers, which leads to the lack of fabrics with artistic patterns. This makes it difficult to satisfy the psychological needs of consumers and develop a strong clothing brand image [2]. However, when combined with weaving technology, pattern design can find more practical applications in production [3,4].

Although some domestic studies focus on pattern design, pattern design of knit woolen clothing is rarely researched. On one hand, design and position printing of knit garment patterns that follows the trend of restoring ancient ways is studied by Zhang Ting in his famous work on design elements [5]; The aesthetic significance and application prospects of the modern abstract art design methods in digital printing are studied by Ding Zhuo [6]. Lulisha, Song Xiaoxia et al in Mexico porcelain as the research object, discussed patterns and color characteristics, and expanded application of knit fabric patterns [7]; Through analysis of Van Gogh's representative works, Wu Yufei found that some artistic elements in Van Gogh painting can be used in knit garment pattern design [8]. All these studies were conducted only from the art design viewpoint - they don't align studies of design and the manufacturing process.

On the other hand, Zhang Dan introduced warp-knit fabric image acquisition and pre-process and then described how to texture image segmentation algorithm knitting for automatic identification [9]; Wu Haihong proposed an algorithm based on color and spatial information of fabric texture. It is based on using color distribution uniformity measurement to judge whether or not a pixel is located at the boundary of a certain part of a garment [10]. These papers focus on extraction and recovery of knitting texture from the fabric and then carrying out the design of the pattern. There are few studies that look into on extraction of patterns from paintings which are then used as a base for knit wool woven fabric texture creation. According to this, abstract expressionism paintings, for example, can be used to extract pattern elements for knit wool fabric design. Additionally, various requirements of knit wool fabric design can be consolidated into a database and later used as a reference in garment production design.

2. Experiment

2.1 The experimental sample pretreatment

The experiment was conducted in the pretreatment stage mainly for pattern selecting and analysis of visual estheticism.

2.1.1 The sample input

Before the experiment, the first action is to select and collect abstract paintings that would be possible to use as patterns in knit wool clothing. The samples can come from two sources: collecting design resources directly from the Internet or using input devices to collect patterns from physical objects [11]. The outlines of post-processing will be more clear and the resulting image will be more detalized if resolution of 300 DPI is used in the experiment.

2.1.2 The sample pretreat

The pre-processing is required because the image may be distorted when scanned or shot with different light or at a different angle [12]. In general, image preprocessing methods will cover image information to some extent, therefore, the original image information should be reserved for further use when dealing with the sample [13].

In the real factory settings, workers collect patterns with their mobile phone, as it is simply convenient, which immensely reduces the accuracy of scanning. The pattern will be distorted because of the light or angle deformation in the shooting (e.g. Fig.1a - the photo doesn't perfectly fit into a rectangular box). In addition, when collecting images from books, smooth surface of coated paper, which is used in numerous publications with pictures, has strong light reflection glare that masks the real figure (see Fig.2a, Fig.2b - the lighting of the image is unbalanced).

(a) Original image (b) Using "cut" to repair (c) Using "deformation" to repair

Fig.1 The correction of distortion sample

(a)Before a high reflective image processing (c) Before shadow difference image processing
(b) After high reflective image processing (d) After Shadow difference image processing

Fig.2 The correction of color interval sample

To avoid these issues, firstly, a correct angle should be selected and the picture should be flat and clear in order to balance the light conditions, avoiding the influence of highlights and shadows during shooting. Then, if it's not possible to eliminate the issues completely, the pictures need to undergo post-processing with graphic software (e.g. Adobe Photoshop). The distortion can be fixed with "deformation" tools for shooting angle issues (such as Fig.1c), or directly adjusted from the maximum deviation of each side of the image with "cut" tool (see Fig.1b); The shadow problem for lighting can be repaired by adjusting image contrast, color balance color interval, which will increase clarity of the image (see Fig.2 - correction of color interval sample).

2.2 Pretreatment and analysis of sample aplication

After pretreatment of samples is complete, the main consideration is how the selected sample painting can be applicated to be used in clothing design [14]. The samples need to be applicated and analyzed, and there are three main ways to do this.

2.2.1 Direct use of abstract paintings as clothing patterns

In pattern development, demand of technology and the beauty must be met simultaneously. In other words, it should express the aesthetics of clothing items, achieve visual harmony and satisfy manufacturing requirements. These patterns were relatively easy to handle in the subsequent process by simply extracting a sample unit of the pattern (as shown in Fig.3a and Fig.3b). Finally, similar color of the sample unit should be merged up to 4 colors by DesignStar, otherwise known as the design system for artificial fur.

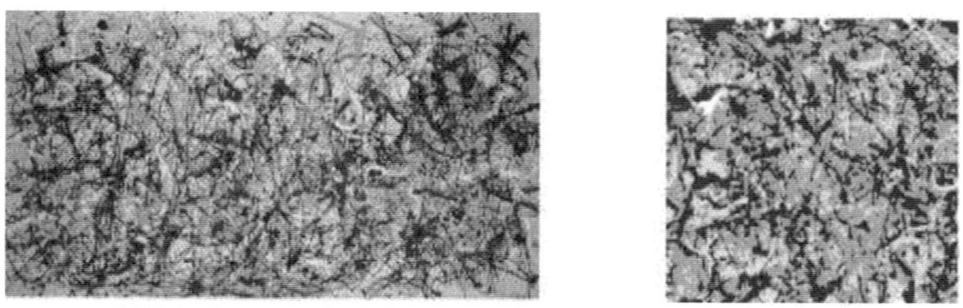

(a)Painting (b)Ab - pink single cycle

Fig.3 Directly extracting painting pattern processing

2.2.2 Partial modification of abstract paintings for creation of clothing patterns

For the next pattern, redesigning is carried out by simplifying, adding, reconstituting, deforming. Introducing new lements and designs each year is popular in different areas of fashion, which can closely unite the garment design industry and the modern society (e.g. Fig.4b is designed based on Fig.4a).

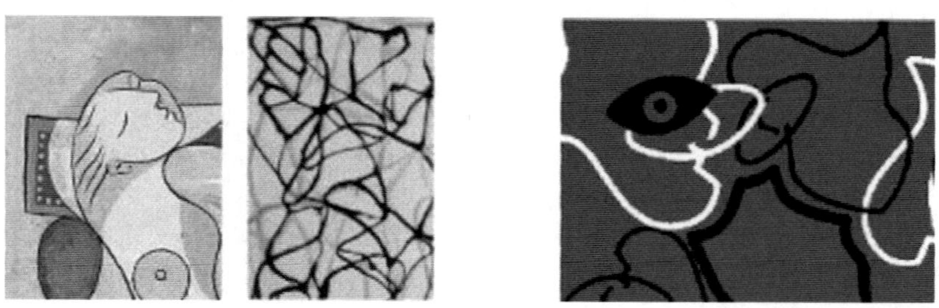

(a) Painting (b) Ab - black single cycle

Fig.4 Extracting painting to do local modification

2.2.3 Extracting elements of abstract painting for pattern design

It is possible to extract painting color combinations and composition to create unique designs, thus using the classical elements to synthesize new clothing patterns (e.g. Fig.5b uses Fig.5a as a reference in the terms of the color scheme).

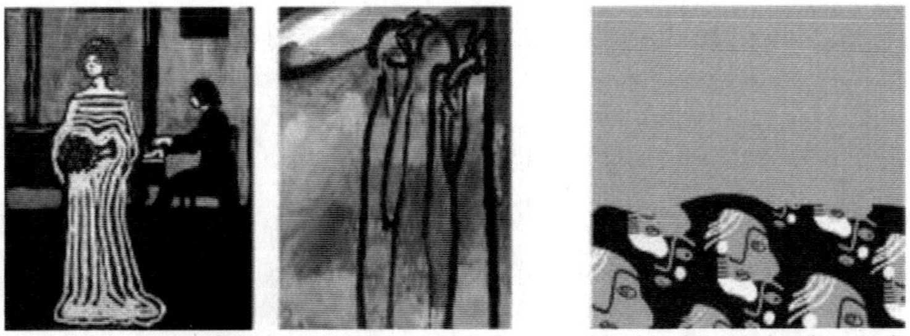

(a)Painting (b) Ab -blue single cycle

Fig.5 Extracting painting element to do pattern design

3. Process of transforming an original pattern into a cycle pattern

If a pattern cannot directly be applied to knit woolen fabrics, it should be converted into into a cycle pattern (a graphic that uses symbols and colors to depict the pattern in small squares [15]). During this modification, issues of size, color types and cycle-splice of the pattern should be taken into consideration. The process of creating cycle pattern mainly utilizes the DesignStar.

3.1 Size processing of transforming an original pattern into a cycle pattern

For patterns that that need to be drawn and modified again, it is possible to utilize common software (such as Adobe Photoshop) and then export to DesignStar design system of artificial fur in*.bmp format to process the cycle pattern. It is possible to draw the cycle pattern in the system of DesignStar. Before drawing, it is important to control pattern size and then calculate the size of its cycle pattern according to the size of the fabric pattern that will be used in production. It is mainly about converting the pin numbers of the machine into a pattern pixel (px) to carry out the calculation. At present, the available industrial fabric sizes are L (2 pins/cm), M (4 pins/cm) and S (8 pins/cm), among which the fabric style in L is the most rough one, the M style is less rough, and style in S is the most exquisite (according to the px of each size):

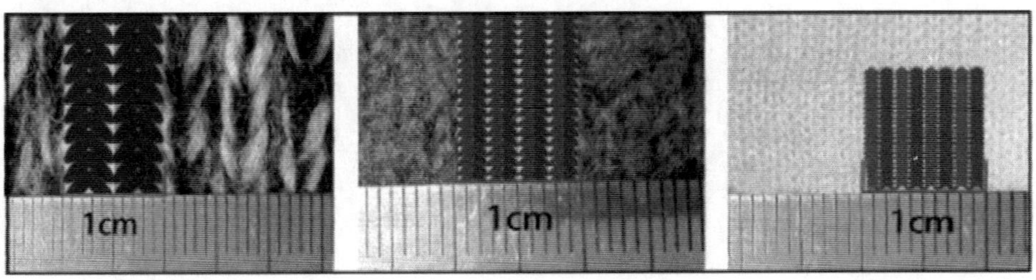

(a) L size (b) M size (c) S size

Fig.6 Sizes of the cycle patterns

According to the formula, it is known that:

$$Pi = Y \times Q_n \quad (1)$$

Here, Pi refers to pixels, Y is the needed pattern length (unit: cm), and Q_n is woven specifications (specifications of the three sizes are respectively L: Q_1=2, M: Q_2=4, S: Q_3=8). Take a single image in 5b as an example. In weaving, width of a single image of the real fabric should be roughly 30cm. If it is produced in S size, then 30*8=240 pixels shall be set in drawing of the cycle pattern. In this way, size of real products can be guaranteed to be appropriate.

3.2 Color processing for transforming an original pattern into a cycle pattern

Design system of this pattern will mainly control the color by dividing all the color into layers through "hopper-feeding picture". A refers to color of the whole pattern, and the hopper-feeding picture consists of several nonempty subsets A_n of A, of which $n \in (0,4]$ and $n \in N+$. In other words, n is the color type of the pattern, and the real weaving color shall be limited to four basic colors.

For example, there are three colors in the pattern Ab-blue (Fig.7b), and they are respectively 15001 light blue, 15075 off-white and 15019 dark angelica green, so it is essential to produce three hopper-feeding pictures (Fig.7a). One color will be extracted from each hopper-feeding picture, and the pattern will be divided into three color layers to weave (substituting green for white to distinguish it from the background).

(a) Ab-blue hopper-feeding picture (b) Ab-blue pattern

Fig.7 Design of hopper-feeding picture

In terms of patterns with four and more colors, it is essential to combine the colors within four. Other bitmap patterns drawn by other software are shown in Fig.8a. For patterns drawn by Photoshop with many similar colors borderline with each other, it is required to be exported to the DesignStar system in order make use of the color combination function to carry out the filling of similar colors of the pixel. After color combination, mixed colors on the edge of the pattern will be combined into pixels of identical color (Fig.8b).

(a) Original pattern (b) Color combination

Fig.8 Color processing of a cycle pattern

3.3 Cycle processing of a cycle pattern

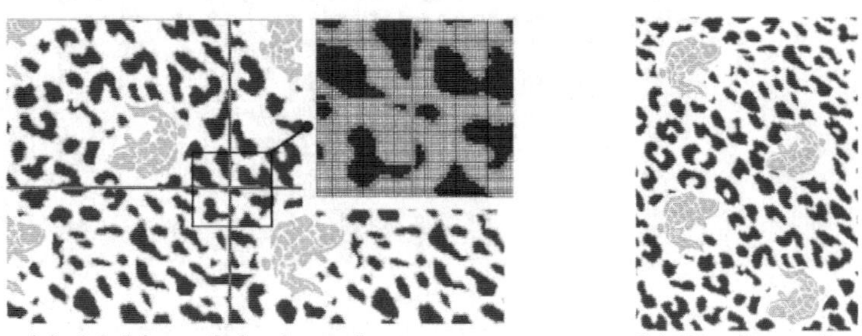

(a) Linking of single cycle pattern (b) Single cycle

Fig.9 Single cycle processing of the cycle pattern

As the size of the finished wool fabric is about 1.5m, control of the cycle pattern unit shall be taken into consideration in the design. While designing the pattern, suppose that a cycle pattern is limited to a rectangular frame with length and width of x and y respectively. It means that x is the length of a single cycle pattern, and y is the width of a single cycle pattern ($x,y > 0$). The length of x determines the splice of longitudinal repeat. If the length of fabric is larger than x, then longitudinal cycle pattern shall be considered. The length of y determines the splice of horizontal repeat. When the value of y is 1184px, the width of a single cycle will be equal to the width of cloth, thus eliminating issues of horizontal repeat splice. When the value of y is less than 1184px, artificial modification of horizontal cycle shall be made, so as to make the whole pattern spread soft and smooth. As is shown in Fig.9a, system of DesignStar is utilized to modify the horizontal and longitudinal cycles of pattern Tr-fish, and finally single cycle that can be directly applied in weaving is generated (Fig.9b).

When the machine can identify the cycle pattern, all the pattern design and processing is finished. Then fabric can be woven, and garments can be produced. The patterns in our experiments and the final garments are shown in Table 1.

Table 1. Test specimens

Name	Single cycle of pattern	Color	Trim size (cm)	Garments pictures
Ab-blue		*15001 lightblue *15075 off-white *15019 dark angelica green	148*200	
Tr-fish	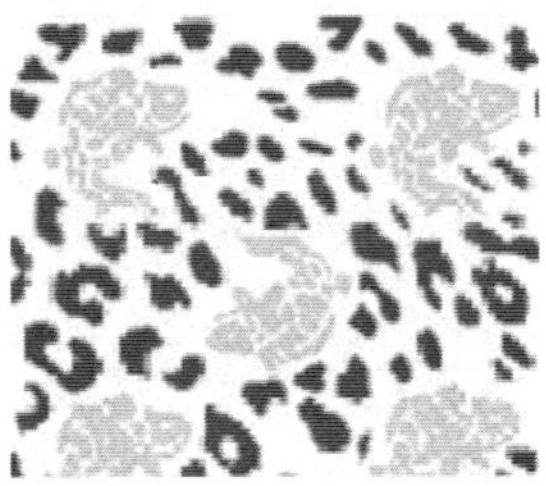	*15047 brown * SX103 yellow *150192# white	45*42	
Ab-black	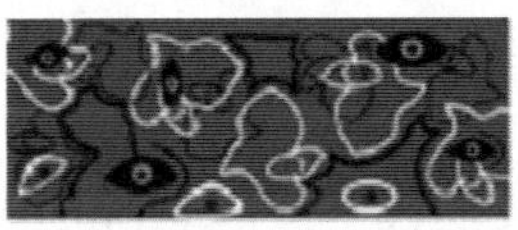	*14104 grey *13020 off white *4F15019 black	148*100	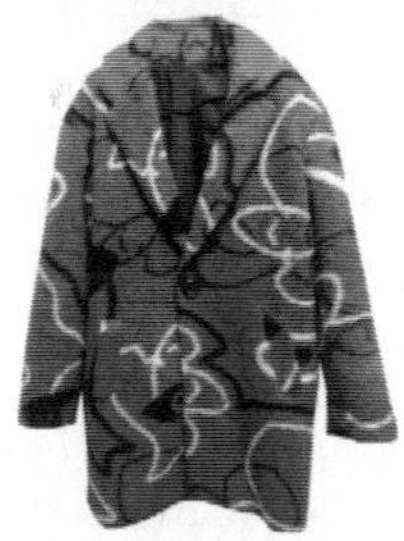
Ab-pink		*SX1410 orange pink *14147 white *14D004B dark grey	28*25	

4. Conclusion

In this study, some abstract paintings were selected as samples, different methods were used in different pattern design, and final patterns identifiable for weaving machine were obtained. Several garments were made at last. Conclusions derived from this research are as follows:

1)When referring to the painting for fashion design creation, characteristics of the garment shall be taken into consideration.

2)Designers should ensure accuracy in fulfillment of each step from collecting patterns when they are designing patterns of knit nylon fabrics. In this way, the number of unnecessary problems will be reduced in post processing which can further improve productivity and enhance fidelity of the pattern.

3)In terms of knit woolen fabrics, it is difficult to preserve fine fashionable patterns. Therefore, pattern design shall produced as clearly as possible and take characteristics of fabrics into consideration before designing.

5. Acknowledgement

We would like to thank the Haoyunlai knitting Co. Ltd. for funding this research project. This project is an important part of the program "Development and application of knitting woolen fabric pattern database system - a case study of abstract painting " of the contract number 2014KJ-048 (HYL201509).

References

[1] Lv H. The kinds and development trend of knitting fabrics. Melliand China: 2007; 04: 71-74.
[2] Hu JL, Li C, Han SG. Analysis of factors affecting clothing brand image and determination of their weights. Journal of Textile Research: 2012; 07: 141-145.
[3] Wang Y, Research on pattern design of knitting wear product. Melliand China: 2011; 05: 28-31.
[4] Li Q, Zhao ZY. CAD system of woolen knitting dress' pattern designing. Wool Textile Journal: 2004; 02: 59-61.
[5] Zhang T. Design and positioning printing of knit garment pattern in the retro fashion. Knitting Industries: 2014; 10: 41.
[6] Ding Z. Design and application of digital printing pattern based on abstract art. Journal of Ningxia University(Humanities & Social Sciences Edition): 2008; 06: 224-226+244.
[7] Lu LS, Song XX. Application of Morocco Porcelain Pattern in Knitting Sweater Fabric Pattern. Journal of Silk: 2015; 11: 60-63.
[8] Wu YF. Research on application of post-impressionism in knitwear-case study of manifestation in Van Gogh's paintings. Zhejiang Sci-Tech University: 2014.
[9] Zhang D, Jiang GM, Cong HL. Automatic identification method of jacquard warp knitted patterns with net. Journal of Textile Research: 2010; 10: 45-49.
[10] Wu HH. Study on key algorithms used in automatic identification system for fabric wave pattern. Zhejiang University: 2006.
[11] Li ZY. The development of computer-aided design system for jacquard patterns. ZheJiang University: 2010.
[12] Hu JF, Lv Y. Text areas segmentation approach for business card images. Computer Engineering: 2008; 21: 169-171.
[13] Pan RR, Gao WD. High precision identification of woven fabric density via image processing. Journal of Textile Research: 2008; 11: 128-131.
[14] Chen XM. Abstract painting in fashion design applied research. Beijing Institute Of Fashion Technology: 2012.
[15] Yang PY. Study on pattern recognition and automatical notation creation of weft jacquard knitted fabric using digital image processing techniques. Donghua University: 2007.

Clothing Technology Research on Vigorous Styles of Zhou Dynasty

Yi-Ping Zhang[1], Jing-Wei Liu[2*]

[1]*School of Fashion & Art Design, Xi'an Polytechnic University, Road jinhuananlu, Xi'an, Shaanxi, 710048,China*

[*]Corresponding author's email: weijingliu@163.com

Abstract

The purpose of this paper is to determine the style of Zhou Dynasty Artifacts by studying the history of Zhou Dynasty Artifacts and identifying adjectives that can be used to describe them. This article integrates the theory of Linguistic Pragmatics and Semantic to explain the designing style. In this paper, take vigorous style as an example, a semantic analysis, is conducted, followed by a description of the design concept. The ultimate goal is to prove that these adjectives can represent vigorous style and can be applied in clothing design.

Keywords: Style; Pragmatics; Vigorous

1. Introduction

One of the main ways in which culture is represented is through language. Language helps put people's thoughts into words, and it also promotes cultural exchange from generation to generation. In this paper, Zhou's artifacts, as a representative of the Zhou culture, some words were collected from intellectual sources such as professional books and documentaries. These words were collected from relevant researcher's evaluation of Zhou's artifacts, and were classified accordingly.

A general picture of the Zhou Dynasty artifacts style is created through questionnaires and interviews with experts. On the basis of traditional aesthetics, modern design elements are blended to better represent a person's personality and style.

2. Style and Pragmatics

2.1 Style

The word "style", originated in the early 16th century from the Latin word "stilus", was first used to represent a form of writing. However, as life becomes more complicated and diversified, "style" can be used in many different contexts. According to Oxford dictionary, the interpretation of the "style" can be divided into a variety of categories and characteristics, showing the relationship between formal language and style features. These features are related to a certain country, a certain group, a certain period, a person. For instance, gypsy style music, baroque clothing and so on [1].

Art historian Gombrich pointed out that one of the most sensitive and subtle scholars Heinrich Wolfflin had a concise discussion on the matter of style in 1988. According to Wolfflin, "style" can be proved the theory, the rest of the mouthpiece with one voice in the historical background at the time, This holistic view is widely accepted by historians and artists. That is, the formation of a period's style features will not only be reflected on a piece of art. It is widely presented in buildings, clothing, music, art, furniture and other aspects;these artifacts together form a complete style group [2].

Style refers to a certain type of artifact as a whole in terms of manifestation and significance, the pursuit of lifestyle, aesthetic and emotional appeal. It reflects an era of collective psychology. Above all, this is the new way of life in the eyes of modern times. People are not satisfied with having the necessities anymore, they hope to achieve spiritual satisfaction through fulfilling their materialistic life. They want to find a new way of life to meet their psychological needs.

2.2 Pragmatics

Pragmatics is a concept introduced by Morris, an American philosopher, in 1938. To determine the accurate meaning and expression of the words used to develop basic principles and guidelines and that these words , various texts and other media of different contexts were studied. Pragmatics in linguistics is a discipline for the study of language meaning. Among the pragmatic definition, a major study is the meaning and other is context. It is specialized to understand the language and use it to communicate appropriately. Words in specific scenarios are examined and to fully understand its usage in different contexts [3].

Language is a symbolic record that exists in the form of knowledge, and represents people's living on abstract form and coding. Knowledge of a language is accumulative and each language presents its own interpretation of the world. Language is an important part of communication that helps people express their feelings and message in art work, but itself can also be an artistic expression in our everyday life.

Previous research has supported the idea mentioned above. For example, Tong JW's paper (2015) "Research of the pragmatic technology of clothing based on Qin culture" [4] talks about how culture is one of the resources used in design. Some special words were identified from the Qin culture that assisted people in expressing their design ideas. Tu SJ's paper (2012) conducted an analysis on the design method of women suit, which suggests that various semantic extension was expressed through different symbolic elements resulted from the collocation between women suit and different clothing [5].

3. Statistics of Zhou style

In this paper, artifacts from the Zhou Dynasty artifacts, such as bronze ware, calligraphy, lacquer and jade,such as Fig.1,Fig.2 and Fig.3, are used as examples to illustrate the impact of culture on language and design. Adjectives that can be used in asthetic evaluation were collected from documentaries and books from the Zhou dynasty. Words that are used by modern scholars to describe existing artifacts from the Zhou dynasty were also collected. Semantic words collected are shown below:

Fig.1 Bronze

Fig.2 calligraphy

Fig.3 lacquer

After collecting these words, manual sorting and lexical semantic analysis of the groups were performed as follows: 1) Sort the collected words and remove words with duplicate meanings. For example, multiple words convey a mysterious feeling and only one of those words is included. 2) Words with similar meanings are grouped together. For example, magnificent and large are put under the same category. 3) Words that share the same characteristics of semantic classification are sorted out to produce the above list.

Table 1. Zhou style semantic group

Zhou style semantic group				
Plump	Natural	Delicate	Bright	Bold
Solemn	Simple	Exquisite	Pretty	Novel
Majestic	Plain	Dexterity	Light	Minute
Domineering	Unadorned	Flexible	Colorful	Complex
Fearful	Classic	Lithe	Sprightly	Enlaced
Mysterious	Honest	Slim	Gorgeous	Overlapped
Weird	Practical	Subtle	Glazed	Ruled
Stiff	Fresh	Accurate	Bonzer	Symmetry
Massiness	Concise	Strange	Ornate	Harmonious
Vigorous	Dignified	Detailed	Luxuriant	Still
Sturdy	Quaint	Rhythm	Artful	Simplified
Rough	Minimalist	Neat	Superb	Suitable
Powerful	Soldierly	Romantic	Sophisticated	Free
Eye-catching	Careful	Winding	Fine	Chic
Tall	Good	Vivid	Beautiful	Unique
Great	Thoughtful	Lively	Scenery	Grave
Burly	Public	Active	Smooth	Magnificent
Imposing	Monotonous	Variety	Perfect	Magical

4. Vigorous style as an example

In this article, vigorous style is used as an example to show how language plays a role in the design of vigorous style.

4.1 Build a semantic database

Contemporary Chinese Corpus at Peking University used "vigorous" as the keyword to screen for passages or publications. Ten or fewer words that are before and after the keyword "vigorous" were selected from these texts and analyzed. The ones that can be effectively used to describe the vigorous state are listed below:

Table 2. Vigorous style semantic group

Vigorous style semantic group				
dense	abundant	robust	simple	mellow
rich	brief	clear	simplicity	unique
exquisite	neat	lasting long	beautiful	delicate
nightlife	steady	long	austere	pure
handsome	dignified	magnificent	kind	flowing
grand	sturdy	vast	sincere	clear
sonorous	implicit	moist	warmth	crystal-clear
concentrated	upright	old and strong	natural	grazioso
red-hot	mature	verdant	childish	soft
gorgeous	harmonization	crude	unadorned	elegant
cool-headed	solemn	powerful	lovely	pretty
experienced	tempestuous	strong	tranquil	graceful
prudent	uninhabited	heavy	heroic	forceful

profound	symmetry	unrestrained	generous	heaviness
steadfast	concise	rude	rustic	rakish
melodious	harmonious	bold	guileless	open-minded
Free and easy	bright	rugged	honest	transparent
luxury	undisturbed	messiness	luminous	full

The collected words have been classified according to the meaning of vigorous.

Firstly, artificial classification was performed by interviewing experts' perspective on lifestyle. Secondly, after classifying the words, we chose the words that describe the vigorous state, and these words were computed and classified by a software that identified similarity in words. These words were then analyzed according to the similarity value and the confidence model of classification was determined, which can be used to improve the classification of work [6]. Word similarity is a software from HowNet that can judge the classification between words.It belongs to the Institute of Computing Technology.

4.2 Vigorous model classification

According to statistics that was produced in the manual classification on the basis of the similarity values, a similarity value of over 0.6 indicates that the words are highly similar, allowing the words to be put under the same category. The similarity value of words associated with the word "simple" and "grazioso" is calculated and shown below.

Table 3. Similarity value of comparison words associated with "simple"

Comparison word	Similarity value
grazioso	0.907407
unique	0.61604
crystal-clear	0.655026
bright	0.655026
flowing	0.907407
luminous	0.655026
clear	0.655026
pretty	0.78836
pure	0.907407
transparent	0.655026
soft	0.749373
graceful	0.78836
full	0.749373

Table 4. Similarity value of comparison words associated with "grazioso"

Comparison word	Similarity value
unique	0.61604
crystal-clear	0.655026
bright	0.655026
flowing	0.907407
luminous	0.655026
clear	0.655026
pretty	0.78836
pure	1
transparent	0.655026
soft	0.749373
graceful	0.78836
full	0.749373

According to the above data, the similarity values of the comparison words are greater than 0.6, meaning that most of the words included in the lists are accurate.

However, the software's results is not comprehensive. After classifying the words by the software, interview with experts and validate the vigorous's style. On the base of these results ,the vigorous style can be described using the following adjectives: free, easy, uninhabited, rude and unreasonable, beautiful and elegant, and open-minded .

4.3 Extract vigorous model design elements

On the basis of these words, we conducted a semantic query again.

If "vigorous and simple "is used as the theme of clothing design, some words from table 5 can be used. People can choose one point, for example "simple", to show the beauty and elegance of the design. Choose the words and homologous visual images,and then analyze the theme words. For example, we chose "transparent" as the key word in the CCL(Center for Chinese Linguistics), and we picked some appropriate words accordingly. These appropriate words were examined using the word similarity software and the semantic words that can be used to describe clothing style, shape, color and design elements. For instance, the clothing will be "transparent" if we select light and thin fabric as the material, or use pure white or pale yellow as the color of choice, emphasizing on pureness and clearness in this occasion.

Table 5. Semantic word description

Key word	Semantic word description						
limpid	rarefied	pure	innocency	azure	pale yellow	clarify	transparent
similarity	0.62406	0.655026	0.61604	0.655026	0.655026	1	1
	Light and thin	white	crystal-clear	neat			
	0.61604	0.68254	1	0.655026			
full	sweety	morbidezza	graceful	symmetrical	plump	massiness	flowing
similarity	0.749373	1	0.749373	0.648243	0.61604	0.61604	0.749373
	guileless	crystal-clear	exquisite	natural	vigorous	brisk	concise
	0.78836	0.61604	0.907407	0.749373	0.78836	0.78836	0.78836
	pure	fresh	gliding	lovely	smooth	soft	pure
	0.749373	0.78836	0.78836	0.907407	0.749373	0.78836	0.749373
transparent	luminous	virid	crystal-clear	transparent	azure	pretty	serene
similarity	0.68254	0.68254	1	1	0.68254	0.655026	0.61604
	cheerful	clarity	vigorous				
	0.61604	1	0.61604				
unique	concise	amazing	pure	charming	subtle		
similarity	0.61604	0.802632	0.61604	0.61604	0.62406		
neutral	morbidezza	charming	dignified	innocency	grazioso	graceful	sweety
similarity	0.78836	1	0.78836	0.765042	0.78836	1	0.78836
	bold	pretty	tranquil	experienced	comfortable	gorgeous	exquisite
	0.775828	1	0.749373	0.749373	0.749373	0.765042	0.78836
	slim	flowing					
	0.78836	0.78836					
pure	innocency	quiet	unadorned	graceful	slim	flowing	exquisite
similarity	0.749373	0.749373	0.809635	0.78836	0.78836	0.907407	0.61604
	natural	active	charming	clarity	beautiful	luminous	stoicism

	1	0.749373	0.78836	0.655026	0.78836	0.749373	0.61604
flowing	serene	free and easy	airy	tranquil	elegant	luminous	loose
similarity	0.749373	1	0.907407	0.78836	1	0.78836	0.749373
	rakish	subtle	comfortable	romantic	stoicism	uninhibited	solemn
	1	0.880952	0.749373	0.656781	0.61604	0.749373	0.749373
	smooth	massiness	strong	pure	untrammelled	implicit	dignified
	0.655026	0.749373	0.775828	0.907407	0.749373	0.61604	0.827202
	soft	gliding					
	0.749373	0.749373					
clear	soft	transparent	virid	crystal-clear	morbidezza	sweety	massiness
similarity	0.62406	1	0.68254	1	0.655026	0.655026	0.61604
	pure	clarity	rustic	plain	full		
	0.655026	1	0.655026	0.68254	0.61604		
crystal-clear	white	colourful	transparent	innocency	transparent	luminous	decently
similarity	0.68254	0.655026	1	0.61604	1	0.68254	0.631708
	soft	full	dazzling	glittering			
	0.61604	0.61604	0.68254	0.68254			
simple	plain	fresh	soft	exquisite	pretty	pure	shipshape
similarity	0.655026	0.78836	0.749373	0.78836	0.78836	0.749373	0.804318
	demure	calm	romantic	natural	graceful	implicit	luminous
	0.880952	0.749373	0.656781	0.907407	0.78836	0.61604	0.749373
	exquisite	rustic					
	0.749373	0.827202					
grazioso	dignified	beautiful	exquisite	romantic	mature	rustic	luxury
similarity	0.880952	0.78836	0.749373	0.656781	0.749373	0.809635	0.814815
	fresh	implicit	massiness	robust	luxury	solemn	luminous
	0.78836	0.61604	0.749373	0.749373	0.907407	0.749373	0.749373
	flowing	charming					
	0.907407	0.78836					
subtle	exquisite	full	implicit	sweety	soft	gliding	pure
similarity	0.78836	1	1	0.749373	0.78836	0.78836	0.78836
	simple	exquisite	brief	elegant			
	0.749373	0.78836	0.78836	0.749373			
graceful	fresh	concise	exquisite	mystical	sexy	full	tranquil
similarity	0.78836	0.749373	0.78836	0.61604	0.669231	0.749373	0.78836
	pure	subtle	neutral	airy	unadorned	gliding	shipshape
	0.78836	0.61604	0.78836	0.78836	0.749373	0.749373	0.804318
	implicit	steady	morbidezza	natural			
	0.61604	0.749373	0.749373	0.78836			

What kind of feelings is associated with the word “simple”? According to this paper, we can find an adjective to describe this style. First, we can find the semantic words of “simple”, which include: plain, fresh, soft, exquisite, pretty, pure, shipshape, demure, calm, romantic, natural, graceful, implicit, luminous, exquisite and rustic. Second, these words can be converted into visual language and used in specific colors, styles, fabrics. These words can be regarded as design elements.

For color, select light colors, such as pure white or lilac, as shown in Fig.4 [11].

For style, use well-fitted, generous brief type like shape-H, as shown in Fig.5 [12].

For fabric, use soft, gauzy, or silky products, as shown in Fig.6 [13].

Fig.4 Color element

Fig.5 Style element

Fig.6 Fabric element

In summary, select the desired design elements that together create the desire themed clothing. Each style has their own combination of design elements. The process of transforming the visual design elements into specific forms is not mentioned here. Therefore, using pragmatics, this paper provides a way to create a vigorous model for fashion design.

5. Conclusion

In conclusion, the vigorous style does not only exist in Zhou Dynasty, but is also a popular style in modern life. Under the guidance of pragmatics, we use corpus to set the vigorous style in five models: the style of free and easy, the style of uninhabited, the style of rude and unreasonable, the style of beautiful and elegant, and the style of open-minded. It also provides the methods of clothing design that can better reflect the style and personality of the person wearing the clothing.

There are some limitations in the article. The content of Center for Chinese Linguistics PKU (CCL) corpus libraries is only available in the People's Daily and the word similarity calculation software is based on the "HowNet", which does not generate enough data. In future studies, we will try to eliminate these limitations.

6. Acknowledgement

This paper is supported by the Innovation Fund Project.

I hereby express my sincere gratitude towards my mentor, Professor Jing-Wei Liu. This paper is finished under her detailed guidance. Her scientific methods of education, acquired professional knowledge and rigorous academic attitude influenced and inspired me deeply, I benefited a lot from her.

Besides, I also want to thank my classmates and my family for their support and assistance while finishing this paper.

References

[1] Information on http: //www.oxforddictionaries.com/definition/learner/style.
[2] Liu JW. Design Thinking: 2014; 95-100.
[3] Charles M. Signs, Language and Behavior: 1989; 261-263.
[4] Liu JW. Design Thinking: 2014;79-82.
[5] Yang K. History of the Western Zhou Dynasty: 1999; 78-100.
[6] Tong JW. Research of the pragmatic technology of clothing based on Qin culture. J.Xian Polytechnic Univ: 2015; 45-47.
[7] ROBERT L. Eastern Zhou and Qin Civilizations. F.E. Quart: 1987; 23-27.
[8] MEY JL. Anticipatory pragmatics.J.Pragmatics: 2012; 63-65.
[9] Tong SJ. Study on Clothing Semantic Extension Expression of Women Suit. J.Dalian Polytechnic Univ: 2014; 57-60.
[10] Zhang ZL, Yu LQ. etc. Measurement of word similarity based on Corpus. Softw. Shanghai JiaoTong Univ: 2006; 03: 638-644.
[11] image from http://www.sjfzxm.com/DESIGN/designer/a/201104/224959_3.htm.
[12] image from http://www.sjfzxm.com/DESIGN/designer/a/201104/221907_6.htm.
[13] image from http://www.sjfzxm.com/DESIGN/designer/a/201104/221907_12.htm.

Design Inspiration from Emotion: A Case Study of Hmong Costume in Longlin Guangxi

Yan-Hui Wang[1], Mao-Qin Yao[2]

[1]*Fashion Design Department, Art and Design Teaching Research Center, College of Art, Guilin University of Technology, No.12 JianGan Road, Guilin, Guangxi, 541004, China*
[2]*Non-material Cultural Heritage Protection Center, Baise City, Guangxi, 533000, China*

1st wyhde@163.com ; 2011019@glut.edu.cn
2nd fW2836915@163.com

Abstract

The traditional Hmong clothing crafts was emphasized in this research object, which makes a profile of fully handcrafted Hmong costume, and demonstrated their primitive emotion behind the costume. The purpose of this study is to provide initial understanding, by using field investigation, documentation, and empirical study, of the link between material, craft, and decoration of costume and emotional and ethnic elements of Longlin Hmong, which can be helpful in revealing the unique cultural value of the Hmong costumes. Furthermore, the paper points out the traditional clothing lifestyle reflect the contemporary trend of slow and eco-friend fashion properly, and the ethnic elements offer an inspiration on the apparel design. Finally, based on the study, a series of related researches on costumes as an non-material cultural heritage in Longlin can be established step by step, and a theoretical framework to design is to be molded connecting the modern and tradition elements.

Keywords: Hmong costume; Emotion; Hemp and indigo; Traditional handcraft; Longlin guangxi; Design inspiration

1. Introduction

People living in Zhuchang Township Longlin depicted themselves as "Hmoob Ntxaug", "MengZou" in Chinese, which means their dress is as beautiful as the flower and exquisite as the brocade (refer with: Fig.1). MengZou's ancestors came from Qian Xinan prefecture across Nanpan River to Longlin during the late Ming and early Qing dynasties [1]. Now the MengZou people mainly live in nine villages distributed in the Zhuchang and Gebu townships, and they are all inhabited on the top of the mountain. The MengZou people of Longlin are in a relatively small population, just over one thousand. Put in a tough condition, they still live in slash and burn even now, and the hemp and indigo they plant are the most important cash-bringing crops. But they keep a high passion for beauty, love, and pursuit of a harmonious living.

Mengzou's costume is a vital intangible cultural heritage, of which the artistic value is largely embodied in unique expression in pattern, color and an primary operating mode. Making the best use of the natural resources, MengZou women adroitly guide their clothing production in accordance with the circumstances, and they specialize in patchwork, edging, goffering, stitching, embroidering, ornamental string and such. It is an important part of the Mengzou's traditional culture that the skills of making clothing reflect the ethnically aesthetic ideas and harmonious ecological views. Hemp plan is used to textile fabrics, indigo is essential material for dyeing, and wax is the most important tool for preventing dyeing, that are all taken from mountains living [2]. The fact that MengZou costumes are the most beautiful and of the brightest colours in all six branches of Hmong in Longlin makes them unmatchedly easy to identify. The paper is to first present an overview of their clothes making process and how the profound emotion from these crafts women is translated into decoration on the costumes; next, it will illustrate that designing garments of ethnic style can receive inspirations from the artistic features and emotional expressions with great stability and inheritance.

Fig.1 MengZou women in Longlin

2. Emotion to Nature

There is a natural optimistism among Mengou, though they are situated in an isolated and tough living conditions where economy is less developed and income is low. All is exploited for out of favourable factors to benefit their production and life. The hemp and indigo are of great significance to Mengzou as the important raw material for making clothes. In this section, a comprehensive record is made in the process of cloth production and the national emotion to nature is revealed.

2.1 Hemp thread

Hemp plant is the gift of nature to Mengzou, which is suitable to grow in the alpine environment in the northwest of Guangxi. The hemp fibers in this area are unique with a very soft feel after treatment that is different from the prickling and rough touch of some other linen textile products. What's more, due to the fact that hemp is of great vitality and its planting season is relatively short, which leads to a result of hemp fiber harvest is about two to three times as much as that of cotton in an acre of land, hemp is planted all over Mengzou villages, contributing to the local's daily wears.

The treatment processes are passed down from generation to generation among Mengzou women. Because of the heavy workload, it is necessary to make villagers work together and complete each processing link with great excellence. Sowing seeds annually in mid-to-end of March and harvesting before the beginning of autumn in the month of Chinese lunar calendar, Mengzou women take up the traditional work to prepare hemp cloth after the growing cycle (refer with: Fig.2). The conventional processing is shown from hemp picking, spinning to weaving on the machine (refer with: Fig.3), which is extremely complex with more than a dozen working procedures. More importantly, there are even some special requirements in the process. The originally made cloth is very plain, but after the primary processing hemp cloth looks classical with different color tones from gray to beige, and after dyed it shows a pure and deep blueness. The characteristics of the hemp fabric construct a surprising beauty in the material, and combining the quality, colour and texture altogether the hemp fabric can be used in modern design to achieve excellent work.

2.2 Indigo dye

Hmong is one of the ethnic groups who uses the vegetable dye first and records the technique in China. Since the Hmong's ancestors found the secret that hemp and cotton fabric could be dyed into blue or brown by indigo plants, indigo has become a popular dye used in women's pleated skirts. Being natural, pro-environment, healthily ecological, and not water polluting, indigo has long been in use by Mengzou women in succession generation after generation. Indigo blue is an original expressive colour of infinite vitality. Due to the difference in dye ratio, weather condition, personal preference and such, the hand-woven cloth can deal with indigo to present slightly different blue colours. Distinguished from other Hmong people, Mengzou women dye hemp fabric after wax painting, before making the pleated skirt.

Fig.2 Harvest Hemp

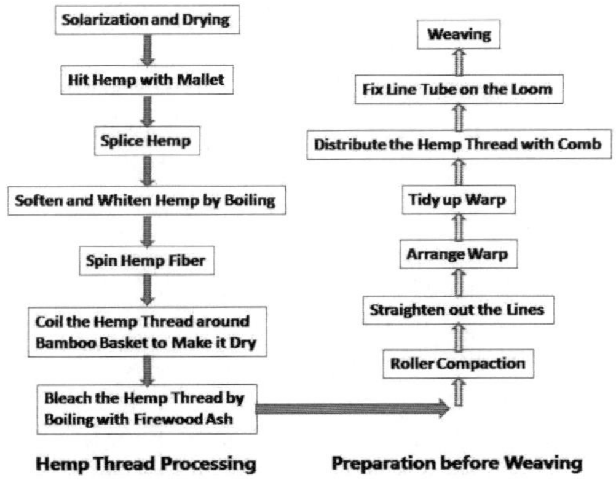

Fig.3 Manufacturing Process

Materials and tools needed in Batik production include hemp cloth, wax, a wax knife, a bottom plate, a pot for melting wax, charcoal, fire-tongs, natural dyeing assistant, indigo, and vat for dyeing, stirring rods and so on (refer with: Fig.4). The main processing are in the following steps:

1) Wax painting, which is the most important technology showing the producers' batik skills and artistic accomplishment. It requires mastering of the wax knife, temperature change in wax liquid, and the wielding angle to achieve smooth drawings, mellow and full to contribute to vivid images. Another key secret is that the wax liquid temperature should be controlled at around 80°C-If too low, wax floats on the cloth unable to penetrate into the fiber but making fuzzy lines, while too high, wax penetrates into the fabric too fast to make good effect on the lines and points, ultimately damaging the fluency and uniformity of drawing [2].

2) Dyeing, with a liquid dye consisting mainly of indigo, soda, alcohol and some other raw materials mixed at a certain ratio. First, dye modulation is successful as the liquid looks pale yellow and the blue foam on the surface thick and sticky accompanied with fragrance of the plant. Then put the batik cloth into the dying liquid with dipping, mixing, and softening, and take it out in 10 minutes. Put the cloth into dye vat again as long as the cloth is not wet. Finally repeat the process several times until the dyed color is satisfying.

3) Dewaxing, which is the finally step. Soak the dried batik cloth in hot water, and keep stirring by hands to make the wax melt off slowly.

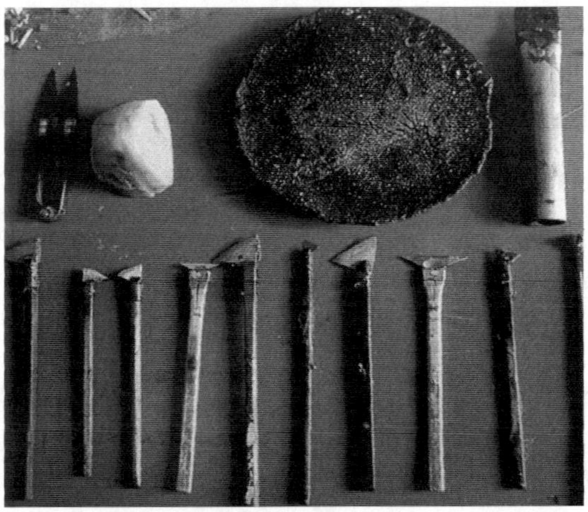

Fig.4 Batik Tools

Fig.5 The Color Distribution in Mother and Daughter's Costume

2.3 Natural color

Due to the natural environment in a high altitude mountainous area, MengZou's color expression in costume is distinctive from that of the other Hmongs. The characteristic of MengZou costume's colour lies in its brilliance and purity. To express their feelings for the totem and nature worship, MengZou people try particularly to create a brilliant sense of order by using a high saturation of colours with great subjectivity. Unlike that of the other Hmong, MengZou' costumes look strongly shining from top to bottom by employing of highly pure colours, making an expression in original colours from the true state of natural images. Apart for their beauty, the costume's function and the matching of colours are not of the same significance. MengZou women, married or childbearing, would rather be dressed up brightly in folk festivals than in black or dark blue, unlike the other Hmong groups. Many rich colours in MengZou's costume get inspired from the colours in the nature. For example, purple is inspired from alpine rhododendron, red from sun, yellow from ripe rice and corn, green from trees, and black from dark nights. These colours, though common in the natural world, are all colours of purity and display strong visual contrasts when matching together. The visual balance recovers when black, white, or gray proportionally [3]. MengZou women's costumes, such as shirt, apron, leggings, and their pleated skirts with batik, are embroidered with red, purple, yellow, and green silk threads, as the critical foil to make them look more gorgeous. Colours in MongZou clothing style have an individual aesthetics character, reflecting the woman's instinct reaction to natural beauty, revealing her idea beauty and life force by using many warm and bright colours in a positive emotion. In the picture of the finished costumes, the authors grasp the Hue colour and its distribution pattern (refer with: Fig.5).

3. Memories to People

As is known, no text about Hmong is recorded; their ethnic costumes are thus regarded as the proof to their historical memory, which includes the history of a migration or family event, by means of abstract patterns in brocade, details in costume form, and decorative figures on the cloth [4]. In this respect, costumes made by MengZhou women are completely different from those ready-made garments, for so many unique feelings are interwoven. Dozens of processes, including hemp planting, thread spinning, cloth weaving, wax printing, pattern decorating, and clothing sewing, being required, MengZou women take a year to make a set of hemp clothes, so such so that a complicated handcraft is a piece of art work for not only its cultural but also emotional value. Making garment is manual labor work never given up in a lifetime. Through the clothing production people show their hope for the coming back of all good things, feeling closer to their families and ethnic groups.

3.1 Loves to Family

Loves for family are embodied in some small things like sewing clothes stitch by stitch. There are two kinds of looms widely used in MengZou's family; one is pedal loom, and the other is simpler and primitive in that it controls warps crossing location up and down by a rope (refer with: Fig.6). Since it doesn't require a large operating space, it is suitable for weaving in any place. MengZou women like to take it to the field creating brocade in their resting time. The authors had inspected on the structure of the other kind of loom found in TongDao HuNan, which, by contrast, is more complex (refer with: Fig.7). The traditional Dong brocade is usually performed by slant rack loom known as "steep machine". The order of the pattern combination is decided by the arrangement of bamboo sticks [5]. But if the simpler loom is used, the warps doesn't need special treatment, and the main varieties depend on wefts, which allows woman to capture many brocade patterns in mind (refer with: Fig.8). In order to tailor new clothes for the family, these women spin and weave all day, knitting time into the costume to express their love for the loved ones. Older MengZou women are passing their skills of sewing and weaving down to the next generation, making the weaving process very sacred. In a sprightly and tidy rhythm, the operators as artists push shuttle back and forth, and move hands up and down and left to right, as if they were not only removing troubles and hardship but also weaving happiness and good health.

In addition to spinning and weaving, MengZou women have also mastered embroidery, cross-stitch, and patchwork, using these handicrafts skilfully to create a wide variety of patterns. In Longlin, all women take the places as daughters, wives and mothers, playing a critical role in the economic survival of their families. It is believed that the degree of decoration delicacy on costume implies a woman's diligence and kindness, which suggests that embroidery demands great care and there mustn't be one single sloppy stitch in the work. The decorative characteristics are mainly displayed in the women's coat, skirt, leg wrappings, and apron that make costumes a medium of emotional communication, satisfying the needs of niceness. By and large, the patterns used for decorating clothes are derived from their loves to family and the longing for a better life. These patterns are mainly based on the religious belief in animism, from which full of natural objects and phenomena are closely related to MengZou people. Some modeling elements are made up of dots and dashes abstracted from natural images, such as the geometric patterns like '卍', '+', '≈', 'o', '△', '井', '…', '≡', '⥱⥲', and '∞' etc., and the geometrical expressions representing natural images include the sun emblem, bronze drum emblem, star emblem, cured grass emblem, eight-rayed flower emblem. Among these elements, designs of flowers, butterflies, sun, dragon, and phoenix are frequently used by MengZou women. If a woman uses the design of eight-rayed flower to decorate the costume, she can either use it alone or together with other patterns to compose a new sense of rhythm, or fill it in other patterns with auspicious designs, all of which show woman's creativity and artistic freedom.

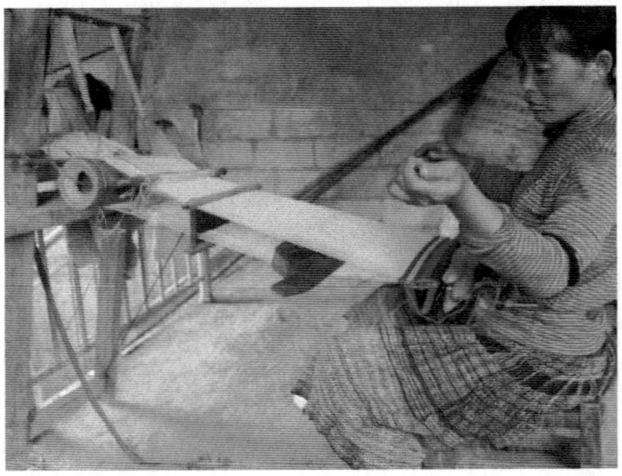

Fig.6 Loom with no Pedal

Fig.7 Brocade Loom in Tongdao

Fig.8 MengZou Brocade

3.2 Remembers to Ancestor

Fig.9 The Seal of Hmong Leader on the Coat Back

The history of the Hmong known in view of migration songs, road guiding songs, and costume designs, may be inaccurate and not detailed, but not illusory [6]. The different kinds of patterns shown in costume are mainly divided according to plants, animals, natural phenomena, and life appliances. The basic pattern used by Hmong is similar, just slightly different in size and composition. The MengZou people want to commemorate their ancestors by using flower belt as a symbol of armor. To remember the history of the migration and inherit culture, the MengZou women put their thoughts of the ancestor's performance into costumes by decorating all kinds of patterns. A MengZou's pleated skirt is made up with two pieces of hemp cloths, on which was decorated batik patterns of four blocks in different width. With basic modelling elements of the diverse permutations and combinations, these patterns are all based on squares and lines, and depict their ancestors' migration over three rivers they crossed and many castles they lived [7]. The MengZou women are clever and deft, they use a needle for a pen, a piece of cloth for that of paper, and thread for colour ink, and finally they keep an account of their remembrance to the past hard road by embroidering every pattern carefully. On the back of the women's coat, there is a piece of square pattern form, which represents the seal of the Hmong leader (refer with: Fig.9). These decorative skills have been passed down from mother to daughter over several generations. Almost all that are woven and embroidered on the costume is to hold a deep memory to the national leaders from the ethnic history legends, customs, aesthetic, emotion, etc. Therefore, it is the clothes instead of the words that have become the ethnic epic. We should find the MengZou's costume is not only exquisite, but also full of emotion and what strengthens the ethnic solidarity and cohesion.

4. Conclusion

This article introduces the three aspects involved in the MengZou costume, including the process of making cloths, color preference, and decorative craft. The people are so keen on natural patterns that suggest the feelings for nature and family, conveyed by the traditional costumes. It can be found that the costume is a critical linkage to natural environment and ethnic emotion, through emotional encoding from cloths, colors, and patterns. By comparison, these products containing real emotion are better than those industrialized beauties. Techniques of cloths making in the MengZou women's traditional costume are unique in that they are from an ethnic culture in the long-term development, which reflects the laborers' wisdom and love. The MengZou women convey the pleasant, cheerful, and sweetheart mood in their own crafts, arousing the positive appreciation of the meaning of life.

Today, many hot issues such as tradition, handcraft, eco-friend, sustainability and so on, have been sprung up in the field of apparel design [8], we need to take account of an inescapable reality, that is, the social change brought by economic development, which poses such a major challenge to the view of nature and people. More and more minorities have abandoned their own traditional cultural and crafts for the pursuit of modern civilization, but in so doing, they lose significant clues about eco-friendliness and sustainability, as well as the value of emotional factors. Now people pay more attention to emotions inspired by costumes which can fully reflect their values and enrich their everyday life. In another word, people want more pleasure experience from a piece of product. Donald Arthur Norman, A famous marketing expert, concludes "Pleasure Experience" in four aspects: Physio-pleasure, Socio-pleasure, Psycho-pleasure, and Ideo-pleasure [9]. We can look for inspirations from apparel fabrics, gorgeous rich colours, unique exaggerated fashion modeling, profound implication of apparel design or even exquisite handicraft technology. Thus, on the basis of collecting and analyzing MengZou's costume crafts, the article is to discuss further on how to use ethnic elements for reference, in order to give some hints to apparel designers on making an ethnic style fashion. The revival of each ethnic culture always comes from the summarization of its heritage.

5. Acknowledgement

The investigation and study is carried out smoothly with the help from Non-material cultural heritage Center of Baise City, culture bureau in Longlin county and local residents in Zhuchang township. The last acknowledgement is to "The Twelfth Five-year Planning Project" of Guangxi Science Education in 2015.

References

[1] Guang MY, etc. Longlin County National Series, Longlin Hmong. Guangxi Nationalities Publishing House. Nanniny: 2013; p: 3-5.

[2] Xiao MH. Printing Manufacture Craft Procedure of Miao Ethnic in Guangxi Longlin. Journal of Guangxi University for Nationalities (Natural Science Edition): 2009; 7; 50-54.

[3] Yan HW, Shu XD. Research on the Application of Ethnic Look for the Textile Visual Design. Advanced Materials Research: Vol. 627 (2013); 229-232.

[4] Xing X, Jun L. Research on Costume of Basha Miao Nationality in Southeast Guizhou. Journal of Nanjing Arts Institute: 2010; 4: 113-116.

[5] Yan HW, Wei C. Pondering the Actuality of Devolvement and Succession of Dong brocade in Tongdao Human Province. Silk Monthly: 2010; 9; 52-55.

[6] Zheng WY. Culture of Hmong Costume. Guizhou National Publishing House. Guiyang: 1998; p: 249.
[7] Zhi XL. The Practicality and Features of Chinese Ethnic Minorities' Crafts and Arts. Ph. D. Thesis : Minzu University of China. 2011; P: 26.
[8] Yu L. An Initial Exploration of Sustainable Development Models in Apparel Design. M. D. Thesis : Beijing Institute of Fashion Technology. 2012: Abstract.
[9] ArthurNorman D. Emotional Design. In: Xiaomei H, Qiuxing O, translators. CITIC Publishing House. Beijing: 2012; P: 86.

A Study on Clothing Structure Design and Children Wearable Devices

Ying-Chun Liang[1*], Hong Dai[2]

[1]*School of Apparel & Art Design, Xi'an Polytechnic University, Xi'an, 710048, China*
[2]*Apparel & Art Design, Xi'an Polytechnic University, Xi'an, 710048, China*

*Corresponding author's email: 420764278@qq.com

Abstract

Along with the age of electronic information, the clothing industry is beginning to become more intelligent. There has been a flourish of new ideas that combines one industry with other industries. With the rapid development of intelligent clothing, it is becoming more and more applicable to fields of military, health-care, professional sportswear and leisure life etc. Children' clothing is a key part of the apparel market and has a higher demand in intelligent clothing development. In this paper, general intelligent clothing will be split into two parts, electronic equipment and clothing section. Then a deeper inquiry is made to improve the wearable structure portion. Electronic intelligent electronic devices are integrated to improve the use of young children's intelligent clothing in regards to the application of life and health.

Keywords: Intelligent clothing design; Children; Structural design; Comfort; Security

1. Development of children intelligent clothing

1.1 Background and development trend of intelligent clothing

The 21st century is an information era. With the rapid development of science and technology, a variety of mutual high-tech products have brought a lot of achievements, such as smart clothing [1]. With the development of the times and technology, people have become increasingly demanding of clothing, it not only need the simple function, such as cold, warm, naked and visual aesthetic. But also need a Smart clothing combines biotechnology, information technology, materials science, traditional textile and apparel technology organically together [2].

Currently there has been a lot of research on wearable devices. Some people used LED technology and photosensitive material to design jerseys some used temperature humidity sensors and fiber optic equipment to monitor clothing microclimate, there are also some who designed a BSN remote healthcare system [3], as shown in Fig.1.

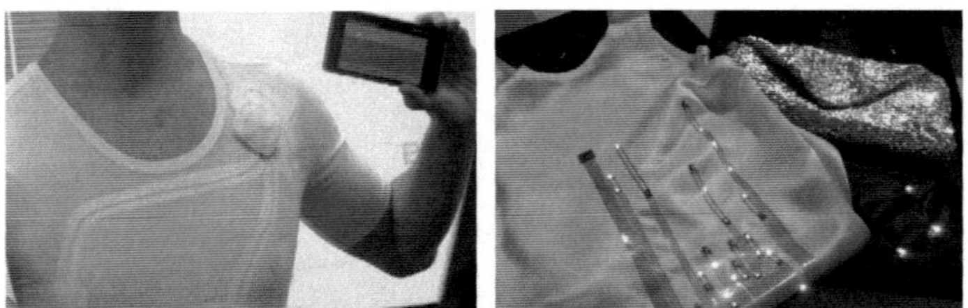

Fig.1 Intelligent clothing products (Picture Source: http://www.image.baidu.com)

1.2 Development Status of children wearable smart devices

The children clothing industry has always had its own unique foothold. With modernization, people demanded for a more comfortable and functional requirement in children's clothing design [5]. Children as a kind of special populations, have higher demanded in clothing functional and useful level, it is critical to look into the development of digital clothing and electronic information [6].

Currently about 70% of the products in the children fashion electronics market are protective and health detection intelligent clothing [7]. For instance, the Owlet baby intelligent socks which can monitor a baby's skin temperature, heart rate, blood oxygen and sleep quality; Mimo baby Kimono leotard clothing can monitor a baby's breathing, heart rate and activities; Sproutling intelligent foot ring can monitor the indoor temperature, the baby's heartbeat and movement, as shown in Fig.2.

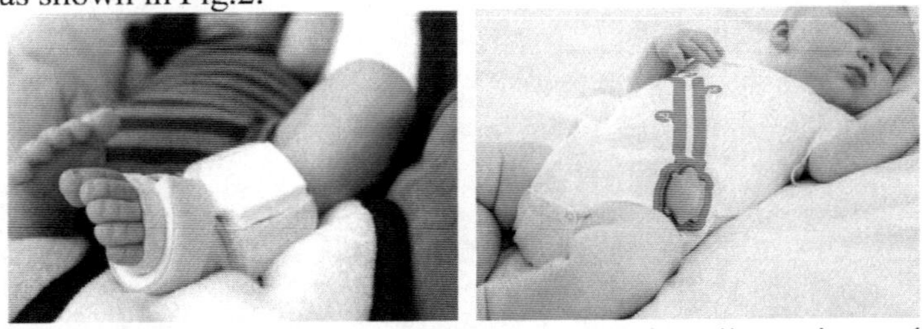

Fig.2 Kids intelligent equipment (Picture Source: http://www.image.baidu.com)

With ergonomic development, children's clothing are beginning to combine with intelligent electronic devices. The purpose is to improve the degree of compatibility between the clothing and electronic products. This paper aims to design a more universal intelligent children's clothing and improve its ability to monitor the children's condition through questionnaires to understand customer satisfaction. There is no need to change clothing type, only an adjustment to the chip can be changed to alter the functionality of the wearable device. Make the cloth device better adapted to the children wearable device improvement.

2. Clothing structure design about children intelligent monitoring equipment

The article through around children class intelligent wearable monitoring equipment design, to make a more comprehensive design of the wearable clothing basic type, fabric selection, structural design, health, safety and other characteristics design. A foundation for the satisfaction assessment of the garment has been established for the following chapter.

2.1 1-3 years old Children intelligent clothing design requirements

From an industry point of view, according to the National Bureau of Technical Supervision, the international harmonization of standards, clothes designed for aged 1-3 years old children are classified under young children clothing. Intelligent monitoring of children clothing design mainly relates to human ergonomics. The wearable device combine with the human, clothing and environment [8].

From the human body itself, during 1-3 years old, the child's height and weight are developing rapidly, the general height between 70 to 90 cm. They are typically characterized with a large head, short neck and thick, narrow shoulders, short limb and so on [9]. Children typically have a strong curiosity about the outside world and generally likes to move. It is around this time that they begin to learn how to walk, run and jump, and theirthinking ability also increases. Although they cannot effectively express their own opinion and perceptions, their subconscious will have already underwent a series of classification, with their likes and dislikes, comfort and discomfort [10]. It is important at this time to allow them to try out new things to discover and broaden their mind.

From the perspective of intelligent clothing, eaten by children by mistake and wear comfortable is most commonly encountered situation. So in the young children clothing design: Firstly, we should to choose design the suitable move, easy to wear and replacement clothing; secondly, we should design the clothes that children can not eaten by mistake; thirdly, we should using durable fabrics to design, improve the intelligence apparatus wearable [11]. From an environmental point of view, the intelligent design of a child's cloth should first avoid harming human health and the environment such as produce radiation or toxic substances [12].

2.2 Fabric selection

Clothing is the carrier of intelligent devices, so fcuntional fabrics are chosen. In this design, polyester spandex fabrics are chosen as the main fabrics, the accessories are butterfly net fabrics (polyester mesh fabric). Table 1 shows the fabric's characteristic. (The fabric source:Jiang sheng knitting co., LTD)

Table 1. Fabric Characteristic

	polyester spandex fabrics	polyester mesh fabric
Fabric		
Tissue type	Warp rib fabric	warp-knitting
Density (e/10cm)	350	----
Fabric density(D)	150	----
Weight (g/m2)	148.6	66.72
Ventilation rate (L/m2.s)	237.7	----
Elasticity	Low bomb	No bomb
Finish	Deodorant, Anti-bacterial	----

Fabric: High density low elastic fabric should be selected that not only has good wear resistance, but also high elasticity so that it can accommodate different children' activities. Knitted fabrics not only meets the thickness requirement, they also have good air permeability that protects children from moisture permeability.

Accessories: Design inner bag, using polyester mesh fabric store the electronic chip device, can make the monitoring data more accurate and improve ventilation.

2.3 Construction design

With the growing popularity of wearable technology equipment, every child can own a wearable intelligent clothing. Different functions can be accommodated through changing the chip, there is no need to change the wearable clothing equipment as shown in Fig.3.

Fig.3 Construction design

Wearable devices can be more convenient to use and dress; the design of the vest is to liberate the child limbs, to reduce constraints that can hinder children's activities. Electornic products are placed under the left armpit. This design has two advantages: First, the armpit is a position that can be easily hidden from the children's secondly, it is one of the key areas to effectively monitor the heartbeat, heart rate, pulsation and temperature, etc.

2.4 Functional design

For functional design, safety and comfort are two fundamental principles of design.The wearable devices back used mesh fabric's to improves moisture permeability; within underarm bag design electronic products, improve the security of wearable devices, to avoid children eating problems; compact wearable device wear off easy, easy washing, but also by replacing the chip, play vital signs monitoring function of different kind as shown in Fig.4. As a health protective garment it it helps with parental control and has easy operation features.

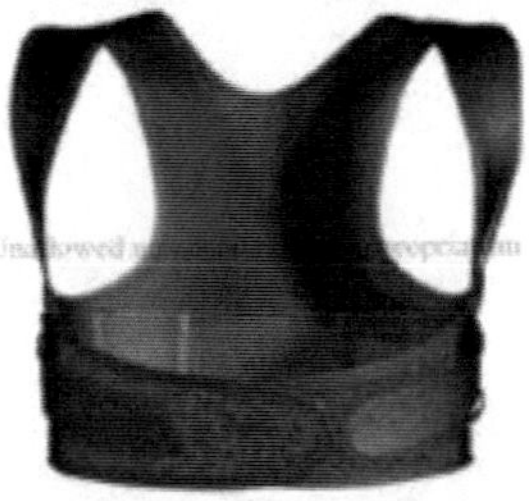

Fig.4 Wearable device show

Through the design of wearable intelligent devices, it can be combined with the actual user's evaluation, to determine whether the user is satisfied with the product..

3. Using semantic analysis for wearable device designed test and evaluation

Conducting an evaluation method through a questionnaire is a semantic analysis, where a semantic differential scale is used to study the meaning of things, constructed by the American psychologist Osgood and his colleagues. The method is conducted with a paper and pencil. Testers were asked to rate by a number scale of seven levels about their opinion of a certain object or a concept (such as cars, neighbor), in order to understand the meaning of the object or a concept and strength in all dimensions. Two endpoints of the level sequence are usually opposing adjectives, such as honesty and dishonesty, strong and weak, important and unimportant. Generally, people do not feel confused when confronted with a scale with no more than 7 levels, so the level of subjective evaluation scale is often based on different objects that are divided into a scale of five or seven [13].

3.1 Questionnaire design

3.1.1 Scoring method design

In the questionnaire , the respondents subjective evaluation is divided into five grades, shown in Figure 5, for example with the adjectives "beautiful - ugly", 1 represents a very beautiful, 2 shows beautiful, 3 represents neither beautiful nor ugly, 4 shows a comparison of ugly, 5 said very ugly.

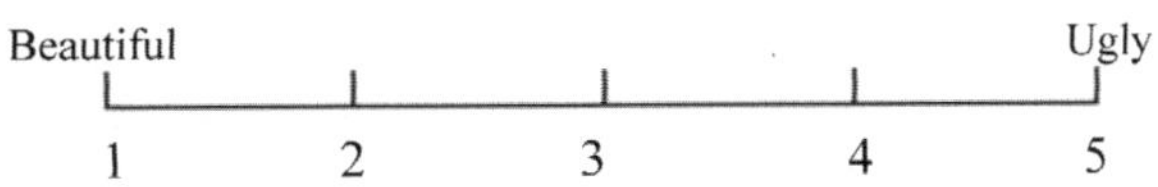

Fig.5 Scoring method

3.1.2 Emotional adjectives set

Some emotional adjectives were selected to reflect the feeling of the user and bystanders, then 50 young children and their mother were selected to try the wearable device. After the final discussion analysis 15 pairs of antisense adjectives were selected to describe the wearable device technology. The 15 pairs of antonyms are shown in Table 2.

Table 2. Selection of antonyms

Antonym	
Touchable - Untouchable	Unhealthy - Health
Comfortable - Uncomfortable	Not practical - Practical
Easy to wash - Wash hard	New - Old
Easy to wear - Difficult to wear	Fine - Rough
Beautiful - Ugly	Strong visual- Weak visual
Smart - Bound	Light - Heavy
Facilitate monitoring - Monitoring hinder	Colorful - Monotonous
Common - Useless	

3.1.3 The survey sample size and selection

The survey is conducted with 50 young children and their parents. Taking into account the specificity and comprehensiveness of the survey population, in order to reflect the evaluation of children and parents, the questionnaire's two point (can be touched - not touch, comfortable - uncomfortable) options can be observed by their parents. But parents should depend on children' wear presentation to predicated and choose the word. The questionnaire includes some basic information such as age, gender, and occupationetc, The main method used is face to face filled out, 50 questionnaires were distributed and 47 were recovered, the recovery rate was 94%.

3.2 Analysis of survey results

3.2.1 Basic Evaluation and Analysis

Table 3. Evaluation and Analysis

Adjectives	Number of observations	Average	Variance
A1	47	4.17	1.04
A2	47	1.95	0.84
A3	47	2.14	0.97
A4	47	2.29	1.46
A5	47	2.72	1.76
A6	47	2.42	1.20
A7	47	2.14	0.97
A8	47	2.74	0.89
A9	47	4.29	1.05
A10	47	2.85	0.99
A11	47	2.55	1.09
A12	47	2.23	1.24
A13	47	2.29	0.99
A14	47	2.65	1.00
A15	47	2.88	1.36

As seen from Table 3,Comfortable, easy to wash, easy to monitor, healthy and practical. From the general description of the design the results are quite good as some of the basic points of the score have been reached.

3.2.2 Factor analysis of the adjectives

Factor analysis is a statistical method used to extract potential common factors of multiple variables. Few potential common factors were synthesized and extracted among the many observable variables to maximize the interpretation of the original observed variables information to reveal the nature of things. Using the factor analysis can effectively achieve dimension reduction and data simplification. The basic idea of factor analysis is to group observed variables according to their correlation sizes, so that the correlation between the observed variables in each group can be higher, and the correlation between different groups can be higher too. Each set of observed variables represents a basic structure and can be represented by the potential common factors of these variables. In simple terms, the contribution rate and influence of the factors can be obtained through analysis.

1) Correlation matrix analysis

Table 4. Correlation matrix

	A1	A2	A3	A4	A5	A6	A7	A8	A9	A10	A11	A12	A13	A14	A15
A1	1.000	.300	.521	-.168	.259	.162	-.287	.308	.824	-.388	.584	.136	.046	-.146	-.196
A2	.300	1.000	.215	-.473	.466	.803	-.809	-.625	-.726	-.849	-.214	-.779	.037	-.416	-.908
A3	.521	.215	1.000	.265	.759	-.369	-.371	.265	-.339	-.518	.221	-.043	-.283	.503	.024
A4	-.168	-.473	.265	1.000	.266	-.351	.205	.081	-.094	.205	-.183	.198	-.141	.714	.474
A5	.259	.466	.759	.266	1.000	-.444	-.759	-.226	-.757	-.761	-.209	-.504	-.208	.278	-.265
A6	.162	.803	-.369	-.351	-.444	1.000	.291	.022	.295	.223	.219	.129	.200	-.520	-.113
A7	-.287	-.809	-.371	.205	-.759	.291	1.000	.468	..027	.891	.207	.717	-.063	.232	.708
sA8	.308	-.625	.265	.081	-.226	.022	.468	1.000	.653	.438	.659	.793	.001	.252	.670
A9	.824	-.726	-.339	-.094	-.757	.295	.027	.653	1.000	.741	.518	.824	.002	.051	.651
A10	-.388	-.849	-.518	.205	-.761	.223	.891	.438	.741	1.000	.165	.664	.071	.156	.709
A11	.584	-.214	.221	-.183	-.209	.219	.207	.659	.518	.165	1.000	.645	-.183	.099	.173
A12	.136	-.779	-.043	.198	-.504	.129	.717	.793	..025	.664	.645	1.000	.039	.269	.710
A13	.046	.037	-.283	-.141	-.208	.200	-.063	.001	.002	.071	-.183	.039	1.000	-.376	-.152
A14	-.146	-.416	.503	.714	.278	-.520	.232	.252	.051	.156	.099	.269	-.376	1.000	.498
A15	-.196	-.908	.024	.474	-.265	-.113	.708	.670	.651	.709	.173	.710	-.152	.498	1.000

Table.4 shows the adjective variable correlation matrix. The highest correlation with 0.891 are A7 and A10 adjectives which indicates that the terms convenient and practical monitoring are veryrelevant. A1 and A9 with 0.824 is also very relevant high, which indicates the properties of healthy and untouchable is also very important. A2 and A6 relevant higher, 0.803 (1 as the limit), it is related to comfort and restraint. Between common variables inherent component significant correlation between explanatory variables, it signifies that there is a continued need for factor analysis.

2) KMO and Bartlett's Test

Table 5. KMO and Bartlett's Test

Bartlett's Test of Sphericity	
Kaiser-Meyer-Olkin Measure of Sampling Adequacy.	.611
Approx. Chi-Square	286.072
df	91
Sig.	.000

As shown in Table 5 for the value of KMO and Bartlett test of sphericity, KMO indicates a partial correlation between variables, the results of the KMO is 0.611. According to the criteria, it belongs to an acceptable range. Bartlett's test of sphericity reaches to a significant level, indicating that there is a significant variable between the clothes structural and relevance thereby rejecting the hypothesis, it is possible to factor analysis.

3) Common variable degree

As can be seen from Table 6, the common reaches 0.7 and above more, so common variable degree is relatively high, it is an ideal state.

Table 6. Communalities Test

Antonym	Initial	Extraction	Antonym	Initial	Extraction
Touchable - Untouchable	1.000	.710	Colorful - Monotonous	1.000	.829
Comfortable - Uncomfortable	1.000	.788	Strong visual- Weak visual	1.000	.814
Easy to wash - Wash hard	1.000	.712	Fine - Rough	1.000	.793
Easy to wear - Difficult to wear	1.000	.605	Common - useless	1.000	.503
Beautiful - Ugly	1.000	.588	Unhealthy - Health	1.000	.796
Smart - Bound	1.000	.641	Not practical - Practical	1.000	.670
Facilitate monitoring - Monitoring hinder	1.000	.843	New - Old	1.000	.856
Light - Heavy	1.000	.687			

Extraction Method: Principal Component Analysis.

4) Total Variance Explained

Table 7. Total Variance Explained

	Initial Eigenvalues			Extraction Sums of Squared Loadings		
Component	Total	% of Variance	Cumulative %	Total	% of Variance	Cumulative %
1	6.200	44.288	44.288	6.200	44.288	44.288
2	3.108	22.202	66.490	3.108	22.202	66.490
3	2.488	17.771	84.261	2.488	17.771	84.261
4	.666	4.758	89.020			
5	.459	3.281	92.301			
6	.349	2.490	94.791			
7	.252	1.797	96.588			
8	.154	1.097	97.684			
9	.130	.926	98.610			
10	.085	.609	99.219			
11	.042	.303	99.523			
12	.037	.264	99.787			
13	.021	.152	99.939			
14	.009	.061	100.000			
15	.187	1.246	100.000			

Extraction Method: Principal Component Analysis.

As shown in Table 7 the first factor has 6.200 eigenvalues with a variance contribution rate of 44.288, which can be interpreted as representing 44.288% of all the variables. The variance contribution rate is one of the largest principal component. The second factor variance contribution rate is 22.202%; the third factor variance contribution rate is 17.771%, the first three factors explained 84.2% of all the variables, the contribution rate is higher, and the eigenvalues are greater than 1, as a result three common factors can be extracted.

3.3 Evaluation

The SPSS17.0 factor analysis has the great influence and contribution rate of a set of words in respondents mind. Through the analysis and classification 15 groups of adjectives can be classified into three parts. Finally the factor investigator assessment and key areas for improvements can be obtained.

3.3.1 Wear clothing basic evaluation

With the overall feeling evaluation, from looking at the average and variance of the objective data, it can be seen that far from the center value score there are six factors that were available, they are to touchable, comfortable, easy to wash, easy to monitor healthy and practical.

Through factor analysis, the molecular correlation conclusion is acquired. In the design facilitate monitoring and practicality has a closely relevant, safety and health have relevant wearer comfort with a strong correlation. These results are in line with our design process purview. KMO and Bartlett test of sphericity only verifies whether these 15 factors are for factor analysis, and whether there is relationship between the various factors which can constitute a subsequent clustering analysis.

Through the joint analysis of variables, we can probably be divided into 15 groups for three partial factor, more than high degree of common 0.8, that investigators cognitive degree of unity are specific factors: convenience monitoring group, drab group, depending on the strong sense of group, new group; greater than 0.7 also reflects a relatively high degree of common, the main factors are not touch, comfortable, easy to wash, fine, healthy set of these adjectives, reflecting the design characteristics and feel to the investigators; of course factor is less than 0.7 are probably not very good design part of the next improved to provide the direction and space. The total variance was out 15 there are three big factors influence and contribution factors, the importance of factors.

3.3.2 Design conclusion and improvement

Through the analysis it can be concluded that wearable design complies with the design requirements and basis, can be better to realize its function, young children can wearing comfort and safety and bring convenient to their parents. The low scoring factors, is also the place where respondents have objections and therefore a change needs to be made.

4. Conclusion

In this article, based on the monitoring of young children's intelligent clothing design it tries to improve the efficiency and convenience of wearable devices by complying with the basic requirements of the design demand and reflect the functional requirements. Through questionnaire and semantic analysis product design was evaluated to find its strengths and weaknesses as well as potential areas for improvement. The intention is to enhance the design of intelligent devices in clothing and make it more convenient to work with electronic equipments.

5. Acknowledgment

The authors will give their great thanks to the Apparel & Design College of Xi'an Polytechnic University.

References

[1] Shen X. Clothing Applications of Computer Intelligence. Technical Discussions: 2009; 41-43.

[2] Huang CY. Application and Prospect of Intelligent Electronic Clothing Outlook. Ninth functional textiles and nanotechnology seminar papers: 2009; 5.

[3] Sharon Baurley. Interactive and Experiential Design in Smart Textile Products and Applications. Personal and Ubiquitous Computing: 2004; 8: 274-281.

[4] Wang SR, Liu C. Now or Future Smart Wearable Devices. Frontier Viewpoint: 2013; 6: 21-23.

[5] Yang Y. Design of Puzzle-based Infant Clothing. Jiangnan Univ: 2007; 6.

[6] Chen H. Smart Clothing for Health Care Systems Research and Development. Donghua Univ: 2008.
[7] Qiu ML. Design and Implementation of The Human Body Wearable Sensor Network-based Remote Healthcare System. Zhejiang Univ: 2013; 1.
[8] Yu XY. Children's Health and Safety Performance of Clothing Design. Nantong Univ: 2005; 4: 55-56.
[9] Wang KX. Infant Clothing Design Health and Safety Research. Northwestern Univ: 2014; 12.
[10] Zhao YX. Relationship of 0-3 year-old Infant Clothing and Body Structure. Beijing Institute of Clothing: 2009; 12.
[11] Zhou LY, Hu XD. Series of Children's Wear Design. China Textile: 2003; 6-7: 37-38.
[12] Jia TT. Safety of Children' Clothing Design Research. Jiangnan Univ: 2009; 6.
[13] Ye YY, XU J. The application of human ergonomics in children's clothing design.GuangXi Textile: 2009; 38: 23-24.

Body Classification of Baiku Yao Women Via 3D Body Scanning

Xiao-Ya Feng, Rong Zheng*, Pan-Pan Li, Hao Zhang, Jing Guo

Aimer Human Engineering Research Centre, Beijing Institute of Fashion Technology, Beijing, 100029, China

*Corresponding author's email: yuki.zheng@gmail.com

Abstract

In the current research on Baiku Yao women's body shape, measurements are conducted manually with seldom involvement of the three dimensional shape of the human body. This study collected three-dimensional data of 20 Baiku Yao women 22-46 years of age in both static and dynamic states and analysed body shape evolution through the comparison of previous data. The body shape evolution showed that the change in results of height measurements is relatively small and that of girth measurements is relatively large. A factor analysis method was used to extract six principal components with the accumulative contribution rate of 75.64% such as weight and height and to analyse the change of data corresponding to body shape in dynamic states as well.

Keywords: Baiku yao; Body shape; 3D body scanning; Factor analysis

1. Introduction

Physical anthropology is a scientific discipline used to investigate physical characteristics and the structure of human beings as well as the latter's formation and development [1]. It also provides a perspective on human evolution and racial categories. Some scholars find that body shapes of Baiku Yao men and women are obviously different from those of ethnic Han Chinese and other ethnic groups in Guangxi. The following studies have collected and analysed data corresponding to body shapes of Baiku Yao men and women.

In 1999 and 2000, Luo and Yang measured 63 body parts including heads, faces, and bodies of 953 Baiku Yao men and women (605 men, 348 women) 22-55 years of age representing three generations of the Baiku Yao ethnic group. Measurement apparatus such as straight pin gauge, spreading caliper, and sitting height chair were used to measure 63 body parts' parameters. They classified such body parts in accordance with relevant standards. Clustering analyses were then conducted from available physical data of 16 ethnic groups in the south to summarise similarities and differences of different ethnic groups. Finally, physical characteristics of the Baiku Yao ethnic group were determined and the main race type was found to be the south China type of southern Asians of Mongolian ancestry [2]. In 2005, clustering analyses were conducted by Yang and Yu for head shape characteristics in accordance with the constituent ratio of three head shape indexes of 14 ethnic groups such as Maonan, Shui, Gelao, and Baiku Yao. Results showed that the genetic distance of head shape characteristics of ethnic groups that had some original relation with ancient tribes was small, and 14 ethnic groups could be divided into three types, whereby the genetic distance of head shape characteristics of Baiku Yao and Miao populations in Guangxi is close [3]. In 2011, Peng et al. observed seven genetic characteristics of hair, Darwin's tubercle, incisor type, lip thickness, chin type, penetration wrinkle, and fingernails of 402 Baiku Yao men and women (202 men, 200 women) in Yao Mountain Yao Village, Libo County, Guizhou using a cluster random sampling method. Findings showed that seven genetic characteristics of the Baiku Yao population were deferentially distributed in different genders and of some correlation [4].

Body measurement of the Baiku Yao population in the aforementioned studies is generally conducted in accordance with Wu et al.'s Body Measurement Method [5] and Shao's Body Measurement Manual [6]. However, the head measurement of men and women is the primary focus of research; few studies focus on body measurement and no research has employed 3D body scanning of human bodies, as manual measurement remains the principal measurement method. In addition, data obtained from human body measurement are time-limited and size measurement results of large-scale samples must be updated about every 10 years [7]. Thus, it is both beneficial and necessary to employ a 3D scanning method to supplement 3D data of human body measurement corresponding to Baiku Yao men and women's body shapes that could be analysed and described [8-10].

This study collected data corresponding to Baiku Yao women's body shapes via 3D scanning and comparatively analysed body-shape characteristics derived from such data. A factor analysis method was employed to extract the principal component factor to describe the characteristics of Baiku Yao women's body shapes and to analyse changes in the data related to body shapes in both static and dynamic states. The use of the 3D scanning method thus allowed us to collect, analyse, and summarise data related to adult Baiku Yao women's body shapes.

2. Body shape data collection

Several field studies conducted in Yao Village in Lihu Country from 2014 to 2015 found that scenes of Baiku Yao women spinning, embroidering, dyeing, and sewing clothes can be observed everywhere in the village. Thus, typical postures of Baiku Yao women in everyday life were selected to collect 3D data, including static standing posture, posture of bending over to work, as well as sitting and resting postures. Certain minor adjustments were made for postures to accommodate the research technology (see Table 1).

Table 1. Setting of 3D scanning postures

	Standing (static)	Bending over (work)	Sitting (rest)
Typical scenes of life			
Posture extracting and setting	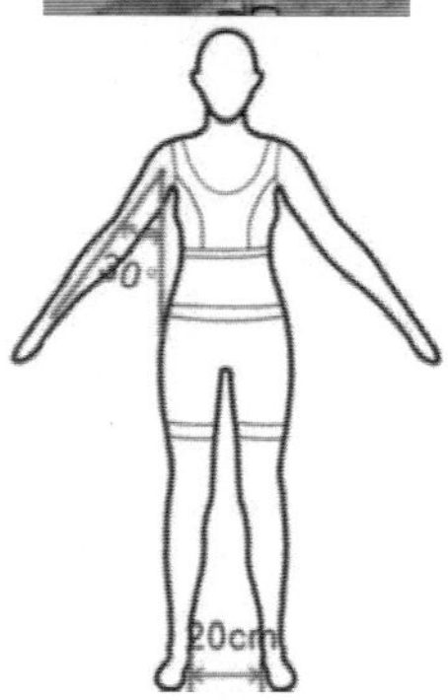		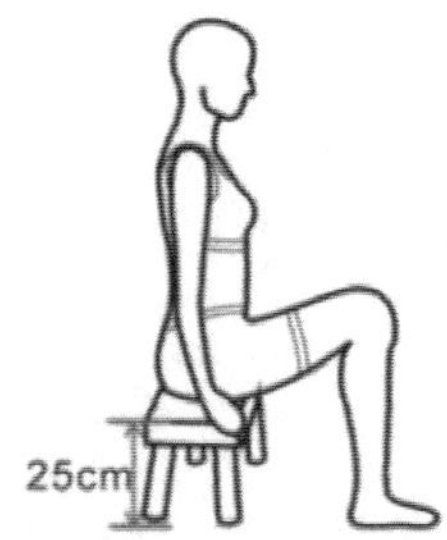

One-dimensional manual measurement, two-dimensional photographing [11], and three-dimensional scanning measurement methods were combined to measure 82 parameters of 20 Baiku Yao women 22-46 years of age in a static standing posture, and to measure 8 parameters of the same sample population in dynamic states (bending over, sitting). The women whom we measured met the following conditions: all three generations belong to Baiku Yao ethnic group; their home towns are located in the region, with a large population in Nandan County (Lihu, Bayu Yao village); they are long-term residents in their home towns. The 82 static measurements include 18 height measurements, 14 girth measurements, 13 width measurements, 9 thickness measurements, 25 length measurements, and 3 other measurements. The 8 dynamic measurements include neck base girth, shoulder breadth, chest girth, waistline, hipline, abdominal girth, maximum thigh girth, and back length.

Due to limitations of the test conditions and the local situation, the maximum recruitment of female volunteers was done to collect the body data. According to the relative error value a and the coefficient of variation CV, we found that the relative error value of height, girth, width, thickness, and length were relatively small and credible with the range of 0~8.7. However, the CV value of angle measurements changed greatly - such as above half body axis, neck tilt, and chest angles - which showed that these data lack credibility. We therefore abandoned the latter data that showed great changes of CV value.

3. Analysis of body shape data

3.1 Evolution of body shape

Using anthropological body shape data from 2000 and Baiku Yao women as the control group, 15 measurements closely related to Baiku Yao costumes were selected from 37 body measurements. The newly collected data were analytically compared to previous data using a T-test method. Table 2 shows the comparative measurements and data.

Table 2. Comparison of differences in data of women's body form

Serial no.	Measurements	Average value 2000 (cm)	Average value 2015 (cm)	Mean difference	Sig. (bilateral)
1	Height	148.6	148.6	0.0	.971
2	Navel point high	92.1	87.7	-4.4	.000
3	Anterior point of iliac crest	80.7	80.9	0.2	.793
4	Middle knee high	41.1	39.1	-2.0	.000
5	Height of malleolus medialis	6.4	6.6	0.2	.117
6	Neck base girth	30.3	38.9	8.6	.000
7	Chest girth	78.0	86.8	8.8	.000
8	Waistline	67.6	75.1	7.5	.000
9	Abdominal girth	71.5	86.3	14.7	.000
10	Maximum biceps girth	22.1	29.8	7.7	.000
11	Maximum shank girth	30.8	35.0	4.2	.000
12	Minimum shank girth	19.6	20.8	1.2	.000
13	Shoulder width	29.6	32.9	3.3	.000
14	Arm length	47.0	49.3	2.3	.000
15	Weight	44.6	53.1	8.5	.000

By contrast, there are significant differences in most measurements (as shown in red in Table 2, the significance level is 0.05) and there are no significant differences in a small number of measurements. By observing average values, the mean difference in Baiku Yao women's height over the 15-year span is small, ranging from 0 to 4.4cm, which shows that body shape factors related to height are stable; the 15-year mean difference of weight and girth is greater, with those for chest girth, waistline, and abdominal girth being over 7cm, which shows an overall increase in girth of women's bodies. This may be related to changes in the ethnic group's living style and diet conditions.

3.2 Data analysis of static body shape

The principal component factor analysis conducted for 82 measurements using SPSS software and the Scree test criterion shows that 4-15 principal component factors are relatively suitable (Fig.1). By further observing results of principal component factor analysis (as shown in Table 3), we found that the analysis result of each component factor is best when the number of principal component factors is six and the accumulative contribution rate is 75.64%; thus we decided to use the six principal component factors to describe body shapes. The contribution rate of the first principal component factor is 32.55% and measurements with high load values are CL girth (0.940), horizontal waist girth (0.934), weight (0.925), abdominal girth (0.917), chest girth (0.906), and other factors related to human body shapes, which are collectively named as fatness factors. The contribution rate of the secondary principal component factor is 21.48% and measurements with high load values are anterior (posterior) axillary point height (0.968), back cervical height (0.964), height (0.921), and other factors related to body height, which are collectively named as height factors. The total contribution rate of the two factors is 54.03%, significantly higher than that of the other four factors.

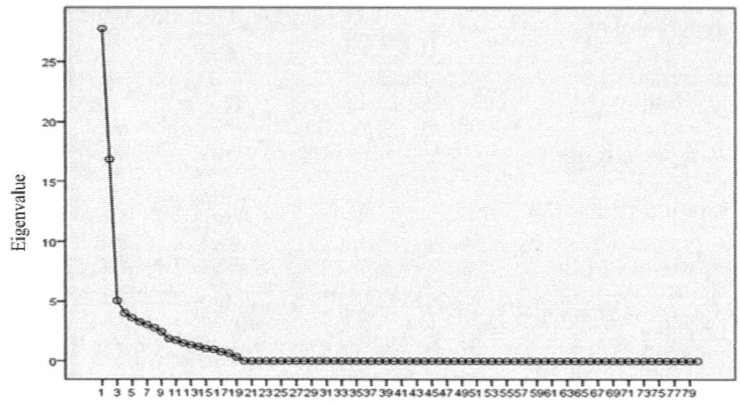

Fig.1 Scree test criterion of characteristic values.

Table 3. Analysis result of principal component factors

Factor	1	2	3	4	5	6
Initial eigenvalue	27.78	16.86	5.06	3.98	3.59	3.25
Proportion of measurement	18	11	5	6	6	5
Rotatory square and loaded	26.04	17.19	5.10	4.19	4.11	3.89
Variance interpretation degree/%	32.55	21.48	6.38	5.24	5.13	4.86
Variance accumulation contribution	32.55	54.03	60.41	65.65	70.79	75.64

Table 4. Terms of principal component factors and description of extreme body shapes

Principal component factor	Description of extreme figures		Variance interpretation degree
Factor 1: fatness factor	(+) Fat	(-) Thin	32.55%
Factor 2: height factor	(+) Tall	(-) Short	21.48%
Factor 3: lower body shape factor	(+) Small angle of inclination of lower body part (-) Large angle of inclination of lower body part		6.38%
Factor 4: upper body shape factor	(+) Small angle of inclination of upper body part (-) Large angle of inclination of upper body part		5.24%
Factor 5: shoulder shape factor	(+) Slope shoulder	(-) Square shoulder	5.13%
Factor 6: chest shape factor	(+) Acute triangle	(-) Equilateral triangle	4.86%

The six principal component factors comprise the primary component factor (fatness factor), the secondary component factor (height factor), the third component factor (lower body shape factor), the fourth component factor (upper body shape factor), the fifth component factor (shoulder shape factor), and the sixth component factor (chest shape factor). Terms of six principal component factors and description of extreme body shape of each factor are shown in Table 4.

3.3 Data analysis of dynamic body shape

Upon continuous observation of data changes of body shape in the dynamic state of the human body, we measured the human body's three-dimensional model based on the reference surface and reference point. Results are shown in Table 5. Upon comparison of the data for the bending-over posture and standing posture, the data change is mainly reflected in the upper body part and the greatest change in body surface size corresponds to back length, which is 2.5cm; shoulder width, chest girth, waistline, abdominal girth, and hipline are also changed. Upon comparison of the data for the sitting posture and standing posture, we found that the change in human body size is mainly reflected in the lower body part (i.e., waistline, abdominal girth, hipline, and the maximum thigh girth). The greatest change in the standing posture is reflected in measurements for hipline, abdominal girth, and waistline, with the change value in hipline being 4.9cm.

Table 5. Average change rate of body size in dynamic posture

Serial no.	Measurements	Sample size	Bending over-Standing posture (cm)	Sitting-Standing posture (cm)
1	Neck base girth	20	-	-
2	Shoulder width SP1	20	2.0	-
3	Chest girth	20	1.8	0.7
4	Waistline	20	1.1	1.5
5	Hipline	20	1.8	4.9
6	Abdominal girth	20	1.2	2.1
7	Maximum thigh girth	20	-	1.8
8	Back length	20	2.5	1.4

4. Conclusion

Our analysis of 3D body shape data of 20 Baiku Yao women 22-46 years of age in static and dynamic states found that the change in results of height measurements is relatively small and that of girth measurements is relatively large. In addition, through the analysis of body shape data corresponding to 82 measurements in the static standing posture, the following six principal component factors were extracted and used to describe Baiku Yao women's body shapes: fatness factor, height factor, lower body shape factor, upper body shape factor, shoulder shape factor, and chest shape factor, with the accumulative contribution rate of 75.64%. Comparison of data for body shapes in static and dynamic states shows that the change in back length is the greatest in the bending-over posture, with a mean value of 2.5cm; the change in hipline was the greatest in the sitting posture, with a mean value of 4.9cm.

Findings from this study's use of the 3D scanning method to collect, analyse, and summarise Baiku Yao adult women's body shapes can enrich the database for research pertaining to Chinese body shapes, and the acquired data related to Baiku Yao people's body shapes can be applied to other research areas, such as those involving the creation of mannequins, museum exhibits, and other virtual displays. The limitation of this research is that only 20 Baiku Yao women's body shapes were collected. We therefore recommend that future research in this field be extended to increase the reliability of the data.

References

[1] Zhang S. Subject system and practices of Chinese anthropology. J of Yunnan Normal University (Philosophy and Social Sciences ed) 2012;3: 88-94.

[2] Luo ZG, Yang JL. An anthropological study of Baiku Yao population. J Anthropol 2003; 2: 150-160.

[3] Yang JH, Yu YS. Clustering analysis of head shape characteristics of fourteen ethnic groups in southern China. J of Qiannan Medical College for Nationalities 2005; 2: 67-68.

[4] Peng HB, Tian JP, Zhu L, Wang B, Wang YM, Li QY, Yu YS. Study on seven human population genetic characters of Baiku Yao nationality. J of Qiannan Medical College for Nationalities 2011; 1: 1-5.

[5] Wu RK, Wu XZ, Zhang ZB. Body measurement method. Beijing: China Science Publishing & Media, 1984.

[6] Shao XQ. Human body measurement manual. Shanghai: Shanghai Lexicographic Publishing House, 1985.

[7] Liu BS, Guo XC, Ma XS. Applicable features of body size measurement of Chinese male aviators. Ergonomics 2002; 8(4): 1-3.

[8] Liu Z, Hu SR. Extraction of the mannequin character parts based on the three-dimensional anthropometry. J of Jiaxing University 2009; 21(3): 117-119.

[9] Cong S, Fang F. Classified study of lower body part shape of female college students in Shanghai. J of Donghua University (Natural Science ed) 2007; 33(5): 668-672.

[10] Ge QG. A study of basic patterns of trousers of women with different body shapes based on 3D body data. Donghua University: 2014.

[11] Zheng R, Zhang H, Huang HQ, Hu XD, Li J, Duan XJ. A human body measurement and analysis method. Beijing: CN1596827, 2005.

Design Research based on the Adjustable Function of Maternity Clothes

Mi-Nong Weng, Xing Zhang*

Xi'an Polytechnic University, 19 Jinhua south road, Xi'an, Shanxi, 710048, China

*Corresponding author's email: Zhxing029@126.com

Abstract

As time progresses and with the improvement of women's social status, a new era of women have higher expectations regarding the requirements for maternity clothes. This paper first studies the status of current existing domestic maternity clothes before providing an adjustable function to the maternity clothes. The aim is to design maternity wear so that it can be used before and after pregnancy, which not introduces an element of fun but also reduces the waste of resources.

Keywords: Adjustable; Clothes design; Maternity clothes

1. The concept of The adjustable function of maternity clothes and introduce the status quo

1.1 The status quo and development of maternity clothes in China

Maternity clothes are essential during pregnancy. With the improvement of women's social status there has been an increasing attention to the choice of clothing during maternity. Professional maternity clothes appeared in China in 1989 through "Gennie's Maternity" and is already considered a very late progression. Even now, the research and development of maternity clothes is still at a semi mature stage, many kinds of brand have a huge difference price and quality,the development level is lower than the world. As the government began to encourage families to have a second childmany policies were introduced favoring the younger generation. People born in the 1980s, are gradually becoming the main target for maternity clothes market [1]. These people are independent, hence have a higher expectation for maternity clothes being not only practical but also fashionable. The maternity clothes in the market at the present are unable to meet the demands of the new generation of pregnant women.

1.2 Introduce the adjustable function of maternity clothes

Maternity clothing is very complex as it the pregnancy period is divided in different stages. It is very hard for designers to create maternity clothes that matches the extreme pace in which the female body's changes which is also why pregnant women are hesitant to buy them. Maternity clothes designed for only the pregnancy period results in a waste of economy and also a waste of resources.

Therefore as short term consumer product, maternity clothes that have an adjustable function will have obvious advantages. It is designed so that it matches the pregnant women's body growth.

At the moment, maternity clothes are designed as simply larger version of an ordinary dress however they are unsuccessful in combining functionality with fashion. However there are high grade maternity clothing that do provide adjustable maternity clothing style however, no brand have been established.

2. Design points for the adjustable function of maternity clothes

2.1 Adjust parts illustrate for maternity clothes

The pregnancy period has a total of 40 weeks (10 months), early pregnancy for 1 ~ 12 weeks (1 ~ 3 months), mid pregnancy is 13 ~ 27 weeks (4 ~ 7months), late pregnancy to 28 ~ 40 weeks (8 ~ 10 months) during which the women's body is always changing. Early pregnancy does not need maternity clothes but simply dress to fit, However, starting from mid pregnancy, there is an obvious uplift to the abdomen, increase in chest circumference waistline and hipline which is when women start wearing maternity clothes.

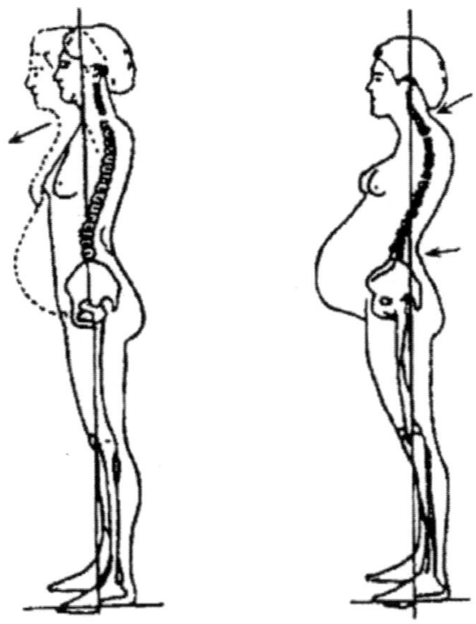

Fig.1 Body changes in early and late pregnancy"have a wrong place, should under the picture which in 8 line. [2]

Body changes in early and late pregnancy is shown in Fig.1 [2], Fig.2 [3]. Chest circumference and abdominal circumference are the biggest change. From the beginning of the third month, abdominal circumference size gradually increases from about average 74cm to about 98cm [4], Seventh month to tenth month abdominal growth grows more quickly, the chest circumference also increases by 2~3 cups [5]. Ordinary clothing are unable to meet such a growth rate. So the chest and abdomen should be adjusted as the key part from the third month of pregnancy.

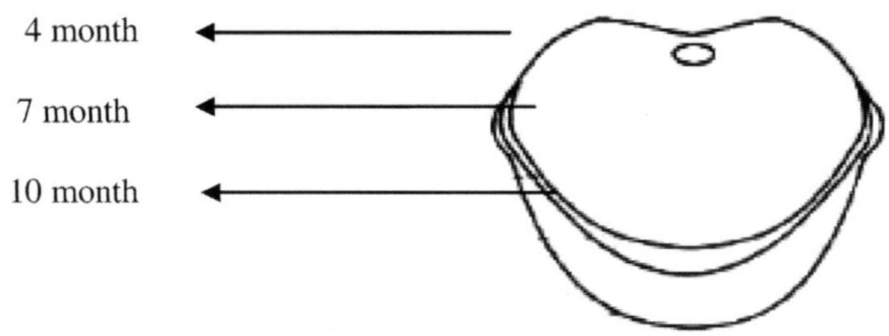

Fig.2 The abdominal section of pregnant [3]

2.2 Garment-tolerance illustrated for the adjustable function of maternity clothes

To calculate the effectiveness in the adjustable function of maternity clothes it is important to measure the body parts primarily the chest and abdomen.The garment-tolerance of the maternity clothes must be looser than ordinary clothing. The garment tolerance required between the period of 4-6 months in the chest area is an increase of 15cm, and in the abdomen area by 15 cm. The garment tolerance required between the period of 7-10 months in the chest area is an increase of 25cm, and in the abdomen area by 35 cm.

According to the needs of the body to divided into several times adjustment, it is "The adjustable function of maternity clothes" is different from the ordinary design method of normal pregnant, measuring the chest circumference in the early pregnancy minimum is"c", the abdominal circumference in the early pregnancy minimum is"h", measure the chest circumference and the abdominal circumference in the late pregnancy maximum are"C", "H". the garment-tolerance of the maternity clothes is"R".So the adjustment of the chest circumference is:

$$(C - c) + R \tag{1}$$

the adjustment of the abdominal circumference is:

$$(H - h) + R \tag{2}$$

assuming that set 3 sub-file of the maternity, it is means adjust 3 times, each chest circumference regulating variable is:

$$(C - c) + R / 3 \tag{3}$$

abdominal circumference variable is:

$$(H -h) + R / 3 \tag{4}$$

This method can be widely used In the "Hidden adjustive method" and "Split and combine method which based on the deconstruction ".

3. Design methods for the adjustable function of maternity clothes

3.1 Adjusted by using the fabric performance

This design method is to use the elastic fabrics, such as knitting fabric, modal fabric to design fitting maternity clothes. In pregnant women sweaters, dress collocation is mainly inside. As each fabric elastic coefficient is different, the degree of how loose the garment is going to be is also different. we shoud ensure that the pregnancy clothes is not to tight for women's body. This kind of design is simple, practical and widely used.

3.2 Adjusted by using drape

Ajusting the drape involves warping fabrics, optional twisting. It uses the inverted pleat so that the maternity cltohes are adjustable.The drape have their own garment ease,it can form different effects according to body size. General in the chest and waist to use drape to design (Fig.3), using mostly in pregnant women dress. Drape method is flexible with a beautiful effect. However, it is important to consider the number of drapes to avoid appearing fat.

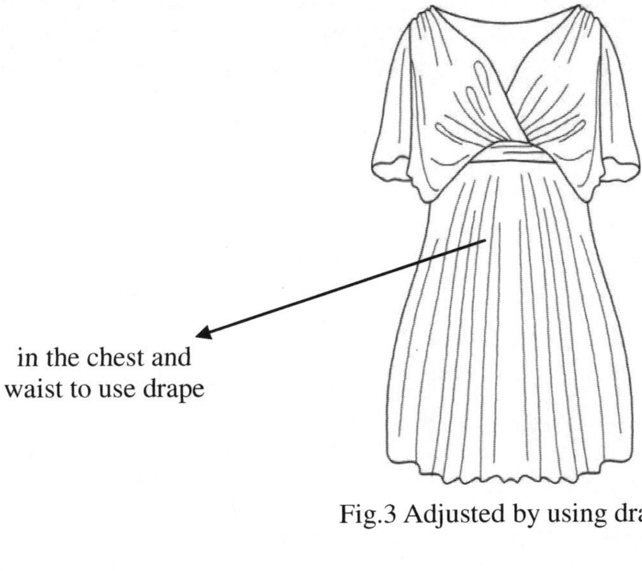

Fig.3 Adjusted by using drape

3.3 Adjusted by clothing modelling

Modeling design method is another way to adjust maternity clothes by using the profile of the garment and its shape characteristics. Some profile of clothes already have a regulation system that can adjust with body size. This model is generally for body shape that is hidden well, such as “A” shape, “O” shape the bat sleeve shape, sweep and side seam use slits (Fig.4). These shapes can appropriately hide the body figureduring pregnancy.

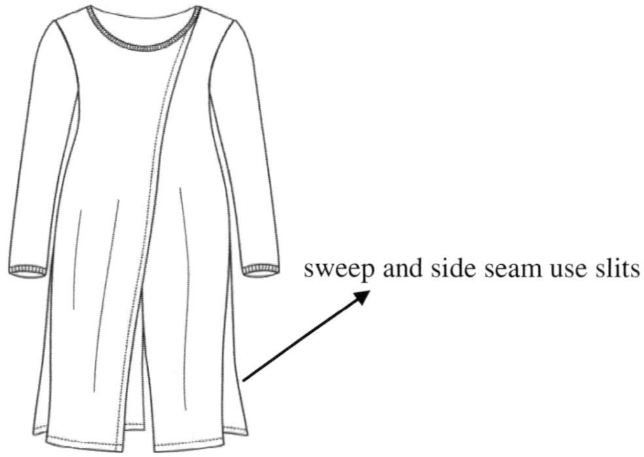

Fig.4 For this shape of clothing the side seams are used as slits

3.4 Adjusted by using accessories

Garment accessories is an addition to the fabric decorative clothing and expand the function of clothing materials with zippers, buttons, lace, cord, fasteners, etc. The use of garment accessories is to diversify the connection between the clothing structure. An appropriate use can allow the maternity clothes to be adjustable but at the same time visually appealing.

3.4.1 Hidden adjustable method

Hidden adjustment method is use the zipper, poppers, buttons,and velcro for adjust. Adjust parts is the abdomen, side seam, shoulder straps. The side seams uses multiple invisible Zipper design,abdommen also use invisible zipper design. (as Fig.5: When the zipper is opened, clothes can be loose,it is a pregnant clothes. When the zipper is closed, it is an ordinary clothes). This kind of design from the external of clothes it is hard to see clear difenrences, This method use mostly on maternity coat, but the design needs to consider the comfort of the accessories.

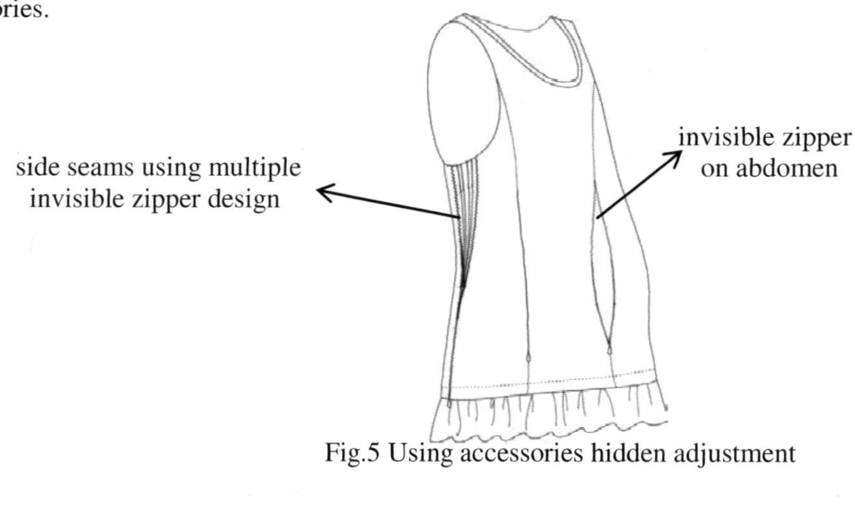

Fig.5 Using accessories hidden adjustment

3.4.2 Adjustable function using rope.

Another method to allow the adjustment of maternity clothing is using rope around, the waist and the sweep. The rope is very comfortable and used mostly on maternity coat (Fig.6: Use a rope in the chest of the coat.) and trousers. While the body is changing during pregnancy the woman is able to adjust the rope to either relax or tighten the clothes to fit their body figure.

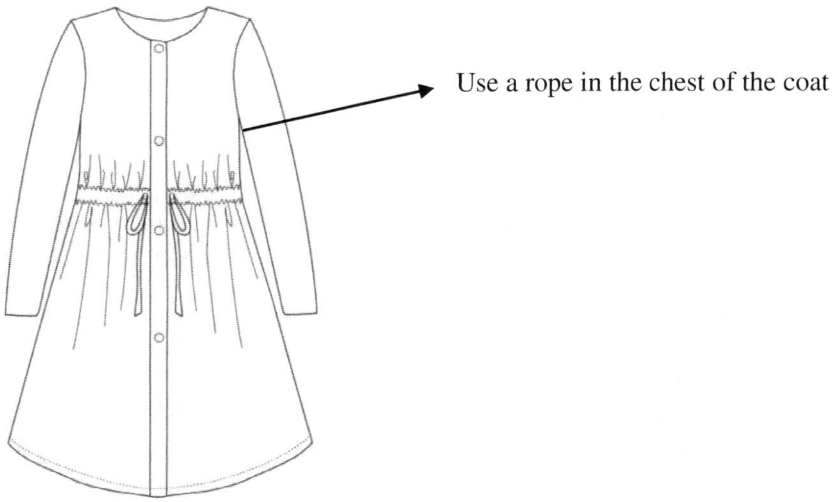

Fig.6 Adjusted by using rope

3.4.3 Split and combine method which based on the deconstruction

Split and combine method is which use szipper, fasteners to connect the clothing together with a variety of different combinations. Generally applied in maternity coat and dress design, adjustable clothes using this method will have a main "body". other garment pieces and parts can be split and put together in a variety of different combination (Fig.7) to change the size of maternity clothes.This adjustable design is the most suitable for maternity clothes design, this method is flexibility and adjustable,but the design of the mthod have a higher demand for designers and pattern makers, should fully consider the comfort of the connection parts, and design expense.

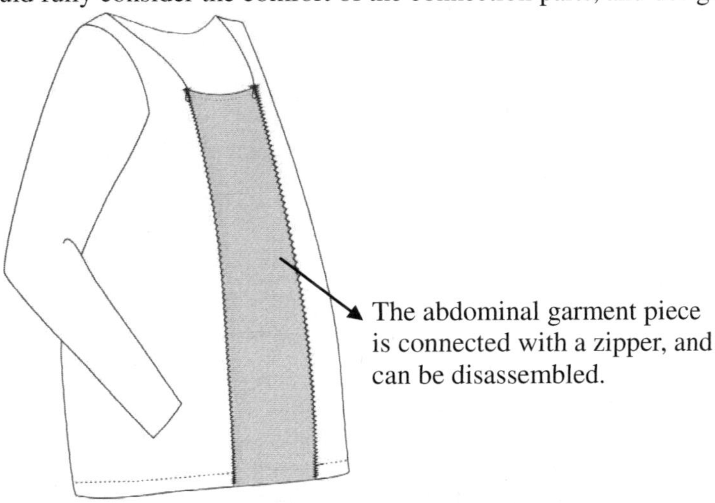

Fig.7 Split and combine method

3.5 Comprehensive application method

Comprehensive application method uses a combination of the above mentioned techniques that allow maternity clothing to be adjustable.

4. Design illustration for the adjustable function of maternity clothes

The adjustable function of maternity clothes is a part of the maternity clothes, and many different from the ordinary clothing, The general rules and basic design methods for adjustable function of maternity clothes are as follows: 1) When designing adjustable function for maternity clothes, there should be a focus in the consideration on the practicality and functionality. 2) When designing the adjustable function for maternity clothes there should be strict references to pregnant women size data in order to make a reasonable sub-file and stitching pieces. 3) Design the adjustable function of maternity clothes so that it is comfortable to wear as pregnancy can cause a lot of inconvenience. 4) Accessories selection for the adjustable function of maternity clothes should be payed more attention to 5)The adjustable function of maternity clothes should also consider the fabric comfort and color fashion. 6) Finally, the adjustable function of maternity clothes for designers put forward higher request, to be both creative and ingenious principle.

5. Applications and future prospects for the adjustable function of maternity clothes

The adjustable function of maternity clothes are designed based on the demand of women during their pregnancy. They are designed so that the clothes are not only suitable to wear but also gives women a fashion choice with lots of flexibility. Such design take care of the pregnant women's body change, not only make a theoretical contribution for maternity clothes, but also provide a theoretical reference for the development of products and brands, Can be used in the clothing company of pregnant woman clothing design,service of society, at the same time produce economic value.We believe this design has lots of potential due to its efficiency and convenience allowing the entire pregnancy experience to be much more enjoyable.

References

[1] Peng GX. Maternity clothes: how to differentiate innovation.Sales and marketing: 2010; 11: 82-84.
[2] Du J, Fu ZW. Maternal health care. Beijing science and Technology Publishing House: 2000: 12-26.
[3] Gao YW. Mmiddle-late pregnancy bottoms design. Knitting industry: 2008; 8: 33-35.
[4] Ye QZ, Chen DS. The data analysis of the pregnant women in the development of maternity dress. Journal of Xi'an Polytechnic University: 2015; 4: 487-491.
[5] Shao XW. Design of maternity health safety based on the physiological characteristics of pregnant Women. Journal of Southwestern Normal University: 2009; 5: 180-185.
[6] Wang XZ. Analysis of the structure of pregnant women's clothing based on Ergonomics.International Textile Leader: 2008; 1: 76-79.
[7] Lv X. Study of functional maternity dress design. Zhejiang University: 2007; 5.
[8] Li D, Xia F. "The clothes wear more application" in the clothing design research. Silk: 2014; 1: 50-54.
[9] Tang HY. Maternity clothes design research.Shanghai Donghua University: 2007; 12.
[10] Xu YJ. The basic theory of maternity dress design. Art and design: 2011; 12: 226-227.
[11] Zhou ZS. Pregnancy childbirth child care book. China Women press: 2010.
[12] Dai GN. Based on the trend of fashion maternity dress design. The China Academy of Art: 2012; 6.
[13] Xiao LQ. Different protruding abdomen body clothing structure design. Clothing accessories: 2012; 2: 67-69.
[14] Zhang WB, Fang F. Clothing ergonomics body. Donghua university press: 2008.
[15] Xu CH. The differences between the maternity and general clothing.Textile industry and technology: 2012; 3: 89-91.
[16] Information on http://corp.octmami.com.

Emotional Cognition of Clothing Color

Qian-Yun Chen[1], Xiao-Feng Jiang[1,2*]

[1]*College of Textile and Clothing Engineering, Soochow University, Suzhou, Jiangsu, 215021, China*
[2]*National Engineering Laboratory for Modern Silk, Soochow University, Suzhou, Jiangsu, 215123, China*

*Corresponding author's email: xfjiangsz@163.com

Abstract

Six priming words with positivity or negativity and eight kinds of hue varied with brightness and saturation were used as stimuli in this paper, and the color cognitive behavioral experiments were conducted in two blocks where whether started with the semantic priming or not. The purpose was to investigate the emotional cognition of the clothing colors, and the cognitive rules of positive or negative color emotions. Results indicated that the high brightness and high saturation color were cognized as positive colors. On the contrary, low brightness and low saturation color were classified as negative colors. There is no significant difference of color emotion evoked between priming words and by no priming words. However, the cognitive activity with the semantic priming was obvious to delay the reaction times. The classification speed of positive emotion color is obviously faster than that of the negative one.

Keywords: Clothing colors; Positive-negative; Emotion; Cognitive

1. Introduction

Color plays a key role in clothing on account of contributing to the first impression and impacting on the decision making of individuals [1]. A large number of literatures show that color can arouse individual emotions and affect individual moods. For example, the color red has been perceived as warmth, and green has been associated refreshment. Colors, in general, have both positive and negative emotions. For instance, blue has been found to have positive impressions such as intelligence, trust and calm, but on the other hand, have negative impressions such as coldness, depression and unfriendliness [2,4].

Colors can be represented by hue, brightness and saturation, and their variety can also affect color emotion. The brightness can be changed by adding black or white, which may evoke dissimilar human feelings. H. J. Suk has discussed the qualitative relationship of the three attributes of color and the three dimensional volume of emotion. Results indicated that there is a positive linear relationship between emotion and the saturation or brightness of color [5]. J. G. Brennan has defined the lightness of color as its attributes that it can be equivalent divided into a set of achromatic series from fuzzy to clear [6]. J. H. Xin has studied the differences of color emotion in diverse culture and geography, and found that the hue was much less predominant for color emotions than brightness and saturation [7]. Li Chen Qu designed the single color psychology experiments, and put forward scientific color emotion model for basic color, the results displayed that there is no gender difference in emotion of color, but between diverse culture, and two groups emotionally metrics (tense-relaxed, like-dislike) differences exist [8]. X. P. Gao has researched the relationship between color attributes and emotions, as well as influence of different cultural backgrounds on the color emotion. Experimental results revealed that there is no significant regional difference. Experimental results revealed that active emotion affected by saturation, and potency factor influenced by brightness, definition coefficient affected by both saturation and brightness [9]. Above, these studies proved the relationship between clothing color and human emotion cognition, but didn't really point out that the laws of clothing color emotional cognitive and the reasons for the formation of this pattern, and our research is mainly aimed at this problem.

In the current study, we will explore the law and its forming reason of positive-negative clothing color emotional cognition, mainly from the three aspects which are the presence or absence of emotional words, positive-negative emotion words and the change of color itself to explore the laws. By investigating subjects' color emotion evaluation ratio and their response time, find out the laws of color emotional cognitive. On this basis, combining with relevant scholars conjecture, explored and analysed the formation reasons of those cognitive laws, to have a better understanding to the positive-negative emotion words and emotional perception of color.

2. Experiment

2.1 Participants

36 undergraduates from Soochow University (18 males, 18 females, aged in 22-24) participated in the experiment, and all participants had normal or corrected-to-normal vision. Participants are all right-handed and have to be tested individually, and have never done a similar experiment.

2.2 Stimuli

1) The choice of emotive priming words. Through the experiment of 60 emotional word-pairs, we selected three word-pairs: gorgeous-unadorned, relaxed-nervous, bright-somber.

2) The choice of the color. The colors of this study came from the research of Palmer [10].We selected the eight kinds of hues (red, orange, yellow, yellow-green, green, blue, ultramarine, violet), each hue with low saturation (LS), low brightness (LB), high saturation (HS) and high brightness (HB) four kinds of tones, thus 32 kinds of color formed the stimulus.

2.3 Procedure

The experiment is conducted on the computer modulated by the E-Prime2.0 software (E - Prime is by the university of Pittsburgh learning research and development centers in the United States and the United States PST companies jointly developed a set of experimental generation system for computerized behavior research). The experiment consisted of two blocks, one block evaluated the color perception with emotive priming words and the other block evaluated the color perception without emotive priming words. Each trail began with the presentation of a fixation point "+" in the center of the screen for 100 ms, after 400 ms, one of six emotive priming words appeared on the central of the computer screen for 300 ms, and then followed by a color image which appeared randomly. After that, the participants must evaluate the emotion of the color. Responses were collected via keyboard, and reaction times recorded automatically by the software system. The subjects pressed "D" in keyboard if he thought the color emotion was in accordance with the word. Otherwise, he should press "K" in keyboard. The whole experiment was about 25 minutes. Before the experiment, 10 trial images were also provided to acquaint with the manipulative method. After the experiment, we will adopt the method of ANOVA (analysis of variance) analysis of the experimental results.

2.4 Results

2.4.1 Color emotional evaluation

The data were fed into multiple-factor repeated-measures ANOVA with Priming (with words and without words) ×Emotion (positive and negative) ×Color (LS, LB, HS and HB). Fig.1 shows the mean categorization accuracy of positive-negative emotions for three word pairs. There was a significant main effect of emotion ($F = 7.957$, $p = 0.000$), implying that different colors arouse diverse emotions. In general, LS and LB colors evoke unadorned and somber emotions. On the contrary, HS and HB colors evoke gorgeous and bright emotions.

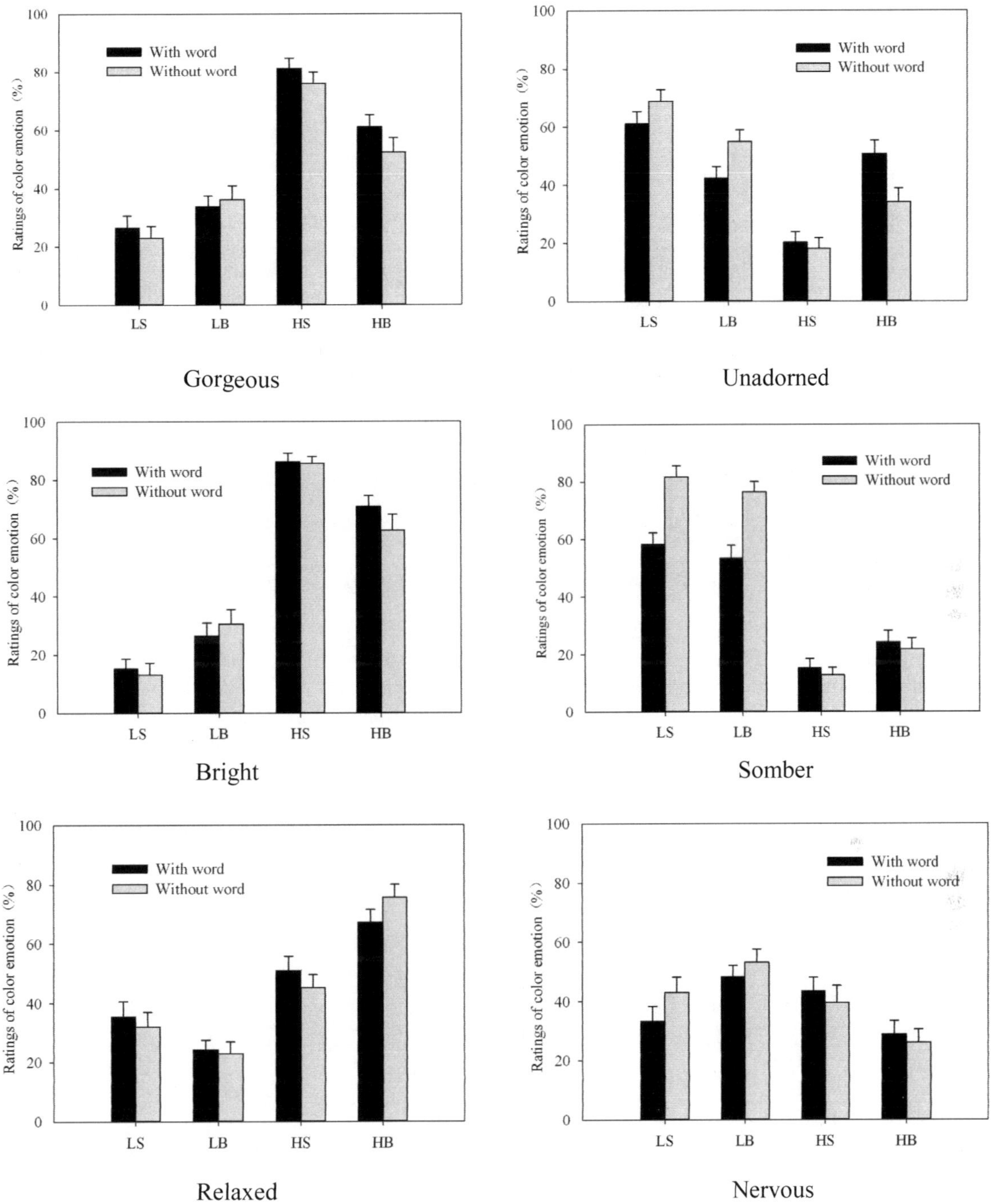

Fig.1 Different color emotional cognition

2.4.2 Color cognition with and without priming words

Viewed from Fig.2, results clearly manifest that the evaluable results of color emotions with priming words are similar to that of without priming words. However, there was no significant main effect of priming [F = 0.543, p=0.466], suggesting that the color emotions triggered with or without priming words are not remarkably different. Namely the emergence of emotional words or not doesn't affect the evaluation results.

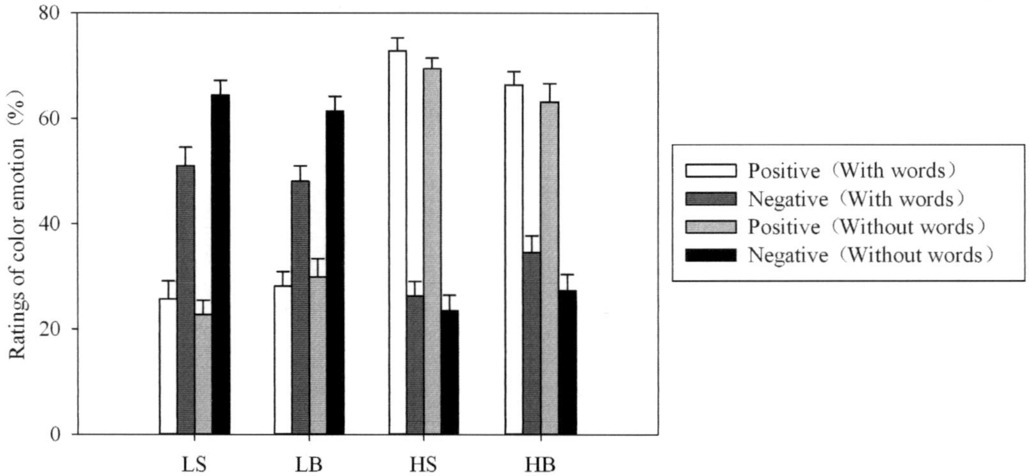

Fig.2 Color cognition with and without priming words

2.4.3 Reaction times (RTs) of color cognition

Table 1 is the reaction time spent in cognition for positive-negative emotion color. It is clearly shown that with the help of semantic priming, the RTs of the subjects during evaluation of color emotion are longer than that with no semantic priming, suggesting that the Stroop effect obviously under the condition of emotional words, namely the color words conflicts make the participants in the evaluation of emotional color delayed time. Table 1 also shows that regardless of whether emotional words existed or not, the time participants took to evaluate positive emotion is less than that to evaluate negative emotion.

In Table 2, the analysis of variance showed: there was significant main effects between the presence of emotional words and the reaction time of color emotional evaluation ($F = 12.336$, $p = 0.001$), indicating that positive-negative emotional priming words have a significant impact on the evaluation of color emotion; It had significant effects between the positive-negative emotional words and reaction time ($F= 5.863$, $p = 0.021$), implying that there was prominent influence of positive-negative emotional words on color emotional evaluation; the interaction of the presence of emotional words and color also has an obvious influence on reaction time of color emotional evaluation.

Table 1. RTs spent in cognition for positive-negative emotion color (ms)

Semantic priming	Positive-negative	RTs	Std. Error	95% Confidence interval	
				Lower limit	Upper limit
With priming words	Positive	965.552	54.119	855.684	1075.403
	Negative	987.563	58.652	868.493	1106.631
Without priming words	Positive	880.613	55.154	768.645	992.580
	Negative	912.695	55.553	799.917	1025.472

Table 2. ANOVA of repeated measures data for RTs of color cognition

Sources	Sum of square	df	Mean squares	F	p	ηp^2
Semantic	919379.917	1.000	919379.919	12.336	0.001	0.261
Positive-negative	105339.681	1.000	105339.680	5.863	0.021	0.143
Color	62262.983	2.336	26650.784	1.620	0.200	0.044
Semantic × Positive-negative	3651.776	1.000	3651.776	0.235	0.631	0.007
Semantic × Color	103639.091	2.385	43462.783	3.141	0.040	0.082
Positive-negative × Color	4251.220	2.385	1782.438	0.180	0.870	0.005
Semantic × Positive-negative × Color	40548.136	2.821	14373.929	1.223	0.305	0.034

3. Discussion

In the cognitive processes of clothing color, this paper focused on the presence or absence of semantic priming, positive-negative emotion and the impact of color change on color emotional cognition by using the behavior experiment, results indicated that the presence and positive-negative emotion words has important implications for the clothing color emotion cognition.

Color emotion can be described as two-dimensional structure space constituted by positive emotion and negative emotion. The experiment divided color priming affection into positive and negative two dimensions, in which researched the cognition trend of this three groups emotional words and the response time to evaluate color emotion [11].

Firstly, it has significant relationship between positive-negative emotional words and color evaluation, when it appears positive emotional words, people tend to choose high purity and brightness colors, and for the negative emotion words, the tendency towards matching the low purity and low brightness color.

Secondly, there is no prominent relation between emotional evaluation of color and whether there is an emotional word, the word appears or not does not affect the results of the evaluation. But we can also see that in the absence of words, evaluation of the effect is more significant for negative emotional words; and in case with words, the effect of evaluation on positive emotional words is more remarkable [12].

Thirdly, with the help of emotional words, the reaction time of subjects are longer than that with no emotional words, which shows that it occurs Stroop effect, the perception time of color emotion would be prolonged for the conflict between words and colors [13,14]. At the same time, regardless of whether there are emotional words, the time participants took to evaluate positive emotional words has less than the time of negative emotional words.

Finally, in terms of color itself, people were no significant differences on the time of color emotional evaluation. Statistics displayed that the change of tone itself has little effect on the cognitive evaluation of color affection, however, with the increase of brightness and purity, the time spent on evaluating is relatively decreased, which is consistent with the judging result of positive emotion words in the third point.

4. Conclusion

Through experiments to explore the classification rules of positive-negative emotion, we obtained the following conclusions:

1) The color with high brightness and saturation is cognized as positive emotion, conversely, low brightness and low saturation color is cognized as negative emotion;

2) In the cognitive activities with the semantic priming, the color-words conflict effect occurred. But there is no remarkable difference between with the semantic priming and without the semantic priming;

3) No matter whether there are priming words, the subjects responded more quickly in cognized the colors as positive emotion than that as negative emotion, which suggest that the color cognition affected by the color-word conflict and related to the color approach or avoidance effects.

5. Acknowledgment

This work was funded by A Project Funded by the Priority Academic Program Development of Jiangsu Higher Education Institutions (PAPD)

References

[1] Crane D, Bovone L. Approaches to material culture: The sociology of fashion and clothing. Poetics 2006; 34(6): 319-333.
[2] Li WH, Jiang ZQ, Liu Y, Wu Q, Zhou ZJ, Jorgensen N, Li X, Li C. Positive and negative emotions modulate attention allocation in color-flanker task processing: Evidence from event related potentials. Motiv Emot 2014; 38: 451-461.
[3] Jiang XF, Bian XP. Positive-negative emotional categorization of clothing color based on brightness. Engineering 2013; 5: 189-194.
[4] Jiang XF, Cai LR. Evaluation of aesthetic response to clothing color combination: a behavioral and electrophysiological study. JFBI 2013; 6(4): 405-414.
[5] Suk HJ, Irtel H. Emotional response to color across media. Color Research & Application 2010; 35(1): 64-77.
[6] Brennan, JG, Burnham RW, Newhall SM. Color terms and definitions. Psychological Bulletin 1948; 45(3): 207-230.
[7] Xin JH, Cheng KM, Taylor G, et al. Cross-regional comparison of color emotions Part I: Quantitative analysis. Color Research & Application 2004; 29(6): 451-457.
[8] Ou LC, Luo MR, Woodcock A, et al. A study of color emotion and color preference. Part I: Color emotions for single colors. Color Research & Application 2004; 29(3): 232-240.
[9] Gao XP, et al. Analysis of cross-cultural color emotion. Color research & application 2007; 32(3): 223-229.
[10] Palmer SE, Schloss KB. An ecological valence theory of human color perference. PNAS 2010; 11:107(19).
[11] Cothran DL, Larsen R. Comparison of inhibition in two timed reaction tasks: the color and emotion Stroop tasks. JRL 2008; 142(4): 373-385.
[12] Liotti M, Woldorff MG, Perez III R, et al. An ERP study of the temporal course of the Stroop color-word interference effect. Neuropsychologia 2000; 38(5): 701-711.
[13] Verhaeghen P, De Meersman L. Aging and the Stroop effect: A meta-analysis. Psychology and Aging 1998; 13(1): 120.
[14] Luo CR. Semantic competition as the basis of Stroop interference: evidence from color-word matching tasks. Psychological Science(PSS) 1999; 10(1): 35-40.

The graphical representation of clothing model design knowledge and its Application in Intelligent Design

Bo-An Ying*, Lin Ma, Xin Zhang

Apparel & Art Design College, Xi'an Polytechnic University, South Jinhua Road 19#, Xi'an, Shaanxi, 710048, China

*Corresponding author's email: 361443916@qq.com

Abstract

In material economy transition and the knowledge economy, knowledge is becoming the most important capital as a factor of production. For the apparel industry, its performance trends on industry collaboration and enterprise technology innovation put forward higher requirements. Male suit companies accumulated a large number of plate-making technology and design experience, but the lack of effective knowledge management and the accumulation of knowledge and innovation cannot be reused, resulting in weaknesses in secondary development and technological innovation capability. This study combines expert knowledge to build knowledge representation, organization, and reorganization, generates and uses data graphs to show related technologies for complex data of male suit plate-making process and to solve the problem in knowledge representation in plate making. Multi-dimensional classification parameters of the male suit Thoughts of plate sorts out knowledge representation in intelligent CAD system. Plate-making process based on technical analysis of charts of potentially valuable relationships for Men in smart suits plate, making process of the main control point, and line discussions and extract give a visual representation. And hierarchy the plate making knowledge, structured and associated plate making integrate knowledge resources. Rules extracted from male suit plate making process control and critical control point for the linear model of intelligent design establishment offer a basis for reference and ideas. This paper aims to provide an effective way for a chart showing plate making efficiency and reuse of knowledge to reduce the waste of intellectual resources. In practical applications, streamlining the repetitive work of product and project development shortens the development cycle and improve secondary development and technological innovation.

Keywords: Men suit; intelligent CAD; graphical representation; knowledge classification; knowledge mining

1. Introduction

In knowledge economy, male suit industry needs to rely on experience from the traditional mode of production design for upgrades, which is a logical, rational, systematic mode of operation. As the importance of knowledge as a factor of production capital becomes more prominent ever since 1986, the concept of knowledge management and knowledge-based economy have been proposed. Knowledge Management becomes a hot topic in management, economics, library and information science, computer science, and other disciplines. and very broad scope of its research, based on the presence of knowledge between different disciplines, the research system and the more entry points there large difference [1]. This article focuses on men suits plate making knowledge, which includes information on plate making, plate-making knowledge to collect, organize, store, produce new knowledge, mutual transformation of explicit knowledge and tacit knowledge. Therefore, this paper introduces the knowledge graph showing related technologies and the association between the characteristics of knowledge and aims to analyze the information required by male suit plate making. This paper also explores the theoretical basis and method for the classification and rules of knowledge representation, establishes theory that applies to male suit model of intelligent design section, better development and utilization of knowledge in male suit plate making information resources such that that version of the division of valuable experience and knowledge accumulated to reuse and sharing.

Intelligent CAD system used to design men suit individual process steps in the knowledge, experience and methods is summarized using a series of modules, prototype package male suit properties and design method, the formation of standardized templates. By setting the template constraints and introducing artificial intelligence algorithms, a combination suit, expansion, rapid deformation design was achieved. In male suit pattern structure design major suit style design, using the template will automatically and quickly expand into a structure diagram pattern [2]. Intelligent CAD technology is currently used to design for men suits fit, using the parameters of predefined and user-interactive input to create a model of the way clothing [3]. The process for the special part due to the lack of a reasonable model builder may cause it to deform so that the model of conventional treatment methods with traditional knowledge pattern difference is relatively large - export to pattern pieces do not actually fit problems [4].This requires intelligent design in a model that uses knowledge to sort out and express. Identifying the main lines and control the critical control points on the template is crucial such that timely optimization of the manual plate is possible.

2. Pretreatment of male suit plate knowledge

For male suit plate, in terms of access to knowledge, explicit knowledge can be identified with a certain nature - it is easily accessible. The tacit knowledge often experiences, word of mouth and other forms exist in the human brain, it is more difficult to obtain. For this article the model of knowledge representation male suit selected template design explicit knowledge, tacit knowledge representation and for the need to establish the corresponding mathematical model and evaluate the rules on this basis.

2.1 Obtaining male suit plate knowledge

There are three methods of Obtaining explicit knowledge: 1) A distributed search. Sources of knowledge are diverse and knowledge of male model suit design is varied; therefore, during the knowledge acquisition, knowledge is gathered together (without regard to source of knowledge), and from which a model of intelligent design knowledge is generated. 2)Smart Agent. (Having certain capabilities include problem solving, and perceptual entity is expected to interact with the outside world). 3)Data mining. (By analyzing each data to find its own rules from large amounts of data, thus revealing the process of implicit, previously unknown and potentially valuable knowledge).

Since the search can be distributed knowledge retrieval and access, regardless of the source of knowledge, this article on the access to knowledge male suit plate uses a distributed search strategy. Knowledge of the structure of the male suit model is shown in Fig.1.

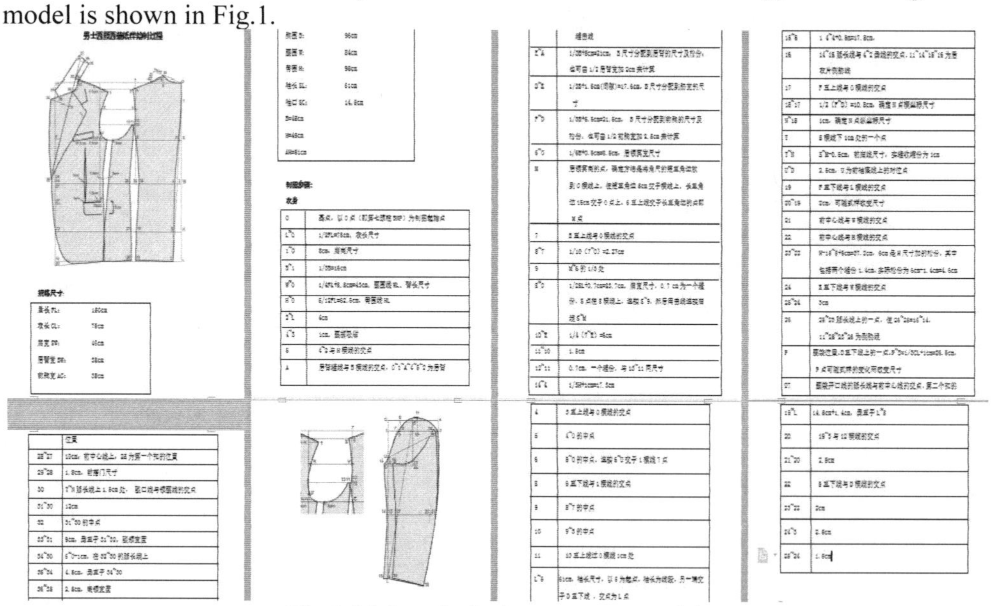

Fig.1 Male suit design process model

2.2 Modular male suit plate processing knowledge

From a macro perspective, modular design is the basis for a model of intelligent design. But also for the microscopic point of parametric model of intelligent design pattern design and provide the basis for generation.

The so-called modular processing system is to divide into several modules, each perform a specific function, and then these are brought together to form a whole (ie system) to complete a specified function [16].

The whole Intelligent design pattern suit system as the overall objectives and functions. However, these objectives are the results of the various components working together and implement functions are interrelated, and in the process of coordination between the various parts of the overall system according to the requirements of. In Intelligent CAD systems, this decomposition and coordination has certain requirements and based on that [17]: 1) According to the function of the system decomposition; 2) The information processing ways and means of decomposition; 3) In accordance with the workflow system decomposition; 4) In accordance with the development, maintenance and modification of convenience exploded. The modular approach provides a series of modules of decomposition - coordination principles and techniques, the whole system into a number of relatively independent modules. Through the coordination of relations between the module and the module design, the function of the whole software system is achieved. As shown in Fig.2.

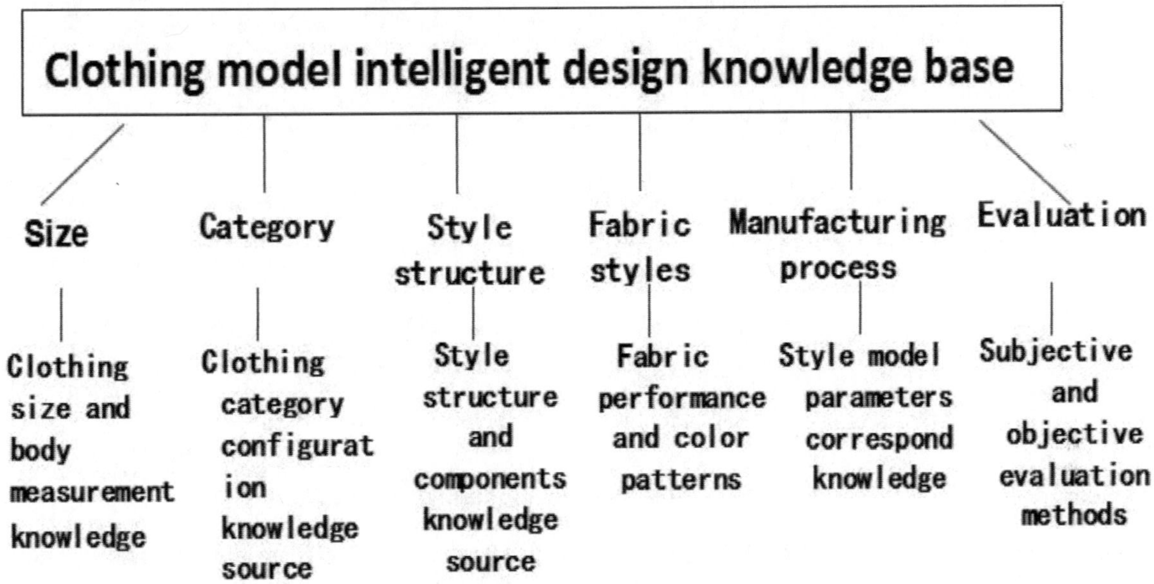

Fig.2 Male suit intelligent design module

3. Chart representation of male suit plate structure knowledge

3.1 Male suit plate structure knowledge Chart representation

Since the structure of knowledge male suit plate belonging multivariate data also has some sequential, the corresponding chart represents selected chart and scatter plot representation combination. The specific illustration is shown in Fig.3.

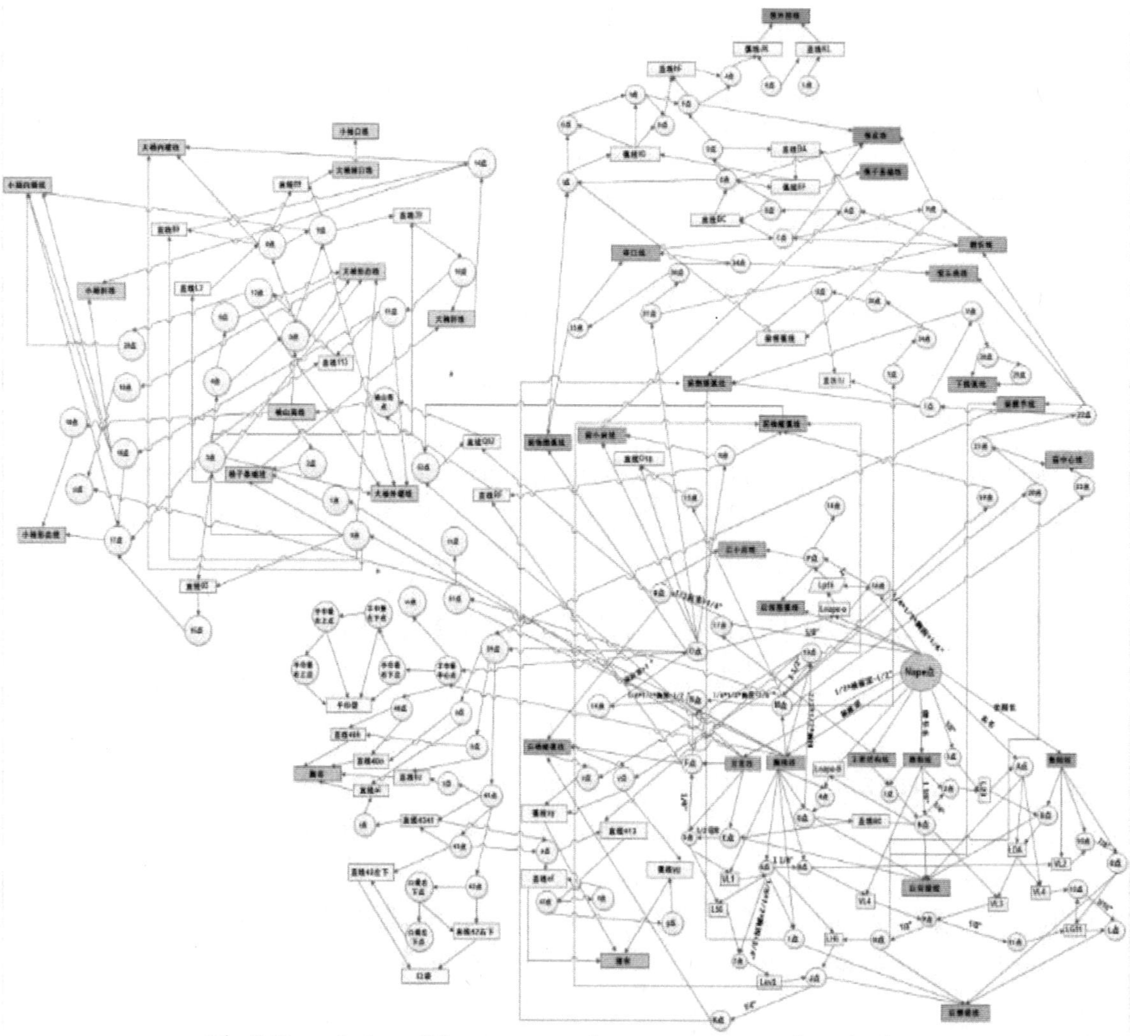

Fig.3 Knowledge of the structure chart represents male suit plate

The chart shown in Fig.3 uses "EDRAW" mapping software. The dotted line represents the relationship and its association with the logical reasoning. The circular bubble point represents the male suit design corresponding point data track involved, rectangular bar graph indicates a model involved in the structural design of the structure lines. Straight-line with the direction of the line connecting the dots corresponds to the structure of the line and build order. Some text corresponding to the text data related to logical inference is further elaborated as shown in Fig.4.

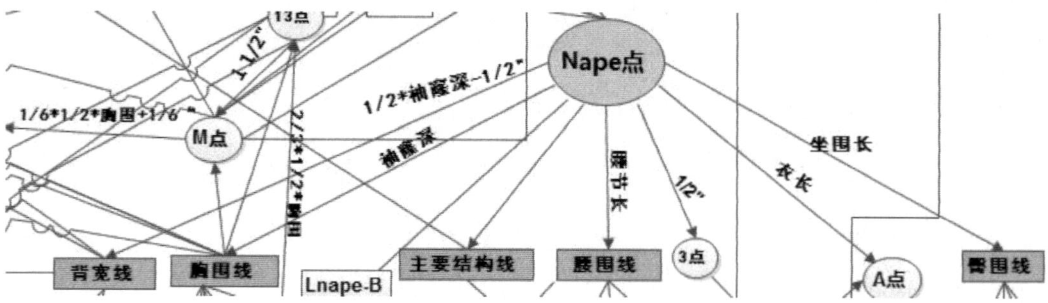

Fig.4 Sample design knowledge representation relations

4. Chart data analysis

4.1 Data processing

Use mapping tools for male suit template design process with charts and scatter, the relationship between the structure of the dotted line with the visual representation. But it is easy to see in Fig.3. a chaotic complex relationship. Using the "EDRAW" Drawing tool structure knowledge male suit plate were modular processing and logic Figure showing leaf reference sequence. As shown in Fig.5, The plate structure of male suit knowledge is divided into back panel, front panel, sleeve films, from films, provincial, and pocket six modules.

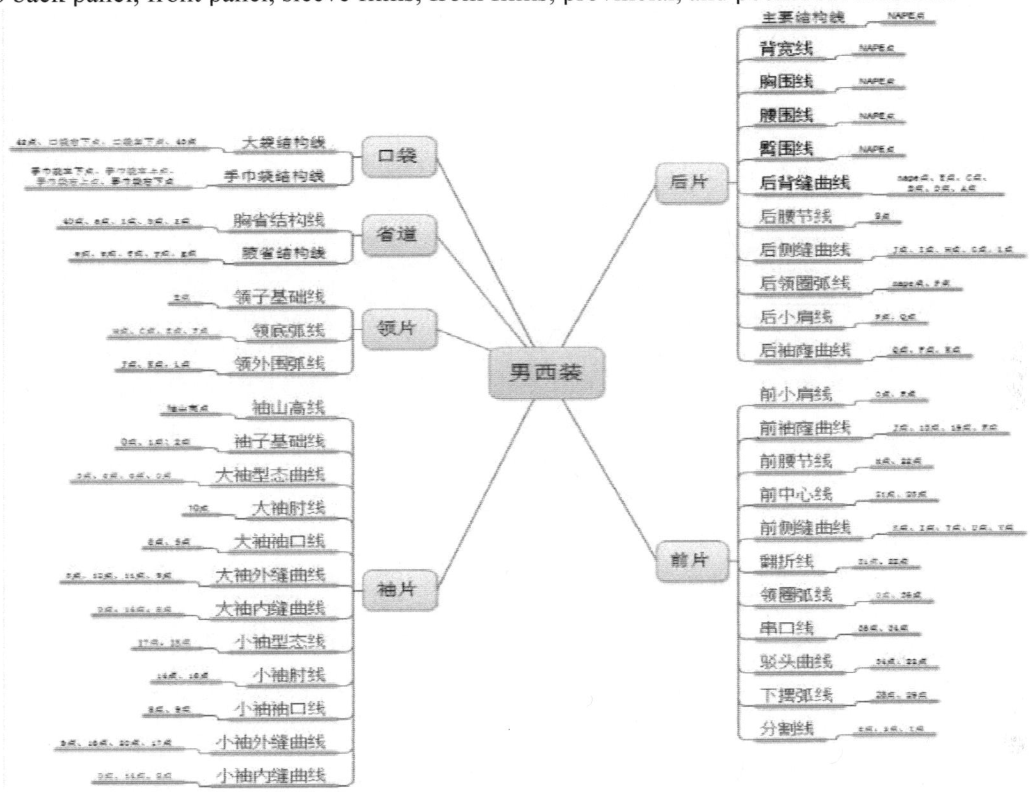

Fig.5 Male suit plate structure knowledge leaf Figure

4.2 Critical control points extracting

Through Fig.3, one can find the plate-making process of the male suit dotted line is not a one to one correspondence relationship, but a many to one relationship. For example, the NAPE point controls the bust line, waist line, hip line and back suture line is from by the point E, point C , point A D together. Extracted these special control line and the accuracy of the control points template for intelligent design template rule generation and model has a vital role.

Using "THE BRAIN" software, front, rear, sleeve films, from films, pockets, and other provincial extracted data extracted for further analysis can be obtained male suit template structure design Critical Control Point, as shown in Fig.6,7,8,9.

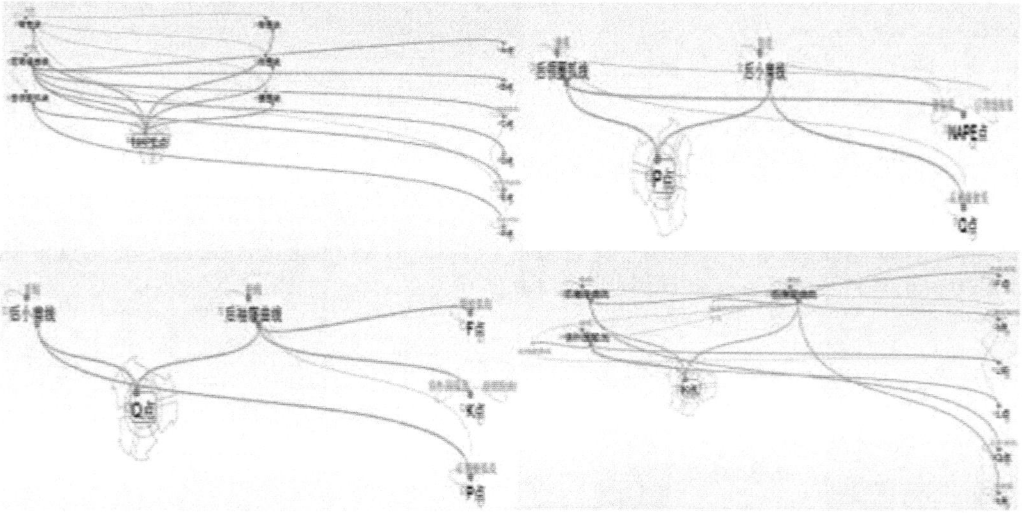

Fig.6 Back garment critical control point

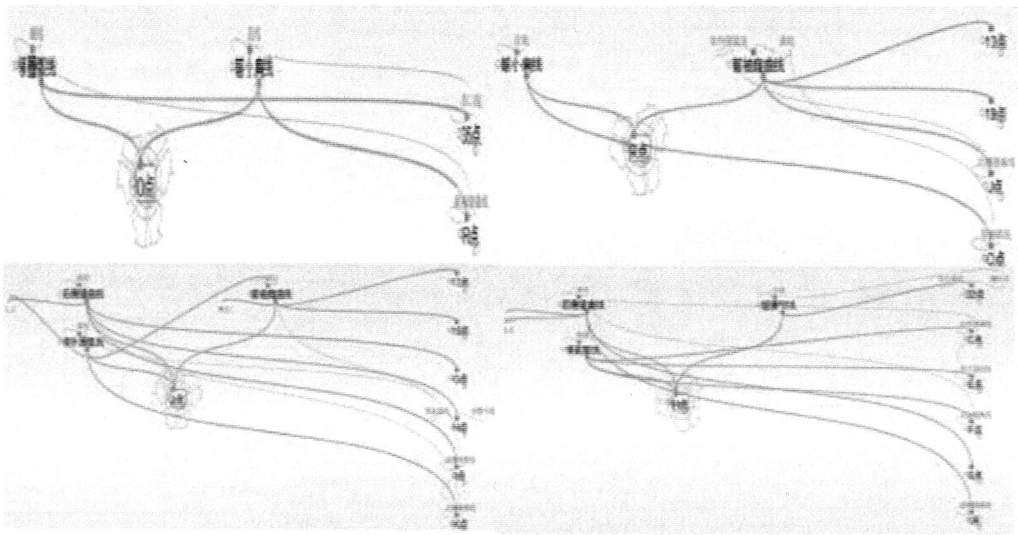

Fig.7 Front garment critical control point

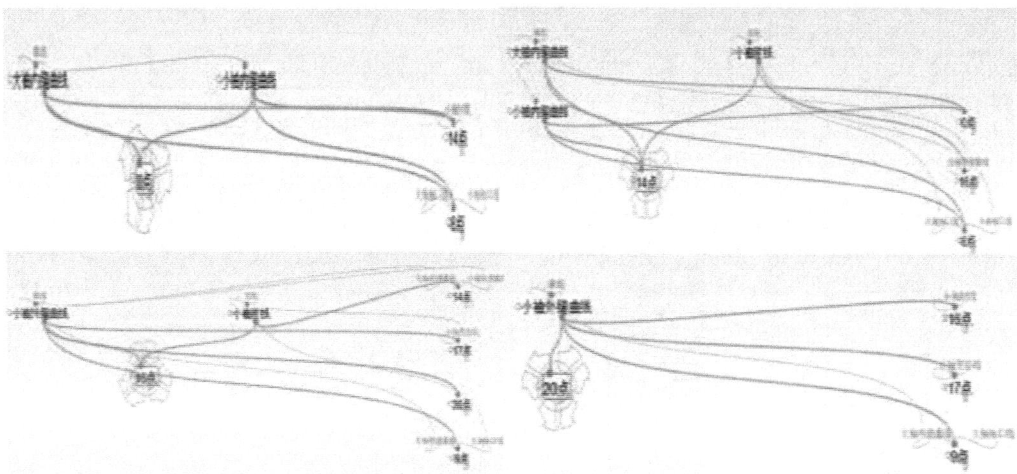

Fig.8 Sleeve critical control point

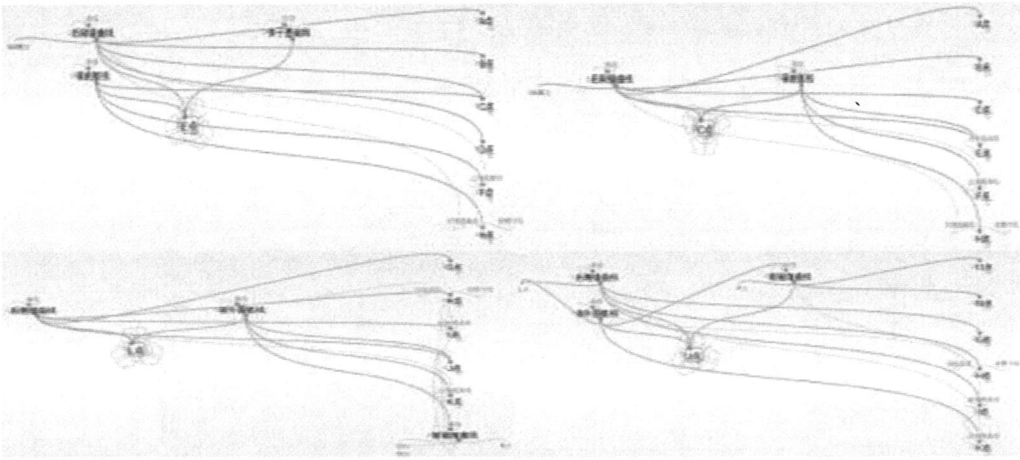

Fig.9 From films critical control point

The mind map template extracted in the structural design of the critical control points are:

E point, B point, C point, D point, A point, J point, I point, H-point, G point, L point, P point, Q points, F point, K point, O point, R-point, J point, H-point, T point, U point, V point, 13-point, 19 points, 21-point, 22 point, 23 point, 28 point, 29 point, 31 point, 34 point, 35 point.

4.3 Critical control lines extracting

Using "THE BRAIN" software will be analyzed according to the plate structure knowledge extracted critical control points available male suit template structure design front, back, sleeve piece, between lower collar piece, pockets, dart modules and modules The key line of control are shown in Fig.10,11,12,13.

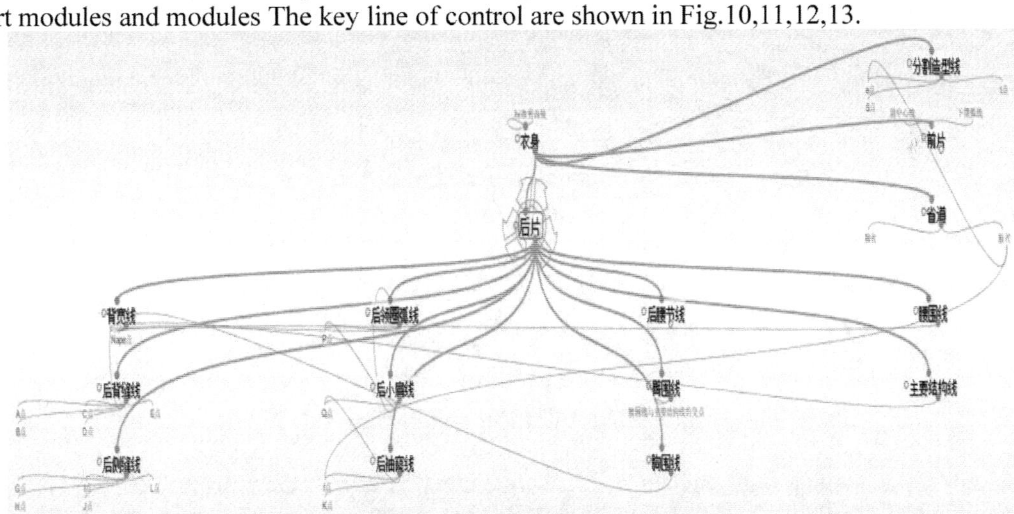

Fig.10 Back garment Critical control lines

Using mind mapping extracted rear panel to control the straight lines: the main structure line, back width line, bust line, waist line, hip line, the lumbar line, after shoulder line;

The back main control curve: back seam curve, the rear seam curve, back curve collar, back armhole curve;

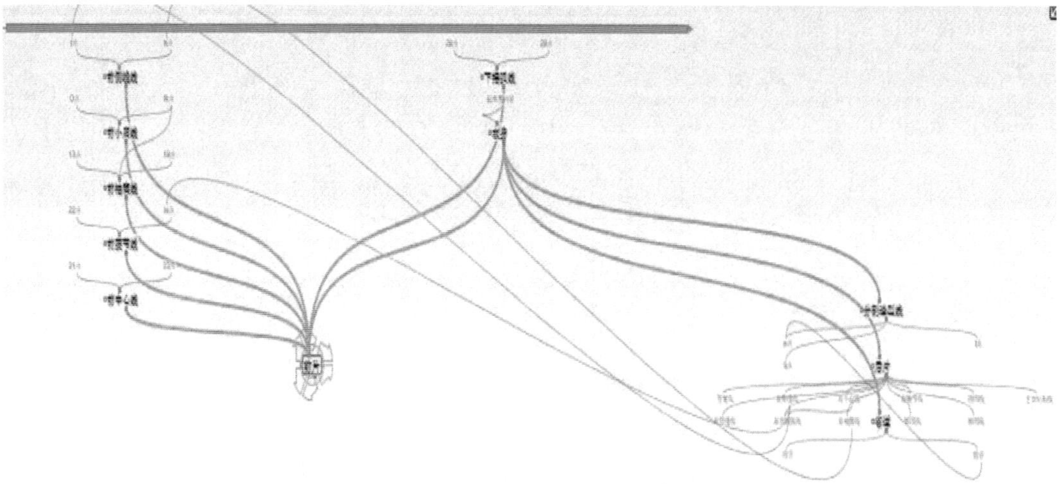

Fig.11 Front garment Critical control lines

The main front panel control straight line: the former shoulder line, the front waist line, the front center line, the folding line; serial line;

The main front panel control curve: front armhole curve, the front side seam curves, arcs collar, lapels curve, hem curve;

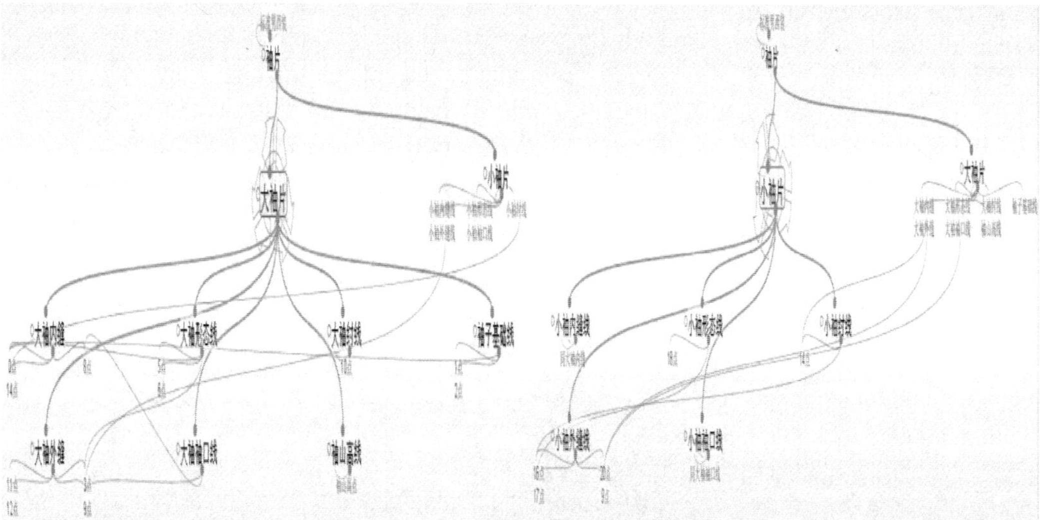

Fig.12 Sleeve Critical control lines

The main control for the sleeve lines: sleeve line, baseline sleeves, elbow sleeves big line, big sleeves cuff line, line patterns small sleeve, small sleeve elbow line, small sleeve cuffs line;

The main control sleeve curves: big sleeves pattern curve, big sleeves stitched curve, big seam curve, small outer seam curve, small seam curve;

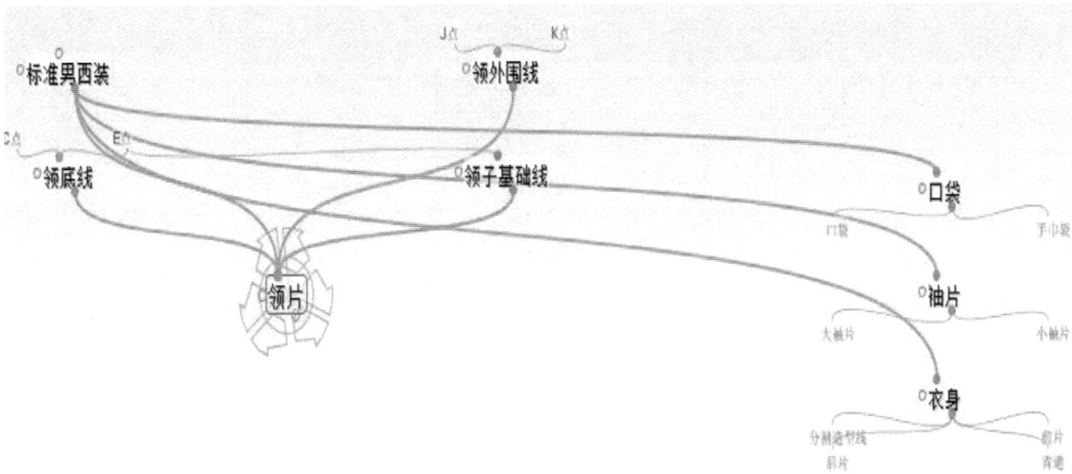

Fig.13 From films Critical control lines

The main control from films line: baseline collar, collar peripheral line;
The main control from films curves：the end of the from curve, and collar peripheral curve.

5. Conclusion

The male suit plate knowledge chart representation includes relational structure, properties, and hidden knowledge male suit plate knowledge nodes or between classes representation. In this paper, the introduction of "The Brain" and "Edraw" tools such as knowledge of graph representation, its semantic relationship between the presence of knowledge into a visual form of knowledge acquisition and improve mining efficiency and depth of knowledge resources to achieve efficient use of a certain practical significance. Simultaneous extraction of rule out male suit plate-making process in a straight line and control of the critical control points set up an intelligent design model that provides a basis for future reference and ideas.

6. Acknowledgements

I would like to thank my tutors: Associate Professor Bo An Ying and Professor Zhang Xin in helping me with the completion of this thesis. Their careful guidance made me improve further in all aspects of basic theory and professional knowledge. In particular, my mentors' rigorous scholarship and science, innovative research ideas, have benefitted me a lot. On this occasion, I would like to extend my heartfelt thanks to my tutors.

As a part of the whole subject, the completion of this thesis has been possible because of the strong support of all staff studio teacher Xiao feng Wang, the smooth experimental part of the paper has been done with great help from Shaanxi Province Engineering Center laboratory, as well as the careful teachings from my teacher Xueying Song; here I would like to deeply thank all the people who helped.

In addition, thanks to my family, for providing support in my process of thesis writing, and for providing a good environment in which I can concentrate on writing.

References

[1] Cheng Y. Study [D] based on uncertain knowledge management body. Chinese Academy of Sciences, Dr: 2007.
[2] Xu ZN, Zhang WM, Chen WW. retrieval [J] Ontology-based Intelligent Information Computer Science: 2001; (6): 21-25.
[3] (US) Drucker Peter F. Knowledge Management [M] Beijing: China Renmin University Press: 2002.

[4] Chu M section. Knowledge management research at home and abroad, the main achievements and future trends [J]. Information and Documentation: 2006 (5).

[5] Zou P. Application and Development of CAD technology in the field of modern clothing [J], Zhejiang Textile & Fashion College: 2007 (03): 15.

[6] Wu YH. Knowledge Classification and Application of Knowledge Management Research [D]. Ocean University of China: 2008.

[7] Xu L, Dai HP, Zhang WB. Application and development of the garment CAD [J], modern textile technology: 2007 (06): 46-49.

[8] Xu MX, Ning XY. parametric design study [J], Computer Aided Design and Computer Graphics: 2002 (11): 57-60.

[9] Liang JJ, Feng YL. Li Ruqin Status and direction of clothing CAD technology research [J], Journal of Textile Research: 24 (06): 612 - 614.

[10] Chen T. Parts Parametric Design Method and System Implementation [D], Nanjing University, master's thesis: 2003.

[11] Qu H. Sun Zheng Shun Construction and policy knowledge management system [J]. Computer Engineering and Applications: 2001 (37): 52 -54.

[12] Liu SS. Constraints parametric clothing plate making anthropometric dimensions [D], Beijing Institute of Clothing, Master's thesis, 2012.

[13] Vail, Edmond F. Knowledge mapping: Getting started with knowledge management[J]. Information Systems Management: 1999 (4): 16~23.

[14] Qin JH. research design and implementation of intelligent combination of style wear [D], Xi'an University of Engineering, Master's Thesis: 2004.

[15] Gong X. Relational Database semi-automatic ontology-based local study [D] Wuhan University, master's degree: 2008.

[16] Gu L. Parametric female body mannequin thoracolumbar modeling methods,research [D] Xi'an Polytechnic University, Master's Thesis: 2006.

[17] Gan RC. Information systems analysis and design [M]. Beijing: Higher Education Press: 2003 (10): 182-201.

[18] Song XY. Molded cup bra designed for Knowledge Management System [D]. Xi'an Polytechnic University Degree: 2011.

The Study on Clothing Design based on Low Carbon Concept

Xu-Lan Zhang[1], Lei Yao[1], Yu Bai[2]

[1]*Business school,Beijing Institute of Fashion Technology, Beijing, 100029, China*
[2]*School of Materials Science and Engineering, Beijing Institute of Fashion Technology, Beijing, 100029, China*

[*]Corresponding author's email: sxyyl@bift.edu.cn

Abstract

With the development of economy, with unceasing enhancement human material life, survival demand rises to developmental demand. Due to improper resource utilization, global warming, environmental degradation and waste pollution are threatening our daily life. Economic development cannot be built at the expense of the environment, and the pursuit of beauty can't treat pollutants as stepping stones, meanwhile, the deterioration of environment stimulates us to adjust developmental mode, low carbon development has become an important direction, which lead people to raise a hot wave of low carbon in all walks of life. As an important industry of our national economy, how to implement low carbon development effectively has became an important topic. It is worthy of studying for the clothing industry. This article mainly analyzes how to realize low carbon development in the clothing design steps by choosing two steps to study specific path of low carbon development: selecting fabrics and the format designing. Hope that, through this article research on low carbon path of clothing design steps, we can make a certain contribution to the development of low carbon for garment industry, and remind designer of paying more attention to low carbon factors, and providing some low carbon design elements in the clothing design.

Keywords: Low carbon; Clothing design; Fabric choice; Format design

1. Introduction

1.1 Background and significance of research

Green design, under the concept of "low carbon development", is an international design trend appearing in the late 1980s. It reflects the thinking of environment and ecological destruction caused by modern civilization of science and technology, but also reflects the return of designers' morality and social responsibility [1]. Low carbon development, is the inevitable orientation of sustainable development, is the important requirement of sustainable consumption, also is the most effective way to resolve new trade barriers. Comparing to sustainable design, it emphasis on the lower calculation of carbon emissions.

In apparel industry, for example, China is the largest producer of clothing over the world, and clothing is one of the main industries gaining foreign trade surplus by export in our country. With the rapid development of economy, purchasing power of the residents has improved significantly, and the demand for textile and garment has been growing quickly. At the same time, with the advance of rural urbanization pace, the development of small and medium-sized cities, and constantly improving of the network coverage, new lifestyle spreads, which promote the development of the clothing consumption.

And textile and garment industry belongs to high-carbon industry, according to a annual meeting report of the blue-sky national energy conservation and environmental protection of printing and dyeing industry in 2009, that energy consumption is roughly 4.84 tce per tons fiber consumption, among them, the garment industry consumes 1.05 tce per ton garment, printing and dyeing industry consumes 2.5-3.2 tce, and the average consumption is 2.84 tce per ton fiber[2]. Thus, low carbonization of clothing is imminent under the strong consumers' demand.

Clothing low carbon development is the stress and trend of the future development of clothing industry in the world. From the procurement of clothing raw materials, to production and processing, to marketing, consumption and recycling process of the whole industry chain, using new energy sources, researching and developing new technology, developing the low carbon economy, completing enterprise model transformation from original extensive high-carbon development pattern to intensive low carbon development, then realizing the sustainable development of the apparel industry is a key way for the future development of clothing enterprises, it reduces greenhouse gas emissions and curb global environmental degradation process.

After low carbon becoming a common topic, building a sustainable development world becomes a consensus. the design breaks away from the commercialization, get out of the situation dominated by consumer. The sustainable development of the design bursts with results constantly at the same time, then "design" and "sustainable development" merge well, becoming the current mainstream design theory - "sustainable design" [3]. This article mainly discusses the low carbon development path of clothing design links.

1.2 Reviewed research

With the concept of low carbon becoming a popular feeling, awareness of environmental protection of consumers enhances. low carbon clothing research and sales also showed vigorous developing trend, and clothing enterprises began to emphasis on low carbon seriously, among them. many well-known clothing enterprises have introduced low carbon clothing as the theme of environmental protection, meanwhile, stars also have put on low carbon clothes, began to promote low carbon living. Scholars have conducted a series of related research, which mainly concentrating on the research of the low carbon economy development appeared after 2010, mainly focused on the research of low carbon economy mode, the construction of low carbon economy evaluation system, or the application in packaging industry, automobile industry, etc. There are some difference among different industries to study low carbon economy patterns, and they are mature in the product packaging industry, automobile industry, etc while garment industry and other basics are still in the theoretical research stage. Most studies focused on the research and development and use of low carbon fabric, formulating relevant carbide execution mechanism, etc. In a word, low carbon economic research based on the link of garment design is relatively deficient. Therefore, this research has very important significance.

2. Relevant basic theories

2.1 Clothing design process

The specific design process for each dress is different, and each stage has overlapping. Therefore, to research low carbon development easily, costume design process can be divided into four stages including fabrics selection, version designing, sample manufacturing and modification on sample. There is a simple introduction about the several stages.

2.1.1 Fabrics selection

Before every garment designed, we should select the fabric and supplementary materials in priority. According to different seasons, different consumer groups, different effect, etc, fabric choice of each dress is different, such as we choose warm kind fabrics in winter, light fabrics in summer.

2.1.2 Version designing

Designers will look for inspiration from nature, society and so on, combine with the fashion trend to do fashion design. Firstly, designers sketch line-art of the clothing styles, and then draw images according to the fabric color and type, finally the pattern maker converts the design to version. and combine with the fashion trend to do fashion design. Firstly, designer sketches line art of the clothing styles, and then maps image according to the fabric color and type, finally, the plate-maker converts the design to version.

2.1.3 Sample manufacturing

According to the version made by pattern maker, tailors the corresponding fabrics and accessories, sews plastic and makes the sample.

2.1.4 Modification on sample

Sample may have deviation with the designer's idea, this requires designer, pattern maker to modify design and version according to the sample, and the revision is repeated many times, ultimately achieving designer ideal effect. After Change completed, the designer's idea can be intuitive. This article mainly study low carbon development path from two links: fabrics selection, design version.

2.2 low carbon relevant concepts

2.2.1low carbon design

Low carbon design named "from cradle to cradle", is based on environmental awareness and the thought of sustainable development, and the product of "design-production-sales-use-recycling" is throughout the life cycle, paying attention to product' environmental properties. The main purpose of low carbon design is to protect the environmental pollution and reduce greenhouse gas emissions and use the best materials.

2.2.2 low carbon products

The low carbon products is these products generating less carbon emission in the process of the production, which are more environmentally friendly products compared to the reform before. low carbon materials tend to be natural, and production process is more simple. Products are easy recycling, and products' life-cycle is long, which can reduce unnecessary energy consumption.

2.3 Green design principles

2.3.1 Technology advanced principle

Technology is the precondition of green design. Green design emphasizes the adoption of advanced technology in the product life-cycle, from the technical guarantee to achieve safe, reliable and economic products of various functions and performance to ensure the whole process of product life-cycle, which has the very good environment coordination [3].

2.3.2 Pollution minimization principle

Green design should completely abandon the traditional end governance way "polluting first, governing later", while fully consider how to make the products in the whole life-cycle with the minimum pollution to the environment in the design, how to eliminate pollution sources, and radically eliminate the pollution. "Environmental pollution to zero" of products in the whole life-cycle is the ideal goal of green design [4].

2.3.3 Safety principle

The green design not only requires the design of the product to cause the least damage to people's mental and physical health during the whole life-cycle, also asks for product producers and users to provide comfortable and pleasant working environment [5].

2.3.4 Best comprehensive benefit principle

Economic rationality is one of the factors which must be considered in green design. Different from traditional design, the green design should not only consider the economic benefits of enterprise itself, but also the viewpoint of sustainable development, considering the behavior of the ecological environment and social environment caused by the life-cycle of the product, namely considering ecological and social benefits of design [6]. To obtain the biggest economic benefits, ecological benefits and social benefits is most important at the lowest cost.

3. The low carbon path design of one of costume design links—the fabrics selection part

For low carbon clothing, these selected fabrics not only meet the general performance of the fabric, but have good environmental benefit, this is, choosing low carbon fabric. In general, recycled materials should be the priority. At the same time, these selected fabrics should be environmentally friendly materials with low energy consumption. We should fully understand the characteristics of each fabric before selecting fabrics, in order to facilitate subsequent applications well. In modern society with rapid development of green-economy, the choice of fabric has been basically achieved low carbon economy. Now, low carbon development of fabric mainly starts from the following two aspects: (1) the selection of the new material; (2) recreation of textile.

3.1 The selection of the new materials

New materials of low carbon clothing can be divided into the followings:

3.1.1 Natural fiber

Existent of nature itself or the textile fiber gained directly from the artificial cultivated plants or captive animals are the natural fiber, mainly including four classes: the cotton, linen, silk and wool fabric. This kind of fiber has excellent environmental protection performance, and the production process is simple, which can be used as good low carbon fabric.

1) The cotton fabric

Cotton fabric feels soft and comfortable, also has good hygroscopicity, but the wrinkle resistance is poor. In addition to the ordinary cotton, organic cotton and the colored cotton is better low carbon fiber.

Organic cotton is used in agricultural production, based on organic fertilizer, biological control of plant diseases and insect pests, natural farming management and not using chemicals, and the production process are natural.

The colored cotton is a kind of new textile materials cultivated by using modern biological engineering technology which has natural colours when cotton boll opens [7]. The color of the colored cotton is natural, so it reduces the carcinogenic substances produced in the process of printing and dyeing, while reduces serious pollution and damages to the environment [8].

2) Linen fabric

Linen fabric used in the clothing surface fabric has a very long history. Linen fabric clothing is comfortable, has good hygroscopicity. All kinds of linen fabric also have good performance, corrosion resistance. White or bleaching linen have natural soft luster, as the fabric have aesthetic feeling of elegant, natural simplicity [9]. Today, under the green economy, linen fabric has get more and more favour from consumers, especially, ambari hemp, apocynum and han ma. Apocynum contains volatile substances, such as hemp sterol on staphylococcus aureus, pseudomonas aeruginosa, escherichia coli, etc, and they have different bacteriostasis, and odor-proof, invigorate the circulation of depressurization. So it have became the hot spot in the market in recent years [6].

3) The silk

The silk is named “material queen” among taking fabric, which has gorgeous appearance, soft luster, can drape, is elegant and whine, also has special silk sounds when fricted, is the ideal of high-end clothing fabric [9]. Silk products are mostly white, without bleaching, reducing the pollution from bleaching process. In addition, there is colored silk in the silk market, which is natural color, without subsequent dyeing process to reduce the pollution to the environment.

4) Wool fabric

Wool fabric has different varieties, and their performance is uneven. According to its production process, it can be divided into pure wool-like fabric and woollen fabric [9]. Wool-like fabric is soft, have clear texture, and cloth body is strong. Pure wool fabric is elastic and qualified for mid-grade or high-grade fabrics. Thick woollen fabric with fuzz surface has good warmth, and it’s smoothly. wool-like fabric has good absorption and permeability [9]. Beyond this, products made of colored wool, organic wool and colored rabbit hair are more low carbon.

Today, several big brands or retailers use organic cotton to make clothing products in the global, for example, Wal-mart and Nike of the United States, Belgium 's C&A, Sweden's H&M, Spain's Zara and so on. Sweden's

H&M promoted clothing made of organic cotton, wool and reused wool and polyester fabrics, and labeled a special carbon emissions. GUESS put organic cotton jeans, trousers with very small amounts of chemicals and simple way of flushing, meanwhile, volume label is all made of recycled paper and soy ink to printed by bean [2].

3.1.2 Regenerated fiber

1) The regenerated cellulose fiber

Regenerated cellulose fiber is the earliest chemical fiber, still occupies an important position in the chemical fiber. Loycell fiber has good hygroscopicity and is comfortable to wear like traditional regenerated cellulose fiber, but improves the strength [9], and fiber production process is no pollution to the environment.

2) The regenerated protein fiber

The common characteristics of regenerated protein fiber is they all have good environmental performance in the use of production, which is a very great low carbon fiber. The current applicated kinds mainly have soybean protein fiber and corn protein fiber. Many clothing enterprises began to use low carbon fabrics, according to introducing, among new products from Fox Spirit Clothing Brand in the spring and summer in 2010, the new environmental protection materials, like bamboo fiber, soybean fiber, milk fiber and so on, has been widely applied to the apparel products. In addition to response to the call of the state, low carbon fabric itself also has its advantages, these fabrics are comfortable and comply with the feel of the spring/ summer outfit clothing requirements well.

3.1.3 Synthetic fiber

Synthetic fiber clothing material have characteristic of high strength, drought, maintenance, durability and other excellent performance. At this stage, synthetic fiber can be used in low carbon fabrics mainly including: ecological cotton, Cleancool fiber, Sorona fiber, Fly-cool, Tactel fiber, superfine fiber and biodegradable polyester fiber, biodegradable Poly fiber, degradable polyethylene fiber, sunlight biodegradable polyolefin fiber and so on.

1) Ecological cotton

Ecological cotton is a kind of great polypropylene melt-blown fiber. The material is usually as a heat preservation layer of clothing or quilt, because it's light, thin, soft, warm, breathable. The content ecological cotton formaldehyde is about 10% less than ordinary cotton fabric's, and 37% of raw materials comes from natural corn sugar reducing 63% carbon dioxide emissions, which is energy conservation and environmental protection [10]. Ecological cotton have excellent stretching and resilience, and after 15%fiber elongation it can still rebound. So ecological cotton is a kind of excellent low carbon fiber material.

2) Sorona fiber

Sorona polymer is a advanced polymer innovation platform introduced by the United States dupont with more than 60 years of innovation to raw material. It endows fibers and fabrics by the unique performance, meanwhile, it leads the use of biotechnology among chemical fiber raw materials to achieve the perfect combination of the function of textile and environmental protection [11].

3) Tactel fiber

Tactel fiber is a nylon fiber producted by the United States dupont, has formed a series of products within the market, such as TactelAquator fiber with good moisture permeability; TactelDiabialo fiber with unique luster drapability; TactelMultisoft fiber, which is soft and have different gloss effect; TactelMicro fiber which is a super thin fiber[12]. Tactel fabric is very light. Its dry time is three times as fast as the cotton fiber, and it's not changeable with outstanding anti-wrinkle ability.

Each garment enterprise also actively develops new fabrics, such as in 2009, Li ning cooperated with Teijin group of Japan, developed the ECOCIRCLE environmental protection material, launched a series of recycled clothing. This ECOCIRCLE fabric is made of recycled coke bottle scrap, old clothes, natural hemp and nontoxic dye. so it can be infinite cycled to realize the cycle of clothing life. Jiangsu Xiake Chardonnay Co., LTD, by processing and manufacturing PET bottle, polyester and other fiber waste, than recycling dyeing by using dope dyeing technology, eventually made green colored fiber [2].

3.2 Recreation of textile

The costume design under the background of low carbon environmental protection covers a form of design, whose shape should be smooth and natural, with the lowest carbon emissions, but clothing products have the highest utilization rate [10]. The most commonly method used in fabric reconstruction is changing the original old clothes into new clothes by clipping, disassembling and restructuring.

Meeting the demand of human psychology, realizing consumer's demands, and attracting consumers is the first condition to generate fabric reconstruction idea. Considered the basis of clothing raw material, by chemical and physical methods to modify the garment fabrics, appearance characteristics, or state, it can make different visual effect compared to the previous clothing, and it's fabrics can avoid single clothing, at the same time, make the fabric with new and different style features, such as terylene/ blended fabric burnt-out, printing, to make the garment's appearance and features produce big change.

Recreation in the clothing design has great advantage in the industry, and it is not blind to design, but on the basis of meeting the aesthetics requirements of the consumers and designer, in accordance with the links among clothing style, color, and the function to generate the design [10].

Major measures of fabric reconstruction have the followings :

3.2.1 Destructive design

Destructive design is based on the fabric cutting, grinding and other methods to change the original basic structure and morphological characteristics of fabrics, and treats the fabrics as a kind of object, which can change with the designer's idea. Fabric can build a kind of incomplete, irregular visual experience after cutting, grinding, adding beauty to the clothing and the hazy beauty of art. it can significantly improve the clothing external form of tension, line with the psychological needs of consumers who pursue clothing external form of beauty by design methods such as grind, tear, shear, hollowing out, drawnwork, at the same time, this design method can create a empty spirit, a unique style, elegant artistic breath, to make the person feel creative.

Fig.1 Fabric' Hollowing out processing

Fig.2 Drawnwork processing of Valentino

3.2.2 Fabric stitching design

Common fabric stitching method is assembling all kinds of fabrics, such as combining different materials, different color fabrics and then dyeing [10]. In general, it joins the same colour fabrics and different materials together to make it present a fault, administrative levels, fold, strong texture contrast effect. Or put a different color and the same material fabric together, to create rich and colorful,gradient visual effect. In addition, the leftover materials can be stitched together to make a new clothing or other items.

Fig.3 Fabric stitching

Fig.4 Geometric stitching shirt

3.2.3 Fabric additional decoration design

Fabric additional decoration design is mainly adding the same or different materials on the finished fabric to change the original appearance of the fabric. There are a lot of adornment gimmick to attach, including embroidery, hanging, sewing, beaded, rope, etc [10], and it also can applicate integrated feather, bead piece, and three-dimensional flower and so on.

Fig.5 Three-dimensional flowers

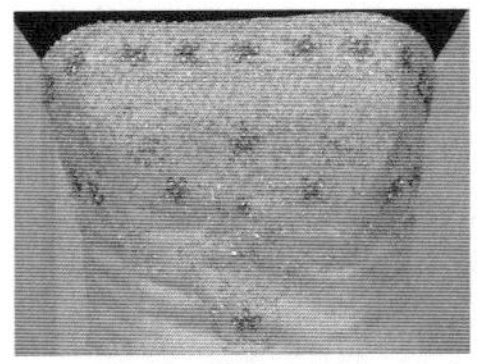

Fig.6 Embroidery, beaded

4. The low carbon path design of one of costume design links---the version designing part

In version of the design, the implementation of low carbon development mainly includes: the market segmentation, the design of fabric reconstruction, the joker design, such as method of transformation o fold clothes.

4.1 Segmenting markets

It was necessary refined the current market before designed. Different style Consumers has different demands, we could effectively avoid the extra clothing waste, when we designed clothing according to the demands of consumers. Targeted design clothing can also make the design more concise, likewise, it avoid changing your design Over and over.

For instance, some consumers maybe focus on the level of comfortable, so the designers should take more time to choose their clothing's fabrics. Then, people wear different style clothing in different occasions, like Sportswear, suits and dresses, it Should not be confused. The style of clothing in Fast fashion brand and Advanced customization was also different. Therefore, refined the market was vitally.

4.2 the recreation of textile

The recreation of textile makes a new clothing from an old one. It was not only reducing carbon emissions in the course of recreation, but also recycling the old clothing, and avoiding the big waste.

In order to recreate clothing more reasonable, the designers must follow the basic rules of formal beauty of clothing. such as symmetry, balance, contrast, reconcile, rhythm, proportion, exaggeration, repeat and so on, but whatever rules should under the designer's control [11].

4.2.1 Partial design

According to the idea of fashion designer, we could design something special in some part of the clothing, this special part will made a clothing unique. We also can recreate this textile which was used on the spacial part.

Collar, shoulder, sleeve edge, waist, chest, clothes edge were needed to "the special part". For example it will be more creative to sew different beads or embroider in the front of collar. Some designers added tassels to the edge of clothing, it made the clothing look more liquidity.

Fig.7 Fabric reconstruction of local application

4.2.2 The application in garments' integral design

The whole of garment fabrics is used usually in simpler clothing style, to. highlight the characteristics of the fabric itself, such as fabric texture, etc. This kind of fabric reconstruction needs fashion designers have deep knowledge of fabrics, and can control the good coordination between clothing design and fabrics.

The representative of the Japanese fashion designer, issey miyake' design directly spread into the fabric design field, and he used the fabric processing craft used by ancient farmers to make clothing have a distinctive appearance. He applied rice paper, white cotton cloth, knitted cotton, linen and other traditional materials to modern technology, combined with his personal philosophy, created a variety of skin texture effect of fabric, and designed unique and amazing clothes, who is known as the "fabric magician". The "fold life" pioneered by him, shows the infinite charm by fabrics' secondary creating, is still a model of the fabric design [13].

4.3 Going with everything

Design systematized means building a systematic wardrobe. It is a considerable issue that designer should make a n integral closet which includes clothes for all seasons. Meanwhile, they also need to pay attention to the match of up-wear with down-wear, inside with outside, different seasons and different series. The style should be stable an d going with other cloth, such as T-shirt could be worn in summer, but also worn inside of coat in winter. This rul e is important for most of consumers.

Besides, one piece going with everything is being sparkplugged. One piece of cloth match with other clothes co uld rub out different spark, which not only reduce purchasing but also reduce carbon emission.

4.4 Clothing transformation

Remaking worn clothing with some physical and chemical methods could maximize clothing utilization. Such as, transform an old suit into a unique handbag, pants into shorts, make some embroidery and beads on damaged clothing or grind holes, and darks with bleach to make tie-dye effects. Therefore, we cloud make worn even dama ged clothing to be a new item with many kinds of ways.The clothes can’t be modified.they could be reformed into some household items. For example, making jeans into tablecloth.Some adornments can be used to intersperse cur tains, and the leftover materials could be made into rags.

A lot of famous brands started advocating to recycle and reconstruct old clothing. The famous Spanish fashion brand ZARA Madrid in cooperation with the European Institute of Design and Development about recycling old c lothes and fabric fashion production. They encourage designers design with a faith of environment friendly and sustainable development. Designers of Burberry make design with low carbon,ventilative and fresh wool.

And Burberry advocates this kind of wool cardigan instead of suitcase for men. Moreover, some other famous brands like Dolce & Gabbana, Versace, bottegaveneta design loosely and produce suitcase with thin material, whi ch in order to reduce use of air-conditioner in summer. Materials recycled from industrial products are used by Vi ctorinox Swiss’design, which make clothes more durable, varied and nice.

5. Conclusion

With the continuous development of the global low carbon economy, the vigorous promotion and support of the national related department, the advancement of university-enterprise cooperation process, the increasing of low carbon environmental protection consciousness of consumers, it's high time for the garment industry to oporate low carbon transformation. Clothing industry is a high pollution, high carbon industry, and to achieve low carbonin the garment industry, we can get down from the production, design, marketing, and other aspects.

This paper, based on the concept of low carbon, researched a path of low carbon among specific segments of the garment design, enriched the academic theory about the low carbon development of the clothing design, provided a theoretical basis for the whole textile clothing industry, the clothing design process of low carbon development.

At present, for the low carbon economy development of China's garment industry, there is still a gap with foreign countries, but with the promotion and popularization of low carbon concept in our country, our clothing low - carbonization can get better development in the future.

References

[1] Wang S, Wang YC. Producer responsibility under the perspective of green design system research [J]. Ecological economy: 2009-11-01.

[2] Zhang YL. Based on the research of low carbon concept of women's clothing design and application [D]. henan:zhongyuan institute of technology: 2013.

[3] Xu YQ. Green design [M]. Second edition. Beijing: Beijing university of science and technology publishing house. 2013: 37, 39.

[4] Su QH, Huang CL, Wang Y, Hu SG. Based on the concept of green product design electrostatic precipitators [J]. Journal of engineering design: 2003-02-28.

[5] Mo XB. Green design based on product configuration method research [D]. Anhui, hefei university of technology: 2007.

[6] Feng DK. Green website and green assessment system [D]. Beijing: Tsinghua university: 2002.

[7] Wei F. Try to talk about low carbon economy environment fashion design innovative development [J]. Journal of arts education research: 2011-10-20.

[8] The global textile net http://www.tnc.com.cn/products/detail/262872.html

[9] Wang GF. Clothing materials science [J]. China textile press, 2006: 20 -28, 161.

[10] Anonymous. [J]. Ecological cotton market modernization: 2013-01-30.

[11] Zhang YS. Sportswear to blow "low carbon wind" dupont aptitude in hand ~ (TM) Sorona ~ at sporting goods [J]. Journal of textile and apparel weekly: 2010-05-31.

[12] Fabricintroducing - Dupont Tactel_ tropic outdoor tourism. Network: http://www.xici.net/d24720783.htm

[13] Fu LN. the recreation of creative technique and its application in clothing design [J]. Journal of hebei textiles. 2010-03-30: 54-55.

The Study of the Pattern Design Model for Female Upper Garment based on Mapping Principle

Qian Ma, Xin Zhang*, Bu-Gao Xu, Bo-An Ying

Apparel & Art Design College, Xi'an Polytechnic University, Xi'an, Shaanxi, 710048, P.R. China

*Corresponding author's email: xianzhangxin@sina.com

Abstract

This paper presents the study of garment pattern design model based on mapping principle in order to better meet the customer's fit and personalized needs. The key point of this model is to establish the mapping relationship between the parameter set and personalized clothing prototype. Firstly, we defined some feature points, feature lines and exert the important characteristic parameters to establish the design model based on female upper mannequin. Then we input the basic pattern model in an Autopattern software developed by our team which generates the clothing template automatically. In the end, by changing the parameters we were able to make different female upper garment pattern samples and evaluate them to prove that the design model is scientific and reasonable.

Keywords: MTM; Pattern Design Model; Mapping Principle; Characteristic Parameter.

1. Introduction

In recent years, as people's living standard are rising, there has been more attention on the fitness and personality of the clothes. Thus there has been a potential market growth for tailor made clothes. Garment pattern design is an important part of the fashion design process. It is an extension of style design and a basis for process design. Therefore, it is necessary to study the pattern relation model of a garment. This study currently focused on the following two aspects:

1) The method of Garment pattern design.

Zhang Wenbin , Zhang Weiyuan of Donghua University, after years of accumulation and refinement of the scientific research and prototype theory was able to establish a prototype for the Chinese people in 1998-Donghua prototype [1,2]. Since then, with the improvement of living standards, the body has changed a lot. Therefore, Liu Bofei of Donghua University conducted a continuous study on the prototype of Donghua, according to the data of human thoracic and lumbar hip analysis results. The original Donghua prototype expanded into a, B two shapes [3]; Mi Tingting of Donghua University made a complementary study of the prototype of Donghua, and made some revision so that it can effectively reflect the characteristics of the current body [4];

Liu Ruipu of Beijing Institute Of Fashion Technology put forward clothing style and pattern design method based on the TPO knowledge system. This method uses "incremental" design idea, namely from the basic pattern to sub basic pattern, to class basic pattern, to achieve basic pattern series pattern design. Basic pattern is the basis and primary plate of garment pattern design, its loose design is in the middle state [5]; Wu Houlin invented a method of structure design for a garment series plate. Its composition is based on new proportions different from traditional clothing pattern design method. This method uses scientific and reasonable design calculation to come up with the formula for the proportion. It is simple, easy to learn, and can better reflect the relationship between the human body and the structure of women's wear [6]; Wang Xiuzhi of Tianjin University of Technology put forward a new kind of garment pattern making method - Basic Law. This method reserves the advantages of the prototype method and the ratio method,gives full consideration to the apparel categories, especially the way to structure classification and takes advantage of classification structure, creates a new method of pattern making [7]; Cui Fangfang of the Xi'an Polytechnic University, uses the method of three-dimensional cutting to establish a jacket type professional prototype, and establish the optimized loose quantity of the jacket type professional tooling, based on the analysis of the distribution of the relaxation capacity and the performance of the fabric, and then establish the final version of the jacket type professional tooling [8].

2) The study of model of automatic generation system.

For the research of sample system automatically generates, domestic colleges and universities have carried out. Such as: Zuo Pingdi focused on the constraints of garment pattern information model and studied the garment pattern design system based on the application of constraint parametric design method. In the sub module of parametric design, the storage, solution and in place editing of the constraint condition of clothing sample are realized, and automatically generating the Multi Size Model. In the special body board change module, realized the special body type changes based on standard plate [9]; Zhang Qi gives the development of garment pattern automatic generation of key technology of expert system from the three aspects of parameterized design thought, expert knowledge acquisition and the establishment of knowledge base, and introduced the concrete realization method of the automatic generation of pattern [10].

Aimed at the present development of pattern design, this paper proposes a new pattern making idea that's called mapping method, which is like a mirror that could reflect the mapping relationship between the structure parameters and physical characteristic parameters. Then it establishes garment pattern's parametric relation model based on physical characteristic according to the above mapping relationship. This study takes a certain type (Aiwa Missy 8) as an example,the relation model is also applicable to the human body with similar types.

This paper's research achievement will offer technical reference to personalized pattern design, and also lay the foundation for the automatic pattern designing system.

2. The parameter extraction of female mannequin for personalized garment pattern design

Body measurement projects are closely linked with the garment form which is the deciding factor of clothing and body fitness. Therefore, correctly determining the feature point, line features, and correctly measuring characteristic parameters for clothing, are crucial to generate basic pattern.

2.1 Female mannequin feature points

The aim of human body measurements is to replicate the charactersitics of the human body which is needed for garment structure drawing. From an anthropometry perspective, body measurements are based on measurements of bone to determine the measuring point, therefore, different parts of bones on the human body are precisely characteristic points. 24 feature points extracted from female mannequin are as shown in the following fig.1.

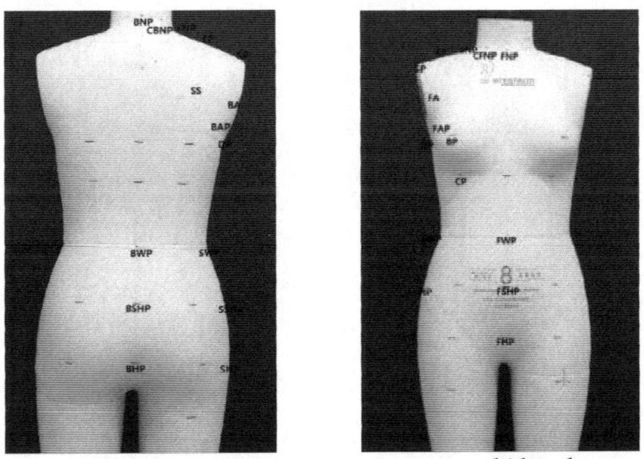

a) front b)back

Fig.1 Feature points of female upper mannequin

2.2 The relationship between female mannequin feature points and human mannequin structure line

The characteristic points of the human mannequin are the reference point of the structure of the human mannequin. The human mannequin structure lines are actually the lines through the human skin surface with two or more marking points. Figure 2 clearly depicts the correspondence between the female upper body trunk feature points and the trunk structure lines.

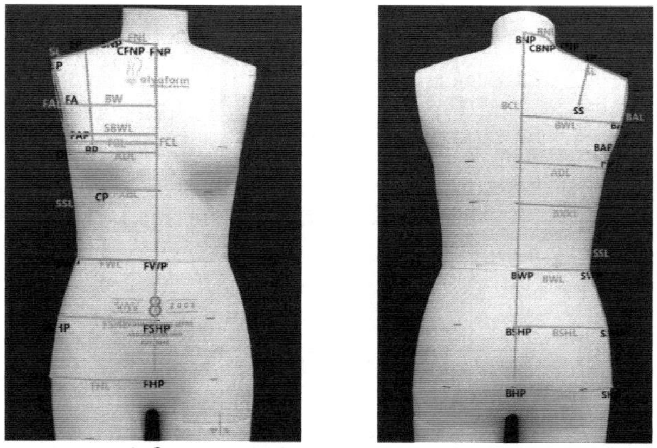

a) front b) back

Fig.2 Characteristic structure line of female upper mannequin

2.3 Extracting characteristic parameters of female upper mannequin

According to the analysis of the female upper mannequin trunk of feature points and structure lines, we have to extract the following 32 female mannequin characteristic parameters to draw the basic pattern of the women's garments.

Table1. characteristic parameters of female upper mannequin

number	Project categories	Project	code	Measurement standard	unit :mm
1		The chest long	L1	Trunk anterior midline in the neck opening FNP to the waistline and W1 intersection curve arc length	
2		Back length	L2	anterior midline in the neck opening FNP to hip line H1 intersection curve arc length	
3		Front armhole height	L3	The trunk midline anterior in the neck opening FNP to the bust line B1 intersection curve arc length	
4	Standing length large class	Posterior shoulder heigh	L4	Curve arc length of the posterior midline of the trunk from the point BNP of the cervical spine to the intersection of the horizontal line of the shoulder blade	
5		Lateral cervical papilla	L5	Curve arc length from the side of the neck to the point of SNP	
6		Lateral neck shoulder length	L6	The curve arc length from the side of the neck to the SNP of the shoulder blade	

7		Anterior axillary point	L7	Trunk anterior midline vertical curve in FNP to the neck opening chest width B4 line arc length
8		Posterior axillary point high	L8	The vertical curve arc length of the posterior midline of the trunk from the cervical point BNP to the B6 of the dorsal line
9		The shoulder point Gallery bottom arc length	L11	From the point of the shoulder to the armhole bottom arm circumference arc length
10		Neck side lumbar segment length	L12	The curve length of SNP from the neck to the waist line
11	Standing body circumference	Anterior axillary arc length	A2	the curve arc length From the shoulder point to anterior axillary point
12		Posterior arc length of shoulder	A3	The curve arc length from the shoulder point to the posterior axillary point
13		Neck circumference	N1	The front neck, neck side point, after a week around the circumference of neck
14		upper bust	B1	The length of chest circumference of the right posterior axillary fossa
15		Bust 1	B2	The level of chest circumference of the nipple
16		Under bust girth	B3	Subpectoral along the chest girth level
17		Hipline	H2	Girth of the hip level at the peak of the hip
18		Fron waist circumference	FW1	The front waist line length of the front waist line is divided into the side seam.
19		Back waist Circumference	BW1	The rear waist curve length with the side seam as the dividing line
20		Front hip	FH2	The front hip arc length with the side seam as the dividing line
21		Back hip	BH2	The posterior hip arc length with the side seam as the dividing line
22		Waist	W2	Waist level around the umbilical point
23	Standing body width	Chest width	B4	Horizontal straight line at the most prominent part of the chest at the point of the nippl
24		Distence between breasts	B5	Linear distance between two milk points
25		Back width	B6	Horizontal straight line distance between left and right axillary points
26		Width of the Shoulder blade	B7	The horizontal straight line distance of the left and right shoulder blades

27		Inclined shoulder length	B8	The curve of the distance from the neck point to the lateral acromion
28		Shoulder width 1	B10	About the distance between the acromion
29		Shoulder width 2	B11	About BNP the arc length of acromion
30		Armhole width	B12	The level distance between the front and the back of the armpit
31	Standing body angle	Posterior shoulder angle	α	/
32		Front shoulder angle	β	/

3. The mapping relationship between mannequin parameters and garment pattern design model

In fact, using mapping method to establish the process of female mannequin basic model is the process of mirroring the corresponding relationship between the clothing and the body. According to the mapping principle, by placing the 3D human body feature points with the structure of the human body and the corresponding size parametrs map to two dimensional clothing, we are able to obtain women's basic model. Figure 3 shows the mapping relationship between the human body and the female basic pattern. It can be seen from the figure that the human body anatomy, the structure, and the size of the human body is an important input data to establish the basis pattern system in mapping method. The mapping relations between the parameter set and the individual garment version and the front piece are discussed respectively.

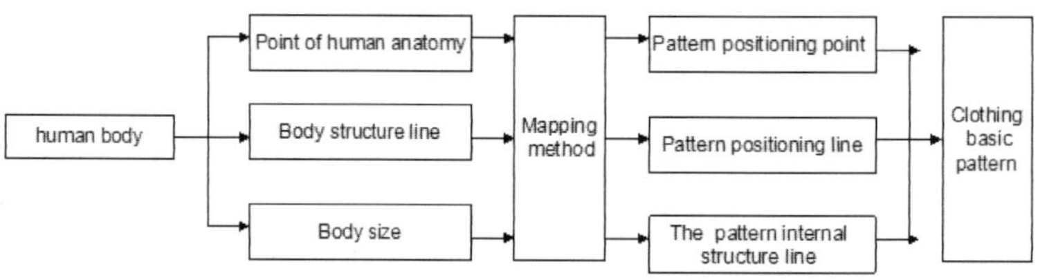

Fig.3 Mapping relation between the human body and the female basic pattern

3.1 The mapping relationship of the back garment piece

In the computer, individual clothing prototypes are composed of a number of lines and curves. The mapping relationship between the parameter sets and the back garment piece are mainly between the points tothe points, the points to the lines, and the lines to the lines. Among them, the plane formation of the prototype of the personalized clothing is composed of 11 feature points and 10 characteristic lines. As shown in Figure 4.

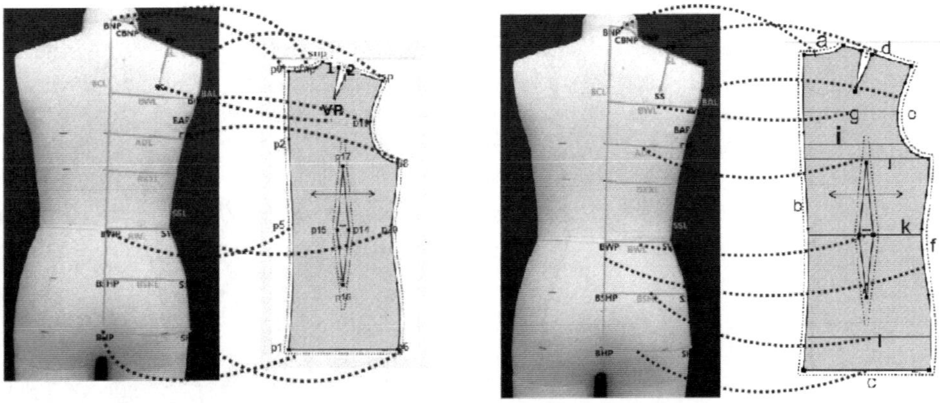

Fig.4 Mapping relation of back garment piece

In the plane formation,feature points p0 and BNP, CBNP and CBNP, p1 and BHP, p6 and SHP, SNP and SNP, SP and SP, SS and SS, BA and p19, DP and p8, BWP and p5, SWP and p20 are corresponding one by one respectively;characteristic lines BNL and a, BCL and b, BAL and e, SL and d, SSL and f, BW and g, ADL and i, BWL and k, BSL and l, BHL and c are corresponding one by one respectively.

3.2 The mapping relationship of the front garment piece

And the same to the former one, The plane formation of the front garment piece of the personalized clothing prototype is composed of 11 characteristic points and 12 characteristic lines. As shown in Figure 5.

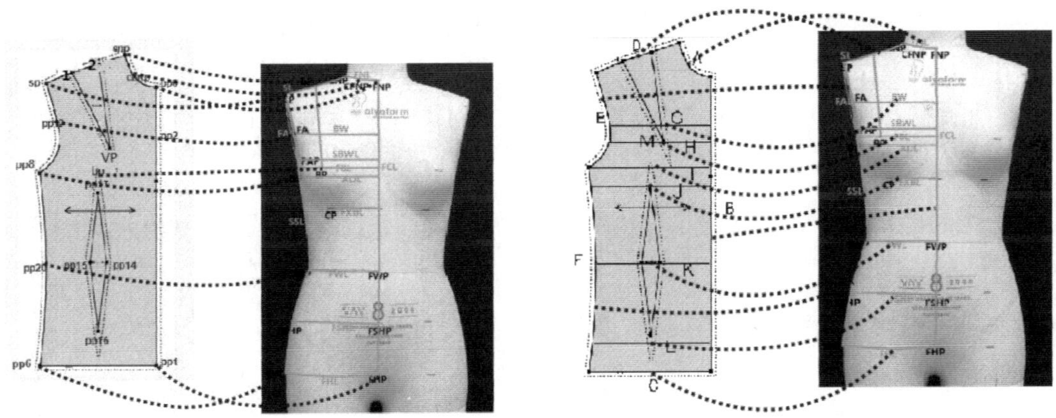

Fig. 5 Mapping relation of back garment piece

In the plane formation, feature points, pp0 and FNP, CFNP and CFNP, PP1 and FHP , pp6 and SHP, SNP and SNP, SP and SP, BP and BP, FA and pp19, DP and pp8, FWP and PP5, SWP and P20 are corresponding one by one respectively; characteristic lines, FNL and a, FCL and B, FAL and E, SL and D, SSL and F, BW and G, SBL and m, ADL and I, FWL and K, FSL and l, FHL and C are in one-to-one correspondence respectively.

3.3 Automatic generation of basic model

Shaanxi Clothing Engineering Center has been committed to the research on automatic model generation. Computer software AUTOPATTERN produces relatively mature research results at the present. In this software we entered the basic pattern model based on the mapping principle of the above research to formulate an automatic generation of the model. The generated template model is shown in Figure 6.

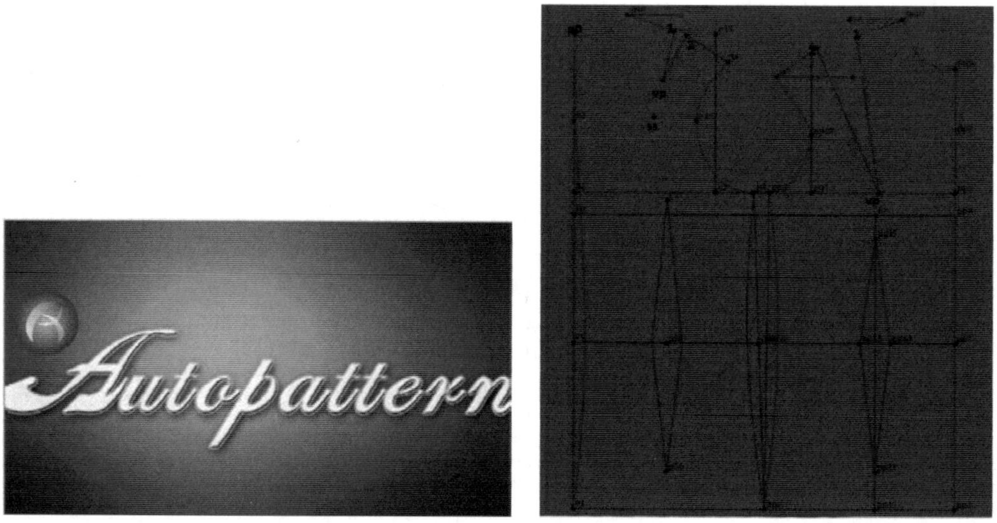

Fig.6 The generated template model

4. Evaluation of the design model for female upper garment pattern

There are two samples resulted from the transformation of the interactivity of the pattern model. Just as the following fig. show, Fig.7 is no ease, and in Fig.8 loose amount is 2cm. Then the comparison of two kinds of patterns will be shown below, observing the alignment of the feature points and feature lines.

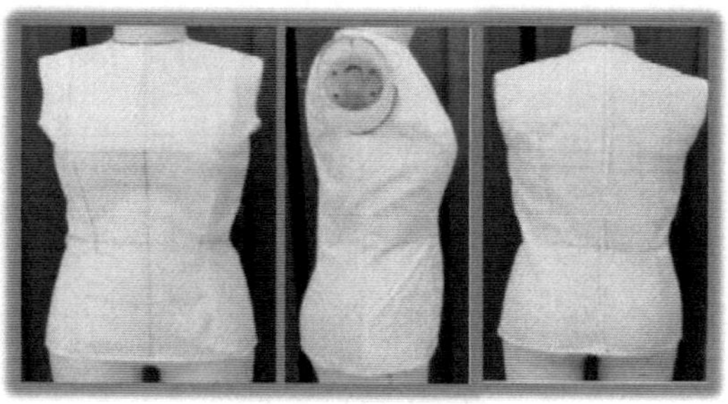

Fig.7 The sample without ease

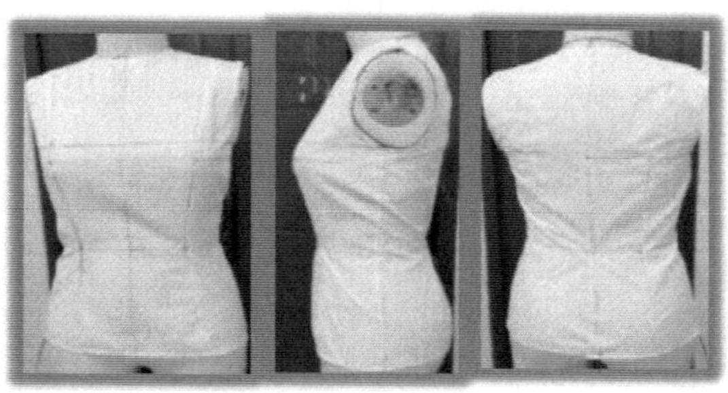

Fig.8 The sample with a certain amount of ease

4.1 Analysis of the effect of clothing wearing

Two samples were found on the models: each feature points on the two models, and the structure lines can be aligned with the marking points on the human stage.

From the lateral view, the basic type of neckline along the neck is round and smooth; armhole deep line, bust line, waist line, line width of chest, back width line were located in the body of the positions that correspond to horizontal; shoulder line located in the lateral neck and shoulder point to connect the shoulder position.

From the vertical perspective, like clothing wove whole lengthwise along straight, front center line, center line, side sea.

From the fitted degree, the dressing effect of the base type for three angles in front and center, the waist line and bust line and hip line are at fully fit.

5. Conclusion

In this paper, we define the feature points and the feature lines that can reflect female mannequin shape well and extract 32 characteristic parameters. In the computer software, Autopattern, the relationship model of basic pattern is established and the automatic generation of the clothing pattern has achieved. In addition, we make two samples based on the basic model, one without ease, another with ease and attempt to evaluate them, to verify whether the design of basic pattern relationship model is scientific and reasonable.

References

[1] Zhang W, Zhang WY. The prototype structure design principles and techniques of Chinese clothing Donghua. Journal of Donghua University (NATURAL SCIENCE EDITION): 2002; (3): 44-49.

[2] Zhang WB. clothing structure design. Beijing: China Textile Press: 2006.

[3] Liu BF. the continuing research on Donghua prototype. Donghua University: 2010.

[4] Mi TT. the complementary research on Donghua prototype, Donghua University: 2009.

[5] Liu RP. the garment pattern design principle and techniques of women's. Beijing: Chinese Textile Press: 2005.

[6] Wu HL. Study on pattern design of dress. Shanghai Textile Science & technology: 2010; (6): 27-32.

[7] Wang XZ. the study of a new design method of garment pattern, Tianjin University of Technology: 2003.

[8] Cui FF. a study on the fit of jacket type professional work, Xi'an Polytechnic University, master's thesis: 2011.

[9] Li J. Study on expert system for automatic generation of shirt customization pattern design Beijing Institute Of Fashion Technology: 2008.

[10] Zuo PD. garment pattern design system based on constraints. Beijing Institute Of Fashion Technology: 1998.

[11] Zhang Q, Zhang WB, Zhang WY. Study on expert system for the automatic generation of garment pattern PDS. Journal of China Textile University: 2000; (5): 62-64.

A study on Balance Between Creativity and Production in Clothing Design

Jie Chen

[1]*China Academy of Art, No.109 Chunxiao Road, Shanghai, 200012, P.R. China*

*Corresponding author's email: jiechenmy@163.com

Abstract

The balance between creativity and production in clothing design was discussed. As a method of creativity in fashion design, the application of structuralism has been discussed and the way of creativity in structuralism summarized. Two cases have been designed and finished in order to analyze the balance between creativity and production, and the look, creativity, fabric, structure, process, and marker were evaluated. Also, the relationship between creativity of design, feasibility of manufacture, and the economic benefit were disclosed.

Keywords: creativity; design; production; balance; structuralism

1. Introduction

Under the trend of economic globalization, the lifestyle and demand of contemporary urban people have become more international and diversified, which makes people's requirement of clothing being personalized and novel. It is not enough that function of clothing is to satisfy people's basic needs. The public daily are more willing to wear both practical and creative costumes.

Although most creative clothing is novel and chic, it is difficult to find the specific rules to follow in fashion design. In order to achieve good effect there are often many time in adjusting samples with multiple correction. Therefore the process of production would be a long time, and the production cost would be higher, so the production mode for creative clothing is often custom-made or small-lot, which cannot meet the demand for general people and mass-production.

This paper was discussed on how to meet the needs of urban mass-fashion and optimize the creative design method as well as production of clothing. The balance between production and creativity in mass-production are researched. The result of research were aimed to meet the urban people's pursuit of fashionable life style, and the result would have good practicability and potential economic value.

There are many creative methods in fashion design; among these methods, we selected the structuralism in our clothing design. In this research work, two cases were designed to analyze the balance between creativity and production. Then, the appearance of clothing, the creativity of design, the selection of fabrics, the manufacturing process, and the marker were evaluated, respectively. Also, the relationship between creativity of design, feasibility of manufacture, and the economic benefit were disclosed.

2. The creative way of structuralism in clothing design

The characteristics of structuralism clothing is based on the human body modelling, emphasizing stereo and expand feeling as well as sense of proportion and hierarchy. It has a strong sense of order, which fully reflects the western aesthetic ideas of traditional costumes. Structuralism clothing pursuit beautiful shape, structure and form, whose three-dimensional space are shaped by arrangement, segmentation and proportion. The simple shape and structure are advocated in structuralism, which is a kind of 'low-key' in the whole and wonderful style. Although the whole look is simple, actually there are rich changes in clothing detail.

2.1 The embodiment of the structuralism on clothing silhouette

Silhouette is the primary factor in the fashion design, not only because of its own rhythm and balance but also the coordinated relationship with the ideal image of the human body. The outline of structuralism clothing are simple and chic, which shows beauty of human body by using simple geometric form.

2.2 The embodiment of the structuralism on clothing fabric

It is important to choose appropriate material to reflect three-dimension of clothing. It is usually that structuralism clothing choose looks being like simple but delicate materials, through whose characteristics interprets the design concept. At the same time, the structuralism clothing stressed the purity of clothing structure, and this kind of art concept makes them more likely to choose simple and elegant fabrics without patterns. These kind of fabrics can avoid pattern being a reversal of the order of host and guest, which would weaken the structure of the visual impact.

2.3 The embodiment of the structuralism on clothing color

Plain fabrics are often chosen in structuralism clothing, and the color could be bold or colorful, characterized by single color area. The three-dimension of clothing is formed through method of structure rather than use of color.

2.4 The embodiment of the structuralism on clothing details

Structuralism clothing is without reservation to the structure of the display, through the decomposition of structure and composition for the interpretation of the structure design.

3. Design of experiment

3.1 The samples design

Due to the consideration of the creative way of structuralism, two samples (i.e, cashmere and leather) were designed with two styles: the big loose coat and the long conjoined shirt.

3.1.1 The style of samples design

S1 is a gray woolen coat, with an A-line silhouette; this coat has a grow-on collar and bust pleats in the front bodice. S2 is a satin and leather shirt, with an H-line silhouette; this shirt has a stand collar and pleats in princess seam (Table 1).

Table 1. The style of samples

samples	Silhouette	Fabric	Color	Details
S1	A line	Heavy woolen fabric	Grey	Grow-on collar, bust pleat and waist pleat
S2	H line	Light satin and leather	White	Stand collar, pleats in princess seam

Both of them will be structuralism as a means of creating elegant, modern atmosphere, makes every effort to show the wearer's personality and status(Fig.1).

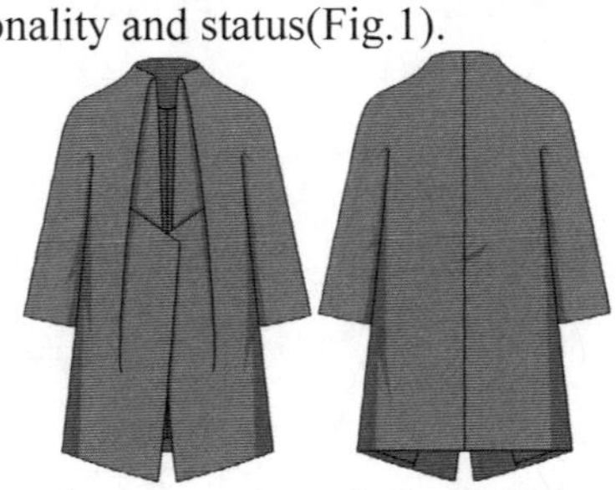

Fig.1 Drawing of sample S1

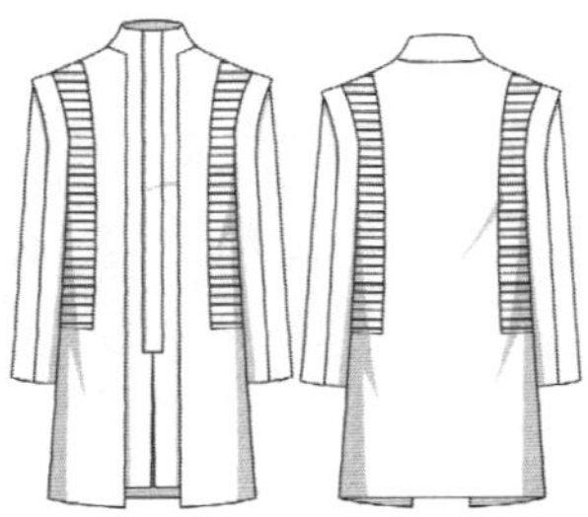

Fig.2 Drawing of sample S2

3.1.2 The specifications of samples

The specifications of samples are as follows (Table 2). The main parts of samples specifications are length, shoulder width, chest, waist, hip, sleeve and cuff.

Table 2. Specifications of samples

Sample Spec. (cm)	Length	Shoulder width	Chest	Waist	Hip	Sleeve	Cuff
S1	84	40	100	108	116	48	24
S2	80	40	92	70	94	60	28

3.2 The Patterns of Samples

According to the drawing of samples, the patterns of samples are made. Patterns S1 shows that the sleeve is raglan sleeves, and the collar is a stand collar joined together with bodice. The main characteristic is the big pleat through down the bodice which put a big dart in line L1(Fig.3). The bodice piece is changed from gray piece (left of Fig.3) to one piece with big pleat added.

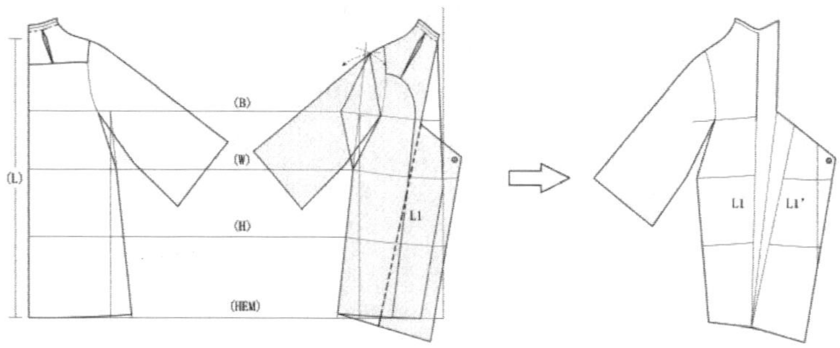

Fig.3 S1 process of patterns making

Through patterns S2 (Fig.4, Fig.5) as you can see, this is a co-joined shirt with bodice and pants together. The collar is stand and the side seams are joined together. The two gray pieces are put together at position of L2 and L2' (Fig.4). The sleeve is broad at top and slim at arm. Seam lines are designed at position of lines L3 and L4. In order to achieve wide shoulder some allowance are added in top of sleeve (Fig.5).

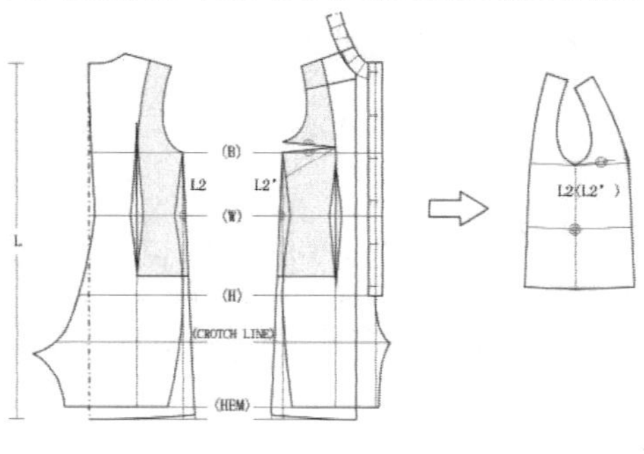

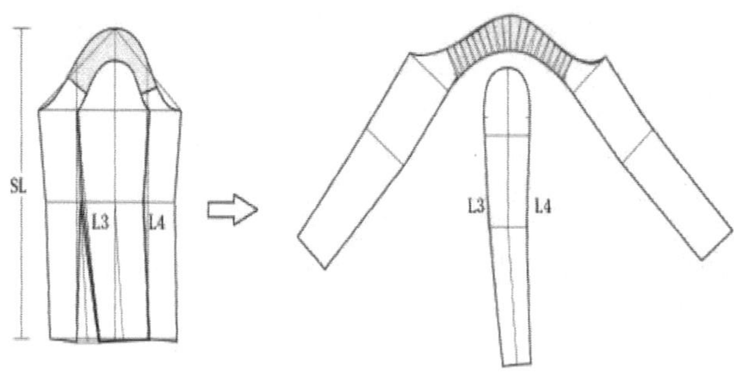

Fig.4 S2 process of bodice patterns making

Fig.5 S2 process of sleeve patterns making

The visual effects of these two clothes are displayed in Fig.6 and 7.

Fig.6 S1 Presentation

Fig.7 S2 Presentation

4. The case analysis of the balance between creativity and production

4.1 The embodiment of creativity

Under the design philosophy of structuralism, the clothes S1 is a gray gabardine coat, the silhouette gives a person to feel with concise atmospheric. The front collar connect with placket together, seemingly simple, but contains a wealth of design points. There is a big pleat straight from neck in front and a bust dart is designed to insert it, which is a main characteristic of structuralism in pattern making. The sleeve is three-quarter length, a little loose, echoing with whole clothing style (Fig.8).

The clothes S2 is used white satin chiffon material, and joined with white leather fabrics in detail. Through the structural segmentation, split, exhibition, it looks likely simple but contains a wealth of design points. There are two big seams in front, and the side pieces with many pleats. The whole clothes S2 is a two-piece suit, with a jump suit inside and long coat outside. The wide shoulders echo with whole clothing style as well, being natural, unrestrained and handsome (Fig.9).

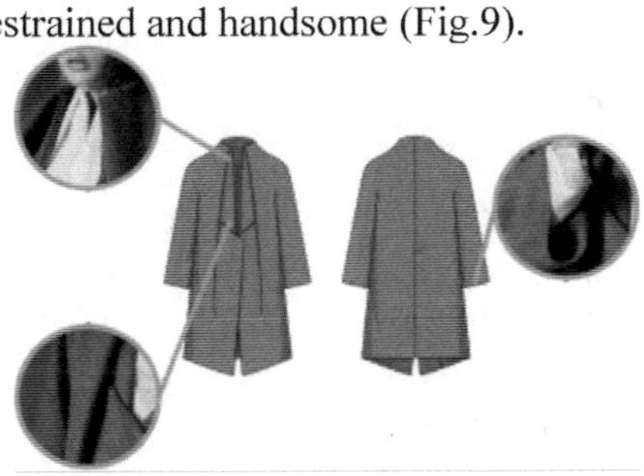

Fig.8 S1 creativity in detail

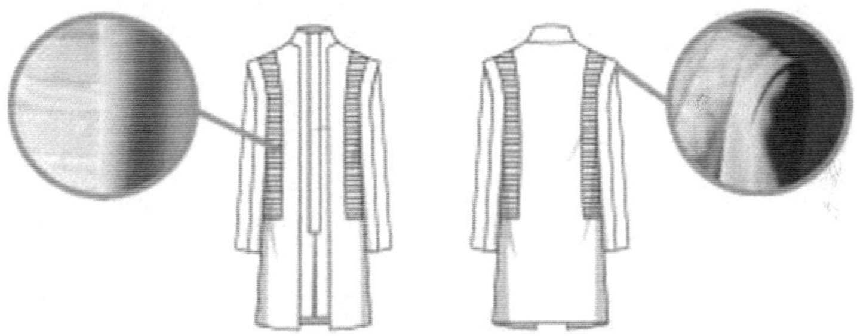

Fig.9 S2 creativity in detail

4.2 The manufacturing feasibility of the experiment case analysis

1) fabric

Sample S1 is suitable for autumn and winter wear. It is easy to find the right fabric. Different fabrics can be chosen according to the need of the production cost. For example, if planned cost is higher, the high-quality woolen fabrics could be chosen; if planned cost is limited, the synthetic fabrics is more suitable. In short, the fabric of the optional scope is bigger.

Sample S2 is a suitable for underwear and summer wear. The main fabric is white satin, matched with white leather. The white color is widely used and suitable for most people. Different fabrics can be chosen according to the need of the production cost. For example, if planned cost is higher, silk and leather could be chosen; if planned cost is limited, man-made fabrics and man-made leather are more suitable. In short, the fabric of the optional scope is bigger as well.

2) structure

Through pattern figures of two samples can be seen, although the details of the garment are more creative, the whole patterns are general. This will bring convenience to the following process of production. Because structure of clothing are not complicated, pattern-making can get the patterns, and do not need to use the draping technique as a supplementary means.

3) craft

Sample S1 has less joined pieces, especially the seam in front piece formed by folded pleats, and craft of making clothing is less difficult. Sample S2 has more joined pieces, especially the pleats in the seam, and craft of making clothing is more difficult, but has more higher creativity. When this style are in bulk production, the production time will be increased, so the structure design can be more optimized.

4) materials using

Seeing from the garment patterns, the two samples are general regular. There is little waste of fabrics in the layout process. Many creative clothing often waste a lot of fabrics because of exaggerated patterns. Both samples success in avoid fabrics wasting as well as high production cost.

4.3 The balance of creativity and production

Through the creativity and manufacturing feasibility analysis, it can be seen that the two samples not only has good creativity but also are easy to wear for customer and produce to manufacturer. The two sample of structuralism have good balance between creativity and production.

5. Conclusion

The pattern making of most daily wear have regulation to obey, such as princess seams. While creative modeling clothing are hard to find the regulation. It is hard to find these kind of clothing in mass market but in high fashion. It is because their creativity will bring difficult to pattern making and production for regular-less. If the creativity and production feasibility can be balanced well, the clothing can be put in mass-production. The need of mass people can be satisfied, not only can meed the personality of custom but also promote industry's economic value.

References

[1] Jiang HY. The structuralism and deconstruction style of structure design. Journal of xi 'an Engineering University: 2011; 25 (2): 32-36

[2] Chen JH. Research on the structure and deconstruction in clothing design style. Decoration, January 2005; p: 14-20.

[3] Dennic Chunman Lo. Pattern Cutting. Beijing: China's Textile Press: 2014; P: 121-145.

[4] Information on http://www.sxxl.com/

[5] Information on http://www.eeff.net/

[6] Information on http://www.wow-trend.com/

Pants Preferences and Body Cathexis Among Senior Women in Northern China

Shi-Yang Yu, Rong Zheng*, Wen Wu

Aimer Human Engineering Research Centre, Beijing Institute of Fashion Technology, Beijing, 100029, China

*Corresponding author's email: yuki.zheng@gmail.com

Abstract

As the world's most populated country, China also has one of the largest populations of middle-aged and elderly people. Chinese aged 45-65 have considerable disposable income, yet are dissatisfied with choices available in the current apparel industry. To date, most domestic studies mainly focused on older adults' consumption status and consumer behaviour, and have overlooked their specific needs and wants with regard to clothing. There is also a dearth of research regarding the relationship between body cathexis and apparel preferences. This study sought to investigate pants preferences and body cathexis of 120 middle-aged and older women in northern China. Survey results provide the apparel industry with more accurate data of pants preferences and psychological factors that could impact design preferences and better cater to this market segment of consumers to maximize industry profits.

Keywords: Pants consumption; Body cathexis; Questionnaire; Pants preferences; Older women

1. Introduction

Like most nations facing industrialization, China is experiencing the phenomenon of an ageing population [1]. One of the fastest growing segments of the Chinese population is middle-aged and senior adults ages 45 to 65, whose sheer number and increasing amounts of disposable income [2] provide opportunities for apparel manufacturers and retailers. Yet, research has shown that this segment of consumers are frustrated with the current apparel market due to its lack of well-fitted, age-appropriate, and desirably styled apparel [3].

Thomas and Peters [4] discovered that although elderly women remain interested in fashion and are physically and socially active, they feel that fashion designers ignore their needs and wants. Lao's [5] study of older people's consumption habits found that the current market does not satisfy their needs and must develop alternative strategies to cater to older people's different income levels. While some domestic studies [5,6] have investigated consumption behaviours of this age cohort, few have considered their actual garment preferences and shopping concerns. Understanding this age group requires in-depth knowledge of factors that affect their clothing purchases.

Body cathexis refers to satisfaction or dissatisfaction with one's body, which could affect women's self-image, as well as their garment-fit assessment and preferences. As women age, physical changes such as a thicker waist, shortened height, sagging breasts, larger abdomens, slumped shoulders, and widened hips [7,8] significantly impact their body cathexis. Only a handful of studies to date have examined the link between body cathexis and clothing preferences [9]. Labat et al [10]. discovered that poor-fitting pants was the most common frustration of senior women, due to the outdated sizing system and complex shape and proportion of the lower body, and that the fit and design of pants for older women needed to be improved. Therefore, this study sought to identify how body cathexis impacts senior women's garment selection and preferences. We chose pants as our target apparel item and investigated pants preferences and body cathexis of women aged 45-65 in northern China.

2. Approach

2.1 Questionnaire design

The study employed an online questionnaire consisting of four sections and 39 questions. The questionnaire was pilot tested and modified to ensure clarity and relevance, and to estimate length of time required to complete it. The final questionnaire took approximately 10 minutes to complete. Each question was unambiguous and appropriate.

The first section of the survey consisted of questions regarding age, marital and fertility status, education background, employment, and monthly disposable income. The second section included four questions regarding location and timing (frequency) of and monthly expenditures on clothing purchases, while the third section explored respondents' satisfaction towards body appearance, especially in regard to the lower body. The fourth section consisted of 14 questions regarding detailed pants preferences and factors impacting consumers' purchases.

The survey was conducted on the Questionnaire Star website. All participants were randomly selected from the website. Data collection began on December 13, 2015 and ended on December 26, 2015. In total, 120 effective questionnaires were acquired.

3. Results

3.1 Demographic information

The demographics of the individuals are as follows.

The age composition of respondents was: 44.2% between 45 and 50 years of age; 35.8% between 50 and 55; 12.5% between 55 and 60; and 7.5% between 60 and 65.

All of the subjects were married and had given birth; 62% still work full-time, over half (55%) hold a college degree, and 33.3% hold a bachelor's or a master's degree.

With regard to personal monthly disposable income, half of the respondents (51%) earn between 2000 to 4000 RMB, 27% have more than 4000 RMB, and 22% have less than 2000 RMB. Fig.1 shows respondents' employment status and personal monthly disposable income.

The results to some extent indicate that the respondents have significant consumption potential.

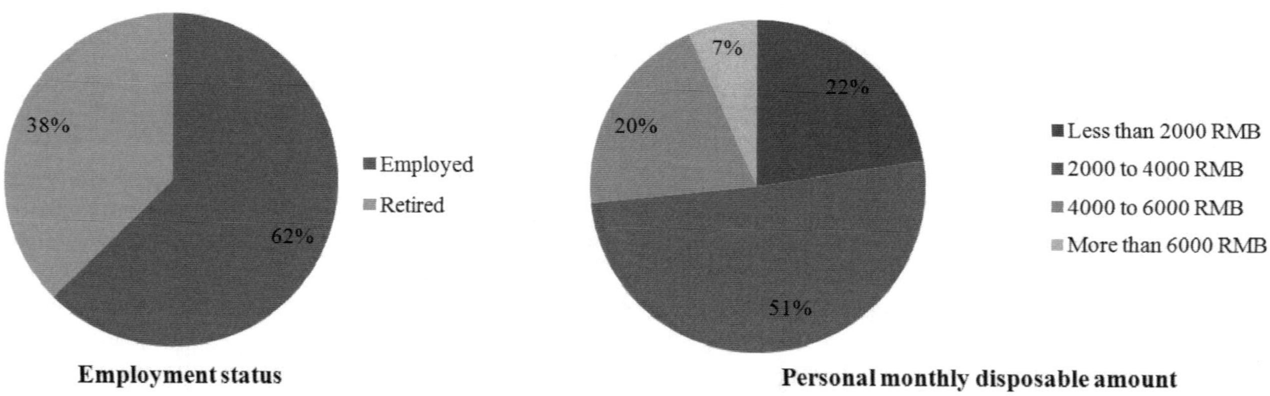

Fig.1 Respondents' employment status and monthly disposable income

3.2 Fashion consumption

With their disposable income, nearly 54% of respondents spend 500 to 1000 RMB on apparel each month, and 18.5% of respondents spend more than 1000 RMB per month. 42.5% of respondents shop twice or more than twice on average each month.

Fig.2 shows the occasions when respondents will shop. Many of the respondents (30.3%) choose to shop when there is a need for them to buy, 25.5% of the respondents will purchase when they see something they really like,

22.2% choose to buy when there is a discount or sales promotion, and 19% of them will purchase for festivals. Only 3% of respondents will purchase the new arrivals. The results confirmed again that the older generation is more rational regarding fashion consumption, yet they are still willing to buy something that they really like.

With regard to consumption places, the most common locations reported by respondents are brand shops (24%) and department stores (28.7%); the least commonly reported location is marketplaces (11%). See Fig.3.

The results of monthly frequency and expenditure on apparel show that the respondents have significant spending power. They are economy-minded but are willing to buy something that caters to their tastes. Through relevance analysis, we also discovered that the employed group spent more on apparel with their disposable income than did the retired group. For the respondents who choose online shopping as a way of buying clothes, 73% of them were from the employed group. The places that respondents indicated they choose to shop also provide retailers and marketers with strategic data to heighten benefits and better cater to the consumers' needs.

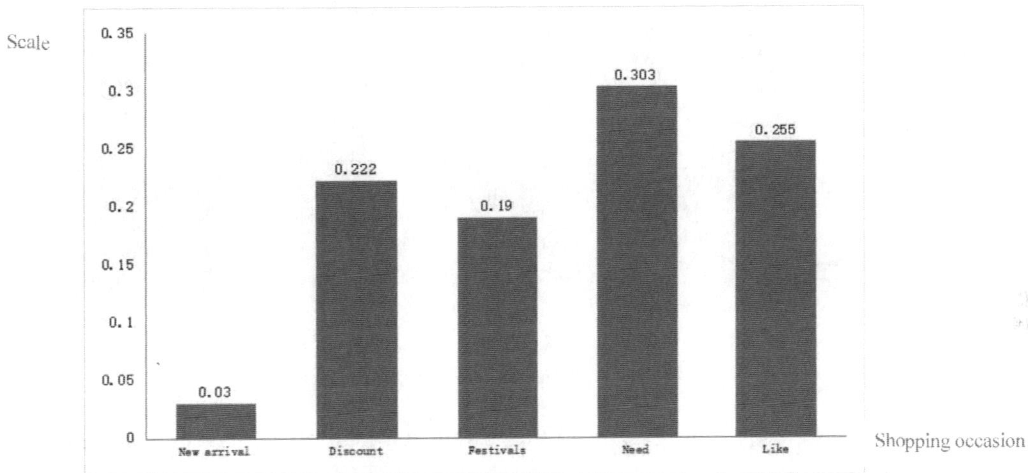

Fig.2 Shopping occasions

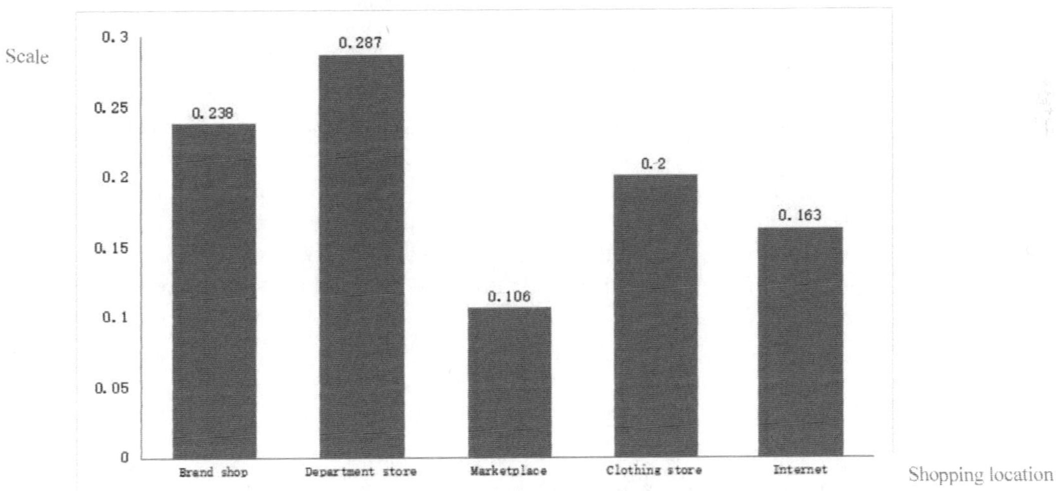

Fig.3 Shopping locations

3.3 Body cathexis

We calculated the Body Mass Index [BMI=Weight (kg)/ $Height^2$ (m)] of the respondents, based on the height and weight they reported. According to the World Health Organization standard, a BMI greater than or equal to 25 is overweight, and a BMI greater or equal to 30 is obesity [11]. It turns out that most of the respondents (91.3%) have the average weight for their body; only 7% were overweight and 1.7% were underweight.

Respondents were asked to evaluate the lower body parts as thin, average, or fat. Results are shown in Fig.4. Very few of the respondents viewed their waist (3.3%) and abdomen (3.3%) as "thin". More than half of the respondents (55.8%) evaluated their abdomen as "fat", and a majority considered their hips and legs as "average". The self-evaluation of their own body can directly or indirectly influence their self-esteem, fashion involvement, and clothing preferences.

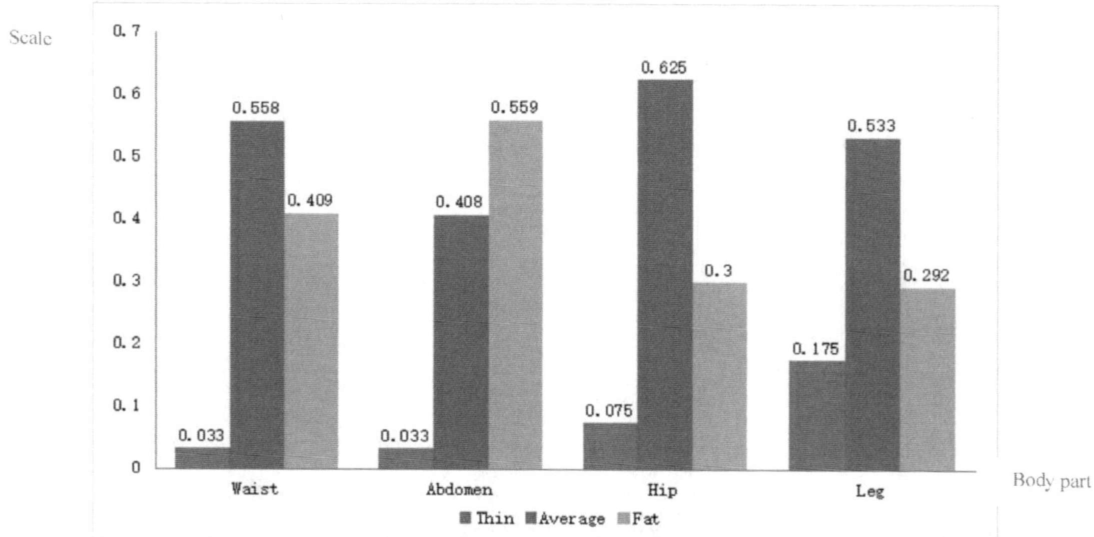

Fig.4 Evaluation of lower body parts

Some of the questions were multiple-choice sorting problems. Respondents were asked to choose three answers and arrange the options in order; the first option was scored as 3 points, the second option as 2 points, and the third option as 1 point. Final scores for options were calculated using Eq.1 shown below; the higher the score, the greater the weight of options, meaning that the respondents were more inclined to agree with that option.

$$\text{score} = \Sigma \text{frequency} * \text{weight} / \text{sample size} \quad (1)$$

Respondents think that it is socially acceptable to show their neck (7.68), lower legs (6.73), and forearms (5.54). Three body parts that they tend to hide are the abdomen (8.05), breasts (6.85), and upper legs (4.75). 95% of respondents mind their physical changes as they grow old; the top three body parts that they care about are the abdomen (7.92), breasts (5.87), and neck (3.96). This indicated that increased abdomen girth, sagging breasts, and neck wrinkles are noted more obviously, and unpleasantly affected the consumers' apparel selection.

With regard to body cathexis of the lower body, respondents were asked to rate their lower body parts using a 5-point scale where 1 = most dissatisfied, 2 = dissatisfied, 3 = neutral, 4 = satisfied, and 5 = most satisfied. The mean values are shown in Table 1.

Table 1. Mean values of satisfaction towards lower body parts

Body part	Mean
Height	3.22
Weight	3.06
Waist	2.81
Abdomen	2.63
Hip	2.97
Upper legs	3.02
Lower legs	3.16
Leg length	3.05
Lower body	3.09

Even though the majority of respondents had average BMI scores, the overall results of body cathexis still show a tendency between neutral and dissatisfied: the upper leg, lower leg, leg length, height, weight, and lower body received neutral evaluation, and the abdomen, waist, and hip received negative evaluation. This could imply that most of the respondents were not satisfied with their lower body, and this slightly low body cathexis is partly due to the ever-changing body, which includes waist and abdominal girth expansion, spinal curvature, coupled with the change of body fat distribution.

3.4 Pants preferences

We investigated detailed pants preferences and factors that impact purchases by using single and multiple-choice questions

Fig.5 shows respondents' detailed preferences of pants. Results indicate that more respondents prefer casual (37.5%) and fashion (34.2%) style, straight-leg pants (44.2%) and tapered pants (44.2%), loose fit (36.7%) and just fit (28.3%) patterns, the use of elastic waistbands (36.7%), and common waistbands (zipper & button) (35.8%).

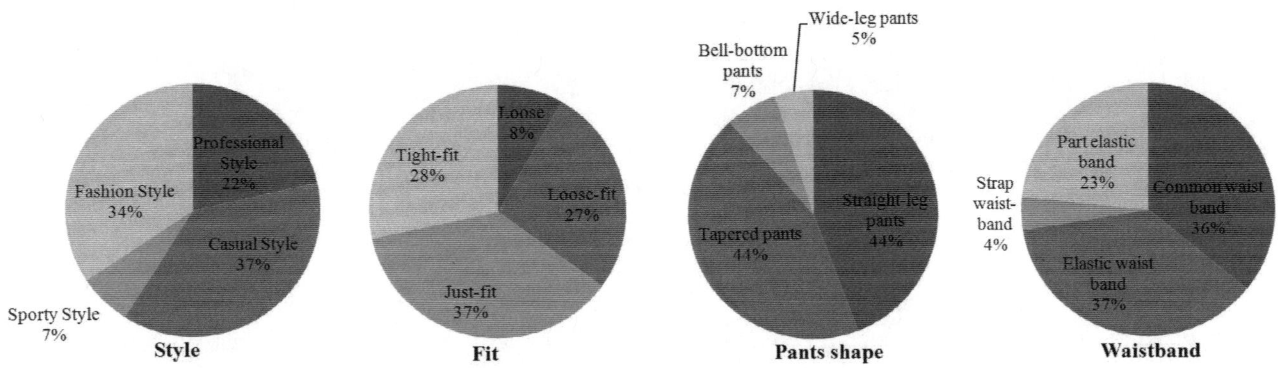

Fig.5 Respondents' style preferences of pants

Most of the respondents (89.2%) favoured dark colours, like black and deep blue, which are stain-resistant and create a slimming effect. A majority of respondents (58.3%) buy pants with a price range of 100 and 200 RMB.

Fabric is extremely important to this age cohort. They expressed a consistent preference for natural fibres (70%), which are perceived as comfortable, practical, breathable, and easy to care for. Moisture permeability & breathability (29.2%), and elasticity (33.3%) of the fabric are the most important aspects when choosing pants. This indicates that fabric manufacturers should develop synthetic and blended fabrics that are more breathable, soft, and comfortable and that could match the natural fabrics. Furthermore, they should increase public awareness of the concept that synthetic fibres do not always equate to bad quality, and that textile science and technology have increased the quality of synthetic fibres.

Table 2. Factors impacting purchase decisions

Factor	Score
Style	3.76
Fabric	3.50
Colour	2.08
Quality	3.58
Comfort	4.51
Fit	6.11
Price	1.65
Brand	0.10
Good mobility	1.83
Easy to match	2.32
Easy to care for	1.11
Flatter the figure	1.39
Hide figure flaws	4.06

With regard to factors that influence respondents' purchase decisions, the scores were also calculated using Eq.1. Results are presented in Table 2. First, the respondents assess whether the pants provide a good fit (6.11), followed by the comfort of the pants (4.51) and the ability to hide their figure flaws (4.06). Style (3.76), quality (3.58), and fabric (3.5) are important considerations as well. Respondents indicated they then check if the pants are easy to match (2.32), and whether the colour (2.08) is desirable or not. They placed the least emphasis on the clothing brand (0.1).

Our analysis of factors that contribute to consumers' dissatisfaction revealed that the most common complaints pertained to: pants that do not suit their body types (4.2), bad fabric quality (3.65), and improper pattern (3.19). Other factors that create dissatisfaction included: uncomfortable to wear (3.17), unreasonable price (2.69), bad sewing quality (2.65), outdated style (2.28), bad fitting (2.23), unable to find the right size (1.75), and restricts movement (1.19) should also to be noted. Scores again were calculated using Eq.1. Results are shown in Table 3.

Table 3. Factors contributing to dissatisfaction

Factor	Score
Outdated style	2.28
Bad fabric	3.65
Bad quality	2.65
Uncomfortable	3.17
Unreasonable price	2.69
Bad fitting	2.23
Does not suit body type	4.20
Unable to find the right size	1.75
Improper pattern	3.19
Restricts movement	1.19

When respondents were asked about their satisfaction levels toward the current pants market, only 20.8% thought the current market satisfies their needs, and even fewer respondents (11.7%) indicated no dissatisfaction with pants currently in the market. These results reiterate that the current market has not yet satisfied most consumers' needs and wants, as the majority still have trouble finding a pair of well-fitted and comfortable pants.

At the end of the questionnaire, an open-ended question asked respondents about any suggestions they may have towards pants design. Some responses included: "I wish designers could pay more attention to the older people, everyone will be old someday"; "We want to have more stylish apparel, not too trendy, not too dowdy"; "Fit will always be the first consideration"; "The fabric should be more soft and more comfortable": "Comfortable, trendy, and fit to my body type"; "I like pants with higher waist position, pockets that can carry many things, and with a zipper to ensure safety. Most importantly, the fabric should be comfortable". From the above, "fit", "comfort", and "fabric" are the most common aspects that respondents value. Companies and designers should be more aware of the aging population's needs and preferences through a choice of comfortable, age-appropriate, stylish, and good quality products.

4. Discussion

Limited research has been undertaken that investigates body cathexis and pants preferences of middle-aged and older women in China. This research addresses some of the psychological concerns towards body image and begins the process of identifying key factors that need to be considered by fashion retailers, designers, companies, and manufacturers.

Survey results provide designers, manufacturers, retailers, and companies with beneficial and applicable data. Converting the research findings pertaining to specific pants preferences - such as fashion/ casual style over professional and sporty style; elastic/ common (zipper & button) waistband over strap waistband and half elastic waistband; loose fit/ just fit pattern over tight fit and loose pants; tapered pants/ straight leg pants over bell-bottom pants - into pants design could result in better fitting and more appropriate styles of pants for these consumers.

Retailers can use these findings to better understand middle-aged and older women's preferences for comfort, fit, and style that flatter their figure types and hide figure flaws. By being aware of specific product attributes, more value will be added to the products and a competitive advantage can be gained over other retailers.

This study not only has implications for apparel manufacturers and retailers who target senior females, but also has social implications related to the well-being and body image satisfaction of seniors as they age. The physical changes significantly affect their body cathexis and impact their shopping preferences. The deterioration of body shape has to be addressed to ensure that garments have appropriate fit and features to improve the quality of the products. Designers should become aware of the aging population's physical changes and psychological needs in order to design better quality products. As this market segment continues to expand, it is imperative that the clothing market environment becomes a more accurate reflection of senior women's needs and wants.

5. Limitation

The sample population in this study was limited to the northern China area, and the online survey approach eliminated a certain group of people who are unable or have no access to use the Internet. This may have limited the application of results to a more diverse middle-aged and senior population located throughout the country. Furthermore, this research is limited to women's lower body (i.e., pants). It would be of interest to look into the possibility of examining the full body, or with the male population or different age groups.

References

[1] Information on http://data.stats.gov.cn/index.htm

[2] Luo M. Clothing market segmentation and marketing strategy of the grey market. Market Modernization 2011; 635: 56-58.

[3] Salusso CJ, Borkowski JJ, Reich N, Goldsberry E. An alternative approach to sizing apparel for women 55 and older. Cloth Text Res J: 2006; 24(2): 96-111.

[4] Thomas JB, Peters CLO. Silver seniors: exploring the self-concept, lifestyles, and apparel consumption of women over age 65. Int J Retail Distrib Manag 2009; 37(12): 1018-1040.

[5] Lao YM. Research on clothing consumption habits of older consumer – a case study of Shaoxing Region. China Textile Leader 2013; 12(4): 89-91.

[6] Dai SJ. Research on analysis of Shaoxing middle-aged and older consumers' clothing purchase mentality and behavior. Journal of Shaoxing University 2013.

[7] Goldsberry E, Shim S, Reich N. Women 55 years and older: part 1 – current body measurements as contrasted to the PS 42-70 data. Cloth Text Res J 1996; 14(2): 108-120.

[8] Katou C, Nakho Y. Shapes of abdomen and hips of elderly women for the drafting pattern of basic blocks. Japan Res Assoc Text End-uses 2001; 42(2): 251-261.

[9] LaBat KL, DeLong MR. Body cathexis and satisfaction with fit of apparel. Cloth Text Res J 1990; 8(2): 43-48.

[10] LaBat KL, Schofield N, Ashdown S, Hethorn J, Salusso C. Improving pant fit for women 55 and older through an exploration of two pant shapes. Cloth Text Res J 2006; 24(2): 147-160.

[11] Information on http://www.who.int/mediacentre/factsheets/fs311/en/

Research on Relational Model for Body Block of Menswear based on Digital 3D Human Model

Jin Duan[1], Xin Zhang[1*], Bo-An Ying[1], Jue Zhang[2]

[1]*Apparel & Art Design College, Xi'an Polytechnic University, Xi'an, Shaanxi, 710048, P.R. China*
[2]*Shenzhen Xingtai Season Wind Fashion Co. Ltd., Guangdong, 518054, P.R. China*

[*]Corresponding author's email: xianzhangxin@sina.com

Abstract

Research on relational models for apparel pattern has been a popular and challenging topic in recent years. In relational models for apparel pattern, individuals are taken as research objects, whereas the idea of apparel pattern classification is applied to the design of apparel patterns. According to apparel pattern classifications, the pattern generating procedure could be divided into three processes, namely body block, garment block and pattern. This ensures stylization of the pattern generating procedure and accuracy of pattern generating results. This paper studies the design of a relational model for apparel pattern for men's tops. Through digital means, the paper explores the body block of menswear, which is the first step in designing the relational model, establishes the relational model and a digital 3D human body model for the body block of menswear and further verifies the relational model.

Keywords: Men's apparel pattern; Body block; Digitizing custom made; Relational model

1. Introduction

As the society and economy develop, clothing is influencing the psychology, physiology and social comfort of individuals as "the environment closest to humans" in the customer-oriented age. Broun (2003) suggested that whether clothes fit well has been widely considered to be the most important factor affecting the appearance and vendibility of clothing, as consumers expect clothing to be tailored to their physical characteristics [1]. However, with the continuous improvement in living standards thanks to scientific and technological innovations, the consumption now tends to be affordable, diversified, hierarchical and personalized. Strong demand for clothing has continued to move the development of clothing design methods away from traditional design to computer-aided design (CAD), and from 2D designs to 3D designs. Similarly, it has constantly changed the development trends in apparel patterns.

In a recent theoretical research study on apparel patterns in relation to clothing fitness, Jiang (2008) defined a new concept in which apparel patterns are directly generated from the human body by analyzing physical characteristics and researching the relationship between the human body and apparel [2]. Zhu and Shi (2010) elaborated on an apparel processing method that automatically generates tailor-made apparel patterns based on digital tailor-made processes [3-4]. In studies of 3D apparel CAD, Yang et al. (2011) outlined a digital design method for women's tailor-made apparel patterns based on 3D human body and the discussion of fitness in women's apparel patterns with different split lines [5]. Hong and Daanen (2008) presented fitness research studies for women's tailor-made dresses based on scanning technology [6]. Huang et al. (2012) put forward the idea of building parameterized models of women's design patterns based on digital 3D human body model. The group went on to validate the claim by establishing relational models for women's dress patterns [7].

From the above studies, most research at present on apparel fitness are women's clothing-oriented. Studies on the theories and digitalization of men's apparel patterns are relatively deficient. Therefore, this paper proposes building relational models for body block of men's tops based on men's physical characteristics and digital 3D human models. The models should also take into consideration existing traditional methods of menswear design and factors which influence the individual needs of consumers and designs of men's apparel patterns.

2. Theoretical Analysis

2.1 Definition of relational models for men's body block

Relational models for patterns of men's tops require male individuals as research objects and regard customization as the design objective of apparel patterns. Jing (2014) and Yue (2014) pointed out that, in order to ensure the convenience and accuracy of apparel pattern design, the design process should be divided into three stages: body block, garment block and pattern [8-9].

Men's body blocks are jointly influenced and generated by the detailed dimensions and characteristic curves of men's upper body. While reserving the physical characteristics of men's upper body in its entirety, body blocks of men's tops needs to ensure the minimum amount of breathing without changes in style and function.

Model M of Alva Form, which is designed and produced by the American Alvanon Ltd., is selected for the research of relational models for basic patterns of men's tops as it is a versatile form for medium-sized Chinese men. This paper mainly analyzes the relationship between physical characteristics of men's upper body and the body block design of men's tops. Furthermore, it establishes and realizes the relationship on 3D digital human models by building, exhibiting and modifying the body block of men's tops based on 3D digital human models. Finally, this paper generates samples to verify the established relational models.

2.2 Physical characteristics of men's upper body and design parameters of men's body block

2.2.1 Corresponding relationship between feature points of men's body block and those of men's upper body

Feature points of men's apparel patterns are the basic elements constituting the patterns of men's tops. Men's body blocks consist of two parts: front and back. In the design of men's body block, the following feature points should be considered: front neck point (P_N), front point of bust width (P_FBF), side point of bust width (P_FBC), front point of front bust circumference (P_B), side point of bust circumference (P_BC), front point of front waist (P_WF), side point of waist circumference (P_WC), front point of armhole depth (P_AHF), armhole side point (P_AHC), shoulder front point (P_SBF), shoulder midpoint (P_SBM), shoulder back points (P_SBB), back neck point (P_NB), front point of back width (P_BBF), side point of back width (P_BBC), back point of armhole depth (P_AHP), bust back point (P_BB), waist back point (P_WB), bust point (P_BP), front waist dart point (P_WM), back waist dart point (P_WMB) and back shoulder dart point (P_SS). There is a corresponding relationship between its feature points and those of men's upper body. (Fig.1)

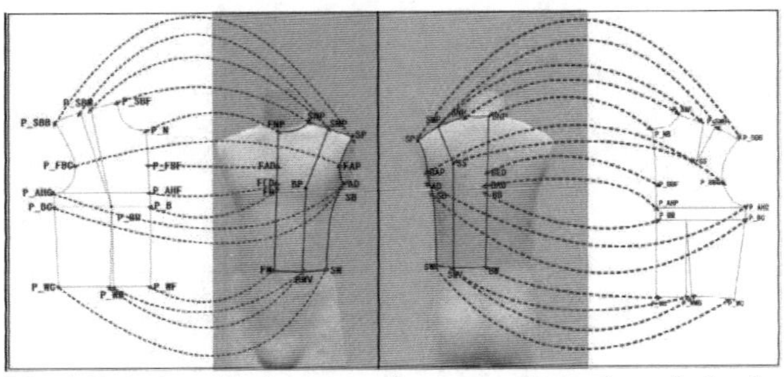

Fig.1 Relationship between feature points of men's body block and those of men's upper body

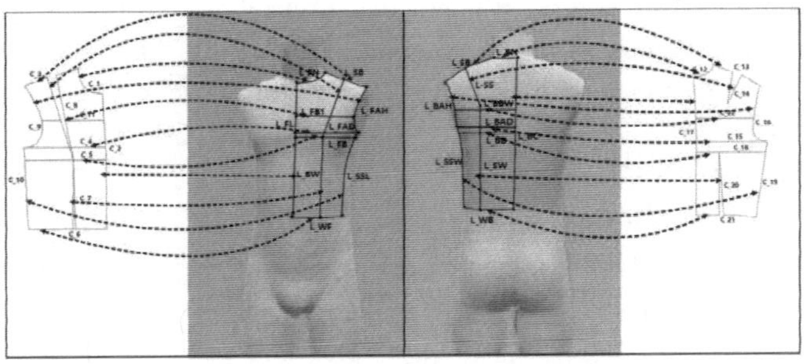

Fig.2 Relationship between feature lines of men's body block and those of men's upper body

2.2.2 Corresponding relationship between feature lines of men's body block and those of men's upper body

Feature lines of men's upper body constitute two or more feature points in men's tops. The feature lines of body block of men's tops also consist of two or more feature points of apparel patterns of men's tops. Since there is a corresponding relationship between feature points of apparel patterns of men's tops and the physical characteristics of men's upper body, feature lines of men's body block correspond to those of men's upper body. (Fig.2, Table 1)

Table 1. Relationship between feature lines of men's body block and those of men's upper body

Feature lines of body block of men's tops	Feature lines of body block of men's upper body	Feature lines of body block of men's tops	Feature lines of body block of men's upper body
Front neck curve (C_1)	Front neck length (L_FN)	Back neck curve (C_12)	Back neck length (L_BN)
Centre front (C_2)	Front line length (L_FL)	Back shoulder (C_13)	Side shoulder length (L_SB)
Front shoulder (C_3)	Side shoulder length (L_SB)	Dart line of back shoulder (C_14)	Dart line of back shoulder (L_SS)
Front armhole depth (C_4)	Length of front armhole depth (L_FAD)	Back armhole depth (C_15)	Length of back armhole depth (L_BAD)
Front bust (C_5)	Front bust length (L_FB)	Back bust (C_16)	Back armhole length (L_BAH)
Front waist (C_6)	Front waist line (L_WF)	Centre back (C_17)	Centre back length (L_BC)
Front waist dart (C_7)	Front waist dart length (L_BW)	Back bust (C_18)	Back bust length (L_BB)
Front shoulder dart (C_8)	Front shoulder dart length (L_BS)	Back side seam (C_19)	Side seam line length (L_SSL)
Front armhole curve (C_9)	Front armhole length (L_FAH)	Back waist dart (C_20)	Back waist dart length (L_SW)
Front side seam (C_10)	Side seam line length (L_SSL)	Back waist (C_21)	Back waist length (L_WB)
Front width line (C_11)	Front width line length (L_FB1)	Back width line (C_22)	Back width line length (L_BBW)

3. Preparation of 3D garment model

3.1 Data acquisition and measurement of 3D digitizing human models

It was determined that the medium male form of Alva Form is to be used as the digital 3D human model. There are a number of marked feature points on the Alva Form. Generally, the calibrated feature points are classified into two categories to facilitate the construction of feature lines after scanning the 3D human body: feature points of dress forms are determined in accordance with the feature points and definition of men's upper body; supplementary feature points are added in places with large curvature. (Fig.3)

The German TecMath non-contact 3D scanner was selected to obtain digital 3D human models and measurements of the Alva M male Form. The scanning image of Alva M male Form is shown in Fig.4.

Table 2. Measurements of the Alva M male form

Measurement	Data (MM)	Measurement	Data (MM)
Front neck length (L_FN)(1/2)	130	Side seam line length (L_SSL)	253
Front line length (L_FL)	405	Front width line length (L_FB1) (1/2)	185
Side shoulder length (L_SB)	148	Back neck length (L_BN) (1/2)	90
Length of front armhole depth (L_FAD)(1/2)	245	Length of back armhole depth (L_BAD) (1/2)	220
Front bust length (L_FB)(1/2)	248	Dart line of back shoulder (L_SS)	175
Front waist line (L_WF)(1/2)	213	Back armhole length (L_BAH)	245
Front waist dart length (L_BW)	205	Centre back length (L_BC)	470
Front shoulder dart length (L_BS)	260	Back bust length (L_BB)(1/2)	235
Front armhole length (L_FAH)	215	Back waist length (L_WB)(1/2)	195

Manual interactive measurement for Alva M male Form was conducted with ScanWorX, the measurement software that comes with the TecMath non-contact 3D scanner of Germany Vitronic Co. Table 2 demonstrates the measured dimensions of Alva M male Form.

3.2 Modeling of 3D digital human models

Data of Alva M male Form obtained from the Vitus Smart 3D scanning system cannot be directly applied to the construction of 3D digital human models. The first reason is that data which is measured through the 3D optical scanning method of non-contact measurement is always accompanied by noise points. This could be caused by the following three reasons: errors caused by the various factors (such as reflectance, color, etc.) on the surface of the dress forms; errors caused by the measurement device itself (such as accuracy, resolution and vibration of the equipment); and errors caused by surrounding environmental factors (such as brightness of lighting). The second reason is that the size of the measured data is large, therefore it is necessary to filter the redundant points. If not, the excessively vast cloud data will slow down the processing efficiency of computers and lead to ruggedness of the reconstructed curves.

In order to achieve the construction of digital 3D human body model, CATIA V5 was adopted to remove noise points in the cloud data in the first step. Since the surface of male dress forms is relatively flat and the data varies evenly, the tolerance ball algorithm was then employed to streamline the point clouds. The experiment shows that when the radius of the tolerance ball is 8mm, the requirements of accuracy and smoothness can be met simultaneously. Finally, a surface model which is able to represent the physical characteristics of men's upper body was constructed via griddization of point clouds. The constructed digital 3D human models are shown in Fig.5.

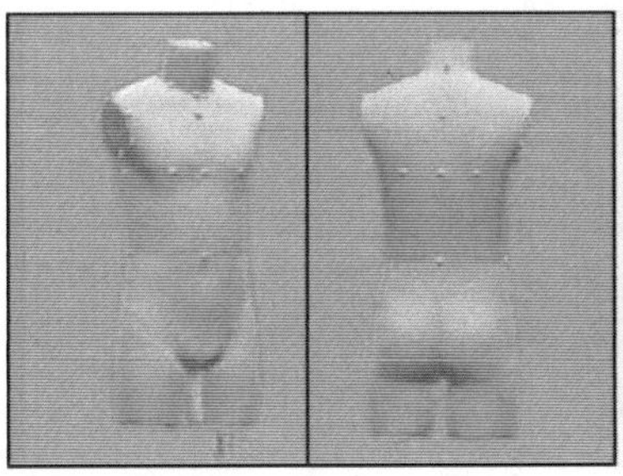

Fig.3 Marked feature points on Alva Form

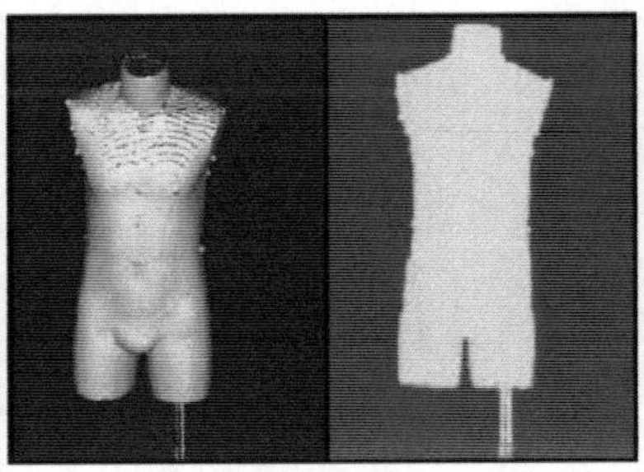

Fig.4 Scanning image of Alva M male Form

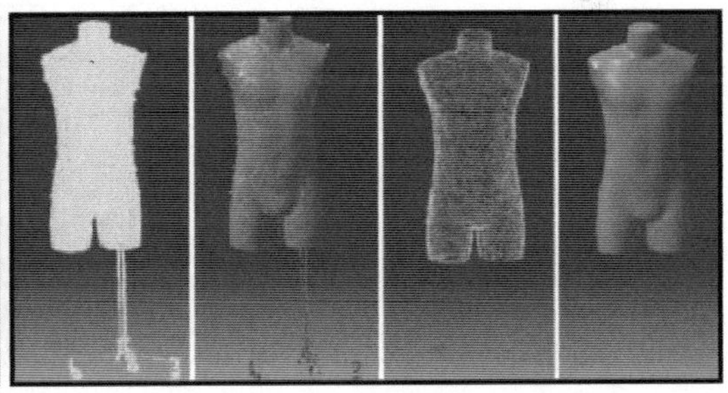

Fig.5 The constructed 3D digitizing human models

4. Design of Men's body block of 3D Digitizing Human Models

4.1 Design of characteristic curves of 3D digitizing human models

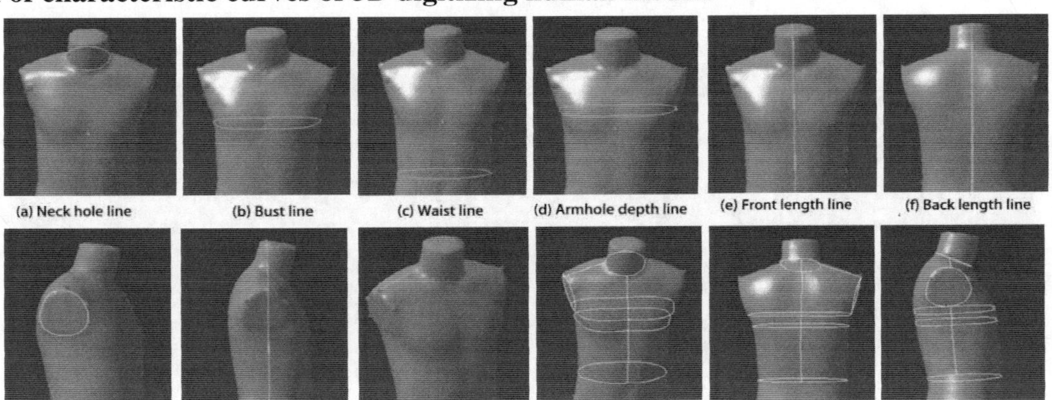

Fig.6 Design of characteristic curves of 3D digitizing human models

Since feature lines of the body block of men's tops correspond to those of men's upper body, the design of characteristic curves of basic patterns of men's tops is required to be performed on the digital 3D human models. These characteristic curves include the neck hole line, the chest line, the waist line, the armhole depth line, the front length line, the back length line, the armhole curve, the side seam line and the side shoulder line (Fig.6). Table 3 shows the design of characteristic curves of based on the digital 3D human models of men's tops.

Table 3. Design of characteristic curves of 3D digitizing human models

Characteristic curves design	**Key point**
Neck hole line	Curve on neck line curved surface，through front neck point、shoulder neck point、back neck point
Bust line	Curve on bust curved surface，though bust point
Waist line	Curve on waist cured surface，though front waist point，side waist point，back waist point
Armhole depth line	Curve on bust curved surface，though armhole depth point
Front length line	Front curve on curved surface of men's upper form, though front neck point、front bust point、front waist point、back neck point、back bust point、back waist point
Back length line	Back curve on curved surface of men's upper form, though front neck point、front bust point、front waist point、back neck point、back bust point、back waist point
Armhole curve	Curve on curved surface of men's upper form, though shoulder point、front depth point、back depth point、armhole depth point
Side seam line	Curve on curved surface of men's upper form, though shoulder point、armhole depth point 、side waist point、side bust point
Side shoulder line	Straight line on curved surface of man's upper form, though shoulder neck point、shoulder point

4.2 Obtaining men's body block from the digital 3D human models

In order to obtain men's body block of the 3D digital human models which can be flattened, there is a need to construct fitting curves on the basis of 3D digitizing human models. The so-called fitting curves are reconstructed from feature lines of the men's body block which are based upon the streamlined point clouds. Firstly, edge contours of the characteristic curves were extracted from the 3D digitizing human models. The contours of the body block's edges must be closed curves. Fig.7 is the contour extraction map for the front and back body blocks of men's tops. Secondly, the body blocks were filled according to the trends of the point clouds indicating the contours of the front and back body blocks. Lastly, the constructed curves were adjusted by virtue of the feature lines of men's body block. Fig. 8 is the fitting surface chart of men's body block.

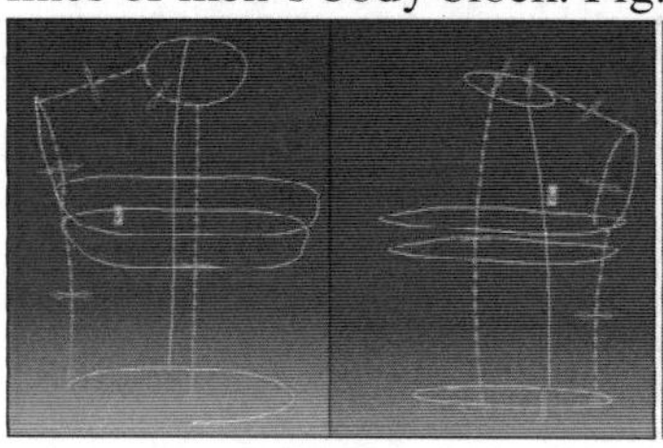

Fig.7 The contour extraction map for body block

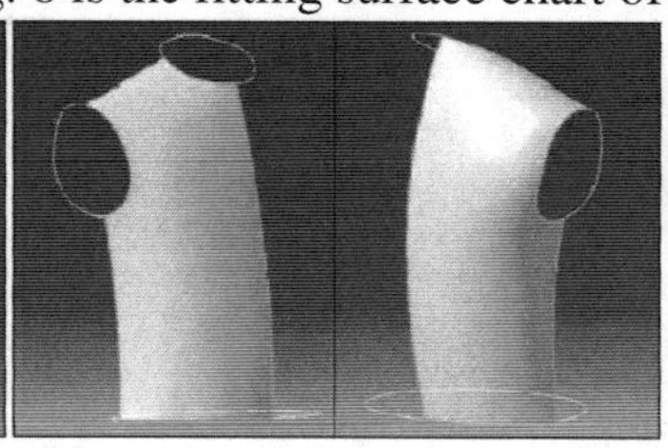

Fig.8 Fitting surface chart of body block

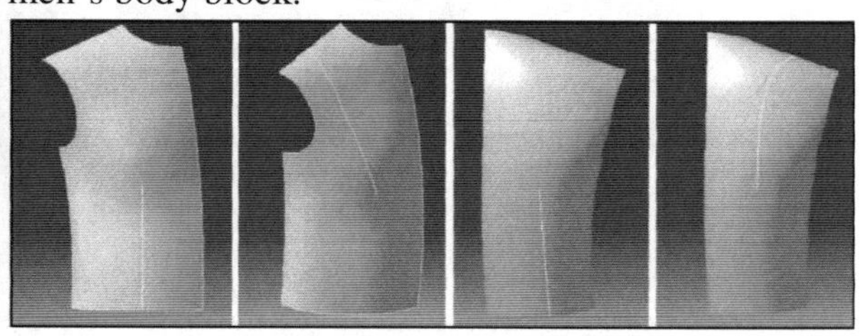

Fig.9 Split lines design in body block for men's tops

4.3 Determination of split lines of men's body block of 3D digital human models

The human body is an irregularly curved surface. Due to the raised male chest and scapula, there is a certain amount of increase in these parts. Such increases causes extra folds in the chest and scapula when the fitting surfaces of the basic patterns of men's tops are developed into two-dimensional ones. Consequently, the apparel patterns cannot fit the surface of human body well. As for the folds, they are called clothe body float remainder.

In order to eliminate the clothe body float remainder, dividing designs are required in the bulge regions. 2D patterns of clothes can better fit the human body by dividing after eliminating the float remainder. While designing the dividing positions, the following two principles should be followed: the dividing sharp points should point to the bumps; the split lines should be easily available. According to the two principles, split lines were drawn on the fitting surfaces of body block of men's tops. Fig.9 shows the split lines design in body block for men's tops.

4.4 Flattening of men's body block of 3D digitizing human models

Since the established surfaces of men's body block are complex and undevelopable, we propose a new flattening algorithm based on the feature points and feature lines. As shown in Fig.10, curved surfaces of the body block of men's tops were divided by the feature lines. The development of divided surfaces was summarized as a problem seeking unconstrained minimization (Su, 2010) [10]. The basic idea is to divide the known surfaces into meshes, map the 3D grid points to 2D surfaces, minimize the changes in distance between grid points both before and after the development, realize approximate development, build spring-mass models on the basis of the initial geometric development, release energy and modify the geometric flattening results.

4.4.1 Initial development of the surface meshes

Lee (1999) and Wang et al. (2002) claimed that curved surfaces of the body block can be flattened first with the triangle method. The flattening process includes unconstrained and constrained development of triangles [11-12]. On the mesh surface in Fig.11 (b), T shares its edges with both T_1 and T_2. Suppose that T_1 and T_2 are developed triangles. If the method of unconstrained development is adopted, the triangle to be developed T will share its edges Q_1Q_2 (P_1P_2 in Fig.b) with the developed triangle T_1. The third point of T, Q_3, can be obtained through the intersection point of the two circles. The two circles, with P_1 and P_2 as their centers respectively, take $R_{1,3}$ and $R_{2,3}$ as their radiuses. $R_{1,3}$ and $R_{2,3}$ are Euclid distances of Q_1Q_3 and Q_2Q_3 in three-dimensional meshes. Since surfaces of the basic patterns of men's tops are undevelopable, the position of Q_3 is P'_3 on the 2D surface. The two points fail to coincide with each other. Consequently, there will be cascading or cracks. To ensure integrity of the topology, it is necessary to constrain the development of triangles. At this time, an average location can be used to solve this conflict preliminarily. Nevertheless, this will lead to distortion of triangles, thus causing errors.

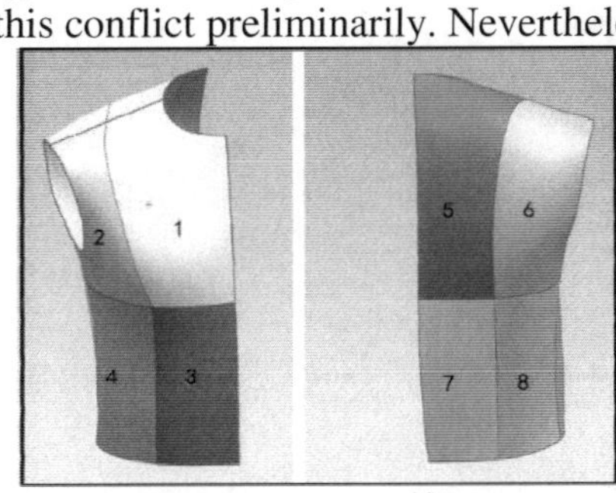

Fig.10 Divided the basic patterns

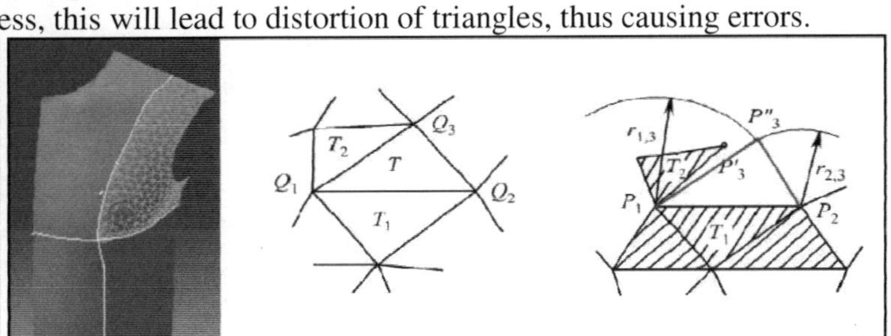

Fig.11 Initial development of the surface meshes

4.4.2 Spring-mass models

In order to reduce errors after the initial flattening of 3D mesh surfaces, spring-mass models were built on the basis of the initial flattening. Points on the meshes were regarded as mass points and the edges were considered as springs. When the edges stretch or compress with the development of curved surfaces, a certain amount of elastic energy will accumulate in the form of elastic deformation energy. In the development process, energy accumulated in the system will be constantly released. Eventually, flattened surfaces can be obtained. As shown in Fig.12, the points are deemed to be connected by springs. Liu et al. (2013) and Han and Liu (2011) stated that the initial edge length of springs on the original curved surface is the original spring length. If the distance between mass points on the surface exceeds the original spring length in the deformation process, it presents attractive forces, otherwise it presents repulsive forces [13-14].

The accumulation mode of energy and power of Pi before and after deformation can be expressed by the following equation:

$$E(P_i)=\sum_{j=1}^{n}\frac{1}{2}C\left(\left|P_iP_j\right|-d_j\right)^2,\quad \vec{f}(P_i)=\sum_{j=1}^{n}(\left|P_iP_j\right|-d_j)\vec{n}_{P_iP_j} \tag{1}$$

Where C is the coefficient of elastic deformation; P_iP_j is the distance between P_i and P_j after the spring deformation; d_j is the original spring length; $\vec{n}_{P_iP_j}$ is the unit vector pointing P_i to P_j.

During the development process of curved surfaces, the movement of mass points can be described by Lagrange's equation, that is:

$$M\ddot{q}+D\dot{q}+Kq=g_q+f_q \tag{2}$$

Where M, D and K are the mass matrix, damping matrix and stiffness matrix of the system respectively; g_q is the internal force of the system caused by the difference between local freedom and global freedom; f_q is the external force of the system. Upon analysis, g_q and f_q in the spring-mass system is 0 since the simulation fabric is flattened.

In addition, the damping is negligible. If Formula (2) is simplified, the following formula can be obtained:

$$M\ddot{q}+Kq=0 \tag{3}$$

In the spring-mass system, the time interval Δt of mass point movement is small. The acceleration of P_i can be considered to be a constant. All mass points in the system are in equilibrium. If Lagrange's equation is solved via the Euler method, the following method can be obtained:

$$m_i=\frac{\xi}{3}\sum A_k,\quad \ddot{q}_{i_t}=\frac{f_{i_t}}{m_i},\quad \dot{q}_{i_{t+\Delta t}}=\dot{q}_{i_t}+\Delta t\ddot{q}_{i_t},\quad q_{i_{t+\Delta t}}=q_{i_t}+\Delta t\dot{q}_{i_t}+\frac{\Delta t^2}{2}\ddot{q}_{i_t} \tag{4}$$

In the formula, m_i is the quality of P_i; ξ is the surface density of the curved surface; A_k is the area of the kth triangle in all triangles containing nodes in space domain; $\ddot{q}_{i_t}$ is the acceleration of node P_i; f_{i_t} is the elastic force of P_i; $\dot{q}_{i_t}$ is the velocity of P_i; q_{i_t} is the location of node P_i in time t. The flattened pattern by the spring-mass models is shown in Fig.13.

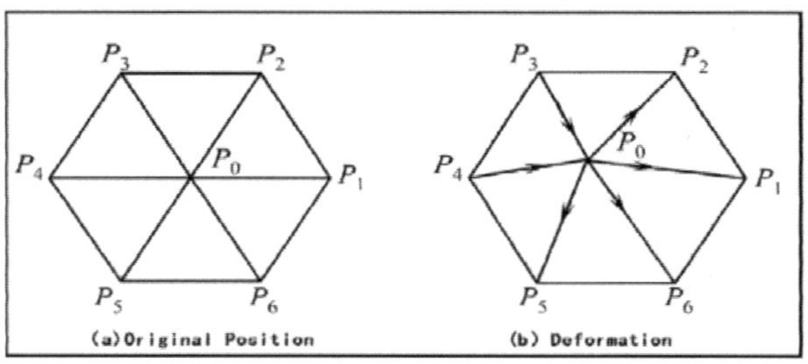

Fig.12 Spring-mass models

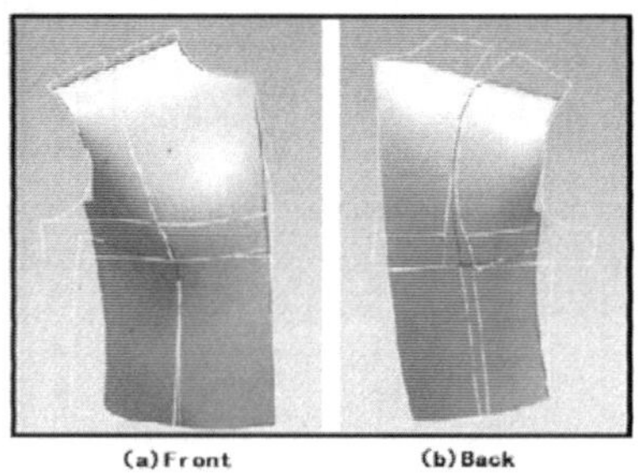

Fig. 13 The flattened pattern

4.4.3 Modification

The patterns flattened by spring-mass models are faulty. The error is usually measured in length. Therefore, it is necessary to modify the flattened results. As shown in Fig.14, the flattened models were imported to the coordinate reference line to align the waist circumference line, chest circumference line, armhole depth line, front center line and back center line with the coordinate system. The curve importing the coordinate reference line was modified into a line. In addition, the measured dimensions of dress forms in Table 2 were utilized to review and modify the flattened patterns. The modified patterns are shown in Fig.15 (a).

The human body is not static, both breathing and movement of human body have demands for apparel patterns on the amount of relaxation. Therefore, the design of body block for men's apparel must ensure sufficient relaxation for normal physiological functions, including allowance for breathing and blood circulation. Generally speaking, body block of men's apparel must guarantee a minimum amount of relaxation allowed. As a result, it is required to add 4cm in the circumference direction of the body block in the first modification to serve as the allowance for respiration (Wu, 2001) [15]. Wang (2005) suggested that the relaxation amount for chest width, back width and armhole should account for 30%, 40% and 30% in the total amount of relaxation respectively [16]. Fig. 15 (b) exhibits the body block of men's tops with the basic amount of relaxation.

5. Validation of relational models for 3D digital human body of men's tops

We performed validation of the relational models for body block of men's apparel on Alva M Form by putting samples of body block on the Alva M Form. The results are shown in Fig.16. Alva M Forms are dress forms without any amount of relaxation. According to Fig.16, samples of body block of men's tops fit the upper body of the dress forms. Characteristic curves of the patterns are basically consistent with the feature lines of men's upper body. The shoulder angle is appropriate without leaving the shoulder empty. They are in equilibrium as a whole.

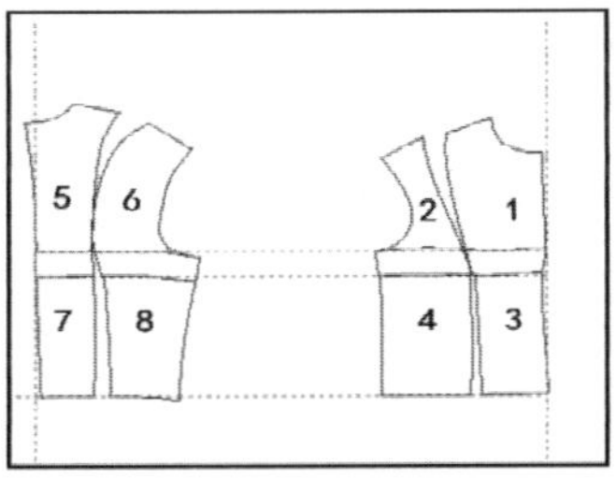

Fig.14 imported the reference line

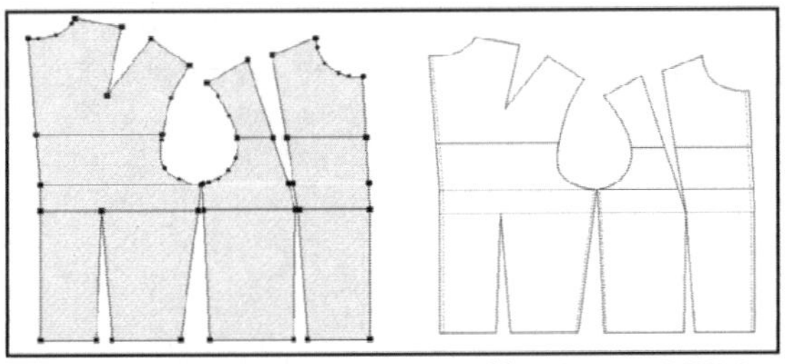

(a) The Firstly Modified (b) The Secondary Modified

Fig.15 The modified patterns

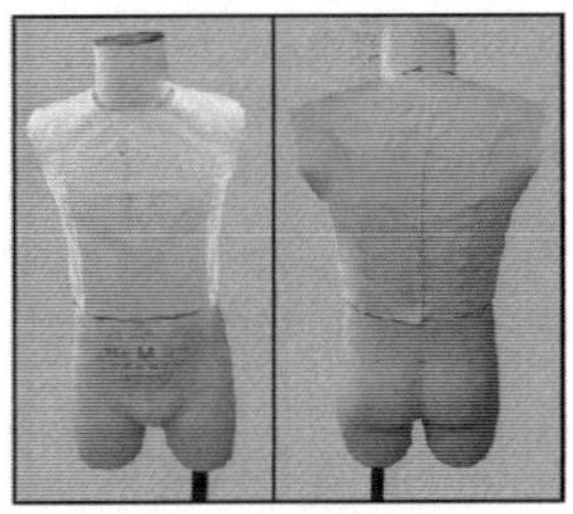

(a) Front (b) Back

Fig.16 samples of body block

6. Conclusion

This paper presents the definition and functions of men's body block, proposing that body block of men's tops is generated and influenced by dimensions and shapes of men's upper body. In addition, this paper analyzes and establishes the corresponding relationships between characteristics of men's upper body and body block of men's tops by utilizing digital design of apparel patterns for men's tops, obtaining body block of men's tops by flattening curved surfaces and making samples of body block of men's tops. Therefore, men's body block is a tight fitting block that incorporates all of the important points, measurements and lines which are based upon men's upper body. Through evaluations on the wear trials, the body block of men's tops fit the human body well. It validates the feasibility of the design methods of body block of men's tops. Moreover, it lays foundation for the next step of menswear design: design of garment block and pattern design of men's clothing.

References

[1] Broun PJ. Consumer choices clothing ideas for people with special needs. Iowa State University Extension: 2003.
[2] Chuan J. Pattern Making System for Men's Upper Wear. Soochow University: 2008.
[3] Yan WS. Establishment of 3D Human Body Database Based on Men's Suits e-MTM Mode of Production. DongHua University: 2010.

[4] Guang YZ. Study on digital Clothing Body Measuring and individual Clothing Prototype Generation Technologies Based on eMTM. DongHua University: 2007.
[5] Yang YC, Zou ZY, Li Z, et al. Development of a Prototype Pattern Based on the 3D Surface Flattening Method for MTM Garment Production. Fibres Text East Eur: 2011.
[6] Hong S, Daanen H. Made-to-measure pattern development based on 3D whole body scans. Int J Cloth Sci Tech: 2008; 20(1): 15-25.
[7] Huang HQ, Mok PY, Kwok YL, et al. Block pattern generation: From parameterizing human bodies to fit feature-aligned and flattenable 3D garments. Comput Ind: 2012; 63(7): 680-691.
[8] Jing Q. Basic Bra Pattern Based on Reflection Theory, Textile Bioengineering and Informatics Symposium Proceedings: 2014:926-933.
[9] Yue XC. Research on Bra Pattern Design and Evaluation Methods for Digital Made to Measure, Xi'an Polytechnic University: 2014.
[10] Bu QS. A Practical Introduction to Differential Geometry. Science press: 2010.
[11] Lee CK. Automatic adaptive mesh generation using metric advancing front approach. Eng Computation: 1999; 16(2): 230-263.
[12] Wang C, Smith S, Yuen M. Surface flattening based on energy model. Comput Aided Design: 2002; 34(11): 823–833.
[13] Liu Q, Xi J, Wu Z. An energy-based surface flattening method for flat pattern development of sheet metal components. Int J Adv Manuf Tech: 2013; 68(5-8): 1155-1166.
[14] Lin H, Bin L. An Algorithm of Triangular Mesh Surface Flattening Based on Spring-Mess Mode and Its Application. J Huaqiao Univ, Nat Sci: 32.6; (2011): 601-606.
[15] Zhi MW. Correct Determination of Garment Tolerance. J Wuxi Univ Light Ind: 2001; 20(6): 649-651.
[16] Xuan W. Analysis Study of the Ease Allowance Needed for Garment. J Text Res: 2005; 26(4): 126-1.

Design and Evaluation of a Full Cup Bra Based on Bra Prototype

Shi-Yang Yu, Rong Zheng*, Jing Zhao, Hai-Jing Liu, Wen Wu

Aimer Human Engineering Research Centre, Beijing Institute of Fashion Technology, Beijing, 100029, China

Corresponding author's email: yuki.zheng@gmail.com

Abstract

The bra is one of the most close-fitting intimate garments worn by women, designed to support and mould the soft tissues of the upper female body. Apart from aesthetic appeal, fitness and comfort are increasingly important in today's bras. This study adopted the 75B bra prototype developed by BIFT-Aimer Human Engineering Research Centre as a starting point to determine ideal design features and material selection of a full cup bra. The study performed market research, made adjustments based on the bra prototype, created a paper pattern, and produced the sample bra. Six female volunteers aged 20-39 with 75B bra size were recruited for the bra fitting experiment which included 2D measurement and 3D scanning under nude and bra-wearing status, stability testing under bra-wearing status, and subjective questionnaire evaluation on shaping effect, stability, and pressure comfort. Test results indicate the full cup bra sample has good shaping effect and stability, positive evaluation on subjective pressure comfort, and thus can be used as a guide for enterprises.

Keywords: Full cup bra; Market research; Shaping effect evaluation; Pressure comfort; Stability evaluation

1. Introduction

The bra is an item of functional intimate apparel designed to support, shape, cover, and protect females' breasts. However with the development of bra technology, consumers are attaching greater importance on a bra's comfort, fitness, and shaping features rather than its protective effect. The ideal comfort-fitting bra should have the following features: contain the breasts; beautify breast curve and bear supporting breast weight; have proper drawing force in binding tape and shoulder strap to prevent concentration pressure formed in artery/ vein and muscle/ skeletal sensitive parts; and ensure stability during bodily movement [1].

The breast is mainly made of fat and breast tissue. Because it has no bone and muscle to support it, displacement is liable to occur during bodily movement, which not only will bring embarrassment to the wearer but may also be detrimental to the health of the female breast [2]. Some research has been conducted on breast displacement: Eden et al.'s [3] study of 15 females with 36C breast size examined breast displacement frequency as subjects walked in free and fixed stride wearing four kinds of sport bras, and found that average displacement in vertical direction was 2.06 to 3.07cm; Li and Wang's [4] study of displacement based on six key points found that horizontal displacement of all points was 0.05 to 0.28cm, and 0.09 to 0.87cm in vertical direction, with displacement trends of cup moving up, wire retraction, and rear sheets moving upward.

A bra's comfort is affected by many factors, such as the bra's style, material, and structure. Duan and Yu found that pressure comfort is a primary aspect influencing physiological comfort of bras [5]. While a bra's pressure can make women feel more comfortable, protect and support breasts, reduce breast ptosis (sagging), and make the breast curve more beautiful, too much pressure can hinder blood circulation of the breast and thus have negative health effects. Wang and Zhao's evaluation test on pressure comfort revealed that pressure is generally high at the most curved body parts - such as the shoulder-strap, under-bust, and back blade positions - and that over-concentrated pressure can create health risks [6]. Miyatsuji et al., who studied pressure effects of two kinds of bras by measuring heartbeat, found that too much pressure will have a great negative effect on the body's internal and circulatory system [7]. Similarly, through a series of studies, Morooka et al. found that extension of material greatly affects pressure comfort, and subjective pressure comfort was different for different volunteers wearing same style bras, and same volunteers wearing same material and different style bras [8]. The aforementioned studies show that breast comfort and health are not only affected by motion intensity but also by physiological conditions of the human body, style of bra, and usage of materials. Thus it is important to improve the pressure comfort of bra.

The full cup bra is the most functional cup; it can hold the whole breast in the cup, and has the effect of supporting, lifting, and gathering, which is appropriate for women with large breasts and soft skin. For the purpose of product design and development, this study investigated full cup bra styles and materials currently used in the market, designed a full cup bra style based on bra prototype 75B developed by BIFT-Aimer Human Engineering Research Centre, and performed evaluations for both subjective and objective shaping effects, stability, and pressure comfort. Findings will not only provide reference data for bra production enterprises, but also give guidance for end consumers when selecting bras.

2. Design of the full cup bra

2.1 Market research

This study focused on full cup bras currently in the market. The 68 samples encompassed five brands - Aimer, Maniform, Gujin, Wacoal, and Triumph - and the investigation considered style features and materials used in each part of the bra. Specific research results are noted below.

2.1.1 Style

The primary full cup bra in the current market has a "T" line cutting cup, and uses an ultrathin lace cup; the height of the junction part is often 5cm to 5.9cm, or 7cm to 7.9cm; the side wing seam heights are between 6.5cm to 11.5cm, most of them (60%) are equal or greater than 9cm; 74% of the full cup bras have a "U" character back wing shape, and 26% of full cups have a "-" character back wing shape.

2.1.2 Fabric

The main research findings pertaining to fabric are as follows:

a) The surface fabric of the cup mainly uses lace fabric, accounting for 74% of total samples. This is because the lace fabric has a decorative effect, and its use as an outer layer can make the bra more beautiful.

b) The lining fabric of the cup mainly uses cotton fabric, accounting for 46% of total samples. This is because the lining of the cup is in direct contact with skin, and natural fibres can meet the requirements of hygiene and comfort.

c) Centre junction part, underband, and side wings are generally viewed as a whole; most full cup bras adopt layered fabrics, with surface fabric using flexible lace fabric and lining using stabilised yarn.

d) Back wings usually adopt double layer fabric, with outer layer using warp knitting fabric, mesh security fabric, double fabric, and lining using mesh lycra.

2.1.3 Auxiliary

Findings regarding bras' auxiliary features are as follows:

a) The elastic band is mainly used in wings; elastic band with 10mm width is used in the underarm band, and elastic band with 13 mm is often used in the underband.

b) 10mm and 12mm elastic band are most common for use in the shoulder strap.

c) 8 hooks and 9 hooks are most commonly used; the hooks mainly rank in three lines.

d) The length of side brace bones and the type of wire are selected according to design requirements.

e) Other auxiliary features include velvet strap, loop and threads, often used in the same standard of full cup bra (except for colour).

2.2 Design procedure

2.2.1 Style and fabric design

Based on the market research results noted above and combined with production requirements of enterprises, the study determined the style features of the bra for use in the experiment: full cup, "T" line cutting cup, high centre junction part, high side hurdle, and "-" character wing shape. The style of the full cup bra is shown in Fig.1.

The selection of fabric and auxiliary features is: tricot lace, stabilised yarn, superfine fabric, mesh fabric, normal wire, 10mm underarm elastic band and 13mm hurdle elastic band, 12mm shoulder strap band, 8 buckles, 10cm-long plastic bones, 9 hooks in three lines. Tables 1 and 2 show fabric and auxiliary usage details.

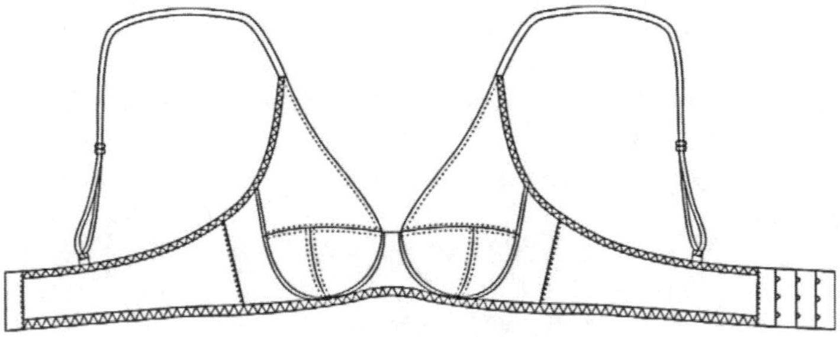

Fig.1 Style of the full cup bra

Table 1. Fabric usage

Part		Fabric name	Composition	Gram weight	Elastic extension
Cup	External clothing	Tricot lace	88.1% nylon 11.9% spandex	17g/m^2	57%
	Inner clothing	Stabilised yarn	100% nylon		
Front part	External clothing	Tricot lace	88.1% nylon 11.9% spandex	17g/m^2	57%
	Inner clothing	Stabilised yarn	100% nylon		
Wing	Outer clothing	Superfine fabric	66% nylon 34% spandex	145g/m^2	Warp direction 224% waft direction142%
	Inner clothing	Mesh fabric	83% nylon 17% spandex	63g/m^2	Warp direction 245%±15% waft direction110%±15%

Table 2. Auxiliary usage

Part	Material name	Specification
Cup and chin connecting parts	Steel ring	Normal ring
Hurdle	Elastic band	10mm
Underband	Elastic band	13mm
Shoulder strap	Elastic band	12mm
Side seam	Plastic bones	10cm
Centre back	buckles	Three line, three buckles
Others	Coarse cotton cloth, multifilament binding type, cotton thread, etc.	

2.2.2 Pattern design

Based on the 75B prototype developed by BIFT-Aimer Human Engineering Research Centre (see Fig.2), we made some adjustments to the clothing pattern according to the style design and features of used material; the modifications are discussed below.

The following sequence of adjustments was made to the bra prototype:

a) Based on shrinkage of the fabric, preshrink the paper pattern of prototype.

b) No adjustment on the cup, as there is little variation in it.

c) The width of underband should be matched with the width of elastic band, so the double pointers of steel ring can cover the edge of elastic band without inside out. Thus the width of the underband is 1.8cm (1.3cm elastic band width plus 0.5 cm folding amount). On the basis of that, make parallel line on bottom surrounding curve and draw an arc at the 0.5cm upper lifting of underband for the purpose of beauty effect.

d) Measure the height of heart shape and side wing on the basis of newly drawn underband, which is 7.4cm and 11.8cm, respectively. Estimate the finished size is 7cm and 10.6cm, which meet the requirements of style design.

e) As the elastic bands selected for upper and lower bust are different, so too is the drawing force, with upper bust being less than lower bust. In order to have a balanced force, move the side seam line 0.5 forward, and connect the two points to draw a straight line

f) As the hook loop selects three line three buckles, with 5.4cm width, draw a smooth line at rear centre line 5.4cm distance away from bottom surrounding.

g) After 20% final shrinkage, we finally get the paper pattern of the bra (shown in Fig.4).

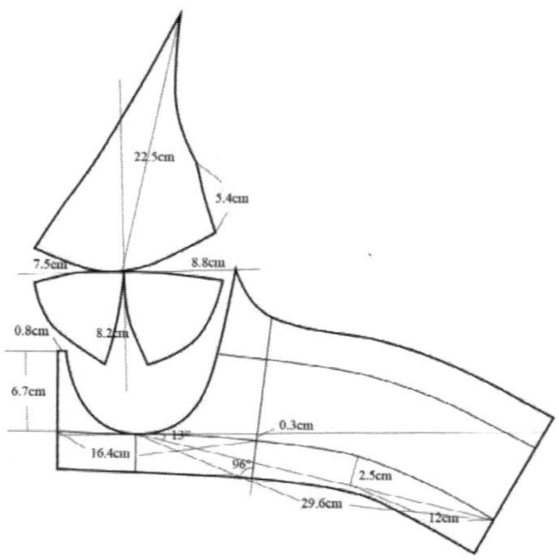

Fig.2 75B bra prototype developed by BIFT-Aimer Human Engineering Research Centre

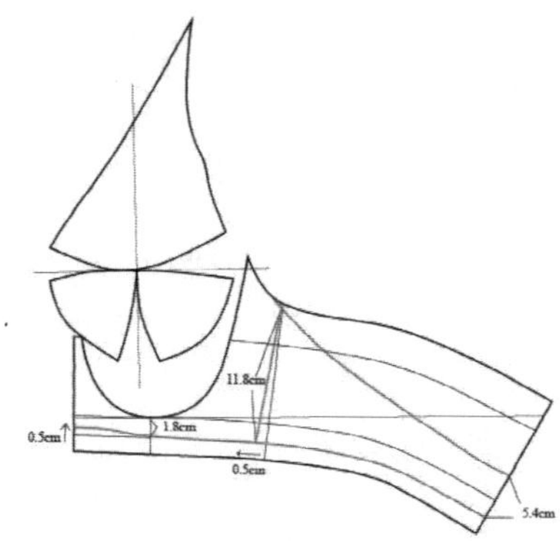

Fig.3 Pattern modification.

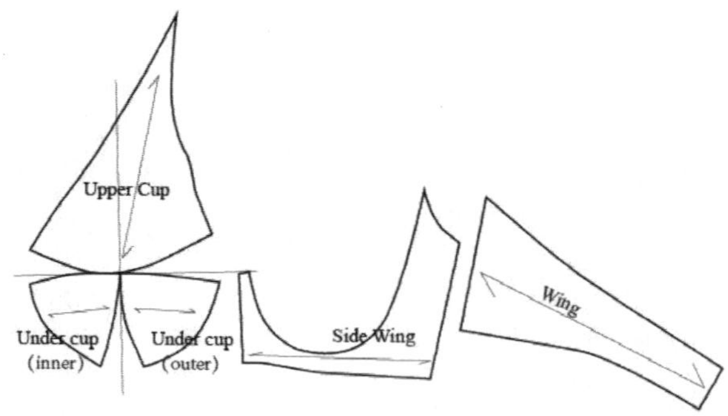

Fig.4 Final pattern.

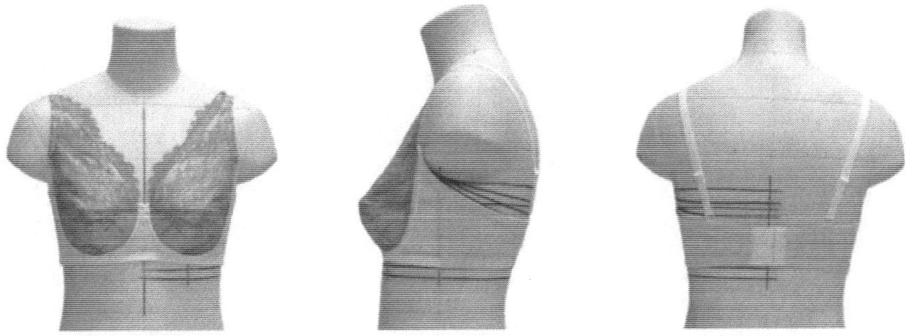

Fig.5 The sample bra

2.2.3 Bra sample

Six sample bras were produced based on the final paper pattern of the bra, shown in Fig.5.

3. Fitting evaluation experiment of the bra

3.1 Design of the experiment

The study recruited six female volunteers for the bra fitting experiment, aged 20 to 39 and with 75B bra size. Evaluation and analysis was performed on bra wearing effect, pressure comfort, and stability. Fig.6 shows the experiment procedure.

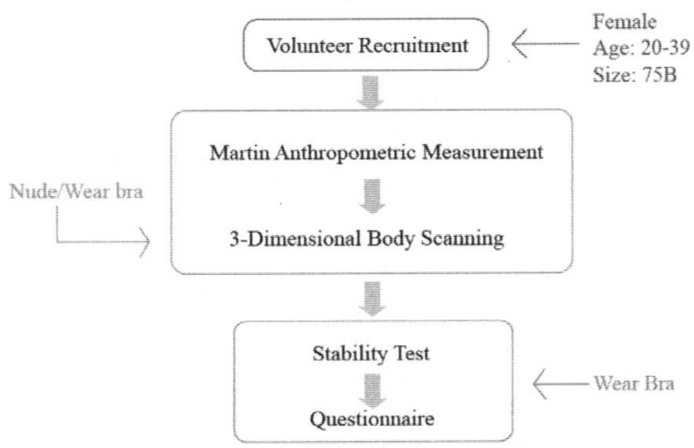

Fig.6 Experiment procedure.

3.1.1 Experiment environment

The experiment took place at the BIFT-Aimer Human Engineering Research Centre lab. The experiment environment included: temperature 25±2℃, air humidity 60%±10%, and wind speed less than 1m/s.

3.1.2 Sample

We recruited six volunteers between 20 to 39 years of age, with a 75B bra size for the fitting evaluation experiment. The basic information of the volunteers is shown in Table 3.

Table 3. Basic information of the subjects

Items	Sample	Range	Max	Min	Mean	SD	Variance
Height(cm)	6	11.70	169.30	157.60	163.17	4.70	22.08
Weight(kg)	6	14.70	63.10	48.40	54.95	4.97	24.74
BL(cm)	6	3.70	88.50	84.80	86.93	1.55	2.39
UBL(cm)	6	3.30	75.80	72.50	74.27	1.43	2.05
BMI(cm)	6	3.04	22.01	18.98	20.61	1.15	1.33

3.1.3 Experiment standard

The experiment used the following standards:

1) Standard posture: stand naturally; look vertically straight with a square body; open the arms at an angle of 30° with the body; maintain natural breathing during measurement.

2) Wearing status: volunteers wear flesh-tone underwear on the lower body, and two statuses of upper body nude and wearing bra. The first volunteer randomly selects a bra from the six bras, the second volunteer randomly selects a bra from the remaining five bras, and so forth for the other four volunteers.

3) Bra wearing standard: Hook loop is buckled on the middle line buttons. Wire is fitted on the bottom of the breast. Adjust shoulder strap to the point where volunteer feels most comfort with shoulder strap length. Keep bottom surrounding level. Adjust other parts until volunteer feels comfort and the bra is smooth and fit.

4) Perform Martin measurement and 3D scanning for the two statuses, respectively.

3.1.4 Measurement

The experiment selected 11 items for measurement of shaping effect of bra (see Fig.7). Martin measurement and 3D scanning measurements were performed under nude and wearing-bra status, respectively. Measurement methods are defined in Table 4.

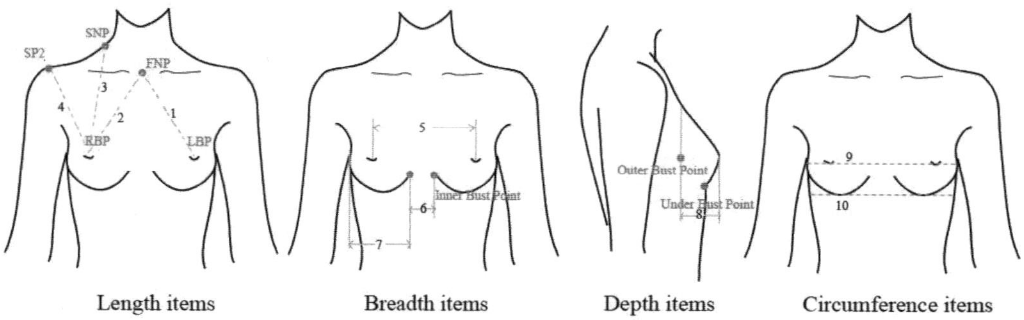

Fig.7 Schematic diagrams of the measurement.

Table 4. Definition of basic measurement of the items

No.	Measurements	Measurement methods
1	FNP~LBP	Length from front neck point to left bust point
2	FNP~RBP	Length from front neck point to right bust point
3	SNP~RBP	Length from side neck point to right bust point
4	SP2~RBP	Length from shoulder point to right bust point
5	Breast pitch	Straight-line distance between breast points
6	Bust inner side diameter	Vertical distance from bust inner points
7	Right bust breadth	Vertical distance from right bust inner point to right bust outer point
8	Right bust depth	Vertical distance from RBP to cross-section of inner bust point to outer bust point
9	Bust circumference	Vertical length through right bust point
10	Under bust circumference	Vertical length through right under bust point
11	Bust difference	Value of bust circumference minus under bust circumference

3.2 Results and discussion

3.2.1 Shaping effect

To verify if the bra had good shaping effects, the study first computed the measuring value of wearing bra, then subtracted the corresponding measuring value of nude. Negative values shown in Table 5 indicate the measuring value of wearing bra decreased compared with the measuring value of nude. After wearing bra: the length of FNP~LBP, FNP~RBP, SNP~BP, and SP2~BP all decreased, indicating the bra has a lifting effect; both breast pitch length and breast breadth decreased, indicating the bra has a gathering effect; the width of breadth decreased but the depth increased, indicating the bra can make the breast more solid; bust girth and under bust girth decreased a little, but the D-value increased, indicating the bra has a good shaping effect.

In addition, vertical cross-section and horizontal cross-section for human body in nude and wearing bra status were extracted from 3D measurement software, and overlaid on each other. As shown in Fig.8, red represents nude status, and blue represents bra wearing status. For BP displacement in horizontal and vertical direction, positive value indicates the bust point moved downward and towards the side of the body. Fig.8 and Table 6 show that, compared with nude status, the bust points of bra wearing status will both move to middle and upper, which verifies the results of Martin that the bra has lift and gathering effects.

The volunteer testers performed a satisfaction evaluation on shaping effects of wearing bra. The evaluation was divided into five grades, with the two ends of the scale being "dissatisfaction" (1) and "satisfaction" (5). After processing the results with normalised analysis, scores higher than 4 indicated the bra is stable during simple movements and, as shown in Table 7, we could determine that the wearing effects of the bra is good.

Table 5. Difference value (measurements of wearing bra minus measurements of nude) (n=6; unit: cm)

Measurement items	Range	Maximum	Minimum	Mean value	Variance	SD
1. FNP~LBP	-2.00	-1.30	-1.80	-2.10	-1.70	-2.00
2. FNP~RBP	-1.50	-1.00	-2.10	-1.00	-2.20	-1.50
3. SNP~RBP	-1.90	-0.20	-1.90	-0.50	-1.80	-0.80
4. SP2~RBP	-1.00	-0.40	-0.30	-0.40	-0.30	0.50
5. Breast pitch	-3.00	-0.90	-4.50	-4.90	-3.70	-3.60
6. Bust inner side diameter	-0.60	-0.80	-1.50	-0.80	-1.20	-1.50
7. Bust breadth-1	-2.50	-1.20	0.60	-1.80	-1.70	-1.20
8. Bust depth-1	0.30	0.50	0.70	0.50	0.50	1.10
9. Bust girth	-1.00	-3.30	-1.00	-1.30	-0.20	-2.50
10. Under bust girth	-3.30	-2.50	-2.30	-3.00	-1.50	-2.00
11. BL-UBL	2.30	-0.80	1.30	1.70	1.30	-0.50

Table 6. BP displacement (unit: cm)

	Range	Maximum	Minimum	Mean	Variance	SD
X axis	3.00	-3.42	-0.42	-1.94	0.86	0.93
Y axis	7.94	-8.62	-0.68	-4.25	9.51	3.08

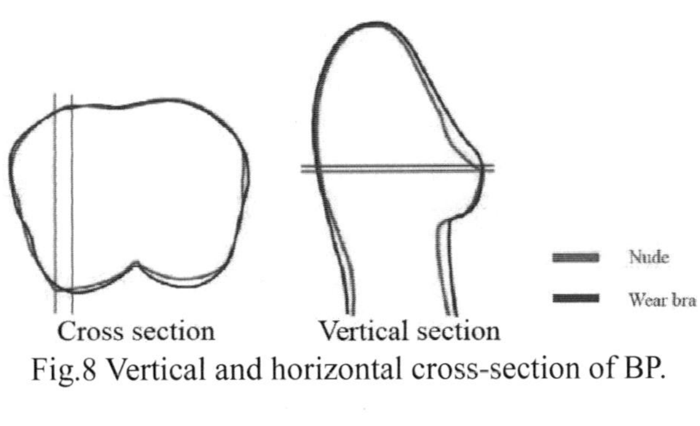

Fig.8 Vertical and horizontal cross-section of BP.

Table 7. Subjective shaping effect evaluation

Shaping effect	Value
Push-up effect	3.85
Push-in effect	3.24
Cup containing effect	4.40
Underarm containing effect	4.49
Overall	4.02

In conclusion, the bra has an overall positive wearing effect, and can lift and gather the female breast, as well as beautify the shape of the breast.

3.2.2 Stability evaluation

The study selected 8 testing points (see Fig.9). We marked the testing points and performed 3D scanning after the volunteer tester donned the bra properly, turned the body to the limit eight times, lifted arms to upright position eight times, and bent down to the limit eight times before recovering to normal status, where we again performed 3D scanning. Finally, we measured and recorded the points displacement in x and y direction in 3D software.

By computing the average displacement of testing points, we determined: displacement of A point on body surface; displacement of B, C, D, F, and G in x and y direction; and displacement of E1 and E2 in y and z direction. We performed an analysis and summary on displacement of each point (see Table 8); after simple movement, the bra is moved upward in a whole, accompanied with certain retraction, but from the value we can see the displacement is very small and close to 0cm.

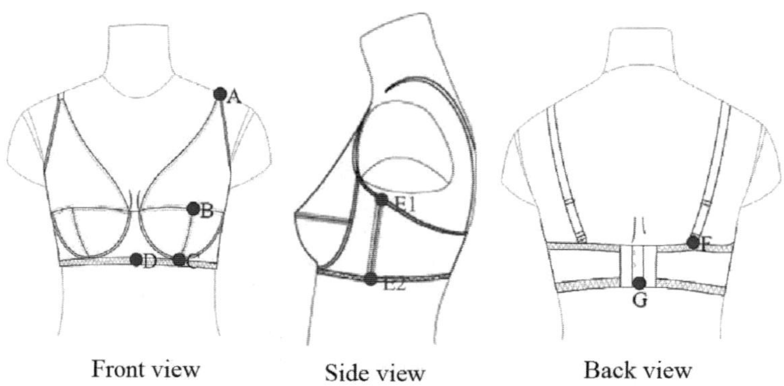

Fig.9 Stability test points

Table 8. Test point displacement (unit: cm)

Part	Number	Moving Direction		Average Displacement
Shoulder strap point	A	Towards shoulder point	↙	0.52
BP area	B	Towards centre front & Up	↳	(-0.11,-0.11)
breast bottom point	C	Towards centre front & Up	↳	(-0.28,-0.35)
Centre underband point	D	Up	↑	0.04
side seam upper point	E1	Up & Front	↳	(-0.23,-0.13)
side seam lower point	E2	Up & Front	↳	(-0.28,-0.57)
Back strap loop	F	Towards centre back & Up	↲	(-0.14,-0.18)
Back hook	G	Up	↑	-0.32

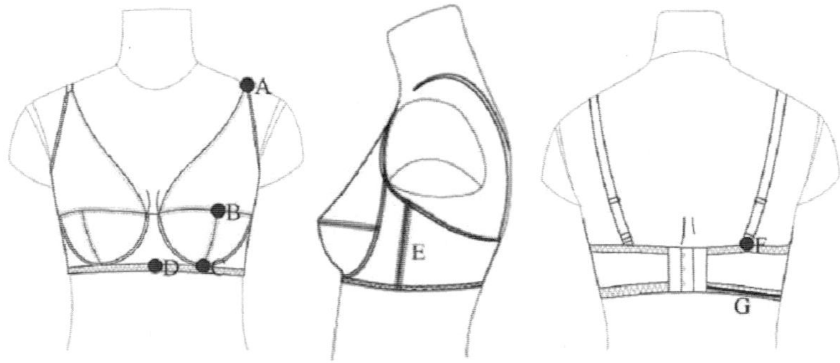

Fig.10 Schematic diagram for subjective stability evaluation

Table 9. Subjective stability evaluation value

Position	Value
Shoulder strap	5.50
BP point	5.44
Wire lowest point	5.17
Underband centre point	5.55
Side seam	5.27
Back strap loop	5.63
Back underband	4.77
Overall	5.16

The volunteer testers performed subjective evaluation on each part during the movement process. There were eight evaluation parts (see Fig.10), and the evaluation standard adopted a 7-point scale, with 1 being "very unstable" and 7 being "very stable." After processing with normalised analysis (see Table 9), scores greater than 4 indicated that the bra is stable during simple movement, showing that the style of the bra has good stability.

In conclusion, the measurement for each part of the bra has some variation. These variations cannot be avoided, and are very small (close to 0cm). Combined with subjective evaluation results, we can say the style of this bra has good stability.

3.2.3 Pressure comfort

The volunteer testers evaluated 11 parts for pressure comfort during the wearing process (see Fig.11). The evaluation was divided into subjective press evaluation and subjective pressure comfort evaluation; the evaluation standard also employed a 7-point scale.

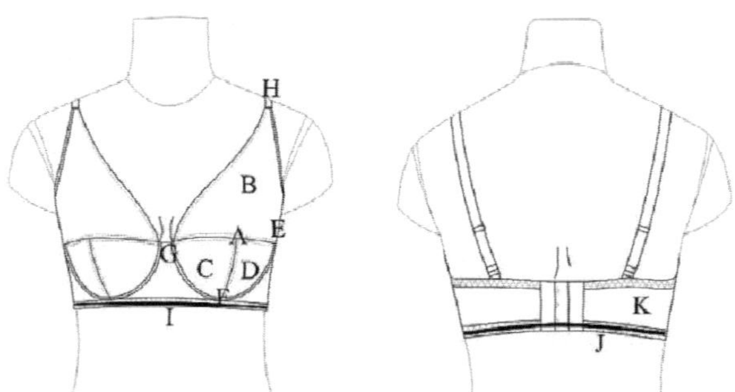

Fig.11 Schematic diagram for subjective pressure evaluation and pressure comfort evaluation

1) Subjective pressure evaluation

The 7-point scale for evaluation of bra subjective pressure ranged from "No pressure" (1) to "overpressure" (7). After processing with normalised analysis (Table 10), scores less than 4 indicated the pressure of the bra is very weak, showing that the bra has a weaker pressure.

Table 10. Subjective pressure evaluation score

Position	Value
BP area	2.45
Upper cup	2.12
Inner lower cup	2.93
Outer lower cup	2.97
Wire edge point	3.13
Wire lowest point	3.63
Junction	3.50
Shoulder strap	3.78
Front underband	3.86
Back underband	4.00
Back wing	3.47
Overall	3.50

2) Subjective pressure comfort evaluation

The 7-point scale for evaluation of bra subjective pressure comfort ranged from "Very uncomfortable" (1) to "Very comfortable" (7). After processing the results with normalised analysis (see Table 11), scores greater than 4 indicated the bra wearing is very comfortable, which shows the pressure comfort of the bra tends to be comfortable.

In conclusion, the evaluation of the subjective pressure comfort for the bra is good, indicating the pressure on the parts is reasonable. From this we can see the style of this bra has good pressure comfort.

Table 11. Subjective pressure comfort evaluation value

Position	Value
BP area	5.99
Upper cup	6.37
Inner lower cup	5.64
Outer lower cup	5.64
Wire edge point	5.32
Wire lowest point	5.56
Junction	4.96
Shoulder strap	5.16
Front underband	4.50
Back underband	5.19
Back wing	5.34
Overall	5.48

4. Conclusions

We arrived at the following conclusions through evaluation and analysis of the bra wearing test:

1) When wearing the bra, the length of FNP~LBP, FNP~RBP, SNP~RBP, SP2~RBP, Breast pitch, Breast inner point distance, and Breast breadth all decreased; in addition, Breast depth increased, and bust point moved toward upper and front centre. These data indicate that this full cup bra has good push-up and push-in effects.

2) Through simple movement after wearing bra, we can see that the bra moves upward and has certain retraction, but the values are all close to 0cm. We also found the overall subjective evaluation values tend to be stable, thus we can say this bra has good stability.

3) After evaluation for pressure and pressure comfort when wearing bra, we can see the pressure of the bra is

very low and at the same time it has good pressure comfort.

References

[1] Hardaker CHM, Fozzard GJW. The bra design process - a study of professional practice. Int J Cloth Sci Tech: 1997; 9(4): 311-325.

[2] Bowles KA, Steele JR, Munro B. What are the breast support choices of Australian women during physical activity? Brit J Sport Med: 2008; 42(8): 670-673.

[3] Eden KB, Valiant GA, Jawson LJ, et al. Three dimensional kinematic evaluation of sport bra design. Med Sci Sport Exer: 1992; 24(1): 81-87.

[4] Li SX, Wang JP. Research on the influence of wing tension on bra pressure and stability. Shanghai: Donghua University, 2013.

[5] Duan XY, Yu WD. On the factors affecting bra wearing comfort. J Jiangsu Teachers U Tech: 2007; 13(4): 63-19.

[6] Wang YP, Zhao P, Gao XS, et al. Objective test and evaluation on pressure comfort for brassiere. J Text Res: 2006; 27(11): 90-93.

[7] Miyatsuji A, Matsumoto T, Mitarai S. Effects of clothing pressure caused by different types of brassieres on autonomic nervous system activity evaluated by heart rate variability power spectral analysis. J Physiol Anthropol Appl Human Sci: 2002; 21(1): 67-74.

[8] Morooka H, Fukuda R, Nakahashi M, et al. Clothing pressure and wear feeling at under-bust part on a push-up type brassiere. Sen-I Gakkaishi: 2005; 61(2): 55-60.

Study on the Application of Zero Waste Design in New Chinese-Style Fashion Brands

Yan Yuan, Chen Yang, Ping-Ping Zhang, Zhu Cong-Tian

Jin Hua Road, Xi'an, Shaanxi, 710048, China

Corresponding author's email: yuanyan2003@126.com

Abstract

Focused on the concept of Zero Waste Design, this article analyzed the origins and aesthetic characteristics of Zero Waste Design. As a kind of creative design approach, it brought the new thinking and new aesthetics into fashion design. For the new Chinese-Style fashion brands, Zero Waste Design provided an impact and designed from the view of minimalism, the Wabi-Sabi and Daoism philosophy. As a result, this article explored and summarized its application in New Chinese-style Fashion Brands by experimental practices of Zero Waste Design.

Keywords: Zero Waste Design, New Chinese-Style, Fashion Brands

1. The concept of Zero Waste Design

Originally, Zero Waste Design is mainly referred to as no waste produced during the process of clothing manufacturing, especially in pattern cutting and marker making. Now, it could also be treated as a creative approach to do research and design. It is a way of thinking that pays attention to both the 2-dimensional and 3-dimensional aspects of fashion design.

2. The origins of Zero Waste Design

2.1 The original way of dressing in ancient time

As an effective way of pattern cutting and marker making, Zero Waste Design can be traced back to its origins in the ancient time. You can find similar clothing's construction in both western and eastern countries.

From the one-piece dress in Ancient Rome and Ancient Greek to the oriental country's clothing, they were all made in a similar way. All of them came from one whole piece of fabric without any or less cutting. The original ways of dressing in ancient time were summarized into three main ways-wrapping (Figure 1), cutting holes (Figure 2) and sewing into a tubular [3]. By these ways, the clothing presented the flat and plain shape in 2-dimensional pattern and the elegant and magic silhouette in 3-dimensional vision at the same time.

Figure 1 showed Ancient Greek's women wear-Chiton and Peplos. It was made by folding a piece of fabric and fixing it on the shoulder and waist. Figure 2 showed the Japanese Kimono. The basic shape of a Kimono's pattern also presented a square piece. Figure 3 showed the Chinese traditional clothing and the corresponding construction. When you flatten all of these clothing, you will find the similar geometric shape with regular cutting and less waste. Above all, these main forms of clothing, regardless of the time or country it is from, presented the basic shape in 2-dimensional pattern and can be considered as the origins of the Zero Waste Design.

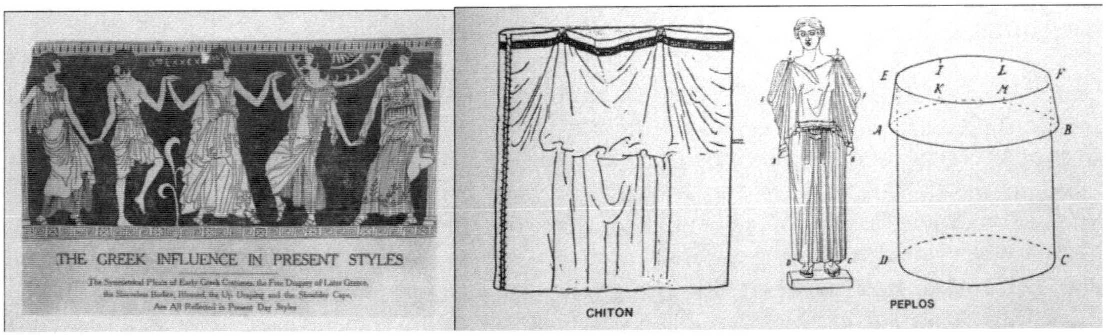

Fig.1 Ancient Greek's women wear-Chiton and Peplos [1]

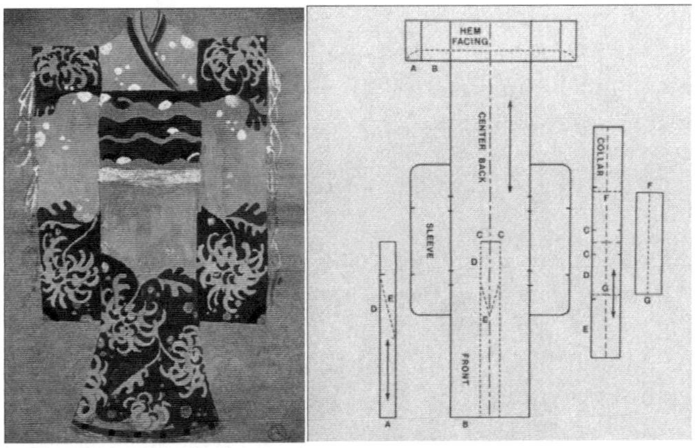

Fig.2 Japanese Kimono and 2D construction [2]

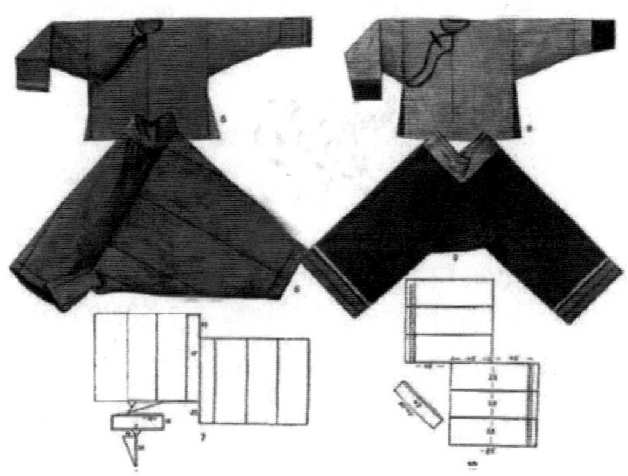

Fig.3 Chinese tradional clothing and 2D construction [3]

2.2 Madeleine Vionnet's Creative Bias Cutting

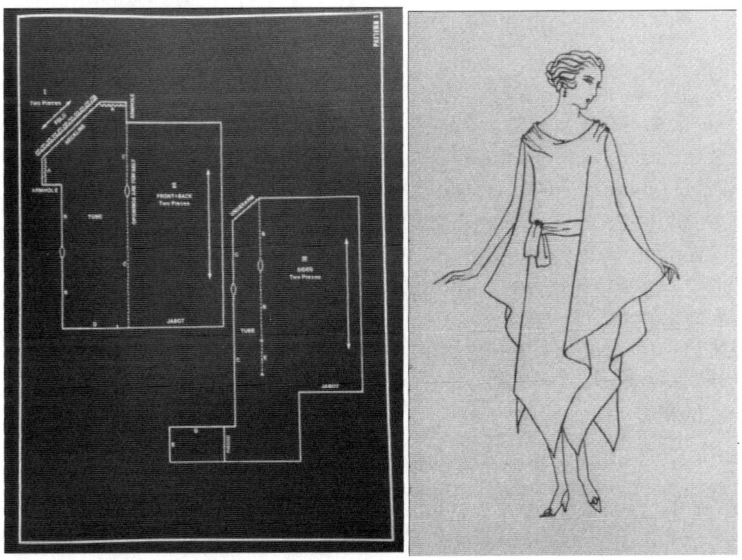

Fig.4 Madeleine Vionnet's dress and 2D construction [3]

Madeleine Vionnet was the creator and inventor of bias cutting and had explored a lot of magical works with the technology of bias cutting. At the same time, almost all her works took on the simple shape in a 2-dimensional pattern but very elegant silhouette in a 3- dimensional vision. When you noticed the bias cutting work of Madeleine Vionnet, you cannot find the normal neckline or armhole line in the 2-dimensional pattern. The normal pattern shapes of the body or sleeves were replaced by the regular or irregular geometric shape.

Madeleine Vionnet's research had moved the original way of dressing further and be full of magic (Figure 4). This method of bias cutting was another higher stage version of origins of Zero Waste Design. This idea has given a strong impact on a lot of fashion designer's work.

3. The Aesthetic Characteristics of Zero Waste Design

How to evaluate the aesthetic of Zero Waste Design? Ordinary clothing would be evaluated when it was dressed on and fit the body shape. That was the normal aesthetic standard for normal clothing. Different from this standard, Zero Waste Design was evaluated by a lot of aspects, from the 2-dimensional pattern to the 3-dimentioanl vision, from the idea to the silhouette, from the style to the sustainable significance. In this sense, the aesthetic characteristic of Zero Waste Design could be summarized into several points as follows.

3.1 The conflict and drama between the plain effect of 2-dimensional pattern and the elegant effect of 3-dimensional vision

The 2-dimensional pattern shapes of Zero Waste Design's clothing were normally geometric shapes such as square, triangle, round and so on. From the biggest part, such as body piece, to the smallest part, such as the pockets or collars, all of them took on the regular or irregular geometric shape. Almost every small piece will not be wasted and every piece has its function. This will be an organic component of the clothing. They just looked like the seven-piece puzzle and were integrated into the clothing under some special rule.

3.2 Minimalism and reduction

Fig.5 Tonysmith, freeride, sculpture [7]

As a modern art movement in sculpture and painting that began in the 1950s, minimalism emphasized on extreme simplification of form and color (Figure 5). "Founded as a backlash against abstract expressionism, minimalism was characterized by simplified, stripped-down forms and materials used to express ideas in a direct and impersonal manner. By presenting objects as simple objects, minimal artists sought to communicate without referring to expressive or historical themes. This critical movement branched out into land art, performance art, and conceptual art, and was still a major influence today." [7]

As a lifestyle and fashion trend, minimalism became more and more acceptable and popular in the 1990s. "Back to basics"was the main slogan at that time. With the lead of this trend, the fashion designer needed to do subtraction from the basics. This subtraction could be mainly referred to the cutting way or the sewing way from the construction. How to make the clothing without too many seams or with less cutting, or less sewing, was a very interesting idea that was worth exploring from different ways. At the same time, minimal color and minimal design, so there will not even be a button in the clothing (Figure 6).

In this sense, minimalism explored the designer's creativity as much as possible and will bring Zero Waste Design endless creative vision.

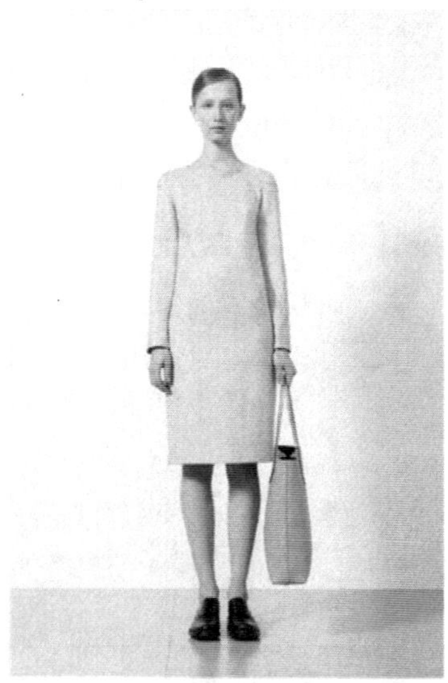

Fig.6 Jil Sander 2011aw [8]

3.3 Sustainable design

As we all know, for a lot of apparel industries, industrial wastes were produced during the process of manufacturing. Inefficiency of pattern cutting and marker making were the main reasons. From a sustainable point of view, Zero Waste Design can solve this problem by smart pattern design and marker making. Figure 7 showed the Zero Waste Design from Timo Rissanen in 2008, which was inspired by the Vionnet's cutting. The pattern of this hoodie was a square and every piece was the component of the hoodie. The construction was very smart and ingenious.

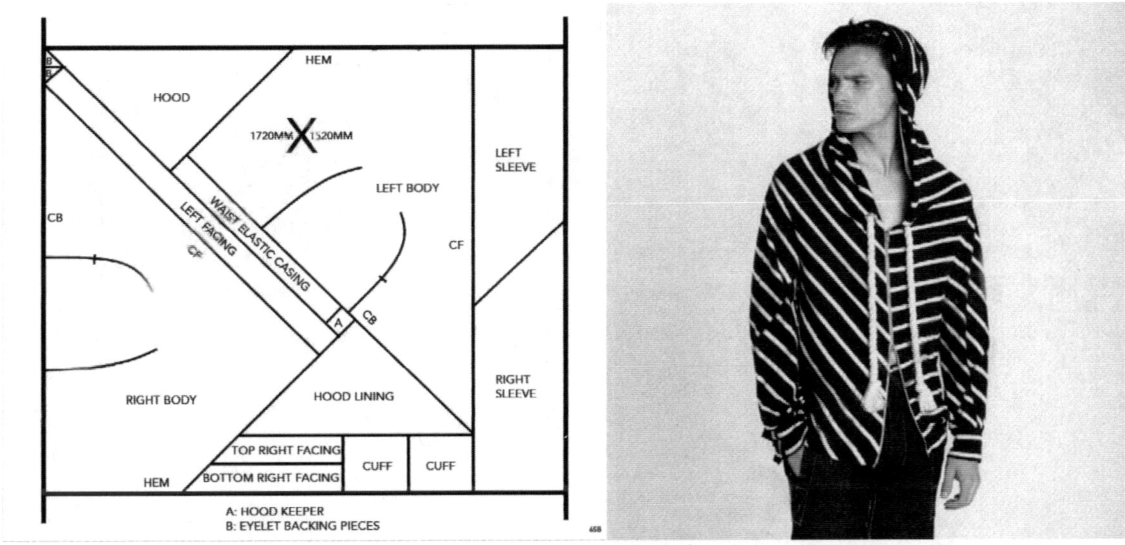

Fig.7 Hoodie and jeans by Timo Rissanen (2008) [9]

4. Application of Zero Waste Design in Fashion Brands

Just as mentioned above, fashion designers were interested in how to apply the method of Zero Waste Design more effectively and its important role in fashion brands. Can we develop smarter designs from Zero Waste Design? Can we help some enterprises treat their industrial waste or make more fashionable things by Zero Waste Design? Can we reuse this method to rebuild the spirit of the New Chinese-Style Fashion Brands? Here are some examples of experimental design from Zero Waste Design in fashion brands.

4.1 Zero Waste Design provided a creative way of thinking to explore the experimental idea and multiple possibilities for fashion brands

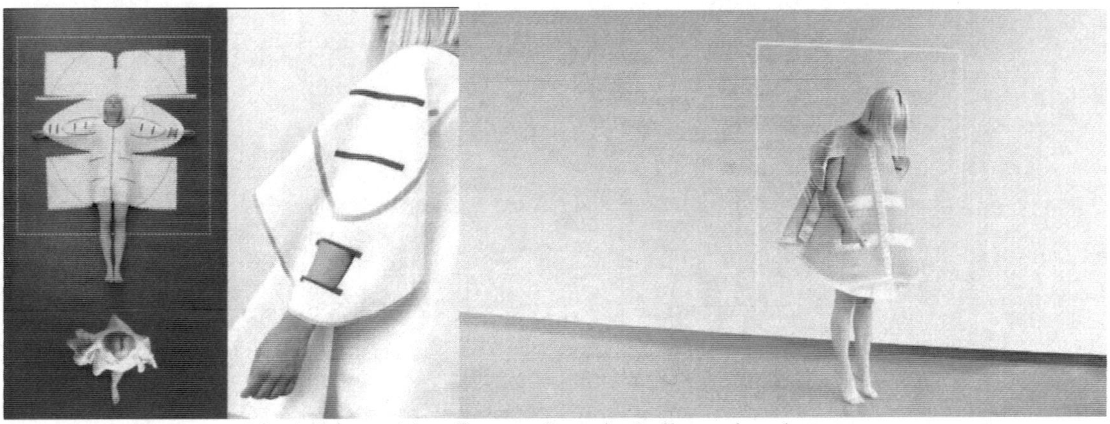

(a) A dress with a flower shape in 2-dimensional pattern

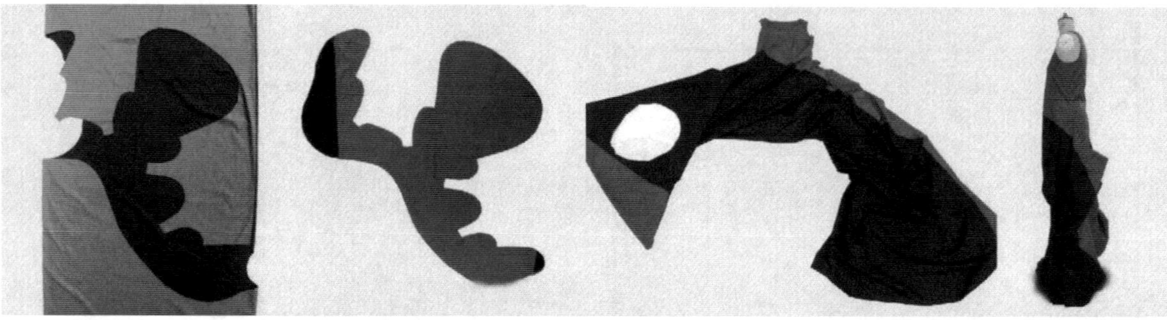

(b)The body pieces of the dress were placed against the normal way and were connected together

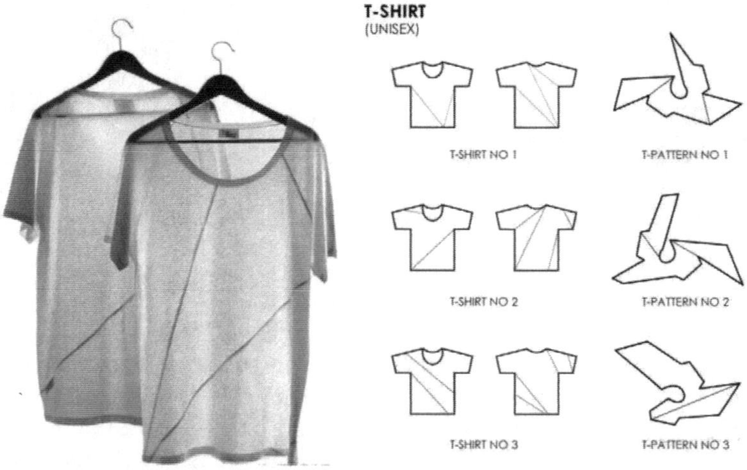

(c) Normal shape T-shirt with very strange 2-dimensional pattern

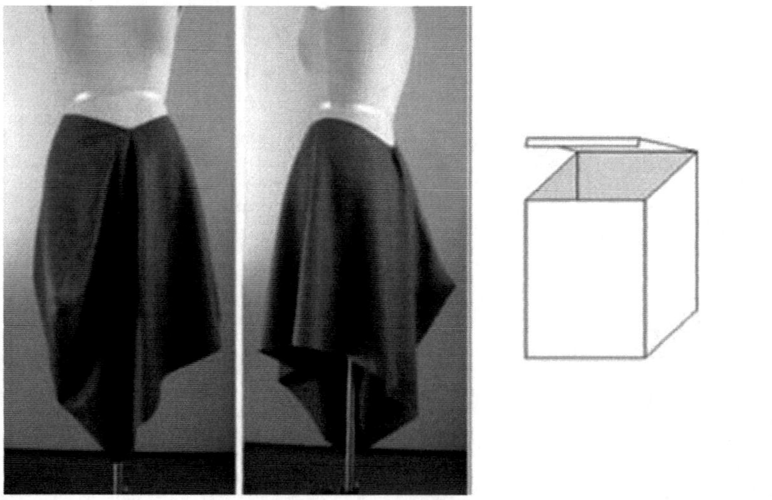

(d) 2005 aw Boundless

Fig.8 Creative way of thinking to do the experimental designs (a) (b) (c) (d)

The designs shown in figure 8 were very interesting and fascinating.

(a) A dress with a flower shape 2-dimensional pattern; The human body resembles that of a small toy with stickers on it.
(b) The body pieces of the dress were placed against the normal way and were connected together. When it was dressed on the body, the silhouette of this dress will be very strange and unique.

(c) Experimental design of T-shirt. "We are fascinated by the triangular polygon, a geometric shape we see as the basic unit of digital substance and wearable form. All our products are made in a new digital environment that allows us to reconstruct every piece of clothing from the core to any degree of complexity. With this new approach we establish a new dimension in which 2D, 3D and the space in between is reconsidered and experimented with." [10]
In doing so, we stir triangular shapes and fine fabrics into basic apparel that subtly exposes its digital descent. As the possibilities in polygonal shaping are unlimited, we also engage in the conception of extensive art pieces to shed light on the extremities. Our ultimate mission is to lift the hurdle that separates the possible and the impossible, and redefine the aesthetics of all things jersey."[10]
(d) This was a skirt design from Boundless 2005aw. The designer of Boundless is Zhang Da. He got the inspiration from a box and used this construction to build the skirt, giving it a more 3-dimensional vision.

All of these wonderful designs provided a unique view to observe clothing construction from the flat way. This was also the main core spirit of Zero Waste Design and could be a way of thinking that lead the designers to construct clothing and to establish the fashion brands' style from an experimental view.

4.2 Zero Waste Design provided magic silhouette and style to the clothing that was helpful for fashion brands to establish its profound philosophy

4.2.1 Zero Waste Design got the inspiration from Daoism and Wabi-Sabi

To be looked from a superficial view, Zero Waste Design could be treated as a kind of technology in construction designing and marker making. Actually, it embodied profound philosophical meaning. It got inspiration from the silhouette from Taoism and got the fabric texture feeling from Wabi-sabi.

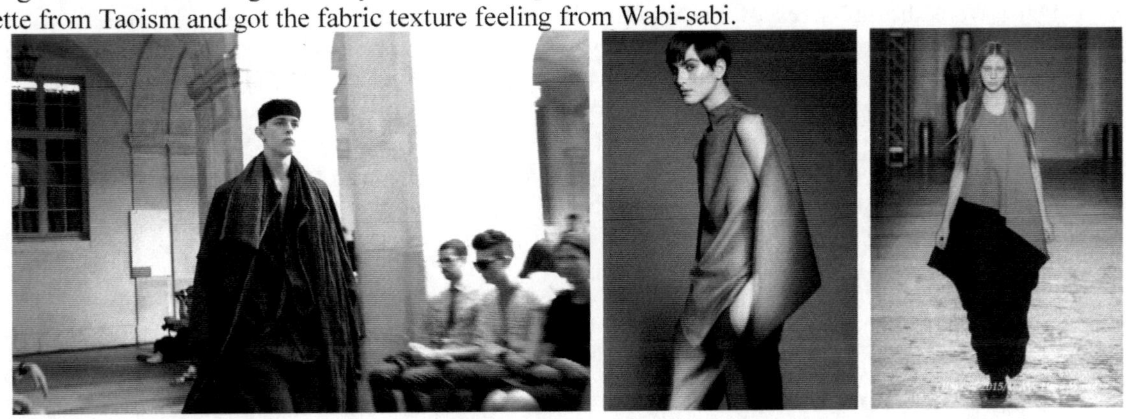

Fig.9 Damir Doma (Left 1/2) and Uma Wang's design(Right)

These two genres of philosophy addressed the different aesthetics standards.

1) Taoism's wu wei and the harmony between heaven and men.

Taoism addressed the emptiness, wu wei, unity of heaven and men and following the nature's rule. "Taoism is a philosophical, ethical or religious tradition of Chinese origin that emphasizes on living in harmony with the Tao." [11]

"The term wu-wei constitutes the leading ethical concept in Taoism. Wei refers to any intentional or deliberated action, while wu carries the meaning of "there is no ..." or "lacking, without". Common translations are "nonaction", "effortless action" or "action without intent" [11].

Taoism does not identify one's will as the root problem. Rather, it asserts that one must place their will in harmony with the universe. Thus, a potentially harmful interference may be avoided, and in this way, goals can be achieved effortlessly." [11]

From this point of view, Taoism's wu wei can be translated into less design and less cutting in clothing. Harmony could be translated into the free feeling of clothing when it was dressed on the body, and the harmony among the people, the clothing and the surroundings.

2) Wabi-sabi's incomplete, imperfect, plain and rough feeling.

"The words wabi and sabi do not translate easily. Wabi originally referred to the loneliness of living in nature, remote from society; sabi meant "chill", "lean" or "withered". Around the 14th century, these meanings began to change, taking on more positive connotations [12]. Wabi now connotes rustic simplicity, freshness or quietness, and can be applied to both natural and human-made objects, or understated elegance. It can also refer to quirks and anomalies arising from the process of construction, which add uniqueness and elegance to the object. Sabi refers to beauty or serenity that comes with age, when the life of the object and its impermanence are evidenced in its patina and wear, or in any visible repairs." [12]

From an engineering or design point of view, wabi may be interpreted as the imperfect quality of any object, due to inevitable limitations in design and construction especially with respect to unpredictable or changing usage conditions; then sabi could be interpreted as the aspect of imperfect reliability, or limited mortality of any object.

The fashion brands, such as the Damir Doma (Left 1/2 in Figure 9) and Uma Wang (right in figure) had explored this philosophy from different points of view and took on a different look. Damir Doma addressed the free and plain design. Uma Wang paid more attention to the texture of the fabric with very loose silhouette. All of these two designers' works have a similar construction of Zero Waste Design.

4.2.2 Zero Waste Design provided valuable insights into the New Chinese-style fashion brands

Just as mentioned above, Chinese-style was characterized by the 2-dimentional method to construct the clothing. The New Chinese-Style fashion brands mainly refer to the fashion brands emerged in recent years and established their brand idea based on the oriental philosophy. We try to seek for ways to apply the Zero Waste Design concept in new Chinese-Style Brands.

Figures 10,11,12 were the practices of Zero Waste Design of virtual new Chinese-Style fashion brands of Edgeviewstudio. These three designs were kept in very plain construction. The dress in Figure 10 addressed its shape from one side of the body and took on an unbalance feeling, just like a distorted and transformed dress. The dress in Figure 10 showed its shape in the front of the body; the dress in figure 11 displayed the three holes in the body piece. It will take on a different style when dressed differently.

These practices explored the possibilities for the new Chinese-style fashion brands by applying the concept of Zero Waste Design into the construction , the style and the silhouette.

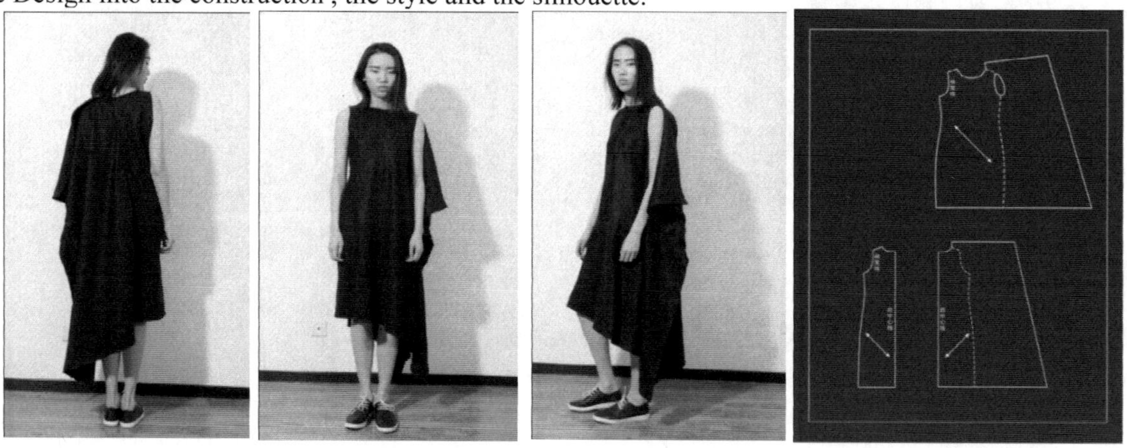

Fig.10 Zero Waste design work I from Fashion Brand Edgeviewstudio (Designed and photographed by author)

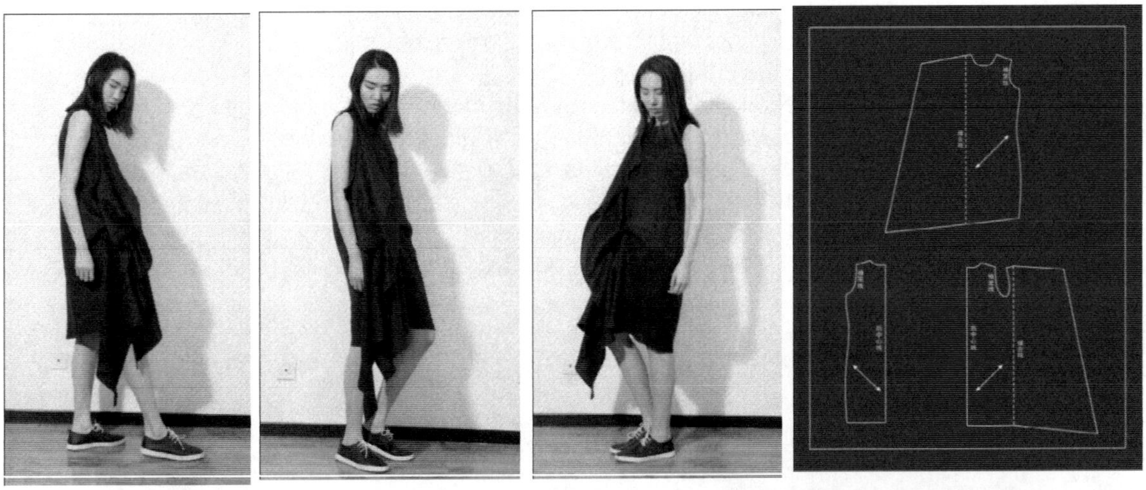

Fig.11 Zero Waste design work II from Fashion Brand Edegeviewstudio (Designed and photographed by author)

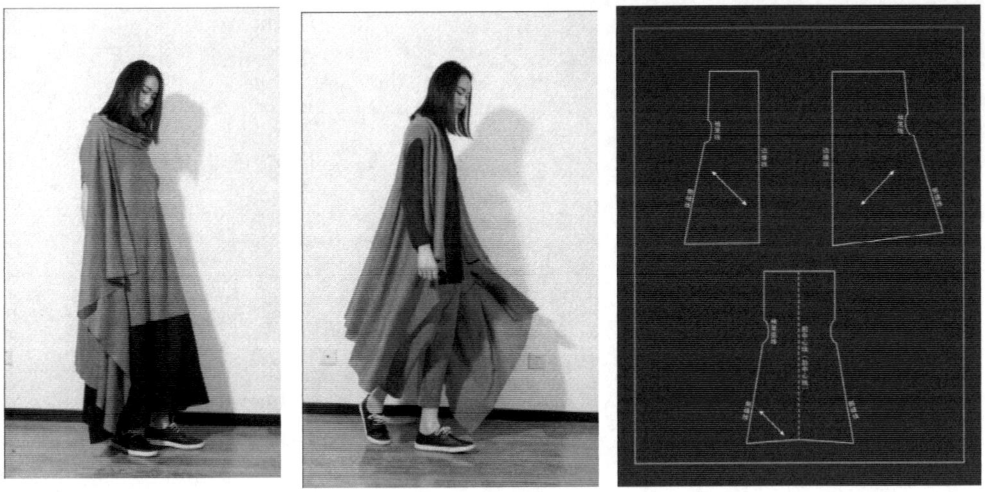

5. Conclusion

Zero Waste Design was an ancient way of dressing in both western and eastern countries.

1) As an ancient way to construct clothing, Zero Waste Design provides a feeling of the oriental style to construct the new Chinese-style aesthetics.

2) Zero Waste Design was developed into further studies by Madeleine Vionnet. Since she invented bias cutting and this technology brought a strong impact on fashion designers, such as Issye Miyake, Yohji Yamamoto, Damir Moma and Uma Wang.

3) As a creative way of thinking to design, Zero Waste Design provided a new point of view to view clothing and to design and redefine the aesthetics of Zero Waste Design.

4) As a sustainable way, Zero Waste Design provided an effective way to reduce resources used during the process of production.

Above all, it was a valuable theme to rebuild the style and construction of clothing, and to redefine the aesthetic value for new Chinese-style brands from Zero Waste Design's point view. At the same time, this technology brought us a new angle to view clothing design and lead us to more experimental and conceptual designs in the future.

Referances

[1] Betty Kirke. Madelein Vionnet [M]. Tkoyo:Kturyudo Art Publishing Co.LTD: 1991(41-42).

[2] Betty Kirke. Madelein Vionnet [M]. Tkoyo:Kturyudo Art Publishing Co.LTD: 1991(46).

[3] Chiyoko Tanaka, Li DQ [Translator]. World Natianality Costume-Exploring the Wisdom of the human dressing way [M]. Beijing: China Textile Publish House: 2001.
[4] Betty Kirke. Madelein Vionnet [M].Tkoyo:Kturyudo Art Publishing Co.LTD: 1991 (46-47).
[5] Timo Rissnen, Holly Mcquillan. Zero Waste Fashion Design [M].Fairchilds books (14).
[6] Daniel Marzona. Minimal Art [M]. TASCHEN America Lic: 2009.
[7] https://en.wikipedia.org/wiki/Minimalism.
[8] www.jilsander.com.
[9] Timo Rissnen, Holly Mcquillan. Zero Waste Fashion Design [M].Fairchilds books (85-86).
[10] https://www.douban.com/note/189745757/?type=like.
[11] https://en.wikipedia.org/wiki/Taoism.
[12] https://en.wikipedia.org/wiki/Wabi-sabi.

Research on the Basic Pattern of Bras on Medium-Sized Mannequin Used for Underwear Design

Jing Zhao[1], Hao Zhang[1*], Rong Zheng[1], Hong-Yu Zhang[2], Xiao-Juan Duan[2], Hai-Jing Liu[1], Ke Liu[2]

[1]*Aimer Human Engineering Research Centre, Beijing Institute of Fashion Technology, Beijing, 100029, China*
[2]*Beijing Aimer Lingerie Co., Ltd., Beijing, 100102, China*

[*]Corresponding author's email: zhhh1cn@126.com

Abstract

Current bra patterns are designed primarily using empirical methods based on standard mannequins. Because of inconsistencies in patterns, bra manufacturers have an urgent need for statistical data on a basic bra pattern that can be applied practically. This study designed a bra based on a medium-sized mannequin used for underwear design; it employed a Japanese paper copying method to obtain the 3D pattern of bra, and convex point cutting to cut the 3D pattern to obtain patterns of three types of cups and two types of under band and wings. Through testing and analysis, an "optimum" pattern was selected and used to make a grey cloth bra sample using current production practices. Upon verification, the basic bra pattern proved to be reasonable and can be used as the basic pattern for manufacturers making the actual bra pattern.

Keywords: Basic bra pattern; Medium-sized mannequin used for underwear design; Japanese paper; Sample

1. Introduction

The bra is an undergarment with a complex structure designed to support a woman's breasts. Currently manufacturers design bra patterns primarily using the empirical method without a standard theoretical basis, rules, or specifications. However, both research and production practices suggest that the most ideal bra pattern should be designed based on structural changes of the basic bra pattern [1,2], and the basic pattern design is very important for the design of the whole bra pattern.

Since 1968, international manufacturers have used a drawing system for bra products [3]; the pattern is mainly drawn based on a customized bodice basic pattern; with margin removed and darts multiplied, the structure that fits women's breasts is formed. In China, there still is not a systematic drawing method of a basic bra pattern, and three methods are generally used - Size Defining Method, Size Defining-Formula Method, and 3D-Plane Method. The former two are the ones mostly used in manufacturing, but both have limitations: it is difficult to use the first one to realize style changes due to indeterminacy of data source and drafting method, while the second one will lead to poor form-fitting effect as it is based on the ordinary shirt basic pattern. By contrast, the 3D-Plane Method remedies deficiencies of the other two methods and should be the mainstream method used for the basic bra pattern design. However, when the method is used, the demand for operator's techniques and use of a mannequin is high, thus it is not popular amongst manufacturers.

The mannequin is an important tool for ensuring the form-fitting effect of clothes during the industrial production process [5] and reasonable clothing patterns can only be tailored based on suitability of the mannequin. Most bra patterns are obtained through tridimensional tailoring based mainly on standard mannequins that in turn are based on data corresponding to the naked body, without any shaping processing. Such mannequins cannot accurately present the form (s) of bodies that wear the bra, thus there are some differences in the sizes of bra patterns.

This study aimed to design and research basic bra patterns using a 3D-Plane Method based on a medium-sized mannequin used for underwear design and to create a basic bra pattern that can be used for style changes and production practice and provide methodological support and theoretical guidance for the design of the basic bra pattern.

2. Design and development of basic bra pattern

In order to obtain an accurate and applicable basic bra pattern, this study's research and design methods were based on the medium-sized mannequin used for underwear design jointly developed by the study team and Beijing Aimer Underwear Co., Ltd. The mannequin was developed based on body size of women aged 20-39 and with breast size of 75B whose information was contained in the Aimer database. The medium-sized mannequin was made via 3D shaping and morphing modeling; it is structurally and aesthetically significant for use in underwear design [6] and has been put into use by Beijing Aimer Underwear Co., Ltd.

2.1 Design of basic bra pattern

As shown in Fig.1, the bra is basically composed of cups, centre gore, under band, wings, and straps [7,8]. The bra pattern in turn is mainly composed of three parts: cup pattern, wings pattern, and under band pattern (generally, centre gore and under band constitute an overall structure, so the under band pattern referred to in this study includes centre gore pattern and under band pattern).

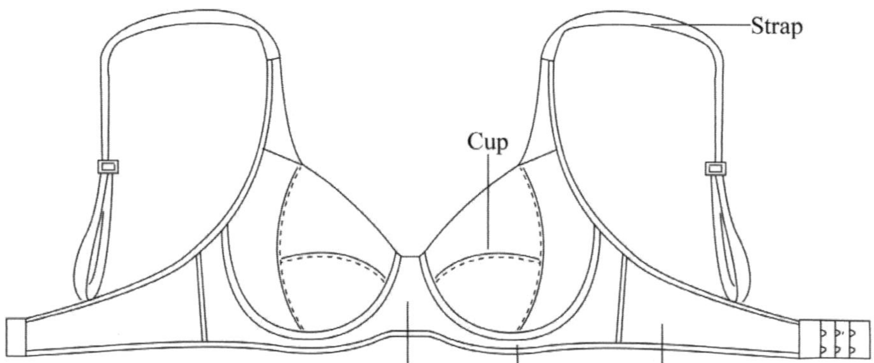

Fig.1 Schematic diagram of basic bra structure

2.1.1 Design of cup

The cup refers to the part that covers the woman's breasts, which are the only organs that are not supported by any bone in the woman's body; generally, they are located between the second and the sixth ribs and the width of one breast is the distance from the inner margin to the front armpit. Based on the shape of the female breast and under the principle that the breast can be maximally covered, the silhouette of cup of the basic bra pattern is designed with the mannequin and marked with 3mm strip (see Fig.2).

2.2.2 Design of under band and wings

Using the same principle applied in the design of cup and on the basis of characteristics of the classic full-cup bra, the maximum silhouettes of centre gore, under band, and wings are determined (see Fig.3).

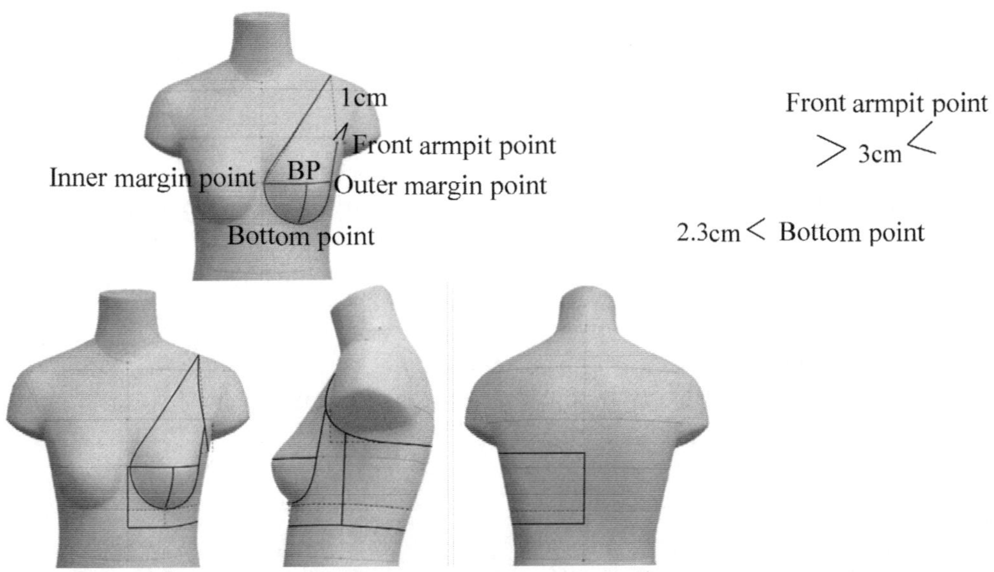

Fig.2 Schematic diagram of cup design Fig.3 Schematic diagram of under band and wings design

2.2 Comparison of bra pattern unfolding methods

Rui employed vertical, horizontal, oblique, and convex point cutting methods to unfold the 3D form of bra cup and evaluate the spread patterns; she found that the pattern unfolding by convex point cutting was the optimum method [9]. This study obtained 3D shape of bra using the Japanese paper copying method [10] and the 3D shape was unfolded in different ways using the convex point cutting method; three cup cutting ways and two under band and wings cutting ways were developed. Three cup patterns and two under band and wings patterns were finally obtained by slightly trimming the patterns after drawing smooth lines along the margin (see Table 1).

Table 1. Cutting method of 3D shape and formation of basic bra pattern

	Cup A-1	Cup A-2	Cup A-3	Under band & wings B-1	Under band & wings B-2
Cutting method*	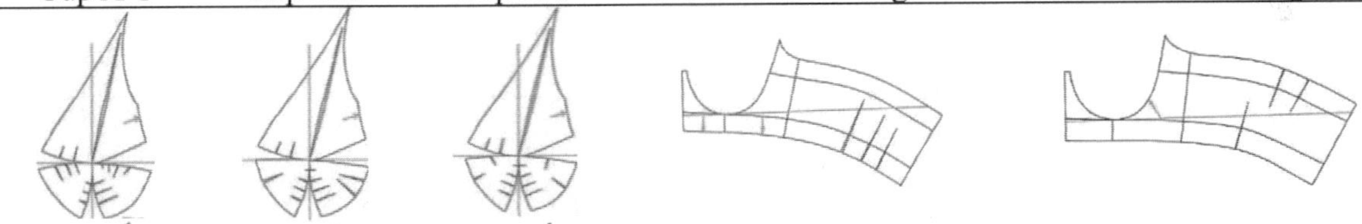				
Unfolded pattern**	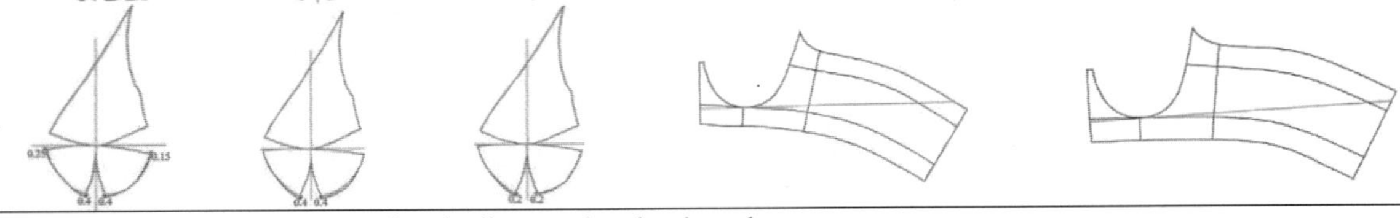				

* Blue line indicates dart is open; green line indicates dart is closed.
** Black line represents smoothing line; red line represents modifying line.

3. Evaluation and analysis of unfolding method of basic bra pattern

Generally speaking, to be transformed into a 2D pattern, the 3D cut piece shall meet the following requirements (i.e., before and after the transformation) [11,12]:

1) The area of 2D pattern is approximately equal to the corresponding 3D body surface area;
2) The boundary length of 2D pattern is basically equal to that of the corresponding 3D body surface;

3) Positions of key points on 2D pattern correspond with those on 3D finished product.
With reference to the above standard, the study evaluated patterns developed with different unfolding methods.

3.1 Evaluation on area change

After inputting the unfolded patterns into clothing CAD software for area measurement, we compared the area of 2D patterns of each bra part and that of the whole 2D pattern with that of corresponding 3D surface of the mannequin. The result showed that the change rates of Cup A-3 and Under Band & Wings B-2 to be close to 0 (see red figures in Tables 2 and 3), meaning the area of these two 2D patterns were closest to 3D surface (see Tables 2 and 3).

Table 2. Area change of each part of cup pattern

		Upper piece	Lower piece (front)	Lower piece (side)
3D	Area (cm^2)	145.83	40.96	46.67
A-1	Area (cm^2)	150.72	41.4	47.16
	Change rate (%)	3.17	1.07	1.05
A-2	Area (cm^2)	150.72	40.17	45.32
	Change rate (%)	3.17	-1.92	2.89
A-3	Area (cm^2)	150.72	40.73	47.11
	Change rate (%)	3.17	-0.56	0.94

Table 3. Area change of each part of under band and wings pattern

		Under band	Wings
3D	Area (cm^2)	113.3	278.56
B-1	Area (cm^2)	113.89	279.75
	Change rate (%)	0.52	0.43
B-2	Area (cm^2)	113.07	279.39
	Change rate (%)	0.2	0.36

3.2 Length change

By comparing the length of each part and side of the bra's 2D pattern before and after the pattern is cut with that of corresponding side of the mannequin, we found that the change rates of length of each side approached 0 (see red figures in Tables 4 and 5), meaning there is no significant change in the length of all sides of 2D patterns compared with corresponding 3D body surface (see Fig.4, Tables 4 and 5).

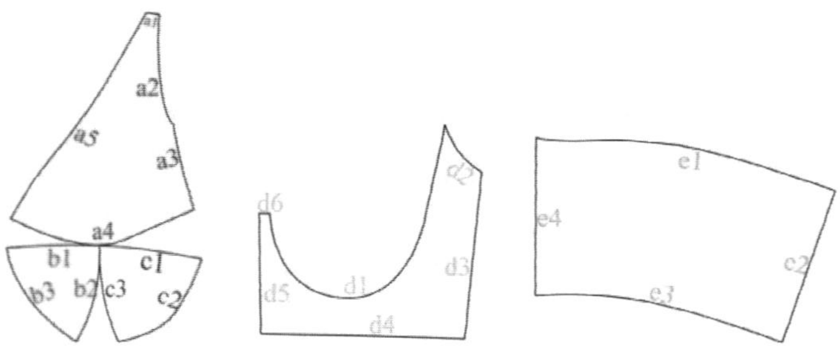

Fig.4 Numbering on each side of bra pattern

Table 4. Area change of each part of cup pattern

		Upper piece					Lower piece (front)			**Lower piece (side)**		
		a1	a2	a3	a4	a5	b1	b2	b3	c1	c2	c3
3D	Length (cm)	1.2	12	5.5	16.16	21.7	7.68	8	9.79	8.48	10.3	8
A-1	Length (cm)	1.2	11.42	5.45	16.81	20.82	8.07	8.04	9.76	8.55	10.38	8.05
	Change rate (%)	0.00	-0.05	-0.01	0.04	-0.04	0.05	0.00	0.00	0.01	0.01	0.01
A-2	Length (cm)	1.20	11.41	5.45	16.33	20.71	7.74	8.04	9.89	8.59	10.29	7.99
	Change rate (%)	0.00	-0.05	-0.01	0.01	-0.05	0.01	0.00	0.01	0.01	0.00	0.00
A-3	Length (cm)	1.20	11.42	5.45	16.81	20.82	7.79	8.04	9.88	8.75	10.27	8.18
	Change rate (%)	0.00	-0.05	-0.01	0.04	-0.04	0.01	0.00	0.01	0.03	0.00	0.02

Table 5. Length change of each side of under band and wings pattern

		Under band						**Wings**			
		d1	d2	d3	d4	d5	d6	e1	e2	e3	e4
3D	Length (cm)	25.59	4.5	12.5	16.12	9.5	0.8	24.32	12.9	19.5	12.5
B-1	Length (cm)	25.59	4.54	12.46	16.5	9.55	0.8	24.35	12.96	19.98	12.46
	Change rate (%)	0.00	0.01	0.00	0.02	0.01	0.00	0.00	0.00	0.02	0.00
B-2	Length (cm)	25.16	4.54	12.50	16.12	9.55	0.80	24.51	12.93	19.60	12.50
	Change rate (%)	-0.02	0.01	0.00	0.00	0.01	0.00	0.01	0.00	0.01	0.00

3.3 Evaluation on paper model

By making the paper model to restore the 3D shape of bra, we could judge whether the positions of key points correspond with the marked positions on the mannequin after the 3D shapes are restored with 2D patterns of three kinds of bras. Upon analysis, the 3D shape developed through the combination of Cup Pattern A-3 and Under Band & Wings Pattern B-2 is optimal.

Table 6. Fitting and evaluation of paper model of bra

	Cup A-1	Cup A-2	Cup A-3	Under band & wings B-1	Under band & wings B-2
Fitting of paper model:	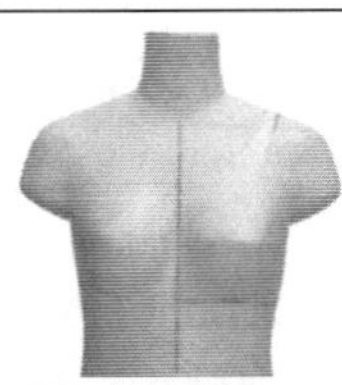	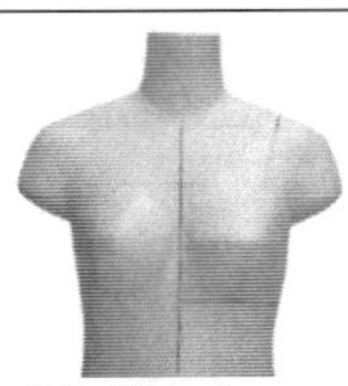	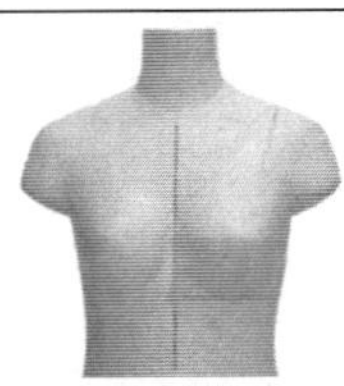		
Evaluation:	BL line is too long; circumference of cup mouth deviates outward and total capacity of cup is too large.	BL line is too short; circumference of cup mouth deviates inward and total capacity of cup is too small.	BL line is moderate and total capacity of cup is moderate.	The part of side wings closing to lower piece is too redundant; centre back line is not identical with centre back, with short upper circumference and long lower circumference.	All parts match with corresponding parts of the mannequin.

3.4 Evaluation on 3D fitting

Upon combination of the three cup patterns and two under band and wings patterns, we developed six bra patterns. We then imported the six patterns into CLO 3D fitting software separately, with the imported medium-sized mannequin used for underwear design as the virtual human body, and the bra patterns were seamed in accordance with the structural design for virtual fitting on the human body. Table 7 illustrates the pressure distribution after six patterns were seamed. Except for the A-3 and B-2 combination, it was found that the greatest pressure would be produced at bottom circumference and back centre in the other five combinations. Thus, the combination of A-3 and B-2 boasts more ideal pressure distribution.

Table 7. Pressure distribution in virtual bra fitting

B-1 B-2 Pressure range

A-1

A-2

A-3

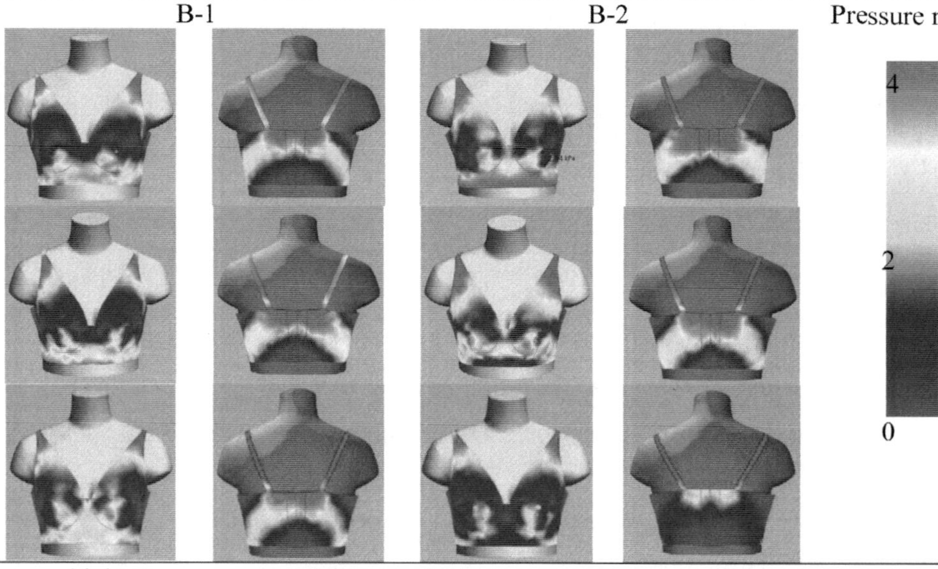

Note: Red areas = highest pressure; purple areas = lowest pressure; pressure ranges from 0kPa to 4kPa

The aforementioned evaluation result shows that the combination of Cup A-3 and Under Band & Wings B-2 is the optimum pattern.

4. Making of basic bra pattern

In order to use the optimum pattern with Japanese paper to make the most comfortable bra, the study fabricated a grey cloth bra sample and modified the unfolded pattern. Fig.5 illustrates the following unreasonable parts of the grey cloth bra sample: 1) the centre gore part is not tight-fitting and the cup circumference at the centre front cannot fit the breast root; 2) there is a large space between the upper piece and the mannequin; 3) there is space between the centre line of the lower piece and the mannequin; 4) necklines are not tight-fitting. The bra sample should be modified under the principle of the bra sample modification (all bra parts shall fit well with the mannequin, with smooth curves and the best visual appearance shall be realized). Fig.6 illustrates the comparison figures before and after the modification.

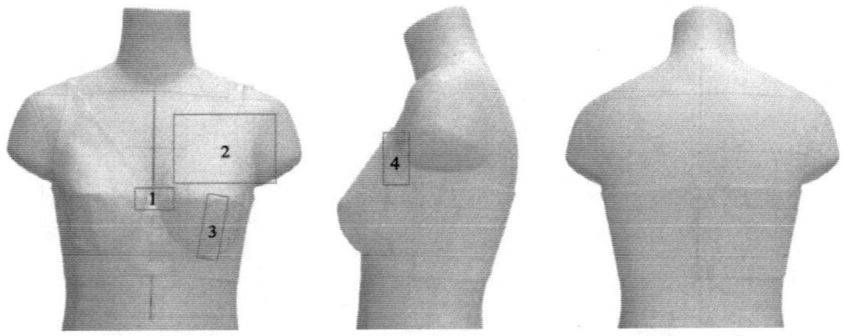

Fig.5 Schematic diagram of grey cloth bra sample

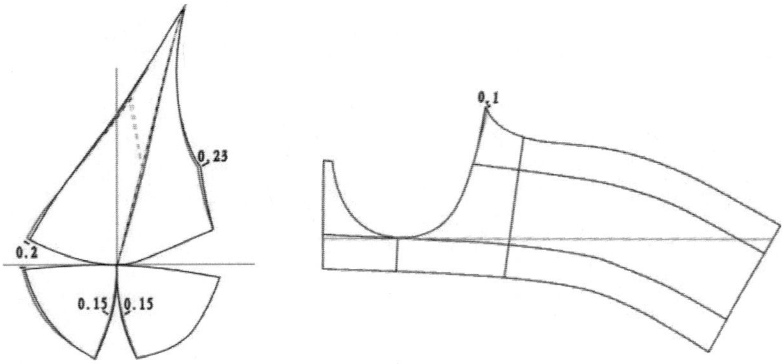

Fig.6 Schematic diagram of modification method of "optimum" pattern
(Red line in figure refers to the modified pattern)

After the grey cloth bra sample was made with the modified pattern, it was found that the bra sample became most form-fitting; thus, we determined that the modified pattern of the bra sample can be used as the pattern of the bra sample (see Fig.7).

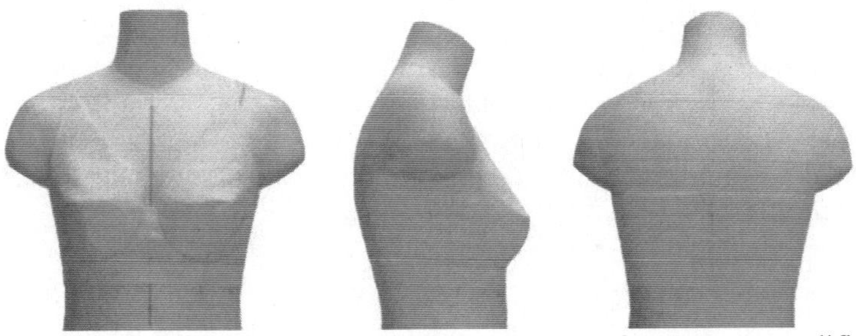

Fig.7 Schematic diagram of grey cloth bra sample after pattern modification

5. Verification of basic bra pattern

The study selected commonly used bra materials to make the bra sample to verify the basic pattern. Before making the bra sample, the basic bra pattern was shrunk. In accordance with data provided by the underwear manufacturer, shrinkage rate was set as 20%. The bra sample was made after the paper pattern was drawn. See Fig.8 for details.

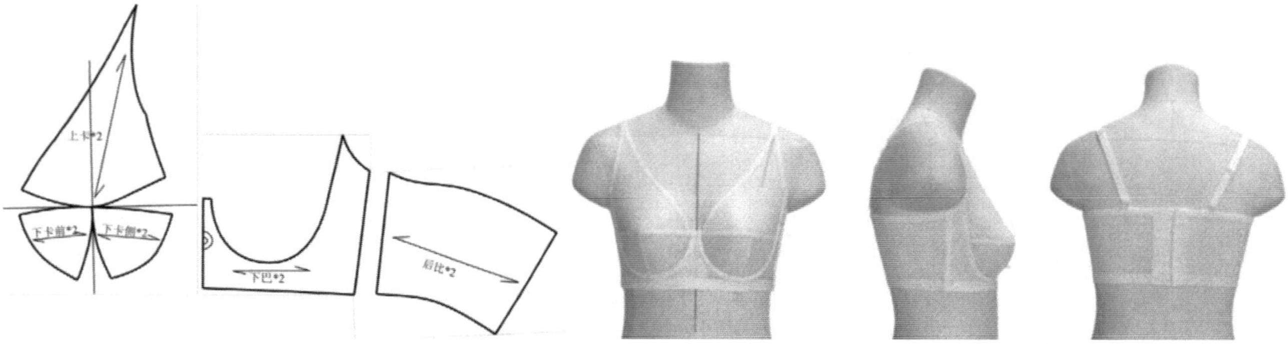

a b

Fig.8 (a) Shrunk bra sample pattern (wings shrunk by 20%); (b) bra sample

As shown in Fig.8(b), the bra sample fit the mannequin well and the designed lines matched with corresponding positions of the mannequin. Thus, it could be determined that the basic bra pattern is reasonably designed and can be applied in production.

6. Conclusions

The research team created the basic bra pattern based on a medium-sized mannequin used for underwear design and arrived at the following conclusions:

1) 3D-Plane Method is the best method to develop the form-fitting basic bra pattern and the mannequin specially used for the underwear design is the key element for obtaining the 3D shape of bra. The study employed the mannequin used for underwear design that has been shaped and beautified, and the bra pattern designed based on it is more form-fitting.

2) When the 3D shape of bra is unfolded, the cutting of different parts of the bra directly influences the change rates of area and side length and pressure distribution at the time of fitting of the bra pattern, thus the cutting method is vitally important;

3) The study determined "optimum" pattern developed from the combination of Cup A-3 and Under Band & Wings B-2. The combination is unfolded by cutting along each side of the pattern, and the final effect is more reasonable than other combinations.

4) Upon further analysis and modification of the "optimum" pattern by making it into the grey cloth bra sample, the final prototype bra pattern was developed. Combined with the actual production practice, the prototype bra was used to make and verify the bra sample and finally the pattern has proven to be reasonable and qualified for use by the manufacturer for bra making.

References

[1] Hager A. Pattern design of underwear, swimsuit, beachwear and sportswear. Beijing: China Textile & Apparel Press: 2001.

[2] Suzhen L, Xin Z, Dongsheng C. Body foundation of major part dimensions in bra patterns. J Text Res 2008; 29: 69-73.

[3] Hardaker CHM, Fozzard GJW. The bra design process: a study of professional practice. Int J Cloth Sci Tech 1997; 9: 311-325.

[4] Xiaopeng C, Junfang S, Yan W. Analysis on design of bra prototype. Silk 2012; 49: 37-40.

[5] Xiaojun D, Lijun W, Fengyuan Z, et al. A study of body form characteristics of young women in Jiangsu and Zhejiang. J of Zhejiang Sci-Tech University 2006; 23: 150-153.

[6] Yunchao Z. Research and development of mannequin used for underwear design with 3D human body data. Master's thesis, Beijing Institute of Fashion Technology: 2013.

[7] Bressler KW, Newman C, Proctor G. A century of lingerie. Beijing: Textile Industry Press: 2000.

[8] Jianrong Y. Course of underwear structure design. Beijing: China Textile & Apparel Press: 2006.

[9] Rui Z. Research on the optimization system of bra pattern based on the underwear design mannequin. Master's thesis, Beijing Institute of Fashion Technology: 2013.
[10] Miyoshi T. Clothing model theory. Theoretical articles. Beijing: China Textile & Apparel Press: 2006.
[11] Hui Z, Jianguo Y, Chaoyun Z. Computer costume design. Shanghai: Fudan University Press: 1996.
[12] Heisey F, Brown P, Johnson RF. Three-dimensional pattern drafting. Part I: Projection. Text Res J 1990; 60: 690-696.

Research On Xilankapu -Let The Traditional And Fashion Develop in Parallel

Jia-Tong Liu*, Li-Li Mao, Xiao-Yan Li

College of Clothing and Art Design, Xi'an Polytechnic University, Xi'an, Shanxi,710048, China

*Corresponding author's email:254013219@qq.com

Abstract

Xilankapu is the first batch of national intangible cultural heritage; it is the treasure that Tujia ancestors left us. However, it has been lagging behind in development due to changes in people's lifestyle. In this paper we are going to describe the Xilankapu weaving technology, color and patterns. We will also analyse Xilankapu's current developments and problems.We suggest that the traditional and fashionable elements should be developed in parallel by utilising the different Xilankapu patterns in apparel and home design. We want the clothing and home designs to meet the modern day pursuit of a high quality of life, and ensure the development of Xilankapu patterns, so that more people can understand Xilankapu and the Tujia culture.

Keywords: Xilankapu; Tujia culture; traditional patterns; cultural heritage

1. Introduction

Cultural inheritance and development can exist through different forms. Even though the languages of the ethnic minorities from Southern China were not adopted as the national language for historical reasons, these people have left behind an extremely rich cultural heritage, Xilankapu being part of it. The Tujia brocade, the Zhuang brocade, the Li brocade and the Dai brocade are collectively known as China's big four folk brocade. Brocade in China has more than 3000 years of history, bringing with it rich cultural connotation and sophisticated technology. Xilankapu is a piece of treasure that is left to us by the Tujia ancestors. It is the most representative and typical of the Tujia brocade, and it interweaves many aspects of the Tujia people, from religious beliefs to the assessment of the Tujia girl attributes in female members of the tribe.

Duetorapiddevelopment of the society and Chinese economy,large-scale machineryandefficient manufacturing operations have been adopted toenrich the people's lives. However, Xilankapu became more and more expensive as a result of its low production efficiency, making it difficult to be a part of people's everyday lives.Xilankapu is now mostly considered only for gifts or collection purposes, which has a devastating impact on the brocade. We believe in the need for both Xilankapu and the society to develop. Hence, in this paper, we are going to focus on Xilankapu's artistic charm its present developments and problems faced by the traditional pattern. Our analyses suggest that the Xilankapu tradition pattern is a popular element that can be integrate into modern life.

2. The artistic charm of Xilankapu

2.1 Weaving technology of Xilankapu

In our daily lives, woven fabric is made in large volumes at high speeds with weaving machines. Xilankapu, however, uses cotton as warp, stranded cotton thread or yarn as weft. Xilankapu is made by hand with a wooden loom based on the skill of "passing the warp thread, cutting the weft thread". The colourful fabric can be divided into two categories, 1) "for oblique" - a series of plain colour and plain weave and 2) "upper and lower oblique" - a coloured twill series. The weaving of Xilankapu fabric is comprised of a series of complicated processes, with a total 12 treatments: 1) twisting spinning; 2) dyeing; 3) back-winding; 4) pulling strings; 5) installing reed; 6) rolling line; 7) picking up fully; 8) artemisia over; 9) cross-stitching; 10) bundle of rods machine; 11) weaving; 12) picking up knitting. The different weaving processes can only be completed manually, no machine is capable of replacing the skilled hands of the Tujia people. And because of this unique weaving process, the Xilankapu pattern is unable to form smooth curves, leaving only combinations of slanted and linear shapes possible, such as the Yang Que flower shape in figure 1. It is this weaving process that makes the patterns mosaic-like, which form its unique artistic charm.

Fig.1 (From Tujia brocade .Beijing Academic Press P24)

2.2 Color of Xilankapu

Color is the first and foremost element of emotional expression. In a good piece of work, people's attention is usually first drawn to the fabric's colour, then followed by its pattern. Color is often an expression of the weaver's inner world. From the use of color, you can perceive the different emotions that the artwork entails, therefore the colours used is particularly important in bringing out the message behind the artwork.

Fig.2 Fig.3

(From http//image.baidu.com)

As recorded in the Journal of the Yong Shun House, "The Yong Shun fabric of five colours was made by the natives… They wove with one hand and cross-stitched using small ox horns with the other, giving the brocade its five colors." The five colors are "… red, black, yellow, green and white". "Indigo herbs grew on the hills and in fields (were) used for dyeing. Herbs that grew near the dyeing shop became sparser as more of them were bought…" We can thus see that a lot of Xilankapu is dyed naturally with plants from the Wuling Mountain area in the Yong Shun county. On the use of colours, the Tujia people are "fond of black and hate white". And as the native sayings go "the deep color supports the light", and "black and white do not match, red and green has no expression; it is the blue and yellow that will shine". The customs of different Tujia areas are quite different, for instance the Dragon Slope Foot, the Indigo Room and the Yong Shun mountain prefers black and indigo; whereas Lao Ce and Cats Coast favour an earthy red. A good example is Tujia's favorite Yang Que flower, where different places adopt different colors. Figure 2 is a production of the Dragon mountain, with similar colors, colour repeats and patterns that exemplifies the beauty of harmony. Dragon mountain Indigo room is depicted in figure 3 with interlacing patterns. Each bird is of a different color, giving a free and lively feeling.

The Tujia people currently use over 20 colours in the production of Xilankapu. The colour matching method is highly sophisticated due to the number of colours used. Some, such as figure 4, are colorful but expresses a sense of freedom, flexibility and uniqueness. Others, such as figure 5, are simpler, but subtly rich and not dull. Each has its own merits.

2.3 Xilankapu pattern and its implications

Fig.4

Fig.5

Fig.6

(From http//:iamge.baidu.com)

There are nearly 160species of Xilankapu's classic traditional grain pattern left, describing the production activities and animals, plants, text, geometry and auspicious words. We can see from figure 2 and figure 3 two Yang bird flower shape is cute, bright and full of spirit. Patterns of Xilankapu are not only exquisite, but the composition is also very particular. Two of the Yang Que flower pattern have two different heads. Figure 2 is made up of repeating patterns, using plain "Monkey hands" as a header, expressing a strong sense of harmony. Figure 3 is uses the "eight-petal flower" as the header, making the original vivid birds more lively, as if they are dancing from flower to flower. The Tujia people's ancestors said, "Patterns must be meaningful, as that will be propitious." Each of the 160 classic patterns has its own catchy song. In Figure 6, a brocade with the pattern of a phoenix flying above a peony is depicted, and its song goes, "Asking any girl under the Wuling Mountains, the beautiful tapestry of thousands of fame, with daylight only without night, the dark fragrance of peony flowers."

The brocade of Tujia people is used to reflect their love of life, mainly via homophonic techniques and the use of symbols to express the good wishes. With the changing of aesthetics, the brocade pattern change from hazy to clear, feelings change from weak to strong, ideas change from casual to intentional creation, colors change from colorful to unique charm, filling the patterns of Xilankapu with endless artistic charm.

3. Development situation and problems of Xilankapu

In this era of rapid modernisation,Xilankapu's development has been devastatingly hampered.Xilankapu had difficulties being integrated into modern life because of its unique weaving process. Firstly, although the Xilankapu fabric is beautiful, the complex weaving process limits the possibility of automation, keeping the yield low and production costs high. Therefore, it is replaced by large quantities of printed fabric. Secondly, Xilankapu is very heavy and is not easy to wash. Modern people pay more attention to healthy habits. As it will take a long time to wash, the native people wear ordinary clothes in their everyday lives. Xilankapu is worn only during major Tujia festivals. This has led to a diminishing range of applications and a whittling list ofskilled traditional manual artists. Thirdly, Xilankapu's quality is mainly determined by the shape of the pattern, hence high elasticity will be considered as defective. However, in the modern day, our clothes, especially sportswear, and beddingare all focused on their comfort.So Xilankapu cannot answer to our everyday needs.Finally, many local people are in the pursuit of a modern life. Traditional skills that are passed down for thousands of years are lost as Tujia girls, who were the main source of productivity, no longer focus on Xilankapu.

As it is difficult for Xilankapu to integrate into the modern life,we think it can only be used forappreciation, collection or as tourist souvenirs in this form. But Xilankapu represents the Tujia culture and is the Tujia people's spiritual sustenance.We cannot just stand aside and watch it disappear.Xilankapu is not only the traditional representative of the Tujia people, but also an outstanding representative of Chinese traditional culture. It has repeatedly represented China abroad in cultural exchanges, and it has been chosen several times as a representative national gift to foreign friends. In 2006, the Tujia brocade was included in the first batch of national intangible cultural heritage. However, modernization is a double-edged sword, on the one hand it promotes progress of the society, on the other hand, it also leads to the destruction of many traditional cultures. In this case, the government has set out to save the Xilankapu. In the Longshan County, the governmentset up a free Xilankapu classroom and organized a Tujia brocade skills competition, in which every participant was offered a subsidy. In order to let the Longshan people and tourists enjoy the essence of Xilankapu, the Longshan County has also placed a special booth dedicated to Xilankapu in its national hotel. Each hotel room is also decorated with Xilankapu works.

National development and the preservation of cultural inheritance is sometimes at odds with each other. However, we cannot allow our inheritance of the exquisite patterns and interesting songs to be lost to social progress and development.

4 .Traditional and popular side-by-side

Xilankapu consists of around 160 species of classic patterns, depicting animals, plants and people's worklife, text, geometric composition, auspicious words. As the weaving process can only be done manually, therefore we can apply these patterns to different places together with its ballads, in a new form to allow more people to appreciate its patterns and songs, and to learn about the Tujia culture, revitalising the Tujia brocade. In addition, we should also gain an in-depth understanding of the morals and meanings behind each kind of pattern. Divided from its meaning and color, we could extract patterns of Xilankapu, make use of its fashion elements and apply them in different aspects.

Fig.7 (From http//:iamge.baidu.com)

Figure 7 is Spring / Summer men series issued by China's young designers Zhou Xiangyu in 2016 Fashion Week. The series is a fine combination of traditional Chinese patterns and modern day fashion. Traditional Patterns and color combinations are used in apparel, creating clothing textures that are similarto that of the traditional brocade. This makes the entire series of clothes gorgeous without losing a sense of fashion. Xilankapu can also be used in a manner such that its patterns can be turned into popular elements used in the design, which brings out strong ties to the Tujia culture and its rich histories.

Because of the particularity of the weaving process of Xilankapu, it could not completed by machine and mass production, so we can choose its pattern, and combine it with different materials, embroidery, knitting and dyeing.For designs of domestic merchandise, wallpaper, pillow, cushion, sofa fabrics and all kinds of bedding could be considered.Depending on different stylesof design patterns different Xilankapu patterns could be selected as ornaments. Thus it could improve cultural connotation and the lifestyle of the inhabitant, and allow the inhabitant to explore the meaning of the patterns in leisure with visitors and learn about culture of Tujia nationality enjoyably. For example, the Yang bird flower with bright colors and a cute style can be used for decoration of a children's room. In the eyes of the Tujia people, the liking of mandarin ducks can be expressed through this song,"Tujia sister love affair, skillfully weaving brocade pattern in turned turret, Yang bird called outside the window, missing best man traveling together." You can also apply it in the marriage room, meaning beautiful love.

The simplest use in clothing design is the wedding garment category, there are many moral and festive Xilankapu patterns meaning love and happiness, like magpies and plum blossoms, Yang bird flower, Lion hydrangea, welcoming relatives, two phoenix and peony, phoenix and horse dragon pattern, auspicious phoenix and dragon, butterflies peony, peony richness, star anise fragrance four color printing, in such bright and peaceful colors. And the colourful Three Lucky Men pattern that can be used on parents' birthday to signify a long life of abundance and respectability. Children's clothing can use adorable patterns like the wild deer with flowers pattern and the lion and Hydrangea pattern. The monkey hand pattern, meaning catching money, can be considered for various parts of daily clothing and bags. Some geometric patterns can be used in large areas and in repetition.

Despite the fast pace of modern life, people have high standards on the quality of life andpay much attention to both material and spiritual needs. Only the combination of traditional and contemporary elements can achieve a win-win situation that meets the needs of people and cultural development.

5. Conclusion

In summary, tradition is the product of people's production and life in a certain period of time. In the new era, as people's living habits change, we must discard the outdated and retain only the essence. Even though Xilankapu's weaving process has its own merits, mass production cannot be carried out. In order to adapt to the rapid development of modern life, we can use a new way to weave Xilankapu, so as to blend into the modern people's lives. We can use its pattern, and then use the embroidery, knitting, printing and dyeing methods to processing fabrics, to recreate its patterns. When its patterns appear in the people's daily lives in this form, it can allow more people to pay attention to the Xilankapu and the Tujia culture. It is fundamental that culture and fashion are enabled to develop together.Xilankapu can only be receive due respect if more people know and appreciate it. In this way we can preserve the Tujia culture,and let more traditional Xilankapu became collections.

References

[1] Wang WY, Tian SX. Hunan brocade, Hunan Fine Arts Publishing House, the 2008; p: 117.
[2] Tian M. Tujia brocade, Beijing, Academic Press: 2008; p: 24-p27.
[3] Zhang XP, Tian Y, Xiang X. Tujia brocade, Changsha, Hunan Normal University publishing house: 2011; p: 201.
[4] Liu JH. Analysis of the transformation of the traditional culture of ethnic minorities in the modern context. Huaihua University: 2006.
[5] Yang Y. Chinese national costume culture "M". Beijing: Popular Literature Press: 1999; p: 144.
[6] Xiong QQ. "Xilankapu" - Beautiful Tujia brocade: 2005; p: 46.
[7] Zhao L, Chen MZ. The overall expression of Xilankapu: 2008; p: 221.
[8] Qi ZJ. A preliminary study on the historical stage of Tujia costumes: 2009; p: 4.
[9] Gao L. An introduction of Xilankapu: 2006; p: 20.

Implementation of Portable Information Extraction from 3D Human Body Data Points in Garment CAD

Yue Wang[1], Shou-Shan Jiang[1,2]

[1]*College of Mechanical and Electrcal Engineering, Northwestern Polytechnical University, Xi'an, Shanxi, 710072, China*

[2]*School of Computer Science, Xi'an Polytechnic University, Xi'an, Shanxi, 710048, China,*

*Corresponding author's emaile: ourfool@163.com

Abstract

The main goal of this study is to improve the rotation of the laser for non-contact human body scanning system, and to get the true and accurate 3D information of the measured human body. In the aspect of processing digital images of human body, the extraction accuracy of laser stripe's pixel coordinates is one of the important factors to affect the reliability of non-contact measurement. Therefore, this paper mainly discussed the characteristics of the human body laser stripe images. Secondly, the principle of the image processing system is brought up. Lastly, the process and method of image processing are analyzed.

Keywords: Human body measurement; Body Scanning; Point cloud processing; the laser stripe center; Pixel coordinates

1. Introduction

Human body measurement is different from dimension measuring of industrial applications. Because the human body is a life body, and its biological characteristics is flexible and complex, so the human body measurement can not be used in the harmful methods and tools [1-3]. Therefore, the human body measurement should be pleasant. So there are special requirements for non-contact measuring equipment of human body. The laser stripe images of human body which is collected by charge coupled device camera is based on structured light vision measurement system. The quality of human body laser stripe images and its processing algorithm directly affect the system scanning precision [4,5].

In order to satisfy the accuracy requirement of the image processing system, laser stripe images have to be done as the following three steps: 1) per-processing; 2) extraction pixel coordinates of the laser stripe center location; 3) Curve fitting the human body laser line. Through the reasonable algorithm, the laser light whose band width more than one pixel in images is developed into a complete, single pixel of the human body laser line [6-9]. These processes met the requirements of the subsequent of three dimensional human body coordinates reconstruction. Because the characteristics of human body laser stripe images, these three steps have higher requirements.

In pixel coordinates calculating, the extraction accuracy of human body laser light center pixel coordinates is one of the important factors that affect measuring precision [10-11], and the camera discrete pixels can cause human body surface originally a smooth band becomes not smooth [12-15]. The main goal of this study is to improve the rotation of the laser for non-contact human body scanning system, and to get the true and accurate 3D information of the measured human body. So this paper from the the second step of the non-contact human body measurement, which means to study the data processing technology of laser stripe images.

Therefore, this paper focuses on the three parts of algorithm, firstly analysis the characteristic of human body laser stripe images; extract pixel coordinates of the laser stripe center location; in the case of the images on the border between the cameras as sensitive unit, by using the bezier curve to fitting human body laser center line, in order to effectively improved the body scan, the smoothness of the center line of the laser light.

2. The working principle of human body 3D reconstruction system

As shown in Fig.1, Using this system for non-contact measurement of human body, we need to place a scanning unit respectively in the front and rear of the human body. Two CCD cameras are installed in the double digital pan tilt, as the stepping motor driving, they begin to rotate scan, and the two scanning pan tilt must be synchronized movement. The rotation angle of the pan tilt at any given moment is one of the important parameters in 3D reconstruction of the human laser stripe of the time. If these two pan tilts have synchronous rotation, we can ensure that the rotation angle at any moment of the two pan tilt are the same, which makes the two scanning units had synchronous scanning, and which make a closed curve by the front and rear formed surface by the human body scanning point cloud data be at the same height, and which can make a convenience for the following calculating of 3D human body reconstruction, otherwise it will cause the error of the measurement system.

The system working principle is: two scanning heads were driven by axis rotate center of pan tilt for synchronous rotation; Line laser that on the top of scanning head casting laser beam passing through a cylindrical lens in the space form a narrow laser plane. The plane bottom-up or top-down scanned the surface of human body, and the intersected with body surface. It was formed the light of the intersection of the body surface by modulation of the surface shape. The dynamic 3D image was placed in each below the laser of the cameras every fixed time interval detection, which can obtain a series of light two-dimensional distortion images. When scanning heads rotated to different azimuth, the CCD cameras obtained two-dimensional images can be transmitted to the computer for analysising.

The system used the establishment of the mathematical model to calculate the 3D reconstruction of each light wire center location pixel coordinates, and the 3D human body coordinates were also calculated through the extraction of the laser light line of each image at the computer. Then it determined the three-dimensional human body data of front and back. 3D human body data point cloud can be unified into the same coordinate system according to the relative position of the relationship between two scanning units. Eventually we got raw data point cloud on the surface of the human body, a human body surface 3D reconstruction in the form of point cloud. The time of the system scanning mainly considered the human body during this period remains relatively motionless, minimizing the effects of motion blur.

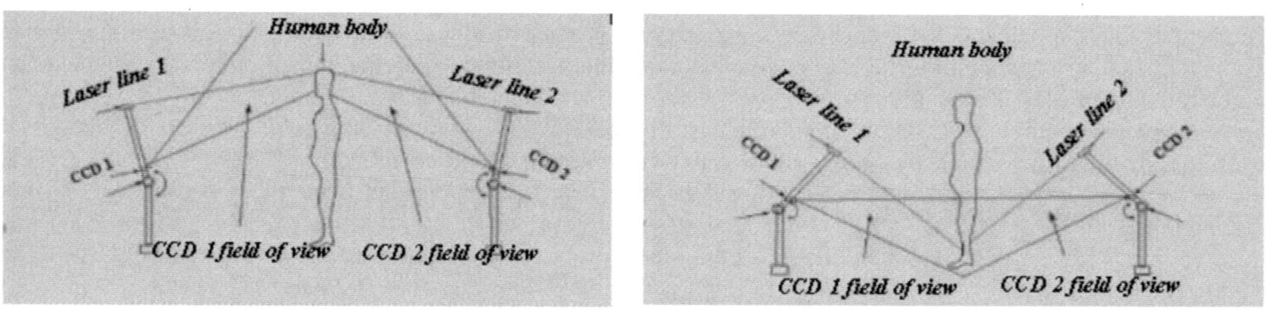

(a) Rotation model in stationary state (b) Rotation model for motion state

Fig.1 The simple description of the measurement principle

3. Steps of image processing experiments

A series of laser stripe images are collected in the experiment of image acquisition.The next work is to process these images, to get the full and single pixel width of laser lines. These laser lines are used for the reconstruction of 3D human body data.

The steps of image processing experiments are shown as follows:

In the VB environment, run the system module named the main program of coordinate calculation. We can get into the interactive interface of image processing experiments, as shown in Fig.2. Click the button on the left side of the interface to load the images, then pop up the search dialog box, which is shown in Fig.3. After finding the target images in the storage path, click the start button, and then enter the images we want to process as shown in Fig.4. Inputs the first frame of image. The specification of the image is BMP file.

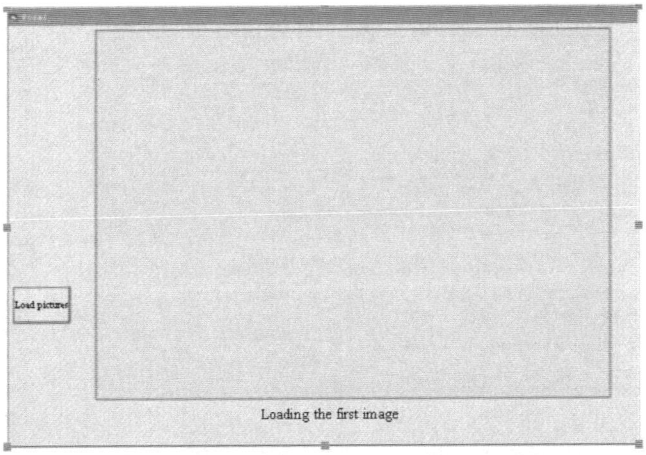

Fig.2 Initial interface for image processing

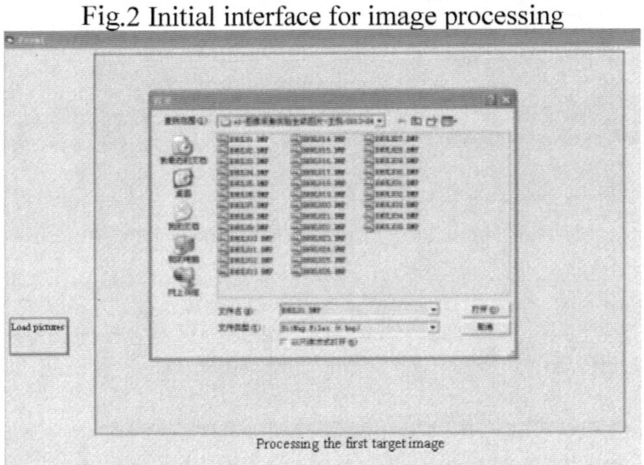

Fig.3 The dialog box of loading images

The system displays the target image, at the interface, through the mode of drop square box operation, visually divided into a target laser line sub image (in Fig.4). This is to reduce the processing area of the target image. After the selection of the sub image is complete, click the button of select area that on the left side of the interface, then enter the interface of the parameter setting of measurement system. Selects a sub-image from the BMP image file, which contains the target light (the human body laser stripe). Through this step, the processing area of the original BMP image is reduced, which becomes more convenient to handle.

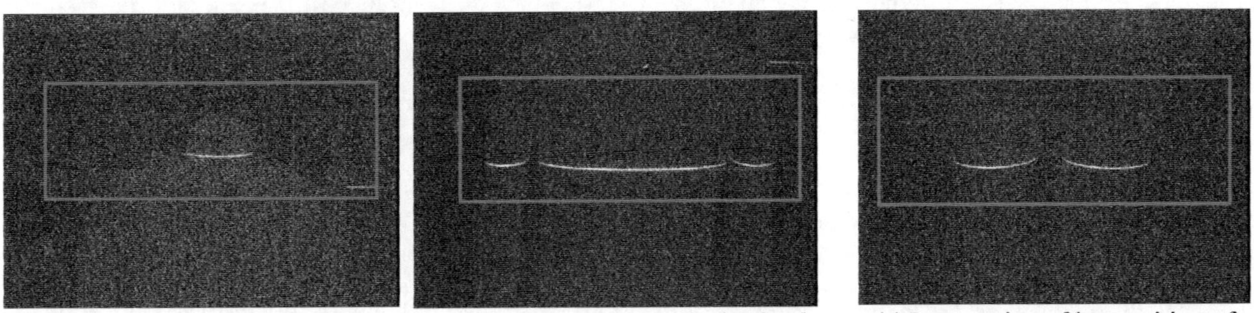

(a) Laser stripe on the neck of the human body

(b) The laser stripe on the back of human body

(c) Laser stripe of leg position of human body

Fig.4 Selection a target sub-image which contains the human body laser stripe

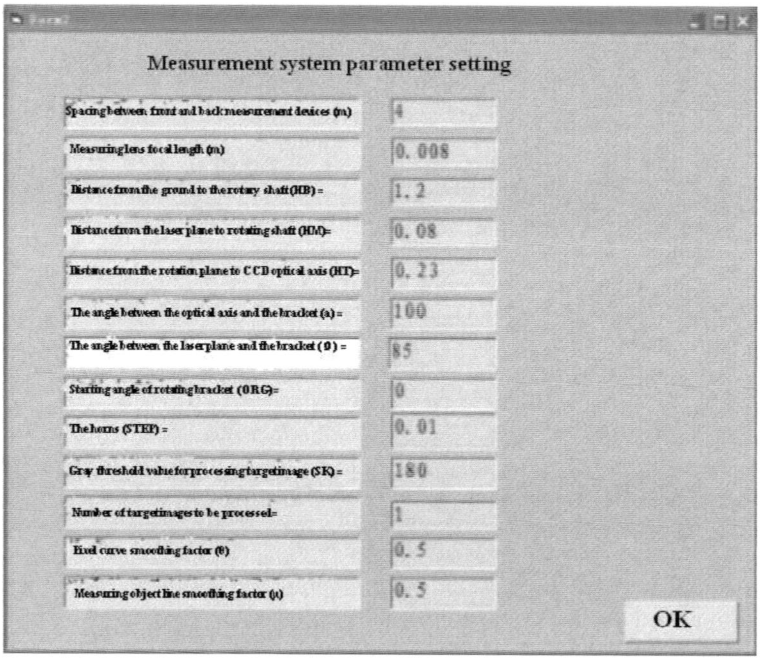

Fig.5 Setting interface for measuring system parameters

In a set to parameters of the measure system interface, the system has already been initialized assignment and displayed on the 13 parameters (in Fig.5). The parameter values of the measurement system can be modified according to different experimental conditions. When the parameter value is modified, click OK, and the system will automatically process the target images.

Read gray value of the sub-image file, which contains the target data information (including the human body laser stripe). That is to say, under the guidance of the target human body laser stripe, the coordinate information data of the crossing curve exists in a TXT file.

Noise removal. The noise of the structure light vision of rotary 3D human body scanning system is random, and it can not repeat.Therefore, the extraction error of the human body laser stripe center pixel coordinates belongs to the random error. In order to improve the extraction accuracy of the human body laser stripe center pixel coordinates, the influence of random noise must be eliminate. It can be seen from the Fig.4, the human body laser stripe is relatively lighter than the other area of the image. The gray value of the human body laser stripe is relatively larger than the gray value of the background. The noise gray value of the background is smaller than the human body laser stripe. So the gray values are easy to distinguish.

In this paper, the median filtering method is used to pre-process the sub-images. Considering the factor of window size of median filtering affects the calculation, the speed of filter processing, the smooth degree of human body laser stripe and the missing gray value information of it, the 3×3 window size of median filtering is selected for pre-processing sub-images. In this paper, noise removal is confirm by the following steps: Setting a gray threshold (I_t); In sub-images, the gray threshold of $I(i,j)$ hasn't a fixed value, and it is greater than 0 and less than 255, that is $0\le I(i,j)\le 255$.

In this process, considering the brightness of the human body laser stripes, and according to many experiments, generally, the value of I_t can be set between 160 and 220, that is $160\le I_t\le 220$. Fig.6 shows a column of gray distribution histogram of a human body laser stripe image. Each $I(i,j)$ of column of gray threshold in sub-image compares with I_t, according to the equation (1) [16-18], can not only filter out noises, but also preserve gray value of human body laser stripe pixels.

$$I(i,j)=\begin{cases} I(i,j), if I(i,j)\geq I_t \\ 0, \quad if I(i,j)<I_t \end{cases} \quad (1)$$

Among them, $i=1,2,3...M$, $j=1,2,3...N$, besides the value of M and N are the number of row and column of a image.

Image gray-scale binary. After de-noising processing, all the pixels of de-noising image are useful. For the convenience of subsequent calculation, in de-noising image, we select the minimum gray threshold as t_k . The processes of binarization conversion use the following equation (2) [19-20]. This step is used to extract the laser stripe image in human body. That is to say, this step is to go to the 0-1 two value of the image processing.

$$I'(i,j)=\begin{cases} 1, if I(i,j)\geq t_k \\ 0, if I(i,j)<t_k \end{cases} \quad (2)$$

Calculate the up and down band position of human body laser stripe. For the binarization image, all the gray value of the human body laser stripe pixels is 1. The gray value of other background pixels is 0. In this case, reading each column of a image, calculate the value of image gradient G . In each column of a image, using the equation (3), the value of G have two values: 1 and 0.

$$G[I(x,y)]\cong |I(x,y)-I(x,y+1)| \quad (3)$$

When the gray value of the odd number is 1, which indicates that these positions are up edges of the human body laser stripe, so we mark these positions as d_k .When the gray value of the even number is 1, which indicates that these position are down edges of the human body laser stripe and mark these position as u_k . So all the up and down edge position of human body laser stripe can be obtained, named $[d_k, u_k]$, while $k=1,2,3...$ is the number of human body laser stripe.

Measure the center of human body laser stripe. For each value of $[d_k, u_k]$, using its up edge position minus its down position, which means calculate the value of t, $t=u_k-d_k$. The position of half of t is the human body laser stripe center.

Determine the number of images that will be disposed. Thus the processing of the extraction of human body laser stripe center pixel coordinates can be finished.

4. Results of image processing experiments

Median filtering of each sub image processing, the purpose is to remove the random clutter point, so that the target information is more clear. The column processing method is used to find the target point in each column. The concept of gray threshold is used here. There is no fixed value of gray threshold in image processing, and the threshold value is between 0 and 255 according to the experience. After determining the number of images to be processed, we need to make a smooth processing of the decision point. The 3D reconstruction software flow chart is shown in Fig.6.

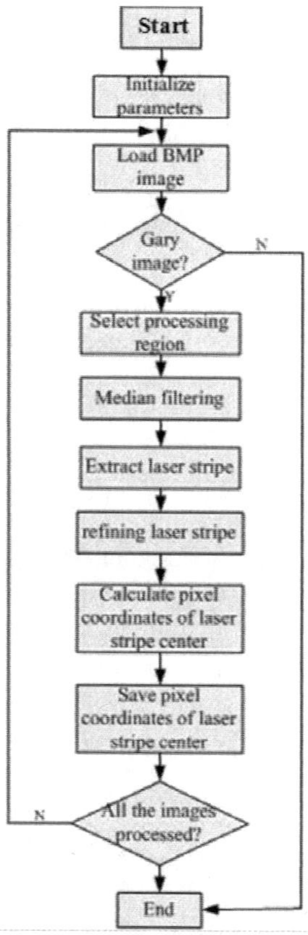

Fig.6 The program flow chart of image processing

5. Characteristic of human body laser stripe images

By using a three-dimensional structured light vision measurement principle as well known [9], the depth parameter of this study is determined by the measured human body laser stripe pixel coordinates in the camera position. Laser stripe shapes that camera received were varied with the transformation of human body's surface situation.

According to the characteristics of laser stripe images, the human body laser stripe images whose acquisition was based on the laser stripe rotating 3D human body scanning system usually has the following characteristics:

1) Most of the human body laser stripe images are simple stripe images. These images usually have one or more smooth stripes. That is to say, there are one or more continuous laser stripes in one human body image. These simple laser stripes appear in the relatively flat surface region of the human body, such as the position of neck region, the position of back or shoulder and so on.

2) Because of the human body are measured by natural standing posture, when the laser stripe is located at the position of the torso, both side of the arms images will be obtained. Usually, these images have one or more segment of laser stripes, and these stripes can not overlap. For example, the laser stripe image of human body's back is shown in Fig.4, it can be seen from the Fig.4 that the optical laser stripe consists with three parts, namely, the left arm laser stripe, torso laser stripe and the right arm laser stripe.

6. Conclusion

The system can be used not only in human body measurement, but also in 3D position measurement. Improve the hardware to gurantee the precision.

1) Rotation Structured-light Scanner Technique makes the volume small and measuring range large.

2) In the structured-light measurement system, special mechanical installation make the calibration simple.

3) The goal pixel region is limited to make the image processing time-saving. In the process of non-contact

measurement of human body, the position of the laser stripe can be estimated through the establishment of the mathematical model. By establishing the position model of the human laser stripe in the image, we can find out the limit position of this object. A new goal pixel searching algorithm is used make the computation effective. High-degree Rational Bezier curve is used in smoothing the measured data.

References

[1] Liu Y, Jiang SS, Zhang X. Body measurement point of monocular vision based on cloud data processing [J]. Journal of Textile. Vol30, No.3. 2009; 03: P: 126-131.

[2] Li XJ. Study on key technologies to enhance the performance of laser 3D body scanning system [D]. Tianjin University. 2010; 01: P: 16.

[3] Shu FF. Study on portable apparel customization system based on monocular vision [D]. Xi'an Polytechnic University. 2008; P: 62.

[4] Huo L, Zhang AJ, Liu WJ. Study on the extraction method of light knife center visual 3D measurement of structured light [J]. Computer applications and software. Vol23, No.10. 2006; 10: P: 100-102.

[5] Ai Y. The new optical measurement application [M]. Wuhan: University of science and technology of Surveying and Mapping Press. 1994.

[6] Xu SL. Numerical analysis and algorithm [M]. Beijing: Mechanical Industry Press. 2007.

[7] Fang J, Ren LH, Ding YS, et al. A 3D body measurement method with a single camera [J]. Journal of DongHua University(Natural Science). 2010; 36(5): P: 536-540.

[8] Chen DD, Yu DM, Xu ZB, et al. Research on three views of 3D measurement on human body with line laser of vision calibration system [J]. Computer Measurement & Control. 2008; 16(9): P: 1249-1251.

[9] Yuan Y. Basing on camera self calibration technique of non-contact body measurement [D]. Da Lian Light Industry Institute. 2005; P: 1-2.

[10] Zhu Z. Research and virtual trying on technology to get the 3D information of human body [D]. Hua Zhong University of Science and Technology. 2004; P: 12-13.

[11] Wu L. Basing on three dimensional body measurement parameter mannequin [D]. Shanxi: Xi'an Polytechnic University. 2006; P: 9-11.

[12] Zhao JM. The non-contact body measurement system of digital image processing and measurement of human information acquisition [D]. Tian Jin University of Technology. 2004; P: 1-3.

[13] Xia L. Three dimensional body measurement technology in apparel industry [J]. 2006; 34(5): P: 76-77.

[14] Yu YW. Three-dimensional body measurement and fitting garment cutting [J]. Journal of Textile. (6), 1996; P: 29-31.

[15] He DQ. Research on automatic measuring system for 3D human body size [D]. Jiao Zuo Institute of Technology. 2001; 4: P: 3.

[16] Zhu Z. Research and virtual trying on technology to get the 3D information of human body [D]. Hua Zhong University of Science and Technology. 2004: 12-13.

[17] Yuan Y. Basing on camera self calibration technique of non-contact body measurement [D]. Da Lian Light Industry Institute. 2005: 1-2.

[18] Di WB. The body measurement based on image processing [D]. Ji Lin University. 2006: 3-4.

[19] Lu XL. System development of automatic measurement of human dimensions [J]. Journal of Beijing Institute of Fashion Technology. 1999.

[20] Yu YW. Three-dimensional body measurement and fitting garment cutting [J]. Journal of Textile. (6), 1996: 29-31.

A Study on Air Thermal Concept Apparel Design

Bing-Jie Chou, Rong Zheng*, Ying Wang

Aimer Human Engineering Research Centre, Beijing Institute of Fashion Technology, Beijing, 100029, China

*Corresponding author's email: yuki.zheng@gmail.com

Abstract

Domestic studies on thermal concept apparel design using knit fabrics with air as insulation medium are scarce. This study sought to design a high-performance sport apparel that can be inflated and deflated by adding air channels in the single-piece basic jacket, thus modifying the knit fabric. As the medium to retain warmth, air is injected into the hollow part of the lining and exterior fabric by means of hot pressing. The study also considered muscular ductility and distribution of human sweat glands based on human ergonomics, aided by research on motion and ductility of skin creases and quantitative aesthetics. Comparative experiments were conducted and analysed to obtain the warmth retention index and assess thermal comfort.

Keywords: Environmental protection; Sustainable design; Aesthetics; Human ergonomics; Aeration; Light warm; Composite technology

1. Introduction

As the high-performance apparel manufacturing system gradually improves, warmth-retaining and light multi-functional clothes are gaining popularity for winter wear. New technology on fabrics and lining is making jackets much lighter and more effective in warmth retention. While the superiority of synthetic versus natural thermal fabric is still contended and environmental protection is regarded as an inevitable element in design, the newly introduced fabric padded with down feathers and synthetic fillings is bringing about a new concept on light, warm, and comfortable wear [1], and keeping warm does not necessarily entail bloated winter wear. The innovative heat insulation solution via down-feather filling and lap-joint design ushered in a new era of thermal wear design. PrimaLoft® introduced Down Blends products combining goose down and synthetic polyester fibre, which features better waterproofing properties, compressibility, and durability, and can be dried four times more quickly than duck down [2]. Meanwhile, Columbia Sportswear introduced TurboDown™, combining goose down and Columbia's Omni-Heat™ thermal insulation technology, thus creating a more effective thermal insulation layer and improving permeability [3].

In the "AirShop" store in Barcelona, Spain, the design and production process of inflatable apparel is displayed, presenting visitors with a new visual impact [4].

Based on existing processing techniques, this study designed air channels with considerations of relevancy between human joints and positions, distribution of sweat glands, as well as distribution of human muscles and extension direction and changes of human skin in stretching movements [5]. The relationship between fabric and clothing pressure has also been taken into account in producing air thermal concept apparel featuring good warmth-retention properties.

2. Approach

2.1 Men's muscle and sweat gland distributions

After analysing movements of muscles in five categories of sport (basketball, running, badminton, football, and heel-and-toe walking), data pertaining to major and secondary muscles in motion and movement extension were obtained, as shown in Fig.1. The yellow sections illustrate the major active muscles, green shows the secondary active muscles, and blue indicates great muscle extension. It can be observed that the yellow section covers major parts of the human body, the blue parts mainly cluster in human limbs, and the green parts are inserted into the yellow part [6]. It can also be observed that traction and compression of clothes are mostly caused by flexion and extension motions, and limb movements. Therefore, while seeking corresponding solutions, such factors as flexion and extension degrees of limbs, and primary and secondary relations of muscular movements should be taken into account.

The basis for the structure line often used in sportswear can be found in the muscle segmentation shown in Fig.1. The following factors can be integrated and applied in designing the structure line. According to the brightness division in the perspiration-area diagram shown in Fig.2, the darker the colour, the higher the perspiration degree [7]. In Fig.2, it can be observed that the area below the chest is the human body's major perspiration area, followed by the shoulder-blades area. The part below the waist usually has less perspiration [8]. However, those body parts featuring less perspiration need more protection against coldness.

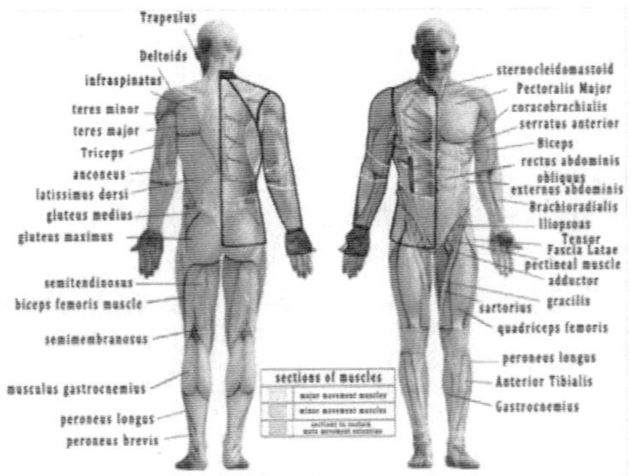

Fig.1 Distribution of men's movement muscles

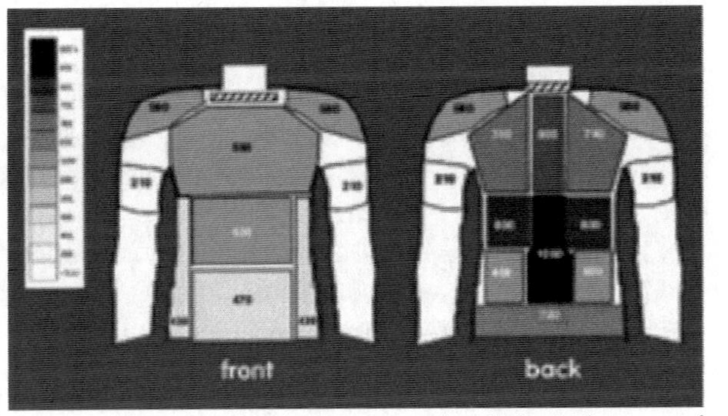

Fig.2 Distribution of men's sweat glands in 10km running

2.2 Skin extension direction and factor analysis of quantitative aesthetics

Infusing air into double-layer fabric will involve aesthetic distribution of air dimension. The study tried to convey the volume and existence of the warmth retention structure visually based on the extension principle of skin wrinkles [9].

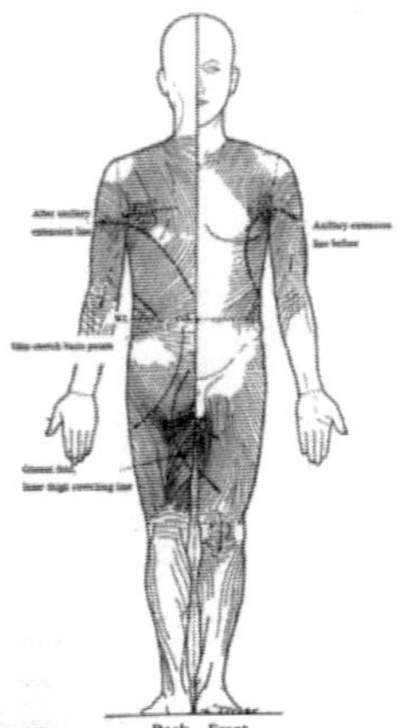

Fig.3 Extension direction of skin wrinkles [8]

Fig.4 Human body's extension and flexion in motion [8] [9]

The inflation area distribution will create a wrinkle prototype in the contour of human body surface. To design the inflation circuit based on the tendency distribution of the wrinkle prototype, the wrinkle principle should be analysed. Human skin can be taken as a piece of clothes to observe its structure. We analysed the extension direction of skin in the human body's breast and back, and clothes' function in sport. As shown in Fig.3, the extension direction of skin wrinkles is in line with the movement direction of connective tissues of skin. From the extension direction of human skin, it can be concluded that the extension direction of the movement paper pattern of the human body is consistent with the extension direction of skin based on the extension function of skin wrinkles shown in Fig.3. A rough linear pattern can be abstracted from the front and back central points of the human body and the extension line of the front and back armpits. As for the design of the back central line, human spine shape should be taken into consideration. As shown in Fig.4, air pressure needs to be distributed rationally [10], to meet the human body's flexion and extension needs in movement, which is directly relevant to the back shape of the inflation channel design. The proper structure design is based on the comprehensive consideration of sweat gland distribution, muscle distribution, human perspiration areas, and men's muscle distribution.

3. Results and discussion

3.1 Style design

This style of air thermal apparel mainly emphasises the design of air inflation to realise warmth retention. Based on current research related to inflation techniques, the study designed air channels with consideration of men's muscle distribution and skin extension direction; different inflation areas (e.g., spine, breast, and back waist areas) were designed that incorporated relatively more inflation areas. In building the morphology aesthetics, a ventilation design was subtly introduced in order to reach a balance between warmth-retention function and comfort, trying to create a microclimate featuring a beneficial cycle [11].

We considered the case of someone jogging outdoors wearing the inflation apparel. The increasing outdoor wind speed can accelerate cross-ventilation and heat exchange in the wearer's microclimate. When the inflation apparel is equipped with a ventilation design, the warm air of the microclimate in the inflation apparel will go up along the body surface. When the inflation apparel has openings in such upward parts as collar or armpits, the rising hot air will flow out from the openings. Therefore, we concluded that the ventilation design could be conducive for heat inside the apparel to be transferred to the outside. Location of openings in the clothes and their sizes will also influence the heat dissipation. The ventilation design can also be helpful to dissipate vapours inside the clothes. This style of inflation clothes needs a cross-ventilation design because the fabric's airtightness should be ensured, all the while maintaining a balance of comfort. Therefore, the armpit part should have a ventilation design. To realise the design, a precise pattern design should be made and a fabric with good airtightness should be selected.

The design of separating line should be based on human skeletal and muscle anatomies. In designing the men's inflation apparel, the study took advantage of thermal conductivity of air to realise warmth retention in parts of the apparel with consideration of human body muscle distribution and men's sweat gland distribution, as shown in Fig.1.

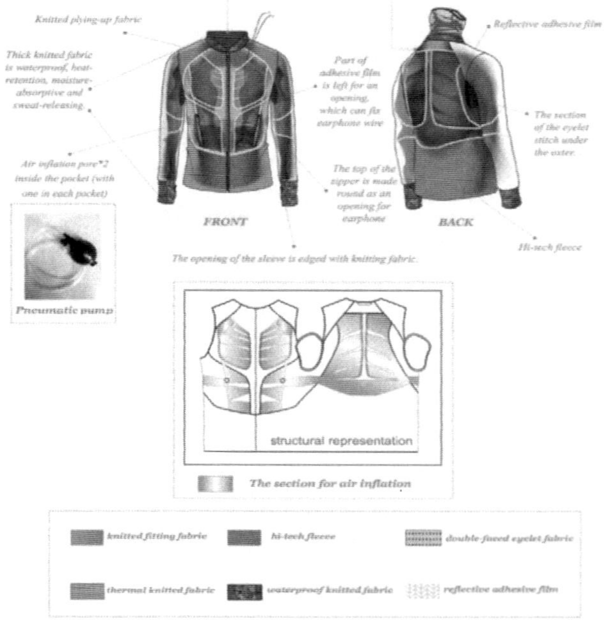

Fig.5 Design effect drawing of men's air thermal jacket

In light of major movement muscle distribution in the upper body, double-layer fabric was used and inflated with the contour lines of area divisions coinciding with the interspaces between muscles. Considering that secondary movement muscles would carry away some heat in perspiration, based on men's sweat gland distribution, single-layer fleece with short fur was used in the waist heat dissipation area of the apparel. In designing areas that will see relatively large extension, elastic-knit laminated fabric with basic warmth retention function was used, with the final effect drawing shown in Fig.5.

3.2 Pattern design

The design of air channels can directly influence microclimate of the human body, thus forming a chaining effect, so the pattern design must take warmth retention as a priority, integrating the front and back parts without lateral lines separating the pattern. In the case of cross-ventilation between the front and back pieces, air heat can be transmitted between them, as shown in Fig.6 and 7.

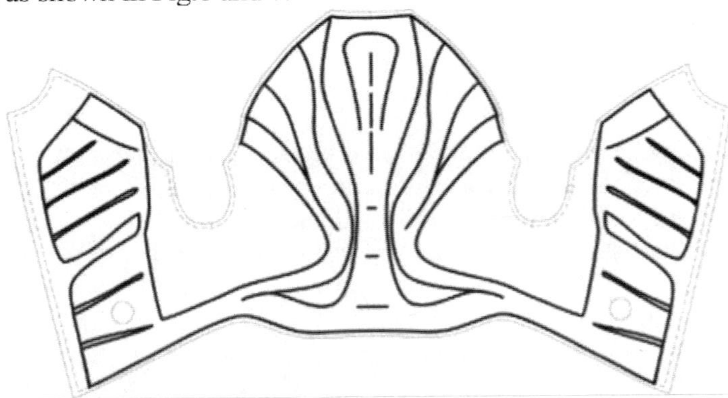

Fig.6 Pattern design of the inflation structure of the single-piece bodice

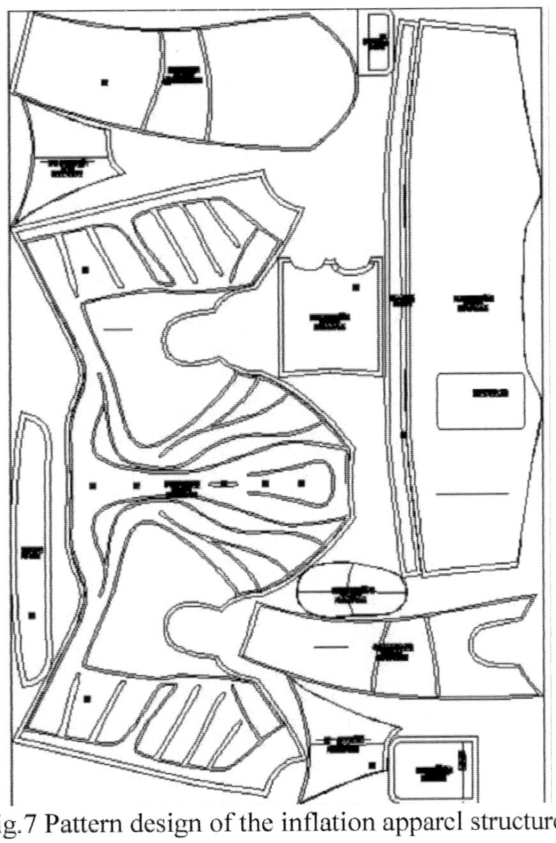

Fig.7 Pattern design of the inflation apparel structure

3.3 Air channel manufacturing technique

Seamless bonding technique was used to splice fabric for the major part of the inflation apparel. Knitted laminated fabric (100% polyester fibre) was used for the major part of the apparel, able to resist moderate rain. Splashing-water-prevention and bonding techniques were used in producing fabric to be used as the apparel surface. Since the knitted laminated fabric is elastic, the industrial membrane to be used in the secondary processing should also be elastic. The knitted laminated fabric spliced with membrane was used as the major fabric. A sample was taken to test its anti-tearing property; the higher the figure, the better the function, as shown in Table 1.

Table 1. Anti-tearing strength property of knitted laminated fabric

Material name	Element	Discolouration in washing	Staining in washing	Dry rubbing	Wet rubbing	Laundering durability	Fracture strength	PH value
Knitting laminated fabric	Polyester fibre	4-5	4-5	4-5	4-5	Qualified	Unable to be ripped	6.4

After testing, ultrasonic welding and hot melting adhesive film bonding techniques were used to realise seamless joints. The following auxiliary materials were used: reinforcing band (seamlessly splicing hems), TPU (thermoplastic polyurethane with good tightness, high strength, high modulus, excellent thermal stability, and wear resistance), and Sewfree206D hot melt adhesive (good adhesion nature and stability) produced by Bemis Associates Inc.

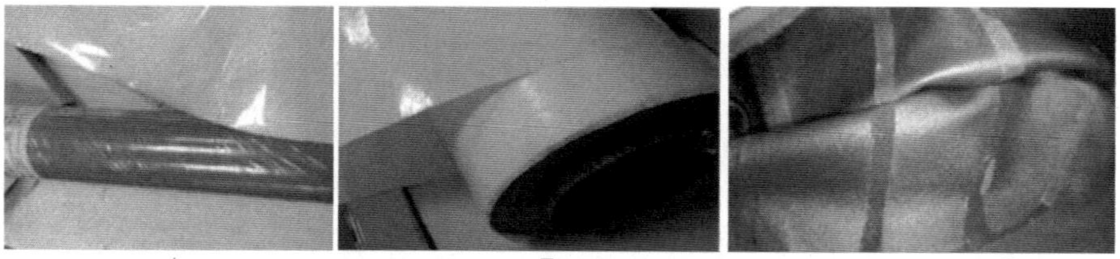

A B C

Fig.8 (A) Hot melt adhesive film, (B) TPU film, and (C) inside view of sample apparel after adhesive film is torn away

Single-layer knitted fabric was used in perspiration and heat dissipation areas as much as possible, except for the inflation part in the bodice. The fabric spliced with film basically solves the tightness issue, but meanwhile disrupts air mobility in the microclimate of the body surface and the comfort property of the fabric. Single-layer knitted net was used in the armpit area to balance ventilation and comfort, and to help discharge heat produced in human body movement.

Each pocket in the left and right areas, respectively, is equipped with an aeration hole for inflation and deflation. The cuff is designed with two openings (for thumb and other four fingers, respectively) with edges covered by elastic fabric. The jacket was externally pressed with reflection adhesive film. The pattern created by the adhesive film is in line with the aesthetic beauty of men's body structure, meeting night runners' need in safety and aesthetics, as shown in Fig.9. TPU was used in earphone punctures in the right and left sides of the collar, convenient for entertainment and communication, as shown in Fig.10. Flannelette was used as interior lining of the collar, enhancing warmth retention. The zipper in the frontal fly area is waterproof, improving tightness. Each side below the frontal waist part is equipped with a cellphone pocket. In addition, considering night exercise safety, the surface of the exterior fabric in the bodice and sleeves of the apparel is adhered with reflection adhesive film in a dynamic design, thus improving the apparel's multiple functions.

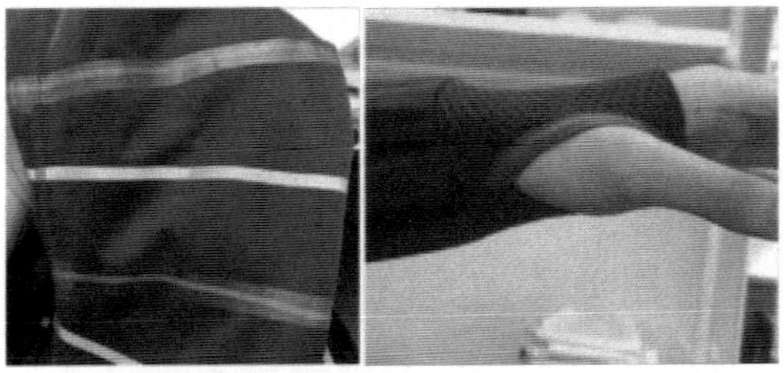

Fig.9 Improved pattern techniques in the elbow, cuff, and pocket areas

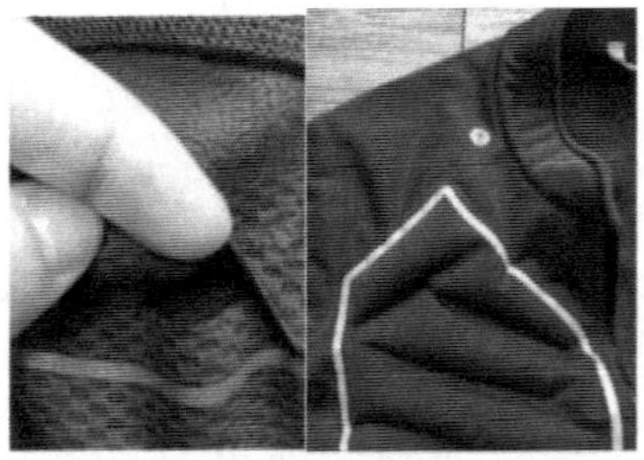

Fig.10 Internally placed pocket and earphone punctures made with TPU

3.4 Inflator design

Modern precision electronic apparatus can be used in seamless laminating. The aeration hole of the air channel is the valve to inflate the jacket. Air tightness of the aeration hole determines whether or not the inflation will be successful. The aeration hole is a plastic puncture, equipped with an inflator; hot lamination technique was used to laminate fabric, ensuring tightness. Light treatment was made from interior to exterior of the apparel, trying to present a sense of technology. The inflator is shown in Fig.11.

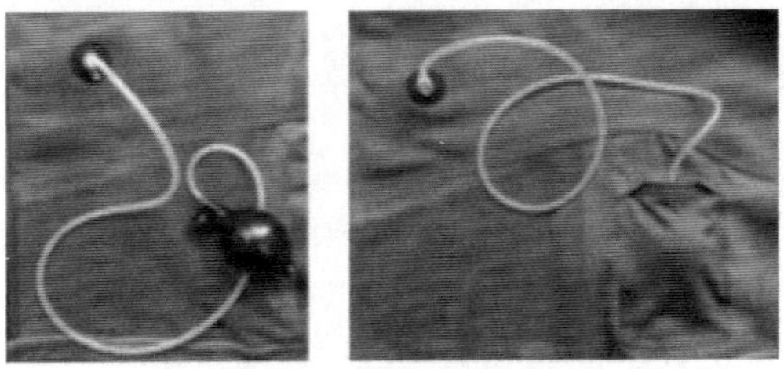

Fig.11 Inflator design

The design has been improved in terms of exterior processing technique and details influencing comfort after two versions. The effect drawing of the final apparel is shown in Fig.12.

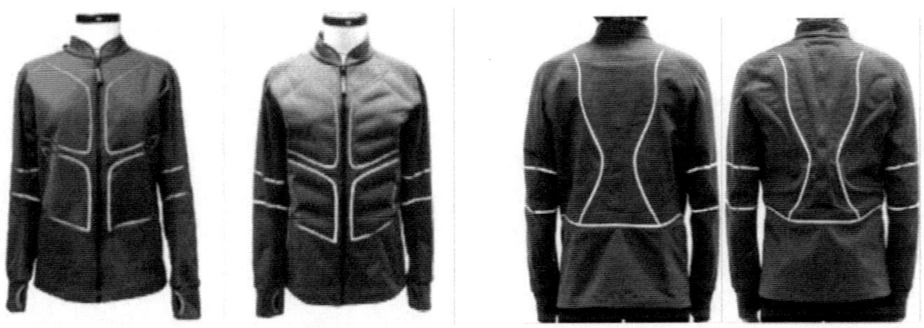

Fig.12 Front and back of the air thermal jacket before and after inflation

Subjective and objective tests of the air thermal apparel were carried out with 15 runners enrolled to participate in the tests. Subjective evaluation encompassed such aspects as overall evaluation of the apparel, comfort degree of the apparel, body feelings of cold or warmth, and skin feelings of dryness and wetness. The objective test included comparative testing of temperature and humidity. Tests proved that the air thermal apparel has a certain degree of warmth retention property and can realise the function of moisture absorption and sweat dissipation. The design realised the pre-set purpose.

4. Conclusion

This study's air thermal concept apparel design was based on the microclimate of the body surface in human ergonomics. Several elements such as human muscle distribution, men's sweat gland distribution, three-dimension quantitative aesthetic distribution, and extension direction of skin creases were taken into consideration and integrated in design; properly placed partial inflation areas were designed with a successful combination of human ergonomics and aesthetics. The jacket pattern is suitable for the human body to have basic stretching movements. The design has been improved after two versions with tightness enhanced. Moreover, the study method and theoretical basis can also be applied in design and development of other sportswear.

References

[1] David M. Insulation (heat) equipment & supplies. Adv Text Technol: 2014; 10.

[2] The fruit of research. Selezione Tessile: 2002; 40(2): 48-55.

[3] Polartec AP. Future Materials Group. Future Materials: 2013; (1)15-16.

[4] White S, Demko J, Tomich A. Flexible aerogel as a superior thermal insulation for high temperature superconductor cable applications. AIP Conference Proceedings: 2010; 1218(1): 788-795.

[5] Ma S, Liu W. The aesthetic deduction of ecological idea in clothing design. China Textile Leader: 2014; 2: 23-25.

[6] Yu L, Duan L. Average body temperature change and thermal sensation forecast as environmental temperature changes suddenly. Refrigeration Air Conditioning & Electric Power Machinery: 2006; 19: 45-47.

[7] Gu M, Song Y, Zhang L. Discussion on relationship between clothes microclimate and human feeling. Hebei Fangzhi: 2009; 1: 20-21.

[8] Machuru N. Human body and clothes: body structure, aesthetic elements and pattern.

[9] Liu L. Mathematical simulation of human body heat and moisture transmission and establishment of thermal comfort feeling model. Shanghai: DongHua University: 2002.

[10] Chen L. Influence factor analysis of waterproof and moisture permeable fabric wearability. Cotton Text Technol 2014; 7: 20-21 [11] Chen D. Clothing hygiene. Beijing: China Textile and Apparel Press: 2000.

[11] Chen S. The research on the designing of modern sporting apparel. Jiangsu, China: Suzhou University: 2008.

A Study on the Current Situation of Zhouzhi Embroidery and its Development

Xi Li, Zhao-Hua Liang*

Apparel & Art Design College, Xi'an Polytechnic University, Xi'an, 710048, China

*Corresponding author's email: liangzhaohua123@126.com

Abstract

Zhouzhi embroidery has varied stitches, novel patterns, soft colors, and rich structures, which fully demonstrates the folk art of creativity. This paper mainly describes the formation and development of Zhouzhi embroidery from aspects of location, resources and cultural customs etc. The aim of this paper is to not only help us understand this particular form of art embroidery and its unique characteristics, it also lays a theoretical foundation for future generations and the development of Zhouzhi embroidery's research in hope to preserve its culture and practices.

Key words: Zhouzhi; Embroidery; Yabai embroidery; Historical development

As a traditional embroidery craft, embroidery originated in a folk, which is the skill that working people beautify production. Embroidery in the Shaanxi region is widely popular in both the rural and urban context. This is particularly true for the making of rural women's dowry, the old's and children's hat, clothes and shoes, etc, which are typically decorated with embroidery.

1. The Basic Situation of Shaanxi Embroidery

From the late Qing Dynasty to the Republic of China, Shaanxi embroidery has not been properly developed, as a result the embroidery industry is mostly agricultural based and part of the sideline family artisanal production. There is no professional commodity production until 1949. After 1949, the government paid much more attention to protect and promote arts and crafts. Due to this change, embroidery had an unprecedented development. The art of embroidery quickly became more refined than ever towards the artistic direction. Many embroidery factories, research institutes, embroidery product factories and others were established in all regions. In 1954, the local government gathered artists specializing in handicraft embroidery artists and toke the road of co-operation. The first embroidery group (ie, Xi'an Jinjiang Embroidery predecessor) and the second embroidery cooperative group (ie, Xi'an Phoenix Embroidery predecessor) were built at that time. Furthermore, the prosperity of the tourist industry encouraged Shanxi folk embroidery to develop for commodity production. To meet the demands of the market model, various forms of embroidery enterprises have been established throughout the entire province, specializing in the production of embroidery. "Regulations on Protection of Traditional Arts and Crafts" was promulgated in 1997, enabling the embroidery industry to flourish under the principles of "protection, development and improvement" and "practical, economic and aesthetic".

2. The Geographical Features of Zhouzhi

68 kilometers away from Xi'an, is a famous Guanzhong plain county called Zhouzhi which is part of the Xi'an municipal county. Siting in the junction of Guanzhong Plain and Zhongnan mountain, north Weihe River, South in accordance with the end of Nanshan. Zhouzhi has a warm temperate continental monsoon climate. It has a beautiful environment, wonderful scenery and abundant resources, and is known as "golden weeks". Zhongzhi is often considered as the rich land in the guanzhong region.

Zhouzhi embroidery is contrastive, vivid and multi-colored mainly featuring red, brilliant green, yellow, and blue.

Full of strong local folk flavor, different embroidery patterns are usually used according to different occasions. For instance multi-mandarin ducks and Kirin sons are often used in a girl's wedding dress whereas longevity figures and crane are used in elderly embroidery. Zhouzhi embroidery has diverse stitches, mainly featuring flat needle jumps, the senate needle, needle cloud, playing mainly seeds and other six categories. The skills are also varied, especially concise style embroidery and clear stitches, which is plain but gorgerous.

The Guanzhong area belonging to the middle region of the Shanxi province is fairly flat. As a result transportation is so convenient that people come here frequently from the outside. Affected by secular culture, folk stories and customs, the embroidery style is full of fineness and rich colors. The Folk embroidery in Zhouzhi region, inherited the fine tradition of ancient Chinese culture, which has maintained strong local characteristics. Under long-term development, it not only became an important part of endemic folklore, beautify the life, it has also improved the quality of life and source of income. The county owns more than 60,000 embroidery workers, and the annual output is more than 500 thousand million yuan, Yabai embroidery market is the largest embroidery market in the Northwest embroidery market.

Fig.1 Rural women were sewing embroidery in Zhouzhi, Zhouzhi embroidery production

Table 1. Comparison of Zhouzhi County do embroidery enterprises and Xi'an Phoenix, Jinjiang embroidery factory

Factory name	Total labor productivity (constant price) yuan \ person	Per capita net worth (constant price)yuan \ person
Xi'an hoenix embroidery factory	8375	2372
Jinjiang Xi'an embroidery factory	6825	2977
Zhouzhi County embroidery company	6491	1437
Zhouzhi County embroidery factory	3442	1083

The Zhouzhi silkworm industry has a long history. The Yabai area located in the Guanzhong Plain has a mild climate which is very suitable for mulberry sericulture and silk woven production. As a result the silk industry is very prosperous and has become a solid foundation to develop Zhouzhi embroidery. In addition, there are also unearthed weeks spinning wheels and bone needles and other artifacts, which we can get a glimpse of the situation that our ancestors mulberry sericulture, silk reeling silk weaving, and embroidery sewing.

3. The Historical Origins of Zhouzhi Embroidery

The capital of Western Zhou Dynasty located in Feng Hao, Zhouzhi is near the capital, as well as Yabai. "Systematic 'Kaogongji" described embroidery, silk and painted practicing, which suggests that embroidery is very popular among Zhouzhi's slave-owning aristocracy in Zhou Dynasty. From Qin to Han Dynasty, in the east of Heihe, there is Shanglinyuan which means splendid palace, including Chang Yang palace, five oak Palace, the

Palace of grapes, and shooting bears Palace. The walls are made of pepper and smear mud, and above the wall is a magnificent and beautifully adorned Embroidery [4]. In the Middle Western of china rich landowners began to adorn large amounts of embroidery which traditionally was only used by the royal families. What's more they began to use embroidery decoration for the the walls of their living room. It is clear that embroidery became prevalent early in Zhouzhi. As Zhouzhi is very close to the capital of Sui and Tang dynasties, embroidery spread to the countryside in Sui and Tang dynasties. The emperor in Tang Dynasty visited Lou Guan Tai many times, which resulted in the influence of royal embroidery on Taoism embroidery. To this day, embroidery in Taoist temples is a continuation of traditional embroidery from the Tang dynasty. From Tang Dynasty to Qing Dynasty, religion was a major instutition, and as a result many palaces and temples in rural and urban areas were decorated with exquisite embroidery, this is particularly true for the clergy's clothes. Zhang set up Shaanxi Opera classes in Ming Dynasty in Zhouzhi, and the drama dresses were all decorated with embroidery and made the from Zhouzhi county. The county also supplied outward drama dress embroidery. In the Qianlong period, Zhouzhi embroidery was very popular and is used mainly for daily necessities. For instance it is made for a girl's dowry and decoration, children's clothing, the old man's cigarette packets and so on. Despite this, the embroidery industry is fading because of serious corruption in the late Qing Dynasty as well as the press and exploitation of Kuomintang. After the liberation, Zhouzhi embroidery resumed production, but the development of embroidery industry was greatly restricted under the ideological influence of the "left", which suggests that the family embroidery is the impact of national industrial production.

4. The Rise of Zhouzhi Embroidery

In early 1970s, there was a huge overcapacity in Zhouzhi's labor force. Yabai was the first town to introduce embroidery machine. Huaihua village of Yabai and Xi'an Phoenix Embroidery worked together, running a machinery factory to process embroidery. Later on, many villiages like Dong Feng, Tuan Jie, and Xia San Tuan also set up the embroidery factories that is linked to Xi'an Jinjiang, Phoenix Embroidery. According to the patteren of Huaihua village, they engaged in processing too. Until 1978, there were 13 village embroidery factories which owned more than 620 sewing machines and 630 embroidery practitioners. Their annual income was 300,000 yuan [1]. At the same time, having recognized the good opportunity to make profit, these embroidery workers disguised their household sewing machines and repurchased Spiraea and embroidered materials to produce embroidery. Some women sellers took the products to the streets, which was very popular way at that time. The task of embroidery workers was producing many embroideries every day and wholesaled them to vendors.

After the Third Plenum of the CPC(the Communist Party of China) in 1978, that the policy on "invigorating the domestic economy, opening up" enabled embroidery production to rapidly develop. As a result foreign trade and embroidery export increased. The production began to spread to the countryside. Meanwhile, many processing sites were set in Zhouzhi. In 1979, the embroidery workers of Zhouzhi increased from 420 in 1978 to 1280 in 1981 because of the national reform. The net income increased from 241,000 yuan to 180 million yuan. Thanks to the demonstration and inspiration of Yabai embroidery, the embroidery workers rapidly increased from 630 in 1978 to 10,600 in 1981, net income of $ 300,000 to $ 2.61 million yuan in the whole county.

In the 1980s, Yabai-centered embroidery production quickly extended to the whole county. In the spring of 1982, the CPC Zhouzhi county, the county people's government held a embroidery forum in Yabai town, making the decision "About developing the production and marketing of Yabai embroidery", which urged the county economic sectors to vigorously support embroidery through engaging funds on raw materials, sewing and embroidery machines and other supplies. Meanwhile, they also released many measures, like simplifying sales accreditation review and other procedures, opening theater dumb Parker, setting up a embroidery materials market, relaxing embroidery sales policy, encouraging farmers salesman to sell embroidery, doing inspection regularly in the plains towns, and guiding embroidery production. Under the policy and support of all parties, Zhouzhi embroidery developed very fast, especially the following towns: Er Qu, Si Tun, Qing Ha, Zhu Yu, Hou Village, Guang Ji, Ma Zhao and so on. Yabai and Situn became professional embroidery towns, and more than ten villages including Huaihua became the professional embroidery villages [1].

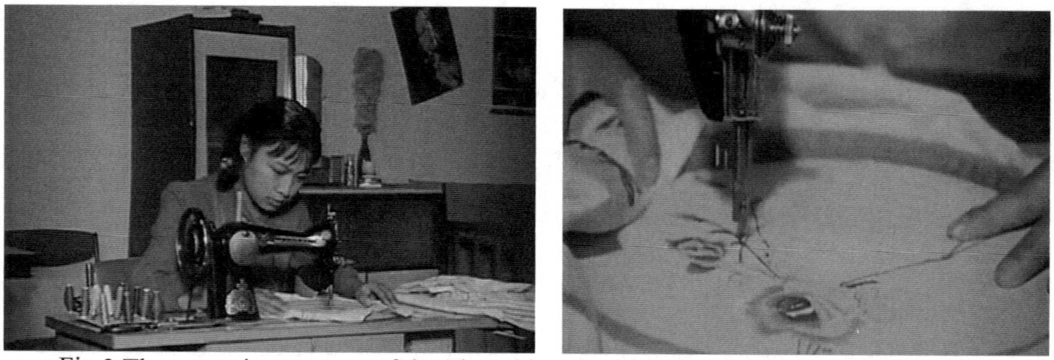

Fig.2 The operating process of the Zhouzhi embroidery by Zhouzhi county annals

Meanwhile, the county government has also set up training courses. Embroidery Factory and the county branch of the embroidery backbone site to teach embroidery workers some techniques, production and marketing of embroidery. Most farmers found it is an excellent opportunity to become rich, so they achieved good results according to the measures. By 1986, there were 41072 embroidery workers in the whole county, and the net income increased to 28.28 million yuan. Due to the large amount of Spiraea, Yabai town began to set up a collective dyeing factory in 1987, and dyeing factory started operation after 3 years. In 1992, the second private line dyeing factory was built. Zhouzhi embroidery was colorful, exquisite and well-designed which expressed the people's good wishes and ideals to life. Coupled with sufficient labor reserve resources and application of machine embroidery, embroidery became more beautiful and practical with a lower price. Zhouzhi embroidery firmly occupied the market, exported embroidery to Gansu, Qinghai, Shanxi, Sichuan, Ningxia, Xinjiang and other provinces.

Fig.3 Yabai embroidery market by Zhouzhi county annals

On December 10, 1991, the Yabai embroidery market was constracted, which was lead and managed by the industry and commerce administration departments. The new Yabai embroidery market consists of 11 319.86 square meters, 5581 cubic meters, 81 square façade and 600 greenhouse boothes, which was the western region's largest distribution center for embroidery. Nowadays, Yabai Embroidery County is the pillar industry of Zhouzhi, which developed from a small collective factory and embroidery supplies at home and abroad every year. Patterned embroidery lines using mechanic operation, delicate, subtle, beautiful color, all kinds of varieties of bedding supplies such as bed sheets, quilt, bedspread, single tables, pillowcases, etc. There were more than 300 kinds varieties. It is a representative of Xi'an exquisite handicraft works. Yabai embroidery market is focused on Zhouzhi's wholesale embroidery, every lunar third, sixth, nineth days, embroidery is put on the market, whereby most of the county embroidery sells to wholesale customers around.

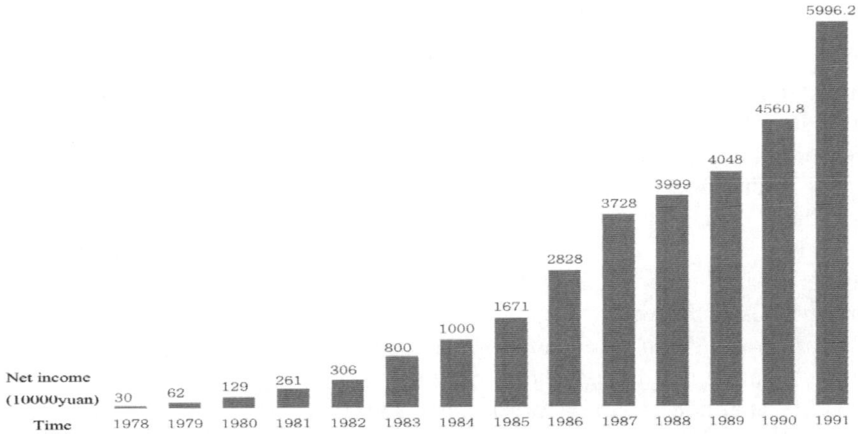

Fig.4 1978~1991 Zhouzhi years embroidery industry income statement.

Table 2. 1978~1991 years of Zhouzhi County embroidery table

Particular year	Embroidered work (person)	Salesman (person)	Net income (10000 yuan)
1978	630	78	30
1979	2 400	190	62
1980	5 000	400	129
1981	10 600	600	261
1982	11 000	800	306
1983	13 500	1 400	800
1984	16 000	2 600	1 000
1985	21 193	4 600	1 671
1986	41 072	5 928	2 828
1987	48 000	5 186	3 728
1988	48 063	5 186	3 999
1989	50 890	4 235	4 048
1990	59 100	4 818	4 560.8
1991	59 800	4 912	5 996.2

5. The present situation and development of Zhouzhi embroidery

In the 1990s, special embroideries were bought from other town by businessmen in Zhouzhi who ran a prosperous embroidery business so that their embroidery business flourished.

However, good times did not last long, in the late 1990s, embroidery in zhouzhi rapidly faded, popular embroidery became unwanted. Some insiders believed that the decline was due to the lack of market economy sense and peasant consciousness. The Malicious competition between farmers conducted to the price cannibalism. In order to reduce the cost and improve batch production many poor quality cloth and silk were used in embroidery which resulted in the poor quality of product and fading colors. In addition, the most important factor was due to the lack of innovation and their single conservative business ideas which made the products fall behind.

The fading of embroidery made local people and the government feel bitter, and as a result they made up their mind to reform the embroidery industry through establishing the concept of integrity and exploring the market actively to restore and further develop Zhouzhi embroidery. With the development of market economy, Zhouzhi embroidery has gradually matured. 'company-farmers-scientific', the three-in-one development mode was gradually adopted by the industry. The original brilliant embroidery industry left abundant resources to Zhouzhi. There are still many sellers in Kunming, Chengdu, Guiyang wholesale markets selling Zhouzhi embroidery as well as Xian Kangfu Road wholesale market.

Some of the locals in Yabai town local are very good at the appreciation of embroidery. They create sales network sell embroidery on the net. Except the merchants who supply the repairing of sewing machines a large number of experienced embroidery rural women also surround the Zhouzhi town. The industrial chain of product-supply-sale are forming which can benefit the recovery of Zhouzhi embroidery. After the baptism of the market economy, Zhouzhi embroidery operators pay more attention to product quality and even more attention to consumer demand. Zhouzhi embroidery gradually adapted to the new ideas of producing for the market.

Superior natural environment encourages Zhouzhi to gather talents. Profound history and culture accumulated precious cultural heritage. The fertile soil recorded Zhouzhi mellow multiple forms of folk art. Zhouzhi embroidery has been developing hundreds years and becoming one of the exotic flowers of Shaanxi folk art. Its strong local flavor and local characteristics are welcomed by the broad masses of people. We will continue to strive for innovation and continuous progress, and further cultivate Zhouzhi embroidery for the better. .

References

[1] Shaanxi local history books, embroidery Zhouzhi county annals. Xi'an: Shaanxi Press: 1993.

[2] Shaanxi Province Local History Compilation Committee, Shaanxi Province Chi. Ethnography. Xi'an: Shaanxi Press: 2000.

[3] Ge F. Qin embroidery village can reproduce CLS. western development: 2008 (09).

[4] Wang AQ. Gold needlepoint embroidery flowers pin Kyushu - Embroidery industry Zhouzhi County History .Chinese industry and commerce: 1993; 36.

[5] Liu SH. Machine embroidery industry way out lies in innovation. Zhejiang Arts and Crafts: 1995 (01).

[6] Wu JX. Craft art history in shaanxi province. shaanxi: people's fine arts publishing house: 1993; 12.

[7] Zheng SS. embroidery . Beijing: China Social Sciences Publishing House: 2007; 10.

[8] Wang X. The ancient Chinese embroidery. Beijing: China Business Press: 2014; 12.

[9] Li HF. Chinese embroidery culture interpretation .Beijing: intellectual property press: 2015; 01.

[10] Li YL, Li YF. Ming and qing dynasties embroidery. Shanghai: donghua university press: 2015.

[11] Yang JZ. The Chinese folk custom series, shaanxi folk. Lanzhou: gansu people's publishing house: 2002; 316.

[12] 30 in xi 'an history data series, the guanzhong folk. Xi 'an: shaanxi people's education press: 2012.

[13] Women volunteers compilation committee, shaanxi province, shaanxi volunteer women volunteers. Xi 'an: shanxi people's publishing house: 2000.

[14] Left in hanzhong city, Chinese folk fine arts modelling. hunan: hunan fine arts publishing house: 1992.

[15] Wang P. Chinese folk art general theory. Nanjing: southeast university press: 2010.

[16] Zhang P, Liang ZH. Artistic nature of traditional embroidery and computer embroidery comparison.art and design: 2010 (07).

[17] Nie YT, Liu F. Inheritance Analysis embroidery and embroidered craft - A Case Study of the Miao Costume Huang County. Guizhou Ethnic Institute: 2011 (05).

[18] Su J. Embroidery products aimed at the general public and tourists Nuggets.Suzhou Daily: 2010.

AUTHOR INDEX

AUTHOR INDEX

AUTHOR INDEX

Textile Bioengineering and Informatics Society (TBIS)
Hong Kong Polytechnic University
ST733
Hung Hom, Hong Kong

ISBN 978-1-5108-2898-8

CURRAN ASSOCIATES INC.
proceedings.com

9th Textile Bioengineering and Informatics Symposium Proceedings (TBIS 2016)

Melbourne, Australia
12-15 July 2016

Volume 2 of 2

Editors:

Yi Li
Rajiv Padhye

ISBN: 978-1-5108-2898-8

Printed from e-media with permission by:

Curran Associates, Inc.
57 Morehouse Lane
Red Hook, NY 12571

Some format issues inherent in the e-media version may also appear in this print version.

Printed by Curran Associates, Inc. (2016)

Textile Bioengineering and Informatics Society (TBIS)
Hong Kong Polytechnic University
ST733
Hung Hom, Hong Kong

Phone: (852) 2766 4206
Fax: (852) 2764 5489

tbis@tbisociety.org

Additional copies of this publication are available from:

Curran Associates, Inc.
57 Morehouse Lane
Red Hook, NY 12571 USA
Phone: 845-758-0400
Fax: 845-758-2633
Email: curran@proceedings.com
Web: www.proceedings.com

PROCEEDINGS OF TBIS 2016

TABLE OF CONTENT

Track 1 Textile Material Bioengineering

Track 2 Textile Biomechanical Engineering

Track 3 Clothing Bioengineering

Track 4 Textile Supply Chain Informatics

Track 5 Textile Thermal Bioengineering

Track 6 Textile Science and Engineering

Track 7 Textile Biomedical Engineering

Establishment of Regression Models of Human Shoulder Arc

Xiao-Yu Xing, Tian Yang , Xiao-Mei Shang*

College of Textile and Clothing Engineering, Soochow University, No.1, Shizi Street, Suzhou City, Jiangsu Province, 215006, China

*Corresponding author's email: shangxiaomei@suda.edu.cn

Abstract

In this paper, an approximate curve of human shoulder arc was obtained by tracing point method, then we draw the circumscribed circle along the curve. By comparing each function fitting effect, we selected the cubic polynomial as the best fitting function. Then we amended to get function relation between length of shoulder arc curve, shoulder wide and shoulder thick. Finally, by zooming the curve, the length of shoulder arc of the same height was calculated and computing algorithms for non-contact automatic measurement were provided. Experimental results show that using this function can curve fitting out small errors, and basically meet the requirements of costume design size error.

Keywords: Shoulder arc; Spline fitting; Least squares fitting; Parametric equation

1. Introduction

With the development of computer technology, non-contact body measurement technology has a wider application and research in the field of clothing. Compared with the traditional manual measurement, its efficiency is greatly improved and the accuracy is higher than before. The non-contact measurement has played a huge role for the apparel mass customization and the census of human dimensions. Automatic three-dimensional non-contact measurement method generally use laser measuring, infrared measurement, moire method and photogrammetry method to collect the the original data of body size, then rebuild the human body to get the body size of each part [1-3]. The equipment developed from these measurements have higher accuracy, but because of the complex system architecture, high cost, expensive, they are not conducive to promote. The method of obtaining the two-dimensional front, side and back of the image from three-dimensional body, and extracting the outline, then fitting the girth using a mathematical model which based on directly measured body data to get the body size, because of its lower costs, occupancy little space, the price is not high, basically meet the needs of accuracy and other characteristics, get widely attention and research [4,5].

Non-contacting human body auto-measurement technology obtains human body size quickly and exactly, to establish human body form database. It is applied widely not only for the study of human body measurement, the standard of clothing size and custom-made, human body modeling for all kinds of person, but also for providing the research resource about bodily form trends of our country and data sustaining by which experts design more comfortable, more customizedly daily necessities [6-8]. In our country, it's very late for the study of Non-contacting human body auto-measurement [9]. We introduce into advanced equipments from American, Europe, and Japan. These equipments are too expensive and not fit for our countries' bodily form. Meanwhile, the study is not applied for clothing manufacture and still for experiments [10]. Aiming at above disadvantages, we develop a non-contacting human body auto-measurement system based on 2D images. The main study in this paper is that: Because of the disadvantage that we can't obtain the human body's circumferences size from 2D images, we sort and analysis for circumferences, mapping the examples' circumference curves approximately. Then we make relationship using functional model between the circumferences and its width, depth, human stature and avoirdupois. Thus, it is easy to estimate the sizes by the models. Otherwise, we present a approach about based example functional model of human trunk for actual human body modeling.

Among the many human dimensions, girth size calculation is the most complex and not easy to control, the error is also one of the biggest challenges. Because of photogrammetry acquired two-dimensional image information of the human body, and the body girth curve where the plane perpendicular to the plane of the image, we can not get information on the body circumference linear curve from a two-dimensional image. Therefore, we must study the human body by girth linear curve, with a reasonable degree of scientific methods to estimate the length of the curve around the use of the existing constraints. Through image processing to the front body, the back side of a two-dimensional, some of the height, width information can be obtained directly, but some very important girth information can not be obtained directly from the image, such as the body's bust, waist, hip circumference which are the essential data for garment production. So, how can we calculate the direct image data obtained in the circumference of the data is the focus of our research. Tianjin Polytechnic University use mapping table section curve to measure the body circumference curve linear, and use logarithmic function with its least squares fit. The experimental results showed the error between the obtained and the true length is basically meet the requirements of costume design size error [11-13]. Experimental procedures in this article found that using a logarithmic function of the measured bust smoother curve fitting results were not good, and use cubic polynomial fitting, the error is very small, suitable for analysis circumference curves. In summary, what is known through the body width, thickness and height and other information, accurately calculate the body size measurements is key to a two-dimensional non-contact body measurement system development [14]. At the same time, due to differences in body size, the computational model interior with more uncertainty, if the human body reasonable classification, were established model, can greatly improve the accuracy girth size [15].

In addition, in traditional manual measurement shoulder often drawn is the arc distance between points around the shoulder,and in the definition of ISO and GB shoulder size is the straight-line distance between two points of the shoulder.The existing variety computer-aided measurement is mostly derived from the straight-line distance, ignoring the thickness and curvature of the human shoulder [16-18]. Clearly, manual measurement Shoulder way to draw more precise dimensions of human shoulder. In order to solve the difficulties of human dimension automatic measuring non-contact two-dimensional image of encounter, the paper file 160 adult women shoulder arc curve for the study, the first acquisition of analog linear bust curve 160/86A, and then manually marked shoulder arc circumcircle curve along the curve to simulate manual measurement, discuss smoother shoulder arc curve linear function type, function and use fitting function to get a smoother relationship with the shoulder arc length between the shoulder and the shoulder thick [19,20].

2. Data and methodology

2.1 Shoulder arc curve collection

According to GB/T 1335.1-2008, the clothing size standard, classify people using the size of height and the difference number of bust and waist as the basis. That is to say, each 5cm height and thoracolumbar difference divide body type. Select the height between 157.5~162.5cm and thoracolumbar difference between 14~18cm in adult women as the experimental samples.

In the experiment, first to measure shoulder linear distance and shoulder thick for the sample's shoulder arc curve. Then find the intersection the of center of the human body and shoulder arc and use this shoulder arc center A_0 as the measurement starting point. Besides, use point of the shoulder B_0 as measuring end point. A_0 as a starting point, take a little every 1.5cm as feature points to performance the trend of shoulder arc curve, remember as A_i (i = 1,2, ... n) until the end point B_0, a total of m points. To make the point distribution, adjust the position of the m-th point by using the midpoint of m-1th point and B_0 as m-th point.

The test tools are: vernier caliper, soft ruler, straightedge and mark lines. When mark shoulder arc curve, pay attention that the feature points of curve should be in the same plane. When recording shoulder thickness as a, and shoulder width as b. When measuring shoulder thick, shoulder point and the midpoint of shoulder arc are projected onto the measurement surface, when both the same height, read out the horizontal distance between the two and record as shoulder thick value. If the height is inconsistent, then adjust the tag line until conditions are met to keep the experimental data accurately.

During the experiment, turn recording the linear distance between A_i and A_0, A_i and B_0, and were recorded as r_{i1} and r_{i2}. Chose the mid-point (O) of left (C_0) and right (B_0) shoulder peak points as the original heart to establish a

Cartesian coordinate system, and the coordinates of A_0 is (0, a), coordinate of B_0 is (b/2,0). Besides, A_i is the intersection of A_0 as the center, r_{i1} radius of the circle and B_0 as the center, r_{i2} radius of the circle in the first quadrant. Using matlab software to quickly find the coordinates of A_i. Then input point coordinate through AutoCAD, connect the various points to generate multi-segment, and generate a spline curve to fit body shoulder arc curve. Figure 1 shows a sample of the shoulder arc curve which is used as a reference curve.

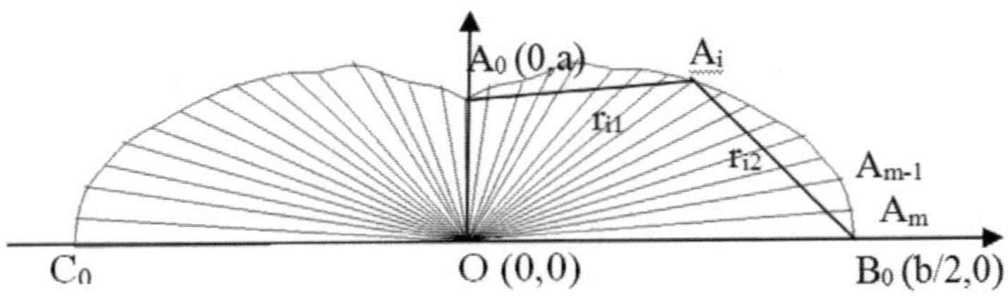

Fig.1 Shoulder arc spline

In the actual measurement, the sample's manual measurement size of shoulder arc is 44.8cm. Besides, the shoulder arc spline length is 44.22 cm, and the actual body-length is 45cm, thus the difference of spline length and the actual body-length is 0.78cm, the error is within 1cm. Thus, the spline fitting the shoulder arc can be approximated as a linear curve to be studied.

2.2 Get a smoother curve

When measuring the body shoulder arc, the body-shoulder long is not the length of the shoulder arc, but the closed curve length using a soft ruler to measure external shoulder arc. And similarly, we can not generate the spline fitting as shoulder arc length. Then we need to simulate soft foot movements to simulate a circle along the external periphery of the spline curve, the circle along the curve must pass through the external outermost point of the shoulder arc curve, namely the scapula points. By measuring the circumscribed circle of the shoulder along the curve to approximate the arc length. Therefore, the simulation of the circumscribed circle along the curve is a convex curve, as shown in Figure 2.

The length of shoulder arc smoother curve is 44.45cm, then compare with the length measured with the manual soft ruler, find the error is 0.35cm which is less than 1 cm. Therefore a smoother curve depicts the basic trend of soft foot when manual measurement, and it can be used to fit the curve and calculate the shoulder arc length.

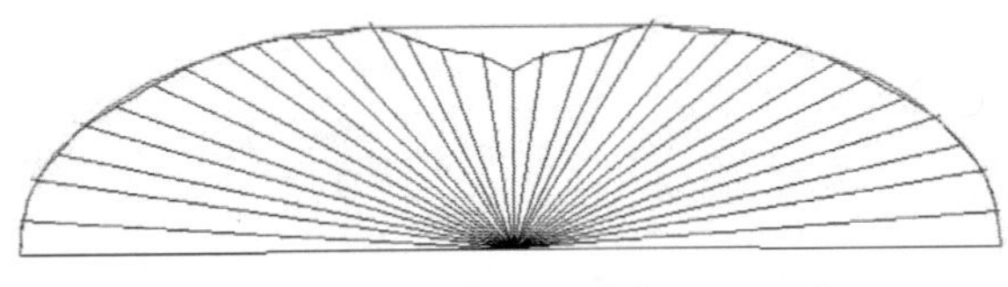

Fig.2 Shoulder arc spline and the smoother curve

3. Results and discussion

The smoother curve of shoulder arc is a curve through shoulder arc curve around and outermost points, so that the shoulder arc curve contains information of shoulder width and thickness of the shoulder, and the two data are only direct access to critical information from the human body two dimensional image.

Specific fitting step is firstly to take a series of equidistant points on a circle along the curve and respectively using quadratic function, a cubic polynomial function, quartic polynomial function, five polynomial function to fit smoother curve. Then choose the appropriate function type as fitting function, and correct the fitted curve as the reference function of other shoulder length of the same type. Therefore, by entering the shoulder width and shoulder thick, we can get estimated value of shoulder arc of the same type.

3.1 Determining fitting function

In smoother curve of 160/ 86A file shoulder arc case, by observing the point set which are equidistantly picked up on the smoother curve, use Robust regression analysis of minimum absolute residuals (LAR) and respectively use quadratic polynomial, cubic polynomial function, four polynomial function, five polynomial function to fit smoother curves. Fitting results is shown in Table 1.

Table 1. Each fitting function expression and the goodness of fit

Fitting function	Expression	Domain	Icon	Goodness of fit
Quadratic polynomial	y= -0.05113*x^2+0.52189*x+8.21925	[0,18]	Figure 3	0.98109
Cubic polynomial	y= -0.00196*x^3+0.0034*x^2+0.13154*x+8.70209	[0,18]	Figure 4	0.98871
Four polynomial	y=-0.000169485*x^4+0.00421*x^3-0.06736*x^2+0.39505*x+8.5456	[0,18]	Figure 5	0.98924
Five polynomial	y= -0.000046246*x^5+0.00193*x^4-0.02907*x^3 +0.14912*x^2-0.09476*x+8.68871	[0,18]	Figure 6	0.99073

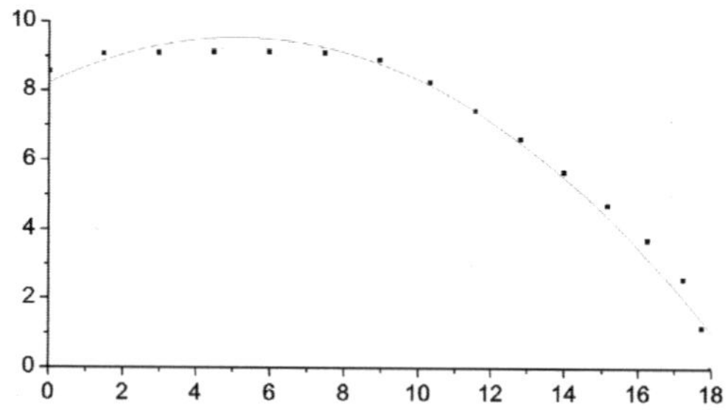

Fig.3 Quadratic polynomial fitting curve

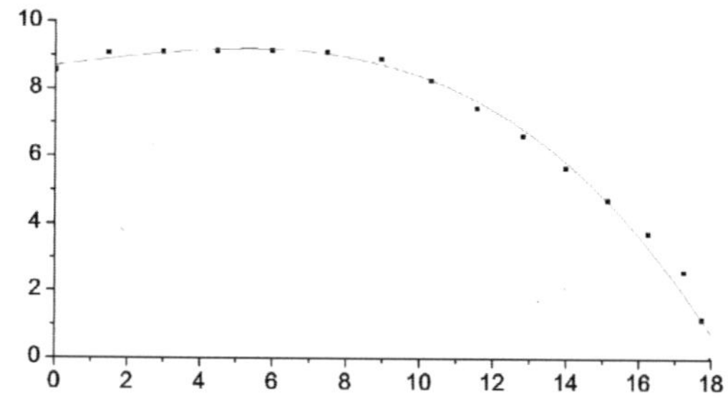

Fig.4 Cubic polynomial fitting curve

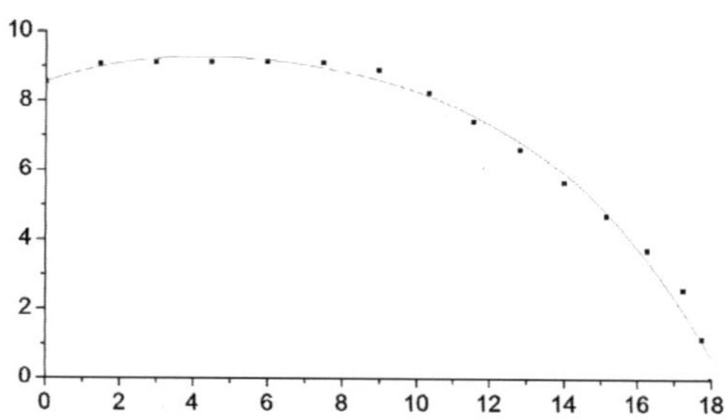

Fig.5 Four polynomial fitting curve

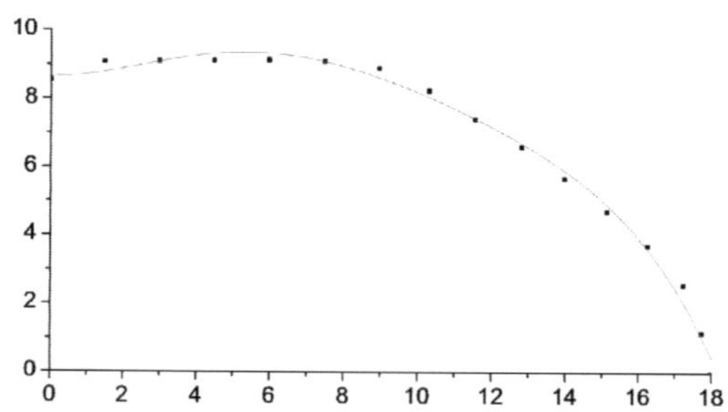

Fig.6 Five polynomial fitting curve

As seen from the figures above, the effect of a cubic polynomial function fitting is better than other functions, and function expressions are more concise and goodness of fit is quite high (see Table 2).

Table 2. Fitting length of different shoulder arc curve's fitting functions

Fitting function	Fitting length of the shoulder arc curve	Difference between smoother curve length	Difference between manual measurement length
Quadratic polynomial	42.655	1.795	2.145
Cubic polynomial	42.800	1.65	2.000
Four polynomial	43.139	1.311	1.661
Five polynomial	42.728	1.722	2.072

3.2 Estimation of shoulder arc curve

In 160/86A shoulder arc curve case, its shoulder arc fitting curve function expressed as:

$$F(x) = f(x) + f(0) \tag{1}$$

Among them, $y = f(x)$ is the shoulder arc fitting function curve, $f(0)$ is the correction value. Use the above 160/86A profile fitting function as a reference function, convert y=f(x) to parametric equations.

$$\begin{cases} x = (\partial / a)t \\ y = (\beta / b) f((\partial / a)t) \end{cases} \text{(t as parameters)} \tag{2}$$

Among them, ∂ and β separately represents the width and thickness of shoulder arc which need to calculate, and a, b correspond to the width and thickness of reference shoulder arc curve. On this basis, through the x, y proportional scaling of parameters representative function integral equation demand curve length, the results is shown in Table 3. Enter the width and thickness of the same type, it can easily calculate shoulder arc length with the indirect method.

Table 3. Shoulder arc curve length obtained from the reference curve (cm)

Type	Shoulder width	Shoulder thick	Calculated shoulder arc length	Manual measurement length	Deviation
160/86A	37.1	9.70	47.857	46.6	1.257
160/86A	34.5	8.80	43.581	42.5	1.081
160/86A	36.3	8.60	46.236	45.0	1.236
160/80A	34.2	8.20	43.337	42.5	0.837
160/84A	36.2	7.81	46.934	46.0	0.934
155/82A	35.1	7.50	43.603	41.8	1.803
155/84A	35.1	6.68	43.606	42.2	1.406
165/86A	35.0	9.57	46.421	45.0	1.421
165/88A	36.0	6.40	43.758	42.5	1.258

4. Conclusion

In this paper we have researched the linear shapes in each sample of the shoulder arc curve, and finally chose a cubic polynomial as fitting function by comparing with others. Furthermore, we got a new function by scaling the corresponding proportion of shoulder width and shoulder thickness of same type of shoulder arc. In this way, we can indirectly acquire the shoulder arc length of same type and can be effectively calculated the human body size information in a non-contact two-dimensional image which provides a most direct way for the development of non-contact 3D body automatic measurement. However, this method used manual measurement data and in each step of the calculation process accumulated a calculation error. Besides, the measurement tools and methods of measurement could be improved.

In accordance with the method described above, we can obtain relatively satisfactory shoulder arc length of the curve with same type. Therefore, under the promotion of other human girth computing, such as waist circumference, hip circumference, neck circumference and other minimum, human dimension data is able to indirectly acquire which can not be obtained in a two-dimensional image but indispensable in the clothing design and production and has a very important value.

5. Acknowledgement

The authors would like to thank all anonymous reviewers for constructive comments. The work described in this paper was funded by the Priority Academic Program Development of Jiangsu Higher Education Institutions. Partial support by Soochow University.

References

[1] Peng Z. Application and progress of anthropometric and digital human modeling techniques [J]. Ergonomics, 2007; 13(4): 67-69.
[2] Huang HQ, Mok PY, Kwok YL. Block pattern generation: From parameterizing human bodies to fit Feature aligned and flat tenable 3D garments [J]. Computers in Industry, 2012; 63(7): 680-691.
[3] Wang YZ, Zhang Y, Zhang T, et al. Study on precision comparasion of 3D scanning and manual measurement [J]. Shandong Textile Science and technology, 2013; 54(3): 46-48.
[4] Pan L, Wang J, Yao T, et al. Comparison of contact with non-contact measuring data in garment [J]. Journal of Dalian Polytechnic University in China, 2013; 32(2): 140-142.
[5] Zhang SJ, Ding XY, Zhou FY. Research on the relationship between data extracted from 3D body scanner and manual measurement data [J]. Journal of Zhejiang Sci-Tech University, 2006; 23(3): 298-301.
[6] Cao B, Jia SL, Ji JH, et al. Study on accuracy affected by the proficiency of body measurement and the characters of points related [C]. Manufacturing and Applied Research, 2014: 405-409.
[7] Liu H, Zhang M, Cheng DS, et al. Study on bust curve of standard female figure [J]. Journal of Textile Research, 2011; 02:117-120.
[8] Ni SM. Research of Young Female Body Shape Recognition Based on Longitudinal Section Curve Form [D]. Journal of Zhejiang Sci-Tech University, 2014.
[9] Zhang JH, Wang HF. Study of womens' shoulder based on 3D body measurement [J]. Journal of Textile Research, 2011; (5): 23-28.
[10] Lu C. The study and implementation of image-based non-contacting human body auto-measurement system [D]. Suzhou: Soochow University, 2009: 10-62.
[11] Wang YX. Non-contact body measurement system - image processing, curve analysis and calculation of body circumference. Tianjin: Tianjin University of Technology master's degree thesis, 2003.
[12] Zhao JM. Non-contact body measurement system - digital image processing and anthropometric access to information.Tianjin: Tianjin University of Technology master's degree thesis, 2003
[13] Wang YX, Li XJ. Establishment of the ridge regression model of human hip girth [J]. Knitting Industries, 2006; (7) : 27-29.
[14] Li XJ, Wang YX. The Hip Girth Calculation in Planar Non-contact Body Measurement System [J]. Journal of Textile Research, 2004; 25(2) : 98-100.
[15] Wang YX, Li XJ, Liu H. Fuzzy clustering analysis of body type in 2D non-contacted body measurement system [J]. Journal of Textile Research, 2007; 28(2): 100- 103.
[16] Huang XL. Research on measurement of young female's bodies based on digital images [D]. Suzhou: Soochow University, 2009: 54-61.
[17] Xu F, Zhang H, Zhang R. Prediction for bust girth using regression analysis from photographic and anthropometric data [J]. Journal of Textile Research, 2006; 27(8) : 49-52.
[18] Cao B, Shang XM. Analysis and optimization of the body measurement program supporting the clothing ergonomics [C]. Advanced Materials Research, 2013: 443-446.
[19] Cai L, Li PW, Hong ZL, et al. Taking the human body measurement analysis of degree of difficult [C]. International Journal of Materials Science and Engineering, 2014: 19-22.
[20] Driessens FCM, Boltong MG, Bermudez O, Planell JA. Formulation and setting times of some calcium orthophosphate cements: a pilot study. J Mater Sci: Mater Med, 1993; 4: 503-508.

Research on Underwear Decoration Design Based on Kansei Engineering

Qiao-Yun Tang, Jie Zhou*, Jiang Wu

School of Apparel & Art Design, Xi'an Polytechnic University, Xi'an, 710048, China

*Corresponding author's email: 1347467333@qq.com

Abstract

This study explored the use of the Reverse Kansei Engineering method with regards to evaluating customer feedback on women's underwear designs. The Reverse Kansei Engineering method is used to study the difference in how designs are perceived. The study was conducted by grading underwear decorative design renderings according to feedback from our sampling group, following which the Semantic Differential (SD) method to quantify the results. Finally, we summed up the degree of sensibility through average value comparison and cluster analysis. The research shows that the perceptual demands of underwear decorative design include gender, occasion, structure factor and target citizens. Combining element of decoration with perceptual demand of consumer can reflect better reflect customer interest, and facilitate the development of an emotional system for consumers who shop for underwear online.

Keywords: Kansei engineering; Underwear design; Interior design

1. Foreword

Kansei engineering is widely used in architectural design, interior design and industrial design, and is now being used in the textile industry [1], cosmetic industry, industrial manufacturing, and other fields. Kansei engineering started late in China, and has been applied exclusively to industry or electronic products development [2], with very little application in the research and development of garment industry. Liu Guolian and others have looked into the relationship between women's style and consumer's perceptual psychology by using the principle of perceptual Engineering [3]. The relative quantification between the discrete design elements and clothing emotional information have been studied by an experiment, as outlined in Lv J and Chen DS, 2014 [4]. Using the perceptual design principle and method of clothes and ornaments, Liu Yaping improved upon women's office suit design [5]. The individual differences of the customer's perception evaluation of clothing design is studied by Gao W and Xiao J, 2014 [6]. Using semantic analysis, nine kinds of woven fabric with different organizational structures were analyzed by French scholar S. Bensaid and others. The research results show that the fabric parameters have significant influence on the customer's touch [7]. However, there are few studies on the field of Kansei engineering in underwear design with parameters such as profile, color, decoration and so on. This study has evaluated the underwear design emotional response and transformed customer's perceptual demands into quantitative data. The author adopts reverse Kansei Engineering to evaluate the user based on design concept, and then sums up the design according to the emotional evaluation of user, gradually enabling the product to achieve a state consistent with the customer's demands. The Study design ideas are shown in Fig.1.

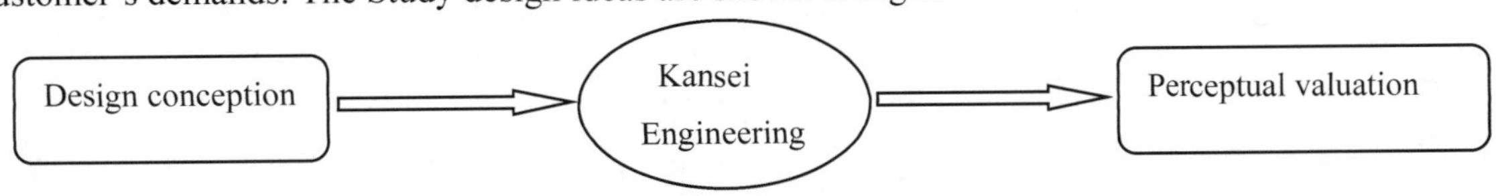

Fig.1 Study design ideas

2. Definitions and glossary established

2.1 Definition of underwear design and decoration

Underwear decorative design is based on Artistics and Aesthetics binding. Appropriate decorative materials or decorative means are used to add design highlights for the underwear and to form underwear style. Demand for underwear varies with each individual. and it is important to take into account the different decorative styles in underwear decorative design [8].

2.2 Definition of Kansei Engineering

Kansei Engineering is a product development technology which is guided by customer needs, and a method by which one might translate feelings and impressions into product parameters [9]. Perceptual design is a method [10] which is based on customer sensory and psychological evaluation. In Japan and other countries, this technology has been widely used in the automotive industry, and in designing home appliances, office supplies, and so on [11,14].

2.3 Determination of the emotional vocabulary

1) Vocabulary summary: The vocabulary was built using the SD method. The SD method is an experimental method of studying psychological intention of testees or measurement technology. It was proposed by American Psychologist Osgood in 1942 [14]. It pairs the vocabulary based on color, texture, structure and workmanship, resulting in with 11 groups of words for the questionnaire method and the selection of experts. These words are:

Female-Neutral	Delicate-Rough	Graceful-Masculine	Introverted-Playful
Modern-Classical	Solemn-casual	Personality- Public	Elegant-Tacky
Fresh-Fancy	Luxury-Simple	Generous-Complex	

2) Vocabulary Assignment: Through the analysis, in this study, each word has been assigned 5 points. The intermediate value is 3 points, adjectives oppositing each other on both sides of the line were 1 point and 5 point, the score closer to the description represents feeling more intense. The decorative style of underwear evaluation is based on its first feeling, and the value should be filled in corresponding form. In the case of the emotional adjective " Female-Neutral ", the method of vocabulary assignment is shown in Fig.2.

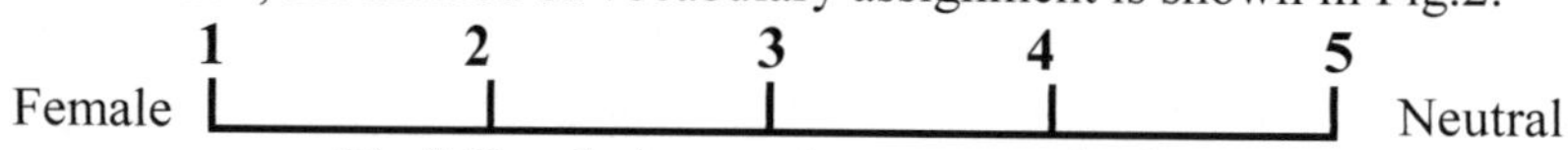

Fig.2 Vocabulary assignment method

3. Design Concept

3.1 Style design of underwear

This research has adopted 1/2 cup bra and inner pants as the test design, shown as Fig.3.

Characteristics of 1/2 cup bra patterns: center front gore height 6.5cm, underwire bra, firm and comfortable; The back band was designed with a U-shape. The closing was designed with three rows of three buckle fastenings to prevent falling and enhance stability of wearing, and to present the perfect effect after dressing; Plastic stay built in wide side with height 11cm, 10cm can gradually push the throatiness into the cup, and makes cups more beautiful.

Characteristics of inner pants patterns: high-waist design, 27.8 cm front length and 18cm side seam length; The segmentation lines were built on both sides, the bottom with the cutting sack for practicality.

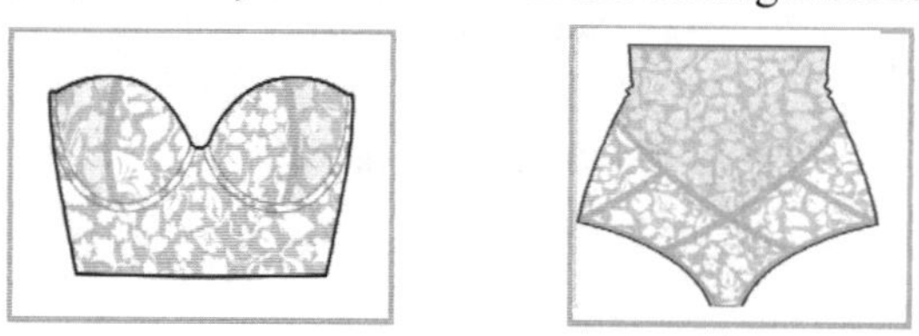

Fig.3 Underwear style

3.2 Decorating design of underwear

In order to avoid interference of variations in the test samples, the same model with different decorative design were selected. The design elements included: Po-phase flowers, Chinese paper-cuts, Blue Calico, Lianzhu pattern, rococo, paisley pattern, waves style pattern, wave point gradient, constitute emotional, rational structure, prominent theme, the theme of assimilation, floral patterns, animal patterns, pop themes, line drawing patterns, painted designs, shading patterns, abstract patterns, deconstruction. Underwear models were be designed to fit renderings as underwear sensual stimulation analysis diagram and were numbered.

20 stimulation pictures are replaced with Arabic numerals 1-20, shown as Fig.4.

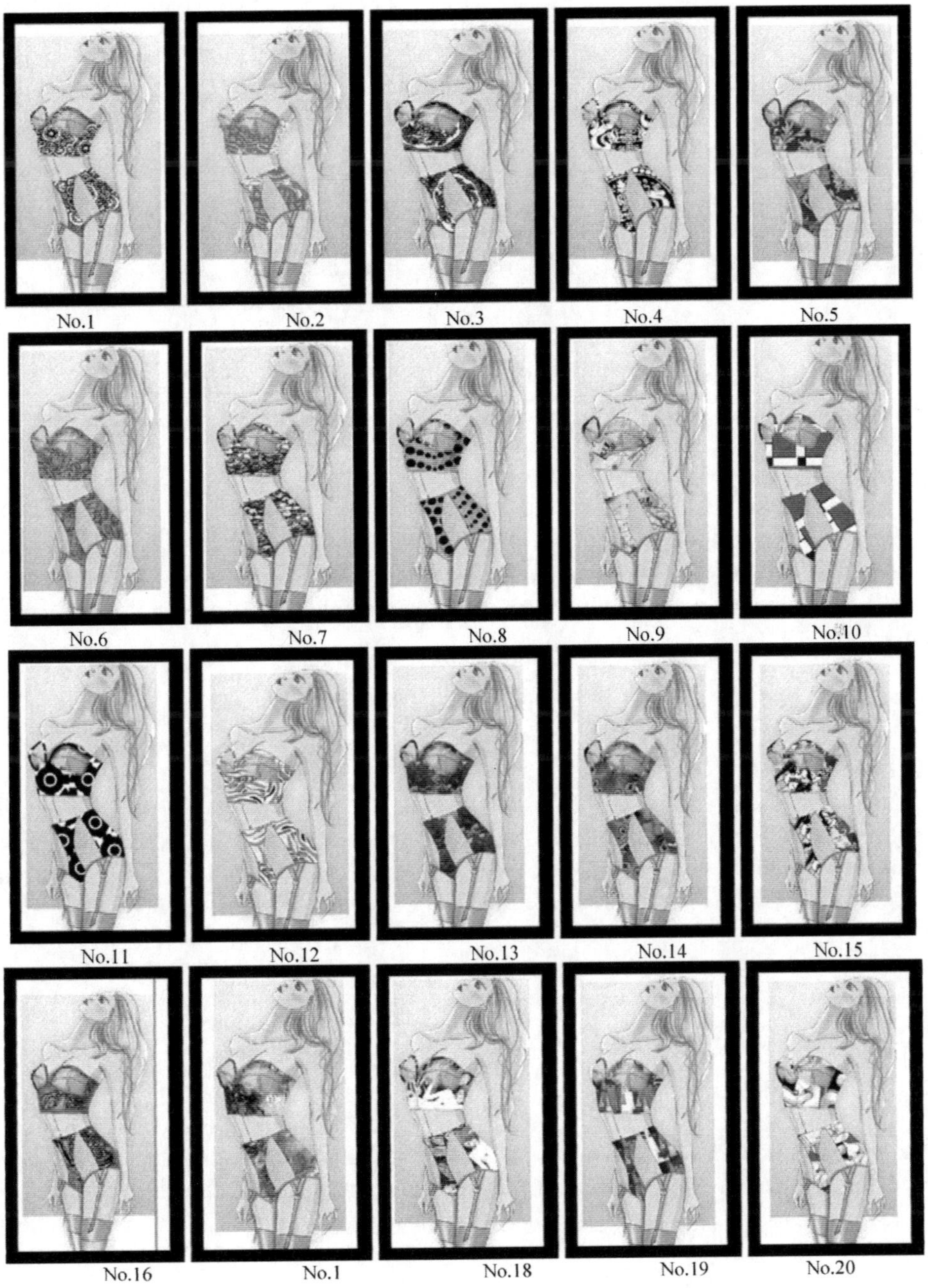

Fig.4 Stimulation pictures

3.3 The establishment of the questionnaire

The underwear decoration design questionnaire adopted five-level scale about the decorative design of underwear. and provided the basic information about respondents. The questionnaire included 20 stimulus diagrams, and each stimulus diagram was corresponded with 11 groups perceptual vocabulary respectively. It has taken "Chinese paper-cuts" of stimulus diagram as an example. Chinese paper-cuts used in underwear design does not seek perspective, anatomy and proportion accuracy, but deliberately pursues deformation exaggeration. The questionnaire design is shown in Fig.5.

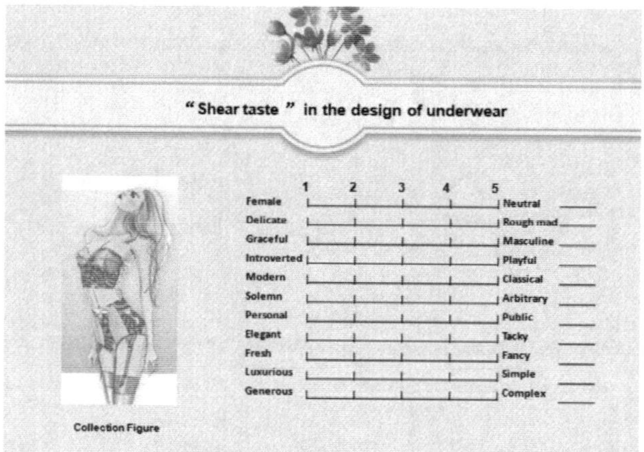

Fig.5 Questionnaire design

3.4 Implementation process

The questionnaire was distributed by combination of field survey and network survey. The target population was dominated by persons aged 20-35 with college students and young white-collar females. Using a random sampling method, a total of 40 questionnaires were sent out. The study was divided into three phases: investigation, statistical analysis, and conclusions.

4. Analysis and results

4.1 Data processing

The taxonomic revision of the experimental data was analyzed by using Excel and IBM Spss Statistics 19. In order to let the subjects understand the investigation object, data was collected through face to face questionnaire survey by one person. According to the relevant literature analysis [14], a total of 40 people were selected to fill the questionnaire survey, 34 questionnaires were returned, making the collecting rate 85%. The data were reliable, even if it had 30 to 40 people who have good sense and can meet the needs of the experiment.

4.2 Analysis of means

In order to facilitate statistical analysis, the data input and mean value were mainly handled and calculated by Microsoft Excel. Calculating the average of each pair of adjectives based on the corresponding stimulation view, indicates that there are still some gaps between the each piece of figure stimulation and the feelings of the people. Therefore, it is necessary to make a deeper and more concrete analysis, which is shown in table 1.

Table1. Stimulation figure sensitivity analysis mean score

Phrase stimulate	Female-Neutral	Delicate-Rough	Graceful-Masculine	Introverted-Playful	Modern-Classical	Solemn-Arbitrary	Personal-Public	Elegant-Tacky	Fresh-Fancy	Luxurious-Simple	Generous-Complex
NO.1	2.46	2.71	1.71	2.54	3.76	2.50	2.00	2.11	4.04	2.21	2.68
NO.2	1.29	1.96	2.04	4.04	2.50	2.64	2.25	2.46	1.96	2.36	4.14
NO.3	2.75	2.46	1.79	1.61	2.36	1.86	2.43	1.89	2.54	2.43	2.07
NO.4	3.87	2.57	4.07	4.79	2.43	2.61	2.11	2.36	3.45	2.61	2.46
NO.5	2.07	2.29	2.36	2.68	4.39	1.82	2.14	4.6	3.86	2.07	2.64
NO.6	2.25	2.39	1.82	2.21	3.46	2.71	2.39	2.46	1.96	2.61	2.36
NO.7	4.07	4.32	2.75	3.89	2.14	4.39	2.54	2.14	4.57	4.11	3.96
NO.8	2.36	2.71	4.04	4.25	1.71	4.07	2.50	2.75	2.61	3.39	2.57
NO.9	1.79	2.00	2.04	2.43	2.46	4.36	2.04	1.86	1.89	1.96	2.11
NO.10	4.18	3.39	3.64	3.25	2.25	1.86	2.54	1.96	1.86	1.25	2.32
NO.11	3.14	3.50	3.36	3.36	2.36	1.93	2.61	1.82	2.71	3.25	2.11
NO.12	1.89	2.29	2.21	3.32	2.25	3.50	4.3	2.50	4.11	3.39	4.04
NO.13	1.46	1.64	1.5	2.61	2.68	2.64	1.93	2.36	3.43	2.14	3.79
NO.14	4.21	2.89	4.04	1.96	2.04	3.32	2.25	3.71	3.68	4.86	3.43
NO.15	2.93	4.18	3.32	4.82	2.36	3.57	2.11	1.96	3.54	2.32	1.89
NO.16	2.32	2.39	2.46	2.18	4.11	3.61	4.07	2.36	2.39	4.04	2.32
NO.17	1.93	2.79	2.68	3.96	2.32	3.21	2.18	2.39	3.36	2.57	4.14
NO.18	2.61	2.82	3.82	3.43	2.00	3.50	2.00	2.64	2.75	3.11	1.82
NO.19	3.96	3.50	4.18	4.68	2.25	3.18	1.93	4.54	4.29	2.71	3.36
NO.20	2.54	2.61	4.41	3.54	2.29	3.21	1.82	2.64	2.57	3.35	2.57

1) The average cross-sectional analysis results are summarized in table 2:

Table 2. Horizontal analysis of stimulation figure

Stimulation fig	Sensitivity
NO.1	Subtle、Classic、Thick
NO.2	Women、Fine、Subtle、Nifty、
NO.3	Subtle、Introverted、Stately、Elegant
NO.4	Masculine、Nifty、Neutral
NO.5	Classic、Stately、Thick、Luxury
NO.6	Subtle、Individuality、Refreshing、Generous
NO.7	Neutral、Rough、Casual、Thick
NO.8	Masculine、Nifty、Modern、Casual
NO.9	Women、Fine、Refreshing、Luxury
NO.10	Neutral、Stately、Fine、Refreshing
NO.11	Stately、Fine、generous
NO.12	Women、Public、Thick、Fancy
NO.13	Neutral、Fine、Subtle、Individuality
NO.14	Neutral、Masculine、Introverted、Plain
NO.15	Ruggedly、Nifty、Elegant、Generous
NO.16	Classic、Public、Plain、Generous
NO.17	Women、Nifty、Fancy
NO.18	Masculine、Modern、Individuality、Generous
NO.19	Individuality、Masculine、Nifty、Individuality
NO.20	Masculine、Individuality、

2) The average of the longitudinal analysis in Fig.6 and Fig.7

The mean of six groups of adjectives, which were analyzed in Fig.6, shows that NO.2, NO.20 obviously give a feminine impression; NO.8, NO.10, NO.14, NO.19 give a neutral impression; NO.2, NO.13 give a delicate impression; NO.7, NO.15 give a rough impression; NO.1, NO.13 give the graceful impression; NO.4, NO.8, NO.14, NO.19 give a masculine impression; NO.3 gives a restrained impression; NO.4, NO.8, NO.15 and NO.19 give a playful impression; NO.8 gives a modern feeling; NO.5, NO.16 and NO.20 give the classical impression; NO.3, NO.5 and NO.10 give the impression of solemnity, and NO.7, NO.8 and NO. 9 give the feeling of arbitrariness.

By extreme value analysis, it can be seen in the figure that there are 20 pairs of stimuli. Taking the previous three groups of words as example to analyze, we can see that NO.15 is the most playful in decorative; NO.21 is the most feminine; NO.9 is the most casual; NO.13 is the most graceful.

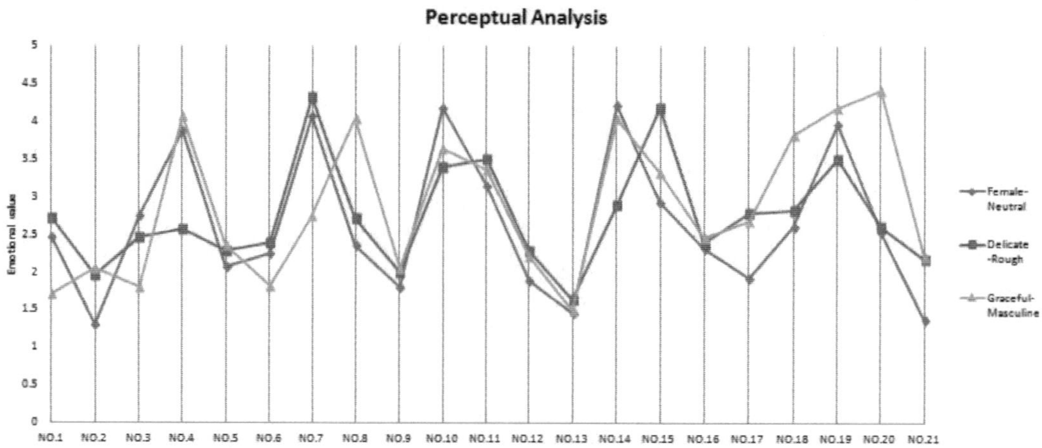

Fig.6 Figure sensibility analysis

Analysis of the mean after five groups Fig.7 The mean analysis of the 3 groups of adjectives shows that NO.13 and NO.20 obviously give the impression that the personality; NO.12 and NO.16 obviously give the impression that the public; NO.3, NO.9, NO.11 obviously gives the feeling is elegance; NO.5 and NO.14 give the obvious impression that is tacky; NO.2, NO.6 and NO.9 give an obvious impression that is refreshing; NO.7 and NO.12 have a clear impression that is messiness; NO.10 gives a clear impression that is luxurious; NO.14 and NO.16 obvious impression is simplicity; NO.15 and NO.18 give the obvious impression of generosity; and NO.2, NO.12, and NO.17 clearly give the impression that implies fanciness.

By extreme value analysis, it can be seen in the figure 20 pairs of stimulation, NO.7 is the most fancy under this decorative design, NO.10 is the most luxurious.

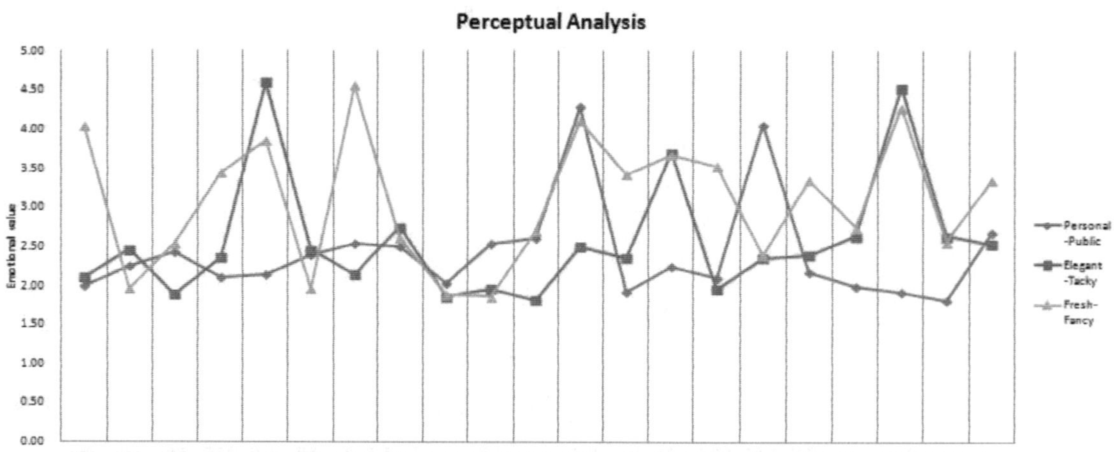

Fig.7 Sensibility analysis

4.3 Figure stimulate clustering analysis

Carrying out cluster analysis of design sketches based on the data about the target group's perceptual knowledge of effect figure which were obtained through the questionnaire survey. The data contains 11 categories of variables, which correspond to 11 pairs of adjectives (the differences between each pair of adjectives only on the degree of difference and the difference in degree is divided into five classes, which is in order to convert the perceptual knowledge to the extent index). Each index is derived from the 25 sample questionnaires, and then the average value is taken. Cluster analysis was not delicate enough ,but te characteristics various types are still analyzed, using clustering analysis of each index in each type. The results are shown in Fig.8.

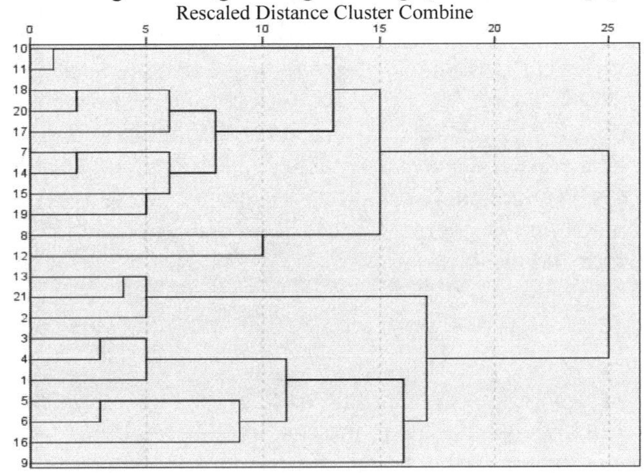

Fig.8 Underwear emotional evaluation index tree

1) The results show that the adjectives can be summarized as gender factor, wearing occasions factor, structure factor, and demographic factor. The gender factor is a comprehensive index to show graceful-masculine, feminine-neutral and delicate-rough; The factor of Wearing occasion is a comprehensive index to explain factor-casual, modern-classical, restrained-playful and luxurious-simple; Structure factor is a comprehensive index that interpreted as refreshing-thick, generous-fancy and elegant-tacky; Audience factor is a comprehensive index that interpreted as personality-public.

2) Fig.8 is aggregated results shown in Table 3 and Table 4:

Table3. Summary stimulate graph clustering

Classification	The first	The second	The third	The fourth	The fifth	The sixth	The seventh
Number	10、11	17、18、20	7、14、15、19	8、12	2、13	3、4、1	5、6、16、9

From Table 3, it is found that adopts the systematic clustering method to classify those stimulation figure and obtains seven categories. Simplified graphics can be divided into four categories shown in Table 4.

Table 4. Summary table of FIG stimulate clustering

Classification	The first	The second	The third	The fourth	The fifth	The sixth	The seventh
Number	10、11、17、18、20、7、14、15、19	8、12	2、13	3、4、1、5、6、16、9	8、12	2、13	3、4、1、5、6、16、9

According to the stimulus graph cluster from Table 4, the 20 types of stimulus figure of this study will be divided into four categories.

The first category: NO.10, NO.11, NO.17, NO.18, NO.20, NO.7, NO.14, NO.15, and NO.19 will be summarized to abstract painting decorative printing. The graphic composition in first category is mainly based on the principle of plane and simple geometry and can convey the abstract concept and aesthetic form. Moreover, they emphasize visual sense and the use of these on underwear design gives a strong sense of shock.

The second category: NO.8 and NO.12 will be summarized as a geometry change and unity decorative design. Obviously, gradient can be seen as strong unity in the underwear, and underwear is created to be smooth, with orderly, rhythmic beauty; The color of underwear design was achieved by the assimilation effect, so that the over all tone of underwear formed a harmonious and unified effect.

The third category: NO.2 and NO.13 will be summarized as a natural theme underwear embroidery designs. The underwear design composition is mainly dominated by plane in the form of composition, to express the actual situation of underwear and the space relationship with the size of the line thickness, density, and image.

The fourth category: NO.3, NO.4, NO.1, NO.5, NO.6, NO.16 and NO.9 will be classified in the traditional Chinese printing and dyeing decorative pattern design. The use of bright blue and white contrast in the underwear makes the whole tone of underwear fresher, clearer and fast-paced. It is an ingenious combination of the silhouette profile and dot short-term that constitute this underwear's unique decorative effect and rhythm. Widespread adoption of curve, arcs, the composition is complex, multi-level superimposed halo of color, making the overall effect of the underwear is more rich and wealthy.

5. Conclusion

The decorative design of underwear plays a very important role in the style of underwear design, which should pay more attention to the influence of individual factors on the customer's perception.

Different methods of underwear decoration to reflect the people's feelings are different - some of them have a high correlation between the adjectives. The main perceptual factors of underwear decorative technique can be summarized as gender factor, wearing occasion factor, structure factor, and audience factor. Starting from the feeling of these four main decorations and designs, we can obtain the orientation of the underwear decoration design, and the underwear design can truly meet the different needs of the consumer's psychological beauty.

Reference

[1] Liang YY. Perceptual design research. Shandong University: 2009; 5: 9-10.

[2] Dong XF. The research and practice of dress design based on kensei engineering. Zhejiang Sci-Tech University: 2012.

[3] Liu GL, Jiang Y. Study on kansei of fashion style based on human sensibility. fzxb: 2007; 28 (11): 101.

[4] Lv J, Chen DS. Fashion perceptual data mining based on clustering algorithm. fzxb: 2014; 35(5): 108-112.

[5] Diu YP. The procedure optimization of woman's office suit design based on kansei engineering. Progress in Textile Science & Technology: 2011; 2: 108-112.

[6] Gao W, Xiao J. Individual differences in perceptual evaluation for fashion design: taking female students as research subjects. fzxb : 2014; 35(5): 137-141.

[7] BENSAID S, OSSELIN JF, SCHACHER L, et al. The effect of pattern construction on the tactile feeling evaluatedthrough sensory analysis. Journal of theTextile Institute: 2006; 97 (2): 137-145.

[8] Xu LP. The decorative design research of modern women underwears. Suzhou University: 2007.

[9] Li YZ. New design concept -kansei engineering.New art: 2003.

[10] KyuSik Kwon. Human sensibility ergonomics in product design.International Journal of Cognitive Ergonomics: 1999; 3(1): 51-62.

[11] Xu J, Zhang X. Based on user preferences imagery product modeling construction law. Chengdu Textile College: 2004; 5 (23): 21-23.

[12] Su JL. Research into kansei engineering and its application in industrial design. Xi'an Jiao tong University: 2004; 1 (38): 60-63.

[13] Xu WF. Use inductive principles of engineering design product modeling. Jiaozuo Institute of Technology: 2003; 5 (22): 406-409.

[14] Li YE, Wang ZY, Xu N. Kansei engineering. Beijing: Hai dian Press: 2009; (10): 77.

[15] Dai J. The fourth value of product-Kansei value. Chongqing University: 2007.

Study of Current Status of Sizhou Opera and Opera Costumes in Suzhou

Meng-Lin Wang, Rong Zheng*, Wen Wu

Aimer Human Engineering Research Centre, Beijing Institute of Fashion Technology, Beijing, 100029, China

*Corresponding author's email: yuki.zheng@gmail.com

Abstract

This study undertook a systematic investigation of the general status of Sizhou Opera in Suzhou using a survey questionnaire and in-depth interviews with local residents and members of Sizhou Opera troupes. Research results show that local residents' awareness and favourability of Sizhou Opera are low, but they are highly interested in traditional opera costumes. Most of Sizhou Opera's audience is comprised of older people. The overall development status of Sizhou Opera is not optimistic. In order to address the problem, recommendations are made for local opera troupes to try to improve plays, costumes, and other opera design features.

Keywords: Sizhou opera; Current status; Awareness; Opera costumes

1. Introduction

Sizhou Opera, known also by its popular name "Lahun Air" originates from the Qianlong-Jiaqing period of the Qing Dynasty and has a history spanning more than 200 years [1]. Sizhou Opera is prevalent mainly in Suzhou and Bengbu in Anhui, covering both shores of the Huaihe River in north Anhui, especially in the Huaibei area [2]. Sizhou Opera is one of four major operas in Anhui and listed in the catalogue of the first batch of China's National Intangible Cultural Heritages [3,4].

Sizhou Opera in Suzhou region has developed from its initial form in which artists wandered about the streets to talk and sing, to the current form involving performances by local professional opera troupes. Similarly, the performance organisation also has been transformed from informal opera folk troupes to amateur opera troupes and finally to today's professional opera troupe. During this development process, Sizhou Opera in Suzhou region reached its pinnacle from the middle of 1970s to the end of 1990s, when both performances and creations were very active in numerous special performance venues such as Suzhou Grand Theatre, Xiao County People's Theater, and Si County People's Theatre [5]. Currently, however, people's cultural appreciation and habits have changed, largely due to the development of technology. Most notably, the popularisation of modern media has greatly impacted local opera, and performance venues have gradually disappeared or are used for other purposes - even Sizhou Opera troupes have folded in some areas [6,7]. During the Sizhou Opera's initial stages, performances used no special opera costumes, and only accoutrements such as a long gown, top hat, and artificial whiskers were worn, but these were replaced by today's professional opera costumes.

Some scholars have started to research multiple factors influencing and restricting the further development of the Sizhou Opera and concluded that the mass consumer base is a key factor that cannot be ignored. Therefore, it is important to investigate local residents' interest in Sizhou Opera and to determine if they want to learn more about relevant cultural institutions such as Sizhou Opera and its traditional opera costumes, and if Sizhou Opera troupes can make any new contributions that might attract and increase audiences.

Using local residents and Sizhou Opera troupes as its research subjects, this study sought to identify and analyse local residents' attitudes towards Sizhou Opera, traditional opera costumes, and the current status of Sizhou Opera troupes; discuss the interaction between residents' attitudes and the development of Sizhou Opera troupes; and provide corresponding basic research data for further research. The study's primary goal was to help local opera troupes improve the current status of local opera and to promote the development, inheritance, and protection of intangible cultural heritages.

2. Study design and implementation

2.1 Research objective and content

This investigation employed research findings from national and international studies along with a survey questionnaire and in-depth interviews with local residents and members of Sizhou Opera troupes to analyse and summarise the overall status of Sizhou Opera in Suzhou.

The study's specific objective was to identify the following: 1) local residents' favourability and awareness for Sizhou Opera; 2) local residents' favourability and awareness for opera roles and costumes; 3) factors influencing residents' favourability and awareness; 4) composition of professionals in Sizhou Opera troupes in Suzhou; and 5) challenges facing opera performances.

2.2 Survey questionnaire

The survey questionnaire was administered to residents in Suzhou, and 240 questionnaires were distributed and collected between December 16, 2015 and December 18, 2015. A total of 236 questionnaires were collected, and the recovery rate was 98.3%; the number of valid questionnaires was 233 and the effective rate of questionnaires was 98.7%.

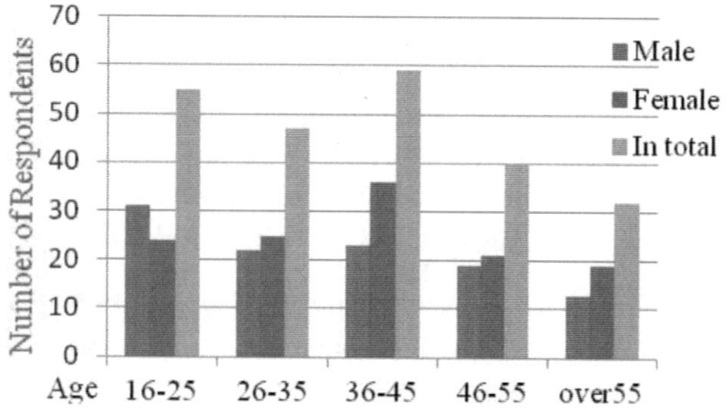

Fig.1 Gender and age distribution.

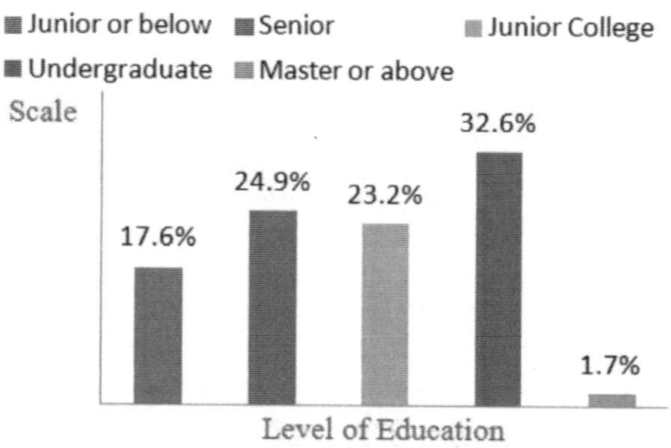

Fig.2 Level of education.

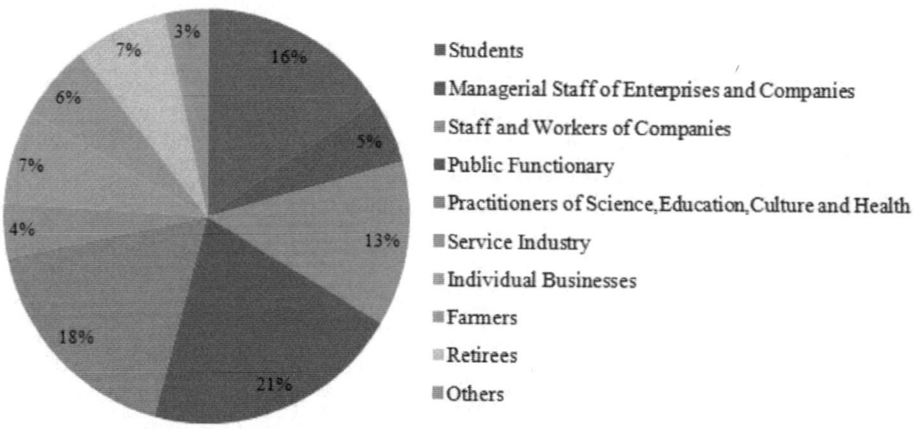

Fig.3 Participating residents' occupations.

Figs.1-3 show the basic information corresponding to participating residents' survey questionnaire results. Overall, samples are well represented in gender and age distribution and different degrees of education, which provides the fundamental guarantee for the survey's coverage rate and validity.

2.3 Exclusive in-depth interviews

In accordance with the research content and interview outline, survey questionnaires were completed by residents but the in-depth interviews were conducted with relevant members of opera troupes. Interviews addressed composition of personnel of Sizhou Opera troupes, performance frequency, plays performed, opera costumes, basic audiences, and problems facing the development of Sizhou Opera. Table 1 shows basic information pertaining to personnel who participated in the in-depth interviews.

Table 1. Basic information of personnel participating in in-depth interviews

No.	Gender	Position
1	Male	Costumer
2	Female	Performer
3	Male	Director
4	Male	Deputy head
5	Male	Relevant expert

3. Current status of local residents' understanding of Sizhou Opera and traditional opera costumes

3.1 Analysis of local residents' favourability and awareness

Favourability for opera, frequency of watching and listening to Sizhou Opera, degree of understanding Sizhou Opera, and intention to understand Sizhou Opera are classified into five grades, ranging from Grade 1 (with 1 being "totally uninterested") to Grade 5 (with 5 being "very interested"). As shown in Table 2, in terms of favourability for operas, the number of persons selecting Grade 3 accounts for 42.5%, which shows that most people's favourability level is average and shows a direct correlation with 33.5% of Grade 2 and 34.8% of Grade 3 responses in terms of the frequency of watching and listening to Sizhou Opera, respectively.

Table 2. Favourability and awareness for Sizhou Opera

	Grade (% of respondents' answers)					
Item	1	2	3	4	5	Mean value
Favourability for Sizhou Opera	15.0	24.0	42.5	15.0	3.4	2.62
Frequency of watching/listening to Sizhou Opera	13.7	33.5	34.8	13.3	4.7	2.68
Degree of understanding Sizhou Opera	24.5	44.2	17.6	10.3	3.4	2.24
Intention to understand Sizhou Opera	15.9	33.0	24.5	22.7	3.9	2.66

In terms of the degree of understanding Sizhou Opera, the number of persons selecting Grade 1 "Never heard before" accounts for 24.5% and that of persons selecting Grade 2 "Heard before" accounts for 44.2%; over half of the sampled population have never heard Sizhou Opera, or only have heard but have not actually listened to Sizhou Opera, and only 30% of persons have listened to Sizhou Opera and understood Sizhou Opera to some degree.

In terms of comments on Sizhou Opera, 15.9% of respondents indicated they have no intention of understanding Sizhou Opera, half of them think Sizhou Opera does not matter, and only 26.6% indicated a strong desire to understand and watch Sizhou Opera. Such results show that awareness and favourability of residents in Suzhou for Sizhou Opera are low. Among the reasons that people do not like Sizhou Opera, 54.5% selected the option "Cannot understand," which becomes a major factor impairing people's favourability for Sizhou Opera; 35.6% think they cannot feel at ease when appreciating Sizhou Opera due to its slow rhythm and obsolete plays; and 8.6% think that the ugliness of opera costumes and stage is also the reason for the unpopularity of Sizhou Opera.

In terms of the survey question regarding suggestions to improve the development of Sizhou Opera from the perspective of audience members, 50% of respondents selected the options "Combination of opera improvement with modern literature" and "development of adolescents' interest"; 14.5% think it is necessary to improve the opera costumes and stage effects; and 41.2% think that it is necessary to strengthen the talent development and enhance advertisements of Sizhou Opera.

People mainly understand the opera by means of TV channels (see Fig.4). As shown in Fig.5, over 90% of respondents are familiar with Peking Opera, Huangmei Opera, and Shaoxing Opera, all of which belong to China's five major opera genres. Peking Opera is the quintessence of Chinese culture and Huangmei Opera is one of the most popular genres of opera in Anhui province; the latter two genres of opera are often performed in TV programs or broadcasted via specialty TV channels. It can thus be seen that TV media plays an active role in the dissemination of opera.

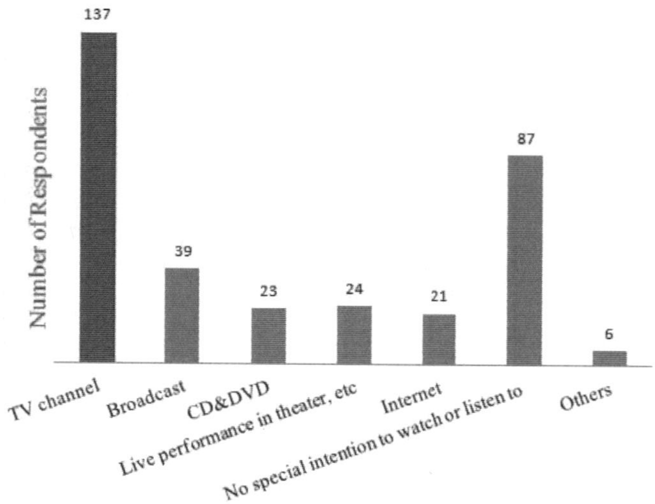

Fig.4 Channel for understanding.

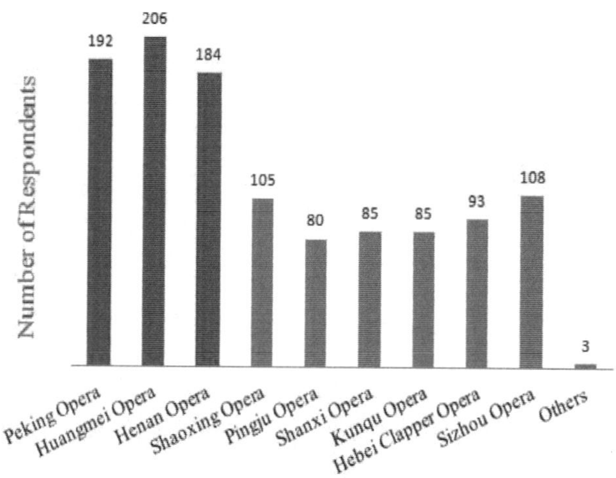

Fig.5 Known genres of opera.

3.2 Analysis of awareness and favourability for traditional opera roles and costumes

In terms of favourability for traditional opera costumes and the intention to understand traditional opera costumes, Grade 1 was set as "Totally uninterested", Grade 2 as "Slightly interested", Grade 3 as "Generally interested", Grade 4 as "Interested/ Intend to understand", and Grade 5 as "Totally interested/Intend to understand". Table 3 shows that the number of persons totally uninterested in traditional opera costumes and without favourability account for less than 20% of responses; 62.7% of persons are interested in and have favourability for traditional opera costumes, so the latter intend to understand traditional opera costumes.

Table 3. Favourability for traditional opera costumes

	Grade (% of respondents' answers)				
Item	1	2	3	4	5
Favourability for traditional opera costumes	19.3	34.8	27.9	15.9	2.1
Intention to understand traditional opera costumes	15.0	33.0	21.0	25.8	5.2

Ten questions were put forward to test awareness of traditional opera costumes, with 1 point assigned for each correctly answered question. Among 233 samples involved in the research, only 27 persons each received 6 points, 28 persons each received 0 points, and only 6 persons each received the highest score - 7 points, as shown in Table 4. This shows that local residents have low awareness of traditional opera roles and costumes, and many persons' awareness of opera costumes is merely that such costumes are opera costumes and they know little about the specific classification of roles.

Table 4. Awareness of traditional opera costumes

Scores:	0	1	2	3	4	5	6	7	8	9	10
Number of people:	28	23	40	43	36	36	21	6	0	0	0

4. Analysis of current status of Sizhou Opera troupes in Suzhou

Using professional Sizhou Opera troupe members as research subjects, this study investigated the current status of two mainstream Sizhou Opera troupes. Table 5 shows the composition of each opera troupe.

Table 5. Composition of personnel in Sizhou Opera troupes

	Troupe no.	
	1	2
Category	Number of people	
Troupe size	28 (18 men, 10 women)	28 (19 men, 9 women)
Performance team	13	14
Playing team	8	6
Creation team	3	3
Other personnel	4	5

4.1 Performance forms and plays

Upon investigation, it is understood that:

1) Currently, Sizhou Opera troupes launch activity forms such as art performance, "performance to countryside," and commercial performance. For example, there are about 200 commercial performances each year given by Opera Troupe 1, creating income of RMB 1,000,000.

2) Performed plays include modern playlets, excerpts, and traditional full-scale operas. Statistics show there are more than 200 plays in the Sizhou Opera repertoire, including traditional full-scale operas and more than 60 playlets and excerpts, but not all plays will be performed. Nowadays, modern playlets are the main plays that are performed; for example, there are 4 modern playlets and 1 costume play frequently performed by Opera Troupe 1. The reason is that modern playlets are closer to life and easier to understand compared with other plays. As Interviewee 3 said, "modern young people are more interested in fashionable and fast-paced things and we will also integrate some popular elements into our creative works to attract more audience".

3) Currently, most of Sizhou Opera's audience are older people. Interviewee 2 indicated that the "opera audience that really watch and listen to the opera are mostly old people, while young people just join in the fun". Data collected from the questionnaires support the latter respondent's viewpoint. In the survey questionnaire, 32 respondents were over 55 years of age; 30 of them will generally watch or listened to the dramatic performance and they show more preference for operas than do other age groups. In order to attract younger people's attention, opera troupes have also made many improvements. For example, Interviewee 3 indicated that Opera Troupe 2 has added many modern pop music elements in its compositions.

4.2 About opera costumes

The study's in-depth interviews revealed that opera costumes correspond with the classification of plays; for example, players wear modern costumes in modern-life playlets and wear traditional opera costumes in traditional full-scale operas. Traditional opera costumes bear the most distinctive characteristics, which are specifically reflected in both structural modelling and durability as follows:

1) Traditional opera costumes are updated slowly and are made of specialised, high-priced materials. Old costumes continue to be used until they are worn out. In the in-depth interviews, all 5 interviewees mentioned that "they would rather wear traditional costumes into holes than wear improper traditional costumes". Opera Troupe 2, for example, still uses opera costumes purchased in the 1980s.

2) The same dramatic costume can be worn in multiple plays. If roles in several plays correspond to similar historic circumstances and/ or character roles, the same costume can be adapted for each role (e.g., players can wear the same dramatic costume if portraying a queen; some small roles such as yamen runner, servant girl, and servant boy can also wear identical opera costumes in different plays; and special costumes are prepared for special roles such as Bao Zheng, Hua Mulan, Mu Guiying, et cetera).

3) Most local opera troupes purchase finished traditional opera costumes directly and few of them prepare

tailor-made traditional opera costumes; ready-made costumes can basically meet the demands of a dramatic troupe. As noted by Interviewee 1, “performers are of similar height and weight, so costumes of the same size can meet their demand. In case of costumes that are too long, the lower inner hem of the costume can be manually stitched below the knees, while too-long sleeves cannot interfere with hand gestures”. Traditional opera costumes can be large-sized for male performers, medium-sized for female performers, and small-sized for child performers.

4) The prices of ready-made costumes are the same as those for tailor-made traditional opera costumes; materials for ready-made costumes are more carefully selected than those of tailor-made ones and most are artificially embroidered.

5) The sustained rate of usage for tailor-made costumes is low: when a performer’s bodily form changes, tailor-made costumes cannot be worn any longer. Most notably, a tailor-made costume for a female performer cannot be worn by the latter if she marries and becomes pregnant, which constitutes a waste of resources for the dramatic troupe.

6) There are no dramatic costume manufacturers in Suzhou, so most local opera troupes purchase costumes at the nearest dramatic costume shop in Zhengzhou, Henan.

In addition to traditional opera costumes, there is another kind of opera costume referred to as ancient costumes. Opera Troupe 2 prepared tailor-made opera costumes for performers of a costume play rehearsed recently and each major performer’s costume cost RMB 100,000. This type of opera costume is different from traditional ones, as it is specially designed by the costume designer in accordance with the script, scene, personal characteristics, director’s requirements, et cetera, and is similar with clothing props used in the costume opera and film.

5. Conclusions

5.1 Findings

This investigation revealed the following findings:

1) Residents’ favourability and awareness for Sizhou Opera are relatively low.

The research results show that local residents are less interested in Sizhou Opera; most notably, people younger than 35 years of age are obviously less interested in Sizhou Opera than middle-aged and older people. Due to low favourability and awareness, residents do not have enough motivation to watch Sizhou Opera; with the impact of modern media, people are more interested in watching TV programs than performances by opera troupes. This calls for improvements in music, costumes, and plays to enhance their attractiveness and residents’ favourability. When residents’ favourability and awareness are promoted, they will actively or passively accept Sizhou Opera and the latter’s audiences will increase, as will the popularity of Sizhou Opera.

2) Residents are interested in traditional opera costumes of Sizhou Opera. Relevant knowledge can be popularised intentionally.

Many local residents are willing to know more about opera costumes. In the overall research on Sizhou Opera, information on costumes is insufficient and even less information is available for residents. As such, in order to promote the development of Sizhou Opera, it is necessary to summarise and analyse the traditional opera costumes and popularise the knowledge of traditional opera costumes from the perspective of clothology.

5.2 Suggestions

The troupes made the following suggestions to improve costumes.

The decreasing audiences and shrinking performance market both restrict the development of Sizhou Opera. In response to this situation, there are three ways to improve the costume:

1) Rely heavily on the director’s requirements for costume design (e.g., Han Dynasty costumes are worn in the play related to the Han Dynasty, and Qing Dynasty costumes are worn in the play related to the Qing Dynasty).

2) On the basis of existing, traditional stylised shape, add only minor adjustments; for example: waist darts and cut lines; a set-in sleeve structure used to beautify bodily form (as most old costumes feature a raglan sleeve structure); the elastic cord used for colourful trousers (lacing is used for old ones and such structure is still used in the military play nowadays).

3) Design costumes mainly for commercial gains (e.g., aesthetic beauty is the focus in terms of costume design

based upon the dramatic demand of commercialised products; whether it is the characteristics of costumes of the Ming Dynasty or those of the Han Dynasty, all beautiful elements can present in costumes as design elements). Ultimately, during the process of costume improvement, there are many problems to be researched and solved.

5.3 Summary

This study explored and summarised the current status of Sizhou Opera in Suzhou from the perspective of residents' understanding of Sizhou Opera and traditional opera costumes and situations of Sizhou Opera troupes; hopefully, the research findings can help more local opera troupes, improve the current status of local opera, and promote the development and protection of intangible cultural heritages.

References

[1] Wan YZ. Brief discussion of Sizhou opera. Bengbu, China: Anhui Art Institute: 1987.
[2] Xue L. Investigation and research of regional cultural characteristics of Lahun Air of operatic circle. Nanjing, China: Nanjing University of the Arts: 2008.
[3] Wang AC. Research on protection and development of Sizhou opera based on concept of "active state." Drama Studies 2011; 1: 161-177.
[4] Chen SG. Suzhou opera. Hefei, China: Anhui People Press: 2009.
[5] Kong DD. A study of current status of Sizhou opera from perspective of development of rural cultural work. Northern Music 2011; 5: 107-108.
[6] Li R. A study of inheritance and protection of intangible cultural heritages of local opera in Anhui. Modern Chinese 2010; 7: 147-148.
[7] Yan CL. Brief analysis of status and measures of protection and utilization of intangible cultural heritages in North Anhui. J of Hubei Normal University 2015; 2: 81-83.
[8] Wang H. Sizhou opera and folk culture in North Anhui. Anhui, China: Anhui University: 2010.
[9] Yang Y. Analysis of Sizhou opera from perspective of folklore. Anhui, China: Anhui University: 2012
[10] Xu J. Discussions of Sizhou opera's development. J of Hubei Correspondence University 2015; 1: 192-194.

The study on the Approach of Conceptual Fashion Design based on 3D printing Technology

Yan Yuan, Meng-Zhu Zhang, Guang-Zu He, Tian-Zhu Cong

Jin Hua Road, Xi'an, Shaanxi, 710048, China

Corresponding author' email: yuanyan2003@126.com

Abstract

With the development of scientific progress and the improvement of fashion design aesthetics, conceptual fashion has become the optimal form to express fashion designers' concepts. In order to meet the public demands for a higher quality of life, emerging high-tech design methods such as 3D printing technology, has replaced both traditional fashion design and production processes, and has become an irresistible trend.

Focused on the combination of 3D printing technology and conceptual fashion, this article analyzed and studied the operational essentials, the design methodology and the production process. Simultaneously, this article discussed and explored methods to use 3D printing technology more efficiently and conveniently, methods to develop conceptual fashion design, and methods to speed up the production period in order to achieve bolder breakthroughs in conceptual fashion design. This article also explored more possibilities and practicability of conceptual fashion based on 3D printing technology. In conclusion, this article tried to find the effective approaches to design the conceptual fashion based on the 3D printing technology.

Keywords: 3D printing, conceptual fashion, design methodology, production processing

With the emergence of "customization" and cloud web manufacturing, 3D printing has become a fashionable design approach and production process in recent years, and has been widely used in the design field. Within several decades of development, 3D printing has reached an unprecedented level of maturity. However, its large development potential has yet to be completely explored. Customization based on individual demand is no longer an impossible dream. With 3D printing technology, designers can print a dye sample within several hours, saving hours of research from marketing to production.

The fashion industry has developed within a very short period of time. For conceptual fashion design, speed, accuracy and a more advanced expression of creativity is what counts. In the past, the development of new conceptual fashion called for a long R&D process (Research & Development) and production. This problem is further augmented with little control over the quality of the final product, where despite repeated production tests, it is near impossible to achieve the desired effect, thus impeding the progress and production of conceptual fashion products. However, upon the arrival of 3D printing technology, enterprises can now design and check products through digital adjustment, whereby only a few samples need to be made. In this way, the time from product design to production can easily be shortened, and enterprises can reap sustainable profits. Designing and making conceptual fashion with 3D printing is cost effective, environmentally friendly, and a unique way to do design. Thus, employing 3D printing into the fashion industry can fully meet the requirements of designers in today's society and save a large amount of raw and occasionally rare materials. The author is convinced that, in the near future, high-tech manufacturing techniques including 3D printing will dominate the fashion industry in terms of design [1].

Currently, most conceptual fashion based on 3D printing both at home and abroad focus on highlighting the complex 3D structure achieved through 3D printing. Though conceptual fashion has alleviated a few traditional shackles, comfort and practicality limit the progression of conceptual fashion into the mainstream due to its pursuit of exaggerating shapes and colors. In terms of manufacturing, 3D printing in the past was solely molded directly through printing, thus lacking innovative combinations with traditional thread-based fabrics. This paper attempts to explore methods to combine 3D printing technology with conceptual fashion design from multiple perspectives.

1. Advantages of 3D printing in conceptual fashion design

1.1 Creating a new design vocabulary that is not bound by traditional processes

Due to its unique design characteristics, conceptual fashion carries a strong importance on the process of manufacturing. During the traditional design process, designers often fail to achieve the expected design. Limited by traditional manufacturing processes, conceptual fashion allows a way for designers to truly express themselves. Compared with other techniques, 3D printing is unique and beautifully complex. It offers an unrivalled advantage over traditional methods of manufacturing in making complicated, sophisticated or special designs - so unique that it goes beyond imitation and transcends other techniques. Therefore, not only can 3D printing make up for the disadvantages of the traditional manufacturing techniques, it also allows conceptual fashion design to flourish, where complex physical structures and exaggerated abstract designs are achievable.

1.2 Providing new ideas for conceptual fashion design

3D printing has provided a creative design stage for designers, rapidly spurring their imagination and capabilities. This is also the greatest advantage of 3D printing. With 3D printing, designers can help their thinking achieve its maximum potential [2]. Besides, the interdisciplinary characteristic of 3D printing also greatly increases its room for creativity. Previously, designers often had to give way to more practical designs when their fanciful inspiration could not find proper techniques to be materialized. In design courses, teachers repeatedly emphasized this as a design taboo. However, there is nothing wrong with a rich imagination. The previous design requirement also reflected the limit of the traditional design thinking. It was not until the application of 3D printing to conceptual fashion design that designers could get rid of the traditional design shackles, broaden their design thinking and express their wide imagination through conceptual fashion design.

1.3 Shortening the design development period

Previously, conceptual fashion design ideas that were hard to realize called for repeated experiments. Such repetition lengthened the design and manufacturing period while inhibiting the development process of conceptual fashion. While conceptual fashion based on 3D printing is made by computer-aided 3D models, problems encountered during the production of conceptual fashion design can be easily solved through quick rectification of these models. The application of 3D printing has practical significance during the design R&D process. It can be seen that 3D printing efficiently combines the design and manufacturing into one linear process, thus eliminating slow experimentation and greatly improving the efficiency for the production of conceptual fashion design.

1.4 Reduce the design and production cost of conceptual fashion

Cost is an important factor influencing conceptual fashion design. Many good design ideas are put aside due to high manufacturing costs. While traditional manufacturing methods also allow manual customization, it is time and cost consuming which calls for multiple rectifications before completion. However, 3D printing is different. Not only does it offer an advantage in making complicated, sophisticated and unique fashion structures, but it also boasts promising application prospects in terms of reducing production cost.

2. Application principles of 3D printing in conceptual fashion

2.1 Design creativity

Conceptual fashion based on 3D printing is different from fashion design of other types. It has no fixed target population. In other words, conceptual design based on 3D printing is used to express personal design concepts of designers. From a deeper spiritual perspective, conceptual fashion design is an external expression of concepts. Its significance has far outweighed the fashion itself. Hussein Chalayan once said that, "design is experimental and conceptual. Concept is, if not being more important than fashion itself, is at least equally important." [3] Thus, it is justified to say that creativity is the source and principle for conceptual fashion based on 3D printing. If conceptual fashion design based on 3D printing loses the support of creative ideas, it will also lose its fundamental attributes.

2.2 Design aesthetics

Conceptual fashion based on 3D printing has unique aesthetic characteristics in terms of its form. During the design phase, it should not only adhere to the fashion concept in terms of form, but, more importantly, use new ideas and theories to challenge traditional aesthetics in fashion so as to redefine the aesthetics of conceptual fashion. This new expression attributed to 3D printing has largely changed the trend in conceptual fashion design, and has developed a new fashion trend out of high-tech manufacturing methods.

2.3 Design functionality

Previously, the function of conceptual fashion was to show personality. Currently, with the support of 3D printing, it has been a possibility for conceptual fashion to embark on the development towards artificial intelligence. The current focus of conceptual fashion design based on 3D printing should be on intelligent functional fashion. The combination of 3D printing technology and artificial intelligence will be highly practical in academic and medical fields. It is necessary to add functional structures into conceptual fashion design based on 3D printing so as to reflect its forward-looking characteristics.

3. Application method of 3D printing in conceptual fashion

3.1 Directly using 3D printing technology

3D printing technology is known as Rapid Prototyping Manufacturing, also referred to as RPM .The biggest advantage of this technology is the molding process for one-time molding. Through the use of 3D printing technology, we can instantly print clothing instead of using the traditional method of pattern making and sewing [4]. Designers can directly take advantage of this special technology to do fashion design, which, to some extent, can change the garment silhouette and structure beyond human control. This design method can put different inspirations together. Not limited or restricted; unrestrained. People who are not familiar with traditional patternmaking can use 3D printing technology to design a unique piece of fashion.

3.1.1 Reverse Engineering (RE)

Using the 3D scanner to scan an existing physical model, one can extrapolate that data and receive the corresponding digital 3D data. We can create the final 3D model through the use of this data combined with model drawing software. This method is simple and can be operated easily because it is not time consuming to create this model. One simple needs to design according to the ordinary method. Three dimensional inversion measuring is suitable for designers who have no contact with the modeling technology of 3D printing.

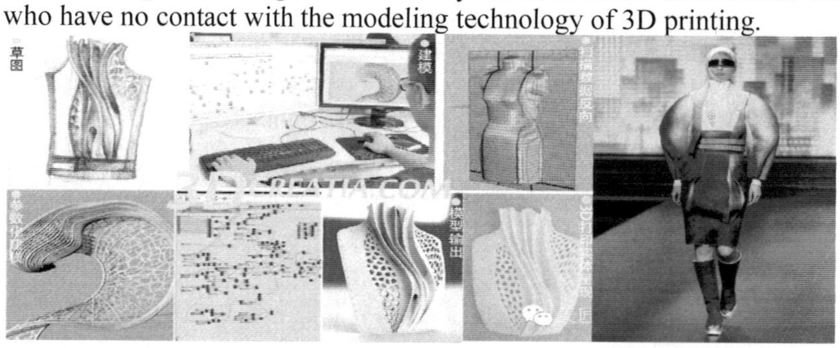

Fig.1 Liang Xiu Dong "Mysterious City"

Fig.1 shows the making process of a final work that was from a design "mysterious city" by a designer who won the 2015 Hanbo Garment Competition.

3.1.2 Splicing

If we want to get a soft and comfortable piece of clothing, we can imitate a way that is originally from the ancient Knight's soft allegations armor. As long as we design a simple model that can be repeated, we can also create it through 3D printing modeling. For example, as shown in Fig.2, the structure is composed of many hexagonal honeycomb pieces through 3D printing sheet splicing. Although these structures are stitched together, they can also be molded using a 3D printer. This simple method that stitches different structures together is suitable for designers who have a basic understanding of building a model.

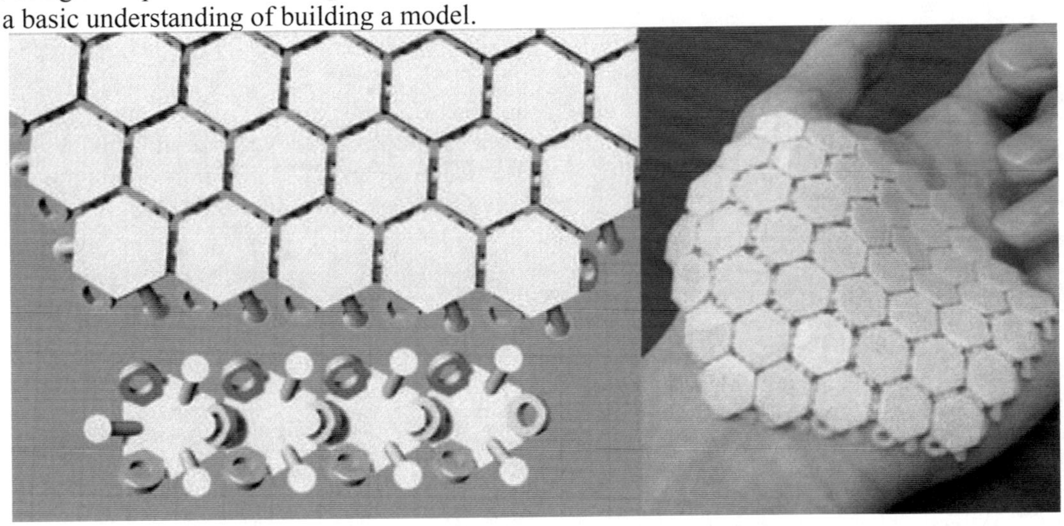

Fig.2 A fabric stitch of a hexagonal honeycomb structure

3.2 Design method of the combination of 3D printing and textile materials

3.2.1 Crocheting

When we use the 3D modeling in combination with knitting materials, the best way to connect the different materials is to use twine. Due to the special structure of the knitted material, the 3D model could directly be wound to the structure of the knitting materials which are highly durable. Mark Beecroft and Laura McPherson, two researchers from the Manchester School of Art, cooperated on a project that combined digital technology with textile materials.

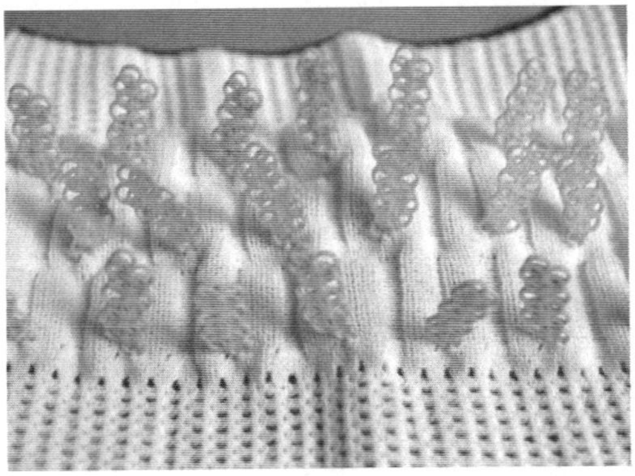

Fig.3 3D model combines with knitting

Fig.4 The experiment with seam fabric and elastic TPU

With regards to Fig.3, this connection method is utilized by extremely skilled designers due to its complexity.

3.2.2 Suture

Suture is the most basic way of connection. This method can also be applied to the combination of 3D printing model and clothing, but it requires special treatments. If TPU elastic or nylon can be sutured with clothing, it would be very much similar to the connection that is made with two normal pieces of clothing. If the materials are harder than 3D printing models, holes will have to be punctured through the fabric prior to the design and manufacturing phase.

Fig.4 illustrates the experimental results of direct stitching using traditional fabrics and elastic TPU.

3.2.3 C. Using adhesive

If we can't make holes on the fabric or we can't suture material and textile materials, the next method is to use adhesive. Because most 3D printing materials for garments are made of thermo softening plastic, nylon, silica gel and rubber, such materials can be glued very well with notions. Put simply, if one were to combine a 3D printed model with traditional clothing accessories, we can adopt this technology using heat-activated adhesive, sticky drill glue and bonding web.

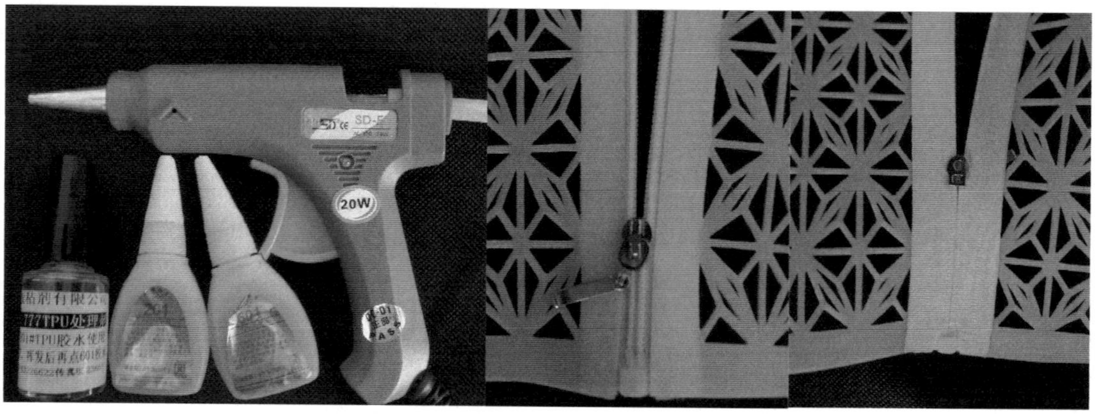

Fig.5 adhesive bonding glue gun and 3D model and the experiment of zipper adhesion

Fig.5 shows a test illustrating the method to connect the 3D printed model with traditional clothing accessories (e.g. zipper) by using the measure with the bonding web.

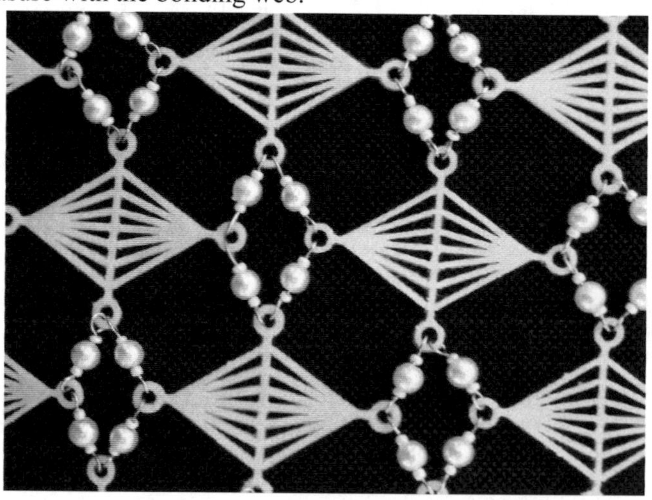

Fig.6 3D printing model of series structure

3.3 The method of the combination of non-textile materials

3.3.1 To stick

The design method of sticking utilizes the same method of combining textile materials. A 3D printed model and other materials can be directly stuck together, which is the most traditional design method. No matter what material the 3D printing model is made of, it can be stuck together with any non-textile materials by using glue, bonding webs and glue guns. This kind of design method is consistent with the traditional industrial processing methods and we do not require many tools or conditions for manufacturing.

3.3.2 To connect the ring structure in series

Utilizing a ring structure as a connection method requires holes that need to be prepared when we are designing models. We have two ways to achieve this method. The first is to reserve the holes on the inner edge of the model, and the second is to add a ring structure on the edge of the model. These connection methods are simple and can be started easily. Figure 6 illustrates an experiment where 3D printing is used to combine needles and rings together.

4. The limitations of designing

Whether the fabric is printed directly, or a combination of both textile and non-textile materials, 3D printing is not always successful in making clothing especially when taking into account the underlying concept of clothing. The materials themselves are often the main cause of this limitation. Material choice is always the key in a successful piece of clothing. Currently many kinds of fabrics and materials can be used for 3D printing, which can easily meet the needs of designing. But sometimes we must make up the shortcoming by changing some of the materials' innate characteristics or structure.

Objectively, with the rapid development of 3D printing technology, it continues to have a variety of technical problems that need to be improved, such as the lack in development and finding new materials. 3D printing could be looked as a new high-tech technology that can be combined with machine manufacturing, art, design and computer numerical control.

Above all, the practices here make use of 3D printing to create a collection of clothing, to verify the feasibility of the design methods that we have mentioned earlier (Fig.7).

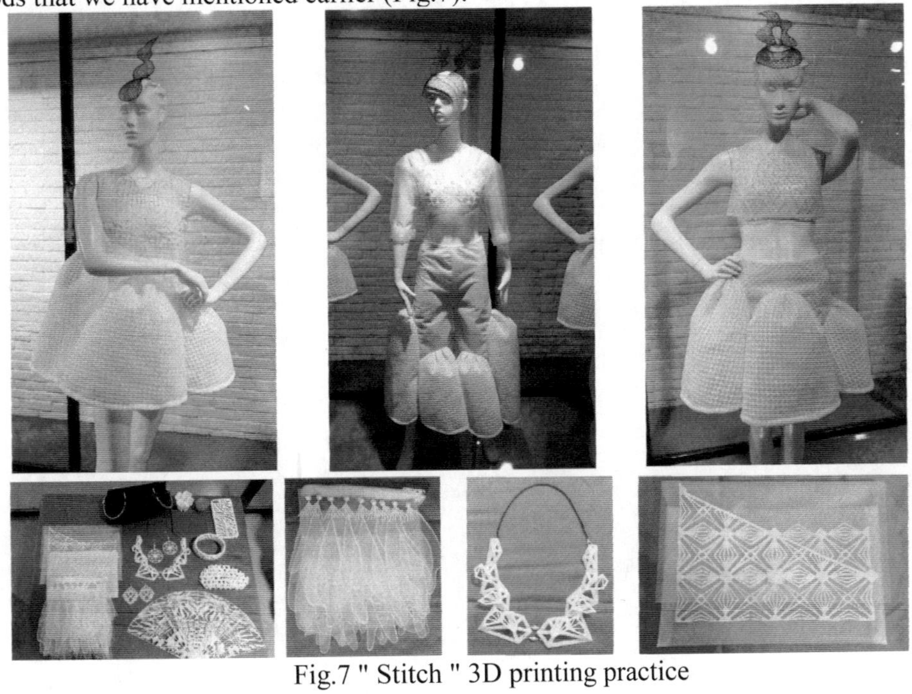

Fig.7 " Stitch " 3D printing practice

5. Conclusion

As mentioned above, this article focused on the combination of 3D printing technology and conceptual fashion with the main points outlined below:

1) By comparison and research, this article studied the principle, process and equipment of 3D printing, and explained the technology that supported 3D customization in conceptual fashion design;

2) From an artistic standpoint and a business and manufacturing point of view, the application of 3D printing was introduced. By analyzing and forecasting the direction of 3D printing, the development of 3D printing is highly regarded to be prevalent in the near future.

3) Explored the methodology to create conceptual fashion based on the technology of 3D printing.

4) Explored the use of 3D printing technology in more efficient and convenient methods,

5) Explored the possibilities and practicality of conceptual fashion based on 3D printing technology.

In conclusion, this article attempted to find the effective approaches to design conceptual fashion based on the 3D printing technology. With 3D printing technology becoming increasingly mature, it has become easier and much more convenient to create complex and unpredictable designs in the future.

References

[1] Wang CC. Analysis of the development condition of 3D printing [J]. Screen printing: 2012 (9): 39 ~ 43.

[2] Chao G. The 3D printing technology is been using in modern clothing art [J]. Art technology, 2015 (1): 11 ~ 68.

[3] Niu YF, the study of the 3D printing in the teaching system [J]. Plastic packaging: 2015 (1): 35 ~ 39.

[4] Feng ZM, Zhao J, Feng C, and some masters: the world's 100 fashion designer quotes [M]. Beijing: Chemical Industry Press: 2008:104.

A Novel Method for Detailed Categorization of Female Buttock Shapes

Xiao-Feng Yao[1], Jian-Ping Wang[1*], Susan P Ashdown[2]

[1]*Donghua University, Yanan Road, Shanghai, 200051, China*
[2]*Cornell University, Thurston Avenue, Ithaca, NY, 14850, America*

*Corresponding author's email: wangjp@dhu.edu.cn

Abstract

The shape of buttocks is one of the most important factors affecting clothing fit issues. The purpose of this paper is to develop a novel method for the classification of female buttock shapes and to identify a range of typical buttock shapes. Characteristics of FBS were analyzed and eight influencing factors, including two length ratios, two area ratios and four angles, were calculated to represent buttock shape features. With measurements extracted from 3D scan data of Northeastern American women, a two-step K-means cluster analysis was used to classify buttock shapes in detail. In this study, we developed a method to locate landmarks on the buttocks, to describe the primary features that describe FBS. A two-step K-means cluster analysis method was used to do in-depth categorize the different shapes. The use of proportional measures allowed us to concentrate on shape variation without considering size. Three groups with ten subgroups of different buttock shapes were finally obtained, and thus this classification method was testified. This study provides a novel method to do categorization of female buttock shapes in detail. These findings could make a supplement to current assist in the development of pants sizing systems and could be used for improving fit of pants and of girdles fit issue. This method also can be used for detailed categorizing other body parts.

Keywords: Female buttock shape; Body shape classification; K-means cluster analysis; 3D scan technology

1. Introduction

The importance of body analysis and classification has been recognized by researchers since ready to wear (RTW) garments began to be popular. To make RTW garments fit various human body shapes and sizes, apparel sizing systems are established based on key dimensions selected from anthropometric data [1]. Current sizing systems are usually based on one body shape only, using an hourglass-shaped fit model with specified key dimensions, which means only those women who are in this figure type can find well-fitted clothing [2]. It has been reported that over 50% of female consumers are unsatisfied with the fit of RTW apparel, and sizing systems developed for very fitted or tight garments particularly need to address a wider range of body types [3,4]. Therefore, deep analysis of body shape is needed for clothing pattern designers and manufacturers.

Buttocks shape is one of the most significant factors impacting people's overall body type and also plays an important role in clothing fit issues [5]. Because most excess fat is stored in the buttocks, hips and thighs in women, female buttock shapes (FBS) may vary even among women of the same size [6,7]. Therefore, to provide a good fit, some apparel companies advertise pants styles in different shapes for their customers. For instance, Levi's [8] developed Levi's Curve ID based on extensive anthropometric data to help customers find jeans fitted to their individual shapes. The Levi's website classified the shape from waist to hip as 'straight hips with a bit of curve', 'some definition from waist to hip', 'much smaller waist than hips' and 'defined waist and fuller hips', while levels of curviness in the seat were categorized as 'not very curvy', 'some curve', 'ample and curvy' and 'very robust and curvy'. Their customers choose their jeans based on their self-judgment of their body shape including buttock shapes based on these designations and line drawings of various buttock curves. Levis has recently modified this sizing system to offer a selection of choices without customer input and recommended shapes, but still provides jeans for a range of body types.

In order to learn more about buttock shapes, a reliable method for buttock shape classification is needed. Two major methods were developed in former studies. The first one, developed by Connell [9] was based on visual analysis. Hip and buttock shapes were evaluated from the front and side views. Front views were analyzed by level of greatest width and degree of curve from waist to the crotch line, and classified into Straight and Nearly Straight hips, High hips, Mid-level hips and Low hips. Based on different shapes and prominence of buttocks as viewed from the side, two shapes, (flat 'Low Hook' and round, curvy 'Even Arch') were identified, but in the study of Connell et al [9], all shapes were then collapsed to three (flat, average, and prominent) for ease of analysis. As detailed shape analysis was usually a compliment for developing a sizing system, sometimes shape characteristics were discussed along with dimensions. McKinney [10] developed a pattern development method by measuring depth at the hip/ buttock level and using the results to create patterns for women with different buttock shapes. Fifty women were finally classified into seven groups. In this method, the buttock and hip shapes were treated separately in the front and side view, so they were only partially evaluated and do not distinguish shapes effectively as a whole. It is also a visual analysis study, which may be influenced by subjective perception issues.

Another method is reported that uses cluster analysis and principal component analysis. This method was developed by Song and Ashdown [11]. Body measurements of 2488 women from SizeUSA body scan data were analyzed based on five principle components (PCs) (PC1: waist to top hip silhouette, PC2: top hip to full hip silhouette, PC3: buttocks prominence) and two additional measures expressed as Z-scores (Z-score 1:abdomen prominence and Z-score 2: slope from abdomen point to front hip point). Three lower body shapes were categorized in this analysis. This method made use of objective measurements and a statistical analysis technique, resulting in a categorization method that was more reliable than the method requiring visual comparisons, but it was designed to address general apparel fit and was limited to three very general lower body types.

Buttock shapes was chosen as example to illustrate how to apply Song and Ashdown's method of identifying body types to classify areas of the body in detail [11]. Measurements with the potential to describe buttock shape's characteristics were converted to angles and proportions. Using two-step K-means cluster analysis, a range of typical buttock shapes were identified.

2. Experiment

2.1 Shape classification considerations:

1) The fixed landmarks on the body for this study needed to be easily identifiable from a 3D scan of the body. Landmarks were used to locate prominent body shape structures. These landmarks must be identifiable on all the subjects at equivalent places. In traditional anthropometry body landmarks were identified at anatomical landmarks, which can be located by visual inspection or palpation at the body's surface. In this study, the landmarks chosen were all located between the waist line and mix-thigh line, and were identified by visual inspection of the 3D scans.

2) Angles and proportions were calculated to define the buttock shapes in order to focus on shape instead of size variation. According to Mosimann [12], "Shape" can be expressed as a set of proportions or ratios that represent dimensionless quantities in a physical space, which is irrelevant to size and has been used in anthropometric analysis to distinguish different species by body types. Therefore, when the buttock shapes are categorized, all the measurements should be converted into proportions or angles [11].

3) Posterior projection, volume, and ptosis (amount sagging or drooping) were included for categorization of the buttock shapes. According to studies by plastic surgeons [13,14], most buttock beauty operations were to improve these three aspects. Tight pants or girdles are designed to wrap the hip and reveal or beautify the curve of buttocks; as a result these three characteristics should be taken into consideration.

2.2 Sample Selection

The goal of creating a size chart is to use as few sizes as possible to represent as many people as possible, but shape classification requires more detail and a decision about the appropriate number of shape categories. To increase efficiency, the study population can be decreased by controlling demographic variables based on the research purpose. For this study we control subject age, hip girth and WHR (Waist to Hip Ratio).

The dataset used for this study was a set of 83 body scans of US females aged 18 to 35, selected randomly from a database of women from the Northeastern part of the United States. All the female participants had hip girths ranging from 86.4cm (34 in) to 121.9 cm (48 in), which is consistent with the standard sizes 2 to 20 of ASTM D5585-95 (the US standard of body measurements for adult female misses figure type sizes).

2.3 Measurement Selection

The original measurements were done using Anthroscan software from Human Solutions software with manual identification of landmarks and manual measurement of the scans. Slice areas and angles were calculated using Geomagic Studio.

All measurements that could define buttock shape characteristics were used for the analysis. In order to focus on the buttocks, but not abdomen, torso or thigh shapes, only measurements relating to the buttocks were used. Girths were not included in the analysis. To focus on the shape characteristics, but not size, all the measurements were calculated as proportions or angles. Three features of the buttock shapes: the volume (the degree of fullness), the proportional position of the maximal gluteal projection point, and the 2D buttock shape from the sagittal plane were considered. All measurements were made on the right side of the body. The landmark locations are shown in Figure 1. Table 1 describes the landmark locations and measurement procedures.

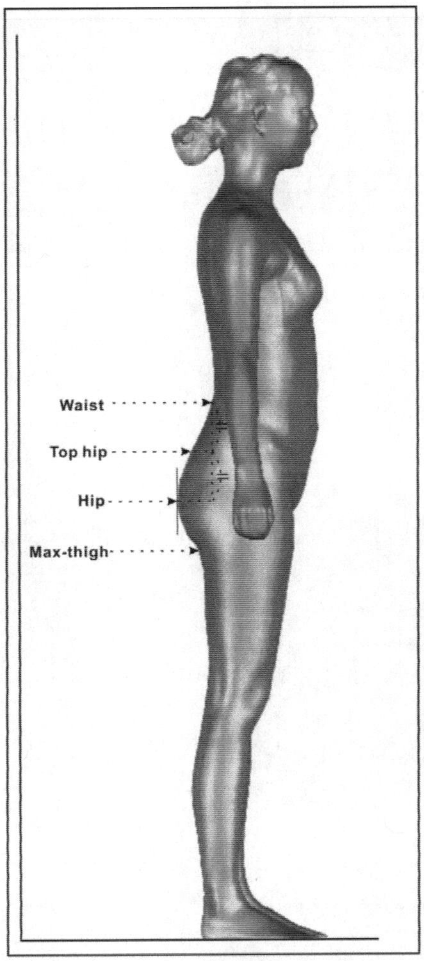

Fig.1 Landmark positions on the human body

Table 1. Measurements procedures and landmarks

	Measurement procedure
Height	A vertical distance taken perpendicular to the floor.
Width (side to side)	A horizontal distance taken parallel to the floor on the frontal plane.
Depth (front to back)	A horizontal distance taken parallel to the floor on the sagittal plane.
Area	A two dimensional measure of a cross section or slice of the body
Angle	A measure of the relative position of two rays on the same plane.
	Landmark location
Waist	In the middle of the lumber vertebrate (L3-L4)
Top hip	A point half the distance between waist and full hip heights.
Full Hip	The furthest posterior extension of the buttocks.
Thigh	One inch below the crotch.

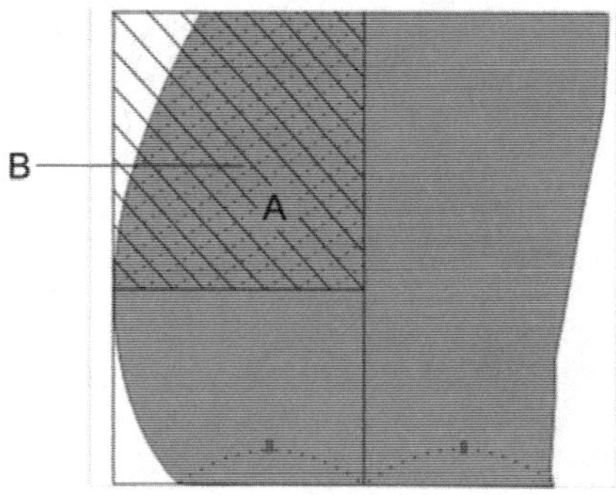

{1} A: Upper buttock area of the slice.
{2}B: Upper rectangle area.

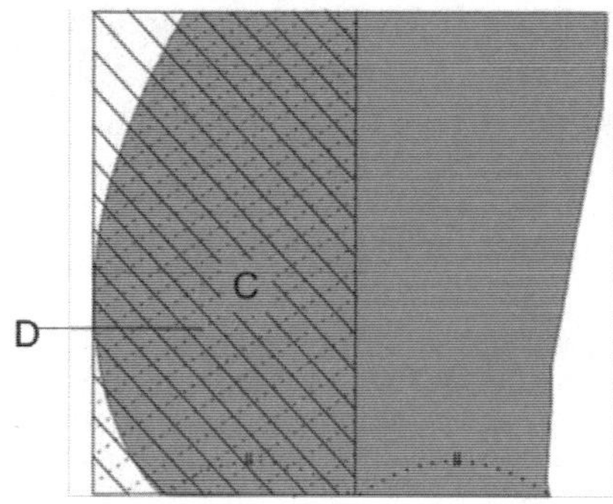

{1}C: Whole buttock area of the slice.
{2}D: Whole rectangle area.

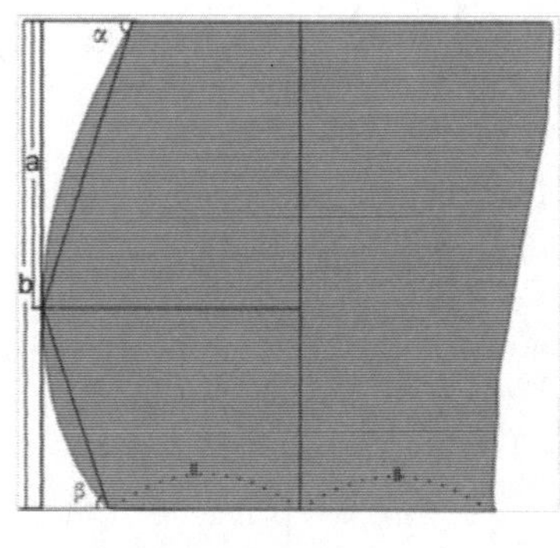

{5} a: Waist hip height.
{6} b: Waist max-thigh height.
{7} angle α: Angle of the full buttock point from the waist.
{8} angle β: Angle of the full buttock point to the max-thigh.

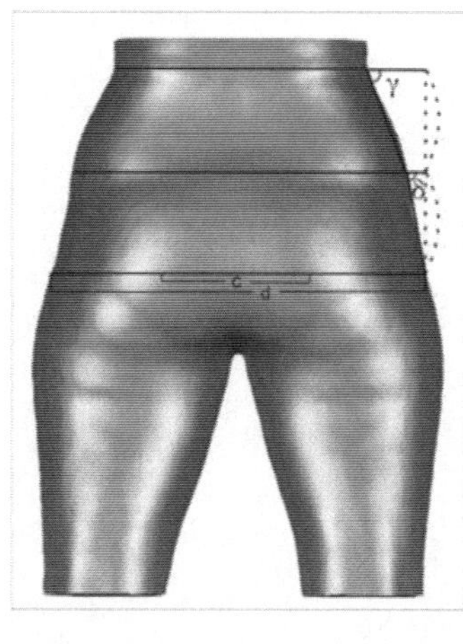

{9} c: Distance between two buttock bumps.
{10} d: Largest width of hip.
{11} angle γ: Angle of the top hip from the waist.
{12} angle δ: Angle of the full buttock point from the top hip

Fig.2 Measurements used in the calculations

A set of measurements were taken from each scan to assess the range of different buttock shapes. In order to focus the analysis on the buttock shape only, a vertical slice was made on the sagittal plane at the location of the furthest posterior extension of the right buttock between the waist level and the max-thigh level. Then the midpoint of the max-thigh line on the sagittal plane was identified and extended vertically to the waist line of the body slice to provide a reliable midline location to segment the buttock area of the slice, thereby excluding the profile of the abdomen. Measurements taken from the resulting slices included ratios of the upper and whole buttock area of the slice (A/B and C/D). These ratios established the volume (the degree of fullness) of the buttock shapes. Waist hip height/ waist max-thigh height (a/b), the angle of the full buttock point from the waist (angle α), and the angle of the full buttock point to the max-thigh (angle β) were used to identify the position of maximal gluteal projection point in the slice of the body. In addition, the distance between two buttock bumps in proportion to the width of hip (c/d), the angle of the top hip from the waist (angle γ) and the angle of the full buttock point from the top hip (angle δ) were used to identify details of the buttock shape from a posterior view. Figure 2 shows the measurements taken from the buttock shapes.

2.4 Body Classification Method

K-means cluster analysis was performed with the subjects. All the variables were standardized to eliminate the dimensional relationship between them. In order to identify as many distinct buttock shapes as possible, the study included two steps. First, four buttock angles (angle α. β. γ. δ), c/d and C/D are used to do the K-means cluster analysis to get the general buttock shape. The analysis was performed starting with different numbers of clusters. The subjects were divided into two, three, four and five groups. Characteristics of each classification method were examined and the three cluster calculation was selected as the best to represent the complex buttock shapes in this study. Then, A/B and a/b are used to K-means cluster analysis to further segment buttock shapes.

3. Results

The participants were divided into different numbers of clusters to determine which cluster number divided the participants into visually distinct groups of relatively equal sizes. When two or three clusters were categorized, the data was judged to be relatively evenly divided into each cluster as follows: in the two-cluster model (cluster 1:n=20, 41.7%, cluster 2:n=28, 58.3%); and in the three-cluster model (cluster 1:n=14, 29.2%, cluster 2:n=13, 27.1%, cluster 3:n=21, 43.7%). However, in the four-cluster model, cluster 3 only represents 10% of the total data (cluster 1:n=10, 20.8; cluster 2:n=18, 37.5%; cluster 3:n=5, 10.4%; and cluster 4:n=15, 31.5%). In the five-cluster model, cluster 2 and cluster 4 only represent 10% and 4% of the total data (cluster 1:n=10, 20.8%; cluster 2:n=5, 10.4%; cluster 3:n=18, 37.5%; cluster 4:n=2, 4.2%; and cluster 5:n=13, 27.1%) Therefore, we concluded that the four or more than four cluster methods were not appropriate for the buttock analysis for this stage of our study.

In the two cluster model, cluster 1 had a straight back silhouette, small distance between buttock bumps and prominent buttocks, while cluster 2 had a straight back silhouette, large distance between buttock bumps and flat buttocks. The third cluster in the three cluster model had a curved back silhouette, large distance between buttock bumps and prominent buttocks, a shape category which did not exist in the two cluster method. In addition, the significance probability (P) of c/d is 0.049 in the two cluster model, very close to 0.05, which means that c/d almost made no contribution to the cluster analysis. Therefore we concluded that the two cluster model does not divide participants into groups that represent all characteristics of the buttocks.

Table2. Mean and deviation of the variables for the three buttock shape group

Variables for cluster analysis	Cluster 1		Cluster 2		Cluster 3		F	Sig.
	mean	Std. Deviation	mean	Std. Deviation	mean	Std. Deviation		
Angle α	69.842	4.056	74.943	2.994	67.182	3.257	20.425	0.000*
Angle β	61.192	3.653	66.600	3.819	64.223	3.526	7.504	0.002*
C/D	0.777	0.046	0.862	0.0352	0.816	0.026	19.557	0.000*
Angle γ	78.632	4.631	80.869	3.479	75.657	4.703	5.911	0.005*
Angle δ	76.786	3.704	82.200	4.640	75.096	4.115	12.062	0.000*
c/d	0.136	0.015	0.179	0.025	0.177	0.016	23.191	0.000*

*P<0.05

In the three cluster model, a one-way AVONA test was performed based on 95% confidence level to compare the mean scores of the six variables used in the cluster analysis. In table 2 it can be seen that overall F for the one-way AVONA test was significantly different ($P<0.05$). These results were similar to the results of the lower body analysis in Song and Ashdown (2011), providing additional validation of this method.

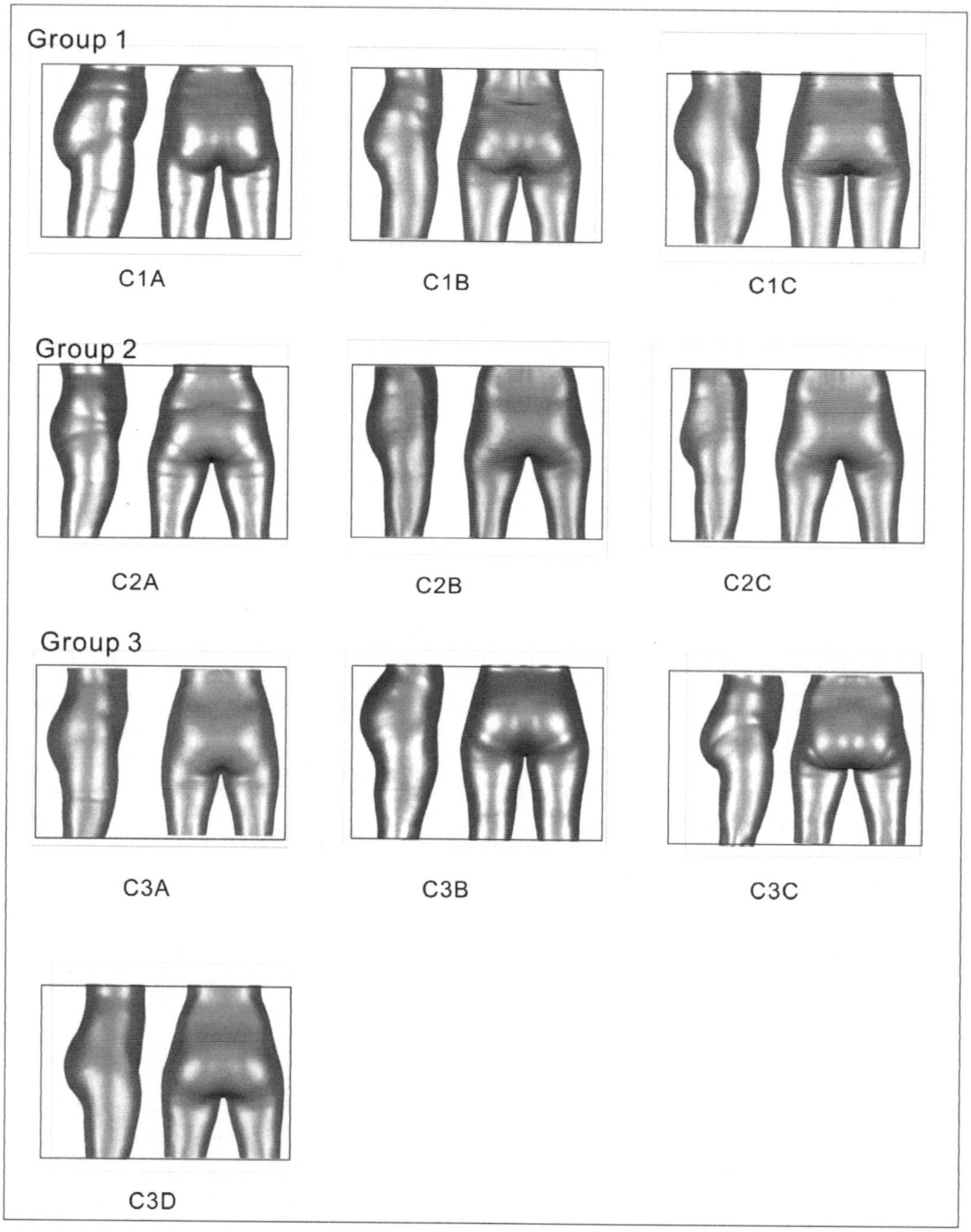

Fig.3 Center figures of the ten clusters

Starting with these basic shape classification results, the variables A/B and a/b were then sorted within each buttock shape group using K-means cluster analysis to further classify the buttock shapes into subgroups. These two factors (A/B and a/b) described the buttock bump position and the fullness of upper and lower buttock area. An analysis process similar to that described above was used to identify visually distinct groups of relatively equal sizes. Once cluster one, two and three were subdivided based on the new variables, the decisions about numbers of subgroup clusters were calculated separately. Three valid subgroups were identified for cluster 1 and 2, and four valid subgroups were identified for cluster three resulting in ten subgroups total.

The three subgroups within cluster 1 were classified as follows: C1A had a low buttock bump position. The buttock bump positions of C1B and C1C were higher than C1A, almost in the middle of the total buttock area, but C1C had plumper upper buttock area than C1B. The three subgroups of the cluster 2 followed a similar pattern. The four subgroups in cluster 3 can be classified as follows: C3B had the highest buttock bump position and also the fullest buttocks, and C3C had the lowest buttock bump position. C3A and C3D had similar medium level buttock bump position, but C3D was plumper. Images of a representative from each subgroup are shown in Figure 3, demonstrating the characteristics of each group.

4. Discussions and conclusion

An analysis method for female's buttock shapes (FBS) was developed and the prominent factors influencing the classification of FBS were investigated using the three-dimensional (3D) scanning technology and 3D image processing software. Three important considerations for shape classification of buttocks were developed in our study, including landmark identification, shape structure description and methods to describe shape features. These three considerations can be used for analysis of body shape for other isolated areas of the body. As a supplement to size analysis, shape classification needs in-depth discussion of can be demonstrated using body ratio features.

In order to concentrate on shape but not size variation, angles and proportions were used to describe the buttock shape characterization. Unlike previous studies [11,15], our study only focused on buttock shape alone, excluding or minimizing the factors of overall lower body shape, abdomen shape, torso shape, or thigh shape as much as possible. From the mid-point of the max-thigh, the lower body was divided into two parts on the frontal plane in order to isolate the buttock area.

Eight factors from three aspects were selected to describe buttock features. There were two stages in the buttock classification process. In the first stage, we did an overall classification of the shape based on angles and overall proportions resulting in clusters which were significantly different from one another. A second cluster analysis was done within each of these three clusters to identify subgroups based on proportions that identified the placement and shape of the buttock bumps, resulting in 10 different buttock shapes. The first two clusters had three subgroup shapes and the last cluster had four subgroup shapes. Our method used a variety of key measurements and differentiates buttock shapes in all dimensions (both side and back view factors).

5. Limitations

Limitations of the study are that the shape analysis was conducted on a subset of the population (women aged 18 to 35 who fit in the size range of 2-20 in the Missy size range, who had a neutral posture and who were not rectangular in body shape). Subjects in the study were also only a subset as they consisted of females from the Northeastern United States. Further study on identification of buttock shapes from a wider range of the population with diverse people will be undertaken in future studies.

References

[1] Zakaria N. Body shape analysis and identification of key dimensions for apparel sizing systems. Anthropometry Apparel Sizing & Design. Cambridge: 2014; p: 95-119.

[2] Connell LJ, Ulrich P, Knox A, Hutton G, Woronka D, Ashdown SP. Body scan analysis for fit models based on body shape and posture analysis. National Textile Center Annual Report: 2003; p: 1-10.

[3] Alexander M, Connell LJ, Presley AB. Clothing fit preferences of young female adult consumers. International Journal of Clothing Science & Technology: 2005; 1: 52-64.

[4] Song HK, Ashdown SP. Development of automated custom-made pants driven by body shape. Clothing & Textiles Research Journal: 2012; 4: 315-329.
[5] Alexander M. Applying three-dimensional body scanning technologies to body shape analysis. Auburn University: 2003.
[6] Nakazawa S. Clothing and Body Shape. Tokyo: Bunka Pub: 1996.
[7] Delavier F. Women's Strength Training Anatomy. Human Kinetics. Pudesy: 2003.
[8] Information on: http://us.levi.com/shop/index.jsp?categoryId=4370093.
[9] Connell LJ, Ulrich PV, Brannon EL, Alexander M, Presley AB, Connell LJ. Body shape assessment scale: instrument development for analyzing female figures. Clothing & Textiles Research Journal: 2006; 2: 80-95.
[10] McKinney EC. Towards a three-dimensional theory of pattern drafting: relationship of body measurements and shapes to pattern measurements and shapes. University Of Minnesota: 2007.
[11] Song HK, Ashdown SP. Categorization of lower body shapes for adult females based on multiple view analysis. Textile Research Journal: 2011; 9: 914-931.
[12] Mosimann JE. Size and shape analysis. Encyclopedia of Statistical Sciences: 1988.
[13] Cuenca-Guerra R, Lugo-Beltran I. Beautiful buttocks: characteristics and surgical techniques. Clinics in Plastic Surgery: 2006; 3: 321-332.
[14] Lee EI, Roberts TL, Bruner TW. Ethnic considerations in buttock aesthetics. Seminars in Plastic Surgery: 2009; 3: 232-243.
[15] Alexander M, Pisut GR, Ivanescu A. Investigating women's plus-size body measurements and hip shape variation based on SizeUSA data. International Journal of Fashion Design Technology and Education: 2012; 1: 3–12.

Research for Development of Digital Personalized Customization

Bing-Bing Liu[1*], Hong Dai[2]

School of Apparel & Art Design, Xi'an Polytechnic University, Xi'an, 710048, China

*Corresponding author's email: 1571684221@qq.com

Abstract

With the development of computer network technology, future apparel industry will inevitably bring about the Personalized Customization of clothing by the network environment. Personalized Customization is the tailoring of products to user requirements, reflecting different user needs and desires. Therefore, there is an urgent need for the development of a personalized customization clothing system based on the network of environment and development. In this paper, theoretical research and practical application of a combination of perspectives, combined with the status of Personalized Customization digital development, analyzes the problems faced by now. This paper discusses the building of the system architecture and development direction of Personalized Customization clothing digital, and establishes the basic ideas for further study in fashion design digitization.

Keywords: Personalized Customization; Digital clothing technology; 3D; Costume design digitizing

1. Background

With the rapid development of the Chinese economy, the garment industry has also ushered into rapid development, despite facing challenges. The traditional garment industry has been unable to meet the structural requirements of development. Information technology and digital technology as the wave of the third industrial revolution, has penetrated into various fields gradually and also provides new opportunities for the garment industry for development. Today, the apparel industry is undergoing unprecedented transformation. Various information systems and digital applications have appeared in the garment industry, resulting in the rapid development of the garment industry, evident in new digitized features.

The digital garment industry is digital information-based, computer technology and network-based. Through the collection of costume design, processing, marketing and other aspects of information integration, storage, transport and application, its ultimate goal is to achieve the optimization configuration of clothing enterprise resources [1]. Digital costume design as an important part of the digital garment industry affects the development process of the whole garment industry directly.

In the development of the current global garment industry, the network environment for development of Personalized Customization of clothing has become an important trend in the development of the world's clothing. Costume design three-dimensional, personalized and fashion has become the trend [2]. The concept of fitting cropped has become a new generation of apparel supply guiding strategy [3]. More and more consumers choose personally customized clothing when they purchase clothes. With the development of computer network technology, the garment industry is bound to realize that the online customization of clothing is the best way to tailor to the needs of consumers. Hence, research and development in a Personalized Customization clothing system in the network is imperative.

2. Fruits and issues for the development of personalized customization clothing

2.1 Development of digital clothing technology

Garment computer aided design technology in China has been developing for nearly 40 years. Since its birth, the production efficiency and product quality of garment processing has improved greatly. It actively promotes the modernization of garment enterprises to improve management level and enhance the ability of garment enterprises to adapt in the markets. It plays an irreplaceable role in the modernization process of the apparel industry. Driven by the garment CAD technology, computer technology is widely used in the traditional garment processing industry, creating a dynamic of their own development.

In recent years, domestic textile and garment industry of computer-aided design technology had matured and become quite commercialized. This includes computer-aided design fabric tissue, clothing grading, marker making and two-dimensional design system software development. Currently, the more influential domestic garment CAD systems are: Gerber (United States), PGM (United States), PAD (Canada), Lectra (France), Toray (Japan), Investronica (Spain), Arisa (China Institute of Aerospace 710, China), Echo (Hangzhou, China) and so on. These systems are widely used in the field of clothing design and processing, reducing the apparel product development cycle greatly.

In analyzing these systems, it is not difficult to find the plane 's design technology is the core of two-dimensional graphics and image processing, and clothing production is a complex, three-dimensional process requiring the continuous advances in computer hardware and software technology. For three-dimensional garment CAD, increasingly urgent demands for improvement have become widespread in recent years. It has become an important research topic for domestic and foreign academics, where basic research has been in full swing.

The so-called three-dimensional garment CAD technology is the realization of three-dimensional body measurements, three-dimensional human modeling, three-dimensional garment design, three-dimensional garment cutting sutures and other aspects of virtual display technology in an interactive fashion design environment. Its ultimate aim is not the production of clothing, but to be able to preview a completed design in virtual reality before production to save time and money. Not only is this conducive to garment production efficiency, but it also improves the quality of design and satisfaction in final products [4]. The technology roadmap of the dimensional garment CAD system is shown in Figure.1 [5].

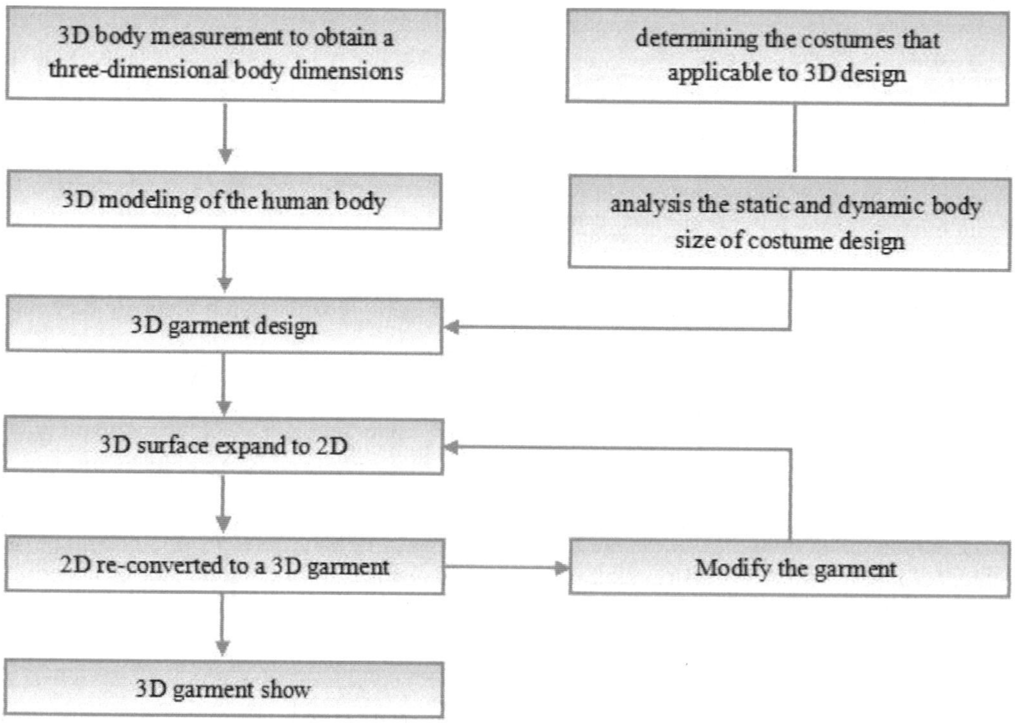

Fig.1 Technology roadmap of dimensional garment CAD system

This system is very helpful for the Personalized Customization of clothing, and personalized clothing design on this basis will improve efficiency and satisfaction greatly.

2.2 Description of personalized customization clothing

In the current era of personalized, customer-centric consumption, fit and sex apparel occupies a very important position in fashion design. The huge population in China and the inevitable differences in geographical distribution create enormous challenges for clothing design, production and sales. China is a garment-producing country. There are nearly 50,000 garment enterprises with an annual output of nearly 14 billion clothing. Visible improvements in garment fit through technology will bring economic and social benefits to the country's garment industry. Thus, personalized customization clothing has become an increasingly important trend in the apparel manufacturing industry.

There are two main methods of traditional clothing customization, one is hand-tailored. Designers measure customers' bodies by hand using simple tools and tailors clothing according to recorded measurements. The second is in accordance with the number of standard production series. This customized approach is to classify the human body, and to develop a series of size standards based on different body characteristics.

Tailoring has existed throughout the world for many years. However, it has several shortcomings: first, it is a lengthy process; secondly, it is based on the experience and judgment of the tailor, that is to say, it lacks quantifiable criteria and is not conducive to the production of large quantities of custom clothing; thirdly, the accuracy of the design can be problematic and may require changes. In producing custom clothing, the first step is to determine what size category customers belong in, and then determine the size of the corresponding customer's type of clothes. This approach requires relatively little time, and is also conducive to the production of large quantities of clothing.

There are many problems in number series in accordance with the standard apparel production. First, the individual's body size on the standard discrete industrialized production of clothing does not meet the "fit" type of clothing demanding individual needs. Secondly, the current number series standard classification method is relatively old, it has been unable to meet current human body features. Hence, the development of a personalized customization clothing digitization system is imminent.

Personalized Customization clothing is custom tailored based on the combination of advanced computer technology and costume design theory, computer graphics technology, computer artificial intelligence, virtual reality technology and computer engineering clothing to achieve a new model of custom clothing system [6-10]. Personalized emphasize bringing about design, production and sales in the Internet environment. The goal of personalized Customization clothing is to use digital technology and network technology digitized anthropometric measurements, style selection, shape analysis, personalized body modeling, virtual try on, costume design, expert guidance, and other aspects of clothing ordered organically combined to achieve effective and fast custom clothing production. Not only to make well-fitted garments, but also to meet individual requirements.

Personalized Customization clothing tailored achieve a digitization and networking garment process from anthropometric measurements-design-processing-sales. Thus, the cycle can be shortened from the original 2-3 months to a week, or even 2-3 days. It can provide clothes tailoring services and personalized services in the apparel buying process to consumers. Personalized Customization clothing's transition to digital garment enterprises provide a strong support in the clothing industrialization, production, design and sale of well-fitted garments to meet the demands of clients. This will have a huge economic and social significance for the Chinese apparel industry which exports garments to 220 countries and represents 20 percent of the country's total exports.

2.3 Issues the development of digital Personalized Customization clothing confronted with

Compared to the application of garment making and grading systems, the development of digital fashion design has not been not given much attention. The existing apparel software modules are mostly concerned with fabric design and is more mature and innovative. But the ideal costume design software should do more than just filling and changing pictures. Ideally, it should provide a truly free adjunct design space to produce highly intelligent, creative designs.

Foreign experts have realized these problems in digital fashion design. Meanwhile, in Europe, Japan and the United States electronic systems tailored clothing have been developed and put into the market test run. But such studies are merely theoretical research, contacts cannot happen with the actual costume design and production. In short, the design of the data base is still full of graphic design software piracy. The key to rectifying this is to organically integrate digitization software design and costume ideas dimensional design.

3. Design of digital personalized clothing custom system

3.1 Concept of system design

Costume design is the service platform that includes research, development andtransformation. It is based on the application of digital electronic information and other high-tech. Its pilot is the development and application of digital technology, based on fashion design talents of digital technology training and promotion, corporate design services, corporate branding, design digital resource specification and application, clothing spread of culture, and user surveys that provide services for the integration of resources, improve efficiency and to achieve leapfrog development of the garment industry.

The current fashion design system is developed for the professional designer. It relies on the designer's drawing skills, expertise and design capability. The intelligent fashion design system for non-professional users. The core idea is to express design knowledge in the way that a computer can accept, process and store in the database. Its operating mode uses intelligent human-computer interaction. Under the guidance of the system, through a series of design command prompt, the user uses the system's resources and knowledge to complete costume design without needing to learn more about specialized design details.

3.2 Structure of the system

The Personalized Customization clothing technology research focused on: (1) a three-dimensional human body measurement and data processing; (2) the establishment of a three-dimensional computer model of the human body; (3) three-dimensional fashion design and simulation on a 3D body model; (4) clothing pieces mapping 2D to 3D, three-dimensional pieces of the plane, and two-dimensional pieces unfold in the corresponding parts of the human body model.

Prior to this, there are a series of preliminary work, including: (1) analysis of the demand for costume design digital network platform; (2) build costume design digital network platform; (3) construct a database of costume design digital network platform; (4) study fashion design digital network platform application models. This paper aims to address these.

3.2.1 The three-dimensional anthropometric

Anthropometric data plays a very important role in the process of custom clothing system. Custom clothing systems tailors clothing based on the user's body measurement data uploaded. With the advent of three-dimensional scanning technology, many manufacturers produce three-dimensional body scanner services for the apparel industry. In fact, three-dimensional anthropometric is an application of three-dimensional imaging technology in the clothing industry. 3D whole system captures an image of the light reflected from a body's surface (laser, white light and infrared). Light is projected onto a photo sensitive device to the three-dimensional characteristics of the human body. Companies that offer this technology include the popular Cyberware from the UK, TC2 from the US, and TechMath from Germany. Three-dimensional scanning systems have a short scan time, high precision, multi-measurements site and other features that are better than traditional measurement techniques and tools.

Three-dimensional body measurement technology can be used for different areas of the body, obtain valid data, and establish a database for characteristics of the human body objectively and accurately. At present, Donghua University, Xi'an Polytechnic University and other institutions have established a database to lay a solid foundation for garment production and mass customization tailoring.

3.2.2 Fabric design

Not only does fabric CAD involve sprinting color processing technology, but it is also related to the plane graphics processing technology. The use of digital simulation technology to simulate the design of clothing fabrics include yarn simulation, structural analog to weave woven fabrics and knitted fabrics, and digitized analog pattern. The main feature of the simulation system is designed to display realistic fabric renderings and organization charts on the display, and are able to cut pieces of fabric on the basis of direct printing or embroidery patterns required in accordance with the design requirements. With this system, customers can choose their favorite fabrics, so the degree of satisfaction is improved greatly.

3.2.3 Fashion design

Adobe Photoshop, CorelDraw, Painter, Illustrator, Freehand and other two-dimensional graphics processing software can be used to draw clothing renderings. Thus form a powerful two-dimensional garment design platform. Using these fashion design software has obvious advantages compared to traditional hand-drawn designs as they can improve design efficiency, accuracy and performance ability greatly.

3.2.4 Garment pattern design

Garment pattern design includes plate, grading, marker making and other aspects. The use of a two-dimensional garment CAD system can easily produce Personalized Customization apparel patterns. Currently, the main use of it is in plating, grading, and nesting modules for pattern design. Clothing playing board, grading and marker CAD systems are mainly related to plane graphics processing technology, and its development and application has been relatively mature. Also, you can use three-dimensional flattening technology clothing model clothing from the three-dimensional shape of the garment finally to get the plane model.

3.2.5 Simulation of 3D garment

Three-dimensional simulation technology is a new trend in the development of clothing apparel CAD systems. Currently, its development is faster than a three-dimensional garment design but there remains many technical. Three-dimensional garment simulation technology can be applied to the following aspects:

1) Garment simulation process

The use of three-dimensional virtual technology can shorten the design time by simulating the production process of the clothing, thus reducing the garment production cycle greatly. At present, Japanese Dressing SIM technology can make two-dimensional patterns to the models by establishing a parameterized three-dimensional electronic model. It can also add animation and background to the dress-model, and set fabric pattern, color and elongation. It shows the clothes in use so that customers can make better selections.

2) Virtual Dressing

Using virtual human models of customers based on inputted scan data or measurements, customers can try the library's clothing by a simple operation. The system can show what the clothing is like while worn, simulate the three-dimensional effect of different fabrics drape, rotate images 360 degrees and other functions. Such applications can also achieve remote control, using the Internet for e-commerce. This technology allows consumers to choose and purchase clothing while sitting at home.

3) Simulation dynamic fashion show

Japanese digital fashion company developed Dressing SIM technology to enable virtual fashion shows, while MiraLAB Research Group at the University of Geneva, Switzerland, has developed a three-dimensional fashion design platform which reflects the effect of apparel being worn on a model while walking.

4) Virtual clothing stores

Build a digital computer simulation of clothing stores by using existing technology. With an e-commerce platform, visitors can choose the right clothing by browsing through the store's website.

4. Conclusion

Personalized Customization clothing is the inevitable product of the development of the information age, but it is new and immature. Currently, information technology is in the development stage, so there are still a lot of problems. It cannot meet the actual demand, and improvements are needed as the process of development continues. Because the spread of any kind of technology cannot be done overnight, it will build awareness and understanding based on people - this is a longer-term process. Therefore, the popularization and application of Personalized Customization clothing will be an important part of our mission and long-term future of the garment industry transformation.

5. Acknowledgement

The authors will give their great thanks to the Apparel & Design College of Xi'an Polytechnic University.

References

[1] L Lei, Liu XG. Present situation and development concept of digital fashion design. Journal of Textile Research 2005; 26: 135-136, 139.
[2] Zhao NP, Zhang WY, Zhang HZ. Development and outlook on dress CAD technology. Journal of Tianjin Institute of Textile Science and Technology 2000; 19: 70-73.
[3] Chen DS, Li HH. The state-of-art and trend of computer aided apparel design. Journal of Jilin Institute of Technology 2000; 21: 34-37.
[4] Li X. 3D body-form and garment modeling pattern flattening CAD techniques for clothing industry. Journal of Zhejing Institute of Science and Technology 2002; 19: 175-180.
[5] Zhu GZ, Li XJ. Research of three-dimensional garment CAD technology. Knitting Industries 2005; 4: 55-57.
[6] Li M, Tang XZ. Research of electronic tailored clothing. Progress in Textile Science & Technology 2007; 12: 97-98.
[7] Li HM, Fang JA, Chen JX. The research on the online eMTM system. Beijing Textile Journal 2005; 26: 45-49.
[8] Ling C, Fang JA, Chen JX. Design and implementation of remote custom-tailor system based on web. Computer Engineering and Design 2005; 26: 375-377, 379.
[9] Yang N. Development of new technologies clothing. Textile Science Research 2005; 26: 45-49.
[10] Wang HX, Zhang HZ. Network virtual fashion design. China Textile Leader 2002; 2: 28-30.

Analysis on Blouse Fashion based on Intelligent Conversion System Between Garment Fashion and Garment Structure Figure

Dan Liu[1*]

School of Apparel & Art Design, Xi'an Polytechnic University, Xi'an, 710048, China

*Corresponding author's email: dd15991092232@163.com

Abstract

Modern clothing structure design has become an important technical method in the second design of clothing. The topic on changing design concept into a structural template is important for structural design technology research development. So the research of an intelligent conversion system between garment fashion to garment structure figure would offer a more convenient way for garment structure. In this paper, based on the intelligent conversion system between garment fashion to garment structure figure, the blouse fashion has been analyzed. Through the analysis, the characteristics and segmentation principle of the blouse in the intelligent conversion system between garment fashion to garment structure figure have been summarized. The theoretical basis of the blouse pattern was automatically generated based on the system.

Keywords: Garment fashion; Garment structure; Analysis; Intelligent conversion

1. Introduction

Clothing CAD is a modern technology people use by taking advantage of computers to finish all steps of apparel design [1]. The first clothing CAD system was the MAR-COM System developed in the United States in 1972. After that, Gerber also from the United States, developed a series of clothing CAD products, opening up the market of such systems [2].

In China, the introduction of Clothing CAD was comparatively late, but the speed of development was very fast. It can be said that there has been world-shaking changes from 1980s until now [3]. At present, many independent R & D enterprises in China have already become developers and providers of Clothing CAD systems with great influences at home and abroad. Especially in recent years, Chinese-made software has become more and more mature in terms of technological development in plate-making, grading, layout and others, which are comparable with foreign CAD software. With regard to price, domestic CAD software' prices are equivalent to one tenth of imported CAD software. They have quickly popularized for this obvious advantage of price [4-7].

Applications of those techniques are primarily focused on 3D transformation. Relatively speaking, the 2D transformation of clothing Garment Fashion and Garment Structure have been paid less attention to in terms of development and application. Zhou Guangjun and others [8] in Donghua University attempted to convert the clothing style presented on the picture into general style plans through technologies such as picture processing. Thus, consumers receive Garment Fashion they wanted at will after combining with database. Chen Zhe and others [9] researched on how to acquire 3D data based on clothing Garment Fashion; Guo Dongmei in Chongqing Normal University put forward the conversion technology of clothing Garment Structure to industrial pattern. These researches are essentially focusing on the design and generation of clothing Garment Fashion, as well as on the generation of clothing Garment Structure separately. However, there's a little research which combines Garment Fashion with Garment Structure.

This paper puts forward a brand new automatic generation concept for paper pattern, that is "converting Garment Fashion directly into Garment Structure", which can further improve the automatic generation technology of clothing paper pattern. This will allow plate-making to become more convenient for structural design personnel in big, small and medium apparel enterprises, as well as save time, reduce cost, simplify processes, offer a technical backup service platform for non-professionals of clothing construction design, and be of great application value in the process of industrial digitization and automatic production of apparel enterprises.

2. Overview of Intelligent Conversion System from Garment Fashion to Garment Structure

2.1 Introduction of Intelligent Conversion System from Garment Fashion to Garment Structure

Smart conversion from Garment Fashion to Garment structure, refers to a series of structural design processes in which clothing pattern makers convert abstract conceptions of designers into specific structure plans after overall inspection of clothing Garment Fashion and optimal configuration of all component factors, in order to further amend and improve industrial Garment Structure through this system.

2.2 Constructing Significance of Intelligent Conversion System from Garment Fashion to Garment Structure

For industrial production and individual manufacturing, clothing construction design consumes a lot of time and resources, as well as requiring people with professional knowledge and experience, which is often very limited. The research about this system has explored the relationships among the characteristics of the human body, style and structure in the design of clothing paper patterns. It establishes systematic models of characteristic relations, and offers basic theory and method for parameterization, automation and intellectualized design of clothing paper patterns. The completion of this system can further improve and transform technologies of clothing CAD software, making it more convenient for structural design personnel of big, small and medium apparel enterprises to make plates and offer a technical backup service platform for non-professionals of clothing construction design.

2.3 Design and Composition of Intelligent Conversion System from Garment Fashion to Garment Structure

The design concept of the system are as follows:

1) The system shall be equipped with input functions of Garment Fashion.

2) The system shall be equipped with a segmentation function for Garment Fashion. That is to say, the system can segment the Garment Fashion into geometric figures which are more beneficial to paper pattern generation in accordance with the requirements of customers.

3) The system shall be equipped with a data measurement function.

4) The system shall be equipped with a revision function of Garment Structure. After the Garment Fashion is segmented into geometric figures in accordance with requirements, the system can adjust the geometric figures in accordance with the drawing requirements of paper pattern so as to enable it meet the requirements of standard paper patterns.

To sum up, the system includes four constituent parts: input of Garment Fashion, segmentation of Garment Fashion, data measurement and Garment Structure generation.

3. Decomposition Principle of Clothing Garment Fashion Based on the System

This part mainly introduces decomposition principle of Garment Fashion, i.e. design reference in the Garment Fashion decomposition module of the second part of the system.

3.1 Necessity and Importance of Garment Fashion Decomposition

This system directly converts to Garment Structure through analysis of Garment Fashion. Therefore to establish of the system, the first step is to analyze Garment Fashion. The first step of the system is to convert clothing styles into specific geometric figures, body pieces and sleeves of the clothing into geometric figures such as squares and triangles based on the features of this system, which is also preparation for latter technological process. Whether the Garment Fashion is segmented reasonably in accordance with norms will directly influence the work in the next step.

3.2 Features of clothing Garment Fashion based on the system

Clothing Garment Fashion is a further distinct expression of clothing styles' design ideas of clothing designers in accordance with effective charts of garments [11]. In the process of clothing production, clothing Garment Fashion involves two steps of design and plate-making, which are an important intermediate link, as well as the basis of clothing plate-making work. Requirements of plotting modes are neat and precise, with distinct and clean graphics, which can be completed by hand-drawing and CAD software.

In the actual process of development and production, the application of clothing Garment Fashions is extensive. It is not just fundamental for pattern designers' plate-making, but also basis for sewing workers to sew clothing. The major features of clothing Garment Fashion are its scientific and accuracy, that is to say, taking body structure as basis, fully considering the ratio and fluctuation relationships of all parts of the body and complying with the disciplines of human motions and principles of clothing structure and clipping [12,13].

At present, there are many statements in the category of clothing Garment Fashion. Here, the conclusions and categories of clothing Garment Fashions of the author of Clothing Garment Fashion Classification and Intensified Training Methods Before Competition will be quoted, namely, to classify common clothing Garment Fashion into the following three kinds [12,14].

1) "Drafting Garment Fashion", as shown in Fig.1, refers to Garment Fashion which is drawn in accordance with the same ratio of outline in the picture, a totally graphic Garment Fashion without perspective feelings [12,14].

Fig.1 "Drafting Garment Fashion"

Fig.2 "Mannequin Garment Fashion"

Fig.3 "T Garment Fashion"

2) "Mannequin Garment Fashion" refers to stereoscopic Garment Fashion which considers the perspective effect of body thickness without body dynamics; the equivalent of the drawing seen from vertical angle of bodies or mannequins when wearing the clothing on human bodies or mannequins, as shown in Fig.2 [12,14]. The designers can only rely on experience to command the ratio and structure of the Garment Fashions when drawing this kind of Garment Fashions.

3) "T Garment Fashion" as shown in Fig.3, refers to stereoscopic Garment Fashion which considers the perspective effect of clothing on bodies with thickness in the drawing process and has body dynamics [12,14]. This kind of Garment Fashion is the one with most features in the clothing effect chart. While it is more applicable to aesthetics, it has weak guidance for production and technologies.

3.2.1 Drawing Requirement of clothing style design

This article chooses the first, "Drafting Garment Fashion" mentioned above as the topic of research. Drawing on strict accordance with the requirements, that is to say, the internal and external contour lines shall be distinct and clear: external contour of clothing, also called external contour lines shall be shown with full thick lines; internal structure shall be shown with thin lines [15]. Drawing of lines shall be smooth and clear-cut with clean pictures. Generally speaking, following tips shall be paid attention to in the process of drawing:

1) Arrangement of Pictures

The size of clothing Garment Fashion shall be decided in accordance with that of the mannequin, generally including front view and back view, taking the front view as the principal view.

2) Drawing Ratio

The drawing ratio of Garment Fashion shall be in accordance with actual wearing form or the appearance result the designers expect.

3) Drawing Principle

① Static and Natural Wearing Principle ② Graphic Principle ③ Principle of Drapes Omission

3.2.2 Standard Graphic Tools in Clothing Garment Fashion

Common professional graphics software can make standard Garment Fashion in accordance with requirements. Adobe Photoshop, Adobe Illustrator and CorelDRAW are commonly used in enterprises to make the result more accurate and of high quality.

3.2.3 Drawing Basis and Methods of clothing style design

The drawing ratio of clothing style design is totally dependent on normal body ratio. As a general rule, body ratio is calculated taking head length as a unit. Adults are generally divided into two kinds, seven heads and eight heads. Seven heads refer to Asian bodies and eight heads, European bodies. However, there may be some differences due to different nationalities, regions, genders and ages, so these ratios are not absolute in practice. This article only analyzes body ratios of Asians; or seven heads.

The division of body ratio of seven heads from top to bottom is successively: head length, chin bottom to ligature of two-papilla, ligature of two-papilla to navel, navel to butt crack, butt crack to patella, patella to middle of shank, middle of shank to pelma [16]. The specific allocations are shown in Fig.4.

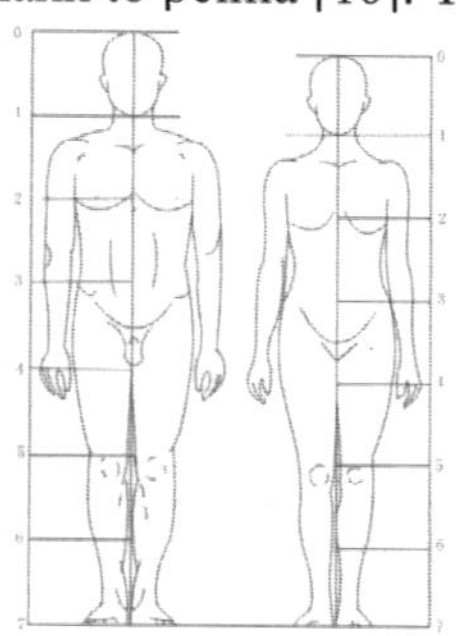

Fig.4 Allocation of 7-Heads Body Ratio

Fig.5 Graphic Design Drawing of Blouses

In seven-head ratio, the straighten length of two arms is also seven-head length with the body in an up-right state. Therein, shoulder breadth is two-head lengths. In other words, the distance between two shoulder points is two-head lengths; the distance from the axilla point (boundary point of chest breadth) to middle finger point is approximately three-head lengths.

The clothing style design can be drawn in accordance with these basic body ratios, such as in Fig.5. However, this is only a basic ratio, and can be seen as the foundation for drawings. To obtain complex and changeable clothing Garment Fashion, the original ratio of clothing shall be adjusted based on the basic body ratio to gain expected Garment Fashion.

3.3 Decomposition Principle for Clothing Garment Fashion of Blouses

This article focuses on the research of decomposition principle of Garment Fashion of blouses, offering theoretical basis for the smart conversion of blouses from Garment Fashion to Garment Structure.

3.3.1 Experiment 1: tiled experiment of body piece paper pattern of common blouses

1) Experimental objective: to analyze the features of tiled body piece of common blouses as shown in Fig.5.

2) Experimental materials

① Tools: ruler, tape, pencil, idler wheel, scissors, camera

② Materials: ivory board, white calico cloth

3) The Contents of Experiment

① Draw basic Garment Structure of front bodice by adopting ratio method with specification of 160/84A, the specific information can be seen in Table 1.

Table 1. Size Information about Front Bodice of Cloth Samples

Serial No.	Drawing parts	Products specification
1	Length	65.0cm
2	Waist Length	40.0cm
3	Chest Measurement	94.0cm
4	Shoulder Breadth	40.0cm
5	Neck Line	40.0cm

② Copy the paper pattern on white calico cloth to clip.

③ Putt the cut pieces flat on the level desk and taking 5cm*5cm profile-paper as back surface and put the cut pieces on the profile-paper, aligning the antemedial line of cut pieces with the perpendicular line on the profile-paper, left and right axilla points with horizontal line on the profile-paper. In this way, it is easier to fix the location of cut pieces and ensures the placement of cut pieces is horizontal and vertical.

④ Aim the camera lens at middle point of paper pattern and take a photo from a distance of 90cm to obtain photo of front bodice of sample dress. Lead the photo into drawing software and draw grids with scale plate in the software in accordance with the grids made in the last step, and adjust the shift caused by camera lens to the right place, as shown in Fig.6. The dark colored lines in the chart represent body pieces' radial drapes in tiled state.

Fig.6 Outlook of Front Bodice of Shirts in Tiled State

When the body pieces are in the tiled state, the measurement size of tiled clothing will be influenced by SHENG, as shown in Table 2.

Table 2 Size Information of Front Part in Tiled State

Serial No.	Drawing parts	Drawing specification	Measurement specification
1	Waist length	40.0cm	39.8cm
2	Half chest breadth	47.0cm	46.5cm
3	Chest breadth	17.3cm	16.9cm

4) Analysis of the Experiment Result

① As shown in Chart 6, the waistline part of blouses' front bodice can be totally flat and there are some drapes in the waistline and lower hem.

② The parts of front bodice which cannot be flat are mainly centralized in the location of BP point (chest salient point). It can be seen from Table 2 that there is a difference of 0.4cm between specification and dimension of breast breadth and measurement size clothing is designed to be well-fitted and hence takes the convexity of females' breast into account. Close breast dart may deform the clothing lining when it is dragged so the breadth of front bodice shall be smaller than the specification and dimension when it is tiled, as shown in Fig.7. The stereoscopic state of pyramid cannot be kept due to flexibility of the fabric and it forms subsided drapes, as shown in Fig.8.

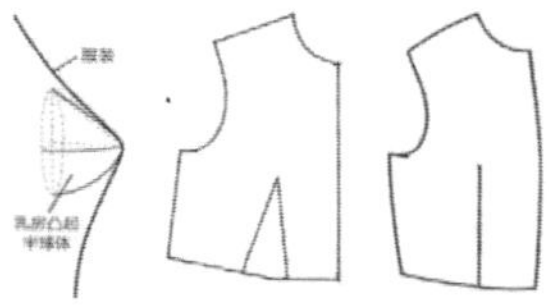

Fig.7

Fig.8

Fig.7 Schematic Diagram of Chest Surface and Dart
Fig.8 Schematic Diagram of Formation of Front Bodice BP Point Drapes

③ The part from bust line to hip line are disciplinary radial drapes. Assuming that from the waistline to the hip line is two connected truncated-cone-type structures, as shown in Fig.9, six darts of front, back and side shall be gathered to suit different body shapes.

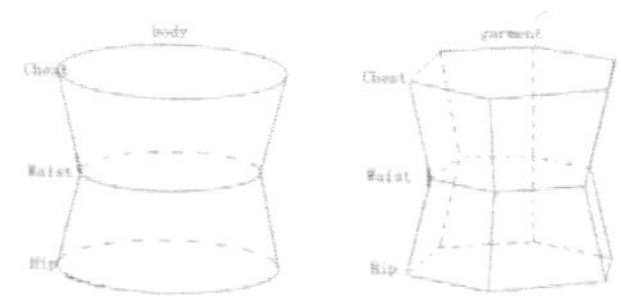

Fig.9 Structural Diagram of Chest, Waist and Hip

It can be seen from aforesaid experiment that there must be a reason for drapes of body pieces. However, the sizes and locations of drapes are unpredictable on the sample dress. Card paper is used to make front bodice of sample dresses due to the flexibility and deformability of the fabric, and to make further analysis.

3.3.2 Experiment of shirt paper pattern's spatiality of common blouses

1) Experimental objective: to analyze and further verify the reason for which clothing cannot be unfold without considering the influence of fabric deformation on unfolding of body piece.

2) Laboratory appliance and materials

① Tools: ruler, pencil, scissors, camera

② Materials: ivory board

3) The Contents of Experiment

① Adopt ratio method to draw on ivory board as the Garment Structure of common blouses shown in 3-3. The specification and dimensions are the same as the first experiment.

② Cut off the paper pattern from ivory board, cohere the darts with cello tape so as to make it close.

③ Take 5cm*5cm profile-paper as back surface and put the paper pattern with closed darts on the profile-paper. Align the antemedial line with the perpendicular line on the profile-paper, left and right axilla points with horizontal line on the profile-paper.

④ Aim the camera lens at the middle point of paper pattern and take a photo from a distance of 90cm to obtain photo of front bodice of sample dress, as shown in Fig.10.

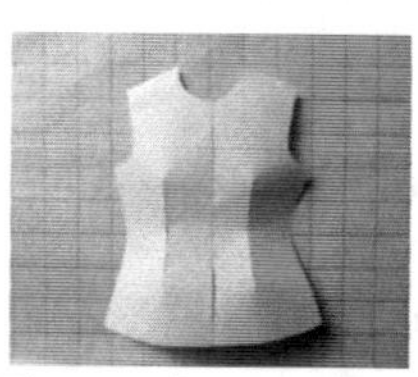

Fig.10 Diagrammatic Figure of Paper Pattern

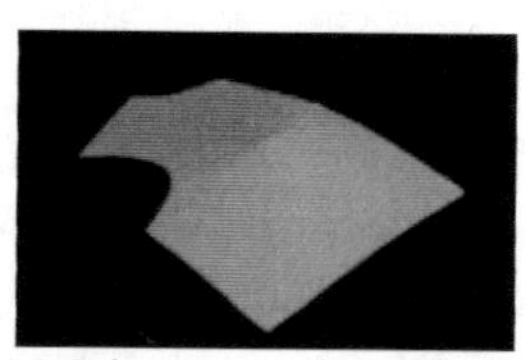

Fig.11 Diagram of Conical Structure of Cut-Part

4) Experiment Result

It can be seen from the chart that the body piece is a stereoscopic space structure and cannot be put flat on a desk without considering fabric deformation.

3.3.3 Cone Structure of Body Piece in Tiled State

It can be seen from Chart 7 and 8 that the breast dart of front bodice can be seen as a sector structure, which is a conic surface naturally formed from cut pieces after closing darts. The aforesaid experiment illustrates that the body piece is a stereoscopic space structure without considering the fabric. Hence, the overall body piece can be seen as a curved surface, taking dart vertex as a conical tip, as shown in Fig.11.

This is applicable to cone surface with one vertex, but it is found in actual operation that cut pieces of some style have several darts and their vertexes are not in the same point, as shown in Chart 12. Then, the situation that two cone vertexes occur on one surface in closed dart may happen. Where two cone vertexes are connected, a straight line may occur with two bevels on both sides. There are four vertexes which form a flat quadrangle if it is seen from an overall front part. The neighboring faces of this quadrangle are bevels and curved. As a result, the difficulty degree of Garment Fashion decomposition under this circumstance is high, and at the same time, requires an enormous amount of mathematical calculations when building a space structure model. The same situation may occur when adopting this method for cut-parts with halving line as shown in Chart 13.

Fig.12 Clothing Cut-Part Chart of Two Darts

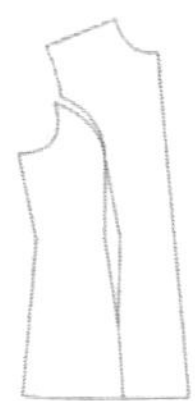

Fig.13 Clothing Cut-Parts with Halving Line

3.3.4 Polygonal Structure of Body Piece in Tiled State

The 3D space establishment with unfolded body piece can be seen as formed by many pyramid surfaces, as shown in Fig.14. Taking vertexes of darts as peaks of pyramids and respectively connecting them with structural feature points to decompose a cut-part into different geometric figures, which are all graphic figures, will have the benefit benefit of establishing model and analytic calculation. This method can be employed to decompose whether it is a cut-part with several darts in Fig.12 or with halving lines in Fig.13.

Fig.14 Diagram of Polygonal Structure of Cut-Parts

4. Conclusion

To segment the Garment Fashion of blouses, the first step is to collect data of Garment Structures, then to vectorize the Garment Fashions, segment the charts and analyze their features. Finally, build up characteristic relations between Garment Fashion and Garment Structure. This article only puts forward a theoretical segmentation basis and will involve knowledge such as image processing technology and the building of mathematical models, which are undoubtedly difficult for personnel of the clothing industry and need to be strengthened.

It can be concluded that the tiled state of half shirt body pieces is a stereoscopic spatial state without considering lining deformation through aforesaid analyses. Furthermore, the stereoscopic spatial state can be seen as many adjoined pyramids. Taking peak of provincial road as the acme of pyramid, the front bodice can be seen as a combination of a rectangular pyramid taking BP point as acme. Most contour lines of this kind of structure are straight, which makes it convenient for calculation and analysis. It is taken as the research basis of space structure when researching an Intelligent Conversion System from Garment Fashion to Garment Structure, and the Garment Fashion of blouses are segmented in accordance with this space structure.

5. Acknowledgement

The authors will give their great thanks to the Apparel & Design College of Xi'an Polytechnic University.

References

[1] Luo WY, Li FX, Gan YJ. An Analysis on the Prospect of Fashion CAD Operation System all Round the World. Journal of Jilin Institute of Technology: 2002; 23: 59-61.
[2] Meng FY, Wang D. Application status and development trend of CAD system in China. Journal of Liaoning University of Technology (Social Science Edition): 2010; 12: 84-86.
[3] Teng HJ. Application status and development of garment CAD model system. Knitting Industries: 2011; 12: 53-54.
[4] Cheng YF. Application of clothing CAD system in textile and garment enterprises. Technology Wind: 2011; 9: 107-109.
[5] Zhang QZ, Huang F. Development and application analysis of CAD technology in China. Light and Textile Industry and Technology: 2007; 36: 51-54.
[6] Xu J, He MJ. Key Technique in Three-dimensional Clothing CAD. Journal of Tianjin Polytechnic University: 2007; 26: 52-57.
[7] Zheng R, Li X. Analysis and Evaluation of Lectra, ET and Richpeace Marking CAD Technology. Journal of Zhejiang Sci-Tech University: 2009; 26: 695-699.
[8] Zhou GJ. Research of Image Feature Extraction in Personal Garment Pattern System. Shanghai: Donghua University: 2009.
[9] Chen Z. Study on Obtaining Clothing 3D Data from Design Drawing. Shanghai: Donghua University: 2013.
[10] Guo DM. The Techniques of Converting the Clothing Structure Drawing in to Industrial Pattern. Shanghai Textile Science & Technology: 2008; 36: 30-31,39.
[11] Guo QH. Research on the Keypoint in the Teaching of Garment Drawing. Decorate: 2010; 212: 129-130.
[12] Dong CX. The Study of Auto-identification of Garment Illustration. Shanghai: Donghua University: 2013.
[13] Weng XQ. Discussion on the importance of fashion design in fashion design. New Arts: 2003; 24: 91-92.
[14] Wang Q. The training method of classification and garment drawing before competition. Art Education Research: 2012; 1: 130-131.
[15] Hu Y, Zhao YQ, Ni JC. Clothing style design and version of the practical manual - shirt. Shanghai: Donghua University press: 2008.
[16] Zhang WB, Fang F. Apparel Somatology. Shanghai: Donghua University press: 2008: 28, 60.

Research on the Functional Design of Kindergarten's Clothing in China

Cun-Xin Liang[1*], Hong Dai[2]

[1]*School of Apparel & Art Design, Xi'an Polytechnic University, Xi'an, 710048, China*
[2]*Apparel & Art Design, Xi'an Polytechnic University, Xi'an, 710048, China*

[*]Corresponding author's email: 519727160@qq.com

Abstract

Kindergarten uniform, as a kind of specific uniform clothing, can reflect the unique cultural development of national and regional education. In this paper, both kindergarten clothing development and cultural characteristics of these clothing are being studied and compared, uncovering many problems. Through interviewing students, parents and school workers in the form of a questionnaire, their view on kindergarten clothing design was collected, which includes clothing style design, color design, fabric selection and related accessories design ideas. Their input provided valuable insights into future design development of kindergarten's uniform.

Key words: Kindergarten; Clothing; Functional style; Color fabric

1. Introduction

1.1 Significance of topic selection

Kindergarten clothing is embedded with the culture, origin and development of both the kindergarten and education in China. Since preschool is the primary stage of compulsory education, proper clothing is crucial to creating a positive and healthy cultural atmosphere for learning. Kindergarten clothing is part of the school's regulations, and should be designed according to factors such as the students' safety concerns, health, comfort, and function, beauty and personality. By satisfying the needs of the students and adapting to their characteristics, it is hoped to create a campus with healthy cultural atmosphere.

1.2 Related concepts

Uniform emphasizes unity in an environment and kindergarten clothing is a type of uniform. Kindergarten suit refers to the uniform, which should be prescribed by the kindergarten. In general, students are expected to wear the school uniform that has the school crest on it during all school events. The uniform can directly affect the image of the school and the school's spirit, making it an important component of campus culture construction. At the country level, school uniform is also an embodiment of national spirit and culture.

1.3 Problems that exist in the kindergarten take present situation in our country

As living standards continue to improve, concerns over school uniforms have also been increasing over the years, gradually uncovering and exposing problems in the development of school uniforms.

School uniform is regarded as clothing that does not have a nice appeal, carries limited cultural meaning and is not as popular as other types of clothing. In 2014, the First Lady of the United States, Michelle, visited the second affiliated middle school of Beijing Normal Univeristy, as shown in Fig.1. Students at the middle school were wearing a set of white and blue loose sports uniform, which is very common in China. The uniform is deemed to be ugly and carries a careless and ugly feeling, in which the gender of the person wearing it cannot be determined by the looks of it.

Fig.1 Second affiliated middle school of Beijing Normal University

With rough sewing effect, people have shown concerns over the quality of kindergarten clothing. Due to the constraints of local protectionism and a variety of other factors, the production of uniforms is more inclined to local production companies, making it difficult to develop the national market.

There is no unified and effective park service quality supervision system. The principle behind the optional kindergarten clothing is "school autonomy, parents voluntarily, unified bidding, the standard management". But in practice, it has been simplified as "school's choice is equivalent to parents' choice".

1.4 Research purpose and meaning

We tried to identify culture embedded in kindergarten clothing. There is more to the function of kindergarten clothing; good kindergarten clothing does not only fulfill the function of protecting the students and representing the school, but also looks aesthetic on the outside, People have just started to realize the deficiencies of current kindergarten uniforms, in which development in this field may be slower. The authors of this paper hope that one day kindergarten students will be able to wear unique kindergarten uniforms that reflect themselves.

1.5 Research methods

In this paper, study methods used mainly include:

1) Collection of relevant knowledge: through access to the library, network retrieval of relevant domestic and overseas papers, literature and books, to understand and study domestic and foreign research on school uniforms.

2) Interview through Questionnaires: to conduct research on kindergarten clothing, three types of questionnaires that ask primary school students, parents and teachers to classify and comment on the design of uniforms were given out. Completed questionnaires were reviewed and a summary of them revealed existing relevant problems in kindergarten uniforms.

2. Analysis of the development of kindergarten clothing development in the past and at present

Kindergarten clothing is the symbol of the kindergarten, which cultivates children's team spirit and strengthens the overall image of the kindergarten. In the history of China, however, history of kindergarten clothing is even longer than that of uniform. In China, kindergarten clothing has existed for one hundred years. while school uniform is only created recently. Regardless of the short history of uniform, it has already undergone earth-shaking changes. [1]

Since the Revolution broke the thousands-year-old restrictions on clothing with "five other people, lint five colours" [1], with the kindergarten clothing change chase the footsteps of history winds its way all the way, each of us carries the youth almost all of the laughter and tears, shine in our memories.

In the 1920s, uniform-style kindergarten suits first appeared; in the 1930s, cheongsam pampered on campus; in the 1940s, the same style from the 1930s was worn; in the 1950s, more casual, the lack of specific kindergarten clothing; in the 1960s and 1970s, old military uniforms. The spate of "movement" obliterated the beautiful land of brilliant colors; in the 1980s, kindergarten clothing reappeared, but more scattered; in the 1990s and the early 21st century, kindergarten sports clothing dominated the world; at present, sports clothing is dominant in kindergarten. Some kindergartens mimic foreign schools and design formal dress clothing for their students [1].

3. Investigation and analysis of the current situation of kindergarten clothing in China

In the form of questionnaires, using Xi'an as an example, research students, parents and teachers were interviewed. Questions regarding the style, color and fabric of kindergarten clothing were asked; questionnaires were distributed to nearly 100 nurseries, ranging from international to public to private kindergartens, in Xi'an by random sampling. 200 students, 100 parents and 100 teachers were chosen to complete the simple questionnaire. These answers were analyzed to discover existing problems in primary school students' previous kindergarten clothing and to provide reference for the design and manufacturing of kindergarten clothing.

The questionnaire covered different aspects of kindergarten clothing, including interviewee's satisfaction towards kindergarten clothing, design of the style, fabric, color collocation and functionality. In the questionnaire design, under the premise of full analysis of research content, a lot of literature was reviewed and referenced. Even though some challenges may arise during the research (such as uncooperative research subject, bad attitude, etc), all questionnaires selected were completed by the research subject to ensure the accuracy and authenticity of the survey data. A total of 200 questionnaires were issued and 186 valid questionnaires were recovered, giving a recovery rate of 93%. 40 questionnaires were given to international kindergartens, which accounts for 20% of the total sampling population; 100 questionnaires were given to public kindergartens, which makes up 50% of the total sampling population; and the remaining 60 were given to private kindergartens, which takes up 30% of the total sampling population.

3.1 Choice of color pattern in kindergarten clothing

Based of survey results, 55% of parents thought that kindergarten clothing lacks aesthetics in overall. Teachers who thought that kindergarten clothing gave a disgruntling feeling account for 30%; and 60% or more of the students are not satisfied with the integral feeling of their existing kindergarten clothing, as shown in Fig.2:

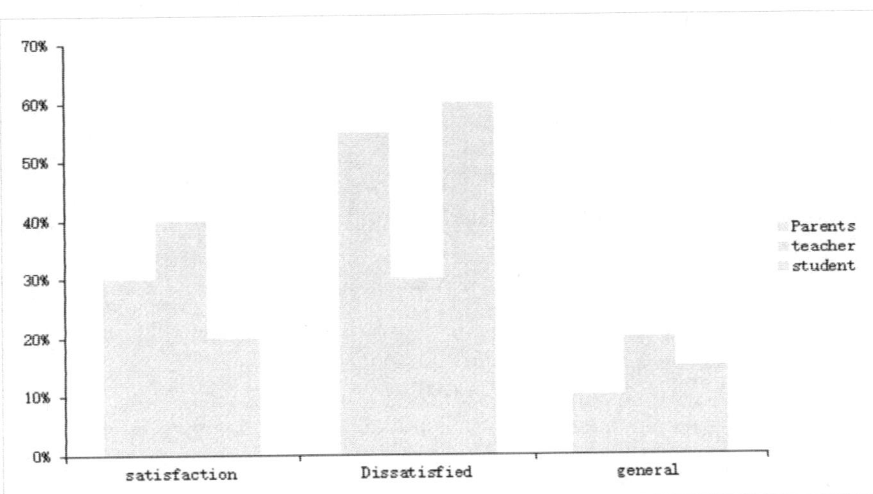

Fig.2 Survey data

3.2 Style design of kindergarten clothing

A set of questions were designed for students and teachers who appeared during the First Lady's visit, as shown in Figure 1. 55% of students like formal kindergarten clothing, in which sportswear accounted for 30% and casual wear accounted for 15%. There are students who want to have a kindergarten school bag with their names on it. From the case study in Xi'an, teachers commented on the design of the kindergarten's uniform jacket pocket. For example, the fake pocket account for 65%; true pocket account for 20%; and 15% of the teacher prefer no pocket. Pockets in kindergarten uniform is another category that needs to be studied.

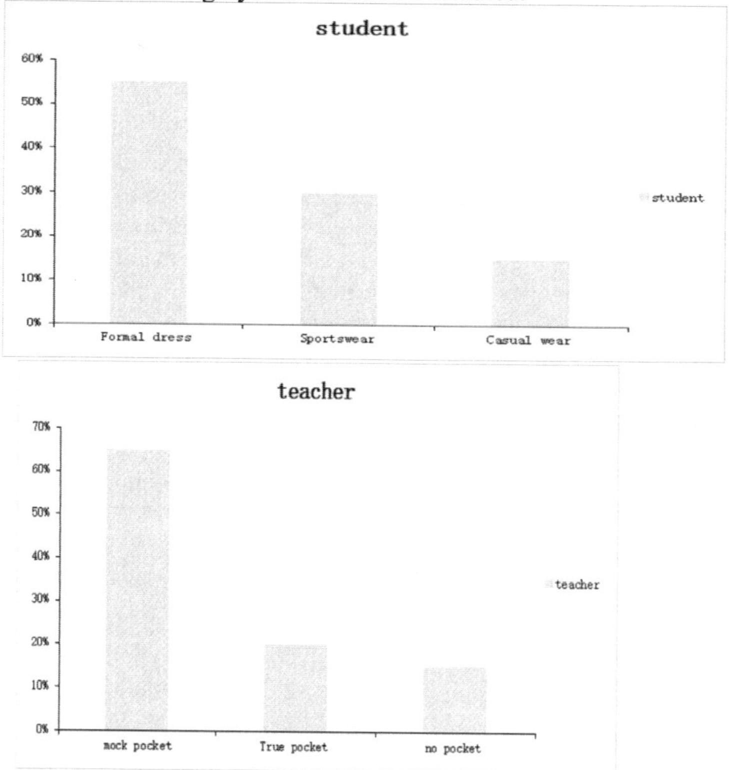

Fig.3 Survey data

3.3 Choice of fabric for kindergarten clothing

Figure analysis, 85% of the parents believe choosing middle-graded fabric, such as polyester-cotton blended fabrics, is more appropriate, as it is more affordable and light in weight. There is also a small proportion of parents who prefer buying kindergarten clothing made of high-graded cotton fabrics. 10% of the parents do not care much about kindergarten uniform, which account for only 5% of the total sampling population. The field survey below indicates that many kindergarten use synthetic fabrics, which did not take environmental protection and wearing comfort into consideration:

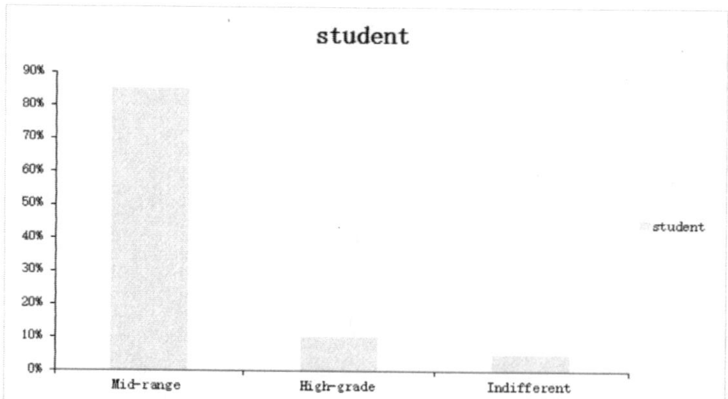

Fig.4 Survey data

4. Research on functional elements of kindergarten clothing design in China

4.1 Analysis of functional color design

Color is one of the three elements in fashion design, which is indispensable [3]. Color has a subtle complex relationship with people. Since the optic nerve of kindergarten students is not fully developed at this stage, a brighter tone can be used [4], such as red, yellow, blue, green and pure white. According to the survey, students in kindergarten are lively and active, eye-catching patterns or cartoon images can be included in the uniform. Some children may prefer decorations like animals or simple objects. In addition, color design of the nursery kindergarten clothing should take into account the number of students at the kindergarten; if there are more students, a darker color can be used, whereas in a smaller school, chroma bright colors can be used.

4.2 Analysis of functional style

Style designed is to primarily increase the aesthetics of clothing, while realizing its function. Kindergarten clothing style design mainly refers to kindergarten clothing silhouette design, which is the main form of uniform design style. Profile of traditional clothing in an alphabetical form is divided into H, A-type, O-type, X-type and T-type [5]. H-type primary school uniforms carry comfortable, relaxed and simple wear characteristics. Kindergarten clothing at home and abroad are mainly classified as H-type, mainly in the category of styles straight onto clothing, straight jeans and shorts, which are comfortable enough to meet the basic physical needs of students on campus.

Fit aesthetics and structure of the kindergarten dress shape is designed to evaluate the pros and cons of important evidence. Students' favorite toys and snacks are served in the pocket of their kindergarten clothing; some children can even fir their favorite pencil in. To adapt to the students' daily activities and to ensure their safety, the design of the kindergarten's uniform jacket pocket, from parents and teachers interviewed in Xi'an, should adopt a reduction design, as shown in Fig.5, 6.

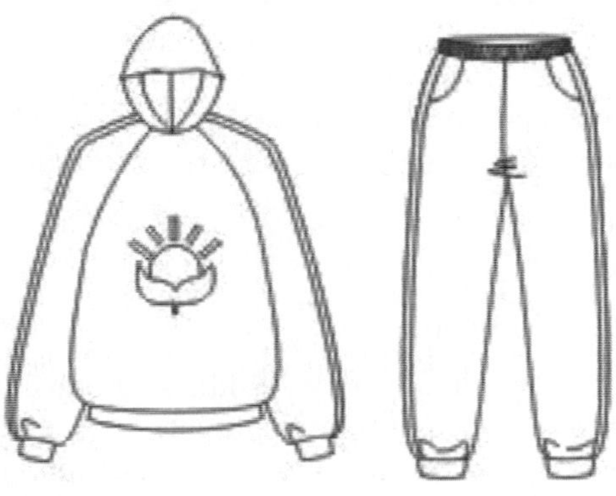

Fig.5 Autumn clothes

Fig.6 Summer clothes

4.3 Analysis of functional fabrics

Fabric is one of the three elements of costume design, which constitutes the material basis of clothing. Considerations for fabric for kindergarten clothing are different from that of our daily clothing, as the fabric does not only have to ensure certain functionality and comfort, but also safety. When choosing fabrics for uniforms, the fabrics should not only be easy to maintain, but also have to be tough, resistant to heat and moisture to prevent bacterial growth.

In current kindergarten clothing design, the fabric should be washable, wrinkle or shrink-proof. Functional fabrics refers to fabrics that have some special properties and uses, mainly through high-tech production of raw materials and high-tech dyeing, in which the feedstock is particularly critical. Main features of functional fabrics include: easy care, anti-UV, anti-bacterial, anti-static, radiation protection, anti-oil and dust retardant, health care and other functions. These features are not only practical, but also have a close relationship with health, making them widely welcomed. Cotton fabrics should be used in athletic park clothing, which does not only have excellent qualities of natural fibers, but also a certain degree of resilience and moisture permeability, allowing students to remain comfortable while exercising.

4.4 Safety requirements of kindergarten clothing

Followed by clothing functionality is the safety requirements of kindergarten clothing. Since kindergarten students are known to have little self-protection or are less aware of danger, safety performance tests on clothing that they wear are very important. Safety index of children's clothing is generally divided into mechanical safety requirements of relevant existing data indicate that: children's clothing business on the mechanical security requirements seriously enough. GB / T 22705-2008 "children's clothing and pull the rope with safety requirements" specified in the standard [3]: Buttons should not have sharp edges to prevent danger. Avoid using slider in clothing. Use pull belts instead of knotted ropes and belts.

5. Development trend of functional kindergarten clothing in our country

Research and development on our uniforms started relatively late in comparison to other countries, in which both theoretical and practical research are not systematic and standardized. However, school uniforms carry unique cultural meaning, in the performance of a national and regional economy, education, culture, and so has a significant role, it displays the spirit of a generation of students at the same time. Thus study of uniform design is imperative.

Kindergarten functional clothing design concept is evolving, with global emphasis on environmental issues; for example, the energy crisis also caused panic in all clothing developers. Kindergarten clothing design design, from the perspective of recycling to improving the wearing comfort of clothing, materials used can be recycled or biodegradable, or minimize chemical contamination dyeing and processing of fabrics produced. There is practical significance to the choice of fabric used, hoping to protect the environment. The spirit of this design concept, the choice of materials used on our uniforms a lot of new technology research and development and environmental protection can be recycled. Although its price is higher than the uniform product, but people pay more attention to environmental protection and its additional functionality as to choose this type of clothing, which will be a big trend in the future development of kindergarten clothing; also the future development the clothing industry in our country.

6. Conclusion

Kindergarten clothing design is a comprehensive art, which concerns over safety, health, comfort, functionality and beauty as a design concept. The nursery kindergarten clothing is an important part of the national dress. Development of kindergarten clothing, when compared with other countries, has a late start and a slow pace in theoretical and functional research on kindergarten clothing design, which are not systematic and standardized. From the side showing the state compulsory education culture, it is one of the components of civilization identity, and thus smaking kindergarten study design imperative. The production of systematic, professional design, marketing standardization, with the apparel market, giving uniforms an advantage in the market.

We have a common dream, hoping students will be able to wear clean, beautiful and well-made kindergarten clothing every morning. From them, we will not see the prosperity gap, nor discrimination, but only equality and dignity. They will say, "I like to wear our kindergarten clothing, because it makes me feel very confident !"

References

[1] Xu Q. students of fashion design and development of China Xianjian: 2010.5.
[2] Fu LN, Xu QQ. Status and Analysis domestic textile science students uniforms: 2006 (18) IV.
[3] Liu YF. costume design [M]. Beijing: China Textile Press: 1997; 153.
[4] Bai Y. student uniforms Reflected Education Management [J] Contemporary Education Forum: 2006; (8):

37-38.
[5] Cui YM. Kids design [M]. Donghua University Press: 2010.
[6] GB / T 22705-2008 "children's clothing and pull the rope with safety requirements" specified in the standard.
[7] Wang M. strategic market of uniforms Sustainable Development of Liaoning silk: 2008 (3).
[8] Liu XG. Shanghai children's clothing design: Donghua University.
[9] Zhang P. Mayenne sportswear and functional fabrics [J] Textile Leader: 2009 (9); 88-89.
[10] Liu YF. costume design [M]. Beijing: China Textile Press: 1997; 153.
[11] Baidu library uniforms: 2014.

Assistance of New Media Technology on the Protection and Inheritance of Traditional Costume Culture

Fen-Fen Ma[1], Yan Jiang[1*], Rui-Liang Guo

[1]*Fen-Fen Ma Beijing Institute of Fashion Technology, No. A2, East Yinghua Street, Chaoyang District, Beijing, P.R, China*

*Corresponding author's email: jsjjy@bift.edu.cn

Abstract

"In the new media age where digital technology has become mainstream, new opportunities and challenges for the protection and transmission of traditional costume culture are coming. This paper describes the impact of new media technology on the traditional costume culture, and the main outlet of traditional costumes and cultural communication in the new media environment was summarized by listing examples of Qing dynasty emperor clothes costume culture and somatosensory 3D display platform, to further explain that the new media technology will effectively expand the way of traditional costumes cultural transmission.And it will add new vitality.

Keywords: New media technology; Traditional costume culture; Somatosensory interaction; Virtual reality

1. Introduction

"New media" is the endorsement of the new media age, and "traditional dress culture" is the cohesion of the Chinese civilization for thousands of years. Now, with digital technology, Internet technology, mobile communications technology becoming the representative of the new media technology ,information has been widely spread digital TV, outdoor high-definition video or other forms. In the 1990s, Nicholas had already predicted that the exchanges of information are able to touch every sense. Obviously, this media era has came. However, with respect to the new media, traditional dress culture is of extraordinary significance to us. China's huge traditional dress culture is a part of traditional culture, it is the crystallization of material, at the same time, it has been given a spiritual connotation.

As a cultural successor, protection and transmission of traditional culture Language is the responsibility that people of this era can not shirk. With the mass media and communications media changing rapidly, the mode of transmission of traditional clothing culture can also escape out of the traditional way of exhibitions like museum's physical static exhibition, reports and other forms from paper. With the help of virtual reality technology to produce "From reality to virtual reality and then 'the three-dimensional virtual presentation, as well as' a new clothing culture of communication in traditional dress walked mobile client" and so on. In the new media age, we urgently need to use the power of the new media technologies, so that the traditional dress culture can get out of the woods before faltering, and ultimately achieve the purpose of protection and inheritance.

2. Dissemination of Effective Models Clothing Culture Under the New Media Environment

Internet and other digital technology make Chinese traditional culture spread more conveniently. Such rich and diverse way of communication expands the transmission range, to meet the individual needs of the audience [1]. Clothing culture propagation is a common majority of all social activities, which the general public again dominated social groups and school groups. Thus, in the face of these two groups need to have time and "differentiated" mode of transmission, but spread the content and form of communication is still required to have a certain appeal.

2.1 Dissemination of New Media Space Occupied Areas, Expanding the Dress Culture

Under the new media environment, mobile phone, tablet, as the carrier of the mobile platform has became the main source of the masses access to information. Therefore,occupying the space of new media spreading areas to expand the dress culture is a top priority. Social masses is a principal part of traditional dress culture spreading.To provide them with some information or a fun interactive platform, it is a very meaningful thing.For example, Yan Jiang a teacher of Beijing Institute of Clothing Technology has made this practical research. She developed cheongsam three-dimensional interactive learning system, the art of embroidery patterns show system, provide a learning platform for the students.

2.2 Explore New Teaching Model that Allows Traditional Dress Culture into Campus Life

Traditional culture is a kind of universal education.And formal education in school is the basic way of traditional culture to propagating [2]. So new teaching mode, is one of the important topics focus on exploration in school. For example, Beijing Institute of Clothing Technology by reforming Chinese clothing history curriculum for the clothing historiography development of educational information, dissemination and sustainable development in the new media environment of traditional culture to explore new ideas. This research and its practical application have multiple meanings in cultural, social and economic [3].

2.3 Through the Fusion of Cultures, Let Traditional Dress Culture Luster the Social Life

The best way of protecting the cultural heritage is making it existed in people's lives like air fall upon everywhere, at same time, making it grow into people's lives and the continuous development and innovation. In a word, traditio nal dress culture have to infiltrate through all areas.And rely on the power of technology, the traditional dress cult ure and social life can make close contact,. For example, clothing culture and the combination of film culture, or cu lture into the clothing online game character modeling and design for young audiences to pay attention to Chinese t raditional costumes, in order to attract them to a deeper understanding of the traditional dress culture, or in social lif e, holding Chinese wedding, pay attention to the main elements of the traditional clothing brand, they are all select directions by these people in their industry [4].

3. The Practice of Traditional Costumes and Cultural Heritage Protection

Under the "cloud era" Background, the evolution of mobile Internet promote the development of national cultural heritage and provide a turning point. The protection of national heritage and traditional culture is rather feasible [5]. Both to protect the "body" and the "soul" of the traditional culture. The following is the example practice under the new media environment and the protection of traditional dress culture heritage.

3.1 Forbidden City Launched the "Emperor of the Qing Dynasty costumes" Apply

"Emperor of the Qing Dynasty costumes" is a platform based IOS application (APP). This application shows the three-dimensional modeling of dresses, Kyrgyzstan clothes, uniforms, Rong Fu, clothing line,this six emperors casual clothing, and also display with clothing design proofing diagram, process the information and material details, fully embodies the Qing Dynasty beauty emperor costumes (Fig.1). This application is for the audience to sort out the system of court costumes, and give users a superior visual and interactive experience.

(a) Qing Dynasty Emperor apparel APP icon

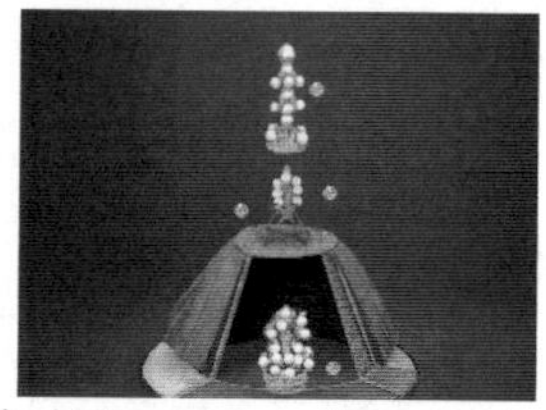

(b) APP content impressions

Fig.1 Qing Dynasty Emperor apparel App (From http://img3.imgtn.bdimg.com)

3.2 Cheongsam Three-dimensional Interactive System and Art Embroidery Patterns System

"Cheongsam three-dimensional interactive learning system" and "Art Embroidery patterns system" have developed already by means of virtual reality technology, both applications have received State Copyright Bureau software copyright certificate, with independent intellectual property rights. "Cheongsam three-dimensional interactive learning systems" dress will be in the form of three-dimensional display, users can interact zoom freely or rotate a three-dimensional character models and to be free to choose cheongsam collar, sleeve type, Length and patterns. So users can get intuitive feelings about the effects of local changes in the overall shape to the dress effect brought about. And apart from this, users can provide through the software system i (information) button to browse the text, images, or video. It provides patterns and embroidery stitch knowledge which can be fully understood. Fig.2 and 3 are APP icon and some display systems interface.

Fig.2 Cheongsam three-dimensional interactive system(From Cheongsam system APP)

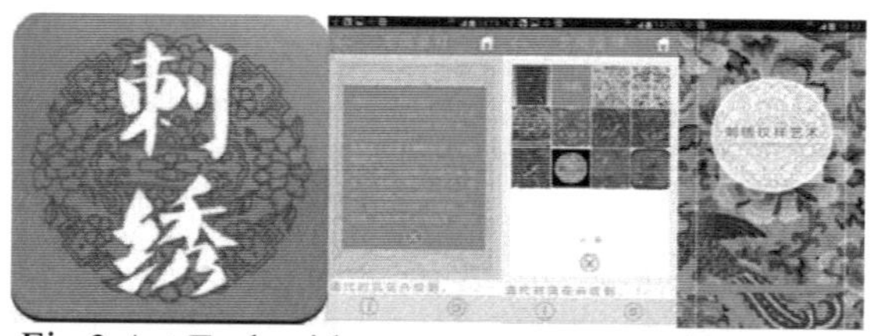

Fig.3 Art Embroidery patterns system(From Art Embroidery patterns system APP)

Dress with embroidery is an important part of traditional Chinese clothing culture. Compared with the traditional graphic material, this two systems provide a more interesting learning platform. Cheongsam three-dimensional interactive system development and Art Embroidery patterns system has a certain significance of other types of clothing cultural development. Used in schools teaching practice, it is possible to improve the student's learning interest and initiative, and promoting teaching effectiveness significantly.

3.3 Somatosensory 3D Clothing Culture Display Platform

Somatosensory 3D clothing culture display platform based on virtual machine in the form of 3D virtual fitting system. This system consists of Beijing Institute of Clothing Technology school, Beijing science development center, the new Guangzhou Intelligent Technology Co., Ltd. It uses cutting-edge science and technology: somatosensory interactive technology and virtual reality technology. System contains ancient Chinese costumes, ethnic costumes, the Olympic apparel, APEC and other apparel sections. Good effects and three-dimensional simulation of virtual clothing follower feature, giving users better interactive experiences. In addition, the system also describes the different costumes corresponding cultural background, knowledge of popular culture clothing of different costumes by means of text, images and other forms. Somatosensory 3D clothing culture display platform enhances cultural heritage innovation capacity of the school, especially provides a strong technical support to the exhibition of new clothing line technical.

For a long time, the forms of research are confined to books, papers, academic reports or other communications, these results spent researcher's lifelong effort, but are not suitable for the general public to make science propaganda. So Chinese clothing culture spread has been greatly limited. At the same time, the museum exhibits precious clothing collections protected by the restrictions, it can only be watched through the windows of the exhibits ,allow visitors to make a simple understanding of the learning process but lack interactive entertainment.So the dress culture propagation is not ideal [6]. Fig.4 is traditional clothing fashion show in FIG. The somatosensory 3D clothing culture display platform, with its good interactivity and realism, to make up for the limitations of traditional display mode.

In September 2014, this platform was meet with the general public at "National Science Day Beijing home activities". The audience responded enthusiastically. Thousands of visitors to experience the 48 sets of clothing of system, Fig.5 is the virtual fitting effect by audience.

Fig.4 Traditional display mode (From http://pic.108tian.com)

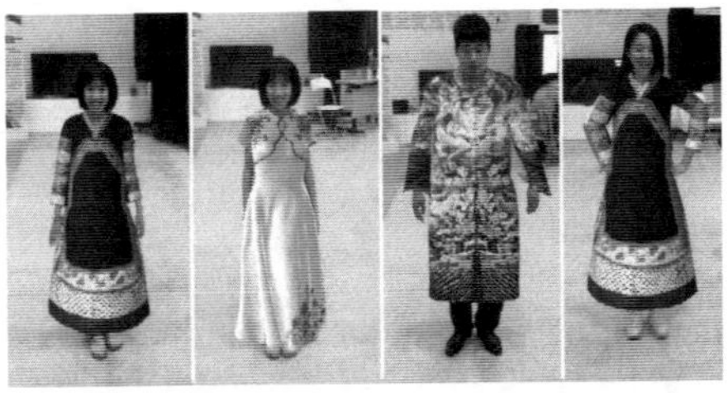

Fig.5 Somatosensory 3D clothing Culture showcase display effect (From http://img1.gtimg.com)

4.Conclusion

New media technologies and diversifying traditional costumes cultural heritage opened the door, at the same time set up a high threshold. So the department has traditional costumes cultural resources and the agency arms master modern technology should strengthen exchanges and cooperation. On the one hand to make the new technology serves for cultural heritage , and on the other hand , the widespread of culture has become the embodies value for new technology itself. In addition, training technical-clothing-complex talent is a viable path. In short, the traditional dress of Chinese culture as ethnic culture has great importance for whole nations and whole world. To stimulate the enthusiasm of their common protection for traditional culture, so, traditional culture can be endless. At the same time, only accompanied by cultural prosperity and national prosperity, ethnic pride is able to enhance essentially.

5. Acknowledgements

This paper is sponsored by the Importation and Development of High-Caliber Talents Project of Beijing Municipal Institutions (CIT&TCD201404025) and Science, Technology Project of Beijing Municipal Education Commission (SQKM201610012008) and Beijing Municipal Education Commission Funded Project (KM201410012006).

References

[1] Peng XN. Promotional Impact of New Media on Traditional Chinese Cultural Transmission.Radio & Television: 2015: 12; 15-17.

[2] Xu H. Play the Advantages of the New Media Culture of Innovation and Popularization of Traditional Ways. Forum on Chinese Culture, 2014: 11; 178-182.

[3] Gao L, Wang ZY.Transmission and Stainable Development of Traditional Culture in the Context of New Media: Taking Chinese Costume Culture as an Example.Hundred Schools In Arts, 2011: S1; 179-182.

[4]Qiu ZY, Song J. Chinese Traditional Costumes Cultural Predicament and Media Propagation Paths.Liaoning Tussah Silk: 2014: 04; 23-24.

[5] Shi YC. Under the "Cloud Era" New Media Applied Ethnic Cultural Studies.Minzu Tribune,2014: 11; 78-81.

[6] Jiang Y, Ma WX, Chen JH, Liu ZD. Kinect-based 3D Costume Culture Display System. China Textile Leader, 2015: 03; 74-76.

Extracting Individual Point Cloud Torso Surface based on Reverse Engineering Technology

Shi-Ning Yuan, Yan-Bo Ji*

No.19 Jinhua Road, Xi'an Polytechnic University, Xi'an, Shaanxi Province, Xi'an China

Corresponding author's email: www.1114043409@qq.com

Abstract

This paper selected a 3D body scanning technology and reverse engineering techniques in order to extract individual torso surface which can further be applied in tight clothing design, fashion design and automatic development of three-dimensional patterns as well as 3D-printing of clothing. In this paper, point cloud data was obtained by scanning a human body, using to the resulting point cloud to plot a reference curve for different parts of the surface feature based on specific classifications, and further consturuction of digital body surface with different methods in order to derive an ideal tight clothing design. The resulting surface can be applied to creation of tight clothing prototypes, 3D priting or used as basis for reconstructing contour of an object.

Keywords: Three-dimensional body; Reverse engineering; Surface

1. Introduction of Background and Present Situation

With the advancement of science and technology to promote social and economic development of society, the traditional standard pattern can't fit the needs of modern consumers. The spending power of different groups of people needs to be intense. At the same time, three-dimensional printing technology matures, making it possible for apparel industry to develop towards speed and personalization.

It is the combination of scientific approach and technology development that has enabled the garment industry to be more efficient, making the processes from design to production run quicker, which in turn spurred development of "MTM" (Made to Measure) tailoring. Clothing enterprises will invest more time and money to deal with the problems of traditional problems.

In order to meet the current market demand for clothing, 3D GCAD (3-Demension Garment Aid Design) employs three-dimensional body measurement, modeling, design, virtual stitching and virtual fitting as well as other tools.

Professor Fang Jingjing studied Energy-based optimal darted pattern for garment design [1]; S Thomassey and P. Bruniaux applied a reversed approach based on the amount of clothing derived from model production methods, and consequently verified the result with a virtual body model [2]; Slavenka Petrak, etc. also proposed a three-dimensional surface model of the human body which was based on the basic pieces of segmentation [3]; Wang Mei and others using 3D scanning and mannequin reconstruction in LookStailorX system generated aim garment surface [4].

According to former studies, three-dimensional garment design will be widely applied in the garments industry in the near future.

2. Research Contents and Methods

2.1 Research Contents

This paper describes a the three-dimensional model that is based on body shape data obtained from point cloud scan and analysis, curvature of which is subsequently analyzed using different methods of data extraction and curve fitting, and rebuilt into a new surface. The use of reverse engineering technology, relying on known human point cloud and the resulting curve, rebuilds the body's surface step by step, restoring the surface of the body shape as realistic as possible.

2.2 Research Methods

This paper applies theoretical analysis combined with experimental research methods, as well as reverse engineering, literature search, experimental operation and data analysis. Finally, a conclusion will be made through the verification process of the entire course of the experiment. Experimental procedure begins with extraction of curve characteristics and size from the three-dimensional virtual human model, based on the scan point cloud data to construct a human body surface. It is followed by a comparison of the original surface and dots cloud, with controlled deviation. Finally, a solid body surface prototype is constructed (Fig.1).

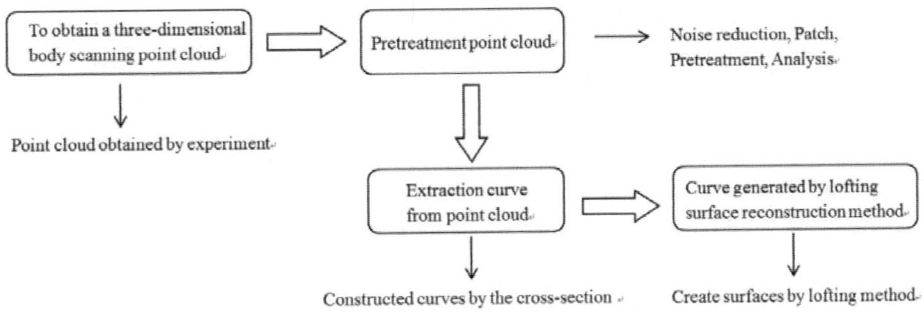

Fig.1 Study method

2.3 The Purpose and Significance of Research

This study aims to develop a substitution for the traditional two-dimensional patterns, in a way that would make it possible for digital platform to be applied in individual three-dimensional clothing design, eliminating the traditional complicated process of "pattern making - prototyping test - pattern changes - prototyping test - to produce the final pattern ". A unique prototype can be used in tight functional clothing or as a basis for designing costume surfaces. It can also be combined with 3D printing technology to generate three-dimensional solid model. These high-technological tools provide the means for improvement the costume design process, making it more efficient, and modernization of the garment industry.

3. The Three-dimensional Scanning and Reverse Engineering

3.1 Three-dimensional Scanning Technology with Human Modeling

The MTM implementation is based on use of non-contact 3D scanning technology and three-dimensional virtual model of the human body. The process of the three-dimensional body scanning measurement can be described with the following steps:

1) Project grating or mechanical movement source onto the body surface;
2) Project the camera sync diagram onto the human body;
3) Extract image data from the body surface information;
4) Construct a three-dimensional visualization of the human body model [5].

The way curve and surface fitting measurement data entry are being combined will directly affect the quality of products and sample detection precision of the post-virtual model, as well as the accuracy of the model fitting for the subsequent reconstruction of curves and surfaces [5]. In this paper, the point cloud data has been auto-repaired by computer after scanning, and the repaired results can be directly applied to the subsequent steps.

3.2 Reverse Engineering

Reverse engineering has originated in precision measurement and quality inspection. It is opposite to the traditional forward engineering concept - “product concept design-a CAD product model–final product” [6].

Reverse engineering aims to achieve the outcomes of the traditional mode of thinking in a reverse process. Traditional design is a process of "feature-principle-structure". Reverse design is "a principle of physical prototypes, a functional three-dimensional reconstruction" work process. After the adoption of digital data processing, three-dimensional reconstruction of the physical prototype has become the key part of the process [7].

This article focuses on two aspects of the reverse engineering: first, the three-dimensional surface data acquisition, and second, curve and surface construction [9].

4. Point Cloud Processing and Surface Generation

Firstly, this work completed the step of measuring a human body with an extraction experiment of the torso surface of standard female body (160-84-64-90: height 160cm, bust 84cm, waist 64cm, hip 90cm). This was followed by analysis of features of point cloud data. Finally, we constructed a prototype of a tight covering surface of the body surface curved according to the curve characteristics.

4.1 Platform introduced

IMAGEWARE is one of the well-known international reverse engineering software, produced by the American company EDS, and is now being incorporated into its NX product line. Its strong point cloud processing power, surface editing capabilities and the ability to build a new surface are widely applied in a number of areas. As used herein, version 13.0 version (Fig.2), support Windows7 operating system.

4.2 Curve Extraction and Surface Reconstruction

4.2.1 Get Three-dimensional Scanning Point Cloud

To obtain a three-dimensional surface of the body at this stage of digitalizations a three-dimensional scanner is used. Germany Tecmath Scanning Equipment Company's non-contact laser scanning three-dimensional facility includes Vitus Smart three-dimensional laser scanning hardware and digital body measurement software. This system has hardware with resolution of 0.5mm; while the software can handle the more than 500,000 data points (Fig.3).

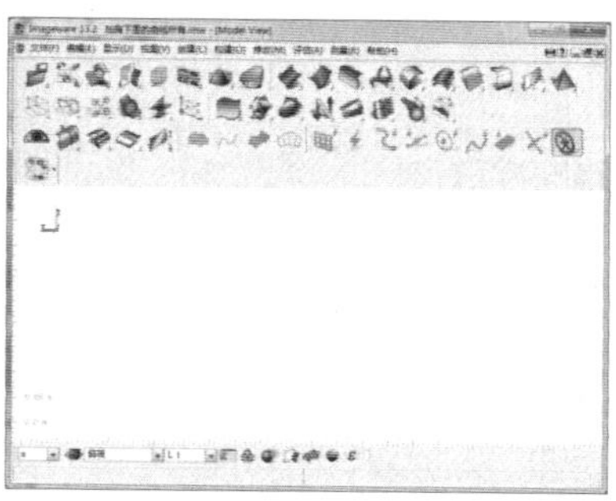

Fig.2 Operating system

Fig.3 Vitus Smart Laser Scanner [9]

Experiment process:

1) Scanner device is turned on and debugged.

2) The mannequin is positioned in the centre of the scan area. The exact positioning is adjusted with a trial scan.

3) The scanned point cloud form is checked for shifts caused by equipment failure missing the point cloud and other issues.

4) The model is fine-tuned and debugged to exclude to human error or malfunction.

5) After the adjustment, the standard mannequin is scanned.

6) Results of the scan are processed and stored. The experiment is complete. The scanner is closed and the model is removed.

4.2.2 Analysis of Point Cloud Model

Point cloud data acquisition scan can help to analyze the basic characteristics of the human body by observation. The known principle of the three views is applied: the outer surface of the object can be determined from the shape of the front view, left view, top view. However, in this particular body surface, there is a problem of left-right asymmetry. Therefore, the experiment requires a multi-view together comparative study to determine the surface characteristics.

The virtual human model (160-84-64-90) is based on a standard female body (height 160cm, Bust 84cm, waist 64cm, hip 90cm).

4.2.3 Pretreatment Point Cloud

In order to obtain a full and correct measurement data to facilitate subsequent modeling work, the need for measurement data preprocessing includes: data format conversion, filtering and smoothing, interpolation and data streamlining, etc [10]. Suikerbuik et al proposed a method of marking the body with a small cube of feature points [11] - the method of point cloud noise reduction. After point cloud data processing, it removes noises interference points, reduces the complexity of data processing, and fixes the vulnerability of the point cloud. Therefore, the model is consistent with the basic requirements (Fig.4).

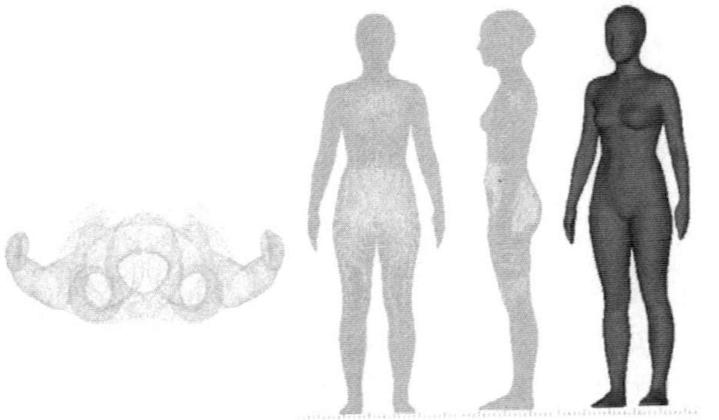

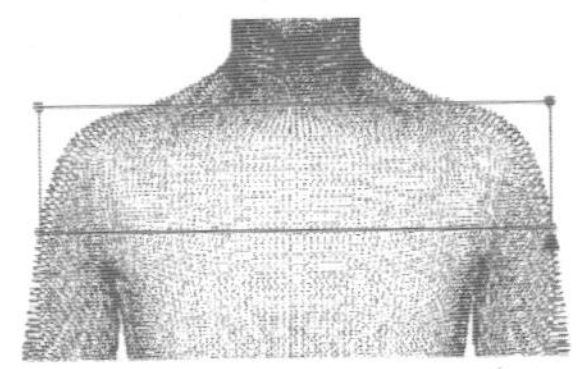

Fig.4 Different views Fig.5 Left shoulder slightly higher than the right shoulder

4.3 Cloud Point Extraction Based Body Curve

4.3.1 Point Cloud Based Extraction Body Curve

Extraction body curve is based on the fluctuation of the surface of the body surface, and the aim is to find the corresponding feature position, and then extract the characteristic curve data. For the prototype of tops, we need to extract feature sizes including: neck line, chest, waist length, armhole line, shoulder line, back center line, and front center line. These dimensions mentioned before constitute the basic elements of the prototype, and the surface of the body is more obvious, so the three-dimensional body can extract the data.

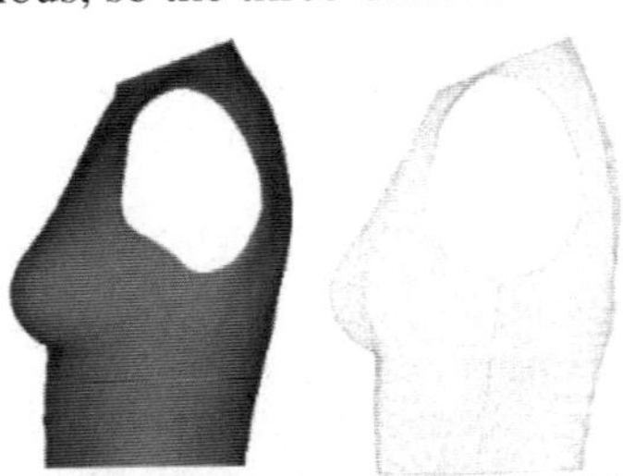

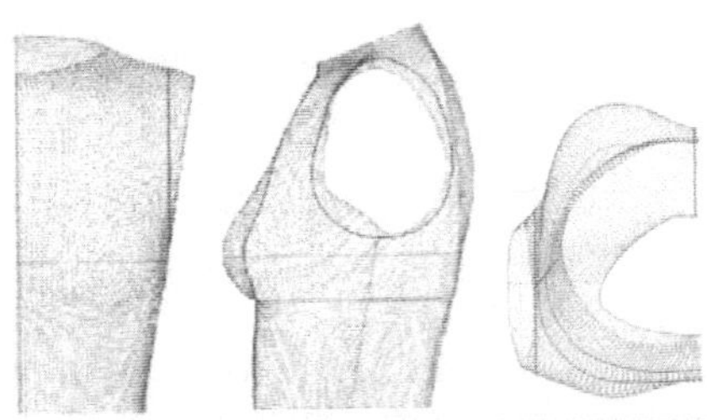

Fig.6 One-half net dot cloud extract Fig.7 After extraction point cloud

Point cloud analysis of this body showed that the left shoulder is slightly higher than the right shoulder. While left shoulder portion of study chosen slightly higher compensates for lack of the right shoulder slightly lower, so the left half physical model was chosen for further studying (Fig.5,6). Fitting characteristic curve is shown in Fig.7.

In order to facilitate comparison with the net post-finished bodies (Table 1).

Extraction curve for bust, under bust, waist, hip, is extracted in the horizontal direction, single direction, and it can be carried out in parallel to the point cloud intercept. After generating an optimal shape of the curve, its parameters are adjusted to fit the cross-sectional shape of the resulting curve point cloud shapes.

For extraction of armhole, shoulder, center front, center back, especially given that curves like the armhole one do not exist in actual body, the position of the shoulder line can be offset. Thus, according to the principle of the special circumstances of special consideration, you should use a different method to extract the data size.

Table 1.Extraction net body feature point cloud data

Name	Original Body Size（m）
1/2 Front NL	0.0993
1/2 Back NL	0.1081
1/2 Front BL	0.2373
1/2 Back BL	0.1820
1/2 Front Under Bust	0.1869
1/2 Back Under bust	0.1699
1/2 Front WL	0.1577
1/2 Back WL	0.1485
Armhole Line	0.4543
Shoulder Line	0.0605
Front Center Line	0.3231
Back Center Line	0.3501

4.3.2 Point Cloud Based on the Significance of the Body Curve Extraction

After several attempts, the scanning showed that the body surface is irregular. Therefore, the extraction process should be conducted in different ways for different parts. Extraction process in accordance with the position of the distribution curve is different for the torso. For bust, waist, hipline and neck the first three are horizontal - there is no displacement in the z-axis direction, so we can use parallel intercept point cloud method to intercept the point cloud. When the point cloud is based on curve fitting, we selected a point cloud elliptical cross-sectional shape similar to curve fitting. To extract the neck circumference, we adjusted the surface in three directions, so that it aligned with the curve in generate from the point cloud.

Armhole long extraction is the same situation. Define the reference feature point, generating curve is still in the point cloud of the curve, but its shape should be consistent with the body's natural physiological structure in order to not become a serious impediment to movement. For length (front and back centerline) extraction: in cross-section view, they are not straight line, while rebuilding the surface can be used as a boundary line. When special circumstances are taken into account and meet after work to extract feature lines is complete.

4.4 Based Extraction Point Cloud Body Surface

4.4.1 Based on the Extracted Human Body Surface Point Cloud

The human body is a complex surface, substantially parallel to within a small interval have changed a lot, so the surface must be divided into many levels, with subsequent modeling, in order to complete the extraction of the real surface.

Extraction of torso below the armhole curve: the torso is parallel to the point cloud intercept with uniform distribution. The larger the change area, the higher the density of intercepts. For intercept point cloud curve fitting, interpolation curve can be chosen to emphasize the curvature as much as possible through all the points on this section; the generated curve will be based on uniform adjustment parameters which will be used to coordinate direction

Resulting curve will vary in length and direction, which is problematic. It is difficult to reach standardized generated surfaces which distinguish between different situations. In Fig.8,9,10, the arc below the bottom of the armhole curve is parallel to the ground, so the resulting curve should be adjusted in the same direction. Both ends of each arc should intersect the center front and side seam, to ensure the smoothness of the finalized surface.

Curves above the armhole are more inclined and follow slowly from parallel with the shoulder line into to the angle of the generated gradient. When the curve lines at the intersecting portions of the collar and armhole are too abrupt, the curve should be trimmed. All curves re-parameterize, first set the number of nodes of a characteristic curve, this curve according to the following curve to re-parameterization.

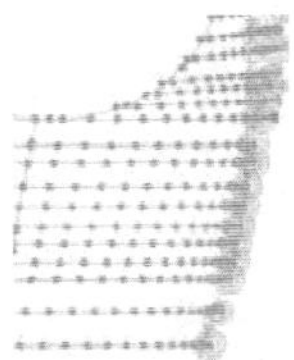

Fig.8 Under the armhole

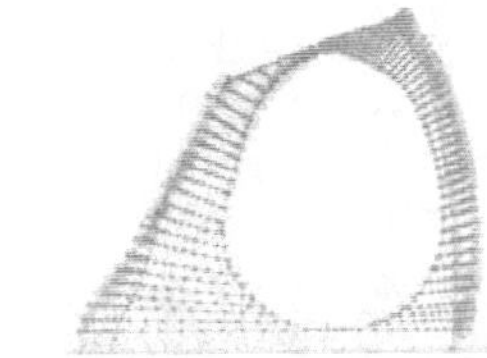

Fig.9 Above the armhole

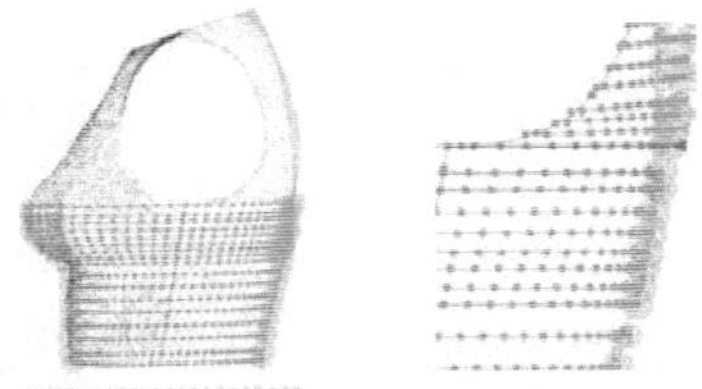

Fig.10 The curve re-parameterize change

Parameter modified two-part distinction: similar to the cylindrical portion of the end under the armhole, compare the rules, during parameterization, the number and distribution of nodes is not used as much change; above the armhole curves, varying of lengths, and the inclination change is not the same, to be individually modified.

Bulk editing can't guarantee that the curves will go in the same direction. The direction of a curve needs to be coordinated and adjusted, the node number and order of the length of the curve to be modified, etc; as far as it is possible to generate curves that fit the standard.

4.4.2 Lofted Surface is Generated Based on The Curve Shape

If a curve was set through consistent direction coordination and after due fulfillment of other steps, the preparatory work can be considered complete. In that case you can rely on the existing curves to build the surface. In reverse engineering, a commonly used method is to create the first approximation of data points by interpolation or fitting into spline (or parametric curve), then to use modeling tools to complete a patch shape, and ,finally, get the complete surface model with extension, trimming and other modifications [12].

Loft surface is curved in accordance with the framework in a certain order, and connects the endpoints of these framework to form a complete frame which contains the surface of these curves. The method of forming such a continuous surface characterized reservations obviously, so it is used to construct the body surface.

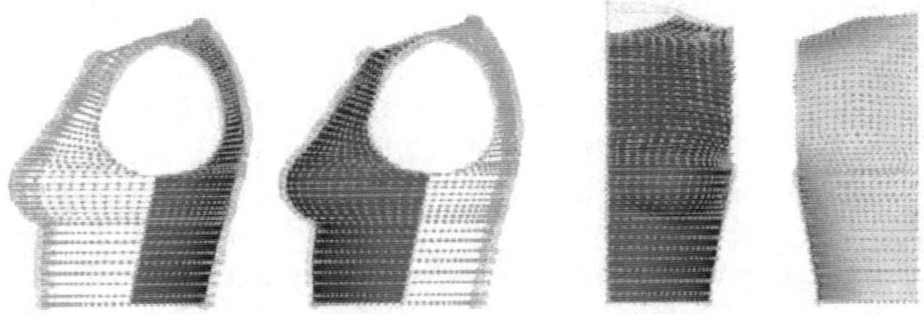

Fig.11 Lofted surface generation

The lofted surface should generated be in a certain order, and it should be selected one by one sequentially to build the curve in order to generate the final smooth surface. In the layout method for surface in IMAGEWARE environment is constructed by a plurality of points in the same direction with approximately parallel curves that link up to the same parameter values on two adjacent curve points with line segments connecting the resulting surface. Therefore, a large array interval distance curve and the degree of uniformity of the closed curve, and called the start point and the alignment direction of the curve to be consistent, to form a uniform surface distribution of the position of the boundary curve, not a large difference in length, otherwise, poor surface quality [13]. A complete surface is shown above in Fig.11.

4.4.3 Surface Check

After the establishment of the surface, in order to check the built surface for errors, it should be verified. Since IMAGEWARE has an unmatched computer-aided surface inspection capabilitiy, surface modeling and Rapid

Prototyping, inspection for analytical errors in curved surfaces can be conducted in the same software environment [14]. Construction of the complete surface point cloud is anazlyzed with the initial errors highlighted in color. Error Analysis Report contains: maximum error, mean and standard error [14].

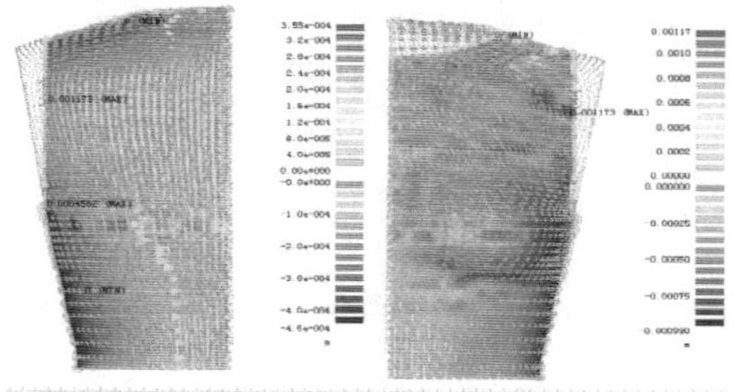

Fig.12 Check result

As shown in Figure 12, the maximum error is less than 2 × 10-3m as most of the surface is colored green, while undulating local places error is relatively large. In general, the reconstruction of the surface and the body surface is consistent. Reconstruction of the original data and the comparison surfaces is shown in Table 2.

Table2. the net volume data and the data reconstruction surface were compared

Name	Body size（m）	New size（m）	Deviation（m）
1/2 Front NL	0.0993	0.0993	0
1/2 Back NL	0.1081	0.1081	0
1/2 Front BL	0.2373	0.2373	0
1/2 Back BL	0.0820	0.1815	0.0005
1/2 Front Under Bust	0.1869	0.1869	0
1/2 Back Under Bust	0.1699	0.1694	0.0005
1/2 Front WL	0.1577	0.1577	0
1/2 Back WL	0.1485	0.1485	0
Armhole Line	0.4543	0.4543	0
Shoulder Line	0.0605	0.0603	0
Front Center Line	0.3231	0.3231	0
Back Center Line	0.3501	0.3501	0

The raw data and reconstruction data were compared, with the deviation falling within 0.0005m. Most parts of the surface didn't change, so it is evident that the reconstruction of the surface condition of the human body is capable of expressing undulating surface, and the scattered points on the surface in the form are presented accurately to rebuild a layer of "skin".

Error analysis: the quality of the surface depends on the distance between two adjacent curves. If the distance is large, errors are likely to occur. Figure 12 shows that errors are common near the bust and armhole areas. This is due to large fluctuations in these areas - the distance between the curves changes more radically then in the other parts and the difference between the number and location of nodes is relatively large. These differences will translate into surface deformation, affecting the surface shape and creating errors.

4.4.4 Cloud Point Extraction Based on Body Surface Meaning

Extraction of nude human torso surface point cloud data is the basis for generation of body surface, analysis, calculation, and deeper design on new costumes surface. Therefore, the extracted net surface is equivalent to a three-dimensional garment design digital mannequins, clothing design subsequent surface, structure lines, decorative lines design - all of these are based on the appearance of this digital model.

This demonstrates the first step towards "one to one" design and manufacturing, which provides a solid foundation for digital fashion and promotes development of 3D printing technology. It may be an indispensable part on clothing line digital technology in the future.

5. Conclusions

Standard body model female torso surface was extracted and, by using reverse engineering technology a reconstructed surface with the same feature was created. Below is the summary of the entire process of the surface extraction:

1) Human point cloud scanning is multi-faceted, with errors causing negative impact on the experiment results, reducing the accuracy of the experiment, but at the same time, improving the efficiency of the experiment.

2) Human body model characteristic curve extraction, is according to different classification needs to discuss its location and the nature of the selection of the most appropriate and most efficient construction methods and point cloud curve.

3) Resulting surface of the original human point cloud needs to to undergo comparative data validation, during which deviations from the surface of the built body built will be detected. Only after its reliability is verified, can it fully represent surface features of the scanned body.

Reconstruction of the human body surface is a copy of the layer of "skin", which is of great significance for the study of special functional clothing. It can be used by designers to apply various different fabrics to develop more special features for functional clothing.

In addition, segmentation analysis and body surface reconstruction experiments provide a theoretical and technical base for establishing a method of surface and costume construction that would allow the process to be conducted with ease. In the future, the use of these construction methods and basic operations will increase ease scanning body surfaces in order to get ideal and realistic costumes.

6. Acknowledgements

First of all, I want thank my mentor for the great support he provided. I took guidance from his recommendations and research. Secondly, I'm very thankful for all the support that came from Xi'an Polytechnic University Institute of Clothing scanning device (VITUS Smart LC-D scanning system) that made this research run smoothly.

References

[1] Fang JJ, Yu D. Energy-based optimal darted pattern for garment design, International Journal of Clothing Science and Technology: 2014; Iss 2: 164-183.

[2] Thomassey S, Bruniaux P. A template of ease allowance for garments based on a 3D reverse methodology, IJIE: 2013; 43.

[3] Slavenka PDR. Systematic representation and application of a 3D computer-aided garment construction method,International Journal of Clothing Science and Technology: 2006; Iss 3.

[4] Wang M, Lu GD, Zhang DL. Three dimensional body garment CAD modeling technology research and application. JEG: 2007; 1.

[5] Li XZ Studies. Based 3D human reconstruction method of non-contact measurement. [PhD thesis]: 2011.

[6] Chen LY. Application Status and Development Trend Analysis of garment CAD [J]. Liaoning silk: 2007; 3.

[7] Peng YJ, Wang cream. PENG Europe.UG, IMAGEWARE applied research in the three-dimensional model reconstruction in reverse engineering. MDM: 2011; 5.

[8] Yang B. Based on the body's point cloud data processing and virtual human modeling. [MS Thesis]: 2011.

[9] Information on http://www.vitronic.de/en/industry-logistics/sectors/other-industries/3d-body-scanner.html.

[10] Zhang JT. Lu ZP. ImageWare achieve defect using point cloud surface reconstruction [Papers]. agricultural research: 2005; 5.

[11] Huang HQ, Mok PY*, Kwok YL, Au JS. Block pattern generation: From parameterizing human bodies to fit feature-aligned and flattenable 3D garments. Computers in Industry: 2012; 63.

[12] Liu WS. Based on surface features Imageware establishment. China Machinery [Papers]: 2014; 14.

[13] Deng XH. On the use of "Imageware software" feel. Technology Information [Papers]: 2014; 31.

[14] Tian ZM, Li G. Reconstruction Technique of law based Imageware loft surfaces. University Journal (NATURAL SCIENCE): 2011; 2; 27.

Personalized Clothing Template Model Design and Research based on Human Body Characteristics

Si-Chen Pan, Bo-An Ying*, Xin Zhang

Apparel & Art Design College, Xi'an Polytechnic University, Xi'an, Shaanxi, 710048, P.R. China

*Corresponding author's email: 546905232@qq.com

Abstract:

With the development of the apparel industry, and increased attention on personalized clothing, individuation, clothing production cycle and tailored clothing, the clothing MTM production mode has become more popular.

This study is based on the key factors of the clothing MTM production mode, using characteristics of the human body as the starting point. The clothing research and development of engineering technology research center in Shanxi assisted in the auto pattern plate making system, combining the physical parameters to complete the model for personalized clothing. The template was made based on a combination of design and experiments that verify the feasibility and rationality of the sample model [1].

Keywords: Body shape characteristics; Garment model; Fitting evaluation; Personalized customization

1. Introduction

In the MTM production mode of modern clothing enterprises, the design of clothing is based on the features of the consumer, which can then meet the needs of people from different size groups. Therefore, it is key to realize the customized and enhance the rapid response capability of the garment industry in China. The analysis and extraction of the human body shape characteristics and its key role in the production of the clothing model are the emphasis of this paper [3.4].

Methods used in this study will be specified. The automatic plate making system that adoptsthe method of preparation for the new cultural prototype of women did not include and combine the human body characteristics of a model; thus, the model has general applicability to clothing of any styles and sizes. After incorporating the combined parameters in the design on the two models, the two models each has its own specific characteristics, in which they are only applicable to certain type or size of clothing. The feasibility and rationality of this model are verified by comparing characteristics of a real person to that of the model.

2. Classification analysis and data extraction from human body shape features

2.1 Analysis, extraction and classification of body shape characteristics

Body size is a reflection of the characteristics of the human body and the proportion of the various components of the logo. Before clothing production, body types are classified based on their features to reasonably allocate the appropriate amount of resources, such as number of clothing batches and amount of garment needed, to the production process. In the sales of clothing, apart from making sure the clothing suits the consumer's height, chest, waist and other body parts, we also need to understand their body, which is commonly known as "checkmark evaluative". The current garment classification can merely reflect the measurements of human body parts and can no longer fit the customer's height requirement. Feature classification, on the other hand, can be used in personalized clothing, producing clothing that better fit the customer.

2.2 Individual feature parameter extraction based on body type classification

If the human body cadre is connected to two similar elliptical (see Fig.1), the characteristic of model C will most resemble that of the human body. A stands for ½ of chest width; B stands for ½ of chest breadth; H represents the distance between the chest cross-section to the waist section; a denotes the width of ½ of the waist; b represents ½ of waist breadth; A' represents ½ of hip width; B' represents half the thickness of the hip; and H' denotes the distance between the waist cross section to the hip cross-section [2].

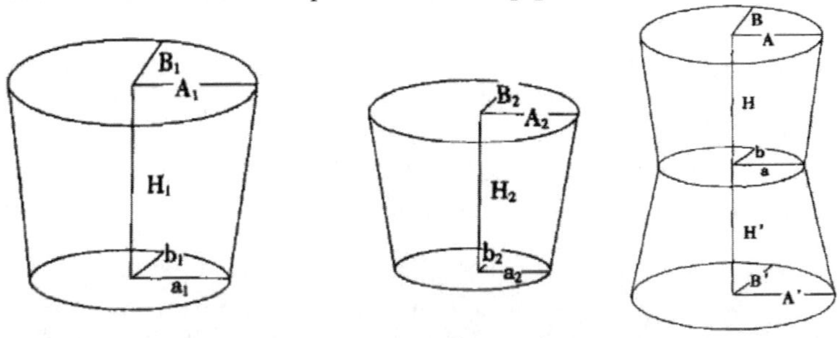

Fig.1

By C1 and C2, C3, C4, C5, C6, C7 can decide the basic shape of the human body cadres.

C= {C1, C2, C3, C4, C5, C6, C7}.

C1=A: B =1/2 Chest width: 1/2 Thick chest.

C2=A: H=1/2 Chest width: The chest cross-section to waist section distance.

C3=A: a=1/2 Chest width: 1/2 waist breadth.

C4=a: b=1/2 Waist width: 1/2Waist thick.

C5=a: H'=1/2 Waist width: Waistline cross section to hip circumference cross section.

C6=a: A'=1/2 Waist width: 1/2 Hip width.

C7= A': B'=1/2 Hip width: 1/2 Hip breadth.

C9, C10, C11 further describes the information of human positive projection in the chest and shoulders, C8, C12, C13 are further described in the side projection of the chest, back, abdomen and buttocks of the information.

C= {C8, C9, 10, C11, C12, C13}.

C8=BG: BH= Under the breast circumference radius: Breast height.

C9=BD/2A= distance between breast: Chest width.

C10=SW: 2a= Shoulder width: Waist width.

C11=α=The shoulder Angle α.

C12= (BN-BW): (FN-FW)= Back length: Center line of the long before.

C13= (FW-CR): (BW-CR)= Length of the arc between attacking midfielder section FW and Gluteal point CR.

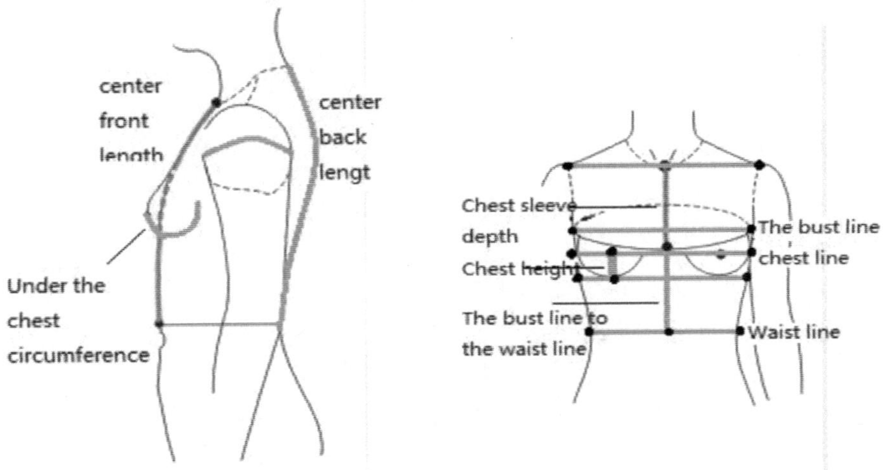

Fig.2

The trunk of the female body can be described by 13 parameters, which are embodied in the body model as shown in Fig.2.

2.3 Classification of personalized clothing pattern design body based on characteristics of the human body

At the garment engineering technology research center of Shanxi Province that provides a platform for the development of auto pattern plate model (Fig.3), a garment model with the desired characteristics of the human body, which are dependent on each other, was made. All the parameters have its own initial set value. A few versions of the model were made: the primary model for not loose quantity of the most basic type of clothing, the latest version is used for personalized clothing design [7].

The body shape characteristics, which are dependent on each other, can then be combined to make the secondary garment model prototype. The secondary model shows pertinence and individuality, and is applicable to individuals with different body shapes.

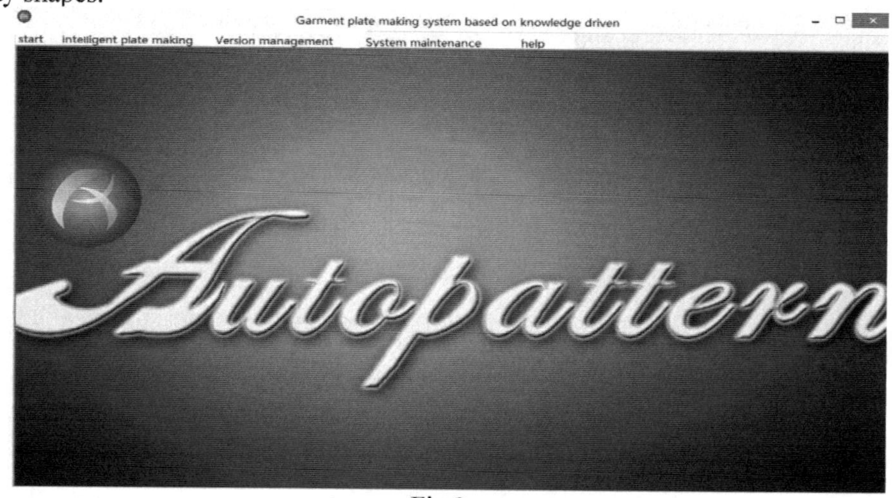

Fig.3

3. Methodology

3.1 The production of first level clothing model prototype

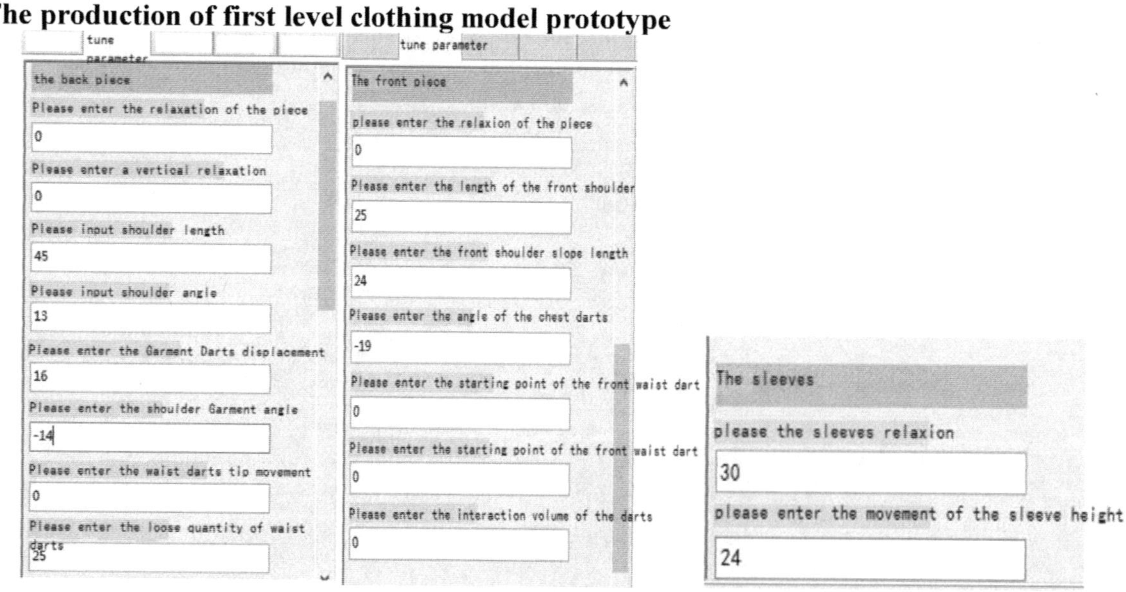

Fig.4

First level clothing model prototype uses 8 standard levels of clothing model prototype, which include the traditional Chinese female body and that of Japanese women to incorporate new cultural elements. The process of combining the dependent variables and setting the amount of interaction to meet the needs of the 8 different models is shown in Fig.4.

As shown in the figure, the effects of addition of relaxation in the chest circumference, chest bottom circumference and waist circumference of a piece of clothing are shown in a before and after comparison: the length of the human body centerline; length of shoulder slope and shoulder angle control in the human body and the change in the slope interaction; amount of dart movement and angle control shoulder to chest slope changes in the interaction; waist province pine and waist dart movement control of the change of value is the difference in thoracic lumbar interaction; and bust dart angle and control over the beginning and the end of the bust dart angle that affects different uplift degree of the change of the chest interaction.

According to the size of codons, the new prototype plate is adopted to make first level clothing model prototype (see Fig.5). In the automatic design system, the first step is to draw after piece, then adjust according to the size of the front and sleeves. Reference feature points of the "after" piece is documented as follows: total of 130, before piece 150, sleeves 80, as shown in the figure. Mobile visual reference points are observed, while most of the auxiliary point are not visible.

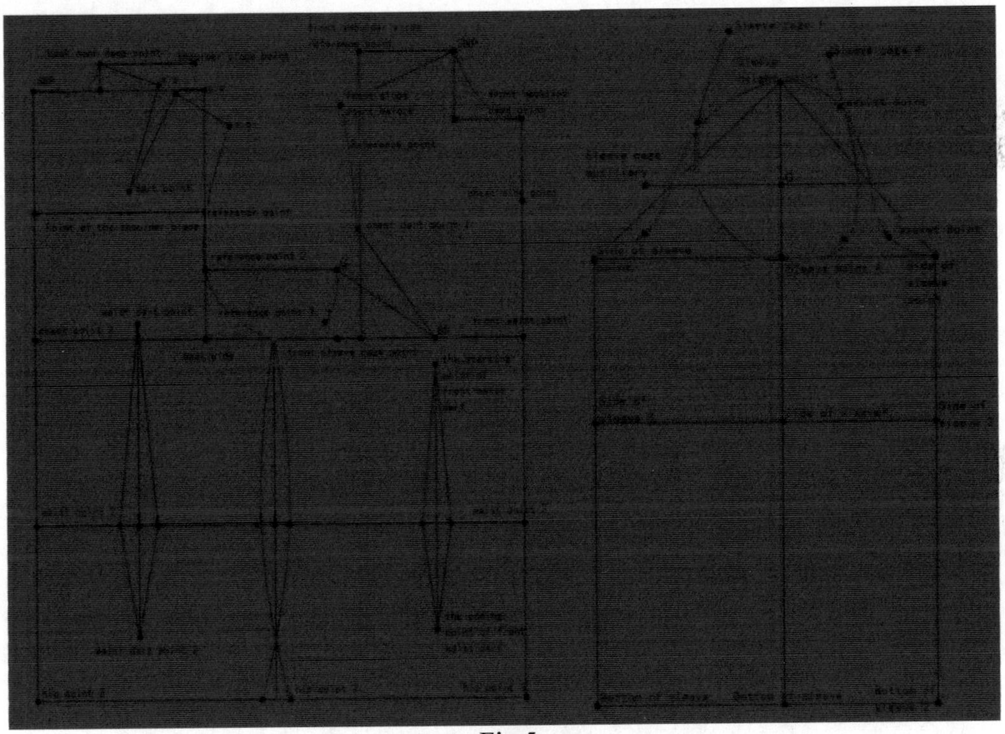

Fig.5

3.2 The production of second level clothing model prototype

Second level clothing model prototype is built upon the first level model to combine dependent characteristic (see Fig.6). In order for the author to try and feel the garment made using the model, data from the author were used to create the model [8,9].

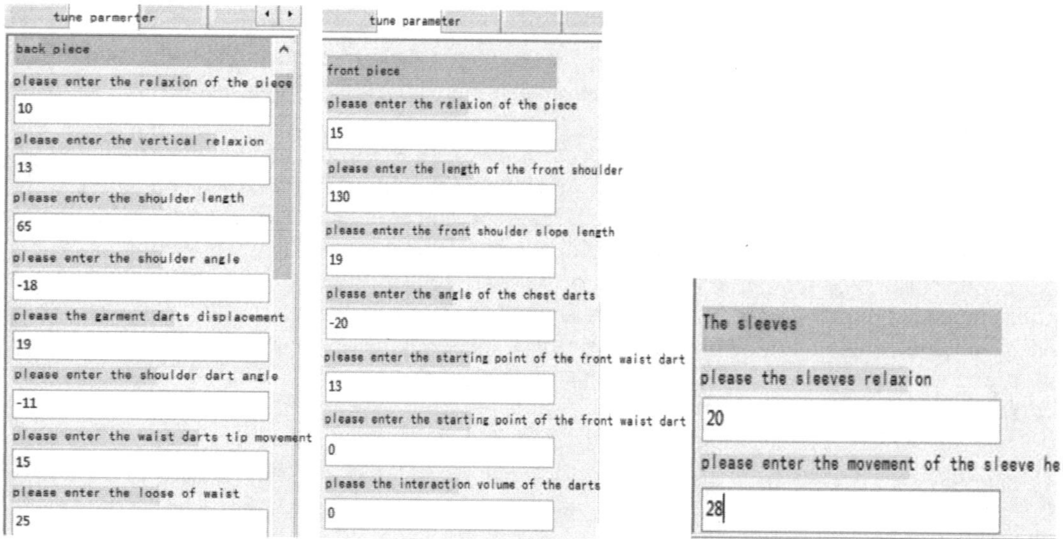

Fig.6

A lot of changes were made to the second interaction model, in which these changes are mainly concentrated in the length, chest girth, the shoulder area, chest waist and the chest area. The difference before and after the change is shown in Fig.7-1-2-3.

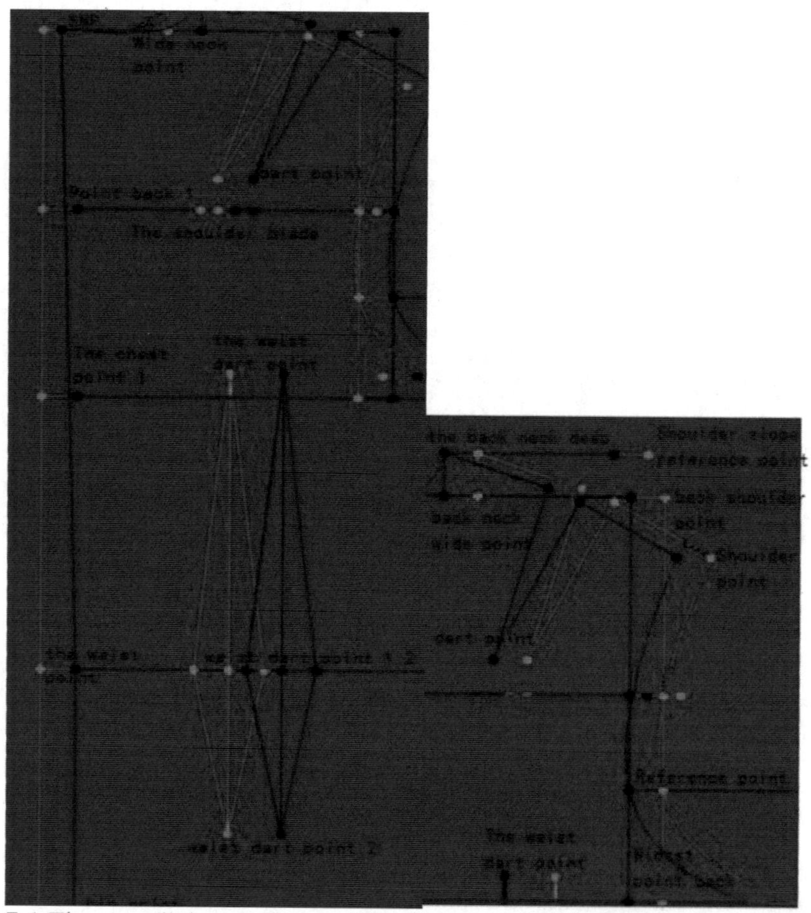

Fig.7-1 The overall degree change. The change in the shoulders and sleeve cage

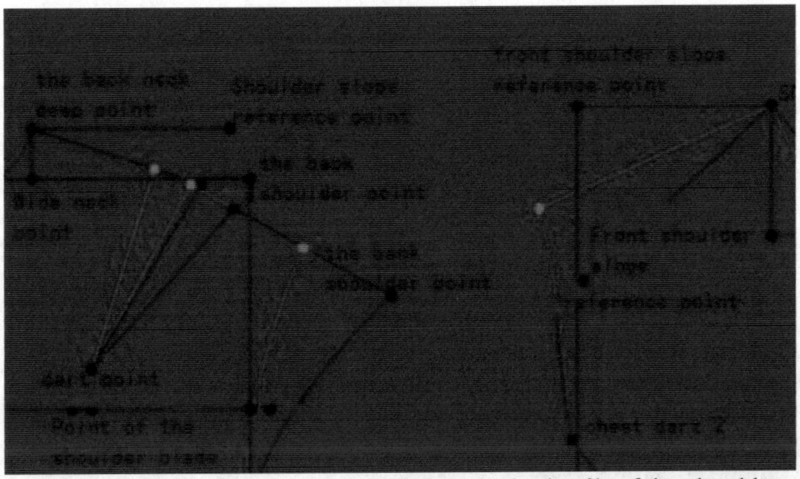

Fig.7-2 Back shoulder ministry and changes in the tilt of the shoulders

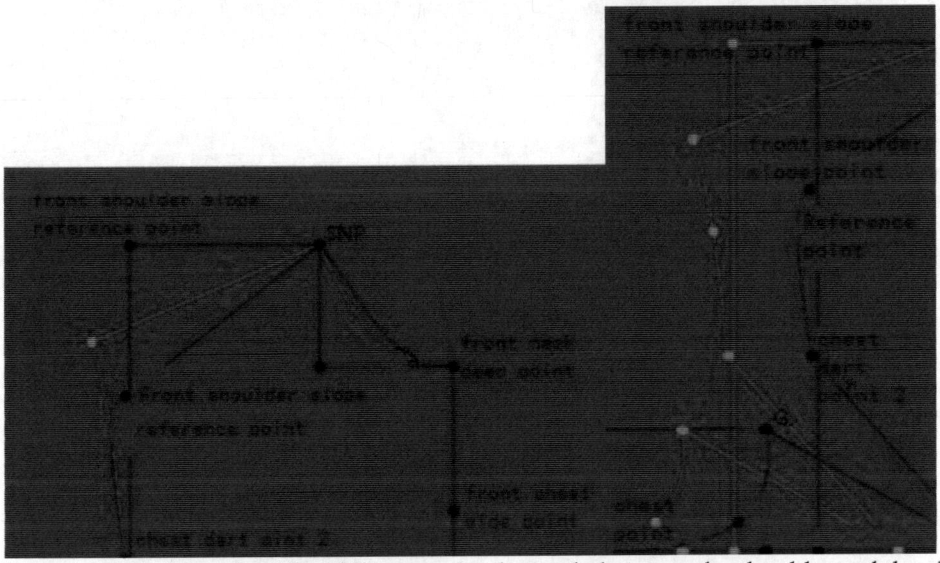

Fig.7-3 The changes in the front shoulder oblique angle, the angle between the shoulder and the chest province

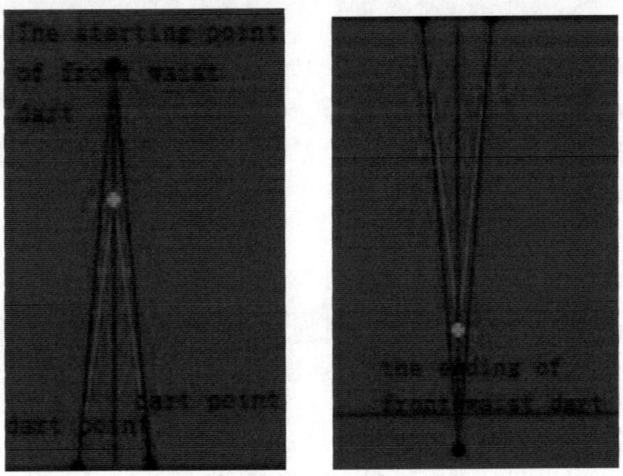

Fig.7-4 Dart size changes

After combining the redundant characteristics and modifying excess or abrupt curve, the sample became more intuitive. The model shown in Fig.8 is made according to the characteristics and measurements of the author, which serves as a specific individual model template.

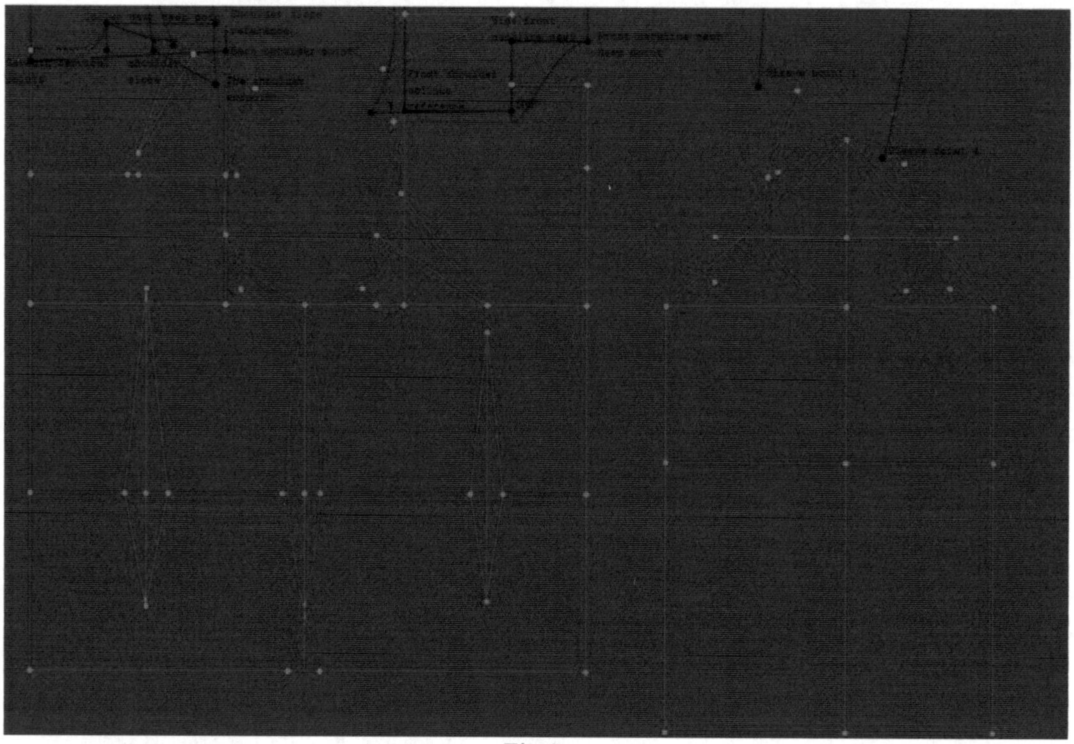

Fig.8

4. Evaluation after garment fitting

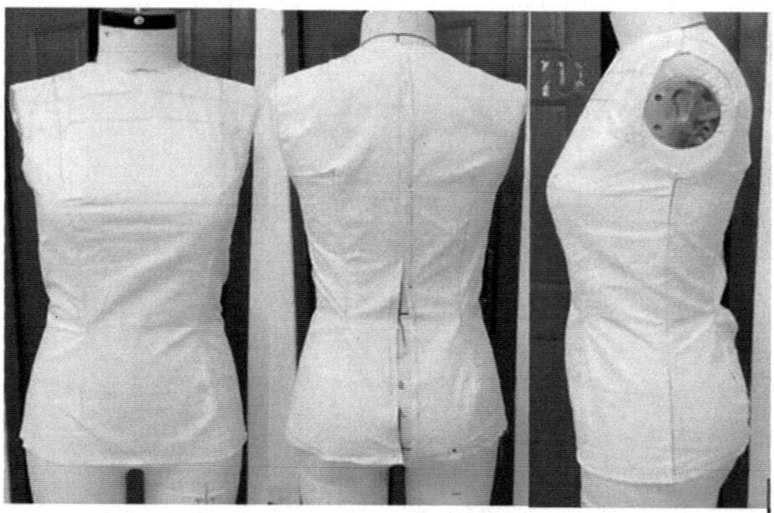

Fig. 8

First the 8 level clothing garment pattern was tried on and evaluated. The effects are shown in Fig.8. The basic fit of clothing can be visualized, in which the chest and waist are well-fitted, while there is some deficiency observed in the back due to the lack of garcon. Basic determination can be used for basic clothing sample [5,6].

After the effects of the first version provided basic affirmation, the characteristics of the model generated using the author's data are shown in Fig.9.

Fitting results of the model generated using the secondary shirt style show that the shape features are larely in line with that of individuals, in which the fitting result is relatively similar and the effect is satisfying. This topic applies to people with height of 160 cm and a chest build of 75B.

The series of experiments proved that the use of human body characteristics enhance the feasibility and rationality of the model.

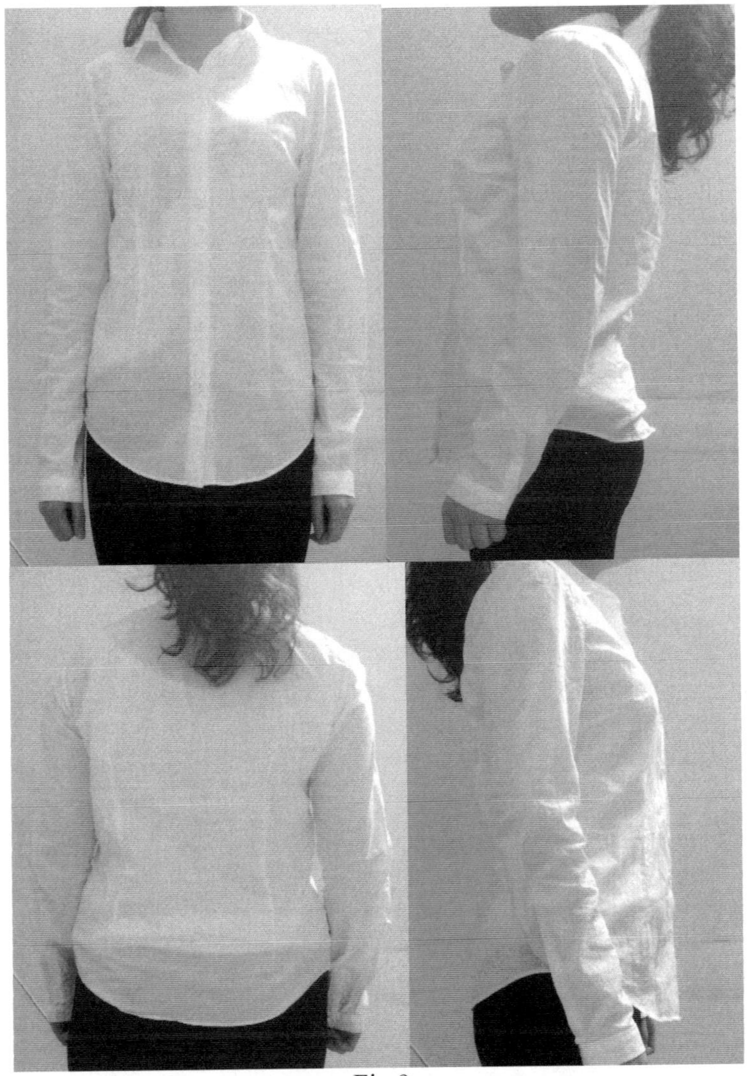

Fig.9

References

[1] Wang H. The characteristic parameters of the body part of the female body. 2005. Journal of Jiaxing University [J].

[2] Zheng Y. Research on the figure classification of female college students in the western and central China. 2005 modern clothing textile high tech Development Seminar [M].

[3] Ning GH. Female body characteristics of the [D]. research on the relationship between the blouses parametric model based on the. 2013. model of Xi'an Polytechnic University.

[4] Zhang WY, Zhang WB. Chinese costume prototype-design principles and techniques of Donghua prototype

structure [J]. Journal of Donghua University (NATURAL SCIENCE EDITION): 2002 (3): 44-49.
[5] Chen YH, Yin GF. Computer aided design technology and application [M]. Science Press: 2000.
[6] Li GS. Body size and garment design [J]. East Textile Science and technology: 002; (3): 45-47.
[7] Chen WF, Wan X. A new strategy for the development of mass customization clothing industry [J]. China Textile: 2002; (8): 34-36.
[8] State Bureau of Technical Supervision issued. The people's Republic of China National Standard GB / T 1335-1997 "garment" [S]. Beijing: China Standard Press: 1998.
[9] Chen WF. Study on the body size of women in clothing fit [D]. Shanghai: Donghua University, doctoral dissertation: 001.

Effect of Support Pants on Body Fat Consumption

Tamaki Mitsuno, Aya Hibino

Faculty of Education, Shinshu University, 6-Ro, Nishinagano, Nagano, 380-8544, Japan

*Corresponding author's email: mitsuno@shinshu-u.ac.jp

Abstract

Physical exercise is important for limiting fat mass and preventing obesity. This study aimed to identify conditions that enhance body fat consumption during aerobic exercise, including the effects of wearing support pants. Twenty-one young Japanese women completed aerobic exercise sessions while wearing each of four types of support pants. Total body fat consumption during aerobic exercise was 16 times greater than at rest. Partial compression garments helped reduce body fat more efficiently than either non-compression or constant compression garments.

Keywords: Support pants; Body fat consumption; Aerobic exercise; Respiratory metabolism; Clothing pressure

1. Introduction

Physical exercise is important for limiting fat mass and preventing obesity [1-8]; in particular, aerobic exercise is known to have health benefits [9,10]. Many factors can influence the body fat utilization that occurs during exercise. For example, in women during the low temperature phase of the menstrual cycle, aerobic exercise before lunch is most effective for maintaining body fat consumption [11,12]. Season has also been clearly shown to affect variations in body fat [13-15]. This study aimed to identify easily achievable conditions that enhance body fat consumption during aerobic exercise. Consumption of body fat was assessed by measuring respiratory metabolism before, during, and after aerobic exercise; the device used measured the amount of inhaled oxygen and exhaled carbon dioxide for every lungful of air. Body fat consumption was then calculated from the energy consumption and oxygen uptake. We also aimed to clarify the effect of support pants on consumption of body fat by comparing lipid metabolism while wearing one of three types of compression pants (one constant compression garment and two partial compression garments) or non-compression pants during aerobic exercise.

2. Experimental method

2.1 Experimental pants

Fig.1 shows the four types of short compression pants used in this study. Subjects remained seated for 1 hour while wearing a short-sleeved 100% cotton t-shirt and one of the following: 100% polyester running pants (A), basic compression pants with 80D (8.9 Tex) yarn (B), 80D (8.9 Tex) yarn pants designed to provide partial compression (C), 110D (12.2 Tex) yarn pants with the same design as C (D). Pants A provided almost no compression, pants B supported the lower abdomen and thighs uniformly, and pants C and D provided varying amounts of partial compression, particularly over the lower abdomen and medial thigh [16,17]. The basic knitting structure of these compression pants were single-rib stitches. The tensile force and strain were measured using by KES (FB1-AUTO-A, Kato Tech, Kyoto, Japan) under the following condition; Speed: 0.2mm/s, Chuck: 5.0cm, Maximum load: 50gf/cm.

2.1.1 Clothing pressure of the experimental pants

The clothing pressures (CP) of pants B, C, and D were measured by the hydrostatic pressure-balance method [18-21]. Fig.2 shows the 33 CP measurement sites, which were all on the right side of the body.

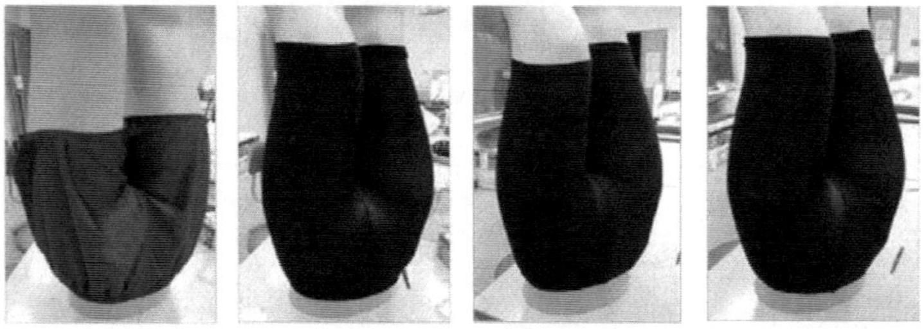

A
Control

B
Full compression pants
(80D: 8.9Tex)

C
Partial compression
(80D: 8.9Tex)

D
Partial compression
(110D: 12.2Tex)

Fig.1 Experimental pants [16,17]

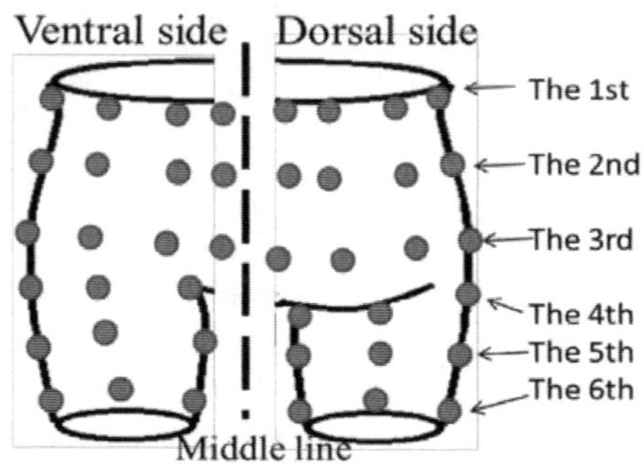

Fig.2 CP measurement regions
1st: waist, 2nd: lower waist, 3rd: hip, 4th: groin, 5th: thigh, 6th: rib lines

2.1.2 Perceived pressure of the experimental pants

Each subject's perception of pressure while wearing the experimental pants was examined by questionnaire using a visual analogue scale [22,23], as shown in Fig.3. The subjects indicated their own perceptions of a loose and a perfect-fitting feeling before wearing the experimental pants, and then recorded the pressure that they felt while wearing the pants. If they felt that the pants were loose, the subjects indicated their score on the lower line, and if they felt strong compression from the pants, they indicated their score on the upper line (Fig.3). The perceived pressures were then converted into numerical values, with not wearing scored as "0" and perfect fitting as "1". "Loose-fitting feeling" and "Tight-fitting feeling" were similarly converted into numerical values, and the subject's perception of the pressure provided by the pants was calculated.

Before wearing experimental pants

Mark the point between not wearing and perfect fitting that you consider to indicate **a loose-fitting feeling**.

Not wearing — Perfect fitting

Mark the point between loose fitting and tight fitting that you consider to indicate **perfect-fitting feeling**.

Loose fitting — Tight fitting

After wearing experimental pants

Choose the most appropriate of the two lines below and mark the point that indicates your perception of the fit of the experimental pants. Please also adjust the formatting of the figure so that the caption can be clearly read.

Perfect fitting — Tight fitting

Perfect fitting — Loose fitting

Fig.3 Questionnaire of the pressure feeling of experimental pants

2.2 Respiratory metabolism assessment

2.2.1 Subjects and experimental schedule

The subjects were 21 young Japanese females, aged in their 20s, who ranged in weight from slightly slender to obese (BMI: 18.4-29.4). Each woman completed four exercise sessions while wearing each of four types of pants (pants A, B, C, and D). Exercise sessions were separated by possible consecutive 4 days in low temperature phase of the equivalence menstrual cycle. Table 1 shows their physical characteristics and the intensity of the exercise loads used in this study, and Fig.4 shows the experimental schedule. They woke at 07:00 after 7 hours of sleep and ate a prescribed meal until 07:30. At 10:30 they entered a climate-controlled room (24.5°C, 50.0%) where they remained sitting in a relaxed position for 1 hour prior to the pre-exercise assessment of respiratory metabolism at 11:30.

Table 1. Physical constitution and given intensity of exercise loads for subjects

Items	Age (Year)	Height (m)	Weight (kg)	BMI ($kg \cdot m^{-2}$)	Waist girth (cm)	Hip girth (cm)	Thigh girth (cm)	Upper knee girth (cm)	Intensity of exercise (W) Step 1	Step 2	Step 3	Srep 4
Mean	21.6	1.57	56.7	23.0	72.3	96.5	54.3	40.1	25.4	39.1	52.3	66.4
SD	1.3	0.09	7.7	2.7	6.9	4.8	3.7	2.3	8.0	8.9	7.4	6.5

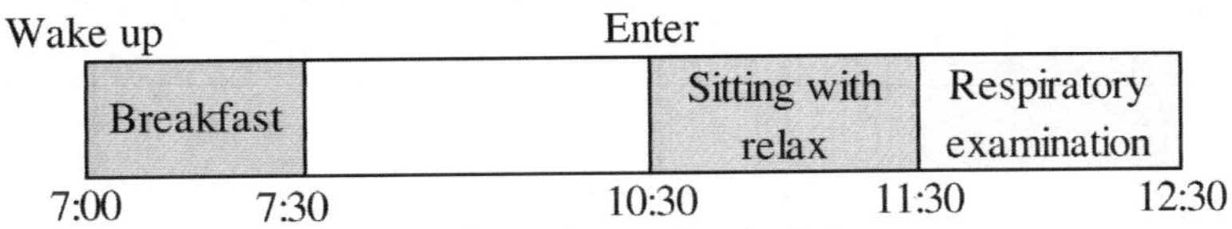

Fig.4 Experimental schedule

2.2.2 Measurement of respiratory metabolism

Fig.5 shows a subject undergoing assessment of respiratory metabolism, and Fig.6 shows a typical exercise load for the subjects. The subjects engaged in aerobic exercise on a bicycle ergometer (Nihonkoden, Tokyo, Japan) at workloads of 30% to 60% of the maximum heart rate for a total of 30 minutes. Respiratory metabolism and heart rate were measured using a VmaxS229 (Sensormedics, CA, USA) before, during, and after exercise. The amounts of energy, adipose, and carbohydrate consumed during exercise were calculated from the respiratory quotients (RQs).

Fig.5 A subject during assessment of respiratory metabolism

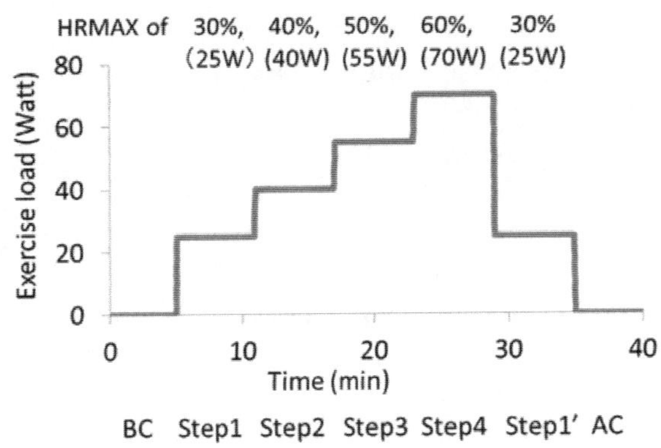

Fig.6 A typical exercises load for subjects

2.3 Calculation of energy/carbohydrate/lipid consumption from RQs

Respiratory metabolism was assessed using the breath-by-breath method [8], which allowed us to obtain the oxygen uptake and carbon dioxide output of every lungful of air. RQs were calculated by Equation (1) as shown below:

$$RQ = \frac{\text{Carbon dioxide output}}{\text{Oxygen uptake}} \tag{1}$$

In the case of short-duration aerobic exercise, such as that performed in this study, it is thought that protein metabolism does not influence the respiratory quotient [22]. Table 2 [24,25] shows the relationships between RQ and proportion of carbohydrate/ lipid, and energy metabolism. As shown in Table 2, if the RQ is 0.7 then the energy source is considered to be lipid only, and if the RQ is 1.0 the energy source is considered to be carbohydrate only. The energy distribution was determined from these percentages and the oxygen consumption (O2), and the amounts of carbohydrate and lipid consumed per minute were calculated based on energy values of 4.1cal/g and 9.3cal/g for carbohydrate and lipid, respectively.

Table 2. Proportion of carbohydrate/lipid, energy, and RQ

RQ	Carbohydrate (%)	Lipid (%)	Energy (kcal/O_2ℓ)
0.70	0.0	100.0	4.686
0.72	4.8	95.2	4.702
0.80	33.4	66.6	4.801
0.82	40.3	59.7	4.825
0.98	82.7	12.8	5.022
1.00	100.0	0.0	5.047

2.4 Statistical analysis

Statistical analyses were performed using SPSS (IBM, Ver.22). The significance of differences between actual and perceived CPs were determined by one-way ANOVA with post-hoc Scheffe tests. The correlations between lipid consumption (LC) and the various factors that potentially influenced LC were assessed using Pearson's correlation coefficients. Differences were considered significant if $p < 0.05$.

3. Results and Discussion

3.1 Compression pants

3.1.1 Actual and perceived clothing pressure

The images of pants in Fig.7 show clothing pressure distributions in support pants B, C, and D. Actual CP is shown on the left side of each image, and perceived CP is shown on the right side. The CP was lesser on the midline and greater on the sides of the body. The actual CP was significantly less in pants B than C, and less in pants C than D, and the perceived followed the same order. Pants B were evaluated as feeling like "a perfect fit" (almost 1.0), while pants C and D were evaluated as "slightly tight" and "tight feeling", respectively (shown in Fig.7).

3.1.2 Elasticity of compression pants

Fig.8 shows a typical example of the tensile test in three compression pants. From this figure, the compression pants were easy to be stretched out to the wale direction in order to Pants B, C, D and more hysteresis was observed.

From this, the knitting structure in Pants B was easy to lengthen, but is hard to return. On the other hand, Pants D was hard to lengthen, and was easy to return.

3.2 Energy, carbohydrate, and lipid consumptions during 40 min of exercise

Fig.9 shows a typical example of energy, carbohydrate, and lipid consumption (EC, CC, and LC) in 10-s increments during 40 min of exercise. EC and CC changed with exercise load intensity, but LC did not. Figure 10 shows the total LC during exercise for each of the four types of pants. The LC during aerobic exercise increased to

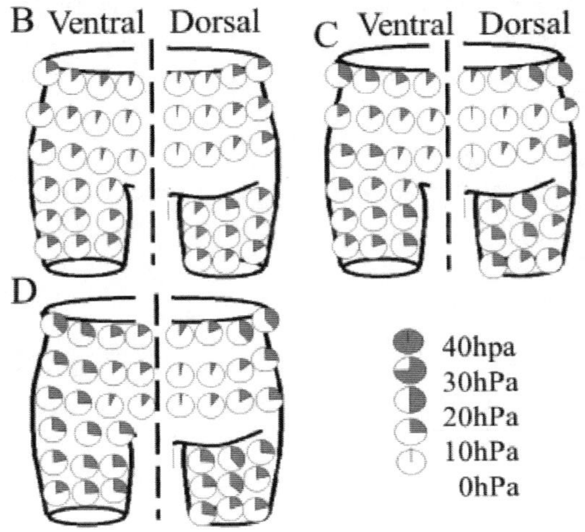

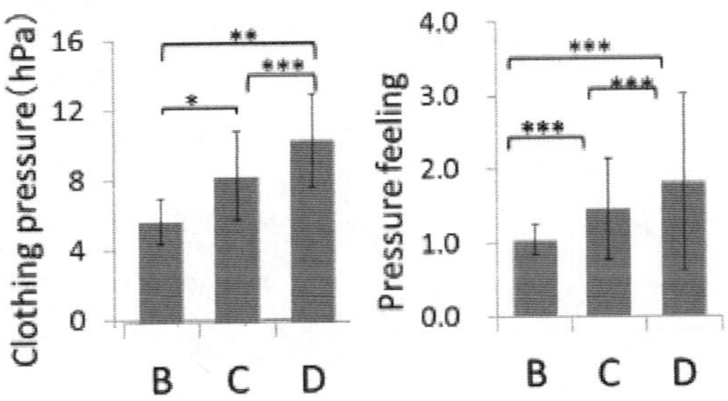

B/C/D: support pants (See 2.1 for more details.)
A value of 1.0 on the vertical axis indicates “a perfect-fitting feeling” and 2.14 indicates a “tight-fitting feeling”.
(*: P<0.05, **: P<0.01, ***: P<0.001)

Fig.7 Actual and perceived clothing pressure distributions and perceptions of fitting feeling wearing the experimental pants

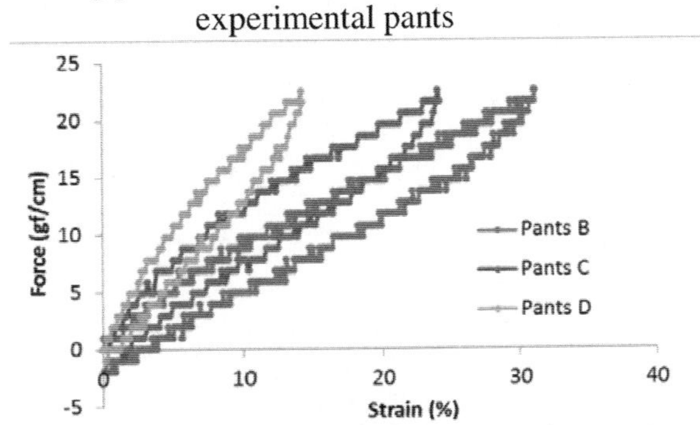

Fig.8 A typical example of the tensile test in three compression pants

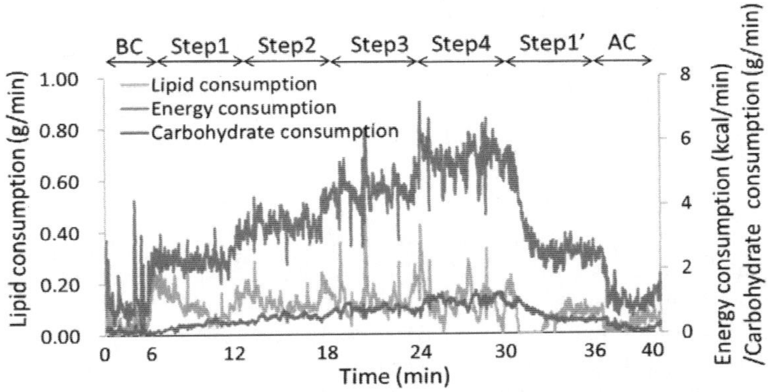

Fig.9 A typical example of EC, CC, and LC during 40 min of exercise

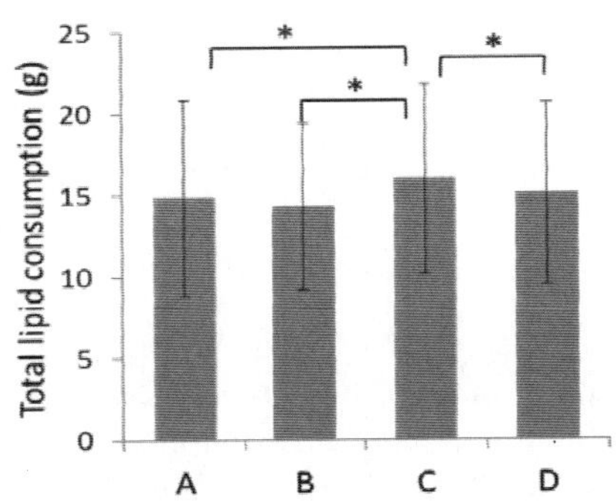

Fig.10 Total LC in four pants

14–16 times greater than before-exercise control values for all of the pants groups. Pants C were the most effective for enhancing body fat consumption during exercise; the lipid consumption for subjects wearing pants C was 16 times greater during exercise than their before-exercise control values.

3.3 Effect of various factors on lipid consumption

The correlation coefficients between the mean LC and the EC, CC, LC, O2, and RHR (relative heart rate based on before-exercise control values at rest) were calculated for each subject during each step of the experiment (Table 2). The LC was significantly correlated with the type of support pants worn, and with BMI/ BF (body mass index/ body fat). There were no strong correlations between LC and EC or RHR, but for pants C wearers, there was a significant correlation between LC and O2. That is, pants C enhanced body fat consumption when oxygen uptake was high.

3.4 Effects of each pants type on lipid consumption and body girth during each experimental step

Table 2. Correlation coefficients between LC and various experimental factors

Conditions		BMI	Body fat	Energy consumption	Carbohydrate consumption	Oxygen uptake	Relative heart rate
Non pressuer pants							
	BC	0.009	0.008	0.295	-0.709 ***	0.385	-0.033
	1	0.201	0.273	0.647 **	-0.770 ***	0.715 ***	-0.125
	2	0.288	0.447 *	0.359	-0.696 ***	0.443 *	-0.222
A	3	0.470 *	0.570 **	0.502 *	-0.536 *	0.561 **	-0.232
	4	0.476 *	0.571 **	0.403	-0.530 *	0.465 *	-0.265
	1'	0.177	0.272	0.488 *	-0.627 **	0.555 **	-0.176
	AC	-0.176	-0.261	0.151	-0.727 ***	0.244	-0.472 *
Supported pants							
Even pressure							
	BC	0.324	0.265	0.484 *	-0.716 ***	0.564 **	-0.230
	1	0.268	0.297	0.626 **	-0.826 ***	0.704 ***	-0.290
	2	0.394	0.399	0.453 *	-0.690 ***	0.531 *	-0.221
B	3	0.656 **	0.636 **	0.399	-0.598 **	0.469 *	-0.313
	4	0.689 ***	0.633 **	0.339	-0.589 **	0.409	-0.432 *
	1'	0.728 ***	0.640 **	0.316	-0.698 ***	0.401	-0.347
	AC	0.475 *	0.361	0.227	-0.728 ***	0.322	-0.584 **
Added to pressure partially							
	BC	0.512 *	0.539 *	0.695 **	-0.816 ***	0.758 ***	-0.124
	1	0.497 *	0.623 **	0.568 **	-0.783 ***	0.648 **	-0.170
	2	0.561 **	0.686 ***	0.413	-0.604 **	0.483 *	-0.313
C	3	0.669 **	0.712 ***	0.467 *	-0.541 *	0.529 *	-0.498 *
	4	0.639 **	0.669 ***	0.443 *	-0.463 *	0.499 *	-0.569 **
	1'	0.532 *	0.585 **	0.386	-0.609 **	0.456 *	-0.500 *
	AC	0.408	0.446 *	0.342	-0.859 ***	0.468 *	-0.579 **
	BC	0.176	0.234	-0.077	-0.883 ***	0.070	-0.187
	1	0.427 *	0.502 *	0.605 **	-0.670 ***	0.669 ***	-0.370
	2	0.582 **	0.608 **	0.205	-0.640 **	0.283	-0.217
D	3	0.686 ***	0.596 **	0.366	-0.549 **	0.432 *	-0.389
	4	0.681 **	0.642 **	0.371	-0.445 *	0.427 *	-0.427 *
	1'	0.486 *	0.507 *	0.158	-0.634 **	0.232	-0.434 *
	AC	0.091	0.094	0.235	-0.793 ***	0.347	-0.152

Judgment of the significance
*: $P<0.05$,
**: $P<0.01$,
***: $P<0.001$

BC/AC: Before/After contrc
1-4,1': 1step-4step,1'step

A/B/C/D: A/B/C/D pants

Table 3 shows the correlation coefficients between nude body girth (waist/ hip/ thigh/ upper knee girths) and LC for each pants group during each step of the experiment (steps 1-4 and BC). Pants A and B, which are non-compression or constant compression garments, respectively, had strong positive correlations with steps 3 and 4; that is, LC increased during slightly heavy aerobic exercise loads. Pants C and D, which are partial-compression garments, helped to reduce body fat more efficiently than pants A or B. Pants C and D were designed with belt-shaped elasticized tapes running from the medial thighs to the buttocks; these effectively enhanced the effects of exercise on body fat. The subjects were 21 healthy Japanese females in their twenties who did not normally participate in regular exercise. As for the elasticized tape sewed on the surface of the pants, Pants D is hard to lengthen in comparison with Pants C. Therefore, the tape of Pants C helped their exercise, but the tape of Pants D rather disturbed their exercise. There are three factors increasing LC which regard as CC, RHR, O2. LC decreased

Table 3. Correlation coefficients between body nude girth and LC in each step

Pants Step	BC	1	2	3	4
A pants					
Waist	0.022	0.183	0.240	0.397	0.438 *
Hip	0.046	0.275	0.296	0.422	0.422
Thigh	-0.020	0.385	0.404	0.491 *	0.436 *
Upper knee	0.095	0.316	0.254	0.434 *	0.404
B pants					
Waist	0.259	0.220	0.403	0.628 **	0.624 **
Hip	0.291	0.256	0.266	0.500 *	0.572 **
Thigh	0.195	0.171	0.197	0.392	0.403
Upper knee	0.281	0.248	0.204	0.375	0.482 *
C pants					
Waist	0.463 *	0.386	0.456 *	0.595 **	0.550 **
Hip	0.532 *	0.507 *	0.473 *	0.517 *	0.493 *
Thigh	0.376	0.425	0.448 *	0.487 *	0.385
Upper knee	0.435 *	0.495 *	0.424	0.426	0.398
D pants					
Waist	0.397	0.593 **	0.606	0.667 ***	0.693 ***
Hip	0.074	0.441 *	0.543 *	0.471 *	0.523 *
Thigh	-0.107	0.292	0.529 *	0.384	0.420
Upper knee	-0.081	0.150	0.307	0.156	0.240

BC: Before control
1-4: 1step-4step

Judgment of the significance;
*: P<0.05,
**: P<0.01,
***: P<0.001

with RHR/ CC increasing, on the other hand, LC increased with O2 increasing. It is necessary to study the relations with CC, RHR, O2 which seems to increase LC. Our findings suggest that partial compression garments may be the most effective for reducing body fat and girth. Pants C were effective from the abdomen to the leg even before exercise; thus, wearing these pants might enhance body fat consumption during normal daily activities.

4. Conclusion

Our findings revealed that partial compression garments (pants C and D) enhanced exercise-associated reductions in body girth and fat more efficiently than non-compression (pants A) or constant compression (pants B) garments. In particular pants C, which were designed with belt-shaped elasticized tapes running from the medial thighs to the buttocks, enhanced body fat consumption even without exercise. Thus, wearing this type of partial compression pant both during exercise and during general daily activities may be an easily achieved strategy for obesity prevention.

5. Acknowledgement

The authors wish to acknowledge Miss Nagata K for cooperating experiment, and Mr. Yoshida A for making prototypic experimental pants. This work was partly supported by a Grant-in-Aid for Scientific Research (C) (no. 24500924) from JSPA, and a Grant of The DESCENTE of Sports Science on "Tool and Clothing" Area from Ishimoto Memorial Foundation.

References

[1] Bouchard C. Physical activity and obesity. Human Kinetic Publishers Inc., USA, Canada, Europe, Australia, NewZealand, 2000; p: 1-389.

[2] Galgani J, Ravussin E. REVIEW, Energy metabolism, fuel selection and body weight regulation. Int J Obesity 2008; 2: S109-S119.

[3] Fujioka K, Seaton TB, Rowe E, Jelinek CA, Raskin P, Lebovitz HE, Weinstein SP. and the Sibutramin/ Diabetes Clinical Study Group, Weight loss with sibutramine improves control and other metabolic parameters in obese patients with type 2 diabetes mellitus. Diabetes, Obesity and Metabolism 2000; 2: 175-187.

[4] Whyte HM, Phil D, Blood Pressure, Obesity. Circulation 1959; XIX: 511-516.

[5] Carr MC, Brunzell JD. Abdominal obesity and dyslipidemia in the Metabolic Syndrome: Importance of Type 2 Diabetes and Familial Combined Hyperlipidemia in Coronary Artery Disease Risk. JCEM 2004; 89: 2601–2607.

[6] Yusuf S, Hawken S, Ounpuu S, Bautista L, Frnzosi MG, Commerford P, Lang CC, Rumboldt Z, Ones CL, Lisheng L, Tanomsup S, Wangai P, Razak F, Sharma AM, Anand SS. Obesity and the risk of myocardial infarction in 27 000 participants from 52 countries: a case-control study. The Lancet 2005; 366: 1640–1649.

[7] Aoyagi K, Kusano Y, Takamura N, Abe Y, Osaki M, Une H. Obesity and cardiovascular risk factors among men and women aged 40 years and older in a rural area of Japan. J Physiol Anthoropol 2006; 25: 371–375, http://www.jstage.jst.go.jp/browse/jpa2 [DOI: 10.2114/jpa2.25.371].

[8] Xiao DY, Shu JW, Yaru X, Jing L. A clinical follow-up study on the risk of cerebral infarction in Chinese aging overweight and obese population. Obes Res Clin Pract 2011; 5: e1-e78. doi: 10.1016/j.orcp. 2010.09.251

[9] Mersy DJ. Health benefits of aerobic exercise. Postgrad Med 1991; 90: 103-110.

[10] Ootani K, Hashimoto S, Hayashi S, Kimura T, Naito M. Effects of aerobic exercise training on lipoprotein metabolism and antioxidant status in sedentary but otherwise healthy young women. J Integrated Study dietary Habits 2011; 22: 114-120.

[11] Shinohara Y, Mitsuno T. Effect of the exercise time of day on respiratory metabolism. Abstract of 2013 annual meetings of Japan Home Economics Society on http://www.jstage.jst.go.jp/browse/kasei/65/0/ contents/-char/ja/ (in Japanese)

[12] Lamina S, Hanif S, Muhammed H. Influence of menstrual cycle on maximal aerobic power of young female adults. AJPARS 2011; 3: 36-41.

[13] Tsumura Y, Makonakwkeyoon L, Limtrakul P, Hirota N, Sone Y. Seasonal variation in amount of unabsorbed dietary carbohydrate from the intestine after breakfast in young female Thai subjects: comparison with that of Japanese subjects. J Physiol Anthropol 2010; 29: 141-147.

[14] Morinaka T, Wozniewicz M, Jeszjka J, Bajerska J, limtrakul P, Makonkawkeyoon L, Hirota N, Kumagai S, Sone Y. Comparison of seasonal variation in the fasting respiratory quotient of young Japanese, Polish and Thai women in relation to seasonal change in their percent body fat. J Physiol Anthropol 2012; 31: 1-12 http://www. jphysiolanthropol.com/content/31/1/10.

[15] Ando Y, Mitsuno T. Various effect factors for lipid consumption under aerobic exercise. Abstract of 2014 annual meeting of Japan Research Association for textile end-uses (in Japanese)

[16] Mitsuno T, Yoshida A. Inner wear. Japan patent pending No. 12600, 2015 (in Japanese) .

[17] Mitsuno T, Morishima M. Development of clothes to promote a body fat burning effect. Descente sports science 2016: 37 (in printing)

[18] Mitsuno T, Makabe H, Momota H, Ueda K. Study on the clothing pressure (Part 1) Measurements by a hydrostatic pressure-balanced method. J Jpn Res Assoc Text End-Uses, 1991: 32: 362-367 (in Japanese).

[19] Mitsuno T, Makabe H, Momota H, Ueda K. Study on the clothing pressure (Part 2) Measurements by a hydrostatic pressure-balanced method. J Jpn Res Assoc Text End-Uses, 1991: 32: 368-367 (in Japanese).

[20] Mitsuno T, Ueda K. Measurement for clothing pressure. J Home Econ Jpn, 1994; 45: 179-188.

[21] Mitsuno T, Ueda K. Time relation among clothing pressure developed at waistband, respiratory movement, and girth of abdomen. SEN'I GAKKAISHI 2010; 66: 74-81.

[22] Bijur PE, Silver W, Gallagher MAE. Reliability of the visual analog scale for measurement of acute pain. Academic Emergency Medicine 2001: 8: 1153-1157.

[23] Mitsuno T, Ueda K. The clothing pressure developed at waistband and the pressure sense estimated in a ratio scale. J Home Econ Jpn, 1997: 11: 989-998 (in Japanese).
[25] Du Bois. Clinical calorimetry, Thirty-fifth paper. A graphic representation of the respiratory quotient and the percentage of calories from protein fat and carbohydrate. J Biol Chem 1924; 59: 43-49.
[26] Lusk G. Animal calorimetry, Twenty-fourth paper. Analysis of the oxidation of mixtures of carbohydrate and fat. J Biol Chem 1924; 59: 41-42.

The Innovative Design and Application based on Fashion Color of Tujia Brocade Pattern

Li-Ping Zhao, Hong Xiao*

College of Apparel & Art Design, Xi'an Polytechnic University, No.19 Jinhua Rd(S), Xi'an, Shaanxi, 710048, China

*Corresponding author`s email:xh309@126.com

Abstract

Tujia brocade is one of the most unique handmade folk art of the Tujia Ethnic minority which is made with variety colors and rich patterns. However, because of the process, pattern, color and other factors, the development of Tujia brocade is in a bottleneck. This paper is written mainly from the perspective of visual art, which analyses the color feature of the pattern of traditional Tujia brocade and the feasibility of modern design. It aims to combine innovative design of Tujia brocade patter color with popular colors and applying it to clothing.

Key words: Tujia brocade; Color; Innovative design; Application

Tujia Nationality in Xiangxi has a long history. They are able to express themselves through the Tujia Brocode which are weaved by Tujia girls and is a artwork that has been passed from generation to generation. There are over 200 different variety of Tujia brocade, all with exquisite workmanship, distinctive theme, colorful, vivid design, rough and simple pattern, and durable etc. In early 1957, Tujia brocade had appeared in the London International Exhibition and its style had become very well known internationally [1].

At present, with the implementation of cultural development strategy, Tujia brocade has resurfaced and has become very popular. Hanging series (It is a decorative fabric hanging on the wall), bag series, interior products, clothing and other products series are in the market. However traditional Tujia brocade that are hand woven has poor practicability and low production efficiency thus they cannot meet the vast number of consumer demand. In addition, because of the absence of the successor of Tujia brocade and the fierce competitive consumer market, there is a chance that it may die out. So this article analyzes Tujia brocade pattern color with unique national characteristics the goal being to combine old popular color with today's popular color so that we are able to preserve the Tujia brocade.

1. The colour analysis of tujia brocade

1.1 The culture analysis of tujia brocade color

Art reflects the culture. When Laiden , an American anthropologist analyzed the function of art within a small society he discovers that art is produced in a social environment and has an association with c specific beliefs and values [2]. There are two main reasons to promote the production of traditional tujia brocade color concept [3]. First of all, tujia people have inherited a very traditional way of living where men plowed the fields and women wove cloth. It's community is very enclosed, undisturbed by the outside. Unique colors and characteristics of the Tujia brocade are inspired from it ssurroundings. Secondly, the brocade color also witholds ancient ethnic color concepts which were accumulated during the history of Tujia. Red and black are the primary colors clearly influenced by the people's favorite, Ba and Chu which directly relates to life in itself. The color concept originate from basic human life instincts therefore the Tujia brocade can be interpreted as moving color forms.

1.2 Characteristics analysis of the Tujia brocade color

Tujia brocade has formed a fixed and unique color style by using the weaving technology of cross junction. Warping commonly uses dark color yarns such as red , blue, black. The weft on the other hand is woven by colorful lines that are refered to as "good colored clothes" [4]. Tujia brocade has its own set of color fomulas: "Nobody matches black and white and nobody chooses to put green and red together, but it will be shining and surprising if y ou combine yellow and blue [1]." Tujia people are very good at choosing complementary colors. Besides, the primary and secondary color the contrast of colors is what makes Tujia Brocade unique and rich. 100 artists of Tujia weave the same brocade pattern with around 100 different variety of colors. Tujia brocade is dyed with natural ore plant. Brocade artists will match color according to the type and number of the existing materials. It is the individualistic expression of colors that makes Tujia brocade a unique visual symbol. the weaving process of pass through the weft.

1.3 Feasibility analysis of Tujia brocade color and color combination

Socialized mass production means that the tujia brocade must keep pace with the current times. The ancient brocade color are not up to date. Only by sustaining innovations can the Tujia brocade create pattern color to meet the demand of modern market. It is necessary that the Tujia brocade work in combination with fashion colors to become involved in the current trend. If tujia brocade is integrated into the modern fashion world the traditional tujia brocade will not only be preserved, but also enriches modern clothing design elements and further expand the international garment industry.

2. Innovative Design and Application of fashion color and Tujia brocade pattern fusion

Fashion color is the benchmark of popularity. Color plays an important role in the fashion consumption sector. Especially Europe, America, Japan, South Korea and other higher consumer level market, color guides the direction of the fashion industry (Fig.1). International trends authorities -Fashion Snoops Release the top ten color in 2017 spring and summer which are: pink orange, maroon, pale cream-colored, pink and purple, light blue, rose pink, purple and blue, malachite green, copper, orange and olive green (as shown Fig.2).

Fig.1 2016 - 2017 spring and summer fashion color forecast
(Picture Source: http: //www.eeff.net/thread-1363476-1-1.html)

Fig.2 2016 - 2017 spring and summer fashion color
(Picture Source: http://www.ylifeng.com/newsshow.php?cid=4&id=34)

These patterns mainly contain classic hook profile, auspicious and folklore patterns. According to the spiritual connotation which Tujia brocade contains, some representative Tujia brocade patterns are selected in this paper, they are combined with the fashion color to achieve secondary innovation and put it into use in clothing design.

2.1 Cultural Connotations analysis of Tujia brocade pattern

In classical patterns, hook patterns were turned from a single eight-hook into forty-eight-hook [5], which symbolizes Tujia people's stubbornness by the use of single-hook plants that are brave in climbing, as shown in picture 3a; and the "Wan" pattern [6] symbolizes eternity and continuous reproduction breed in an endless succession, as shown in picture 3b; and there are also serpent patterns [3], the totem worshiped by Tujia people, as shown in picture 3c.

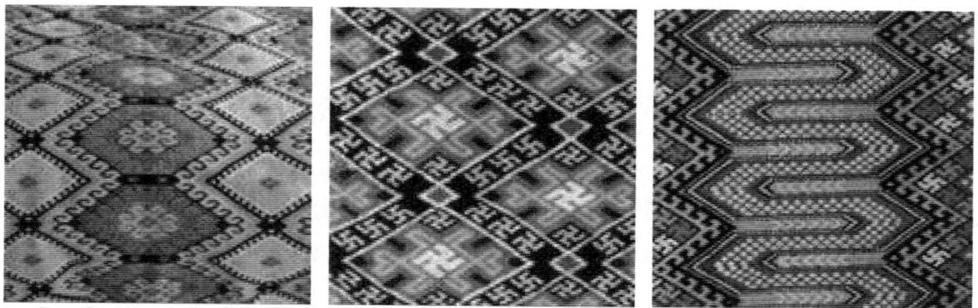

3a. forty-eight hook 3b. Words'Wan'pattern 3c. serpentine

Fig.3 Patterns on hook (Photos taken by the author)

In auspicious patterns, nine plum blossom [5] implies good fortune and joy, as shown in picture 4a; and lotus pattern [5] is a metaphor which means that the more sons you have the more happiness you would gain, as well as the continuous reproduction of descendants, as shown in picture 4b; and azalea pattern [7] implies the meaning of good fortune due to cuckoo is regarded as bird of good fortune, as shown in picture 4c.

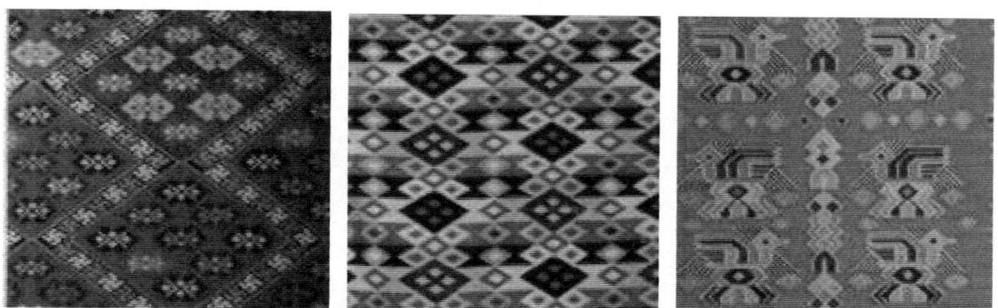

4a. nine plum pattern 4b. lotus pattern 4c. sun bird pattern

Fig.4 Auspicious patterns (Photos taken by the author)

In folklore patterns, there are ancient folk Tujia Waving Dance theme pattern, shown in Fig.5a. Tujia "White Tiger rush" reflects a folklore associated with Tujia people that avoids disaster brought by evil and blessing of good represented by the Tai Tai flower [5] shown in Fig.5b. The Tujia "drought dragon boat" story has evolved boats flower [8], shown in Fig.5c.

5a. Waving hands Dance 5b. Tai-Tai flower 5c. Boats flower

Fig.5 National story pattern (Photos taken by the author)

2.2 Color Innovation Design and Application of Tujia brocade pattern

The design breaks through the Tujia brocade's strong visual sense of color contrast and conforms to the fashion trend of 2016-2017 spring and summer color. It displays soft warm color tone giving a sense of comfort and harmony. According to the above description in this section, from the cultural connotation brocade pattern, the design divided into classic hook profile series, folk auspicious series and the series of stories.

2.2.1 The classic hook profile series

Based on the concept tof timelessness, the series selects the classic coat collection and classic hook pattern of fashion style. Pink-orange (Fig.2) is the main tone as it gives off a gentle and harmonious feel which drives away the winter spirit.

Coat Design 1: Forty-eight hook were selected as design elements with common Tujia brocade patterns. 2016-2017 SP printing fashion has swept the world, large lapel costumes with printing performance practices, the style is fashionable, as Fig.6; Coat Design 2: The Wan-words Symbols were used as a decorative pattern, combined with continuous Wan-word quilted hook profile. Fashion design elements are created by bounding points and lines of form, coupled with the rich cultural temperament on the collar of the suitshown in Fig.6b; Coat Design 3: Serpentine pattern continuously arranged in a stripe shape, with a broken line hook pattern, giving a staggered visual appeal. There areno collar cardigan design and rustic ash tones. The atmosphere is quite literary style, as shown in Fig.6c.

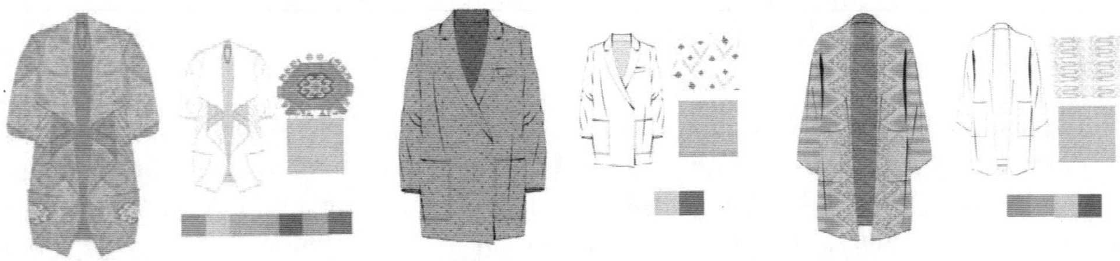

6a. 6b. 6c.

Fig.6 Classic hook theme profile design (Designed by the author)

2.2.2 Auspicious series

Design ideas: In Chinese tradition, the color Red is an auspicious symbol, and for traditional Tujia brocade, behind the patterns displayed on the clothes are usually a red background, as shown in above picture nine plum flower pattern and a male bird flower. The pink and purple are higher are usually brighter which, represents a soft, soothing and sweet style. to The T-shirt is designed with a romantic feel to reflect on the passion of summer giving people good wishes.

T-shirt design 1: combines the nine plum and divided Wan-words hook profile as a small unit which are continuously arranged in the clothing tiled., The loose feature expresses lightness and elegance as shown in Fig.7a; T-shirt Design 2: the lotus profile pattern appears in the form of four consecutive (Which is a organizational method of Patterning) with ice blue lotus pattern which are extremely eye-catching. This combined with the shirring collar, creates freshness and vitality, as shown in Fig.7b; T-shirt Design 3: Pink and purple colors combined with sling style design expresses a sweet girl style. Blue tones Yang bird forms an irregular pattern arrangement which gives off a playful and cute look. , shown in Fig.7c .

7a. 7b. 7c.

Fig.7 Auspicious theme design (Designed by the author)

2.2.3 Folklore Series

Design Ideas: folklore legend is only known from generation to generation. Casual style is seen mostly in everyday clothing. The Tujia brocade pattern with ethnic story applies to everyday clothing design, combined with a friendly and stable olive color design wild casual clothing style, will play a promotional role to brocade Tujia culture .

Short-sleeved design with a single abutment floral pattern with a simple gray green short-sleeved, brings a cool sense of lightness in summer ,as shown in Fig 8a; Vest design: there are distinct cut-off rule and crisp model design, and if it is added waving dance patterns in certain parts it looks handsome and naughty as well, as shown in picture 8b. Trousers design: in style, loose harem pants style is chosen; in patterns, geometrical boat flower is used; in colors, the popular olive color is selected, which make the garment look simple and unique as well, as shown in picture 8c.

8a. 8b. 8c.

Fig.8 Folklore theme design (Designed by the author)

3. Conclusion

The paper based on 2016-2017 spring and summer fashion color analysis, breaks the consistent color matching style of Tujia brocade pattern, using unique design ideas, combining fashion color with Tujia brocade. It redefines the look and providesa reference for the development of new Tujia brocade. In addition, Pattern is an indispensable language to express the art of clothing, and it plays an important role in fashion design. Tujia brocade has rich patterns and various colors. However,the development of Tujia clothing is at a bottleneck because of the condition of production engineering and heavy minority style. So the attempt of innovative design is combining its design concept with the modernization, internationalization of costume design concept. This will not only enrich the cultural connotation of the modern fashion design but also perserve the old Tujuia brocade culture. In addition to clothing design, we can expand the application areas of Tujia brocade to home furnishing, decoration, advertising, packaging, labeling and so on. We hope that the Tujia brocade can be more and more popularized through continuous innovation and change.

4. Acknowledgement

This work was supported by the Social science fund of Shaanxi Province (11K118), which is the Research on the m ining and innovative design arts of Han and Tang Dynasties weavingembroidering arts.

References

[1] Li K. Xilankapu color study in Modern Advertisement Design. HBUT: 2011; 15-19.
[2] Robert L. Anthropology of Art . Literature and Art Publishing House: 1990: 51.
[3] Xin YH, Luo B. Aesthetic Characteristics of Tujia Brocade. CCNU: 2001; (3): 74.
[4] Xiang KY. On the Xiangxi Tujia brocade Characteristics and Inheritance. HUNNU: 2008: 6.
[5] Chen H. Xiangxi Xilankapu traditional patterns and research in Modern Garment Design. GXAU: 2014; 7-9: 11.
[6] Liu L. Xiangxi Tujia brocade pattern aesthetic characteristics and its application. HNUST: 2013: 17.
[7] Wang ZM. Xiangxi Tujia brocade pattern art research. HUNNU: 2007: 4.
[8] Ran HF. Study on the change of Tujia brocade culture[D]. SCUEC: 2007: 36.

Study on Inheritance of Tujia Ethnic Brocade based on Digitization Design

Jia Wang, Gao-Yun Pan, Peng Zhou, Yan Yuan*, Zhi-Dong Li

Jin Hua Road, Xi'an, Shaanxi, 710048, China

Corresponding author' email: 424787619@qq.com

Abstract

Tujia Ethinic Brocade is a historical precipitation for Tujia people, especially for the Tujia women who have unique aesthetic sentiments and pursuits. With the rapid development of contemporary society, it is difficult and challenging for the Tujia minority to protect and pass on their Brocades. In view of the seriousness of this phenomenon, we must adopt effective measures to defend them. It is essential to explore innovative ways by which to collect data to study Tujia Brocades. By exploring digitization design of Tujia Ethinic Brocades, we try to find effective ways to establish the basis for reusing and redesigning this handcraft in the future.

Keywords: Tujia ethnic brocades; Inheritance; Digitization design

1. The origin and overview of Tujia Brocade

Tujia Brocade is one of the traditional handcrafts of the Tujia, the 6th largest ethnic minority in People's Republic of China. It is one of the Four Excellent Brocades of ethnic minorities in China, which consist of Tujia Brocade, Zhuang Brocade, Li Brocade, and Dai Brocade. Traditional Tujia Brocade is a kind of hand-knitted fabric, using organic materials such as linen, cotton, and silk. Its artistic prestige is not only due to its aesthetic value, but also based on its exquisite and unique craft. Tujia brocade, called "Xi Lan Ka Pu" in Tujia language, is known as "Tu flower cloth" or "bedding flower" in Chinese. "Xi Lan" means bedding, and "Ka Pu" is flower in the Tujia language. There is a saying in the folk custom that "Xi lan" is the name of a girl, and "Ka Pu" means the flower on the quilt. Tujia brocades are mainly produced at the Maoer shoal, Laoche, Dianfang and slope toe of Xiche River Basin in Longshan County in western part of Hu Nan province, in Duishan and Fengqi of Yongshun County, and in such valley and river belts in high mountains in Purong and Qingshui in Baojing County. The traditional application customs of brocade are maintained in these areas. Brocade is named "counting yarn flower" along Yongshun, that is, during the cross-stitch work, it is woven according to the longitude and latitude of linen finishing, and the black linen finish is often seen on the brocade surface while line-adorning seems like peach blossom [1].

In folk, Tujia brocade is mainly used to make quilts, suck as the quilt for children, the foot quilt, andthe skirt It is alsoused as the shawl for a girl's dowry, as a pledge of love, and even for waving dance in the folk activity. It was also used as a wall hanging in the front of the hall for worship. Generally, Tujia girls begin to learn to weave from their mothers from 9 to 10 years of age, with pattern design and color turning from simple to complex. After they come of age, they are made to memorize the common dermatoglyphic patterns, leaving no need for models. In stockaded villages along Xiche River of Mountain Long, nearly every household has a loom. It is a common phenomenon that a family may has several looms, the number of which [1] Tujia brocade tutorial. Hainan publishing housecorresponds to the number of girls in the family. A girl's intelligence is determined by the number of dermatoglyphic patterns she can weave, flat woven articles, and skillful color matching. Flowery quilts for dowry are woven and sewed by the girl herself. Some girls may weave up to 20 quilts with different design and color. Well-woven dowries are never transferred, and start to be prepared and accumulated several years before a girl gets married. Articles for children before and after the marriage are imbued with the Tujia women's pure wishes for their children, and are meant to show their maternal love.

Up till now, it originates from people's most basic production and life, with very rich and colorful ways of expression. The main characteristic is metaphor that feelings are conveyed through things. That is, phenomena related to sexual love and reproduction during the process of growth of every thing in the world, such as animals and plants in pairs. or which breed multiple descendents, are used to imply love which is reflected in brocade line-adorning.

Fig.1 Tujia Brocades(Photogtaphied by author)

The name of "Tujia Brocade" itself reflects a beautiful story amongst the Tujia people. The story goes that Western Ran, abeautiful and clever Tujia girl from ancient times, weaves all the flowers in the world except for the maidenhair flower, "which blossoms during the Yin period and fades during the Mao period". Western Ran stays alone under the maidenhair tree thoughout the night, unknowing that her jealous sister-in-law is slandering her name and questioning her character. In a night lighted by the moon, the maidenhair suddenly blossoms. Western Ran picks a flower, whispering to it. At this time, her older brother comes back home heavily drunk, and becomes furious with her, believing in the slander perpetuated by her sister-in-law. He takes a cloth beating bar,and beats Western Ran to death under the maidenhair tree. After awaking from a drunken sleep, he sees that Western Ran has become a bird, crying, "My brother feels pain at mistakenly killing me," the older brother feels terrible grief. He wears Western Ran's brocade every day, and finally, the bird becomes placated and leaves. Every year, after the spring rain of Qingming, the bird flies back to the village, urging the Tujia people to plough for spring sowing [2]. This bird is called Yang bird in folk, which is depicted with beautiful posture and unique shape in Tujia brocade of "Yang bird flower". In memory of Western Ran, people hand weave brocades inspired by Western Ran, and hand them down from generation to generation as the quilt cover. Western Ran has become the weaving goodness in The hearts of the Tujia people's.

At present, Tujia brocade has hundreds of patterns of different appellations, and quite a part of appellations are transliterated. These patterns can be classified into six categories: animal patterns, plant patterns, daily utensil patterns, abstract geometric patterns, character patterns, and astronomical phenomenon patterns. Animal patterns include Mabi (pony and other small animals), Yangque, swallow, big dragon, small dragon, snake, cat footprints, phoenix stripes, butterfly and peonies, marriage of daughter of mouse, etc. Plant patterns include lotus, leek flower, nine plum flowers, power on the dyke, trumpet flower, flower on the cane, pumpkin handle,and peonies. Daily utensil patterns includechaisr, desks, a box with eight angles, tofu bracket, baba bracket, buckets, ships, saws, shuttles, and the back-cage. Abstract geometric patterns include the single-side eight hook, double-side eight hook, 48 hook, 12 hook, 24 hook and [2] Tujia brocade tutorial. Hainan publishing house Other variations of the hook. Lucky character patterns include the Chinese characters of Fulushouxi, Wangziliushui, Yipindangchao, Tian, and Shou. Astronomical phenomena patterns include the sun, the full moon, the half moon, shuttle, and a sky full of stars. Clustered flowers, broken branches, and lucky patterns are obviously influenced by Han Culture of Ming and Qing Dynasties and peach flower patterns, but most of these patterns are recombined and shaped by the Tujia people. For example, peonies and butterfly and peonies are different from the image of the peony in the embroidery and carving of Ming and Qing Dynasties. "Marriage of daughter of mouse" is a new pattern derived from the peach flower pattern, which shows the inclusion, absorption, and creation of foreign cultures.

 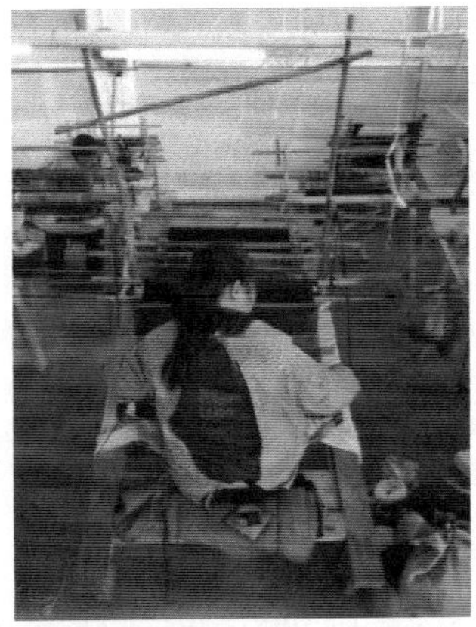

Fig.2 Tujia Brocades handcraft (Photogtaphied by author)

2. The analysis of the development factors of Tujia Brocades

Since the reform and opening policy, the roads in the mountains have been in use, and is open to the public, leading to the Tujia ethnic groups's economy to gradually improve. However, a large number of families have abandoned this DIY clothing tradition, instead opting to go shopping in clothing outlets. More and more people go out to work and have no time to make Brocades at home, except only a few old women and some local people who still have a strong interest in Tujia Brocades. Due to their rarity in modern times, Tujia Brocades are particularly valuable, attracting a lot of scholars and researchers to study the art. With the acceleration of the process of the protection of intangible cultural heritage, Tujia Brocades that were hiding in the lonely mountain have been unearthed, attracting much attention.

3. The problems to be faced in the process of inheritance

The rapid development of society brings various kinds of goods and materials to the common man. This will lead to change people's esthetics and criteria, which influences our folk culture and modern civilization. Tujia Brocades are ancient handicraft products that are a family inheritance, and can meet people's demands of life and their own esthetics even in modern times. However, this phenomenon threatens the inheritance and development of our Tujia Brocades misleading them easily in the wrong way, which will destroy traditional Brocades someday.

Old Tujia artists who made traditional Brocades died, leaving only a few people to inherit this type of art. What's more, the technology of the Tujia minority makes producing the brocades very complex and time consuming. And the high prices and bad markets are also the key factors in causing the Tujia Brocade to lose favour:

1) The market impacts the inheritance of Tujia Brocades patterns causing the loss of valuable crafts.

2) The Tujia Brocade market is sluggish, and few people work for it, so it is hard for the apprentice and trainers to concentrate on inheriting the Tujia Brocades. In addition to this, the process of making Tujia brocades is extremely complex, and the cost of technical requirements is high, thus not allowing for a high revenue stream.

3) The materials of Brocades are thick, told noting its use to a specific product. Furthermore, the patterns used are too traditional, making it difficult to adapt for different products.

4. The inheritance of Tujia Brocades

With the rapid development of society, Tujia Brocades are facing near extinction. Some traditional brocades have already been destroyed by human beings, which ras resulted in the loss of a large number of Brocades. The older artists who made traditional Brocades have passed on, so that this kind of craftwork, and the only way to inherit the traditional culture, is soon to be lost forever. Therefore, we should work out relevant methods to rescue Tujia Brocades - a special traditional folk art.

4.1 To Collect important information about Tujia Brocade and protect from the original way

Many traditional brocades have been destroyed, threatening the inheritance and development of the Tujia Brocades. Ifwe want to protect and improve the handcraft technology of Tujia Brocades, the first thing we must do is to collect all the aspects of information and handcraft technology about Tujia Brocades canst as to establish a database museum to broadcast this handcraft. This will be a brand new endeavor, allowing more of those who ar interested to inherit this traditional art.

4.2 To develop the all kinds of design by utilizing Tujia Brocades from defferent lever to meet the people's increasing aesthetic demands

Up until now, there has already been a lot of different forms of production developed from the Tujia Brocade in the market. However, some of them are out of fashion and look very old; while some of them are the best quality of Tujia Brocades, but are sold at a very cheap price for young people, who cannot find a better product, leaving them to use poorly redesigned products stemming from this traditional craft. Considering the modern market, it is worth considering how to integrate folk art handcraft with modern design concepts.

4.2.1 The vectorization of Tujia Brocades pattern

Tujia Brocade patterns are closely related to people's daily lives. The pattern communicates the Tujia ancestors' understanding of nature, and contains a wealth of national and cultural information.

The patterns here are explore the vectorization of Tujia Brocade patterns.

1) Ma Bi Pattern

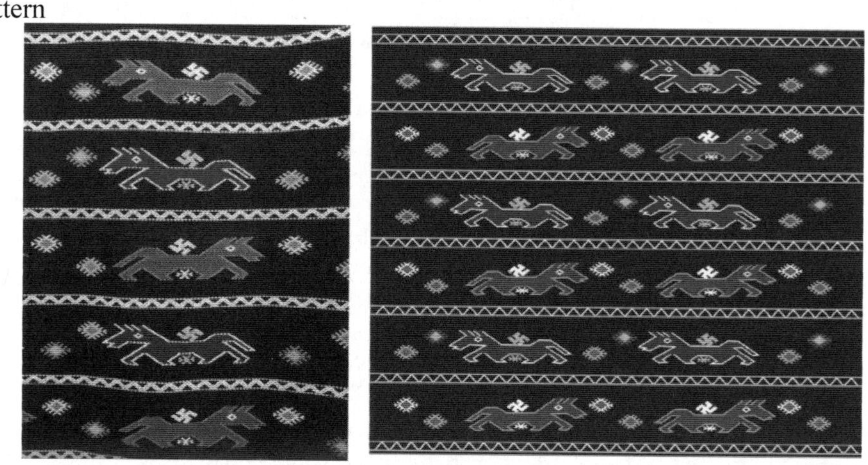

Fig.3 Ma Bi Pattern (Photogtaphied by author) and vectorization design by coreldraw

The Ma Bi Pattern is typical of natural image. "Bi" means small, and "Ma Bi" means the 'little horse' in the Tujia language. The design of the "Ma Bihua"pattern is modeled after the running horse. The horse symbolizes the majesty in Tujia traditional culture.

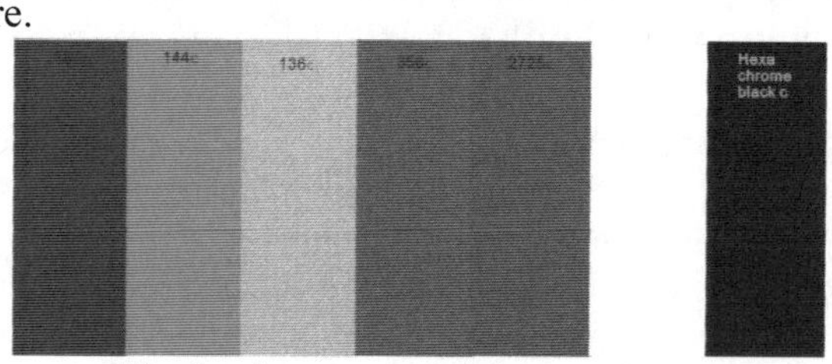

Fig.4 color extraction by coreldraw

Based on the panton color card, color piece for the color label. On the bottom of the black with the adornment style of rainbow colors.Due to the organic use of black and light colors thought intense color contrast of changes in temperature, so, image characteristics is more outstanding.

2) Male Bird Pattern

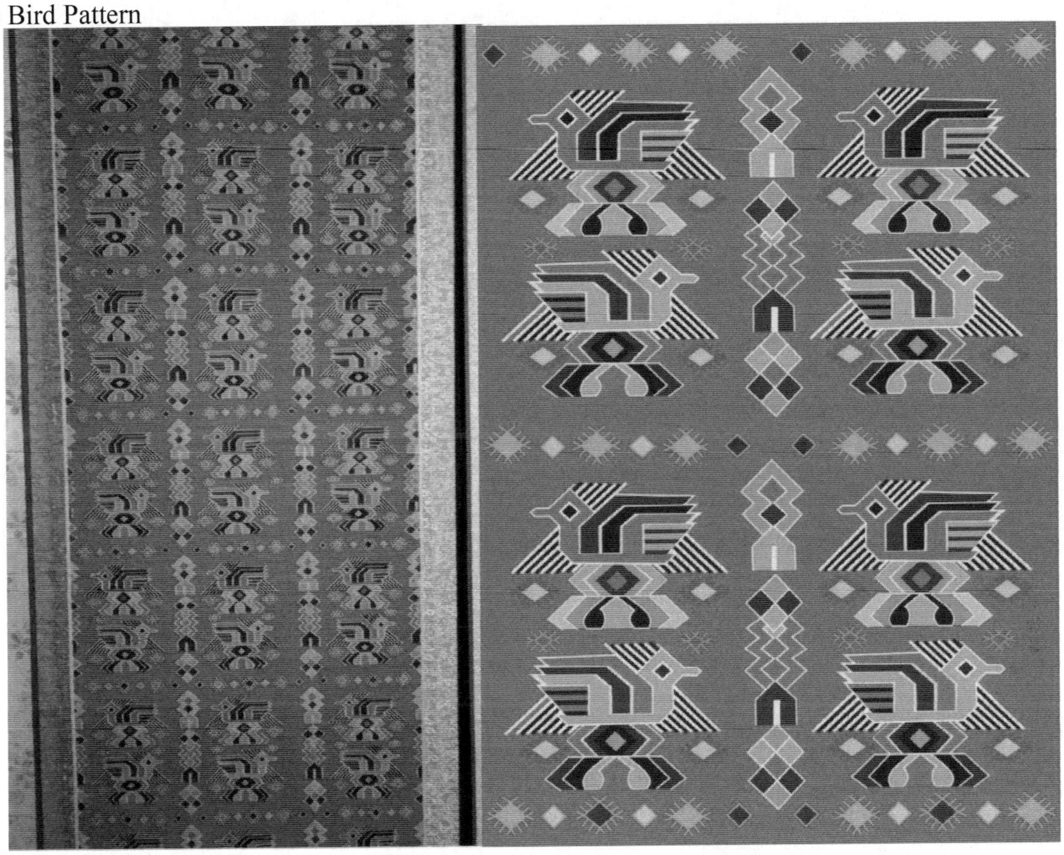

Fig.5 Male Bird Pattern (Photogtaphied by author) and vectorization design by coreldraw

The Yang bird flower patterns use the traditional "one oblique flower" production process, the use of both ends of the lace "on the ramp flower" production, lace a "crab legs" as a two-party continuous pattern. "Crab" is the protection of God Tujia night. "Yang bird flower" is the whole shape of the middle of the division andby a combination of geometry, the entire image has both lively and features depicting the image of birds. The origin of the Male Bird Pattern of Tujia brocade has its own legends. On the one hand it is said that the origin and the ancients observe the change of climate by the seasonal migratory birds , and told the people for farming. On the other hand said that the legendary Xi Lan turned into a bird when he died, and year after year heralds spring for the Tujia people. The Tujia people not only called the brocade "Xi Lan Ka Pu", but also make the male bird into the Tujia brocade in order to better appreciate him.

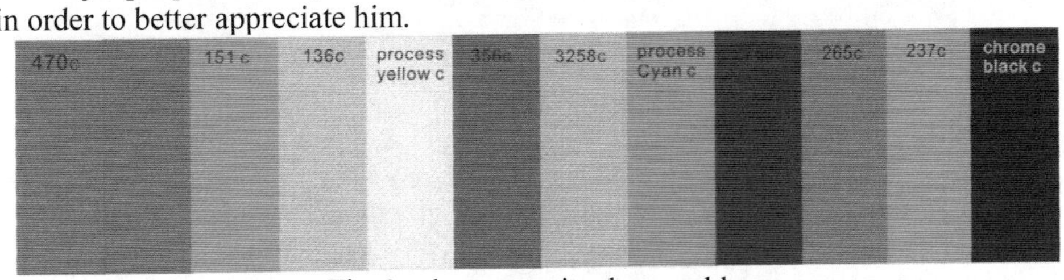

Fig.6 color extraction by coreldraw

Based on panton color card, color piece for the color laber. This design mainly gives priority to warm colors. It is an expression of the meaning of auspiciousness. This kind of collocation method which uses warm and cool colors can strengthen the administrative levels of design, giving the wearer a warm and bright look.

3) Eight hook pattern

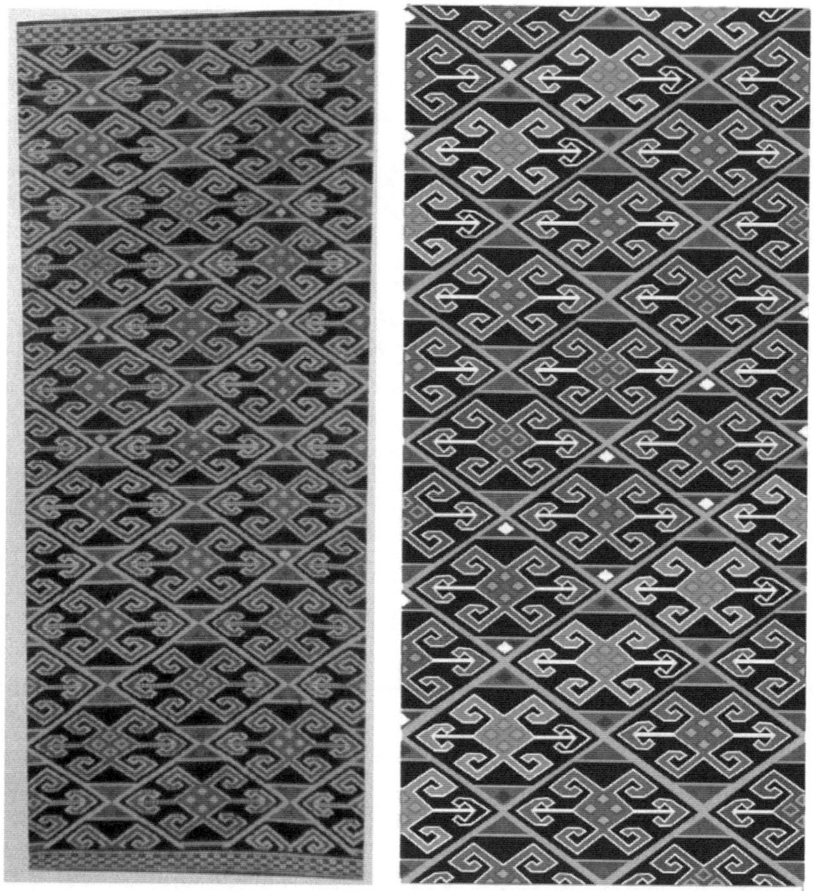

Fig.7 Eight Hook Pattern (Photogtaphied by author) and vectorization design by coreldraw

The Eight hook pattern is an abstract pattern. This pattern is from the four corners or edge of the central body extending eight hook pattern. Hook pattern reflects a typical pattern of fire, It is like the leaping flames. The design seemed to originate in the spring mountains.

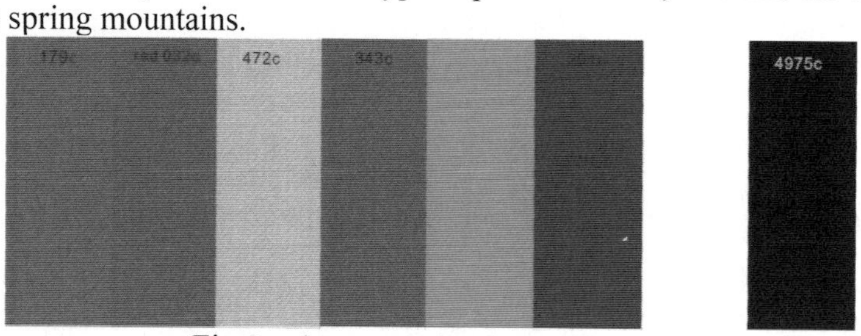

Fig.8 color extraction by coreldraw

Based on panton color card, color piece for the color laber. This design mainly gives priority to warm colors, with a small area dedicated to cool colors. The picture contrast brings out the strong adornment effect.

4.2.2 Try to develope various forms of the production by redesigning and reuseing this vectorization design from Tujia Brocades

This vectorization design from Tujia Brocades has some advantages:

1) To change colors more easily and more flexibly.
2) To be convenient to be saved.

3) To be reused and redesigned into different forms of production in the printing and dyeing process.

Therefore, we can explore various forms of the production on the basis of redesigning and reusing it. And at same time, develop new production methods to meet the peoples' increasing aesthetic demands.

5. Conclusion

Facing the threat of losing the Tujia Brocades art and technology, we should explore some effective ways by which to protect, broadcast, and inherit this traditional craft. What we could do is concluded in several ways in this paper. First, we should collect as much information as possible to establish a database for the Tujia Brocades. Secondly, we could analyze and make vectorization designs of the patterns. Lastly, we can reuse and redesign this vectorization design into more fashionable ways to meet the people's increasing aesthetic demands. Above all, this exploration will be developed further with more and more people realizing the importance of protecting the traditional handcraft.

References

[1] Zuo H. Folk Brocades [M]. Hunan Fine Art Publish: 1994.

[2] Qiao YC, Ye HG. The revelation of the integrated marketing of broadcasting the Tujia Brocades's travel production [J]: 2013 (04).

[3] Yang B, Xiao G. Tujia Brocades of the Western Hunan. Fine Art [J]: 2012 (4).

[4] Tan ZM, Huo XL. The study on the prctection and inheritance of Tujia Brocads from the culture space pointview. Journal of Western Nationality University: 2013 (10).

[5] Ran HF. The study on the inheritance and innovation of the Tujia Brocases. The Journal of Yangtze Normal University [J]: 2008 (01).

[6] Luo Y. The study onthe inheritance of the Tujia Brocades technology. Central Univeristy for Nationality: 2013 (05).

[7] Wang ZM. The study on the art of Tujia Brocades of the Western Hunan. Hunan Normal Univeristy: 2007 (10).

[8] Hu WJ. Research on the Art Symbol of Tujia Brocade. Hunan university of technology: 2008.

[9] Li YF. The current research of tujia brocade productive research. Minzu university of China: 2012 (5)

[10] Shao CB. Tujia brocade contemporary development research. South-central university for nationalities: 2013 (5)

[11] Ran HF. Tujia brocade cultural change research. The Journal of Yangtze Normal University: 2007 (5)

Study of the Model and Structure of Collar Edge

Shi-Yi Li[1], Ming Zhao[2*]

[1]*Li Shiyi, Yinghua east street No.2, Chaoyang district, Beijing, 100029, China*
[2]*ZhaoMing, Yinghua east street No.2, Chaoyang district, Beijing, 100029, China*

[*]Corresponding author's email: mingzhao21@163.com

Abstract

In traditional Chinese costume, collar has become an important structure in clothing since the appearance of ancient Chinese robes, and it not only remains the position of human visional focus, but also undertakes the role of decoration. With a combination of the conclusion of the trimming collar decoration's evolution in history and the fixing role the trimming collar decoration plays in the various clothing patterns from the surplice to the garment, this research focuses on the arts and cultural implications of collars' function and decoration by means of illustrations and images describing the connectivity between the fixing way of the traditional collars and the arts so as to provide the references to the use of modern clothing structure.

Keywords: Clothing structure; Collar edge; Functionality; Cloting culture; Thoughts of creation

1. Introduction

The notion, "Clothing for embroidery, brocade for edge" emerged in the Pre-Qin period when the custom systems had first established its roots. And the thick and rigid brocade was used for making the collar edges. In the depiction of the unearthed clothe quills in the Chu Dynasty Tomb of Jiangling by Shen Chongwen: "In terms of the edge decorations, brocade was used for collar, sleeves and clothe edges, which corresponded to the system of Clothing for embroidery, brocade for edge. The heaviness and thickness he brocade edges formed the framework of the clothing, which benefits the stability and fitness of the custom formats" [1]. Therefore, the edge of collar is made out of thick and wide brocade originally due to its good fixing function. The decoration of the collar is constantly changed as times pass by.

It can be learned from historical records that as early as the periods of Yellow Emperor and Emperor Yao and Emperor Shun, dresses had appeared, and as one of the modes of dealing with the dresses-collar edge also emerged, and along with people's living and the development of times, they have seen relevant changes. In the classic work of Interpretation of names-Interpretation of dresses, the author described a top band as: "The women used 'Jiang (the color of red)' for dress-making, with the upper and lower parts connected and edges were made on all surroundings, and they were also known as robes. Through such a glimpse, it can be learned that by then the making of edges on the dresses; edges had been quite wide-spread. By Tang Dynasty, it became popular; originating from Northern Zhou Dynasty, the round-collar robes that fused the elements of the robe dresses of the nationalities of the barbarian tribes began to run parallel with the cross-collar robe dresses (cross-collar mean the left and right foreparts of the clothing are overlapped before the breast, neckline usually have wide edge as a collar.), and there had been fundamental distinctions in the patterns of their collar edge. By the Qing Dynasty, the elements of the ethnic minorities were again fused, and the "厂" character unlined upper garments had replaced the round-collar robe dresses, and by this period, the techniques, colors and decoration extents of the edge decorations have reached the extreme of this period.

This Paper will focus on the top band and their patterns, and it starts from the Shenyi (ancient Chinese robe) in the form of crossed collars in the Han Dynasty, to the round-collared gowns, as well as the "厂(like Chinese word)" character unlined upper garment of the Qing Dynasty. Taking into consideration of the abovementioned historical backgrounds and cultural connotations over different times, the research is carried out in the fixing function of clothing necklines from various patterns and ways of the edge collar decoration.

2. Fixation Modes of the Collar Edges of Fabric Edge

2.1 The formats of Shenyi

The format of upper dress and lower lap was a dressing format since ancient times, and even the upper dress and the lower laps are joined, there is a need to apply a horizontal band to distinguish the up and lower parts [2]. Shenyi was just one of the robe dresses that had joined upper and lower parts, and the dress and the lower laps were cut separately and they were sewed at the waist in pursuit of an effect of the integrity, and the upper dress and the lower lap correspond to the harmony of yin and yang. A dress is a medium dignifying the levels of status, and naturally the edge ornaments became one of the symbolic elements, and the edge ornaments were often described as "pure" in the classical literature. Zheng Xuan, a Confucian of the Eastern Han Dynasty said: "The colors of the dress edges of people of different classes also differ". The colors and materials of purity featured explicit dressing regulations to dignify the status [4].

2.2 Fixations of the Parallel Edges-Pure Gauze Unlined Dress in the Western Han Dynasty（about 2nd century B.C.）

The Pure Gauze Unlined Dress (Fig.1) unearthed in Tomb 1 at Mawangdui [12]. It had crossed collar, composed of the upper and lower pats, right overlapping, no inner lining, and it was covered on the outwear or served as an innerwear, and the luster of the collar, sleeves and lappet collar edge formed the edge decorations of the edges, which were also the parts that weigh most in this dress that is renowned for its light weight, then why had luster been used for the making of the collar edge for such a dress that had "almost weightless light gauze"?

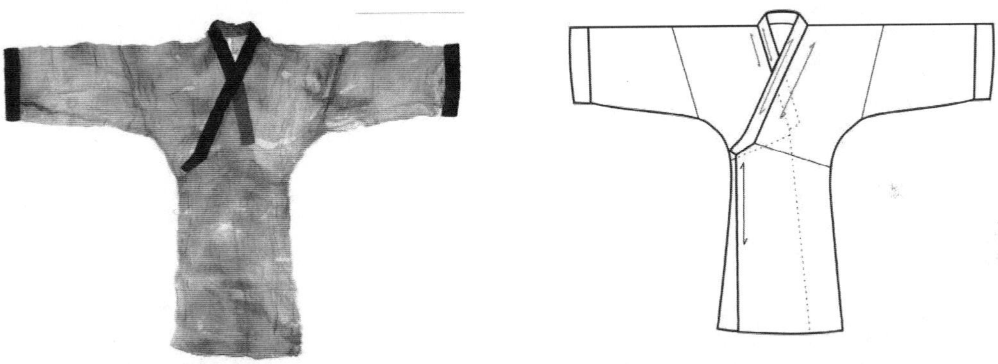

Fig.1 Pure Gauze Unlined gown of the actual object and style pattern

During the Han Dynasty with highly distinctive ritual systems, this conformed to the costume system of "Clothing for embroidery, brocade for edge". Juan (hick but loosely woven silk) was a fabric of the raw silk, easy to take off yarn, which was thick and sparsely woven. During the Han Dynasty, the gauze clothing was not only a supreme manifestation of the social status, and it also carried the splendid craftsmanship of the dress-makers for the service of the rulers. Seen from the Fig.1, the gauze that were surrounded by the collar edges were in the straight gauze direction but are not vertical to the ground; in other words, they had the same gauze direction as the Juan that covered the edges, and at this time, as the collar edge, the Juan played a role of wrapping the rough edges and fixing the cloth edges, and it also featured further consolidation for the stability at the neckline of the dresses.

2.3 Fixations of the clothe slanting silk - the ochen-colored letter embroidered silk robe of the Western Han Dynasty

Similarly, another ochen-colored letter embroidered silk robe of the Western Han Dynasty [11] (Fig.2) unearthed in Tomb 1 at Mawangdui, Changsha City featured curved garment front, cross collars, and right overlapping parts, and the white pure silk of the wide collar edge was spread to the back from the neckband as a front and then return to the lower lap, and this conformed to the recording of "extending the overlapping and hooking the edges" in "Li Ji, Shen Yi", and the collar edge take a big area in the entire dress, and what distinguished it from the straight front of Pure Gauze Unlined Dress was that such a silk robe could be used as a formal and make people to the etiquette of "walking without the showing of the feet". The transformation of the roles of the collar edge from collars to the extensions of the front and then to the lower lap accomplished the visual unity, and it also became a style that is unique to the costumes of the Han nationality.

In this brocade gown, the wide collar edge occupy large areas, and starting from the amount of extension at the collar, it enabled the wearer to retain his due posture. In terms of the fixation forms, it also differed from the Pure Gauze Unlined Dress, and seen from the slanted patterns of the fabric of the brocade gown. The patterns from the neck side to the lower lap were basically vertical to the ground (with no consideration given to the fact that when the predecessors were extending the overlapping and hooking the edges), some slanting were generated by stretching the cloth edge to the waist in order to increase the coverage amount of between the closures, while the directions of the cloth of the silk clothing generates an included angle with the cloth of the silk clothing that were vertical to the ground, and the fabric for the dress bodies at the sewed parts of these two must be slanted cloth and can easily be stretched, and meanwhile, the wide edges that served as the high collars at the slanted positions of the cross collars also played the role of fixing the slanted cloth of the dress body, and such garment examples were numerous in the Han Dynasty.

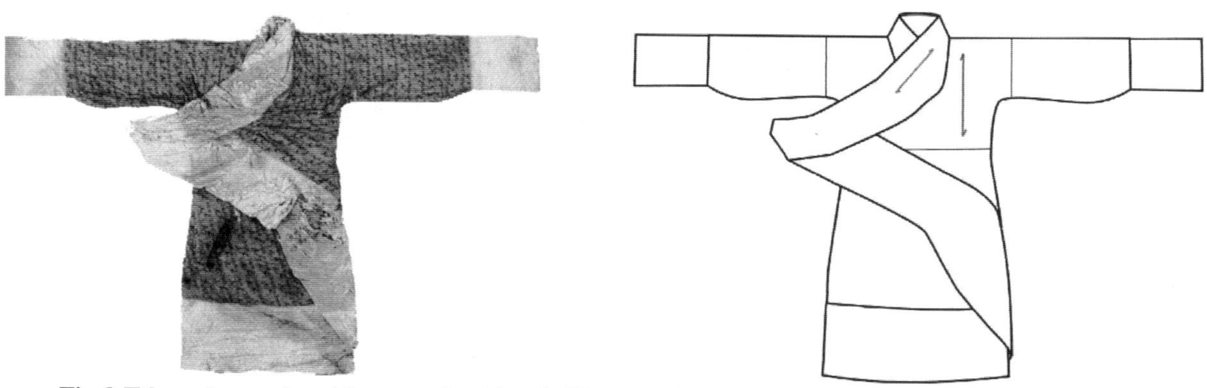

Fig.2 Tthe ochen-colored letter embroidered silk robe of the Western Han Dynasty and style pattern

The fixation modes of the collar edges on the cloth of silk clothing have also been constantly changing amid the alterations of the dynasties, and starting from the time when the pure-colored "brocade as edges", they played the role of the power of the ruling class. In the Tang Dynasty (618-907 A.C.) [3], only high-ranking official could use the black wide collar edges (Fig.3), and use them to manifest their steady bearing; the cross collars of the Song Dynasty were coupled with black or red color edges, which were mainly aimed to manifest the rigorousness and dignities of the royal family members and the officials. On the red collar edge of the seated portrait of Empress of Emperor Gaozong of Song Dynasty (about 1110 A.C.) was also embroidered with the three-jaw dragon design (Fig.4), manifesting the supreme hierarchy of the empress. With the progress of people's decoration techniques, decoration detail more and more,the space on the surfaces of the collar edges became less and less, and became a location for the manifestation of the refined craftsmanship, and they were constantly filled with such decoration modes as embroidery, gold patching and golden threading, as shown in Fig.5, and the matches with the fabric colors, models and patterns were constantly strengthened.Collar edges with the fixation roles becoming increasingly important along with the rich and expensive ornaments [5].

Fig.3 Officer of the Tang Dynasty

Fig.4 Empress of Emperor Gaozong of Song Dynasty

Fig.5 The Qing Dynasty Mandarin jacket of Bright yellow silk [10]

3. Round-collar robes and the unlined upper garments the collar edge-fixations on fabric patches

3.1 Formats of the Round-collar Robes

As the most prosperous dynasty in Chinese history, the Tang Dynasty (about 700 A.C.)manifested its tolerance to either trading and commerce or the costumes, and the diversities of the garment models and the merging and fusions of the decorative methods all represent a free era. Although the round-collar robes were only an insignificant portion of the multiple dresses in the Tang Dynasty, they became an important model that has impacted the sequence of thoughts and development of the garments in later generations. In the Sui Dynasty, such dresses, which originated from the Zhou Dynasty and feature round collars, medium splitting and have edge decorations in the collars, sleeves and the lower parts had become a common dress mode.

In the Tang Dynasty Princess Tomb of Yongtai's murals, two dress models were rather typical [3]: one kind was the robe dress that fused the style of the nationalities of the barbarian tribes but retained the opening in Fig.6. and closure of the cross-collar, as shown in the maidservant in Fig.7. Another kind was a round-collar robes that were extended into the Song Dynasty, as shown in the waitress in Fig.8. Such two kinds of robe dresses had been detached from the Shenyi with crossed collars and the extension of the edging into the lower lap, and only one ring of edges was applied in the neck ring [10]. there were no obvious collar edge below the collar edges, and the treatment method was reversed to the addition of an inner lining. Regarding whether the opening of the flap of the latter's round-color robes was the turndown collar was not known. However, the edging at the neckbands of the round-collar robes and the buttoning on the part of the right neck should have been a very well-fitting position around the neck. And this is also a point of positioning after the entire dress is worn, and the neckband is bound to be dug downward, and after unbuttoning, the collar edge with the flipped collar surface would not be a straight line, as the opened front collar edge in Fig.7. No matter whether the collar surface is flipped and folded, these two dresses had seams in the middle, and the straight or curved wide edges were provided on the necks of these two models of dresses, thus becoming the secondary fixation after the sewing of the front and middle cloth patches.

Fig.6 Maid of turndown collar Fig.7 Maid of close collar Fig.8 Ming Dynasty robe of round collar

3.2 Influences of the Cutting Modes

Prior to modeling, a dress may be subjected to the influences by a multitude of factors, and the width of the cloth has determined the cutting modes since ancient times. During the long course of Chinese history, the nation has long had self-sufficiency, with men engaged in farming and women in weaving; usually, with families as units, the women were responsible for cloth-weaving and dress-making, and the primitive loom involved the waist loom with the weaver seated on the ground and the operation could be done by just one person and the fabric width was usually only about 30cm; later, it was developed into a loom that needed to be operated by two or more people in cooperation, with a fabric width of about 50cm; During the periods of Ming and Qing Dynasty(about 1640A.C.), in the region south of the Yangtze river, imperial textile bureaus were created for the royal families and the uses by the Imperial Courts, thus forming an entire weaving system, and naturally, the fabric width was widened, but still not as wide as those of modern times.

Under the cloth with preset widths, two modes of material cutting were employed on this Red color Kesi (a type of weaving done by the tapestry method in fine silks and gold thread) brocade colored drawing on pottery eight group s of plum flowers, orchids, bamboos and chrysanthemum collar fastening up the front of imperial or ceremonial robes gown of Daoguang period of the Qing Dynasty(about 1820A.C.) [9]. The first kind featured no medium seam, as shown in Fig.9, and in this way, under the limited fabric width, the whole body could be cut; in case of the big front, there will be seam elements. And the edge ornaments would be used to supplement them, and the first cloth edge strengthening mode was conducted using such decoration techniques as the inlaying of the bindings, and it was necessary to supplement the small fronts and the sleeves otherwise; the second kind was cutting with a seam-breaking in the front, middle and rear, as shown in Fig.10, and the sleeves can be cut at the limited fabric width, and even there is no need for joined sleeves (the dress was a model with deliberate sleeve joining, and it belonged to a fashionable model of that time), and on the blank space in the right, such trivial component as the large front can be arranged, and in comparison to the first cutting mode, it makes more sufficient use of the materials, and this also complies with the creation view of “valuing substances a gold” and “Adroitly guiding action according to circumstances” [6].

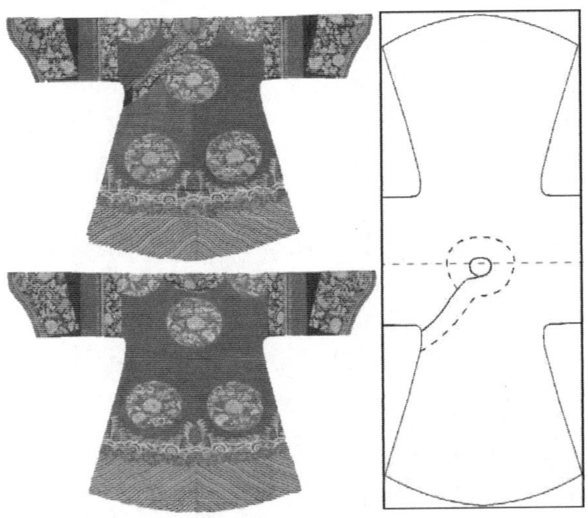

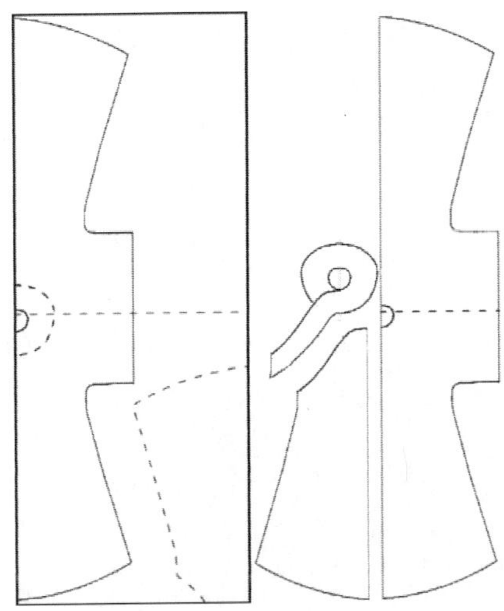

Fig.9 The Kesi Clip robe and one of pattern cutting illustration Fig.10 The other one of pattern cutting illustration

The fixation mode for the collar edge of this Kesi silk brocade collar fastening up the front of imperial or ceremonial robe gown and the aforementioned two models of the robes of the Tang Dynasty was the same; firstly, it played the role of a collar and it even fixes the easily deformed cloth edges in the joined body and even the tight collar ring; secondly, it conducts another fixation and shaping for the front, rear and middle broken seams with the covering areas of the wide edges at the collar. After top band were added into the ribbons of the Qing Dynasty, their parallelism with the wide edge of the collar sides also played the same roles, and the ornaments and ribbons on the wide edges of the collar edge also became one of the most gorgeous ornament methods of the Qing Dynasty.

4. Unlined Upper Garments the Collar Edge-Fixations on the Temporary Structures

4.1 The Emergence of Stand Collars

In Chinese costumes, the elements of stand collars emerged in mid-Ming Dynasty (about 1550-1600 A.C.), and they became quite popular during the late Ming Dynasty (about 1630 A.C.). The emergence and popularity were also associated with the freezing weather which was resulted from the Little Ice Age encountered in 1550. It can be seen from (Fig.11) that the stand collars of the Ming Dynasty had square corners and were tightly snapped at the neck. Later, after the powerful Qing Dynasty treaded on the land of Central Plains, it began to establish the costume system with its national characteristics. There are also stand collars in its national costumes, with the only difference being that those in the Ming Dynasty had square corners, whose those of the Qing Dynasty had round corners (Fig.12 and 13), and, along with the "厂" character unlined upper garment, they became the basic patterns of the Costumes of the Qing Dynasty.

Fig.11 The portrait of Empress Zhang of the Emperor Hsiao Tsung of the Ming Dynasty

Fig.12 Collar of the Qing Dynasty robe

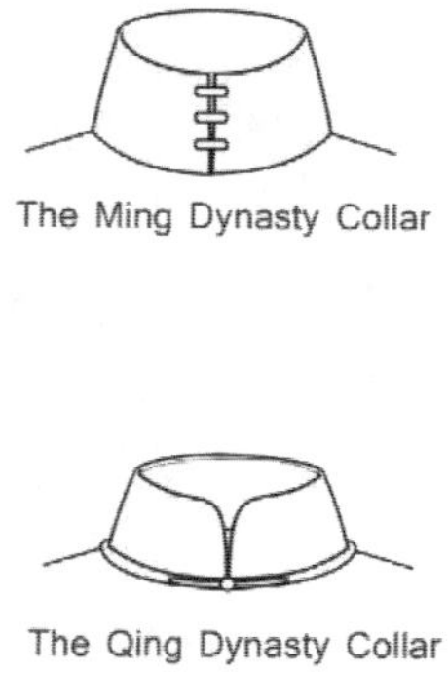

Fig.13 Different between the collars

4.2 Fixations between the various components

The traditional round-collar gowns usually had a small neck circumference, which clapped person's neck to ensure the dress will not move back; therefore, they could be easily stained by sweat. However, most of the fabrics of such dresses are too precocious to be washed easily; in addition, during the dress-making process of the dress bodies, usually such traditional techniques as starch finishing were used, so after washing, the original styles will be lost, thus making it hard for the dress bodies to be washed at ease. This gave rise to the detachable collars, and during the development of Cheongsam in the period of the Republic of China, the dedicated boxes for containing the stand collars so as to contain multiple stand collars (Fig.14) [7].

The stand collars of the Changyi Gowns of the late Qing Dynasty (about 1800 A.C.) (Fig.15) is a representative of the independent sewing. A circle of piped edge was tapped on the collar ring [8], and it was sewed together with the dress body. Consequently, most of the overcoats that we can see at present are the collarless "厂" character unlined upper garment, and the bright colors and the patterns of the original fabric are still retained, but without collar. Such a way of having the detachable components was not only popular during the late Qing Dynasty and the period of the Republic of China, it was also widely adopted in the costumes of the ethnic minorities (Fig.16). The detachable collars not only retained personal hygiene, it also provided the garments with the collars surfaces that can be matched by a number of configurations; with adjustable heights and being bright colored, they make the collars become a fashionable element with rich changes in the traditional Chinese dresses. The collars and the dress bodies are joined with bindings, a method that is labor-saving and time-saving; not only does it cover the stitches between the dress bodies and the collars, it also deepening the stability of the collar.

Fig.14 Removable collar of traditional Chinese cheongsam

Fig.15 Changyi Gowns of the Qing Dynasty[10]

Fig.16 Collar edge of Shes ethnic group Traditional costume

5. Conclusion

In Chinese traditional costumes, collar edge have long assumed an important task, no matter with regard to the use of pure-colored and thick brocade as a collar edge to supplement the extension amounts resulted from the fabric restrictions, or the constant overlapped ornament technique of the dresses and the collars during late Qing Dynasty, their fixed functions have never disappeared, but, instead have been changing along with the costume changes amid the alterations of the dynasties. The roles of the collar edge in the collar edges were further substantiated, from the basic wrapping of the fabric edge to make the dress edges smooth and integral; to the fixations of the fabric strands so that the inclined strand of different arch degrees at the collars will not be tilted; meanwhile, they all became the recruitment of the closures; to the covering on the broken seam between the front, read and middle so as to fix the two fabrics; and then to the fixation of the dress body and the collar in the form of bindings, all these taxation modes have far surpassed the collar edges significance as collars. Similarly, the fabrics that carry such different fixation modes are also expressing the voices of the power hierarchies and nobility or humbleness via such techniques as different colors, textures, embroidery and attachments of golden patches.

Without consideration of colors, materials and fine techniques, it is quite significant concept of design to explore the gentle relationship between clothing structures and the development and changes in terms of people's lifestyle as well as discover the interaction of clothing structures. Amid such an era characterized by constant changes, it is necessary to have some creation concept that originally existed to arouse people reminiscence and valuing on the nature of dresses.

References

[1] Shen CW. China ancient costume research in the Chu Dynasty Tomb of Jiangling. M Shanghai Bookstore Publishing House, 2002; 22-27.

[2] Qi Y. Technique and folk costume culture of side opening clothing in Qilu area in modern times. J Journal of Textile Research, 2010; 31(11) : 116-121.

[3] Li CC. The Comparative Research of Collar Development and Evolution in Tang and Song Dynasties. D Shaanxi University of Science & Technology, 2013; 8-35

[4] Wei Na. Research on Trimmings of Chinese traditional Costume. [D], Jiangsu: Soochow University, 2014; 19-33

[5] Zhao M. Line Cutting & Doublet Structure: A Study Of Chinese Ethnic Costume's Structure J Art&Design: Folks Crafts,2012\01; 110-112

[6] Zhao M. Cultural Collision and Fusion-Traditional Chinese Cheongsam Research J Art&Design Research:Apparel Research, 2013 03.Autumn; 019-023

[7] Lu HX. A Study on Pattern and Shape Evolution of Ttaditional Chinese Cheongsam D Beijing: Beijing Institute of Fashion Technology, 2012; 52-62

[8] Yu Y. Study in Tranditional Chinese Cheongsam Style and Structure in Republic of China D Beijing: Beijing Institute of Fashion Technology, 2014; 60-62

[9] Wang SH. Study on Shapes and Patterns of Ttaditiona Manchu Costume D Beijing: Beijing Institute of Fashion Technology,2012;28-31

[10] Fang W. The "Gown" Collar Style of Hunni Burial Puppet in Tang Dynasty J Directory of Open Access Journals, Asian Social Science, 2009; Vol.5 (10)

[11] Information on http://www.dpm.org.cn/shtml/1/@/9057.html

[12] Information on http://www.hnmuseum.com/hnmuseum/index_gb.jsp

Research on Business Suit Lapel based on Kansei Engineering

Dong-Hui Wang, Jing-Wei Liu*

Apparel & Art Design College, Xi'an Polytechnic University, Xi'an, 710048, China

*Corresponding author's email: weijinglui@163.com

Abstract

In the business environment, it is important to wear a suit to establish a professional image. There is a wide variety of collars for business suits on the market which includes closure collar, flat collar and shawl collar. The difference can be distinguished by collar size, collar locations and so forth. This study aims to reflect on how different collars are suited to different business styles. In this survey, market mainstream suit collar models are first collected, then these models are classified and summarized to create a representative suit collar working sketch. Perceptual words related with suit are gathered by using the perceptual semantic method to form suit lapel's semantic space. Based on the above collar shape and the collar semantic space, to design the experiment of relationship between collar shape and collar semantics, through the analysis of the data in spss 13.0 to get the relationship between the two. From the emotional point of view of what the suit lapel style to better express the business style for designer and consumer, suitable for business occasions and to a establish a method for suit collar structure design.

Keywords: Kansei engineering; Business style; Suit lapel

1. Introduction

Clothing are typically products of perception where every detail and line are constrained by the designer's aesthetic experience. Due to the potential conflictbetween designers' expressions and customers' requirements, it is necessary to research how to connect customer's demands and the designer's intentions together. The original intention of this research is to develop product semiotics in Business Suit Lapel as in recent years, product semiotics has been efficiently used in many fields [1].

There are many studies based on Kansei Engineering at home and abroad, where the theory of Kansei Engineering is applied to clothing research, mostly for the clothing fabric or color, but also on clothing outline and structure.

When consumers chooses a business suit, they will give a business suit a psychological position. There is a presence of some kind of subconscious emotional factors that determines what kind of suit collar better reflectsbusiness style. It will be able to describe the business style and set out a semantic differential method of Kansei Engineering to more clearly comprehend the consumer's subconscious emotional factors, given by respondents for a business suit collar stimulation figure that is specially made, selected, analyzed. Through this survey, business suits and design are selected to provide a theoretical basis. This paper will analyze what kind of lapel best expresses the business style and the best suitable business environment by choices for the business suit collar stimulation figure as respondents. The analysis will provide a certain theoretical basis for the choice of business suits and suit collar design.

2. Concept of Kansei Engineering

Kansei Engineering is first proposed by the Japanese design community in the 1980s. It quantifies the past can not be quantified and only qualitative, non-logical, irrational emotional concept under conditions of modern technology, and then adapted to develop products and technology. Kansei Engineering is oriented to the needs of customers to implement product development technology, the customer's images and feelings into fashion design elements of technology [2].

3. Research on business suit lapel based on Kansei Engineering

3.1 Research object

In this paper, men's business suit lapel is selected as the research object. In the business environment, where people wear business suits, more and more people will start thinking about what kind of suit style is more suitable for a business environment. This article is for a detail of business suit, collar, to study which as the most important part of the suit to show the views of the suit style. Therefore, this paper studies what kind of suit collar style better reflects the business style of clothing.

3.2 Research method

This paper uses the semantic difference method. When designing the Semantic Differential Scale, it is important to first determine the set of properties that is related to the object to be measured, where a pair of opposite adjectives is chosen to establish a scale for defining the property. The scale is divided into five, seven or nine consecutive grades and the scale is divided into five successive levels in the research. Respondents were asked to mark at the right level position based on their evaluation of the measured object view. The basic flow of the semantic difference method is shown in Fig.3.2.1.

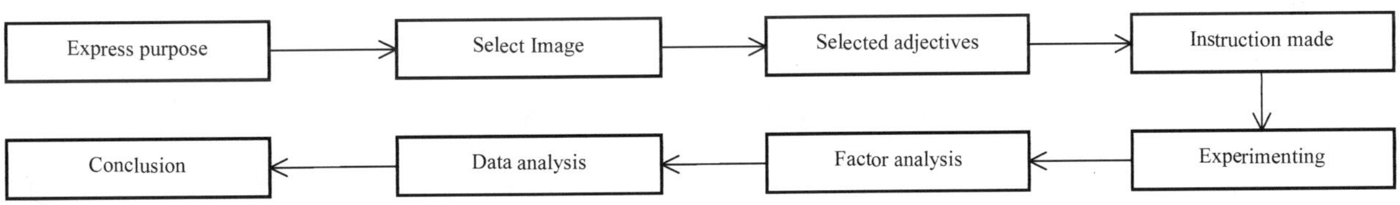

Fig.1 Flow chart

3.3 Research process

First, the existing business suit style on the market were collected and sorted into a selection of representative styles. Then a working sketch was developed by selecting word pairs to represent the business style. This is then followed by the design of a questionnaire to a target population. Finally the data is analyzed and the appropriate conclusions are drawn.

3.3.1 Determined the research sample

In this sample selection, the men's business suits style were collected by means of a network and summarized. After the collection of suit styles, it was possible to classify men's business suit lapel under the following categories of closure collar, flat collar, shawl collar and so on(Fig.2,Fig.3,Fig.4).

Fig.2 Closure collar

Fig.3 Flat collar

Fig.4 Shawl collar

In order to control unwanted variable in suit style that only changes the construction of the lapel while maintaining the length of the collar and the position the same and avoid the influence of color on the senses, the study provides no color style figure as the stimulation Figure (Fig.5-26). These samples are divided into three categories which are closure collar, flat collar and shawl collar. The style changes are based on the three collar, such as adding a dividing line and changing the split mouth angle. There are altogether 22 samples in the experiment.

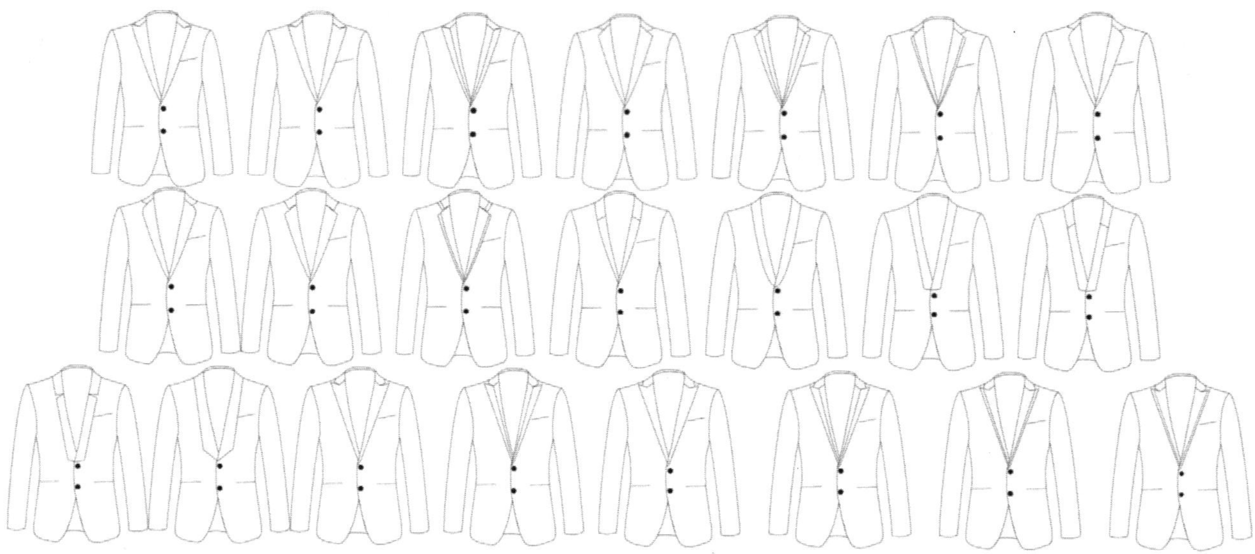

Fig.5-26 The stimulation FIG

3.3.2 Selected adjectives

The study collected and summarized to reflect the business style of vocabulary and the corresponding words, and finally thirteen word pairs were carefully selected. The paper collected and summarized 13 adjectives that reflect business style adjectives and found their corresponding word after a review of relevant professional and clothing books, magazines, newspapers and consult professionals and discuss with the students. The carefully selected thirteen pair words are the following:serious - casual, sober - lively, mature - inexperienced, elegant - inelegant, quiet - fickle, formal - optional, implicit - unrestrained, cheesy - restrained, starched - weak, simple - cumbersome, modest - arrogant, delicate - rough, confident - unconfident.

3.3.3 Design and distribution of questionnaire

The adjectives are divided into five levels in the questionnaire which are 1, 2, 3, 4,5 (Fig.27) where 1 indicates a more serious style, and 5 a more casual style. Respondents of the questionnaire are mainly graduate students who specialize in garment knowledge.

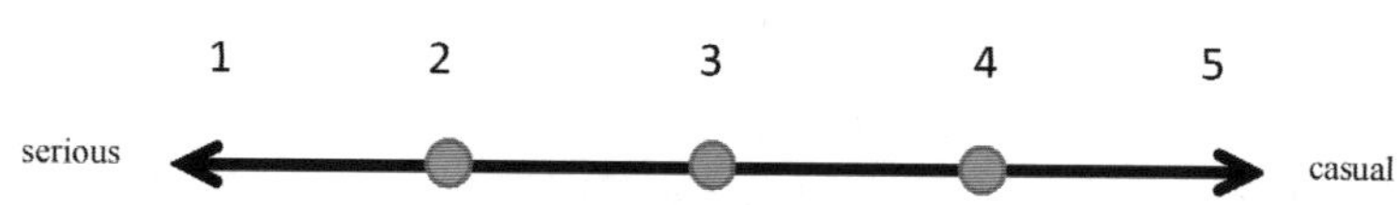

Fig.27 Emotional level

3.3.4 Data collection and analysis

After the questionnaire was conducted, the average scores indicated that there are about 22 suit collar samples. (Table 1)

Preliminary analysis can be made to the above sample mean that the mean of the sample 1, sample 3, sample 5, the sample 7 and the sample 15 is less than three. So the study conclude that the five samples more have business style.

Table 1. Average scores of sample

	serious - casual	sober - lively	mature - inexperienced	elegant - inelegant	quiet - fickle	formal - optional	implicit - unrestrained	cheesy - restrained	starched - weak	simple - cumbersome	modest - arrogant	delicate - rough	confident - unconfident
1	***2.36***	***2.32***	***2***	***2.84***	***2.2***	***2.28***	***2.28***	***2.84***	***2.28***	***2.12***	***2.28***	***2.64***	***2.28***
2	3.04	2.92	2.6	2.72	2.72	2.76	2.96	2.84	2.56	3.36	2.8	2.52	2.84
3	***2.92***	***2.76***	***2.4***	***2.16***	***2.32***	***2.52***	***2.88***	***2.76***	***2.6***	***2.56***	***2.76***	***2***	***2.2***
4	3.08	2.96	2.76	2.52	2.72	2.92	2.52	2.88	2.68	3.2	2.76	2.28	2.72
5	***2.56***	***2.44***	***2.48***	***2.76***	***2.16***	***2.32***	***2.64***	***2.72***	***2.68***	***1.96***	***2.56***	***2.68***	***2.28***
6	3.08	3	2.84	2.52	2.68	3.32	2.8	3.08	2.56	3.2	2.8	2.36	2.56
7	***2.24***	***2.36***	***2.6***	***2.52***	***2.2***	***2.04***	***2.76***	***2.52***	***2.24***	***1.84***	***2.56***	***2.68***	***2.32***
8	2.96	2.96	2.56	2.72	2.92	2.84	3	2.84	2.64	3.52	2.76	2.48	2.68
9	2.96	2.8	3.08	2.48	2.56	2.76	3.04	2.88	2.52	3.12	2.72	2.4	2.4
10	3.2	2.6	2.8	2.92	2.28	2.48	2.8	2.52	2.8	1.64	2.8	2.64	2.28
11	3	2.96	2.92	2.72	3.04	2.92	2.84	2.96	2.76	3.4	3.24	2.56	2.92
12	2.8	2.64	2.56	2.48	2.68	2.52	2.68	2.8	2.72	3.16	2.84	2.76	2.52
13	3.2	3.16	3.08	2.84	2.64	3.2	2.68	2.48	3.44	2.04	2.8	2.56	2.48
14	3.48	2.96	2.76	2.96	2.4	2.76	2.72	2.48	3.52	2	2.56	2.44	2.48
15	***2.92***	***2.64***	***2.2***	***2.48***	***2.04***	***2.56***	***2.64***	***2.68***	***2.16***	***2.16***	***2.68***	***2.68***	***2.04***
16	3.4	3.29	3.12	2.24	2.6	3.16	2.96	2.52	2.36	3.36	2.6	2.28	2.08
17	3.36	2.6	3	2.72	2.48	3.04	2.44	3.12	3.04	2.2	2.36	2.68	2.32
18	3.56	3.28	3	2.48	2.6	3.16	3.04	2.4	3.2	2	2.52	2.48	2.28
19	3.12	2.8	2.64	2.72	2.64	3.08	2.72	2.76	3	2.08	3.04	2.8	2.72
20	3.04	3.12	2.72	3	2.88	3.44	2.8	2.76	3.12	2.68	3	2.72	2.56
21	3.4	2.96	2.8	2.8	2.52	3.08	3.16	2.6	2.8	2.64	3	2.76	2.52
22	3.2	2.76	2.64	2.48	3.04	3.16	3.12	2.36	2.92	1.92	3	2.84	2.16

First, the data derived from the study is examined to determine whether it is suitable for factor analysis. The data were analyzed with SPSS13.0 software and 13 Kansei image variable as a variable factor analyzed. The paper obtained a correlation matrix of 13 variables. The KMO and Bartlett's test are shown in Table 2 and Table 3.

Table 2. Correlation matrix

	serious - casual	sober - lively	mature - inexperienced	elegant- inelegant	quiet - fickle	formal - optional	implicit - Unrestrained	cheesy - restrained	starched - weak	simple - cumbersome	modest - arrogant	delicate - rough	confident - unconfident
serious - casual	1.000	.743	.654	.042	.369	.739	.414	-.227	.624	.058	.200	-.147	.044
sober - lively	.743	1.000	.648	-.079	.589	.813	.507	-.179	.454	.425	.349	-.365	.268
mature - inexperienced	.654	.648	1.000	-.027	.434	.604	.396	-.079	.473	.204	.198	-.179	.174
Elegant - inelegant	.042	-.079	-.027	1.000	.034	.071	-.260	.011	.486	-.284	.077	.517	.393
quiet - fickle	.369	.589	.434	.034	1.000	.685	.438	.134	.346	.553	.643	.035	.591
formal - optional	.739	.813	.604	.071	.685	1.000	.353	.034	.524	.277	.419	-.028	.253
implicit - unrestrained	.414	.507	.396	-.260	.438	.353	1.000	-.384	.074	.259	.493	-.097	.043

cheesy - restrained	-.227	-.179	-.079	.011	.134	.034	-.384	1.000	-.276	.548	.004	-.164	.457
starched - weak	.624	.454	.473	.486	.346	.524	.074	-.276	1.000	-.295	.180	.134	.243
simple - cumbersome	.058	.425	.204	-.284	.553	.277	.259	.548	-.295	1.000	.320	-.417	.529
modest - arrogant	.200	.349	.198	.077	.643	.419	.493	.004	.180	.320	1.000	.184	.534
delicate - rough	-.147	-.365	-.179	.517	.035	-.028	-.097	-.164	.134	-.417	.184	1.000	.029
confident - unconfident	.044	.268	.174	.393	.591	.253	.043	.457	.243	.529	.534	.029	1.000

Table 3. KMO and Bartlett's test

Adequate sampling of the Kaiser-Meyer-Olkin measure		.537
Bartlett test of sphericity	Chi-square approximation	193.387
	df	78
	Sig	.000

KMO value was greater than 0.5, so this study is suitable for analysis factor.

The appropriate method was then selected for factor analysis. Principal component analysis and common factor analysis are the two basic factor analysis methods. Among them, the principal component analysis is mainly used with the least possible number of variables to explain variance. Common factor analysis is mainly used to identify common factors related to the main dimension. Therefore, the principal component analysis was chosen for the study. the results of the common factor variance derived from the principal component analysis method is shown in Table 4:

Table 4. The common factor variance

	initial value	extraction value
serious-casual	1.000	.834
sober-lively	1.000	.879
mature-inexperienced	1.000	.657
Elegant-inelegant	1.000	.820
quiet-fickle	1.000	.820
formal-optional	1.000	.788
implicit-unrestrained	1.000	.832
cheesy-restrained	1.000	.842
starched-weak	1.000	.852
simple-cumbersome	1.000	.904
modest-arrogant	1.000	.816
delicate-rough	1.000	.760
Confident-unconfident	1.000	.837

Analysis of variance table as shown in Table 5:

Table 5. Variance analysis table

Composition	initial eigenvalue			Extraction quadratic sum	
	total	variance %	accumulated %	total	variance %
1	*4.783*	*36.789*	*36.789*	*4.783*	*36.789*
2	*2.418*	*18.597*	*55.387*	*2.418*	*18.597*
3	*2.110*	*16.231*	*71.618*	*2.110*	*16.231*
4	*1.331*	*10.238*	*81.856*	*1.331*	*10.238*
5	.600	4.617	86.473		
6	.458	3.523	89.996		
7	.372	2.861	92.857		
8	.325	2.501	95.358		
9	.234	1.799	97.157		
10	.169	1.301	98.458		
11	.111	.852	99.310		
12	.076	.584	99.894		
13	.014	.106	100.000		

The characteristic value of certain factors represents the variance of the size related to the factors, therefore only factors with eigenvalues greater than 1 were retained. As a result only keep eigenvalues greater than 1 factor, since the characteristic value of less than 1 factor is not better than the original variables. The four factors are retained of which eigenvalues greater than 1 as shows in table and it is determined 4 factors.

Table 6. The rotation components matrix table

	Composition			
	1	2	3	4
serious - casual	***.896***	.091	-.147	-.008
sober - lively	***.852***	.309	.087	-.222
formal - optional	***.816***	.306	.161	.055
mature - inexperienced	***.795***	.126	.029	-.094
starched - weak	***.732***	-.036	-.108	.550
modest - arrogant	.141	***.855***	.188	.175
implicit -unrestrained	.316	***.739***	-.305	-.307
quiet - fickle	.465	***.663***	.400	.068
cheesy - restrained	-.162	-.176	***.882***	-.080
confident - unconfident,	.155	.377	***.753***	.322
simple - cumbersome	.122	.372	***.719***	-.483
elegant - inelegant	.095	-.095	.146	***.883***
delicate - rough	-.271	.251	-.219	***.759***

The paper adopts orthogonal rotation that gained the rotation components matrix table as shown in Table 6. All Kansei variables can be eventually attributed to four factors: factor 1. serious - casual, sober - lively, formal - optional, mature - inexperienced, starched - weak; factor 2, modest - arrogant, implicit - unrestrained, quiet - fickle; factor 3, cheesy - restrained, confident - unconfident, simple - cumbersome; and factor 4, legant - inelegant, delicate - rough.

The chart considers retaining all variables because the load of all the variables are so large. The explanatory variables is explain the properties of the factor which by factor analysis ,then named the factor. It is a very important step in factor analysis. By factor analysis the paper has learned that consumers obtain the perception evaluation of business suit from four aspects.

Factor 1-Temperament factor: serious - casual, sober - lively, formal - optional, mature - inexperienced, starched - weak;

Factor 2-Personality factor: modest - arrogant, implicit - unrestrained, quiet - fickle;

Factor 3-Confidence factor: cheesy - restrained, confident - unconfident, simple - cumbersome;

Factor 4-Charm factor: elegant - inelegant, delicate - rough.

Reprocessing data of these four factors and the average as shows in Table 7.

Table 7. Average of four factors

	Factor 1	Factor 2	Factor 3	Factor 4
1	***2.04***	***2.25***	***2.41***	***2.64***
2	2.78	2.83	3.01	2.52
3	***2.64***	***2.65***	***2.51***	***2***
4	2.88	2.67	2.93	2.28
5	***2.50***	***2.45***	***2.32***	***2.68***
6	2.96	2.76	2.95	2.36
7	***2.30***	***2.51***	***2.23***	***2.68***
8	2.79	2.89	3.01	2.48
9	2.82	2.77	2.80	2.4
10	2.78	2.63	2.15	2.64
11	2.91	3.04	3.09	2.56
12	2.65	2.73	2.83	2.76
13	3.22	2.71	2.33	2.56
14	3.10	2.56	2.32	2.44
15	***2.50***	***2.45***	***2.29***	***2.68***
16	3.07	2.72	2.65	2.28
17	3.01	2.43	2.55	2.68
18	3.24	2.72	2.23	2.48
19	2.93	2.80	2.52	2.8
20	3.09	2.89	2.67	2.72
21	3.01	2.89	2.59	2.76
22	2.94	3.05	2.15	2.84

that the results of the data analysis are the following:

1) The determinants of the business style business suits are grouped into the following four categories temperament factor, personality factors, self-confidence factor and charm factor.

2) Regarding the factor of determining business style, temperament factor is the highest and charm factor is relatively low.

3) The paper ignores average figures which are close to three and have little contribution rate to the data analysis. The left arrow points business style vocabulary, and the right arrow points the opposition words and the measured value is smaller while the test has a business style suit collar;

Fig.28 Business style suits

4) The paper reprocesses the date of four types factors that the results obtained and the conclusions of mean method are similar to No. 1, No. 3,No. 5, No.7,No. 15 (Fig.28)are very business styles.

4. Conclusion

The most important business suits style is serious, sober, formal, mature and starched. Consumers will choose the suit by judging whether it is serious, sober, formal, mature and starched. The study results show that No.1 No.3 No. 5 No.7 and No.15 were considered the most business style. There is a commonality in characteristic between the suit styles: a straight line lapel mouth; few dividing lines and simple appearance. This conclusion can be used for business suit collar design and the language of the text is converted to visual language.

References

[1] Qiao C. The Collar Structure Research Based on Product Semiotics: 2012; 1.
[2] Jonathan C, Craig MV. Creating breakthrough products-innovation from product planing to program approval. NJ: 2002; 1-31.
[3] Yamamoto K. Kansei engineering - the art of automotive development at Mazda.Ann Arbor: The university of Michigan: 1986; 1-23.
[4] Su JN. Research on Kansei engineering and it application product design.Journal of xi'an jiaotong university: 2004; 1: 60-63.
[5] Liang YY. Research on the method of Kansei design.
[6] Shinya N. Present state of Kansei engineering in Japan. IEEE International conference on systems, man and cyberneties: 2004; 333-338
[7] Mitsuo N. Kansei Engineering in Consumer produet Design. The Magazine of Human Factors applications: 2002; vol.10: 5-9.
[8] Gao C. The study on emotional design and kansei engineering system's development of office suit.
[9] Achmad S, Taufiq I. Design of Innovative Alarm Clock Made from Bamboo with Kansei Engineering Approach. Agriculture and Agricultural Science Procedia: 2015; vol. 3: 184–188.
[10] Ngip KC, Ashok S, Mizhanim MS. Kansei Engineering for e-commerce Sunglasses Selection in Malaysia. Procedia - Social and Behavioral Sciences: 2013; vol. 97: 707–714.

A Component-Based 3D Garment Modeling Algorithm

Ruo-Mei Wang, Gong-Hai Xie, Bing-Guang Zhou, Fan Zhou*

National Engineering Research Center of Digital Life, State-Province Joint Laboratory of Digital Home Interactive Applications, School of Data Science & Computer Technology, Sun Yat-sen University Guangzhou, 510006, China

*Corresponding author's email: isszf@mail.sysu.edu.cn

Abstract

virtual garment has a wide range of application in different areas. However, it is difficult to construct the soft garments with rich styles, complex components. In order to generate virtual garment easily, an easy-to-use 3D garment generation method is proposed in this paper. The main contribution of our method is use the garment component assemble 3D garment according to the requirement of consumer or designer.

Keywords: Garment; Component; Computer software; assembly

1. Introduction

With the fast development of the advanced modeling and visualization techniques of virtual garment, 3D garments have been more and more popular type of digital media. Virtual 3D garment is important in the application of computer aid garment design, character-based games environment and virtual reality.

Virtual garment is the result of a combination of many techniques which concentrate on the simulation and visualization features. Researchers have made tremendous efforts to create a virtual garment where designers are able to preview their design conceptions accurately and effectively before the garment is actually made or the consumers are able to try on the garment and to preview fitting effect and even animate the clothing in a virtual scenario[1,2].

Some landmark research results can be reviewed. Stylios reported the dynamic drape of garments modeling. Real garments made from cotton/ polyester and polyester lightweight fabrics have been modeled and have been used on synthetic human in a virtual fashion show [3]. Nadia Magnenat-Thalmann stated the reach a truly 3D Virtual Try On which is the possibility of any one to give dimensions on her predefined body and obtain her own sized shape body, select a 3D cloth and see oneself animated in real-time, walking along a catwalk [4]. Other systems existed today such as Swivel (Facecake) overlay a still image of the garment on the user's Figure to view the garment fitting effect, and Fitnect 2011 provides the simple cloth animation when the user moves [5].

Alought some significant researches have been reported, but the application of 3D garment is lack.The research of virtual garment remains a challenging problem. The main problems to construct a virtual garment is time cost and complex construction process. The first, garment is made by soft textile material with low bending stiffness rather than rigid solid objects, the main character of 3D garment is large deformation. Therefore, the garment construction and simulation is complex and high time cost. Secondly, the garment structure is very complex because of the garment style is very rich, and the garment components such as collar and sleeve, are assembled together following specific pattern making rules which are totally different from the conventional assembly methods [6].

The purpose of this paper is to present an efficient algorithm for 3D garment modeling based on the component assemble to solve the low time cost problem. This is more practical to design a CAD system to preview virtual 3D garment effect in the daily life.

2. RelatedWork

Physically-based mechanical simulation is often used in 3D garment modeling for reproducing adequately the shape and the motion of the garment on the virtual body.

In 1995, Provot [7] proposed a mass-spring system to model clothing. Breen and House [8,9] developed a non-continuum particle model for fabric drape that explicitly represents the micro-mechanical structure of a fabric via a particle system.Then the realistic-looking 3D garments are generated according to fabric physical properties in the virtual environment. Baraff et al [3]. also developed the implicit integral cloth simulation method based on spring-particle system.

The research of 3Dgarment is attractive propositions in emerging markets, has wide application in virtual clothing, digital customizing, clothing industry, and e-commerce. Virtual fitting is a systematic and quantitative way of previewing the garment simulation effect before the garment is actually made. The significative researches include the dynamic drape of garments modeling reported by Stylios [10], the realistic simulation of comprehensive garment developed by the research group of Prof. Nadia Thalmann [2,11], a truly 3D Virtual Try On was reported in which any one can select a 3D cloth and see oneself animated in real-time, walking along a catwalk [4]. Recently, in order to meet the requirement of application, some research results such as preserving garment transfer, dressing for any person are reported [12-14]. Hongyan Liu et al. reported the research of reusability of geometric model [15]. Ei Chaw Hlaing et al. presented the methods of creating 3D garment design as well as grading in 3D, based on scalable virtual models [16]. Other systems existed today such as Swivel (Facecake) overlay a still image of the garment on the user's Figure to view the garment fitting effect, and Fitnect 2011 provides the simple cloth animation when the user moves [5].

However, the research of virtual fitting remains a challenging problem. It has to address some special issues. The first, garment is made by soft textile material with low bending stiffness rather than rigid solid objects, the main character of 3D garment is large deformation. Therefore, the garment construction and simulation is complex and high time cost. Secondly, the garment structure is very complex because of the garment style is very rich, and the garment components such as collar and sleeve, are assembled together following specific pattern making rules which are totally different from the conventional assembly methods [6]. Thirdly, as we known, in the real world, dressing up is a layered manner. People would like various combinations of garments to be tried to reach maximum satisfaction [17]. But the efficient and reality in the layered garment construction modeling and simulation numerical computation are key problems.

Assembly-based modeling is the technology that the separate components are combined together to build a new one. One of earliest assemblybased modeling is presented by Funkhouser et al [18]. They propose a Modeling by Example system for composing models using text or shape retrieval of existing parts of models. The target model is generated by merging components of different source models. Kreavoy et al [19]. use compatible parts of similar class models to assemble novel model. Their method is based on the assumption that same class models have the same number of components, so that they don't need to concern about the correlation of components.

We use the principle of assembly for garment modeling and the simple garment components are assembled together according to the requirement of garment modeling.

3. Garment Modeling Methodology

3.1 Garment Components

Garment components are generated by 2D patterns, which can be easily done by physical simulation and the garment components are represented by triangular meshes.

Based on the requirement of garment modeling, some components are selected. The we use a mesh merging algorithm to assemble them together. Like how people assemble things, we regard the synthesis constituted by former selected components as target object and the latter components as source objects. The source objects are merged into the target one by one, then generate a new target object. So each time we merge two meshes, a target and a source.

Mean Value Coordinates (MVC) is first proposed by Floater [20]. He used it to simplify mesh parameterization and morphing by simulating harmonic functions with MVC method. Subsequent research [21] generalizes MVC in

R3. Hormann and Floater [22] presented application examples with MVC in computer graphics and geometric modelling. Our goal is to merge source mesh and target mesh together seamlessly. Thus we choose MVC method since it can reduce computation complexity and then speed up our merging process.

3.2 Boundary Ajustment

The MVC merging algorithm is transforming the source mesh to a new mesh which has the same boundary with the target mesh. The origin boundary points are regarded as weight points, playing an important role in the merging process. So the function of boundary ajustment is used to keep the original information of the boundary lines of both meshes and make them consistent between source and target objects. Two new boundary lines should have the same number of points and points need to be distributed with the same proportion.

For the source boundary line $Ls = Ps_i$ and target boundary line $Lt = Pt_j$, we calculate the length of them (lens and lent). Then we find distribution ratio set of the two line using the equation (1).

$$\begin{cases} dr_i = \dfrac{\sum_{k=1}^{i+1} dist_k}{len}, \quad i = 1,2,...,m-1 \\ dist_k = \|\mathbf{P}_k - \mathbf{P}_{k-1}\|_2, \quad k = 2,3,...,m \end{cases} \tag{1}$$

Where m is the number of the points on the boundary, len is lens for the source line or lent for the target line. Finally we insert new points into the boundary lines in appropriate positions to make them consistent with each other. For example, if we get the dr_s = {0.25,0.5,0.75,1} and dr_t = {0.17,0.33,0.5,0.67,0.83,1}. Two points are inserted at ratios {0.25,0.75} for the target line, and four points are inserted at ratios {0.17,0.33,0.67,0.83} for the target line.

3.3. MVC Merging Algorithm

The coordinates of the points on the mesh are needed to define first .

Considering $\partial B = (p_1, p_2, \cdots, p_m, p_1), p_i \in R^3$. The mean value coordinates of point x of the mesh is defined as $\lambda(x)$:

$$\lambda_i(\mathbf{x}) = \omega_i(\mathbf{x}) \Big/ \sum_{j=1}^{n} \omega_j(\mathbf{x}) \tag{2}$$

Figure 1 shows the mean value coordinates of point.

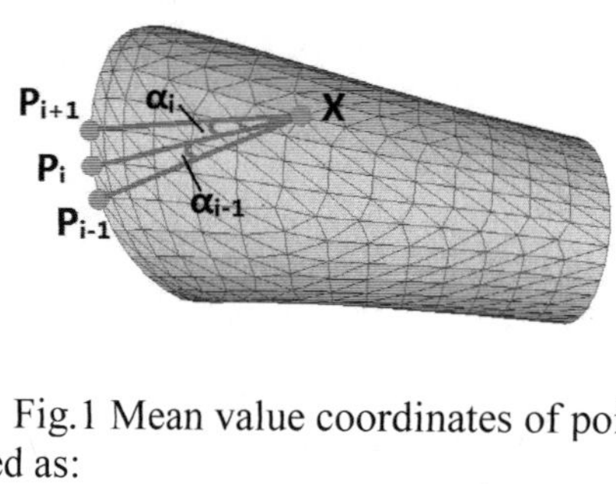

Fig.1 Mean value coordinates of point

Where the weight function is defined as:

$$\omega_i(\mathbf{x}) = \frac{1}{\|\mathbf{P}_i - \mathbf{x}\|}(\tan\frac{\alpha_i}{2} + \tan\frac{\alpha_{i-1}}{2}) \tag{3}$$

Here, α_i is $\angle p_i x p_{i+1}$. Then for source mesh $S = s_i$ and target mesh $T = t_i$, we call the boundary line of them ∂S and ∂T. The distance between corresponding points on ∂S and ∂T can be computed by the equation (4).

$$f_d(i) = s_i - t_i \tag{4}$$

Then source mesh S is transformed into a new mesh $S' = \{s_i'\}$ by the equation (5).

$$\begin{cases} S'(x) = S(x) + R(x) \\ R(x) = \sum_{i=0}^{m-1} \lambda_i(x) f_d(x) \end{cases} \tag{5}$$

Now the new source S' is seamless matched the target mesh along the boundary lines. Then we get the new target mesh by merging S' and T together.

3.4 Garment Manufacturing

Manufacturability is an important factor for garment modeling, especially in Computer Aid Garment Design (CAGD) area. In the case when modeler need to make the real garment, the corresponding 2D patterns are required. Other researches such as transferring method [12] ahave to do 3D-to-2D extraction. Benefit from the origin sources are generated by 2D pattern simulation, it is easy for us to get the patterns (ignoring the distortion caused by mesh merging). Figure2 shows the corresponding components used for the garment modeling.

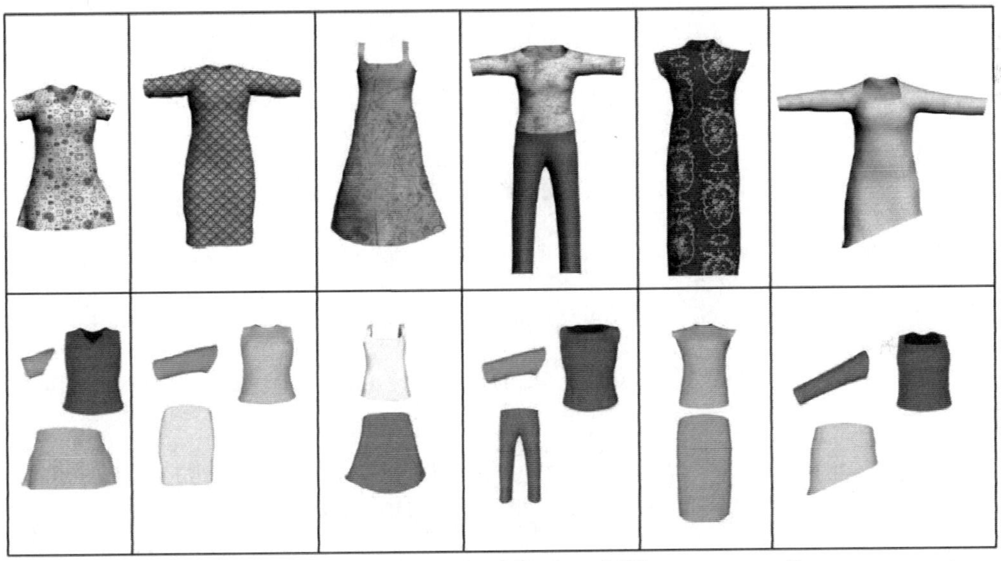

Fig.2: Top) The 3D garments modeled by our method. Bottom) The corresponding components used for the garment modeling.

4. Conclusion

In this paper, a new approach for 3D garment modeling based on component assembly is reported. This is an easy way to construct the real garment. The main contribution is to decrease the time cost and solve the problem of complex construction process about the garment modeling. So the components database can be designed to store different types of simple components, these components can be used to generate different types of garment style. Once these 3D garment is constructed, they can be simulated by different material and texture mapping to visulaze the realism virtual garment.

5. Acknowledgements

We would like to thank the anonymous reviewers for their valuable comments. We were supported by the National Natural Science Foundation of China (61379112, 61272192). Fan Zhou is the corresponding author.

References

[1] Yue Q, Zhong BX. Three-dimensional Garment Dressing Simulation. Textile Research Journal: 2012; 79 (9); p: 792-803.

[2] Carignan M YY, Magnenat-Thalmann N, Thalmann D. Dressing Animated Synthetic Actors with Complex Deformable Clothes. Comput graph (Proc ACM SIGGRAPH 92). ACM: 1992; p: 99-104.

[3] Stylios GK., Wan TR, Powell NJ. Modelling the dynamic drape of garments on synthetic humans in a virtual fashion show. International Journal of Clothing Science and Technology: 1996; 8 (3); p: 95-112.

[4] Magnenat-Thalmann N. Modeling and simulating bodies and garments: 2010: Springer.

[5] Li L, R W, Zhuo S. Mesh-based anisotropic cloth deformation for virtual fitting. Multimed Tools Applications: 2014; 71; p: 411-433.

[6] Liu Y, Z D, Yuen M. A survey on CAD methods in 3D garment design,. Computers in Industry: 2010; 61 (6); p: 576-593.

[7] Provot X. Deformation Constraints in a Mass-Spring Model to Describe Rigid Cloth Behavior. Proceedings of Graphics Interface'95: 1995; p: 147-154.

[8] Breen DE, House DH, Wozny MJ. A Particle-Based Model for Simulating the Draping Behavior of Woven Cloth. Textile Res. J.: 1994; 64 (11); p: 663-685.

[9] Breen DE, House DH, Getto PH. A Physically Based Particle Model of Woven Cloth. The Visual Computer: 1992; 8 (6); p: 264-277.

[10] Stylios GK, Wan TR, Powell NJ. Modelling the dynamic drape of garments on synthetic humans in a virtual fashion show. International Journal of Clothing Science and Technology: 1996; 8 (3); p: 95-112.

[11] Volino P, Courchesne M, Thalmann NM. Versatile and Efficient Techniques for Simulating Cloth and Other Deformable Objects. Proceedings of SIGGRAPH 95, computer graphics proceedings, annual conference series. ACM: 1995; p: 137-144.

[12] Brouet R, S A, Boissieux L, Cani M. Design Preserving Garment Transfer. ACM Transactions on Graphics (TOG): 2012; 31 (4); p: 36: 1-11.

[13] Guan P, R L, Hirshberg D, Weiss A, Black M. DRAPE: Dressing any person. ACM Transactions on Graphics (TOG): 2012; 31 (4); p: 35:1-10.

[14] Jituo Li JY, Wang YS, Bai L, Lu GD, Fitting 3D garment models onto individual human models. Computers & Graphics: 2010; 34 (6); p: 742-755.

[15] Liu HY, Y Z, Wang SY. Converting to reusable garment model. International Journal of Clothing Science and Technology: 2012; 24 (5); p: 308-316.

[16] Ei Chaw Hlaing S. Krzywinski,Hartmut Roedel, Garment prototyping based on scalable virtual female bodies. International Journal of Clothing Science and Technology: 2013; 25 (3); p: 184-197.

[17] Zhong Y. Fast penetration resolving for multi-layered virtual garment dressing. Textile Research Journal: 2009; 79 (9); p: 815 - 821.

[18] Funkhouser T, K M, Shilane P, Min P, Kiefer W, Tai A, Rusinkiewcz S, Dobkin D. Modeling by example. ACM Trans. Graph: 2004; 23 (3); p: 652-663.

[19] Kreavoy V, J D, Sheffer A. Model composition from interchangeable components. PG'07. 15th Pacific Conference; 2007; p: 129-138.

[20] Floater SM. Mean value coordinates. Computer Aided Geometric Design: 2003; 20 (1); p: 19-27.

[21] Floater Loater S, K G, Reimers M. Mean value coordinates in 3D. Computer Aided Geometric Design: 2005; 22 (7); p: 623-631.

[22] Hormann K, F M S. Mean value coordinates for arbitrary planar polygons. ACM Trans. Graph: 2006; 25 (4); p: 1424-1441.

Analysis on the Change of Adult Body Shape in Xi'an

Jing Qi*, Long Wu, Xin Zhang, Bo-An Ying

Apparel & Art Design College, Xi'an Polytechnic University, Xi'an, Shaanxi, 710048, P.R. China

*Corresponding author's email: qijing0928@126.com

Abstract

The changing dietary structure and lifestyle of a modern citizen, has resulted in the change of human body measurements. It has a significant impact on all aspects of clothing designing from production to product development and so forth. With the comparison of the data regarding standard sizing systems for garments with the data we measured in 2009 and 2015, the paper analyzed the change of adult body size data. The results showed that great changes have taken place in both human body distribution proportion and body measurements. The distribution of the number and the mean of main body measurements divided by age stage and by both age stage and body type were analyzed in samples 2015, which can support enterprises in building garment size systems, production and garment sales.

Keywords: Human body measurements; Adult body shape; Human body type; Sizing systems for garments; Body measurements distribution

1. Introduction

The present Chinese adult standard sizing systems for garments, GB1335-2008, was developed based on the statistics of a large sample of data which were measured in 1987 [1]. With the change of the dietary structure and lifestyle of the average modern citizen, a change of human body measurements have taken place. Many results show that an average person's height and circumference size has increased [2-7]. Foreign researches were also focused on the change of human body shape [8-11]. Using advanced non-contact 3D body scanner and manual measuring tool, Research Center of Clothing Engineering and Technology of Shaanxi Province in Xi'an Technology University, has continuously gathered human body measurements and conducted research on the human body shape from 2002 [12-14]. They cooperated with the China National Institute of Standardization (CNIS), and as the representative of western China, we conducted two large-scale activities of gathering human body measurements in xi'an in 2009 and 2015. Some adults aged from 18 to 75 were measured by 3D scanning and manual measuring methods. In this paper, body sizes were measured in 3D images who were randomly selected from body measurements database in 2009 and 2015. Then the changes of adult body shape were analyzed by comparing them with the data of GB.

2. Measuring items and methods

Male and female samples were randomly selected from the database, then were measured using their 3D images to get their body measurements. In order to compare with the data of GB1335-2008, the same measuring items and methods were adopted from GB/T 16160 [15]. (See Table 1).

3. Distribution of the data

3.1 Distribution by body tape

Based on the value of chest and waist, Chinese body shape was divided into four types: Y, A, B and C [16,17]. Y represents thin, and A represents normal, and B represents a little fat, and C represents fat. On this basis, we added two extra kinds of body type, named X and D. If someone's data of breast-waist difference is bigger than type A, its body type will be X. And if the data is smaller than type C, its body type will be D. The proportion of each body type of male and female samples were compared with the data of whole country and Middle Western areas in GB (see Table 2 and Table 3).

Table 1. Measure items and methods

Items	Method
Body height	Distance from the head to the ground vertically
Cervical height	Distance for the seven cervical point to the ground vertically
Arm length	Length from the elbow to the wrist point（right side）
Waist height	Distance from the waist line to the ground vertically
Chest girth	Chest girth through the breast nipples
Neck base girth	Neck root Girth
Whole should width	Length of the curve through the left shoulder point to the right point by way of the back neck point
Waist girth	Natural waist curve girth between the top of hipbone and the low edge of the rib
Hip girth	Curve girth around the most plump part of hip

Table 2. Comparison of body types frequency in female

Body type	Breast-waist difference（cm）[16]		GB [16]		Samples in 2009		Samples in 2015	
	GB	Calculated Value	Whole country	middle western areas	N	%	N	%
X	—	>=24.5	—	—	2	0.39%	5	1.62%
Y	19~24	18.5~24.4	14.82%	17.50%	102	19.69%	62	20.06%
A	14~28	13.5~18.4	44.13%	46.79%	233	44.98%	138	44.66%
B	9~13	8.5~13.4	33.72%	30.34%	118	22.78%	86	27.83%
C	4~8	3.5~8.4	6.45%	4.52%	16	3.09%	17	5.50%
D	—	<=3.4	—	—	47	9.07%	1	0.32%
Total			99.12%	99.15%	518	100%	309	100.00%

Table 3. Comparison of body types frequency in male

Body type	Breast-waist difference（cm）[17]		GB [17]		Samples in 2009		Samples in 2015	
	GB	Calculated Value	Whole country	middle western areas	N	%	N	%
X	—	>=22.5	—	—	—	—	1	0.40%
Y	17~22	16.5~22.4	20.98%	19.66%	10	1.61%	35	14.00%
A	12~16	11.5~16.4	39.21%	37.24%	155	25.00%	88	35.20%
B	7~11	6.5~11.4	28.65%	29.97%	239	38.55%	89	35.60%
C	2~6	1.5~6.4	7.92%	9.50%	149	24.03%	32	12.80%
D	—	<=1.4	—	—	67	10.81%	5	2.00%
Total			96.76%	96.37%	620	100.00%	250	100.00%

Table 2 showed the same values as GB, Type A is still the largest proportion in female body. But, the proportion in male body has greatly changed (see Table 3). The result of data analysis is, male samples that were both measured in 2009 and 2015, type B has become the largest number, and the numbers of type C and D are both showing a tendency to increase.

In order to understand the change of human body, the mean of main measurements of type A, B, C and D between GB/T 1335.2-2008 and the samples were compared. (see Table 4 and Table 5) The comparison shows that regardless of whether male or female, the girths (bust, waist and hips) have increased significantly, but the change of the heights are not obvious. It is not consistent with many other research results. The result also shows that the method of only using the value of chest and waist to divide human body shape is not reasonable.

Table 4. Comparison of the mean of main body measurements in female (cm)

Body type		Body height	Cervical height	Arm length	Waist height	Chest girth	Neck base girth	Whole should width	Waist girth	Hip girth
Y	GB	160.0	136.2	50.4	98.2	84.0	33.4	39.9	63.6	89.2
	2009	158.5	133.1	53.6	98.4	88.9	36.0	38.6	68.6	91.4
	2015	160.9	136.7	51.9	100.3	89.9	36.5	39.2	69.7	91.2
A	GB	160.0	136.0	50.4	98.1	84.0	33.7	39.9	68.2	90.9
	2009	156.8	131.6	53.1	97.4	92.6	37.1	38.9	76.7	93.6
	2015	157.7	134.1	51.0	98.0	92.0	36.9	39.4	76.1	92.5
B	GB	160.0	136.3	50.5	98.0	88.0	34.7	40.3	76.6	94.8
	2009	157.9	132.6	54.2	98.0	97.8	39.4	39.7	86.3	98.3
	2015	157.7	134.2	51.2	97.9	96.7	38.2	40.0	84.9	96.0
C	GB	160.0	136.5	50.5	98.2	88.0	34.9	40.5	81.9	96.0
	2009	157.1	131.9	53.0	96.9	100.7	40.5	43.0	94.6	102.4
	2015	157.3	134.2	51.4	97.3	103.3	39.9	40.4	96.5	104.8

Table 5. Comparison of the mean of main body measurements in male (cm)

Body type		Body height	Cervical height	Arm length	Waist height	Chest girth	Neck base girth	Whole should width	Waist girth	Hip girth
	GB	170.0	144.8	55.4	102.6	88.0	36.3	43.6	69.1	87.9
Y	2009	174.6	147.6	56.9	107.8	88.7	41.4	43.8	70.6	87.9
	2015	172.3	147.0	56.4	106.7	91.8	40.5	44.3	73.4	92.0
	GB	170.0	145.1	55.3	102.3	88.0	37.0	43.7	74.1	90.1
A	2009	170.0	144.0	55.2	104.1	85.1	41.9	44.2	71.9	86.9
	2015	170.7	145.7	56.3	106.3	92.7	42.1	43.7	79.3	92.8
	GB	170.0	145.4	55.3	101.9	92.0	38.2	44.5	82.8	94.1
B	2009	168.9	143.2	54.4	103.3	85.8	42.3	44.1	76.8	88.9
	2015	170.2	145.7	56.3	106.3	95.4	42.5	43.3	86.2	95.2
	GB	170.0	146.1	55.4	101.6	96.0	39.5	45.3	92.6	98.1
C	2009	169.4	143.9	55.0	102.8	90.3	43.5	44.1	86.1	92.9
	2015	169.5	145.6	56.5	106.7	100.4	44.0	43.8	95.2	97.8

3.2 Distribution by age stage

According Chinese Human Body Measurement project organized by China National Institute of Standardization (CNIS), the adults are divided into four groups by age. Table 6 and Table 7 shows the mean of main body measurements of female and male by age stage in samples 2015. The result shows that with the increase of age, the adult's height measurements decreases, and their girth measurements increases.

Table 6. The mean of main body measurements of female by age group in samples 2015 (cm)

Age stage	N	%	Body height	Cervical height	Arm length	Waist height	Chest girth	Neck base girth	Whole should width	Waist girth	Hip girth
18~24	60	19.40%	160.7	136.7	51.9	100.5	87.4	36.0	39.0	69.0	90.8
25~34	51	16.50%	160.0	135.8	51.4	99.0	87.8	36.1	39.0	70.3	90.3
35~59	153	49.50%	157.7	134.2	51.1	97.9	96.2	38.0	39.8	82.1	94.9
60~75	45	14.60%	156.1	132.9	51.0	97.2	99.2	38.1	40.1	87.1	98.3
Total	309	100.00%	158.4	134.8	51.3	98.5	93.5	37.3	39.6	78.3	93.9

Table 7. The mean of main body measurements of male by age group in samples 2015 (cm)

Age stage	N	%	Body height	Cervical height	Arm length	Waist height	Chest girth	Neck base girth	Whole should width	Waist girth	Hip girth
18~24	59	23.60%	171.1	146.0	55.9	106.3	90.6	41.3	43.5	76.4	92.6
25~34	67	26.80%	172.3	147.0	56.5	107.1	92.8	42.3	44.3	79.7	94.2
35~59	101	40.40%	169.3	145.1	56.4	106.0	97.6	42.9	43.5	88.5	95.1
60~75	23	9.20%	168.2	144.3	56.0	105.4	97.0	42.4	42.7	88.6	95.4
Total	250	100.00%	170.4	145.8	56.3	106.3	94.6	42.3	43.6	83.3	94.3

3.3 Distribution by age stage and body type

Fig.1 showed the distribution by age and body type in samples 2015. For example, female aged from 35 to 59, type A is the largest number, and type B is second largest.

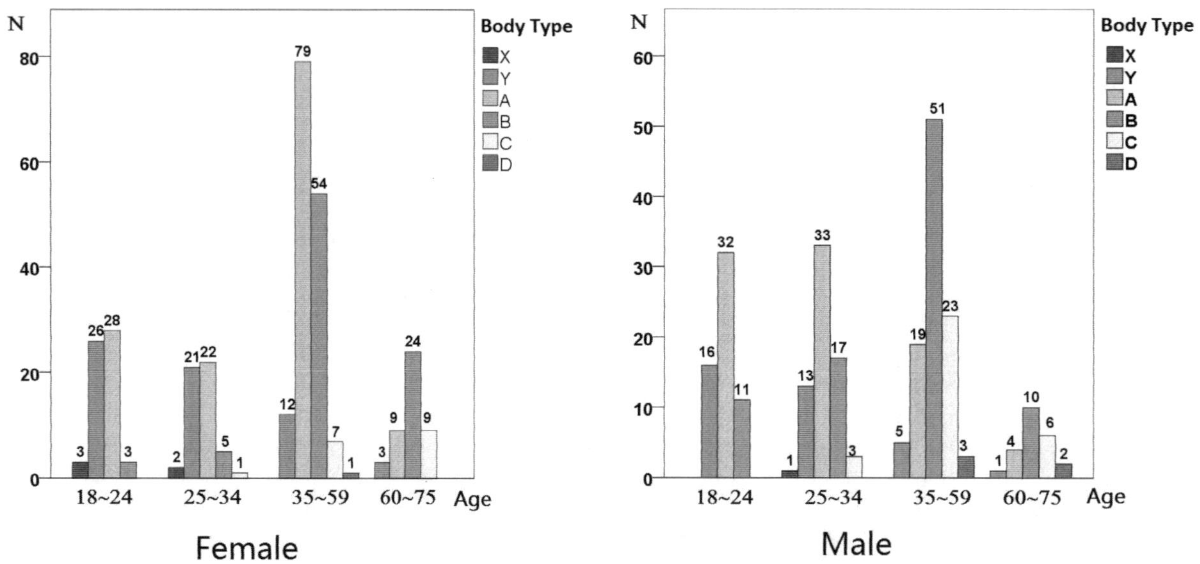

Fig.1 Distribution by age stage and body type in samples 2015

Table 8. The mean of main body measurements divided by both age stage and body type in samples 2015 (cm)

Age stage		Female						Male					
		X	Y	A	B	C	D	X	Y	A	B	C	D
18~24	Body height	166.7	160.4	160.2	160.9				171.4	170.7	171.8		
	Cervical height	141.8	136.3	136.4	137.1				146.1	145.8	146.7		
	Arm length	53.8	51.7	52.0	51.3				56.0	56.0	55.6		
	Waist height	103.5	100.3	100.4	99.6				105.9	106.3	106.8		
	Chest girth	92.7	88.8	85.6	87.3				91.4	90.0	90.9		
	Neck base girth	37.0	36.0	35.9	36.8				40.7	41.3	41.9		
	Whole should width	38.3	38.8	39.2	39.0				46.1	46.1	45.2		
	Waist girth	67.3	68.3	69.1	75.1				72.9	76.5	81.2		
	Hip girth	93.6	90.6	90.5	93.6				92.0	92.2	94.6		
25~34	Body height	159.1	163.9	156.9	157.4	159.7		161.3	174.1	171.7	172.7	172.3	
	Cervical height	133.6	139.2	133.1	133.9	135.0		136.9	148.6	146.3	147.7	147.6	
	Arm length	51.8	52.5	50.3	51.4	50.0		54.0	56.9	56.2	56.8	57.7	
	Waist height	97.6	101.8	97.0	96.7	96.5		99.0	107.8	106.7	108.0	107.1	
	Chest girth	92.8	88.6	86.7	85.5	92.9		94.9	91.6	92.4	93.1	101.4	
	Neck base girth	35.6	36.4	35.8	36.1	38.0		41.4	40.4	42.8	42.6	43.0	
	Whole should width	38.0	39.7	38.5	38.1	40.1		47.8	47.1	46.5	46.9	49.0	
	Waist girth	67.3	68.7	70.5	74.8	85.7		68.7	73.6	78.9	83.4	96.1	
	Hip girth	87.0	90.7	89.6	92.3	94.2		87.9	92.5	93.4	96.6	99.4	
35~59	Body height		157.9	157.6	157.9	157.6	159.0		170.3	169.9	169.4	169.1	165.1
	Cervical height		134.4	134.1	134.4	134.4	136.3		145.5	145.2	145.0	145.5	142.0
	Arm length		51.1	51.0	51.3	51.6	53.2		56.0	57.2	56.2	56.6	54.2
	Waist height		98.2	97.8	97.9	97.6	98.1		105.6	106.0	105.8	106.7	104.4
	Chest girth		93.6	94.7	98.1	101.5	100.0		93.9	97.6	96.8	100.8	94.9
	Neck base girth		37.7	37.5	38.7	38.8	41.2		40.4	42.4	42.6	44.2	43.3
	Whole should width		39.1	39.5	40.2	40.5	44.3		43.5	47.5	46.5	47.7	46.5
	Waist girth		73.9	79.2	86.1	94.6	97.6		75.5	84.6	87.7	95.5	94.3
	Hip girth		92.3	93.7	96.2	103.8	95.0		90.9	92.9	95.2	97.5	94.9
60~75	Body height		156.6	153.3	156.9	156.8			175.0	166.2	168.2	169.6	165.2
	Cervical height		132.7	130.4	133.5	133.9			149.8	143.1	144.4	144.9	141.8
	Arm length		51.5	50.5	51.0	51.3			57.4	56.2	56.6	55.3	54.7
	Waist height		98.0	94.9	98.0	97.2			108.4	104.1	105.5	106.5	103.2
	Chest girth		92.4	100.3	97.0	106.0			89.6	93.1	97.5	98.7	100.7
	Neck base girth		36.0	37.4	37.5	41.0			38.5	41.8	41.8	43.8	44.5
	Whole should width		39.2	40.5	40.0	40.3			48.3	43.7	45.6	47.4	48.9
	Waist girth		73.2	84.2	85.4	99.1			68.4	79.6	89.0	93.2	101.4
	Hip girth		95.2	94.9	96.7	106.9			89.5	92.9	94.2	98.1	101.6

Table 8 showed the mean of main measurements divided by both age stage and body type in samples 2015. It showed more detail that should be considered in building garment size system and producing and sale garments.

5. Conclusion

By comparing the body measuring data of GB1335-2008 with the samples data measured in 2009 and 2015, the distribution of body type and the measurements of Mediate Body of each type were analyzed. The result shows that people's body shape has changed a lot, and the girths significantly increased. Furthermore, the number of type B greatly increased, and has a tendency of over type A. So, the method from GB/T 1335-2008 which only using the value of chest and waist to divide human body type can no longer fit for the requirement of garments production. Some advanced methods of dividing human body type, building garments sizing systems and standards should be presented through organizing large scale survey activities, measuring body sizes, and comparing and analyzing the measurements. The mean of main body measurements of samples 2015 were computed by age stage. The cross table of age stage and body type was listed, which can support enterprises considering the integrated factors of consumer's age and body type in garment design, production and sale.

6. Acknowledgement

The authors wish to acknowledge the National natural science of China, Controlled surface geometry constraint conditions can change continuously the female breasts digital model research (Grant No. 61303120) providing support to this research.

References

[1] <Size designation of clothes> Standard Editor Groups. Instruction and Application of Chinese Standard <Size Designation of Clothes>. Beijing: Chinese Standard Publication 1992.
[2] Huang Y, Cui ZY, Xia M. Research on lower body shape of adult female in east China. Journal of Donghua University 2012; 38(3): 297-302.
[3] Wang HJn, Wang HF. Garment sizes for old aged people in the north part of Shaanxi. Journal of Textile Research 2009; 30(2) : 108-112.
[4] Song XX, Sun X. An analysis on the body forms of the middle and old aged men in the Songjiang area of Shanghai. Knitting Industries 2009; 4: 26-29.
[5] Gao Y, XIie H. Segmentation of women' s garment sizes based on age differences. Journal of Shanghai University of Engineering Science 2012; 26(2): 163-167.
[6] Nie YY. The study on optimizing male garment sizing systems in the lower yangtse river region. Shanghai: Donghua University 2007.
[7] Shang L, Xu YY, Du XH. Study on geographical difference of male youth somatotypes among different areas in China. Acta Anthropologica Sinica 2004; 23(1): 55-60.
[8] Lee JY, Istook CL, Nam YJ, Park SM. Comparison of body shape between USA and Korean women. International Journal of Clothing Science and Technology 2007; 19(5): 374-391.
[9] Park J, Nam Y, Choi K, L, Y, Lee K. Apparel consumers' body type and their shopping characteristics. Journal of Fashion Marketing and Management 2009; 13(3): 372-393.
[10] Song HK, Ashdown SP. Categorization of lower body shapes for adult females based on multiple view analysis. Textile Research Journal 2011; 81(9): 914-931.
[11] Shin JY, Nam YJ. A Study on Body Shape for 3D Virtual Body Shape. Transformation - Focusing on the Women with age of forties -. Fashion & Textile Research Journal 2015; 17(2): 265-277.
[12] Qi J. Research of MTM for sweater based on body measurement. Xi'an: Xi'an University of Engineering Science & Technology 2005.
[13] QI J, Li Y, Zhang X. Classification and high speed inquiry of body type of Chinese males. Textile Bioengineering and Information Symposium Proceedings 2008; 876-882.
[14] Qi J, Li Y, Zhang X, Description and classification of young men's body type in the west China. Journal of Textile Research 2010; 31(5): 107-111.

[15] GB/T 16160-2008, Location and method of anthropometric surveys for garments 2008.
[16] GB/T 1335.2-2008, Standard sizing systems for garments-Women 2008.
[17] GB/T 1335.1-2008, Standard sizing systems for garments-Men 2008.

Study on Free Curve Control Mode in Clothing Parametric Design

Xiao-Feng Wang*, Xue-Ying Song, Xin Zhang, Bo-An Ying

Apparel & Art Design College, Xi'an Polytechnic University, Xi'an, Shaanxi, 710048, P.R. China

*Corresponding author's email: Mr_wxf@163.com

Abstract

Parameterized clothing is when a whole pattern can be derived by changes of the relative position of a single point. As the point changes, the corresponding straight line and curve also changes. In this process, creating a curve line is more difficult than a straight line for parameterization and to create a parametric curve line design to satisfy the intention of designers more easily. As a result establishing a curve model of garment for supporting interactive design has become one of the top research studies. The study on curve controlling model is a key component of garment parametric design.

The purpose of this paper is to research a universal curve controlling model for meeting the needs of apparel design, which is faster, more flexible and more accurate for users to control the changes of curves.

In this paper, free curve and applicability of clothing were studied because different scribble gives different control modes. The apparels designed with curve modes have smooth curves and can meet the requirements of parametric design. The users can modify the curve parameters directly by this method, and if it is still not ok after that, they can modify the curve modes, which is very convenient and flexible to put into effect.

Keywords: Parametric clothing design; Curve model; Clothing curve model.

1. Introduction

Parametric design is used to define geometric size and size relations by a set of numerical parameters. The result of the design is driven by the changes of parameters. Parametric design of apparel pattern is to use a set of numerical parameters for restricting the sequence of structure size due to there are some corresponding connections between them, meaning it is possible to project an apparel pattern by designing a model once and transferring to new patterns by changing the number of parameters. With this technique it is possible to facilitate the change of coordinated type, which enables easy modification to the pattern, and avoids a lot of unnecessary repetitive work.

Parameterized garment model is essential to make patterns through the changes of the relative position of the points, corresponding straight or curve line can be changed by the changes of the points. In this process, a curve line is more difficult to construct than a straight line for parameterization and creating a parametric curve line design to satisfy the intention of designers more easily. As a result establishing a curve model of garment for supporting interactive design has become one of the top research studies. The study of curve controlling model is the key point of garment parametric design.

The purpose of this paper is to research a universal curve controlling model for meeting the needs of apparel design, which is faster, more flexible and more accurate for the users to control the changes of curves.

In recent years, with the continuous expansion of the domestic apparel market, more and more attention has been focused on intelligent clothing CAD garment industry research. The research in this paper focuses on clothing curve digitals and has obtained some preliminary results [1-7].

For example, Li Jingfang used Bezier spline curve, surface structure in his studies and conducted research in garment CAD applications [1]. Huang Jiang gave us some new costume pattern design methods such as cubic Bezier spline curves and T-Bezier curves [2]. Xu Xiaohui (2011) studied the generation prototype based on human characteristics tight collars, sleeves curve rule for clothing model provides a method for the automatic generation of research and technical reference [3]. Zhang qiao (2004) studied the garment CAD representation of the curve, the Bezier curve is the main technical analysis [4], and so on.

The above-mentioned study [1-7] benefited the promotion and application of intelligent clothing CAD, but after a long-term study, it was discoverd that the studies above have the following problems:

1) Some mathematical researches did not consider the nature of fashion design and costume design habits;

2) Some fashion design did not combine mathematical research;

3) Study on the one-side expression of a curve was defined in one type of curve, one type curve mode did not meet the requirements of many types of clothing design curve;

4) Some did not design curves from parametric design.

The purpose of this paper is to research the curves of a general control model, which can create the shape of the curve in a more flexible manner, with greater accuracy and speed which allows more costume design during the parametric clothing design process.

2. The characteristics and applicability of common free curve

At present, the cubic hermit curve, cubic parametric spline curves, second order or third order Bezier curve and B-spline curve, interpolation spline curve and nurbs curve are commonly used in apparel designing and modeling. This topic analyzes mathematical models and its properties as follows:

1) cubic hermit curve.

The general expression: Q (t) =At3+Bt2+Ct+D 0=<t=<=1; the initial conditions is: the position of the two points and the initial vector and terminal vector is determined, then the smooth curve according to the given conditions are calculated, as shown in Fig.1.

Fig.1 Cubic hermit curve

2) cubic parameters spline.

The general expression: Qi(t): =Ait3+Bit2+Ci+Di 0≤t≤l, i=l, 2, ··· , n ; A, B, C, D is Vector coefficient which can be determined by specific conditions.

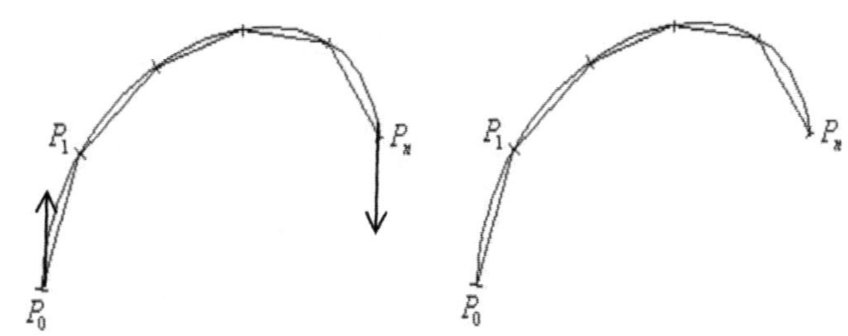

(a) the free end curve (b) control end free curve

Fig.2 Cubic parameters spline

As shown in Fig.2, when multiple Hermit curves realize smooth link, the cubic parameter spline curve are constituted, according to the difference of the condition of initial vector and terminal vector of the curve, the spline can be divided into control end free curve and free end control curve.

The initial condition: For control end free curve, as shown in Fig.2(a), each point is first coordinated, the direction and magnitude of vector of the start and end points are determined. Then the curve passes through all data points, and the final shape of the curve can be controlled by each point. The curve tangent vector at each point is influenced by coordinate transformation of adjacent control points. Above all, it is flexible and is easy to change the shape of the curve. Just remember to make sure that the coordinate and the tangent line are in the same direction.

For the free end curve, as shown in Fig.2(b), the initial condition is to give the coordinate of the end points. This curve is most suited when the end point vector is not required.

3) Bezier curve.

Bezier curve is based on a kind of approximation which is determined by a controlling polygon. Bezier curve is intuitive for combining functional approximation and the method of geometric representation. As shown in Fig.3, the route of second order Bezier curve is tracked by the function B (t) of given point P1,P2,P3: Quadratic Bezier, the general expression: $B(t) = (1-t)^2 P_0 + 2t(1-t)P_1 + t^2 P_2, t \in [0,1]$

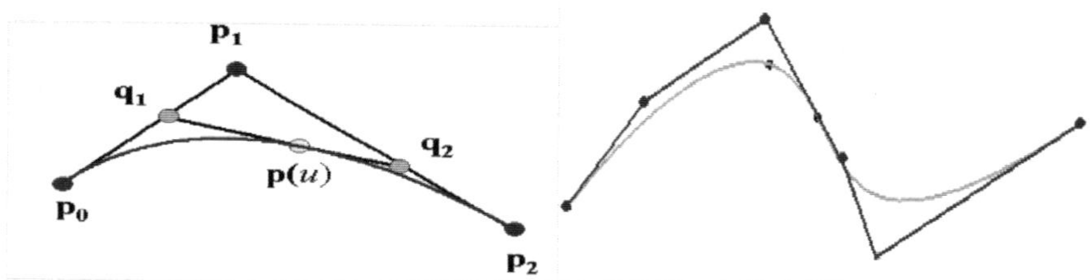

Fig.3 Bezier curve

Cubic Bezier, The general expression:

$$B(t) = P_0(1-t)^3 + 3P_1 t(1-t)^2 + 3P_2 t^2(1-t) + P_3 t^3, t \in [0,1]$$

It means that the curve is located in the “convex hull” of its control vertex. The “convex hull” is a convex range that is formed by the whole control points or part of it, which is called the convexification of Bezier curve.

4) B-spline.

The general expression is:

$$P(t) = \sum_{t=0}^{n} P_i F_{i,n}(t), t \in [0,1]$$

i=1, 2, …,

B-spline has similar feature with Bezier curve except for their basic function. It maintains not only the intuitive, convenience and convexity of Bezier curve, but also has the advantages of having local adjustability, the shape of the curve is more close to the controlling polygon, and can be calculated quite easily, so it has been widely used nowadays, Quadratic B-spline can be expediently used in the design of the hem of clothing, as shown in Fig.4:

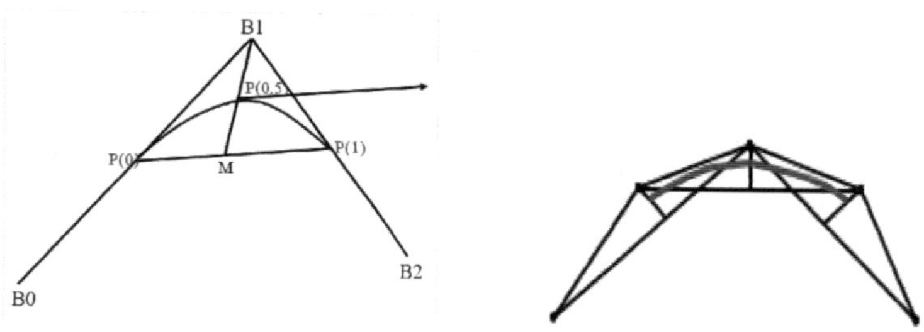

Fig.4 B-spline

3. Analysis of the applicability of the main clothing curve

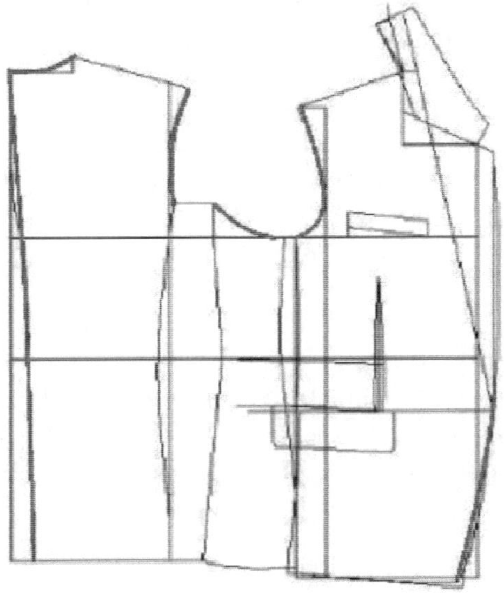

Fig.5 Curves in clothing

In the process of pattern making, there are numerous factors we should consider: human body structure, design factors, fabrics factors, technology factors and so on. The control of the size of each part, the aesthetic degree of the cutting seam, the connection of each pattern and the complexity of the sewing technology expresses the clothing curve line. The ever-changing clothing curve line is crucial for the connotation of clothing design and the quality of clothing products. As shown in Fig.5.

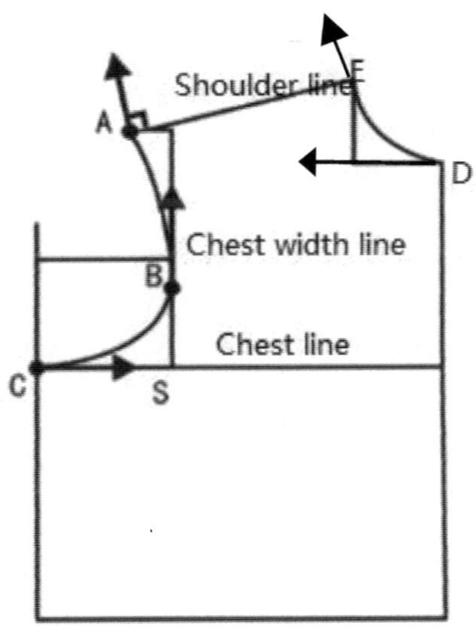

Fig.6 Control point and direction of collar nest line and Armhole line

For example, when drawing the neckline, two points can make a curve line, the direction of the curve line are the normal direction of shoulder line and front center line respectively. Hermit curve line is very suitable for drawing necklines. As shown in Fig.6.

A neckline is drawn through three points, as shown in Fig.6, the three points' direction on the curve line are the normal direction of bust line, chest breadth line and shoulder line respectively. Tangential direction of the end point of the first half of the armhole curve line should be subject to shoulder line, tangential direction should ensure vertical to shoulder line, so that the smooth transition of neckline at neck point when the front and behind cloths are pieced together can be ensured. This curve line is suitable for drawing with multi hermit curve line for these three points that has a vector requirement.

When drawing the back line of a suit, the following key points must be determined and then connected in turn. The requirements of the points are not high, it is suitable for using free end free curve line to draw the line.

With the change of the garment, there are different requirements for curve line. The curve line in this topic can basically meet the design requirements of clothing curve. However, these curves are used for not only meeting the requirements of parametric design but also to easily control the focus of our research.

4. The research on the control model for clothing curve line

The analysis above illustrates that different parts should use different curve models. In this topic, in order to use the curve line more conveniently, the following aspects for the design of the curve pattern must be considered: 1.accurate control to meet the needs of parametric design; 2.easy control; 3: use the name of points as parameters for obtaining the coordinates of points. Clothing curve control model is developed as follow.

4.1 Collar nest line control mode

As shown in the last section, the basic logic of drawing a neckline is similar to hermit curve line. The characteristics is that the start direction is in the horizontal direction, vector size defaults to the distance of beginning point to the end point, and the size is controlled by times of modulus. The terminal direction is vertical to shoulder line. In this passage, the vector size is defined same as the shoulder line and the size is controlled by times of modulus. It not only meets the requirement of the curve line itself, but also meets the requirement of designing a curve line. Meanwhile, users can control it well with this process. The basic mode is that when users input a curve mode, they only need to input the name of start point [ ], end point [ ] and control point [ ], if the line does not meet the requirement, they can control it by modulus multiple of start point [1] and end point [1].

Table 1. Collar nest line control mode

Curve	Collar nest line	start point [ ],end point [ ],control point [ ],Modulus multiple of start point [1], Modulus multiple of end point [1]

4.2 Armhole line control mode

It is possible to draw a neckline through three points, like fig 6, the three points' direction on the curve line are the normal direction of bust line, chest breadth line and shoulder line respectively. Tangential direction of the end point of the first half of the armhole curve line should be subject to shoulder line, tangential direction should ensure vertical to shoulder line, so that the smooth transition of neckline at neck point when the front and behind cloths are pieced together can be ensured. The basic mode is that when users input a curve mode, they only need to input the name of start point [ ], intermediate point [ ] and control point [ ], if the line does not meet the requirement, they can control it by modulus multiple of start point [1] intermediate point [1] and end point [1]

Table 2. Armhole line control mode

Curve	Armhole line	Start point [ ], intermediate point [ ],end point [ ],control point [ ], Modulus multiple of start point [1],Modulus multiple of intermediate point [1], Modulus multiple of end point [1]

4.3 Control end curve control mode

Control end free curves apply to curves that should pass through a few points and the direction of the two endpoints has to be controlled strictly. In this topic, the direction of starting vector and end points passes the control points, and the size of the vector can be controlled by the extent and angle.

Table 3. Control end curve control mode

Curve	Control end curve	Start point [ ], intermediate point [ ], end point [ ], start control point [ ], long [ ], angle [ ], start control point [ ], long [ ], angle [ ]
Curve	Control end curve	Start point [ ], intermediate point [ ], intermediate point [ ], end point [ ], start control point [ ], long [ ], angle [ ], start control point [ ], long [ ], angle [ ]

4.4 The free end curve control mode

The free end curves are made by connecting a few points in turn, and the number of points can be adjusted by users themselves.

Table 4. The free end curve control mode

Curve	The free end curve	Point [ ], point [ ], point [ ]
Curve	The free end curve	Point [ ], point [ ], point [ ], point [ ]

4.5 The multi-point hermit curve control mode

When clothing curves are more complex, it is important to control a curve with two points, as well as inputting the start and end vectors. When the curves pass through multiple points, and the direction of each point are strictly controlled, it can also be used in this mode.

Of course, the parameter of this curve model is the largest when compared with others, and users should choose the last few curve modes.

Table 5. The multi-point hermit curve control mode

Curve	The multi-point hermit curve	Point [ ], Horizontal modulus [1], Vertical modulus [1]
Curve	The multi-point hermit curve	Point [ ], Horizontal modulus [1], Vertical modulus [1], point [ ], Horizontal modulus [1], Vertical modulus [1]

5. Verification of clothing curve control mode

Having designed the curve control mode above, a parametric CAD model costume design was applied, Xi'an Polytechnic University self-developed an automatic pattern making software AutoPattern. The suit pattern was developed, and the main results of the experiment are as follows. As shown in fig.7. The black line represents the quantity of bust ease as 160, the black line represents bust ease as 180.

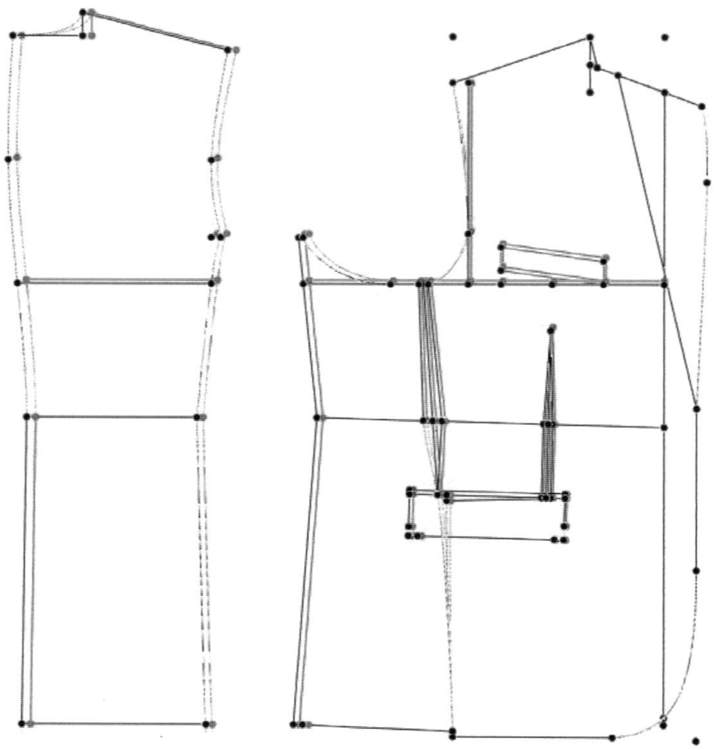

Fig.7 results of experiment

The smoothness of the plate curve is determined by two factors: The first factor iswhether the curve data points (key points) are good or not, the second factor is whether the curve mode is suitable. In Fig.5 and 6, the three curve modes have very smooth curves, which suggests that the three curve modes have an ideal effect when the position of data points is suitable. In this case, the collar nest line, armhole curve line, back stitching, these three curve modes had comply with the designer's ideas and requirements which were different. Different designers can also pick other curve modes. When the users are not satisfied with the results after parameterization, they can modify the curve parameters. If it is still not ok, they can modify the curve modes directly, which is very convenient and flexible to manage. In this experiment, parameterization is applied to modify the bust line of the suit, and the plates change with the changes of the parameters. As shown in the last figure, the curve lines change with the changes of the parameters, and the more smooth the curves, the more convenient the application is for users.

6. Conclusion

In this paper, free curve and the applicability of clothing were studied because different Scribble gives different control modes, in order to meet the requirements of the fashion design free curve model. When these models were applied in clothing parametric design, smooth clothing curve can meet the parametric design requirements, and can repair the curve parameters first for controlling the curve, if modified curve parameters cannot meet the requirements, you can modify the curve model that is convenient, flexible and easy to use.

7. Acknowledgment

This research is supported by the Shaanxi science and technology innovation engineering by overall planning project (2012KTCQ01-42).

References

[1] Fang LJ. Parametric curves and surfaces modeling in the garment CAD, Zhejiang University of Technology: 2010.

[2] Huang J, Some issues in garment CAD modeling of curves and surfaces. Zhejiang University of Technology: 2008.

[3] Xu XH. Study on rule curve of tight collar and sleeves based on the characteristics of the human prototyp, Soochow University: 2011; 2: 59-67.

[4] Zhao NP, Zhang WY, Zhang HZ. Development and prospects of apparel CAD technology development and prospects [J], Tianjin Institute of Textile Technology: 2006; 9: 120-116.

[5] Ling HY. Research on method of Intelligent garment pattern design [D]. Tianjin University of Technology: 2008.

[6] Cao WL. Research on aided garment parametric system [D]. Beijing Institute of Clothing: 2007.

[7] Tan ZH. Object-oriented clothing hit / push system [D]. Xiangtan University: 2007.

Artificial Intelligence: Technology and Application in Apparel Manufacturing

Rajkishore Nayak[1], Kanesalingam Sinnappoo[1*], Lijing Wang[1], Rajiv Padhye[1]

[1]School of Fashion and Textiles, RMIT University, 25 Dawson street, Brunswick, VIC 3056, Australia

*Corresponding author's email: sinnappoo.kanesalingam@rmit.edu.au

Abstract

Apparel manufacturing is labor intensive, which is characterized by low fixed capital investment; a wide range of product designs and hence input materials; variable production volumes; high competitiveness and often high demand on product quality. To cater these demands, the labor intensive processes should be converted into automated processes accomplished by the use of computers, models, digital components and artificial intelligence. Artificial intelligence is the field of study that deals with the synthesis and analysis of computational agents that act intelligently. In this review, a systematic study has been done to explore various applications of artificial intelligence in apparel manufacturing. This also includes different types of artificial intelligence such as expert systems, neural network, Fuzzy logic, genetic algorithm and other approaches used in garment manufacturing.

Keywords: Artificial Intelligence; Apparel Manufacturing; Supply Chain Management; Garment Inspection; Retailing

1. Introduction

In this era of information technology, Artificial Intelligence (AI) has revolutionized the field of engineering, physics, medicine and management [1]. Significant results like improving quality, increasing productivity and lowering production cost can be achieved with the help of AI [2,3]. Around a decade ago, the main emphasis on apparel production was solely on the volume of production, which has now changed a lot. The recent focus is on achieving good quality, reduced cost, appropriate information management, statistical process control, just-in-time manufacturing and computer integrated manufacturing in addition to the volume of production [4,5].

AI is a field of computer science that can simulate characteristics of human intelligence and human sensory capabilities [6]. Conventional computers lack the ability to learn and this restricts them to operate only under the conditions for which they are programmed. In essence the limitations of computers can be summarized that:

- They are not intelligent,
- They have no self-awareness, and
- They have no understanding of the environment.

In contrast although humans are slow and unreliable compared to computers, they possess the intelligence and understanding that allow them to solve infinitely varied problems. AI is the creation of machines that can simulate human intelligence and work accordingly. The heuristic and intelligent nature of AI assists in providing superior solutions over the classical techniques.

Apparel manufacturing is labour intensive, which is characterised by low fixed capital investment; a wide range of product designs and hence input materials; variable production volumes; high competitiveness and often high demand on product quality. Although the manufacturing process is associated mainly with apparels and household linens, it is also used in a variety of industries and crafts such as upholstery, shoe-making, sail-making, book-binding and the production of varieties of sporting goods. Sewing is the fundamental process with ramifications into a variety of textile arts and crafts, including tapestry, quilting, embroidery, appliqué and patchwork. The apparel manufacturing process evolved as an art and underwent several technical changes. The technological advancements in the apparel industry include the use of computerized equipment (especially in design, patternmaking and cutting), 3D scanning technology, integration of wearable technology, advanced material transport systems, automation and use robotics or artificial intelligence. The recent important development involves increasing use of robotics or AI to transport components and materials within the plant or facilitate the production, which helps in improving the production efficiency. However, the apparel industry especially the sewing technology has remained significantly less automated compared to many other manufacturing industries.

Apparel manufacturers have to produce a diverse product mix as consumers are difficult to understand and predict [4]. Their choice is unstable and unpredictable, and there is wide variation in their demographics and physiographics. The product quality depends on several factors related to yarn manufacturing, fabric preparation, chemical processing and garment manufacturing. Hence, all these factors can be better controlled by the application of AI in apparel manufacturing. Although there is some automation, the apparel industries are still far behind the other sectors and rely on manual intervention. AI is gaining impetus over the last two decades, in the apparel industry in different areas. The automation of various instruments by the application of AI in spreading, cutting, sewing and material handling can reduce the production cost and minimize faults.

This review focusses on various applications of AI in the apparel industry such as pattern design, production planning and control (PPC), marker making, sewing automation, sales forecasting, fashion recommendation and supply chain management (SCM). A number of AI techniques are employed to handle these problems, such as expert system (ES), neural network (NN), fuzzy logic (FL) and genetic algorithm (GA).

2. Types of artificial intelligence

AI can be broken into several different disciplines. Each of them is unique, but often they intermix to accomplish a specific task. The different disciplines, mainly used are:

- Expert system,
- Neural network,
- Fuzzy logic,
- Genetic algorithm.

2.1 Expert System

Expert system is a knowledge intensive computer program, which solves a problem that normally requires human expertise. It operates in the same manner as a human expert. Expert systems are made of two parts: a knowledge base and an inference engine. The program in the expert system incorporates the knowledge of the expert consisting of facts and heuristics by storing it in the form of rules. The inference engine is the software used to search and interpret the rules in the knowledge base by logical deduction. Facts are the information that is widely available and generally agreed upon by the experts. Heuristics are the rule of thumb derived from personal experience and judgment of individual experts. Production rules of IF/ THEN form are the easiest way of coding the knowledge. 'IF' portion contains the conditions and 'THEN' portion contains the actions. By prompting the right questions and considering the user's reply, an expert system decides which element of knowledge to use as the basis for further questioning until the intended goal is reached. This goal is the information, which the expert is expected to provide.

Expert systems can be built in two ways. One way is to develop the computer code by using programming languages like PROLOG and LISP, or by using any of the high level programming languages such as BASIC and FORTRAN. The other way is to use expert system building tools called expert system shells. These are specially designed programs by specialists in computer science. MYCIN is the first expert system developed in the world. This expert system has been used by physicians to diagnose certain forms of bacterial blood infections and to determine suitable treatments for various diseases.

2.2 Neural Network

Mankind is distinguished from other animals by their ability to think and the intelligence. Animals react adaptively to the changing environment by their nervous system. Similar responses are produced by the appropriate simulation of the nervous system. The nervous system is built up by simple units: The Neurons. Human brain consists of 1011 to 1014 neurons. These neurons receive any process and transmit electrochemical signals via synaptic connections. Artificial Neural Network (ANN) behaves similar to a human brain in a digital computer. ANN may be applied where traditional techniques have failed to give satisfactory results, or where small improvement in modeling performance can make a significant difference in operational efficiency.

ANN is the representation of biological neural network composed of many artificial neurons, which are linked together by special network architecture and have the ability to learn from examples. Though artificial neural network was established before the advent of computers, it has evolved into a powerful tool for solving complex problems only recently. In AI, the input signal is converted to output signal by the neurons obeying some transfer function. This output becomes an input to the subsequent neuron. A simple neuron is shown in Fig.1.

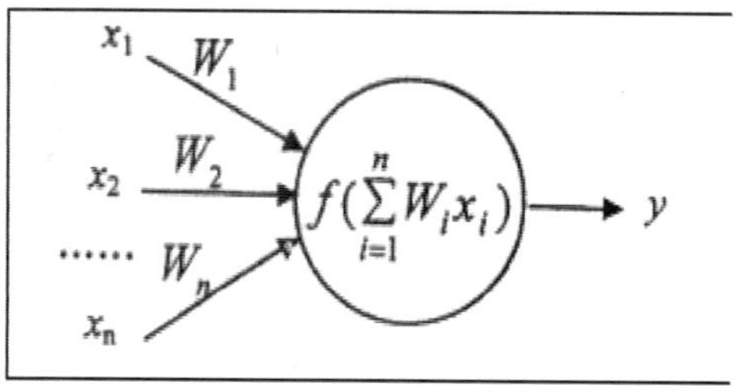

Fig.1 A simple neuron

In the above figure x1, x2,....., xn are inputs; W1, W2,....., Wn are the weights and y is the Output. The single neuron converts the input units to output units according to the weights and the transfer function. The single neuron acts in a linear behavior. This is also known as Single Layer Perception (SLP). All most all the neural networks are based on Multi-layer Perception (MLP) consisting of an agglomerate of neurons interconnected in some specific architectural way. Fig.2 (a) shows the typical neural network with one hidden layer and Fig.2 (b) with multiple hidden layers between the input and the output. The selection of the number of hidden layers is important. AI may follow a linear function or a non-linear (sigmoid) function.

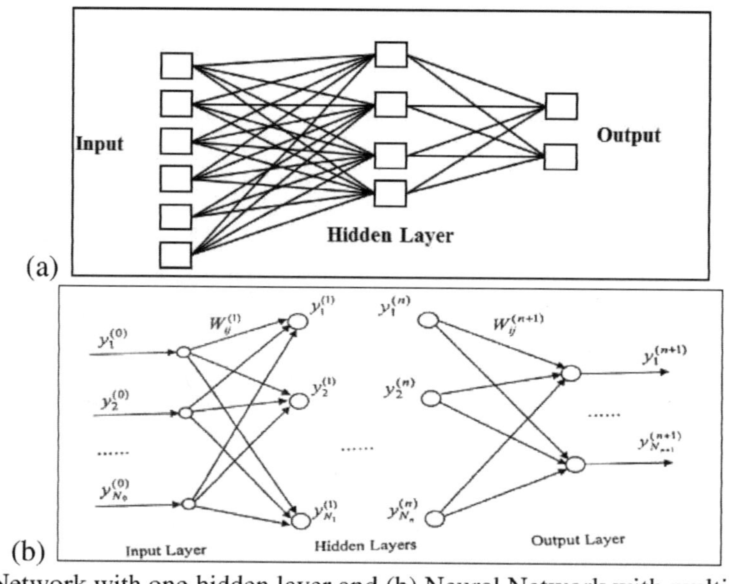

Fig.2 (a) Neural Network with one hidden layer and (b) Neural Network with multiple hidden layers.

In MLP the units are arranged in distinct layers and each unit receives weight input from each unit in previous layer. Initially the output is calculated with initial set of weights. Then weights are modified according to the expected output by reducing the difference between the actual and expected output values. Each input and output pairs would be presented to the network in turn and small changes made to weights every time. The presentation of all the training pairs to the network is known as an epoch. Training may require many hundreds of thousands of epochs and could take many hours even in a fast computer.

When the training is complete the values of the weights are fixed. During the recall phase the weights are not modified but the network can be used to predict the output values when set of input values are given. If the network has been trained correctly then it will be able to do this accurately for all the training data and also for other input values that were not included in the training set. This important property is called generalization. The training of the neural networks is a complex activity and it can be done by Back Propagation, Counter Propagation, Standard Conjugate Gradient, Dynamic LVQ etc. Among these the Back Propagation is widely used.

2.3 Fuzzy Logic

Fuzzy logic is used where the behavior is not linear (e.g. the trend in today's dyeing industry is to use highly concentrated dyeing liquor that do not behave in a linear fashion due to interaction between the dye molecules). In fuzzy logic IF/ THEN reasoning is used when applying a Fuzzy Inference System (FIS). Adaptive-Neuro-Fuzzy Inference Systems (ANFIS) are used where the back-propagation method is implemented like work for modeling or fitting of spectroscopic data. The number of Membership Functions (MFs) has a clear influence on the ANFIS training results. The root-mean-square (RMS) error is determined as the difference when the number of MFs increases.

2.4 Genetic algorithm

Both the constrained and unconstrained optimization problems can be solved by GA, which is a method based on a natural selection process that mimics biological evolution. In GA, a population of individual solutions is repeatedly modified by the algorithm. GA is a part of the larger class of evolutionary algorithms (EA), which generate solutions for optimization problems using techniques inspired by natural evolution, such as inheritance, mutation, selection and crossover. GA is an iterative procedure that maintains a population of chromosomes representing different possible solutions to an optimization problem. Each individual iteration is called a generation and in each generation, the fitness of each chromosome is evaluated by a suitable fitness function. By this approach, the optimal solution can be obtained in various fields of apparel manufacturing such as apparel design, marker planning, PPC, production scheduling, fabric and other materials location and replenishment decision making in apparel supply chain.

2.5 Other AI approaches

Various other approaches such as artificial immune system (AIS), genetic programming (GP), multi-agent system (MAS) and simulated annealing (SA), have also been used in various section of apparel manufacturing. AIS is inspired by theoretical immunology and observed immune functions, which is an adaptive system based on principles and models. GP is an evolutionary algorithm, which is inspired by biological evolution and based on computer programs that perform a user-defined task. MAS is composed of multiple interactive intelligent agents, those possessing the characteristics of interaction and cooperation and can help the agents execute different tasks with autonomous functions on receipt, transfer and delivery of information simultaneously. SA approach is based on a generic probabilistic meta-algorithm, which is inspired by the annealing process in used in metallurgy.

3. Applications of AI

Out of the three main classes of Artificial Intelligence, Neural Networks are mostly used in apparel production in following fields:

1) Prediction of mechanical properties,
2) Classification and grading,
3) Identification and analysis of faults,
4) Process control and online monitoring,
5) Supply chain management and retailing.

Some of the applications of AI are discussed here.

3.1 Fabric analysis, grading, selection and sewability

- AI can be applied for fault inspection and grading of incoming fabrics to a garment industry. The image of the faults are stored in the system and compared with the captured image. The grading of fabrics can be done on the basis of frequency and severity of the faults by the application of AI. Fabric quality is dependent of the fibre type, yarn properties and fabric structure. For different types of clothing, the fabric quality requirements are different. AI can be used for appropriate fabric selection depending on the endue application of the clothing.
- Furthermore, during the fabric manufacturing a pre-trained neural network can be used to detect and mark the fabric faults in the loom or knitting machine by charge coupled device (CCD) cameras. This can be used as a process control tool and save the time and cost of fabric checking at later stages.
- AI system can be used to find the sewability of different fabrics during garment production [7,8]. Fabric mechanical properties affect the performance during spreading, cutting and sewing. For example Chen et al [9]. used fabric mechanical properties as input parameter to evaluate their performance during garment manufacturing.
- Fabric can be classified according to their hand value by the application of AI. This can be used for selecting right type of fabrics according to the season. A fuzzy neural network can predict the total hand value of outer fabrics [10]. It can be applied for process control, online monitoring, process planning sales and other industrial issues.
- The performance properties of apparel fabrics like seam puckering, seam slippage, yarn severance, drapability and crease retention can be predicted by AI. Seam puckering is detected by back-propagation algorithm by taking fabric GSM, thickness and bending stiffness in warp and weft way as input. Also two networks can be used for the prediction of puckering. The process involves pattern reorganization by a laser beam followed by learning.
- AI can be applied for intelligent manufacturing of clothing. The network can predict the clothing properties after a particular process. So it can be used for fabric engineering and suitable garment designing for a given fabric.

3.2 Colour solutions

- Colour is the first element of design to which the consumers respond. People select the clothing on the basis of the colour appeal. AI system can be used for selecting favorite colours for a target group of customers. For example Hiu et al[11]. designed a fuzzy system that can predict the color for a group depending on the age, gender, height and skin colour.
- AI can be used for color matching of fabrics and shade sorting. It can be used for the true color production by predicting the concentration of dyes from their spectrophotometric absorbance.

3.3 CAD systems

- CAD systems are used in garment manufacturing for creating designs, pattern making and grading operations. Several attempts have been made by researchers to integrate AI with CAD systems to generate designs automatically. For example an intelligent system was developed by Zeng et al [12]. to assist the designers in decision making process. The objectives were to predict the fashion messages conveyed by the design elements in the garment and to extract elements concrete design elements from fashion design elements for real garment design.
- Experienced personal are needed for appropriate pattern design for different clothing styles. However, the AI system can be used to provide the expert knowledge of experienced designers [13]. Hu et al [14]. designed an expert system that combines the knowledge of experts and designers. It provided valuable suggestion on specific design by interactive design functionalities.

- The other area is the creation of virtual 3D garment designs. Fan et al [15]. developed an intelligent system to predict and display the drape images for women's using different fabrics and styles. The major problem here lies in exact prediction of the drape.
- To set-up a huge anthropometrical database for designing the garments that fit the wide ranges of variation in body type is time-consuming and uneconomical. To avoid that AI can be remarkably used for prediction of shirt pattern from 3D body measurements by multiple linear regression.

3.4 Production planning and control

- Production planning and control (PPC) coordinates between various departments of production so that delivery dates are met and the customer's orders are delivered on time. Various research activities focused on the problems related to PPC and avoid the bottlenecking [16,17] .
- AI can be used to solve or optimize the problem of machine layout, operation assignment and sewing line balancing. This can help in achieving the objectives of PPC objectives.

3.5 Sewing room problems

- An AI can be used to produce the required sewability, tailorability, tensile properties, surface properties, shear and bending properties in a fabric, which can be used for optimum selection of sewing room parameters.
- The judicious selection of fiber, yarn and fabric structure to achieve the required garment quality can be done by back-propagation algorithm. Selection of right fabric quality can reduce the sewing room problems and improve the productivity and efficiency.
- AI can be used to identify the relationships among clothing parameters, task requirements, environmental conditions and worker characteristics, and their effects on the performance of the worker. It can be used to predict negative effects of protective clothing on worker's performance.
- The sewing parameters for a particular type of fabric can be set by giving the input parameters of needle penetration force and needle heating to the AI.

3.6 Final garment inspection

- The final quality inspection of finished garments is still done by experts during garment manufacturing, which is very time consuming and subjective in nature. The final quality of garments depends on the sewing quality and other faults present in the finished garment. Automated inspection can be performed by the use of AI and image processing for inspection of these faults [18].
- In manufacturing of seamless garments [19] AI can be used to detect faults on line. The Image of the final garment is captured and compared with the standard and any variation from the standard is reported as a fault which can be mended at that time or a marking is done where the fault occurs.
- Using the AAATC wrinkle rating, garments can be evaluated and rated by AI. In wrinkle rating a laser sensor is used for evaluation of the quality and sharpness of the wrinkles and the garment is being rated.

3.7 Application in supply chain

- Supply chain management (SCM) integrates various business processes, activities, information and resources for creating value for customers. Appropriate SCM can manage the cost and business competitiveness. AI based models can be used for integrating and sharing information at any point in SCM [20,21].
- Various AI approaches can be used for effective inventory management, reorder point and inventory replenishment [22,23].
- Although SCM is a wide area in garment manufacturing, limited research has been done to address the issues in SCM. The future research should focus on the application of AI to optimize SCM network for solving decision making and inventory management problems.

3.8 Application in retailing

- AI can also be applied in retailing such as sales forecasting, style suggestion to consumers and fashion coordination. Historical data on sales in combination with the style, colour and garment size can be used for sales forecasting [24-26]. AI suggestion systems can be used for selection of appropriate style and design combination for consumers [27,28]. In addition it can be also used to select the right clothing combination depending on the occasion. In fashion coordination, the AI can be used to provide improves service to customers and increase sales. Professional advice can be given to consumers on style selection based on price.
- Today's consumers are more aware on the comfort features than before. AI can also be used for selection of right type of garment for providing necessary comfort including the appearance, which can be used by the customers [3,29,30].

4. Conclusion

In this modern era artificial intelligence is being used in many areas to solve various problems with intelligence similar to human being. Artificial intelligence is now making its path in apparel industry starting from material sourcing till retailing. Developed countries have already started using the AI to improve customer service and hence sales. Much progress is undergoing in this rapidly and in near future it will become an important tool for the export houses for enhancing quality, increasing production, lowering operating costs and exercising in house control over production leading to quick response and just-in-time concept.

References

[1] Vas P. Artificial-intelligence-based electrical machines and drives: application of fuzzy, neural, fuzzy-neural, and genetic-algorithm-based techniques. Cambridge, 1999. Oxford University Press.

[2] Shamey R, Hussain T. Artificial intelligence in the colour and textile industry. Rev of Progress in Color and Related Topics, 2003; 33(1): 33-45.

[3] Wong A, Li Y, Yeung P. Predicting clothing sensory comfort with artificial intelligence hybrid models. Text Res J, 2004; 74(1): 13-19.

[4] Nayak R, Padhye R. Introduction: the apparel industry. In Garment Manufacturing Technology, Nayak R, Padhye R, Editors. Elsevier, NY, 2014; p: 1-18.

[5] Guo Z, Wong WK, Leung SYS, Li M. Applications of artificial intelligence in the apparel industry: a review. Text Res J, 2011: 1-22.

[6] Eckert CM, Cross N, Johnson JH. Intelligent support for communication in design teams: garment shape specifications in the knitwear industry. Design Studies, 2000; 21(1): 99-112.

[7] Nayak R, Padhye R, Gon DP. Sewing performance of stretch denim. J of Text and Apparel, Technol and Manag, 2010; 6(3): 1-9.

[8] Nayak R, Padhye R. Sewability of air-jet textured sewing threads in denim. J of Text and Apparel, Technol and Manag, 2013; 8(1): 1-11.

[9] Chen Y, Zhao T, Collier B. Prediction of fabric end-use using a neural network technique. J of the Text Inst, 2001; 92(2): 157-163.

[10] Nayak R, Padhye R. Garment Manufacturing Technology. Elsevier, NY: 2014.

[11] Hui CL, Lau TW, Ng SF, Chan CC. Learning-based fuzzy colour prediction system for more effective apparel design. International J of Clothing Sci and Technol, 2005; 17(5): 335-348.

[12] Zeng X, Ge X, Bruniaux P. Developing an intelligent system for supporting fashion design decisions. In Proceedings of the 8th International FLINS Conference. 2008. World Scientific.

[13] Fang JJ, Ding Y. Expert-based customized pattern-making automation: Part I. Basic patterns. International J of Clothing Sci and Technol, 2008; 20(1): 26-40.

[14] Hu ZH, Ding YS, Zhang WB, Yan Q. An interactive co-evolutionary CAD system for garment pattern design. Computer-Aided Design, 2008; 40(12): 1094-1104.

[15] Fan J, Newton E, Au R, Chan S. Predicting garment drape with a fuzzy-neural network. Text Res J, 2001; 71(7): 605-608.

[16] Wong W, Leung S, Au K Real-time GA-based rescheduling approach for the pre-sewing stage of an apparel manufacturing process. The International J of Advanced Manufacturing Technol, 2005; 25(1-2): 180-188.

[17] Guo Z, Wong WK, Leung S, Fan J, Chan S. Genetic optimization of order scheduling with multiple uncertainties. Expert Systems with Applications, 2008; 35(4): 1788-1801.

[18] Yuen C, Wong WK, Qian S, Chan L, Fung E. A hybrid model using genetic algorithm and neural network for classifying garment defects. Expert Systems with Applications, 2009; 36(2): 2037-2047.

[19] Nawaz N, Nayak R. Seamless garments, in Garment Manufacturing Technology, Nayak R, Padhye R, Editors. 2015: Cambridge, UK. 373-383.

[20] Lo WS, Hong TP, Jeng R, A framework of E-SCM multi-agent systems in the fashion industry. International J of Production Economics, 2008; 114(2): 594-614.

[21] Au K, Wong W, Zeng X. Decision model for country site selection of overseas clothing plants. The International J of Advanced Manufacturing Technol, 2006; 29(3-4): 408-417.

[22] Dong A, Leung S. A simulation-based replenishment model for the textile industry. Text Res J, 2009; 79(13): 1188-1201.

[23] Pan A, Leung S, Moon K, Yeung K. Optimal reorder decision-making in the agent-based apparel supply chain. Expert Systems with Applications, 2009; 36(4): 8571-8581.

[24] Sztandera LM, Frank C, Vemulapali B. Prediction of women's apparel sales using soft computing methods. In Knowledge-Based Intelligent Information and Engineering Systems. 2004. Springer, NY; 37-48.

[25] Thomassey S, Happiette M, Castelain JM. A short and mean-term automatic forecasting system-Application to textile logistics. European J of Operational Res, 2005; 161(1): 275-284.

[26] Thomassey S, Fiordaliso A. A hybrid sales forecasting system based on clustering and decision trees. Decision Support Systems, 2006; 42(1): 408-421.

[27] Ding Y, Xu Y. Intelligent optimal selection of garment sizes by using immune algorithm and AHP method. J of the Text Inst, 2008; 99(3): 281-286.

[28] Hsu C, Lee T, Kuo H. Applying fuzzy theory based data mining to establish the female sizing systems for garment production. In Proceedings of the 8th WSEAS International Conference on Applied Computing and Applied Computational Science, Hangzhou, China, World Scientific and Engineering Academy and Society. 2009.

[29] Luo X, Hou W, Li Y, Wang Z. A fuzzy neural network model for predicting clothing thermal comfort. Computers and Mathematics with Applications, 2007; 53(12): 1840-1846.

[30] Nayak R, Punj S, Chatterjee K, Behera BK. Comfort properties of suiting fabrics. Indian J of Fibre and Text Res, 2009; 34: 122-128.

Development of Digitally Camouflaged Uniforms for Ethiopian Defense Forces

Biruk Ketema, Saminathan Ratnapandian*

Ethiopian Institute of Textile and Fashion Technology (EiTEX), Bahir Dar University, PO Box 26 Bahir Dar, Ethiopia

*Corresponding author's email: drsnrtex@gmail.com

Abstract

Military camouflage is the art and science of making defense personnel and equipment difficult to detect by enemy forces. Digital camouflage uses colors and digitally developed patterns to achieve low degree of detection. Ethiopian defense forces are using an obsolete US camouflage pattern that does not relate to the deployment area. In this research three digital camouflage designs have been developed considering the diverse topography of Ethiopia. The new patterns were found to be at least 20% more effective than the patterns in use at present when evaluated using NATO workshop SCI - 012 photo simulation procedures. Field trials are recommended to enhance the effectiveness.

Keywords: Digital camouflage; Ethiopia; Woodland pattern; Arid pattern; Desert pattern; Detectability

1. Introduction

Camouflage is a form of deceptive concealment that allows animals, people and objects to remain unnoticed by blending with their environment (crypsis) or by resembling something else (mimicry) [1]. Military camouflage is the use of camouflage by a military force to protect personnel and equipment from observation by enemy forces [2]. In the beginning, the design of camouflage patterns was in the hands of artists and designers, who, during the decades of the twentieth century, improved their work in close collaboration with the army [3].

The advancement in target detecting technology like radar, heat sensing and night vision devices that do not depend on visual detection, has led to the development other means of camouflaging technique which collectively are known as "stealth technology" [4,5]. Many modern camouflage textiles address visibility not only to visible light but also near infrared, for concealment from night vision devices. However visual concealment retains priority in ground forces [6]. Today, camouflage uniforms are an indispensable part of military equipment and serve not only as a means of disguising an object in a certain environment but also as a national symbol. Digital camouflage is the latest camouflage design used widely in most countries.

1.1 Digital Camouflage

Digital camouflage provides a disruptive effect through the use of pixilated patterns that helps to defeat observation at a range of distances. The "digital" refers to the coordinates of the pattern, which are digitally defined. The term is also used for computer generated patterns like the non-pixilated Multicam and the Italian Fractal Vegetato pattern. Contemporarily, the most accurate definition of digital camouflage would be a camouflage pattern that has been designed using computer algorithms that are programmed to create micro-patterns for effective disruption [7].

The theory behind micro-patterns is that large blotches of color with sharp outlines are easier to see, while "blurring" or "dithering" the edges of the colored patches makes the outlines more difficult to discern. In its common usage, however, the term digital has come to refer to any camouflage design that incorporates pixels rather than organic shapes to create the design. Although the term "pixilated" camouflage is more accurate, digital has become a part of the common vocabulary amongst military and collector communities [2,8].

Most of the current digital patterns incorporate both a macro pattern and a digital micro pattern matrix (Fig.1). Micro pattern matrix or pixel digitizing is required to conceal the target when the focal (recognition) region of the eye, which is only a few degrees wide, is looking directly at the target. While the macro pattern, overall larger part of the pattern, is designed to conceal the target from the ambient (detection) portion of the view which is about 10 times wider than the focal region of our eyesight [9].

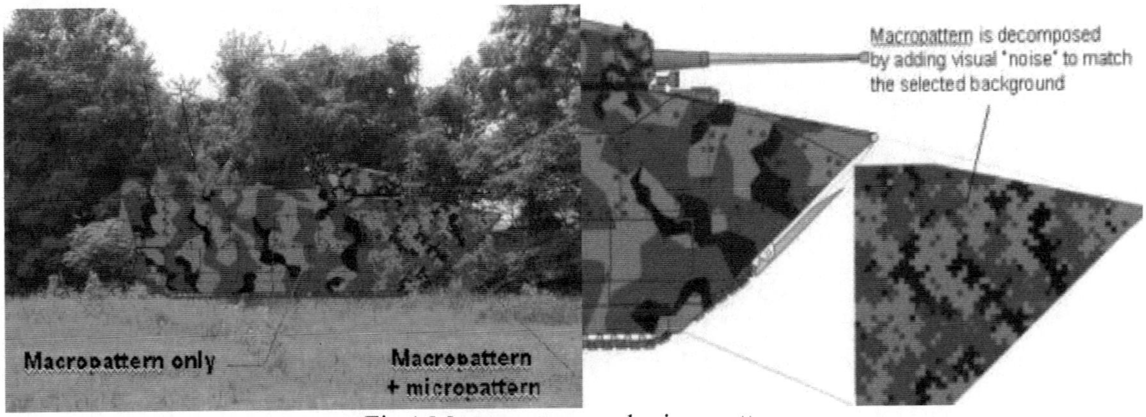

Fig.1 Macro pattern and micro pattern

Field studies in the United States and Canada show that a micro pattern addition to even the best macro patterns dramatically improves the concealment factor. Digital patterns can not only be used for uniforms but also enlarged in scale for applications to vehicles, aircraft, naval vessels and buildings. In such cases both the macro patterns and micro patterns become larger in scale

Recently the new Canadian Camouflage pattern CADPAT (Left) and the U.S. Marines pattern MARPAT (Right), shown in Fig.2, based on the Canadian developed pattern, have garnered a lot of attention as the pattern is made up of a digitized image using four colors. Canadian CADPAT is 30% more effective than Olive Drab in field testing. This digital effect generates a dithering effect between colors (no solid lines) and works well within 100 yards of an adversary. However, this advance in camouflage is minimal as the colors tend to blend into each other at farther distances. If this blended color of the uniform is different from the background the human shape is revealed [10].

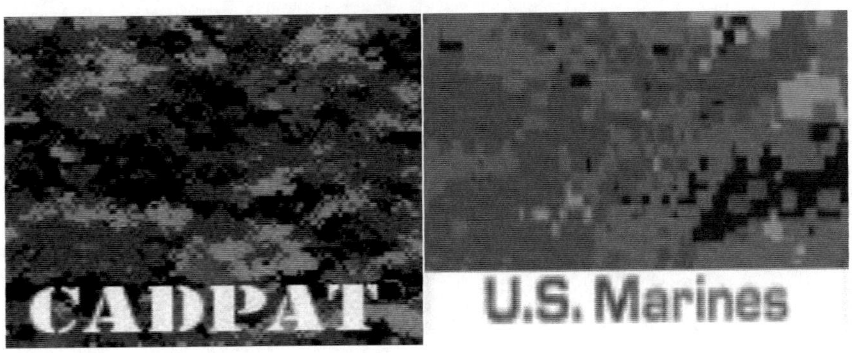

Fig.2 CADPAT and MARPAT digital camouflage

1.2 Procedure for Digital camouflage development

The procedure for designing a digital camouflage pattern can be generally described as computer aided processing of digital photos acquired in a specific environment. Some developers of camouflage use math based algorithms and graphics like Camouflage Designated Enhanced Fractal Geometry (C2G) shown in Fig.3 Other developers use technical and artistic principles and visual aspects to design camouflage. For instance Friškovec [6] has developed camouflage for Slovenian urban environments using the theory of artistic and scientific design principles like Gasalt psychology and fractal geometry [11].

Fig.3 Fractals pattern by C2G

According to Cramer [12], the C2G patterns that have been developed using fractals that are found in most global regions can be used around the world under different color schemes without modifying the initial pattern. He states that Fractal is found around the globe in both natural and populated areas. This may be broadly classified into temperate and tropical (tree branches) / desert - (arid ground and salt flats) / urban - cured, aged concrete.

Camouflage uniforms need to be made and distributed to a large number of soldiers. The design of camouflage uniforms therefore involves a tradeoff between camouflaging effect, recognizability, cost, and manufacturability. Four-color combinations camouflage patterns are common because as the number of color increases the cost also increases. Recent development in US military shows that the following terrain colors are appropriate for different environmental locations [12].

Table 1. Camouflage colors of different environments

Environment	Color
Woodland	Tan, Brown, Green, Black
Desert	Tan, Brown, Khaki, Dark Tan
Urban	Tan, Light Gray, Medium Gray, Black
Desert/Urban	Tan, Brown, Light Gray, Dark Tan

1.3 Camouflage Effectiveness Testing

The measures of effectiveness for a camouflage system are [13,14]

- Target detectability - the time at which the observer first correctly pointed to the location of the target (Initial Detection).This is almost always evaluated by a number of observers, either in the field or in laboratory.
- Recognition - describe the classification of target and
- Identification - identify the target specifically.

A laboratory model of assessment uses standardized procedures and apparatus (photo simulations) can provide some indication of field performance. Different models such as Janus, Castforem, Deep Target Acquisition Model and NATO Workshop SCI-012 have been developed to evaluate the effect of camouflage [15].

2. Analysis of the Existing Ethiopian Army Camouflage Dresses

According to the data provided by the Ethiopian Army Foundation, the Ethiopian Defense force has developed material, product and quality specification of camouflage trousers, shirt and cap for its soldier and ground force uniforms. However no scientific analysis, documentation and specification are available for the camouflage design patterns (Fig.4) used currently. The authors identified the specification by comparing it with world camouflages and it is found to be an imitation of US woodland pattern 81 camouflage [16]. This pattern was the default camouflage pattern issued to United States defense forces from 1981, until its replacement around 2006 by the digital MARPAT camouflage pattern. The use of Woodland pattern is now obsolete in the United States.

Ethiopian defense force is neither using a camouflage pattern designed according to its geographical environment nor based on the science and principles of digital camouflage. Hence this work aims to develop new set of camouflage based on modern science and art in order to improve its effectiveness during military duties.

This was achieved in three stages:

1) Classifying Ethiopian geographical environment for camouflage application
2) Developing digital camouflages
3) Experimentally assessing the effectiveness

Fig.4 Photo of Ethiopian woodland camouflage (Courtesy Ethiopian army foundation)

3. Results and Discussion

3.1 Classification of Ethiopian geographical environments for camouflage applications

Ethiopia and its surrounding geographical terrain can be classified into three main regions for the purpose of effective military camouflages. These are

- Woodland- where forests of large trees are available, mostly located in south west region.
- Arid lands- located in the majority of central and north western regions
- Desert lands - located in eastern and few of southern regions.

3.2 New design development for different terrains

The authors developed three digital camouflages for the three major terrains of Ethiopia and its neighboring environments namely woodland, arid and desert. The combination of Adobe Illustrator and Adobe Photoshop was used for the function of vector and raster image analysis and design developments.

Adobe illustrator's image trace, live paint, pixilation filter and art color editing tools were employed to extract colors and shapes of terrain image. Adobe Photoshop's tools were mainly used to convert the developed design into seamless pattern for applying at different scales for various military objects, photo model simulation.

1) Existing woodland camouflages designs were modified using CAD software to develop new pixilated digital woodland camouflage with the same shape and color. For abbreviation and identification, the existing Ethiopian woodland 81 camouflage was named Original Woodland Pattern with Woodland Colour (OWPWC) and the newly developed digital camouflage as Digital Woodland Pattern with Woodland Colour (DWPWC) shown in Fig 5a and 5b. The component colors and their hexagonal codes are listed in Table 2.

2) The digital images of sample geographical areas of arid areas (Fig.6a) were digitally analyzed to identify the common shapes and color composition (Fig.6b). Based on this, new digital arid camouflage design was developed (Fig.6c). As the pattern was subjectively developed using artistic and scientific principles of digital camouflage design and CAD parameters, there may be other design alternatives. Table 3 contains details on the colors and their codes.

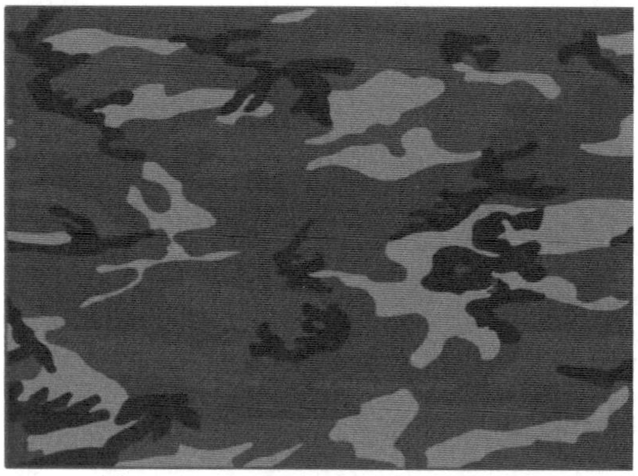

Fig.5a OWPWC (original pattern)

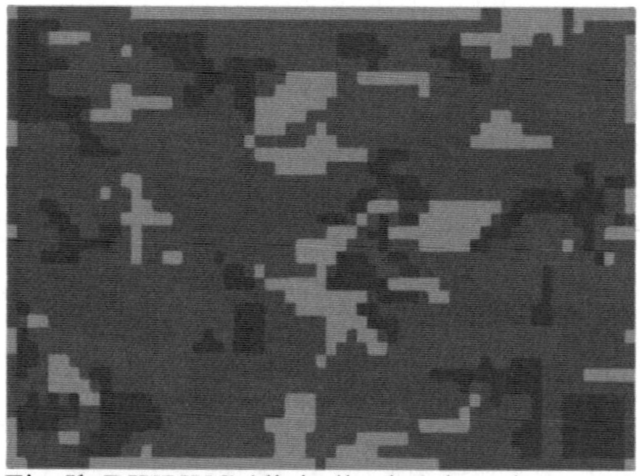

Fig.5b DWPWC (digitally developed pattern)

Table 2. Woodland Hexagonal color code

Colour	RGB colour mode
Green	4A5A5E
Brown	555334
Black	40404A
Grey	868A8B

Fig.6a Sample of selected geographic area (Courtesy Panoramyo photo explorer)

Fig.6b Shape pattern analysis using Adobe Illustrator

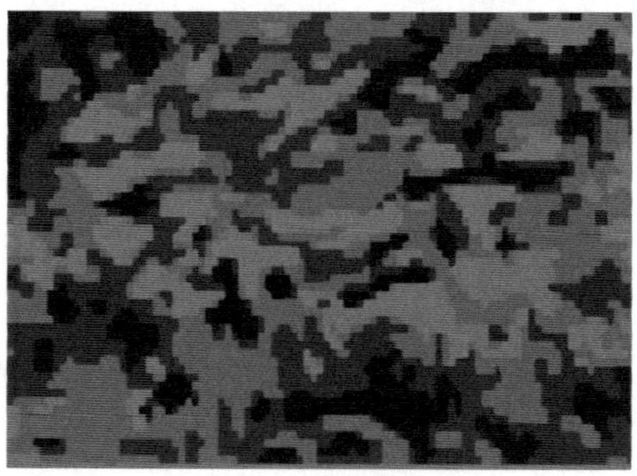

Fig.6c DAPAC (Digital arid pattern arid color)

Table 3. Arid Hexagonal color code

Colour	RGB colour mode
Green	585A3C
Grey	3D4034
Black	232623
Brown	786C47

3) The third was developing new digital desert camouflage design based on the digital camouflage shapes developed for arid pattern and using desert colors widely found in modern desert camouflage colors. The colors and developed pattern are shown in Table 4 and Fig.7.

Table 4. Desert Hexagonal color code

Colour	RGB colour mode
Khaki	E0C0AA
Brown	95664E
Tan	CDAB8A
Dark Tan	AE8868

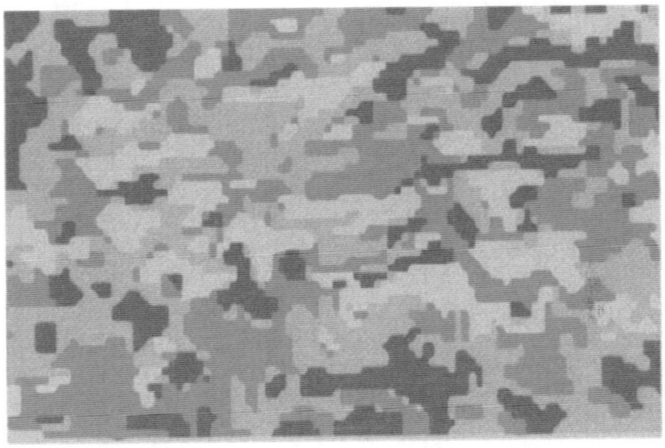

Fig.7 DAPDC (Digital Arid Pattern Desert Color)

3.2 Seamless repeat pattern and model simulation

Fig.8a and b show the seamless patterns of the four designs as applied to soldier and jeep. These were used for model simulation and digital photo simulation camouflage effectiveness assessment.

Fig.8a OWPWC Seamless repeat pattern, soldier and jeep car model simulation

Fig.8b Seamless repeat pattern, soldier and jeep car model simulation

3.3 Camouflage Effectiveness Assessment Testing Procedure and Experiments

The NATO workshop SCI-012 digital photo simulation camouflage assessment technique was adapted to compare the relative effectiveness of the newly developed and existing camouflage pattern. In this model a photo simulation experiment using a power point projected to a white screen in its full image view was used (Fig.9).

A power point presentation with alternate background images at different ranges is necessary to evaluate the relative effectiveness of the camouflages under comparison. Targets were placed at short, medium and far ranges and also at most possible camouflaging effect in each background.

Observers viewed each image for 14 seconds and the power point was adjusted to advance automatically. A data sheet for recording detection time and target identification was prepared. Trial tests and procedures were briefed in detail. The observers were given practice using two sets of trial digitized power point containing a different background area before the actual test was performed. Around 18-25 observes were enough to obtain statistically significant data on detection time.

Fig.9 NATO Workshop SCI 012 photo simulation camouflage assessment

Power point presentations with 14 slides consisting of 3 different background images were selected. Two of the background images were arid area backgrounds from Ethiopian Amhara region, Addis Zemen town with specific location around tall nosed mountain. One of the background images with a desert area background was selected.

The experiment was done in CAD lab of EiTEX using 22 students as observers. Three camouflages namely DAPAC, OWPWC and DAPDC were tested. The targets selected were military camouflaged soldier, minibus and jeep and did not include any civil person, animals and equipment. As seen in Fig.10, the targets were placed so as to obtain best camouflage effect in each background area.

Fig.10 Example of PPT slides DAPAC jeep in arid area image background

The test was done to compare the effectiveness of the newly developed DAPAC and OWPWC in arid area and DAPDC and OWPWC in desert area. The objective was to compare the existing Ethiopian military camouflage design, OWPWC, with newly developed camouflage for desert and arid areas.

3.4. Data analysis

Data were analyzed after finalizing the effectiveness assessment experiment. Data with wrong identification were discarded and only those detection times with perfect identification were used for analysis. The average of detection time is calculated and the results are shown in Table 5.

Table 5. Average detection times

Environment	Arid		Desert	
	Camouflage type	Average Detection Time (seconds)	Camouflage type	Average Detection Time (seconds)
	OWPWC	3.88	OWPWC	3.88
	DAPAC	5.57	DAPDC	4.99
	Difference	1.69	Difference	1.09

The 1.69 second (43.5%) increase in detection time indicates that in arid area the newly developed DAPAC is more effective than OWPWC. Similarly in desert environment the developed DAPDC is more effective with an increase in detection time by 1.09 seconds (a 27.3% increase).

4. Conclusions

It may be concluded that digital camouflage can achieve low degree of detection and could be adopted by defense forces. Specific patterns and colors have to be developed according to the area of deployment. Extensive field trials would enhance camouflage effectiveness.

5. Acknowledgements

The authors are thankful to EITEX, Bahir Dar University, for funding this research and allowing the use of required facilities. They also acknowledge the cooperation extended by Mr. Israel Mezgebe from EiTEX, Bahir Dar University and officers of Ethiopian Defense Foundation.

References

[1] Stevens M, Merilaita SAnimal camouflage. current issues and new perspectives. Philosophical transactions, 2009.

[2] Cho WJ, et al. Making Pixel Patterns Automatically for Camouflage - Using Color Information from Their Background, in HCI International 2011 - Posters' Extended Abstracts: International Conference, HCI International 2011, Orlando, FL, USA, July 9-14, 2011, Proceedings, Part II, C. Stephanidis, Editor. 2011, Springer Berlin Heidelberg: Berlin, Heidelberg. P: 98-101.

[3] Puzikova NP, et al. Principles of an approach for coloring military camouflage. Fibre Chemistry, 2008. 40(2); p: 155-159.

[4] Rao G. Mahulikar SIntegrated review of stealth technology and its role in airpower. Aeronautical Journal, 2002.

[5] Bahret W. The beginnings of stealth technology. IEEE Transactions on aerospace and electronic systems, 1993: 29(4).

[6] Friskovec M. Gabrijelcic H and Simoncic BDesign and evaluation of a camouflage pattern for the Slovanian urban environment. Journal of Imaging Science and Technology, 2010: 54(2).
[7] Yu J, et al. Research of digital camouflage generation algorithm [J]. Opto-Electronic Engineering 2010.
[8] Gabrijelcic H, Friskovec M, Dimitrovski K. The use of image analysis for defining the shapes of urban camouflage pattern. 2009.
[9] Santos L, et al. Camouflage US marine corps combat utility uniform fabric and design, USPTO, Editor. 2004: USA.
[10] Storey E. Canadian battlegroup badges. Canadian Military History, 2015: 19(1).
[11] Crilly A, Earnshaw R, Jones H, eds. Application of fractals and chaos, the shape of things. 2013, Springer Science and Business Media.
[12] Cramer G. Dual texture - US army digital camouflage. 2008.
[13] BirkemarkC CAMEVA a methodology for computerized evaluation of camouflage effectiveness and estimation of target detectability. in AeroSense'99 International society for optics and photonics. 1999.
[14] Gretzmacher, Ruppert, Nyberg. Camouflage assessment considering human perception data. in Aerospace/Defense sensing and controls international society of optics and photonics. 1998.
[15] Rubeziene, et al. Evaluation of camouflage effectiveness of printed fabrics in visible and NIR spectral ranges. Materials Science, 2008: 14(4).
[16] Newark, Camouflage. 2007: Thames & Hudson.

Modelling of Carbon Footprint of Polyester Sports Shirt

Shadia Moazzem[1], Fugen Daver[2], Lijing Wang[1*]

[1]*School of Fashion & Textiles, RMIT University, Brunswick, Victoria 3056, Australia*
[2]*School of Engineering, RMIT University, Bundoora, VIC 3083, Australia*

[*]Corresponding author email: lijing.wang@rmit.edu.au

Abstract

Sustainability is a vital issue in clothing manufacturing industry due to the large percentage of greenhouse gas emissions from textile production. Polyester is the most widely used synthetic fibres in the world which is produced from non-renewable resources. Demand of the polyester fibre is growing day by day specially for sportswear as most of the sportswear is manufactured by polyester fibre. In this study a preliminary assessment of carbon foot print (CFP) of polyester sportswear has been done through life cycle analysis. This life cycle assessment reveals that polyester production stage is the most energy consuming part that is causing negative environmental impact, followed by the use stage and spinning stage. Production of synthetic fibres is causing significant green house gas emissions. Hence, it is necessary to explore other fibres to produce high performance sports apparel and reduce environmental impact.

Keywords: Sustainability; Greenhouse gas emission; Life cycle assessment; Synthetic fibre

1. Introduction

Currently sportswear is playing a significant role in the apparel industry due to the high volume of demand and production. Demand of the sportswear is increasing day by day by the general consumer for specific performances such as comfort, wicking and dryness [1]. According to a statement from Euromonitor International, sportswear apparel market in Australia grows rapidly than other general apparel. It has been stated that from 2008 to 2013 the sportswear market consumption has grown by more than 10% and at the same time the general apparel market consumption has grown by 5.9% [2]. Currently most of the sportswear produced from synthetic fibres, such as polyester (PET), polyester-spandex and nylon [1]. Not only sportswear, currently half of the world's apparel are produced from synthetic fibres. In 2011 synthetic fibres occupied 61.3% and cotton fibre occupied 31.2% of the total production of textile fibres [3]. Synthetic fibres have negative impact to the environment because they are manufactured from non-renewable resource, crude oil. Production of synthetic fibres and yarns requires more energy and produces higher greenhouse gas than natural fibres. Synthetic fibres are non-biodegrable [3]. It has been observed from a recent study that 20kg CO_2 has been emitted from one kilogram polyester fibre production whereas only 3.3kg CO_2 emits from one kilogram cotton fibre production [3].

Nowadays sustainability is a vital issue in textile and clothing industry because buyers are more concern about sustainability issues of their clothing products specially purchasing imported apparel. Now fashion organisations are willing to develop their policies for sustainable environmental practice by giving attention to the carbon footprint and energy used in the whole supply chain [4]. Sustainability can be maintained by several ways such as reducing energy consumption, improving productivity, reducing waste generation and recycling waste and by using less material in production. Energy consumption is the vital issue for textile production as it directly related to various environmental releases. Therefore, producing energy saving textile products can be the most important environmental practice. Life cycle assessment (LCA) is the best option to access the sustainability of a product or process as it involved the environmental impacts associated with the product or process, raw material acquisition, energy, resources, waste, and emissions to air, water and soil, etc. Sportswear is mainly manufactured with light weight knit synthetic fabrics, though some winter sportswear is made with heavy weight fabrics. High performance comfort property is essential for sportswear. Most of the research done on sportswear has concentrated on the performance and comfort characteristics of the fabrics used to manufacture sportswear. Production of synthetic fibres causes significant green house gas emissions (GHG). Therefore, it is necessary to consider other fibres to produce high performance sports apparel to reduce green house gas emissions. In this study life cycle of a general polyester sport shirt is investigated to identify the carbon footprint (CFP) throughout its life cycle.

2. Carbon footprint of textile products

Several researches have investigated carbon footprint of textile products. Most of the studies investigated life cycle assessment of supply chain, raw material growing, spinning, weaving, knitting, clothing manufacturing, transportation, end use and disposal, energy, material, equipment and so on. A raw material preparation process depends on the types of fibre. The production process of natural fibre starts with cultivation and farming and manmade fibre production process starts with crude oil manufacturing. Several life cycle assessment studies have been done on different fibres to identify the global warming potential. A research study on comparison of the energy consumption of different fibres indicated that the total energy required to manufacture polyester fabric is significantly greater than other fibres. As carbon footprint is related to the consumed energy, so carbon footprint from polyester production is obviously greater than other fibre production. Different conversion factors are used to relate the energy and CO_2 according to the geographical location of production. An LCA study on a pair of jeans conducted by Levi Strauss & Company in 2007 showed that about 58% of the total energy was consumed in the use stage. Fabric manufacturing phase and cutting/sewing phase consumed 21% and 10% energy respectively of the total energy consumption. Another study on khaki trouser showed the use phase consumed 63% energy of the total [3]. Another life cycle study on a polyester knit blouse conducted by Franklin Associates showed that 82% of the total energy consumed at the consumer use phase [5]. A similar study on polyester jacket used functional unit of 100 days life time and six times wash during its life time. During the whole life time the selected jacket consumed total 156.39 MJ of energy [6]. A study on cotton T-shirt and viscose blouse showed that cotton T-shirt consume total 112 MJ energy over the whole life cycle, of which 65 MJ was only for use phase. Production, material acquisition and transport phase consumed 24 MJ, 16 MJ and 7 MJ energy, respectively. Energy profile of viscose blouse is completely different from cotton T-shirt. Material acquisition phase of viscose consumed 65% of the total energy [3]. Another study on cotton long sleeve shirt conducted by Shah & Sinha [7] illustrates that material production phase contributed 33% carbon foot print then followed by consumer use stage.

3. Methodology

Life cycle assessment is an important tool to identify the environmental impact of any textile product. The stages of life cycle include raw materials production/ cultivation, yarn production, fabric production, transport, use and disposal. Specific procedures and standards are followed to assess the environmental impact of a textile product through life cycle assessment. The ISO14040 is a specific standard used in this life cycle assessment which includes goal and scope, functional unit, system boundary, inventory analysis and impact assessment. The first step of life cycle assessment is to determine the goal. The goal of this study is to access the carbon foot print of polyester sportswear through life cycle analysis as a preliminary assessment. Fig.1 shows the system boundary of this study which includes fibre extraction to disposal. Secondly, a standard functional unit has been chosen to calculate the materials throughout the system boundary. Then energy and emissions of each individual sub process have been calculated on the basis of functional unit.

The functional unit for this study is as follows:

50 days' use of polyester knitted sports T- shirt over one year life time. It has been assumed that the sportswear is washed after each wearing. In this study we assumed that sportswear weight is 180 g/m^2 [8]. Energy requirements of each individual process have been determined according to the fuel consumption and electricity unit, kWh (kilowatt hour). Two types of data are required to calculate the carbon footprint of a textile product, activity data and emission factor data. Activity data involved the amount of all input and output materials, transportation, use phase and energy. Emission factor data refers to the link between activity data and emission data that convert the amount to greenhouse gas emission. Activity data and energy/ emission factor data can be collected from primary and secondary sources. In this study literature source has been used for data collection.

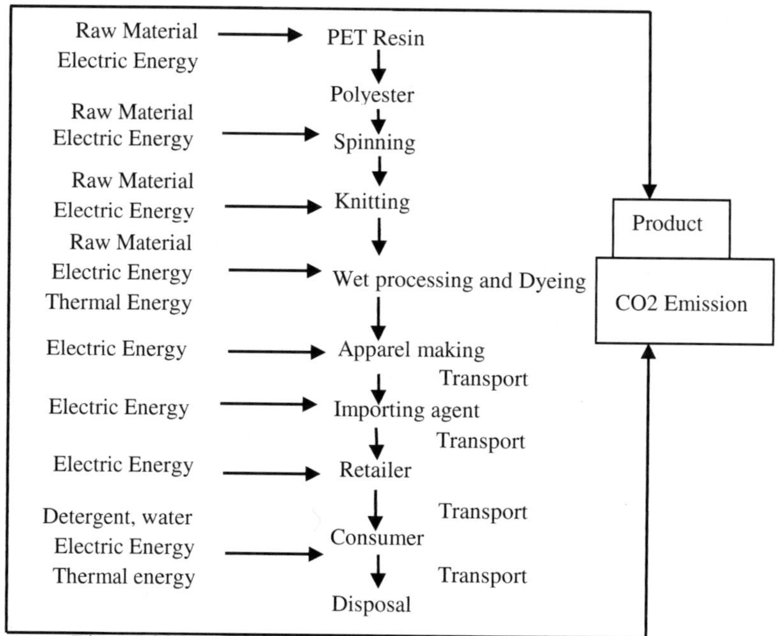

Fig.1 Life Cycle Assessment system boundary

The carbon footprint of each sub process of supply chain was calculated from the input of materials, input of energy and CO_2 conversion factor for energy. The specific equation used to calculate the carbon footprint of each sub process is shown in Equation 1. Transportation energy can be calculated through estimating the transport distances and transport modes for the whole supply chain. Transport mode can be a combination of several transports such as truck, ship and car. Conversion factor is used to determine transport fuel energy and CO_2 emission. Equation 2 illustrates the carbon foot print from transport.

$$CO2\ emission\ from\ process(kgCO2equvalent) = \left[\{M + (M * loss\%)\} * E_m\right] * W_m \quad (1)$$

Where, CFP_m is carbon footprint of material M, M is material input; E_m is energy conversion factor for material M; W_m is CO_2 conversion factor for energy.

$$\text{CO2}\ Emissions\ from\ transport(kgCO2equavalent) = \sum\left[M_{mt} * D_t * E_t * W\right] \quad (2)$$

Where, M_{mt} is mass of material transported; D_t is travel length; E_t is mass of transport fuel; W is CO_2 conversion factor for fuel.

Carbon Footprint of knitting sub process for one functional unit is estimated as follows,

$$CFP_{knitting} = [[\{204.97 + (204.97 * 1.5\%)\}/1000] * 8.08\ ] * 0.8529 = 1.4\ Kg\ CO_2 \quad (3)$$

3.1 Raw material extraction phase/ PET production

Polyester is produced from crude oil and natural gas. One kg PET production consumes 0.614Kg crude oil and 0.251kg natural gas. The form of these raw materials during production is chemicals, namely xylene and ethylene. Polymer is produced through polymerisation between terephthalic acid and ethylene glycol [9]. Energy requirements for PET resin production were estimated by comparing different literatures. A comparatively recent study on two Marks & Spencer products, cotton briefs and polyester trousers, has been conducted in 2002 [10]. This study has been done based on the information provided by Franklin Associates [11]. It has been observed from the literature review that most of the study on polyester was based on the Franklin Associates data. Franklin Associates shows a detailed study on polyester blouse. In this study we adopted PET fibre modified data based on

Franklin Associates study [10]. It was assumed that PET fibre was produced in Zhejiang province. Then it was transported to Wuhan for material production.

3.2 Production phase

The main raw materials for PET yarn production are PET chips, spin finish oil, anti-microbial agent, caustic soda, sodium hydrosulphite, acetic acid (98% Glacial) and retarding agent [12]. Many studies have been done to identify the energy demand of yarn spinning. Spinning is the vital energy consuming stage of textile manufacturing. Many studies have been done to work out the energy demand of yarn spinning. Electric energy and thermal energy are required to operate spinning industry. Electrical energy is used to operate all of the machines involved in spinning, air conditioning and illumination and compressors. Thermal energy is used for the fixation of yarn [13]. About 78% of energy is required to operate all machines, 3% for illumination, 3% for compressors and 16% for air conditioning and humidification [3]. Energy consumption for ring and open end yarn is closer. Yarn spinning energy is depend on fibre type (natural or synthetic), yarn count and yarn end use. Yarn used for weaving requires more twist than knitting yarn. Due to this spinning energy is comparatively lower for knitting yarn [14].

According to a study performed by Collins and Aumônier [9], the total energy consumption of a spinning process is 22.23 kWh/kg and the fibre preparation process consumes 7.13 kWh/kg and this estimation was based on British Textile Technology Group [15]. We consider this conversion factor to calculate energy consumption for this study. We also assume that the loss percentage of spinning process and fibre preparation process are 15% and 5%, respectively.

The environmental impact from fabric manufacturing depends on the manufacturing process, namely knitting and weaving. Knitting is a comparatively simple process and produces less emission as it uses only single machines to convert yarn to fabrics, whereas weaving involves several pre-processes that consume more energy and emit more emissions to the environment. In this study we assume that the knitted sportswear fabric is polyester single jersey 165 g/m^2. Energy consumption was assumed 8.08 kWh/kg and yarn loss during knitting is 1.5% [10,15]. Washing, boiling, bleaching and dyeing processes are involved in wet processing. Before dying, the fabric is pre-treated to remove wax and any residue. All of the processes of wet processing are usually carried out in the same machine except the dyeing process. Bleaching process has been done according to the required colour. Dyeing process of knitted fabrics is carried out in jet or winch dyeing machine. Caustic liquor and acidic liquor are used during washing and boiling. Acidic liquor is used to remove alkali metal compounds and caustic liquor is used to remove waxes and grease from raw fabric. Reactive dyes and disperse dyes are used to dye cotton and polyester fibres, respectively. In case of blend fabric suitable dyestuffs are used to dye blend fabrics. Knitted fabrics are also finished chemically by softener, antistatic, enzyme, piling resistant or crease resistant [16]. Different dyeing machine consume different energy such as winch dyeing machine, jet dyeing machine and stenter consume 6.0-17.0 GJ, 3.5-16.0 GJ and 2.5 GJ respectively per ton material output. Dyeing and pre-treatment process consumes 90% fuel energy and 10% electric energy. In this study we assume 1% fabric loss during dyeing and pre-treatment and energy consumption 7.11 kWh/kg [10,15].

Apparel manufacturing process involves design, laying, cutting, pattern making, sewing, ironing and finishing. According to a recent study [3], apparel production process has relatively low impact to the environment among other sub processes. An apparel production process requires 2.472 MJ energy of which 49.8% energy is used for sewing, 29.6% energy is used for cutting and 20.06% is used for packaging. In this paper, we assume 11.6% material loss during inspection, lying, cutting and sewing and energy consumption 2.38 kWh/kg was based on study carried out by Collins and Aumônier [10]. Data for energy requirement in each production stage are assumed based on British Textile Technology Group (BTTG) and study by Collins and Aumônier [10], where energy refers to the process energy and raw material extraction energy. It includes electricity and fuel used by the machine and product. The energy data are based on medium voltage electricity supply and oil as the fuel [10].

3.3 Transportation phase

In this study, it is assumed that the whole manufacturing process (from raw materials production to apparel production) of sportswear takes place in China and then this product is shipped to Melbourne, Australia. Transportation energy has been calculated through estimating the transport distances and transport modes for the whole supply chain. It is a combination of several transport modes such as truck, ship and car. Fuel required for each transport mode was calculated based on Table 1. Transport fuel energy content was calculated based on the fuel required for each transport mode shown in Table 2 [17]. Travel distance was calculated based on the calculator of sea-distances.org [18]. Then Transport fuel energy was calculated based on the conversion factor of energy content per gallon fuel in MJ [17]. The following conversion factors were used to calculate the fuel amount and CO_2 emission from transport (Table 3). In this study we used conversion factor of CO_2 emission for transport (kg CO_2/Mt ton-mile) from Table 3.

Table 1. Fuel required for each transport mode (ton-miles /gallon or kg-km/gallon) [17]

Transport mode	Distance in ton-miles per gallon of fuel	Distance in kg-km per gallon of fuel
Automobile	19	5636.98
Truck	59	33258.2
Train	386	217588
Boat	522	294251

Table 2. Energy Content in the fuel (MJ/ gallon) [17]

Transport Mode	Fuel	Energy Content in MJ
19 mpg Car	Gasoline	132
Truck, train, boat	Diesel	146

Table 3. Conversion factor of CO_2emission (kg CO_2 / Mt ton-mile) [17]

Transport Mode	Kg CO_2/ Metric Ton-Mile
19 mpg Automobile	0.4305
Truck	0.0937
Train	0.0260
Boat	0.0132

3.4 Product use and customer care

The use phase of any apparel starts from the purchase of apparel by consumer from the retail shop and it is controlled by consumer choice. Energy required and environmental impact from this phase depend on some factors such as frequency of washing, washing and drying temperature, detergent type and ironing. It is assumed that sportswear (180 g/m^2) washed 50 times in its one year lifetime. This means 9kg polyester sportswear will be washed in its life-time. Washing machine load was assumed 7kg. House hold washing with cold machine wash (20 °C temperature) with mild powdered detergent, without softener and line dry was assumed according to the recommended fabric care instructions of polyester sportswear [19]. It was stated from Energy Star that about 90% of the total washing machine energy consumed for water heating. So cold water washing saves lots of energy than hot or warm water. Cold water detergent is required for cold water wash. For this study, 66 litre water per machine wash of 7kg load [20] was assumed and water consumption 84.5 litres was calculated for its life time. Based on LCA studies of Marks and Spencer apparel, we assume that 0.001 kWh of electricity is consumed for per litre of water supply, treatment and pumping. This energy includes waste water and clean water supply and pumping [10]. CO_2 emission from water treatment and supply was calculated based on the Melbourne greenhouse gas emissions from water related energy demand [21]. Two scoop detergents are assumed to be required per wash which equates to 30g. Detergent extraction energy was assumed 0.28 KWh for 30g detergent [10]. The emission factor for detergent production is assumed to be 0.11 ton CO_2/GJ according to Chinas Manufacturing subsector emission [22]. It was assumed that washing machine energy for cold wash is 0.3 KWh per machine load (7kg) [23]. Carbon footprint of washing machine has been calculated based on the emission factors of purchased electricity from the grid of different states in Australia [24].

3.5 End of life

End of life of any textile product depends on the consumer choice. It can be reuse, recycle, incineration and landfill [3]. It has been assumed that sportswear will be discarded for reuse after one year through SATC (Salvation Army Trading Company). Based on a SATC report, collection, processing, and distribution of second hand clothing consume 1.7 kWh energy per kg clothing [10]. We adapted the data to calculate total energy for reuse in this study.

Victories greenhouse gas emission factor (1.18kg CO_2 /kWh) of purchased electricity was used as a conversion factor for energy consumption during reuse [24].

3.6 Conversion factor for energy and carbon footprint

The energy source involved in this CFP assessment is electricity and fuel. Energy required for material production throughout the supply chain was calculated based on the previous study as shown in the Table 4.

Table 4. Energy required throughout the supply chain [10,15]

Process	Energy (KWh/kg)	Assumption
Fibre preparation (opening, cleaning, drawing, roving)	7.13	100% Electricity
Spinning	22.23	100% Electricity
Knitting	8.08	100% Electricity
Scouring, dyeing and finishing	7.11	90% Fuel,10% Electricity
Cutting and sewing	2.38	100% Electricity

Electric power for Zhejiang province and Hubei province are supplied from China east power grid and central grid respectively. A carbon emission factor for electricity consumption was calculated based on CO_2 emission factor of different power grids in China [22]. CO_2 emission factor 0.7826 ton CO_2/MWh for electricity consumption of Zhejiang province was used in this study. Accordingly emission factor 0.8529t CO_2/MWh was used for Hubei Province in this study. Carbon emission factor 77 ton CO_2/TJ (0.25 kg/kWh) for fuel oil was estimated based on IPCC [22]. Carbon footprint of consumer stage and end of life stages have been calculated based on the emission factor (1.18kg CO_2 /kWh) of purchased electricity from the power grid of Victoria, Australia [24].

4. Results and discussion

The total carbon footprint of polyester shirt is 21.50kg CO_2 for one functional unit which includes 5.95kg CO_2 for raw material manufacturing, 8.79kg CO_2 for product manufacturing, 0.89kg CO_2 for transportation, 5.51kg CO_2 for consumer use and 0.361kg CO_2 for disposal as shown in Table 5. Fig.2 illustrates the comparison of energy and CFP of life cycle stages. Spinning is the highest energy consuming part in product manufacturing stage due to high electric energy demand. Dyeing, finishing and pre-treatment stages are affected by diverse range of chemicals and energy sources.

Table 5. Estimated Carbon Footprint of Life Cycle Stages

Product phase	Estimated carbon footprint (Kg)
Disposal	0.361
Use	5.51
Transport	0.89
Product manufacturing	
Fabric inspection, lying, cutting and sewing	0.4
Dyeing and finishing	0.45
Pre-treatment	0.45
Knitting	1.44
Spinning	4.53
Fibre preparation	1.52
Raw material manufacturing	5.95
Total Energy = 36.86 KWh / functional unit	
CO_2 emission = 21.50 kg CO_2/functional unit	

The CFP contribution analysis of each life cycle process is shown in Fig.3. The PET fibre production process contributes the highest CFP 27.67%. Within the product manufacturing process, each industrial sub-production process contributes the following: fibre preparation 7.06%, spinning 21.06%, knitting 6.7%, pre-treatment 2.1%, dyeing and finishing 2.09%, inspection, cutting and sewing 1.86%. Transportation and disposal stage contributes 4.13% and 1.67%, respectively. Consumer use phase is the second largest CFP contributing part of life cycle which is 25.62%. CFP of transportation stage depends on transportation mode and distance. In this study we assumed that all of the production was occurred in the same place which reduces CFP from transportation stage. Raw material for the main production was transported from another province.

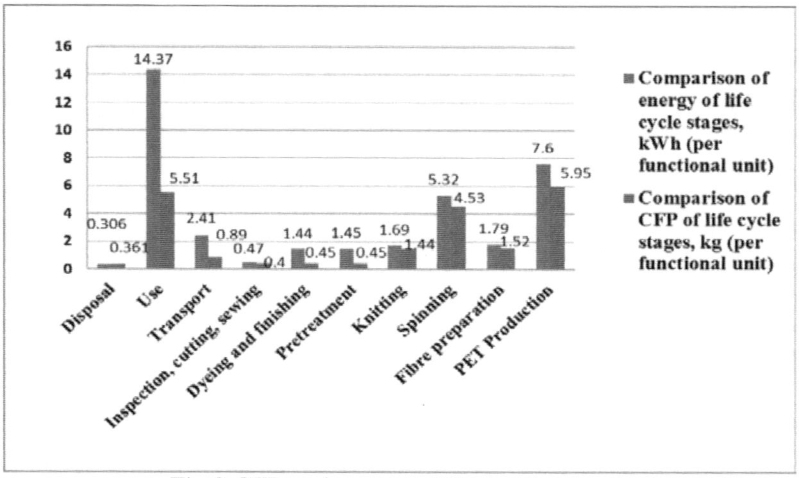

Fig.2 CFP and energy of life cycle stages

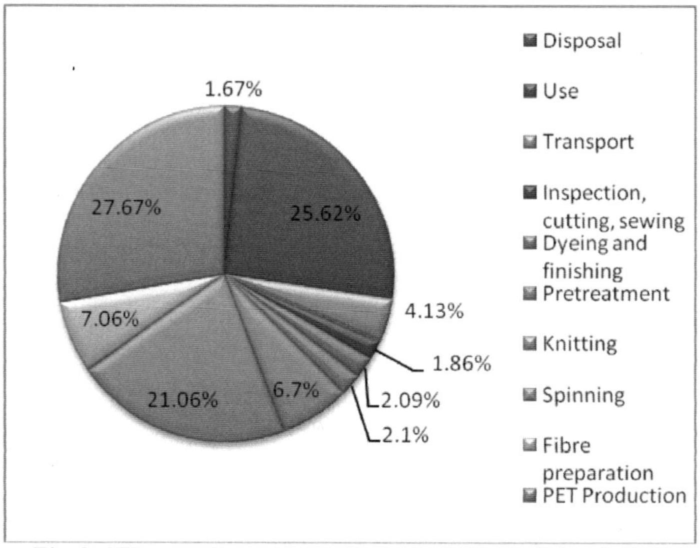

Fig.3 CFP contribution analysis throughout the life cycle

The CFP of consumer stage completely depends on consumer habit. Washing frequently with warm or hot water will result in high CFP. Washing machine load and machine selection according to energy rating also influence CFP. In this study we followed the care instruction of polyester dress which was cold water wash with no drying and ironing. Due to this our study shows comparatively less CFP than previous studies. As most of the studies shows that the use phase is the main CFP contributing phase due to the consumer wash habit with hot water, which results in high energy use. Generally the use phase contributes 80% of environmental impact among the whole chain of the product [3]. Detergent use is another contributing factor due to its manufacturing energy.

Users have a number of choices at the end step of any apparel such as discard, sell, or donate to charity or municipalities. Depending on the quality, the donated textiles are sent to the op-shop for resell, recycling or incineration plant. In this study we assumed the product will be ended up through reuse which causing minimum CFP from the use phase.

5. Conclusion and recommendation

In this study, we have included the most industrial sub production processes and other life cycle stages involved in sports shirt manufacturing. Assessment data were collected from secondary sources due to the unavailability of primary sources. However, it is very difficult to collect more detailed information regarding production process, input and output from previous studies. Due to this we used only the specific information of the production process in this study. Furthermore, the detailed information and data are necessary for complete CFP modelling of any product. From this study we can conclude that polyester production process accounted for the highest energy use among all the stages of life cycle. The next energy intensive stages are use stage and spinning process. Improving energy efficiency and implementing renewable energy sources in production stages can reduce the energy demand which ultimately result in less CFP and improved environmental performance.

This study could partially present the current situation of environmental sustainability of polyester sports shirt and provide some basic data with limited resources to guide more sustainable production of sports shirt. Fabric selection for the sports apparel should be addressed both for the performance requirements of sports apparel and environmental sustainability. Further study is required on the environmental performance of sportswear made with different fibres that possess the same required performance and quality.

Polyester raw materials, crude oil and natural gas, are non-renewable, whereas wool is available all the year round in Australia and other countries of the world. Wool is a natural fibre that has naturally breathability and active moisture and temperature management properties. Wool is naturally biodegradable and renewable. Energy and carbon footprint from wool production is comparatively less than polyester production. Wool can be considered as an alternative fibre for sports fabric manufacturing for sustainability. Cotton can be another alternative. Other options for reducing environmental burden from polyester sportswear are reuse and recycling. Applications of recycled polyester to produce different products are growing day by day. Recycling process involved mechanical and chemical treatment processes which also have environmental burden. Product reuse is indeed a good option for sustainability. But the aspects of recycling process to reduce environmental footprint should be studied in more detail to find out the best environmental practice.

References

[1] Skomra E. A comparative study of athletic apparel made from cotton. Masters Theses and Doctoral Dissertations. 2006 Jan 1: 21.

[2] Sportswear, why is it so expensive. https://www.choice.com.au/health-and-body/diet-and-fitness/sportswear-and-shoes/articles/sportswear-why-is-it-so-expensive, [Viewed January 2016].

[3] Muthu SS. 2015. Handbook of Life Cycle Assessment (LCA) of Textiles and Clothing. Burlington, Elsevier Science: 2015.

[4] Beamon BM. Designing the green supply chain. Logistics information management. 1999; 12(4): 332-42.

[5] Smith GG, Barker RH. Life cycle analysis of a polyester garment. Resour. Conserv. Recy. 1995; 14(3): 233-49.

[6] Steinberger JK, Friot D, Jolliet O, Erkman S. A spatially explicit life cycle inventory of the global textile chain. The International Journal of Life Cycle Assessment. 2009; 14(5): 443-55.

[7] Shah R, Sinha P. Examine the Role of Different Actors Across the Textile Fashion Supply Chain to Understand the Issues Regarding Growth of Eco-labelled Sustainable Textile Products: University of Manchester: 2010.

[8] DRI-Power Sports Shirt: 2016, http://www.jerzees.com/21MR.shtml. [Viewed January, 2016].

[9] By PU, Kuczenski B, Geyer R. Life Cycle Assessment of Polyethylene Terephthalate (PET) Beverage Bottles Consumed in the State of California: 2004.

[10] Collins M, Aumônier S. Streamlined life cycle assessment of two Marks & Spencer plc apparel products. Environmental Resources Management, Oxford: 2002.

[11] Franklin Associates. Resource and Environmental Profile Analysis of a Manufactured Apparel Product: Woman's knit polyester blouse, Washington DC: American Fiber Manufacturers Association.1993, pp3-4.

[12] Profile on the Production of Polyester Yarn. http://www.ethioembassy.org.uk/trade_and_investment/Investment%20Profiles%20EIA/Textile%20Industry%20Profiles/polyster%20yarn.pdf. [Viewed January, 2016].
[13] Koç E, Kaplan E. An investigation on energy consumption in yarn production with special reference to ring spinning. Fibres & Textiles in Eastern Europe. 2007(4 (63)):18-24.
[14] Kaplan E. The Cost Factor in Textile Industry and General Utilization of Energy Cost (Doctoral dissertation, MsC Thesis, Cukurova University, Institute of Natural and Applied Science, Textile Engineering Department, Adana 2004 (in Turkish).
[15] BTTG. Waste Minimisation and Best Practice, British Textile Technology Group. 1999.
[16] Ahonen E, Hansen J. Bat for Textile Industry: Pre-treatment, Dyeing, Printing and/or Finishing of Textile Products. Nordic Council of Ministers; 1996.
[17] Patagonia. Patagonia's common threads garment recycling program: A detailed analysis. https://www.patagonia.com/pdf/en_US/common_threads_whitepaper.pdf. [Viewed January 2016].
[18] Sea-distances.org. http://www.sea-distances.org. [Viewed January, 2016].
[19] How To Wash Nike Dri-Fit: 2016, http://help-en-us.nike.com/app/answers/detail/a_id/649/~/how-to-wash-nike-dri-fit. [Viewed on January, 2016].
[20] Samsung washing machine: 2016, http://www.samsung.com/au/washingmachinerecall/. [Viewed January 2016].
[21] Kenway SJ, Priestley A, Cook S, Seo S, Inman M, Gregory A, Hall M. Energy use in the provision and consumption of urban water in Australia and New Zealand. Water Services Association of Australia (WSAA): Sydney, Australia: 2008.
[22] Lu H. China's industrial carbon dioxide emissions in manufacturing subsectors and in selected provinces. ECEEE Industrial Summer Study, Arnhem, the Netherlands, September 11-14: 2012.
[23] Price per load (electricity), based on water temperature. http://michaelbluejay.com/electricity/laundry.html. [Viewed January 2016].
[24] National Greenhouse Accounts Factors, Australian National Greenhouse Accounts, August, 2015. https://www.environment.gov.au/system/files/resources/3ef30d52-d447-4911-b85c-1ad53e55dc39/files/national-greenhouse-accounts-factors-august-2015.pdf. [Viewed January 2016].

Decomposition Analysis for Assessing Driving Forces of CO_2 Emissions in Hangzhou's Textile and Apparel Industry

Li-Rong Sun[1], Lai-Li Wang[1, 2*], Yi Li[3]

[1]*School of Fashion Design & Engineering, Zhejiang Sci-Tech University, Hangzhou, 310018, China*
[2]*Zhejiang Sci-Tech University (Engineering Research Center of Clothing of Zhejiang Province), Hangzhou, 310018, China*
[3]*Zhejiang Provincial Key Research Institute of Philosophy and Social Sciences for Ecological Civilization, Hangzhou, 310018, China*

*Corresponding author's email: wangll@zstu.edu.cn

Abstract

Reducing CO_2 emissions of industry sectors is one of the key issues in the 13th Five-Year Plan of Hangzhou as it is establishing low carbon city. In order to figure out the driving factors influencing CO_2 emissions in Hangzhou's textile and apparel industry, a decomposition analysis based on the LMDI method was carry out to investigate how changes in industrial scale, energy intensity, industrial structure and energy mix influenced the CO_2 emissions from 1998 to 2014. The results showed that industrial scale effect and energy intensity effect were the main determinants of the change in energy-related CO_2 emissions. The industrial scale effect contributed the most significantly to increase the CO_2 emissions, especially from 1999 to 2007. Energy intensity effect was the only cause for the reduction of CO_2 emissions till 2014. Industrial structure and energy mix had limited effect contributed to increase emissions.

Keywords: decomposition analysis; driving forces; CO_2 emissions; textile and apparel industry

1. Introduction

Global climate change, in a trend of warming, poses serious threats on environment, ecology and socio-economic systems. The increasing anthropogenic emissions of greenhouse gases (GHGs) is one of the most important factors in the global climate change [1]. Among the anthropogenic GHGs, CO_2 which mainly comes from the fossil energies consumption is the main component [2]. In China, CO_2 emissions have increased tremendously as a result of the rapid process of industrialization. China has overtaken the United States as the world's top annual emitter of CO_2 according to a report [3].

China now is the world leading country for the textile and apparel production and export. Textile and apparel industry is Chinese traditional superiority industry. It has been equipped with the most complete industry chain and much higher processing level in the world. However, textile and apparel industry also accounts for a considerable share of the total CO_2 emissions in China's industries for the reason of its energy consumption structure which mainly rely on fossil energies such as coal, oil and natural gas [4].

Zhejiang province occupies an important place in China's textile and apparel industry. The production and export of textile and apparel products of Zhejiang province ranked top three in recent ten years. Hangzhou's textile and apparel industry is a representative in Zhejiang province. Its total industrial output value (TIOV) generally occupied 20 percent of the total TIOV of Zhejiang's textile and apparel industry. However, Hangzhou's textile and apparel industry also contributed a lot to the large quantities of Zhejiang's CO_2 emissions. Actually, the CO_2 emissions of Hangzhou's textile and apparel industry increased largely in recent fifteen years [5].

There have been several literatures about analysis of CO_2 emissions in China's industries, such as iron and steel industry, cement industry, power industry, food industry, chemical industry, textile industry, etc [6-11]. Decomposition analysis is often used to quantify the impact of different factors on the change of CO_2 emissions. They are generally two decomposition categories, the structural decomposition analysis (SDA) and the index decomposition analysis (IDA) [12]. Among the various SDA and IDA methods, Log Mean Divisia Index (LMDI) was proposed as the preferred method because this technique makes it possible to present decomposition without residuals, a very desirable property in decomposition analysis [13-14]. Then, many studies with regard to change and influencing factors of CO_2 emissions in China's industries have been done using LMDI method. The existing research results in this area showed that main influencing factors of the change of CO_2 emissions were energy intensity, energy mix, industrial structure and industrial scale [15]. For example, Wang et al, Lin and Moubarak all found that industrial scale contributed to the increase of CO_2 emissions in China's textile industry, energy intensity was the inhibiting factors of the increase of CO_2 emissions [16-17]. However, researches on the CO_2 emissions in China's textile industry are still limited in the national level. Few literatures address on the main factors and the degrees of influence impact of the CO_2 emissions in regional textile industry. Reducing CO_2 emissions of industry sectors is one of the key issues in the 13th Five-Year Plan of Hangzhou as it is establishing low carbon city. As prerequisite, it is important to figure out the driving factors influencing the CO_2 emissions. In this context, we applied LMDI method to explore influencing factors of CO_2 emissions in Hangzhou's textile and apparel industry during 1998~2014 period. It will provide information contributes to define strategies for mitigating CO_2 emissions. Furthermore, this paper will also give useful insight and orientation for policy interventions on CO_2 emissions reduction in other regional industries.

2. Methodology and data

2.1 Decomposition model of CO_2 emissions

In general, CO_2 emissions from stationary sources can be calculated by multiplying energy consumption by the corresponding emission coefficient. The method is summarized as:

$$C_{total} = \sum_{i,j} C_{i,j} = \sum_{i,j} E_{i,j} \times f_j \tag{1}$$

where C_{total} is the total CO_2 emissions of textile and apparel industry (Mt (10^6 ton) CO_2). $C_{i,j}$ is the CO_2 emissions of sub-industry i caused by the consumption of energy j (Mt CO_2). $E_{i,j}$ is the consumption of energy j in sub-industry i (Mt SCE). f_j is the emission coefficient of energy j (t CO_2/ t SCE).

In order to investigate influencing factors of C_{total}, it should be decomposed. Eq.1 can be further expressed as follows:

$$C_{total} = \sum_{i,j} C_{i,j} = \sum_{i,j} V_{total} \times \frac{V_i}{V_{total}} \times \frac{E_i}{V_i} \times \frac{E_{i,j}}{E_i} \times f_j = \sum_{i,j} V_{total} \times P_i \times I_i \times S_{i,j} \times f_j \tag{2}$$

where V_{total} is the TIOV of the textile and apparel industry (10^8 yuan), represents the industrial scale effect. V_i is the TIOV of sub-industry i (10^8 yuan). P_i is the share of economic output in sub-industry i, represents the industrial structure effect. E_i is the total energy consumption in sub-industry i (Mt SCE). I_i is energy intensity of sub-industry i (t SCE/10^4 yuan), represents the energy intensity effect. $E_{i,j}$ is the consumption of energy j in sub-industry i (Mt SCE). $S_{i,j}$ is energy mix ratio for energy j in sub-industry i, represents the energy mix effect.

The change quantity of CO_2 emissions between a base year 0 and a target year t can be expressed with ΔC_{total}. This parameter can be decomposed into its components in additive form according to the LMDI method, as illustrated as follows:

$$\begin{aligned}\Delta C_{total} &= C^t_{total} - C^0_{total} \\ &= \sum_{i,j} V^t_{total} \times P^t_i \times I^t_i \times S^t_{i,j} \times f^t_j - \sum_{i,j} V^0_{total} \times P^0_i \times I^0_i \times S^0_{i,j} \times f^0_j \\ &= \Delta C(V_{total}) + \Delta C(P_i) + \Delta C(I_i) + \Delta C(S_{i,j}) + \Delta C(f_j)\end{aligned} \tag{3}$$

where $\Delta C(X)$ represents the contribution values to ΔC_{total} of component X. It can be calculated as follows:

$$\Delta C(X) = \sum_{i,j} \frac{C^t_{i,j} - C^0_{i,j}}{ln\, C^t_{i,j} - ln\, C^0_{i,j}} \times ln \frac{X^t}{X^0} \tag{4}$$

If $\Delta C(X) > 0$, the component X has drove up effect on the increase of CO_2 emissions. If $\Delta C(X) < 0$, the component X has drove down effect on the increase of CO_2 emissions. If $\Delta C(X)$=0, the component X has zero impact on the change of CO_2 emissions.

Furthermore, the change of CO_2 emissions can be also expressed with multiplicative mode. D_{total} represents the change rate of CO_2 emissions, it is illustrated as follows:

$$D_{total} = \frac{C^t_{total}}{C^0_{total}} = \frac{\sum_{i,j} V^t_{total} \times P^t_i \times I^t_i \times S^t_i \times f^t_j}{\sum_{i,j} V^0_{total} \times P^0_i \times I^0_i \times S^0_i \times f^0_j} = D(V_{total}) \times D(P_i) \times D(I_i) \times D(S_i) \times D(f_j) \tag{5}$$

$$D(X) = exp\left(\sum_{i,j} \frac{\left(C^t_{i,j} - C^0_{i,j}\right) \times \left(ln\, C^t_{total} - ln\, C^0_{total}\right)}{\left(C^t_{total} - C^0_{total}\right) \times \left(ln\, C^t_{i,j} - ln\, C^0_{i,j}\right)} \times ln \frac{X^t}{X^0} \right) \tag{6}$$

If $D(X) > 1$, the component X has drove up effect on the increase of CO_2 emissions. If $D(X) < 1$, the component X has drove down effect on the increase of CO_2 emissions. If $D(X)=1$, the component X has zero impact on the change of CO_2 emissions.

2.2 Data sources

The sub-sectors of Hangzhou's textile and apparel industry are manufacture of textile (MT) sector, manufacture of textile wearing apparel, footware, and caps (MTWAFC) sector and manufacture of chemical fibers (MCF) sector according to GB/T 4754-2002 Industrial Classification for National Economic Activities. Data used in this paper cover the period from 1998 to 2014. The energy consumption data and economic output came from Hangzhou Statistical Yearbook (1999~2015). The TIOV is measured in constant 1998 price. CO_2 emissions coefficients (Table 1) in China were referred to previous work by Cao and Xie [18].

Table 1. CO_2 emissions coefficients of energies (Unit: t CO_2/t SCE (standard coal equivalent))

Energy	CO_2 emissions coefficients	Energy	CO_2 emissions coefficients
Raw Coal	2.492	Kerosene	2.051
Cleaned Coal	2.631	Diesel Oil	2.167
Other Washed Coal	2.492	Fuel Oil	2.219
Coke	2.977	Liquefied petroleum gas	1.828
Coke Oven Gas	1.288	Other Petroleum Products	2.126
Other Gas	1.288	Natural Gas	2.162
Crude Oil	2.104	Heat	3.212
Gasoline	1.988	Electricity	6.113

3. Results and discussion

According to Wang and Li's research, the CO_2 emissions of Hangzhou's textile and apparel industry increased from 3.32 Mt CO_2 in 1998 to 11.69 Mt CO_2 in 2014 [5]. The CO_2 emissions of the three sub-sectors differed quite a bit with each other. MT sector occupied more than 64 percent of the total CO_2 emissions, followed by MCF sector and MTWAFC sector (less than 3 percent).

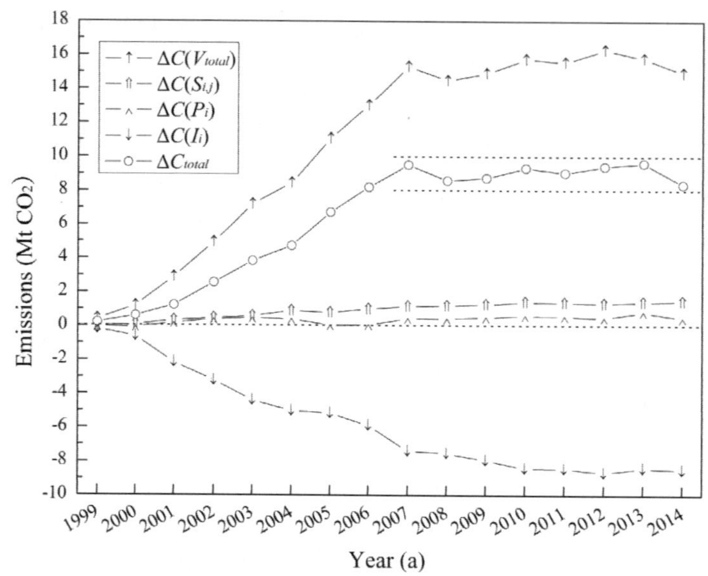

Fig.1 The degree of influence determinants of change quantity of CO_2 emissions.

Figure 1 shows the decomposed results of change quantity of CO_2 emissions of Hangzhou's textile and apparel industry. It can be seen that there has been significant increase of CO_2 emissions from 1999 to 2007. The increase of industrial scale, in the form of TIOV growth, was the most important factor that led to the increase of CO_2 emissions during this period. The energy mix effect and industrial structure effect also drove up the CO_2 emissions, but their contribution was much less significant than industrial scale effect. The energy intensity effect was the only factor that drove down the CO_2 emissions. However, the degree of driving up effect was larger than that of driving down effect. From 2007 to 2014, the industrial scale effect showed a volatile trend interchanging intervals of increase and decrease though it continually contributed mostly in increasing CO_2 emissions. The energy intensity effect still increased throughout this period. As a result, the change quantity of CO_2 emissions fluctuated between 8 Mt CO_2 and 10 Mt CO_2. The emissions coefficients of energy type had zero impact on the change quantity as they were fixed.

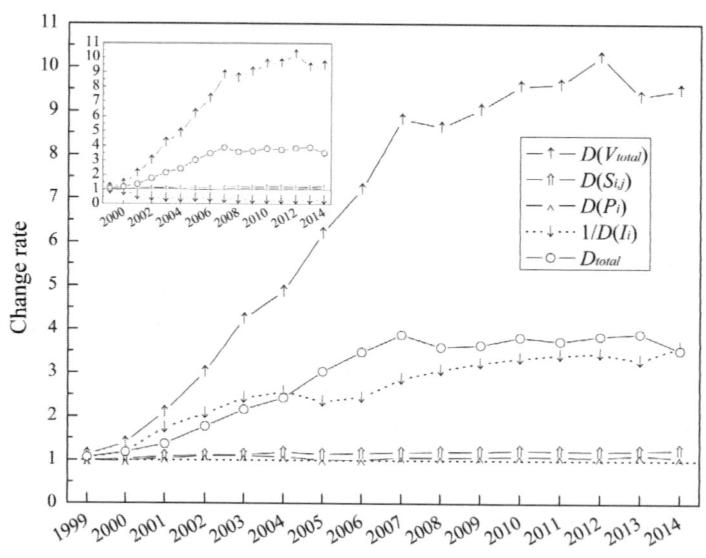

Fig.2 The degree of influence determinants of change rate of CO_2 emissions.

The decomposed results calculated according to equation (6) are shown in Figure 2. In order to enhance the comparability of factors' contribution ratios, the reciprocal of contribution ratio caused by energy intensity effect was expressed here. From Figure 2 it can be seen that industrial scale was mainly responsible for the increase of CO_2 emissions. On the contrary, energy intensity effect had negative impact on the increase of CO_2 emissions. This was consistent with the decomposed results with additive mode showed in Figure 1.

From 1998 to 2007, the scale of Hangzhou's textile and apparel industry extended rapidly benefited from China's booming export of textile and apparel products. For example, the TIOV of Hangzhou's textile and apparel industry increased approximately 6.5 times from 1998 to 2007. The industrial production of textile and apparel products consumed large quantity of energy, causing the increase of CO_2 emissions. With the improving of industrial production technology and the implementation of energy saving policies, energy intensity decreased continually. This led to decrease the CO_2 emissions. The world financial crisis in 2008 and 2009 caused the shrink of international demands on textile and apparel products. Consequently, the industrial production scale of Hangzhou's textile and apparel industry decreased in 2008. In the latter years until 2014, the driving up effect on CO_2 emissions contributed by industrial scale was not so significant as before. However, energy intensity reduced continually and the driving down effect on CO_2 emissions was enhanced during that period. Change of industry structure can promote the transformation and upgrading of Hangzhou's textile and apparel industry to gain a further and faster development. Unfortunately, this measure did not decrease the CO_2 emissions as MT sector occupied so significant proportion of CO_2 emissions. Generally, adjusting energy mix to reduce the share of fossil energies can mitigate the increase of CO_2 emissions. However, increasing consumption of electricity, mainly from heat-engine plant, undermined the expectation.

4. Conclusion

In this paper we investigated the change of energy-related CO_2 emissions in Hangzhou's textile and apparel industry. Application of decomposition analysis based on the LMDI method provided appreciable quantitative analysis on how changes in industrial scale, energy intensity, industrial structure and energy mix influenced the CO_2 emissions from 1998 to 2014. The results showed that industrial scale effect and energy intensity effect were the main determinants of the change in energy-related CO_2 emissions. The industrial scale effect contributed the most significantly to increase the CO_2 emissions level, especially from 1999 to 2007. Energy intensity effect was the only cause for the reduction of CO_2 emissions till 2014. Industrial structure and energy mix had limited effect contributed to increase emissions.

According to the results, several policies and measures should be implemented by enterprises and government in order to efficiently reduce CO_2 emissions in Hangzhou's textile and apparel industry. Energy intensity reduction must be taken up on a priority basis. Generally, technology upgrading is always an effective method to resolve industrial issues. For example, through allocating more investment for energy conservative technologies to improve the energy efficiency can eventually reduce energy intensity. Technological improvement will not only reduce the energy intensity, but will also improve the industrial productivity. Measures of shifting from fossil energies consumption to less polluting energy sources, such as hydroelectric generation and wind electricity, can also bring positive results for the mitigation of CO_2 emissions. Then, energy mix will have driving down effect on CO_2 emissions. Moreover, industrial structure adjustment should be continued. The industrial production of emissions intensive sectors should be decreased and promote the production of high value-added textile and apparel products with less emissions intensity.

5. Acknowledgement

The authors are grateful to Hangzhou Federation of Humanities and Social Sciences Circles for providing funding supports to this research through Hangzhou philosophy and social sciences project (Z16JC093).

References

[1] Wang HF. Economic costs of CO_2 emissions reduction for non-Annex I countries in international shipping. Energy Sustain Dev 2010; 14: 280-286.

[2] Zhao M, Tan LR, Zhang WG, Ji MH, et al. Decomposing the influencing factors of industrial carbon emissions in shanghai using the LMDI method. Energy 2010; 35: 2505-2510.
[3] Information on http://www.nature.com/news/2007/070618/full/news070618-9.html
[4] Wang LL, Du C, Wu XY. Analysis of carbon emission of China's textile and apparel industry. China Text Lea 2011; 19-22.
[5] Wang LL, Li Y. Investigation on the carbon emissions of Hangzhou's textile and apparel industry. Journal of Silk, in process.
[6] Wang XL, Lin BQ. How to reduce CO_2 emissions in China's iron and steel industry. Renew Sust Energ Rev 2016; 57: 1496-1505.
[7] Lin BQ, Zhang ZH. Carbon emissions in China's cement industry: A sector and policy analysis. Renew Sust Energ Rev 2016; 58: 1387-1394.
[8] Yang LS, Lin BQ. Carbon dioxide-emission in China's power industry: evidence and policy implications. Renew Sust Energ Rev 2016; 60: 258-267.
[9] LIN BQ, XIE X. CO_2 Emissions of China's food industry: an input–output approach. J Clean Prod 2016; 112: 1410-1421.
[10] Lin BQ, Lou HY. Emissions reduction in China's chemical industry – based on LMDI. Renew Sust Energ Rev 2016; 53: 1348-1355.
[11] Lin BQ, Moubarak M. Mitigation potential of carbon dioxide emissions in the Chinese textile industry. Appl Energ 2014; 113: 781-787.
[12] Chang CC. A multivariate causality test of carbon dioxide emissions, energy consumption and economic growth in China. Appl Energ 2010; 87: 3533-3537.
[13] Ang BW, Liu FL, Chew EP. Perfect decomposition techniques in energy and environmental analysis. Energ Policy 2003; 31: 1561-1566.
[14] Ang BW. Decomposition analysis for policy making in energy: which is the preferred method. Energ Policy 2004; 32: 1131-1139.
[15] Rre SG, Yin HY, Chen XH. Using LMDI to analyze the decoupling of carbon dioxide emissions by China's manufacturing industry. Environ Dev 2014; 9: 61-75.
[16] Lin BQ, Moubarak M. Decomposition analysis: change of carbon dioxide emissions in the Chinese textile industry. Renew Sust Energ Rev 2013; 26: 389-396.
[17] Wang LL, Wu XY, Ding XM, Li M. Analysis of carbon emission for energy consumption in China's textiles and clothing sectors: based on LMDI model. Environ Sci Technol 2013; 36: 201-205.
[18] Cao SX, Xie GD. Tracking analysis of carbon footprint flow of China's industrial sectors. Resour Sci 2010; 32: 2046-2052.

Application of "Crushable Elements" in Clothing Brand Extension

Ya-Juan Wang[1], Zhao-Hui Gu[2*]

Xi'an Polytechnic University, 19 Golden Road, Xi'an, Shaanxi, 710048, China

*Corresponding author's email: lightgu2008@126.com

Abstract

In the age of overflowing information and proliferating media, it became increasingly difficult to get consumers' attention. Therefore, it is very important for apparel enterprises to develop distinct characteristics to stand out. In this study, market investigation and market investigation are used to analyze and study the effect of "crushable elements" in brand extension, aiming to provide a more effective method. The research shows that the effect of "crushable elements" in brand extension is effective and a feasible method is put forward to define the "crushable elements".

Keywords: New media; Crushable elements; Clothing; Brand extension

1. Preface

The concept of "crushable elements" first came from the Coco-Cola bottle. In 1915, the requirements for bottle design features are the use of distinctive features, in which every part of the bottle should represent Coca-Cola. Even when the bottle shatters, the person who picks up the debris should be able to link it to Coca-Cola. In short, even without the company LOGO, consumers should still be able to identify the brand [1].

1.1 Research Status of "crushable elements"

Relatively speaking, "crushable elements" are more widely studied in foreign countries than in China. The famous international brand marketing expert Martin Lindstrom is the main researcher of "crushable elements". He analyzed a large number of cases to determine ways to define the "crushable elements" [2]. He reveals the common factor of success in top global brands: most of them use sensory brand marketing means, to create a new world using the "five senses" [3]. In the meantime, the case analyses revealed how to define the "crushable elements": crushing color (maintenance exclusive rights), crushing language (consistence), crushing behaviour (how to make a deep impression), crushing shape and name [4]. In overall, although there are a lot of research performed on "crushable elements", its application in the garment enterprise brand extension remain under-developed.

Technological advances allow the free flow of information in the media, in which vast amount of information is disseminated. Information dissemination virtualization is one of the main features of the new media environment, in which information continues to be fragmented, diverging customers' attention. Clearly, the traditional way of brand extension in the new environment is no longer effective, affecting apparel brand extension. Therefore, it is urgent to find a new brand extension method to remove current limitations on the development of apparel enterprises.

This study analyzes and summarizes the successful experience of predecessors, which include the role and significance of dialectical analysis in "crushable elements" that shed light on deriving a new method, thus reducing the cost for apparel enterprises in brand extension.

1.2 The influence of new media environment on brand extension

1.2.1 Definition of New Media and Brand Extension

1) New Media

New media is compared with traditional media. New media is based on digital technology, network technology and other support. For example, provide information and entertainment services to the users of spread form and media form through the Internet, wireless communication network terminal channels, mobile phone and computer, to. Examples of new media include virtual community, podcasts, microblog, SNS and outdoor LED [5].

2) Brand Extension

Brand extension uses a well-known brand as the core and starting point, in which the enterprise already has a high reputation in brand extension, taking advantage of the original well-known brand to launch new products and new market processes. Brand extension should follow proper procedures as the way and method have significant influence over the brand image [6]. Brand extension in a broad sense can be divided into two categories: product line extension and product species extension. Product line extension refers to the use of mother brand product line to introduce new daughter products under the same project; the new products usually have different characteristics, such as composition, tastes, forms and sizes, when compared to products from the mother brand. For instance, an extension of crest toothbrush products include crest multiplay partner toothbrush, crest white partner, kids' crest cavity protection toothpaste. Product range extension refers to the use of mother brand to launch different kinds of new products, such as the famous French brand Chanel that has an extensive range of products from hats, clothing to perfume.

1.2.2 Benefit analysis

Application and popularization of new media, which creates a more complex and hierarchical environment that encourages sharing and free flow of information, allow major publishing companies to conduct market research and increase publicity through a variety of ways. These companies can also incorporate brand extension to create more publicity.

1.2.3 Drawbacks analysis

1) Raising the cost of brand extension.

Under the new media environment, the limitations of traditional advertising in brand extension are gradually revealed. In the case of advertising effectiveness, the traditional way of advertising has shown to be less effective, but some companies still choose to adhere to the nature way of innovation. The new method and the traditional way of advertising should be combined to constantly changing environment.

2) Traditional brand extension mode can not meet the practical needs.

In general, corporate brand extension is usually carried out in the form of advertising, but excessive advertising among the massive amount of information shared in the new media is inevitable. Companies invest a lot of money on advertising, but the effect is not ideal. The new media environment makes traditional brand extension mode can not meet the practical needs, so it is important for the garment enterprises to make use of the new media environment to find a new way to brand extension by merging with the traditional way [7].

2. Brand extension in Apple and Coca-Cola

2.1 Application of "crushable elements" on Apple

Apple is a "crushable" brand as consumers can easily identify its products without the company logo. When Apple releases another device, regardless of the name or design style, consumers will think with no doubt that this is another product of Apple. One of the "crushable elements" of Apple is its unique navigation system, which distinguishes itself from other products in the market. This unique navigation system has been successfully extended to other various kinds of Apple products, which has paved the way for brand extension and product promotion. The system will remain one of Apple's inherent elements, which is undoubtedly a powerful promotional tool [8].

In general, the retail prices of Apple products are relatively high, but Apple is still able to attract a lot of costumers. In order to understand the charm of Apple products, this study launched a consumer survey, which was designed using the Likert scale (feeling quantification), on Apple's iPhone. The survey evaluated 5 level sensory quantization, the conciseness and complexity of iPhone case was ranked from 1 to 5: 1 = concise; 2 = relatively concise; 3 = neither simple nor complicated; 4 = relatively complex; 5 = very complex [9]. Since the design method of the questionnaire is 5 level sensory quantization, such as "mature - immature", the more closer to 1 the score, the more intense the "mature" feeling, the more closer to 5 the score, the more intense the "immature" feeling, so, 3 is a watershed, equivalent to coordinate origin, the farther its score from the 3 the feeling more obvious.Thus, each adjective score is calculated as follow: if the average score is smaller than 3, 3 subtract the average and the final number named after the adjectives on the left; if the average score is greater than or equal to 3, the average score subtract 3, named after the right of the adjectives. case in"concise - complex", if the average score is 2, 3-2 = 1, 1 represents the adjective score of "concise", if the average score is 4, 4-3 = 1, 1 represents the adjective score of "complex". The survey issued a total of 150 copies of the questionnaire, of which 135 questionnaires were recovered, and out of the 135 recovered questionnaires, 130 valid responses were received. The questionnaire was mainly divided into three parts: appearance, function and service. Ten adjectives were chosen to describe each aspect, which in total produced 30 adjectives. In the end, three adjectives with the highest score in each of the three aspects were picked out, and is shown in Table 1 with the corresponding scores.

Table 1. Adjectives and its score

Number	Classification	Adjective	Score
1	Appearance	Succinct line	1.20
		Elegant colors	1.10
		Fashionable and generous	1.10
2	Function	Responsive	1.40
		Diverse functions	1.35
		High resolution	1.25
3	Service	Warm and welcoming	1.65
		Thoughtful after-sales service	1.40
		Guaranteed	1.35

The adjectives identified in Table 1 are considered the “crushable elements” of Apple that have attracted a vast number of consumers.

2.2 Application of “crushable elements" on Coca-Cola

The concept of “crushable elements” first came from the Coco-Cola bottle in 1915, in which the intrinsic characteristics of the bottle, which is a big red and unique streamline bottle, are still in use today. As required by the Coca-Cola Company, the bottle should be designed in a way that consumers are still able to recognize a Coca-Cola product even when it is shed into pieces. In order to understand the consumer's perception of the Coca-Cola bottle, this study conducted an investigation by handing out surveys to 150 people.

The questionnaire included abstract images of the Coca-Cola bottle in the questions, and all the questions are related to the image attached. There were five types of questionnaires, and each questionnaire only had one abstract figure (as shown in Fig.1). In Figure 1, five kinds of abstract figure, named (1) to (5), are shown on the basis of five different graphic designs. The same questions were asked, regardless of the picture attached, to reduce errors caused by improper picture selection. A total of 150 copies of the questionnaire were issued, of which 148 questionnaires were recovered, and out of the 148 recovered questionnaires, 145 valid responses were received.

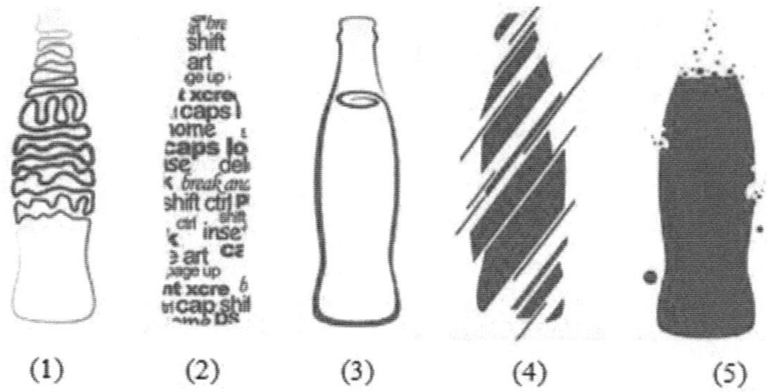

Fig.1 Coca-Cola bottles
(This figure is taken from http://www.zcool.com.cn/work/ZMTA1OTg4.html)

The analysis shows that, over 80% of those surveyed can quickly identify the Coca-Cola bottle. Around 20% of those surveyed could not recognize the bottle or did not know about the brand Coca-Cola, in which the majority of them were elderly. Clearly, the application of "crushable elements" on Coca-Cola is very effective in customer recognition (as shown in Fig.2).

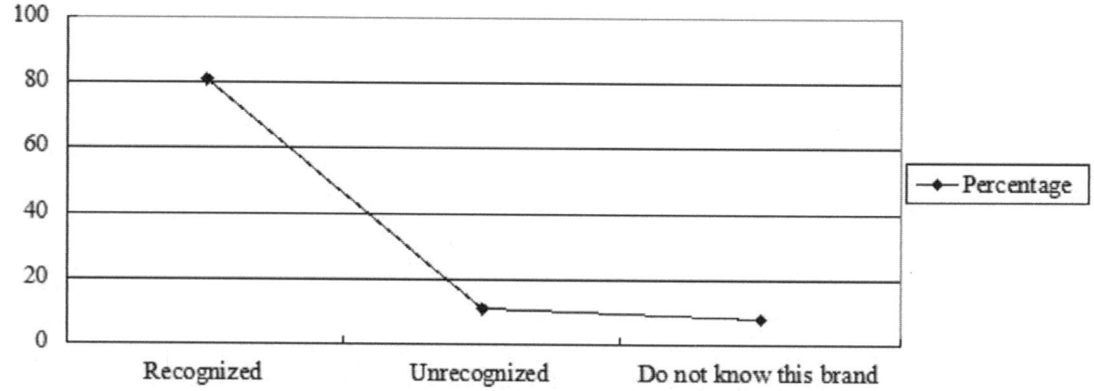

Fig.2 The degree of recognition ratio

At the same time, the data analysis shows that the color and streamline of the bottle were the major factors that consumers recognize, which accounted for 92% of the ones surveyed. 88% and 86% of those surveyed recognized the bottle using its contour and color, respectively. 78% of the people relied on feeling and impression. This shows that Coca-Cola's brand has been deeply rooted among the people, in which the color and bottle body lines are important characteristics, and thus they are also important "crushable elements" for Coca-Cola (as shown in Fig.3).

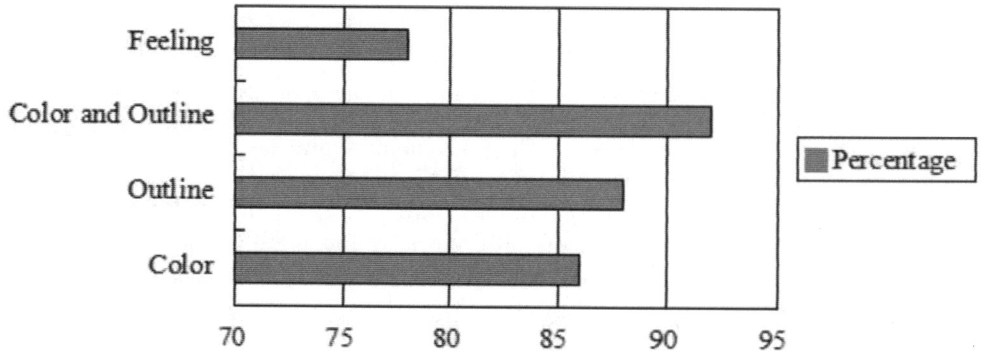

Fig.3 The degree of judgment basis

Although the Coca-Cola bottles were broken down into pieces that no longer resemble the usual complete bottle, its usual color and streamline were still recognized by most consumers. Therefore, "crushable elements" plays a very important role in corporate brand extension and there are ways to identify them. For example, Coca-Cola conducted brand extension from the visual angle such as color and outline. These elements have great significance on the brand extension of garment enterprises, and when garment enterprises look for "crushable elements", they can begin from the human senses such as sight, hearing, smell, taste and touch.

3. Analysis of the function of "crushable elements" in apparel brand extension

3.1 Case analysis: Chanel

Chanel was created by Coco Chanel in 1910 in Paris, France. Brand products range from clothing, jewelry and accessories, cosmetics, skin care products to perfume. All the products, particularly its perfume and fashion, are famous. Chanel is a famous brand with a history of one hundred years, and its fashion design has always remained a decorous, concise and elegant style [10]. Brand extension of Chanel has maintained its own unique style. Where there are many unique features, a few are selected and elaborated on.

1) Clothing: classic little black dress

In the 20th century, Gabrielle Chanel created a simple and luxurious little black dress for women in the upper class and has successfully shaped unique female temperament. Until today, Chanel's little black dress is still the dream choice for women in the world. At a ball, women dressed in black or white will always become the centre of attention, and there are constantly changes in fabric, design details and production to make the little black dress a unique fashion masterpiece. This dress has always been synonymous with the term "modern classic" [11].

2) Perfume

In Chanel, all the perfume is not made simply by following the production procedures, but to fully mobilize the scent of the perfume coordinator through the human senses. The connection and experience when using the product are key to the production process, striving to create perfumes that leave remarkable impressions on the users.

3) Glasses

Chanel glasses also inherited the tradition of Chanel, in which the fine details are carefully crafted. Designed Chanel style glasses can be clearly seen from the carved details, innovative technology and excellent lines, contours. In addition to the elegant signs, glasses themselves have a distinctive contour, avant-garde designs and colors to attract the public eye. The Chanel trademark can be found on the frame of each pair of Chanel glasses, reflecting the user's taste and identity [12].

4) The national flower of the Chanel kingdom: Camellia

The national flower of the Chanel kingdom is camellia. Its delicacy, pureness, elegance and unobtrusive form' have made the flower Ms. Chanel's love her whole life, but also condensed her expectations of beauty, charm and simple. Camellia initially appears in the front decoration of noble man in fashion design. Today, the craftsmen emphasize on the unique skill on presenting camellia flower's wonderful nature, embodying the perfect Chanel noble, elegant and unique taste [13].Chanel aims for the high road, creating a simple, stylist, elegant, pure and beautiful youth. "Pop fleeting, style forever", is still the motto of the brand. Chanel designs its products from the male point of view, which dominated the trend of fashion design in female beauty.

Chanel is not the only brand that has successfully extended the range of products, Versace is also one of them. Versace used to only produce products for women at first, with the extraordinary glamorous Greek goddess Medusa as a logo to confuse everyone. Versace is designed for glamorous and attractive people. Versace was founded in 1991, and since then its product line has been extended to men's clothing, watches, fragrances, glasses, scarves, neckties, underwear, bags and high-end resort hotel [14].

3.2 The importance of "crushable elements" to apparel brand extension

3.2.1 Simplification of the process of apparel brand extension

Identifying "crushable elements" is very simple. In general, as long as the brand has distinctive features, then the brand is crushable and it will be easy to find the elements. Compared with advertising, this way is simpler and less time-consuming. Chanel's unique characteristic lasted for over one hundred years and is still outstanding in the garment industry. As long as the enterprise maintains the brand's unique style and innovation, brand extension will become relatively simple as well.

3.2.2 Reduction in clothing brand extension publicity expenses

At present, the cost of advertising is an important factor in many clothing enterprises. Meanwhile, advertising costs soared in recent years; a second of television advertising costs more than tens of thousands of yuan, more importantly, these expenditures may not even be effective in promoting the products or the brand. As consumers live in an environment of advertising information overflow, consumers automatically filter out information that they deem less useful to them as they are unable to process all of them, which make advertising ineffective. Effects of "crushable elements", on the other hand, once been established, is stable and permanent. Consumers will automatically be able to link the products out in the market with the parent brand. Therefore, “crushable elements” are need in apparel brand extension to effectively make up for the defect in advertising.

Aside from its role in brand extension, “crushable elements” may play other roles with varying degrees of significance, depending on the brand.

4. Methods to identify "crushable elements"

4.1 Characteristics of "crushable elements"

1) "Crushable elements" have obvious features after being crushed

Like the human genome, some elements of the parent brand features will be extended to the “daughter brand” or “daughter company”, it must have very distinctive traits to be identified. Due to their obvious characteristics, these elements can still be identified after being crushed. Take eggs as an example, when an egg is broken into pieces, people are still able to identify the pieces as an egg’s, due to the shell’s features such as arc, thickness and color.

2) Characteristics of the parent brand are retained after crushing

When a leaf is torn apart, people can still tell the type of tree leaves using characteristics such as texture, cntext, taste and color. These characteristics can be used to confirm their identity. Generally speaking, it will be difficult to distinguish between the parent brand and its sub-brands if the crush elements are too similar. Therefore, crush elements of each brand must be distinctive [15].

3) "Crushable elements" can imply some association to the target market of the brand

Brand is valuable in that it can provide consumers certain physical and psychological services. Therefore, the first task before brand extension is to conduct market research,the purpose is to understand the “crushable elements” can make the consumers starting up what association. LILANZ, for example, from a psychological point of view can make the person produces a sense of nobility, rigorous, smart, from the features of the product can make the person produces a feeling of good quality, style, fashionable, lenovo plays a vital role in consumers accepting the new brand .

4) Relative concept of crushable elements and elements that cannot be crushed

"Crushable elements" and elements that cannot be crushed is a relative concept. “Crushable elements” are not always created in brand extension. For example, a brand cannot be crushed in the extension of product A, but “crushable elements” can be identified in the extension to the product B, it is not affirmatory, anyhow should according to the actual situation to judge [16].

4.2 Ways to crush elements

"Crushable elements" can be tangible but can also be invisible. There can be many ways to identify “crushable elements” in clothing; for example, these elements can be identified from different human senses.

1) Starting from visual

Humans have five senses: visual, hearing, smell, taste and touch. The eyes are the window to the soul. In visual terms, crushed elements of clothing can be color, style, size and so on. For example, some fresh clothing brand designed to be wind-proof, although the style is different, the overall style comes with a fresh feeling.as the new brand inherits the parent brand-specific elements.

2) Starting from hearing

Starbucks create their own music and when customers walk into the store, they will listen to music that they have never heard before. After a while, the customers will naturally be familiar with Starbucks' unique music. The auditory aspect of "crushable elements", can be music, a word, or any unique sound, as long as it carries some relevance from the parent brand elements.

3) Starting from tactile

Clothing brand in the search for "crushable elements", can begin from touch and degree of comfortableness, for instance silk gives a feeling of smoothness and softness and linen gives a rough texture. Sometimes, "crushable elements" can be a unique touch or in the shape of a button. "Crushable elements", regardless of its size, should be capable of embodying that of the parent brand.

4) Starting from olfactory

Starbucks coffee selection is very particular about the quality, each coffee has its own unique taste and smell, which can be used to classify them. For the types of coffee a café sell, it produces a unique scent. Therefore, When Starbucks coffee grinding out, there will always be a special aromatic smell spread to every corner of the cafe, it has become a unique signature of Starbucks. In addition, of course, there are a lot of scents or fragrances that can be used and perfume is a typical example.

5) Starting from an idea

"Crushable elements" are not necessarily visible, sometimes an invisible element carry greater significance than the tangible elements of influence, such as some thoughts or ideas, both of which will influence the minds of consumers in an invisible way, making brand extension more effective and easier.

6) Starting from behavior

Behavior can be a smile, can be a greeting, can be a service, and so on. Certainly, the above-mentioned methods are not the only ways to search for "crushable elements", there is need to constantly explore and innovate. The methods are not independent of each other and can be combined to develop "crushable elements" for brand extension.

5. Conclusion

The new methods of brand extension can be merged with the traditional means to reduce costs and improve the effects of brand extension.

"Crushable elements" is a new means for apparel enterprises to develop their brand under the new media environment, it is a supplement to the traditional brand extension ways. The combination and application of the two to clothing brand extension can reduce costs and enhance the effects of brand extension. There are so many methods for identifying "crushable elements" of a clothing brand, and different brands can use different search methods. Not only can "crushable elements" be applied to clothing enterprises in brand extension, but can also be applied to other types of brands too. Therefore, "crushable elements" have a wide range of applications and potential.

6. Acknowledgement

The authors would like to express their gratitude towards the professors at the Apparel & Design College of Xi'an Polytechnic University.

References

[1] Martin L, Lin ZY. Crushable brand. 21st CBR: 2011; 6: 36-38.
[2] Martin L. Buyology. United States: Doubleday: 2008; 58-88 .
[3] Martin L. Brand sense. Tianjin: TJEPH, 2011; 23-55.
[4] Martin L. Smash your brand. Business(Review): 2011; 8: 53-56.
[5] Deng XL. Information propagation characteristics of the new media environment. Journal of Editing: 2011; 3: 2-3.

[6] Liu GQ. About brand extension and the focus of contrast research. Science Tribune (ten-day): 2014; 1: 181, 239.
[7] Cai JJ. Brand design methodology. SHU, 2013: 18-30.
[8] Wang SZ. Brand extension theory and strategic management research. XTDX, 2005: 11-20.
[9] Zhao QF, Wang ZY, Fan BT. Kansei engineering and its study in Japan. Art and Design: 2007; 7: 32-34.
[10] Karen G. Unique Chanel. National New Books Information: 2010; 19: 58-59.
[11] Information on http://www.chanel.com/zh CN/
[12] Zhao W. Crossover,the winner-takes-all plan of brands. China Apparel: 2013; 11: 22-30.
[13] Li YT. Extension strategy of luxury brand. ZGSC: 2008; 18: 11-14.
[14] Shao YM, Yuan AF. From Versace to see our brand of high-end clothing. Jiang Su Commercial Forum: 2010; 8: 147-149.
[15] Martin L, Lin ZY. Crushable brand. 21st CBR: 2011; 6: 36-38.
[16] Liu Y. Theory of brand extension strategy. Guide of Sci-tech Magazine: 2013; 8: 90,162.

Analyses of Aimer's Investment Climate In Tailand

Lu Wang[1], Xiao Hong[*2], Xuan Liu[3]

[1,3]*School of Fashion, Shaanxi institute of international trade&commerce*
[3,5] *tongyi west Road,XianYang, Shaanxi, 712046, P.R. China*
[2]*School of Fashion & Art Design, Xi'an Polytechnic University*
[1,9] *Jinhua South Road, Xi'an, Shaanxi, 710048, P.R. China*

[*]Corresponding author's email: xh309@126.com

Abstract

The investment climate is highly competitive in the 21st century, and so many countries provide different preferential policies to attract foreign investors. On the other hand, some companies want to expand globally, in order to create value of the company. So based on the research about Thailand and Aimer, the paper divided into three parts. At first, the introduction of Aimer will be shown and explained why choose Aimer as a background. And in the second part there is a description about investment climate of Thailand. After that, the third part will present the overview of Thailand's lingerie market. The opportunities and challenges of Aimer's entering Thailand explored in detail.

Keywords: Investment climate; Tailand; Marketing

1. Introduction

1.1 The Introduction of Aimer

During the recent decade, Aimer has developed quickly and become one of biggest lingerie brands in China. And it has had the highest market share and the largest sales volume for several years according to the records of Chinese General Chamber of Commerce. Aimer company has professional production bases and subsidiary fabric factories in Beijing and Suzhou, with a capacity over 10 million pieces per year. Table 1 lists the branches of Aimer brands.

Table 1. Aimer Brand's branches

	Aimer 爱慕	LA CLOVER	im's	Shine love	AIMER men	Aimer kids	MODELAB
Price Arrange (RMB)	190 ~1380	780 ~ 1600	145 ~ 399	129 ~ 259	168 ~ 380	76 ~ 356	460 ~ 980
Brands Position	Middle-upper Class	Slight luxury	Young Target	Sexy style	men underwear	Kid underwear	Functional, Shapewear

And the following graph shows Aimer's product lines matrix.

Table 2. Aimer's product lines matrix

Bras	CamiBra	Push-up	Convertible	Front Close	T-shirt	Padded
	Seamless	Underwire	Wirefree			
Panties	Thongs	Briefs	Boxers			
Shape wear	Tops	Bodysuits	Control Panties	Control Slips	Thigh Slimmers	Waist Cinchers
Thermal wear	Tops	Bottoms				
Sleep wear	Camisoles	Chemises	Gowns	Pyjamas	Nighties	
Swim wear	Swimsuits	Bikinis	Tankinis	Accessories		
Men	T-shirst	Tank Tops	Briefs	Boxers	Trunks	
Kids	Thermal Wear	Tops	Girl's Bras	Underwears	Swim	Sleepwears
	Parent-child Matching					
Accessories	Bra Extenders	Lingerie Wash Bag	Bust Pads	Nipple Covers		

1.2 The Mature Time to develop overseas market

Aimer has large numbers of fans looking for high-quality goods in China, including an increasing number of foreign customers, the company's luxury lingerie products such as Aimer and La Clover are sold across Europe, the Middle East and most recently, Singapore. It is high time for Aimer to expand business globally. Since Aimer opened their online shopping mall, they have received a high quantity of orders from overseas buyers paying with PayPal.

Aimer Group also aims to expand their business to Southeast Asia. This paper aims to study Thailand is a proper choose.

2. Economic Environment

2.1 Key Economic Indicators

Table 3. Key Economic Indicators

Indicator	In 2015
Nominal GDP	US$317.7 billion
Increase rate of Price Index (manufacturer)	9.60%
Economic Growth Rate	7.80%
Unemployment Rate	0.90%
Per Capita GNI	US$4,719

2.2 Economic Performance

The National Economic and Social Development Board (NESDB) reported that the Thai Economy expanded by 7.8 percent in 2010, supported by global economic recovery and strong domestic demand. Private investment and consumption grew by 13.8 and 4.8 percent respectively, while the value of Thai exports expanded by 28.5 percent to set a new record of US$19538 billion.

Table 4. Thailand's GDP & Exports statistics

	2014	2015
GDP(%)	-2.3	+7.8
Exports(%)(US$ basis)	-14	+28.5
Imports	-25.2	+36.8
Current Account to GDP (%)	+8.3	+4.6
Inflation	-0.8	+3.3
Unemployment	+1.5	+1.0

Besides, Bank of Thailand indicates that the financial policy maintain the stability of the financial system, in order to adapt to the changing of external situation and push the long-term stability and development of Thailand economy.

Table 5. Exchange Rate

Year	2008	2009	2010	2011	2012	2013	2014
USD & Baht Exchange Rate	34.56	33.36	33.34	31.66	30.76	30.46	30.14

2.3 Bilateral Trade between Thailand and China

Thailand and China signed a Free Trade Agreement in 2003. And now China is Thailand's second largest export market. Besides, China has planned to create China City Complex in Thailand to boost trade and get around trade barriers in the ASEAN region. China takes advantage of the ASEAN-China Free Trade Area which came into effect in 2014. This allows its goods to be exported through ASEAN countries with zero or reduced trade barriers.

Table 6. Thailand's main trading partners to export

	2014		Unit: million USD
Nations and Regions	Amount	Year Basis (%)	Occupy (%)
Total	92,647	36.8	100
China	**10,059**	**47.8**	**10.9**
Japan	9,357	32.2	10.1
Amercia	9,123	24.8	9.9
Hong Kong	5,936	39.9	6.4
Malaysia	5,168	62.4	5.6

Table 7. Thailand's main trading partners to import

201	4	unit:million USD	
Nations and Regions	Amount	Year Basis(%)	Occupy(%)
Total	87,319	52	100
Japan	17,834	76.1	20.4
China	**11,501**	**58.5**	**13.2**
Malaysia	5,341	43.3	6.1
Amercia	5,118	32.3	5.9
Korea	3,986	63.9	4.6

2.4 Special Economic Zones in Thailand

Thailand government provides Export Processing Zones (EPZs) and Free Zones (FZs) to promote international trade in the country by implementing various tax privileges. Companies in EPSs are granted tax and non-tax privileges offered by the Industrial Estate Authority of Thailand. Currently, there are 10 EPZs in Thailand.

Machinery, equipment, tools and supplies including parts transported into the EPZ for the purpose of manufacturing goods, and constructing or assembling a factory or a building in the EPZ are not subject to special surcharges under the Investment Promotion Act, import duty, excise tax and VAT.

Table 8. Export Processing Zone

Export Processing Zone	Location
Lard Krabang Industrial Estate	Lard Krabang, Bangkok
Banpoo Industrial Estate	Bangpoo Mai, Muang, Samut Prakan
Bangpa-In Industrial Estate	Bangpa-In, Ayutthaya
Hi-Tech Industrial Estate	Ban Wha, Bangpa-In, Ayutthaya
Gateway city Industrial Estate	Plaengyao, Chachoengsao
Hemaraj Chonburi Industrial Estate	Bo-Win, Sriracha, Chonburi
Leam Chabang Industrial Estate	Leam Chabang, Sriracha, Chonburi
Northern Industrial Estate	Banklang, Muang, Lamphun
Southern Industrial Estate	Chalung, Hat Yai, Songkhla
Phichit Industrial Estate	Nhong Lum, Wachirabarami, Phichit

The Free Zone Program has been implemented to promote competitiveness by encouraging companies to maintain and expand their operations in Thailand. Foreign and local goods transported into FZs are eligible for tax and duty privileges as well.

3. Textile and Apparel Industry

Textile and apparel industry is one of the most important manufacture industries in Thailand which involved more than 4,500 companies, 1.2 million staff and accounts for 17% of GDP. For production facilities, Thai factories hold 130,000 looms (50% to 60% of them are shuttleless loom), 110,000 modern knitting equipments and 758,000 sewing machines. However, we should not ignore that the labor shortage problem which lacks of more than tens of thousands workers, especially for those professional senior textile engineers.

4. Thailand Lingerie Market

4.1 Lingerie sales grew in 2014 amid increasing competition

Demand for lingerie is determined by demographic factors, such as age, gender, and economic factors, including disposable income and the level of unemployment. In addition, product accessibility is driving a new market opportunity as the existing market is close to maturity. The role of modern trade sales outlets has been increasing over the past decade in Thailand. Besides, lingerie sellers have increased the level of market penetration to service new customer groups including young and trendy customers.

4.2 Customer Behavior

Young customers are mostly attracted by the desire to follow the latest fashion trends, which in turn encourages manufacturers to continuously introduce new collections. For working women, a target group with higher brand loyalty than younger customers, fit and suitable sizes are the major factors influencing the purchase decision. In addition, the cup size of Thai women is a little smaller than Chinese or European.

Colorful designs are increasingly popular, and swimwear is moving towards more fashionable styles. For example, two pieces swimwear like the tankini are growing in popularity.

4.3 China Dominated Lingerie Import

Competition in the lingerie market remains high. In 2011, brassieres imported from China and Hong Kong accounted for 86% of total brassiere imports into Thailand. The remaining porting were imported from Taiwan, Vietnam, Korea, Germany, and Japan.

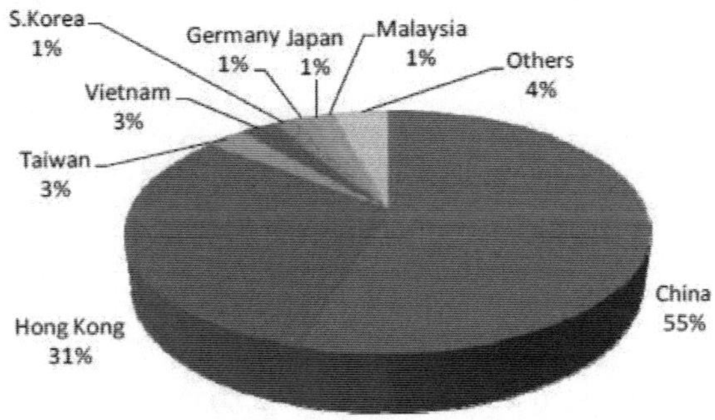

Fig.1 Thailand's Brassiere Imports by Country of Origin in 2014

4.4 Lingerie market continues to grow

In 2014, the value of major branded lingerie sold in department stores in Thailand totaled approximately Bt4,600 million, a rise of 12.4% from 2013, according to a survey conducted by I.C.C. International PLC (ICC). In 2015, lingerie sales are expected to grow at moderate rate due to the economic recovery. Demand may also rise because the provincial market segment is expanding, companies are instigating more marketing activities and campaigns, and firms are launching new innovative products.

4.5 Thai Consumers

4.5.1 Thai Consumers' lifestyle

Shopping/ hanging out around in shopping area in a very important weekend activity for consumers in Thailand, especially city dwellers in Bangkok as there are not that many alternative entertainment activities.

4.5.2 Rising Affluence of Middle Class Concentrated in Bangkok

The increasingly affluent middle class in Thailand is looking to branded fashions for identity and a better lifestyle. It is estimated that currently around 3% of the Thai population (about 2 million) are -upper middle class earning 20,000-30,000 baht (about US$500-700) per month , who are highly concentrated in Bangkok and nearby cities.

4.5.3 Demanding for underwear and swimwear in holidays

Demand for women's swimwear tends to peak during summer time, from March to May, which is the period when there are school breaks and parents often take their children for long holidays. The preferred destinations are on the coast. Therefore, promotions and campaigns, as well as new collection releases, tend to occur during those months.

4.5.4 Young People are Increasingly Fashion Target

More frequent marketing activities and more rapid launches of fashion-styled lingerie are crucial tools to reach consumers in the young segment and to counter aggressive competition. As aforementioned, young customers at 20s-30s are the major target of department stores and shopping complexes in Thailand, which are also the Aimer's target customer. As a matter of fact, over one-third of its population is within the age group of 19-39 - the segment with the spending power and willingness to spend.

Table 9. Thai Population by Age Group 2014

60 & Above	8.70% 10.30%	
40 - 49	21.90%	22.50%
20 - 39	35%	35.30%
14 - 19	17.60%	16.40%
0 - 14	16.80%	15.30%

0.00% 10.00% 20.00% 30.00% 40.00% 50.00% 60.00% 70.00% 80.00%

■ Males ■ Females

5. Challenges that Aimer Might Face

The challenges are divided into 6 parts: competition, political uncertain, grey market, repositioning problem, unknown market demand, logistics issue.

5.1 Fierce competition from foreign brands

Aimer has to compete with large numbers of lingerie brands in Japan and Europe. For instance, in underwear market, Wacoal PCL held the leading position due to strong brand loyalty to Wacoal, who dominated the market with around 56% market share, followed by Sabina (13%) and Triumph (11%).

Brands in swimwear favoured by local consumers in Thailand include Triumph, Jantzen and Speedo. Trends in women's swimwear have seen a shift beyond functionality to include aesthetics and fashion, which require brands provide more varieties products.

Hypermarkets/ convenience stores compete on price; they are highly price sensitive and look for -cost-savings suppliers.

5.2 Domestic political uncertainty

Although Thailand remains top spot for work, its domestic political uncertainty and the global economic situation remain potential threats to demand for women's underwear, nightwear and swimwear. If another bout of political instability occurs, it will definitely cause many tourists to avoid visiting Thailand, and Thai people will be less inclined to go travelling or shopping. Due to Aimer would take Thailand into consideration for the first globally expansion, a stable environment would be preferred.

5.3 Grey Market Issue

With the ASEAN-China FTA, tariff-free entry of clothing and lingerie items of mainland origin to Thailand. Aimer might face grey market issues relative to other foreign brands.

5.4 Market positioning

Thailand's businesses and consumers are extremely price-conscious, often favoring lower prices over product quality or other benefits. Exporters with products that are competitive for reasons other than price should plan to work with their local partner to undertake an extensive marketing strategy. So competitors who are highly price sensitive like hypermarkets and convenience stores. Wacoal and Triumph retail prices are really low in Bangkok (compared to China), it's essential for Aimer to consider its market position and price strategy. The original position of Aimer will directly compete with Wacoal and Triumph, which will increase risk for entering a new market.

5.5 Not enough information about Chinese upper brands

Several Mainland Chinese brands have entered in Thailand market. It's still a question whether Thailand customers can accept Chinese design culture and characteristics.

5.6 Non-Normal Distributions

Thailand transportation system lacks of efficiency, the infrastructure is not enough and the total logistics cost is a little high.

In quantitative portfolio management, the company absolutely thinks of achieving the highest possible expected return for some level of risk. Thailand customers prefer variety and fashionable products; it would add difficulties for retailers to get economy scale and control logistics system effectively. In addition, Aimer Group's factories located in Beijing and Suzhou, whose distances to Thailand are long. The Selection of distribution center and transportation is easy to increase cost if managers handle it inappropriately.

6. Conclusion

Compared opportunities with challenges, it can be concluded that Thailand deserves to be invested as a first global country that Aimer will enter. Thailand's open polices, pleasant climate, smile culture, emerging economy, mature tourism and huge domestic lingerie demand constitute a great investment environment.

References

[1] Anna Strutt, Thomas W. Hertel, Susan Stone, "Chapter 8 Exploring Poverty Impacts of ASEAN Trade, Liberalization for Cambodia, Lao PDR, Thailand and Vietnam", John Gilbert, in (ed.) New Developments in Computable General, Equilibrium Analysis for Trade Policy (Frontiers of Economics and Globalization, Volume 7), Emerald Group Publishing Limited: 2013; pp: 217-245.

[2] Doing Business in Thailand: 2014 Country Commercial Guide for U.S. Companies, U.S. Commercial Service: 2014.

[3] Economic and Trade Relations between China and Thailand Kingdom, http://www.chinagate.cn/english/365.htm

[4] Euromonitor International Ltd. 60-61 Britton Street, EC1M 5UX, World Consumer. Income and Expenditure Patterns 2014.

[5] Johansson JK.. Global Marketing: Foreign Entry, Local Marketing, and Global Management, 4th ed., McGraw -Hill Co., New York, NY: 2012.

[6] Joséde Sousa, Thierry Mayer, Soledad Zignago(2014). Market access in global and regional trade, Regional Science and Urban Economics.

[7] Kawai M, Wignaraja G. ASEANt3 or ASEANt6: which way forward? ADBI Discussion Paper No.77, Asian Development Bank Institute, Tokyo: 2011.

[8] Key Economic Indicators, Statistics Office, Bank of Thailand, 0ctober31: 2014

[9] Kittichai (Tu) Watchravesringkan, Elena Karpova, Nancy Nelson Hodges, Raedene Copeland. "The competitive position of Thailand's apparel industry: Challenges and opportunities for globalization", Journal of Fashion Marketing and Management, Vol. 14 Iss: 4: 2013; pp: 576 -597.
[10] Kunz GI, Garner MB, Going Global. The Textile and Apparel Industry, Fairchild Publications Inc., New York, NY: 2011.
[11] The Government of The Grand Duchy of Luxembourg, Ministry of the Economy and Foreign Trade, Market Entry Guide Thailand, Luxembourg for Business: February 2013
[12] Will Martin. Ashington, Free trade area membership as a stepping stone to development : the case of ASEAN / Emiko Fukase, D.C. World Bank: c2009.

Design and Realization of Middle and Small-Scale Clothing Enterprises Evaluation System through Social Media

Yan-Chao Wei[1,2], Zhao-Hui Gu[1*], Jin Duan[1]

[1]*Apparel & Art Design College, Xi'an Polytechnic University, Xi'an, Shaanxi, 710048, P.R. China*
[2]*Apparel & Art Design College, Minnan University of Science and Technology, Quan Zhou, Fu Jian, P.R. China*

[*]Corresponding author's email: lightgu2008@126.com

Abstract

With the rapid development of technology and the widespread use of internet, marketing through social medial will soon become the main channel for marketing. Especially, the middle and small-scale clothing enterprises can greatly improve marketing efficiency through utilizing the social media. As a new emerging marketing channel, there is no guideline to evaluate the effect of marketing through social media. Through studying the social media marketing strategies of small- and middle-scale clothing enterprises, the article made a classification on social media marketing to discuss the advantages and disadvantages of marketing through social marketing in these clothing enterprises. Solutions to existing or current problems are also discussed, in hopes of developing an evaluation system that includes basic modules and direction of data flow.

Keywords: Middle and small-sized enterprises; Socialized media; Clothing marketing; Evaluation system

1. Introduction

With the rapid development of internet, the global economy has entered an information era. According to the 34 Times China Internet Network Development State Statistic Report published by China Internet Network Information Center (CNNIC) in 2014, as of June 2014, our country's net citizens scale reached 632 million, with internet popularizing rate reaching up to 46.9% [1]. The rise of informatization changes the public's way and efficiency of obtaining information, which also changes the ways and means of marketing. Marketing through social media rises in response to the era of rise of informatization.

The seasonality and uncertainty of fashion trends of the clothing industry shorten the lifespan of the clothing market in comparison to other markets, as the clothing market changes more quickly. The short lifespan creates both opportunities and challenges for complying with the trends of informatization era, and for adapting to the huge change brought by marketing activities through social media.

2. Social Media Marketing Definition

2.1 The Concept of Social Media

Relatively speaking, social media is a new concept. People from different fields are doing a lot of research on the concept of social media. Different opinions exist in its definition and utilization. In order to correctly understand the concept of social media, we analyzed definitions of social media published in different papers (Table 1).

In Table 1, we know that the core essence of social media is users' participation, which relies on the internet technology. It emphasizes change in the way of thinking, in which media is no longer a traditional and single mechanical spreading, but a kind of polyhedral communication spreading. Although, since the proposal of the social media concept in 2007, different specialists and scholars have presented different researches and definitions, and I still believe that the definition of Antony Mayfield to social media is the most concise and accurate, in which social media is a new kind of online media that gives users a great interacting space.

Table 1. The Concept of Socialized Media

Time	Author	Literature	Definition
2007	Antony Mayfield	What is social media	Social media is a kind of new online media which gives users a great participating space [2].
2008	Jim Tobin	Social Media is a Cocktail Party	Social media is a kind of advanced online space. It allows people to share, comment, communicate and even spread everything based on the network techniques, such as internet, mobile phone or mobile applications (APP) [3].
2008	Public Relations Society of America	The Point of Connection Social Media Discussion. PRSA 2008 International Conference	From the perspective of use, social media is a kind of decentralized network established by people and takes people as basis. From the perspective of technology, social media shall mean the web services or tools of supporting "taking people as basis" [4].
2010	Jesse Grainger	Social Media and The Fortune 500	Social media refers to the interaction between individuals through the high-tech internet technology of wide contact surface, which is a kind of online new media. The social media utilizes internet and internet technology, and changes the original one-to-many media dialogue to a many-to-many media dialogue [5].
2011	David Meerman Scott	The New Rulers of Marketing and PR	Contents of social media can be established, made and generated by anyone [6].
2011	Wang Xiaoguang, Guo Shujuan	The First Analysis of Sociality Media	Social media is a communication tool generated by relying on internet technology, and social media is a kind of digital media [7].
2011	Peng Lan	Double Challenges Integrated by Socialized Media and Medium	Social media is a platform for users' social communication and information sharing [8].
	Baidu Baike		It is also known as social media, which refers to the websites and technology where people can write, share, evaluate, discuss and intercommunicate. The so-called socialized media should be the process of a large number of net citizens' voluntary contributions, extraction, creating news, and intercommunicating.
2013	Li Liubang	The Research of Social Media Marketing of the Enterprises-A FMCG Enterprises Perspective	Social media is a new kind of online media that gives users a great interacting space [9].
2013	Xu Yixin	The Precision Marketing Research Based on Social Media	Social media contains the content generated by promoting and supporting the communication between users [10].
2014	Lin Bei	Social Media Marketing Strategy Research of Hongkong I.T Apparel Group	It is a new-type online media of high interactivity, and a platform which gives users a great interacting space. Through utilizing this platform and tool, it allows people to share their own information, opinions, experience and ideas with the many-to-many mode, and enables users to maintain the relationship network between each other [11].

2.2 The Marketing Concept of Social Media

Social media marketing is a marketing mode of utilizing the social media or its way of thinking. Comparing to traditional marketing, social media marketing possesses the properties of interactive communication, fairness and openness, and media connectivity. Therefore, it can realize the precise localization to the target audience, bringing the long tail effect to enterprises' marketing. It can also greatly reduce marketing cost and maximise the enterprises' marketing profits.

3. Social Media Marketing and Middle and Small-Scale Clothing Enterprises

3.1 The Current Situation of China's Clothing Industry

China's clothing industry, which is mainly comprised of middle and small-scale enterprises, possesses its own characteristics. In 2013, China had more than 4000 clothing enterprises, and among which, state-owned enterprises accounted for 2.7%, the collectively-owned enterprises accounted for 26.8%, while the three kinds of investment enterprises occupied 19.2%, and the private enterprises making up 53.5% of all enterprises. Most of the clothing enterprises are middle and small-scale , meaning that they have fewer than 500 staff and their operation revenue is under 40 million Yuan [1].

3.2 Advantages of Middle and Small-Scale Clothing Enterprises under the Social Media Marketing

Due to the one-way information flow of traditional marketing, the marketing process depends greatly on the occupation to information and the control to resources. Comparing to the large-scale enterprises, the middle and small-scale clothing enterprises have congenital deficiencies and weaknesses. The middle and small-scale clothing enterprises are inferior to large enterprises from the aspects of material resources, financial resources, human resources, information resources, market occupation, and customer resources, etc. Under the context of traditional media, the limited resources possessed by the middle and small-scale clothing enterprises have created difficulties in presenting information adequately, thus leading to the dilemma of the strong becoming stronger and the weak becoming weaker.

Social media marketing relies on digitization and informatization, which provide the middle and small-scale clothing enterprises with new marketing ideas, new marketing means, and new marketing methods. Under the context of social media marketing, the middle and small-scale clothing enterprises can conduct brand communication by depending on new media. The accuracy, timeliness, universality and interactivity of flow of information possessed by social media can realize the brand spread's precision spread, wide spread, real-time transmission and interaction. For middle and small-scale clothing enterprises, it can furthest save costs, and invest more money into product quality or brand culture construction, obtaining better and quicker development.

3.3 Disadvantages of Middle and Small-Scale Clothing Enterprises under Social Media Marketing

As a new concept, social media marketing has its own significant advantages. However, as the saying goes that every opportunity can also be seen as a crisis, under its advantages, it hides the incredible crisis. Firstly, social media can realize the precise localization to the target audience, and it can enhance publicity by limiting the costs. Unfortunately, in the marketing process, it only targets a single audience, which loses a lot of potential audience in the publicity process. Secondly, social media depends on digitization and informatization. In delivering information, it can realize information's real-time transmission and wide spread, bringing the long tail effect for enterprises' marketing. This marketing effect is divided into two outcomes: positive and negative. The former can promote brand value, while the latter means that if the social media marketing method operates wrongly, the small mistakes will be amplified, resulting in devastating effects. Moreover, because China still has not formed the laws of specification on public's opinion on social media, information on social media is diversified, with some of them being true and the rest being faulty, making consumers judge the authenticity of the information based on their own experience. Consequently, it influences the marketing effect of social media.

The largest blank of social media marketing at the current stage is the evaluation of its marketing effect. Every enterprise shall know the outcome before choosing a marketing mean, but there is no platform integrating the data formed by the influence of social media marketing. Enterprises cannot judge its effect by simply relying on the data generated from the amount of searches, comments, reprints, and other miscellaneous data. They also cannot measure the influence on the products' reputation.

As middle and small-scale clothing enterprises have limited recourses, an evaluation system is urgently warranted to utilize the social media platforms for marketing, improve marketing efficiency and reduce marketing risks.

4. Construction of an Evaluation System for Social Media Marketing

4.1 Principles of Social Media Marketing Evaluation System

Evaluation of social media marketing involves multiple targets, which can be divided into two aspects: qualitative evaluation and quantitative evaluation. The first one is evaluation based on quantitative measurements, including the exposure rate of spreading content, users' visiting rate, sales conversion rate. In each marketing activity, the data before and after the advertisement or information is released will be compared; these data include the amount of followers and participation ratio such as the enrolment of online activities, the number of participants, website page view, sales data, number of fans following the official account, area, content forward and comments. The second measurement approach is qualitative marketing evaluation through reviewing the comments from audiences and customers. These comments can be found on social media, which include the evaluation ratios of the public's opinion.

4.2 Construction of a Marketing Evaluation System for Clothing Enterprises

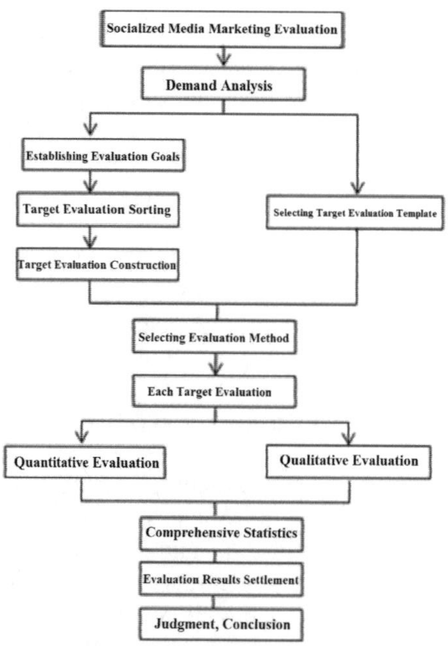

Fig.1 Social Media Marketing Effect Evaluation Flow Diagram

As a multiple-target design evaluation system, the social media marketing evaluation system can be divided into two parts, namely "tool" and "template". "Tool" means that the middle and small-scale clothing enterprises' valuers conduct data target construction through inputting the mastered spread contents' exposure rate, users clicking rate, sales conversion rate, and other fragmentation data, as well as various kinds of operating and operations. "Template" means the knowledge carrier of design evaluation. The system possesses the database system of complete functions, and strong query, maintenance, and operation functions, which guarantees the efficiency and accuracy of the evaluation process. The social media marketing evaluation system adopts the object-oriented system analysis method to solve problems of the social media marketing effect evaluation. The system flow is shown in Fig.1.

The social media marketing evaluation system is an interactive system. Users conduct comparative analysis through inputting data group, and it draws conclusions towards the end. There are a lot of sources for data for social media marketing effect evaluation, including the data inputted by users, the systems built-in database, and the data generated by systems; therefore, its data flow direction is shown in Fig.2.

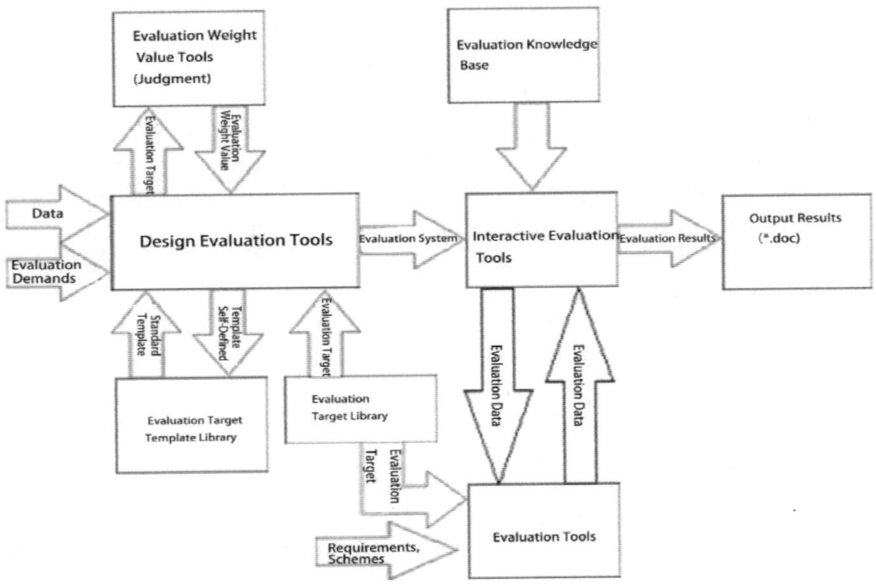

Fig.2 Social Media Marketing Evaluation System Data Flow

It is shown in Fig.2 that users can make a comparative analysis to the system's built-in database through the input data, evaluation demands, design evaluation tools and interactive tools, and at last, it outputs evaluation results. Therefore, for system's design, it designs the operation interface with the principles of good interactivity and simple operation interface (Fig.3).

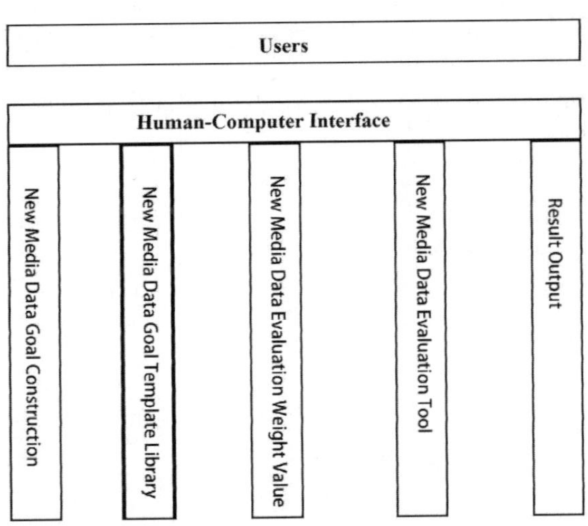

Fig.3 Social Media Marketing Evaluation System Operation Interface

As shown in Fig.3, we input the existed fragmentation data and they are divided into the new media data target building modules, in which the fragmentation data is reconstructed. The new media data target module can set up the new media data target template, and new media's data evaluation weight can be set in new media data evaluation weight module according to small- and medium-scale clothing enterprises' different marketing targets and expectations.Then, it selects the evaluation method through evaluation tools, and at last, it produces evaluation results.

5. Conclusion

Starting from the characteristics of the social media, this paper concludes the advantages and disadvantages of applying social media marketing in middle and small-scale clothing enterprises. This paper puts forward solutions by aiming at the current stage's greatest gap in social media marketing, and the problems which are difficult to be evaluated by social media marketing.The article first summarized and concluded the definition of social media marketing, followed by an analysis of the advantages and disadvantages of small- and medium-scale clothing enterprises who adopt social medial marketing. Thirdly, the basic module and data flow direction are analyzed in small- and medium-scale clothing enterprises' social media marketing system to provide new solutions and methods for reluctant puzzles. Lastly, heoritical foundation is provided to establish a social marketing evaluation system for small- and medium-scale enterprises.

References

[1] China Internet Network Information Center. The 34 Times China Internet Network Development State Statistic Report. China Internet: 2014(7).

[2] A Mayfield, What is social media, Cibmtr Org: 2008.

[3] Tobin J, Braziel L. Social Media is a Cocktail Party: Why You Already Know the Rules of Social Media Marketing. Create Space, 2008. The Point of Connection Social Media Discussion. PRSA 2008 International Conference: 2008.

[4] Public Relations Society of America.

[5] Grainger, Jesse. Social media and the Fortune 500: How the Fortune 500 uses, perceives and measures social media as a marketing tool. PQDT: 2010.

[6] Scott DM. The New Rules of Marketing and PR: 2015.

[7] Wang X, Guo S. The First Analysis of Sociality Media, China New Media Communication Association Annual Conference: 2008.

[8] Peng L. Double Challenges Integrated by Socialized Media and Medium, Press Cir.: 2012(1).

[9] Li L. The Research of Social Media Marketing of the Enterprises-A FMCG Enterprises Perspective, ShangHai University: 2013.

[10] Xu Y. The Precision Marketing Research Based on Social Media, Dalian Maritime University: 2013.

[11] Lin B. Social Media Marketing Strategy Research of Hongkong I.T Apparel Group, Shanghai International Studies University: 2014.

A Case Study on the Influence of Regional Ecological Environment based on Fishing Woman Dress in Xun Pu Village

Xin-Yan Lu[1*], Chen Huang[2], Juan Chen[3]

[1]Xiamen Academy of Arts and Design Fuzhou University, XiaMen, Fujian, 36000, China
[2]Economic Institute of Technology Anhui Agricultural University, Anhui, Hefei, 230000, China
[3]College of Arts and Media Anhui university Anhui, Hefei, 230000, China

* Corresponding author's email: luxyart@126.com

Abstract

Taking fisher-woman dress in the coastal environment as an example, to analyze the effect of the natural ecological environment on dress, the functional effect of Marine natural environment on headwear and dress, the effect of the environment on clothing material and color, and the effect of humanistic ecological environment on clothing decoration and accessories under the background of Marine culture. it is concluded that regional dress features, from the structure to the color, occurred, developed and evolved in a particular ecological environment. It has close relationship with the environment.the difference of the natural environment is one of the important factors of diversification of clothing design. It comes from nature, and also beautify the natural. This is the charm of Oriental clothing art and the clothing art beauty important features.

Keywords: Clothing Design; The ecological environment; The cultural connotation

1. Introduction

Xun Pu female is the fisher-women living at Xun Pu village Fengze District quanzhou city fujian province.XunPu female, huian female and meizhou female are called three fisher-women in fujian province.xunpu village is located in the estuary of jinjiang, quanzhou bay, and the port of quanzhou bay. according to legend, there were many red sturgeon holes on the beach (" redxun "is a kind of crab), the name is from "Xun"and"Pu".XunPu female living in on the coast of East China Sea has a unique clothing shape, "ZanHua wai" headdress called "walking garden", the living fossil of spicule fixed hair, the traditional right placket coat and the wide leg trousers of 7/10 length with region characteristic and hook earrings, this particular dresses have became their "marks".Xun Pu is the starting point of "Marine silk road". Xun Pu women dress culture has been included in the national nonmaterial cultural heritage list blending Marine culture and central plains culture. Xun Pu female clothing is influenced by the ecological environment in two aspects: one is the natural ecology of the oceanic climate, annual average temperature and other natural elements, as well as the working mode of production, the second is the humanities ecological Marine silk road, the influence of Xun Pu is located in quanzhou, the starting point of sea silk road, the blend of Chinese and western culture, the influence of cultural elements such as history.

2. The natural ecological environment of Xun Pu female hairstyles, the influence of the tire

Xun Pu is located in the subtropical Marine climate, annual average temperature between 17.2 ℃ to 20.8 ℃, the differences were small, evergreen all the year round, so Xun Pu female clothing shape regardless of the four seasons, unique flowers for Xun Pu female headdress, provide the guarantee of the four seasons of flowers. Xun Pu female flowers tire is also called "ZanHua wei" (Fig.1), is to use the four seasons small flower into neat wreath, the main material of flowers are usually seasonal flowers, a smiling, magnolia, grapefruit flowers quietly elegant, full of class also has a colourful peach flower buds, small roses, carnations, etc. ZanHua circumference is wan hair into a bun hairstyle, hair with ivory plastic chopsticks, also have to put a red chopsticks as adornment, bun was the most peripheral 2-3 bunch of flower garlands, known as the "ZanHua wei", young and old, regardless of seasons, ooze incense day throughout the year.

Fig.1 ZanHua wei

2.1 Production mode determines the style characteristics

Xun Pu female headdress dish hair not bang hair comes from the need of their life working way, as the saying goes, near the sea eat sea, tidal flats farming is a major way of working, they their homework all the year round in the seaside natural environment, ebb and flow of the sea at 5-6 hours a day, the sea is mostly bend the bow, in order not to make the hair dipped into water, hung broken fishing nets, they are all the hair dish in, twist has been Xun Pu female continuation of hair, in the process of work, in order not to make the hair down, and conveniently forward to eat with chopsticks or chopstick even smooth branches, fishbone, etc., over time, practical and decorative, give attention to two or morethings have now ivory chopsticks, gold and silver chopsticks, due to the sun, the sea of time need to put on rain to block the sun hat, fixed go ivory chopsticks have played an important role in supporting hat, Xun Pu female hairdo "ivory material fixed hair ornaments" or "spicule as fixed hair ornaments" lasted for one thousand years of Chinese "with spicule hair", become a "living fossil" of The Times.

2.2 The complementary of the colorful flowers headdress and the blue ocean

Loving beauty is every woman's nature, the Xun Pu female living on the seashore of the four seasons of flowers was found at a glance while working on the vast blue sea,and they are attached to the hat with a red scarf inclined, because the Xun Pu female live in subtropical Marine climate, flowers bloom all the year round, in order to make it more visible, brilliant under the background of blue ocean,they choose the four seasons of flowers to decorate the bun, in order to make the flowers orderly decorative knots, they will be the same color of flowers with hemp rope strung into wreaths, surrounding the bun, not enough 1 ring, 2, 3 rings, holidays up to seven rings, flowers varies with the seasons. the old women in Xun Pu village are also used in ZanHua outsourcing again on hair scarf, which is originated from the coastal climate is hot in summer, rainy season and the rainy and wet, the old women resist the wind blowing, formed the habit, wrapped head. Different regions clothing formed in a certain mode of production and lifestyle, and shows the geographical environment adaptation. The differences of natural environments created the different regions in cultural differences, Xun Pu female headdress is characterized in the specific natural environment of the occurrence, development and evolution.

3. Influence of natural ecological environment on xunpu women clothing shape

3.1 Influence of natural environment on the structure of clothing

Natural environment not only decides the complicated and simple， width, length of dress shape, etc., but also restricts the people's production, life style and the clothing materials. [2] Xun Pu woman clothing shape continues the ancient Chinese han nationality traditional clothing collar right surplice, cuffs tighten, sleeve length above the wrist, easy to work in the sea, called "cloth unlined upper garment" (Fig.2), bottom is wide leg trousers, the length between the ankle and calf, called 7 minutes of pants, waist head use white, blue Mosaic, folded in half with a rope around the waist (Fig.3).

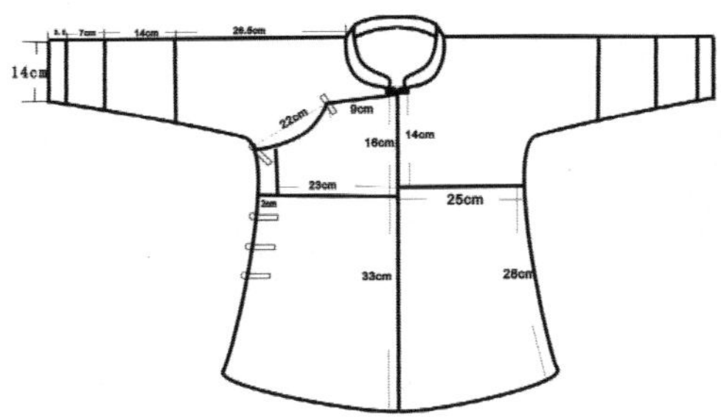

Fig.2 Cloth unlined upper garment

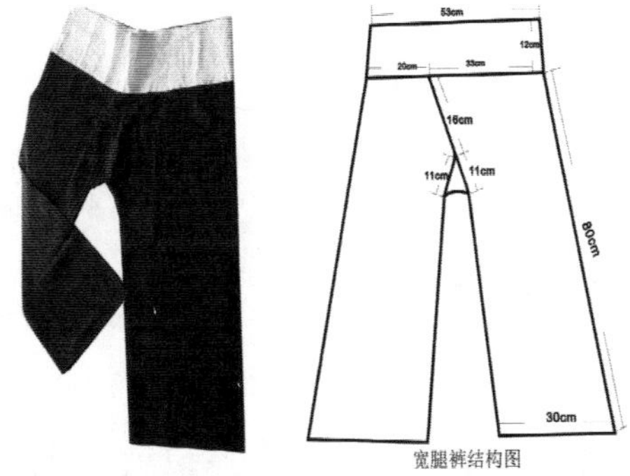

Fig.3 Folded in half with a rope around the waist

The relationship between human and the natural environment is inseparable, people not only get the livelihood from geographical environment information, but also make spiritual relationship and geographical environment. The kind of geographical environment decides the kind of folk custom [3].Wide leg is closely connected with the local Marine life environment, Xun Pu female wide-legged pants perennial work in sea water, wide-legged pants convenient work, rolled up the legs not easily wet by water, even wet because of the wide easily by the wind blow dry, pedal-driven vehicles is more at ease when walking, especially pants crotch widen assumes the isosceles triangle, easy to bend down. Xun Pu women and is called a "zhui zai" shorts is the embodiment of the mode of production, pants crotch insert a big triangle (Fig.4), to increase the level of activity crotch, designed to squat labor under the sea in summer, such as digging non, etc.

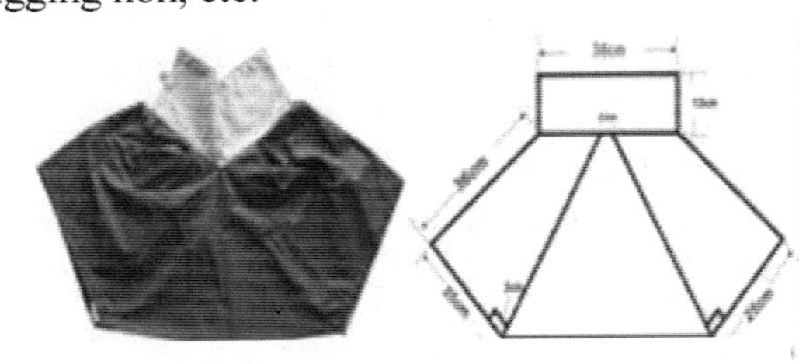
Fig.4 "zhui zai"

Xun Pu women mainly engage in the breeding of tidal flats, maritime transport trade loading and unloading of farm work and housework. In ancient times had bound feets Xun Pu female not bound feet was referred to as the "coarse foot's" crown. Due to the work in sea for a long time, pay more attention to feet is the common characteristic of the coastal fisher girl, the convenient plastic shoes become their favorite, four seasons can wear a pair of plastic slippers, even married chicken male shoes approximate slippers, fisher girl plain the good quality.

3.2 The influence of regional material on garment dyeing

In order to adapt to the impatient dirty, abrasion proof and working environment, they use litchi bark processing in the local as a stain, which belongs to the environmental protection of pure plant dyestuff, make clothes purple color, and use a kind of plant known as "potato Lang" , rolling out the red juice, dyed yellow tonal, known as "red clothes" or "potato Lang garment".Local plants cloth dyeing technology is squeezing plant into juice (fig.5), after adding water, put the cotton clothes in soaking, heating to remove dried, repeated several times, to make deeper color, better luster, and add pig blood or egg white, it is said that this kind of coat is more suitable for maritime Labour, being difficult of entangling in fishing nets, not fear water, corrosion resistance, not easy dirty, this is an early Xun Pu female clothing[4]. Now young Xun Pu female clothing color is not affected by the limitations of printing and dyeing, it is colorful. different color piece together is a typical decoration Xun Pu female clothing style, plain-coloured and bright-coloured, plain-coloured and bright-coloured, the joining together of different fabrics, not only the coat but also the waist head, which also uses different tone color matching design.

Fig.5 Local plants cloth dyeing

4. Xun Pu female clothing is influenced by regional humanities ecological environment

4.1 Xun Pu female flowers tire association with the origin of the Marine silk road

Quanzhou Citong port in the tang dynasty is one of China's four major foreign trade port. Xun Pu village was next to Citong port. In song and yuan period Citong port and Alexandria are known as the Oriental first port, "Marine silk road" went to the world from here. Xun Pu "flowers surround" hair woman, is not only look beautiful, also marks grown-up, Xun Pu woman keeped long hair, and combed "flowers surround" at the age of 12, 3, "flowers surround" became Xun Pu symbol of female adults. Xun Pu female earrings are also a sign of age and seniority, unmarried girls are wearing is not add eardrop like earrings, married woman can under like earrings with various eardrop, known as "clove pattern earrings"; But grandmother wears the earrings called "mom clove pattern pendant earrings". the jasmine, jasmine, such as exotic flowers that Xun Pu women are wearing now came from western regions in Ming dynasty and continues today, There was a village named Yunli near Xun Pu village, is the southern song dynasty Arab PuShouGeng' "Yunli Garden" in the late Yuan and the early Ming, quanzhou was the "Oriental first port" at that time, Arab PuShi descendants changed there family name "Bu" later , at present, more than half of the village residents family name are "Bu", because they are Arab descent. the old woman in Xun Pu village was used to wound turban on his head, like the Arab style of "towel", flowers not only make Xun Pu, the four seasons fragrance, also is permeated with rich Marine culture breath.

4.2 Xun Po female jewelry modeling and the origin of the maritime silk road

Xun Pu female earrings and gold hairpin geometric patterns reflect the modeling of exotic culture, Xun Pu woman earrings, fluctuations in the curve of rotating pattern hair pin, This kind of the earrings have not been found in the Chinese ancient, in the five dynasties wearing earrings is a sign of difference between Fan and Han. the geometric shape ornaments of Xun Pu female coincide wavy curve fit the Arab decorative pattern, geometry features, affected by the Arab decoration pattern of the trade. the other kind of hairpin looks like a stick of western missionaries, wave rotation curve pattern, rich rhythm of abstract lines nested mutually, affected by the Arab decoration pattern of the trade. From Xun Pu female clothes we can find Marine culture inclusive in the form of Minyue culture, central plains culture and foreign culture.

5. Complementation of Xun Pu female clothing color and ecological environment

Different geographical environment makes different aesthetic temperament and the feelings of color. Xun Pu female live near the sea all year around. blue sky and the sea became their background color, the dress has its corresponding habit color, long ago because of the lacking of dyeing technology, The main Xun Pu female clothing color are blue, black and brown red, using a headful of colorful flowers to ornament. Now Xun Pu female clothing color is rich, in stark contrast and the natural environment. Red, yellow, blue and green, all sorts of color stitching. Now the main Xun Pu female clothes color is warm color, for example red and yellow. especially the festival season, the old woman wear more red.The main Xun Pu female headdress color are red, yellow, purple and white, with golden hair and green leaves, with the awaken of spring, colorful, hat and headdress are colorful, black trousers has the effect of stable harmony.

Modern Xun Pu female various trousers replaced wide leg trousers, jacket is still big fringing unlined upper garment, it is connected with their labor way. mudflat aquaculture production mode has gradually exited Xun Pu female life, but the color of the pants is still black. Xun Pu female clothing has different color tendency in different period, developing in inheritance and change, and settling into the style of dress color now. Natural colors cultivate people's aesthetic consciousness and aesthetic ability. people learn the beauty of nature to beautify themselves, which embodies the natural son of feelings and aesthetic creation that people depict nature and be good use of nature [5].

6. The conclusion

Natural environment is the important base and prerequisite, the humanities environment is developed with it. Bath of them create and nourish the local national art of ecological environment. as a non-material cultural heritage, Xun Pu female clothing is not exist as the material culture, but exist as human activities. the survival and development environmental change will inevitably lead to the culture change[6]. The Emergence and development of Xun Pu female clothing are closely related to the ecological environment. We should protect the ecological environment and cultural ecological environment of good Xun Pu female clothing, coordinated and optimize the various ecological factors, treat all kinds of cultural phenomenon on the premise of ecological protection, in order to inherit and carry forward the traditional culture better.

7. Acknowledgement

The authors wish to acknowledge the national social science foundation of China (15BG104) for the financial support.

References

[1] Wang QL. Clothing and geographical environment on the second folk tourism resources .J Hebei Normal University Journal:1991; 04: 95-97.
[2] Chen J. Dress with the natural environment [J]. Journal of fujian institute of education: 2005; 01: 109-113.
[3] Chen QY. Chinese traditional folk custom and the geographical environment [J]. Journal of Hainan Normal University: 2002; 12: 59-62.
[4] Li JC. Saying aunt "Xun Pu". The door to all the wonderful - maritime silk road and xun Pu folk culture research album (internal): 2004; 12: 48-49.
[5] Chen LQ. Ring beibu gulf in guangxi ethnic costumes and natural ecology [J]. Nanning vocational and technical college: 2012; 03: 17-20.
[6] Song SG. National art and cultural ecology-economic globalization under the background of the development of national art aesthetic thinki [J]. Inner Mongolia social science: 2002; 01: 82-86.

Research on Digital Evaluation of Commercial Circle, based on the Clothing Brand

Ren Hui[1], Liang-Jian Fang[1*]

[1]*Apparel & Art Design College, Xi'an Polytechnic University, Xi'an, Shaanxi ,710048, China*

[*]Corresponding author's email: liangjianfang69@163.com

Abstract

The main purpose of this study is to make the site selection more data and scientific in the clothing field. According to analysis of the factors affecting the clothing brand location, this study constructs an evaluation index system using the AHP (Analytic Hierarchy Process) method to build a judgment matrix which calculates weight of index and sets up the evaluation method of index to assess index digitally. This study concludes with empirical evaluation through case research around the Xi'an commercial district, and verifies the scientificity and rationality of the evaluation index system constructed above.

Keywords: Clothing brand; Commercial district; Digital; Evaluation index; Analytic Hierarchy Process

1. Introduction

In the increasingly competitive fashion retail industry, it is very important to evaluate and analyze the location of the physical store. Many scholars have studied this topic; the earliest research was concerning classic business model theories [1,2,3,4] such as experience judgment method, the law of attraction in retail, the Huff Model, retail saturation theory and so on. However, due to lack of effective tools to support research and difficulty in obtaining data, the result is that these theories and methods are very difficult to be applied in the store location in practice. This is followed by the site selection of technical research such as bitmap analysis, GIS analysis [5], geographic information system software analysis [6]; the latest research is about evaluation methods such as AHP analysis [7], fuzzy clustering method [8], AHP- fuzzy comprehensive evaluation method [9], and multilayer fuzzy comprehensive evaluation [10].

Throughout the current situation, clothing stores location selection is mainly based on market research and qualitative analysis, at home or abroad, and it is short of a clear indicator system and site demonstration. Hence, based on the current problems faced in the industry, this study aims to improve the evaluation system, quantify the evaluation index and ensure the scientific value and rationality of methodologies in combination with cases studies.

2. Determination of evaluation index system

Based on previous literature, field research of clothing retail enterprises and telephone interviews of professional information research institutions, and considering the operability of enterprises' site selection, cost and other factors, finally, the factors affecting the site selection of clothing brands were determined, including elements of the market economy, location, traffic, customer flow, and competition.

2.1 Factors affecting the site selection

1) Market Economy

Market economy is the evaluation of the economic prosperity of the district. It mainly contains economic indexes about the current assessment of industry development of different cities and business districts. The evaluation is mainly based on the economic analysis of cities and regions.

2) Location

Location is mainly aimed to evaluate the architectural pattern of business districts and the surrounding environment, malls' hardware facilities and entertainment services, etc. Assessment of the location factors of this study includes the scale of the mall, hardware facilities, store visibility, catering support, and entertainment service facilities.

3) Traffic

In judging whether a place is developed and convenient depends mainly on investigating the convenience of traffic. The relationship is very delicate between traffic and business. Traffic can promote the development of a place of business, but it can also impede business development. Therefore, traffic for commercial development has very important influence. In this study, assessment of the traffic factors includes assessing the convenience of transportation, the north and south of the road, parking facilities, the surrounding geographical location and environment construction planning.

4) Passenger Flow

Passenger flow is calculated by using the Pedestrian Flow Index to count passengers passing the store at certain periods. It has important influence on business orientation and business strategy in all kinds of retail business enterprises in the business district. Therefore, this study is centered on studying passenger flow, including the study of district population size, the proportion of effective passenger flow in business district, proportion of passenger flow in store, the structure of passenger flow in store, and store transaction rate, etc.

5) Competition

Enterprises participate in market competition not only to understand who their customers are, but also to know who their competitors are. The survey of competitive factors is mainly about the target area proportion, the average distance to the competitors and the number of competitors.

2.2 Using AHP to construct evaluation index system

AHP (Analytic Hierarchy Process) [7] mainly divides the index into three layers: the first is the target layer, to solve the problem and name the total target; the second is the criterion layer, to achieve the overall goal of the intermediate links as a measure of the overall goal of the criterion and index; the third is the scheme layer, to implement the overall goal of the alternatives. AHP transforms a complex decision-making system into a hierarchical process. Through the layer by layer comparison of the importance of various related factors for analysis, it can provide quantitative basis for decision-making.

This section uses AHP as an analysis method to build the evaluation index system shown in Table 1.

3. Construction of the judgment matrix and determination of the weight of index

3.1 Construction of judgement matrix

According to describe the importance of the subject, constructs the judgment matrix by the factors of pairwise compared. The judgment matrix reflects the importance of this factor in relation to other relevant factors. It is the key to the Analytic Hierarchy Process and it usually takes an integer between 0-9 to scale. As Table 2 shows, the index comparative scores in Table 3 and 4 and 5 and 6 and 7 and 8 were considered by experienced pundits such as business owners, industry experts, professors and other professionals.

3.2 Determination of the weight of index [7]

According to the judgment matrix, using the power method or the root method approximates the maximum eigenvalue of the matrix and the single sort weight vector, and calculates the consistency index CI and random consistency ratio CR. This paper takes the shopping mall as an example for the final weight calculation, and results are shown in Table 9.

Table 1. The evaluation index system of site selection

Target layer A	Criterion layer B	Scheme layer C
Index system of district location	Market Economy B1	Urban economic level C11 Regional economic level C12
	Location B2	Scale level C21 Hardware facilities C22 (goods ladder configuration, straight flow ladder, indoor comfort, etc.) Store visibility C23 Catering support and entertainment C24
	Traffic B3	Convenience of transportation C31 Parking facilities C32 The north and south of the road C33 Surrounding geographical location and environment construction planning C34
	Passengers flow B4	District population size C41 Proportion of effective passenger flow in business district C42 Proportion of passenger flow in store C43 Structure of passenger flow in store C44 Store transaction rate C45
	Competition B5	Target area proportion C51 Average distance to the competitors C52 Number of competitors C53

4. Empirical case study

The main purpose of this case study is to verify the scientificity and rationality of the evaluation index system. The case study is divided into two parts: the evaluation of the application of the index, and the evaluation of the index. This case uses Master Sportsman as the research subject, a mid-range underwear brand. As a case study, this choice has certain comprehensiveness and effectiveness.

Target area includes the KaiYuan shopping mall, XiaoZhai SEG International, WAL-MART supermarket in Li Feng International, Century Ginwa walking street, and Jiaotong University mall community store, making a total of five research sites in three categories.

Table 2. Location score description

Scaling	Meaning	Scaling	Meaning
1	Compared to the two elements, "equal" is important.	9	The former is extremely important compared to the latter.
3	The former is slightly more important than the latter.	2、4、6、8	Intermediate values of the adjacent judgments
5	The former is significantly more important than the latter.	The inverse of 1-9	The latter is more important than the former.
7	The former is extremely important compared to the latter.		

According to the scoring rules of Table 2, the following results are through expert making:

Table3. Criterion layer B judgment matrix

B	Market Economy	Location	Traffic	Passenger Flow	Competition
Market Economy	**1**	**1/3**	**1/4**	**1/7**	**1/8**
Location	**3**	**1**	**1/2**	**1/4**	**1/5**
Traffic	**4**	**2**	**1**	**1/3**	**1/4**
Passenger Flow	**7**	**4**	**4**	**1**	**1/2**
Competition	**8**	**5**	**5**	**2**	**1**

Table 4. Scheme layer C1 judgment matrix

C1	Urban economic level	Regional economic level
Urban economic level	**1**	**1/3**
Regional economic level	**3**	**1**

Table 5. Scheme layer C2 judgment matrix

C2	Scale level	Hardware facilities	Catering support and entertainment
Scale level	**1**	**3**	**4**
Hardware facilities	**1/3**	**1**	**2**
Catering support and entertainment	**1/4**	**1/2**	**1**

Table 6. Scheme layer C3 judgment matrix

C3	Convenience of Transportation	Parking Facilities	The north and south of the road	Surrounding geographical location and environment construction
Convenience of Transportation	**1**	**4**	**3**	**1/4**
Parking facilities	**1/4**	**1**	**1/2**	**1/7**
The north and south of the road	**1/3**	**2**	**1**	**1/5**
Surrounding geographical location and environment construction	**4**	**7**	**5**	**1**

Table 7. Scheme layer C4 judgment matrix

C4	District population size	Proportion of effective passenger flow in business district	Proportion of passenger flow in store	Structure of passenger flow in store	Store transaction rate
District population size	**1**	**1/4**	**1/5**	**1/3**	**1/6**
Proportion of effective passenger flow in business district	**4**	**1**	**1/2**	**3**	**1/3**
Proportion of passenger flow in store	**5**	**2**	**1**	**3**	**1/3**
Structure of passenger flow in store	**3**	**1/3**	**1/3**	**1**	**1/5**
Store transaction rate	**6**	**3**	**3**	**5**	**1**

Table 8. Scheme layer C5 judgment matrix

C5	Target area proportion	Average distance to the competitors	Number of competitors
Target area proportion	**1**	**1/5**	**1/3**
Average distance of the competitors	**5**	**1**	**3**
Number of competitors	**3**	**1/3**	**1**

Table 9. Distribution of weight value of shopping mall

Two level index			Three level index				
Criterion layer B	B relative to the weight of A	CR	Scheme layer C	C relative to the weight of B	CR	C relative to the weight of A	Index ranking
B_1	0.038	0.024	C_{11}	0.25	0	0.010	
			C_{12}	0.75		0.029	
B_2	0.084		C_{21}	0.625	0.017	0.053	6
			C_{22}	0.238		0.020	
			C_{23}	0.137		0.012	
B_3	0.131		C_{31}	0.230	0.04	0.030	
			C_{32}	0.064		0.008	
			C_{33}	0.105		0.014	
			C_{34}	0.601		0.079	4
B_4	0.299		C_{41}	0.034	0.05	0.010	
			C_{42}	0.136		0.040	
			C_{43}	0.199		0.060	5
			C_{44}	0.068		0.020	
			C_{45}	0.365		0.108	3
B_5	0.448		C_{51}	0.105	0.04	0.047	
			C_{52}	0.637		0.285	1
			C_{53}	0.258		0.116	2

4.1 Data collection

Table10.The evaluation method of index

Index	Evaluation Method
C11 C12	Access to the relevant documents, assignment in accordance with the GDP ranking and the city's internal regional economic ranking
C21	Set rating criteria in accordance with the mall level
C22	Consider the goods ladder and straight ladder facilities through calculation and observation methods; Consider temperature and humidity through the degree of comfort by five-level scale
C23	Through the research method and observation method, assess consumer experience and their degree of satisfaction
C31	Consider traffic jam and vehicle flow by taking spot photos in at certain times
C32	Measure the ratio of parking area and shopping area and calculate the number of parking spaces by measuring method. Through the interview method, consider the availability of parking spaces in relation to consumer traffic analysis in the peak period of shopping malls
C33	Consider the north and south of the road through the observation method. In general, the north side is better than the south .
C34	Through observation of the surrounding environment, understand urban planning and whether or not rebuild because of the social policies, then set rating criteria.
C41	Enter the population of the various communities in the district, and as the attributes of the Tyson polygon storage, then divide by the area of each polygon to get the Tyson polygon as the boundary of the population density: The first level of low density area：0-5000 people/km ^ 2 The second level of Medium density region：5000-10000 people/km ^ 2 The third level of high density region：>10000 people/km ^ 2
C41 C42 C43 C44 C45	The installation of camera equipment in key areas of brand counters and shops facilitates image acquisition to collect passenger flow data. There are three sections in the selection period of the passenger flow measurement: Monday to Friday; Saturday; Sunday and holidays. According to the monitoring of customer flow at different times every day, set up rating criteria by calculating the proportion of passenger flow.
C51	According to the observation method, set up the scoring standard by the proportion of the area
C52	With a certain distance as the benchmark, the value decreased with the increase of distance
C53	According to the increase of the number of competing stores, the value decreased

According to different categories of business circleS in different cities, the index scoring criteria are also different. Our team is objective to evaluate the type of score, score, rating scale, and weight value which correspond to the type of commercial circle according to the evaluation method [11] of Table 10. then By carrying out the survey in the form of research tables, we summarize and analyze data. This study presents all indicators' scores out of 10. According to the division of different levels for different proportions, different numbers are assigned as index score. The final standard points are set to 6 points, that is to say, through the research, the total score of more than 6 points of the business district is suitable for location.

4.2 Result and discussion

Table11. The score summary table of three types of business circle

Period Type	Shopping mall		Supermarket	Community	
	Kaiyuan shopping mall in Bell Tower	SEG International in Xiaozhai	Wal-Mart in Li Feng International	Century Ginwa pedestrian street	Jiaotong University mall walking street
Weekday	5.873	4.539	6.485	6.823	6.444
Weekend	5.950	4.539	6.823	5.703	6.328
Average	5.912	4.539	6.654	6.411	6.386

Table 11 indicates that the Wal-Mart supermarket in three types of business district looks the most suitable as the location for master sportsman, when each item score is compared to the average.

The reasons for the scores are as follows: For shopping malls that are not suitable store locations for Master Sportsman, the two most important indicators are the proportion of the target area, and the number of competitors. The small underwear area is in the large construction of Kaiyuan and SEG area, so that the proportion of the target area tends to be zero in accordance with the scoring rules. And this small area is distributed in 20 different grades of brands, making the competition fierce; The community score comparison can be seen that weekday total scores were 6 points or more, even Jinhua pedestrian street can reach 7 points, however, Jinhua scores plummet at the weekend. The main cause for the decline in the value of the index is the store transaction rate and the number of competitors. Tracking cause is found to be due to some unexpected cause: in Jinhua street, a lingerie store was relocating, resulting in the store turnover rate plummeting. For community stores, there should not be too much competition in the shop, but these two community stores have five competitive brands, resulting in lower scores.

Research results are consistent with the Master Aportsman location orientation; it illustrates that the evaluation index system and data evaluation method based on the clothing brand building is scientific and reasonable.

5. Conclusion

In this paper, the evaluation index system, the weight distribution, and the evaluation method of index are obtained in the clothing brand field. The specific conclusions are as follows:

1) Factors affecting the site selection of clothing brand are market economy, location, traffic, passenger flow and competition. Of those, competition, passenger flow and traffic are the main considerations .

2) According to the weight distribution, the main evaluation indicators are the average distance to the competitors, the number of competitors, the store transaction rate, the proportion of passenger flow in store, the surrounding geographical location and environment construction planning, the scale level, etc.

3) According to the evaluation method, each index is assigned by the division of different levels, for different proportion and different numbers to make the site selection more data and scientific.

Finally, the case study proves that this study not only enriches the district location theory, but also has guiding and practical significance.

References

[1] Louise C. Geographies of retailing and consumption. Prog Hum Geog, 2000; 24(2): 45-47.

[2] Patrick J, Kaufmann, Naveen D, Charles M. Multi-unit retail site selection processes: incorporating opening

delays and unidentified competition. J Retail, 2003: Vol. 76(1).
[3] Zeng Q. Retail attraction: an empirical study based on reilly rule and huff model. Fin&Trd Econ, 2010; 12(04): 142-144.
[4] W J. Relly, Methods for the study of retail relationships. Austin: Bureau of business research, Universtiy of Texas: 1929.
[5] Song GF. The application research of shopping mall location based on GIS. Guangxi: Guilin Institute of Technology: 2007.
[6] Song XD. Geographic information system course. Beijing: Science Press, 2004: 211-239.
[7] Sun J. Research on the business level community site and the layout planning of Xian High-Tech Zone. Xi'an University of Architecture &Technology: 2012.
[8] Zhao K. Louis Vuitton optimal strategy of stores location planning in China. Lanzhou University: 2013.
[9] Han L, Mei Q. Analysis and research of AHP-- fuzzy comprehensive evaluation. Chin Saf Sci. J: 2004: 7(7).
[10] Xie SM, Research on index system of chain hypermarkets' Site selection - take chain hypermarkets in Xi'an as an example. Xi'an Polytechnic University: 2011.
[11] Fan SS. Study on the specialty store location based on commercial circle. Xi'an University of Architecture & Technology: 2013.

Empirical Analysis of Beijing Textile and Clothing Industry between International Market Share and Carbon Emission

Yi-Lun Wu[1], Lei Yao[1], Chun-Ying Ge[1]

[1]*Business School, Beijing Institute of Fashion Technology, Beijing, 100029, China*

Corresponding author's email: sxyyl@bift.edu.cn

Abstract

Textile and clothing industry is the mainstay industry of Beijing and makes numerous contribution to the export market. In recent years, with the development of the society, people pay more attention on environmental issues, how to ensure that the textile and clothing industry keep contributing to the economic development of Beijing while protecting the environment and reducing its carbon emissions at the same time has become a focal issues, which is also a main question about how to make the industrial upgrading come true. Based on the data form 2003 to 2012 of Beijing textile industry energy consumption, exports value and other related data, the author analyzing and building econometric model, aimed at studying the relation between international competitiveness and carbon emissions of Beijing textile and clothing industry. According to the result of the analysis, suggestions will be given to reduce the energy emission and build a sustainable model of textile and clothing industry.

Keywords: Carbon emissions; Textiles and clothing; International competitiveness; International market share; Energy-intensity

1. Introduction

Under the background of global climate change, China has become the world's leading energy consumer and carbon dioxide emission country and made a positive contribution to climate change. In the "Twelfth Five Year Plan" of the textile industry, the relevant leading departments have put forward a series of quantitative indicators. For example, energy consumption per unit of industrial added value decreased to 80% of 2010, industrial carbon dioxide emissions intensity decreased to 80% of 2010 and emissions of major pollutants decreased by 10% compared with 2010. To speed up the establishment of recycling system of textile fibers, we have to reuse textile fibers to reach about 8 million tons.

Under strict environmental regulation, Beijing textile and garment export is under great pressure. Beijing labor costs continue to rise. The consumer market always emerge product quality problems. All of the above phenomenon lead to the textile factories to move to Southeast Asia where the labor force is cheaper, meanwhile cause the weakening of the competitiveness of Beijing textile trade. From the perspective of textile and garment industry in Beijing, based on the statistical yearbook, data of China's textile and garment industry and environmental data, the study evaluated the carbon emission of textile and garment industry in Beijing, analysed the main factors influencing the carbon emission. It can provide the basis for the relevant departments to provide emission reduction strategies. In addition, based on China's statistical yearbook data and trade data released by the United Nations, measuring the international competitiveness of textile and garment industry in Beijing. To provide policy recommendations and scientific basis for the development of low carbonization of the textile and garment industry in Beijing, it analysis the relationship between the international competitiveness of economic indicators and carbon emissions.

2. Development status

2.1 Development status of Beijing's textile and garment industry

Textile and garment industry is the urban industry of Beijing, it makes full use of the advantages of Beijing, provides jobs opportunities and meets the needs of the consumer market in Beijing. According to the Beijing's statistical yearbook data, Beijing textile and garment industry exports increased from $3164.54 million in 2004 to $4333.16 million in 2008, then a rapid decline to $3545.32 million in 2009, and finally picked up to $4354.06 million in 2012. But the proportion of Beijing's textile and garment industry exports to the total amount of Beijing is declining year by year, especially between 2005 and 2008, from 16.3% in 2003 to 7.53% in 2012, shown in Figure 1. This proportion of the reduction, on the one hand due to the downturn of international export market and the export squeeze of other regions, on the other hand due to the industrial relocation policy issued in Beijing.

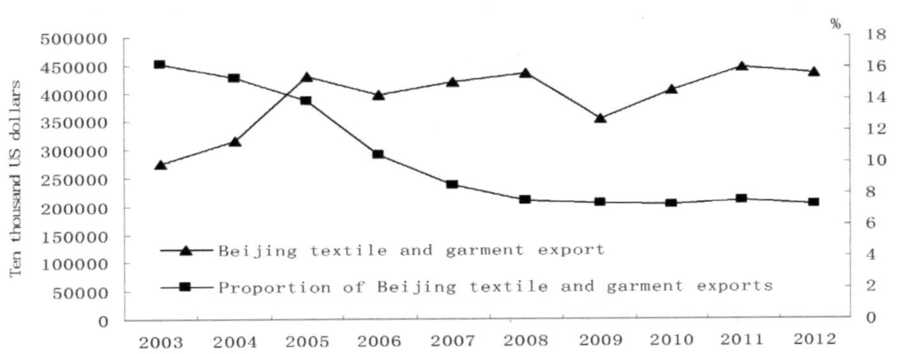

Data source: the author gathered the data from "Beijing Statistical Yearbook" and "Beijing Commerce Committee data statistics"

Fig.1 2003-2012 Beijing textile and garment industry export volume and the proportion of total exports of Beijing

2.2 Policy development status of low carbonization of textile industry in Beijing

In 2004, China promulgated the "Planning framework of Beijing fashion industry development and the fashion capital construction". If Beijing wants to become the world's fifth largest fashion capital of the world, it's not enough to just rely on super high productivity, Beijing should accelerate the transformation of industrial structure for further. It required Beijing to transform from processing production to R&D of the high-end product by setting up R&D Center, Information Center and Fashion Exhibition Center. In this way, it cloud maintain the export competitiveness of textile and clothing and accomplish carbon emissions abatement.

In 2007, the State Council issued "The First National Census Program of Pollution Sources". It said that textile and garment industry is one of the heavy polluting industries we should focus on. Eighteen Chinese Communist Party put forward the concept of "Beautiful China". Energy saving and emission reduction is an essential measure to realize "Beautiful China" vision. It require the textile and garment industry to play a bigger role in the national economy and build a sustainable development pattern, and turn the "textile big country" into a "textile powerful country".

2.3 Carbon emission of textile industry in Beijing

The textile and garment industry has made great contribution to the national economy, but also bears the great pressure of greenhouse gas emission reduction. At present, the growth mode of the textile and garment industry is still extensive form, which means that the use of resources in the industry is not reasonable, resulting in waste of resources and a serious pollution of the environment. This is the reason why measurement of carbon emission in textile industry is necessary.

3. Concept definition and index calculation

3.1 Concept and index calculation of carbon emission

3.1.1 Concept and calculation formula

Carbon emission is a general term for greenhouse gas emission. Carbon emissions from the consumption of the energy sector are mainly due to the burning of fossil fuels, so we calculate carbon emissions according to the amount of fuel consumption and the carbon emission coefficient of different fuels. Carbon emission calculation used the statistical data of textile and garment industry terminal energy consumption. Using carbon emission coefficient method to calculate the carbon emissions of the industry, the formula is as follows:

$$c=\sum_{i=1}^{n} E_i \times u_i \tag{1}$$

Among them, C for carbon emissions, E_i for energy consumption of i fossil fuels, which is measured in tons. And u_i for the energy carbon emission coefficient of i fossil fuels. The actual value of the carbon emission coefficient is equal to the product of the theoretical value and the oxidation fraction.

3.1.2 Sources and processing of data

According to the textile and garment industry's energy consumption statistical data published by "Beijing Statistical Yearbook", It can be known that the carbon emission of textile and garment industry in Beijing mainly comes from coal, coke, gasoline, kerosene, diesel oil, fuel oil, liquefied natural gas, heat and electricity. Electricity and heat is the secondary energy, they should not be included in the carbon emissions calculation. As the emission coefficient is mainly determined by the carbon content of the fuel, calculation is based on the IPCC data in this paper. In the process of calculating carbon emission, the effects of changes in the combustion conditions on the carbon emissions of the entire combustion process are not considered. Carbon emission coefficients of various types of energy are shown in table 1.

Table1. Carbon emission coefficients of various types of energy

Types of energy	Carbon emission coefficient(kg CO_2/kg)	Types of energy	Carbon emission coefficient(kg CO_2/kg)
Raw coal	2.46	Kerosene	2.56
Hard coke	3.14	Diesel oil	2.73
Fuel oil	2.219	Liquefied petroleum gas	1.75
Gasoline	1.988	Natural gas	2.33

Data source: the author calculated according to the IPCC data. Among them, the unit of natural gas is: kg CO2/m3

3.1.3 Estimation results and analysis

The textile and garment industry's total carbon emissions from energy consumption fell by 27.53%, decreased from 653.2 thousand tons CO_2 in 2003 to 473.4 thousand tons CO_2 in 2012, shown in Table 2.

Table 2. Carbon emission of Beijing textile and garment industry in 2002-2012

Year	Carbon emission of textile and garment industry (ten thousand tons of CO_2)
2003	65.318
2004*	65.518
2005	64.825
2006	65.506
2007	64.730
2008*	58.220
2009	51.709
2010	49.076
2011	48.298
2012	47.331

Notice: As the data of the year marked with start is missing, the result is the average value for upper and lower two years

The study found that from 2003 to 2012, the energy consumption of textile industry mainly concentrated in the consumption of coal, gasoline, diesel oil, which accounted for about 98%. The coal consumption brought the highest greenhouse gas emissions. But CO_2 emissions of coal consumption decreased from 627.7 thousand tons in 2003 to 446 thousand tons in 2012. CO_2 emissions of gasoline consumption increased from 13.7 thousand tons in 2003 to 25.3 thousand tons in 2009, then decreased to 18.3 thousand tons in 2012. The variation trend of diesel oil CO_2 emissions is similar to the CO_2 emissions of gasoline, rise before 2007 and fall after 2007. The detail of the main CO_2 emissions of energy consumption. The proportion of greenhouse gases emitted by coal in greenhouse gas emission is not significantly decreased, shown in Figure 2.

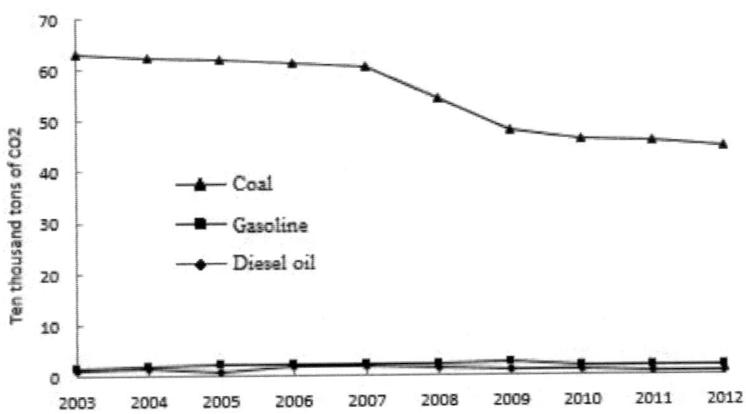

Fig.2 2002-2012 Beijing textile and garment industry major energy carbon emissions

In recent years, China's textile industry developed rapidly. In order to control the pollution, Beijing will focus on the development of high-tech industry based on the relocation policies of industrial enterprises. So the energy consumption of the textile industry in Beijing does not increase.

3.2 Indicators of international competitiveness

3.2.1 Concept and calculation formula of international market share

International market share index is a quantitative indicator used in the analysis of international competitiveness. According to the object of study in this paper, this index is used to measure the trade competitiveness of textile and garment industry in Beijing. The index can be expressed as a regional industry exports value accounted for the world's total exports value. The formula is as follows:

International market share = Total exports value of a certain region / World total exports value of the industry

This index is between 0 and 1. Index closer to 0 indicates lower competitiveness. While, index closer to 1 indicates that the region's international market competitiveness is higher.

3.2.2 Sources and processing of data

This paper selects data from 2003 to 2012. Export value of textile and garment industry derived from the statistics of Beijing Business Council. World total export value of the industry derived from the UN Comtrade database. Data is shown in Table 3.

Table 3. Beijing textile and garment industry international share in 2003-2012

Year	Beijing textile and garment export(ten thousand US dollars）	Word textile and garment export（ten thousand US dollars）	The international share of Beijing textile and garment industry（%）
2003	275195	42076203.5793	0.654
2004	316454	47050286.9138	0.673
2005	429181	50016564.2350	0.858
2006	396827	54917605.8120	0.723
2007	418770	60631593.9223	0.691
2008	433316	63555251.6134	0.682
2009	354532	54871131.3427	0.646
2010	402964	62799030.4890	0.642
2011	445658	73217694.6146	0.609
2012	435406	69050002.7272	0.631

Data source: Beijing Statistical Yearbook, Beijing Business Council, UN Comtrade database

3.2.3 Estimation results and analysis

As we can see from the above table, the world's textile and garment industry's exports have been showing an upward trend since 2003. As the outbreak of the financial crisis in 2008, the world's textile and garment industry's exports decreased from $635.55 billion to $548.71 billion. Due to the international export market downturn, the textile industry exports in Beijing is affected. Beijing's textile and garment industry's exports fell from $4.333 billion in 2008 to $3.545 billion in 2009.

Overall, Beijing textile and garment industry has a downward trend in the international share. It may be due to the tremendous pressure of textile exports Beijing was facing, for example, labor costs are no longer cheap, consumers are more critical about the product quality. The development of other countries in textile industry and the transfer of the textile factory caused the weakening of the trade competitiveness of Beijing.

4. Empirical analysis of model

4.1 Basic theoretical assumption

The international market share index reflects the competitiveness of international trade in the textile industry. Increase in exports will lead to an increase in international competitiveness. Therefore it could assume that the international share index is positively related to carbon emissions.

Energy intensity is the ratio of total industrial energy consumption and industrial gross output value. It reflects the energy consumption of industry output value per unit. Therefore, the higher the total amount of carbon emissions are, the higher the energy intensity will be. These two index have a positive relation.

4.2 Variables definition and data description

In this paper, the total amount of carbon emission in the textile and garment industry of Beijing was chosen as the dependent variable. The international market share index and the energy intensity were selected as the explanatory variables. After compiling, as shown in Table 4.

Table 4. Final variables and data sources

Type of the variable	Variable name	Data source
Dependent variable	Carbon emissions of textile and garment industry in Beijing Y	Beijing Statistical Yearbook: Industry energy consumption IPCC Database: Carbon emission coefficient of various types of energy consumption
explanatory variables	International market share index X_1	Beijing Statistical Yearbook, Beijing Business Council, UN Comtrade database
	Energy intensity X_2	Beijing Statistical Yearbook

According to the variables definition and data description, the data is collected in Table 5.

Table 5. Main factors of carbon emission in Beijing textile industry from 2003 to 2012

Year	Carbon emissions of textile and garment industry in Beijing Y (ten thousand tons of CO_2)	International market share index of Beijing textile and garment industry X_1 (%)	Energy intensity of Beijing X_2 (ten thousand tons of standard coal per ten thousand US dollars)
2003	65.318	0.654	0.000134
2004	65.518	0.673	0.000152
2005	64.825	0.858	0.000115
2006	65.506	0.723	0.000111
2007	64.730	0.691	0.000104
2008	58.220	0.682	0.000101
2009	51.709	0.646	0.0000984
2010	49.076	0.642	0.0000831
2011	48.298	0.609	0.0000881
2012	47.331	0.631	0.0000952

Data source: Beijing Statistical Yearbook, Beijing Business Council, UN Comtrade database

4.3 Model building

The variables selection and data finishing lay the foundation of modeling. The measurement model of this paper is as follows:

$$\ln(Y)=C_0+C_1\ln(X_1)+C_2\ln(X_2)+\mu \quad (2)$$

Among them, Y represents the total carbon emissions of Beijing's textile and garment industry. C_0 is a constant. X_1 represents international market share index of Beijing textile and clothing industry, X_2 represents the energy intensity of the textile and garment industry in Beijing, μ is a random error.

4.4 Results and analysis of the model

4.4.1 Regression results of the model

According to the data of Table 5, using statistical software Eviews 7.2 to conduct regression analysis, we get the following regression results:

Table 6. Regression results

Dependent Variable: LOG(Y)				
Variable	Coefficient	Std. Error	t-Statistic	Prob.
C	8.910589	1.259931	7.072282	0.0002
LOG(X_1)	0.632419	0.270352	2.339246	0.0519
LOG(X_2)	0.504154	0.141138	3.572066	0.0091
R-squared	0.793475	Mean dependent var		4.052344
Adjusted R-squared	0.734468	S.D. dependent var		0.142789
S.E. of regression	0.073579	Akaike info criterion		-2.137591
Sum squared resid	0.037897	Schwarz criterion		-2.046815
Log likelihood	13.68795	Hannan-Quinn criter.		-2.237171
F-statistic	13.44708	Durbin-Watson stat		1.267997
Prob(F-statistic)	0.004003			

By the regression results, the regression equation is:

$$\ln(Y)=0.6324\ln(X_1)+0.5042\ln(X_2)+8.9105 \quad (3)$$

4.4.2 Interpretation of regression results

The regression results of Table 6. show that the regression coefficients are significant and the coefficients are obviously not zero. R^2 value is 0.793475, indicated that all variables are explained by 79.35%. And the model was established with good regression result.

It can be seen from the regression results that the two explanatory variables X_2 and X_1 are positively correlated with the dependent variables Y, which consistent with the expected assumptions. The coefficient of ln (X_1) is 0.63. It means that when the energy intensity remains with the international market share index increased by 1 percentage, carbon emissions of Beijing textile and garment industry will increase 0.63 percentage. The coefficient of ln (X_2) is 0.5. It means that when the international market share index remains with the energy intensity increased by 1 percentage, the carbon emissions will increase 0.5 percentage.

4.4.3 Model checking

The calculation results of the model show that adjusted R^2 is 0.73, which means that the model was established with good regression result. Through the White test, we can know that there is no heteroscedasticity.

When the world textile and clothing exports remain the same, Beijing textile and garment has a greater exports volume. Then the International market share index will increase and the carbon emissions will also increase. The results of coefficient of X_1 and X_2 are positive, which consistent with the economics hypothesis.

4.4.4 Regression analysis

Based on the research results, it is found that the carbon emission of Beijing textile and garment industry has a positive relationship with the international market share index and energy structure coefficient. The increase of the international market share index will lead to an increase of carbon emissions of Beijing textile industry and the increase in the proportion of energy consumption. And the increase of the industrial output in the industry will lead to an increase of the carbon emissions of Beijing textile industry.

The data show that the carbon emissions of Beijing textile industry is decreasing year by year, as shown in Figure 3. From the year of 2005, the international market share index showed a downward trend. Until 2011, the index has some recovery, suggesting that the export competitiveness of textile and garment industry in Beijing is decreasing. During this period, carbon emissions and international market share index changes in the same direction.

Fig.3 2002-2012 Beijing textile industry carbon emissions and international market share index trends

On the one hand, developed countries have a strong comprehensive national strength, technological innovation and brand competitiveness, so it leads the world's textile and garment industry's technological innovation and high-end market. On the other hand, developing countries develop rapidly in recent years, especially China. But as the gradual abolition of textile quotas, export volume increased year by year. And the price of textile products has shown a downward trend. In addition to the "special protection clause" issued by the western countries, Beijing should focus on the development of technology innovation ability of textile industry to avoid threats from developed countries, improve the trade competitiveness of the textile industry and reduce carbon emissions.

5. Policy proposals

5.1 Improve trade competitiveness

5.1.1 Developing the Green textile brand and image

Beijing is the capital of China and a cultural metropolis. According to Beijing's unique characteristics among textile and garment industry, it should explore new kinds of environmental protection material. Then textile enterprises in Beijing should develop independent "green" brand. With the environmental performance of the special textile materials and the increasing awareness of Beijing textile, Beijing could create brand image of the city and send it out to the global market. In this way, we can make use of the advantages of "green and environmental protection" image to break through some trade barriers and expand the export of the textile and garment products.

5.1.2 Industrial chain technological innovation

In order to promote the export of products, Beijing textile and garment industry should focus on the development of low carbonization of the whole industry chain. Beijing should improve technological innovation ability among design, producing, delivery and waste recycling, provide environmental protection training to all of the employees and optimize information system which could ensure sufficient information to reduce waste. In this way, Beijing is not only famous for its green products, but also a green city known for its green industry.

5.2 Reduce energy intensity

5.2.1 Improving energy management

At present, coal is the main energy. Carbon emissions from coal consumption is much higher than that of oil and natural gas, which also caused a lot of carbon emissions. A large number of carbon emissions caused by energy structure, which is the reason why the textile industry is filthy industry. Therefore, the use of coal fuels should be restricted to a certain extent and vigorously develop renewable energy sources. The energy structure changed from the coal based to the renewable resources based will reduce the pollution of the environment.

5.2.2 Encouraging technological innovation

Since 2003, carbon emissions of Beijing's textile and garment industry have been decreasing year by year, which is related to the technological innovation of enterprises. Technological innovation of textile enterprises is embodied in the technological innovation of equipment, technology, managerial philosophy and personnel management. Survey shows that the operating efficiency of motor drive system in developed country is nearly 20% higher than the average level of the industry. With an average annual operation of 4000 hours, if the total capacity of the motor increased by 10%, the power saving rate will be up to 25%. Therefore, equipment renewal is an effective way to reduce carbon emission in Beijing's textile and garment industry. In addition, technology research and development can also improve the out-put per unit energy and reduce energy intensity effectively.

6. Acknowledgments

[1] National Social Science Fund (Fund Code: 14BJL056)
[2] Beijing Institute of Fashion Technology 2014 Scientific Research Promote Project "Beijing used clothing recycling system built and suggestion research" (Fund code: 2014AL-05)
[3] Beijing Excellent Talents Cultivating Funding Projects (Fund Code: 2014000026833ZS06)
[4] The Project of Construction of Innovative Team and Selection and Development of Excellent Talents for Beijing Institute of Fashion Technology (Fund Code: PTTBIFT_TD_007)
[5] The key project of social science program for Beijing Municipal Education Commission 2016, Study on recycling and reuse of waste textiles from the perspective of urban waste reduction

References

[1] Du YS, Zhang WF. Nanjing University of Finances and Economics. Study on the growth and driving factors of China's export trade implied carbon emissions. J Journal of International Trade: 2012; 3: 97-107.
[2] Wang LL, Wu XY, Ding XM. Donghua University. Analysis of carbon emission factors of energy consumption in China's textile and garment industry. J Environmental Science & Technology: 2013; 5: 201-205.
[3] Wang HM. National Engineering Research Center for Dyeing and Finishing of Textiles; Rong-Li Donghua University; Lin-Lin BIK (Shanghai) Trading Co.,Ltd..China Dyeing and Printing Association J. Dyeing & Finishing: 2015; 3: 50-53
[4] Wang BC. Tianjin University of Commerce, Problems and Countermeasures of sustainable development of China's textile and garment export J. China Textile: 1998; 10: 7-9
[5] Fu JX, Jiao FY, Zhang YF. Beijing Institute Of Fashion Technology. Thoughts on the sustainable development of Beijing garment and textile industry J. Market Modernization, 2006; 5: 277-279
[6] Wang HT, Shu-ru Mu SR. Central University of Finance and Economics. Study on the trend of total energy consumption structure and carbon emission in Beijing City J. Urban Studies, 2010; 9: 55-61
[7] Bahareh Zamani, Magdalena Svanstom, Gregory Peters, Tomas Rydberg. A Carbon Footprint of Textile Recycling J. Journal of Industrial Ecology,2014; 11: 1-12

[8] Jun Wu Department of economy and management Huaiyin Normal University, Analysis on the international competitiveness of China's textile export in the ten years of China's WTO entry J. Prices Monthly, 2013; 2: 46-48
[9] Wu GS. Hunan University of Arts and Science. On the difference of clothing design in our country from the traditional elements J. Intelligence, 2009; 3: 203
[10] Chen ZG. Science and technology innovation to speed up the textile industry in Beijing J. KEJICHAO, 2009; 2: 12-13

Analysis on the Characteristics of Middle-aged Women Bra Consumer Behaviours

Shao-Di Liu, Chi Liu*, Li-Zhen Wang

Apparel & Art Design College, Xi'an Polytechnic University, Xi'an, 710048, China

*Corresponding author's email: liuchi646691@126.com

Abstract

Middle-aged women is a special consumer group and their behavior of buying bras is different from young people. We have carried out a research on the middle-aged women bra consumer behavior and summarized their favorite bra fabric, color, select cup style in order to discover the sources of discomfort as it is an important factor which can influence the popularity and amount of purchases. We found that many middle-aged women likes to select bras that can remove the straps, with steel ring, cotton fabric, three rows of clasps, maintain breast shape, simple style, 3/4 cup, the price between 50-100 yuan, incarnadine, decorative of lace. We hope to provide some useful information for the underwear production enterprise based on our analysis.

Keywords: Middle-aged women; Bra; Choice preference

1. Introduction

With the continuous improvement of people's living standards, the understanding of clothing and requirements has also become higher and higher, gradually from the stress of the coat to the attention of underwear. Chinese underwear consumption concept is constantly upgrading, promoting the development of the underwear industry continuously [1]. China's underwear market capacity and growth potential is huge, annual sales exceeds more than 100 billion growing by nearly 20% a year, In the garment industry, woman's underwear accounted for 60%, become the underwear industry's top priority [2]. However, most women are unable to find a bra which can meet their basic requirements. An issue for women is that high quality bras are unaffordable therefore to resolve this it is important to question how can we improve the manufacturing process in factories to ensure quality control. Underwear enterprises should find a balance between the price and the quality.

Middle-aged women as a special consumer group, occupy an important position in the underwear market. It will play a vital role in the development of the potential consumer market of underwear enterprises. While there are lack of reach of middle-aged women bra consumer behaviors, we can find lots of article analysis on young girl and college students' behaviors. This article intends to discover the middle aged woman's underwear consumption behavior through a questionnaire survey involving different design elements of the bra . The research aims to grasp the characteristics of middle-aged woman's underwear consumption behaviors and provide a reference for underwear business plans to produce underwear for middle-aged women.

2. Questionnaire

I made a preliminary investigation of 150 people before the distribution of the formal questionnaire. During this process some people have been eliminated. Then I distributed 120 questionnaires, these people were targeted. In total 109 questionnaires were received, the effective rate is 90.8%.

2.1 Respondents

109 middle-aged women of different monthly income in Xi'an of Shaanxi province were investigated. According to

the survey target income level, the wealth gap is divided into five levels, Below 1000 yuan, 2000-3000 yuan, 3000-4000 yuan, 4000-5000 yuan, above 5000. The chosen women are between the age of 30-50 years old, to avoid the investigation of repeatability and chance.

2.2 Survey method

Different income levels have different consumption demands [3], When choosing sample, according to the cluster sampling method, unit in a different class of people group, a level as a group, each group of Numbers 20 people. The questionnaires were carried out face to face, to guide the target groups and ensure the integrity and objectivity of the questionnaire.

2.3 Questionnaire design and survey content

The questionnaire includes four parts. The first part contains basic information of the respondents, mainly including: Age. The second part is the design elements of the bra, mainly including: is the shoulder strap detachable, whether there is a steel ring, bra fabrics, latch type, function of bra, the bra cup style, style, price, surface decoration, color of bra. The third part is the discomfort of wearing a bra, mainly including: wear ability, shoulder strap slide, shoulder straps and bra, decorative lace worn skin, steel leakage tingling of the skin, joint friction of the skin, lower edge band make the skin produce strangulation, skin allergy, tickling feeling. The fourth part is the key factors affecting the purchase behaviors, mainly includes: price, style, fabric, color, functional, comfortable degree.

3. Analysis of survey results

3.1 Shoulder strap

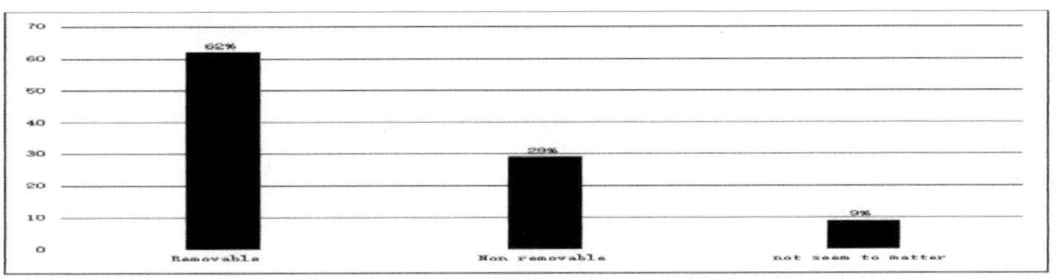

Fig.1 The choice tendency of the shoulder strap

The figure 1 shows that 62% of the middle-aged women will choose the bra that have detachable straps. First of all, this bra is convenient for unpicking and washing. Secondly, detachable shoulder strap can match different styles of garment as it is possible to randomly take off [4].

29% of middle-aged women tend to choose straps that make the bra not removable, this kind of bra integrity is strong, can increase the shoulder belt bearing force improving the overall figure.

9% of middle-aged women said that the bra straps are removable or not can be accepted.

3.2 Steel ring

Figure2 shows that, 52% of women chose the bra with steel ring. Middle-aged women wear this in order to reduce the sagging breast phenomenon Furthermore, the bra steel ring plays an important role in supporting the chest which gives a more beautiful figure [5]. Most women desire curvaceous figure which is why middle age women appreciate rim bra. However at the same time, the blunt mechanical force used to shape the breast can cause discomfort which is very bug concern.

42% of the middle-aged women choose bra without steel ring. Although this bra is less effective at keeping chest shape, it's very comfortable and will not affect the health of the human body. Therefore, the quantity of consumer is relatively large.

2% of the middle-aged women believe that women wear a bra that with or without rims are the same.

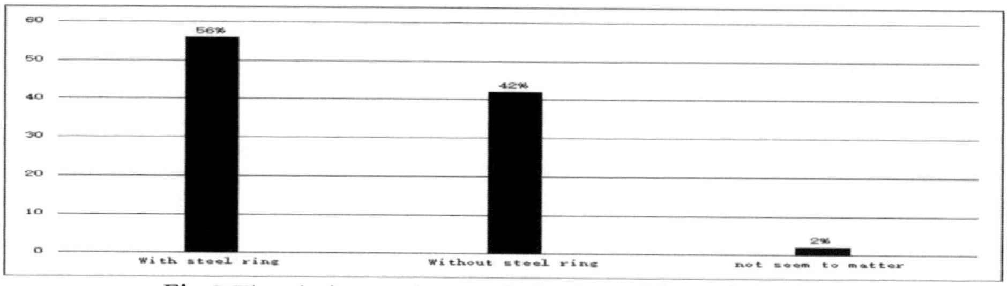

Fig.2 The choice tendency of whether with steel ring bra

3.3 Fabric

Fig.3 The choice tendency of the bra fabric

Figure 3 shows that,86% of people would choose cotton bra. The cotton itself has excellent permeability and hygroscopicity, giving a different feel from other fabrics. From an aesthetic perspective, cotton cloth has a natural and simple characteristic, and the price moderate, therefore most people like this kind of material. Fabric is not only the base for the style of clothing but is also one of the essential elements in the form of artistic expression [6]. 6% of people would choose silk bra, because the silk fabric has elegant and luxurious appearance, adding different embroidery to the fabric enhancing the visual appeal of the bra. This effect is hard to accomplish with other fabrics, but it comes at a cost. So the choice of it is relatively few. Nylon ,spandex belong to chemical fiber fabrics which a total of 8% people choose. The data indicate that the majority of middle-aged women prefer bra made out of natural fabric.

3.4 Hook

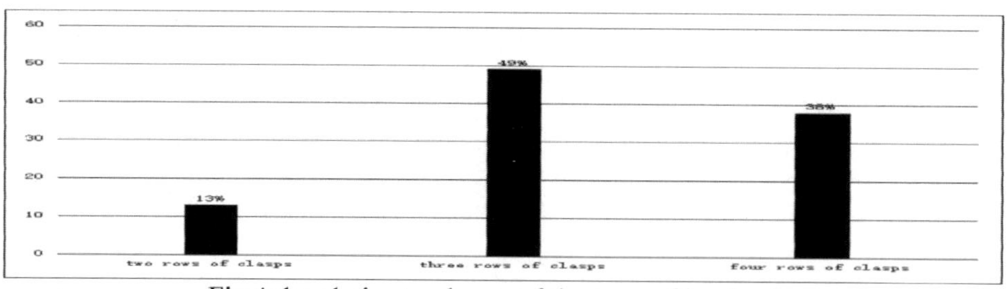

Fig.4 the choice tendency of the type of bra hook

Many Bra hook buckles are not convenient to wear with less influence on the shape of the chest. Two rows of hook bra is usually a relatively light bra, suitable for women who are small and thin. Wearing this kind of bra in the summer does not only feel cool but is also very convenient. However most of the middle-aged women figure are plump, so generally will select bras with more rows of hook. Demand for the two breasted bra accounted for 13%. The bra of three rows buttons are the most conventional as they account for the highest proportion, compared with the bra that has two rows of buckles. The number of clasp rows increase, to widen the armpit part of the bra, enhance the stability of the bra.

If the chest is too big, wearing four-breasted bra helps gather fat and control it. By matching of the rims, it can effectively adjust the chest sagging phenomenon ensuring the chest to become upright. From figure 4,three rows of buttons and the four-breasted bra is the most popular type of bra.

3.5 Functional

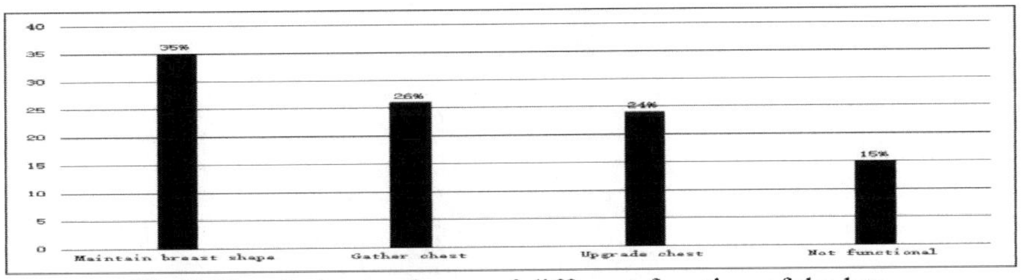

Fig.5 The choice tendency of different function of the bra

Middle aged female breast's will generally sag overtime which is a huge concern to them as well as many underwear enterprises [7]. Due to the influence of gravity, body fat will gradually move down with age. A functional bra can help middle-aged women improve the chest figure or correct the phenomenon of outside expansion. 35% of women choose to the bra that maintains the best chest figure to prevent the breasts continue sagging or outside the expansion;

Gathered bra accounted for 26%. The pushup bra is able to make the breasts appear more full by pushing the breasts from both sides towards the center.

The type of Lifting bra can improve the sagging breasts phenomena. Middle-aged women demand for the type of ascension bra is relatively high and accounts for 24%;

15% of middle-aged women choose the bra that has no function a tall as they care more about bra wearing comfort.

3.6 Style

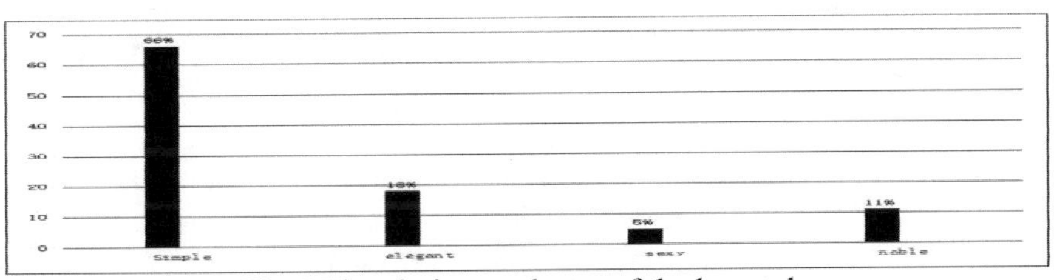

Fig.6 The choice tendency of the bra style

Bra for women not only protect the breast but also, adjusts the chest form and, is a symbol of personal life taste and emotional appeal. The different bra style shows middle-aged women with different concept of life and aesthetic taste. Chinese culture influenced by Confucianism points to the direction of moral spirit and culture grade in elements of fashion [8].The middle-aged women are mostly reserved, so generally a functional but simple bra can meet their requirements. There is highest demand for bra with a as it accounts for 66%.

The bra that focuses on style and color with an elegant look is called tie in, This bra style is concise in accordance with the characteristics of middle-aged woman's pursuit of clean, concise and convenience. This bra accounts for.

Middle-aged women are introverted. There are the least demand for bra's with a sexy style as it only accounts for 5%.

Bra with style that highlight a women's elegant noble features attracts middle-aged female with an intellectual pursuit. The demand is higher and accounts for 11%.

3.7 Cup style

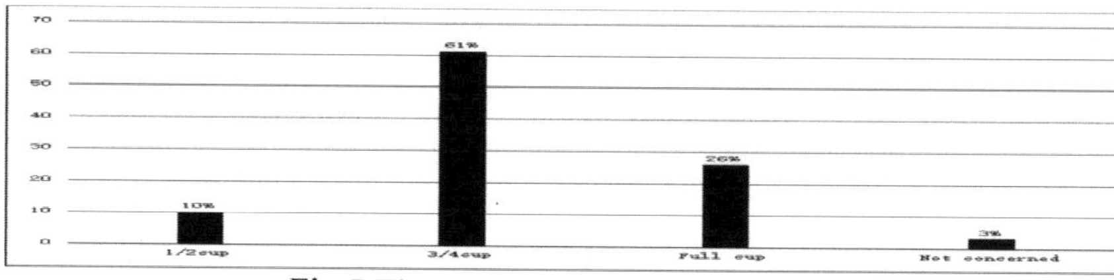

Fig.7 The tendency of cup style choice

Bra in accordance with the cup size is divided into 1/2 cup, 3/4 cup and full cup [9]. Woman's breast shape is different from person to person. 1/2 cup bra size fits smaller people, in general. Bra designed for Middle aged female body with a plump figure has fewer demand accounting for 10%. 3/4 cup is bigger than 1/2 cup and its smaller than full cup, if you wear this type of bra which can highlight the curve of the breast. It's especially suitable for these woman who has large breast, the amount of demand accounted for 61%.

Full cup bra can make the breast appear full. This bra has the function of support, enhance, concentrated female chest, which have a very big action to support large breast. This type of bra is proper for these woman who has a sagging breasts, the amount of demand accounted for 26%.From this data we can find that its commonly phenomenon in middle-aged woman whose breast has been sag .

3% middle-aged women doesn't concerned about the bra's cup style.

3.8 Price

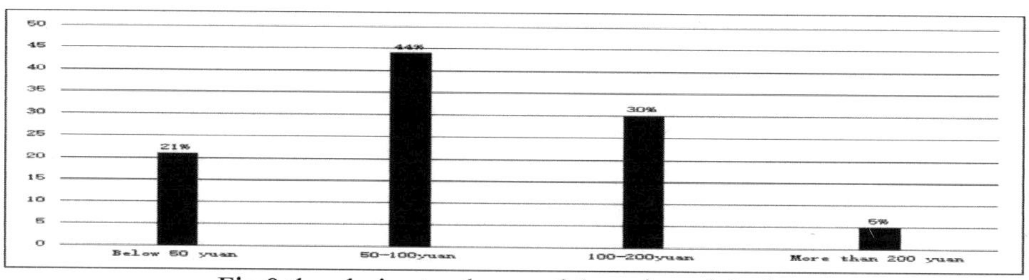

Fig.8 the choice tendency of the price of the bra

In this survey it was discovered that the price and income are proportional when it comes to purchasing bra. Lower income level are willing to accept bra price under 50 yuan. This accounts for21%. 0-200 yuan bra cost is the ideal price for middle-aged women, 74% of women said they were able to accept it.

Middle aged women with high income do not care about the price of the bra and can afford bra costing over 200 yuan. They account for 5%.

3.9 Decoration

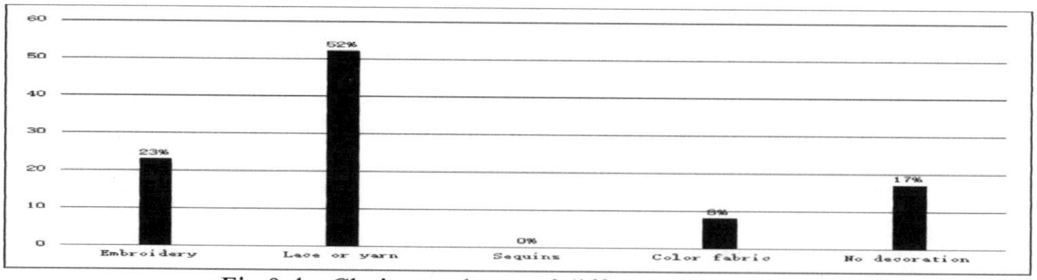

Fig.9 the Choice tendency of different decorative bra

Bra decoration goes past the functional element of the bra and deals with aesthetic and outlook [10]. The use of decoration is so that the use of the bra is more flexible., Different decoration provides different style. Embroidery and clothing are inextricably linked, from the past to the present. Embroidery is one of the most original clothing decoration in both Chinese and Foreign clothing [11]. Embroidery bra has an elegant temperament, delicate and not rough. General embroidery bra style is also more classical. The demand for this kind of bra isn. 23%;

In fashion's long history, lace has always expressed a variety of beauty of women in different times [12]. Lace or gauze fabrics is very commonly applied in bra design. the combination of the two kinds of material and the bra can meet the psychological needs of middle-aged women in their pursuit of natural, simplicity. This accounts for the highest proportion 52%;

Middle-aged women generally is relatively simple hence the use of sequins is not suitable for bra design Non decorated bra has softer characteristics as too much decoration can shorten the service life and cause discomfort. The demand for non-decorative bra is 17%.

3.10 Color

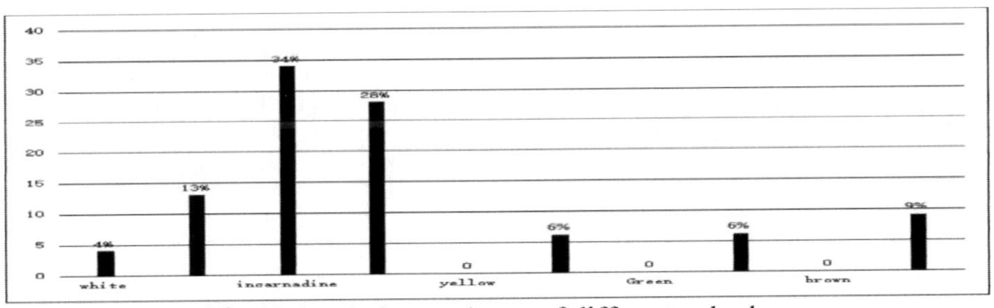

Fig.10 the choice tendency of different color bra

Color is an art of expressing emotions and is the soul of the whole garment [13], bra's basic function is to protect, support the breasts but now has gradually evolved into a kind of aesthetic decorations. Color of the bra has attracted more and more attention as, different colors of the bra portray a different meaning. The type of color choice reflects upon people's living habits.

34% of the middle-aged women tend to choose bra that has a natural skin tone as it draws less attention.

White, pink, light purple, light blue bra, these color are soft and suitable to wear in the summer. Pink has the highest proportion of 28%.

Yellow and green color draws too much attention whereas brown give off an unclean feel which makes it undesired.

Black bra is easy to clean and maintain, red symbolizes of good fortune Hence they have a higher demand.

4. conclusion

In view of the above investigation and analysis, middle aged woman like the bra of shoulder straps can be remove, because they washes this kind of bra is convince and they can change different style of shoulder strap. They needs bra with steel ring which can effectively support the shape of the breast. They need the cotton fabric bra because this type of bra will make them feel comfortable. In addition, they like the bra of 3/4 cup, this type of cup can meet most middle aged women's need and suit for their shape. The price of bra should not be too high in best, low price of bra will attract more women to buy more bras. The surface of bra's decoration should be used yarn, because most middle aged woman like this decorations. The bra's color should used light color ,for instance pink. Production enterprises should cate to the preferences of the middle-aged women, to enhance the grade of the bra design, meet the consumer's purchase desire.

5. Recommendations

5.1 Fabric

The fabric is the key factor that determines the bra comfort. Chemical fiber bra has a poor permeability which likely leads to itchy skin [17]. As cotton have very beneficial properties such as good air permeability, moister permeability it is the favourite amongst consumers [18], but the elasticity of the cotton is poor. Bra's that are made 100% out of cotton can oppress the chest.t. To balance this dilemma, production enterprises should find a proper blending ratio between chemical fiber and natural fiber. To obtain a comfortable elastic bra fabric, different fabrics can be used in different parts of the bra. The cups with breathable cotton, the side strip using good elasticity of chemical fiber fabrics. In addition, the bra can use double fabrics. The side touching the skin can be made out of natural fiber fabrics whereas the opposite side can be made out of chemical fiber lace or yarn to give a more aesthetic appeal.

5.2 Rims

Steel ring for the lower edge of the bra, is an important part of the shape of the brassiere as it can gather and support the chest. It also plays the role of beautifying female bosom curve [19]. Proper use of steel ring provides bra comfort. First of all, Instead of choosing excellent materials for the steel ring it can be substituted with material that has good elasticity and stability. Secondly, if steel with nickel coating is used it should not have direct contact with the skin as it may cause allergic reaction to the body. To ensure that the steel ring outer edge is smooth, make the rims at both ends oval, to reduce the leakage situation of steel abroad.

Through improving the bra structure the rims with steel rings will not be necessary making it not only more convenient but eliminating the feeling of confinement.

5.3 Shoulder straps

Suitable elastic straps can effectively prevent the shoulder straps from slipping. If there is too much elasticity, it will cause the Shoulder straps to lose their support. However if the elasticity is too small, it will cause a lot of pressure on the shoulder and back. To ensure that the shoulder strap has a moderate elasticity is very important.

5.4 Decorative lace

The decorative lace is divided into two kinds, one such will not touch the skin Only play the role of decoration, another type will touch the skin. Some underwear production enterprises use of chemical fiber material lace in the bra while ignoring how comfortable it is when worn. Decorative lace that is not exposed to the surface can only focus on its aesthetic appeal but if it does have contact with the skin, the manufactures must ensure that the comfort is not comprised.

References

[1] Luo Y. Intimate fashion underwear design. Chin Tex Press: 1999; 113-127.
[2] Ma XH, Liu JW. Researchon early warning system of underwear industry. XA Pol Un: 2007; 68.
[3] Zhao XL. The analysis of the characteristics of Chinese residents' consumption behaviours and the policy choice of starting up the consumption demand. Chin Ec: 2000; 4: 18-19.
[4] Wang X, Xu D. Investigation and analysis of woman's underwear knitting Market. Kn Ind: 2007; 12: 25-28.
[5] Li LL, Liu C. Study on analysis of middle-aged female breast shape bra steel ring based on the central and Western Regions. XA Pol Un: 2012; 87.
[6] Che WD, Wang JJ, Tang Y. The relationship between the material and the style of the clothes. Tex Lead: 2013; 5: 94-96.
[7] Shen L, Wang SM. Comparison of Chinese and Western women's underwear's style. J Tex Res: 2009; 5: 114-117.
[8] Qi Q, Liu C, Dong XY. Optimization of bra structure in the case of breast sagging in middle-aged women. XA Pol Un: 2014; 97.
[9] Wang CC. Research and Analysis on the characteristics of adjustable bra.Kn Ind: 2012; 9: 53-56.
[10] Cao YY, Wu ZM. Design method of bras decoration. Hun Sch Art: 2011; s2: 155-157.
[11] Ye S. Embroidery and garment decoration.Silk Mon: 2004; 8; 16-17.
[12] Cao YY, Wu ZM, Yang X, eta. An application of laces to bra.Hun Sch Art: 2011; s1: 161-164.

[13] Xi X. On the color of clothing and its significance.Wr Mag: 2009; 18: 241-242.
[14] Yu XK, Cai LG, Xiao P, eta. Requirements of Wearing Comfort for Bra Straps and Its Influence Factors [J]. J DH Un (Nat Sc): 2014; 1: 48-53,63.
[15] Chen XN, Wang JP. Bra pattern making and its affecting factors. J Tex Res: 2013; 11;155-160.
[16] Han L, Yu WD. Research on the Mechanism of Fabric Prickle. BJ Tex J: 2011; 4: 51-53.
[17] Zhang HY. Good bra to make breast more healthy. Heal Life: 2007; 3:16.
[18] Zhu SW, Liu JW. Clothing materials science. Chin Tex Press: 2015; 113-127.
[19] Wang SH, Ding YR. The analysis of human factors affecting the bra steel ring design. SX Tex Gar: 2013; 3: 9-14.

Survey of Single Color Preference and Multi-color Emotion

Yuan Xue[1*], Man-Yi Zhang[1] , Bing-Jie Guo[1], Haruki Imaoka[2]

[1]Apparel and Art Design College, Xi'an Polytechnic University, No.19 Jinhua Rd(S), Xi'an, Shaanxi, 710048, P.R. CHINA
[2]Department of Life Science and Human Technology, Nara Women's University, Kitauoyahigashi-machi, Nara, 630-8506, JAPAN

[*]Corresponding author's email: xyuan98@hotmail.com

Abstract

In this study, forty-two colors were selected from INTERCOLOR of the Spring-Summer color trends, and twenty-four kinds of color combinations were made according to the six color concepts. The study included two parts: one was single color preference investigation; the other was color emotion survey of multi-color combinations. In the second part, fifteen color emotion scales, including "modern-classic", "vulgar-elegant", "formal-casual", "fresh-stale", "gaudy-tasteful", "lively-stiff", "implicit-flaunty", "gorgeous-plain", "fashionable-dated", "individual-popular", "powerful-powerless", "rural-urban", "warm-cool", "intellectual-nonintellectual", "like-dislike" were selected according to the Semantic Differential Method. The survey was conducted among the undergraduate college students from Xi'an Polytechnic University. Color histogram was used to express the result of single color preference, and it was clear that the popularity of these released colors was not even. Factor Analysis and cluster analysis were used to analyse the data from the second part of the survey. Three factors were extracted and labeled "personality", "taste", and "fashion". Twenty-four multi-color combinations were classified into three groups. It indicated that comprehensive color emotions were more important than other factors.

Keywords: Color preference; Color emotion; Multi-color combination; Semantic Differential Method; Factor analysis

1. Introduction

The International Commission of Color held its first meeting in 1963. From then on, countries began to establish their own association of color one after another. These societies were responsible for forecasting and releasing fashion colors. After several decades of development, the trend reports were referred to by more and more companies and designers. Series of fashionable colors are released in themes twice each year (Spring-Summer and Autumn-Winter) [1]. This study seeks to confirm if these colours are genuinely popular.

In our previous study, color psychology for each single color was investigated, and was compared between the Chinese and Japanese. It has been concluded that the respondents were inconsistent between the abstract meaning of color and color preference, but consistent between natural color and custom [2-4]. Uravis Tangkijviwat researched the relationship between color preference and the mode of color appearance. The results showed that color preference is dominated not only by color attributes but also by the mode of color appearance [5]. Iris Bakker studied color preferences on different topics in connection to personal characteristics. It indicated that color preferences were dependent upon the topic and personal characteristics [6].

Color preference is a kind of color emotion, which shows more complicated psychology and has been extensively researched upon.

Li-Chen Ou et al. had studied color emotions for single-color and two-color, and a color-appearance-based model was established for color-combination emotions. With this model one can predict color emotions for a color pair if color-appearance attributes of the component colors in that pair are known [7,8].

In this study, one set of colors released by China Fashion Color Association is used as color samples. Single color preference and multi-color emotion are investigated respectively. The aim is to test the reliability of color trends and to propose better ways of using color trends.

2. Method

2.1 Stimuli and questions

In this study, forty-two released colors were selected from six themes of International Textile color trades. The six themes were sublimation, filtering, gradation, natural blue tone, roadway and hanging. Fig.1 shows the forty-two color samples in the CIE 1976 (L^*, a^*, b^*) color space (CIE is an abbreviation of the International Commission on Illumination). In Fig.1, most of these released colors are concentrated around the origin of the axis where achromatic colors and colors similar to achromatic colors are located. The color names and the RGB values for each color are shown in Table 1. Each color name came from the theme number, for example, the first color of "sublimation" is named "T1-1". It indicated that achromatic colours and other similar colours are more likely to become popular.

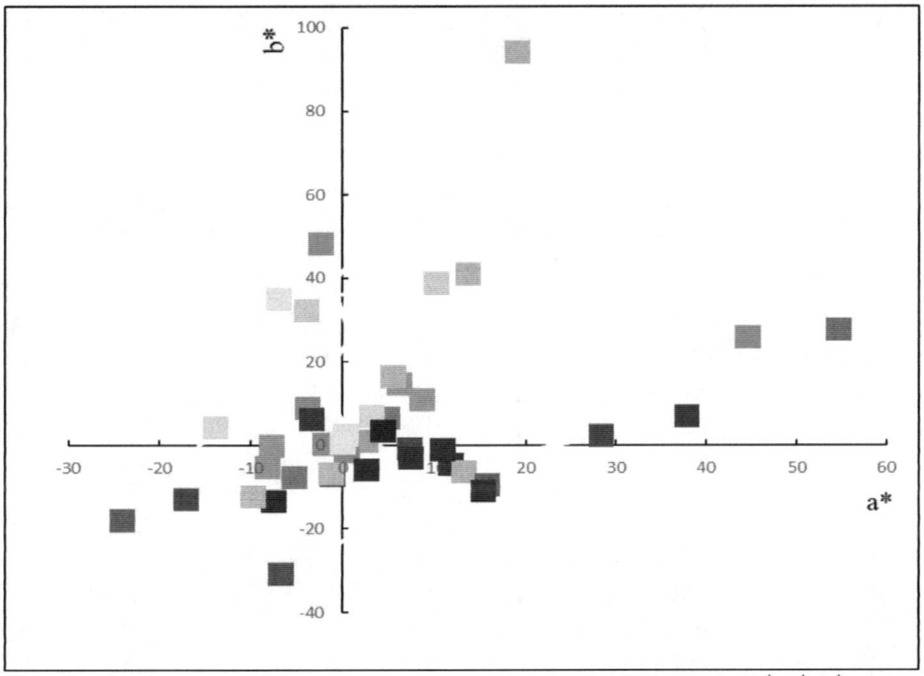

Fig.1 The distribution map of forty-two color samples in the CIE 1976 (L^*, a^*, b^*) color space

In the six released themes, there are several kinds of color combinations. Based upon to these color combinations, twenty-four multi-color combinations were presented in the form of Chinese Jiugongge as shown in Fig.2, which allows for the expression of the main color and secondary color. In the aesthetics of fashion color combination, contrast must exist in one or more aspects, such as lightness, saturation, hue or area.

Fifteen pairs of color emotion words were used in the second part of survey: "modern-classic", "vulgar-elegant", "formal-casual", "fresh-stale", "gaudy-tasteful", "lively-stiff", "implicit-flaunty", "gorgeous-plain", "fashionable-dated", "individual-popular", "powerful-powerless", "rural-urban", "warm-cool", "intellectual-nonintellectual" and "like-dislike".

Table 1. The value of selected colors in different color systems

Color Item	Number in CNCS	R	G	B	L	a*	b*
T1-1	008 70 00	182	174	175	71.84	2.54	0.84
T1-2	024 65 07	189	160	147	67.99	8.69	10.61
T1-3	128 60 00	147	148	151	61.22	0.61	-1.76
T1-4	152 35 02	99	84	89	37.21	7.42	-1.03
T1-5	032 65 07	183	157	136	66.61	6.27	14.53
T1-6	032 75 07	214	187	162	77.53	5.58	16.21
T1-7	088 60 00	148	152	151	62.60	-1.78	0.04
T2-1	048 90 00	233	230	226	91.51	0.41	2.32
T2-2	024 55 02	150	133	126	56.87	4.90	6.31
T2-3	024 85 02	234	220	210	88.56	3.25	6.63
T2-4	048 60 17	170	148	61	61.84	-2.19	47.91
T2-5	048 80 12	224	209	148	83.88	-3.76	32.11
T2-6	056 60 02	150	151	134	62.06	-3.68	8.75
T2-7	056 30 02	73	76	65	31.72	-3.20	5.94
T3-1	152 25 02	71	58	67	26.02	7.45	-3.15
T3-2	152 25 07	76	55	69	25.90	11.79	-4.70
T3-3	160 35 17	126	65	81	35.52	28.51	2.18
T3-4	002 30 27	126	43	64	30.74	37.95	6.64
T3-5	007 50 32	215	75	80	51.92	54.82	27.56
T3-6	010 60 27	233	115	108	62.26	44.71	25.63
T4-1	120 20 02	43	52	61	21.09	-0.97	-7.35
T4-2	104 35 12	24	96	108	37.19	-16.97	-13.25
T4-3	088 65 02	149	169	165	67.55	-7.53	-0.41
T4-4	000 90 00	233	232	232	92.20	0.14	0.37
T4-5	120 75 07	186	196	207	78.60	-1.16	-6.99
T4-6	112 55 07	116	137	147	55.74	-5.12	-7.83
T4-7	112 25 12	30	66	83	26.25	-7.33	-13.69
T4-8	119 40 22	26	105	150	41.90	-6.59	-31.08
T5-1	144 75 12	211	184	204	77.50	13.28	-6.50
T5-2	144 45 12	129	101	126	46.30	15.91	-9.52
T5-3	160 25 07	76	55	63	25.68	10.95	-1.33
T5-4	144 25 12	76	53	78	25.57	15.42	-11.14
T5-5	024 20 02	60	49	47	21.50	4.44	3.16
T5-6	112 75 12	161	202	218	78.95	-9.64	-12.57
T5-7	039 75 37	254	172	0	76.29	19.18	93.94
T6-1	080 85 07	198	231	215	88.98	-13.77	3.97
T6-2	103 45 17	0	121	138	45.77	-24.04	-18.30
T6-3	104 65 07	143	170	174	67.69	-8.06	-5.39
T6-4	136 25 02	59	58	68	24.93	2.72	-6.06
T6-5	048 90 12	241	232	163	91.24	-6.78	34.68
T6-6	032 85 17	252	201	137	83.99	10.25	38.57
T6-7	032 75 17	236	178	114	76.74	13.76	40.75

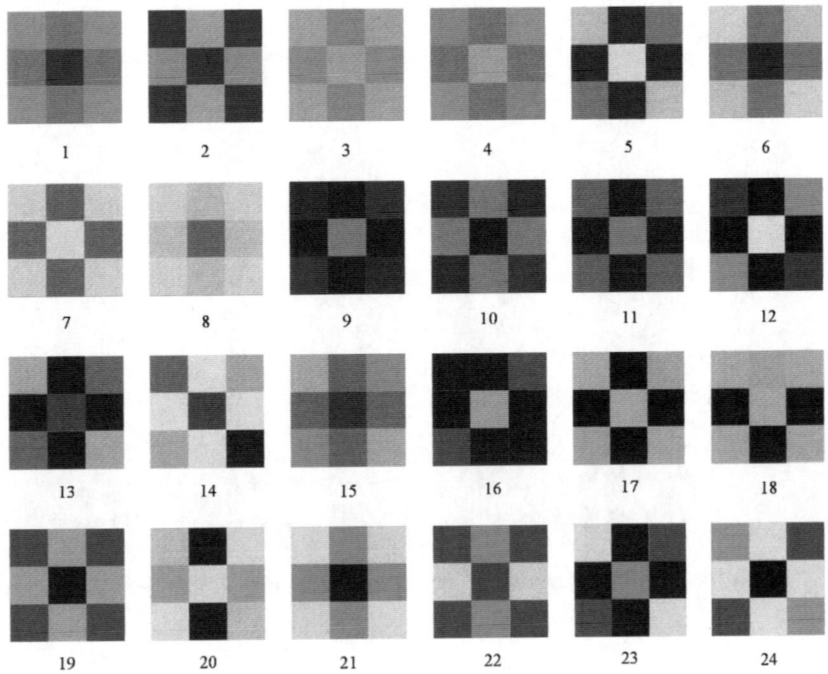

Fig.2 Color stimuli for the survey of multi-color emotion

These were the most frequently used words in these studies. For each pair of words, a scale of one to five was used. These color emotion words and a scale of one to five were printed in answer sheets.

The stimuli were generated by a computer and displayed by a projector in 3500 lm with a resolution of 1024×768 pixels. The screen was shown to the respondents at visual angles of 13° (vertical direction) and 17.5° (horizontal direction) from a distance of 10 meters. The respondents with common vision were gathered in a dark room with just enough light for reading the questionnaire. The respondents were asked to watch the stimuli and answer the questionnaire.

2.2 Respondents and survey period

The survey was conducted in 2015. A total of 204 undergraduate students were surveyed at Xi'an Polytechnic University. All the students were aged between 20 to 24, and has normal color vision.

2.3 Methods of analysis

Color histogram and the Semantic Differential Method were used in this study, and the data from the survey was analysed by using factor analysis and cluster analysis.

3. Results and discussion

3.1 Results of single color preference

In the first part of this study, forty-two single colors were used as samples to investigate color preference. Each respondent was asked to choose five favorite colors from the forty-two samples. The results were expressed by the color histogram shown in Fig.3. In Fig.3, the most popular colors were T5-6, T6-1, T6-5 and T5-1, which belonged to the fifth and sixth themes. They were pink-purple, soft blue, light green and pale yellow. They had several similar features: chromatic, soft and bright. Some colors were not chosen at all. They were T2-2, T3-4, T5-3, T5-4,

T5-5 and T6-7. Half of them came from the fifth theme. These unpopular colors also had some common characters, such as very low saturation and dull.

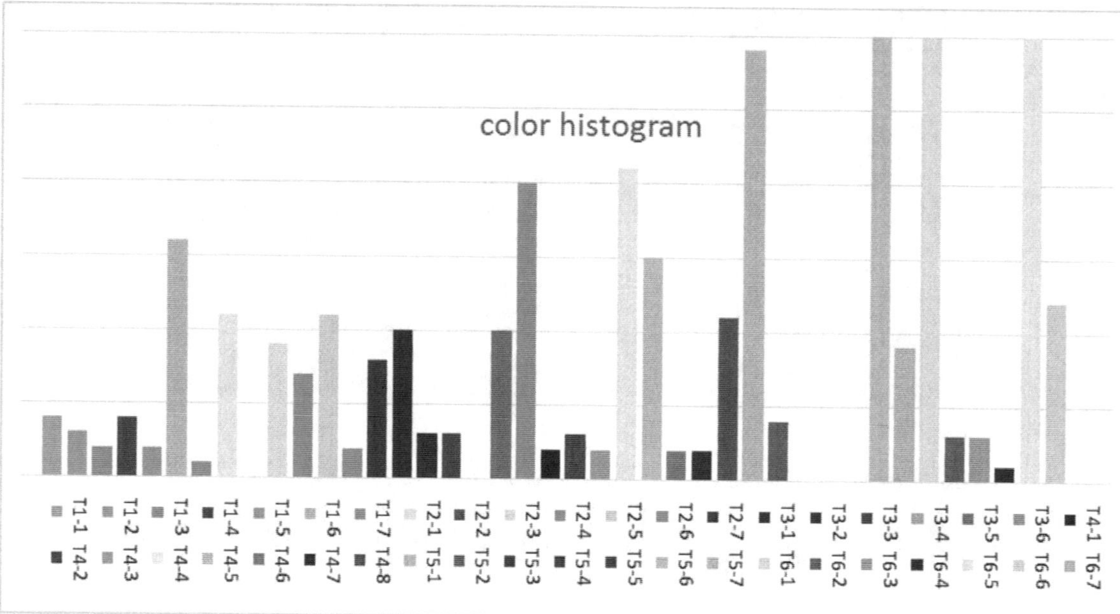

Fig.3 Results of single color preference in color histogram

Looking at the colours by themes, colours of the first theme are relatively balanced in popularity. There was no clear favoritism or dislike for a particular colour, and each colour had its own supporters, though limited in numbers. The feature of the first theme was its tone. All the colors had the lowest saturation. Their saturation were smaller than 10 in the CNCS (China National Color System) color system, and three colors had 0 saturation. The fifth and sixth themes were extreme. The most popular and unpopular color existed side by side. It can be inferred that there may be strong contrast.

3.2 Basic statistics of multi-color combinations

The second part of this study was multi-color emotion. According to the released color trend, twenty-four multi-color combinations were made and used as color stimuli. There were three to five kinds of color combinations in each theme. Semantic Differential Method was used in this part. The average score of each color stimuli for each pair of adjectives is first obtained, which is shown in Table 2. "Multi-color" was abbreviated as "MC". Each adjective pair was expressed by capitalizing their first letters.

From the averages in Table 2, color emotions of each multi-color combination was clear. MC-1 was a combination of T1-1, T1-4 and T1-7, and it is associated with the following emotions: stale, tasteful, flaunty, plain, and intellectual. MC-2 was a combination of T1-1, T1-4 and T1-6, and the associated emotions were modern, urban and elegant. MC-3 was a combination of by T1-2, T1-5 and T1-6, and the associated emotions were stale, stiff, flaunty, plain and powerless. MC-7 was a combination of T2-1, T2-2 and T2-3, and the associated emotions were more elegant, tasteful, and intellectual. MC-8 was a combination of T2-2, T2-3 and T2-5, and the associated emotions were casual and lively. MC-9 was a combination of T3-1, T3-3, T3-4 and T3-6, and the associated emotions were stale, gaudy, gorgeous, powerful and urban. MC-10 was also a combination of T3-1, T3-3, T3-4 and T3-6, but the main color was T3-6, and the associated emotions were gaudy, implicit, gorgeous, powerful and warm. MC-11 was a combination of T3-1, T3-5 and T3-6, and the associated emotions were gaudy, implicit, gorgeous, powerful, urban and warm. MC-20 was a combination of T6-4, T6-5, T6-6 and T6-7, and the emotions associated were modern, casual, lively, implicit, fashionable, individual and warm. MC-21 was a combination of T6-1, T6-3, T6-4 and T6-5, and the associated emotions were casual and tasteful. MC-22 was a combination of T6-1, T6-2 and T6-3, and the associated emotions were tasteful and flaunty. MC-23 was a combination of T6-2, T6-3,

T6-4 and T6-5, and the emotions it was associated to was relatively balanced across the adjective pairs. MC-24 was a combination of black, white and gray, the associated emotions were showed elegant, tasteful, flaunty and urban.

Table 2. The average score of each multi-color stimuli for each adjective pair

Items	M-C	V-E	F-C	F-S	G-T	L-S		I-P	P-P	R-U	W-C	I-N	L-D
MC-1	3.13	3.51	2.81	4.01	4.04	3.90	......	3.34	3.21	3.23	2.96	2.99	2.10
MC-2	2.19	3.80	2.89	3.27	3.63	2.94	......	2.60	2.29	3.96	2.66	2.39	2.26
MC-3	3.37	3.01	3.53	4.19	3.64	3.76	......	3.39	3.89	2.93	2.47	3.70	3.20
......	......	......	......	......	......	......	......	......	......	......	......	......	......
MC-7	2.57	3.96	2.40	2.66	4.10	3.16	......	3.14	2.56	3.79	3.47	2.04	1.93
MC-8	2.44	3.16	3.77	2.64	3.53	2.36	......	2.53	3.00	3.11	2.54	2.73	3.03
MC-9	3.36	3.13	2.93	4.20	1.90	3.04	......	2.41	2.29	3.83	2.47	3.04	3.11
MC-10	2.89	3.53	2.93	3.40	2.11	2.67	......	2.54	2.16	3.83	2.21	2.61	2.71
MC-11	2.60	3.33	2.90	3.29	1.87	2.45	......	2.50	2.10	3.91	2.23	2.83	2.86
......	......	......	......	......	......	......	......	......	......	......	......	......	......
MC-20	2.00	3.14	4.00	2.36	2.70	1.89	......	1.97	2.41	3.06	2.11	2.37	3.06
MC-21	2.34	3.39	3.84	2.41	3.79	2.41	......	2.53	3.00	3.06	3.00	2.43	3.01
MC-22	3.14	3.53	2.91	3.20	3.91	3.49	......	3.13	2.97	3.51	3.74	2.91	2.63
MC-23	2.40	2.87	3.26	3.37	3.10	2.81	......	2.53	2.53	3.67	2.84	3.21	3.26
MC-24	2.57	3.83	2.60	2.80	4.03	3.27	......	3.29	2.89	3.79	3.70	2.30	2.33

Fig.4 shows the scatter plots of color emotion of MC-7, MC-9 and MC-20 as examples.

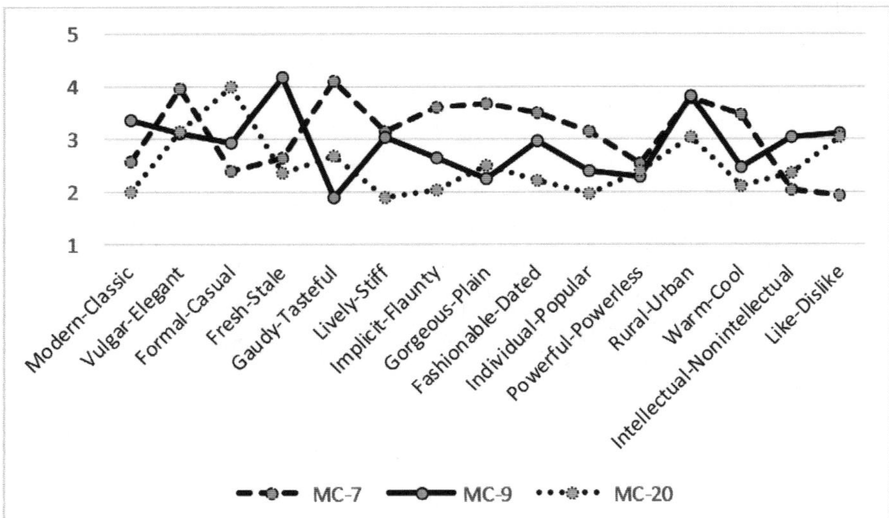

Fig.4 The scatter plots of color emotion of MC-7, MC-9 and MC-20

3.3 Multi-statistics and discussion

In order to investigate the emotion space of consumers in relation to the released color and color coordination, factor analysis was used to deduce the dimensions of consumer's impression. In the software SPSS (version 15.0), PCA (principal component analysis) and cluster analysis were selected. The KMO test was first performed with a result of 0.696, implying factor analysis was possible. Eigenvalues were set over 1, and three factors were extracted as shown in Table 3. The variance of the first factor was 37.566 and the total variance was 88.900. The factor rotation method was varimax, and the rotated results are also shown in Table 3. The factor plots of multi-color emotion scales in rotated space are shown in Fig.5.

From Table 3 and Fig.5, the first factor included "Gorgeous-Plain", "Gaudy-Tasteful", "Powerful-Powerless", "Implicit-Flaunty", "Individual-Popular" and "Warm-Cool". These adjectives were connected to characters, so this factor was named "personality". The second factor included "Fresh-Stale", "Modern-Classic", "Intellectual-Nonintellectual" and "Lively-Stiff". These words represented aspects related to education, so it was named "taste". The third factor included "Vulgar-Elegant", "Like-Dislike", "Formal-Casual", "Rural-Urban" and "Fashionable-Dated". These words expressed the feeling of time, so it was named "fashion".

It indicated that color emotion of multi-color coordination was relative to personality, taste and fashion. Fashion is always the main point of consumer demand. Personality is used to distinguish one person from another, which can also be expressed by clothings. Taste of clothings can show artistic appreciation.

Table 3 Rotated factor matrix of color emotions

Phase	Component1	Component2	Component3
Gorgeous-Plain	0.953	0.121	0.080
Gaudy-Tasteful	0.952	-0.118	0.174
Powerful-Powerless	0.867	0.292	-0.318
Implicit-Flaunty	0.856	0.387	0.293
Individual-Popular	0.789	0.476	0.313
Warm-Cool	0.692	-0.016	0.357
Fresh-Stale	-0.055	0.971	-0.048
Modern-Classic	0.230	0.835	0.103
Intellectual-Nonintellectual	-0.005	0.753	-0.592
Lively-Stiff	0.602	0.738	0.268
Vulgar-Elegant	0.369	-0.213	0.836
Like-Dislike	-0.385	0.083	-0.830
Formal-Casual	-0.100	-0.508	-0.827
Rural-Urban	-0.496	0.140	0.700
Fashionable-Dated	0.526	0.574	0.596
% of variance	37.566	26.003	25.331
Cumulative (%)		88.900	

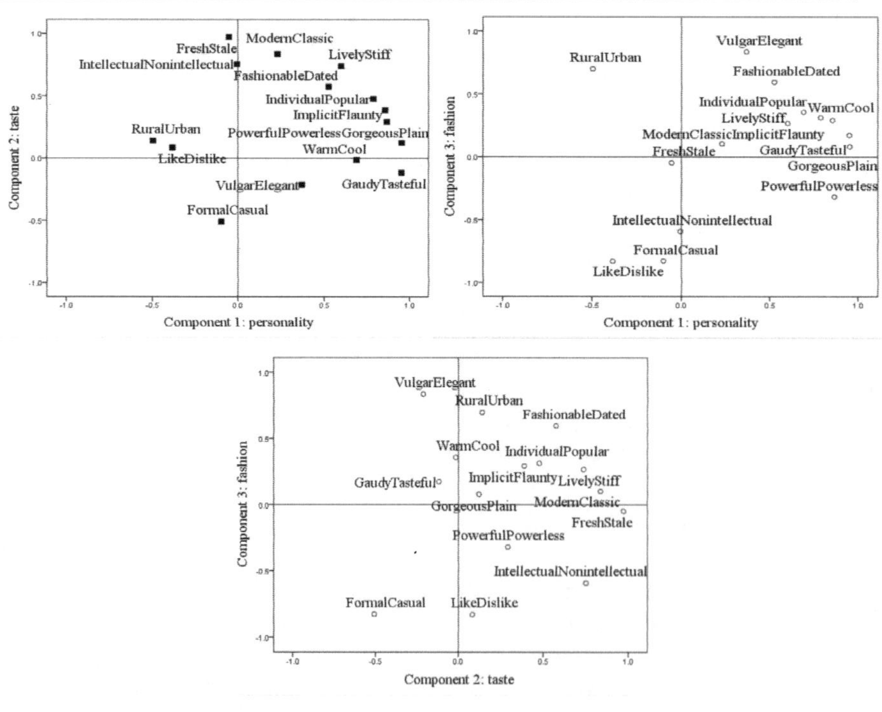

Fig.5 Factor plots of ten color emotion scales in rotated space

In order to obtain the response of consumers to the released color coordination, cluster analysis was used to conduct the data analysis. The factor scores of all the stimuli to each factor were regarded as variables, and a hierarchical cluster analysis was performed. A tree graph of cluster analysis was shown in Fig.6. From the graph, twenty-four stimuli can be divided into three groups.

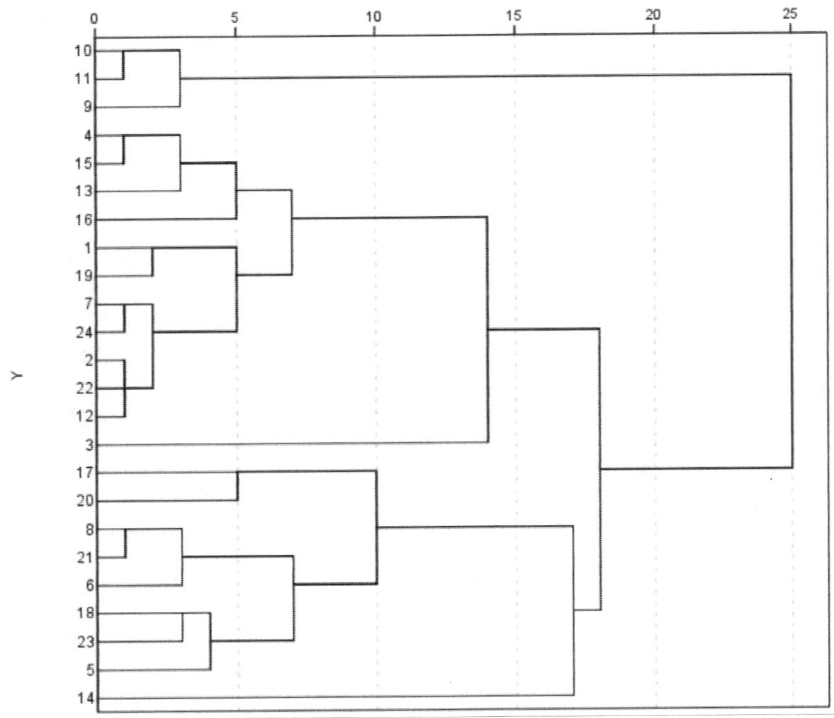

Fig.6 The tree graph of cluster analysis

One group included MC-9, MC-10 and MC-11. The three color coordination were of the wine tone, which is a combination of different red and dark colors. Red is striking, powerful and warm. Different red colours occurring at the same time can easily bring out the feelings of gaudy, gorgeous and implicit.

The second group included MC-4, MC-15, MC-13, MC-16, MC-1, MC-19, MC-7, MC-24, MC-2, MC-22, MC-12 and MC-3. These color combinations are closely related to color. For example, MC-3 was combined by T1-2, T1-5 and T1-6. The hue number in CNCS of T1-5 and T1-6 were both 032. MC-7 was a combination of T2-1, T2-2 and T2-3. The hue number in CNCS of T2-2 and T2-3 were both 024. It is very easy to achieve good harmony with this kind of color combination, which are naturally related to intellectual, flaunty and tasteful. In particular, MC-24 was a combination of black, white and gray which are forever popular. The achromatic tone showed more intellectual and tasteful emotions.

The third group included MC-17, MC-20, MC-8, MC-21, MC-6, MC-18, MC-23, MC-5 and MC-14. Most of these color combinations had strong color contrast or a strong color. For instance, MC-17 and MC-18 had the same color combination which were comprised of T5-5, T5-6, T5-7. Amongst the colours, T5-5 was close to black, T5-6 was grayish blue, and T5-7 was vivid orange. Black and blue are a classic, intellectual, soft color combination. Orange added power and energy to black-blue, and it made them younger, more fashionable. MC-5, MC-6 and MC-8 projected similar color emotions and shared a same color, T2-5. T2-5 is pale yellow, with a CNCS value of 80. This color with high lightness are included in these color combinations at different proportion. It was of the spring tone which brought vitality, hope, and a light-soft feeling.

Through the survey of single color preference, it was observed that the popularity of the forty-two single colors were not very even. Some colors were very popular, but some colors were not. This meant that designers could not select the single colors on their own. These single colors need to appear in groups with other colors. In other words, colors should be released in groups.

When two or three or more colors are combined, color property is not the only factor influencing the effect of the color combination. Location, area and proportion are all the important factors. From the above results, although the color components were the same, associated color emotions changes as the color proportion is adjusted.

After factor analysis and cluster analysis, three factors were extracted and twenty-four multi-color stimuli were classified into three groups which were different from six released themes. It indicated that multi-color emotion and the way of color construction were more important than single color property. Multi-color emotion creates a comprehensive feeling, which relates to the color property of the main color, color area, color construction and so on.

4. Conclusion

In this study, a two-part survey was performed. One was single color preference, and the other was color emotion of multi-color combination. Forty-two single colors were selected and twenty-four multi-color combinations were prepared. Color histogram was adopted to process single color preference. Semantic Differential Method was used to conduct the survey of multi-color emotion, and factor analysis and cluster analysis were used to analyze the data. From these results, some conclusions could be drawn.

1) The popularity of the released colors was not even, so designers cannot propose design plans based merely on particular single colors.

2) Three factors were extracted from the survey of multi-color emotion, and these factors were labeled "personality", "taste", and "fashion".

3) Multi-color emotion and the way of color construction were more important than single color property.

4) Designers must focus on the comprehensive color emotion, which entails main color property, color area, and color construction.

5. Acknowledgement

The project is supported by project BS0908 of Xi'an Polytechnic University, and by project JK0973 of Shaanxi Education Department. The authors would like to thank all the respondents of Xi'an Polytechnic University.

References:

[1] History of Intercolor. http://www.intercolor.nu/history.html.

[2] Xue Y, Imaoka H. Comparison of color psychological aspects for single color between Chinese and Japanese. J. Jpn. Res. Assn. Text. End-Uses: 2009; 50: 141-147.

[3] Xue Y, Imaoka H. Comparison of color preferences for two-color combinations between Chinese and Japanese. J. Jpn. Res. Assn. Text. End-Uses: 2009; 50: 309-316.

[4] Xue Y, Imaoka H. Comparison of color preferences for three-color combinations between Chinese and Japanese. J. Jpn. Res. Assn. Text. End-Uses: 2009; 50: 687-694.

[5] Uravis T, Kitirochna R, Hiroyuki S. Color preference affected by mode of color appearance. Col. Res. Appl: 2010; 35: 50-61.

[6] Iris B, Theo van der V, Peter V, Jan de B, Conne B. Color preferences for different topics in connection to personal characteristics. Col. Res. Appl: 2015; 40: 62-71.

[7] Ou LC, Luo MR, Woodcock A, Wright A. A study of colour emotion and colour preference. Part I: Colour emotions for single colours. Col. Res. Appl: 2004; 29: 232-240.

[8] Ou LC, Luo MR, Woodcock A, Wright A. A study of colour emotion and colour preference. Part II: Colour emotions for two-colour combinations. Col. Res. Appl: 2004; 29: 292-298.

Study on Online Fabric Pictures

Meng Yuan[1], Wei Tian[2*]

Xi'an Polytechnic University, 19 Golden Road, Xi'an, Shaanxi, 710048, China

*Corresponding author's email: 569790315@qq.com

Abstract

This paper mainly analyzes garment material web display of some famous e-shopping platforms, from shooting technique and photoshop sides to finding the limitations of garment material web display. This paper puts forward a correct way of displaying pictures online. Firstly, pictures produced from the shoot then post-processed are analyzed. Secondly, using the above mentioned method 3 kinds of different affection pictures are made with 5 groups in total. Thirdly, questionnaire survey designed for customers was conducted to find out which pictures they accept. Finally, the following conclusions are drawn: 1) garment material display is an important part of website display, an independent module should be considered for the verbal description and the pictures show on the display. 2) For the display, details on fabric material is important, aesthetics is not the first factor to be considered. 3) For female customer groups, the pictures should be slightly modifed (e.g. adjusting the hue-saturation). For male customer groups, the pictures should directly represent original photo as much as possible.

Keywords: Online shopping platform; Fabric web display; Post-processing; Authentic extent

1. Foreword

In recent years, with the rapid development of electronic commerce, and the adjustment of business development pattern, the traditional shopping model has been changed to an online platform. Especially in the last few years, with the continuous development and proliferating breakthroughs in the business model of various online shopping platforms, there is a huge market in the online shopping website. In particular, the daily business transaction was 57.1 billion on "double 11" in 2014 at taobao website, and after a year, the "double 11" day's transaction volume stopped at 91.2 billion on taobao. Online shopping consumer behavior has become an essential part of People's Daily consumption behavior.

However, we can't touch and feel the product during online shopping. For garment design, the web page is a window for consumers to know the product, which may include the quality, colors, materials, performance and other aspects. Consumers will then combine the experience and knowledge to make a decision whether to buy or not. Thus, online display has a high effect on accessing information and to make the purchase decisions for consumers [1].

Psychological research shows that out of all theinformation that are accepted, 83% is from the visual channel, 11% are from the auditory channel, the other 6% comes from the sense of smell, touch and gustation [2]. Some other Studies show that according to the relationship of the stimulus character with attention and memory perspective, the general image can bring about 78% of the attention, and words only retains 22% attention [3]. In general, the information pickup points of clothing products are the following: price, style, fabric, color, quality, style, collocation and brand, etc [4].

Fabric is a key factor to evaluate the quality of the clothing products, however, on many network platforms, they do not put the fabric at a primary position in online display. Therefore, the products may return back because the goods is not consistent with the pictures on the website. As time goes by, consumers will not believe the seller which may harm the reputation of the network platform. This paper mainly studies the display of fabric module on online shopping platform which aims at the problem in online shopping platform.

2. The fabric display analysis on some famous online shopping platform

2.1 The analysis of taobao and tmall shopping platform

According to the investigation research on a taobao website, it shows that some sellers can display the fabric of a product, and add some exaggerated literal interpretation beside the picture. This way it can stimulate people's consumption to some extent. But the fabric image pixel processing are too fuzzy to reflect the real degree (as show in Fig.1). At the same time, some sellers do the explanation of the fabric in the garment details, and have no explanation on the feeling and material composition of the fabric. Most sellers do not understand the principle of how to use pictures and ,verbal description and also do not consider the customers' needs and concerns from their perspective.

Fig.1 Fabric show

Compared to the taobao website, the display fabric on the mall website are more normal, the quality of the image pixel are also more clear, magnified details are calso shown on the website (as show in Fig.2). Meanwhile, the same problem on the mall website also exist, some garment brands' display did not show the fabric in a clear way.

Fig.2 Fabric show

2.2 The analysis of other online shopping website

The research of this part also has 3 online shopping platforms which is included in jumeiyoupin, vip-shop and meilishuo website.

The display of garment in jumeiyoupin website mainly divides into commodities parameter and details. The quality of the picture is more beautiful than the picture in taobao website while the quantity is less than taobao website. Many of the displays lack fabric pictures which can explain the material of the garment.

Vip-shop is a shopping platform for some high level brands. Regarding the garment, the level of detail of their display depends on the brand itself, different brand has different level of details. Above all, the display is limited into the quantity of the pictures.

Meilishuo includes fabric display in many stores, but it has the same problem as taobao as it includes over word processing and the over process of the images. Thus the representation is very different to the actual product.

3. The process of image generation on online shopping platform

3.1 The analysis of photography

The equipment used was a pro-camera. Before the shooting, the fabric and garment must be ironed and pressed. Regarding the place and light conditions of the shooting, whether it is indoors and outdoors, using a natural light or artificial light source must be considered. For the background, the first requirement is a flat background, the other request is the color as it should not be a contrasting color with the cloth.

3.2 The analysis of image on post-processing

Post processing mainly involves photoshop. With photoshop the three tools that are mainly used for post-processing are Spot Heading Brush, Clone Stamp and Adjust Tools. The first and the second repairs the drawbacks of the original image. The Hue/ Saturation tool which includes the Adjust Tools can adjust the color a which enables to make it look like the real product.. The Curve Tool is mainly used to strengthen the color of picture stereo and increases the color contrast, thus it will highlight the brightness of the products.

4. Questionnaire

4.1 The design of the questionnaire

This paper made 3 different kinds of effect pictures with 5 groups of fabric. Then questionnaire was conducted for the consumers.

The original pictures (as show in Fig.3) were shot with a profession SLR camera, nikon 5300, in a condition of natural light, indoors. photo shop was used in the first picture (as show in Fig.4) to adjust the correct size of the image and to deal with defects and flaws. The second picture (as show in Fig.5) is based on the first picture and was adjusted based on the saturation and contrast of the color. The purpose of this step is to highlight the brightness of the fabric and texture. The third picture (as show in Fig.6) on the online shopping platform is the most common way to display fabric. The use of bending of the fabric crease, shadows and highlights on photo shop which is to deal with fabric. At the same time, it also increases the fuzzy degree of partial shadow parts to highlight the quality of fabric.

Fig.3 The original picture

Fig.4 The first picture

Fig.5 The second picture

Fig.6 The third picture

In order to avoid random errors being produced by a set of images, 5 groups of images were designed for the investigation. The color of the 5 groups are different and the fabric texture aspects also differ from each other.

4.2 The process of questionnaire survey

The objections of the questionnaire are college students and the people who have rich shopping experience, a total of 40, among 20 girls and 20 boys. The investigation place was indoors, it was quiet and peaceful. In order to avoid the random error caused by different displays, the display of the research was shown on the Y470 lenovo laptop. The questionnaire information such as the pictures are also involved in the same word document.

4.3 The the results analysis of the questionnaire survey

The statistical results of questionnaire survey in the following table (as show in Table 1).

Table1. The statistical results of questionnaire survey

		Group1	Group2	Group3	Group4	Group5	The average of percentage
The number of Fig.1 choice	female	10	10	12	13	7	52%
	male	17	16	14	18	15	80%
The number of Fig.2 choice	female	9	8	8	6	9	40%
	male	3	2	4	2	4	15%
The number of Fig.3 choice	female	1	2	0	1	4	8%
	male	0	2	2	0	1	5%

From analyzing the statistical data, the average percentage of females who chose Fig.1 is 52%, the average percentage of the female who chose Fig.2 is 40%, the average percentage of the female who chose Fig.3 is only 8%. While the average percentage of the male who chose Fig.1 is 80%, the second average percentage is 15% and the third average percentage is 5%. In conclusion, the average percentage of all the participants is Fig.1 which amounts to 66%, the second average percentage is 27.5%, the third average percentage is 6.5%.

Through the analysis of the data, it can be concluded that consumers in the shopping process is more focused on whether the pictures can accurately reflect the fabric texture and color. In regards to requirements of the aesthetics, it is not too clear. According to the statistics data, the average percentage of the people who chose Fig.2 are female which is higher than their male counterpart. It shows that contrasted with the male participants, the female participants have ore demand on the aesthetic measures of the pictures while men may pay more attention to the reality of the pictures. As a result for the sells on the website, in order to attract more consumers, they should put the pictures which are merely close to the truth on the online display. The picture with which matches to the real situation will result in more customer satisfaction. At the same time, for female consumers, a little bit embellish can be made in the pictures; for male consumers, it is better for the pictures to reflect the original product.

4.4 The suggestions to the online sellers

While sellers do display for garment on website, a separate module should be made for fabric display and quality, and the pictures must be close to the actual product in the simplest manner (such as tiled display). Decorating the pictures too much may cause the harm to the reputation of the online store. Meanwhile, to avoid the aberration caused by the display, sellers can use pantone color number to have further annotation in regards to the color of the garment. All of these methods would bring better customer satisfaction.

5. Conclusion

Through the research and the analysis of some famous online shopping website and the questionnaire survey analysis, for the garment of fabric display modules, the following conclusions can be made: garment material display is an important part on website display, an independent module should be considered for the verbal description and the pictures shown on the display. For the display, details should be as accurate as possible to reflect the real fabrics, aesthetic is not the first factor to be considered. For female customergroups the pictures ccan be mofidifed slightly (e.g. adjusting the hue-saturation). For male customer groups, the pictures should be extremely close to the original product.

6. Acknowledgement

The author would like to express their gratitude to the teacher and the Apparel & Design College of Xi'an Polytechnic University.

References

[1] Chen HX, Jiang L, Wang BL. Online clothing display and relationship between the risk perception of online clothing. J. Journal of Beijing Institute of Fashion Technology: 1001-0564 (2010) 01-0008-05: 8-12.
[2] "Lewd" god - visual effect on the terminal consumers. http://www.boraid.com/article/html/13/1309.asp.
[3] MU Y. The secrets of visual marketing on fast fashion. J. Journal of Sales and marketing: 2011 (1) : 16.
[4] Zhao SS. The research of picture communication of products on Tmall with women's clothing boutique. HuNan: Xiang Tan University: 2014; 12-23.
[5] Shen DQ. The impact between online store pictures and the intention of the consumes-the adjustment of product type. HangZhou: Zhejiang Gongshang university: 2014; 11-15.

Operating Mechanism and Suggestions of Waste Textiles Recycling and Reusing System in France

Lei Yao[1], Yi-Lun Wu[1], An-Yi Gen[1]

[1]*Business school, Beijing Institute of Fashion Technology, Beijing, 100029, China*

Corresponding author's email: sxyyl@bift.edu.cn

Abstract

The paper carries out an analysis on the legal system of recycling of waste textiles and introduces the composition and the implementation method of the recycling system of waste textiles in France. Based on the present situation of China's waste recycling method and using France as a reference, some suggestions are put forward to address existing issues of recycling and reusing waste textiles in China. This includes the development of appropriate laws and regulations, a clear unified regulations, the implementation of recycling system construction, etc.

Keywords: France; Waste textiles; Recycling and reusing; Operating mechanism

1. Introduction

Textile industry is an important pillar industry in China. According to the data from Chinese first textile network, adding up the weight of the scrap which textile enterprises produced in the process of production and the household waste are more than 20 million tons every year. Only 5% of them can make it true to recycle the resource. The other has been dumped in landfills as garbage [1]. The recycling rate of waste textiles in China is relative to 16% of France and 17% of the UK [2]. There is still a big gap between China and the developed countries. The development of recycling of waste textiles can significantly reduce carbon dioxide emissions. The reduction of environmental pollution in the textile industry is in line with the goal of building a circular economy in China and an important measure of the country.

France is the second country in Europe to put forward the classification of domestic waste, recycling and reuse, second only to Germany [2]. It has paid great attention to the recycling of waste textiles. Back in 2000, the French companies began to work on the recovery of polyester fiber and widely use its products in the production of automotive textiles. Due to the establishment of regulations and recycling system, the recovery rate of waste textiles in France is 16%. Every year at least 2.5 million tons of waste textiles are recycled. (Data from the French waste textiles recycling agency Eco TLC)

2. Legal basis of recycling system of waste textiles in France

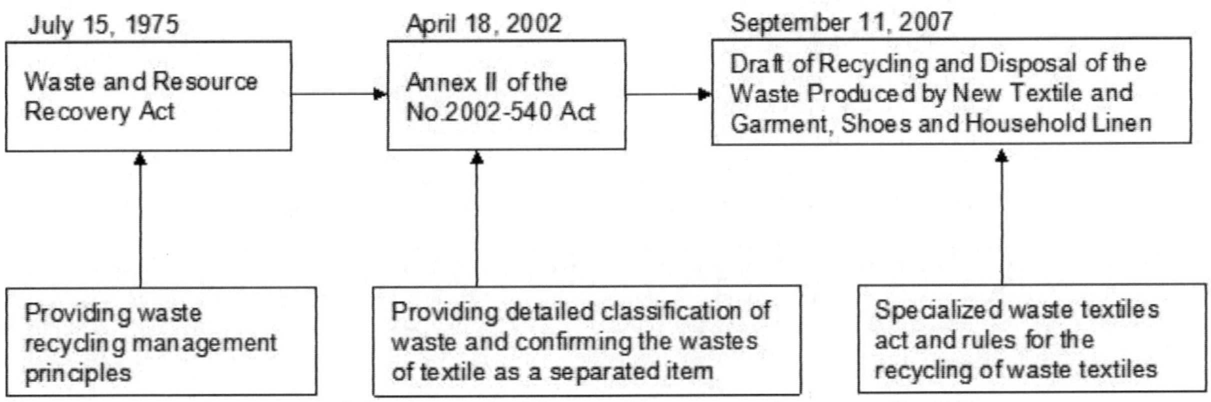

Fig.1 The legal development of the recycling system of waste textiles in France

As a member of the European Union, the law of recycling of waste textiles in France must be in line with the relevant laws and regulations of the European Union, France has developed a law accords with the national conditions, including the following three components. The first one is the Waste and Resource Recovery Act (No.75-633), promulgated in 1975. The second one is Annex II of the No.2002-540 Act promulgated in April 2002. The last one is Draft of Recycling and Disposal of the Waste Produced by New Textile and Garment, Shoes and Household Linen (No.G/TBT/N/FRA/66), promulgated in September 2007 [3]. From the basic waste recycling, to the classification of waste textiles recycling, and then to the promulgation of the law on the recycling of waste textiles, the French government has gradually realized the importance of recycling of waste textiles. And the gradual promulgation of laws and regulations has made a solid legal basis for the recycling system of waste textiles, shown in Fig.1.

2.1 Waste and Resource Recovery Act

Waste and Resource Recovery Act (No.75-633) came into effect on July 15, 1975. The act defines waste in detail, proposes that we should recycle the waste to make it value added, put forward some principles of waste management and specifies the resources recycling and the removal of waste residues.

This is the first time that the French government provides waste recycling management principles in the form of law and proposes waste should be recycled as a resource.

2.2 Annex II of the No.2002-540 Act

In April 18, 2002, the French government in Annex II of the No.2002-540 Act provides a unified list of waste and detailed classification of waste. The wastes from the leather, fur and textile industries are set up separately in the fourth item. Cloth and other unknown special protective clothing and packaging waste is classified in the fifteenth item, shown in Table 1.

Table 1. Waste classification list in Annex II of the No.2002-540 Act

1	Wastes from physical and chemical treatment of exploration and mining
2	Wastes from agriculture, horticulture, aquaculture, forestry, hunting, fishing and food preparation and processing
3	Wastes from the processing of trees, wood, furniture production, pulp, paper and paperboard
4	Wastes from leather, fur and textile industry
5	Wastes from petroleum refining, natural gas purification, coal pyrolysis treatment
6	Wastes from mineral chemical methods
7	Wastes from organic chemistry
8	Wastes from production, preparation, distribution and use of coatings, such as paints, lacquers and enamels vitreous body, and sealant and printing ink
9	Wastes from the photographic industry
10	Wastes from the heat process
11	Wastes from surface chemical treatment, metal, other materials and wet coating of nonferrous metal
12	Wastes comes from surfaces physical, mechanical processing and formatting of metal and plastic
13	Use of petroleum and liquid fuels
14	Wastes from organic solvents, refrigerants and Propellants
15	Packaging and packaging waste, absorbent, wipes, filter materials and other unspecified special protective clothing
16	Other wastes that are not described in the list.
17	Wastes from construction and demolition, including dredging of contaminated sites.
18	Wastes from medical or veterinary services and research
19	Wastes from waste management facilities, sewage and garbage field of water treatment and preparation of water for human consumption and industrial water waste
20	Municipal waste

Annex II of the No.2002-540 Act distinguished the wastes of textile industry from the waste of life. Although there are still defects in classification standards, it is enough to show that the government has begun to attach importance to the recycling of waste textiles.

2.3 Draft of Recycling and Disposal of the Waste Produced by New Textile and Garment, Shoes and Household Linen

On September 11, 2007, the French Ecology and Sustainable Development Ministry released the Draft of Recycling and Disposal of the Waste Produced by New Textile and Garment, Shoes and Household Linen. The draft aims to ensure the sustainability of the waste textiles industry, increase the number of selected waste textiles recycling and reduce the cost of waste of clothing, footwear and household linen.

Draft have been developed for the plan and the organization procedures about the "extension responsibility of textile waste producer" and clear the recovery target of each system and institution, which specific to classification, reuse, recycling and recycling. At the same time, the arrangement and utilization of scientific research personnel are also made detailed provisions. If the unauthorized textiles were put into the French market, the state will impose fines on producers.

This draft is specifically applicable to the recycling of waste textiles in France so far. It not only specifies the sorting, recycling and recycling of waste textile, but also carries on the reorganization plan to the entire textile industry's industrial chain. The concept of recycling runs through the whole process of textile production, such as sale and recycling.

3. Analysis the operation mechanism of waste textiles recycling system in France

3.1 Main operation of the waste textiles recycling system

The recycling system of waste textiles in France is made up of three main operation: Local government, waste textiles recycling agency Eco TLC, trade association and charitable organizations. Among them, the local government is dominant. Eco TLC is responsible for the specific implementation of recycling. And industry associations and charitable organizations play a role of auxiliary and propaganda.

Eco TLC co. was established in December 2008. According to the environmental code file, No.D. 543-214 and No.L 541-10-3, this non profitable private companies are dedicated to the recycling of waste textiles. Based on French extended responsibility system of textile wastes producer, all producers are responsible for the recycling of the textile products they produce. They can choose to carry out recycling by their own or commission third-party. TLC Eco is such a third party company. It is authorized by the local government responsible for the recycling of waste textiles, while responsible for tracking and confirm the logistics of waste textiles, including the source of the upstream and downstream whereabouts. It relies on the environmental taxes paid by the producer to maintain the operation.

3.2 Pattern of the waste textiles recycling system

Through the study of the three main bodies of the recycling system of waste textiles in France in this paper, French recycling system of waste textiles can be divided into two aspects: macro and micro.

3.2.1 Macro Pattern

The main body of the macro recycling system includes local government, Eco TLC recycling companies, industry associations, charities and the public, shown in Fig.2:

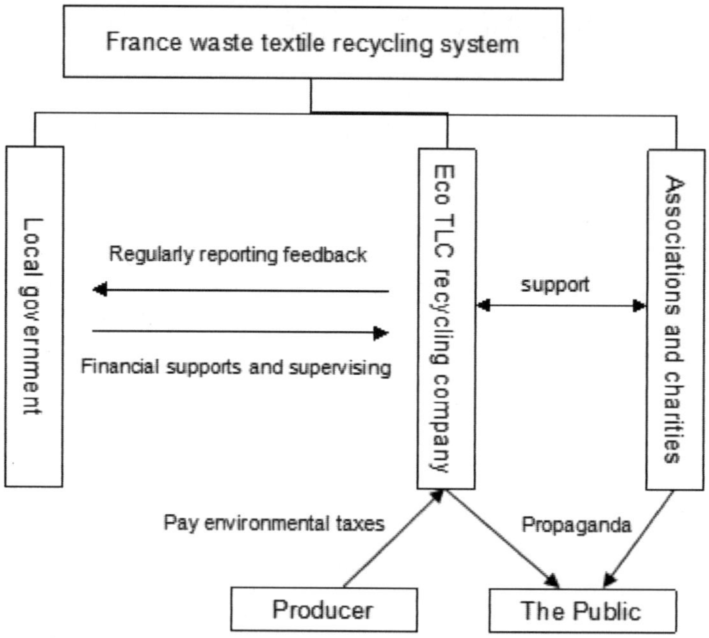

Fig.2 Macro pattern of the waste textiles recycling system

Local government is the dominant of the recycling system, plans and guides for the whole recycling industry chain. The government's functional departments are responsible for the overall planning, laws and regulations, carry out the supervision and financial support for Eco TLC company. Part of the Eco TLC's funds comes from the government's financial support, and the other part comes from the environmental tax paid to the Eco TLC by the producers, used for recycling of waste textiles. Recycling companies are bound by law, operate on a self-financing basis and perform the specific business, such as recycling, removal and disposal of waste textiles. It reports to the local government regularly. TLC Eco company provides financial support for the garbage classification association and R & D department of new fabrics and product annually, in order to encourage the development of environment-friendly products, promote garbage classification and improve people's awareness of the classification and recycling of waste textiles. Charity and Industry Association is responsible for assisting in the collection of waste, providing support to the Eco TLC company and promoting awareness education to citizens.

The system effectively separated the enterprise from the government. The efficient operation of Eco TLC company is an important factor to ensure the operation of the whole macro system.

3.2.2 Micro Pattern

Eco TLC is a typical representative of the micro recycling companies and relies on the following 3 ways to recycle waste textiles: First, House Call Collecting by the staff. Second, personal donations to charitable organizations or associations, and then transferred to the recycling company. Third, residents put the white trash to the designated location. Operation pattern is shown in Fig.3.

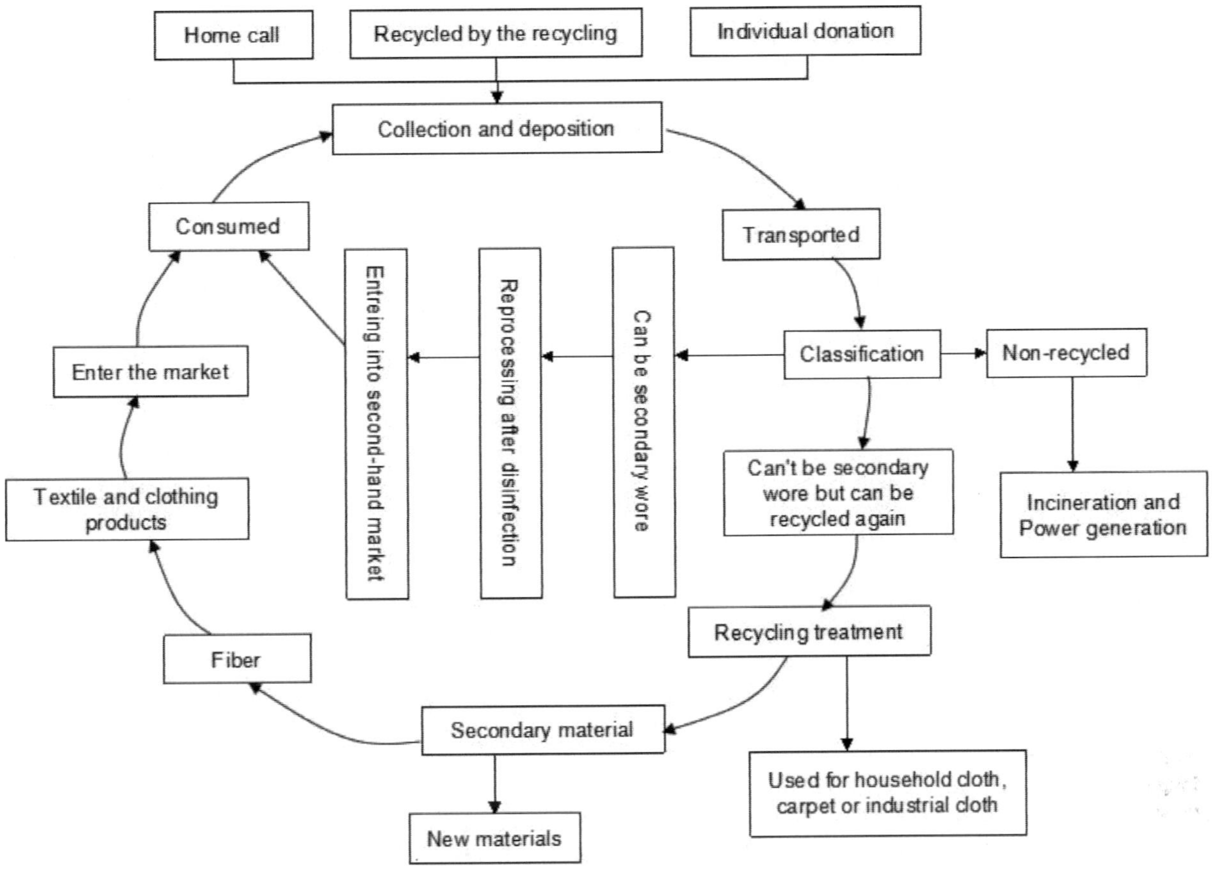

Fig.3 Operation Pattern of Eco TLC company

The waste textiles will be collected and stored uniformly, and then go to the Eco TLC company for transportation and professional classification. There are three kinds of treatment for the classified waste textiles. First, non recyclable waste can be burned directly to power generation. Second, the waste textiles which can be worn second time can be sold into the secondary market trading after disinfection and reprocessing. Third, the waste textiles which is recyclable but can't be worn for second time can be processed into secondary materials, and then used for the production of clothing, shoes and etc., and finally sold into the consumer market.

Eco TLC recovered more than 2.5 million pieces of clothing from the French market each year, and this number is still expanding.

4. Suggestions on the recycling of waste textiles for China

At present, the China government has been aware of the importance of recycling of waste textiles recycling, and experimented recovery facilities in Beijing, Shanghai, Guangzhou and other first tier cities. For example, in 2009 Shanghai set up many waste recycling facilities in the community. However, facing the excellent prospects of waste textiles recycling market, whether from the perspective of national legislation or from the perspective of industry development, China's waste textiles recycling market is still facing many constraints and problems. Learning from the experience of the French recycling system, several suggestions are put forward in this paper.

4.1 Strengthening the construction of legal basis

China promulgated the Circular Economy Promotion Law of the People's Republic of China in 2009, although it has made it clear that we should establish and improve the recycling system of renewable resources, urban life waste management is not mentioned. At present, there is no law aimed at the classification, recycling and reuse of municipal waste in China. Meanwhile local governments have different regulations.

Based on this situation, first of all, we should be clear that on the legal level urban life waste is a recyclable resource, which should be recycled. Second, the government should be definite about classification standards of the waste and sort waste processing. Finally, aiming at different types of waste, the government should develop suitable special recycling act to unite the names of recycling wasted, the color of the classified garbage bins etc. If someone's behavior does not conform to the provisions of the classification, there should be a specific legal punishment. The government should also establish a strict supervision system to build the legal basis of urban life waste and improve recycling of waste textiles.

4.2 Improving the construction of the operation system

At present, there is no lack of waste recycling companies in China, for example Shanghai Yuan Yuan Industrial Co., Ltd. Up to 2013, the waste recycling bins set up by Yuan Yuan company have reached 1800 units in the community, schools and other places in Shanghai. However, some of the recycling companies just sort, regardless of the treatment. And some enterprises don't have normative collection process. These companies do not reuse the waste resources.

From the perspective of improving the construction of the system, the government should play a leading role in the whole system. In Beijing recycling work of waste textiles is led by the Civil Affairs Bureau, while in Shanghai the Greening and City Management Bureau has led the work about clothing recycling station since 2010. This shows that the government has been aware of the role they should play in the recycling industry, but it is far from enough to do just that. The government also needs to supervise the whole recycling chain, listen to the feedback of the enterprises, deal with the new policy issues and etc. The recycling companies should establish a set of standardized audit system to ban the companies which don't have a license for recycling. And this system should be reviewed regularly.

Recycling companies are the most important part of the recycling system so that the extent of regulation and efficiency of recycling companies affect the operation of the entire recovery system. China should expand the types of waste recycling companies, not only by encouraging the charitable and environmental organizations to increase the recycling of waste textiles, but also mobilizing market forces. For example, the permission for the private enterprises to enter the market of recycling waste textiles. It requires enterprises to operate independently, but provides them with financial support to encourage their innovative R&D and to improve their operational efficiency.

4.3 Strengthening the construction of the whole environment

Recycling of waste can reduce greenhouse gas emission and protect the environment. This is not just what the government and the enterprise should do, every citizen has the obligation to participate in it.At present, our citizens are lack of awareness of waste classification and recycling, while in developed countries, like Germany and France, education of garbage classification start from kindergarten, and every citizen regarded it as his own responsibility and obligation. Charities and industry associations can make propaganda film and posters to guide the public to deal with garbage classification. To enhance public awareness, the whole environment can be improved so that it will promote the recycling of waste textiles.

5. Acknowledgment

[1] National Social Science Fund (Fund Code: 14BJL056)
[2] Beijing Institute of Fashion Technology 2014 Scientific Research Promote Project "Beijing used clothing recycling system built and suggestion research" (Fund code:2014AL-05)
[3] Beijing Excellent Talents Cultivating Funding Projects (Fund Code: 2014000026833ZS06)
[4] The Project of Construction of Innovative Team and Selection and Development of Excellent Talents for Beijing Institute of Fashion Technology (Fund Code:PTTBIFT_TD_007)
[5] The key project of social science program for Beijing Municipal Education Commission 2016, Study on recycling and reuse of waste textiles from the perspective of urban waste reduction

Reference

[1] Chinese first textile network: http://www.chinairn.com/news/20140810/103157736.shtml.
[2] Oakdene Hollins. 2013: http://www.oakdenehollins.co.uk/reports.php.
[3] The official website of Eco TLC: http://www.ecotlc.fr/index.php.
[4] Guo Y. Present situation and suggestions on recycling and reusing of waste textiles in China [J].Cotton Textile Technology.2013; 41 (4): 59-61.
[5] Luo YH, Pu ZY, Huang YH. The present situation and development trend of recycling and reuse of waste textiles [J]. Progress in Textile Science & Technology.2012 (3): 9-12.
[6] Bai J. Preliminary study on the development of recycling and reuse of waste textiles in foreign countries [J].Shandong Textile Economy.2012 (12): 54-56.

[7] Xu H, Tang SJ. Analysis on the development status of recycling and utilization of waste textiles in foreign countries [J]. China Textile Leader.2012 (7): 31-32.
[8] Wang LL. Discussion on recycling and reusing of waste textiles [J]. China Textile Leader.2009 (4): 26-28.
[9] Yang XJ, Wang XL. Preliminary study on the development of recycling and reuse of waste and old clothing at home and abroad [J].Shandong Textile Economy.2013 (7): 11-13.
[10] Li P, Ye HW, Chen YD, Zhang Y. Present situation of recycling and utilization of waste textiles at home and abroad [J]. Synthetic Fiber in China.2014; 43 (4): 41-45.

Current Status of the UK Textile Industry and its Sustainable Development

Si-Nuo Li[1], Yi Li[2*], Ming-Liang Cao[1]

[1]*Institute of Textiles and Clothing, The Hong Kong Polytechnic University, Hung Hom, Kowloon, Hong Kong, China*
[2]*School of Materials, The University of Manchester, Manchester M13 9PL, UK*

[*]Corresponding author's email: henry.yili@manchester.ac.uk

Abstract

The textile industry in the UK is under the period of resources re-allocation and technology upgradation for revival. The author introduced and adopted a novel strategic tool - Industry Technology Roadmap to draw long-term plans for the repatriation of the textile industry to the UK. This article firstly conducted an in-depth research on current status of the UK textile industry and then proposed feasible plans for its sustainable development. Through data mining of industrial resource distribution and academic research strength, industrial supply chain and innovation chain of the UK textile industry were investigated. Based on the results, SWOT analysis and a comprehensive industrial technology roadmap were created for strategic planning, which might contribute its originality and innovativeness to the literatures.

Keywords: Industry technology roadmap; Supply chain; Innovation chain; Sustainability

1. Introduction

Alliance Project, a not-for-profit organization has been established to support the repatriation of textile manufacture to the UK, for which two reports "Repatriation of the Textile Industry to the UK" and "The Revival of the UK Textiles Industry" were published [1-2]. According to the reports, the UK textile industry had its strength, such as solid traditional textile foundation and strong academic resources, and also had weakness like fragile supply chain and outdated infrastructure. To inject new energy into this traditional industry, a sustainable plan to re-allocate and re-integrate existing resources and guide the industry on a customized development route was what the UK needed urgently.

Industry technology roadmapping has become a world-wide used strategic planning tool, in both single contexts of company or institute and collective contexts of industry or economy [3-6]. Technology roadmaps created a full image of near future development with clear key areas, important breakthrough projects and resources flow, which were based on supply chains to integrate different industrial sectors in vertical levels and competitors/cooperative partners in longitudinal levels. This would be an effective and scientific method for industrial mid-term or long-term targets, especially for those in survival and leaping stages.

This paper used the technology roadmapping methodology to make a long-term plan for the revival of textile industry to the UK, which might contribute original and innovative ideas to relevant literatures. Firstly, research backgrounds and methods are introduced. Secondly, basic status of supply chain and innovation chain of the UK textile is investigated through data mining and analysis. Last but not least, SWOT analysis and a comprehensive and customized industrial technology roadmap are presented.

2. Research Background

To evaluate an industry's status and development, two key components including supply chain management and innovation chain maintenance should be investigated first and foremost. Different organizations and experts proposed various definitions of supply chain [7-9]. In this paper, Ayers' definition of supply chain as "product life cycle processes comprising physical, information, financial and knowledge flows whose purpose is to satisfy end-user requirements with physical products and services from multiple, linked suppliers" [10] was adopted. The status of the industrial supply chain revealed the strength and weakness of the industry directly and had crucial influences on its survival and future development. As for the concept of innovation chain, both general and narrow definitions had been made. Malerba defined Sectoral Systems of Innovation and Production as a system comprising products, marketing agents and interactions for the purposes of users' satisfaction, which included generally the whole industrial chain [11]. Other researchers, like Edquist and Zeng gave their narrow definitions to Industrial Innovation System as activities aggregation of the production, distribution and application of innovation which comprised ability construction, knowledge creation, applying transformation, entrepreneurial behavior, market establishment and institution reform and so on [12-13]. The industrial innovation chain indicated the core competition and development power of the industry in different area contexts.

Industry technology roadmap that created firstly for Motorola's business development blueprint and plans has become a novel type of industrial technology management and innovation practices in different areas by varied entities, including traditional strategic planning tool for a certain company or organization, and innovative scientific methodology for the survival and progress of an entire industry [3,14-15]. Industrial technology roadmapping had been adopted in company business plans and industry long-term strategy projecting [3-4,13-15]. A number of studies from both academia and industry claimed the effectiveness and advantages of technology roadmaps in technology prediction, linkages between innovation and business strategies as well as long-term livable planning in dynamic contexts and so on [5-6,19]. Some pioneering researchers conducted studies on the theory and practices of how to evolve a successful roadmap [15,20-21]. Basing on the status of industrial supply chain and innovation chain, technology roadmapping could be applied to the UK textile industry in order to have effective strategies and feasible methods.

3. Methods

In this paper, data mining, SWOT analysis and industry technology roadmapping were adopted. To collect information and data for supply chains and innovation chains of the UK textile industry, the author searched data from the database SCIFinder and Web of Science, with which relevant patents and SCI index papers were analyzed and illustrated by figures. SWOT analysis, the world-wide-use methodology, was applied to explore the strength and the weakness of the UK textile industry itself and the opportunities and the threats from outside environments. With the integration of these four factors, corresponding strategies - SO, WO, ST and WT were presented. Last but not least, industry technology roadmapping was conducted to draw the whole picture of the future development guidelines for the UK textile industry.

4. Results and Discussion

The UK textile industry was analyzed from three different aspects, including supply chain circumstances of selected companies and overall status of innovation chains, which were followed by a comprehensive SWOT analysis.

4.1 Supply Chains in the UK Textile Industry

An information database of 94 textile-relating companies that were collected from The Revival of the UK Textiles Industry of NBrown National Textiles Growth Programme [1] had been established for illustration. These 94 companies were categorized by locations and supply chain sectors.

Fig.1 showed that 35 companies in the funded companies of Alliance Project belonged to West Yorkshire, 26 belonged to Great Manchester and 17 belonged to Lancashire, where traditional textiles had strong industrial foundations and still have great impact on today's textile industry in the UK. From Figure 2, we founded that the companies mainly covered four sectors as spinning, weaving and knitted cloth, apparel and clothing, home textiles and dyeing and finishing. All these four sectors were relatively traditional and crucial parts in textile industry. In addition, there were 5 companies in fabric design and industrial textiles respectively, which represented the new trend in textile industry in global. Last but not least, none funded companies were listed in the fashion design sector, suggesting that resources had not been well integrated in the textile supply chain, as for the fact that the UK fashion design was one of the best in the world.

From the analysis results, it showed that the funded companies were geographically clustered in districts with strong textile foundation and rich history, while complete supply chains with collaborated sectors and integrated resources had not been established.

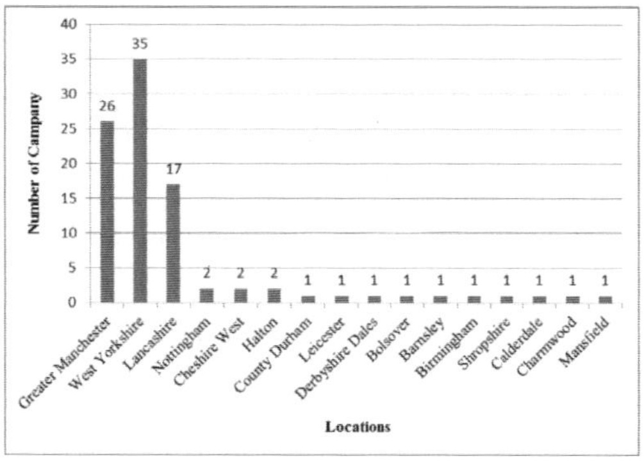

Fig.1 Geographical Distribution of the Funded Companies

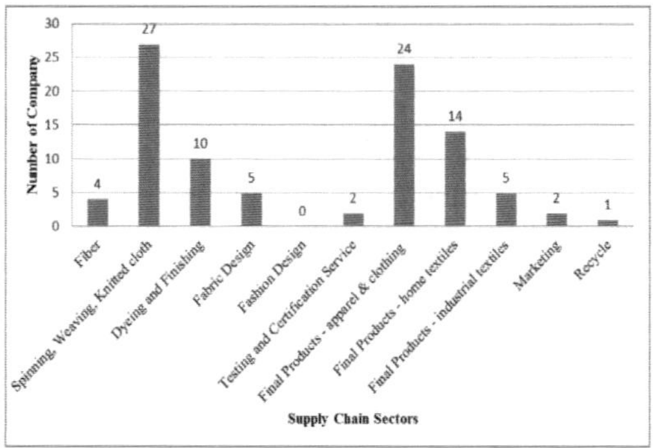

Fig.2 Supply Chain Sectors Distribution of the Funded Companies

4.2 Innovation Chains in the UK Textile Industry

Innovation plays a more and more important role in the textile industry, especially in the period of global upgradation at this moment. To examine the innovation chain in the industry, previous research works had been done through the investigation of patent and academic paper research [22]. In this section, two different aspects would be analyzed, including patent filling status and academic paper index. Both international and domestic circumstances in the UK would be investigated as for the full image of innovation chains in the UK textile industry.

4.2.1 Patent Search

Through the SCIFinder, the author selected Textile and/or Clothing as key words to search for relevant patents in English from 2000 to 2015. As illustrated in Figure 3, the top three countries that possessed most patents published in English in these 15 years were the United States, the UK and Italy. Comparing with other countries, the United States had 1366 pieces of patent relating to textile and clothing, much more than the second one the UK with 160 pieces. However, the data could not explain the whole developing status globally, since some countries might publish their patents in their own languages. For example, Japanese textile industry continued to develop its strength of advanced technology and innovation, especially in the fields of technical textiles with numbers of patents in Japanese. In addition, the author founded that the integrated strength of Chinese textile industry had been increasing due to its domestic industrial upgradation, and also performed well in textile innovation with patents published in Chinese.

The following Figure 4-5 presented the status of textile and clothing patents in the UK, which may provide us a picture of its industrial performance in textile innovation chains. In Figure 4 from the perspective of patent filling, 2002-2011 were the prosperous ten years for technology and innovation development in textile industry in the UK, and the number sharply decreased from 2011 to 2015, which was the booming period of technical textiles in the whole world [22-23]. The traditional development pattern of the UK textile industry might explain it, and the way how to upgrade its core competitive power and develop novel products should be the exact concern for the nation. From Figure 5, we could see that Unilever Plc with 52 patents listed top of organizations. The followed organization P2i Ltd and The Secretary of State for Defense only had 5 and 4 pieces respectively. Unilever Group was playing a dominant position in the UK.

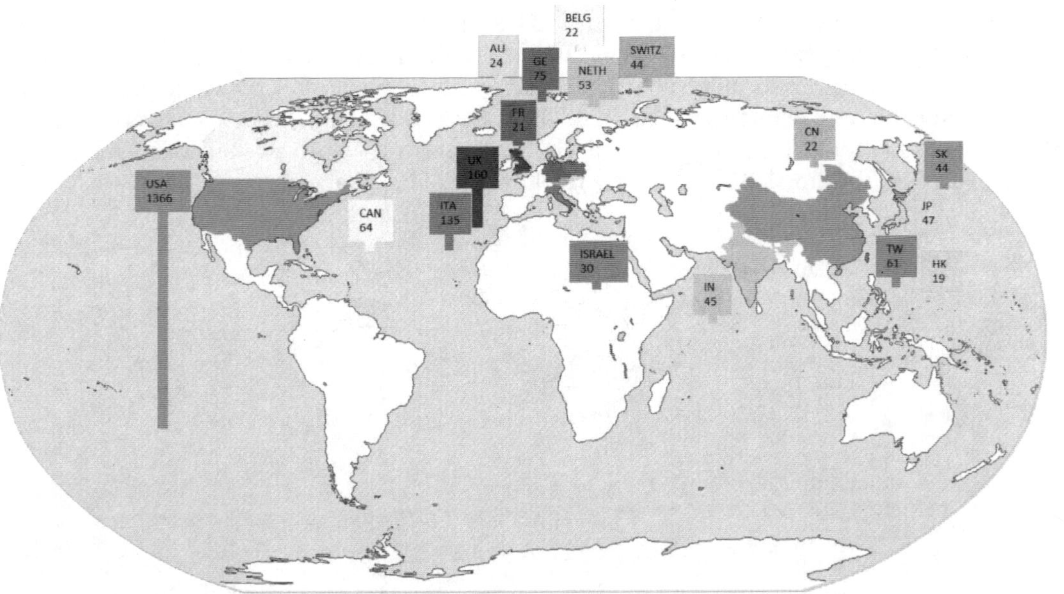

Fig.3 Geographical Distribution of Searched Patents (English, 2000-2015)

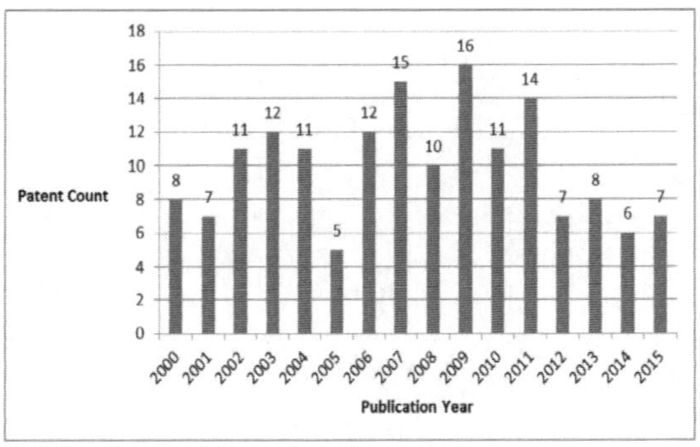

Fig.4 Patent Count by Publication Year in the UK

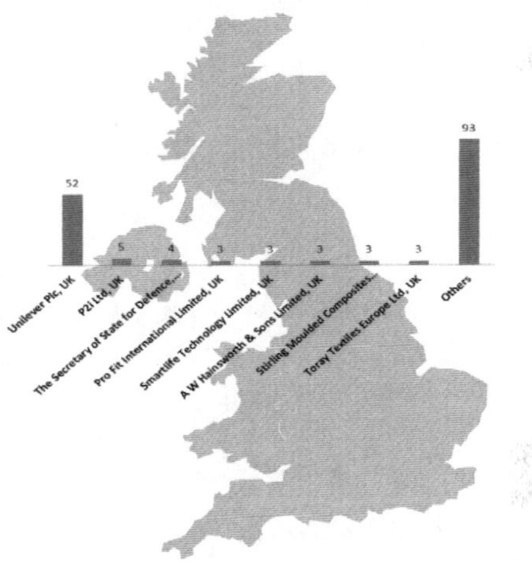

Fig.5 Patent Count by Organization in the UK

4.2.2 SCI Index Paper Search

By searching from Web of Science, SCI index papers in key words of Textile and/or Clothing from 2000 to 2015 were calculated. As presented in Figure 6, the United States possessing 7478 pieces ranked top in the world, which was followed by China with 6930 pieces. The UK, Germany and India listed from the third to fifth places, with 2843, 2575 and 2531 pieces respectively. Different from the status of English patents filling, the number of SCI papers published by Chinese organizations was relatively close to that of the United States and far more than the other countries. The reason of it might be that SCI index paper publication had been attached much importance in Chinese academic areas in recent years, and technology and innovation of textile industry developed rapidly and vigorously under the upgradation circumstances. The UK ranked the third place but only took up a little more of one third of that in the United States. India was the other developing country that performed outstandingly in this field, which might be the results of development of domestic biomedical textiles. In addition, Turkey, Japan and South Korea also had great performance of SCI paper publication.

Fig.7-9 showed the SCI paper publication status of the UK textile academia. From 2000-2014, it was increasing in publication number year by year, especially from 2012 to 2014, in which the data of 2015 was continuing to be updated. Regarding the research areas, material science and textile engineering were listed top two. The UK textile industry had its innovative competition and strength from the academia. Figure 9 indicated that University of Manchester, University of Nottingham and University of Southampton with many other universities/institutes had been publishing updated innovation in textiles to the whole world.

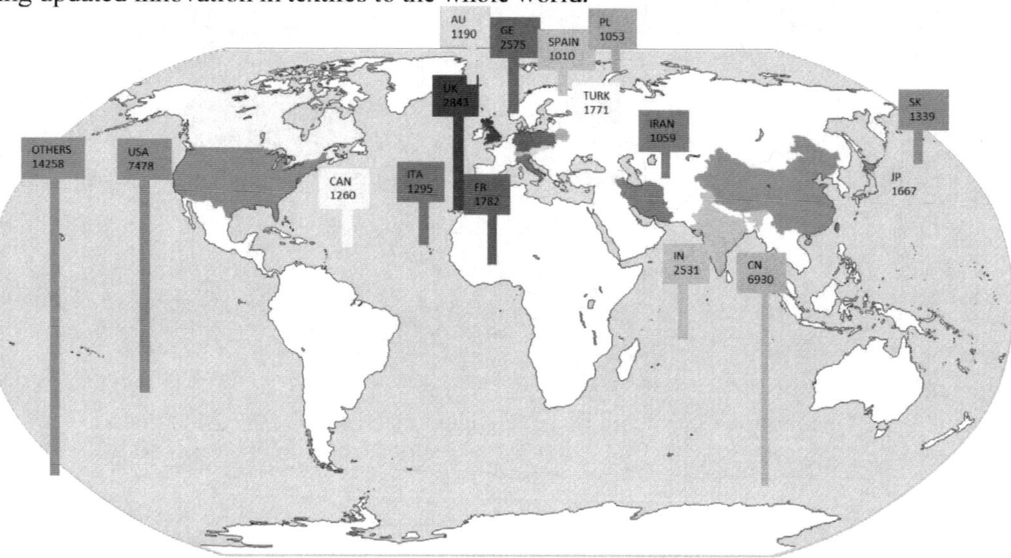

Fig.6 Geographical Distribution of Searched SCI Paper (2000-2015)

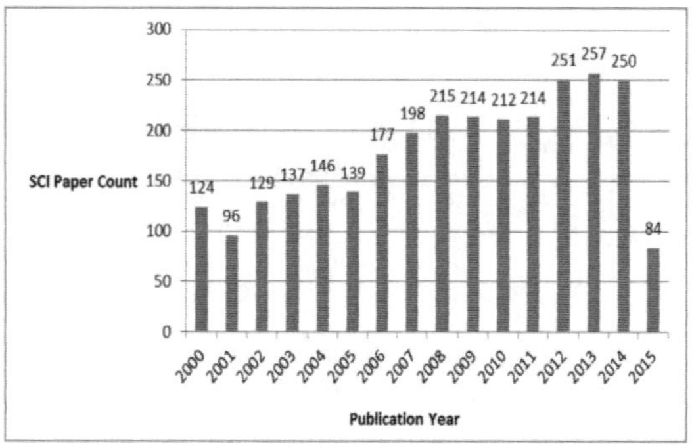

Fig.7 SCI Paper Count by Publication Year in the UK

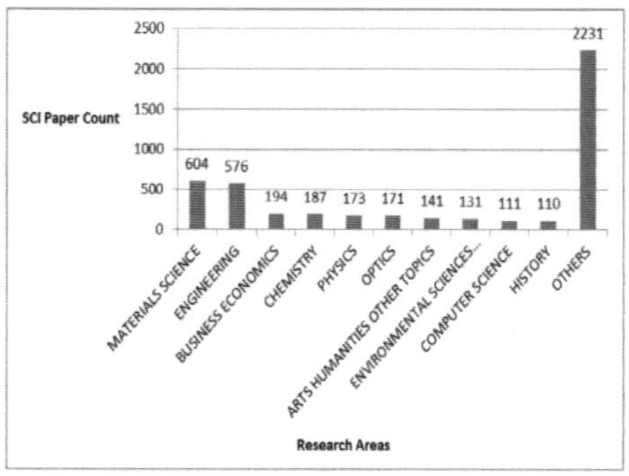

Fig.8 SCI Paper Count by Research Area in the UK

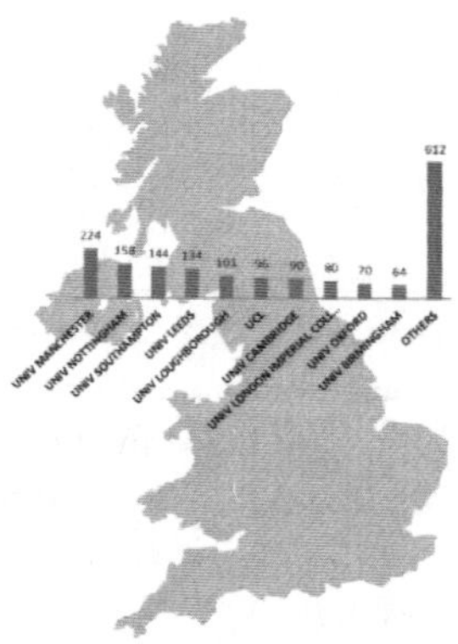

Fig.9 SCI Paper Count by Publication Organization in the UK

4.3 SWOT Analysis

Table 1. SWOT Analysis of the UK Textile Industry

Internal Factors / External Factors	STRENGTH (S)	WEAKNESS (W)
	1. Solid traditional textile and clothing industrial foundation 2. Advanced technology and innovation 3. Strong educational resources 4. High quality and reputation products 5. Famous fashion brands 6. Innovative development of performance apparels	1. Fragmented supply chains 2. Small-scale companies 3. Insufficient labor force 4. Shortage of professionals 5. Aging and weak infrastructure 6. Lack of international marketing strategies 7. Inconvenient financial transfer 8. Lack of technology platform 9. E-business
OPPORTUNITIES (O)	**SO STRATEGY**	**WO STRATEGY**
1. Rapid market expansion within the world 2. Continuous development of technology 3. Opportunities in E-commerce 4. Development of E-based manufacturing and supply chain management 5. Development of E-finance 6. Development of technical textiles 7. Increase of health care du tot global population aging	Upgrade internal strength and exploit external opportunities 1. Increase core competence of traditional techniques/processing 2. Develop functional products with high added value 3. Develop industry and university cluster innovations 4. Exploit international markets 5. Brand management and strategic marketing 6. Participate in establishment of global industrial standards	Exploit external opportunities to overcome internal weakness 1. Establish database for global cooperation in supply chains 2. Develop industrial clusters 3. Introduce new positions in novel textile industry employments 4. Introduce updated worldwide information to the industry 5. Set up public technology service platforms to integrate different sectors in the new supply chains 6. Add new-trend areas into traditional textiles & clothing educations
THREATS (T)	**ST STRATEGY**	**WT STRATEGY**
1. Traditional raw materials shortage and price increase 2. Competitions in novel materials 3. Competitions in functional and innovative design 2. Increase of labor cost 3. Carbon emission control 4. Rapid development of developing countries 5. Impacts of E-business on traditional store-shopping mode 6. Globalization	Take advantage of internal strength and solve external threats 1. Research on functional materials 2. Research on fashion and function design 3. Research on carbon footprint and develop low-carbon economy 4. Attract high level professionals 5. Promote E-Fashion business 6. Establish market entrance standards for online products	Improve internal weakness and reduce external threat impact 1. Set up the UK special style in strong sectors in industrial chains 2. Large company leading or small-scale company alliance modes in innovation research 3. Establish training bases for technicians and management talents 4. Rebuild industrial image of textiles in domestic economy 5. Upgrade domestic infrastructure 6. Renew global marketing strategies

5. Conclusion

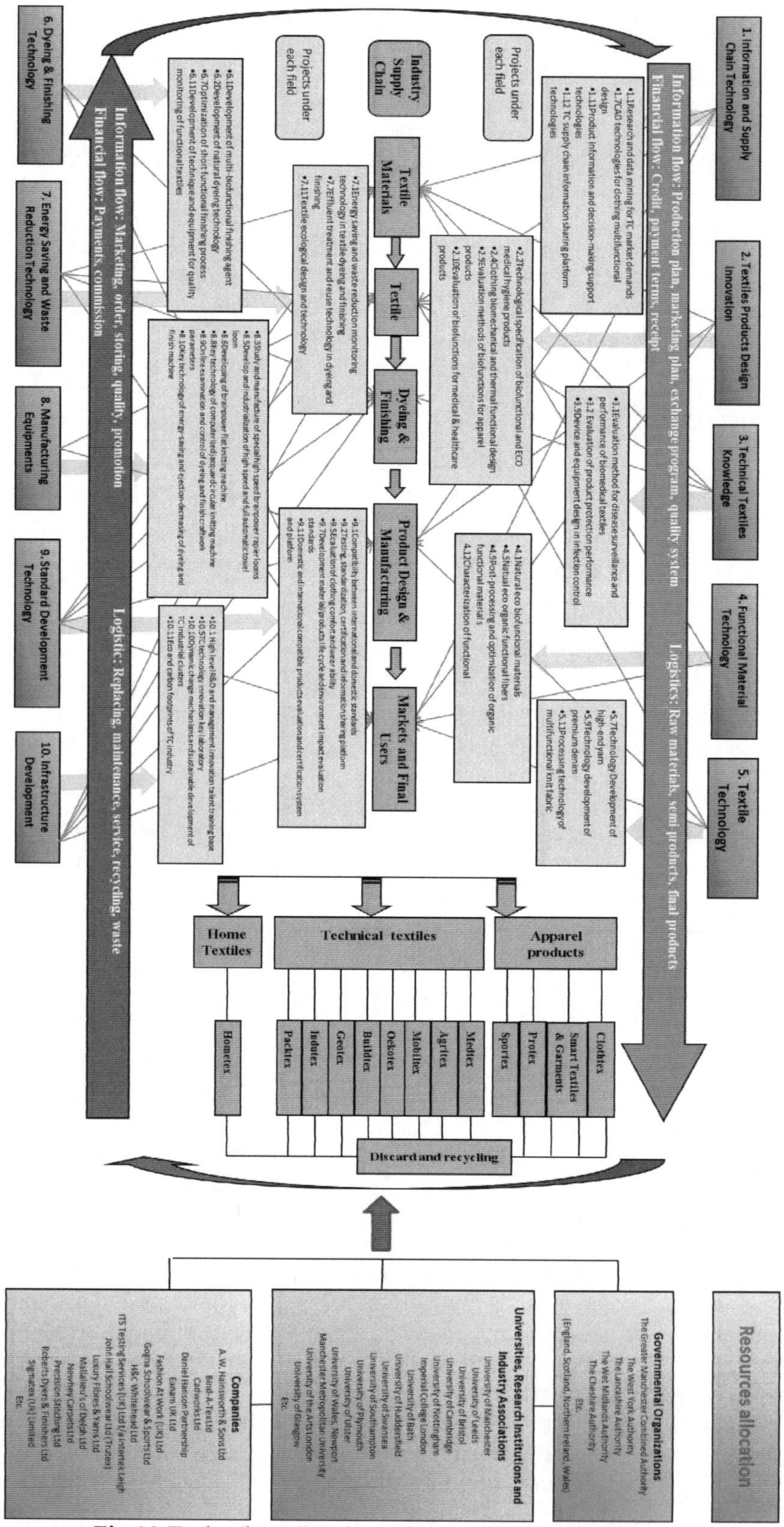

Fig.10 Technology Roadmap for the UK Textile Industry

To make a sustainable plan for the UK textile industry, it was crucial to integrate resources in different aspects. By the mean of roadmapping, a comprehensive technology roadmap had been created to provide feasible strategies for the 10 to 20 years' development. The supply chain comprised Textile Materials, Textile, Dyeing and Finishing, Product Design and Manufacturing, Markets and Final Users (Apparel Textiles, Technical Textiles and Home Textiles) and Discard and Recycling, which involved distribution and integration of resources from governmental organizations, universities/institutes and companies. Ten technical areas had been developed - Information and supply chain technology, Textile product design innovation, Technical textile knowledge, Functional material technology, textile technology, Dyeing and finishing technology, Energy saving and waste reduction technology, Manufacturing equipment, Standard development technology and Infrastructure development. A number of specific projects relating to different areas had also been proposed, whose relations with corresponding industrial sectors were also indicated. Detailed information and full planning route are illustrated in Figure 10.

6. Acknowledgement

The authors wish to acknowledge the research team of Prof. Yi LI.

References

[1] NBrown National Textiles Growth Programme. The Revival of the UK Textiles Industry. Report: 2014.
[2] NBrown National Textiles Growth Programme. Repatriation of the UK Textiles Industry. Report: 2015.
[3] Li Y, Xiong XY. Technology Roadmap of Textile and Clothing Industry-Guangdong Textile and Clothing Industry Technology Management Innovations. Beijing: China Textile Press: 2010.
[4] Garcia ML. Introduction to technology roadmapping: the semiconductor industry association's technology roadmapping process. Report SAND97-0666, Sandia National Laboratories: 1997.
[5] Phaal R, O'Sullivan E, Routley M, et al. A framework for mapping industrial emergence. Technological Forecasting and Social Change: 2011; 78(2); p: 217-230.
[6] Yasunaga Y, Watanabe M, Korenaga M. Application of technology roadmaps to governmental innovation policy for promoting technology convergence. Technological Forecasting & Social Change: 2009; 76; p: 61-79.
[7] Panetta K. Supply Chain. ECN: Electronic Component News: 2014; 58(13); p: 30.
[8] Gibson BJ, Mentzer JT, Cook RL. Supply chain management: the pursuit of a consensus definition. Journal of Business Logistics: 2005; 26; p: 17–25.
[9] Ton G, De Kok AG, Jan CF. Planning Supply Chain Operations: Definition and Comparison of Planning Concepts. In Handbooks in Operations Research and Management Science, edited by Graves SC, De KoK AG: 2003; 11; p: 1-765.
[10] Ayers JB. Handbook of supply chain management, Second Edition. New York: Auerbach Publications: 2006.
[11] Malerba F. Sectoral systems of innovation and production. Research Policy: 2002; 31(2); p: 247–264.
[12] Edquist C. Systems of innovation: perspectives and challenges. In: Fagerberg J, Mowery D, Nelson N. Oxford Handbook of Innovation. Oxford: Oxford University Press: 2004.
[13] Zeng L, Tang YL, Li CD, et al. Industry technology roadmap - exploring strategical innovation industry cultivation patterns. Beijing: Science Publications: 2014; p: 48.
[14] Li SN, Li Y, Li MW, et al. Development of Technology Roadmap for Guangdong Biomedical Textile Industry. TBIS2012, Ueda: Sep 2012.
[15] Jeffrey H, Sedgwick J, Robinson C. Technology roadmaps: an evaluation of their success in the renewable energy sector. Technological Forecasting & Social Change: 2013; 80; p: 1015-1027.
[16] Albright RE, Kappel TA. Roadmapping the corporation, Res. Technol. Manag: 2003; 46 (2); p: 31–40.
[17] Richey JM, Grinnell M. Evolution of roadmapping at Motorola. Res. Technol. Manag: 2004; 47 (2); p: 37–41.
[18] Lee SK, Mogi G, Kim JW. Energy technology roadmap for the next 10 years: the case of Korea. Energy Policy: 2009; 37 (2); p: 588-596.
[19] Ronald NK, Robert B, Gene RS. Disruptive technology roadmaps. Technological Forecasting & Social Change: 2004; 71; p: 141-159.

[20] Phall R, Farrukh C, Probert D. Tehcnology roadmapping: linking technology resources to business objectives. University of Cambridge: 2014.
[21] Nathasit G, Ronald SV, Sasawat D. Dealing with the dynamics of technology roadmapping implementation: a case study. Technological Forecasting & Social Change: 2009; 76; p: 50-60.
[22] Li SN, Li Y, Li MW. Analysis of Industrial Value Chains for Biomedical Textile Industry Technology Roadmap. TBIS2014, Hong Kong: Aug 2014.
[23] Li SN, Li Y, Li MW, at el. Development of Technology Roadmap for Guangdong Biomedical Textile Industry. TBIS2013, Xi'an: Sep, 2013.

Evaporative Cooling of Wet Fabrics used for Clothing

Uwe Reischl[1*], Kylie Pace[1], Conrad Colby[1], Ravindra Goonetilleke[2]

[1]*Boise State University, 1910 University Ave, Boise, Idaho 83725, USA*
[2] *The Hong Kong University of Science and Technology, Clear Water Bay, Hong Kong*

*Corresponding author's email: ureischl@boisestate.edu

Abstract

Protective clothing is most often used to provide safety and comfort for persons exposed to hot or cold environments. To assess the impact of clothing moisture and wetness on cooling, a series of wind-tunnel tests was conducted to quantify the evaporative cooling capacity of selected fabric samples. Single-layer cotton, polyester, nylon and silk were evaluated. The results showed that the onset and magnitude of evaporative cooling is determined by the amount of water contained in the fabric. The results also showed that the "skin" exhibited more cooling when covered with a fabric than when it was not. The information gained can assist in the selection of garment materials to optimize comfort and safety.

Keywords: Wet Clothing; Evaporative Cooling; Comfort; Safety

1. Background

Evaporation of water (sweat) from the skin and clothing provides a powerful physiological cooling mechanism for persons exposed to hot environments. However, clothing can complicate the evaporation process by creating a barrier to metabolic heat loss and generating a humid microclimate that may reduce the overall cooling effects of sweat evaporation [1]. In general, measuring the evaporation efficiency in clothing systems is complex because sweat may evaporate on the skin or wick into the clothing where evaporation takes place on the surface of the clothing. Additionally, excessive sweat may roll off the skin or remain trapped in the clothing. It is a common assumption that sweat vaporized on the skin surface delivers better cooling than when sweat evaporates from clothing because the location of the phase change is removed from the skin and much of the heat is then extracted from the environment [2]. Assessment of cooling by sweat evaporation for clothed persons has been carried out in the past on human subjects and thermal manikins [3-6]. However, the results have been difficult to incorporate into new clothing design since the outcomes are confounded by heat loss through latent heat of vaporization which is always difficult to quantify [7]. Cooling dynamics of wet clothing, independent of body metabolic heat loss, and the associated latent heat of evaporation have not been investigated separately. To address this topic, a test protocol was developed to measure the direct temperature drop created by evaporative cooling that is not influenced by metabolic heat production and sweating.

2. Methods and Procedures

Wind-tunnel tests were carried out on cotton, polyester, nylon and silk fabric samples to determine the cooling generated by water evaporation from the fabric samples under controlled temperature, humidity, and air velocity conditions. A "control" configuration that did not include a fabric sample was used as a reference. The temperature drop due to water evaporation was measured using a thermocouple temperature probe imbedded in the surface of the fabric sample mounting platform. All tests were performed three times and the data averaged.

2.1 Wind Tunnel

A negative pressure laminar air flow wind-tunnel was used in this study. The wind-tunnel measured 2.3 meters in length, 40.5 cm in height and 30.5 cm in width. Fabric samples were placed onto a mounting platform which was positioned in the center of the wind-tunnel 1.2 meters from the inlet. The air velocity was maintained at 1.5 m/sec. Air temperature was maintained at 22^{0}C (±2^{0}C) and relative humidity maintained at 15% (±5%). A side view of the wind tunnel is illustrated in Figure 1.

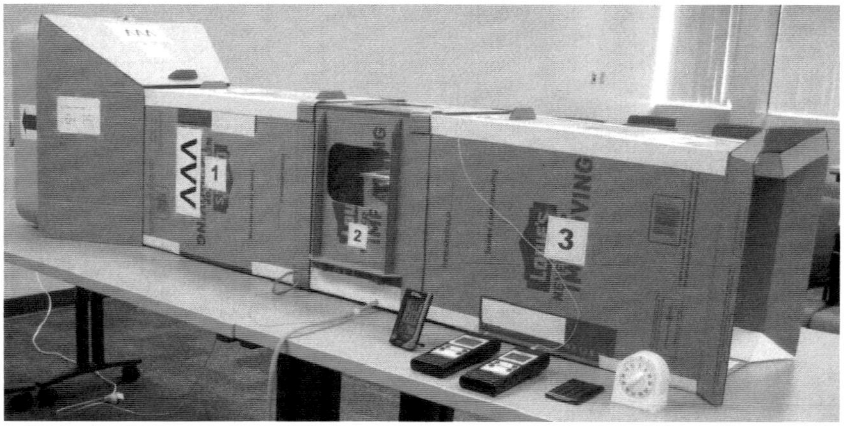

Fig.1 Illustration of wind-tunnel used for the assessment of evaporative cooling of fabric samples

2.2 Mounting Platform

Fabric samples were placed onto a convex surface which allowed a fabric sample to lay smoothly on the platform under its own weight. The platform shape provided for a gravity run-off of excess water. The device was made of Styrofoam which was impermeable to water. Each fabric sample was secured to the platform by four pins to prevent displacement of the sample during exposure to air flow. The fabric mounting platform is illustrated in Figure 2.

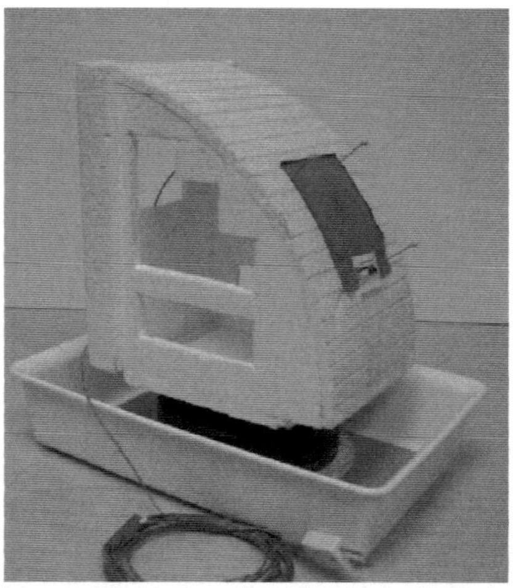

Fig.2 Illustration of the mounting platform which was placed inside the wind-tunnel. The platform was not heated and the fabric sample irrigation water was maintained at ambient air temperature. The platform was oriented towards the air flow

2.3 Fabric Samples

Four types of fabric materials were tested including 100% cotton, 100% polyester, 100% nylon, and 100% silk. Each sample dimension was 5.0 cm wide and 6.0 cm long.

2.4 Test Protocol

The test protocol included the following:

- Evaluation of a "control" configuration that established the magnitude of evaporative cooling for the mounting -platform alone, without a fabric cover.
- Evaluation of the cotton sample for a saturated condition and for a "moist" condition. The saturated condition was created by applying excessive water spray to the sample for a 10-second period and until the irrigation water dripped freely from the mounting platform. The non-saturated (moist) condition was created by compressing the saturated cotton sample until all excess water was removed from the fabric.
- The moist condition and the saturated fabric samples including the cotton, polyester, nylon and silk samples where each tested separately three times using a randomized selection protocol.

10 seconds prior to each test, the fabric sample was sprayed with irrigation water which was maintained at room-temperature. The water was applied until the fabric sample was saturated and the water was observed to drip freely from the mounting platform. The fabric sample tested for the "moist" condition was first saturated and then pressed to remove all excess water from the fabric. No water was added at to any sample at a later time. The temperature drop created by each sample was measured for 15 minutes. The temperature was recorded at 0.5 minutes, 1.0 minute, 1.5 minutes, 2 minutes, 5 minutes, 10 minutes and 15 minutes.

3. Results

3.1 Control

Table 1 summarizes the evaporative cooling observed for the "control" condition and the saturated cotton fabric configurations. The temperature drop seen for the control condition was 5.1^{0}C for the first five minutes and then declined to 4.4^{0}C afterwards. However, the temperature drop observed for the cotton sample was 8.6 ^{0}C during the first 10 minutes but showed no decline afterwards.

Table 1. Data summary of evaporative cooling observed for the cotton sample and the temperature drop observed for the "control" configuration

Exposure Time (Minutes)	"Control" Configuration (^{0}C)	Cotton Sample (^{0}C)
0	0	0
0.5	3.4	4.2
1.0	4.4	6.2
1.5	4.7	7.3
2.0	4.8	7.9
5.0	5.1	8.5
10.0	4.5	8.6
15.0	4.4	8.6

3.2 Water Content

Table 2 summarizes the evaporative cooling for the "moist" and the "saturated" cotton samples. During the first 1.5 minutes of exposure, cooling by the "moist" sample was substantially greater for the cooling observed for the "saturated" sample. However, after two minutes of exposure, the "saturated" cotton sample exhibited substantially greater cooling than the "moist" cotton sample.

Table 2. Summary of evaporative cooling observed for the "moist" and "saturated" conditions

Exposure Time (Minutes)	Cotton Sample (Moist) (^{0}C)	Cotton Sample (Saturated) (^{0}C)
0	0	0
0.5	6.7	5.7
1.0	7.8	7.4
1.5	8.3	8.2
2.0	8.4	8.5
5.0	8.5	8.8
10.0	8.5	8.8
15.0	8.5	8.8

3.3 Fabric Materials

Table 3 summarizes the evaporative cooling for the cotton, polyester, nylon and silk saturated fabric samples. During the first 1.5 minutes the temperature drop measured was greater for the Nylon and silk samples in comparison to the temperature drop measured for the cotton and polyester samples. However, this relationship was reversed after five minutes when the cotton and polyester cooling was observed to be greater than that for the nylon and silk samples.

Table 3. Data summary of the evaporative cooling observed for the saturated cotton sample, polyester sample, nylon sample, and the silk sample

Exposure Time (Minutes)	Cotton Sample (^{0}C)	Polyester Sample (^{0}C)	Nylon Sample (^{0}C)	Silk Sample (^{0}C)
1.5	8.5	8.3	9.1	9.2
2.0	9.1	8.9	9.2	9.3
5.0	9.6	9.5	9.3	9.4
10.0	9.7	9.6	9.2	9.4
15.0	9.7	9.6	9.0	9.4

4. Analysis

Comparing the temperature drop of the cotton sample to the control condition suggests that the improved cooling for the saturated cotton fabric sample was due to the water in the fabric sample. Since the testing platform was made of non-absorbing material, water applied to the platform flowed off the surface freely with only a surface layer of water remaining as a result of surface tension. This provided only a limited availability of water for evaporation. As shown in Fig.3, the "control" cooling decreased over time while the cotton fabric sample exhibited no such decrease. This difference in cooling suggests that the total amount of water retained by the fabric sample determines the overall cooling capacity of the fabric sample.

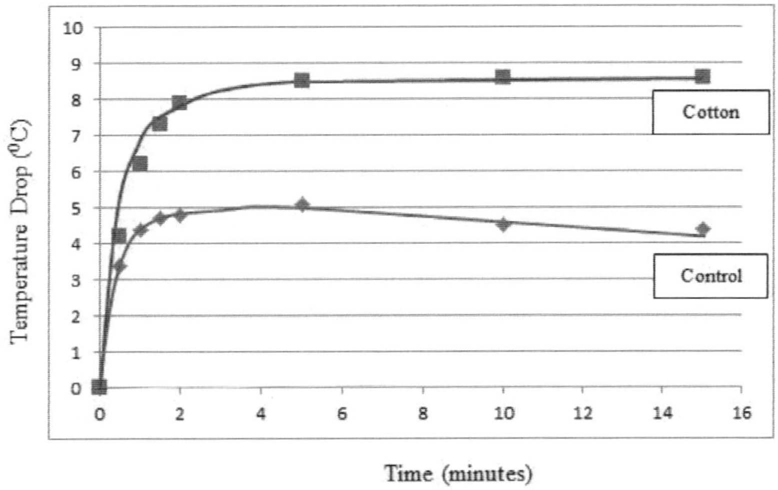

Fig.3 Illustration of the evaporative cooling observed for the saturated cotton sample and the evaporative cooling of the "control" configuration. The platform covered by the cotton fabric sample exhibited greater cooling than when the platform was not covered by the fabric sample

The relationship between water retention in a fabric sample and the cooling capacity of the sample is illustrated in Fig.4. It can be seen that the saturated (high water content) cotton sample exhibited greater cooling overall than the moist (low water content) sample. However, the moist sample performed much better during the first 1.5 minutes while the saturated sample performed better afterwards. The dynamics of the relative cooling advantage offered over time by each of the water levels is illustrated in Fig.5. This suggests that the maximum cooling does not always occur when the water retained in a sample is high

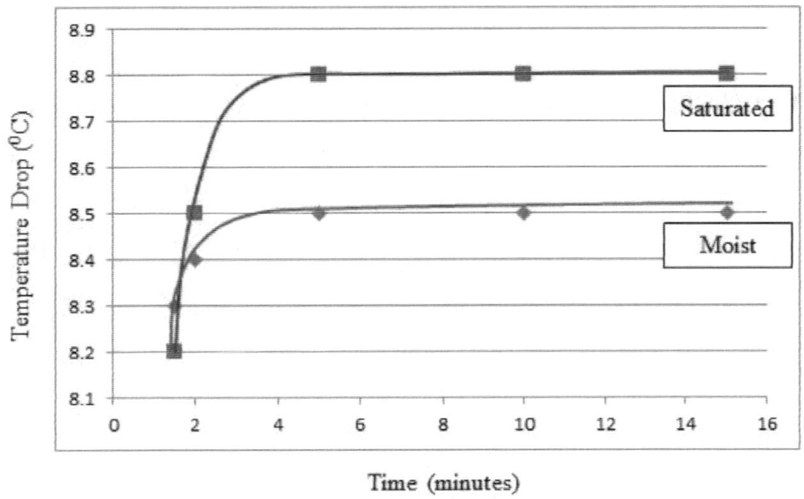

Fig.4 Illustration of the evaporative cooling observed for the saturated and the moist cotton sample. The moist sample generated greater cooling during the first 1.5 minutes while the saturated sample generated greater cooling afterwards

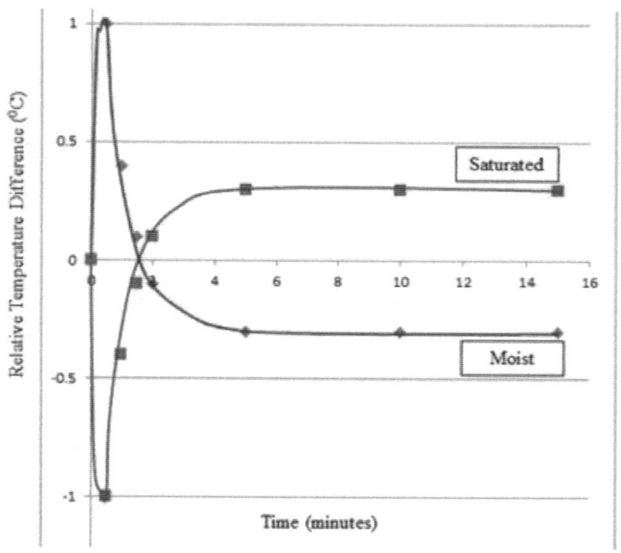

Fig.5 Illustration of the relative evaporative cooling advantage offered by a moist sample and a saturated sample during a 15-minute exposure period. The moist sample provided an immediate cooling advantage (red curve) while the saturated sample provided a cooling advantage later (blue curve)

A comparison of the cooling dynamics exhibited by the cotton, polyester, nylon, and silk samples is illustrated in Fig.6. It can be seen that the cotton and polyester samples offered greater evaporative cooling overall than the silk and nylon samples. However, the nylon and silk samples initially provided greater evaporative cooling. This finding again supports the idea illustrated in Fig.4 and Fig.5 where the moist sample exhibited greater initial evaporative cooling than the saturated sample.

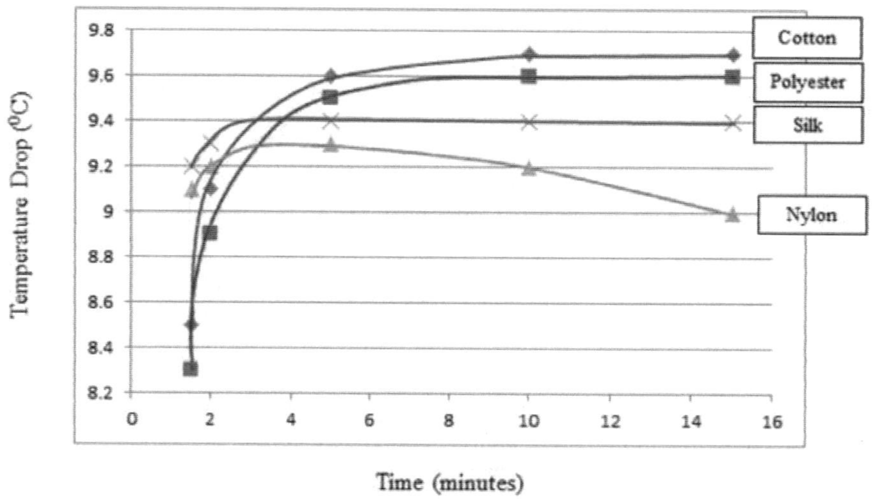

Fig.6 Illustration of the evaporative cooling observed for the cotton, polyester, silk and nylon fabric samples. The silk and nylon samples created greater cooling during the first 1.5 minutes while the cotton and polyester samples created greater cooling afterwards

5. Conclusions

Based on the findings obtained in this study, the following conclusions can be reached:

- Wet clothing can create cooling that is significantly greater than the cooling generated by water evaporating directly from the skin.
- The level of cooling created by wet clothing over time is determined by the water content in the fabric. The greater the amount of water, the longer the cooling effect over time.
- Initial cooling by wet fabrics is greatest when a fabric contains a minimal amount of water (moist) while cooling is optimal over a longer period of time when the fabric contains high levels of water (saturated).

The results explain why persons report improved comfort in hot environments when wearing single-layered clothing rather than no shirts or blouses. The findings also explain why clothing materials that absorb low amounts of water such as nylon and silk are often preferred over materials that absorb larger amounts of water such as cotton. Application of these findings can help in the future selection and development of clothing systems that can provide both immediate and long-term comfort and safety for wearers in hot and cold environments.

References

[1] Nunnaly S. Heat Stress in Protective Clothing, Scan. J. Work Environ Health 1989;15(suppl 1):52-57.

[2] Craig FN, Moffit JJ. Efficiency of Evaportive Cooling from Wet Clothing. J Appl Physiol 1974;36:313-316.

[3] Lotens WA, Havenith G, Van de Linde FJG. Effects of Condensation in Clothing on Heat Transfer. Ergonomics 1995;38:1114-1131.

[4] Lotens WA, Havenith G. Effects of Moisture Absorption in Clothing on the Human Heat-Balance, Ergonomics 1995;38:1095-1113.

[5] McLellan TM, Pope JL, Cain JB, Cheung SS. Effects of Metabolic Rate and ambient Vapor Pressure on Heat Strain in Protective Clothing, Eur J Appl Physiol1996;74:518-527.

[6] Mitchell D, Wyndham CH, Atkins AR, Vermeulen AJ, Hofmeyer HS, Strydom NB, Hodgson T. Direct Measurement of Thermal Responses of Nude Men Resting in Dry Environments. Arch Ges Physiol 1968;74:324-343.

[7] Havenith G, Brode P. Den Hartog, E, Kuklane, K., Homer, I, Rossi RM, Richards, M, Farnworth, B, and Wang, X. Evaporative Cooling: Effective Latent Heat of Evaporation in Relation to Evaporation Distance from the Skin, J Appl Physiol 2013;114:778-785.

Investigation on the Performance of Four Cooling Systems under Heatwave Conditions through a Thermal Manikin

Cheng-Jiao Zhang[1, 2], Peng Zheng[1, 2], Wen-Fang Song[1, 2], Fa-Ming Wang[1, 2*]

[1]*Laboratory for Clothing Physiology and Ergonomics (LCPE), National Engineering Laboratories for Modern Silk, Soochow University, 199 Ren'ai Road, Suzhou, 215123, China*
[2]*Department of Apparel Design and Engineering, Soochow University, Suzhou, 215021, China*

[*]Corresponding author's email: dr.famingwang@gmail.com

Abstract

The objective of current study was to investigate the effectiveness of four personal cooling systems (i.e., electric fan [FAN], evaporative cooling vest [ECG], liquid cooling vest [LCG] and using evaporative cooling vest together with electric fan [ECV&FAN] in mitigating heat strain under heatwave conditions by a thermal manikin. The thermoregulatory model control mode was selected, and a metabolic rate of 1.2 METs was used to simulate a person resting quietly. Two testing environments, i.e., 36 °C and 33% RH (hot-dry), 36 °C and 67% RH (hot-humid), were used. It was found that LCG and ECG&FAN could provide better cooling than the ECG and FAN in the two studied testing environments. The ECG was more effective in mitigating body heat strain compared with the FAN in the hot-dry environment, whereas the FAN was superior to the ECG in removing body heat in the hot-humid environment. Thus it was recommended that both FCG and ECG&FAN could be used to alleviate body heat strain during heateaves.

Keywords: Personal cooling system; Electric fan; Evaporative cooling garment; Liquid cooling garment; Heat strain; Heatwave

1. Introduction

Severe heatwaves have become a more frequent occurrence worldwide in the last few decades [1]. Heat stress caused by hot conditions has been widely recognized as an underlying cause for many work incidents and heat-related injuries. This is especially true for occupants involving activities under high radiation or hot environments [2-5]. In hot environments, heat stress resulted from heat accumulation could cause failure of human thermoregulatory system, eventually resulting in heat-related illnesses such as heat cramps, heat syncope, or even fatal heat stroke [1,4]. An average of 658 heat-related deaths was reported annually in the USA between 1999 and 2009 [4]. Generally, human dissipates heat to its surrounding environment via dry (i.e., convection) and latent (i.e., evaporation) avenues of heat transfer. However, when the ambient temperature is higher than that of body surface, the heat exchange from the body to the surrounding environment is negative [5]. Thus the physiological defence against heat stress depends on the evaporation of perspiration [1,5]. Unfortunately, clothing confines the evaporation of perspiration, and leads to heat accumulation in the body, raises core temperature and induces heat strain [2]. To reduce these potential risks caused by the hostile hot environments, thermal management systems are required to manage the heat strain in human body.

During the last few decades, many technologies in protective fabrics, clothing systems and cooling systems have been developed to mitigate the body heat strain, and they could be divided into two categories: macroenvironment cooling systems and personal cooling system (PCSs) [5]. Macroenvironment cooling is achieved by the cooling of the interior space or a room using mainly air conditioning or ventilation fans. In contrast, microenvironment cooling is normally designed to cool the wearer by wearing personal cooling system, and thus more practical, energy saving compared to the macroenvironment cooling systems [5]. During the last decades, various types of PCSs have been developed and they could be classified into four basic categories according to their cooling mechanism: fluid/gas cooling garments (FCGs), evaporative cooling garments (ECGs), phase change material garments (PCMs) and ventilation cooling garments (VCGs) [5,6]. FCGs are usually constructed by embedding fine tubes into clothing and a refrigerated coolant tank is required for cooling purpose. The coolant could be fluid, gas or non-toxic aqueous solutions. The coolant circulates through the inside of tubes and removes body heat away via conduction and convection [5,6]. ECGs harness the large latent heat of moisture evaporation to achieve the cooling effect. The substantial latent heat of water evaporation promises the effectiveness of ECGs using a small amount of water [6]. For PCMs, heat is drawn away from the body during the phase change from solid to liquid state [5-8]. For VCGs, the cooling effect is obtained by the increased heat transfer via convection, as well as the heat transfer by promoted moisture evaporation [6].

The current study investigated the actual cooling performance of different cooling systems (i.e. FCG, ECG, FAN) under both hot dry (HD) and hot humid (HH) environments by means of a thermal manikin. Besides, the cooling effect of a hybrid cooling system (i.e., ECG&FAN, combined by the ECG and FAN) was examined and compared. In addition, the effect of relative humidity on the cooling performance was evaluated. It was believed that the findings in this study could provide customers useful suggestions on how to choose cooling systems based on intended exposure environments.

2. Methodology

2.1 Thermal manikin

A 34-zone 'Newton' type thermal manikin (Thermetrics LLC, Seattle, WA, USA) was used in this study. The thermoregulatory model control mode was selected to investigate the actual cooling performance of these four cooling systems. This model program was developed by ThermoAnalytics Inc. (Calumet, MI) [9], and the manikin surface temperature and sweating rate were controlled in analogy to the prediction of the Fiala multi-segmental thermophysiological model [10]. Prior to using the thermoregulatory model control mode, the manikin was preheated to a defined thermoneutral condition. The segmental temperatures of the manikin under the thermoneutral condition are presented in Table 1. The thermophysiological parameters (i.e., the mean skin and hypothalamic temperatures) were recorded by the ThermDAC® (version 8.2.26.0). The prediction of subjective perceptions was based on the UC Berkeley Comfort Model [11].

Table 1. the segmental surface temperatures of the manikin under defined thermoneutral condition

Segments	Temperature (°C)	Segments	Temperature (°C)
Face	35.50	Waist	34.49
Head	35.24	Low back	34.70
Right upper forearm-front	33.48	Right pelvis front	34.25
Right upper forearm-back	33.48	Right pelvis back	34.81
Left upper forearm-front	33.32	Left pelvis front	33.91
Left upper forearm-back	33.32	Left pelvis back	35.00
Right upper arm-front	33.27	Right thigh front	33.86
Right upper arm-back	33.27	Right thigh back	33.95
Left upper arm-front	33.33	Left thigh front	33.79
Left upper arm-back	33.33	Left thigh back	34.05
Right hand	34.69	Right shin	33.78
Left hand	34.58	Right calf	33.66
Upper chest	34.79	Left shin	34.10
Shoulders	34.85	Left calf	33.49
Stomach	34.48	Right foot	30.73
Mid back	34.70	Left foot	30.77

2.2 Cooling systems

A set of clothing ensemble (included a short-sleeved polyester shirt, cotton briefs and shorts, the thermal insulation: 0.25 clo) was selected as the basic clothing (i.e., CON), and was used in all tests. Four different cooling systems were selected (shown in Fig.1) and their characteristics are briefly described as below:

Fluid cooling garment (FCG). The FCG used in this study was an ice-chilled cooling system (COMP-IBPO2-B, Compcooler Technology Co., Shenzhen, China). It was comprised of a cooling vest and a backpack (in Fig.1). The cooling vest was constructed by incorporating the zigzag flexible micro tubes into two elastic mesh polyester fabrics. The backpack was used to locate the coolant reservoir (i.e., a sealed bag filled with 1500 ml ice bottle and water), micro pump and battery. On application, the coolant reservoir was connected to the inlet and outlet of the tubes, and the cold water was pumped from the coolant reservoir and circulated in the tubes. The vest should be in close contact with the human body by adjusting the straps to size of the up torso. The total weight of the FCG used was 7.0 kg and the maximum output power of the pump was 15 W.

Evaporative cooling garments (ECG). The evaporative cooling vest (HyperkewlTM, TechNiche, Vista, California, USA) used in this study was composed of a three-layer structure main fabric and two elastic mesh fabric along the body sides. The three-layer main fabric was comprised of a nylon outer-layer, a water repellent nylon inner-layer, and a water storing layer (i.e., a kind of flocculent materials) filled in between. The ECG was prepared according to the following steps before use: soaked it in tap water (around 23±2 °C) for 2 min, gently squeezed it to remove the excess water (activated). The total weight of the activated ECG was 0.88 kg.

Electric fan (FAN). In this study, an electric fan (SF-45MV-1V, Suiden, Tokyo, Japan) was placed at the front of the manikin at a distance of 1.0 m. A mean constant air velocity of 1.0 m/s was generated. The air velocity was measured by a hot wire anemometer (Fluke 923, Fluke, Washington, America) placed at the manikin's chest region.

Hybrid cooling system (ECG&FAN). A hybrid cooling system was developed by combining the abovementioned ECG and FAN.

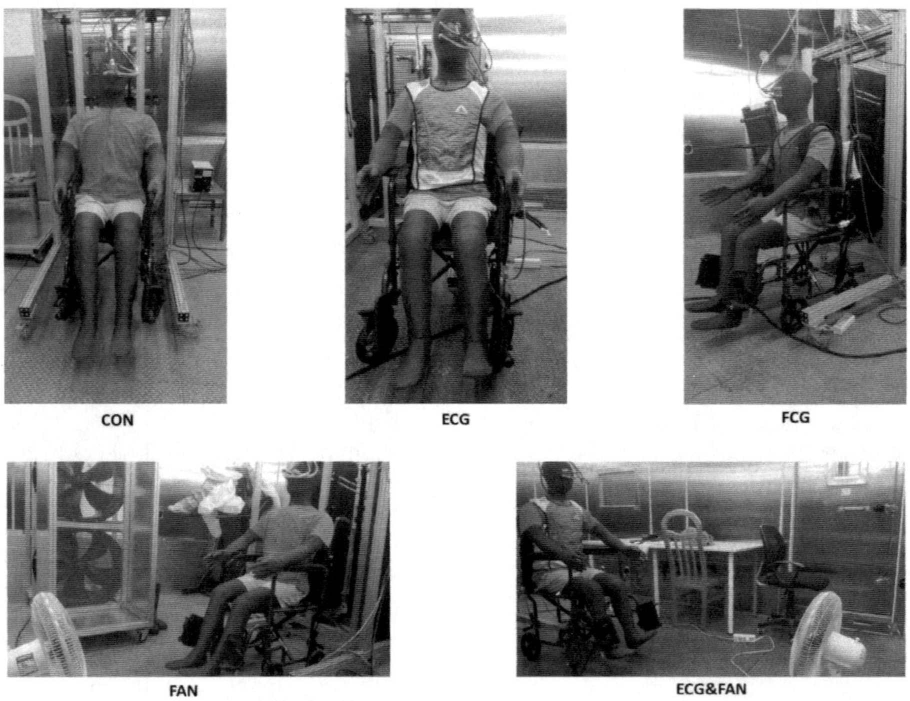

Fig.1 Test garment systems: CON, ECG, FCG, FAN, ECG&FAN.

2.3 Test conditions

All tests were performed in a climate chamber. Two environmental conditions were selected, i.e., hot-dry environment (HD, 36±0.5 °C and RH = 33±5%, the partial water vapour pressure was 2000 Pa) and hot-humid environment (HH, 36±0.5 °C and RH = 67±5%, the partial water vapour pressure was 4000 Pa). A constant air velocity of 0.15±0.1 m/s was applied to all tests except the FAN scenario. A metabolic rate of 1.2 METs was used to simulate a person resting quietly.

2.4 Test protocol

During preparation, the manikin was placed outside the chamber and dressed with the basic ensemble, (i.e., a short-sleeved shirt, cotton briefs and shorts), and then was preheated to the defined thermoneutral condition (i.e., the segmental surface temperatures listed in Table 1) at a room temperature of 20±2 °C. It was then transferred into the climate chamber with the pre-set environmental condition, and randomly assigned to a cooling system. The thermoregulatory model control mode was activated, and the metabolic rate was set to 1.2 METs. The testing duration was 60 min, and the hypothalamic temperature (T_{hy}) and the mean skin temperature (T_{sk}) were recorded throughout the 60 min test with a constant interval of 1 min. A total of 30 tests were performed, (i.e., five testing scenarios [i.e., four cooling systems and one CON] × two environments × three times).

2.5 Statistical analysis

All data were presented as mean ± standard deviation. A mixed model two-factor ANOVA was performed to examine whether there were significant differences in the mean skin and hypothalamic temperatures between the four cooling systems (i.e., ECG, FCG, FAN and ECG&FAN) and CON. If there were significant differences, paired-sample *t*-tests were conducted to ascertain the significant effect between scenarios at different time points separated by 5 min interval. All statistical analysis was conducted by the SPSS v.20.0 (IBM Inc., Armonk, NY) and the alpha significance level, *p*, was set to 0.05.

3. Results and discussion

3.1 Cooling effect of FCG

Time course changes in the mean skin (T_{sk}) and hypothalamic temperatures (T_{hy}) are presented in Fig.2. It could be clearly seen that the FCG significantly mitigated the increase in T_{sk} and T_{hy} in both two testing environments. Significant lower T_{sk} and T_{hy} were detected from 35th min to the end of the test and from the 45th min till the end of the test in the HD (i.e., 36 °C and 33% RH), and from 20th min to the end of the test and from the 30th min till the end of the test in the HH (i.e., 36 °C and 67% RH), respectively ($p<0.05$). FCG presented significantly lower T_{sk} and T_{hy} than FAN (i.e., during the 20-60th min for T_{sk} and the 45-60th min for T_{hy}. $p<0.05$), and ECG (i.e., from 30th min to the end of the test and from the 50th min to the end of the test, respectively. $p<0.05$) in HH. Thus it could be concluded that the FCG was more effective in reducing heat strain than ECG and FAN in both two environment, which was in good agreement with the statement of Yang et al. [6], who described that the FCG could bring a large cooling effect and has been recognized as one of the most effective personal cooling strategies [6].

3.2 Cooling effect of FAN

It was found that the FAN had almost no effect on the mitigation of heat strain in HD. This was in line with the suggestion of the U.S. Environmental Protection Agency (EPA), that the highest air temperature limits for air fans operation were between 35.6-37.2 °C [12]. In contrast, significantly lower T_{sk} and T_{hy} values were observed in FAN compared with those in CON in HH, i.e., from 40th min to the end of the test and from the 45th min to the end of the test, respectively ($p<0.05$). The difference in the finding between the two environments might be due to the cooling mechanism of the fan. The cooling effect of the fan mainly depended on the convection and evaporation induced by the forced-air movement. When the ambient temperature was higher than the body surface temperature, the air movement caused by the fan would cause negative convective heat flown into the body [13]. In the HH, the reduction in body heat strain may be an indicator that the amount of heat loss induced by the fan due to moisture evaporation exceeded the convective heat gain [13].

3.3 Cooling effect of ECG

Significant lower T_{sk} and T_{hy} values were detected in ECG compared with those in CON in HD, i.e., from the 40th min to the end of the test and from the 45th min to the end of the test, respectively ($p<0.05$). However, no cooling effect was detected in ECG compared with CON in the HH. This decrement in cooling effect was attributed to the narrowed water vapour pressure gradient between the body surface and the ambient. For evaporative cooling garment, the cooling effect was obtained via heat absorbed from the human body surface during water evaporation, and water evaporation was greatly depended on the water vapour pressure gradient between the body surface and the ambient [13]. The moisture evaporation was facilitated by the large water vapour pressure gradient between the body surface and the ambient in HD, whilst it was greatly restricted in the HH [13].

3.4 Cooling effect of ECG&FAN

ECG&FAN showed a positive effect on the reduction of heat strain in both two testing environments compared with CON. In HD, ECG&FAN displayed significant lower T_{sk} and T_{hy} from the 40th min till the end of the test compared to CON ($p<0.05$). Moreover, the T_{hy} was significantly lower in ECG&FAN from the 45th min till the end of the test compared with that in ECG ($p<0.05$). Significant lower T_{sk} and T_{hy} in ECG&FAN in HH compared to CON were detected from 25th min to the end of the test and from the 30th min to the end of the test, respectively ($p<0.05$). In addition, the differences between ECG&FAN and FAN or ECG were effective. It was also discovered that ECG&FAN induced significantly lower T_{sk} and T_{hy} compared with FAN (i.e., from the 25th min to the end of the test and at the 60th min of the test, respectively) and ECG (i.e., from the 35th min to the end of the test and from the 50th min to the end of the test, respectively) ($p<0.05$). Hence, it could be concluded that the ECG&FAN had a better cooling performance compared to ECG and FAN. The cooling benefit induced by ECG&FAN may be an indicator that the evaporative heat loss promoted by the hybrid cooling system was much greater than the enhanced convective cooling heat gain. This finding was consistent with the observation of Chan et al., who described that the heat transfer by evaporation promoted by the electric fan could provide more cooling benefit overweighed heat transfer due to enhanced convection in hot environments [14].

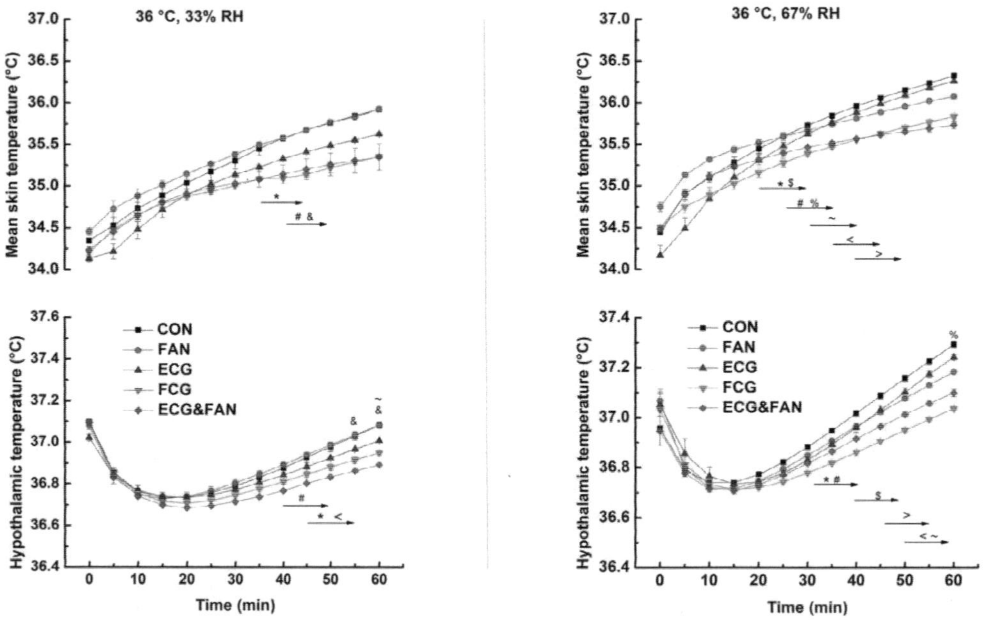

Fig.2 Time course changes in the mean skin and hypothalamic temperatures under HD and HH environments. *, significances between CON and FCG; #, significances between CON and ECG&FAN; $, significances between FAN and FCG; %, significances between FAN and ECG&FAN; &, significances between CON and ECG; ~, significances between ECG and FCG; <, significances between ECG and ECG&FAN; >, significances between CON and FAN.

Finally, our study has clearly demonstrated that the FCG and ECG&FAN brought the best cooling effect among the four cooling systems in both two studied environments. The cooling effectiveness of ECG and FAN differed in different environments. We must acknowledge that the investigation and comparison were only confined to the thermal manikin, which was a limitation of this study. Besides, only one level of metabolic rate was simulated. Therefore, further studies focusing on human trials with different labour intensities should be conducted in order to get a comprehensive understanding of the actual effectiveness of different cooling systems.

4. Conclusions

In this study, the effectiveness of four cooling systems (i.e., the electric fan, the evaporative cooling garment, fluid cooling garment, and cooling system combined by the evaporative cooling vest and the electric fan) in mitigating the heat strain was investigated and compared in HD and HH environments by means of a thermal manikin. The fluid cooling ferment and the combined cooling system were found to have the best cooling effect in both two studied environments. It was also found that the evaporative cooling garment was superior to the electric fan in the reduction of body heat strain in HD environment. In contrast, FAN showed better cooling performance in HH environment compared to ECG and thus ECG is more suitable for applications in HD environments, whereas FAN is recommended for applications in HH environments.

References

[1] Li M, Gu S, Bi P, Yang J, Liu Q. Heat waves and morbidity: current knowledge and further direction-a comprehensive literature review. Int J Environ Res Public Health 2015; 12: 1231-1250.

[2] Elson J, Eckels S. An objective method for screening and selecting personal cooling systems based on cooling properties. Appl Ergon 2015; 48: 33-41.

[3] Guo T, Shang B, Duan B, Luo X. Design and testing of a liquid cooled garment for hot environments. J Therm Biol 2015; 49-50: 47-54.

[4] Miniño AM, Murphy SL, Xu J, Kochanek KD. Deaths: final data for 2008. Natl Vital Stat Rep 2011; 59: 1-126.

[5] Yang Y, Stapleton J, Diagne BT, Kenny GP, Lan CQ. Man-portable personal cooling garment based on vacuum desiccant cooling. Appl Therm Eng 2012; 47: 18-24.

[6] Sun Y, Jasper WJ. Numerical modeling of heat and moisture transfer in a wearable convective cooling system for human comfort. Build Environ 2015; 93: 50-62.

[7] Lu Y, Wei F, Lai D, Shi W, Wang F, Gao C, Song G. A novel personal cooling system (PCS) incorporated with phase change materials (PCMs) and ventilation fans: an investigation on its cooling efficiency. J Therm Biol 2015; 52: 137-146.

[8] Song W, Wang F. The hybrid personal cooling system (PCS) could effectively reduce the heat strain while exercising in a hot and moderate humid environment. Ergonomics 2015. DOI: 10.1080/00140139.

[9] Fiala D, Lomas KJ, Stohrer M. A computer model of human thermoregulation for a wide range of environmental conditions: the passive system. J Appl Physiol 1999; 87: 1957-1972.

[10] Wang F, Ferraro SD, Molinaro V, Morrissey M, Rossi R. Assessment of body mapping sportswear using a manikin operated in constant temperature mode and thermoregulatory model control mode. Int J Biometeorol 2014; 58: 1673-1682.

[11] Huizenga C, Zhang H, Arens E. A model of human physiology and comfort for assessing complex thermal environments. Build Environ 2000; 36: 691-699.

[12] Information on http://emergency.cdc.gov/disasters/extremeheat/heat_guide-page-3.asp.

[13] Jay O, Cramer MN, Ravanelli NM, Hodder SG. Should electric fans be used during a heat wave ?. Appl Ergon 2015; 46: 137-143.

[14] Chan AP, Song W, Yang Y. Meta-analysis of the effects of microclimate cooling systems on human performance under thermal stressful environments: potential applications to occupational workers. J Therm Biol 2015; 49-50: 16-32.

Thermo-physiological Comfort and Frictional Characteristics of Running Socks in different Moisture Conditions

Diyana Tasron[1*], Siti Hana Nasir[2], Olga Troynikov[2], Roger Lewis[1], Matt Carré[1]

[1]*Department of Mechanical Engineering, The University of Sheffield, Mappin St., Sheffield, S1 3JD, UK*
[2]*School of Fashion and Textiles, RMIT University, Melbourne Vic 3056, Australia*

[*]Corresponding author's email: d.tasron@sheffield.ac.uk

Abstract

During prolonged walking and running, the plantar aspect of the foot is subjected to repetitive pressure and cycles of shearing, along with high levels of moisture due to perspiration and heat. The combination of all these elements not only affects comfort, but also provides an ideal condition for friction blister formation. The current study explored the physical properties of a predominantly nylon anti-blister sock and a cotton-rich sock along with their characteristics relevant to thermo-physiological comfort. The anti-blister fabric exhibited a slightly higher sliding friction than the cotton-rich fabric in dry condition. However, in the presence of moisture, the anti-blister fabric produced lower sliding friction than the cotton-rich fabric-which could be attributed to the low water vapour resistance and good one-way transport ability of anti-blister fabric. It is hoped that these results could be used to inform the actual frictional performance of the socks in different moisture settings, taking into account the propensity for sweating to occur.

Keywords: Foot-sock interface; Biotribology; Moisture conditions; Running sock; Friction blister

1. Introduction

Comfort is related to interaction of materials with the human body and can be classified into three inter-related aspects which are physiological comfort, psychological comfort and physical comfort [1]. Branson and Sweeney [2], defined clothing comfort as "a state of satisfaction indicating physiological, psychological and physical balance among a person with his or her clothing, and his/ her environment". The researchers emphasized that "balance" was important to achieve satisfaction or clothing comfort. Wearer comfort can enhance or suppress their performance and efficiency. Clothing comfort is a complex concept influenced by various factors such as thermo-physiological, sensorial and ergonomic. Thermo-physiological comfort relates to heat and moisture interaction between the human body and clothing worn. Sensorial comfort is a function of the feel of garment against the skin. Ergonomic comfort relates to the fit of the garment and may be influenced by factors such as a tendency to stick to the skin [1].

Moisture build-up within a shoe environment is primarily due to high perspiration rate of the feet and warm or humid environment generated in enclosed shoe. It has been reported that an athlete may induce sweat losses of nearly three litres per hour during a long run in a humid environment [3]. Sweat from the body will vaporise into the atmosphere much quicker compared to the sweat accumulated within an enclosed shoe environment. This will increase the presence of moisture at the sock fabric - plantar skin interface and in return influence the available friction [4-8]. It is therefore important for sock fabrics to have the capability to "wick" and transport the moisture away from the plantar skin into the external environment [9,10]. Furthermore, accumulated moisture at the sock fabric - plantar skin interface has potential to bridge air gaps between fibres which subsequently increases the contact area between these two surfaces. This could contribute to an increase in the available friction and/ or cause lubrication [11], in addition to influencing the thermal resistance and thermal conductivity of the sock fabrics [12]. The socks should not generate additional shear force at sock fabric - plantar skin interface since this could have a negative impact on the range of movement and could even potentially lead to friction blisters [11,13], which obviously would introduce discomfort to the wearer [13].

The purpose of the current study was to assess the properties of two commercially-available running socks influencing the thermo-physiological comfort of the wearer and relate the findings to the frictional performance of the socks obtained from an earlier study [8].

2. Experimental design

2.1 Sock fabrics and physical characteristic tests

Two types of newly-purchased running socks of differing material compositions and fabric construction were selected for the purpose of this study [8,14]. In order to remove any contaminants trapped in the sock fibres as well as to maintain their dimensional stability, all socks were hand-washed using water and a mild liquid detergent and left to air-dry at room temperature for at least 72 hours prior to the test [8].

For the purpose of the physical characteristic tests, five specimens were obtained from each sock and the mean values of each measured parameter were tabulated in Table 1 below. The experimental fabric samples were conditioned for 24 hours under standard conditions of 20 ± 2°C and 65 ± 2% relative humidity to eliminate the influence of atmospheric moisture content according to AS2001.1-1995 [15].The number of wales and courses per unit length of fabric sample was measured according to AS 2001.2.6-2001 [16], whereas the thickness of fabric samples was measured as the distance between the reference plate and parallel presser foot of the thickness tester as per AS 2001.2.15-1989 standard [17]. The mass per unit area was also calculated as the mean mass per unit area of five specimens according to AS 2001.2.13-1987 standard [18]. All the experiments were carried out under similar standard conditions for conditioning of the fabrics.

Table 1. The physical characteristics of the running socks used in this study

Sock type	Material compositions	Fabric construction	Mean thickness ± SD (mm)	Wales per cm	Courses per cm	Mean fabric mass per unit area ± SD (g/m^2)
Anti-blister "ABS"	99% nylon and 1% elastane	Single jersey	1.18 ± 0.04	9	14	304.00 ± 8.49
Cotton-rich "CRS"	70% cotton, 29% nylon, and 1%elastane	Single jersey, plush terry	2.62 ± 0.08	7	11	540.00 ± 5.66

The cotton-rich fabric is 44% heavier than that of the predominantly nylon anti-blister fabric despite having lower stitch density. It is also noted that the cotton-rich fabric is significantly thicker than the anti-blister fabric due to the plush knitted structure on the next-to-skin side of the fabric.

2.2 Tests relevant to thermo-physiological comfort

The thermo-physiological tests include obtaining the thermal resistance, water vapour resistance and moisture management properties of both sock fabrics. A sweating guarded hot-plate was used for this purpose in accordance to the ISO 11092 [19] standard, with introduced modifications due to the limited size of the fabric samples. The sweating guarded hot-plate has been widely used to simulate both heat and moisture transfer from a body surface through the fabric layers to the environment.

The moisture management tester (MMT) was used to measure and record the liquid water transfer and distribution of each fabric sample in multiple directions according to the American Association of Textile Chemists and Colorists (AATCC) test method 195-2009 [20]. Five samples with the size of 80 mm × 80 mm were obtained from each sock fabric for testing. The fabric was in contact with the sensor rings, allowing the liquid content and the liquid moisture transfer behaviour between the fabric surfaces to be determined. The top surface on the instrument is normally the surface in contact with the skin (next-to-skin side) and bottom surfaces of the fabric is the surface closest to the neighbouring environment (away-from-skin side).

2.3 Friction experiments

The friction experiments were conducted at The University of Sheffield, UK as part of a wider study [8]. Twenty-six healthy participants (18 males and 8 females; average age in years, 25.4 ± 4.3 SD) were recruited from the same institution for the purpose of this study. Approval was obtained from the Ethics Committee at the University of Sheffield and all participants provided their written consent prior to study after being well-informed of the entire testing procedure. Exclusion criteria include having any acute or chronic wounds on the feet, or a history of skin disorders. The plantar aspect of the first metatarsal head (1MTH) was selected as the most appropriate test region and therefore all friction tests were performed in this particular region.

The 1MTH hydration level was measured and monitored at specific intervals throughout testing using a capacitance-based device, the Corneometer® CM825 (Courage and Khazaka Electronic, Cologne, Germany) and an established test protocol [8]. The Corneometer® is capable of differentiating between dry and well-hydrated skin [21]. The measurement depth of the device is between 10 to 20 μm, which only reflects the amount of moisture within the stratum corneum layer without any influence from the deeper skin layers (dermis and subcutaneous tissue). All measurements are reported in "arbitrary units" (AU) ranging from 0 to 120, as in other studies that use this device [7,8,22]. The cleaning and acclimatisation of the foot was performed according to the standard protocol as highlighted in previous studies [7,8]. All testing was performed in a controlled-laboratory setting with a temperature of between 20 to 22°C and a relative humidity of 40 to 60%.

All newly-purchased sock materials were prepared in the exact same way as described for the physical characteristic tests to ensure consistency in the fabric properties. In order to achieve three sock moisture conditions that were significantly different from one another: "dry", "low moisture" and "wet", a standard moisture control protocol was employed [22] using the Corneometer® CM825. The intended Corneometer® range and the measured Corneometer® readings of both sock types are included in Table 2. It should be noted, however, that the maximum reading that the Corneometer® was capable of was 120 AU and that some of the "wet" condition samples were measured to be at this level although the amount of moisture present may have been slightly higher than indicated.

Table 2. The Corneometer® measurements of each moisture condition [8]

Moisture condition	Intended Corneometer range (AU)	Measured Corneometer reading ± SD (AU)	
		ABS socks	**CRS socks**
Dry	15-17	15.81 ± 0.27	15.98 ± 0.19
Low Moisture, "LMo"	40-50	46.70 ± 4.39	46.13 ± 5.04
Wet	110-120	119.35 ± 1.66	18.64± 1.76

A custom-built friction rig developed at the University of Sheffield, adapted from a previous study [5] was used to carry out the friction experiments. This approach had been compared with other methods and chosen as the most suitable for this type of study [14]. Friction tests were performed on the next-to-skin side of the sock fabrics within the marked area of 102 × 54 mm. Participants pressed their 1MTH against the sock fabric and then pushed their foot forwards, maintaining the initial level of normal load and a relatively consistent, self-monitored, sliding velocity. The same procedure was repeated for five different ranges of normal load for each moisture condition. This test protocol was based on previous studies [12].

3. Results and discussion

3.1 Sock characteristics relevant to thermo-physiological comfort

The experimental data showed that the anti-blister fabric has a lower thermal resistance and water vapour resistance than the cotton-rich fabric by 27% and 41% respectively. The t-test analysis of the data also revealed that both anti-blister and cotton-rich fabrics behaved significantly differently in terms of their thermal resistance and water vapour resistance ($p < 0.005$). The higher value of thermal resistance and water vapour resistance is attributed to the thickness of the cotton-rich sock fabric. Since the cotton-rich fabric is made from plush jersey, it increases the air volume contained in the knitted structure within the thickness of the sock which increases the passage of heat through the fabric from the skin to the environment, thus increasing its thermal resistance. Fabric thickness has been reported by many researchers as one of the paramount fabric parameters influencing the thermal and water vapour resistance of a fabric [12,23].

The moisture management capability of the fabric is mainly assessed on three indices which include the average moisture absorption rate at the top and bottom surfaces, accumulative one-way transport index (AOTI) and overall moisture management capacity (OMMC) [9]. Fig.1 illustrates the results for the main indices of the moisture management properties of the sock fabrics.

It can be seen from Fig.1(a) that the absorption rate of the anti-blister fabric is almost twice as high as the cotton-rich fabric for both the top and bottom surfaces. According to Özdil et al [24]. a thinner and finer fabric will have a higher absorption rate compared to a thicker fabric due to its comprising finer yarn. The absorption rate on the bottom surface is slightly higher than the top surface for both fabrics which indicates that the liquid (sweat) diffuses efficiently from the next-to-skin side to the away-from-skin side and spreads on this side. Therefore, the top surface which would be in contact with the wearer's skin will remain drier than the bottom surface which is ideal for the wearer.

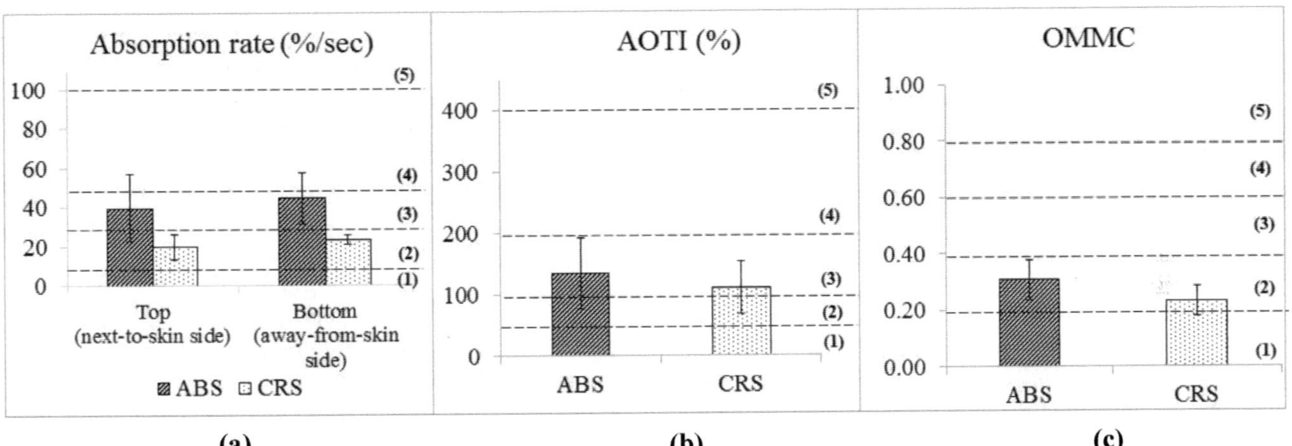

Fig.1 (a) Average moisture absorption rate at the top and bottom surfaces; (b) accumulative one-way transport index (AOTI); (c) overall moisture management capacity (OMMC) of both anti-blister "ABS" and cotton-rich "CRS" fabric samples
(Grading scale provided by the AATCC Test Method 195-2009 [19]: (1) - poor; (2) - fair; (3) - good; (4) - very good; and (5) - excellent)

Fig.1(b) demonstrates the comparison of the AOTI values obtained from both fabrics, where fabric with higher AOTI value could transport the liquid sweat quicker from next-to-skin side to the external environment. The AOTI of anti-blister fabric is slightly higher than that of the cotton-rich fabric, although the difference is insignificant ($p > 0.05$). When the values of AOTI were converted into a grading scale, it showed that the anti-blister fabric has "good" one-way transport ability (3/5) whilst the cotton-rich fabric has "fairly-good" (2.5/5) one-way transport ability.

The OMMC values of the fabrics are plotted in Fig.1(c). According to the grading scale, both fabrics are in the "fair" category in terms of their moisture management capacity. For active sportswear, the moisture management properties of fabrics are critical in order to keep the skin dry and provide maximum comfort to the wearer. Factors such as fibre, yarn, fabric construction, and fabrics finishes are important to improve the moisture management performance of the fabrics.

3.2 Relating thermo-physiological properties to frictional performance

The dynamic coefficient of friction (DCOF) values were computed by dividing the predicted sliding friction force data by the 100N normal force for all participants. The mean DCOF values for all six fabric-moisture combinations are presented in Table 3, along with the respective standard deviations and the level of significance from an ANOVA test. Apart from when comparing the low moisture cotton-rich fabric with wet anti-blister fabric, all other comparisons were statistically significant ($p < 0.05$). The anti-blister fabric exhibits a slightly higher sliding friction than the cotton-rich fabric in dry condition. However, in the presence of moisture, the anti-blister fabric produced lower sliding friction than the cotton-rich fabric. This could be attributed to the fact that the anti-blister fabric has slightly better wicking properties compared with the cotton-rich fabric. Moreover, the anti-blister fabric demonstrates better water vapour transmission and due to its low resistance to water vapour and good one-way transport ability, allowing the water vapour from the added moisture to be transported through the air spaces and interstices between the sock fabric (see Fig.1).

It should be noted that although the feet were well-ventilated throughout the friction testing, some of the water may still be absorbed by the 1MTH skin when it is in contact with the "low moisture" and "wet" fabrics during sliding. This could increase the area of contact between the skin and the sock fabric due to increased skin tissue flexibility, as seen in previous studies [4,25]. By monitoring the foot hydration level at specific intervals in accordance to the established test protocol [8] ensured that this effect could be considerably minimised.

In real-world scenarios moisture in the sock is dictated by the tribological properties [26] and hydrophobicity of the fabrics. During vigorous physical activity, the amount of perspiration produced could exceed the absorption capacity of any sock fabrics. In order for the dry in-shoe environment to be maintained, efficient wicking process is required to transport the moisture away from the skin surface into the external environment. Based on the findings of this current study, the anti-blister sock would more likely to remain dry within a shoe environment, hence producing lower sliding friction than the cotton-rich sock. This would not only reduce the probability of blistering but also help to ensure comfort to the wearer.

Table 3. Comparisons of sock frictional performance in different moisture conditions including the values of mean DCOF ± SD (▲: highest DCOF;; =: similar DCOF level)
The level of significance is indicated by * ($p < 0.05$)

	ABS_Dry (DCOF = 0.44 ± 0.10)	**ABS_LMo (DCOF = 0.61 ± 0.08)**	**ABS_Wet (DCOF = 0.69 ± 0.07)**
CRS_Dry (DCOF = 0.33 ± 0.07)	ABS (▲) p=0.000*	ABS (▲) p=0.000*	ABS (▲) p=0.000*
CRS_LMo (DCOF = 0.67 ± 0.08)	CRS (▲) p=0.000*	CRS (▲) p=0.011*	ABS = CRS p=0.448
CRS_Wet (DCOF = 0.74 ± 0.08)	CRS (▲) p=0.000*	CRS (▲) p=0.000*	CRS (▲) p=0.028*

4. Conclusions

The current study demonstrated the physical properties of two different running sock fabrics, which are predominantly-nylon anti-blister sock and cotton-rich sock, along with their frictional performance in "dry", "low moisture" and "wet" conditions. These parameters are relevant to thermo-physiological comfort of the wearer. Based on the findings, if either sock was chosen to be worn, the anti-blister sock is likely to be dry and low moisture whereas the cotton-rich sock will likely to be low moisture or wet. This could be attributed to the fact that the anti-blister fabric possesses low water vapour resistance and good one-way transport ability. By employing similar thermo-physiological and friction test protocols highlighted in this study, other sock fabrics could be tested and assessed in the future to gain a much better understanding on how fabric properties influence its frictional performance in different moisture settings.

References

[1] Fan J. Physiological comfort of fabrics and garments, in Engineering apparel fabrics and garments. vol. 1, Fan J. and Hunter L. Eds., ed FL, USA: Woodhead Publishing, 2009; p: 201-250.

[2] Branson DH, Sweeney M. Conceptualization and measurement of clothing comfort: toward a metatheory, in Critical linkages in textiles and clothing: Theory, method and practice, S. Kaiser and M. L. Damhorst, Eds., ed Monument, Colorado: International Textile and Apparel Association, 1991; p: 94 - 105,

[3] Costill DL. Sweating: its composition and effects on body fluids. Ann. N. Y. Acad. Sci., 1977; 301: 160-174.

[4] Tomlinson SE, Lewis R, Liu X, Texier C, Carré MJ. Understanding the friction mechanisms between the human finger and flat contacting surfaces in moist conditions. Tribol Lett 2011; p: 283-294.

[5] Liu X, Lu Z, Lewis R, Carré MJ, Matcher SJ. Feasibility of using optical coherence tomography to study the influence of skin structure on finger friction. Tribol Int 2013; 63: 34-44.

[6] Troynikov O, Ashayeri E, Fuss FK. Tribological evaluation of sportswear with negative fit worn next to skin. Proc. Inst. Mech. Eng. J J. Eng. Tribol 2012; 226: 588-597.
[7] Tasron DN, Thurston TJ, Carré MJ. Frictional behaviour of running sock textiles against plantar skin. Procedia Eng 2015; 112: 110-115.
[8] Tasron D, Maiti R, Hemming M, Lewis R, Carré MJ. Frictional interaction between running sock fabrics and plantar aspect of first metatarsal head in different moisture conditions. Proceedings of the 11th Conference of the International Sports Engineering Association 2016 (in press) 2016.
[9] Konopov I, Oggiano L, Chinga-Carrasco G, Troynikov O, Sætran L, Alam F. Aerodynamic and comfort characteristics of a double layer knitted fabric assembly for high speed winter sports. Procedia Eng 2010; 2: 2837-2843.
[10] Richie DH. Athletic socks, in Athletic Footwear and Orthoses in Sports Medicine, ed: Springer, 2010; p: 69-78.
[11] Baussan E, Bueno MA, Rossi RM, Derler S. Analysis of current running sock structures with regard to blister prevention. Text. Res. J. 2013; 83: 1-13.
[12] Van Amber RR, Wilson CA, Laing RM, Lowe BJ, Niven BE. Thermal and moisture transfer properties of sock fabrics differing in fiber type, yarn, and fabric structure. Text. Res. J 2015; 85: 1269-1280.
[13] Bertaux E, Derler S, Rossi RM, Zeng X, Loehl L, Ventenat V. Textile, physiological, and sensorial parameters in sock comfort. Text. Res. J. 2010; 80: 1803-1810.
[14] Carré MJ, Tasron DN, Lewis R, Hashmi F. Investigating foot-sock friction: a comparison of two different methodologies. Proceedings of the 11th Conference of the International Sports Engineering Association 2016 (in press), 2016.
[15] Australian standard, Method of test for textiles; Conditioning procedures. 1995
[16] Australian standard, Method of test for textiles; Method 2.6 physical test; Determination of the number of wales and courses per unit length of knitted fabric. 2001.
[17] Australian standard, Method of test for textiles; Method 2.15 physical test; Determination of thickness of textile fabrics. 1989.
[18] Australian standard, Method of test for textiles; Physical test; Determination of mass per unit area and mass per unit length of fabrics. 1987.
[19] ISO standard, Textiles - Physiological effects - Measurement of thermal and water-vapour resistance under steady-state conditions (sweating guarded-hotplate test). 2014.
[20] AATCC Test Method 195-009, Liquid Moisture Management Properties of Textile Fabrics. 2009.
[21] Fluhr JW, Gloor M, Lazzerini S, Kleesz P, Grieshaber R, Berardesca E. Comparative study of five instruments measuring stratum corneum hydration (Corneometer CM 820 and CM 825, Skicon 200, Nova DPM 9003, DermaLab). Part II. In vivo. Skin Res Technol 1999; 5: 171-178.
[22] Tasron DN, Maiti R, Hemming M, Lewis R, Carré MJ. Investigating a methodology to measure moisture in skin–textile friction experiments. Footwear Sci 2015; 7: S15-S16.
[23] Cimilli C, Nergis BU, Candan C, Özdemir M. A Comparative Study of Some Comfort Related Properties of Socks From Different Fiber Types. Text. Res. J. 2009.
[24] Özdİl N, Süpüren G, Özçelik G, Pruchova J. A study on the moisture transport properties of the cotton knitted fabrics in single jersey structure. TEKSTİL ve KONFEKSİYON 2009; 3: 218-223.
[25] Gerhardt LC, Strässle V, Lenz A, Spencer ND, Derler S. Influence of epidermal hydration on the friction of human skin against textiles. J. R. Soc. Interface 2008; p: 1317-1328.
[26] Baussan E, Bueno MA, Rossi RM, Derler S. Experiments and modelling of skin-knitted fabric friction. Wear 2010; 268: 1103-1110.

Objective Evaluation of Mattress Materials by using a Sweating Thermal Manikin

Jin Li, Xiao-Qun Dai*, Jiang-Yuan Qiu, Guang-Yu Su, Jie-Xian Huang

Soochow University, College of Textile and Clothing Engineering, Suzhou, 215006, China

*Corresponding author's email: daixqsz@suda.edu.cn

Abstract

A mattress plays an important role in the bed system. However, it is often difficult to evaluate the mattress materials by the conventional textile tests because of their 3D structure. In this study a sweating thermal manikin is used to evaluate three types of mattress materials objectively. The commonly used polyurethane foam material is found not good for summer use, because of its poor heat and moisture transfer properties. For the 3D fiber transverse laying material, although its moisture transfer property has been improved, it is still not good enough for summer mattress due to its high thermal insulation. The 3D knitted spacer fabric material which is superior in heat and moisture transfer, can maintain a comfort bed climate while avoiding much water loss through sweating.

Keywords: Mattress; Sweating thermal manikin; Bed climate; Heat and moisture transfer

1. Introduction

We human beings spend one-third of our lifetime on beds, and an integration of psychological, physiological, and physical conditions affects the quality of sleep. A typical sleeping system is defined as mattress + support structure + head cushion. It is supposed to provide the body sufficient support, avoiding spinal distortion, and generate comfort bed climate. Mattress plays an important role in this system. However, as Parsons pointed out over 40 years ago [1], it remains true to this day that the literature regarding the objective evaluation of mattresses is sparse. Most of the identified studies that have dealt with objective measurements were concerned with the healing of wounds or the prevention of bed sores for those who are bedridden. Since pressure is a major factor in the development of such sores [2], those papers evaluated mattresses on the basis of pressure levels generated by the person lying on the bed or by the number or severity of bed sores [3-5]. There is very few report available about evaluating mattresses for normal daily use.

The thermal environment is a key determinant of sleep because thermoregulation is strongly linked to the mechanism regulating sleep [6]. Providing a thermally comfortable sleeping environment is important for sleep maintenance and contributes positively to human health and their daytime activities. In general, the bed climate is mainly affected by the body biological character, the ambient environment and the bedding system conditions. Bischof et al. stated that thermal comfort during sleep is directly connected to the microclimate in the bed and only secondly to ambient conditions and found a correlation between the quality of sleep and the variations of temperature and that larger fluctuations of temperature in the bed are connected to a shorter deep-sleep time [7]. Although it was found very high or low air temperatures decreased slow wave sleep, and increased the frequency and duration of wakefulness [8-10], Wang et al. stated higher or lower indoor temperature does not fully address sleep thermal comfort. Compared with the indoor operative temperature, the bedding micro climate has a greater effect on sleep thermal comfort with a body-covering quilt [11].

Sleep is influenced by the micro thermal environment around human body. When asleep, energy expenditure is reduced due to the supine position and lack of physical activity, and the core temperature is kept low [12]. The appropriate surrounding temperature yields an advantage for attaining and stabilizing the level of skin blood flow, and more importantly, maintaining physiological homeostasis through the thermoregulatory mechanisms [13]. There are just a few papers about the bedding system conditions available. Okamoto et al. investigated the effects of a specially designed bed sheet on bed climate and physiological responses and found that it increased temperature and humidity, and the comfort sensation and subjective sleep evaluation [14]. Ikeda and Fukai investigated how bed climate changes when disposable waterproof sheets are used in addition to usual ones and found that the disposable waterproof sheets significantly increased the relative humidity in bed [15].

Park and Lee investigated the effects of mattress types on sleep quality, and found that mean skin temperature, deep sleep (III and IV), sleep efficiency, sleep latency and subjective ratings were significantly affected with mattress types. When subjects slept in comfortable beds, their skin temperature of the lower body, sleep efficiency and the percentage of deep sleep were higher [16]. Okamoto et al. investigated the effects of various mattresses upon bed climate, and found that the relative humidity was higher in an air mattress, which is of the impermeable structure, than a regular cotton Futon(F); the rectal temperature was significantly lower in a truss mattress(T) than F, because of lower thermal insulation of T, resulting from its high air permeability [17,18]. They further compared polyurethane foam mattresses and camel mattresses in mild humid environment, and found that the foam mattress material could increase the subjective humid sensation and the requirement for decreasing mattress temperature by increasing the bed climate and microclimate temperature and humidity, and even changing the rapid eye movement sleep (REM) distribution [19].

Lin and Deng measured the total insulation values for a wide range of bedding systems (through different combinations of bed, bedding and its percentage coverage over a human body, and sleepwear) commonly used in the subtropical regions using a thermal manikin. The total insulation values of the measured bedding systems varied greatly from 0.90 to 4.89 clo, depending upon bedding, bed and mattress, the type of sleepwear, and percentage coverage of body surface area by bedding and bed. However, as they stated, the sweating and moisture permeability of the quilts, were not considered on the measurement of the total insulation values for the bedding systems [20]. In this paper, three types of typical mattress materials were evaluated by using a sweating thermal manikin, the properties of heat and moisture transfer were quantified by dry and wet tests respectively, and the bed climate was further generated by the comfort test.

2. Methods

2.1 Materials

Three types of typical mattress materials were evaluated in this study, polyurethane foam(PF), 3D fiber transverse laying (3DT), and 3D knitted spacer fabric (3DK). Fig.1 shows the structure of these materials, and their physical parameters are listed in Table 1.

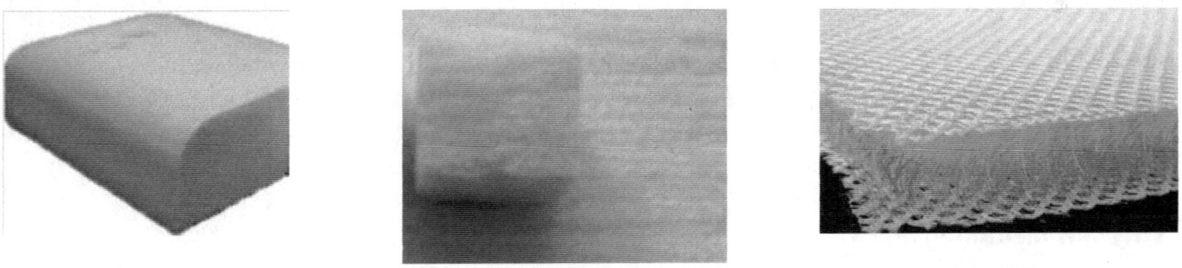

(a) PF (b) 3DT (c) 3DK

Fig.1 Mattress materials

Table 1. Physical parameters of mattress materials

Type	PF	3DT	3DK
Density (g/m^3)	0.025	0.05	0.065
Thickness (cm)	4.0	4.0	2.0

2.2 Measurement

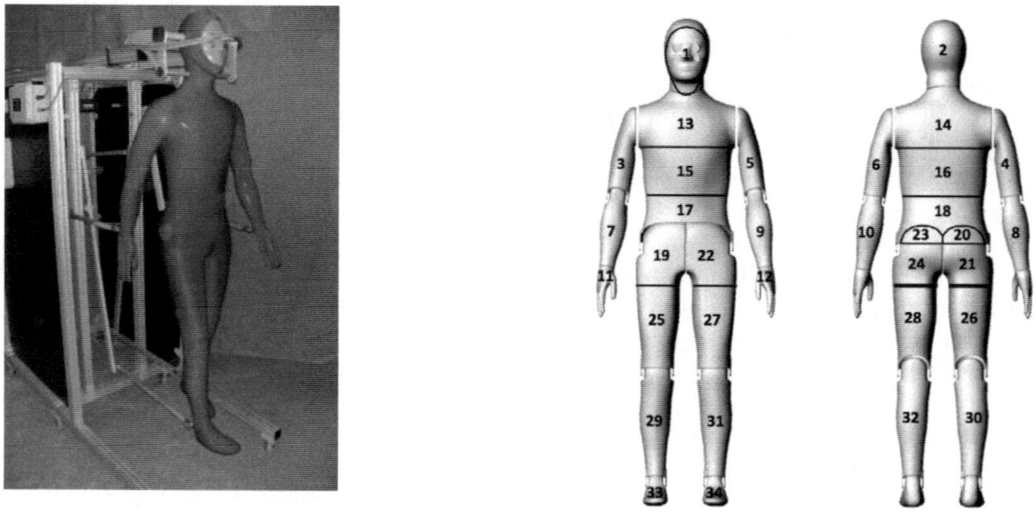

(a) Sweating manikin (b) Manikin zones

Fig.2 Sweating thermal manikin

A 34-zone Newton type sweating thermal manikin (MTNW, Seattle, WA, USA) was used in this study. Each mattress material was placed on a wooden board, and was covered with a 100% cotton sheet. The manikin was then placed with a supine position over the sheet, and the whole body except the head was covered with a thin 100% polyester blanket for summer use. The constant temperature mode of the manikin was used to measure the thermal insulation and evaporative resistance of this bedding system. All tests were carried out in a climatic chamber. The chamber was set as 20°C, 50%RH for the thermal insulation test and 34°C, 50%RH for the vapor resistance test. The air velocity was 0.4±0.1 m/s.

The physiological control mode of the manikin (comfort test) was used to assess the bed climate. Before running the physiological control mode, the manikin was preheated to a defined thermal neutral condition. Then a 1 MET metabolism level was assigned to the manikin, to simulate the heat generation of a human body during resting and the manikin skin temperature and sweating rate were regulated by the Fiala human thermoregulatory model [21]. Lan et al. proposed that the comfortable indoor temperature was 23°C in winter and 26°C in summer during the sleep period [10]. Since we simulated sleeping condition in summer, the temperature was set as 26°C, and the relative humidity was 55% for the comfort test. The bed climate was monitored by placing two paired temperature and humidity sensors at the shoulder and the chest of the manikin surface respectively.

3. Results and discussion

3.1 Thermal insulation and vapour resistance

Table 2 shows the thermal insulation and vapour resistance of the mattress materials obtained from manikin tests. The "entire" obtained by grouping the whole body zones excepted the head with the parallel method, and "back" is the result by grouping only the back zones (shoulder 14, mid back 16 and lower back 18 shown in Fig.1 (b)) of the upper body which directly touches the mattress. The "entire" thermal insulation ranges from 0.24 to 0.364°C·m^2/W, and the "back" one ranges from 0.261 to 0.928°C·m^2/W; the "entire" vapour resistance ranges from 38.13 to 43.88 Pa·m^2/W, and the "back" one from 97.94 to 162.98 Pa·m^2/W. Compared to the "entire" values, the "back" ones showed much bigger difference among materials. Except the thermal insulation for "3DK", the thermal insulation and vapour resistance for "mid back", "lower back" zones were far higher than the other body zones. It was confirmed that the mattress was the biggest barrier for heat and moisture diffusion in this bed condition. Since the back parts directly touches the mattress, and the heat and moisture transfer from this body part is mainly through the mattresses. Thus, the "back" values are more representative to the thermal properties of the mattress materials.

Table 2. Thermal insulation and vapour resistance of the materials

Materials	Thermal insulation (°C·m²/W)		Vapour resistance(Pa·m²/W)	
	Entire	Back	Entire	Back
PF	0.364	0.639	43.88	162.98
3D-T	0.373	0.928	39.79	122.82
3D-K	0.24	0.261	38.13	97.94

The thermal insulation for 3DT is the highest (0.928°C·m²/W) among three materials because of its high mass density. However, the thermal insulation for 3DK is the lowest while its mass density is the highest among the three. It is not just because that its thickness is only half of the other two, but also because of its special 3D space structure as shown in Fig.1(c). The thermal insulation for the "back" is quite close to the "entire" one. It confirmed that the 3D space formed channels to allow air to flow through, resulted in low thermal insulation. The vapor resistance for 3DK was also far lower than the other two, benefiting from its space structure. However, the "back" one is far higher than the "entire" one, indicating that the 3DK mattress is still a big barrier for moisture diffusion.

3.2 Results of comfort test

Fig.3 shows the results from the comfort test. Tsk is the mean skin temperature for the manikin and Thy is the virtual temperature of the hypothalamus for the manikin and is used as the core body temperature for the manikin. Swa denotes the amount of sweat generated by active thermoregulation in grams/min. Since the values of Swb, which denotes the amount of sweat generated to maintain basal skin wetness, were close to each other, they were not shown here. Qm displays current amount of metabolic power being generated by the manikin in Watts. Although the same rest condition (1MET) was simulated, Qm still varied within a small range (107.75~108.1W) and the difference among materials was up to 0.51W. The difference of Thy among materials was up to 0.08°C, PF the highest and 3DK the lowest. The Tsk also showed similar difference among materials, while the amount was bigger, up to 0.32°C. The big difference among materials was found in Fig.3(b), the difference between PF and 3DK was up to 0.69 g/min. The onset time of sweating were 53, 60 and 79 min. for 3DK, 3DT and PF respectively.

Although for 3DK the active sweat amount was the lowest, the Tsk kept the lowest through the whole period. Moreover, although the thermal insulation of 3DT was higher than PF, the manikin on it sweated less and later. It indicated that the more part of the basal and active sweat amount in materials of lower vapor resistance evaporated and took away body heat. For PF, although its sweat amount was the highest, most of them just flowed away as liquid and was not helpful for body cooling. It was confirmed that in summer, the ability to release moisture of the mattress is very critical.

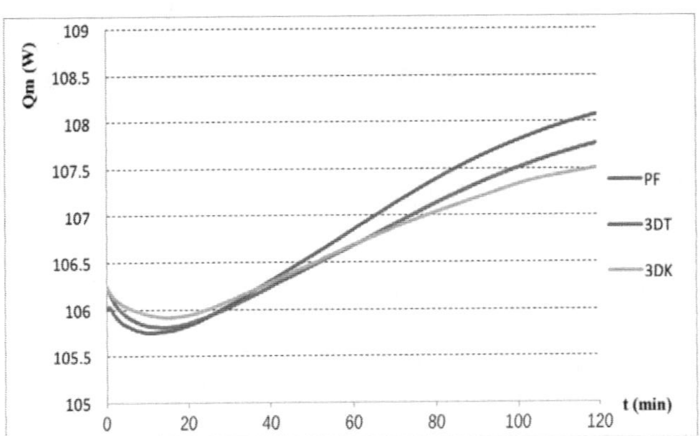

(a) Metabolic power (Qm)

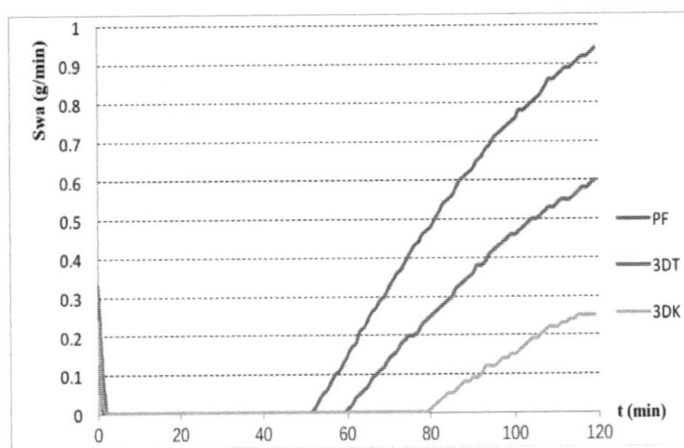

(b) Active sweat rate (Swa)

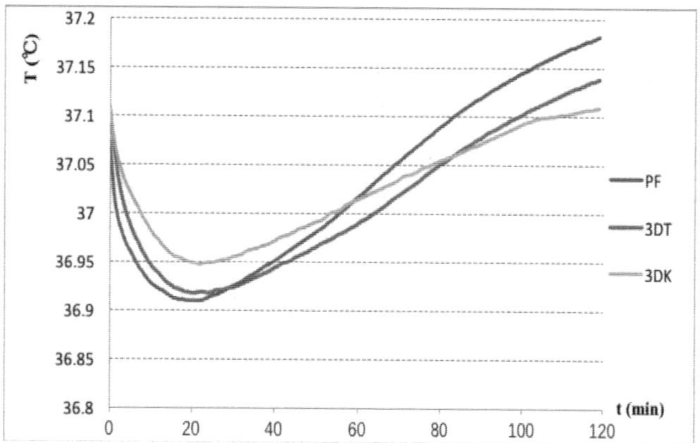

(c) Core temperature (Thy)

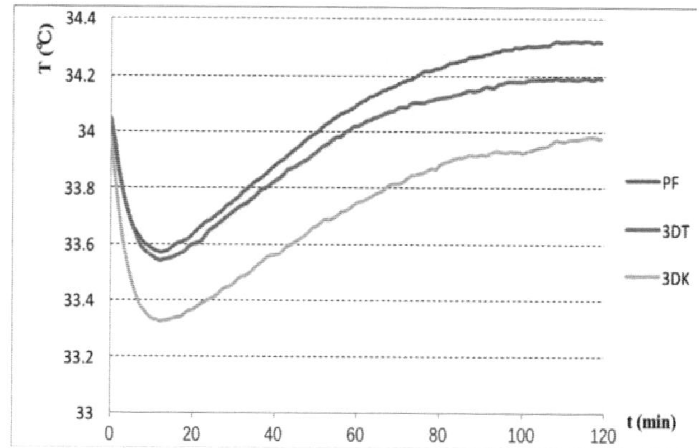

(d) Mean skin temperature (Tsk)

Fig.3 Comfort test

3.3 Bed climate

In Fig.4(a) shows the local skin temperature of manikin at chest and shoulder, (b) is the sweat rate for these two zones, (c) and (d) are the temperature and relative humidity measured at the manikin surface as at chest and shoulder. For all three mattresses, the skin temperature at chest is higher than at shoulder, the difference is up to 1.8°C. This confirmed that the manikin skin temperature varies among zones under the physiological control mode. There might be two reasons for such variation, one is that the hot air goes upward, then the temperature for upper zones is higher than the lower zones; another is that the sweat rate for the back is usually higher than the front of the body torso. Fig.4(b) shows the the active sweat rate for shoulder is far higher than that for chest, explaining the big temperature drop at shoulder shown in Fig.4(a). One interesting thing is for PF material, the sweat rate at shoulder is far higher than the other two, but the shoulder temperature only dropped by 0.65°C; the sweat rate at shoulder for 3DK material is the lowest, but the shoulder temperature dropped by 1.52°C. It confirmed again the 3DK material is effective in heat and moisture diffusion.

Since the skin temperature sensors in the manikin system are embedded in the manikin body and covered by the skin layer, the skin temperature reported by the manikin system is not really the surface temperature. Thus the data shown in Fig.4(c) and (d) are more representative to the bed climate. Although Fig.4(c) shows similar trend with Fig.4(a), the difference between shoulder and chest is as big as 4.9°C and the difference among materials is also much bigger, up to 4.1°C. Since the manikin lying over the PF material started to sweat first, the relative humidity increased rapidly first, around 7 min later, the relative humidity over the 3DT material began to increase quickly. The relative humidity kept to be lower than 60% for over one hour and then started to increase gradually. It has been reported that the relative humidity needs to be lower than 60% for a comfort microclimate around human body. For three materials, the relative humidity at chest kept to be around 40%RH. It can be concluded that the 3DK material helps to maintain a relatively comfort bed climate.

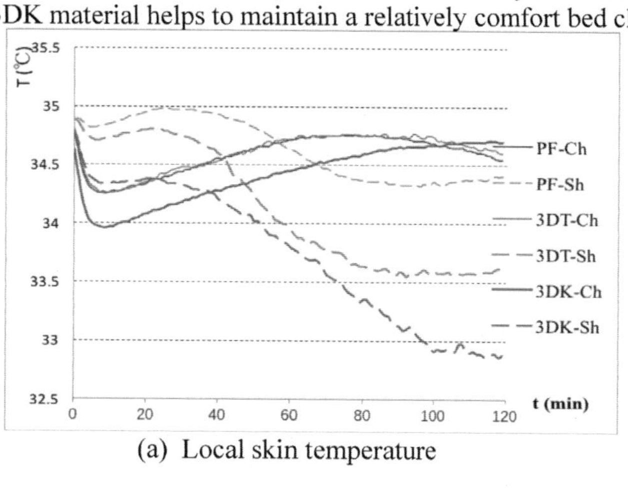

(a) Local skin temperature

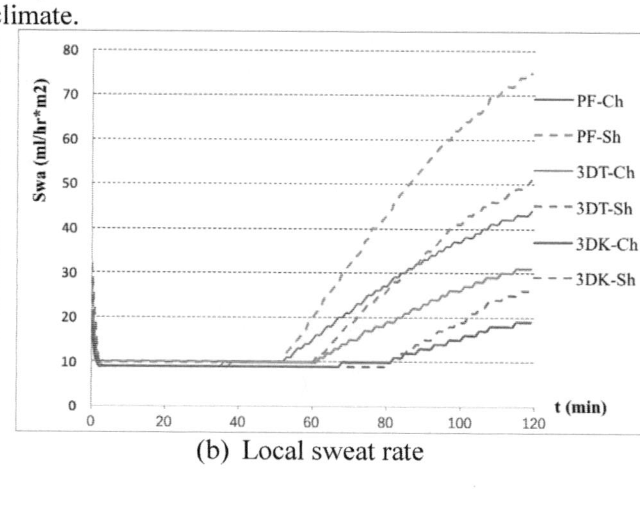

(b) Local sweat rate

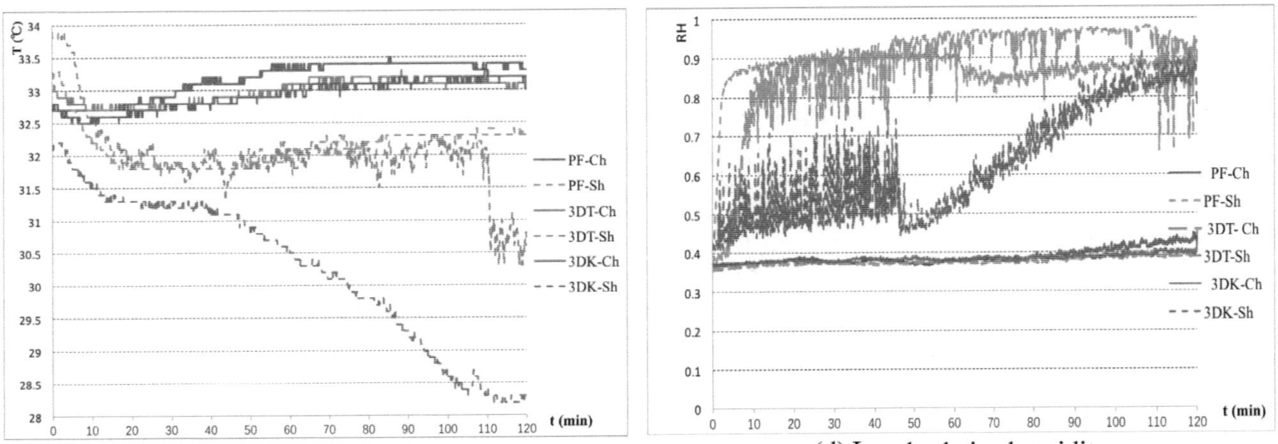

(c) Temperature at manikin surface (d) Local relative humidity

Fig.4 Local data at chest and shoulder

It is difficult to measure the mattress materials by the conventional textile tests because of their 3D structure. This study confirmed that manikin test is an effective way to evaluate mattress materials objectively. The heat and moisture transfer properties can be measured by the manikin dry and wet tests, the local thermal insulation and vapor resistance at the back which contacts the mattress can be used to quantify such properties. Moreover, the bed climate can be simulated by the comfort test although the comparison with real human trial need to be carried out in future. The commonly used PF material is not good for summer, because of its poor heat and moisture transfer properties. For 3DT material, the moisture transfer property is improved, due to its high thermal insulation, it is still not good enough for summer use. The 3DK material which has 3D space structure, thus is of low thermal insulation and vapor resistance, can maintain a comfort bed climate in summer while avoiding much water loss through sweating.

4. Conclusion

In this paper, three typical mattress materials were evaluated by using a sweating thermal manikin. The local thermal insulation and vapor resistance at the back which contacts the mattress were confirmed to be useful to quantify the heat and moisture transfer properties. Through the comfort test, it was further made clear that the 3DK material which has 3D space structure, thus is of low thermal insulation and vapor resistance, can maintain a comfort bed climate in summer.

5. Acknowledgement

This is a project funded by the priority academic program development (PAPD) of Jiangsu higher education institutions. The project was also partially funded by Nantong Textile Institute of Soochow University.

References

[1] Parsons HM. The bedroom. Hum Factors 1972; 14 (5): 421-450.
[2] Defloor T, De Schuijmer JD. Preventing pressure ulcers: an evaluation of four operating-table mattresses. Appl Nursing Res 2000; 13(3): 134-141.
[3] Shelton F, Barnett R, Meyer E. Full-body interface pressure testing as a method for performance evaluation of clinical support surfaces. Appl Ergon 1998; 29 (6): 491-497.
[4] Bain D, Ferguson-Pell M, McLeod A. Evaluation of mattresses using interface pressure mapping. J Wound Care 2003; 12 (6): 231-235.
[5] Bale S, Price P, Crook H, Morgan T, Harding KG. 1999. Clinical evaluation of a new pressure-relieving mattress. J. Wound Care 8 (10): 520-524.

[6] Gilbert SS, van den Heuvel CJ, Ferguson SA, Dawson D. Thermoregulation as a sleep signaling system. Sleep Med Rev 2004, 8: 81-93.
[7] Bischof W, Madsen TL, Clausen J, Madsen PL, Wildschiodtz G. Sleep and the temperature field of the bed. J Therm Biol 1993: 18(5/6): 393-398.
[8] Okamot-Mizuno K, Mizuno K. Effects of thermal environment on sleep and circadian rhythm. J Physiol Anthropol 2012; 31:14.
[9] Muzet A, Libert JP, Candas V. Ambient temperature and human sleep. Experientia 1984; 40: 425-429.
[10] Lan L, Pan L, Lian ZW, Huang HY, Lin YB. Experimental study on thermal comfort of sleeping people at different air temperatures. Build Environ 2014; 73: 24-31.
[11] Wang YY, Liu YF, Song C, Liu JP. Appropriate indoor operative temperature and bedding micro climate temperature that satisfies the requirements of sleep thermal comfort. Build Environ 2015; 92: 20-29.
[12] Krauchi K. The human sleep-wake cycle reconsidered from a thermoregulatory point of view. Physiol Behav 2007: 90 (2-3): 236-245.
[13] Parmeggiani PL. Interaction between sleep and thermoregulation: an aspect of the control of behavioral states. Sleep 1987; 10 (5): 426-435.
[14] Okamoto K, Matuo K, Nakabayashi K, Okudaira N. Effects of a specially designed bed sheet on sleep and bed climate. J Home Econ Jpn 1997; 48(12): 1077-1082.
[15] Nakamura-Ikeda R, Fukai K. Effects of different bed sheets on bed climate and thermal response.Jpn J Nurs Sci 2005; 2: 51–55.
[16] Lee HJ, Park SJ. Quantitative effects of mattress types (comfortable vs. uncomfortable) on sleep quality through polysomnography and skin temperature. Int J Ind Ergonom 2006; 36: 943-949.
[17] Okamoto K, Mizuno K, Okudaira N. The effect of a newly designed air mattress upon sleep and bed climate. Appl Human Sci 1997; 16(4): 161-166.
[18] Okamoto K, Nakabayashi K, Mizuno K, Okudaira N. Effects of truss mattress upon sleep and bed climate. Appl Human. Sci. 1998, 17(6): 233-237.
[19] Okamoto K, Nakabayashi K, Mizuno K, Okudaira N. Effects of bed mattress material on sleep onset period under mild humid heat exposure(in Japanese).J Home Econom Jpn 2012; 63(7): 391-397.
[20] Lin ZP, Deng SM. A study on the thermal comfort in sleeping environments in the subtropics - Measuring the total insulation values for the bedding systems commonly used in the subtropics. Build Environ 2008; 43: 905-916.
[21] Fiala D, Lomas KJ, Stohrer M. Computer prediction of human thermoregulatory and temperature responses to a wide range of environmental conditions. Int J Biometeorol 2001; 45(3): 143-159.

A Study on the Thermal Properties of Polysulfonamide Fiber

Xin-Ping Lu[1], Ying-Qi Xu[1,2*], Jian-Qing Ye[2]

[1]*Fashion School, Shanghai University of Engineering Science, Shanghai, 201620, China*
[2]*The Key Lab of Textile Science & Technology Ministry of Education, Donghua University, Shanghai, 200051, China*

*Corresponding author's email: xyqq911@hotmail.com

Abstract

Polysulfonamide (PSA) fiber belongs to polyamide aromatic polymer materials. In order to study the termal stability of PSA, thermo gravimetric analysis was used to obtain the thermogravimetric progress of the fiber. The pyrolysates of the fiber were identified by pyrolysis-gas chromatography/mass spectrometry, and the thermal behavior was analyzed using differential scanning calorimetry. The results obtained from thermal decomposition process indicate that the existence of the conjugated aromatic rings and the additional sulfone group structure in its molecular main chain enable PSA fiber has high degradation temperature and can be applied to various fields in our daily life.

Keywords: PSA fiber; thermal degradation; pyrolysis; TG; DSC; Py-GC/MS

1. Introduction

Polysulfonamide (PSA) fiber is a high performance fiber with great thermal stability and has been widely used as high-temperature filtration materials, insulation materials, thermal protective textiles, composite materials, membrane materials, etc for its excellent mechanical properties, high-temperature resistance performance, flame-retardant properties, and easy dyeing properties [1-5]. As is known, PSA was first developed by Shanghai Textile Research Institute and Shanghai Synthetic Fiber Research Institute in 1974 and owns China's independent intellectual property rights. PSA, prepared by wet or dry spinning with copolymer of terephthaloyl chloride [6],3,3'-diaminodiphenylsulfone and 4,4'-diaminodiphenylsulfone(3:1:4) dissolved in dimethylacetamide solvent, is one kind of aromatic polyamide fibers with 75% para linked and 25% meta linked compounds. The molecular structure of PSA fiber cannot be easily destroyed at high temperature [7, 8]. The limiting oxygen index (LOI) of PSA fiber may reach up to 33, whereas that of Nomex (polyisophthaloyl metaphenylene diamine), and Kevlar (Poly (p-phenylene terephthamide)) are 29 and 26 respectively.

The thermal degradation of aramid fibers including Kevlar and Nomex has been fairly well investigated in the last few decades. Many characteristic temperature of their degradation have been obtained [9-15]. But there are few information in the literature on PSA fibers.

In the present study, the thermal studies were carried out to assess the relative stabilities of PSA by the combined thermo analytical techniques of thermogravimetry (TG), pyrolysis-gas chromatography/mass spectrometry (Py-GC/MS) and differential scanning calorimetry (DSC).

2. Problem formulation

2.1 Materials

PSA fiber was provided by Shanghai Tanlon Fiber Co., Ltd. Kevlar® 149 and Nomex® were purchased from DuPont Inc., USA. All of these fibers are 2.0dtex in linear density and 51mm in length.

2.2 TG Analysis

The thermal degradation of PSA was measured using a thermal gravimetric analyzer TG2050 over the temperature range of 50 to 800 ℃ with a heating rate of 20℃/min in nitrogen atmosphere. In this experiment, the nitrogen flow was maintained at 60 mL /min. The sample weight around 15 mg in primary mass was held in an alumina crucible, and then the loss of the sample mass was measured under a temperature program. Thermal response curves in the form of mass change and heat flow as a function of temperature and time were recorded using the SetSys software (V 4.3). In this software, the derivative of the primary mass change (DTG) was utilized, thereby extended the capability and scope of the analysis.

2.3 Py-GC/MS experiment

The gas chromatography/mass spectra analysis of PSA fiber was performed by GC/MS-QP2010 gas chromatograph and mass spectrometer. Powder sample (3 mg) was put into the platinum cup and dropped into the pyrolysis quartz capillary. The pyrolysis was carried under helium carrier gas at a flow rate of 3 mL/ min, the temperature is 570℃, and the pyrolysis time is 12secs. The forward press of GC column was 34.2kPa. The instrument was programmed for a heating rate of 10℃ from 40℃ for 3 mins to 300℃. The temperature of GC/MS interface was set at 300℃. The MS detector was scanned from 30 to 650 m/z with 70eV ionization energy at 230℃.

2.4 DSC measurement

The thermal decomposition experiments on PSA, Kevlar and Nomex were carried out with a NETZSCH DSC404 differential scanning calorimeter from 200 to 600℃ at a heating rate of 10℃/min under argon environment with a flow rate of 20mL/min. The weight of all specimens was maintained around 10 mg. Test samples were placed in sample pans with perforated lids which allowed any volatile degradation products from the sample to escape into the surrounding chamber and be swept away by the gas flow which purges the system. All samples were allowed to come to equilibrium within the chamber before heating. The testing methodology employed in this work fully adhered to the ASTM methods [16].

3. Problem solution

3.1 TGA/DTG Analysis

Thermo gravimetric analysis is useful to determine thermal stability and degradation behavior of a polymer or composite [17]. TGA quantifies the mass/change of mass due to different processes including out gassing, dehydration, devolatilization, decomposition, oxidation, reduction, and other chemical reactions as a function of temperature and/or time. DTG (first derivation of TGA) which describe the mass loss rate (dm/dt) is used to understand various decomposition temperatures as well as the number of steps involved in the thermal degradation of the PSA fiber.

TG–DTG curves presented in Fig.1 show the thermal behavior of PSA under nitrogen. As can be seen from the plot that a small weight loss of 1.801% occurred at the initial area from a (49℃) to b(123℃) mainly attributed to the evaporation of free water in the sample. Subsequently no significant weight loss took place. The total weight loss in the range 123-373°C is only 0.840％, possibly via dehydration of crystal water or some additive in fiber. From then on weight loss increased rapidly above 400°C to the end.

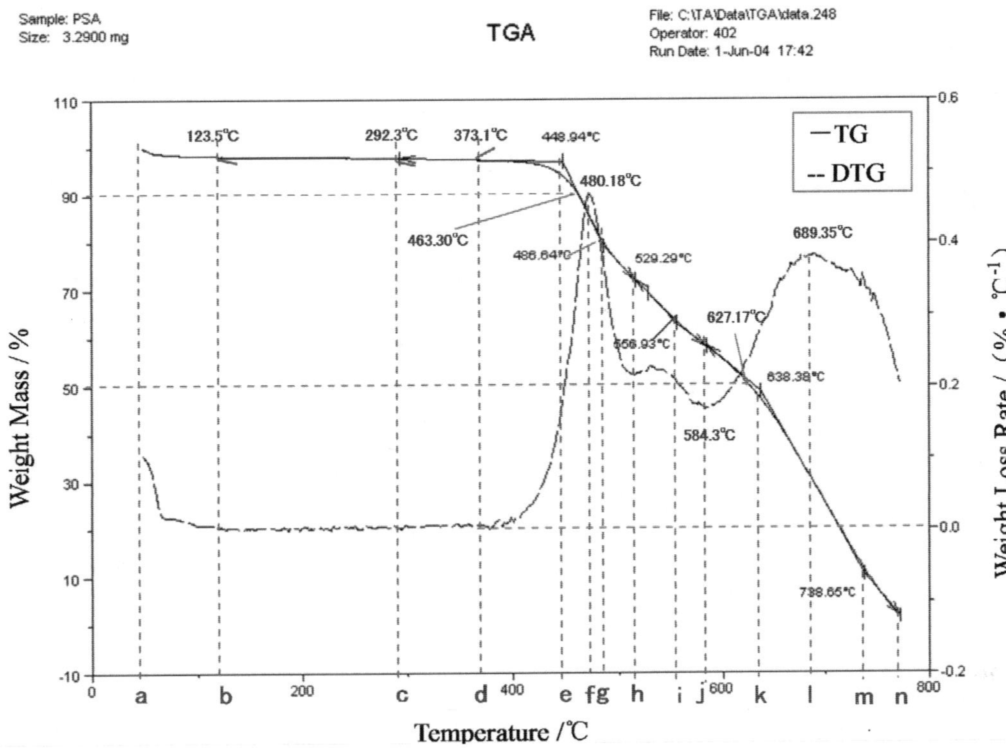

Fig.1 TG–DTG plot of PSA tested at a heating rate of 10℃/min in N2 atmosphere

According to the differential mass loss (DTG) thermograms, PSA fiber has three regions of weight loss, implying at least three stages of degradation from373 to 772℃. The degradation at first stage (ranging from 373-529℃), at second stage (ranging from 529-584℃) and at third stage (ranging from 584-772℃) are reflected in three peaks in the DTG curves at around 480℃, 541℃ and689℃, respectively.

The first stage corresponds to the temperature span of 156.19℃ showing 28.06 % mass loss attributed to the rupture of the molecule chains or the release of low vaporizing components. As for PSA, the change in the base-line slope near 373℃ is probably due to the physical changes in fibers. The extrapolated onset temperature of this region te, calculated by construction method, is 448.94℃. Since te is mostly reproducible, it is often used to characterize the stability of polymer.[18] The weight loss raised up to 10% at 463.30℃.And the maximum mass loss rate 0.430%·℃-1 can be seen as the peak in the DTG curve at approximately 480℃.

The weight loss in the second stage which called transition decomposition is 9.92 % .When the temperature increased those materials with small molecular decomposed further. The breakage of molecular chain increased as well [19].

The third region of DTG curve is evident spanning the temperature range of 584.3-772℃. The extrapolated onset temperature of this region is 638.38℃.At 627.17℃， the mass loss is 50 % .When the temperature rises to 689.35℃,the mass loss rate arrives to the second peak of DTG curve at 0.383%·℃-1. And rapid weight loss occurs up to a residual weight of about 2% at 780°C.This mass loss is representative of the amount of the volatilization of highly volatile matter such as moisture, residual solvents, or other low melting components, and moderately volatile matter.

3.2 Thermal pyrolysis process analysis

Pyrolysis is one of the first steps of all thermochemical processes occurring in an inert atmosphere. It is a thermal decomposition process based on a series of complex reactions that are influenced by many factors, such as heating rate, temperature, pressure, residence time, moisture etc [20]. Fig.2 presents the total ion chromatogram of the pyrolysis products of PSA fiber under He atmosphere at 570℃. It can provide useful information on the mechanism of the thermal degradation in polymers.

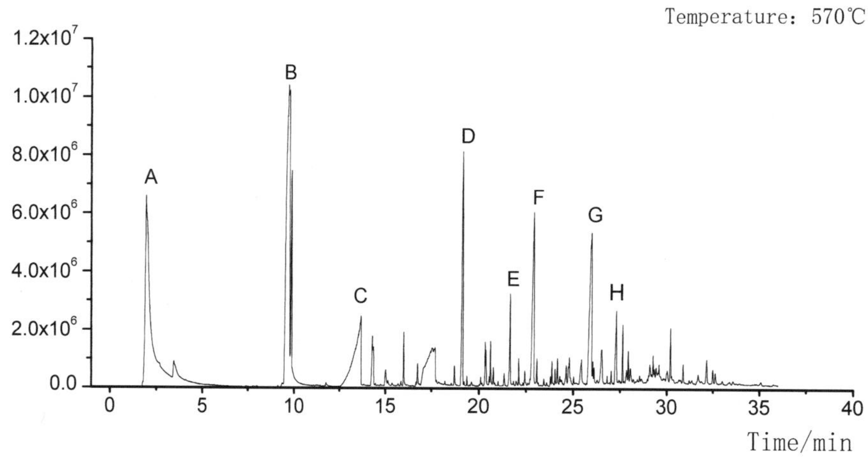

Fig.2 Total ion spectrum of the Pyrolysates of PSA fiber in Helium at 570℃

Py-GC/MS is taken at a temperature point and affords the advantages of the separation power of gas-chromatography coupled with the identification power of mass spectrometry. As a result, volatile thermal products, temporarily non-reactive, thermally stable and fragments can be detected [15]. The MS results of eight chromatographic peaks A~H in Fig.2 are shown in Fig.3 a-h, respectively.

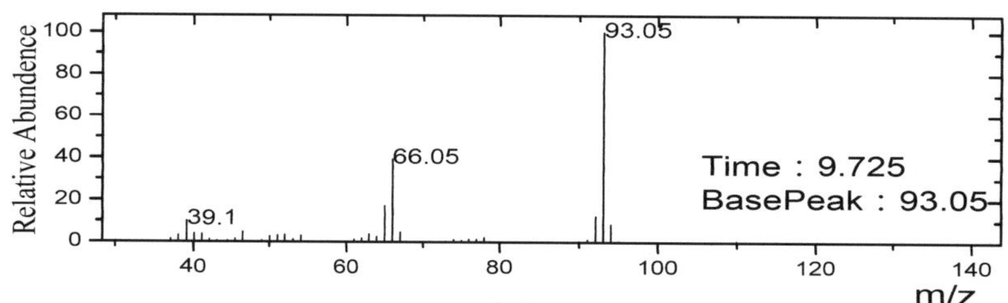

(b)

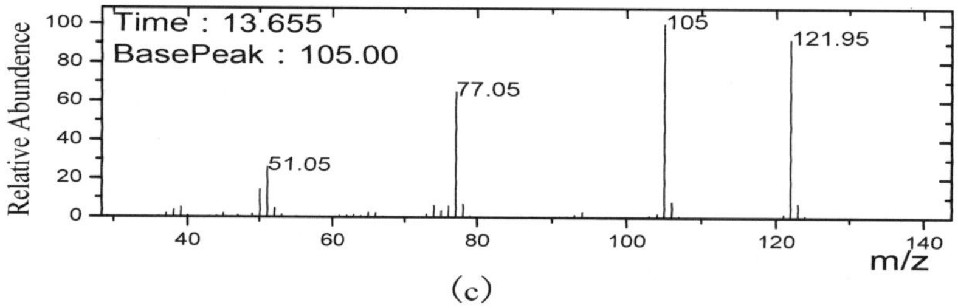

(c)

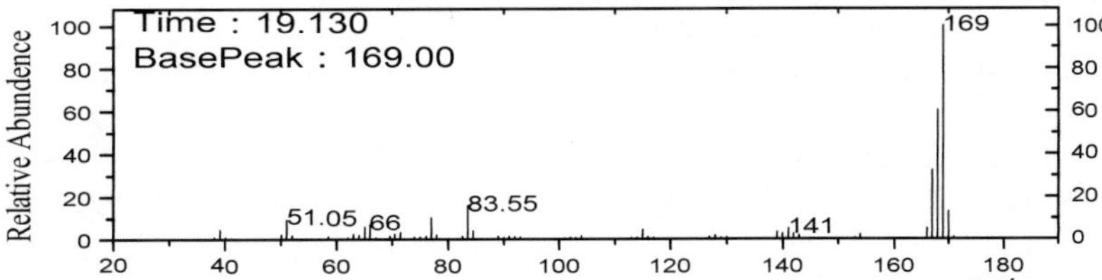

(d)

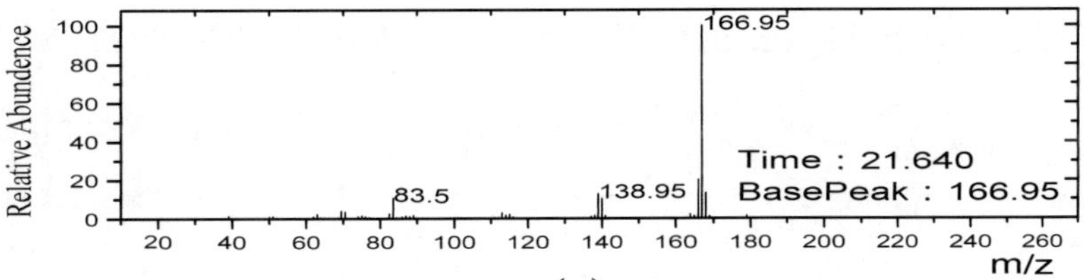

(e)

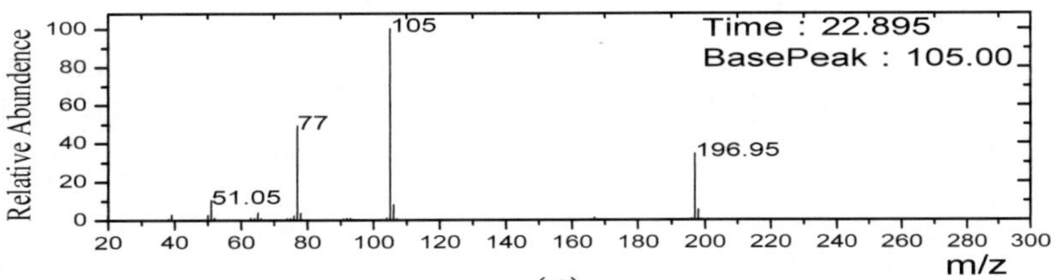

(f)

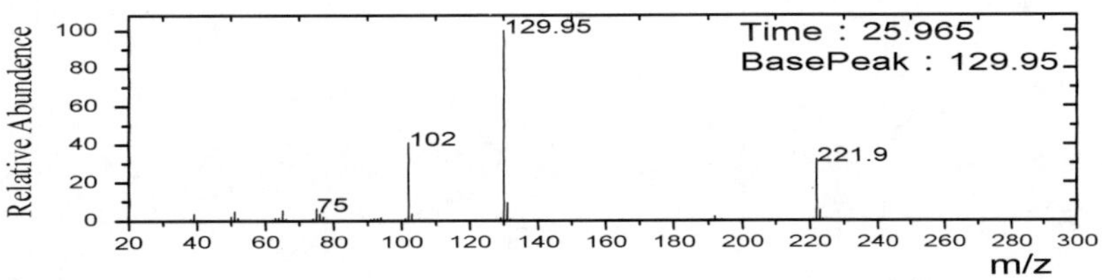

(g)

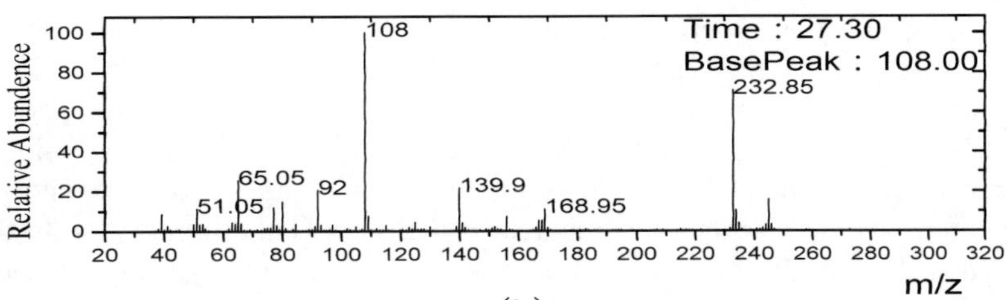

(h)

Fig.3 The mass spectrum of chromatographic peaks

There is a signature representative of the formation and liberation of different organic compounds due to the presence the characteristic peaks. Identification of the pyrolysis gaseous products is achieved by comparing the observed mass spectra to those in the NIST mass spectral library [21].

Gaseous compounds that evolved during thermal decomposition, identified using the MS spectra, are listed in the order of retention times in table1. Of course, GC/MS only gives partial products in the pyrolysates, but it is still useful to study the pyrolysis process of fibers. As shown in table 1 the pyrolysates of PSA fiber under helium at 570°C mainly consist of Sulfur dioxide, Aniline, Benzoic Acid, Diphenylamine, N-Phenyl-benzamide, 3-Cyano-benzoic acid, 4-Aminodiphenyl sulphone and other compounds, corresponding to the molecular structure of PSA fiber 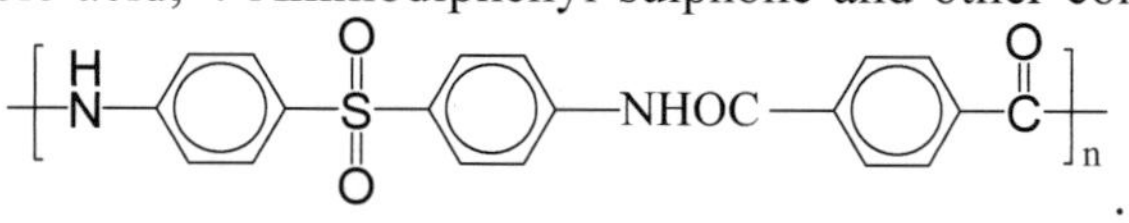 .

Table1. Pyrolysates of PSA during thermal decomposition under helium at 570°C

Peak	Molecular formula	Molecular mass	Constitutional formula	Index of similarity (%)
A	SO_2	64	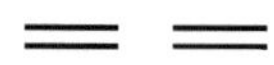	87
B	C_6H_7N	93	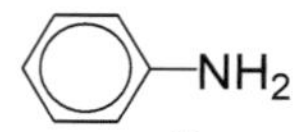	96
C	$C_7H_6O_2$	122	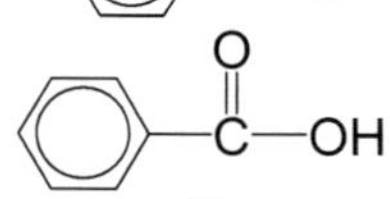	92
D	$C_{12}H_{11}N$	169	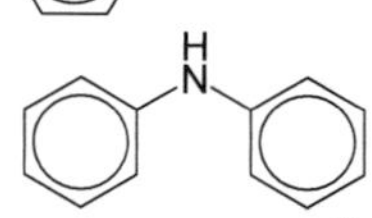	91
E	$C_{16}H_{29}NO_8$	363	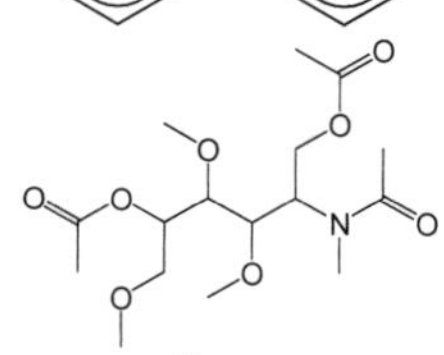	95
F	$C_{13}H_{11}NO$	197	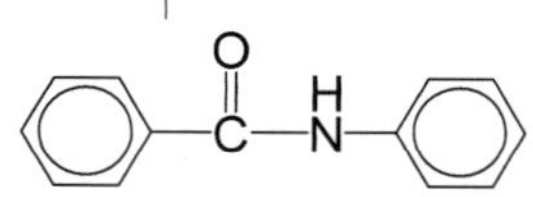	97
G	$C_8H_5NO_2$	147	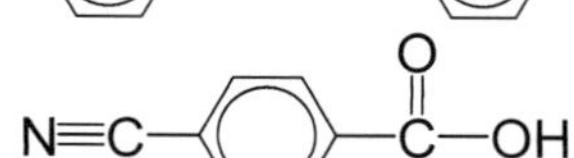	81
H	$C_{12}H_{11}NO_2S$	233	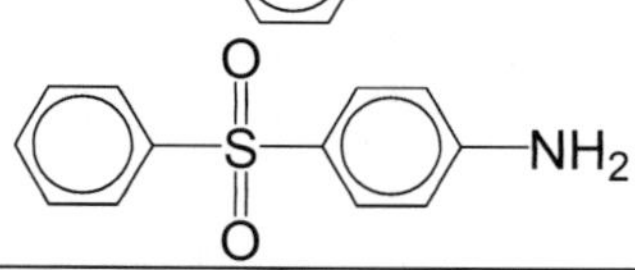	88

The pyrolysis may consist o f one or more pyrolytic reactions occurring simultaneously or sequentially. The path of a pyrolytic process depends on the experimental conditions. During pyrolysis, polymeric chain scissions take place randomly or at the weak link as a successive removal of the monomer units from the polymeric chain. C-S bridge linkage is ruptured; a series of compounds of sulfur dioxide, benzamide and 4- benzoylamino benzamide are formed. Then, benzonitrile and N-(4-cyanophenyl) benzamide are produced from benzamide and 4- benzoylamino benzamide by the elimination of water. Pyrolysis of polysulphoneamide could be shown as. Of course, the recomposition and polycondensation of fragments occur with hydrolys is and homolysis. In this temperature range, most fragments appeared in the results of TGA-DTA/MS and Py- GC/MS can be interpreted by the routes of hydrolys is and homolysis.

3.3 DSC calorimetric analysis

Differential scanning calorimetry is an efficient technique for quantitative thermal analysis. It examined the heat evolution from/absorption by a sample in a controlled environment and provides essential information regarding phase transitions, chemical reactions, precipitation, and dissolution. DSC heating–cooling curve is the curve of a simultaneous differential thermal analysis and represents the heat flow (endothermic or exothermic reactions) during a measurement. DSC can be used to identify a polymer based on its melting point. It also provides essential information on the phase transition characteristics [22].

The heat flow curves generated from DSC for the PSA investigated fibers in argon are shown in Fig.4.

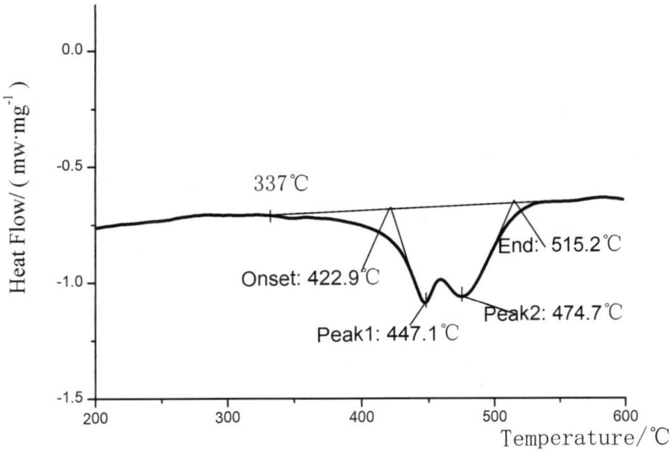

Fig.4 DSC plot of PSA tested at a heating rate of 10 °C min-1 in argon from 200 to 600℃

When temperature rises to 337, endothermic base-line shifts associated with glass transitions. Then two small endothermic peaks with maximum at 447 and 474°C are observed respectively. A decomposition endotherm over a narrow temperature range may easily be mistaken for a melting endotherm, unless the criteria of reversibility and heating rate independence of first-order transitions are applied. In the absence of such evidence, similar endotherm should be treated as decomposition endotherm [9].The onset of the first peak in the DSC curve is at approximately 422.9°C and represents the starting point of the thermal degradation of the PSA fiber. At this temperature, however, no significant mass loss is visible yet in the TGA curve. The area under the second peak in the DSC curve at 474.7 °C represents the energy required for thermal degradation and evaporation of released gases. This second peak of the DSC curve almost coincides with the maximum mass loss rate which can be seen as the peak in the DTG curve at approximately 480.18°C.By 530°C the DSC curve returns to the baseline.

In order to make further comparison analyses with other polyamide aromatic fibers, Nomex and Kevlar are chosen to do the DSC measurements in the same condition. Fig.5 displays the representative DSC heating curves of PSA, Nomex and Kevlar. It can be seen that there is only one strong exothermic peak in DSC traces of both Nomex and Kevlar instead of two in that of PSA.

The characteristic temperature of fibers that obtained from the curve of DSC by the method of Ref [23] indicated that the initial degradation temperature of the Kevlar is the highest. The initial degradation temperature of Nomex is around 370°C, quite similar with that of PSA. However it clearly shows that the end decomposition temperature of Nomex is much lower than PSA fiber. The relative stabilities may perhaps be explained by comparing the general chemical structures. Due to the existence of the conjugated aromatic rings and the additional sulfone (-SO2-) group structure in its molecular main chain, the PSA fiber performs better in thermal stability compared with other high-performance products for the conjugated double bond.

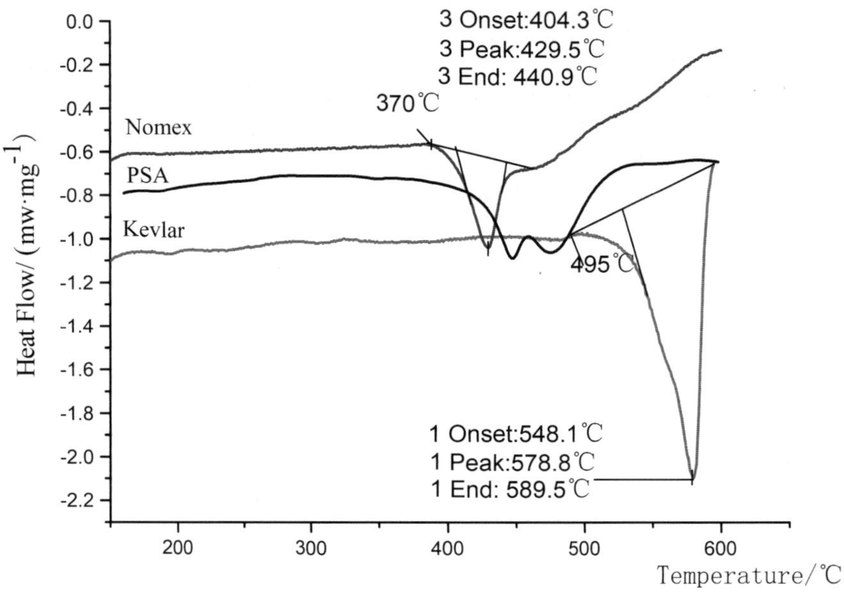

Fig.5 DSC heating curves of PSA, Nomex and Kevlar

4. Conclusion

In this work a study of the thermal decomposition behavior of PSA fiber is investigated with a combination of TG–Py-GC/MS–DSC. Finally, the following conclusions can be obtained:

1) Thermal decomposition of the PSA occurs in several stages. At the start, when heated to a temperature of 400℃, the weight of the sample is reduced by 2.85%. Then the temperature of 10% decomposition is 463.30℃.When the temperature rises to 500℃, the weight loss is equal to 24.17%. On heating to 600℃, the weight of the PSA is reduced by 44.01%. The temperature of half decomposition is 627.17℃.

2) Mass loss rate comes to the maximum at approximately 480.18℃.

3) The decomposition volatiles of PSA fiber at 570°C are mainly SO_2, C_6H_7N, $C_7H_6O_2$, $C_{12}H_{11}N$, $C_{16}H_{29}NO_8$, $C_{13}H_{11}NO$, $C_8H_5NO_2$, and $C_{12}H_{11}NO_2S$.

4) The degradation temperature of PSA fiber is 420℃, lower than Kevlar but a little higher than Nomex. The initial degradation temperature of PSA fiber is 373℃.The first onset temperature is 449℃.

References

[1] Ren JR, Wang XF, Zhang YH. Development for market and application of polyphenylene sulphona-mide fiber[J]. TechnicalTextiles,2007; 25(5): 1-6. (in Chinese)

[2] Liu JX, Tang ZY, Zhang DR, Sun JL. Preparation and performance of polysulfonamide nano-composites and it's fiber[J]. TechnicalTextiles,2007; 2: 14-20.(in Chinese)

[3] Xin BJ, Chen ZM, Wu XJ. Preparation and characterization of PSA/nano-TiO2 composites and fibers [J].Journal of the textile institute, 2013; 104(2): 164-169.

[4] Chen ZM, Xin BJ, Wu XJ, Wang XF, Du WP. Preparation and characterization of PSA/CNT Composites and fibers[J]. Fibres&Textiles in Eastern Europe.2013; 21(1): 24-28.

[5] Zhang LP, Jiang YY, Ma H. Functionalization of polysulfonamide (PSA) fabrics with AI(OH)(3) thin films by electrostatic self-assembly method[J].Industrial & engineering chemistry research. 2013; 52(10): 3607-3612.

[6] Banduryan SI, Ivanova NA, Efimova SG, *et al.* Polysulfonamide fibers produced by dry and wet spinning processes[J].KhimicheskieVolokna.1977: 5,59.
[7] Wang ZH, Shen HG. Tensile Behavior of Polysulfonamide Fiber after Chemical Treatment， International Conference on Future Energy, Environment , and Materials (FEEM), Hong Kong, CHINA, 2012: 432-437.
[8] Ding X, Chen Y, Chen S, *et al.* Copolymer Structure and Properties of Aromatic Polysulfonamides[J].Journal of Macro molecular Science,,PartB:Physics,2012; 51: 1199- 1207.
[9] Brown JR, Ennis BC. Thermal analysis of Nomex and Kevlar fibers [J]. Text Res J, 1977; 47(1):62-6.
[10] Brown JR, Hodgeman DKC. An ESR study of the thermal degradation of Kevlar 49 aramid [J]. Polymer, 1982; 23:365-8.
[11] Xiao CF, Zhang YF, Wu SZ. Investigation on the thermal behaviors of polyarylate /polyester blend fibers [J]. Journal of Tianjin Institute of Textile Science and Technology, 1998; 17(5): 1-5.
[12] Villar-Rodil S, Martı0nez-Alonso A. Studies on pyrolysis of Nomex polyaramid fibers [J]. J Anal Appl Pyrolysis, 2001; 58(2): 105-15.
[13] Brown JR, Power AJ. Thermal degradation of aramids: part I: pyrolysis/gas chromatography/mass spectrometry of poly (1, 3-phenyleneisophthalamide) and poly (1, 4-phenylene terephthalamide) [J].Polym Degrad Stab, 1982; 4: 379-92.
[14] Brown JR, Power AJ. Thermal degradation of aramids–part II:pyrolysis/gas chromatography/mass spectrometry of some modelcompounds of poly(1,3-phenyleneisophthalamide) and poly(1,4-phenylene terephthalamide) [J]. Polym Degrad Stab, 1982; 4: 479-90.
[15] Cai GM, Yu WD. Thermal degradation of high performance fibers by TG/FTIR and Py-GC/MS [J]. J Therm Anal Calorim, 2011; 104: 757-763.
[16] ASTM D3417 99 Standard Test Method for Enthalpies of Fusion and Crystallization of Polymers by Differential Scanning Calorimetry (DSC).
[17] Bodirlau R, Teaca C A, I. Spiridon, Thermal investigation upon various composite materials [J]. Rev. Roum. Chim, 2007; 52: 153-158.
[18] Lin ML. Applications of Thermal Analysis on Determination of Compound Composition [J]. Guangzhou Chemistry, 2001; 29(1): 32-34.(in Chinese)
[19] Zhang XY, Cheng J, Zhao JB. Polymer chemistry [M]. Beijing: China Light Industry Press, 2000: 243. (in Chinese)
[20] Laranci P, Fantozzi F, Bidini G. CFD simulation of biomass pyrolysis syngas vs.natural gas in a microturbine annular combustor [J]. Proceedings of ASME turbo Expo 2010: power for land, sea and air, June 14-18, Glasgow, UK.
[21] NIST/EPA/NIH Mass Spectral Library 2011[M]. U. S: Wiley.
[22] Zhai W, Wang WL, Geng DL, Wei B. A DSC analysis of thermodynamic properties and solidification characteristics for binary Cu–Sn alloys[J]. Acta Materialia 2012; 60: 6518-6527.
[23] Zhang MZ, Liu BJ, Gu XX. Study on the method of polymer [M]. Beijing: China Light Industry Press,2000: 125.(in Chinese)

Influence of Wet Cooling Shirt on Firefighters Clothing

Shadi Houshyar[1*], Daniela Zavec Pavlinic[2], Rajiv Padhye[1]

[1]*School of Fashion and Textiles, RMIT University, Dawson Street, Brunswick, 3056, Australia*
[2] *Titera ltd., Obrtna ulica 40, Murska Sobota, 9000, Slovenia*

[*]Corresponding author's email: shadi.houshyar@rmit.edu.au

Abstract

Protective clothing is designed to protect the firefighter against different threats and it must be comfortable and should not restrict motion and enable heat transfer from the wearer to the environment. The possibility of cooling a person at the microenvironment level must be considered while working in a hot environment to prevent overheating. In this article, passive system for cooling has been investigated. The passive system is in a form of a non-woven fleece material, which was distributed across the surface of a T-shirt on the upper front chest and backside of the T-shirt.

Non-woven fleece with hydro-crystals on contact with cold water swells and in that state lowers the body temperature or keep the body temperature at normal level. In this study, we tested a thermal insulation of the wet cooling T-shirt within the firefighter clothing using a thermal manikin.

Keywords: Cooling shirt; Firefighters clothing; Manikin; Thermal insulation; Hydro-crystals

1. Introduction

Firefighters' protective clothing has to comply with European Standards like EN 469 and AU 4967-2009 to assure their protective properties. Such garments are made by combining 3 or 4 layers of various fabrics with good wicking and comfort properties. This multi-layered fabric system constrains heat dissipation from the human body which leading to un-compensable heat stress. Consecutively accumulation of sweat arises often as well. This reduces thermal and wearing comfort, limits performance at working activities and even becomes life threatening in an extreme hot environment.

We have investigated cooling T-shirt as an under layer of firefighter's protective clothing. The aim of this study was to investigate the influence of cooling T-shirt on overall thermal insulation of firefighter's clothing.

1.1 Thermal parameters of firefighter's clothing

Many researchers worldwide have studied the influence of firefighter's protective clothing characteristics on fire safety improvement and optimization, heat production and injury prevention, like Havenith and Heus [1], Groeller and Taylor [2], Raimundo and Figueiredo [3]. These studies rely on the environmental working conditions to which firefighters are exposed, on simulations of thermoregulatory responses and on parameters of firefighter's clothing. Latter have to be investigated in details as accumulation of heat in the human body and sweat in the microenvironment of clothing mostly depend on the garment characteristics. Heat and sweat are captured between the human skin and the outer shell of the firefighter's clothing. As this clothing is made of few layered materials with high thermal insulation, heat and sweat transfer is restricted. It is obvious that clothing insulation impacts physiological and behavioral reactions of individuals and can cause important, undesired reactions, namely introversion (violent sweating, loss of judgment, amnesia), superficial skin damage (pain and first degree burns), heat stroke (fainting, cessation of sweating, central nervous system alteration) and permanent injuries (greater than first degree burn, brain damage or, in more serious cases, death) [3,4], Fig.1.

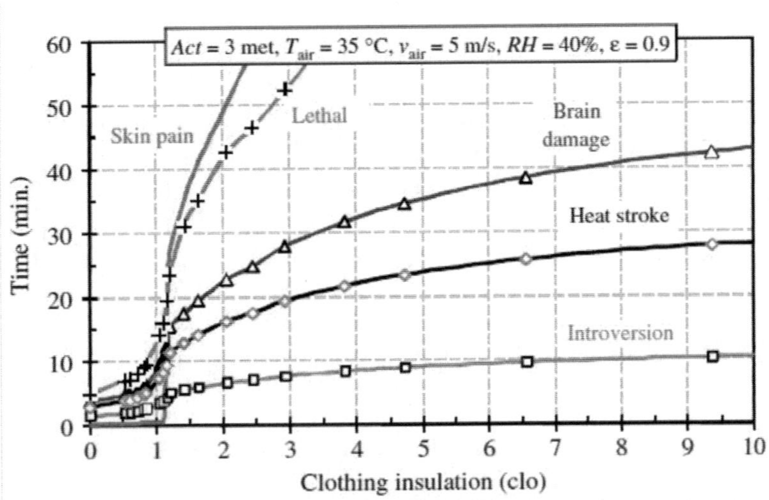

Fig.1 Time for the beginning of skin pain, of introversion, of heat stroke, of brain damage and of death with clothing insulation [3]

It is known that firefighters working activities time is related to the amount of the air in the breathing apparatus, which is around 30 minutes. With reference to the data shown in Fig.1, this is already a time interval, where one can get different undesired reactions and permanent injuries.

2. Experimental

There are different systems are being used for cooling humans during work out, e.g. Drager, Arctic Heat, Cool comfort and Flexi Ice Vest. They are different in their performances over time and functionality. This paper deals with the passive system for cooling of humans using a T-shirt. The passive system in a form of non-woven fleece material was distributed across the surface of a T-shirt on the upper chest and back parts. Non-woven fleece with hydro-crystals on contact with cold water swells and in turn cools down a person or keeps their skin temperature at normal levels. The lower part of the shirt was made of a polyester air-netted material. We assume that this cooling T-shirt can decrease the thermal insulation of firefighter's clothing. The cooling T-shirt can be used in different protocols such as while performing high intensive work, like fire fighting and can be used as standalone product or in combination with other breathable or non- breathable clothing. Furthermore, it can be used as pre-cooling system before the working activity or as inter-cooling over working activities and as well as post- cooling after high intensive work. The user should know his working conditions, like humidity and intensity of work.

During the study thermal insulation of the cooling T-shirt and combination of cooling T-shirt with firefighter's clothing under dry, semi wet and wet conditions were investigated. Both samples were covered with PVC to prevent water vapours escaping from the system. Therefore the changes in the thermal insulation were measured as a result of changes in conductivity not evaporation of the system. The test assemblies were shown in Fig.2.

The Firefighter jacket was made of three layers, inner layer of Nomex (DuPont™ Nomex® Comfort), and middle layer is a thermal barrier and the outer layer is protective made of PBI/Kevlar. The ensembles include the cooling T-shirt with and without jacket. In the first set of experiment the cooling T-shirt was tested wet and dry and in the second set, the cooling T-shirt was covered with the firefighter's jacket. All the ensembles were then covered by impermeable PVC film.

Fig.2 Thermal manikin clothed with the cooling T-shirt, firefighter's ensemble and PVC membrane

Thermal resistance of the ensemble (R_{ct} ($m^2.°C/w$)) was measured using thermal manikin, Newton P357. The manikin consisted of 20 independently controlled thermal zones. Cooling T-shirt was fully wetted before the start of the experiment by immersing the T-shirt into the warm water for 2 minutes. T-shirt was then placed on the towel and rinsed. Ensembles with cooling T-shirts and fire-fighter's jacket on the manikin were then covered with an external impermeable PVC film. To avoid any loss of vapours, manikin was sealed at wrists, neck and the top of the thigh. The external impermeable PVC membrane was used to prevent moisture and liquid evaporation from the system to the environment. All tests were performed in the climate chamber under controlled environmental temperature of 23± 1 °C and RH of 50 %, in accordance with ASTM F1291. The humidity of the room did not affect the testing conditions for the ensemble covered by PVC as the test run on the closed environment. To measure thermal insulation of semi-wetted cooling T-shirt, the cooling T-shirt immersed into the warm water for around 1 minutes and the same procedure as mentioned above (for fully wetted ensemble) was followed.

The clothing cover factor was considered constant at 1.48 for jacket (ASTM F1291); the measurements performed with the standing manikin.

In all experiments "parallel-thermal resistance" was measured, due to the uniform skin temperature, which is closer to the real situation in firefighting. Manikin skin temperature was set to 35 °C and all manikin results were calculated for the area covered by the T-shirt/Jacket ($A=0.978\ m^2$). Generally the thermal/steady-state condition on manikin, in which heating power changes over time to less than 2%, was reached within 1-2 hours. All measurements were taken for 45 minutes after the system has reached steady-state, with coefficients of variation (CV %) < 5 on three replicates.

3. Results and discussion

To investigate the effects of wet cooling T-shirt on the real insulation and heat loss of the protective clothing, the thermal manikin study was done. The study mimicked the real situation during firefighting. The manikin test results of cooling T-shirt are presented in the Table 1 and it can be observed that, by wetting the cooling T-shirt, the thermal insulation of ensemble dropped dramatically. There are few reasons for this phenomenon. Hydro-crystals inside the T-shirt swelled in wet state; therefore it cooled down the wearer and helped to keep the skin temperature at normal levels, by trapping in the evaporated water in the hydro-crystals. Water present inside the cooling T-shirt was evaporated by absorbing energy from the skin of the manikin and led to lowering down the temperature of the skin.

On the other hand the air layer inside the system was gradually replaced by the evaporated moisture which facilitated the heat transfer from cooling T-shirt to the environment due to the higher thermal conductivity of this layer in comparison with air. The amount of the water content inside the system also has an effect on the thermal insulation of the ensembles.

As it can be seen from the Table 1, reduction in thermal insulation of the wet T-shirt was increased significantly from 42% to 65%, when the moisture content of the system was increased from 50% (semi-wet) to 100% (fully wet), respectively. These results were confirmed when compared jacket, cooling T-shirt and their ensembles as shown in Fig.3. Dry jacket has high thermal resistance but it dropped when the cooling T-shirt was placed under the jacket. Even with covered PVC film a thermal resistance of the ensemble was lower than the dry jacket by itself.

Table 1.Thermal resistance of the cooling T-shirt ensembles

	R_{ct} [m^2.°C/w]	
	Average	Std
Wetted cooling T-shirt no cover	0.04	0.00
Wetted cooling T-shirt covered with PVC	0.05	0.02
Wetted cooling T-shirt covered with jacket and PVC	0.13	0.01
Semi-wetted (50%) cooling T-shirt covered with PVC	0.08	0.01
Semi-wetted (50%) cooling T-shirt covered with jacket and PVC	0.26	0.01
Dry cooling T-shirt covered with PVC	0.14	0.01
Dry cooling T-shirt covered with jacket and PVC	0.38	0.02

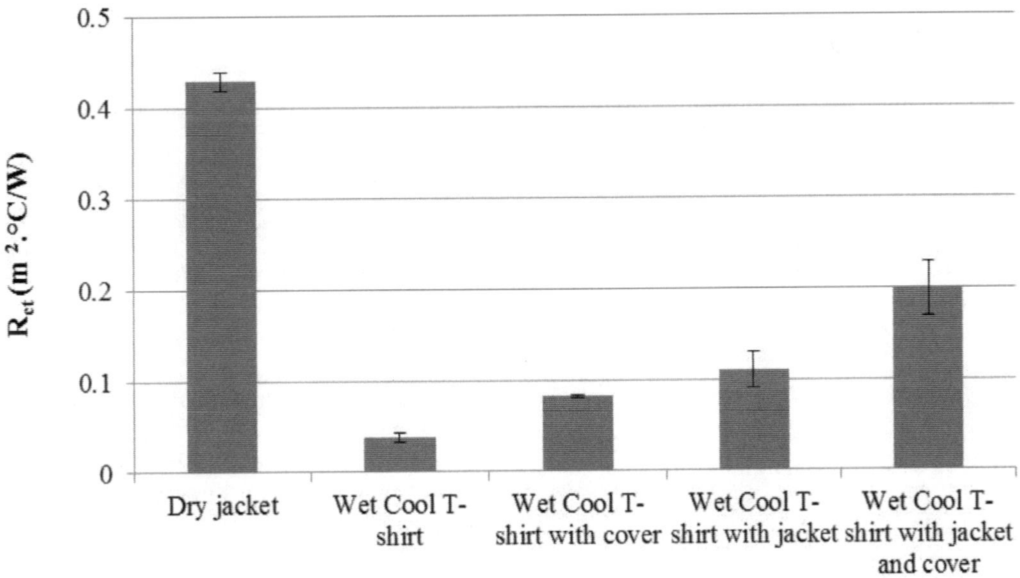

Fig.3 Thermal resistance of the jacket, cooling T-shirt and ensemble

Cooling T-shirt was covered by the impermeable PVC to prevent evaporation to the surrounding environment. Therefore, part of the moisture evaporated from the T-shirt, which taking up the latent heat of the evaporative cooling has again condensed inside the PVC film, which led in releasing the heat and rising the temperature of the film. Dry heat loss to the environment from the system was increased due to the raised in temperature of the film. Therefore, it resulted in more moisture condensation inside the membrane that has been transferred from the cooling T-shirt to this layer.

In the second set of experiment the cooling shirt was covered by the jacket and impermeable PVC, Table 1. Wetting cooling T-shirt in the ensemble was resulted in reduction of thermal insulation of the system by 32% and 65%, when the moisture content of the system was 50% (semi-wet) and 100% (fully-wet) respectively. This can be explained by the fact that evaporation occurred close to the skin resulted in cooling down the skin and reducing the skin temperature.

Wicking is important factor that helps in transferring the moisture from the cooling T-shirt to the PVC membrane. However, wicking was stopped just before reaching semi permeable layer, (thermal barrier), inside the jacket, which allowed the passage of the moisture only to the outer layer and led to the moisture accumulation and elevating temperature of the fabric membrane. This mechanism helped in greater moisture evaporation and heat loss to the environment inside the system and even elevating the rate of condensation inside the PVC film.

Several studies by Havenith et al[9]. Wang et al[10]. and Brode et al[11]. confirmed this mechanism. It means by pre wetting the garment close to the skin, the chances of heat strain can be reduced due to the higher thermal conductively of the wetted garment and the location of the wet layer in the system.

The produced cooling T-shirt with hydro-crystal provides the conditions which resulted in reducing the heat strain and improving the comfort of the wearer. However, further investigation is required to consider steam burn during activity at high temperature.

4. Conclusion

Wet under layers inside firefighter's clothing, like cooling T-shirt, helps in reducing thermal insulation of the system. However, further study needed to investigate the mechanism of the moisture transfer from under layer to the environment. Possibly, the best way would be, to combine the protocols of intermittent working periods; pre-, inter- and post cooling; while investigating the feasibility of developing and producing the thinner and lighter clothing for harsh environments.

5. Acknowledgements

The authors wish to acknowledge to the company PreventDeloza Ltd. from Slovenia who provided the cooling T-shirt for the testing.

References

[1] Havenith G, Heus R. Ergonomics of protective clothing. In: Kuklane K, Holmer I, editors. Proceedings of Nokobetef 6 and First European Conference on Protective Clothing. Stockholm, Sweden, 2000.

[2] Groeller H, Taylor NAS. Physical Performance optimisation, A handbook for Firefighters. Human Performance Laboratories, University of Wollongong, Australia. ISBN: 978-1-74128-152-1.

[3] Raimundo AM, Figueiredo AR. Personal protective clothing and safety of firefighters near a high intensity fire front. Fire Safety Journal 2009; 44: 514 – 521.

[4] EN 469 norm, Protective clothing for firefighters, European Standard, CEN, Brussels, 1995.

[5] Dräger. Dräger comfort vest CVP 5220: http: //www.draeger.com/sites/enus_us/Pages/Fire-Services/DraegerComfort-Vest.aspx.

[6] Personal Climate Systems- The Challenge: http: //www.security-technologynews.com/article/personal-climatesystems-challenge.html.

[7] Arctic heat body cooling products: http: //www.arcticheat.com.au/.

[8] Cooling vests arctic heat USA: http: //www.arcticheatusa.com/.

[9] Havenith G,Wang X, Richards M, Candas V, Meinander H, Broede P, den Hartog EA, Holmer I, Nocker W. Apparent and real cooling efficiency of moisture evaporation from the skin while wearing protective clothing. J Physiol Anthropol 2007; 26: 272-273.

[10] Wang F, Shi W, Lu Y, Song G, Rossi RM, Anaheim S. Effect of moisture content and clothing fit on clothing apparent "wet" thermal insulation: A thermal manikin study. TRJ 2015: 1-7.

[11] Brode P, Havenith G, Wang X, Candas V, Hartog EA, Griefahan B, Holmer I, Kuklane K, Meinander H, Nocker W, Richards M. Non-evaporative effect of a wet mid layer on heat transfer through protective clothing. Eur J Appl Physiol 2008; 104: 341-349.

Effect of Air Gap volumes on Thermal Insulation and Evaporative Resistance of Body Heat Underwear

Xing-Yuan Duan[1*], Yuen-Shing Wu[2]

[1*]*School of Art Design, Jiangsu University of Technology, Jiangsu Changzhou, 213001*
[2]*Human Ecology Department, Cornell University, United State New York, 14850*

Corresponding author's email: dxy0126@163.com

Abstract

Evaporative resistance and thermal insulation of clothing are important factors influencing clothing thermal comfort and engineering design. In this study, a standing sweating fabric manikin was used to measure the thermal insulation and the evaporative resistance of men's body heat underwear (size from M to XXL), and the air volumes between the body heat underwear and the manikin surface were obtained by applying a three-dimensional human body scanner and reverse engineering software. On the basis of above, two polynomial regression equations were established separately to predict the thermal insulation and the evaporative resistance from air gap volumes. The results proved that the thermal insulation and the evaporative resistance all increase first and then decrease when air volumes reach some value. Further study shows the moisture permeability index of the body heat underwear is the largest in size M, and the corresponding air volume is 1888cm^3. That's to say, the thermal comfort of body heat underwear is the best when the underwear size is M.

Keywords: Body heat underwear; Air gap volumes; Thermal insulation; Evaporative resistance; Moisture permeability; 3D human body scanner

1. Introduction

Air gap volume between a garment and a human body is an important factor affecting the garment's thermal comfort due to the good thermal resistance of static air. Air gaps caused by garment fit. On account of the specific geometry of human body shape and unique fabric properties, the space developed between these two surfaces and the overall contour formed along the body represents the garment fit [1]. Some experimental investigations on air gap and its dimensions for clothing thermal comfort have been reported. Psikuta et al [2]. reported that the air gap of loose-fitting clothing was higher than that of tight-fitting clothing. McCullough et al [3,4]. used a standing manikin to measure the thermal insulation of two pairs of loose-fitting and two pairs of corresponding tight-fitting long trousers, and found that the loose-fitting trousers have much greater thermal insulation than the corresponding tight-fitting long trousers, and the air volume in microclimate is an important factor for determining the insulation value of clothing ensemble. Havenith et al [5]. also measured the thermal insulation of loose fit and tight fit clothing ensembles and found the similar results with McCullough h et al [3,4]. Y. S. Chen et al [6]. investigated the effects of jackets fit on clothing thermal insulation and moisture vapor resistance by using a manikin, and found both of which increased with the thickness of the air gap between the garment and the body when the air gap is small. Yejin Lee et al [7]. also measured the insulation of vests over shirts as a layered clothing system, and found the insulation of the clothing system was increased as the pattern size increased, but beyond a certain limit, the insulation started to decrease due to convection and ventilation.

With regard to the effect of air gap sizes on garment thermal comfort, much of the past work has focused on the measurement of thermal resistance of regular clothing and fabric, without considering the influence of functional fabric, for example, clothing making of far-infrared fabric, and there are no work to report the relationship between the air gap sizes and the body heat functional garment. In fact, different fabric has some influences on thermal properties [7]. In this study, the thermal insulation and evaporative resistance of body heat underwear were all tested, and the air gaps between the underwear and the manikin surface were also measured. The purpose of the study is to ascertain the relationship between air gap sizes and the thermal insulation and evaporative resistance of body heat underwear.

2. Methods

2.1 Experimental garments

To investigate the relationship between air gaps, thermal insulation and evaporative resistance, four men's body heat underwear made of solar touch performance fabric that can generate heat by special material's vibration were selected to be measured. These fabrics respond to infrared ray from sun or the human body. Size M is close-fitting, size L is fitted, and size XL and XXL are loose-fitting.

The physical characteristics of fabrics for the experimental underwear were shown in Table 1. The FAST-1 was used to test the thickness of knit fabrics at a low load (196 Pa), the FAST-2 was used to test bending rigidity, and the PERMETEST was used to test the thermal insulation of fabric.

Table 1. Physical characteristic of fabrics for the experimental underwear

garment	Fiber content(%)	Knit	Thicknes s(mm)	Weight (g/m^2)	Bending rigidity (μN·m)	Fabric thermal insulation (℃·m^2/W)
Body heat underwear	Arcylic 40 Rayon 20 Polyester 35 Polyurethane 5	Plain weft	0.53	131.9	36.4	0.0233

All the underwear was made in the same style of long sleeved and zip up at the back center. They gradually increased at chest girth, waist girth and shoulder width with the increase of sizes, but the sleeve length, neck girth and underwear length keep the same to build the different air gap between the underwear and manikin skin. The measurements were shown in Table 2.

Table 2. Measurements of experimental underwear

Underwear size	Chest girth(cm)	Waist girth(cm)	Shoulder width(cm)	Neck girth(cm)	Underwear length(cm)	Sleeve length(cm)
M	88	86	39.5	40.5	72	56
L	92	90	40.5	40.5	72	56
XL	96	94	41.5	40.5	72	56
XXL	100	98	42.5	40.5	72	56

2.2 Manikin

In this study, a standing sweating fabric manikin Walter [8-10] was used to measure the thermal insulation and evaporative resistance of underwear directly. The manikin wore experiment garments with different sizes but the same form-fitting pants to measure.

2.3 3 D body scanning

A 3D body scanner VITUS was used for this investigation, which can generate highly precise 3D images of the human body according to ISO20685: 2005 [11]. Before test, The 3D scanner was calibrated with a cylindrical tube (diameter 110 mm, height 2100 mm, measured at a constant temperature 25℃±2) and was less than 1 mm of average maximal girth error.

A standing fabric manikin (same size with the manikin Walter) was used to scan. It was fixed in the top of the head and suspended from a support, and it can move up and down and turn around the fixed position. The manikin was scanned nude and dressed with experimental underwear using the 3D canner. In order to get reliable measurement data, each item was scanned three times with the manikin random turn around. First, the manikin is bared and scanned to get the surface images of naked bodies, and then put on the experimental underwear to scan to get the surface images of the clothed bodies.

2.4 Determination of thermal insulation and evaporative resistance as well as moisture permeability index

For thermal insulation (I_t), it is determined according to ASTM 1291-10(ASTM 2010a) [12]:

$$I_t = \frac{A_s\,(T_s - T_e)}{H_d} \tag{1}$$

where I_t is thermal insulation (°C m^2/ W); A_s is the total area of manikin (1.79m^2); T_s is the area weighted mean skin temperature (°C); T_e is the environment temperature (°C); H_d is the dry heat loss of the manikin (W); H_d is equal to total heat loss subtracted by evaporative heat loss (Fan and Chen, 2002; Wu et al., 2011).

Evaporative resistance (R_e) is determined by (Fan and Chen [8], ASTM 2010b [13], Wu et al [14].):

$$R_e = \frac{A_s\,(P_{ss} - P_{as} RH_a)}{H_e} - R_{es} \tag{2}$$

Where R_e is evaporative resistance (Pa m^2/W); P_{ss} is the saturated moisture vapor pressure at skin temperature (P_a); P_{as} is the saturated moisture vapor pressure at environment temperature (P_a); RH_a is the relative humidity of the environment (%); R_{es} is the evaporative resistance of the skin (8.6 Pa m^2/W) [8,14]; H_e is the evaporative heat loss (W). $H_e = \lambda Q$, where Q is the perspiration rate and λ is heat of evaporation.

With the measured thermal insulation and evaporative resistance, the permeability index (i_m) is calculated by ISO 9920: 2007(ISO 2007) [15]:

$$i_m = 60.6 \times \frac{I_t}{R_e} \tag{3}$$

2.5 Calculating air gap sizes

The air gap sizes can be obtained by calculating the volumes of naked and clothed bodies with Rapidform software.

The air gaps between the underwear and the manikin surface can be calculated by following formula [16]:

$$V_{air} = V_{cl} - V_{body} \tag{4}$$

Where V_{air} is the volume of air gaps under clothing; V_{cl} is the volume of the clothed manikin; V_{body} is the volume of the naked manikin.

2.6 Thermal insulation (I_t) and evaporative resistance (R_e) experiment procedures

I_t and R_e of body heat underwear were measured under the following conditions and procedures:

1) The experiment was conducted in a climate chamber at a constant environment temperature and relative humidity , and the environment temperature, relative humidity were set as 25℃±2, 65%±5%, respectively.

2) The manikin was maintained at a constant mean skin temperature of 35℃ during the measurement.

3) All experimental underwear was hung in a climate chamber for 24 hours before the test.

4) All experimental underwear was tested three times and they were taken off and put on again between repetitions.

5) Measurements of It and Re were taken only after the moisture accumulation rate within the underwear was stabilized, and the measurements were taken after at least 8 hours.

Fig.1 shows the measurement of four body heat underwear. From left to right: M, L, XL, XXL.

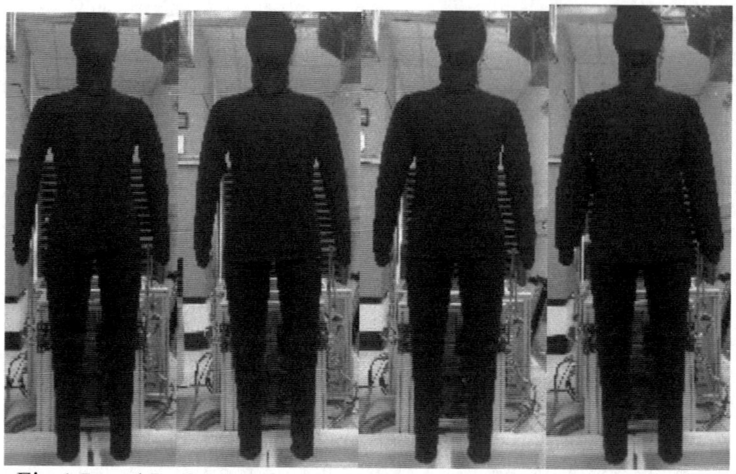

Fig.1 It and Re Measurements of four body heat underwear

3. Result and discussion

3.1 Thermal insulation and evaporative resistance of the experimental underwear

Table 3. Thermal insulation and evaporative resistance of underwear with different air gap volumes

Size	I_t (℃m²/W)	R_e (Pam²/W)	Air volumes (cm³)
M	0.144	21.7	1888
L	0.148	23.0	3115
XL	0.156	23.5	4531
XXL	0.157	23.8	6234

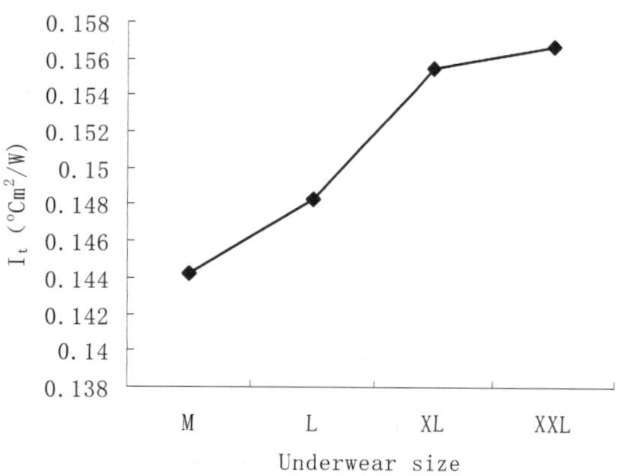

Fig.2 Thermal insulation of underwear

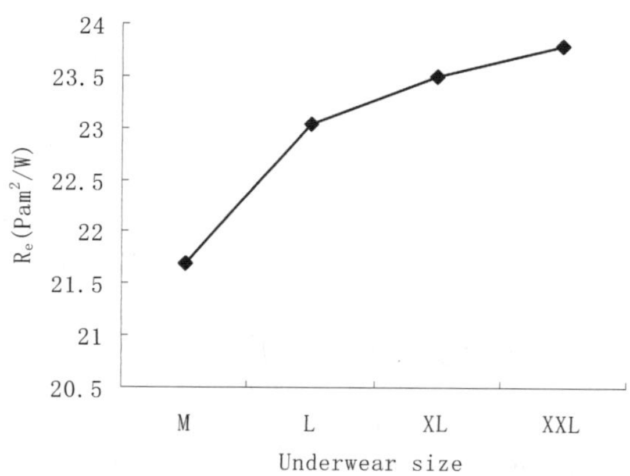

Fig.3 Evaporative resistance of underwear

The values of thermal insulation and evaporative resistance of manikin measurements are listed in Table 3. Fig.2 and Fig.3 shows the change of I_t and R_e. From the two figures we can see the thermal insulation and the evaporative resistance of body heat underwear increase gradually with the underwear size. This is probably because loose fitting garment tend to entrap more still air which has a lower conductivity than fiber and causes some increase in thermal insulation and evaporative resistance. We also can see the thermal insulation increases obviously from size M to size XL, but there is no significant difference from size XL to size XXL. While the evaporative resistance increases obviously from size M to size L, but a little increase from size L to size XXL.

In order to research thermal comfort of body heat underwear further, it is necessary to calculate and analysis the values of Permeability index of underwear.

3.2 Moisture permeability index (i_m) of body heat underwear

Fig.4 shows the moisture permeability of body heat underwear. From it we can see the value of moisture permeability index decreases first and then increases. It reaches maximum when the underwear size is M, and minimum when the underwear size is L, which means the thermal comfort is the best when the size is M, and the corresponding air gap volume is 1888cm³.

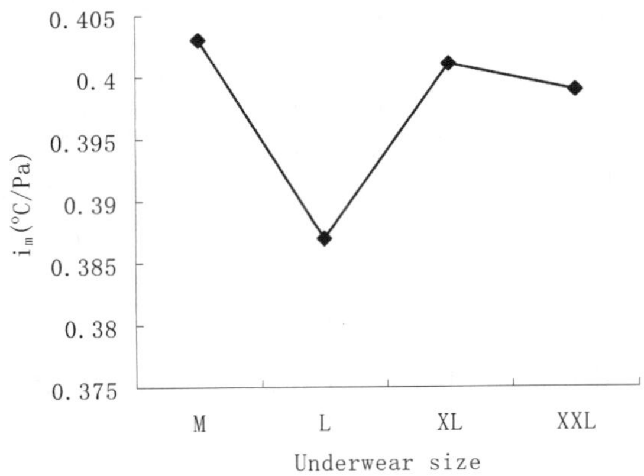

Fig.4 Permeability index of underwear

3.3 Effect of air gap volumes on thermal insulation and evaporative resistance

3.3.1 Relationship between air gap volumes and thermal insulation

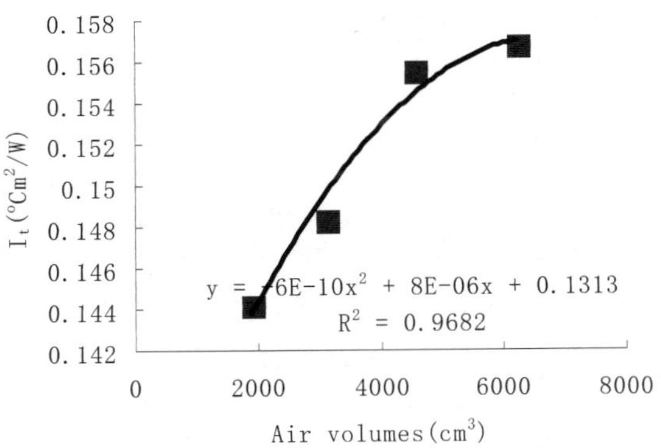

Fig.5 Relationship between air volumes and thermal insulation

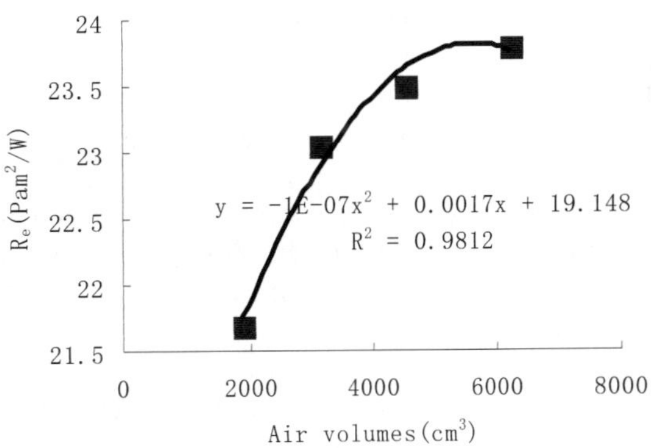

Fig.6 Relationship between air volumes and evaporative resistance

Fig.2 shows that the thermal insulation of body heat underwear continues to increase with the underwear size, but the rate of increase gradually decreases from size XL to size XXL.

To further analyze the effect of air volumes on the thermal insulation of underwear, the polynomial regression equation is established to predict the thermal insulation of body heat underwear from air gap volumes, which is shown in Fig.5. The relationship between air gap volumes and thermal insulation can be described by the following polynomial regression equation (R^2=0.97, $p < 0.05$):

$$y=0.1313 - 6E\text{-}10x^2+8E\text{-}06x \quad (5)$$

Where y is the thermal insulation of underwear (℃m^2/W); x is the air volumes between the underwear and manikin surface (cm^3).

From the polynomial regression equation (5) we can see the thermal insulation of body heat underwear increases within some air volumes, but decreases gradually when air volumes reaches some volumes.

3.3.2 Relationship between air gap and evaporative resistance

As shown in Fig.3, the evaporative resistance of body heat underwear keeps increasing with the underwear size, and it increases obviously from size M to size L, but a little increase from size L to size XXL. That to say, the increase is not clear from size L to size XXL.

To further investigate the relationship between air volumes and evaporative resistance, the polynomial regression equation is also established to predict the evaporative resistance of body heat underwear from air gap volumes, which is shown in Fig.6. The relationship for the body heat underwear can be described by the following polynomial regression equation (R^2=0.98, $p < 0.05$):

$$y=19.148 - 1E\text{-}07x^2+0.0017x \quad (6)$$

Where y is the evaporative resistance of underwear (Pam^2/W); x is the air volume between the underwear and manikin surface (cm^3).

From the polynomial regression equation (6) we can see the evaporative resistance of body heat underwear also increases first, and decreases gradually when air volumes reaches some volumes.

4. Conclusions

In this study, a standing sweating fabric manikin was used to measure the thermal insulation and the evaporative resistance of body heat underwear and to investigate the relationship between the air gap volumes and the thermal insulation and the evaporative resistance and the thermal comfort of body heat underwear. The polynomial regression equation was established separately to predict the thermal insulation and the evaporative resistance from air gap volumes. It was found that the thermal insulation and the evaporative resistance continue to increase with the underwear size, but the rate of increase of the thermal insulation gradually decreases from size XL to size XXL, while the rate of increase of the evaporative resistance decreases gradually from size L to size XXL. From the polynomial regression equation (5) and (6) we can see the thermal insulation and the evaporative resistance all increase first and then decrease when air volumes reach some value. Further, it is found that the moisture permeability of body heat underwear decreases first and then increases, and it reaches maximum when the underwear size is M, which means the thermal comfort is the best when the size is M, and the corresponding air gap volume is 1888cm^3.

The subsequent of this topic is to compare the thermal comfort of body heat underwear with regular underwear, and make a subjective evaluation of comfort and aesthetics on body heat underwear, so as to determine the appropriate allowance in body heat underwear design.

References

[1] Lu YH, Song GW, Li J. A Novel Approach for Fit Analysis of Thermal Protective Clothing Using Three-dimension Body Scanning. Applied Ergonomics, 2014; 45: 1439-1446.

[2] Psikuta A, Frackiewicz-Kaczmarek J, Fridrych I, Rossi R. Quantitative Evaluation of Air Gap Thickness and Contact Area Between Body and Garment. Textile Research Journal, 2012; 82: 1405-1413.

[3] McCullough EA, Jones BW, Huck J. A Comprehensive Data Base for Estimating Clothing Insulation. ASHRAE Trans.1985; 91: 29-47.

[4] McCullough EA, Hong S. A Data Base for Determining the Decrease in Clothing Insulation Due to Body Motion. ASHRAE Trans. 1994; 100: 765-775.

[5] Havenith G, Heus R, Lotens WA. Resultant Clothing Insulation: A Function of Body Movement, Posture, Wind, Clothing Fit and Ensemble Thickness. Ergonomics 1990; 33(1): 67-84.
[6] Chen YS, Fan JT, Qian X. Effect of Garment Fit on Thermal insulation and Evaporative Resistance. Textile Research Journal, 2004; 74(8): 742-748.
[7] Lee YJ, Hong K, Hong S. 3D Quantification of Microclimate Volume in Layered Clothing for the Prediction of Clothing Insulation. Applied Ergonomics, 2007; 38: 349-355.
[8] Fan JT, Chen YS. Measurement of clothing thermal insulation and moisture vapour resistance using a novel perspiring fabric thermal manikin. Measurement Science and Technology, 2002; 13: 1115–1123.
[9] Fan JT, Qian XM. New functions and applications of Walter, the sweating fabric manikin. European. Journal of Applied Physiology, 2004; 92 (6): 641–644.
[10] Wan XF. Fan JT. A novel test method for measuring the thermal properties of clothing ensembles under dynamic conditions. Measurement Science and Technology, 2008; 19: 065702 (8pp).
[11] ISO20685:2005(ISO 2005). 3-D scanning methodologies for internationally compatible anthropometric databases.
[12] ASTM 1291-10(ASTM 2010a). Standard Test Method for Measuring the Thermal Insulation of Clothing Using a Heated Manikin.
[13] ASTM F 2370-10(ASTM 2010b). Standard Test Method for Measuring the Evaporative Resistance of Clothing Using a Sweating Manikin.
[14] Wu YS, Fan JT, Yu W. Effect of positions on the evaporative resistance and thermal insulation of clothing. Ergonomics, 2011; 54(3): 301-313.
[15] ISO 9920:2007(ISO 2007). Ergonomics of the thermal environment - Estimation of thermal insulation and water vapour resistance of a clothing ensemble.
[16] Li J, Zhang ZH,Wang YY. The Relationship between Air Gap Sizes and Clothing Heat Transfer Performance. The Journal of the Textile Institute, 2013; 104(12): 1327-1336.

Influence of Fabric Stitching on Thermal Comfort of Sports Vest

Tong-Hui Zhang*, Yan-Bo Ji

Xi`an Polytechnic University, Jinhua Road, Xi`an, Shan Xi, 710048, China

Corresponding author's email: m18392856087@163.com

Abstract

In recent years, people engage in exercise increasingly often, and sport vest, as popular type of close-fitting clothing, is a particularly important object of research in terms of thermal and moisture comfort that it provides. Firstly, this work selected common sports fabrics available in the market. Then, thermal and moisture properties of these fabrics went through objective testing. Furthermore, spliced vests were designed according to distribution of heat and moisture in the human body. In different experiments, various performance fabrics were applied to different parts of human body in order to determine a combination that provides the best thermal comfort in a sports vest. Finally, subjective wearing experiments were used to evaluate the sports vest, which verified that different heat and moisture distribution resulting from particular splicing designs can enhance thermal and moisture comfort of a vest.

Keywords: Fabric stitching; Sports vest; Thermal comfort; Subjective evaluation

1. Introduction

At present, the research on the fabric heat moisture comfort in China and abroad mainly focuses on the following aspects: Fabric moisture transfer performance [1]; Heat and moisture transfer properties of textile and clothing and its influence on clothing comfort [2]; Research and development of comfortable fabric for sportswear [3]; research of new test methods and equipment applied in this field. Most of works are focused on studying clothing comfort function at this stage. Various types of thermal mannequins were developed to simulate human body, clothing and environment systems [4,5]. In recent years, most scholars have produced a lot of research on thermal and moisture comfort of clothing, but most of them are based on a single fabric. In this paper, different properties of various fabrics were allocated to different parts of sports vest. And a new type of experiment clothing was made to evaluate thermal and moisture comfort of sports vest during exercise. On one hand, this meets the aesthetic demand. On the other hand, it enables hot and wet comfort of sports vest to be increased to top performance levels and provide appropriate guidance for the wearing and selection of sports vests. Research methods of this work are as follows:

1) According to distribution of temperature and humidity in the process of human body movement, five different temperature and moisture distribution regions were identified. 5 different parts of distribution of heat and moisture were upper arm, lower arm, waist to hip line, chest and back raphe respectively. However the upper and lower part of the arm were omitted in this paper because the research object is a sports vest. In this paper, experiment material was chosen from selection available in the market, such as cotton fiber fabrics, modal, bamboo fiber, Odell; and the experiment materials were assessed based on their thermal properties, air permeability, moisture permeability and propensity to wick effect. Three different parts of a sports vest were matched and spliced based on objective experiment data in order to achieve an optimum thermal and moisture comfort.

2) This research selected six objects of 160/84A; experiment clothing was sewed. The subjects were dressed in laboratory clothing in an artificial climate chamber. Subjective evaluation of thermal and moisture comfort of the experiment clothing was conducted during different levels of exercise [6].

3) The sports vest was evaluated comprehensively according to the subjective experiment data combined with its air permeability and moisture permeability.

2. Testing of thermal and moisture comfort properties of fabrics

The thermal and moisture comfort of clothing means the ability to transfer heat and moisture through its area, which results in different body comfort levels in changing environments. This particularly includes heat transfer performance of a fabric (heat insulation, thermal insulation, thermal conductivity, heat) and moisture transfer properties (water resistance, breathability, etc).

2.1 Determination of experiment fabrics

A 10×10 factorial experiment design was used for testing. Fabrics were conditioned for 24 hours prior to testing in a Standard environment in accordance with GB/T 6529-2008 (i.e. 20.0±2 °C, 65±4% RH) [7]. Basic structural parameters of fabric is shown in Table 1.

Table 1. basic structural parameters of fabrics

Sample number	Fiber	Structure	Thickness (mm)	Weight (g/m^2)	Vertical density/10cm	Horizontal density/10cm
1#	95% cotton 5% spandex	Plain	1.07	211.33	280	224
2#	100% modal	Plain	0.57	119.17	188	310
3#	70% bamboo fiber 30% cotton	Plain	0.64	131.03	302	210
4#	95%Odell 5% spandex	Plain	0.53	169.47	278	214

2.2 Testing of thermal and moisture comfort properties of fabrics

This paper focuses on studying the influence of fabric splicing on thermal and moisture comfort of a sports vest. Therefore, the physical properties of fabric, including heat and moisture performance (insulation, air permeability, moisture permeability, moisture transfer) were tested. In order to ensure accuracy of the experiment, the sample involved in the experiment was pre-conditioned under standard pressure regulating temperature and humidity for at least 24 hours.

2.2.1 Thermal insulation properties

Table2. Test results of fabric thermal insulation

Sample number	Thermal resistance (10^{-3})	Thermal conductivity	Clo (10^{-3})	Warmth Retention (%)
1#	41.44	24.18	267.3	41.03
2#	26.27	38.18	169.43	30.98
3#	26.89	37.32	173.4	31.77
4#	23.60	42.86	152.2	28.19

1) Experiment instrument: YG606E textile thermal resistance tester, specified in GB/T 11048-2008 [8].

2) Experiment environment: Temperature: 202 degrees Celsius; humidity: 65%±2%; wind speed of less than 0.1m/s.

3) Experiment procedures: Firstly, set the data of the experiment instrument. Secondly, implement the experiment. Open the glass door above the instrument, then put the sample on the test board so that the test board is completely covered with the sample. The edges of the sample were fixed with adhesive tape in order to ensure the smoothness of the sample. Press the "Yes" button - the instrument will start automatically. The test is completed

after a beep, then the instrument automatically stops.

4) Experiment results: The experiment results are shown in Table 2.

2.2.2 Air permeability

1) Experiment instrument: YG461E-III full automatic air permeability tester

2) Experiment environment: Temperature: 202 degrees Celsius; humidity: 65%±2%; wind speed is less than 0.1m/s.

3) Experiment procedures: Set the data of the experiment instrument in accordance with GB/T 5453-1997 [9]. The fabrics were clamped on the sample circle with enough tension to keep the fabric smooth and not deformed. Padding gasket prevented leakage during experiment.

4) Experiment results: The results are shown in Table 3.

Table 3. Air permeability of fabric

Sample number	1#	2#	3#	4#
Results（mm/s）	147.7	2177.3	1198.9	1262.4

2.2.3 Moisture permeability

1) Experiment instruments: YG601H computer type fabric moisture meter. The dish. Balance. Cylinder. Dryer. Anhydrous calcium chloride is moisture absorption agent in this experiment.

2) Experiment environment: Temperature: 202 degrees Celsius. Humidity: 65%±3%. Wind speed was less than 0.1m/s.

3) Experiment procedures: The specimen tails were placed on a permeation cup, and installed with a washer and a pressure ring. Then the nut was tightened and then sealed with tape from the side of the dish. It was taken out after 30 minutes in the test box. At last, balanced for 30 minutes in the silica gel dryer.

4) Experiment results: The results are shown in Table 4.

Table 4. The experiment results of fabric moisture permeability

Sample number	1#	2#	3#	4#
WVT($g/(m^2 \cdot d)$)	4926	7159	8484	5648

2.2.4 Wick effect

1) Experiment instruments: The tester is composed of a rectangular container and a sample holder. The sample holder has three scales, and the upper end of the scale is provided with a sample holder. Its minimum scale is 0.1cm; Thermometer; Several tension clamps.

2) Experiment environment: Temperature: 202 degrees Celsius. Humidity: 65%±3% . Wind speed is 0.1m/s.

3) Experiment procedures: The end of samples were vertically fixed on the specimen shelf. Clipping tension clamp (about 3g) in the lower 8-10mm specimen were ensures to remain vertical. The level of the beam was lowered to the zero and measure of height when sample gets in touch with water was 30min.

4) Experiment results: The results are shown in Table 5.

Table 5. Sample core suction height after 30 minutes

Sample number	Vertical （cm）	Horizontal（cm）
1#	10.8	10.3
2#	10.3	8.5
3#	15.5	12.2
4#	0	0

3. Design of sports vest

This paper studies the influence of fabric splicing on the thermal and moisture comfort properties of sports vests. In line with the principle that the experiment is accurate, this experiment chose the straight tubular sports vest.

3.1 Reference of dividing line design

The temperature and moisture distribution in different parts of human body is different when people exercise, so the demand for the thermal and moisture comfort of clothing is not same. The body surface is divided into 16 parts by H Y, Wu [10]. According to the temperature and humidity values of clustering and statistical analysis, he got 5 different temperature and moisture parts as shown in Fig.1 and Fig.2. Both are references from Wu's paper.

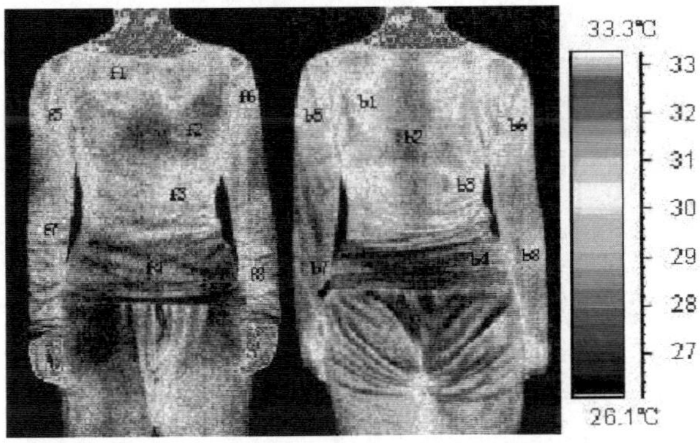

Fig.1 Temperature distribution in different parts of the human body

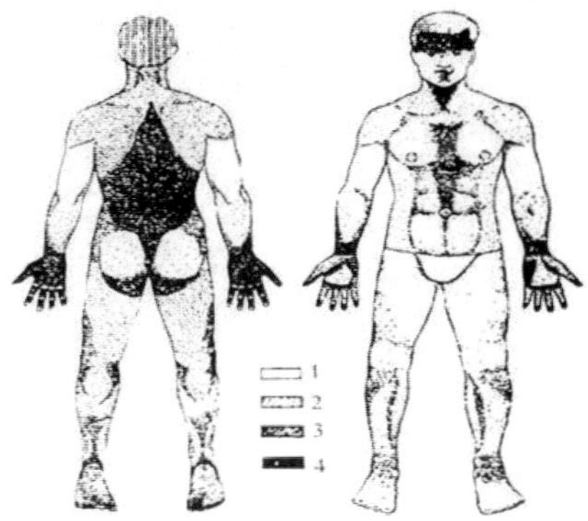

Fig.2 Moisture distribution in different parts of human body

As can be seen in Fig.1, The five parts of the hot wet difference are the upper arm (f5,f6,b5,b6),From waist to hip line (f4,b4), Under the arm (f7,f8,b7,b8), The front and back of the raphe (f2,f3,b2), From the collar and the lower part of the back of the scapula to the back waist (b1,b3,f1) respectively.

As can be seen in Fig.2, 1 stands for the minimum sweating area, 2 for the weak sweating area, 3 for the middle area, 4 for the strong sweating area.

The part going from neckline and back scapula to waist parts was most similar to the human body, so this site is a high calorie area. However, the moisture levels of this part are not the highest, because the movement of sweat mainly accumulates in the chest and back, therefore the humidity of those part is exceeds that of other areas.

3.2 Design of dividing line and the selection of the various parts of the fabric

Based on different parts of the heat and moisture, we made the following design for the splicing line of the sports vest, as shown in Fig.3.

1) Site A is a high calorie area. Vest was designed by using low thermal resistance fabrics 2# and 3# and 4# so that the site can quickly dissipate heat, and improve the overall comfort vest.

2) Site B is a high humidity region. Sweat is mostly gathered in the front and back, when the person is moving. Although back scapular region has high temperature, sweat is usually gathered and condensed in the center back. So 2# and 3# fabric that has better thermal resistance and moisture transfer properties were applied in the high humidity region of this sports vest, making moisture transfer and heat emission in this part quicker. This improved the overall comfort of the vest.

3) The temperature of site C is not high. We use 2#, 3# and 4# fabrics that can quickly absorb moisture, making this part maintain low moisture concentration.

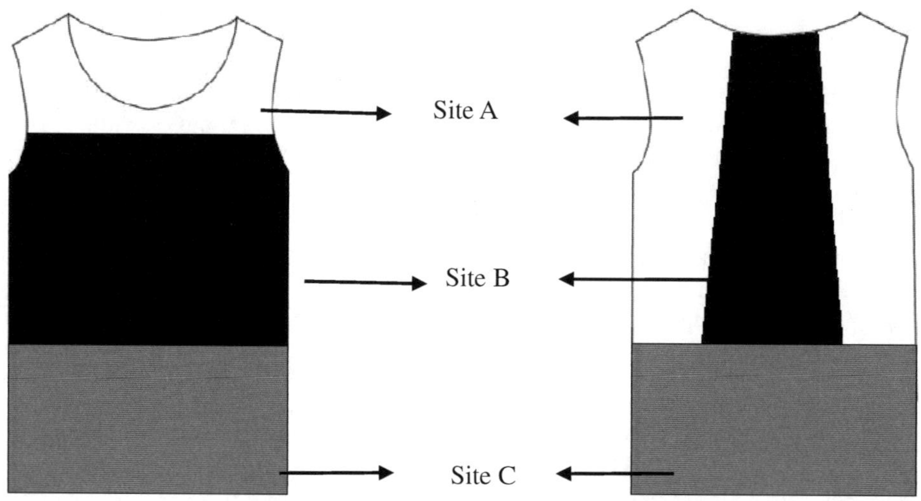

Fig.3 Design of the split line of sports vest

The specific splicing method is shown in Table 6:

Table 6. Method of sport vest splicing

Dress code/Fabric number	Site A	Site B	Site C
1#	2#	3#	1#
2#	3#	2#	1#
3#	4#	2#	3#
4#	4#	2#	1#
5#	4#	3#	2#
6#	4#	3#	1#
7#	1#		
8#	2#		
9#	3#		
10#	4#		

3.3 Specifications of sports vest

The sizes of sports vest are shown in Table 7.

Table 7.The sizes of the sports vest

Parts	Length	Bust	Front neck depth	Front neck width	Back neck depth	Back neck width
Size(cm)	54	84	11	10	3.5	10

4. Design of subjective experiment

The performance of thermal and moisture comfort of fabric can be reflected by objective evaluation of fabric insulation, moisture permeability and other methods. However, comfort is a subjective psychological quality and can not be measured directly. Therefore, it is necessary to evaluate thermal and moisture comfort of sports vests by subjective evaluation in spite of objective evaluation [11].

Firstly, the clothing was made according to certain rules; secondly, a certain number of test subjects were selected; after that, a subjective evaluation test was carried out under certain conditions; Again, performance of personal subjective feeling can use descriptive words in accordance with the requirements of the experiment, or a descriptive vocabulary, or a score or given symbol [12]; Finally, evaluation results were analyzed and conclusions were made.

4.1 Experiment clothing

The subjects were wearing experiment clothes, specifications and design of which are shown in Table 7 and Fig.3. Fabric parameters were shown in Table 1. Lower body part of subjects was covered in unified sports shorts and light sports shoes. All the clothes were cleaned before the experiment, and the temperature and humidity of laboratory have been balanced for at least 12 hours.

4.2 The choice of subjects

This experiment selected 6 female college students aged 21-24 years, height 160±2cm, perimeter of chest 84±1cm. The subjects were required to have good health and sweat easily. The subjects were trained to be familiar with and understand the meaning of the term and grade of the subjective sensation and the physiological characteristics before the experiment. In order to reduce the effects of individual differences, the subjects were not allowed to do a large amount of exercises before the experiment, consume spicy food, alcohol, caffeine and other stimulating food. The experiment was carried out after an hour after a meal.

4.3 The determination of subjective evaluation index and grade

The experiment mainly aimed to evaluation the thermal and moisture comfort of the 10 sport vests. The four indexes closely related to thermal and moisture comfort of clothing are "wet", "thermal", "stuffy" and "sticky". Wet sense means a moist feeling that was made by the sweat stuck on the surface of the skin, without rapid evaporation. This can occur because the clothes absorb sweet slowly. Ability of transferring moisture and the drying rate of the liquid water in the fabric are related to the wet feeling. Thermal sensation refers to ambient temperature, which is related to the thermal resistance and the rate of heat preservation. The stuffy feeling refers to airflow. Sticky sense is the feeling of stickiness on the human body. The grades of the four subjective evaluation indicators are shown in Table 8.

Table 8. The rating scale of subjective sensation

Sensation/ level	1	2	3	4	5
Wet sense	Not at all	Less	Medium	More	Most
Thermal sense	Not at all	Less	Medium	More	Most
Stuffy sense	Not at all	Less	Medium	More	Most
Sticky sense	Not at all	Less	Medium	More	Most

4.4 Activity level

The experiment was divided into four stages: meditation, mild exercise, jogging, and meditation sequentially. The subjects sat for 10 minutes in the first stage. and then carried out the subjective evaluation. The second stage was the mild movement stage. The subjects were evaluated after running for 10 minutes at a speed of 3Km/h on atreadmill. The third stage is the stage of motion. The subjects evaluated again after running for 10 minutes at the speed of 6.5Km/h . The fourth stage is the stage of meditation. and the subjects carried out the subjective evaluation once every 5 minutes. Subjects sat for 15 minutes to adapt to the environment in constant temperature and humidity of the chamber before the experiment . Every experiment only allowed only one subject into the laboratory.

4.5 Experiment procedure

1) Every experiment only allowed one subject into the laboratory. The subjects sat for 15 minutes to adapt to the environment before the experiment, while the testers asked subjects to conduct the experiment process. This stage

did not allow being noisy or walking around spontaneously.

2) The subjects randomly selected clothing for an experiment after 15min. Then they sat down for 10 minutes while wearing the clothing. It is the first time to carry out the subjective evaluation of its heat moisture comfort.

3) The subjects were evaluated after running for 10 minutes on the treadmill at the speed of 3Km/h.

4) The subjects were evaluated after running for 10 minutes on the treadmill at the speed of 6.5Km/h.

5) The treadmill was slowly stopped, and the subjects began to evaluate the clothing respectively after the rest of the meditation at 5 minutes, 10 minutes and 15 minutes .

6) Repeated steps as described above.

7) The experiment clothing was washed in order to facilitate the next use in accordance with the GB/T829 - 2001.

5. Results and analysis

Evaluation scores about the feeling of "wet", "thermal", "stuffy" and "sticky" in four different stages were obtained from the wearer during experiment as it is shown in Fig.4. In this paper, the average value of each experiment service in different stages of subjective score was studied.

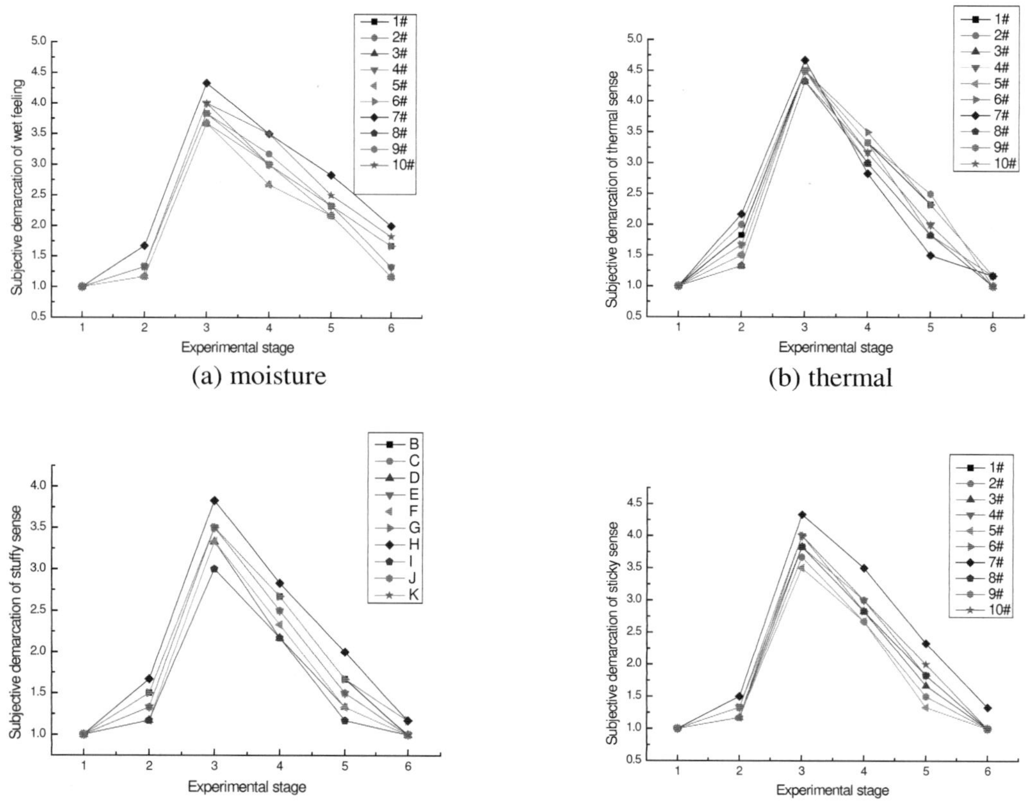

(a) moisture (b) thermal

(c) stuffy (d) sticky

Fig.4 Trends in various types of subjective evaluation

The results show that 7# is higher in six different experiment stages of subjective evaluation. That is to say that 7# is the most uncomfortable in four indicators. That is because 7# fabric is thicker, and possesses the maximum thermal resistance. Therefore, wearing 7# experiment clothing is the mostly uncomfortable in summer periods. 3#, 5#, 8# and 9# are more comfortable than 7# in terms of the selected four senses and are most suitable for people doing a lot of exercises in summer. In general, the spliced vest based can improve the overall comfort based on ergonomics.

6. Conclusions

This research is based on the physical properties of 4 kinds of common fabrics. This work matched fabric with different body areas in order to optimize thermal and moisture comfort of the whole vest according to different properties of 4 kinds of fabrics that relate to sweat and heat transfer. Specific conclusions combined with the results of subjective evaluation are as follows:

1) The neckline and back below scapula to waistline are parts of the clothing that are closest to human body. So the temperature of these part is very high. Based on properties of 2#, 3# and 4# fabrics such as thermal resistance and heat transfer characteristics, this part of the body should use fabrics with high temperature, not high humidity resistance when spliced. Therefore, thermal and moisture comfort of the sports vest is improved and the any discomfort is reduced.

2) Moisture levels in the front and the back parts of the human body increase during movement. Although the temperature at the back part of the scapula is very high, the most sweat accumulates and condenses on the back. So 2# and 3# fabrics should be used to balance the high temperature and high humidity.

3) 3#, 5#, 8# and 9# provide a higher level of overall comfort. So these are most suitable to be worn by people doing large amounts of exercising in summer periods. There is no obvious difference in comfort of different splicing of a sport vest. But these fabrics perform bettes than a 7# sports vest.

The results of this study show that ,according to distribution of body surface heat and moisture, it is feasible to design splicing vest based on the difference of heat and moisture properties. It is evident that there is practical significance of innovative design a sports vest.

7. Acknowledgment

We thank Chun-xiang Yuan for assisting in preparation of this text.

References

[1] Zhang WY, Han JY. A study on the human body dressing test under artificial climatic conditions. J of China Text University: 1991; 2: 72-80.

[2] Wu JH, Yu WD. Fuzzy evaluation of thermal and moisture comfort of knitted sports fabrics. J of Wuhan University of Sci and Eng: 2008; 9: 9-11.

[3] Wu HY. Optimization design of fabric splicing for improving thermal and moisture comfort of sports clothing. Shanghai Text Sci and tech: 2010; 4: 39-41.

[4] Li FF. Research on clothing thermal and moisture comfort based on sweating fabric thermal manikin.Master's degree thesis of Zhejiang Sci-Tech University: 2014; 3-5.

[5] Lei ZX, Qiqn XM. Research status and development trend of sweating manikin. Silk: 2015; 9: 32-36.

[6] Ralph FG. Environmental Ergonomics: Whence-What-Wither. The 11th ICEE in Sweden: 2005.

[7] GB/T 6529-2008, Textiles-standard atmospheres for conditioning and testing: 2008.

[8] GB/T 11048-2008, Textiles--Physiological effects-Measurement of thermal and wate-vapour resistance under steady-state conditions: 2008.

[9] GB/T 5453-1997, Textiles-Determination of the Permeability of Fabrics to Air: 1997.

[10] Wu HY, Zhang WY, Li J. The thermal and moisture comfort of innovational natural hygroscopic fibres. in Pro. IWTC: 2006; 67-72.

[11] Gao MG, Fu JF. Application of subjective sensory analysis in textile and garment industry. Guangxi Text Sci and tech: 2007; 2: 21-25.

[12] Liu LY, Liu L. Study on the relationship between subjective evaluation of clothing comfort and subjective and objective indexes. J of Qingdao University: 2013; 3: 60-63.

The Effect of Air Gap Thickness on Fabric Thermal and Evaporative Resistance

Zhao-Hua Zhang, Ping Xiao*, Shi-Jin Zhai, Si-Yuan Yin

[1]*Fashion and Art Design School, Donghua University, Shanghai, 200051, China*
[2]*Key Laboratory of Clothing Design & Technology, Donghua University, Shanghai, 200051, China*

*Corresponding author's email: xiaoping@dhu.edu.cn

Abstract

Air gaps located between human skin and fabric influence the heat and mass transfer of clothed human body. To quantify the influence of air gap thickness on thermal and evaporative resistance of fabric system, the sweating hot plate was reconstructed with air gap setup to simulate different air gap thickness, and the heat and moisture transfer through human skin, air gaps and fabric was tested as well. The experimental results show that the thermal and evaporative resistance of fabric system increase with the air gaps firstly, and then go down when the air gaps reaches 12mm, after that they go up again. Although the fabric thickness influence the thermal and evaporative resistance of fabric system, the influence of air gaps under clothing is more significant. Taking into account of the fit and thermal comfort of clothing, the local ease allowance for cold protective clothing is suggested to be within 10mm. This paper is helpful to understand the relationship between clothing ease design and heat transfer property, which provides an instruction to clothing comfort design.

Keywords: Fabric system; Thermal insulation; Evaporative resistance; Air gaps; Sweating hot plate

1. Introduction

The air gaps located between human skin and inner surface of clothing influence both the exercise space and heat transfer of human body. Different clothing pattern design has different three-dimensional air gap distributions, and forms different clothing fitness [1-2]. On the other hand, the static air gaps between skin and clothing have better thermal insulation than fibers, especially under windy conditions air ventilation through clothing openings is benefit for heat loss from human body. Chen [3] found that the thermal insulation and moisture vapor resistance increased with the thickness of air gap when the air gap was small, but the rate of increase gradually decreased as the air gap became thicker. Havenith's study [4] showed that when a wearer was sitting or walking in the wind, the thermal insulation of tight clothing was lower by 6%-31% than that of loose one. Recently, the relationship between air gaps under thermally protective garment and skin burn injury patterns has gained increasing attention. Tannie [5-6] measured the air gaps between a female mannequin and protective coveralls using a three-dimensional body scanner, investigated the effect of garment style and fit on thermal protection using a flash-fire instrumented female mannequin evaluation system and proved that air gap size had a positive correlation to burn injury.

Above researches demonstrate that the air gaps influence the heat transfer and thermal protective properties of clothing, which is a vital factor in functional clothing design. But few researches have systematically simulated different air gap thickness and quantified its effect on thermal and evaporative resistance of fabric system. By constructing the air gaps between skin and fabrics, this study analyzed the influence rule of air gaps on heat and moisture transfer of fabric system by measuring the thermal and evaporative resistance with the help of sweating hot plate.

2. Experimental

2.1 Material

Five fabrics with different thickness and air permeability were chosen as the experimental fabrics. As shown in Table 1, the fabric 1# and 5# have different thickness but similar air permeability, while fabric 2#, 3# and 4# have different air permeability but similar thickness. Evaporative resistance of fabric was cut into three testing samples with size 50×50cm², and pre-conditioned in an chamber of temperature 25±0.5 ⁰C, relative humidity 65±4% and wind speed 1±0.1m/s for 24 hours.

2.2 Experimental setup

2.2.1 Air gap setups

To simulate the air gaps between human skin and fabric, five acrylic plates with the same size as the testing samples but increasing thickness of 3mm were used to simulate the air gaps of 3mm, 6mm, 9mm, 12mm and 15mm. As shown in Fig.1, the acrylic spacer was hollowed out the middle and left surrounding sides 5cm wide to make the air go through freely from testing plate through fabric to outside. At the same time, small holes were punched every 5cm alongside the space edge, and the nylon thread was woven through the holes to form a mesh to support the fabric and avoiding fabric collapse by its weight. In such way, the evenly distributed thickness of air gaps can be achieved. During the experiments, the testing samples were pasted to the spacer with adhesive tapes and each sample was tested once under five air gap thickness conditions.

Table 1. Fabric specification parameters

Code	Fiber content/ %	Air permeability/ (mm·s⁻¹)	Density/(10 cm⁻¹)		Weight/ (g·m⁻²)	thickness/ mm
			Warp	Weft		
#1	34% Polyester/65% linen	301.57	210	200	181	0.43
#2	45% Polyester /55% linen	587.19	210	200	121	0.30
#3	100% Silk	213.28	430	360	135	0.30
#4	55% Wool/45% Polyester	317.90	280	250	148	0.31
#5	60% wool/40% Polyester	304.85	140	110	421	2.49

2.2.2 Sweating hot plate

As shown in Fig.2, the experimental setup contains sweating hot plate (SGHP 306-200/400, Northwestern testing technology co., LTD, USA), acrylic spacer and testing sample to test the thermal and evaporative resistance of fabric system (simulated skin-air gaps-fabric). The testing plate can simulate temperature and sweating of human skin, and the guarded plate is used to avoid the heat loss from side and bottom of testing plate. The whole testing equipment was put inside a chamber with controlled temperature and humidity.

2.3 Testing method

The ASTM-F1868 standard (part A) was used to test the thermal insulation of fabric system, the temperature of hot plate was controlled at 35±0.5℃; the climate chamber was controlled with temperature of 25±0.5℃, humidity 65±4% and wind speed 1±0.1m/s. Then the total thermal insulation of fabric system can be calculated by the following formula:

$$R_{ct} = (T_s - T_a)A/H_c \tag{1}$$

Where is the total thermal insulation of fabric system and air layer ($^0C \cdot m^{2} \cdot W^{-1}$); is the temperature of hot plate (^0C); is the temperature of environment (^0C); *A* is area of the plate test section (m^2); is power input (W).

To eliminate the influence of dry heat, the Isothermal testing method (Part B) was used to test the evaporative resistance of fabric system. The temperature of hot plate and chamber was maintained at the same value of 35±0.5^0C, then the total evaporative resistance of fabric system can be calculated as follows:

$$R_{et} = (P_s - P_a)A/H_E \tag{2}$$

Where is the evaporative resistance of fabric system and air layer ($kPa \cdot m^{2} \cdot W^{-1}$); is water vapor pressure in the air (kPa); is water vapor pressure at the plate surface (kPa); *A* is area of the plate test section (m^2); is power input (W).

Then the effective thermal and evaporative resistance of fabric system are the insulation exclude the effect of boundary air layers of fabric:

$$R_{cf} = R_{ct} - R_{co} \tag{3}$$

$$R_{ef} = R_{et} - R_{eo} \tag{4}$$

Where, is effective thermal insulation of fabric system ($^0C \cdot m^{2} \cdot W^{-1}$); is thermal Insulation of the air layer on the surface of fabric ($^0C \cdot m^{2} \cdot W^{-1}$); is effective evaporative resistance of fabric system ($kPa \cdot m^{2} \cdot W^{-1}$); is evaporative resistance of the air layer on the surface of fabric ($kPa \cdot m^{2} \cdot W^{-1}$).

2.4 Data analysis

The statistical analysis of experimental data was processed with Origin 9.0. Liner regression was used to establish the relationship between air gap thickness and fabric effective thermal and evaporative resistance, the significant level was set at $p < 0.01$.

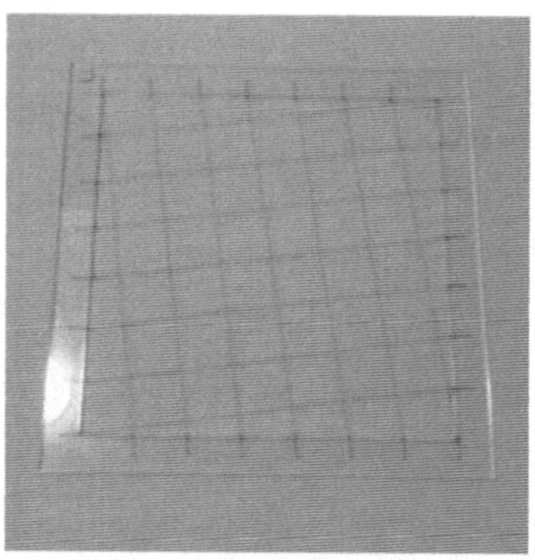

Fig.1 Acrylic spacer to simulate the air gaps

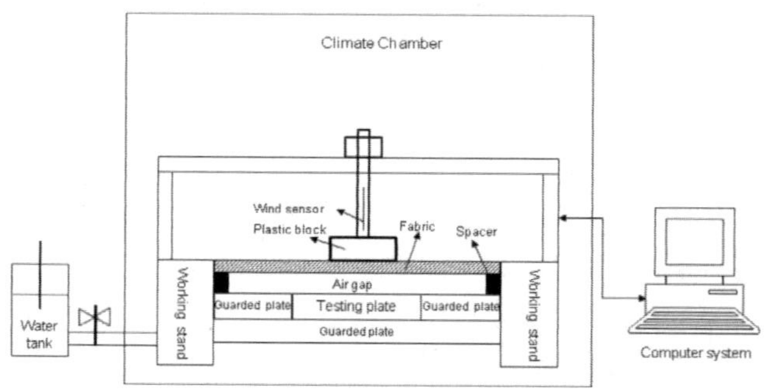

Fig.2 The experimental setup

3. Results and discussion

3.1 Thermal insulation

When air gaps increase from 0mm to 15mm, the effective thermal insulation of fabric system is shown in Fig.3. It can be concluded the following results:

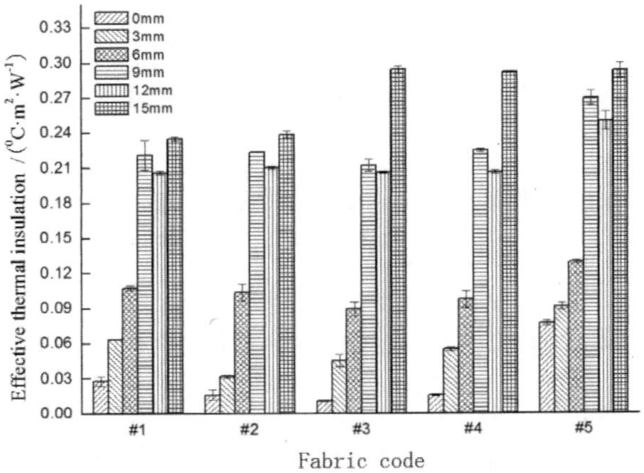

Fig.3 The fabric effective thermal insulation under different air gaps

3.1.1 The thermal insulation increases with the air gap when its thickness is within 9mm

When the air gap thickness is zero, the thicker fabric has better thermal insulation ($3^{\#}<4^{\#}<2^{\#}<1^{\#}<5^{\#}$). But with the increase of air gaps to 9mm, the thermal insulation of five fabric increases gradually with air gap. Linear regression analysis shows the significant linear relationship between air gap thickness and thermal insulation of fabric system (Ad. R^2=0.9; Pearson's r=0.94, p<0.01).

$$y=0.0024+0.0172x \tag{5}$$

Where, is effective thermal insulation of fabric system (^{0}C·m^2/W); is thickness of air gaps under clothing (mm).

3.1.2 The thermal insulation begins to decrease when the air gap increases to 12mm

With the further increase of air gaps, the testing found the effective thermal insulation begins to decrease when air gap thickness reaches 12mm. To testify if natural convection emerges in the air gaps, the heat transfer coefficient was calculated according to heat transfer theory. As the experimental setup can be regarded as a horizontal cavity, the heat transfer within air gaps changes to natural convection when Rayleigh number is bigger than 1700 [9]. The Rayleigh number (Ra) can be expressed as:

$$R\alpha = g\beta(T_s-T_{in})L^3/\alpha\gamma \tag{6}$$

Where, β is coefficient of volumetric (^{0}C $^{-1}$); L is air gap thickness (m); and are temperature of hot plate and inner surface of fabric respectively (^{0}C);g is acceleration of gravity (m·s^{-2}); α and are thermal diffusivity and kinematic) viscosity (m^2·s^{-1}).

Under current experimental conditions, the calculated Ra of 1724 shows natural convection will emerge in air gaps with thickness 12mm. Natural convection increases heat loss and decreases the thermal insulation of fabric system. Thus the threshold of air gap thickness when thermal insulation of fabric system begins to decrease is 12mm, when the heat transfer mode changes from conduction to natural convection as well. Assuming human body as a cylinder, then the air gaps thickness*2π is the ease allowance of clothing, so the clothing ease allowance can be designed as 12mm*2π=7.5cm to get best thermal insulation and fitting beauty .

3.1.3 The effect of fabric thickness on thermal insulation is not significant when air gaps located between skin and fabric

As shown in Table 2, the thickness of fabric $5^{\#}$ is 82.7 % bigger than fabric $1^{\#}$, which results in thermal insulation of fabric system $5^{\#}$ 63.6 % bigger than fabric $1^{\#}$ when there are no air gaps between hot plate and fabric. After inserting 9mm air gaps between hot plate and fabric, the thermal insulation of fabric system $5^{\#}$ is only 18% bigger than fabric $1^{\#}$. It shows that introduction of air gaps weakens the influence of fabric thickness on thermal insulation. Take fabric $2^{\#}$ for example, compared with no air gap condition, the thermal insulation of fabric system increases 49.5% after 3mm air gaps is introduced, and continually increases 84.4% after 6mm air gaps is introduced.

As can be seen from above, although fabric thickness is important for thermal insulated property, it's only the case when the air gaps located under clothing is zero or very thin. Under actual dressing situation of human body, air gaps between skin and clothing represents not only ease allowance design of clothing, but also influence heat transfer from human body.

3.1.4 The influence of fabric air permeability on thermal insulation is not as significant as fabric thickness

Fabrics 3#, 4# and 2# have the same thickness but increasing air permeability, their thermal insulation under different air gaps are shown in Fig.4. Thermal insulation of fabric is increasing with the increase of fabric air permeability when the air gaps is 0mm. Compared with fabric $3^{\#}$, air permeability of fabric $2^{\#}$ increases 63.7 %, while thermal insulation only increases 34.7 %. It demonstrates that the influence of air permeability is not so significant as fabric thickness.

As the air gaps increase to 12mm, the effect of air permeability on thermal insulation of fabric is not significant any more, especially under 15mm air gaps, although fabric $2^{\#}$ has the biggest air permeability, the thermal insulation of fabric $2^{\#}$ is the smallest. This is because airflow in the air gaps is easier to transfer heat by bigger air permeability.

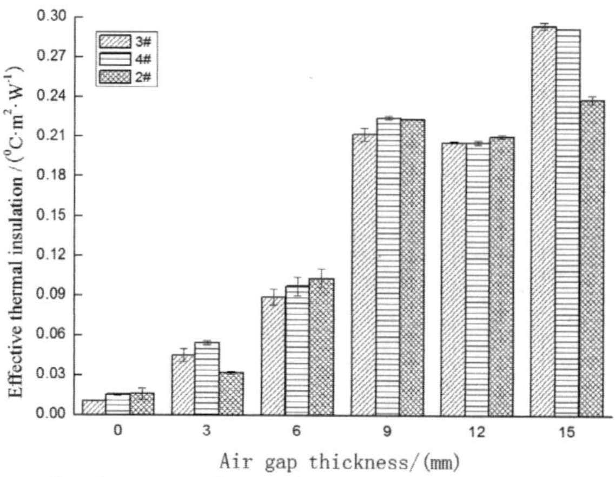

Fig.4 Comparing between three fabrics with different air permeability

3.2 Evaporative resistance

When air gaps increase from 0mm to 15mm, the effective evaporative resistance of fabric system is shown in Fig.5. It can be concluded the following results:

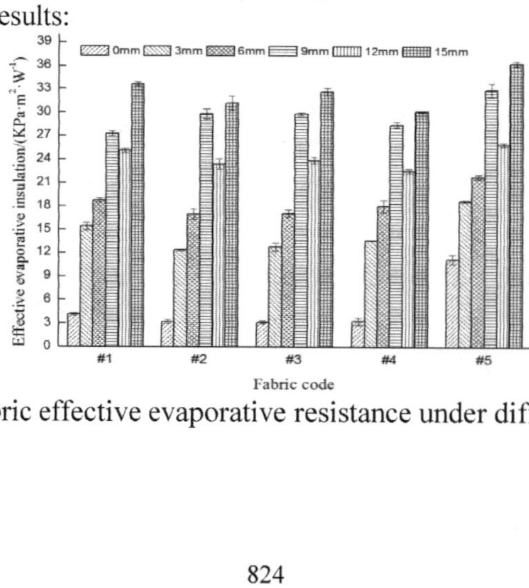

Fig.5 The fabric effective evaporative resistance under different air gaps

3.2.1 Evaporative resistance increases with air gaps, but begins to decrease when thickness of air gaps reaches 12mm

There are similar trend between thermal and evaporative resistance, the critical thickness for evaporative resistance begins to drop is also 12mm. As shown in Figure 5, the evaporative resistances of five fabrics increase gradually when the air gap thickness increases from 0mm to 9mm, this is because vapor transfers through air gaps by molecular diffusion. But evaporative resistance begins to decrease as convective mass transfer instead of molecular diffusion increases the heat loss from fabric system when air gap thickness reaches 12mm.

Linear regression analysis shows the significant linear relationship between air gap thickness and evaporative resistance of fabric system (Ad. R^2=0.82; Pearson's r=0.91, p<0.01). The regression equation can be expressed as:

$$y = 8.0175 + 1.7024x \tag{7}$$

Where, is effective evaporative resistance of fabric system ($kPa \cdot m^2 \cdot W^{-1}$); is thickness of air gaps (mm).

Experimental results show that thermal and evaporative resistances begin to increase again while air gaps increase to 15mm. The thermal and evaporative resistance of 15mm air gaps are even bigger than those of 9mm air gaps. One possible reason is that under 15mm air gaps the conductive heat transfer is more than natural convective heat transfer because of the good closure of experimental setup. Currently none of literatures are available about this phenomenon. Further experiments are still needed to check what will happen under bigger air gap thickness.

3.2.2 The effect of fabric thickness on evaporative resistance is not as significant as air gaps

Similar to thermal insulation of fabric system, when there is air gaps located between hot plate and fabric, the influence of fabric thickness on evaporative resistance is smaller than that of none air gaps situation. It can be seen from Table 2, the thickness of fabric $5^{\#}$ is 82.7 % bigger than fabric $1^{\#}$, which results in evaporative resistance of fabric system $5^{\#}$ is 63 % bigger than fabric $1^{\#}$ when there is no air gaps between hot plate and fabric. But when the air gaps between hot plate and fabric is 9mm, the evaporative resistance of fabric system $5^{\#}$ is only 17% bigger than fabric $1^{\#}$. It shows that introduction of air gaps weakens the influence of fabric thickness on evaporative resistance. Correspondingly, for the same fabric, air gaps can increase the evaporative resistance of fabric system dramatically. Take fabric $2^{\#}$ for example, compared with no air gap condition, the evaporative resistance of fabric system increases 74.2% after 3mm air gaps is introduced, and continually increases 81.2% after 6mm air gaps is introduced.

Table 2. The change of fabric thermal insulation

Name	Fabric thickness / (mm)	Thermal insulation with air gaps of 0mm /($^{0}C \cdot m^2 \cdot W^{-1}$)	Thermal insulation with air gaps of 9mm /($^{0}C \cdot m^2 \cdot W^{-1}$)	Evaporative resistance with air gaps of 0mm /($pa \cdot m^2 \cdot W^{-1}$)	Evaporative resistance with air gaps of 9mm /($pa \cdot m^2 \cdot W^{-1}$)
$1^{\#}$	0.431	0.028	0.221	4.14	27.295
$5^{\#}$	2.488	0.077	0.27	11.193	32.888
$(5^{\#}-1^{\#})/5^{\#}\times\%$	82.7%	63.6 %	18 %	63 %	17 %

3.2.3 The relationship between air permeability and evaporative resistance of fabric

It's a pity that in our current study, no linear relationship is found between air permeability and evaporative resistance of fabric because of the little sample numbers. But some relative researches [7-8] have found negative correlated between them, that is, better air permeability of fabric usually has smaller evaporative resistance.

4. Conclusions

This research based on the testing theory of sweating hot plate, simulated the heat and moisture transfer through skin- air gaps-fabric system by adding acrylic spacer between hot plate and fabric, and analysed the influence of air gap thickness on thermal and evaporative resistance of fabric system.

The experimental results demonstrate that thermal and evaporative resistance increase with air gap thickness and the thermal and evaporative resistance of fabric system will increase 1 times by adding 3mm air gaps. But the increase of insulation is not linear, and begins to decrease because of natural convection when air gaps increase to 12mm. The textile science has known that there is linear relationship between fabric thickness and thermal insulation, but this research found the air gaps have more influence than fabric thickness on insulated property of clothing. This research can be used in designing protective clothing with better ease allowance.

It is worth noting that this experiment simulated air gaps as a total closure space, while under practical dressing situation, air gaps are not fully enclosed and convection through clothing openings such as neck, cuff, helm et. al. will enhance heat transfer from human body to climate. In future, the sweating manikin should be used to analyse the actual effect of clothing fitness on thermal and evaporative resistance of clothing system by changing environmental temperature and wind speed.

5. Acknowledgement

The authors wish to acknowledge the support from Doctoral Program of Chinese Higher Education for New Teacher (ZX201311000009) and Fundamental Research Funds for the Central Universities for this research (16D110709).

References

[1] Song HK, Ashdown SP. An exploratory study of the validity of visual fit assessment from three-dimensional scans. Clothing and TRJ 2010; 4: 263-278.

[2] Loker S, Ashdown S, Schoenfelder K. Size-specific Analysis of body scan data to improve apparel fit. J of Textile and Apparel, Technology and Management 2005; 4: 1-15.

[3] Chen Y, Fan J, Qian X. Effect of garment fit on thermal insulation and evaporative resistance [J]. Textile Research Journa: 2004; 74(8): 742-748.

[4] Havenith G, Heus R, Lotena WA. Clothing ventilation, vapour resistance and permeability index: changes due to posture, movement and wind. Ergonomics 1990; 8: 989-1005.

[5] Mah T, Song G. Investigation of the contribution of garment design to thermal protection. part 1: characterizing air gaps using three-dimensional body scanning for women's protective clothing. TRJ 2010; 13: 1317-1329.

[6] Mah T, Song G. Investigation of the contribution of garment design to thermal protection. part 2: instrumented female mannequin flash-fire evaluation system. TRJ 2010; 14: 1473-1487.

[7] Ding D, Tang T, Song G, et al. Characterizing the performance of a single-layer fabric system through a heat and mass transfer model - part II: thermal and evaporative resistances. TRJ 2011; 9: 945-958.

[8] Jiang P, Yan H. Research on the forecast of heat resistance and moisture resistance of fabrics. Shanghai Textile Science & Technology 2004; 4: 63-64.

[9] Mei F, Zhang X, Pei R. Heat transfer theory. 5th ed. Beijing: China Architecture & Building Press: 2007; 168-175.

Functional Benefits of Introduction of Merino Wool into Cotton Denim for Cold Weather Applications

Nazia Nawaz[1*], Chris Watson[1], Olga Troynikov[1]

[1]*School of Fashion and Textiles, RMIT University, 25 Dawson Street, Brunswick, Melbourne VIC, 3056, Australia*

[*]Corresponding author's email: nazia.nawaz@rmit.edu.au

Abstract

The present study aimed to determine the functional performance benefits of merino wool/ cotton blended denim in comparison to 100% cotton denim when tested under dynamic ambient conditions in a simulated cold weather environment typical of European conditions in late autumn and early spring. The results demonstrated that 66 /34% and 83/17% cotton/ wool jeans performed better in comparison to 100% cotton jeans as relevant to thermal comfort attributes of the wearer at 17°C and different relative humidities due to moisture buffering properties of wool fibres. Furthermore, 66/34% cotton/ wool and 83/17% cotton/ wool jeans demonstrated higher vapour permeability index than 100% cotton jeans due to hygroscopic nature of wool fibres indicating their higher ability to transfer water vapour from next to skin to environment.

Keywords: Merino wool; Cotton; Denim; Cold weather; Functional materials; Moisture buffering; Permeability index

1. Introduction

Human thermophysiological comfort is associated with thermal balance of the human body, which is highly dependent on metabolic rate, physical activity, ambient temperature and relative humidity, and the heat and moisture transport properties of the worn clothing [1].

Thermal and water vapour resistance of clothing are the attributes relevant to the thermal comfort of the wearer. [2]. Water vapour transport properties of textile fabrics are a critical property of clothing systems which assist the maintenance of the thermal equilibrium of the wearer's body [2,3].

Previous studies demonstrated that hygroscopicity and moisture buffering properties of fibres are considered as the most important parameters influencing the water vapour resistance of clothing and resultant thermophysiological responses of the wearer under dynamic conditions [4]. The hygroscopicity of wool fibre is the highest among all natural and synthetic fibres, capable of absorbing up to 35% of its weight in moisture. Due to this higher amount of heat of sorption is released when moisture is absorbed by wool fibres in comparison to natural and synthetic fibres. When sweating starts, highly hygroscopic fibres absorb considerable amounts of sweat and their temperature rises due to the heat of sorption. The elevated fabric temperature interacts with the wearer's body, maintaining higher skin temperature [4, 5]. Hence, it is believed that the greater heat of sorption would reduce the heat loss from the body to the environment in cold conditions, when clothing is made of wool fibres [4, 6]. Both fibre hygroscopicity and ambient relative humidity influence the moisture exchange processes between the fabric and the ambient air and between the fabric and the skin during contact. The ambient relative humidity influences the rate of evaporation of moisture from the fabric [7]. In a study, Li et al. investigated the influence of fibre hygroscopicity on thermoregulatory responses by conducting subjective trials using wool and polyester garments during exercise at 20 °C and 35RH%. Subjects exercised by peddling on a cycle. It was found that humidity measured in the microclimate of clothing increased more quickly when wearing polyester than when wearing wool. The difference between the two humidities was significant in the period between ten and twenty minutes after exercise [8]. This study provides a basis for investigating the effect of moisture buffering properties of wool fibres on thermal and water vapour resistance due to change in ambient relative humidity.

In contrast to the use of steady ambient conditions [9-14] as per standard methods or under extreme hot/cold ambient conditions [15-19], some authors attempted to investigate the effect of different set of ambient conditions on water vapour resistance of fabrics. For example, in 2009, Skenderi Z et. al [19] investigated water vapour transfer through knitted fabrics under three different ambient conditions and standard ambient conditions which were 27°C and 43%RH, 20°C and 58%RH, 16°C and 65%RH and 35°C and 40%RH simulating environmental conditions in Athens, Zagreb, Helinski and the conditions for Standard method ISO 11092. For this investigation four fabrics were knitted from cotton, cotton/ modal, viscose and Tencel® yarn of 20 tex comprising same knit construction and differing in raw material and mass per unit area. Their study demonstrated that water vapour transfer through textile structures is essentially governed by the vapour pressure difference. The pressure difference is larger in climatic conditions with a lower temperature and higher humidity (16 °C and 65% RH) because in such conditions, textile fabrics allow the passage of a higher amount of water vapour. Further, the authors concluded that the viscose and Tencel® assemblies pose a lower barrier to water vapour transfer, which would result in a drier feeling than in the case of 100% cotton or cotton/modal fabric. By contrast, a cotton and cotton/modal fabric might be perceived as wetter. Considering the climatic conditions, viscose and Tencel® fabrics should be preferred in hot and humid climates, such as Athens, where the body is required to sweat heavily to assist in heat transfer [19]. A limitation of this study was that fabrics made of synthetic and highly hydroscopic wool fibre were not included.

Later in 2010, Huang J [20] investigated the effects of air temperature, relative humidity, and wind speed on water vapour resistance of air permeable fabrics; micro porous membrane laminated fabrics; hydrophilic membrane laminated or coated fabrics; and hybrid laminated or coated fabrics. The air temperatures were varied from 10°C to 30°C in an interval of 5°C dynamically. The relative humidity conditions ranged from 0% to 50% and the wind speed was varied from 0.1 m/s to 0.4 m/s. In this study a new method was developed for the measurement of water vapour resistance of experimental materials. The results indicated that the water vapour resistance of four types of fabrics did not vary at the various experimental temperatures. For microporous membrane laminated fabrics and air permeable fabrics, the water vapour resistance grew gradually with the increase in the relative humidity. Hydrophilic membrane laminated fabrics and hybrid laminated fabrics exhibited concentration dependent transport behaviour, i.e. the water vapour resistance was low as the relative humidity was high and thus it would create a drier feeling for the wearer. In addition, high wind speed led to low water vapour resistance [20]. This study identified the effect of different ambient relative humidity on water vapour resistance of laminated fabrics and membranes but lacks the analysis of fabrics made of natural or synthetic fibres without membranes. Further, in this study the authors developed a new method for the measurement of water vapour resistance without providing validation of their method. Therefore, the results of this study needs to be cautiously interpreted.

It should be noted, however, that all the previous studies that are reported in literature are subjected to few limitations. To the best of this author's knowledge, the case of investigating thermal and water vapour resistance of cotton garments versus wool garments objectively at different ambient relative humidity has not been given attention by previous researches which have provided a novel basis for this study. Although some past researchers [5-7] have conducted subjective trials and examined the physiological responses of the wearers in changing ambient relative humidity and activity levels, these studies lack findings on water vapour resistance due to these changes.

Therefore, the present study aimed to address this knowledge gap and determine the functional performance benefits of merino wool/cotton denim in comparison to 100% cotton denim when tested under dynamic ambient conditions in a simulated cold weather environment typical of European conditions in late autumn and early spring.

2. Experimental

In the present study, functional performance benefits of introducing Merino Wool into Cotton Denim were investigated under steady state conditions. The study was conducted in a cold environment to simulate cold weather of European conditions in late autumn and early spring with different relative humidity (Table 1). Attributes relevant to the thermal comfort of the wearer such as thermal and water vapour resistance of denim jeans were determined through steady state testing using 20 zone thermal manikin Newton manufactured by Thermetrics when manikin was operating in constant skin temperature mode (CST). For CST mode skin temperature of each zone was set at 35 °C and the heat flux was regulated by the manikin Thermal software to retain the set skin temperature near constant.

Table 1. Environmental conditions for steady state manikin experimental studies

Steady state ambient conditions		
1	Temperature 17 °C ± 1°C ,	RH: 50%±5%
2	Temperature 17 °C ± 1°C,	RH: 80%±5%

2.1 Materials and Methods

2.1.1 Materials

Three experimental specimens of sample denim jeans of the same construction, style and fit were produced to fit the thermal manikin. All experimental garments then were subjected to denim garment washing process and all of each resultant specimens of each sample were tested as part of the full experimental clothing ensemble (Fig.1).

The fabrics used in experimental jeans samples were of the same twill weave in different fibre compositions: 66/34% cotton/ wool, 83/17% cotton/ wool, 100% cotton.

For each test manikin was dressed in full clothing ensemble consisting of commercial shell jacket, polypropylene long-sleeve 1x1 rib knitted base-layer shirt, men's brief, socks, running shoes and experimental denim jeans constructed of denim twill fabric of each of fibre-blends, where to reduce a number of variables the shirt, underwear, socks, jacket and footwear remained the same for all tests.

2.1.2 Methods

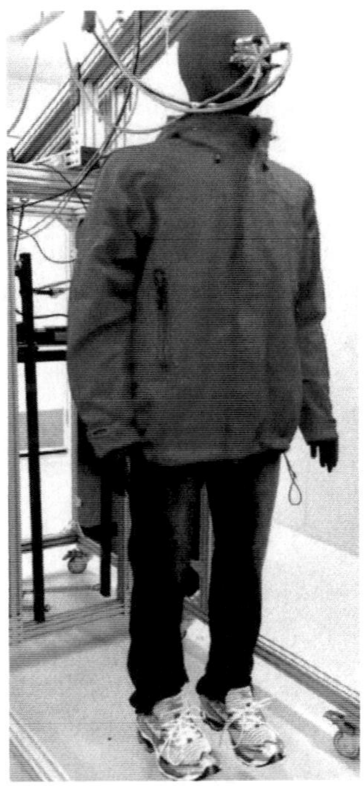

Fig.1 Manikin dressed in experimental ensemble

For determination of physical parameters standard test methods were used to measure fabric mass per unit area, fabric thickness, fabric thread density and air permeability [21-23] with details of the methods described elsewhere [9]. The air permeability for the sample fabrics were determined according to the ISO 9237 [24] where all the tests were conducted at pressure drop of 100Pa. For the determination of fabric density (g/cm^3) mass of the experimental fabrics was measured per cubic centimetre. All physical parameters of fabrics were determined after the manikin tests to account for the variation of the garment washing process applied to all experimental garments.

Thermal and water vapour resistance of jeans were determined by standard test methods for measuring heat flux, thermal and evaporative resistance of clothing using a heated sweating manikin, ASTM F1291-2010 and ASTM F2370-2010 [25, 26] at the set environmental conditions (Table 1) where the whole manikin system was placed in a controllable environmental chamber. The data for skin surface temperature and the heat flux of each zone were obtained directly from MTNW's ThermDAC software. Furthermore, the special "Jeans" group of zones relevant to the areas covered by the jeans was defined (Fig.2).

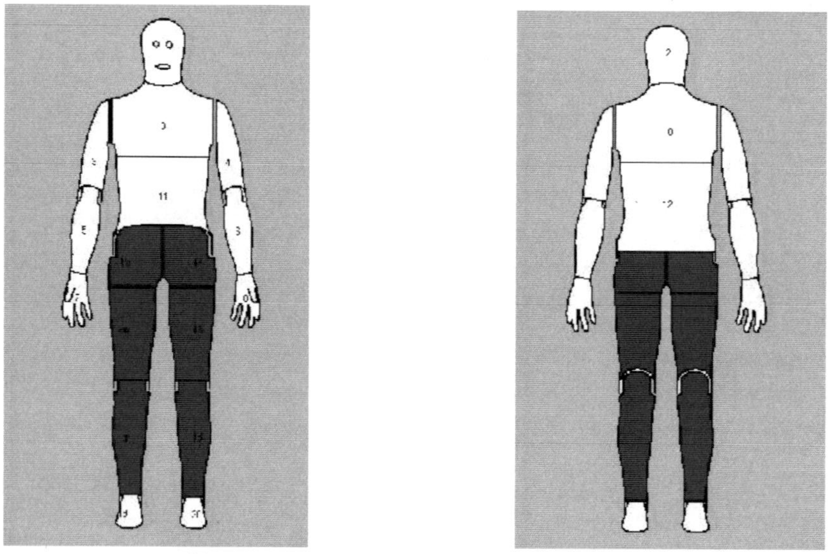

(a) (b)

Fig.2 Jeans Group: a) Manikin front b) Manikin back

Total thermal resistance $I_{T(tr)}$, and the total evaporative resistance $R_{eT(tr)}$ for the Jeans group were calculated and analysed.

The total thermal resistance $I_{T(tr)}$ of jeans according to the defined group was measured using a standard method and calculated by the following formula for the jeans group using parallel method [27-29].

$$I_{T(tr)} = \frac{(T_{skin} - T_{amb})}{Q/A} \quad \text{[Eq.1]}$$

Where,

$I_{T(tr)}$ – Total thermal resistance of jeans including the boundary air layer in m ^{2} °C/W

T_{skin} - Zone average temperature in °C;

T_{amb} - Ambient air temperature in °C;

Q/A - Area weighted heat flux in W/m^2.

The total water vapour, $R_{eT(tr)}$ of jeans was measured and calculated by the following formula, using parallel method [27-29].

$$R_{eT(tr)} = \frac{(P_{sat} - P_{amb})}{Q/A - [(T_{skin} - T_{amb})]/I_{T(tr)}} \quad \text{[Eq.2]}$$

Where,

$R_{eT(tr)}$ - Total evaporative resistance of jeans including the boundary air layer in m^2 Pa/W;

P_{sat} - Saturation vapour pressure at the manikin skin surface in Pa;

P_{amb} - Vapour pressure in the ambient air in Pa;

Q/A - Area weighted heat flux in W/m^2;

T_{skin} - Zone average temperature in °C;
T_{amb} - Ambient air temperature in °C;
$I_{T(tr)}$ - Total thermal resistance of jeans including the boundary air layer in m ^{2} °C/W.

The parallel method is preferred when the manikin is set on a uniform surface temperature, as in present study, rather than the serial method.

The permeability index I_m, developed by Woodcock (1962), was used as an indicator of the evaporative performance of a fabric relative to its thermal resistance. The permeability index is dimensionless, and has values between 0 and 1. A value of 0 implies that the fabric is water-vapour impermeable, that is, it has infinite water-vapour resistance, and a fabric with a value of 1 has both the thermal resistance and water-vapour resistance of an air layer of the same thickness [30]. It is given as,

$$I_{m(t_r)} = \frac{K.I_{T(t_r)}}{R_{eT\,(t_r)}} \qquad \text{[Eq.3]}$$

Where,
$I_{m(tr)}$ - Permeability Index of jeans;
$I_{T(tr)}$ - Total thermal resistance of jeans including the boundary air layer in m ^{2} °C/W;
$R_{eT(tr)}$ - Total evaporative resistance of jeans including the boundary air layer in m^2 Pa/W;
K - Constant (60.6515 Pa/°C).

In addition to account for variation in fabric physical parameters, both $I_{T(tr)}$ and $R_{eT(tr)}$ were also normalised by the mass/unit area of comprising fabrics to arrive at $I_{T(tr)}$ and $R_{eT(tr)}$ for each experimental ensemble. Three experiments were carried out for each experimental ensemble and mean results of each ensemble were presented using bar charts with errors bars indicating the standard deviations of the results.

3. Results and discussion

From Table 2, it is evident that fabric constructed of 100% cotton has the highest mass per unit area and higher thread density, both in warp and weft in comparison to the fabrics constructed of 66/34% cotton/ wool and 83/17% cotton/ wool fibres. Due to higher density, 100% cotton fabric demonstrates lower air permeability in comparison to the fabrics of 66/34% cotton/ wool and 83/17% cotton/ wool.

Table 2. Physical parameters of fabrics comprising experimental garments

Material composition of each jeans	**Mean fabric thickness (mm)**	**Mean fabric mass per unit area (g/m^2)**	**Mean fabric air permeability (l/m^2/s)**	**Mean No. of warp/cm**	**Mean No. of weft /cm**	**Mean fabric density g/cm^3**
100% cotton	1.804	400	24.98	24	30	0.2217
66/34% cotton/wool	1.939	384	25.90	23	25	0.1988
83/17% cotton/wool	1.931	364	26.65	22	25	0.1885

Furthermore, it can also be seen from Table 2 that fabric of 83/17% cotton/ wool composition is one of the lightest fabrics among all, with the similar thread density to 66/34% cotton/ wool. This variation in fabric parameters could have an influence on the thermal and vapour resistance of the experimental garments.

In Fig.4, $I_{T(tr)}$ of experimental jeans is presented.

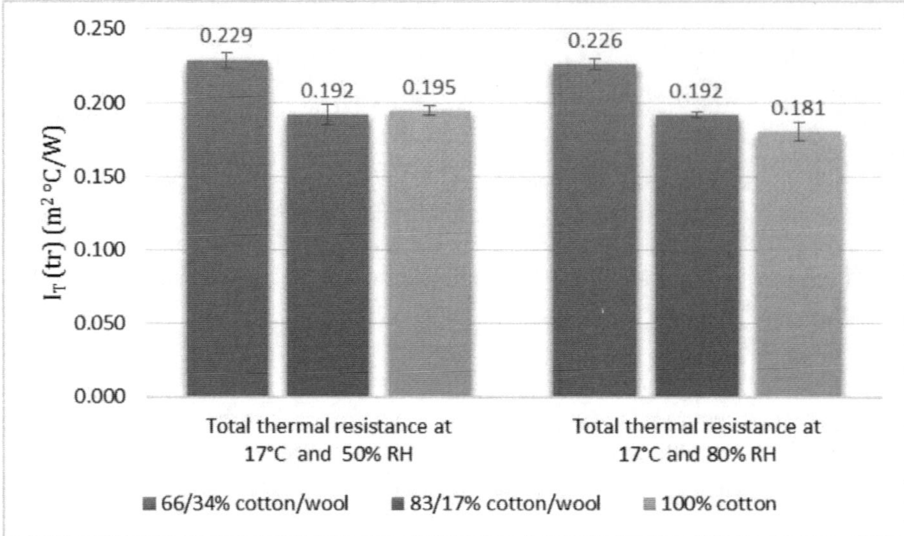

Fig.4 $_{T(tr)}$ at 17°C and different relative humidities

Fig.4 demonstrates that at 17°C and 50% RH $I_{T(tr)}$ is the highest for the jeans made from 66/ 34% cotton/ wool blended fabric where these are approximately 17% higher as compared to the jeans made of 100% cotton fabric and 83/ 17% cotton/ wool blended fabric at the same environmental conditions. Thus, these results suggest that the addition of wool to the cotton fibres increased $I_{T(tr)}$ for 66/ 34% cotton/ wool jeans. However, in case of jeans made from 83/ 17% cotton/ wool blended fabric, $I_{T(tr)}$ is similar to $I_{T(tr)}$ the of 100% cotton jeans for the same environmental conditions. The variation in fabric physical parameters has to be given consideration here (Table 2) where the wool containing denims have lower density and mass/ unit area in comparison to that of 100% cotton.

In Fig.5, $I_{T(tr)}$ of jeans are presented. From fig.5 it is evident that if variation in mass/ unit area is eliminated in each case, 100% cotton jeans will demonstrate the lowest normalised $I_{T(tr)}$ among all sample jeans.

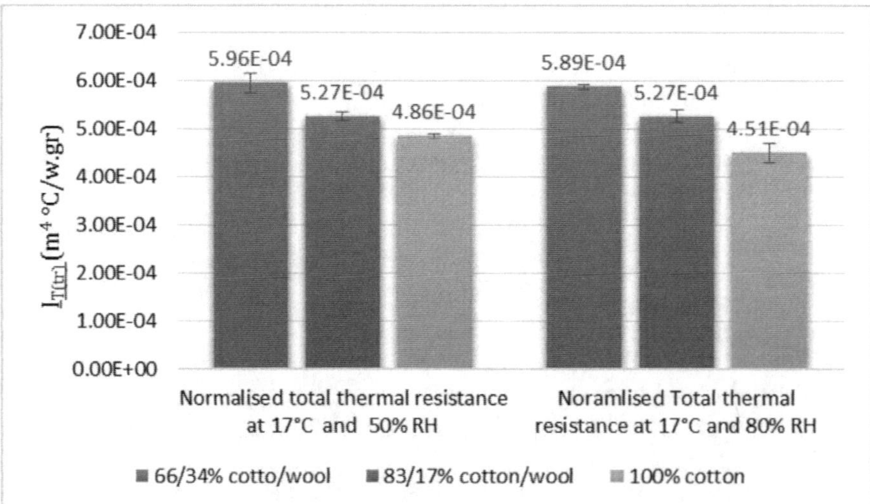

Fig.5 Normalised $\boldsymbol{I_{T(tr)}}$ at 17°C and different relative humidities

From Fig.4-5, it is also evident that for the steady state conditions there is no significant impact of increased environmental humidity on $I_{T(tr)}$ of jeans, except for 100% cotton jeans which exhibit the slight drop in thermal resistance in humid conditions (Fig.5). This is due to the fact that for the calculation of steady state $I_{T(tr)}$ the temperature at the skin of the manikin and the ambient temperature is taken into account after the system reaches steady state (Eq.1).

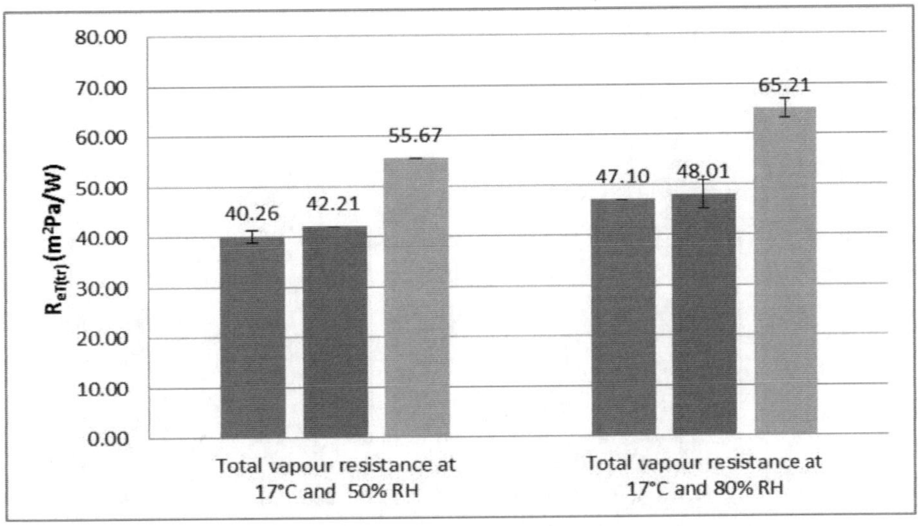

Fig.6 $R_{eT(tr)}$ of jeans at 17°C and different relative humidities

Fig.6 demonstrates that water vapour resistance $R_{eT(tr)}$ is the highest for 100% cotton jeans as compared to 66/34% cotton/ wool and 83/17% cotton/ wool jeans at 17°C and 50% RH. $R_{eT(tr)}$ of 100% cotton jeans is 27% higher than 66/34% cotton/wool jeans, and 23% higher than 83/17% cotton/ wool jeans. These lower $R_{eT(tr)}$ values are the result of addition of wool to the cotton fibres due to the significantly higher hygroscopicity of wool fibre.

Furthermore, from Fig.6 it can also be seen that when relative humidity increased from 50%RH to 80% RH, $R_{eT(tr)}$ values for all sample jeans also increased. At 80% RH, $R_{eT(tr)}$ of 100% cotton jeans increases by 15% and $R_{eT(tr)}$ of 66/34% cotton/ wool and 83/17% cotton/ wool jeans increased by 13% and 15% respectively. Increase in relative humidity results in increase in ambient vapour pressure. Due to this higher ambient vapour pressure at 80%RH (being less than saturated vapour pressure at skin), the ability of a material to transfer water vapour from next to skin microclimate to the environment decreases. Thus, there is a consistent increase in water vapour resistance of all the experimental jeans at 80% RH. Form Fig.7, it is evident that $R_{eT(tr)}$ is higher for 100% cotton jeans. Thus, addition of wool fibre to cotton fibre reduced water vapour resistance of cotton/ wool blended denim, in comparison to 100%cotton denim jeans due to moisture buffering properties and hygroscopic nature of wool fibres.

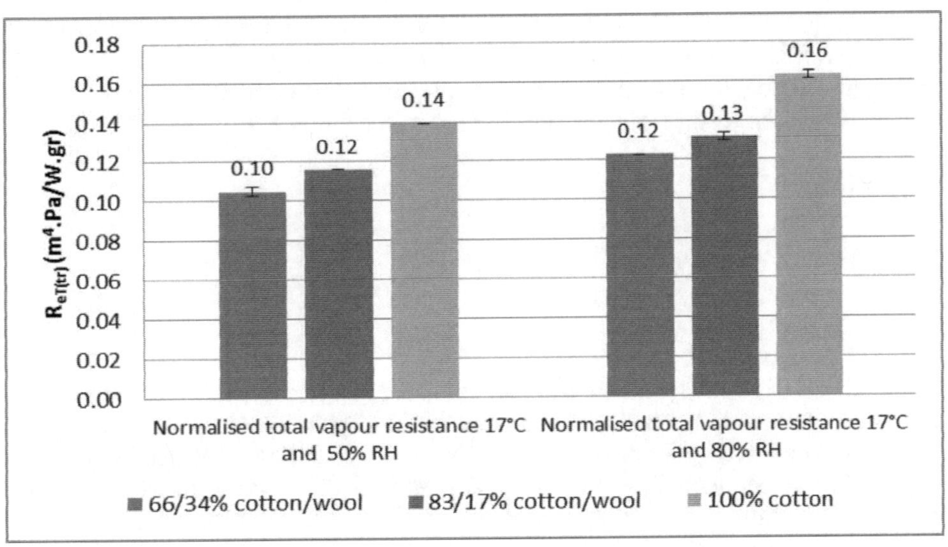

Fig.7 Normalised $R_{eT(tr)}$ of jeans at 17°C and different relative humidities

In Fig.8, permeability index $I_{m(tr)}$ of each sample jeans is presented at 17°C, 50% and 80% RH.

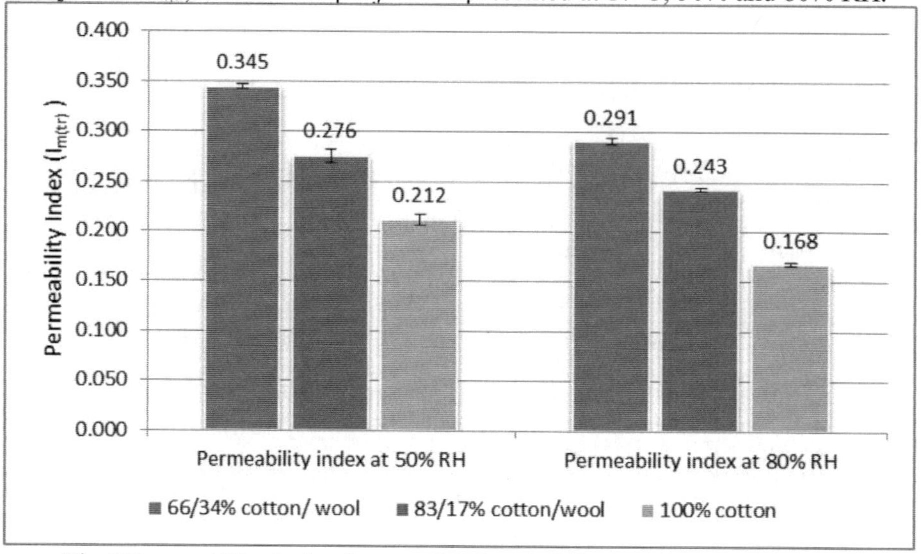

Fig.8 Permeability index $I_{m(tr)}$ at 17°C and different relative humidities

Fig.8, reveals that 66/ 34% cotton/wool jeans demonstrate highest $I_{m(tr)}$ in comparison to the 83/ 17% cotton/ wool and 100%cotton jeans at both relative humidities indicating its highest ability to transfer vapour from skin to the environment and is an attribute of highly hygroscopic nature of wool fibre.

4. Conclusion

From the results, it is evident that, due to presence of wool fibres, 66/ 34% cotton/ wool jeans exhibit highest thermal resistance ($I_{T(tr)}$) and lowest vapour resistance ($R_{eT(tr)}$) in comparison to 83/ 17% cotton/ wool and to 100% cotton jeans. Results for normalised $I_{T(tr)}$ and $R_{eT(tr)}$ of jeans demonstrate that if the mass per unit area of fabrics for each experimental jeans is the same, 66/ 34% cotton/wool and 83/ 17% cotton/ wool jeans will perform better in comparison to 100% cotton jeans as relevant to thermal comfort attributes of the wearer at 17 °C and different relative humidities. This is due to moisture buffering properties of wool fibres. Furthermore, 66/ 34% cotton/ wool and 83/ 17% cotton/ wool jeans demonstrate higher vapour permeability index than 100% cotton jeans due to the hygroscopic nature of wool fibres indicating their higher ability to transfer water vapour from next to skin to the environment and keep the wearer drier in humid conditions. However, to comprehensively assess the performance of clothing in various climatic conditions it is important to not only consider their behaviour in steady state but also in dynamic state which would better replicate the real-life use of the garments.

5. Acknowledgement

This research was funded by Australian Wool Innovation Limited.

References

[1] Havenith G. Interaction of clothing and thermoregulation. Exogenous Dermotology: 2002; 1(5): 221-230.
[2] Holmer I. Protective clothing in hot environments. Industrial Health: 2006; 44: 404-413.
[3] Rossi R. Comfort and thermoregulatory requirements in cold weather clothing. In: Williams JT editor. Textiles for Cold Weather. Woodhead Publishing Series in Textiles, Woodhead Publishing: 2009; 3-18.
[4] Li Y. The science of clothing comfort Textile progress:2001; 31 (1-2): 1-135.
[5] Li Y, Plante AM, Holcombe BV. Fiber Hygroscopicity and Perceptions of Dampness: Part II: Physical Mechanisms. Textile Research Journal:1995; 65: 316-324.

[6] Y Li, Wang ASW editors. Clothing biosensory engineering. The Textile institute: 2006. CRC Press, Woodhead Publishing Limited.
[7] Plante AM, Holcombe BV, Stephens LG. Fiber Hygroscopicity and Perceptions of Dampness: Part I: Subjective Trials. Textile Research Journal: 1995; 65: 293-298.
[8] Li Y, Holcombe BV. Fiber Hygroscopicity and Thermoregulatory Responses During Exercise, presented at 2nd Asian Textile Cotif, Seoul, South Korea: 1993.
[9] Nawaz N, Troynikov O, Watson C, "Thermal Comfort Properties of Knitted Fabrics Suitable for Skin Layer of Protective Clothing Worn in Extreme Hot Conditions. Advanced Materials Research :2011; 331: 184-189.
[10] Varga K, Schädel U, Nilsson H, Persson O, Schuster K Ch. Measuring the Heat of Wetting of Textile Fibres by Reaction Calorimetry. Fibres & Textiles in Eastern Europe: 2007; 15(5-6): 64-65.
[11] Onofrei E, Rocha AM, Catarino A. The Influence of Knitted Fabrics' Structure on the Thermal and Moisture Management Properties. Journal of Engineered Fibers and Fabrics: 2011; 6 (4).
[12] McCullough E, Huang J, Kim CS. An Explanation and Comparison of Sweating Hot Plate Standards Journal of ASTM International: 2004.
[13] ISO 11092:1993 Textiles - Physiological effects - Measurement of thermal and water-vapour resistance under steady-state conditions (sweating guarded-hotplate test).
[14] ASTM F 1868-XX, Standard Test Method for Thermal and Evaporative Resistance of Clothing Materials Using a Sweating Hot Plate.
[15] Mark GM, Richardsa, Rossia R, Meinander H, Broedec P, Candasd V, Hartoge ED, Ingvar Holmér I, Nockerg W, Havenith G. Dry and Wet Heat Transfer Through Clothing Dependent on the Clothing Properties Under Cold Conditions. International Journal of Occupational Safety and Ergonomics: 2008; 14(1).
[16] Osczevski RJ. Water vapour transfer through a hydrophilic film at Subzero temperatures. Textile Research Journal: 1996; 66 (1): 24-29.
[17] Salopek I, Skenderi Z. Thermophysiological comfort of knitted fabrics in moderate and hot environment. Proceedings of 3rd International Ergonomics Conference, Ergonomics: 2007; 287-293.
[18] Osczevski RJ. Water vapor transfer through a hydrophilic film at subzero temperatures. Textile Research Journal: 1996; 66 24-29.
[19] Skenderi Z, Čubrić IS, rdjak M. Water Vapour Resistance of Knitted Fabrics under Different Environmental Conditions. FIBRES & TEXTILES in Eastern Europe: 2009; 17(2): 72-75.
[20] Huang J, Chen Y. Effects of Air Temperature, Relative Humidity, and Wind Speed on Water Vapor Transmission Rate of Fabrics. Textile Research Journal March: 2010; 80: 422-428.
[21] AS/NZS 2001.2.13(1987), Methods of test for textiles - Method 2.13: Physical tests -Determination of Mass Per Unit Area. Standards Association of Australia.
[22] AS 2001.2.15 (1989). Methods of test for textiles - Method 2.15: Physical tests -Determination of thickness of textile fabrics. Standards Association of Australia.
[23] AS2001.2.5-1991-Methods of Test for Textiles Physical Tests - Method 2.5 - Determination of the Number of Threads Per Unit length, 1991: Standards Association of Australia.
[24] ISO, EN ISO 9237 - 1995 Textiles - Determination of the permeability of fabrics to air. International Organisation of standardisation.
[25] ASTM, F1291 - 10 Standard Test Method for Measuring the Thermal Insulation of Clothing Using a Heated Manikin, 2010: American society for testing and materials, West Conshohocken, USA.
[26] ASTM, F2370 - 10 Standard Test Method for Measuring the Evaporative Resistance of Clothing Using a Sweating Manikin, 2010: American society of testing and materials, West Conshocken, USA.
[27] Wang F, Gao C, Kuklane K, Holmér I. A comparison of three different calculation methods of clothing evaporative resistance: 2012; Proceedings of the 5th European Conference on Protective Clothing (ECPC), Valencia, Spain.
[28] Kuklane K, et al. Calculation of clothing insulation by serial and parallel methods: effects on clothing choice by IREQ and thermal responses in the cold. Int J Occup Saf Ergon: 2007; 13(2): 103-116.
[29] Troynikov O, Ashyeri E, 2011. Thermoregulatory evaluation of triathlon suits in regards to their physiological comfort properties, Procedia Engineering: 13; 357-362.
[30] Woodcock AH. Moisture transfer in textile systems, part 1. Textile Research Journal 1962; 32: 628-633.

Design of Underwear based on Kansei Engineering

Li-Zhen Wang, Chi Liu*, Shao-Di Liu

Apparel & Art Design College, Xi'an Polytechnic University, Xi'an, 710048, China

*Corresponding author's email: liuchi646691@126.com

Abstract

Lingerie appearance is one of the factors affecting consumers when buying underwear. In terms of different underwear, different consumers have different feelings. This paper, based on Kansei Engineering theory, analyzes people's emotional evaluation of underwear appearance using questionnaires. The emotional evaluation of consumers was translated to specific data, which was related to underwear design elements. With the help of SPSS software, the emotional evaluation of underwear is attributed to structure, style and material factor.

Key Words: Underwear; Emotional evaluation; Kansei engineering; Factor

1. Introduction

Kansei Engineering was first proposed in the 1980s by the Japanese design community. In the past, products were assessed qualitatively, based on emotions but not logic. The use of modern computer technology has rectified this by making it possible to conduct quantitative analysis to develop a new generation of products and technology.

Traditional product development follows the process of "concept design, detailed design, process design, manufacturing, test verification, and design changes". Since the front and rear of the critical design-related subtle design elements cannot be perfect convergence ignored, which causes the situation to improve cost and waste of resources, etc[1]. Deng Xiao-Rong applied Kansei Engineering to fashion design [2]. By surveying consumers and conducting practical clothing Kensei evaluations, Dong Xiang-Fang obtained effective design guidance elements through analyzing quantified data of consumers' evaluation of different clothing. She also practiced and explored design methods based on Kensei Engineering Theory, and put forward a complete design concept [3]. With the help of questionnaires, Wang Yu-Hong analyzed people's Kansei evaluation on the appearance of different kinds of T-shirts. He also analyzed young T-shirt consumers' consumption behaviors and differences among different groups of people [4]. Wang Ying researched the style of woman's overcoats by the method of fore-quantitive deducing sensory engineering. Firstly, the style of woman's overcoat was analyzed by the shape analytical method, in which 25 design elements were found to influence the perception judgment of consumers. Secondly, 6 word-pairs which were used to describe the perception of women's overcoat were obtained through collection, selection and classification. Then, the mathematical models between perception and design elements of women's overcoat were established by the quantification theory, and were applied to predict the value of perception evaluation. Finally, these mathematical models were applied to a system with two functions - perception analysis and style research, which were developed by visual basic 6.0 [5].

Hou Xiao-Peng discussed the relationship between material texture and consumers' psychology images. The Kensei Engineering Method is used to achieve the visual and tactile sense of quantity. Through experiments and statistical analysis of results, the materials image database is created. The mapping tree is constructed by the Top-Down method, and products' texture CAID system is constructed according to consumer demands and the emotional habits of the designers [6]. Chen Yi-Ru researched UI interface color design based on Kansei engineering [7]. Liang Xuan's discussion framework can be summed up as the basis in the design of interactive experience, perceptual research, and cognitive habits for research relying on Kansei engineering principles from the perspective of cognitive science and psychology, translation emotional elements, and virtual product design engineering [8]. In Li Yong-Feng's paper, product design method based on Kansei Engineering was adopted. The first objective was to clearly define the design orientation. Then, Kansei words and design properties were identified and typical products were evaluated with Kansei words. Finally, the results were analyzed by using mathematical methods, and the relationship that guided the product design between Kansei words and design properties was established. The method was proved to be valid and feasible by the design practice of an eye protection table lamp [9]. With “keep fit” as the primary requirement, perceptual factors such as sense of comfort, sense of safety, and sense of beauty are taken into consideration and specified in designing school uniforms to satisfy the perceptual requirements of students from the selection of fabric, right through to the garment structure and sewing quality, Wang Ying wrote [10].

From the present state of research on Kansei Engineering, it can be seen that Kansei Engineering is oriented towards the needs of customers in implementing product development technologies. The customer's perception and feelings become elements of costume design technology. More than ever, importance is place in seizing the psychological needs of consumers, and a scientific methodology makes the design process simple and accurate.

Although many researchers have explored the Kansei used in clothing fields, most of these studies focused on apparel and marketing. Since its application in respect of underwear is relatively rare, this paper discusses the application of Kansei Engineering in intimate apparel.

2. Defining the scope of the study

2.1 Defining the research object

In this paper, the research object will eventually locate 13, 14, and 15 graduate students in Xi'an Polytechnique University, They are 18-25 years old women.

Women were chosen as research subjects because women are more emotional and have experience wearing underwear. Compared to 18-year-old women, 18-25 years old women have a more mature concept of consumption, stronger purchasing power, and higher sensitivity to fashion trends. On the other hand, compared to 35 or older women, 18-25 years old women with no children have less household expenses and tend to form a “buy what they want” concept of consumption.

2.2 Research Content

In this paper, the main research is lingerie; mainly bras and panties, because bras and panties are every woman's wardrobe essential and is the best in showing female physical beauty. At the same time, intimate apparel plays an important part in women’s daily lives as they are one of the most functional clothing showing women’s emotional and psychological needs.

3. Emotional vocabularies and underwear models database

3.1 Emotional vocabulary collection of underwear

According to the study, this paper collected a variety of vocabularies for describing intimate apparel from network, magazines, dictionaries, posters, TV, and so on.

3.2 Sample collection and collation of underwear models

Collecting samples is very critical. Therefore, in order to establish an effective sample database, and gain emotional needs of consumer, interviews and questionnaire surveys have been conducted. We learned from the interviews with these women who usually have more concern to fashion cycle, that they are often affected by popular information from network. Therefore, we selected a number of popular models from network, posters, magazines and other channels to collect a total of 30 samples.

With the preferences and needs of women aged 18-25 (this paper’s sample database) in mind, 15 samples were chosen. To ensure consumers are not affected by the color of the pictures in the evaluation process, the color of the samples were removed through software and the pictures were numbered, as shown in Fig.1, 15 sets of underwear models included.

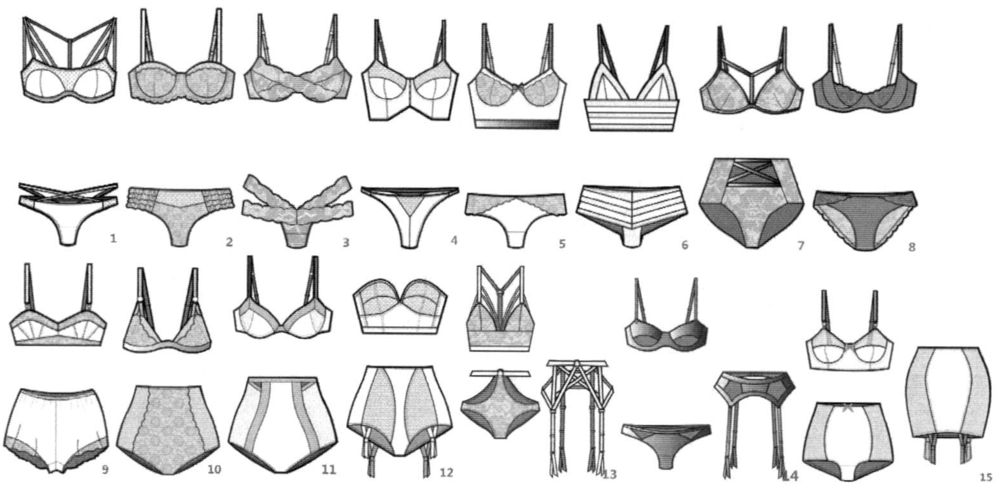

Fig.1 15 sets of underwear models

3.3 Building emotional vocabulary rating scale

3.3.1 Adjective vocabulary sorting

According to the semantic differential method, it is reasonable to sort out adjectives into 5 to 15 pairs. In combination with a special study of a Kansei Engineering teacher along with the help of 10 students, a professional clothing emotional vocabulary was completed. Eventually, 13 pairs of adjectives were sorted: romantic - realistic, comfortable - uncomfortable, soft - tough, classic - fashionable, delicate - rough, elegant - vulgar, simple - complex, mature - cute, plump - light, open - conservative, retro - novelty, lively - dull, natural - contrived.

3.3.2 Vocabulary assignment

According to the theory of semantic differential method, the rating scale's assigned value should be an odd number. In this study, each group assigned each vocabulary 5 points and the median is 3 points. Taking the first pair of adjectives for example: 1 means feeling close to a "romantic" adjective side, 5 means feeling close to a "realistic" adjective side, 2 points indicates relatively close to the "romantic" adjective side, and 4 points indicates feel closer to "reality" adjective side.

Table 1. Vocabulary assessment scale

adjective	1	2	3	4	5	adjective
romantic						realistic

4. Intimate apparel design elements extract

4.1 Questionnaire survey

The research uses questionnaires to sort the 13 pairs of emotional words, 26 kinds of feeling, the sample database of 15 sets of underwear pictures and vocabulary evaluation scale binding. This is to enable consumers to evaluate the different styles of underwear on a subjective feeling. The same or similar design elements of underwear will be extracted.

4.2 Data Analysis and design elements of underwear

4.2.1 The emotional and psychological evaluation scores of consumers for underwear

300 questionnaires were distributed and a total of 260 valid questionnaires were collected. Using EXCEL statistical analysis software for data averaged, the emotional and psychological evaluation scores of consumers for 15 sets of underwear were obtained.

4.2.2 Emotional dimension analysis of underwear figure

By analyzing the questionnaires collected, it was clear that different underwear styles provoked different feelings. The 13 pairs of adjectives embodied 13 kinds of psychological feeling, which we further analyzed below. The emotional adjectives were simplified to allow us to further understand how images influenced consumer psychology in the purchase of underwear.

Using EXCEL software for statistical survey results, we obtained psychological evaluation score of each underwear, then score data were imported using SPSS software (spssstatistics 17.0). Finally, by using factor analysis, perceptual evaluation scores were analyzed.

4.2.2.1 The concept of Factor Analysis

Factor Analysis is a branch of multivariate statistical analysis techniques; its main purpose is to concentrate data. Through the study of many internal dependencies between variables, Factor Analysis also explores the basic structure of the observed data, and uses a few hypothetical variables to represent the basic data structure. These hypothetical variables reflect key information represented by a large number of observed variables and explain these interdependencies between variables. We call these hypothetical variables underlying variables. Factor Analysis is the analysis of a large number of observed variables and condensing these into a few key factors with a minimal loss of information.

Factor Analysis correspond to the contents of this study that different respondents' emotional and psychological evaluation of 15 different underwear can be expressed in several factors. That is, people's emotional evaluation to underwear has been converted to design elements. In this way, more suitable underwear will be designed.

4.2.2.2 Factor Analysis Model

Any one variable, after $z = \frac{x - \bar{x}}{\sigma}$ Transformation ($\bar{x}$ is the mean of x, σ is the standard deviation of x) became standardized variables.

Factor Analysis Model is similar to Multivariate Regression Models in form. Each observed variable represents a linear combination of a factor. It assumes K observed variables, respectively $x_1, x_2, ..., x_k$. x_i is with zero mean, unit variance standardized variable. The below is general expression of the Form Factor Model.

$$x_i=a_{i1}f_1+ a_{i2}f_2+\ldots+ a_{im}f_m+ u_i \qquad (i=1,2,\ldots, k)$$

In this model, f_1, f_2, ... f_m are called common factor. They are common factors of each observed variables to explain the correlation between variables. u_i is called special factor, which is unique to each observation variable factor, and the equivalent of multiple regression residuals indicates that the variable is not part of the common factors explained. a_{ij} is called load factor, which is the i variable on the j common factor of the load, the equivalent of multiple regression analysis of standardized regression coefficients (i = 1, ..., k; j = 1, ..., m). This model assumes that k is independent of each other between the special factor. The relation between common factor and these special factors is also independent.

4.2.2.3 Preparation of factor analysis

Table 2. KMO and Bartlett's Test

Kaiser-Meyer-Olkin Measure of Sampling Adequacy.		.548
Bartlett's Test of Sphericity	Approx. Chi-Square	170.909
	df	78
	Sig.	.000

Before the Factor Analysis method was used to analyze the score of the perceptual evaluation, the KMO (Kaiser-Meyer-Olkin) test was carried out. The measure is based on the simple correlation coefficient between the observed variables and the relative size of the partial correlation coefficient, and its value is from 0 to 1. When the value of KMO is greater than 0.5, it is suitable for Factor Analysis. As can be seen from Table 2, KMO between variables is 0.548, which means that it is possible to do Factor Analysis.

4.2.2.4 Using Factor Analysis to analyze perceptual reviews

1) Common variance

Table 3. Communalities

	adj.1	adj.2	adj.3	adj.4	adj.5	adj.6	adj.7	adj.8	adj.9	adj.10	adj.11	adj.12	adj.13
Initial	1.000	1.000	1.000	1.000	1.000	1.000	1.000	1.000	1.000	1.000	1.000	1.000	1.000
Extraction	.594	.826	.596	.917	.618	.807	.856	.669	.884	.927	.918	.861	.897

Common factor variance is also called common degree, also known as common variance. It refers to the proportion of variance in the observed variables determined by common factors. Variance variable consists of two parts: one part is determined by a special factor; determining factor by the public. Common factor variance represents the common factor of variance, which can be explained in part. The greater common factor variance is, the higher the degree of variable factors can be described. Using principal component analysis, from the common factor variance table, we can see that every adjective extracting common factor variance are greater than 0.5, as shown in Table 3. Therefore, 13 pairs of adjectives can be explained by common factors, which contained the main information.

2) Total variance explained

As shown in Table 4, the total variance explained by the extraction of three factors, which are taken as the main factors with eigenvalues greater than 1, from the 13 components. Thereby, emotional adjectives converted into three factors. The first characteristic value factor is 6.401, explaining 49.235% of the original 13 variables total variance. Eigenvalues' second factor is 2.777, explaining 21.361% of the original 13 variables total variance. Eigenvalues' third factor is 1.194, 9.182% of the original 13 to explain the total variance of variables. These three factors contribute to the total cumulative variance of 79.778%. These three factors summed up most of the information in regards to adjectives on 15 sets of underwear's sensual style. It is able to explain the perceptual psychology.

Component	Initial Eigenvalues			Extraction Sums of Squared Loadings			Rotation Sums of Squared Loadings		
	Total	% of Variance	Cumulative %	Total	% of Variance	Cumulative %	Total	% of Variance	Cumulative %
1	6.401	49.235	49.235	6.401	49.235	49.235	6.288	48.366	48.366
2	2.777	21.361	70.596	2.777	21.361	70.596	2.754	21.181	69.547
3	1.194	9.182	79.778	1.194	9.182	79.778	1.330	10.231	79.778
4	.969	7.457	87.235						
5	.649	4.991	92.226						
6	.307	2.365	94.591						
7	.236	1.812	96.403						
8	.195	1.502	97.905						
9	.134	1.030	98.934						
10	.071	.544	99.479						
11	.054	.417	99.895						
12	.009	.070	99.965						
13	.004	.035	100.000						

Table 4. Total Variance Explained

3) Scree plot

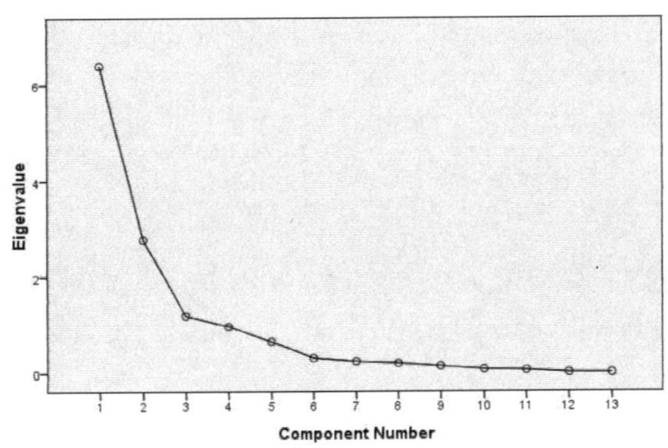

Fig.2 Scree plot

Fig.2 "Scree Plot" shows the number of factors and the ordinate eigenvalue. We can see that the first factor eigenvalues gave the greatest contribution to the interpretation of the original variable maximum. In Fig.2, starting from the fourth factor, the line slowly leveled off last eigenvalues almost close to zero. That is, from the beginning of the fourth factor, the contribution factor behind the global variable gets smaller and smaller. So selecting "scree plot" near the origin, the first three factors (three factors have eigenvalues greater than 1), is more appropriate.

4) Rotated Component Matrix

Table 5. Rotated Component Matrix

Component	adj.13	adj.4	adj.7	adj.2	adj.6	adj.10	adj.11	adj.8	adj.12	adj.1	adj.5	adj.3	adj.9
1	**.944**	**.923**	**.897**	**.894**	**.871**	**-.813**	**.750**	**-.619**	-.391	.038	.384	.423	.308
2	.075	-.100	.052	.154	.210	.504	-.585	-.153	**.826**	**.770**	**.636**	**.606**	-.004
3	-.026	.236	.221	.057	.070	-.108	.115	.512	-.158	.025	.256	-.222	**.889**

As shown in Table 5, the maximum variance method used the factor loading matrix obtained by applying orthogonal rotation factor loading matrix, which makes the explanatory factors named. After rotated, the amount of rotation of load factor reflects the difference between factors and variables; the greater the absolute value, the closer the relationship between the variables and the corresponding factor, and the greater the contribution factor. So we can see from Table 5: "the 13th adjective pair (natural - artificial), the 4th adjective pair (Classic - fashionable), the 7th adjective pair (simple - complex), the 2nd adjective pair (comfortable - uncomfortable), the 6th adjective pair (elegant - vulgar), the 10th adjective pair (open - conservative), the 11th adjective pair (retro - novelty), the 8th adjective (pair mature - lovely)" scored highest on the first factor; namely a major factor explaining these variables can be named " structure factor". "The 12th adjective pair (lively - rigid), the 1st adjective pair (romantic - realistic), the 5th adjective pair (fine - rough), the 3rd adjective pair (soft - tough)" carries a greater load on the second factor; the main explanation of these variables can be named as the "style factor". "The 9th adjective pair (plump - thin)" has a large load on the third factor; a variable that can be named as the "material factor".

As can be seen from the above analysis, consumers' perceptual evaluation of underwear appearance consists of three factors: structure, style, and material.

5. Conclusion and expectation

This is the underwear Kansei Engineering Design Theory. It is aimed at female consumers through emotional or feeling the need for research and analysis, to explore consumer feelings and blend of sensual underwear design of adaptive methods to meet the emotional needs of today's women's underwear.

This paper analyzes consumers' subjective emotional evaluations when they are faced with lingerie products. According to the theory of Kansei Engineering, and questionnaires using SPSS software classified by factor analysis of the emotional dimension of consumer evaluations, finally we extracted three factors: structure, style and material factor.

References

[1] Li YE, Wang ZY, Xu N. Kansei Engineering. Beijing: Haidian Press: 2009; 10: 18-19.

[2] Deng XR. Apparel design application for Kansei Engineering. Light and Textile Industry and Technology: 2014; 5: 80-81.

[3] Dong XF. The research and practice of dress design based on Kensei Engineering. Zhejiang: Zhejiang University of Technology: 2012.

[4] Wang YH. The Research on emotional evaluation and consumption behavior of T-shirt. Jiangsu: Suzhou University: 2006.

[5] Wang Y. The sensory research and application on the style of woman's overcoat. Jiangsu: Suzhou University: 2007.

[6] Hou XP, Yu SH, Zhao W. Research on the material texture CAID system of product based on Kensei Engineering. Modern Manufacturing Engineering: 2008; 12: 44-47.

[7] Chen YR. The research and application of UI color design based on Kansei Engineering. Anhui: Anhui University: 2014.

[8] Liang X. Based on Kansei Engineering product interactions. Shaanxi: Xi'an Polytechnic University: 2013.

[9] Li YF, Zhu LP. Research on product design method based on Kansei Engineering. Packaging Engineering: 2008; 11: 112-114.

[10] Wang Y, Chen Y. Application of Kansei Engineering to student uniform design. Journal of Textile Research: 2007; 3: 96-99.

A Study on Cycling Comfort of Non-competitive Cyclists

Jing Guo[1,2], Rong Zheng[1*], Heng-Chan Yin[2]

[1]*Aimer Human Engineering Research Centre, Beijing Institute of Fashion Technology, Beijing, 100029, China*
[2]*College of P.E and Sports, Beijing Normal University, Beijing, 100875, China*

[*]Corresponding author's email: yuki.zheng@gmail.com

Abstract

This study investigated non-competitive cyclists' comfort during cycling activities. An online survey of 92 non-competitive cyclists revealed their levels of comfort when cycling for commuting/non-commuting purposes. Study findings show: 1) inconvenience is the main factor preventing non-competitive cyclists from wearing cycling pants; 2) during commuting cycling, cyclists in certain groups feel more discomfort than those in other groups; 3) during non-commuting cycling, cyclists' discomfort initially increases but decreases as the cycling duration increases. The study provides practical guidance for consumers selecting cycling garments and for designers developing cycling apparel for non-competitive cyclists. Findings are also theoretically significant for assessment of cycling comfort and cycling apparel design for non-competitive cyclists.

Keywords: Non-competitive cycling; Cycling comfort; Cycling pant; Survey

1. Introduction

Non-competitive cycling is strongly appealing because of its convenience, environmental friendliness, and adaptability for different user groups. However, cycling requires cyclists to sit on a bike saddle with their torsos leaning forward and a repetitive foot-pedaling motion that could adversely affect users' physical well-being and cycling comfort. Comfort has become one of the top considerations for consumers selecting bicycles and for designers developing cycling-related products. Face-to-face interviews and a review of the literature have revealed that commuting and non-commuting cycling are two major occasions for non-competitive cyclists, and comparable studies of non-competitive cyclists' comfort during commuting and non-commuting cycling as well as those related to cycling pants are equally limited.

Investigators in this study designed a questionnaire administered in an online survey regarding the cycling comfort of 92 non-competitive cyclists engaged in commuting and non-commuting activities. More specifically, the study sought to identify and compare features of cycling pants that affect non-competitive cyclists' comfort during commuting and non-commuting cycling, and to propose suggestions to improve comfort. This study provides practical guidance for consumers when selecting cycling garments and for designers developing cycling apparel for non-competitive cyclists. Findings are also theoretically significant for assessment of cycling comfort and apparel design for non-competitive cyclists.

2. Approach

The study used a snowball sampling method, with non-competitive cyclists as subjects. We define non-competitive cyclists as amateur adult cyclists who self-determine cycling duration and distance. The three-part questionnaire was designed to address: 1) cyclists' basic characteristics (age and gender); 2) cyclists' comfort during commuting cycling and factors related to cyclists' use of cycling pants (most comfortable cycling pants, cyclists' decision to wear or not wear cycling pants, and reasons for not wearing them); specific cycling features (experience, frequency, duration, bicycle, saddle, body position, and environment); comfort criteria (degrees of overall and partial discomfort and corresponding sources for discomfort, measured by a 5-point scale devised to identify degrees of discomfort); and 3) cyclists' comfort during non-commuting cycling and related factors corresponding to commuting cycling noted above. After the questionnaire was formulated, it first was used in a trial to ensure its feasibility. Its subsequent revision mainly addressed the logic of the questionnaire's content and language use, and the study used the questionnaire's final draft.

Via an online survey, a total of 92 valid copies of the questionnaire were collected. SPSS18.0 was used for statistical analysis of the collected data, covering descriptive analysis, T-test of independent samples, single-factor analysis of variance, and univariate analysis of variance. The significance level was set as $P<0.05$.

Table 1 shows the distribution of the samples.

Table 1. Distribution of the samples

	Age				
	25 and below	26-35	36 and above	Male	Female
Frequency	42	34	16	53	39
Scale	45.7%	37.0%	17.4%	57.6%	42.4%

3. Results and analysis

3.1 Descriptive analysis of current cycling

3.1.1 Analysis of current cycling pant usage

All of the surveyed cyclists ride bicycles for non-commuting activities, while 58 of them also ride bicycles when commuting. The commuting and non-commuting groups are overlapped. As shown in Fig.1, only 29.35% of the surveyed cyclists chose to wear cycling pants during non-commuting cycling; during commuting cycling, 87.93% of surveyed cyclists chose not to wear them. Therefore, it can be concluded that most cyclists do not wear cycling pants when cycling.

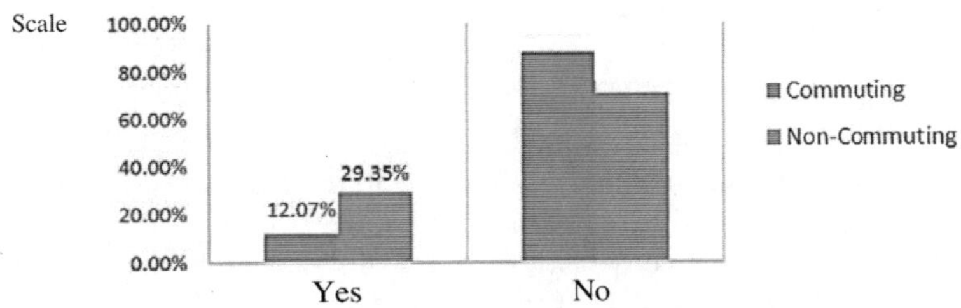

Fig.1 Percentage of cyclists wearing/ not wearing cycling pants during commuting/ non-commuting cycling

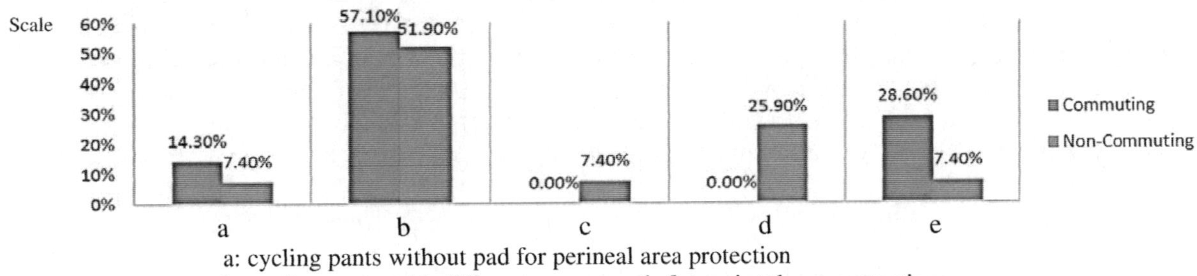

a: cycling pants without pad for perineal area protection
b: cycling pants with different sponge pads for perineal area protection
c: cycling pants with sponge pad for perineal area protection
d: cycling pants with different silica gel pads for perineal area protection
e: cycling pants with silica gel pad for perineal area protection

Fig.2 Distribution of cyclists' choices of what constitutes most comfortable cycling pants

Seven of the surveyed cyclists have tried cycling pants when commuting, while 27 of them have tried cycling pants for non-commuting activities. Fig.2 shows that cyclists who wear cycling pants during both commuting and non-commuting cycling shared a similar opinion regarding which cycling pants are most comfortable. Half of the surveyed cyclists (57.10% in commuting, and 51.90% in non-commuting) regarded the cycling pants with different sponge pads for perineal area protection as the most comfortable. For non-commuting, 25.90% of cyclists regarded the cycling pants with different silica gel pads for perineal area protection as the most comfortable.

Fig.3 shows that 49.30% of the surveyed cyclists indicated that inconvenience was the main reason they did not wear cycling pants during commuting cycling; 27.50% of the cyclists who did not wear cycling pants during non-commuting cycling also shared this opinion. Another important reason cited by respondents for not wearing cycling pants during both commuting and non-commuting cycling was "no need to wear," which was expressed by more than one-third of the surveyed cyclists (36.20% in commuting, and 33.30% in non-commuting).

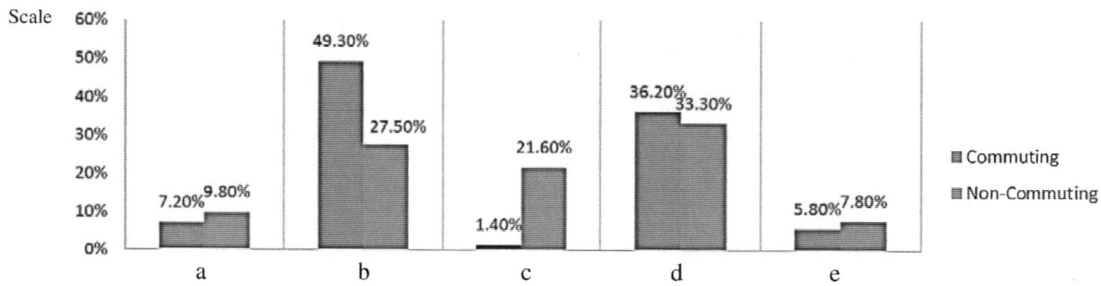

a: Too expensive
b: Troublesome to wear
c: Pants diminish wearer's image
d: No need to wear
e: Worrying about other people's opinion

Fig.3 Reasons for not wearing cycling pants.

3.1.2 Analysis of current cycling features

The surveyed cyclists indicated similar cycling experiences among commuting and non-commuting groups, mainly with 1 year or less cycling experience (41.40% in commuting, and 45.70% in non-commuting). In the commuting group, 44.80% of cyclists ride a bike 1-4 times per week, and 15.50% cycle more than 11 times per week. In the non-commuting group, 70.7% of respondents cycle 1-4 times per month, but 18.5% cycle more than 11 times per month. In the commuting group, 79.30% of respondents cycle less than 1 hour each time; in the non-commuting group, 59.80% cycle about 1-4 hours each time, but 4.30% cycle for more than 8 hours each time.

Whether commuting or non-commuting, cyclists indicated similar choices with regards to bike, saddle, and cycling position: 1) nearly half of the surveyed cyclists mainly ride a mountain bike (50.00% in commuting, and 44.60% in non-commuting), while the second largest group is cyclists who ride a road bike (17.20% in commuting, and 28.30% in non-commuting); 2) nearly half of the surveyed cyclists choose the narrow, long but hard saddle (34.50% in commuting, and 43.50% in non-commuting), while a considerably large population prefer a wide and soft saddle (29.30% in commuting, and 31.50% in non-commuting); 3) the position of cycling a mountain bike dominates in terms of cycling position (50.00% in commuting, and 51.10% in non-commuting), while about 10% of respondents (17.20% in commuting, and 7.60% in non-commuting) reported that they do not pay attention to cycling position.

3.1.3 Analysis of discomfort in key parts of the human body

The questionnaire examined cycling comfort from six aspects relating to the overall feeling and specifically to the head, shoulders, waist, buttocks, and knees. Considering that the human body is quite sensitive to discomfort, the questionnaire used a 5-point scale to indicate the level of discomfort with the higher scale number indicating a higher degree of discomfort. Table 2 shows the following: 1) the most uncomfortable body part is the buttocks (scoring 2.67 in commuting, and 2.63 in non-commuting), nearly equal to the medium discomfort; 2) scores of other body parts in terms of discomfort did not show great difference, basically falling into the category of slight discomfort; 3) except for the head and buttocks, scores for cycling discomfort in commuting were all slightly lower than those in non-commuting.

By further examining sources for discomfort in key body parts, it can be concluded that basically the same sources cause discomfort in key body parts in commuting and non-commuting: 1) pressure (40.60% in commuting, and 38.30% in non-commuting) and moist heat (37.50% in commuting, and 36.20% in non-commuting) are the major sources of discomfort in the head during cycling, which is in line with results of related studies; 2) pressure (45.80% in commuting, and 48.60% in non-commuting) and vibration (27.10% in commuting, and 33.30% in non-commuting) are the major sources of discomfort in the shoulders; 3) pressure (50.00% in commuting, and 54.20% in non-commuting) and vibration (30.60% in commuting, and 23.70% in non-commuting) are the major sources of discomfort in the waist; 4) pressure (31.40% in commuting, and 34.10% in non-commuting) and abrasion (32.60% in commuting, and 30.10% in non-commuting) are the major sources of discomfort in the buttocks; 5) pressure (38.20% in commuting, and 39.10% in non-commuting), friction (35.30% in commuting, and 26.60% in non-commuting), and vibration (23.50% in commuting, and 25.00% in non-commuting) are the major sources of discomfort in the knees.

Table 2. Discomfort scores of key parts of the human body

	N	Minimum	Maximum	M±SD
Overall discomfort while commuting	58	1	4	1.98±0.783
discomfort in head	58	1	4	1.62±0.834
discomfort in shoulders	58	1	4	1.78±0.773
discomfort in waist	58	1	4	1.72±0.812
discomfort in buttocks	58	1	5	2.67±1.114
discomfort in knees	58	1	5	1.72±1.005
Overall discomfort while non-commuting	92	1	4	2.11±0.943
discomfort in head	92	1	4	1.59±0.814
discomfort in shoulders	92	1	5	1.96±0.96
discomfort in waist	92	1	4	1.89±0.988
discomfort in buttocks	92	1	5	2.63±1.307
discomfort in knees	92	1	5	1.93±1.057

3.2 Difference analysis of relevant factors regarding cycling comfort

The study summarized scores for discomfort in six different body parts corresponding to cyclists' commuting and non-commuting cycling, respectively, and labelled them as discomfort in commuting and discomfort in non-commuting based on their features and values. Then a comparative analysis was made, including T-test of independent samples, single-factor analysis of variance, and univariate analysis of variance. Total scores for each category could be as high as 30; 1-6 implies no discomfort, 7-12 represents slight discomfort, 13-18 means medium discomfort, 19-24 stands for severe discomfort, and 25-30 means extreme discomfort.

3.2.1 Difference analysis of cycling comfort among cyclists with different basic features

1) Degree of discomfort among cyclists of different age groups in both commuting and non-commuting populations ranged between 8-12 (categorized as slight discomfort). 2) As cyclists' age increases, degree of discomfort gradually decreases in both commuting and non-commuting groups. 3) Only the 25-and-below age group saw little difference in terms of cycling discomfort in commuting and non-commuting; the other two groups both indicated that degree of discomfort is higher in non-commuting cycling. The single-factor analysis of variance found that degree of discomfort in commuting cycling among different age groups varies greatly with $F(2,55)=3.906$ and $p=0.026<0.05$, and degree of discomfort among different age groups in non-commuting cycling showed no significant difference with $F(2,89)=1.021$ and $p=0.364>0.05$. LSD multiple comparisons of discomfort during commuting cycling among different age groups showed that degree of discomfort among the 25-and-below age group is significantly higher than for other groups (as shown in Table 3).

In both commuting and non-commuting cycling, scores for degree of discomfort for the surveyed male and female populations were generally 8-12, again categorized as slight discomfort. T-tests of independent samples found the following results: 1) no significant difference in discomfort among different gender groups in cyclists' commuting cycling, t (56)=-0.738, p=0.463; and 2) no significant difference in discomfort among different gender groups in non-commuting cycling, t (90)=1.758, p=0.082.

Table 3. Variance analysis of cycling discomfort among different age groups

Dependent variables	Age（I）	Age（J）	Mean difference（I-J）	Standard error	Significance
Discomfort in commuting cycling	25 and below	25-35 (including 35)	2.25714*	1.08740	**0.043**
		above 35	3.35714	1.36828	**0.017**
	25-35 (including 35)	above 35	1.10000	1.43849	0.448
Discomfort in non-commuting cycling	25 and below	25-35 (including 35)	0.44258	1.09868	0.688
		above 35	1.99405	1.39913	0.158
	25-35(including 35)	above 35	1.55147	1.44383	0.285

* P<0.05

3.2.2 Difference analysis of discomfort in buttocks in different cycling pants

Considering the intimate contact between cycling pants and the human body, the study analyzed the discomfort difference in the buttocks in different cycling pants.

Fig.4 shows the following: 1) scores for discomfort in buttocks during commuting and non-commuting cycling with and without cycling pants were all around 2-4, categorized as medium discomfort; 2) in non-commuting cycling, scores for discomfort in buttocks in cycling pants were slightly lower than without cycling pants; 3) the score for discomfort in buttocks in commuting cycling was 3.29, which probably related to the small number of samples (n=7). T-tests of independent samples further identified the following results: 1) cycling pants did not lead to a significant difference of discomfort in buttocks in commuting cycling, t (56)=-1.573, p=0.121; 2) cycling pants did not cause significant difference of discomfort in buttocks in non-commuting cycling, t (90)=-1.056, p=0.294.

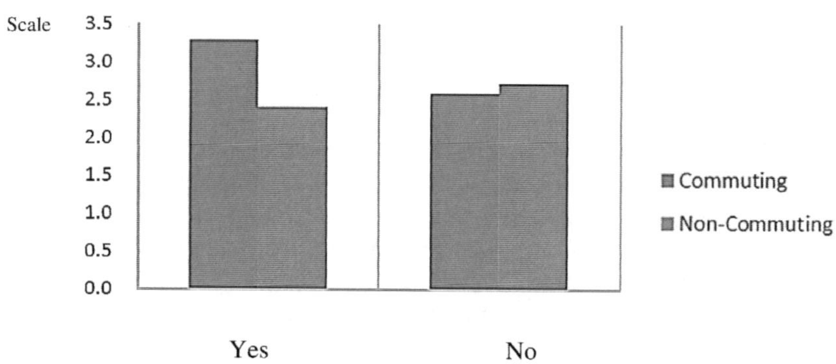

Fig.4 Buttocks discomfort scores in commuting and non-commuting cycling with or without cycling pants.

Due to the small number of samples (n=7) in cycling pants in commuting cycling, the study mainly examined buttocks discomfort in non-commuting cycling. Fig.5 illustrates the following: The score for buttocks discomfort while wearing cycling pants with different sponge pads for perineal area protection was the lowest, followed by the score for buttocks discomfort while wearing cycling pants with different silica gel pads for perineal area protection. The results conform to the results and analyses of studies on the most comfortable cycling pants. Single-factor analysis of variance found no significant difference in buttocks discomfort in non-commuting cycling with different cycling pants, F (4,22)=1.195, P=0.341.

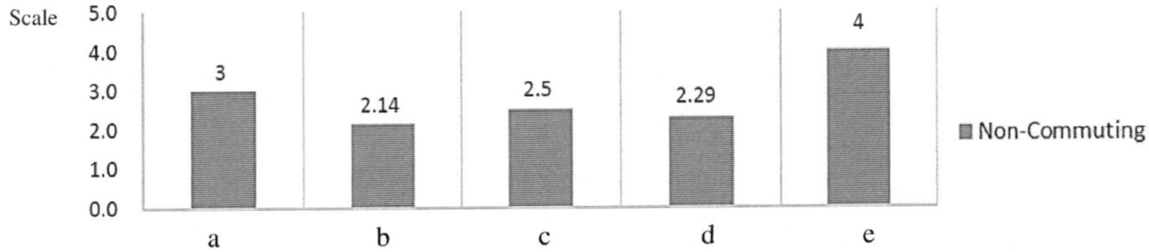

a: cycling pants without pad for perineal area protection
b: cycling pants with different sponge pads for perineal area protection
c: cycling pants with sponge pad for perineal area protection
d: cycling pants with different silica gel pads for perineal area protection
e: cycling pants with silica gel pad for perineal area protection

Fig.5 Scores of bottom discomfort in non-commuting cycling in different cycling pants

3.2.3 Difference analysis of discomfort among groups with different features in cycling

Table 4 shows the following: 1) in commuting cycling, the group with a cycling experience of 1 year or less reported significantly higher discomfort than the group with a cycling experience of 5 years or more; 2) in commuting cycling, the group who cycle 1-4 times per week reported significantly higher discomfort than corresponding groups; 3) in commuting cycling, the group who cycle for less than 1 hour each time showed significantly lower discomfort than corresponding groups; 4) in non-commuting cycling, significant difference of cycling comfort exists between the group who cycle for less than 1 hour each time and the group who cycle for 1-4 hours each time, and significant comfort difference was also found between the group who cycle for 1-4 hours and the group who cycle for more than 8 hours each time. A basic trend can be observed: the discomfort degree first increases and then decreases as the cycling duration increases; 1-4 hours could be a turning point, and discomfort could reach the lowest point when the cycling duration is more than 8 hours.

Table 4. Comfort difference among groups with different cycling experience, frequencies, and time

	Discomfort in commuting	Discomfort in non-commuting
Cycling experience:		
1 year and below (a)	12.6667±4.3306	12.6667±5.2946
2-4 years (b)	11.2727±3.9058	142.1563±4.5516
5 years and above (c)	9.5833±1.8320	10.7222±3.6430
Tamhane test	**Pac***	no significant difference
Cycling frequency:		
1-4 times (weekly in commuting; monthly frequency non-commuting)	13.1154±4.3481	12.3077±4.9998
5-10 times (weekly in commuting; monthly in non-commuting)	10.3913±3.1723	11.1000±4.2019
above 11 times (weekly in commuting; monthly in non-commuting)	9.6667±3.8989	11.9412±4.2789
S-n-K test	subset 1: 1-4 times subset 2: 5-10 times, 11 times and above	no significant difference
Cycling duration each time:		
less than 1 hour (a)	10.7391±3.5617	10.2083±4.0215
1-4 hours (b)	14.4167±3.8954	13.2727±4.8284
4-8 hours (c)		11.6667±5.0745
more than 8 hours (d)		8.5000±2.0817
T-test (commuting) LSD test (non-commuting)	t=-3.126,**P=0.003**	**Pab*** 、 **Pbd***

* P<0.05

To further examine related cycling factors' effect on comfort, we undertook a univariate analysis of variance. The following results were obtained: In commuting cycling, the interaction of cycling experience, cycling frequency, and cycling duration each time is very significant ($F=2.761$, $P=0.007<0.01$). Further analysis of mean values indicated that the group with a cycling experience of 2-4 years who cycle 5-10 times weekly with each cycling period lasting for less than 1 hour showed the lowest discomfort degree (scoring 8.7778), falling into the category of slight discomfort, and the group with a cycling experience of 1 year or less who cycle 1-4 times weekly with each cycling period lasting for 1-4 hours showed a higher degree of discomfort (scoring 17.2500), categorized as medium discomfort. In non-commuting cycling, the interaction of cycling experience, frequency, and duration was insignificant.

In both commuting and non-commuting cycling, different bikes, saddles, positions, and environment did not create a significant difference in terms of cycling comfort.

4. Conclusions and recommendations

4.1 Conclusions

1) The group categorized at age 25 and below in commuting cycling showed significantly higher discomfort than other age groups.

2) As for the use of cycling pants, in both commuting and non-commuting cycling, most of the surveyed cyclists chose not to wear cycling pants. The majority of cyclists who wear cycling pants regarded cycling pants with different sponge pads for perineal area protection as the most comfortable. The two major reasons for the surveyed cyclists not to wear cycling pants in commuting and non-commuting cycling are "inconvenience caused by wearing" and "no need to wear." Discomfort in the buttocks without cycling pants is higher than that with cycling pants, but not significantly different. Cyclists who wear cycling pants with different sponge pads for perineal area protection reported lowest discomfort in the buttocks, followed by cyclists who wear cycling pants with different silica gel pads for perineal area protection.

3) As for features of cycling behaviours, survey respondents basically have similar experience in commuting and non-commuting cycling, mostly 1 year and below. In commuting cycling, most of the respondents cycle 1-4 times weekly, and in non-commuting cycling most people cycle 1-4 times monthly. In commuting, the cycling duration for most respondents is less than 1 hour each time. In non-commuting, the cycling duration is mostly 1-4 hours each time. In commuting and non-commuting cycling, respondents basically showed no significant difference in terms of their selection of bikes, saddles, cycling positions, and environment; mountain bikes equipped with a long, narrow, and hard saddle and the position of cycling a mountain bike are their major choices, and they mainly cycle on urban roads. In commuting, the group with a cycling experience of 1 year or less reported a significantly higher discomfort than the group with a cycling experience of 5 years or more. In addition, the group who cycled 1-4 times weekly showed significantly higher discomfort than other groups; the group who cycle for less than 1 hour reported significantly lower discomfort than other groups. Significant differences in terms of cycling comfort exist among different combinations of cycling features. In non-commuting cycling, the basic trend is that the degree of discomfort first increases and then decreases as the cycling duration increases.

4) As for discomfort in key body parts, in commuting and non-commuting cycling, most discomfort was found in the buttocks (nearly reaching the medium discomfort level) with the discomfort in commuting slightly lower than the discomfort in non-commuting. The major discomfort source in commuting and non-commuting cycling is pressure, with other sources of moist heat, friction, and vibration also contributing greatly.

4.2 Recommendations

1) Designers and researchers should pay more attention to cyclists' comfort in the buttocks, which is affected by pressure in non-commuting cycling, and therefore design related products for non-competitive cycling, also taking into consideration comfort related to factors such as moist heat, friction, and vibration. It is recommended that designers focus on and improve the cycling features of mountain bikes, such as the latter's narrow, long, and hard saddle, as well as the cycling position of mountain bikes in relation to urban-road cycling. Considering that inconvenience related to the wearing of cycling pants is non-competitive cyclists' major reason for not wearing

cycling pants, designers and researchers should take convenience in wearing cycling pants into account. Restricted by the limited study time and the small sample population in the present study, we recommend that subsequent studies expand their samples in order to delve deeper into the field.

2) Regarding the selection of cycling equipment, we recommend that non-competitive cyclists pay more attention to comfort related to pressure in non-commuting cycling. To counter cycling discomfort in the buttocks, consumers should select cycling pants with different sponge pads for perineal area protection. Male cyclists in young age groups in particular should pay greater attention to cycling comfort. Given that the lowest of disagree of discomfort in commuting cycling was noted among the group with cycling experience of 2-4 years who cycle 5-10 times weekly with each cycling lasting less than 1 hour, it is recommended that commuting cyclists with cycling experience of 1 year or less who cycle infrequently (about 1-4 times weekly) should pay more attention to cycling comfort. Cyclists also should take into account the cycling duration in non-commuting cycling.

5. Acknowledgement

This study has been financed by the 2013A-01 and BKSP 02150202/001 programs.

References

[1] Shen WQ, Parsons KC. Validity and reliability of rating scales for seated pressure discomfort. Int J Ind Ergonom 1997; 20(6): 441-461.

[2] Liu YS, Tsay TS, Chen CP, Pan HC. Simulation of riding a full suspension bicycle for analyzing comfort and pedaling force. Procedia Eng 2013; 60(11): 84-90.

[3] Petrone N, Giubilato F. Development of a test method for the comparative analysis of bicycle saddle vibration transmissibility. Procedia Eng 2013; 60(2): 288-293.

[4] Ghosh AK, Baere ID, Walleghem JV, et al. Effect of flax fibre reinforcement on the cycling comfort of composite racing bicycle frames. Procedia Eng 2010; 2(2): 34-35.

[5] Li ZB, Wang W, Liu P, et al. Physical environments influencing bicyclists' perception of comfort on separated and on-street bicycle facilities. Transport Res D-Tr E 2012; 17(3): 256-261.

[6] Ayachi FS, Dorey J, Guastavino C. Identifying factors of bicycle comfort: an online survey with enthusiast cyclists. Appl Ergon 2015; 46: 124-136.

[7] Alam F, Chowdhury H, Elmir Z, et al. An experimental study of thermal comfort and aerodynamic efficiency of recreational and racing bicycle helmets. Procedia Eng 2010; 2(2): 2413-2418.

[8] Bruyne GD, Aerts JM, Sloten JV, et al. Quantification of local ventilation efficiency under bicycle helmets. Int J Ind Ergonom 2012; 42(3): 278-286.

[9] Marcolin G, Paoli A, Panizzolo FA, et al. A method for the analysis of cyclist shorts with different pads for perineal area protection: comparison between drum and road tests. Procedia Eng 2010; 2(2): 2831-2835.

[10] Bressel E, Bliss S, Cronin J. A field-based approach for examining bicycle seat design effects on seat pressure and perceived stability. Appl Ergonom 2009; 40(3): 472-476.

Influence of Weave Factor and Soil Slope on the Erosion Control Performance of Coir Geomesh

Vinay Kumar Midha*, Suresh Kumar S

Department of Textile Technology
Dr B R Ambedkar National Institute of Technology, Jalandhar, Punjab, 144011, India.

*Corresponding author's email: midhav@rediffmail.com

Abstract

Soil erosion can be effectively controlled by proper soil management techniques using geomeshes. Commercial plain woven coir geomeshes suffers from high flexural rigidity due to their higher interlacements. In this study coir geomeshes of 400, 700 and 900 g/m² were structurally modified with 2/1 twill and 4 end broken twill weaves, to reduce their flexural rigidity. The performances of modified coir geomesh were then evaluated. Rainsplash erosion control tests were carried out according to ASTM D 7101, at different soil slopes of 15°, 30° and 45°. Germination tests were performed according to ASTM D 7322. It was observed that 2/1 twill woven geomesh with 1.5 weave factor showed better erosion control at the lower and medium slope angles of 15° and 30°, whereas the commercial plain woven geomeshes with 1.0 weave factor showed better performance at higher slope angle of 45°. It is due the higher weft contact of modified geomesheswith higher float lengths. 2/1 twill woven geomeshalso showed highest total primary rooting length, whereas 4 end broken twill showed highest value of vegetation (%). Hence, at lower and medium angle of soil slopes, modified 2/1 twill woven coir geomesh is recommended, whereas plain woven geomesh is recommended for higher slope angles.

Keywords: Soil Erosion; Geomesh; Weave Factor; Soil Slope; Erosion Control; Germination Test

1. Introduction

Removal of top layer of soil, along with organic matter and vegetation known as soil erosion is one of the major and silent threat to our planet, followed by pollution and population. Air, water and other human activities are the general mediums of soil erosion; among those major erosion occurs due to the water and rainsplash [1-3]. This also reduces soil stability which is important for the construction of roads and embankments, underlays at hill slopes, and seashores [4-6]. Such problems of soil erosion are prevalent over 53% of Indian land area and lead to major disasters like land sliding, floods, chocking down the bridges, sedimentation of dams, desertification, etc. North-Western Himalayan regions including the states of Jammu Kashmir, Himachal Pradesh, Uttarakhand, Punjab and Haryana are the frequent victims of such disasters.Soil erosion can be effectively controlled by proper soil management systems or bio-engineering techniques (utilization of vegetation) using geomeshes [7].

Geomeshes are woven or nonwoven geotextile products made of different natural or synthetic fibres, or combination of both. They are widely used in controlling rainsplash soil erosion, until the permanent vegetation grows sufficiently to reinforce the soil, and are also known as rolled erosion control products (RECPs). Among various geomeshes, geomeshes made from natural fibres perform better in environmentally sensitive regions, due to their biodegradability. Also, the protruding fibre surface of natural fibre geomesh reduces the water flow velocity and has effective control on soil erosion [8-10]. The better water retaining capacity of natural fibres offers better microclimate for the growth of vegetation by retaining water in the soil along with the controlled lighting during germination [6,11-14]. Biodegradable geomeshes made of coir and jute fibres are widely used in India for erosion control, due to their abundance in southern and eastern part of the subcontinent. Though jute geomeshes perform better in erosion control, but faster degradation of jute in damp conditions has restricted its usage [15-17]. Hence the coir geomeshes with better durability (3 to 5 years) are observed to be a better alternative to resolve the soil erosion problems of Himalayan regions.

The commercially available coir geomeshes suffer from the problem of high flexural rigidity, due to the high lignin content in the coir fibres. Moreover, the plain woven structure of commercially available coir geomeshes with higher interlacement of yarns increases their flexural rigidity and therefore limits their erosion control performance. Chemical modification of coir geomeshes to reduce their flexural rigidity results in the removal of lignin content from coir fibres and therefore reduces their durability. Therefore, in this study coir geomeshes were structurally modified by increasing their weave float lengths (2/1 twill and 4 end broken twill weave), and weave factors were calculated. Coir geomeshes of 400, 700 and 900 g/m^{2}were used for the study. Since Himalayan region has narrow to very steep slopes, the rainsplash erosion tests were carried out at different slope angles of 15°, 30° and 45°.

2. Materials and Methods

The performances of three different grade (400, 700 and 900 g/m^2) plain woven commercial coir geomeshes along with their modified woven structures (2/1 twill and 4 end broken twill weave) were studied for rainsplash erosion control. Soil from the lower foothills of the Himalayas/ Shiwalik ranges, having 40% sand, 40% silt and 20% clay particles, was used in the study. The North-western lower Himalayan foothills receive an average annual rain fall of 950 mm/year. During heavy rainy day, there will be an average rainfall of 100 mm in a single day with 3 rain spells of each 20 to 30 minutes [18-21]. The E_r ratio (Ratio between the average rainfall per day to the average rainfall per year) of these regions is observed to be 0.1 [7]. Same geographical conditions were considered for the study. Table 1 shows the characteristics of coir geomeshes used in the study.

Table 1. Characteristics of coir geomeshes

Geomesh mass/unit area (g/m^2)	Weave	% cover	Warp density (ends/ m)	Weft density (picks/ m)	Warp linear density (Tex)	Weft linear density (Tex)	Mesh opening size (mm x mm)
	Plain	21	52	40			
400	2/1 twill	21	52	40	5895	4646	19x21
	4 end broken twill	21	52	40			
	Plain	33	58	46			
700	2/1 twill	33	58	46	4942	4942	17x19
	4 end broken twill	33	58	46			
	Plain	25	52	42			
900	2/1 twill	25	52	42	5895	4942	19x19
	4 end broken twill	25	52	42			

Woven geomesh made from different weaves cannot be compared with regard to their physical and mechanical properties unless the weave effect is normalized. Hence weave factor of different woven geomeshes were calculated [22-23]. The weave factor is a number that accounts for the number of interlacement of warp and weft in a given repeat. It is also equal to the average float and expressed as in Eq.1.

$$\text{Weave factor (M)} = \frac{\text{Number of threads per repeat (E)}}{\text{Number of intersections per repeat of the cross thread (I)}} \tag{1}$$

The performance of modified coir geomeshes for rainsplash erosion control were tested according to ASTM D 7101 using a bench-scale setup with some modifications (Fig.1) [24]. Since the coir geomeshes are open structures with mesh opening size of 17-21 mm, the test core of 20 cm diameter specified in ASTM D7101 were replaced by a test tray 75 cm in length, 50 cm in width, and 25 cm in depth. Test trays were filled with 2 cm of sand at the bottom and with 20 cm of soil to be tested (Shiwalik soil) at the top and allowed to settle for 120 days [25]. Testing was carried out at three different slope angles of 15°, 30° and 45°, so the settled test trays were placed in the required angle of slope and covered with and without modified coir geomeshes. Rainsplash based on the geographical conditions (100 mm/hr rainfall intensity) was simulated on the trays, and the eroded soil was collected for 2 minutes and weighed. The percentage difference of eroded soil from the trays with geomesh and without geomesh was calculated to evaluate the erosion control performance of geomesh (Eq.2).

$$\text{Erosion control \%} = \frac{(\text{eroded soil without geomesh} - \text{eroded soilwith geomesh})}{\text{eroded soil without geomesh}} \times 100 \quad (2)$$

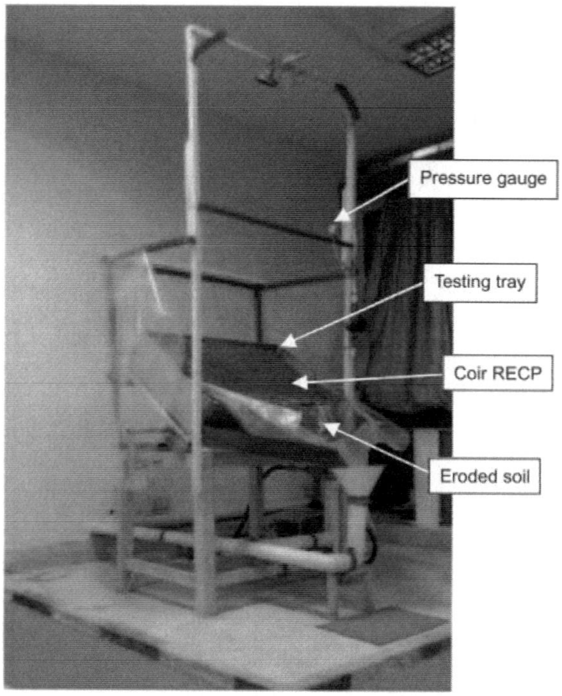

Fig.1 Rainsplash erosion control test to evaluate the erosion control performance of geomesh.

Germination tests were performed according to ASTM D 7322 [26]. Germination tests wereperformed using earthen pots filled with soil, and sown with fixed number of wheat seeds (90 seeds/ pot). The pots were covered with modified coir geomeshes at surface of the soil as shown in the Fig.2a, and one pot was left uncovered for the control test. The test pots were kept at uniform temperature and lighting, and watered until the 21st day (Fig.2b). At the end of the 21stday, percentage of vegetation in each pot was calculated using Eq.3 with reference to the number of plants germinated in the control test pot.

$$\text{\% of vegetation} = \left(\frac{\text{Number of plants germinated in the pot with geomesh}}{\text{Number of plants germinated in the pot without geomesh}}\right) \times 100 \quad (3)$$

Other important factors that affect the soil stability and germination of plants are the number and length of roots, so total rooting was also calculated at the end of 21st day. Primary roots in wheat plants germinate in 10 days and secondary roots start appearing after 20 days [15]. After 21 days, the average primary root length and average number of roots/plant were measured in each pot by harvesting 10 plants randomly. Total rooting after 21 days was calculated using Eq.4 [11].

$$\text{Total rooting (cm)} = N \times n \times L \quad (4)$$

Where, N is the vegetation density (total number of plants germinated after 21 days), n is the average number of roots/ plant, and L is the average root length (cm).

a b

Fig.2 Germination test; a. 1st day; b. 21st day.

The coir geomeshes were also tested for various index tests, including tensile strength, elongation, and flexural rigidity [27]. Tensile strength and elongation were tested according to ASTM D4595 on 10 cm wide samples at 30 cm gauge length. Wet tensile strength was also tested after dipping the samples in water for 2 minutes until they were completely wet. Flexural rigidity was tested in Shirley bending length tester using ASTM D1388, with 15 cm wide samples. Shirley bending length tester works on 'cantilever principle', in which a horizontal strip of fabric is clamped at one end and the rest of the strip is allowed to hang under its own weight [28-29].Ten specimens for each sample were tested. Four bending length measurements were taken on each specimen; two on the face side and two on back side, and flexural rigidity was calculated using Eq.5 [30].

$$\text{Flexural rigidity } (\mu\text{Nm}) = \text{Fabric weigth in g/m}^2 \times c^3 \times 9.807 \times 10^{-6} \quad (5)$$

Where, c is the bending length (mm).

3. Results and discussion

3.1 Rainsplash erosion test

Table 2. shows the performance test (rainsplash erosion and germination test) results of modified coir geomeshes.

Table 2. Performance test results of modified coir geomeshes

Geomesh mass/unit area (g/m²)	Weave	Weave factor	Erosion control percentage			% of vegetation at 21st day (%)	Average number of primary roots/ plant	Average primary root length (cm)	Total primary rooting length (cm)
			15°	30°	45°				
400	Plain	1.0	59.1	47.6	42.2	81.8	5.0	12.0	4320.0
	2/1 twill	1.5	69.9	44.7	30.2	96.6	4.5	12.0	4590.0
	4 end broken twill	2.0	41.6	35.8	22.7	97.7	4.0	11.0	3784.0
700	Plain	1.0	60.2	49.1	43.1	80.7	5.0	12.5	4437.5
	2/1 twill	1.5	71.1	66.3	36.9	86.4	5.0	12.0	4560.0
	4 end broken twill	2.0	79.1	45.0	31.5	89.8	4.5	11.3	4017.2
900	Plain	1.0	60.2	51.1	48.8	84.1	5.0	11.8	4366.0
	2/1 twill	1.5	78.5	53.6	33.4	89.8	5.0	11.6	4582.0
	4 end broken twill	2.0	71.9	46.5	31.1	98.9	4.0	11.0	3828.0

It was observed that the 2/1 twill woven geomesh with 1.5 weave factor shows better rainsplash erosion control at the lower and medium angle of 15° and 30° slopes, while the plain woven coir geomesh with weave factor 1.0 showed better performance at the higher slope angle of 45° (Table 2). This is due to the higher float length and structural stability of 2/1 twill structure. The higher weft float at the back side of the twill structure lies against the direction of water flow, which offers obstruction to the water flow and reduces its velocity. Moreover, higher weft floats at the back side of 2/1 twill woven coir geomesh offer better contact with the soil and increase the stability of soil. During rainfall and soil erosion, the soil surface develops contours (channel or path of water flow in the soil) because of it the contact between soil and geomesh continues to reduce. The depth and number of contours formed depends on various factors including angle of slope, water flow velocity, and soil characteristics. Hence geomesh with lower flexural rigidity can follow the contours easily as compared to those with higher flexural rigidity

However, the broken twill structure geomeshes with maximum float length and weave length of 2.0was observed to have reduced erosion control performance at higher angle of slope, due to the structural disorientation of the geomesh. It increases the direct exposure of soil to the rainsplash and increases the rate of erosion. For the same reason the 2/1 twill woven geomesh was also observed to have limited performance at higher angle of slope, where the increased water velocity disorients the structure of geomesh and induces the rate of erosion. When combined effect of weave factor and slope angle was observed, plain woven geomesh with weave factor 1.0 was observed to have higher erosion control percentage at the lower slope angle of 15°, which was followed by 30° and 45° (Fig.4). Similar trend was observed in all 400, 700 and 900 g/m^2 geomeshes. 2/1 twill woven and broken twill woven geomesh with weave factor 1.5and 2.0 was also observed to have better erosion control at lower angle of slope due to the reduced water flow velocity at lower angles.

At 15° slope angle increase in weave factor (float length) of 900 g/m^2 geomeshes from 1.0 to 1.5 was observed to have an increased erosion control percentage, but further increase in weave factor to 2.0 was observed to have decrease in erosion control percentage. It is due to the disorientation of structure at increased float length. However, at 15° increase in weave factor of 700 and 900 g/m^2 geomeshes from 1.0 to 2.0 was observed to have a continuous increase in erosion control percentage due to the higher weft yarn contact with soil and better stability of geomesh structure. But at higher angle of slope (30° and 45°) with higher water flow velocity 700 and 900 g/m^2 geomeshes fail to control erosion due to the disorientation of their structures (Fig.3).

Among coir geomeshes with different masses, it was observed that 700 g/m^{2}geomesh showed higher erosion control in most of the slopes. This is due to the higher yarn density (Ends or warp yarns per meter, and Picks or weft yarns per meter) and percentage cover of 700 g/m^{2}geomesh when compared to other geomeshes (Table 1). Higher yarn density leads to higher number of interlacements per unit area and influences the stability of the structure. At higher interlacements per unit area, yarn offers higher restrictions to disorientation even at the higher velocity of water flow at steeper angles. Further the higher % cover of the 700 g/m^2 grade coir geomesh reduces the direct exposure of soil to the rainfall and provides better stability to the soil.

3.2 Germination test

Table 2 shows the results of % vegetation and total primary rooting length grown after 21 days using different coir geomeshes. It was observed that the total rooting length of plain woven structure is low when compare to twill woven structure, due its lower % of vegetation. On the other hand, the broken twill stucture with higher % of vegetation results in lower total rooting length, due to the lower number of roots per plant and lower average root length. In twill woven stucture, the moderate root length, number of roots/ plant, and % of vegetation results in the highest total root length. The average root length and number of roots per plant are much influenced by the water retaining capacity and air circulation in the soil [31], which depends on the woven structure of geomeshes. Higher float lengths in 4 end broken twill weave reduces the contact between soil and geomesh at certain places andincreases the contact at floats. This non-uniform contact of broken twill woven structure reduces the air circulation and water retention capacity of the soil, which results in the reduced root length and number of roots per plant. Whereas in the plain and 2/1 twill woven structures the uniform contact of geomeshes to the soil increases the water retention capacity and air circulation, which increases the total primary rooting length. However, the % of vegetation has influence on erosion control, but the total primary rooting has a major influence on erosion control due to its influence on soil stability. The improved total rooting can highly improve the soil stability and reduce the soil erosion. Hence the twill woven coir geomeshes are prferable for rainsplash erosion control.

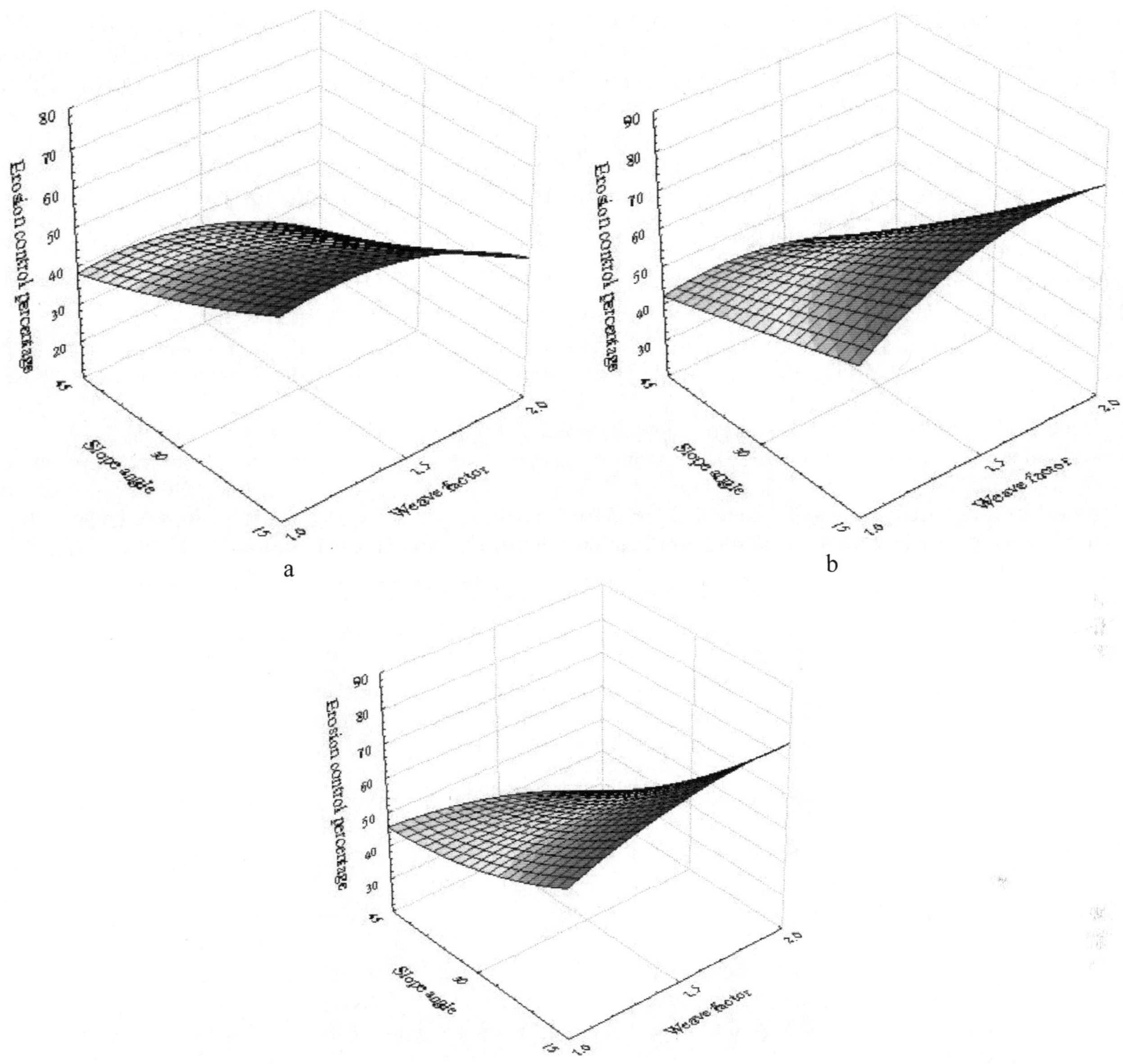

Fig.3 Influence of weave factor and slope angle on erosion control percentage; a. 400 g/m^2 geomesh; b. 700 g/m^{2}geomesh; c. 900 g/m^2.

3.3 Other Index tests

Tensile strength and elongation performance of modified coir geomeshes were calculated. It was observed that woven structure of geomeshes have less influence on its tensile strength and elongation, both in dry and wet conditions due to open structures. The warp wise tensile strength of different woven structures of 400 g/m^2 grade coir geomeshes is in the range of 181.9 to 121.9 kg force and its elongation ranges from 35.2 to 33.6%.

4. Conclusion

From the study it was observed that the limitations of coir geomesh for erosion control can be overcome by proper structural modifications. Along with the weave structure, soil slope angle also has greater influence on the performance of geomesh. 2/1 Twill woven geomeshes with weave factor 1.5 shows better soil erosion control at lower and medium slopes; whereas, plain woven structures with weave factor 1.0 shows better results at higher slope angles. In the case of germination of plants, it was observed that the germination of plants improves with the structural modification of commercial geomesh. The percentage of germination is highest for the 4 end broken twill structure with weave factor 2.0, and lowest for plain woven structure. Total rooting length was found to be higher for 2/1 twill structures of all masses per unit area. Finally, 2/1 twill woven coir geomeshes are preferable for the lower and medium slopes, whereas plain woven coir geomeshes are preferable for higher slopes of lower Himalayan regions, for the effective control on soil erosion.

References

[1] Sutherland RA. A critical assessment of the research conducted at the hydraulics and erosion control laboratory: a focus on rolled erosion control systems applied to hill slopes. Geotext Geomembranes 1998; 16: 87-118.

[2] Pimentel D. Soil erosion: a food and environment threat. J Environ Dev Sustain 2006; 8: 119-137.

[3] Guidelines for Installing Rolled Erosion Control Products in Slope, Channel and Shoreline applications. Information on http://www.researchgate.net/publication/237563570_Guidelines_for_Installing_Rolled_Erosion_Control_Products_in_Slope_Channel_and_Shoreline_Applications. Accessed on 14th April 2014.

[4] Giroud JP, Williams ND, Pelte T, Beech JF.Stability of geosynthetic: soil layered systems on slopes. Geosynth Int 1995; 2: 1115-1148.

[5] Naemura S, Miki H. Design and construction of geotextile reinforced soil structures for road earthworks in Japan. Geosynth Int 1996; 3: 49-62.

[6] Mathai PM. Coir. In: Frank RR, editor. Bast and other plant fibres. Cambridge: WP, 2005; p: 274-314.

[7] Kothyari UC. Erosion and sedimentation problem in India. In: Exeter Symposium.Erosion and sediment yield: global and regional perspectives. Proceeding, 1996; p: 531-540.

[8] John NWM. Geotextiles. London: BA&P, 1996; p: 78-85.

[9] Ogbobe O, Essien KS, Adebayo A. A study of biodegradable geotextiles used for erosion control. Geosynth Int 1998; 5: 545-553.

[10] BhatiaSK, Smith JL, Lake D, Walowsky A. Technical and economic evaluation of geosynthetic rolled erosion control products in highway drainage channels. Geosynth Int 2002; 9: 125-148.

[11] Midha VK, Suresh KS. Influence of woven structure on coir rolled erosion-control products. Geosynth Int 2013; 6: 396-407.

[12] Balan K. Coir Geotextiles: an eco-friendly engineering material. In: Roa GV, Banerjee PK, Shahu JT, Ramana GV, editors. Geosynthetics: new horizons. New Delhi: ABP, 2004; p: 365-376.

[13] Soil Erosion control products. Information on http://www.erosionpollution.com/soil-erosion-control.html. Accessed on 14th April 2014.

[14] Coir Bhoovastra some thoughts on market promotion. Information on http://coirboard.gov.in/News/Articles/bvast-mktg.pdf. Accessed on 11th August 2012.

[15] MidhaVK, Suresh KS, Sharma A. Performance of transesterified jute rolled erosion control products. Geosynth Int 2014; 21: 301-309.

[16] Thomson JC, Ingold, TS. Use of Jute Fabrics in Erosion Control. In: Report to the Jute Market Promotion Project: International Jute Organization. Geneva: UNCTAD/ GATT: 1986.

[17] Sarsby RW. Use of limited life geotextiles (LLGs) for basal reinforcement of embankments built on soft clay. Geotext Geomembranes 2002; 25: 302-310.

[18] State of Environment report on Himachal Pradesh. In: DEST (Department of Environment, Science & Technology). Shimla, 2008; p: 30-37.

[19] Guidelines on erosion control and drainage of railway formation. In: GED/RDSO (Geo-technical Engineering Directorate Research Design and Standards organization). Lucknow, 2005; p: 8-12.

[20] State of Environmental Reports. Information on http://www.soeatlas.org/Reports.aspx. Accessed on 14th June 2011.

[21] Rainfall Calculator. Information on http://www.lenntech.com/calculators/rain/rainfall-precipitation.html. Accessed on 21st January 2012.
[22] Grosicki ZJ. Watson's Textile Design & Colour. Cambridge: WP, 2004; p: 4-61.
[23] Behera BK, Hari PK. Woven textile structure: theory and applications. Cambridge: WP, 2010; p: 6-8.
[24] ASTM D 7101. Standard index test method for determination of unvegetated rolled erosion control product (RECP) ability to protect soil from rain splash and associated runoff under bench-scale conditions. Pennsylvania: ASTM.
[25] Fan JC, Wu MF. Effect of Soil Strength, Texture, Slope Steepness and Rain Fall Intensity on Interril Erosion of Soils in Taiwan. In: Sustaining the Global Farm. International Soil Conservation Organization. Indiana, 2001; p: 588-593.
[26] ASTM D 7322. Standard test method for determination of rolled erosion control product (RECP) ability to encourage seed germination and plant growth under bench-scale conditions. Pennsylvania: ASTM.
[27] Li MH, Khanna S. Aging of rolled erosion control products for channel erosion control. Geosynth Int 2008; 15: 224-231.
[28] Booth JE. Principle of textile testing. New Delhi: CBS, 1996; p: 353-380.
[29] Allen SR. Evaluation and standardization of rolled erosion control products. Geotext Geomembranes 1996; 14: 207- 221.
[30] Saville BP. Physical Testing of Textiles. Cambridge: WP, 2002; p: 256-260.
[31] Blum A. Plant Breeding for water- limited Environments. New York: SP, 2010; p: 12-44.

Dependence of Friction-induced Vibration on Constructional Parameters of Woven Fabric

Ji-Yong Hu[1,2*], Rui-Tao Jiang[2], Xu-Dong Yang[2], Qian Zhang[2], Bao-Zhong Sun[2], Yu-Ling Li[2], Xin Ding[2]

[1]*Key Lab of Textile Science & Technology, Ministry of Education, Donghua University, Shanghai, 201620, China*
[2]*College of textiles, Donghua University, Shanghai, 200237, China*

[*]Corresponding author's email: hujy@dhu.edu.cn

Abstract

Normal vibration features have been recognized to play an important role in the tactile qualities of textiles, whereas the tangential components are commonly neglected. Here, we explored the dependence of different vibrational components on constructional parameters of woven fabrics. This was achieved by measuring the friction-induced vibration, and analyzing the relationship of extracted vibration features with constructional properties of woven fabrics. The results showed that the yarn weave densities majorly affect the normal vibration intensity, the yarn fineness and the weave pattern determine the tangential vibration intensity, and the fundamental frequency of vibration characterizes the periodic textures of woven fabric. It is concluded that the normal and tangential components of friction-induced vibration represent different constructional specifications of woven fabrics.

Keywords: Vibration; Fabric; Construction; Fingerprint

1. Introduction

To facilitate the research on tactile texture of textiles, a key question is what it would take for us to easily manipulate the feel of surface textures. Controlling fabric tactile textures requires a means to "measure" tactile textures. Only if fabric tactile textures can be measured quantitatively, accurately, and reproducibly is it possible to identify the fundamental physical characteristics of the fabric sheet that impact its tactile textures. And then, the potential exists to develop a model that can be used to optimize the production process through a variety of means, including weave design, wet end chemistry, machine operations, creping, fiber selection and surface treatments.

Previous researchers developed models relating various measured parameters to panel handfeel results [1]. However, these models do not make it possible to identify the specific sheet characteristics that cause one sheet to be more or less than another sheet or to identify how to improve fabric sheet's tactile textures. A key factor in overcoming this deficiency is to base the model on fundamental, measureable fabric properties that the fabric manufacturer has the ability to alter, and just the constructional properties of woven fabric are sufficient.

The other motivation for our research is in the area of haptic virtual reality and tactile quality quantification of textiles, where tactual texture plays an important role in realism. Interestingly, the synthesis of rich, artificial tactual textures have tended to resemble audio rendering techniques, being based on generating a spatial waveform that is played at the speed of exploration. The artificial waveform can be created either from measurements [2], or procedurally from stochastic or fractal models. All these synthesis processes are based on the assumption that the contact is made by a rigid probe that obeys an elastic linear law, and on simplified solid friction models to generate stimuli. Unfortunately, these assumptions break down completely in the case of a bare finger [3].

When the human finger slides over various surfaces, vibrations are produce from the sliding friction [4,5]. Friction develops between sliding surfaces regardless of the magnitude of relative motion between them. Since the early works of Katz, as well as with more recent studies [6,7], it is generally accepted that these vibrations play a determinant role in the tactile perception of textures, or absence thereof. The vibration produced are related to the perceived roughness [8,9], softness [8,10], and slippery sensation [8].

The human sense of touch excels at sensing and interpreting the friction-induced vibration to gather information about the physical contacts taking place, as is ascribed to the specialized receptors in human finger to detect these vibrations, e.g, the Pacinian corpuscle with a peak sensitivity of 250 Hz and Meissner corpuscle with a peak sensitivity of 100Hz. Information about stimulus frequency and amplitude, depending on the roughness and stiffness of the surface tested, is multiplexed within a single population of neurons[6,7]. The frequency content of skin vibrations over above 50Hz by natural textures is conveyed through the timing of the response with millisecond precision [7,11], and the vibration amplitude alone is only marginally informative about texture identity [12].

One factor behind the generation of the oscillations by the sliding skin is the distribution and nature of asperities on object surfaces which result from manufacturing and fabric morphology. When human fingertip touches a rigid surface, the frequency spectrum distribution of the induced vibration is mainly determined by the relative geometry of the two contact surfaces and by the contact parameters (i.e. velocity) [4]. As a typical object of fabrics with deformable surfaces, the vibration signals produced by the relative motion between human fingertip and the touched fabrics are representative of both the constructional details of fabrics and the materials properties. Fagiani et al has observed that the structure and deformation of the textiles itself largely affect the spectrum of the induced vibration [4,13]. And also, the excitation frequency from the texture pattern of the fabric by the relative motion between a rigid tip and textiles is nearly recognized from the amplitude of friction force responses, whereas the viscoelastic nature of the textiles can lead to the shift of the friction response frequency [5]. In addition, previous studies focus on the contribution of the harmonic components of normal vibration signals to the content of tactile textures [14-16], whereas the tangential vibration was observed to play a significant role in generating tactile sensation [3, 17]. The human tactile system can indeed discriminate between vibrations with the same fundamental frequency but different spectral characteristics. Therefore, it needs to learn a low dimensional representation of the vibration data. This low-dimensional space should capture the vibrations caused by fabric surfaces, but exclude dimensions containing the additional vibrations that are irrelevant to the tactile sensing task.

The finger also contributes peculiar geometrical and material properties that determine the tribology of the contact [18-20]. When sliding on woven fabric surfaces, human fingers generate audible noise [21]. However, the friction between fabric and metal plate was less sensitive to fabric morphology and rubbing direction [22]. In this sense, only the fabric properties in the contact area can't represent the stimulus features pertinent to cutaneous mechanoreceptors responsible for tactile textures.

In the present study, we will ask whether the vibrations generated during the slip of a finger on woven fabrics could be characterized in terms of their normal and tangential spectrum contents. In other words, we wondered whether these vibrations encode the underlying surface geometry and material property. Fabrics are inherently complex, and their surfaces textures have both stochastic indiscernible features like sandpaper and regular distinguishable features of patterned textures like grooves. A simple plain woven fabric (with a uniform structure) can provide valuable understanding of the spectra that is produced in simplified systems. For this reason, four sets of plain woven fabrics as well as a set of fabrics with complex weave patterns were investigated.

Friction vibration was produced by human fingertip sliding across a fabric on aluminum platform, and the subsequent vibration spectra were analyzed to generate total vibration and frequency, and were compared to the frequencies one would expect according to the fabrics' microstructure, specially the space between adjacent yarns and the relationship between the spectra and woven structure parameters. The analysis of the spatial frequency content of these vibrations gave some insights into the mechanisms that can possibly be at play.

2. Materials and Methods

2.1 Stimuli

Five sets of fabrics were considered, and each set consists of five fabric samples. The construction properties of five sets are: the warp and weft yarn weave density (Dj, Dw), the warp and weft yarn count (Nj, Nw), the adjacent yarn spacing (Pj, Pw), and the weave patterns, respectively. Details of these fabrics are given in Table 1. The yarn diameter is calculated as followed: $d = 1.1284 / \sqrt{N_{j,w} \times \delta_y}$. δy is the linear density of yarn (0.8 g/cm^3).

Table1. Construction details of the woven fabrics studied

set	Fab	DjxDw (ends/cm)	NjxNw (tex)	Pattern	d(μm)	Pw(mm)	Pj(mm)
1	1	280x160	37x37	plain	240	0.691	1.404
	2	280x180	37x37		240	0.645	1.08
	3	280x200	37x37		240	0.644	1.007
	4	280x220	37x37		240	0.643	0.835
	5	280x240	37x37		240	0.663	0.810
2	6	280x240	37x59	plain	305	0.652	0.905
	7	280x240	37x48		278	0.605	0.911
	8	280x240	37x42		258	0.657	0.973
	9	280x240	37x37		240	0.67	0.992
	10	280x240	37x28		210	0.708	1.024
3	11	200x200	37x37	plain	240	0.828	0.958
	12	220x200	37x37		240	0.739	1.043
	13	240x200	37x37		240	0.684	1.053
	14	260x200	37x37		240	0.645	0.973
	15	280x200	37x37		240	0.649	0.993
4	16	280x200	59 x37	plain	305	0.722	0.832
	17	280x200	48x37		278	0.708	0.828
	18	280x200	42x37		258	0.690	0.890
	19	280x200	37x37		240	0.715	0.827
	20	280x200	28x37		210	0.694	0.865
5	21	280x220	37x37	plain	240	0.647	0.97
	22	280x220	37x37	3/1 ↗ twill	240	1.248	1.957
	23	280x220	37x37	5/1 ↗ twill	240	1.811	2.731
	24	280x220	37x37	8/5warp-faced satin	240	2.421	2.515
	25	280x220	37x37	8/5 weft-faced satin	240	2.430	3.947

2.2 Apparatus

To characterize vibrations transmitted to the finger during a test of tactile perception, a system showed in figure 1 was developed. The human finger plays the role of a resonator and the roughness/ adhesion of tested material as a source of excitation [23]. The stimulus, i.e the fabric sample is fixed on the upper center part of the apparatus. The vibrations created by the scanning of the fabric surfaces were recorded by means of a lightweight, triaxial displacement acceleration sensor (PCB 356A01, USA) securely bonded to the index finger nail. The other experimental conditions used with the human finger are summarized as follows: normal force: 0.2–0.4 N, sliding speed: 16 mm/s, sliding distance: 20 mm, scanning direction: from left to right side. Note, a large preload flattens the yarn surface and diminishes the pattern effect. Koc et al has observed that the contact area converges to a constant value with increasing preload [24].

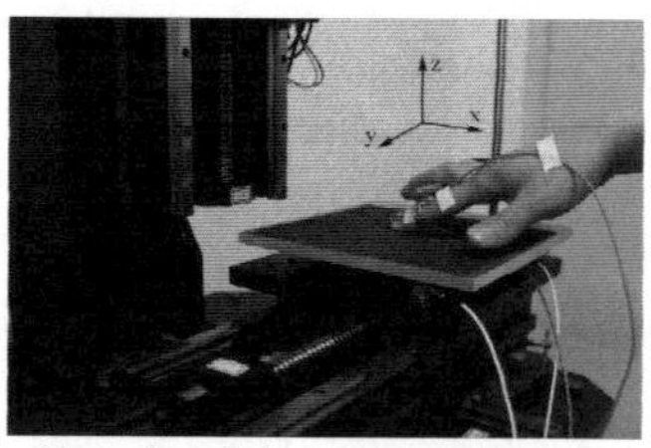

Fig.1 Sketch of the measurement setup

Considering the sliding direction effect in the friction measurement of fabrics [24], the sliding direction was from left to right side and was corresponding to the weft direction or warp direction depending on the changing fabric construction parameters.

2.3 Experimental Procedure

The subject was sitting comfortably in front of the apparatus. The fabric was scanned from left to right on the right index fingertip of the receptor-bearing digit. These strokes will produce a time based waveform applicable for signal processing algorithms. Each participant made 5 satisfactory recordings with each fabric, called one recording block. Each recording has different starting sliding position and sliding path. Recordings lasted less than one second, so that slowly varying effects due to viscoelastic behavior of a fingertip would not interfere with the results.

To quantify the vibration noise introduced by the measurement system when moving the table, measurements without friction between the finger and the surface, i.e the called null recordings, were carried out.

For each subject, there was one recording block per surface. A set of null recordings was also obtained in the same experimental session as the set of fabric surface recordings for which it served as a control. There were 2-3 blocks per session, one or two with fabric surfaces, and one with the null surface. The acceleration signals were acquired with a digital-to-analog, analog-to-digital data acquisition system (NI9234, National Instruments, Austin, TX, USA), which is linked by standard microcomputer. The sampling period was set at 10 kHz in order to have a comfortable sampling margin and reduce distortion during post-filtering. However, other frequency peaks can be seen within each vibration spectrum, and may relate to multiple harmonics of the construction.

2.4 Data Processing

Considering primarily frequency and amplitude, but not acceleration of the vibrations were perceived [25], the acceleration data was transformed into vibration displacement by an integral algorithm. The followed data preprocessing method is similar to Bensmaia's [26]. For each subject, 50 noise-corrected power spectra were obtained for each surface, the median noise-corrected spectrum for each stimulus was then computed for each subject (medians were used, rather than means, to minimize the influence of outliers). The surface spectra obtained from each subject were then normalized by the mean spectral power across surfaces to account for (across-subject) differences in recording gain.

To identify characteristic features of the sample explored, signals are analyzed in the spectral domains. The peak of intensity in each window of the spectrogram was estimated and its corresponding frequency was extracted. For the simple woven fabric sample with periodic surface textures, the energy was extracted to be precise located at the mechanical fundamental frequency. The fundamental frequency in a specified sliding velocity is estimated by $f_0(Hz) = v_{slid}(\mathrm{mm/s}) / \mathrm{P}_{j,w}(\mathrm{mm})$,where v_{slid} is the relative sliding speed. For the variation of yarn count, the adjacent warp or weft yarn spacing is calculated by: $P = 2\times(2\times10^5 / D - d)$. Meanwhile, the vibration wavelength of the dominant frequency was estimated by the sliding velocity. The space and the estimated wavelength were compared.

3. Results and Discussions

The frequency content of the measurements made on the texture revealed multiple features of the transformation between the simple woven fabric and the generated vibration amplitude fluctuations. From the spectrum analysis, we could thus extract relevant information for the discrimination of period stimuli such as woven fabric. The vibration signal contained a fundamental frequency component corresponding to the finger velocity divided by the spatial period of the stimulus. This regularity was found for a wide range of textural length scales, and the intensity of the vibrations could be related to the microgeometry of the scanned surfaces.

3.1 Consistency

The ranking consistency coefficients (*R*c) between fabric construction parameters and the extracted spectral features, as well as the 2-side confidence intervals, were listed in table 2, and Rc no less than 0.7 was indicated in the bold italic text. Generally, the vibration amplitude component in the z direction nearly represent the relative magnitude to the weave density of these fabric samples, and the vibration wave components in the tangential direction was the change of yarn count and the number of yarns in a complete repeat. In the following sections, these vibration features with high ranking consistency coefficient were further analyzed.

Table 2. Ranking consistency between fabric construction parameters and vibration features

Features		f_0	A_0x	A_0y	A_0z	f_1	A_1x	A_1y	A_1z
Weft density	Rc	-0.300	-0.400	-0.300	***-0.700***	-0.300	0.400	0.300	***-0.800***
	95%CI	-0.894	-0.915	-0.894	-0.964	-0.894	-0.609	-0.676	-0.977
		0.676	0.609	0.676	-0.258	0.676	0.915	0.894	0.033
Weft count	Rc	***0.700***	***-0.700***	***-0.700***	-0.500	-0.00800	***-0.700***	***-0.700***	-0.500
	95%CI	-0.258	-0.964	-0.964	-0.933	-0.042	-0.964	-0.964	-0.933
		0.964	0.258	0.258	0.524	-0.025	0.258	0.258	0.524
Warp density	Rc	-0.100	0.100	***0.700***	***0.900***	-0.300	0.100	0.200	-0.200
	95%CI	-0.843	-0.774	-0.258	0.328	-0.894	-0.774	-0.73	-0.87
		0.774	0.843	0.964	0.989	0.676	0.843	0.870	0.730
Warp count	Rc	-0.400	***-0.700***	-0.600	-0.100	-0.100	***-0.700***	***-0.700***	-0.200
	95%CI	-0.915	-0.963	-0.949	-0.843	-0.843	-0.963	-0.963	-0.870
		0.609	0.258	0.412	0.774	0.774	0.258	0.258	0.730
Repeat	Rc	0.600	***-0.900***	***-0.900***	-0.600	0.600	***-0.800***	-0.600	-0.600
	95%CI	-0.412	-0.984	-0.984	-0.949	-0.412	-0.977	-0.949	-0.949
		0.949	-0.328	-0.328	0.412	0.949	0.033	0.412	0.412

Note: f and A is the frequency and amplitude of vibration wave components, the subscript digit "0" and "1" represent the fundamental and the first harmonic wave, and x, y and z the vibration direction, respectively.

3.2 Yarn density

Fig.2 also plotted the components of normal vibration amplitude against weave density. Observably, there was a turning point of the changing curve of vibration amplitude against an increasing weave density. From the microscopic image of fabric surface, the weft yarn spacing was about 417~625μm, and the fingerprint width of the subject was about 518μm. It has been observed in Zahouani's work [27] that the friction-induced vibration intensity during fingertip across fabric surfaces was affected by the relative ratio of fingerprint to yarn spacing. And also, Fagiani et al mentioned that the not perfectly sinusoidal profile of human fingerprint and the sample roughness was response for the upper harmonic components of the response [28]. Here, when the weft density is 180 fillings per centimeter, the weft yarn spacing is about 555μm. Therefore, when the fingerprint width is less than the yarn spacing, the yarn density determines the vibration intensity. With an increase of yarn density, the increasing number of yarn intersections in a unit contact area results in higher surface roughness at larger yarn densities. Thus, human fingertip confronts more resisting points on the surface of the tested fabrics and yields higher vibration frequency and intensity as increasing weaving yarn densities, and vice versa.

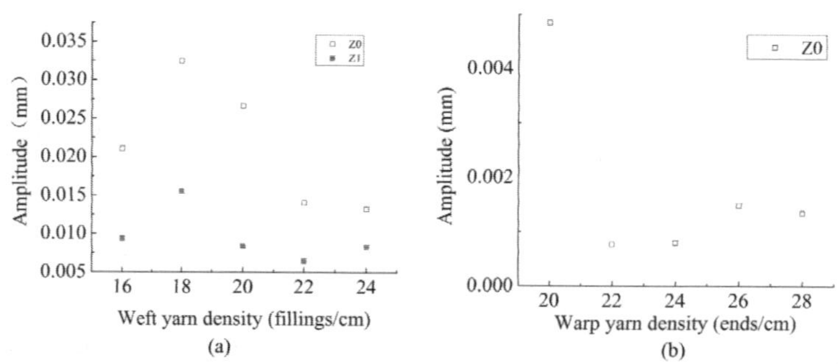

Fig.2 Relationship between vibration amplitude (Z) and yarn weaving density

3.3 Yarn fineness

From the consistency analysis results, the yarn fineness (i.e. count) majorly impacted the fundamental frequency of normal vibration and the amplitude of tangential vibration. When the weft yarn fineness increased from 28 tex to 59 tex, the theoretical yarn spacing increased from 210μm to 305 μm. Apparently, these yarn spacing was less than that of fingerprint width of the subject. The yarn spacing dominated the friction-induced vibration frequency. Fig.3(a) and 3(b) showed those frequencies predicted and the experimentally measured frequencies against yarn fineness. As weft yarn fineness (or diameter) of fabrics with same patterns and yarn densities decreased, the vibration frequency decreased as well due to reduction in contact and yarn intersection area and increased in smoothness. As the yarn fineness of these woven fabrics was increased, the fundamental frequency was decreased. The frequencies predicted of most of fabrics were found to exist within the vibration spectra produced, showing that the calculated frequencies and the predicted frequencies nearly matched and therefore this related to the micro-structure of the plain woven fabrics. Therewith, it was possible to create a fingerprint of the frequency spectra according to the microstructure of the woven fabric, particularly in the case of plain woven fabrics.

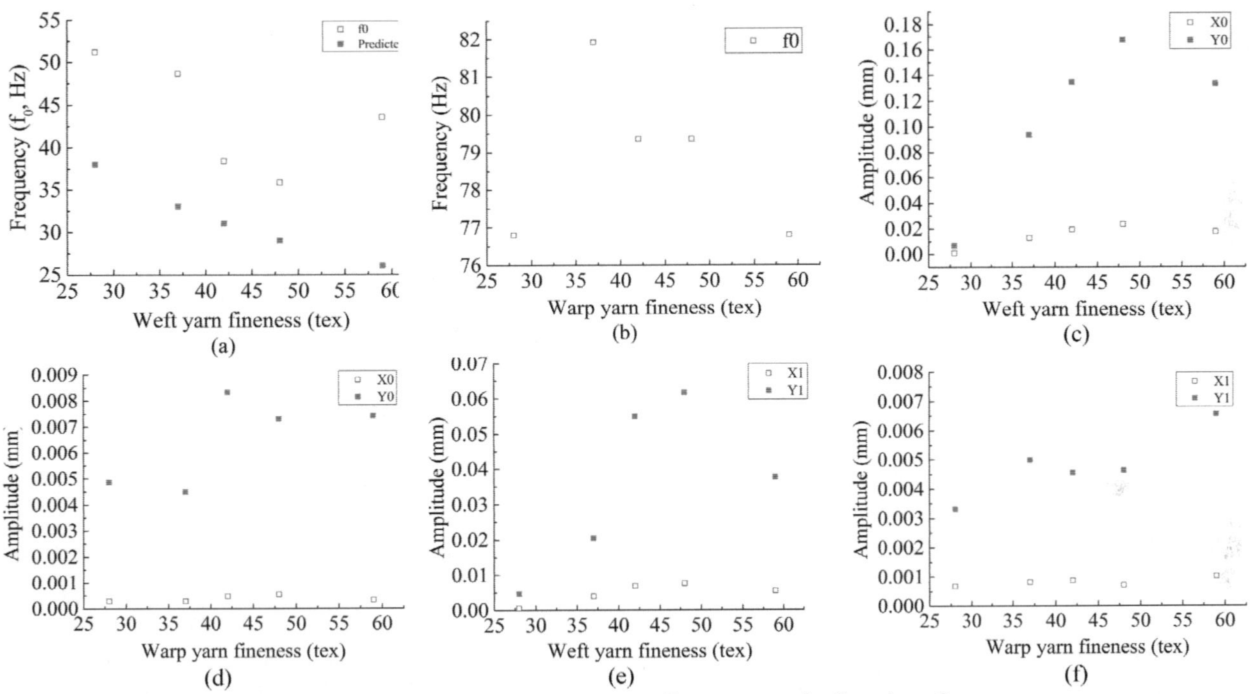

Fig.3 Relationship between yarn fineness and vibration features

On the other hand, fig.3(e)-3(f) showed some similar features to those in the case of weave density. The fundamental tangential vibration amplitude increased with weft yarn fineness over the range from 28 tex to 59 tex, and the main component of the generated vibration intensity was corresponding to the sliding direction. Considering the same weave density and weaving pattern, an increase of the yarn fineness flatter yarns in a fabric with a constant weaving density, and results in a higher lateral compression resistance. Fontaine et al has observed that the contact hardness between the sensor and surfaces were expected to influence the sensor response [29]. In the view of the mechanical interaction, the yarn count can affect the mechanical interlock between fingertip and fabric surface textures, so that the transverse vibration was enhanced. However, the change of vibration amplitude with weft yarn fineness was significantly different from that with warp yarn fineness, as showed in fig.3.

3.4 Pattern effect

The structure of the fabric and its appearance are affected by the weaving pattern to a large extent. The effect of pattern on friction-induced vibration holds the contribution of yarn densities and diameter of weft and warp yarns together with weaving pattern of the yarns. Considering the equivalent consistency with respect to the number of yarns in unit pattern (NYUP), the crossing-over firmness factor and the floating yarn factor [30], only the effect of NYUP on the friction-induced vibration was further discussed. Fig.4 showed the fundamental tangential vibration amplitude against NYUP. It can be seen from figure, the fundamental tangential vibration amplitude decreased with an increasing NYUP.

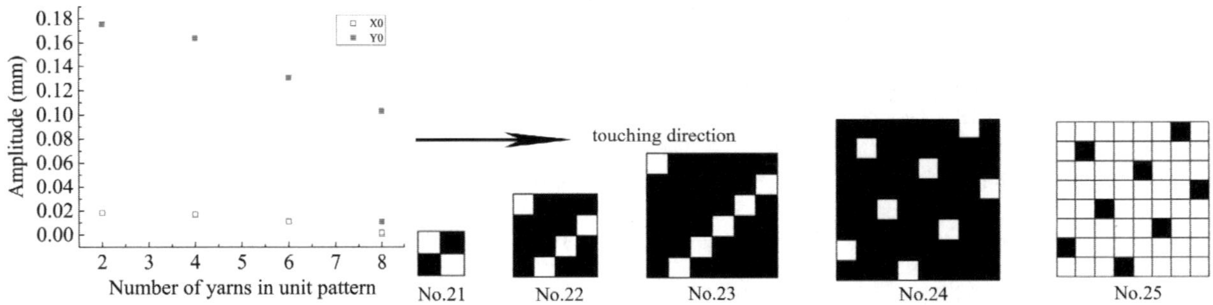

Fig.4 Relationship between NYUP and the tangential fundamental vibration amplitude. The illustrations of weaving pattern of five fabrics in set No.5 is on the right

In fact, the floating length increases with an increase of NYUP, and the floating yarns occupy a large part of the unit surface area. When human fingertip slides across these fabric surfaces, the weaving pattern illustrated in fig.4 will lead to a low intermittent interaction between human fingerprint and fabric texture elements. Koc's studies showed that the friction coefficient decreases with an increase of the floating yarn length or the proportion of floating yarn over unit surface area. As seen in figure 4, the unit length of warp yarns floating oven weft yarns in the size of 8x8 unit^2 fabric sample are 4 unit length for plain, 6 unit length for 3/1 twill, 7 unit length for 5/1 and 7 unit length for satin and 1 unit length for sateen. Observably, increase in floating yarn length and larger numbers of floating yarn pieces per unit area in warp direction results in smoother contact between fingertip and fabric, and decreases the vibration intensity. Hence, pattern effect is more influential parameter on vibration than yarn densities and yarn fineness.

4. Conclusions

The above study identified a peculiar dependence of the main harmonic components of the vibration response, both amplitude and frequency, on the constructional parameters of the woven fabric samples. The FFT analysis confirms that the accelerator is capable of identifying dominant frequencies of clean consistent textured surfaces. The relationship between fabric constructional parameters and friction-induced vibration investigated in this research primarily depends on the number of intersection points and total area of floating yarns in a unit pattern area in the scanning direction. And also, there are relative effects of the normal and tangential characteristics of scanned fabric surfaces on friction-induced vibration. These results provide the grounding for further development towards a realtime algorithm that could be integrated with the current tactile sensing algorithms of textile qualities.

5. Acknowledgement

This work is supported by the National Natural Science Foundation of China through project (No. 51405079), and the Fundamental Research Funds for the Central Universities.

Reference

[1] Hu J, Ding X, Wang R. A review on the cognitive study of fabric hand. J Donghua Univ 2007; 33: 677-681, 686.

[2] Romano JM, Kuchenbecker KJ. Creating realistic virtual textures from contact acceleration data. IEEE T Haptics 2012; 5:109-119.

[3] Wiertlewski M, Lozada J, Hayward V. The spatial spectrum of tangential skin displacement can encode tactual texture. IEEE T Robot Autom 2011; 27: 461-472.
[4] Fagiani R, Massi F, Chatelet E, Costes JP, Berthier Y. Contact of a finger on rigid surfaces and textiles friction coefficient and induced vibrations. Tribol Lett 2012; 48: 145-158.
[5] Sancakdar K, Sumer B. An experimental study of the frictionally induced vibrations on rough textile fabrics. 40th Leeds-Lyon Symposium on Tribology & Tribochemistry Forum 2013, Lyon, France, 2013.
[6] Harvey MA, Saal HP, Dammann JF, Bensmaia SJ. Multiplexing stimulus information through rate and temporal codes in primate somatosensory cortex. PLoS Biol 2013; 11: e1001558.
[7] Weber AI, Saal HP, Lieber JD, Cheng JW, Manfredi LR, Dammann III JF, et al. Spatial and temporal codes mediate the tactile perception of natural textures. Proc Natl Acad Sci USA 2013; 110: 17107-17112.
[8] Muramatsu Y, Niitsuma M. Correspondence relationships between vibrotactile stimuli and tactile sensations determined by semantic differential. IEEE RO-MAN: The 22nd IEEE International Symposium on Robot and Human Interactive Communication. Gyeongju, Korea: IEEE; 2013; p: 668-673.
[9] Hollins M, Bensmaïa SJ. The coding of roughness. Can J Exp Psy 2007; 61: 184-195.
[10] Tanaka M, Tanaka Y, Chonan S. Measurement and evaluation of tactile sensations using a PVDF sensor. J Intel Matl Syst Str 2008; 19: 35-42
[11] Mackevicius EL, Best MD, Saal HP, Bensmaia SJ. Millisecond precision spike timing shapes tactile perception. J Neurosci 2012; 32: 15309-15317.
[12] Manfredi LR, Saal HP, Brown KJ, Zielinski MC, Dammann III JF, Polashock VS, et al. Natural scenes in tactile texture. J Neurophysiol 2014; 111: 1792-17802.
[13] Fagiani R, Massi F, Chatelet E, Berthier Y, Sestieri A. Experimental analysis of friction-induced vibrations at the finger contact surface. Proc Inst Mech Eng, Part J: J Eng Tribol 2010; 224: 1027-1035.
[14] Picinali L, Feakes C, Mauro DA, Katz BFG. Spectral discrimination thresholds comparing audio and haptics for complex stimuli. Lect Notes Comput Sc 2012; 7468: 131-140
[15] Park G, Choi S. Perceptual space of amplitude-modulated vibrotactile stimuli. IEEE World Haptics Conference (WHC), 2011; p: 59-64.
[16] Cholewiak SA, Kwangtaek K, Tan HZ, Adelstein BD. A frequency-domain analysis of haptic gratings. IEEE T Haptics 2010; 3: 3-14.
[17] Smith A, Chapman C, Deslandes M, Langlais J, Thibodeau M. Role of friction and tangential force variation in the subjective scaling of tactile roughness. Exp Brain Res 2002; 144: 211-223.
[18] Wiertlewski M, Hayward V. Mechanical behavior of the fingertip in the range of frequencies and displacements relevant to touch. J Biomech 2012; 45: 1869-1874.
[19] Adams MJ, Briscoe BJ, Johnson SA. Friction and lubrication of human skin. Tribol Lett 2007; 26: 239-53.
[20] Tomlinson SE, Lewis R, Carré MJ. Review of the frictional properties of finger-object contact when gripping Proc Inst Mech Eng, Part J: J Eng Tribol 2007; 221: 841-50.
[21] Cooper C, Norton J, Marshman C, Norton I. The acoustics of friction and microstructure of model and apparel fabrics. Text Res J 2014; 84: 626-637.
[22] Das A, Kothari VK, Vandana N. A study on frictional characteristics of woven fabrics. AUTEX Res J 2005; 5: 133-140.
[23] Zahouani H, Vargiolu R, Boyer G, Pailler-Mattei C, Laquieze L, Mavon A. Friction noise of human skin in vivo. Wear 2009; 267: 1274-1280.
[24] Koc IM, Aksu C. Tactile sensing of constructional differences in fabrics with a polymeric finger tip. Tribol Int 2013; 59: 339-349.
[25] Pongrac H. Vibrotactile perception examining the coding of vibrations and the just noticeable difference under various conditions. Multimedia Syst 2008;13 : 297-307.
[26] Bensmaia SJ, Hollins M. The vibrations of texture. Somatosens and Mot Res 2003; 20: 33-43.
[27] Zahouani H, Mezghani S, Vargiolu R, Hoc T, Mansori ME. Effect of roughness on vibration of human finger during a friction test. Wear 2013; 301: 343-52.
[28] Fagiani R, Barbieri M. Modelling of finger-surface contact dynamics. Tribol Int 2014; 74: 130-137.
[29] Fontaine S, Marsiquet C, Nicoletti N, Renner M, Bueno MA. Development of a sensor for surface state measurements using experimental and numerical modal analysis. Sensors Actuat A: Phys 2005; 120: 507-517.
[30] Milasius V. An integrated structure factor for woven fabrics Part I: estimation of the weave. J Text Inst 2000; 91: 268-76.

The Innovative Application of Modern Design Elements to Shaanxi Homespun Pattern

Xiao-Yan Li, Ai-Xiang Gao*

School of Apparel & Art Design, Xi'an Polytechnic University, Xi'an, 710048, China

*Corresponding author's email: 362759945@qq.com

Abstract

This paper describes the aesthetics of the traditional homespun patterns in Shaanxi and compares modern and traditional design elements. This article focuses on the similarities and differences between traditional homespun patterns and modern graphic elements. It explores innovative applications of modern design elements to Shaanxi homespun patterns. In this way people can get a better understanding of the traditional homespun patterns, as well as modern design elements. This article explores how modern design elements could be mixed with traditional patterns. In this text, some aspects about fabrics are discussed, focusing on the performance of the plain, twill and satin fabrics as well as their innovation applications.

Keywords: Soil weaving traditional patterns; Innovative contemporary elements; Modern graphic design

Fig.1 Process of weaving

Fig.2 Cushion

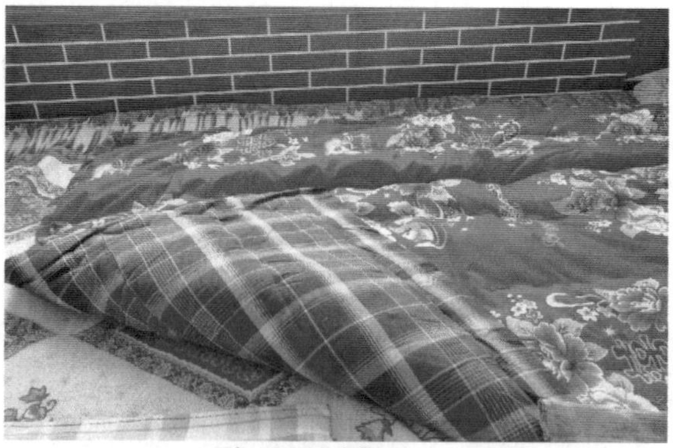

Fig.3 Bedding

Fig.4 Apron

Fig.5 Cotton trousers

Shaanxi homespun has become one of the most useful forms of fabrics in people's daily life. It had a pivotal role in China's textile history due to its richness and distinctiveness of local characteristics. Its diverse application also means that it affects every aspect of local people's lives (Fig.2, Fig.3, Fig.4 and Fig.5). Shaanxi's weaving industry is famous for homespun weaving. Using different colored threads, through the mutual weaving process between the threads of longitude and latitude, Shaanxi homespun is known for its unique geometric design. Unfortunately, with rapid industrialization and the advent of new fabrics, the homespun had gradually faded in popularity. Recently, because of the heightened awareness in environmental protection and a surge in the search for comfortable fabrics, interest in the Shaanxi homespun has returned once again. There has been an unprecedented level of attention to those unique textiles due to their breathable fabrics, colorful patterns, artistic characteristics and practical value.

Shaanxi homespun was complimented for a period of time, offering a level of comfort hardly to be found anywhere else. The aesthetic characteristics of the homespun aligned perfectly with the beauty-appreciation of modern design principles. Such aesthetics are not only reflected in the simple beauty of its material but also in the unique combination of patterns, colors and shapes. Besides, it is said that the current design is not only a reflection of temporary lifestyles but also a transcendent and revolutionary lifestyle. The patterns design should use elements in modern daily life as basis, since design styles come in a variety of ways. In Shaanxi, the traditional homespun has been ongoing with its typical representations throughout history.

1. Relationship between modern design elements and pattern

According to the dictionary <Lexicon>, the word 'pattern' refers to processing some kind of shape or structure, color, grain of the vessel, preparing the plan of construction in advance which has to evolve into a specific design. Contemporary elements can be understood like this, i.e. they can reflect modern concepts and aesthetic appeals, which are the combination of people's visual sense and their lifestyle changes in modern society.

Innovation doesn't mean we should be irrespective of the origin or pursue the novelties blindly. Instead, it is the inheritance of traditional ideal aesthetics and its development. There is a legacy of weaving traditions from previous generations that made its way into modern practices. New ideas are actively pursued with a simultaneous respect for historic traditions. This is the essence of design as well as the requirements of the current time. Furthermore, in today's society, innovation becomes more and more necessary to attract the people's eyes. This is the same for traditional patterns as they are closely related to modern designs but still retains ancient culture and aesthetic psychology of the Chinese nation. To redesign traditional patterns, people should also add some contemporary elements. Only by doing like this can people reach the combination of tradition and modernity.

To design the traditional patterns of the homespun, the most important thing is the focus on the completeness of the pattern as well as ensuring specific shapes in a pattern. Moreover, current traditional homespun patterns can no longer meet the aesthetic needs of the consumers. For this reason, the urgent problem is to ensure that traditional Chinese design features involved in the textile industry should seek to adapt to satisfy contemporary people's taste and aesthetic demands. Not only does modern design focus on solid and positive forms, but it also focuses on negative forms and use of several approaches like reconfiguration. Rearranging various forms to create an array of graphics and interesting patterns has some implied meanings as well. In short, it is for the sake of improving existing patterns and pursuing new visual imagery. Through innovating homespun patterns, it is possible to bring fresh blood into the textile market, stimulating design thoughts, in order to sustain and innovate homespun patterns. On the basis of the original homespun patterns, even more modern fashionable elements join in order to innovate. As well as raising homespun's popularity among the people because of the patterns in the sense of multi-changes and sense of modernity.

There are three components which are the baseline or even the core of modern design. The comparison between beauty in the design itself and beauty in the composition as a whole is suitable for homespun pattern design which adopts modern design elements. In fact, beauty of composition pays more attention to the works' inherited performance, unique visual form and ways of composition, which arouse people's strong visual aesthetic feelings. Paying attention to visual feelings and logical thoughts is partial to rational thinking. However, the traditional patterns of homespun stem from life and are the result of natural abstraction. Adding great importance to intuitive thinking and emanating a vivid perceptual beauty shows the richness of free beauty. Although traditional homespun patterns stress an exquisite sense of form and possess some certain abstract elements, it is primarily based on materialization, representing a significant degree of abstraction of thoughts and ideas. For the abstraction of modern elements, the most important thing is the abstraction of images. Such a perception of abstract images causes a delicate momentum as well as dynamism, rhythm and shape, resulting in a strong visual sense of composition.

Innovating modern design elements in homespun pattern designs should use all kinds of creative techniques. At the same time, encourage serious thinking to use modern design elements within the homespun pattern design Fig.6, Fig.7 and Fig.8 are all designed by myself. In this way, it can reflect the theme and gain consumer's popularity. In order to achieve such a goal, firstly, the design objectives and market positioning should be made; uniqueness and originality should be acquired. Through the distinctive artistic language, the theme should be highlighted, and the form, innovation and contents should be united. Being fit for people in the transmission is one of the principles that transmit the creativity of the pattern in a certain theme; it means the designers have to think about all the situations of objects when it comes to creativity and design by pointing out the targeted object and considering the effective demands. Only in this way can one truly tap into the psychological acceptance of desired people. Therefore, modern design elements in the traditional homespun patterns are an ingenious approach. Both of them become a new way to design homespun pattern that must have far reaching significance and get consumers' satisfaction easily as well.

Fig.6 1st design pattern

Fig.7 2nd design pattern

Fig.8 3rd design pattern

The application of modern design elements to the homespun pattern design should be on the premise of respect for traditional pattern. Making the traditional patterns under the modern design theory, it could endow traditional patterns in a new form due to the use of artistic skills of modern pattern design. Since the newly created graphics must be something the consumers like and did not expect, it is not the ultimate goal to combine all kinds of modern design elements with a variety of methods that can reduce the traditional pattern. To combine the modern design elements and traditional pattern designs, as a famous Hong Kong graphic designer, Master Kan, cites three essential principles to achieve the beauty: conception - using image instead of sprit; innovation - breaking the stereotype and establishing the new situation; application - being flexible and vivid. In order to achieve this, one has to view things from an overall perspective.

2. Reflection of modern design elements in pattern

Analyzing and refining modern design elements through the association and analysis of emotional cognition to rational analysis, the visual image can fully express the outstanding trait of visual elements, looking for the connection and showing the theme. During this process, the unique thought method became the connecting bond between elements and the theme. As a result, creative graphics give new connotation or new visual images to the visualization, which also means that it can add new meanings to representation.

From the application of modern design elements to the process of designing homespun patterns, the integral effect is the key parameter. During the design process, simplification and generalization should be highly prioritized, also should select and simplify both of them. Traditional patterns and modern design elements can be simplified in order to achieve beauty at its purest state. In order to make the pattern much richer, we should also fortify detailed depictions with ethnic flavor on the basis of the simplification and exaggeration, whilst including Chinese culture to reach a well-balanced combination of patterned art modernism and symbolism.

By adding and considering modern design elements in the process of traditional pattern design, integrated effects of the works often determine their styles. The entire pattern embodies the works' main ideology and the artistic style of the homespun. By applying the modern design elements, we can make the pattern of the homespun conform to modernism. Conveying spirit through appearance and showing the beauty of artistic conception can not only reveal sufficient hints of Chinese culture and make room for reflection, but also make it become fashion-conveyed forms that people are likely to accept and satisfy their spiritual needs.

In the process of pattern design, it is important to find out one's greatest interest as a basic model, and then separate this fundamental form. On the basis of arrange, increase and decrease, or conclusion of the geometric shape, using the modern design method, such as superposition and iterative techniques so as to get a new unique design. This form is the model of the structure that should be recognized and recreated. In this way, homespun patterns could transfer imagination into visual sensation with specific connotation. Meanwhile, the design of homespun patterns should exert as much as possible the fabric weaving, which is a unique art language. The combination of the traditional pattern and modern design elements plays an important role in expressing the common people's spiritual world with the inner charm of the combined works.

3. Influence of modern design elements on homespun skills

Design of the homespun pattern and manual skills will finally meet the requirements of homespun pattern design. Design patterns without modern elements perhaps are very aesthetically pleasing, but cannot be created using manual skills. Since homespun with this pattern cannot be weaved, the modern elements for manual weaving process are very crucial. As connotation of element is rich and flexible, weaving staff should think twice before they weave in terms of the homespun pattern design requirements, combining the different weave design methods and manual skills, and implementing how to accomplish it. Before starting to weave homespun, manual calculations for weaving selvedge and total root number that the pattern needs should be meticulously considered, and will be equipped with longitude in color configuration order, with the proper density intertwine on the special frame work. In terms of modern homespun by hand, the three basic textile weaves are mainly Sanyuan tissue: plain weave, twill weave and satin weave.

Bedding materials on the market can be roughly divided into tabby, tweed, satin weave and various fabrics. In plain weave, the warp and weft are aligned so they form a simple criss-crossing pattern. Each weft thread crosses the warp threads by going over one, then under the next, and so on. The next weft thread goes under the warp threads that its neighbor went over, and vice versa. In plain weave, each wire is woven alternately over and under the warp wires through the cloth at 90 degree angles. The shedding cycle for plain weave (which is the simplest weave) will repeat on two picks or two weaving cycles (Fig.9).

Twill weave, which has more warps than the wefts floating on the face of the fabric is called warp faced and those with wefts predominating are known as weft faced (Fig.10). Twill is done by passing the weft thread over one or more warp threads and then under two or more warp threads and so on, with a "step" or offset between rows to create the characteristic diagonal pattern. Because of this structure, twills generally drape well. Examples of twill fabric are chino, drill, denim, gabardine, tweed and serge.

A third form of weave is the satin weave, the new kind of fabrics and the most complicated one (Fig.11). The satin weave is characterized by four or more cool fill or weft yarns floating over a warp yarn or vice versa, four warp yarns floating over a single weft yarn. This explains the even sheen, as unlike in other weaves, the light reflecting is not scattered as much by the fibers, which have fewer tucks. Therefore, it is also the latest one among the three elementary weaves without crossing basically, extraordinary level and fabrics luster. A satin fabric may have as few as five harnesses to form the repetition (in this case it is called a 5- shaft satin), or it may have more. The plain woven fabric has the best shape - memory properties, but there are no obvious differences among other woven structures. It is a crisp, smooth, and a sheen plain-woven fabric. The twill fabric, compared with the tabby in the same sub-fabrics density of feeling is more comfortable than tabby. On the folded fabric, always plan the layout on the wrong side with the right sides together. Uneven plain and twill fabrics have a right and a wrong side and therefore are not considered reversible. Khaki is a color but is now synonymous with military twill pants because soldiers returned to civilian life from World War II continued to wear their military chinos, especially in college. Chino doesn't have to be twill, but is an often firm weave of cotton. Most premium jeans' cotton denim fabric comes from the primary maker of high-end denim fabric used in the U.S. and Europe: Greensboro, N.C. - based Cime Denim, a unit of the International Textile Group. Wherever you had an old pair of chinos, now you have a bona fide catwalk look. The great white cotton twill pants with white gauze cowboy shirts is an interesting way to use the trend and makes it approachable. The research concluded that the moisture permeability, UV resistance and wrinkle recovery of the fabrics with satin weave are better than that of fabrics with plain and twill weave. But the floats in satin and sateen fabrics tend to snag and abrade easily, thus, such fabrics are not as durable as plain or twill weave fabrics. Processes of dyeing and coating, in terms of fabric appearance satin design, colorful, people pleasing, to create a living environment of hope. Superior satin fabric is only for high- rating hotels.

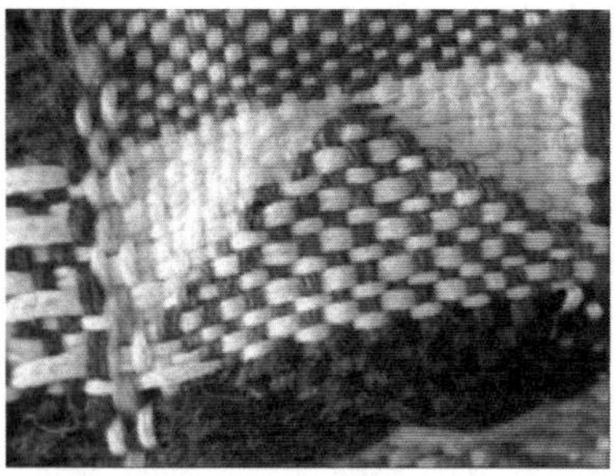
Fig.9 Plain weave homespun

Fig.10 Twill weave homespun

Fig.11 Satin weave homespun

The all kinds of fiber materials can also be used as weft thread that weave in single, double or multi-warp. These constitute a variety of colorful cloth texture, so that we can make the effect of homespun design richer. In the process of weave by hand, in order to keep the cloth and cloth cover different, generally plain weave or basket weave on the cloth edge is used. Most of the pattern can be weaved using the plain weave or the changes of plain weave tissue. The patterns that need to be outstanding in the tissue to increase the weft tissue point (such as weft faced twilled grain or the weft faced satined grain) as its basic tissue. During the process, something that can best match with the warp faced tissue, by contrast using different fabric texture tissue, makes the weaving patterns' gradation more distinct and well- proportioned.

Of course you can also use the k' o-ssu method, which is a type of weaving done by the tapestry method with fine silks and gold thread. It is also a kind of distinctive traditional manual weave skill. With antique loom, Cross-knife to take off by Wei, choose the opposite approach woven by hand woven by pick. This kind of manual technology is particularly suitable for various colors and the homespun patterns of shade levels. By applying the method of k 'o-ssu manual soil weaving cloth, we can not only smooth the surface of the whole cloth but also the color composition of the whole picture. Knotted pile and tufted fabric are applied to the patterns that need designs full of thick style manual weaving, and both of them are for the complex structure and the sense of space.

The homespun pattern uses the knotted pile method to weave, and the cloth surface scatters the dense of fine hair. While the homespun pattern uses the tufted fabric method to weave, the cloth surface spread all over the loop pile that should be curved. It must make the colored weft threads meet the requirements tightly in different ways, and make the color weft pile or loop pile erect on the surface of the fabric. After completion of the weaving, we should use scissors for appropriate dressing. Using the method of knotted pile or tufted fabric, we can significantly increase the richness and strength of the homespun pattern. At the same time, we can also enhance the artistic expression of pattern, which is required to highlight the dynamic effects of patterns.

4. Conclusion

In a word, modern homespun is all manually produced, with primitive simplicity, full of rural style, and caters to modern city folks who pursue a carefree and comfortable way of life. However, with the rapid development and progress of the whole society, continuous innovation on the basis of maintenance of the traditional essence can make the market much more stable and would not be engulfed by the era tide. In such an industrialized and highly developed modern society, homespun is seen as a spiritual symbol of the contains a strong national spirit; the perfect combination of modern design elements and classical elements. With the heightened awareness of environmental protection and innovation in this field, the homespun pattern is creating a new style in the fashion industry and is bound to be outstanding in the market.

References

[1] Gao AX. Modern indoor textile design. Journal of Xi'an University of Engineering Science and Technology 2002; 03: 16.

[2] Gao AX. Analysis on the combination of figure design and weave design in modern manual wall-hanging. Journal of Xi'an University of Engineering Science and Technology 2004; 18: 69.

[3] Li YH. Weixian folk spinning and weaving research. Hebei Normal University 2007.

[4] Kathleen FJ. A weaver looks at Tinguain blankets. Textile Society of America/ Newsletter 2011; 3: 23.

[5] Sandy B. Eco-chic: The Fashion Paradox. Black Dog Publishing 2008.

[6] Luo M, Liu H. Homespun application research in modern household decoration design. The mass of literature and art 2013.

[7] Liu YZ. Pucheng traditional homespun textile technology research. Xi 'an: xi 'an academy of fine arts 2006.

[8] Xu XW. History of Jiangnan homespun. Shanghai: Shanghai academy of social sciences publishing house 1992.

[9] Du XC. The historical status of the rural weaving industry in modern China. Guizhou academy of social sciences 2012; 10: 53-60.

[10] Song YH. The beauty of the cloth with fashion. The Chinese manual 2009; 9: 24-25.

Preparation and Properties of Core-and-Effect Air-jet Textured Yarns Blended by Nylon6 and Poly(ethylene terephthalate) Filaments

Li-Qian Huang[1,2*], Xue-Li Wang[3], Feng Chen[1], Rui-Yun Zhang[1,2], Jian-Yong Yu[3]

[1]*College of Textiles, Donghua University, Shanghai, 201620, China*
[2]*Key Laboratory of Education Ministry of Textile Technology, Donghua University, Shanghai, 201620, China*
[3]*Modern Textile Research Institute, Donghua University, Shanghai, 201620, China*

*Corresponding author's email: hlqian@dhu.edu.cn

Abstract

In this paper, core-and-effect composite yarns blended by Nylon6 and Poly (ethylene terephthalate) (PET) filaments were prepared by air-jet texturing process. The influence of over-feed rate, filament combination of core and effect on structure and properties of prepared textured yarn were mainly studied. The results show that the tenacity and unevenness of textured yarn decrease with the rise of overfeeding rate, while linear density, instability and loop ratio of textured yarn are increased with the increased in overfeeding rate. The textured yarn composed completely by Nylon6 filaments has more large loops with highest loop ratio than the textured yarns blended by Nylon6 and PET filaments, and a typical core-and-effect yarn structure is formed. For the blended textured yarns, the yarns with PET filaments as core component exhibit higher initial modulus, lower instability and lower unevenness, indicating more stable and even structures are formed.

Keywords: Air-jet textured yarn; Core-and-effect; Overfeeding rate; Nylon6; PET; Properties

1. Introduction

As one member of Polyamides family, Nylon6 fiber has been widely used in areas such as apparel, carpet and tire cords, attributed to its excellent properties in tenacity, elasticity, moisture absorption, abrasion resistance and so on. However, compared to Poly(ethylene terephthalate) (PET) fibers, the shape retention as well as the sun light resistance of Nylon6 fiber and its textiles are inferior because of its relatively flexible molecular chain, which limits its further development to great extent [1]. Traditionally, the simplest way to solve above problems, is interweaving PET yarns and Nylon yarns periodically in the fabric. However, for both Nylon and PET filament yarn, such processing as false-twist texturing, air-jet texturing are usually adopted before weaving, namely, air-jet textured yarn(ATY) and drawn-textured yarn (DTY) are produced and thereafter used for fabric weaving. Therefore, it is reasonable to believe that the properties of Nylon6 fabrics can be improved by producing Nylon/ PET composite filament yarn in these texturing processes before weaving. Among these texturing technologies, air-jet texturing is an effective way to produce composite filament yarns by mechanical interlocking.

Air-jet texturing technology was firstly invented by DuPont Company in the early of 1950s, aiming to produce newly filament yarns with similar appearance and physical characteristics to spun yarns [2]. By this technology, the original parallel arrangement of parent filament yarn is converted into a disordered and entangled structure with numbers of loops and arcs on the yarn surface under the action of supersonic turbulent air flow in texturing nozzle. Core-and-effect composite yarn can be produced by feeding two or more parent filament tows into nozzle. The filament tow which is processed with a higher overfeeding rate than the other tows becomes the skin of yarn, providing loopy, bulky morphology and soft hand to the textiles, while the core filaments fed at comparatively lower overfeeding rate locates in the inner part of the yarn with a compact structure and plays the major role on load-bearing and determines the mechanical properties of the yarn. Many research works [3-10] have been done to study the effect of texturing process parameters on structures and properties of air-jet textured yarns. Texturing process and properties of air-jet textured yarn blended by diacetate and PET filament were studied by Jing Zhang et al [11,12]. The properties of bilateral air-textured yarn blended by polyester and nylon fibers were investigated by

M.Y.Gudiyawar et al [12]. However, few works have been reported on core-and-effect air-jet textured yarn composed by Nylon6 and PET fibers.

In present work, Nylon6 and PET filaments are used to prepare core-and-effect composite yarns by air-jet texturing technology. The influence of overfeeding rate and fiber combination on structure and properties of textured yarn have been mainly discussed.

2. Experiments

2.1 Raw Materials

Nylon6 and PET filaments used as the parent yarn to prepare core-and-effct air-jet textured yarns were supplied by Zhejiang Taihua New Materials Co. Ltd. The specification and properties of the filaments are shown in Table 1.

Table 1. Specification and Properties of Parent Filaments

Filament	Specification	Tenacity /cN.dtex^{-1}	Elongation at Breaking /%
Nylon6	70D/68F	4.49	41.15
PET	75D/72F	3.69	26.34

2.2 Preparation of Air-jet Textured Yarns

The samples of air-jet textured yarns were processed on AIKI RIOTECH air-jet texturing machine with JTC V3069C jet nozzle, with Nylon6 and PET filaments acting as the core or effect component respectively. The core component was wetted by water jet at 17℃ before entering the jet nozzle, the whole process is completed at temperature of 20~25℃. In order to explore the influence of overfeeding rate on the properties of textured yarns, 3 overfeeding rates of 15/10, 25/10, 35/10 were designed. Core-and-effect air-jet textured yarn composed completely by Nylon6 filaments was also prepared as reference. The code and detailed texturing conditions are illustrated in Table 2.

Table 2. Processing Parameters of Air-jet Textured Yarns

Yarn code	Effect component	Core component	Overfeeding rate(effect/core)	Texturing speeding /m.min^{-1}	Air pressure /MPa
NP-1			15/10		
NP-2	Nylon6	PET	25/10		
NP-3			35/10		
PN-1			15/10	420	0.9
PN-2	PET	Nylon6	25/10		
PN-3			35/10		
NN	Nylon6		35/10		

2.3 Characterization of the Yarns

The linear density of air-jet textured yarns was calculated by weighing a yarn with the length of 100m, average of 20 data is used for each sample. Tensile tests were carried out on YG061-1500 Tensile Tester under standard condition with the gauge length of 500mm and loading speed of 500mm/min, the pretension load is (0.05±0.005)cN/dtex, 30 readings for each sample. Yarn unevenness is tested on YG135G evenness tester at speed of 400m/min for 1min, 10 times were tested. Instability of air-jet textured yarn was measured by a modified Heberlein Method [13]. The procedure is illustrated in Fig.1 and explained as follows:

1) The yarn is first tensioned with a light load of 0.01cN/dtex for 1 min, then the mark is made at length a of

50cm;
2) After removal of the light load, a heavy load of 0.5cN/dtex is then applied for 1min and the length *b* is measured;
3) 1 min after removing the heavy load, the light load is reapplied to the sample, and after a further 1 min, the length *c* is measured;
4) The instability index is then calculated by: Instability(%) = [(c-a)/a]×100.

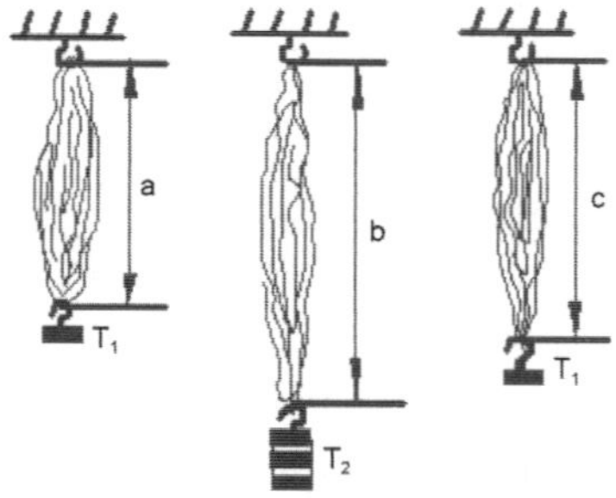

Fig.1 Instability of textured yarn [12]

The photos of air textured yarns were taken using JJ7701-0151 Digital Microscopy by winding the yarns on black velvet board. The loop rate is defined as the ratio of loop structure over the whole yarn structure, and is calculated from the yarn photos by image processing method [14].

3. Results and Discussion

The properties of air textured yarns are summarized in Table3.

Table 3. Properties of Air Textured Yarn

Yarn code	Linear density /dtex	Unevenness CV /%	Tenacity/ cN.dtex^{-1}	Elongation at Breaking /%	Initial modulus/ N.tex^{-1}	Instability /%	Loop rate /%
NP-1	170.02	3.88	3.36	23.75	2.02	2.26	1.28
NP-2	175.31	3.45	2.77	28.34	2.06	2.37	2.88
NP-3	180.56	3.40	2.53	31.73	2.14	2.41	19.34
PN-1	169.55	4.32	3.61	33.53	1.80	3.10	2.38
PN-2	175.17	3.76	3.28	33.76	1.64	4.25	16.44
PN-3	180.81	3.60	3.11	34.68	1.56	4.73	21.58
NN	182.33	3.04	2.80	23.63	1.93	5.39	25.88

3.1 Effect of Overfeeding Rate on Structure and Properties of Textured Yarn

From Fig.2, it can be seen that with increasing of overfeed rates, both NP yarns and PN yarns have more loops present on the surface of yarn, that is the yarns become more bulky. By comparing PN yarns to NP yarns, it is found that NP yarns are more straight and denser than PN yarns. In PN yarns, kind of thick-and-thin structure can be seen, and the filaments are packed more loosely and less oriented along the yarn axis.

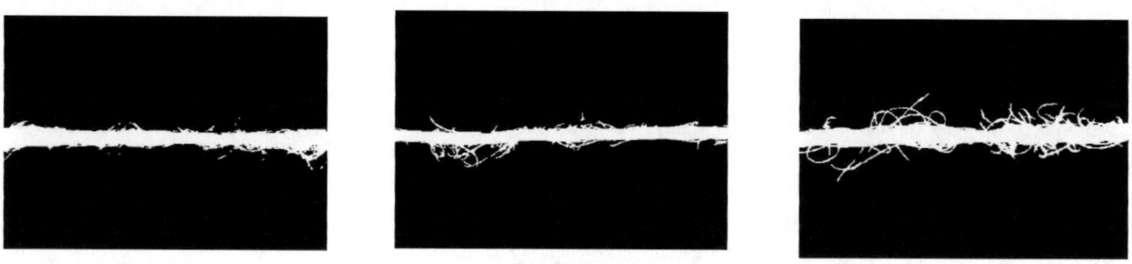

(a)NP-1 (b)NP-2 (c)NP-3

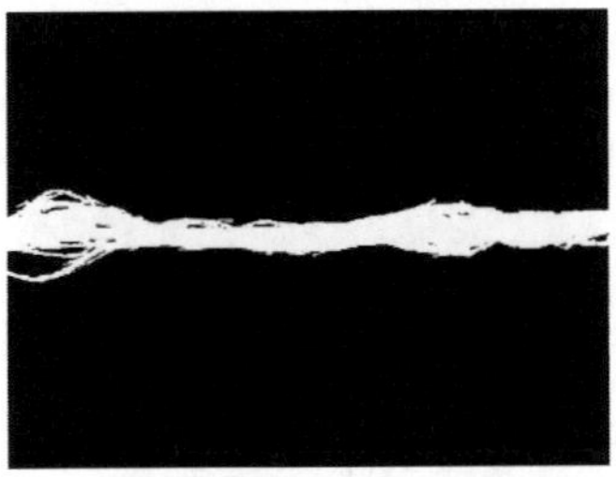

(d)PN-1

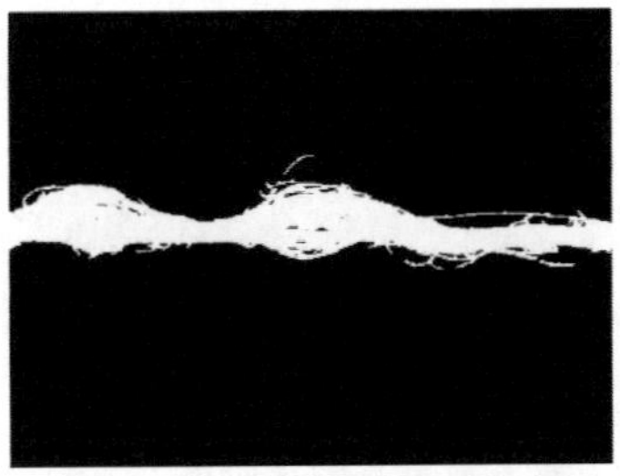

(e)PN-2

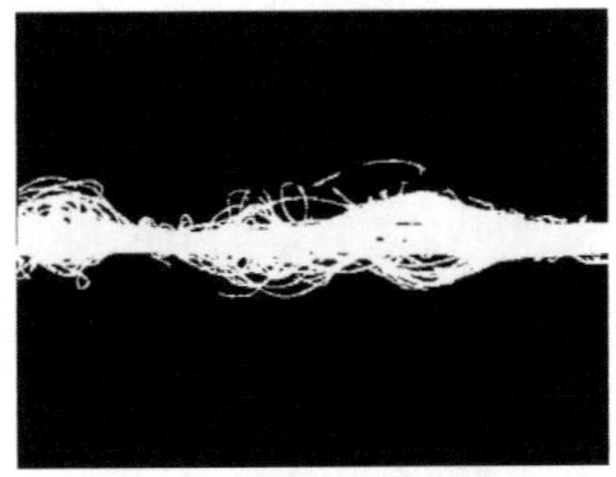

(f)PN-3

Fig.2 Images of air textured yarn at different overfeed rates (×160)

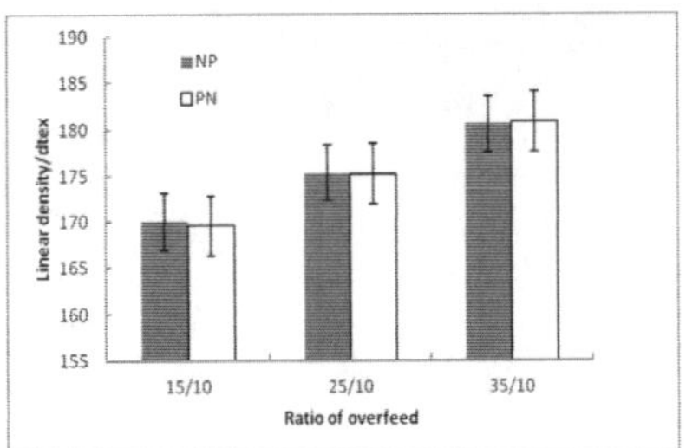

(a)

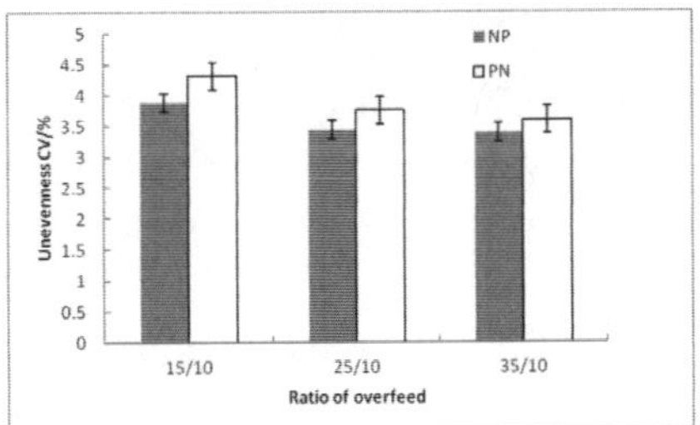

(b)

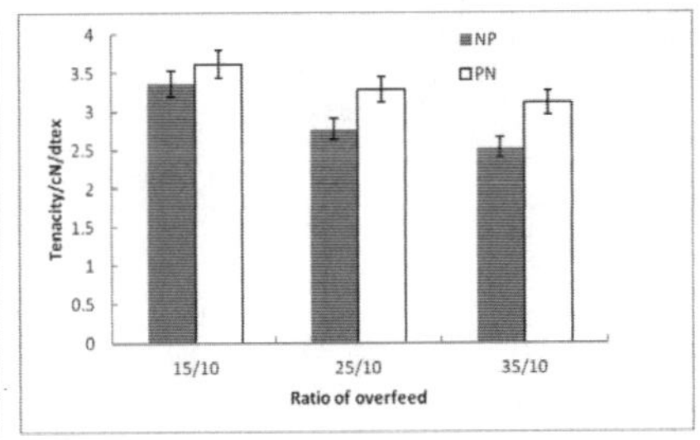

(c)

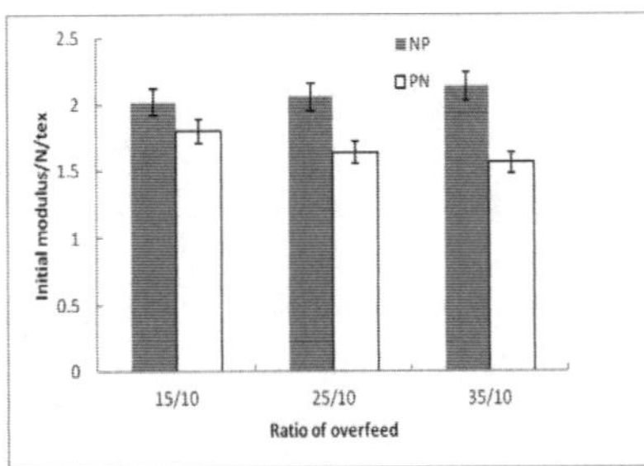

(d)

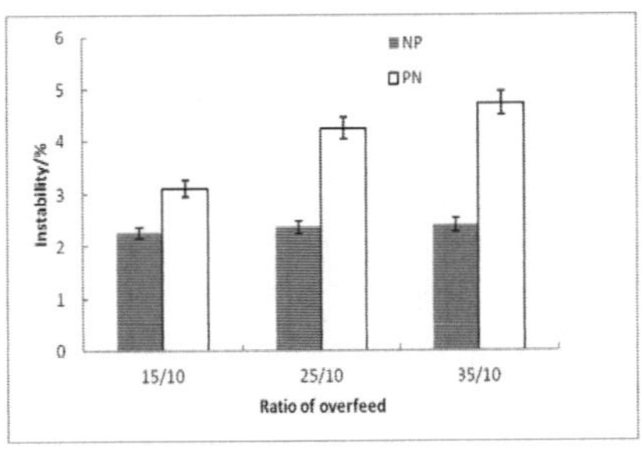

(e)

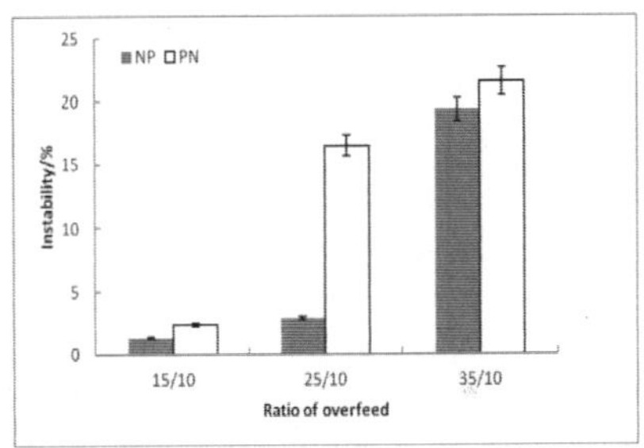

(f)

Fig.3 Properties of air textured yarn at different overfeed rates

From Fig.3(a) and Fig.3(b), it can be seen that the linear density of textured yarns increases slightly at high overfeed rates while the unevenness of yarn decreases with the rise of overfeed rate. At a higher overfeed rate, more excess length of filaments participates in loop formation and fiber entanglement, which results in a higher linear density of the textured yarn. The smaller unevenness of yarns at a higher overfeed rate indicates the variation of yarn mass or linear density along yarn axis is lessened, which provides a better yarn quality.

Fig.3(c) shows that there is an obvious decrease in the tenacity of textured yarn with the rise of the overfeed rate. It is well-known that the tenacity of filament yarn is mainly decided by the breakage of straight filaments in a yarn [2]. For core-and-effect air textured yarn, with the increase of overfeed rate of core to effect filaments, more excess length of filaments participates in the loop formation in the yarn, which disturbs the fiber orientation along yarn axis, leading to a greater unequal strain distribution as well as more step-by-step rupture, and eventually brings with the decline of yarn breaking strength.

In Fig.3(d), it is observed that the initial modulus of NP yarns at high overfeed rate is greater than that at low overfeed rate, while for PN yarns, lower initial modulus at high overfeed rate. The linkage between stretched fibers is the key factor to determine the initial modulus of multi-filaments yarn. For NP yarns, the higher modulus at high overfeed rate means more entanglements formed among filaments. By contrast, for PN yarns, lower modulus at high overfeed rate indicates less entanglements formed filaments.

As seen in Fig.3(e), the instability both of NP and PN yarns increase with the rise of overfeed rate. At higher overfeed rate, more loops are formed , which enhance the chance of fiber slippage. Especially for PN yarns, the instability of yarn is greater than that of NP yarns, indicating a less stable yarn structure.

From Fig.3(f), the loop ratio of yarns increase with increasing overfeed rate, which is easy to understand that at a higher overfeed rate, more excess length of filaments participates in loop formation, thus forming a bulkier yarn structure.

3.2 Effect of Core/ Effect Components on Structure and Properties of Textured Yarn

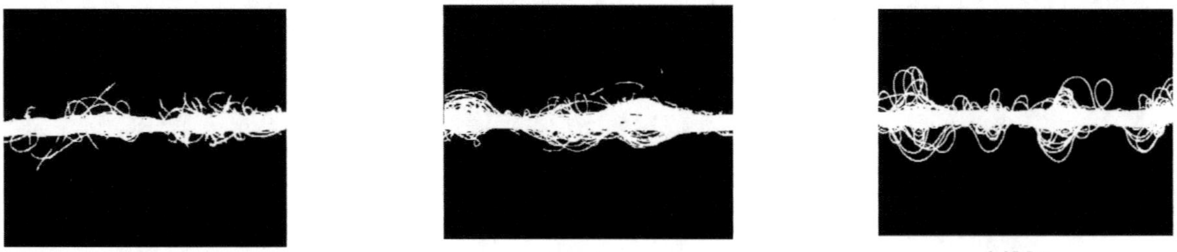

(a)NP-3 (b)PN-3 (c)NN

Fig.4 Images of air-jet textured yarns (×160)

Three samples, NP-3, PN-3 and NN, which were produced at the same textured processing condition except for the different filament combination for core and effect components are chosen.

From Fig.4, it can be observed that the morphology of textured yarns varies obviously with different core/ effect filament combination. For NN yarn, there are a great number of large loops on the yarn surface and a typical core-and-effect yarn structure can be observed. But for NP-3, a few fine and small fiber loops protrude on the yarn surface, which makes the yarn looks more like the spun yarn with numbers of hairiness on the surface. In PN-3, the yarn is more bulky than NP-3 and a thick and thin structure can be found. From Table3, it is found that the loop rate of NN is highest followed by PN-3, and the loop rate of NP-3 is least. The higher loop rate, the bulkier the yarn. This result is accordance with the longitudinal appearance as illustrated in Fig.4.

As shown in Table3, the unevenness of NN is the lowest, that is, the variation of mass for NN yarn along the yarn axis is least. It can be explained that in NN, the same fiber, Nylon6, is used as core as well as effect components, which is helpful to minimize such processing uncertainty as the processing tension fluctuation, thus reducing the variation of yarn mass along yarn axis. The unevenness of NP-3 is lower than that of PN-3, indicating that when PET is used as core component, the variation of yarn mass along yarn axis is less. Furthermore, it should be pointed out that the mass variation along yarn axis is not always accordance with the yarn morphology observed, especially for core-and –effect yarns, because the packing density in the yarn is not a constant.

The tenacity of NP-3 is lower than NN and PN-3. In the core-and-effect structure yarn, the core filaments bear the loading and determine the mechanical properties of yarn. In NP-3, PET filaments act as the core component, while both in PN-3 and NN, Nylon6 filaments are the core component. From Table 1 we know that the tenacity of PET filaments is lower than that of Nylon6, which results in the lower strength for NP-3. However, the initial modulus of NP-3 is higher than NN and PN-3 because NP-3 is packed denser than NN and PN-3 as observed in Fig.4, and the modulus of PET core fiber is higher than that of Nylon6. It is generally believed that the higher initial modulus for fibers or yarns, the less deformation they produces, consequently, the better shape retention the textiles have. Therefore, it is beneficial to improve the shape retention ability of Nylon6 fabrics by blending PET filaments with Nylone6 and choosing PET filaments as core component in an air-jet textured yarn.

The instability of NP-3 is lower than NN and PN-3, which demonstrates that taking the PET filaments as core component is helpful to improve the structure stability of air-jet textured yarn. From the data in Table3, we can know that the structure stability of NN is lowest. It may be caused that after removing the heavy load, the great numbers of entanglements of Nylon6 fibers in the yarn prevent the drawback of the fibers to its original position.

4. Conclusions

According to present study, it can be concluded that:

1) With the increasing overfeed rate of effect-to-core filaments, the textured yarns become more bulky and even with higher linear density, loop ratio and evenness. But the tenacity and yarn stability reduce with the rise of overfeed rate. The initial modulus of NP yarns increase with the rise of overfeed rate, while the initial modulus of PN yarns decrease with the rise of overfeed rate, which indicates the different entangled structure inter-filaments in the yarns.

2) By comparing the yarns with the varied core and effect components, the textured yarn composed completely by Nylon6 filaments has more large loops with highest loop ratio than the textured yarns blended by Nylon6 and PET filaments, and a typical core-and-effect yarn structure is formed. For the blended textured yarns, the yarns with PET filaments as core component exhibit higher initial modulus, lower instability and lower unevenness, indicating more stable and even structures are formed.

5. Acknowledgement

The authors wish to acknowledge the National Natural Science Foundation project (No.51303021), Natural Science Foundation of Shanghai (13ZR1400400), National Key Technology R&D Program project (2013BAE01B02).

References

[1] Yu WD. Textile Materials. Beijing: China Textile & Apparel Press, 2006; P: 26.

[2] Wang SY, Yu XY. New Textile Yarns. Shanghai: Donghua University Press, 2007; P: 334-336.

[3] Demir A, Acar M, et al. Air-Jet textured yarns: the effects of process and supply yarn parameters on the properties of textured yarns. Textile Research Journal: 1988; 58(6): 318-328.

[4] Rengasamy RS, Kothari VK, et al. Effect of process variables and feeder yarn properties on the properties of core-and-effect and normal air-jet textured yarns. Textile Research Journal: 2004; 74(3): 261-263.

[5] Baldua RK, Kothari VK., et al. Effect of feed filament fineness and process parameters on air-jet textured yarns properties. Journal of Textile Institute: 2014; 105(11): 1186-1192.

[6] Sengupta AK, Kothar VK., et al. Effects of filament modulus and linear density on the properties of air-jet textured yarns. Textile Research Journal: 1996; 66(7): 452-455.

[7] Acar M, Demir A. Yarn wetting mechanism and suitable supply yarns for air-jet texturing. Proc. 2nd Int. Conf. on Air-jet Texturing and Mingling/ Interlacing: Loughborough, UK, 1989: 111-130.

[8] Kothar VK, Sengupta AK., et al. Influence of interfilament friction on structure and properties of air-jet textured yarns. Textile Research Journal: 1989; 59(6): 317-323.

[9] Rengasamy RS, Kothar VK, et al. Effect of specific characteristics on structure and properties of air-jet textured yarns: part I effect of filament fineness. Melliand Textilber: 1990; 71((9): 655-659.

[10] Rengasamy RS, Kothar VK, et al. Effect of raw material characteristics on structure and properties of air-jet textured yarns: part II effect of filament cross-sectional shape. Melliand Textilber: 1990; 71(12): 935-937.

[11] Zhang J, Zhang ZL, et al. Investigation on air texturing process for diacetate blending with polyester filaments. Fibers and Polymers: 2007; 8(4); 427-431.

[12] Zhang J, Zhang ZL, et al. Properties of core-and-effect air textured yarns blended by diacetate and polyester filaments. Fibers and Polymers: 2007; 8(1): 84-88.

[13] Demir A, Acar M, et al. Instability tests for air-jet textured yarns.Textile Research Journal: 1986; 56(2): 191-198.

[14] Chen F, Huang LQ, et al. Study on structure and evenness of air textured yarn by image processing method . Journal of Donghua University: 2015; 41(6): 739-742, 749.

Plant Dyeing Using Che Yellow: The Once Exclusive Costume Colour for Chinese Emperors

Min Shao*

Min Shao, Shanghai Theatre Academy, 630# Huashan Road, Shanghai, 200040, P.R. China

Corresponding author's email: shtengsheng@126.com

Abstract

As an important part of Chinese ritual system, costume colour in ancient China was strictly enforced in accordance with ritual law. Yellow is one of China's original traditional colours and part of the five elements. It represents "Earth" and occupies the centre position. As the once exclusive costume colour for Chinese emperors, yellow conferred special status in ancient China.

Plant dyeing technology has been developed for thousands of years in China. From the structure of Chinese calligraphy "染" ("dye") we find that dyeing white fabric is a difficult and highly technical skill. The sad thing is, about a hundred years after chemical dyeing replaced plant dyeing, the great ancient techniques basically disappeared from people's view in China.

Ancient Chinese documents recorded more than a dozen plants which could dye yellow, and Che (cudrania tricuspidata) was a very important one. But this kind of valuable timber grows very slowly, and over-harvesting in order to obtain economic benefits made matters worse. Currently, existing wild Che trees are fewer and fewer. Therefore, the goals of the Che dyeing project included three parts: first, obtain the traditional dyeing materials and process; second, obtain dyeing fabric samples and transfer them into CMYK colour data; third, dye some apparel goods with waste Che wood and minimize the waste of Che resources.

Keywords: Che; Plant Dyeing; Yellow; CMYK Data

1. Introduction

1.1 Yellow as an exclusive costume colour

As an important part of the Chinese ritual system, costume colour in ancient China was strictly enforced in accordance with ritual law. The ritual system was based on three origins: The origin of nature (conforming to the world), the origin of culture (respecting the ancestors), and the origin of class (separating the ruler from the ruled) [1]. Five original colours, including blue, red, yellow, white and black, represented the five greatest gods [2]. As the offspring of the gods, the Chinese emperor was charged with the responsibility to worship heaven and earth [3]. These five original colours, the important visual elements for worship, ancestor and rule, were strictly limited.

At the same time, the five original colours correspond to five orientations and five elements [4]. Yellow represents "Earth" and occupies the centre position [5]. Yellow was a symbol of the emperor for a long historical period in China; as the once exclusive costume colour for the emperor, yellow conferred a special status in ancient China.

In order to highlight their lofty status, the emperor had repeatedly banned ordinary folk wearing yellow dress [6]. Dyeing materials were kept in the most important royal warehouse, and only the court workers could dye yellow [7].

1.2 Che tree and Che yellow

Although in ancient China, the colour Yellow represents nobility and was once exclusive to the emperor, it is not very difficult to find yellow pigmentation in nature. Chinese traditional documents, such as 本草纲目(Compendium of Materia Medica), 齐民要术(Qi Min Yao Shu), 天工开物(Heavenly Creations), 急就篇(Ji Jiu Pian), 玉篇(Yu Pian), 正字通(Zheng Zi Tong), recorded about a dozen plants which could be used to dye yellow. The following are the most important ones: the root of rehmannia glutinosa, the fruit of gardenia jasminoides, the flower of sophora japonica, the root of curcuma aromatica, the branch of berberis thunbergii, the flower of koelreuteria paniculata, the leaf of arthraxon, the branch of cudrania tricuspidata (Che), the branch of cotinus coggygria, the branch of phellodendron amurense, the root of curcuma longa, and so on. Different plant can be dyed different kind of yellow, and Che yellow looks more like golden, which represents glory and virtue. So Che was regarded as a crucial plant for dyeing yellow in ancient China.

Che (cudrania tricuspidata), also known as Chinese mulberry or silkworm thorn, is a Chinese member of the moraceae family. As a kind of common mountain tree, Che was easily found in Jiangsu, Guangxi, Henan and other provinces. Che wood is hard and dense, perfect material for crossbows since ancient times; Che leaves are used to feed silkworms, just slightly harder than mulberry; Che branches can be dyed yellow and dedicated to the emperor [8]; Che root is one ingredient of TCM (Traditional Chinese Medicine), focusing on cough and phlegm; Che fruits are edible and can be used to make wine. It can be said that in ancient China, Che was regarded as a treasure tree, especially the magical quality whereby its leaves can feed silkworm silk and its wood can be used for dyeing silk. Therefore, according to traditional seasonal law, it was forbidden to cut Che in spring [9].

Che grows very slowly, and a Che tree takes a hundred years to reach a diameter of 20 centimetres. Every 3 to 5 years, the Che tree presents a clear beautiful gold thread, which would be lighter and thinner if in South China, and darker and thicker if in North China (Fig.1-2). Che wood is hard and the colour is beautiful, so it is perfect for high-quality furniture. Over-harvesting in order to obtain economic benefits depleted existing wild Che tree supplies. Che wood is still sold in Chinese pharmacies as TCM, but it is from the artificial fast-growing Che tree.

Fig.1 Che slices from Henan province

Fig.2 Che trunk from Guangxi province

2. Experimental

2.1 Dyeing material

In order to restore the ancient colour as closely as possible, wild Che wood and the traditional dyeing process will be accepted.

Two kinds of wild Che wood was found in Guangxi and Henan provinces, finally, Henan Che was choesd as dyeing sample because it was thicker and older. In order to reduce the waste of wildwood resources, waste Che material (block and powder) after making woodware (such as Che bracelet) was used (Fig.3).

Fig.3 Wasted wood material from wild Che trunk

2.2 Dyeing process

For most of dyeing plants, the traditional process is decoction, similar to TCM. I could not find any detailed dyeing record from ancient documents, so I tried normal decoction and successfully created dye. The specific steps are as follows:

1) The first dye: soak Che material in cold water for 60 minutes, heat with strong fire until boiling and then switch to mild fire for 30 minutes, turn off the fire and cool down, then filter the dregs (Fig.4a)

2) The second dye: add cold water again and soak for 30 minutes, heat with strong fire until boiling and then switch to mild fire for 40 minutes, turn off the fire and cool down, then filter the dregs (Fig.4b).

Fig.4a The first dye

Fig.4b The second dye

The basic process of Che dyeing is from light to dark, one layer over another. The dyeing colour will appear slightly different on different natural fabrics (cotton, linen, silk and wool), and in order to restore the traditional Chinese colour for the emperor's costume, I chose 30 momme silk plain satin (Fig.5) as the fabric sample for Che dyeing experiments. The specific steps are as follows (Fig.6):

1) Silk fabric sample: cut silk into small pieces 50*80MM, soak in cold water for 10 minutes, take out and squeeze the water out.

2) The first dyeing sample: heat the second dye until boiling, put undyed wet silk samples into the liquid for 3 minutes, take out, rinse and dry in the shade.

3) The second dyeing sample: heat the second dye until boiling, put the first dyeing samples into the liquid for 5 minutes, take out, rinse and dry in the shade.

4) The third dyeing sample: heat the second dye until boiling, put the second dyeing samples into the liquid for 8 minutes, take out, rinse and dry in the shade.

5) The fourth dyeing sample: heat the first dye until boiling, put the third dyeing samples into the liquid for 3 minutes, take out, rinse and dry in the shade.

6) The fifth dyeing sample: heat the first dye until boiling, put the fourth dyeing samples into the liquid for 5 minutes, take out, rinse and dry in the shade.

7) The sixth dyeing sample: heat the first dye until boiling, put the fifth dyeing samples into the liquid for 8 minutes, take out, rinse and dry in the shade.

Fig.5a 30 momme silk plain satin   Fig.5b Silk fabric sample

Fig.6a The first dyeing sample   Fig.6b The second dyeing sample   Fig.6c The third dyeing sample

Fig.6d The fourth dyeing sample   Fig.6e The fifth dyeing sample   Fig.6f The sixth dyeing sample

2.3 More sample

After successfully completing the dyed fabric samples, I continued to dye a bigger sample Che yellow. I chose a 200s white wool scarf, so that it could be directly used as apparel.

I designed three colour effects: flat dyeing, hanging dyeing (Fig.7) and clip dyeing. In order to increase colour fastness, I soaked the scarf in alum (Fig.8) water for 10 minutes after dyeing, and the result satisfied me (Fig.9-11).

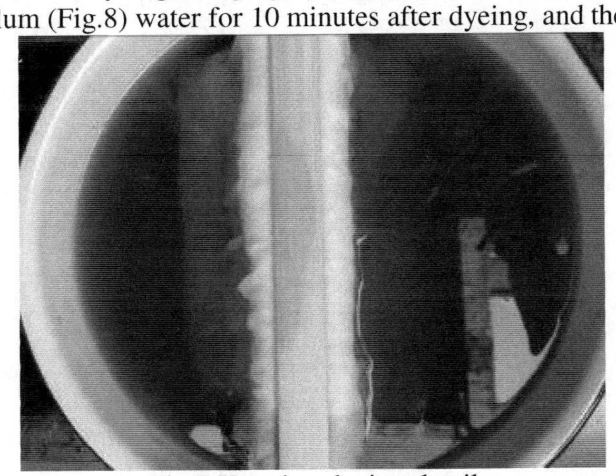

Fig.7 Hanging dyeing detail

Fig.8 Natural alum is helpful for colour fastness

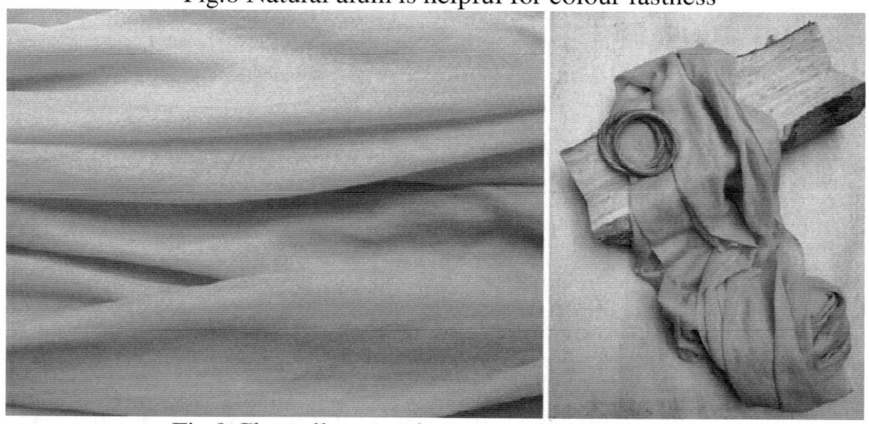

Fig.9 Che yellow wool scarves with flat dyeing

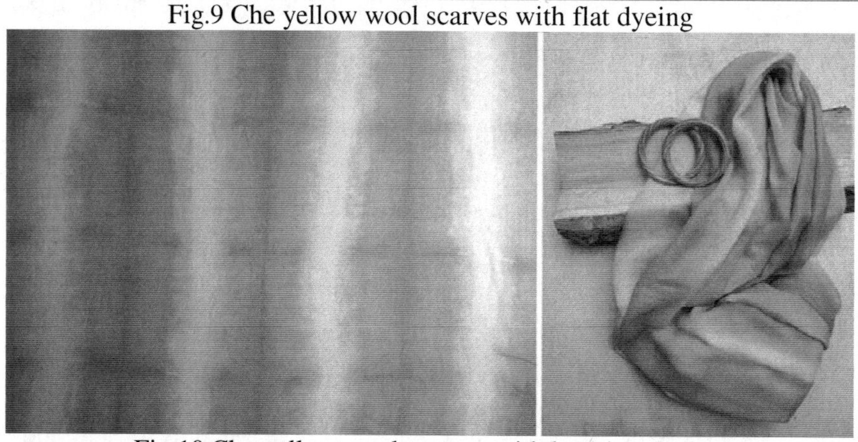

Fig.10 Che yellow wool scarves with hanging dyeing

Fig. 11. Che yellow wool scarves with clip dyeing.

3. Results

When samples were prepared, I used a Sucolor 3NH SC-10 colorimeter (Fig.12) for data acquisition and changed colour samples into LAB digital samples. In order to get digital data more accurately, I collected data from five different locations of each fabric, and averaged the final colour data. The results are as follows (Table 1):

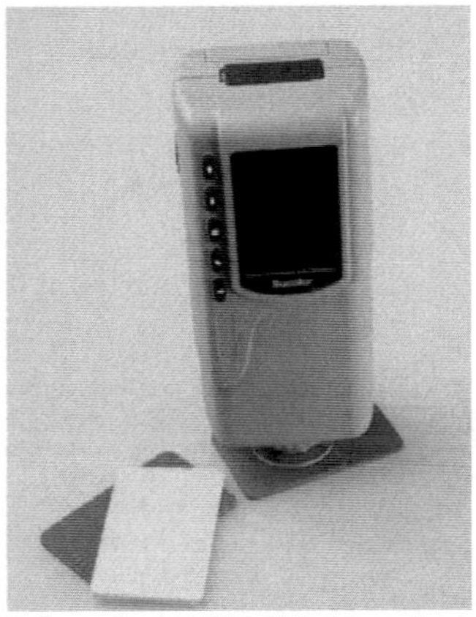

Fig.12 Sucolor 3NH SC-10 colorimeter

Table 1. Che yellow samples with LAB value

No.	Sample	L value	A value	B value
Che-1	The first dyeing	84.99	1.76	36.27
Che-2	The second dyeing	80.26	6.00	53.11
Che-3	The third dyeing	76.25	8.17	50.07
Che-4	The fourth dyeing	71.74	12.83	54.69
Che-5	The fifth dyeing	69.81	14.52	61.81
Che-6	The sixth dyeing	66.13	15.57	54.25

Finally, with Adobe Photoshop software, I transferred the LAB data into a CMYK digital colour sample, the results are as follows (Table 2. and Fig.13):

Table 2. Che yellow samples with CMYK value

No.	C value	M value	Y value	K value
Che-1	12	20	50	0
Che-2	14	28	69	0
Che-3	18	33	70	0
Che-4	19	40	78	0
Che-5	20	44	87	0
Che-6	25	48	84	0

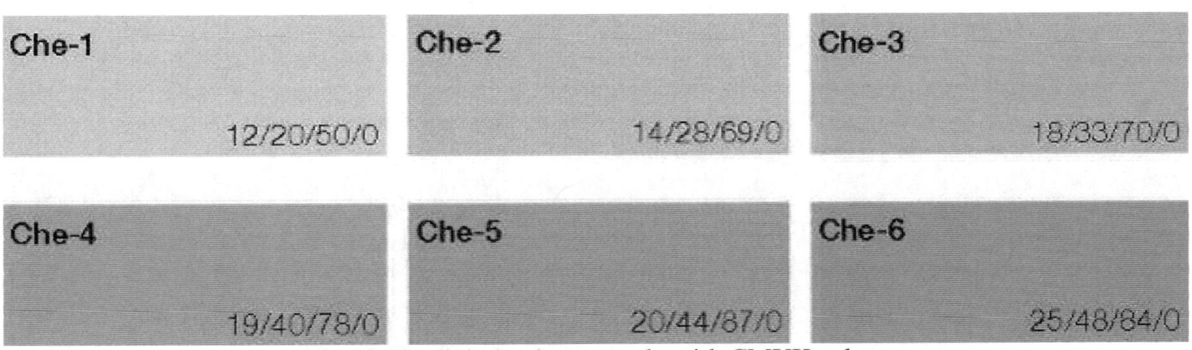

Fig.13 Six digital colour sample with CMYK values

4. Conclusion

In this paper, a traditional Chinese fabric color was been restored, based on the traditional plant dyeing technique. With the same dyeing material and process as in ancient China, emperor yellow appeared in visual way, instead of a word in document. In addition, the digital data has been determined so that the color can be be defined as CMYK value.

References

[1] Dai D. 四部丛刊初编·大戴礼记 (Dai's Rites, from Four Series). Jia Qu Tang edition. Wu Xi: Ming Dynasty; 10.

[2] Ji H, Cao RH. 钦定四库全书·钦定续通典(Qing Ding Xu Tong Dian, from Qing Ding Sikuquanshu). Beijing: Qing Dynasty; 47: 2.

[3] Zheng X. 礼记注疏(Li Ji Zhu Shu). Xi An: Han Dynasty; 12: 1.

[4] Yang S. 钦定四库全书·丹铅总录(Dan Qian Zong Lu, from Qing Ding Sikuquanshu). Beijing: Qing Dynasty; 1: 15.

[5] Xu ZQ. 考工记通(Kao Gong Ji Tong). Beijing: Ming Dynasty; 1: 65.

[6] Shen SX. 大明会典(Law of Ming Dynasty). Nan Jing: Ming Dynasty; 61: 13.

[7] Liu RY. 酌中志(Zhuo Zhong Zhi) . Beijing: Ming Dynasty; 16: 34.

[8] Li SZ. Compendium of Materia Medica. Hu Chenglong edition. Beijing: Ming Dynasty, 1596; 36: 7.

[9] Zheng X. 礼记(Li Ji). Beijin : Qing Dynasty, 1723-1735; 3: 46.

Characterization of Motion Artifacts from the Interfacial Instability of Textile Electrodes and Skin using a Simulated Method

Peng-Jun Xu[1*], Fa-Ming Wang[2], Hao Liu[3], Xiao-ming Tao[4]

[1]*Faculty of Clothing and Design, Minjiang University, 1 Wenxian Rd., Minhou County, Fuzhou, Fujian Province, 350108, China*
[2]*Laboratory for Clothing Physiology and Ergonomics (LCPE), 199 Ren'ai Rd., Soochow University, Suzhou, Jiangsu Province, 215123, China*
[3]*School of Textiles, Tianjin Polytechnic University, 399 West Binshui Rd., Xiqing District, Tianjin, 300387, China*
[4]*Institute of Textiles and Clothing, 11 Hong Chong Rd., The Hong Kong Polytechnic University, Hung Hom, Kowloon, Hong Kong*

[*]Corresponding author's email: xupj@mju.edu.cn

Abstract

An objective method to evaluate the motion artifacts of textile electrodes was presented using a self-made apparatus. The apparatus simulates skin-electrode mechanical interaction by controlling electrodes moving on a volume conductor with various speed and contact pressure. Three different types of electrodes was characterized and the motion induced electrical noise (E_{ocp}) of a pair of the same structural electrodes were recorded, peak-to-peak of the noise (ΔE_{ocp}) was calculated for evaluating the amplitude of motion artifacts. The results shown that motion artifacts decreased with increased movement speed and contact pressure. For textile electrodes, woven plain structure exhibits the minimum motion artifacts.

Keywords: Electrocardiogram (ECG); Textile Electrodes; Motion Artifacts; Interfacial Instability; Open Circuit Potential (OCP)

1. Introduction

Wearable systems for ubiquitous health monitoring have received much attention in recent years. The prevailing scenario is to integrate electronic devices or sensors into textiles and garments in easy to use, comfortable to wear, and minimally obtrusive fashion [1]. One of the typical examples in wearable healthcare is for Electrocardiogram (ECG) monitoring using wearable electrodes fabricated from textile structures which also called textile electrode, or textrode. These electrodes are flexible, air and moisture permeable, easy integrated in to clothes, suitable for long-term health monitoring application [2].

The state-of-the-art wearable electrodes can acquire good ECG signal in static state while suffered by interferences in dynamic state. The interference, called motion artifacts, mainly attributes to skin-electrode interfacial instability and skin deformation induced noise [3]. In dynamic state, there is skin electrode relative movement which generates streaming potential at the skin-electrode interface. Considering the clothing pressure, skin deformation or stretching occurs which is another main source of motion artifacts. Moreover, textile structural electrodes are flexible which easily deformed by external force from skin-electrode relative movement and compression, thus affecting skin-electrode interfacial contact and stability. Plenty of work has been done on the motion artifacts of traditional gelled metal plate electrode [3-9] which revealed that the skin-gel-electrode streaming potential is in the order of several micro-Volts (μV) much smaller than that from skin deformation induced noise of several milli-volt (mV) [7]. However, for the dry textile electrodes where no gel presented at the skin-electrode interface, the noise level from skin-electrode interfacial instability and influence of textile structures on motion artifacts yet remain unknown which is the main concern of this paper.

Under dynamic state, textile electrode structural deformation, skin deformation and skin-electrode interfacial contact area variation are concurrently making the evaluation of motion artifacts from different noise sources even more complicated. To characterize skin-electrode interfacial noise, one of the approaches is using simulated method to prevent the influence of skin deformation induced noise. For one thing, this can avoid human factors which may cause variations from subject to subject, or within subject over time [10]. For the other, it has the potential to standardize the characterization of wearable electrodes. Westbrook [11] evaluated the quality of textile electrodes using electrochemical impedance spectroscopy by putting textile electrodes into a plastic tube filled with sodium chloride (NaCl) electrolyte. This method is similar to that described in ANSI/ AAMIEC12:2000 standard by measuring the impedance of a pair of gelled disposal electrode putting face to face. Beckmann [12] characterized textile electrodes using standardized measurement setups using a simulated skin dummy, the impedance of a pair of electrode with different structures were measured under varied contact pressures. In these studies, the textile electrodes were in static state and characterized using impedance spectroscopy. Whereas, motion artifacts do not necessarily depends on skin-electrode impedance variation [6]. Actually, motion artifacts is the noise superimposed on the measured bioelectric signal, from the electrochemical point of view, which is the open circuit potential (E_{ocp}) of a pair of electrodes. Therefore, we adopted E_{ocp} for direct characterization of the motion artifacts of wearable electrodes.

In this paper, a dynamic evaluation apparatus was developed to characterize electrode-skin motion artifacts from their mechanical interaction. The apparatus comprises three main parts: a simulator to mimic skin-electrode contact, a motion controller to control skin-electrode relative movement and a data acquisition unit to record mechanical and electric signals during experiment. The simulator is the core part of this equipment, which is made from Teflon tube with two pieces of membrane fastened at each end to form a container filled with electric conductive gel. A pair of electrodes contacted and moved on the membrane, open circuit potential (E_{ocp}) of the electrode-gel-electrode cell was simultaneously recorded. The E_{ocp} can be assumed to be the motion artifacts solely from skin-electrode interfacial instability as no skin stretch and deformation induced potential arouse like the measurement carried out on living skin [9]. By comparing E_{ocp} variation of different wearable electrodes, we can evaluate the influence of fabric structures on motion artifacts, thus optimizing fabric structures.

2. Experimental

2.1 Measurement setup

To characterize the motion artifacts of wearable electrode, we developed a dynamic measurement setup which comprises a human skin simulator, an electrode motion controller and a data acquisition unit. The skin simulator, as shown in Fig.1, is the core part of this measurement setup made from Teflon tube (height 92mm, inner diameter 60mm). Two filtration membrane (Millipore, USA, pore size 0.22 μm) fastened at each end of the tube to simulate skin. The tube and two membranes formed a container which was filled with electric conductive gel (Lectron II, USA). The motion controller is assembly of two linear stages (New port, USA) placed in perpendicular directions, driving electrodes moving in a plane along a designed trajectory. The linear stage has a travel range of 50mm with a resolution of 0.1 μm and 100mm/sec of maximum speed. The data acquisition unit comprises a computer, an electrochemical Interface 1252A (Solartron, UK) and a frequency Response Analyzer 1287 (Solartron, UK), an auxiliary signal conditioning modules and two force sensors. The accuracy for the E_{ocp} measurement is as high as 1 μV , and for force detection is 0.1 cN . All measurements were done in an ambient condition of 20 and R.H 65%.

Fig.1 Illustration of electrode testing setup

2.2 Electrode fabrication

As shown in Fig.2, three kinds of electrodes, two textile electrodes woven and embroidery terry structures, and one commercial metal plate electrode, were prepared in this study. The woven structural electrode were made from commercial silver coated multifilament (140D, 2-ply, 40 filaments) on a sample loom (SL 8900, CCI TECH INC, Taiwan) which has the same weft and warp density of 50 ends per inch. The terry structural electrodes were made on an embroidery machine (TAJIMA, TMCE-61202, TOKAI Industrial sewing machine Co, LTD, Japan) with a terry height of 2mm, ling spacing of 2mm, and stitch width of 0.5mm. The textile electrodes were stuck on a layer of foam substrate (4mm thick) and a snap button was punched through the centre for lead wire connection. For the commercial metal plate electrode, the gel was removed. The diameters of the three electrode types are 30mm, 30mm and 10mm with an effective area of 628 mm^2, 628 mm^2, and 78.5 mm^2, respectively.

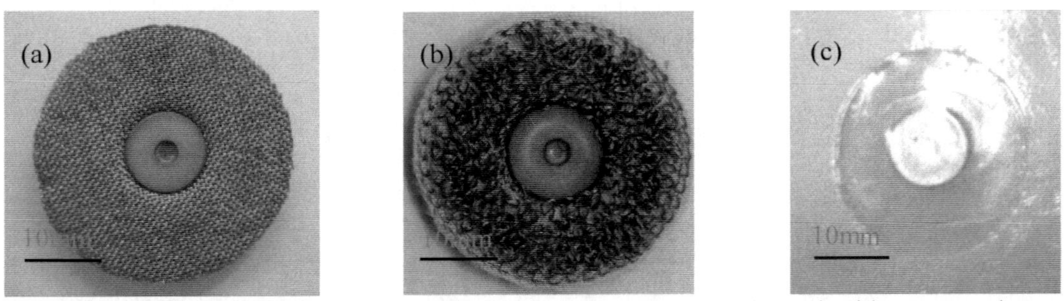

Fig.2 Comparison of different structural electrodes: (a) woven electrode, (b) embroidery terry electrode and (c) commercial metal plate electrode

2.3 Electrode characterization

The skin-electrode mechanical interaction was defined as horizontal relative movement and vertical compression. To characterize the mechanical interaction induced motion artifacts, textile structures and skin-electrode relative movement speed as well as their contact pressure were the main factors taken into consideration, as listed in Table 1. The electrodes were moved along a circular trajectory (dia. 3mm) passing through the centre of the membrane (Fig.1) under various electrode-membrane contact pressure at various linear velocities according to the experimental parameters in Table 1. Moreover, to study the influence of the skin-electrode compression on motion artifacts, the electrodes were put at the centre of the membrane without cyclic movement while stepwise changing the contact pressure. During these experiments, the open circuit potential E_{ocp} of the upper and lower electrodes was simultaneously measured.

Table 1. Parameters of electrode movement during testing

Sample	Electrode type	Speed(mm/sec)	Pressure(kPa)	Radius(mm)
1-5	woven	1,3,5,7,9	2	3
6-12	woven	1	0.25,0.5,1,2,3,4,5	3
13-17	embroidery	1,3,5,7,9	2	3
18-24	embroidery	1	0.25,0.5,1,2,3,4,5	3
25-29	metal plate	1,3,5,7,9	2	3
30-36	metal plate	1	0.25,0.5,1,2,3,4,5	3

2.3.1 Speed variation

To characterize the motion artifact from in-plane movement, a pair of electrode with a 2KPa preloaded pressure contacting with the membrane to measure the E_{ocp} variation. The lower electrode was stationary while the upper one moved along a circumferential trajectory with a diameter of 3mm at various linear speeds of 1mm/sec, 3mm/sec, 5mm/sec, 7mm/sec and 9mm/sec. During the movement, the electrode-membrane contact pressure and E_{ocp} signals were recorded in the sample rate of 50Hz and 5Hz, respectively, latter of which is limited by the electrochemical instrument.

2.3.2 Pressure variation

Different from the above measurement, the electrode membrane relative movement speed was set to 1mm/sec, while varying their contact pressure under 0.25KPa, 0.5KPa, 1KPa, 2KPa, 3KPa, 4KPa and 5KPa, respectively, for each measurement.

2.3.3 Skin-electrode compression

Apart from the horizontal movement (speed and pressure variation), the electrode contact pressure is another influential factor of the motion artifacts. The lower electrode is stationary with a 2KPa preloaded pressure while the upper one is subjected to stepwise pressure variation in the sequence of 0.25-0.5KPa, 0.5-1KPa, 1-2KPa, 2-3KPa, 3-4KPa, and 4-5KPa, each step held for approximately 20sec.

3. Results and Discussion

Motion artifacts induced by the skin-electrode mechanical interaction was studied using simulated method from horizontal relative movement and perpendicular pressure variation aspects. As the diameter of woven and terry electrodes are different from that of metal plate electrode, to compare the noise level of these electrodes, the noise level of should be normalized to the same electrode area. According to Huigen's investigation [7], the noise level (n) is proportion to reciprocal of square root of the electrode contact area (A) as shown in Eq.1.

$$n \propto \frac{1}{\sqrt{A}} \tag{1}$$

3.1 Influence of electrode pressure

The motion artifacts of different electrode types under different preloaded pressures were shown in Fig.3 (a-c). It is obvious that the motion artifacts decreased with the increased contact pressure because of the stable electrode membrane interfacial contact under higher pressure. As shown in Fig.3(d), woven structural electrodes have the minimum motion artifacts, while terry structure and commercial metal plate are much higher. The overall trend of motion artifacts decreased with pressure, which may attributed to the enhanced contact stability of electrode and membrane.

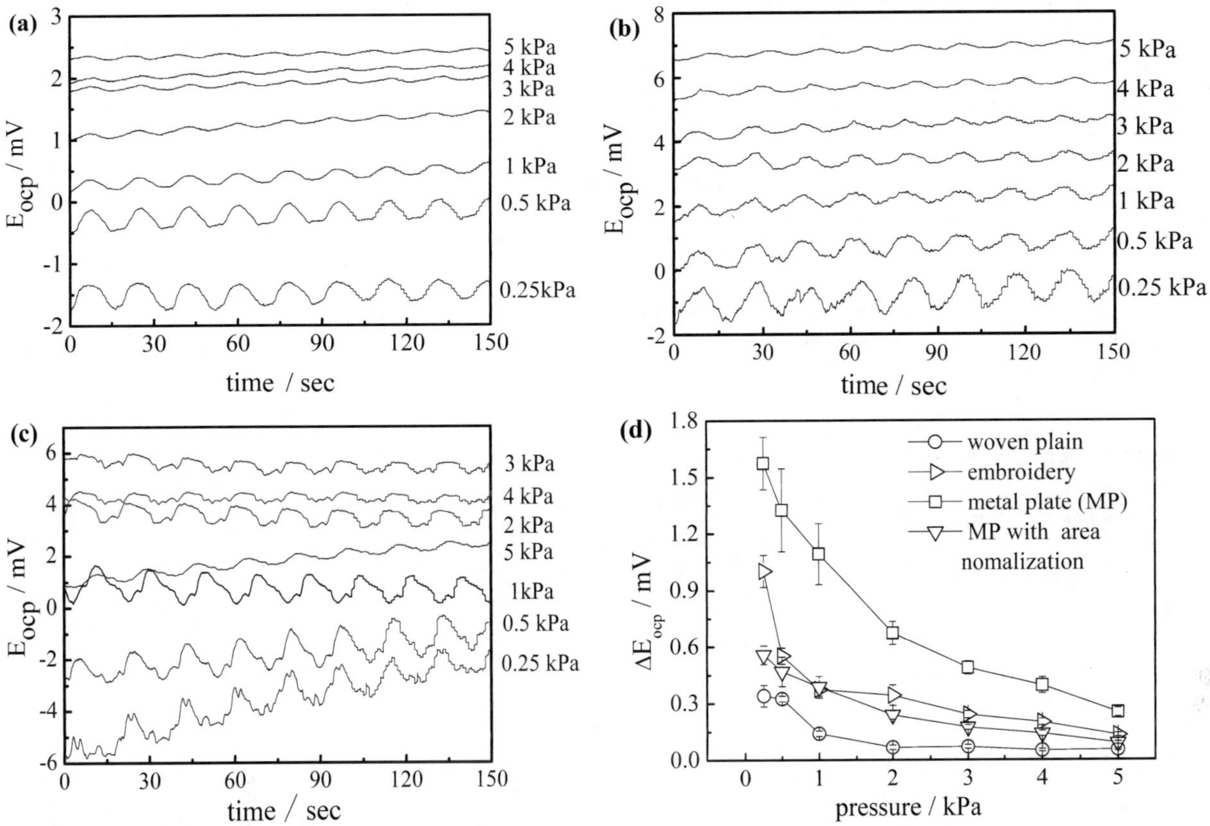

Fig.3 Motion artifacts of different structural electrodes moving at constant speed under various preload pressures: (a) woven plain, (b) embroidery terry, (c) metal plate (d) comparison of three type electrodes

3.2 Influence of moving speed

As shown in Fig.4, E_{ocp} variation of embroidery terry structural electrodes have the maximum variation while woven textile electrode is the minimum, and commercial metal plate electrodes are the intermediate. Motion artifacts of woven electrode are smaller than that of embroidery terry one due to its simple and stable structure, and small contact surface variation during movement. Moreover, E_{ocp} variation of three types of electrode has the same trendy, which is decreased with the increasing movement speed. This can be explained as the interface moisture effect, the quicker speed, the shorter duration of electrodes moved to a specific position, thus the minimum equilibrium potential differences. Ödman S [6]. assumed the electrode movement as a newly applied electrode that needs time for stabilization.

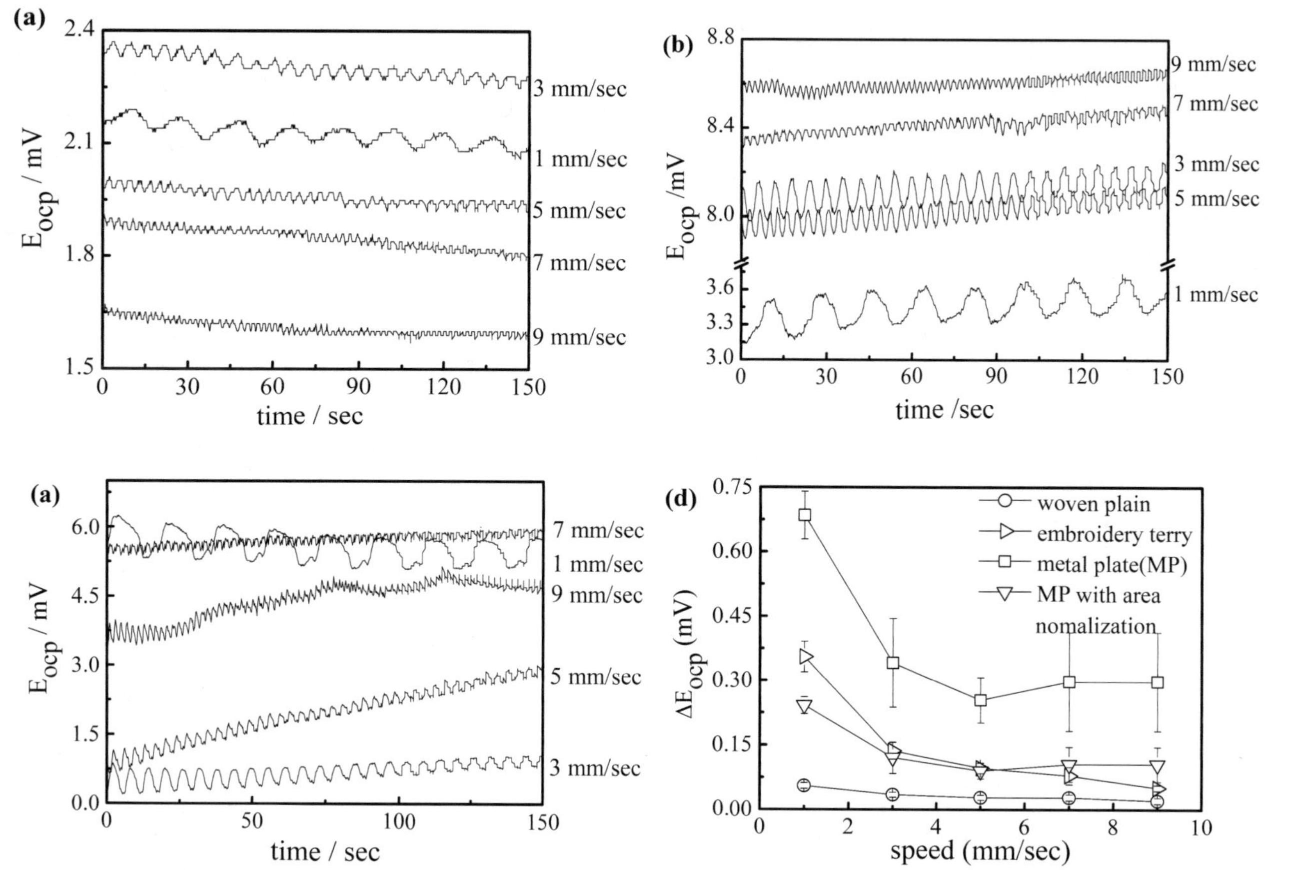

Fig.4 Motion artifacts of different structural electrodes moving at constant speed under various preload pressures: (a) woven plain, (b) embroidery terry, (c) metal plate (d) comparison of three type electrodes

3.3 Influence of vertical compression

The instant change of electrode-membrane contact pressure also causes considerable motion artifacts. As shown in Fig.5, the abrupt shift of E_{ocp} gradually decreased as increased electrode-membrane contact pressure. Terry structural electrodes have the highest E_{ocp} shift while the commercial metal plate has the lowest which could be attributed to the compressive Young's modulus of electrode, the higher modulus, the less influence of pressure on the interfacial contact area. The metal plate electrode has the highest modulus exhibiting no obvious E_{ocp} shift. For the woven electrode, when the electrode contact pressure is higher than 2kPa, the E_{ocp} decreased to a minimum which can be assumed that the interfacial contact came to stable state. Therefore, a pressure higher than 2kPa is recommended for wearable electrodes in practical application.

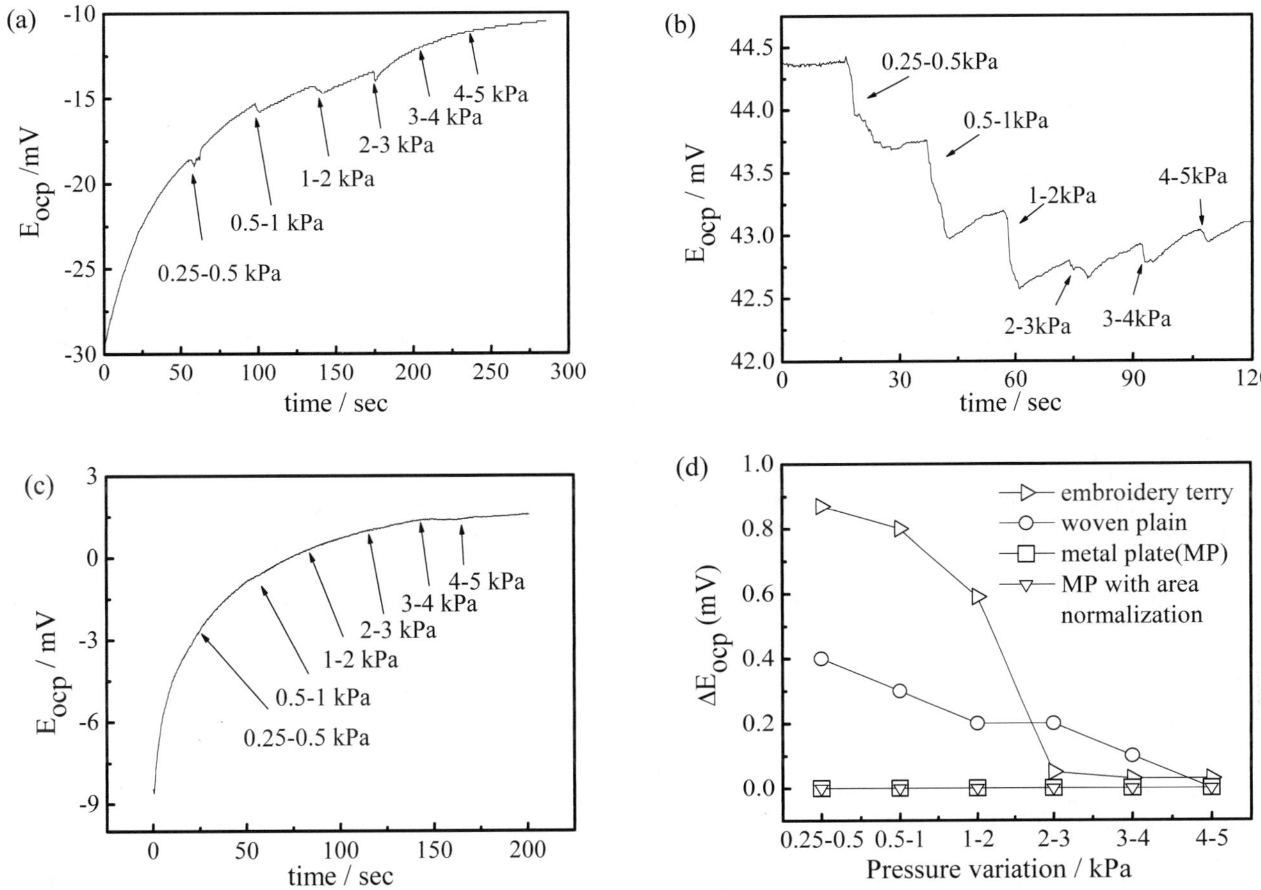

Fig.5 Motion artifacts of different structural electrodes moving at constant speed under various preload pressures: (a) woven plain, (b) embroidery terry, (c) metal plate (d) comparison of three type electrodes

3.4 Influence of electrode structures

Textile structures have a significant influence on the motion artifacts. In Fig.3 and 4, the electrodes moved in the horizontal plane parallel with the membrane under varied contact pressures and moving speeds. The results demonstrated that the woven structural electrodes exhibited the smallest motion artifacts, while the terry structural electrodes were the highest, and the metal plate electrodes were in between. In Fig.5, the electrode-membrane contact pressure was stepwise increased, inducing obvious motion artifacts for the woven and terry structures. From these results, we can infer that textile electrodes made from simpler and stiffer textile structures may result in a better dynamic performance in terms of motion artifacts.

4. Conclusion

In this paper, the motion artifacts of textile electrodes induced by skin electrode mechanical interaction were evaluated using simulated method. The mechanical interaction was defined as a horizontal relative movement and a vertical compression of the electrode versus skin. The motion artifacts, indicated by the open circuit potential (E_{ocp}) of a pair of electrode, which decreased with the increased electrode-skin contact pressure and movement speed. Abrupt variation of electrode-skin contact pressure also caused significant motion artifacts which decreased as the pressure increased and came to a stable state when pressure higher than 2kPa. Textile structures were also an important contributor to the motion artifacts suggesting that the simpler and stiffer textile structural electrodes have a stable electrode-skin contact, thus introducing less motion artifacts in the dynamic state. To lower the motion artifacts of wearable electrodes, the interfacial stability of electrode and skin should be enhance by carefully designing of electrode structure and properly setting the electrode-skin contact pressure. However, there are properties differences between membrane and real human skin, a comparison study will be carried out to verify the effectiveness of using simulated method.

5. Acknowledgement

The authors wish to acknowledge the The Hong Kong Polytechnic University (1-BB99), Minjiang University (YKQ13004 and MJUA2013003), the Scientific Research Foundation for Talents of Minjiang University, the National Natural Science Foundation of China (51403098), the Natural Science Foundation of Fujian Province (2015J05106), the Educational Commission of Fujian Province (JA14258), and the Program for New Century Excellent Talents in Fujian Provincial Universities (NCETFJ) for their financial support.

References

[1] Bonfiglio A, Rossi DD. Wearable Monitoring Systems, Springer Science & Business Media: 2010.
[2] Xu P, Zhang H, Tao X. Textile-structured electrodes for electrocardiogram, Textile Progress: 2008; 40 (4); pp: 183-213.
[3] Ödman S. On the spread of deformation potentials in the skin, Medical and Biological Engineering and Computing: 1982; 20(4); pp: 451-456.
[4] Ödman S. Potential and impedance variations following skin deformation, Medical and Biological Engineering and Computing: 1981; 19(3); pp: 271-278.
[5] Ödman S. Changes in skin potentials induced by skin compression, Medical and Biological Engineering and Computing: 1989; 27(4); pp: 390-393.
[6] Ödman S, Öberg PÅ. Movement-induced potentials in surface electrodes, Medical and Biological Engineering and Computing: 1982; 20(2); pp: 159-166.
[7] Huigen E, Peper A, Grimbergen CA. Investigation into the origin of the noise of surface electrodes, Medical and Biological Engineering and Computing: 2002; 40(3); pp: 332-338.
[8] Talhouet Hde, Webster JG. The origin of skin-stretch-caused motion artifacts under electrodes, Physiological Measurement: 1996; 17(2); pp: 81.
[9] Vos WK, Bergveld P, Marani E. Low Frequency Changes in Skin Surface Potentials by Skin Compression: Experimental Results and Theories, Archives of physiology and biochemistry: 2003; 111(4); pp: 369-376.
[10] Gaumond RP, Arellano P. Simultaneous variation of inter-electrode potential and impedance, Proceedings of the 1989 Fifteenth Annual Northeast Bioengineering Conference: 1989; pp: 89-90.
[11] Westbroek P, et al. Quality control of textile electrodes by electrochemical impedance spectroscopy, Textile research journal: 2006; 76(2); pp: 152-159.
[12] Beckmann L, et al. Characterization of textile electrodes and conductors using standardized measurement setups, Physiological measurement: 2010; 31(2); pp: 233.

A Study on the Operating Mechanism of Waste Textile's Recycling and Reusing System in Germany

Yi-Lun Wu[1], Lei Yao[1], An-Yi Geng[1]

[1]*Business school, Beijing Institute of Fashion Technology, Beijing, 100029, China*

Corresponding author's email: sxyyl@bift.edu.cn

Abstract

This article has carried out a detailed record on the German waste textile recycling system and introduced the composition and the concrete method of Germany's textile recycling system. Take Germany as an example, combining with the status and problems of textile recycling we are now facing in China, we put forward some suggestions for the recycling of waste textile, such as making corresponding laws and regulations, clearing unified provisions, implementing the plan of building the recycling system, etc.

Keywords: Germany; Waste textiles; Recycling and reusing; Suggestions

1. Introduction

Textile industry is an important pillar industry in China. Data from the China textile network shows that each year, offcuts produced in the processes involved in textile and clothing enterprises and waste textiles produced by households exceeds the 20 million tons, of which only 5% are recycled. The other 95% of them end up in the land fill as garbage. Compared to France and UK, there is still a large gap between China and developed countries. The recycling of waste textiles can significantly reduce carbon emission. The plans of the construction of circular economy are also an important issue for our country and people.

Germany is the first country putting forward to sorting, recycling and reusing the living garbage in Europe, and takes the recycling of waste textiles very seriously. As early as 1914, family-owned second-hand clothing recycling companies were formed in Germany. Germany is the first country formulating garbage sorting and recycling system. In 2008, a study done by Germany's federal association of secondary raw materials and waste organization (bvse) shows that in Germany, about 90% of the old clothes can be recycled. Now about 750,000 tons of second-hand clothes are recycled every year.

2. Legal basis of Germany's waste textiles recycling system

In the 1980s, German government called on reducing, decomposing and reusing the waste. The development process of its legal system can be summarized by three important law. First, "Federal waste management" in 1972 proposed the garbage classification for the first time. Second, "The packaging waste management act" in 1991 followed the circular economy idea of "Resources - Products - Resources". Third, "Circular economy and waste management" in 1996 accorded to the idea of "Avoid - Recycling - Final disposition". In this way, German developed and perfected their own law on garbage classification, and its new philosophy and economic practice for waste disposal, which has a great impact on the world.

2.1 Federal waste management

In 1972, the federal republic of Germany issued the "Federal waste management", setting waste collection and disposal in the form of law for the first time. This issue changed the living habits of Wessi, for instance, some general life rubbish, plastic containers and paper beginning to be classified and recycled. Germans sort out their garbage according to the classification shown in Table 1.

Table 1. Germany's garbage classification

Types of garbage	Instructions
Glasses	None
Paper	Newspapers, magazines, waste paper, paper bags, etc., Kleenex excluded
Organic waste	Organic garbage of green plants , remnants of the fruits and vegetables, etc. (processed excluded)
Special(risk) garbage	Acids, medicine bottle containing residual tablets, etc
Electronic waste	Daylight lamp, energy-saving lamps, batteries, etc
Large waste	Old furniture, TV, refrigerator, etc waste which can not be contained by 120L trash
Garbage not belonging to the above categories	none

As for the "Federal waste management", the environmental protection consciousness of the west public has been improving constantly. in 1986, the federal republic of Germany proposed this concept of "avoid waste generation" developed principles on avoiding wasting, recycling and safety treatment and fixed them in the form of law. The understanding of waste rose from the level of "how to deal with waste" up to "how to avoid the generation of waste". German government gradually established the concept of circular economy development.

2.2 Packaging waste management regulations

The import of the idea of circulation economic in Germany was manifested in "Packaging waste management regulations" which was enacted in 1991. This is the first management regulation about reducing and recycling packaging waste in the world. By means of controlling the packaging waste, Germany wants to promote the sustainable development of packaging industry and reduce resource waste and environmental pollution caused by the large number of packaging waste.

Because of "polluter pays" principle, the regulations cleared the obligation of commodity production and circulation industry to recovering and recycling of packaging waste. Commodity packaging of the catering industry must be recycled by the manufacturer. Packaging waste produced by purchasing in the store must be retrieved by the seller. Packaging waste produced in the process of transportation must be recycled by the manufacturer. Packaging waste recycled must be circled.

In 1992, in line with better implementation of the law, Germany began to implement the famous recovery system called "Green Dot". This "Green Dot" indicates that this commodity producers have paid for the recovery of the goods after using it. Residents should throw commodity packaging printed with "Green Dot" into recycling bins in yellow or yellow bag, which will be collected by the staff who work for "Green Dot" company. Concrete waste should be dumped as shown in Table 2.

Table 2. Garbage sorting in Germany

Types of garbage	Color of recycling bins
Glasses	White or green
Paper	Blue
Waste labeled "green dot"	**yellow**
Organic waste	brown
Special(risk) garbage	Special collecting points
Electronic waste	Special collecting points (such as large cosmetic chain)
Large waste	Special collecting points
Garbage not belonging to the above categories	black

Effects from this meticulous and thorough garbage classification are obvious. According to the German federal statistics office's data, in 2007 the amount of German business packaging waste recycled is up to 3.2 million tons, which accounts for 88% of all quantity of packing manufactured for the whole year. From 1996 to 2007, Germany's net reduction of total waste amount went beyond 37.7 million tons.

2.3 Circular economy and waste management

In October 1996 "Circular economy and waste management" went into effect and became the cardinal law of environmental policy in Germany. The law made a fundamental change of environmental protection movement in the world, which transformed the circular economy from "controlling the end" to "controlling the source".

The "waste" in the regulations not only refers to the users' waste, but also means all remnants in the productive process. Producers must be responsible for its treatment. It requires that the dosage of all resources must be reduced package waste, all waste products and waste carpet and fabric.

Textiles is separately listed in the appendix of this law, and divided it into two categories as cotton and jute.

Table 3. The explanation to textiles in appendix of "Circular economy and waste management"

materials	abbreviation	number
cotten	TEX	60
jute	TEX	62

Source: The appendix of "the circular economy and waste management"

3. Operating mechanism analysis of German textile waste recycling system

3.1 Main operational bodies of German textile waste recycling system

Germany's textile waste recycling system mainly consists of three main operational bodies: local government, textile recycling association FTR and textile waste recycling companies which make their own management decisions and take full responsibility for their own profits and losses. Among them, the local government are the guider, then FTR association and textile waste recycling companies are responsible for specific work like the recovery, classification and recycling of the waste textiles, and the promotion and development of recycling consciousness and ideas to the public.

Textile recycling association (Fachverbands Textile Recycling, FTR) was founded in 1948, which is responsible for the collecting, sorting and recycling of the old clothes, shoes and other textiles. In 2011, the association became a subordinate department of secondary raw materials and waste organization of the Federal Association (hereinafter referred to as bvse), which specially recycles waste textiles.

Germany has a lot of companies which recycles waste textiles, such as the East - West Kursun textiles recycling Co. Limited, k. & a. Wenkhaus GmbH textiles recycling company, etc. They are mostly family-run operations, self-financing. Some of them even have more than 90 years old experience for textile recycling (such as k. & a.

Wenkhaus GmbH). They design recycling containers, collect and empty containers with old clothes regularly. The Bureau of International Recycling (BIR) and other waste textile recycling organizations in the world is the important cooperative members of FTR association.

3.2 The operation mode of German textile waste recycling system

After studying three main operational bodies of Germany's textile waste recycling' system separately, the author believes that Germany's textile waste recycling system can be divided into two aspects, macro and micro. From the macro perspective the local government, FTR association, textile waste recycling companies and the public formed a virtuous circle of operating system. At the micro level, within FTR association and textile waste recycling companies there is an efficient and complete operation system. Running efficiently at the micro level ensured the integrity and efficiency of the macro system.

3.2.1 The operation mode of Germany textile macro waste recycling system

Main bodies of the macro recycling system include local government, FTR association, textile waste recycling companies and the public. The operation mode is shown in figure 1.

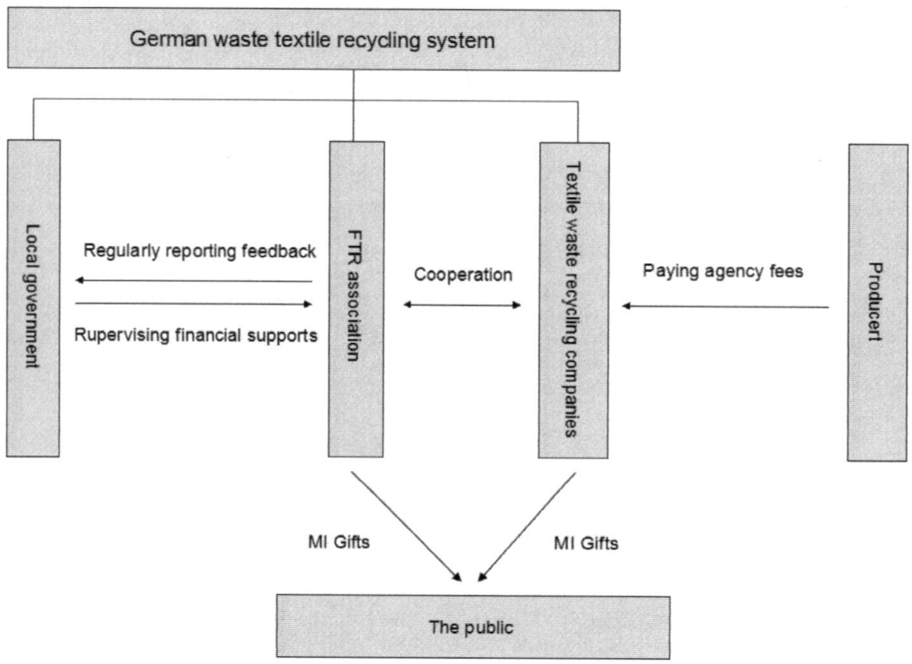

Fig.1 Germany waste textile recycling system

The local government is the guider of recycling system, planning and leading for the chain of recycling industry. Government functional departments are responsible for the overall planning, regulations formulating, association supervising, financial supporting, etc. Operating cost of FTR association and recycling organizations is partially financed by the state, and the other part is from agency fees paid by polluter. Textile producers in Germany pay agency fees to recycling companies, which are independent and self-financing, to make concrete business including recycling, cleaning and disposing within the legal framework. At the same time, FTR association and recycling organizations promote public awareness education, help people improve the consciousness of recycling and classify the waste textiles etc.

3.2.2 The operation mode of Germany textile micro waste recycling system

Source: Official website of k. & a. Wenkhaus GmbH
Fig.2 A white recycling bin designed by k. & a. Wenkhaus GmbH

Micro recycling system mainly includes FTR and waste textile companies. Recycling of waste textiles mainly relies on the following two methods. First, the recycling team buy or charge directly by house call, such as k. & a. Wenkhaus GmbH textiles recycling company. Second, the residents put in the recycling bins of various companies at the specified position. This bins are emptied by the corresponding recycling companies regularly. Other ways of recycling include individual donation etc.

In addition to the traditional ways of recycling, the recycling companies are also constantly innovating recycling methods. K. & a. Wenkhaus GmbH textiles recycling company made all of its recycling trucks equipped with an electronic balance system. The digital scanner weighing system are developed by the company to provide preciseness and transparent of per kilogram for clients. This is the only digital recycling way of waste textiles in Germany so far, greatly reducing the workload of recycling team, and is suitable for recycling of large quantities of waste textiles.

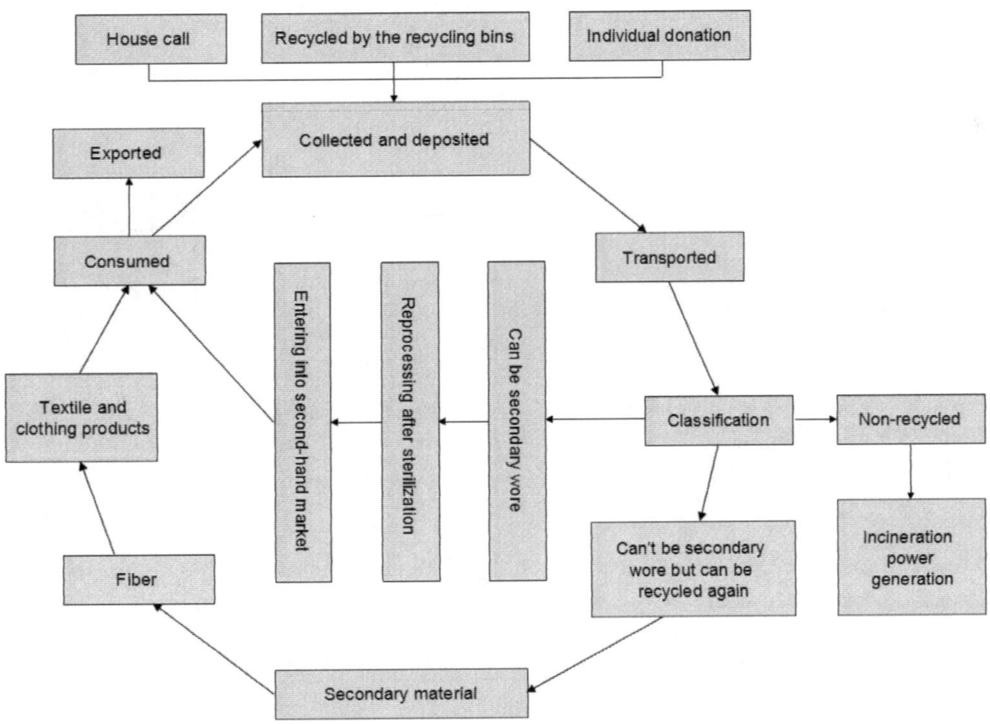

Fig.3 The operation mode of disposing waste textiles

After old textiles recycled, transported to the recycling companies, they will carry on the classification and arrangement. All employees of these multiple recycling companies, cooperating with FTR association, have accepted higher education and professional skills training. They know how to categorize waste textiles recycled, and all work of the classification is finished by hand. This ensures a high-quality sorting, and the classification link is the best link that German old textile recycling system did.

Sources: Official website of East-West Kursun textile recycling co., LTD

Fig.4 A workshop of waste textiles sorting

Sources: Official website of Texaid

Fig.5 Staff of Texiad cooperating with FTR association are sorting second-hand clothes manually

However, to achieve efficient manual sorting, waste textiles recycled must be clean and protected. FTR association put forward, old clothes are collected mainly for one purpose: wearing. Before textile products were recycled, therefore, they should be well packed, for example, packed in a plastic bag. According to this principle and textile recycling guidelines of bvse, FTR association launched "Quality Seal" in 2013, claiming to establish a unified seals standards and airtight container, to stamp, register, sealed package strictly for qualified waste textiles recycled, and providing recycling track and record of recycling business to the association, which gave external independent experts to examine regularly. This unique measure ensured the hygiene of the recycled clothes. What is more important, it's more conducive to quality controlling and ensuring the public transparency. In order to ensure transparency of the recycling, FTR provides corresponding name and seal number for the recycling companies certified, being responsible for sealing.

Fig.6 A sealing seal of old textile recycling company Mustermann co., LTD
Image: Official website of FTR association

Fig.7 A sealing seal of K., & a. Wenkhaus GmbH
Image: Official website of k. & a. Wenkhaus GmbH

After sorting, waste textiles are divided into three parts. Frist, those are non-recycled, disposed as waste which are sent to generate power or other ways of using. Second, those can be secondary wore by reprocessing after sterilization. Those can be sent to textile recyclers or a charity. Most of them are exported to Asia or Africa market. Third, part of those can't be secondary wore but can be recycled again to make secondary material such as fiber used to produce clothing, shoes and other products again. The third one finally entered the consumer market.

4. Suggestions to Chinese textile waste recycling system

4.1 Strengthen the construction of legal basis

"The circular economy promotion law of the People's Republic of China" issued in 2009 clearly put forward that recycling system of renewable resources shall be established and perfected, but did not mention the management of urban waste. For waste recycling, it can only play a guide role. At present, there is no law to unify regulations for sorting, recycling and reusing of urban life waste in China, and regulations produced by local governments of various levels have some differences.

Based on the present situation, we should firstly clear form the law-making view that city life waste is a recyclable resource, then we should deal with the waste by recycling. Second, subdividing the waste and clearing classification standard of waste are important. Finally, aiming at the different types of waste, special law suitable for its recycling and more meticulous planning, after recycling, on classification, reusing, recycling, should be produced, then unifying its name after recycling of the waste, classify color of the bin, etc. It also should clear the punishment measures with specific legal provisions for the behaviors that do not conform to the classification rules, and establish a strict system of supervision. Gradually we build the legal basis for urban living waste guides and make recycling utilization of waste textiles in order.

4.2 Perfect the construction of system

At present, our country is not lack of old textile recycling companies, such as Shanghai Yuanyuan industrial co., LTD. In 2013, recycling bins of old clothes set in the communities, schools and other places in Shanghai by Yuanyuan, have reached 1800. Some recycling companies, however, in the name of collecting old clothes, partially just care about classification, and they didn't actually do the recycling of resources. As the most important part of the recycling processes, whether recycling companies are specific highly effects the efficiency of the operation of the recycling system.

From the point of developing system construction, the government should play a leading role in the whole system. The work of current waste textile recycling in Beijing is dominated by the civil affairs bureau, and Greenery and Public Sanitation Bureau of Shanghai also began lead pilot of the old clothes recycling in the whole city in 2010. This suggests that the government has realized they should play the right role in this industry. However, only doing these are not enough. At the same time of leading recycling, the government also needs to supervise the whole chain of recycling, listen to the feedback information in time and deal with the emergence of new policy problems timely. For the companies being responsible to recycling, they should establish a set of normative examination systems, which can be reviewed time to time. Recycling companies which are substandard should be banned without licenses.

As one of the most important part of the recycling system, recycling companies' specification and efficiency highly affect the operation of the recycling system. Take Germany for instance, our country should expand the type of old textile recycling companies, not only to encourage charitable institutions, environmental groups and other organizations to increase recycle waste textiles, but also transfer the market forces, allowing more private enterprises into the market of waste textile recycling. While asking for enterprises being independent, self-financing. The government should also give them the corresponding financial support, encourage them to innovate and research and improve their efficiency.

4.3 Strengthen the construction of peripheral operation environment

Recycling of waste can greatly reduce greenhouse gas emissions and protect the environment. This is not only what the government and enterprises should do, but also an obligation of all the citizen. At present, citizens in our country are not known of waste recycling and have no consciousness of garbage sorting. In Germany, France and other developed countries, they conduct the education of garbage classification to the children when they were in kindergarten, and each citizen treats the garbage classification as their responsibility and obligation. According to the survey, in Germany, 85% of respondents are expected their wastes to get further used. Citizens, who have strong consciousness of recovery and actively participate in the recycling behavior, are the important force to promote the waste textile recycling going on smoothly in Germany.

Charities and industrial associations can give the education about waste classification to the public by making propaganda, posters and so on. Strengthening the public awareness and reducing phenomena of throwing litter disorderly, are bound to promote the recycling work of waste textiles.

5. Conclusion

Taking Germany as an example, this paper introduced the legal basis of federal waste management, packaging waste management regulations and circular economy and waste management. The main operational bodies of German textile waste recycling system are local government, textile recycling association FTR and textile waste recycling companies. They operate rigorously macro and micro in the macro and micro levels. At present, China has already realized the significance of waste textiles recycling. Recycling work in Beijing, Shanghai, Guangzhou and other first-tier cities have had experimental units, such as in 2009 "recycling point of old clothes walks into the housing estate" in Shanghai. However, facing excellent prospect of waste recycling in textile market, whether from the perspective of national legislation or from the perspective of industrial development, the market of waste textiles recycling in China still faces many constraints and problems. Several suggestions are given aimed at legal basis, construction of system and operation environment.

6. Acknowledgments

[1] National Social Science Fund (Fund Code: 14BJL056)

[2] Beijing Institute of Fashion Technology 2014 Scientific Research Promote Project “Beijing used clothing recycling system built and suggestion research” (Fund code: 2014AL-05)

[3] Beijing Excellent Talents Cultivating Funding Projects (Fund Code: 2014000026833ZS06)

[4] The Project of Construction of Innovative Team and Selection and Development of Excellent Talents for Beijing Institute of Fashion Technology (Fund Code: PTTBIFT_TD_007)

[5] The key project of social science program for Beijing Municipal Education Commission 2016, Study on recycling and reuse of waste textiles from the perspective of urban waste reduction

References

[1] K. & a. Wenkhaus GmbH's: http://www.wenkhaus.de/

[2] Official website of FTR association: http://www.bvse.de/15/Textil

[3] Official website of DSD green dot:http://www.gruener-punkt.de/en/duales-system-deutschland-gmbh.html

[4] Li P, Ye HW, Zhang Y. The present situation of waste textiles recycling at home and abroad [J]. Journal of synthetic fiber, 2014 (4): 41-45.

[5] Guo Y.The present situation and suggestions of recycling and reusing waste textiles in our country [J]. Journal of cotton manufacturing technology, 2013(4): 59-61.

[6] Yang XJ, Wang XL. The development old clothes recycling at home and abroad [J]. Journal of shandong textile economy, 2013 (7): 11-13.

[7] Lou YH, Pu ZY, Huang YH. Present situation and development trend of waste textiles recycling [J]. Journal of Progress in Textile, 2012 (3): 9-12.

[8] Bai J. Preliminary studying on the development of foreign waste textile recycling [J]. Journal of Shandong textile economy, 2012 (12): 54-56.

[9] Xu H. Tang SJ. 2012. The analysis to the development of foreign waste textile recycling [J]. Journal of textile guides.

[10] Wang LL.The discussion on waste textile recycling. [J]. Journal of textile guides, 2009 (4): 26-28.

[11] Xiang J. German legislation time of developing circular economy [J]. Journal of cooperative economy and technology, 2008(3): 122-123.

Adsorption Capacity of Pure Chitosan and Chitosan/Titania Composites for Removal of C.I. Reactive Red 35 Dye Solution

Siriwan Kittinaovarat[1*], Prangsiri Maneenoun[1]

[1]*Department of Materials Science, Faculty of Science, Chulalongkorn University, Bangkok, 10330, Thailand*

[*]Corresponding author's email: siriwan.k@chula.ac.th

Abstract

This study investigated the adsorption capacity to remove C.I. Reactive Red 35 by three different adsorbent beads, consisting of pure chitosan and two different chitosan/ titania (TiO_2) composite beads with the TiO_2 being obtained either from the sol-gel method or from added commercial TiO_2 (A100-TiO_2). From the obtained dye adsorption capacities, the chitosan/ TiO_2 composite beads with A100-TiO_2 had the highest adsorption ability to remove reactive dye solution up to 97%, compared to the chitosan/ TiO_2 composite beads with TiO_2 obtained from the sol-gel method (adsorption capacity of 70-45%). Pure chitosan adsorbent also showed a good adsorption capacity (95%) due to the electrostatic interactions between chitosan and C.I. Reactive Red 35.

Keywords: Chitosan; Titania; Adsorption; Isotherm

1. Introduction

This study investigated the chemical treatment of wastewater by combining adsorption with photocatalytic oxidation. Among adsorbents with the capacity to remove dye effluents, activated carbon, chitin, chitosan and clay minerals have all been shown to effectively adsorb or remove organic compounds from textile effluents. Of these, chitosan [1-8] was selected to use as the adsorbent in this study because it is a waste biodegradable polysaccharide obtained from the deacetylation of chitin, the second most plentiful natural polymer after cellulose. Moreover, chitosan has many interesting characteristics, such as it is hydrophilic, biocompatible, biodegradable, and has an antibacterial property and a flocculating regeneration ability. These properties of chitosan have led to its widespread applications. Chitosan has been regarded as a useful material to remove transition metal ions and organic substances from waste water because the amino and hydroxyl groups on the chitosan chains serve as reaction sites. The photocatalytic oxidation process has been widely studied for degradation of several organic pollutants and has been established as having a high potential for wastewater treatment [9-11]. Titania (TiO_2) is currently the most promising photo catalyst employed to treat textile wastewater. In recent years, the use of TiO_2 as a photocatalyst has gained more attention in textile applications due to its self-cleaning and UV-protection properties. Under ultraviolet light and in the presence of water, TiO_2 forms hydroxyl radicals (OH^-) and superoxide anions (O_2^-) and these two oxidation reactants decompose (oxidize) the organic compounds to carbon dioxide and water. However, since TiO_2 is a powder it is problematic to retrieve following treatment of the textile wastewater to degrade the organic pollutants. Therefore, the use of TiO_2 in the form of a composite would make its handling much easier. It was in this regard that the techniques of photodegradation and adsorption were combined to prepare chitosan/TiO_2 composite beads for use in reactive dye removal from an aqueous solution. Therefore, the objective of our research involved chitosan/TiO_2 composite beads for discoloration and degradation of textile dye aqueous solutions similar to previous research articles [12-14], but the idea to prepare chitosan/ TiO_2 composite beads for removal the dye solution differently. In this study, the chitosan/ TiO_2 composite beads were prepared from both the in TiO_2 formation using titanium tetraisopropoxide (TTIP) as the precursor in a sol-gel technique and also by the addition of commercial TiO_2 (A100-TiO_2). The effect of different chitosan: TiO_2 (w/w) ratios in the chitosan/ TiO_2 composite beads on the level of reactive dye removal obtained were also studied, including the adsorption isotherm of the dye molecules onto the chitosan/ TiO_2 composite beads and pure chitosan beads.

2. Experimental

2.1 Materials

Commercial grade high molecular weight chitosan (480 kDa) with a 90% degree of deacetylation was supplied from Bio-Line Co., Ltd. The TTIP, used as precursor for preparing TiO_2 in the sol-gel method, was purchased from A.C.S. Xenon Limited Partnership, Thailand. Glacial acetic acid from Mallinckrodt Baker, lnc. was used as a solvent for chitosan. Sodium hydroxide (NaOH) and ethanol, used to coagulate chitosan beads, and nitric acid were from the T.C. Sathaporn Group Ltd. All chemical reagents were used as received and were A.R. grade. Commercial TiO_2 (A100-TiO_2) and C.I. Reactive Red 35 were supplied from Dystar Co., Ltd. The chemical structure of the C.I. Reactive Red 35 used in this research is shown in the Fig.1.

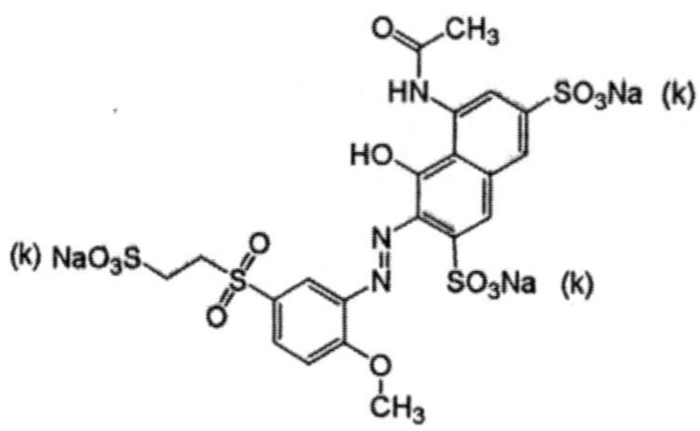

Fig.1 The chemical structure of C.I. Reactive Red 35

2.2 Preparation of chitosan beads

The 2% (w/v) chitosan solution was prepared by dissolving 2g of chitosan flakes in 100 ml of 2% (v/v) aqueous acetic acid under magnetic stirrer for 2h at ambient temperature (30 °C) until the chitosan dissolved completely. This chitosan solution was then dropped into neutralization solution (9:1 (v/v) ratio of 8% (w/v) NaOH and ethanol) and the chitosan beads that had formed were left in the solution for 2d prior to being washed with deionized water until the solution was neutral and then dried at 100 °C for 3h before use.

2.3 Preparation of TiO_2-sol by the sol-gel method

The TiO_2-sol was prepared by mixing TTIP at a concentration of 20% (v/v) in 5% (v/v) acetic acid and 0.1% (v/v) nitric acid and then heating to 60 °C for 1h until the TTIP dissolved completely. After that the TTIP solution was stirred for 2h at ambient temperature (30 °C) to obtain the clear TTIP solution.

2.4 Preparation of chitosan/ TiO_2-sol composite beads

Two ratios of chitosan/ TiO_2-sol (80:20 and 60:40 (v/v), respectively) were used to make the chitosan/ TiO_2-sol composite beads. These were prepared by adding 20 ml or 40 ml of 20% (v/v) TTIP solution (from section 2.3) into 80 ml or 60 ml, respectively, of 2% (v/v) chitosan solution, mixing and then stirring until a homogeneous solution and then dropped into the neutralization solution, washed and dried as detailed in section 2.2. The chitosan/ TiO_2-sol composites prepared from an 80:20 and 60:40 (v/v) ratio of chitosan: TiO_2 are referred to hereafter as chitosan/ TiO_2-sol-20 and chitosan/ TiO_2-sol-40, respectively. To evaluate the effect of curing on the TiO_2 crystalline phase in the chitosan/ TiO_2-sol composite beads, after drying for 3h at 100 °C the beads were heated at 150 °C for 5 min. The TiO_2 crystalline phases were then examined (section 2.9).

2.5 Preparation of the chitosan/ A100-TiO_2 composite beads

Two ratios of chitosan: A100-TiO_2 were used to prepare the chitosan/ A100-TiO_2 composite beads. To 100 ml of 2% (v/v) chitosan solution was added either 0.798 or 1.597 g of A100-TiO_2, selected so as to give the equivalent amount of chitosan: TiO_2 as in the chitosan/ TiO_2-sol-20 and chitosan/ TiO_2-sol-40 composite beads formed by the sol-gel method (section 2.4), respectively. Thereafter, the synthesis was performed as described in section 2.4.

2.6 Adsorption procedure

C.I. Reactive Red 35 was dissolved in distilled water to 60 mg/l. This dye concentration was used to study the effect of the different adsorbent beads on the reactive dye removal. The dry adsorbent beads (2g) were immersed in 10 ml dye solution in a 150 ml beaker for 30 min in the dark and then exposed to a UV light lamp in a color assessment cabinet for 8 h. An aliquot of the aqueous solution was removed every hour and analyzed for its absorbance at a wavelength of 512 nm (A_{512}) using an UV/Visible spectrometer (Specord S 100). Each experiment was performed in triplicate and the average absorbance was converted to the concentration of dye using the linear absorbance versus concentration standard calibration curve. The converted concentration was then used to calculate the percentage dye removal.

2.7 Re-adsorption of adsorbents

Each of the different adsorbents were evaluated for their re-adsorption ability by transferring the adsorbents from the first adsorption to a fresh dye solution and proceeding as reported in section 2.6 thereafter.

2.8 Adsorption isotherms

Adsorption isotherms were performed in a set of five beakers containing 10 ml of reactive dye solutions at an initial concentration of 60, 80, 100, 120 and 140 mg/l. Adsorbent beads (2g) were then added to each dye solution and kept in a beaker for 8h. After that the final concentration of dye in the solution was determined from the A_{512} (section 2.6) and the amount of adsorption at equilibrium q_e (mg/g) was calculated from Eq.1 [13-14].

$$q_e = (C_0-C_e)V/W \qquad (1)$$

where C_0 and C_e (mg/l) are the liquid-phase concentrations of dye at initial and equilibrium, respectively, V is the volume of the solution (l) and W is the mass of dry adsorbent used (g). The adsorption data, obtained from the different initial dye concentrations using Eq.1, were used to determine the type of adsorption isotherm in terms of their fit to the Langmuir or Freundlich isotherm.

2.9 X-ray diffraction (XRD) analysis

The chitosan and TiO_2 polymorphs in the different chitosan/ TiO_2 composite beads and their crystalline phases were examined by XRD analysis using a XRD D8 Advance Bruker AXS with the scan mode operating at 40 kV and 40 mA and a scan range 2θ equal to 5-60°.

3. Results and discussion

3.1 XRD pattern of the chitosan and chitosan/ TiO_2 composite beads

The XRD pattern of pure chitosan beads (Fig.2) showed a diffraction pattern including the two major characteristic peaks of chitosan at 2θ values of 10.6° and 20.6°, which implied that this chitosan was in a crystalline state.

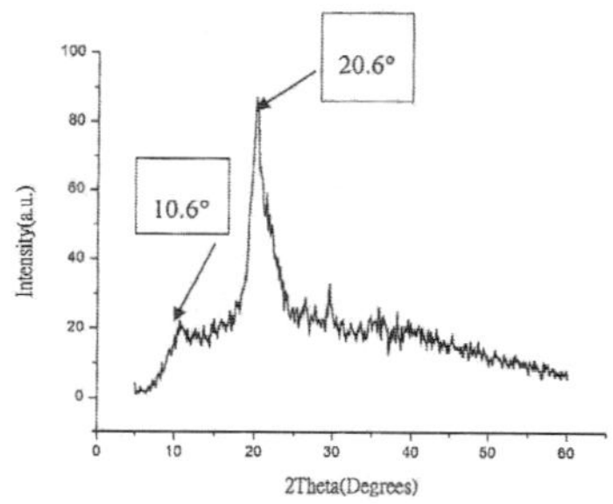

Fig.2 Representative XRD pattern of the pure chitosan beads

The XRD patterns of the chitosan/ A100-TiO_2-20 and chitosan/ A100-TiO_2-40 composite beads (Fig.3) showed similar strong diffraction peaks at a 2θ value of 25.3°, 37.8°, 48.0° and 55.0°, which were regarded as an attributive indicator of the anatase crystalline phase of TiO_2 [15]. That the two major characteristic diffraction peaks of chitosan (2θ values of 10.6° and 20.6°) did not appear in the XRD patterns of either type of chitosan/ A100-TiO_2 composite beads might be because the high intensity peaks of the anatase crystalline phase of TiO_2 concealed the diffraction peaks of chitosan in the composite beads.

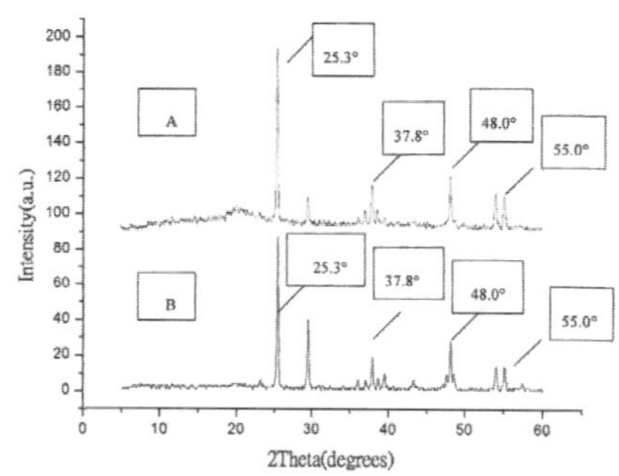

Fig.3 Representative XRD patterns of (A) chitosan/ A100-TiO_2-20 and (B) chitosan/ A100-TiO_2-40 composite beads

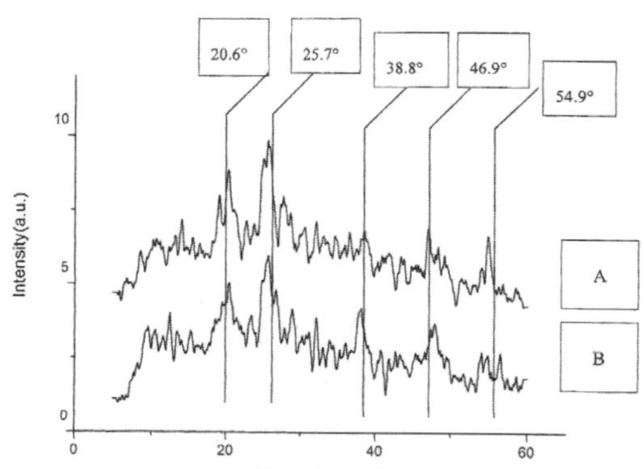

Fig.4 Representative XRD patterns of chitosan/ TiO_2-sol-20 composite beads (A) without and (B) with curing at 150 °C for 5 min

In contrast, the XRD pattern of the chitosan/TiO_2-sol-20 composite beads (Fig.4) did not show any strong diffraction peaks like those of the two different types of chitosan/ A100-TiO_2 composite beads. This implied that the TiO_2 in the chitosan/ TiO_2-sol-20 composite beads was not in the crystalline phase. The diffraction peaks exhibited at 2θ values of 25.7°, 38.8°, 46.9° and 54.9° were broad and blunt and of a low intensity and so correspond to anatase TiO_2 that was not in the crystalline phase. In addition, the diffraction peak of chitosan at a 2θ value of 20° also existed in the XRD patterns of the chitosan/ TiO_2-sol composite beads, but the intensity of this peak was quite low compared with the diffraction peak at a 2θ value of 20.6° for the pure chitosan beads. This might imply that the TiO_2 prepared by the sol-gel method affected the crystalline phase of chitosan.

With respect to the effect of curing the composite beads, the XRD patterns of the chitosan/ TiO_2-sol-20 composite beads with or without curing showed broadly similar diffraction peaks without any high intensity peaks (Fig.4). Thus, curing at 150 °C for 5 min was not strong enough to induce the formation of any more crystalline phase of anatase TiO_2 in the chitosan/ TiO_2 composite beads.

The representative XRD patterns of the chitosan/ TiO_2-sol-40 composite beads with and without curing (Fig.5) showed similar characteristic diffraction peaks to those from the chitosan/ TiO_2-sol-20 composite beads (Fig.4) without any strong diffraction peaks of a high intensity. Thus, using a higher proportion of TTIP (and so TiO_2) during the preparation of the chitosan/ TiO_2-sol composite beads did not enhance the level of the crystalline phase of anatase TiO_2. In contrast, the chitosan/ TiO_2-sol-40 composite beads affected the crystalline peaks of chitosan by shifting the 2θ to a lower value at 10°, which could imply that the increased proportion of TiO_2 in the composite impaired the crystalline state of chitosan.

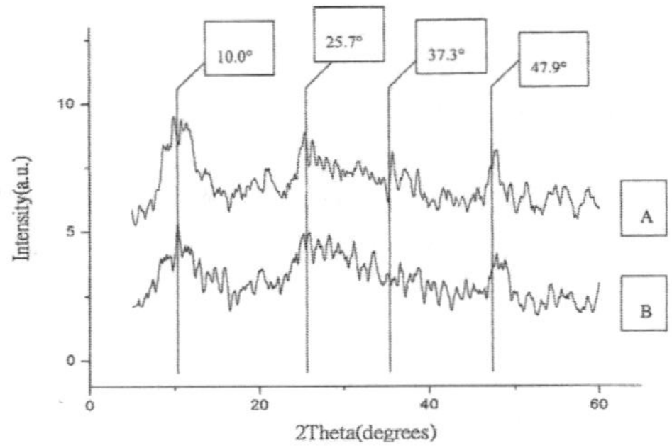

Fig.5 Representative XRD patterns of the chitosan/ TiO_2-sol-40 composite beads (A) without and (B) with subsequent curing at 150 °C for 5 min

3.2 Comparison of the adsorption capacity to remove C.I. Reactive Red 35 by three different adsorbents

The adsorption capacities (% dye removal) of C.I. Reactive Red 35 of the chitosan beads and the chitosan/ A100-TiO_2-20 and chitosan/ A100-TiO_2-40 composite beads are shown in Fig.6a, while that for the chitosan/ TiO_2-sol-20 and chitosan/ TiO_2-sol-40 composite beads, along with chitosan beads for reference, are shown in Fig.6b.

The maximum dye removal by the chitosan/ A100-TiO_2-20 and chitosan/ A100-TiO_2-40 composite beads exposed to UV light for 8 h was close to 97%, while the maximum dye removal obtained with the pure chitosan was up to 95% (Fig.6a). In contrast, the maximum dye removal obtained with the chitosan/ TiO_2-sol-20 composite beads either with or without curing was about 70%, while that for the chitosan/ TiO_2-sol-40 composite beads was lower at 50% after curing and 45% without curing. That the adsorption capacity of the chitosan/ TiO_2-sol composite beads was much lower than those of chitosan and the chitosan/ A100-TiO_2 composite beads may reflect that the higher crystalline phase of anatase increased the adsorption capacity of the chitosan/ A100-TiO_2 composite beads.

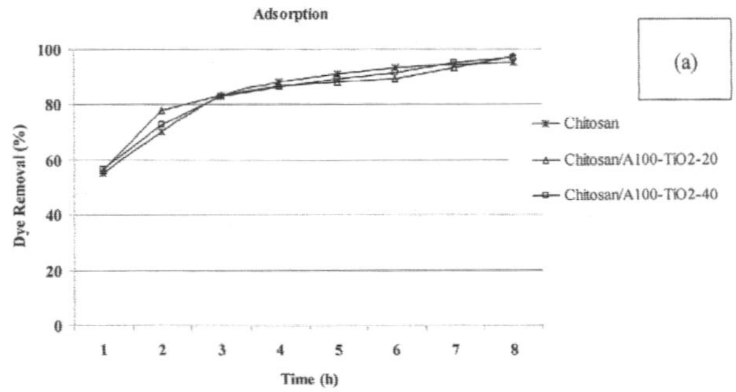

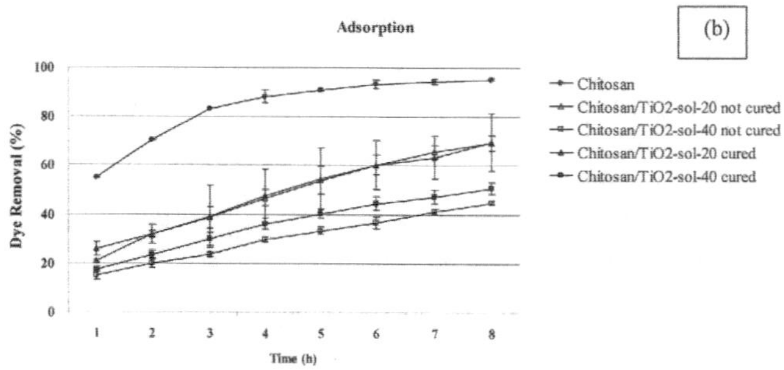

Fig.6 Adsorption capacity of C.I. Reactive Red 35 of the (a) chitosan and chitosan/ A100-TiO_2 composite beads and (b) chitosan and chitosan/ TiO_2-sol composite beads

3.3 Re-adsorption capacity of the different adsorbents

The re-adsorption capacities for C.I. Reactive Red 35 of the chitosan and chitosan/ A100-TiO_2 composite beads are shown in Fig.7a, while those for the chitosan/ TiO_2-sol composite beads, along with chitosan beads for reference, are shown in Fig.7b.

The re-adsorption level of C.I. Reactive Red 35 on the chitosan and chitosan/ A100-TiO_2-20 composite beads was similar to each at other approximately 90%. In contrast, the re-adsorption on the chitosan/ A100-TiO_2-40 composite beads was unexpectedly markedly lower at nearly 76%. It was expected that a higher proportion of A100-TiO_2 in the composite beads would enhance the re-adsorption of the chitosan/ A100-TiO_2 composite beads due to the photocatalytic property of TiO_2. That the pure chitosan showed a high re-adsorption level of this dye may reflect that the 8h adsorption period was insufficient to reach saturation, and so the residual adsorption capacity then appeared as a re-adsorption capacity.

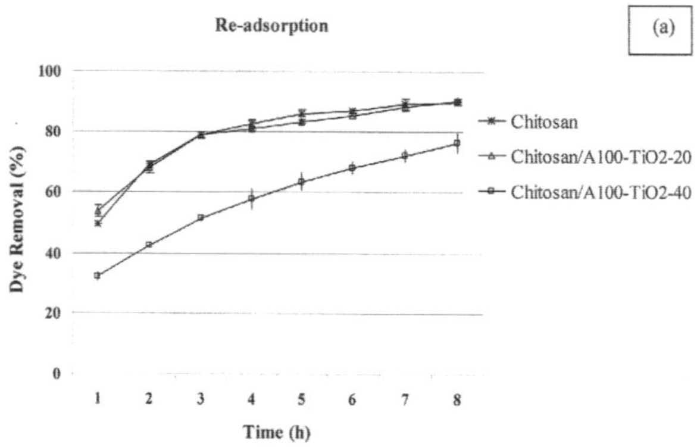

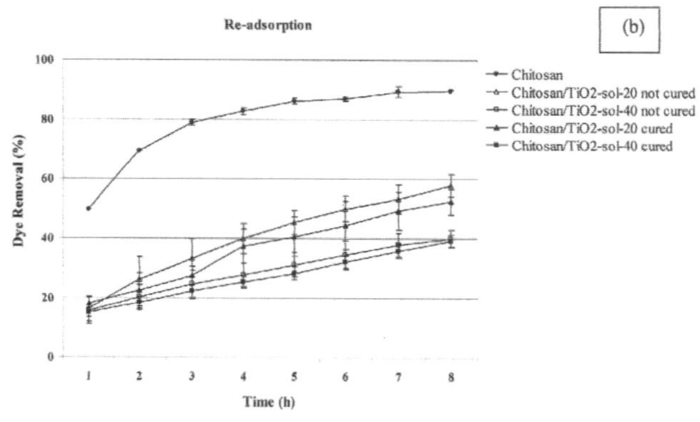

Fig.7 Re-adsorption capacity for C.I. Reactive Red 35 of the (a) chitosan and chitosan/ A100-TiO_2 composite beads and (b) chitosan and chitosan/TiO_2-sol composite beads with or without curing

However, the re-adsorption of C.I. Reactive Red 35 on the chitosan/ TiO_2-sol composite beads at either proportion of TiO_2 and with or without curing was much lower than those of chitosan and the respective chitosan/ A100-TiO_2 composite beads. The maximum re-adsorption of the chitosan/ TiO_2-sol-20 composite beads without or with curing was 58% and 52%, respectively, whereas that for the chitosan/ TiO_2-sol-40 composite beads without or with curing were both close to 40%. Thus, the chitosan/ A100-TiO_2 composite beads showed much better adsorption and re-adsorption capacities than the chitosan/ TiO_2-sol composite beads. This may be because of the crystalline phase of A100-TiO_2 in the chitosan/ A100-TiO_2 composite beads.

3.4 Adsorption Isotherms

Adsorption isotherms can describe how solutes interact with the adsorbent in terms of how the adsorbate molecules distribute between the liquid and solid phases at an equilibrium state. In this study they were evaluated using the Langmuir and Freundlich isotherms. The applicability of the respective isotherm model was evaluated by judging the correlation coefficient (R^2) of the fit between the experimental data and that predicted from the respective isotherm model. To optimize the design of an adsorption system for the adsorption of adsorbates, it is important to establish the most appropriate correlation for the equilibrium curves. The linear form of Langmuir isotherm is expressed as shown in Eq.2 [13-14];

$$q_e = X_m b C_e /(1+bC_e) \tag{2}$$

where q_e is the amount of dye adsorbed per unit weight of adsorbent (mg/g) and C_e is the equilibrium concentration of dye in solution (mg/l). The constant X_m signifies the adsorption capacity (mg/g) and b is calculated as the Langmuir constant.
The Freundlich isotherm can be generally described as shown in Eq.3 [13-14];

$$q_e = KC_e^{1/n} \tag{3}$$

where q_e is the adsorption capacity (mg/g) and C_e is the equilibrium concentration (mg/l). By plotting log q_e as the y-axis and log C_e as the x-axis, K and 1/n were calculated as Freundlich constants. The results of parameters obtained from the Langmuir and Freundlich isotherm models and the correlation coefficient between the predicted and experimental results are reported in Table 1.

Table 1. Values of parameters obtained from two different isotherms and the relative correlation coefficient on three different adsorbents

Type of adsorbent	Langmuir equation			Freundlich equation		
	X_m (mg/g)	b	R^2	K	1/n	R^2
Chitosan	0.064	0.018	0.989	0.062	0.454	0.963
Chitosan/A100 TiO_2-20	0.200	0.010	0.463	0.040	0.838	0.967
Chitosan/TiO_2-sol-20	0.039	0.001	0.946	0.015	0.411	0.949

From the obtained correlation coefficients, the adsorption isotherm of the pure chitosan and chitosan/ TiO_2-sol-20 composite beads could be described reasonably well by both the Freundlich and Langmuir models. However, the isotherm of the pure chitosan beads was slightly better fitted by the Langmuir model, while the chitosan/ A100-TiO_2-20 composite beads were best fitted by the Freundlich model.

4. Conclusion

With respect to the dye adsorption capacities, the chitosan/ A100-TiO_2 composite beads had an adsorption capacity for C.I. Reactive Red 35 dye removal of up to 97%, whereas that for the chitosan/ TiO_2-sol-20 composite beads with and without curing at 150 °C was lower at close to 70% and the maximum C.I. Reactive Red 35 dye removal of the chitosan/ TiO_2-sol-40 composite beads was lower still at close to 50% with curing and 45% without curing. That the chitosan/ A100-TiO_2 composite beads had a much better adsorption capacity than the chitosan/ TiO_2-sol composite beads was because the commercial TiO_2 powder had a much higher anatase crystalline phase content than that obtained from the sol-gel method in the chitosan/ TiO_2-sol composite beads. Pure chitosan showed an excellent dye adsorption capacity, up to 95%, due to the electrostatic interactions between chitosan and C.I. Reactive Red 35. All these different adsorbent beads had a good re-adsorption capacity but the efficiency of such differed amongst them. The maximum re-adsorption capacity was found with chitosan/ A100-TiO_2 composite beads followed by pure chitosan and chitosan/ TiO_2-sol composite beads, respectively. Moreover, the adsorption isotherm of reactive dye adsorption on pure chitosan beads and chitosan/ TiO_2-sol composite beads fitted reasonably well with both the Langmuir and Freundlich models, whereas that for the chitosan/ A100-TiO_2 composite beads followed the Freundlich model only.

5. Acknowledgement

The authors thank the Chulalongkorn University graduate school thesis grant, Chulalongkorn University for financial support.

References

[1] Ngah WSW, Teong LC, Hanafiah MAKM. Adsorption of dyes and heavy metal ions by chitosan composites: A review. Carbohydr Polym 2011; 83: 1446-1456.
[2] Cheung WH, Szeto YS, McKay G. Enhancing the adsorption capacities of acid dyes by chitosan nano particles. Bioresour Technol 2009; 100: 1143-1148.
[3] Salehi R, Arami M, Mahmoodi NM, Bahrami H, Khorramfar S. Novel biocompatible (chitosan-zinc oxide nanoparticle): Preparation, characterization and dye adsorption properties. Colloids Surf B Biointerfaces 2010; 80: 86-93.
[4] Soltani RDC, Khataee AR, Safari M, Joo SW. Preparation of bio-silica/chitosan nanocomposite for adsorption of a textile dye in aqueous solutions. Int Biodeterior Biodegradation 2013; 85: 383-391.
[5] Annadurai G, Ling L, Lee JF. Adsorption of reactive dye from an aqueous solution by chitosan: isotherm, kinetic and thermodynamic analysis. J Hazard Mater 2008; 152: 337-346.
[6] Zainal Z, Hui LK, Hussein MZ, Abdullah AH, Hamadneh IMKR. Characterization of TiO_2-chitosan/glass photocatalyst for the removal of a monoazo dye via photodegradation-adsorption process. J Hazard Mater 2009; 164: 138-145.
[7] Kittinaovarat S, Kansomwan P, Jiratumnukul N. Chitosan/modified montmorillonite beads and adsorption Reactive Red 120. Appl Clay Sci 2010; 48: 87-91.
[8] Nawi MA, Sabar S, Sheilatina. Photocatalytic decolourisation of Reactive Red 4 dye by an immobilised TiO_2-chitosan layer by layer system. J Colloid Interface Sci 2012; 372: 80-87.
[9] Han F, Kambala VSR, Srinivasan M, Rajarathnam D, Naidu R. Tailored titanium dioxide photocatalysts for the degradation of organic dyes in wastewater treatment. Appl Catal A-Gen 2009; 359: 25-40.
[10] Saquib M, Tariq MA, Faisal M, Muneer M. Photocatalytic degradation of two selected dye derivatives in aqueous suspensions of titanium dioxide. Desalination 2008; 219: 301-311.
[11] Colpini LMS, Alves HJ, Santos OAAD, Costa CMM. Discoloration and degradation of textile dye aqueous solutions with titanium oxide catalysts obtained by the sol-gel method. Dyes Pigm 2008; 76: 525-529.
[12] Farzana MH, Meenakshi S. Photo-decolorization and detoxification of toxic dyes using titanium dioxide impregnated chitosan beads. Int J Biol Macromolec 2014; 70: 420-426.
[13] Zubieta CE, Messina PV, Luengo C, Dennehy M, Pieroni O, Schulz PC. Reactive dyes remotion by porous TiO_2-chitosan materials. J Hazard Mater 2008; 152: 765-777.

[14] Zhu H, Jiang R, Fu Y, Guan Y, Yao J, Xiao L, Zeng G. Effective photocatalytic decolorization of methyl orange utilizing TiO_2/ ZnO/ chitosan nanocomposite films under simulated solar irradiation. Desalination 2012; 286: 41-48.
[15] Xie Y, Yuan C. Visible-light responsive cerium ion modified titania sol and nanocrystallites for X-3B dye photodegradation. Appl Catal B-Environ 2003; 46: 251-259.

An Investigation of Color Analysis Method for Printed Fabrics Based on K-means Clustering

Bin-Jie Xin[1]

[1]*College of Fashion, Shanghai University of Engineering Science, Shanghai, China*

Corresponding author's email: xinbj@sues.edu.cn

Abstract

In this paper, a novel method based on a self-adaptive K-means algorithm is developed for the color classification of the printed fabrics. The color image of printed fabrics has been digitalized through an imaging system with a self-developed sample holder based on flat scanner. The image of the printed fabrics could be decomposed into three sub-images in red, green and blue channels separately. The median filters with different template sizes have been applied to process the three sub-images respectively. Subsequently, the filtered image in *RGB* color space, reconstructed from the three sub-images, could be converted into *LAB* color format. After that, a self-adaptive K-means algorithm has been used to carry out the color clustering of the printed fabric image for the purpose of color classification. Our experimental results indicate that the proposed method could achieve the automatic color clustering of printed fabric in *LAB* color space, much better than the conventional methods in *RGB* color space and *HSV* color space.

Keywords: Color clustering; Printed fabrics; Self-adaptive K-means algorithm; *LAB* color space; Image analysis

1. Introduction

The objective evaluation technology based on image analysis and artificial intelligence develops very rapidly due to the advances of computer technology, which can be used for the detecting and identification of the fabric texture or related properties. Quite a lot of researchers in textile fields have attempted to present image analysis methods to identify the fabric structural or color parameters, such as fabric density [1-3], fabric pattern [4-6] and fabric colors [7-11].

As well known, the printed fabrics have been extensively prevalent among people since invented, for its unique gorgeous color and extremely abundant pattern. Hence, the color segmentation of printed fabrics plays an essential role during the fabric manufacturing and analysis processes, which is considered as the most important issue for the quality of subsequent plate molding and printing. Conventionally, color segmentation has been accomplished manually and subjectively by means of optical microscope, which is time-consuming and labor-intensive. To accurately realize the color clustering and avoid redundant calculation, many researchers have presented some automatic color clustering methods for printed fabrics. Generally speaking, the suggested methods could be categorized into two varieties: unsupervised based methods and supervised based methods.

One way of the color segmentation in printed fabrics relies on the unsupervised analysis method, as pointed out by Kuo et al [12]. an unsupervised analysis method using a fuzzy C-means (FCM) clustering algorithm and a specific cluster-validity criterion was developed. It was reported that the difficulties of manual recognition could be solved by using this method, it was effective for the analyzing the color characteristics of printed fabrics in RGB color space. The color analysis method is difficult to be achieved for the printed fabrics, which have different colors expressing the similar intensity but with different hue values, although it is suitable for the printed fabrics showing the variation in the image intensity in the RGB color space. In this situation, Xin et al [13]. separated the polychromatic image into monochromatic regions based on watershed transformation in CIELAB color space and selected hierarchical histogram analysis to eliminate the inherent drawbacks of over-segmentation in the watershed algorithm. Furthermore, an innovated Printed Fabric Computerized Automatic Color Separation System was introduced to achieve fast and accurate color separation [14]. The digitalized color image of printed fabric was converted from RGB (red, green, blue) mode to HSI (hue, saturation, intensity) mode which could be suitable for human vision. By using the Genetic Algorithm, smaller sub-images with the same color distribution could be identified to replace the original fabric image so as to shorten the separation process. Ultimately, the printed fabric image was automatically segmented by employing the FCM-based (Fuzzy C-Means) Region Splitting Method.

On the other hand, the neural network has the advantage of acquiring information from external environment or other artificial neural, and make simple computation, and output results to external environment or other artificial neuron. Accordingly, it could be trained well to do the color separation. Papamarkos et al [15]. proposed a multi-thresholding technique to segment both gray-level and color images, which was implemented by a Principle Component Analyzer (PCA) and a Self-Organized Feature Map (SOFM) neural network. On the basis of this method, a Self-Organizing Map (SOM) was used to map patterns in a three-dimensional color space to a two-dimensional space [16]. Subsequently, a problem of identifying major clusters in input patterns could become a problem of finding peaks and valleys in the density map. The number of the colors included in an image corresponding to the number of the neurons with distinct values in the density map. Nevertheless, the maximum number of "main" colors in an image is limited with respect to a certain SOM neural network and it could not be very bigger in practical using. Therefore, a supervised RGB computerized color separation system based on backward-propagation (BP) neural network was proposed [17]. After Gene Algorithm processing, the BP neural network was applied to conduct separation of printed fabrics RGB sub-image. In addition, the PANTONE® standard color ticket had been utilized to do the color matching. It was reported that the supervised method could differentiate the color on printed fabric image.

Nevertheless, most of the researches are still principally focused on solid colored fabrics. In order to complete the color clustering for printed fabrics, we attempt to employ an innovative method based on a self-adaptive K-means algorithm to do the color clustering analysis in LAB color space. Since the color variation, caused by the uneven light reflection, may have a negative impact on the recognition of the printed fabric color. A series of median filters having different template sizes could be used to process the three sub-images in red, green and blue channels separately. After reconstructing from the three sub-images, the filtered image has been converted from RGB color space to LAB color space. Ultimately, the color clustering of printed fabric is realized with the aid of a self-adaptive K-means algorithm.

2. Methodology

2.1 Image Acquisition System Set-up

Before completing the image acquisition, the materials and structures of printed fabric samples should be analyzed using the manual method with the aid of traditional tools. Simultaneously, the densities along both the warp and weft directions could be measured. Specifically, six printed fabric samples have been selected for this investigation and the particular information is listed in Table 1.

Table 1. Specifications of fabric samples

Sample ID	Weave structure	Composition	Yarn density (yarns per inch)	
			Warp	Weft
1	Plain 1/1	100%Cotton	109.2	78.2
2	Twill 4/1	100%Polyester	137.7	81.3
3	Plain 1/1	68%Polyester+ 32%Rayon	147.3	81.3
4	Twill 2/1	100%Cotton	138.5	72.6
5	Plain 1/1	98%Cotton+ 2%Spandex	154.9	73.7
6	Twill 4/1	68%Polyester+ 32%Rayon	134.3	78.3

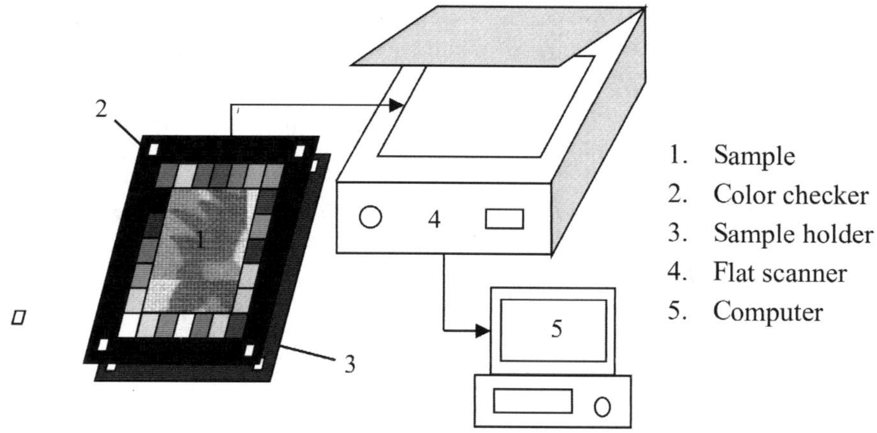

Fig.1 Overview of the image acquisition system

A set of image acquisition system (Fig.1) has been established for the digitalization of the printed fabrics, which includes three main components: 1) a high-resolution flat scanner, which is used to capture the reflective surface images of printed fabric samples; 2) a personal computer with software, which is employed to process and analyze the captured images; 3) a self-developed sample holder, which is not only used to grid the sample with one pair of plates, jointed by four rectangle magnets; but also is used to do color calibration with the color checker, consisting of 24 squares of color blocks.

This study use Microsoft Visual Studio 2010 as a software tool to develop the image acquisition system. The Epson scanner V700 PHOTO with the resolution of 1200 dpi is employed to digitalize the printed fabric image in full-color space (RGB space). The personal computer used is an INTEL Xeon E5 (CPU 3.6GHz). In order to facilitate the computer operation, the original image is first decomposed into several small sub-images. With the aid of the self-adaptive K-means algorithm, the sub-image could be divided into different clusters. After reconstruction, color clustering of the original image will be done based on the clusters of the sub-images. Six images of the printed fabrics have been cropped into small sub-images as shown in Fig.2.

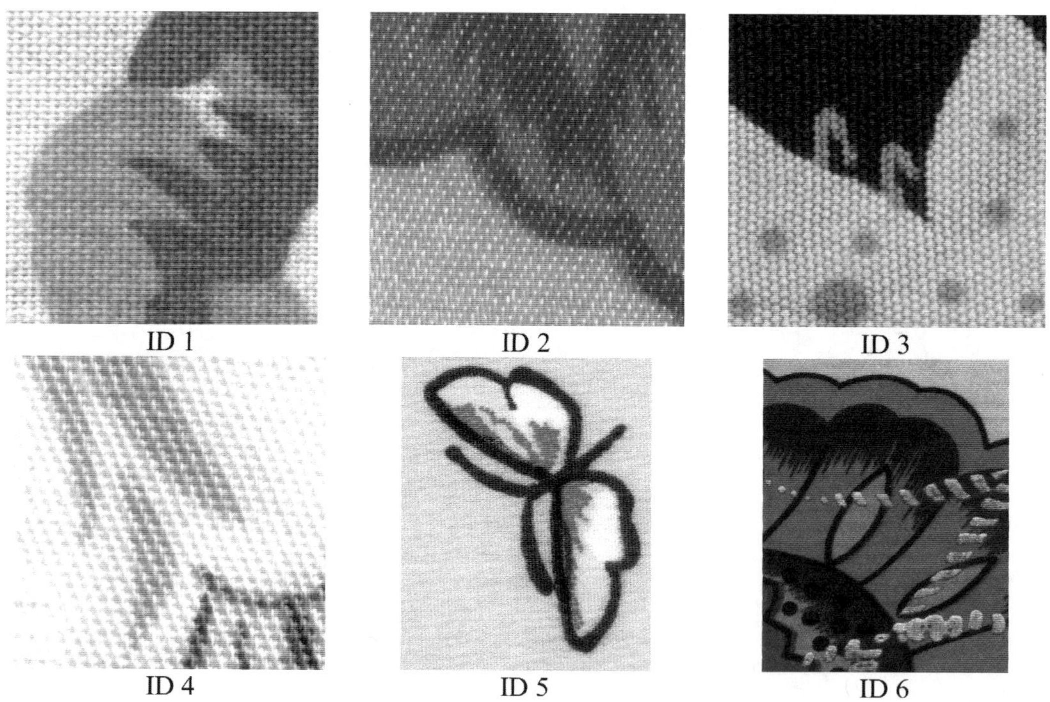

Fig.2 Sub-images of the printed fabrics (6 samples)

2.2 Development of Color Clustering Algorithm

To substantially complete the color clustering analysis, this study uses the median filter to process the three sub-images in red, green and blue channels respectively so as to eliminate the noises. After reconstructing from the three sub-images, the conversion from RGB color space to LAB color space could be done for the subsequent color separation algorithm. An adaptive K-means algorithm is also used to carry out the color separation procedure in LAB color space. In the end, different color could be divided into disjoint clusters and the particular processes are shown in Fig.3.

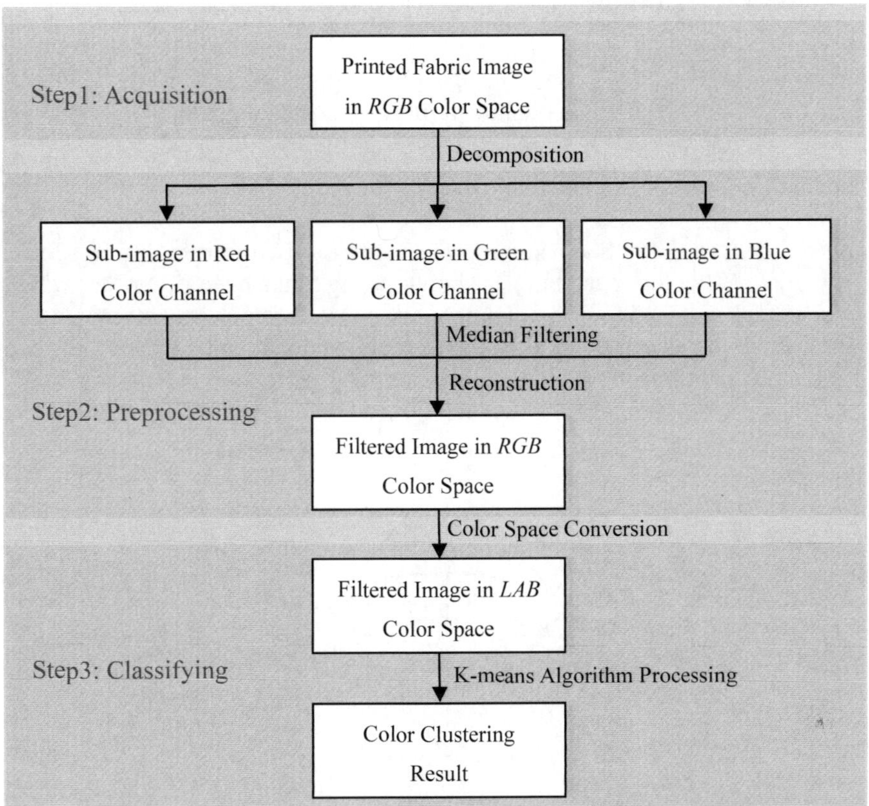

Fig.3 Flowchart of the color clustering processing for the printed

2.2.1 Median Filtering

The principal idea of the median filter is to eliminate the noises caused by the hairiness or uneven light reflection. Thus, a special domain median filter has been applied to enhance the image quality for the subsequent color analysis.

Assuming the pattern of neighbors has an odd number of entries, it is just the middle value after all the entries in the window could be sorted numerically [18]. Here, a 3×3 template has been selected as the example (Fig.4). Supposing $f(x,y)$ and $g(x,y)$ represents the original image and the filtered image separately, 0,32,64,96,128,160,192,225 and 255 are the original grayscale values of each pixel. Since the filter has been utilized to process all the pixels of the original image point by point, 128 could be reserved as the value of the same position in the filtered image.

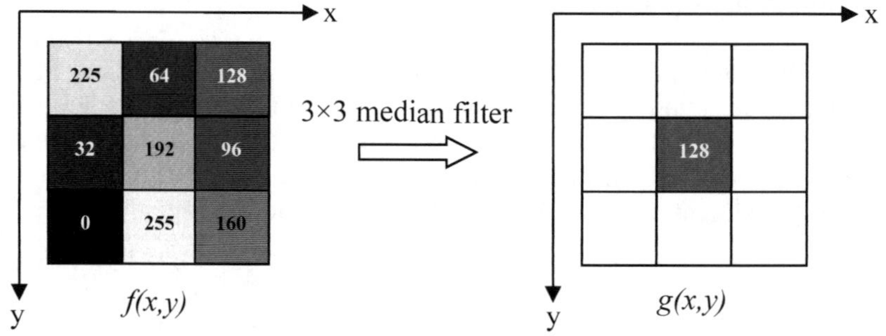

Fig.4 Establishment of the 3×3 median filter

2.2.2 Color Space Conversion

Conventional color space is expressed by hue, saturation and intensity while it is difficult for users to recognize a color precisely. Accordingly, LAB color space [19] as shown in Fig.5 has been proposed, which is more similar to human vision for its perceptual uniformity. The L component approximately accommodates human perception of lightness. Therefore, the output curves can be modified to achieve accurate color balance corrections in the A and B components. Furthermore, the L component could be used to adjust the lightness contrast as well.

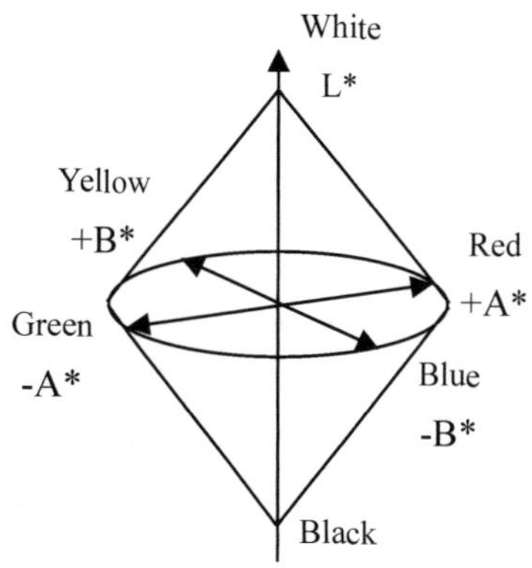

Fig.5 Diagram of *LAB* color space

To accomplish the conversion from RGB color space to LAB color space, the fabric image in RGB color space should be transferred to XYZ color space. The color correction matrix is introduced to do the color calibration. Since the three color components of the original image have been multiplied by the color correction matrix, the calibrated image could be obtained. The specific conversion steps are illustrated as the following equations:

$$r = R/255 \tag{1}$$

$$\text{if } (r > 0.04045) \quad \text{then } r = \left(\frac{r+0.055}{1.055}\right)^{2.4}, \text{ else } r = \frac{r}{12.92} \tag{2}$$

Similarly, g and b can be calculated from the G and B components in the same way.

$$\begin{bmatrix} X \\ Y \\ Z \end{bmatrix} = \begin{bmatrix} 0.4125 & 0.3576 & 0.1805 \\ 0.2126 & 0.7152 & 0.0722 \\ 0.0193 & 0.1192 & 0.9505 \end{bmatrix} \bullet \begin{bmatrix} r \\ g \\ b \end{bmatrix} \tag{3}$$

Then, the image can be converted to *LAB* color space from *RGB* color space by the aid of *XYZ* color space.

$$\text{if } (X > 0.008856) \quad x = X^{\frac{1}{3}} \tag{4}$$

$$\text{else} \quad x = 7.787 \times X + \frac{16}{116} \tag{5}$$

Analogously, y and z can be calculated from the Y and Z components using the above equations.

$$\begin{cases} L = 116 \times y - 16 \\ A = 500 \times (x - y) \\ B = 200 \times (y - z) \end{cases} \tag{6}$$

Once the conversion has been finished, the L, A, B values will range from -128 to +128. Therefore, the clustering method could be used to classify the printed fabrics with LAB color space.

2.2.3 Self-adaptive K-means Clustering Algorithm

Clustering analysis is the process of grouping a provided series of patterns into disjoint clusters, which has been long used for image segmentation. Among the clustering algorithms, the K-means algorithm has been widely used for many practical applications, such as statistics, neural network and colors. In this section, the self-adaptive K-means algorithm could divide the colors into different blocks. Generally, the clustering algorithm is used to divide one type into one clustering and classify into clustering whose degree of membership is maximum. These procedures are illustrated as the following steps [20-22].

1) Choose K initial cluster centers $Z_1(1), Z_2(1), \ldots, Z_k(1)$.

2) At the k-th iterative step, distribute the samples $\{x\}$ among the K clusters using the relation,

$$x \in C_j(k) \ \text{ if } \ \|x - z_j(k)\| < \|x - z_i(k)\| \tag{7}$$

for all $i=1, 2, \ldots, K;\ i \neq j;$ where $C_j(k)$ denotes the set of samples whose cluster center is $z_j(k)$.

3) Compute the new cluster centers $z_j(k+1), j=1, 2, \ldots, K$ such that the sum of the squared distances from all points in $C_j(k)$ to the new cluster center is minimized. The measure which minimizes this is simply the sample mean of $C_j(k)$. Therefore, the new cluster center is given by

$$z_j(k+1) = \frac{1}{N_j} \sum_{x \in C_j(k)} x, \quad j=1, 2, \ldots, K \tag{8}$$

Where N_j is the number of samples in $C_j(k)$.

4) If $z_j(k+1)=z_j(k)$ for $j=1, 2, \ldots, K$ then the algorithm has converged and the procedure is terminated. Otherwise go to Step (2).

The value of K is variable and different values are appropriate for different images. Since the value of K is determined, the self-adaptive K-means algorithm could divide the colors into the corresponding K clusters.

3. Experimental Results

Obviously, the image filtering processes in R, G and B three parallel channels could be illustrated in Fig.6 (a1-a3), (b1-b3) and (c1-c3). Different median filter templates have been used to process the sub-images, incorporating a 3×3 template and a 13×13 template. Meanwhile, these filtered sub-images could be presented using three dimensional (3D) images as exhibited in Fig.6 (d1-d3), (e1-e3) and (f1-f3) so that the changes after different conducts could be revealed distinctly. At last, the filtered images of printed fabric have been displayed in Fig.6 (g1-g3). From Fig.6 (g3), it can be seen that the hairiness or fibers spreading over the fabric surface have been eliminated successfully and the color information is also effectively reserved.

(a1) Sub-image in red color channel

(a2) 3×3 median filtered image of (a1)

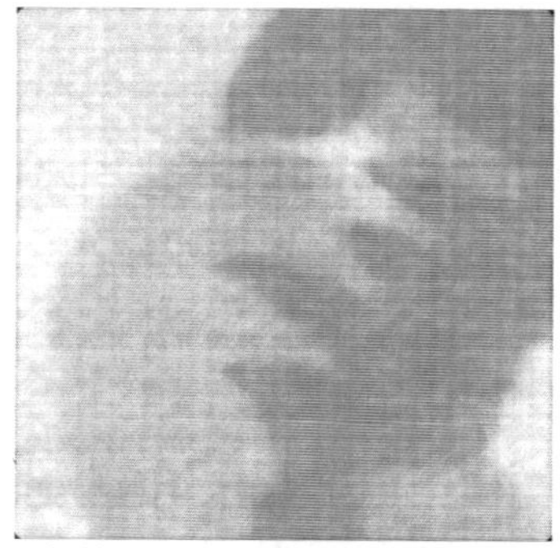

(a3) 13×13 median filtered image of (a1)

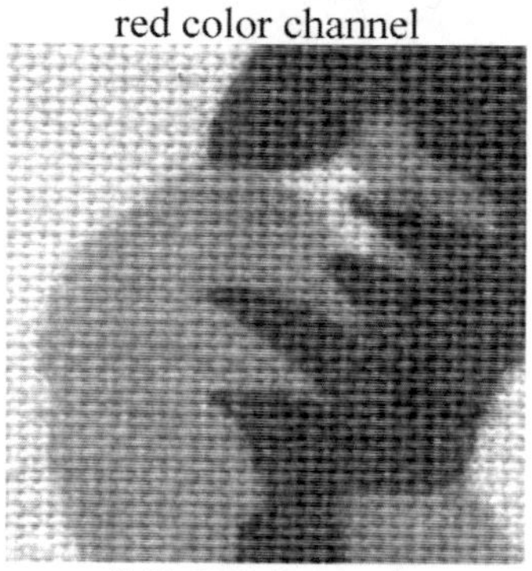

(b1) Sub-image in green color channel

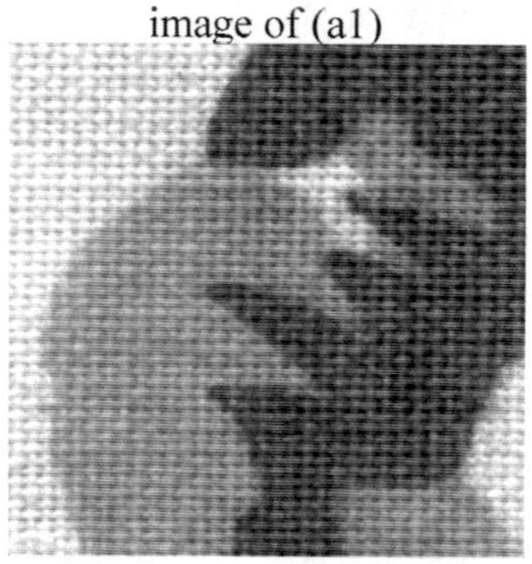

(b2) 3×3 median filtered image of (b1)

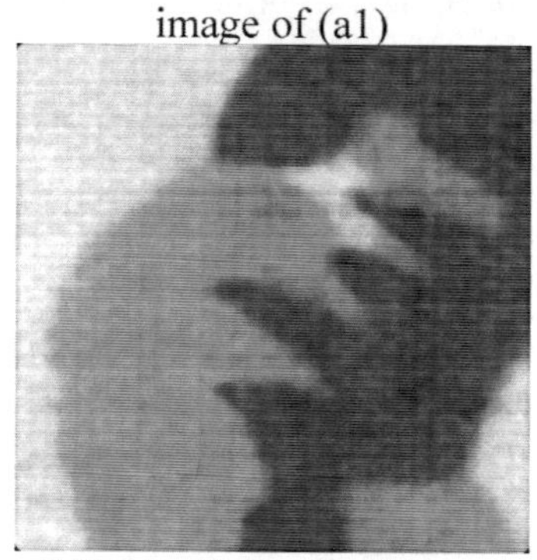

(b3) 13×13 median filtered image of (b1)

(c1) Sub-image in blue color channel

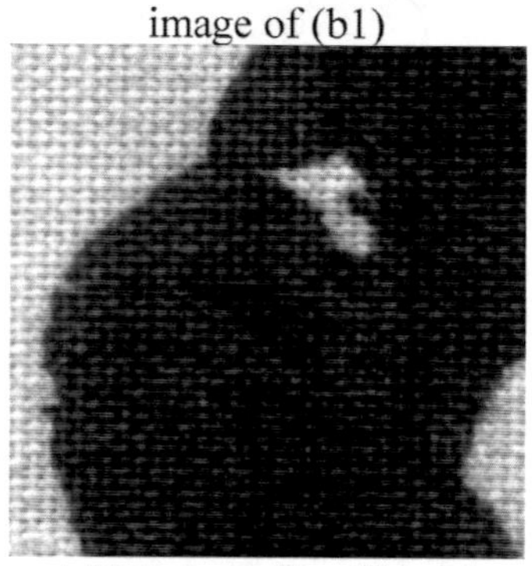

(c2) 3×3 median filtered image of (c1)

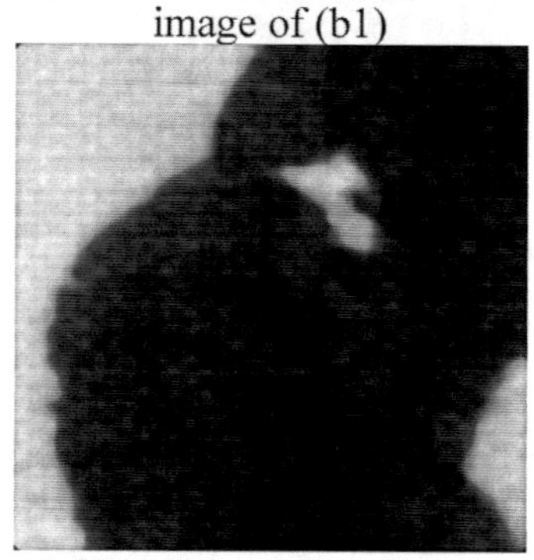

(c3) 13×13 median filtered image of (c1)

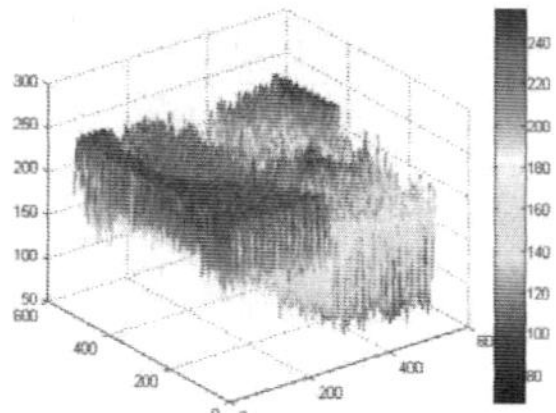

(d1) 3D image of (a1)

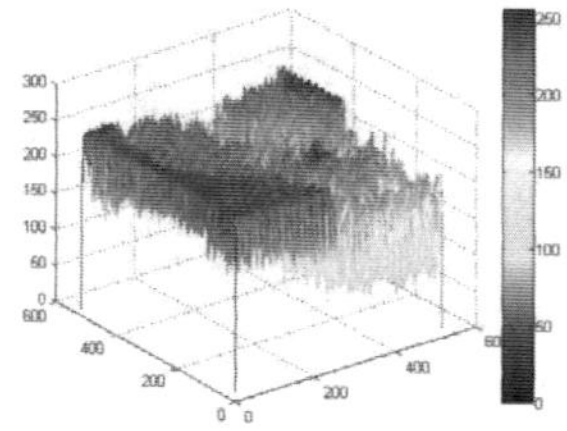

(d2) 3D image of (a2)

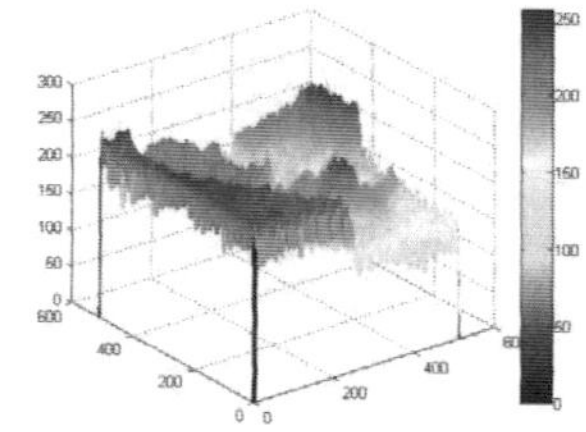

(d3) 3D image of (a3)

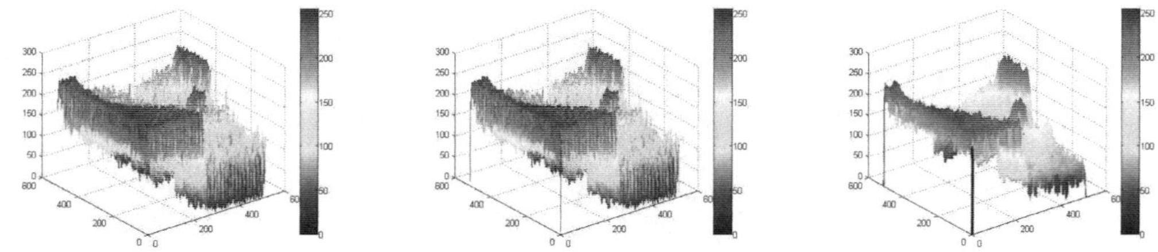

(e1) 3D image of (b1) (e2) 3D image of (b2) (e3) 3D image of (b3)

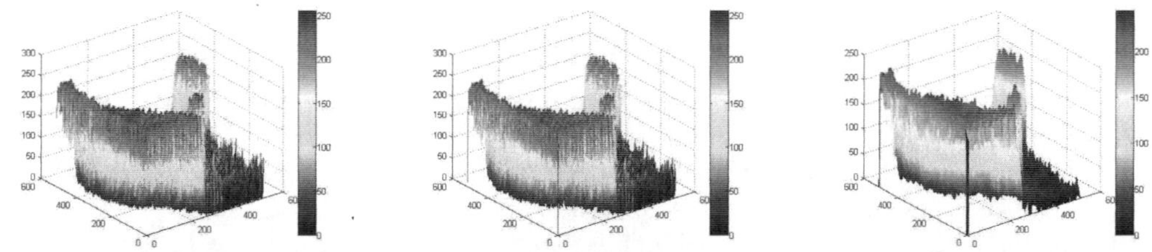

(f1) 3D image of (c1) (f2) 3D image of (c2) (f3) 3D image of (c3)

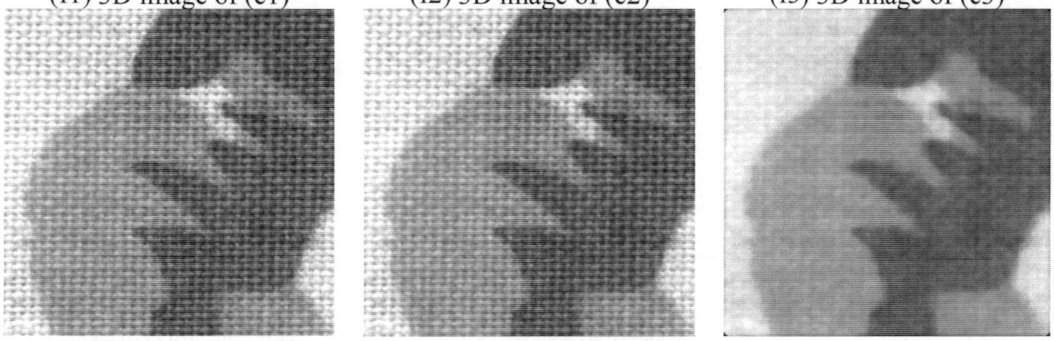

(g1) The printed fabric image of ID 1 (g2) 3×3 median filtered image of (g1) (g3) 13×13 median filtered image of (g1)

Fig.6 Demonstration of the image filtering process

To enhance the precision of the color separation, a self-adaptive K-means algorithm based on LAB color space has been used to achieve the goal, and the segmentation results are shown in Fig.7. It can be found that three different clusters have been identified as illustrated in Fig.7 (a2-a4), although there are some isolated noise points. Thus, a series of median filters own different templates have been utilized to solve the problem.

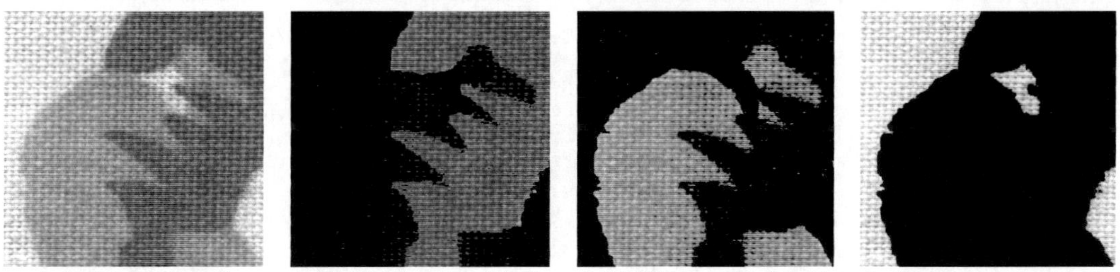

(a1) The printed fabric image of ID 1 (a2) Cluster 1 of (a1) (a3) Cluster 2 of (a1) (a4) Cluster 3 of (a1)

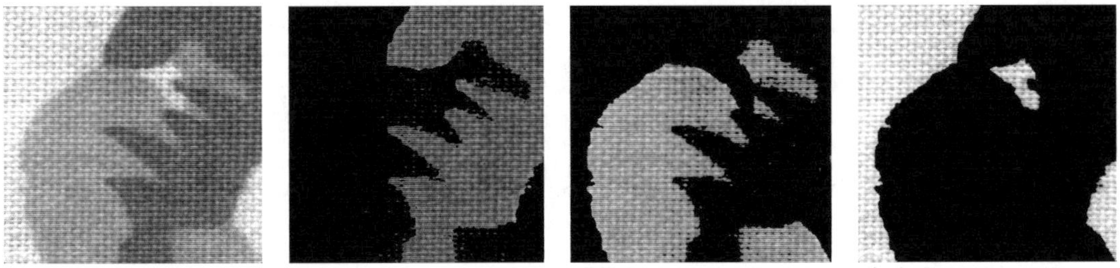

(b1) 3×3 median filtered image of (a1) (b2) Cluster 1 of (b1) (b3) Cluster 2 of (b1) (b4) Cluster 3 of (b1)

Fig.7 Segmentation results of ID 1 in *LAB* color space

In the previous section, it was declared that the printed fabric image was composed of three different colors. In order to remove the isolated noise points, different templates have been used and the 13×13 template shows the best quality as shown in Fig.8 (a). Subsequently, the classification results based on the 13×13 template have been described in Fig.8 (b1-b3), (c1-c3) and (d1-d3). As a contrast, it could be seen that the classification results in *LAB* color space exhibit better performance than the classification results in *RGB* color space and *HSV* color space. Dramatically, when the colors have the similar intensity, hue values but with different saturations, the colors could not be segmented well in *RGB* or *HSV* color space.

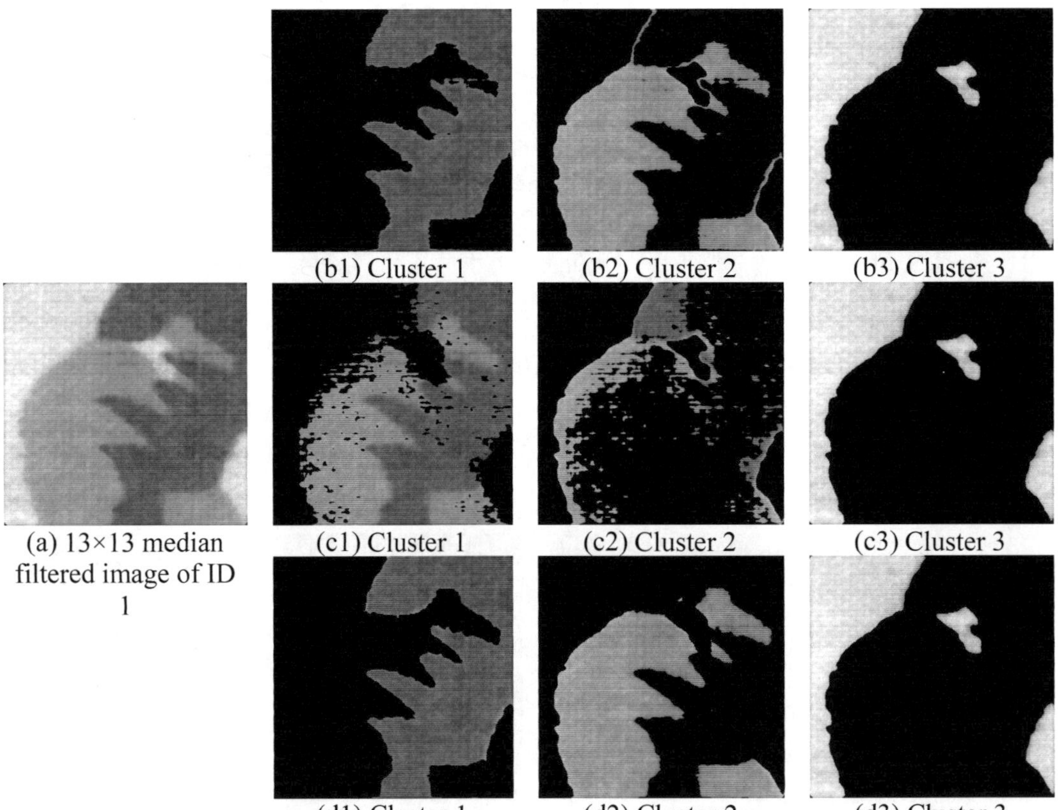

(b1) Cluster 1 (b2) Cluster 2 (b3) Cluster 3

(a) 13×13 median filtered image of ID 1 (c1) Cluster 1 (c2) Cluster 2 (c3) Cluster 3

(d1) Cluster 1 (d2) Cluster 2 (d3) Cluster 3

Fig.8 Segmentation results with K-means algorithm for the printed fabric: (b1-b3) segmentation results in the *RGB* color space (k=3); (c1-c3) segmentation results in the *HSV* color space (k=3); (d1-d3) segmentation results in the *LAB* color space (k=3).

With the proposed method, three separated clusters have been classified and defined as cluster 1, cluster 2 and cluster 3 respectively. The histograms of the three clusters along axis *R*, *G* and *B* could be described in Fig.9. Using the analogously method based on *LAB* color space, the color segmentation results when K=3 have been illustrated in Fig.10, the color segmentation results when K=5 have been illustrated in Fig.11, the color segmentation results when K=7 have been illustrated in Fig.12.

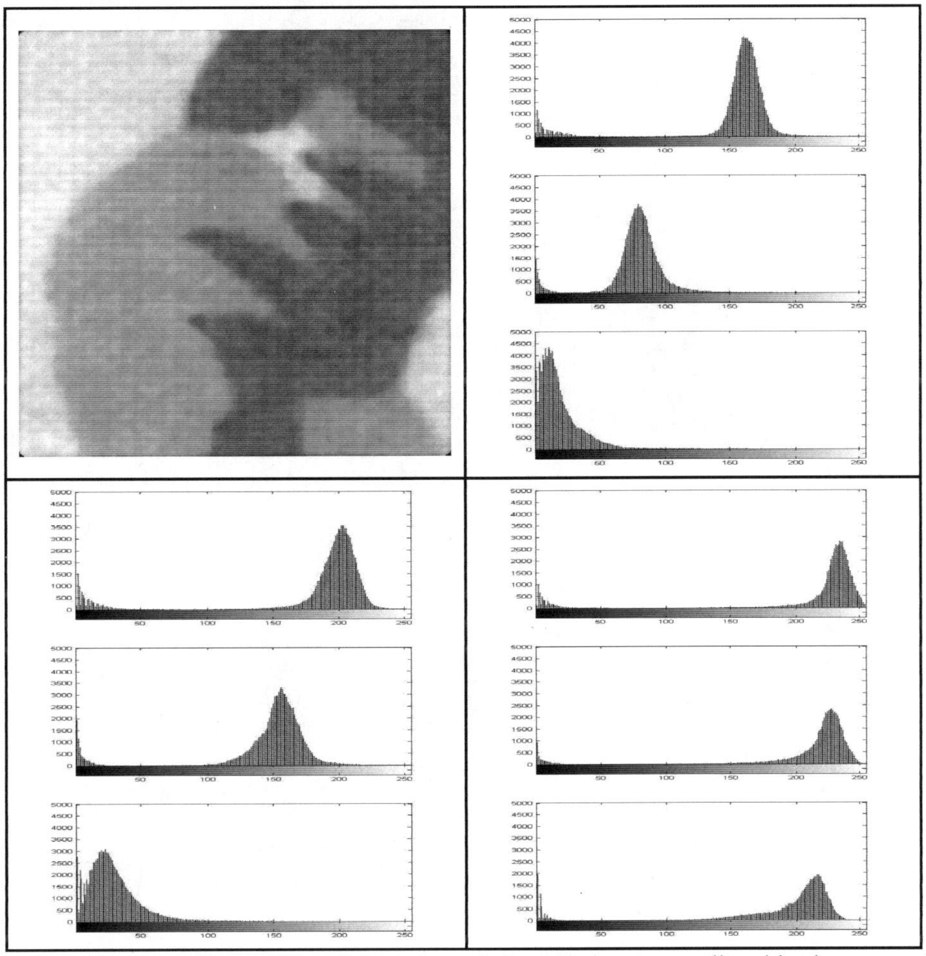

Fig.9 Illustration of color clustering results of ID 1. Three divided clusters are listed in the upper right (cluster 1), the bottom left (cluster 2) and the bottom right (cluster 3) directions separately. The first row is the histogram along axis Red; the center row is the histogram along axis Green; the bottom row is the histogram along axis Blue.

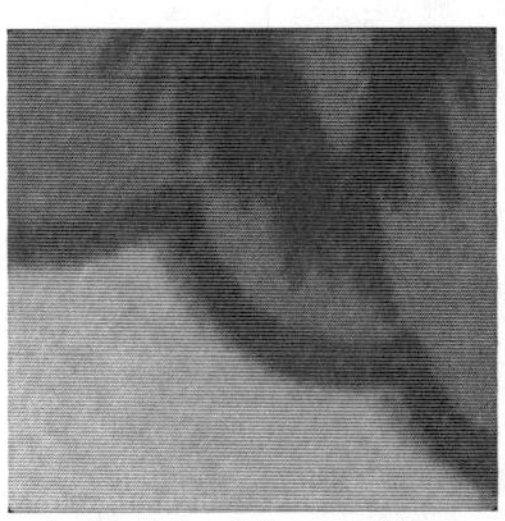

(a1) 13×13 median filtered image of ID 2

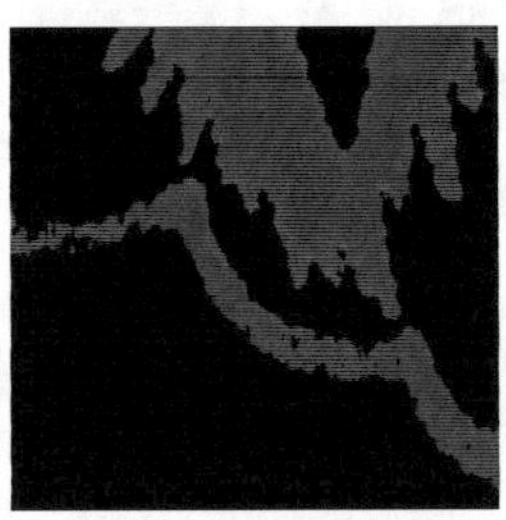

(a2) Cluster 1 of (a1)

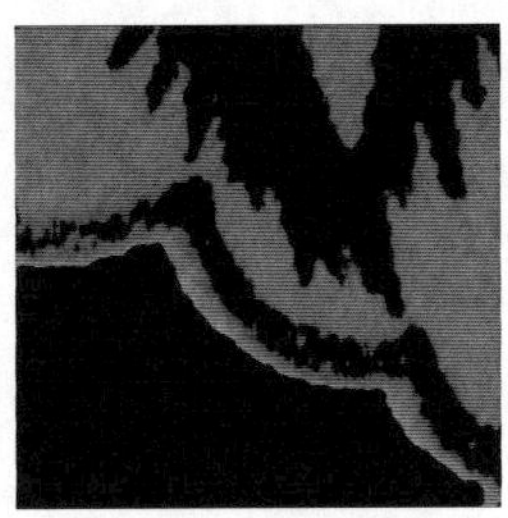

(a3) Cluster 2 of (a1)

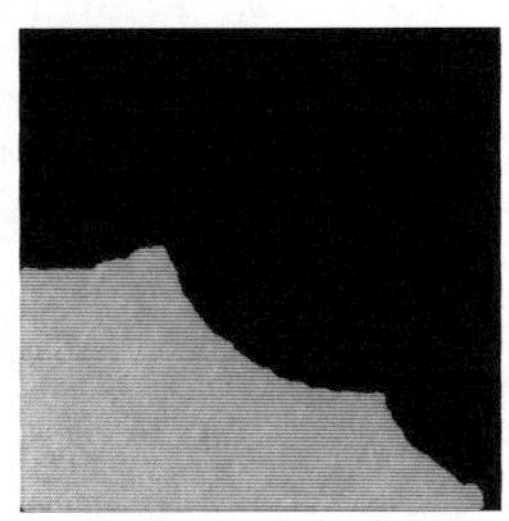

(a4) Cluster 3 of (a1)

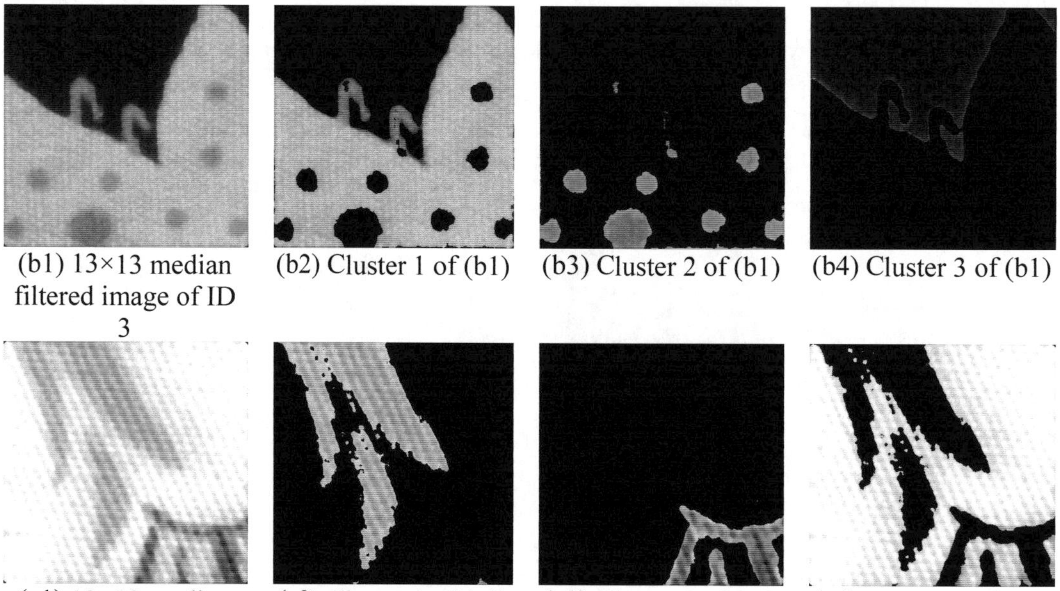

(b1) 13×13 median filtered image of ID 3 (b2) Cluster 1 of (b1) (b3) Cluster 2 of (b1) (b4) Cluster 3 of (b1)

(c1) 13×13 median filtered image of ID 4 (c2) Cluster 1 of (c1) (c3) Cluster 2 of (c1) (c4) Cluster 3 of (c1)

Fig.10 K=3, the 13×13 median filtered images and the classified three clusters

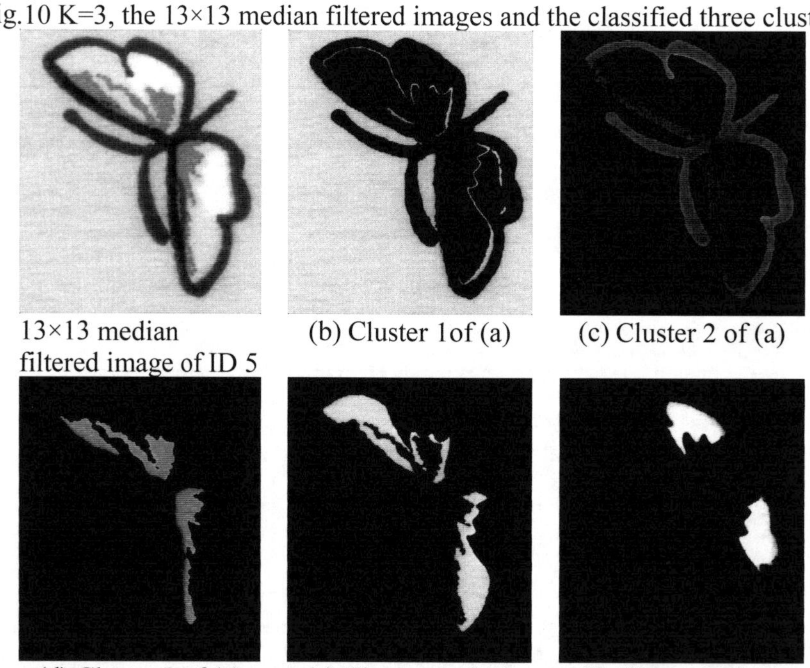

13×13 median filtered image of ID 5 (b) Cluster 1of (a) (c) Cluster 2 of (a)

(d) Cluster 3 of (a) (e) Cluster 4 of (a) (f) Cluster 5 of (a)

Fig.11 K=5, the 13×13 median filtered images and the classified five clusters

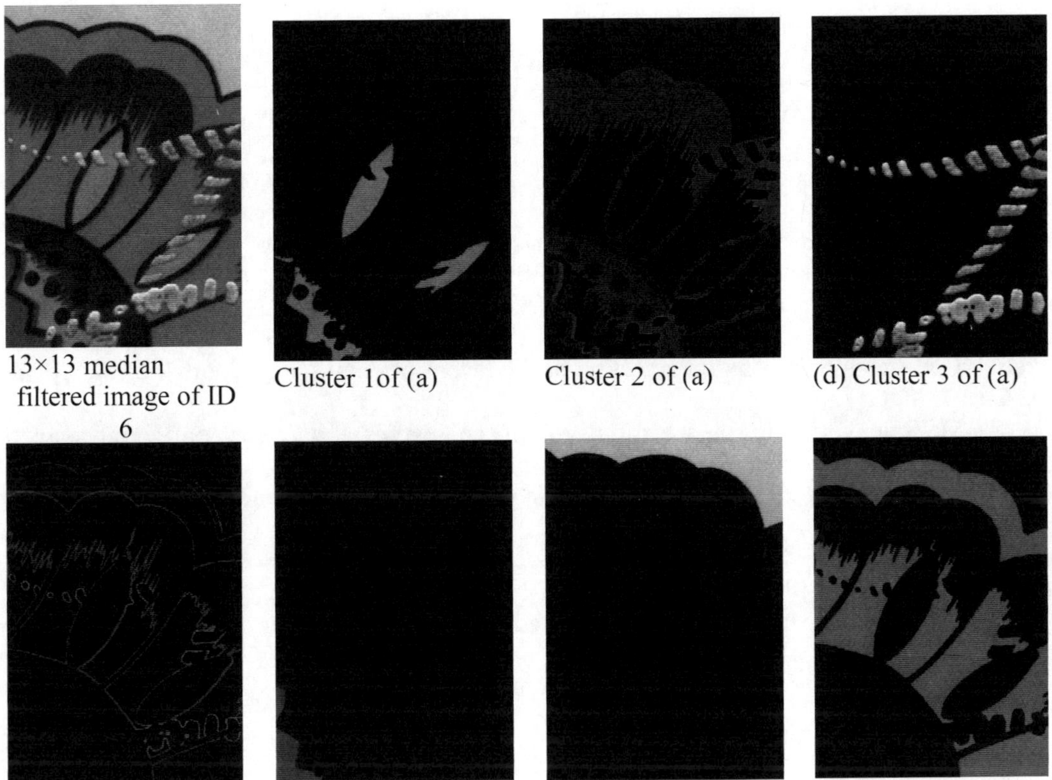

13×13 median filtered image of ID 6 | Cluster 1of (a) | Cluster 2 of (a) | (d) Cluster 3 of (a)

(e) Cluster 4 of (a) | (f) Cluster 5 of (a) | (g) Cluster 6 of (a) | (h) Cluster 7 of (a)

Fig.12 K=7, the 13×13 median filtered images and the classified seven clusters

4. Discussion and Conclusions

In this study, an innovative color clustering method for printed fabrics in LAB color space has been put forward. The fabric image with the resolution of 1200dpi, captured by a flat scanner with a self-developed sample holder, is decomposed into three sub-images in red, green and blue channels separately. After median filtering, the three sub-images can be reconstructed into one filtered image in RGB color space. By means of conversion from RGB color space to LAB color space, the fabric image could be segmented and classified into three different clusters with the self-adaptive K-means algorithm. According to our experimental results, the LAB color space based method could differentiate the disjoint colors better than the RGB color space based method and the HSV color space based method.

One of the limitations of our work is that the border colors of different clusters may be influenced not only by the original reflection light itself, but the diffuse reflection light which generated by the disjoint yarns. The foundation of the color blended model and diffuse reflection model will be the emphasis of our next discussion.

5. Acknowledgement

This project is supported by the National Natural Science Foundation of China under grant No.61271419 and the Graduate Innovation Fund of Shanghai University of Engineering Science under grant No.A-0903-13-01116.

References

[1] Xu BG. Identifying fabric structures with Fast Fourier Transform techniques. Textile Research Journal, 1996; 66(8): 496-506.

[2] Pan RR, et al. Automatic inspection of double-system-mélange yarn-dyed fabric density with color-gradient

image. Fibers and Polymers, 2011; 12(1): 127-131.
[3] Tunák M, Linka A, Volf P. Automatic assessing and monitoring of weaving density. Fibers and Polymers, 2010; 10(6): 830-836.
[4] Kang TJ, Kim CH, Oh KW. Automatic recognition of fabric weave patterns by digital image analysis. Textile Research Journal, 1999; 69(2): 77-83.
[5] B JX, et al. Investigation on the Classification of Weave Pattern Based on an Active Grid Model. Textile Research Journal, 2009; 79(12): 1123-1134.
[6] Huang CC, Liu SC, Yu WH. Woven Fabric Analysis by Image Processing. Part1: Identification of Weave Patterns. Textile Research Journal, 2000; 70(6): 481-485.
[7] Pan RR, et al. Automatic detection of the layout of color yarns with logical analysis. Fibers and Polymers, 2012; 13(5): 664-669.
[8] Pan RR, et al. Genetic algorithm-based detection of the layout of color yarns. Journal of the Textile Institute, 2011; 102(2): 172-179.
[9] Dong G, Xie M. Color clustering and learning for image segmentation based on neural networks. Neural Networks, IEEE Transactions on, 2005; 16(4): 925-936.
[10] Shen HL, Xin JH. Analysis and synthesis of multicolored objects in a single image. Optics letters, 2005; 30(18): p: 2378-2380.
[11] Gorji Kandi S, Amani Tehran M, Rahmati M. Colour dependency of textile samples on the surface texture. Coloration Technology, 2008; 124(6): p: 348-354.
[12] Kuo CFJ, et al. Color and pattern analysis of printed fabric by an unsupervised clustering method. Textile Research Journal, 2005, 75(1): 9-12.
[13] Kuo CFJ, Shih CY. Printed fabric computerized automatic color separating system. Textile Research Journal, 2011; 81(7): 706-713.
[14] Xu B, Lin S.. Automatic color identification in printed fabric images by a fuzzy-neural network. AATCC Review, 2002; 2(9): 42-45.
[15] Xin J, Shen H. Recolouring digital textile printing design with high fidelity. Coloration Technology, 2004; 120(1): p: 6-13.
[16] Kuo CFJ, Su TL, Huang YJ. Computerized color separation system for printed fabrics by using backward-propagation neural network. Fibers and Polymers, 2007; 8(5): 529-536.
[17] Kuo CFJ, Kao CY, Chiu CH, Kuo. Integrating a genetic algorithm and a self-organizing map network for an automatically separating color printed fabric system. Textile Research Journal, 2009; 79(13): 1235-1244.
[18] Papamarkos NC. Strouthopoulos, and I. Andreadis, Multithresholding of color and gray-level images through a neural network technique. Image and Vision Computing, 2000; 18(3): p: 213-222.
[19] Bao XM, et al. Textile image segmentation based on semi-supervised clustering and Bayes decision. in Artificial Intelligence and Computational Intelligence, 2009. AICI'09. International Conference on. 2009. IEEE
[20] Lucchese L and Mitra S. Unsupervised segmentation of color images based on k-means clustering in the chromaticity plane. in Content-Based Access of Image and Video Libraries, 1999.(CBAIVL'99) Proceedings. IEEE Workshop on: 1999. IEEE.
[21] Ray S, Turi RH. Determination of number of clusters in k-means clustering and application in colour image segmentation. in Proceedings of the 4th international conference on advances in pattern recognition and digital techniques: 1999.
[22] Tomasi C, Manduchi R. Bilateral filtering for gray and color images. in Computer Vision, 1998. Sixth International Conference on: 1998. IEEE.

Dyeing and Ultraviolet Protection of Cotton Fabric using Natural Dyes Extracted from Windmill Palm Fiber

Shu Wang, Li-Mei Fang, Xing-Xing Liu, Jia-Li Li, Ying Jiang, Min Guo, Tong-Hua Zhang*

Southwest University, Tiansheng Street, Beibei District, Chongqing, 400715, China

*Corresponding author's email: zhtonghua@aliyun.com

Abstract

The ultraviolet protective properties of cotton fabrics using natural dyes extracted from windmill palm fiber were investigated. To achieve the highest absorbency in the ultraviolet region, the optimum extraction conditions were obtained for windmill palm fiber. The optimum conditions were extraction temperature of 100 °C, extraction time of 120 min, and material-to-liquor (water) ratio of 1:7. The extraction process was repeated twice. The UPF and *T*(UVA) values for the cotton fabrics dyed by the optimum dye solution meet the American Society for Testing and Materials Standard (UPF=29, *T*(UVA)=2.96). According to the standard, the cotton fabrics belong to the "very good" protection category. The relationship between the UPF and *K/S* values was also studied.

Keywords: Windmill palm fiber; Ultraviolet protection; Natural dye; Aqueous extract; Dyeing; Fabric

1. Introduction

Environmental pollution is gradually damaging the ozone layer. Harmful ultraviolet (UV) rays can easily penetrate the atmosphere and damage human health. One third of cancers worldwide are skin-related. Many factors are involved in the development of skin cancers, and cumulative ultraviolet exposure is a critical variable. An effective remedy is the use of suitable clothes, hats, and sunscreen [1]. The range of UV wavelength is further subdivided into UVA (320-400 nm), UVB (280-320 nm), and UVC (100-280 nm) radiation. The effects of UV radiation on human health depend on the UV wavelength. UVA radiation is believed to contribute to premature ageing and wrinkling of the skin and has been recently implicated as a cause of skin cancer. UVB radiation has been implicated as the major cause of skin cancers and sunburn. UVC radiation is extremely dangerous, but much of the dangerous UV radiation is absorbed in the atmosphere's ozone layer before it reaches the ground. Ultraviolet protection factor (UPF) is utilized as an indicator of fabric protection capacity [2]. The characteristics of suitable clothes identified to affect the protective properties include color [3-5], fiber content [3,6], and structure [6]. Although synthetic fibers with high concentrations of dyes possess good ultraviolet protection properties, fabrics made of synthetic fibers are uncomfortable to wear [7]. It is known that fibers with high cellulose content are vulnerable to UV light. Cotton, whose cellulose content is approximately 100%, is a type of cellulose fiber. Flax is also a kind of cellulose fiber. Grifoni et al dyed cotton and flax fabrics with multifunctional plant extracts to achieve better UV-protective properties. Flax fabrics mordanted with alum and dyed with Lavandula and Rubia, and flax fabrics mordanted with tannin and dyed with Rubia and Helychrysum reached the minimum protection level. A very good protection level was reached by flax mordanted with alum and dyed with Helychrysum[8]. The mordants and dyes used in that study are natural.

The concept "natural life" has introduced a new trend in the use of natural fibers and natural dyes [9-10]. Several natural dyes not only provide unique and elegant colors, but also impart functional properties, such as antibacterial and ultraviolet protection, to fabrics. These natural dyes have been applied to wool, flax, and cotton [11-17]. Windmill palm leaf sheath fiber is known for its excellent ultraviolet protection properties [18]. Lignin is one of the main components of palm fiber [19]. Lignin is an aromatic polymer with abundant benzene structures, which can absorb UV rays well [20]. Palm fiber is a multicellular fiber with a "lacuna" from the transverse section, and this unique structure is similar to that of hemp fiber [20]. In short, palm fiber has the potential to reflect and absorb UV rays. Wang et al. investigated the UV absorption behavior of windmill palm fiber and confirmed the excellent UV absorption of palm fiber [18].

The combination of cotton and palm fiber aqueous extract should be ideal for the manufacture of ultraviolet protective clothes with excellent color qualities. However, only a few studies have investigated this area. The objective of this study is to investigate whether the aqueous extract obtained from palm fiber, either in the form of a liquid dye or after application to cotton fabrics, provides some level of ultraviolet protection.

2. Experimental

2.1 Materials and chemicals

Windmill palm fiber samples were obtained from a local plantation in Honghe County, Yunnan Province, China. No other treatment was applied to the fiber before the specimen was ground into powder form (approximately 200 um in size). The scientific name is windmill palm leaf sheath fiber, belonging to Trachycarpus genus and fortune species. The characteristics of the fabrics obtained from a local company in Chongqing are presented in Table 1. The chemical reagents used as mordants during dyeing are presented in Table 2. These chemicals were obtained from Shanghai Macklin Biochemical Co., Ltd., Shanghai, China.

Table 1. Characteristics of fabrics used in the study

Fabric type	Weaves	Thickness (mm)		Mass per unit area (g/m2)	
		Mean	s.d.	Mean	s.d.
Cotton	Plain	0.30	0.02	95	0.04

2.2 Extraction of dyes

The colored matter in the windmill palm fiber was extracted in distilled water. The steps of the extraction were as follows: boiling in water (three times, 120 min each time), cooling to room temperature, filtration (several times), combining the filtered liquid, and concentration and extraction of liquid dye solution [22].

2.3 Dyeing

According to traditional Chinese dyeing methods using natural dyestuff, the dyeing process of fabrics can be conducted in four different ways: direct, pre-mordant, simultaneous, and post-mordant dyeing. For the last three methods, a mordanting bath containing metal ion is necessary. Therefore, alum, ferrous sulfate, and potassium dichromate were used in this study (shown in Table 2) to obtain the excellent ultraviolet protection performances of the cotton fabrics. The recipes for dyeing and the steps in the dyeing process are shown in Table 3. The common parameters of fabric pretreatment before dyeing were as follows: the fabric-to-liquor ratio for scouring was 1:50, the proportion of scouring agent (sodium hydroxide) was 2.5%, and boiling for 1 h at 100 °C under continuous stirring [23]. After washing, the fabrics were dried in an oven at 60 °C. Before dyeing, the dried fabrics were immersed in distilled water for 10 min. Afterward, direct, pre-mordant, simultaneous, and post-mordant dyeing were conducted to dye the cotton fabrics. After dyeing, the fabrics were oven-dried at 60 °C. The fabrics were then tested and characterized.

Table 2. Chemical reagents used in this study

No.	Substances	Molecular formula	Origin	Purity (%)
1	Aluminum Sulfate	$Al_2(SO_4)_3 \cdot H_2O$	Shanghai	99.00
2	Ferrous Sulfate	$FeSO_4 \cdot 7H_2O$	Shanghai	99.99
3	Potassium Dichromate	$K_2Cr_2O_7$	Shanghai	99.99

Table 3. Recipe for dyeing the fabrics with palm fiber extracts and steps in the dyeing process

Dyeing method	Mordanting		Dyeing		Process
	Mordants % (owf.)	Temp. (°C)	Bath Ratio	Temp. (°C)	
Direct dyeing	-	-	1:50	90	Dyeing-washing-drying
Pre-mordanting dyeing	3	90	1:50	90	Mordanting-dyeing-washing-drying
Simultaneous dyeing	3	90	1:50	90	Dyeing & mordanting-washing-drying
Post-mordanting dyeing	3	90	1:50	90	Dyeing-washing-mordanting-washing-drying

2.4 Measurement

The dyeing performance and ultraviolet protective property of the fabric samples were measured to identify the characteristic of the extracts obtained from palm fiber.

The color yield of the dyed fabrics was calculated through light reflectance measurements using Datacolor 650 (Datacolor, Lawrenceville, NJ, USA). The color strength (*K/S*) value was assessed with the Kubelka-Munk equation (1). It represents the amount of dye yielded by fabrics.

$$K/S = (1-R)^2 / 2R, \tag{1}$$

where *K/S* is the Kubelka-Munk ratio, K is the absorption coefficient depending on the colorant concentration, *S* is the light-scattering coefficient caused by the fabric, and *R* is the decimal fraction of the reflectance of the dyed fabrics.

According to EN 13758-1, the UPF of the fabric was determined from the total spectral transmittance as follows:

$$UPF = \frac{\sum_{\lambda=290}^{\lambda=400} E(\lambda)\varepsilon(\lambda)\Delta\lambda}{\sum_{\lambda=290}^{\lambda=400} E(\lambda)T(\lambda)\varepsilon(\lambda)\Delta\lambda} \tag{2}$$

where E(λ) is the solar irradiance expressed in Wm-2nm-1, ε(λ) is the erythema action spectrum, λ is the wavelength interval, and T(λ) is the spectral transmittance at wavelength λ.

UPF is actually the measure of blocked UV radiation (UVA and UVB). UPF is measured by a spectrophotometer. The UPF of fabric specimens were also measured in a dry state. The UVR transmittance of the fabrics was measured over the spectrum (290–400 nm) with 5 nm intervals. The fabrics are classified according to UV protection in Table 4.

Table 4. Protection categories of fabrics according to ASTM D 6603

Protection category	UPF range	Rating
Excellent	40-50, 50+	40, 45, 50, 50+
Very good	25-39	25, 30, 35
Good	15-24	15, 20

Determination of the UV absorbance for aqueous extracts obtained from palm fiber was conducted after the aqueous extracts were diluted 5, 10, and 15 times using distilled water. The UV absorbance of the extracts was investigated with a UV-vis absorption spectrometer (TU 1901, Pgeneral, China) in transmission mode.

3. Results and discussion

3.1 UV absorbance of dyes in solution

Fig.1 shows the absorbance curves of aqueous extraction under different dilutions. The absorption of the palm fiber extracts in water is strong in the UVB (from 280 nm to 320 nm) and UVC (from 100 nm to 280 nm) regions. In the ultraviolet region, the absorbance curve of the aqueous extracts diluted five times has three peaks, which are at approximately 207, 209, and 280 nm. However, the strongest peaks of the absorbance curves diluted 10 and 15 times appear at 201 and 200 nm, respectively, which might be due to the effect of the diluent's (distilled water) large polarity. However, the peak of curves at 280 nm could weaken with the increase in diluted multiples. This condition may also be attributed to the large polarity of distilled water. Notably, the absorbance curve of aqueous extraction obtained from undiluted palm fiber has three distinct peaks at 215, 237, and 280 nm (not shown in this study). Absorption in this region is expected to offer good protection from harmful UV radiation.

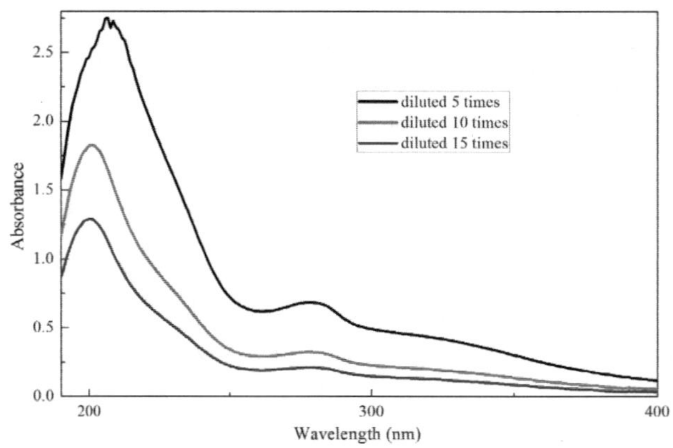

Fig.1 Absorbance curves of aqueous extracts of palm fiber under different dilutions

3.2 UPF of fabrics dyed with colored matter

The dye solution extracted was applied to the fabric samples to obtain the best ultraviolet protection. To achieve this, the properties of the fabric samples dyed with different mordants were compared. The properties of the fabric samples dyed by the four different methods were also compared. The results are presented below.

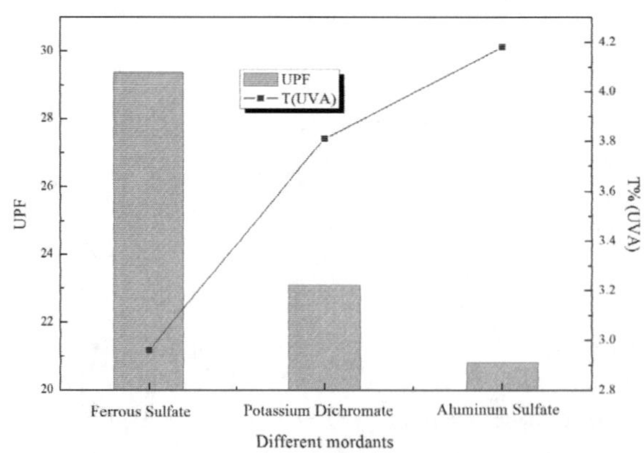

Fig.2 Ultraviolet protection performances of three samples dyed with different mordants

Fig.2 shows the ultraviolet protection performances of three samples dyed with different mordants using post-mordant dyeing. Evidently, the maximum values for UPF and minimum *T*(UVA) values were obtained when ferrous sulfate was used as mordant in the dyeing process. *T*(UVA) pertains to the transmittance in the UVA region (320–400 nm). However, all three samples dyed with palm fiber aqueous extract using ferrous sulfate, potassium dichromate, and aluminum sulfate reached or exceeded the protection category "good" according to ASTM D 6603 in Table 4. Particularly, the cotton fabrics dyed by palm fiber aqueous extract using ferrous sulfate as mordant reached the protection category of "very good". According to the increase in UPF value and decrease in *T*(UVA), the sequence of increase in the ultraviolet protection of dyed cotton fabrics using different mordants is as follows: aluminum sulfate mordanted cotton fabrics, potassium dichromate mordanted fabrics, and ferrous sulfate mordanted fabrics. Thus, we selected ferrous sulfate as the mordant when we dyed fabrics using aqueous extracts from palm fiber.

Fig.3 shows the amount of ultraviolet protection provided by the four samples. We dyed the cotton fabrics using palm fiber aqueous extract. We selected ferrous sulfate as the mordant. Different methods, including direct, pre-mordant, simultaneous, and post-mordant dyeing, were conducted under the above conditions. Fig.3 shows that all dyed samples had a significantly higher UPF value than the undyed reference sample and reached or exceeded the protection category "good" according to ASTM D 6603 in Table 4. In addition, *T*(UVA) decreased dramatically after dyeing, thereby offering better protection against harmful UV radiation than the undyed reference sample. Given that the post-mordant dyed sample had the highest UPF value and the lowest *T*(UVA) among all four dyed samples, we selected the post-mordant dyeing method to perform the experiment.

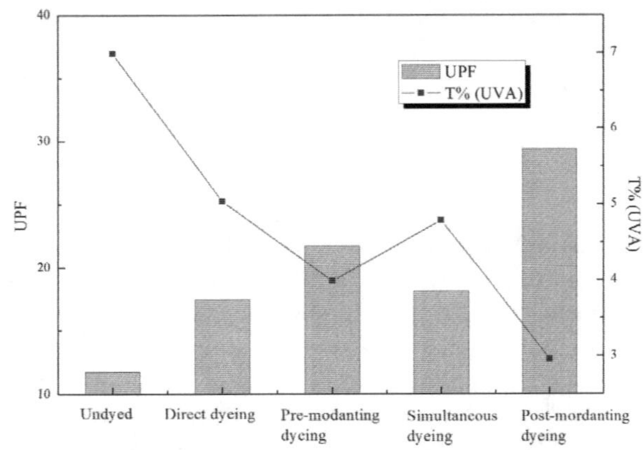

Fig.3 Ultraviolet protection performances of four dyed samples

3.3 *K/S* value of fabrics dyed with colored matter

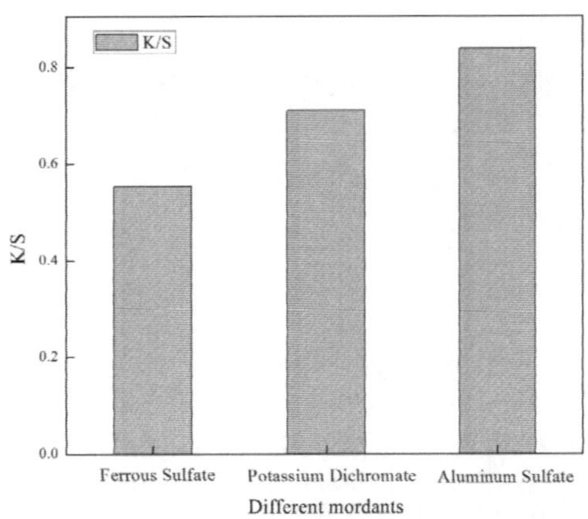

Fig.4 *K/S* values of three samples dyed with different mordants

Fig.4 shows the *K/S* values of three samples dyed by palm fiber extract with different mordants. The *K/S* values of the dyed fabrics increased in the order of ferrous sulfate, potassium dichromate, and aluminum sulfate. By combining Figures 2 and 4, we found that a deep color of the dye solution corresponds to low UV protection, which is contrary to the results of Wang et al [22]. Therefore, both the chemistry of the dye and the type of mordant determine the ultraviolet protection performance.

Fig.5 shows the *K/S* values of four samples dyed through different methods. The *K/S* values of the dyed fabrics decreased in the order of direct, pre-mordanting, simultaneous, and post-mordanting dyeing, which was inconsistent with the increased UPF values. Therefore, the ultraviolet protection performance is determined by the chemistry of the dye.

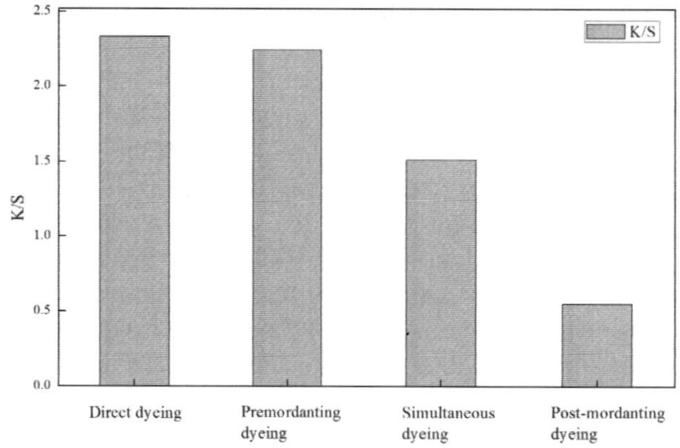

Fig.5 *K/S* values of four dyed samples

4. Conclusion

The dye solution extracted from palm fiber has excellent ultraviolet absorption even under different dilutions.

Among the three mordants utilized in this study, the post-mordanting dyed sample using ferrous sulfate resulted in the highest UPF value and possessed the lowest transmittance in the UVA region. Comparison of the four different dyeing methods employed in this study also showed that the post-mordanting dyed samples obtained the highest UPF value and the lowest transmittance in the UVA region. The increase in UPF values was inconsistent with the increase in *K/S* values. Under these conditions, the UPF value of the dyed cotton fabric was 29 (>25), and the transmittance in the UVA region, *T*(UVA), was 2.96% (<5%), which reached the "very good" category.

The aqueous extracts obtained from windmill palm fiber have the potential to provide fabrics with ultraviolet protective performance.

5. Acknowledgement

This work was supported by the Fundamental Research Funds for the Central Universities (XDJK2015D006, XDJK2013A021), National College Students Innovation and entrepreneurship training program (No.201510635047), and the Postgraduate Scientific Research Innovation Project in Chongqing, China (CYB14057).

References

[1] Hoffmann K, Laperre J, Avermaete A, et al. Defined UV protection by apparel textiles[J]. Archives of Dermatology, 2001; 137(8): 1089.

[2] Standards Australia/Standards New Zealand, AS/NZS 4399: 1996, "Sun Protective Clothing – Evaluation and Classification," Standards Australia, Homebush NSW and Wellington, and Standards New Zealand, Wellington, 1996.

[3] Davis S, Capjack L, Kerr N, et al. Clothing as protection from ultraviolet radiation: which fabric is most effective?[J]. International journal of dermatology, 1997; 36(5): 374-379.

[4] Srinivasan M, Gatewood BM. Relationship of dye characteristics to UV protection provided by cotton fabric[J]. Textile Chemist & Colorist & American Dyestuff Reporter, 2000: 32(4).

[5] Wang SQ, Kopf AW, Marx J, et al. Reduction of ultraviolet transmission through cotton T-shirt fabrics with low ultraviolet protection by various laundering methods and dyeing: clinical implications[J]. Journal of the American Academy of Dermatology, 2001; 44(5): 767-774.

[6] Crews PC, Kachman S, Beyer AG. Influences on UVR transmission of undyed woven fabrics[J]. Textile Chemist & Colorist, 1999: 31(6).

[7] Sarkar AK. An evaluation of UV protection imparted by cotton fabrics dyed with natural colorants[J]. BMC dermatology, 2004; 4(1): 1.

[8] Grifoni D, Bacci L, Di Lonardo S, et al. UV protective properties of cotton and flax fabrics dyed with multifunctional plant extracts[J]. Dyes and Pigments, 2014; 105: 89-96.

[9] Borland VS. Natural Resources: Animal and vegetable fibers for the 21st century[J]. America's Textile Industries, 2000; 29: 66-70.

[10] Glover B. Are natural colorants good for your health? Are synthetic ones better?[J]. Textile Chemist & Colorist, 1995: 27(4).

[11] Gupta D, Jain A, Panwar S. Anti-UV and anti-microbial properties of some natural dyes on cotton[J]. Indian Journal of Fibre & Textile Research, 2005; 30(2): 190-195.

[12] GUPTA D. UPF characteristics of natural dyes and textiles dyed with them[J]. Colourage, 2007; 54(4): 75-80.

[13] Kim S. Dyeing characteristics and UV protection property of green tea dyed cotton fabrics[J]. Fibers and Polymers, 2006; 7(3): 255-261.

[14] Feng XX, Zhang LL, Chen JY, et al. New insights into solar UV-protective properties of natural dye[J]. Journal of Cleaner Production, 2007; 15(4): 366-372.

[15] Iqbal J, Bhatti IA, Adeel S. Effect of UV radiation on dyeing of cotton fabric with extracts of henna leaves[J]. Indian J Fibre Text Res, 2008; 33: 157-162.

[16] Sha X, Wang L, Zhang X, et al. Research on dyeing and antimicrobial activities of vegetable dyes extracted from W-CT on wool fabric[C]//Proceedings of conference, Wool Sci. Technol. 2007; 35(1): 5-9.

[17] Hou X, Zhou Q, Zhang X, et al. Dyeing behavior of natural Flos Sophorae buds dyes and its dyeing process for wool fabrics[J]. Journal of Textile Research, 2007; 28(8): 58.

[18] Wang S, Zhang T, Li J, et al. Exploration of the Origin of the UV Absorption Performance of Windmill Palm Fiber[J]. BioResources, 2016; 11(1): 2607-2616.

[19] Zhang T, Guo M, Cheng L, et al. Investigations on the structure and properties of palm leaf sheath fiber[J]. Cellulose, 2015; 22(2): 1039-1051.

[20] Zhu, P. (2006). Functional Fiber and Textiles, China Textile and Apparel Press, Beijing, China.

[21] Cheng L, Zhang TH, Guo M, et al. Structural characteristics and properties of windmill palm leaf sheath fiber[J]. Wood Fiber Sci, 2014; 46: 270-279.

[22] Wang L, Wang N, Jia S, et al. Research on dyeing and ultraviolet protection of silk fabric using vegetable dyes extracted from Flos sophorae[J]. Textile Research Journal, 2009.

[23] Lee YH, Hwang EK, Baek YM, et al. Deodorizing function and antibacterial activity of fabrics dyed with gallnut (Galla Chinensis) extract[J]. Textile Research Journal, 2015; 85(10): 1045-1054.

Study on Ultra High Molecular Weight Polyethylene Fibers with Disperse Dyes

Jin-Long Yan[1], Xiao-Chun Wang*[1,2], Li-Ping Zhang[1], Guo-Liang Zhao[1]

[1]*Beijing Institute of Fashion Technology, Chaoyang District, Beijing, 100029, China*
[2]*Tianjin Polytechnic University, Xiqing District, Tianjin, 300387, China*

* corresponding author's email: clywxch@bift.edu.cn

Abstract

Two disperse dyes with high planarity and super hydrophobic structure are studied to dye ultra high molecular weight polyethylene (UHMWPE) fiber which has extreme hydrophobicity and high level of crystallinity. The influences of dyeing temperature, dye time, the dispersant and pH value of dyeing bath were researched. Compared with 1,4-Bis (methylanimo) anthraquinone, the results show that 4- (Diethylamino) azobenzene has good dyeability for UHMWPE fiber with high K/S value and high fastness. The optimal dyeing process was achieved at the concentration of dye 5% owf, the concentration of dispersant 0.2%, 130℃ for 60min. The K/S value reached 11.61 and the color fastness to rubbing and light was higher than rating 4.

Keywords: Ultra high molecular polyethylene fiber; Super hydrophobicity; Dispersion dye; Dyeability

1. Introduction

Ultrahigh molecular weight polyethylene (UHMWPE) fiber and carbon fiber, aramid fiber are known as the three major high-performance fiber material. UHMWPE fiber has many excellent features, such as high strength, high modulus, low specific gravity, abrasion resistance, cut resistance, light and soft. It plays an important role in many areas such as military defense, aviation and maritime, civil and other. With the expansion of its applications, the demand for colored UHMWPE fiber is increasing strongly [1,2].

The UHMWPE fiber, which contains only C-H bond and no any functional groups in molecular chain, shows super hydrophobicity and high level of crystallinity. Therefore, it has been considered that UHMWPE fiber can't be dyed with general dyes with any dyeing systems.

Based on similar dissolve mutually theory, dyes with high planarity, super hydrophobic structure and low molecular weight may have affinity toward UHMWPE fiber. In this study, therefore, in order to dye UHMWPE fiber, the dyeability of 4- (Diethylamino) azobenzene yellow disperse dye and 1,4-Bis (methylanimo) anthraquinone blue disperse dye were researched [3,4].

2. Experimental

2.1 Materials and equipment

UHMWPE fiber with 1800D/ 1380F and the melting point of 147.8℃ supplied by Sinopec Yizheng Chemical Fibre Company Limited was used. 4- (Diethylamino) azobenzene yellow dye and 1,4-Bis (methylanimo) anthraquinone blue dye supplied by Alfa Aesar were used. Primary Alcobol Ethoxylate (AEO-9) was used as dispersion agent.

SDM2-12-140 high temperature oil bath dyeing machine (Fong's National Engineering (Shenzhen)CO.,LTD) was used to dye UHMWPE fiber.

The color strength (K/S) and the light fastness of the dyed fabrics were examined respectively by ORINTEX colorimeter (Italia) and Xenon Test Chamer.

2.2 Experimental methods

2.2.1 The dispersion of dyes

The dye was dissolved in fatty alcohol polyoxyethylene ether (AEO-9). A certain amount of AEO-9 and dye were mixed in dye solution. In order to make uniform and stable dye solution, it was exposed in the ultrasonic apparatus at 50℃ for 20min.

2.2.2 Dyeing of UHMWPE fiber with the method of single factor experiment

The factors of temperature, dyeing time and the concentration of AEO-9 were discussed.

As the following dyeing curve Fig.1, the dyeing of fibers with red dye and blue dye was carried out at 5% (on weight of fabric), at 1:100 (material to liquor ratio) for particular time, at given temperature, at given pH. The dyed samples were reduction-cleared and dried in shade at room temperature.

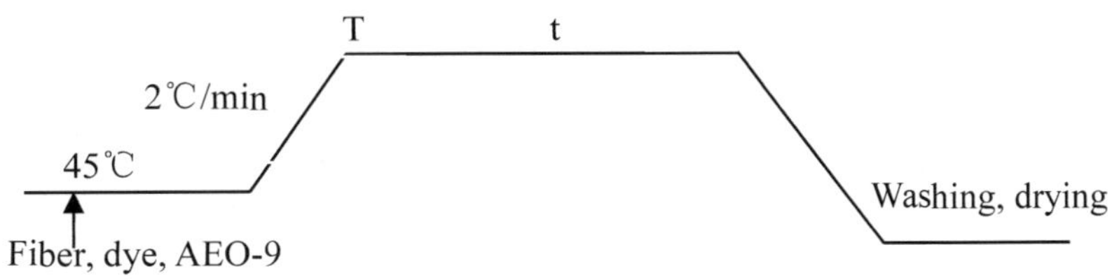

Fig.1 Dyeing curve

2.2.3 Dyeing of UHMWPE fiber with the method of Orthogonal experiment

Orthogonal experiment for optimization of the dyeing processing with 4- (Diethylamino) azobenzene yellow dye was designed as Table 1.

Table 1. The parameters of Orthogonal experiment

Experimen \ Factor	Temperature /℃	Dispersant /%	Time /min
1	130	0.2	60
2	130	0.3	90
3	125	0.2	90
4	125	0.3	60

2.2.4 Build-up property of two disperse dyes

The dyeing process was the concentration of dispersant 0.3%, 130℃ dyeing for 60min at 1:100 (material to liquor ratio).

2.2.5 Measurement of dyed fibers

The color strength (K/S) of the dyed fibers were assessed using ORINTEX colorimeter (Italia). D_{65} lamp-house and 10° visual angel were adopted.

The rubbing fastness and light fastness of the dyed samples were examined by the procedures of GB/T3920-2008 and GB/T 8426-1998, respectively.

3. Results and discussion

3.1 Effect of dyeing temperature

For the diffusivity of dispersion dye and the size of instant free volume of UHMWPE fiber, temperature is very significant. The effect of dyeing temperature was investigated firstly and the results showed in Fig.2, Fig.3 and Table 2. The dyeing process was the concentration of dye 5% owf, the concentration of dispersant 0.3%, dyeing for 60 min with pH value 5.3 at 1:100.

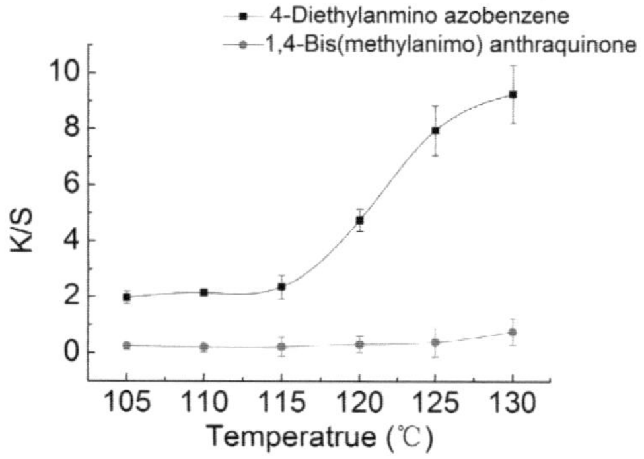

Fig.2 Effect of dyeing temperature on K/S

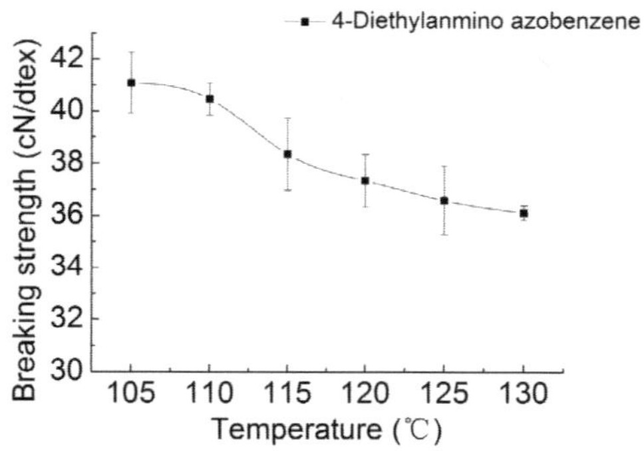

Fig.3 Effect of temperature on breaking strength

Table 2. Effect of dyeing temperature to fastness of dyed fiber

Temperature℃	130		125		115		110		105	
Rubbing	Dry	Wet	Dry	Wet	Dry	Wet	Dry	Wet	Dry	Wet
4-(Diethylamino) azobenzene	4	4-5	3-4	4	2-3	3	4	3-4	2-3	3
1,4-Bis(methylanimo) anthraquinone	2	2-3	2	2-3	2	3	2-3	4	2	3

Although both 4- (Diethylamino) azobenzene yellow dye and 1,4-Bis (methylanimo) anthraquinone blue dye are free from -OH, $-NH_2$ polar groups and have good structure planarity with simple molecular structure. As shown in Fig.2 and Table 2, the color strength and fastness of UHMWPE fiber dyed by 4- (Diethylamino) azobenzene yellow dye was obviously higher than 1,4-Bis (methylanimo) anthraquinone blue dye. It revealed that the compatibility and affinity between the monoazo yellow dye and UHMWPE fiber were better than those of the blue anthraquinone dye.

From Fig.2, when the dyeing temperature exceeded more than 115℃, the K/S values of 4- (Diethylamino) azobenzene yellow dye increased remarkably,but no significant change in 1,4-Bis (methylanimo) anthraquinone. This is attributed to that, only in high temperature, the macromolecular segments of UHMWPE fiber moves violently and forms a larger free volume instantly, which is beneficial to reduce the resistance of the dye diffusion. Meanwhile, high temperature makes diffusion kinetic energy and solubility of dye increase dramatically. However, the dyeing temperature need to be decided very prudently since the heat resist property of UHMWPE fiber is not enough. When the temperature rised, the thermal motion of the molecular chain and the relaxation process accelerated, and the modulus and strength decreased [5]. Fig.3 showed the breaking strength loss of fiber by high temperature.

Table 2 showed that the rubbing fastness of UHMWPE fiber dyed in low temperature was poor relatively. It indicated that more surface adsorption of dye occurred in low dyeing temperature. Moreover, compared with 4-(Diethylamino) azobenzene yellow dye, rubbing fastness of UHMWPE fiber dyed with anthraquinone blue dye was very poor.

3.2 Effect of dispersant

The effect of the concentration of dispersant on color strength was discussed, the results showed in Fig.4. The dyeing process was the concentration of dye 5% owf, 130℃ dyeing for 60min with pH value 5.3 at 1:100.

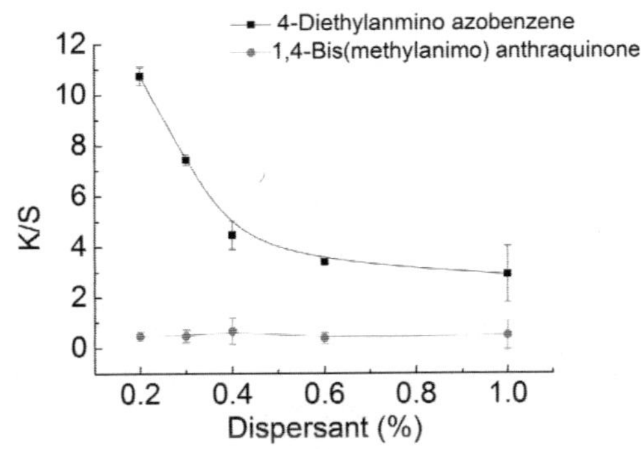

Fig.4 Effect of the concentration of dispersant on K/S value

Because of small affinity between 1,4-Bis (methylanimo) anthraquinone with fiber, the influence of dispersant on 1,4-Bis (methylanimo) anthraquinone showed in Fig.4 was not obvious. However, with the concentration of dispersant increased, the color strength of fiber dyed with 4- (Diethylamino) azobenzene yellow dye decreased continuously. This was attributed to that the interaction between the dispersant and dye would impede the diffusion of dye on to fiber. It also attributed to that AEO-9 played a spreading agent which would occupy the free volume space of UHMWPE fiber and reduced the dye uptake on balance. Therefore, the concentration of dispersant AEO-9 0.2% is appropriate.

3.3 Effect of dyeing time

Dyeing time is the important factors to color strength, fastness and retention of mechanical properties of the fiber and the results showed in Fig.5 and Table 3. The dyeing process was the concentration of dye 5% owf, the concentration of dispersant 0.3%,130℃ dyeing with pH value 5.3 at 1:100.

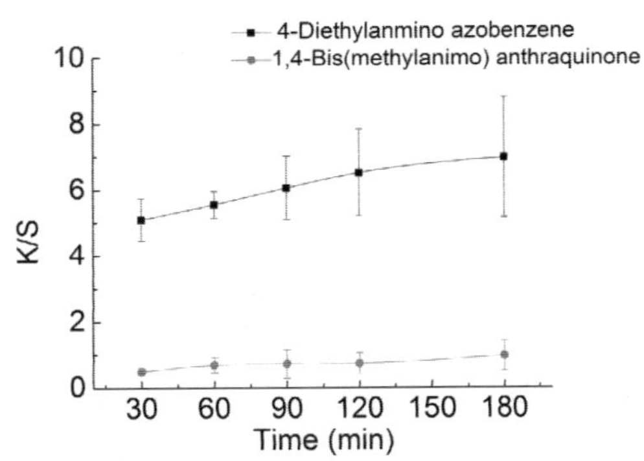

Fig.5 Effect of dyeing time on K/S value

From Fig.5 and Table 3, with the dyeing time increasing, the color strength of fiber dyed with 4- (Diethylamino) azobenzene gradually increased and the rubbing fastnesses was very good showing higher than rating 4. This is due to extending the time in favor of dye adsorption, dye diffusion and dye penetrability.

Due to the obvious damage on breaking strength of UHMWPE fiber by high temperature, it is important to note that the dyeing time could not be prolong more than 90 min for protecting mechanical property of fiber.

Table 3. Effect of time to fastness of dyed fiber

Time/min	30		60		90		120		180	
Rubbing	Dry	Wet	Dry	Wet	Dry	Wet	Dry	Wet	Dry	Wet
4-(Diethylamino) azobenzene	4	4-5	4-5	4-5	4	4-5	4	4-5	4-5	4-5
1,4-Bis(methylanimo) anthraquinone	2	2-3	2	3	2	3	2	3	2	3

3.4 Effect of pH

When the dye bath is at high temperature, some dyes may be damaged by pH value of dye bath. It would lead to reducing of uptake rate, causing color changes in hue, poor dyeing reproducibility. Therefore, the influence of dye bath pH value was discussed and the results showed in Fig.6 and Table 4. The dyeing process was the concentration of dye 5% owf, the concentration of dispersant 0.3%,130℃ dyeing for 60 min at 1:100.

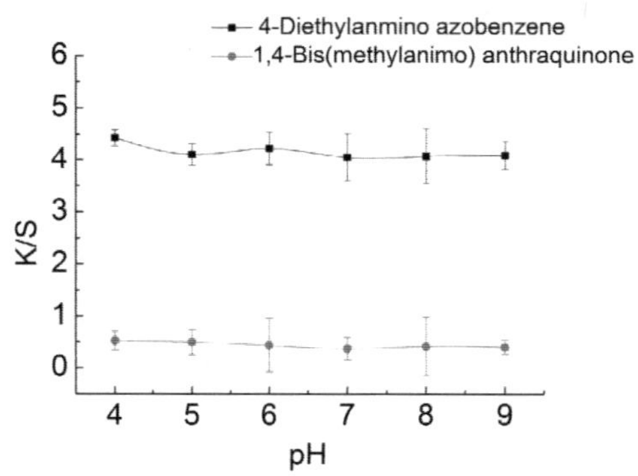

Fig.6 Effect of pH on K/S value

Table 4. Effect of pH to fastness of dyed fiber

pH value	4		5		6		7		8		9	
Rubbing	Dry	Wet	Dry	Wet	Dry	Wet	Dry	Wet	Dry	Wet	Dry	Wet
4-(Diethylamino) azobenzene	4	4	4	4-5	4	4	4-5	4	4	4-5	4	4
1,4-Bis(methylanimo) anthraquinone	2	2-3	2	2-3	2	3	2-3	3	2	3	2-3	3

As shown in Fig.6 and Table 4, when the pH value of dye bath was increasing from 4 to 9, the color strength and rubbing fastness of fiber dyed with 4- (Diethylamino) azobenzene yellow dye were stable. It indicated that 4-(Diethylamino) azobenzene dye had good acid and alkali resistance, and had a wide adaptability for dye bath pH value.

3.5 Multi-factor orthogonal experiment

On the basis of the single factor, orthogonal experiment for the optimization of the dyeing process was researched and evaluated by color strength, rubbing fastness, light fastness and mechanical property retention of UHMWPE fibers. The results showed in Table 5.

Table 5. Results of orthogonal experiment

Experiment	Results							
Dyes	No.	K/S	Rubbing		Light Fastness	Breaking strength (cN/dtex)	Initial modulus (cN/dtex)
			Dry	Wet			
4-(Diethylamino) azobenzene	1	7.583	4	4-5	3-4	35.39	257.21
	2	11.223	4	4-5	4-5	43.39	367.45
	3	5.261	4-5	4-5	4	42.96	375.71
	4	11.610	4	4-5	4-5	39.26	332.39
1,4-Bis(methylanimo) anthraquinone	1	0.581	2	3	3	37.70	302.57
	2	0.707	2	3	3	44.10	349.97
	3	0.539	2	3	3	39.18	353.56
	4	0.751	2	3	3	36.52	289.19

Table 6. Analysis of orthogonal experiment results

Dyes	Factors	Analysis	Temperature/℃	Time/min	Dispersant/%
4-(Diethylamino) azobenzene	K/S	Mean1	9.596	9.403	6.422
		Mean 2	8.242	8.435	11.416
		Range	1.354	0.968	4.994
	Breaking strength (cN/dtex)	Mean 1	37.325	39.392	39.177
		Mean 2	43.174	41.107	41.322
		Range	5.849	1.715	2.145
1,4-Bis(methylanimo) anthraquinone	K/S	Mean 1	0.666	0.644	0.560
		Mean 2	0.623	0.645	0.729
		Range	0.043	0.001	0.169
	Breaking strength (cN/dtex)	Mean 1	37.110	40.899	38.439
		Mean 2	41.643	37.854	40.314
		Range	4.533	3.045	1.875

Table 6 showed that the concentration of dispersant and the dyeing temperature were significantly factors for dyeing. The 4- (Diethylamino)azobenzene yellow dye has good dyeability and good fastness for UHMWPE fiber, and the optimum dyeing procedure is carried out at 130℃ for 60 minutes with 0.3% AEO-9.

1,4-Bis (methylanimo)anthraquinone blue dye showed poor dyeability and poor fastness for UHMWPE fiber. It indicated that 1,4-Bis (methylanimo)anthraquinone was not suit to dye UHMWPE fiber.

3.6 Build-up property of two disperse dye

From Fig.7, as the concentration of dyes increased, the color strength of fiber dyed with 4- (Diethylamino) azobenzene increased continuously. When the concentration of dye up to 10% owf, the maximum was still not occur. Therefor, 4- (Diethylamino) azobenzene showed good build-up property for UHMWPE fiber.

Meanwhile, due to the low affinity between 1,4-Bis (methylanimo)anthraquinone blue dye and UHMWPE fiber, the increasing concentration of dye was not favor of the build-up of color strength.

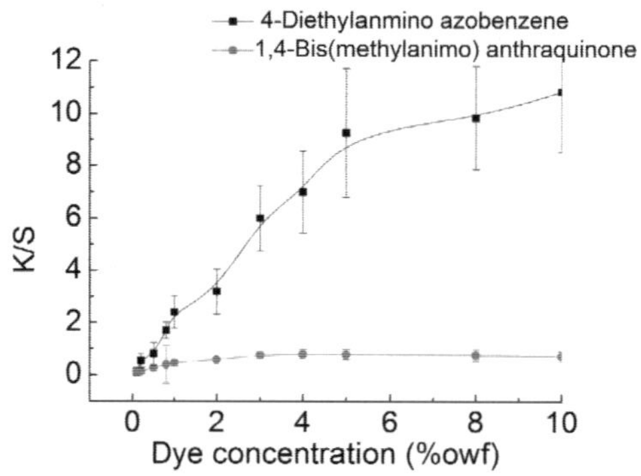

Fig.7 Build-up property of two disperse dye

4. Conclusion

The 4- (Diethylamino) azobenzene yellow dye shows good dyeability and good fastness for UHMWPE fiber, and the optimum dyeing procedure is carried out at 130℃ for 60 minutes with 0.3% AEO-9. Therefore, the dyeing of UHMWPE fiber with the single-azo dye with super hydrophobicity is possible.

Dye to the poor dyeability of UHMWPE fiber with 1,4-Bis (methylanimo) anthraquinone, 1,4-Bis (methylanimo) anthraquinone blue dye dose not suit to dye UHMWPE fiber.

5. Acknowledgement

The author wishes to acknowledge support for this work via two projects “Study on dyeing technology of high performance polyethylene made by melt spinning” (2015A-02) and “Graduate Innovative Projects” (120301990122) under Beijing Institute of Fashion Technology.

References

[1] Yuan JX. Property and modification of UHMWPE and its application. NEW CHEMICAL MATERIALS: 2003; 31(3): 19-21.
[2] Wang JM. Ultra-high Molecular Weight Polyethylene Fiber Industry Status and Market Prospect. CHEMICAL INDUSTRY: 2014; 08: 32-38.
[3] Taekyeong K, Seonhee J, Dongsup K, Yuri C. Coloration of Ultra High Molecular Weight Polyethylene Fibers Using Alkyl-substituted Anthraquinoid Blue Dyes. Fibers and Polymers: 2012; 13(2): 212-216.
[4] Taekyeong Kim, Seonhee Jeon. Coloration of Ultra High Molecular Weight Polyethylene Fibers Using Alkyl-substituted Monoazo Yellow and Red Dyes. Fibers and Polymers: 2013; 14(1): 105-109.
[5] Wu QY, Zhang P, Yang WJ, Lin RX. Polymer Physics. Beijing: HIGHER EDUCATION PRESS: 2011: 225-227.

Density Measurement for Yarn Dyed Woven Fabrics based on Wavelet Transform Fusion Method

Rui Zhang, Bin-Jie Xin*

College of Fashion, Shanghai University of Engineering Science, Shanghai, 201620, China

Corresponding author's email: xinbj@sues.edu.cn

Abstract

The yarn density is always considered as the fundamental structural parameter used for the quality evaluation of woven fabrics. Conventional yarn density measurement method is based on one-side analysis. In this paper, a novel density measurement method is developed for yarn dyed woven fabrics based on dual-side fusion technique. Firstly, a lab-used dual-side imaging system is established to acquire both face-side and back-side image of woven fabric and the affine transform is used for the alignment and fusion of dual-side images. Then, the color images of woven fabrics are transferred from RGB to CIE-Lab color space , the intensity information of image extracted from L component is used for the texture fusion and analysis. Subsequently, the Wavelet transform fusion method is developed and utilized to merge the dual-side images. Finally, yarn density of the fused image is measured based on fast Fourier transform, the yarn alignment image could be reconstructed using inverse fast Fourier transform. Our experimental results show that the accuracy of density measurement by using proposed method is close to 99.44% compared with traditional method and the robustness of this new proposed method is better than that of conventional analysis methods.

Keywords: Density measurement; Yarn dyed woven fabric; Image fusion; Image processing; Dual-side scanning

1. Introduction

The yarn density is an important parameter of woven fabric and plays a decisive role for the fabric appearance evaluation and the determination of its physical properties, so fabric density measurement is an indispensable part in textile quality testing. The traditional method of yarn density is manually operated in the textile laboratory based on human eyes with the aid of pins. Although it has the advantages of easy and practical, this method suffers the time consuming and inefficient, and it is also subjectively affected by the knowledge and experience of inspectors. With the rapid development of image processing and artificial intelligence, textile researchers have already presented many digital methods to measure density of woven fabrics automatically.

In 1990, Wood employed the Fourier transform and autocorrelation function to model the spatial periodicity of yarn structure and fabric pattern [1]. Later on, in 1992, Ravandi S. A. Hosseini used Fourier transform to analyze appearance of plain weave fabric [2]. In Xin's researches, The peak filtering was utilized to remove random frequency points in power spectrum after fast Fourier transform (FFT), the selected peak points represent periodicity information of woven fabric [3]. In 2001, one set of fabric structure analysis system was developed based on the Fourier spectrum by Jaume Escofet and his colleagues. The minimum structural unit of fabric texture was extracted through convolution and template matching. The frequency points corresponding to yarn arrangement were identified and located to determine the density of warp and weft yarns [4]. Similar methods have been reported in many references [5-9]. In addition to Fourier transform, the wavelet transform is another effective method to measure yarn density. In 2001, Feng and Li employed to perform a multi-scale decomposition and measure the woven fabric density [10]. In 2007, Shen recommended a novel approach based on Wavelet and Radon transform for fabric weave pattern and structural pattern detection. It was reported that fabric texture directions, weave pattern and yarn density could be determined through proposed method [11]. In 2015, Jing proposed a method combining the multi-scale wavelet transform and morphology processing, via successive operations of dilation and erosion, to detect the density of the warp and weft of woven fabric automatically [12]. In same year, Kang utilized Gabor wavelet and graylevel co-occurrence matrix (GLCM) to extract the texture features of fabric images. Then Probabilistic Neural Network (PNN) was applied to classify the three basic woven fabrics [13].

Although the methods based on spectrum analysis were effective for those fabrics with regular textures, it seems that irregular texture has effects on the performance of the spectrum based methods. The methods based on spatial domain analysis usually could be defined as those methods using statistical or geometrical analysis to characterize the fabric image in spatial domain, extracting the geometrical or statistical parameter of fabric structure. In 1991, Nishimatsu utilized a charge coupled device (CCD) camera and a frame grabber board to digitalize the image of fabric at 50x magnification. It was reported that the recognition accuracy of this method for plain fabrics was as high as 92.5%, while the accuracy for twill fabrics was only 79.1% [14]. In 2002, Gao and his colleagues proposed a correlation function-based approach to calculate the weft and warp repeats [15]. Lin J.J. attempted to use the Gray-level Co-occurrence Matrix to measure yarn density, and the author reported that this method was effective for the density measurement of twill fabrics [16], however, no related works reported for those fabrics with complex textures.

From the previous works described above, it is obvious that both spectrum based methods and space domain based methods have its advantages and disadvantages for the density measurement and texture analysis due to the diversity of woven fabrics. It was found that one-side imaging system and the corresponding algorithms was used in most of previous references. However, the yarn information acquired only from one side is not complete and insufficient for the texture analysis, because there are some interlaced yarns can not be observed from the face-side surface or the back-side surface, it means only parts of one single yarn is visible from one side. In this case, it is reasonable to merge all the image information to improve the accuracy of image analysis.

In this paper, a new dual-side modeling and measurement method for the yarn density based on the dual-side scanning and fusion technique is proposed to digitize more complete information about the yarn structure or fabric textures, the related algorithms for the image alignment and fusion are developed and utilize to determine the yarn density.

2. Methodology

2.1 Dual-side image acquisition system set-up

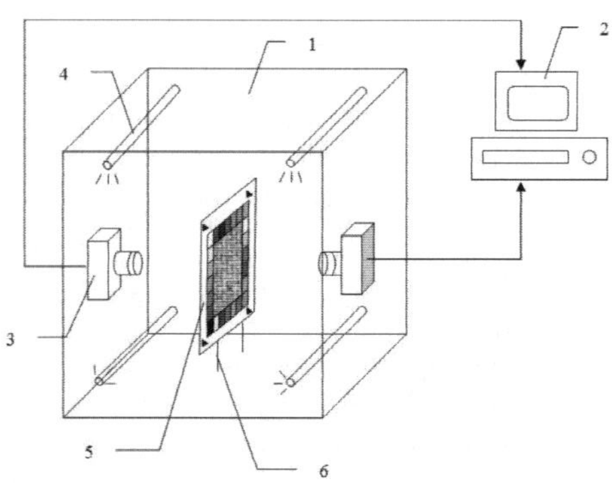

1——Dark box ;2——Computer ;3——CCD cameras ;
4——Illumination ;

Fig.1 Overview of the dual-side imaging system

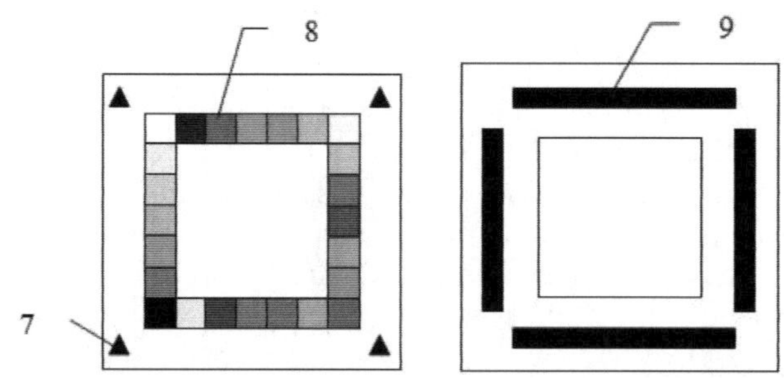

7——Reference marker ;
8——Color blocks

9——Rectangle magnets

Fig.2 The diagram of sample holder

In this research, a new imaging system is set up to acquire dual-side reflective image of fabric samples as illustrated in Fig.1. It consists of five components: 1) Two high resolution CCD cameras, which are used to capture the dual-side reflected images of yarn dyed woven fabric. 2) A self-developed sample holder as shown in Fig.2, which consists of four rectangle magnets and 24 squares of color blocks, can not only used to grid the sample with one pair of plates, but also do color calibration with the color checker. Four black triangular markers have been designed and fixed on each plate, located at the four corners of the sampling windows, which are used as reference points for the alignment and calibration of the face-side and back-side images. 3) Four standard light sources which are put symmetrically on the left and right sides of sample holder, are used to generate bi-directional illumination of both face-side and back-side plates. 4) A closed dark box, which is used to reduce influence of external light in process of image acquisition. 5) A personal computer with software, which is employed to process and analyze the captured images.

2.2 Density measurement algorithms for yarn dyed woven fabrics

Once the dual-side images of woven fabric have been captured, the yarn density can be calculated through the process illustrated in Fig.3. The analysis algorithm mainly consists of four steps: 1) image alignment; 2) color removal; 3) dual-side image fusion; 4) density analysis. The image alignment is used to match face-side image with back-side image at pixel level based on affine transform. The color removal is used to eliminate the color information and only remain the luminance information of fabric images based on color space conversion. The Wavelet transform fusion method is used to merge the face-side image and the back-side image. The density analysis is used to calculate the yarn density eventually based on Fast Fourier Transform and Inverse Fast Fourier Transform (IFFT).

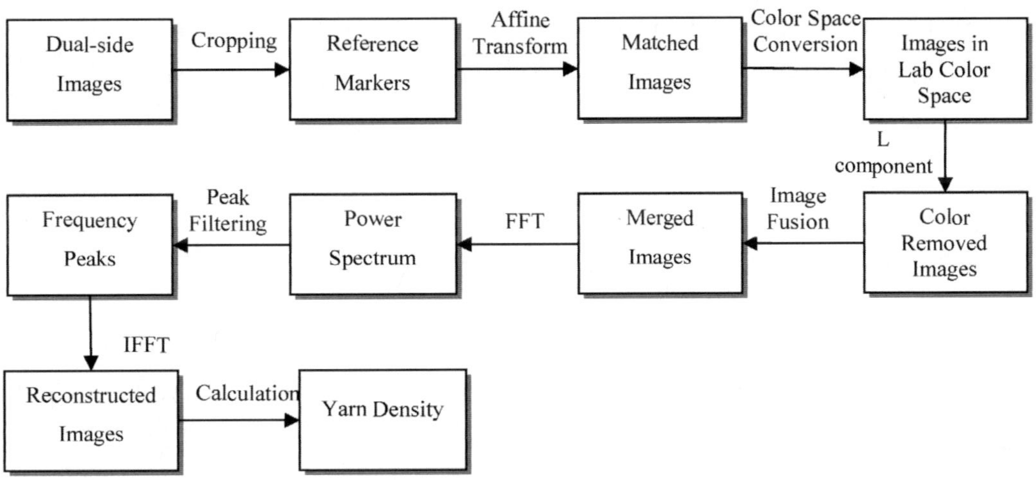

Fig.3 Flowchart of the dual-side image analysis algorithm

2.2.1 Image Alignment

Once the dual-side images have been digitalized, the sub-image of the reference markers could be cropped for the positioning. The face-side and back-side images could be corresponded with each other accurately by using affine transform [17]. The edge of the triangle can be extracted after image segmentation and Sobel edge detection. the Radon transform was adopted to achieve the equations of three lines: $y = kx + b$. The coordinates of the three vertices of the triangle can be computed respectively. Then, the center of gravity of the triangle can be determined and the location of the reference markers of the dual-side images can be obtained. Ultimately, the images inside the sampling region could be selected for further yarn density analysis.

2.2.2 Color removal

Conventional image captured by digital cameras could be expressed in three color value: red, green and blue, which is called RGB color space. However, the periodical structure information of yarn dyed fabric could be affected by the color information of the yarn. In order to eliminate the influence of color information, the original images could be transferred from RGB color space to Lab color space, the luminance information (L component) of image is selected for the image fusion and analysis.

To accomplish the conversion from RGB color space to Lab color space, the original fabric images should be transferred to XYZ color space firstly. The specific conversion procedures [18] are described as following:

$$r = R/255 \tag{1}$$

$$\text{if } (r > 0.04045) \text{ then } r = \left(\frac{r+0.055}{1.055}\right)^{2.4}, \text{ else } r = \frac{r}{12.92} \tag{2}$$

Similarly, g and b can be calculated from the G and B components in the same way.

$$\begin{bmatrix} X \\ Y \\ Z \end{bmatrix} = \begin{bmatrix} 0.4125 & 0.3576 & 0.1805 \\ 0.2126 & 0.7152 & 0.0722 \\ 0.0193 & 0.1192 & 0.9505 \end{bmatrix} \bullet \begin{bmatrix} r \\ g \\ b \end{bmatrix} \tag{3}$$

Then, the image can be converted to Lab color space from RGB color space by the aid of XYZ color space.

$$\text{if } (X > 0.008856) \quad x = X^{\frac{1}{3}} \tag{4}$$

$$\text{else } x = 7.787 \times X + \frac{16}{116} \tag{5}$$

Analogously, y and z can be calculated from the Y and Z components using the above formula.

$$\begin{cases} L = 116 \times y - 16 \\ a = 500 \times (x - y) \\ b = 200 \times (y - z) \end{cases} \tag{6}$$

The color space could be converted eventually through above steps, and the L values could range from 0 to 100, the a, b values could range from -128 to +128. Subsequently, the L component is extracted to do further density analysis, and a component and b component can be used for color texture analysis.

2.2.3 Dual-side image fusion

Once the color space conversion has been accomplished, dual-side image is merged based on Wavelet transform fusion methods to enhance the periodical structure information of fabric image After that, the fused images could be used for density measurement.

Wavelet transform is the inheritance and development of traditional Fourier transform and is defined as follows [19]:

$$W_a(a,b) = \int_{-\infty}^{+\infty} f(t)\Psi_{a,b}(t)dt = \int_{-\infty}^{+\infty} f(t)a^{-\frac{1}{2}}\Psi(\frac{t-b}{a})(t)dt \tag{7}$$

where a is the dilation factor, b is the translation vector. Wavelet transform can be divided into Continuous Wavelet Transform (CWT) and Discrete Wavelet Transform (DWT). Because the digital signal of image is discrete, DWT is adopted. DWT coefficient can be described in equation (8):

$$C_{a,b} = \int_{-\infty}^{+\infty} f(t)\Psi_{a,b}(t)dt \tag{8}$$

while the reconstruction function can be shown as following:

$$f(t) = c\sum_{-\infty}^{+\infty}\sum_{-\infty}^{+\infty} C_{a,b}\Psi_{a,b}(t) \tag{9}$$

A fabric image at level s can be decomposed into four components: the approximation at level s+1 and the details in three orientations (horizontal, vertical, and diagonal) via two-dimension (2D) DWT. The same decomposition could be done for the low frequency image and the same results are obtained, which can continue unlimitedly in theory.

In our research, the value of s is 2. For s=2, fabric images could get the tree of data structure as shown in Fig. 4 via two scales wavelet decomposition. S is the input fabric image. SA is the low frequency, SH, SV and SD are horizontal high frequency, vertical high frequency and diagonal direction high frequency respectively, which are the results from the first decomposition. SSA, SSH, SSV and SSD are the consequence from the second decomposition.

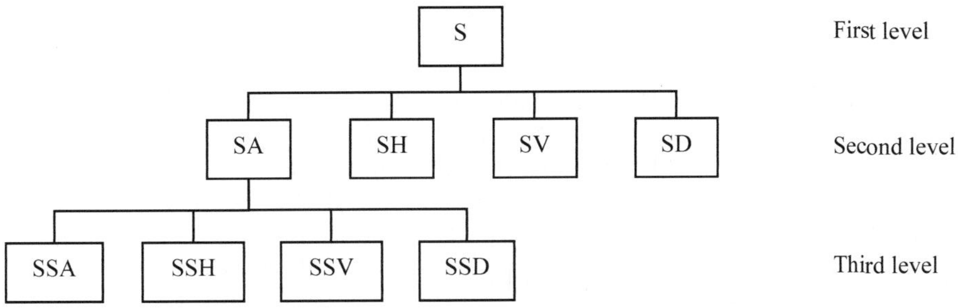

Fig.4 Wavelet decomposition image structure ($s = 2$)

Supposing $S1(m,n)$ and $S2(m,n)$ represent the color removal image of face-side and back-side respectively, $S(m,n)$ represent the fused image. The specific procedures of image fusion based on Wavelet Transform method are described as following:

1) the two-scales DWT is utilized to decomposed original dual-side fabric image, and the data structure tree of S1 and S2 can be obtained.

2) extract the horizontal high frequency of S1 at third level (SSH1) and corresponding frequency of S2 (SSH2). Then choose each pixel as the center, calculate the variance value of 9 pixels in 3*3 area. Supposing D_{SSH1} and D_{SSH2} represent the variance value calculated from SSH1and SSH2 respectively, the horizontal high frequency of at third level (SSH) can be determined by following formula:

$$SSH(m,n) = \begin{cases} SSH1(m,n), & D_{SSH1}(m,n) \geq D_{SSH2}(m,n) \\ SSH2(m,n), & D_{SSH1}(m,n) < D_{SSH2}(m,n) \end{cases} \tag{10}$$

similarly, SSV and SSD can be determined from SSV1 and SSV2, SSD1 and SSD2 in the same way.

3) the low frequency of S at second level (SA) can be reconstructed through Inverse Discrete Wavelet Transform (IDWT) based on SSA, SSH, SSV and SSD.

4) similar method with step2 is used to calculate the horizontal high frequency (SH), vertical high frequency (SV) and diagonal direction high frequency (SD) at second level.

5) the fusion image (S) can be obtained through IDWT based on SA, SH, SV, SD.

2.2.4 Density analysis

Once the fused images of woven fabrics have been acquired, subsequently, the Fourier transform and its inverse transform are utilized to analyze the fabric density since the woven fabric images contain periodical structure.

Supposing two-dimensional function $f(x,y)$ represent the fused image, then the Fourier transform and the inverse Fourier transform could be described as following equations [20]:

$$F(u,v) = \int_{-\infty}^{+\infty}\int_{-\infty}^{+\infty} f(x,y)e^{-i2\pi(ux+vy)}dxdy \tag{11}$$

$$f(x,y) = \int_{-\infty}^{+\infty}\int_{-\infty}^{+\infty} F(u,v)e^{i2\pi(ux+vy)}dudv \tag{12}$$

where u and v are frequency variables, $u = 0,1,\cdots,M-1$, $v = 0,1,\cdots,N-1$. Since the digital signal of image is discrete, the Discrete Fourier transform (DWT) is adopted. The DWT and its inverse are defined as following:

$$F(u,v) = \frac{1}{MN}\sum_{x=0}^{M-1}\sum_{y=0}^{N-1} f(x,y)\exp\left[-i2\pi\left(\frac{ux}{M}+\frac{vy}{N}\right)\right] \tag{13}$$

$$f(x,y)=\sum_{u=0}^{M-1}\sum_{v=0}^{N-1}F(u,v)\exp\left[i2\pi\left(\frac{ux}{M}+\frac{vy}{N}\right)\right] \tag{14}$$

where $u=0,1,\cdots,M-1$, $v=0,1,\cdots,N-1$. The calculation of DFT is relatively slow especially when the image data is huge. Therefore, the fast Fourier transform (FFT) has been developed to reduce the calculating cost.

The woven fabric image can be transformed to frequency domain by using FFT technique, and correspondingly power spectrum can be generated. In power spectrum of woven fabric, periodic structures information could be represented by peak points. Furthermore, two brightest peak points located on the principal horizontal line, which passes through the origin, correspond to the warp yarn repeat. Similarly, two brightest peak points located on the principal vertical line, which also passes through the origin, correspond to the weft yarn repeat [17]. Hence, images could be reconstructed by the inverse fast Fourier transform (IFFT) of the two selected peak points corresponding to the warp yarn repeat and weft yarn repeat respectively. Eventually, the number of warp yarns and the number of weft yarns can be calculated based on the two reconstructed images.

The density of the warp yarns and the weft yarns can be calculated from following equations:

$$\begin{cases}\dfrac{R}{1}=\dfrac{x}{s_w}\\[2ex]\dfrac{R}{1}=\dfrac{y}{s_h}\end{cases}\Rightarrow\begin{cases}s_w=\dfrac{x}{R}\\[2ex]s_h=\dfrac{y}{R}\end{cases} \tag{15}$$

$$\begin{cases}d_j=\dfrac{n_j}{s_w}\\[2ex]d_w=\dfrac{n_w}{s_h}\end{cases} \tag{16}$$

Where, R is the image resolution in dpi; s_w is the width of the original image (unit: inch) and s_h is the height of the original image. x is the size of original image in width direction and y is the size of original image in height direction. d_j and d_w are the density of the warp yarns and the weft yarns, respectively.

3. Experiment Results and Discussion

3.1 Sample preparation and image acquisition

The dual-side imaging system was employed to acquire the images of yarn dyed woven fabrics. The image resolution to the physical fabric size in our experiment is around 1200 pixels per inch. Subsequently, the affine transform was used for image alignment and 20 dual-side images of the yarn dyed woven fabrics have been obtained as illustrated in Fig.5. The size of matched images is 512 pixels in width direction and 512 pixels in height direction.

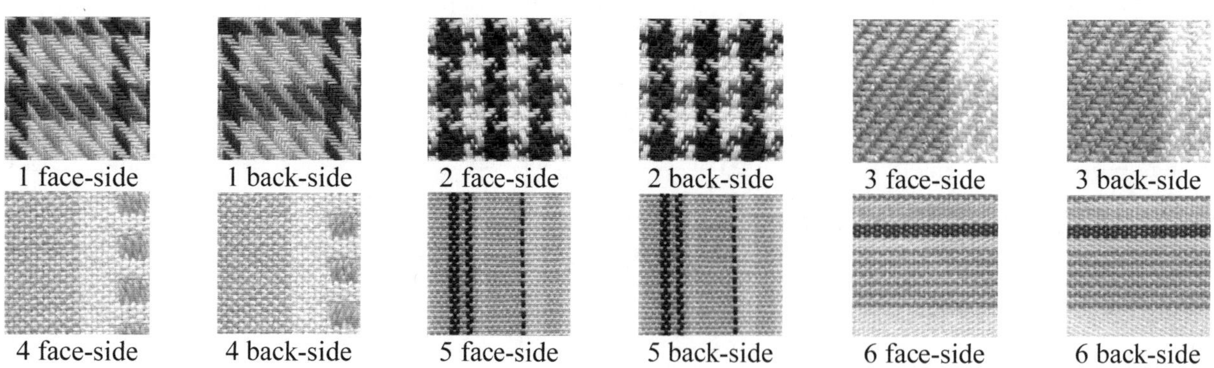

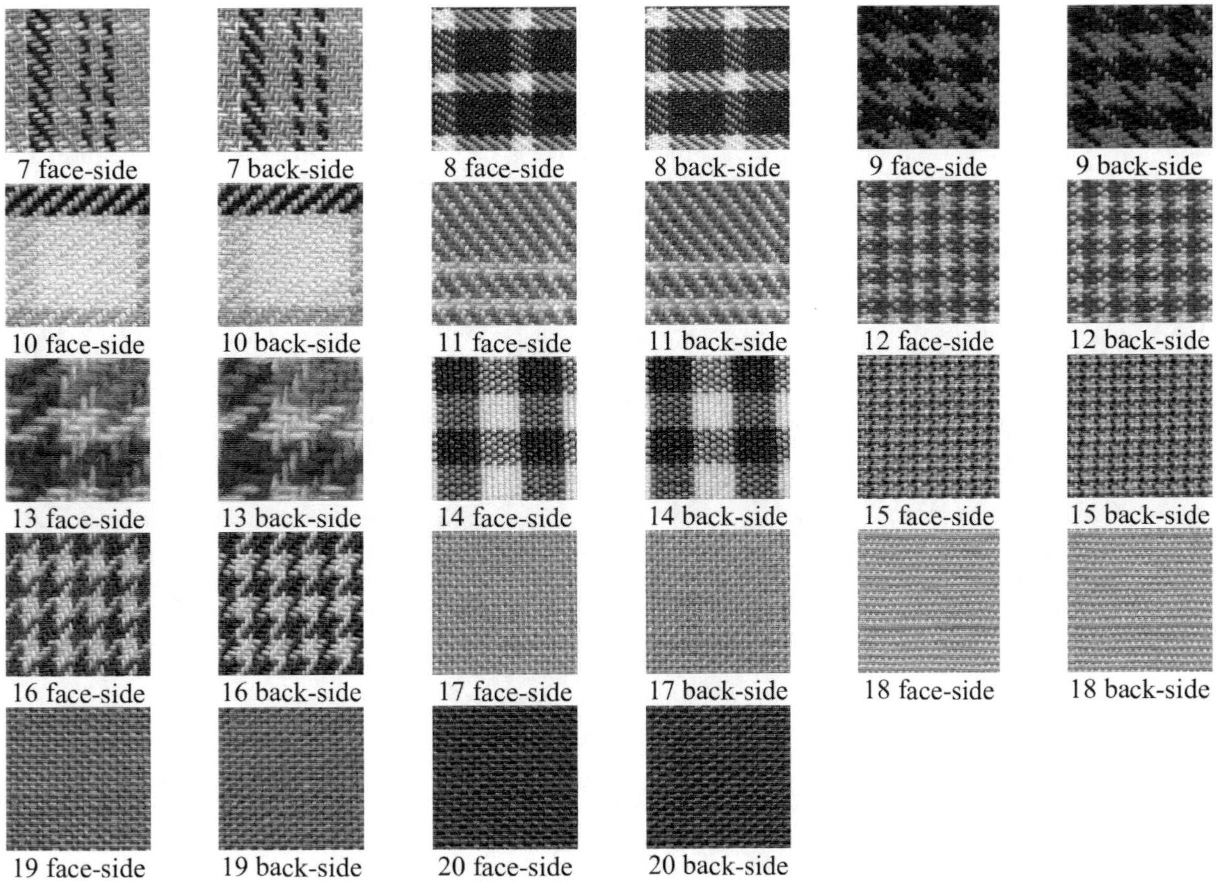

Fig.5 Matched dual-side images of 20 samples

3.2 Dual-side image fusion

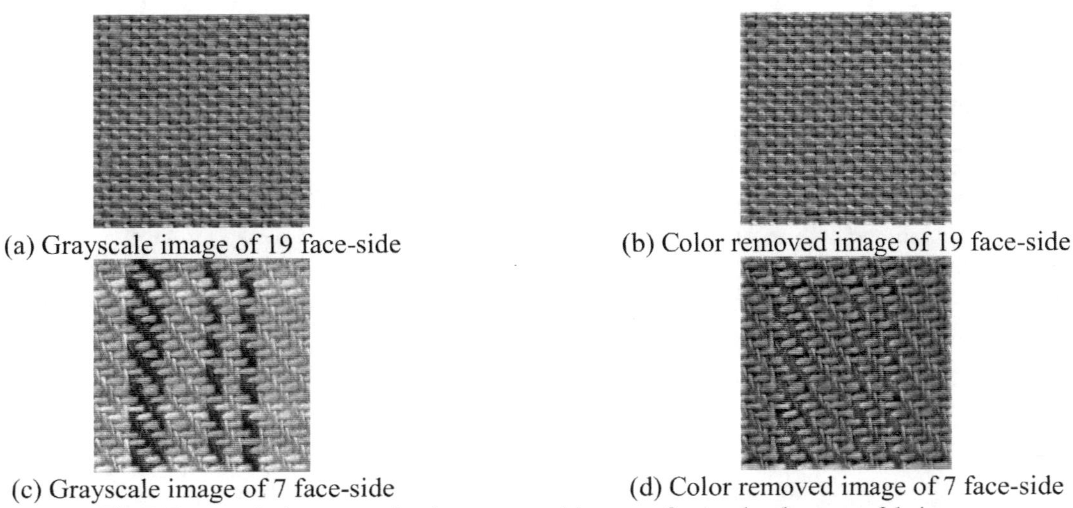

Fig.6 Grayscale image and color removed image of yarn dyed woven fabrics

Once the dual-side images have been matched, the color space of them was transferred to CIE-Lab color space from RGB color space. Then, only luminance information (L component) of image was remained, it is found that in single colored woven fabrics, the color removed images are same with grayscale images transformed from RGB color space. However, in multicolored woven fabrics, the color removed images have more concentrate gray levels which different with them of grayscale images transformed from RGB color space as shown in Fig.6. Therefore, the color removed images are more suitable for periodical structure information analysis of yarn dyed woven fabrics.

The color removed dual-side image of 20 samples have been fused based on the Wavelet transform fusion method. For example, the dual-fused images of sample 3 are exhibited in Fig.7. It is found that the discontinued yarn floats, both warp yarns and weft yarns, in one-side image could be merged continuously to form a complete texture of the yarn alignment in dual-side fusion image. In other words, the yarn repeat could be enhanced in dual-side fusion image.

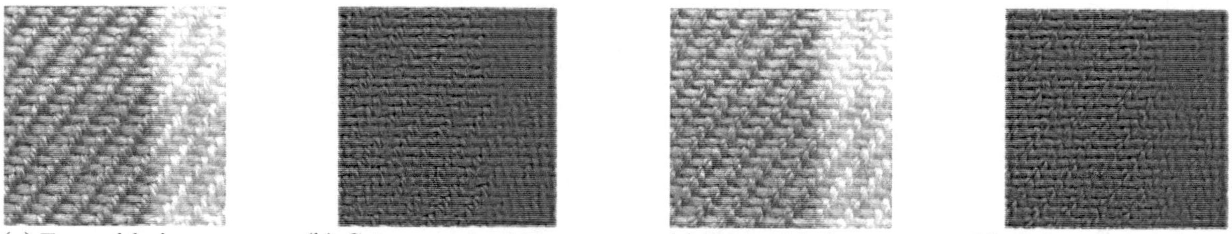

(a) Face-side image (b) Color removed image of (a) (c) Back-side image (d) Color removed image of (c)

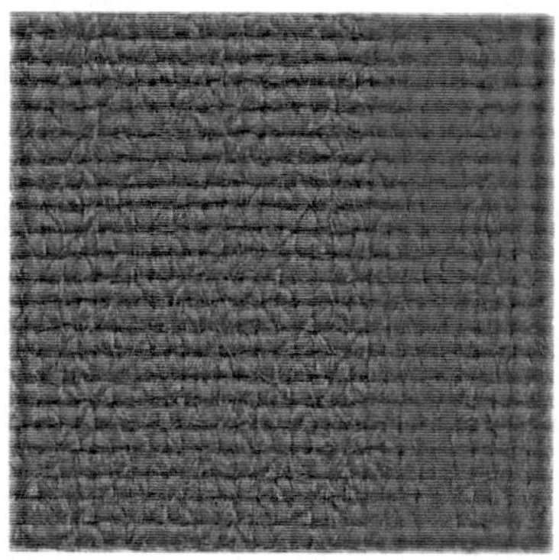

(e) Dual-fused image of sample 3

Fig.7 The dual-fused images of sample 3

3.3 Density measurement

3.3.1 Based on traditional method

Based on the traditional method, although the yarn density of single colored woven fabrics can be accurately computed, some errors may still occur when applying the FFT techniques in multicolored woven fabrics as shown in Fig.8. Since the color periodical information is blended with the yarn repeat inside the fabric image, the peak points in power spectrum generated by FFT techniques may not only corresponding to the yarn repeat, but also contained the color periodical information. Therefore, the yarn density calculated form reconstructed image based on IFFT techniques may not accurate.

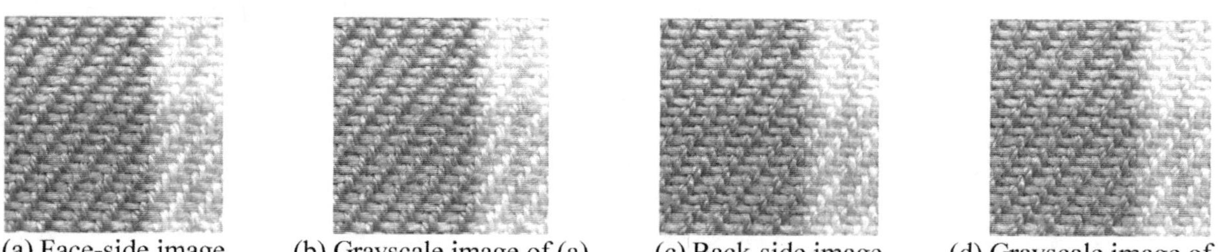

(a) Face-side image (b) Grayscale image of (a) (c) Back-side image (d) Grayscale image of (c)

(e) Power spectrum of (b)

(f) Power spectrum of (d)

Fig.8 The error selected peaks in the power spectrum based on the traditional method

3.3.2 Based on dual-side fusion method

(a) Face-side image

(b) Back-side image

(c) Fusion image based on the Wavelet transform fusion method

(d) Power spectrum of (c)

(e) Two horizontal selected peaks of (d)

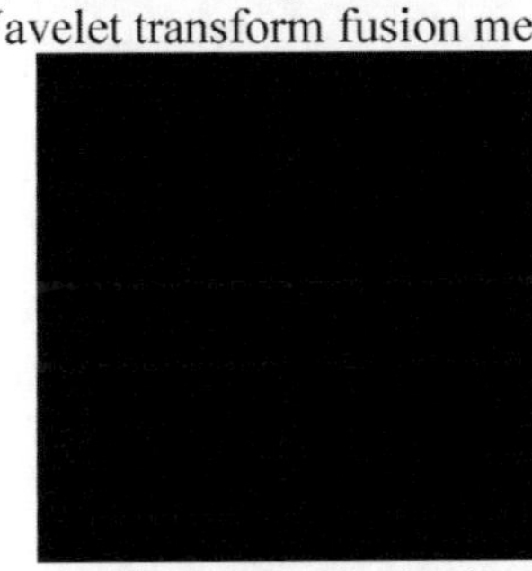

(f) Two vertical selected peaks of (d)

(g) Reconstructed image of (e)

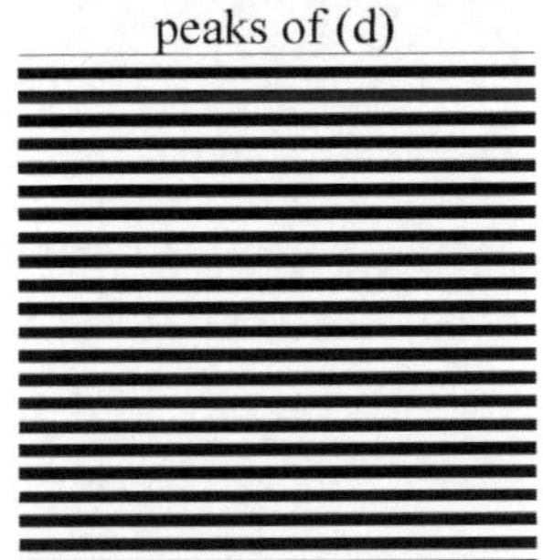

(h) Reconstructed image of (f)

Fig.9 The fusion image and the reconstructed images

The dual-side fusion image of woven fabrics based on the Wavelet transform fusion method is adopted for yarn density analysis in our research. Since the yarn repeat is enhanced and more dominant than the color information in the merged image, the brightest peak points corresponding the yarn repeat information in power spectrum of fusion image are located on the principal lines passes through the origin, which are different in power spectrum of one-side image. Then the yarn density can be calculated accurate. All the fusion image and the reconstructed images of sample 3 are shown in Fig.9.

The yarn density of 20 samples have been measured and the specification are summarized in Table 1. The results obtained from density meter (DM) are also exhibited in table to compared with the results calculated by our methods. According to experiment results, the maximum value of deviation between fusion method and DM method is 1.86 and the average value is 0.56, it indicates that the dual-side fusion technique presented in this research is effective for density measurement of yarn dyed woven fabrics.

Table 1. The yarn density measurement of 20 samples

		Warp yarn density (yarns/inch)			Weft yarn density (yarns/inch)		
Sample	Composition	Fusion method	DM method	Deviation (%)	Fusion method	DM method	Deviation (%)
1	80% wool + 20% polyester	119.5	118.9	0.50	93.8	93.1	0.75
2	100% cotton	119.5	118.9	0.50	70.7	70.1	0.85
3	100% cotton	70.3	70.1	0.28	51.6	50.9	1.37
4	55% linen + 45% cotton	60.9	61.2	0.49	56.3	55.8	0.89
5	68% polyester + 32% rayon	107.8	107.5	0.28	65.6	65.4	0.30
6	55% linen + 45% cotton	128.9	128.1	0.62	70.3	70.1	0.28
7	80% wool + 20% polyester	72.7	71.9	1.11	56.3	56.1	0.36
8	70% cotton + 30% tencel	138.3	138.3	0	98.4	98.8	0.40
9	65% polyester + 35% rayon	70.3	70.1	0.28	56.3	56.7	0.70
10	68% polyester + 32% rayon	68.0	67.8	0.29	58.6	58.2	0.69
11	20% wool + 80% polyester	70.3	70.7	0.56	56.3	56.0	0.53
12	100% cotton	70.3	70.3	0	51.6	51.8	0.39
13	100% cotton	32.8	32.2	1.86	30.5	30.9	1.29

14	80% wool + 20% polyester	84.4	84.3	0.12	56.3	56.1	0.36
15	65% polyester + 35% rayon	103.1	103.6	0.48	91.4	91.4	0
16	65% polyester + 35% cotton	81.0	81.2	0.25	63.3	62.6	1.12
17	100% cotton	56.3	56.6	0.53	53.9	54.5	1.10
18	100% cotton	103.1	103.2	0.09	51.6	50.7	1.77
19	100% linen	58.6	58.9	0.51	51.6	51.6	0
20	100% linen	58.6	58.7	0.17	49.2	49.6	0.81

4. Conclusion

A novel density measurement method based on dual-side fusion technique has been proposed for yarn dyed woven fabric. A lab-used dual-side imaging system has been established to acquire both face-side and back-side image of fabric appearance. The color information of acquired image has been removed by color space transform and only the lightness information has been remained. A self-developed dual-side image fusion method based on Wavelet transform have been utilized to enhance the periodical structure information of warp yarns and weft yarns. Eventually, the yarn density can be measured based on the fused image by using FFT and IFFT techniques.

Twenty samples of woven fabrics have been selected for the validation experiments, including both single colored yarn dyed fabrics and multicolored yarn dyed fabrics. Our experimental results show that the accuracy of density measurement by using dual-side fusion method is close to 99.44% compared with density meter method. It indicates that the presented fusion methods are much more effective than the traditional methods for density analysis, even with the impact of color information. Furthermore, the location and classification of yarn interlacing could be carried out using dual-side methods. Thus, the woven pattern recognition based on the dual-side fusion technique could be the next focus of our research.

Acknowledgments

This project is supported by the National Natural Science Foundation of China under grant no. 61271419.

References

[1] Wood EJ. Applying fourier and associated transforms to pattern characterization in textiles, Textile Research Journal, 60,1990: 212-220.

[2] Ravandi SA, Hosseini, Torium K. Fourier transform analysis of plain weave fabric appearance, Textile Research Journal, 65,1995: 676-683.

[3] Xin BJ, Yu XF, Wu ZP. Density measurement of woven fabrics based on image processing technique, Journal of China Textile University, 3, 1999: 34-37.

[4] Escofet J, Millan MS, Rallo M. Modeling of woven fabric structures based on fourier image analysis, Applied Optics, 40, 2001: 6170-6176.

[5] Rallo M, Escofet J, Millan MS. Weave-repeat identification by structural analysis of fabric images, Applied Optics, 42, 2003: 3361-3372.

[6] Jeong YJ, Jang J. Applying Image Analysis to Automatic Inspection of Fabric Density for Woven Fabrics, Fibers and Polymers, 6, 2005: 156-161.

[7] Tunak M, Linka A, Volf P. Automatic Assessing and Monitoring of Weaving Density, Fibers and Polymers, 10, 2009: 830-836.
[8] Pan RR. Automatic Inspection of Double-system-melange Yarn-dyed Fabric Density with Color-gradient Image, Fibers and Polymers, 12, 2011: 127-131.
[9] Lachkar A. Textile Woven-fabric Recognition by Using Fourier Image-analysis Techniques: Part I: A Fully Automatic Approach for Crossed-points Detection, The Journal of The Textile Institute, 94, 2014: 194-201.
[10] Feng YL, Li RQ. Automatic Measurement of Weave Count with Wavelet Transfe, Journal of Textile Research, 22, 2001: 30-31.
[11] Shen JQ, Geng ZF, Zou X, Pan YB. Detection Method for Fabric Weave Patterns and Structural Parameters Based on Wavelet Transform, Chinese Journal of Scientific Instrument, 28, 2007: 357-362.
[12] Jing JF, Deng QY, Li PF. Automatic Inspection of Woven Fabric Density Based on Digital Image Analysis. Journal of Fiber Bioengineering and Informatics, 8(2), 2015: 259-266.
[13] Kang XJ, Xu MM, Jing JF. Automatic Classification of Woven Fabric Structure Based on Computer Vision Techniques, Journal of Fiber Bioengineering and Informatics, 8(1), 2015: 69-79.
[14] Nishimatsu T. Automatic recognition of the interlacing pattern of woven fabrics part 1: Determination of the pattern by using image information analysis, Sen'i Kikai Gakkaishi, 44,1991: 51-57.
[15] Gao WD, Liu JH. Automatic Identification of Warps Arrangement Parameters in Fabric, Cotton Textile Technology, 3, 2002: 31-34.
[16] Lin JJ. Applying a co-occurence matrix to automatic inspection of weaving density for woven fabrics, Textile Research Journal, 72, 2002: 486-490.
[17] Zhang J, Xin BJ, Wu XJ. Density measurement of yarn dyed woven fabrics based on dual-side scanning and the FFT technique, Measurement Science and Technology, 25, 2014: 1-18.
[18] Pan RR. Automatic detection of the layout of color yarns for yarn-dyed fabric via a FCM algorithm, Textile Research Journal, 80, 2010: 1222-1231.
[19] Feng YL, Li RQ. Automatic Measurement of Weave Count with Wavelet Transfer, Journal of textile research, 2001: 30, 94-95.
[20] Hosseini RS, Toriumi K. Fourier transform analysis of plain weave fabric appearance, Textile Research Journal, 65, 1995: 676-676.

Evaluation of Pressure Sensors Based on Piezoresistive Films for Integration in Smart Medical Textiles

Suresh Parmar[1,2], Ehsan Bafekrpour[1,2], Franz Konstantin Fuss[2,3], Olga Troynikov[1,2*]

[1]*School of Fashion & Textiles, RMIT University, 25 Dawson St, Brunswick, Vic, 3056, Australia*
[2]*Wound Management Innovation Cooperative Research Centre, Brisbane Australia*
[3]*School of Aerospace, Mechanical and Manufacturing Engineering, RMIT University, Melbourne, Vic, 3001, Australia*

*Corresponding author's email address: olga.troynikov@rmit.edu.au

Abstract

In present research, pressure sensors based on piezoresistive films for integration in smart medical textiles are evaluated using rigid electrodes forming rigid experimental pressure sensors. A systematic study was carried out to investigate the effect of sensing area and number of layers of piezoresistive films on experimental pressure sensor's performance parameters such as pressure sensitivity, hysteresis, and accuracy of output. At 10 kPa pressure, these experimental pressure sensors with a sensing area of 100 mm^2 demonstrated pressure sensitivity of 0.05 kPa^{-1}. The pressure sensitivity of sensors increased with an increase in sensing area and with the number of layers of the piezoresistive film. Hysteresis remained in a range of 23% to 25% and demonstrated negligible variations with change in sensing area from 100 mm^2 to 900 mm^2. The accuracy of all the sensors was found to be above 80%, however no pronounced corelation between the sensing area and the number of layers of the piezoresistive film was observed. It was found that the piezoresistive sensing film can be sucessfully used to fabricate a pressure sensor operating in low-pressure range below 50 kPa.

Keywords: Rigid Pressure Sensors; Smart Medical Textiles; Piezoresistive Sensors; Pressure Sensing Films

1. Introduction

Research in 'smart textiles' with applications of sensors and actuators have grown exponentially over the last decade [1-3]. Taking this concept one step further, we envision the emergence of 'smart textiles' with low-pressure sensing capabilities. Such quantitative measurement of pressure has significant importance in applications such as sports, health monitoring and artificial skin sensing. The emergence of these textile-based products will enhance performance of a user by determining and communicating pressure through the clothing itself. In the past few years, several commercial pressure sensors became available on the market for medical and sports applications such as, for example, compression therapy [4] and gait analysis [5-7]. However, these commercial pressure sensors are expensive, uncomfortable to wear against the skin and usually operate in high pressure. Hence, a low cost and simple pressure sensor that is soft and comfortable against human body is desired. Textiles are considered to be the preferred option for incorporation of sensors as they can remain comfortably in continuous contact with body of the wearer. However, the current state of knowledge of flexible and low-cost textile-based pressure sensors is limited [8]. It is observed that amongst the various transduction techniques attempted by researchers, piezoresistance has gained wide acceptance in pressure sensors mainly due to its simplicity and cost-effectiveness [9]. Goy et al [10]. and Donselaar et al [11]. attempted to develop these textile-based wearable pressure sensors from piezoresistive films. The electrical resistance of these piezoresistive films decreases with application of pressure as the conductive particles in the piezoresistive film come in closer contact with each other. However, these flexible piezoresistive sensors developed for applications such as sport and biomedical have not been analysed in depth in terms of their performance: there is a lack of knowledge about the key performance parameters such as sensitivity, hysteresis and accuracy of piezoresistive pressure sensors.

Therefore, in the present research, the performance parameters of a rigid prototype pressure sensor made from a low-cost and simple piezoresistive material that consists of a polyolefin based polymeric film impregnated with carbon black and rigid electrodes were analysed. Effect of the sensing area of the pressure sensor and the number of layers of piezoresistive film on performance parameters such as sensitivity, hysteresis and accuracy were investigated to determine the pressure measurement ability of the piezoresistive film based sensors. In addition, the knowledge gained through this feasibility study will be used in future to develop flexible low-pressure range pressure sensors for a range of applications such as sports, health monitoring and artificial skin sensing.

2. Method

In order to develop a piezoresistive pressure sensor, two main parts, electrodes and piezoresistive film are required. The piezoresistive sensing film changes its conductivity with application of pressure. This film was sandwiched between two electrodes to construct a sensor. Each electrode is positioned to exactly overlap each other and so the size of the electrodes equals to the 'sensing area' (Fig.1) when the piezoresistive film is sandwiched between them. The electrodes were then connected to the pressure measurement device using wires. Electrodes with three different sizes were prepared to investigate the effect of the sensing area on the pressure sensing performance parameters such as sensitivity, hysteresis and accuracy. Similarly, the effect of the number of layers of piezoresistive films on the performance parameters of the pressure sensor was analysed. A known amount of pressure was applied on the sensors through Instron 5565A in compression mode and the output signals such as voltage, conductivity and resistance were recorded by a proprietary pressure measurement device. The recorded raw data was then analysed to obtain the performance parameters. Each experiment was repeated five times after changing the piezoresistive sensing film specimens to allow for the variability within the samples. The results are reported as mean values with standard deviations.

2.1 Sensor

Pressure sensors were constructed using rigid copper PCB plates as electrodes and commercially available piezoresistive film, Velostat™ as sensing material. The piezoresistive film is supplied by 3M Company and has a thickness of 0.1 mm and volume resistivity of less than 500 ohm-cm. In order to investigate the influence of sensing area of electrodes on the resultant performance parameters such as hysteresis, accuracy, and sensitivity, electrodes with effective sensing areas of 10×10 mm^2, 20×20 mm^2, and 30×30 mm^2 were cut from PCB plates using a water jet cutter (Fig.1a). The copper wire connecting the electrodes to the pressure sensor measurement hardware was soldered using soldering wire on the excess part of the copper PCB electrode. The two soldered joints of the electrodes were placed opposite to each other so that they did not emanate from the same side in the assembled sensor (Fig.1b). Piezoresistive film specimens used in the sensor were cut from different locations of the A4 size Velostat™ film and were 5 mm larger than each side of the electrodes for the different electrode sizes. In this multi-layered assembly of the sensor, the effective sensor resistance is equal to the sum of the resistance of the copper plate, resistance of the Velostat™ and the resistance generated due to the difference in area of contact of different layers of the sensor. The resistance of the copper plate does not change during compression.

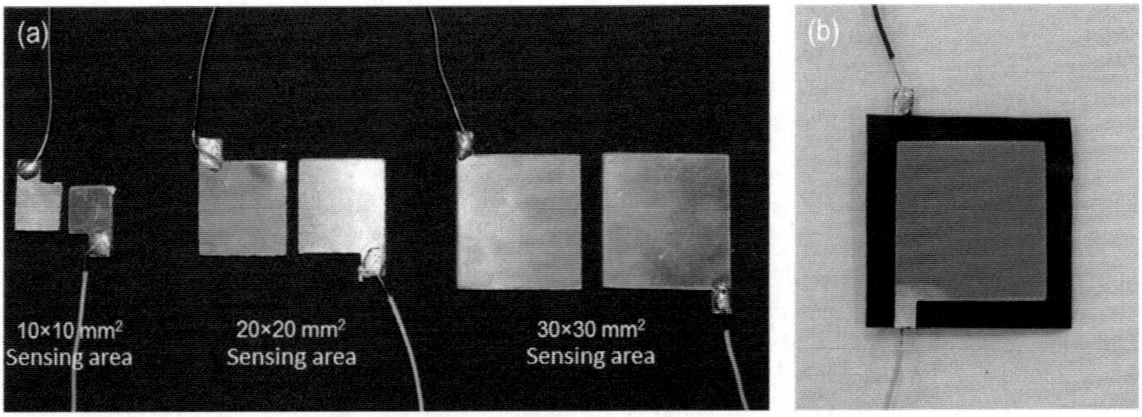

Fig.1 (a) Sensor electrodes; (b) Sensor assembly

2.2 Mechanical loading and pressure measurement

The pressure was applied using an Instron 5565A with a cross head speed of 0.1 mm/s. Cycling loading and unloading test profile with a maximum pressure up to 40 kPa for each sensor was applied through a user-defined program developed using Bluehill 3 software. Each sensor was placed on a platform equipped with a calibrated load cell which was used as a reference for the pressure applied to the experimental sensor. The reference load cell is supplied by Meltrons and works with an error of less than of 0.02 (based on supplier's certificate) to measure force up to 500 N. The reference load cell was calibrated using Instron with a load cell of 2N - 100N force range and its accuracy was verified before committing the experiment. The platform was placed on the compression rig plate (Fig.2b). The electrodes of the sensors were connected to our in-house proprietary pressure detection circuit instrumented on LabVIEW by National Instruments. This electrical circuit measured the change in output voltage on application of pressure and sent it to the software after converting the analog signal into digital signal. The electrical resistivity of the sensing material was calculated in real-time using Eq.1.:

$$R_s = (V_{in}\,R_{ref} - V_{out}R_{ref}) / V_{out} \qquad (1)$$

Where V_{in} is the input voltage and 5.5 V, V_{out} and R_{ref} are the voltage and resistance of the reference resistance in the circuit shown in Fig.2a, respectively. R_{ref} =100 Ω and V_{out} was measured during the experiment. The labVIEW software is programmed to simultaneously collect and analyse the data.

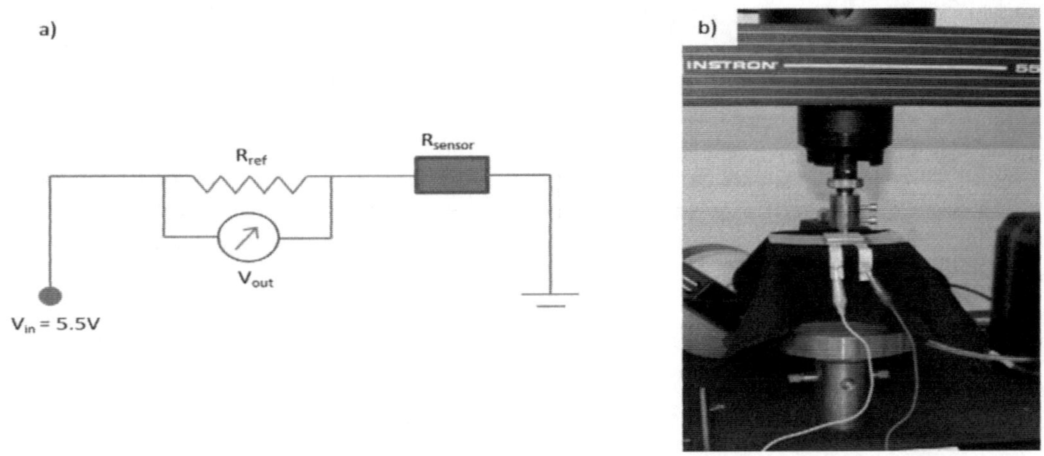

Fig.2 (a) Circuit diagram of pressure measurement assembly; (b) Instron and the platform equipped with a calibrated load cell.

2.3 Sensitivity

Sensitivity is a sensor performance parameter that determines the effectiveness of the output of a sensor. In case of a piezoresistive pressure sensor, it is defined as the slope of the curve of relative change of resistance and pressure of the sensor. It can be calculated at each pressure as (Eq.2):

$$S = (\Delta R/R_0)/P \tag{2}$$

Where, ΔR is the change in resistance, R_0 is the initial resistance before pressure application, and P is the applied pressure. In order to investigate the effect of sensing area and number of layers of piezoresistive film on the sensors sensitivity, each pressure sensor was compressed to a pressure of 40 kPa and the change in the resistance was dynamically recorded. The units of sensitivity were obtained as kPa^{-1}. Each experiment was repeated five times after changing the piezoresistive sensing film to minimize variability of the samples. These sensitivity values were then plotted to get a graphical curve and the mean of relative resistance change - pressure curve of these five tests was obtained by using OriginPro software.

2.4 Hysteresis

Hysteresis (H%) is another significant performance parameter that affects the output of piezoresistive pressure sensors. It is defined as the difference in the output signal (resistance) at the same level of input pressure during loading and unloading on the pressure sensor (Eq.3):

$$H = (R_{loading}-R_{unloading}/R_{max}) \text{ x}100 \tag{3}$$

Where, R_{max} is the resistance value of the sensor at the maximum pressure of the loading-unloading cycle and $R_{loading}$ and $R_{unloading}$ is the resistance values of the sensor at the same pressure (half of the maximum pressure of the cycle) during loading and unloading of the pressure sensor, respectively. To carry out the hysteresis tests, each sample was subjected to 10 cycles of cyclic loading and unloading with pressure of up to 11 kPa. The hysteresis in each sensor was then calculated at 5kPa during loading and unloading on the sensor in Eq.3. Furthermore, each experiment was repeated five times to minimize variability and the mean hysteresis was reported for each sensor. The hysteresis values were then plotted to get a graphical curve.

2.5 Accuracy

Accuracy demonstrates how close the pressure measurements from a sensor are to the actual applied pressure (Eq.4):

$$\text{Accuracy} = 1 - [(P_l-P_s)/P_l] \text{ x}100 \tag{4}$$

Where P_l and P_s are the pressures measured from reference load cell and sensor, respectively. This accuracy in the output of the sensor readings can be expressed either as a percentage of full scale load or in absolute terms. However, in this study, the accuracy of the pressure sensors was calculated at about 5 kPa during the loading cycle. At this pressure, the measured pressures from the reference load cell and sensor were put in Eq.4 and corresponding values of accuracy were obtained. The mean accuracy of five test sensors was then obtained for each sensing area and number of layers of piezoresistive film during 10 cycles. These accuracy values were then plotted to get a graphical curve.

3. Results and discussion

3.1 Sensitivity

Influence of the sensing area and number of piezoresistive films on sensitivity of the sensor is reported in Fig.3. The mean values of relative change in output signal of five sensors are reported (Fig.3) with the standard deviation error bars. The sensitivity of the sensors was calculated by obtaining the derivatives of the mean values of relative change in output signal for each sensing area (Fig.3b and d). It can be seen from Fig.3a and b that for a sensor with a single layer of piezoresistive film, the sensitivity increased with increase in the sensing area. For example, at 10 kPa, for a size of 100 mm^2, the pressure sensitivity of 0.05 kPa^{-1} increased to 0.06 kPa^{-1} and 0.07 kPa^{-1} for a sensing area of 400 mm^2 and 900 mm^2, respectively. However, the difference in sensitivity due to the sensing area decreased as the pressure increased. For example, at pressures above 40 kPa all the pressure sensors with sensing areas of 100 mm^2, 400 mm^2 and 900 mm^2 exhibited 0.02 kPa^{-1} of pressure sensitivity. Therefore, it can be concluded that these pressure sensors generally exhibited bilinear sensitivity curve with a higher sensitivity at lower pressure range below 15 kPa and lower sensitivity at higher pressure levels.

Similarly, the relative change in output signal and average sensitivity of pressure sensors with different number of layers of piezoresistive sensing film is reported in Fig.3c and 3d. It is observed that for a sensor with sensing area of 900 mm^2, the pressure sensitivity values increased from 0.06 kPa^{-1} to 0.08 kPa^{-1} when the number of layers of piezoresistive film increased from 1 to 2. However, on further increasing the number of layers of piezoresistive film from 2 to 4, there was negligible effect on pressure sensitivity of the sensor. In addition, the effect of change in number of piezoresistive films of the sensors also became negligible at pressures above 40 kPa.

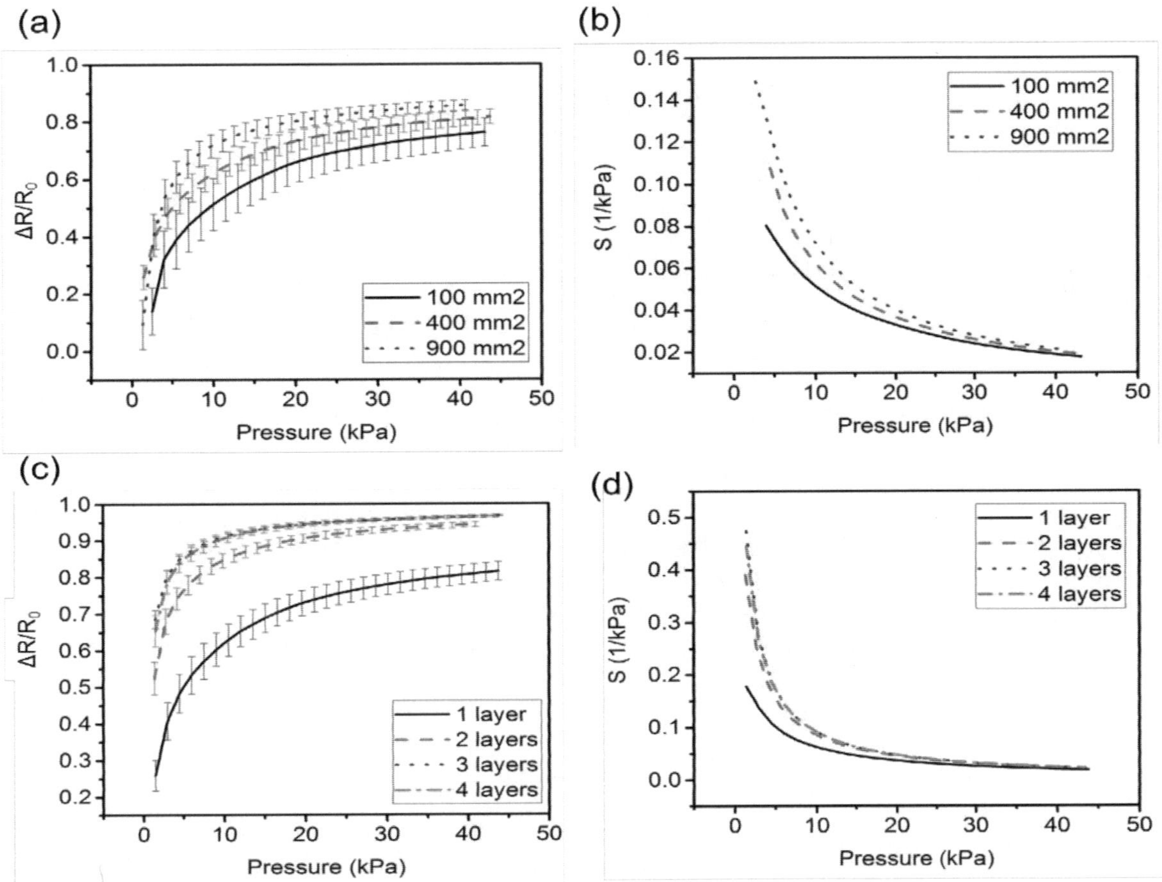

Fig.3 Influence of sensing area on (a) relative change in output signal at different pressure levels; (b) sensitivity at different pressure levels. Influence of number of piezoresistive layers on(c) relative change in output signal at different pressure levels; (d) on sensitivity at different pressure levels.

3.2 Hysteresis

It can be seen that when only one layer of piezoresistive film was used in experimental pressure sensors (Fig.4), then the change in sensing area had negligible effect on the values of hysteresis. For example, it was observed that the hysteresis values remained in a range of 23% to 25% for sensors with sensing area of 100 mm^2 to 900 mm^2 at an applied pressure of 5kpa.

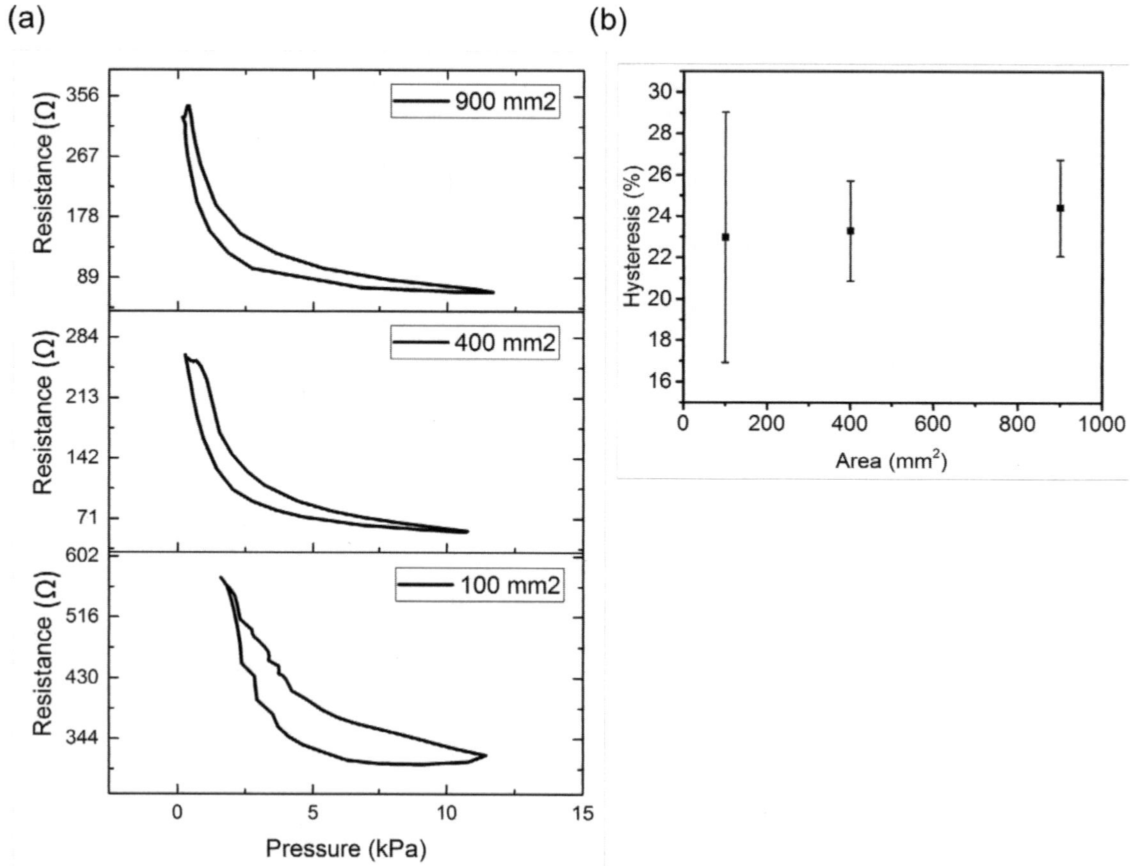

Fig.4 (a) Hysteresis of the sensors with a sensing area of 100 mm^2, 400 mm^2, and 900 mm^2; (b) Effect of sensing area on hysteresis at 5 kPa. Graphs have different scales on the vertical axis.

Similarly, the hysteresis plots of sensors having a sensing area of 900 mm^2 and different number of piezoresistive films is reported in Fig.5. It is observed that for a pressure sensor developed from a single layer of piezoresistive film, the hysteresis values were nearly 25%. However, the pressure sensors with more than one layer of piezoresistive films showed a high hysteresis value of about 45%. Hysteresis, in case of piezoresistive pressure sensor based on polymer materials is usually attributed to viscoelastic nature of the polymer [10], yet more investigation is required to explain the exact phenomenon.

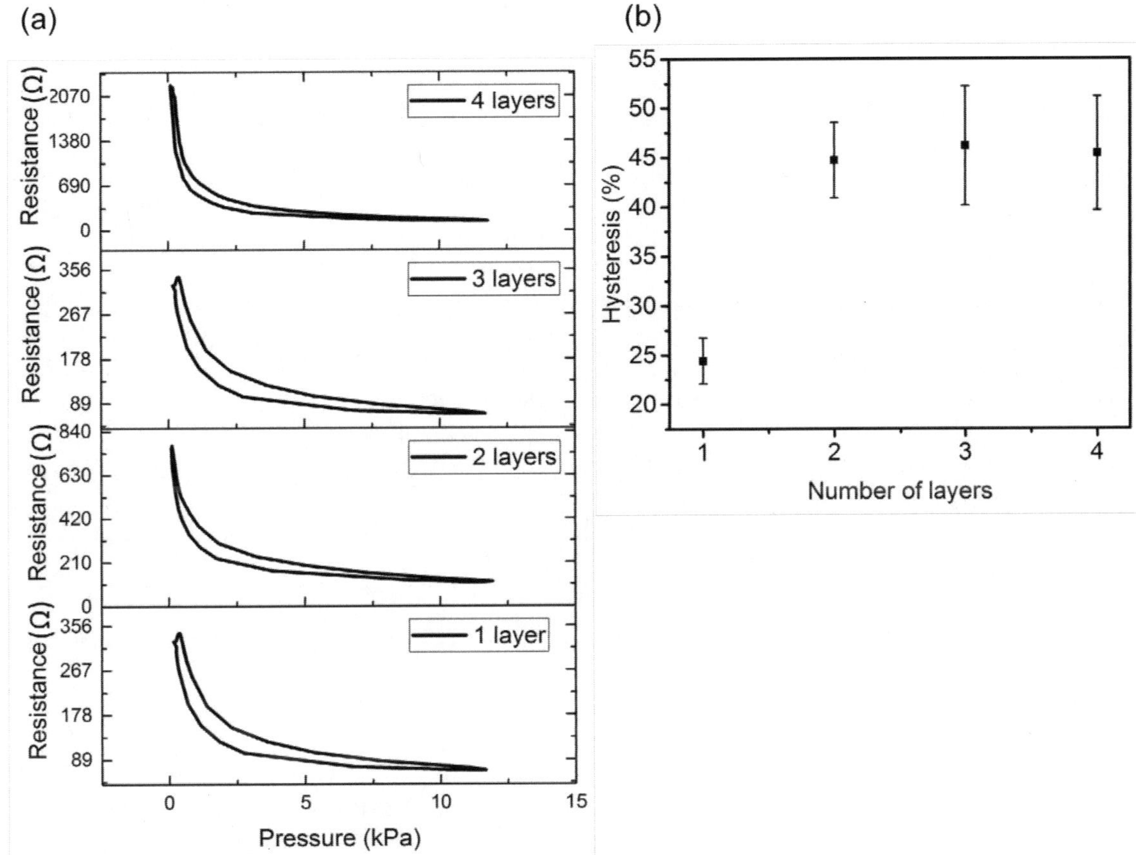

Fig.5 (a) Hysteresis of pressure sensors with 1, 2, 3, and 4 layers of piezoresistive films (b) Effect of number of piezoresistive films on hysteresis at 5 kPa.

3.3 Accuracy

The influence of sensing area on the accuracy of the pressure sensors having a single layer of piezoresisitve film is shown in Fig.6a. It is observed that although the sensors showed high accuracy above 80% in their pressure measurements, yet there was no direct correlation between an increased sensing area and accuracy of the output of the sensors. Similar results were observed when the number of layers of piezoresistive film was increased for a sensor with sensing area of 900 mm^2.

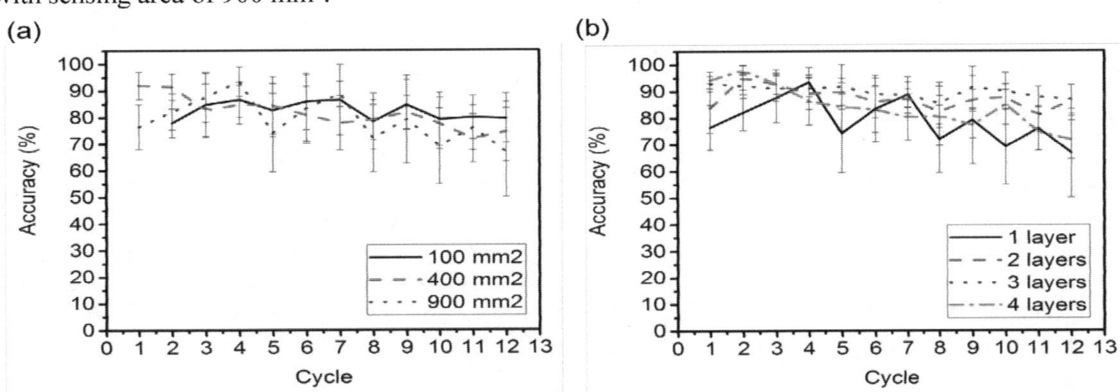

Fig.6 (a) Effect of the sensing area on accuracy of output of the sensors; (b) Effect of the number of the piezoresistive films on accuracy of output of the sensors.

4. Conclusion

In the present research, the performance parameters of a rigid pressure sensor developed from a piezoresistive sensing material consisting of polyolefin film embedded with carbon black were analysed. The effect of sensing area and number of piezoresistive layers on performance output of pressure sensors were presented. At 10 kPa pressure, these experimental pressure sensors with a sensing area of 100 mm^2 demonstrated a high pressure sensitivity 0.05 kPa^{-1}. This sensitivity further increased with increase in sensing area and the number of layers of the piezoresistive film. Hysteresis values remained in a range of 23% to 25% and demonstrated negligible variations with change in sensing area. However, while overall accuracy of the sensors remained above 80%, yet they did not demonstrate any pronounced corelation with sensing area and the number of piezoresistive film layers. This study provides insights into factors affecting performance parameters such as sensitivity, hysteresis and accuracy which will enable improvement in future pressure sensors. In addition, based on the knowledge gained through this study on a rigid experimental pressure sensor, new and flexible piezoresistive sensors that have better performance properties along with the ability to measure pressure in low-pressure range could be developed. These flexible pressure sensors will then find critical applications ranging from sports and health monitoring to artificial skin sensing.

5. Acknowledgement

The authors would like to acknowledge the support of the Australian Government's Cooperative Research Centres Program.

References

[1] Langenhove LV. Smart textiles for medicine and healthcare: materials, systems and application. Elsevier: 2007.
[2] Capineri L. Resistive sensors with smart textiles for wearable technology: from fabrication processes to integration with electronics. Procedia Engineering: 2014; 87: 724-727.
[3] Stoppa M, Chiolerio A. Wearable electronics and smart textiles: a critical review. Sensors (Basel): 2014; 14(7): 11957-92.
[4] Tyler D. Application of pressure sensors in monitoring pressure. In: Hayes SG, Venkatraman P. Materials and technology for sportswear and performance apparel: CRC Press: 2015; p: 289-310.
[5] Das R, Kumar N. Investigations on postural stability and spatiotemporal parameters of human gait using developed wearable smart insole. J Med Eng Technol: 2015; 39(1): 75-78.
[6] Holleczek T, Ruegg A, Harms H, Troster G. Textile pressure sensors for sports applications. IEEE Sensors: 2010; 732-737.
[7] Shu L, Hua T, Wang Y, Li Q, Feng DD, Tao X. In-shoe plantar pressure measurement and analysis system based on fabric pressure sensing array. IEEE Transactions on Information Technology in Biomedicine, 2010; 14(3): 767-775.
[8] Tiwana MI, Redmond SJ, Lovell NH, A review of tactile sensing technologies with applications in biomedical engineering. Sensors and Actuators A: Physical: 2012; 179: 17-31.
[9] Stassi S, Cauda V, Canavese G, Pirri CF. Flexible tactile sensing based on piezoresistive composites: a review. Sensors (Basel): 2014; 14(3): 5296-5332.
[10] Goy CB, Menichetti V, Yanicelli LM, Lucero JB, López MA, Parodi NF, Herrera MC. Design, fabrication and metrological evaluation of wearable pressure sensors. J Med Eng Technol: 2015; 39(3): 208-215.
[11] Donselaar RV, Chen W. Design of a smart textile mat to study pressure distribution on multiple foam material configurations. Proceedings of the 4th International Symposium on Applied Sciences in Biomedical and Communication Technologies: 2011; 129: 1-5.

Design Innovation for Wellbeing: Project Examples Using Smart Materials for the Body and Built Environment

Joan Farrer[1*], Petar Goulev[2], Carolyn Watt[1], Harriet Parr[3], Cressida Bowyer[1], Iain Allan[4]

[1]*University of Portsmouth, Winston Churchill Avenue, Portsmouth, Hampshire, PO1 2DJ, UK*
[2]*Targovia Ltd, Hartley Road, Exmouth, Devon, EX8 2SG, UK*
[3]*Osbourne Energy Ltd, Marine Court, St Leonards-on-Sea, TN38 0DN, UK*
[4]*University of Brighton, Huxley Building, Lewes Road, Brighton, BN2 4GJ, UK*

[*]Corresponding author's email: joan.farrer@port.ac.uk

Abstract

Expertise in design, science, technology, engineering, mathematics and business (D-STEM-B) are combined successfully in transdisciplinary Design Thinking groups to produce innovations to societal problems with wellbeing at the core. Three research proposals currently in progress are presented here: 1. Barrier Solutions-melanoma prevention; 2. DERMA-ageing well; 3. SHAKES-earthquake early warning systems. By their transdisciplinary nature the projects are expected to produce new types of research using mixed methods for textiles and biomedical applications in the design, science and innovation arena.

Keywords: Design; Innovation; Smart Materials; Transdisciplinarity; Wellbeing; Environment

1. Introduction

The creative and cultural industries, in conjunction with science, technology, engineering, mathematics (STEM) and business subjects have developed Piaget's [1] transdisciplinary model where designers, scientists and commerce transfer their skills to produce social and environmental wellbeing solutions; developing smart materials, innovative technologies and applied research for societal good. This is a positioning paper presenting three transdisciplinary research and innovation projects using Design Thinking methodology and includes an in-depth case study on smart materials (conductive textiles) used in the environment.

1.1 Innovation through transdisciplinarity

Transdisciplinarity was introduced to the public in the 1970s by the Swiss philosopher and psychologist Piaget, who addressed the issue of relationships between disciplines in an academic context, purporting the notion of transdisciplinarity to create synergies between unrelated disciplines that enable investigations and opportunities for collaboration in a third space beyond their respective contributions. Transdisciplinarity goes beyond interdisciplinary (combining knowledge and methods from diverse disciplines), crossdisciplinary (viewing one discipline from the perspective of another discipline) and multidisciplinary (people from different disciplines working together, drawing on their respective disciplinary knowledge) [2].

The Department of Business, Innovation and Skills, UK [3, p.16] confirm the necessity of working outside of disciplinary silos in their recent parliamentary white paper:

'The challenges facing the world are complex, and increasingly require multi- or inter-disciplinary approaches. Our ambition is to ensure that our research and innovation system is sufficiently strategic and agile to meet these challenges, and to deliver national capability for the future that drives discovery and growth.'

Working in a transdisciplinary team can prove challenging. However, this approach to dialogue, exchange of skills, knowledge transfer unusual collaboration can lead to holistic and unexpected successful solutions.

Collaborations between design, science, technology, engineering, mathematics and business (D-STEM-B) and non-academic partners in order to create a community of learners, can play an integral and influential role in developing innovative applied research proposals for real-world solutions.

Lang et al [4, p. 3]. confirm the complexity in understanding and application of the model where:

'[T] ransdisciplinary research needs to comply with the following requirements: (a) focusing on societally relevant problems; (b) enabling mutual learning processes among researchers from different disciplines (from within academia and from other research institutions), as well as actors from outside academia; and (c) aiming at creating knowledge that is solution-oriented, socially robust [...] and transferable to both the scientific and societal practice'.

Lang et al [4]. evoke the importance of transdisciplinary research as a method to begin to solve societal issues, described in the following project examples. Mutual learning between experts moves beyond simple knowledge exchange and sharing of best practice, to interaction with concepts, philosophies and blue-sky thinking to enable new knowledge creation, application and true impact.

Cronin [5, p. 1] describes this method as 'a new field of research emerging in the "knowledge society"', where a knowledge society embodies the concept of a community of learners and is directly linked to the creative economy:

'[A] knowledge society should be able to integrate all its members and to promote new forms of solidarity involving both present and future generations. Nobody should be excluded from knowledge societies, where knowledge is a public good, available to each and every individual.' [6, p. 18]

When combined with the discipline of design, transdisciplinary working leads to innovation. Cope and Kalantzis [7, p. 5] suggest design 'is an engine of change' and argue that 'the principles and practices of Design may also become a central concern of every discipline'. Brown [8, p. 88] describes the design process in metaphorical terms:

'[A]s a system of spaces rather than a predefined series of orderly steps. The spaces demarcate different sorts of related activities that together form the continuum of innovation. Design thinking can feel chaotic to those experiencing it for the first time. But over the life of a project participants come to see [...] that the process makes sense and achieves results, even though its architecture differs from the linear, milestone base process typical of other kinds of activities'.

2. How to address real-world societal issues such as health, wellbeing and the environment through design innovation?

2. 1 Smart materials and Design Thinking

'Design Thinking is an approach to designing products and services that takes into account not just needs of the consumers, but also their social and cultural infrastructure'. [9]

Design Thinking enables discussion towards innovative development and amalgamation of practical science, and complimentary systems, which can make more efficient and effective use of technology, resources and materials for new products, services and processes that permeate the entire range of innovation activities with a human-centred approach. The development and application of new technology and smart material projects discussed here create solution pathways to address societal issues and health and wellbeing for the body and the environment. Inspired by the outcomes of collaborative thinking with colleagues from D-STEM-B striving to solve an identified problem, results have produced solutions to enhance lives, economies and societies, adding value and novel benefits to the outcomes. Design Thinking in a transdisciplinary collaboration is a powerful agent for innovation for personal and social transformation, where applications are developed for real-world contexts and can be and a route to social sustainability where products are relevant for the less wealthy.

Smart materials and new technologies are ever more apparent in the healthcare and construction industries, and are increasingly shaping many aspects of society, having been embraced by producers and consumers. Ridell [10, p. 14] describes the active consumer as one 'who engages in innovation and who co-creates value together with other consumers, and with producers, in relation to products'. This audience is often composed of early adopters who are familiar with advances in information and communications technology (ICT), which has made it increasingly possible for these consumers to actively engage in innovation and co-creation.

'The term Smart Textiles refers to a broad field of studies and products that extend the functionality and usefulness of common fabrics. Smart Textiles are defined as textile products such as fibers and filaments, yarns together with woven, knitted or non-woven structures, which can interact with the environment/ user. The convergence of textiles and electronics (e-textiles) can be relevant for the development of smart materials that are capable of accomplishing a wide spectrum of functions, found in rigid and non-flexible electronic products nowadays' [11, p. 11958].

Adaptive and responsive materials and textiles have been combined with ICT, to produce cheaper innovative health and wellbeing applications with a range of functions including data transmission, data harvesting, processing, powering, generating, storing, sensory functions and communication abilities. In a modern hospital, for example, textiles appear in many applications, including implants, tissue engineering, hygiene and healthcare products, packaging, dressings, uniforms and personal protective equipment; while in construction they may be used to monitor stress and strains in civil engineering projects such as bridges, in the sensing and actuation of integrated materials in tower blocks to detect movement and, in piezoceramics, shape memory materials for real-time vibration and shape control. Smart materials and innovative applications are continuing to emerge through transdisciplinary research and Design Thinking for both the body and the built environment.

3. Problem solution: project examples: healthcare

3.1 Barrier Solutions: melanoma prevention

The Barrier Solutions project, under development at the University of Portsmouth, draws on technological, creative and scientific tools to foster innovative and cost-effective solutions for melanoma prevention. Barrier Solutions' focus is on 'smart' textiles combining computer intelligence and digital communication to provide 'early warning systems' of over-exposure to the sun. With a focus on design for behavioral change, the co-creation of smart, healthy, and sustainable products and environments, the project will foster a holistic approach to develop new models and tools for health promotion and ill health prevention. Project partners are experts in: gaming and application program (app) development; ICT solutions; behavior analysis; textile and fashion design; virtual reality and motion capture technology; pharmacy and bimolecular science; and dermatology.

An opportunity for fashion and textiles to contribute to improved health outcomes is seen in the prevention of serious skin cancer and melanoma. Sun protection factors (SPF) were first created by chemist, Milton Blake [12] in his kitchen in Australia; he went on to create Hamilton Laboratories in 1932. However, despite the millions of dollars that are spent by consumers on SPF creams each year [13], the use of SPF creams has not prevented increases in cases of melanoma. In adolescence and young adulthood, between the ages of 15 and 24, our bodies are exposed to 80% of the total harmful radiation that we will encounter during our lifespan. This is an age group where the fashion is to wear t-shirts and shorts suited to an outdoorsy antipodean's lifestyle, and, for teenage boys, going without a shirt is normal and is a particular issue for UV exposure in sensitive areas. Typically, the darker the colour of the fabric, the higher the SPF rating; but these dense colours are less appealing from a fashion angle on sunny days, so designed textiles that consumers want, with intelligent protective systems, are in development, where the fabric can protect and alert against skin cancer or premature ageing of the skin. These are key concerns in parts of the world where protection from harmful UV radiation is required. Currently, fabrics which have a UV protection factor (UPF) still absorb a percentage of UV rays rather than totally blocking them (a UPF rating of 25 means that only 1/25th of UV radiation can penetrate the fabric, and increase the fabrics' unique value proposition [or USP]); for example, polyester and certain polyamides are particularly effective at absorbing dangerous rays. In the area of fabric finishes and coating, chromic inks, which change colour when affected by temperature, will be an early warning indicator to the wearer or, with the inclusion of ceramic molecules into a synthetic fibre or surface in coatings, this technology can assist in deflecting harmful rays. In addition, tighter and denser weaves give increased protection, whereas standard cotton weaves and knitted structures allow harmful rays through the cloth to the skin when the fabric is stretched, yet the wearer may think they are protected.

An increasing public health and wellbeing issue, which impacts a variety of outdoor user situations, is exposure to the sun. Those directly affected are outdoor workers and the general public who follow government advice and guidance which, on the one hand encourages a 'healthy' lifestyle (e.g. enjoying sport and leisure activities), but on the other may put the human body at risk due to UV radiation exposure, potentially leading to health problems that are both serious (e.g. skin cancer) and aesthetically detrimental (e.g. premature ageing of the skin). Thus, public health messages concerning sun safety are mixed, with exposure being both beneficial and risky. Sunlight makes us feel good due to the complex relationship of sunlight, serotonin and melatonin, which affect mood and sleep. Crucially, vitamin D is produced when UV rays from sunlight hit the skin and trigger synthesis; high levels of vitamin D help maintain serotonin levels in the body, which is the chemical targeted by a class of anti-depressants called selective serotonin reuptake inhibitors (SSRIs). A lack of vitamin D is also one of the most common causes of rickets; vitamin D is crucial for the development of healthy strong bones in children. Yet prolonged exposure to solar radiation in humans may result in skin damage and, in extreme cases, malignant melanoma which if untreated could be fatal [14]. Understanding and communicating individual risk, due to lifestyle, family history or skin type, is therefore a critical factor in the personal management of sun safety and, although the background research for Barrier Solutions began in New Zealand and Australia where there is significant longitudinal data relating to northern European population exposure to sunlight and its detrimental effects, the fieldwork and testing will take place mainly in the northern hemisphere where the research team is located. The Barrier Solutions project will engage in dialogue with the most vulnerable target group (young males) in selected European regions using a toolbox of technical interventions and product development to instigate behavioural change and test its efficacy on the user group. Computer intelligence and digital communications combined with functional and aesthetically designed communicative prototypes such as mobile devices, serious games technology and integrating interactive textiles in aesthetically designed clothing and accessories will form 'early warning systems' to alert the user group to over-exposure to the sun. Communication of the toolbox findings from the target group, if behavioural change has taken place, would protect this group moving forward and will be a key and impactful outcome.

3.2 DERMA: Design of Enabling Regenerative Materials for Ageing

DERMA is a European (Interreg 2-Seas funding stream) proposal with an expected start date in late 2016. Diverse partners from the UK, France, Belgium and the Netherlands include University of Brighton, Ghent University, Eurosante and the University of Portsmouth working in a transdisciplinary STEM, Design and commercial partnership.

The DERMA (Design of Enabling Regenerative Materials for Ageing) project engages the development of new healthcare technologies and smart materials to improve the quality of life of the ageing population. The project will address the common challenge of improving the quality of life of our increasingly elderly citizens in northern France, southern England, Belgium and the Netherlands, south coastal area. In Europe, the ageing population is burgeoning and, in 2016, the European Commission [15] stated: 'by 2025 more than 20% of Europeans will be 65 or older, with a particularly rapid increase in the number of over 80s'. The DERMA research cluster will tackle a problem for some elderly people brought on by advanced age-inflammation, infection and ulceration of the skin-which costs the EU €6.5 billion per annum [16]. This transdisciplinary project will utilise the expertise of clinicians, designers and creative technologists from the outset to develop a community of learners, to transfer, communicate and advance our existing novel laboratory technologies to more refined, close-to-market products, and encourage adoption beyond the project in response to market demands of active consumers (end-user clinicians and patients).

The objective of DERMA is to develop, through co-design, new smart material and technological dressings, garments and application models for the management and treatment of dermal ulcers and related skin conditions, including stomas. By addressing market and patient needs, as identified by the project's Observer Partners, the project will deliver advanced new technologies (incorporating marine-sourced biopolymers) ready for exploitation by industry to manufacture improved products for the treatment of patients. It is expected that these enabling technologies will provide cost savings to the EU by improving the efficiency and success of wound management by healthcare providers.

Dermal ulcers are a source of physical discomfort which can lead to further medical complications and can release offensive odours that are distressing to the patient and may lead to social isolation and depression [17]. In response to clinical demand and market knowledge, Observer Partners have identified a need for novel wound dressings to treat chronic dermal ulcers and peristomal inflammation. The DERMA project will develop a designed antimicrobial wound dressing with odour-absorbing agents using marine polymers to manage the release of offensive odours. This diagnostic dressing will indicate the state of healing of the underlying wound, giving a visual indication to the patient and/or carer of whether the dressing needs changing. The technologies to be developed will move beyond the current state-of-the-art in wound care in order to realise real-world changes. A continual dialogue between the end-users, the care providers and the wound dressing technologists will ensure that the product is fit for purpose.

A crucial part of this transdisciplinary research project is to create a narrative and explain ideas and methodology to the non-expert audience, especially in the form of public outreach, behavioural change and the UK's Impact agenda [18]. Creative technology partners will visualise the wound dressing design and application in the form of 3D computer animations. The odour-absorbing and antimicrobial functions will be animated in an easy to understand format in order to demystify the research and communicate the smart material to the general public. Product prototypes will be supported with instructional computer animations and films for end-users and design progress and route-to-market will be documented using video diaries as an educational tool for other developers and users. Innovative communication methods will be used by the partners to collaborate on large scale public events to showcase D-STEM-B research in action leading to pathways for change.

DERMA is an evolution of sister project Biocare Marine, which concluded in 2014. Biocare Marine focused on the development of porous scaffolds based on fish gelatin and exopolysaccharides using cryogelation. Biocare Marine results enabled the optimisation of the polymer processing conditions which form the starting point to develop highly advanced novel wound dressings with antimicrobial and odour adsorbing features to be developed in project DERMA. Biocare Marine isolated, characterised and sustainably utilised marine biomolecules applicable to human healthcare for the social and economic benefit of the coastal regions: France (Nord-Pas de Calais), England (southwest, southeast and east), Belgium (Flanders) and the Netherlands (south coastal area).

3.3 Case study: environment: SHAKES: Structural Health Autonomic Kinetics for Earth Shifts

3.3.1 Introduction

It was postulated that the reduction of human and economic impacts of earth tremor damage may be achieved by the development of an intelligent structural health monitoring (SHM) technology that, in real-time, could collect seismic data and alert users to changes in a structure's integrity.

The proof-of-concept prototype for a reusable intelligent nanotechnology-incorporating textile was developed by the authors. Based upon Professor Joan Farrer's experience with colleagues developing a smart textiles-based vital signs monitoring garment for Zephyr Technologies Ltd., New Zealand, in 2009, the macro textile sensor for the built environment is a conceptual development of the micro body monitor smart textile. The development for the built environment uses a conductive carbon polymer developed by NanoForce [19] of Queen Mary University, London, UK; a flexible textile was developed by woven textile designer Kirsty MacDougal, currently of the Royal College of Art, London, UK, which incorporates the nanopolymer monofilament nylon with additional silicone coating. The theoretical framework for the textile to be used and tested further in the built environment was developed by architect Kate Cheyne, currently of the University of Brighton.

The final intelligent material was selected from the development of a woven structure sample range, tested for best conductivity throughout the material development under laboratory conditions. The research collaborators acknowledge that there is potential for improvement of the textiles resulting in better data gathering and better analysis frameworks which could lead to improved monitoring in earthquake prone areas, better indication of irreparably damaged and unstable structures, and to inform and improve civil engineering in academia and industry. The potential of the research could further enhance the position of the textiles/ electrical engineering/ design/ computing disciplines inside a growing field of advanced materials specifically developing intelligent materials for the built environment. This research is evidence of transdisciplinary engagement between institutes, countries and commerce for societal good.

3.3.2 Research questions and aims

The questions were:

1) Is it possible to create an earthquake damage detection system for buildings from an intelligent reusable textile?

2) Is it possible to use textile nanofibre and computing technology to create a 'mapping system' patch for radio-frequency identification (RFID) data gathering from buildings affected by earth tremors?

The aim was to develop and test a woven textile which incorporated a strain-sensitive conductive carbon nanotube polymer to protect its performance as a tailor-made structural health monitoring system (SHMS) for buildings and produce a proof-of-concept experimental kit. The applied textile research was practice-based, providing architects, weavers, designers and engineers the opportunity to prototype, test, re-evaluate, recreate and retest samples of woven cloth containing nanofibres and nylon in various structures. These woven fabrics also had to be assessed for their strength and durability with regard to possible weathering.

3.3.3 Background and context

Settlements in geographic areas prone to seismic shifts remain conscious of the effects of these on population and property; however, in such areas with low socio-economic development, environmental and financial recovery from a seismic incident may be both costly and lengthy. The average number of quakes globally each year of a magnitude greater than 7.0 is 16 [20]. The 2010 Haitian disaster was recorded at magnitude 7.0, leading to 316,000 deaths and more than 1.3 million people displaced [21]. New Zealand's earthquake of 2011, with a magnitude 6.3 in the Christchurch region, has an estimated rebuild of NZ$20 billion (US$15.1 billion) [22].

Damage detection technology is minimal in the built environment and is often visible bolted-on technology, unlike some sophisticated seismic prediction tools, which are embedded. Following a seismic incident, there may be no visible signs of damage to buildings, and so people return to their homes or places of work believing them to be safe. However, the structure has often been irreparably damaged by seismic activity and is unsafe for use. Should aftershocks occur, the affected structure may collapse unexpectedly, possibly with a loss of human life. To counteract this situation, buildings in a seismic shift region are demolished as a preventative measure, without knowing how unsafe the structure actually is. Therefore, the development of a cost effective, durable, versatile and easy to apply external monitor, which could also work in less affluent world regions, is novel and overdue. It is proposed that a conductive polymer, woven with a moisture-resistant nylon fibre, could be applied to structural joints of buildings and other structures (e.g. bridges or dams) as a capital expense. The importance of regularly monitoring the structure could justify the extra cost of incorporating a SHMS in a new build or retrofitting one to an existing structure to ensure an appropriate lifespan and maintenance, and to protect users at home or at work.

3.3.4 Beneficiaries of the research

The proof-of-concept experiment provides a platform for dialogue with institutions with expertise in SHM to begin to assimilate this conceptual model. There would be benefits in developing collaborations with businesses with experience of intelligent textile and digital development, from which research and development funds may be sought. Centres such as the Pacific Earthquake Engineering Research Center (PEER) in California, US, with existing knowledge and understanding of earth shift patterns and the current methods for tracking and tracing these, may become affiliated. Collaborators may include small and medium-sized enterprises from atypical disciplines and/or locations, with a potentially significant impact on cross-industry technical innovation, bridging gaps between D-STEM-B disciplines.

The transdisciplinary nature of the research means the project could be disseminated through, and be relevant to, three different industries: textile; construction; and material engineering. Geographically, this may eventually incorporate various testing grounds and field studies in the real built environment. Contacts with the construction industry in urban earthquake zones will provide opportunities for installation of the prototype textile in new buildings and existing unstable structures to monitor the extent of structural instability.

Advancement in the field of 'intelligent textiles' for use in structural monitoring will enhance knowledge exchange and development of a second generation 'intelligent textile' SHMS. Socio-economic relevance and benefits include the reduction of humanitarian and financial losses. Advancement in the development and analysis of SHMS for buildings and structures is critical to understand the data to inform how structures behave in earthquakes.

3.3.5 Methodology

The methodology combined case study and practice-based applied research. The case study for the concept development was the earthquake zone in Christchurch New Zealand, used as the theoretical 'test-bed'. The study explored Christchurch post-earthquake, examining first-hand the impact of such a disaster on a city and community, where optimising seismic data results from such a conductive textile may have been useful in analysing which structures should be demolished.

Part one of the experiment was to produce textile samples in various yarns and constructions, which were woven and tested for strength and conductivity. The second part of the experiment was to select the best swatch from the samples to gather data from the textile through a bluetooth device, which was monitored by RFID using a laptop computer. What is important is that the equipment (bluetooth, conductivity textile and recording device) can enable recording, transmitting and data gathering while the textile was both motionless and moving, before, during and after shocks, denoting the 'proof-of-concept'.

Firstly, a series of hand woven experiments to develop the textiles took place. The initial sample textiles were analysed for conductivity and strain testing at NanoForce laboratories, Queen Mary University, where quantitative technical data was produced; this proved inaccessible for some non-scientific collaborators, although all partners realised some samples were more effective than others at possible transmission. The first part of the experiment resulted in the development of a conductive textile with strain-sensing ability and data transmission possibilities towards self-monitoring structures. NanoForce developed the innovative strain sensitive fibre used in the samples, for which they hold the intellectual property. The fibre was composed of thermoplastic polyurethane (TPU) fibres containing multi-walled carbon nanotubes (MWCNT) extruded as a yarn. NanoForce had already proved that it is possible to predict the conductivity of the nanocomposite fibres which have a 'tune-able' level of electrical conductivity. The same extruded TPU/ MWCNT fibres showed interesting strain-sensing abilities under static and dynamic loading conditions and could provide a step change in SHMS.

3.3.6 Technical detail

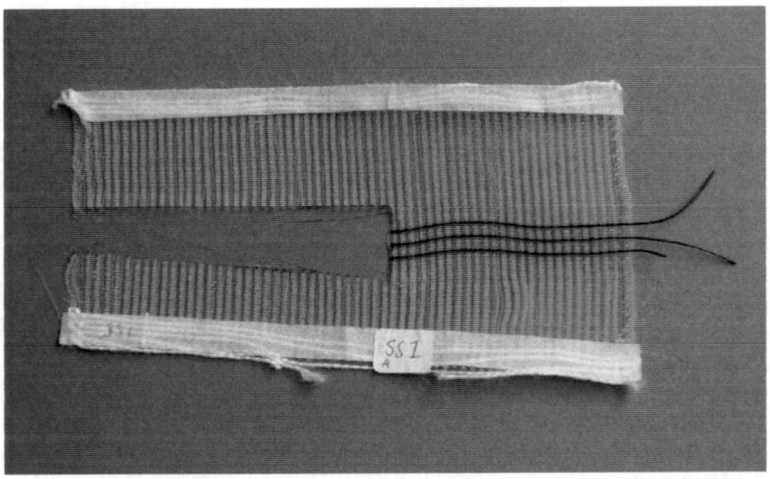

Fig.1 Conductive thermoplastic (black thread) and nylon (white thread) woven textile patch prototype

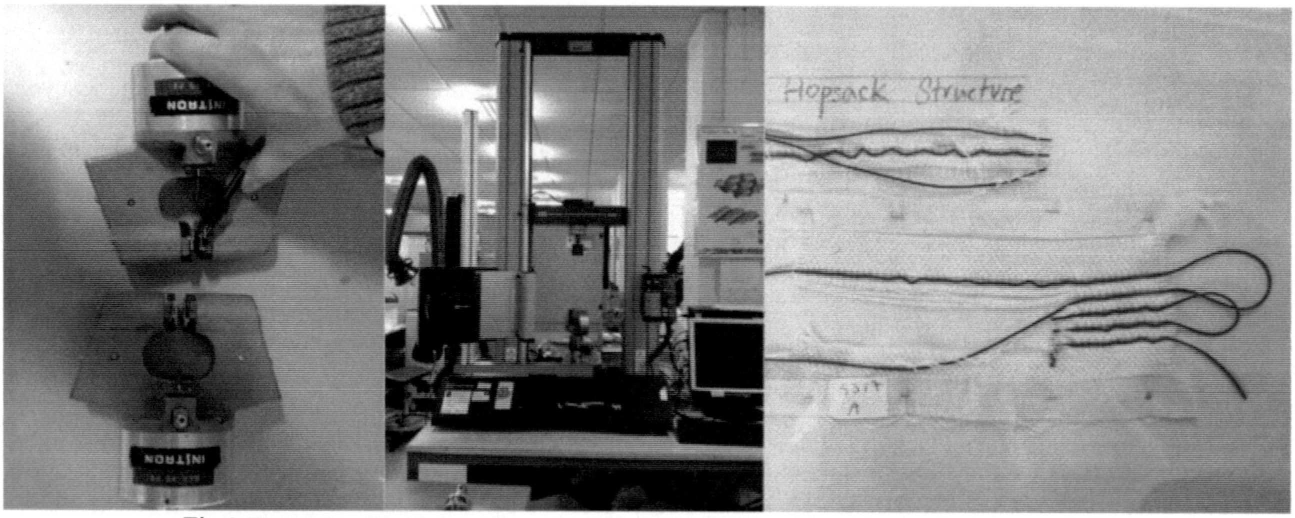

Fig.2 Example of initial strain and conductivity testing kit and sample constructions

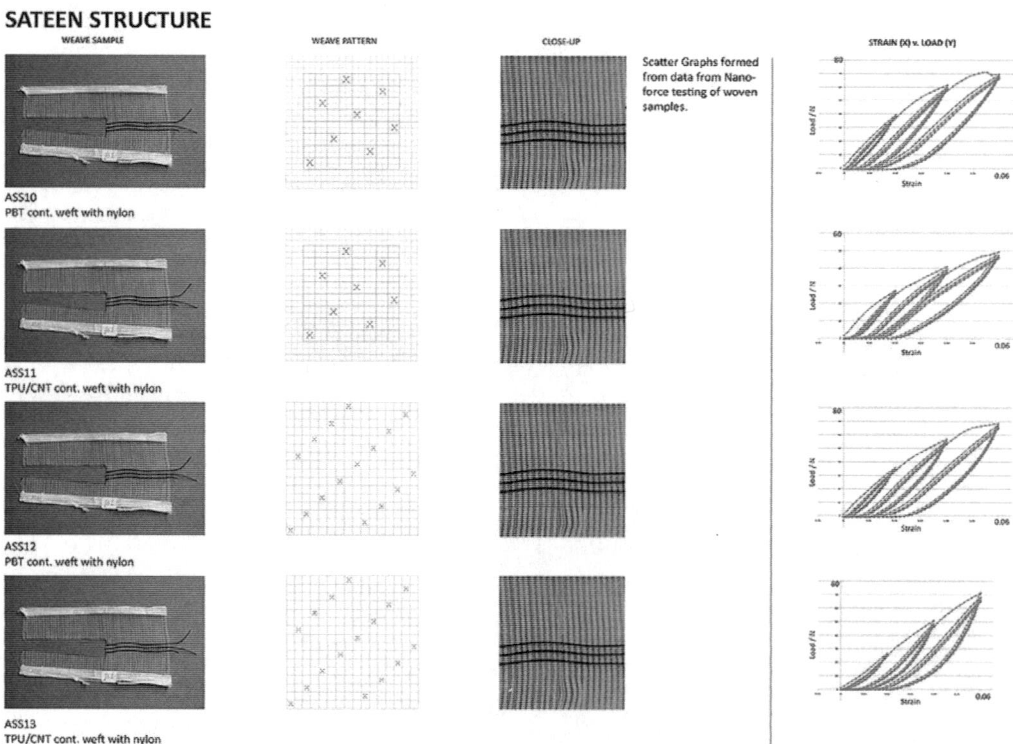

Fig.3 Examples of textile samples, weave plans and testing results

For the purposes of the research, NanoForce provided thermoplastic yarns to be combined for the first time into a woven textile, the yarns are made up of thermoplastic polyurethane (TPU) fibres containing multi-walled carbon nanotubes (MWNT) extruded as a yarn, Zhang at al. [23, p.9449].

The polymer was interwoven on a dobby loom, with moisture-resistant fibres of nylon and lycra in various flexible constructions, which were used for conductivity testing as seen in Fig.1. Samples (25.4 cm × 25.4 cm) were produced in plain weave, twill, leno variations and sateen constructions to allow for a range of tests for accuracy in the transmission of digital signals in relation to the fabric density and stability. The plain weave construction using nylon held the interspersed polymer thread tightly within the textile, restricting possible movement of the nanothread in the material. Twill weave variations produced more movement of the polymer within the fabric, which was unstable. Sateen weave structures placed more of the polymer yarn on the surface of the fabric. A 'crammed and spaced' warp put different tensions on the weft materials and exposed varying degrees of polymer on the fabric surfaces. The polymer was used in different densities of weft stripes with varieties of water resistant yarns in the warp and weft.

Initially, eight different textile samples were produced, combining a variety of weaves, materials and proportions of polymer yarn. The amount and quality of information that the samples were able to transmit then went on to be tested by NanoForce using the fabric's electrical signals. The samples were refined on the basis of the results, with the specific aim of optimising data transfer at the next stage. Further testing was used to test these revisions to achieve the most effective structural monitoring system possible. From these findings an optimised textile was then produced for testing, with a structural element as explained below.

The strain sensitivity of the polymer/ carbon nanotube composites mono filaments in the textiles underwent a measurement of the conductivity change of the materials, while subjected to a given strain profile. The conductivity was measured by a two-point method: a combination of a picoameter (Keithley 6485) and a DC voltage source (Agilent 6614C). The strain profile was set by a computer-controlled Instron tensile test machine. The data were visualised as clearly readable graphs. Input from a software-programming engineer would provide the potential for further research to be visualised in 3D form, representing the structure being analysed. This area of expertise was not part of this research, however.

Fig.2 shows an example of initial sample for testing and the Nanoforce laboratory, whilst Fig.3 is an illustration of the results of the strain sensing experiments on the fabrics: NanoForce tested at least 3 specimens from each fabric samples but were not able to test one of the fabrics as there was too much pretension on the elastic TPU fibres and they could not get a flat specimen. The charts in Fig.3 refer to one specimen, labelled with the fabric name followed by a number indicating the repetition, showing Load [N] Vs Strain [-]. The complete results are market sensitive, still under analysis and confidential therefor this is an indication of our methods rather than publishing findings.

Testing the textiles in the first stage addressed the following issues:

1) Whether the strength of the warp relative to the weft distorts the readings.

2) Whether some of the woven samples hold the strain sensitive polymer in only one direction, thus restricting the kind of information transmitted.

3) The accuracy with which complex structural movement can be read when the textile is integrated into a structure.

4) The degree to which the design of the textile is able to reduce the proportion of polymer to ensure the textile is financially viable while maximising the accuracy and complexity of the structural analysis. Understanding the ways in which the polymer behaves and transmits in a woven structure is identified as an integral component to the research. Following this analysis, the polymer could be modified by NanoForce to particular specifications.

3.3.7 Data collection

The second stage of the experiment was to code an application capable of transmitting data from nanofibre woven in a particular density in a particular structure. This kit successfully produced data via RFID technology in a sedentary state, seen on a computer screen as a gently pulsating horizontal line. When the woven textile moved the signal produced an erratic line on the screen, depending on the agitation. Therefore, the authors hypothesise that, if the textile were incorporated into a structure, when the building was motionless the signal from the sensor would remain constant or, in this case, a horizontal line. When the textile on a building was affected by seismic waves, the textile would transmit to a screen on a computer or mobile device differently to inform those with access to the screen. The proof-of-concept experiment was carried out under laboratory conditions in which the textile was placed on the corner of a washing machine representing a key structural joint in a building. The transmission while relatively motionless on the wash cycle was represented as a horizontal line appearing on a laptop two rooms (16 m) away by bluetooth. When the washing machine moved to spin cycle and the textile patch vibrated, the line became erratic peaks and troughs, which appeared on the laptop screen visualising movement of the textile on a vibrating structure. This could represent vibration of an individual dwelling or commercial building. The experiment was repeated 10 times in the laboratory and results were recreated and recorded. In terms of 'mapping' and making sense of the collected data captured by SHMS, it must be noted that this would be part of a future long term project which could benefit the design and consent process for buildings, structures, civil engineering and land choice when building in earthquake zones.

Textiles, architecture, nanotechnology and engineering developers collaborated successfully to develop this intelligent textile mock-up prototype to demonstrate that it could be possible to embed a smart monitoring textile into a structural system to sense movement and warn of potential structural failure.

This prototype has the potential to gather more structural monitoring information for analysis, for use when deconstructing a building, or to provide information for architects on materials and methods to use when designing a building in earthquake zones. The common electrical problem for monitoring in earthquake zones is that infrastructure is destroyed, disabling effective communication. The unique selling point of SHAKES is that the telecommunication infrastructure is built into the sensor within the fabric. Therefore, providing the joint is 'alive', it is possible to transmit information. This differs from the 'smart communicative' materials already on the market, such as fibre-optic glass which is easily broken, limiting its communication capabilities. Here, the nanofibre textile material will continue to transmit as it is difficult to 'break'. However, additional power is needed within this textile to act as an autonomous power source so, if power is lost, the textile will maintain transmission for at least a few hours until the data is collected remotely or a power supply is reconnected. An inbuilt battery may be solar powered. Another 'smart' element is an internet connection, which is why the textile runs with a bluetooth component and so it can continue to transmit, collect and store data during power disruption. When creating the prototype, these factors of power source and supply were imperative to the design process. The prototype contains technology similar to that seen in mobile telephones, but highly adapted for critical situations. Emergency and first response personnel use hand-held mobile technology in disaster situations; the possibility exists for them to interface with SHAKES retrieving real-time relevant information to assist in evacuation, rescue and recovery.

3.3.8 Conclusion

The prototype is new 'combination technology' created through Design Thinking and transdisciplinary innovation for earthquake disaster zones. The prototype's textile fibres are always 'alive' with power. The bluetooth module has a sequence of passwords to control access to information, enabling engineers/ technicians to monitor the textile independently, when needed. Variable memory storage allows real-time or historical data to be recorded and retrieved; the size of the storage unit would depend on the intended deployment of a SHAKES unit, and would impact on production and retail costs. The SHAKES prototype used a small memory (1GB). The bigger the memory and, therefore, the more data collected, the more comprehensive the results from the prototype will be.

The technology provides opportunity for employment in areas of analysis and maintenance; through the use of mobile technology this may be done remotely and there exists potential to automate parts of the process. The imperative is to retrieve information from a unit with as short a delay as possible following an event to allow engineers to decide on a structure's future-repair or demolition.

Civil engineering training could develop electrical technology for sensor data conversion; as an emerging field this may require a significant amount of time to perfect. However, the critical factor in this research, at this time, is the prototype bluetooth and smart material technology, not its analysis. Eventually, civil engineers will be able to analyse the data collected in the same way that clinicians and medical technicians were able to analyse the vital signs data gathered by the intelligent textiles previously discussed upon which SHAKES was based. In stating this, the tool to start recording and collecting earthquake movement is the proof-of-concept prototype acknowledged in this research as SHAKES. This type of kitchen table design activism research [24] has the potential to influence governmental policy and public opinion in order to incorporate cost effective monitoring which could prevent further disaster by aiding the understanding of building safety and its behaviour during earthquakes.

Currently there are sensors which study the phenomenon of earthquakes, as in the Pacific. These sensors monitor earth movement, not necessarily building movement, making the SHAKES prototype unusual. Existing technology employs fibreglass cables, and engineers determine whether light is passing through these cables or not, acknowledging if the material is alive or dormant and if there has been damage when the light flow stops. However, this technology is one-use only and unsustainable as it has to be re-installed every time there is earth movement. In contrast, smart textile technology may be reused, may cover multiple earthquakes and large regions, does not break, and is more environmentally friendly and financially viable.

As the prototype textile sensors provide realistic and workable analysis readings, the next stage is to test the fabric in combination with compressive and tensile building materials to explore the conductive potential of the textile within a structural system at key points. Research partner Cheyne proposed that section of a post-and-beam structure is fabricated to experiment with the sensing textile. The construction materials of concrete, masonry and steelwork have so far been prohibitively expensive to test the textile in this research project therefore funding and new collaborators will be sought to progress the project.

4. Conclusion

The paper presents funded research in development with unusual partnerships between textiles, biomedical science, business, engineering and design collaborators focusing on transdisciplinarity in conjunction with Design Thinking methodologies for innovative human and environmental wellbeing solutions. The Barrier Solutions project shows there may be an alternative multifaceted approach to excessive sun exposure to address the burgeoning situation of melanoma damage using ICT, fashion and oncology teams. Similarly, DERMA proposes an international collaboration framework between science and communication, business and design partnerships for new materials for ageing well. SHAKES, too, combines design and technology-enhanced fabrications and construction with geological monitoring for an early warning system in earthquake regions of the globe. The projects' research teams synergistically focus on a societal issue then develop together a series of solutions and tests using smart materials, creative technologies, science and applied research for societal good using a transdisciplinary and Design Thinking approach for new outcomes.

5. Acknowledgement

We would like to acknowledge support from the EU Interreg 2-Seas programme, Santander Bank, Innovate UK and Grand Challenges.

SHAKES research project contributions from Kate Cheyne, University of Brighton and Kirsty McDougall, Royal College of Art.

References

[1] Piaget J. The epistemology of interdisciplinary relationships. In: Apostel L, Berger G, Briggs A, Michaud G, editors. Interdisciplinarity. Problems of teaching and research in universities. Paris: OECD, 1972; p: 127-139.
[2] Information on http://www.arj.no/2012/03/12/disciplinarities-2/
[3] The Department of Business, Innovation and Skills. Success as a Knowledge Econonmy: Teaching Excellence, Social Mobility and Student Choice. London 2016; p: 16
[4] Lang D, Wiek A, Bergmann M, Stauffacher M, Martens P, Moll P, Swilling M, Thomas C. Transdisciplinary research in sustainability science: practice, principles, and challenges. Sustainability science: bridging the gap between science and society. Published online 04 February 2012. DOI: 10.1007/s11625-011-0149-x
[5] Cronin K. Transdisciplinary research (TDR) and sustainability overview. Report prepared for the Ministry of Research, Science and Technology (MoRST), 2008; p: 1-36.
[6] United Nations Educational, Scientific and Cultural Organisation. Towards knowledge societies. UNESCO World Report. Condé-sure-Noireau, France: UNESCO, 2005; p: 1-220.
[7] Cope B, Kalantzis M. 'Design' in principal and practice: a reconsideration of the terms of design engagement. Design J 2011; 14: 45-63.
[8] Brown T. Design thinking. Harvard Bus Rev June 2008; 84-92.
[9] Information on https://www.innocentive.com/design-thinking-real-world-solutions
[10] Ridell OP. Who is the active consumer? Insight into contemporary innovation and marketing practices. PhD thesis, Department of Business Studies, Uppsala University, Sweden 2013; 1-95.
[11] Stoppa M, Chiolerio A. Wearable electronics and smart textiles: a critical review. Sensors 2014; 14: 11957-11992.
[12] Information on http://www.hamiltonsun.com.au/about/

[13] Information on http://business.time.com/2013/05/30/5-reasons-youre-paying-too-much-for-sunscreen/
[14] Jain SK, Jain NK. Multiparticulate carriers for sun-screening agents. J Cosm Sci 2010; 32: 89-98.
[15] Information on http://ec.europa.eu/health/population groups/elderly/index en.htm
[16] Harding K. Innovation and wound healing. J Wound Care 2015; 24: 7-13.
[17] Augustin M. Cumulative life course impairment in chronic wounds. Curr Probl Dermatol 2013; 44: 125-129.
[18] Information on http://www.rcuk.ac.uk/innovation/impacts/
[19] Information on http://www.nanoforce.co.uk/
[20] Information on http://earthquake.usgs.gov/earthquakes/eqarchives/year/eqstats.php
[21] Information on http://earthquake.usgs.gov/earthquakes/world/world deaths.php
[22] Information on http://www.bloomberg.com/news/articles/2011-02-28/new-zealand-quake-damages-may-cost-as-much-as-15-1-billion
[23] Zhang R, Bilotti E, Deng H, Baxendale M, Peijs T. Fabrication and Property Prediction of Conductive and Strain Sensing TPU/ CNT Nanocomposite Fibres. Journal of Materials Chemistry 2010; 20: 9449 – 9455.
[24] Farrer J, Goulev P. Interactive design activism. In: Pit J, editor. The computer after me: awareness and self-awareness in automatic systems. London: Imperial College Press, 2014; p:147-158.

Physiological Comfort in State Administrative Institution Employees' Clothing (Case Study)

Iveta Ābele[1], Inese Šroma[2*], Inese Ziemele[3], Ausma Viļumsone[4]

[1, 2, 3, 4] *Institute of Design Technologies, Faculty of Material Science and Applied Chemistry, Riga Technical University, Kipsalas Str.6, Riga, LV-1048, Latvia*

[*]Corresponding author's email: inese.sroma@edu.rtu.lv

Abstract

The thermal resistance of the Republic of Latvia state administrative employees' functional daily clothing basic set layering has been determined with the aim to predict the level of wearer's physiological comfort, body cooling or possible heat stroke wearing particular clothing in its guaranteed period of use. For the study by means of a thermal manikin according to ISO 15831, four experimental functional daily clothing sets from waredrobe of the State Police and State Border Guard were chosen. The clothing sets studied were found to have correspond largely to the officers' specific job assignments at the winter, spring, summer and autumn seasons of the climatic conditions of Latvia. The factors causing body cooling or heatstroke have been identified which may pose a threat to the officer's health and life during performance of his/ her job duties. For evaluation of the experiment's results and the prediction of the clothing functionality the calculation methods of ISO 11079 and ISO 7933 have been used.

Keywords: Functional clothing; Prediction; Thermal resistance; IREQ; Physiological comfort

1. Introduction

Upon analysing the Republic of Latvia (hereinafter - Latvia) state enactments related to state employees' uniforms, special clothing and personal protective equipment [1-4] as well as upon carrying out a study on the satisfaction level among officers with the state-provided clothing [5] it has been concluded that satisfaction with functional clothing is fair. It can be explained by the fact that functional clothing is not designed and completed as one complete layering of clothing. State structural units do not procure and maintain functional clothing and personal protective equipment as one system of clothing ensemble. It has also been determined that there is no uniform documentation on basic and functional requirements for daily functional clothing and protective equipment. Thus, it is not possible to predict the properties of these sets under various environmental conditions and use intensity conditions and the officer performance, as well as functional clothing life cycle monitoring can not be carried out.

Basic requirements for daily uniforms are determined by the officer's official duties and tasks under different environmental conditions (rain, wind, cold, heat etc.). The functional properties (threats reduction or prevention under different ambient conditions) of the clothing are determined by the elements of the clothing sets and variants of their layering. These properties can be described by physiological comfort of the officer, which is assessed subjectively basing on sensations. The wearer receives sensations through his/ her sensory system, and different factors and their combinations influence them.

According to Y. Li, thermophysiological comfort means achieving a comfortable heat and moisture balance, which includes transport of heat and moisture through fabric [6]. It has been determined that sensations depend on the clothing fabrics air-permeability and their layering, thermoregulation and capacity to absorb, transfer or remove moisture. Therefore, upon designing clothing it is essential to predict its level of physiological comfort in use. It has been determined that microclimate under clothing is essential for physiological comfort of the wearer, as the air between the skin and fabric or fabric layers has thermal insulation function [7]. In turn, thermal regulation is a complex of the physiological reactions of organism ensuring constant body temperature. The so-called human core body temperature's norm T_{ref} is 37 °C and its changes within the limits 2-2.5 °C induce disorders of ongoing life processes [8].

2. Problem formulation

It is predicted that the metabolic rate M of the Latvia state administrative institution employees' work intensity can vary from 65 Wm^2 to 520 Wm^2 [9] / 522 W/m^2 [10] (see Table 1). Wearing of inadequately completed functional clothing may result in decrease of the officer's body core temperature - hypothermia (< 26.5 °C) or a heat stroke - hyperthermia (> 37.8 °C), which can be lethal for a human [11].

Table 1. Classification of metabolic rate M in accordance with ISO 8996 and ISO 7243 [9,10]

Class	Average metabolic rate, W/m^2	Max metabolic rate, W/m^2	Expected officer activities
0 – resting	65 (55-70)	115 (100-125)	relaxation or rest state
1 – low metabolic rate	100 (70-130)	180 (125-235)	sedentary office work; checking of persons crossing the state border, vehicles cargo documents; moving of light items; driving under normal conditions; patrolling at speed of up to 2.5 km/h (ISO 7243 up to 3.5 km/h)
2 – moderate metabolic rate	165 (130-200)	295 (235-360)	driving of a truck or vehicle in difficult conditions; moving of medium heavy items; patrolling at speed of 2.5-5.5 km/h (ISO 7243 3.5-5.5 km/h)
3 – high metabolic rate	230 (200-260)	415 (360-465)	intensive work outside the office – arresting of criminals, violators or trespassers, convoy missions; patrolling in full gear at speed of 5.5-7 km/h; moving of heavy items
4 – very high metabolic rate	290 (> 260)	520 (> 465)	extremely intensive work at a quick to maximum tempo outside the office – climbing stairs, ramps; running in full gear; active arresting of criminals, violators or trespassers; patrolling at speed of over 7 km/h

The optimum thermal insulation value of the Latvian state Ministry of Interior institutions system employees' basic sets of daily functional clothing for performance of various duties in the climatic conditions of Latvia has not been determined neither documentary nor practically. To predict the compliance of thermal insulation values it is advisable to use NATO's requirements for thermal insulation properties of military uniforms for respective climate zoning [12]. In accordance with the document AECTO-230, winter, summer, spring and autumn climatic conditions of Latvia correspond to zoning from A2 to C2 or from - 46 °C to + 44 °C [12].

To evaluate the influence of hot, moderate or cold environment on individual, his/her work ability, level of comfort in clothing, ergonomics standards ISO 7730 [13], ISO 7243 [10], ISO 7933 [14], ISO 9886 [15], ISO 10551 [16] and ISO 11079 [17] are used [18]. Hot environment is assessed according to ISO 7243, using wet bulb globe temperature index WBGT, or according to ISO 7933 standard method, determining the required sweat rate Sw_{req} [18]. Assessment of moderate environment is made according to ISO 7730 by calculating the predicted mean vote PMV and the predicted percentage of dissatisfied PPD values [18]. For assessment of cold environment ISO 11079 is applied calculating the required clothing insulation values $IREQ_{neutral}$, $IREQ_{min}$ and $IREQ_{max}$ [18]. ISO 9886 contains references for assessment of physiological measurements of individual subjects and particular groups of occupants, and in ISO 10551 a method can be found for subjective assessment of thermal environment [18]. In order to assess any thermal environment, each of the above mentioned standards may be used separately or together with the following ergonomics standards - ISO 7726 [19], ISO 8996 [9], ISO 9920 [20] and ISO 15265 [21] [18].

Special attention should be paid to the risk of heat stroke in case of heavy workload in summer. Occurrence of heat stroke may be provoked by:

- Natural hazards of the working environment - high air temperature Ta (°C), air velocity v (m/s) and relative humidity RH (%);

- High physical load affecting officers during execution of official duties (arresting of criminals, violators or trespassers in woods, swampy areas, on roads and water);
- The weight of the officer's outfit, personal protective equipment, arms, ammunition, communication means and uniform m (kg) (convoy missions and patrolling in full gear);
- Limited heat exchange with the environment through the functional clothing and equipment.

These factors have a simultaneous and continuous impact on the officer carrying out work-related tasks thus creating a specific hazard for his/ her health and life. All the aforementioned factors are represented as particular values (wet bulb globe temperature index WBGT, metabolic rate M, effective mechanical power W, required sweat rate Sw_{req}, basic clothing insulation Icl etc.) in the heat stroke calculation models. In order to predict the heat stroke the calculation model for estimation of the core temperature (T_{ref}) is used; additionally the PHS2 calculation model [22] elaborated by Lund University (Sweden) can be applied, as well as the calculation model provided in ISO 7933 [14].

The above calculation systems allow determining the following heat stroke parameters:

- Core temperature T_{ref} and the possibility of heat stroke, %;
- Core temperature variation curve as a function of time, min;
- Volume of water loss of organism, required sweat rate Sw_{req}, W/m^2;
- Long term risk assessment, time, min.

Upon determination of the above values it is possible to predict the potential work ability loss of the personnel due to heat stroke and the time of occurring of the heat stroke in order to make corrections in the layering of the functional clothing or the work hours timely.

3. Experimental part

Clothing of seven levels (level one, two and three underwear, level 4-7 jacket and pants, summer and winter footwear, headgear and gloves) is planned for the basic structure of everyday working functional clothing of the Latvian Ministry of Interior system administrative institution employees. By mutually combining selected uniform items, versions of clothing ensembles and their structure have been determined in accordance with the specific character of the service (tasks and work environment conditions) for summer, winter, autumn and spring.

Table 2. State administrative institution SP and SBG employees' clothing system fabrics

Set clothing layer	Fabric description	Fabric mass per unit area, g/m^2
Level 1 underwear	100 % polyester plaited jersey	125 ± 12.5
Level 2 underwear	97 % polyester, 3 % silver coated fiber, face-jersey mesh, back – brushed jersey	115 ± 8,625
Level 3 underwear	93 % polyester, 7 % spandex circular knit plaited jersey	224 ± 22
Level 4 jacket and pants	Pants: 65 % polyester, 35 % cotton	–
	Outer fabric: 60 % cotton, 40 % nylon	160 – 200
Special coverall	Inner fabric - lining: 100 % polyamide	
	Insulation material	
Level 5 jacket and pants	100% polyester circular knit terry, doublesided, heavyweight velour	363 ± 36.3
Level 7 jacket and pants	Outer fabric: face fabric material - 100 % polyester, woven fabric.	108 ± 10.8
	Functional layer - membrane based on ePTFE	
	Inner fabric - lining: 100 % polyamide	55 ± 5
	Insulation material - polyester:	
	jacet body	220 ± 22
	sleeves and pants	145 ± 14.5
Cap/head/neck wear	Breathable, thin, flexible, wear-resistant fabric	–
Winter cap	100% polyester circular double velour	207 ± 20.7
Level 2 gloves	100 % polyester	329 ± 32.9

Table 3. The studied everyday functional clothing sets of the SP and the SBG

Set No	Level 1 underwear	Level 2 underwear	Level 3 underwear	Level 4 jacket and pants	Special coverall	Level 5 jacket and pants	Level 7 jacket	Level 7 pants	Cap / head / neck wear	Winter cap	Level 2 gloves	Socks	Winter boots	Summer boots	Safety helmet	Season	Set abbreviation
1	X	X	X			X	X	X	X	X	X	X	X			cold winter	CW
2	X			X								X		X	X	summer	SM
3		X		X			X		X	X	X	X	X			spring/autumn	SpA
4	X				X				X			X		X	X	all-season (combat uniform)	AS

Thermal resistance of the fabric layering in the most characteristic variants of the most characteristic clothing sets has been determined to predict the level of employees' physiological comfort and its duration or possible heat stroke while wearing particular clothing when performing official duties under particular environmental conditions. Classification of clothing items by levels according to their fabric properties is summarized in Table 2.

Four experimental functional daily clothing sets of the State Police (SP) and and State Border Guard (SBG) of Latvian Ministry of Interior were completed for the research (see Table 3). They correspond to the specifics of the officers' tasks and the seasonality of environment - winter, summer, autumn and spring wearing seasons.

Upon elaboration of the layer sets of officers' functional clothing the clothing layering principle has been followed. Fabrics having good thermal resistance properties (e.g., Level 3 underwear of SP and SBG Rct > 0.58 m^2K/W) and vapour permeability properties have been selected.

Thermal insulation and water-vapour resistance tests were made in the climate chamber of Lund University, Design Science, EAT, Thermal Environment Laboratory by means of a thermal manikin according to ISO 15831 [23]. In the Thermal Environment Laboratory climate chamber using thermal manikin the following values have been determined (see Table 4):

Itot - total thermal insulation of the clothing ensemble with the manikin stationary;

Icle - effective thermal insulation of the clothing ensemble with the manikin stationary;

Icl - basic clothing insulation with the manikin stationary;

Itr - resultant total thermal insulation of the clothing ensemble with the manikin moving;

Iclr - resultant basic thermal clothing insulation with the manikin moving;

Icler - resultant effective insulation with the manikin moving;

Ret - total clothing evaporative resistance and boundary air layer;

Recl - clothing evaporative resistance.

In the experiment air velocity v is 0.4 m/s and walking speed w 0.98 m/s. In the manikin test the air layer thermal resistance between the skin and the lower layer of clothing ensemble layering is 0.076 m^2K/W. To describe the thermal resistance characteristics of clothing layering the unit clo (1clo = 0.155 m^2K/W) was used as well.

4. Results and Discussion

The heat resistance of clothing layering Rct (m^2K/W) has been measured according to ISO 15831:2004 [23] and to LVS EN ISO 11092:2014 [24]. Upon analysing the obtained data metabolic rates associated with the load employees are exposed to were determined and calculations of duration limited exposure Dlim were made according to ISO 11079 [17].

The functional clothing basic set should ensure the officer's physiological comfort in its guaranteed period of use taking into account the climatic conditions in Latvia. According to ISO 11079 there exists a hypothermia risk for an officer wearing clothing which has thermal insulation Iclr < $IREQ_{min}$, and it is necessary to have a break during the work [17]. If clothing thermal insulation Iclr > $IREQ_{neutral}$, hypothermia of an officer is possible [17]. The clothing having thermal insulation Iclr in the interval between $IREQ_{min}$ and $IREQ_{neutral}$ has the regulatory function [17]. The higher the IREQ value at any officer activity level, the higher the environmental influence on cooling of the body [17].

According to ISO 11079 upon modelling employee metabolic rate M (W/m^2) the required clothing insulation IREQ and duration limited exposure Dlim (see Table 5, 6 and 7) have been predicted for SP and SBG officers' functional clothing sets designed for wearing in a cold winter, summer, spring and autumn, and for the all-season special task set. Performing calculations following indices have been defined: ambient air temperature Ta (°C), mean radiant temperature Tr (-25 °C winter, -5 °C and +5 °C spring/autumn), air permeability p (5 l/m^{2}s), relative air velocity v (0.4 m/s), relative humidity RH (65 %), walking speed w (m/s) and resultant basic thermal clothing insulation with the manikin moving Iclr (clo). The employee metabolic rates used in the calculations M (W/m^2) and walking speed w (m/s) are assumed according to ISO 8996, ISO 7243 and ISO 11079 [9,10,17]. The air temperatures Ta correspond to the moderate climate conditions in Latvia in winter -25 °C, in spring and autumn -5 °C and +5 °C.

Upon modelling employee metabolic rate M (W/m^2) for winter season (-25 °C) the summer functional clothing set SM is not analyzed as it is only devised for indoor wear and under summer climatic conditions in Latvia, corresponding to zoning from A2 to B3 or from 23 °C to 44 °C [12]. Upon modelling employee metabolic rate M (W/m^2) for autumn and spring seasons (-5 °C / +5 °C) the functional clothing winter set CW is not analyzed as it is devised for winter climatic conditions in Latvia, corresponding to zoning from C0 to C2 (from -6 °C to -46 °C) [12].

Table 4. Manikin tests results of SP and SBG sets

Set	Itot, m^{2}K/W	Itot, clo	Icle, m^{2}K/W	Icle, clo	Icl, m^{2}K/W	Icl, clo	Itr, m^{2}K/W	Itr, clo	Iclr, m^{2}K/W	**Iclr, clo**	Icler, m^{2}K/W	Icler, clo	Ret, m^{2}Pa/W	Recl, m^{2}Pa/W
CW	0.508	3.28	0.432	2.79	0.455	2.94	0.453	2.92	0.417	**2.69**	0.401	2.59	63.5	57.3
SM	0.190	1.22	0.114	0.73	0.124	0.80	0.153	0.99	0.109	**0.70**	0.102	0.66	-	-
SpA	0.371	2.39	0.295	1.90	0.313	2.02	0.306	1.98	0.267	**1.72**	0.266	1.64	-	-
AS	0.310	2.00	0.234	1.51	0.250	1.61	0.256	1.65	0.215	**1.39**	0.205	1.32	46.6	39.6

Basing on the calculated required clothing insulation $IREQ_{neutral}$ values it is concluded that SP and SBG officers are at hypothermia risk outside premises in winter conditions (Ta = -25 °C, v = 1.2 m/s) wearing functional clothing set cold winter CW only when relaxing or in rest state (M < 80 W/m^2) for longer than 0.6 hours, as thermal insulation of functional clothing Iclr < $IREQ_{min}$, i.e., 2.69 < 5.0 clo (see Table 5). The risk of developing hyperthermia exists already in case of moderate metabolic rate (M > 175 W/m^2), performing official duties longer than for 8 hours as thermal insulation of functional clothing Iclr > $IREQ_{neutral}$, i.e., 2.69 > 2.2 clo (see Table 5).

Table 5. IREQ and Dlim calculations for functional clothing sets in winter season, Ta = Tr = -25 °C [17]

Metabolic rate, W/m^2		80		80		140		140		175		230		290	
Walking speed, m/s		0.1		0.7		0.7		1.2		1.2		1.2		1.2	
Set	Iclr, clo	IREQ, clo		IREQ, clo		IREQ, clo		IREQ, clo		IREQ, clo		IREQ, clo		IREQ, clo	
		min	neutral	min	neutral	min	neutral	min	neutral	min	neutral	min	neutral	min	neutral
CW	**2.69**														
SpA	**1.72**	5.0	5.3	5.0	5.4	2.6	2.9	2.6	2.9	1.9	2.2	1.3	1.6	0.9	1.2
AS	**1.39**														
Set	Iclr, clo	Dlim, h		Dlim, h		Dlim, h		Dlim, h		Dlim, h		Dlim, h		Dlim, h	
		min	max	min	max	min	max	min	max	min	max	min	max	min	max
CW	**2.69**	0.6	0.7	0.6	0.6	2.1	> 8	1.5	3.1	> 8	> 8	> 8	> 8	> 8	> 8
SpA	**1.72**	0.3	0.4	0.3	0.3	0.5	0.7	0.4	0.6	0.7	1.1	4.1	> 8	> 8	> 8
AS	**1.39**	0.3	0.3	0.2	0.3	0.3	0.4	0.3	0.4	0.4	0.6	0.9	3.6	> 8	> 8

Table 6. IREQ and Dlim calculations for functional clothing sets in spring/autumn season, Ta = Tr = -5 °C [17]

Metabolic rate, W/m²		80		80		140		140		175		230		290	
Walking speed, m/s		0.1		0.7		0.7		1.2		1.2		1.2		1.2	
Set	Iclr, clo	IREQ, clo		IREQ, clo		IREQ, clo		IREQ, clo		IREQ, clo		IREQ, clo		IREQ, clo	
		min	neu tral	min	neu tral	min	neu tral	min	neu tral	min	neu tral	min	neu tral	min	neu tral
SM	**0.70**														
SpA	**1.72**	3.0	3.4	3.1	3.4	1.4	1.7	1.5	1.8	1.0	1.3	0.6	0.9	0.3	0.6
AS	**1.39**														
Set	Iclr, clo	Dlim, h		Dlim, h		Dlim, h		Dlim, h		Dlim, h		Dlim, h		Dlim, h	
		min	max	min	max	min	max	min	max	min	max	min	max	min	max
SM	**0.70**	0.3	0.3	0.2	0.3	0.3	0.5	0.3	0.4	0.4	0.7	0.8	> 8	> 8	> 8
SpA	**1.72**	0.7	0.9	0.6	0.8	3.7	> 8	2.1	> 8	> 8	> 8	> 8	> 8	> 8	> 8
AS	**1.39**	0.5	0.6	0.5	0.5	1.2	3.5	0.9	2.1	3.8	> 8	> 8	> 8	> 8	> 8

Table 7. IREQ and Dlim calculations for functional clothing sets in spring/autumn season, Ta = Tr = +5 °C [17]

Metabolic rate, W/m²		80		80		140		140		175		230		290	
Walking speed, m/s		0.1		0.7		0.7		1.2		1.2		1.2		1.2	
Set	Iclr, clo	IREQ, clo		IREQ, clo		IREQ, clo		IREQ, clo		IREQ, clo		IREQ, clo		IREQ, clo	
		min	neu tral	min	neu tral	min	neu tral	min	neu tral	min	neu tral	min	neu tral	min	neu tral
SM	**0.70**														
SpA	**1.72**	2.0	2.4	2.1	2.4	0.9	1.2	0.9	1.2	0.6	0.9	0.3	0.5	0.1	0.3
AS	**1.39**														
Set	Iclr, clo	Dlim, h		Dlim, h		Dlim, h		Dlim, h		Dlim, h		Dlim, h		Dlim, h	
		min	max	min	max	min	max	min	max	min	max	min	max	min	max
SM	**0.70**	0.4	0.5	0.4	0.4	0.7	1.7	0.6	1.3	1.2	> 8	> 8	> 8	> 8	> 8
SpA	**1.72**	1.7	3.3	1.3	2.2	> 8	> 8	> 8	> 8	> 8	> 8	> 8	> 8	> 8	> 8
AS	**1.39**	1.0	1.5	0.8	1.2	> 8	> 8	> 8	> 8	> 8	> 8	> 8	> 8	> 8	> 8

Thus, while performing intensive official duties, breaks for rest and physiological needs should be provided. If official duties are to be performed outside premises in winter conditions (Ta = -25 °C, v = 1.2 m/s) for a prolonged period and intensively, then hyperthermia risk can be reduced by not putting on Level 2 or Level 3 underwear.

There exists officer hyperthermia risk while performing official duties for longer than 4.1 hours at metabolic rate M > 230 W/m² (patrolling/ running in full gear) wearing functional clothing set spring/ autumn SpA in winter conditions (Ta = -25 °C, v = 1.2 m/s) (see Table 5). An officer at high metabolic rate (M = 230 W/m²) wearing all-season combat uniform AS will feel comfortable in winter conditions (-25 °C) for 0.9-3.6 hours, as $IREQ_{min}$ < Iclr < $IREQ_{neutra}$ [17], i.e., 1.3 < 1.39 < 1.6 clo (see Table 5).

It should be pointed out that in case of low (M < 140 W/m²) and average (M < 175 W/m²) officer's metabolic rate when performing passive job duties (patrolling on the street) outside premises in winter conditions (-25 °C) even for a short 0.3-0.8 hour period, hypothermia is possible as well when wearing functional clothing sets spring/autumn SpA and all season AS (see Table 5), as Iclr < $IREQ_{min}$. Performing job duties in case of low and moderate metabolic rate, the hypothermia risk can be reduced by putting on Level 1-4 garments supplementary.

Functional clothing summer set SM of SP and SBG officers is not suitable for performing official duties outdoors in winter, autumn and spring conditions not wearing supplementary Level 2 and/or Level 3 underwear, Level 5 or Level 7 jacket and winter boots, as it is designed only indoor wear in summer under the climatic conditions of Latvia. When due to job specifics it is necessary to perform official duties outside premises and wearing the office functional summer clothing set, then officer prognosis of hyperthermia risk is as follows.

There is officer hypothermia risk when wearing functional clothing summer set SM while performing official duties over 0.2-0.7 hours outside premises in autumn and spring seasons (-5 °C) in case of low and average metabolic rate (M < 175 W/m²) (see Table 6). Outside premises in autumn and spring seasons (+5 °C), in case of low metabolic rate (M < 140 W/m²) activities performing passive official duties longer than 0.4-1.7 hours (see

Table 7). It can be explained by the fact that SP and SBG functional clothing summer set SM thermal insulation Iclr is only 0.70 clo, as Level 4 shirt and trousers thermal resistance approaches zero value (2.93 mK*m^2/ W for shirt and 1.66 mK*m^2/ W for trousers, thermal resistance has been determined by Permetest measuring instrument).

There exists officer hyperthermia risk when wearing functional clothing summer set SM in autumn and spring conditions (-5 °C and +5 °C) only while performing intensive (M > 230 W/m^2) and very intensive (M > 290 W/m^2) official duties outside the office longer than for 8 hours, inclusive of breaks for rest and physiological needs (see Table 6 and 7).

According to ISO 7730, upon modelling employee metabolic rate M (W/m^2) the predicted mean vote PMV and the predicted percentage of dissatisfied PPD VP and VRS have been predicted for the officer functional clothing summer set SM, anticipating wearing it under the environmental conditions during summer (see Table 8). Performing calculations the following parameters have been defined: air temperature Ta (+15 °C, +25 °C and 30 °C outside the premises; +15 °C in the office), air velocity v (1 m/s outdoors, 0.1 m/s in the office), relative humidity RH (65 %) and functional clothing summer set SM Iclr = 0.70 clo, where PPD from -3 cold to +3 hot.

Table 8. Assessment of emploee physiological comfort in summer wearing functional clothing summer set SM (Iclr = 0.70 clo) according to ISO 7730 [13]

Summer set SM	Seasonal temperature Ta = +15 °C				Seasonal temperature Ta = +20 °C, Ofiss			
	Metabolic level, W/m^2				Metabolic level, W/m^2			
	80	140	175	230	80	140	175	230
PMV	-2.83	-0.68	0.02	1.05	-0.43	0.86	1.46	2.42
PPD, %	98.1	14.8	5	28.1	8.8	20.8	48.9	91.7
Summer set SM	Seasonal temperature Ta = +25 °C				Seasonal temperature Ta = +30 °C			
	Metabolic level, W/m^2				Metabolic level, W/m^2			
	80	140	175	230	80	140	175	230
PMV	0.14	1.35	1.99	3.03	1.69	2.42	3.02	4.08
PPD, %	5.4	43.0	76.2	99.3	61.2	91.5	99.2	100

In accordance with the calculation results (see Table 8) it will be possible to wear the functional clothing summer set SM in cool summer and autumn/spring (+15 °C, 1 m/s) when performing tasks at moderate workload (M = 175 W/m^2) out of the office. In the office at +20 °C the officer will feel himself/herself comfortable wearing the functioal clothing set SM when carrying out passive official duties or relaxing and resting (M < 80 W/m^2).

There exists the officer's hyperthermia risk in the office (+20 °C) and out of premises in hot summer (+25 °C) at metabolic rate M < 140 W/m^2. When performing work of a heavier workload (M > 140 W/m^2), officers will feel themselves warm and hot as PMV > 0.5 [21]. In a very hot summer (+30 °C,1 m/s) officer functional clothing summer set SM, which includes a long sleeve shirt and Level 1 T-shirt, is not suitable for officer work as a great risk to get too hot can be predicted.

Based on the calculated $IREQ_{neutral}$ values it is concluded that hyperthermia risk for officers wearing functional clothing sets spring/autumn SpA and all season AS there exists hyperthermia risk while performing official duties longer than for 3.8-8 hours (see Table 6) outside premises in autumn and spring (-5 °C) at an average metabolic rate (M = 175 W/m^2, w = 1.2 m/s). At low metabolic rate (M = 140 W/m^2, w = 0.7-1.2 m/s) activities outside premises in autumn and spring seasons (+5 °C) while performing passive official duties longer than for 8 hours, inclusive of breaks for rest and physiological needs (see Table 7).

In turn, in the above mentioned functional clothing sets spring/autumn SpA and all season AS hypothermia risk in officers already exists outside premises in autumn and spring seasons (-5 °C and +5 °C) at low metabolic rate (M < 80 W/m^2) while performing quite passive official duties (checking of documents, sorting of light items, inspection) or relaxing and staying in rest state longer that for 0.5-0.8 hours (-5 °C, see Table 6) and longer than for 0.8-3.3 hours (+5 °C, see Table 7). If official duties are to be performed outside premises in autumn or spring conditions (-5 °C to +5 °C) for a prolonged period and passively, hypothermia risk can be reduced by putting on Level 2 or Level 3 underwear. The results of the study are summarized in Table 9.

Table 9. Ensured physiological comfort in experimental functional clothing of state administrative officers

Sett	Season, temperature	Metabolic rate, W/m^2	Walking speed, m/s	Comfort duration, h
CW	winter, -25 °C	140	0.7 – 1.2	1.5 – 8
	spring/autumn, -5 °C	230	1.2	0.8 – 8
	spring/autumn, +5 °C	175	1.2	2.1 – 8
SM	Ta = +15 °C, v = 1 m/s	175	0.7 – 1.2	< 8
	Ta = +20 °C, ofice, v = 0.1 m/s	80	< 0.7	< 8
	Ta = +25 °C, v = 1 m/s	80	< 0.7	< 8
SpA	spring/autumn, -5 °C	140	1.2	2.1 – 8
AS	winter, -25 °C	230	1.2	0.9 – 3.6

5. Conclusion

Under present-day conditions the state guaranteed clothing of the state administrative officers has not only the function of identification, but it serves also as a means of individual protection under specific work and environmental conditions. Targeted selection of physical and mechanical properties of the different garment layers of functional clothing sets and their assessment, and designing of clothing layering taking into account particular environmental conditions and employee metabolic rates allows to prognosticate and influence the officer's physiological comfort in clothing when carrying out his/ her official duties.

The officer's physiological comfort in clothing has been found to be influenced by: the environment (wind, air temperature, moisture, wind rate); intensity of official duties and tasks; the mass of officer's equipment, individual means of protection, weapons, ammunition, means of communication and the uniform; the limited heat exchange of the functional clothing and equipment with the environment and the physiological state of an individual wearer's body.

The level of wearer physiological comfort, body cooling or potential heat stroke has been prognosticated when wearing the SP and SBG experimental functional clothing sets within their guaranteed period of use. In order to determine officers' metabolic rate ISO 8996 and ISO 7243 have been referred to. In order to prognosticate cold and moderate environment influence on officers ISO 11079 has been referred to when calculating the values of required clothing insulation $IREQ_{neutral}$ and $IREQ_{min}$.

It has been established that for the Latvian state Ministry of Interior institutions system employees' functional daily clothing basic sets for carrying out of various tasks under the climatic conditions of Latvia the optimum thermal insulation value has not been determined neither documentary nor practically. No particular requirements have been set for the thermo-physiological comfort of functional clothing and its levels.

The thermal resistance of state institutions employees' basic sets of daily functional clothing layering of typical fabrics has been determined by means of a thermal manikin according to ISO 15831. Upon studying four SP and SBG of Latvia experimental daily functional clothing sets and processing the obtained experiment results by ISO 11079 and ISO 7933 calculation methods, the experimental functional clothing sets have been found to largely correspond to the specifics of the officers' tasks and the seasonality of environment in Latvia's wearing seasons.

It has been concluded that it is possible to reduce the officer's hyperthermia risk by stopping work, not putting on Level 2 or 3 underwear, e.g. of the cold winter set, or increasing the number of openings (ensuring moisture removal). It is possible to reduce the officer's hypothermia risk wearing functional clothing sets Summer SM, spring/autumn Spa and All Season AS in winter, autumn and spring outdoor environment by mandatory wearing of Level 1-4 and/ or Level 5 and/or 7 garments additionally according to seasonal environmental conditions.

In order to assess officers' hypothermia risk in various environments at high and very high metabolic rate (M > 230 W/m^2), the study should be continued according to ISO 7933, determining the core body temperature Tref, required sweat rate Sw_{req} and other parameters.

References

[1] Regulations on Officials with Special Service Ranks Working in Institutions of the System of the Ministry of the Interior and the Prisons Administration of the Description of Uniforms and Distinguishing Marks, of Special Clothing, of Personal Protective Equipment, equipment and Inventory Types and of their Issue Norms:

Cabinet of Ministers Regulation No. 137. Adopted on 10 February 2009. [Online] Information on http://likumi.lv/doc.php?id=188255.
[2] Regulations on Officials of Customs Authorities and Financial Police of the State Revenue Service Uniforms, Badges and Ranks Distinguishing Marks: Cabinet of Ministers Regulation No. 705. Adopted on 3 September 2013. [Online] Information on http://likumi.lv/doc.php?id=259555.
[3] On State Revenue Service: Latvia Law. Adopted on 28 October 1993. [Online] Information on http://likumi.lv/doc.php?id=59902.
[4] Law on Police: Latvia Law. Adopted on 4 June 1991. [Online] Information on http://likumi.lv/doc.php?id=67957
[5] Šroma I, Ziemele I, Rudzika S. The Survey of Latvian State Revenue Service Employees' Self - assessment of their Uniforms. Material Science: Textile and Clothing Technology: 2014; 9: 34-44. ISSN: 1691-3132. e-ISSN: 2255-8888. doi: 10.7250/mstct.2014.006.
[6] Y. Li. THE SCIENCE OF CLOTHING COMFORT. Textile Progress: 2001; 31: 1-2: p. 1-135. [Online] Information on http://www.tandfonline.com/loi/ttpr20.
[7] Bivainytė A, Mikučionienė D. Investigation on the Air and Water Vapour Permeability of Double - Layered Weft Knitted Fabrics. FIBRES & TEXTILES in Eastern Europe: 2011; 19; 3 (86): p. 69-73. [Online] Information on fibtex.lodz.pl/2011/3/69.pdf.
[8] Malaja medicinskaja enciklopedija. M.: Medicinskaja enciklopedija: 1991-96. [Online] Information on http://dic.academic.ru/dic.nsf/enc_medicine/30951.
[9] Ergonomics of the thermal environment - Determination of metabolic rate. ISO 8996, 2004.
[10] Hot Environments - Estimation of the heat stress on working man, based on the WBGT-index (wet bulb globe temperature). ISO 7243, 1989.
[11] Choudhury AKRoy, Majumdar PK, Datta C. Factors affecting comfort: human physiology and the role of clothing. In: Guowen Song, ed. Improving comfort in clothing. Woodhead Publishing Limited: Cambridge, 2011; p: 3-60.
[12] Climatic conditions: Allied environmental conditions and test publication: NATO International Staff - Defence investment division. AECTP-230 (Edition 1), May 2009.
[13] Ergonomics of the thermal environment - Analytical determination and interpretation of thermal comfort using calculation of the PMV and PPD indices and local thermal comfort criteria. ISO 7730, 2005.
[14] Ergonomics of the thermal environment - Analytical determination and interpretation of heat stress using calculation of the predicted heat strain. ISO 7933, 2004.
[15] Ergonomics - Evaluation of thermal strain by physiological measurements. ISO 9886, 2004.
[16] Ergonomics of the thermal environment - Assessment of the influence of the thermal environment using subjective judgement scales. ISO 10551, 1995.
[17] Ergonomics of the thermal environment - Determination and interpretation of cold stress when using required clothing insulation (IREQ) and local cooling effects. ISO 11079, 2007. [Online] Information on http://wwwold.eat.lth.se/Forskning/Termisk/Termisk_HP/Klimatfiler/IREQ2002alfa.htm.
[18] Ergonomics of the thermal environment - Principles and application of relevant International Standards. ISO 11399, 1995.
[19] Ergonomics of the thermal environment - Instruments for measuring physical quantities. ISO 7726, 1998.
[20] Ergonomics of the thermal environment - Estimation of the thermal insulation and evaporative resistance of a clothing ensemble. ISO 9920, 2007.
[21] Ergonomics of the thermal environment - Risk assessment strategy for the prevention of stress or discomfort in thermal working conditions. ISO 15265, 2004.
[22] The PHS2 calculation model of Lund University (Sweden). [Online] Information on http://www.eat.lth.se/fileadmin/eat/Termisk_miljoe/PHS/PHS.html30.
[23] Clothing - Physiological effects - Measurement of thermal insulation by means of a thermal manikin. ISO 15831, 2004.
[24] Textiles - Physiological effects - Measurement of thermal and water-vapour resistance under steady-state conditions (sweating guarded-hotplate test) (ISO 11092:2014). LVS EN ISO 11092, 2014.

Fabrication and drug releasing property of Silk Fibroin-based fibrous membrane for the treatment of Anal Fistula

Lei Liu[1], Xu-Sheng Xie[1], Qing Liu[1], Min Zhi[2], Ze-Kun Liu[1], Chao-Heng He[1], Gang Li[1*], Xiao-Qin Wang[1]

[1]*National Engineering Laboratory for Modern Silk, College of Textile and Clothing Engineering, Soochow University, Suzhou, 215123, China*
[2]*Department of Gastroenterology, The Sixth Affiliated Hospital of Sun Yat-Sen University, Guangzhou, 510655, China*

[*]Corresponding author's email: tcligang@suda.edu.cn

Abstract

In the present study, a new compound material system with both drainage property and drug delivery function was developed as alternative materials for the treatment of crohn's fistula disease. Curcumin and 5-Aminosalicylic (5-ASA) acid were loaded into the mixture solution of silk fibroin (SF) and polyethylene glycol (PEG). In order to improve the drug loading and stability, core-shell fibrous membranes were made from the SF aqueous solution (shell layer) and curcumin/ 5-ASA acid (core layer) using coaxial electrospinning method. The morphology structure, secondary structure, drug release and their interaction effects of fibrous films were examined. The sectional images of fibrous clearly demonstrated the core-shell fibrous structures of the drug-loaded SF membranes. Fourier transform infrared spectroscopy (FTIR) results revealed that the membranes were predominantly in random coil and silk I conformation. The experimental result of *in-vitro* drug release showed a stable and effective drug releasing property. In conclusion, the stability and utilization ratio of curcumin/ 5-ASA were improved by loading them into SF/ PEG films, which can be used as a novel material for the treatment of crohn's fistula disease.

Key words: Coaxial Electrospinning; Silk fibroin;Anal Fistula; Curcumin; 5-Aminosalicylic Acid

1. Introduction

Crohn's disease (CD) is a chronic inflammatory bowel disease, the incidence rate is 100/100000 in America, in recent years, incidence rate is gradually increasing in china [1]. Anal fistula is one of the most common complications of Crohn's disease, Anal fistula incidence rate is about 6% in china, or about 1/4 of anorectal disease [2,3]. The clinical manifestation of anal fistula is anal itching associated with discomfort such as anal pain and recurrent infections, it gives patients great suffering [4].

At present, the treatment of Crohn's disease is on the basis of the protection of anal function, 5-aminosalicylic acid (5-ASA) is effective and less adverse reaction [5,6]. However, its oral effect is not ideal. The local surgical procedures in the treatment of complicated Crohn's disease fistula, need to meet the preoperative pelvic magnetic resonance (MR). The purpose of hanging plug in primary anal fistula place is to achieve continuous drainage and prevent anal fistula mouth closes, so as to achieve successful control of inflammation. For different types of anal fistula, commonly used materials and devices are rubber tube, silicone tube, rubber bands, surgical suture, antibacterial Vicryl absorbable sutures (a kind of adding antibacterial surgical suture) and so on [3]. However, the above material is not conducive to promoting the healing of the fistula, and repeated infection rate is 20% - 40%, and easily lead to incontinence and other adverse sequelae, furthermore 8%-13% of the patients appeared different degree of leakage dung complications and increase the pain of the patient [7]. Recent studies have reported that using biological glue fistula for the treatment of Crohn's disease fistula, including the injection of fibrin glue, anal fistula plug (AFP) and acellular dermal matrix (AEM), in which the fibrin glue for complex anal fistula healing rate of is only 10% -67% [8,9]. Anal fistula plug can be treated the low anal fistula, and the healing rate can reach at

70% -100%. But for the complex anal fistula, particularly the poor long-term efficacy is more than 1 year, the cure rate of most studies of less than 50%, or even less than 30% [10-13], but has higher prices. Acellular dermal matrix is a new anal fistula material, it was reported that the effect is good for the treatment of low anal fistula, but lacking of long-term observation for complex anal fistula [14]. Therefore, it is important to develop a new materials for the treatment of Crohn's disease fistula , the material should possess drainage function, with the role of drug delivery, and fistula healing, so as to achieve the effect for the treatment of Crohn's disease.

The latest studies have shown that curcumin has a good anti-inflammatory, anti-proliferative, anti-bacterial properties and security. And combined with the use of 5-ASA, to play the role of coordination and anti-inflammatory. Silk fibroin is an excellent natural polymer material, silk fibroin has rich proteins, and it contains 18 kinds of amino acids, including leucine accelerate cell metabolism, serine, threonine can delay skin aging, and silk fibroin has excellent processing properties, mechanical properties and biological compatibility and biological degradation, it has been attracted much attention of scientists [15-18]. It can be made into a particulate, colloidal and fibrous and other forms, for example, it is used in tissue engineering [24], enzyme immobilization [20,21] and drug delivery [22], as well as other biomedical applications [23,24]. The latest research also proved that silk fibroin can carry the curcumin effectively, and release drug molecules through the role of the enzyme [25]. Drugs-loaded fibrous are main filed, drugs add to spinning solution, it can achieve drug delivery effect [26,27]. Currently, the ideal drug delivery system is that drug or growth factors can be loaded into the fiber wrap, and form core-shell structures [28,29]. An new electrospinning method known as coaxial electrospinning was used to fabricate fibers with have core-shell structures, it has the advantages of high drug loading and high efficiency [28].

In our study, by exploiting the advantages of coaxial electrospinning, we attempted to develop a simple method for fabricating drug-loaded silk fibroin fibrous membranes. We used silk fibroin aqueous solution as outer solution, Polyethylene glycol (PEG, Molecular weight:10000) as a drug carrier, and we chose curcumin and 5-Aminosalicylic acid as the model drug. We fabricated fibrous scaffold with shell-core structured as the outer layer of composite anal fistula plug. On the basis of traditional anal fistula plug, our prototyped anal fistula plug not only has drainage property, but also drug delivery function. The morphology structure, secondary structure, drug delivery and their interaction effects of fibrous films will be examined.

2. Materials and methods

2.1 Materials

Silk was supplied by Shengzhou Silk Ltd., Co., China. Curcumin was purchased from Sigma USA. 5-Aminosalicylic acid (5-ASA) used for this study was purchased from Macklin biochemical Co.,Ltd, China. Polyethylene glycol (PEG) was supplied by Sinopharm chemical reagent Co.,Ltd, China. PBS buffer (PH = 7.4) was supplied by our laboratory. All of the chemicals and reagents are analytical grade.

2.2 Experimental method

2.2.1 Preparation of fibrous membranes

7% concentration of spinning solution of silk fibroin was prepared by degumming silk, dissolution, dialysis, centrifugation and assay concentration. Then concentrating to 27% silk fibroin concentration in 15% PEG, Molecular weight, 20000. We chose 27% silk fibroin aqueous solution as outer spinning solution. 20% PEG as inner solution, on the basis of outer and inner solutions, curcumin and 5-ASA (M_{CUR}:$M_{5\text{-}ASA}$=1:1) were added to inner spinning solution, then to coaxial electrospinning. We chose core-shell ratios and drug percentage as a variable, core-shell ratio is 1:3, 1:6 and 1:12 and drug percentage is 0.4%, 0.9% and 1.9%, respectively.

2.2.2 Surface morphology and sectional structure characterization

The area part of fibrous membranes was randomly selected, specimens were coated with gold. The surface morphology of the electrospun fibrous membranes were examined under scanning electron microscope (SEM, Hitachi S-4800). Wc randomly selected area part of fibrous, and put it into ethanol solution for 1 hour. In order to dissolve inner layer, we put fibrous membranes into ultra-pure water for 2days, then remove dried Liquid nitrogen interrupted fibrous membranes. Finally, the sectional structure of the electrospun fibrous membranes were examined under a scanning electron microscope (SEM, Hitachi S-4800).

2.2.3 Secondary structure characterization

The area part of fibrous membranes with or without ethanol treatment was randomly selected. These samples were grinded into powder. The secondary structure of the membranes was examined using fourier transform infrared spectroscopy (FTIR, Thermo Nicolet 5700). FT-IR wave numbers were set from 4000 to 400 cm^{-1} during 32 scans, with 2 cm^{-1} resolution. The major vibration bands were attributed to specific chemical groups.

2.2.4 Drug releasing characterization

The nanofibrous membranes with different drug percentages were treated by water vapor for 6h. SF/ PEG/ Drugs fibrous membranes were transferred to sealed centrifuge tube, 1ml of 0.01-M PBS (PH=7.4) were added to the centrifuge tubes. Then put them into thermostatic shaker, set temperature is 37 °C, rotation speed is 100 r/min. At regular intervals, 0.5ml of solution was taken from centrifuge tube and its absorbance at 330nm and 425nm was measured using Bio-Tek synergy H1 microplate reader. After measuring the absorbance, the solution was returned to the centrifuge tube to ensure constant volume of the release solution. The release percentage (equation 1) and accumulative releasing (equation 2) were calculated by using the following equations [25,30].

$$RPi(\%) = \frac{Ci * V}{W0} * 100\% \tag{1}$$

$$ARi(\%) = \sum_{i=1}^{i} RPi \tag{2}$$

Where: *RPi* represent release percentage; *ARi* represent accumulative releasing; *Ci* represent drug concentration; *V* represent volume; W_0 represent total entrapped drug.

3. Results and discussion

3.1 The morphology and structure of different core-shell ratios and drug percentage drug-loaded fibrous membranes

Table 1. The group of different core-shell ratios and drug percentage

Group	SF concentration (%)	PEG concentration (%)	Core-shell ratios	Drug percentage (w/w)
a	27	20	1:3	0.4%
b	27	20	1:6	0.4%
c	27	20	1:12	0.4%
d	27	20	1:3	0.9%
e	27	20	1:6	0.9%
f	27	20	1:12	0.9%
g	27	20	1:3	1.9%
h	27	20	1:6	1.9%
i	27	20	1:12	1.9%

In coaxial electrospinning, core-shell ratios and drug percentage are key factors for successful preparation of shell-core structured fibrous membranes. We set core-shell ratio at 1:3, 1:6 and 1:12 and drug percentage at 0.4%, 0.9% and 1.9%, respectively (Table 1).The SEM images of the drug-loaded SF/PEG fibrous membranes at different drug percentages are shown in Fig.1. When core-shell ratios were set at 1:3 and 1:6, the membranes showed linear morphology with a uniform structure and smooth surface. The mean diameter of fibrous was ranged at 5.621±1.320 um. When core-shell ratio was set at 1:12, it's difficult to fabricate fibrous with shell-core structure.

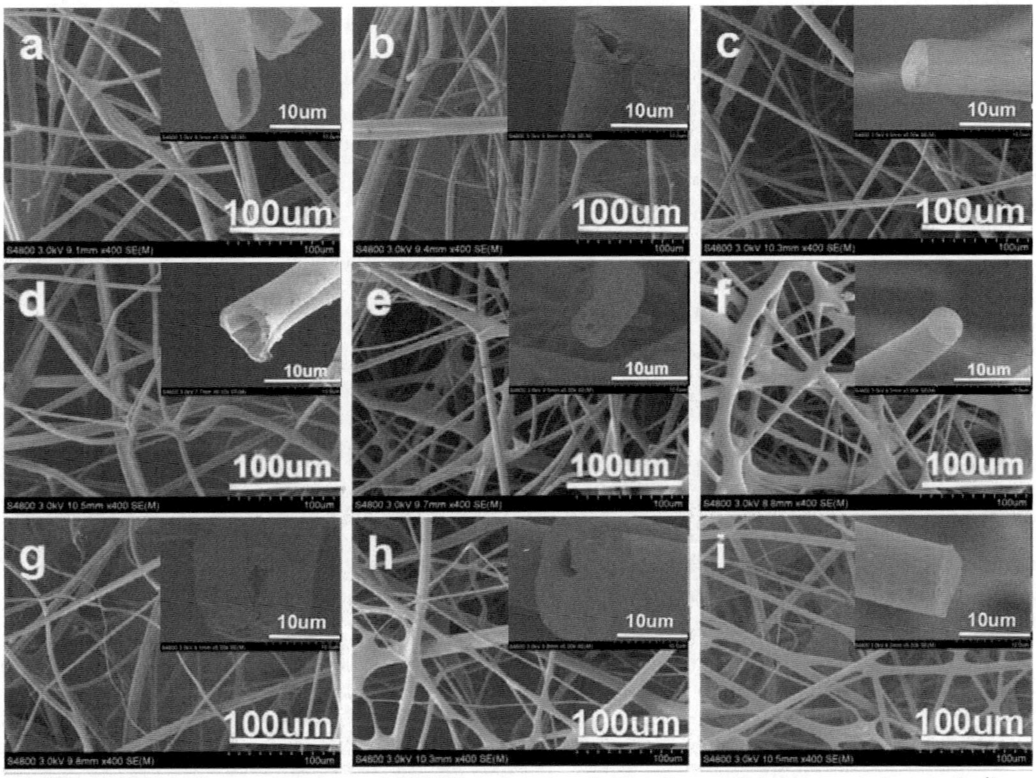

Fig.1 Morphology of fibrous membranes with different core-shell ratios and drug percentages; (a-i) represent different samples as given in Table 1

3.2 Secondary Structure of the fibrous membranes

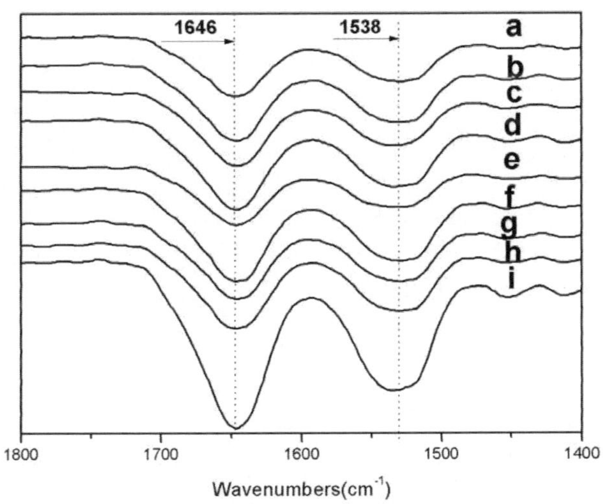

Fig.2 FTIR spectrums of different fibrous membranes, (a-i) represent different samples as given in Table 1

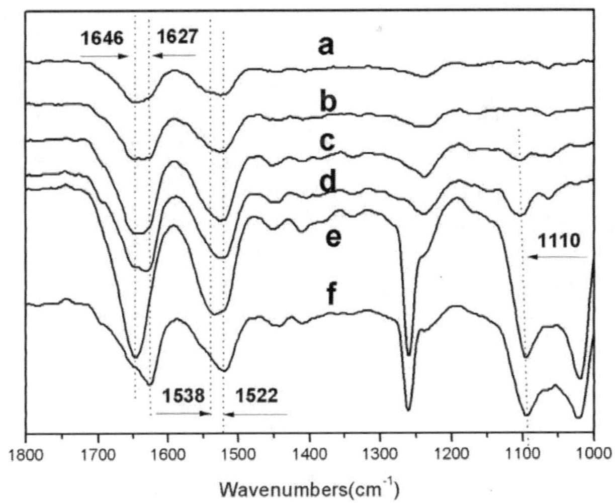

Fig.3 FTIR spectrums of different fibrous membranes after ethanol treated; (a-f) SF, SF/PEG and 0.9% SF/ PEG/ drug fibrous membranes before and after ethanol treated

The FTIR spectrums of fibrous samples with different core-shell ratios and different drug percentages are presented in Fig.2. $1646cm^{-1}$ and $1538cm^{-1}$ are characteristic peaks of silk I (Amide I and II), and $1627cm^{-1}$ and $1522cm^{-1}$ are characteristic peaks of silk II (Amide I and II). The results indicates that different core-shell ratios and different drug percentage didn't change silk fibroin form in fibrous membranes, and the silk fibroin was predominantly in random coil and silk I conformation.

The FTIR spectrums of ethanol treated fibrous membranes are presented in Fig.3. The results indicates that the silk fibroin was predominantly in random coil and silk I conformation. The bands of silk fibroin had an obvious change after ethanol treated fibrous, the silk I conformation changed to silk II conformation mostly. In curves c and d, characteristic peaks at $1110cm^{-1}$ (C-O stretching vibrations) were observed. Before ethanol treated, the membranes were predominantly in random coil and silk I conformation. With the addition of ethanol, characteristic peak occurred significant shift. Curves e and f indicates that silk fibroin was mainly silk I form in drug-loaded fibrous films after ethanol treated [31,32].

3.3 Drug delivery of the fibrous membranes

The drug release profile of the curcumin/5-ASA/ SF/ PEG fibrous membrane at 37 °C was characterized in PBS pH=7.4 as the release medium (Fig.4). The drug release curves showed that the drug release rate and accumulative release rate increase while the drug loading percentage increases. The experimental result of *in-vitro* drug release showed a stable and effective drug releasing property. A burst release of 5-ASA occurred within the 24h as shown in Fig.4(B). The accumulative release rate is above 80%. It happened due to the increased solubility of 5-ASA in PBS.

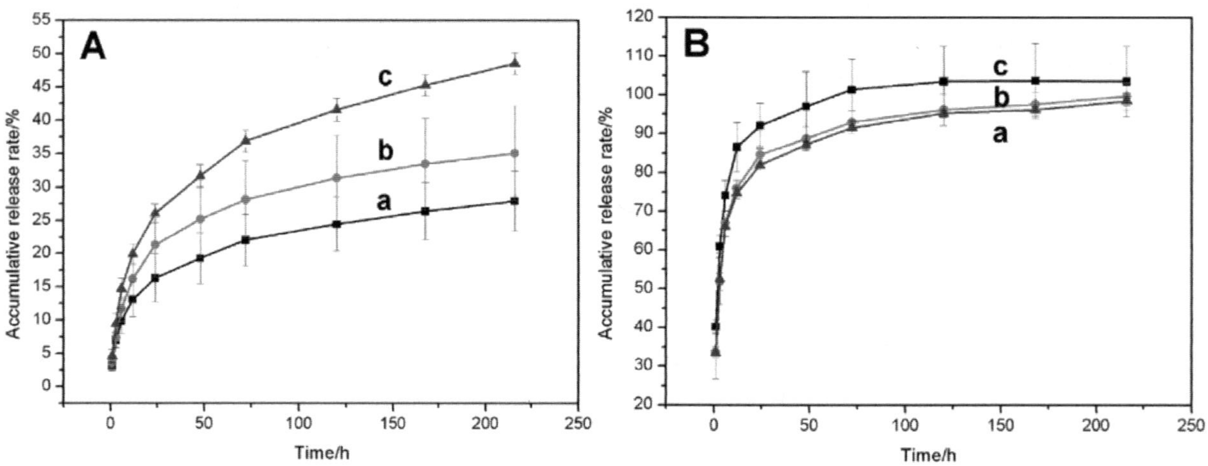

Fig.4 Drug release curves of fibrous membranes with different drug percentages; (A) and (B) represent release curves of curcumin and 5-ASA in the samples; a-c represent the fibrous membranes with different drug percentages at 0.4%,0.9% and 1.9%, respectively

4. Conclusion

In the present work, we successfully fabricated the drug-loaded SF fibrous membranes using coaxial electrospinning as an anal fistula plug material. The morphology, secondary structure, drug release and their interaction effects of fibrous films were examined. When shell-core ratio are 1:3 and 1:6, it successfully fabricated fibrous membranes with shell-core structure. Random coil and silk I conformation were predominantly found in the drug-loaded SF fibrous membranes. The experimental result of *in-vitro* drug release showed a stable and effective drug releasing property. In conclusion, the stability and utilization ratio of curcumin/ 5-ASA were improved by loading them into SF/ PEG films, which can be used as a novel material for the treatment of crohn's fistula disease.

5. Acknowledgement

This work was funded by the National Natural Science Foundation of China (NSFC, 51273138). We would like to thank the support of State Key Laboratory for Modification of Chemical Fibers and Polymer Materials, Donghua University (LK1421) and Natural Science Foundation of Jiangsu Province (BK20150372) and Natural Science Foundation for Young Scholars of Soochow University (SDY2014A06).

References

[1] Wen YL, Niu JK, Miao YL. Non-drug Treatment of Inflammatory Bowel Disease. Shijie Huaren Xiaohua Zazhi: 2015; 23: 3720-3728.
[2] Schwartz DA, Pemberton JH, Sandborn WJ. Diagnosis and Treatment of Perianal Fistulas in Crohn Disease. Annals of Internal Medicine: 2001; 135: 906-918.
[3] He XS, Chen ZX, Lan P. Advances on the Carcinogenesis of Inflammatory Bowel Disease. Chinese Journal of Practical Surgery: 2013; 33: 604-606
[4] Zhu R, Zhang PS, Shen Lin, et al. Advances on the Diagnosis and Treatment of Anal Fistula. Research of Integrated Traditional Chinese and Western Medicine 2011; 3(3): 156-161.
[5] Prosberg MV, Vester-Andersen MK, Andersson M, et al. Long-term Compliance with Oral 5-aminosalicylic Acid Therapy and Risk of Disease Recurrence in Patients with Ulcerative Colitis: A Population-based Cohort Study. Inflammatory bowel diseases: 2016; 22: 925-932.
[6] Wang S, Zhou GH. Progress on Maintenance Treatment for Ulcerative Colitis by Aminosalicylic Acids Combined with Probiotics: 2015; 18: 463-466.
[7] Steele SR, Kumar R, Feingold DL, et al. Practice Parameters for the Management of Perianal Abscess and Fistula-in-ano. Diseases of the Colon & Rectum: 2011; 54: 1465-1474.
[8] Sentovich SM. Fibrin Glue for Anal Fistulas: Long-term Results. American Journal of Gastroenterology: 2003; 46: 498-502.
[9] C Neal E, Rostas JW, Greiner FG. Long-term Outcomes with the Use of Bioprosthetic Plugs for the Management of Complex Anal Fistulas. Diseases of the Colon & Rectum: 2010; 53: 798-802.
[10] Ky AJ, Sylla P, Steinhagen R, et al. Collagen Fistula Plug for the Treatment of Anal Fistulas. Diseases of the Colon & Rectum: 2008; 51: 838-843.
[11] Christoforidis D, Etzioni DA, Goldberg SM, et al. Treatment of Complex Anal Fistulas with the Collagen Fistula Plug. Diseases of the Colon & Rectum: 2008; 51: 1482-1487.
[12] Garg P, Song J, Bhatia A, et al. The Efficacy of Anal Fistula Plug in Fistula-in-ano: a Systematic Review. Colorectal Disease the Official Journal of the Association of Coloproctology of Great Britain & Ireland: 2010; 12: 965-970.
[13] O'Riordan J, Datta I, Johnston C, et al. A Systematic Review of the Anal Fistula Plug for Patients with Crohn's and Non-Crohn's Related Fistula-in-ano. Diseases of the Colon & Rectum: 2012; 55: 351-358.
[14] Song WL, Wang ZJ, Zheng Y, et al. A Anorectal Fistula Treatment with Acellular Extracellular Matrix: A New Technique. World Journal of Gastroenterology: 2008; 14: 4791-4794.
[15] Li G, Li Y, Chen G, et al. Silk-based Biomaterials in Biomedical Textiles and Fiber-based Implants. Advanced Healthcare Materials: 2015; 4: 1134-1151.
[16] Shao Z, Vollrath F. Materials: Surprising Strength of Silkworm Silk. Nature: 2002; 418: 741.
[17] Fan S, Zhang Y, Shao H, et al. Electrospun Regenerated Silk Fibroin Mats With Enhanced Mechanical Properties. International Journal of Biological Macromolecules: 2013; 56: 83-88.
[18] Lin N, Liu XY, Diao YY, et al. Switching on Fluorescent Emission by Molecular Recognition and Aggregation Dissociation. Advanced Functional Materials: 2012; 22: 361-368.
[19] Zoccola M, Aluigi A, Vineis C, et al. Study on Cast Membranes and Electrospun Nanofibers Made from Keratin/ Fibroin Blends. Biomacromolecules: 2008; 9: 2819-2825.
[20] Huang J, Liu L, Yao JM. Technology and Properties of Regenerated Bombyx Mori Silk Fibroin Nano-fiber Membrane by Electrospinning Method. Silk: 2011; 48: 20-23.
[21] Liu Y, Xue HY, Dai LX. Study on SF/ PVA Blending Nanofibers by Electrospinning. S FC: 2006; 35: 13-16.
[22] Cao Y, Wang BC, Chi SP, Zhang W. Silk Fibroin Used As A Drug Delivery Material. Journal of Clinical

Rehabilitative Tissue Engineering Research: 2009; 13: 1533-1536.
[23] Zhou YS, Yang HJ, Liu X, et al. Electrospinning of Carboxyethyl chitosan/ Poly (vinyl alcohol)/ Silk Fibroin Nanoparticles for Wound Dressings. Int J Biol Macromol: 2013; 53: 88-92.
[24] Wei K, Kim BS, Kim IS. Fabrication and Biocompatibility of Electrospun Silk Biocomposites. Membranes (Basel): 2011; 1: 75-98.
[25] Yucel T, Lovett ML, Kaplan DL. Silk-Based Biomaterials for Sustained Drug Delivery. Journal of Controlled Release: 2014; 190: 381-397.
[26] Li G, et al. A 5-Fluorouracil-loaded Polydioxanone Weft-knitted Stent for the Treatment of Colorectal Cancer. Biomaterials: 2013; 34: 6451-6461.
[27] Chen Y, Lin J, Wang HB, et al. Preparation and Anticoagulation Evaluation of Electrospun PLA/Curcumin Composite Membranes. New Chemical Materials: 2011; 39: 50-53.
[28] Wang JG, Liu JJ, Fan CY, et al. Preparation of Coaxial Electrospinning Nanofibers for Nerve Growth Factor Delivery. Journal of Clinical Rehabilitative Tissue Engineering Research: 2008; 12: 4440-4444.
[29] Zhang HT, Luo JJ, Xing TH, et al. Preparation and Slow Release Properties of Nanofibrous Scaffolds with SF/ PCL-BSA Sheath-core Structure. Journal of Zhejiang Sci-Tech University (Natural sciences): 2015; 33: 766-770.
[30] He T, et al. Electrospinning Polyvinylidene Fluoride Fibrous Membranes Containing Anti-bacterial Drugs Used As Wound Dressing. Colloids & Surfaces B Biointerfaces: 2015; 130: 278-286.
[31] Li G, et al. Silk Microfiber-reinforced Silk Composite Scaffolds: Fabrication, Mechanical Properties, and Cytocompatbility. Journal of Materials Science: 2016; 51: 3025-3035.
[32] Li G, et al. Structural Mimetic Silk Fiber-reinforced Composite Scaffolds Using Multi-angle Fibers. Macromolecular Bioscience: 2015; 15: 1125-1133.

Prediction of Sub Bandage Pressure for High Compression Bandages Using ANN

M P Sikka[1*], A Mukhopadhyay[2]

[1, 2]*Dr B R Ambedkar National Institute of Technology, Jalandhar, Punjab, 144011, India*

[*]Corresponding author's email: sikkamonica@yahoo.co.in

Abstract

In this paper artificial neural network (ANN) model has been designed to predict the sub bandage pressure of high compression bandages during wrapping on the limb. Two different types of bandage fabrics (knitted and woven) were taken for the study. The other input parameters include bandage fabric type, circumference of limb, number of bandage layers, stretch% and width of the bandage. In order to reduce the dependency of the results on a specific partition of the data into training and testing sets, a three-way cross validation tests were performed, i.e. total data was divided into training and testing set in three different ways. The predicted sub bandage pressure was correlated to the experimental sub bandage pressure and correlation coefficient between the actual and predicted values obtained. It was observed that the neural network system is able to predict the sub bandage pressure after application with good correlation and less average error.

Keywords: Artificial Neural Network; High Compression Bandage; Sub Bandage Pressure; Limb

1. Introduction

External compression is given during compression therapy using compression materials like bandage, stockings or pressure garments etc. The efficacy of the compression treatment is dependent on the pressure generated between the bandage and the skin which is called the sub-bandage pressure. The sub bandage pressure exerted by the bandage depends on several factors like limb shape or size, application technique, physical and structural properties of the bandage, physical activities taken by the patient etc The pressure generated by a bandage just after its application is a function of the bandage tension, the number of layers applied, the bandage width and the limb circumference, and this relation is based upon the Laplace's law for measuring pressure across a cylindrical thin film[1]. It is important that the bandage should apply the right amount of pressure to the affected part to assist the healing process. Prediction of sub bandage pressure applied by a bandage to a leg facilitates the efficiency of compression bandage. But the use of Laplace's law to determine sub bandage pressure remains a controversial issue. The modified form of Laplace's equation described by Thomas[1] does not consider the increase in radius because of the preceding layers of the bandage. Also several mathematical models have been proposed in the literature since then to predict the sub bandage pressure on a cylindrical limb [1-5] Dale et al. [6] compared four different four-layer bandaging system and observed that the final pressure achieved by a multilayer bandaging system is not equal to the sum of the pressure exerted by each individual layer as predicted by Laplace's law. Schuren and Mohr [7] found that the interface pressure based on theoretical mathematical equations is not supported by experimental results. Al Khaburi et al [4]. have proposed a model based of thick wall cylinder theory to predict interface pressure by multiple layers of medical compression bandages.

Neural network is a knowledge based system that acquires, represents and uses knowledge for a specific purpose. It is based on the theory that intelligence emerges through a large number of processing elements connected together, each performing simple computation. It is widely being used to predict the physical and mechanical properties; identification or classification of defects; process optimization or online monitoring; marketing and planning [8, 9]. Further, in various studies [10, 11] the statistical models have been compared with the ANN models and it has been reported that the ANN models are better in their prediction performance as compared to statistical models. So a neural network model has been applied for predicting sub bandage pressure in the following study.

2. Experimental

2.1 Materials

Two different types of bandage fabrics, i.e woven and knitted bandages are used. Table 1 shows the characteristics of the fabrics. The bandage fabric was given a numeric value of 1 and 2 (Table 2) for using in the neural network.

Table 1. Characteristics of fabrics

Sample	Structure	GSM	Thickness (cm)	EPI/ wales per cm	PPI/ courses per cm	Elastic yarn	
						Number of elastic yarns/inch	Count of elastic yarns (tex)
1	Woven	288	0.187	14	24	10	75
2	Knit	336	0.112	6	28	15	200

2.2 Methods

Three different circumferences of rigid tubes with a layer of foam were used for applying the bandage (Fig.1). The sub bandage pressure was measured at three different stretch% (30, 40 & 50) for two different widths of the bandages (8cm,10cm) at spiral 50% and 66% overlap generating two and three layers respectively with Kikuhime pressure monitoring device after calibration. Ten measurements were taken for each sample for each combination and the error was found to be less than 3% at 95% confidence level.

Fig.1 Rigid plastic tube with layer of foam

The widespread applicability of neural networks comes from the fact that the multilayer feed forward networks are universal approximators[12]. The problem of predicting the sub bandage pressure of high stretch bandages from the bandage and application parameters can be viewed as one of the function approximation. The sub bandage pressure is a function of the application parameters, and the goal is to approximate that function from a set of measurements. The feed forward neural network is used for this study. A feed forward network is a network which receives inputs only from previous layers i.e. the output of one layer is the input for next layer and there is no feedback from the output to the input.

3. Network architecture

In order to train a feed forward neural network properly, various network parameters have to be optimized. One or two hidden layers are able to map any response to a good degree of accuracy and increasing the number of hidden layers does not give a significant increase in the prediction performance of the network [13]. In the present study, two hidden layers are used and different number of neurons in the hidden layers is tried to optimize the network. The starting point for number of neurons in the hidden layer was chosen by a thumb rule Eq.1, i.e. [13].

$$n_{hidden} > 2 \times [\max\ (input\ neurons,\ output\ neurons] \qquad (1)$$

Using too few neurons starves the network of the resources it needs to solve the problem. Using too many will increase the training time and may cause a problem called over fitting. The error on the training set is driven to a very small value, but when new data is presented to the network, the error becomes very large. Although the network is able to map the training set, but it cannot generalize to new situations. Hence, some of the test data points give very high errors. One method for improving network generalization is to use a network that is just large enough to provide an adequate fit. But it is difficult to know beforehand how large a network should be for a specific application. There are two other methods for improving generalization viz. regularization and early stopping. In the present study regularization was carried out to avoid over fitting. This is done by modifying the performance function of mean square error or '*mse*'. The new function called '*msereg*' is given by Eq.2.

$$msereg = \gamma\ mse + (1-\gamma)msw \qquad (2)$$

Where *msw* is mean square weights given by Eq.3, γ is the performance ratio (default value 0.5).

$$msw = \frac{1}{n}\sum_{j=1}^{n} W_j^2 \qquad (3)$$

Where n is the number of neurons in a layer and wj is the weight matrix. The value of 'msereg' becomes lower than'mse' hence the total error will be reduced.

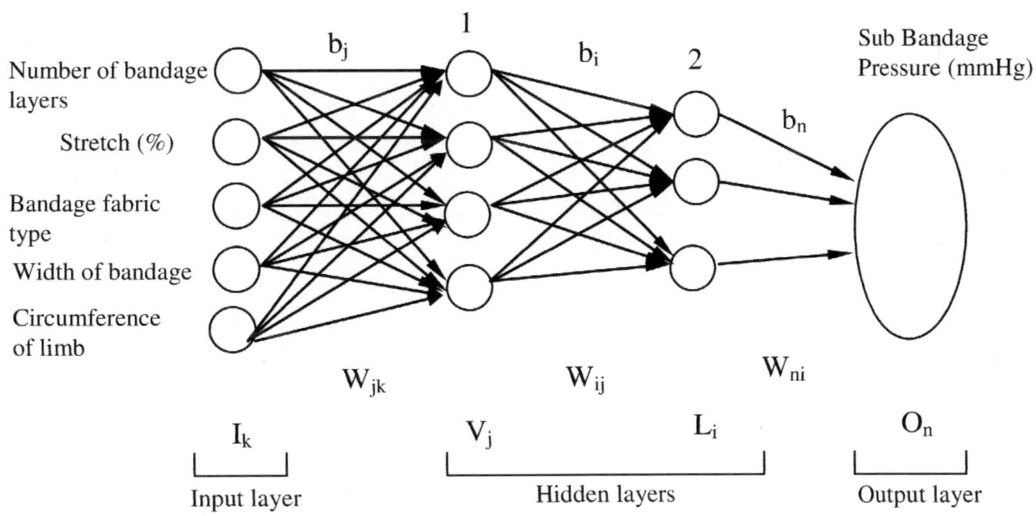

Fig.2 Architecture of a three-layered neural network used in the study: W_{jk} is the weight matrix, b_j is the bias, and V_j and L_i are the number of nodes in the hidden layers

Fig.2 shows the feed forward neural network using five nodes in the input layer (equal to the number of input parameters), two hidden layers and one node in the output layer (corresponding to the output, i.e. sub bandage pressure).

Back-propagation is used for the network. It is an iterative gradient descent algorithm that minimizes the sum of squared error between the desired output and the actual output of a set of patterns [14]. In addition, it is capable of handling noisy data. In order to train a feed forward neural network properly, various network parameters have to be optimized. These parameters include the number of hidden layers, the number of neurons in the hidden layers, the learning rate and the number of training cycles (also known as epochs). In the present study, two hidden layers are used and different number of neurons in the hidden layers is tried to optimize the network.

3.1 Network training and cross validation

A total of 72 values of sub bandage pressure were taken for the study, of which 54 (about 75%) samples were used for training and 18 for testing the network. Table 2 shows the training and test data sets. The prediction performance of the neural network was evaluated by testing with eighteen sample data which were not used during the training. In order to reduce the dependency of the results on a specific partition of the data into training and testing sets, a three-way cross validation test was performed, i.e. total data was divided into training and testing set in three different ways.

Table 2. Specifications of the training and test data set

S.No.	Bandage fabric type	Circumference (cm)	Stretch%	Number of bandage layers	Width (cm)	Pressure (mmHg)	Partition		
							1	2	3
1	Woven(1)	23	30	2	8	13		*	
2	Woven(1)	23	30	3	8	20	*		
3	Woven(1)	23	40	2	8	17	*		
4	Woven(1)	23	40	3	8	23			*
5	Woven(1)	23	50	2	8	20			
6	Woven(1)	23	50	3	8	26			*
7	Woven(1)	23	30	2	10	12			*
8	Woven(1)	23	30	3	10	19			
9	Woven(1)	23	40	2	10	15			
10	Woven(1)	23	40	3	10	22		*	
11	Woven(1)	23	50	2	10	18	*		
12	Woven(1)	23	50	3	10	25		*	
13	Woven(1)	26	30	2	8	11		*	
14	Woven(1)	26	30	3	8	18			
15	Woven(1)	26	40	2	8	14			
16	Woven(1)	26	40	3	8	21			*
17	Woven(1)	26	50	2	8	16			
18	Woven(1)	26	50	3	8	25	*		*
19	Woven(1)	26	30	2	10	10	*		*
20	Woven(1)	26	30	3	10	17			
21	Woven(1)	26	40	2	10	13			
22	Woven(1)	26	40	3	10	20	*	*	
23	Woven(1)	26	50	2	10	15			
24	Woven(1)	26	50	3	10	23		*	
25	Woven(1)	32.5	30	2	8	9		*	
26	Woven(1)	32.5	30	3	8	15	*		
27	Woven(1)	32.5	40	2	8	11	*		
28	Woven(1)	32.5	40	3	8	17			*

29	Woven(1)	32.5	50	2	8	14			
30	Woven(1)	32.5	50	3	8	20			*
31	Woven(1)	32.5	30	2	10	7			*
32	Woven(1)	32.5	30	3	10	14			
33	Woven(1)	32.5	40	2	10	10			
34	Woven(1)	32.5	40	3	10	16		*	
35	Woven(1)	32.5	50	2	10	13	*		
36	Woven(1)	32.5	50	3	10	19		*	
37	Knit(2)	23	30	2	8	41		*	
38	Knit(2)	23	30	3	8	52	*		
39	Knit(2)	23	40	2	8	48	*		
40	Knit(2)	23	40	3	8	60			*
41	Knit(2)	23	50	2	8	56			
42	Knit(2)	23	50	3	8	67			*
43	Knit(2)	23	30	2	10	39			*
44	Knit(2)	23	30	3	10	50			
45	Knit(2)	23	40	2	10	45			
46	Knit(2)	23	40	3	10	58		*	
47	Knit(2)	23	50	2	10	53	*		
48	Knit(2)	23	50	3	10	66		*	
49	Knit(2)	26	30	2	8	38		*	
50	Knit(2)	26	30	3	8	51			
51	Knit(2)	26	40	2	8	45			
52	Knit(2)	26	40	3	8	59			*
53	Knit(2)	26	50	2	8	52			
54	Knit(2)	26	50	3	8	65	*		*
55	Knit(2)	26	30	2	10	36	*		*
56	Knit(2)	26	30	3	10	49			
57	Knit(2)	26	40	2	10	44			
58	Knit(2)	26	40	3	10	58	*	*	
59	Knit(2)	26	50	2	10	51			
60	Knit(2)	26	50	3	10	64		*	
61	Knit(2)	32.5	30	2	8	32		*	
62	Knit(2)	32.5	30	3	8	46	*		
63	Knit(2)	32.5	40	2	8	42	*		
64	Knit(2)	32.5	40	3	8	53			*
65	Knit(2)	32.5	50	2	8	48			
66	Knit(2)	32.5	50	3	8	58			*
67	Knit(2)	32.5	30	2	10	31			*
68	Knit(2)	32.5	30	3	10	44			
69	Knit(2)	32.5	40	2	10	40			
70	Knit(2)	32.5	40	3	10	51		*	
71	Knit(2)	32.5	50	2	10	46	*		
72	Knit(2)	32.5	50	3	10	57		*	

* indicates that the sample was used in the test set for that partition, the other samples were used in the training set

When the neural network was trained by training set, the sub bandage pressure was predicted for both training and testing sets. The percent error of each data set and the mean square error were calculated. Since mean square error (MSE) is the average of the square of the difference between the actual and the predicted sub bandage pressure, this value can be used to estimate the prediction power of the ANN model. The predicted sub bandage pressure was correlated to the experimental sub bandage pressure and correlation coefficient between the actual and predicted sub bandage pressure obtained.

4. Results and discussion

The input parameters viz. bandage fabric type, circumference of the limb, number of bandage layers, stretch% and width of the bandage are considered to predict the sub bandage pressure in high stretch bandages. Table 3 shows the experimental and predicted sub bandage pressure of different data partitions. It show the results from three different experiments of three way cross-validation. The measured and predicted sub bandage pressure values are shown along with the ANN performance error (difference between the experimental and predicted sub bandage pressure). It can be seen that the proposed neural network approximates the sub bandage pressure values with a very low output error, regardless of the specific data partition into training and test sets.

Table 3. Experimental and predicted sub bandage pressure for different partitions

S. No.	Partition1			Partition2			Partition3		
	Actual	Predicted	Error (%)	Actual	Predicted	Error (%)	Actual	Predicted	Error (%)
1	20	19.0288	4.86	13	12.0073	7.64	23	22.8046	0.85
2	17	15.9656	6.08	22	21.9719	0.13	26	27.9014	-7.31
3	18	18.2464	-1.37	25	25.0961	-0.38	12	13.6664	-13.89
4	25	23.1583	7.37	11	10.7049	2.68	21	21.1319	-0.63
5	10	10.3877	-3.88	20	20.1211	-0.61	25	26.1092	-4.44
6	20	19.4112	2.94	23	24.3707	-5.96	10	11.6874	-16.87
7	15	14.98	0.13	9	9.6213	-6.90	17	16.6392	2.12
8	11	10.0095	9.00	16	17.1055	-6.91	20	21.5945	-7.97
9	13	11.1093	14.54	19	20.9526	-10.28	7	6.6242	5.37
10	52	52.4579	-0.88	41	40.643	0.87	60	62.1849	-3.64
11	48	46.5067	3.11	58	60.8851	-4.97	67	64.5211	3.70
12	53	53.2909	-0.55	66	63.6542	3.55	39	35.9984	7.70
13	65	64.1596	1.29	38	38.8301	-2.18	59	57.457	2.62
14	36	39.616	-10.04	58	58.1887	-0.33	65	63.9914	1.55
15	58	56.5356	2.52	64	63.1491	1.33	36	33.3408	7.39
16	46	42.9513	6.63	32	32.9242	-2.89	53	50.0009	5.66
17	42	40.9111	2.59	51	50.6427	0.70	58	56.3248	2.89
18	46	45.0766	2.01	57	60.0071	-5.28	31	27.2008	12.26

4.1 Prediction performance

Table 4 shows the prediction performance of the network for three ways cross validation data.

The mean square errors in the training set ranges from 1.30 to 1.57 in all three partitions. The maximum and minimum errors in the training set are 13.11 and 0.0002 %, respectively. The coefficient of determination is 0.99. In the test set, the mean square error ranges from 1.87 to 2.23 and the maximum and minimum errors are 14.43 and 0.0004 %, respectively. The coefficient of determination in the three partitions is 0.99.

Table 4. Performance parameters of the neural network

	Partition 1	Partition 2	Partition 3
Network architecture	5-6-6-1	5-6-6-1	5-6-6-1
Performance goal	0.01	0.01	0.01
Epochs	73	51	49
Performance ratio	0.95	0.95	0.95
TRAINING SET			
Mean square error (MSE)	1.57	1.36	1.30
Root means square error (RMSE)	1.25	1.16	1.14
Maximum error	8.60	13.11	9.02
Minimum error	0.007	0	0.0002
Coefficient of determination (r^2)	0.99	0.99	0.99
TEST SET			
Mean square error (MSE)	2.23	1.87	1.88
Root means square error (RMSE)	1.49	1.36	1.37
Maximum error	13.07	9.04	14.43
Minimum error	0.0004	0.0007	0.017
Coefficient of determination (r^2)	0.9942	0.9954	0.9920

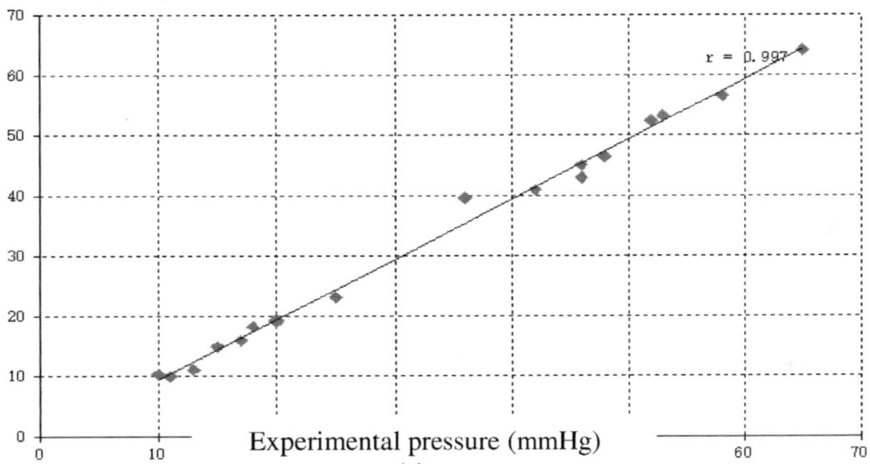

(a)

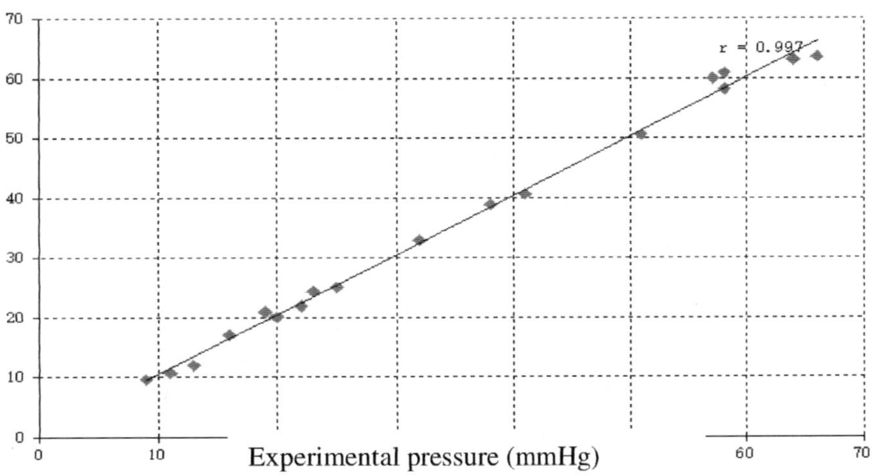

(b)

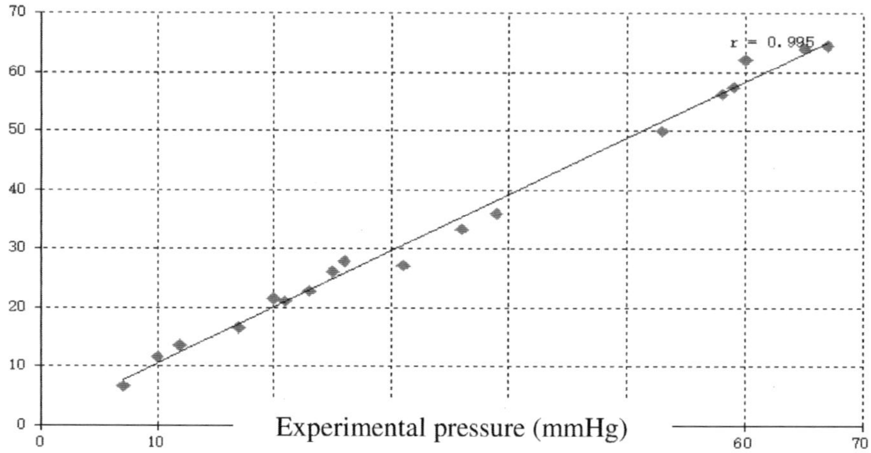

(c)

Fig.3 Correlation between actual values and predicted values for sub bandage pressure for test sets using ANN: (a) Partition 1, (b) Partition 2, (c) Partition 3

Fig.3 shows the correlation between the experimental and predicted sub bandage pressure of bandage fabrics. The values of correlation coefficient for three partitions are 0.997, 0.9971 and 0.995 respectively. A good correlation between the measured and predicted sub bandage pressure is obtained. The neural network system is able to predict the sub bandage pressure after wrapping with good correlation and low error.

5. Conclusions

An ANN model is proposed and employed to predict the sub bandage pressure of high compression bandages during wrapping on the limb. Five input parameters like bandage fabric type, circumference of limb, number of bandage layers, stretch% and width of the bandage are used. The results show that neural networks provide an effective method to predict the sub bandage pressure using different bandage types using a single model. The performance of the proposed neural network is tested on experimental data and cross validation of data is done by dividing the training and test data in three different ways. A reasonably good correlation between the experimental and predicted tenacity loss is obtained for each partition. The neural network system is able to predict the sub bandage pressure of high compression bandages after application on the limb with good correlation and low MSE.

References

[1] Thomas S. The use of the Laplace equation in the calculation of sub-bandage pressure. EWMA J 2003; 3: 21-3.

[2] KhaburiJAl, Nelson EA, Hutchinson J, Dehghani-Sanij AA. Impact of multilayered compression bandages on sub-bandage pressure: a model. *Phlebology* 2011; 26(1): 75-83.

[3] Yildiz N. A novel technique to determine pressure in pressure garments for hypertrophic burn scars and comfort properties. *Burns* 2007; 33:59-64.

[4] Al Khaburi J, Dehghani-Sanij AA, Nelson EA, Hutchinson J. Effect of bandage thickness on interface pressure applied by compression bandages. *Med EngPhys*2012; 34(3):378-85.

[5] Kumar B, Das A, Alagirusamy R. An approach to determine pressure profile generated by compression bandage using quasi-linear viscoelastic model. *J BiomechEng*2012; 134(9).

[6] Dale JJ, Ruckley CV, Gibson B, Brown D, Lee AJ, Prescott RJ. Multi-layer compression: comparison of four different four-layer bandage systems applied to the leg. *Eur. J. Vasc. Endovasc. Surg* 2004; 27: 94-99.

[7] Schuren J, Mohr K. The efficacy of laplace's equation in calculating bandage pressure in venous leg ulcers. *Wounds UK*2008; 4(2): 38-47.

[8] Mukhopadhyay A. Application of Artificial Neural Networks in Textiles. *Text Asia 2002*; 33(4): 35-39.

[9] Chattopadhyay R, Guha A.Artificial Neural Networks: Applications to Textile.*Text Prog* 2003: 35(1).
[10] Debnath S, Madhusoothanan M, Srinivasamoorthy VR. Prediction of Air Permeability of Needlepunched Nonwoven Fabrics using Artificial Neural Network and Empirical Models. *Indian J Fbr Text Res* 2000; 25(4): 251-255.
[11] Zurada JM. Introduction to Artificial Neural Systems. Mumbai:Jaico Publishing House, 1997.
[12] Hornik K, Stinchcombe M. White H.Multilayer Feedforward Networks are Universal Approximators. *Neural Networks* 1989; 2(5): 359.
[13] Guha A. Ph. D Dissertation. Indian Institute of Technology, Delhi, 2002.
[14] Bhattacharjee D, Kothari VK. A Neural Network System for Prediction of Thermal Resistance of Textile Fabrics. *Text Res J* 2007; 77(1): 4-12.

Body Posture Change of Pregnant Women During Pregnancy

Hui-Wei Cao, Rong Zheng*, Xiao-Jiao Sun, Hao Zhang, Hai-Jing Liu

Aimer Human Engineering Research Centre, Beijing Institute of Fashion Technology, Beijing, 100029, China

*Corresponding author's email: yuki.zheng@gmail.com

Abstract

Abdomen-supporting underwear and maternity pants support women's belly weight during pregnancy. The contents of this study include a market evaluation of existing abdomen-supporting products, analysis of pregnant women's body shape, and research on the structural properties of abdomen-supporting underwear. It was revealed that there is a shortage of abdomen-supporting pants in the Chinese market as well as a lack of new data and research regarding body shape analysis of pregnant women. The study conducted a two-year market survey on abdomen-supporting underwear and performed statistical analysis of body shape data of 26 pregnant women. We concluded that body posture of women during pregnancy can be divided into three stages - early pregnancy (8-12 weeks), second trimester (16-24 weeks), and late pregnancy (28-36 weeks) - with each stage having particular maternity garment requirements. Study findings can play a guiding role in structural design of abdomen-supporting underwear and establishment of dress forms for pregnant women.

Keywords: Abdomen-supporting underwear; Pregnant women; Body posture analysis; Descriptive analysis

1. Introduction

Abdomen-supporting maternity garments help support women's belly weight during pregnancy. The abdomen-support functions of pregnant women's underwear can be designed according to body posture changes during the pregnancy cycle [1]. However, in China there are few brands of abdomen-supporting pants available to pregnant women, and research and development related to such garments is somewhat obsolete [2]. According to the Reports on China Population and Labour Issues, there will be a baby boom in China from 2005 to 2020, and the population will see a growth rate of 13 to 15 million each year. Thus, the market for products for pregnant women will have a strong development potential, large base, and stable growth [3].

The body posture of pregnant women has no great change in early pregnancy. With fetal development, pregnant women's bellies become large during the second trimester of pregnancy and late pregnancy. In order to maintain body balance during the latter stages of pregnancy, women must lean backward, causing lumbar lordosis that aggravates the lower back ligament, leading to back and waist pains [4]. While an abdomen-supporting belt can play an important role for women with back and waist pain during indoor or outdoor activities [5], pregnant women are concerned that the pressure caused by this type of belt may have adverse effects on the fetus [6]. We thus recommend that future studies should focus on the biomechanics and physiology fields [7].

The placenta is in the upper-half part of the uterus for 90% of pregnant women, and the whole or a part of the placenta is above the uterus in 31% of pregnant women; during the ninth month, such condition exists in 40% of pregnant women - that is to say, the abdominal distension is apparent in women during pregnancy [8]. The growth of the fetus increases intra-abdominal pressure, and portal venous obstruction causes the upper and lower venous congestion of the rectum, at times resulting in internal or external hemorrhoids. The enlarged uterus pressing against the vessels also may cause dizziness in pregnant women [9]. Therefore, the increasing size of women's belly during the second trimester and late pregnancy creates problems and poses health risks during the later stages of pregnancy. In addition, pregnant women have difficulty choosing pants; pants with appropriate abdominal circumference but large waist and hip circumference may easily fall, while those with appropriate waist and hip circumference but small abdominal circumference may have an adverse effect on fetal growth and development [10]. As such, there is an urgent need for abdomen-supporting underwear that conforms to ergonomics and has both abdominal and waist supporting functions as well as a high comfort level.

In this study, we conducted a two-year survey pertaining to the requirements of abdomen-supporting underwear in pregnant women. We collected and tracked data on changes in women's body posture during pregnancy. The purpose of the study was to perform analyses on the body posture of pregnant women and to provide data support for the structural design of abdomen-supporting underwear with a strong theoretical basis and application prospects that fully consider body posture changes during the whole pregnancy.

2. Questionnaire survey

A pre-market survey was made based on the existing abdomen-supporting products for pregnant women. A group of questionnaires were designed to identify market conditions of clothing for pregnant women, and to provide a research basis for body posture changes of pregnant women.

2.1 Questionnaires

The survey content encompassed the product brand, satisfaction, comfort, and structure of abdomen-supporting products for pregnant women as well as recommendations for improvement. The survey was administered in Beijing Baodao Healthcare, Beijing Chaoyang Women & Children Hospital, and Beijing Maternity Hospital. A total of 38 questionnaires were distributed and all of them were valid.

2.2 Analysis of results

2.2.1 Use of brands of abdomen-supporting underwear

As shown in Fig.1, among the respondents there are more women wearing non-branded products, accounting for 37.8%, suggesting that pregnancy-related abdomen-supporting underwear has not gained enough attention. Among the branded products surveyed, OCTMAMI occupies the largest market share, accounting for 35.1%, suggesting that there is a development potential of abdomen-supporting underwear in China.

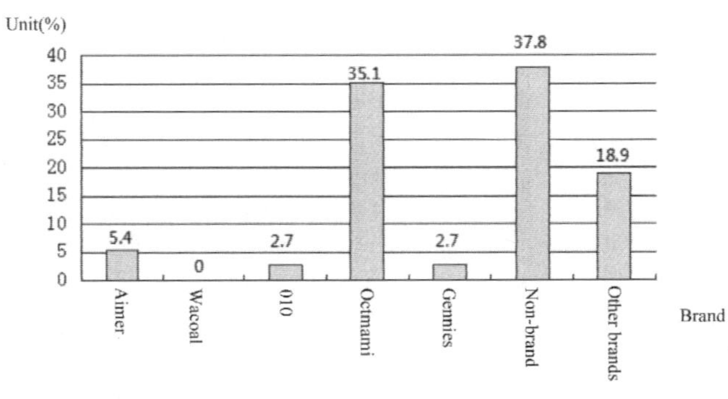

Fig.1 Popularity of different brand

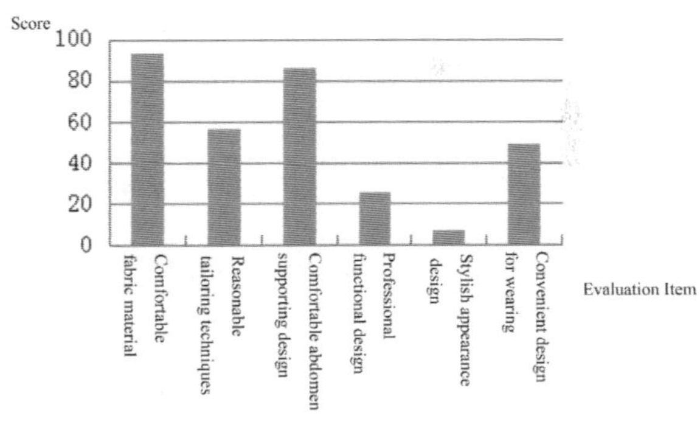

Fig.2 Characteristics of satisfactory abdomen-supporting underwear

2.2.2 Survey of characteristics of satisfactory abdomen-supporting underwear among pregnant women

Characteristics of abdomen-supporting underwear were ranked in an order of descending importance; those ranked as most important were scored as 6, second-most important scored as 5, and so forth. Results were as follows: 93 for comfortable fabric material; 56 for reasonable tailoring techniques; 86 for comfortable abdomen-supporting design; 27 for professional functional design; 7 for stylish appearance design; and 49 for convenient design for wearing (as shown in Fig.2). Thus, the products recognized by pregnant women usually feature comfortable fabric material, reasonable abdomen-supporting design, reasonable tailoring techniques, et cetera. We suggest that in addition to fabric comfort, structural design conforming to ergonomics is also a key factor for the design of abdomen-supporting products.

2.2.3 Characteristics of unsatisfactory abdomen-supporting underwear among pregnant women

The satisfaction survey pertaining to existing abdomen-supporting underwear indicated that 51.4% of pregnant women believed that product satisfaction needs to be improved. Unsatisfactory factors mainly included appearance, fabric, touch, difficulty to don and doff, appearance after wearing, heat and moisture comfort, pressure, and wearing comfort. As shown in Fig.3, our scoring method revealed that the top four factors were: fabric (score of 48), wearing comfort (41), touch (36) and heat and moisture comfort (31). Thus, it can be seen that the fabric used for abdomen-supporting products available in the marketplace needs to be improved; touch and heat as well as moisture comfort were mainly associated with the fabric, while the wearing comfort was associated with the fabric, structure, and comfortable abdominal pressure.

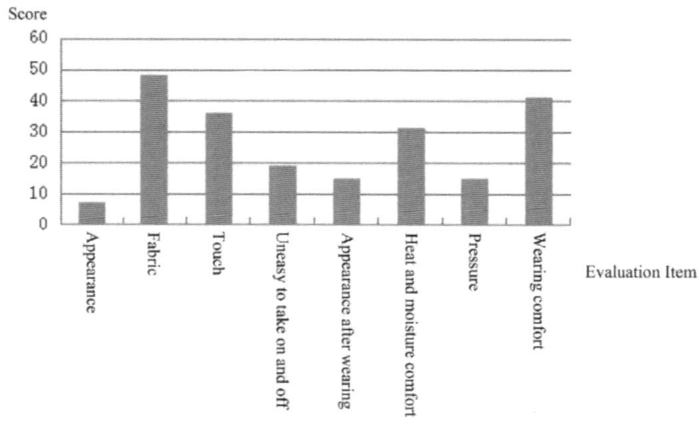

Fig.3 Characteristics of unsatisfactory abdomen-supporting underwear

2.2.4 Preference of style for abdomen-supporting underwear

The survey asked consumers to choose the preferred styles for the front and backs of abdomen-supporting underwear for pregnant women (as shown in Figs.4 and 5). The first preference was scored 4, the second preference was scored 3, and so forth. It can be seen that among the styles of the front part of abdomen-supporting underwear, styles B and C had the highest scores (65 each), followed by style A (53) and finally style D (41). For the back part, style F had the highest score (54), followed by style E (47) and style G (40). The results showed that for the front part of abdomen-supporting underwear, a straight waist and long legs, and a high waist with adjustable abdomen-supporting structure were the most popular styles; while for the back part, a straight and lateral support structure of back waist was most popular. Thus, special attention should be paid to the changes in women's waist, abdominal, hip, and thigh circumferences during pregnancy.

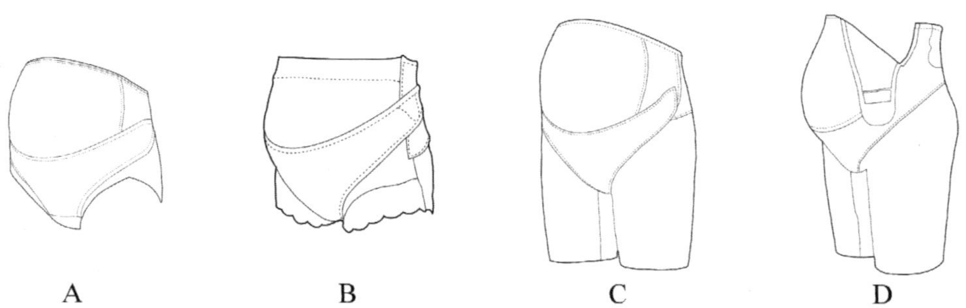

Fig.4 Style of front part of abdomen-supporting underwear for pregnant women

Fig.5 Style of back part of abdomen-supporting underwear for pregnant women

2.3 Summary of survey results

1) People do not pay much attention to abdomen-supporting underwear for pregnant women; most respondents wear non-branded products, indicating that the development potential of abdomen-supporting underwear is great in China.

2) Products recognized by pregnant women feature comfortable fabric material, reasonable and comfortable abdomen-support design, and reasonable tailoring techniques, suggesting that structural design conforming to ergonomics is a key factor in addition to fabric comfort for the design of abdomen-supporting products.

3) The wearing comfort was associated with the fabric, structure, and comfortable abdominal pressure.

4) A straight waist and long legs, and high waist with adjustable abdomen-supporting structure are the most popular styles for the front part of abdomen-supporting underwear; while a straight and lateral support structure for the back waist is the most popular style for the back part. The styles should have a structural design suitable for the changes of body shape of women during pregnancy.

3. Data collection

3.1 Selection of measurement items

Data collection involved three steps: first, manual measurement with a Martin anthropometric instrument; second, 2D-picture plotting measurement using Shape Anthropometric System 1.2; and third, extraction of key sections of human body through Sliding Gauge.

3.2 Data collection procedures

We recruited a total of 26 pregnant women volunteers from Beijing Region and collected their basic body posture data. Data collection and measurement items were determined based on the market survey and previous studies. The body posture change trend of examinees from early pregnancy to the pre-natal was monitored and collected once every 4 weeks before the 24th week of pregnancy, and once every 2 weeks after the 24th week. The measurement items included height, body weight, body axial angle, waist circumference, abdominal circumference, and hip circumference. The total of 60 items included 6 for circumference, 10 for height, 10 for width, 7 for thickness, 8 for angle, 14 for proportion, and 3 for actual length.

4. Data analysis

After data of six secondary pregnant women were excluded, we conducted an exploratory analysis of 60 measurement items of 20 pregnant women, and a descriptive analysis of change in the body posture at all stages of the pregnancy, including BMI, upper and lower body axial angle, as well as rear horizontal waist circumference, and abdominal, umbilical, and hip circumferences associated with the lower body posture of pregnant women.

4.1 Changes in BMI

As seen from the statistical analysis of BMI data during the pregnancy (Fig.6), BMI of women during pregnancy has always shown an overall increase with the deepening of pregnancy, and the increased value of BMI can be up to the maximum (1.38 kg/m^2) at 24-28 weeks of pregnancy, with a changing rate of 24.78%, as shown in Table 1.

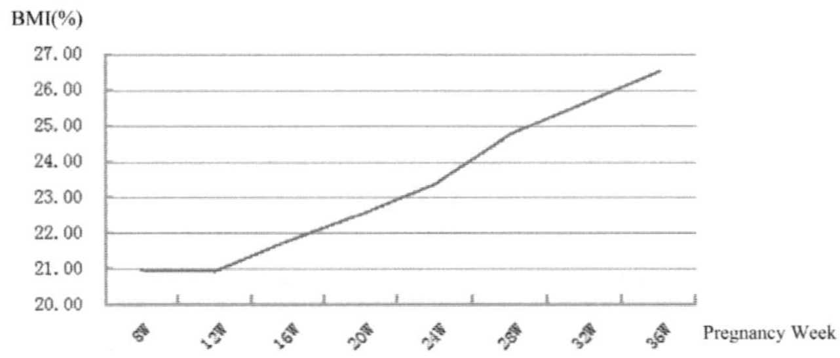

Fig.6 Women's BMI during pregnancy

Table 1. Women's BMI increase trend change during pregnancy (Unit: kg/m^2)

Pregnancy weeks:	8-12	12-16	16-20	20-24	24-28	28-32	32-36
BMI:	(0.01)	0.82	0.77	0.85	1.38	0.90	0.84
Change rate:	0.18%	14.72%	13.82%	15.26%	24.78%	16.16%	15.08%

4.2 Body axial angle change

Based on data analysis of body axial angle of pregnant women's upper and lower body, body axial angle of upper body (angle of the vertical line that runs from the midpoint of the waist thickness to the midpoint of the base of neck thickness, viewed in profile) during pregnancy showed an overall rising trend (as shown in Fig.7), and achieved the maximum rate of change of 41.01% at 8-12 weeks, while body axial angle of lower body (angle of the vertical line that runs from the midpoint of the smallest part of the calf thickness to the midpoint of the waist thickness, viewed in profile) also showed an overall rising trend (as shown in Fig.8), and achieved the maximum rate of change of 34.62% at 24-28 weeks. As shown in Fig.9, pregnant women's body centre of gravity gradually moved forward to form a unique posture of pregnant women. Table 2 shows how body axial angle of pregnant women's upper and lower body changed greatly at 24-28 weeks.

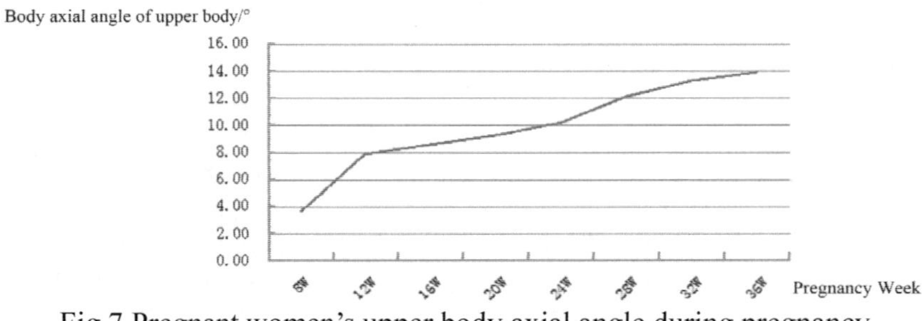

Fig.7 Pregnant women's upper body axial angle during pregnancy

Table 2. Pregnant women's upper and lower body axial angle increasing trend change during pregnancy (unit: °)

Pregnancy weeks:	8-12	12-16	16-20	20-24	24-28	28-32	32-36
Upper body axial angle:	4.22	0.66	0.74	0.94	1.95	1.16	0.62
Upper body axial angle rate of change:	41.01%	6.41%	7.19%	9.14%	18.95%	11.27%	6.03%
Lower body axial angle:	(0.21)	0.15	0.07	0.01	0.45	0.13	0.28
Lower body axial angle rate of change:	16.15%	11.54%	5.38%	0.77%	34.62%	10%	21.54%

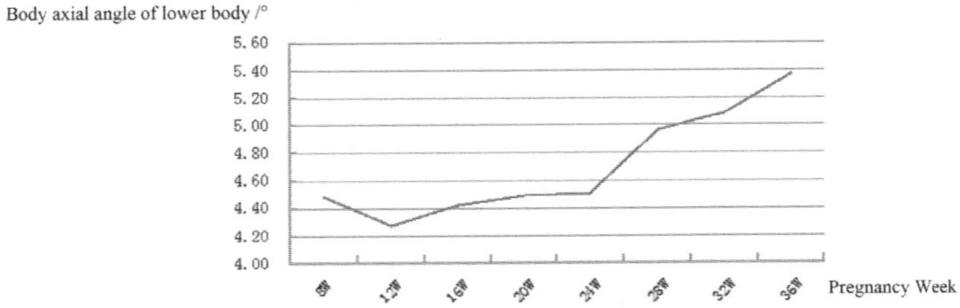

Fig.8 Pregnant women's lower body axial angle during pregnancy

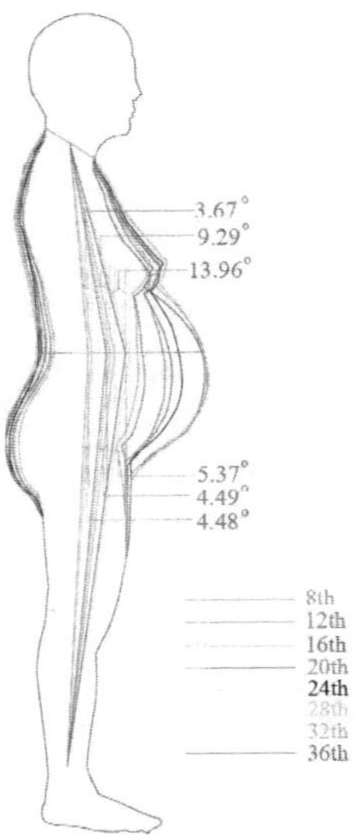

Fig.9 Pregnant women's upper and lower body axial angle changes during pregnancy

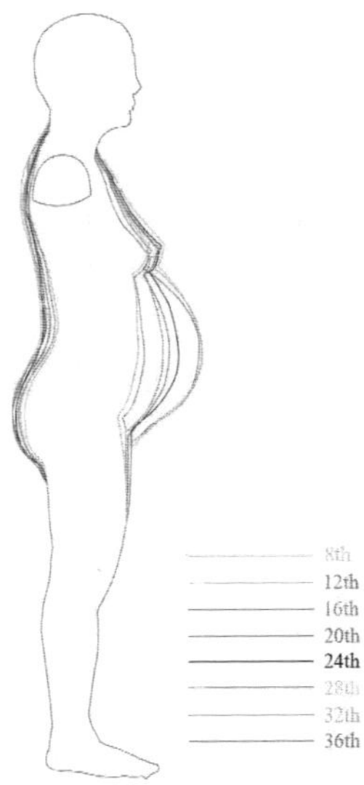

Fig.10 Phases of pregnant women's body posture changes during pregnancy

As shown in a vertical cross-sectional diagram (Fig.10), women's body posture during pregnancy can be divided into three distinct phases: the first phase (early pregnancy, 8-12 weeks), shows a slightly protruding abdomen; the second phase (second trimester of pregnancy, 16-24 weeks), shows a significantly protruding abdomen; and the third phase (late pregnancy, 28-36 weeks), shows a significantly protruding upper and lower abdomen.

4.3 Hip and abdomen data change

As shown in Fig.11, compared to rear horizontal waist circumference, abdominal circumference, and umbilical circumference, the hip circumference changing trend was relatively small, and the above data showed greater change at 24-28 weeks (Table 3). As shown in Fig.12 comparing rear horizontal waist circumference thickness, abdominal circumference thickness, and hip circumference thickness, hip circumference changing trend of thickness was also relatively small, and the rear horizontal waist circumference and abdominal circumference changing trend of thickness were consistent.

As seen from the descriptive analysis of upper and lower body axial angle, rear horizontal waist circumference, abdominal circumference, umbilical circumference, and hip circumference of pregnant women, all sites of pregnant women's bodies have shown an increasing trend (Figs.6-12), with a total increase of 25.7cm in rear horizontal waist circumference, total increase of 22.32cm in abdominal circumference, total increase of 26.45cm in umbilical circumference, and a relatively small increase of only 8.09cm in hip circumference. Maximum increase was achieved in all the above-described sites at 24-28 weeks, respectively; that is, the lower body posture change of pregnant women at 24-28 weeks was more obvious than during other periods of time. As shown in Table 4, the average increase in umbilical circumference was the largest, and the increase in hip circumstance was the smallest.

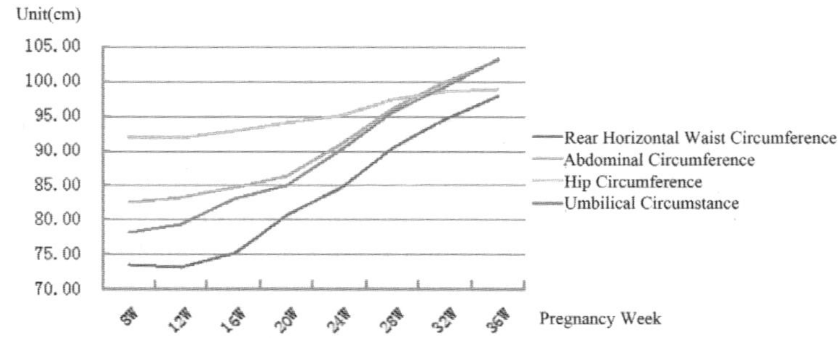

Fig.11 Women's rear horizontal waist, abdominal, hip, and umbilical circumferences during late pregnancy

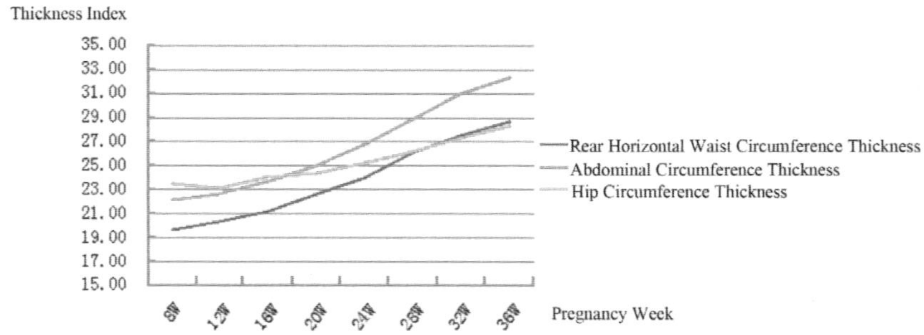

Fig.12 Women's rear horizontal waist, abdominal, and hip circumference thicknesses during late pregnancy

5. Conclusion

Analysis on body posture data obtained from 20 pregnant women shows that BMI increased with the progression of the pregnancy and eventually reached a maximum at 24-28 weeks. The greatest change took place in the rear horizontal waist circumference, umbilical circumference, and abdominal circumference; hip circumference showed relatively small change, and rear horizontal waist circumference and abdominal circumference changing trend of thickness were consistent. Lower body posture change of pregnant women at 24-28 weeks was more obvious than at other periods of time. The average increase in umbilical circumference was the largest, while the average increase in hip circumference was the minimum in terms of average increase of all sites. Axial angle of pregnant women's lower and upper body gradually moved forward to form the unique posture of pregnant women. Body posture of women during pregnancy can be divided into three distinct phases: early pregnancy at 8-12 weeks; second trimester of pregnancy at 16-24 weeks; and late pregnancy at 28-36 weeks. Enterprises can design abdomen-supporting underwear according to the measured posture of pregnant women, and produce dress forms for pregnant women according to the measurement data at the same time.

Table 3. Increase in trend change per pregnant women's body site (unit: cm)

Pregnancy weeks:	8-12	12-16	16-20	20-24	24-28	28-32	32-36
Rear horizontal waist circumference:	(0.39)	2.09	5.53	3.86	5.85	4.12	3.40
Rear horizontal waist circumference rate of change:	1.55%	8.27%	21.92%	15.28%	23.18%	16.33%	13.47%
Rear horizontal waist circumference thickness:	0.65	0.82	1.49	1.29	2.11	1.49	1.07
Rear horizontal circumference thickness rate of change:	7.25%	9.18%	16.68%	14.46%	23.63%	16.75%	12.05%
Abdominal circumference:	0.71	1.41	1.74	4.59	5.15	3.85	3.18
Abdominal circumference rate of change:	3.43%	6.81%	8.44%	22.27%	24.95%	18.65%	15.44%
Abdominal circumference thickness:	0.49	1.04	1.33	1.66	2.17	2.05	1.38
Abdominal circumference thickness rate of change:	4.83%	10.31%	13.17%	16.42%	21.42%	20.20%	13.64%
Hip circumference:	(0.03)	1.04	1.12	1.13	2.37	1.08	0.39
Hip circumference rate of change:	0.42%	14.54%	15.69%	15.74%	33.11%	15.11%	5.40%
Hip circumference thickness:	(0.46)	0.97	0.33	0.81	0.95	1.15	0.95
Hip circumference thickness rate of change:	8.19%	17.30%	5.88%	14.43%	16.96%	20.42%	16.82%
Umbilical circumference:	1.18	3.53	2.16	4.93	5.55	3.71	3.94
Umbilical circumference rate of change:	4.72%	14.14%	8.65%	19.71%	22.20%	14.83%	15.75%

Table 4. Average increase of each site of pregnant women's body during pregnancy (unit: cm)

Rear horizontal waist circumference increase	Abdominal circumference increase	Hip circumstance increase	Umbilical circumference increase
25.70	22.32	8.09	26.45

References

[1] Tang HY. Design research of maternity clothes. Shanghai: Donghua University: 2006.

[2] Shao XW. Health and safety design of pregnant women's clothes based on maternity physiological characteristics. J of Southwest China Normal University (Natural Science) 2009; 34: 5.

[3] Lu X. Design research of functional maternity clothes. Hangzhou, China: Zhejiang University: 2007.

[4] Kalus SM, Kornman LH, Quinlivan JA. Managing back pain in pregnancy using a support garment: a randomised trial. Brit J Obstet Gynaec 2008; 115: 68-75.

[5] Ho SSM, Yu WWM, Lao TT, et al. Garment needs of pregnant women based on content analysis of in-depth interviews. J Clin Nurs 2009; 18(17): 2426-2435.

[6] Moorcroft D, Stitzel J, Duma S, Duma G. The effects of uterine ligaments on fetal injury risk in frontal automobile crashes. J Auto Eng 2003; 217(12): 1049-1055.

[7] Ho SSM, Yu WWM, Lao TT, et al. Effectiveness of maternity support belts in reducing low back pain during pregnancy: a review. J Clin Nurs 2009; 18(11): 1523-1532.

[8] Fei XH, Li YM. Body posture analysis and couture during the pregnancy. J of Hangzhou Normal College (Natural Science) 2003; 2(4); 66-69.

[9] Ma LK, Yu Q. Prenatal care—Union Medical College Hospital Famous Doctors' Health Advice Series. Marrickville, Australia: Science Press: 2009.

[10] Zheng R, Zhang H, Han JY (trans). Costumes science: theory chapter. Yunnan, China: China Textile Press: 2008.

Durable Antibacterial Activity onto Galangal Essential Oil/Polyurethane-Urea Microcapsules Treated Cotton Fabric

Sireerat Charuchinda[1*], Areerat Pramsiri[1]

[1]*Faculty of Science, Chulalongkorn University, Phyathai Rd, Patumwan, Bangkok, 10330, Thailand*

[*]Corresponding author's email: sireeratc@gmail.com

Abstract

In this study, Galangal essential oil/Polyurethane-urea (GEO/PUU) microcapsules were treated onto knitted cotton fabrics using nano-polyurethane binder with and without fixing agent via impregnation and spray-coating. It was found that spray-coating with 10% GEO/PUU microcapsules, 3% nano-polyurethane binder and fixing agent exhibited excellent antibacterial activity against *Staphylococcus aureus* (> 99% reduction of bacteria) onto cotton fabrics. Moreover, the antibacterial activity was durable to washing up to 25 wash cycles, to light, to hot-press and to perspiration. However, the antibacterial activity durability to acid perspiration is lower when compared to alkali perspiration. Nevertheless, the bursting strength after 25 wash cycles was slightly decreased. Moisture management (MMT)-grading of the treated fabric was poor, whereas the untreated fabric was fair.

Keywords: Antibacterial activity; Galangal essential oil; Polyurethane-urea; Cotton fabric; Microcapsules

1. Introduction

In recent years, functional textiles are the new dimension of "textiles" which are not only for the aesthetic apparels but also for the protective textiles especially in terms of health and hygiene. Generally, antibacterial activity can be imparted to textile materials by incorporating antibacterial agents onto fibres or fabrics. Most of antibacterial agents for textiles are organo-metallics, phenolic compounds, quaternary ammonium salts, triclosan, halamine precursor compounds and organo-silicones which are chemical substances and expensive [1]. Some chemical substances may be toxic and hazardous to human health and environment. Consequently, safety, non-toxicity and biodegradability are required for antibacterial agents. Herbs are naturally prosperous and abundant in Thailand. They are known for their antibacterial activity such as curcumin, lemongrass, Indian gooseberry and galangal [2-4]. Galangal, *Alpinia galangal Linn.,* essential oil enriched with 1,8-cineol, eugenol and flavonoids are acclaimed to exhibit antibacterial activity resulting in cell's growth inhibition (e.g. against *Staphylococcus aureus)* [5]. Thus, galangal essential oil (GEO) is viewed as one of the new potential alternative natural, non-toxic source and as a potential replacement for chemical antibacterial agents. However, its antibacterial activity on textile fabrics quickly disappears after washing due to its quick release characteristic because of its high volatility [6]. Since high concentration of herbs is needed to inhibit bacterial growth, this could make eye and skin irritation. To solve these problems, encapsulation of GEO in alginate or polyurethane-urea (PUU) microcapsule was undertaken to reduce irradiation control and also to increase the longevity and efficiency of antibacterial finishing [7,8] 。

The microcapsules can be applied on any fabric by conventional method such as impregnation, dipping and coating [9,10]. The main advantage of coating is the larger quantity of microcapsules can be added. This is especially important for encapsulated galangal essential oil as their quantity is strongly related to their antibacterial activity. In addition, durability of the microcapsules on the fabric in relation to their controlled release properties in some end uses such as washing, light exposure, ironing and perspiration is challenging [11]. The major advantages of the spray coating are the high speed of coating, low cost of equipment, the large area coatings. In addition, this technique is efficiency and has high feasibility in an industrial scale up compared to other techniques. Moreover, the spray coating could maintain the microcapsules shape and protect them from breaking during coating. Hence, in this study, different technique: impregnation and spray coating were used to fix galangal essential oil/polyurethane-urea (GEO/PUU) microcapsules prepared from our previous study onto cotton fabric. Quantitative analyses of antibacterial activity of the treated and washed (light exposed, hot-pressed, perspiration soaked) fabrics were determined by AATCC-100 test method. Morphology of such fabrics was also examined by scanning electron microscope. The bursting strength and moisture management (MMT) of the fabrics was also determined.

2. Experimental

2.1 Materials

The galangal essential oil/polyurethane-urea (GEO/PUU) microcapsules used in this study were prepared via interfacial polymerization in our laboratories. The microcapsules which prepared by using stirring rate at 14000 rpm were selected to coat onto cotton fabrics due to their small particle size of 6.72 μm and quite narrow size distribution. Moreover, these microcapsules still contained high loading capacity (16.87%). Ethanol 99.8% (AR grade, Merck) was purchased from T.C. Sathaporn Group Ltd., Part. Nano-polyurethane binder (Baypret Nano PU, particle size <100 nm) was kindly provided by Tanatex Chemicals. The chemical materials were also used as received without any further purification. A bleached single jersey-knit 100% cotton fabric (a weight of 202.22 g/m^2 and a thickness of 0.62 mm) was were kindly supplied by August Chem Co., Ltd.

2.2 Fabric treatment with GEO/PUU microcapsules

In this study, impregnation and spray-coating were used to apply onto knitted cotton fabric.

2.2.1 Impregnation

Cotton fabric was treated with a mixture of 10% of selected GEO/PUU microcapsules, 3% nano-polyurethane binder, 0.2% wetting agent and 1% dispersing agent using a pad mangle with a pressure nip at 80-90 % wet pick up. The treated fabric was then dried at 60 °C for 5 min and cured at 170°C for 30 second.

2.2.2 Spray-coating

Cotton fabric was treated with a mixture of 10% of selected GEO/PUU microcapsules, 3% nano-polyurethane binder, 0.2% wetting agent, 1% dispersing agent with and without fixing agent using spray-gun (Model F-75 and PC-4 C 400 cc, Somax, Thailand). Then, the treated fabric was dried at 60 °C for 5 min and cured at 170°C for 30 second.

2.3 Antibacterial activity determination

The antibacterial activity against Staphylococcus aureus (S. aureus) (gram-positive bacteria) of treated fabrics was determined quantitatively using agar diffusion method (AATCC 100-1999). The untreated and treated cotton fabrics were UV-sterilized for 20 minutes and incubated for 24 hours. S. aureus was chosen for antibacterial test because it is capable of causing various infections of the skin and other organ. S. aureus infections are common in people with frequent skin injury, particularly if the skin is dry. Staphylococcal skin infections are seen most commonly in certain occupational groups such as healthcare workers [12].

2.4 Durability of antibacterial activity

The durability of antibacterial activity of the cotton fabrics to 5, 10, 15, 20 and 25 washing cycles (ISO105-E40) using Gyrowash 815/20: James H. Heal Co. Ltd., United Kingdom, to light (AATCC 16E-2004) using Xenon Solar Box 1500e RH: CO.FO.ME.GRA, Italy, to perspiration (AATCC 15-2002) using Perspirometer M231: SDL Atlas Inc., United Kingdom, and to hot-press (AATCC 125-2004) using Scorch Tester M247A: SDL Atlas Inc., United Kingdom were determined.

2.5 Morphology of microcapsules anchored to the surface of treated cotton fabrics

Surface characteristic of treated cotton fabrics with microcapsules were investigated by scanning electron microscopy (SEM, a JEOL JSM 5410 LV, Japan) under accelerated of 15 kV at a magnification of 1500X.

2.6 Bursting strength

Bursting strength of treated cotton fabrics with microcapsules were examined by bursting strength testing equipment (CY-6103A1, Chun Yen Testing Machine Co., Ltd., Taiwan) according to ASTM D 3786 standard test method.

2.7 Moisture management

Fabric comforts after treatment with microcapsules in term of moisture management were conducted by moisture management tester (M290, Technical Color Solution, Pakistan) according to AATCC 195-2010 standard test method.

3. Results and discussion

3.1 Antibacterial activity of cotton fabrics via impregnation and spray-coating

Table 1 shows the antibacterial activity against S. aureus of cotton fabrics treated with 10% GEO/PUU microcapsules and 3% nano-polyurethane binder by different techniques. The untreated fabric exhibited 0% reduction of S. aureus, whereas, the treated fabric via impregnation and spray-coating presented 43.54% and 99.89%, respectively. This clearly indicated that galangal essential oil/polyurethane-urea microcapsules exhibited antibacterial activity. This can be elucidated that the galangal essential oil which enriched of eugenol and 1,8-cineol denatured of protein inside the cell and also disrupted the cytoplasmic membrane function of S. aureus cells which resulted in a loss of cytoplasmic constituents and ions [5].

Moreover, treatment of cotton fabric via spray-coating is more effective than impregnation. It is attributed that pressure from pad mangle via impregnation could break and damage microcapsules which led essential oil or active agents inside evaporate during drying and curing process. Furthermore, the amount of microcapsules anchored to the fiber was less than that of spray-coating. This is in agreement with SEM images shown in Fig. 1 that a good shape and a high quantity of microcapsules was detected on the fabric when spray-coating was used. Thus, spray-coating was then selected to treat onto cotton fabric to further investigate the durability of antibacterial activity to wash, to light, to perspiration and to hot-press.

Table 1. Antibacterial activity against S. aureus of cotton fabrics treated with 10% GEO/PUU microcapsules and 3% nano-polyurethane binder by impregnation and spray-coating

Technique	Add-on [%]	Reduction of bacteria [%]
Untreated	0	0
Impregnation	30.15	43.54
Spray-coating	33.84	99.89

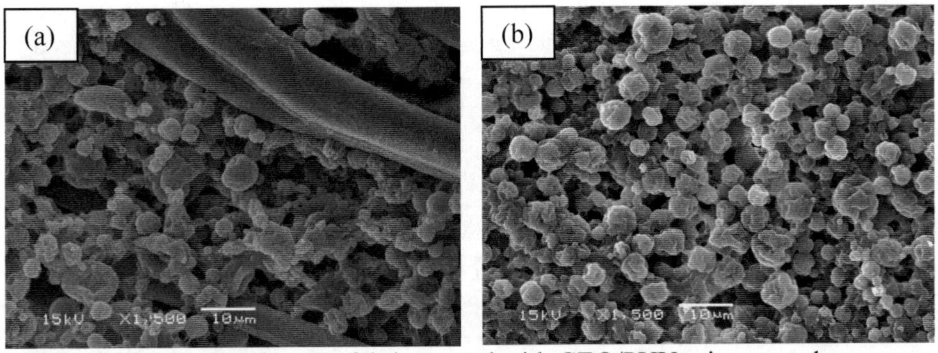

Fig.1 SEM images of cotton fabric treated with GEO/PUU microcapsules
(a) Impregnation (b) Spray-coating at a magnification of 1500X.

3.2 Durability of antibacterial activity of cotton fabrics

The company recommended the formulation of coatings using nano-polyurethane binder without any fixing agent, however, the effect of fixing agent on the durability of antibacterial activity of treated cotton fabric should be considered. Thus, the durability of antibacterial activity to wash, to light, to perspiration and to hot-press of the cotton fabric treated with 10% GEO/PUU microcapsules and 3% nano-polyurethane binder with and without fixing agent via spray-coating was investigated.

3.2.1 Wash durability

Durability of antibacterial activity against S. aureus to washing of cotton fabrics treated with 10% GEO/PUU microcapsules and 3% nano-polyurethane binder with and without fixing agent was shown in Table 2. It was found that untreated cotton fabric was absolutely attacked by S. aureus. Since GEO inside PUU microcapsules release and play an important role to inhibit S. aureus cell growth, cotton fabric treated with GEO/PUU microcapsules displays antibacterial activity. Table 2 shows that nano-polyurethane binder could anchor microcapsule to the fabric and are durable to washing. Although both cotton fabrics treated with and without fixing agent exhibited % reduction of bacteria more than 97%, fixing agent plays an important role to fix microcapsules onto fabrics better than that of without fixing agent. Addition of fixing agent led to fewer amounts of microcapsules leached out after washing (see the % microcapsule leached out). It is seen that the % reduction of bacteria of treated cotton fabric with fixing agent after 10-25 wash cycles was 99.99% which was more than that of without fixing agent.

Table 2. Durability of antibacterial activity against S. aureus to washing (up to 25 cycles) with 10% GEO/PUU microcapsules and 3% nano-polyurethane binder with and without fixing agent

Wash cycles	Microcapsules leached out [%]		Reduction of bacteria [%]	
	without fixing agent	with fixing agent	without fixing agent	with fixing agent
Untreated	0	0	0	0
0	0	0	99.89	98.97
5	6.51	12.63	97.78	99.80
10	28.30	8.87	97.77	99.99
15	21.99	9.87	98.07	99.99
20	20.08	8.29	99.82	99.99
25	19.76	4.83	99.82	99.99

3.2.2 Light, hot-press and perspiration durability

Durability of antibacterial activity against *S. aureus* to light, hot-press and perspiration of cotton fabrics treated with 10% GEO/PUU microcapsules and 3% nano-polyurethane binder with and without fixing agent was shown in Table 3.

Table 3. Durability of antibacterial activity against S. aureus to light, hot-press and perspiration of cotton fabric treated with 10% GEO/PUU microcapsules and 3% nano-polyurethane binder with and without fixing agent

	Reduction of bacteria [%]							
	Light		Hot-press		Perspiration			
					acid [pH 5.5]		alkali [pH 8]	
	before	after	before	after	before	after	before	after
Untreated	0	0	0	0	0	0	0	0
No fixing agent	99.89	99.99	99.89	92.58	99.89	0	99.89	0
Fixing agent	98.97	99.99	98.97	99.25	98.97	72.57	98.97	99.42

It was found that both cotton fabrics treated with GEO/PUU microcapsules with and without fixing agent shows % reduction of bacteria before and after light exposure, hot-press exposure more than 90%. This indicated that PUU walls are resistant to light and heat which led to protect GEO and could release GEO to inhibit growth of bacteria. This is in agreement with SEM images shown in Fig.2 that a high quantity of microcapsules was detected on the fabric when fixing agent was used and still remained after light exposure. However, it is seen that fixing agent plays an important role in perspiration durability. Treated cotton fabric without fixing agent shows 0% reduction of bacteria after soaking in both acid perspiration (pH 5.5) and alkali perspiration (pH 8), whereas, treated cotton fabric with fixing agent still exhibited 72.57 % and 99.42% reduction of bacteria after soaking in acid perspiration and alkali perspiration, respectively. It is elucidated that microcapsules could anchor to the fabric better than that of without fixing agent. Hence, the quantity of microcapsules leached out when soaking in perspiration is less than that of without fixing agent. The remained GEO could release antibacterial activity compounds to inhibit S. aureus cell growth. Interestingly, the antibacterial activity of galangal essential oil against S. aureus decreased when the fabric was soaked in an acid perspiration (pH 5.5). This may be demonstrated that the cytoplasm of S. aureus cell was acidified which led to protein denaturation and coagulation of cell contents [5].

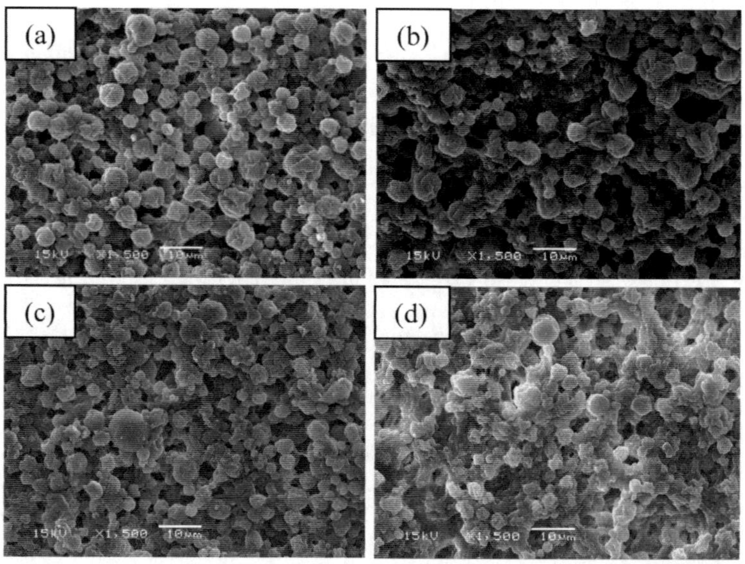

Fig.2 SEM images of cotton fabric treated with GEO/PUU microcapsules:
(a) without fixing agent before light exposure (b) without fixing agent after light exposure
(c) with fixing agent before light exposure (d) with fixing agent after light exposure
at a magnification of 1500X.

3.2.3 Bursting strength

In this study, we selected the cotton fabric treated with 10% GEO/PUU microcapsules and 3% nano-polyurethane binder with fixing agent via spray-coating which shows very good antibacterial activity and also very good durable antibacterial activity to washing, light, hot-press and perspiration compared to other formulation to determine bursting strength after treatment. Bursting strength of untreated and treated cotton fabric is shown in Table 4. It is seen that bursting strength of fabric after treatment slightly increase due to crosslinked structure of film layer anchored to fabric.

Table 4. Bursting strength of cotton fabric treated with 10% GEO/PUU microcapsules and 3% nano-polyurethane binder with fixing agent

Fabric	Bursting strength [KPa]
Untreated (1 wash)	822.5
Treated (1 wash)	845.0

3.2.4 Moisture management (MMT)

Although fabric treated with GEO/PUU microcapsules using nano-polyurethane binder and fixing agents via spray-coating show high efficiency and high durability of antibacterial activity, fabric comforts is also important for their application. In this study, the liquid moisture management properties of fabrics were evaluated as fabric comforts.

Table 5 displays the grading scale for each of the nine MMT test parameters as well as the composite score representing overall moisture management capability (OMMC). It is seen that the OMMC of untreated fabric is grading 2, whereas the treated one is grading 1. This means that the OMMC of untreated fabric is fair, whereas the treated one is poor. This may be due to the film layer of binder cover fabric surface as a barrier.

Table 5. Moisture management (MMT) grading of cotton fabric treated with 10% GEO/PUU microcapsules and 3% nano-polyurethane binder with fixing agent

Index	Grade			
	Untreated cotton fabric		Treated cotton fabric	
	Top	Bottom	Top	Bottom
Wetting time	3.5	2.5	3.5	2
Absorption rate	2.5	5	3.5	1.5
Maximum wetted radius	4	3	2	3
Spreading speed	2	2.5	1	1
One-way transport capability	(2)		(1)	
Overall moisture management capability	(2)		(1)	

4. Conclusion

The antibacterial treatment of knitted cotton fabric was successfully conducted by using GEO/PUU microcapsules prepared via interfacial polymerization in our laboratories. Different technique: impregnation and spray coating were used to treat cotton fabric with GEO/PUU microcapsules, nano-polyurethane binder, with and without fixing agent. When spray-coating was used, good shape and a high quantity of microcapsules was detected on the fabric and enough to exhibit excellent antibacterial activity against Staphylococcus aureus (> 99% reduction of bacteria). Fixing agent plays an important role to anchor microcapsules onto fabric. PUU walls are resistant to light and heat and can protect GEO and could release GEO to inhibit growth of bacteria. This leads the antibacterial activity was durable to washing up to 25 wash cycles, to light, to hot-press and to perspiration. However, the antibacterial activity durability to acid perspiration is lower when compared to alkali perspiration. Nevertheless, the bursting strength and fabric comforts of the treated fabric have to be further improved.

5. Acknowledgement

The authors gratefully acknowledge the 90th Anniversary of Chulalongkorn University Fund (Ratchadaphiseksomphot Endowment Fund), the Ratchadaphiseksomphot Endowment Fund (2015), Chulalongkorn University (CU-58-034-AM) and the National Science and Technology Development Agency

(NSTDA) (Grant No. P-10-10934) for their financial supports.

References

[1] Gupta D, Antimicrobial treatments for textiles. Indian J Fibre Text 2007; 32: 254-263.
[2] Han S, Yang Y. Antimicrobial activity of wool fabrics treated with curcumin. Dyes Pigments 2005; 64: 157-161.
[3] Leimann FV, Gonçalves OH, Machado RAF, Bolzan A. Antimicrobial activity of microencapsulated lemongrass essential oil and the effect of experimental parameters on microcapsules size and morphology. Mat Sci Eng C 2009; 29: 430-436.
[4] Mayachiew P, Devahastin S. Antimicrobial and antioxidant activities of Indian gooseberry and galangal extracts. LWT-Food Sci Technol 2008; 41: 1153-1159.
[5] Oonmetta-aree J, Suzuki T, Gasaluck P, Eumkeb G. Antimicrobial properties and action of galangal (*Alpinia* galangal Linn.) on Staphylococcus aureus. LWT-Food Sci Technol 2006; 39: 1214-1220.
[6] Chaliewsak J, Charuchinda S. Improvement of antimicrobial property of cotton fabric with galangal essential oil and galangal extracts. Proceedings of the 10th Asian Textile Conference (ATC-10) September 7-9, 2009 Ueda, Japan.
[7] Jiamrungraksa T, Charuchinda S. Preparation and characteristics of galangal essential oil/alginate microcapsules. JMMM 2010; 20: 89-92.
[8] Podshivalov AV, Bronnikov S, Zuev VV, Jiamrungraksa T, Charuchinda S. Synthesis and characterization of polyurethane-urea microcapsules containing galangal essential oil. The statistical analysis of encapsulation Process. J Microencap 2013; 30: 198-203.
[9] Rodrigues SN, Martins IM, Fernandes IP, Gomes PB, Matac VG, Barreiro MF, Rodrigues AE. Scentfashion®: Microencapsulated perfumes for textile application. Chem Eng J 2009; 149: 463-472.
[10] Sathianrayanan MP, Bhat NV, Kokate SS, Walunj VEW. Antibacterial finish for cotton fabric from herbal products. Indian J Fibre Text 2010; 35: 50-58.
[11] Van Parys M. Smart textiles using microencapsulation technology. In: Ghosh SK, editor. Functional Coatings: by Polymer Microencapsulation. Weinheim: Wiley-VCH, 2006. p. 221-258.
[12] Information on http://www.dermnetnz.org/bacterial/staphylococci.html

Variation in the Interface Pressure Profile and Gradient of Commercial Medical Compression Stockings

Yong-Rong Wang[1,2*], Li-Su Zhu[1], Yin-Xing Liu[1], Yi Liu[1]

[1] *College of Fashion and Design, Donghua University, 1882 Yanan(w) Road, Shanghai, 200051, China*
[2] *Key Laboratory of Clothing Design & Technology, Ministry of Education, Shanghai, 200051, China*

[*]Corresponding author's email: yrwang@dhu.edu.cn

Abstract

This study is aimed to assess the variability of interface pressure profile and gradient among multiple volunteers and examine the actual compression provided by medical stockings from different manufacturers. The interface pressure distribution and gradient for two brands of medical compression stockings were examined on five healthy volunteers' right lower limb, four height levels(ankle, calf, knee, thigh) along leg, and four directions(anterior, lateral, posterior, medial) at each transverse section, totally 16 measured locations are enclosed. The pressure profiles at varied directions and height levels are discussed, and the variation in interface pressure and gradient among multiple volunteers is analyzed.

Keywords: Medical compression stockings; Pressure gradient; Pressure distribution; Lower limb

1. Introduction

Compression therapy is the major component of non-invasive management of venous leg ulcers, varicose veins, deep vein thrombosis and chronic venous insufficiency, etc [1]. Compression may be achieved by different modalities, such as bandages, compression hosieries, and external pneumatic compression pumps, etc. Medical stockings can provide external and controlled gradient pressure (The pressure is highest at ankle and gradually reduces as it approaches thigh.) to the skin and underlying tissues of lower limbs. Medical stockings are one of the most widely used mechanical compression approaches as a result of their convenience, ease of use, and sustained, ambulant therapeutic function [2]. Although the exact physiological mechanisms are not entirely understood, compression stockings have been demonstrated to be effective in reversing increased venous hypertension, augmenting the calf muscle pump, and facilitating venous return [3-5].

The mechanism of compression therapy are complex, but it is verified that the interface pressure and gradient on the skin correlates with therapeutic effect, and more severe conditions require application of higher pressure garments [6,7]. If the pressure applied during compression therapy is insufficient or not sustained, the efficacy of compression therapy will not be realized [8], however, excessive compression can lead to tissue damage and necrosis [9]. That is the challenges of correct use of compression in everyday practice: applying sufficient compression to achieve therapeutic benefit while minimizing risks associated with excessive local compression.

Direct pressure measurement is rarely performed, most of the published studies, and all randomized trials still used exclusively manufacturers provided box labels for determining interface pressure, which is determined in vitro, although European guidelines for studying compression encouraged in vivo pressure measurements [10]. Measurements of the interface pressure under compression stockings in volunteers showed the pressure variability among different brands and types of compression stockings [11]. However, the variations in pressure profile among volunteers are still unknown, since the actual interface pressure distribution and gradient are subjects of variability due to the differences in shape of the leg and mechanical properties of soft tissues among the patients.

This study is aimed to assess the variability of interface pressure profile and gradient among multiple volunteers and examine the actual compression provided by medical stockings from different manufacturers.

2. Experimental

2.1 Subjects and stockings

Two brands of commercial medical compression stockings (Brand 1 and 2) from different manufacturers with approximately the same quality were chosen to be studied, which are in the style of thigh length and especially designed for Asian patients (Table 1), were measured on five healthy female volunteers (mean age: 22±2 years, height:168±2 cm,weight:55±4kg). Table 2 shows the anthropometric parameters of subjects' right lower limb. The experimental protocol was informed, and consent was obtained from all volunteers prior to undertaking the study.

Table 1. Characteristics of medical compression stockings

No.	Specified Ankle Pressure		Fiber content	Description
	(mmHg)	(Pa)		
Brand 1	23-32	3066-4266	70% Nylon, 30% Lycra	Thigh high compression stockings
Brand 2	23-32	3066-4266	73% PA, 27% EA	Thigh length w/silicone lace-open toe

* Pressure Unit: 1 mmHg = 133.32 Pa;

* PA: Poliammidica-Polyamid- Polyamide; EA: Elastan-Elasthan-Elasthanne

Table 2. Anthropometric parameters of subjects' right lower limb (Unit:cm)

Location	Dimensions	Student 1	Student 2	Student 3	Student 4	Student 5	Mean	STDEV
Ankle	Girth	21.14	19.72	22.03	19.76	22.52	21.03	1.28
	Height	12	11	10	12	11	11.20	0.84
Calf	Girth	36.68	35.24	35.97	33.52	36.57	35.60	1.29
	Height	35	32	34	33	34	33.60	1.14
Knee	Girth	39.17	37.77	39.91	36.27	39.63	38.55	1.52
	Height	48	47	49	48	48	48.00	0.71
Thigh	Girth	45.13	45.81	44.17	41.09	45.34	44.31	1.90
	Height	58	57	59	58	58	58.00	0.71

2.2 Testing equipment

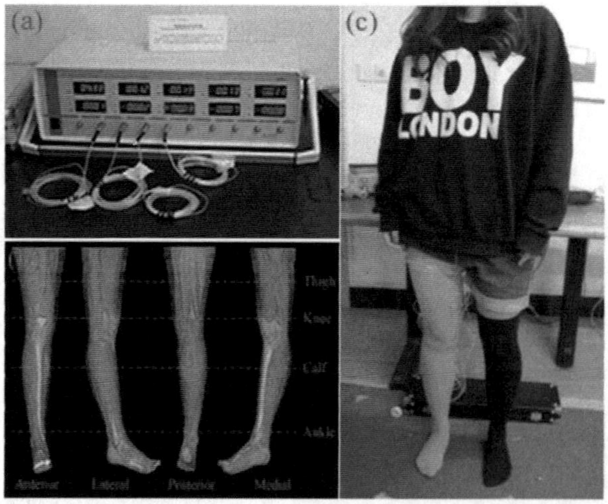

Fig.1 The compression measurements of medical stockings: (a) the pressure measurement system; (b) sensors distributed locations; (c) testing state example.

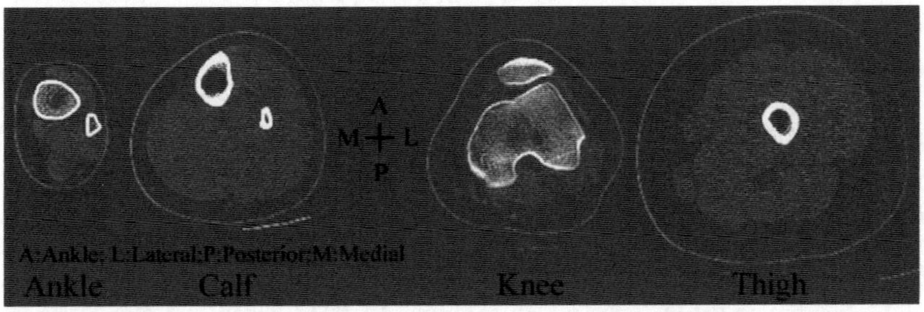

Fig.2 CT images of transverse sections of a healthy female's right leg

Compression was examined using Air-pack type contact pressure sensor(AMI TECHNO CO.,LTD, Japan) and a multichannel measuring system. The pressure sensor has a circular, flexible probe (pressure measurement range: 0-35kPa; precision: ±0.1kPa) at one end. The individual pressure sensors were placed at 16 different locations on the skin surface distributed at four height levels (ankle, the widest circumference of the calf, knee and mid-thigh) and in four directions (anterior, lateral, posterior, and medial) around the leg, shown in Fig.1, CT images of transverse sections of a healthy female's right leg is shown in Fig.2.

3. Result and discussion

The interface pressure profiles for two brands of medical compression stockings were examined on five healthy volunteers' right lower limbs, four height levels(ankle, calf, knee, thigh) along leg, and four directions(anterior, lateral, posterior, medial) at each transverse section, totally 16 measured locations are enclosed. The pressure profiles at varied directions and height levels are discussed, and the variation in interface pressure among different volunteers is analyzed.

3.1 The pressure profile at multiple directions around human lower limb

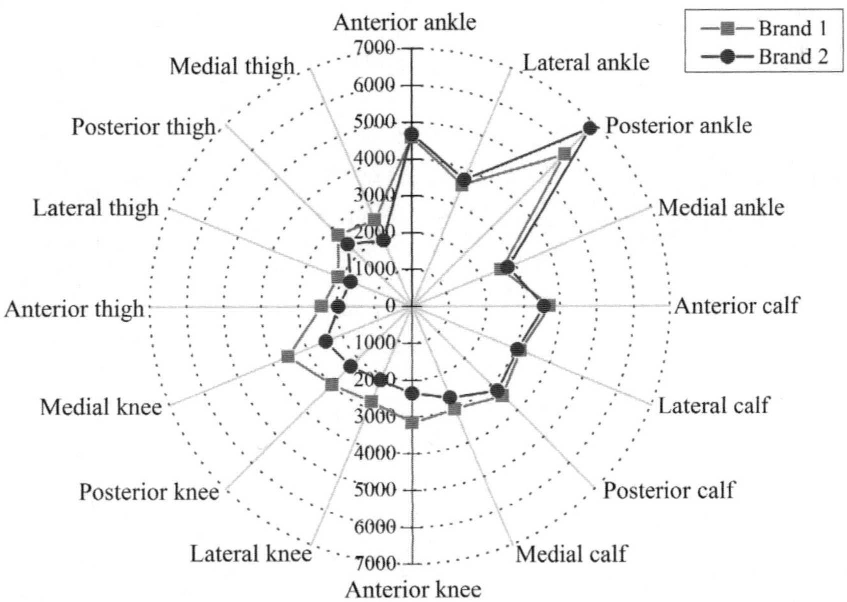

Fig.3 Interface pressure profile at different directions around human lower limb

Fig.3 shows the interface pressure profile at four directions around human lower limb, the pressure magnitude is the mean value which is measured on five volunteers. There were significant pressure differences between directions on height of ankle level, which is more than 3000Pa. The maximal pressure value is measured at posterior direction for both brands of stockings, 5825Pa and 6828Pa, respectively; the minimum is measured at medial direction, 2589Pa and 2779Pa, respectively. According to Fig.2, the anatomical structure and its mechanical properties mainly effect the pressure distribution, such as the posterior of ankle with less soft tissue and larger curvature, in contrast, the medial of ankle with thicker soft tissue and minor curvature. On height of calf level, the maximal pressure magnitudes are 3700Pa and 3544Pa at anterior direction for each brand stockings, the minimum are 3025 Pa and 2680Pa at medial direction, its pressure difference range is 600 to 900Pa. The next-to-skin tibia may relate to the higher pressure value at anterior direction. On the height of knee level, the maximal pressure magnitudes are 3538 Pa and 3544 Pa at its medial direction, the minimum are 2789Pa and 2156Pa, the pressure difference is 300 to 800Pa. On the height of thigh level, the maximal pressure magnitudes are 2747Pa and 2393Pa, minimum is 2105Pa and 1748Pa, the pressure difference is around 650Pa. In summary, other than ankle level, the pressure difference is less than 900Pa and pressure distribution tends to homogeneous.

3.2 The pressure profile at multiple height levels

$$P_{average}=(P_{anterior}+P_{lateral}+P_{posterior}+P_{medial})/4 \quad (1)$$

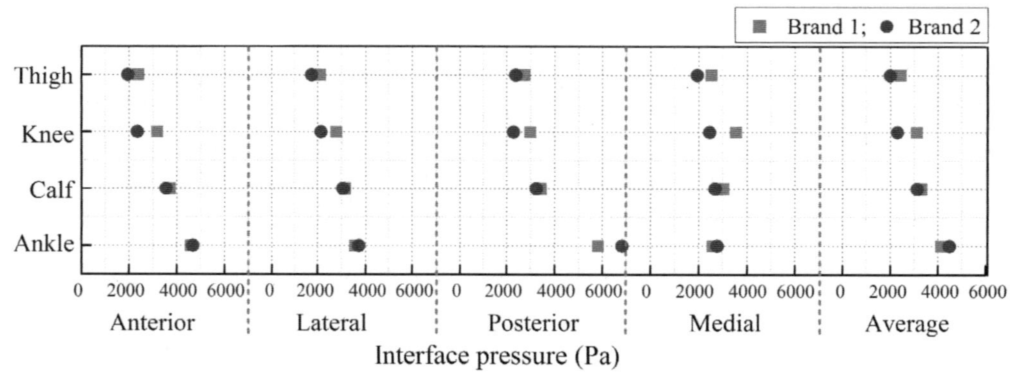

Fig.4 Interface pressure distributions at different heights along the leg

Table 3. The pressure gradient along human lower limb

No.	Specified pressure gradient Ankle:Calf:Thigh	Calculated pressure gradient Ankle:Calf:Knee:Thigh	
Brand 1	100:70:50	Anterior	100:81:69:52
		Lateral	100:88:78:59
		Posterior	100:59:51:47
		Medial	100:117:137:98
		Average	100:80:75:59
Brand 2	100:70:40	Anterior	100:76:50:42
		Lateral	100:82:58:47
		Posterior	100:47:34:35
		Medial	100:96:89:70
		Average	100:70:51:45

Fig.4 shows the pressure profile at four height levels along human leg, the pressure magnitude is the mean value which is measured on five volunteers, and the average pressure is calculated by Eq.1. Table 3 lists the specified pressure gradient and calculated pressure gradient through tested results. The each calculated gradient at four directions is different from the specified ones for both of stockings. Other than the medial direction for brand 1, the pressure magnitude reduces as the height from ankle to thigh. At the medial direction for brand 1 stockings, the maximal pressure is measured on knee level, and value on thigh and calf also more than ankle. At the posterior direction, the pressure on calf, knee and thigh gradually reduction with small amplitude, but the pressure on ankle is around 2 times than other locations. For the average pressure for each height level, the pressure gradient is inconsistent with the specified ones, which is 100:80:75:59,100:70:51:45, respectively.

3.3 The compression performance between different brands of stockings

Fig.5 shows the pressure difference between two brands of stockings on each measured locations. There is significant variability among stockings, both in the absolute pressures and in the pressure gradients generated from the ankle to the calf, thought to be functionally important for venous flow. On ankle level, the pressure measured from Brand 1 is more than Brand 2, especially at posterior direction with difference of 1003Pa. On calf, knee and thigh level, the pressure measured from Brand 2 is more than Brand 1. The pressure difference is least on calf level, it is less than 350Pa, most on knee level, and it is 1076Pa at medial direction. The pressure gradients are listed in table 3, the Brand 2 has greater pressure gradient than Brand 1. For the average pressure gradient, the Brand 2 is more close to its specified value.

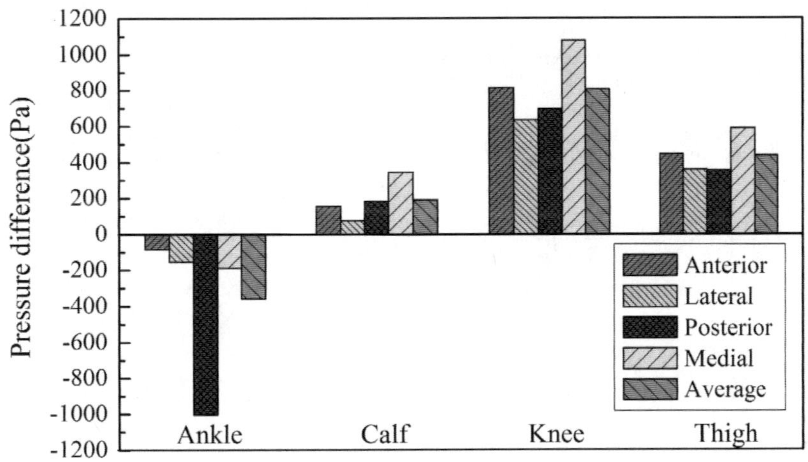

Fig.5 The compression performance comparison between two brands of stockings

3.4 The variation in the pressure of different volunteers

Fig.6 and Fig.7 shows the variation in the interface pressure among volunteers, the pressure difference and pressure percentage are calculated by Eq.2 and Eq.3, respectively. The compression pressures generated by stockings are almost all outside its specified range. For brand 1, the pressure magnitude is 11.2%, 14.4% more at calf and thigh level, respectively. For brand 2, the magnitude is 5.4%, 4.8%, 17.6% more at ankle, calf and thigh, respectively. For each direction, the pressure at lateral ankle is in its range, the medial ankle is less than specified range, and others are more than specified ranges.

There are significant discrepancies in pressure magnitudes among five volunteers, especially at the posterior ankle. For Brand 1, the most pressure difference is at posterior ankle, comes to 4081Pa, the least is at medial calf with 310Pa. Compared to Brand 1, the pressure difference is minor at all measured locations, most is at lateral ankle with 1416Pa, and least is at medial knee with 209Pa. The measured results suggest that mechanical properties of soft tissues and the shape of the leg can significantly impact the magnitude of the interface pressure. And the mechanical properties, structure and design of stockings are also should be considered to the pressure variations.

$$P_{difference}=P_{maximum}-P_{minimum} \quad (2)$$

Where, $P_{maximum}$ and $P_{minimum}$ is the maximal and minimal pressure magnitude measured among five volunteers, respectively.

$$P_{percentage}=(P_{maximum}-P_{minimum})/P_{mean} \quad (3)$$

Where, P_{mean} is the mean pressure magnitude measured among five volunteers.

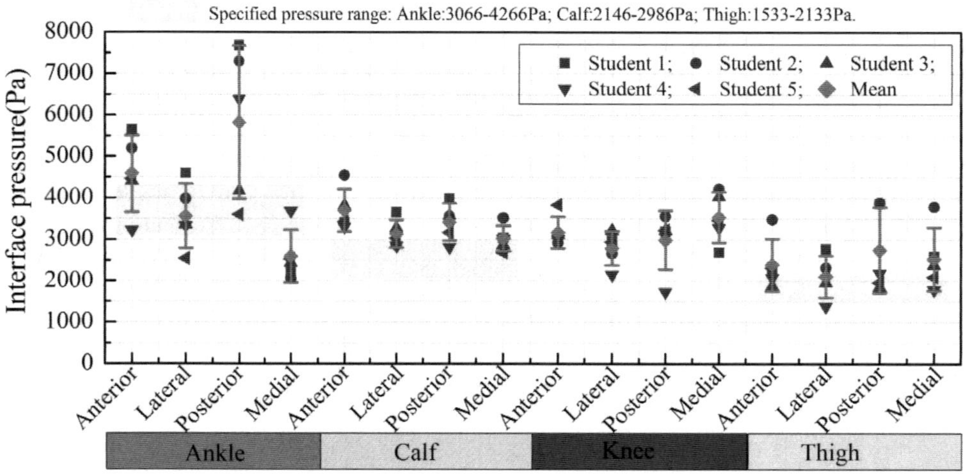

(1)

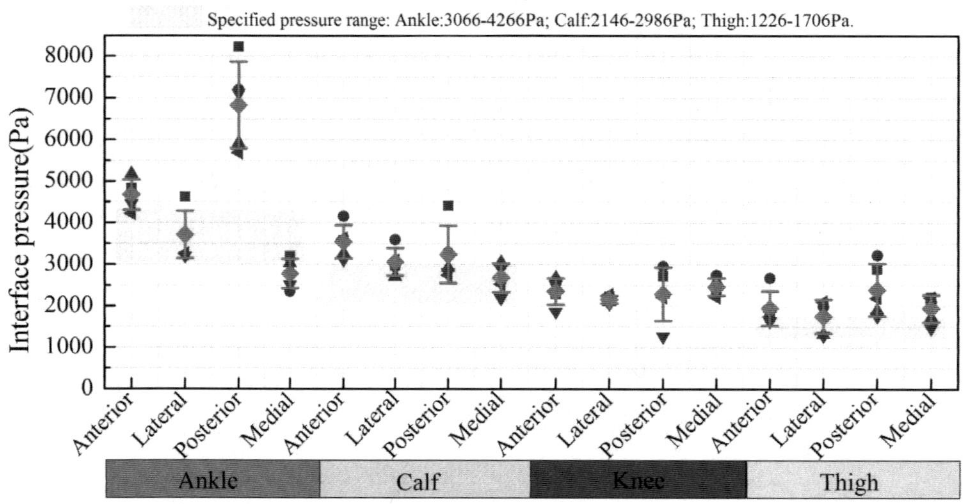

(2)

Fig.6 The variation in the interface pressure among volunteers: (1) Brand 1; (2) Brand 2.

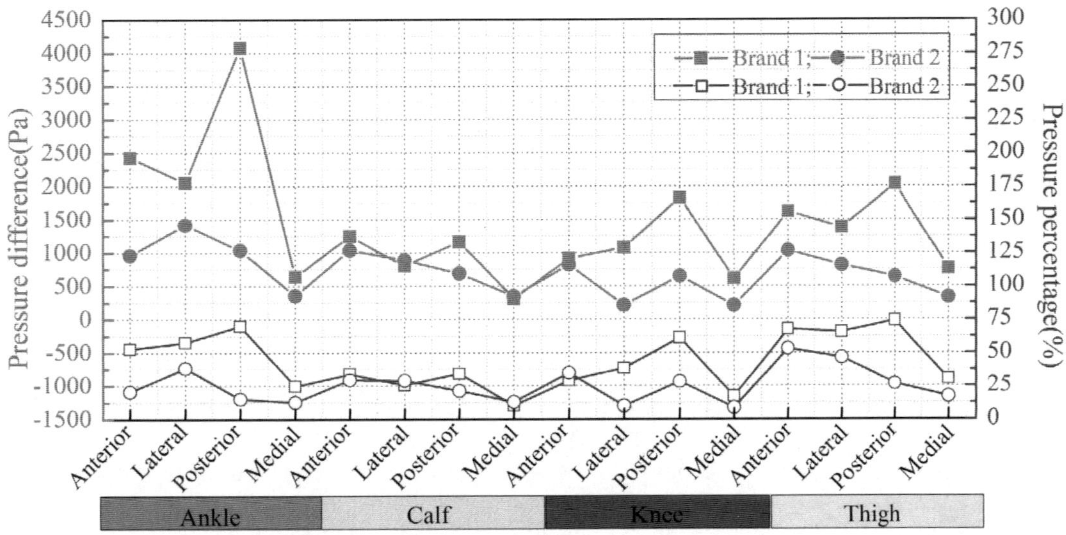

Fig.7 The pressure difference and percentage among volunteers

4. Conclusion

The interface pressure distribution and gradient for two brands of medical compression stockings were examined on five healthy volunteers' right lower limb, four height levels(ankle, calf, knee, thigh) along leg, and four directions(anterior, lateral, posterior, medial) at each transverse section, totally 16 measured locations are enclosed. The pressure profiles at varied directions and height levels are discussed, and the variation in interface pressure and gradient among multiple volunteers is analyzed. The following conclusions were drawn:

1) There are significant pressure discrepancies at different directions of multiple transverse sections around human lower limb, and the pressure distribution is non-uniform.

2) From ankle to thigh level, the pressure magnitude is gradually reduced at anterior, lateral and posterior directions, which keep consistent gradient trend with specified on each brand of stockings, But at medial direction, the pressure magnitude generated by Brand 1 at ankle is less than both of calf and knee.

3) Compared with two brands of stockings, which are specified with the same pressure level and medical function, however, the pressure profiles are significantly difference, the calculated pressure gradients also deviated from its specified, and the pressure magnitudes are outside of ranges at mostly measured locations.

4) The variations in the interface pressure among multiple volunteers are significant, and the variation varies with type of stockings. Which demonstrated that the mechanical properties of soft tissue and stockings, shape of leg, structure and design of stockings can significantly impact the magnitude of interface pressure, and that will directly affect the medical treatment.

The results of this study suggest that in vivo measurements of interface pressure and gradient should be a requirement for clinical studies of medical compression stockings, and with personal customization of medical compression stockings may be reasonable for ensuring appropriate pressure level in clinical practice. Also, it is one of the greatest challenges for development of medical compression stockings in future which can achieve the ability to adapt to different patient characteristics and limb shapes while ensuring effective compression.

5. Acknowledgement

The authors wish to thank the National Natural Science Foundation of China（Grant number：51505080）for funding the research.

References

[1] Moffatt C. Variability of pressure provided by sustained compression. Int Wound J 2008; 5: 259-265.

[2] Liu R, Kwok YL, Li Y, Lao TT, Zhang X. Skin pressure profiles and variations with body postural changes beneath medical elastic compression stockings. Int J Dermatol 2007; 46(5): 514-523.

[3] Holford CP. Graded compression for preventing deep venous thrombosis. Br Med J 1976; 2: 969-970.

[4] Sarin S, Scurr JH, Coleridge SPD. Mechanism of action of external compression on venous function. Br J Surg 1992; 79: 499-502.

[5] European Committee for Standardization (CEN). Non-active Medical Devices. Working Group 2 EVN 12718: European Prestandard "Medical Compression Hosiery." CEN TC 205. Brussels: CEN, 2001.

[6] Cullum N, Nelson EA, Fletcher AW, Sheldon TA. Compression for venous leg ulcers. Cochrane Database Syst Rev 2001; 2: CD000265.

[7] Olin JW, Beusterien KM, Childs MB, Seavey C, McHugh L, Griffiths RI. Medical costs of treating venous stasis ulcers: evidence from a retrospective cohort study. Vasc Med 1999; 4: 1-7.

[8] Thomas S, Fram P. An evaluation of a new type of compression bandaging system. World Wide Wounds 2003; 1-14.

[9] Callam MJ, Ruckley CV, Dale JJ, Harper DR. Hazards of compression treatment of the leg: an estimate from Scottish surgeons. BJM (Clin Res Ed) 1987; 295: 1382.

[10] Rabe E, Partsch H, Jünger M, Abel M, Achhammer I, Becker F, et al. Guidelines for clinical studies with compression devices in patients with venous disorders of the lower limb. Eur J Vasc Endovasc Surg 2008; 35: 494-500.

[11] Harry M, John B, Rafael DM, Kevin ET. Variability in leg compression provided by gradient commercial stockings. Journal of vascular surgery: venous and lymphatic disorders 2015; 431-437.

Design and Characterization of Flexible Pressure Sensors for Integration in Smart Medical Textiles

Suresh Parmar[1,2], Ehsan Bafekrpour[1,2], Franz Konstantin Fuss[2,3], Olga Troynikov[1,2*]

[1]*School of Fashion & Textiles, RMIT University, 25 Dawson St, Brunswick, Vic, 3056, Australia*
[2]*Wound Management Innovation Cooperative Research Centre, Brisbane Australia*
[3]*School of Aerospace, Mechanical and Manufacturing Engineering, RMIT University, Melbourne, Vic, 3001, Australia*

*Corresponding author's email address: olga.troynikov@rmit.edu.au

Abstract

This paper is a feasibility study of using conductive fabric as electrode for flexible pressure sensors. A new low-cost textile-based pressure sensor was developed using piezoresistive carbon loaded polyolefin film as the sensing element. The sensor design as well as its manufacturing is described in detail. The performance parameters of the sensor such as hysteresis, accuracy, and sensitivity were investigated experimentally. The sensor exhibited a bilinear pressure-sensitivity curve with a sensitivity values of 0.1 kPa^{-1} at a lower range of pressure (<10 kPa) and 0.025 kPa^{-1} in a higher pressure range (>10 kPa). The sensor displayed an average overall accuracy of about 88% and hysteresis of 13% at 5 kPa. The results obtained from the proposed pressure sensor configuration demonstrate that it can be successfully used for the development of flexible textile-based low-pressure sensors for biomedical and other similar applications.

Keywords: Smart Medical Textiles; Piezoresistive Sensors; Flexible Pressure Sensors; Pressure Sensing Films

1. Introduction

The continuous and comfortable contact of clothing worn next to the skin of a human body makes textile materials an ideal choice for wearable integrated sensor systems. Textile materials with added functionalities such as sensing are termed "smart textiles", and have gained significant attention in the last decade for application in wearable technologies [1,2]. A considerable amount of work in the area of textile-based sensing has been carried out, particularly in health monitoring and sports applications [3,4]. In these applications physiological parameters such as blood pressure, breathing rate and skin temperature have been widely studied. In addition, emerging function of sensors for these applications is measurement of pressure applied to the human body at hands, legs and feet. Currently, this pressure sensing is mainly performed using rigid electronic sensors [5,6]. However, as the comfort properties and flexibility of the electronic sensors are not comparable with textile materials, it is highly desirable that the textile material itself performs the function of pressure sensing. Recent research works focused on the development of these flexible textile-based pressure sensors for foot plantar pressure measurement [7,8]. Despite significant developments in flexible textile-based pressure sensing, the performance attributes relevant to low-pressure measurement in applications such as compression bandages or pressure garments for venous leg ulcer treatment have not been reported. In this scenario, the sensors are required to measure interface pressure between the bandage or garment and the leg of the wearer in a low-pressure range of 0-10 kPa. In scientific literature and market, there are currently several sensors available whose electrical properties change on application of pressure. This change in electrical properties is then used to quantify the applied pressure. A promising type of pressure sensor for pressure measurement applications is the piezoresistive sensor [9,10]. These sensors are based on the change in electrical resistance of the sensing material in the sensors in response to an applied pressure. It should be noted that they are different from strain gauge sensors that are based on the variation in electrical resistance in response to a change of length and width of the conductor. The piezoresistive sensors have the advantage of being simple, robust, low-cost and easy to manufacture.

In this paper, we present the design and fabrication process of a textile-based flexible pressure sensor for medical application. We also discuss the design of a pressure measurement device incorporating this textile-based sensor and its calibration technique using a reference load-cell. Its performance is evaluated at a low-pressure range (<10kPa) required to be applied for the venous leg ulcer treatment. The pressure sensing performance of the flexible pressure sensor such as pressure sensitivity, hysteresis and accuracy are reported and analysed. The structure and properties of the main components of the flexible piezoresistive pressure sensor such as piezoresistive materials and conductive textile materials are also characterized in detail.

2. Method

The basic design of a piezoresistive pressure sensor consisted of two electrodes that are separated by a commercial piezoresistive sensing material whose properties vary on application of pressure. Woven conductive fabric was used as electrodes for passage of current through the piezoresistive material. The pressure sensing properties of the pressure sensors were then evaluated using a pressure measurement assembly. The operation of pressure measurement assembly was based on measurement of change in voltage across the pressure sensor on application of pressure. In order to minimise the effect of variability in properties of sensors each experiment was conducted five times by developing the pressure sensor from the same fabric electrodes but after changing the piezoresistive sensing material specimen. Further in order to attribute any change in the output voltage signal (change in resistance) to the piezoresistive material, it is important that the conductive fabric electrode does not change its electrical properties on application of pressure. Therefore, the suitability of conductive fabrics as sensor electrode was investigated by measuring the change in output voltage on application of pressure.

The performance parameters of the pressure sensors such as sensitivity, hysteresis and accuracy were evaluated. Pressure sensitivity was defined as the slope of the curve of relative change of resistance and pressure of the sensor. It can be calculated at each pressure as (Eq.1):

$$S = (\Delta R/R_0)/P \quad (1)$$

Where, ΔR is the change in resistance, R_0 is the initial resistance before pressure application, and P is the applied pressure. Hysteresis was defined as the difference of the output signal at the same level of input pressure during loading and unloading on the sensor. Whereas, accuracy was defined as the maximum difference that will exist between the actual input values and the indicated output values of the sensor. Average values for five samples and the standard deviations as error bars were reported.

2.1. Materials

A flexible polyolefin film uniformly loaded with conductive carbon particles, Velostat™ was used as the piezoresistive sensing material. It is commercially manufactured by 3M and is available as 30cm x 30cm sheets of film with the volume resistivity of <500 ohm.cm and thickness of 0.1 mm. The woven conductive fabric, made of metal coated nylon fibers, for the experimental sensor electrodes was sourced from Microwear Australia (www.microwear.com.au).

2.2 Sensor assembly

The experimental pressure sensor was developed by placing a single Velostat™ specimen between two conductive L-shaped fabric electrode sheets with an overlap area of 30 mm × 30 mm (Fig.1). The two fabric electrodes were placed so that their ends of an L-shape are located on the same side of the pressure sensor but do not touch each other. After sandwiching them along with the sensing material, the layers were firmly secured by an adhesive tape that covered the electrodes.

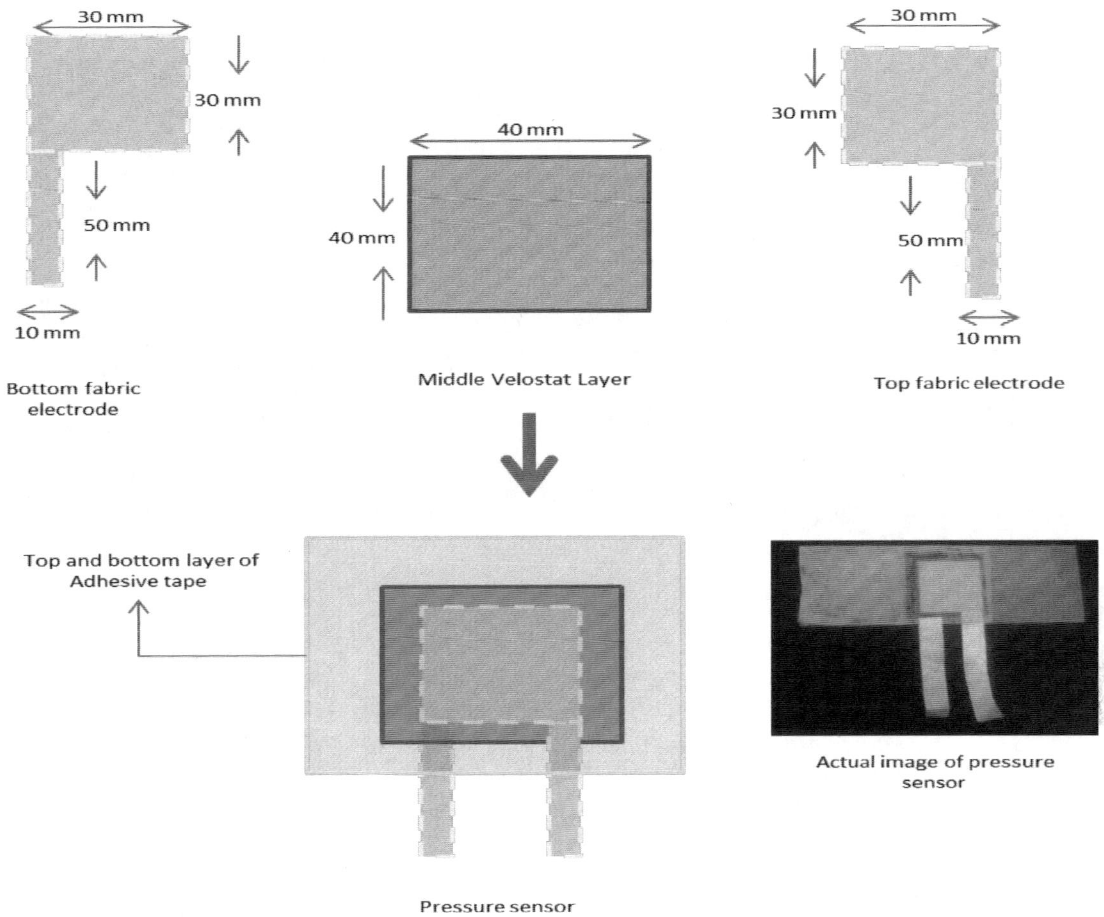

Fig.1 Pressure sensor preparation using conductive fabric and Velostat^{TM}

2.3 Experimental setup

The measurement and verification of pressure applied to the sensor was carried out by placing each experimental pressure sensor sample on a calibrated reference load cell which was further placed on compression rig plate of Instron 5565A as shown in Fig.2. The reference load cell was supplied by Meltrons and works with an error of less than of 0.02 (based on supplier's certificate) to measure force up to 500 N. The reference load cell was calibrated using Instron with a load cell at 2N - 100N force range and its accuracy was examined and verified before committing the experiment. The Instron with a load cell of 2N - 100N force range was also used to compress the sensors with a loading rate of 1 mm/s. The electrodes of the pressure sensor were connected to the developed proprietary pressure detection circuit instrumented on National Instrument's LabVIEW. The change in voltage output during the application of pressure on the sensor was measured and recorded by proprietary pressure reading software. To facilitate the analysis of sensor performance, this pressure reading software collected a reading every 0.1 seconds and simultaneously plotted graphs of output voltage (V_f), read pressure of reference load cell (kPa), and time (second).

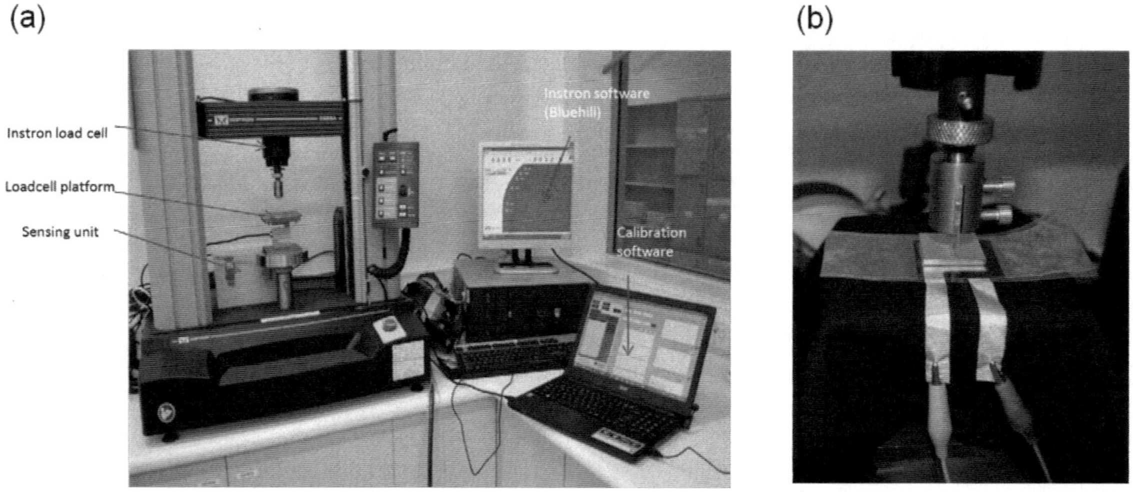

Fig.2 (a) Experimental setup including Instron for loading and calibrated reference load cell for measurement; (b) Designed pushing square pin and the sensor on the load cell platform

2.4 Scanning electron microscopy

The micro-level morphological structure of the piezoresistive Velostat™ film was determined by performing scanning electron microscopy (SEM) using FEI Quanta SEM on random specimens. SEM images were obtained at accelerating voltage of 30kV in high vacuum mode. Furthermore, in order to achieve an enhanced stability, reduced charging and increased secondary electron emission to visualise at higher resolution, SEM images were also obtained after gold sputter coating for 1 min. These were then observed under Philips XL30 SEM in high vacuum.

2.5 Thermogravimetric analysis

The thermogravimetric analysis (TGA) was conducted in the presence of oxygen on Perkin Elmer TGA instrument to determine the percentage of metal content in the conductive nylon fabric as well as the percentage of carbon content in the Velostat™ film. Initially, the sample was heated to 50°C and maintained at this temperature for 5 min and then heated to 800°C at a heating rate of 20°C/min and maintained for 3 min before cooling down to room temperature.

The degradation onset and offset temperatures and the temperature of the highest weight loss rate were calculated using the first derivatives of the TGA curve. The onset and offset temperatures is found from the derivative curve where its value changes from zero to non-zero and from non-zero to zero, respectively. Temperature of the highest weight loss rate is found from the main weight loss derivative peak.

3. Results and discussion

3.1 Electrode performance

Fig.3 shows the resulting output voltage signal when the fabric electrode is compressed under pressure of 40 kPa. This output voltage in Fig.3 is not a voltage drop between two conductive electrodes, but it is the voltage between two ends of a single piece of silver fabric under a compression test as shown by the inset photo of Fig.3. It is seen that the standard variation in the output voltage over the applied pressure range is negligible at $\pm 5.34e^{-5}$ V. The electrodes do not change their resistance under pressure, therefore, it was inferred that any change in the output voltage signal during application of pressure could be attributed only to the change in resistance of the piezoresistive material.

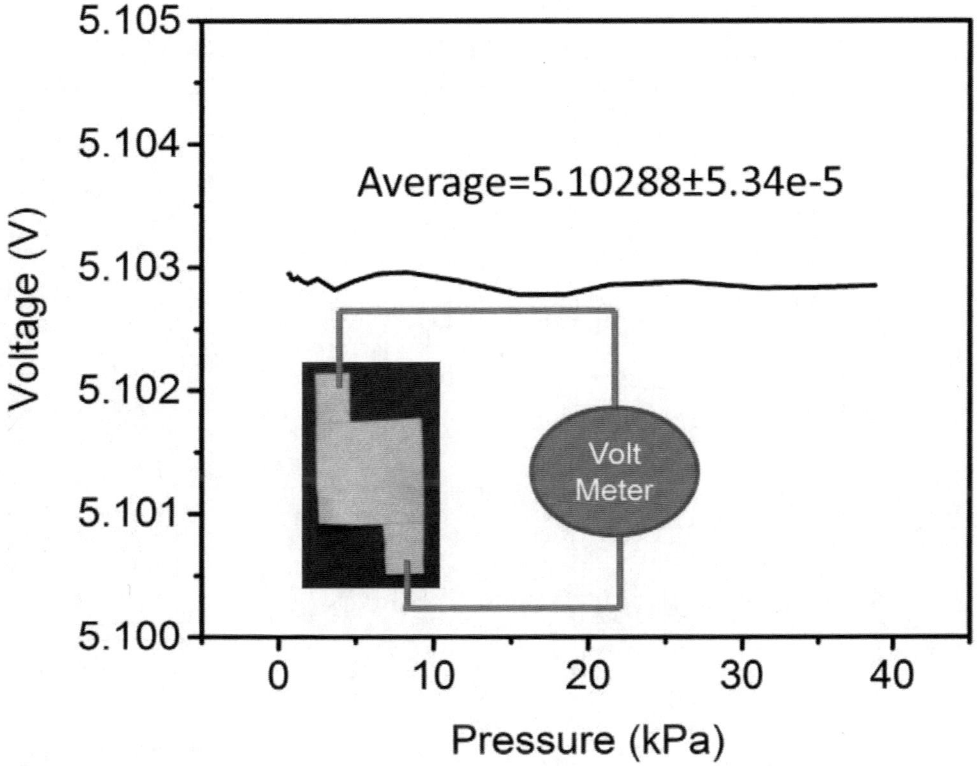

Fig.3 Change in the output voltage signal of the conductive fabric electrode under pressure. The inset photo shows the shape of the electrode

3.2 Sensor performance

The ability of the sensor to measure pressure in low-pressure range under 10 kPa during a loading-unloading cycle is exhibited in Fig.4a. It is observed that the calibrated output of this pressure sensor exhibited a very good agreement with the ones measured by the reference load cell during both loading and unloading period. In addition, it shows that the pressure sensor has a very quick response time in the order of milliseconds. The average sensitivity of five experimental sensors under a compression of up to 40 kPa is presented in Fig.4b. The pressure sensitivity is the relative change in output sensor signal per pressure unit at each pressure. It can be observed that this sensor showed a bilinear sensitivity with 0.1 kPa^{-1} sensitivity value below an applied pressure of 10 kPa, whereas it reduced to 0.025 kPa^{-1} sensitivity value when the applied pressure was increased to 40 kPa. On analysis of accuracy of the output of the sensor in comparison to a reference load cell (Fig.4c), it is observed that the average accuracy of five experimental sensor samples increased from 80% at 2 kPa to about 95% when the pressure level was increased to 6 kPa and then decreased to 85% at 12 kPa. Over the measured pressure range the sensors exhibited an average accuracy of 87%. Fig.4d shows the hysteresis of the five samples. The sensors exhibited hysteresis values of 13% at 5 kPa which can be attributed to the viscoelastic nature of the piezoresistive polymer.

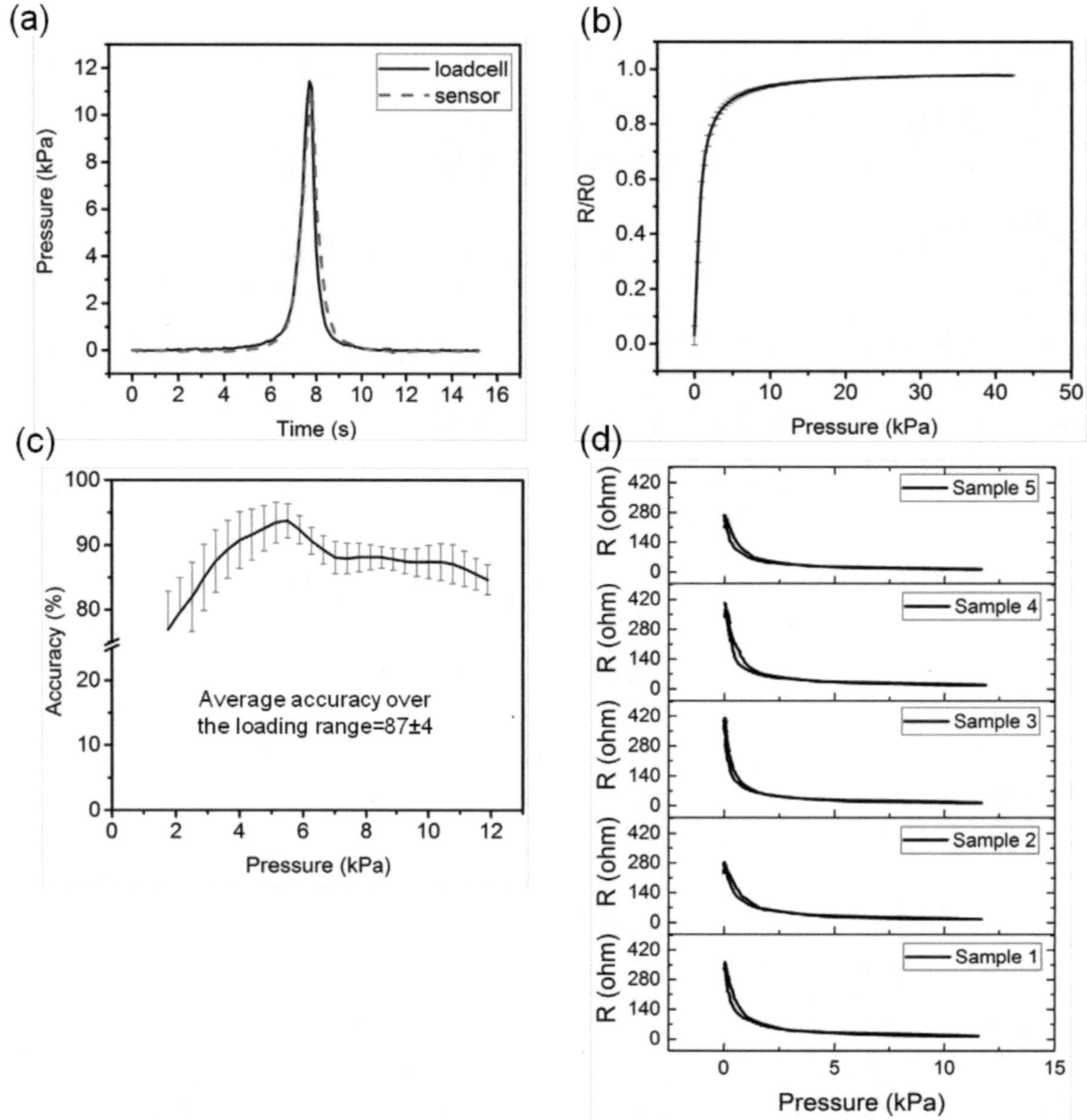

Fig.4 (a) Comparison between the sensor and the load cell; (b) Sensitivity; (c) Accuracy; (d) Hysteresis.

3.3 SEM

At 10000X magnifications and with 30 kV accelerating voltage (Fig.5a) uniformly loaded round spheres is observed in the Velostat^{TM} structure. These round spheres are separated by small gaps between them. It is to be noted that these round spheres are dispersed carbon particles that form large connected colonies rather than being situated as individual particles. The interspatial distance between the colonies of such carbon particles is observed to be nearly 10 times larger in comparison to the individual carbon particles themselves which are observed to be of dimensions less than 100 nm diameter each.

Fig.5 Scanning Electron microscopy image of Velostat™ sheet at (a) uncoated Velostat™ sheet at 10000X with silver conductive base; and (b) coated Velostat™ sheet at 15000X with carbon tab conductive base.

The presence of the gap between each carbon black colony overall seems to result in a porous conductive structure. Under compression, carbon black particle in these colonies come in closer contact with each other, thereby resulting in an enhanced conductivity of the Velostat™ sheet. In addition, these carbon black colonies are expected to increase the conductivity of the piezoresistive film on the basis of percolation threshold theory, However, when these films were gold coated and observed under SEM, then it is seen (Fig.5b) that there is a significant change in the surface features of the Velostat™ . The large colonies of carbon particles are not seen and a plane surface of polymer with carbon particles embedded in it is observed. It could be that this change in features occurred due to the application of heat during gold coating. It may also have happened that the surface features of the Velostat™ film were covered by thicker gold coating. However, we believe that this is the surface of the polymer itself, which may not have been visible during the previous SEM image, since the high energy electron beam may have penetrated polyolefin polymer and was reflected only from carbon black particles.

3.4 TGA

Fig.6a shows the thermal degradation profile of conductive nylon fabric in presence of oxygen. It is observed that the degradation process of conductive nylon fabric started at about 361°C with a main weight loss derivative peak at 423°C. This sudden degradation stopped at 463°C and resulted in presence of conductive metal and burnt residual oxides in the sample. It was further observed that at a temperature of 800°C, the sample still contained about 45% of its weight fraction which demonstrates the metal content along with residual matter in the conductive woven fabric. Similar to the TGA curve of conductive fabric, Fig.6b demonstrates the thermal degradation profile of Velostat™ piezoresistive film in presence of oxygen. It is observed that the Velostat™ also started losing weight at 430°C with the highest weight loss rate at 495°C and then finally stopping at 530°C. As Velostat™ contains polyolefin loaded with conductive carbon particles, it can be inferred that the initial sudden fall below 500°C corresponds to the degradation of polyolefin. A 10% loss at about 400°C can also be attributed to the presence of additives in the sample. Finally, a weight fraction of 30% at the end of the TGA cycle is attributed to presence of about 30% conductive particles and residual matter in the Velostat™ film.

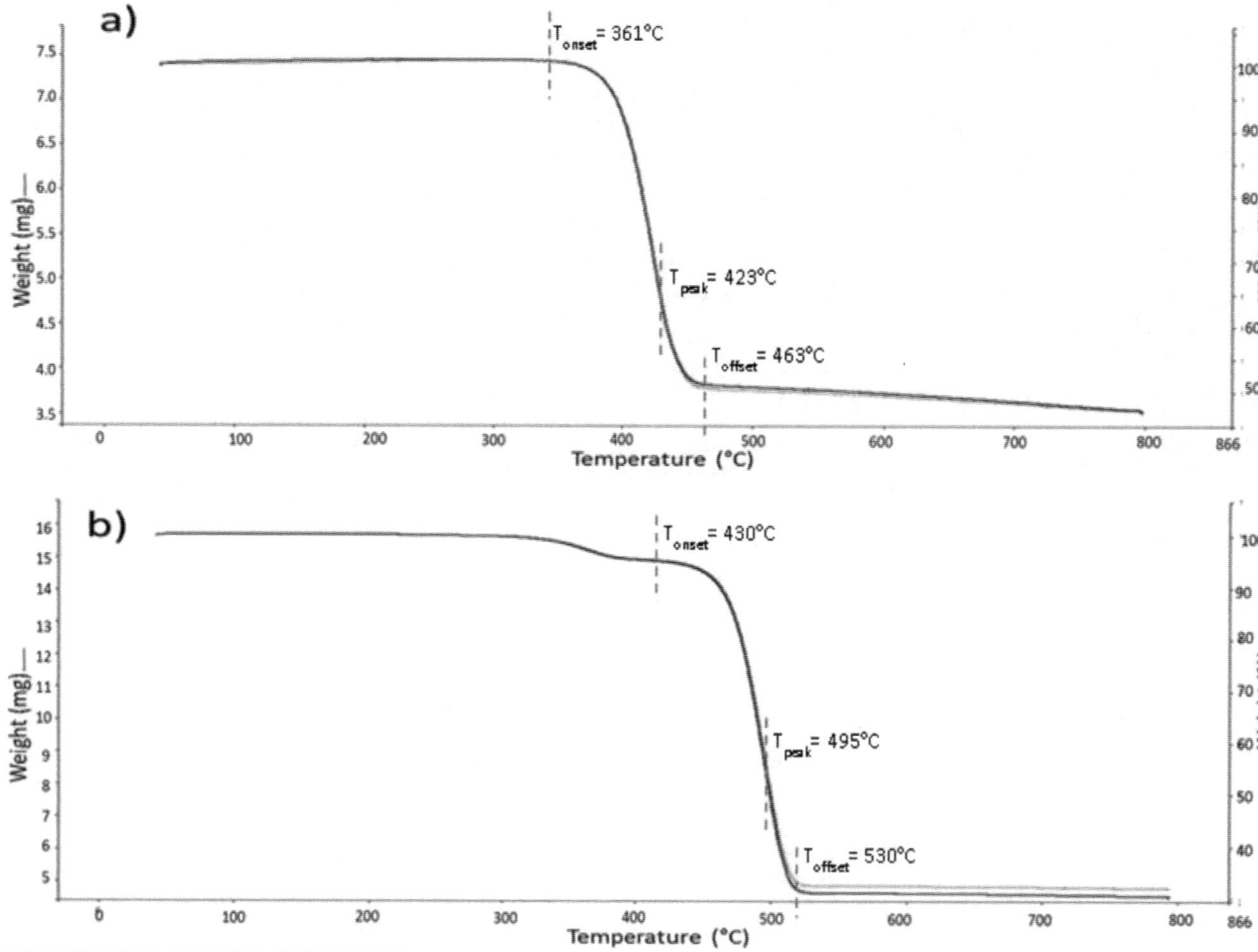

Fig.6 (a) Thermal degradation profile of conductive nylon fabric; (b) Thermal degradation profile of Velostat™ piezoresistive film

4. Conclusion

In this paper, a preliminary study on a new low-cost textile-based pressure sensor was presented and its pressure sensing performance properties with reference to a load cell were analysed. It was concluded that the flexible woven conductive fabric could be used for development of a flexible pressure sensor. The characteristics of the different components of this flexible piezoresistive pressure sensor were also investigated and it was seen that this sensor exhibited a bilinear pressure-sensitivity curve with a sensitivity values of 0.025 kPa^{-1} and 0.1 kPa^{-1} at high and low pressure ranges, respectively. The sensor also displayed an average overall accuracy of about 88% and hysteresis of 13% at 5 kPa. Thus, the pressure measurements obtained from this pressure sensor were found to be in good agreement with the requirements of pressure sensors for biomedical applications. In our future work, we will aim to further develop and improve the properties of this low-cost and simple textile-based pressure sensor.

5. Acknowledgement

The authors would like to acknowledge the support of the Australian Government's Cooperative Research Centres Program. We would also like to acknowledge Dr. Muthu Pannirselvam for supporting our use of facilities of RMIT Rheology and Materials Characterisation laboratory for TGA experiments.

References

[1] Langenhove LV. Smart textiles for medicine and healthcare: materials, systems and applications. Elsevier: 2007.

[2] Stoppa M, Chiolerio A. Wearable electronics and smart textiles: a critical review. Sensors (Basel): 2014; 14(7): 11957-11992.

[3] Pantelopoulos A, Bourbakis NG. A survey on wearable sensor-based systems for health monitoring and prognosis. IEEE transactions on systems, man, and cybernetics, Part C: Applications and Reviews: 2010; 40(1): 1-12.

[4] Rai P, Kumar PS, Oh S, Kwon H, Mathur GN, Varadan VK, Agarwal MP. Smart healthcare textile sensor system for unhindered-pervasive health monitoring. Proceedings of SPIE smart structures and materials+ nondestructive evaluation and health monitoring: 2012; 8344: 1-10.

[5] Martínez-Nova A, Cuevas-García JC, Pascual-Huerta J, Sánchez-Rodríguez R. BioFoot® in-shoe system: Normal values and assessment of the reliability and repeatability. The Foot: 2007; 17(4): 190-196.

[6] Xu W, Huang MC, Amini N, Liu JJ, He L, Sarrafzadeh M. Smart insole: a wearable system for gait analysis. Proceedings of the 5th international conference on pervasive technologies related to assistive environments: 2012; 18: 1-4.

[7] Rashidi FRM, Hussein O, Hasan W. Investigation on developing of a piezoresistive pressure sensor for foot plantar measurement system. Proceedings of IEEE regional symposium on micro and nanoelectronics (RSM): 2015.

[8] Abdul Razak AH, Zayegh A, Begg RK, Wahab Y. Foot plantar pressure measurement system: a review. Sensors: 2012; 12(7): 9884-9912.

[9] Salibindla S, Ripoche B, Lai DTH, Maas S. Characterization of a new flexible pressure sensor for body sensor networks. Proceedings of IEEE eighth international conference on intelligent sensors, sensor networks and information processing: 2013.

[10] Capineri L. Piezoresistive sensors fabricated with conductive textiles for monitoring the step rate with read-out electronics and wireless connection to a smart watch. J Fashion Technol Textile Eng 2015; 3(3): 1-6.

AUTHOR INDEX

AUTHOR INDEX

AUTHOR INDEX

Textile Bioengineering and Informatics Society (TBIS)
Hong Kong Polytechnic University
ST733
Hung Hom, Hong Kong

ISBN 978-1-5108-2898-8

CURRAN ASSOCIATES INC.
proceedings.com